FEATURES AND BENEFITS

ALGEBRA 1: Integration, Applications, Connections

1. **Integration** Integration helps students to view mathematics as a whole and not as compartmentalized areas of instruction. *Lessons* that connect the various areas of mathematics to algebra are included as they are appropriate (see Lesson 3-4). This integration also occurs *within* algebra lessons. *Lesson Openers* often reflect integration (see page 173), as do *Examples* (see Example 2 on page 169) and *Exercises* (see Exercises 37–38 on page 171).

2. **Applications** Because the ability to grasp concepts and skills is greatly enhanced when they are tied to applications, most lessons open with a real-world application (see page 150). This is strengthened by including *Applications and Problem Solving* in every set of exercises (see page 149), as well as in *Examples* (see Example 2 on page 145).

3. **Connections** Connections to *interdisciplinary areas*, such as science, geography, history, music, and so on, enhance learning while increasing student interest. These connections appear as *Lesson Openers* (see page 119), *Examples* (see Example 1 on page 179), and as *Exercises* (see Exercise 29 on page 183).

4. **Technology** Although this program is *not* dependent upon graphing calculators, the graphing calculator is integrated throughout the program in various ways.

Graphing Technology Lessons	See Lesson 7-8A.
Graphing Calculator Explorations	See page 157.
Graphing Calculator Programs	See Exercise 39 on page 171.
Graphing Calculator Exercises	See Exercises 50–55 on page 284.

   Spreadsheets also play a role in the program, particularly with the integration of statistics (see the Exploration on page 297). *Technology Tips* are provided as appropriate.

5. **Modeling** Students have the opportunity to bridge the gap between the concrete and the abstract through *hands-on experiences* in the *Modeling Mathematics* activities.

Modeling Mathematics Lessons	See Lesson 3-1A.
Modeling Mathematics Activities	See page 164.
Modeling Mathematics Exercises	See Exercise 5 on page 165.

6. **Problem Solving** Problem-solving strategies are *integrated* within lessons (see *work backward* in Example 2 on page 157). *Applications and Problem Solving* and *Critical Thinking* in *every* set of exercises illustrate the ongoing attention to problem solving (see Exercises 46–49 on pages 160–161).

7. **Functions** The concept of function is a unifying theme, facilitating the introduction of graphing early on in the text.

Introduction to *Functions*	See Lesson 1-9.
Relations and *Functions*	See Lesson 5-5.
Functions and Patterns	See Lesson 5-6.
Graphing Linear *Functions*	See Lesson 6-5.
Graphing Quadratic and Exponential *Functions*	See Lessons 11-1 and 11-4.

TEACHER'S WRAPAROUND EDITION

Glencoe

Algebra 1

Integration
Applications
Connections

GLENCOE

McGraw-Hill

New York, New York   Columbus, Ohio   Mission Hills, California   Peoria, Illinois

Visit the Glencoe Mathematics Internet Home Page at...

glencoe.com

For teachers, you will find classroom resources and information about *Algebra 1* and the entire Glencoe Mathematics Series. You'll also find ordering information, toll-free numbers for customer service, and links to information about NCTM conferences.

For students and parents, there are challenging activities such as problems of the week and group activities.

Come see us at http://www.glencoe.com/sec/math

Glencoe/McGraw-Hill

A Division of The ***McGraw-Hill*** *Companies*

Send all inquiries to:
Glencoe/McGraw-Hill
936 Eastwind Drive
Westerville, OH 43081-3329

ISBN: 0-02-825326-4 (Student Edition)
0-02-825328-0 (Teacher's Wraparound Edition)

3 4 5 6 7 8 9 10 071/043 05 04 03 02 01 00 99 98 97

TEACHER'S WRAPAROUND EDITION

Table of Contents

Algebra 1

INTEGRATION ● APPLICATIONS ● CONNECTIONS

Glencoe Brings Algebra 1 to Life!

INTEGRATION ● APPLICATIONS ● CONNECTIONS

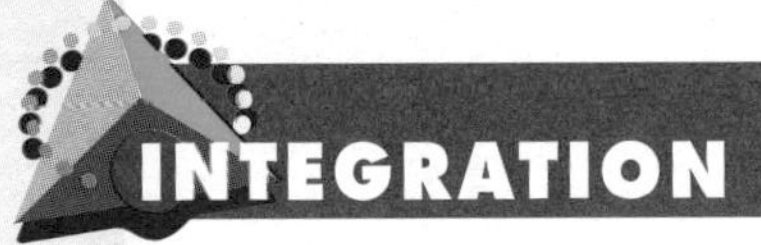

Students relate and apply algebraic concepts to geometry, statistics, data analysis, probability, and discrete mathematics.

Realistic and relevant applications help answer the question "When am I ever going to use this stuff?" Sports, space, world cultures, and consumerism are just a few of the real-life problem settings that students explore.

Students connect mathematics to other topics they are studying, like biology, geography, art, history, and health, through problems that are rich in algebraic content.

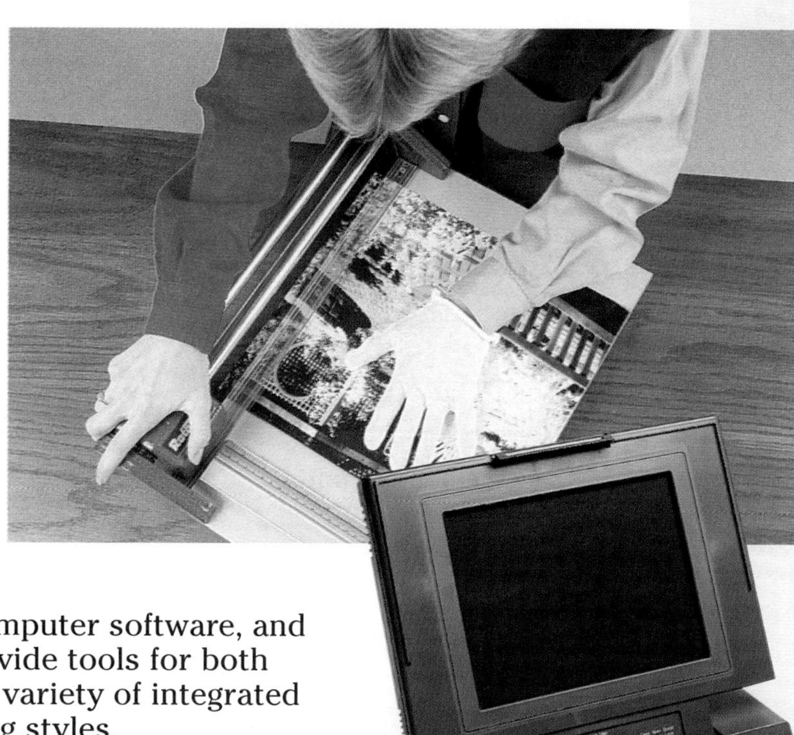

TECHNOLOGY
Graphing calculators, interactive computer software, and CD-ROM multimedia technology provide tools for both problem solving and discovery. The variety of integrated technology supports diverse learning styles.

HANDS-ON ACTIVITIES
Modeling Mathematics activities and multi-day Investigations engage students with manipulatives and mathematical models. Both college-bound and tech-prep students find meaningful hands-on activities that relate to life and work.

ASSESSMENT
A broad spectrum of evaluation and assessment tools, ranging from short-response questions to open-ended problems, lets all students demonstrate what they have learned.

ACCESSIBILITY
Practical strategies meet the needs of students with differing ability levels, rates of learning, and learning styles. Modeling, journal writing, cooperative learning projects, and technology tools help all students experience success.

NCTM STANDARDS
The sequence of topics, the content emphasis, integration with other mathematics areas, hands-on activities, and technology combine to form a program that reflects the NCTM Curriculum, Teaching, and Assessment Standards.

Glencoe's Dynamic Lesson Structure

Each lesson follows a straightforward format —

What You'll Learn/Why It's Important
Application/Connection/Integration
Examples
Check for Understanding
Exercises

— that focuses on concepts and applications.

WHAT YOU'LL LEARN/WHY IT'S IMPORTANT

Students quickly see what concepts are covered in the lesson and why it is beneficial to learn them.

APPLICATION/CONNECTION/INTEGRATION

A motivating lesson opener introduces the concept by posing a real-world application, an interdisciplinary connection, or a mathematical integration.

EXAMPLES

A set of completely worked-out examples with clear explanations supports the objectives and mirrors the exercises in the Guided Practice and Practice sections that follow.

MODELING MATHEMATICS/ EXPLORATION

Activities with hands-on manipulatives and technology are fully integrated into appropriate lessons.

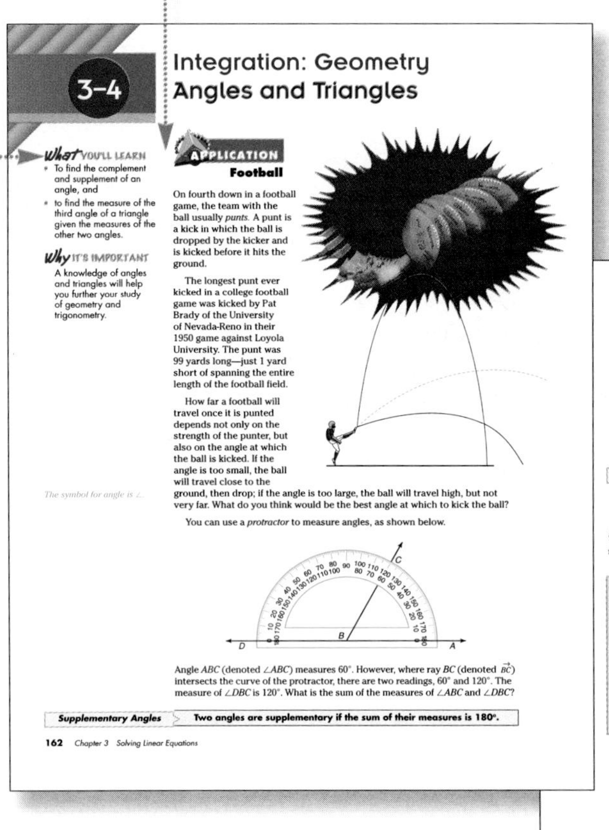

3–4 Integration: Geometry
Angles and Triangles

What YOU'LL LEARN
- To find the complement and supplement of an angle, and
- to find the measure of the third angle of a triangle given the measures of the other two angles.

Why IT'S IMPORTANT
A knowledge of angles and triangles will help you further your study of geometry and trigonometry.

APPLICATION **Football**

On fourth down in a football game, the team with the ball usually *punts.* A punt is a kick in which the ball is dropped by the kicker and is kicked before it hits the ground.

The longest punt ever kicked in a college football game was kicked by Pat Brady of the University of Nevada-Reno in their 1950 game against Loyola University. The punt was 99 yards long—just 1 yard short of spanning the entire length of the football field.

How far a football will travel once it is punted depends not only on the strength of the punter, but also on the angle at which the ball is kicked. If the angle is too small, the ball will travel close to the ground, then drop; if the angle is too large, the ball will travel high, but not very far. What do you think would be the best angle at which to kick the ball?

The symbol for angle is ∠.

You can use a *protractor* to measure angles, as shown below.

Angle ABC (denoted $\angle ABC$) measures 60°. However, where ray BC (denoted $\overrightarrow{BC}$) intersects the curve of the protractor, there are two readings, 60° and 120°. The measure of $\angle DBC$ is 120°. What is the sum of the measures of $\angle ABC$ and $\angle DBC$?

Supplementary Angles	Two angles are supplementary if the sum of their measures is 180°.

162 Chapter 3 Solving Linear Equations

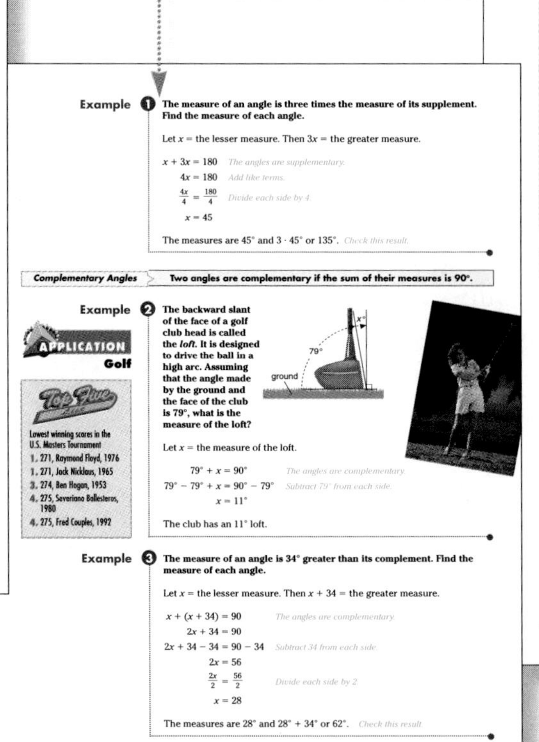

Example 1 **The measure of an angle is three times the measure of its supplement. Find the measure of each angle.**

Let x = the lesser measure. Then $3x$ = the greater measure.

$x + 3x = 180$ *The angles are supplementary.*
$4x = 180$ *Add like terms.*
$\frac{4x}{4} = \frac{180}{4}$ *Divide each side by 4.*
$x = 45$

The measures are 45° and $3 \cdot 45°$ or 135°. *Check this result.*

Complementary Angles	Two angles are complementary if the sum of their measures is 90°.

Example 2 APPLICATION **Golf**

The backward slant of the face of a golf club head is called the *loft.* It is designed to drive the ball in a high arc. Assuming that the angle made by the ground and the face of the club is 79°, what is the measure of the loft?

Top Five
Lowest winning scores in the U.S. Masters Tournament
1. 271, Raymond Floyd, 1976
1. 271, Jack Nicklaus, 1965
3. 274, Ben Hogan, 1953
4. 275, Severiano Ballesteros, 1980
4. 275, Fred Couples, 1992

Let x = the measure of the loft.

$79° + x = 90°$ *The angles are complementary.*
$79° - 79° + x = 90° - 79°$ *Subtract 79° from each side.*
$x = 11°$

The club has an 11° loft.

Example 3 **The measure of an angle is 34° greater than its complement. Find the measure of each angle.**

Let x = the lesser measure. Then $x + 34$ = the greater measure.

$x + (x + 34) = 90$ *The angles are complementary.*
$2x + 34 = 90$
$2x + 34 - 34 = 90 - 34$ *Subtract 34 from each side.*
$2x = 56$
$\frac{2x}{2} = \frac{56}{2}$ *Divide each side by 2.*
$x = 28$

The measures are 28° and 28° + 34° or 62°. *Check this result.*

A **triangle** is a polygon with three sides and three angles. What is the sum of the measures of the three angles of a triangle?

Lesson 3–4 INTEGRATION Geometry Angles and Triangles 163

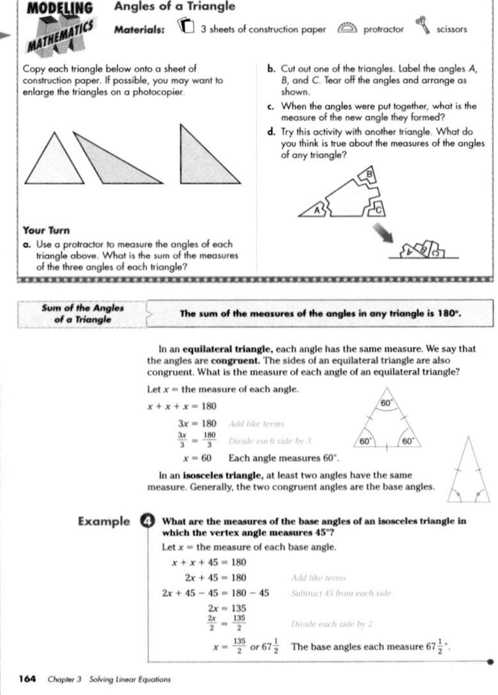

MODELING MATHEMATICS **Angles of a Triangle**

Materials: 3 sheets of construction paper, protractor, scissors

Copy each triangle below onto a sheet of construction paper. If possible, you may want to enlarge the triangles on a photocopier.

Your Turn

a. Use a protractor to measure the angles of each triangle above. What is the sum of the measures of the three angles of each triangle?
b. Cut out one of the triangles. Label the angles *A*, *B*, and *C*. Tear off the angles and arrange as shown.
c. When the angles were put together, what is the measure of the new angle they formed?
d. Try this activity with another triangle. What do you think is true about the measures of the angles of any triangle?

Sum of the Angles of a Triangle	The sum of the measures of the angles in any triangle is 180°.

In an **equilateral triangle,** each angle has the same measure. We say that the angles are **congruent.** The sides of an equilateral triangle are also congruent. What is the measure of each angle of an equilateral triangle?

Let x = the measure of each angle.

$x + x + x = 180$
$3x = 180$ *Add like terms.*
$\frac{3x}{3} = \frac{180}{3}$ *Divide each side by 3.*
$x = 60$ Each angle measures 60°.

In an **isosceles triangle,** at least two angles have the same measure. Generally, the two congruent angles are the base angles.

Example 4 **What are the measures of the base angles of an isosceles triangle in which the vertex angle measures 45°?**

Let x = the measure of each base angle.

$x + x + 45 = 180$
$2x + 45 = 180$ *Add like terms.*
$2x + 45 - 45 = 180 - 45$ *Subtract 45 from each side.*
$2x = 135$
$\frac{2x}{2} = \frac{135}{2}$ *Divide each side by 2.*
$x = \frac{135}{2}$ or $67\frac{1}{2}$ The base angles each measure $67\frac{1}{2}°$.

164 Chapter 3 Solving Linear Equations

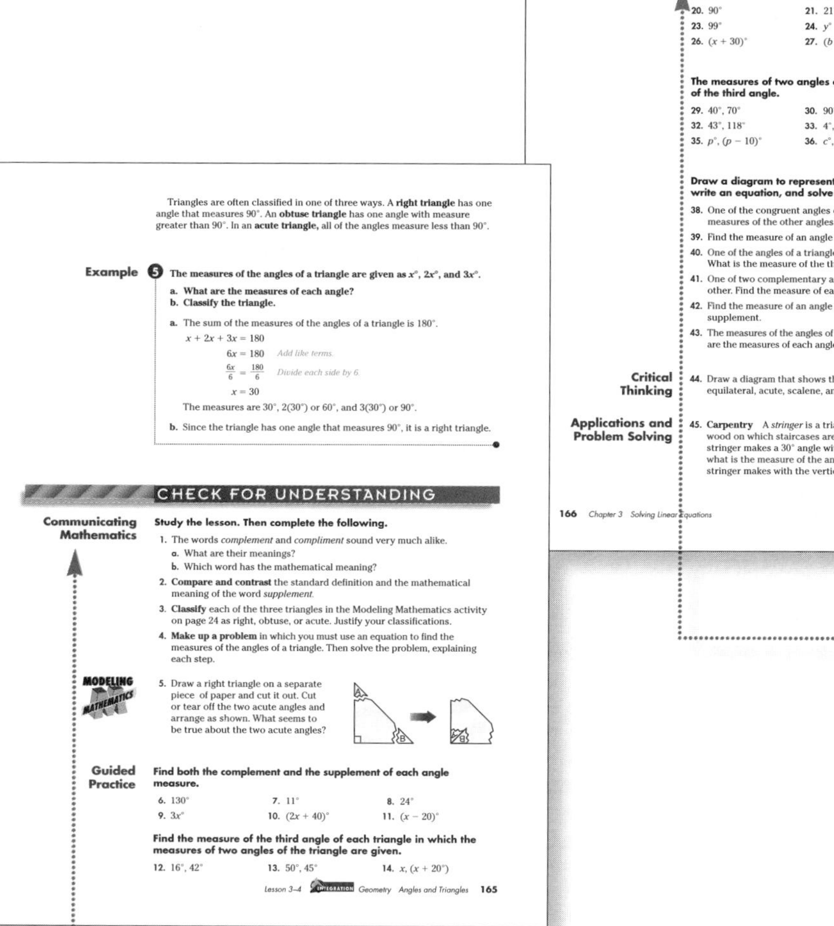

Triangles are often classified in one of three ways. A **right triangle** has one angle that measures 90°. An **obtuse triangle** has one angle with measure greater than 90°. In an **acute triangle,** all of the angles measure less than 90°.

Example 5 **The measures of the angles of a triangle are given as $x°$, $2x°$, and $3x°$.**

a. What are the measures of each angle?
b. Classify the triangle.

a. The sum of the measures of the angles of a triangle is 180°.

$$x + 2x + 3x = 180$$
$$6x = 180 \quad \text{Add like terms.}$$
$$\frac{6x}{6} = \frac{180}{6} \quad \text{Divide each side by 6.}$$
$$x = 30$$

The measures are 30°, 2(30°) or 60°, and 3(30°) or 90°.

b. Since the triangle has one angle that measures 90°, it is a right triangle.

CHECK FOR UNDERSTANDING

Communicating Mathematics

Study the lesson. Then complete the following.

1. The words *complement* and *compliment* sound very much alike.
   a. What are their meanings?
   b. Which word has the mathematical meaning?
2. **Compare and contrast** the standard definition and the mathematical meaning of the word *supplement.*
3. **Classify** each of the three triangles in the Modeling Mathematics activity on page 24 as right, obtuse, or acute. Justify your classifications.
4. **Make up a problem** in which you must use an equation to find the measures of the angles of a triangle. Then solve the problem, explaining each step.

MODELING MATHEMATICS

5. Draw a right triangle on a separate piece of paper and cut it out. Cut or tear off the two acute angles and arrange as shown. What seems to be true about the two acute angles?

Guided Practice

Find both the complement and the supplement of each angle measure.

6. 130°	7. 11°	8. 24°
9. $3x°$	10. $(2x + 40)°$	11. $(x - 20)°$

Find the measure of the third angle of each triangle in which the measures of two angles of the triangle are given.

12. 16°, 42°	13. 50°, 45°	14. x, $(x + 20°)$

Lesson 3–4 INTEGRATION Geometry Angles and Triangles 165

Make a diagram to represent each situation. Then define a variable, write an equation, and solve each problem. Check your solution.

15. An angle measures 38° less than its complement. Find the measures of the two angles.
16. The measures of the angles of a triangle are given as $x°$, $(x + 5)°$, and $(2x + 3)°$. What are the measures of each angle?

EXERCISES

Practice

Find both the complement and the supplement of each angle measure.

17. 42°	18. 87°	19. 125°
20. 90°	21. 21°	22. 174°
23. 99°	24. $y°$	25. $3a°$
26. $(x + 30)°$	27. $(b - 38)°$	28. $(90 - z°)$

The measures of two angles of a triangle are given. Find the measure of the third angle.

29. 40°, 70°	30. 90°, 30°	31. 63°, 12°
32. 43°, 118°	33. 4°, 38°	34. $x°$, $y°$
35. $p°$, $(p - 10)°$	36. $c°$, $(2c + 1)°$	37. $y°$, $(135 - y)°$

Draw a diagram to represent each situation. Then define a variable, write an equation, and solve each problem. Check your solution.

38. One of the congruent angles of an isosceles triangle measures 37°. Find the measures of the other angles.
39. Find the measure of an angle that is 30° less than its supplement.
40. One of the angles of a triangle measures 53°. Another angle measures 37°. What is the measure of the third angle?
41. One of two complementary angles measures 30° more than three times the other. Find the measure of each angle.
42. Find the measure of an angle that is one-half the measure of its supplement.
43. The measures of the angles of a triangle are given as $x°$, $(3x)°$, and $(4x)°$. What are the measures of each angle?

Critical Thinking

44. Draw a diagram that shows the relationships among right, isosceles, equilateral, acute, scalene, and obtuse triangles.

Applications and Problem Solving

45. **Carpentry** A *stringer* is a triangular piece of wood on which staircases are based. If the stringer makes a 30° angle with the floor, what is the measure of the angle that the stringer makes with the vertical?

stringer
30°

166 Chapter 3 Solving Linear Equations

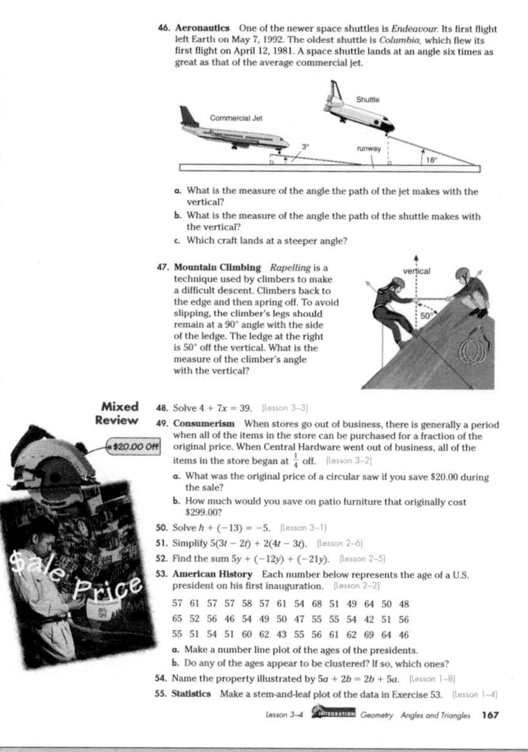

46. **Aeronautics** One of the newer space shuttles is *Endeavour.* Its first flight left Earth on May 7, 1992. The oldest shuttle is *Columbia,* which flew its first flight on April 12, 1981. A space shuttle lands at an angle six times as great as that of the average commercial jet.

   a. What is the measure of the angle the path of the jet makes with the vertical?
   b. What is the measure of the angle the path of the shuttle makes with the vertical?
   c. Which craft lands at a steeper angle?

47. **Mountain Climbing** *Rapelling* is a technique used by climbers to make a difficult descent. Climbers back to the edge and then spring off. To avoid slipping, the climber's legs should remain at a 90° angle with the side of the ledge. The ledge at the right is 50° off the vertical. What is the measure of the climber's angle with the vertical?

Mixed Review

48. Solve $4 + 7x = 39$. (Lesson 3–3)
49. **Consumerism** When stores go out of business, there is generally a period when all of the items in the store can be purchased for a fraction of the original price. When Central Hardware went out of business, all of the items in the store began at $\frac{1}{4}$ off. (Lesson 3–2)
   a. What was the original price of a circular saw if you save $20.00 during the sale?
   b. How much would you save on patio furniture that originally cost $299.00?
50. Solve $h + (-13) = -5$. (Lesson 3–1)
51. Simplify $5(3t - 2t) + 2(4t - 3t)$. (Lesson 2–6)
52. Find the sum $5y + (-12y) + (-21y)$. (Lesson 2–5)
53. **American History** Each number below represents the age of a U.S. president on his first inauguration. (Lesson 2–2)

    57 61 57 57 58 57 61 54 68 51 49 64 50 48
    65 52 56 46 54 49 50 47 55 55 54 42 51 56
    55 51 54 51 60 62 43 55 56 61 62 69 64 46

   a. Make a number line plot of the ages of the presidents.
   b. Do any of the ages appear to be clustered? If so, which ones?
54. Name the property illustrated by $5a + 2b = 2b + 5a$. (Lesson 1–8)
55. **Statistics** Make a stem-and-leaf plot of the data in Exercise 53. (Lesson 1–4)

Lesson 3–4 INTEGRATION Geometry Angles and Triangles 167

CHECK FOR UNDERSTANDING

This section ensures that all students are engaged in the lesson and understand the concepts.

Communicating Mathematics

Students work in small groups or as a whole class to define, explain, describe, make lists, draw models, use symbols, or create graphs.

Modeling Mathematics and **Math Journal** activities, which further strengthen communication skills, appear frequently.

Guided Practice

Keyed to the lesson objectives, the Guided Practice exercises present a representative sample of the exercises in the Practice section.

EXERCISES

The Exercises always include **Practice, Critical Thinking, Applications and Problem Solving,** and **Mixed Review.**

Practice

The Practice exercises are separated into A, B, and C sections, indicated only in the Teacher's Wraparound Edition. The A and B sections generally match the Guided Practice exercises in a 3:1 ratio. The C section offers challenging, thought-provoking questions.

Critical Thinking

Each lesson contains critical thinking exercises, in which students explain, evaluate, and justify mathematical concepts and relationships.

Applications and Problem Solving

Students find numerous opportunities to apply concepts to both real-life and mathematical problem situations.

Mixed Review

This spiraled, cumulative review comprises about 15% of the total number of exercises in each lesson and includes at least one application, connection, or integration problem.

Glencoe's Algebra 1 Exemplifies the NCTM Standards

CONTENT EMPHASIS

Algebra 1: Integration, Applications, and Connections implements the shift from just manipulative skills to algebra as a means of representation.

> *"The proposed algebra curriculum will move away from a tight focus on manipulative facility to include a greater emphasis on conceptual understanding, on algebra as a means of representation, and on algebraic methods as a problem-solving tool." NCTM Standards, page 150, 1989.*

In each chapter, students explore the language of algebra in verbal, tabular, graphical, and symbolic forms. Problem-solving activities and applications encourage students to model patterns and relationships with variables and functions.

> *"The primary role of algebra at the school level is to develop confidence and facility in using variables and functions to model numerical patterns and quantitative relations." NCTM Statement on Algebra, February 1994.*

Each lesson opener motivates students to master the content they need to solve the application, connection, or integration. Additional applications, connections, and integration in the exercises enable students to apply what they have learned.

Glencoe's ***Algebra 1*** incorporates graphing-calculator and computer-software activities for discovery, problem solving, and modeling. The program deemphasizes paper-and-pencil methods of transforming, simplifying, and solving symbolic expressions, while retaining those procedures that students need for algebraic understanding.

> *"Development of algebra should use technology to explore ideas and methods from at least three connected perspectives—graphic, numeric, and symbolic." NCTM Statement on Algebra, February 1994.*

Examples, exercises, and assessment activities in Glencoe's ***Algebra 1*** integrate algebra topics with statistics, geometry, and discrete mathematics.

> *"Refocusing algebra on modeling...serves to break down artificial barriers between algebra and statistics, between algebra and geometry, and between algebra and discrete mathematics." NCTM Statement on Algebra, February 1994.*

Applications, modeling activities, and open-ended projects encourage a diversity of approaches and engage today's students in Algebra 1.

ASSESSMENT

The wide and varied collection of assessment tools and materials that accompany Glencoe's ***Algebra 1*** support the NCTM Assessment Standards. *[Note: The bulleted headings and quotations below are taken from the Assessment Standards for School Mathematics, NCTM, 1995, pages 11, 13, 15, 17, 19, and 21.]*

- **The Mathematics Standard —** "Assessment should reflect the mathematics that all students need to know and be able to do."
  Skills, procedural knowledge, factual knowledge, reasoning, applying mathematics, and problem solving are assessed through a broad range of short-response exercises and longer-term, open-ended activities.

- **The Learning Standard —** "Assessment should enhance mathematics learning."
  Assessment is an integral part of the Communicating Mathematics, Guided Practice, and Exercises sections of each Glencoe's ***Algebra 1*** lesson. Students are able to evaluate their own learning by writing in their math journals and selecting portfolio items.

- **The Equity Standard —** "Assessment should promote equity."
  The wide variety of assessment tools enable all students to demonstrate what they know and can do. Performance Assessment tasks for each chapter include a scoring guide.

- **The Openness Standard —** "Assessment should be an open process."
  Learning objectives tell students what they are expected to learn in each lesson. The examples, activities, and assessment match those objectives. Guided Practice prepares students to successfully complete the Exercises.

- **The Inferences Standard —** "Assessment should promote valid inferences about mathematics learning."
  The assortment of assessment tools provide many sources of evidence. Exercises, Mixed Review, and Chapter Tests are all correlated to the learning objectives and practice activities.

- **The Coherence Standard —** "Assessment should be a coherent process."
  The numerous, varied assessment tools are carefully designed to complement each other. They enable you to easily construct the assessment program that best fits your needs.

Learning Objectives & NCTM Standards Correlation

Lesson	Lesson Objectives	NCTM Standards
1-1	Translate verbal expressions into mathematical expressions and vice versa.	1-5
1-2	Solve problems by looking for a pattern.	1-5
1-3	Use the order of operations to evaluate real number expressions.	1-5, 8
1-4	Display and interpret data on a stem-and-leaf plot.	1-5, 10
1-5	Solve open sentences by performing arithmetic operations.	1-5
1-6	Recognize and use the properties of identity and equality. Determine the multiplicative inverse of a number.	1-5
1-7A	Model the distributive property.	1-5
1-7	Use the distributive property to simplify expressions.	1-5, 8
1-8	Recognize and use the commutative and associative properties when simplifying expressions.	1-5
1-9	Interpret graphs in real-world settings. Sketch graphs for given functions.	1-6, 8
2-1	State the coordinate of a point on a number line. Graph integers on a number line. Add integers by using a number line.	1-5, 14
2-2	Interpret numerical data from a table. Display and interpret statistical data on a line plot.	1-4, 10
2-3A	Use counters to add and subtract integers.	1-5
2-3	Find the absolute value of a number. Add and subtract integers.	1-5
2-4	Compare and order rational numbers. Find a number between two rational numbers.	1-5
2-5	Add and subtract rational numbers. Simplify expressions that contain rational numbers.	1-5
2-6A	Use counters to multiply integers.	1-5
2-6	Multiply rational numbers.	1-5
2-7	Divide rational numbers.	1-5
2-8A	Use base-ten tiles to model square roots.	1-5
2-8	Find square roots. Classify numbers. Graph solutions of inequalities on number lines.	1-5, 14
2-9	Explore problem situations. Translate verbal sentences and problems into equations or formulas and vice versa.	1-5
3-1A	Solve equations by using cups and counters to model variables and integers.	1-5
3-1	Solve equations by using addition and subtraction.	1-5
3-2	Solve equations by using multiplication and division.	1-5
3-3A	Solve equations by using cups and counters to model variables and integers.	1-5
3-3	Solve equations involving more than one operation. Solve problems by working backward.	1-5
3-4	Find the complement and supplement of an angle. Find the measure of the third angle of a triangle given the measures of the other two angles.	1-5, 7

Lesson	Lesson Objectives	NCTM Standards
3-5	Solve equations with the variable on both sides. Solve equations containing grouping symbols.	1-5, 7
3-6	Solve equations and formulas for a specified variable.	1-5, 7
3-7	Find and interpret the mean, median, and mode of a set of data.	1-5, 10
4-1A	Collect data to determine ratios.	1-5
4-1	Solve proportions.	1-5
4-2	Find the unknown measures of the sides of two similar triangles.	1-5, 7
4-3	Use trigonometric ratios to solve right triangles.	1-5, 9
4-4	Solve percent problems. Solve problems involving simple interest.	1-5
4-5	Solve problems involving percent of increase or decrease. Solve problems involving discounts or sales tax.	1-5
4-6	Find the probability of a simple event. Find the odds of a simple event.	1-5, 11
4-7	Solve mixture problems. Solve problems involving uniform motion.	1-5
4-8	Solve problems involving direct and inverse variation.	1-5
5-1	Graph ordered pairs on a coordinate plane. Solve problems by making a table.	1-5, 8
5-2A	Use a graphing calculator to graph relations.	1-5, 8
5-2	Identify the domain, range, and inverse of a relation. Show relations as sets of ordered pairs, tables, mappings, and graphs.	1-5
5-3A	Use a graphing calculator to investigate relations and determine the ranges when the domains are given.	1-5
5-3	Determine the range for a given domain. Graph the solution set for the given domain.	1-5
5-4A	Use a graphing calculator to graph linear relations and functions.	1-5, 8
5-4	Graph linear equations.	1-5, 8
5-5	Determine whether a given relation is a function. Find the value of a function for a given element of the domain.	1-6
5-6	Write equations to represent relations, given some of the solutions for the equation.	1-6
5-7A	Use a graphing calculator to calculate measures of variation.	1-4, 10
5-7	Calculate and interpret the range, quartiles, and interquartile range of sets of data.	1-4, 10
6-1A	Use a geoboard to model a line segment and calculate its slope.	1-5, 8
6-1	Find the slope of a line, given the coordinates of two points on the line.	1-5, 8
6-2	Write linear equations in point-slope form. Write linear equations in standard form.	1-5, 8
6-3	Graph and interpret points on a scatter plot. Draw and write equations for best-fit lines, and make predictions by using those equations. Solve problems by using models.	1-5, 10
6-4	Determine the x- and y-intercepts of linear graphs from their equations. Write equations in slope-intercept form. Write and solve direct variation equations.	1-5
6-5A	Use a graphing calculator to determine whether a group of graphs forms a family.	1-5
6-5	Graph a line given any linear equation.	1-5
6-6	Determine if two lines are parallel or perpendicular by their slopes. Write equations of lines passing through a given point, parallel or perpendicular to the graph of a given equation.	1-5, 8

Lesson	Lesson Objectives	NCTM Standards
6-7	Find the coordinates of the midpoint of a line segment in the coordinate plane.	1-5, 8
7-1	Solve inequalities by using addition and subtraction.	1-5
7-2A	Use cups and counters to solve inequalities.	1-5
7-2	Solve inequalities by using multiplication and division.	1-5
7-3	Solve linear inequalities involving more than one operation. Find the solution set for a linear inequality when replacement values are given for the variables.	1-5
7-4	Solve problems by making a diagram. Solve compound inequalities and graph their solution sets. Solve problems that involve compound inequalities.	1-5
7-5	Find the probability of a compound event.	1-4, 11
7-6	Solve open sentences involving absolute value and graph the solutions.	1-5
7-7	Display and interpret data on box-and-whisker plots.	1-5, 10
7-7B	Use a graphing calculator to compare two sets of data by using a double box-and-whisker plot.	1-5, 10
7-8A	Use the "Shade C" command on a graphing calculator to graph inequalities in two variables.	1-5
7-8	Graph inequalities in the coordinate plane.	1-5
8-1A	Use a graphing calculator to solve systems of equations.	1-5, 8
8-1	Solve systems of equations by graphing. Determine whether a system of equations has one solution, no solutions, or infinitely many solutions by graphing.	1-5, 8
8-2	Solve systems of equations by the substitution method. Organize data to solve problems.	1-5
8-3	Solve systems of equations by the elimination method using addition or subtraction.	1-5
8-4	Solve systems of equations by the elimination method using multiplication and addition.	1-5
8-5	Solve systems of inequalities by graphing.	1-5, 8
9-1	Multiply monomials. Simplify expressions involving powers of monomials. Solve problems by looking for a pattern.	1-5
9-2	Simplify expressions involving quotients of monomials. Simplify expressions containing negative exponents.	1-5
9-3	Express numbers in scientific and standard notation. Find products and quotients of numbers expressed in scientific notation.	1-5
9-4A	Use algebra tiles to model polynomials.	1-5
9-4	Find the degree of a polynomial. Arrange the terms of a polynomial so that the powers of a variable are in ascending or descending order.	1-5
9-5A	Use algebra tiles to add and subtract polynomials.	1-5
9-5	Add and subtract polynomials.	1-5
9-6A	Use algebra tiles to model the product of simple polynomials.	1-5
9-6	Multiply a polynomial by a monomial. Simplify expressions involving polynomials.	1-5
9-7A	Use algebra tiles to find the product of binomials.	1-5
9-7	Use the FOIL method to multiply two binomials. Multiply any two polynomials by using the distributive property.	1-5
9-8	Use patterns to find (a + b)2, (a - b)2, and (a + b)(a - b).	1-5

Lesson	Lesson Objectives	NCTM Standards
10-1	Find the prime factorization of integers. Find the greatest common factors (GCF) for sets of monomials.	1-5
10-2A	Use algebra tiles to factor binomials.	1-5
10-2	Use the GCF and the distributive property to factor polynomials. Use grouping techniques to factor polynomials with four or more terms.	1-5
10-3A	Use algebra tiles to factor simple trinomials.	1-5
10-3	Solve problems by using guess and check. Factor quadratic trinomials.	1-5
10-4	Identify and factor binomials that are the differences of squares.	1-5
10-5	Identify and factor perfect square trinomials.	1-5
10-6	Use the zero product property to solve equations.	1-5
11-1A	Use a graphing calculator to graph a quadratic function and find the coordinates of its vertex.	1-6
11-1	Find the equation of the axis of symmetry and the coordinates of the vertex of a parabola. Graph quadratic functions.	1-6, 13
11-1B	Use a graphing calculator to study the characteristics of families of parabolas.	1-6
11-2	Use estimation to find roots of quadratic equations. Find roots of quadratic equations by graphing.	1-6, 13
11-3	Solve quadratic equations by using the quadratic formula.	1-6
11-4A	Use a graphing calculator to graph exponential functions.	1-6, 13
11-4	Graph exponential functions. Determine if a set of data displays exponential behavior. Solve exponential equations.	1-6, 13
11-5	Solve problems involving growth and decay.	1-6
12-1	Simplify rational expressions. Identify values excluded from the domain of a rational expression.	1-5
12-1B	Use a graphing calculator to check the simplification of rational expressions.	1-6
12-2	Multiply rational expressions.	1-5
12-3	Divide rational expressions.	1-5
12-4	Divide a polynomial by a binomial.	1-5
12-5	Add and subtract rational expressions with like denominators.	1-5
12-6	Add and subtract rational expressions with unlike denominators. Solve problems by making an organized list of possibilities.	1-5
12-7	Simplify mixed expressions and complex fractions.	1-5
12-8	Solve rational equations.	1-5
13-1A	Use a geoboard to find areas by using the Pythagorean theorem.	1-5, 7
13-1	Use the Pythagorean theorem to solve problems.	1-5, 7
13-2	Simplify square roots. Simplify radical expressions.	1-5
13-3A	Use a graphing calculator to simplify and approximate values of expressions containing radicals.	1-5
13-3	Simplify radical expressions involving addition and subtraction.	1-5
13-4	Solve radical equations.	1-5
13-5	Find the distance between two points in the coordinate plane.	1-5, 8
13-6A	Use algebra tiles as a model for completing the square.	1-5
13-6	Solve problems by identifying subgoals. Solve quadratic equations by completing the square.	1-5

Planning Your Algebra 1 Course

40- TO 50-MINUTE CLASS PERIODS

One Year

The chart below gives suggested pacing guides for two options, Standard and Honors, for four 9-week grading periods. The Standard option covers Chapters 1-12 while the Honors option covers Chapters 1-13. *A pacing guide that addresses both of these options is provided in each lesson of the Teacher's Wraparound Edition.*

The total days suggested for each option is about 160 days. This allows for teacher flexibility in planning due to school cancellation or shortened class periods.

Grading Period	Standard		Honors	
	Chapter	Days	Chapter	Days
1	1 2 3	15 15 12	1 2 3	14 15 12
2	4 5 6	15 14 13	4 5 6	8 7 7
3	7 8 9	14 9 15	7 8 9 10-1 to 10-3	13 8 14 5
4	10 11 12	12 11 14	10-4 to end 11 12 13	6 10 13 11

Two Years

The suggested pacing guide for teaching Glencoe's ***Algebra 1*** in two years is based on teaching Chapters 1-7 the first year preceded by a 2-week review of pre-algebra. The second year begins with a 6-week review of the first year and then Chapters 8-13. *The pacing guide in each lesson of the Teacher's Wraparound Edition also addresses teaching Algebra 1 in two years.*

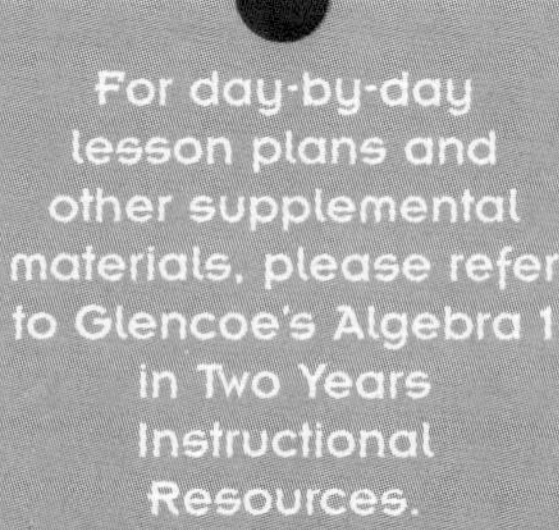
For day-by-day lesson plans and other supplemental materials, please refer to Glencoe's Algebra 1 in Two Years Instructional Resources.

Grading Period	Year One		Year Two	
	Chapter	Days	Chapter	Days
1	Review 1 2-1 to 2-3	10 22 8	Review 8-1 to 8-4	30 9
2	2-4 to end 3	20 19	8-5 to end 9 10-1 to 10-3	5 26 10
3	4 5	20 21	10-4 to end 11 12-1 to 12-3	12 19
4	6 7	18 21	12-4 to end 13	18 22

BLOCK SCHEDULE, 90-MINUTE CLASS PERIODS

The charts below give suggested pacing guides for teaching Glencoe's ***Algebra 1*** using block scheduling. The guides are based on teaching Chapters 1-13. *The pacing guide in each lesson of the Teacher's Wraparound Edition also addresses block scheduling.*

90 Days *(1 Semester or 1 Year)*

The suggested pacing guide below can be used for block scheduling in 1 semester (classes meet every day) or 1 year (classes meet about 85 days).

Chapter	Class Periods
1	7
2	9
3	7
4	7
5	7
6	7
7	7
8	5
9	8
10	6
11	5
12	6
13	7

180 Days *(1 or 2 Years)*

The suggested pacing guide below can be used for block scheduling in 1 year (classes meet every day) or 2 years (classes meet about 85 days per year).

One Year	
Chapter	**Class Periods**
Review	7
1	12
2	15
3-1 to 3-3	4
3-4 to end	6
4	11
5	12
6	10
7	12
8	8
9	14
10-1 to 10-3	5
10-4 to end	7
11	10
12	14
13	12

Two Years	
Chapter	**Class Periods**
Review	5
1	11
2	14
3	10
4	10
5	11
6	9
7	11
Review	15
8	7
9	13
10-1 to 10-4	6
10-5 to end	5
11	10
12	14
13	11

For day-by-day lesson plans, please refer to Glencoe's Algebra 1 Block Scheduling Booklet.

Classroom Vignettes

Glencoe's unique **Classroom Vignettes** are classroom-tested teaching suggestions made by experienced mathematics educators. Each vignette was written by a teacher who uses Glencoe's ***Algebra 1*** or by a Glencoe reviewer, consultant, or author. Glencoe thanks these outstanding mathematics educators for their unique contributions.

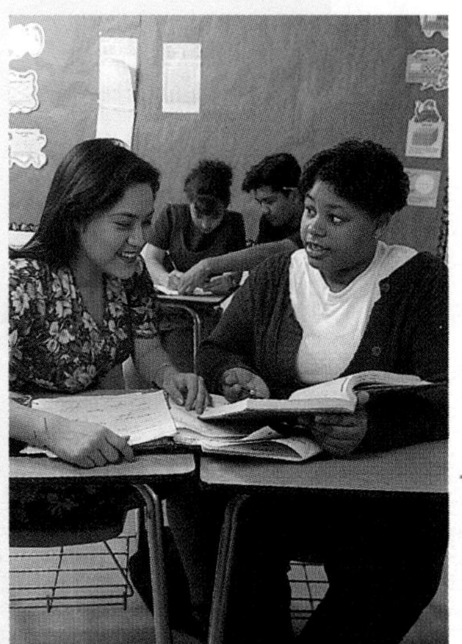

Frank Arcuri (p. 508)
Croton Harmon High School
Croton-on-Hudson, New York

Betsy Barron (p. 74)
Rochester High School
Rochester, Indiana

Cindy Boyd (p. 465)
Abilene High School
Abilene, Texas

Jenny Brawner (p. 87)
Lee-Scott Academy
Auburn, Alabama

William Brinkman (p. 235)
Paynesville High School
Paynesville, Minnesota

John Cline (p. 241)
Greenfield Central High School
Greenfield, Indiana

Emma Lou Crawford (p. 223)
Roanoke High School
Robersonville, North Carolina

Linda Crawford (p. 121)
Brown County Junior High School
Nashville, Indiana

Lara Dillinger (p. 219)
Rock Creek High School
St. George, Kansas

Barbara Ferguson (p. 243)
Enumclaw High School
Enumclaw, Washington

Eva Gates (pp. 143 & 346)
Independent Mathematics Consultant
Pearland, Texas

Joan Gell (p. 637)
Palos Verdes High School
Palos Verdes Estates, California

Linda Glover (p. 537)
Conway Junior High School
Conway, Arkansas

Harvey A. Johnecheck (p. 505)
Gaylord High School
Gaylord, Michigan

Nancy Kinard (p. 25)
Palm Beach Gardens High School
Palm Beach Gardens, Florida

Barbara Lasko (p. 130)
Roosevelt-Wilson Middle School
Clarksburg, West Virginia

Joyce Anne MacKenzie (p. 662)
Camdenton R-III High School
Camdenton, Missouri

Bea Moore-Harris (p. 56)
Educational Specialist, Region IV
Education Service Center
Houston, Texas

Stan Musgrave (p. 254)
Rochester High School
Rochester, Indiana

Barbara Owens (p. 722)
Davidson Middle School
Davidson, North Carolina

Karen C. Painter (p. 585)
Smoky Mountain High School
Sylva, North Carolina

Deborah Phillips (pp. 258 & 272)
St. Vincent-St. Mary High School
Akron, Ohio

Donna Rehling (p. 145)
Elder High School
Cincinnati, Ohio

Carolyn Stanley (p.128)
Tidewater Academy
Wakefield, Virginia

Linda Weissinger (p. 421)
Rochester High School
Rochester, Indiana

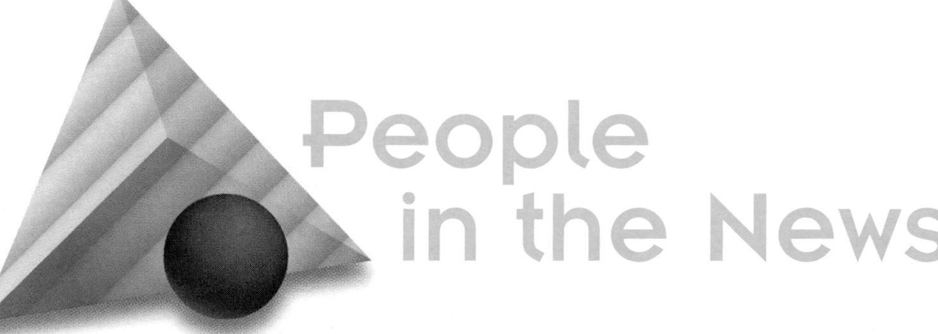

People in the News

Glencoe's unique **People in the News** features young people from across the United States who have appeared "in the news" because of a noteworthy accomplishment. Glencoe congratulates these exceptional citizens.

Internet Connections

Glencoe's unique **Internet Connection** is a source for more information related to the **Chapter Project** and the interests of the person featured in **People in the News.**

The sites referenced in Glencoe's **Internet Connections** are not under the control of Glencoe. Therefore, Glencoe can make no representation concerning the content of these sites. Extreme care has been taken to list only reputable links by using educational and government sites whenever possible. Internet searches have been used that return only sites that contained no content apparently intended for mature audiences. The following list of mathematics Internet sites may prove helpful. Useful search tools include http://webcrawler.com/, http://www.yahoo.com/search.html, and http://www.microsoft.com/search.

Internet Site	Comments
http://www.yahoo.com/Science/Mathematics	An ideal starting point that has links to other math sites.
http://www.tc.cornell.edu/Edu/MathSciGateway	Provides links to math and science resources; for grades 9-12.
http://forum.swarthmore.edu/dr.math/	Students in grades K-12 can ask Dr. Math their own questions.
http://www.enc.org/	This is the Eisenhower clearinghouse for K-2 math and science instructional materials.
http://forum.swarthmore.edu/mathmagic	K-12 telecommunications project in Texas that uses computers to increase problem-solving and communication skills.

Glencoe's Algebra 1 Research Activities

Glencoe's ***Algebra 1: Integration, Applications, and Connections***, as well as the entire Glencoe Mathematics Series, is the product of ongoing, classroom-oriented research that involves students, teachers, curriculum supervisors, administrators, parents, and college-level mathematics educators.

The programs that make up the Glencoe Mathematics Series are currently being used by millions of students and tens of thousands of teachers. The key reason that Glencoe Mathematics programs are so successful in the classroom is the fact that each Glencoe author team is a mix of practicing classroom teachers, curriculum supervisors, and college-level mathematics educators. Glencoe's balanced author teams help ensure that Glencoe Mathematics programs are both practical and progressive.

Prior to publication of a Glencoe program, typical research activities include:

- a review of educational research and recommendations made by groups such as NCTM
- mail surveys of mathematics educators
- discussion groups involving mathematics teachers, department heads, and supervisors
- focus groups involving mathematics educators
- face-to-face interviews with mathematics educators
- telephone surveys of mathematics educators
- in-depth analyses of manuscript by a wide range of reviewers and consultants
- field tests in which students and teachers use pre-publication manuscript in the classroom

Feedback from teachers, curriculum supervisors, and even students who currently use Glencoe mathematics programs is also incorporated as Glencoe plans and publishes new and revised programs. For example, Classroom Vignettes, which are printed in the Teacher's Wraparound Editions, are one result of this feedback.

All of this research and information is used by Glencoe's authors and editors to publish the best instructional resources possible.

WHY IS ALGEBRA IMPORTANT?

Why do I need to study algebra? When am I ever going to have to use algebra in the real world?

Many people, not just algebra students, wonder why mathematics is important. ***Algebra 1*** is designed to answer those questions through **integration, applications,** and **connections.**

Geometry

Did you know that algebra and geometry are closely related? Topics from all branches of mathematics, like geometry and statistics, are integrated throughout the text.

You'll learn how to find the complement and supplement of an angle and how to find the measure of the third angle of a triangle. (Lesson 3–4, pages 162–164)

Nutrition

I can't believe that a double cheeseburger has that much fat! Real-world uses of mathematics are presented.

The number of grams of fat in a double cheeseburger is determined by solving an open sentence. (Lesson 1–5, page 32)

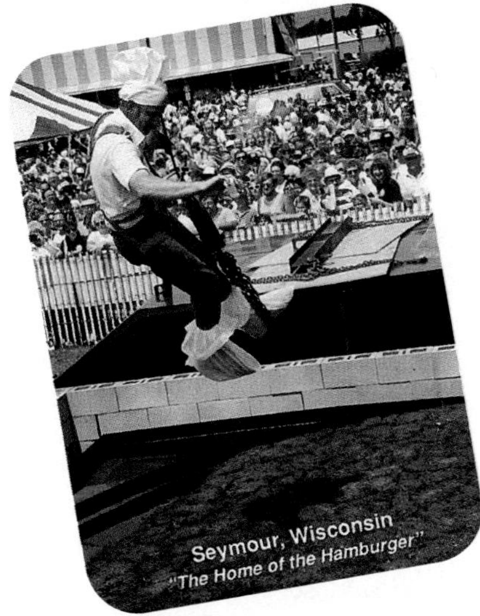

Biology

What does biology have to do with mathematics? Mathematical topics are connected to other subjects that you study.

Punnett squares, which are models that show the possible ways that genes combine, are connected to squaring a binomial. (Lesson 9–8, page 542)

Authors

WILLIAM COLLINS teaches mathematics at James Lick High School in San Jose, California. He has served as the mathematics department chairperson at James Lick and Andrew Hill High Schools. Mr. Collins received his B.A. in mathematics and philosophy from Herbert H. Lehman College in Bronx, New York, and his M.S. in mathematics education from California State University, Hayward. Mr. Collins is a member of the Association of Supervision and Curriculum Development and the National Council of Teachers of Mathematics, and is active in several professional mathematics organizations at the state level. He is also currently serving on the Teacher Advisory Panel of the *Mathematics Teacher.*

"In this era of educational reform and change, it is good to be part of a program that will set the pace for others to follow. This program integrates the ideas of the NCTM Standards with real tools for the classroom, so that algebra teachers and students can expect success every day."

GILBERT CUEVAS is a professor of mathematics education at the University of Miami in Miami, Florida. Dr. Cuevas received his B.A. in mathematics and M.Ed. and Ph.D., both in educational research, from the University of Miami. He also holds a M.A.T. in mathematics from Tulane University. Dr. Cuevas is a member of many mathematics, science, and research associations on the local, state, and national levels and has been an author and editor of several National Council of Teachers of Mathematics (NCTM) publications. He is also a frequent speaker at NCTM conferences, particularly on the topics of equity and mathematics for all students.

ALAN G. FOSTER is a former mathematics teacher and department chairperson at Addison Trail High School in Addison, Illinois. He obtained his B.S. from Illinois State University and his M.A. in mathematics from the University of Illinois. Mr. Foster is a past president of the Illinois Council of Teachers of Mathematics (ICTM) and was a recipient of the ICTM's T.E. Rine Award for Excellence in the Teaching of Mathematics. He also was a recipient of the 1987 Presidential Award for Excellence in the Teaching of Mathematics for Illinois. Mr. Foster was the chairperson of the MATHCOUNTS question writing committee in 1990 and 1991. He frequently speaks and conducts workshops on the topic of cooperative learning.

BERCHIE GORDON is the mathematics/science coordinator for the Northwest Local School District in Cincinnati, Ohio. Dr. Gordon has taught mathematics at every level from junior high school to college. She received her B.S. in mathematics from Emory University in Atlanta, Georgia, her M.A.T. in education from Northwestern University in Evanston, Illinois, and her Ph.D. in curriculum and instruction at the University of Cincinnati. Dr. Gordon has developed and conducted numerous inservice workshops in mathematics and computer applications. She has also served as a consultant for IBM, and has traveled throughout the country making presentations on graphing calculators to teacher groups.

"Using this textbook, you will learn to think mathematically for the 21st century, solve a variety of problems based on real-world applications, and learn the appropriate use of technological devices so you can use them as tools for problem solving."

Beatrice Moore-Harris is an educational specialist at the Region IV Education Service Center in Houston, Texas. She is also the Southwest Regional Director of the Benjamin Banneker Association. Ms. Moore-Harris received her B.A. from Prairie View A&M University in Prairie View, Texas. She has also done graduate work there, at Texas Southern University in Houston, Texas, and at Tarleton State University in Stephenville, Texas. Ms. Moore-Harris is a consultant for the National Council of Teachers of Mathematics (NCTM) and serves on the Editorial Board of the NCTM's *Mathematics Teaching in the Middle School.*

"This program will bring algebra to life by engaging you in motivating, challenging, and worthwhile mathematical tasks that mirror real-life situations. Opportunities to use technology, manipulatives, language, and a variety of other tools are an integral part of this program, which allows all students full access to the algebra curriculum."

James Rath has 30 years of classroom experience in teaching mathematics at every level of the high school curriculum. He is a former mathematics teacher and department chairperson at Darien High School in Darien, Connecticut. Mr. Rath earned his B.A. in philosophy from The Catholic University of America and his M.Ed. and M.A. in mathematics from Boston College. He has also been a Visiting Fellow in the mathematics department at Yale University in New Haven, Connecticut.

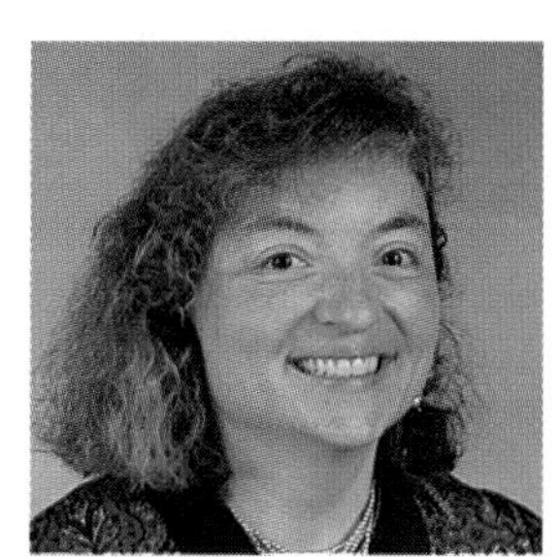

Dora Swart is a mathematics teacher and department chairperson at W.F. West High School in Chehalis, Washington. She received her B.A. in mathematics education at Eastern Washington University in Cheney, Washington, and has done graduate work at Central Washington University in Ellensburg, Washington, and Seattle Pacific University in Seattle, Washington. Ms. Swart is a member of the National Council of Teachers of Mathematics, the Western Washington Mathematics Curriculum Leaders, and the Association of Supervision and Curriculum Development. She has developed and conducted numerous inservices and presentations to teachers in the Pacific Northwest.

"Glencoe's algebra series provides the best opportunity for you to learn algebra well. It explores mathematics through hands-on learning, technology, applications, and connections to the world around us. Mathematics can unlock the door to your success—this series is the key."

Leslie J. Winters is the former secondary mathematics specialist for the Los Angeles Unified School District and is currently supervising student teachers at California State University, Northridge. Mr. Winters received bachelor's degrees in mathematics and secondary education from Pepperdine University and the University of Dayton, and master's degrees from the University of Southern California and Boston College. He is a past president of the California Mathematics Council-Southern Section, and received the 1983 Presidential Award for Excellence in the Teaching of Mathematics and the 1988 George Polya Award for being the Outstanding Mathematics Teacher in the state of California.

Consultants, Writers, and Reviewers

Consultants

Cindy J. Boyd
Mathematics Teacher
Abilene High School
Abilene, Texas

Gail Burrill
National Center for Research/
Mathematical & Science Education
University of Wisconsin
Madison, Wisconsin

David Foster
Glencoe Author and Mathematics
Consultant
Morgan Hill, California

Eva Gates
Independent Mathematics
Consultant
Pearland, Texas

Joan Gell
Mathematics Department Chairman
Palos Verdes High School
Palos Verdes Estates, California

Daniel Marks
Consultant, Real-World Applications
Associate Professor of Mathematics
Auburn University at Montgomery
Montgomery, Alabama

Melissa McClure
Consultant, Tech Prep
Mathematics Consultant
Teaching for Tomorrow
Fort Worth, Texas

Dr. Luis Ortiz-Franco
Consultant, Diversity
Associate Professor of Mathematics
Chapman University
Orange, California

Writers

David Foster
Writer, Investigations
Glencoe Author and Mathematics
Consultant
Morgan Hill, California

Jeri Nichols-Riffle
Writer, Graphing Technology
Assistant Professor
Teacher Education/Mathematics and
Statistics
Wright State University
Dayton, Ohio

Reviewers

Susan J. Barr
Mathematics Department Chairperson
Dublin Coffman High School
Dublin, Ohio

William Biernbaum
Mathematics Teacher
Platteview High School
Springfield, Nebraska

John R. Blickenstaff
Mathematics Teacher
Martinsville High School
Martinsville, Indiana

Wayne Boggs
Mathematics Supervisor
Ephrata High School
Ephrata, Pennsylvania

Donald L. Boyd
Mathematics Teacher
South Charlotte Middle School
Charlotte, North Carolina

Judith B. Brigman
Mathematics Teacher
South Florence High School
Florence, South Carolina

William A. Brinkman
Mathematics Department Chairperson
Paynesville High School
Paynesville, Minnesota

Louis A. Bruno
Mathematics Department Chairperson
Somerset Area Senior High School
Somerset, Pennsylvania

Luajean Nipper Bryan
Mathematics Teacher
McMinn County High School
Athens, Tennessee

Kenneth Burd, Jr.
Mathematics Teacher
Hershey Senior High School
Hershey, Pennsylvania

Todd W. Busse
Mathematics/Science Teacher
Wenatchee High School
Wenatchee, Washington

Esther Corn
Mathematics Department Chairperson
Renton High School
Renton, Washington

Kimberly C. Cox
Mathematics Teacher
Stonewall Jackson High School
Manassas, Virginia

Janis Frantzen
Mathematics Department Chairperson
McCullough High School
The Woodlands, Texas

Vicki Fugleberg
Mathematics Teacher
May-Port CG School
Mayville, North Dakota

Sabine Goetz
Mathematics Teacher
Hewitt-Trussville Junior High
Trussville, Alabama

Dee Dee Hays
Mathematics Teacher
Henry Clay High School
Lexington, Kentucky

Ralph Jacques
Mathematics Department Chairperson
Biddeford High School
Biddeford, Maine

Dixie T. Johnson
Mathematics Department Chairperson
Paul Blazer High School
Ashland, Kentucky

Jane Housman Jones
Mathematics/Technology Teacher
du Pont Manual High School
Louisville, Kentucky

Rebecca A. Luna
Mathematics Teacher
Austin Junior High School
San Juan, Texas

Sandra A. Nagy
Mathematics Resource Teacher
Mesa Schools
Mesa, Arizona

Linda Palmer
Mathematics Teacher
Clark High School
Plano, Texas

Cynthia H. Ray
Mathematics Teacher
Ravenswood High School
Ravenswood, West Virginia

Hazel Russell
Mathematics Teacher
Riverside Middle School
Fort Worth, Texas

Bernita Jones
Mathematics Department Chairperson
Turner High School
Kansas City, Kansas

Nancy W. Kinard
Mathematics Teacher
Palm Beach Gardens High School
Palm Beach Gardens, Florida

Frances J. MacDonald
Area Mathematics Coordinator, K–12
Pompton Lakes School District
Pompton Lakes, New Jersey

Roger O'Brien
Mathematics Supervisor
Polk County Schools
Bartow, Florida

Deborah Patonai Phillips
Mathematics Department Head
St. Vincent-St. Mary High School
Akron, Ohio

Carrol W. Rich
Mathematics Teacher
High Point Central
High Point, North Carolina

Judith Ryder
Mathematics Teacher
South Charlotte Middle School
Charlotte, North Carolina

Cheryl M. Jones
Mathematics Department Coordinator
Winnetonka High School
Kansas City, Missouri

Margaret M. Kirk
Mathematics Teacher
Welborn Middle School
High Point, North Carolina

George R. Murphy
Mathematics Department Chairperson
Pine-Richland High School
Gibsonia, Pennsylvania

Barbara S. Owens
Mathematics Teacher
Davidson Middle School
Davidson, North Carolina

Kristine Powell
Mathematics Teacher
Lehi Junior High School
Lehi, Utah

Mary Richards
Mathematics Teacher
Clyde A. Erwin High School
Asheville, North Carolina

Donna Schmitt
Reviewer, Technology
Mathematics and Computer Science Teacher
Dubuque Senior High School
Dubuque, Iowa

Mabel S. Sechrist
Mathematics Teacher
Mineral Springs Middle School
Winston-Salem, North Carolina

Diane Baish Stamy
Mathematics Chairperson
Big Spring School District
Newville, Pennsylvania

Robert L. Thompson
Mathematics Teacher
Raines High School
Jacksonville, Florida

John J. Turechek
Mathematics Department Chairperson
Bunnell High School
Stratford, Connecticut

Jan S. Wessell
Mathematics Supervisor
New Hanover County Schools
Wilmington, North Carolina

Kathy Zwanzig
Middle School Algebra Resource Teacher
Jefferson County Public Schools
Louisville, Kentucky

Richard K. Sherry
Mathematics Department Chairman
Roosevelt Junior High School
Altoona, Pennsylvania

Carolyn Stanley
Mathematics Department Chairperson
Tidewater Academy
Wakefield, Virginia

Linda K. Tinga
Mathematics Teacher
Laney High School
Wilmington, North Carolina

Tonya Dessert Urbatsch
Mathematics Curriculum Coordinator
Davenport Community Schools
Davenport, Iowa

Ted Wiecek
Mathematics Department Chairperson
Wells Community Academy
Chicago, Illinois

Catherine C. Sperando
Mathematics Teacher
Shades Valley High School
Birmingham, Alabama

Richard P. Strausz
Math/Computer Coordinator
Farmington Schools
Farmington, Michigan

Kimberly Troise
Mathematics Teacher
A. C. Reynolds High School
Asheville, North Carolina

Anthony J. Weisse
Mathematics Teacher
Central High School
LaCrosse, Wisconsin

Rosalyn Zeid
Mathematics Director
Union High School
Union, New Jersey

Table of Contents

CHAPTER 2 Exploring Rational Numbers 70

CHAPTER 3 Solving Linear Equations 140

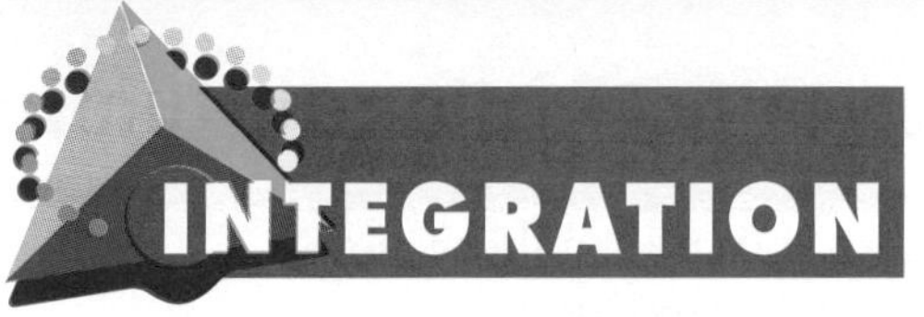

Content Integration

What does geometry have to do with algebra? Believe it or not, you can study most math topics from more than one point of view. Here are some examples.

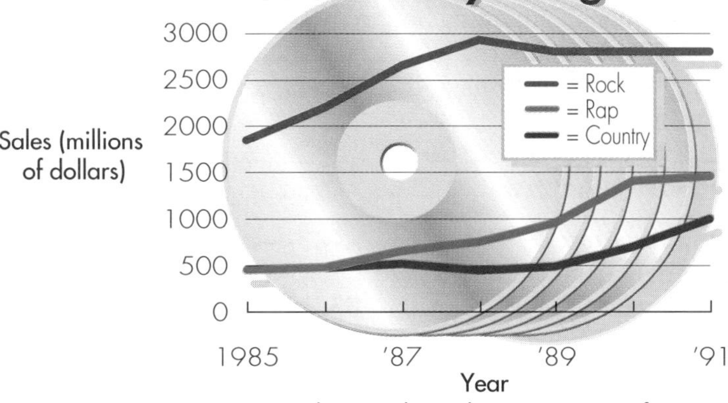

Source: The Recording Industry Association of America

▲ **Problem Solving** You'll interpret graphs and learn how to sketch graphs of real-world situations. (Lesson 1–9, page 62)

◀ **Probability** You'll model the roll of two dice with relations and line plots. (Lesson 5–2, Exercise 45)

LOOK BACK

You can refer to Lesson 1-7 for information on the distributive property.

Source: Lesson 9–6, page 529

Look Back features refer you to skills and concepts that have been taught earlier in the book.

Discrete Mathematics ▶ You'll investigate patterns and sequences and use them to solve a variety of problems. (Lesson 1–2, pages 12–18)

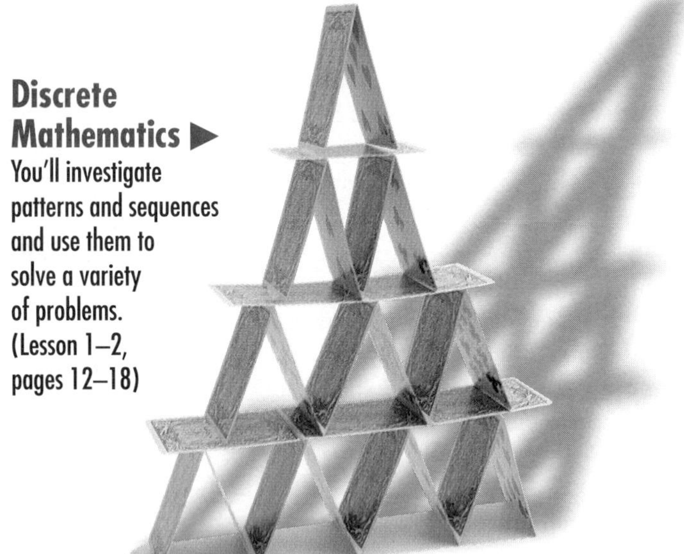

Geometry You'll use your algebra skills to find the missing measures of similar triangles. (Lesson 4–2, pages 201–205)

Statistics You'll learn how to ▶ display data about the speeds of the fastest animals on a line plot. (Lesson 2–2, page 80)

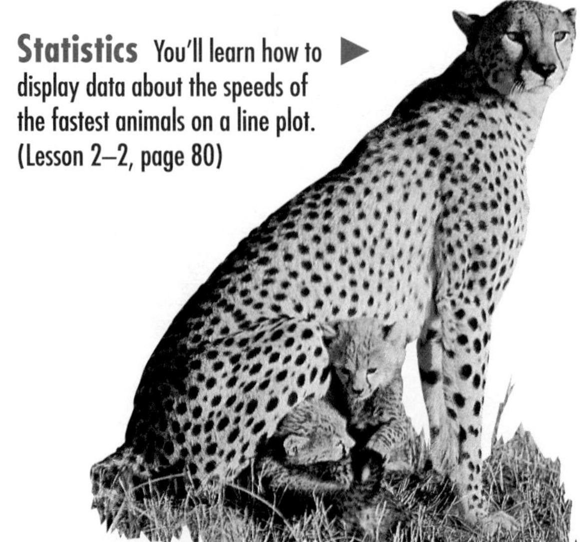

CHAPTER 4 Using Proportional Reasoning 192

CHAPTER 5 Graphing Relations and Functions 252

CHAPTER 6 Analyzing Linear Equations 322

CHAPTER 7 Solving Linear Inequalities 382

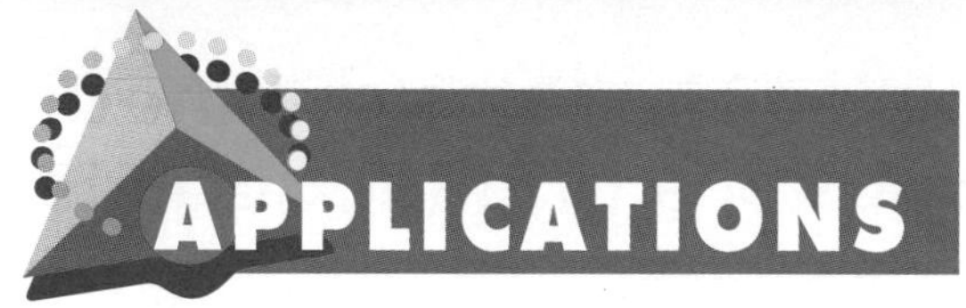

Real-Life Applications

Have you ever wondered if you'll ever actually use math? Every lesson in this book is designed to help show you where and when math is used in the real world. Since you'll explore many interesting topics, we believe you'll discover that math is relevant and exciting. Here are some examples.

Top Five List, FYI, and **Fabulous Firsts** contain interesting facts that enhance the applications.

Selling Prices of Paintings

1. *Portrait du Dr. Gachet* by van Gogh, $75,000,000
2. *Au Moulin de la Galette* by Renoir, $71,000,000
3. *Les Noces de Pierrette* by Picasso, $51,700,000
4. *Irises* by van Gogh, $49,000,000
5. *Yo Picasso* by Picasso, $43,500,000

Source: Lesson 9–4, page 514

Football You'll see how punting a football is related to angle measure. (Lesson 3–4, page 162)

Carpentry You'll study an industrial technology application that involves computing with rational numbers. (Lesson 2–7, Exercise 47)

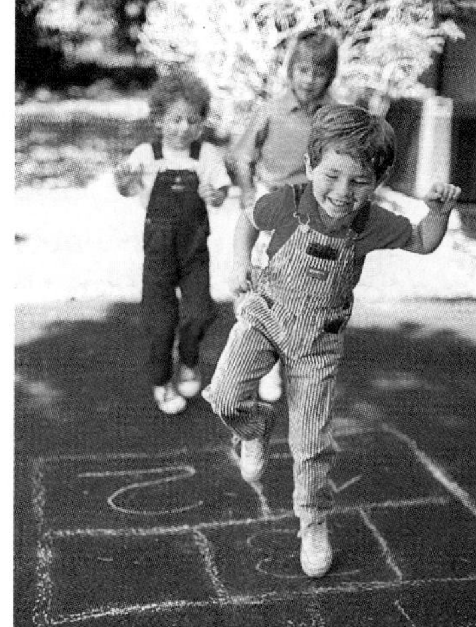

◄ Recreation Games from several world cultures that are similar to hopscotch will illustrate multiplying a polynomial by a monomial. (Lesson 9–6, page 529)

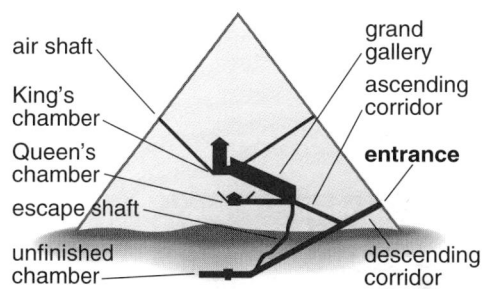

F Y I

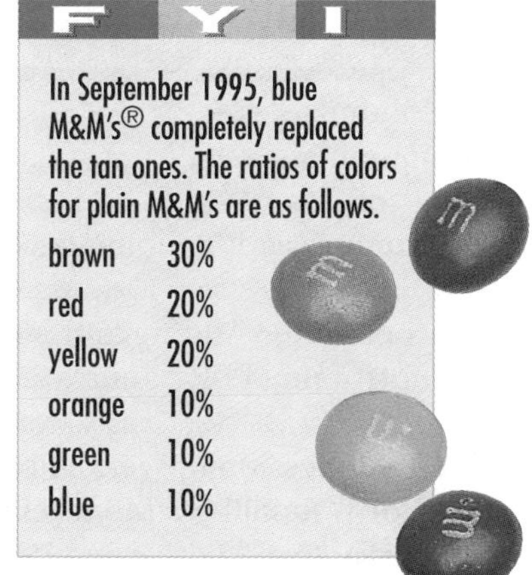

In September 1995, blue M&M's® completely replaced the tan ones. The ratios of colors for plain M&M's are as follows.

brown	30%
red	20%
yellow	20%
orange	10%
green	10%
blue	10%

Source: Lesson 2–2, page 79

air shaft
grand gallery
King's chamber
ascending corridor
Queen's chamber
entrance
escape shaft
unfinished chamber
descending corridor

▲ World Cultures The Great Pyramid of Khufu is the setting for an application involving systems of linear equations. (Lesson 8–3, Exercise 42)

fabulous FIRSTS

Elmer Simms Campbell (1906–1971)

Elmer Simms Campbell was the first African-American cartoonist to work for national publications. He contributed cartoons and other art work to *Esquire, Cosmopolitan,* and *Redbook,* as well as syndicated features in 145 newspapers.

Source: Lesson 8–3, page 469

Space Science You'll discuss the extreme temperatures encountered during a walk in space as you learn about absolute value. (Lesson 2–3, page 85)

Teen Talk Barbies

What do Barbie dolls have to do with mathematics? Actual reprinted articles illustrate how mathematics is a part of our society. (Lesson 3–7, page 183)

CHAPTER 8 Solving Systems of Linear Equations and Inequalities 450

CHAPTER 9 Exploring Polynomials 494

CHAPTER 10 Using Factoring 556

CHAPTER 11 Exploring Quadratic and Exponential Functions 608

Interdisciplinary Connections

Did you realize that mathematics is used in biology? in history? in geography? Yes, it may be hard to believe, but mathematics is frequently connected to other subjects that you are studying.

GLOBAL CONNECTIONS

About 15,000 years ago, an important application of the lever appeared as hunters used an *atlatl*, the Aztec word for spear-thrower. This simple device was a handle that provided extra mechanical advantage by adding length to the hunter's arm. This enabled the hunter to use less motion to get greater results when hunting.

Source: Lesson 7–2, page 392

◀ **Global Connections** features introduce you to a variety of world cultures.

Health You'll write inequalities to model target heart rates and exercise. (Lesson 7–8, Exercise 47)

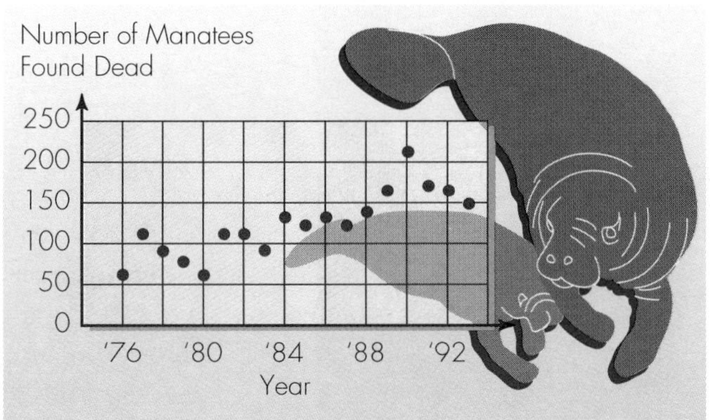

▲ **Biology** You'll study the mortality rates of Florida's manatees, an endangered species. (Lesson 5–2, page 262)

CAREER CHOICES

A **zoologist** studies animals, their origin, life process, behavior, and diseases. They are usually identified by the animal group in which they specialize, for example herpetology (reptiles), ornithology (birds), and ichthyology (fish).

Zoology requires a minimum of a bachelor's degree. Many college positions in this field require a Ph.D. degree.

For more information, see:

Occupational Outlook Handbook, 1994–95, U.S. Department of Labor.

Source: Lesson 5–2, page 264

◀ **Career Choices** features include information on interesting careers.

Geography In 1994, the population of New York was surpassed by Texas. This situation is modeled as a system of linear equations. (Lesson 8–2, page 462)

▲ **Art** A Mondrian painting entitled *Composition with Red, Yellow, and Blue* is used to model polynomials. (Lesson 9–3, page 514)

Math Journal exercises give you the opportunity to assess yourself and write about your understanding of key math concepts. (Lesson 7–1, Exercise 6)

6. Sometimes statements we make can be translated into inequalities. For example, *In some states, you have to be at least 16 years old to have a driver's license* can be expressed as $a \geq 16$, and *Tomás cannot lift more than 72 pounds* can be translated into $w \leq 72$. Following these examples, write three statements that deal with your everyday life. Then translate each into a corresponding inequality.

CHAPTER 12 Exploring Rational Expressions and Equations 658

CHAPTER 13 Exploring Radical Expressions and Equations 710

TECHNOLOGY

Do you know how to use computers and graphing calculators? If you do, you'll have a much better chance of being successful in today's high-tech society and workplace.

GRAPHING CALCULATORS

There are several ways in which graphing calculators are integrated.

- **Introduction to Graphing Calculators** On pages 2–3, you'll get acquainted with the basic features and functions of a graphing calculator.
- **Graphing Technology Lessons** In Lesson 5–2A, you'll learn how to plot points using a graphing calculator.
- **Graphing Calculator Explorations** You'll learn how to use a graphing calculator to find the mean and median in Lesson 3–7.
- **Graphing Calculator Programs** In Lesson 3–5, Exercise 39 includes a graphing calculator program that can be used to solve equations of the form $ax + b = cx + d$.
- **Graphing Calculator Exercises** Many exercises are designed to be solved using a graphing calculator. For example, see Exercises 40–42 in Lesson 7–8.

COMPUTER SOFTWARE

- **Spreadsheets** On page 297 of Lesson 5–6, a spreadsheet is used to help write an equation that models a relation.
- **BASIC Programs** The program on page 8 of Lesson 1–1 is designed to evaluate expressions.
- **Graphing Software** The graphing software Exploration on page 469 in Lesson 8–3 involves graphing and solving systems of linear equations.

Technology Tips, such as this one on page 216 of Lesson 4–4, are designed to help you make more efficient use of technology through practical hints and suggestions.

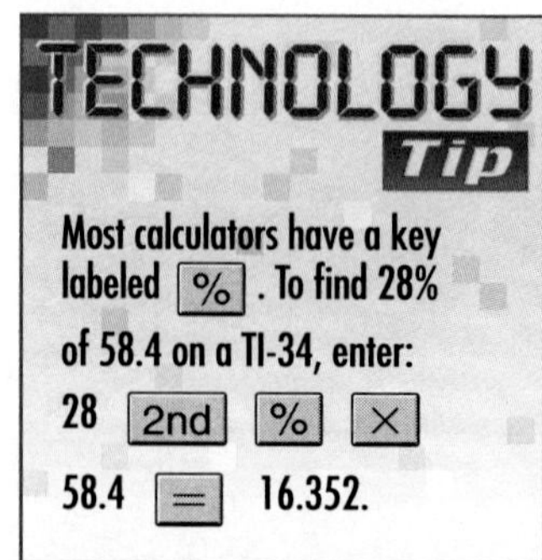

TECHNOLOGY
Tip
Most calculators have a key labeled % . To find 28% of 58.4 on a TI-34, enter:
28 2nd % ×
58.4 = 16.352.

SYMBOLS AND MEASURES

Symbols

$=$	is equal to
$\neq$	is not equal to
$>$	is greater than
$<$	is less than
$\geq$	is greater than or equal to
$\leq$	is less than or equal to
$\approx$	is approximately equal to
$\sim$	is similar to
$\times$ or $\cdot$	times
$\div$	divided by
$-$	negative or minus
$+$	positive or plus
$\pm$	positive or negative
$-a$	opposite or additive inverse of a
$\lvert a \rvert$	absolute value of a
$a \stackrel{?}{=} b$	Does a equal b?
$a : b$	ratio of a to b
$\sqrt{a}$	square root of a
$P(A)$	probability of A
O	origin
π	pi
%	percent
$0.1\overline{2}$	decimal 0.12222...
$°$	degree
$f(x)$	f of x, the value of f at x
(a, b)	ordered pair a, b
$\overline{AB}$	line segment $\overline{AB}$
$\overset{\frown}{AB}$	arc AB
$\overrightarrow{AB}$	ray AB
$\overleftrightarrow{AB}$	line AB
AB	measure of $\overline{AB}$
$\angle$	angle
Δ	triangle
$\cos A$	cosine of A
$\sin A$	sine of A
$\tan A$	tangent of A
( )	parentheses; *also* ordered pairs
[ ]	brackets; *also* matrices
{ }	braces; *also* sets
$\varnothing$	empty set

Measures

mm	millimeter
cm	centimeter
m	meter
km	kilometer
g	gram
kg	kilogram
mL	milliliter
L	liter
in.	inch
ft	foot
yd	yard
mi	mile
in^2 or sq in.	square inch
s	second
min	minute
h	hour

GETTING ACQUAINTED WITH THE GRAPHING CALCULATOR

What is it?
What does it do?
How is it going to help me learn math?

These are just a few of the questions many students ask themselves when they first see a graphing calculator. Some students may think, "Oh, no! Do we *have* to use one?", while others may think, "All right! We get to use these neat calculators!" There are as many thoughts and feelings about graphing calculators as there are students, but one thing is for sure: a graphing calculator *can* help you learn mathematics.

So what is a graphing calculator? Very simply, it is a calculator that draws graphs. This means that it will do all of the things that a "regular" calculator will do, *plus* it will draw graphs of simple or very complex equations. In algebra, this capability is nice to have because the graphs of some complex equations take a lot of time to sketch by hand. Some are even considered impossible to draw by hand. This is where a graphing calculator can be very useful.

But a graphing calculator can do more than just calculate and draw graphs. You can program it, work with matrices, and make statistical graphs and computations, just to name a few things. If you need to generate random numbers, you can do that on the graphing calculator. If you need to find the absolute value of numbers, you can do that, too. It's really a very powerful tool—so powerful that it is often called a pocket computer. But don't let that intimidate you. A graphing calculator can save you time and make doing mathematics easier.

As you may have noticed, graphing calculators have some keys that other calculators do not. The Texas Instruments TI-82 will be used throughout this text. The keys located on the bottom half of the calculator are probably familiar to you as they are the keys found on basic scientific calculators. The keys located just below the screen are the graphing keys. You will also notice the up, down, left, and right arrow keys. These allow you to move the cursor around on the screen and to "trace" graphs that have been plotted. The other keys located on the top half of the calculator access the special features such as statistical and matrix computations.

There are some keystrokes that can save you time when using the graphing calculator. A few of them are listed below.

- Any light blue commands written above the calculator keys are accessed with the [2nd] key, which is also blue. Similarly, any gray characters above the keys are accessed with the [ALPHA] key, which is also gray.
- [2nd] [ENTRY] copies the previous calculation so you can edit and use it again.
- Pressing [ON] while the calculator is graphing stops the calculator from completing the graph.
- [2nd] [QUIT] will return you to the home (or text) screen.
- [2nd] [A-LOCK] locks the [ALPHA] key, which is like pressing "shift lock" or "caps locks" on a typewriter or computer. The result is that all caps will be typed and you do not have to hold the shift key down. (This is handy for programming.)
- [2nd] [OFF] turns the calculator off.

Some commonly used mathematical functions are shown in the table below. As with any scientific calculator, the graphing calculator observes the order of operations.

Mathematical Operation	Examples	Keys	Display
evaluate expressions	Find 2 + 5.	2 [+] 5 [ENTER]	2+5 7
exponents	Find 3^5.	3 [^] 5 [ENTER]	3^5 243
multiplication	Evaluate 3(9.1 + 0.8).	3 [×] [(] 9.1 [+] .8 [)] [ENTER]	3(9.1+.8) 29.7
roots	Find $\sqrt{14}$.	[2nd] [√] 14 [ENTER]	√ 14 3.741657387
opposites	Enter −3.	[(−)] 3	−3

Graphing on the TI–82

Before graphing, we must instruct the calculator how to set up the axes in the coordinate plane. To do this, we define a **viewing window.** The viewing window for a graph is the portion of the coordinate grid that is displayed on the **graphics screen** of the calculator. The viewing window is written as [left, right] by [bottom, top] or [Xmin, Xmax] by [Ymin, Ymax]. A viewing window of [−10, 10] by [−10, 10] is called the **standard viewing window** and is a good viewing window to start with to graph an equation. The standard viewing window can be easily obtained by pressing [ZOOM] 6. Try this. Move the arrow keys around and observe what happens. You are seeing a portion of the coordinate plane that includes the region from -10 to 10 on the x-axis and from -10 to 10 on the y-axis. Move the cursor, and you can see the coordinates of the points for the current position of the cursor.

Any viewing window can be set manually by pressing the WINDOW key. The window screen will appear and display the current settings for your viewing window. First press ENTER. Then, using the arrow and ENTER keys, move the cursor to edit the window settings. Xscl and Yscl refer to the x-scale and y-scale. This is the number of tick marks placed on the x- and y-axes. Xscl=1 means that there will be a tick mark for every unit of one along the x-axis. The standard viewing window would appear as follows.

Xmin = -10
Xmax = 10
Xscl = 1
Ymin = -10
Ymax = 10
Yscl = 1

Graphing equations is as simple as defining a viewing window, entering the equations in the Y= list, and pressing GRAPH. It is often important to view enough of a graph so you can see all of the important characteristics of the graph and understand its behavior. The term **complete graph** refers to a graph that shows all of the important characteristics such as intercepts or maximum and minimum values.

Example: Graph $y = x - 14$ in the standard viewing window.

Enter: Y= X,T,θ − 14 GRAPH

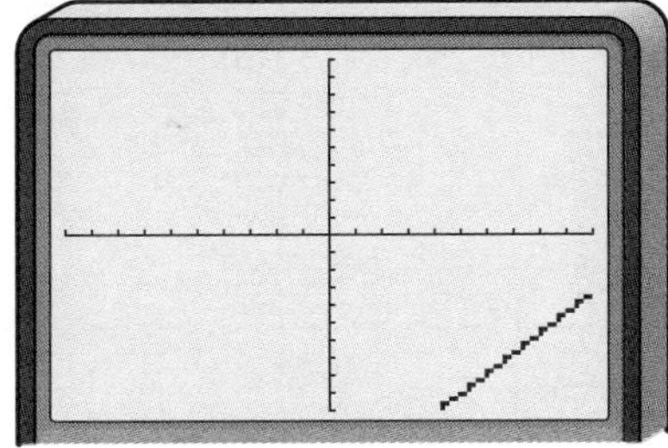

We see only a small portion of the graph. Why?

The graph of $y = x - 14$ is a line that crosses the x-axis at 14 and the y-axis at -14. The important features of the graph are plotted off the screen and so the graph is not complete. A better viewing window for this graph would be $[-20, 20]$ by $[-20, 20]$, which includes both intercepts. This is considered a complete graph.

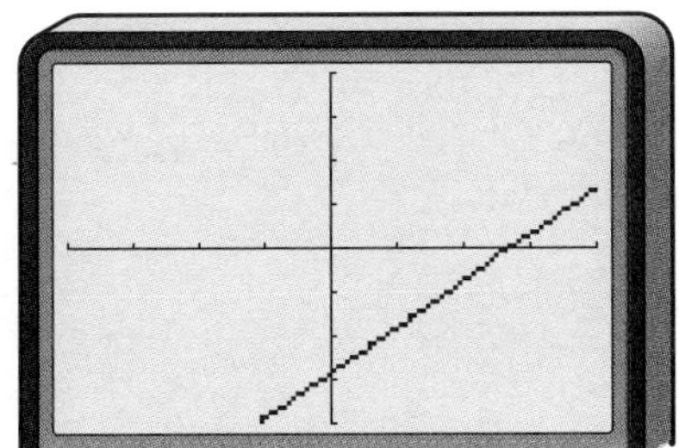

Programming on the TI–82

The TI–82 has programming features that allow us to write and execute a series of commands to perform tasks that may be too complex or cumbersome to perform otherwise. Each program is given a name. Commands begin with a colon (:), followed by an expression or an instruction. Most of the features of the calculator are accessible from program mode.

When you press PRGM, you see three menus: EXEC, EDIT, and NEW. EXEC allows you to execute a stored program by selecting the name of the program from the menu. EDIT allows you to edit or change an existing program and NEW allows you to create a new program. To break during program execution, press ON. The following example illustrates how to create and execute a new program that stores an expression as Y and evaluates the expression for a designated value of X.

1. Enter PRGM ▶ ▶ ENTER to create a new program.
2. Type EVAL ENTER to name the program. (Make sure that the caps lock is on.) You are now in the program editor, which allows you to enter commands. The colon (:) in the first column of the line indicates that this is the beginning of the command line.
3. The first command lines will ask the user to designate a value for x. Enter PRGM ▶ 3 2nd A-LOCK " ENTER THE VALUE FOR X " ALPHA ENTER PRGM ▶ 1 X,T,θ ENTER.
4. The expression to be evaluated for the value of x is $x - 7$. To store the expression as Y, enter X,T,θ − 7 STO▸ ALPHA Y ENTER.
5. Finally, we want to display the value for the expression. Enter PRGM ▶ 3 ALPHA Y ENTER.
6. Now press 2nd QUIT to return to the home screen.
7. To execute the program, press PRGM. Then press the down arrow to locate the program name and press ENTER, or press the number or letter next to the progam name. The program asks for a value for x. You will input any value for which the expression is defined and press ENTER. To immediately re-execute the program, simply press ENTER when Done appears on the screen.

While a graphing calculator cannot do everything, it can make some things easier. To prepare for whatever lies ahead, you should try to learn as much as you can. The future will definitely involve technology, and using a graphing calculator is a good start toward becoming familiar with technology. Who knows? Maybe one day you will be designing the next satellite, building the next skyscraper, or helping students learn mathematics with the aid of a graphing calculator!

CHAPTER

1 Exploring Expressions, Equations, and Functions

PREVIEWING THE CHAPTER

This chapter provides the necessary introduction and practice to prepare students for the successful study of algebra. Students apply the concept of variables as they write and simplify algebraic expressions. Patterns and sequences are introduced to enhance problem-solving skills, and stem-and-leaf plots are integrated into the chapter to provide students with another tool for data representation. Students connect relevant number rules and properties to the organization and structure of algebra. The distributive property is emphasized because of its vital role in performing algebraic functions. The chapter concludes with students using graphs to interpret and organize information.

Lesson (Pages)	Lesson Objectives	NCTM Standards	State/Local Objectives
1-1 (6–11)	Translate verbal expressions into mathematical expressions and vice versa.	1–5	
1-2 (12–18)	Extend sequences.	1–5	
1-3 (19–24)	Use the order of operations to evaluate real number expressions.	1–5, 8	
1-4 (25–31)	Display and interpret data on a stem-and-leaf plot.	1–5, 10	
1-5 (32–36)	Solve open sentences by performing arithmetic operations.	1–5	
1-6 (37–43)	Recognize and use the properties of identity and equality. Determine the multiplicative inverse of a number.	1–5	
1-7A (44)	Model the distributive property.	1–5	
1-7 (45–50)	Use the distributive property to simplify expressions.	1–5, 8	
1-8 (51–55)	Recognize and use the commutative and associative properties to simplify expressions.	1–5	
1-9 (56–62)	Interpret graphs in real-world settings. Sketch graphs for given functions.	1–6, 8	

A complete, 1-page lesson plan is provided for each lesson in the ***Lesson Planning Guide***. Answer keys for each lesson are available in the ***Answer Key Masters.***

ORGANIZING THE CHAPTER

You may want to refer to the **Course Planning Calendar** on page T12 for detailed information on pacing.
PACING: Standard—15 days; **Honors**—14 days; **Block**—7 days; **Two Years**—22 days

LESSON PLANNING CHART

			Blackline Masters											
Lesson (Pages)	**Materials/ Manipulatives**	**Extra Practice (Student Edition)**	**Study Guide**	**Practice**	**Enrichment**	**Assessment and Evaluation**	**Modeling Mathematics**	**Multicultural Activity**	**Tech Prep Applications**	**Graphing Calculator**	**Science and Math Lab Manual**	**Real-World Applications**	**Interactive Mathematics Tools Software**	**Teaching Transparencies**
1-1 (6–11)	graphing calculator	p. 756	p. 1	p. 1	p. 1								1-1	1-1A 1-1B
1-2 (12–18)	string scissors calculator	p. 756	p. 2	p. 2	p. 2									1-2A 1-2B
1-3 (19–24)	graphing calculator	p. 756	p. 3	p. 3	p. 3	p. 16	p. 72		p. 1				1-3	1-3A 1-3B
1-4 (25–31)		p. 757	p. 4	p. 4	p. 4									1-4A 1-4B
1-5 (32–36)		p. 757	p. 5	p. 5	p. 5	pp. 15, 16		p. 1			pp. 1–4	1		1-5A 1-5B
1-6 (37–43)		p. 757	p. 6	p. 6	p. 6									1-6A 1-6B
1-7A (44)	algebra tiles* product mat*						p. 19						1-7A	
1-7 (45–50)	algebra tiles*	p. 758	p. 7	p. 7	p. 7	p. 17		p. 2						1-7A 1-7B
1-8 (51–55)		p. 758	p. 8	p. 8	p. 8									1-8A 1-8B
1-9 (56–62)		p. 758	p. 9	p. 9	p. 9	p. 17			p. 2	p. 1		2		1-9A 1-9B
Study Guide/ Assessment (63–67)						pp. 1–14, 18–20								

*Included in Glencoe's Student Manipulative Kit and Overhead Manipulative Resources.

ORGANIZING THE CHAPTER

OTHER CHAPTER RESOURCES

Student Edition
Chapter Opener, pp. 4–5
Mathematics and Society, p. 11

Teacher's Classroom Resources
Investigations and Projects Masters, pp. 25–28
Algebra and Geometry Overhead Manipulative Resources, p. 1

Technology
Test and Review Software (IBM and Macintosh)
CD-ROM Interactions (Windows and Macintosh)

Professional Publications
Block Scheduling Booklet
Glencoe Mathematics Professional Series

OUTSIDE RESOURCES

Books/Periodicals
A Gebra Named Al, Free Spirit Publishing
Assessment Standards for School Mathematics, NCTM
Curriculum and Evaluation Standards for School Mathematics, NCTM
Everybody Counts: A Report to the Nation on the Future of Mathematics Education, National Academy Press
Mathematics: The Science of Patterns, W. H. Freeman and Co.

Software
Algebraic Patterns, Sunburst/Wings for Learning
Algebra Concepts, Volume 1, Ventura Educational Systems

Videos/CD-ROMs
Getting Acquainted with Algebra/Order of Operations, The Wisconsin Foundation for Vocational, Technical, and Adult Education, Inc.
How to Take the Mystery Out of Algebra, Learning Forum

ASSESSMENT RESOURCES

Student Edition
Math Journal, pp. 28, 34, 41, 53, 59
Mixed Review, pp. 18, 24, 31, 36, 43, 50, 55, 62
Self Test, p. 36
Chapter Highlights, p. 63
Chapter Study Guide and Assessment, pp. 64–66
Alternative Assessment, p. 67
 Portfolio, p. 67

Teacher's Wraparound Edition
5-Minute Check, pp. 6, 12, 19, 25, 32, 37, 45, 51, 56
Check for Understanding, pp. 9, 15, 22, 28, 34, 40, 48, 52, 59
Closing Activity, pp. 11, 17, 23, 30, 36, 43, 50, 55, 61
Cooperative Learning, pp. 27, 34

Assessment and Evaluation Masters
Multiple Choice Tests, Forms 1A (Honors), 1B (Average), 1C (Basic), pp. 1–6
Free Response Tests, Forms 2A (Honors), 2B (Average), 2C (Basic), pp. 7–12
Calculator-Based Test, p. 13
Performance Assessment, p. 14
Mid-Chapter Test, p. 15
Quizzes A–D, pp. 16–17
Standardized Test Practice, p. 18
Cumulative Review, pp. 19–20

ENHANCING THE CHAPTER

Examples of some of the materials for enhancing Chapter 1 are shown below.

Multicultural Activity Masters, pp. 1, 2

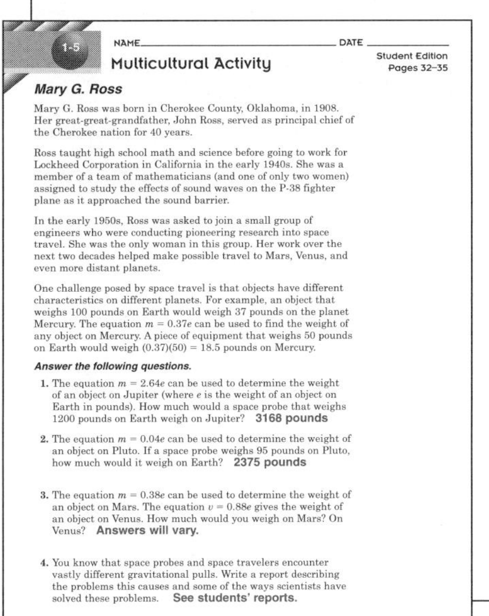

1-5 NAME__________ DATE ________

Multicultural Activity — Student Edition Pages 32–35

Mary G. Ross

Mary G. Ross was born in Cherokee County, Oklahoma, in 1908. Her great-great-grandfather, John Ross, served as principal chief of the Cherokee nation for 40 years.

Ross taught high school math and science before going to work for Lockheed Corporation in California in the early 1940s. She was a member of a team of mathematicians (and one of only two women) assigned to study the effects of sound waves on the P-38 fighter plane as it approached the sound barrier.

In the early 1950s, Ross was asked to join a small group of engineers who were conducting pioneering research into space travel. She was the only woman in this group. Her work over the next two decades helped make possible travel to Mars, Venus, and even more distant planets.

One challenge posed by space travel is that objects have different characteristics on different planets. For example, an object that weighs 100 pounds on Earth would weigh 37 pounds on the planet Mercury. The equation $m = 0.37e$ can be used to find the weight of any object on Mercury. A piece of equipment that weighs 50 pounds on Earth would weigh $(0.37)(50) = 18.5$ pounds on Mercury.

Answer the following questions.

1. The equation $m = 2.64e$ can be used to determine the weight of an object on Jupiter (where e is the weight of an object on Earth in pounds). How much would a space probe that weighs 1200 pounds on Earth weigh on Jupiter? **3168 pounds**
2. The equation $m = 0.04e$ can be used to determine the weight of an object on Pluto. If a space probe weighs 95 pounds on Pluto, how much would it weigh on Earth? **2375 pounds**
3. The equation $m = 0.38e$ can be used to determine the weight of an object on Mars. The equation $v = 0.88e$ gives the weight of an object on Venus. How much would you weigh on Mars? On Venus? **Answers will vary.**
4. You know that space probes and space travelers encounter vastly different gravitational pulls. Write a report describing the problems this causes and some of the ways scientists have solved these problems. **See students' reports.**

Real-World Applications, 1, 2

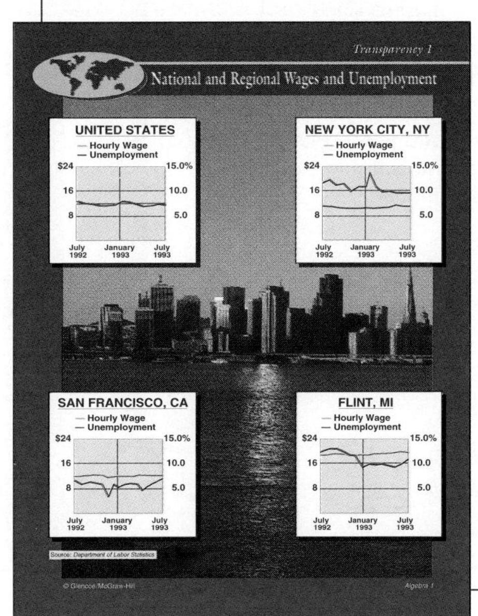

Graphing Calculator Masters, p. 1

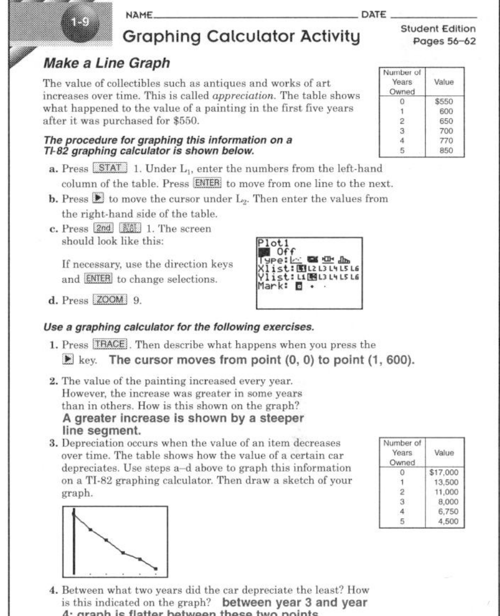

1-9 NAME__________ DATE ________

Graphing Calculator Activity — Student Edition Pages 56–62

Make a Line Graph

The value of collectibles such as antiques and works of art increases over time. This is called *appreciation.* The table shows what happened to the value of a painting in the first five years after it was purchased for $550.

Number of Years Owned	Value
0	$550
1	600
2	650
3	700
4	770
5	850

The procedure for graphing this information on a TI-82 graphing calculator is shown below.

a. Press STAT 1. Under L_1, enter the numbers from the left-hand column of the table. Press ENTER to move from one line to the next.
b. Press ▶ to move the cursor under L_2. Then enter the values from the right-hand side of the table.
c. Press 2nd STAT PLOT 1. The screen should look like this:

```
Plot1
On Off
Type:
Xlist:L1 L2 L3 L4 L5 L6
Ylist:L1 L2 L3 L4 L5 L6
Mark: ▪ + ·
```

If necessary, use the direction keys and ENTER to change selections.

d. Press ZOOM 9.

Use a graphing calculator for the following exercises.

1. Press TRACE. Then describe what happens when you press the ▶ key. **The cursor moves from point (0, 0) to point (1, 600).**
2. The value of the painting increased every year. However, the increase was greater in some years than in others. How is this shown on the graph? **A greater increase is shown by a steeper line segment.**
3. Depreciation occurs when the value of an item decreases over time. The table shows how the value of a certain car depreciates. Use steps a–d above to graph this information on a TI-82 graphing calculator. Then draw a sketch of your graph.

Number of Years Owned	Value
0	$17,000
1	13,500
2	11,000
3	8,000
4	6,750
5	4,500

4. Between what two years did the car depreciate the least? How is this indicated on the graph? **between year 3 and year 4; graph is flatter between these two points**

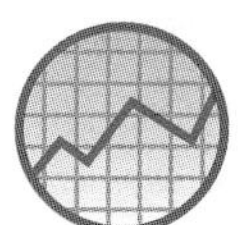

TECH PREP

Tech Prep Applications Masters, pp. 1, 2

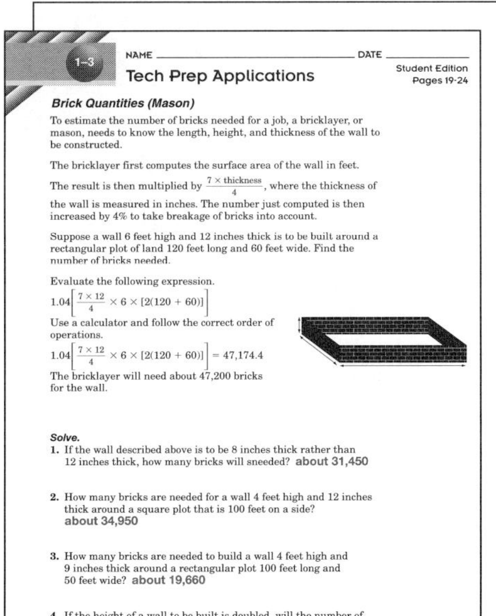

1-3 NAME__________ DATE ________

Tech Prep Applications — Student Edition Pages 19-24

Brick Quantities (Mason)

To estimate the number of bricks needed for a job, a bricklayer, or mason, needs to know the length, height, and thickness of the wall to be constructed.

The bricklayer first computes the surface area of the wall in feet. The result is then multiplied by $\frac{7 \times \text{thickness}}{4}$, where the thickness of the wall is measured in inches. The number just computed is then increased by 4% to take breakage of bricks into account.

Suppose a wall 6 feet high and 12 inches thick is to be built around a rectangular plot of land 120 feet long and 60 feet wide. Find the number of bricks needed.

Evaluate the following expression.

$1.04\left[\frac{7 \times 12}{4} \times 6 \times [2(120 + 60)]\right]$

Use a calculator and follow the correct order of operations.

$1.04\left[\frac{7 \times 12}{4} \times 6 \times [2(120 + 60)]\right] = 47{,}174.4$

The bricklayer will need about 47,200 bricks for the wall.

Solve.

1. If the wall described above is to be 8 inches thick rather than 12 inches thick, how many bricks will sneeded? **about 31,450**
2. How many bricks are needed for a wall 4 feet high and 12 inches thick around a square plot that is 100 feet on a side? **about 34,950**
3. How many bricks are needed to build a wall 4 feet high and 9 inches thick around a rectangular plot 100 feet long and 50 feet wide? **about 19,660**
4. If the height of a wall to be built is doubled, will the number of bricks needed to build it double? **yes**

Science and Math Lab Manual, pp. 1–4

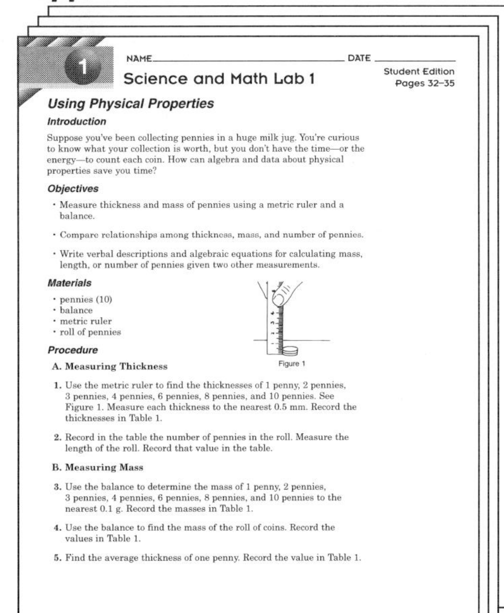

1 NAME__________ DATE ________

Science and Math Lab 1 — Student Edition Pages 32–35

Using Physical Properties

Introduction

Suppose you've been collecting pennies in a huge milk jug. You're curious to know what your collection is worth, but you don't have the time—or the energy—to count each coin. How can algebra and data about physical properties save you time?

Objectives

- Measure thickness and mass of pennies using a metric ruler and a balance.
- Compare relationships among thickness, mass, and number of pennies.
- Write verbal descriptions and algebraic equations for calculating mass, length, or number of pennies given two other measurements.

Materials

- pennies (10)
- balance
- metric ruler
- roll of pennies

Figure 1

Procedure

A. Measuring Thickness

1. Use the metric ruler to find the thicknesses of 1 penny, 2 pennies, 3 pennies, 4 pennies, 6 pennies, 8 pennies, and 10 pennies. See Figure 1. Measure each thickness to the nearest 0.5 mm. Record the thicknesses in Table 1.
2. Record in the table the number of pennies in the roll. Measure the length of the roll. Record that value in the table.

B. Measuring Mass

3. Use the balance to determine the mass of 1 penny, 2 pennies, 3 pennies, 4 pennies, 6 pennies, 8 pennies, and 10 pennies to the nearest 0.1 g. Record the masses in Table 1.
4. Use the balance to find the mass of the roll of coins. Record the values in Table 1.
5. Find the average thickness of one penny. Record the value in Table 1.

Problem of the Week Cards, 1, 2, 3

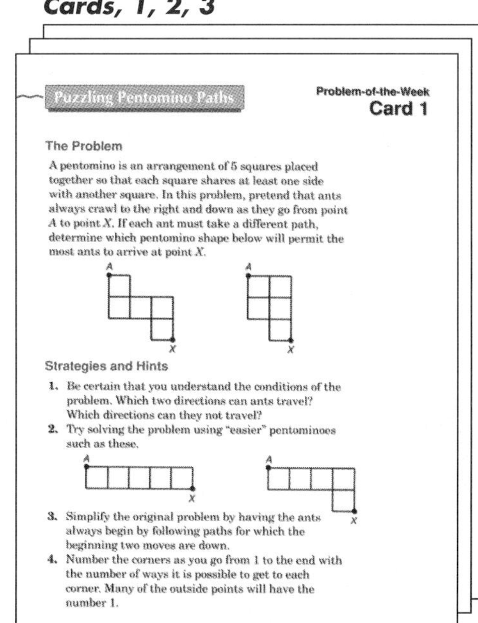

Puzzling Pentomino Paths — Problem-of-the-Week **Card 1**

The Problem

A pentomino is an arrangement of 5 squares placed together so that each square shares at least one side with another square. In this problem, pretend that ants always crawl to the right and down as they go from point A to point X. If each ant must take a different path, determine which pentomino shape below will permit the most ants to arrive at point X.

Strategies and Hints

1. Be certain that you understand the conditions of the problem. Which two directions can ants travel? Which directions can they not travel?
2. Try solving the problem using "easier" pentominoes such as these.
3. Simplify the original problem by having the ants always begin by following paths for which the beginning two moves are down.
4. Number the corners as you go from 1 to the end with the number of ways it is possible to get to each corner. Many of the outside points will have the number 1.

MAKING MATHEMATICS RELEVANT

This two-page introduction to the chapter provides students with an opportunity to explore contemporary topics and their applications to mathematics.

Background Information

Computer Poetry Many think that the sciences and the arts are worlds apart. But even computer scientists have contributed to poetry writing. For years, they have tried to develop computer programs that will write poetry. Such programs are no match for human poets, but they can be fun. A program called MicroWorld Language Arts, manufactured by LOGO Computer Systems, assists students in writing advertising slogans and simple forms of poetry such as haiku.

CHAPTER 1

Exploring Expressions, Equations, and Functions

Objectives

In this chapter, you will:

- translate verbal expressions into mathematical expressions,
- solve problems by looking for a pattern,
- use mathematical properties to evaluate expressions,
- solve open sentences, and
- use and interpret stem-and-leaf plots, tables, graphs, and functions.

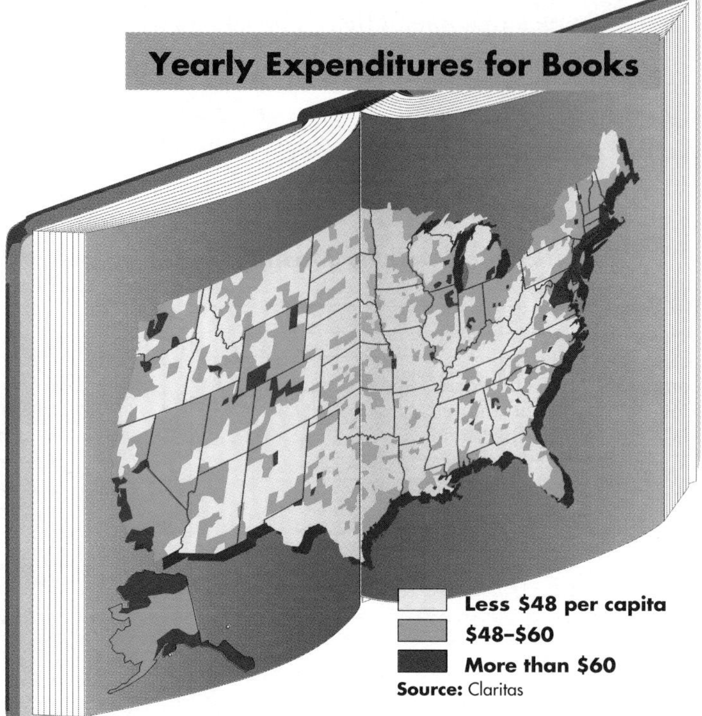

The writing and publishing of books is an important part of our economy. Do you like to express yourself? Do you share your thoughts by writing poems or short stories? Would you like to eventually make a living putting words on paper?

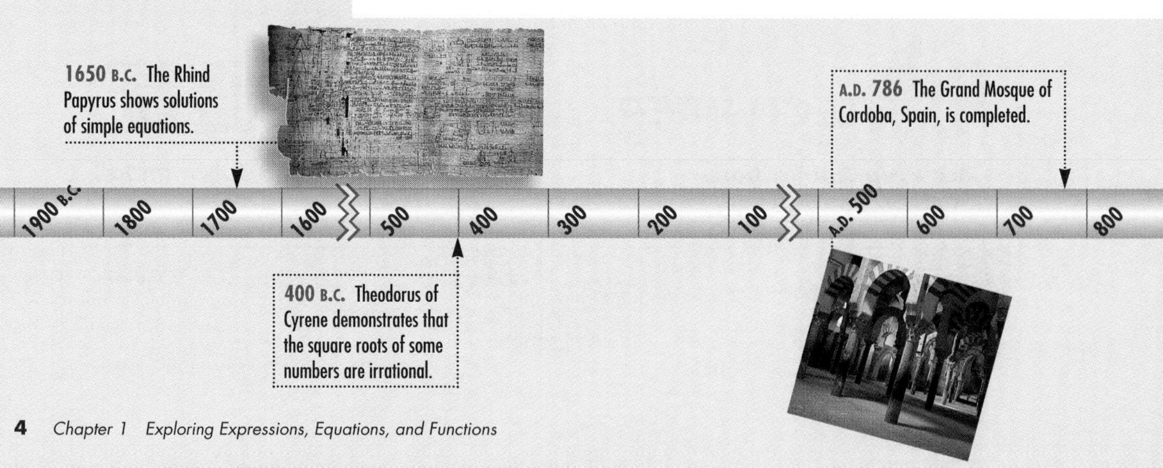

Ask students to conduct research and write a one-page report about both Egyptian and Greek mathematics. Have students compare and contrast the two.

interNET CONNECTION

Resources for writers (career planning, advice from professional writers, writing interactive multimedia, and more) are available on-line!

World Wide Web
http://www.seanet.com/Users/warlock/writers1.html

PEOPLE IN THE NEWS

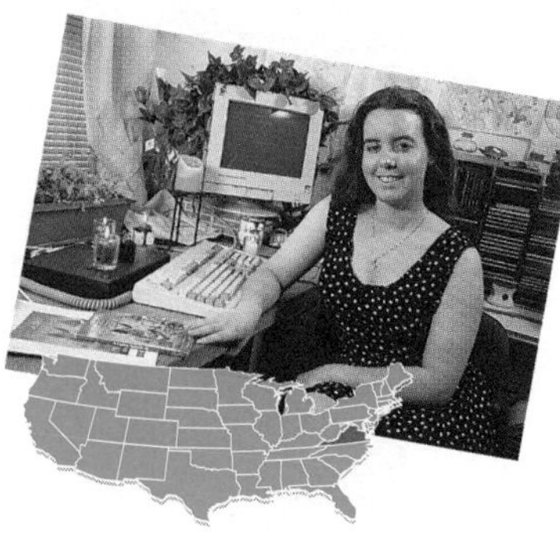

When **Wendy Isdell** was in her eighth grade Algebra 1 class, she got an idea for a short story. Later, she wrote an outline and completed a first draft. The following year, she entered the finished story in the Virginia Young Authors' Contest and won first place in the state competition. Wendy sent her story to Free Spirit Publishing in Minneapolis, and her book *A Gebra Named Al* was published during her senior year in high school.

Wendy's Algebra 1 and 2, geometry, trigonometry, chemistry, Earth science, and physical science classes have provided background for the information in her book. She hopes to make a living as a professional writer.

Chapter Project

Choose a poet and read 10 of his or her poems.

- Describe some patterns that you see in these poems.
- Count the number of words in each poem. Make a stem-and-leaf plot of the number of words in the poems. Does the poet seem to write poems with about the same number of words? Are one or more of the poems exceptionally long or short for this poet?
- Make a graph that models one of the poems. Write a paragraph that describes the relationship between the graph and the poem.

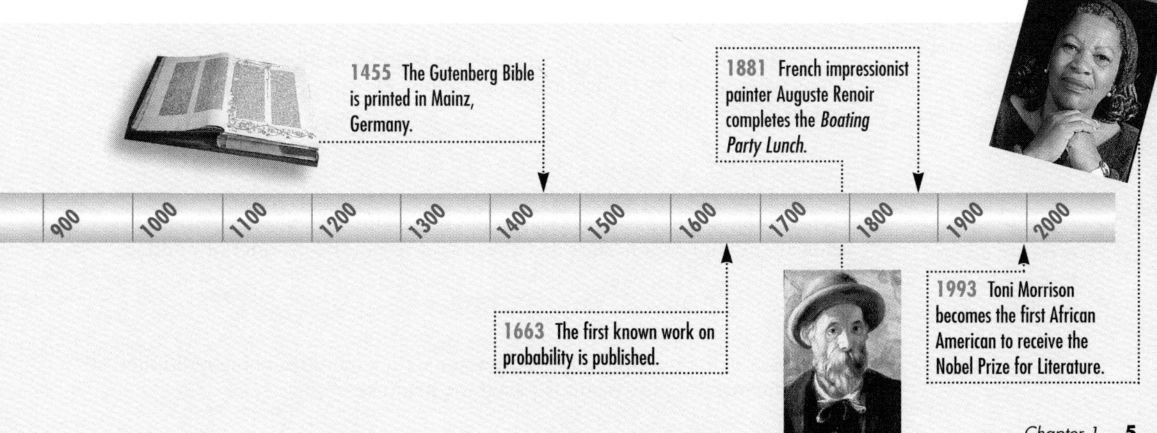

1455 The Gutenberg Bible is printed in Mainz, Germany.
1663 The first known work on probability is published.
1881 French impressionist painter Auguste Renoir completes the *Boating Party Lunch.*
1993 Toni Morrison becomes the first African American to receive the Nobel Prize for Literature.
900
1000
1100
1200
1300
1400
1500
1600
1700
1800
1900
2000

PEOPLE IN THE NEWS

Mathematical ideas offer a wealth of ideas for creative fiction. Working imaginatively with the concept of space and time has led to many science fiction stories. One is the book, *Flatland,* by A. Square (real name: Edwin A. Abbott) in 1884. In this story, the characters are points, lines, and two-dimensional figures like squares, triangles, and so on. Abbott illustrates the concept of dimension through this tale. It also contains biting social commentary about the class structure of the time.

Chapter Project

Cooperative Learning You may choose to have students pair off or arrange them in cooperative groups to create the graph. Students need to set up the horizontal and vertical axes. Since there are decreases as well as increases, the vertical axis must extend below the horizontal axis. One student can construct the axis system, being careful to allow for an appropriate range of values. The other student(s) can plot the data on the graph.

Alternative Chapter Projects

Two other chapter projects are included in the *Investigations and Projects Masters.* In Chapter 1 Project A, pp. 25–26, students extend the topic in the chapter opener. In Chapter 1 Project B, pp. 27–28, students analyze sports statistics.

Investigations and Projects Masters, p. 25

NAME______________ DATE__________

1 Chapter 1 Project A — Student Edition Pages 6–62

The "Write" Way

1. Conduct a survey in which you ask at least 20 working people to rank the importance of good writing skills from 1 to 3, with 1 being "very important for my job," 2 being "somewhat important for my job," and 3 being "not important for my job."
2. Using the data you gathered in step 1, calculate the percentage of responses in each category.
3. Go to the library and find a book that contains descriptions of various jobs and the skills that are necessary to do them. Make a list of jobs that require good writing skills.
4. Research the kind of writing that is involved in each career on your list. You might want to interview people who do a lot of writing on the job. Find out about the purpose of the writing each person does. Is it meant to inform? To entertain? To persuade? You might also want to ask about the audience for each piece of writing.
5. Make a poster or brochure that helps younger students understand the importance of having strong writing skills.

1-1 LESSON NOTES

NCTM Standards: 1–5

Instructional Resources

- Study Guide Master 1-1
- Practice Master 1-1
- Enrichment Master 1-1

Transparency 1-1A contains the 5-Minute Check for this lesson; **Transparency 1-1B** contains a teaching aid for this lesson.

Recommended Pacing	
Standard Pacing	Days 1 & 2 of 15
Honors Pacing	Day 1 of 14
Block Scheduling*	Day 1 of 7
Alg. 1 in Two Years*	Days 1 & 2 of 22

*For more information on pacing and possible lesson plans, refer to the *Block Scheduling Booklet* and *Algebra 1 in Two Years.*

1 FOCUS

5-Minute Check

Write an arithmetic expression that represents each situation. Then find its value.

1. the sum of 18, 23, and 51 **$18 + 23 + 51$; 92**
2. the quotient of 2,910,000 and 550,000 **$2{,}910{,}000 \div 550{,}000$; 5.291**
3. the square of 9 **9^2; 81**
4. the product of 4, 5, and 6 **$4 \times 5 \times 6$; 120**
5. 10 decreased by the cube of 2 **$10 - 2^3$; 2**

Motivating the Lesson

Questioning Ask students the following questions concerning the protection of the skin from the sun.

1. Do you use sunscreen when you go to the beach or pool?
2. If you use sunscreen, do you know its SPF number?
3. Do you know how long you can be in the sun without burning?

1-1 Variables and Expressions

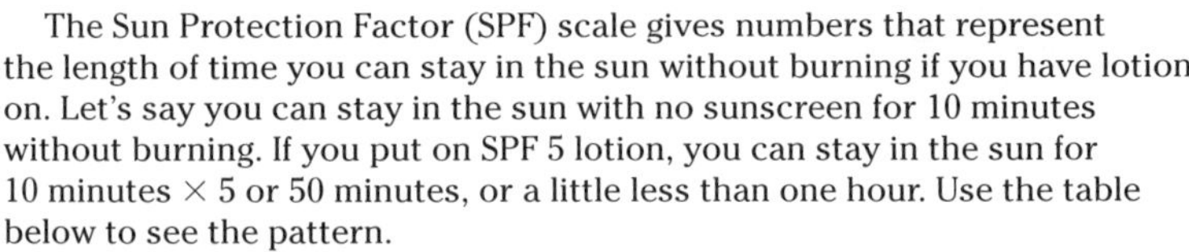

APPLICATION
Health

What YOU'LL LEARN
- To translate verbal expressions into mathematical expressions and vice versa.

Why IT'S IMPORTANT
You can use expressions to solve problems involving health, recycling, and geometry.

Many people enjoy going to the beach in the summer. Unfortunately, the ultraviolet radiation in direct sunlight causes skin cancer. In recent years, the number of skin cancer cases has increased. Many sunbathers use sunscreen lotions to protect themselves in the sun.

The Sun Protection Factor (SPF) scale gives numbers that represent the length of time you can stay in the sun without burning if you have lotion on. Let's say you can stay in the sun with no sunscreen for 10 minutes without burning. If you put on SPF 5 lotion, you can stay in the sun for 10 minutes $\times$ 5 or 50 minutes, or a little less than one hour. Use the table below to see the pattern.

10×5 and 10×15 are numerical expressions.

No Sunscreen	With Sunscreen	
Minutes in Sun Without Burning	SPF Number	Minutes in Sun Without Burning
10	5	10×5
10	15	10×15
5	4	5×4
15	8	15×8
m	s	$m \times s$

CAREER CHOICES

A **registered nurse (RN)** cares for the injured and sick, administers medications, and provides for the physical and emotional needs of his or her patients.

Nursing requires graduation from an accredited nursing school and passing a national licensing examination. About 20% of the 3.8 million new jobs in health care will be for RNs.

For more information, contact:

National League for Nursing
250 Hudson St.
New York, NY 10014.

The letters m and s are called **variables,** and $m \times s$ is an **algebraic expression.** In algebra, variables are symbols that are used to represent unspecified numbers. Any letter may be used as a variable. We selected m because it is the first letter of the word "minutes" and s because it is the first letter of "SPF".

An algebraic expression consists of one or more numbers and variables along with one or more arithmetic operations. Here are some other examples of algebraic expressions.

$$x - 2 \qquad \frac{a}{b} + 3 \qquad t \times 2s \qquad 7mn \div 3k$$

The symbol "×" as shown in $x \times y$ is often avoided because it may be confused with the variable "x."

In algebraic expressions, a raised dot or parentheses are often used to indicate multiplication. Here are ways to represent the product of x and y.

$$xy \qquad x \cdot y \qquad x(y) \qquad (x)(y) \qquad (x)y \qquad x \times y$$

In each of the multiplication expressions, the quantities being multiplied are called **factors,** and the result is called the **product.**

CAREER CHOICES

Encourage students to discuss the possible uses of mathematics in nursing and other related health care fields. For example, standard dosages are given for a certain body weight.

It is often necessary to translate verbal expressions into algebraic expressions.

Verbal Expression	Algebraic Expression
7 less than the product of 3 and a number x	$3x - 7$
the product of 7 and s divided by the product of 8 and y	$7s \div 8y$
four years younger than Sarah (S = Sarah's age)	$S - 4$
half as big as last night's crowd (c = size of last night's crowd)	$\frac{c}{2}$

Example  **Write an algebraic expression for each verbal expression.**

a. three times a number x subtracted from 24

$24 - 3x$

b. 5 greater than half of a number t

$\frac{t}{2} + 5$ or $\frac{1}{2}t + 5$

Another important skill is translating algebraic expressions into verbal expressions.

Example 2 Write a verbal expression for each algebraic expression.

a. $(3 + b) \div y$

the sum of 3 and b divided by y

b. $5y + 10x$

the product of 5 and y plus the product of 10 and x

An expression like x^n is called a **power.** The variable x is called the **base,** and n is called the **exponent.** The exponent indicates the number of times the base is used as a factor.

x^2 means $x \cdot x$. 12^4 means $12 \cdot 12 \cdot 12 \cdot 12$.

Symbols	Words	Meaning
5^1	5 to the first power	5
5^2	5 to the second power or 5 squared	$5 \cdot 5$
5^3	5 to the third power or 5 cubed	$5 \cdot 5 \cdot 5$
5^4	5 to the fourth power	$5 \cdot 5 \cdot 5 \cdot 5$
$3a^5$	three times a to the fifth power	$3 \cdot a \cdot a \cdot a \cdot a \cdot a$
x^n	x to the nth power *An expression of the form x^n is read as "x to the nth power."*	$\underbrace{x \cdot x \cdot x \cdot \ldots \cdot x}_{n \text{ factors}}$

2 TEACH

Teaching Tip Letters in algebra are called *variables* because their values can *vary.*

Teaching Tip Note that for a product of a number and a variable, the number is written first. For example, the product of x and 5 is $5x$, not $x5$.

In-Class Examples

For Example 1
Write an algebraic expression for each verbal expression.

a. m increased by 5 **$m + 5$**
b. the difference of x and 9 **$x - 9$**
c. 7 times the product of x and t **$7xt$**

For Example 2
Write two different verbal expressions for each algebraic expression.

a. $9t$ **9 times t, product of 9 and t**
b. $8 + a$ **8 plus a, sum of 8 and a**
c. $7 - 3y$ **difference of 7 and 3 times y, 7 minus the product of 3 and y**

GLENCOE Technology

 **Interactive Mathematics Tools Software**

This multimedia software provides an interactive lesson for finding the relationship between the number of cricket chirps per minute and the degrees Fahrenheit. A **Computer Journal** gives students an opportunity to write about what they have learned.

For Windows & Macintosh

Alternative Teaching Strategies

Student Diversity Review lesson vocabulary with students. If necessary, write a variety of expressions on the board and have students identify the variables, factors, products, bases, and exponents within the expressions.

In-Class Examples

For Example 3
Write a power that represents the number of small squares in the large square. 5^2

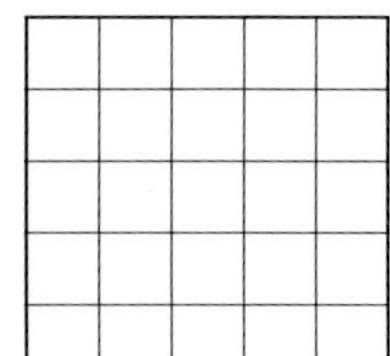

For Example 4
Evaluate each expression.

a. 3^5 **243** **b.** 5^3 **125**

Teaching Tip Students may get a different result in Example 4, such as 64 (4^3) because they entered 4 and then 3. Emphasize the need for making a mental estimate when evaluating expressions using a calculator. This helps verify that keys were used in the proper order.

EXPLORATION

Students should note that a numeric variable in BASIC is represented by a capital letter or a capital letter followed by a numeral. Also, point out that the BASIC multiplication symbol, *, may never be omitted, which is why the algebraic expression *ab* is written as *A * B* in BASIC.

Example 3 **Write a power that represents the number of smallest squares in the large square.**

Geometry

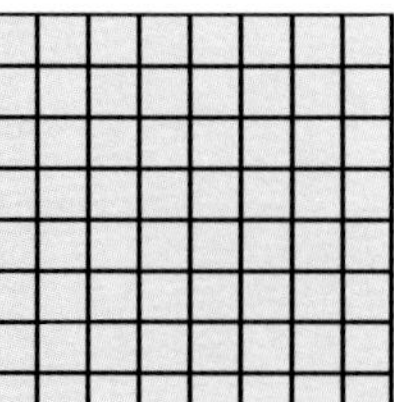

There are 8 squares on each side.

The total number of squares is $8^2 = 8 \cdot 8$ or 64.

You can use the x^2 key on a calculator to square a number.

Enter: 8 [x^2] *64*

To **evaluate** an expression means to find its value.

Example 4 **Evaluate 3^4.**

$$3^4 = 3 \cdot 3 \cdot 3 \cdot 3$$
$$= 81$$

You can use the [y^x] key on a calculator to raise a number to a power.

Enter: 3 [y^x] 4 [=] *81*

EXPLORATION PROGRAMMING

BASIC is a computer language. The symbols used in BASIC are similar to those used in algebra.

+ means add.
– means subtract.
* means multiply.
/ means divide.
↑ means exponent.

Numerical variables in BASIC are represented by capital letters. The program at the right can be used to add, subtract, multiply, divide, and find powers.

```
10 INPUT A,B
20 PRINT "A + B ="; A + B
30 PRINT "A - B ="; A - B
40 PRINT "A * B ="; A * B
50 PRINT "A / B ="; A / B
60 PRINT "A ↑ B ="; A ↑ B
```

Your Turn

a. Let $a = 10$ and $b = 3$. Use the program to find $10 + 3$, $10 - 3$, 10×3, $10 \div 3$, and 10^3. **13; 7; 30; 3.333; 1000**

b. Let $a = 9$ and $b = 3$. Use the program to find $9 + 3$, $9 - 3$, 9×3, $9 \div 3$, and 9^3. **12; 6; 27; 3; 729**

c. Let $a = 5.2$ and $b = 2$. Use the program to find $5.2 + 2$, $5.2 - 2$, 5.2×2, $5.2 \div 2$, and 5.2^2. **7.2; 3.2; 10.4; 2.6; 27.04**

Alternative Learning Styles

Auditory Read the following verbal expressions to students and instruct them to write the algebraic expression.

- three-fourths times the cube of a number $\frac{3}{4}x^3$ or $\frac{3x^3}{4}$
- two times the sum of *x* and *y*, decreased by three times *b* $2(x + y) - 3b$
- one-half the sum of 23 and *s* $\frac{1}{2}(23 + s)$ or $\frac{23 + s}{2}$

CHECK FOR UNDERSTANDING

Communicating Mathematics

Study the lesson. Then complete the following.

1. **Describe** how you would compute the length of time you could sunbathe using an SPF 20 sunscreen lotion if you can stay in the sun for 15 minutes without burning. **Multiply 15 times 20.**
2. Can you find the volume of a cube if you only know the length of a side, s? How? **Yes; evaluate s^3 where s is the measure of a side.**
3. **Describe** what x^7 means. **x used as a factor 7 times**
4. **Explain** the difference between numerical expressions and algebraic expressions. **Algebraic expressions include variables.**
5. **You Decide** Darcy and Tonya are studying for an exam. Darcy says, "The square of a number is always greater than the number." Tonya disagrees. Who is correct? Explain your answer.
Tonya; Sample explanation: $0.5^2 = 0.25$ and $0.5 > 0.25$.

Guided Practice

Write an algebraic expression for each verbal expression.

6. the product of the fourth power of a and the second power of b **$a^4 \cdot b^2$**
7. six less than three times the square of y **$3y^2 - 6$**

Write a verbal expression for each algebraic expression.

8. 7^5 **7 to the fifth power**
9. $3x^2 + 4$ **3 times x squared increased by 4**

Write each expression as an expression with exponents.

10. $4 \cdot 4 \cdot 4$ **4^3**
11. $a \cdot a \cdot a \cdot a \cdot a \cdot a \cdot a$ **a^7**

Evaluate each expression.

12. 6^2 **36**
13. 2^5 **32**
14. **Geometry** The area of a circle can be found by multiplying the number π times the square of the radius. If the radius of the circle is r, write an expression that represents the area of the circle. **πr^2**

EXERCISES

Practice

Write an algebraic expression for each verbal expression.

15. the sum of k and 20 **$k + 20$**
16. the product of 16 and p **$16p$**
17. a to the seventh power **a^7**
18. 49 increased by twice a number **$49 + 2x$**

19. two thirds of the square of a number **$\frac{2x^2}{3}$ or $\frac{2}{3}x^2$**
20. the product of 5 and m plus half of n

20. $5m + \frac{n}{2}$ or $5m + \frac{1}{2}n$

21. 8 pounds heavier than his brother (b = brother's weight) **$b + 8$**
22. 3 times as many wins as last season (w = last season's wins) **$3w$**

Write a verbal expression for each algebraic expression.

23. $4m^5$ **4 times m to the fifth power**
24. $\frac{x^2}{2}$ **x squared divided by 2**
25. $c^2 + 23$ **c squared plus 23**
26. $\frac{1}{2}n^3$ **one half the cube of n**
27. $2(4)(5)^2$
28. $3x^2 - 2x$

27. 2 times 4 times 5 squared

28. 3 times x squared minus 2 times x

Reteaching

Translating Expressions Have students answer each question with an algebraic expression. Let d represent Dawn's age now.

1. How old will Dawn be in two years? $d + 2$
2. How old is Ellen now if she is five years younger than Dawn? $d - 5$
3. How old is Frank now if he is three times as old as Dawn? $3d$

3 PRACTICE/APPLY

Check for Understanding

Exercises 1–14 are designed to help you assess your students' understanding through reading, writing, speaking, and modeling. You should work through Exercises 1–5 with your students and then monitor their work on Exercises 6–14.

Error Analysis

Students often translate expressions involving subtraction in the wrong order. Point out that "the difference of 5 and t" and "5 decreased by t" are translated as $5 - t$ while "5 less than t" and "5 subtracted from t" are translated as $t - 5$.

Assignment Guide

Core: 15–49 odd
Enriched: 16–46 even, 47–50

For **Extra Practice,** see p. 756.

The red A, B, and C flags, printed only in the Teacher's Wraparound Edition, indicate the level of difficulty of the exercises.

Study Guide Masters, p. 1

NAME ______________ DATE ______________

Study Guide — Student Edition Pages 6–11

Variables and Expressions

Any letter used to represent an unspecified number is called a variable. You can use variables to translate verbal expressions into algebraic expressions.

Words	Symbols
4 more than a number	$x + 4$
a number decreased by 8	$b - 8$
the product of 5 and a number	$5c$
a number divided by 8	$h \div 8$ or $\frac{h}{8}$
a number squared	y^2

The algebraic expression x^n represents a product in which each factor is the same. The small raised n is the exponent and it tells how many times the base, x, is used as a factor.

Example: Evaluate 3^4.

$3^4 = 3 \cdot 3 \cdot 3 \cdot 3$
$= 81$

Answers may vary. Sample answers are given.

Write a verbal expression for each algebraic expression.

1. $w - 1$ **one less than w**
2. $\frac{1}{3}a^3$ **one third the cube of a**
3. $81 + 2x$ **eighty-one increased by twice x**

Write an algebraic expression for each verbal expression.

4. a number decreased by 5 **$n - 5$**
5. four times a number **$4n$**
6. 8 less than a number **$n - 8$**
7. a number divided by 6 **$n \div 6$**
8. a number multiplied by 37 **$37n$**
9. the sum of a number and 9 **$n + 9$**
10. 3 less than 5 times a number **$5n - 3$**
11. twice the sum of 15 and a number **$2(15 + n)$**
12. 7 more than the product of 6 and a number **$6n + 7$**
13. 30 increased by 3 times the square of a number **$30 + 3n^2$**

Write each expression as an expression with exponents.

14. $7 \cdot 7 \cdot 7$ **7^3**
15. $3 \cdot p \cdot p$ **$3p^2$**
16. $9(b)(b)(b)(b)(b)$ **$9b^5$**

Evaluate each expression.

17. 2^3 **8**
18. 10^5 **100,000**
19. 4^4 **256**

Write each expression as an expression with exponents.

29. $8 \cdot 8$ **8^2**
30. $10 \cdot 10 \cdot 10 \cdot 10 \cdot 10$ **10^5**
31. $4 \cdot 4 \cdot 4 \cdot 4 \cdot 4 \cdot 4 \cdot 4$ **4^7**
32. $t \cdot t \cdot t$ **t^3**
33. $z \cdot z \cdot z \cdot z \cdot z$ **z^5**
34. $d \cdot d \cdot d \cdot d \cdot d \cdot d \cdot d \cdot d \cdot d \cdot d$ **d^{10}**

Evaluate each expression.

35. 7^2 **49**
36. 9^2 **81**
37. 4^3 **64**
38. 5^3 **125**
39. 2^4 **16**
40. 3^5 **243**

Write an algebraic expression for each verbal expression.

41. triple the difference of 55 and the cube of w **$3(55 - w^3)$**
42. four times the sum of r and s, increased by twice the difference of r and s

42. $4(r + s) + 2(r - s)$

43. the sum of a and b, increased by the quotient of a and b **$a + b + \frac{a}{b}$**
44. the perimeter of a square if the length of a side is s **$4s$**
45. the amount of money in Karen's savings account if she has y dollars and adds x dollars per week for 10 weeks **$y + 10x$**

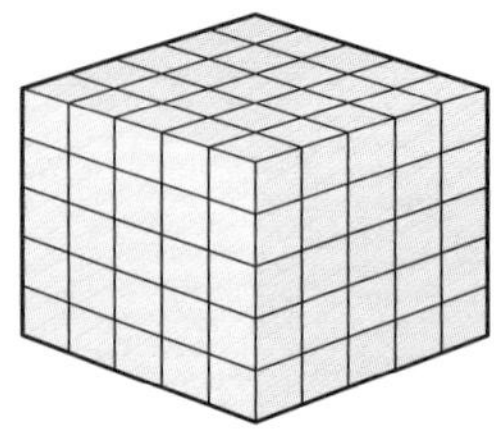

46. **Geometry** Write an expression that represents the total number of small cubes in the large cube shown at the right. Then evaluate the expression. **5^3; 125 cubes**

Critical Thinking

47. Evaluate 4^2 and 2^4. **16; 16** **47a. They are equal; no; no.**
   a. What do you notice? From this example, can you say that the same relationship exists between 2^5 and 5^2? What about between 3^4 and 4^3?
   b. In general, do you think that $a^b = b^a$ for positive whole numbers a and b? Support your answer. **no; for example, $2^3 \neq 3^2$**

Applications and Problem Solving

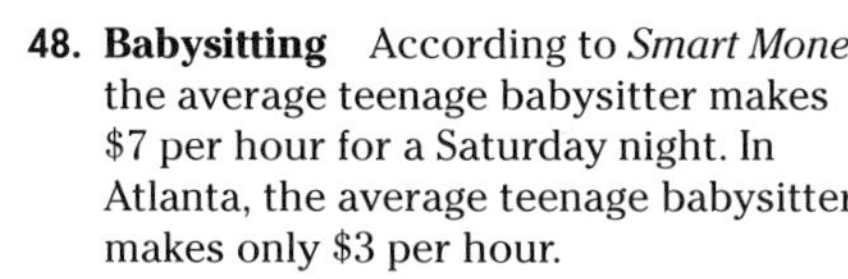
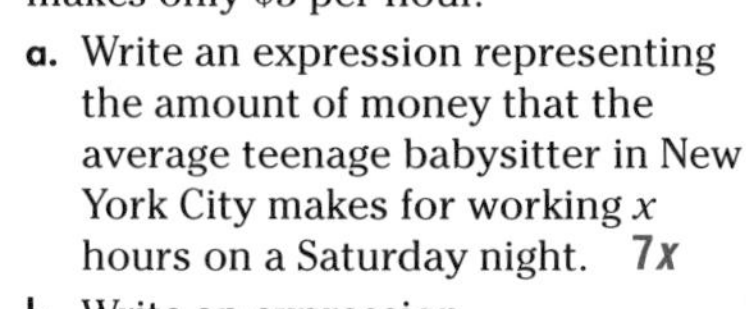
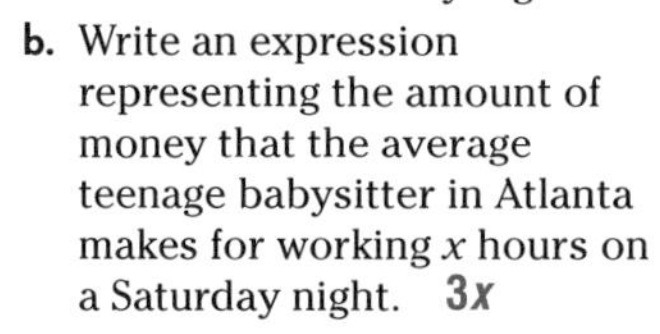
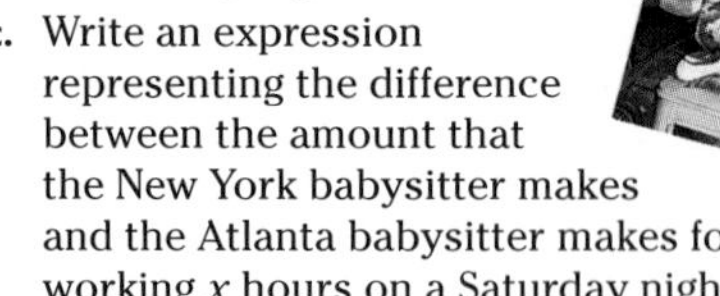

48. **Babysitting** According to *Smart Money* magazine, in New York City, the average teenage babysitter makes \$7 per hour for a Saturday night. In Atlanta, the average teenage babysitter makes only \$3 per hour.
   a. Write an expression representing the amount of money that the average teenage babysitter in New York City makes for working x hours on a Saturday night. **$7x$**
   b. Write an expression representing the amount of money that the average teenage babysitter in Atlanta makes for working x hours on a Saturday night. **$3x$**
   c. Write an expression representing the difference between the amount that the New York babysitter makes and the Atlanta babysitter makes for working x hours on a Saturday night. **$7x - 3x$**

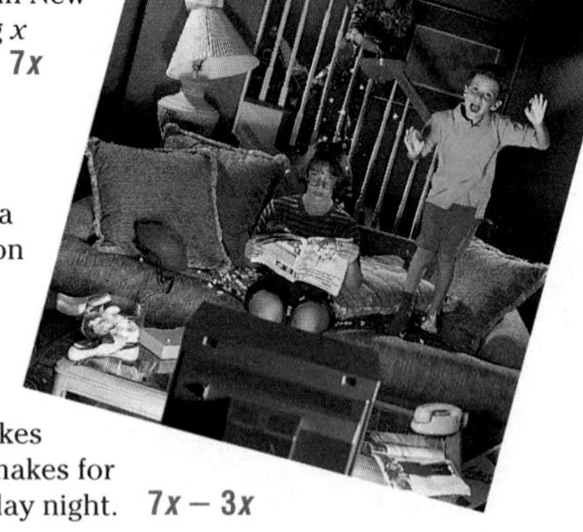

Practice Masters, p. 1

1-1 NAME________ DATE ______

Practice

Student Edition Pages 6–11

Variables and Expressions

Answers may vary. Sample answers are given.

Write a verbal expression for each algebraic expression.

1. $c + 4$ the sum of c and 4
2. $5x$ the product of 5 and x
3. z^2 z squared
4. $\frac{a}{b}$ the quotient of a and b
5. $6y^5$ six times y to the fifth power
6. $k - 3$ 3 less than k
7. $5 - t$ five decreased by t
8. 10^3 ten cubed

Write an algebraic expression for each verbal expression.

9. six times a number $6n$
10. a number to the fourth power n^4
11. eleven increased by a number $11 + n$
12. a number decreased by twenty $n - 20$
13. the quotient of a number and 2 $\frac{n}{2}$
14. the product of 8 and a number $8n$
15. three times the square of a number $3n^2$
16. a number to the fifth power n^5
17. the sum of 9 and a number $9 + n$
18. a number divided by 6 $\frac{n}{6}$
19. a number multiplied by 10 $n \cdot 10$
20. two less than a number $n - 2$
21. a number plus 12 $n + 12$
22. five times a number $5n$
23. a number subtracted from 12 $12 - n$
24. one-half the cube of a number $\frac{1}{2}n^3$
25. five increased by three times a number $5 + 3n$
26. 20 less than eight times a number $8n - 20$

Extension

Reasoning Have students solve: Carolyn received d dollars on Sunday. She spent half of this money on Monday, one-third of what was left on Tuesday, and one-fourth of what was left on Wednesday. Write an algebraic expression to represent how much money she spent each day.

a. Monday $\frac{d}{2}$

b. Tuesday $\frac{1}{3}\left(d - \frac{d}{2}\right)$

c. Wednesday $\frac{1}{4}\left[\frac{d}{2} - \frac{1}{3}\left(d - \frac{d}{2}\right)\right]$

49. **Recycling** The alarming amount of garbage produced by our society requires that everyone work to recycle. According to *Vitality* magazine, each person in the United States produces an average of 3.5 pounds of trash a day.

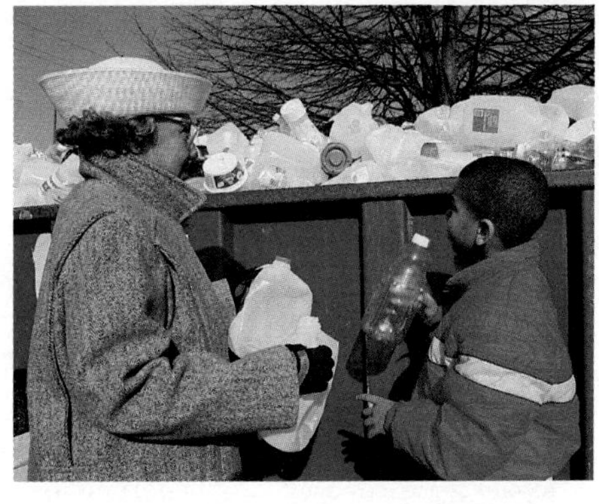

a. Write an expression representing the pounds of trash produced in a day by a family that has x members. **$3.5x$**

b. Write an expression representing the pounds of trash produced in a day by a family that has y members. **$3.5y$**

c. Write an expression representing the pounds of trash produced in one day by both of the families in parts a and b. **$3.5x + 3.5y$**

50. **Geometry** The surface area of a rectangular prism is the sum of the product of twice the length ℓ and the width w and the product of twice the length and the height h and the product of twice the width and the height. Write an expression that represents the surface area of this type of prism. **$2\ell w + 2\ell h + 2wh$**

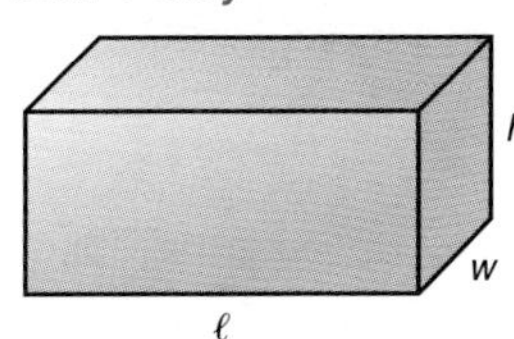

International Mathematical Olympiad

The excerpt below appeared in an article in *TIME* magazine on August 1, 1994.

If the United States sends six kids to the International Mathematical Olympiad in Hong Kong, where a perfect individual score is 42, and together they score 252, does the country have reason to cheer? If you can't answer that one, then you desperately need a remedial course in arithmetic (or perhaps just new batteries in your calculator). The U.S. team members, all public high-school students, started out by competing against 350,000 of their peers on the American High School Mathematics Examination, aced two tougher exams, and prepped for a month at the U.S. Naval Academy. Only then did they board a plane and became the first squad in the Math Olympiad's 35-year history to get perfect scores across the board, outstripping 68 other nations to win the competition. . . . "They showed the world!" suggests the justifiably proud U.S. coach, Walter Mientka, a math professor at the University of Nebraska at Lincoln. ■

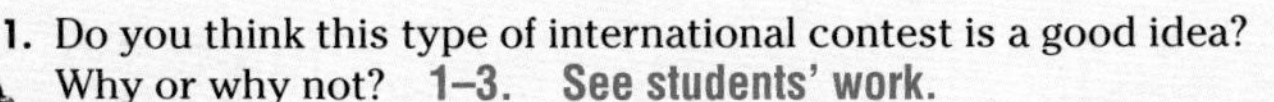

1. Do you think this type of international contest is a good idea? Why or why not? **1–3. See students' work.**
2. Would you like to represent your country in this competition? Why or why not?
3. How do you feel about the math courses you have had in school? What factors have influenced the way you feel?

Mathematics and SOCIETY

Encourage students to continue their discussion with the following question. Because these six students won the competition, beating students from 68 other nations, does it follow that American students are better at mathematics than students from these other countries? Students should support their answers with mathematical reasoning.

4 ASSESS

Closing Activity

Writing Have students list phrases that imply addition, subtraction, multiplication, and division. Then have them make generalizations as to whether the order in which terms are written makes a difference.

Enrichment Masters, p. 1

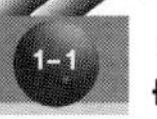

NAME______________________ DATE ________

Student Edition Pages 6–11

1-1 Enrichment

The Four Digits Problem

One well-known problem in mathematics is to write expressions for consecutive numbers from 1 upward as far as possible. On this page, you will use the digits 1, 2, 3, and 4. Each digit is used only once. You can use addition, subtraction, multiplication (not division), exponents, and parentheses in any way you wish. Also, you can use two digits to make one number, as in 12 or 34.

Answers will vary. Sample answers are given.

Express each number as a combination of the digits 1, 2, 3, and 4.

$1 = (3 \times 1) - (4 - 2)$	$18 = (2 \times 3) \times (4 - 1)$	$35 = 2^{(4+1)} + 3$
$2 = (4 - 3) + (2 - 1)$	$19 = 3(2 + 4) + 1$	$36 = 34 + (2 \times 1)$
$3 = (4 - 3) + (2 \times 1)$	$20 = 21 - (4 - 3)$	$37 = 31 + 2 + 4$
$4 = (4 - 2) + (3 - 1)$	$21 = (4 + 3) \times (2 + 1)$	$38 = 42 - (3 + 1)$
$5 = (4 - 2) + (3 \times 1)$	$22 = 21 + (4 - 3)$	$39 = 42 - (3 \times 1)$
$6 = 4 + 3 + 1 - 2$	$23 = 31 - (4 \times 2)$	$40 = 41 - (3 - 2)$
$7 = 3(4 - 1) - 2$	$24 = (2 + 4) \times (3 + 1)$	$41 = 43 - (2 \times 1)$
$8 = 4 + 3 + 2 - 1$	$25 = (2 + 3) \times (4 + 1)$	$42 = 43 - (2 - 1)$
$9 = 4 + 2 + (3 \times 1)$	$26 = 24 + (3 - 1)$	$43 = 42 + 1^3$
$10 = 4 + 3 + 2 + 1$	$27 = 3^2 \times (4 - 1)$	$44 = 43 + (2 - 1)$
$11 = (4 \times 3) - (2 - 1)$	$28 = 21 + 3 + 4$	$45 = 43 + (2 \times 1)$
$12 = (4 \times 3) \times (2 - 1)$	$29 = 2^{(4+1)} - 3$	$46 = 43 + (2 + 1)$
$13 = (4 \times 3) + (2 - 1)$	$30 = (2 \times 3) \times (4 + 1)$	$47 = 31 + 4^2$
$14 = (4 \times 3) + (2 \times 1)$	$31 = 34 - (2 + 1)$	$48 = 4^2 \times (3 \times 1)$
$15 = 2(3 + 4) + 1$	$32 = 4^2 \times (3 - 1)$	$49 = 41 + 2^3$
$16 = (4 \times 2) \times (3 - 1)$	$33 = 21 + (3 \times 4)$	$50 = 41 + 3^2$
$17 = 3(2 + 4) - 1$	$34 = 2 \times (14 + 3)$	

Does a calculator help in solving these types of puzzles? Give reasons for your opinion.

Answers will vary.

1-2 LESSON NOTES

NCTM Standards: 1–5

Instructional Resources

- Study Guide Master 1-2
- Practice Master 1-2
- Enrichment Master 1-2

Transparency 1-2A contains the 5-Minute Check for this lesson; **Transparency 1-2B** contains a teaching aid for this lesson.

Recommended Pacing	
Standard Pacing	Day 3 of 15
Honors Pacing	Day 2 of 14
Block Scheduling*	Day 2 of 7 (along with Lesson 1-3)
Alg. 1 in Two Years*	Days 3 & 4 of 22

*For more information on pacing and possible lesson plans, refer to the *Block Scheduling Booklet* and *Algebra 1 in Two Years.*

1 FOCUS

5-Minute Check
(over Lesson 1-1)

1. Write a verbal expression for $2x - x^2$. **twice a number less the square of the number**
2. Evaluate 3^4. **81**

Write an algebraic expression for each verbal expression.

3. a number increased by 17 $n + 17$
4. twice a number decreased by 45 $2n - 45$
5. the square of a number decreased by twice the number $n^2 - 2n$

Motivating the Lesson

Questioning Write the following sequence on the board: *a, abb, cabbc, cabbcdd,* . . . . Have students continue the sequence. Then have them explain the steps they followed to continue the sequence.

1-2 Patterns and Sequences

What YOU'LL LEARN

- To extend sequences.

Why IT'S IMPORTANT

You can use patterns to extend sequences and solve problems.

APPLICATION

Handicrafts

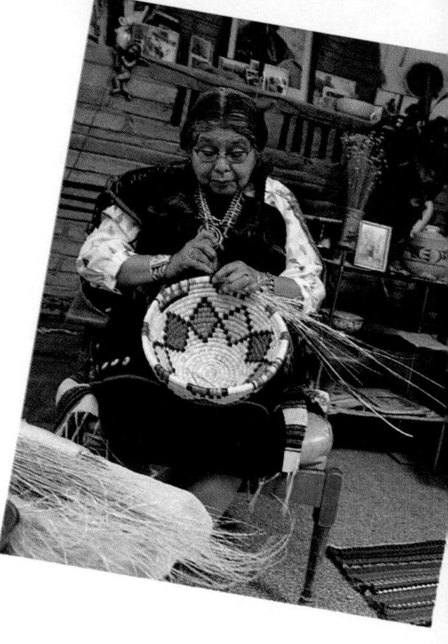

The photograph at the right shows a Hopi woman completing a basket. Baskets are important in the Hopi culture since they can serve either a functional or a spiritual purpose. Looking at the inside of the basket, we can see a pattern. Following this pattern in a clockwise direction, what color would you expect to find on the first diamond shape that cannot be seen in the photo? **green**

GLOBAL CONNECTIONS

Many American Indians play traditional games passed down through the generations. In one such game, markers made from bone, pottery, or shell are marked and tumbled in a small basket. The array of the markers is the basis for scoring. Five or six of a kind would win.

Example 1 **Study the pattern below.**

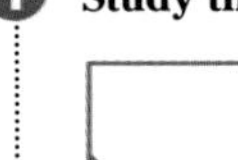
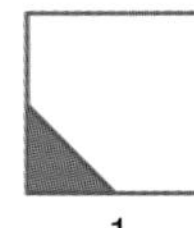

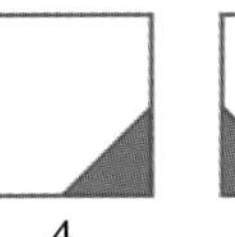

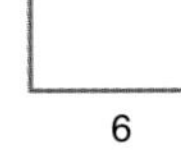

1 2 3 4 5 6

a. Draw the next three figures in the pattern.
b. Draw the 35th square in the pattern.

a. The pattern consists of squares with one corner shaded. The corner that is shaded is rotated in a clockwise direction. The next three figures are drawn below.

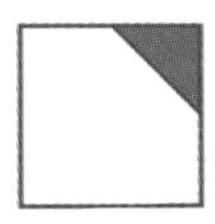
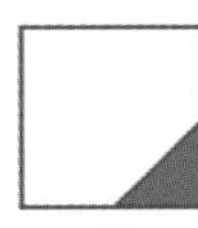

7 8 9

You may find a different pattern. For example, the seventh square may be the same as the fifth square, the eighth square may be the same as the fourth square, and the ninth square may be the same as the third square.

b. The pattern repeats after every 4 designs. Since 32 is the greatest number less than 35 that is divisible by 4, the 33rd square will be the same as the first square.

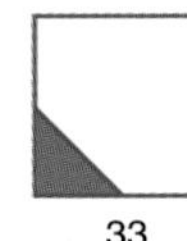
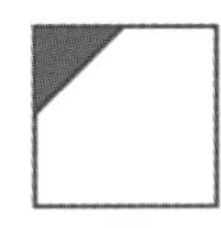
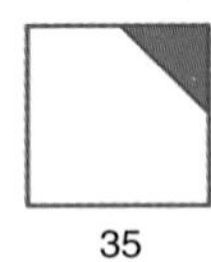

33 34 35

The 35th square will have its upper right corner shaded.

GLOBAL CONNECTIONS

These traditional games required an intuitive understanding of probability. Six of a kind is less likely than five of a kind. Why do three spotted and two striped beat three of a kind? **Because the former is less likely to occur than the latter.**

The numbers 2, 4, 6, 8, 10, and 12 form a pattern called a **sequence.** A sequence is a set of numbers in a specific order. The numbers in the sequence are called **terms.**

Example  **Find the next three terms in each sequence.**

a. 7, 13, 19, 25, . . .

Study the pattern in the sequence.

7, 13, 19, 25,...

+6 +6 +6

Each term is 6 more than the term before it.

$25 + 6 = 31$

$31 + 6 = 37$

$37 + 6 = 43$

The next three terms are 31, 37, and 43.

b. 243, 81, 27, 9, . . .

Study the pattern in the sequence.

243, 81, 27, 9,...

$\times\frac{1}{3}$ $\times\frac{1}{3}$ $\times\frac{1}{3}$

Each term is $\frac{1}{3}$ of the term before it.

$9 \times \frac{1}{3} = 3$

$3 \times \frac{1}{3} = 1$

$1 \times \frac{1}{3} = \frac{1}{3}$

The next three terms are 3, 1, and $\frac{1}{3}$.

One of the most-used strategies in problem solving is **look for a pattern.** When using this strategy, you will often need to make a table to organize the information.

Example  **What is the number of diagonals in a 10-sided polygon?**

Geometry

Drawing a 10-sided polygon and all its diagonals would be difficult. Another way to solve the problem is to study the number of diagonals for polygons with fewer sides and then look for a pattern.

triangle
3 sides
0 diagonals

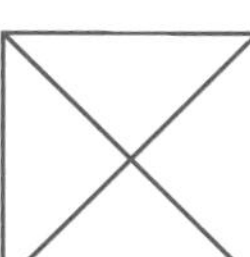

quadrilateral
4 sides
2 diagonals

pentagon
5 sides
5 diagonals

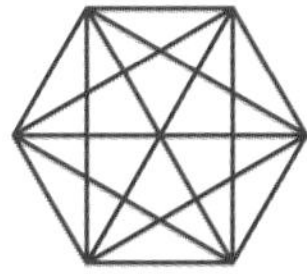

hexagon
6 sides
9 diagonals

(continued on the next page)

Alternative Learning Styles

Visual Share with students that most numerical sequences can be represented by points with lines joining the points. The sequence 3, 6, 12. . . can be visualized as follows.

The sequence can also be visualized as follows.

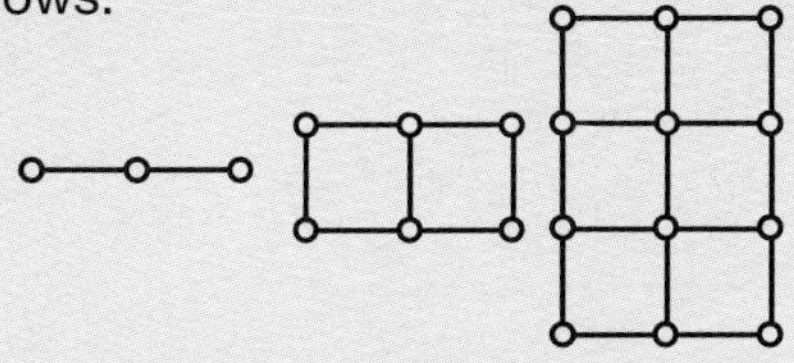

Since the visual representation is not unique, have students use grid paper and creativity to draw their own representations for this sequence.

2 TEACH

In-Class Examples

For Example 1
Study the pattern below.

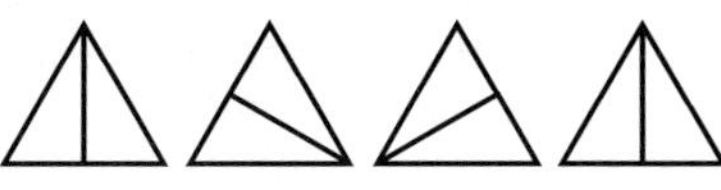

a. Draw the next three figures in the pattern.

b. Draw the 26th triangle in the pattern.

For Example 2
Find the next three numbers in each sequence.

a. 51, 43, 35, 27 **19, 11, 3**
b. 3072, 768, 192 **48, 12, 3**

For Example 3
How many points does the sixth figure in this sequence have? **63**

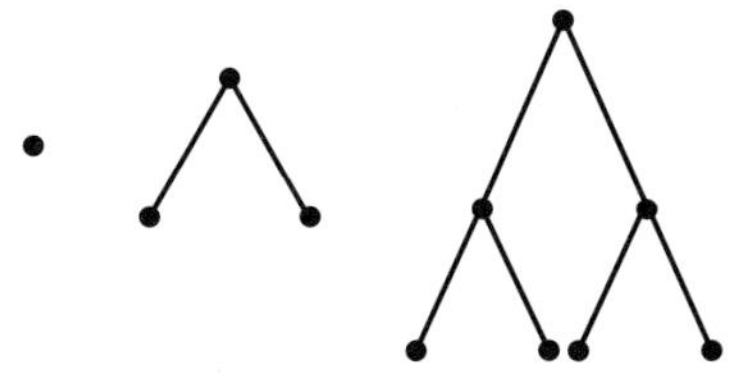

Teaching Tip In Example 2, suggest that students begin looking for a pattern by determining if the second term is derived from the first term by addition, subtraction, multiplication, or division.

Teaching Tip Point out to students that in some cases it looks as if there is a pattern, but later on the pattern is broken. For example, show how the pattern is broken in the eighth digit of $111{,}111{,}111 \div 9 = 12{,}345{,}67\underline{9}$.

In-Class Example

For Example 4

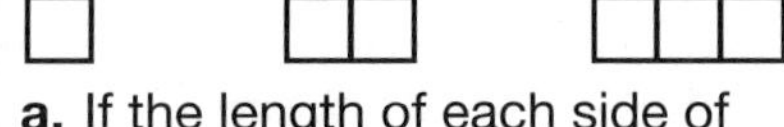

a. If the length of each side of the square is 1 unit, what is the perimeter of each figure in the pattern above? **4, 6, 8**

b. What is the perimeter of the *n*th figure in the pattern? **$2n + 2$**

Use a chart to see the pattern.

Number of Sides	3	4	5	6	7	8	9	10
Number of Diagonals	0	2	5	9	?	?	?	?

+2 +3 +4

Use the pattern to complete the chart.

Number of Sides	3	4	5	6	7	8	9	10
Number of Diagonals	0	2	5	9	14	20	27	35

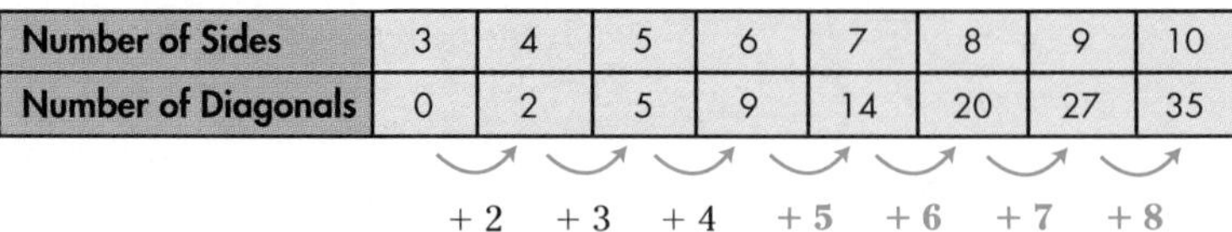

+2 +3 +4 +5 +6 +7 +8

A 10-sided polygon has 35 diagonals.

Recognizing patterns is important, but it does not always guarantee finding the correct answer. You may wish to check this answer by drawing a 10-sided polygon and its diagonals.

Sometimes a pattern can lead to a general rule that can be written as an algebraic expression.

Example  **Study the following pattern. Write an algebraic expression for the perimeter of the pattern consisting of *n* trapezoids.**

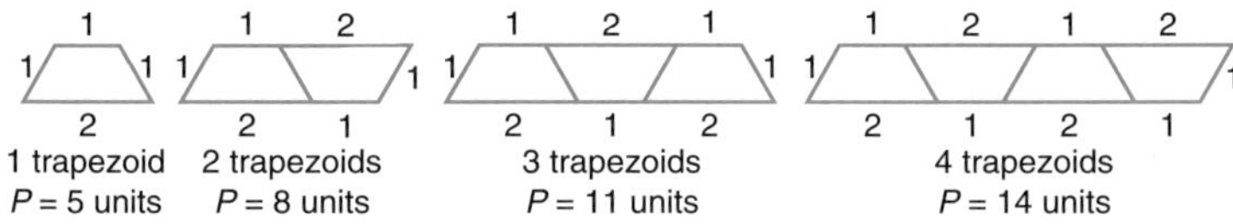

Use a chart to look for a pattern.

Number of Trapezoids	1	2	3	4	5	6	n
Perimeter	5	8	11	14	?	?	?

+3 +3 +3

Notice that each trapezoid adds 3 units to the perimeter.

For 1 trapezoid, the perimeter is $3 \cdot 1 + 2$, or 5 units.
For 2 trapezoids, the perimeter is $3 \cdot 2 + 2$, or 8 units.
For 3 trapezoids, the perimeter is $3 \cdot 3 + 2$, or 11 units.
For 4 trapezoids, the perimeter is $3 \cdot 4 + 2$, or 14 units.

Extend this pattern.

For 5 trapezoids, the perimeter should be $3 \cdot 5 + 2$, or 17 units.
For 6 trapezoids, the perimeter should be $3 \cdot 6 + 2$, or 20 units.
For n trapezoids, the perimeter should be $3 \cdot n + 2$, or $3n + 2$ units.

The algebraic expression $3n + 2$ represents the perimeter of this pattern with n trapezoids.

Looking for Patterns

Materials:  string scissors

If you use a pair of scissors to cut a piece of string in the normal way, you will have 2 pieces of string. What happens if you loop the string around one of the cutting edges of the scissors and cut?

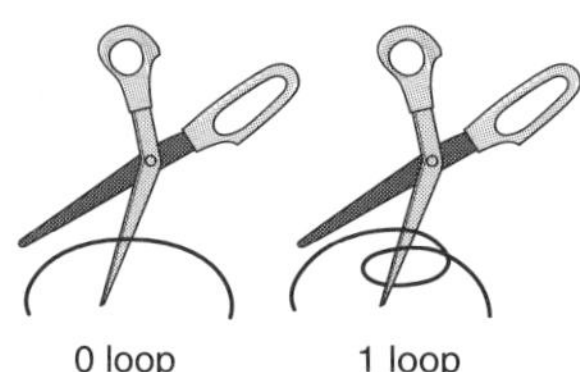

Your Turn

a. Make a piece of string loop once around your scissors as shown above. Cut the string. How many pieces do you have? **3 pieces**

b. Make 2 loops and cut. How many pieces do you have? **4 pieces**

c. Add 2 to the number of loops; 3, 4, 5, 6, 7, ...

c. Continue making loops and cutting until you see a pattern. Describe the pattern and write the sequence.

d. How many pieces would you have if you made 20 loops? **22 pieces**

e. Now tie the ends of the string together before you loop the string around the scissors. Investigate to determine how many pieces you would have if you made 10 loops with this string. **22 pieces**

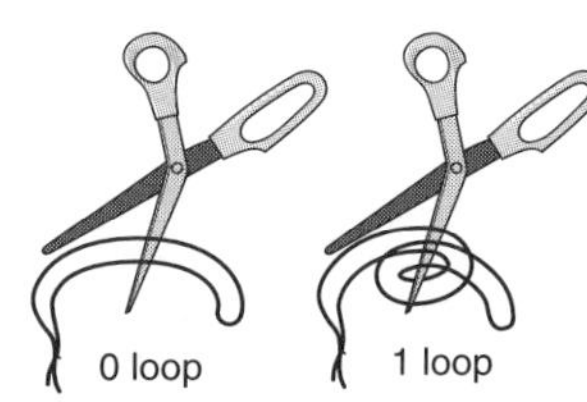

CHECK FOR UNDERSTANDING

Communicating Mathematics

1. See margin.
2. Sample answer: 7, 12, 17, 22

Study the lesson. Then complete the following.

1. **Explain** how looking for a pattern can help to solve some problems.
2. **Write** a sequence that has 22 as its fourth term.
3. **You Decide** Chi-Yo studies the sequence 1, 2, and 4. She notices that the second number is 1 more than the first, and the third number is 2 more than the second. She concludes that the next number in the sequence is 4 + 3 or 7. Alonso disagrees. He notices that each number is twice the number before it. He says the next number in the sequence is 8. Who is correct? Explain. **See margin.**

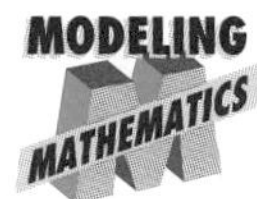

4. Refer to the Modeling Mathematics activity above. Suppose the ends of a string are *not* tied together.
   a. The string is looped around the scissors 50 times, and the string is cut. How many pieces will you have? **52 pieces**
   b. The string is looped around the scissors y times, and the string is cut. How many pieces will you have? $y + 2$

Guided Practice

5–7. Sample answers are given.

7. $5x + 1$, $6x + 1$

Give the next two items for each pattern.

5. 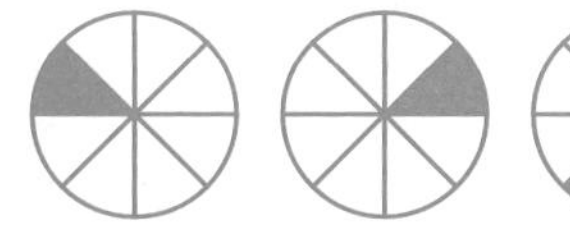  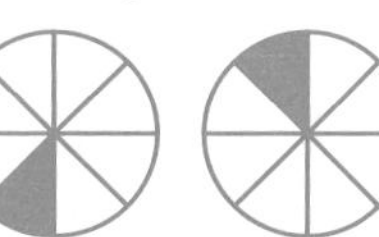  

6. 85, 76, 67, 58,... **49, 40**

7. $1x + 1, 2x + 1, 3x + 1, 4x + 1,...$

 Instead of counting the number of loops, students can consider the number of cuts the scissors make when the string is cut. The pattern is different: number of pieces = number of cuts + 1.

3 PRACTICE/APPLY

Check for Understanding

Exercises 1–9 are designed to help you assess your students' understanding through reading, writing, speaking, and modeling. You should work through Exercises 1–4 with your students and then monitor their work on Exercises 5–9.

Error Analysis
Students will sometimes find a pattern in the first three terms and not notice that the fourth term does not fit the pattern.

Additional Answers

1. **When the problem is difficult, you can solve simpler problems and look for a pattern that could help you solve the more difficult problem.**
3. **Either person could be right. You would need to know the fourth term to decide which pattern is correct.**

Reteaching

Using Patterns Find the next two terms of this sequence.

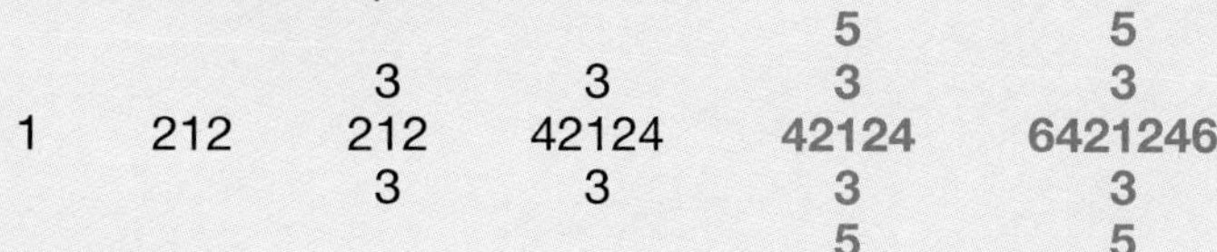

				5	5
		3	3	3	3
1	212	212	42124	42124	6421246
		3	3	3	3
				5	5

Assignment Guide

Core: 11–25 odd, 26, 27, 29–35
Enriched: 10–24 even, 26–35

For **Extra Practice,** see p. 756.

The red A, B, and C flags, printed only in the Teacher's Wraparound Edition, indicate the level of difficulty of the exercises.

Additional Answers

8b.

12.

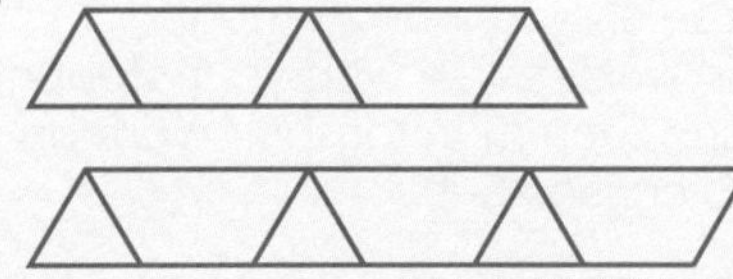

Study Guide Masters, p. 2

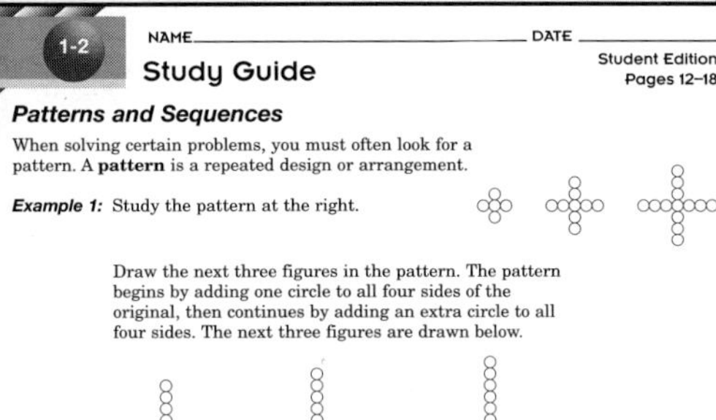

1-2 NAME________________ DATE________

Study Guide — Student Edition Pages 12–18

Patterns and Sequences

When solving certain problems, you must often look for a pattern. A **pattern** is a repeated design or arrangement.

Example 1: Study the pattern at the right.

Draw the next three figures in the pattern. The pattern begins by adding one circle to all four sides of the original, then continues by adding an extra circle to all four sides. The next three figures are drawn below.

The numbers 1, 3, 5, 7, 9 and 11 form a **sequence.** A sequence is a set of numbers in a specific order. The numbers in a sequence are called **terms.**

Example 2: Find the next three numbers in each sequence.

a. 2, 6, 18, 54, ⋯ Study the pattern in the sequence. Each term is 3 times more than the term before it. The next three terms are 162, 486, and 1,458.

b. 128, 64, 32, ⋯ Study the pattern in the sequence. Each term is $\frac{1}{2}$ of the term before it. The next three terms are 16, 8, and 4.

Give the next two items for each pattern.

1. 

2. 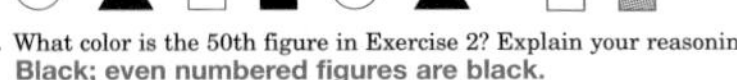

3. What color is the 50th figure in Exercise 2? Explain your reasoning. Black; even numbered figures are black.
4. 2, 12, 72, 432, ⋯ 2,592 and 15,552
5. $b + 1, b + 4, b + 9, b + 16, \cdots$ $b + 25$ and $b + 36$
6. 10, 7, 11, 8, 12, 9, 13, ⋯ 10 and 14

8. Consider the following pattern.

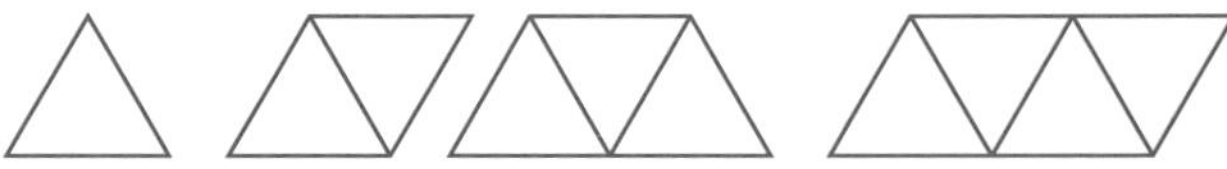

a. Suppose the length of each side of the triangles is 1 unit. What is the perimeter of each figure in the pattern? **3 units; 4 units; 5 units; 6 units**

8b. See margin; 7 units.

b. Draw the next figure in the pattern. What is the perimeter of this figure?

c. What is the perimeter of the tenth figure in this pattern? **12 units**

d. What is the perimeter of the *n*th figure in this pattern? **$n + 2$ units**

9. a. Copy and complete the following table.

4^1	4^2	4^3	4^4	4^5
4	16	**64**	**256**	**1024**

b. If you found the value of 4^6, what number do you think will be in the ones place? Check your conjecture. **6; $4^6 = 4096$**

c. Find the number in the ones place for the value of 4^{225}. Explain your reasoning. **4; When the exponent is odd, 4 is in the ones place.**

EXERCISES

Practice **Give the next two items for each pattern.**

10.

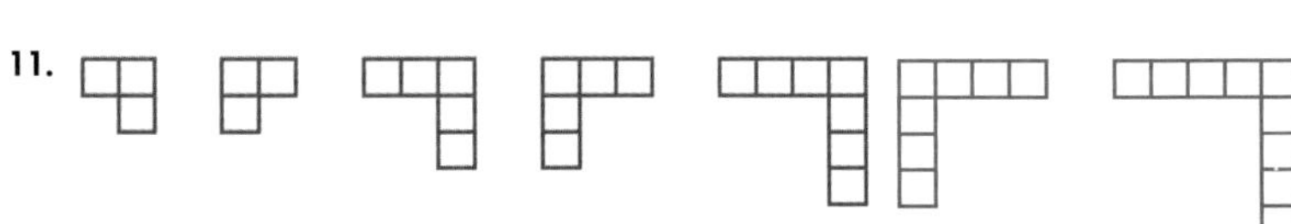

11.

12. See margin.

12.

13. 3, 6, 12, 24, ... **48, 96**

14. 4, 5.5, 7, 8.5, ... **10, 11.5**

15. 1, 4, 9, 16, ... **25, 36**

16. 9, 7, 10, 8, 11, 9, 12, ... **10, 13**

18. $x - 4y, x - 5y$

17. $a + 1, a + 3, a + 5, \ldots$ **$a + 7, a + 9$**

18. $x - 1y, x - 2y, x - 3y, \ldots$

19. a. Draw the next three figures in the following pattern.

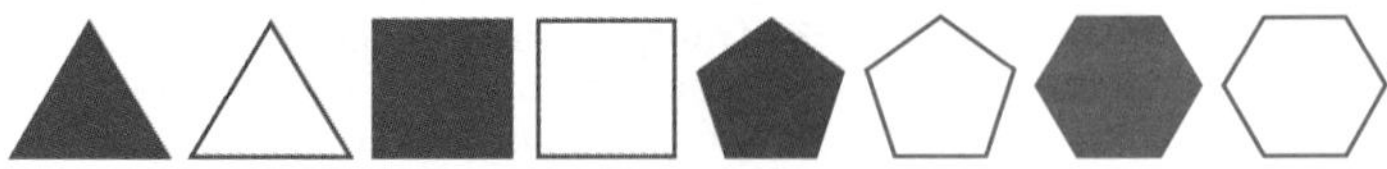

19b. White; even-numbered figures are white.

19c. 12 sides; For reasoning, see margin.

b. What is the color of the 38th figure? Explain your reasoning.

c. How many sides will the 19th figure have? Explain your reasoning.

20. a. Copy and complete the following table.

3^1	3^2	3^3	3^4	3^5	3^6
3	9	**27**	**81**	**243**	**729**

b. Write the sequence of numbers representing the numbers in the ones place. **3, 9, 7, 1, 3, 9, ...**

c. What are the next six numbers in this sequence? **7, 1, 3, 9, 7, 1**

d. Find the number in the ones place for the value of 3^{100}. Explain your reasoning. **1; See margin for reasoning.**

21. a. Copy and find each sum.

$1 = 1$
$1 + 3 = ?$ **4**
$1 + 3 + 5 = ?$ **9**
$1 + 3 + 5 + 7 = ?$ **16**
$1 + 3 + 5 + 7 + 9 = ?$ **25**

21b. See margin.

b. Pythagoras is credited with a discovery about the sum of consecutive odd numbers such as those shown in part a. What did he discover?

c. What is the sum of the first 100 odd numbers? **10,000**

d. What is the sum of the first x odd numbers? x^2

22. If y represents the counting numbers 1, 2, 3, 4, ..., then the algebraic expression $2y$ represents the terms of a sequence. To find the first term of the sequence, replace y with 1 and find $2 \cdot 1$.

a. How could you find the second term of the sequence?

b. Write the first ten terms of the sequence.

c. Describe the terms of this sequence. **even numbers**

22a. Replace y with 2 and find $2 \cdot 2$.

22b. 2, 4, 6, 8, 10, 12, 14, 16, 18, 20

23. a. Use a calculator to find each product.

i. $999{,}999 \times 2$ **1,999,998**
ii. $999{,}999 \times 3$ **2,999,997**
iii. $999{,}999 \times 4$ **3,999,996**
iv. $999{,}999 \times 5$ **4,999,995**

b. Without using a calculator, find the product of 999,999 and 9. **8,999,991**

24. a. Copy and complete the following table.

10^1	10^2	10^3	10^4	10^5
10	100	**1000**	**10,000**	**100,000**

b. Use the pattern to find the value of 10^0. **1**

25. If n represents the counting numbers 1, 2, 3, 4, ..., then the algebraic expression $3n + 1$ represents the terms of a sequence. Write the first five terms of the sequence. **4, 7, 10, 13, 16**

Critical Thinking

26. A special sequence is called the *Fibonacci sequence.* It is named after Leonardo Fibonacci from Italy who presented it in 1201. This sequence is very intriguing because the numbers in the sequence are often found in nature. The first six terms of the sequence are listed below. Give the next six numbers in the sequence. 1, 1, 2, 3, 5, 8, ... **13, 21, 34, 55, 89, 144**

Applications and Problem Solving

27. **Transportation** Olga has part of a bus schedule. She wishes to take the bus to go to the mall, but she cannot leave until after 1:00 P.M. What is the earliest time Olga can catch the bus? **1:13 P.M.**

Bus Schedule
Departures
8:25 A.M.
9:13 A.M.
10:01 A.M.
10:49 A.M.

4 ASSESS

Closing Activity

Modeling Have students work in pairs. One student should use toothpicks to create a pattern. The other student should determine the pattern and continue it.

Additional Answers

19c. The shapes come in pairs. Since $19 \div 2 = 9.5$, the 19th figure will be part of the 10th pair. The 10th pair will have 10 + 2 or 12 sides.

20d. 100 is divisible by 4. According to the pattern, all powers with exponents divisible by 4 have 1 in the ones place.

21b. The sums are perfect square numbers; that is $1^2, 2^2, 3^2, 4^2, 5^2, \ldots$.

Practice Masters, p. 2

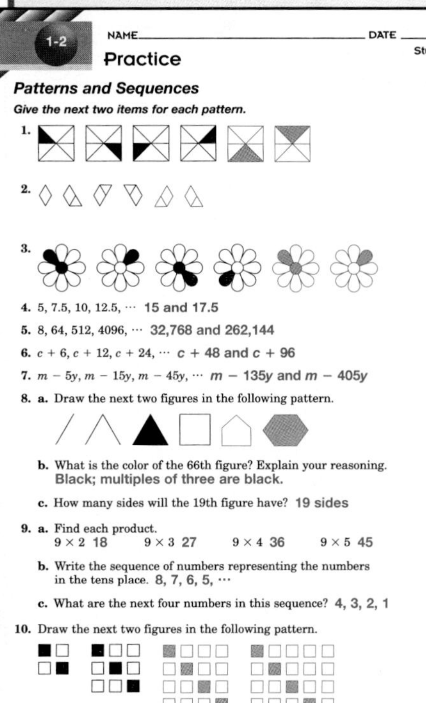

1-2 NAME ________ DATE ________

Practice — Student Edition Pages 12–18

Patterns and Sequences

Give the next two items for each pattern.

1.

2.

3.

4. 5, 7.5, 10, 12.5, ··· 15 and 17.5

5. 8, 64, 512, 4096, ··· 32,768 and 262,144

6. $c + 6, c + 12, c + 24, \cdots$ $c + 48$ and $c + 96$

7. $m - 5y, m - 15y, m - 45y, \cdots$ $m - 135y$ and $m - 405y$

8. a. Draw the next two figures in the following pattern.

b. What is the color of the 66th figure? Explain your reasoning. Black; multiples of three are black.

c. How many sides will the 19th figure have? 19 sides

9. a. Find each product.
9×2 18 9×3 27 9×4 36 9×5 45

b. Write the sequence of numbers representing the numbers in the tens place. 8, 7, 6, 5, ···

c. What are the next four numbers in this sequence? 4, 3, 2, 1

10. Draw the next two figures in the following pattern.

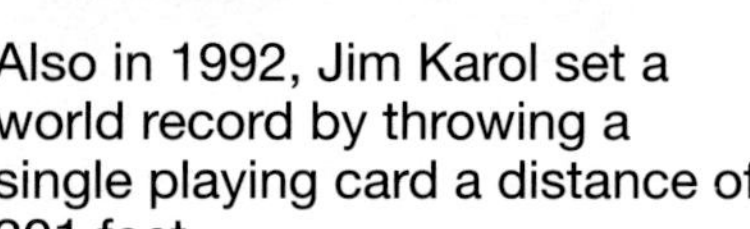

Also in 1992, Jim Karol set a world record by throwing a single playing card a distance of 201 feet.

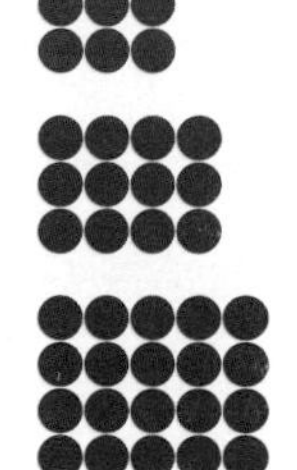

top layer

second layer

third layer

28. Sales Paula needs to make a tower of soup cans as a display in a supermarket. Each layer of the tower will be in the shape of a rectangle as shown at the left. The length and width of each layer will be one less than the layer below it.

a. How many cans will be needed for the fourth layer? **30 cans**

b. What is the total number of cans needed for an 8-layer tower? **328 cans**

29a. 100 cards

29b–d. See students' work.

FYI

In 1992, Bryan Berg broke a world record by building the tallest house of cards. He used 208 decks to build 75 stories.

29. Recreation The house of cards at the right used 26 cards to build four stories.

a. How many cards are needed to build a similar house that is eight stories high?

b. Plan another way to construct a house of cards. Make a drawing showing four stories of your house.

c. Write a sequence that represents the number of cards needed to build 1, 2, 3, and 4 stories of your house.

d. How many cards will you need to build eight stories of your house?

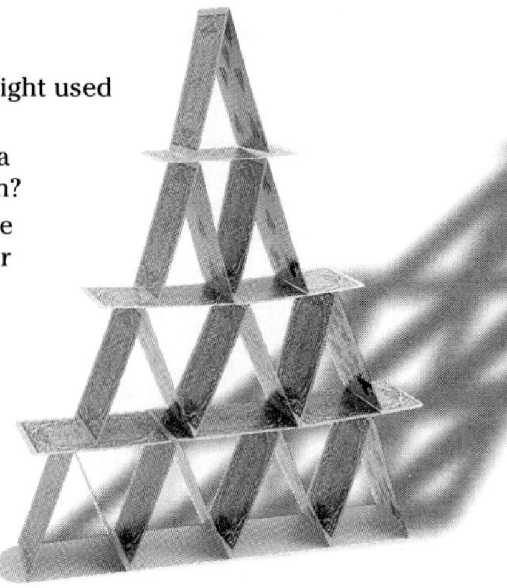

Mixed Review

30. Write an algebraic expression for *eight less than the square of q.* (Lesson 1–1) $q^2 - 8$

31. *x* cubed divided by 9

31. Write a verbal expression for $\frac{x^3}{9}$. (Lesson 1–1)

32. Write $m \cdot m \cdot m \cdot m \cdot m \cdot m \cdot m \cdot m \cdot m$ as an algebraic expression with exponents. (Lesson 1–1) m^9

33. **Geometry** Write an expression that represents the total number of small cubes in the large cube at the right. Then evaluate the expression. (Lesson 1–1) **4^3; 64 cubes**

34. Write an algebraic expression for the amount of mileage on Seth's car if the car initially has 20,000 miles and Seth adds an average of x miles per month for 2 years. (Lesson 1–1) **20,000 + 24*x***

35. $x + \frac{1}{11}x$

35. **Science** When water freezes, its volume is increased by one eleventh. In other words, the volume of the ice would equal the sum of the volume of the water and the product of $\frac{1}{11}$ and the volume of the water. Suppose x cubic inches of water is placed in a freezer. Write an expression for the volume of the ice that is formed. (Lesson 1–1)

Enrichment Masters, p. 2

NAME______________________ DATE __________

Enrichment

Student Edition Pages 12–18

The Tower of Hanoi

The Tower of Hanoi puzzle has three pegs, with a stack of disks on peg a. The object is to move all of the disks to another peg. You may move only one disk at a time. Also, a larger disk may never be put on top of a smaller disk.

Peg a Peg b Peg c

A chart has been started to record your moves as you solve the puzzle.

Another way to record the moves is to use letters. For example, the first two steps in the chart can be recorded as 1c, 2b. This shows that disk 1 is moved to peg c, and then disk 2 is moved to peg b.

Peg a	Peg b	Peg c
1 2 3		
2 3		1
3	2	1
3	1 2	
	1 2	3
1	2	3
1		2 3
		1 2 3

Solve each problem.

1. Finish the chart to solve the Tower of Hanoi puzzle for three disks.

2. Record your solution using letters.
1c, 2b, 1b, 3c, 1a, 2c, 1c

3. On a separate sheet of paper, solve the puzzle for four disks. Record your solution.
1c, 2b, 1b, 3c, 1a, 2c, 1c, 4b, 1b, 2a, 1a, 3b, 1c, 2b, 1b

4. Solve the puzzle for five disks. Record your solution.
1c, 2b, 1b, 3c, 1a, 2c, 1c, 4b, 1b, 2a, 1a, 3b, 1c, 2b, 1b, 5c, 1a, 2c, 1c, 3a, 1b, 2a, 1a, 4c, 1c, 2b, 1b, 3c, 1a, 2c, 1c

5. If you start with an odd number of disks and you want to end with the stack on peg c, what should be your first move?
1c

6. If you start with an even number of disks and you want to end with the stack on peg b, what should be your first move?
1c

Extension

Reasoning For Exercise 29a, describe the number of cards needed to construct level n when levels are counted from top to bottom and n is not the bottom level. **$3n$** Describe the number of cards needed when n is the bottom level. **$3n - n$**

1–3 Order of Operations

What YOU'LL LEARN

- To use the order of operations to evaluate real number expressions.

Why IT'S IMPORTANT

You use the order of operations to evaluate expressions and solve equations.

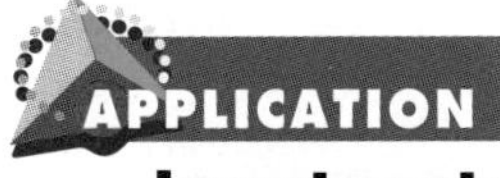

APPLICATION

Investments

Bobbie Jackson is investing money in stocks to help pay for her child's college education. She buys one share of Nike stock at $16. She also purchases five shares of Disney stock at $35 each. The expression below represents the amount of money Ms. Jackson spends for the stock purchase.

cost of 1 share of Nike stock → 16; *cost of 1 share of Disney stock* → 35

$$16 + 5 \cdot 35$$

number of Disney stocks → 5

Numerical and algebraic expressions often contain more than one operation. A rule is needed to let you know which operation to perform first. This rule is called the **order of operations.**

To find the total amount of money Ms. Jackson invests, evaluate the expression $16 + 5 \cdot 35$. Which of the following methods is correct?

Method 1		**Method 2**	
$16 + 5 \cdot 35 = 16 + 175$	*Multiply first.*	$16 + 5 \cdot 35 = 21 \cdot 35$	*Add first.*
$= 191$	*Then add.*	$= 735$	*Then multiply.*

The answers are not the same because a different order of operations was used in each method. Since numerical expressions must have only one value, the following order of operations has been established.

Order of Operations

1. **Simplify the expressions inside grouping symbols, such as parentheses, brackets, and braces, and as indicated by fraction bars.**
2. **Evaluate all powers.**
3. **Do all multiplications and divisions from left to right.**
4. **Do all additions and subtractions from left to right.**

Based on the context of the problem and the order of operations, Method 1 is correct. Therefore, Ms. Jackson invested $16 + 5 \cdot 35$ or $191 in the stock market.

1-3 LESSON NOTES

NCTM Standards: 1–5, 8

Instructional Resources

- Study Guide Master 1-3
- Practice Master 1-3
- Enrichment Master 1-3
- Assessment and Evaluation Masters, p. 16
- Modeling Mathematics Masters, p. 72
- Tech Prep Applications Masters, p. 1

Transparency 1-3A contains the 5-Minute Check for this lesson; **Transparency 1-3B** contains a teaching aid for this lesson.

Recommended Pacing	
Standard Pacing	Day 4 of 15
Honors Pacing	Day 3 of 14
Block Scheduling*	Day 2 of 7 (along with Lesson 1-2)
Alg. 1 in Two Years*	Days 5 & 6 of 22

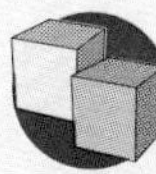

*For more information on pacing and possible lesson plans, refer to the *Block Scheduling Booklet* and *Algebra 1 in Two Years.*

1 FOCUS

5-Minute Check
(over Lesson 1-2)

Write the next three numbers in each sequence.

1. 2, 4, 10, 28 . . . **82, 244, 730**
2. 5, 15, 35, 75 . . . **155, 315, 635**
3. 10, 90, 890, 8890 . . . **88,890; 888,890; 8,888,890**
4. 5, 17, 53, 161 . . . **485, 1457, 4373**
5. 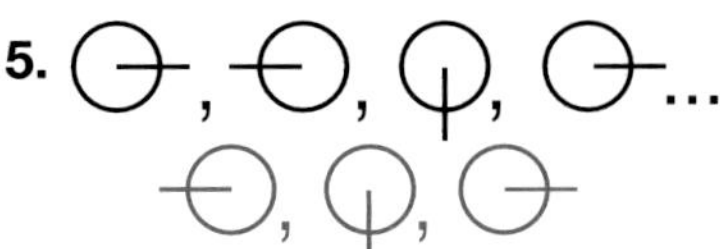

Motivating the Lesson

Hands-On Activity Have students take apart and then reassemble a simple mechanism such as a flashlight. Ask students if there was a specific sequence of steps necessary to reassemble the item or if it could have been done in a number of ways.

2 TEACH

Teaching Tip Ask students to explain what it means for a numerical expression to have a unique value. **The expression must be equal to "one and only one" value.**

In-Class Examples

For Example 1
Evaluate each expression.

a. $3(3 + 7)^2 \div 5$ **60**
b. $27 \div 3(5 - 3)^2$ **36**

For Example 2
Evaluate $\frac{2}{3}\left[8(a - b)^2 + 3b\right]$ if $a = 6$ and $b = 3$. **54**

For Example 3
In the rectangle below, $a = 12$ mm.

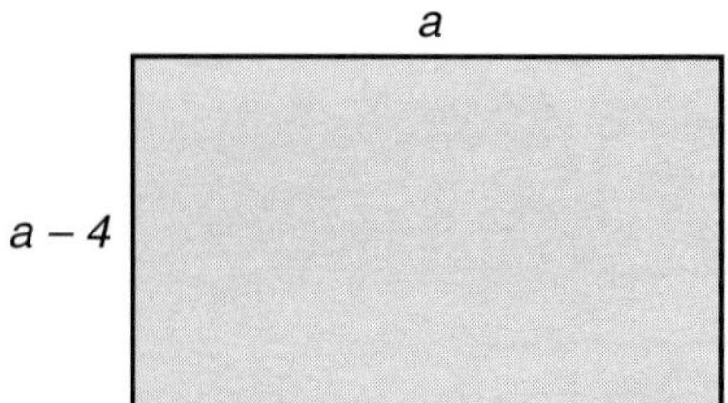

a. Find the perimeter of the rectangle. **40 mm**
b. Find the area of the rectangle. **96 mm²**

Other expressions can be evaluated by using the order of operations.

Example  **Evaluate $5 \times 7 - 6 \div 2 + 3^2$.**

Evaluate $5 \times 7 - 6 \div 2 + 3^2 = 5 \times 7 - 6 \div 2 + 9$ *Evaluate 3^2.*
$= 35 - 6 \div 2 + 9$ *Multiply 5 by 7.*
$= 35 - 3 + 9$ *Divide 6 by 2.*
$= 32 + 9$ *Subtract 3 from 35.*
$= 41$ *Add 32 and 9.*

In mathematics, grouping symbols such as parentheses ( ), brackets [ ], and braces { } are used to clarify or change the order of operations. They indicate that the expression within the grouping symbol is to be evaluated first. When more than one grouping symbol is used, start evaluating within the innermost grouping symbols.

Example  **Evaluate $8[6^2 - 3(2 + 5)] \div 8 + 3$.**

$8[6^2 - 3(2 + 5)] \div 8 + 3 = 8[6^2 - 3(7)] \div 8 + 3$ *Add 2 + 5, the innermost group.*
$= 8[36 - 3(7)] \div 8 + 3$ *Evaluate 6^2.*
$= 8[36 - 21] \div 8 + 3$ *Multiply 3 by 7.*
$= 8[15] \div 8 + 3$ *Subtract 21 from 36.*
$= 120 \div 8 + 3$ *Multiply 8 by 15.*
$= 15 + 3$ *Divide 120 by 8.*
$= 18$ *Add 15 and 3.*

Algebraic expressions can be evaluated when the values of the variables are known. First, replace the variables by their values. Then, calculate the value of the numerical expression.

Example  **The figure at the right is a rectangle.**

INTEGRATION Geometry

a. Find the perimeter of the rectangle when $s = 5$.
b. Find the area of the rectangle.

a. The perimeter of the rectangle is the sum of 2 times the measure of the width (s) and 2 times the measure of the length ($s + 3$).

$P = 2s + 2(s + 3)$
$= 2(5) + 2(5 + 3)$ *Replace the variable s with 5.*
$= 2(5) + 2(8)$ *Add 5 and 3.*
$= 10 + 2(8)$ *Multiply 2 and 5.*
$= 10 + 16$ *Multiply 2 and 8.*
$= 26$ *Add 10 and 16.*

The perimeter is 26 mm.

GLENCOE Technology

Interactive Mathematics Tools Software

This multimedia software provides an interactive lesson that uses real-world experiences, such as shopping, as a tool for learning the order of operations. A **Computer Journal** gives students an opportunity to write about what they have learned.

For Windows & Macintosh

b. The area is the product of the width (s) and the length ($s + 3$).

$A = s(s + 3)$
$= 5(5 + 3)$ *Replace the variable s with 5.*
$= 5(8)$ *Add 5 and 3.*
$= 40$ *Multiply 5 and 8.*

The area is 40 mm^2. *Area is expressed in units squared.*

EXPLORATION GRAPHING CALCULATORS

You can use a graphing calculator to evaluate algebraic expressions. Use a calculator to evaluate $\frac{0.25x^2}{7x^3}$ when $x = 0.75$.

Enter: .75 [STO▸] [X,T,θ] [2nd] [:] [(] .25 [X,T,θ] [x^2] [)] [÷] [(] 7 [X,T,θ] [^] 3 [)] [ENTER] 0.0476190476

Your Turn

a. Evaluate the expression when $x = 24.076$. **0.0014833978**

b. Evaluate $\frac{2x^2}{(x^2 - x)}$ when $x = 27.89$. **2.074377092**

c. Work through some of the examples in this lesson using a graphing calculator.

The fraction bar is another grouping symbol. It indicates that the numerator and denominator should each be treated as a single value.

$$\frac{2 \times 3}{1 + 2} \text{ means } (2 \times 3) \div (1 + 2) \text{ or } 2.$$

Example  **Evaluate $\frac{x^3 + y^3}{x^2 - y^2}$ when $x = 4.2$ and $y = 1.8$.**

$$\frac{x^3 + y^3}{x^2 - y^2} = \frac{(4.2)^3 + (1.8)^3}{(4.2)^2 - (1.8)^2}$$

Estimate: $\frac{4^3 + 2^3}{4^2 - 2^2} = \frac{64 + 8}{16 - 4} = \frac{72}{12}$ *or 6*

Use a scientific calculator.

Enter: [(] 4.2 [y^x] 3 [+] 1.8 [y^x] 3 [)] [÷] [(] 4.2 [x^2] [−] 1.8 [x^2] [)] [=] 5.55

Is the answer reasonable?

EXPLORATION

Point out to students that some calculators may require parentheses to perform the operations. Without parentheses, the operations will be performed in the order entered.

In-Class Example

For Example 4
Use a calculator to evaluate $\frac{a^6 - 22}{a^3 + x^7}$ if $a = 12$ and $x = 9$. Round to the nearest hundredth. **0.62**

Teaching Tip Have students analyze the necessity of grouping symbols in Example 4 by working the example without using the parentheses keys and then interpreting the resulting value of 71.178612.

Alternative Teaching Strategies

Reading Algebra Emphasize the importance of looking for key words when translating verbal expressions into algebraic expressions. For example, the word *quantity* usually indicates that parentheses or other grouping symbols are to be used.

3 PRACTICE/APPLY

Check for Understanding

Exercises 1–12 are designed to help you assess your students' understanding through reading, writing, speaking, and modeling. You should work through Exercises 1–3 with your students and then monitor their work on Exercises 4–12.

Error Analysis

Students may tend to evaluate expressions from left to right, ignoring the correct order of operations. For example, they may evaluate $24 - 8 \div 2$ as $16 \div 2$ or 8 and not $24 - 4$ or 20. Emphasize the importance of looking at the entire expression to determine the correct order of operations before evaluating.

Assignment Guide

Core: 13–41 odd, 42, 43, 45–53
Enriched: 14–40 even, 42–53

For **Extra Practice,** see p. 756.

The red A, B, and C flags, printed only in the Teacher's Wraparound Edition, indicate the level of difficulty of the exercises.

Study Guide Masters, p. 3

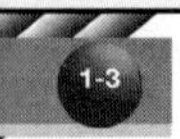

NAME________________ DATE ________

Study Guide

Student Edition
Pages 19–24

Order of Operations

Ian said Shelley won first prize and I won second prize. Without punctuation, this sentence has three possible meanings.

Ian said, "Shelley won first prize and I won second prize."
or
"Ian," said Shelley, "won first prize and I won second prize."
or
Ian said Shelley won first prize and I [that is, the speaker] won second prize.

In mathematics, in order to avoid confusion about meaning, an agreed-upon order of operations tells us whether a mathematical expression such as $15 - 12 \div 4$ means $(15 - 12) \div 4$ or $15 - (12 \div 4)$. That order is shown at the right.

Order of Operations
1. Simplify expressions inside grouping symbols. 2. Evaluate all powers. 3. Do all multiplications and divisions from left to right. 4. Do all additions and subtractions from left to right.

You can evaluate an algebraic expression when the value of each variable is known. Replace each variable with its value and then use the order of operations to perform the indicated operations. Remember to do all operations within grouping symbols first.

Example 1: Evaluate $15 - 12 \div 4$.

$$15 - 12 \div 4 = 15 - 3 = 12$$

Example 2: Evaluate $x^3 + 5(y - 3)$ if $x = 2$ and $y = 12$.

$$x^3 + 5(y - 3) = 2^3 + 5(12 - 3) = 2^3 + 5(9) = 8 + 5(9) = 8 + 45 = 53$$

Evaluate each expression.

1. $10 + 8 \cdot 1$ 18
2. $3^2 \div 3 + 2^2 \cdot 7 - 20 \div 5$ 27
3. $12(20 - 17) - 3 \cdot 6$ 18
4. $\frac{15 + 60}{30 - 5}$ 3

Evaluate each expression when $x = 2$, $y = 3$, $a = \frac{4}{5}$, and $b = \frac{3}{5}$.

5. $x + 7$ 9
6. $3x - 5$ 1
7. $6a + 8b$ $\frac{48}{5}$, or $9\frac{3}{5}$
8. $a^2 + 2b$ $\frac{46}{25}$, or $1\frac{21}{25}$
9. $\frac{5a^2b}{y}$ $\frac{16}{25}$
10. $(10x)^2 + 100a$ 480
11. $23 - (a + b)$ $\frac{108}{5}$, or $21\frac{3}{5}$
12. $\frac{x^4 - y^2}{3ay}$ $\frac{35}{36}$

CHECK FOR UNDERSTANDING

Communicating Mathematics

Study the lesson. Then complete the following. 1. Multiply 7 and 2.

1. When evaluating the expression $4 + 7 \cdot 2$, what would you do first?
2. **Name** two types of grouping symbols and explain how you may use them. 2–3. See margin.
3. **Explain** how you would evaluate the expression $2[5 + (30 \div 6)]^2$.

Guided Practice

Evaluate each expression.

4. $15 + 3 \cdot 2$ 21
5. $5^3 + 3(4^2)$ 173
6. $\frac{38 - 12}{2 \cdot 13}$ 1
7. $12 \div 3 \cdot 5 - 4^2$ 4

Evaluate each expression when $w = 12$, $x = 5$, $y = 6$, and $z = 4$.

8. $wx - yz$ 36
9. $2w + x^2 - yz$ 25
10. Use two or more of the variables in Exercises 8–9 and their assigned values to write an expression whose value is 18.
11. If $a = 1.5$, which is less, $a^2 - a$ or 1? $a^2 - a$
12. **Geometry** Find the area of the rectangle when $t = 6$ inches. 78 in^2

2t + 1

t

10. Sample answer: $xz - \frac{w}{y}$

EXERCISES

Practice

Evaluate each expression.

A

13. $3 + 2 \cdot 3 + 5$ 14
14. $(4 + 5)7$ 63
15. $29 - 3(9 - 4)$ 14
16. $50 - (15 + 9)$ 26
17. $15 \div 3 \cdot 5 - 4^2$ 9
18. $4(11 + 7) - 9 \cdot 8$ 0

B

19. $\frac{(4 \cdot 3)^2 \cdot 5}{(9 + 3)}$ 60
20. $[7(2) - 4] + [9 + 8(4)]$ 51
21. $\frac{6 + 4^2}{3^2(4)}$ $\frac{11}{18}$
22. $(5 - 1)^3 + (11 - 2)^2 + (7 - 4)^3$ 172
23. $\frac{2 \cdot 8^2 - 2^2 \cdot 8}{2 \cdot 8}$ 6
24. $7(0.2 + 0.5) - 0.6$ 4.3

Evaluate each expression when $v = 5$, $x = 3$, $a = 7$, and $b = 5$.

25. $(v + 1)(v^2 - v + 1)$ 126
26. $\frac{x^2 - x + 6}{x + 3}$ 2
27. $[a + 7(b - 3)]^2 \div 3$ 147
28. $v^2 - (x^3 - 4b)$ 18
29. $(2v)^2 + ab - 3x$ 126
30. $\frac{a^2 - b^2}{v^3}$ $\frac{24}{125}$

Reteaching

Using Guess and Check Have students insert parentheses so that each expression has the indicated value.

1. $56 \div 6 + 2 \div 1$; 7
   $56 \div (6 + 2) \div 1$
2. $2 \cdot 8 - 2 + 6 \div 3$; 14
   $2 \cdot (8 - 2) + 6 \div 3$
3. $2 \cdot 8 - 2 + 6 \div 3$; 0
   $2 \cdot (8 - (2 + 6)) \div 3$
4. $20 - 2 \cdot 3^2 + 12 \div 6$; 0
   $20 - (2 \cdot 3^2 + 12 \div 6)$

Additional Answers

2. parentheses, brackets, braces, fraction bar; to show what operation should be completed first
3. First, divide 30 by 6, which equals 5. Second, add 5 and 5, which equals 10. Third, square 10, which equals 100. Fourth, multiply 100 times 2, which equals 200.

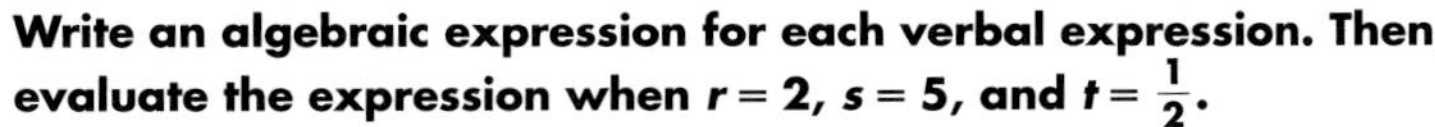

Write an algebraic expression for each verbal expression. Then, evaluate the expression when $r = 2$, $s = 5$, and $t = \frac{1}{2}$.

31. the square of r increased by $3s$ **$r^2 + 3s$; 19**
32. t times the sum of four times s and r **$t(4s + r)$; 11**
33. the sum of r and s times the square of t **$(r + s)t^2$; $\frac{7}{4}$**
34. r to the fifth power decreased by t **$r^5 - t$; $31\frac{1}{2}$**

Integration: Geometry

A formula for the perimeter of each figure is given. Find the perimeter when $a = 5$, $b = 6$, and $c = 8.5$.

35. triangle
$P = a + b + c$

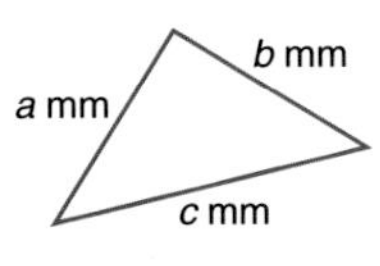

19.5 mm

36. square
$P = 4c$

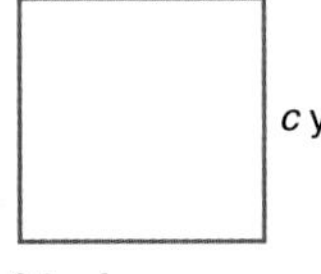

34 yd

37. parallelogram
$P = 2(a + b)$

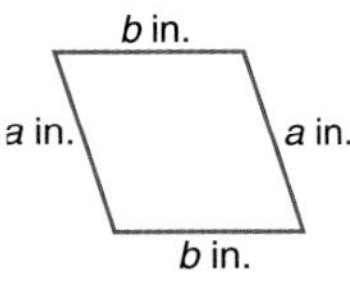

22 in.

Graphing Calculator

Use a graphing calculator to evaluate each expression to the nearest hundredth.

38. $5(2)^4 + 3$ **83**
39. $\frac{(5 \cdot 7)^2 + 5}{(9 \cdot 3^2) - 7}$ **16.62**
40. $4.79\,(0.05)^2 + 0.375\,(6.34)^3$ **95.58**
41. $1 - 2u + 3u^2$ when $u = 1.35$ **3.77**

Critical Thinking

42. **Patterns** Consider the value of the expression $\frac{2x - 1}{2x}$ for various values of x.
   a. Copy and complete the chart below.

x	$\frac{1}{2}$	1	10	50	100
$\frac{2x - 1}{2x}$	**0**	**$\frac{1}{2}$**	**$\frac{19}{20}$**	**$\frac{99}{100}$**	**$\frac{199}{200}$**

b. As the value of x gets larger, the value of the expression approaches 1.

   b. Describe the pattern that is emerging.
   c. What would the approximate value of the expression be if x were a very large number? **1**

43. Consider the following sequence of numbers.

   4 2 5 3 2

   a. Insert operation symbols and parentheses in the sequence so that its value is 2. **Sample answer: $(4 - 2)5 \div (3 + 2)$**
   b. Insert operation symbols and parentheses in the sequence so that the answer is the largest possible number. **$4 \times 2 \times 5 \times 3 \times 2$**

Applications and Problem Solving

44. **Accounting** Alicia and Travis are selling tickets for a school talent show. Bleacher seats cost $3.00, and floor seats cost $4.00. Alicia sells 30 bleacher seat tickets and 25 floor seat tickets. Travis sells 65 bleacher seat tickets.
   a. Write an expression to show how much money Alicia and Travis have collected for the tickets. **30(3) + 25(4) + 65(3)**
   b. How much money have they collected? **$385**

4 ASSESS

Closing Activity

Writing Have students work as teams writing numerical expressions using only the four digits from the four-digit number that represents the current year. For example, 1997 can be represented as

$$1 \cdot (9 - 9) \cdot 7 = 0.$$

Compare answers to see what values were generated and how many different expressions were written for particular values.

Chapter 1, Quiz A (Lessons 1-1 through 1-3), is available in the *Assessment and Evaluation Masters,* p. 16.

Practice Masters, p. 3

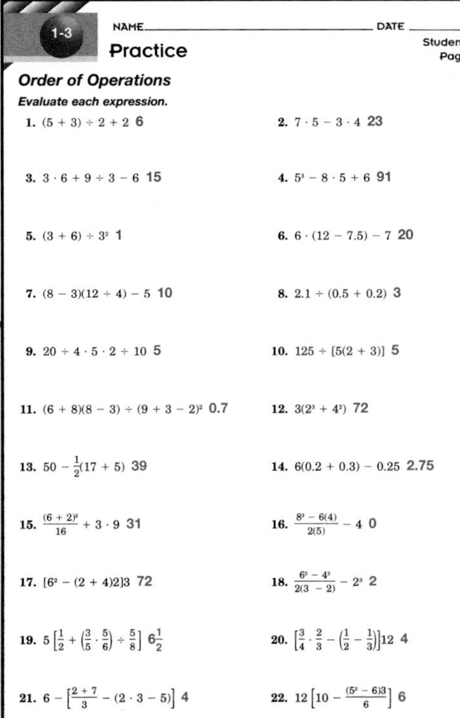

1-3 NAME ____________ DATE ______

Practice — Student Edition Pages 19–24

Order of Operations

Evaluate each expression.

1. $(5 + 3) \div 2 + 2$ 6
2. $7 \cdot 5 - 3 \cdot 4$ 23
3. $3 \cdot 6 + 9 \div 3 - 6$ 15
4. $5^3 - 8 \cdot 5 + 6$ 91
5. $(3 + 6) \div 3^2$ 1
6. $6 \cdot (12 - 7.5) - 7$ 20
7. $(8 - 3)(12 \div 4) - 5$ 10
8. $2.1 \div (0.5 + 0.2)$ 3
9. $20 \div 4 \cdot 5 \cdot 2 \div 10$ 5
10. $125 \div [5(2 + 3)]$ 5
11. $(6 + 8)(8 - 3) \div (9 + 3 - 2)^2$ 0.7
12. $3(2^3 + 4^2)$ 72
13. $50 - \frac{1}{2}(17 + 5)$ 39
14. $6(0.2 + 0.3) - 0.25$ 2.75
15. $\frac{(6 + 2)^2}{16} + 3 \cdot 9$ 31
16. $\frac{8^2 - 6(4)}{2(5)} - 4$ 0
17. $[6^2 - (2 + 4)2]3$ 72
18. $\frac{6^2 - 4^2}{2(3 - 2)} - 2^3$ 2
19. $5\left[\frac{1}{2} + \left(\frac{3}{5} \cdot \frac{5}{6}\right) \div \frac{5}{8}\right]$ $6\frac{1}{2}$
20. $\left[\frac{3}{4} \cdot \frac{2}{3} - \left(\frac{1}{2} - \frac{1}{3}\right)\right]12$ 4
21. $6 - \left[\frac{2 + 7}{3} - (2 \cdot 3 - 5)\right]$ 4
22. $12\left[10 - \frac{(5^2 - 6)3}{6}\right]$ 6

45. **World Landmarks** The Great Pyramid of Cheops in Egypt is considered to be one of the Seven Wonders of the World. It is also the largest pyramid in the world. The area of its base is 4050 square meters. The volume of any pyramid is one-third the product of the area of the base B and its height h.

Pyramid	Height (meters)
Great Pyramid of Cheops in Egypt	147
Bent Pyramid in Egypt	101
Inca Pyramid in Peru	75
Pyramid of the Sun in Mexico	60
Step Pyramid of Djoser in Egypt	60
Pyramid of the Sun in Peru	50

a. Write an expression that represents the volume of a pyramid. $\frac{1}{3}Bh$

b. Find the volume of the Great Pyramid of Cheops. **198,450 m^3**

Mixed Review

46. Find the next two terms in the sequence 2, 4, 8, 16, ... (Lesson 1–2)

47. Find the next two terms in the sequence 2, 5.5, 9, 12.5, ... (Lesson 1–2) **16, 19.5**

48. Find the next two items in the pattern a, a^2b, a^3b^2c, $a^4b^3c^2d$, ... (Lesson 1–2)

49. Ishi wants to save her money to buy a drum set from her uncle. Her uncle is willing to sell her the drums for \$525. She has already saved \$357, and she plans to add \$1 on June 1, \$2 on June 2, \$3 on June 3, and so on. On what day will Ishi be able to buy the drums? (Lesson 1–2) **June 18**

50. Write an algebraic expression for *h to the fifth power*. (Lesson 1–1) h^5

51. The price of tickets for the spring concert is \$3 more than last year. If last year's tickets cost t dollars, write an expression for the cost of the tickets this year. (Lesson 1–1) $t + 3$

52. Evaluate 11^2. (Lesson 1–1) **121**

53. Write a verbal expression for $9 + 2y$. (Lesson 1–1) **9 more than two times y**

46. 32, 64

48. $a^5b^4c^3d^2e$, $a^6b^5c^4d^3e^2f$

Extension

Reasoning Determine all the possible values for the numerical expression $8 - 2 \cdot 3 + 1$ if you ignore the rules for order of operation.

There are five possible values: 0, 1, 3, 19, and 24.

Enrichment Masters, p. 3

1-3 NAME________ DATE ________

Enrichment — Student Edition Pages 19–24

Puzzling Primes

A *prime number* has only two factors, itself and 1. The number 6 is not prime because it has 2 and 3 as factors; 5 and 7 are prime. The number 1 is not considered to be prime.

1. Use a calculator to help you find the 25 prime numbers less than 100.

   2, 3, 5, 7, 11, 13, 17, 19, 23, 29, 31, 37 41, 43, 47, 53, 59, 61, 67, 71, 73, 79, 83, 89, 97

Prime numbers have interested mathematicians for centuries. They have tried to find expressions that will give all the prime numbers, or only prime numbers. In the 1700s, Euler discovered that the expression $x^2 + x + 41$ will yield prime numbers for values of x from 0 through 39.

2. Find the prime numbers generated by Euler's formula for x from 0 through 7.

   41, 43, 47, 53, 61, 71, 83, 97

3. Show that the expression $x^2 + x + 31$ will not give prime numbers for very many values of x.

   It works for $x = 0, 2, 3, 5$, and 6 but not for $x = 1, 4$, and 7.

4. Find the largest prime number generated by Euler's formula. 1601

Goldbach's Conjecture is that every nonzero even number greater than 2 can be written as the sum of two primes. No one has ever proved that this is always true. No one has disproved it, either.

5. Show that Goldbach's Conjecture is true for the first 5 even numbers greater than 2.

   $4 = 2 + 2$, $6 = 3 + 3$, $8 = 3 + 5$, $10 = 3 + 7$, $12 = 5 + 7$

6. Give a way that someone could disprove Goldbach's Conjecture.

   Find an even number that cannot be written as the sum of two primes.

1-4 Integration: Statistics Stem-and-Leaf Plots

What YOU'LL LEARN

- To display and interpret data on a stem-and-leaf plot.

Why IT'S IMPORTANT

Stem-and-leaf plots are a useful way to display data.

APPLICATION

Consumerism

Carmela Perez sells artificial joints to hospitals. She does a lot of highway driving and frequently works with the doctors who perform operations to replace joints. Her present car is old, and she needs to buy a new one. Before deciding which car to buy, she wants to find out the miles-per-gallon (MPG) ratios of the cars she likes. This information will help her decide which cars to investigate in detail before she chooses one to buy. The MPG ratios for 25 cars are listed below.

31	30	28	26	22	31	26	34	47
32	18	33	26	23	18	29	13	40
31	42	17	22	50	12	41		

Which MPG ratio occurs most frequently? Which is the highest MPG ratio? Which is the lowest?

Each day when you read newspapers or magazines, watch television, or listen to the radio, you are bombarded with numerical information about food, sports, the economy, politics, and so on. Interpreting this numerical information, or **data,** is important to your understanding of the world around you. A branch of mathematics called **statistics** helps provide you with methods of collecting, organizing, and interpreting data.

Graphs are often used to display data. The misuse of graphs can lead to false assumptions. One way that graphs are often used to mislead the reader is by labeling the vertical and horizontal scales inconsistently. All of the interval marks should represent the same units. If either of the scales does not begin at zero, this should be indicated by a broken or jagged line. For example, suppose Ana wants to convince her parents that her math grades are improving. Which graph below seems to show the greatest improvement in Ana's grades?

Graph A

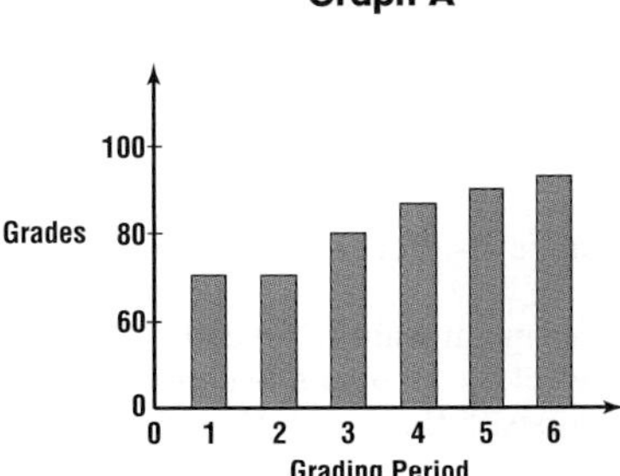

Graph B

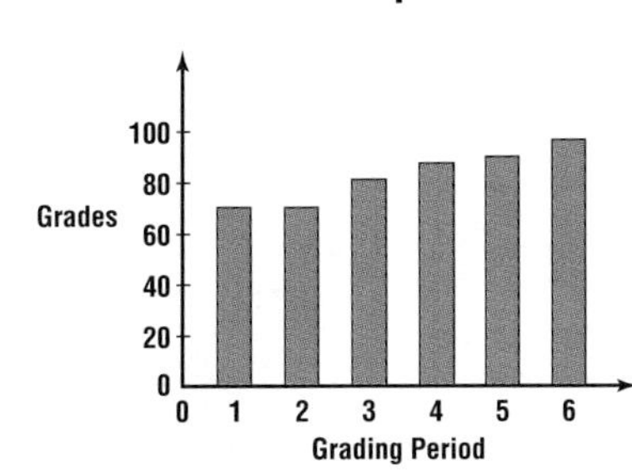

Graph A seems to show the greatest improvement. However, it is misleading. Notice the vertical axis. The distance from 0 to 60 is the same as the distance from 80 to 100. This is an incorrect representation of the data and can lead to the wrong conclusions.

Classroom Vignette

"I use a version of stem-and-leaf plots all the time to show grades from each test."

Example:

10	0
9	5 6 9
9	0 3 4
8	9
8	0 3 3

Nancy Kinard
Palm Beach Gardens High School
Palm Beach Gardens, Florida

1-4 LESSON NOTES

NCTM Standards: 1–5, 10

Instructional Resources

- Study Guide Master 1-4
- Practice Master 1-4
- Enrichment Master 1-4

Transparency 1-4A contains the 5-Minute Check for this lesson; **Transparency 1-4B** contains a teaching aid for this lesson.

Recommended Pacing	
Standard Pacing	Days 5 & 6 of 15
Honors Pacing	Days 4 & 5 of 14
Block Scheduling*	Day 3 of 7
Alg. 1 in Two Years*	Days 7 & 8 of 22

*For more information on pacing and possible lesson plans, refer to the *Block Scheduling Booklet* and *Algebra 1 in Two Years.*

1 FOCUS

5-Minute Check
(over Lesson 1-3)

Evaluate each expression.

1. $5 + 6 \cdot 3 + 7$ **30**
2. $3(10 + 6) - 8 \cdot 6$ **0**
3. $5(3^2 + 2^3)$ **85**

Evaluate if $x = 4$, $y = 2$, and $z = 5$.

4. $3xz - y^3$ **52**
5. $6x + y^2 - z^2$ **3**

Motivating the Lesson

Hands-On Activity Have students pair off and measure one another's height, recording the data in inches rounded to the nearest whole number. Students should write their partner's height on the board in no particular order. Then ask students to think of ways to organize and present the data for the entire class.

In-Class Example

For Example 3

Measurements were taken of a stand of oak and maple trees. Below is a table summarizing heights rounded to the nearest foot. Construct a stem-and-leaf plot and answer the questions.

Oaks		Maples	
35	51	28	27
42	28	35	28
38	35	38	31
35	48	27	22
47	42	22	32
28	38	41	35
38	36	32	29
30	42	35	27

Oak	Stem	Maple
1	5	
22278	4	1
05556888	3	122558
88	2	22777889

3 | 1 = 31

a. What is the height of the shortest oak? The shortest maple? **28, 22**
b. What is the difference in height between the shortest oak and the shortest maple? **6**
c. What does 3 | 5 represent in each plot? **35 feet**
d. What is the greatest number of oaks that are the same height? What is the greatest number of maples that are the same height? **3, 3**
e. How many maples are over 35 feet tall? How many oaks are over 40 feet tall? **2, 6**

Teaching Tip Notice that all of the leaves in the examples are in numerical order. Encourage students to automatically order the leaves.

3 PRACTICE/APPLY

Check for Understanding

Exercises 1–10 are designed to help you assess your students' understanding through reading, writing, speaking, and modeling. You should work through Exercises 1–4 with your students and then monitor their work on Exercises 5–10.

In this case, the heights of the boys and the heights of the girls are to be compared. To compare these numbers most effectively, use a back-to-back stem-and-leaf plot.

Boys	Stem	Girls
3 2 1 1 0	7	1 2
9 6 5 3 0	6	0 1 1 1 2 3 4 5
9 8	5	7 9 *5 | 7 = 57*

a. What is the height of the shortest boy? the shortest girl?
The shortest boy is 58 inches tall. The shortest girl is 57 inches tall.

b. What is the difference in height between the shortest boy and the tallest girl?
72 − 58 or 14 inches

c. What does 6 | 3 represent in each plot?
63 inches

d. What is the greatest number of boys who are the same height? What is the greatest number of girls who are the same height?
There are 2 boys who are 71 inches tall and 3 girls who are 61 inches tall.

e. What patterns, if any, do you see in the data?
The boys appear to be slightly taller than the girls. Seven boys are 65 inches or taller, while only 2 of the girls are that tall.

CHECK FOR UNDERSTANDING

Communicating Mathematics

Study the lesson. Then complete the following.

1. **Describe** the information you can determine by looking at a stem-and-leaf plot. **See margin.**
2. **Name** two or three ways in which a graph can be misleading.
3. **List** the steps used to make a stem-and-leaf plot. **See margin.**
4. **Assess Yourself** Write a paragraph about your use of stem-and-leaf plots. Begin the paragraph by completing this sentence: "I can use stem-and-leaf plots to ____________." Think of the ways in which you deal with numbers: earnings from a job, number of assignments turned in, test scores, attendance at club meetings, and so on. **See margin.**

2. not beginning an axis at 0, not using uniform spacing on scales, not indicating that the scale does not begin at 0 with a broken line

MATH JOURNAL

Guided Practice

Suppose the number 25,678 is rounded to 25,700 and plotted using stem 25 and leaf 7. Write the stem and leaf for each number below if the numbers are part of the same set of data. **7. stem 126, leaf 9**

5. 12,221 **stem 12, leaf 2** **6.** 6323 **stem 6, leaf 3** **7.** 126,896

8. Write the stems that would be used for a plot of the following set of data.

57, 43, 34, 12, 29, 8 **0, 1, 2, 3, 4, 5**

Reteaching

Using Discussion Have students discuss the advantages and disadvantages of stem-and-leaf plots. Have them compare a stem-and-leaf plot to a simple list of data. Emphasize that while original data must be reconstructed from a stem-and-leaf plot, it is a more concise representation of the data and displays the data in order.

Additional Answers

1. **Answers will vary. Responses should include least and greatest values, frequency in categories, distribution, clusters, and gaps.**

3a. **Decide on stems.**
3b. **Place stems and leaves on a chart.**
3c. **Arrange leaves in numerical order.**

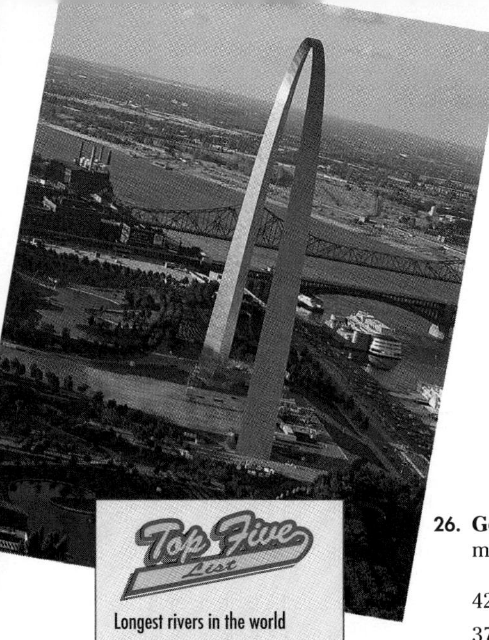

Top Five List

Longest rivers in the world

1. Nile, 4145 miles
2. Amazon, 4007 miles
3. Yangtze-Kiang, 3915 miles
4. Mississippi-Missouri-Red Rock, 3710 miles
5. Yenisey-Angara-Selenga, 3442 miles

25. **Economics** The table at the right shows the average acreage per farm for six western states in 1980 and 1993 according to the *Universal Almanac.*

State	1980	1993
Arizona	5080	4557
Colorado	1358	1286
Montana	2601	2445
Nevada	3100	3708
New Mexico	3467	3274
Wyoming	3846	3742

a. Round each number to the nearest hundred. Then make a back-to-back stem-and-leaf plot of the average farm sizes for 1980 and 1993 for these six states. **See margin.**

b. Which interval shows the most common farm size found in 1980 for these six states? in 1993? **3000 to 3900; same**

c. Write a statement about the information the stem-and-leaf plot presents. **Sample answer: Farm sizes seem to be shrinking.**

26. **Geography** The numbers below show the lengths (in miles) of 30 major rivers in North America as listed in the *World Almanac.*

424	313	444	301	659	652	314	538
377	800	883	525	360	512	500	722
865	360	390	425	309	336	430	692
540	610	800	350	300	420		

a. Round each number to the nearest ten. Then make a stem-and-leaf plot of the data. **See margin.**

b. What is the difference in length between the shortest and longest rivers? **580 miles**

c. How many rivers are less than 400 miles long? **11**

Mixed Review

27. Evaluate $3 \cdot 6 - \frac{12}{4}$. (Lesson 1–3) **15**

28. Evaluate $9a - 4^2 + b^2 \div 2$ when $a = 3$ and $b = 6$. (Lesson 1–3) **29**

29. **Weather** The time between seeing lightning flash and hearing the bang of thunder it produces can be used to approximate the distance from the lightning. The distance in miles from lightning can be approximated by dividing the number of seconds between seeing the lightning and hearing the thunder by 5. Lightning within three miles can be dangerous and is a warning to take shelter.

a. Suppose there are s seconds between seeing the lightning and hearing the thunder. Write an algebraic expression for the distance from the lightning. (Lesson 1–1) $\frac{s}{5}$

b. Fernando counted 10 seconds between seeing lightning and hearing thunder. How far away was the lightning? Is Fernando in danger? (Lesson 1–3) **2 miles; yes**

30. Find the next two terms in the sequence $\frac{1}{2}, \frac{3}{4}, \frac{5}{8}, \frac{7}{16}, \frac{9}{32}, \ldots$ (Lesson 1–2) $\frac{11}{64}, \frac{13}{128}$

31. Give the next two items for the following pattern. (Lesson 1–2)

The Nile, the world's longest river, flows through four countries: Egypt, Sudan, Ethiopia, and Uganda.

Additional Answers

25a.

1980	Stem	1993
4	1	3
6	2	4
8 5 1	3	3 7 7
	4	6
1	5	

$3|7 = 3700$

26a.

Stem	Leaf
3	0 0 1 1 1 4 5 6 6 8 9
4	2 2 3 3 4
5	0 1 3 4 4
6	1 5 6 9
7	2
8	0 0 7 8

$6|5 = 650$

Extension

Reasoning The table below shows the percent of households in two major cities that own the appliances listed. Explain why a stem-and-leaf plot of the data would not be an adequate representation.

Appliance	City A	City B
TV	88	94
Dishwasher	28	43

Even though a back-to-back stem-and-leaf plot would distinguish between the data for City A and City B, it would not distinguish between the data for TVs and dishwashers.

Enrichment Masters, p. 4

NAME ______ DATE ______

Enrichment

Student Edition Pages 25–31

Latin Squares

In designing a statistical experiment, it is important to try to randomize the variables. For example, suppose 4 different motor oils are being compared to see which give the best gasoline mileage. An experimenter might then choose 4 different drivers and four different cars. To test-drive all the possible combinations, the experimenter would need 64 test-drives.

To reduce the number of test drives, a statistician might use an arrangement called a **Latin Square**.

For this example, the four motor oils are labeled A, B, C, and D and are arranged as shown. Each oil must appear exactly one time in each row and column of the square.

The drivers are labeled D(1), D(2), D(3), and D(4); the cars are labeled C(1), C(2), C(3), and C(4).

Now, the number of test-drives is just 16, one for each cell of the Latin Square.

	D(1)	D(2)	D(3)	D(4)
C(1)	A	B	C	D
C(2)	B	A	D	C
C(3)	C	D	A	B
C(4)	D	C	B	A

Create two 4-by-4 Latin Squares that are different from the example. Answers will vary. Sample answers are given.

1.

	D(1)	D(2)	D(3)	D(4)
C(1)	B	C	A	D
C(2)	D	A	C	B
C(3)	C	D	B	A
C(4)	A	B	D	C

2.

	D(1)	D(2)	D(3)	D(4)
C(1)	D	C	B	A
C(2)	A	B	C	D
C(3)	B	D	A	C
C(4)	C	A	D	B

Make three different 3-by-3 Latin Squares.

3.

	D(1)	D(2)	D(3)
C(1)	A	B	C
C(2)	B	C	A
C(3)	C	A	B

4.

	D(1)	D(2)	D(3)
C(1)	A	C	B
C(2)	C	B	A
C(3)	B	A	C

5.

	D(1)	D(2)	D(3)
C(1)	B	C	A
C(2)	C	A	B
C(3)	A	B	C

1-5 LESSON NOTES

NCTM Standards: 1–5

Instructional Resources

- Study Guide Master 1-5
- Practice Master 1-5
- Enrichment Master 1-5
- Assessment and Evaluation Masters, pp. 15–16
- Multicultural Activity Masters, p. 1
- Real-World Applications, 1
- Science and Math Lab Manual, pp. 1–4

Transparency 1-5A contains the 5-Minute Check for this lesson; **Transparency 1-5B** contains a teaching aid for this lesson.

Recommended Pacing	
Standard Pacing	Day 7 of 15
Honors Pacing	Day 6 of 14
Block Scheduling*	Day 4 of 7 (along with Lesson 1-6)
Alg. 1 in Two Years*	Days 9 & 10 of 22

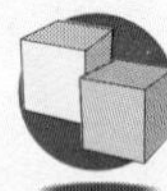

*For more information on pacing and possible lesson plans, refer to the *Block Scheduling Booklet* and *Algebra 1 in Two Years.*

1 FOCUS

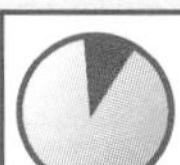

5-Minute Check
(over Lesson 1-4)

1. Use a stem-and-leaf plot to represent the following test data: 56, 75, 73, 89, 92, 87, 64, and 82.

Stem	Leaf
9	2
8	2 7 9
7	3 5
6	4
5	6

$9|2 = 92$

2. Based on the plot, in what interval did the most grades fall? **80–89**
3. Given the data 125, 128, 134, 158, 164, 144, 168, 172, and 165, what stems would you use to represent the data? **12, 13, 14, 15, 16, 17**

1-5 Open Sentences

What YOU'LL LEARN

- To solve open sentences by performing arithmetic operations.

Why IT'S IMPORTANT

You can use open sentences to solve problems involving nutrition, biology, and weather.

APPLICATION
Nutrition

The grams of fat found in cheeseburgers sold by various fast food chains are listed in the stem-and-leaf plot at the right.

Stem	Leaf
1	3 3 3 6 9
2	0 5 8 9
3	6 9
4	6
6	3

$4|6 = 46$

Obviously, some of the fast food chains are selling burgers with more than the usual amount of fat. Some students at Middletown High School have formed a consumer awareness club and are concerned about the high content of fat in this particular cheeseburger. They decide to write a letter to the company recommending a way to reduce the fat of their cheeseburger to 37 grams.

FYI

The largest hamburger in the world was made in Seymour, Wisconsin, on August 2, 1989. It was 21 feet in diameter and weighed 5520 pounds.

The students know that a one-ounce slice of American cheese has about 7 grams of fat and that ground beef has about 6 grams of fat per ounce. The students need to determine the number of ounces of ground beef that can be used to make a cheeseburger with a fat content of 37 grams.

Seymour, Wisconsin
"The Home of the Hamburger"

Let b represent the number of ounces of beef that will go into the cheeseburger. This problem can be represented by the equation below.

$$6b + 7 = 37$$

The number 6 represents the amount of fat in one ounce of ground beef. The 7 represents the number of grams of fat in one slice of cheese. The 37 represents the total number of grams of fat for a cheeseburger.

Mathematical statements with one or more variables, or unknown numbers, are called **open sentences.** An open sentence is neither true nor false until the variable has been replaced by a value. Finding a replacement for the variable that results in a true sentence is called **solving the open sentence.** This replacement is called a **solution** of the open sentence.

A little arithmetic will show that each square foot of the hamburger weighed 15.9 lbs. Have students use the current price of hamburger to determine the cost of this giant burger.

Replace b in $6b + 7 = 37$ with the values 3, 4, 5, and 6. Then see whether each replacement results in a true or false sentence.

Replace b with:	$6b + 7 = 37$	True or False?
3	$6(3) + 7 \stackrel{?}{=} 37 \rightarrow 25 \neq 37$	false
4	$6(4) + 7 \stackrel{?}{=} 37 \rightarrow 31 \neq 37$	false
5	$6(5) + 7 \stackrel{?}{=} 37 \rightarrow 37 = 37$	true
6	$6(6) + 7 \stackrel{?}{=} 37 \rightarrow 43 \neq 37$	false

Since $b = 5$ makes the sentence $6b + 7 = 37$ true, the solution for $6b + 7 = 37$ is 5. The students can write a letter suggesting that the chain reduce the amount of fat to 37 grams by using 5 ounces of ground beef in their cheeseburger.

A set of numbers from which replacements for a variable may be chosen is called a **replacement set.** A **set** is a collection of objects or numbers. Sets are often shown by using braces { }. Each object or number in a set is called an **element,** or member. Sets are usually named by capital letters. Set A has three elements; they are 1, 3, and 5.

$A = \{1, 3, 5\}$ $\qquad B = \{2, 4, 5\}$ $\qquad C = \{1, 2, 3, 4, 5\}$

The **solution set** of an open sentence is the set of all replacements for the variable that make the sentence true.

Example 1 **Find the solution set for $y + 5 \leq 7$ if the replacement set is {0, 1, 2, 3, 4}.**

Replace y with:	$y + 5 \leq 7$	True or False?
0	$0 + 5 \stackrel{?}{\leq} 7 \rightarrow 5 \leq 7$	true
1	$1 + 5 \stackrel{?}{\leq} 7 \rightarrow 6 \leq 7$	true
2	$2 + 5 \stackrel{?}{\leq} 7 \rightarrow 7 \leq 7$	true
3	$3 + 5 \stackrel{?}{\leq} 7 \rightarrow 8 \not\leq 7$	false
4	$4 + 5 \stackrel{?}{\leq} 7 \rightarrow 9 \not\leq 7$	false

The symbol ≤ means "less than or equal to." The symbol ≥ means "greater than or equal to."

Therefore, the solution set for $y + 5 \leq 7$ is {0, 1, 2}.

A sentence that contains an equals sign, =, is called an **equation.** A sentence having the symbols $<$, $\leq$, $>$, or $\geq$ is called an **inequality.** Which of the open sentences below are equations? Which ones are inequalities?

Mathematical Sentence	Equation or Inequality?
$2x + 10 = 50$	equation
$3a \leq 43$	inequality
$y - 6 > 12$	inequality

Sometimes you can solve an equation by simply applying the order of operations.

Motivating the Lesson

Situational Problem Ramon, Kesha, Larry, and Maria are going to the movies together. After buying her ticket, Kesha has $1.75. She wants to buy a bucket of popcorn, which costs $2.90. Ramon has $1.25, Larry has $1.10, and Maria has $1.18. How can Kesha determine which of her friends does not have enough money to help her buy the popcorn? **She should add the money that she has ($1.75) to the money that each of the friends has ($1.25, $1.10, and $1.18) and determine which sum is less than $2.90. Since $1.75 + $1.10 < $2.90, Larry is the one who does not have enough money.**

2 TEACH

Teaching Tip In mathematics, sentences that can be classified as true or false are called *statements.* In general, *open sentences* are true for some variable replacements and false for others. The original sentence is neither true nor false.

In-Class Example

For Example 1
Find the solution set for each open sentence if the replacement set is {0, 2, 4, 6, 8}.

a. $8 - x < 7$ **{2, 4, 6, 8}**

b. $\frac{x + 3}{4} > 3$ **{} or 0**

Teaching Tip Point out to students that an open sentence in mathematics, like that in Example 1, is like an English sentence with a blank in it. The sentence ______ *is a color* is true if *Blue* is placed in the blank, false if *Hard* is placed in the blank, and neither true nor false if the blank is left empty.

Alternative Learning Styles

Kinesthetic Have students work in groups of four. Provide each group with two pennies, two nickels, and two dimes. Ask the groups which pair of the same coins would add up to more than 15 cents. Students should realize that only the pair of dimes adds up to more than 15 cents. Explain that they just solved the open sentence $2c > 15$ where c is the value of a coin. Provide each group with two quarters and have them construct and answer similar problems.

In-Class Examples

For Example 2

a. Solve $m = \dfrac{7 \cdot 7 + 2}{10 + 7}$. 3

b. Solve $\dfrac{4.2 + 2(3.9)}{3.1 + 2.9} = k$. 2

For Example 3

If McDonald's wanted to offer a sandwich with two quarter-pound patties and one slice of cheese, how many grams of fat would this sandwich contain? 55

3 PRACTICE/APPLY

Check for Understanding

Exercises 1–12 are designed to help you assess your students' understanding through reading, writing, speaking, and modeling. You should work through Exercises 1–5 with your students and then monitor their work on Exercises 6–12.

Study Guide Masters, p. 5

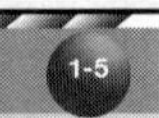

NAME ______________________ DATE ____________

Study Guide — Student Edition Pages 32–36

Open Sentences

Mathematical statements with one or more variables are called **open sentences.** Open sentences are **solved** by finding a replacement for the variable that results in a true sentence. The replacement is called a **solution.**

Example 1: Replace a in $3a + 12 = 39$ with the value 9.

$$3a + 12 = 39$$
$$3(9) + 12 = 39$$
$$27 + 12 = 39$$
$$39 = 39 \quad \text{true}$$

Since $a = 9$ makes the sentence $3a + 12 = 39$ true, 9 is a solution.

A set of numbers from which replacements for a variable may be chosen is called a **replacement set.** The set of all replacements for the variable in an open sentence that results in a true sentence is called the **solution set** for the sentence.

A sentence that contains an equals sign, =, is called an **equation** and sometimes may be solved by simply applying the order of operations. A sentence having the symbols < or > is called an **inequality.**

Example 2: Solve $\dfrac{2(3 + 1)}{3(2 + 1)} = b$.

$$\frac{2(3 + 1)}{3(2 + 1)} = b$$
$$\frac{2(4)}{3(3)} = b$$
$$\frac{8}{9} = b$$

State whether each equation is true or false for the value of the variable given.

1. $y + \frac{1}{2} = \frac{1}{4} + \frac{1}{2}, y = \frac{7}{2}$ False
2. $x^4 = 2^8, x = 4$ True
3. $a^2 + a^3 + 2 < 10, a = 2$ False
4. $\frac{2^3 - 2d}{3^2 - 1} \leq 2, d = 2$ True

Find the solution set for each inequality if the replacement sets are $x = \{\frac{1}{2}, 2, 3, \frac{1}{4}\}$ and $y = \{2, 4, 6, 8\}$.

5. $x + 4 < 6$ $\{\frac{1}{2}, \frac{1}{4}\}$
6. $3y \geq 18$ {2, 4, 6}
7. $\frac{y}{3} > 1$ {4, 6, 8}

Solve each equation.

8. $s = \frac{15 - 6}{27 - 24}$ 3
9. $w = 6^2 - 3^3$ 9
10. $c = 3\frac{1}{2} + 2\frac{1}{4}$ $5\frac{3}{4}$

Example  Solve $\dfrac{5(3 + 5)}{3 \cdot 2 + 2} = d$.

$$\frac{5(3 + 5)}{3 \cdot 2 + 2} = d$$
$$\frac{5(8)}{6 + 2} = d$$
$$\frac{40}{8} = d \quad \textit{Evaluate the numerator and denominator.}$$
$$5 = d \quad \textit{Divide.}$$

The solution is 5.

Nutrition

Example 3 **Refer to the application at the beginning of the lesson. Solve for the number of grams of fat f in a McDonald's Quarter Pounder® with Cheese. This cheeseburger has one slice of cheese and four ounces of ground beef. Assume that the bun and toppings contain no fat.**

$f = 6b + 7$ — *f = grams of fat and b = ounces of beef*

$f = 6(4) + 7$ — *Replace b with 4.*

$f = 24 + 7$ — *Evaluate using the order of operations.*

$f = 31$

A Quarter Pounder® with Cheese has about 31 grams of fat.

CHECK FOR UNDERSTANDING

Communicating Mathematics

1–4. See Solutions Manual.

Study the lesson. Then complete the following.

1. **Explain** why an open sentence always has at least one variable.
2. **Explain** the difference between an expression and an open sentence.
3. **Define** in your own words the phrase *solution set of an open mathematical sentence.*
4. **Explain** how to find the solution set for $7 + 2n > 31$ if the replacement set for n is {10, 11, 12, 13}.

5. Make up an inequality with a replacement set. Find the solution set and explain how you obtained it. See student's work.

Guided Practice

State whether each equation is *true* or *false* for the value of the variable given.

6. $7(x^2) - 15 \div 5 = 25, x = 2$ true
7. $\dfrac{7a + a}{(9 \cdot 3) - 7} = 4, a = 5$ false

Find the solution set for each inequality, given the replacement set.

8. $3x + 2 > 2$; {0, 1, 2} {1, 2}
9. $2y^2 - 1 > 0$; {1, 3, 5} {1, 3, 5}

Solve each equation for *y* if *x* is replaced with 6.

10. $x - 2 = y$ 4
11. $2x^2 + 3 = y$ 75

Reteaching

Decision Making State whether each sentence is *true, false,* or *open.* If the sentence is open, replace the variable with a value to make the sentence true.

1. $2 + 3 \cdot 5 = 30$ false
2. $2 + 3 \cdot 5 = x$ open; 17
3. $40 - 8 \div 2^2 = 38$ true
4. $x - 8 \div 2^2 = 24$ open; 26

Cooperative Learning

Group Discussion Separate the class into small groups. Have them fill in the blanks in the open sentence, "The green ____ is bigger than the blue ____" in as many ways as possible. Discuss how students might have filled in this open sentence to make it false. For more information on the group discussion strategy, see *Cooperative Learning in the Mathematics Classroom,* one of the titles in the Glencoe Mathematics Professional Series, page 31.

12. **Diet** During a lifetime, the average American drinks 15,579 glasses of milk, 6220 glasses of fruit juice, and 18,995 glasses of soft drinks.
   a. Write an equation for the total number of glasses of milk, juice, and soft drinks that the average American drinks in a lifetime.
   b. What is the total number of glasses of milk, juice, and soft drinks consumed by the average American in a lifetime? **40,794 glasses**
   12a. $g = 15{,}579 + 6220 + 18{,}995$

EXERCISES

Practice

State whether each equation is *true* or *false* for the value of the variable given.

13. $a + \frac{3}{4} = \frac{3}{2} + \frac{1}{4}, a = \frac{1}{2}$ **false**
14. $\frac{3 + 15}{x} = \frac{1}{2}(x), x = 6$ **true**
15. $y^6 = 4^3, y = 2$ **true**
16. $3x^2 - 4(5) = 6, x = 3$ **false**
17. $\frac{5^2 - 2y}{5^2 - 6} \leq 1, y = 3$ **true**
18. $a^5 \div 8 \div a^2 \div a < \frac{1}{2}, a = 2$ **false**

Find the solution set for each inequality if the replacement sets are $x = \left\{\frac{1}{2}, \frac{3}{4}, 1, \frac{5}{4}\right\}$ and $y = \{5, 10, 15, 20\}$.

20. {10, 15, 20}
21. $\left\{\frac{1}{2}, \frac{3}{4}\right\}$

19. $y - 2 < 6$ **{5}**
20. $y + 2 > 7$
21. $8x + 1 < 8$
22. $2x > 1$ $\left\{\frac{3}{4}, 1, \frac{5}{4}\right\}$
23. $\frac{y}{5} \geq 2$ **{10, 15, 20}**
24. $3x \leq 4$ $\left\{\frac{1}{2}, \frac{3}{4}, 1, \frac{5}{4}\right\}$

Solve each equation.

25. $y = \frac{14 - 8}{2}$ **3**
26. $4(6) + 3 = a$ **27**
27. $\frac{21 - 3}{12 - 3} = x$ **2**
28. $d = 3\frac{1}{2} \div 2$ $1\frac{3}{4}$
29. $s = 4\frac{1}{2} + \frac{1}{3}$ $4\frac{5}{6}$
30. $x = 5^2 - 2^3$ **17**

Critical Thinking

31. Find five pairs of values for p and q such that the open sentence $q + 2 > 3p$ is true. **See margin.**

Applications and Problem Solving

32. **Weather Forecasting** The graph at the right shows the states with the most tornadoes.

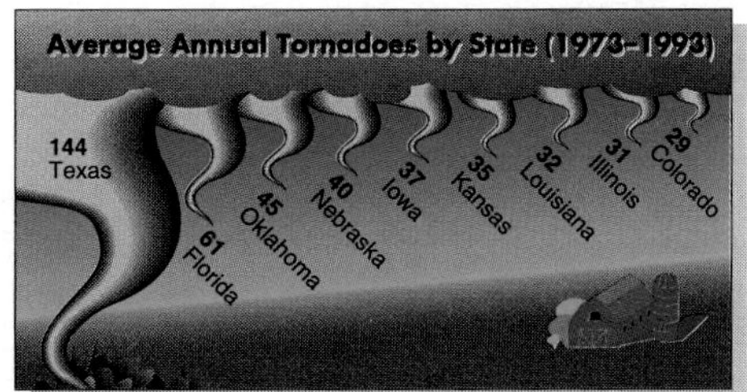

Source: National Severe Storm Forecast Center

   a. Write an equation that estimates the number of tornadoes Texas will have in the next three years. Justify why your equation will provide a good estimate. **See margin.**
   b. At this rate, how many tornadoes will Texas have in the next three years? **432 tornadoes**
   c. Make up another problem using the data in the graph. **See margin.**

33. **Nutrition** A person must burn 3500 calories to lose one pound of weight.
   a. Write an equation that represents the number of calories a person would have to burn a day to lose 4 pounds in two weeks.
   b. How many calories would the person have to burn each day?

33a. $c = \frac{3500 \cdot 4}{14}$
33b. 1000 calories

Assignment Guide

Core: 13–33 odd, 35–40
Enriched: 14–30 even, 31–40
All: Self Test, 1–10

For **Extra Practice,** see p. 757.

The red A, B, and C flags, printed only in the Teacher's Wraparound Edition, indicate the level of difficulty of the exercises.

Additional Answers

31. Sample answer: $p = 1$ and $q = 2$, $p = 2$ and $q = 10$, $p = 3$ and $q = 8$, $p = 4$ and $q = 20$, and $p = 5$ and $q = 15$

32a. $t = 3(144)$; Texas should have about 3 times their yearly average

32c. Sample answer: Write an equation that estimates the number of tornadoes that Oklahoma will have in the next five years. $t = 5(45)$

Practice Masters, p. 5

NAME__________ DATE __________

Practice — Student Edition Pages 32–36

Open Sentences

State whether each equation is true or false for the value of the variable given.

1. $\frac{2 + 12}{y} = \frac{1}{4}y, y = 7$ false
2. $5x^2 - 3(4) = 8, x = 2$ true
3. $n^3 \div 4 \div n^2 \div 2 \leq \frac{1}{2}, n = 4$ true
4. $\frac{2^3 - 3y}{4^2 - 2} \geq 12, y = 2$ false
5. $n^3 - n^2 + n < 130, n = 6$ false
6. $\frac{0.16 - 0.08}{x - 0.2} > 0.04, x = 0.3$ true

Find the solution set for each inequality if the replacement sets are $x = \left\{3\frac{1}{2}, 2, 3, 1\frac{1}{4}\right\}$ and $y = \{3, 5, 6, 8\}$.

7. $y + 6 < 10$ {3}
8. $x - 1 > 3$ ∅
9. $4x - 2 < 5$ $\left\{1\frac{1}{4}\right\}$
10. $6y \geq 36$ {6, 8}
11. $\frac{y}{3} \leq 2$ {3, 5, 6}
12. $2x - 1 \leq 2$ $\left\{1\frac{1}{4}\right\}$

Solve each equation.

13. $s = 0.87 - 0.33$ 0.54
14. $v = 2\frac{1}{4} \div 9$ $\frac{1}{4}$
15. $k = \frac{15 - 3}{0.6 + 0.6}$ 10
16. $w = 140 \div [2(8 - 3)]$ 14
17. $d = 10^3 - 5^4$ 375
18. $p = \frac{16 - 4[2(3 - 2)]}{18 - 2[5 + 1]}$ $1\frac{1}{3}$
19. $r = \frac{3^3 - 1^3}{8^2 - 12}$ $\frac{1}{2}$
20. $x = \frac{5^3 - 2^5}{4^3 - 2^2}$ $1\frac{11}{20}$
21. $a = \frac{0.6 - 0.18}{0.02 - 0.01}$ 42
22. $b = \frac{0.76 - 0.52}{0.75 - 0.63}$ 2

Tech Prep

Meteorologist Meteorologists must often look to data like that in Exercise 32 as they prepare long-range weather forecasts. Students who are interested in meteorology may wish to find out more about how meteorologists use that information in the science of weather forecasting. For more information on tech prep, see the *Teacher's Handbook.*

Extension

Communication Write a verbal statement for each open sentence.

1. $2 + 6x = 26$ **Two more than six times a number is equal to 26.**
2. $5(r - 4) = 20$ **Five times the difference of a number and 4 is 20.**
3. $a^3 > 10$ **The cube of a number is greater than 10.**
4. $7(1 + 3z)^2 \leq \frac{z}{2}$ **Seven times the quantity one plus three times z squared is less than or equal to half of z.**

4 ASSESS

Closing Activity

Modeling Have students explore the classroom for pairs of objects that they can handle such that each pair satisfies the open sentence "_____ is heavier than _______." Lay the objects out in such pairs.

Chapter 1, Quiz B (Lessons 1-4 and 1-5), is available in the *Assessment and Evaluation Masters,* p. 16.

Mid-Chapter Test (Lessons 1-1 through 1-5) is available in the *Assessment and Evaluation Masters,* p. 15.

Answers to the Self Test

7.

Stem	Leaf
4	8
5	4
6	7
7	7
8	5 9

$6|7 = 67$

8.

Stem	Leaf
1	0
2	4 5
3	5 9
4	5
7	5 6

$3|5 = 350$

Enrichment Masters, p. 5

NAME______________________ DATE ____________

Enrichment

Student Edition Pages 32–35

Solution Sets

Consider the following open sentence.

It is the tallest building in the world.

You know that a replacement for the variable *It* must be found in order to determine if the sentence is true or false. If *It* is replaced by either the Empire State Building or the Sears Tower, the sentence is true.

The set {Empire State Building, Sears Tower} is called the *solution set* of the open sentence given above. This set includes all replacements for the variable that make the sentence true.

Write the solution set of each open sentence.

1. It is the name of a state beginning with the letter A. 1. {Alabama, Alaska, Arizona, Arkansas}
2. It is a primary color. 2. {red, yellow, blue}
3. Its capital is Harrisburg. 3. {Pennsylvania}
4. It is a New England state. 4. {Maine, New Hamp., Vermont, Mass., Rhode Is., Conn.}
5. $x + 4 = 10$ 5. {6}
6. It is the name of a month that contains the letter *r*. 6. {Jan, Feb, Mar, Apr, Sept, Oct, Nov, Dec}
7. During the 1970s, she was the wife of a U.S. President. 7. {Pat Nixon, Betty Ford, Rosalyn Carter}
8. It is an even number between 1 and 13. 8. {2, 4, 6, 8, 10,12}
9. $31 = 72 - k$ 9. {41}
10. It is the square of 2, 3, or 4. 10. {4, 9, 16}

Write an open sentence for each solution set.

11. {A, E, I, O, U} 11. It is a vowel.
12. {1, 3, 5, 7, 9} 12. It is an odd number between 0 and 10.
13. {June, July, August} 13. It is a summer month.
14. {Atlantic, Pacific, Indian, Arctic} 14. It is an ocean.

34. **Biology** Insects are the largest class of animals in the world with at least 800,000 species. The fastest-moving insect is the large tropical cockroach. It scurries at speeds of up to 3.36 miles per hour, which is about 2.3 feet per second.
    a. Write an equation that represents how many miles the roach can travel in 1.5 hours. $d = (3.36)(1.5)$
    b. How many miles can the roach travel in 1.5 hours? **5.04 mi**
    c. Can the roach travel 100 feet in less than a minute? Explain. **Yes; it can travel 100 ft in 100 ÷ 2.3 or 43.5 s**

Mixed Review

35. Write the stems that would be used to form a stem-and-leaf plot for the following set of data. (Lesson 1–4)
    2.7, 5.9, 2.0, 7.7, 5.2, 6.0, 5.4, 9.9, 5.4 **2, 5, 6, 7, 9**

36. **Hockey** The stem-and leaf plot at the right shows the greatest number of goals scored by a National Hockey League player during any one season. Mario Lemieux is credited with the fourth greatest number of goals in a season. How many goals did Lemieux score during that season? (Lesson 1–4) **85**

Stem	Leaf
7	1 2 3 6 6 6
8	5 6 7
9	2

$7|2 = 72$

37. Evaluate $5(13 - 7) - 22$. (Lesson 1–3) **8**

38. Write the next two expressions for the pattern $2a + 1, 4a + 3, 6a + 5, 8a + 7, \ldots$. (Lesson 1–2) $10a + 9, 12a + 11$

39. Write a verbal expression for $x^5 - 5$. (Lesson 1–1) **5 less than *x* to the fifth power**

40. **Geometry** Write an algebraic expression for the perimeter of the triangle at the right. (Lesson 1–1) $a + b + c$

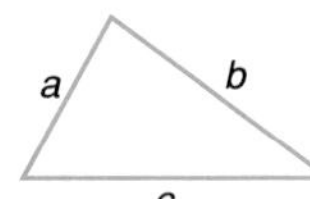

SELF TEST

Write an algebraic expression for each verbal expression. (Lesson 1–1) **1.** $3a + b^2$

1. the sum of three times *a* and the square of *b*
2. *w* to the fifth power minus 37 $w^5 - 37$
3. **Patterns** Each Tuesday morning at Central High School, the first seven bells ring at 8:00, 8:43, 8:47, 9:30, 9:34, 10:17, and 10:21 A.M. Give the times of the other morning bells. (Lesson 1–2) **11:04, 11:08, 11:51, 11:55**

Evaluate each expression. (Lesson 1–3)

4. $5(8 - 3) + 7 \cdot 2$ **39**
5. $6(4^3 + 2^2)$ **408**
6. $(9 - 2 \cdot 3)^3 - 27 + 9 \cdot 2$ **18**

Make a stem-and-leaf plot for each set of data. (Lesson 1–4) **7–8. See margin.**

7. 67, 85, 54, 48, 89, 77
8. 236, 450, 748, 254, 755, 347, 97, 386

Find the solution set for each open sentence if the replacement set is {4, 5, 6, 7, 8}. (Lesson 1–5)

9. $x + 2 > 7$ **{6, 7, 8}**
10. $9x - 20 = x^2$ **{4, 5}**

SELF TEST

The Self Test provides students with a brief review of the concepts and skills in Lessons 1-1 through 1-5. Lesson numbers are given to the right of exercises or instruction lines so students can review concepts not yet mastered.

1-6 Identity and Equality Properties

What YOU'LL LEARN

- To recognize and use the properties of identity and equality, and
- To determine the multiplicative inverse of a number.

Why IT'S IMPORTANT

You can use the identity and equality properties to evaluate expressions and solve equations.

APPLICATION

Football

According to the graph at the right, the price of Super Bowl tickets increased by \$0 from 1975 to 1977. Also, from 1981 to 1983, ticket prices stayed at \$40. The open sentences below represent each situation.

Super Bowl Ticket Prices 1967–1995

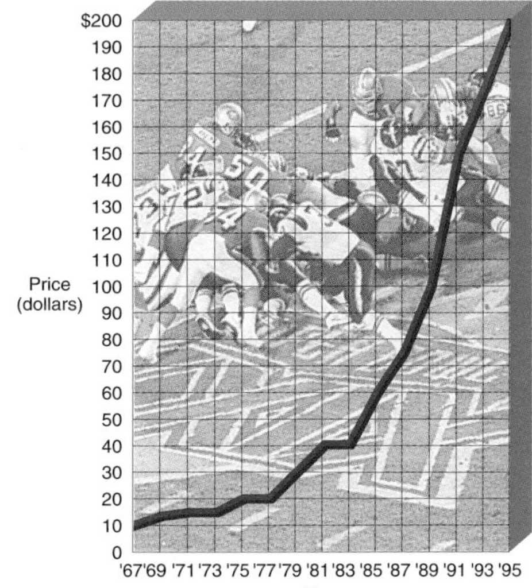

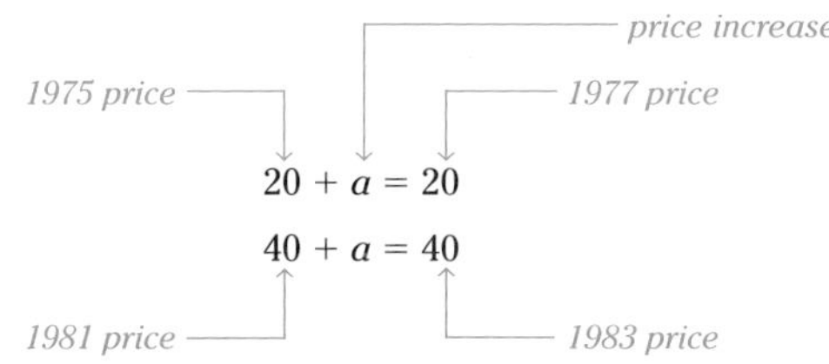

The solution for each equation is 0. Super Bowl ticket prices increased \$0 during each time period.

Equations such as these can be summarized in algebraic terms. The sum of any number and 0 is equal to that number. Zero is called the **additive identity.**

Additive Identity Property	**For any number a, $a + 0 = 0 + a = a$.**

The equations below also represent the ticket prices from 1975 to 1977 and from 1981 to 1983. The variable m represents the number of times of increase.

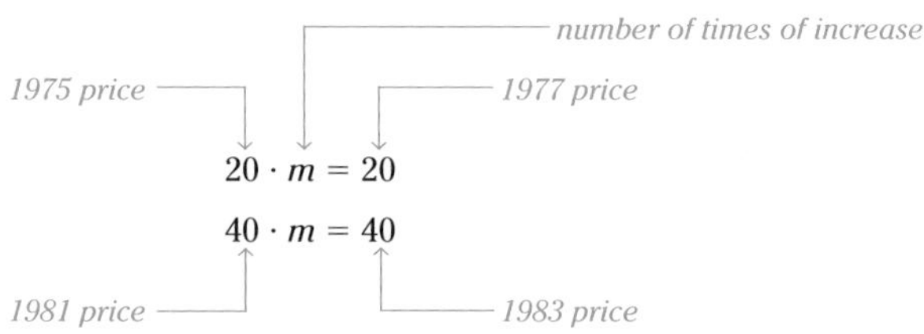

1-6 LESSON NOTES

NCTM Standards: 1–5

Instructional Resources

- Study Guide Master 1-6
- Practice Master 1-6
- Enrichment Master 1-6

Transparency 1-6A contains the 5-Minute Check for this lesson; **Transparency 1-6B** contains a teaching aid for this lesson.

Recommended Pacing	
Standard Pacing	Day 8 of 15
Honors Pacing	Day 7 of 14
Block Scheduling*	Day 4 of 7 (along with Lesson 1-5)
Alg. 1 in Two Years*	Days 11 & 12 of 22

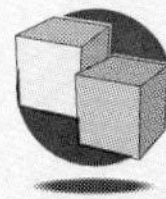

*For more information on pacing and possible lesson plans, refer to the *Block Scheduling Booklet* and *Algebra 1 in Two Years.*

1 FOCUS

5-Minute Check
(over Lesson 1-5)

Solve each equation.

1. $b = \frac{10 + 6}{4}$ **4**
2. $15.2 - 4.25 = s$ **10.95**
3. $k = \frac{8.6 - 1.16}{(1.2)(2)}$ **3.1**
4. $t = \frac{3 \cdot 9 - 5^2}{2^3 + 3^2}$ $\frac{2}{17}$
5. Find the solution set for $3x + 1 \geq 3(x + 1)$ if the replacement set is {2, 3, 4, 5, 6}. **{ } or ∅**

Motivating the Lesson

Situational Problem To attract customers to a car dealership, the manager decides to sell one of the models at cost so she will make \$0.00 in profit on each car sold. How much profit will she make if she sells 2 cars? If she sells 10 cars? if she sells n cars? **\$0.00, \$0.00, \$0.00**

2 TEACH

Teaching Tip To help explain why it must be assumed that no variable equals zero in Example 1, have students explore division by zero on a calculator.

Enter: 6 [÷] 0 [=] ERROR

Why does the display show "ERROR"?
Division by zero is undefined.

> **In-Class Example**
>
> ***For Example 1***
> **Name the multiplicative inverse of each number or variable. Assume that no variable equals zero.**
>
> **a.** 3 $\frac{1}{3}$
>
> **b.** $\frac{x}{3}$ $\frac{3}{x}$
>
> **c.** xy $\frac{1}{xy}$

The solution for each equation is 1. Since the product of any number and 1 is equal to the number, 1 is called the **multiplicative identity.**

Multiplicative Identity Property	**For any number a, $a \cdot 1 = 1 \cdot a = a$.**

Suppose you purchase a number of Super Bowl tickets for $200 each. If you sell three of them at face value ($200), your profit is $0. The following equation describes the situation.

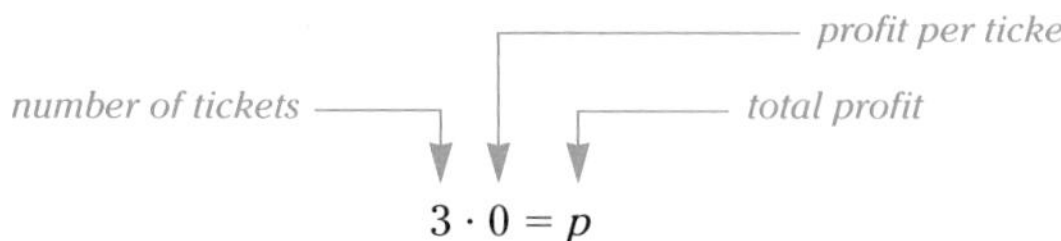

In this equation, one of the factors is 0 and the value of p is 0. This equation suggests the following property.

Multiplicative Property of Zero	**For any number a, $a \cdot 0 = 0 \cdot a = 0$.**

Two numbers whose product is 1 are called **multiplicative inverses** or **reciprocals.** Zero has no reciprocal because any number times 0 is 0.

Multiplicative Inverse Property	**For every nonzero number $\frac{a}{b}$, where $a, b \neq 0$, there is exactly one number $\frac{b}{a}$ such that $\frac{a}{b} \cdot \frac{b}{a} = 1$.**

Example 1 **Name the multiplicative inverse of each number or variable. Assume that no variable equals zero.**

a. 5

Since $5 \cdot \frac{1}{5} = 1$, $\frac{1}{5}$ is the multiplicative inverse of 5.

b. *x*

Since $x \cdot \frac{1}{x} = 1$, the multiplicative inverse is $\frac{1}{x}$.

$x \neq 0$; why?

c. $\frac{2}{3}$

Using the property, the multiplicative inverse is $\frac{3}{2}$.

At the beginning of algebra class, Ms. Escalante gave each of her students a single strip of paper 8 inches long. She instructed the students to divide their paper strips any way they wished.

Staci left her strip as one 8-inch strip. | ← 8 in. → |

Amad cut his strip to form a 6-inch strip and a 2-inch strip. | ← 6 in. → | 2 in. |

Liam cut his strip to form a 5-inch strip and a 3-inch strip. | ← 5 in. → | ←3 in.→ |

Using the strips of paper, we know the following to be true.

$8 = 8$ $\quad$ $6 + 2 = 6 + 2$ $\quad$ $3 + 5 = 3 + 5$

The **reflexive property of equality** says that any quantity is equal to itself.

Reflexive Property of Equality	**For any number a, $a = a$.**

Using the paper strips, we can show the following statements are true.

If $8 = 6 + 2$, then $6 + 2 = 8$.

If $3 + 5 = 6 + 2$, then $6 + 2 = 3 + 5$.

The **symmetric property of equality** says that if one quantity equals a second quantity, then the second quantity also equals the first.

Symmetric Property of Equality	**For any numbers a and b, if $a = b$, then $b = a$.**

A third property can also be shown using the paper strips.

If $3 + 5 = 8$ and $8 = 6 + 2$, then $3 + 5 = 6 + 2$.

If $8 = 3 + 5$ and $3 + 5 = 6 + 2$, then $8 = 6 + 2$.

The **transitive property of equality** says that if one quantity equals a second quantity and the second quantity equals a third quantity, then the first and third quantities are equal.

Transitive Property of Equality	**For any numbers a, b, and c, if $a = b$ and $b = c$, then $a = c$.**

We know that $5 + 3 = 6 + 2$. Since $5 + 3$ is equal to 8, we can substitute 8 for $5 + 3$ to get $8 = 6 + 2$. The **substitution property of equality** says that a quantity may be substituted for its equal in any expression.

Substitution Property of Equality	**If $a = b$, then a may be replaced by b in any expression.**

Teaching Tip Point out to students the differences between the transitive and substitution properties of equality. Stress the important use of the substitution property to simplify expressions.

In-Class Example

For Example 2
Suppose that next month the pep club at Roosevelt High School decides to sell apples for \$0.35 each. Write an expression that represents the profit for 50 sandwiches, 100 glasses of lemonade, and 50 apples. Evaluate the profit.
50(3.00 – 2.00) + 100(1.00 – 0.25) + 50(0.35 – 0.25); \$130

3 PRACTICE/APPLY

Check for Understanding

Exercises 1–19 are designed to help you assess your students' understanding through reading, writing, speaking, and modeling. You should work through Exercises 1–5 with your students and then monitor their work on Exercises 6–19.

Error Analysis
In such large and complex expressions, students must be careful to follow the rules of the order of operations. One suggestion is to number the operations first and then evaluate the expression.

Example:

$3\,(8 \bullet 2^3 - 7 \bullet 3)$

You can use the properties of identity and equality to justify each step when evaluating an expression.

Example 2 APPLICATION **Sales**

The pep club at Roosevelt High School is selling submarine sandwiches, lemonade, and apples at the district swim meet. Each sandwich costs \$2.00 to make and sells for \$3.00. Each glass of lemonade costs \$0.25 to make and sells for \$1.00. Each apple costs the club \$0.25, and the members have decided to sell apples for \$0.25 each. Write an expression that represents the profit for 80 sandwiches, 150 glasses of lemonade, and 40 apples. Evaluate the expression, indicating the property used in each step.

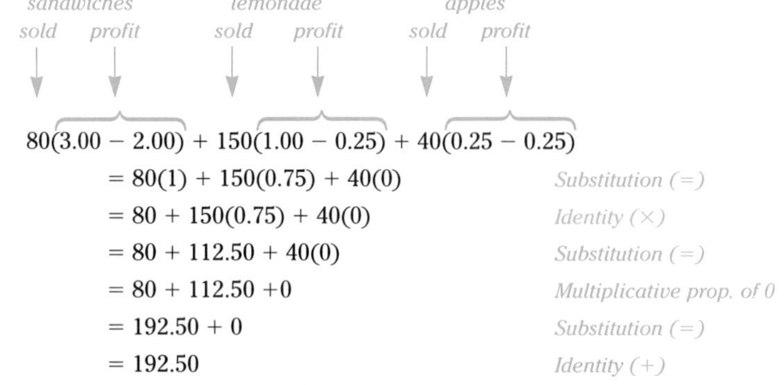

$$80(3.00 - 2.00) + 150(1.00 - 0.25) + 40(0.25 - 0.25)$$
$$= 80(1) + 150(0.75) + 40(0) \quad \text{Substitution } (=)$$
$$= 80 + 150(0.75) + 40(0) \quad \text{Identity } (\times)$$
$$= 80 + 112.50 + 40(0) \quad \text{Substitution } (=)$$
$$= 80 + 112.50 + 0 \quad \text{Multiplicative prop. of 0}$$
$$= 192.50 + 0 \quad \text{Substitution } (=)$$
$$= 192.50 \quad \text{Identity } (+)$$

The club would make a profit of \$192.50.

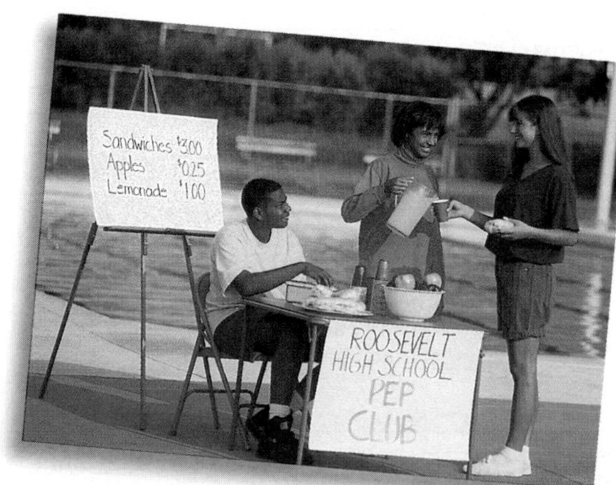

1. two or more things that are exactly alike
2. No; because $a + 1 \neq a$.
3. You cannot divide by 0.

CHECK FOR UNDERSTANDING

Communicating Mathematics

Study the lesson. Then complete the following.

1. **Define** the term *identity* in your own words.
2. **Explain** whether or not 1 can be the additive identity.
3. **Explain** why 0 does *not* have a multiplicative inverse.
4. **Name** the multiplicative inverse of 1. **1**

Reteaching

Using Guess and Check Replace the variable to make each open sentence true.

1. $(2 + x)^2 = 5^2$ **3**
2. $(2 + x)^2 = 2^2$ **0**
3. $3(2 - a) = 3$ **1**
4. $3(2 - a) = 0$ **2**
5. $m(1) + m(0) = 5$ **5**

5. Write a paragraph explaining how to determine the multiplicative inverse of a number.
Reverse the numerator and the denominator.

Guided Practice

Name the multiplicative inverse of each number or variable. Assume that no variable represents zero.

6. 7 $\frac{1}{7}$ 7. $\frac{9}{2}$ $\frac{2}{9}$ 8. c $\frac{1}{c}$

Match the expressions in the left-hand column with the properties in the right-hand column.

9. $0 \cdot 36 = 0$ c
10. $1(68) = 68$ b
11. $14 + 16 = 14 + 16$ e
12. $(9 - 7)(5) = 2(5)$ g
13. $\frac{3}{4} \times \frac{4}{3} = 1$ d
14. $0 + g = g$ a
15. If $8 + 1 = 9$, then $9 = 8 + 1$. f

a. Additive identity property
b. Multiplicative identity property
c. Multiplicative property of 0
d. Multiplicative inverse property
e. Reflexive property (=)
f. Symmetric property (=)
g. Substitution property (=)

Name the property used in each step.

16. $(14 \cdot \frac{1}{14} + 8 \cdot 0) \cdot 12 = (1 + 8 \cdot 0) \cdot 12$ multiplicative inverse
$= (1 + 0) \cdot 12$ multiplicative prop. of 0
$= 1 \cdot 12$ additive identity
$= 12$ multiplicative identity

17–18. For properties, see Solutions Manual.

Evaluate each expression. Name the property used in each step.

17. $6(12 - 48 \div 4) + 9 \cdot 1$ 9 18. $3 + 5(4 - 2^2) - 1$ 2

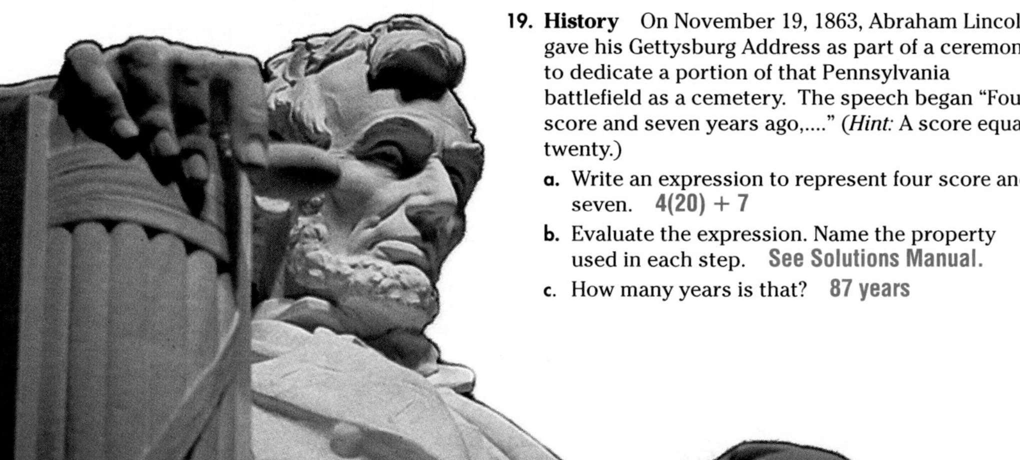

19. **History** On November 19, 1863, Abraham Lincoln gave his Gettysburg Address as part of a ceremony to dedicate a portion of that Pennsylvania battlefield as a cemetery. The speech began "Four score and seven years ago,...." (*Hint:* A score equals twenty.)
   a. Write an expression to represent four score and seven. $4(20) + 7$
   b. Evaluate the expression. Name the property used in each step. See Solutions Manual.
   c. How many years is that? 87 years

Study Guide Masters, p. 6

NAME______________ DATE ______

Study Guide

Student Edition Pages 37–43

Identity and Equality Properties

The identity and equality properties in the chart below can help you solve algebraic equations and evaluate mathematical expressions.

Additive Identity Property	For any number a, $a + 0 = 0 + a = a$.
Multiplicative Identity Property	For any number a, $a \cdot 1 = 1 \cdot a = a$.
Multiplicative Property of Zero	For any number a, $a \cdot 0 = 0 \cdot a = 0$.
Substitution Property	For any numbers a and b, if $a = b$ then a may be replaced by b.
Reflexive Property	$a = a$
Symmetric Property	If $a = b$, then $b = a$.
Transitive Property	If $a = b$ and $b = c$, then $a = c$.

Example: Evaluate $24 \cdot 1 - 8 + 5(9 \div 3 - 3)$. Indicate the property used in each step.

$24 \cdot 1 - 8 + 5(9 \div 3 - 3) = 24 \cdot 1 - 8 + 5(3 - 3)$ Substitution (=)
$= 24 \cdot 1 - 8 + 5(0)$ Substitution (=)
$= 24 - 8 + 5(0)$ Multiplicative identity
$= 24 - 8 + 0$ Multiplication property of zero
$= 16 + 0$ Substitution (=)
$= 16$ Additive identity

Solve each equation.

1. $a(9) = 0$ 0
2. $15 \cdot m = 15$ 1
3. $0 + p = 3$ 3
4. $7(0) = y$ 0

Name the property or properties illustrated by each statement.

5. $0 + 21 = 21$ Add. identity
6. $(0)15 = 0$ Mult. Prop. of Zero
7. If $4 + 5 = 9$, then $9 = 4 + 5$. Symmetric Prop.
8. $(1)94 = 94$ Mult. identity
9. If $3 + 3 = 6$ and $6 = 3 \cdot 2$, then $3 + 3 = 3 \cdot 2$. Transitive Property
10. $(14 - 6) + 3 = 8 + 3$ Substitution
11. $23 \cdot 1 = 23$ Mult. identity
12. $4 + 3 = 4 + 3$ Reflexive Prop.

Evaluate each expression. Name the property used in each step.

13. $10 \div 5 - 2^2 \div 2 + 13$
$= 10 \div 5 - 4 \div 2 + 13$ Sub. (=)
$= 2 - 4 \div 2 + 13$ Sub. (=)
$= 2 - 2 + 13$ Sub. (=)
$= 0 + 13$ Sub. (=)
$= 13$ Add. identity

14. $3(5 - 5 \cdot 1^2) + 21 \div 7$
$= 3(5 - 5 \cdot 1) + 21 \div 7$ Sub. (=)
$= 3(5 - 5) + 21 \div 7$ Mult. identity
$= 3(0) + 21 \div 7$ Sub. (=)
$= 0 + 21 \div 7$ Mult. Prop. of Zero
$= 0 + 3$ Sub. (=)
$= 3$ Add. identity

Assignment Guide

Core: 21–45 odd, 46, 47, 49–57
Enriched: 20–44 even, 46–57

For **Extra Practice**, see p. 757.

The red A, B, and C flags, printed only in the Teacher's Wraparound Edition, indicate the level of difficulty of the exercises.

EXERCISES

Practice

Name the multiplicative inverse of each number or variable. Assume that no variable represents zero.

20. 9 $\frac{1}{9}$ **21.** $\frac{1}{9}$ 9 **22.** $\frac{1}{4}$ 4

23. p $\frac{1}{p}$ **24.** $\frac{2}{a}$ $\frac{a}{2}$ **25.** $1\frac{1}{2}$ $\frac{2}{3}$

Name the property or properties illustrated by each statement.

26. If $7 \cdot 2 = 14$, then $14 = 7 \cdot 2$. **27.** $8 + (3 + 9) = 8 + 12$

28. $(10 - 8)(5) = 2(5)$ **29.** $mnp = 1mnp$

30. $\left(\frac{3}{4}\right)\left(\frac{4}{3}\right) = 1$ **31.** $3\left(5^2 \cdot \frac{1}{25}\right) = 3$

32. $0 + 23 = 23$ **33.** If $6 = 9 - 3$, then $9 - 3 = 6$.

34. $5(0) = 0$ **35.** $32 + 21 = 32 + 21$

36. If $4 \cdot 2 = 8$ and $8 = 6 + 2$, then $4 \cdot 2 = 6 + 2$. transitive (=)

26. symmetric (=)
27. substitution (=)
28. substitution (=)
29. multiplicative identity
30. multiplicative inverse
31. multiplicative inverse, multiplicative identity
32. additive identity
33. symmetric (=)
34. multiplicative property of 0
35. reflexive (=)

Name the property used in each step.

37. $2(3 \cdot 2 - 5) + 3 \cdot \frac{1}{3} = 2(6 - 5) + 3 \cdot \frac{1}{3}$ substitution (=)
$= 2(1) + 3 \cdot \frac{1}{3}$ substitution (=)
$= 2 + 3 \cdot \frac{1}{3}$ multiplicative identity
$= 2 + 1$ multiplicative inverse
$= 3$ substitution (=)

38. $26 \cdot 1 - 6 + 5(12 \div 4 - 3) = 26 \cdot 1 - 6 + 5(3 - 3)$ substitution (=)
$= 26 \cdot 1 - 6 + 5(0)$ substitution (=)
$= 26 - 6 + 5(0)$ multiplicative identity
$= 26 - 6 + 0$ multiplicative prop. of 0
$= 20 + 0$ substitution (=)
$= 20$ additive identity

39. $7(5 \cdot 3^2 - 11 \cdot 4) = 7(5 \cdot 9 - 11 \cdot 4)$ substitution (=)
$= 7(45 - 44)$ substitution (=)
$= 7 \cdot 1$ substitution (=)
$= 7$ multiplicative identity

Evaluate each expression. Name the property used in each step.

40. $4(16 \div 4^2)$ 4 **41.** $(15 - 8) \div 7 \cdot 25$ 25

42. $(8 \cdot 3 - 19 + 5) + (3^2 + 8 \cdot 4)$ 51 **43.** $(2^5 - 5^2) + (4^2 - 2^4)$ 7

44. $8[6^2 - 3(11)] \div 8 \cdot \frac{1}{3}$ 1 **45.** $5^3 + 9\left(\frac{1}{3}\right)^2$ 126

40–45. See Solutions Manual for properties.

Practice Masters, p. 6

NAME __________ DATE ________

1-6 Practice

Student Edition Pages 37–43

Identity and Equality Properties

Name the property illustrated by each statement.

1. $0 + b = b$
Additive identity prop.

2. If $x + y = 3$, then $3 = x + y$.
Symmetric prop.

3. $x = x$
Reflexive prop.

4. $4 \cdot 1 = 4$
Multiplicative identity prop.

5. $1 \cdot y = y$
Multiplicative identity prop.

6. $6 = 6$
Reflexive prop.

7. $0 = 0 \cdot 12$
Multiplicative prop. of zero

8. $5 = 5 + 0$
Additive identity prop.

9. If $12 = 17 - 5$, then $17 - 5 = 12$.
Symmetric prop.

10. $7(8 - 3) = 7(5)$
Substitution prop.

11. $w + (4 + 6) = w + 10$
Substitution prop.

12. $x + 2 = x + 2$
Reflexive prop.

13. $(6 + 9)x = 15x$
Substitution prop.

14. $(7 - 4)(6) = 3(6)$
Substitution prop.

15. $xyz = 1xyz$
Multiplicative identity prop.

16. $8 + 5 = (4 + 4) + 5$
Substitution prop.

17. If $6 + 3 = 9$ and $9 = 3(3)$, then $6 + 3 = 3(3)$.
Transitive prop.

Evaluate each expression. Name the property used in each step.

18. $2 + 6(9 - 3^2) - 2$
$= 2 + 6(9 - 9) - 2$ Substitution (=)
$= 2 + 6(0) - 2$ Substitution (=)
$= 2 + 0 - 2$ Mult. prop. of zero
$= 2 - 2$ Additive identity
$= 0$ Substitution (=)

19. $5(13 - 39 \div 3) + 7 \cdot 1$
$= 5(13 - 13) + 7 \cdot 1$ Substitution (=)
$= 5(0) + 7 \cdot 1$ Substitution (=)
$= 0 + 7 \cdot 1$ Mult. prop. of zero
$= 0 + 7$ Multiplicative identity
$= 7$ Additive identity

20. $7(16 \div 4^2)$
$= 7(16 \div 16)$ Substitution (=)
$= 7(1)$ Substitution (=)
$= 7$ Multiplicative identity

Critical Thinking

46. Think about the relationship "is less than," represented by the symbol <. Does < work with each of the following? Explain why or why not. Give examples that support your reasoning.
   a. reflexive property **no**
   b. symmetric property **no**
   c. the transitive property **yes**

See margin for explanations.

Applications and Problem Solving

47. **Entertainment** The students in Mr. Toshio's class are planning an ice cream party. They have selected the ice cream flavors shown in the table below. The prices listed are for $\frac{1}{2}$ cup servings. A survey of the students shows that 12 students want vanilla, 15 want chocolate, and 10 want chocolate chip cookie dough.

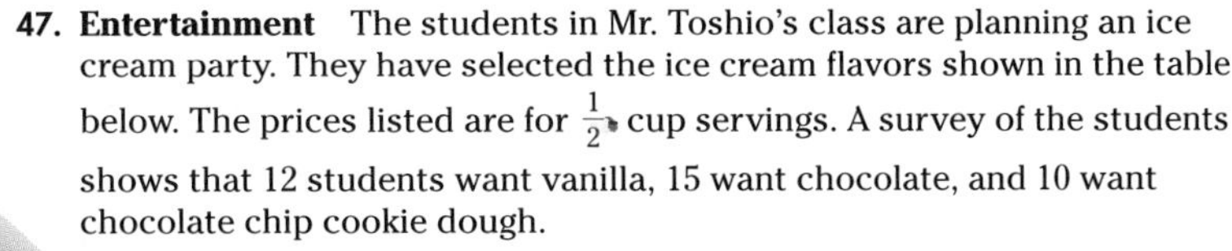

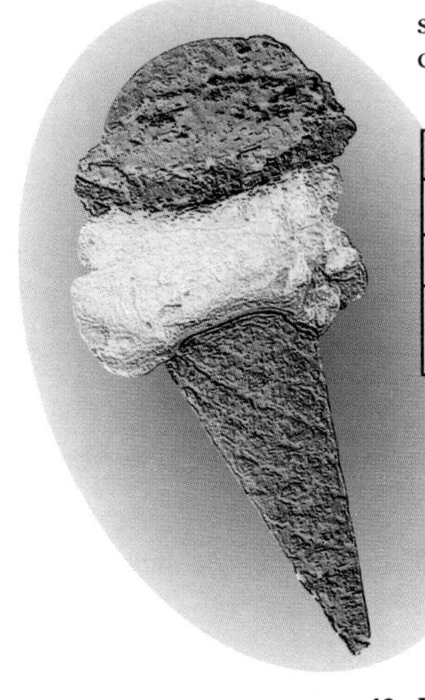

Flavor	Brand	Price
Vanilla	Breyers	21¢
Chocolate	Edy's/Dreyer's Grand	23¢
Chocolate Chip Cookie Dough	Ben & Jerry's	67¢

   a. Write an expression that represents how much it will cost to purchase ice cream for the class if each student gets a 1-cup serving. **See margin.**
   b. Evaluate the expression. Name the property used in each step. **See margin.**
   c. What is the total cost? **$25.34**

48. **Postal Rates** Patricia wants to mail a package to her cousin in Los Angeles. The cost of first class mail is \$0.32 for the first ounce and \$0.23 for each additional ounce or fraction of an ounce. Patricia's package weighs 14.4 ounces.
   a. Write an expression that represents the cost to mail the package.
   b. Evaluate the expression. Name the property used in each step.
   c. How much will Patricia need to pay in postage? **$3.54**

48a. 0.32 + 0.23(14)
48b. See margin.

Mixed Review

49. *True or False*: $15 \div 3 + 7 < 13$. (Lesson 1–5) **true**
50. Solve $m = (18 - 3) \div (3^2 - 2^2)$. (Lesson 1–5) **3**
51. *True or False*: $(2n^2 + 6) \div 4 < 5$, when $n = 3$. (Lesson 1–5) **false**
52. When is a back-to-back stem-and-leaf plot used? (Lesson 1–4)
53. Write the first and last names of ten of your classmates. Make a stem-and-leaf plot showing the number of letters in these ten names. (Lesson 1–4)
54. Evaluate $5(7 - 2) - 3^2$. (Lesson 1–3) **16**
55. Evaluate $xy - 2y$ when $x = 6$ and $y = 9$. (Lesson 1–3) **36**
56. Find the next two terms in the sequence 1, 4, 7, 10, ... (Lesson 1–2) **13, 16**
57. Write an algebraic expression for the number of months in y years. (Lesson 1–1) **12y**

52. when there are 2 sets of data
53. See students' work.

4 ASSESS

Closing Activity

Speaking Discuss the following relationships between two people to determine which ones work with the

a) reflexive property
b) symmetric property, or
c) transitive property.

1. is taller than **no, no, yes**
2. sits next to **no, no, yes**
3. has same color eyes as **yes, yes, yes**
4. is the father of **no, no, no**
5. lives within one mile of **yes, yes, no**

Additional Answers

46a. A number is never less than itself. Sample example: 4 is *not* less than 4.
46b. If one number is less than a second number, then the second can never be less than the first number. Sample example: 4 is less than 5, but 5 is *not* less than 4.
46c. If one number is less than a second number and a second number is less than a third number, then the first number will be less than the third number. Sample example: 4 < 5 and 5 < 6, so 4 < 6.

Enrichment Masters, p. 6

NAME ______ DATE ______

1-6 Enrichment — Student Edition Pages 37–43

The Conditional Statement

If p and q represent statements, the compound statement "if p then q" is called a *conditional*.

symbol: $p \rightarrow q$ read: either "if p then q" or "p only if q"

The statement p is called the *antecedent*, and the statement q is called the *consequent*.

For each conditional statement identify the antecedent (A) and the consequent (C).

1. If it is nine o'clock, then I am late.
   A: It is 9:00.
   C: I am late.
2. If Karen is home, then we will ask her to come.
   A: Karen is home.
   C: We will ask Karen to come.
3. The fish will die if we don't feed them.
   A: We don't feed the fish.
   C: The fish will die.
4. There will be no school if it snows.
   A: It snows, or it is snowing.
   C: There will be no school.
5. There will be no school only if it snows.
   A: There will be no school.
   C: It is snowing.
6. If $y + 2 = 6$ then $y = 4$.
   A: $y + 2 = 6$
   C: $y = 4$

Extension

Problem Solving Dentmaster Car Rental charges \$21.95 per day and \$0.11 a mile to rent one of its cars. Bryan rented a car for one day to visit friends in Altton, 85 miles away, Latton, 80 miles away, or Talton, 75 miles away. If he can spend only \$40.00 on his car rental, which friends could he visit? **the friends in Latton or Talton**

Additional Answers

47a. $[21(12 \cdot 2)] + [23(15 \cdot 2)] + [67(10 \cdot 2)]$

47b. $[21(12 \cdot 2)] + [23(15 \cdot 2)] + [67(10 \cdot 2)]$
$= [21(24)] + [23(30)] + [67(20)]$ Substitution (=)
$= 504 + 690 + 1340$ Substitution (=)
$= 2534$ Substitution (=)

48b. $0.32 + 0.23(14) = 0.32 + 3.22$ Substitution (=)
$= 3.54$ Substitution (=)

1-7A LESSON NOTES

NCTM Standards: 1–5

Objective
Model the distributive property.

Recommended Time
Demonstration and discussion: 15 minutes; Exercises: 30 minutes

Instructional Resources
For each student or group of students
Student Manipulative Kit
- algebra tiles

Modeling Mathematics Masters
- p. 15 (product mat)
- p. 19 (worksheet)

For teacher demonstration
Algebra and Geometry Overhead Manipulative Resources

1 FOCUS

Motivating the Lesson
Addition can be modeled by the length and width of a rectangle, and multiplication can be modeled by its area. The distributive property relates these two operations. How could the relationship between adding and multiplying be represented geometrically? **Answers will vary.**

2 TEACH

Teaching Tip Watch for students who may try to give the x-tile a numerical value by comparing the length of the x-tile. Reinforce the idea that the tiles are models, not exact quantities, and that x represents an unknown quantity.

3 PRACTICE/APPLY

Assignment Guide

Core: 1–7
Enriched: 1–7

MODELING MATHEMATICS

A Preview of Lesson 1–7

1–7A The Distributive Property

Materials:  algebra tiles, product mat

Throughout your study of mathematics, you have used rectangles to model multiplication. For example, the figure below shows the multiplication $2(3 + 1)$ as a rectangle that is 2 units wide and $3 + 1$ units long. The model shows that the expression $2(3 + 1)$ is equal to $2 \cdot 3 + 2 \cdot 1$. The sentence $2(3 + 1) = 2 \cdot 3 + 2 \cdot 1$ illustrates the distributive property.

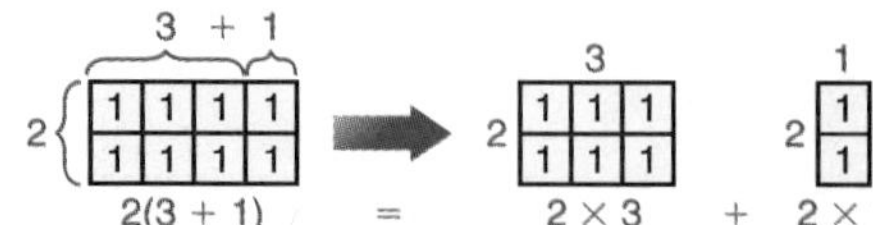

You can use special tiles called **algebra tiles** to form rectangles that model multiplication. A 1-tile is a square that is 1 unit long and 1 unit wide. Its area is 1 square unit. An x-tile is a rectangle that is 1 unit wide and x units long. Its area is x square units.

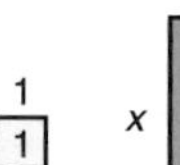

Activity **Find the product $3(x + 2)$ by using algebra tiles. First, model $3(x + 2)$ as the area of a rectangle.**

Step 1 The rectangle has a width of 3 units and a length of $x + 2$ units. Use your algebra tiles to mark off the dimensions on a product mat.

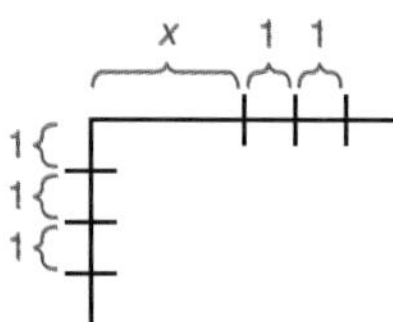

Step 2 Using the marks as a guide, make the rectangle with algebra tiles.

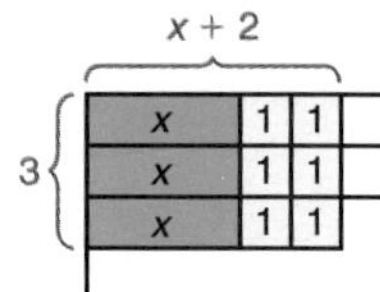

The rectangle has 3 x-tiles and 6 1-tiles. The area of the rectangle is $x + 1 + 1 + x + 1 + 1 + x + 1 + 1$, or $3x + 6$. Thus, $3(x + 2) = 3x + 6$.

Model **Find each product by using algebra tiles.**

1. $2(x + 1)$ **$2x + 2$**
2. $5(x + 2)$ **$5x + 10$**
3. $2(2x + 1)$ **$4x + 2$**
4. $2(3x + 3)$ **$6x + 6$**

Draw **Tell whether each statement is true or false. Justify your answer with algebra tiles and a drawing.** **5–6. See Solutions Manual for drawings.**

5. $3(x + 3) = 3x + 3$ **false**
6. $x(3 + 2) = 3x + 2x$ **true**

Write

7. **You Decide** Helen says that $3(x + 4) = 3x + 4$, but Adita says that $3(x + 4) = 3x + 12$.
   a. Which classmate is correct? **Adita**
   b. Use words and/or models to give an explanation that demonstrates the correct equation. **See Solutions Manual.**

4 ASSESS

Observing students working in cooperative groups is an excellent method of assessment.

GLENCOE Technology

Interactive Mathematics Tools Software

In this interactive computer lesson, students use algebra tiles and a product mat to model the distributive property. A **Computer Journal** gives students an opportunity to write about what they have learned.

For Windows & Macintosh

1–7

The Distributive Property

What YOU'LL LEARN

- To use the distributive property to simplify expressions.

Why IT'S IMPORTANT

You can use the distributive property to evaluate expressions and solve equations.

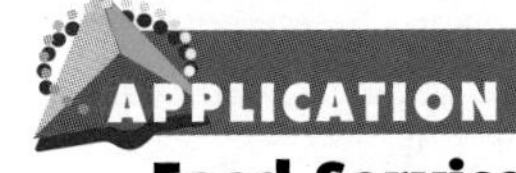

Food Service

In the school cafeteria, each student can pick one hot or cold entree and one salad from the following in order to get a "Meal Deal" lunch at a special low price.

Hot Lunch	Cold Lunch	Salads
• Tacos • Meat loaf and mashed potatoes • Steamed veggies and rice • Spaghetti	• Submarine sandwich • Turkey club sandwich • Tuna salad with tomato	• Tossed salad • Fruit salad

Jenine wants to know how many different lunches are possible. She can make a table to represent all the possible combinations.

Hot Lunch				
	Entree			
Salads	Tacos	Meat Loaf	Veggies	Spaghetti
Tossed	x	x	x	x
Fruit	x	x	x	x

Cold Lunch			
	Entree		
Salads	Submarine	Club Sandwich	Tuna Salad
Tossed	x	x	x
Fruit	x	x	x

There are $2 \cdot 4$ or 8 possible hot lunches.
There are $2 \cdot 3$ or 6 possible cold lunches.
There are $(2 \cdot 4) + (2 \cdot 3)$ possible lunches from which students can choose.
Note that $(2 \cdot 4) + (2 \cdot 3) = 8 + 6$ or 14.

1-7 LESSON NOTES

NCTM Standards: 1–5, 8

Instructional Resources

- Study Guide Master 1-7
- Practice Master 1-7
- Enrichment Master 1-7
- Assessment and Evaluation Masters, p. 17
- Multicultural Activity Masters, p. 2

Transparency 1-7A contains the 5-Minute Check for this lesson; **Transparency 1-7B** contains a teaching aid for this lesson.

Recommended Pacing	
Standard Pacing	Day 10 of 15
Honors Pacing	Day 9 of 14
Block Scheduling*	Day 5 of 7
Alg. 1 in Two Years*	Days 14 & 15 of 22

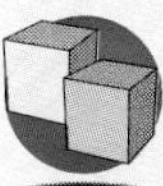

*For more information on pacing and possible lesson plans, refer to the *Block Scheduling Booklet* and *Algebra 1 in Two Years.*

1 FOCUS

5-Minute Check
(over Lesson 1-6)

Solve each equation.

1. $0 + r = 5$ **5**
2. $b \cdot 1 = 3$ **3**

Name the property illustrated by each statement.

3. If $3 + 0 = 3$, then $3 = 3 + 0$. **symmetric (=)**
4. $3(0) = 3(0)$ **reflexive (=)**
5. If $8 + 7 = 15$, then $15 = 8 + 7$. **symmetric (=)**

Motivating the Lesson

Questioning Which of the following expressions represent the area, in square meters, of the large rectangle at the left? **2, 3, and 7**

6 m
5 m
x m

1. $30 + x$	2. $30 + 6x$
3. $6(5 + x)$	4. $6(5x)$
5. $30 + x^2$	6. $11x$
7. $(5 + x)6$	8. $30x$
9. $5 = 6x$	10. $22 + 2x$

2 TEACH

Teaching Tip A common mistake is for students to evaluate an expression such as 2(5)(8) as 2(5) + 2(8), getting 26 instead of 80.

In-Class Example

For Example 1
Evaluate each expression in two different ways.

a. $11(8 + 2)$
$11(8 + 2) = 11(10) = 110$
$11(8 + 2) = 11(8) + 11(2) = 110$

b. $6(7 - 3)$
$6(7 - 3) = 6(4) = 24$
$6(7 - 3) = 6(7) - 6(3) = 24$

Jenine can also represent this problem by using the following table.

Salads	Entree						
	Hot Lunch				Cold Lunch		
	Taco	Meat Loaf	Veggie	Spaghetti	Sub	Club Sand.	Tuna
Tossed	×	×	×	×	×	×	×
Fruit	×	×	×	×	×	×	×

According to this table, there are 2 types of salads times the total number of entrees, 4 + 3, or 7.

$$2(4 + 3) = 2 \cdot 7 \text{ or } 14$$

Either way you look at it, there are 14 possible lunches. That's because the following is true.

$$2(4 + 3) = 2 \cdot 4 + 2 \cdot 3$$

This is an example of the **distributive property.**

Distributive Property	**For any numbers *a*, *b*, and *c*,** $a(b + c) = ab + ac$ **and** $(b + c)a = ba + ca$; $a(b - c) = ab - ac$ **and** $(b - c)a = ba - ca$.

Notice that it doesn't matter whether a is placed on the right or the left of the expression in parentheses.

The symmetric property of equality allows the distributive property to be written as follows.

If $a(b + c) = ab + ac$, then $ab + ac = a(b + c)$.

Example 1

Geometry

The Oak Grove High School Spirit Club wants to make a banner for the championship football game. The students have two large pieces of paper to use for the banner. One piece is 5 feet by 13 feet, and the other is 5 feet by 10 feet. They plan to use both pieces for the banner. Find the total area of the banner.

The total area of the banner can be found in two ways.

13 ft, 5 ft; 10 ft, 5 ft

Method 1: Add the areas of the smaller rectangles.

$$A = w\ell_1 + w\ell_2$$
$$= 5(13) + 5(10)$$
$$= 65 + 50$$
$$= 115$$

Use a rectangular piece of paper as a geometric model to illustrate the distributive property.

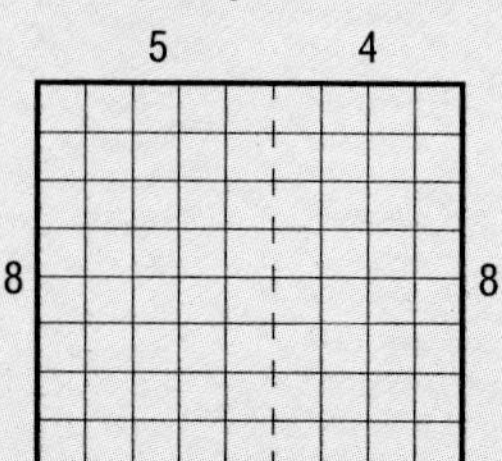

(1) $A = 8(5 + 4) = 8(9)$ or 72
(2) $A = 8(5) + 8(4) = 40 + 32$ or 72

Method 2: Multiply the width by the length.

$A = w\ell$
$= 5(13 + 10)$ *$w = 5, l = 13 + 10$*
$= 5(23)$
$= 115$

13 ft   10 ft
5 ft

The area is 115 square feet.

You can use the distributive property to multiply in your head.

Example  **Use the distributive property to find each product.**

a. $\mathbf{7 \cdot 98}$

$7 \cdot 98 = 7(100 - 2)$
$= 700 - 14$
$= 686$

b. $\mathbf{8(6.5)}$

$8(6.5) = 8(6 + 0.5)$
$= 48 + 4$
$= 52$

A **term** is a number, a variable, or a product or quotient of numbers and variables. Some examples of terms are x^3, $\frac{1}{4}a$, and $4y$. The expression $9y^2 + 13y^2 + 3$ has three terms.

Like terms are terms that contain the same variables, with corresponding variables having the same power. In the expression $8x^2 + 2x^2 + 5a + a$, $8x^2$ and $2x^2$ are like terms, and $5a$ and a are also like terms.

We can use the distributive property and properties of equality to show that $3x + 8x = 11x$. In this expression, $3x$ and $8x$ are like terms.

$3x + 8x = (3 + 8)x$ *Distributive property*
$= 11x$ *Substitution (=)*

The expressions $3x + 8x$ and $11x$ are called **equivalent expressions** because they denote the same number. An expression is in **simplest form** when it is replaced by an equivalent expression having no like terms and no parentheses.

Example  **Simplify** $\frac{1}{4}x^2 + 2x^2 + \frac{11}{4}x^2$.

In this expression, $\frac{1}{4}x^2$, $2x^2$, and $\frac{11}{4}x^2$ are like terms.

$\frac{1}{4}x^2 + 2x^2 + \frac{11}{4}x^2 = \left(\frac{1}{4} + 2 + \frac{11}{4}\right)x^2$ *Distributive property*
$= 5x^2$ *Substitution (=)*

The **coefficient** of a term is the numerical factor. For example, in $23ab$, the coefficient is 23. In xy, the coefficient is 1 since, by the multiplicative identity property, $1 \cdot xy = xy$. Like terms may also be defined as terms that are the same or that differ only in their coefficients.

In-Class Examples

For Example 2
Use the distributive property to find each product.

a. 16(101) **1616**
b. 9(10.6) **95.4**

For Example 3
Simplify $\frac{6xy^2}{5} + \frac{7xy^2}{5}$. $\mathbf{\frac{13xy^2}{5}}$

Teaching Tip When the text refers to a coefficient, it will be referring to the numerical coefficient of the term.

Example 4 **Name the coefficient in each term.**

a. $\mathbf{145x^2y}$ The coefficient is 145.

b. $\mathbf{ab^2}$ The coefficient is 1 since $ab^2 = 1ab^2$.

c. $\mathbf{\frac{4a^2}{5}}$ The coefficient is $\frac{4}{5}$ because $\frac{4a^2}{5}$ can be written as $\frac{4}{5} \cdot a^2$.

Example 5 **Simplify each expression.**

a. $\mathbf{4w^4 + w^4 + 3w^2 - 2w^2}$

Remember that $w^4 = 1w^4$.

$$4w^4 + w^4 + 3w^2 - 2w^2$$
$$= (4 + 1)w^4 + (3 - 2)w^2$$
$$= 5w^4 + 1w^2$$
$$= 5w^4 + w^2$$

b. $\mathbf{\frac{a^3}{4} + 2a^3}$

Remember that $\frac{a^3}{4} = \frac{1}{4}a^3$.

$$\frac{a^3}{4} + 2a^3 = \frac{1}{4}a^3 + 2a^3$$
$$= \left(\frac{1}{4} + 2\right)a^2$$
$$= 2\frac{1}{4}a^3$$

CHECK FOR UNDERSTANDING

Communicating Mathematics

1. See margin.

Study the lesson. Then complete the following.

1. **Explain** why the equation $2(a - 3) = 2a - 3$ is *not* a true sentence.
2. **Write** an expression that meets the following conditions.
   a. It has four terms, Sample answer: $5a + 3a + a + 2b$
   b. three terms are like terms, and
   c. one has a coefficient of 1.
3. **Describe** how you would simplify $3(2x - 4)$. See margin.

4. **Draw** a rectangular model or use tiles to represent $4(x + 1)$. See margin.

Guided Practice

Match an expression in the left-hand column with the equivalent expression in the right-hand column.

5. $8(10 + 4)$ b
6. $(12 - 3)6$ e
7. $(4 \cdot 2) + (x \cdot 2)$ a
8. $2(x + 5)$ d
9. $2(x - 1)$ c

a. $(4 + x)2$
b. $8 \cdot 10 + 8 \cdot 4$
c. $2x - 2$
d. $2x + 10$
e. $(12)(6) - (3)(6)$

Use the distributive property to rewrite each expression without parentheses.

10. $3(2x + 6)$ $6x + 18$
11. $2(a - b)$ $2a - 2b$

Use the distributive property to find each product.

12. $15 \cdot 99$ 1485
13. $28\left(2\frac{1}{7}\right)$ 60

In-Class Examples

For Example 4
Name the coefficient in each term.

a. y^2 1
b. $\frac{r}{4}$ $\frac{1}{4}$
c. $\frac{5x^2}{7}$ $\frac{5}{7}$

For Example 5
Simplify.

a. $13a^2 + 8a^2 + 6b$ $21a^2 + 6b$
b. $\frac{5}{6m} - \frac{m}{6} + 4m^2$ $4m^2 + \frac{5 - m^2}{6m}$

Teaching Tip Emphasize that in Example 5a, $5w^4$ and w^2 cannot be combined into one term because their exponents are different.

Teaching Tip Emphasize that in Example 5b, the coefficient of $\frac{a^3}{4}$ is $\frac{1}{4}$. This is often difficult for students to understand.

3 PRACTICE/APPLY

Check for Understanding

Exercises 1–23 are designed to help you assess your students' understanding through reading, writing, speaking, and modeling. You should work through Exercises 1–4 with your students and then monitor their work on Exercises 5–23.

Study Guide Masters, p. 7

NAME__________ DATE ______

Study Guide

Student Edition
Pages 45–50

The Distributive Property

When you find the product of two integers, you find the sum of two partial products. For example, you can write

$$\begin{array}{r} 58 \\ \times\ 5 \\ \hline 290 \end{array} \quad \text{as} \quad \begin{array}{r} 50 + 8 \\ \times \quad\ 5 \\ \hline 250 + 40 \end{array} \leftarrow (50 \times 5) + (8 \times 5)$$

The statement $(50 + 8) \times 5 = (50 \times 5) + (8 \times 5)$ illustrates the distributive property. The multiplier 5 is distributed over the 50 and the 8.

Distributive Property
For any numbers a, b, and c, $a(b + c) = ab + ac$ and $(b + c)a = ba + ca$; $a(b - c) = ab - ac$ and $(b - c)a = ba - bc$.

You can use the distributive property to simplify algebraic expressions.

Example: Simplify $4(a^2 + 3ab) - ab$.

$4(a^2 + 3ab) - ab = 4(a^2 + 3ab) - 1ab$	Multiplicative identity
$= 4a^2 + 12ab - 1ab$	Distributive property
$= 4a^2 + (12 - 1)ab$	Distributive property
$= 4a^2 + 11ab$	Substitution

Name the coefficient of each term. Then name the like terms in each list of terms.

1. $3x, 3x^2, 5x$ 3, 3, 5; $3x$, $5x$
2. $2mn, 10mn^2, 12mn^2, mn^2$ 2, 10, 12, 1; $10mn^2$, $12mn^2$, mn^2

Use the distributive property to find each product.

3. $4 \cdot 315$ 1260
4. $3 \cdot 24$ 72

Use the distributive property to rewrite each expression.

5. $5(4x - 9)$ $20x - 45$
6. $9r^2 + 9s^2$ $9(r^2 + s^2)$

Simplify each expression, if possible. If not possible, write in simplest form.

7. $7b + 3b$ $10b$
8. $4(5ac - 7)$ $20ac - 28$
9. $21c + 18c + 31b - 3b$ $39c + 28b$
10. $10x^2 - 6x^3$ in simplest form
11. $10xy - 4(xy + xy)$ $2xy$
12. $0.2(0.8 + 7y) + 0.36y$ $0.16 + 1.76y$

Reteaching

Using Alternative Methods Explain how the multiplication algorithm is just an example of the distributive property. Students should demonstrate this by using an example like the following.

$$\begin{array}{rl} 14 & \\ \times\ 76 & \\ \hline 84 & = 14 \times 6 \\ +980 & = 14 \times 70 \\ \hline 1064 & = 14(70 + 6) \end{array}$$

Additional Answers

1. According to the distributive property, $2(a - 3) = 2a - 6$.
3. Multiply 3 times $2x$ and 3 times 4. Subtract the product of 3 and 4 from the product of 3 and $2x$.
4.

	$x + 1$	
4	x	1
	x	1
	x	1
	x	1

Name the coefficient of each term.

14. $2.5cd$ **2.5**
15. $7a^2b$ **7**
16. $\frac{3b}{5}$ **$\frac{3}{5}$**

Name the like terms in each expression.

17. $4y^4 + 3y^3 + y^4$ **$4y^4, y^4$**
18. $3a^2 + 4c + a + 3b + c + 9a^2$ **$3a^2, 9a^2; 4c, c$**

Simplify each expression, if possible. If not possible, write *in simplest form*.

19. $t^2 + 2t^2 + 4t$ **$3t^2 + 4t$**
20. $25x^2 + 5x$ **in simplest form**
21. $16a^2b + 7a^2b + 3ab^2$ **$23a^2b + 3ab^2$**
22. $7p + q - p + \frac{2q}{3}$ **$6p + 1\frac{2}{3}q$**
23. **Employment** Maria and Mark are salesclerks at a local department store. Each earns $5.35 per hour. Maria works 24 hours per week, and Mark works 32 hours per week. Write two expressions to represent the amount the two of them earn per week. **23. 5.35(24) + 5.35(32); 5.35(24 + 32)**

EXERCISES

Practice

Use the distributive property to rewrite each expression without parentheses.

24. $2(4 + t)$ **$8 + 2t$**
25. $(g - 9)5$ **$5g - 45$**
26. $5(x + 3)$ **$5x + 15$**
27. $8(3m + 6)$ **$24m + 48$**
28. $28\left(y - \frac{1}{7}\right)$ **$28y - 4$**
29. $a(5 - b)$ **$5a - ab$**

Use the distributive property to find each product.

30. $5 \cdot 97$ **485**
31. $\left(3\frac{1}{17}\right) \times 17$ **52**
32. $16(102)$ **1632**
33. $24(2.5)$ **60**
34. $999 \cdot 6$ **5994**
35. 3×215 **645**

Simplify each expression, if possible. If not possible, write *in simplest form*.

36. $15x + 18x$ **$33x$**
37. $14a^2 + 13b^2 + 27$ **37. in simplest form**
38. $10n + 3n^2 + 9n^2$ **$10n + 12n^2$**
39. $5a + 7a + 10b + 5b$ **$12a + 15b$**
40. $7(3x^2y - 4xy^2 + xy)$ **40. $21x^2y - 28xy^2 + 7xy$**
41. $13p^2 + p$ **in simplest form**

C

42. $5(6a + 4b - 3b)$ **$30a + 5b$**
43. $3(x + 2y) - 2y$ **$3x + 4y$**
44. $\frac{2}{3}\left(c - \frac{3}{4}\right) + c(1 + b)$ **44. $\frac{5}{3}c - \frac{1}{2} + cb$**
45. $a + \frac{a}{5} + \frac{2}{5}a$ **$1\frac{3}{5}a$**
46. $4(3g + 2) + 2(g + 3)$ **$14g + 14$**
47. $3(x + y) + 2(x + y) + 4x$ **$9x + 5y$**

Programming

48. The program at the right tests values of *A*, *B*, and *C* to determine if the distributive property holds for division. That is, does $\frac{A + B}{C} = \frac{A}{C} + \frac{B}{C}$?

```
Program: DISTPROP
: Prompt A, B, C
: If(A + B)/C = A/C + B/C
: Then
: Disp "YES, IT HOLDS."
: Else
: Disp "TRY AGAIN."
: END
```

a. Run the program for 10 different sets of values for *A*, *B*, and *C*. Does the property hold? **yes, if $C \neq 0$**
b. How could you change the program to test if $(A + B)^C = A^C + B^C$? Does the distributive property hold for powers? **48b. Change the / in the If statement to ∧; only for $C = 1$.**

Error Analysis
Students may express $4(x - y)$ as $4x - y$. Remind them that in the first expression, both *x* and *y* are multiplied by 4 while in the second, only *x* is multiplied by 4. Students may also apply the distributive property to multiplication or division problems. They might express $5(2 \cdot x)$ as $10 \cdot 5x$. Emphasize that multiplication only distributes over addition or subtraction.

Assignment Guide

Core: 25–51 odd, 52–60
Enriched: 24–48 even, 49–60

For **Extra Practice**, see p. 758.

The red A, B, and C flags, printed only in the Teacher's Wraparound Edition, indicate the level of difficulty of the exercises.

Using the Programming Exercises The program given in Exercise 48 is for use with a TI-82 graphing calculator. For other programmable calculators, have students consult their owner's manuals for commands similar to those printed here.

Practice Masters, p. 7

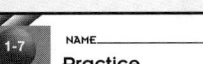

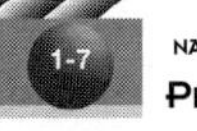

NAME ____________ DATE ________

Practice — Student Edition Pages 45–50

The Distributive Property

Use the distributive property to rewrite each expression.

1. $5(2x + 7)$ $10x + 35$
2. $6x - 6y$ $6(x - y)$
3. $3(m + n)$ $3m + 3n$
4. $3p - 3q$ $3(p - q)$

Simplify each expression, if possible. If not possible, write in simplest form.

5. $2x + 8x$ $10x$
6. $17j + j$ $18j$
7. $4ab + ab$ $5ab$
8. $2x^2 + 6x^2$ $8x^2$
9. $16m - 10m$ $6m$
10. $5t^3 - 3t^3$ $2t^3$
11. $15i^2 - 15i^2$ 0
12. $3(5 + 2h)$ $15 + 6h$
13. $5(r + 2) + 7r$ $12r + 10$
14. $a + b + a + c$ $2a + b + c$
15. $6x + 7(y + x)$ $13x + 7y$
16. $3(r + 2s) - 3r$ $6s$
17. $10c + 5c + 6d + d$ $15c + 7d$
18. $(x + 5)y + 2y$ $xy + 7y$
19. $7rs + 2rs + 3rs$ $12rs$
20. $w + 14w - 6w$ $9w$
21. $4.5v + 23v + v$ $28.5v$
22. $3r^2 + 6r + 2s^2$ in simplest form
23. $c^2 + 4d^2 + d^2 + 3d$ $c^2 + 3d + 5d^2$
24. $\frac{1}{3}n + \frac{2}{5}n$ $\frac{11}{15}n$
25. $a + \frac{1}{3}b + \frac{4}{3}a + \frac{4}{3}b$ $\frac{7}{3}a + \frac{5}{3}b$
26. $35a + 5n + 2(n - 1)$ $35a + 7n - 2$

4 ASSESS

Closing Activity

Writing Have students list other possible "distributive" properties and state whether these properties are always true. Examples might include addition distributing over multiplication, division distributing over subtraction, and raising to a power distributing over multiplication. They should include counterexamples to show properties that are not always true.

Additional Answer

56.

Stem	Leaf
2	9
3	6 6 7
4	5 5
5	1 5 8

$3|7 = 37$

Chapter 1, Quiz C (Lessons 1-6 and 1-7), is available in the *Assessment and Evaluation Masters,* p. 17.

Enrichment Masters, p. 7

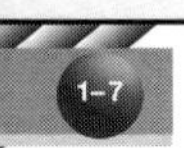

NAME______________________ DATE ____________

Enrichment — Student Edition Pages 45–50

Closure

A *binary operation* matches two numbers in a set to just one number. Addition is a binary operation on the set of whole numbers. It matches two numbers such as 4 and 5 to a single number, their sum.

If the result of a binary operation is always a member of the original set, the set is said to be *closed* under the operation. For example, the set of whole numbers is not closed under subtraction because 3 − 6 is not a whole number.

Is each operation binary? Write yes or no.

1. the operation ↵, where a ↵ b means to choose the lesser number from a and b **yes**
2. the operation ©, where a © b means to cube the sum of a and b **yes**
3. the operation sq, where $sq(a)$ means to square the number a **no**
4. the operation exp, where $exp(a, b)$ means to find the value of a^b **yes**
5. the operation ⇑, where a ⇑ b means to match a and b to any number greater than either number **no**
6. the operation ⇒, where $a \Rightarrow b$ means to round the product of a and b up to the nearest 10 **yes**

Is each set closed under addition? Write yes or no. If your answer is no, give an example.

7. even numbers **yes**
8. odd numbers **no; 3 + 7 = 10**
9. multiples of 3 **yes**
10. multiples of 5 **yes**
11. prime numbers **no; 3 + 5 = 8**
12. nonprime numbers **no; 22 + 9 = 31**

Is the set of whole numbers closed under each operation? Write yes or no. If your answer is no, give an example.

13. multiplication: $a \times b$ **yes**
14. division: $a \div b$ **no; 4 ÷ 3 is not a whole number**
15. exponentation: a^b **yes**
16. squaring the sum: $(a + b)^2$ **yes**

Critical Thinking

49. If $2(b + c) = 2b + 2c$, does $2 + (b \cdot c) = (2 + b)(2 + c)$? Choose values for b and c to show that these may be true or find *counterexamples* to show that they are not. **No; sample counterexample: $2 + (4 \cdot 5) \neq (2 + 4)(2 + 5)$**

Applications and Problem Solving

50. **Economics** According to *American Demographics,* the average 13- to 15-year-old male received \$16.15 per week for allowance in 1993. The average 16- to 19-year-old female received \$32.45 per week. Toni was 17 years old, and her brother Carlos was 14 in 1993. Assume that they got the average amount for their weekly allowance.
    a. Write two expressions to represent the amount their parents paid in allowance during the month of February.
    b. What was the amount their parents paid in allowance during that month? **\$194.40**

50a. 4(16.15 + 32.45); 4(16.15) + 4(32.45)

Keong Emas Imax Theater

51. **Geometry** The largest permanently installed theater screen has an area of 6768 square feet. It is at the Keong Emas Imax Theater in Jakarta, Indonesia. The movie screen below represents a typical movie screen in U.S. movie theaters.
    a. Write an expression for the perimeter of the movie screen below and then simplify the expression. **$2[x + (x + 14)] = 4x + 28$**

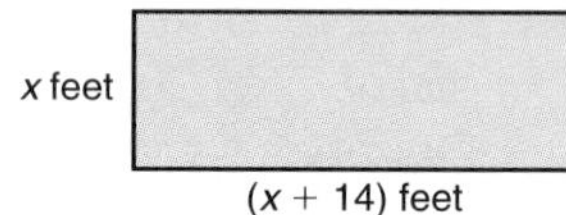

    b. Evaluate the expression if $x = 17$. **96**
    c. Find the area of the movie screen if $x = 17$. **527 ft^2**

Mixed Review

52. Name the property illustrated by the following statement: *If 19 − 3 = 16, then 16 = 19 − 3.* (Lesson 1–6) **symmetric (=)**
53. Name the property illustrated by $9 \times 0 = 0$. (Lesson 1–6)
54. Find the solution set for inequality $3x - 5 > 7$ if the replacement set is {2, 3, 4, 5, 6}. (Lesson 1–5) **{5, 6}**
55. **Physics** Sound travels 1129 feet per second through air. (Lesson 1–5)
    a. Write an equation that represents how many feet sound can travel in 2 seconds when it is traveling through air. **$d = (1129)(2)$**
    b. How many feet can sound travel in 2 seconds when traveling through air? **2258 ft**
56. Make a stem-and-leaf plot for the following set of data. (Lesson 1–4)
    37, 45, 36, 51, 55, 29, 45, 58, 36 **See margin.**
57. Evaluate $\frac{4^2 - 2^3}{24 - 2(10)}$. (Lesson 1–3) **2**
58. What are the next two expressions for the pattern $5a, 10a^2, 15a^3, 20a^4, \ldots$? (Lesson 1–2) **$25a^5, 30a^6$**
59. **Culture** Each year in the Chinese calendar is named for one of 12 animals. Every 12 years, the same animal is named. If 1992 was the Year of the Monkey, how many years in the 20th century were Years of the Monkey? (Lesson 1–2) **8 years**
60. Write an algebraic expression for *37 less than 2 times a number k.* (Lesson 1–1) **$2k - 37$**

53. multiplicative prop. of 0

Extension

Connections Carita placed six 2-cm by 2-cm squares flat on her desk so that each square had one side in common with at least one other square. What are the possible perimeters for all the figures she can make? (*Hint:* Make drawings for all the possible figures.) **20 cm, 24 cm, 28 cm**

1-8 Commutative and Associative Properties

What YOU'LL LEARN

- To recognize and use the commutative and associative properties to simplify expressions.

Why IT'S IMPORTANT

You can use the communitative and associative properties to evaluate expressions and solve equations.

Networks

The map below shows the location of Leticia's house and school. It also shows the time in minutes that it takes Leticia to walk around her neighborhood.

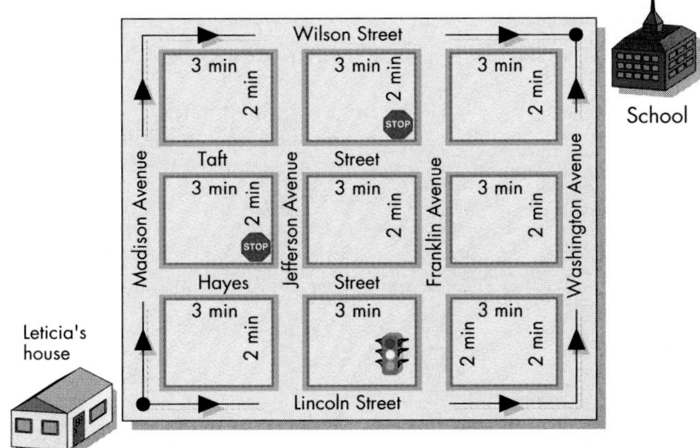

Yesterday, Leticia walked to school by way of Lincoln Street and Washington Avenue. She took 3 + 3 + 3 + 2 + 2 + 2 or 15 minutes to walk to school. Today, Leticia walked to school by way of Madison Avenue and Wilson Street. She took 2 + 2 + 2 + 3 + 3 + 3 or 15 minutes. The time it took Leticia to walk to school is the same for both days.

$$3 + 3 + 3 + 2 + 2 + 2 = 2 + 2 + 2 + 3 + 3 + 3$$

In this equation, the addends are the same, but their order is different. The **commutative property** says that the order in which you add or multiply two numbers does not change their sum or product.

Commutative Property	**For any numbers a and b, $a + b = b + a$ and $a \cdot b = b \cdot a$.**

One easy way to find the sum or product of numbers is to group or *associate* the numbers. The **associative property** says that the way you group three numbers when adding or multiplying does not change their sum or product.

Associative Property	**For any numbers a, b, and c, $(a + b) + c = a + (b + c)$ and $(ab)c = a(bc)$.**

1-8 LESSON NOTES

NCTM Standards: 1–5

Instructional Resources

- Study Guide Master 1-8
- Practice Master 1-8
- Enrichment Master 1-8

Transparency 1-8A contains the 5-Minute Check for this lesson; **Transparency 1-8B** contains a teaching aid for this lesson.

Recommended Pacing	
Standard Pacing	Day 11 of 15
Honors Pacing	Day 10 of 14
Block Scheduling*	Day 6 of 7 (along with Lesson 1-9)
Alg. 1 in Two Years*	Days 16 & 17 of 22

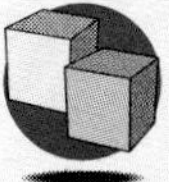

*For more information on pacing and possible lesson plans, refer to the *Block Scheduling Booklet* and *Algebra 1 in Two Years.*

1 FOCUS

5-Minute Check
(over Lesson 1-7)

Simplify each expression.

1. $2a(4a^2 - 2ab)$ $\mathbf{8a^3 - 4a^2b}$
2. $a(2a + 3b) - ab$ $\mathbf{2a^2 - 2ab}$
3. $3m + 2n + 4p + 3n - m$ $\mathbf{2m + 5n + 4p}$
4. $4(3x + 2y) + 6(3y - 2x)$ $\mathbf{26y}$
5. $2(7x + y) + 4(3x + 2y)$ $\mathbf{26x + 10y}$

Motivating the Lesson

Situational Problem Have students list the steps they would follow, in order, when wrapping a gift. Then have them indicate on the list what steps cannot come after other steps. For example, tissue paper cannot be placed in the package after ribbon has been wrapped around the package.

2 TEACH

In-Class Example

For Example 1
Which of the following is the greatest?

a. $6 \times 4\frac{1}{2} \times 3$
b. $5\frac{1}{2} \times 5 \times 4$
c. $7 \times 3 \times 2\frac{1}{2}$
$b = 110$

For Example 2

a. Write an algebraic expression for the verbal expression, *the difference of x cubed and three increased by the difference of 5 and x cubed.* $(x^3 - 3) + (5 - x^3)$
b. Simplify the algebraic expression in part a, indicating all of the properties used. 2, associative, commutative

Teaching Tip Ask students to explain the difference between the commutative and associative properties. The commutative properties allow a different ordering of numbers while the associative properties allow a different grouping of numbers.

3 PRACTICE/APPLY

Check for Understanding

Exercises 1–15 are designed to help you assess your students' understanding through reading, writing, speaking, and modeling. You should work through Exercises 1–5 with your students and then monitor their work on Exercises 6–15.

Error Analysis

Students often confuse the symmetric property of equality and the commutative properties. Emphasize that expressions are interchanged with respect to an equal sign for the symmetric property, while expressions are interchanged with respect to either an addition or a multiplication sign for the commutative properties.

You can group 11 + 12 + 7 + 9 + 7 + 6 to make mental addition easy.

$$\begin{aligned} 11 + 12 + 7 + 9 + 7 + 6 &= 11 + 9 + 7 + 6 + 7 + 12 && \text{Commutative } (+) \\ &= (11 + 9) + (7 + 6 + 7) + 12 && \text{Associative } (+) \\ &= 20 + 20 + 12 \\ &= 52 \end{aligned}$$

The commutative and associative properties can be used with the other properties you have studied when evaluating and simplifying expressions.

Example 1

Marketing

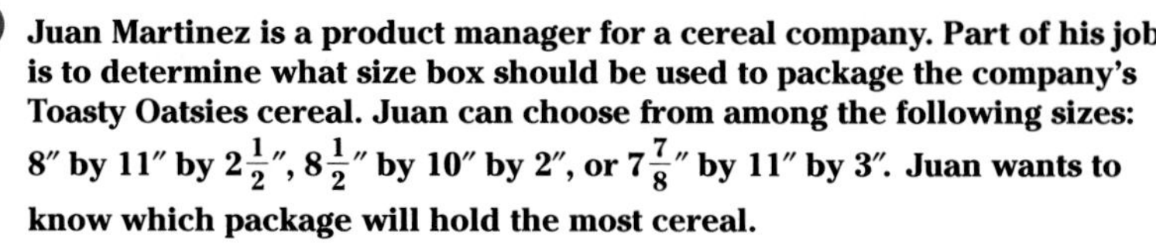

Juan Martinez is a product manager for a cereal company. Part of his job is to determine what size box should be used to package the company's Toasty Oatsies cereal. Juan can choose from among the following sizes: 8″ by 11″ by $2\frac{1}{2}$″, $8\frac{1}{2}$″ by 10″ by 2″, or $7\frac{7}{8}$″ by 11″ by 3″. Juan wants to know which package will hold the most cereal.

The box with the greatest volume will hold the most cereal. To find the volume of each box, multiply the length times the width times the height.

Why has the order of the factors changed?

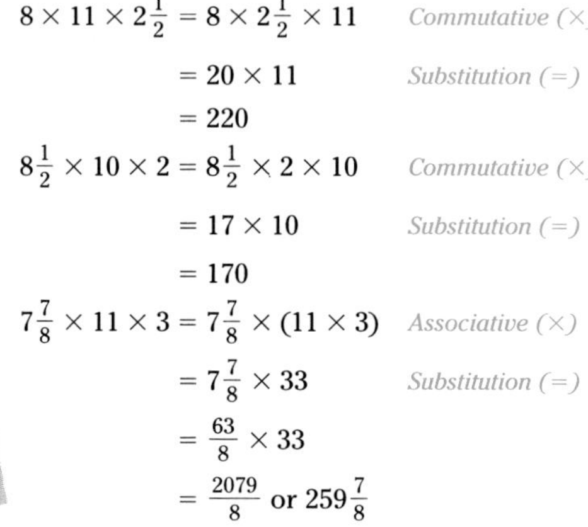

$$\begin{aligned} 8 \times 11 \times 2\tfrac{1}{2} &= 8 \times 2\tfrac{1}{2} \times 11 && \text{Commutative } (\times) \\ &= 20 \times 11 && \text{Substitution } (=) \\ &= 220 \\ 8\tfrac{1}{2} \times 10 \times 2 &= 8\tfrac{1}{2} \times 2 \times 10 && \text{Commutative } (\times) \\ &= 17 \times 10 && \text{Substitution } (=) \\ &= 170 \\ 7\tfrac{7}{8} \times 11 \times 3 &= 7\tfrac{7}{8} \times (11 \times 3) && \text{Associative } (\times) \\ &= 7\tfrac{7}{8} \times 33 && \text{Substitution } (=) \\ &= \tfrac{63}{8} \times 33 \\ &= \tfrac{2079}{8} \text{ or } 259\tfrac{7}{8} \end{aligned}$$

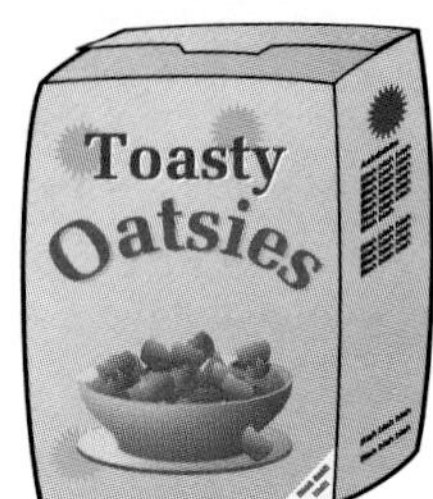

The box that is $7\frac{7}{8}$″ by 11″ by 3″ has the greatest volume.

The chart below summarizes the properties you can use to simplify expressions.

The following properties are true for any numbers *a*, *b*, and *c*.		
	Addition	**Multiplication**
Commutative	$a + b = b + a$	$ab = ba$
Associative	$(a + b) + c = a + (b + c)$	$(ab)c = a(bc)$
Identity	0 is the identity. $a + 0 = 0 + a = a$	1 is the identity. $a \cdot 1 = 1 \cdot a = a$
Zero		$a \cdot 0 = 0 \cdot a = 0$
Distributive	$a(b + c) = ab + ac$ and $(b + c)a = ba + ca$	
Substitution	If $a = b$, then a may be substituted for b.	

Example 2 a. **Write an algebraic expression for the verbal expression *the sum of two and the square of t increased by the sum of t squared and 3.***

b. **Then simplify the algebraic expression, indicating all of the properties used.**

a.

the sum of two and the square of t	*increased by*	*the sum of t squared and 3*
$2 + t^2$	$+$	$t^2 + 3$

b.
$$
\begin{aligned}
2 + t^2 + t^2 + 3 &= t^2 + t^2 + 2 + 3 && \text{Commutative } (+)\\
&= (t^2 + t^2) + (2 + 3) && \text{Associative } (+)\\
&= (1 \cdot t^2 + 1 \cdot t^2) + (2 + 3) && \text{Multiplicative identity}\\
&= (1 + 1)t^2 + (2 + 3) && \text{Distributive property}\\
&= 2t^2 + 5 && \text{Substitution } (=)
\end{aligned}
$$

CHECK FOR UNDERSTANDING

Communicating Mathematics

Study the lesson. Then complete the following. 1–4. See margin.

1. **Illustrate** the associative property of multiplication with a numerical equation.
2. **Explain** how a property or properties can be used to find the product of 5(6.5)(2) without using a calculator or paper and pencil.
3. **Explain** the difference between the commutative and associative properties.
4. **Write** a short explanation as to whether or *not* there is a commutative property of division. Consider these examples.
   $5 \div 3 = 1\frac{2}{3}$ $\quad\quad$ $3 \div 5 = \frac{3}{5}$

5. **Assess Yourself** Do you believe that you understand the properties described in this chapter? Which properties were easiest for you to understand? List any properties that give you difficulty and explain why. See students' work.

Guided Practice

Name the property illustrated by each statement. 6–9. See margin.

6. $(6 + 4) + 2 = 6 + (4 + 2)$
7. $7 + 6 = 6 + 7$
8. $(2 + 5) + x = 7 + x$
9. $7(ab) = (7a)b$
10. Name the property used in each step.
    a. $ab(a + b) = (ab)a + (ab)b$ — distributive property
    b. $= a(ab) + (ab)b$ — commutative (×)
    c. $= (a \cdot a)b + a(b \cdot b)$ — associative (×)
    d. $= a^2b + ab^2$ — substitution (=)

Simplify.

11. $4a + 2b + a$ $\quad$ $5a + 2b$
12. $3p + 2q + 2p + 8q$ $\quad$ $5p + 10q$
13. $3(4x + y) + 2x$ $\quad$ $14x + 3y$
14. $6(0.4x + 0.2y) + 0.5x$ $\quad$ $2.9x + 1.2y$
15. Write an algebraic expression for the verbal expression *the product of six and the square of z, increased by the sum of seven, z^2, and 6.* Then simplify, indicating the properties used. See margin.

Reteaching

Logical Thinking For each situation, determine whether "followed by" is commutative.

a. Putting mustard on your hamburger, followed by putting ketchup on your hamburger. yes

b. Opening the car door, followed by starting the car's engine. no

c. Buying tickets for the homecoming dance, followed by buying an outfit to wear to the homecoming dance. yes

d. Have students make up their own example of a situation in which "followed by" is commutative and a situation in which "followed by" is not commutative. Answers may vary.

Additional Answers

1. Answers will vary. Sample answer: $(2 \cdot 3) \cdot 5 = 2 \cdot (3 \cdot 5)$
   $6 \cdot 5 = 2 \cdot 15$
   $30 = 30$
2. Use the commutative property of multiplication to rewrite the problem as $5 \cdot 2 \cdot 6.5$. Since $5 \cdot 2 = 10$, the answer 65 can be determined without a calculator or paper and pencil.
3. The commutative properties allow a different *ordering* of numbers while the associative properties allow a different *grouping* of numbers.
4. Division is *not* commutative since the examples have different quotients.
6. associative (+)
7. commutative (+)
8. substitution (=)
9. associative (×)
15. $6z^2 + (7 + z^2 + 6)$
    $= 6z^2 + (z^2 + 7 + 6)$ commutative (+)
    $= (6z^2 + z^2) + (7 + 6)$ associative (+)
    $= (6 + 1)z^2 + (7 + 6)$ distributive property
    $= 7z^2 + 13$ substitution (=)

Study Guide Masters, p. 8

NAME________________ DATE ________

Study Guide

Student Edition Pages 51–55

Commutative and Associative Properties

The commutative and associative properties can be used to simplify expressions.

Commutative Properties
For any numbers a and b, $a + b = b + a$ and $a \cdot b = b \cdot a$.

Associative Properties
For any numbers a, b, and c, $(a + b) + c = a + (b + c)$ and $(ab)c = a(bc)$.

Example: Simplify $8(y + 2x) + 7y$.

$$
\begin{aligned}
8(y + 2x) + 7y &= (8y + 16x) + 7y && \text{Distributive property}\\
&= (16x + 8y) + 7y && \text{Commutative property of addition}\\
&= 16x + (8y + 7y) && \text{Associative property of addition}\\
&= 16x + (8 + 7)y && \text{Distributive property}\\
&= 16x + 15y && \text{Substitution property of equality}
\end{aligned}
$$

Simplify.

1. $4x + 3y + x$ $\quad$ $5x + 3y$
2. $8r^2s + 2rs^2 + 7r^2s$ $\quad$ $15r^2s + 2rs^2$
3. $6(2x + 4y) + 2(x + 9)$ $\quad$ $14x + 24y + 18$
4. $3a^2 + 4b + 10a^2$ $\quad$ $13a^2 + 4b$
5. $4xy + 7x^2y + xy$ $\quad$ $5xy + 7x^2y$
6. $3ab + 4a^2b + 5(2a^2b)$ $\quad$ $14a^2b + 3ab$
7. $6(a + b) - a + 3b$ $\quad$ $5a + 9b$
8. $0.5(18x + 16y) + 13x$ $\quad$ $22x + 8y$
9. $\frac{2}{3} + \frac{1}{2}(x + 10) + \frac{4}{3}$ $\quad$ $7 + \frac{1}{2}x$
10. $5(0.3x + 0.1y) + 0.2x$ $\quad$ $1.7x + 0.5y$
11. $5(2y + 3x) + 6(y + x)$ $\quad$ $16y + 21x$
12. $z^2 + 9x^2 + \frac{4}{3}z^2 + \frac{1}{3}x^2$ $\quad$ $\frac{7}{3}z^2 + \frac{28}{3}x^2$, or $2\frac{1}{3}z^2 + 9\frac{1}{3}x^2$

Name the property illustrated by each statement.

13. $6x + 2y = 2y + 6x$ Commutative property of addition
14. $15(a + 4) = 15a + 15(4)$ Distributive property
15. $1 \cdot b^3 = b^3$ Multiplicative identify
16. $(2c + 6) + 10 = 2c + (6 + 10)$ Associative property of addition

Assignment Guide

Core: 17–47 odd, 48–56
Enriched: 16–44 even, 45–56

For **Extra Practice**, see p. 758.

The red A, B, and C flags, printed only in the Teacher's Wraparound Edition, indicate the level of difficulty of the exercises.

Additional Answers

16. commutative (+)
17. multiplicative identity
18. distributive property
19. associative (×)
20. multiplicative property of zero
21. distributive property
22. substitution (=)
23. commutative (×)
24. associative (+)
25. associative (+)
39. $2(s + t) - s$
 $= 2s + 2t - s$ distributive property
 $= 2t + 2s - s$ commutative (+)
 $= 2t + (2s - s)$ associative (+)
 $= 2t + (2s - 1s)$ multiplicative identity
 $= 2t + (2 - 1)s$ distributive
 $= 2t + 1s$ substitution (=)
 $= 2t + s$ multiplicative identity

EXERCISES

Practice

Name the property illustrated by each statement. 16–25. See margin.

16. $67 + 3 = 3 + 67$
17. $1 \cdot b^2 = b^2$
18. $10x + 10y = 10(x + y)$
19. $(5 \cdot m) \cdot n = 5 \cdot (m \cdot n)$
20. $(3x^2) \cdot 0 = 0$
21. $4(a + 5b) = 4a + 20b$
22. $(2 + 3)a + 7 = 5a + 7$
23. $fh + 2g = hf + 2g$
24. $7a + \left(\frac{1}{2}b + c\right) = \left(7a + \frac{1}{2}b\right) + c$
25. $(5x^2 + x + 3) + 15 = (5x^2 + x) + (3 + 15)$
26. Name the property used in each step.
 a. $3c + 5(2 + c) = 3c + 5(2) + 5c$ distributive property
 b. $= 3c + 5c + 5(2)$ commutative (+)
 c. $= (3c + 5c) + 5(2)$ associative (+)
 d. $= (3 + 5)c + 5(2)$ distributive property
 e. $= 8c + 10$ substitution (=)

Simplify. 37. $\frac{3}{4} + \frac{5}{3}m + \frac{4}{3}n$ 38. $\frac{23}{10}p + \frac{6}{5}q$

27. $4x + 5y + 6x$ $10x + 5y$
28. $8a + 3b + a$ $9a + 3b$
29. $5x + 3y + 2x + 7y$ $7x + 10y$
30. $4 + 6(ac + 2b) + 2ac$ 30. $4 + 8ac + 12b$
31. $2(3x + y) + 4x$ $10x + 2y$
32. $4y^4 + 3y^2 + y^4$ $5y^4 + 3y^2$
33. $16a^2 + 16 + 16a^2$ $32a^2 + 16$
34. $3.2(x + y) + 2.3(x + y) + 4x$ 34. $9.5x + 5.5y$
35. $\frac{1}{4}x + 2x + 2\frac{3}{4}x$ $5x$
36. $0.5[3x + 4(3 + 2x)]$ $5.5x + 6$
37. $\frac{3}{4} + \frac{2}{3}(m + 2n) + m$
38. $\frac{3}{5}\left(\frac{1}{2}p + 2q\right) + 2p$

Write an algebraic expression for each verbal expression. Then simplify, indicating the properties used. 39–42. See margin.

39. twice the sum of s and t decreased by s
40. half of the sum of p and $2q$ increased by three-fourths q
41. five times the product of x and y increased by $3xy$
42. four times the sum of a and b increased by twice the sum of a and $2b$
43. Find the following product.
 $\frac{1}{2} \cdot \frac{2}{3} \cdot \frac{3}{4} \cdot \ldots \cdot \frac{98}{99} \cdot \frac{99}{100}$
 Recall that an ellipsis ". . ." means "to continue the pattern." Write a sentence or two explaining how you found the product. See margin.

Critical Thinking

44. Is subtraction commutative? Write a short explanation that includes examples to support your answer. See margin.
45. Suppose the operation ✱ is defined for all numbers a and b as $a ✱ b = a + 2b$. Is the operation ✱ commutative? Give examples to support your answer. See margin.

Applications and Problem Solving

46. **Chemistry** Chemists use water to dilute acid. One important rule of chemistry is to always pour the acid into the water. Pouring water into acid could produce spattering and burns.
 a. Would you say that combining acid and water is commutative? no
 b. Give an example from your life that is commutative. See margin.
 c. Give an example from your life that is not commutative. See margin.

Additional Answers

40. $\frac{1}{2}(p + 2q) + \frac{3}{4}q = \frac{1}{2}p + q + \frac{3}{4}q$ distributive property
 $= \frac{1}{2}p + \left(q + \frac{3}{4}q\right)$ associative (+)
 $= \frac{1}{2}p + \left(1q + \frac{3}{4}q\right)$ multiplicative identity
 $= \frac{1}{2}p + \left(1 + \frac{3}{4}\right)q$ distributive property
 $= \frac{1}{2}p + \frac{7}{4}q$ substitution (=)

41. $5xy + 3xy = (5 + 3)xy$ distributive property
 $= 8xy$ substitution (=)

42. $4(a + b) + 2(a + 2b)$
 $= 4a + 4b + 2a + 4b$ distributive property
 $= 4a + 2a + 4b + 4b$ commutative (+)
 $= (4 + 2)a + (4 + 4)b$ distributive property
 $= 6a + 8b$ substitution (=)

Practice Masters, p. 8

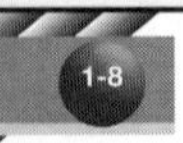

NAME ______ DATE ______

Practice — Student Edition Pages 51–55

Commutative and Associative Properties

Name the property illustrated by each statement.

1. $x + y = y + x$ Commutative prop. of add.
2. $5(m \cdot n) = (5 \cdot m)n$ Associative prop. of mult.
3. $k + 0 = k$ Additive identity prop.
4. $3t + 2r = 2r + 3t$ Commutative prop. of add.
5. $6(u + 2v) = 6u + 12v$ Distributive prop.
6. $0 = 100 \cdot 0$ Multiplicative prop. of zero
7. $(2a + 3b) + 4c = 2a + (3b + 4c)$ Associative prop. of add.
8. $pq + n = qp + n$ Commutative prop. of mult.
9. $gx = xg$ Commutative prop. of mult.
10. $15(c + d) = 15(d + c)$ Commutative prop. of add.

Simplify.

11. $2x + 5y + 9x$ $11x + 5y$
12. $a + 9b + 6a$ $7a + 9b$
13. $2p + 3q + 5p + 2q$ $7p + 5q$
14. $\frac{3}{4}x + z + \frac{1}{4}x$ $x + z$
15. $3j + 4k + 2l + 6k$ $3j + 10k + 2l$
16. $\frac{2}{5}m^2 + 3m + m^2$ $1\frac{2}{5}m^2 + 3m$
17. $6 + 3xz + 4(xz + y)$ $7xz + 4y + 6$
18. $3(2pq + 3r) + 4pq$ $10pq + 9r$
19. $\frac{1}{3}x + \frac{1}{5}y + \frac{1}{2}z + \frac{4}{5}y$ $\frac{1}{3}x + y + \frac{1}{2}z$
20. $0.5x + 0.2y + x$ $1.5x + 0.2y$
21. $2(2e + f) + 3f + 1$ $4e + 5f + 1$
22. $3 + 2(r + s) + 2r$ $3 + 4r + 2s$
23. $r + 3s + 5r + s$ $6r + 4s$
24. $6k^2 + 6k + k^2 + 9k$ $7k^2 + 15k$
25. $4p + 3(3p + 5) + 5$ $13p + 20$
26. $9 + 3(xz + 2y) + xz$ $9 + 4xz + 6y$
27. $5(0.6m + 0.4n) + m$ $4m + 2n$
28. $4(2a + b) + 2(a + 2b)$ $10a + 8b$

FYI

In 1994, the roller coaster with the greatest and fastest drop was the Steel Phantom at the Kennywood Amusement Park in West Mifflin, PA. It has a vertical drop of 225 ft with a design speed of 80 mph.

47. **Physics** The pressure you experience as you ride a roller coaster is produced by the fast changes in direction. This pressure is what pins you down to the back or bottom of your seat. This pressure is known as *g-force.* The formula used to evaluate the g-force is, $G = \frac{(speed)^2}{32.2 \times radius\ of\ the\ curve}$, where the speed is in feet per second, and the radius of the curve is in feet. The acceleration of gravity in $\frac{ft}{sec^2}$ is 32.2.

    a. If you are riding the Steel Phantom at 60 feet per second (about 40 mph) and the car is turning a curve with a radius of 30 feet, what g-force will you experience? **$G = 3.73$**

    b. The g-force has an effect on your weight. For example, if you weigh 100 pounds and experience a g-force of 2 g's, your weight at that moment feels like 2×100 or 200 pounds. What will your weight feel like while taking the curve at 60 feet per second? **See students' work.**

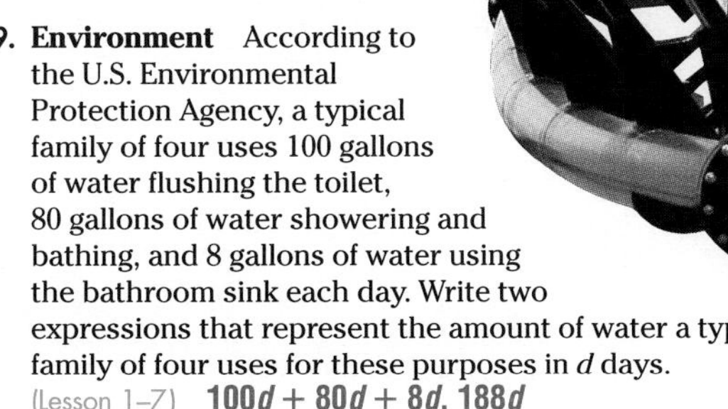

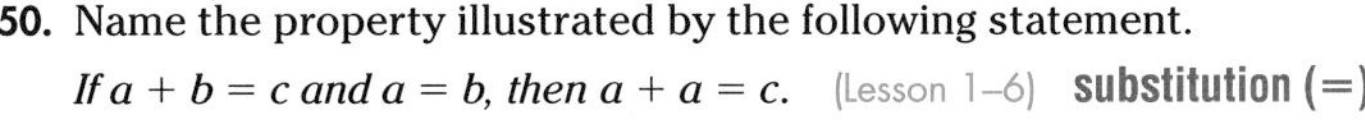

Mixed Review

48. Name the coefficient of $\frac{4m^2}{5}$. (Lesson 1–7) **$\frac{4}{5}$**

49. **Environment** According to the U.S. Environmental Protection Agency, a typical family of four uses 100 gallons of water flushing the toilet, 80 gallons of water showering and bathing, and 8 gallons of water using the bathroom sink each day. Write two expressions that represent the amount of water a typical family of four uses for these purposes in *d* days. (Lesson 1–7) **$100d + 80d + 8d$, $188d$**

50. Name the property illustrated by the following statement. *If $a + b = c$ and $a = b$, then $a + a = c$.* (Lesson 1–6) **substitution (=)**

51. Name the multiplicative inverse of $\frac{2}{3}$. (Lesson 1–6) **$\frac{3}{2}$**

52. *True or False:* $\frac{3k-3}{7} + 13 < 17$, when $k = 8$. (Lesson 1–5) **true**

53. **Weather** The following data represents the average daily temperature in Fahrenheit for Santiago, Chile, for two weeks during the month of February. Organize the data using a stem-and-leaf plot. (Lesson 1–4)

    101, 95, 100, 100, 99, 101, 97,
    99, 97, 100, 103, 101, 101, 99 **See margin.**

54. Evaluate $(25 - 4) \div (2^2 - 1^3)$. (Lesson 1–3) **7**

55. Find the next two items in the sequence. (Lesson 1–2)

56. Write an algebraic expression for *five times p to the sixth power.* (Lesson 1–1) **$5p^6$**

Extension

Reasoning Define the operation × as follows: For any numbers *a* and *b*, $a \times b = a + b + ab$. Use the commutative and associative properties to show that the operation × is commutative.

$$a \times b = b \times a$$
$$a + b + ab = b + a + ba$$
$$a + b + ab = a + b + ba$$

FYI

The Magnum XL-200 at Cedar Point Park in Sandusky, Ohio, is one of the fastest roller coasters in the world. It has a vertical drop of 195 ft and a speed of 72 mph.

4 ASSESS

Closing Activity

Writing Have students list the different properties discussed in Lessons 1-6, 1-7, and 1-8 along with an example of each property.

Additional Answers

43. $\frac{1}{100}$; Each denominator and the following numerator represent the number 1. The resulting expression is 1 in the numerator and 100 in the denominator multiplied by numerous 1s. Since 1 is the multiplicative identity, the product is $\frac{1}{100}$.

44. No; answers will vary. Sample answer: $5 - 3 \neq 3 - 5$.

45. No; Sample example: Let $a = 1$ and $b = 2$; then $1 * 2 = 1 + 2(2)$ or 5 and $2 * 1 = 2 + 2(1)$ or 4.

46b. Sample answer: practicing piano and doing homework

46c. Sample answer: putting clothes into the washer and dryer

53.

Stem	Leaf
10	0 0 0 1 1 1 1 3
9	5 7 7 9 9 9

10|3 = 103

Enrichment Masters, p. 8

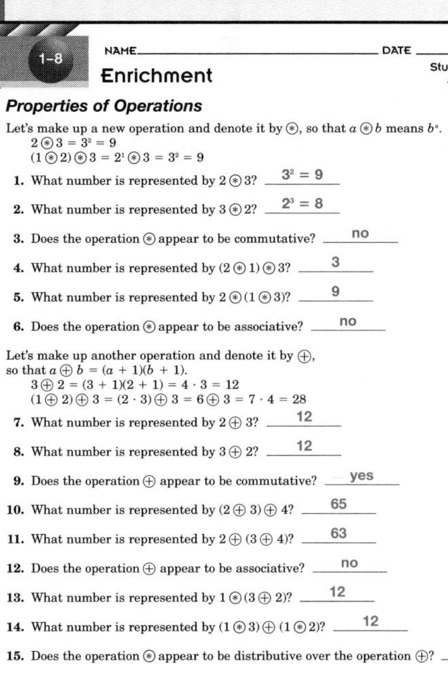

1-8 NAME________ DATE________

Enrichment Student Edition Pages 51–55

Properties of Operations

Let's make up a new operation and denote it by $\circledast$, so that $a \circledast b$ means b^a.

$2 \circledast 3 = 3^2 = 9$

$(1 \circledast 2) \circledast 3 = 2^1 \circledast 3 = 3^2 = 9$

1. What number is represented by $2 \circledast 3$? $3^2 = 9$
2. What number is represented by $3 \circledast 2$? $2^3 = 8$
3. Does the operation $\circledast$ appear to be commutative? no
4. What number is represented by $(2 \circledast 1) \circledast 3$? 3
5. What number is represented by $2 \circledast (1 \circledast 3)$? 9
6. Does the operation $\circledast$ appear to be associative? no

Let's make up another operation and denote it by $\oplus$, so that $a \oplus b = (a + 1)(b + 1)$.

$3 \oplus 2 = (3 + 1)(2 + 1) = 4 \cdot 3 = 12$

$(1 \oplus 2) \oplus 3 = (2 \cdot 3) \oplus 3 = 6 \oplus 3 = 7 \cdot 4 = 28$

7. What number is represented by $2 \oplus 3$? 12
8. What number is represented by $3 \oplus 2$? 12
9. Does the operation $\oplus$ appear to be commutative? yes
10. What number is represented by $(2 \oplus 3) \oplus 4$? 65
11. What number is represented by $2 \oplus (3 \oplus 4)$? 63
12. Does the operation $\oplus$ appear to be associative? no
13. What number is represented by $1 \circledast (3 \oplus 2)$? 12
14. What number is represented by $(1 \circledast 3) \oplus (1 \circledast 2)$? 12
15. Does the operation $\circledast$ appear to be distributive over the operation $\oplus$? yes
16. Let's explore these operations a little further. What number is represented by $3 \circledast (4 \oplus 2)$? 3375
17. What number is represented by $(3 \circledast 4) \oplus (3 \circledast 2)$? 585
18. Is the operation $\circledast$ actually distributive over the operation $\oplus$? no

1-9 LESSON NOTES

NCTM Standards: 1–6, 8

Instructional Resources

- Study Guide Master 1-9
- Practice Master 1-9
- Enrichment Master 1-9
- Assessment and Evaluation Masters, p. 17
- Graphing Calculator Masters, p. 1
- Real-World Applications, 2
- Tech Prep Applications Masters, p. 2

Transparency 1-9A contains the 5-Minute Check for this lesson; **Transparency 1-9B** contains a teaching aid for this lesson.

Recommended Pacing	
Standard Pacing	Days 12 & 13 of 15
Honors Pacing	Days 11 & 12 of 14
Block Scheduling*	Day 6 of 7 (along with Lesson 1-8)
Alg. 1 in Two Years*	Days 18 & 19 of 22

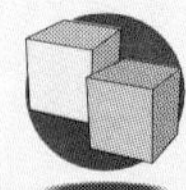

*For more information on pacing and possible lesson plans, refer to the *Block Scheduling Booklet* and *Algebra 1 in Two Years.*

1 FOCUS

5-Minute Check
(over Lesson 1-8)

Name the property illustrated by each statement.

1. $by + 3a = yb + 3a$ **commutative (×)**
2. $4b + 5c = 5c + 4b$ **commutative (+)**
3. $(2 \cdot a) \cdot b = 2 \cdot (a \cdot b)$ **associative (×)**

Simplify.

4. $4x + 5y + 3z + 6x$ $\mathbf{10x + 5y + 3z}$
5. $4(3a + b) + 3a$ $\mathbf{15a + 4b}$

1-9 A Preview of Graphs and Functions

What YOU'LL LEARN

- To interpret graphs in real-world settings, and
- to sketch graphs for given functions.

Why IT'S IMPORTANT

You need to know how to sketch and interpret graphs in order to study trends and display data.

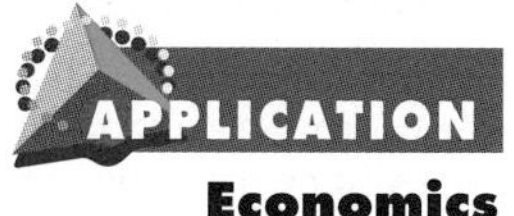

Economics

The dollar value of a car begins to decrease immediately after it is sold. This is called *depreciation.* Mathematically, depreciation can be defined by the following open sentence.

depreciation = original cost of car − value of car when sold

The table below shows how a typical $15,000 car depreciates over a period of five years. From the table, you can see that as the age of the car increases, the value decreases. Why do you think this happens?

Age of Car (year)	Approximate Value of Car
0	$15,000
1	11,000
2	8000
3	6000
4	5000
5	4000

This information can also be presented in a graph. The graph below shows the relationship between the current value of the car and the age of the car.

The value of the car is said to be a **function** of the age of the car. A function is a relationship between input and output. In a function, the output depends on the input. There is exactly one output for each input.

There are two number lines on the graph. The line that represents the current value of the car is the **vertical axis.** The line that represents the age of the car is the **horizontal axis.**

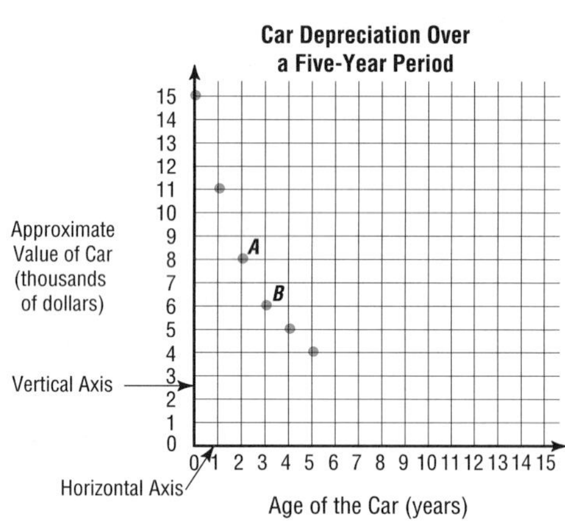

Classroom Vignette

"Ask students to research (using Blue Books, newspaper ads, etc.) the price of several makes of cars to determine the better buy. Insurance costs, maintenance, and fuel efficiency should also be considered. They should then prepare graphs to show the depreciation of the cars researched as part of their findings. This can serve a dual role of addressing interdisciplinary instruction and performance assessment."

Bea Moore-Harris
Educational Specialist,
Region IV Education Service Center
Houston, Texas

Notice that there are six points shown on the graph. Each point represents the dollar value of the car (in thousands) at the end of each year. For example, after year 1, the value is \$11,000. You can express this relationship as (1, 11). What do points A and B represent?

$A(2, 8)$ means that at the end of year 2 the value of the car is about \$8000.

$B(3, 6)$ means that at the end of year 3 the value of the car is about \$6000.

(1, 11), (2, 8), and (3, 6) are called **ordered pairs.** The order in which the pair of numbers is written is important. Ordered pairs are used to locate points on the graph. The first number in an ordered pair corresponds to the horizontal axis, and the second number in an ordered pair corresponds to the vertical axis. The ordered pair (0, 0) corresponds to the **origin**.

Many real-world situations can be modeled using functions.

Example 1

Sales

Shim owns a farm market. The amount a customer pays for sweet corn depends on the number of ears that are purchased. Shim sells a dozen ears of corn for \$3.00.

a. Make a table showing the price of various purchases of sweet corn.

Number of Dozen	Ears of Corn	Price
$\frac{1}{2}$	6	\$1.50
1	12	\$3.00
$1\frac{1}{2}$	18	\$4.50
2	24	\$6.00

b. Write four ordered pairs that represent the number of ears of corn and the price of the corn.

Four ordered pairs can be determined from the chart. These ordered pairs are (6, 1.50), (12, 3.00), (18, 4.50), and (24, 6.00).

c. Describe a set of axes that could be used to graph the number of ears of corn and the price of the corn.

Let the horizontal axis represent the number of ears of corn, and let the vertical axis represent the price of the corn.

d. Draw a graph that shows the relationship between the number of ears of corn and the price.

Graph the four points described by the ordered pairs.

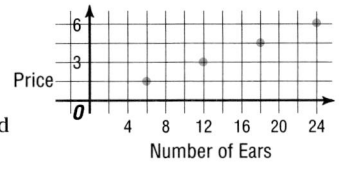

e. As you read the graph from left to right, describe the trend you see. Explain.

The graph goes upward, because the price increases as the number of ears increases. The price is a function of the number of ears purchased.

Motivating the Lesson

Hands-On Activity Have students bring to class examples of graphs. For each graph, have them identify the meaning of the horizontal axis and the vertical axis. They should also identify the range of values that occur on each axis and determine why the axes stop at the maximum and minimum values at which they stop.

2 TEACH

In-Class Example

For Example 1

A shoe store is having a sale. The first pair of shoes sells for \$40. The second pair sells for half price, or \$20. The next pair sells for half of that, and so on.

a. Make a table showing the cost of 1 to 5 pairs of shoes.

Number of Pairs	Total Cost
1	**\$40**
2	**\$60**
3	**\$70**
4	**\$75**
5	**\$77.50**

b. Write the data as a list of ordered pairs.
(1, 40), (2, 60), (3, 70), (4, 75), (5, 77.5)

c. Draw the graph that shows the relationship between the number of pairs of shoes and the total price.

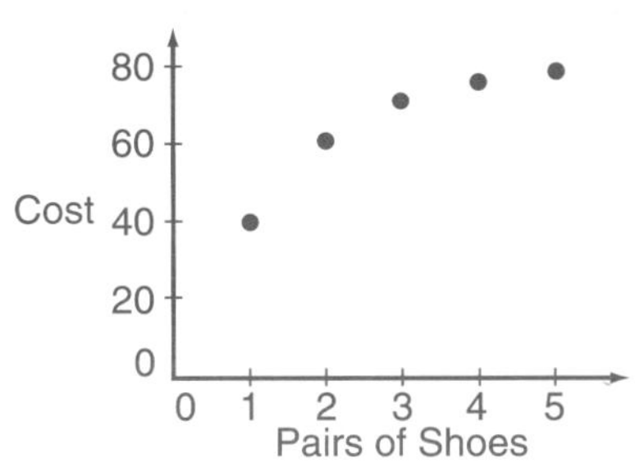

Teaching Tip Emphasize to students the idea that graphs are used to show the relationship between two different sets of data, such as the number of ears of corn and the price of corn in Example 1.

GLENCOE Technology

CD-ROM Interaction

A multimedia simulation connects graphing and equations with softball. A blackline master activity with teacher's notes provides a follow-up to the CD-ROM simulation.

For Windows & Macintosh

In-Class Examples

For Example 2
A balloon is being inflated with air. As more and more air is pumped into the balloon, the volume increases until it finally bursts. Make a sketch of a graph of the situation.

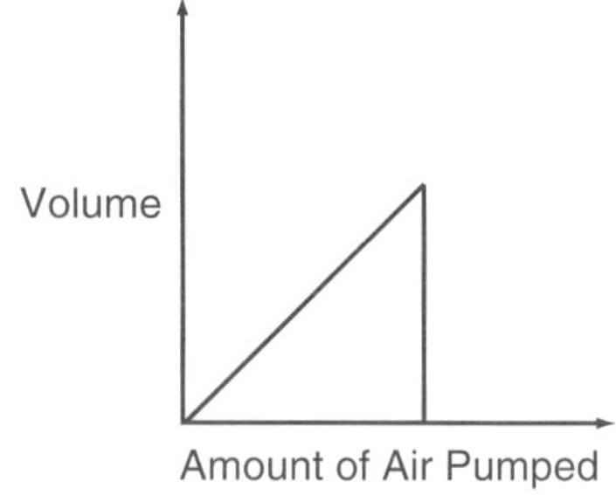

For Example 3
The price of gasoline at the pump is dependent on the availability of fuel oil. Write a story about the relationship in the following graph.

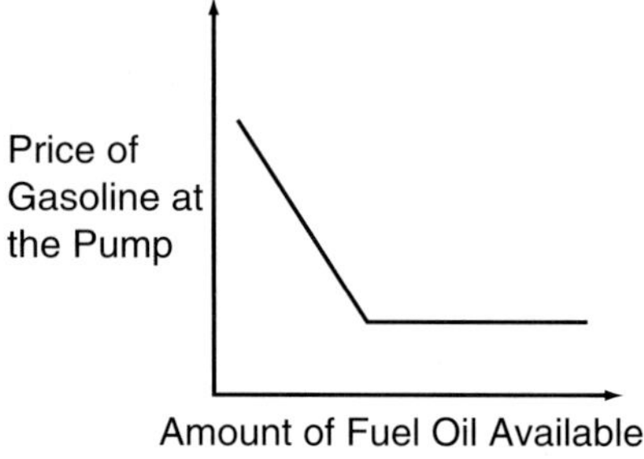

Sample story: When fuel oil is scarce, the prices are high. As more becomes available, the prices drop. There is, however, a price below which no gas station will set its price.

In Example 1, the price a customer pays depends on the number of ears purchased. Therefore, the number of ears purchased is called the **independent variable** or **quantity,** and the price is called the **dependent variable** or **quantity.** Usually the independent variable is graphed on the horizontal axis and the dependent variable is graphed on the vertical axis.

You can use a graph without scales on either axis to show the general shape of the graph that models a situation.

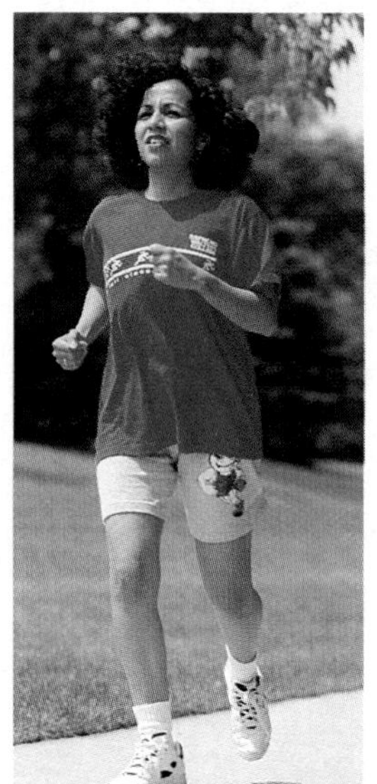

Example 2 **For a certain time, Lucinda jogs up a hill at a steady speed. Then she runs down the hill and picks up her speed.**

a. What happens to her speed when Lucinda jogs at a steady pace? when Lucinda runs down the hill?

The speed remains the same when her jogging pace is steady. The speed increases when she runs downhill.

b. Identify the independent and dependent quantities.

Time is the independent quantity, and speed is the dependent quantity.

c. Match the graph to the situation.

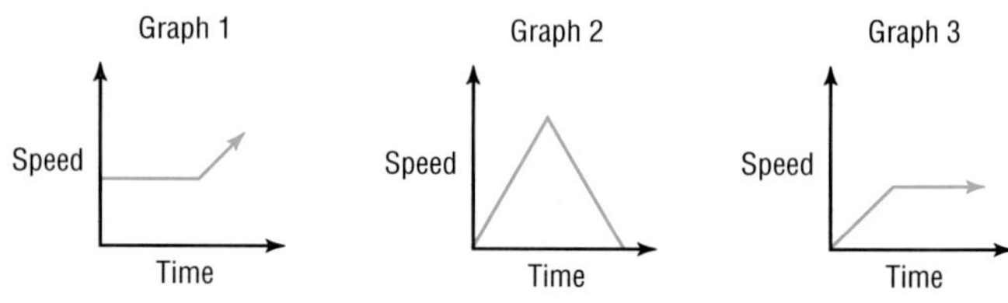

Graph 1 shows a steady speed (a horizontal line), followed by an increase in speed (a line that goes up) the longer she exercises.

A **relation** is a set of ordered pairs. The set of first numbers of the ordered pairs is the **domain** of the relation. The domain contains values of the independent variable. The set of second numbers of the ordered pairs is the **range** of the relation. The range contains values of the dependent variable.

Example  **Ryan is not feeling well and has missed a day of school. The graph shows Ryan's temperature as a function of the time of day.**

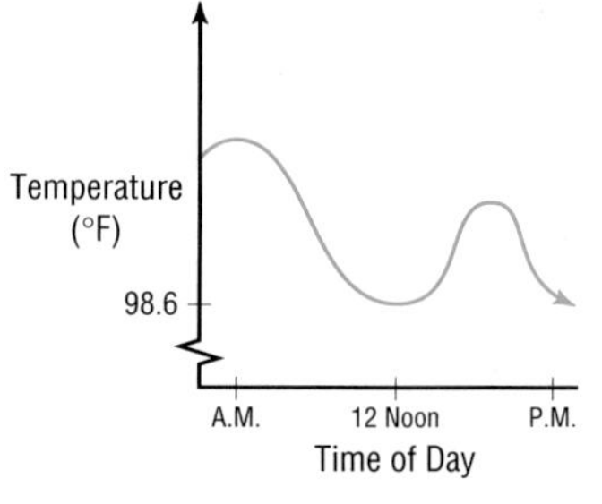

a. Write a story about Ryan's illness.

When Ryan got up in the morning, he had a high fever. As the morning went on, his temperature went down. By noon it was almost normal, but his temperature went up again after lunch. Then his mom gave him some medication. By evening his fever was gone, and his temperature was back to normal.

b. Identify a reasonable domain and range for this problem.

The domain includes the 24 hours in a day. A reasonable range would include all of the temperatures between the highest and lowest temperature a person can have and still be alive.

CHECK FOR UNDERSTANDING

Communicating Mathematics

1. See margin.
2. Sample answers: $3000; $800

Study the lesson. Then complete the following.

1. **Explain** why the order of the numbers in an ordered pair is important.
2. Refer to the application at the beginning of the lesson. Extend the graph to predict the value of the car after 6 years and after 10 years.
3. Use the graph at the right to answer the following questions. **a–e. See margin.**

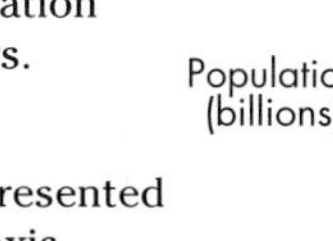

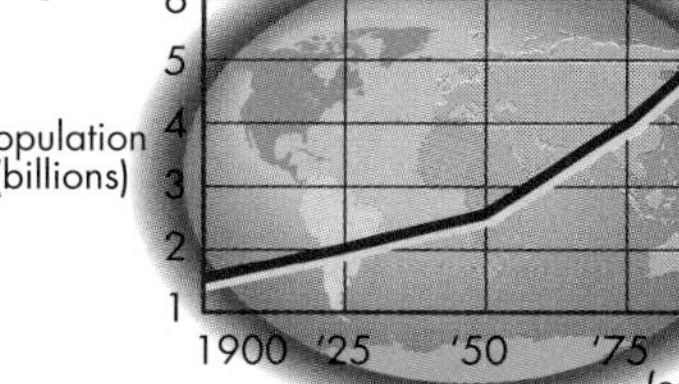

   a. **Describe** the information that the graph shows.

   b. **Identify** the variable represented along each axis.

   c. **Identify** two points on the graph and write the ordered pair for each point. Explain what each point represents.

   d. **Explain** what the ordered pair (2000, 6.2) represents.

   e. **Write** a statement that explains the function in the graph.

HERMAN

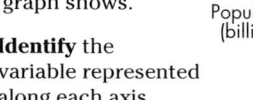

"I really look forward to your cheery little visits."

4. **Write** a story that explains what the graph in the cartoon at the left might represent. Identify the independent and dependent variables in your story. **Sample answer: Every day, the patient's condition becomes worse. Time is the independent variable and the patient's condition is the dependent variable.**

Guided Practice

5. Match each description with the most appropriate graph.

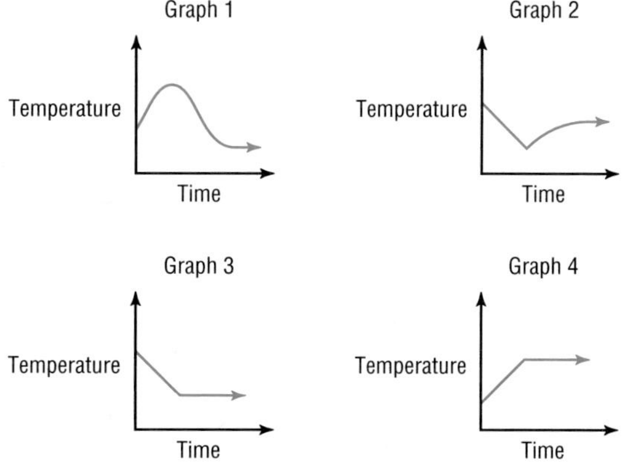

5a. Graph 3
5b. Graph 4
5c. Graph 2
5d. Graph 1

   a. In August, you enter a hot house and turn on the air conditioner.

   b. In January, you enter a cold house and turn up the thermostat to 68° F.

   c. You put ice cubes in your fruit punch and then drink it slowly.

   d. You put a cup of water in the microwave and heat it for one minute.

3 PRACTICE/APPLY

Check for Understanding

Exercises 1–7 are designed to help you assess your students' understanding through reading, writing, speaking, and modeling. You should work through Exercises 1–4 with your students and then monitor their work on Exercises 5–7.

Error Analysis
Students may mix up the axes. Remind them that it is customary to represent the independent variable along the horizontal axis and the dependent variable along the vertical axis.

Additional Answers

1. The numbers represent different values. The first number represents the number on the horizontal axis and the second represents the number on the vertical axis.

3a. world population in billions of people from 1900 to the year 2000

3b. horizontal axis: time in years; vertical axis: population in billions

3c. Sample answer: (1925, 2): In 1925 the world's population was about 2 billion people. (1950, 2.5): In 1950 the world's population was about 2.5 billion people.

3d. The estimated world population for the year 2000 is 6.2 billion.

3e. Sample answer: The world's population has increased since 1900.

Reteaching

Using Data Have groups of students think of data that could be portrayed using a graph and construct their own graph using their data. Also ask students to write five questions that pertain to the graph. Make copies of each group's graph and questions and distribute them to the other groups to answer the questions.

Assignment Guide

Core: 9–15 odd, 16, 17, 19–23
Enriched: 8–14 even, 16–23

For **Extra Practice**, see p. 758.

The red A, B, and C flags, printed only in the Teacher's Wraparound Edition, indicate the level of difficulty of the exercises.

Additional Answers

6.

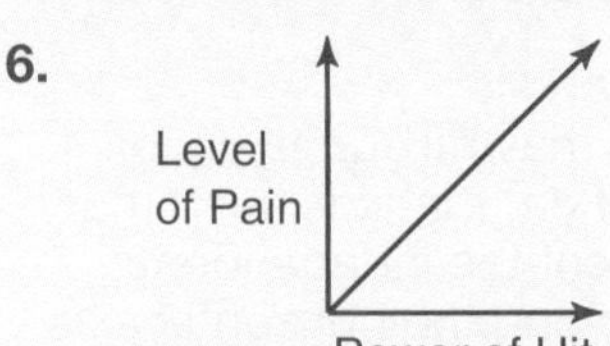

The independent variable is how hard you hit your thumb; the dependent variable is how much it hurts.

7a. *A* is the younger player, but *B* runs the mile in less time.

7b. *A* made more 3-point shots, but *B* made more 2-point shots.

7c. *B* is the older player and *B* made more 2-point shots.

7d. *A* is the younger player and *A* made more free-throw shots.

8. b; The bus gains speed, and then travels at a level speed. Later, it slows to a stop. Then the pattern continues.

Study Guide Masters, p. 9

NAME ______________________ DATE ________

Study Guide Student Edition Pages 56–62

A Preview of Graphs and Functions

A plane is landing from an altitude of 35,000 ft. The plane descends at a rate of 3,200 ft/min. Mathematically, the plane's altitude can be defined by the following open sentence.

altitude = 35,000 − 3,200*t*, where *t* is the time of descent

The table at the right shows the plane's altitude over a period of 3 minutes.

Time of Descent	Plane's Altitude
0 min	35,000
1 min	31,800
2 min	28,600
3 min	25,400

This information can also be represented in a graph. The graph shows the relationship between the plane's altitude and its time of descent.

Example 1:

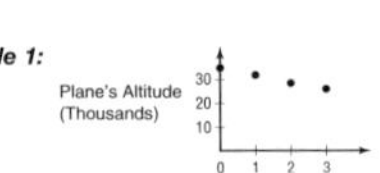

You can use a graph without scales on either axes to show the general shape of the graph that represents a situation.

Example 2: Kelley walks up a hill at a steady pace. Then she runs down the hill. Make a sketch of the graph of the situation.

a. Speed / Time b. Speed / Time c. Speed / Time

To select the graph that best represents this situation, find the graph that shows a steady pace, followed by an increase in speed. Graph c matches this description.

1. Identify the graph that matches the following statement. Explain your answer. Your body temperature rises as you exercise. **b. You start out with a normal temperature and after time it increases.**

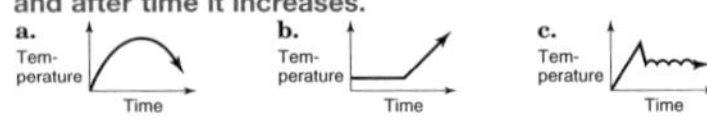

Sketch a reasonable graph for each situation.

2. Tim is riding his bike at a steady pace. Then he has a flat tire. He walks up a hill to a gas station to have the tire fixed. When the tire is fixed, he continues his ride.

Speed / Time

3. Ken likes to trade marbles. During the first week of the month, he added lots of marbles to his collection. Then during the next three weeks of the month, he lost several others.

No. of marbles / Time

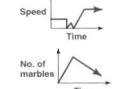

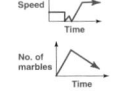

6. Sketch a reasonable graph for the following statement. *How hard you hit your thumb with a hammer impacts on how much it hurts.* Name the independent and dependent variables. **See margin.**

7. The basketball coach at City College is recruiting basketball players. The graphs below describe two of the players being recruited.

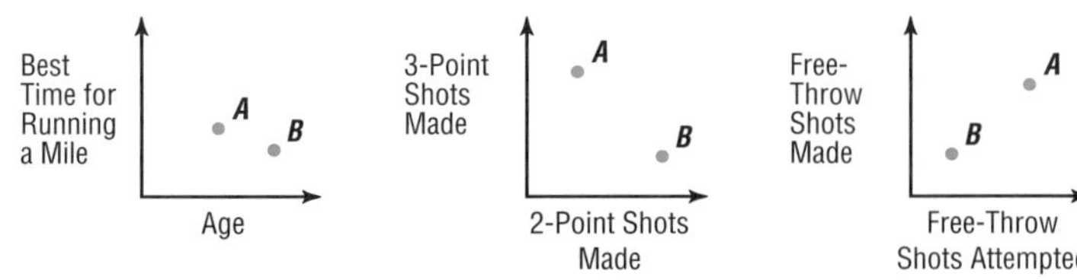

Determine whether each statement is *true* or *false*. Explain your reasoning. **a–d. See margin for explanations.**

a. The younger player is faster than the older player. **false**

b. One player made more 2-point and more 3-point shots than the other.

7b. false

c. The older player made more 2-point shots. **true**

d. The younger player tried and made more free-throw shots. **true**

EXERCISES

Practice

A

8. Identify the graph that matches the following statement. Explain your answer. **See margin.**

A bus frequently stops to pick up passengers.

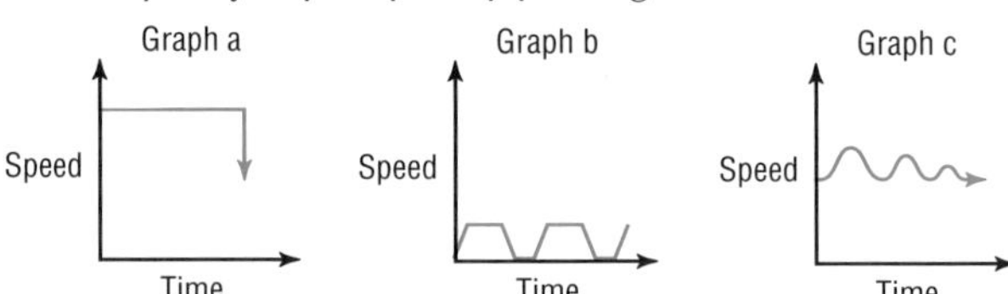

9. Identify the graph that represents annual income as a function of age. Explain your answer. **See margin.**

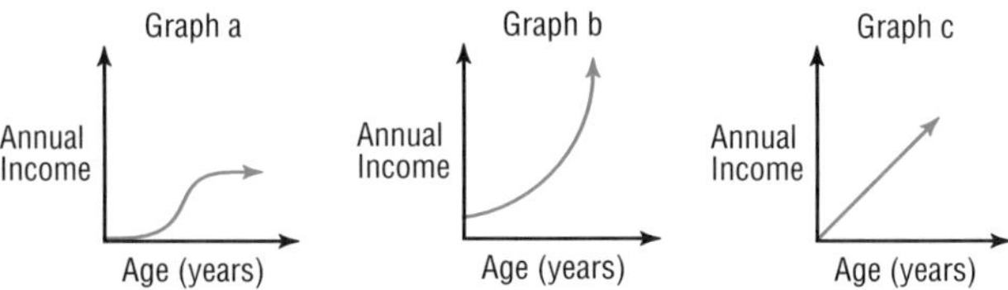

10. The graph at the right shows the amount of money in Jorge's savings account as a function of time. **See margin.**

a. Describe what is happening to Jorge's money. Explain why the graph rises and falls at particular points.

b. Describe the elements in the domain and range.

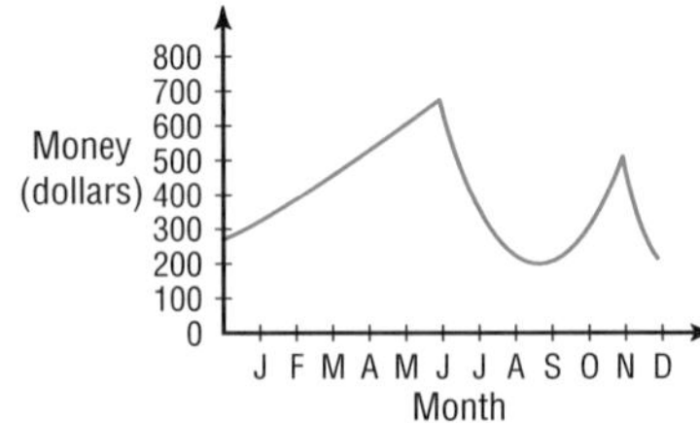

Additional Answers

9. a; An average person makes no money as a child, then his or her income rises for several years, and finally levels off.

10a. Sample answer: Jorge saved steadily from January to June. In July, he withdrew money to go on vacation. He started saving again in September. Then in November he withdrew money for holiday presents.

10b. The domain includes the months of the year; the range is the amount of money.

11. Match each table of data to the most appropriate graph.

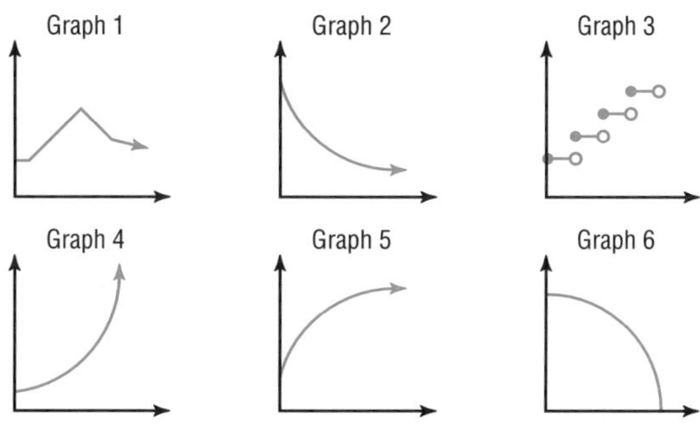

a. Primary Seismic Wave **Graph 5**

Distance from Epicenter (km)	1000	2000	3000	4000	5000	6000	7000
Time (minutes)	2	4	5.8	7.2	8.5	9.7	10.8

b. Cost of First-Class Postage Stamps **Graph 3**

Year	1983	1985	1987	1989	1991	1993	1995
Cost of Stamp	22¢	25¢	25¢	25¢	29¢	29¢	32¢

c. Immigration into the United States **Graph 1**

Year	1987	1988	1989	1990	1991	1992	1993
Legal Admissions (millions)	0.6	0.6	1.1	1.5	1.8	1.0	0.9

d. Development of Human Fetus **Graph 4**

Time (weeks)	4	8	12	16	20	24	28
Mass (grams)	0.5	1	28	110	300	650	1200

e. Height of Falling Object **Graph 6**

Time (seconds)	0	0.5	1.0	1.5	2.0	2.5	3.0
Height (feet)	200	196	184	164	136	100	56

f. Cooling of Boiling Water **Graph 2**

Time (minutes)	0	1	2	3	4	5	6
Temperature (°F)	100	50	25	12.5	6.25	3.125	1.5625

Sketch a reasonable graph for each situation. **12–15. See margin.**

12. Rashaad likes to trade basketball cards. During the first two weeks of the month, he added lots of cards to his collection. Then, he lost some cards and sold others.

13. A radio-controlled car moves along and then crashes against a wall.

C 14. Phyllis is riding her bike at a steady pace. Then she has a flat tire. She walks to a gas station to have the tire repaired. When the tire is fixed, she continues her ride.

15. The height of a football that was thrown into the end zone for a touchdown compared to the distance the football is from the end zone.

4 ASSESS

Closing Activity

Speaking Encourage a discussion of the main terms introduced in this lesson: *vertical axis, horizontal axis, origin, ordered pair, independent variable,* and *dependent variable.* Encourage students to find a sentence definition for each term.

Chapter 1, Quiz D (Lessons 1-8 and 1-9), is available in the *Assessment and Evaluation Masters,* p. 17.

Teaching Tip The graphs in Exercise 11 are previews of the step functions, quadratic functions, and exponential functions students will study in this text.

Practice Masters, p. 9

1-9 NAME______ DATE______
Practice
Student Edition Pages 56–62

1. Identify the graph that matches the following statement. Explain your answer. *A train stops to let off passengers.*

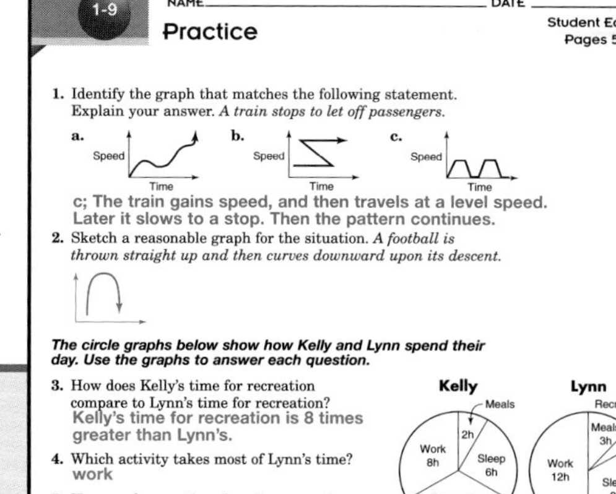

c; The train gains speed, and then travels at a level speed. Later it slows to a stop. Then the pattern continues.

2. Sketch a reasonable graph for the situation. *A football is thrown straight up and then curves downward upon its descent.*

The circle graphs below show how Kelly and Lynn spend their day. Use the graphs to answer each question.

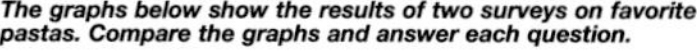

3. How does Kelly's time for recreation compare to Lynn's time for recreation? Kelly's time for recreation is 8 times greater than Lynn's.

4. Which activity takes most of Lynn's time? work

5. How much more time does Lynn spend working than Kelly does? 4 hours

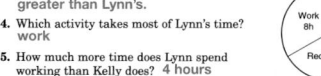

The graphs below show the results of two surveys on favorite pastas. Compare the graphs and answer each question.

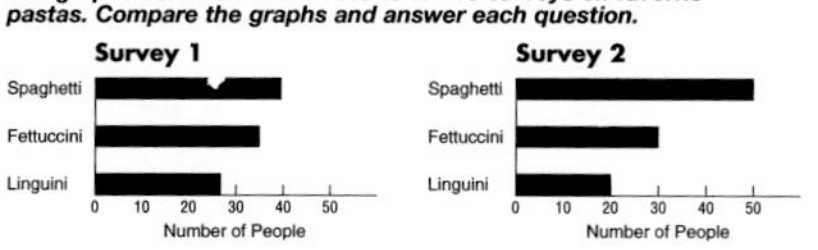

6. In Survey 1, the number of votes for spaghetti is twice the number of votes for which pasta in Survey 2? Linguini

7. How many more people preferred spaghetti in Survey 2 than preferred spaghetti in Survey 1? 10 people

8. In Survey 1, which pasta received the most votes? the fewest votes? spaghetti; linguini

Additional Answers

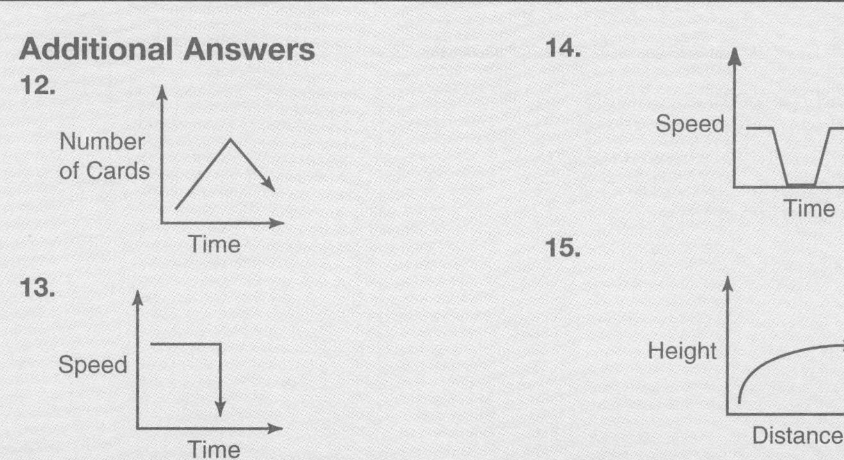

Additional Answers

17a. independent variable: time in years; dependent variable: millions of dollars.
17b. In 1991 the telethon raised 45 million dollars.
17c. Money raised in the telethon decreased, then increased steadily.
17d. The money raised increased.
17e. Answers will vary. See students' work.
18. Sample answer: Rock music was the best seller. Rap music sold better than country music.

Critical Thinking

16. Which of the sports listed will produce a graph like the one below? Explain why you chose the sport you did and give reasons why you did not choose the other sports. **See students' work.**

Archery	High Jumping
Biking	Javelin Throwing
Fishing	Pole Vaulting
Golf	Skateboarding
High Diving	Sky Diving

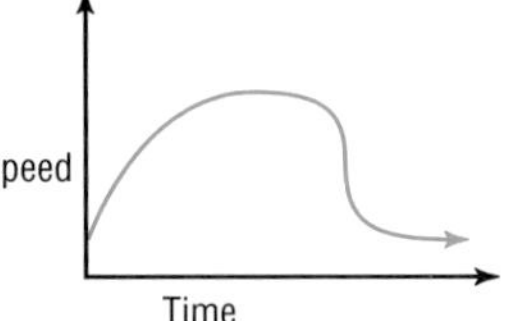

Applications and Problem Solving

17a–e. See margin.

17. **Charities** The graph at the right shows the amount of money raised by the Muscular Dystrophy Association over a 14-year period.

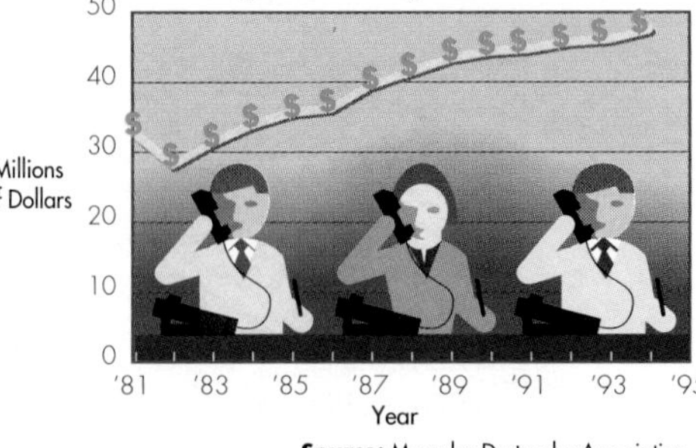

a. Name the independent and dependent variables.
b. What does the ordered pair (1991, 45) represent?
c. Write a statement about the amount of money raised from 1981 to 1986.
d. Write a statement about the amount of money raised from 1987 to 1994.
e. Use the information to predict the amount of money raised each year since 1994. Research to see if your prediction is correct.

18. **Music Business** The graph at the right shows the sales in millions of dollars for different music categories. Write two conclusions you can make from the information presented in the graph. **See margin.**

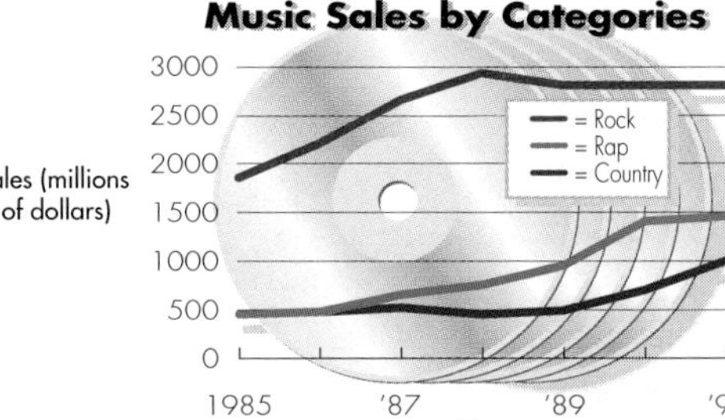

Mixed Review

19. Simplify $5p + 7q + 9 + 4q + 2p$. (Lesson 1–8) $\mathbf{7p + 11q + 9}$

20. Simplify $9a + 14(a + 3)$. (Lesson 1–7) $\mathbf{23a + 42}$

21. Evaluate $14(1) - 27(0)$. (Lesson 1–6) **14**

22. *True or False:* $5m + 6^2 = 56$, when $m = 4$. (Lesson 1–5) **true**

23. **Geometry** The volume of a cone equals one-third the area of the base B times the height h. Write an algebraic expression for the volume of a cone. (Lesson 1–1)

23. $\frac{1}{3}Bh$

Enrichment Masters, p. 9

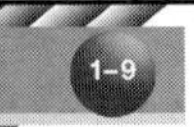

NAME_______________ DATE_______

Enrichment — Student Edition Pages 56–62

Figurate Numbers

The numbers below are called **pentagonal numbers.** They are the numbers of dots or disks that can be arranged as pentagons.

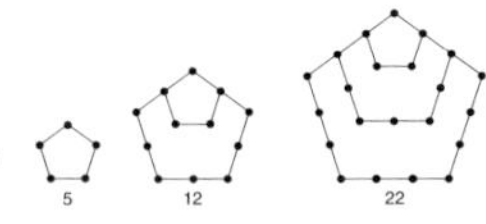

1. Evaluate the expression $\frac{1}{2}n(3n - 1)$ for values of n from 1 through 4. **1, 5, 12, 22**
2. What do you notice? **They are the first four pentagonal numbers.**
3. Find the next six pentagonal numbers. **35, 51, 70, 92, 117, 145**
4. Evaluate the expression $\frac{1}{2}n(n + 1)$ for values of n from 1 through 5. On another sheet of paper, make drawings to show why these numbers are called the triangular numbers. **1, 3, 6, 10, 15**
5. Evaluate the expression $n(2n - 1)$ for values of n from 1 through 5. Draw these hexagonal numbers. **1, 6, 15, 28, 45**
6. Find the first 5 square numbers. Also, write the general expression for any square number. **1, 4, 9, 16, 25; n^2**

The numbers you have explored above are called the plane figurate numbers because they can be arranged to make geometric figures. You can also create solid figurate numbers.

7. If you pile 10 oranges into a pyramid with a triangle as a base, you get one of the tetrahedral numbers. How many layers are there in the pyramid? How many oranges are there in the bottom layer? **3 layers; 6**
8. Evaluate the expression $\frac{1}{6}n(n + 1)(n + 2)$ for values of n from 1 through 5 to find the first five tetrahedral numbers. **1, 4, 10, 20, 35**

Extension

Reasoning Markham starts running quickly, slows down, stops, runs quickly again, then stops. Draw a rough graph of speed against time, and a graph of distance against time.

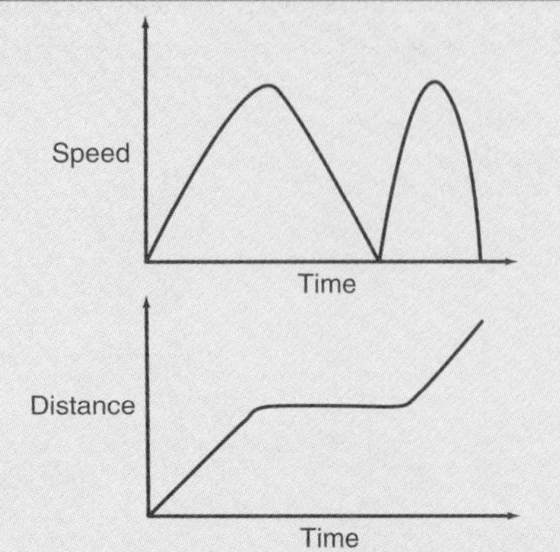

CHAPTER 1 HIGHLIGHTS

Using the CHAPTER HIGHLIGHTS

The Chapter Highlights begins with a listing of the new terms, properties, and phrases that were introduced in this chapter. Have students define each term and provide an example or two of it, if appropriate.

VOCABULARY

After completing this chapter, you should be able to define each term, property, or phrase and give an example or two of each.

Algebra

additive identity (p. 37)
algebraic expression (p. 6)
algebra tiles (p. 44)
associative property (p. 51)
base (p. 7)
coefficient (p. 47)
commutative property (p. 51)
dependent variable (p. 58)
distributive property (p. 46)
domain (p. 58)
equation (p. 33)
equivalent expressions (p. 47)
evaluate (p. 8)
exponent (p. 7)
factors (p. 6)
function (p. 56)
horizontal axis (p. 56)
independent variable (p. 58)
inequality (p. 33)
like terms (p. 47)
multiplicative identity (p. 38)
multiplicative inverse (p. 38)
multiplicative property of zero (p. 38)
open sentence (p. 32)
order of operations (p. 19)
ordered pairs (p. 57)
power (p. 7)
product (p. 6)
range (p. 58)
reciprocal (p. 38)
reflexive property of equality (p. 39)
relation (p. 58)
replacement set (p. 33)
simplest form (p. 47)
solution (p. 32)
solution set (p. 33)
solving the open sentence (p. 32)
substitution property of equality (p. 39)
symmetric property of equality (p. 39)
term (pp. 13, 47)
transitive property of equality (p. 39)
variable (p. 6)
vertical axis (p. 56)

Discrete Mathematics

element (p. 33)
set (p. 33)
sequence (p. 13)

Statistics

data (p. 25)
statistics (p. 25)
stem-and-leaf plots (p. 25)

Problem Solving

look for a pattern (p. 13)

UNDERSTANDING AND USING THE VOCABULARY

Choose the letter of the term that best matches each statement or phrase.

1. For any number a, $a + 0 = 0 + a = a$. **a**
2. For any number a, $a \cdot 1 = 1 \cdot a = a$. **c**
3. For any number a, $a \cdot 0 = 0 \cdot a = 0$. **e**
4. For any number a, there is exactly one number $\frac{1}{a}$ such that $a \cdot \frac{1}{a} = \frac{1}{a} \cdot a = 1$. **d**
5. For any number a, $a = a$. **f**
6. For any numbers a and b, if $a = b$, then $b = a$. **h**
7. For any numbers a and b, if $a = b$, then b may be substituted for a. **g**
8. For any numbers a, b, and c, if $a = b$ and $b = c$, then $a = c$. **i**
9. For any numbers a, b, and c, $a(b + c) = ab + ac$. **b**

a. additive identity property
b. distributive property
c. multiplicative identity property
d. multiplicative inverse property
e. multiplicative property of zero
f. reflexive property
g. substitution property
h. symmetric property
i. transitive property

Instructional Resources

Three multiple-choice tests and three free-response tests are provided in the *Assessment and Evaluation Masters*. Forms 1A and 2A are for honors pacing, and Forms 1B, 1C, 2B, and 2C are for average pacing. Chapter 1 Test, Form 1B is shown at the right. Chapter 1 Test, Form 2B is shown on the next page.

Assessment and Evaluation Masters, pp. 3–4

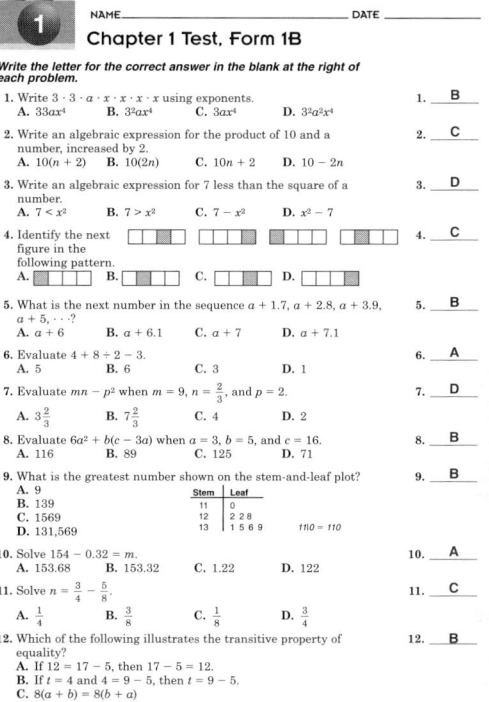

1 NAME________________ DATE ________

Chapter 1 Test, Form 1B

Write the letter for the correct answer in the blank at the right of each problem.

1. Write $3 \cdot 3 \cdot a \cdot x \cdot x \cdot x \cdot x$ using exponents. 1. **B**
   A. $33ax^4$ B. 3^2ax^4 C. $3ax^4$ D. $3^2a^2x^4$
2. Write an algebraic expression for the product of 10 and a number, increased by 2. 2. **C**
   A. $10(n + 2)$ B. $10(2n)$ C. $10n + 2$ D. $10 - 2n$
3. Write an algebraic expression for 7 less than the square of a number. 3. **D**
   A. $7 < x^2$ B. $7 > x^2$ C. $7 - x^2$ D. $x^2 - 7$
4. Identify the next figure in the following pattern. 4. **C**
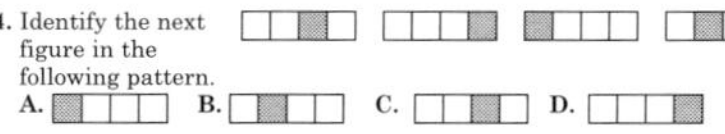
   A. B. C. D.
5. What is the next number in the sequence $a + 1.7$, $a + 2.8$, $a + 3.9$, $a + 5, \cdots$? 5. **B**
   A. $a + 6$ B. $a + 6.1$ C. $a + 7$ D. $a + 7.1$
6. Evaluate $4 + 8 \div 2 - 3$. 6. **A**
   A. 5 B. 6 C. 3 D. 1
7. Evaluate $mn - p^2$ when $m = 9$, $n = \frac{2}{3}$, and $p = 2$. 7. **D**
   A. $3\frac{2}{3}$ B. $7\frac{2}{3}$ C. 4 D. 2
8. Evaluate $6a^2 + b(c - 3a)$ when $a = 3$, $b = 5$, and $c = 16$. 8. **B**
   A. 116 B. 89 C. 125 D. 71
9. What is the greatest number shown on the stem-and-leaf plot? 9. **B**
   A. 9
   B. 139
   C. 1569
   D. 131,569

Stem	Leaf
11	0
12	2 2 8
13	1 5 6 9

$11|0 = 110$

10. Solve $154 - 0.32 = m$. 10. **A**
    A. 153.68 B. 153.32 C. 1.22 D. 122
11. Solve $n = \frac{3}{4} - \frac{5}{8}$. 11. **C**
    A. $\frac{1}{4}$ B. $\frac{3}{8}$ C. $\frac{1}{8}$ D. $\frac{3}{4}$
12. Which of the following illustrates the transitive property of equality? 12. **B**
    A. If $12 = 17 - 5$, then $17 - 5 = 12$.
    B. If $t = 4$ and $4 = 9 - 5$, then $t = 9 - 5$.
    C. $8(a + b) = 8(b + a)$
    D. If $5 + 6 = 11$, then $6 + 5 = 11$.

1 NAME________________ DATE ________

Chapter 1 Test, Form 1B (continued)

13. What is the simplest form of $7rs + 2rs + 3$? 13. **A**
    A. $9rs + 3$ B. $12rs$ C. $7r^2s^2 + 3$ D. $9rs^2 + 3$
14. What is the simplest form of $6b + 3(2a^2 + b)$? 14. **D**
    A. $2a^2 + 10b$ B. $2a^2 + 9b$ C. $6a^2 + 7b$ D. $6a^2 + 9b$
15. Which property is illustrated by $7(6 \cdot 2) = (7 \cdot 6)2$? 15. **D**
    A. distributive property
    B. reflexive property of equality
    C. commutative property of multiplication
    D. associative property of multiplication
16. Which of the following illustrates the distributive property? 16. **C**
    A. $5(xy + 4) = (xy + 4)5$ B. $5(xy + 4) = 5(4 + xy)$
    C. $5(xy + 4) = 5xy + 20$ D. $5(xy + 4) = 5(xy) + 4(xy)$
17. What is the simplest form of $4[5t + 2(3t + 5)]$? 17. **C**
    A. $4(11t + 10)$ B. $26t + 10$ C. $44t + 40$ D. $84t$
18. Which property is illustrated by $4n^2 + (6 + 2n) = 4n^2 + (2n + 6)$? 18. **C**
    A. distributive property
    B. associative property of addition
    C. commutative property of addition
    D. reflexive property of equality
19. Which of the following best describes the graph? 19. **C**

    A. The price of a share of the company's stock increased more in the morning than in the afternoon.
    B. The price of a share of the company's stock decreased more in the morning than in the afternoon.
    C. The price of a share of the company's stock increased more in the afternoon than in the morning.
    D. The price of a share of the company's stock decreased more in the afternoon than in the morning.
20. Identify the graph that matches the following statement: The accident rate for middle-aged automobile drivers is lower than the rate for younger and older drivers. 20. **C**
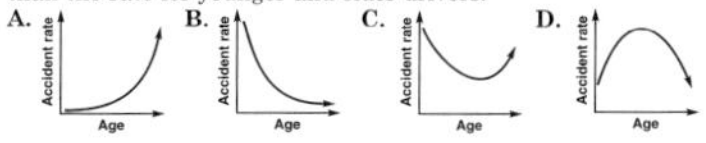

Bonus In some leagues, the equation $f = \frac{4}{5}(200 - m) + a$ is used to find a bowler's final score (f). In this equation, m = the bowler's average score and a = the actual score. Find the final score if: Bonus a. **204** b. **187**
a. Peter averages 110 but bowled a 132 this game.
b. Rhonda averages 160 but bowled a 155 this game.

Using the STUDY GUIDE AND ASSESSMENT

Skills and Concepts Encourage students to refer to the objectives and examples on the left as they complete the review exercises on the right.

Assessment and Evaluation Masters, pp. 9–10

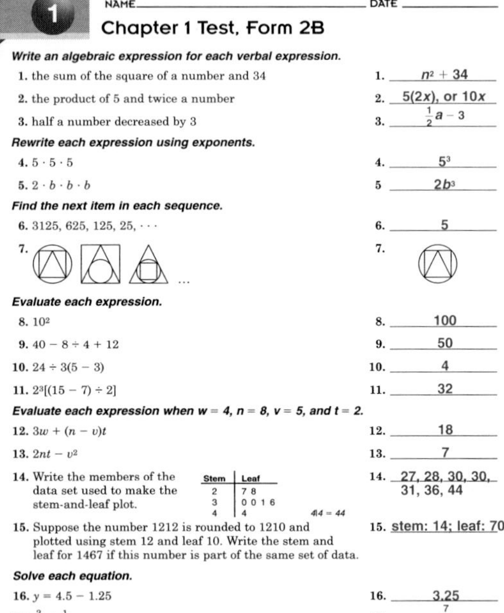

NAME ______________ DATE ______

Chapter 1 Test, Form 2B

Write an algebraic expression for each verbal expression.

1. the sum of the square of a number and 34 — 1. $n^2 + 34$
2. the product of 5 and twice a number — 2. $5(2x)$, or $10x$
3. half a number decreased by 3 — 3. $\frac{1}{2}a - 3$

Rewrite each expression using exponents.

4. $5 \cdot 5 \cdot 5$ — 4. 5^3
5. $2 \cdot b \cdot b \cdot b$ — 5 $2b^3$

Find the next item in each sequence.

6. 3125, 625, 125, 25, · · · — 6. 5
7. ... — 7.

Evaluate each expression.

8. 10^2 — 8. 100
9. $40 - 8 \div 4 + 12$ — 9. 50
10. $24 \div 3(5 - 3)$ — 10. 4
11. $2^3[(15 - 7) \div 2]$ — 11. 32

Evaluate each expression when $w = 4$, $n = 8$, $v = 5$, and $t = 2$.

12. $3w + (n - v)t$ — 12. 18
13. $2nt - v^2$ — 13. 7
14. Write the members of the data set used to make the stem-and-leaf plot. — 14. 27, 28, 30, 30, 31, 36, 44

Stem	Leaf
2	7 8
3	0 0 1 6
4	4

4|4 = 44

15. Suppose the number 1212 is rounded to 1210 and plotted using stem 12 and leaf 10. Write the stem and leaf for 1467 if this number is part of the same set of data. — 15. stem: 14; leaf: 70

Solve each equation.

16. $y = 4.5 - 1.25$ — 16. 3.25
17. $\frac{3}{8} + \frac{1}{2} = c$ — 17. $\frac{7}{8}$
18. $m = 8 \div 4\frac{1}{4}$ — 18. $\frac{32}{17}$, or $1\frac{15}{17}$
19. $\frac{6 + 3^2 \cdot 4}{7 - 1} = y$ — 19. 7

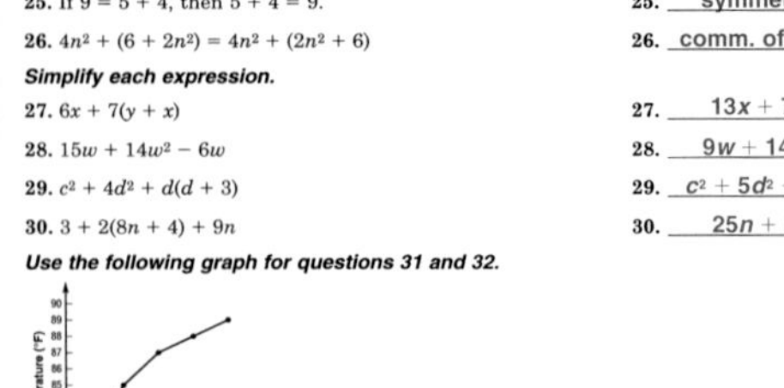

NAME ______________ DATE ______

Chapter 1 Test, Form 2B (continued)

State the property illustrated by each statement.

20. $5 + 0 = 5$ — 20. add. ident.
21. $w + (4 + 6) = w + 10$ — 21. subst.
22. $7(8 - 3) = 7 \cdot 8 - 7 \cdot 3$ — 22. distributive
23. $4(1) = 4$ — 23. multi. ident.
24. $7(6 \cdot 2) = (7 \cdot 6)2$ — 24. assoc. of mult.
25. If $9 = 5 + 4$, then $5 + 4 = 9$. — 25. symmetric
26. $4n^2 + (6 + 2n^2) = 4n^2 + (2n^2 + 6)$ — 26. comm. of add.

Simplify each expression.

27. $6x + 7(y + x)$ — 27. $13x + 7y$
28. $15w + 14w^2 - 6w$ — 28. $9w + 14w^2$
29. $c^2 + 4d^2 + d(d + 3)$ — 29. $c^2 + 5d^2 + 3d$
30. $3 + 2(8n + 4) + 9n$ — 30. $25n + 11$

Use the following graph for questions 31 and 32.

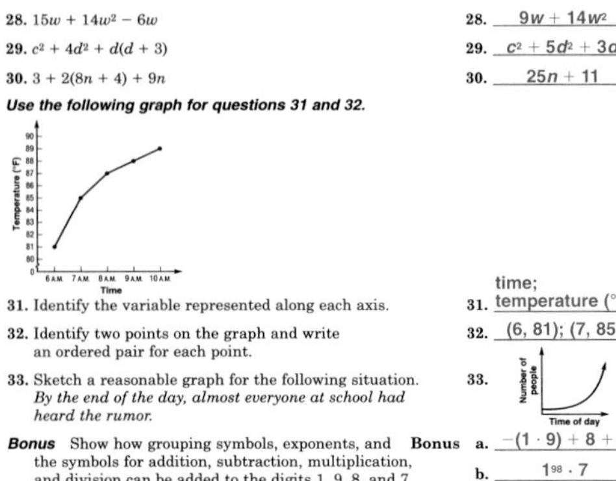

31. Identify the variable represented along each axis. — 31. time; temperature (°F)
32. Identify two points on the graph and write an ordered pair for each point. — 32. (6, 81); (7, 85)
33. Sketch a reasonable graph for the following situation. *By the end of the day, almost everyone at school had heard the rumor.* — 33. (Number of people vs. Time of day)

Bonus Show how grouping symbols, exponents, and the symbols for addition, subtraction, multiplication, and division can be added to the digits 1, 9, 8, and 7 (in that order) to form expressions that will yield the following values when evaluated.
a. 6
b. 7
c. 9

Bonus a. $-(1 \cdot 9) + 8 + 7$ b. $1^{98} \cdot 7$ c. $1 + (9 - 8) + 7$

CHAPTER 1 STUDY GUIDE AND ASSESSMENT

SKILLS AND CONCEPTS

OBJECTIVES AND EXAMPLES

Upon completing this chapter, you should be able to:

- translate verbal expressions into mathematical expressions and vice versa (Lesson 1–1)

Write an algebraic expression for the sum of twice a number x and fifteen.

$$2x + 15$$

Write a verbal expression for $4x^2 - 13$.
four times a number x squared minus thirteen

15. three times a number m to the fifth power

16. the sum of one half a number x and 2

17. the difference of four times m squared and twice m

REVIEW EXERCISES

Use these exercises to review and prepare for the chapter test.

Write an algebraic expression for each verbal expression. **12. $3x - 8$**

10. the sum of a number x and twenty-one **$x + 21$**
11. a number x to the fifth power **x^5**
12. the difference of three times a number x and 8
13. five times the square of a number x **$5x^2$**

Write a verbal expression for each algebraic expression. **14. twice a number p squared**

14. $2p^2$
15. $3m^5$
16. $\frac{1}{2}x + 2$
17. $4m^2 - 2m$

OBJECTIVES AND EXAMPLES

- solve problems by extending sequences (Lesson 1–2)

Give the next two items for each sequence.

(dot pattern: 1 dot over 2 dots, 2 dots over 2 dots, 2 dots over 3 dots ..., next: 2 dots over 2 dots and 2 dots over 2 dots)

3, 7, 15, 31, . . .

3 7 15 31 63 127
+4 +8 +16 +32 +64

The next two terms are 63 and 127.

REVIEW EXERCISES

Give the next two items for each pattern.

18. 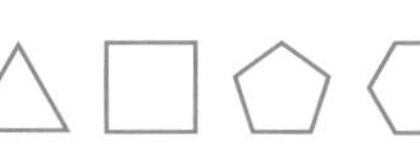 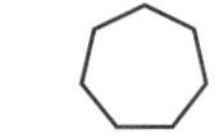 

19. 2, 4, 8, 16, . . . **32, 64**
20. 1, 1, 2, 3, 5, . . . **8, 13**
21. $x + y, 2x + y, 3x + y, \ldots$ **$4x + y$, $5x + y$**

22. 100,000, 1,000,000

Consider the following sequence.

10, 100, 1000, 10,000, . . .

22. Give the next two terms of the sequence.
23. Express the first four terms in the sequence as a power of 10. **10^1, 10^2, 10^3, 10^4**
24. What are the 9th and 13th terms? **10^9, 10^{13}**

OBJECTIVES AND EXAMPLES

- use the order of operations to evaluate real number expressions (Lesson 1–3)

$5 \times 2^2 - 15 \div 3 + 4$	$[3(6) - 4^2]^3 \times 15 \div 5$
$= 5 \times 4 - 15 \div 3 + 4$	$= [3(6) - 16]^3 \times 15 \div 5$
$= 20 - 15 \div 3 + 4$	$= [18 - 16]^3 \times 15 \div 5$
$= 20 - 5 + 4$	$= [2]^3 \times 15 \div 5$
$= 15 + 4$	$= 8 \times 15 \div 5$
$= 19$	$= 120 \div 5$
	$= 24$

REVIEW EXERCISES

Evaluate each expression.

25. $3 + 2 \times 4$ **11**
26. $(10 - 6) \div 8$ **$\frac{1}{2}$**
27. $18 - 4^2 + 7$ **9**
28. $0.8\,(0.02 + 0.05) - 0.006$
29. $(3 \times 1)^3 - (4 + 6) \div (9.1 + 0.9)$ **26**

28. 0.05

Evaluate each expression when $x = 0.1$, $t = 3$, and $y = 2$.

30. $t^2 + 3y$ **15**
31. xty^3 **2.4**
32. $ty \div x$ **60**
33. the product of y and x plus t raised to the second power **9.2**

GLENCOE Technology

Test and Review Software

You may use this software, a combination of an item generator and item bank, to create your own tests or worksheets. Types of items include free response, multiple choice, short answer, and open ended.

For IBM & Macintosh

OBJECTIVES AND EXAMPLES

- display and interpret data on a stem-and-leaf plot (Lesson 1–4)

Stem	Leaf
2	6 8
3	3 5 6 8 9
4	0 2 3 4 4 4 5 6 7 7 8 8 9
5	0 1 2 2 2 3 3 4 6 7
6	1 3 5 9
7	0 3 2|6 = 26

The highest entry is 73. The lowest entry is 26. The total number of entries is 36. Most of the data is in the interval 40–49. **36. 46 38. 60–70**

REVIEW EXERCISES

The ages (in years) at the time of death for the presidents of the United States are listed below.

67, 90, 83, 85, 73, 80, 78, 79, 68, 71, 53, 65, 74, 64, 77, 56, 66, 63, 70, 49, 56, 67, 71, 58, 60, 72, 67, 57, 60, 90, 63, 88, 78, 46, 64, 81 **34. See margin.**

34. Make a stem-and-leaf plot of the data.
35. How old was the oldest president at death? **90**
36. How old was the youngest president at death?
37. Does the total number of entries match the number of presidents? Explain. **See margin.**
38. In which interval did most of the ages fall?

- solve open sentences by performing arithmetic operations (Lesson 1–5)

State whether each sentence is true or false.

$x + 3 \geq 5, x = 1$	$5^2 - 3y \leq 1, y = 8$
$1 + 3 \geq 5$	$5^2 - 3(8) \leq 1$
$4 \geq 5$	$25 - 24 \leq 1$
false	$1 \leq 1$
	true

State whether each sentence is *true* or *false* for the value of the variable given.

39. $x + 13 = 22, x = 8$ **false**
40. $2b + 2 < b^3, b = 2$ **true**
41. $(y + 4) \div (y + 2) \leq 2, y = 2$ **true**
42. Find the solution set for $3x + 1 \leq 13$ if the replacement set is {2, 4, 6, 8}. **{2, 4}**

Solve each equation.

43. $y = 4\frac{1}{2} + 3^2$ **$13\frac{1}{2}$** 44. $y = 5[2(4) - 1^3]$ **35**

- recognize and use the properties of identity and equality (Lesson 1–6)

$36 + 7 \times 1 + 5(2 - 2)$

$= 36 + 7 \times 1 + 5(0)$ *Substitution (=)*

$= 36 + 7 + 5(0)$ *Multiplicative identity*

$= 36 + 7$ *Multiplicative prop. of zero*

$= 42$ *Substitution (=)*

48. multiplicative identity 49. additive identity

Name the multiplicative inverse of each number or variable.

45. 3 **$\frac{1}{3}$** 46. y **$\frac{1}{y}$** 47. 0.2 **5**

Name the property illustrated by each statement.

48. $xy = 1xy$ 49. $0 + 22x = 22x$
50. Evaluate $2[3 \div (19 - 4^2)]$ and name the property used in each step. **2; See margin.**

- use the distributive property to simplify expressions (Lesson 1–7)

$5(t + 3)$	$2x^2 + 4x^2 + 7x$
$= 5(t) + 5(3)$	$= (2 + 4)x^2 + 7x$
$= 5t + 15$	$= 6x^2 + 7x$

51. $2 \times 4 + 2 \times 7$ 52. $4x + 4$ 53. $1 - 3p$

57. $8m + 8n$ 58. $2p + 32pr$

Use the distributive property to rewrite each expression without parentheses.

51. $2(4 + 7)$ 52. $4(x + 1)$ 53. $3\left(\frac{1}{3} - p\right)$

Use the distributive property to find each product.

54. 6×103 **618** 55. 3×98 **294** 56. $12(1.5)$ **18**

Simplify each expression.

57. $3m + 5m + 12n - 4n$ 58. $2p(1 + 16r)$

Additional Answers

34.

Stem	Leaf
4	6 9
5	3 6 6 7 8
6	0 0 3 3 4 4 5 6 7 7 7 8
7	0 1 1 2 3 4 7 8 8 9
8	0 1 3 5 8
9	0 0 7|2 = 72

37. No, because some presidents are still alive.

Applications and Problem Solving Encourage students to work through the exercises in the Applications and Problem Solving section to strengthen their problem-solving skills.

OBJECTIVES AND EXAMPLES

- when simplifying expressions recognize and use the commutative and associative properties (Lesson 1–8)

$3x + 7xy + 9x$

$= 3x + 9x + 7xy$ *Commutative (+)*

$= (3 + 9)x + 7xy$ *Distributive*

$= 12x + 7xy$ *Substitution (=)*

64–65. See Solutions Manual for algebraic expressions and properties used.

REVIEW EXERCISES

Name the property illustrated by each statement.

59. $7(p + q) = 7(q + p)$ **commutative (+)**

60. $3 + (x + y) = (3 + x) + y$ **associative (+)**

61. $(b + a)c = c(b + a)$ **commutative (×)**

62. $2(mn) = (2m)n$ **associative (×)**

63. Simplify $2x + 2y + 3x + 3y$. **$5x + 5y$**

Write an algebraic expression for each verbal expression. Then simplify, indicating the properties used. **64. $3x + 5y$ 65. $3pq$**

64. five times the sum of x and y decreased by $2x$

65. twice the product of p and q increased by pq

- interpret graphs in real-world settings (Lesson 1–9)

Sketch a graph for the following statement.

As a band gets more popular, their prices increase to a point and then level off.

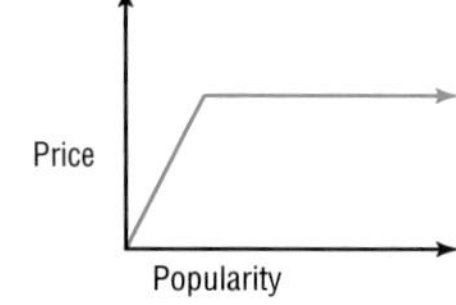

Match each description with the most appropriate graph.

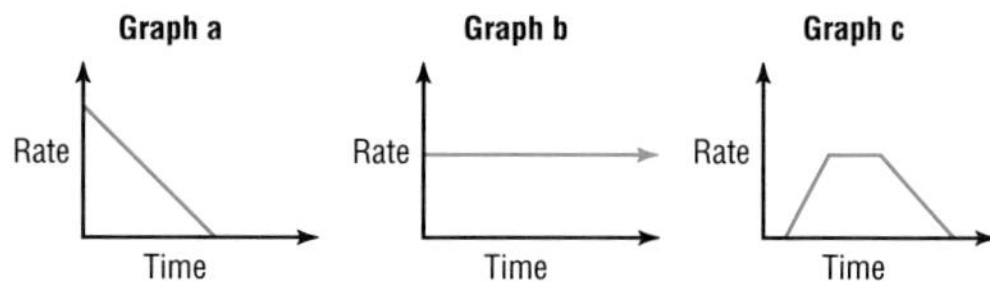

66. the speed of light **Graph b**

67. an airplane taking off, then landing **Graph c**

68. a car approaching a stop sign **Graph a**

71a. $3 + 4 > 5$, $3 + 5 > 4$, $4 + 5 > 3$ 71b. $x > 3$ ft

APPLICATIONS AND PROBLEM SOLVING

69. Food and Nutrition There are 80 Calories in one serving of skim milk and 8 servings in a half gallon. (Lesson 1–1)

a. Write an expression that describes how many Calories you get if you drink s servings of skim milk. **$80s$**

b. How many Calories do you consume if you drink 4 servings of skim milk a day? **320**

c. How many Calories are in a half gallon of skim milk? **640**

70. Investments The equation $I = prt$ describes simple interest on a savings account, where I is the amount of interest, p is the amount deposited, r is the annual interest rate, and t is the time in years. (Lesson 1–3)

a. Solve for I if $p = 100$, $r = 0.05$, and $t = 2$. **\$10**

b. If you deposit \$200 in a savings account that earns 6% interest (6% = 0.06), how much money will be in your account after the first year? **\$212**

71. Geometry The triangle inequality states that if you have a triangle, then the sum of any two of its sides must be greater than its third side. (Lesson 1–5)

a. Verify that a triangle with sides of length 3, 4, and 5 units satisfies the triangle inequality.

b. Suppose you want to construct a triangle such that two of the sides are 3 feet and 6 feet long. What is the minimum length of the third side? (Use x for the length of the third side.)

A practice test for Chapter 1 is provided on page 787.

ALTERNATIVE ASSESSMENT

COOPERATIVE LEARNING PROJECT

Operating a Music Store In this chapter, you learned how to model verbal sentences with algebraic sentences. The basic principles of modeling verbal sentences involve being able to determine whether or not you have an equality or an inequality and determining what your variables are.

In this project, imagine that you and your friends plan to open a new music store that will sell only compact discs. After some research, you find that the monthly rent for your store will be \$500 and that you can expect to spend about \$200 a month for utilities. Assume that your inventory of compact discs must be at least 400 different titles with 4 discs per title and that the wholesale cost for compact discs is \$8.00 per disc.

Outline the costs for starting your business and plan how much in CD sales you would need to show a profit each month.

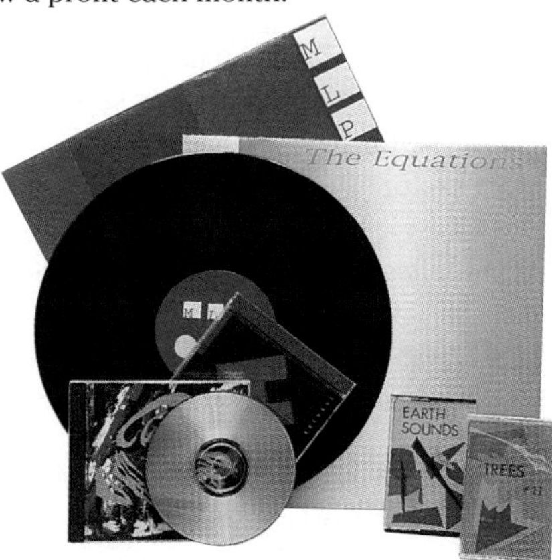

Follow these steps to organize your business.

- Determine the cost to open your new business.
- Determine your monthly overhead costs.
- Find a price for which you will sell your CDs.
- Write an algebraic model that describes the number of CDs you have to sell in order to break even each month.
- Draw a graph to show that if sales increase over time, the profits will also increase.
- Write several paragraphs describing your plan and incorporate your algebraic models and graphs to help support it.

THINKING CRITICALLY

- Does the distributive property also apply to multiplication? In other words, does the product $a(bc) = (ab)(ac)$? Why or why not?
- Create a sequence of numbers in which there is more than one pattern. Explain why this is possible.

PORTFOLIO

Do you organize your work so that you and anyone else reading it can tell what you were thinking? Find an example of your work that you have done that is well organized and list the qualities that make it so. Then find an example of your work that isn't as well organized and list what you could have done to make it more so. Place both of these in your portfolio.

SELF EVALUATION

Are you confident with your knowledge of mathematics? Most math students would rather not try to solve a math problem than try to and get the wrong answer.

Assess yourself. How confident are you? Would you rather not attempt a problem, or are you willing to give your best effort and learn from the outcome? Describe how you can become more confident in your problem solving both in mathematics and in your daily life.

Assessment and Evaluation Masters, pp. 14, 25

NAME __________ DATE __________

Chapter 1 Performance Assessment

Instructions: ***Demonstrate your knowledge by giving a clear, concise solution to each problem. Be sure to include all relevant drawings and justify your answers. You may show your solution in more than one way or investigate beyond the requirements of the problem.***

1. Some important area formulas are given below.

   Area of a circle $= \pi r^2$, where r is the radius of the circle

   Area of a trapezoid $= \frac{(b_1 + b_2)}{2} \cdot h$, where b_1 and b_2 are the lengths of the bases and h is the height of the trapezoid

   a. Write a formula for the area of the figure shown at the right.

   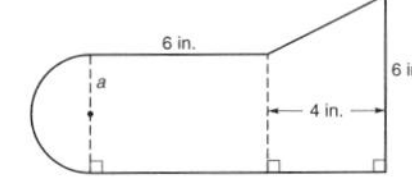

   b. Simplify the formula in part a. Justify each step in your simplification by naming the property used.

   c. State the order of operations indicated by the formula in part b. Then find the area when $a = 2$.

2. In each sentence, circle key words or phrases that indicate a mathematical operation and write the corresponding mathematical symbol above each. Then write an equation for each sentence.

   a. Three times a number decreased by four is equal to twenty-five.

   b. The sum of the square of a number and a second number is eighteen.

   c. A number multiplied by 7 and divided by three gives five more than the number.

   d. One-half of a number added to itself equals twice the difference of the number and five.

3. Think of a situation that could be modeled by this graph. Then label the axes of the graph and write several sentences describing the situation.

Scoring Guide
Chapter 1
Performance Assessment

Level	Specific Criteria
3 Superior	• Shows thorough understanding of the concepts of *evaluating and simplifying mathematical expressions, translating verbal expressions into mathematical expressions,* and *interpreting graphs.* • Uses appropriate strategies to solve problems. • Computations are correct. • Written explanations are exemplary. • Goes beyond requirements of some or all problems.
2 Satisfactory, with Minor Flaws	• Shows understanding of the concepts of *evaluating and simplifying mathematical expressions, translating verbal expressions into mathematical expressions,* and *interpreting graphs.* • Uses appropriate strategies to solve problems. • Computations are mostly correct. • Written explanations are effective. • Satisfies all requirements of problems.
1 Nearly Satisfactory, with Serious Flaws	• Shows understanding of most of the concepts of *evaluating and simplifying mathematical expressions, translating verbal expressions into mathematical expressions,* and *interpreting graphs.* • May not use appropriate strategies to solve problems. • Computations are mostly correct. • Written explanations are satisfactory. • Satisfies most requirements of problems.
0 Unsatisfactory	• Shows little or no understanding of the concepts of *evaluating and simplifying mathematical expressions, translating verbal expressions into mathematical expressions,* and *interpreting graphs.* • May not use appropriate strategies to solve problems. • Computations are incorrect. • Written explanations are not satisfactory. • Does not satisfy requirements of problems.

Alternative Assessment

The Alternative Assessment section provides students with the opportunity to assess their own work by thinking critically, working with others, keeping a portfolio, and honestly evaluating their own progress. For more information on alternative forms of assessment, see *Alternative Assessment in the Mathematics Classroom,* one of the titles in the Glencoe Mathematics Professional Series.

Performance Assessment

Performance Assessment tasks for this chapter are included in the *Assessment and Evaluation Masters.* A scoring guide is also provided.

In•ves•ti•ga•tion

TEACHER NOTES

NCTM Standards: 1–5

This Investigation is designed to be completed over several days or weeks. It may be considered optional. You may want to assign the Investigation and the follow-up activities to be completed at the same time.

Objective

Use mathematics to analyze aspects of the greenhouse effect and report your findings.

Mathematical Overview

This Investigation will use the following mathematical skills and concepts from Chapters 2 and 3.

- making charts and graphs
- adding and subtracting integers
- looking for patterns

Recommended Time

Part	Pages	Time
Investigation	68–69	1 class period
Working on the Investigation	83, 111, 154, 177	20 minutes each
Closing the Investigation	184	1 class period

Instructional Resources

Investigations and Projects Masters, pp. 1–4

A recording sheet, teacher notes, and scoring guide are provided for each Investigation in the *Investigations and Projects Masters.*

1 MOTIVATION

News reports often refer to the threat of global warming. Ask students what is meant by *global warming* or the *greenhouse effect* and what dangers the situation poses for Earth.

LONG-TERM PROJECT

In•ves•ti•ga•tion

the Greenhouse Effect

MATERIALS NEEDED

- lamp with a 100-watt bulb
- stopwatch
- thermometer
- plastic zipper-style sandwich bag
- ruler

The *greenhouse effect* refers to the warming of Earth by the sun. The main culprit in the greenhouse effect is carbon dioxide (CO_2). Carbon dioxide in the atmosphere works like the panes of glass on a greenhouse. Glass is transparent to visible light, allowing the sun's rays to warm Earth's surface. But when the surface gives off excess heat, the hot air stays in the greenhouse, which continues to keep the air warm. Similarly, CO_2 in the atmosphere absorbs the sun's infrared rays, allowing some of the excess heat to stay in the atmosphere rather than escaping into space. How much heat is retained depends on how much CO_2 is in the air.

Over the last 200 years, the amount of CO_2 in our atmosphere has increased, raising Earth's average temperature. Some scientists predict that if global warming continues and Earth's average temperature goes up an additional 3° to 8°F, we could see a marked increase in the number of weather-related disasters like heat waves, droughts, floods, and hurricanes.

In this Investigation, you will use mathematics to analyze aspects of the greenhouse effect and report your findings. As a scientist commissioned to explore the greenhouse effect, you will conduct two experiments. The first experiment is a *control experiment* that will be used to compare data with the second experiment, called the *greenhouse experiment.* Both experiments will involve temperature as a *function* of distance.

Make an Investigation Folder in which you can store all of your work on this Investigation for future use.

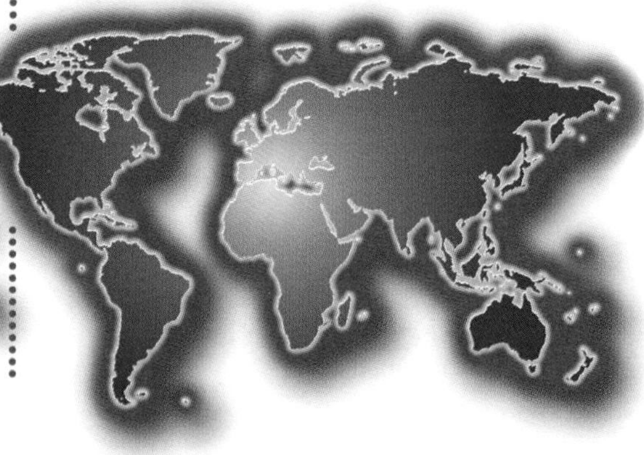

Cooperative Learning

This Investigation offers an excellent opportunity for using cooperative learning groups. For more information on cooperative learning strategies and group management, see *Cooperative Learning in the Mathematics Classroom,* one of the titles in the Glencoe Mathematics Professional Series.

CONTROL EXPERIMENT				
Test	Distances Between Thermometer and Bulb	First Temperature	Second Temperature	Difference in Temperatures
1				
2				
3				
4				

CONTROL EXPERIMENT

1. Begin by copying the chart above onto a sheet of paper.
2. Set your lamp about 5 inches away from the thermometer. Make sure the lamp is turned off.
3. Measure and record the temperature and the distance between the thermometer and the bulb.
4. Turn the lamp on and leave it on for five minutes. Use the stopwatch to keep accurate time.
5. Record the new temperature and find the difference of the temperatures.
6. Let the thermometer and bulb cool to room temperature and repeat the process, moving the lamp a little farther from the thermometer. Conduct the test a total of four times at four different distances. Make sure you let the thermometer cool to room temperature between each test.

GREENHOUSE EXPERIMENT

7. Begin by making a chart like the one you made for the Control Experiment.
8. Make sure that the thermometer has cooled to room temperature. Then record the temperature.
9. Put the thermometer in the plastic bag and close the bag. Make sure that the bulb of the thermometer does not touch the bag. Then place the thermometer the same distance from the lamp as you did in the control experiment (about 5 inches). Record the distance between the thermometer and the lamp bulb. Then turn the lamp on and leave it on for five minutes.
10. Record the new temperature. Then find the difference of the temperatures.
11. Conduct four tests, using the same distances you used in the control experiment. Make sure you take the thermometer out of the bag and let it cool to room temperature between each test. Be sure to take accurate readings and measurements.

You will continue working on this Investigation throughout Chapters 2 and 3.

Be sure to keep your chart and materials in your Investigation Folder.

The Greenhouse Effect Investigation

2 SETUP

You may wish to have a student read the first two paragraphs of the Investigation to provide information about why the greenhouse effect is an important issue. You may then wish to read the third paragraph, which introduces the activity. Discuss the activity with the class. Then separate the class into groups of four.

3 MANAGEMENT

Each group member should be responsible for a specific task.

Recorder Collects data.
Measurer Sets and measures the distances between the lamp and the thermometer.
Timer Times different aspects of the activity and turns the lamp on and off at the appropriate times.
Weatherperson Reports temperature readings to the recorder and makes sure that the thermometer is adequately prepared for each aspect of the activity.

At the end of the activity, each member should turn in his or her respective equipment.

Investigations and Projects Masters, p. 4

2, 3

NAME__________ DATE ________

Investigation, Chapters 2 and 3 — Student Edition Pages 68–69, 83, 111, 154, 177, 184

The Greenhouse Effect

Work with your group to add to the list of questions to be considered.

- What was the difference in the way the two experiments were designed?
- Do you think life could exist on Earth without CO_2 ?
- Do you think life could exist on Venus with CO_2 ?
- How much CO_2 was in the atmosphere in 1800 and in 1950?
- How many degrees on the Celsius scale would 15°F be?
- What is the difference between the boiling point and the freezing point on each of the scales?

Reread the information about the greenhouse effect on pages 68–69 to find significant information. You may wish to look in some other sources. List your findings below.

Please keep this page and any other research in your Investigation folder.

CHAPTER

2 Exploring Rational Numbers

PREVIEWING THE CHAPTER

This chapter explores the basic operations as applied to integers and rational numbers. The number line is used as a mathematical model to develop the rules for the addition of integers. Students apply their knowledge of the rules for addition and the additive inverse property to develop rules for the subtraction of integers. Then the connection is established between these rules for integers and the rules for rational numbers. Rules for multiplying and dividing rational numbers are also explored. Finally, students examine the information presented in verbal problems and translate that information into algebraic expressions, as well as reversing the process and writing descriptions of the problems from algebraic expressions.

Lesson (Pages)	Lesson Objectives	NCTM Standards	State/Local Objectives
2-1 (72–77)	State the coordinate of a point on a number line. Graph integers on a number line. Add integers by using a number line.	1–5, 14	
2-2 (78–83)	Interpret numerical data from a table. Display and interpret statistical data on a line plot.	1–4, 10	
2-3A (84)	Use counters to add and subtract integers.	1–5	
2-3 (85–92)	Find the absolute value of a number. Add and subtract integers.	1–5	
2-4 (93–99)	Compare and order rational numbers. Find a number between two rational numbers.	1–5	
2-5 (100–104)	Add and subtract rational numbers. Simplify expressions that contain rational numbers.	1–5	
2-6A (105)	Use counters to multiply integers.	1–5	
2-6 (106–111)	Multiply rational numbers.	1–5	
2-7 (112–117)	Divide rational numbers.	1–5	
2-8A (118)	Use base-ten tiles to model square roots.	1–5	
2-8 (119–125)	Find square roots. Classify numbers. Graph solutions of inequalities on number lines.	1–5, 14	
2-9 (126–132)	Explore problem situations. Translate verbal sentences and problems into equations or formulas and vice versa.	1–5	

A complete, 1-page lesson plan is provided for each lesson in the ***Lesson Planning Guide***. Answer keys for each lesson are available in the ***Answer Key Masters***.

ORGANIZING THE CHAPTER

You may want to refer to the **Course Planning Calendar** on page T12 for detailed information on pacing.
PACING: Standard—15 days; **Honors**—15 days; **Block**—9 days; **Two Years**—28 days

LESSON PLANNING CHART

			Blackline Masters											
Lesson (Pages)	**Materials/ Manipulatives**	**Extra Practice (Student Edition)**	**Study Guide**	**Practice**	**Enrichment**	**Assessment and Evaluation**	**Modeling Mathematics**	**Multicultural Activity**	**Tech Prep Applications**	**Graphing Calculator**	**Science and Math Lab Manual**	**Real-World Applications**	**Interactive Mathematics Tools Software**	**Teaching Transparencies**
2-1 (72–77)	graphing calculator	p. 759	p. 10	p. 10	p. 10						pp. 5–8		2-1	2-1A 2-1B
2-2 (78–83)	small packages of plain M & Ms	p. 759	p. 11	p. 11	p. 11							3		2-2A 2-2B
2-3A (84)	counters* integer mat*						p. 20						2-3A	
2-3 (85–92)	computer calculator	p. 759	p. 12	p. 12	p. 12	p. 44	p. 73		p. 3			4		2-3A 2-3B
2-4 (93–99)	calculator	p. 760	p. 13	p. 13	p. 13					p. 2			2-4	2-4A 2-4B
2-5 (100–104)		p. 760	p. 14	p. 14	p. 14	pp. 43, 44		p. 3					2-5	2-5A 2-5B
2-6A (105)	counters* integer mat*						p. 21							
2-6 (106–111)		p. 760	p. 15	p. 15	p. 15				p. 4			5		2-6A 2-6B
2-7 (112–117)	calculator	p. 761	p. 16	p. 16	p. 16	p. 45	pp. 36–38							2-7A 2-7B
2-8A (118)	base-ten tiles*						p. 22							
2-8 (119–125)	calculator	p. 761	p. 17	p. 17	p. 17									2-8A 2-8B
2-9 (126–132)	rectangular box scissors	p. 761	p. 18	p. 18	p. 18	p. 45	pp. 39–41	p. 4				6	2-9	2-9A 2-9B
Study Guide/ Assessment (133–137)						pp. 29–42, 46–48								

*Included in Glencoe's Student Manipulative Kit and Overhead Manipulative Resources.

ORGANIZING THE CHAPTER

Other Chapter Resources

Student Edition
Investigation, pp. 68–69
Chapter Opener, pp. 70–71
Mathematics and Society, p. 77
Working on the Investigation, pp. 83, 111

Teacher's Classroom Resources
Investigations and Projects Masters, pp. 29–32
Algebra and Geometry Overhead Manipulative Resources, pp. 2–5

Technology
Test and Review Software (IBM and Macintosh)
CD-ROM Interactions (Windows and Macintosh)

Professional Publications
Block Scheduling Booklet
Glencoe Mathematics Professional Series

Outside Resources

Books/Periodicals
Gray, Virginia, *The Write Tool to Teach Algebra,* Dale Seymour Publications
Taylor, Harold and Loretta Taylor, *Developing Skills in Algebra One,* Dale Seymour Publications

Software
Tobbs Learns Algebra, WINGS for Learning, Sunburst

Videos/CD-ROMs
Algebra for Everyone, NCTM

Assessment Resources

Student Edition
Math Journal, pp. 97, 122
Mixed Review, pp. 77, 83, 92, 99, 104, 111, 117, 125, 132
Self Test, p. 99
Chapter Highlights, p.133
Chapter Study Guide and Assessment, pp. 134–136
Alternative Assessment, p. 137
Portfolio, p. 137
Cumulative Review, pp. 138–139

Teacher's Wraparound Edition
5-Minute Check, pp. 72, 78, 85, 93, 100, 106, 112, 119, 126
Check for Understanding, pp. 75, 80, 89, 96, 102, 109, 115, 122, 129
Closing Activity, pp. 76, 82, 91, 99, 104, 111, 117, 125, 132
Cooperative Learning, pp. 113, 127

Assessment and Evaluation Masters
Multiple-Choice Tests, Forms 1A (Honors), 1B (Average), 1C (Basic), pp. 29–34
Free-Response Tests, Forms 2A (Honors), 2B (Average), 2C (Basic), pp. 35–40
Calculator-Based Test, p. 41
Performance Assessment, p. 42
Mid-Chapter Test, p. 43
Quizzes A–D, pp. 44–45
Standardized Test Practice, p. 46
Cumulative Review, pp. 47–48

ENHANCING THE CHAPTER

Examples of some of the materials for enhancing Chapter 2 are shown below.

Multicultural Activity Masters, pp. 3, 4

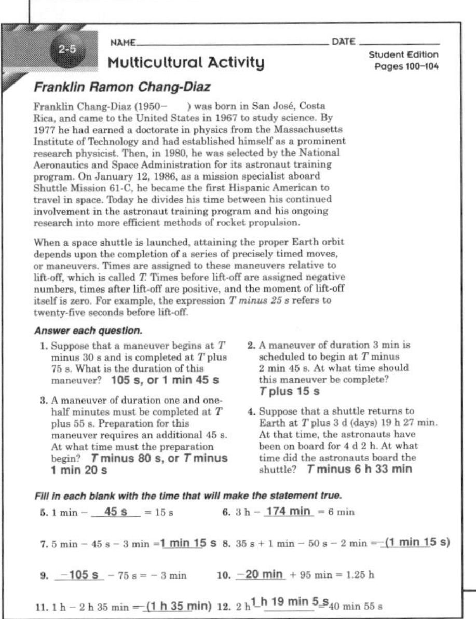

NAME ______ DATE ______

2-5 **Multicultural Activity**

Student Edition Pages 100–104

Franklin Ramon Chang-Diaz

Franklin Chang-Diaz (1950–) was born in San José, Costa Rica, and came to the United States in 1967 to study science. By 1977 he had earned a doctorate in physics from the Massachusetts Institute of Technology and had established himself as a prominent research physicist. Then, in 1980, he was selected by the National Aeronautics and Space Administration for its astronaut training program. On January 12, 1986, as a mission specialist aboard Shuttle Mission 61-C, he became the first Hispanic American to travel in space. Today he divides his time between his continued involvement in the astronaut training program and his ongoing research into more efficient methods of rocket propulsion.

When a space shuttle is launched, attaining the proper Earth orbit depends upon the completion of a series of precisely timed moves, or maneuvers. Times are assigned to these maneuvers relative to lift-off, which is called T. Times before lift-off are assigned negative numbers, times after lift-off are positive, and the moment of lift-off itself is zero. For example, the expression *T minus 25 s* refers to twenty-five seconds before lift-off.

Answer each question.

1. Suppose that a maneuver begins at T minus 30 s and is completed at T plus 75 s. What is the duration of this maneuver? **105 s, or 1 min 45 s**
2. A maneuver of duration 3 min is scheduled to begin at T minus 2 min 45 s. At what time should this maneuver be complete? ***T* plus 15 s**
3. A maneuver of duration one and one-half minutes must be completed at T plus 55 s. Preparation for this maneuver requires an additional 45 s. At what time must the preparation begin? ***T* minus 80 s, or *T* minus 1 min 20 s**
4. Suppose that a shuttle returns to Earth at T plus 3 d (days) 19 h 27 min. At that time, the astronauts have been on board for 4 d 2 h. At what time did the astronauts board the shuttle? ***T* minus 6 h 33 min**

Fill in each blank with the time that will make the statement true.

5. 1 min − **45 s** = 15 s
6. 3 h − **174 min** = 6 min
7. 5 min − 45 s − 3 min = **1 min 15 s**
8. 35 s + 1 min − 50 s − 2 min = **−(1 min 15 s)**
9. **−105 s** − 75 s = − 3 min
10. **−20 min** + 95 min = 1.25 h
11. 1 h − 2 h 35 min = **−(1 h 35 min)**
12. 2 h − **1 h 19 min 5 s** = 40 min 55 s

Real-World Applications, 3, 4, 5, 6

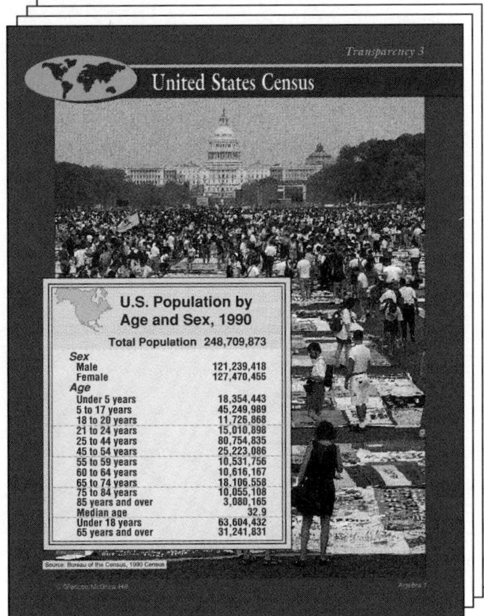

U.S. Population by Age and Sex, 1990

Total Population	248,709,873
Sex	
Male	121,239,418
Female	127,470,455
Age	
Under 5 years	18,354,443
5 to 17 years	45,249,989
18 to 20 years	11,726,868
21 to 24 years	15,010,898
25 to 44 years	80,754,835
45 to 54 years	25,223,086
55 to 59 years	10,531,756
60 to 64 years	10,616,167
65 to 74 years	18,106,558
75 to 84 years	10,055,108
85 years and over	3,080,165
Median age	32.9
Under 18 years	63,604,432
65 years and over	31,241,831

Source: Bureau of the Census, 1990 Census

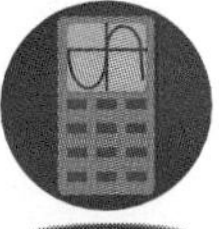

TECHNOLOGY

Graphing Calculator Masters, p. 2

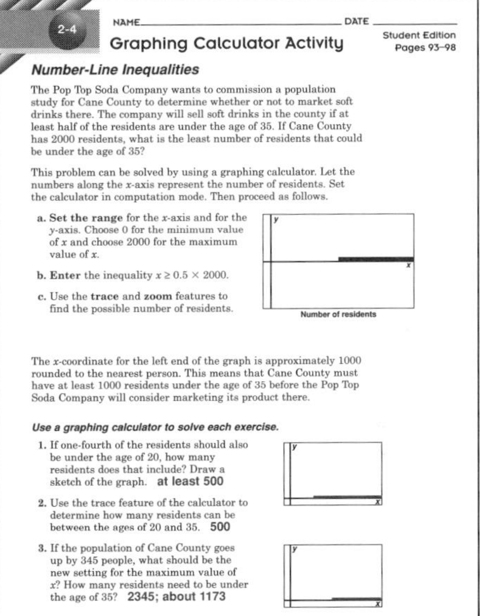

NAME ______ DATE ______

2-4 **Graphing Calculator Activity**

Student Edition Pages 93–98

Number-Line Inequalities

The Pop Top Soda Company wants to commission a population study for Cane County to determine whether or not to market soft drinks there. The company will sell soft drinks in the county if at least half of the residents are under the age of 35. If Cane County has 2000 residents, what is the least number of residents that could be under the age of 35?

This problem can be solved by using a graphing calculator. Let the numbers along the x-axis represent the number of residents. Set the calculator in computation mode. Then proceed as follows.

a. **Set the range** for the x-axis and for the y-axis. Choose 0 for the minimum value of x and choose 2000 for the maximum value of x.

b. **Enter** the inequality $x \geq 0.5 \times 2000$.

c. Use the **trace** and **zoom** features to find the possible number of residents.

Number of residents

The x-coordinate for the left end of the graph is approximately 1000 rounded to the nearest person. This means that Cane County must have at least 1000 residents under the age of 35 before the Pop Top Soda Company will consider marketing its product there.

Use a graphing calculator to solve each exercise.

1. If one-fourth of the residents should also be under the age of 20, how many residents does that include? Draw a sketch of the graph. **at least 500**
2. Use the trace feature of the calculator to determine how many residents can be between the ages of 20 and 35. **500**
3. If the population of Cane County goes up by 345 people, what should be the new setting for the maximum value of x? How many residents need to be under the age of 35? **2345; about 1173**

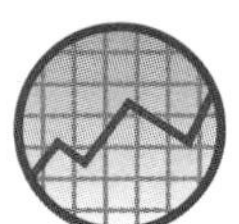

TECH PREP

Tech Prep Applications Masters, pp. 3, 4

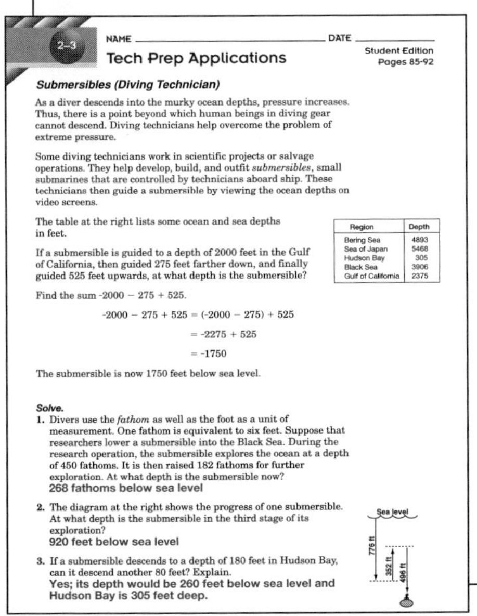

NAME ______ DATE ______

2-3 **Tech Prep Applications**

Student Edition Pages 85-92

Submersibles (Diving Technician)

As a diver descends into the murky ocean depths, pressure increases. Thus, there is a point beyond which human beings in diving gear cannot descend. Diving technicians help overcome the problem of extreme pressure.

Some diving technicians work in scientific projects or salvage operations. They help develop, build, and outfit *submersibles*, small submarines that are controlled by technicians aboard ship. These technicians then guide a submersible by viewing the ocean depths on video screens.

The table at the right lists some ocean and sea depths in feet.

Region	Depth
Bering Sea	4893
Sea of Japan	5468
Hudson Bay	305
Black Sea	3906
Gulf of California	2375

If a submersible is guided to a depth of 2000 feet in the Gulf of California, then guided 275 feet farther down, and finally guided 525 feet upwards, at what depth is the submersible?

Find the sum -2000 − 275 + 525.

$$-2000 - 275 + 525 = (-2000 - 275) + 525$$
$$= -2275 + 525$$
$$= -1750$$

The submersible is now 1750 feet below sea level.

Solve.

1. Divers use the *fathom* as well as the foot as a unit of measurement. One fathom is equivalent to six feet. Suppose that researchers lower a submersible into the Black Sea. During the research operation, the submersible explores the ocean at a depth of 450 fathoms. It is then raised 182 fathoms for further exploration. At what depth is the submersible now? **268 fathoms below sea level**
2. The diagram at the right shows the progress of one submersible. At what depth is the submersible in the third stage of its exploration? **920 feet below sea level**
3. If a submersible descends to a depth of 180 feet in Hudson Bay, can it descend another 80 feet? Explain. **Yes; its depth would be 260 feet below sea level and Hudson Bay is 305 feet deep.**

Sea level
776 ft
352 ft
496 ft

Science and Math Lab Manual, pp. 5–8

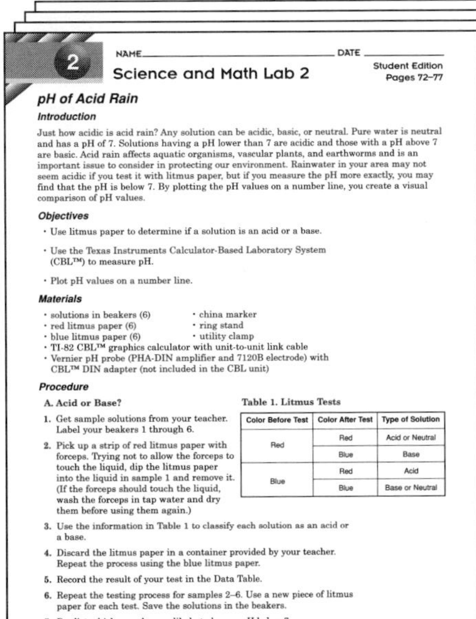

NAME ______ DATE ______

2 **Science and Math Lab 2**

Student Edition Pages 72–77

pH of Acid Rain

Introduction

Just how acidic is acid rain? Any solution can be acidic, basic, or neutral. Pure water is neutral and has a pH of 7. Solutions having a pH lower than 7 are acidic and those with a pH above 7 are basic. Acid rain affects aquatic organisms, vascular plants, and earthworms and is an important issue to consider in protecting our environment. Rainwater in your area may not seem acidic if you test it with litmus paper, but if you measure the pH more exactly, you may find that the pH is below 7. By plotting the pH values on a number line, you create a visual comparison of pH values.

Objectives

- Use litmus paper to determine if a solution is an acid or a base.
- Use the Texas Instruments Calculator-Based Laboratory System (CBL™) to measure pH.
- Plot pH values on a number line.

Materials

- solutions in beakers (6)
- red litmus paper (6)
- blue litmus paper (6)
- china marker
- ring stand
- utility clamp
- TI-82 CBL™ graphics calculator with unit-to-unit link cable
- Vernier pH probe (PHA-DIN amplifier and 7120B electrode) with CBL™ DIN adapter (not included in the CBL unit)

Procedure

A. Acid or Base?

1. Get sample solutions from your teacher. Label your beakers 1 through 6.
2. Pick up a strip of red litmus paper with forceps. Trying not to allow the forceps to touch the liquid, dip the litmus paper into the liquid in sample 1 and remove it. (If the forceps should touch the liquid, wash the forceps in tap water and dry them before using them again.)
3. Use the information in Table 1 to classify each solution as an acid or a base.
4. Discard the litmus paper in a container provided by your teacher. Repeat the process using the blue litmus paper.
5. Record the result of your test in the Data Table.
6. Repeat the testing process for samples 2–6. Use a new piece of litmus paper for each test. Save the solutions in the beakers.
7. Predict which samples are likely to have a pH below 7. ______

Table 1. Litmus Tests

Color Before Test	Color After Test	Type of Solution
Red	Red	Acid or Neutral
	Blue	Base
Blue	Red	Acid
	Blue	Base or Neutral

Problem of the Week Cards, 4, 5, 6

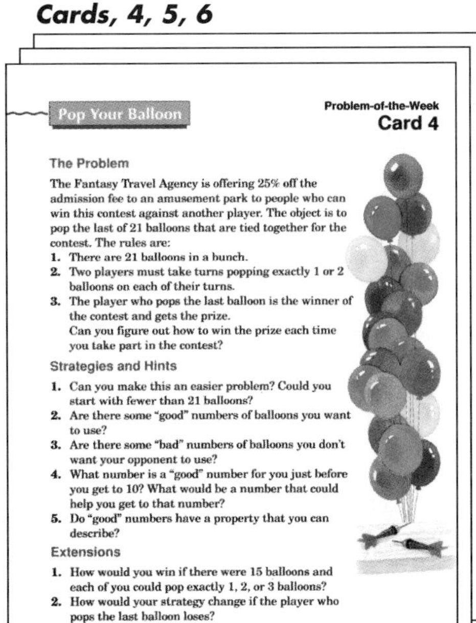

Pop Your Balloon

Problem-of-the-Week **Card 4**

The Problem

The Fantasy Travel Agency is offering 25% off the admission fee to an amusement park to people who can win this contest against another player. The object is to pop the last of 21 balloons that are tied together for the contest. The rules are:

1. There are 21 balloons in a bunch.
2. Two players must take turns popping exactly 1 or 2 balloons on each of their turns.
3. The player who pops the last balloon is the winner of the contest and gets the prize.

Can you figure out how to win the prize each time you take part in the contest?

Strategies and Hints

1. Can you make this an easier problem? Could you start with fewer than 21 balloons?
2. Are there some "good" numbers of balloons you want to use?
3. Are there some "bad" numbers of balloons you don't want your opponent to use?
4. What number is a "good" number for you just before you get to 10? What would be a number that could help you get to that number?
5. Do "good" numbers have a property that you can describe?

Extensions

1. How would you win if there were 15 balloons and each of you could pop exactly 1, 2, or 3 balloons?
2. How would your strategy change if the player who pops the last balloon loses?

MAKING MATHEMATICS RELEVANT

This two-page introduction to the chapter provides students with an opportunity to explore contemporary topics and their applications to mathematics.

Background Information

Veggie Lovers Unite! There are many types of vegetarians. Vegetarians eat only vegetables, fruits, nuts, seeds, grains, and legumes. A lacto-vegetarian will also consume milk and milk products. Lacto-ovo-vegetarians include eggs in their diets. Pesco-vegetarians will eat fish, while pollo-vegetarians will include poultry in their diets. All vegetarians abstain from eating red meat.

CHAPTER 2

Exploring Rational Numbers

Objectives

In this chapter, you will:

- display and interpret statistical data on line plots,
- add, subtract, multiply, and divide rational numbers,
- find square roots, and
- write equations and formulas.

Veggie Lovers Unite!

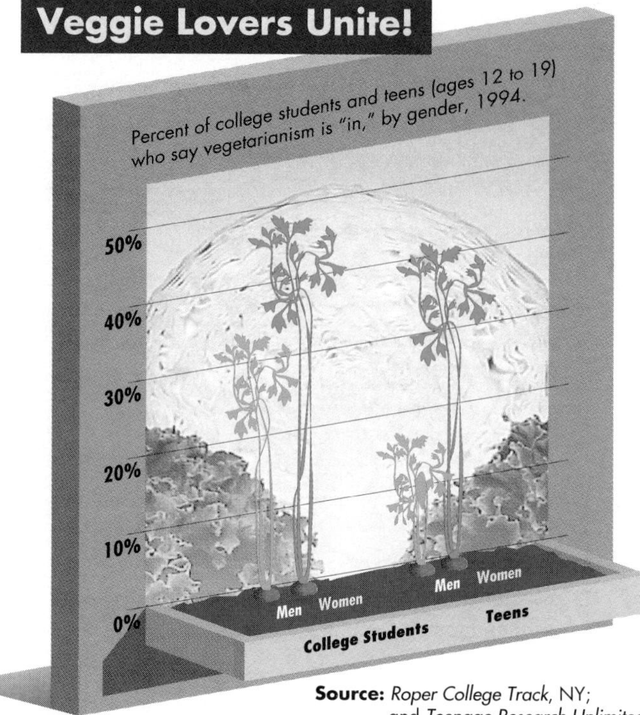

Percent of college students and teens (ages 12 to 19) who say vegetarianism is "in," by gender, 1994.
50%
40%
30%
20%
10%
0%
Men
Women
College Students
Men
Women
Teens

Source: *Roper College Track*, NY; and *Teenage Research Unlimited*

Hamburgers and pepperoni pizzas have been teen diet staples for decades. But lately there is a growing trend toward meatless meals. Why do teens switch to "veggie values" and go for tofu and eggplant? Are there benefits to a diet without meat? Are there dangers to a diet without meat?

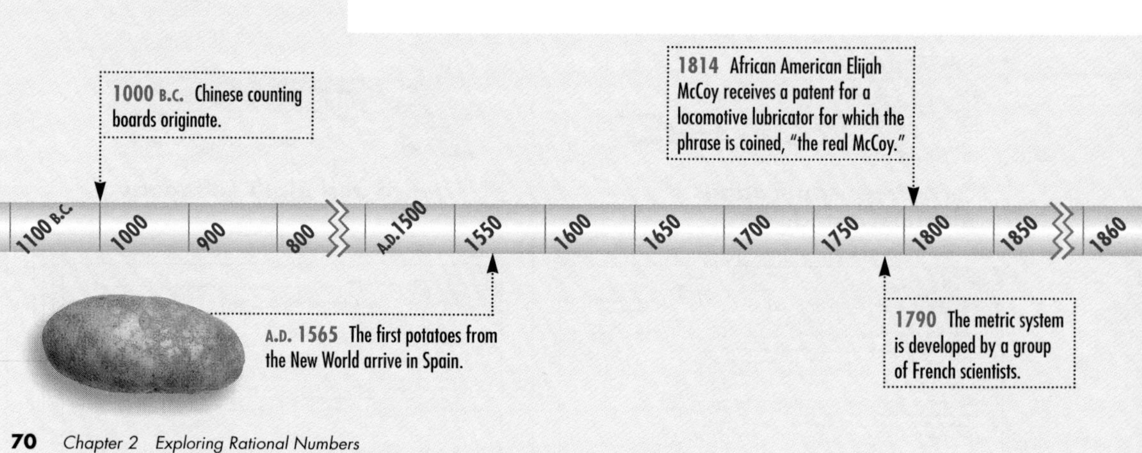

TIME Line
1000 B.C. Chinese counting boards originate.
1814 African American Elijah McCoy receives a patent for a locomotive lubricator for which the phrase is coined, "the real McCoy."
1100 B.C.
1000
900
800
A.D. 1500
1550
1600
1650
1700
1750
1800
1850
1860
A.D. 1565 The first potatoes from the New World arrive in Spain.
1790 The metric system is developed by a group of French scientists.

TIME Line

The process of counting primarily deals with the positive integers. Students might find it interesting to investigate and report on the Chinese counting boards that were developed as far back as 1000 B.C.

interNET CONNECTION

You can learn about vegetarian news, recipes, and environmental and biological concerns on-line.

World Wide Web
http://www.cs.unc.edu/~barman/vegetarian.html

PEOPLE IN THE NEWS

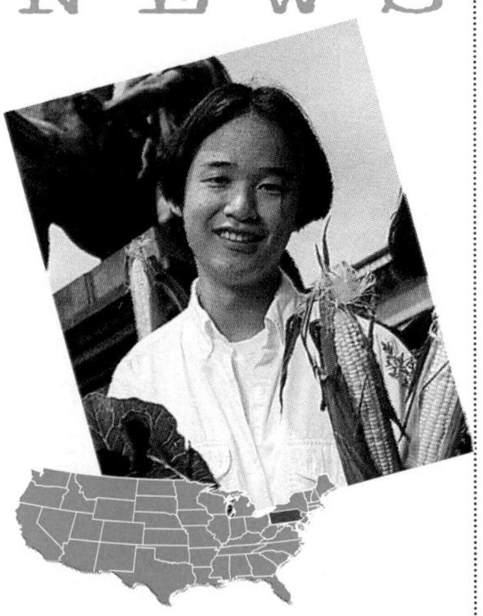

The concern for animals is one reason teenagers give up eating meat. Groups like Earth 2000, a teens-only advocacy group founded by **Danny Seo,** want the focus of their generation to be "no animal cruelty, no meat." Danny and members of Earth 2000 cite teen celebrities like Michael Stipe of R.E.M., Eddie Vedder of Pearl Jam, and Jennie Garth of *Beverly Hills 90210* as vegetarian role models.

Danny also cites health reasons for maintaining a meatless diet, such as cutting back on the intake of fat. In a balanced vegetarian diet, only about one third of the calories come from fat.

Chapter Project

What would you eat if you didn't eat meat? Research nutrition and dietary resource materials at the library to help you devise a healthy vegetarian diet and a healthy meat diet for an average teenager.

- Plan two diets. One should be for a no animal meat vegetarian, and one should be for a meat-eating person.
- Devise a seven-day menu for each diet, listing all nutrition facts such as calories, fat, cholesterol, sodium, carbohydrates, proteins, and vitamins.
- Can a healthy diet be maintained without animal products? Is an increase in fruit and vegetable consumption healthy? Explain your reasoning.
- Draw some conclusions about individual diets and food production from your research. Share your findings with the class.

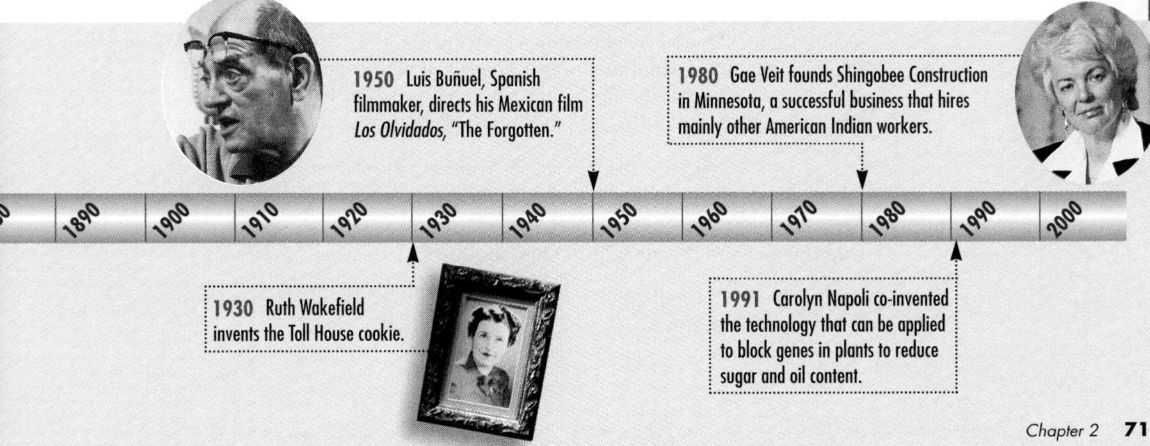

1950 Luis Buñuel, Spanish filmmaker, directs his Mexican film *Los Olvidados,* "The Forgotten."
1980 Gae Veit founds Shingobee Construction in Minnesota, a successful business that hires mainly other American Indian workers.
1890
1900
1910
1920
1930
1940
1950
1960
1970
1980
1990
2000
1930 Ruth Wakefield invents the Toll House cookie.
1991 Carolyn Napoli co-invented the technology that can be applied to block genes in plants to reduce sugar and oil content.

Alternative Chapter Projects

Two other chapter projects are included in the *Investigations and Projects Masters.* In Chapter 2 Project A, pp. 29–30, students extend the topic in the chapter opener. In Chapter 2 Project B, pp. 31–32, students design and produce a board game.

PEOPLE IN THE NEWS

The following menu lists what a healthy 13-year-old vegetarian might eat in a day. Of the 2260 calories, 32% come from fat.

Breakfast orange; oatmeal with soy milk; toast with jam

Lunch miso soup with tofu; pita-bread sandwich with hummus, tomato, and spinach; glass of carrot juice; dried figs

Dinner rice, bean, squash, onion, and broccoli casserole with safflower oil; cooked kale

Snacks slice of watermelon; strawberries; trail mix

Chapter Project

Cooperative Learning Have students work in groups of three to investigate levels of fat, protein, and carbohydrates in cereal. One group member should investigate a sugary children's cereal, another should look at a cereal that is promoted as healthful, and the third group member should investigate a cereal that does not obviously fall into either category. Have group members compare their findings and then discuss them with the class.

Investigations and Projects Masters, p. 29

NAME__________ DATE________

Chapter 2 Project A

Student Edition Pages 72–132

Good for You!

1. Choose at least five different kinds of breakfast cereal. Copy the information given in the "Nutrition Facts" chart on each package and note the price. (Be sure each package contains the same amount of cereal.)
2. For each cereal find the total number of vitamins with a Percent Daily Value of at least 25% in one serving. (Use the information given for the cereal alone—without milk.)
3. Rank the cereals in order from least sugar to most sugar per serving.
4. Rank the cereals in order from lowest price to highest price.
5. Present your findings to the class by doing one of the following activities.
   - Make a poster that summarizes what you discovered about the cereals.
   - Write an article for the school newspaper about nutrition in breakfast cereals.
   - Create a magazine or television ad for the cereal you think is most nutritious.

2-1 LESSON NOTES

NCTM Standards: 1–5, 14

Instructional Resources

- Study Guide Master 2-1
- Practice Master 2-1
- Enrichment Master 2-1
- Science and Math Lab Manual, pp. 5–8

Transparency 2-1A contains the 5-Minute Check for this lesson; **Transparency 2-1B** contains a teaching aid for this lesson.

Recommended Pacing	
Standard Pacing	Day 1 of 15
Honors Pacing	Day 1 of 15
Block Scheduling*	Day 1 of 9 (along with Lesson 2-2)
Alg. 1 in Two Years*	Days 1, 2, & 3 of 28

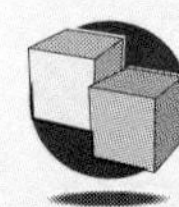

*For more information on pacing and possible lesson plans, refer to the *Block Scheduling Booklet* and *Algebra 1 in Two Years.*

1 FOCUS

5-Minute Check
(over Chapter 1)

Simplify.

1. $6r + 2s + 3r + s$ **$9r + 3s$**
2. $9(a + b) + 3a$ **$12a + 9b$**

Evaluate each expression if $a = 5$, $b = 4$, and $c = 3$.

3. $3ac - bc$ **33**
4. $b - c + 2ab$ **41**

Name the property shown.

5. $5 + (x + y) = 5 + (y + x)$ **commutative property (+)**

Motivating the Lesson

Situational Problem An outdoor thermometer shows a temperature of 10°C. Several hours later, the temperature has dropped 13 degrees. What temperature does the thermometer show now? **−3°C**

2–1

Integers and the Number Line

What YOU'LL LEARN

- To state the coordinate of a point on a number line,
- to graph integers on a number line, and
- to add integers by using a number line.

Why IT'S IMPORTANT

You can use number lines to add and subtract integers and to display data.

APPLICATION
Entertainment

Have you ever played Monopoly®? Most likely you have at one time or another. Parker Brothers has sold over 100 million sets of the game of Monopoly since it was invented in 1933 by Charles Darrow.

In order to play Monopoly, you must know how to move playing pieces around the board. To do this, you add the numbers rolled on the dice and then move your playing piece the corresponding number of spaces forward. You also need to know how to move your playing piece *backward* when directed to do so. Understanding how to add and subtract integers can help you to play.

Top Five List

Most Landed-on Squares in Monopoly®
1. Illinois Ave.
2. Go
3. B&O Railroad
4. Free Parking
5. Tennessee Ave.

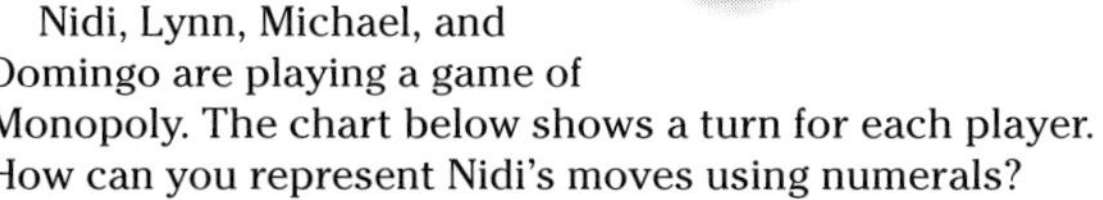

Nidi, Lynn, Michael, and Domingo are playing a game of Monopoly. The chart below shows a turn for each player. How can you represent Nidi's moves using numerals?

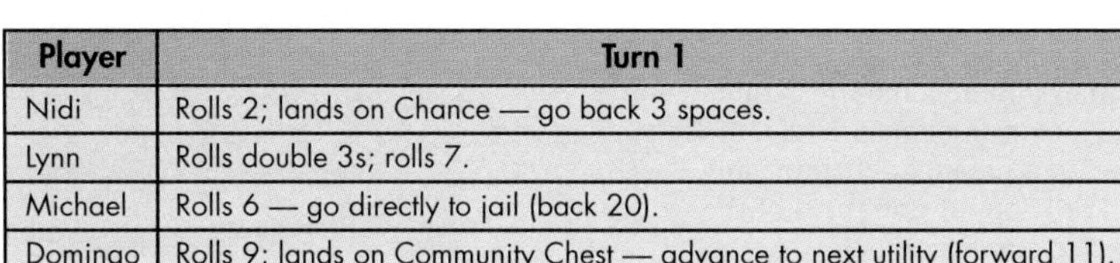

Player	Turn 1
Nidi	Rolls 2; lands on Chance — go back 3 spaces.
Lynn	Rolls double 3s; rolls 7.
Michael	Rolls 6 — go directly to jail (back 20).
Domingo	Rolls 9; lands on Community Chest — advance to next utility (forward 11).

You can solve problems like the one above by using a **number line.** A number line is drawn by choosing a starting position, usually 0, and marking off equal distances from that point. The set of **whole numbers** is often represented on a number line. This set can be written {0, 1, 2, 3, ...}, where "..." means that the set continues indefinitely.

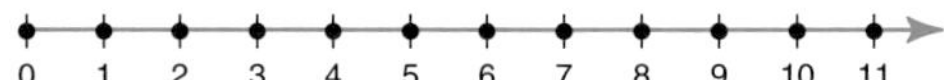

Although only a portion of the number line is shown, the arrowhead indicates that the line and the set of numbers continue.

We can use the expression $2 - 3$ to represent Nidi's moves.

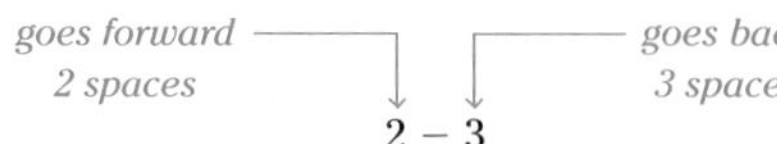

Mr. Darrow's Monopoly became so popular among his family and friends that he began making them their own sets. He then began selling sets for $4.00 each. When first offered the game, Parker Brothers turned it down, believing the game would never be accepted by the public.

The number line below shows that the value of 2 − 3 should be 1 less than 0. However, there is no whole number that corresponds to 1 less than 0. You can write the number 1 *less than* 0 as −1. This is an example of a **negative number.**

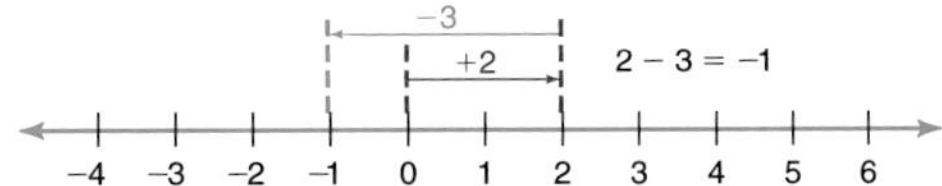

Any nonzero number written without a sign is understood to be positive.

To include negative numbers on the number line, extend the number line to the left of zero and mark off equal distances. The points to the right of zero are named using the *positive sign* (+). The points to the left of zero are named using the *negative sign* (−). Zero is neither positive nor negative.

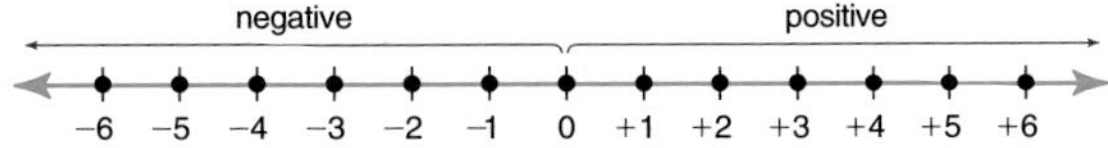

Read "−5" as *negative* 5. Read "+5" as *positive* 5.

Some of the most popular board games around the world are:

Pachisi—India
Go—Japan
Mancala—Africa
Alquerque—Spain
Senet—Egypt
Backgammon—Middle East
Checkers—England
Dominoes—Korea

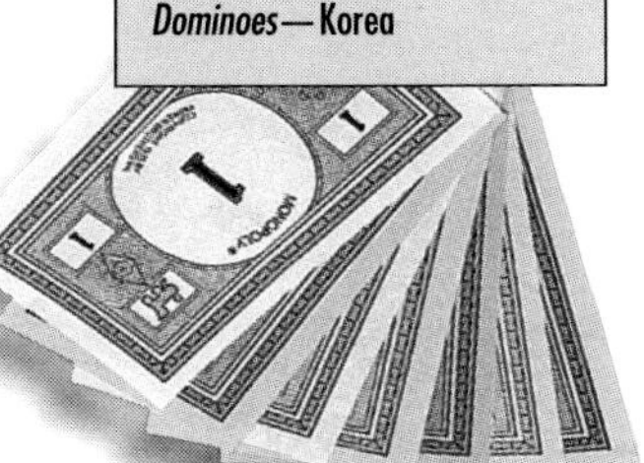

The set of numbers used on the number line above is called the set of **integers.** This set can be written {..., −6, −5, −4, −3, −2, −1, 0, 1, 2, 3, 4, 5, 6, ...}.

Venn diagrams are figures often used to represent sets of numbers.

Sets	Examples
Natural numbers	1, 2, 3, 4, 5, ...
Whole numbers	0, 1, 2, 3, 4, ...
Integers	..., −2, −1, 0, 1, 2, ...

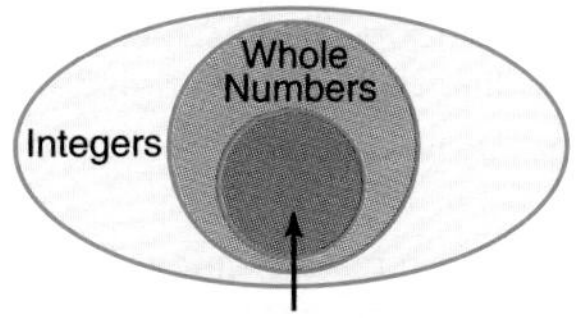

Notice that the natural numbers are a subset of the whole numbers and the whole numbers are a subset of the integers.

To **graph** a set of numbers means to draw, or plot, the points named by those numbers on a number line. The number that corresponds to a point on a number line is called the **coordinate** of that point.

Example  **Name the set of numbers graphed.**

a. 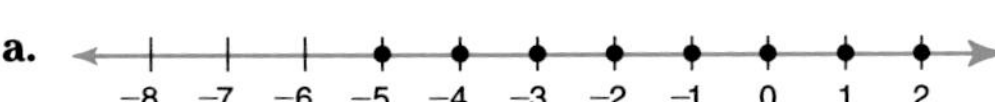

The bold arrow means that the graph continues indefinitely in that direction.

The set is {−5, −4, −3, −2, −1, 0, 1, 2, ...}.

b. 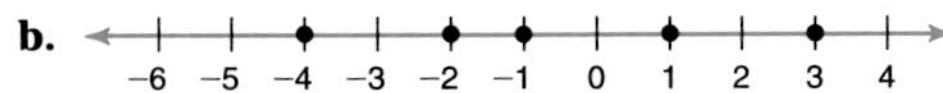

The set is {−4, −2, −1, 1, 3}.

2 TEACH

GLOBAL CONNECTIONS

There are many varieties of Domino games. In China, one finds Tiu-ü (fishing), Tau Ngau (bull fighting), Pai Kow, Tien Kow, and Mah Jong. In Europe, one finds the Block Game, Cyprus, Matador, Bergen, Tiddlywinks, and Blind Hughie.

Teaching Tip In a Venn diagram, ovals, circles, and squares or rectangles are used to represent the universal, or largest, set and its subsets. However, the size and shape of a Venn diagram is unimportant.

In-Class Example

For Example 1
Name the set of numbers graphed.

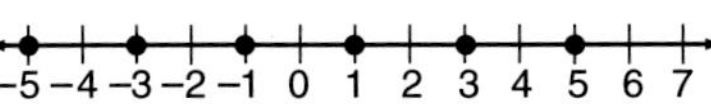

{−5, −3, −1, 1, 3, 5}

GLENCOE Technology

 **Interactive Mathematics Tools Software**

This multimedia software provides an interactive lesson that uses a number line to determine the rules for adding and subtracting positive and negative integers. A **Computer Journal** gives students an opportunity to write about what they have learned.

For Windows & Macintosh

Alternative Teaching Strategies

Reading Algebra After studying the concept of positive and negative numbers, ask students to discuss what business terms such as *negative worth* and *negative cash flow* might mean.

MODELING MATHEMATICS When showing addition on a number line, emphasize that negative numbers are indicated by arrows pointing to the left, and positive numbers are indicated by arrows pointing to the right. The first arrow must start at the origin. The second arrow starts at the arrowhead of the first arrow. The sum is at the head of the second arrow.

Answers for Modeling Mathematics

a. $2 + (-5) = -3$

b.

c.

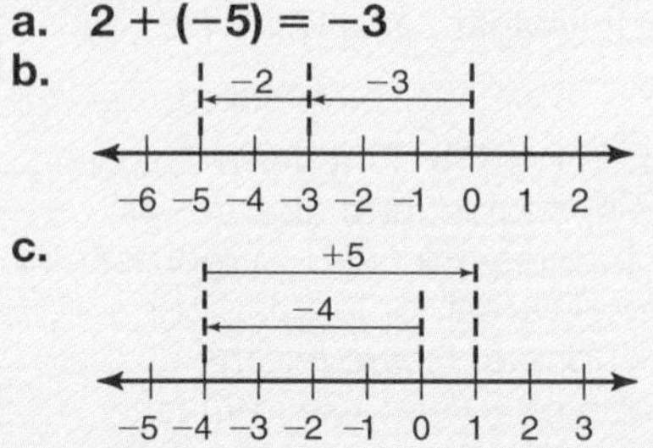

Teaching Tip Point out that a number line is not always horizontal. A thermometer is an instrument with a vertical number line.

> **In-Class Example**
>
> ***For Example 2***
> The next day in Casper, Wyoming, the low temperature was −8°F. The high was 21° higher than the low temperature. What was the high temperature? **13°F**

Teaching Tip Point out that since addition is commutative, the sum of (2) + (−4) can be found either by finding +2 on the number line and then moving 4 units left, or by first finding −4 on the number line and then moving 2 units to the right.

Additional Answers

1. **An arrow would be drawn, starting at 0 and going to −4. Starting at −4, an arrow would be drawn to the right 6 units long. The arrow ends at the sum of 2.**
2.

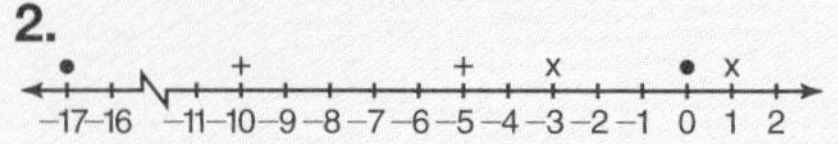

You can use a number line to add integers.

MODELING MATHEMATICS — Adding Integers

To find the sum of −7 and −5, follow the steps below.

Step 1
Draw an arrow starting at 0 to −7, the first addend.

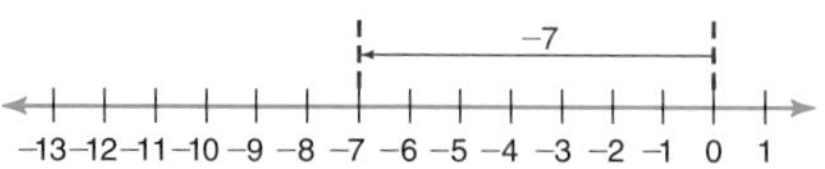

Step 2
Starting at −7, draw an arrow and go to the left 5 units long. This represents the second addend, −5.

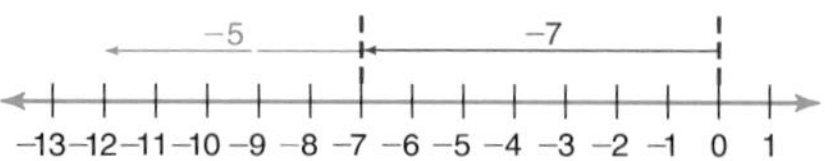

Step 3
The arrow ends at the sum, −12.

So, $-7 + (-5) = -12$.

Notice that parentheses are used in the equation so that the sign of the number is not confused with a subtraction sign.

Your Turn **a–c. See margin.**

a. What addition sentence is modeled on the number line shown at the right?

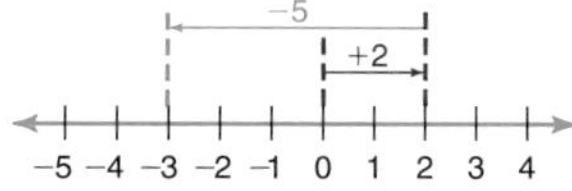

b. Draw a number line to show the addition sentence $-3 + (-2) = -5$.
c. Use a number line to find $-4 + 5$.

Example 2

APPLICATION
Meteorology

At 4:08 A.M., the temperature in Casper, Wyoming, was −10°F. By 1:30 P.M., the temperature had risen 17° to the daytime high. What was the high temperature?

Draw and label a number line. Draw an arrow from 0 to −10. Then draw an arrow to the right 17 units long.

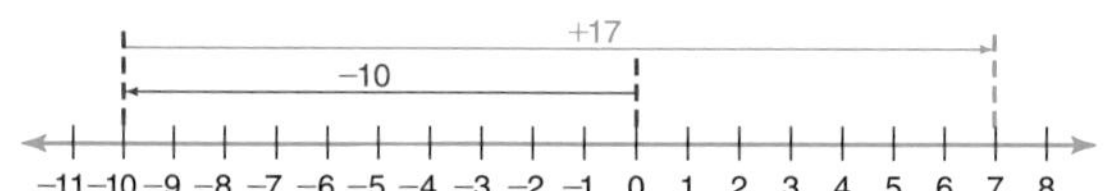

The high temperature of the day was 7°F.

CHECK FOR UNDERSTANDING

Communicating Mathematics

Study the lesson. Then complete the following. **1–2. See margin.**

1. **Explain** how you would find the sum $-4 + 6$ on a number line.
2. **Draw** a number line to show that $-17 < 0$, $-3 < 1$, and $-5 > -10$.
3. Refer to the application at the beginning of the lesson. Draw number lines to show the first Monopoly® moves for Lynn, Michael, and Domingo. **3. See Solutions Manual.**
4. **Research** instances where integers are used in real-life situations by collecting newspaper clippings that show integers. **See students' work.**

Classroom Vignette

"We actually put a number line on the floor and left it there for weeks. We used it to demonstrate problems throughout this chapter. It really seemed effective, and the students enjoyed physically adding and subtracting on the number line."

Mrs. Betsy A. Barron

Betsy Barron
Rochester High School
Rochester, Indiana

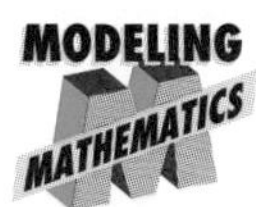

5. Use a number line to find $5 + (-6)$. **See margin.**

Guided Practice

Name the set of numbers graphed.

6. −6 −5 −4 −3 −2 −1 0 1 2 3
{−3, −2, −1}

7. −7 −6 −5 −4 −3 −2 −1 0 1 2
{−1, 0, 1, 2,...}

Graph each set of numbers on a number line.

8. {0, 2, 4, 6}
9. {−1, 0, 1, 2, 3,...}
10. {integers greater than −3} **8–10. See Solutions Manual.**

Write a corresponding addition sentence for each diagram.

11. 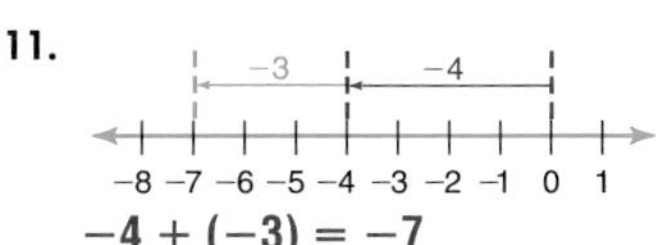

$-4 + (-3) = -7$

12. 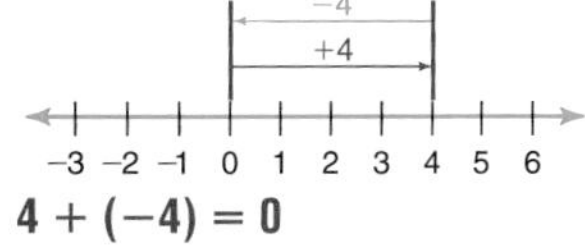

$4 + (-4) = 0$

Find each sum. If necessary, use a number line.

13. $-8 + 3$ **−5**
14. $-7 + (-15)$ **−22**

15. **Football** The Barnesville Bruins' offense lined up on their 20-yard line. They gained 5 yards on the first down, lost 3 yards on the second down, gained 7 yards on the third down, and punted on the fourth down. What was the net gain or loss after the third down? **9-yard gain**

EXERCISES

Practice

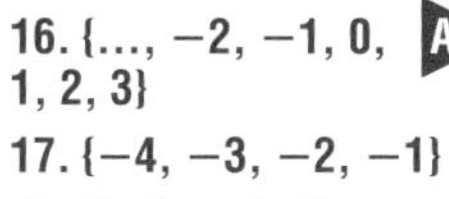

16. {..., −2, −1, 0, 1, 2, 3}
17. {−4, −3, −2, −1}
18. {0, 2, 5, 6, 8}
19. {−7, −3}
20. {−2, −1, 1, 2}
21. {..., −5, −4, −3, −2, −1, 0}

22–30. See Solutions Manual.

A

Name the set of numbers graphed.

16. 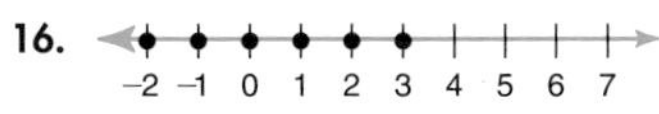

17. −6 −5 −4 −3 −2 −1 0 1 2 3

18. 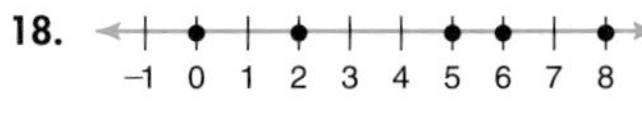

19. −8 −7 −6 −5 −4 −3 −2 −1 0 1

20. 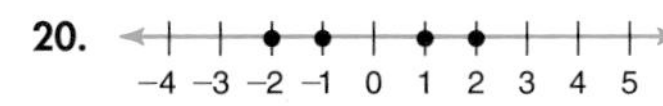

21. −5 −4 −3 −2 −1 0 1 2 3 4

Graph each set of numbers on a number line.

B

22. {−4, −3, −1, 3}
23. {..., −2, −1, 0, 1}
24. {integers between −6 and 10}
25. {integers less than 0}
26. {integers between −2 and 5}
27. {integers less than $-3 + (-1)$}
28. {whole numbers greater than $-4 + 4$}

C

29. {integers less than 0 but greater than −6}
30. {integers less than or equal to −3 and greater than or equal to −10}

Find each sum. If necessary, use a number line.

31. $8 + 5$ **13**
32. $-6 + (-7)$ **−13**
33. $7 + (-12)$ **−5**
34. $-4 + (-9)$ **−13**
35. $0 + (-12)$ **−12**
36. $5 + (-3)$ **2**
37. $-3 + 9$ **6**
38. $-13 + 13$ **0**
39. $-14 + (-9)$ **−23**

3 PRACTICE/APPLY

Check for Understanding

Exercises 1–15 are designed to help you assess your students' understanding through reading, writing, speaking, and modeling. You should work through Exercises 1–5 with your students and then monitor their work on Exercises 6–15.

Error Analysis

The "−" is used in three ways in math. Students may interchange "minus," "negative," and "opposite of."

Expression	Meaning
$-x$	opposite of x (relationship)
−3	negative (direction)
5 − 4	5 minus 4 (operation)

The context determines how the "−" is being used. Consider $-x - 3 - (-2) - y$. This means the opposite of x, minus 3, minus negative 2, minus y.

Assignment Guide

Core: 17–39 odd, 40, 41, 43, 45–52
Enriched: 16–38 even, 40–52

For **Extra Practice,** see p. 759.

Study Guide Masters, p. 10

NAME ______ DATE ______

Study Guide — Student Edition Pages 72–77

Integers and the Number Line

The figure at the right is part of a number line. On a number line, the distances marked to the right of 0 are named by members of the set of **whole numbers.**

The set of numbers used to name the points marked on the number line at the right is called the set of **integers.**

To graph a set of numbers means to locate the points named by those numbers on the number line. The number that corresponds to a point on the number line is called the **coordinate** of the point.

Name the coordinate of point G.
The coordinate of G is −2.

A number line is often used to show addition of integers. For example, to find the sum of 3 and −5, follow the steps at the right.

Step 1 Draw an arrow, starting at 0 and going to 3.
Step 2 Start at 3. Draw an arrow 5 units to the left.
Step 3 The second arrow points to the sum, −2.

Name the coordinate of each point.

1. M **5** 2. Q **12** 3. H **−2** 4. E **−7** 5. J **0** 6. A **−12**
7. G **−4** 8. P **11** 9. F **−6** 10. N **8** 11. O **9** 12. D **−9**

Graph each set of numbers on a number line.

13. {−3, −1, 1, 3} 14. {−5, −2, 1, 4} 15. {integers less than 0}
16. {0, 1, 3, 5} 17. {−3, −2, 2} 18. {..., −2, −1, 0, 1}

Find each sum. If necessary, use a number line.

19. $2 + 3$ **5** 20. $9 + 1$ **10** 21. $-5 + (-1)$ **−6**
22. $-10 + 6$ **−4** 23. $9 + (-9)$ **0** 24. $0 + (-4)$ **−4**
25. $-8 + (-3)$ **−11** 26. $6 + (-10)$ **−4** 27. $-6 + 6$ **0**

Reteaching

Act It Out "Walk out" addition problems by dividing a large strip of paper into squares labeled −9 to 9, with positives on the right and negatives on the left. To "walk out" $2 + (-5)$, start at zero and walk 2 places to the right (+2). Then walk 5 places to the left (−5) ending on −3. Have students "walk out" $2 + 5$, $(-2) + 5$, and $(-2) + (-5)$.

Additional Answer

5. 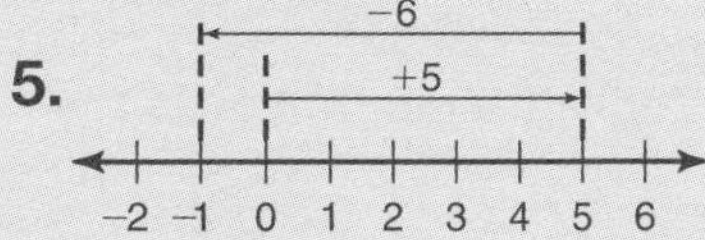

4 ASSESS

Closing Activity

Writing Have students determine the relationship that exists between two addends with the same sign and the sign of their sum. Then have students identify the relationship between addends with different signs and the sign of their sum. Students should write a short paragraph describing their findings.

Additional Answers

42a. 11 P.M. the same day
42b. 7 A.M. the same day
42c. 12 noon the same day
42d. 8:30 P.M. the same day

Practice Masters, p. 10

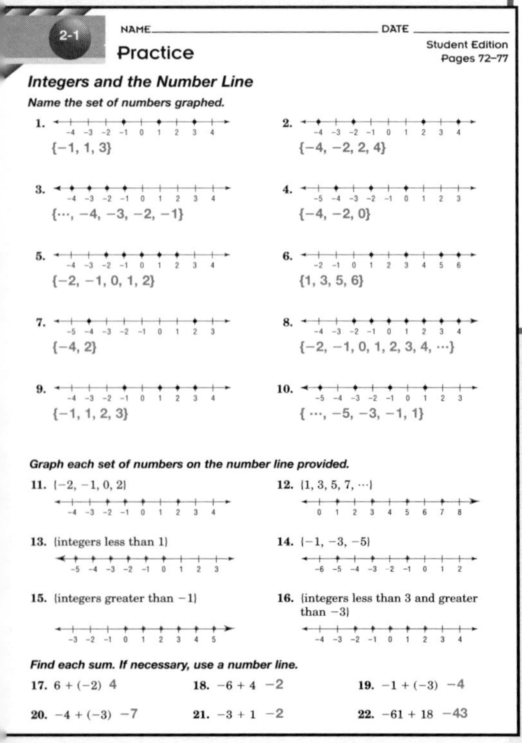

2-1 NAME________ DATE ________

Practice

Student Edition Pages 72–77

Integers and the Number Line

Name the set of numbers graphed.

1. {−1, 1, 3}
2. {−4, −2, 2, 4}
3. {···, −4, −3, −2, −1}
4. {−4, −2, 0}
5. {−2, −1, 0, 1, 2}
6. {1, 3, 5, 6}
7. {−4, 2}
8. {−2, −1, 0, 1, 2, 3, 4, ···}
9. {−1, 1, 2, 3}
10. { ···, −5, −3, −1, 1}

Graph each set of numbers on the number line provided.

11. {−2, −1, 0, 2}
12. {1, 3, 5, 7, ···}
13. {integers less than 1}
14. {−1, −3, −5}
15. {integers greater than −1}
16. {integers less than 3 and greater than −3}

Find each sum. If necessary, use a number line.

17. $6 + (-2)$ 4
18. $-6 + 4$ −2
19. $-1 + (-3)$ −4
20. $-4 + (-3)$ −7
21. $-3 + 1$ −2
22. $-61 + 18$ −43

Critical Thinking

40. Write three different equations using integers that have a sum of −14.
See students' work.

Applications and Problem Solving

For Exercises 41–42, write an open sentence using addition. Then solve each problem.

41. Meteorology On February 10, the low temperature in Houlton, Maine, and Fort Yukon, Alaska, was −17°. On the same date, the high temperature in Lajitas, Texas, was 82° higher. What was the temperature in Lajitas, Texas, on February 10? **$-17 + 82 = 65$; 65°F**

42. Geography Use the map below to determine what time it is in each city if it is 10:00 A.M. in New York City. **a–d. See margin.**

a. Hong Kong **b.** Los Angeles
c. Rio de Janeiro **d.** Bombay

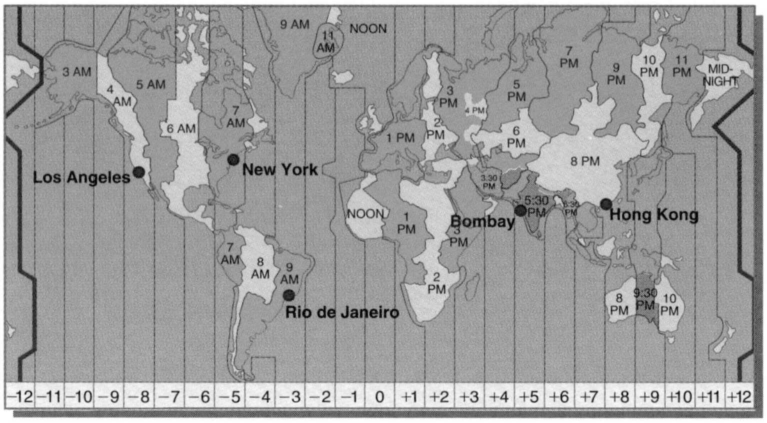

43. Meteorology *Windchill factor* is an estimate of the cooling effect the wind has on a person in cold weather. Meteorologists use a chart like the one below to predict the windchill factor. Use the chart below to find each windchill factor.

a. 20°F, 10 mph
b. 0°F, 15 mph
c. −10°F, 5 mph
d. 30°F, 15 mph

43a. 4°F
43b. −31°F
43c. −15°F
43d. 9°F

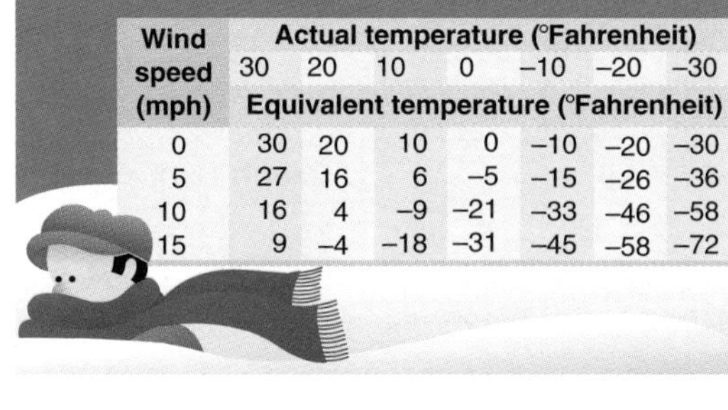

Wind speed (mph)	Actual temperature (°Fahrenheit)						
	30	20	10	0	−10	−20	−30
	Equivalent temperature (°Fahrenheit)						
0	30	20	10	0	−10	−20	−30
5	27	16	6	−5	−15	−26	−36
10	16	4	−9	−21	−33	−46	−58
15	9	−4	−18	−31	−45	−58	−72

44. Baseball Sometimes baseball team owners develop incentive clauses for hitters in an effort to win more games. Suppose an owner decided to pay each hitter \$500 for each run scored, \$400 for each hit, and \$500 for each run batted in. Look in the sports section of a newspaper. Choose a team's box score and calculate the value of the incentive clause for the first five hitters in a game. **See students' work.**

Extension

Reasoning Have students study the following number pattern.

−15, −12, −9, −6, . . .

Ask students to identify the pattern that exists in this sequence. **Each number is 3 more than the number that preceded it.** Have students identify the tenth term in the number pattern. **12**

Mixed Review

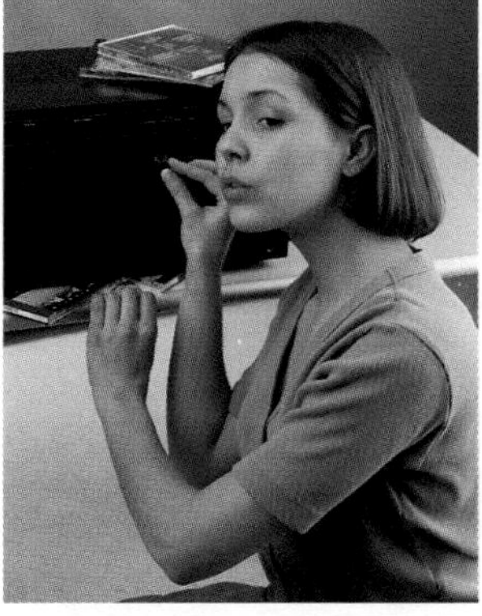

45. **Entertainment** Juanita has the volume on her stereo turned up because her parents are not home. Suddenly, she sees her father's car pull into the driveway. She runs to her room to turn down the volume. Juanita's father comes in, grabs an umbrella, and leaves, so Juanita returns the volume to its previous level. Sketch a reasonable graph to show the volume of Juanita's stereo during this time. (Lesson 1–9) **See margin.**

46. **Physical Fitness** Mitchell likes to exercise regularly. On Mondays, he likes to walk two miles, run three miles, sprint one-half of a mile, and then walk for another mile. Identify the graph that best represents Mitchell's heart rate as a function of time. (Lesson 1–9) **b**

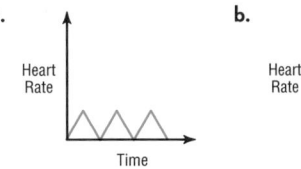

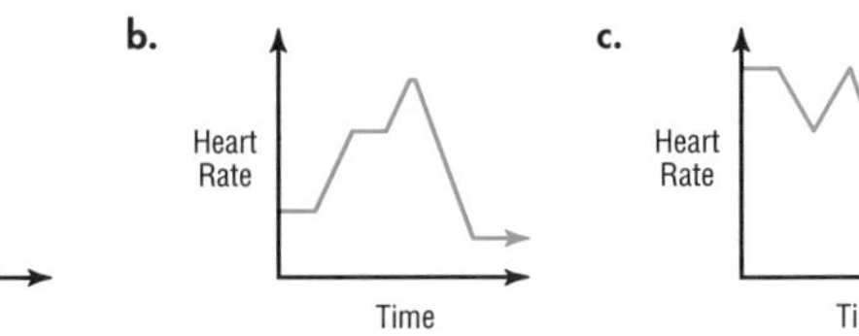

47. commutative (×)

47. Name the property illustrated by $8(2 \cdot 6) = (2 \cdot 6)8$. (Lesson 1–8)

48. Simplify $16a + 21a + 30b - 7b$. (Lesson 1–7) $\mathbf{37a + 23b}$

49. substitution (=)

49. Name the property illustrated by $(12 - 9)(4) = 3(4)$. (Lesson 1–6)

50. Solve $5(7) + 6 = x$. (Lesson 1–5) **41**

51. Evaluate $9(0.4 + 1.2) - 0.5$. (Lesson 1–3) **13.9**

52. **Patterns** Find the next three numbers in the pattern 5, 6.5, 8, 9.5, ... (Lesson 1–2) **11, 12.5, 14**

Mathematics and SOCIETY

Mayan Mathematics

The excerpt below is from an article that appeared in *The Columbus Dispatch* on January 15, 1995. It describes the number system developed by the Mayan Indians, an ancient civilization that lived in southern Mexico and Central America until the 16th century.

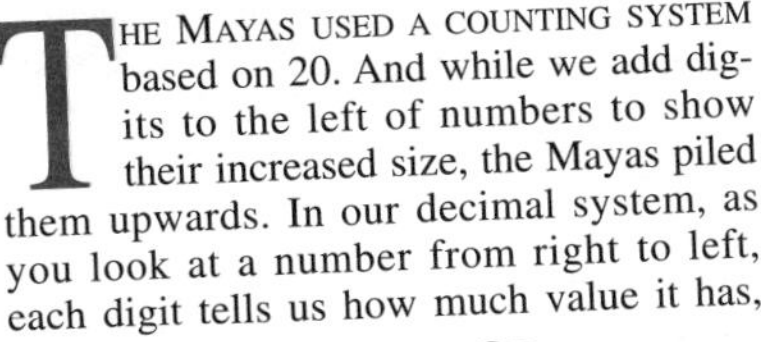

THE MAYAS USED A COUNTING SYSTEM based on 20. And while we add digits to the left of numbers to show their increased size, the Mayas piled them upwards. In our decimal system, as you look at a number from right to left, each digit tells us how much value it has, based on the number 10. The number 326, for instance, is really 3 hundreds, 2 tens and 6 ones....The Mayas used the same principle, but used the vigesimal system (based on 20). Units were represented by dots. Five dots were represented by a horizontal line. ■

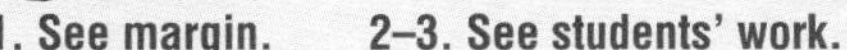

Upper Level
= 1
= 5
= 5
= 5
16 × 20 = 320
Lower Level
= 1
= 5
6 × 1 = 6

1. See margin. 2–3. See students' work.

1. What combinations of dots and horizontal lines (bars) are needed to write the number 252 using the Mayan system? (*Hint:* Two levels are required.)
2. What uses would the Mayas and other ancient civilizations have had for the number systems they developed?
3. Would a number system based on a number other than 10 be as accurate as the decimal system? Would it be as easy to use? Explain your answers.

Mathematics and SOCIETY

The Maya civilization spanned more than 2000 years, from 1000 B.C. to A.D. 1542. The final downfall of the society came in 1542 with the conquest by Francisco de Martejo. Today, more than 2 million descendants of the Maya live in a fashion relatively unchanged since earlier times.

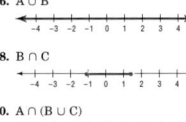

Additional Answer

45.

Volume

Time

Answer for Mathematics and Society

1. Upper Level

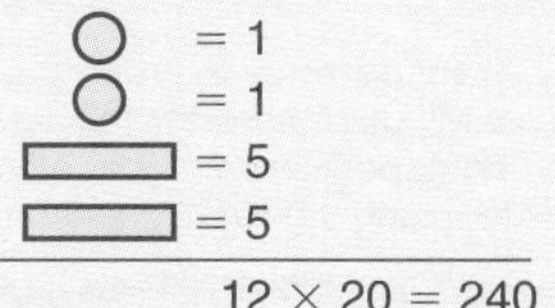

Lower Level

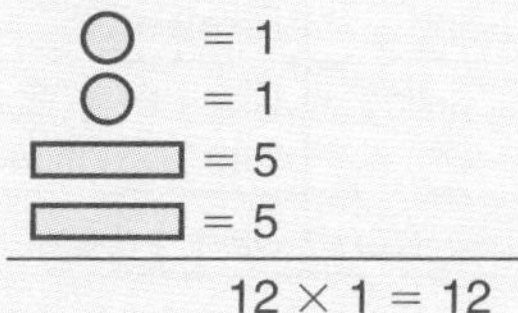

Enrichment Masters, p. 10

NAME________________ DATE ________

Enrichment

Student Edition Pages 72–77

Intersection and Union

The **intersection** of two sets is the set of elements that are in both of the sets. The intersection of sets A and B is written $A \cap B$. The **union** of two sets is the set of elements in either A, or in B, or in both. The union is written $A \cup B$.

In the drawings below, suppose A is the set of points inside the circle and B is the set of points inside the square. Then, the shaded areas show the intersection and union.

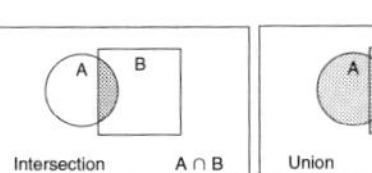

Write $A \cap B$ and $A \cup B$ for each of the following.

1. A = {p, q, r, s, t} B = {q, r, s} $A \cap B = \{q, r, s\}$ $A \cup B = \{p, q, r, s, t\}$
2. A = {the integers between 2 and 7} B = {0, 3, 8} $A \cap B = \{3\}$ $A \cup B = \{0, 3, 4, 5, 6, 8\}$
3. A = {the states whose names start with K} B = {the states whose capital is Honolulu or Topeka} $A \cap B$ = {Kansas} $A \cup B$ = {Hawaii, Kansas, Kentucky}
4. A = {the positive integer factors of 24} B = {the counting numbers less than 10} $A \cap B = \{1, 2, 3, 4, 6, 8\}$ $A \cup B = \{1, 2, 3, 4, 5, 6, 7, 8, 9, 12, 24\}$

Suppose A = {numbers x such that $x < 3$}, B = {numbers x such as $x \geq -1$}, and C = {numbers x such that $x \leq 1.5$}. Graph each of the following.

5. $A \cap B$
   −4 −3 −2 −1 0 1 2 3 4
6. $A \cup B$
   −4 −3 −2 −1 0 1 2 3 4
7. $A \cup B$
   −4 −3 −2 −1 0 1 2 3 4
8. $B \cap C$
   −4 −3 −2 −1 0 1 2 3 4
9. $(A \cap C) \cap B$
   −4 −3 −2 −1 0 1 2 3 4
10. $A \cap (B \cup C)$
   −4 −3 −2 −1 0 1 2 3 4

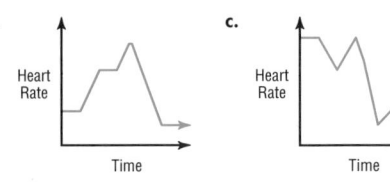

2-2 LESSON NOTES

NCTM Standards: 1–4, 10

Instructional Resources

- Study Guide Master 2-2
- Practice Master 2-2
- Enrichment Master 2-2
- Real-World Applications, 3

Transparency 2-2A contains the 5-Minute Check for this lesson; **Transparency 2-2B** contains a teaching aid for this lesson.

Recommended Pacing	
Standard Pacing	Day 2 of 15
Honors Pacing	Day 2 of 15
Block Scheduling*	Day 1 of 9 (along with Lesson 2-1)
Alg. 1 in Two Years*	Days 4 & 5 of 28

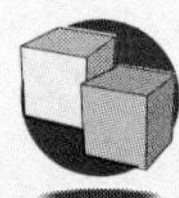

*For more information on pacing and possible lesson plans, refer to the *Block Scheduling Booklet* and *Algebra 1 in Two Years.*

1 FOCUS

5-Minute Check
(over Lesson 2-1)

1. Graph {2, 3, 4} on a number line.

   –2 –1 0 1 2 3 4 5 6

2. Show the addition 4 + (–7) on a number line. **–3**

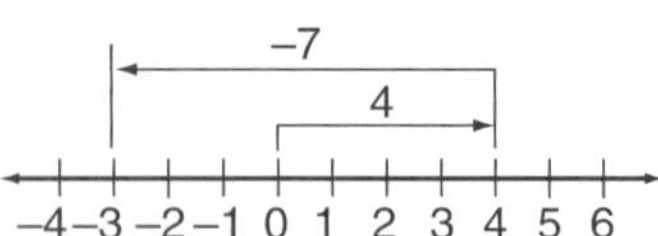

Find each sum.

3. 9 + (–24) **–15**
4. (–18) + (–47) **–65**
5. –27 + 45 **18**

Teaching Tip Have students discuss and write what information can be gained from the line plot in the Example. You might ask, in general, how many wins were necessary to make the playoffs? **between 40 and 60**

2-2 Integration: Statistics Line Plots

What YOU'LL LEARN

- To interpret numerical data from a table, and
- to display and interpret statistical data on a line plot.

Why IT'S IMPORTANT

Line plots are a useful way to display data.

APPLICATION
Television

Are you a fan of Will on the television show *Fresh Prince of Bel-Air*? The Nielsen ratings are used to determine which shows are popular and which shows are not. They are also used to determine which shows should be continued or canceled. The chart at the right shows the Nielsen ratings for Monday, February 6, 1995.

Nielsen Ratings

Monday, February 6, 1995

TIME	PROGRAM	VIEWERS (rounded to nearest million)
8:00	Fresh Prince of Bel-Air (NBC)	22
	The Nanny (CBS)	19
	Melrose Place (Fox)	14
	Coach (ABC)	15
	Star Trek: Voyager (UPN)	14
8:30	Dave's World (CBS)	21
	Blossom (NBC)	19
	Sneakers (ABC)	16
9:00	Murphy Brown (CBS)	22
	Serving in Silence (NBC)	19
	Models, Inc. (Fox)	10
	Platypus Man (UPN)	5
9:30	Cybill (CBS)	19
	Pig Sty (UPN)	4
10:00	Chicago Hope (CBS)	19

In some cases, data can be presented on a number line. Numerical data displayed on a number line is called a **line plot.** The data in the table above can be presented in a line plot as follows.

Step 1 Draw and label a number line. You can see that the data in the table ranges from 4 million to 22 million viewers. In order to represent the data on a number line, a *scale* must be used that includes this range of values. You can use a scale from 0 to 25 with *intervals* of five.

Step 2 Draw the line plot. Write a "v" for each TV show above its share of viewers. A completed line plot for the Nielsen ratings is shown below.

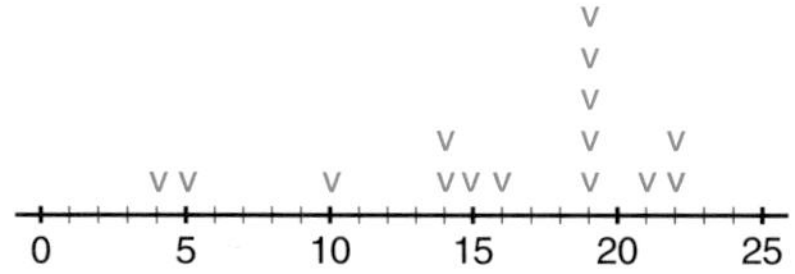

Notice that some data values are located between marked intervals on the number line.

Example **1**

APPLICATION
Basketball

In 1994 and 1995, the Houston Rockets won back-to-back NBA championships. The table at the right shows the standings for each of the 27 NBA teams for the 1994–1995 season on April 28, 1995.

a. Make a line plot to show the number of wins by each playoff team.

Final Regular-Season Standings

EASTERN CONFERENCE			WESTERN CONFERENCE		
ATLANTIC	**W**	**L**	**MIDWEST**	**W**	**L**
y-Orlando	57	25	y-San Antonio	62	20
x-New York	55	27	x-Utah	60	22
x-Boston	35	47	x-Houston	47	35
Miami	32	50	x-Denver	41	41
New Jersey	30	52	Dallas	36	46
Philadelphia	24	58	Minnesota	21	61
Washington	21	61			
CENTRAL	**W**	**L**	**PACIFIC**	**W**	**L**
z-Indiana	52	30	z-Phoenix	59	23
x-Charlotte	50	32	x-Seattle	57	25
x-Chicago	47	35	x-L.A. Lakers	48	34
x-Cleveland	43	39	x-Portland	44	38
x-Atlanta	42	40	Sacramento	39	43
Milwaukee	34	48	Golden State	26	56
Detroit	28	54	L.A. Clippers	17	65

x–clinched playoff berth; y–clinched conference; z–clinched division

Source: *USA Today*, 4-28-95

The scale ranges from 35 to 65 with intervals of 5.

The teams with an x, y, or z next to their names made the playoffs. The number of wins by the playoff teams ranges from 35 to 62. You can use a "w" to represent the number of each team's wins.

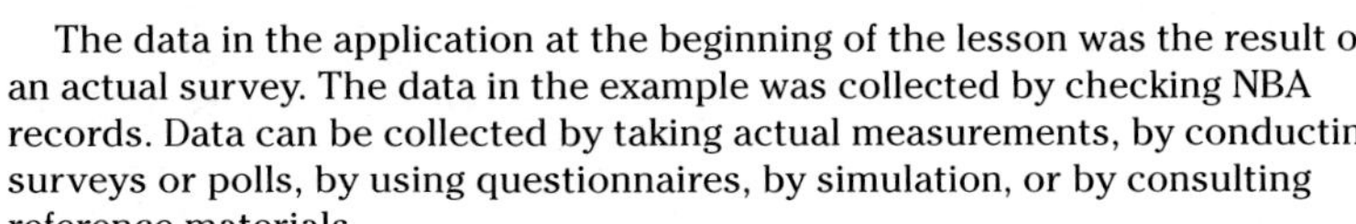

b. How many playoff teams won fewer than 50 games?
From the line plot, you can see that 8 teams won fewer than 50 games. These teams were Boston, Chicago, Cleveland, Atlanta, Houston, Denver, L.A. Lakers, and Portland.

c. Which playoff team had the best record? the worst record?
San Antonio had the best record and Boston had the worst record of the playoff teams.

d. How many teams made the playoffs?
Count the number of w's on the line plot. Sixteen teams made the playoffs.

FYI

In September 1995, blue M&M's® completely replaced the tan ones. The ratios of colors for plain M&M's are as follows.

brown	30%
red	20%
yellow	20%
orange	10%
green	10%
blue	10%

The data in the application at the beginning of the lesson was the result of an actual survey. The data in the example was collected by checking NBA records. Data can be collected by taking actual measurements, by conducting surveys or polls, by using questionnaires, by simulation, or by consulting reference materials.

It is important that you know how the data were obtained. For example, would you want to draw conclusions about changing the name of your school mascot based on a result of a survey of seniors only? Why or why not?

MODELING MATHEMATICS **Line Plots**

Materials: small packages of plain M&M's®

In 1994 and 1995, the Mars Company conducted surveys to determine whether a new color should be added to M&M's®. The choices were blue, pink, purple, or leave them as they are. The color blue was chosen.

Your Turn **a–e. See students' work.**

a. Open a package of M&M's. Separate the candies by color. Find the total number of each color.
b. Make a line plot to show the number of each color. Use b for brown, r for red, y for yellow, o for orange, g for green, and bl for blue.
c. Do the colors cluster around any number?
d. Make a class line plot of your data. Are the data in the class line plot different or the same as yours? Explain.
e. Make a class line plot showing the total number of M&M's in each of your packages. Do the packages have the same number in them?

Alternative Learning Styles

Auditory Read the following problem to the class. Have students write the equation and solve it.
The ladder reaches 11 feet from the top of the building. The building is 67 feet tall. How high does the ladder reach?
$x - 11 = f$, $x = 67$, $67 - 11 = 56$

MODELING MATHEMATICS Most likely students will see that the class line plot will be closer to the Mars Company's specifications. Point out to students that the larger sample size tends to provide more accurate predictions. However, any sample can be far outside the distribution for the entire population.

Motivating the Lesson

Situational Problem Ask students what factors would be important to them when buying a car. Ask them how they would obtain information about these factors. Write the following data on the chalkboard.

Fuel Economy	Model A	Model B	Model C
EPA estimate city/highway	19/29	18/26	19/24
195-mile trip	29	27	25

Have students interpret these data to identify the model that gets the best fuel economy in expressway driving. Ask students to identify the car that has the lowest miles per gallon average for overall performance.

2 TEACH

In-Class Example

For the Example
Garrick's Department Store kept a record of the number of transactions that transpired for each month in 1997. Make a line plot of the data.

Jan.	– 10,243	July	– 15,327
Feb.	– 10,268	Aug.	– 15,489
March	– 11,450	Sept.	– 15,635
April	– 12,676	Oct.	– 15,687
May	– 15,422	Nov.	– 19,213
June	– 15,318	Dec.	– 21,955

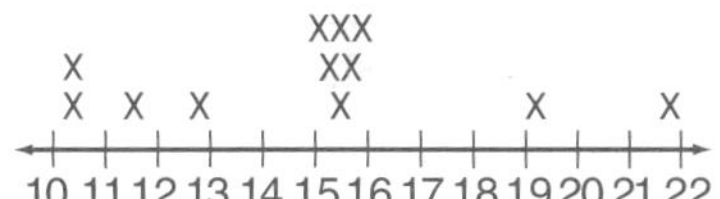

The graph shows the number of 1000s.

FYI

Students might find it helpful to visualize this distribution with a circle graph.

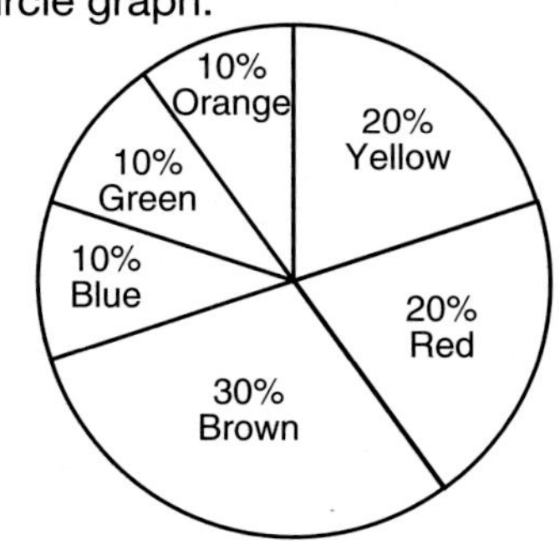

3 PRACTICE/APPLY

Check for Understanding

Exercises 1–6 are designed to help you assess your students' understanding through reading, writing, speaking, and modeling. You should work through Exercises 1–3 with your students and then monitor their work on Exercises 4–6.

Error Analysis
In order to answer questions using a line plot, the scales should be accurate enough to read right off the plot. They should also be as precise as possible.
Example:

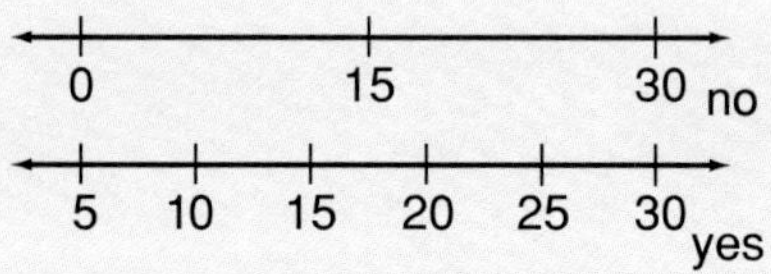

Additional Answers

4. from 0 to 70

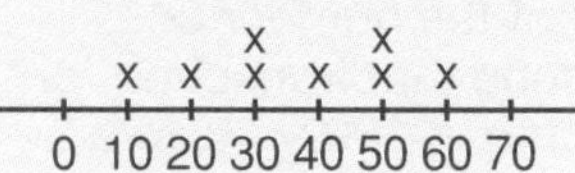

5. from 35 to 80

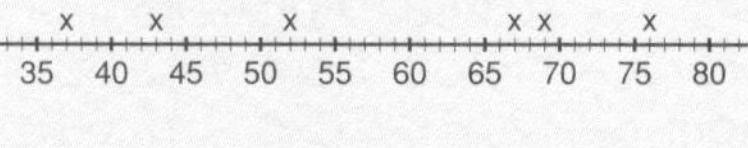

6a.

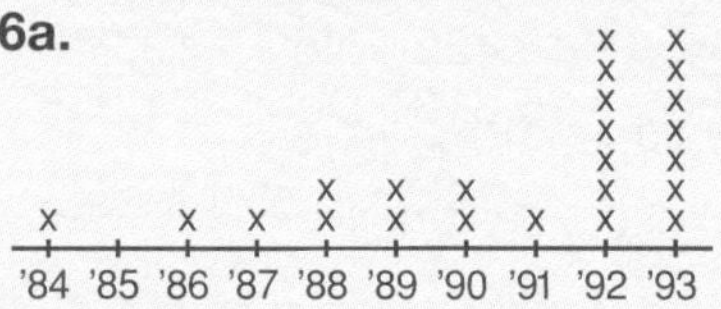

6b. $1950 to $11, 475

6c.

2 3 4 5 6 7 8 9 10 11

7a.

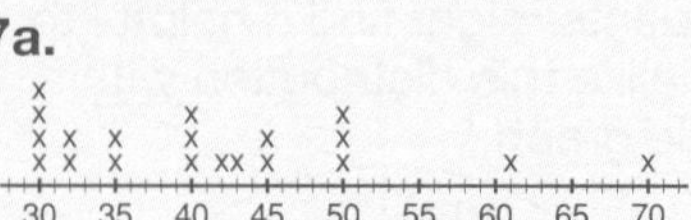

CHECK FOR UNDERSTANDING

Communicating Mathematics

Study the lesson. Then complete the following.

1. **Describe** a situation in which it would be valuable to obtain data in order to make a decision. **1–3. See students' work.**
2. **Compare and contrast** line plots and tables as a means of reporting data. What are the advantages and disadvantages of each?
3. **Develop** a survey and collect data on the Monday night television viewing habits of your classmates. Make a line plot of the data. Compare your class findings to those of another class and to a Nielsen ratings chart from a magazine or newspaper. What conclusions can you draw?

MODELING MATHEMATICS

Guided Practice

State the scale you would use to make a line plot for the following data. Then draw the line plot. **4–5. See margin.**

4. 50, 50, 30, 30, 20, 10, 60, 40
5. 52, 43, 67, 69, 37, 76

6. See margin.

6. **Retail Sales** The advertisement below lists the cars for sale at Texas Motors.

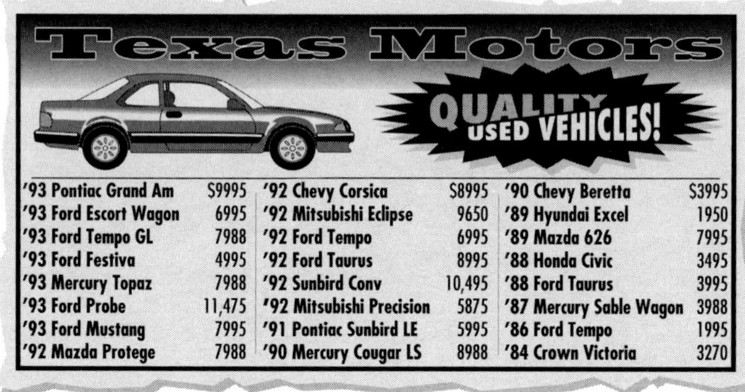

'93 Pontiac Grand Am	$9995	'92 Chevy Corsica	$8995	'90 Chevy Beretta	$3995
'93 Ford Escort Wagon	6995	'92 Mitsubishi Eclipse	9650	'89 Hyundai Excel	1950
'93 Ford Tempo GL	7988	'92 Ford Tempo	6995	'89 Mazda 626	7995
'93 Ford Festiva	4995	'92 Ford Taurus	8995	'88 Honda Civic	3495
'93 Mercury Topaz	7988	'92 Sunbird Conv	10,495	'88 Ford Taurus	3995
'93 Ford Probe	11,475	'92 Mitsubishi Precision	5875	'87 Mercury Sable Wagon	3988
'93 Ford Mustang	7995	'91 Pontiac Sunbird LE	5995	'86 Ford Tempo	1995
'92 Mazda Protege	7988	'90 Mercury Cougar LS	8988	'84 Crown Victoria	3270

a. Make a line plot to show how many cars from each year are in stock.
b. What is the price range for the used cars?
c. Make a line plot to show the costs of the cars. Round each price to the nearest thousand dollars. Let each number on the number line represent thousands of dollars.

EXERCISES

Applications and Problem Solving

7b. 70 mph; cheetah 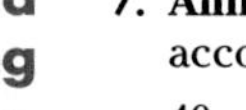

7. **Animals** The speeds of 20 of the fastest animals in miles per hour according the *The World Almanac*, 1995, are listed below.

40	61	50	50	32	70	35	30	50	45
43	40	30	30	35	45	42	32	40	30

a. Make a line plot of the data. **See margin.**
b. What is the fastest speed of any animal? Research to find the name of the fastest animal.
c. What is the slowest speed of the 20 animals? **30 mph**
d. Which speed occurred most frequently? **30 mph**
e. How many animals had a speed of at least 40 mph? **12**
f. How many animals had speeds greater than 30 and less than 40 mph? **4**

Reteaching

Using Discussion Have students compare and contrast how data are presented in a table and on a line plot.

8. History The table below lists the fifty United States and the years they entered the Union.

Years States Entered the Union

State	Year	State	Year	State	Year	State	Year
AL	1819	IN	1816	NE	1867	SC	1788
AK	1959	IA	1846	NV	1864	SD	1889
AZ	1912	KS	1861	NH	1788	TN	1796
AR	1836	KY	1792	NJ	1787	TX	1845
CA	1850	LA	1812	NM	1912	UT	1896
CO	1876	ME	1820	NY	1788	VT	1791
CT	1788	MD	1788	NC	1789	VA	1788
DE	1787	MA	1788	ND	1889	WA	1889
FL	1845	MI	1837	OH	1803	WV	1863
GA	1788	MN	1858	OK	1907	WI	1848
HI	1959	MS	1817	OR	1859	WY	1890
ID	1890	MO	1821	PA	1787		
IL	1818	MT	1889	RI	1790		

Source: *The World Almanac,* 1995

a, c. See margin.

a. What is the range of years for states entering the Union?

b. Make a line plot that shows the year each state joined the Union.

8b. See Solutions Manual.

c. Do the years cluster around a certain number? If so, which years are those? What would explain this?

9. Weather The average rainfall per year, rounded to the nearest inch, for ten cities around the world is shown on the map below.

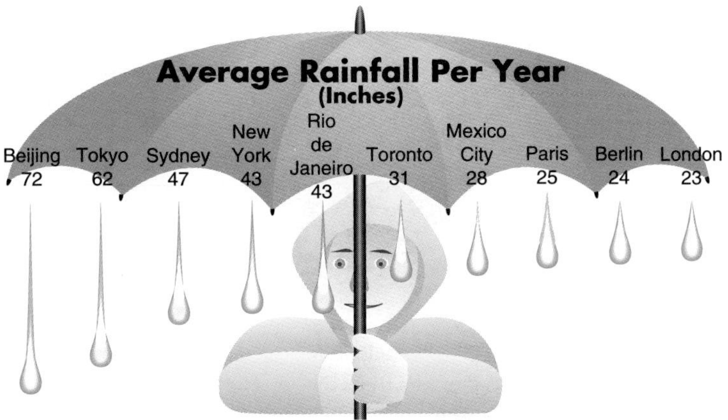

a. Make a line plot of the data. What scale did you use? **See margin.**

b. Do the average amounts of rain cluster about a certain number?

9b. yes, 25 and 42

c. Do you think that the average rainfall for your community is close to that of any of the cities above? Why or why not? **c–d. See students' work.**

d. Look up the average rainfall for your community. How does it compare to the average rainfall for London? for Tokyo?

Assignment Guide

Core: 7–11 odd, 13–18
Enriched: 8–12 even, 13–18

For **Extra Practice,** see p. 759.

The red A, B, and C flags, printed only in the Teacher's Wraparound Edition, indicate the level of difficulty of the exercises.

Additional Answers

8a. from 1787 to 1959

8c. Yes, 1788; states ratified constitution and became official United States of America.

9a. from 20 to 75 by 5s

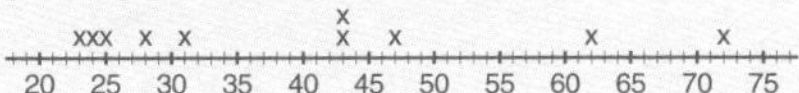

Study Guide Masters, p. 11

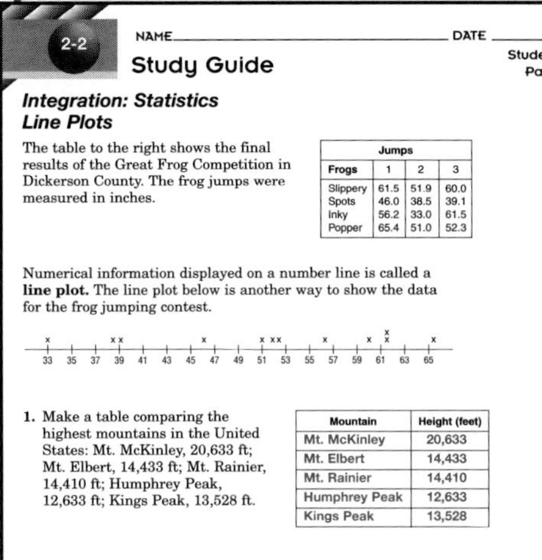

2-2 NAME______ DATE ______

Study Guide

Student Edition Pages 78–84

Integration: Statistics
Line Plots

The table to the right shows the final results of the Great Frog Competition in Dickerson County. The frog jumps were measured in inches.

	Jumps		
Frogs	1	2	3
Slippery	61.5	51.9	60.0
Spots	46.0	38.5	39.1
Inky	56.2	33.0	61.5
Popper	65.4	51.0	52.3

Numerical information displayed on a number line is called a **line plot.** The line plot below is another way to show the data for the frog jumping contest.

1. Make a table comparing the highest mountains in the United States: Mt. McKinley, 20,633 ft; Mt. Elbert, 14,433 ft; Mt. Rainier, 14,410 ft; Humphrey Peak, 12,633 ft; Kings Peak, 13,528 ft.

Mountain	Height (feet)
Mt. McKinley	20,633
Mt. Elbert	14,433
Mt. Rainier	14,410
Humphrey Peak	12,633
Kings Peak	13,528

Use the table at the right for Exercises 2–4.

2. Make a line plot representing the weights of wrestlers. Insert labels and "x"'s on the number line below.

Weights of Junior Varsity Wrestlers (pounds)				
170	160	135	135	160
122	188	154	108	135
140	122	103	190	154

3. How many wrestlers are in weight classes over 140 lb? 7

4. What is the largest weight class? 190 lb

4 ASSESS

Closing Activity

Modeling Have students obtain data from the newspaper that would be good to use on a line plot. Have them cut out the data, make a line plot, and then present it to the class.

Additional Answer

11b. No; in Ms. Martinez's class, the hours that the students talked on the phone were close to each other. (4 ± 2 hours). In Mr. Thomas's class, the students had a wider range of hours spent talking on the phone. The range was from 0 to 8 hours, with no hours clustered around a certain number.

Practice Masters, p. 11

2-2 NAME______________________ DATE __________

Practice

Student Edition Pages 78–83

Integration: Statistics Line Plots

1. The average number of participants per day in certain activities at Carl's Country Club are listed by age groups. Use the table to answer each question.

	Age			
	12–24	25–39	40–59	60 and over
Swimming	79	65	41	18
Bicycling	55	37	22	7
Jogging	51	31	13	2
Tennis	32	20	10	1
Golf	16	13	13	7

a. What is the most popular activity for people ages 60 and over? swimming

b. What is the least popular activity for people ages 40–59? tennis

c. What is the least popular activity for all ages? golf

d. Which age group has the most participants in all activities? ages 12–24

e. Which age group has the fewest participants in all activities? ages 60 and over

f. If you were age 60 or over and wanted to participate in an activity with a partner in your age group, which activity would you want to avoid? tennis; accept other answers with reasonable justification.

2. The income of a private investigator for each month for a recent calendar year is listed in the table at the right.

Month	Income in Dollars
January	6390
February	6390
March	7800
April	6390
May	6000
June	6000
July	5600
August	5130
September	7800
October	7800
November	7080
December	5130

a. Make a line plot of the data on the number line provided.

Income of a Private Investigator (hundreds of dollars)

50 55 60 65 70 75 80 85

b. In which months did the investigator earn the least? August, December

c. In which months did the investigator earn the most? March, September, October

d. Identify the heaviest cluster of data in any 500-unit span on the scale. 6000 to 6500

10. **Tornadoes** A tornado is a powerful, twisting windstorm. The winds of a tornado are the most dangerous winds that occur on Earth, with speeds of more than 200 miles per hour. The table below lists the average number of tornadoes that occurred per year in each state for a 30-year period.

TORNADO OCCURRENCE BY STATE, 1962–1991

STATE	AVG.	STATE	AVG.	STATE	AVG.	STATE	AVG.
AL	22	IN	20	NE	37	SC	10
AK	0	IA	36	NV	1	SD	29
AZ	4	KS	40	NH	2	TN	12
AR	20	KY	10	NJ	3	TX	139
CA	5	LA	28	NM	9	UT	2
CO	26	ME	2	NY	6	VT	1
CT	1	MD	3	NC	15	VA	6
DE	1	MA	3	ND	21	WA	2
FL	53	MI	19	OH	15	WV	2
GA	21	MN	20	OK	47	WI	21
HI	1	MS	26	OR	1	WY	12
ID	3	MO	26	PA	10		
IL	27	MT	6	RI	0		

Source: National Severe Storm Forecast Center

10a. See Solutions Manual.

10b. yes; 1, 2, and 3

a. Make a line plot of the average number of tornado occurrences.

b. Do any of the numbers appear clustered? If so, which ones?

c. Which state had the highest average? Why do you think it had the most tornadoes? **Texas, because of weather patterns and geographical location.**

11. **School** Mr. Thomas and Ms. Martinez each asked 10 students from their algebra classes how many hours they spent talking on the telephone last week. The results are shown in the line plot below. Use this plot to answer each question.

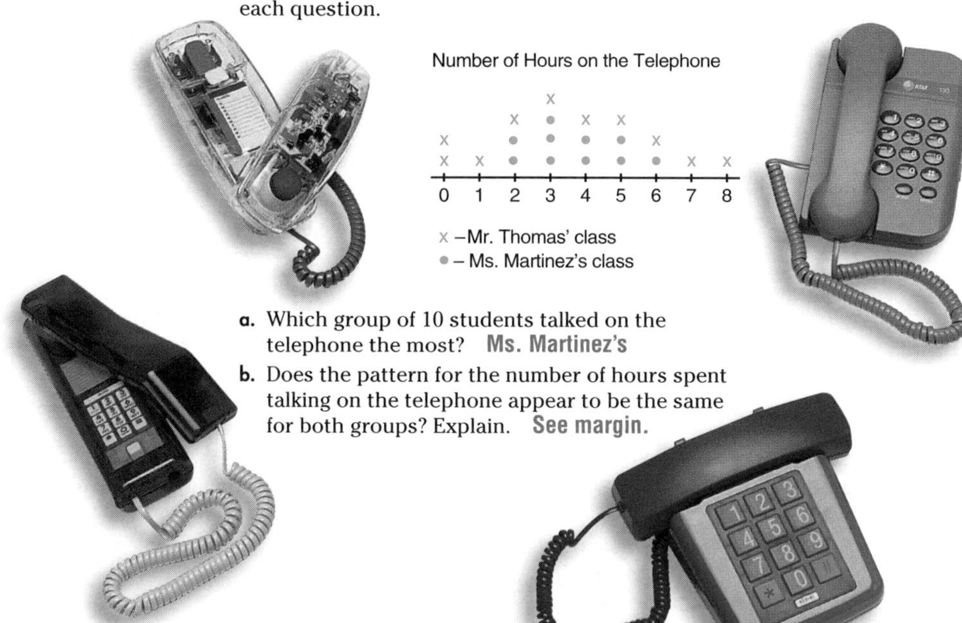

a. Which group of 10 students talked on the telephone the most? **Ms. Martinez's**

b. Does the pattern for the number of hours spent talking on the telephone appear to be the same for both groups? Explain. **See margin.**

Extension

Problem Solving The enrollment per class in each of the three high schools in Brunswick is given in the table. Using the enrollment for each high school as a separate set of data, make a line plot that differentiates between each set.

	9th	10th	11th	12th
North	500	450	432	405
Central	648	640	563	552
South	425	405	372	354

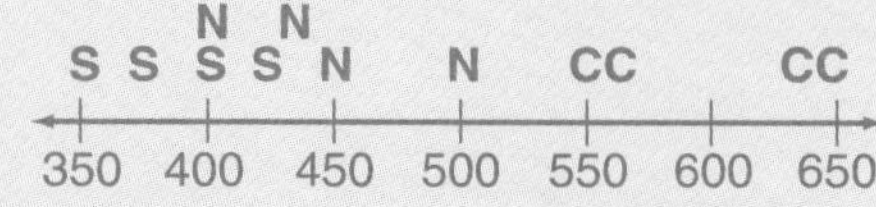

12. **Winter Olympics** The table below lists the medal standings for the 1994 Winter Olympics.

Winter Olympics, Medal Standings

Nation	Gold	Silver	Bronze	Total
Norway	10	11	5	26
Germany	9	7	8	24
Russia	11	8	4	23
Italy	7	5	8	20
United States	6	5	2	13
Canada	3	6	4	13
Switzerland	3	4	2	9
Austria	2	3	4	9
South Korea	4	1	1	6
Finland	0	1	5	6

a. Make a line plot that shows the number of gold medals won by the countries. **a–b. See margin.**

b. Write three questions that can be answered using the line plot.

Critical Thinking

13. Use the line plot in Exercise 11. Find the average number of hours that the students in Mr. Thomas and Ms. Martinez's classes spent on the telephone. Do the values support your answer to Exercise 11b? Explain. **See margin.**

Mixed Review

15. associative (×)

14. Graph {..., −5, −4, −3} on a number line. (Lesson 2–1) **See margin.**

15. Name the property illustrated by $6(jk) = (6j)k$. (Lesson 1–8)

16. Use the distributive property to find 15(124). (Lesson 1–7) **1860**

17. Solve $m = \frac{22 - 8}{7}$. (Lesson 1–5) **2**

18. Evaluate $np + st$ when $n = 7, p = 6, s = 4$, and $t = 5$. (Lesson 1–3) **62**

WORKING ON THE **In·ves·ti·ga·tion**

Refer to the Investigation on pages 68–69.

the Greenhouse Effect

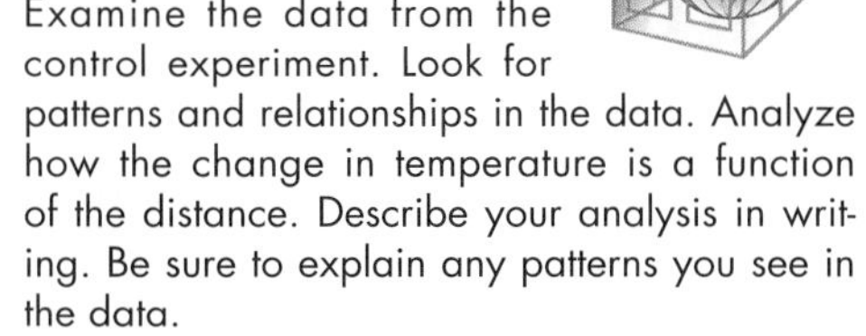

1 Examine the data from the control experiment. Look for patterns and relationships in the data. Analyze how the change in temperature is a function of the distance. Describe your analysis in writing. Be sure to explain any patterns you see in the data.

2 Examine the data from the greenhouse experiment. Look for patterns and relationships in the data. Analyze how the change in temperature is a function of the distance. Describe your analysis in writing. Use the terms *function, independent, dependent, domain,* and *range.*

3 Compare the two experiments. List the similarities and differences in the data and explain the factors that might cause them.

4 Think about the design of the two experiments. How were they different? Why do you think they were designed the way they were? How do they relate to the greenhouse effect?

Add the results of your work to your Investigation Folder.

Additional Answers

12a.

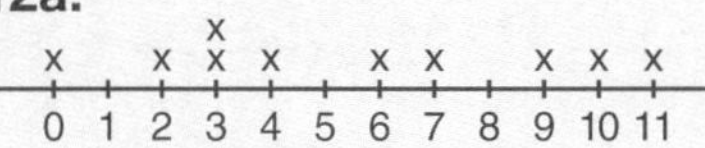

12b. Answers will vary. Sample answer:
1. Did any countries win the same number of gold medals?
2. How many countries won more than three gold medals?
3. How many gold medals did the countries win altogether?

13. Mr. Thomas's, 3.6 hours; Ms. Martinez's, 3.7 hours; yes

14. −5 −4 −3 −2 −1 0 1 2

In·ves·ti·ga·tion

Working on the Investigation

The Investigation on pages 68–69 is designed to be a long-term project that is completed over several days or weeks. Encourage students to keep their materials in their Investigation Folder as they work on the Investigation.

Enrichment Masters, p. 11

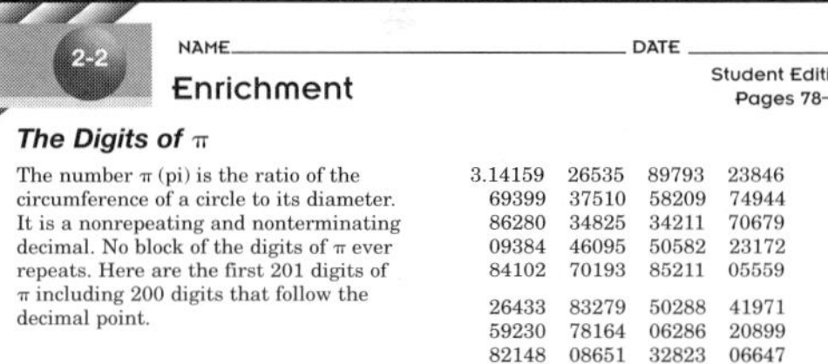

2-2 NAME ______ DATE ______

Enrichment — Student Edition Pages 78–83

The Digits of π

The number π (pi) is the ratio of the circumference of a circle to its diameter. It is a nonrepeating and nonterminating decimal. No block of the digits of π ever repeats. Here are the first 201 digits of π including 200 digits that follow the decimal point.

3.14159 26535 89793 23846
69399 37510 58209 74944
86280 34825 34211 70679
09384 46095 50582 23172
84102 70193 85211 05559

26433 83279 50288 41971
59230 78164 06286 20899
82148 08651 32823 06647
53594 08128 34111 74502
64462 29489 54930 38196

Solve each problem.

1. If each of the digits appeared with equal frequency, how many times would each digit appear in the first 200 places following the decimal point? 20

2. Complete this frequency distribution table for the first 200 digits of π that follow the decimal point.

Digit	Frequency (Tally Marks)	Frequency (Number)	Cumulative Frequency
0	𝍸 𝍸 𝍸 ////	19	19
1	𝍸 𝍸 𝍸 𝍸	20	39
2	𝍸 𝍸 𝍸 𝍸 ////	24	63
3	𝍸 𝍸 𝍸 𝍸	20	83
4	𝍸 𝍸 𝍸 𝍸 //	22	105
5	𝍸 𝍸 𝍸 𝍸	20	125
6	𝍸 𝍸 𝍸 /	16	141
7	𝍸 𝍸 //	12	153
8	𝍸 𝍸 𝍸 𝍸 ////	24	177
9	𝍸 𝍸 𝍸 𝍸 ///	23	200

3. Explain how the cumulative frequency column can be used to check a project like this one. The last number should be 200, the number of items being counted.

4. Which digit(s) appears most often? 2 and 8

5. Which digit(s) appears least often? 7

2-3A LESSON NOTES

NCTM Standards: 1–5

Objective
Use counters to add and subtract integers.

Recommended Time
Demonstration and discussion: 15 minutes; Exercises: 30 minutes

Instructional Resources
For each student or group of students
Student Manipulative Kit
- 20 red/yellow 2-sided counters

Modeling Mathematics Masters
- p. 2 (counters)
- p. 13 (integer mat)
- p. 20 (worksheet)

For teacher demonstration
Algebra and Geometry Overhead Manipulative Resources

1 FOCUS

Motivating the Lesson
Ask students how much money they would really have if they just cashed a check for $40, but owe their parents $25.

2 TEACH

Teaching Tip Caution students that $-1 - (-4)$ presents a problem. Having put down one negative counter, four counters cannot be removed. This can be solved by first putting down four zero pairs.

3 PRACTICE/APPLY

Assignment Guide

Core: 1–13
Enriched: 1–13

Additional Answer
13. Answers should include using the number line.

MODELING MATHEMATICS

A Preview of Lesson 2–3

2–3A Adding and Subtracting Integers

Materials:  counters ▭ integer mat

You can use counters to help you understand addition and subtraction of integers. In these activities, yellow counters represent positive integers, and red counters represent negative integers.

Rules for Integer Models

Rules for Integer Models	
A zero pair is formed by pairing one positive counter with one negative counter.	
You can remove or add zero pairs to a set because removing or adding zero does not change the value of the set.	

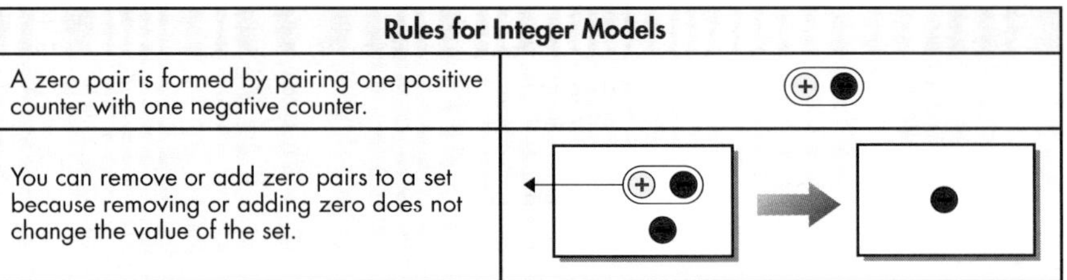

Activity 1 **Use counters to find the sum $-2 + 3$.**

Step 1 Place 2 negative counters and 3 positive counters on the mat.

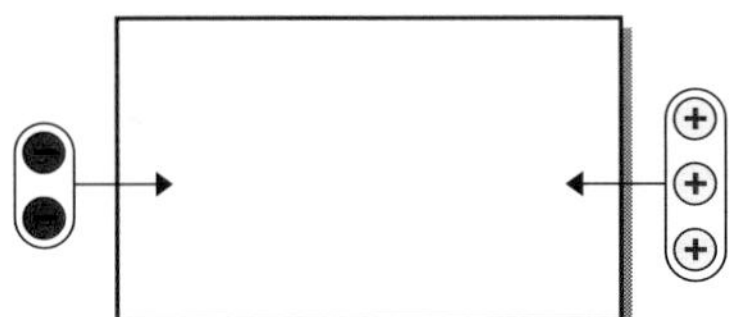

Step 2 Remove the 2 zero pairs. Since 1 positive counter remains, the sum is 1. Therefore, $-2 + 3 = 1$.

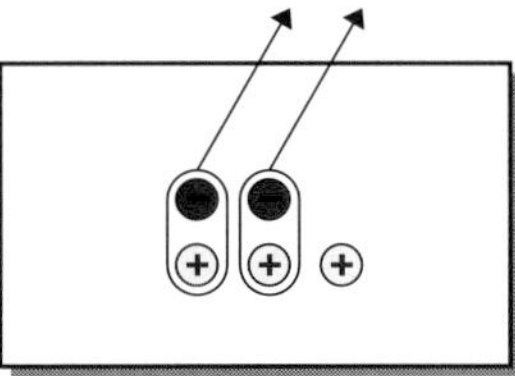

Activity 2 **Use counters to find the difference $3 - (-2)$.**

Step 1 Place 3 positive counters on the mat. There are no negative counters, so you cannot remove 2 negatives. Add 2 zero pairs to the mat. Adding zero pairs does not change the value of the set.

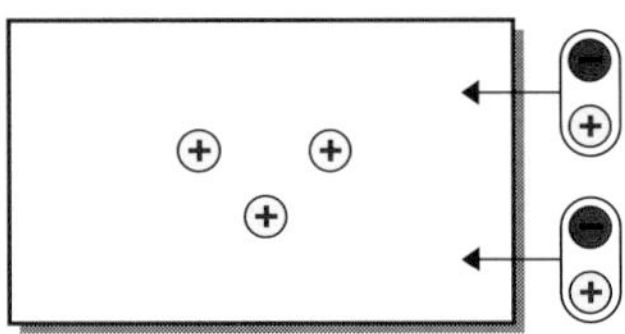

Step 2 Now remove 2 negative counters. Since 5 positive counters remain, the difference is 5. Therefore, $3 - (-2) = 5$.

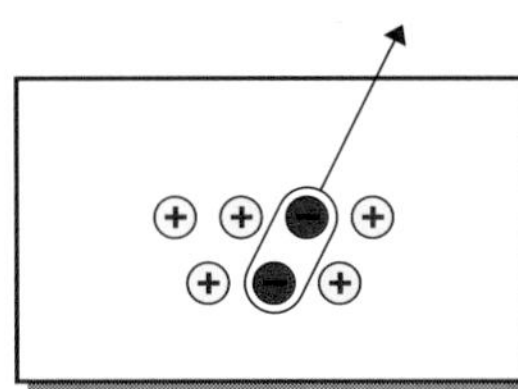

Model **Find each sum or difference by using counters.**

1. $4 + 2$ **6**
2. $4 + (-2)$ **2**
3. $-4 + 2$ **−2**
4. $-4 + (-2)$ **−6**
5. $4 - 2$ **2**
6. $-4 - (-2)$ **−2**
7. $4 - (-2)$ **6**
8. $-4 - 2$ **−6**

Draw **Tell whether each statement is *true* or *false*. Justify your answer with a drawing.**

9. $5 - (-2) = 3$ **false**
10. $-5 + 7 = 2$ **true**
11. $2 - 3 = -1$ **true**
12. $-1 - 1 = 0$ **false**

Write
13. Write a paragraph explaining how to find the sum of two integers without using counters. Be sure to include all possibilities. **See margin.**

4 ASSESS

Observing students working in cooperative groups is an excellent method of assessment.

***GLENCOE* Technology**

Interactive Mathematics Tools Software

This multimedia software provides an interactive lesson using a game in which students find the differences between two positive and/or negative integers. A **Computer Journal** gives students an opportunity to write about what they have learned.

For Windows & Macintosh

2–3 Adding and Subtracting Integers

What YOU'LL LEARN

- To find the absolute value of a number, and
- to add and subtract integers.

Why IT'S IMPORTANT

You can add and subtract integers to help you solve problems involving weather, business, and golf.

Space Science

On February 10, 1995, Dr. Bernard A. Harris, Jr. became the first African-American to walk in space. During the walk, he and Dr. Michael Foale, a British-born astronaut who is a U.S. citizen, were exposed to temperatures as cold as $-125°F$. The scheduled five-hour spacewalk to test thermal improvements to NASA spacesuits was cut short by 30 minutes because the astronauts experienced icy fingers.

Drs. Michael Foale and Bernard Harris in space

Usually, spacewalkers spend some of their time basking in the sun's rays where temperatures can be as high as 200°F. But throughout the spacewalk, Drs. Harris and Foale remained in the shadow of the shuttle Discovery and Earth, the coldest possible spot.

As you can see, spacewalkers must be able to survive in extreme temperatures. You can use a number line to determine the range of the temperature extremes. -125 and 200 are graphed on the number line below. Notice that $-125°F$ is 125 units from 0 and 200°F is 200 units from 0. So, the total number of units from -125 to 200 is $125 + 200$ or 325.

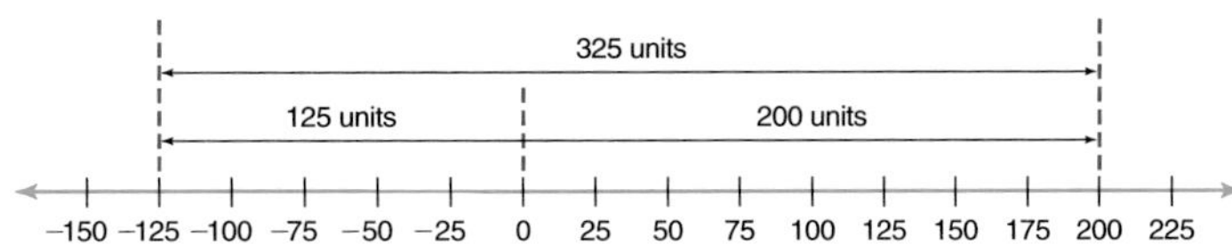

The temperature range from $-125°F$ to 200°F is 325°F.

You used the idea of **absolute value** to find the range from $-125°F$ to 200°F.

Definition of Absolute Value	**The absolute value of a number is its distance from zero on a number line.**

Since distance cannot be less than zero, absolute values are always greater than or equal to zero.

The symbol for the absolute value of a number is two vertical bars around the number.

Note that the absolute value bars can serve as grouping symbols.

$|-125| = 125$ is read *The absolute value of −125 equals 125.*

$|200| = 200$ is read *The absolute value of 200 equals 200.*

$|-7 + 6| = 1$ is read *The absolute value of the quantity −7 + 6 equals 1.*

2-3 LESSON NOTES

NCTM Standards: 1–5

Instructional Resources

- Study Guide Master 2-3
- Practice Master 2-3
- Enrichment Master 2-3
- Assessment and Evaluation Masters, p. 44
- Modeling Mathematics Masters, p. 73
- Real-World Applications, 4
- Tech Prep Applications Masters, p. 3

Transparency 2-3A contains the 5-Minute Check for this lesson; **Transparency 2-3B** contains a teaching aid for this lesson.

Recommended Pacing	
Standard Pacing	Day 4 of 15
Honors Pacing	Day 4 of 15
Block Scheduling*	Day 2 of 9
Alg. 1 in Two Years*	Days 7 & 8 of 28

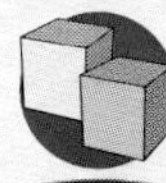

*For more information on pacing and possible lesson plans, refer to the *Block Scheduling Booklet* and *Algebra 1 in Two Years.*

1 FOCUS

5-Minute Check
(over Lesson 2-2)

True or False:

1. A line plot helps you see the distribution and clustering of data. **true**
2. A line plot would be more useful than a table when reviewing your test grades in algebra over the past semester. **false**

Use the line plot below to answer each question.

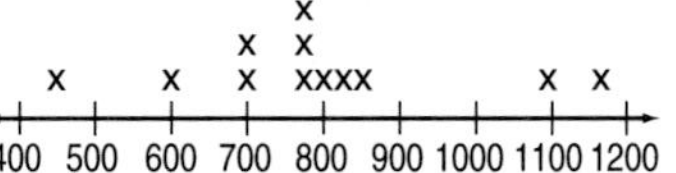

3. What was the greatest attendance? **almost 1200**
4. What appears to be the average attendance, rounded to the nearest hundred? **800**

Motivating the Lesson

Hands-On Activity A scuba diver swimming at a depth of 30 meters descends another 10 meters. Draw a diagram that shows the path of the diver. At what depth is the diver now swimming? **40 meters**

2 TEACH

Teaching Tip Emphasize that the absolute value of a number is always nonnegative. Absolute value will be used in the rules for adding signed numbers.

In-Class Examples

For Example 1
Evaluate each expression.

a. $|r + 2|$ if $r = -5$ **3**
b. $-|4 + t|$ if $t = -8$ **−4**
c. $-|7 + p|$ if $p = 3$ **−10**

For Example 2
Find each sum.

a. $-6 + (-2)$ **−8**
b. $-56 + 42$ **−14**
c. $38 + (-36)$ **2**

Teaching Tip Students may not remember the meaning of the term *addend.* Remind them that an addend is one of the numbers or terms being added.

You can evaluate expressions involving absolute value.

Example 1 **Evaluate $-|x + 6|$ if $x = -10$.**

$-|x + 6| = -|-10 + 6|$ *Substitution property of equality*
$= -|-4|$
$= -(4)$ *The absolute value of −4 is 4.*
$= -4$

You can use absolute value to add integers.

Same Signs

a. $4 + 3 = 7$ *Notice that the sign of each addend is positive. The sum is positive.*

b. $-4 + (-3) = -7$ *Notice that the sign of each addend is negative. The sum is negative.*

Different Signs

c. $6 + (-8) = -2$ *Notice that 8 − 6 = 2. Since the integer −8 has the greater absolute value, the sign of the sum is also negative.*

d. $-6 + 8 = 2$ *Notice that 8 − 6 = 2. Since the integer 8 has the greater absolute value, the sign of the sum is also positive.*

These examples suggest the following rules.

Adding Integers

To add integers with the *same sign,* add their absolute values. Give the result the same sign as the integers.

To add integers with *different signs,* subtract the lesser absolute value from the greater absolute value. Give the result the same sign as the integer with the greater absolute value.

Example 2 **Find each sum.**

a. $-10 + (-17)$

$-10 + (-17) = -(|-10| + |-17|)$ *Both numbers are negative,*
$= -(10 + 17)$ *so the sum is negative.*
$= -27$

b. $39 + (-22)$

$39 + (-22) = +(|39| - |-22|)$ *Subtract the absolute values. Since the*
$= +(39 - 22)$ *number with the greater absolute value,*
$= 17$ *39, is positive, the sum is positive.*

c. $-28 + 16$

$-28 + 16 = -(|-28| - |16|)$ *Subtract the absolute values. Since the*
$= -(28 - 16)$ *number with the greater absolute value,*
$= -12$ *28, is negative, the sum is negative.*

Notice that both +2 and −2 are 2 units away from 0. That is, −2 and 2 have the same absolute value.

Every positive integer can be paired with a negative integer. These pairs are called **opposites.** For example, the opposite of +2 is −2.

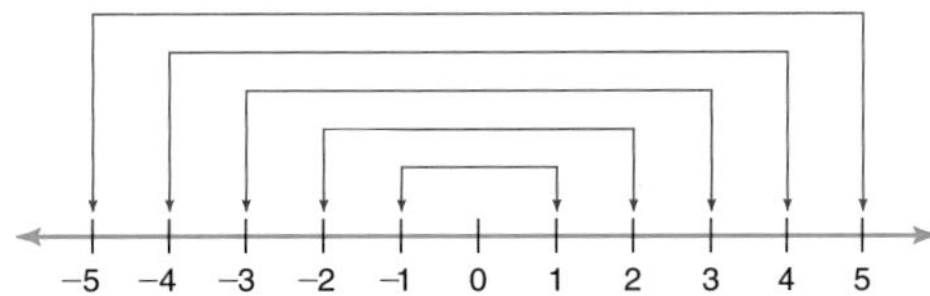

A number and its opposite are called **additive inverses** of each other. Look at the sum of each number and its additive inverse below. What pattern do you notice?

$2 + (-2) = 0 \qquad -67 + 67 = 0 \qquad 409 + (-409) = 0$

These examples suggest the following rule.

Additive Inverse Property	**For every number *a*, $a + (-a) = 0$.**

Additive inverses can be used when you subtract numbers.

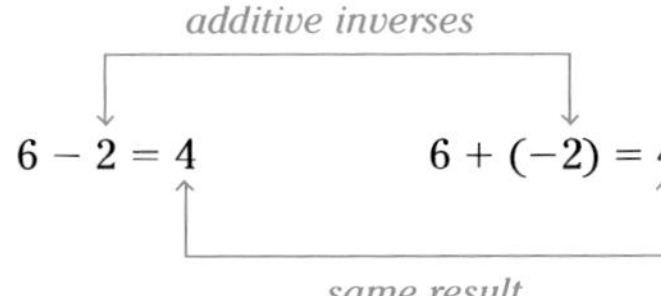

$6 - 2 = 4 \qquad 6 + (-2) = 4$

same result

It appears that subtracting a number is equivalent to adding its additive inverse.

Subtracting Integers	**To subtract a number, add its additive inverse.** **For any numbers *a* and *b*, $a - b = a + (-b)$.**

Example 3 **Find each difference.**

a. 6 − 14

Method 1

Use the rule.

To subtract 14, add −14.

$6 - 14 = 6 + (-14)$

$= -8$

Method 2

Use models.

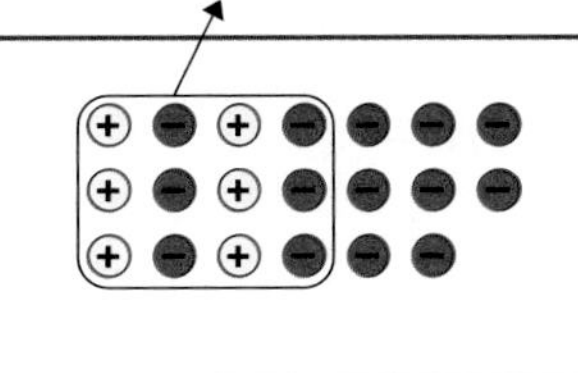

Place 6 positive counters on the mat. Add 14 negative counters. Form and remove six zero pairs.

$6 - 14 = -8$

Teaching Tip Point out to students that in $a + (-a)$, $-a$ can be read as "the opposite of a."

In-Class Example

For Example 3
Find each difference.

a. $-9 - (-4)$ **−5**
b. $-3 - 7$ **−10**
c. $12 - (-48)$ **60**

Classroom Vignette

"I put integers in the five triangles at the bottom of the large triangle and ask students to add or subtract integers as they progress up the triangle. The example shows an addition triangle."

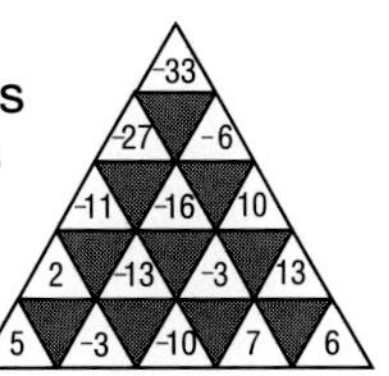

Jenny Brawner
Lee-Scott Academy
Auburn, Alabama

In-Class Examples

For Example 4

a. Add the matrices.

$$\begin{bmatrix} 800 & 650 & 240 \\ 1200 & 880 & 310 \end{bmatrix} + \begin{bmatrix} 650 & 410 & 170 \\ 800 & 600 & 200 \end{bmatrix}$$

$$\begin{bmatrix} 1450 & 1060 & 410 \\ 2000 & 1480 & 510 \end{bmatrix}$$

b. Subtract the matrices.

$$\begin{bmatrix} 800 & 650 & 240 \\ 1200 & 880 & 310 \end{bmatrix} - \begin{bmatrix} 650 & 410 & 170 \\ 800 & 600 & 200 \end{bmatrix}$$

$$\begin{bmatrix} 150 & 240 & 70 \\ 400 & 280 & 110 \end{bmatrix}$$

For Example 5

Simplify $4a - 9a$. $-5a$

3 PRACTICE/APPLY

Check for Understanding

Exercises 1–24 are designed to help you assess your students' understanding through reading, writing, speaking, and modeling. You should work through Exercises 1–6 with your students and then monitor their work on Exercises 7–24.

Error Analysis

Students often interpret $-x$ to be a negative number. Use a table of values to show otherwise.

x	$-x$	$\lvert x \rvert$	$-\lvert x \rvert$
3	−3	3	−3
−2	2	2	−2
−1.5	1.5	1.5	−1.5
0	0	0	0
2	−2	2	−2
−7	7	7	−7

On a TI-82, enter the following keystrokes.

(−) 12 −

(−) 8 ENTER −4

b. $-12 - (-8)$

Method 1

Use the rule.

To subtract −8, add its inverse, +8.

$$-12 - (-8) = -12 + (+8)$$
$$= -4$$

Method 2

Use a scientific calculator.

The +/− key on a scientific calculator, called the *change-sign key*, changes the sign of the number on the display. You can find $-12 - (-8)$ by using this key.

Enter: 12 +/− − 8 +/− = −4

c. $43 - (-26)$

$$43 - (-26) = 43 + 26$$
$$= 69$$

Check by using a calculator.

Enter: 43 − 26 +/− = 69 ✓

The plural of matrix is matrices.

A **matrix** is a rectangular arrangement of elements in rows and columns. Although matrices are sometimes used as a problem-solving tool, their importance extends to another branch of mathematics called **discrete mathematics.** Discrete mathematics deals with finite or discontinuous quantities. The distinction between continuous and discrete quantities is one that you have encountered throughout your life. Think of a staircase. You can slide your hand up the banister, but you have to climb the stairs one by one. The banister represents a continuous quantity, like the graph of {all numbers greater than 3}. However, each step represents a discrete quantity, like an element in a matrix, or a point in the graph of {2, 4, 7, 12}.

When entries in corresponding positions of two matrices are equal, the matrices are equal.

$$\begin{bmatrix} 4 & 2 \\ -2 & 5 \end{bmatrix} = \begin{bmatrix} 4 & 2 \\ -2 & 5 \end{bmatrix} \qquad \begin{bmatrix} 4 & 2 \\ -2 & 5 \end{bmatrix} \neq \begin{bmatrix} 2 & 4 \\ -2 & 1 \end{bmatrix}$$

You can add or subtract matrices only if they have the same number of rows and columns. You add and subtract matrices by adding or subtracting corresponding entries.

Example 4

The Paw Print at Gahanna Lincoln High School in Gahanna, Ohio, is a school store run by seniors in a yearlong marketing education class. The store is stocked with school items such as sweatshirts, T-shirts, school supplies, and snacks. The sales, cost of goods, and expenses for each semester in 1994 and 1995, rounded to the nearest dollar, are shown below.

Semester 1	Sales	Cost of Goods	Expenses
1994	\$47,981	\$32,627	\$12,999
1995	\$70,018	\$49,013	\$12,705

Semester 2	Sales	Cost of Goods	Expenses
1994	\$31,988	\$21,752	\$8666
1995	\$30,008	\$21,005	\$5445

FYI

When the Paw Print opened in 1976, sales totaled $7800. In 1995, sales hit $100,000. The profits are used to buy new equipment and merchandise. Donations are also made to charities.

a. Find the total sales, cost of goods, and expenses for each year at The Paw Print.

To find the total sales, cost of goods, and expenses for each year, add the matrices for semester 1 and semester 2.

$$\begin{bmatrix} \$47{,}981 + \$31{,}988 & \$32{,}627 + \$21{,}752 & \$12{,}999 + \$8666 \\ \$70{,}018 + \$30{,}008 & \$49{,}013 + \$21{,}005 & \$12{,}705 + \$5445 \end{bmatrix}$$

$$= \begin{bmatrix} \$79{,}969 & \$54{,}379 & \$21{,}665 \\ \$100{,}026 & \$70{,}018 & \$18{,}150 \end{bmatrix}$$

b. Find the store's gross income, or profit, by year. The gross income is the difference between the sales and the sum of the cost of goods and expenses.

Profit for 1994 $= \$79{,}969 - (54{,}379 + 21{,}665)$
$= \$3925$

Profit for 1995 $= \$100{,}026 - (70{,}018 + 18{,}150)$
$= \$11{,}858$

The rule for subtracting integers and the distributive property can be used to combine like terms.

Example 5 **Simplify $2x - 5x$.**

$2x - 5x = 2x + (-5x)$ *To subtract 5x, add its additive inverse, −5x.*
$= [2 + (-5)]x$ *Use the distributive property.*
$= -3x$

CHECK FOR UNDERSTANDING

Communicating Mathematics

1. See margin.
2–3. See margin for sample answers.

Study the lesson. Then complete the following.

1. **Explain** how a zero pair of counters illustrates the additive inverse property.
2. **Write** an integer on a piece of paper. Have a friend explain how to find its absolute value without a number line and without knowing the integer.
3. **Explain** to a friend how you would find each sum or difference below.
   a. $-3 + (-1)$ b. $5 + (-7)$ c. $4 - 9$ d. $6 - (-5)$
4. a. Find a value for x if $x = -x$. **0**
   b. Find a value for x if $|x| = -|x|$. **0**

MODELING MATHEMATICS

5. Write a statement that represents the model shown below. **−3 + 5 = 2**

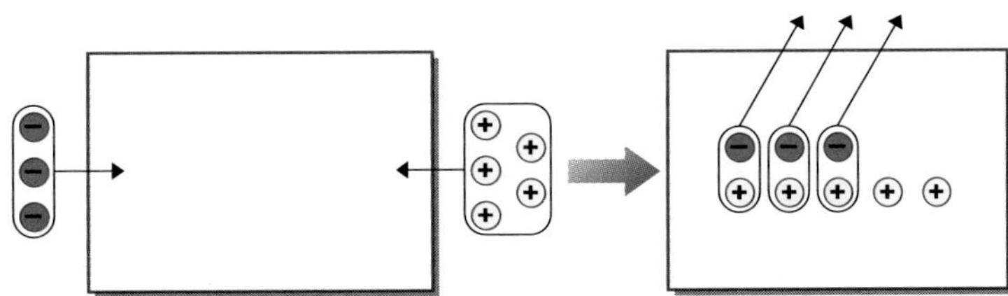

6. Find $-9 - (-4)$ by using counters. **−5**

Additional Answers

1. A zero pair of counters has equal amounts of positive and negative counters, therefore there are equal pairs of opposites and they will add to zero, which shows the additive inverse property.
2. Sample answer: −2; Start at zero. The absolute value is two places to the right.
3a. Sample answers: On a number line, start at zero. Move left three spaces to −3. Then move left one space to add −1. The sum is −4.
3b. On a number line, start at zero. Move right five spaces to +5. Then move left seven spaces to add −7. The sum is −2.
3c. On a number line, start at zero. Move right four spaces to +4. Then move nine spaces to the left to subtract 9. The difference is −5.
3d. On a number line, start at zero. Move right six spaces to +6. Subtracting −5 is the same as adding +5. So move right five more spaces to add +5. The sum is 11.

FYI

This is an increase of over twelve times in nineteen years. Investing in the stock market rarely produces this type of return.

Reteaching

Using Discussion Ask students the following questions.

1. What integer is its own additive inverse? **0**
2. A number x is added to the absolute value of x. If the sum is 0, what conclusion can you draw about the number x? **It is a negative integer.**

Assignment Guide

Core: 25–61 odd, 62, 63, 65, 67–78
Enriched: 26–60 even, 62–78

For **Extra Practice,** see p. 759.

The red A, B, and C flags, printed only in the Teacher's Wraparound Edition, indicate the level of difficulty of the exercises.

Additional Answers

22. $\begin{bmatrix} 7 & 0 \\ 4 & 6 \end{bmatrix}$

23. $\begin{bmatrix} -1 & 4 \\ 6 & -6 \end{bmatrix}$

58. $\begin{bmatrix} 3 & 7 \\ 1 & -5 \end{bmatrix}$

59. $\begin{bmatrix} 0 & 3 \\ 2 & 2 \end{bmatrix}$

60. $\begin{bmatrix} 4 & 12 \\ -6 & -1 \end{bmatrix}$

61. $\begin{bmatrix} -4 & -5 \\ -4 & -5 \\ 10 & -3 \end{bmatrix}$

Study Guide Masters, p. 12

NAME______________________ DATE ____________

Study Guide

Student Edition
Pages 85–92

Adding and Subtracting Integers

Use the following definitions, rules, and properties when adding or subtracting integers.

Definition, Rule, or Property		Example
Definition of Absolute Value	For any real number a: if $a > 0$, then $\|a\| = a$, and if $a < 0$, then $\|a\| = -a$. $\|0\| = 0$	$\|2\| = 2$ $\|-2\| = 2$
Adding Integers with the Same Sign	To add integers with the same sign, add their absolute values. Give the sum the same sign as the addends.	$3 + 2 = 5$ $-3 + (-2) = -5$
Adding Integers with Different Signs	To add integers with different signs, subtract the lesser absolute value from the greater absolute value. Give the result the same sign as the addend with the greater absolute value.	$-7 + 6 = -1$ $8 + (-4) = 4$
Additive Inverse Property	For every number a, $a + (-a) = 0$.	$-9 + 9 = 0$
Subtraction Rule	To subtract a number, add its additive inverse. For any numbers a and b, $a - b = a + (-b)$.	$8 - (-2) = 8 + 2 = 10$

You can use the distributive property and the addition and subtraction rules for integers to simplify expressions with like terms.

Example: Simplify $-6x - x + 9x$.

$$\begin{aligned} -6x - x + 9x &= -6x + (-1x) + 9x \\ &= [-6 + (-1) + 9]x \\ &= (-7 + 9)x \\ &= 2x \end{aligned}$$

Find each sum or difference.

1. $-17 + (-16)$ **−33**
2. $107 + (-40)$ **67**
3. $75 + 86$ **161**
4. $11 - 41$ **−30**
5. $15 - (-21)$ **36**
6. $-33 - (-17)$ **−16**
7. $3m + (-15m) - 11m$ **−23m**
8. $-6a + 15a + (-11a)$ **−2a**
9. $-9y + 20y - (-6y)$ **17y**

Evaluate each expression if x = −4, y = 3, and z = −7.

10. $456 + |z|$ **463**
11. $z + (-71) + |y|$ **−75**
12. $-11 - |x|$ **−15**
13. $31 - y - |x|$ **24**

Guided Practice

State the additive inverse and absolute value of each integer.

7. $+7$ **−7, 7**
8. -20 **20, 20**
9. 0 **0, 0**

Find each sum or difference.

10. $-12 + (-8)$ **−20**
11. $-11 - (-7)$ **−4**
12. $-36 + 15$ **−21**
13. $18 - 29$ **−11**
14. $|-6 - 4|$ **10**
15. $-|5 - 5|$ **0**

Simplify each expression.

16. $-16y - 5y$ **−21y**
17. $32c - (-8c)$ **40c**
18. $-9d + (-6d)$ **−15d**

Evaluate each expression if $x = -5$, $y = 3$, and $z = -6$.

19. $y - 7$ **−4**
20. $16 + z$ **10**
21. $|8 + x|$ **3**

Let $A = \begin{bmatrix} 3 & 2 \\ 5 & 0 \end{bmatrix}$ and $B = \begin{bmatrix} 4 & -2 \\ -1 & 6 \end{bmatrix}$. **22–23. See margin.**

22. Find $A + B$.
23. Find $A - B$.
24. **Space Science** Refer to the application at the beginning of the lesson. The Discovery astronauts were working in temperatures ranging from −125°F to 200°F.
    a. Which is colder, −125°F or 200°F? **−125°F**
    b. Write an equation that uses addition or subtraction to show how much colder one temperature is than the other. **200 − (−125) = 325**

EXERCISES

Practice

State the additive inverse and absolute value of each integer.

25. $+12$ **−12, 12**
26. -45 **45, 45**
27. -302 **302, 302**

Find each sum or difference.

28. $18 + (-16)$ **2**
29. $47 + (-47)$ **0**
30. $-9 + 32$ **23**
31. $0 - 32$ **−32**
32. $16 - (-23)$ **39**
33. $-5 + (-11)$ **−16**
34. $-104 + 16$ **−88**
35. $-9 + (-61)$ **−70**
36. $-18 - 4$ **−22**

B

37. $9 - 24$ **−15**
38. $-21 - (-24)$ **3**
39. $-13 - (-8)$ **−5**

Simplify each expression.

40. $29t - 17t$ **12t**
41. $17b - (-23b)$ **40b**
42. $-6w + (-13w)$ **−19w**
43. $6p + (-35p)$ **−29p**
44. $54y - 47y$ **7y**
45. $-5d + 31d$ **26d**

Evaluate each expression if $b = -4$, $d = 2$, and $p = -5$.

46. $b + 14$ **10**
47. $d - 6$ **−4**
48. $15 + p$ **10**
49. $d + (-17)$ **−15**
50. $d + 7$ **9**
51. $|p|$ **5**
52. $|6 + b|$ **2**
53. $|p - 3|$ **8**
54. $-|-22 + p|$ **−27**

Find each sum or difference. **57. −59** **58–61. See margin.**

C

55. $|-58 + (-41)|$ **99**
56. $|-93| - (-43)$ **136**
57. $-|-345 - (-286)|$

INTEGRATION
Discrete Mathematics

58. $\begin{bmatrix} 3 & 2 \\ 4 & -1 \end{bmatrix} + \begin{bmatrix} 0 & 5 \\ -3 & -4 \end{bmatrix}$
59. $\begin{bmatrix} 2 & 4 \\ -1 & 0 \end{bmatrix} + \begin{bmatrix} -2 & -1 \\ 3 & 2 \end{bmatrix}$
60. $\begin{bmatrix} 1 & 6 \\ -4 & -5 \end{bmatrix} - \begin{bmatrix} -3 & -6 \\ 2 & -4 \end{bmatrix}$
61. $\begin{bmatrix} 2 & 3 \\ 4 & 2 \\ 5 & -5 \end{bmatrix} - \begin{bmatrix} 6 & 8 \\ 8 & 7 \\ -5 & -2 \end{bmatrix}$

Alternative Learning Styles

Visual Use the graph below to complete the following table.

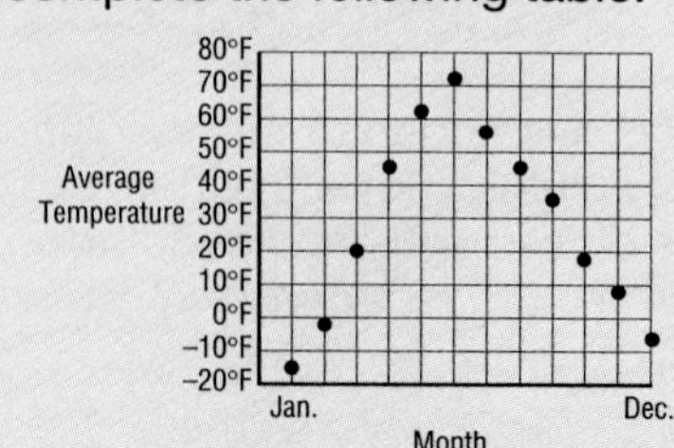

Month	Average Temp.	Increase from Previous Month
Jan.	−15	—
Feb.	−3	+12
March	+20	+23
April	+46	+26
May	+63	+17
June	+72	+9
July	+56	−12
Aug.	+46	−10
Sept.	+37	−9
Oct.	+18	−19
Nov.	+8	−10
Dec.	−7	−15

Critical Thinking

62. **Geometry** Use the number line below to find the length of segment QS if segment RT has a length of 15 units. **10 units**

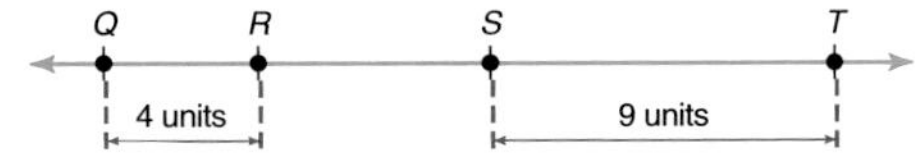

Applications and Problem Solving

Yokohama Landmark Tower

63. **Elevators** The 70-floor Yokohama Landmark Tower houses the world's fastest passenger elevators. Passengers travel from the second floor to the 69th floor in 40 seconds at an average speed of 28 miles per hour. If an employee rode the elevator up to his office on the 67th floor, then came down 43 floors to deliver a report, which floor is he on now? **24th floor**

64. **Puzzles** The sum of the numbers in each row, column, and diagonal of a magic square is the same. In the magic square at the right, the sum is 0. New magic squares can be formed by adding the same number to each entry. Complete each magic square.

1	2	−3
−4	0	4
3	−2	−1

a.

3	4	−1
−2	2	6
5	0	1

b.

−2	−1	−6
−7	−3	1
0	−5	−4

c.

−6	−5	−10
−11	−7	−3
−4	−9	−8

65. **Business** The manager of The Best Bagel Shop keeps a record of each type of bagel sold each day at their two stores. Two days of sales are shown below.

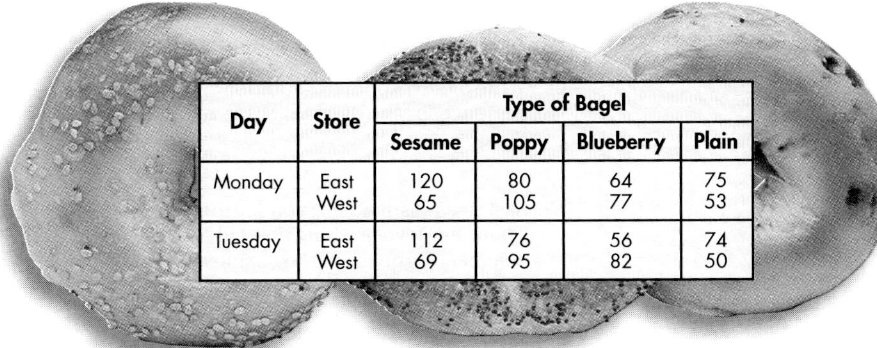

Day	Store	Type of Bagel			
		Sesame	Poppy	Blueberry	Plain
Monday	East West	120 65	80 105	64 77	75 53
Tuesday	East West	112 69	76 95	56 82	74 50

a. Write a matrix for each day's sales. **a–c. See margin.**

b. Find the sum of the two days' sales using matrix addition.

c. Subtract Tuesday's sales from Monday's sales by using matrix subtraction. What does this matrix represent?

66. **Golf** In golf, scores are based on *par.* Par 72 means that a golfer should hit the ball 72 times to complete 18 holes of golf. A score of 68, or 4 under par, is written as −4. A score of 2 over par is written as +2. At the Oldsmobile Classic LPGA tournament in June 1995, at a par-72 course in Michigan, Helen Alfredsson shot 65, 74, 69, and 71 during four rounds of golf.

a. Use integers to write Ms. Alfredsson's score for each round as over or under par. **−7, +2, −3, −1** **b. −7 + 2 + (−3) + (−1) = −9**

b. Add the integers to find Ms. Alfredsson's overall score.

c. Was Ms. Alfredsson's score under or over par? Would you want to have her score? Explain.

66c. Under; yes, it is better than par 72.

Tech Prep

Retail Manager Students who are interested in a career in retail management may wish to do further research into the use of mathematics in that occupation, as mentioned in Exercise 65. For more information on tech prep, see the *Teacher's Handbook*.

Additional Answers

65a. Monday

Columns: Sesame, Poppy, Blue, Plain; rows: East Store, West Store

$$\begin{bmatrix} 120 & 80 & 64 & 75 \\ 65 & 105 & 77 & 53 \end{bmatrix}$$

Tuesday

Columns: Sesame, Poppy, Blue, Plain; rows: East Store, West Store

$$\begin{bmatrix} 112 & 76 & 56 & 74 \\ 69 & 95 & 82 & 50 \end{bmatrix}$$

65b. Monday + Tuesday

Columns: Sesame, Poppy, Blue, Plain; rows: East Store, West Store

$$\begin{bmatrix} 232 & 156 & 120 & 149 \\ 134 & 200 & 159 & 103 \end{bmatrix}$$

65c. Monday − Tuesday

Columns: Sesame, Poppy, Blue, Plain; rows: East Store, West Store

$$\begin{bmatrix} 8 & 4 & 8 & 1 \\ -4 & 10 & -5 & 3 \end{bmatrix}$$

This matrix represents the difference between Monday's and Tuesday's sales in each category.

Practice Masters, p. 12

2-3 NAME________ DATE________

Practice

Student Edition Pages 85–92

Adding and Subtracting Integers

Find each sum or difference.

1. $-94 + (-27)$ **−121**
2. $47 + (-28)$ **19**
3. $38 + (-38)$ **0**
4. $31 - 12$ **19**
5. $8 - 79$ **−71**
6. $11 - 46$ **−35**
7. $-6 + 39$ **33**
8. $-15 - 65$ **−80**
9. $14 - (-8)$ **22**
10. $-19 + (-17)$ **−36**
11. $12 - (-63)$ **75**
12. $-61 - (-5)$ **−56**
13. $17 + (-23)$ **−6**
14. $47 + (-58)$ **−11**
15. $-7 - (-31)$ **24**
16. $69w - 18w$ **$51w$**
17. $12x + (-15x)$ **$-3x$**
18. $-9k + 7k$ **$-2k$**
19. $18p - 25p$ **$-7p$**
20. $-47y - 28y$ **$-75y$**
21. $8h + (-5h)$ **$3h$**
22. $-19z - 5z$ **$-24z$**
23. $-11l + (-6l)$ **$-17l$**
24. $z - 31z$ **$-30z$**
25. $31z - z$ **$30z$**
26. $-16m - (-38m)$ **$22m$**
27. $-8a + (-9a)$ **$-17a$**
28. $6 + (-8) + (-5)$ **−7**
29. $9 + (-3) + (-10)$ **−4**
30. $-2 + (-6) + (-7)$ **−15**

Evaluate each expression if $a = -2$, $b = 4$, and $c = -5$.

31. $a + 12$ **10**
32. $b - 8$ **−4**
33. $13 + c$ **8**
34. $b + (-12)$ **−8**
35. $c + 3$ **−2**
36. $a + b$ **2**
37. $b - c$ **9**
38. $a - c$ **3**
39. $a - b$ **−6**
40. $|a|$ **2**
41. $|3 + c|$ **2**
42. $-|b - 4|$ **0**

4 ASSESS

Closing Activity

Modeling Have students "walk out" addition and subtraction problems, using a large strip of squares labeled −9 to 9 with positives to the right and negatives to the left. Adopt the conventions shown below.

Direction	Face
positive negative	right (R) left (L)

Operation	Walk
addition subtraction	forward (F) backward (B)

Problem	Start	Face	Walk	Steps	End
5 − (+2)	5	R	B	2	3
5 − (−2)	5	L	B	2	7
5 + (+2)	5	R	F	2	7
−4 − (−6)	−4	L	B	6	2

Chapter 2, Quiz A (Lessons 2-1 through 2-3), is available in the *Assessment and Evaluation Masters,* p. 44.

Enrichment Masters, p. 12

NAME__________ DATE __________

Enrichment

Student Edition
Pages 85–92

Equivalent Sets

Two sets are **equal,** or identical, if they contain exactly the same elements. The order in which we name the elements is unimportant. Thus, $\{a, b, c, d\}$ and $\{c, a, d, b\}$ are equal sets. Two sets are **equivalent** if for every element of one set there is one and only one element in the other set; that is, there exists a one-to-one matching between the elements of the two sets. The one-to-one matchings below show that the sets are equivalent.

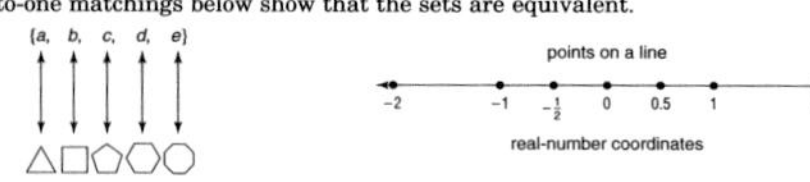

Consider these equivalent sets.

Set of whole numbers = $\{0, 1, 2, 3, 4, 5, \cdots\}$

Set of even whole numbers = $\{0, 2, 4, 6, 8, 10, \cdots\}$

Are there more whole numbers or more even whole numbers? Or might there be the same number of each, even though we have no counting number to tell how many?

A one-to-one matching of the whole numbers and the even whole numbers appears at the right. Each whole number n is matched with the even number $2n$, and each even number $2n$ is matched with the whole number n. Therefore, the two sets are equivalent. This means that there are as many even whole numbers as there are whole numbers.

$\{0, 1, 2, 3, 4, 5, \cdots, n, \cdots\}$
↕ ↕ ↕ ↕ ↕ ↕ ↕
$\{0, 2, 4, 6, 8, 10, \cdots, 2n, \cdots\}$

Use a one-to-one matching to show that the two sets are equivalent.

1. {Amy, Betsy, Carol, Dorothy} and {Al, Bob, Carl, David} {Amy, Betsy, Carol, Dorothy} ↕ {Al, Bob, Carl, David}
2. $\{1, 2, 3, 4, 5, \cdots\}$ and $\{3, 6, 9, 12, 15, \cdots\}$ $\{1, 2, 3, 4, 5, \cdots\}$ ↕ $\{3, 6, 9, 12, 15, \cdots\}$
3. $\{-1, -2, -3, -4, \cdots\}$ and $\{1, 2, 3, 4, \cdots\}$ $\{-1, -2, -3, -4, \cdots\}$ ↕ $\{1, 2, 3, 4, \cdots\}$
4. $\{1, 2, 3, 4, 5, \cdots\}$ and $\{1, 3, 5, 7, 9, \cdots\}$ $\{1, 2, 3, 4, 5, \cdots\}$ ↕ $\{1, 3, 5, 7, 9, \cdots\}$

Mixed Review

67. **Trees** The table below lists twenty different species of American trees and their highest recorded height in feet. (Lesson 2–2)

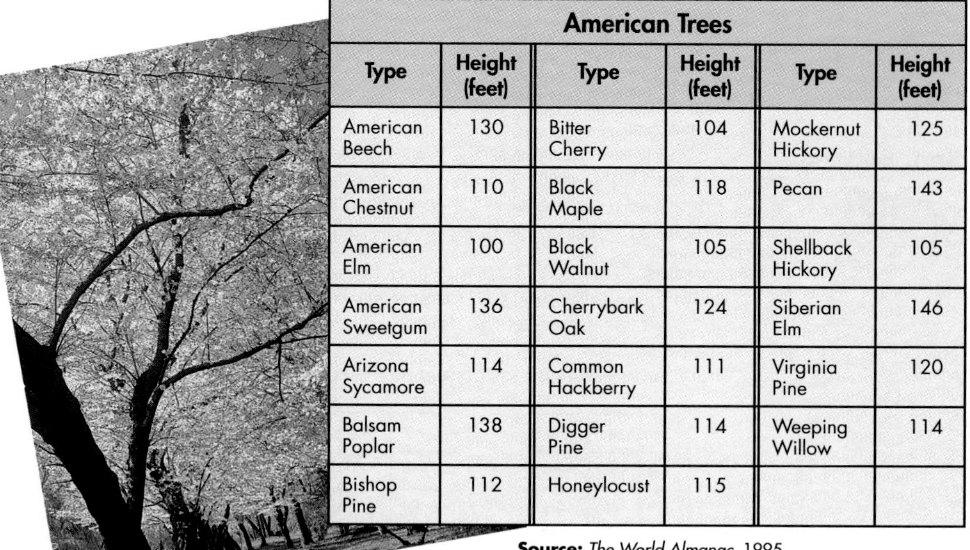

American Trees					
Type	**Height (feet)**	**Type**	**Height (feet)**	**Type**	**Height (feet)**
American Beech	130	Bitter Cherry	104	Mockernut Hickory	125
American Chestnut	110	Black Maple	118	Pecan	143
American Elm	100	Black Walnut	105	Shellback Hickory	105
American Sweetgum	136	Cherrybark Oak	124	Siberian Elm	146
Arizona Sycamore	114	Common Hackberry	111	Virginia Pine	120
Balsam Poplar	138	Digger Pine	114	Weeping Willow	114
Bishop Pine	112	Honeylocust	115		

Source: *The World Almanac,* 1995

a. What is the range of heights for the trees? **from 100 to 146**
b. Make a line plot showing the height of the trees. **See Solutions Manual.**
c. Do the heights cluster around a certain number? **yes; 114**
d. Which tree is the tallest? the shortest? **See margin.**

68. Write a corresponding addition sentence for the diagram at the right. (Lesson 2–1) **$4 + (-6) = -2$**

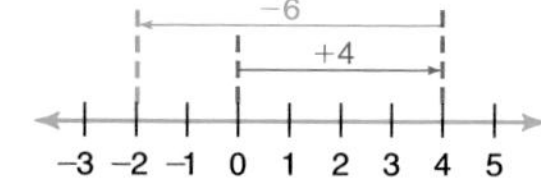

69. Sketch a reasonable graph for the following situation.
The water level in Craig Pond is extremely high at the beginning of the summer. Throughout the summer, the water level lowers steadily. By the end of the summer, the pond is almost dry. (Lesson 1–9) **See margin.**

70. Name the property illustrated by $3 + 4 = 4 + 3$. (Lesson 1–8)

70. commutative (+)

71. Simplify $23y^2 + 32y^2$. (Lesson 1–7) **$55y^2$**

Name the property illustrated by each statement. (Lesson 1–6)

72. $3abc \cdot 0 = 0$

73. If $12xy - 3 = 4$, then $4 = 12xy - 3$.

72. multiplicative property of 0

73. symmetric (=)

74. Solve $p = 6\frac{1}{4} + \frac{1}{2}$. (Lesson 1–5) **$6\frac{3}{4}$**

75. **Statistics** Write the stems that would be used for a plot of the following data: 29.1, 34.7, 20.05, 74.9, 64, 14.2, 84.9, 38.26, 20.9, 34, 59.42, 37.107, and 43.676. (Lesson 1–4) **See margin.**

76. Evaluate $19 + 5 \cdot 4$. (Lesson 1–3) **39**

77. **Patterns** Find the sixth term in the sequence 4, 8, 12, 16, ... (Lesson 1–2)

77. 24

78. Write an algebraic expression for the sum of n and 33. (Lesson 1–1)

78. $n + 33$

Extension

Reasoning Is $|a| + |b|$ always equal to $|a + b|$? Prove your answer with an example. **No. Examples will vary. Sample is $|5| + |-2|$ and $|5 + (-2)|$.**

Additional Answers

67d. Siberian Elm; American Elm

69.

Water Level

Time

75. 14, 20, 29, 34, 37, 38, 43, 59, 64, 74, 84

2–4 Rational Numbers

What YOU'LL LEARN

- To compare and order rational numbers, and
- to find a number between two rational numbers.

Why IT'S IMPORTANT

You need to know the values of rational numbers to solve problems involving consumerism and archaeology.

APPLICATION

Recycling

Many people all over the world recycle their aluminum cans in order to help our environment. The graph below shows the fraction of aluminum cans that have been recycled over the years.

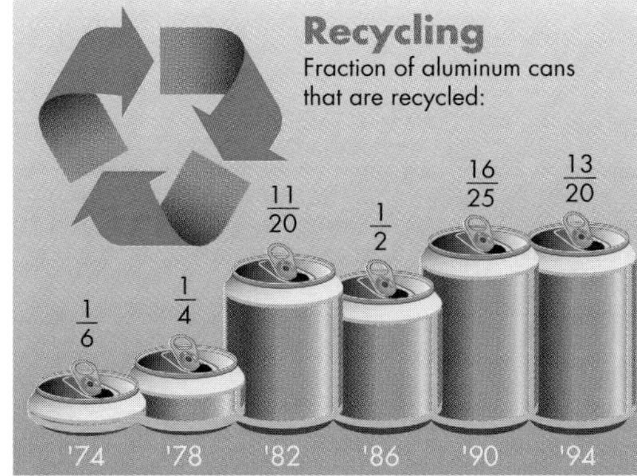

Source: The Aluminum Association

The numbers shown in the graph are examples of **rational numbers.**

Definition of a Rational Number	**A rational number is a number that can be expressed in the form $\frac{a}{b}$, where a and b are integers and b is not equal to 0.**

Examples of rational numbers expressed in the form $\frac{a}{b}$ are shown below.

Notice that all integers are rational numbers.

Rational Numbers	4	$-3\frac{3}{4}$	0.250	0	$0.33\overline{3}$
Form $\frac{a}{b}$	$\frac{4}{1}$	$-\frac{15}{4}$	$\frac{1}{4}$	$\frac{0}{1}$	$\frac{1}{3}$

Rational numbers can be graphed on a number line in the same manner as integers. The number line below is separated into fourths to show the graphs of some common fractions and decimals.

You can compare rational numbers by graphing them on a number line. Recall that a mathematical sentence that uses $<$ and $>$ to compare two expressions is called an *inequality*.

2-4 LESSON NOTES

NCTM Standards: 1–5

Instructional Resources

- Study Guide Master 2-4
- Practice Master 2-4
- Enrichment Master 2-4
- Graphing Calculator Masters, p. 2

Transparency 2-4A contains the 5-Minute Check for this lesson; **Transparency 2-4B** contains a teaching aid for this lesson.

Recommended Pacing	
Standard Pacing	Day 5 of 15
Honors Pacing	Day 5 of 15
Block Scheduling*	Day 3 of 9
Alg. 1 in Two Years*	Days 9 & 10 of 28

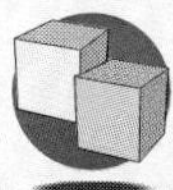

*For more information on pacing and possible lesson plans, refer to the *Block Scheduling Booklet* and *Algebra 1 in Two Years.*

1 FOCUS

5-Minute Check
(over Lesson 2-3)

Find each sum or difference.

1. 18 − 25 **−7**
2. −6 + 29 **23**
3. 9 + (−24) **−15**
4. (−18) + −(47) **−65**

Evaluate each expression if $b = -4$, $j = 2$, and $n = -7$.

5. $b + 12$ **8**
6. $j - n$ **9**
7. $-|b + (-9)|$ **−13**

Motivating the Lesson

Questioning Display a picture of an appliance such as a refrigerator from a sales flyer. Note the sale price on the back of the picture. Have students take turns asking questions about the sale price of the item using the terms "greater than" and "less than." Students should use responses from one another's questions to guess the price.

2 TEACH

Teaching Tip Students may benefit from making concrete comparisons involving height, weight, or length to prove that exactly one of the three comparison sentences is true. For example, use three students in the classroom whose heights are different to demonstrate the comparison sentences.

Teaching Tip The comparison property is sometimes called the *trichotomy property.*

In-Class Example

For Example 1
Replace each __?__ with <, >, or = to make each sentence true.

a. -11 __?__ $-20 + 9$ $=$
b. $-15 + 6$ __?__ 0 $<$
c. 5 __?__ $8 + (-12)$ $>$

Teaching Tip You may wish to introduce the symbols for "is not less than" ($\nless$) and "is not greater than" ($\ngtr$) to your students.

The following statements can be made about the graphs of -5, -2, $1\frac{1}{2}$ and 3.5 shown on the number line below.

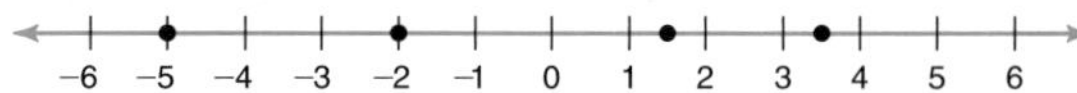

a. The graph of -2 is to the left of the graph of 3.5. $-2 < 3.5$
b. The graph of $1\frac{1}{2}$ is to the right of the graph of -5. $1\frac{1}{2} > -5$

These examples suggest the following rule.

Comparing Numbers on the Number Line	If a and b represent any numbers and the graph of a is to the left of the graph of b, then $a < b$. If the graph of a is to the right of the graph of b, then $a > b$.

If <, >, and = are used to compare two numbers, then the following property applies.

Comparison Property	For any two numbers a and b, exactly one of the following sentences is true. $a < b$ $a = b$ $a > b$

The symbols ≠, ≤, and ≥ can also be used to compare numbers. The chart below shows several inequality symbols and their meanings.

Symbol	Meaning
$<$	is less than
$>$	is greater than
$\neq$	is not equal to
$\leq$	is less than or equal to
$\geq$	is greater than or equal to

Example 

Replace each __?__ with <, >, or = to make each sentence true.

a. -75 __?__ 13

Since any negative number is less than any positive number, the true sentence is $-75 < 13$.

Note that the closed ends of < and > always point to the lesser number.

b. -14 __?__ $-22 + 9$

-14 __?__ $-22 + 9$

-14 __?__ -13 *Simplify.*

Since -14 is less than -13, the true sentence is $-14 < -22 + 9$.

c. $\frac{3}{8}$ __?__ $-\frac{7}{8}$

Since 3 is greater than -7, the true sentence is $\frac{3}{8} > -\frac{7}{8}$.

You can use **cross products** to compare two fractions with different denominators. When two fractions are compared, the cross products are the products of the terms on the diagonals.

GLENCOE Technology

Interactive Mathematics Tools Software

This multimedia software provides an interactive lesson that uses a number line to compare positive and negative rational numbers. A **Computer Journal** gives students an opportunity to write about what they have learned.

For Windows & Macintosh

Comparison Property for Rational Numbers	For any rational numbers $\frac{a}{b}$ and $\frac{c}{d}$, with $b > 0$ and $d > 0$: 1. if $\frac{a}{b} < \frac{c}{d}$, then $ad < bc$, and 2. if $ad < bc$, then $\frac{a}{b} < \frac{c}{d}$.

This property also holds if $<$ is replaced by $>$, $\leq$, $\geq$, or $=$.

Example  **Replace each _?_ with <, >, or = to make each sentence true.**

Alberto Salazar (1958–)

Alberto Salazar won the first marathon he entered, the 1980 New York Marathon. His track career included a world record and six U.S. records.

a. $\frac{7}{13}$ _?_ $\frac{4}{15}$

$7(15)$ _?_ $13(4)$

$105 > 52$

The true sentence is $\frac{7}{13} > \frac{4}{15}$.

b. $\frac{7}{8}$ _?_ $\frac{8}{9}$

$7(9)$ _?_ $8(8)$

$63 < 64$

The true sentence is $\frac{7}{8} < \frac{8}{9}$.

Every rational number can be expressed as a terminating or repeating decimal. You can use a calculator to write rational numbers as decimals.

Example 3 **Use a calculator to write the fractions $\frac{3}{8}$, $\frac{4}{5}$, and $\frac{1}{6}$ as decimals. Then write the fractions in order from least to greatest.**

It is helpful to know these commonly used fraction-decimal equivalencies by memory.

$\frac{1}{2} = 0.5$, $\frac{1}{3} = 0.\overline{3}$,

$\frac{1}{4} = 0.25$, $\frac{1}{5} = 0.2$,

$\frac{1}{8} = 0.125$

$\frac{3}{8} = 0.375$ — This is a terminating decimal.

$\frac{4}{5} = 0.8$ — This is also a terminating decimal.

$\frac{1}{6} = 0.16666\ldots$ or $0.1\overline{6}$ — This is a repeating decimal.

In order from least to greatest, the decimals are $0.1\overline{6}$, 0.375, 0.8. So, the fractions in order from least to greatest are $\frac{1}{6}$, $\frac{3}{8}$, $\frac{4}{5}$.

You can use a calculator to compare the cost per unit, or **unit cost,** of two similar items. Many people comparison shop to find the best buys at the grocery store. The item that has the lesser unit cost is the better buy.

unit cost = total cost ÷ number of units

Example 4 **Erica and Gabriella are running in a 5-kilometer race to raise money for The Multiple Sclerosis Foundation. They want to buy a sports beverage to drink after they finish the race. Which is the better buy?**

Consumerism

32 oz. $1.69

20 oz. $1.09

Use a calculator to find the unit cost of each brand. In each case, the unit cost is expressed in cents per ounce.

unit cost of 20-ounce bottle: 1.09 [÷] 20 [=] 0.0545

unit cost of 32-ounce bottle: 1.69 [÷] 32 [=] 0.0528

Since 0.053 < 0.055, the 32-ounce All Sport® drink is the better buy.

In-Class Examples

For Example 2
Replace each _?_ with <, >, or = to make each sentence true.

a. $\frac{21}{30}$ _?_ $\frac{7}{10}$ **=**

b. $\frac{6}{7}$ _?_ $\frac{9}{10}$ **<**

c. $\frac{-7}{5}$ _?_ $\frac{-11}{6}$ **>**

For Example 3
Write the fractions as decimals. Then write the fractions in order from least to greatest.

a. $\frac{3}{11}$, $\frac{4}{25}$, $\frac{5}{16}$ **$0.\overline{27}$, 0.16, 0.3125; $\frac{4}{25}$, $\frac{3}{11}$, $\frac{5}{16}$**

b. $\frac{7}{18}$, $\frac{1}{6}$, $\frac{9}{25}$ **$0.3\overline{8}$, $0.1\overline{6}$, 0.36; $\frac{1}{6}$, $\frac{9}{25}$, $\frac{7}{18}$**

For Example 4
If the prices on both drinks were lowered by $0.10, which would be the better buy? **Gatorade**

Teaching Tip To further emphasize the ease of using the comparison property, change the fractions to equivalent fractions with a common denominator.

$\frac{6}{11} = \frac{72}{132}$ $\quad$ $\frac{7}{12} = \frac{77}{132}$

Notice that $\frac{72}{132} < \frac{77}{132}$. Thus, $\frac{6}{11} < \frac{7}{12}$.

fabulous FIRSTS

The New York Marathon first took place in 1970. The Boston Marathon first took place in 1897. A woman finished the Boston Marathon for the first time in 1972.

Alternative Learning Styles

Kinesthetic Use items available at students' desks, such as pencils, paper, or backpacks, to demonstrate the comparison and transitive properties by comparing lengths of pencils, weights of backpacks, and thicknesses of paper. Point out to students that the transitive property applies to inequalities also. If $a \geq b$ and $b \geq c$, then $a \geq c$.

In-Class Example

For Example 5

Find a rational number between $\frac{1}{9}$ and $\frac{3}{5}$. **Sample answer: $\frac{16}{45}$**

3 PRACTICE/APPLY

Check for Understanding

Exercises 1–13 are designed to help you assess your students' understanding through reading, writing, speaking, and modeling. You should work through Exercises 1–6 with your students and then monitor their work on Exercises 7–13.

Additional Answers

3. **One way is to represent these numbers as decimals and then find three decimal values between 0.2 and 0.25, such as 0.21 or 0.24, and then convert these back into fraction form. Another way is to use the calculator to find the mean of the two decimal equivalents. That is one in-between value. Then find the mean of the first decimal and the new value for the second value and the mean of the second decimal and the new value for the third value.**
4. **A fraction cannot have zero in the denominator because that would make the fraction undefined.**
5. **Yes; since $\frac{a}{b} < \frac{c}{d}$ then what needs to be shown is that $\frac{a}{b} < \frac{a+c}{b+d} < \frac{c}{d}$.**
   If $\frac{a}{b} < \frac{c}{d}$, then $ad < bc$ and then adding ab to both sides, $ad + ab < bc + ab$.
   Then $a(d + b) < b(c + a)$ and $\frac{a}{b} < \frac{a+c}{b+d}$.
   The same method can be done to show the other side also. Have students show that side.

Can you always find another rational number between two rational numbers? One point that lies between any two points is their midpoint. Consider $\frac{1}{3}$ and $\frac{1}{2}$.

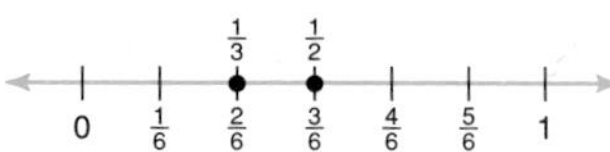

To find the average of two numbers, add the numbers and then divide by 2. Multiplying by $\frac{1}{2}$ is the same as dividing by 2.

To find the coordinate of the midpoint for $\frac{1}{3}$ and $\frac{1}{2}$, find the average, or mean, of the two numbers.

average: $\frac{1}{2}\left(\frac{1}{3} + \frac{1}{2}\right) = \frac{1}{2}\left(\frac{5}{6}\right)$ or $\frac{5}{12}$ ($\frac{1}{3}$, $\frac{5}{12}$, $\frac{1}{2}$)

This process can be continued indefinitely. The pattern suggests the **density property.**

Density Property for Rational Numbers	**Between every pair of distinct rational numbers, there are infinitely many rational numbers.**

Example 5 **Find a rational number between $-\frac{2}{5}$ and $-\frac{3}{8}$.**

Find the average of the two rational numbers.

Method 1

Use a pencil and paper.

$$\frac{1}{2}\left[-\frac{2}{5} + \left(-\frac{3}{8}\right)\right] = \frac{1}{2}\left[-\frac{16}{40} + \left(-\frac{15}{40}\right)\right]$$
$$= \frac{1}{2}\left(-\frac{31}{40}\right)$$
$$= -\frac{31}{80}$$

Method 2

Use a scientific calculator.

Enter: (2 ÷ 5 +/− + 3 ÷ 8 +/−) ÷ 2 = −0.3875

A rational number between $-\frac{2}{5}$ and $-\frac{3}{8}$ is $-\frac{31}{80}$ or −0.3875.

CHECK FOR UNDERSTANDING

Communicating Mathematics

Study the lesson. Then complete the following.

1. **Write** three examples of rational numbers.
2. **State** any assumptions you can make about y and z if $x < y$ and $x < z$.
3. **Show** two ways to find three rational numbers between $\frac{1}{5}$ and $\frac{1}{4}$.
4. **Explain** why the definition of a rational number states that b cannot be zero.
5. **You Decide** To find a number between $\frac{3}{4}$ and $\frac{6}{7}$, John added as follows. $\frac{3+6}{4+7} = \frac{9}{11}$; $\frac{9}{11}$ is between $\frac{3}{4}$ and $\frac{6}{7}$. He claimed that this method will always work. Do you agree? Explain.

1. Sample answers: $\frac{1}{2}$, $\frac{10}{3}$, 0.32
2. none
3–5. See margin.

Reteaching

Translating Expressions Write an algebraic expression for each verbal expression.

1. The cost is under \$195. **$c < 195$**
2. x is positive. **$x > 0$**
3. x is nonnegative. **$x \geq 0$**
4. Tom is at least 18 years old. **$T \geq 18$**
5. The stream is no more than 50 feet across. **$s \leq 50$**

6. **Assess Yourself** Describe two situations in which you would prefer to use decimals instead of fractions. **See students' work.**

Guided Practice

Replace each ? with <, >, or = to make each sentence true.

7. $-4 \underline{\ ?\ } 8$ $<$
8. $-5 \underline{\ ?\ } 0 - 3$ $<$
9. $\frac{5}{14} \underline{\ ?\ } \frac{25}{70}$ $=$

Write the numbers in each set in order from least to greatest.

10. $\frac{2}{3}, \frac{1}{6}, \frac{1}{2}$ $\mathbf{\frac{1}{6}, \frac{1}{2}, \frac{2}{3}}$
11. $2.5, \frac{3}{4}, -0.5, \frac{7}{8}$ $\mathbf{-0.5, \frac{3}{4}, \frac{7}{8}, 2.5}$
12. Which is a better buy: a 6-ounce can of tuna for \$1.59 or a package of 3 three-ounce cans for \$2.19?
13. Find a number between $\frac{1}{2}$ and $\frac{5}{7}$. **See students' work.**

EXERCISES

Practice

Replace each ? with <, >, or = to make each sentence true.

14. $-3 \underline{\ ?\ } 5$ $<$
15. $-1 \underline{\ ?\ } -4$ $>$
16. $-6 - 3 \underline{\ ?\ } -9$ $=$
17. $5 \underline{\ ?\ } 8.4 - 1.5$ $<$
18. $4 \underline{\ ?\ } \frac{16}{3}$ $<$
19. $\frac{8}{15} \underline{\ ?\ } \frac{9}{16}$ $<$

20. $\frac{14}{5} \underline{\ ?\ } \frac{25}{13}$ $>$
21. $\frac{4}{3}(6) \underline{\ ?\ } 4\left(\frac{3}{2}\right)$ $>$
22. $\frac{0.4}{3} \underline{\ ?\ } \frac{1.2}{8}$ $<$

Write the numbers in each set in order from least to greatest.

23–28. See margin.

23. $\frac{6}{7}, \frac{2}{3}, \frac{3}{8}$
24. $-\frac{4}{15}, -\frac{6}{17}, -\frac{3}{16}$
25. $\frac{4}{14}, \frac{3}{23}, \frac{8}{42}$

26. $6.7, -\frac{5}{7}, \frac{6}{13}$
27. $0.2, -\frac{2}{5}, -0.2$
28. $\frac{4}{5}, \frac{9}{10}, 0.7$

Which is the better buy?

29. a 16-ounce drink for \$0.59 or a 20-ounce drink for \$0.89
30. a 32-ounce bottle of shampoo for \$3.59 or a 64-ounce bottle for \$6.99
31. a package of 48 paper plates for \$2.39 or a package of 75 paper plates for \$3.29

Find a number between the given numbers.

32–37. Sample answers are given.

32. $\frac{2}{5}$ and $\frac{7}{2}$ $\mathbf{\frac{1}{2}}$
33. $\frac{19}{30}$ and $\frac{31}{45}$ $\mathbf{\frac{2}{3}}$
34. $-\frac{2}{15}$ and $-\frac{6}{3}$ $\mathbf{-1}$
35. Name a fraction between $\frac{1}{4}$ and $\frac{1}{2}$ whose denominator is 20. $\mathbf{\frac{7}{20}}$
36. Name a fraction between $\frac{1}{3}$ and $\frac{5}{6}$ whose denominator is 12. $\mathbf{\frac{6}{12}}$
37. Name a fraction between -0.5 and $\frac{1}{3}$ whose denominator is 6. $\mathbf{\frac{1}{6}}$

Assignment Guide

Core: 15–41 odd, 43–49
Enriched: 14–36 even, 38–49
All: Self Test, 1–10

For **Extra Practice,** see p. 760.

The red A, B, and C flags, printed only in the Teacher's Wraparound Edition, indicate the level of difficulty of the exercises.

Additional Answers

23. $\mathbf{\frac{3}{8}, \frac{2}{3}, \frac{6}{7}}$
24. $\mathbf{-\frac{6}{17}, -\frac{4}{15}, -\frac{3}{16}}$
25. $\mathbf{\frac{3}{23}, \frac{8}{42}, \frac{4}{14}}$
26. $\mathbf{-\frac{5}{7}, \frac{6}{13}, 6.7}$
27. $\mathbf{-\frac{2}{5}, -0.2, 0.2}$
28. $\mathbf{0.7, \frac{4}{5}, \frac{9}{10}}$

Study Guide Masters, p. 13

NAME ________________ DATE ________

Study Guide

Student Edition Pages 93–99

Rational Numbers

Definition of a Rational Number	A rational number is a number that can be expressed in the form $\frac{a}{b}$, where a and b are integers and $b \neq 0$.

You can compare rational numbers by graphing them on a number line.

Comparing Numbers on the Number Line	If a and b represent any numbers and the graph of a is to the left of the graph of b, then $a < b$. If the graph of a is to the right of the graph of b, then $a > b$.
Comparison Property	For any two numbers a and b, exactly one of the following sentences is true. $a < b$ $a = b$ $a > b$

Example 1: $-3\frac{1}{2} < -\frac{1}{2}$ The graph of $-3\frac{1}{2}$ is to the left of the graph of $-\frac{1}{2}$.

Example 2: $-2\frac{1}{4} > -3\frac{1}{4}$ The graph of $-2\frac{1}{4}$ is to the right of the graph of $-3\frac{1}{4}$.

Example 3: Replace ? with <, >, or = to make the sentence true.
$-15 \underline{\ ?\ } -3$
$-15 < -3$ Since -15 is to the left of -3 on a number line, -15 is less than -3.

The symbols $\neq$, $\leq$ and $\geq$ can also be used to compare numbers and are called **inequality symbols.**

You can use **cross products** to compare two fractions with different denominators.

Comparison Property for Rational Numbers	For any rational numbers $\frac{a}{b}$ and $\frac{c}{d}$, with $b > 0$ and $d > 0$: 1. if $\frac{a}{b} < \frac{c}{d}$ then $ad < bc$, and 2. if $ad < bc$, then $\frac{a}{b} < \frac{c}{d}$.

This property also holds if < is replaced by >, $\leq$, $\geq$, or =.

A property that is true for rational numbers but is not true for integers is the **density property.**

Density Property for Rational Numbers	Between every pair of distinct rational numbers, there is another rational number.

Replace each ? with <, >, or = to make each sentence true.

1. $-4 \underline{\ ?\ } 10$ $<$
2. $\frac{-29}{2} \underline{\ ?\ } -28.5 + 14$ $=$
3. $-5 - 6 \underline{\ ?\ } -12 - 1$ $>$

Write the numbers in each set in order from least to greatest.

4. $3\frac{1}{3}, \frac{5}{8}, 0.4$ $0.4, \frac{5}{8}, 3\frac{1}{3}$
5. $-\frac{3}{2}, \frac{1}{4}, 0.2$ $-\frac{3}{2}, 0.2, \frac{1}{4}$

Find a number between the given numbers. See students' work.

6. $\frac{1}{2}$ and $\frac{7}{9}$ $\frac{23}{36}$
7. $\frac{7}{6}$ and $\frac{9}{8}$ $\frac{55}{48}$
8. $\frac{9}{17}$ and $\frac{2}{5}$ $\frac{79}{170}$

Critical Thinking

38. Three numbers a, b, and c satisfy the following conditions: $b - c < 0$, $a - b > 0$, and $c - a < 0$. Which one is the greatest? **a**

39. Find the coordinates of E, G, and H if the distances between the points are equal.

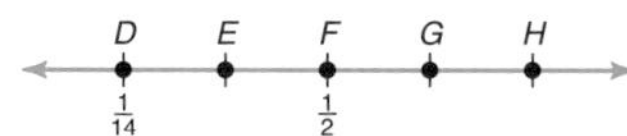

39. $E = \frac{4}{14}$ or $\frac{2}{7}$; $G = \frac{10}{14}$ or $\frac{5}{7}$; $H = \frac{13}{14}$

Applications and Problem Solving

40. **Personal Computers** Study the graph at the right. Write an inequality that compares the total number of PCs shipped for home use in 1995 to those shipped in 1996. **$8.2 < 9.8$**

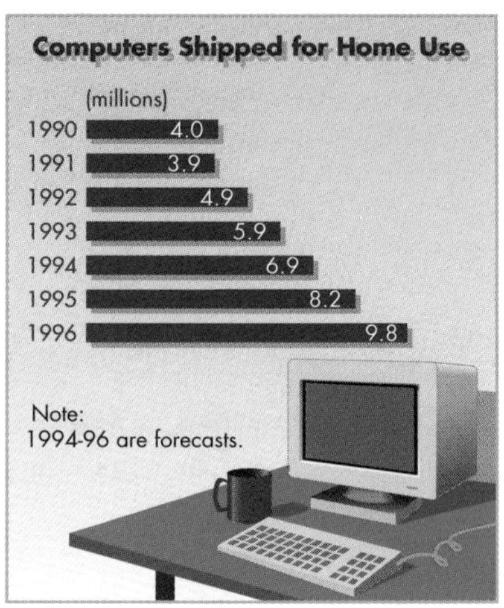

Sources: Dataquest, LINK

41. **Aviation** When building or servicing airplanes, an airplane mechanic sometimes needs to drill holes. Drill bit sizes are given in fractions, yet many blueprints list hole diameters in decimals. One blueprint calls for a hole of 0.391 inches. Find the fractional drill size required if the hole must be drilled $\frac{1}{64}$-inch undersize. **0.375 inch**

42. **Archaeology** To calculate a person's height, archaeologists measure the lengths of certain bones. The bones measured are the femur or thigh bone F, the tibia or leg bone T, the humerus or upper arm bone H, and the radius or lower arm bone R. When the length of one of these bones is known, scientists can use one of the formulas below to determine the person's height h in centimeters. **a. 172.4 cm** **b. 179.6 cm**

Males	Females
$h = 69.089 + 2.238F$	$h = 61.412 + 2.317F$
$h = 81.688 + 2.392T$	$h = 72.572 + 2.533T$
$h = 73.570 + 2.970H$	$h = 64.977 + 3.144H$
$h = 80.405 + 3.650R$	$h = 73.502 + 3.876R$

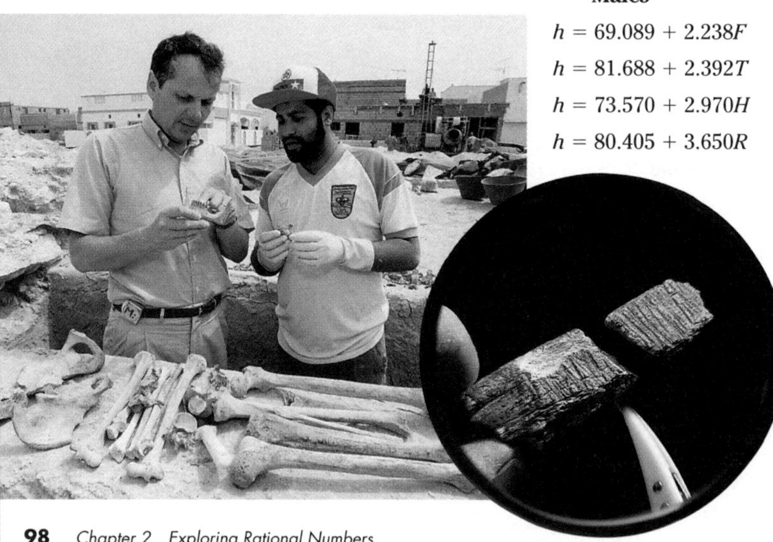

a. The femur of a 37-year old woman measured 47.9 cm. Use a calculator to find the height of the woman to the nearest tenth of a centimeter.

b. The humerus of a 49-year old man measured 35.7 cm. Use a calculator to find the height of the man to the nearest tenth of a centimeter.

Teaching Tip You may wish to have students measure their own tibia or radius and apply the appropriate formula to determine if it provides a reasonably accurate measurement of their height.

Practice Masters, p. 13

2-4 NAME__________ DATE ________

Practice — Student Edition Pages 93–99

Rational Numbers

Replace each ? with <, >, or = to make each sentence true.

1. $\frac{2}{3}$? $\frac{5}{8}$ >
2. $\frac{3}{4}$? $\frac{5}{7}$ >
3. $\frac{4}{2}$? $\frac{5}{2}$ <
4. $\frac{7}{11}$? $\frac{2}{3}$ <
5. $\frac{3}{14}$? $\frac{15}{70}$ =
6. $\frac{4}{9}$? $\frac{1}{2}$ <
7. $\frac{0}{1}$? $\frac{0}{2}$ =
8. $\frac{10}{10}$? $\frac{100}{100}$ =
9. $\frac{9}{7}$? $\frac{7}{4}$ <
10. $\frac{5}{7}$? $\frac{7}{9}$ <
11. $\frac{33}{40}$? $\frac{17}{24}$ >
12. $\frac{12}{6}$? $\frac{18}{9}$ =
13. $\frac{2}{3}$? $\frac{9}{16}$ >
14. $\frac{3}{16}$? $\frac{5}{12}$ <
15. $\frac{1}{5}$? $\frac{1}{2}$ <
16. $\frac{7}{8}$? $\frac{15}{16}$ <
17. $\frac{100}{10}$? $\frac{40}{4}$ =
18. $\frac{3}{4}$? $\frac{11}{16}$ >
19. $\frac{1.1}{3}$? $\frac{3.3}{9}$ =
20. $\frac{1}{8}$? $\frac{2}{3}$ <
21. $\frac{10}{29}$? $\frac{15}{43}$ <

Write the numbers in order from least to greatest.

22. $\frac{18}{29}, \frac{24}{39}, \frac{3}{5}$ $\frac{3}{5}, \frac{24}{39}, \frac{18}{29}$
23. $\frac{10}{29}, \frac{5}{14}, \frac{15}{43}$ $\frac{10}{29}, \frac{15}{43}, \frac{5}{14}$
24. $\frac{3}{4}, \frac{5}{8}, \frac{5}{6}$ $\frac{5}{8}, \frac{3}{4}, \frac{5}{6}$
25. $0.25, \frac{7}{8}, \frac{3}{8}$ $0.25, \frac{3}{8}, \frac{7}{8}$
26. $0.85, 0.84, \frac{19}{22}$ $0.84, 0.85, \frac{19}{22}$
27. $\frac{16}{19}, \frac{18}{21}, \frac{17}{20}$ $\frac{16}{19}, \frac{17}{20}, \frac{18}{21}$
28. $\frac{5}{6}, 0.8, \frac{2}{3}$ $\frac{2}{3}, 0.8, \frac{5}{6}$
29. $\frac{6}{6}, 0, \frac{1}{2}$ $0, \frac{1}{2}, \frac{6}{6}$

Find a number between the given numbers. Sample answers are given.

30. $\frac{1}{5}$ and $\frac{2}{3}$ $\frac{13}{30}$
31. $\frac{9}{10}$ and $\frac{1}{4}$ $\frac{23}{40}$
32. $\frac{18}{5}$ and $\frac{6}{7}$ $\frac{78}{35}$
33. $\frac{1}{2}$ and $\frac{1}{9}$ $\frac{11}{36}$
34. $\frac{2}{3}$ and $\frac{3}{2}$ $\frac{13}{12}$
35. $\frac{6}{1}$ and $\frac{2}{2}$ $\frac{7}{2}$

Tech Prep

Airplane Mechanic Students who are interested in a career in aviation mechanics may wish to do further research into the use of mathematics in that occupation, as mentioned in Exercise 41. For more information on tech prep, see the *Teacher's Handbook*.

Mixed Review

43. State the additive inverse and absolute value of +9. (Lesson 2–3) **−9, 9**

44. **Statistics** State the scale you would use to make a line plot for the following data. Then draw the line plot. (Lesson 2–2)
145, 130, 135, 150, 145, 145, 140, 130, 145, 150, 130

45. Graph {−3, −2, 2, 3} on a number line. (Lesson 2–1) **44–45. See margin.**

46. **Soccer** Bryce, Maria, and Holly are three offensive players for the Spazmatics soccer team. The graph at the right describes the players' offensive skills in their last season. (Lesson 1–9)

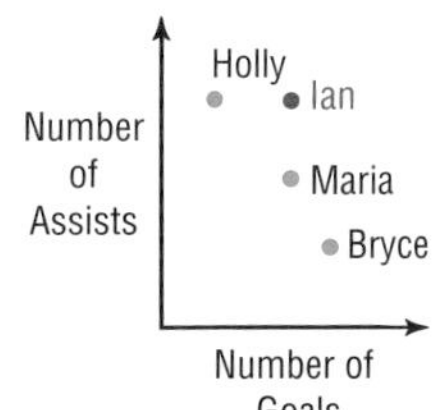

a. Who scored the most goals? **Bryce**

b. Who helped the team most with goals and assists overall? **Maria**

c. Another player, Ian, had the same number of goals as Maria and the same number of assists as Holly. Describe where on the graph a point should go that would show his offensive skills. **See students' work.**

47. Simplify $\frac{2}{5}m + \frac{1}{5}(6n + 3m) + \frac{1}{10}(8n + 15)$. (Lesson 1–6)
$m + 2n + \frac{3}{2}$

48. **Patterns** Find the next three terms in the sequence 1, 3, 9, 27, ... (Lesson 1–2) **81, 243, 729**

49. Write a verbal expression for $2x^2 + 6$. (Lesson 1–1)
two times *x* squared plus six

SELF TEST

Graph each set of numbers on a number line. (Lesson 2–1)

1. {−3, 0, 2}
2. {numbers greater than −2}

1–2. See margin.

Find each sum or difference. (Lesson 2–3)

3. −9 + (−8) **−17**
4. 6 + (−15) **−9**
5. 23 − (−32) **55**
6. −12 − 4 **−16**

7. **Mountain Bikes** The table at the right shows prices of several mountain bikes. (Lesson 2–2)

a. Make a line plot of the data. **See Solutions Manual.**

b. Do the numbers cluster around any certain price? **yes; $270**

Bike	Price
Trek 820	$325
Mongoose Threshold	270
Giant Yukon	370
Giant Rincon	300
Bianchi Timber Wolf	270
Schwinn Clear Creek	320
Schwinn Sidewinder	260
Specialized Hardrock	270
Raleigh M40GS	290
Trek 800	240

Replace each _?_ with <, >, or = to make each sentence true. (Lesson 2–4)

8. $-7 - 5 \;\underline{?}\; -11$ **<**
9. $\frac{7}{16} \;\underline{?}\; \frac{8}{15}$ **<**
10. $\frac{5}{4}(4) \;\underline{?}\; 8\left(\frac{1}{2}\right)$ **>**

SELF TEST

The Self Test provides students with a brief review of the concepts and skills in Lessons 2-1 through 2-4. Lesson numbers are given to the right of exercises or instruction lines so students can review concepts not yet mastered.

Extension

Connections Draw two rectangles with equal dimensions. Then show that $\frac{3}{5} > \frac{1}{3}$.

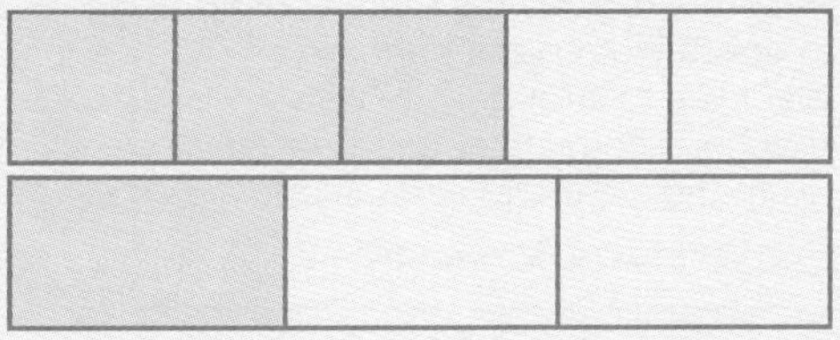

4 ASSESS

Closing Activity

Speaking Have students determine how many whole numbers satisfy the inequality $x < 5$. **5** Then ask students how many rational numbers satisfy the inequality. **an infinite number**

Additional Answers

44. from 130 to 150 by 5s

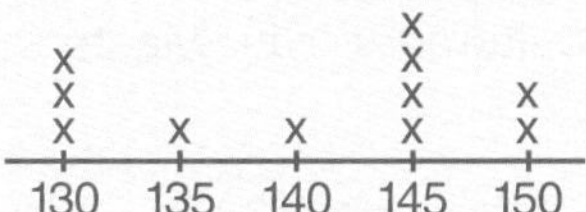

45. [number line: −4 −3 −2 −1 0 1 2 3 4]

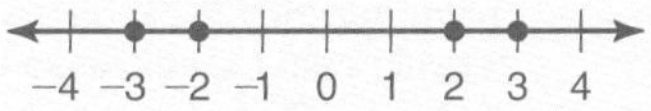

Answers for the Self Test

1.

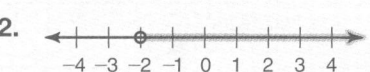

2. [number line: −4 −3 −2 −1 0 1 2 3 4]

Enrichment Masters, p. 13

NAME__________ DATE______

Enrichment Student Edition Pages 93–98

Rounding Fractions

Rounding fractions is more difficult than rounding whole numbers or decimals. For example, think about how you would round $\frac{4}{9}$ inches to the nearest quarter-inch. Through estimation, you might realize that $\frac{4}{9}$ is less than $\frac{1}{2}$. But, is it closer to $\frac{1}{2}$ or to $\frac{1}{4}$? Here are two ways to round fractions. Example 1 uses only the fractions; Example 2 uses decimals.

Example 1:

Subtract the fraction twice. Use the two nearest quarters.

$\frac{1}{2} - \frac{4}{9} = \frac{1}{18}$ $\quad \frac{4}{9} - \frac{1}{4} = \frac{7}{36}$

Compare the differences.

$\frac{1}{18} < \frac{7}{36}$

The smaller difference shows you which fraction to round to.

$\frac{4}{9}$ rounds to $\frac{1}{2}$.

Example 2:

Change the fraction and the two nearest quarters to decimals.

$\frac{4}{9} = 0.4\overline{4}, \frac{1}{2} = 0.5, \frac{1}{4} = 0.25$

Find the decimal halfway between the two nearest quarters.

$\frac{1}{2}(0.5 + 0.25) = 0.375$

If the fraction is greater than the halfway decimal, round up. If not, round down.

$0.4\overline{4} > 0.3675$. So, $\frac{4}{9}$ is more than half way between $\frac{1}{4}$ and $\frac{1}{2}$.

$\frac{4}{9}$ rounds to $\frac{1}{2}$.

Round each fraction to the nearest one-quarter. Use either method.

1. $\frac{1}{3}$ $\frac{1}{4}$
2. $\frac{3}{7}$ $\frac{1}{2}$
3. $\frac{7}{11}$ $\frac{3}{4}$
4. $\frac{4}{15}$ $\frac{1}{4}$
5. $\frac{7}{20}$ $\frac{1}{4}$
6. $\frac{31}{50}$ $\frac{1}{2}$
7. $\frac{9}{25}$ $\frac{1}{4}$
8. $\frac{23}{30}$ $\frac{3}{4}$

Round each decimal or fraction to the nearest one-eighth.

9. 0.6 $\frac{5}{8}$
10. 0.1 $\frac{1}{8}$
11. 0.45 $\frac{1}{2}$
12. 0.85 $\frac{7}{8}$
13. $\frac{5}{7}$ $\frac{3}{4}$
14. $\frac{3}{20}$ $\frac{1}{8}$
15. $\frac{23}{25}$ $\frac{7}{8}$
16. $\frac{5}{9}$ $\frac{1}{2}$

NCTM Standards: 1–5

Instructional Resources

- Study Guide Master 2-5
- Practice Master 2-5
- Enrichment Master 2-5
- Assessment and Evaluation Masters, pp. 43–44
- Multicultural Activity Masters, p. 3

Transparency 2-5A contains the 5-Minute Check for this lesson; **Transparency 2-5B** contains a teaching aid for this lesson.

Recommended Pacing	
Standard Pacing	Day 6 of 15
Honors Pacing	Day 6 of 15
Block Scheduling*	Day 4 of 9
Alg. 1 in Two Years*	Days 11, 12, & 13 of 28

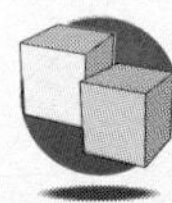

*For more information on pacing and possible lesson plans, refer to the *Block Scheduling Booklet* and *Algebra 1 in Two Years.*

1 FOCUS

5-Minute Check
(over Lesson 2-4)

Write the fractions in order from least to greatest.

1. $\frac{3}{15}, \frac{5}{17}, \frac{2}{7}$ **$\frac{3}{15}, \frac{2}{7}, \frac{5}{17}$**
2. $\frac{7}{12}, \frac{6}{11}, \frac{8}{15}$ **$\frac{8}{15}, \frac{6}{11}, \frac{7}{12}$**

Find a rational number between the given numbers.

3. $-\frac{3}{4}, \frac{5}{6}$ **0**
4. $\frac{1}{2}, \frac{6}{7}$ **$\frac{19}{28}$**
5. Find the unit price if 8 ounces of coffee sells for \$3.84. **\$0.48**

2-5 Adding and Subtracting Rational Numbers

What YOU'LL LEARN

- To add and subtract rational numbers, and
- to simplify expressions that contain rational numbers.

Why IT'S IMPORTANT

You can add and subtract rational numbers to help you solve problems involving the stock market and track and field.

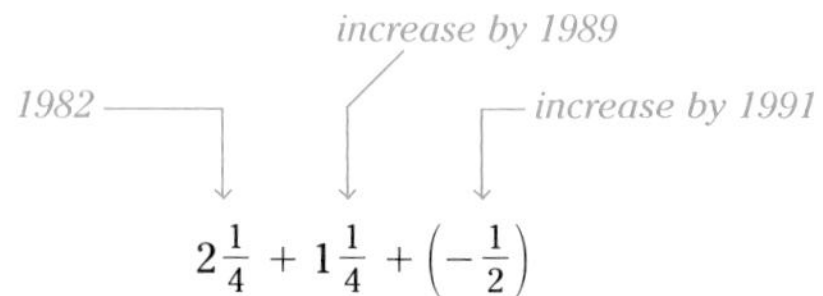

APPLICATION
Employment

In 1982, the average number of weekly overtime hours worked by a U.S. worker was $2\frac{1}{4}$ hours. This average increased $1\frac{1}{4}$ hours by 1989 and then decreased $\frac{1}{2}$ hour by 1991. What was the average number of weekly overtime hours worked in 1991?

To answer this question, you need to find the following sum.

$$2\frac{1}{4} + 1\frac{1}{4} + \left(-\frac{1}{2}\right)$$

(1982 → $2\frac{1}{4}$; increase by 1989 → $1\frac{1}{4}$; increase by 1991 → $\left(-\frac{1}{2}\right)$)

LOOK BACK
You can refer to Lesson 1-8 for information on the commutative and associative properties.

To add rational numbers such as these, use the same rules you used to add integers. When adding three or more rational numbers, you can use the commutative and associative properties to rearrange the addends.

Method 1: Use the rules.

$2\frac{1}{4} + 1\frac{1}{4} + \left(-\frac{1}{2}\right) = 3\frac{2}{4} + \left(-\frac{1}{2}\right)$ *Add the first two addends.*

$= 3\frac{1}{2} + \left(-\frac{1}{2}\right)$ *Simplify.*

$= +\left(\left|3\frac{1}{2}\right| - \left|-\frac{1}{2}\right|\right)$ *$3\frac{1}{2}$ has the greater absolute value,*

$= +\left(3\frac{1}{2} - \frac{1}{2}\right)$ *so the sign of the sum is positive.*

$= 3$

Method 2: Use a number line.

As with integers, you can also add rational numbers by using a number line.

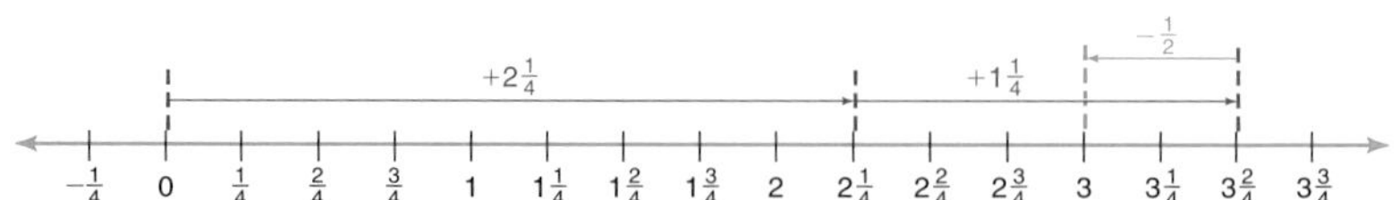

The average number of weekly overtime hours worked in 1991 was 3.

GLENCOE Technology

CD-ROM Interaction

A multimedia simulation allows students to see how math is used in operating a pizza parlor. A blackline master activity with teacher's notes provides a follow-up to the CD-ROM simulation.

For Windows & Macintosh

Example 1 **Find each sum.**

a. $4\frac{1}{8} + \left(-1\frac{1}{2}\right)$ *Estimate:* $4 + (-2) = 2$

$4\frac{1}{8} + \left(-1\frac{1}{2}\right) = 4\frac{1}{8} + \left(-1\frac{4}{8}\right)$ *The LCD is 8. Replace* $-1\frac{1}{2}$ *with* $-1\frac{4}{8}$.

$= +\left(\left|4\frac{1}{8}\right| - \left|-1\frac{4}{8}\right|\right)$ *The sum is positive. Why?*

$= +\left(4\frac{1}{8} - 1\frac{4}{8}\right)$

$= +\left(3\frac{9}{8} - 1\frac{4}{8}\right)$ *Replace* $4\frac{1}{8}$ *with* $3\frac{9}{8}$.

$= 2\frac{5}{8}$ *Compare to the estimate.*

b. $-1.34 + (-0.458)$ *Estimate:* $-1 + (-0.5) = -1.5$

$-1.34 + (-0.458) = -(|-1.34| + |-0.458|)$ *The numbers have the same sign. Their sum is negative.*

$= -(1.34 + 0.458)$

$= -1.798$

c. $-\frac{2}{5} + 1\frac{1}{2} + \left(-\frac{2}{3}\right)$ *Estimate:* $-\frac{1}{2} + 1\frac{1}{2} + \left(-\frac{1}{2}\right) = \frac{1}{2}$

$-\frac{2}{5} + 1\frac{1}{2} + \left(-\frac{2}{3}\right) = \left[-\frac{2}{5} + \left(-\frac{2}{3}\right)\right] + 1\frac{1}{2}$ *Group the negative numbers together.*

$= \left[-\frac{12}{30} + \left(-\frac{20}{30}\right)\right] + \frac{45}{30}$ *The LCD is 30.*

$= -\frac{32}{30} + \frac{45}{30}$

$= \frac{13}{30}$ *Compare to the estimate.*

To subtract rational numbers, use the same process you used to subtract integers.

Example 2 **Find each difference.**

a. $-\frac{3}{5} - \left(-\frac{4}{7}\right)$

$-\frac{3}{5} - \left(-\frac{4}{7}\right) = -\frac{3}{5} + \frac{4}{7}$ *To subtract* $-\frac{4}{7}$*, add its inverse,* $+\frac{4}{7}$.

$= -\frac{21}{35} + \frac{20}{35}$ *The LCD is 35.*

$= -\frac{1}{35}$

b. $-6.24 - 8.52$

$-6.24 - 8.52 = -6.24 + (-8.52)$

$= -14.76$

Motivating the Lesson

Questioning Have students read the stock report from a recent newspaper. Discuss the meanings of such headings as "High," "Low," and "Chg." Have each class member select and monitor a particular stock during a one-week period.

2 TEACH

Teaching Tip Students are often confused with the various sets of numbers. Emphasize that a rational number is a ratio of two integers.

Teaching Tip In Example 1, although the positive sign is not necessary in the step that reads $+\left(\left|4\frac{1}{8}\right| - \left|-1\frac{4}{8}\right|\right)$, encourage students to, at first, place the sign outside the parentheses. Emphasize this step in analyzing what sign the sum will have.

In-Class Examples

For Example 1
Find each sum.

a. $5\frac{3}{7} + \left(-6\frac{1}{2}\right)$ $-1\frac{1}{14}$

b. $-8.007 + (-5.755)$ **−13.762**

For Example 2
Find each difference.

a. $-\frac{5}{9} - \left(-\frac{3}{5}\right)$ $\frac{2}{45}$

b. $-\frac{1}{2} - \left(-\frac{6}{7}\right)$ $\frac{5}{14}$

c. $6.32 - (-3.42)$ **9.74**

GLENCOE Technology

Interactive Mathematics Tools Software

This multimedia software provides an interactive lesson that uses a number line to subtract positive rational numbers. A **Computer Journal** gives students an opportunity to write about what they have learned.

For Windows & Macintosh

In-Class Examples

For Example 3

Find the net change in a stock that gained $3\frac{1}{8}$ points one day and then lost $\frac{3}{4}$ point the next day. **gain of $2\frac{3}{8}$ points**

For Example 4

Evaluate.

a. $p - 7.1$, if $p = 5.2$ **−1.9**

b. $\frac{8}{7} - x$, if $x = \frac{11}{14}$ **$\frac{5}{14}$**

3 PRACTICE/APPLY

Check for Understanding

Exercises 1–13 are designed to help you assess your students' understanding through reading, writing, speaking, and modeling. You should work through Exercises 1–4 with your students and then monitor their work on Exercises 5–13.

Additional Answer

3.

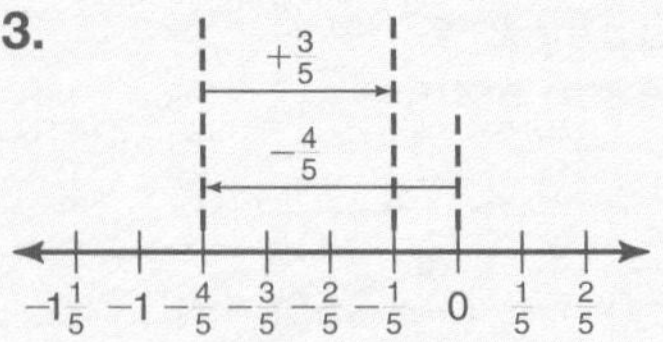

Study Guide Masters, p. 14

NAME________________ DATE ________

Study Guide

Student Edition Pages 100–104

Adding and Subtracting Rational Numbers

The rules for adding and subtracting integers also apply to adding and subtracting rational numbers.

Rational Number	Form $\frac{a}{b}$
3	$\frac{3}{1}$
$-2\frac{3}{4}$	$-\frac{11}{4}$
0.125	$\frac{1}{8}$

Example 1: Add $\left(-2\frac{1}{3}\right) + 5\frac{2}{3}$.

$$\left(-2\frac{1}{3}\right) + 5\frac{2}{3} = +\left(\left|5\frac{2}{3}\right| - \left|-2\frac{1}{3}\right|\right)$$
$$= +\left(5\frac{2}{3} - 2\frac{1}{3}\right)$$
$$= 3\frac{1}{3}$$

Example 2: Subtract $-3.42 - 5.82$.

$$-3.42 - 5.82 = -3.42 + (-5.82)$$
$$= -9.24$$

Previously you have added pairs of numbers. To add three or more numbers, first group the numbers in pairs. Use the commutative and associative properties to rearrange the addends if necessary. Study the example at the right.

Example 3: Add $-\frac{2}{3} + \frac{4}{5} + \left(-\frac{5}{3}\right)$.

$$-\frac{2}{3} + \frac{4}{5} + \left(-\frac{5}{3}\right) = \left[-\frac{2}{3} + \left(-\frac{5}{3}\right)\right] + \frac{4}{5}$$
$$= -\frac{7}{3} + \frac{4}{5}$$
$$= -\frac{35}{15} + \frac{12}{15}$$
$$= -\frac{23}{15} \text{ or } -1\frac{8}{15}$$

Find each sum or difference

1. $-\frac{9}{11} + \left(-\frac{13}{11}\right)$ −2
2. $\frac{5}{8} + \left(-\frac{1}{12}\right)$ $\frac{13}{24}$
3. $-0.005 + 0.0043$ −0.0007
4. $\frac{3}{8} - \left(-\frac{1}{8}\right)$ $\frac{1}{2}$
5. $4.59 - 2.31$ 2.28
6. $-\frac{7}{5} - \frac{2}{7}$ $-\frac{59}{35}$

Evaluate each expression if x = −4, y = 3, z = −7.

7. $x + 16$ 12
8. $0 + y$ 3
9. $27 - (x - z)$ 24
10. $100 + (x + y)$ 99

Find each sum.

11. $-36.4 + 29.15 + (-14.2)$ −21.45
12. $6.5x + 12.3x + (-14.9x)$ 3.9x
13. $0.85 + 13.6 + (-3.01)$ 11.44
14. $-9y + (-20y) + 6y$ −23y
15. $\frac{3}{5} + \left(-\frac{5}{8}\right) + \frac{1}{4}$ $\frac{9}{40}$
16. $12p + 11p + (-23p)$ 0

Example 3 APPLICATION **Stock Market**

The Intel Corporation produces the Pentium™ processor for personal computers. In November of 1994, Intel announced that there was a flaw in the processor that could cause errors in complex mathematical calculations. In one week, Intel's stock dropped $2\frac{1}{4}$ points. By December 13, the stock had dropped another $2\frac{1}{8}$ points. How many points did the stock drop in this time period?

We can represent the drops in the price of the stock with negative numbers.

$$-2\frac{1}{4} - 2\frac{1}{8} = -2\frac{1}{4} + \left(-2\frac{1}{8}\right) \quad \textit{To subtract } 2\tfrac{1}{8}\textit{, add its additive inverse.}$$
$$= -2\frac{2}{8} + \left(-2\frac{1}{8}\right) \quad \textit{The LCD is 8.}$$
$$= -4\frac{3}{8}$$

Intel's stock dropped $4\frac{3}{8}$ points during this time period.

FYI

In the 18th century, the U.S. dollar was equal to the Spanish silver dollar. This coin was often divided into eight parts, so when the stock market opened, the practice of quoting prices in eighths of a dollar began.

You can also evaluate expressions involving the addition and subtraction of rational numbers.

Example 4 **a. Evaluate $p - 3.5$ if $p = 2.8$.**

$$p - 3.5 = 2.8 - 3.5 \quad \textit{Replace p with 2.8.}$$
$$= -0.7$$

b. Evaluate $\frac{11}{4} - x$ if $x = \frac{27}{8}$.

$$\frac{11}{4} - x = \frac{11}{4} - \frac{27}{8} \quad \textit{Replace x with } \tfrac{27}{8}.$$
$$= \frac{22}{8} - \frac{27}{8} \quad \textit{The LCD is 8.}$$
$$= -\frac{5}{8}$$

CHECK FOR UNDERSTANDING

Communicating Mathematics

Study the lesson. Then complete the following.

1. **Describe** two ways to find $-2.4 + 5.87 + (-2.87) + 6.5$. **1. See Solutions Manual.**
2. **Find** an example of addition or subtraction of rational numbers in a newspaper or magazine. **See students' work.**

3. **Draw** and use a number line to show the sum of $-\frac{4}{5}$ and $\frac{3}{5}$. **See margin.**
4. **Use** a number line to solve the problem in Example 3. **See Solutions Manual.**

Guided Practice

Find each sum or difference.

5. $\frac{7}{9} - \frac{8}{9}$ **$-\frac{1}{9}$**
6. $-\frac{7}{12} + \frac{5}{6}$ **$\frac{1}{4}$**
7. $-\frac{1}{8} + \left(-\frac{5}{2}\right)$ **$-2\frac{5}{8}$**
8. $-69.5 - 82.3$ **−151.8**
9. $4.57 + (-3.69)$ **0.88**
10. $-\frac{2}{5} + \frac{3}{15} + \frac{3}{5}$ **$\frac{2}{5}$**

FYI

The trading volume of the New York Stock Exchange is greater than that of all the other American exchanges put together. Other exchanges in the United States include the American Stock Exchange, the Midwest Stock Exchange, the Pacific Stock Exchange, and exchanges in Boston, Philadelphia, and Cincinnati.

Reteaching

Working Backward Determine whether addition or subtraction signs belong in each equation.

1. $\frac{7}{8} \square \left(-\frac{3}{4}\right) \square \frac{1}{2} = \frac{5}{8}$ **+**
2. $\frac{5}{8} \square \frac{3}{10} \square \left(-\frac{1}{4}\right) = \frac{23}{40}$ **−**
3. $\frac{1}{6} \square \left(-\frac{5}{8}\right) \square \frac{7}{12} = \frac{1}{8}$ **+**
4. $\frac{5}{6} \square \frac{7}{9} \square \frac{1}{2} = -\frac{4}{9}$ **−**

Evaluate each expression.

11. $a - (-5)$, if $a = 0.75$ **5.75**

12. $\frac{11}{2} - b$ if $b = -\frac{5}{2}$ **8**

13. **Business** For the week of July 22, 1995, the following day-to-day changes were recorded in the Dow Jones Industrial Average:

Monday, $+2\frac{3}{8}$; Tuesday, $-\frac{1}{4}$; Wednesday, $-\frac{3}{8}$;

Thursday, $+2\frac{1}{4}$; Friday, $-3\frac{1}{2}$.

a. What was the net change for the week? $+\frac{1}{2}$

b. Draw a number line that shows the change each day. **See Solutions Manual.**

EXERCISES

Practice

Find each sum or difference.

14. $\frac{5}{8} + \left(-\frac{3}{8}\right)$ $\frac{1}{4}$

15. $-\frac{7}{8} - \left(-\frac{3}{16}\right)$ $-\frac{11}{16}$

16. $-1.6 - 3.8$ **−5.4**

17. $3.2 + (-4.5)$ **−1.3**

18. $\frac{1}{2} + \left(-\frac{8}{16}\right)$ **0**

19. $\frac{2}{3} + \left(-\frac{2}{9}\right)$ $\frac{4}{9}$

20. $-38.9 + 24.2$ **−14.7**

21. $-0.0007 + (-0.2)$ **−0.2007**

22. $-\frac{3}{7} + \frac{1}{4}$ $-\frac{5}{28}$

23. $-5\frac{7}{8} - 2\frac{3}{4}$ $-8\frac{5}{8}$

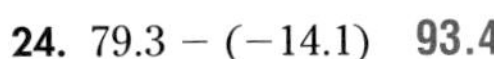

24. $79.3 - (-14.1)$ **93.4**

25. $-0.0015 + 0.05$ **0.0485**

B

26. $-9.16 - (-10.17)$ **1.01**

27. $-5.6 + (-9.45) + (-7.89)$ **−22.94**

28. $\frac{1}{4} + 2 + \left(-\frac{3}{4}\right)$ $1\frac{1}{2}$

29. $-0.87 + 3.5 + (-7.6) + 2.8$ **−2.17**

30. $\frac{3}{4} + \left(-\frac{4}{5}\right) + \frac{2}{5}$ $\frac{7}{20}$

31. $-4\frac{1}{4} + 6\frac{1}{3} + 2\frac{2}{3} + \left(-3\frac{3}{4}\right)$ **1**

Evaluate each expression.

32. $-3\frac{1}{3} - b$, if $b = \frac{1}{2}$ $-3\frac{5}{6}$

33. $r - 2.7$, if $r = -0.8$ **−3.5**

34. $n - 0.5$, if $n = -0.8$ **−1.3**

35. $t - (-1.3)$, if $t = -18$ **−16.7**

C

36. $\frac{4}{3} - p$, if $p = \frac{4}{5}$ $\frac{8}{15}$

37. $-\frac{12}{7} - s$, if $s = \frac{16}{21}$ $-\frac{52}{21}$

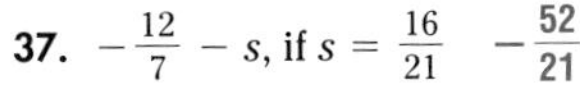

INTEGRATION Discrete Mathematics

Find each sum or difference. **38–42. See margin.**

38. $\begin{bmatrix} 3 & -2 \\ 5 & 6.2 \end{bmatrix} + \begin{bmatrix} 2.4 & 5 \\ -4 & 1 \end{bmatrix}$

39. $\begin{bmatrix} -1.3 & 4.2 \\ 0 & -3.4 \end{bmatrix} - \begin{bmatrix} 2.5 & 4.3 \\ -1.7 & -6.3 \end{bmatrix}$

40. $\begin{bmatrix} \frac{1}{2} & -2 \\ 7 & \frac{1}{4} \end{bmatrix} + \begin{bmatrix} 3 & -5 \\ \frac{2}{3} & 6 \end{bmatrix}$

41. $\begin{bmatrix} \frac{1}{2} & 6 \\ -1 & 5 \\ 4 & -7 \end{bmatrix} + \begin{bmatrix} -\frac{1}{4} & 5 \\ -4 & 3 \\ -3\frac{1}{2} & -5 \end{bmatrix}$

42. $\begin{bmatrix} 3.2 & 6.4 & -4.3 \\ 5 & -4 & 0.4 \end{bmatrix} - \begin{bmatrix} 1.4 & 3.7 & 5 \\ -1 & 5 & 0.4 \end{bmatrix}$

Critical Thinking

43. **Discrete Mathematics** Create four 2 × 2 matrices. Use the matrices to support or disprove the following statement: *Matrix addition is associative and commutative.* **See margin.**

Additional Answer

43. Answers will vary. Sample answer:

$\begin{bmatrix} 5 & 3 \\ 1 & 7 \end{bmatrix}, \begin{bmatrix} 4 & -3 \\ 1 & 2 \end{bmatrix}, \begin{bmatrix} 9 & 0 \\ 2 & 9 \end{bmatrix}, \begin{bmatrix} 18 & 0 \\ 4 & 18 \end{bmatrix}$

$$\begin{bmatrix} 5 & 3 \\ 1 & 7 \end{bmatrix} + \begin{bmatrix} 4 & -3 \\ 1 & 2 \end{bmatrix} = \begin{bmatrix} 9 & 0 \\ 2 & 9 \end{bmatrix}$$

$$= \begin{bmatrix} 4 & -3 \\ 1 & 2 \end{bmatrix} + \begin{bmatrix} 5 & 3 \\ 1 & 7 \end{bmatrix}$$

$$\left(\begin{bmatrix} 5 & 3 \\ 1 & 7 \end{bmatrix} + \begin{bmatrix} 4 & -3 \\ 1 & 2 \end{bmatrix}\right) + \begin{bmatrix} 9 & 0 \\ 2 & 9 \end{bmatrix} = \begin{bmatrix} 18 & 0 \\ 4 & 18 \end{bmatrix}$$

$$= \begin{bmatrix} 5 & 3 \\ 1 & 7 \end{bmatrix} + \left(\begin{bmatrix} 4 & -3 \\ 1 & 2 \end{bmatrix} + \begin{bmatrix} 9 & 0 \\ 2 & 9 \end{bmatrix}\right)$$

Assignment Guide

Core: 15–45 odd, 47–54
Enriched: 14–42 even, 43–54

For **Extra Practice,** see p. 760.

The red A, B, and C flags, printed only in the Teacher's Wraparound Edition, indicate the level of difficulty of the exercises.

Additional Answers

38. $\begin{bmatrix} 5.4 & 3 \\ 1 & 7.2 \end{bmatrix}$

39. $\begin{bmatrix} -3.8 & -0.1 \\ 1.7 & 2.9 \end{bmatrix}$

40. $\begin{bmatrix} \frac{7}{2} & -7 \\ \frac{23}{3} & \frac{25}{4} \end{bmatrix}$

41. $\begin{bmatrix} \frac{1}{4} & 11 \\ -5 & 8 \\ \frac{1}{2} & -12 \end{bmatrix}$

42. $\begin{bmatrix} 1.8 & 2.7 & -9.3 \\ 6 & -9 & 0 \end{bmatrix}$

Practice Masters, p. 14

NAME__________ DATE______

2-5 Practice

Student Edition Pages 100–104

Adding and Subtracting Rational Numbers

Find each sum or difference.

1. $\frac{5}{14} + \left(-\frac{3}{14}\right)$ $\frac{1}{7}$
2. $-\frac{19}{6} + \left(-\frac{5}{6}\right)$ -4
3. $-\frac{6}{11} - \left(-\frac{10}{11}\right)$ $\frac{4}{11}$
4. $-6.2 + 4.6$ -1.6
5. $-27.9 + 36$ 8.1
6. $-3.4 - 4.8$ -8.2
7. $-\frac{3}{4} + \frac{5}{8}$ $-\frac{1}{8}$
8. $\frac{7}{4} + \left(-\frac{1}{3}\right)$ $\frac{17}{12}$
9. $-\frac{3}{5} + \frac{3}{4}$ $\frac{3}{20}$
10. $126.2 + (-200)$ -73.8
11. $-164 + (-8.29)$ -172.29
12. $164.7 - 186.1$ -21.4
13. $\frac{5}{16} - \frac{1}{2}$ $-\frac{3}{16}$
14. $-\frac{5}{6} - \frac{4}{15}$ $-\frac{11}{10}$
15. $-\frac{7}{9} + \frac{5}{12}$ $-\frac{13}{36}$
16. $-9.26 - 38.5$ -47.76
17. $-0.006 + (-0.5)$ -0.506
18. $-0.09 + 0.009$ -0.081
19. $-\frac{1}{2} - \frac{2}{3}$ $-\frac{7}{6}$
20. $-\frac{1}{2} + \left(-\frac{2}{3}\right)$ $-\frac{7}{6}$
21. $-\frac{2}{3} - \frac{1}{2}$ $-\frac{7}{6}$
22. $65.6 - (-13.3)$ 78.9
23. $-6\frac{3}{8} - 2\frac{7}{8}$ $-9\frac{1}{4}$
24. $-1\frac{2}{3} + 1\frac{1}{2}$ $-\frac{1}{6}$
25. $\begin{array}{r} -12x \\ +\ 9x \\ \hline -3x \end{array}$
26. $\begin{array}{r} -15z \\ -\ 39z \\ \hline -54z \end{array}$
27. $\begin{array}{r} 7.9 \\ +(-9.7) \\ \hline -1.8 \end{array}$

Evaluate each expression.

28. $k - (-2)$, if $k = 1.6$ 3.6
29. $d - (-3.1)$, if $d = -15$ -11.9
30. $m - 2.8$, if $m = -1.7$ -4.5
31. $\frac{9}{4} - n$, if $n = \frac{11}{4}$ $-\frac{1}{2}$
32. $\frac{5}{3} - a$, if $a = \frac{5}{6}$ $\frac{5}{6}$
33. $-\frac{13}{9} - c$, if $c = \frac{2}{3}$ $-\frac{19}{9}$

4 ASSESS

Closing Activity

Writing Ask students to write a problem involving a real-life situation in which rational numbers are either added or subtracted. Have students exchange problems with classmates and solve each other's creation.

Chapter 2, Quiz B (Lessons 2-4 and 2-5), is available in the *Assessment and Evaluation Masters,* p. 44.

Mid-Chapter Test (Lessons 2-1 through 2-5) is available in the *Assessment and Evaluation Masters,* p. 43.

Additional Answer

49a.

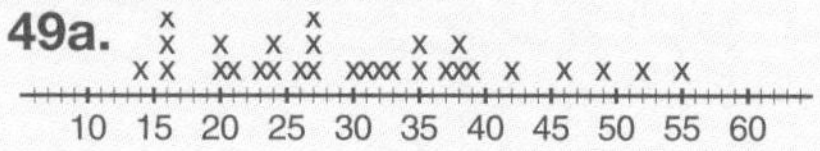

Enrichment Masters, p. 14

2-5 NAME________________ DATE________ Student Edition Pages 100–104

Enrichment

Tangram Puzzles

These seven geometric figures are called **tans.** They are used in a very old Chinese puzzle called **tangrams.**

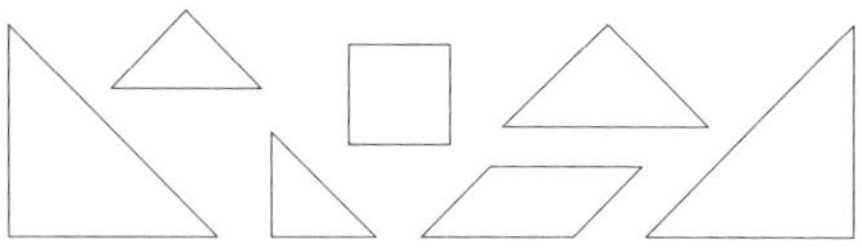

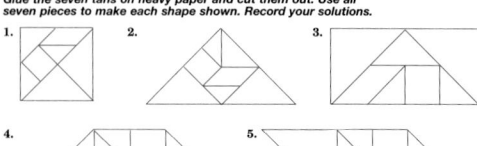

Glue the seven tans on heavy paper and cut them out. Use all seven pieces to make each shape shown. Record your solutions.

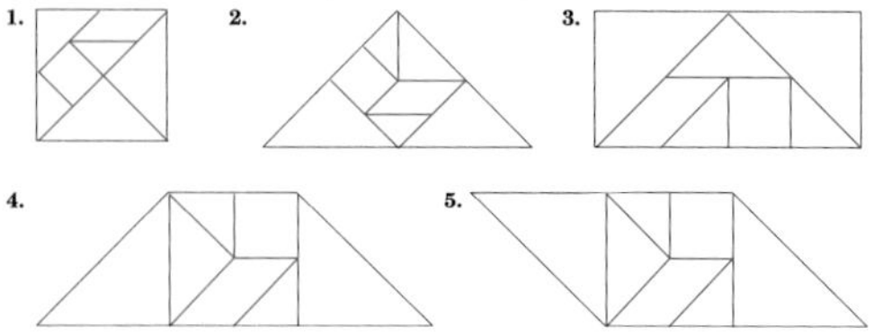

6. Each of the two figures is made from all seven tans. They seem to be exactly alike, but one has a foot and the other does not. Where does the second figure get his foot?
In the left figure, the body is made from 4 pieces rather than 3. The extra piece becomes the foot in the right figure.

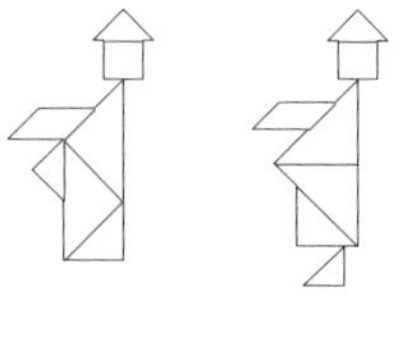

Applications and Problem Solving

44. **Book Binding** A textbook like this one is made up of sets of 32 pages called *signatures*. A signature is made from one piece of large paper that is folded several times. Signatures are bound together to make a book. Suppose when this book was printed, there were $1\frac{1}{2}$ rolls of paper available in the warehouse. The printer had three jobs to do that utilized this type of paper. This textbook required $\frac{1}{4}$ roll, another book required $\frac{2}{5}$ roll, and the third book needed $\frac{1}{2}$ roll. Is there enough paper available for all three jobs? Explain. **Yes; $\frac{7}{20}$ of a roll is left over.**

45. **Discrete Mathematics** An arithmetic sequence is a sequence in which the difference between any two consecutive terms is the same.
 a. Write the next three terms in the arithmetic sequence 4.53, 5.65, 6.77, 7.89, ... **9.01, 10.13, 11.25**
 b. What is the common difference? **1.12**
 c. Write the first five terms of an arithmetic sequence in which the first term is -2 and the common difference is $\frac{3}{4}$. Find the sum of the first five terms. **$-2, -\frac{5}{4}, -\frac{1}{2}, \frac{1}{4}, 1; -2\frac{1}{2}$**

46. **Track and Field** In 1995, Sheila Hudson-Strudwick set the U.S. women's triple jump record. Suppose her first jump was 46 feet 9 inches and her next jump was $1\frac{3}{4}$ inches longer. Her record-setting jump was 1 foot $2\frac{1}{2}$ inches longer than her second jump. What was the length of her record-setting jump? **48 feet $1\frac{1}{4}$ inches**

Mixed Review

47. Replace _?_ with <, >, or = to make -3.833 _?_ -3.115 true. (Lesson 2–4) **<**

48. Evaluate $-|-9 + 2x|$ when $x = 7$. (Lesson 2–3) **−5**

49. **Statistics** The points scored by the winning teams in the first 29 Super Bowls are listed below. (Lesson 2–2)

35	33	16	23	16	24	14	24
16	21	32	27	35	31	27	26
27	38	38	46	39	42	20	55
20	37	52	30	49			

 a. Make a line plot of the data. **See margin.**
 b. Do the numbers cluster around any number? **no**

50. **Restaurant Sales** Rosa's Deli made 3 cheesecakes, each having 12 slices, on Sunday. On Monday, they sold 3 slices. On Tuesday, they sold one entire cheesecake and 2 additional slices. The baker made another cheesecake on Wednesday and sold nine slices. If Rosa's Deli sold 11 slices of cheesecake on Thursday, how many slices were left on Friday? (Lesson 2–1) **11 slices**

51. Name the property illustrated by $7(0) = 0$. (Lesson 1–6) **mult. prop. of 0**

52. Evaluate $7[4^3 - 2(4 + 3)] \div 7 + 2$. (Lesson 1–3) **52**

53. **Patterns** Find the next term in the sequence $\frac{1}{3}, \frac{3}{6}, \frac{5}{12}, \ldots$. (Lesson 1–2) **$\frac{7}{24}$**

54. Evaluate 8^4. (Lesson 1–1) **4096**

Extension

Reasoning Insert one set of parentheses so that the value of the expression shown becomes $-\frac{37}{60}$.

$$\left(-\frac{3}{4}\right) - \left(-\frac{1}{5}\right) - \frac{3}{5} - \left(-\frac{2}{3}\right)$$

$$\left(-\frac{3}{4}\right) - \left(\left(-\frac{1}{5}\right) - \frac{3}{5} - \left(-\frac{2}{3}\right)\right)$$

A Preview of Lesson 2–6

2–6A Multiplying Integers

Materials:  counters integer mat

You can use counters to model multiplication of integers. Remember that yellow counters represent positive integers and red counters represent negative integers. When multiplying whole numbers, 2×4 means *two sets of four items.* When multiplying integers with counters, $(+2)(+4)$ means *to put in two sets of four positive counters.* $(-2)(+4)$ means to take out two sets of four positive counters.

Activity 1 Use counters to find the product (+2)(−4).

Step 1 Place, or *put in,* two sets of four negative counters on the mat.

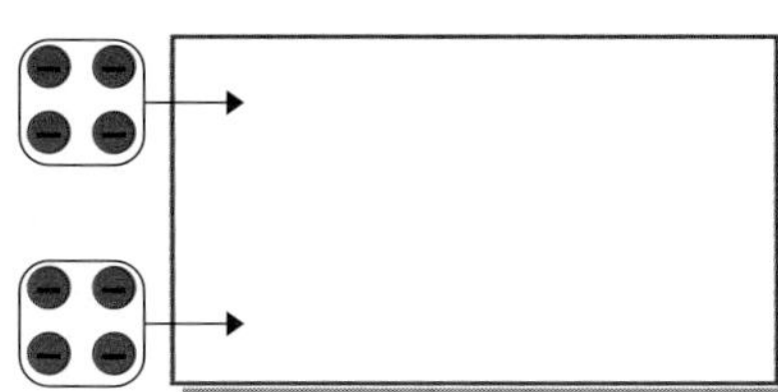

Step 2 Since there are eight negative counters on the mat, the product is -8. Thus, $(+2)(-4) = -8$.

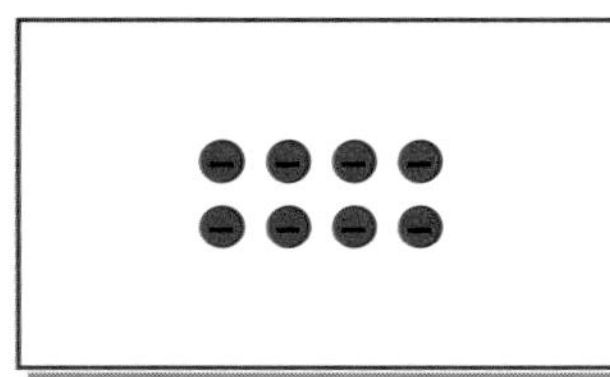

Activity 2 Use counters to find the product (−2)(+4).

Step 1 Add enough zero pairs so that you can *take out* two sets of four positive counters.

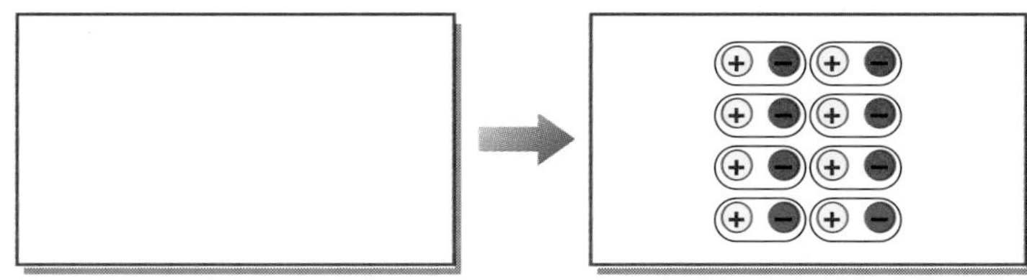

Step 2 Now take out two sets of four positive counters from the mat.

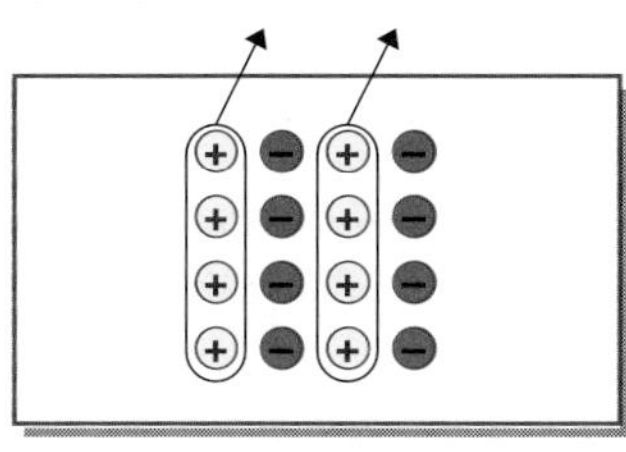

Step 3 Since there are eight negative counters on the mat, the product is -8. Thus, $-2(+4) = -8$.

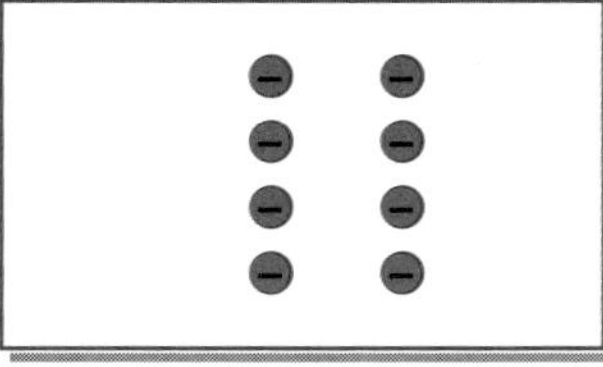

Write
1. Model $-2(-4)$. Write a short paragraph explaining what $(-2)(-4)$ means. **See students' work; take out two sets of negative counters.**

Model
Use counters to find each product. **2–7. See Solutions Manual for models.**

2. $2(-5)$ **−10**
3. $-2(5)$ **−10**
4. $-2(-5)$ **10**
5. $5(-2)$ **−10**
6. $-5(2)$ **−10**
7. $-5(-2)$ **10**

Write
8. How are the operations $-2(5)$ and $5(-2)$ the same? How do they differ? **See margin.**

2-6A LESSON NOTES

NCTM Standards: 1–5

Objective
Use counters to model multiplication of integers.

Recommended Time
Demonstration and discussion: 15 minutes; Exercises: 30 minutes

Instructional Resources
For each student or group of students
Student Manipulative Kit
- 20 red/yellow 2-sided counters

Modeling Mathematics Masters
- p. 2 (counters)
- p. 13 (integer mat)
- p. 21 (worksheet)

For teacher demonstration
Algebra and Geometry Overhead Manipulative Resources

1 FOCUS

Motivating the Lesson
Remind students that multiplication is simply repeated addition. For example, $3 \times 6 = 6 + 6 + 6$. 3×6 is 6 added together 3 times, or 3 added together 6 times. Ask students to solve $3(-6)$ in the same way. What is the sign of the answer? **negative**

2 TEACH

Teaching Tip Point out that in this activity, the − symbol can be viewed in two different ways. When it occurs outside the parentheses, it symbolizes the act of "taking out" counters. When it occurs inside the parentheses, it symbolizes negative counters.

3 PRACTICE/APPLY

Assignment Guide
Core: 1–8
Enriched: 1–8

4 ASSESS

Observing students working in cooperative groups is an excellent method of assessment. You may wish to ask a student from each group to explain how to model and solve a problem. Also watch for and acknowledge students who are helping others understand the concept being taught.

Additional Answer
8. The product is the same. The order of multiplication using counters is different.

2-6 LESSON NOTES

NCTM Standards: 1–5

Instructional Resources

- Study Guide Master 2-6
- Practice Master 2-6
- Enrichment Master 2-6
- Real-World Applications, 5
- Tech Prep Applications Masters, p. 4

Transparency 2-6A contains the 5-Minute Check for this lesson; **Transparency 2-6B** contains a teaching aid for this lesson.

Recommended Pacing	
Standard Pacing	Day 8 of 15
Honors Pacing	Day 8 of 15
Block Scheduling*	Day 5 of 9
Alg. 1 in Two Years*	Days 15 & 16 of 28

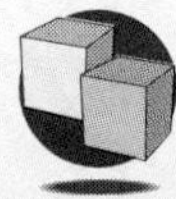

*For more information on pacing and possible lesson plans, refer to the *Block Scheduling Booklet* and *Algebra 1 in Two Years.*

1 FOCUS

5-Minute Check
(over Lesson 2-5)

Find each sum or difference.

1. $3.1 + (-4.56)$ **−1.46**
2. $-\frac{23}{3} + 2\frac{1}{5}$ $\mathbf{-5\frac{7}{15}}$
3. $9.16 - 10.17$ **−1.01**

Evaluate each expression.

4. $-1\frac{2}{3} - b$, if $b = \frac{1}{3}$ **−2**
5. $k - (-5.8)$, if $k = 6.3$ **12.1**

Motivating the Lesson

Questioning Ask students to name examples from everyday life in which they use fractions. They might also suggest instances requiring the use of fractions from their science classes.

2-6 Multiplying Rational Numbers

What YOU'LL LEARN
- To multiply rational numbers.

Why IT'S IMPORTANT

You can multiply rational numbers to help you solve problems involving geometry and business.

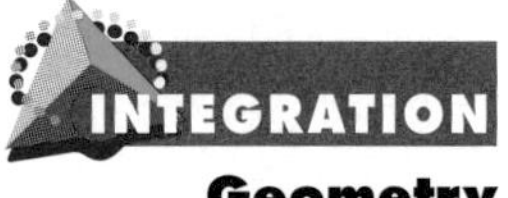

INTEGRATION
Geometry

Once called "the world's largest office building," the Pentagon, the five-story, five-sided defense building in Washington, D.C., was designed by G.E. Bergstrom to maximize space and efficiency. No two offices are more than a seven-minute walk from one another.

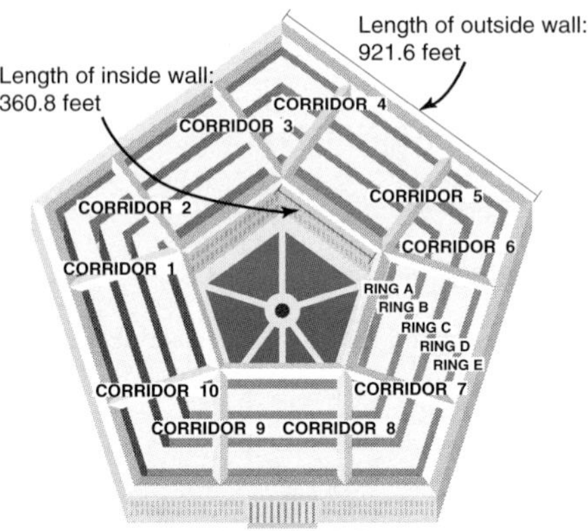

THE PENTAGON

The building is made up of 10 corridors and five rings. There are 230 restrooms, 150 stairways, and 7748 windows. Each outside wall of the building is 921.6 feet in length, slightly longer than three football fields. The inner walls, also in the shape of a pentagon, are each 360.8 feet in length. What is the outer perimeter of the Pentagon?

One way to solve this problem is to use repeated addition.

$$921.6 + 921.6 + 921.6 + 921.6 + 921.6 = 4608$$

An easier method would be to multiply 921.6 by 5.

$$5(921.6) = 4608$$

The perimeter of the Pentagon is 4608 feet.

Since this method would not work if you wanted to find the product of $\frac{2}{3}$ and $-\frac{2}{5}$ or the product of -5 and -0.3, you can use the following patterns to discover a rule for multiplying rational numbers.

$\frac{2}{3} \times \frac{2}{5} = \frac{4}{15}$	$-5 \cdot 0.3 = -1.5$
$\frac{2}{3} \times \frac{1}{5} = \frac{2}{15}$	$-5 \cdot 0.2 = -1.0$
$\frac{2}{3} \times \frac{0}{5} = \frac{0}{15}$	$-5 \cdot 0.1 = -0.5$
$\frac{2}{3} \times \left(-\frac{1}{5}\right) = -\frac{2}{15}$	$-5 \cdot 0 = 0$
$\frac{2}{3} \times \left(-\frac{2}{5}\right) = -\frac{4}{15}$	$-5 \cdot (-0.1) = 0.5$
	$-5 \cdot (-0.2) = 1.0$
	$-5 \cdot (-0.3) = 1.5$

The examples on the previous page suggest the following rules.

Multiplying Two Rational Numbers	The product of two numbers having the *same sign* is positive. The product of two numbers having *different signs* is negative.

Example 1 **Find each product.**

a. $(-9.8)4$ — *Estimate:* $(-10)4 = -40$

$(-9.8)4 = -39.2$ — *Since the factors have different signs, the product is negative.*

b. $\left(-\frac{3}{4}\right)\left(-\frac{2}{3}\right)$ — *Estimate:* $(-1)\left(-\frac{1}{2}\right) = \frac{1}{2}$

$\left(-\frac{3}{4}\right)\left(-\frac{2}{3}\right) = \frac{6}{12}$ or $\frac{1}{2}$ — *Since the factors have the same sign, the product is positive.*

Sometimes you need to evaluate expressions that contain rational numbers.

Example 2 **Evaluate $a\left(\frac{5}{6}\right)^2$ if $a = 2$.**

$a\left(\frac{5}{6}\right)^2 = 2\left(\frac{5}{6} \cdot \frac{5}{6}\right)$ — *Replace a with 2.*

$= 2\left(\frac{25}{36}\right)$ — *Multiply.*

$= \frac{\overset{1}{\cancel{2}}}{1}\left(\frac{25}{\underset{18}{\cancel{36}}}\right)$ or $\frac{25}{18}$

You may need to simplify expressions by multiplying rational numbers.

Example 3 **Simplify each expression.**

a. $(2b)(-3a)$

$(2b)(-3a) = 2(-3)ab$ — *Commutative and associative properties of multiplication*

$= -6ab$

b. $3x(-3y) + (-6x)(-2y)$

$3x(-3y) + (-6x)(-2y) = -9xy + 12xy$ — *Multiply.*

$= 3xy$ — *Combine like terms.*

Notice that multiplying a number or expression by -1 results in the opposite of the number or expression.

$-1(4) = -4$ $\quad (1.5)(-1) = -1.5$ $\quad (-1)(-3m) = 3m$

Multiplicative Property of -1	The product of any number and -1 is its additive inverse. $-1(a) = -a$ and $a(-1) = -a$

To find the product of three or more numbers, you may want to first group the numbers in pairs.

2 TEACH

In-Class Examples

For Example 1
Find each product.

a. $20\left(-\frac{4}{5}\right)$ **-16**

b. $(-1.4)7$ **-9.8**

For Example 2
Evaluate $a\left(\frac{2}{3}\right)^3$ if $a = 3$. **$\frac{8}{9}$**

For Example 3
Simplify each expression.

a. $(-2a)(3b) + (4a)(-6b)$ **$-30ab$**

b. $(5x)(-3y) + (-7x)(4y)$ **$-43xy$**

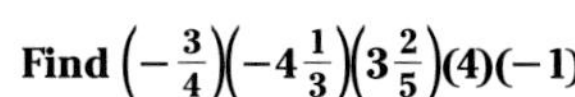

Example 4 Find $\left(-\frac{3}{4}\right)\left(-4\frac{1}{3}\right)\left(3\frac{2}{5}\right)(4)(-1)$.

$$\left(-\frac{3}{4}\right)\left(-4\frac{1}{3}\right)\left(3\frac{2}{5}\right)(4)(-1) = \left[\left(-\frac{3}{4}\right)(4)\right]\left[\left(-4\frac{1}{3}\right)(-1)\right]\left(3\frac{2}{5}\right) \quad \text{Commutative and associative properties}$$
$$= (-3)\left(4\frac{1}{3}\right)\left(3\frac{2}{5}\right)$$
$$= \left(-\frac{3}{1}\right)\left(\frac{13}{3}\right)\left(\frac{17}{5}\right)$$
$$= -\frac{221}{5} \text{ or } -44\frac{1}{5}$$

You can multiply any matrix by a constant. This is called **scalar multiplication.** When scalar multiplication is performed, each element is multiplied by that constant and a new matrix is formed.

Scalar Multiplication of a Matrix

$$m\begin{bmatrix} a & b & c \\ d & e & f \end{bmatrix} = \begin{bmatrix} ma & mb & mc \\ md & me & mf \end{bmatrix}$$

Example 5

APPLICATION Airlines

From July 3 through July 6, 1995, American Airlines quoted round-trip coach fares, in dollars, to and from the selected cities below.

	Chicago	Dallas	Las Vegas
Atlanta	**\$198.00**	**\$198.00**	**\$1214.00**
New York	**\$246.00**	**\$1224.00**	**\$1342.00**

Suppose the major airlines lower fares and begin an airfare war. To increase their air travel, American Airlines decides to lower their fares by 30% or $\frac{3}{10}$. Find the new fares for travel to the cities above.

To find the new fares, you can use scalar multiplication. If fares are reduced by $\frac{3}{10}$, travelers will pay $1 - \frac{3}{10}$ or $\frac{7}{10}$ of the original price. Multiply to find the discount prices.

$$\frac{7}{10}\begin{bmatrix} 198.00 & 198.00 & 1214.00 \\ 246.00 & 1224.00 & 1342.00 \end{bmatrix} = \begin{bmatrix} 138.60 & 138.60 & 849.80 \\ 172.20 & 856.80 & 939.40 \end{bmatrix}$$

The new fares are shown below.

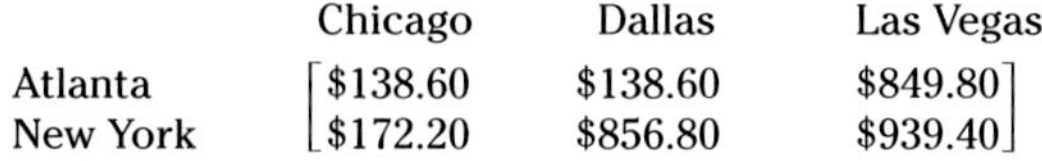

	Chicago	Dallas	Las Vegas
Atlanta	\$138.60	\$138.60	\$849.80
New York	\$172.20	\$856.80	\$939.40

CHECK FOR UNDERSTANDING

Communicating Mathematics

Study the lesson. Then complete the following. 1–2. See margin.

1. **List** the conditions under which each statement is true. Then give an example that verifies your conditions.
   a. ab is positive. b. ab is negative. c. ab is equal to 0.
2. **Explain** what you can conclude about a if each statement is true.
   a. a^2 is positive. b. a^3 is positive. c. a^3 is negative.
3. If $a = -2$, which is greater, $a + a^2$ or -4? $a + a^2$

Teaching Tip The product of any pair of negative numbers is positive. An even number of negative factors can be paired exactly to produce a positive result. An odd number of negative factors cannot be paired exactly, so a negative result is produced.

In-Class Examples

For Example 4
Multiply.

a. $\left(-\frac{2}{3}\right)\left(-\frac{1}{6}\right)\left(\frac{9}{5}\right)\left(-\frac{1}{2}\right)$ $-\frac{1}{10}$

b. $\left(-\frac{5}{6}\right)(-4)\left(7\frac{1}{5}\right)\left(\frac{11}{12}\right)$ 22

For Example 5
Find the new fares if American lowers its fares by 40% or $\frac{4}{10}$.

$$\begin{bmatrix} \$118.80 & \$118.80 & \$728.40 \\ \$147.60 & \$734.40 & \$805.20 \end{bmatrix}$$

Additional Answers

1a. ab is positive if a and b have the same signs, either both positive or both negative.
1b. ab is negative if a and b have opposite signs, one is negative and the other is positive.
1c. ab is equal to 0 if either a or b is 0, or both a and b are 0.
2a. If a^2 is positive, then a is either negative or positive.
2b. If a^3 is positive, then a is positive.
2c. If a^3 is negative, then a is negative.

Alternative Teaching Strategies

Student Diversity Many countries use the Celsius system for measuring temperatures. The United States still uses the Fahrenheit system for weather reports for the general public. The formula used to convert degrees Fahrenheit to degrees Celsius is $C = \frac{5}{9}(F - 32)$.

4. **Write** a multiplication sentence for the model at the right. $2 \times (-3) = -6$

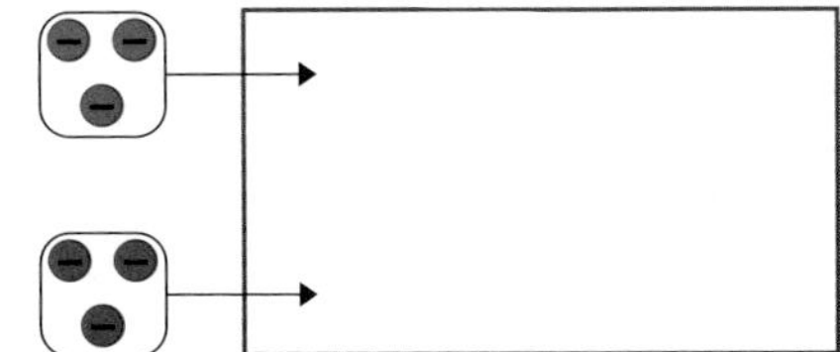

Guided Practice

Find each product.

5. $6(-3)$ **−18**
6. $(-4)(-8)$ **32**
7. $(-4)(2)(-3)$ **24**
8. $\left(\frac{7}{3}\right)\left(\frac{7}{3}\right)$ $\frac{49}{9}$
9. $\left(-\frac{4}{5}\right)\left(-\frac{1}{5}\right)(-5)$ $-\frac{4}{5}$
10. $\left(\frac{3}{5}\right)\left(-\frac{4}{7}\right)$ $-\frac{12}{35}$

Evaluate each expression if $x = \frac{1}{2}$ and $y = -\frac{2}{3}$.

11. $3y - 4x$ **−4**
12. x^2y $-\frac{1}{6}$

Simplify.

13. $5s(-6t) + 2s(-8t)$ **$-46st$**
14. $6x(-7y) + (-3x)(-5y)$ **$-27xy$**

Find each product. **15–16. See margin.**

15. $3\begin{bmatrix} -2 & 4 \\ -1 & 5 \end{bmatrix}$
16. $-5\begin{bmatrix} -1 & 0 \\ 4.5 & 8 \\ 3.2 & -4 \end{bmatrix}$

17. **Business** Employees of Glencoe/McGraw-Hill are reimbursed for the "wear and tear" that occurs to their car while on company business. One employee's odometer read 19,438.6 at the beginning of the day and 19,534.1 at the end of the day. How much should the employee be reimbursed if the rate is $0.30 per mile? **$28.65**

EXERCISES

Practice

Find each product.

18. $6(13)$ **78**
19. $(-5)(12)$ **−60**
20. $(-7)(-6)$ **42**
21. $\left(-\frac{8}{9}\right)\left(\frac{9}{8}\right)$ **−1**
22. $-\frac{5}{6}\left(-\frac{2}{5}\right)$ $\frac{1}{3}$
23. $(-5)\left(-\frac{2}{5}\right)$ **2**
24. $(-100)(-3.6)$ **360**
25. $(-2.93)(-0.003)$
26. $(-5)(3)(-4)$ **60**
27. $\left(\frac{2}{3}\right)\left(\frac{3}{5}\right)(-3)$ $-\frac{6}{5}$
28. $\left(-\frac{7}{12}\right)\left(\frac{6}{7}\right)\left(-\frac{3}{4}\right)$ $\frac{3}{8}$
29. $\frac{6}{11}\left(-\frac{33}{34}\right)$ $-\frac{9}{17}$
30. $(-0.075)(-5.5)$
31. $(-5.8)(-6.425)(2.3)$
32. $(-4)(0)(-2)(-3)$ **0**
33. $\frac{3}{5}(5)(-2)\left(-\frac{1}{2}\right)$ **3**
34. $(3)(-4)(-1)(-2)$
35. $\frac{2}{11}(-11)(-4)\left(-\frac{3}{4}\right)$

25. **0.00879**
30. **0.4125**
31. **85.7095**
34. **−24**
35. **−6**

Evaluate each expression if $m = -\frac{2}{3}$, $n = \frac{1}{2}$, $p = -3\frac{3}{4}$, and $q = 2\frac{1}{6}$.

B

36. $6m$ **−4**
37. nq
38. $2m - 3n$
39. $pq - m$ $-\frac{179}{24}$
40. $m^2\left(-\frac{1}{4}\right)$ $-\frac{1}{9}$
41. $n^2(q + 2)$ $\frac{25}{24}$

37. $\frac{13}{12}$ 38. $-\frac{17}{16}$

Reteaching

Using Reasoning

1. Write the sign of each product.
   a. $(-)(-)$ +
   b. $(-)(-)(-)(+)(+)$ −
   c. product of seven negative and three positive factors −
2. Find each product.
   a. $(-1)^8$ **+1**
   b. $(-1)^{19}$ **−1**
   c. $\left(\frac{3}{4}\right)\left(\frac{2}{3}\right)\left(-\frac{1}{2}\right)$ $-\frac{1}{4}$
   d. $(-2)(-12)(0)(-6)$ **0**

3 PRACTICE/APPLY

Check for Understanding

Exercises 1–17 are designed to help you assess your students' understanding through reading, writing, speaking, and modeling. You should work through Exercises 1–4 with your students and then monitor their work on Exercises 5–17.

Assignment Guide

Core: 19–57 odd, 59–68
Enriched: 18–52 even, 54–68

For **Extra Practice,** see p. 760.

The red A, B, and C flags, printed only in the Teacher's Wraparound Edition, indicate the level of difficulty of the exercises.

Additional Answers

15. $\begin{bmatrix} -6 & 12 \\ -3 & 15 \end{bmatrix}$

16. $\begin{bmatrix} 5 & 0 \\ -22.5 & -40 \\ -16 & 20 \end{bmatrix}$

Study Guide Masters, p. 15

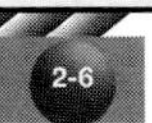

NAME ______________ DATE ______

2-6 Study Guide

Student Edition Pages 106–111

Multiplying Rational Numbers

You can use the rules below when multiplying rational numbers.

Rule or Property		Example
Multiplying Two Numbers with Different Signs	The product of two numbers that have different signs is negative.	$(-2x)5y = (-2)(5)xy$ $= -10xy$
Multiplying Two Numbers with the Same Sign	The product of two numbers that have the same sign is positive.	$(-4)(-7) = 28$
Multiplicative Property of −1	The product of any number and −1 is its additive inverse. $-1(a) = -a$ and $a(-1) = -a$	$(-3)(-4)(-1)(2) = 12(-1)(2)$ $= -12(2)$ $= -24$

To find the product of two or more numbers, first group the numbers in pairs.

Example: Multiply $(-2.3)(5.6)(-0.7)(-0.2)$

$(-2.3)(5.6)(-0.7)(-0.2) = [(-2.3)(5.6)][(-0.7)(-0.2)]$ **Associative property (×)**
$= (-12.88)(0.14)$ **Substitution property (=)**
$= -1.8032$

Find each product.

1. $(-24)(-2)$ 48
2. $(6.0)(-0.3)$ −1.8
3. $(-2)(-3)(-4)$ −24
4. $\left(\frac{1}{2}\right)(-10)(5)$ −25
5. $(-22)(-3)\left(-\frac{2}{3}\right)$ −44
6. $\left(\frac{4}{5}\right)(-5)(0)(4)$ 0

Simplify.

7. $(-6)(5) + (2)(4)$ −22
8. $\left(-\frac{3}{4}\right)\left(\frac{1}{8}\right) - \left(\frac{1}{4}\right)\left(\frac{1}{8}\right)$ $-\frac{1}{8}$
9. $-5(2x + x) - 3(-xy)$ $-15x + 3xy$
10. $(-5ab)6 - ab(4 + 1)$ $-35ab$
11. $2.3(4c - d) - 0.1(-0.5c + 8d)$ $9.25c - 3.1d$
12. $(-0.9)(1.1) - (-5.1)(0.6)$ 2.07

Additional Answers

48. $\begin{bmatrix} 7 & -57.4 & 0 \\ -28 & -39.2 & 7 \\ -22.4 & -49 & -49 \end{bmatrix}$

49. $\begin{bmatrix} 2 & 6 & 3 \\ \frac{5}{2} & 5 & 1 \end{bmatrix}$

50. $\begin{bmatrix} 5.2 & -8 & -16 \\ 2 & -1.2 & 20 \\ 26.4 & 8.4 & -32 \end{bmatrix}$

51. $\begin{bmatrix} -9 & 22.68 \\ -22.4 & -10 \\ 28.8 & -11.12 \end{bmatrix}$

52. $\begin{bmatrix} -1.6 & -36 \\ 11.2 & 24 \\ -24 & -19.2 \\ 56 & 25.6 \end{bmatrix}$

53. $\begin{bmatrix} 6 & 18 & 4 \\ 0 & 2 & \frac{8}{3} \end{bmatrix}$

56a. $\frac{1}{9}, \frac{1}{27}, \frac{1}{81}$

56b. $\frac{1}{3}$

56c. −6, −3, −1.5, −0.75, −0.375, −11.625

Practice Masters, p. 15

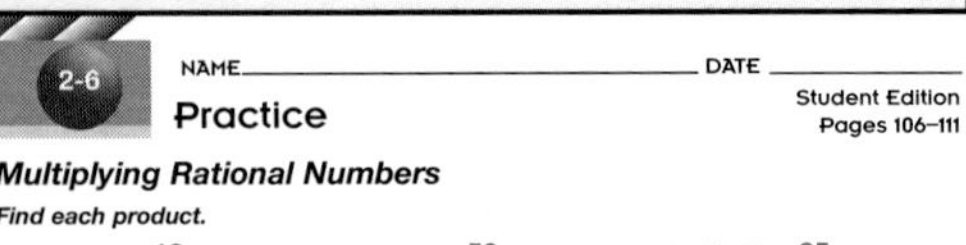

NAME________________ DATE ________

2-6 Practice

Student Edition Pages 106–111

Multiplying Rational Numbers

Find each product.

1. $(-3)(-4)$ 12
2. $-8(7)$ −56
3. $5(-7)$ −35
4. $2\left(-\frac{5}{7}\right)$ $-\frac{10}{7}$
5. $\left(-\frac{3}{4}\right)\left(\frac{7}{8}\right)$ $-\frac{21}{32}$
6. $\left(-\frac{4}{5}\right)\left(-\frac{5}{6}\right)$ $\frac{2}{3}$
7. $(6.8)(-1.3)$ −8.84
8. $(-5)(7)(-3)$ 105
9. $\left(-\frac{2}{3}\right)(-9)$ 6
10. $\frac{3}{5}(-6)(-15)$ 54
11. $\frac{2}{3}(-4)(9)$ −24
12. $(-5)(7)(-12)(0)$ 0
13. $\frac{3}{8}(-24)(5)(3)$ −135
14. $(-3.6)(0.057)$ −0.2052
15. $(-1.32)(-5)(2.1)$ 13.86
16. $\frac{5}{6}\left(-\frac{5}{6}\right)(-9)$ $\frac{25}{4}$
17. $-\frac{1}{2}\left(\frac{7}{12}\right)\left(-\frac{4}{5}\right)$ $\frac{7}{30}$
18. $\frac{8}{9}\left(\frac{3}{5}\right)\left(-\frac{1}{4}\right)$ $-\frac{2}{15}$
19. $\frac{7}{8}(-16)(-7)\left(-\frac{1}{14}\right)$ −7
20. $(5)(-3)(-1)(-4)$ −60
21. $\frac{5}{16}(-2)(-4)\left(-\frac{4}{5}\right)$ −2

Simplify.

22. $5(-7) + (-3)(6)$ −53
23. $7x(-3y) + 3x(-4y)$ $-33xy$
24. $(-2.1)(1.02) - (-3.4)(-10.1)$ −36.482
25. $(-5a)(-2b) - (-8a)(6b)$ $58ab$
26. $-\frac{1}{4}\left(\frac{8}{3}\right) - \left(\frac{5}{6}\right)\left(\frac{2}{5}\right)$ −1
27. $6(3m - m) + 3(m + 2m)$ $21m$
28. $(-2x)(-4y) - (-9x)(10y)$ $98xy$
29. $4[3x + (-2x)] - 5(3x + 2x)$ $-21x$
30. $9(2ab - 3c) - (4ab - 6c)$ $14ab - 21c$
31. $\frac{5}{6}(-24a + 36b) + \left(-\frac{1}{3}\right)(60a - 42b)$ $-40a + 44b$

Simplify.

42. $-2a(-3c) + (-6y)(6r)$ **42. $6ac - 36ry$**
43. $(5t)(-6r) - (-4s)$ $-30rt + 4s$
44. $7m(-3n) + 3m(-4n)$ $-33mn$
45. $5(2x - x) + 4(x + 3x)$ $21x$
46. $(-6b)(-3c) - (-9a)(7b)$ $18bc + 63ab$
47. $3.2(5x - y) - 0.3(-1.6x + 7y)$ $16.48x - 5.3y$

INTEGRATION Discrete Mathematics C

Find each product. 48–53. See margin.

48. $-7\begin{bmatrix} -1 & 8.2 & 0 \\ 4 & 5.6 & -1 \\ 3.2 & 7 & 7 \end{bmatrix}$

49. $\frac{1}{2}\begin{bmatrix} 4 & 12 & 6 \\ 5 & 10 & 2 \end{bmatrix}$

50. $4\begin{bmatrix} 1.3 & -2 & -4 \\ 0.5 & -0.3 & 5 \\ 6.6 & 2.1 & -8 \end{bmatrix}$

51. $-4\begin{bmatrix} 2.25 & -5.67 \\ 5.6 & 2.5 \\ -7.2 & -2.78 \end{bmatrix}$

52. $-8\begin{bmatrix} 0.2 & 4.5 \\ -1.4 & -3 \\ 3 & 2.4 \\ -7 & -3.2 \end{bmatrix}$

53. $\frac{2}{3}\begin{bmatrix} 9 & 27 & 6 \\ 0 & 3 & 4 \end{bmatrix}$

Critical Thinking

54. If a product has an even number of negative factors, what must be true of the product? **It is positive.**

55. If a product has an odd number of negative factors, what must be true of the product? **It is negative.**

Applications and Problem Solving

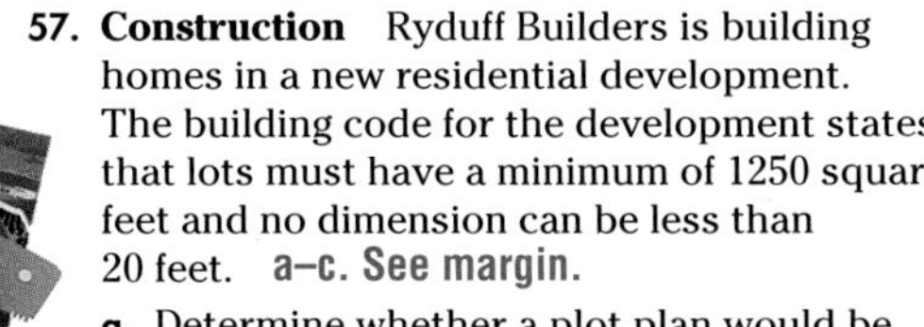

56. **Discrete Mathematics** A *geometric sequence* is a sequence in which the ratio of any term divided by the term before it is the same for any two terms.
    a. Write the next three terms in the geometric sequence 9, 3, 1, $\frac{1}{3}$, ...
    b. What is the common ratio? **a–c. See margin.**
    c. Write the first five terms in a geometric sequence in which the first term is −6 and the common ratio is 0.5. Then find the sum of the five terms.

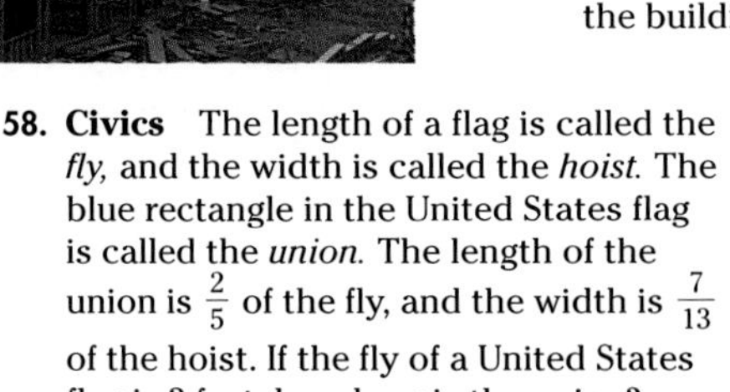

57. **Construction** Ryduff Builders is building homes in a new residential development. The building code for the development states that lots must have a minimum of 1250 square feet and no dimension can be less than 20 feet. **a–c. See margin.**
    a. Determine whether a plot plan would be approved if it measured 32 feet by 38 feet. Explain your reasoning.
    b. Determine whether a plot plan would be approved if it measured 19 feet by 70 feet. Explain your reasoning.
    c. If one dimension of a plot plan was 42 feet, find another dimension that would satisfy the building code conditions.

58. **Civics** The length of a flag is called the *fly,* and the width is called the *hoist.* The blue rectangle in the United States flag is called the *union.* The length of the union is $\frac{2}{5}$ of the fly, and the width is $\frac{7}{13}$ of the hoist. If the fly of a United States flag is 3 feet, how long is the union? **58. $1\frac{1}{5}$ feet**

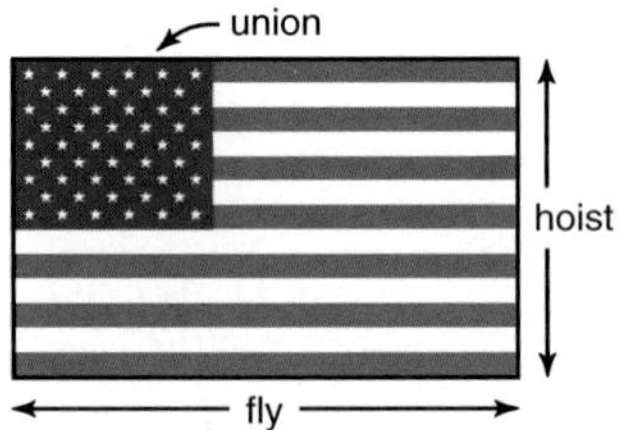

Additional Answers

57a. No; it meets the requirement that no dimensions can be less than 20 feet, but the minimum of 1250 square feet is not met because this yard would have 1216 square feet.

57b. No; it meets the requirement that yards must have a minimum of 1250 square feet because it would have 1330 square feet, but the requirement that no dimension can be less than 20 feet is not met because it has a side of length 19 feet.

57c. Answers will vary. Sample answer: 30 feet

Mixed Review

59. Find $5.7 + (-7.9)$. (Lesson 2–5) **−2.2**

60. Replace _?_ with <, >, or = to make 12 _?_ 15 − 7 true. (Lesson 2–4) **>**

61. Evaluate $a - 12$ if $a = -8$. (Lesson 2–3) **−20**

62. **Statistics** State the scale you would use to make a line plot for the following data. Then draw the line plot. (Lesson 2–2) **See margin.**

    4, 9, 2, 2, 15, 7, 6, 6, 9, 12, 1, 3, 2, 11, 10, 2, 6, 4, 12, 13

63. Find $6 + (-13)$. If necessary, use a number line. (Lesson 2–1) **−7**

64. Write an algebraic expression for *five times the sum of x and y decreased by z.* (Lesson 1–8) **$5(x + y) - z$ or $5(x + y - z)$**

65. Find the solution set for $y + 3 > 8$ if the replacement set is {3, 4, 5, 6, 7}. (Lesson 1–5) **{6, 7}**

66. **Statistics** Suppose the number 27,878 is rounded to 27,900 and plotted using stem 27 and leaf 9. Write the stem and leaf for each number below if the numbers are part of the same set of data. (Lesson 1–4) **See margin.**
    a. 13,245 b. 35,684 c. 153,436

67. Evaluate $6(4^3 + 2^2)$. (Lesson 1–3) **408**

68. **Geometry** Write an expression that represents the total number of small cubes in the large cube shown at the right. Then evaluate the expression. (Lesson 1–1) **4^3; 64 cubes**

4 ASSESS

Closing Activity

Speaking Without finding each product, have students indicate whether the product will be positive, negative, or 0.

1. $(-4)(-3)$ **positive**
2. $(-1.3)(4)(-2.54)$ **positive**
3. $\left(-\frac{3}{4}\right)\left(-\frac{1}{2}\right)(0)\left(-\frac{7}{10}\right)$ **0**
4. $\left(\frac{2}{5}\right)\left(-\frac{3}{8}\right)\left(\frac{5}{9}\right)$ **negative**

Additional Answers

62. from 1 to 15 by 1s

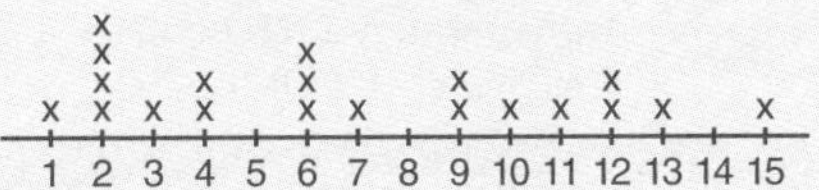

66a. stem 13 and leaf 2
66b. stem 35 and leaf 7
66c. stem 153 and leaf 4

WORKING ON THE

In·ves·ti·ga·tion

Refer to the Investigation on pages 68–69.

the Greenhouse Effect

Without a heat-trapping blanket of naturally occurring CO_2, it is believed that Earth would have an average surface temperature of −17°C, instead of its current average, 15°C. There is evidence that Mars has little CO_2 in its atmosphere, and its temperature never exceeds −31°C. At the other extreme, Venus, with lots of CO_2, has an average temperature of 454°C.

1 What would the difference in temperature be between Earth with CO_2 and Earth without CO_2? Do you think life could exist on Earth without CO_2? Explain.

2 What is the difference in temperature between Mars and Earth without CO_2? Explain what other factors contribute to the difference in temperature.

3 What is the difference in temperature between Venus and Mars? Explain what other factors contribute to the difference in temperature.

4 Make a chart that shows the three planets and their temperatures with CO_2 in their atmospheres and with little or no CO_2. Assume that the differences in the temperatures for the other planets would be the same as the difference in Earth's temperatures with CO_2 and without CO_2. You will use this information later on as you work on the Investigation.

Add the results of your work to your Investigation Folder.

Extension

Problem Solving Negative three is added to a certain number 27 times. If the result is −29, what was the original number? **52**

In·ves·ti·ga·tion

Working on the Investigation

The Investigation on pages 68–69 is designed to be a long-term project that is completed over several days or weeks. Encourage students to keep their materials in their Investigation Folder as they work on the Investigation.

Enrichment Masters, p. 15

NAME______________ DATE ________

Enrichment

Student Edition Pages 106–111

Convergence, Divergence, and Limits

Imagine that a runner runs a mile from point A to point B. But, this is not an ordinary race! In the first minute, he runs one-half mile, reaching point C. In the next minute, he covers one-half the remaining distance, or $\frac{1}{4}$ mile, reaching point D. In the next minute he covers one-half the remaining distance, or $\frac{1}{8}$ mile, reaching point E.

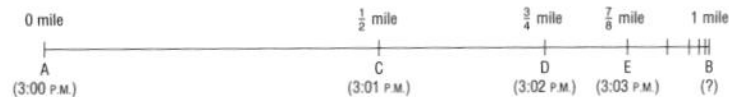

In this strange race, the runner approaches closer and closer to point B, but never gets there. However close he is to B, there is still some distance remaining, and in the next minute he can cover only half of that distance.

This race can be modeled by the infinite sequence $\frac{1}{2}, \frac{3}{4}, \frac{7}{8}, \frac{15}{16}, \cdots$.

The terms of the sequence get closer and closer to 1. An infinite sequence that gets arbitrarily close to some number is said to **converge** to that number. The number is the limit of the sequence.

Not all infinite sequences converge. Those that do not are called **divergent.**

Write C if the sequence converges and D if it diverges. If the sequence converges, make a reasonable guess for its limit.

1. 2, 4, 6, 8, 10, ⋯ **D**
2. 0, 3, 0, 3, 0, 3, ⋯ **D**
3. $1, \frac{1}{2}, \frac{1}{3}, \frac{1}{4}, \frac{1}{5}, \cdots$ **C, 0**
4. 0.9, 0.99, 0.999, 0.9999, ⋯ **C, 1**
5. −5, 5, −5, 5, −5, 5, ⋯ **D**
6. 0.1, 0.2, 0.3, 0.4, ⋯ **D**
7. $2\frac{1}{4}, 2\frac{3}{4}, 2\frac{7}{8}, 2\frac{15}{16}, \cdots$ **C, 3**
8. $6, 5\frac{1}{2}, 5\frac{1}{3}, 5\frac{1}{4}, 5\frac{1}{5}, \cdots$ **C, 5**
9. 1, 4, 9, 16, 25, ⋯ **D**
10. $1, -\frac{1}{2}, \frac{1}{3}, -\frac{1}{4}, \frac{1}{5}, -\frac{1}{6}, \cdots$ **C, 0**
11. Create one convergent sequence and one divergent sequence. Give the limit for your convergent sequence. **Answers will vary.**

2-7 LESSON NOTES

NCTM Standards: 1–5

Instructional Resources

- Study Guide Master 2-7
- Practice Master 2-7
- Enrichment Master 2-7
- Assessment and Evaluation Masters, p. 45
- Modeling Mathematics Masters, pp. 36–38

Transparency 2-7A contains the 5-Minute Check for this lesson; **Transparency 2-7B** contains a teaching aid for this lesson.

Recommended Pacing	
Standard Pacing	Day 9 of 15
Honors Pacing	Day 9 of 15
Block Scheduling*	Day 6 of 9
Alg. 1 in Two Years*	Days 17, 18, & 19 of 28

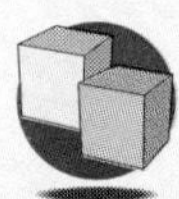

*For more information on pacing and possible lesson plans, refer to the *Block Scheduling Booklet* and *Algebra 1 in Two Years.*

1 FOCUS

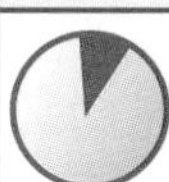

5-Minute Check
(over Lesson 2-6)

Find each product.

1. $\left(-\frac{2}{3}\right)\left(\frac{5}{4}\right)$ $-\frac{5}{6}$

2. $(0.1)(-1.1)(-0.01)$ 0.0011

Evaluate each expression if $x = -3$ and $y = \frac{4}{3}$.

3. $2xy + 4x$ -20

4. $-2(x - 6xy)$ -42

5. Find $\frac{3}{4}\begin{bmatrix} -\frac{3}{4} & 4 & 16 \\ 2 & 0 & -1\frac{1}{3} \end{bmatrix}$.

$\begin{bmatrix} -\frac{9}{16} & 3 & 12 \\ \frac{3}{2} & 0 & -1 \end{bmatrix}$

2–7 Dividing Rational Numbers

What YOU'LL LEARN

- To divide rational numbers.

Why IT'S IMPORTANT

You can divide rational numbers to help you solve problems involving drafting, nursing, and economics.

Aviation

Vicki Van Meter had a lot to write in her report about her summer vacation of 1993. At 12 years old, inspired by Amelia Earhart, she became the youngest female to pilot a plane across the Atlantic Ocean. Vicki did all of the flying, navigation, and communication for her 3200-kilometer flight aboard a single-engine Cessna 210. She also had to be concerned about how supplies and fuel were loaded onto the plane.

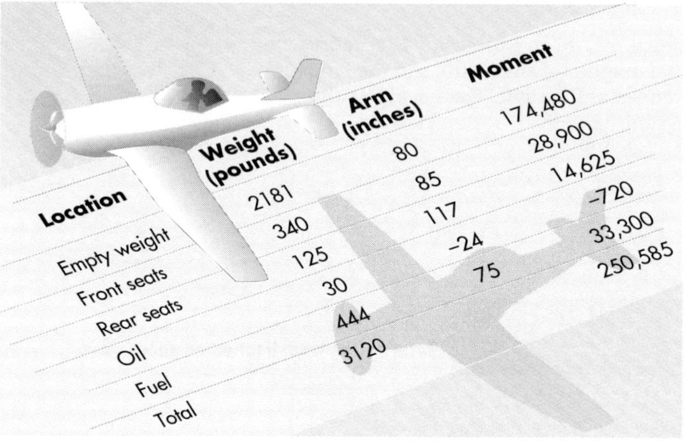

Location	Weight (pounds)	Arm (inches)	Moment
Empty weight	2181	80	174,480
Front seats	340	85	28,900
Rear seats	125	117	14,625
Oil	30	−24	−720
Fuel	444	75	33,300
Total	3120		250,585

Before any small aircraft can take off, the pilot must be sure that the aircraft is loaded so that the center of gravity is within certain safe limits. If it is, the pilot is ready for takeoff. If not, the weight must be rebalanced.

You can use the table at the left to find the center of gravity for a particular aircraft. The safe limit for this aircraft is 82.1. For each location, the weight and the arm are multiplied to find the moment.

Now add the moments, divide by the total weight, and round to the nearest tenth to find the center of gravity.

Total Moments		*Total Weight*		*Center of Gravity*
250,585	÷	3120	≈	80.3

Since 80.3 is less than 82.1, the aircraft is safe and ready for takeoff.

You already know that the quotient of two positive numbers is positive. But how do you determine the sign of the quotient when negative numbers are involved? Since multiplication and division are inverse operations, the rule for finding the sign of the quotient of two numbers is similar to the rule for finding the sign of the product. Study these patterns.

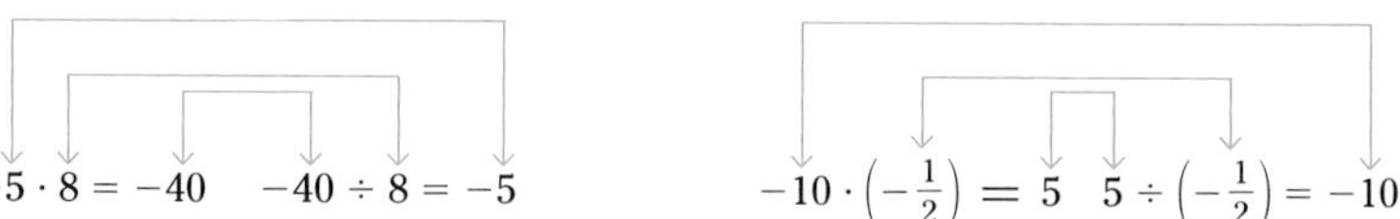

$-5 \cdot 8 = -40 \quad -40 \div 8 = -5$
$-10 \cdot \left(-\frac{1}{2}\right) = 5 \quad 5 \div \left(-\frac{1}{2}\right) = -10$

These examples suggest the following rules.

Dividing Two *Rational Numbers*	The quotient of two numbers having the *same sign* is positive. The quotient of two numbers having *different signs* is negative.

Teaching Tip To find the center of gravity of an aircraft, the pilot multiplies the weight of all of the items on the plane by its *arm,* or the distance from a certain point called the *datum.* This product is called the *moment* of each item. The moments are then added together and divided by the total weight of the items. The quotient equals the distance of the center of gravity from the datum. The pilot then checks this number to see if the center of gravity is within safe limits.

Example 1 **Find each quotient.**

a. $\mathbf{-75 \div (-15)}$

This division problem may also be written as $\frac{-75}{-15}$.

$\frac{-75}{-15} = 5$ *The fraction bar indicates division.*

Since the signs are the same, the quotient is positive.

b. $\mathbf{\frac{72}{-8}}$

$\frac{72}{-8} = -9$ *Since the signs are different, the quotient is negative.*

Recall that you can change any division expression to an equivalent multiplication expression. To divide by any nonzero number, multiply by the reciprocal of that number.

Example 2 **Find each quotient.**

a. $\mathbf{\frac{1}{2} \div 5}$

$\frac{1}{2} \div 5 = \frac{1}{2} \cdot \frac{1}{5}$ *Multiply by $\frac{1}{5}$, the reciprocal of 5.*

$= \frac{1}{10}$ *The signs are the same, so the product is positive.*

b. $\mathbf{-\frac{6}{7} \div 3}$

$-\frac{6}{7} \div 3 = -\frac{6}{7} \cdot \frac{1}{3}$ *Multiply by $\frac{1}{3}$, the reciprocal of 3.*

$= -\frac{6}{21}$ or $-\frac{2}{7}$ *The signs are different, so the product is negative.*

Example 3

Nursing

Three of the measurements nurses commonly use are cubic centimeters (cc), drops, and grains. Use this information to solve the following.

a. A doctor orders $\frac{1}{400}$ of a grain of medicine to be given to a patient. The nurse has a vial labeled $\frac{1}{200}$ grain per cc. How many cc of the medicine should the nurse give the patient?

b. The doctor also prescribes a 1000-cc intravenous (IV) pouch of fluid to be given to the patient over an 8-hour period. If there are 15 drops in 1 cc, for how many drops per minute should the nurse set the IV?

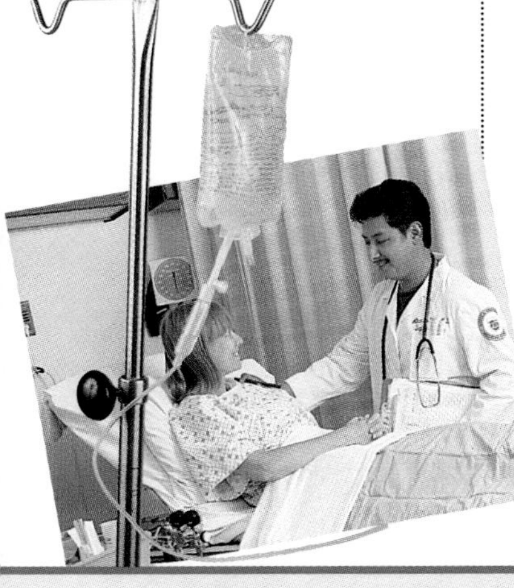

a. How many cc are in $\frac{1}{400}$ of a grain? Divide $\frac{1}{400}$ grain by $\frac{1}{200}$ grain/cc.

$$\frac{1}{400} \text{ grain} \div \frac{1}{200} \text{ grain/cc} = \frac{1 \text{ grain}}{400} \cdot \frac{200 \text{ cc}}{1 \text{ grain}}$$

$$= \frac{\overset{1}{\cancel{1 \text{ grain}}}}{\underset{2}{\cancel{400}}} \cdot \frac{\overset{1}{\cancel{200}} \text{ cc}}{\underset{1}{\cancel{1 \text{ grain}}}}$$

$$= \frac{1}{2} \text{ cc}$$

The nurse should give the patient $\frac{1}{2}$ cc of the medicine.

(continued on the next page)

Motivating the Lesson

Hands-On Activity Have students look up the term *inverse* in the dictionary. Develop student understanding of the term by having a volunteer mime a sequence of three actions. Ask another volunteer to mime the inverse of these actions. Then discuss the meaning of a multiplicative inverse, or reciprocal.

2 TEACH

In-Class Examples

For Example 1
Find each quotient.

a. $-32 \div -8$ 4
b. $\frac{81}{-9}$ -9

For Example 2
Find each quotient.

a. $\frac{5}{6} \div 3$ $\frac{5}{18}$
b. $-\frac{3}{4} \div \frac{5}{8}$ $-1\frac{1}{5}$
c. $-\frac{1}{5} \div \frac{3}{10}$ $-\frac{2}{3}$

For Example 3

a. If the doctor orders $\frac{1}{50}$ of a grain to be given to a patient, how many cc should the nurse administer? 4cc
b. If the doctor prescribes an 800-cc intravenous bottle to be given over a 6-hour period, for how many drops per minute should the nurse set the IV? $33\frac{1}{3}$

Teaching Tip Remind students that division by 0 is not possible. In addition, $\frac{0}{n} = 0$ regardless of the sign of n if $n \neq 0$.

Cooperative Learning

Trade-A-Problem Have students work in pairs. Each pair should develop ten problems dealing with division of fractions, with both students agreeing on the answers. Pairs of students should trade sets of problems, solve them, and then join to check answers.

For more information on the trade-a-problem strategy, see *Cooperative Learning in the Mathematics Classroom,* one of the titles in the Glencoe Mathematics Professional Series, pages 25–26.

In-Class Examples

For Example 4
Write each fraction in simplest form.

a. $\frac{\frac{3}{9}}{4}$ $\frac{4}{3}$

b. $-\frac{6}{\frac{3}{2}}$ -4

For Example 5
Simplify.

a. $\frac{-8x-30}{2}$ $-4x-15$

b. $\frac{-25y-4}{5}$ $-5y-\frac{4}{5}$

3 PRACTICE/APPLY

Check for Understanding

Exercises 1–13 are designed to help you assess your students' understanding through reading, writing, speaking, and modeling. You should work through Exercises 1–4 with your students and then monitor their work on Exercises 5–13.

Error Analysis
Since division is not commutative, determining the major division bar in a complex fraction is crucial.

Consider $\frac{\frac{3}{4}}{8}$. Is this $\frac{\frac{3}{4}}{8}$ or $\frac{3}{\frac{4}{8}}$?

Their values are $\frac{3}{32}$ and 6, respectively. The author of the fraction should indicate the major fraction bar by some emphasis (bolder, longer, or wider). If no major fraction bar is indicated, the lowest one is the major one by default.

Example: $\frac{\frac{64}{8}}{4}$ defaults to $\frac{\frac{64}{8}}{4}$, which is $64 \div 8 \div 4$ or 2.

b. $\frac{1000 \text{ cc}}{8 \text{ h}} = \left(\frac{\overset{125}{\cancel{1000 \text{ cc}}}}{\underset{1}{\cancel{8 \text{ h}}}}\right)\left(\frac{\overset{1}{15 \text{ drops}}}{\cancel{1 \text{ cc}}}\right)\left(\frac{\cancel{1 \text{ h}}}{\underset{4}{\cancel{60 \text{ min}}}}\right)$ *There are 15 drops in 1 cc and 60 minutes in 1 hour.*

$= \frac{125 \text{ drops}}{4 \text{ min}}$

$= 31\frac{1}{4}$ drops/min

The nurse should set the IV for $31\frac{1}{4}$ drops per minute.

If a fraction has one or more fractions in the numerator or denominator, it is called a **complex fraction.** To simplify a complex fraction, rewrite it as a division sentence.

Example 4 **Write each fraction in simplest form.**

a. $\frac{\frac{2}{3}}{8}$

Rewrite the fraction as $\frac{2}{3} \div 8$, since fractions indicate division.

$\frac{2}{3} \div 8 = \frac{2}{3} \cdot \frac{1}{8}$ *Multiply by $\frac{1}{8}$, the reciprocal of 8.*

$= \frac{2}{24}$ or $\frac{1}{12}$

b. $\frac{-5}{\frac{3}{7}}$

Rewrite the fraction as $-5 \div \frac{3}{7}$.

$-5 \div \frac{3}{7} = -5 \times \frac{7}{3}$ *Multiply by $\frac{7}{3}$, the reciprocal of $\frac{3}{7}$.*

$= -\frac{35}{3}$ or $-11\frac{2}{3}$ *The signs are different, so the product is negative.*

You can use the distributive property to simplify fractional expressions.

Example 5 **Simplify $\frac{-3a+16}{4}$.**

Method 1

$\frac{-3a+16}{4} = (-3a+16) \div 4$

$= (-3a+16)\left(\frac{1}{4}\right)$ *To divide by 4, multiply by $\frac{1}{4}$.*

$= -3a\left(\frac{1}{4}\right) + 16\left(\frac{1}{4}\right)$ *Distributive property*

$= -\frac{3}{4}a + 4$

Method 2

$\frac{-3a+16}{4} = -\frac{3a}{4} + \frac{16}{4}$

$= -\frac{3}{4}a + 4$

CHECK FOR UNDERSTANDING

Communicating Mathematics

Study the lesson. Then complete the following. 1, 4. See margin.

1. **Compare** multiplying rational numbers to dividing rational numbers. How are the two operations similar?
2. **Complete** the sentence: Dividing by any number, except zero, is the same as __?__. multiplying by its reciprocal
3. **Find** a value for x if $\frac{1}{x} > x$. Sample answer: $x = \frac{1}{2}$
4. **You Decide** Simone says that $-\frac{4}{5}$ is equal to $\frac{-4}{-5}$. Miguel says that $-\frac{4}{5}$ is equal to $\frac{-4}{5}$ or $\frac{4}{-5}$. Which one is correct, and why?

Guided Practice

Simplify.

5. $\frac{32}{-8}$ -4
6. $\frac{-77}{11}$ -7
7. $-\frac{3}{4} \div 8$ $-\frac{3}{32}$
8. $\frac{2}{3} \div 9$ $\frac{2}{27}$
9. $\frac{\frac{-5}{6}}{8}$ $-\frac{5}{48}$
10. $\frac{54s}{6}$ $9s$
11. $\frac{-300x}{50}$ $-6x$
12. $\frac{6b + 12}{6}$ $b + 2$
13. **Money** On October 10, 1994, *USA Today* reported that the United States government spends about \$168 million per hour. Stanley Newberg, who came to the United States from Austria in 1906, died in 1994 at age 81 and left the government \$5.6 million. How long did it take the government to spend Mr. Newberg's money? about 2 minutes

EXERCISES

Practice

Simplify.

14. $\frac{-36}{4}$ -9
15. $\frac{-96}{-16}$ 6
16. $-9 \div \left(-\frac{10}{17}\right)$ $\frac{153}{10}$
17. $-\frac{2}{3} \div 12$ $-\frac{1}{18}$
18. $-64 \div (-8)$ 8
19. $-\frac{3}{4} \div 12$ $-\frac{1}{16}$
20. $-18 \div 9$ -2
21. $78 \div (-13)$ -6
22. $-108 \div (-9)$ 12
23. $-\frac{2}{3} \div 8$ $-\frac{1}{12}$
24. $\frac{-1}{3} \div (-4)$ $\frac{1}{12}$
25. $-9 \div \left(-\frac{10}{27}\right)$ $\frac{243}{10}$

26. $\frac{\frac{5}{6}}{-10}$ $-\frac{1}{12}$
27. $-\frac{7}{\frac{3}{5}}$ $-\frac{35}{3}$
28. $\frac{-5}{\frac{2}{7}}$ $-\frac{35}{2}$
29. $\frac{-650m}{10}$ $-65m$
30. $\frac{81c}{-9}$ $-9c$
31. $\frac{8r + 24}{8}$ $r + 3$

33. $-20a - 25b$

32. $\frac{6a + 24}{6}$ $a + 4$
33. $\frac{40a + 50b}{-2}$
34. $\frac{-5x + (-10y)}{-5}$ $x + 2y$

37. $-a - 4b$

35. $\frac{42c - 18d}{-3}$ $-14c + 6d$
36. $\frac{-8f + (-16g)}{8}$ $-f - 2g$
37. $\frac{-4a + (-16b)}{4}$

Evaluate if $a = 5$, $b = -6$, and $c = -1.5$.

38. $\frac{b}{c}$ 4
39. $b \div a$ -1.2
40. $(a + b) \div c$ $0.\overline{6}$

41. $-0.8\overline{3}$

41. $(a + b + c) \div 3$
42. $\frac{c}{a}$ -0.3
43. $\frac{ab}{ac}$ 4

Reteaching

Using Alternative Methods Here is an easy way to use signed numbers to find an average A of a set of n numbers. Guess an average G. Subtract G from each number to find its deviation. Find the total of the deviations and then divide by n to get the average deviation D.
$A = G + D$

Scores	Deviations for G = 80	Deviations for G = 75
92	12	17
60	−20	−15
84	4	9
73	−7	−2
81	1	6
TOTAL	−10	15

$D = -2$ $\quad D = 3$
$A = 80 + (-2)$ $\quad A = 75 + 3$

Additional Answers

1. Both multiplying and dividing rational numbers involve multiplication. In division, you multiply by the reciprocal of the second number.
4. Miguel is correct.
$$-\frac{4}{5} = (-1)\frac{4}{5} = \frac{(-1)4}{5} = \left(\frac{1}{-1}\right)\frac{4}{5} = \frac{4}{-5}$$
$$\frac{-4}{-5} = \frac{(-1)4}{(-1)5} = \left(\frac{-1}{-1}\right)\left(\frac{4}{5}\right) = (1)\frac{4}{5} = \frac{4}{5}$$

Assignment Guide

Core: 15–43 odd, 44, 45, 47, 48–55
Enriched: 14–42 even, 44–55

For **Extra Practice,** see p. 761.

The red A, B, and C flags, printed only in the Teacher's Wraparound Edition, indicate the level of difficulty of the exercises.

Study Guide Masters, p. 16

NAME______________________ DATE__________

Student Edition Pages 112–117

Study Guide

Dividing Rational Numbers

Use the following rules to divide rational numbers.

Rule or Property		Example
Dividing Rational Numbers	The quotient of two numbers is positive if the numbers have the same sign. The quotient of two numbers is negative if the numbers have different signs.	$-60 \div (-10) = 6$ $-48 \div 4 = -12$
Multiplicative Inverse Property	For every nonzero number a, there is exactly one number $\frac{1}{a}$, such that $(a)\frac{1}{a} = \frac{1}{a}(a) = 1$.	$\frac{1}{3} \div \frac{3}{4} = \frac{1}{3} \cdot \frac{4}{3}$ $= \frac{4}{9}$
Division Rule	For all numbers a and b, with $b \neq 0$, $a \div b = \frac{a}{b} = a\left(\frac{1}{b}\right) = \frac{1}{b}(a)$.	$6 \div 2 = \frac{6}{2}$ $= (6)\frac{1}{2}$ $= \frac{1}{2}(6) = 3$

Since the fraction bar indicates division, you can use the division rules and the distributive property to simplify rational expressions.

Example: Simplify $\frac{-20a + 15}{5}$.

$$\frac{-20a + 15}{5} = (-20a + 15)\left(\frac{1}{5}\right) = (-20a)\left(\frac{1}{5}\right) + (15)\left(\frac{1}{5}\right) = -5a + 3$$

Simplify.

1. $\frac{-44a}{4}$ $-11a$
2. $\frac{16x}{2}$ $8x$
3. $\frac{80}{5}$ 16
4. $\frac{81}{-27}$ -3
5. $\frac{-144a}{6}$ $-24a$
6. $\frac{-30}{-10} \div \frac{30}{10}$ 1
7. $\frac{57y}{3}$ $19y$
8. $-\frac{1}{2} \div 8$ $-\frac{1}{16}$
9. $\frac{18a - 6b}{-3}$ $-6a + 2b$
10. $\frac{12x}{3} \div \frac{1}{12x} + xyz$ $48x^2 + xyz$
11. $\frac{36a - 12}{12}$ $3a - 1$
12. $\frac{-\frac{5}{8}}{5}$ $-\frac{1}{8}$

Additional Answers

44a. You would get an error since you are asking the calculator to divide by 0 and this is undefined.

44b. You will end up with the original number since you reciprocated and then reciprocated again. If n is even, you will get the original number. If n is odd, you will get the reciprocal of the number.

44c. 1; You multiplied a number by its reciprocal, which always results in a product of 1.

Practice Masters, p. 16

2-7 NAME__________ DATE ________

Practice Student Edition Pages 112–117

Dividing Rational Numbers

Simplify.

1. $\frac{-63}{9}$ -7
2. $\frac{36}{-6}$ -6
3. $\frac{-42}{-7}$ 6
4. $\frac{-144x}{12}$ $-12x$
5. $\frac{-54z}{-9}$ $6z$
6. $\frac{56a}{-8}$ $-7a$
7. $-36 \div (-4)$ 9
8. $-24 \div 6$ -4
9. $75 \div (-15)$ -5
10. $-\frac{2}{3} \div 6$ $-\frac{1}{9}$
11. $-\frac{5}{6} \div (-20)$ $\frac{1}{24}$
12. $-8 \div \left(-\frac{9}{14}\right)$ $\frac{112}{9}$
13. $\frac{\frac{3}{4}}{-6}$ $-\frac{1}{8}$
14. $\frac{6}{-\frac{2}{9}}$ -27
15. $\frac{-7}{\frac{2}{5}}$ $-\frac{35}{2}$
16. $\frac{5x + 25}{5}$ $x + 5$
17. $\frac{18t + 12r}{-3}$ $-6t - 4r$
18. $\frac{5a - 10}{-5}$ $-a + 2$
19. $\frac{18x + 12}{-6}$ $-3x - 2$
20. $\frac{-4c + (-16d)}{4}$ $-c - 4d$
21. $\frac{8k - 12h}{-4}$ $-2k + 3h$
22. $-323 \div (-17)$ 19
23. $-5 \div \frac{3}{7}$ $-\frac{35}{3}$
24. $-\frac{7}{4} \div (-8)$ $\frac{7}{32}$

Critical Thinking

44. The [1/x] key on a scientific calculator is called the *reciprocal key*. When this key is pressed, the calculator replaces the number on the display with its reciprocal. **a–c. See margin.**
   a. Enter 0 and then press the reciprocal key. What happened? Explain.
   b. Enter a number and then press the reciprocal key twice. What happened? Predict what will happen if you press the key n times.
   c. Enter 6.435 [1/x] [×] 6.435 [=]. What is the result? Why?

Applications and Problem Solving

45. **Drafting** Sofia Fernandez is making a dresser for her little sister's nursery. The dresser will be 30 inches high and will have a 4-inch-thick base and a $1\frac{1}{2}$-inch-thick top. Four equal-sized drawers are to fit in the remaining space, with $\frac{3}{4}$ inch between each drawer.
   a. Sketch the dresser, labeling each part. **See margin.**
   b. What is the height of each drawer? **$5\frac{9}{16}$ inches**

46. **Economics** Kim Lee's shoe store has a buy one, get one half-price sale. You must pay full price for the more expensive pair of shoes in order to receive half off the second pair. If you buy a single pair of shoes, you get $\frac{1}{5}$ off. Together, Sharon and LaShondra find five pair of shoes that cost \$45.99, \$23.88, \$36.99, \$19.99, and \$14.99. How should they pay for the shoes in order to get the best buy? **See margin.**

47. **Science** *Precision* is the degree of exactness to which a measurement can be reproduced. It is determined by subtracting the least measurement from the greatest measurement, and dividing by 2. Suppose you conducted an experiment to determine the length of a piece of lumber. After measuring several times, you have recorded measurements ranging between 17.239 cm and 17.561 cm, inclusive.
   a. Use the plus-or-minus symbol (±) to describe your measurements. **17.4 ± 0.161 cm**
   b. What was the precision of your measurement? **0.161 cm**

Additional Answers

45a.

30"
$1\frac{1}{2}$"
$\frac{3}{4}$"
$\frac{3}{4}$"
$\frac{3}{4}$"
4"

46. If they buy them each individually, the total cost will be \$113.47 ($\frac{1}{5}$ off each pair). If Sharon buys the two most expensive pairs and LaShondra buys the next two most expensive pairs and then one of them buys the fifth pair by itself, the total cost will be \$110.37. Therefore, the latter is how they should buy the shoes.

Mixed Review

50. Sample answer: $\frac{19}{24}$

51a. $\begin{bmatrix} -2 & 4 \\ 3 & 12 \end{bmatrix}$

51b. $\begin{bmatrix} 4 & 4 \\ 7 & 2 \end{bmatrix}$

48. Find $\left(-\frac{1}{5}\right)\left(\frac{3}{2}\right)(-2)$. (Lesson 2–6) $\frac{3}{5}$

49. Evaluate $\frac{9}{4} + \frac{x}{6}$ if $x = -7$. (Lesson 2–5) $\frac{13}{12}$

50. Name a fraction between $\frac{2}{3}$ and $\frac{7}{8}$ whose denominator is 24. (Lesson 2–4)

51. Let $A = \begin{bmatrix} 1 & 4 \\ 5 & 7 \end{bmatrix}$ and $B = \begin{bmatrix} -3 & 0 \\ -2 & 5 \end{bmatrix}$. (Lesson 2–3)

a. Find $A + B$.　　b. Find $A - B$.

52. Statistics The table below shows the mean scores for the mathematics portion of the Scholastic Assessment Test (SAT) by state for 1994. (Lesson 2–2)

SAT Mean Mathematics Scores, 1994

State	Score	State	Score	State	Score	State	Score	State	Score
AL	529	HI	480	MA	475	NM	528	SD	548
AK	477	ID	508	MI	537	NY	472	TN	535
AZ	496	IL	546	MN	562	NC	455	TX	474
AR	518	IN	466	MS	528	ND	559	UT	558
CA	482	IA	574	MO	532	OH	510	VT	472
CO	513	KS	550	MT	523	OK	537	VA	469
CT	472	KY	523	NE	543	OR	491	WA	488
DE	464	LA	530	NV	484	PA	462	WV	482
FL	466	ME	463	NH	486	RI	462	WI	557
GA	446	MD	479	NJ	475	SC	443	WY	521

Source: College Entrance Examination Board

a. What is the range of scores for the fifty states? 131

b. Make a line plot of the data. See Solutions Manual.

c. Do the scores cluster around a certain number? If so, which ones? yes, 472

53. Simplify $8b + 12(b + 2)$. (Lesson 1–7) $20b + 24$

54. Name the multiplicative inverse of $\frac{4}{5}$. (Lesson 1–6) $\frac{5}{4}$

55. Statistics The list below shows the prices of several 1995 models of sport utility vehicles, rounded to the nearest hundred dollars. (Lesson 1–4)

\$22,000	\$19,400	\$29,000	\$13,000	\$22,200	\$17,300
\$25,400	\$25,100	\$33,000	\$20,000	\$30,700	\$15,400
\$27,900	\$34,600	\$30,400	\$24,500	\$52,500	\$17,200

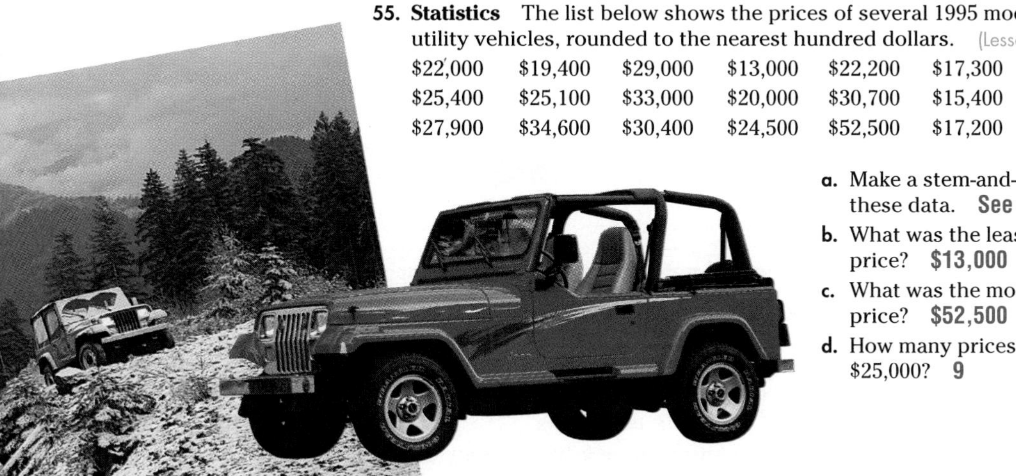

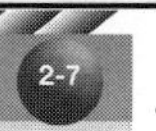

a. Make a stem-and-leaf plot of these data. See margin.

b. What was the least expensive price? \$13,000

c. What was the most expensive price? \$52,500

d. How many prices were over \$25,000? 9

Extension

Reasoning Matt had an average of 72 on three 100-point tests. What score would he need on his next test to raise his average to 75? 84

4 ASSESS

Closing Activity

Speaking Have students explain the steps they would follow when comparing $\frac{3}{-5}$ and $\frac{8}{-9}$ by finding cross products. Encourage students to justify each step in the procedure.

Chapter 2, Quiz C (Lessons 2-6 and 2-7), is available in the *Assessment and Evaluation Masters,* p. 45.

Additional Answer

55a.

Stem	Leaf
13	0
15	4
17	2 3
19	4
20	0
22	0 2
24	5
25	1 4
27	9
29	0
30	4 7
33	0
34	6
52	5

13|0 = 13,000

Enrichment Masters, p. 16

2-7 NAME ______ DATE ______

Enrichment

Student Edition Pages 112–117

Counterexamples

Some statements in mathematics can be proven false by **counterexamples.** Consider the following statement.

For any numbers a and b, $a - b = b - a$.

You can prove that this statement is false in general if you can find one example for which the statement is false.

Let $a = 7$ and $b = 3$. Substitute these values in the equation above.

$$7 - 3 \stackrel{?}{=} 3 - 7$$
$$4 \neq -4$$

In general, for any numbers a and b, the statement $a - b = b - a$ is false. You can make the equivalent verbal statement: subtraction is *not* a commutative operation.

In each of the following exercises a, b, and c are any numbers. Prove that the statement is false by counterexample. Sample answers are given.

1. $a - (b - c) \stackrel{?}{=} (a - b) - c$
   $6 - (4 - 2) \stackrel{?}{=} (6 - 4) - 2$
   $6 - 2 \stackrel{?}{=} 2 - 2$
   $4 \neq 0$

2. $a \div (b \div c) \stackrel{?}{=} (a \div b) \div c$
   $6 \div (4 \div 2) \stackrel{?}{=} (6 \div 4) \div 2$
   $\frac{6}{2} \stackrel{?}{=} \frac{1.5}{2}$
   $3 \neq 0.75$

3. $a \div b \stackrel{?}{=} b \div a$
   $6 \div 4 \stackrel{?}{=} 4 \div 6$
   $\frac{3}{2} \neq \frac{2}{3}$

4. $a \div (b + c) \stackrel{?}{=} (a \div b) + (a \div c)$
   $6 \div (4 + 2) \stackrel{?}{=} (6 \div 4) + (6 \div 2)$
   $6 \div 6 \stackrel{?}{=} 1.5 + 3$
   $1 \neq 4.5$

5. $a + (bc) \stackrel{?}{=} (a + b)(a + c)$
   $6 + (4 \cdot 2) \stackrel{?}{=} (6 + 4)(6 + 2)$
   $6 + 8 \stackrel{?}{=} (10)(8)$
   $14 \neq 80$

6. $a^2 + a^2 \stackrel{?}{=} a^4$
   $6^2 + 6^2 \stackrel{?}{=} 6^4$
   $36 + 36 = 1296$
   $72 \neq 1296$

7. Write the verbal equivalents for Exercises 1, 2, and 3.
   1. Subtraction is not an associative operation.
   2. Division is not an associative operation.
   3. Division is not a commutative operation.

8. For the distributive property $a(b + c) = ab + ac$ it is said that multiplication distributes over addition. Exercises 4 and 5 prove that some operations do not distribute. Write a statement for each exercise that indicates this.
   4. Division does not distribute over addition.
   5. Addition does not distribute over multiplication.

2-8A LESSON NOTES

NCTM Standards: 1–5

Objective
Use base-ten tiles to model square roots.

Recommended Time
Demonstration and discussion: 15 minutes; Exercises: 30 minutes

Instructional Resources
For each student or group of students
Modeling Mathematics Masters
- p. 16 (base-ten models)
- p. 22 (worksheet)

For teacher demonstration
Algebra and Geometry Overhead Manipulative Resources

1 FOCUS

Motivating the Lesson
Ask students to suggest a number whose square root is greater than 2 but less than 3. Point out that $2^2 = 4$ and $3^2 = 9$, so 5, 6, 7, and 8 are all correct answers to the question. Explain that students can use base-ten tiles to find such estimates of squares roots.

2 TEACH

Teaching Tip One strategy is to arrange the first four tiles into a two-by-two square. Fill in two adjacent sides to form a three-by-three square, and so on.

3 PRACTICE/APPLY

Assignment Guide
Core: 1–6 **Enriched:** 1–6

MODELING MATHEMATICS

A Preview of Lesson 2–8

2–8A Estimating Square Roots

Materials:  base-ten tiles

You can use base-ten tiles to model square roots. A **square root** is one of two identical factors of a number. For example, a square root of 144 is 12 since $12^2 = 144$.

Activity 1 Use base-ten tiles to find the square root of 121.

Step 1 Model 121 with base-ten tiles.

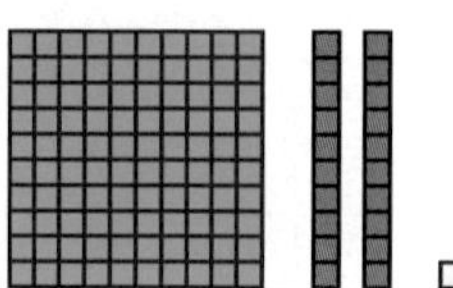

Step 2 Arrange the tiles into a square. The square root of 121 is 11 because $11^2 = 121$.

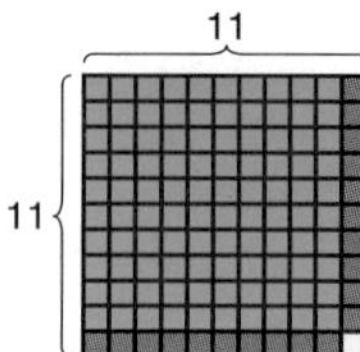

Activity 2 Use base-ten tiles to estimate the square root of 151.

Step 1 Model 151 with base-ten tiles.

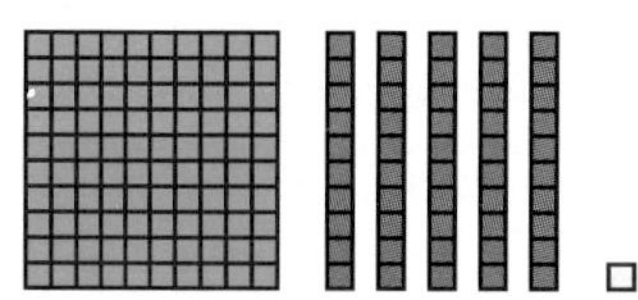

Step 2 Arrange the tiles into a square. The largest square possible has 144 tiles, with 7 tiles left over.

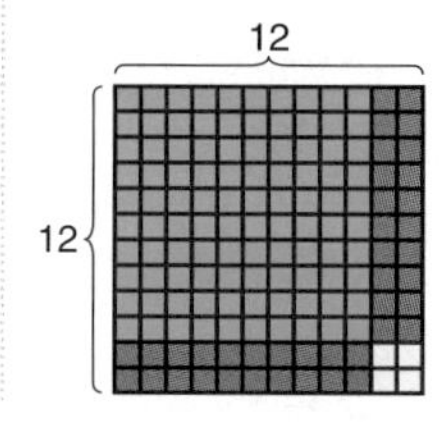

Trade a 10-tile for 10 1-tiles when necessary.

Step 3 Add tiles until you have the next larger square. You need to add 18 tiles. Since 151 is between 144 and 169, the square root of 151 is between 12 and 13.

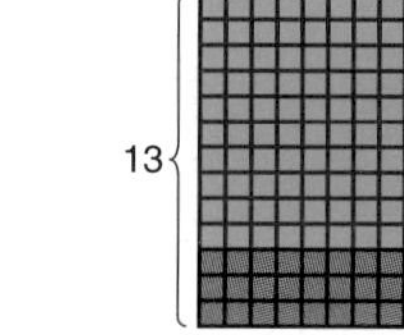

Model Use base-ten tiles to estimate the square root of each number.

1. 20 **4–5**
2. 450 **21–22**
3. 180 **13–14**
4. 200 **14–15**
5. 2 **1–2**

Write

6. If you list all of the factors of a number in numerical order, the square root of the number either is the middle number or lies between the two middle numbers. How can you use tiles to show this? **See margin.**

4 ASSESS

Observing students working in cooperative groups is an excellent method of assessment. You may wish to ask a student from each group to explain how to model and solve a problem. Also watch for and acknowledge students who are helping others understand the concept being taught.

Additional Answer

6. **Begin with a $1 \times n$ rectangle made of blocks, where n is the number being factored. As you form new rectangles with the same blocks, the first factor increases in value while the second factor decreases. When you find the square root, both dimensions are the same.**

2-8 Square Roots and Real Numbers

What YOU'LL LEARN
- To find square roots,
- to classify numbers, and
- to graph solutions of inequalities on number lines.

Why IT'S IMPORTANT

You need to know the values of irrational numbers to solve problems involving traffic safety and flying.

CONNECTION

Biology

Swedish botanist Carolus Linnaeus (1707–1778) developed the system we use today to classify every kind of living thing according to common characteristics. For example, an African elephant is from the Mammalia Class and also from the Animalia Kingdom.

Kingdom: Animalia
Phylum: Chordata
Class: Mammalia
Order: Proboscidea
Family: Elephantidae
Genus: Loxodonta
Species: Loxodonta africana
African elephant

In mathematics, we classify numbers that have common characteristics. So far in this text, we have classified numbers as natural numbers, whole numbers, integers, and rational numbers.

Finding a square root of 81 is the same as finding a number whose square is 81.

The square roots of perfect squares are classified as rational numbers. A **square root** is one of two equal factors of a number. For example, one square root of 81 is 9 since $9 \cdot 9$ or 9^2 is 81. A rational number, like 81, whose square root is a rational number, is called a **perfect square**.

It is also true that $-9 \cdot (-9) = 81$. Therefore, -9 is another square root of 81.

$$9^2 = 9 \cdot 9 = 81$$ *9^2 is read "nine squared" and means that 9 is used as a factor two times.*

$$(-9)^2 = (-9)(-9) = 81$$ *-9 is used as a factor two times.*

Definition of Square Root	**If $x^2 = y$, then x is a square root of y.**

The symbol $\sqrt{}$, called a **radical sign**, is used to indicate a nonnegative or **principal square root** of the expression under the radical sign.

$\sqrt{81} = 9$ — $\sqrt{81}$ indicates the *principal* square root of 81.

$-\sqrt{81} = -9$ — -81 indicates the *negative* square root of 81.

$\pm\sqrt{81} = \pm 9$ — $\pm\sqrt{81}$ indicates *both* square roots of 81.

$\pm\sqrt{81}$ is read "plus or minus the square root of 81."

2-8 LESSON NOTES

NCTM Standards: 1–5, 14

Instructional Resources

- Study Guide Master 2-8
- Practice Master 2-8
- Enrichment Master 2-8

Transparency 2-8A contains the 5-Minute Check for this lesson; **Transparency 2-8B** contains a teaching aid for this lesson.

Recommended Pacing	
Standard Pacing	Day 11 of 15
Honors Pacing	Day 11 of 15
Block Scheduling*	Day 7 of 9
Alg. 1 in Two Years*	Days 21 & 22 of 28

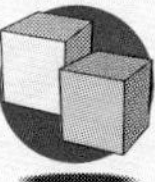

*For more information on pacing and possible lesson plans, refer to the *Block Scheduling Booklet* and *Algebra 1 in Two Years.*

1 FOCUS

5-Minute Check
(over Lesson 2-7)

Simplify.

1. $-\frac{2}{3} \div 6$ $-\frac{1}{9}$
2. $1\frac{1}{5} \div -36$ $-\frac{1}{30}$
3. $\frac{\frac{5}{8}}{4}$ $\frac{5}{32}$
4. $\frac{12 - 8b}{2}$ $6 - 4b$
5. $\frac{-70x + (-15)}{5}$ $-14x - 3$

Motivating the Lesson

Hands-On Activity Open the lesson by allowing students to experiment with the square root key on their calculators. Encourage students to experiment with both positive and negative numbers. Ask students to define *square root* in their own words.

2 TEACH

Teaching Tip Encourage students to check square roots by squaring.

In-Class Examples

For Example 1
Find each square root.

a. $\sqrt{36}$ **6**

b. $-\sqrt{169}$ **−13**

c. $\pm\sqrt{0.49}$ **± 0.7**

For Example 2
Use a calculator to evaluate each expression if $a = 441$, $b = 600$, and $c = 424$.

a. $\sqrt{a}$ **21**

b. $\sqrt{b + c}$ **32**

Example 1 **Find each square root.**

Some square roots can be found mentally.

a. $\sqrt{25}$

The symbol $\sqrt{25}$ represents the principal square root of 25.

Since $5^2 = 25$, you know that $\sqrt{25} = 5$.

b. $-\sqrt{144}$

The symbol $-\sqrt{144}$ represents the negative square root of 144.

Since $12^2 = 144$, you know that $-\sqrt{144} = -12$.

c. $\pm\sqrt{0.16}$

The symbol $\pm\sqrt{0.16}$ represents both square roots of 0.16.

Since $(0.4)^2 = 0.16$, you know that $\pm\sqrt{0.16} = \pm 0.4$.

Most scientific calculators have a *square root* key labeled [√] *or* [√x]. When you press this key, the number in the display is replaced by its principal square root.

Example 2 **Use a scientific calculator to evaluate each expression if $x = 2401$, $a = 147$, and $b = 78$.**

a. $\sqrt{x}$

$\sqrt{x} = \sqrt{2401}$ *Replace x with 2401.*

Enter: 2401 [2nd] [√x] 49

Therefore, $\sqrt{2401} = 49$.

b. $\pm\sqrt{a + b}$

$\pm\sqrt{a + b} = \pm\sqrt{147 + 78}$ *Replace a with 147 and b with 78.*

Enter: [(] 147 [+] 78 [)] [2nd] [√x] 15

Therefore, $\pm\sqrt{a + b}$ is $\pm$ 15.

Numbers such as $\sqrt{5}$ and $\sqrt{13}$ are the square roots of numbers that are *not* perfect squares. Notice what happens when you find these square roots with your calculator.

Enter: 5 [2nd] [√x] 2.236067978...

Enter: 13 [2nd] [√x] 3.605551275...

These numbers continue indefinitely without any pattern of repeating digits. These numbers are not rational numbers since they are not repeating or terminating decimals. Numbers like $\sqrt{5}$ and $\sqrt{13}$ are called **irrational numbers.**

Definition of an Irrational Number	**An irrational number is a number that *cannot* be expressed in the form $\frac{a}{b}$, where a and b are integers and $b \neq 0$.**

The set of rational numbers and the set of irrational numbers together form the set of **real numbers**. The Venn diagram at the right shows the relationships among natural numbers, whole numbers, integers, rational numbers, irrational numbers, and real numbers.

Real Numbers

Rationals
Integers
Whole Numbers
Natural Numbers
Irrationals

Example 3 **Name the set or sets of numbers to which each real number belongs.**

a. 0.8333333... This repeating decimal is a rational number since it is equivalent to $\frac{5}{6}$. *This number can also be expressed as $0.8\overline{3}$.*

b. $-\sqrt{16}$ Since $-\sqrt{16} = -4$, this number is an integer and a rational number.

c. $\frac{14}{2}$ Since $\frac{14}{2} = 7$, this number is an integer, a natural number, a whole number, and a rational number.

d. $\sqrt{120}$ Since $\sqrt{120} = 10.95445115\ldots$, which is not a repeating or terminating decimal, this number is irrational.

The solutions to many real-world problems are irrational numbers.

Example 4 **The area of a square is 325 square inches. Find its perimeter to the nearest hundredth.**

INTEGRATION
Geometry

First find the length of each side. Since the area of a square is the length of the side squared, find the square root of 325.

325 square inches

Enter: 325 [2nd] [$\sqrt{x}$] 18.02775638

The length of each side is about 18.02775638 inches. Use the formula $P = 4s$ to find the perimeter.

$P = 4s$
$= 4 \cdot 18.02775638$ *Replace s with 18.02775638.*

Enter: 18.02775638 [×] 4 [=] 72.11102655

The perimeter is about 72.11 inches.

You have graphed rational numbers on number lines. Yet, if you graphed all of the rational numbers, the number line would still not be complete. The irrational numbers complete the number line. The graph of all real numbers is the entire number line. This is illustrated by the **completeness property.**

Completeness Property for Points on the Number Line	**Each real number corresponds to exactly one point on the number line. Each point on the number line corresponds to exactly one real number.**

In-Class Examples

For Example 3
Name the set or sets of numbers to which each number belongs. Use *N* for natural numbers, *W* for whole numbers, *Z* for integers, *Q* for rational numbers, and *I* for irrational numbers.

a. -9 ***Z, Q***
b. 0.101101110... ***I***
c. $\sqrt{81}$ ***N, W, Z, Q***
d. 0.33333... ***Q***

For Example 4
The area of a square is 827 square inches. Find its perimeter.
approximately 115 inches

Classroom Vignette

"To show the relationship between the various sets of numbers (whole, natural, rational, integers, irrational, and real), I made a set of nested boxes. In each box, I included small slips of paper with representative numbers of that set."

Linda Crawford
Brown County Junior High School
Nashville, Indiana

Linda L. Crawford

Teaching Tip Point out that a dot [•] on the number line means that the value at which the dot occurs is part of the solution set. A circle [o] means that the value is not part of the solution set.

Teaching Tip Any point on the number line can be assigned a value from the set of real numbers. Between any two real numbers there exists another real number.

In-Class Example

For Example 5
Graph each solution set.

a. $x < 3$

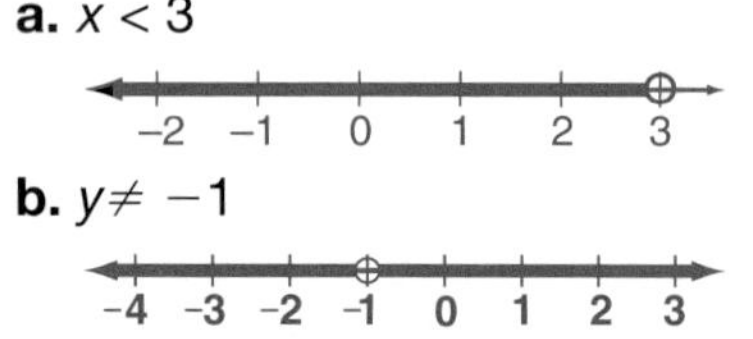

b. $y \neq -1$

3 PRACTICE/APPLY

Check for Understanding

Exercises 1–20 are designed to help you assess your students' understanding through reading, writing, speaking, and modeling. You should work through Exercises 1–6 with your students and then monitor their work on Exercises 7–20.

Error Analysis
Students sometimes mistake approximations of irrational numbers for the numbers themselves. Thus, they may think that $\sqrt{2}$ is *equal* to its decimal approximation 1.4142136. To show that these values are not equal, have students find $\sqrt{2}$ using a calculator and square it. The result is 2. Now, have them enter 1.4142136 and square it. The result is 2.00000001, not 2.

Additional Answers

1. Sample answer:

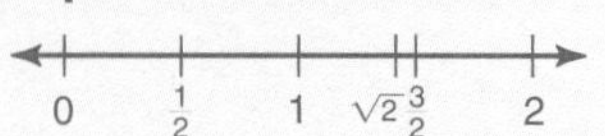

2. The comic is funny because you usually count off by natural numbers {1, 2, 3, 4, ...} not square roots.

Recall that equations like $x - 5 = 11$ are open sentences. Inequalities like $x < 6$ are also considered to be open sentences. To solve $x < 6$, determine what replacements for x make $x < 6$ true. All numbers less than 6 make the inequality true. This can be shown by the solution set {real numbers less than 6}. Not only does this include integers like 3, 0, and -4, but it also includes all rational numbers less than 6 such as $\frac{1}{2}$, $-5\frac{3}{8}$, and -3 and all irrational numbers less than 6 such as $\sqrt{5}$, $\sqrt{3}$, and π.

Example  **Graph each solution set.**

a. $y \geq -7$

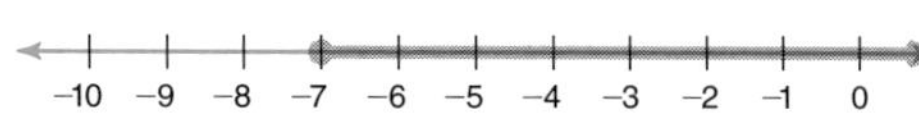

The heavy arrow indicates that all numbers to the right of -7 are included. The *dot* indicates that the point corresponding to -7 is included in the graph of the solution set.

b. $p \neq \frac{3}{4}$

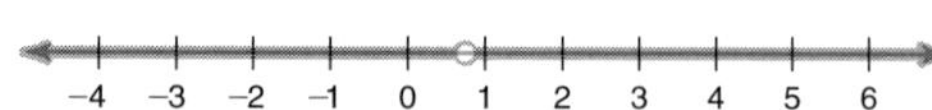

The heavy arrows indicate that all numbers to the left and to the right of $\frac{3}{4}$ are included in the graph of the solution set. The *circle* indicates that the point corresponding to $\frac{3}{4}$ is not included in the graph.

CHECK FOR UNDERSTANDING

Communicating Mathematics

Study the lesson. Then complete the following. 1–2. See margin.

Deep down inside, Coach Knott had always wanted to be a math teacher.

1. You have studied the following sets of numbers: integers, irrational, natural, rational, real, and whole numbers. Draw a number line and label at least one number from each set of numbers. Indicate which number is from each set.
2. Study the comic at the left. Explain why it is humorous.
3. **Determine** whether 36 is a perfect square. If it is, to what set of numbers does $\sqrt{36}$ belong? **Yes, because its square root is a rational number, 6; rationals, integers, whole numbers, natural numbers.**
4. **Explain** why 3 and -3 are both square roots of 9. **$3(3) = 9$ and $(-3)(-3) = 9$. Therefore, 3 and -3 are both square roots of 9.**
5. **Write** an inequality for the graph below. **$x \leq 0$**

6. **Write** a paragraph explaining the difference between rational and irrational numbers to a classmate. **See margin.**

Reteaching

Using Alternative Methods Have students convert repeating decimals to ratios of integers. For example, consider if

$$\begin{aligned} x &= 0.3636..., \text{ then} \\ 100x &= 36.3636... \\ (-)x &= -0.3636... \\ \hline 99x &= 36.0000 \\ x &= \frac{36}{99} \text{ or } \frac{4}{11} \end{aligned}$$

This method cannot be used with nonrepeating decimals. They are irrational numbers.

Guided Practice

10. ± 0.28

Find each square root. Use a calculator if necessary. Round to the nearest hundredth if the result is not a whole number.

7. $\sqrt{64}$ **8** 8. $-\sqrt{36}$ **−6** 9. $\sqrt{122}$ **11.05** 10. $\pm\sqrt{0.08}$

Evaluate each expression. Use a calculator if necessary. Round to the nearest hundredth if the result is not a whole number.

11. $\sqrt{x}$, if $x = 256$ **16** 12. $\sqrt{y}$, if $y = 151$ **12.29**

Name the set or sets of numbers to which each real number belongs. Use N for natural numbers, W for whole numbers, Z for integers, Q for rational numbers, and I for irrational numbers.

13. $-\frac{3}{4}$ **Q** 14. $\frac{8}{4}$ **Q, Z, W, N** 15. 0.6666... **Q** 16. $\sqrt{13}$ **I**

Graph the solution set of each inequality on a number line.

17–19. See margin.

17. $p < 7$ 18. $r \geq -3$ 19. $x \neq 2$

20. **Discrete Mathematics** In the geometric sequence 5, 15, _?_, 135, 405, the missing number is called the *geometric mean* of 15 and 135. It can be found by evaluating $\sqrt{ab}$, where a and b are the numbers on either side of the geometric mean. Find the missing number. **45**

EXERCISES

Practice

Find each square root. Use a calculator if necessary. Round to the nearest hundredth if the result is not a whole number or simple fraction.

21. $\sqrt{169}$ **13** 22. $\sqrt{0.0049}$ **0.07** 23. $\sqrt{\frac{4}{9}}$ $\frac{2}{3}$

24. $-\sqrt{289}$ **−17** 25. $\sqrt{420}$ **20.49** 26. $\sqrt{\frac{25}{64}}$ $\frac{5}{8}$

27. $\sqrt{225}$ **15** 28. $\pm\sqrt{1158}$ **±34.03** 29. $-\sqrt{625}$ **−25**

30. $\sqrt{1.96}$ **1.4** 31. $\sqrt{\frac{9}{25}}$ $\frac{3}{5}$ 32. $-\sqrt{5.80}$ **−2.41**

Evaluate each expression. Use a calculator if necessary. Round to the nearest hundredth if the result is not a whole number.

33. $\sqrt{x}$, if $x = 87$ **9.33** 34. $\pm\sqrt{t}$, if $t = 529$ **±23**

35. $-\sqrt{m}$, if $m = 2209$ **−47** 36. $\sqrt{c + d}$, if $c = 23$ and $d = 56$ **8.89**

37. $-\sqrt{np}$, if $n = 16$ and $p = 25$ **−20** 38. $\pm\sqrt{\frac{a}{b}}$, if $a = 64$ and $b = 4$ **±4**

Name the set or sets of numbers to which each real number belongs. Use N for natural numbers, W for whole numbers, Z for integers, Q for rational numbers, and I for irrational numbers.

39. $-\sqrt{49}$ **Z, Q** 40. 0 **W, Z, Q** 41. 0.4583 **Q**

42. $0.\overline{3}$ **Q** 43. $-\frac{1}{2}$ **Q** 44. $\sqrt{49}$ **N, W, Z, Q**

45. 0.6666 **Q** 46. $\frac{10}{5}$ **N, W, Z, Q** 47. $\sqrt{37}$ **I**

48. 3.14 **Q** 49. $\frac{3}{5}$ **Q** 50. 5 **N, W, Z, Q**

Assignment Guide

Core: 21–65 odd, 67–78
Enriched: 22–60 even, 61–78

For **Extra Practice,** see p. 761.

The red A, B, and C flags, printed only in the Teacher's Wraparound Edition, indicate the level of difficulty of the exercises.

Additional Answers

6. **Sample answer: A rational number is a number that can be expressed as a common fraction. An irrational number is a number that cannot be expressed as a common fraction.**

17. 2 3 4 5 6 7 8 9 10

18. −6 −5 −4 −3 −2 −1 0 1 2

19. −2 −1 0 1 2 3 4 5 6

Study Guide Masters, p. 17

2-8 NAME ____________ DATE ________

Study Guide

Student Edition Pages 119–125

Square Roots and Real Numbers

The chart below illustrates the various kinds of real numbers.

Counting or Natural Numbers, N	{1, 2, 3, 4, ···}
Whole Numbers, W	{0, 1, 2, 3, 4, ···}
Integers, Z	{··· −3, −2, −1, 0, 1, 2, 3, ···}
Rational Numbers, Q	{all numbers that can be expressed in the form $\frac{a}{b}$, where a and b are integers and $b \neq 0$}
Irrational Numbers, I	{numbers that cannot be expressed in the form $\frac{a}{b}$, where a and b are integers and $b \neq 0$}
Real Numbers, R	{rational numbers and irrational numbers}

The square roots of perfect squares are classified as rational numbers. A **square root** is one of two equal factors of a number. For example, the square root of 36 is 6 and −6 since $6 \cdot 6$ or 6^2 is 36 and $(-6)(-6)$ or $(-6)^2$ is also 36. A rational number like 36, whose square root is a rational number, is called a **perfect square.**

The symbol $\sqrt{}$ is called a **radical sign.** It indicates the nonnegative, or **principal,** square root of the expression under the radical sign.

Example: Find $\pm\sqrt{49}$. The symbol $\pm\sqrt{49}$ represents both square roots. Since $7^2 = 49$, we know that $\pm\sqrt{49} = \pm 7$.

Numbers such as $\sqrt{2}$ and $\sqrt{3}$ are not perfect squares. Notice what happens when you find these square roots with your calculator. The numbers continue indefinitely without any pattern of repeating digits. These numbers are not rational numbers since they are not repeating or terminating decimals and they are classified as **irrational numbers.**

Find each square root. Use a calculator if necessary. Round to the nearest hundredth if the result is not a whole number or simple fraction.

1. $\sqrt{81}$ 9 2. $\sqrt{0.00025}$ 0.02 3. $-\sqrt{\frac{25}{16}}$ $-\frac{5}{4}$ 4. $-\sqrt{3600}$ −60 5. $\pm\sqrt{\frac{121}{100}}$ $\pm\frac{11}{10}$

Evaluate each expression. Use a calculator if necessary. Round to the nearest hundredth if the result is not a whole number.

6. $\sqrt{a}$, if $a = 39$ 6.24 7. $\pm\sqrt{s + t}$, if $s = 30$ and $t = 19$ ±7

8. $-\sqrt{\frac{a}{b}}$, if $a = 169$ and $b = 4$ $-\frac{13}{2}$ 9. $\sqrt{cd}$, if $c = 12$ and $d = 15$ 13.42

Name the set or sets of numbers to which each real number belongs. Use N for natural numbers, W for whole numbers, Z for integers, Q for rational numbers, and I for irrational numbers.

10. 3.145 Q 11. $\sqrt{11}$ I 12. $\sqrt{25}$ Z; Q

Additional Answers

51. 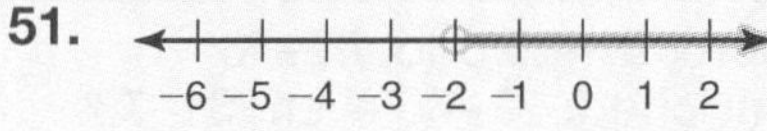
−6 −5 −4 −3 −2 −1 0 1 2

52. −3 −2 −1 0 1 2 3 4 5

53. −7 −6 −5 −4 −3 −2 −1 0 1

54. 2 3 4 5 6 7 8 9 10

55. −15 −14 −13 −12 −11 −10 −9 −8 −7

56. 0 1 2 3 4 5 6 7 8

57. −5.6 −5.5 −5.4 −5.3 −5.2 −5.1 −5.0 −4.9 −4.8

58. $\frac{1}{4}$ 0 $\frac{1}{4}$ $\frac{1}{2}$ $\frac{3}{4}$ 1 $1\frac{1}{4}$ $1\frac{1}{2}$ $1\frac{3}{4}$

59. $4\frac{1}{2}$ $4\frac{3}{4}$ 5 $5\frac{1}{4}$ $5\frac{1}{2}$ $5\frac{3}{4}$ 6 $6\frac{1}{4}$ $6\frac{1}{2}$

61. Yes, it lies between 27 and 28.

62a. $\sqrt{10}$, $\sqrt{11}$, $\sqrt{12}$, $\sqrt{13}$, $\sqrt{14}$, and $\sqrt{15}$

62b. $\sqrt{28}$, $\sqrt{29}$, and $\sqrt{30}$

66a. 103 or more cases

66b. 98 99 100 101 102 103 104 105 106

Practice Masters, p. 17

2-8 NAME______ DATE______

Practice — Student Edition Pages 119–125

Square Roots and Real Numbers

Find each square root. Use a calculator if necessary. Round to the nearest hundredth if the result is not a whole number or a simple fraction.

1. $\sqrt{144}$ 12
2. $\sqrt{225}$ 15
3. $-\sqrt{0.06}$ −0.24
4. $-\sqrt{\frac{100}{49}}$ $-\frac{10}{7}$
5. $\sqrt{\frac{1}{16}}$ $\frac{1}{4}$
6. $\pm\sqrt{1444}$ ±38
7. $\pm\sqrt{2.2}$ ±1.48
8. $-\sqrt{1.44}$ −1.2
9. $\sqrt{0.0625}$ 0.25
10. $-\sqrt{\frac{4}{289}}$ $-\frac{2}{17}$
11. $\sqrt{3.06}$ 1.75
12. $\pm\sqrt{2401}$ ±49
13. $-\sqrt{4900}$ −70
14. $\sqrt{\frac{3136}{25}}$ $\frac{56}{5}$
15. $-\sqrt{11.4}$ −3.38
16. $\sqrt{3364}$ 58
17. $\sqrt{0.906}$ 0.95
18. $-\sqrt{22.58}$ −4.75
19. $\pm\sqrt{8376}$ ±91.52
20. $-\sqrt{\frac{1}{196}}$ $-\frac{1}{14}$
21. $-\sqrt{59{,}312}$ −243.54

Evaluate each expression. Use a calculator if necessary. Round to the nearest hundredth if the result is not a whole number.

22. $-\sqrt{a}$, if $a = 400$ −20
23. $\sqrt{b}$, if $b = 313$ 17.69
24. $\sqrt{st}$, if $s = 10$ and $t = 22$ 14.83
25. $-\sqrt{m+n}$, if $m = 12$ and $n = 4$ −4
26. $\sqrt{c+d}$, if $c = 33$ and $d = 51$ 9.17
27. $\sqrt{de}$, if $d = 19$ and $e = 6$ 10.68
28. $-\sqrt{\frac{16}{a}}$, if $a = 16$ −1
29. $\pm\sqrt{b-a}$, if $b = 6$ and $a = 2$ ±2
30. $\sqrt{mn}$, if $m = 45$ and $n = 3$ 11.62
31. $-\sqrt{d+e}$, if $d = 0$ and $e = 9$ −3

Name the set or sets of numbers to which each real number belongs. Use N for natural numbers, W for whole numbers, Z for integers, Q for rational numbers, and I for irrational numbers.

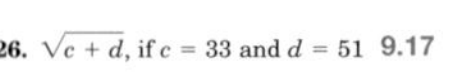

32. $-\sqrt{1849}$ Z; Q
33. $\sqrt{29}$ I
34. $-\frac{2}{5}$ Q
35. 0.676767 ⋯ Q
36. −2 Z; Q
37. $\frac{0}{2}$ W; Z; Q
38. 0.333333 ⋯ Q
39. $\sqrt{93}$ I
40. −20 Z; Q

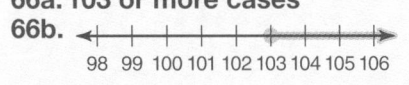

51–59. See margin.

Graph the solution set of each inequality on a number line.

51. $y > -2$
52. $x < 1$
53. $p \geq -4$
54. $n \neq 6$
55. $c > -12$
56. $r \leq 4.5$
57. $b \geq -5.2$
58. $y \neq \frac{3}{4}$
59. $s \leq 5\frac{1}{2}$

60. **Geometry** The volume of a rectangular solid is 100 cm^3. Its height is the product of its length and width. The base of the solid is a square.
   a. Draw the solid from two different perspectives. **See students' work.**
   b. Find the dimensions of the solid. Use a calculator as needed and round decimal answers to the nearest hundredth.

60b. 3.16 cm × 3.16 cm × 10 cm; 10 cm × 10 cm × 1 cm

Critical Thinking

61–62. See margin.

61. Determine whether $\sqrt{733}$ is an irrational number. If it is, name two consecutive integers between which its graph lies on the number line.

62. Find all numbers of the form $\sqrt{n}$ such that n is a natural number and the graph of $\sqrt{n}$ lies between each pair of numbers on the number line.
   a. 3 and 4
   b. 5.25 and 5.5

Applications and Problem Solving

63. **Military** The formula to determine the distance d in miles that an object can be seen on a clear day on the surface of the ocean is $d = 1.4\sqrt{h}$, where h is the height in feet the viewer's eyes are above the surface of the water. About how many miles can the pilot of a U.S. Coast Guard plane see if he is flying at 1275 feet?
   a. Write an equation to represent the distance. Then approximate the distance mentally.
   b. Use a calculator to find the exact distance. **49.989999**

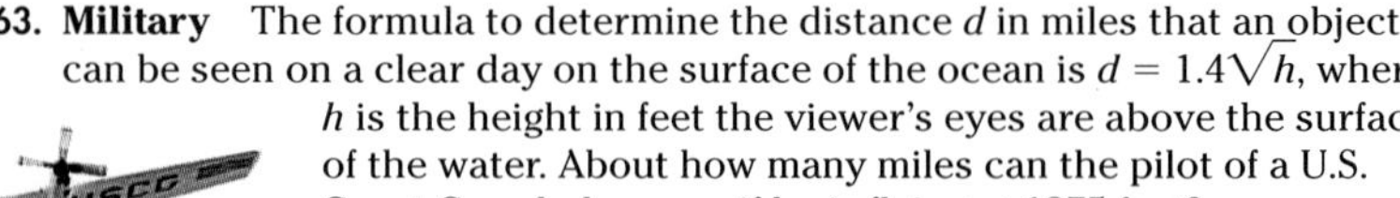

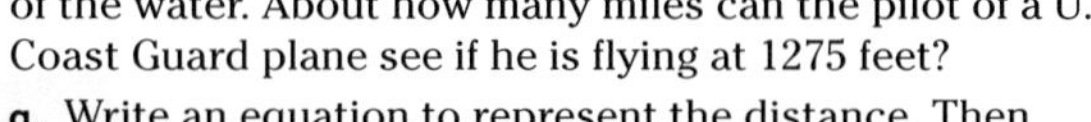

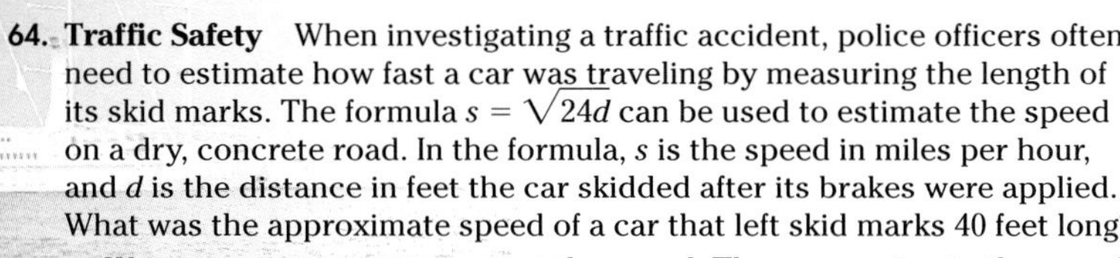

64. **Traffic Safety** When investigating a traffic accident, police officers often need to estimate how fast a car was traveling by measuring the length of its skid marks. The formula $s = \sqrt{24d}$ can be used to estimate the speed on a dry, concrete road. In the formula, s is the speed in miles per hour, and d is the distance in feet the car skidded after its brakes were applied. What was the approximate speed of a car that left skid marks 40 feet long?
   a. Write an equation to represent the speed. Then approximate the speed mentally. **$s = \sqrt{24(40)}$ or $\sqrt{960}$; Estimates should be close to 30.**
   b. Use a calculator to find the exact speed. **30.98386677**

63a. $d = 1.4\sqrt{1275}$; Estimates should be close to 50.

65. **Math History** One of the great mathematical challenges in early mathematics was to find an exact value for π. Today, 3.14 and $\frac{22}{7}$ are two frequently used approximations for π. In the third century, Chinese mathematician Liu Hui used the number $3\frac{7}{50}$ as an approximation of π. The Italian mathematician Leonardo Fibonacci used $\frac{864}{275}$ as an approximation of π. Which mathematician's approximation was closer to the generally-accepted approximation of π, 3.141592654...? **Fibonacci**

66. **Business** The Spanish Club at Alexander Manor High School is selling chocolate bars to raise money for a trip to Mexico. The bars are sold only by the case, with 24 bars per case. Each bar costs \$1.00. The club's goal is to raise at least \$2450 in gross sales.
   a. How many cases of chocolate bars must the club sell in order to reach or exceed their goal? **a–b. See margin.**
   b. Graph the solution set.

Mixed Review

67. Evaluate $(-3a)(4)\left(\frac{2}{3}\right)\left(\frac{1}{6}a\right)$ if $a = 3$. (Lesson 2–6) **−12**

68. Find $-\frac{5}{12} + \frac{3}{8}$. (Lesson 2–5) $-\frac{1}{24}$

69. Which is the better buy, a 1-pound package of lunch meat for $1.95 or a 12-ounce package for $1.80? (Lesson 2–4)

70. Find $-5 - (-6)$. (Lesson 2-3) **1**

71. **Statistics** State the scale you would use to make a line plot for the following data: 504, 509, 520, 500, 513, 517, 517, 508, 502, 518, 509, 520, 504, 509, 517, 511, 520. Then draw the line plot. (Lesson 2–2) **See margin.**

72. Graph {0, 2, 6} on a number line. (Lesson 2–1) **See margin.**

73. Suppose a basketball was dropped from a tall building onto a sidewalk. Identify the graph below that best represents this situation. (Lesson 1–9) **b**

a.
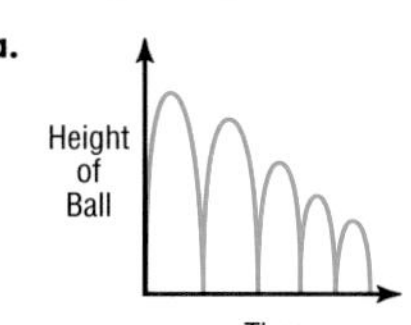

b.
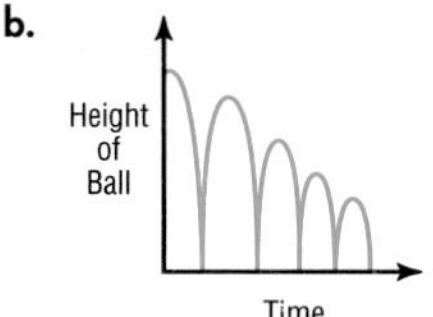

c.
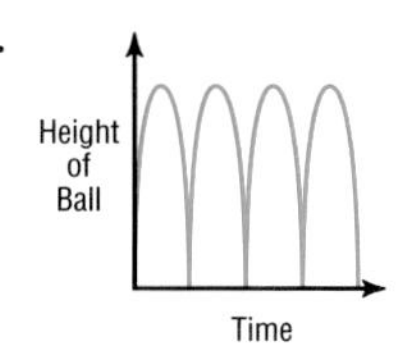

74. Name the property illustrated by $a + (2b + 5c) = (a + 2b) + 5c$. (Lesson 1–8) **associative (+)**

75. **Golf** The stem-and-leaf plot below shows the earnings of the top 25 U.S. women professional golfers for 1994 according to the *Sports Almanac.* (Lesson 1–4)

Stem	Leaf
6	6 9
5	0
4	0 1 2 3 7
3	2 3 4 5 9
2	0 0 1 1 3 4 4 5 6 7 7 8 *4|3 = $430,000*

a. to the nearest ten-thousand

a. How have these numbers been rounded?

b. How many of these golfers made between $300,000 and $490,000? **10**

c. Laura Davies had the highest earnings of all women golfers in 1994. How much money did she earn? **$690,000**

Laura Davies studies her putt.

76. Evaluate $12(19 - 15) - 3 \cdot 8$. (Lesson 1–3) **24**

77. Find the next two terms in the sequence 2, 8, 14, 20,... (Lesson 1–2) **26, 32**

78. Write an algebraic expression for p to the sixth power. (Lesson 1–1) p^6

Extension

Connections The area of a triangle is 38 m². If the base is twice as long as the height, what is the base of the triangle? (Use the formula $A = \frac{1}{2}bh$.)

$2\sqrt{38}$ **m or 12.3 m**

4 ASSESS

Closing Activity

Speaking Have students give examples of natural numbers, whole numbers, integers, rational numbers, and irrational numbers.

Additional Answers

71. from 500 to 520 by 5s

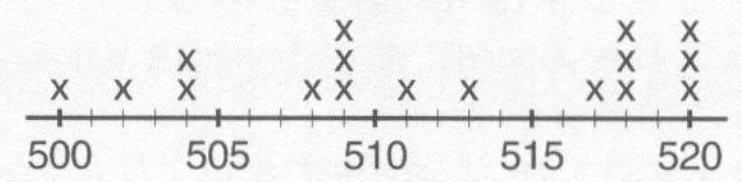

72.
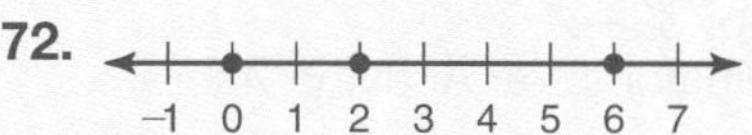

Enrichment Masters, p. 17

NAME______________ DATE______

Enrichment — Student Edition Pages 119–125

Squares and Square Roots From a Graph

The graph of $y = x^2$ can be used to find the squares and square roots of numbers.

To find the square of 3, locate 3 on the x-axis. Then find its corresponding value on the y-axis.

The arrows show that $3^2 = 9$.

To find the square root of 4, first locate 4 on the y-axis. Then find its corresponding value on the x-axis. Following the arrows on the graph, you can see that $\sqrt{4} = 2$.

A small part of the graph at $y = x^2$ is shown below. A 1:10 ratio for unit length on the y-axis to unit length on the x-axis is used.

Example: Find $\sqrt{11}$.

The arrows show that $\sqrt{11} = 3.3$ to the nearest tenth.

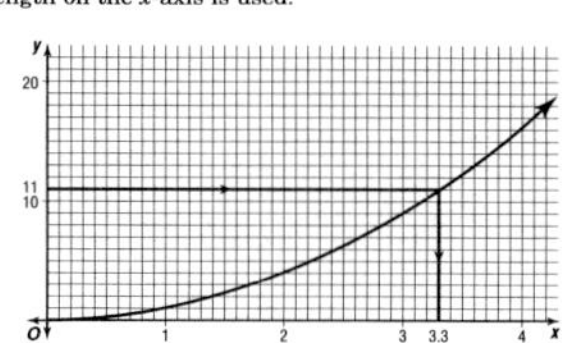

Use the graph above to find each of the following to the nearest whole number.

1. 1.5^2 2	**2.** 2.7^2 7	**3.** 0.9^2 1
4. 3.6^2 13	**5.** 4.2^2 18	**6.** 3.9^2 15

Use the graph above to find each of the following to the nearest tenth.

7. $\sqrt{15}$ 3.9	**8.** $\sqrt{8}$ 2.8	**9.** $\sqrt{3}$ 1.7
10. $\sqrt{5}$ 2.2	**11.** $\sqrt{14}$ 3.7	**12.** $\sqrt{17}$ 4.1

2-9 LESSON NOTES

NCTM Standards: 1–5

Instructional Resources

- Study Guide Master 2-9
- Practice Master 2-9
- Enrichment Master 2-9
- Assessment and Evaluation Masters, p. 45
- Modeling Mathematics Masters, pp. 39–41
- Multicultural Activity Masters, p. 4
- Real-World Applications, 6

Transparency 2-9A contains the 5-Minute Check for this lesson; **Transparency 2-9B** contains a teaching aid for this lesson.

Recommended Pacing	
Standard Pacing	Days 12 & 13 of 15
Honors Pacing	Days 12 & 13 of 15
Block Scheduling*	Day 8 of 9
Alg. 1 in Two Years*	Days 23, 24, & 25 of 28

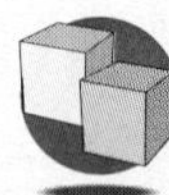

*For more information on pacing and possible lesson plans, refer to the *Block Scheduling Booklet* and *Algebra 1 in Two Years*.

1 FOCUS

5-Minute Check
(over Lesson 2-8)

Find each square root to the nearest hundredth.

1. $\sqrt{0.0256}$ **0.16**
2. $-\sqrt{280}$ **−16.73**

Name the set of numbers to which each real number belongs.

3. $\sqrt{7}$ **irrational**
4. $\sqrt{0.04}$ **rational**
5. Graph the solution set of $x \geq -3\frac{1}{2}$ on a number line.

−5 −4 −3 −2 −1 0 1 2 3 4 5

2-9 Problem Solving Write Equations and Formulas

What YOU'LL LEARN

- To explore problem situations, and
- to translate verbal sentences and problems into equations or formulas and vice versa.

Why IT'S IMPORTANT

In the world of work, exploring problems and translating them into formulas that can be solved are valuable skills.

CONNECTION

Biology

Bugs, bugs, everywhere! That could be the slogan for James Fujita, an avid bug collector. Fujita has been fascinated with bugs since he was bitten by a cricket at the age of three. His collection includes hundreds of bugs—beetles, spiders, millipedes, scorpions, just to name a few—both living and dead.

Fujita has studied bugs as far away as Japan and Costa Rica, as well as in his own backyard in Oxnard, California. He has given bug presentations and workshops, donated some of his bugs to museums and zoos, and even appeared on a television talk show with some of his larger bugs. Fujita is a big believer in reading and exploring all you can about bug collecting—if you get the itch!

You can explore problems by asking and answering questions. In this text, we will use a four-step plan to solve problems. All of the steps are listed below.

Problem-Solving Plan

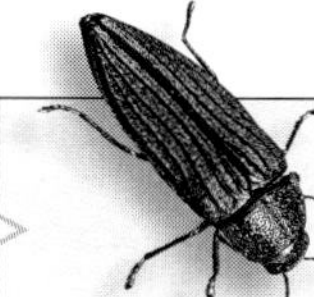

1. **Explore the problem.**
2. **Plan the solution.**
3. **Solve the problem.**
4. **Examine the solution.**

CAREER CHOICES

An **entomologist** is a scientist who studies the origin, development, anatomy, and distribution of insects and their effects on humans.

An inquisitive mind, and a knowledge of biology, chemistry, and mathematics are necessary are required for this career.

For more information, contact:

The Entomological Society of America
9301 Annapolis Road
Lanham, MD 20706

The first step in solving a problem is to read and explore it until you completely understand the relationships in the given information.

Step 1: Explore the Problem
To solve a verbal problem, first read the problem carefully and explore what the problem is about.

- Identify what information is given.
- Identify what you are asked to find.

Questions	Answers
a. How old was James when he became interested in bugs?	**a.** 3 years old
b. In what countries other than the U.S. has James studied bugs?	**b.** Japan and Costa Rica
c. If James is n years old now, for how many years has he been a collector of bugs?	**c.** $n - 3$
d. If James has b bugs and finds 14 new bugs, how many bugs are in his collection now?	**d.** $b + 14$

CAREER CHOICES

Entomology contains many subfields: anatomy, physiology, insect behavior, bionomics and ecology, systematics, economic entomology, and medical entomology. One of the main concerns of medical entomology is the role insects play in the transmission of diseases.

GLENCOE Technology

Interactive Mathematics Tools Software

This multimedia software provides an interactive lesson that allows students to use guess-and-check as a means of finding the formulas for converting degrees Fahrenheit to Celsius and degrees Celsius to Fahrenheit. A **Computer Journal** gives students an opportunity to write about what they have learned.
For Windows & Macintosh

Other strategies are:
- *look for a pattern*
- *solve a simpler problem*
- *act it out*
- *guess and check*
- *draw a diagram*
- *make a table or chart*
- *work backward*

Step 2: Plan the Solution
One strategy you can use to solve a problem is to write an equation. Choose a variable to represent one of the unspecified numbers in the problem. This is called **defining the variable.** Then use the variable to write expressions for the other unspecified numbers in the problem.

Step 3: Solve the Problem
Use the strategy you chose in Step 2 to solve the problem.

Step 4: Examine the Solution
Check your answer within the context of the original problem. Does your answer make sense? Does it fit the information in the problem?

Example 1

APPLICATION Computers

Gregory Arakelian of Herndon, Virginia, set a speed record for typing the most words per minute, with no errors, on a personal computer in the KeyTronic World Invitational Type-Off on September 24, 1991. Suppose his closest competitor typed 10 fewer words per minute than Arakelian. If his closest competitor typed 148 words per minute, how many words did Arakelian type per minute?

Explore
- How many words can Arakelian type in relation to his closest competitor? 10 more words
- How many words per minute did his competitor type? 148 words

Plan Write an equation to represent the situation. Let w represent the words per minute that Arakelian typed.

$$\underbrace{w}_{\text{words Arakelian typed}} - 10 = \underbrace{148}_{\text{words competitor typed}}$$

Solve

$w - 10 = 148$ *Solve mentally by asking,*
$w = 158$ *"What number minus 10 is 148?"*

Arakelian typed 158 words per minute.

Examine The problem asks how many words Arakelian typed per minute. His competitor typed 10 fewer words per minute. Since $158 - 10 = 148$, the answer makes sense.

Gregory Arakelian

In Lesson 2–4, you studied meanings for the symbols $<$, $>$, $\neq$, $\leq$, and $\geq$.

When solving problems, many sentences can be written as equations. Use variables to represent the unspecified numbers or measures referred to in the sentence or problem. Then write the verbal expressions as algebraic expressions. Some verbal expressions that suggest the *equals sign* are listed below.

- is
- equals
- is equal to
- is the same as
- is as much as
- is identical to

Motivating the Lesson

Situational Problem Pose the following problem to students. Pedro is older than his sister, Claudia. Since the age of 21, the sum of the digits in his age has been equal to the sum of the digits in her age. How much older is Pedro than Claudia? **9 years** After students have solved the problem, discuss the strategies they used to find the answer.

2 TEACH

Teaching Tip Point out to students that the variable x is not always used. Encourage them to choose appropriate variables, such as w for weight, when writing a verbal expression.

In-Class Example

For Example 1
Suppose Winata ran a 25-mile race 3 minutes faster than Joanna. Joanna ran the race in 254 minutes.

a. Express Winata's time as an algebraic expression. $s = J - 3$

b. How long did Winata take to run the race? **251 minutes**

Teaching Tip Point out that variables can often be defined in more than one way. In the connection at the beginning of the lesson, if n were equal to the number of years that James was interested in bugs, then his age would be $n + 3$.

Cooperative Learning

Numbered Heads Together Exercises 14–23 can be done in pairs. One student should read the problem aloud to the other student, who should write down the translation. They should alternate roles after each problem. For more information on the numbered heads together strategy, see *Cooperative Learning in the Mathematics Classroom,* one of the titles in the Glencoe Mathematics Professional Series, pages 9–12.

In-Clas

For Exan
Translat
sentenc
One num
a second
these tw
$n + (n +$

For Exar
Translat
a formul

a. The di
produ
time t.

b. The vo
equal
times
$V = \frac{a}{3}$

For Exa
Translat
verbal s

a. $(x + 5$
The s
numb

b. $\frac{n^2 - 1}{3}$
of a r
fiftee
great

For Exa
Write a
given in
w = wei
packed
$w - 30$
partially
$3w + (w$
Sample
packed
less tha
The tot
packed
packed
How m
box we

out of pa
cut-out.
see phys
the diffe

Using the CUMULATIVE REVIEW

These two pages review the skills and concepts presented in Chapters 1–2. This review is formatted to reflect new trends in standardized testing.
A more traditional cumulative review is provided in the *Assessment and Evaluation Masters*, pp. 47–48.

Assessment and Evaluation Masters, pp. 47–48

2 NAME________ DATE________

Chapter 2 Cumulative Review

Write an algebraic expression for each verbal expression. (Lesson 1-1)

1. 7 more than the square of a number — 1. $n^2 + 7$
2. 5 less than 4 times a number — 2. $4n - 5$

Evaluate each expression. (Lesson 1-3)

3. $6 - 2^2 + 14 \cdot 3 \div 7$ — 3. 8
4. $3a^2(a^2 - b^2)$ if $a = 4$ and $b = 3$ — 4. 336

Solve each equation. (Lesson 1-5)

5. $a = \frac{3}{5} \cdot 15$ — 5. 9
6. $\frac{84 \div 14}{27 \div 9} = m$ — 6. 2

Name the property illustrated by each statement. (Lessons 1-6, 1-7, and 1-8)

7. If $x = 7 + 2$ and $7 + 2 = w$, then $x = w$. — 7. transitive
8. $6 + (7 \cdot 5) = 6 + 35$ — 8. substitution
9. $3(4a + b) = 12a + 3b$ — 9. distributive
10. $(7 + x) + a = (x + 7) + a$ — 10. commutative of add.

Simplify. (Lesson 1-7)

11. $6a^2 + 11a^2 - 3$ — 11. $17a^2 - 3$
12. $2(2x + y) + 6(x + 2y)$ — 12. $10x - 14y$

Translate each sentence into an equation, inequality, or formula. (Lesson 2-9)

13. y is the product of a and b, increased by c. — 13. $y = ab + c$
14. The number m equals the square of the difference of n and p. — 14. $m = (n - p)^2$

Replace each _?_ with <, >, or = to make each sentence true. (Lesson 2-4)

15. -3 _?_ -6 — 15. >
16. $\frac{4}{13}$ _?_ $\frac{3}{11}$ — 16. >

2 NAME________ DATE________

Chapter 2 Cumulative Review (continued)

17. Name the set of numbers graphed below. (Lesson 2-1) — 17. {integers less than 3}
18. Graph $\{-3, -1, 0, 2\}$ on the number line provided. (Lesson 2-1) — 18. [number line graph]
19. Graph the solution set of $w \le 3$ on the number line. (Lesson 2-1) — 19. [number line graph]

Find each sum or difference. (Lessons 2-3 and 2-5)

20. $-15 + a$ if $a = -10$ — 20. -25
21. $-13 + 6 + (-8)$ — 21. -15
22. $4m + (-3m) + (-12m)$ — 22. $-11m$
23. $\begin{bmatrix} -3 & -\frac{2}{7} \\ 2.7 & 12 \end{bmatrix} + \begin{bmatrix} -18 & \frac{11}{14} \\ -5.4 & 6 \end{bmatrix}$ — 23. $\begin{bmatrix} -21 & \frac{1}{2} \\ -2.7 & 18 \end{bmatrix}$
24. Make a line plot for the following data. (Lesson 2-2) 21, 20, 29, 29, 22, 20, 24, 27, 29, 26. — 24. [line plot]

Find each scalar product. (Lesson 2-6)

25. $4\begin{bmatrix} -3 & -\frac{3}{4} \\ 2.6 & 12 \end{bmatrix}$ — 25. $\begin{bmatrix} -12 & -3 \\ 10.4 & 48 \end{bmatrix}$
26. $-\frac{2}{3}\begin{bmatrix} 9.6 & \frac{9}{4} & -12 \end{bmatrix}$ — 26. $\begin{bmatrix} -6.4 & -\frac{3}{2} & 8 \end{bmatrix}$
27. Find $\sqrt{38}$ to the nearest hundredth. (Lesson 2-8) — 27. 6.16

Simplify. (Lessons 2-6 and 2-7)

28. $5(-7) + (-3)(6)$ — 28. -53
29. $3(5t - t) + 4(2t + t)$ — 29. $24t$
30. $\left(-\frac{1}{3}\right)\left(\frac{3}{7}\right) + \left(\frac{5}{6}\right)\left(\frac{12}{35}\right)$ — 30. $\frac{1}{7}$
31. $\frac{5a - 10}{5}$ — 31. $a - 2$

Define a variable, then write an equation for each problem. Do not try to solve. (Lesson 2-9)

32. Wayne bought a suit on sale and saved \$15.35. The regular price was \$139.29. What was the sale price? — 32. Let s = sales price; $s = 139.29 - 15.35$
33. Lisa is 5 years older than Susan. The sum of their ages is 25. How old is Susan? — 33. Let s = Susan's age; $s + (s + 5) = 25$

CUMULATIVE REVIEW

CHAPTERS 1–2

SECTION ONE: MULTIPLE CHOICE

There are eight multiple-choice questions in this section. After working each problem, write the letter of the correct answer on your paper.

1. **Patterns** Find the sixth term in the sequence $-6, -3\frac{1}{2}, -1, \ldots$. **D**
   - A. 0
   - B. $2\frac{1}{2}$
   - C. 4
   - D. $6\frac{1}{2}$

2. **Travel** Travel agencies often give information on a state's temperature during the year in order to prepare their customers for their vacation. The stem-and-leaf plot shows daily high temperatures recorded for April in Ohio. Which statement best describes the data in order to prepare travelers for their visit to Ohio in April? **B**

Stem	Leaf
8	0 2 2 5
7	0 3 3 4 5 7 7 7 8 9 9
6	1 2 3 5 5 6 7 8 8 9
5	4 7 8 8
4	9 5|7 = 57

   - A. April temperatures can range from 49 to 85 degrees.
   - B. April is mild, with temperatures usually between 60 and 80 degrees.
   - C. April's temperatures can soar as high as 85 degrees.
   - D. April is a cold month with temperatures as low as 49 degrees.

3. Evaluate $3t^2 - g(w - t)$ if $t = 3$, $g = 5$, and $w = -2$. **D**
   - A. -110
   - B. 20
   - C. 43
   - D. 52

4. Tyrone is 3 years older than Bill. In 5 years, twice the sum of their ages will equal 5 times Bill's age now. Choose the equation that should be used to solve this problem. **C**
   - A. $b + (b + 3) = 5b$
   - B. $2b + (b + 3) + 5 = 5b$
   - C. $2[(b + 5) + (b + 3 + 5)] = 5b$
   - D. $2[(b + 3) + (b - 3 + 5)] = 5b$

5. Quan is working as a mailroom clerk in a busy downtown office building for the summer. His morning route takes him from the mailroom up two floors to the payroll office, up four more floors to the attorney's office, down three floors to the cafeteria, up seven floors to the executive offices, and down twelve floors to the security desk on the first floor. On what floor is the mailroom? **B**
   - A. second floor
   - B. third floor
   - C. sixth floor
   - D. top floor

6. Stephen bought a 5-foot-long submarine sandwich to serve at his Super Bowl party. How many pieces would he have to serve if he cut the sub into $1\frac{1}{2}$-inch pieces? **D**
   - A. 3 pieces
   - B. 8 pieces
   - C. 30 pieces
   - D. 40 pieces

7. Simplify $10a^3 + 7 - 2(2a^3 + a) + 3$. **C**
   - A. $-2a^4 + 10a^3 + 10$
   - B. $8a^3 + 2a + 10$
   - C. $6a^3 - 2a + 10$
   - D. $6a^3 + a + 10$

8. Choose the expression that has a value of 28. **D**
   - A. $4 + 3 \cdot 4$
   - B. $\frac{75}{3} - 2$
   - C. $8 \cdot 4 - 8$
   - D. $(5 + 3) \cdot 7 \div 2$

Standardized Test Practice Questions are also provided in the *Assessment and Evaluation Masters*, p. 46.

SECTION TWO: FREE RESPONSE

This section contains seven questions for which you will provide short answers. Write your answer on your paper.

9. Rewrite the expression $7 + 15 \div 2 + 4 - 2$ using grouping symbols so that the value of the expression is $6\frac{1}{2}$. **$7 + 15 \div 2 + 4\,(-2)$**

10. Three integers x, y, and z satisfy the following conditions:

$y - z < 0$, $x - y > 0$, and $z - x < 0$.

Arrange the integers in order from least to greatest. ***y, z, x***

11. Graph the solution set of $k \neq 1$ on a number line. **See margin.**

12. Geometry Find the perimeter of the figure below if $x = 7$, $y = 3$, and $z = 1\frac{1}{2}$. **16 ft**

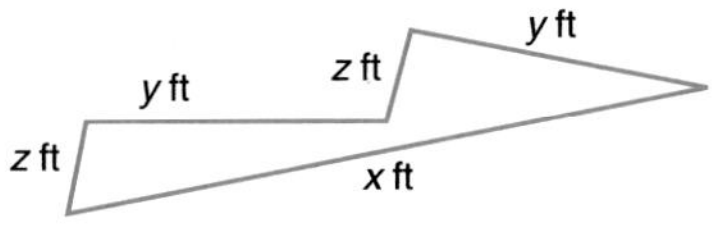

13. School The students in Miss Stickler's algebra class scored the following points on their last test: 88, 89, 85, 92, 91, 86, 90, 95, 91, 86, 90, 92, 91, 89, 91, and 90. Make a line plot of the scores and write a sentence to describe the data shown. **See margin.**

14. The average hourly rate of a teenage baby-sitter on a Saturday night is \$5 in Los Angeles and \$2 in Pittsburgh. Corey, who lives in Pittsburgh, baby-sits his little brother every Saturday night for 5 hours. Ella lives in Los Angeles and baby-sits her cousin for 3 hours every other Saturday night. How many Saturday nights will each have to baby-sit to earn the same amount of money? **Corey, 3; Ella, 2**

15. What is the absolute value of a if $-a > a$? **$-a$**

16. Write an algebraic expression to represent *nine decreased by the product of y and 3.* **$9 - 3y$**

17. Patterns Find the next five terms of the sequence 3, 4.5, 6, ... **Sample answer: 7.5, 9, 10.5, 12, 13.5**

18. Replace __?__ with >, <, or = to make a true statement.

$-2 - 1$ __?__ $-2(-1)$ **<**

SECTION THREE: OPEN-ENDED

This section contains two open-ended problems. Demonstrate your knowledge by giving a clear, concise solution to each problem. Your score on these problems will depend on how well you do the following.

- Explain your reasoning.
- Show your understanding of the mathematics in an organized manner.
- Use charts, graphs, and diagrams in your explanation.
- Show the solution in more than one way or relate it to other situations.
- Investigate beyond the requirements of the problem.

19. Latisha and Joia are planning to purchase a CD player for their dorm room. Latisha works at the campus book store and earns \$5.35 per hour. Joia has a job off campus as a waitress and earns \$2.00 per hour, plus tips. Each student works 20 hours per week.

A. How long will they have to work to earn enough money to buy a CD player worth \$350? **about $2\frac{1}{2}$ weeks, not counting tips**

B. What other factors may affect when they can buy the CD player? **other expenses they have, amount of tips Joia makes**

20. If $2(b + c) = 2b + 2c$, does $2 + (b \cdot c) = (2 + b)(2 + c)$? Use values for b and c to justify your answer. **No; see students' work.**

Additional Answers

11.

13.

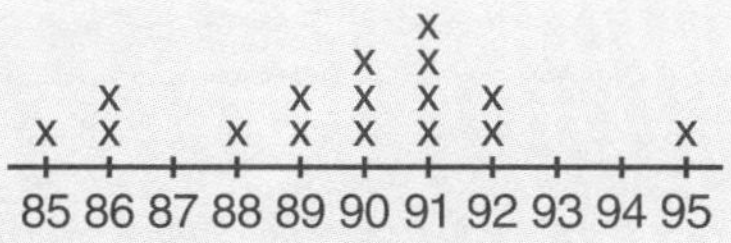

The scores ranged from a low of 85 to a high of 95 with a cluster around the score of 91.

CHAPTER

3

Solving Linear Equations

PREVIEWING THE CHAPTER

This chapter addresses the crucial skills and concepts involved in solving equations. Cups and counters are used as models to develop an understanding of equations, and then, in carefully sequenced lessons, the addition, subtraction, multiplication, and division properties of equality are presented and applied to solving equations. Then students use their knowledge of equations to integrate geometry skills by studying properties of angles and triangles. They also learn to use more than one property to solve equations with the variable on either side or on both sides. The chapter concludes with students learning to find the mean, median, and mode.

Lesson (Pages)	Lesson Objectives	NCTM Standards	State/Local Objectives
3-1A (142–143)	Solve equations by using cups and counters to model variables and integers.	1–5	
3-1 (144–149)	Solve equations by using addition and subtraction.	1–5	
3-2 (150–154)	Solve equations by using multiplication and division.	1–5	
3-3A (155)	Solve equations by using cups and counters to model variables and integers.	1–5	
3-3 (156–161)	Solve equations involving more than one operation. Solve problems by working backward.	1–5	
3-4 (162–167)	Find the complement and supplement of an angle. Find the measure of the third angle of a triangle given the measures of the other two angles.	1–5, 7	
3-5 (168–172)	Solve equations with the variable on both sides. Solve equations containing grouping symbols.	1–5, 7	
3-6 (173–177)	Solve equations and formulas for a specified variable.	1–5, 7	
3-7 (178–183)	Find and interpret the mean, median, and mode of a set of data.	1–5, 10	

A complete, 1-page lesson plan is provided for each lesson in the ***Lesson Planning Guide***. Answer keys for each lesson are available in the ***Answer Key Masters.***

ORGANIZING THE CHAPTER

You may want to refer to the **Course Planning Calendar** on page T12 for detailed information on pacing.
PACING: Standard—14 days; **Honors**—13 days; **Block**—6 days; **Two Years**—19 days

LESSON PLANNING CHART

Lesson (Pages)	Materials/ Manipulatives	Extra Practice (Student Edition)	Blackline Masters: Study Guide	Practice	Enrichment	Assessment and Evaluation	Modeling Mathematics	Multicultural Activity	Tech Prep Applications	Graphing Calculator	Science and Math Lab Manual	Real-World Applications	Interactive Mathematics Tools Software	Teaching Transparencies
3-1A (142–143)	cups and counters* equation mat*						p. 23						3-1A	
3-1 (144–149)	calculator cups and counters* equation mat*	p. 762	p. 19	p. 19	p. 19		p. 74							3-1A 3-1B
3-2 (150–154)	cups and counters* equation mat*	p. 762	p. 20	p. 20	p. 20	p. 72			p. 5	p. 3		7	3-2	3-2A 3-2B
3-3A (155)	cups and counters* equation mat*						p. 24							
3-3 (156–161)	graphing calculator cups and counters* equation mat*	p. 762	p. 21	p. 21	p. 21		pp. 42–44	p. 5	p. 6			8		3-3A 3-3B
3-4 (162–167)	construction paper protractor* scissors	p. 763	p. 22	p. 22	p. 22	pp. 71, 72		p. 6			pp. 9–12		3-4	3-4A 3-4B
3-5 (168–172)		p. 763	p. 23	p. 23	p. 23									3-5A 3-5B
3-6 (173–177)		p. 763	p. 24	p. 24	p. 24	p. 73								3-6A 3-6B
3-7 (178–183)	graphing calculator	p. 764	p. 25	p. 25	p. 25	p. 73						9	3-7	3-7A 3-7B
Study Guide/ Assessment (185–189)						pp. 57–70, 74–76								

*Included in Glencoe's Student Manipulative Kit and Overhead Manipulative Resources.

ORGANIZING THE CHAPTER

Other Chapter Resources

Student Edition
Chapter Opener, pp. 140–141
Mathematics and Society, p. 183
Working on the Investigation, pp. 154, 177
Closing the Investigation, p. 184

Teacher's Classroom Resources
Investigations and Projects Masters, pp. 33–36
Algebra and Geometry Overhead Manipulative Resources, pp. 5–12

Technology
Test and Review Software (IBM and Macintosh)
CD-ROM Interactions (Windows and Macintosh)

Professional Publications
Block Scheduling Booklet
Glencoe Mathematics Professional Series

Outside Resources

Books/Periodicals
Codes, Ciphers, and Secret Writing, Dover
Mathematics and Humor, NCTM

Software
Data Insights, Sunburst
Geometric Connectors: Coordinates, Sunburst
Teasers by Tobbs with Whole Numbers, Sunburst

Videos/CD-ROMs
Becoming Successful Problem Solvers, HRM Video
Math ... Who Needs It?! FASE Productions

Assessment Resources

Student Edition
Math Journal, pp. 170, 175, 181
Mixed Review, pp. 149, 154, 161, 167, 172, 177, 183
Self Test, p. 161
Chapter Highlights, p. 185
Chapter Study Guide and Assessment, pp. 186–188
Alternative Assessment, p. 189
Portfolio, p. 189

Teacher's Wraparound Edition
5-Minute Check, pp. 144, 150, 156, 162, 168, 173, 178
Check for Understanding, pp. 148, 152, 159, 165, 170, 175, 181
Closing Activity, pp. 149, 154, 161, 167, 172, 177, 183
Cooperative Learning, pp. 164, 180

Assessment and Evaluation Masters
Multiple-Choice Tests, Forms 1A (Honors), 1B (Average), 1C (Basic), pp. 57–62
Free-Response Tests, Forms 2A (Honors), 2B (Average), 2C (Basic), pp. 63–68
Calculator-Based Test, p. 69
Performance Assessment, p. 70
Mid-Chapter Test, p. 71
Quizzes A–D, pp. 72–73
Standardized Test Practice, p. 74
Cumulative Review, pp. 75–76

ENHANCING THE CHAPTER

Examples of some of the materials for enhancing Chapter 3 are shown below.

Multicultural Activity Masters, pp. 5, 6

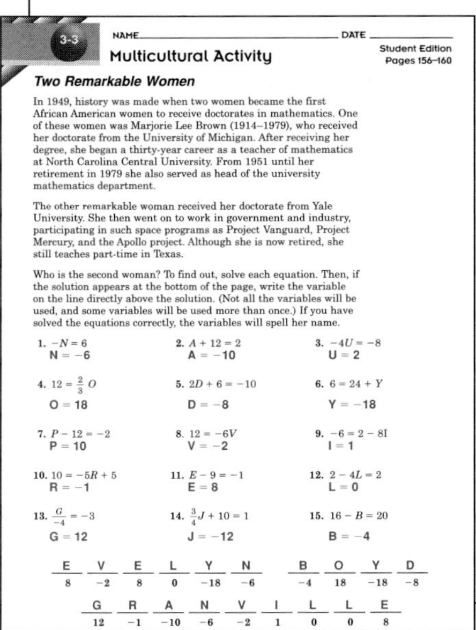

3-3 NAME ______ DATE ______

Multicultural Activity — Student Edition Pages 156–160

Two Remarkable Women

In 1949, history was made when two women became the first African American women to receive doctorates in mathematics. One of these women was Marjorie Lee Brown (1914–1979), who received her doctorate from the University of Michigan. After receiving her degree, she began a thirty-year career as a teacher of mathematics at North Carolina Central University. From 1951 until her retirement in 1979 she also served as head of the university mathematics department.

The other remarkable woman received her doctorate from Yale University. She then went on to work in government and industry, participating in such space programs as Project Vanguard, Project Mercury, and the Apollo project. Although she is now retired, she still teaches part-time in Texas.

Who is the second woman? To find out, solve each equation. Then, if the solution appears at the bottom of the page, write the variable on the line directly above the solution. (Not all the variables will be used, and some variables will be used more than once.) If you have solved the equations correctly, the variables will spell her name.

1. $-N = 6$ $N = -6$
2. $A + 12 = 2$ $A = -10$
3. $-4U = -8$ $U = 2$
4. $12 = \frac{2}{3}O$ $O = 18$
5. $2D + 6 = -10$ $D = -8$
6. $6 = 24 + Y$ $Y = -18$
7. $P - 12 = -2$ $P = 10$
8. $12 = -6V$ $V = -2$
9. $-6 = 2 - 8I$ $I = 1$
10. $10 = -5R + 5$ $R = -1$
11. $E - 9 = -1$ $E = 8$
12. $2 - 4L = 2$ $L = 0$
13. $\frac{G}{-4} = -3$ $G = 12$
14. $\frac{3}{4}J + 10 = 1$ $J = -12$
15. $16 - B = 20$ $B = -4$

E	V	E	L	Y	N		B	O	Y	D
8	−2	8	0	−18	−6		−4	18	−18	−8

G	R	A	N	V	I	L	L	E
12	−1	−10	−6	−2	1	0	0	8

Real-World Applications, 7, 8, 9

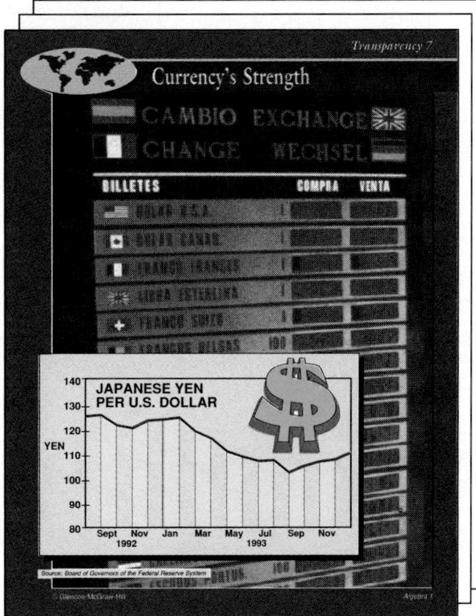

Graphing Calculator Masters, p. 3

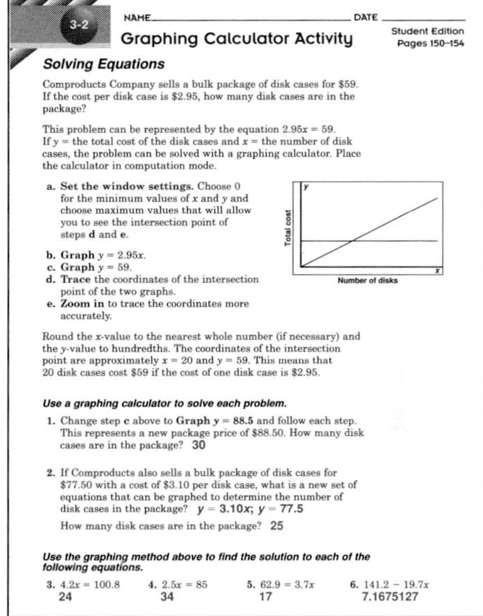

3-2 NAME ______ DATE ______

Graphing Calculator Activity — Student Edition Pages 150–154

Solving Equations

Comproducts Company sells a bulk package of disk cases for \$59. If the cost per disk case is \$2.95, how many disk cases are in the package?

This problem can be represented by the equation $2.95x = 59$. If y = the total cost of the disk cases and x = the number of disk cases, the problem can be solved with a graphing calculator. Place the calculator in computation mode.

a. **Set the window settings.** Choose 0 for the minimum values of x and y and choose maximum values that will allow you to see the intersection point of steps **d** and **e**.
b. **Graph** $y = 2.95x$.
c. **Graph** $y = 59$.
d. **Trace** the coordinates of the intersection point of the two graphs.
e. **Zoom in** to trace the coordinates more accurately.

Total cost / Number of disks

Round the x-value to the nearest whole number (if necessary) and the y-value to hundredths. The coordinates of the intersection point are approximately $x = 20$ and $y = 59$. This means that 20 disk cases cost \$59 if the cost of one disk case is \$2.95.

Use a graphing calculator to solve each problem.

1. Change step **c** above to **Graph** $y = 88.5$ and follow each step. This represents a new package price of \$88.50. How many disk cases are in the package? 30
2. If Comproducts also sells a bulk package of disk cases for \$77.50 with a cost of \$3.10 per disk case, what is a new set of equations that can be graphed to determine the number of disk cases in the package? $y = 3.10x$; $y = 77.5$
   How many disk cases are in the package? 25

Use the graphing method above to find the solution to each of the following equations.

3. $4.2x = 100.8$ 24
4. $2.5x = 85$ 34
5. $62.9 = 3.7x$ 17
6. $141.2 = 19.7x$ 7.1675127

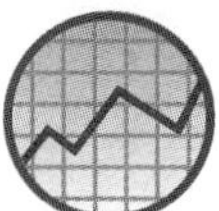

TECH PREP

Tech Prep Applications Masters, pp. 5, 6

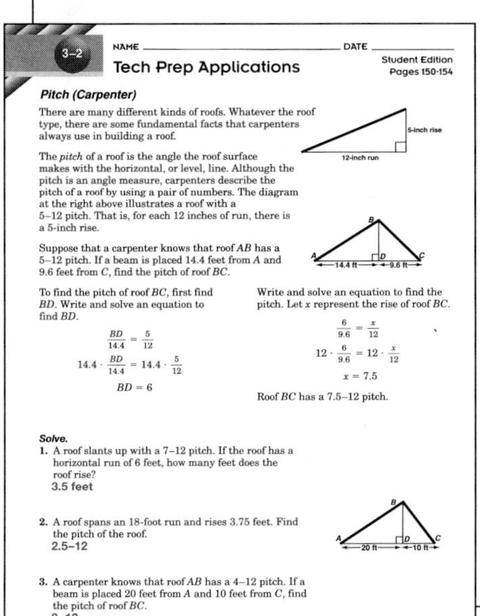

3-2 NAME ______ DATE ______

Tech Prep Applications — Student Edition Pages 150–154

Pitch (Carpenter)

There are many different kinds of roofs. Whatever the roof type, there are some fundamental facts that carpenters always use in building a roof.

The *pitch* of a roof is the angle the roof surface makes with the horizontal, or level, line. Although the pitch is an angle measure, carpenters describe the pitch of a roof by using a pair of numbers. The diagram at the right above illustrates a roof with a 5–12 pitch. That is, for each 12 inches of run, there is a 5-inch rise.

5-inch rise / 12-inch run

Suppose that a carpenter knows that roof AB has a 5–12 pitch. If a beam is placed 14.4 feet from A and 9.6 feet from C, find the pitch of roof BC.

To find the pitch of roof BC, first find BD. Write and solve an equation to find BD.

$$\frac{BD}{14.4} = \frac{5}{12}$$
$$14.4 \cdot \frac{BD}{14.4} = 14.4 \cdot \frac{5}{12}$$
$$BD = 6$$

Write and solve an equation to find the pitch. Let x represent the rise of roof BC.

$$\frac{6}{9.6} = \frac{x}{12}$$
$$12 \cdot \frac{6}{9.6} = 12 \cdot \frac{x}{12}$$
$$x = 7.5$$

Roof BC has a 7.5–12 pitch.

Solve.

1. A roof slants up with a 7–12 pitch. If the roof has a horizontal run of 6 feet, how many feet does the roof rise? 3.5 feet
2. A roof spans an 18-foot run and rises 3.75 feet. Find the pitch of the roof. 2.5–12
3. A carpenter knows that roof AB has a 4–12 pitch. If a beam is placed 20 feet from A and 10 feet from C, find the pitch of roof BC. 8–12

Science and Math Lab Manual, pp. 9–12

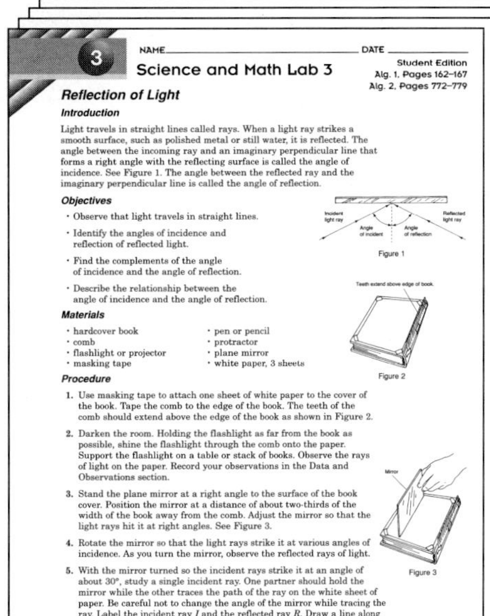

3 NAME ______ DATE ______

Science and Math Lab 3 — Student Edition Alg. 1, Pages 162–167; Alg. 2, Pages 772–779

Reflection of Light

Introduction

Light travels in straight lines called rays. When a light ray strikes a smooth surface, such as polished metal or still water, it is reflected. The angle between the incoming ray and an imaginary perpendicular line that forms a right angle with the reflecting surface is called the angle of incidence. See Figure 1. The angle between the reflected ray and the imaginary perpendicular line is called the angle of reflection.

Objectives

- Observe that light travels in straight lines.
- Identify the angles of incidence and reflection of reflected light.
- Find the complements of the angle of incidence and the angle of reflection.
- Describe the relationship between the angle of incidence and the angle of reflection.

Materials

- hardcover book
- comb
- flashlight or projector
- masking tape
- pen or pencil
- protractor
- plane mirror
- white paper, 3 sheets

Procedure

1. Use masking tape to attach one sheet of white paper to the cover of the book. Tape the comb to the edge of the book. The teeth of the comb should extend above the edge of the book as shown in Figure 2.
2. Darken the room. Holding the flashlight as far from the book as possible, shine the flashlight through the comb onto the paper. Support the flashlight on a table or stack of books. Observe the rays of light on the paper. Record your observations in the Data and Observations section.
3. Stand the plane mirror at a right angle to the surface of the book cover. Position the mirror at a distance of about two-thirds of the width of the book away from the comb. Adjust the mirror so that the light rays hit it at right angles. See Figure 3.
4. Rotate the mirror so that the light rays strike it at various angles of incidence. As you turn the mirror, observe the reflected rays of light.
5. With the mirror turned so the incident rays strike it at an angle of about 30°, study a single incident ray. One partner should hold the mirror while the other traces the path of the ray on the white sheet of paper. Be careful not to change the angle of the mirror while tracing the ray. Label the incident ray *I* and the reflected ray *R*. Draw a line along the edge of the back of the mirror. Label the sheet of paper *Trial A*.

Figure 1 — Incident light ray; Angle of incident; Angle of reflection; Reflected light ray

Figure 2 — Teeth extend above edge of book.

Figure 3 — Mirror

Problem of the Week Cards, 7, 8, 9

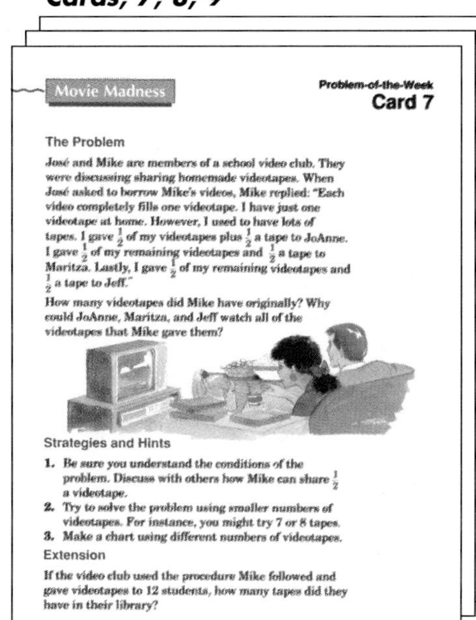

Movie Madness — Problem-of-the-Week Card 7

The Problem

José and Mike are members of a school video club. They were discussing sharing homemade videotapes. When José asked to borrow Mike's videos, Mike replied: "Each video completely fills one videotape. I have just one videotape at home. However, I used to have lots of tapes. I gave $\frac{1}{2}$ of my videotapes plus $\frac{1}{2}$ a tape to JoAnne. I gave $\frac{1}{2}$ of my remaining videotapes and $\frac{1}{2}$ a tape to Maritza. Lastly, I gave $\frac{1}{2}$ of my remaining videotapes and $\frac{1}{2}$ a tape to Jeff."

How many videotapes did Mike have originally? Why could JoAnne, Maritza, and Jeff watch all of the videotapes that Mike gave them?

Strategies and Hints

1. Be sure you understand the conditions of the problem. Discuss with others how Mike can share $\frac{1}{2}$ a videotape.
2. Try to solve the problem using smaller numbers of videotapes. For instance, you might try 7 or 8 tapes.
3. Make a chart using different numbers of videotapes.

Extension

If the video club used the procedure Mike followed and gave videotapes to 12 students, how many tapes did they have in their library?

MAKING MATHEMATICS RELEVANT

This two-page introduction to the chapter provides students with an opportunity to explore contemporary topics and their applications to mathematics.

Background Information

The Rising Cost of College It's never too early to start saving for college. MoneySense, of Dallas, Texas, determined in 1994 that if the parent of a newborn started placing $214 each month into a savings plan earning 8% interest, he or she would accumulate the projected $115,615 required for four years of college in 18 years.

Students should be aware that there are many kinds of grants, loans, and scholarships available for college. There are currently seven federal programs:

- Federal Pell Grant—up to $2300
- Federal Supplemental Educational Opportunity Grant (FSEOG)—up to $4000
- Subsidized Stafford Loan—up to $5500
- Unsubsidized Stafford Loan—up to $5500
- Perkins Loan
- Parent Loan for Undergraduate Students (PLUS)
- Federal Work Study

In addition, thousands of organizations, foundations, and businesses offer scholarships. Almost every college also offers academic, art, athletic, drama, music, and science scholarships. Finally, most colleges offer on-campus part-time employment, regardless of financial status.

Solving Linear Equations

Objectives

In this chapter, you will:

- solve equations using one or more operations,
- solve problems that can be represented by equations,
- work backward to solve problems,
- define and study angles and triangles, and
- find measures of central tendency.

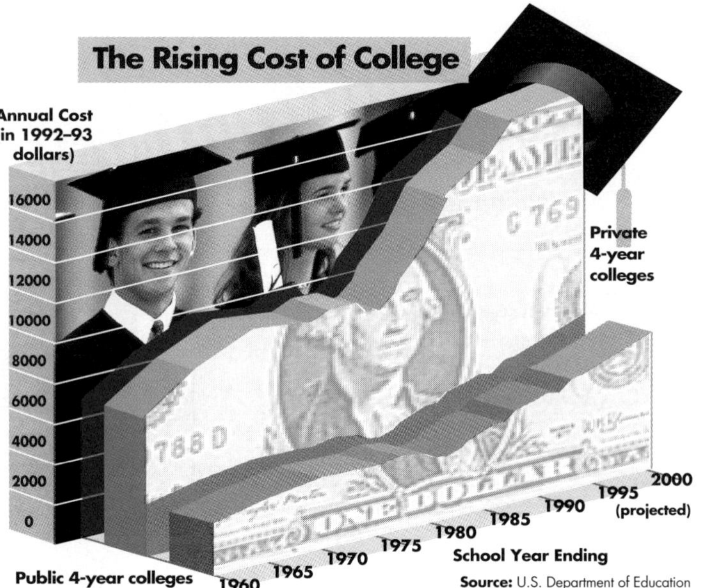

Over the last thirty years, tuition at public 4-year colleges has increased by about 50%, and tuition at private 4-year colleges has increased by a whopping 110%! How do you and your family intend to fund your college experience?

A.D. 240 In *Arithmetica,* Diophantus obtains negative numbers as solutions to equations, but dismisses them as "absurd."

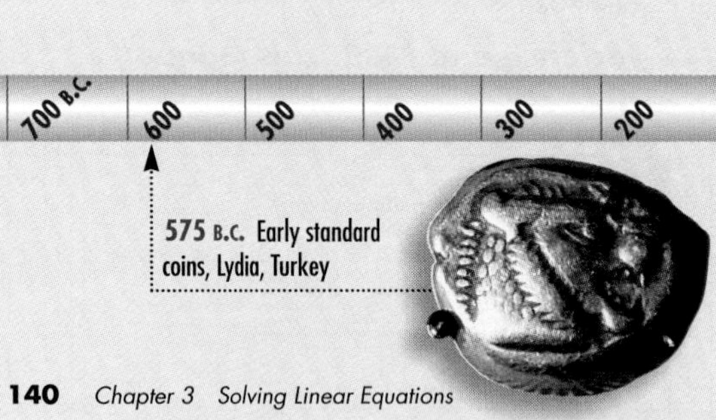

575 B.C. Early standard coins, Lydia, Turkey

Ask students to conduct research and write a one-page report about the use of currency in history. Point out that they could focus on early currency, methods of exchange, coinage, and the advent of paper money.

interNET CONNECTION

Use Fast WEB (a searchable database) to find information about financial aid available for college expenses.

World Wide Web
http://www.cs.cmu.edu/afs/cs/user/mkant/Public/FinAid/finaid.html

PEOPLE IN THE NEWS

When **Marianne Ragins** was a high-school student in her hometown of Macon, Georgia, she applied for scholarship funds from several sources so she could go to college. By the time Marianne was a freshman at Florida A & M University, she had won more than $400,000 in scholarship offers.

When she graduates from college, Marianne will have used $120,000 of those scholarships for her college expenses. Marianne, who is now 21, is the author of *Winning Scholarships for College,* and she now conducts The Scholarship Workshop across the country.

Chapter Project

Juan Ramirez has been offered two college scholarships. Under the terms of the first scholarship, University A will pick up one-half of all his college expenses. Under the terms of the second scholarship, University B will pick up $30,000 of his college expenses. There is $27,500 in the college fund his parents set up for him when he was born.

- Suppose college fees for one year at University A, a state university, are $9500 per year and fees for one year at University B, a private university, are $19,000.
- Write an equation to represent the cost for Juan to attend University A for four years. Write another equation to represent the cost for Juan to attend University B for four years.
- Which school should Juan attend, and why? Be sure to completely explain your reasoning.

780 Muhammed ibn Musa Al-Khowarizmi, Arabian mathematician, born; from his name we get the word *algebra.*
1848 Maria Mitchell is the first woman elected to the American Academy of Science.
800
900
1000
1100
1200
1300
1400
1500
1600
1700
1800
1900
1158 The University of Bologna, Italy, is founded.
1989 Ileana Ros-Lehtinen becomes the first Latina ever elected to Congress.

PEOPLE IN THE NEWS

Marianne began her research by gathering information at her local library. She contacted several corporations and applied for all the scholarships for which she was eligible. The first two she received were from Coca-Cola Corporation ($20,000) and from Rhodes College ($60,000).

Marianne has three tips that she shares with students in her workshop.

1. Don't let your parents do the footwork; you have to really want the scholarships to get them.
2. Write to local banks, as well as to radio and television stations, for information on available scholarships.
3. Attend college fairs. Contact your guidance counselor about fair dates and locations.

Chapter Project

Cooperative Learning You may have students pair off or arrange them in cooperative groups to discuss possible solutions for this activity. This activity allows students to make comparisons between two beneficial plans and choose the one that best fits the situation.

Investigations and Projects Masters, p. 33

NAME________________ DATE ________

Chapter 3 Project A — Student Edition Pages 142–184

Higher Learning

1. What do you think the ideal college would be like? Would it be large or small? Would it be located in a big city or a small town? Would the weather be warm or cold? What would the professors, classes, and sports teams be like? Think about your answers to these questions. Make up your own questions. Then write a description of your ideal college.
2. Go to the library and find a book with descriptions of colleges. Make a list of schools that fit the description that you wrote for exercise 1.
3. Find the estimated annual cost of attending each school.
4. Research the financial aid that is available to students attending three of the schools on your list. What scholarships do the colleges offer? Will the schools assist you in finding a part-time job? How much money can you earn on such a job? Can you get a loan from the colleges, the federal government, or some other agency?
5. Draw up a financial plan for meeting your college expenses.
6. Work with a partner to complete one of the projects listed below, based on the information you gathered in exercises 1–5.
   - Conduct an interview.
   - Write a book report.
   - Write a proposal.
   - Write an article for the school paper.
   - Stage a debate.
   - Write a report.
   - Make a display.
   - Make a graph or chart.
   - Plan an activity.
   - Design a checklist.

Alternative Chapter Projects

Two other chapter projects are included in the *Investigations and Projects Masters.* In Chapter 3 Project A, pp. 33–34, students extend the topic in the chapter opener. In Chapter 3 Project B, pp. 35–36, students research numeration in other civilizations.

Sample Answers

$f_A = \frac{4(9500)}{2}$; $f_B = 4(19{,}000) - 30{,}000$

University A would be more economical and his parents could easily pay the remainder of the fees. University B may provide other benefits such as a smaller campus, more individualized instruction, and so on. Juan's family would have to come up with the remaining fees.

3-1A LESSON NOTES

NCTM Standards: 1–5

Objective

Solve equations by using cups and counters to model variables and integers.

Recommended Time

Demonstration and discussion: 30 minutes; Exercises: 30 minutes

Instructional Resources

For each student or group of students
Student Manipulative Kit
- 20 red/yellow 2-sided counters
- cups

Modeling Mathematics Masters
- pp. 2–3 (cups and counters)
- p. 13 (equation mat)
- p. 23 (worksheet)

For teacher demonstration
Algebra and Geometry Overhead Manipulative Resources

1 FOCUS

Motivating the Lesson

Since this is the first time students will see the process used to solve equations using cups and counters, you may want to allow students some time for exploration and discovery. Introduce the terms and manipulatives and invite students to try to work out the methodology.

2 TEACH

Teaching Tip Another way to solve the equation in Activity 1 is to add 3 positive integers to each side of the mat and then remove the zero-pairs.

A Preview of Lesson 3–1

3–1A Solving One-Step Equations

Materials:  cups & counters  equation mat

You can use cups and counters as a model for solving equations. After you model the equation, the goal is to get the cup by itself on one side of the mat by using the rules stated below.

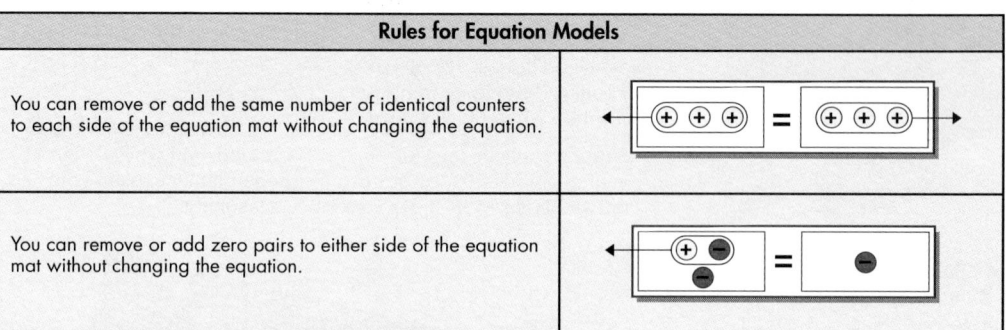

Rules for Equation Models	
You can remove or add the same number of identical counters to each side of the equation mat without changing the equation.	
You can remove or add zero pairs to either side of the equation mat without changing the equation.	

Activity 1 Use an equation model to solve $x + (-3) = -5$.

Step 1 Model the equation $x + (-3) = -5$ by placing 1 cup and 3 negative counters on one side of the mat. Place 5 negative counters on the other side of the mat. The two sides of the mat represent equal quantities.

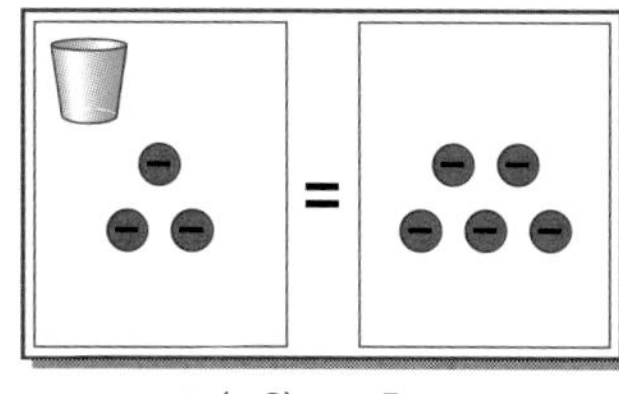

$x + (-3) = -5$

Step 2 Remove 3 negative counters from each side to get the cup by itself.

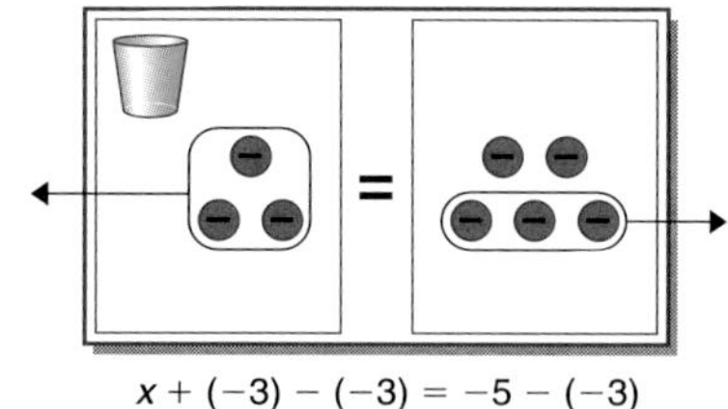

$x + (-3) - (-3) = -5 - (-3)$

Step 3 The cup on the left side of the mat is matched with 2 negative counters. Therefore, $x = -2$.

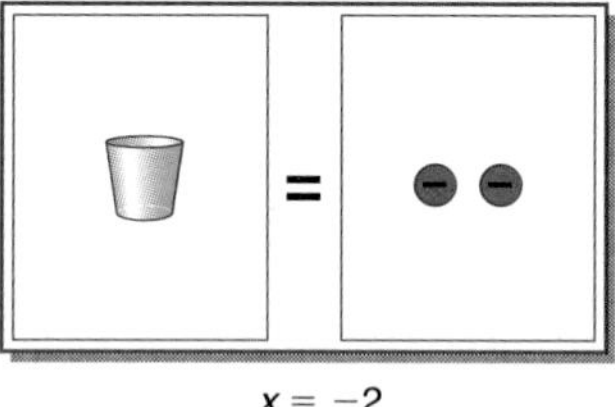

$x = -2$

Cooperative Learning

This lesson offers an excellent opportunity for using cooperative learning groups. For more information on cooperative learning strategies and group management, see *Cooperative Learning in the Mathematics Classroom*, one of the titles in the Glencoe Mathematics Professional Series.

GLENCOE Technology

Interactive Mathematics Tools Software

This multimedia software provides an interactive lesson that uses counters to solve one-step equations with addition or subtraction. A **Computer Journal** gives students an opportunity to write about what they have learned.

For Windows & Macintosh

Activity 2 **Use an equation model to solve $2p = -6$.**

Step 1 Model the equation $2p = -6$ by placing 2 cups on one side of the mat. Place 6 negative counters on the other side of the mat.

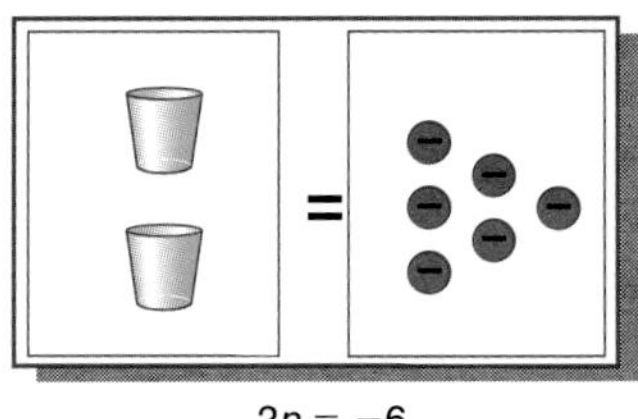

$2p = -6$

Step 2 Separate the counters into 2 equal groups to correspond to the 2 cups. Each cup on the left is matched with 3 negative counters. Therefore, $p = -3$.

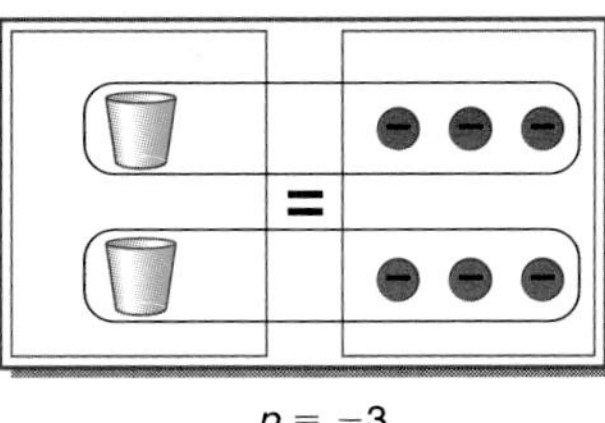

$p = -3$

Activity 3 **Use an equation model to solve $r - 2 = 3$.**

Step 1 Write the equation in the form $r + (-2) = 3$. Place 1 cup and 2 negative counters on one side of the mat. Place 3 positive counters on the other side of the mat. Notice that it is not possible to remove the same kind of counters from each side. Add 2 positive counters to each side.

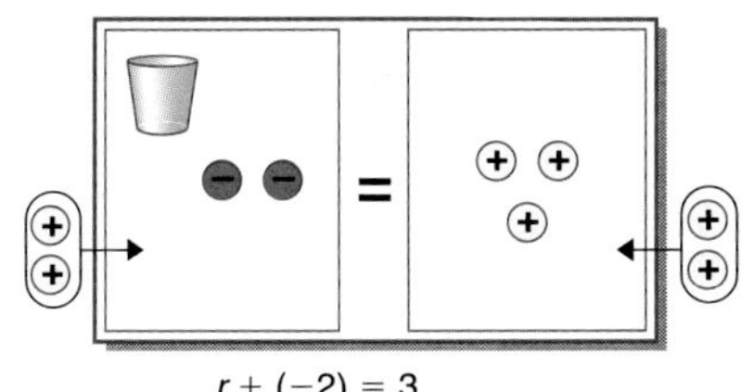

$r + (-2) = 3$

Step 2 Group the counters to form zero pairs. Then remove all the zero pairs. The cup on the left is matched with 5 positive counters. Therefore, $r = 5$.

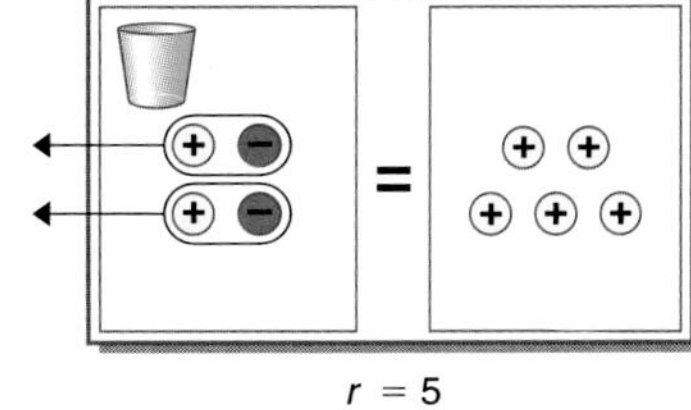

$r = 5$

Model **Use equation models to solve each equation.**

1. $x + 4 = 5$ **1**
2. $y + (-3) = -1$ **2**
3. $y + 7 = -4$ **−11**
4. $3z = -9$ **−3**
5. $m - 6 = 2$ **8**
6. $-2 = x + 6$ **−8**
7. $8 = 2a$ **4**
8. $w - (-2) = 2$ **0**

Draw **Tell whether each number is a solution of the given equation. Justify your answer with a drawing.**

9. -3; $x + 5 = -2$ **no**
10. -1; $5b = -5$ **yes**
11. -4; $y - 4 = -8$ **yes**

Write

12. Write a paragraph explaining how you use zero pairs to solve an equation like $m + 5 = -8$. **See margin.**

Teaching Tip In Activity 3, remind students that subtracting a number is the same as adding its opposite. So subtracting 2 from x is the same as adding -2 to x.

Teaching Tip Remind students to always check their solutions algebraically.

3 PRACTICE/APPLY

Assignment Guide

Core: 1–12
Enriched: 1–12

Additional Answer

12. **Add 5 negative counters to each side. Group and remove the counters that form zero pairs. The cup will be matched with 13 negative counters. Thus, $m = -13$.**

4 ASSESS

Observing students working in cooperative groups is an excellent method of assessment.

Classroom Vignette

"Before I use tiles or cups and counters to demonstrate solving equations, I use a balance scale, some #2 unsharpened pencils, and centimeter cubes to model the concept of equality and transformation of equations. By building an equation using a visual model, students have a conceptual understanding of equality."

Eva Gates
Independent Mathematics Consultant
Pearland, Texas

3-1 LESSON NOTES

NCTM Standards: 1–5

Instructional Resources

- Study Guide Master 3-1
- Practice Master 3-1
- Enrichment Master 3-1
- Modeling Mathematics Masters, p. 74

Transparency 3-1A contains the 5-Minute Check for this lesson; **Transparency 3-1B** contains a teaching aid for this lesson.

Recommended Pacing	
Standard Pacing	Days 2 & 7 of 14
Honors Pacing	Day 2 of 13
Block Scheduling*	Day 1 of 6
Alg. 1 in Two Years*	Days 2 & 3 of 19

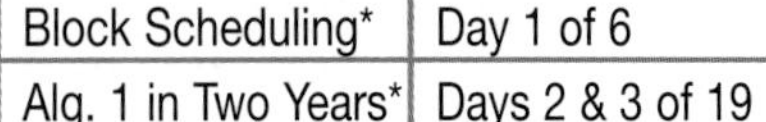

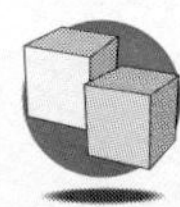

*For more information on pacing and possible lesson plans, refer to the *Block Scheduling Booklet* and *Algebra 1 in Two Years.*

1 FOCUS

5-Minute Check
(over Chapter 2)

Simplify.

1. $-\frac{38}{2}$ **−19**

2. $\frac{30}{-5}$ **−6**

3. $\frac{-200x}{50}$ $\mathbf{-4x}$

Write an equation for each problem.

4. A number decreased by 14 is 72. What is the number? $\mathbf{x - 14 = 72}$

5. Jon is twice as old as Peter. The sum of their ages is 51. Find their ages. $\mathbf{x + 2x = 51}$

3-1 Solving Equations with Addition and Subtraction

What **YOU'LL LEARN**

- To solve equations by using addition and subtraction.

Why **IT'S IMPORTANT**

You can use equations to solve problems involving sports, telecommunications, and economics.

APPLICATION

Football

According to the graph below, the Oakland Raiders and the Dallas Cowboys have each appeared on *Monday Night Football* a total of 41 times through the 1992–1993 season.

Most *Monday Night Football* Appearances

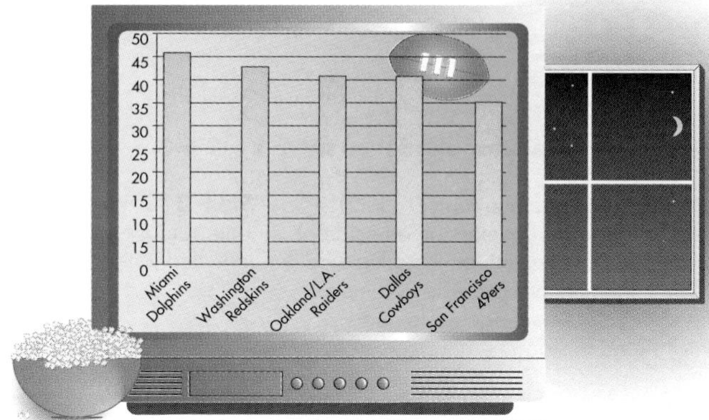

Source: Capital Cities/ABC

FYI

As winners of the Super Bowl in 1993 and 1994, the Dallas Cowboys became the fifth team to win the Super Bowl for two consecutive years. The Pittsburgh Steelers did it twice, winning in 1975, 1976, 1979, and 1980.

If during the next five seasons, each team appears an average of two times per season, the two teams would still have an equal number of *Monday Night Football* appearances.

$41 = 41$

$41 + 5(2) = 41 + 5(2)$ *Two appearances per season for five seasons can be represented as 5(2).*

This example illustrates the **addition property of equality.**

Addition Property of Equality	For any numbers a, b, and c, if $a = b$, then $a + c = b + c$.

Note that c can be positive, negative, or 0. In the equation on the left below, $c = 5$. In the equation on the right, $c = -5$.

$18 + 5 = 18 + 5$ $\qquad$ $18 + (-5) = 18 + (-5)$

If the same number is added to each side of an equation, then the result is an **equivalent equation.** Equivalent equations are equations that have the same solution.

$y + 7 = 13$ *The solution to this equation is 6.*

$y + 7 + 3 = 13 + 3$ *Using the addition property of equality, add 3 to each side.*

$y + 10 = 16$ *The solution to this equation is also 6.*

The Dallas Cowboys were defeated in their quest to become the first team to win the Super Bowl for three consecutive years. After defeating the Cowboys in the NFC championship game, the San Francisco 49ers won the Super Bowl in 1995. The Buffalo Bills, on the other hand, are the only team to have lost the Super Bowl for four consecutive years.

Recall that an equation mat is a model of an equation. The sides of the mat represent the sides of an equation. When you add counters to each side, the result is an equivalent equation, as shown below.

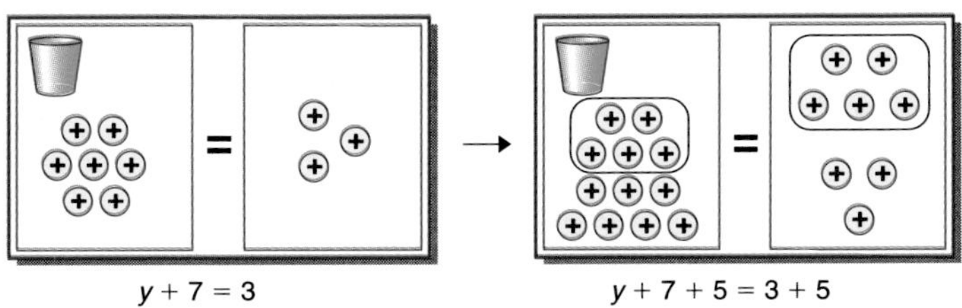

$y + 7 = 3$ $\qquad$ $y + 7 + 5 = 3 + 5$

Remember, x means 1 · x. The coefficient of x is 1.

To **solve an equation** means to isolate the variable having a coefficient of 1 on one side of the equation. You can do this by using the addition property of equality.

Example 1 **Solve 23 + t = −16.**

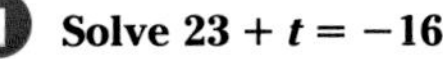

$$23 + t = -16$$
$$23 + t + (-23) = -16 + (-23) \quad \textit{Add } -23 \textit{ to each side.}$$
$$t + 0 = -39 \quad \textit{The sum of 23 and } -23 \textit{ is 0.}$$
$$t = -39$$

To check that −39 is the solution, substitute −39 for t in the original equation.

Check:
$$23 + t = -16$$
$$23 + (-39) \stackrel{?}{=} -16$$
$$-16 = -16 \checkmark$$

The solution is −39.

You can and will use equations to solve many real-world problems.

Example 2

APPLICATION Football

Refer to the application at the beginning of the lesson. The San Francisco 49ers appeared fewer times than the Washington Redskins through the 1992–1993 season. How many more times did the Redskins appear?

Explore Read the problem to find out what is asked. Then define a variable.

The problem asks how many more times the Washington Redskins appeared on *Monday Night Football* than the San Francisco 49ers through the 1992–1993 season. From the chart, we can see that the Redskins appeared 43 times.

Let n = the difference in the number of times each team appeared. Then $43 - n$ = the number of times the 49ers appeared.

LOOK BACK
You can refer to Lesson 2-10 for information on solving problems by writing and solving equations.

Plan Write an equation.

You know that $43 - n$ represents the number of times the 49ers appeared. From the chart, we can see that the 49ers appeared 35 times. Thus, $43 - n = 35$.

(continued on the next page)

Classroom Vignette

"I ask one student to represent a parent and two other students to represent children. The parent stands between the two children and acts like an equals sign. The parent then divides a box or bag of candy between the two children. Whatever the parent does to one side, he or she must do to the other side to keep things fair."

Donna Rehling

Donna Rehling
Elder High School
Cincinnati, Ohio

Motivating the Lesson

Hands-On Activity Display a scale to the class. Place two classroom objects such as a pencil and a piece of chalk on the sides of the scale. Have students explain the result. Develop the idea that the scale is balanced when its sides hold equal weights. Challenge students to find two different objects that are equal in weight. Check the validity of student predictions by placing the objects on the scale.

2 TEACH

In-Class Examples

For Example 1
Solve.

a. $21 + r = -2$ **−23**
b. $5 + p = -10$ **−15**
c. $13 + t = -14$ **−27**

For Example 2
Jeff Simons sold 27 cars this month. He sold 63 cars last month. How many fewer cars did he sell this month? **36 cars**

Teaching Tip Emphasize the importance of utilizing the answer check. For Example 1, show how $t = 39$ produces a false statement in the last line of the check.

$$23 + t = -16$$
$$23 + 39 = -16$$
$$62 \neq -16$$

Teaching Tip Explain the symbol $\stackrel{?}{=}$ to students. For example, $a \stackrel{?}{=} b$ means "Is a equal to b?"

Teaching Tip Review the four steps in the problem-solving plan: explore, plan, solve, and examine. Emphasize the importance of checking the answers to all problems as in Example 2.

Teaching Tip It is helpful to learn a concept with basic examples and then extend to more complex ones.

In-Class Examples

For Example 3
Solve.

a. $5 - t = 8$ **−3**
b. $16 - y = 7$ **9**
c. $190 - x = 215$ **−25**

For Example 4
Solve.

a. $y + 9 = -53$ **−62**
b. $b + 15 = -32$ **−47**
c. $m + 18 = -57$ **−75**

Teaching Tip Point out that the minus sign preceding the variable in Example 3 can be read as "the opposite of." For example, $-k = 21$ means "the opposite of k is positive 21." Therefore, k is negative 21; that is, $k = -21$.

Solve Solve the equation and answer the problem.

$$43 - n = 35$$
$$43 + (-43) - n = 35 + (-43) \quad \text{Add } -43 \text{ to each side.}$$
$$-n = -8 \quad \text{The opposite of } n \text{ is negative 8.}$$
$$n = 8 \quad \text{Therefore, } n \text{ is positive 8.}$$

Thus, the Washington Redskins appeared 8 times more than the San Francisco 49ers.

Examine Check to see if the answer makes sense.

If the Redskins appeared 43 times and the 49ers appeared 35 times, then the difference is 43 − 35 or 8 times.

In addition to the addition property of equality, there is a **subtraction property of equality** that may also be used to solve equations.

Subtraction Property of Equality	**For any numbers a, b, and c, if $a = b$, then $a - c = b - c$.**

Example 3 **Solve $190 - x = 215$.**

$$190 - x = 215$$
$$190 - x - 190 = 215 - 190 \quad \text{Subtract 190 from each side.}$$
$$-x = 25 \quad \text{The opposite of } x \text{ is 25.}$$
$$x = -25$$

Check:
$$190 - x = 215$$
$$190 - (-25) \stackrel{?}{=} 215$$
$$190 + 25 \stackrel{?}{=} 215$$
$$215 = 215 \ \checkmark$$

The solution is −25.

Most equations can be solved in two ways. Remember that subtracting a number is the same as adding its inverse.

Example  **Solve $a + 3 = -9$ in two ways.**

Method 1: Use the subtraction property of equality.

$$a + 3 = -9$$
$$a + 3 - 3 = -9 - 3 \quad \text{Subtract 3 from each side.}$$
$$a = -12$$

Check:
$$a + 3 = -9$$
$$-12 + 3 \stackrel{?}{=} -9$$
$$-9 = -9 \ \checkmark$$

Alternative Learning Styles

Auditory Read the following problem to your students. Have them solve the problem by writing and solving an equation. You may have to read the problem more than once.

Todd went to the store and bought three rolls of tape for \$0.89 each, two packs of notebook paper for \$1.39 each, and a notebook. If he spent \$9.34, not including tax, how much did he spend on the notebook? **\$3.89**

Method 2: Use the addition property of equality.

$$a + 3 = -9$$
$$a + 3 + (-3) = -9 + (-3) \quad \text{Add } -3 \text{ to each side.}$$
$$a = -12 \quad \text{The answers are the same.}$$

The solution is -12.

Sometimes equations can be solved more easily if they are first rewritten in a different form.

Example  **Solve** $y - \left(-\frac{3}{7}\right) = -\frac{4}{7}$.

This equation is equivalent to $y + \frac{3}{7} = -\frac{4}{7}$. *Why?*

$$y + \frac{3}{7} = -\frac{4}{7}$$
$$y + \frac{3}{7} - \frac{3}{7} = -\frac{4}{7} - \frac{3}{7} \quad \text{Subtract } \tfrac{3}{7} \text{ from each side.}$$
$$y = -\frac{7}{7} \text{ or } -1$$

Check:
$$y - \left(-\frac{3}{7}\right) = -\frac{4}{7}$$
$$-\frac{7}{7} - \left(-\frac{3}{7}\right) \stackrel{?}{=} -\frac{4}{7}$$
$$-\frac{7}{7} + \frac{3}{7} \stackrel{?}{=} -\frac{4}{7}$$
$$-\frac{4}{7} = -\frac{4}{7} \checkmark$$

The solution is -1.

EXPLORATION — CALCULATORS

You can use a scientific or graphing calculator to solve equations that involve decimals.

 If you're using a scientific calculator like the TI-34, you can use the *plus/minus* key to input negative numbers or to change the sign of any number.

 If you're using a graphing calculator like the TI-81 or TI-82, you can use the *negative* key to input negative numbers.

 On all calculators, use the *subtraction* key to subtract two numbers.

Your Turn a–d. See margin.

a. Take some time to work through the examples in this lesson using your calculator.

b. Describe the difference between the key you use to indicate a negative number on your calculator and the subtraction key.

c. Does the order in which you use these keys matter? Why or why not?

d. How would you use your calculator to solve $x + (-9.016) = 5.14$?

In-Class Example

For Example 5
Solve.

a. $a - (-6) = 3$ **−3**
b. $r - (-8) = 14$ **6**
c. $m - (-9) = 1$ **−8**

Teaching Tip The equation in Example 5 can be solved by adding −11 to each side, that is, by performing the inverse operation. However, most students can solve it more easily by first writing the equation in a simpler form, as shown. That is also the case with Example 6, which could be solved by adding 7.2 to each side.

EXPLORATION

After you give students some time to explore how their own calculators work, have several students demonstrate solving an equation that involves a negative number on their calculators. Make sure they understand that the order in which they use the keys does matter, as well as the difference between a negative or plus/minus key and the subtraction key.

Study Guide Masters, p. 19

 NAME________ DATE________

Study Guide — Student Edition Pages 144–149

Solving Equations with Addition and Subtraction

You can use the addition and subtraction properties of equality to solve equations. To check, substitute the solution for the variable in the original equation. If the resulting sentence is true, your solution is correct.

Addition Property of Equality	For any numbers a, b, and c, if $a = b$, then $a + c = b + c$.
Subtraction Property of Equality	For any numbers a, b, and c, if $a = b$, then $a - c = b - c$.

Example 1: Solve $r - 6 = -11$.
$$r - 6 = -11$$
$$r - 6 + 6 = -11 + 6$$
$$r = -5$$

Check:
$$r - 6 = -11$$
$$-5 - 6 = -11$$
$$-11 = -11 \checkmark$$

Example 2: Solve $k + 18 = -9$.
$$k + 18 = -9$$
$$k + 18 - 18 = -9 - 18$$
$$k = -27$$

Check:
$$k + 18 = -9$$
$$-27 + 18 = -9$$
$$-9 = -9 \checkmark$$

Sometimes an equation can be solved more easily if it is rewritten first. Recall that subtracting a number is the same as adding its inverse. For example, the equation $g - (-5) = 18$ may be rewritten as $g + 5 = 18$.

Solve each equation. Then check your solution.

1. $b - 17 = -40$ −23
2. $x + 12 = 6$ −6
3. $z + 2 = -13$ −15
4. $-17 = b + 4$ −21
5. $s + (-9) = 7$ 16
6. $v - (-12) = 10$ −2
7. $19 + h = -4$ −23
8. $73 = 29 - q$ −44
9. $-3.2 = l + (-0.2)$ −3
10. $-25 - r = \frac{4}{36}$ $-25\frac{1}{9}$
11. $-\frac{3}{8} + x = \frac{5}{8}$ 1
12. $\frac{5}{9} = -y + \frac{2}{15}$ $-\frac{19}{45}$

Answers for the Exploration

b. The plus/minus or negative key is used to indicate a negative number. For example, you would use the plus/minus key to add −3 and 5. The subtraction key is used only for subtraction. You would use both keys to subtract −7 and 2.

c. Yes, the order in which you use the keys does matter. If you are using the plus/minus key, you press it *after* entering the number. If you are using the negative key, you press it *before* entering the number.

d. 5.14 [−] 9.016 [+/−] [=]

In-Class Example

For Example 6
Solve.

a. $x + (-6.4) = -11.5$ **−5.1**
b. $y + (-2.7) = -9.2$ **−6.5**
c. $p + (-10.1) = -15.3$ **−5.2**

3 PRACTICE/APPLY

Check for Understanding

Exercises 1–13 are designed to help you assess your students' understanding through reading, writing, speaking, and modeling. You should work through Exercises 1–5 with your students and then monitor their work on Exercises 6–13.

Error Analysis

Students may be tempted to solve word problems without translating them into open sentences. With simple word problems, students may solve and "undo" the problem as they read and interpret it. To help students focus on translation, assign some word problems with these instructions: "Translate into open sentences. Do not solve the problems."

Practice Masters, p. 19

NAME ______________________ DATE __________

Practice — Student Edition Pages 144–149

Solving Equations by Using Addition and Subtraction

Solve each equation. Then check your solution.

1. $13 + c = -38$ −51
2. $w + 14 = -8$ −22
3. $4.2 = n + 7.3$ −3.1
4. $y + (-10) = 6$ 16
5. $y - 7 = 8$ 15
6. $-13 = 5 + x$ −18
7. $19 - k = 32$ −13
8. $x - 4 = 6$ 10
9. $-12 = 7 - h$ 19
10. $-15 = u + (-5)$ −10
11. $j - (-17) = 36$ 19
12. $-10 - a = -17$ 7
13. $y - (-1.5) = 0.5$ −1
14. $k + (-4.21) = -19.3$ −15.09
15. $52 = 41 - u$ −11
16. $s + (-28) = 0$ 28
17. $\frac{9}{10} = k + \frac{7}{10}$ $\frac{1}{5}$
18. $-\frac{7}{8} - x = -\frac{7}{12}$ $-\frac{7}{24}$

Define a variable, write an equation, and solve each problem. Then check your solution.

19. A number decreased by 14 is −46. Find the number. −32
20. Thirteen subtracted from a number is −5. Find the number. 8
21. A number increased by −56 is −82. Find the number. −26
22. The sum of a number and 67 is −34. Find the number. −101

Example 6 Solve $b + (-7.2) = -12.5$.

This equation is equivalent to $b - 7.2 = -12.5$. *Why?*

$$b - 7.2 = -12.5$$
$$b - 7.2 + 7.2 = -12.5 + 7.2 \quad \textit{Add 7.2 to each side.}$$
$$b = -5.3$$

Method 1

Check: $b + (-7.2) = -12.5$
$-5.3 + (-7.2) \stackrel{?}{=} -12.5$
$-12.5 = -12.5$ ✔

Method 2

Use a calculator.
5.3 [+/−] [+] 7.2
[+/−] [=] −12.5 ✔

The solution is −5.3.

CHECK FOR UNDERSTANDING

Communicating Mathematics

Study the lesson. Then complete the following.

1. Choose the equation that is equivalent to $2.3 - w = 7.8$. **b**
   a. $w - 2.3 = 7.8$ **b.** $w - 2.3 = -7.8$ **c.** $-w = -5.5$
2. **Write** three equivalent equations. **See students' work.**
3. **Explain** how you can prove that 4.5 is the solution of $x - 1.2 = 3.3$. **3. See margin.**
4. **Complete:** If $x - 6 = 21$, then $x + 7 =$ _?_ . Explain your reasoning. **34**

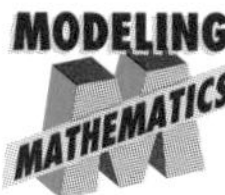

5. Write an equation for the model at the right. Then use cups and counters to solve the equation. **$x - 5 = -10$; $x = -5$**

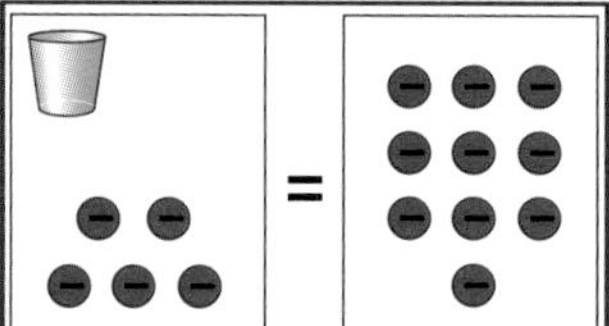

Guided Practice

Solve each equation. Then check your solution.

6. $m + 10 = 7$ **−3**
7. $a - 15 = -32$ **−17**
8. $5.7 + a = -14.2$ **−19.9**
9. $y + (-7) = -19$ **−12**
10. $\frac{1}{6} - n = \frac{2}{3}$ **$-\frac{1}{2}$**
11. $d - (-27) = 13$ **−14**

Define a variable, write an equation, and solve each problem. Then check your solution. **12. $n - 13 = -5$; 8 13. $n + (-56) = -82$; −26**

12. Thirteen subtracted from a number is −5. Find the number.
13. A number increased by −56 is −82. Find the number.

EXERCISES

Practice

Solve each equation. Then check your solution.

14. $k + 11 = -21$ **−32**
15. $41 = 32 - r$ **−9**
16. $-12 + z = -36$ **−24**
17. $2.4 = m + 3.7$ **−1.3**
18. $-7 = -16 - k$ **−9**
19. $0 = t + (-1.4)$ **1.4**
20. $r + (-8) = 7$ **15**
21. $h - 26 = -29$ **−3**
22. $-23 = -19 + n$ **−4**
23. $-11 = k + (-5)$ **−6**
24. $r - 6.5 = -9.3$ **−2.8**
25. $t - (-16) = 9$ **−7**

B

26. $-1.43 + w = 0.89$ **26. 2.32**
27. $m - (-13) = 37$ **24**
28. $-4.1 = m + (-0.5)$ **28. −3.6**
29. $-\frac{5}{8} + w = \frac{5}{8}$ **$1\frac{1}{4}$**
30. $x - \left(-\frac{5}{6}\right) = \frac{2}{3}$ **$-\frac{1}{6}$**
31. $g + \left(-\frac{1}{5}\right) = -\frac{3}{10}$ **$-\frac{1}{10}$**

Reteaching

Using Substeps State the operation and the number you would apply to each side to solve each equation. Then solve. Check each solution.

1. $p + 12 = -4$ **+12 or +(−12); −16**
2. $-12 + b = 12$ **+12 or −(−12); 24**
3. $-14 - a = -21$ **+14 or −(−14); 7**
4. $k + (-13) = 21$ **+13 or −(−13); 34**
5. $b - (-8) = 23$ **+(−8) or −8; 15**
6. $z + (-17) = 0$ **+17 or −(−17); 17**

Additional Answer

3.
$$x - 1.2 = 3.3$$
$$4.5 - 1.2 \stackrel{?}{=} 3.3$$
$$3.3 = 3.3$$

Define a variable, write an equation, and solve each problem. Then check your solution.

32. Twenty-three minus a number is 42. Find the number. $23 - n = 42; -19$
33. A number increased by 5 is equal to 34. Find the number. $n + 5 = 34; 29$
34. What number decreased by 45 is -78? $n - 45 = -78; -33$

35. The difference of a number and -23 is 35. Find the number.
36. A number increased by -45 is 77. Find the number. $n + (-45) = 77; 122$
37. The sum of a number and -35 is 98. Find the number. $n + (-35) = 98; 133$

35. $n - (-23) = 35; 12$

Critical Thinking

38. Suppose a solution of an equation is a number n. Is it possible for $-n$ to also be a solution of this equation? If not, explain why not. If so, give an example of such an equation and its solutions n and $-n$.
Yes; a typical answer is $n^2 = 25$; $n = \pm 5$.

Applications and Problem Solving

Write two different equations to represent each situation. Then solve the problem.

39. **Telecommunications** The graph at the right shows the growth in the number of people who own and use cellular telephones.
    a. How many more cellular phone subscribers were there in 1994 than there were in 1985?
    b. Predict how many cellular phone subscribers there will be in the year 2000. Write a justification for your prediction.

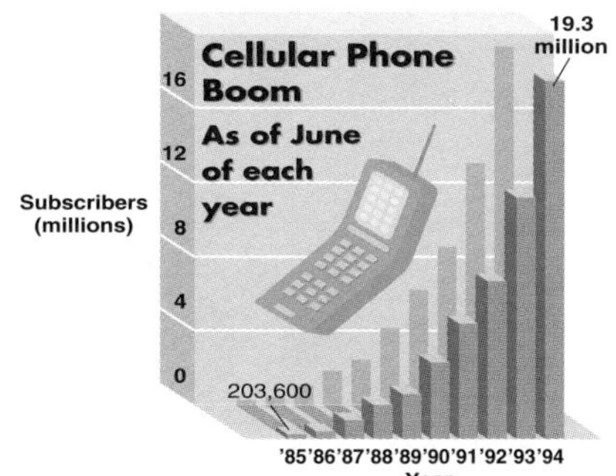

Source: Cellular Telecommunications Industry Association

40. **Consumerism** According to Runzheimer International, in 1994, a tube of lipstick cost \$5.94 in Los Angeles, California. The same tube cost \$4.89 more than that in London, England. In Sao Paulo, Brazil, the lipstick cost \$26.54.
    a. How much did the lipstick cost in London?
    b. How much more did the lipstick cost in Sao Paulo than in Los Angeles?

Mixed Review

41. Alejandra Salazar said, "I am 24 years younger than my mom and the sum of our ages is 68 years." (Lesson 2–9)
    a. How old was Mrs. Salazar when Alejandra was born? 24 years old
    b. How much older than Alejandra is Mrs. Salazar now? 24 years
    c. What will be the sum of their ages in 5 years? 78 years
    d. How old was Mrs. Salazar when Alejandra was 10 years old? 34 years old
    e. In ten years, how much younger than Mrs. Salazar will Alejandra be?
42. Find the principal square root of 256. (Lesson 2–8) 16
43. Simplify $65 \div (-13)$. (Lesson 2–7) -5
44. **Statistics** Odina's algebra average dropped three fourths of a point for each of eight consecutive months. If her average was originally 82, what was her average at the end of eight months? (Lesson 2–6) 76
45. Find the sum. $-0.23x + (-0.5x)$ (Lesson 2–5) $-0.73x$
46. Simplify $6(5a + 3b - 2b)$. (Lesson 1–7) $30a + 6b$
47. Evaluate $12 \div 4 + 15 \cdot 3$. (Lesson 1–3) 48

39a. $203{,}600 + s = 19{,}300{,}000$; 19,096,400 subscribers

39b. Sample answer: 25 million

40a. $t - 5.94 = 4.89$; \$10.83

40b. $5.94 + t = 26.54$; \$20.60

41e. 24 years

Assignment Guide

Core: 15–37 odd, 38, 39, 41–47
Enriched: 14–36 even, 38–47

For **Extra Practice**, see p. 762.

The red A, B, and C flags, printed only in the Teacher's Wraparound Edition, indicate the level of difficulty of the exercises.

4 ASSESS

Closing Activity

Writing Have students make up a problem for which they can write an equation and use addition or subtraction to solve. Then have them write an equation using addition and one using subtraction to solve their problem. Finally, have them solve both equations and compare their answers.

Extension

Reasoning Have students work through the following problem in groups.

Would you rather have 100 pounds of dimes or a ton of pennies?

Students will need to do some research to determine the weights of a dime and a penny to solve the problem.

Since there are 11 dimes per ounce and 9 pennies per ounce, 100 pounds of dimes equals \$1760 while 2000 pounds of pennies equals \$2880.

Enrichment Masters, p. 19

NAME ______________________ DATE __________

Enrichment

Student Edition Pages 144–149

Equations With No Solutions

Not every equation has a solution. Watch what happens when we try to solve the following equation.

$$8 - (3 - 2x) = 5x - 3x$$
$$8 - 3 + 2x = 2x$$
$$5 + 2x = 2x$$
$$5 = 0$$

Since the equation is equivalent to the false statement $5 = 0$, it has no solution. There is no value of x that will make the equation true.

Write a false statement that shows each equation has no solution. Answers may vary. Sample answers are given.

1. $1 - 2t = 2(1 - t)$ $1 = 2$
2. $11y - 7y = 5 + 4y - 6$ $0 = -1$
3. $-7x^2 + 5 + 6x^2 = 12 - (2 + x^2)$ $5 = 10$
4. $2(2 - y^2) = 5 - (5 + 2y^2)$ $4 = 0$
5. $\frac{3}{2} + \frac{2}{3}p - 1 = \frac{1}{3}(1 + 2p)$ $\frac{1}{2} = \frac{1}{3}$
6. $0.5(1 + 3m) = 1.05 - (1 - 1.5m)$ $0.5 = 0.05$

Solve each equation if possible.

7. $5(3 - m) = 15m + 15$ $m = 0$
8. $-9x + 12x = 3(2 - x)$ $x = 1$
9. $10(0.2 + 0.4c) = 10c + 0.2 - 6c$ no solution
10. $13 - (3 - n) = 5(n + 2)$ $n = 0$
11. $2(1 + 4t) = 8 - (3 - 8t)$ no solution
12. $3(d - 1) + 2 = 3(d + 2) - 5$ no solution

3-2 LESSON NOTES

NCTM Standards: 1–5

Instructional Resources

- Study Guide Master 3-2
- Practice Master 3-2
- Enrichment Master 3-2
- Assessment and Evaluation Masters, p. 72
- Graphing Calculator Masters, p. 3
- Real-World Applications, 7
- Tech Prep Applications Masters, p. 5

Transparency 3-2A contains the 5-Minute Check for this lesson; **Transparency 3-2B** contains a teaching aid for this lesson.

Recommended Pacing	
Standard Pacing	Day 3 of 14
Honors Pacing	Day 3 of 13
Block Scheduling*	Day 2 of 6
Alg. 1 in Two Years*	Days 4 & 5 of 19

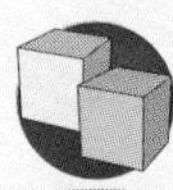

*For more information on pacing and possible lesson plans, refer to the *Block Scheduling Booklet* and *Algebra 1 in Two Years.*

1 FOCUS

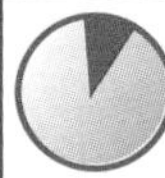

5-Minute Check
(over Lesson 3-1)

Solve.

1. $m + 10 = 7$ **−3**
2. $18 + j = 54$ **36**
3. $t - 11 = 46$ **57**
4. $x - (-24) = -37$ **−61**
5. $-24 - g = -74$ **50**
6. The sum of two integers is −42. One integer is 12. Find the other integer. **−54**

3-2 Solving Equations with Multiplication and Division

What YOU'LL LEARN

- To solve equations by using multiplication and division.

Why IT'S IMPORTANT

You can use equations to solve problems involving construction, sociology, and demographics.

APPLICATION

Construction

Rick Hansen, wheelchair athlete

In 1990, Congress passed into law the Americans with Disabilities Act. One of the provisions of that law has to do with ramps installed on buildings to give people with disabilities access to those buildings. The law states that for a *rise* of 1″, there should be at least 12″ of *run.* The maximum rise is 30″.

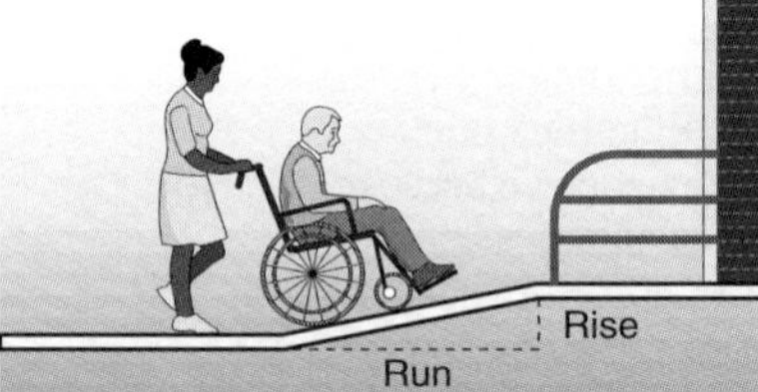

If a contractor wants to build a ramp that has a 180-inch run, what is the greatest rise that the ramp can have?

Since run is dependent on rise, run is a function of rise. Rise is the independent quantity, and run is the dependent quantity. Only positive values for rise and run make sense.

In the table at the right, the pattern suggests that the run is always 12 times the rise. Let x represent the number of inches in the rise. Then $12x$ represents the number of inches in the run. Write an equation to represent the situation.

Rise	Run
1	12
2	24
3	36
4	48
x	$12x$

$12x = 180$ *This equation will be solved in Example 3.*

To solve equations with multiplication and division, you will need new tools. Equations of the form $ax = b$, where a and/or b are fractions, are generally solved by using the **multiplication property of equality**.

Multiplication Property of Equality	For any numbers a, b, and c, if $a = b$, then $ac = bc$.

Example 1 Solve $\frac{g}{24} = \frac{5}{12}$.

$$\frac{g}{24} = \frac{5}{12}$$

$$24\left(\frac{g}{24}\right) = 24\left(\frac{5}{12}\right) \quad \textit{Multiply each side by 24.}$$

$$g = 2(5) \text{ or } 10$$

Check: $\frac{g}{24} = \frac{5}{12}$

$\frac{10}{24} \stackrel{?}{=} \frac{5}{12}$ *Replace g with 10.*

$\frac{5}{12} = \frac{5}{12}$ 

The solution is 10.

Example 2 **Solve each equation.**

a. $\left(3\frac{1}{4}\right)p = 2\frac{1}{2}$

$\left(3\frac{1}{4}\right)p = 2\frac{1}{2}$

$\frac{13}{4}p = \frac{5}{2}$ *Rewrite the mixed numbers as improper fractions.*

$\frac{4}{13}\left(\frac{13}{4}\right)p = \frac{4}{13}\left(\frac{5}{2}\right)$ *Multiply each side by $\frac{4}{13}$, the reciprocal of $\frac{13}{4}$.*

$p = \frac{20}{26}$ or $\frac{10}{13}$ *Check this result.*

The solution is $\frac{10}{13}$.

b. $40 = -5d$

$40 = -5d$

$-\frac{1}{5}(40) = -\frac{1}{5}(-5d)$ *Multiply each side by $-\frac{1}{5}$.*

$-8 = d$ *Check this result.*

The solution is -8.

GLOBAL CONNECTIONS

From 1985 to 1987, Rick Hansen, a wheelchair athlete, went around the world in a wheelchair. He covered 34 countries on four continents, logging 24,000 miles and raising money for wheelchair sports and research on spinal cord injuries.

The equation in Example 2b, $40 = -5d$, was solved by multiplying each side by $-\frac{1}{5}$. The same result could have been obtained by dividing each side by -5. This method uses the **division property of equality**. It is often easier to use than the multiplication property of equality.

Division Property of Equality	**For any numbers a, b, and c, with $c \neq 0$, if $a = b$, then $\frac{a}{c} = \frac{b}{c}$.**

Example 3 **Refer to the application at the beginning of the lesson. Solve $12x = 180$.**

APPLICATION

Construction

$12x = 180$

$\frac{12x}{12} = \frac{180}{12}$ *Divide each side by 12.*

$x = 15$

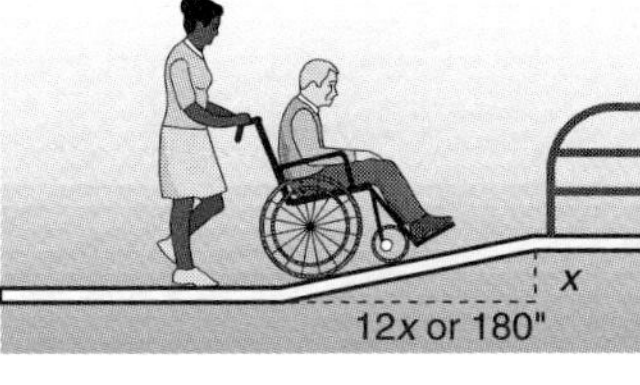

Check: $12x = 180$

$12(15) \stackrel{?}{=} 180$

$180 = 180$ ✓

The rise of the ramp could be at most 15 inches.

Motivating the Lesson

Situational Problem Read the following to the class.

Mr. Kwan bought 15 computer disks and a carrying case for \$28.50. If the carrying case cost \$6.75, what was the cost of each disk?

Then have students identify the procedure they would use to solve the problem. Determine the procedure identified that involves the least number of steps.

2 TEACH

In-Class Examples

For Example 1

Solve.

a. $\frac{3}{4} = \frac{a}{12}$ **9**

b. $\frac{2}{7} = \frac{b}{14}$ **4**

c. $\frac{3}{5} = \frac{c}{30}$ **18**

For Example 2

Solve.

a. $\left(3\frac{1}{2}\right)t = 5\frac{1}{4}$ **$1\frac{1}{2}$**

b. $\left(1\frac{3}{4}\right)m = 4\frac{1}{2}$ **$2\frac{4}{7}$**

c. $\left(-5\frac{2}{5}\right)y = -4\frac{1}{20}$ **$\frac{3}{4}$**

For Example 3

Solve.

a. $-5t = 60$ **-12**

b. $15 = 6n$ **$2\frac{1}{2}$**

c. $-3v = -129$ **43**

Teaching Tip Point out that mixed numbers are less difficult to work with if they are changed to improper fractions or decimals, as in Example 2.

Teaching Tip Be sure to help students develop the habit of checking their solutions, as in Examples 2 and 3.

Teaching Tip You may wish to show Example 3 using the division property of equality.

$$40 = -5d$$
$$\frac{40}{-5} = \frac{-5d}{-5}$$
$$-8 = d$$

The solution is -8.

GLENCOE Technology

Interactive Mathematics Tools Software

This multimedia software provides an interactive lesson that uses counters to solve one-step equations with multiplication or division. A **Computer Journal** gives students an opportunity to write about what they have learned.

For Windows & Macintosh

GLOBAL CONNECTIONS

Rick Hansen was paralyzed as the result of a car accident in 1973, when he was 15 years old. He went on to become the first disabled person at the University of British Columbia to earn a physical education degree. On his journey around the world, Rick saw good times and bad. He had 100 flat tires and was robbed four times.

In-Class Example

For Example 4
Solve.

a. $-12p = 48$ **−4**
b. $-17r = -51$ **3**
c. $24x = -144$ **−6**

3 PRACTICE/APPLY

Check for Understanding

Exercises 1–13 are designed to help you assess your students' understanding through reading, writing, speaking, and modeling. You should work through Exercises 1–5 with your students and then monitor their work on Exercises 6–13.

Additional Answers

1. **Doralina is correct. There is no value for x that would make the statement $0x = 8$ true.**
2. **Rise is the number of units in a vertical direction, and run is the number of units in a horizontal direction.**
4. **17 m; x m; $A = 51\text{ m}^2$**
   Let x = the width. Then $17x = 51$, and $x = 3$.

Study Guide Masters, p. 20

3-2 NAME__________ DATE __________

Study Guide — Student Edition Pages 150–154

Solving Equations with Multiplication and Division

You can solve equations in which a variable has a coefficient by using the multiplication and division properties of equality.

Multiplicative Property of Equality	For any numbers a, b, and c, with $c \neq 0$, if $a = b$, then $ac = bc$.
Division Property of Equality	For any numbers a, b, and c, with $c \neq 0$, if $a = b$, then $\frac{a}{c} = \frac{b}{c}$.

Example 1: Solve $\frac{1}{4}n = 16$.

$\frac{1}{4}n = 16$
$4\left(\frac{1}{4}n\right) = 4(16)$
$n = 64$

Check: $\frac{1}{4}n = 16$
$\frac{1}{4}(64) \stackrel{?}{=} 16$
$16 = 16$ ✔

Example 2: Solve $8n = 64$.

$8n = 64$
$\frac{8n}{8} = \frac{64}{8}$
$n = 8$

Check: $8n = 64$
$8(8) \stackrel{?}{=} 64$
$64 = 64$ ✔

Solve each equation. Then check your solution.

1. $-3r = -24$ **8**
2. $8s = -64$ **−8**
3. $-3t = 51$ **−17**
4. $\frac{1}{4}w = -16$ **−64**
5. $6x = \frac{3}{4}$ **$\frac{1}{8}$**
6. $1\frac{1}{4}y = -3\frac{3}{4}$ **−3**

Define a variable, write an equation, and solve each problem. Then check your solution.

7. Twelve times a number is 96. What is the number? **$12n = 96$; 8**
8. One half of a number is fifteen. What is the number? **$\frac{1}{2}n = 15$; 30**
9. Negative four times a number is −112. What is the number? **$-4n = -112$; 28**
10. Regina paid \$53.50 for 5 basketball tickets. What is the cost of each ticket? **$5t = 53.50$; \$10.70**

Complete.

11. If $4x = 100$, then $8x =$ **200**.
12. If $6y = 36$, then $3y =$ **18**.
13. If $-10a = 53$, then $-5a =$ **26.5**.
14. If $2g + h = 12$, then $4g + 2h =$ **24**.

Example  **Solve $3x = -9$.**

$3x = -9$
$\frac{3x}{3} = \frac{-9}{3}$ *Divide each side by 3.*
$x = -3$

Recall that an equation mat is a model of an equation. You can group each cup on the left side of the mat with the same number of counters on the right side of the mat.

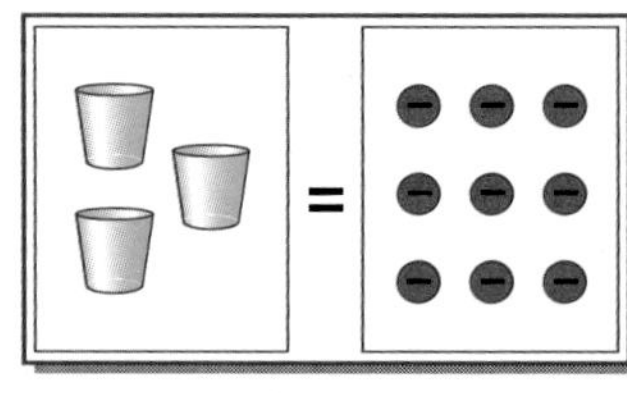

$3x = -9$

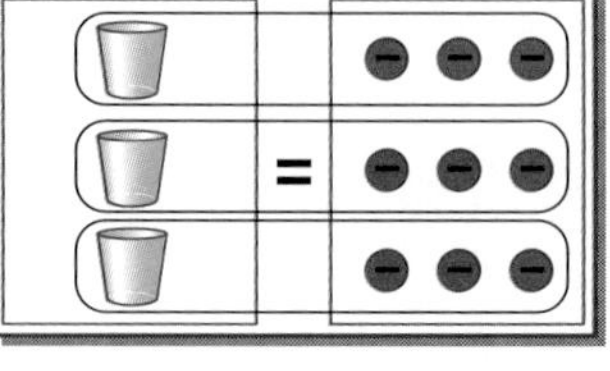

$x = -3$

The solution is −3.

CHECK FOR UNDERSTANDING

Communicating Mathematics

Study the lesson. Then complete the following. **1–2. See margin.**

1. **You Decide** Kezia says that the equation $0x = 8$ has a solution. Doralina says it doesn't. Which one is correct, and why?
2. **Define** the terms *rise* and *run* in your own words.
3. **Refer** to the application at the beginning of the lesson. If the maximum rise is 30″, what is the maximum run? **360 inches or 30 feet**
4. **Draw** a model and explain how to find the width of a rectangle if its area is 51 m² and its length is 17 m. **See margin.**

5. Use cups and counters to model $4x = 16$. Then solve the equation. **See margin.**

Guided Practice

Solve each equation. Then check your solution.

6. $-8t = 56$ **−7**
7. $-5s = -85$ **17**
8. $42.51x = 8$ **0.19**
9. $\frac{k}{8} = 6$ **48**
10. $-10 = \frac{b}{-7}$ **70**
11. $-5x = -3\frac{2}{3}$ **$\frac{11}{15}$**

Define a variable, write an equation, and solve the problem. Then check your solution.

12. Eight times a number is 216. What is the number? **$8n = 216$; 27**
13. The product of −7 and a number is 1.477. What is the number? **$-7n = 1.477$; −0.211**

Reteaching

Using Substeps State the operation and the number you would apply to each side to solve each equation. Then solve. Check each solution.

1. $4n = -20$ **$\times \frac{1}{4}$ or $\div 4$; −5**
2. $12 = -3a$ **$\times \left(-\frac{1}{3}\right)$ or $\div (-3)$; −4**
3. $1.2x = 6$ **$\times \left(\frac{1}{1.2}\right)$ or $\div 1.2$; 5**
4. $\frac{x}{3} = 6$ **$\times 3$ or $\div \frac{1}{3}$; 18**
5. $\frac{m}{-6} = 12$ **$\times (-6)$ or $\div \left(-\frac{1}{6}\right)$; −72**
6. $\frac{c}{2.4} = 12$ **$\times 2.4$ or $\div \frac{1}{2.4}$; 28.8**

Additional Answer

5. 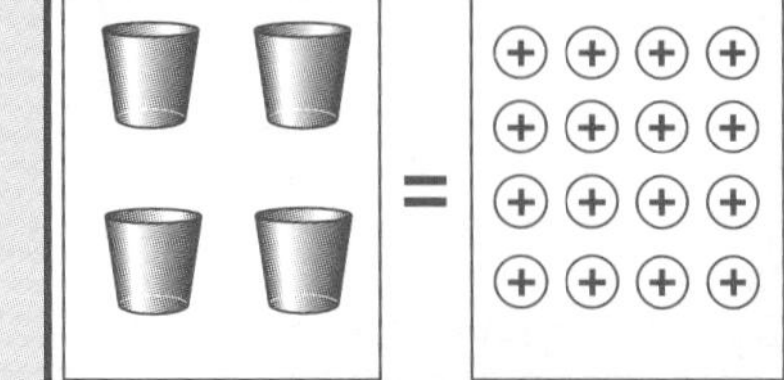

$x = 4$

EXERCISES

Practice

Solve each equation. Then check your solution.

14. $-4r = -28$ **7**
15. $5x = -45$ **−9**
16. $9x = 40$

16. 4.44 or $4\frac{4}{9}$

17. $-3y = 52$
18. $3w = -11$
19. $434 = -31y$ **−14**
20. $1.7b = -39.1$ **−23**
21. $0.49x = 6.277$ **12.81**
22. $-5.73c = 97.41$ **−17**

23. $-0.63y = -378$ **600**
24. $11 = \frac{x}{5}$ **55**
25. $\frac{h}{11} = -25$ **−275**

17. -17.33 or $-17\frac{1}{3}$

26. $\frac{c}{-8} = -14$ **112**
27. $\frac{2}{5}t = -10$ **−25**
28. $-\frac{11}{8}x = 42$ **$-30\frac{6}{11}$**

18. -3.67 or $-3\frac{2}{3}$

29. $-\frac{13}{5}y = -22$ **$8\frac{6}{13}$**
30. $3x = 4\frac{2}{3}$ **$1\frac{5}{9}$**
31. $\left(-4\frac{1}{2}\right)x = 36$ **−8**

Define a variable, write an equation, and solve each problem. Then check your solution.

32. Six times a number is −96. Find the number. **$6n = -96$; −16**
33. Negative twelve times a number is −156. What is the number?
34. One fourth of a number is −16.325. What is the number?
35. Four thirds of a number is 4.82. What is the number? **$\frac{4}{3}n = 4.82$; 3.615**
36. Seven eighths of a number is 14. What is the number? **$\frac{7}{8}n = 14$; 16**

33. $-12n = -156$; 13
34. $\frac{1}{4}n = -16.325$; −65.3

Complete.

37. If $3x = 15$, then $9x =$ __?__. **45**
38. If $10y = 46$, then $5y =$ __?__. **23**
39. If $2a = -10$, then $-6a =$ __?__. **30**
40. If $12b = -1$, then $4b =$ __?__. **$-\frac{1}{3}$**
41. If $7k - 5 = 4$, then $21k - 15 =$ __?__. **12**

Critical Thinking

42. If $-x$ and $-1 \cdot x$ always represent the same number, does $-x$ always stand for a negative number? Explain your reasoning. **No; if x is negative, then $-x$ is positive.**

Applications and Problem Solving

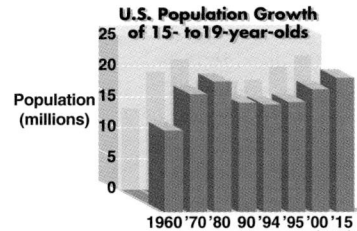

Write an equation to represent each situation. Then solve the problem.

43. **Sociology** Research conducted by *USA Today* has shown that about 1 in 7 people in the world are left-handed.
    a. About how many left-handed people are there in a group of 350 people? 583 people? **$7\ell = 350$, 50 people; $7\ell = 583$, ≈ 83 people**
    b. If there are 65 left-handed people in a group, about how many people are in that group? **$\frac{p}{7} = 65$; 455 people**
44. **Demographics** In the year 2000, there will be about $1\frac{1}{2}$ times the number of 15- to 19-year-olds in the U.S than there were in 1960. If the U.S. Census Bureau projects that there will be 19,819,000 15- to 19-year-olds in the year 2000, about how many were there in 1960?

U.S. Population Growth of 15- to19-year-olds

Source: U.S.Census Bureau

44. $1\frac{1}{2}t = 19{,}819{,}000$; 13,212,667

45. **Telecommunications** In 1995, the long-distance company Sprint introduced Sprint Sense, a plan in which long-distance calls placed on weekends cost only \$0.10 per minute.
    a. How long could you talk for \$2.30? **$0.10m = 2.30$; 23 minutes**
    b. What would be the cost of an 18-minute call? **$0.10(18) = c$; \$1.80**

Assignment Guide

Core: 15–41 odd, 42, 43, 45–53
Enriched: 14–40 even, 42–53

For **Extra Practice,** see p. 762.

The red A, B, and C flags, printed only in the Teacher's Wraparound Edition, indicate the level of difficulty of the exercises.

Practice Masters, p. 20

NAME________ DATE________

3-2 Practice — Student Edition Pages 150–154

Solving Equations by Using Multiplication and Division

Solve each equation. Then check your solution.

1. $-6d = -42$ 7
2. $-7t = 49$ −7
3. $32z = 896$ 28
4. $8k = 36$ $\frac{9}{2}$
5. $18e = -216$ −12
6. $0.26f = 0.0312$ 0.12
7. $\frac{c}{4} = 16$ 64
8. $-18 = \frac{m}{6}$ −108
9. $31 = -\frac{1}{6}m$ −186
10. $6 = \frac{4}{-7}t$ $-\frac{21}{2}$
11. $-\frac{5}{8}w = -9$ $\frac{72}{5}$
12. $\frac{3}{5}x = 15$ 25
13. $-12 = \frac{2}{3}z$ −18
14. $\frac{c}{-7} = -13$ 91
15. $\frac{5}{9}k = -10$ −18
16. $-\frac{3}{5}s = -40$ $\frac{200}{3}$
17. $5n = 2\frac{3}{4}$ $\frac{11}{20}$
18. $-3x = 1\frac{1}{2}$ $-\frac{1}{2}$

Define a variable, write an equation, and solve each problem. Then check your solution.

19. Eight times a number is 216. What is the number? 27
20. Negative seven times a number is 1.477. Find the number. −0.211
21. One third of a number is 8.5. What is the number? 25.5
22. Four thirds of a number is 4.82. What is the number? 3.615

4 ASSESS

Closing Activity

Writing Have students identify and write about specific instances in which algebra is used in geometry. As they write, have them create problems to illustrate their answers.

Chapter 3, Quiz A (Lessons 3-1 and 3-2), is available in the *Assessment and Evaluation Masters,* p. 72.

Additional Answer

50. 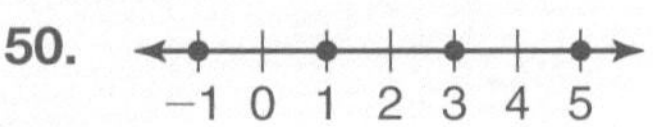

Enrichment Masters, p. 20

NAME ________________ DATE ________

Enrichment — Student Edition Pages 150–154

Identities

Any equation that is true for every value of the variable is called an **identity.** When you try to solve an identity, you end up with a statement that is always true. Here is an example.

$8 - (5 - 6x) = 3(1 + 2x)$

$8 - 5 + 6x = 3 + 6x$

$3 + 6x = 3 + 6x$

State whether each equation is an identity. If it is not, find its solution.

1. $2(2 - 3x) = 3(3 + x) + 4$ — $x = -1$
2. $5(m + 1) + 6 = 3(4 + m) + (2m - 1)$ — identity
3. $(5t + 9) - (3t - 13) = 2(11 + t)$ — identity
4. $14 - (6 - 3c) = 4c - c$ — no solution
5. $3y - 2(y + 19) = 9y - 3(9 - y)$ — $y = -1$
6. $3(3h - 1) = 4(h + 3)$ — $h = 3$
7. Start with the true statement $3x - 2 = 3x - 2$. Use it to create an identity of your own. Answers will vary.
8. Start with the false statement $1 = 2$. Use it to create an equation with no solution. Answers will vary.

Mixed Review

47. $17 + 1\frac{1}{2}y = 33\frac{1}{2}$; 11 years

51a. The number of bags increased during the day.

51b. The machine was refilled.

51c. See students' work.

46. Solve $-11 = a + 8$. Then check your solution. (Lesson 3–1) **−19**
47. Define a variable, then write an equation for the problem below. (Lesson 2–9) Ponderosa pines grow about $1\frac{1}{2}$ feet each year. If a pine tree is now 17 feet tall, about how long will it take for the tree to become $33\frac{1}{2}$ feet tall?
48. **World Records** The longest loaf of bread ever baked was 2132 feet $2\frac{1}{2}$ inches. If this loaf were cut into $\frac{1}{2}$-inch slices, how many slices of bread would there have been? (Lesson 2–7) **51,173 slices**
49. Find a number between $\frac{1}{2}$ and $\frac{6}{7}$. (Lesson 2–4) $\frac{19}{28}$
50. Graph $\{-1, 1, 3, 5\}$ on a number line. (Lesson 2–1) **See margin.**
51. The graph at the right shows the number of bags of potato chips in the snack machine at Lee High School two times during an average day. (Lesson 1–9)
    a. What can you learn from looking at the graph?
    b. What do you think happened between the times indicated by the dots?
    c. If lunch at Lee is between 11:45 A.M. and 1:15 P.M., estimate the number of bags you think would be in the machine at 1:30 P.M. Justify your estimate.

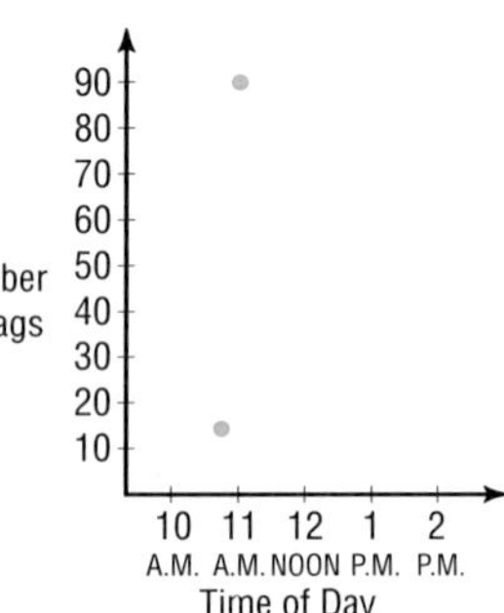

52. Simplify $5(3x + 2y - 4y)$. (Lesson 1–7) **$15x - 10y$**
53. Evaluate $a + b^2 + c^2$ if $a = 6$, $b = 4$, and $c = 3$. (Lesson 1–1) **31**

WORKING ON THE In•ves•ti•ga•tion

Refer to the Investigation on pages 68–69.

the Greenhouse Effect

The changes in the levels of carbon dioxide (CO_2) on Earth happened naturally until nearly 200 years ago. During the Industrial Revolution of the early 1800s, a new factor was thrown into the equation. When wood and fossil fuels like coal, oil, and natural gas are burned, CO_2 is released in large quantities. Oceans and vegetation absorb the gas. But due to the widespread cutting of trees to produce goods in the 1900s, few remained to soak up excess CO_2. In 1920, atmospheric CO_2 stood at about 280 parts per million (ppm). By 1996, it had risen to 356 ppm. **1–4. See margin.**

1. Make a chart that shows the changes in the levels of CO_2 in our atmosphere. Assume that the levels of CO_2 grow at a constant rate over the years. Use the headings *Year, Amount of CO_2 (ppm),* and *Change.* Start the chart with 1920, and end with the present year.
2. Write a paragraph about the patterns you see in your chart. Use the terms *function, independent, dependent, domain,* and *range.*
3. Based on the relationships you found in the previous question, determine how much CO_2 was in the atmosphere in 1800 and 1950. Explain how you found those amounts.
4. Predict the amounts of CO_2 that will be in the atmosphere in the years 2000, 2020, 2050, and 3000. Explain how you made your predictions.

Add the results of your work to your Investigation Folder.

Extension

Problem Solving JoAnn bought 18 cupcakes for a party. She had writing put on seven of the cupcakes at an extra cost of 20¢ each. If the total cost of the cupcakes was \$11.30, how much did each plain cupcake cost? **55¢**

In•ves•ti•ga•tion

Working on the Investigation

The Investigation on pages 68–69 is designed to be a long-term project that is completed over several days or weeks. Encourage students to keep their materials in their Investigation Folder as they work on the Investigation.

A Preview of Lesson 3–3

3–3A Solving Multi-Step Equations

Materials: cups & counters equation mat

You can use an equation model to solve equations with more than one operation or equations with a variable on each side.

Activity 1 Use an equation model to solve $2x + 2 = -4$.

Step 1 Model the equation by placing 2 cups and 2 positive counters on one side of the mat. Place 4 negative counters on the other side of the mat. Add 2 negative counters to each side to form zero pairs on the left.

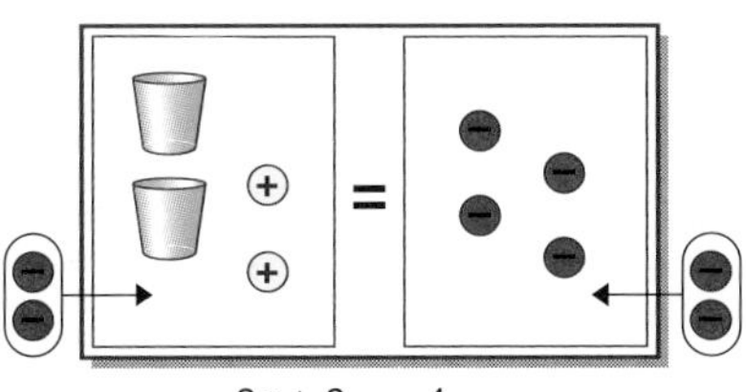

$2x + 2 = -4$

Step 2 Group the counters to form zero pairs and remove the zero pairs. Separate the remaining counters into 2 equal groups to match the 2 cups. Each cup is paired with 3 negative counters. Therefore, $x = -3$.

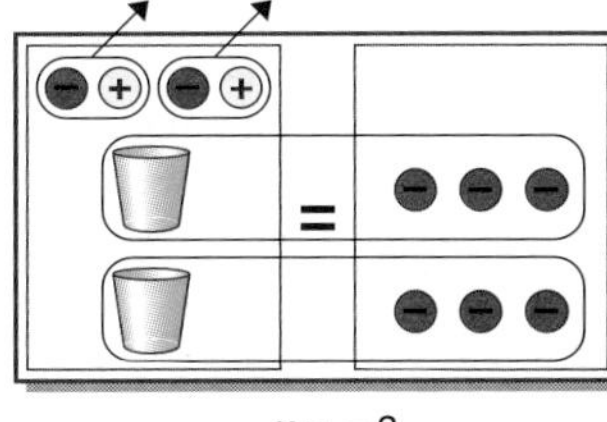

$x = -3$

Activity 2 Use an equation model to solve $w - 3 = 2w - 1$.

Step 1 Model the equation by placing 1 cup and 3 negative counters on one side of the mat. Place 2 cups and 1 negative counter on the other side of the mat. Remove 1 negative counter from each side of the mat.

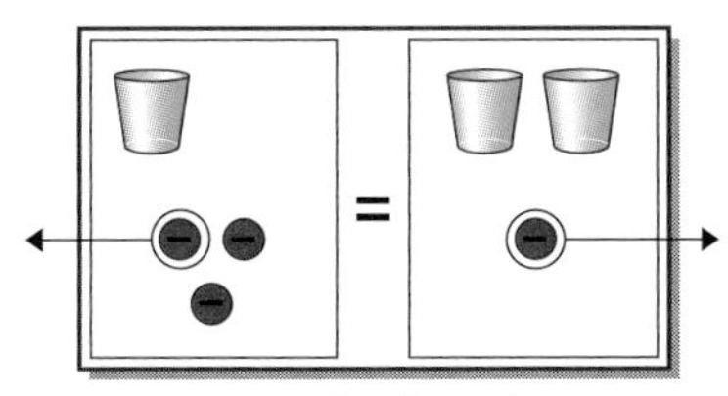

$w - 3 = 2w - 1$

Step 2 You can remove the same number of cups from each side of the mat. In this case, remove 1 cup from each side. The cup on the right is matched with 2 negative counters. Therefore, $w = -2$.

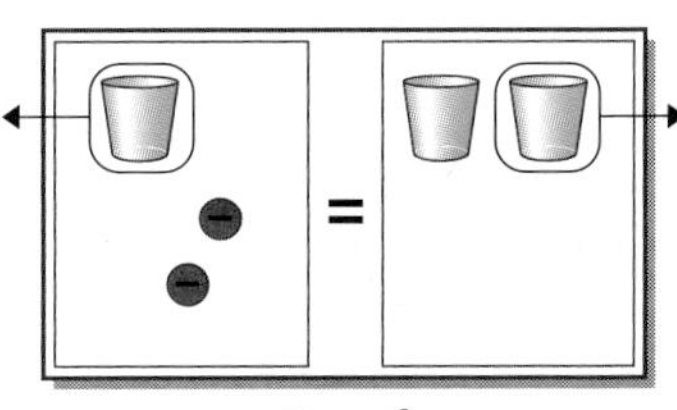

$w = -2$

Model Use equation models to solve each equation.

1. $2x + 3 = 13$ **5**
2. $2y - 2 = -4$ **−1**
3. $-4 = 3a + 2$ **−2**
4. $3m - 2 = 4$ **2**
5. $3x + 2 = x + 6$ **2**
6. $3x + 7 = x + 1$ **−3**
7. $3x - 2 = x + 6$ **4**
8. $y + 1 = 3y - 7$ **4**
9. $2b + 3 = b + 1$ **−2**

3-3A LESSON NOTES

NCTM Standards: 1–5

Objective
Solve equations by using cups and counters to model variables and integers.

Recommended Time
Demonstration and discussion: 15 minutes; Exercises: 30 minutes

Instructional Resources
For each student or group of students
Student Manipulative Kit
- 20 red/yellow 2-sided counters
- cups

Modeling Mathematics Masters
- pp. 2–3 (cups and counters)
- p. 13 (equation mat)
- p. 24 (worksheet)

For teacher demonstration
Algebra and Geometry Overhead Manipulative Resources

1 FOCUS

Motivating the Lesson
Remind students that this is not a new procedure, but a combination of the two types of equations they solved in Lesson 3-1A. Help students to see the connection between the processes used to solve one-step equations and those used to solve multi-step equations.

2 TEACH

Teaching Tip Encourage students to check their solutions algebraically by replacing the variable in the original equation with the solution to see if the equation holds.

3 PRACTICE/APPLY

Assignment Guide
Core: 1–9
Enriched: 1–9

4 ASSESS

Observing students working in cooperative groups is an excellent method of assessment.

3-3 LESSON NOTES

NCTM Standards: 1–5

Instructional Resources

- Study Guide Master 3-3
- Practice Master 3-3
- Enrichment Master 3-3
- Modeling Mathematics Masters, pp. 42–44
- Multicultural Activity Masters, p. 5
- Real-World Applications, 8
- Tech Prep Applications Masters, p. 6

Transparency 3-3A contains the 5-Minute Check for this lesson; **Transparency 3-3B** contains a teaching aid for this lesson.

Recommended Pacing	
Standard Pacing	Days 5 & 6 of 14
Honors Pacing	Day 5 of 13
Block Scheduling*	Day 3 of 6
Alg. 1 in Two Years*	Days 7 & 8 of 19

*For more information on pacing and possible lesson plans, refer to the *Block Scheduling Booklet* and *Algebra 1 in Two Years.*

1 FOCUS

5-Minute Check
(over Lesson 3-2)

Solve.

1. $\frac{k}{8} = -24$ **−192**
2. $39 = -3b$ **−13**
3. $\left(1\frac{1}{3}\right)p = 2\frac{1}{6}$ **$1\frac{5}{8}$**
4. $-65.8 = 8.27t$ **−7.96**
5. $\frac{4x}{5} = -32$ **−40**
6. Suzanne bought 7 books for $55.23. If all the books cost the same amount, what was the price of 1 book? **$7.89**

3-3 Solving Multi-Step Equations

What YOU'LL LEARN

- To solve equations involving more than one operation, and
- to solve problems by working backward.

Why IT'S IMPORTANT

You can use equations to solve problems involving home repair, health, and world cultures.

APPLICATION
World Cultures

On June 9, 1993, Japan's Crown Prince Naruhito wed former diplomat Masako Owada. It took Ms. Owada about $2\frac{1}{2}$ hours to put on her wedding kimono, a 12-layered silk garment that weighed about 30 pounds. The prince wore a kimono of bright orange—the "sunrise color"—that can only be worn by the heir to the throne.

After the wedding, the crown prince changed into a tuxedo and the princess into a bridal gown to formally announce their marriage to the Emperor and Empress. To change into the bridal gown, Princess Masako and her attendants had to remove the wedding kimono, a process that was similar to putting on the garment, only in reverse order. You will solve certain kinds of problems by working in reverse order, or **working backward.** Work backward is one of many *problem-solving strategies* that you can use to solve problems. Here are some other problem-solving strategies.

FYI

The *kimono* has been worn by Japanese men and women since the 7th Century A.D. It is an ankle-length gown with long wide sleeves and is secured by a sash called an *obi.*

Problem-Solving Strategies	
draw a diagram	solve a simpler (or a similar) problem
make a table or chart	eliminate the possibilities
make a model	look for a pattern
guess and check	act it out
check for hidden assumptions	list the possibilities
use a graph	identify subgoals

Example 1 **Due to melting, an ice sculpture loses one-half its weight every hour. After 8 hours, it weighs $\frac{5}{16}$ of a pound. How much did it weigh in the beginning?**

Make a table to show weight as a function of time. Work backward to find the original weight.

Recall that the sculpture loses one-half its weight every hour. Multiply the current weight of the sculpture by 2 to find its weight an hour before. Continue multiplying until you reach the original weight.

The original weight of the sculpture was 80 pounds.

Hour	Weight
8	$\frac{5}{16}$
7	$2\left(\frac{5}{16}\right) = \frac{5}{8}$
6	$2\left(\frac{5}{8}\right) = \frac{5}{4}$
5	$2\left(\frac{5}{4}\right) = \frac{5}{2}$
4	$2\left(\frac{5}{2}\right) = 5$
3	$2(5) = 10$
2	$2(10) = 20$
1	$2(20) = 40$
0	$2(40) = 80$

FYI

The wedding of Crown Prince Naruhito and Masako Owada was the first to take advantage of high-definition television (HDTV). In keeping with tradition, the groom's parents did not attend the wedding ceremony. They were, however, able to watch a computer simulation of the private ceremony on their HDTV set.

To solve equations with more than one operation, often called **multi-step equations,** you will undo the operations by working backward.

Example 2

Home Repair

Mrs. Guzman needs a repairperson to fix her washing machine. Since her machine is pretty old, she doesn't want to spend more than $100 for repairs. The owner of Albie's Appliances told her that a service call would cost $35 and the labor would be an additional $20 per hour. What is the maximum number of hours that the repairperson can work and keep the total cost at $100?

Explore Read the problem and define the variable.
Let h represent the maximum number of hours the repair can take.

Plan Write an equation.

$35 plus $20 per hour for h hours equals $100.

$35 + 20h = 100$

Solve Work backward to solve the equation.

$$35 + 20h = 100$$
$$35 - 35 + 20h = 100 - 35$$
$$20h = 65$$

Undo the addition first. Use the subtraction property of equality.

$$\frac{20h}{20} = \frac{65}{20}$$
$$h = \frac{13}{4} \text{ or } 3\frac{1}{4}$$

Then undo the multiplication. Use the division property of equality.

The repairperson has up to $3\frac{1}{4}$ hours to fix the washing machine.

Examine Check to see if the answer makes sense.

$3\frac{1}{4} \times \$20 = \$65$ $\quad$ $\$35 + \$65 = \$100$

EXPLORATION — GRAPHING CALCULATORS

You can use a graphing calculator, like the TI-82, to solve multi-step equations. The *solve(* function will solve an equation if it is rewritten as an expression that equals zero, or most commonly, in the form $ax + b - c = 0$. This function also requires that you include a guess about the solution to the equation. The format is as follows.

solve(*expression,variable,guess*)

To solve the equation $3x - 4 = 2$, access the math function by pressing the MATH key. Then choose 0, for solve. Enter the expression $3x - 4 - 2$, the variable x, and a guess about its solution. If we guess -1, then this appears on the calculator screen: solve(3X-4-2,X,-1). When you press ENTER, the solution, 2, appears on the screen.

Your Turn

a. Work through the examples in this lesson by using a graphing calculator.

b. Describe the process of using a graphing calculator to solve equations in your own words.

c. Why did we input the equation $3x - 4 = 2$ as $3x - 4 - 2$?

d. How would you use a graphing calculator to solve the equation $3x - 4 = -2x + 6$?

b–d. See margin.

Motivating the Lesson

Situational Problem Read the following problem to students. Have them identify various ways of solving the problem. Then have students use a method described to find the answer.

Each week for 7 weeks, a store reduced the price of a VCR by $8.99. The original price of the VCR was $329.00. Tom wants a VCR but has saved only $295.00. How long does he have to wait to purchase the VCR? **4 weeks**

2 TEACH

Teaching Tip As the work-backward strategy is being introduced, you may wish to have students explore other methods of problem solving, such as guess and check.

In-Class Examples

For Example 1

A school cafeteria sells more garden salads than soup but not as much as pizza. Fewer sandwiches are sold than soup. Order the lunch items from the most popular to the least popular. **pizza, garden salad, soup, sandwiches**

For Example 2

Heather bought a winter coat for $6 less than half its original price. Heather paid $65 for the coat. What was the original price? **$142**

Teaching Tip Emphasize the importance of using the four-step problem-solving plan, as in Example 2, to properly analyze each verbal problem.

EXPLORATION

Students can be fairly creative in their guesses, but this is a good time for them to check the answers in their homework. Point out that if they include the answers they got, they will be able to tell if they were correct or not.

Answers for the Exploration

b. Input an expression of the form $ax + b - c$, a comma, the variable x, another comma, and a guess. Close with a right parenthesis and press ENTER.

c. Because $3x - 4 = 2$ can be written as $3x - 4 - 2 = 0$. There is no need to enter the "= 0" part in the calculator.

d. Rewrite as $5x - 4 - 6 = 0$; then follow the same steps.

You have seen a multi-step equation in which the first, or *leading,* coefficient is an integer. You can use the same steps if the leading coefficient is a fraction.

Example 3 **Solve each equation.**

a. $\frac{y}{5} + 9 = 6$

$\frac{y}{5} + 9 = 6$ *The leading coefficient is $\frac{1}{5}$.*

$\frac{y}{5} + 9 - 9 = 6 - 9$ *First, subtract 9 from each side. Why?*

$\frac{y}{5} = -3$

$5\left(\frac{y}{5}\right) = 5(-3)$ *Then, multiply each side by 5.*

$y = -15$

Check: $\frac{y}{5} + 9 = 6$

$\frac{-15}{5} + 9 \stackrel{?}{=} 6$

$-3 + 9 \stackrel{?}{=} 6$

$6 = 6$ ✔

The solution is -15.

b. $\frac{d-2}{3} = 7$

$\frac{d-2}{3} = 7$

$3\left(\frac{d-2}{3}\right) = 3(7)$ *Multiply each side by 3. Why?*

$d - 2 = 21$

$d - 2 + 2 = 21 + 2$ *Add 2 to each side.*

$d = 23$

Check: $\frac{d-2}{3} = 7$

$\frac{23-2}{3} \stackrel{?}{=} 7$

$\frac{21}{3} \stackrel{?}{=} 7$

$7 = 7$ ✔

The solution is 23.

Consecutive integers are integers in counting order, such as 3, 4, 5. Beginning with an even integer and counting by two will result in *consecutive even integers.* For example, $-6, -4, -2, 0,$ and 2 are consecutive even integers. Beginning with an odd integer and counting by two will result in *consecutive odd integers.* For example, $-1, 1, 3,$ and 5 are consecutive odd integers.

The study of numbers and the relationships between them is called **number theory.** Number theory involves the study of odd and even numbers.

In-Class Examples

For Example 3
Solve.

a. $\frac{t}{3} + 11 = 29$ 54

b. $\frac{x}{5} - 17 = 3$ 100

c. $\frac{2}{3}a - 14 = 61$ $112\frac{1}{2}$

d. $12 = \frac{2-3x}{4}$ $-15\frac{1}{3}$

e. $\frac{3b+1}{2} = -25$ -17

f. $\frac{5+2x}{8} = 14$ $53\frac{1}{2}$

Teaching Tip You may wish to show an alternate method for solving Example 3a.

$\frac{y}{5} + 9 = 6$

$5\left(\frac{y}{5} + 9\right) = 5(6)$

$y + 45 = 30$

$y + 45 - 45 = 30 - 45$

$y = -15$

The solution is -15.

Additional Answers

2. **Let $2n + 2$ represent consecutive even numbers and let $2n + 1$ and $2n + 3$ represent consecutive odd numbers.**
   $2n + 2n + 2 = (2n + 1) + (2n + 3)$
   $4n + 2 = 4n + 4$
   $2 = 4$
   Since this is not a true statement, the sum of two consecutive even numbers can never equal the sum of two consecutive odd numbers.
4. $4x + 3x - 5 = 27$
   $4(-2) + 3(-2) - 5 \stackrel{?}{=} 27$
   $-8 + (-6) - 5 \stackrel{?}{=} 27$
   $-19 \neq 27$
8. **Add 5 to each side, then divide each side by 4.**
9. **Subtract 7 from each side, then divide each side by 3.**
10. **Add 6 to each side, then multiply each side by $\frac{9}{2}$.**
11. **Subtract 3 from each side, then multiply each side by -7.**
12. **Multiply each side by 7, then add 3 to each side.**
13. **Rewrite numerator as $p + 5$, multiply each side by -2, and then subtract 5.**

Alternative Learning Styles

Kinesthetic Have students work in groups of four. Give each group a set of cards numbered 1 through 8. Each student should hold two cards. The group's task is to arrange its members in the formation shown at the right so that no two consecutive integers occupy neighboring squares horizontally, vertically, or diagonally.

	3	5	
7	1	8	2
	4	6	

Once all of the groups have reached a solution, have one member from each group record and explain the group's solution to the class. **There are several solutions. One possible solution is provided.**

Example 4 Find three consecutive odd integers whose sum is −15.

Number Theory

Let n = the least odd integer.
Then $n + 2$ = the next greater odd integer,
and $n + 4$ = the greatest of the three odd integers.

$$n + (n + 2) + (n + 4) = -15$$
$$3n + 6 = -15 \quad \textit{Simplify.}$$
$$3n + 6 - 6 = -15 - 6 \quad \textit{Subtract 6 from each side.}$$
$$3n = -21$$
$$\frac{3n}{3} = \frac{-21}{3} \quad \textit{Divide each side by 3.}$$
$$n = -7$$

$n + 2 = -7 + 2$ $\qquad$ $n + 4 = -7 + 4$

$n + 2 = -5$ $\qquad$ $n + 4 = -3$

The consecutive odd integers are −7, −5, and −3.

Explain why this answer makes sense.

CHECK FOR UNDERSTANDING

Communicating Mathematics

Study the lesson. Then complete the following.

1. a. **Explain** how you would solve $2p + 10 = 42$ if you had to undo the multiplication first. **Divide each term by 2.**
   b. **Explain** why undoing the multiplication first would be inconvenient for solving the equation $7x - 4 = 24$. **Fractions would be introduced.**
2. **Determine** whether the sum of two consecutive even numbers can ever equal the sum of two consecutive odd numbers. Justify your answer. *2. See margin.*
3. If n is an even integer, explain how to find the even integer just before it. *3. Subtract 2; $n - 2$.*
4. **Explain** why −2 is not the solution of the equation $4x + 3x - 5 = 27$. *4. See margin.*
5. **List** three consecutive even integers if the greatest one is −18. *5. −22, −20, −18*
6. **Complete:** If $2x + 1 = 5$, then $3x - 4 =$ __?__ . Explain your reasoning. **2**

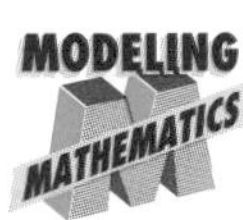
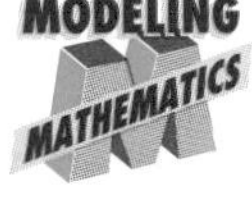

7. Write an equation for the model at the right. Then use cups and counters to solve the equation.
   $3x + 2 = 8$; 2

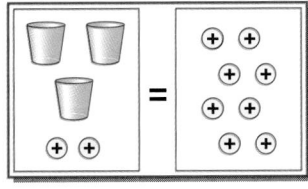

Guided Practice

Explain how to solve each equation. Then solve.

8–13. See margin for explanations.

8. $4a - 5 = 15$ **5**
9. $7 + 3c = -11$ **−6**
10. $\frac{2}{9}v - 6 = 14$ **90**
11. $3 - \frac{a}{7} = -2$ **35**
12. $\frac{x - 3}{7} = -2$ **−11**
13. $\frac{p - (-5)}{-2} = 6$ **−17**

Define a variable, write an equation, and solve each problem. Then check your solution. 15. **$n + (n + 2) + (n + 4) = 21$; 5, 7, 9**

14. Find two consecutive integers whose sum is −31. *14. $n + (n + 1) = -31$; −16, −15*
15. Find three consecutive odd integers whose sum is 21.
16. Twenty-nine is 13 added to 4 times a number. Find the number. *16. $29 = 13 + 4n$; $n = 4$*

In-Class Example

For Example 4
Find four consecutive odd integers whose sum is 72.
15, 17, 19, 21

Teaching Tip In Example 4, be sure students understand that zero is even and that both positive and negative integers can be odd or even.

3 PRACTICE/APPLY

Check for Understanding

Exercises 1–16 are designed to help you assess your students' understanding through reading, writing, speaking, and modeling. You should work through Exercises 1–7 with your students and then monitor their work on Exercises 8–16.

Error Analysis
Students may choose to undo multiplication before undoing addition. Emphasize that the whole side (all terms) must be divided. The distributive property shows why using the multiplicative property of equality is equivalent to multiplying each term by the number.

Study Guide Masters, p. 21

NAME__________ DATE______
3-3 Study Guide — Student Edition Pages 156–161

Solving Multi-Step Equations

When solving some equations you must perform more than one operation on both sides. First, determine what operations have been done to the variable. Then undo these operations in the reverse order.

Example 1: How would you solve $\frac{n}{3} - 7 = 28$?

$\frac{n}{3} - 7 = 28$ — First, n was divided by 3. Then 7 was subtracted. } To solve, first add 7 to each side. Then multiply each side by 3.

Procedure for Solving a Two-Step Equation	1. Undo any indicated additions or subtractions. 2. Undo any indicated multiplications or divisions involving the variable.

Example 2: $5x + 3 = 23$ — Addition of 3 is indicated.
$5x + 3 - 3 = 23 - 3$ — Therefore, subtract 3 from each side.
$5x = 20$ — Multiplication by 5 is also indicated.
$\frac{5x}{5} = \frac{20}{5}$ — Therefore, divide each side by 5.
$x = 4$

Check:
$5x + 3 = 23$
$5(4) + 3 \stackrel{?}{=} 23$
$20 + 3 \stackrel{?}{=} 23$
$23 = 23$ ✔

Solve each equation. Then check your solution.

1. $5z + 16 = 51$ 7
2. $14n - 8 = 34$ 3
1. $0.6x - 1.5 = 1.8$ 5.5
4. $\frac{4b + 8}{-2} = 10$ −7
5. $16 = \frac{d - 12}{14}$ 236
6. $8 + \frac{3n}{12} = 13$ 20
7. $\frac{7}{8}p - 4 = 10$ 16
8. $\frac{g}{-5} + 3 = -13$ 80
9. $-4 = \frac{7x - (-1)}{-8}$ $\frac{31}{7}$, or $4\frac{3}{7}$

Define a variable, write an equation, and solve each problem. Then check your solution.

10. Find three consecutive integers whose sum is 96.
    $n + (n + 1) + (n + 2) = 96$; 31, 32, 33
11. Find two consecutive odd integers whose sum is 176.
    $n + (n + 2) = 176$; 87, 89

Reteaching

Using Alternative Methods Solve the equations by undoing the multiplication or division first. Be sure to multiply or divide each side completely (all terms) by the number.

1. $42 = 10 + 2p$ **16**
2. $\frac{x}{4} + 9 = 6$ **−12**
3. $\frac{d - 4}{3} = 5$ **19**
4. Two germs live in a tin can. At 4:00 P.M. they begin to double in number every second. At 5:00 P.M. the can is completely full of germs.
   a. When was the can
      i. $\frac{1}{2}$ full? **4:59**
      ii. $\frac{1}{4}$ full? **4:58**
      iii. $\frac{1}{8}$ full? **4:57**
   b. How many germs does the can hold? **2^{61}**

Assignment Guide

Core: 17–47 odd, 49–56
Enriched: 18–44 even, 45–56
All: Self Test, 1–10

For **Extra Practice,** see p. 762.

The red A, B, and C flags, printed only in the Teacher's Wraparound Edition, indicate the level of difficulty of the exercises.

Additional Answers

36. $n + (n + 1) + (n + 2) = -33$; $-10, -11, -12$

37. $n + (n + 1) + (n + 2) + (n + 3) = 86$; 20, 21, 22, 23

46b.

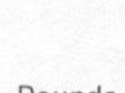

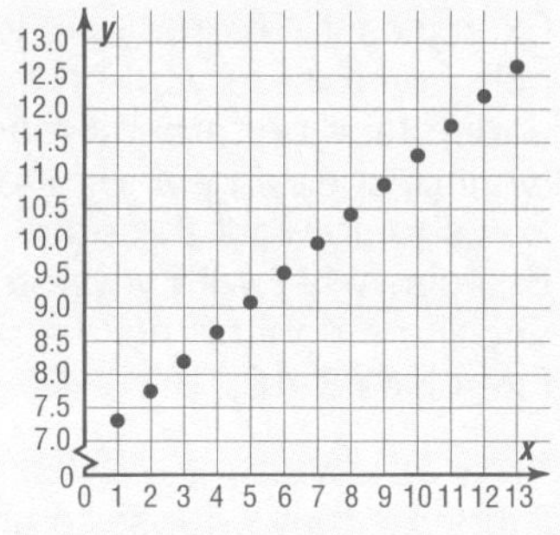

Practice Masters, p. 21

NAME__________ DATE ________

Practice — Student Edition Pages 156–161

Solving Multi-Step Equations

Solve each equation. Then check your solution.

1. $5x + 3 = 23$ 4	**2.** $4 = 3a - 14$ 6
3. $19 = 3y - 5$ 8	**4.** $6 - 5c = -29$ 7
5. $8 - 5w = -37$ 9	**6.** $42 = 18 - 4v$ −6
7. $0.4m - 3 = -1$ 5	**8.** $3.2e + 2.6 = -23$ −8
9. $\frac{n}{3} - 8 = -2$ 18	**10.** $\frac{x}{-4} + 5 = 1$ 16
11. $13 = \frac{w}{-3} - 4$ −51	**12.** $-7 = \frac{c}{-6} + 12$ 114
13. $\frac{3}{4}x - 7 = 8$ 20	**14.** $-12 = 6 + \frac{2}{3}y$ −27
15. $-17 = -32 - \frac{3}{5}f$ −25	**16.** $8 - \frac{3}{8}k = -4$ 32
17. $-14 = \frac{s + 12}{-6}$ 72	**18.** $\frac{u + 12}{-4} = 5$ −32

Define a variable, write an equation, and solve each problem. Then check your solution.

19. Find two consecutive odd integers whose sum is 116. 57, 59

20. Find two consecutive even integers whose sum is 126. 62, 64

21. Find three consecutive odd integers whose sum is 117. 37, 39, 41

22. Find two consecutive even integers whose sum is 217. no solution

23. Find four consecutive odd integers whose sum is 8. −1, 1, 3, 5

24. Find three consecutive even integers whose sum is 396. 130, 132, 134

EXERCISES

Practice

Solve each equation. Then check your solution.

17. $3x - 2 = -5$ **−1**
18. $-4 = 5n + 6$ **−2**
19. $5 - 9w = 23$ **−2**
20. $17 = 7 + 8y$
21. $-2.5d - 32.7 = 74.1$
22. $0.2n + 3 = 8.6$ **28**
23. $\frac{3}{2}a - 8 = 11$ **$12\frac{2}{3}$**
24. $5 = -9 - \frac{p}{4}$ **−56**
25. $7 = \frac{c}{-3} + 5$ **−6**
26. $\frac{m}{-5} + 6 = 31$ **−125**
27. $\frac{g}{8} - 6 = -12$ **−48**
28. $6 = -12 + \frac{h}{-7}$ **−126**
29. $\frac{b + 4}{-2} = -17$ **30**
30. $\frac{z - 7}{5} = -3$ **−8**
31. $-10 = \frac{17 - s}{4}$ **57**

B

32. $\frac{4t - 5}{-9} = 7$ **−14.5**
33. $\frac{7n + (-1)}{6} = 5$ **$4\frac{3}{7}$**
34. $\frac{-3j - (-4)}{-6} = 12$ **$25\frac{1}{3}$**

20. 1.25 or $1\frac{1}{4}$
21. −42.72

INTEGRATION Number Theory

Define a variable, write an equation, and solve each problem. Then check your solution. 36–37. See margin.

35. Twelve decreased by twice a number is −7. $12 - 2n = -7$; $9\frac{1}{2}$

36. Find three consecutive integers whose sum is −33.

C

37. Find four consecutive integers whose sum is 86.

38. Find two consecutive odd integers whose sum is 196. $n + (n + 2) = 196$; 97, 99

INTEGRATION Geometry

Make a diagram to represent each situation. Then define a variable, write an equation, and solve each problem. Check your solution.

39. The lengths of the sides of a triangle are consecutive odd integers. The perimeter is 39 meters. What are the lengths of the sides?

39. $n + (n + 2) + (n + 4) = 39$; 11 m, 13 m, 15 m

40. The lengths of the sides of a quadrilateral are consecutive even integers. Twice the length of the shortest side plus the length of the longest side is 120 inches. Find the lengths of the four sides. $2n + (n + 6) = 120$; 38 in., 40 in., 42 in., 44 in.

Graphing Calculator

Solve each equation. Then use a graphing calculator to verify your solution.

41. $0.2x + 3 = 8.6$ **28**
42. $4.91 + 7.2x = 39.4201$ **4.793**
43. $\frac{3 + x}{7} = -5$ **−38**
44. $\frac{-3n - (-4)}{-6} = -9$ **−16.67 or $-16\frac{2}{3}$**

Critical Thinking

45. Write an expression for the sum of three consecutive even integers if $3n - 1$ is the least integer. $(3n - 1) + (3n + 1) + (3n + 3)$

Applications and Problem Solving

For Exercises 46–47, write an equation to represent each situation. Then solve the problem.

46. Health According to the American Medical Association, the average birth weight of a baby in the United States is 7.5 pounds. After birth, most babies lose an average of 1 ounce a day for the first 5 days and then gain an average of 1 ounce a day for the next 13 weeks.

a. Suppose a baby weighs 7.5 pounds at birth. After how many days will the baby weigh 12 pounds? $(120 - 5) + (d - 5) = 192$; 82 days

b. Draw a graph that shows the expected weight of a 7.5-pound baby for each of its first 13 weeks of life. See margin.

47. World Cultures Hawaii became our 50th state in 1959. It's the only state that is not on the North American continent. The English alphabet contains 2 more than twice as many letters as the Hawaiian alphabet. How many letters are there in the Hawaiian alphabet? $2a + 2 = 26$; 12 letters

648. Wilson, 16 bags; Martinez, 8 bags; Brightfeather, 4 bags; Wimberly, 2 bags

48. **Work Backward** Four families went to a baseball game. A vendor selling bags of popcorn came by. The Wilson family bought half of the bags of popcorn plus one. The Martinez family bought half of the remaining bags of popcorn plus one. The Brightfeather family bought half of the remaining bags of popcorn plus one. The Wimberly family bought half of the remaining bags of popcorn plus one, leaving the vendor with no bags of popcorn. If the Wimberlys bought 2 bags of popcorn, how many bags did each of the four families buy?

Mixed Review

49a–b. See students' work.

49. **Health** One way to calculate your suggested daily intake of fat grams is to multiply your ideal weight in pounds times 0.454. (Lesson 3–2)
    a. Write an equation to represent your suggested daily intake.
    b. What is your suggested daily intake of fat grams?
50. Complete: If $x + 4 = 15$, then $x - 2 = \underline{?}$. (Lesson 3–1) **9**
51. Name the set or sets of numbers to which $\sqrt{11}$ belongs. Use N for natural numbers, W for whole numbers, Z for integers, Q for rational numbers, or I for irrational numbers. (Lesson 2–8) ***I***
52. Simplify $\frac{-200x}{50}$. (Lesson 2–7) $-4x$
53. Simplify $4(7) - 3(11)$. (Lesson 2–6) **−5**
54. Name the property illustrated by $9 + (2 + 10) = 9 + 12$. (Lesson 1–6) **substitution (=)**

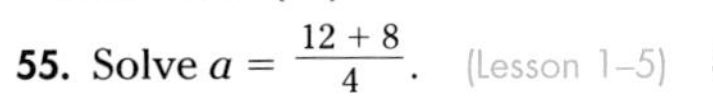

55. Solve $a = \frac{12 + 8}{4}$. (Lesson 1–5) **5**
56. Evaluate $\frac{3}{4}(6) + \frac{1}{3}(12)$. (Lesson 1–3) $8\frac{1}{2}$

SELF TEST

Solve each equation. (Lessons 3–1, 3–2, and 3–3)

1. $-10 + k = 34$ **44**
2. $y - 13 = 45$ **58**
3. $20.4 = 3.4y$ **6**
4. $-65 = \frac{f}{29}$ **−1885**
5. $-3x - 7 = 18$ $-8\frac{1}{3}$
6. $5 = \frac{m - 5}{4}$ **25**

Write an equation and solve. Then check your solution.

7. Twenty-three minus a number is 42. Find the number. (Lesson 3–1) $23 - n = 42; -19$
8. **Geometry** The length of one side of a rectangle is 34 cm. The area of the rectangle is 68 cm^2. Find the width of the rectangle. (Lesson 3–2) $34w = 68$; 2 cm
9. **Number Theory** Find two consecutive even integers whose sum is 126. (Lesson 3–3) $n + (n + 2) = 126$; 62, 64
10. **Economics** According to the *Baltimore Sun,* if inflation continues at the current rate, the price of a movie ticket in 40 years will be \$3.50 less than 10 times the current average price of \$5.05. What will be the price of a movie ticket in 40 years? (Lesson 3–3) $p = 10(5.05) - 3.50$; \$47

Extension

Connections A rectangular garden is fenced on all four sides. If the garden is 6 feet longer than it is wide, what are its dimensions if 156 feet of fencing is needed to enclose it?
42 feet by 36 feet

SELF TEST

The Self Test provides students with a brief review of the concepts and skills in Lessons 3-1 through 3-3. Lesson numbers are given to the right of exercises or instruction lines so students can review concepts not yet mastered.

4 ASSESS

Closing Activity

Writing Have students explain in writing whether the sum of two consecutive even numbers can ever equal the sum of two consecutive odd numbers.

Enrichment Masters, p. 21

NAME__________ DATE_______

3-3 Enrichment Student Edition Pages 156–160

Division by Zero?

You may remember being told, "division by zero is not possible" or "division by zero is undefined" or "we never divide by zero." Have you wondered why this is so? Consider the two equations below.

$\frac{5}{0} = n \qquad \frac{0}{0} = m$

Because multiplication is the inverse of division, these lead to the following.

$0 \cdot n = 5 \qquad 0 \cdot m = 0$

There is no number that will make the first equation true. Any number at all will satisfy the second equation.

For each expression, give the values that must be excluded from the replacement set in order to prevent division by zero.

1. $\frac{x + 1}{x - 1}$ $x = 1$
2. $\frac{2(x + 1)}{2x - 1}$ $x = \frac{1}{2}$
3. $\frac{(x + 1)(x - 1)}{(x + 2)(x - 2)}$ $x = -2$ or $x = 2$
4. $\frac{x + y + 3}{(3x - 1)(3y - 1)}$ $x = \frac{1}{3}, y = \frac{1}{3}$
5. $\frac{x^2 + y^2 + z^2}{2xyz}$ $x = 0, y = 0, z = 0$
6. $\frac{(x + y)^2}{x - y}$ values where $x = y$

Many demonstrations or "proofs" that lead to impossible results include a step involving division by zero. Explain what is wrong with each "proof" below.

7. $0 \cdot 1 = 0$ and $0 \cdot 2 = 0$.
   Therefore, $\frac{0}{0} = 1$ and $\frac{0}{0} = 2$.
   Therefore, $1 = 2$.
   The second step involves division by zero.
8. Assume that $a = b$.
   Then $ab = a^2$
   Therefore, $ab - b^2 = a^2 - b^2$.
   Next it is shown that $a^2 - b^2 = (a + b)(a - b)$.
   $(a + b)(a - b) = (a + b)a - (a + b)b$
   $= a^2 + ba - ab - b^2$
   $= a^2 + 0 - b^2$
   $= a^2 - b^2$
   Therefore, $ab - b^2 = (a + b)(a - b)$.
   Also, $b(a - b) = ba - b^2 = ab - b^2$.
   Therefore, $b(a - b) = (a + b)(a - b)$.
   Therefore, $b = a + b$.
   Therefore, $b = 2b$.
   Therefore, $1 = 2$.
   In moving to the third from the last step, each side is divided by $(a - b)$. Because $a = b$, that is dividing by zero.

3-4 LESSON NOTES

NCTM Standards: 1–5, 7

Instructional Resources

- Study Guide Master 3-4
- Practice Master 3-4
- Enrichment Master 3-4
- Assessment and Evaluation Masters, pp. 71–72
- Multicultural Activity Masters, p. 6
- Science and Math Lab Manual, pp. 9–12

Transparency 3-4A contains the 5-Minute Check for this lesson; **Transparency 3-4B** contains a teaching aid for this lesson.

Recommended Pacing	
Standard Pacing	Day 7 of 14
Honors Pacing	Day 6 of 13
Block Scheduling*	Day 4 of 6 (along with Lesson 3-5)
Alg. 1 in Two Years*	Days 9 & 10 of 19

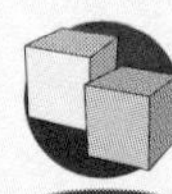

*For more information on pacing and possible lesson plans, refer to the *Block Scheduling Booklet* and *Algebra 1 in Two Years.*

1 FOCUS

5-Minute Check
(over Lesson 3-3)

Solve.

1. $\frac{y}{4} + 7 = -4$ **−44**
2. $-3 + \frac{2r}{3} = 13$ **24**
3. $\frac{(t-5)}{4} = 19$ **81**
4. A rectangle is 5 times as long as it is wide. The difference of its length and width is 9 inches. Find its perimeter. **27 in.**
5. Find three consecutive integers whose sum is 54. **17, 18, 19**

3-4 Integration: Geometry Angles and Triangles

What **YOU'LL LEARN**

- To find the complement and supplement of an angle, and
- to find the measure of the third angle of a triangle given the measures of the other two angles.

Why **IT'S IMPORTANT**

A knowledge of angles and triangles will help you further your study of geometry and trigonometry.

APPLICATION

Football

On fourth down in a football game, the team with the ball usually *punts.* A punt is a kick in which the ball is dropped by the kicker and is kicked before it hits the ground.

The longest punt ever kicked in a college football game was kicked by Pat Brady of the University of Nevada-Reno in their 1950 game against Loyola University. The punt was 99 yards long—just 1 yard short of spanning the entire length of the football field.

How far a football will travel once it is punted depends not only on the strength of the punter, but also on the angle at which the ball is kicked. If the angle is too small, the ball will travel close to the ground, then drop; if the angle is too large, the ball will travel high, but not very far. What do you think would be the best angle at which to kick the ball?

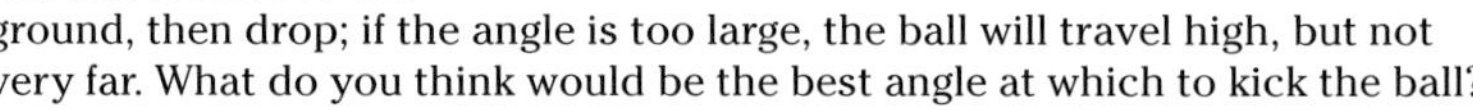

The symbol for angle is ∠.

You can use a *protractor* to measure angles, as shown below.

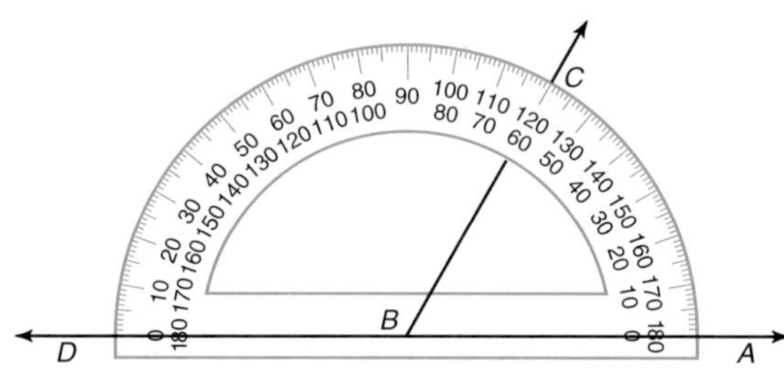

Angle ABC (denoted $\angle ABC$) measures 60°. However, where ray BC (denoted $\overrightarrow{BC}$) intersects the curve of the protractor, there are two readings, 60° and 120°. The measure of $\angle DBC$ is 120°. What is the sum of the measures of $\angle ABC$ and $\angle DBC$?

Supplementary Angles	Two angles are supplementary if the sum of their measures is 180°.

Example 1 **The measure of an angle is three times the measure of its supplement. Find the measure of each angle.**

Let x = the lesser measure. Then $3x$ = the greater measure.

$x + 3x = 180$ *The angles are supplementary.*

$4x = 180$ *Add like terms.*

$\frac{4x}{4} = \frac{180}{4}$ *Divide each side by 4.*

$x = 45$

The measures are 45° and $3 \cdot 45°$ or 135°. *Check this result.*

Complementary Angles — **Two angles are complementary if the sum of their measures is 90°.**

Example 2

Golf

Lowest winning scores in the U.S. Masters Tournament

1. 271, Raymond Floyd, 1976
1. 271, Jack Nicklaus, 1965
3. 274, Ben Hogan, 1953
4. 275, Severiano Ballesteros, 1980
4. 275, Fred Couples, 1992

The backward slant of the face of a golf club head is called the *loft*. It is designed to drive the ball in a high arc. Assuming that the angle made by the ground and the face of the club is 79°, what is the measure of the loft?

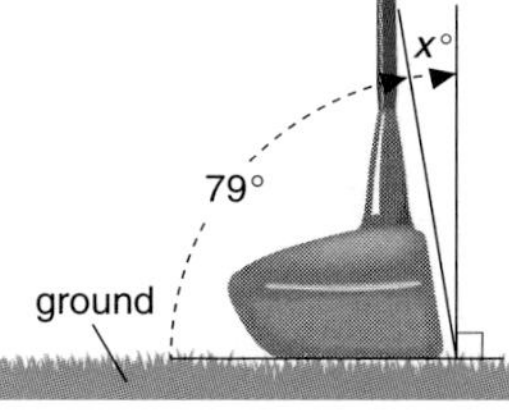

Let x = the measure of the loft.

$79° + x = 90°$ *The angles are complementary.*

$79° - 79° + x = 90° - 79°$ *Subtract 79° from each side.*

$x = 11°$

The club has an 11° loft.

Example 3 **The measure of an angle is 34° greater than its complement. Find the measure of each angle.**

Let x = the lesser measure. Then $x + 34$ = the greater measure.

$x + (x + 34) = 90$ *The angles are complementary.*

$2x + 34 = 90$

$2x + 34 - 34 = 90 - 34$ *Subtract 34 from each side.*

$2x = 56$

$\frac{2x}{2} = \frac{56}{2}$ *Divide each side by 2.*

$x = 28$

The measures are 28° and 28° + 34° or 62°. *Check this result.*

A **triangle** is a polygon with three sides and three angles. What is the sum of the measures of the three angles of a triangle?

Motivating the Lesson

Questioning Draw the angles below on the chalkboard. Ask students to name each angle in two ways.

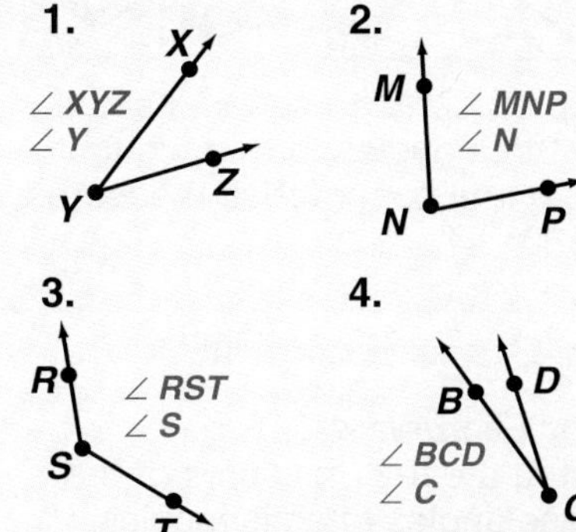

2 TEACH

Teaching Tip Point out that on the protractor shown, we can see that angle *ABC* is the supplement of angle *DBC* and vice versa.

In-Class Examples

For Example 1
The measure of an angle is 20° less than three times its supplement. Find the measure of each angle. **130°, 50°**

For Example 2
The measure of an angle is 26° greater than its complement. Find the measure of each angle. **32°, 58°**

For Example 3
An angle measures 42° less than its complement. Find the measure of each angle. **24°, 66°**

Point out that the winner of a golf game is the player who has the lowest score. The next lowest scores on the list were shared by many players: 276, by Arnold Palmer (1964), Jack Nicklaus (1975), and Tom Watson (1977); 277, by Bob Goalby (1968), Johnny Miller (1975), Gary Player (1978), Ben Crenshaw (1984), Ian Woosnam (1991), Raymond Floyd (1992), and Bernhard Langer (1993).

Alternative Teaching Strategies

Reading Algebra Knowing the Latin derivation of a nonmathematical English word may help students understand and remember a mathematical definition of the same word. For example, the word *complement* comes from the Latin word *complere*, which means to "make full" or "to complete." A nonmathematical meaning in music is an interval that completes a given octave.

Have students look up the meanings of the Latin roots of the following words, as well as a nonmathematical definition.

a. supplement
b. acute
c. congruent

MODELING MATHEMATICS Instead of tearing the corners of the triangle, students can fold the vertices together, as shown below.

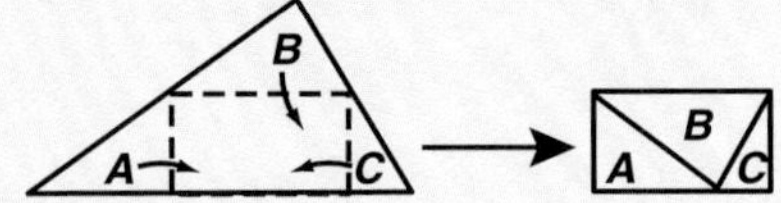

Help students to see that since the three angles form a straight angle, the sum of their measures is 180°.

In-Class Example

For Example 4
What are the measures of the base angles of an isosceles triangle in which the vertex angle measures 20°? **80°**

Teaching Tip Remind students that the word *congruent* means "equal measures."

Angles of a Triangle

Materials:  3 sheets of construction paper  protractor  scissors

Copy each triangle below onto a sheet of construction paper. If possible, you may want to enlarge the triangles on a photocopier.

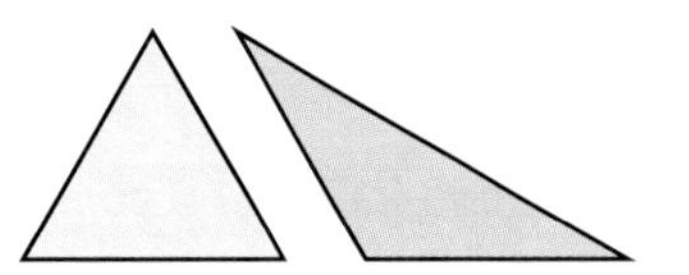

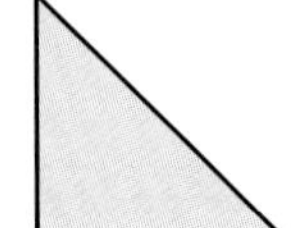

Your Turn

a. Use a protractor to measure the angles of each triangle above. What is the sum of the measures of the three angles of each triangle? **180°**

b. Cut out one of the triangles. Label the angles *A*, *B*, and *C*. Tear off the angles and arrange as shown.

c. When the angles were put together, what was the measure of the new angle they formed? **180°**

d. Try this activity with another triangle. What do you think is true about the measures of the angles of any triangle? **The sum of the interior angles is 180°.**

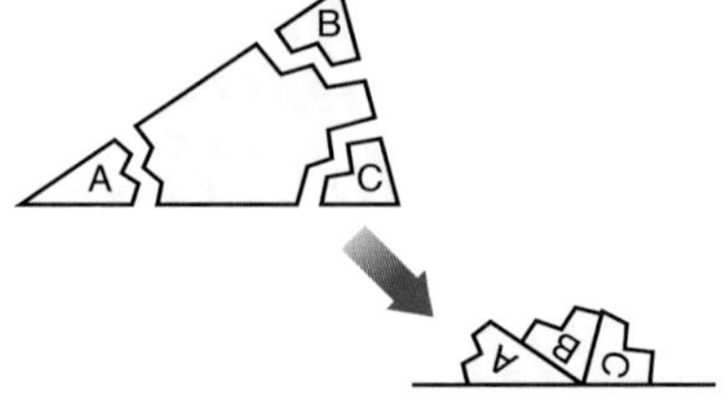

Sum of the Angles of a Triangle	The sum of the measures of the angles in any triangle is 180°.

In an **equilateral triangle,** each angle has the same measure. We say that the angles are **congruent.** The sides of an equilateral triangle are also congruent. What is the measure of each angle of an equilateral triangle?

Let x = the measure of each angle.

$x + x + x = 180$

$3x = 180$ *Add like terms.*

$\frac{3x}{3} = \frac{180}{3}$ *Divide each side by 3.*

$x = 60$ Each angle measures 60°.

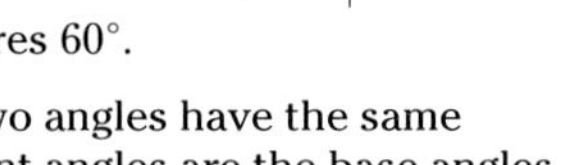

In an **isosceles triangle,** at least two angles have the same measure. Generally, the two congruent angles are the base angles.

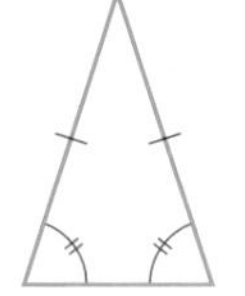

Example 4 **What are the measures of the base angles of an isosceles triangle in which the vertex angle measures 45°?**

Let x = the measure of each base angle.

$x + x + 45 = 180$

$2x + 45 = 180$ *Add like terms.*

$2x + 45 - 45 = 180 - 45$ *Subtract 45 from each side.*

$2x = 135$

$\frac{2x}{2} = \frac{135}{2}$ *Divide each side by 2.*

$x = \frac{135}{2}$ or $67\frac{1}{2}$ The base angles each measure $67\frac{1}{2}$°.

GLENCOE Technology

Interactive Mathematics Tools Software

This multimedia software provides an interactive lesson by helping students find the sum of the interior angles of any triangle by manipulating a triangle into various shapes and sizes. A **Computer Journal** gives students an opportunity to write about what they have learned.

For Windows & Macintosh

Cooperative Learning

Formations You may wish to have students work in cooperative groups to model the various types of triangles. For more information on the formations strategy, see *Cooperative Learning in the Mathematics Classroom*, one of the titles in the Glencoe Mathematics Professional Series, pages 30–31.

Triangles are often classified in one of three ways. A **right triangle** has one angle that measures 90°. An **obtuse triangle** has one angle with measure greater than 90°. In an **acute triangle,** all of the angles measure less than 90°.

Example 5 **The measures of the angles of a triangle are given as $x°$, $2x°$, and $3x°$.**

a. What are the measures of each angle?
b. Classify the triangle.

a. The sum of the measures of the angles of a triangle is 180°.

$$x + 2x + 3x = 180$$
$$6x = 180 \quad \textit{Add like terms.}$$
$$\frac{6x}{6} = \frac{180}{6} \quad \textit{Divide each side by 6.}$$
$$x = 30$$

The measures are 30°, 2(30°) or 60°, and 3(30°) or 90°.

b. Since the triangle has one angle that measures 90°, it is a right triangle.

In-Class Example

For Example 5
The measures of the angles of a triangle are $x°$, $x°$, and $2x°$. What are the measures of each angle?
45°, 45°, 90°

CHECK FOR UNDERSTANDING

Communicating Mathematics

Study the lesson. Then complete the following. 1–2. See margin.

1. The words *complement* and *compliment* sound very much alike.
   a. What are their meanings?
   b. Which word has the mathematical meaning?
2. **Compare and contrast** the standard definition and the mathematical meaning of the word *supplement.*
3. **Classify** each of the three triangles in the Modeling Mathematics activity on page 164 as right, obtuse, or acute. Justify your classifications.
4. **Make up a problem** in which you must use an equation to find the measures of the angles of a triangle. Then solve the problem, explaining each step. See students' work.

3. acute, obtuse, and right, respectively

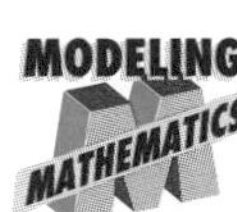

5. Draw a right triangle on a separate piece of paper and cut it out. Cut or tear off the two acute angles and arrange as shown. What seems to be true about the two acute angles? The sum of their measures is 90°.

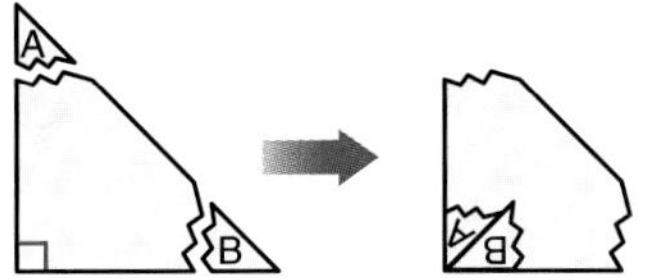

Guided Practice

Find both the complement and the supplement of each angle measure.

6. 130° none, 50°
7. 11° 79°, 169°
8. 24° 66°, 156°
9. $3x°$
10. $(2x + 40)°$
11. $(x - 20)°$

9. $(90 - 3x)°$, $(180 - 3x)°$
10. $(50 - 2x)°$, $(140 - 2x)°$
11. $(110 - x)°$, $(200 - x)°$

Find the measure of the third angle of each triangle in which the measures of two angles of the triangle are given.

12. 16°, 42° 122°
13. 50°, 45° 85°
14. $x°$, $(x + 20)°$

14. $(160 - 2x)°$

3 PRACTICE/APPLY

Check for Understanding

Exercises 1–16 are designed to help you assess your students' understanding through reading, writing, speaking, and modeling. You should work through Exercises 1–5 with your students and then monitor their work on Exercises 6–16.

Additional Answers

1a. A compliment is an expression of praise. A complement is something that completes or fills.

1b. The word *complement* has the mathematical meaning.

2. A supplement is something added to complete something, make up for a deficiency, or extend or strengthen the whole. The supplement of an angle completes a 180° angle, which is a straight line.

Study Guide Masters, p. 22

NAME______________________ DATE ________

Study Guide

Student Edition Pages 162–167

Integration: Geometry
Angles and Triangles

Supplementary Angles	Two angles are supplementary if the sum of their measures is 180°.
Complementary Angles	Two angles are complementary if the sum of their measures is 90°.
Sum of the Angles of a Triangle	The sum of the measures of the angles in any triangle is 180°.

Example 1: The measure of an angle is twice the measure of its complement. Find the measure of each angle.

Let x = the lesser measure. Then $2x$ = the greater measure.

$x + 2x = 90$ **The sum of the measures is 90°.**
$3x = 90$ **Add x and $2x$.**
$\frac{3x}{3} = \frac{90}{3}$ **Divide each side by 3.**
$x = 30$ The measures are 30° and $2 \cdot 30°$ or 60°.

Example 2: The measures of two angles of a triangle are 26° and 77°. Find the measure of the third angle.

Let x = the measure of the third angle.

$26 + 77 + x = 180$ **The sum of the measures of the angles is 180°.**
$103 + x = 180$ **Add 26 and 77.**
$103 - 103 + x = 180 - 103$ **Subtract 103 from each side.**
$x = 77$ The measure of the third angle is 77°.

Find both the complement and the supplement of each angle measure.

1. 39° 51°; 141°
2. 85° 5°; 95°
3. 13° 77°; 167°
4. $t°$ $(90 - t)°$; $(180 - t)°$
5. $(a - 6)°$ $(96 - a)°$; $(186 - a)°$
6. $(35 - x)°$ $(55 + x)°$; $(145 + x)°$

Find the measure of the third angle of each triangle in which the measures of two angles of the triangle are given.

7. 120°, 45° 15°
8. 55°, 55° 70°
9. t, $3t$ $(180 - 4t)°$

Write an equation and solve. Then check your solution.

10. The measure of an angle is 74° greater than its supplement. Find the measure of each angle. 53°, 127°
11. One of two complementary angles is 24° more than twice the other. Find the measure of each angle. 22°, 68°

Reteaching

Using Substeps The measure of an angle is 5 times its complement. If x is the lesser measure, then $5x$ is the greater measure. Fill in the blanks to solve the problem.

1. The angles are complementary.
   _?_ + _?_ = _?_
   $5x + x = 90°$
2. Simplify.
   _?_ = _?_
   $6x = 90°$
3. Divide each side by the same number.
   $\frac{?}{?} = \frac{?}{?}$ $\frac{6x}{6} = \frac{90°}{6}$
4. What are the measures of the angles?
   x = ?, $5x$ = ?
   $x = 15°$, $5x = 75°$

Assignment Guide

Core: 15–43 odd, 44, 45, 47–55
Enriched: 16–42 even, 44–55

For **Extra Practice,** see p. 763.

The red A, B, and C flags, printed only in the Teacher's Wraparound Edition, indicate the level of difficulty of the exercises.

Make a diagram to represent each situation. Then define a variable, write an equation, and solve each problem. Check your solution.

15. An angle measures 38° less than its complement. Find the measures of the two angles. **$(c - 38) + c = 90$; 26°, 64°**
16. The measures of the angles of a triangle are given as $x°$, $(x + 5)°$, and $(2x + 3)°$. What are the measures of each angle?

16. $x + (x + 5) + (2x + 3) = 180$; 43°, 48°, 89°

EXERCISES

Practice

Find both the complement and the supplement of each angle measure.

17. 42° **48°, 138°**
18. 87° **3°, 93°**
19. 125° **none, 55°**
20. 90° **none, 90°**
21. 21° **69°, 159°**
22. 174° **none, 6°**
23. 99° **none, 81°**
24. $y°$
25. $3a°$
26. $(x + 30)°$ **$(60 - x)°$, $(150 - x)°$**
27. $(b - 38)°$ **$(128 - b)°$, $(218 - b)°$**
28. $(90 - z°)$ **$z°$, $(90 + z)°$**

24. $(90 - y)°$, $(180 - y)°$
25. $(90 - 3a)°$, $(180 - 3a)°$

The measures of two angles of a triangle are given. Find the measure of the third angle.

29. 40°, 70° **70°**
30. 90°, 30° **60°**
31. 63°, 12° **105°**
32. 43°, 118° **19°**
33. 4°, 38° **138°**
34. $x°$, $y°$

35. $p°$, $(p - 10)°$ **$(190 - 2p)°$**
36. $c°$, $(2c + 1)°$ **$(179 - 3c)°$**
37. $y°$, $(135 - y)°$ **45°**

34. $180 - (x + y)°$ or $(180 - x - y)°$

Draw a diagram to represent each situation. Then define a variable, write an equation, and solve each problem. Check your solution.

38–43. See students' diagrams.

38. One of the congruent angles of an isosceles triangle measures 37°. Find the measures of the other angles. **$37 + 37 + x = 180$; 37°, 106°**
39. Find the measure of an angle that is 30° less than its supplement.
40. One of the angles of a triangle measures 53°. Another angle measures 37°. What is the measure of the third angle? **$53 + 37 + x = 180$; 90°**
41. One of two complementary angles measures 30° more than three times the other. Find the measure of each angle. **$x + (30 + 3x) = 90$; 15°, 75°**

42. Find the measure of an angle that is one-half the measure of its supplement. **$x + 2x = 180$; 60°**
43. The measures of the angles of a triangle are given as $x°$, $(3x)°$, and $(4x)°$. What are the measures of each angle? **$x + 3x + 4x = 180$; 22.5°, 67.5°, 90°**

39. $x + x - 30 = 180$; 75°

Critical Thinking

44. Draw a diagram that shows the relationships among right, isosceles, equilateral, acute, scalene, and obtuse triangles. **See students' work.**

Applications and Problem Solving

45. **Carpentry** A *stringer* is a triangular piece of wood on which staircases are based. If the stringer makes a 30° angle with the floor, what is the measure of the angle that the stringer makes with the vertical? **60°**

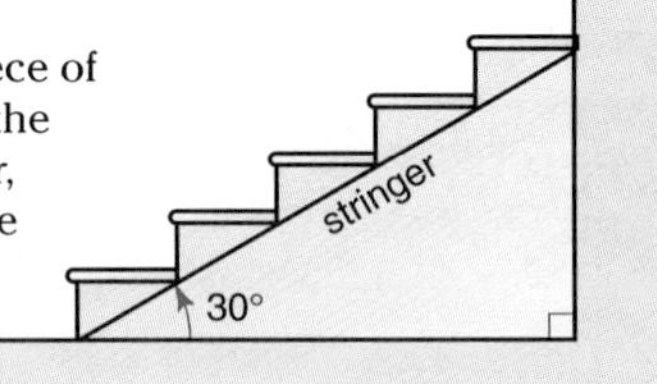

Practice Masters, p. 22

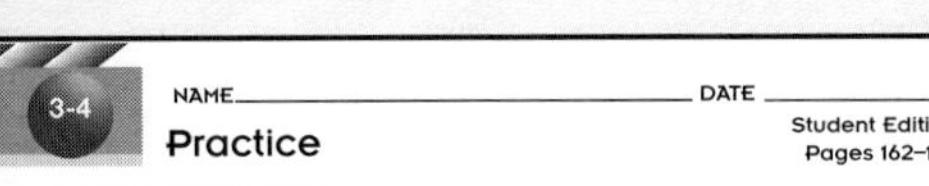

3-4 NAME ______ DATE ______
Practice
Student Edition Pages 162–167

Integration: Geometry
Angles and Triangles

Find both the complement and the supplement of each angle measure.

1. 40° **50°, 140°**
2. 79° **11°, 101°**
3. 116° **none, 64°**
4. 150° **none, 30°**
5. 9° **81°, 171°**
6. 65° **25°, 115°**
7. 100° **none, 80°**
8. 90° **0°, 90°**
9. $c°$ **$(90 - c)°$, $(180 - c)°$**
10. $(x - 30)°$ **$(120 - x)°$, $(210 - x)°$**
11. $2x°$ **$(90 - 2x)°$, $(180 - 2x)°$**
12. $(90 - y)°$ **$y°$, $(90 + y)°$**

The measures of two angles of a triangle are given. Find the measure of the third angle.

13. 10°, 64° **106°**
14. 41°, 72° **67°**
15. 62°, 74° **44°**
16. 18°, 46° **116°**
17. 36°, 45° **99°**
18. 11°, 125° **44°**
19. 35°, 129° **16°**
20. 86°, $x°$ **$(94 - x)°$**
21. $y°$, 110° **$(70 - y)°$**
22. $x°$, $(x + 10)°$ **$(170 - 2x)°$**
23. $y°$, $(y - 20)°$ **$(200 - 2y)°$**
24. $r°$, $(2r - 1)°$ **$(181 - 3r)°$**

Draw a diagram to represent each situation. Then define a variable, write an equation, and solve each problem. Check your solution. **25-28. See students' diagrams and equations.**

25. An angle measures 15° more than its complement. Find the measures of the two angles. **37.5°, 52.5°**
26. One of the congruent angles of an isosceles triangle measures 51°. Find the measures of the other angles. **51°, 78°**
27. One of the angles of a triangle measures 35°. Another angle measures 108°. What is the measure of the third angle? **37°**
28. One of the two supplementary angles measures 30° more than twice the other. Find the measure of each angle. **50°, 130°**

Alternative Learning Styles

Visual Students will need grid paper to solve the following problem. Have students draw the path described in the problem and label their drawings appropriately. Then have them draw and describe three different paths Marta could have taken. Have them identify the most direct path.

Marta walked 8 blocks north and 3 blocks east from her home to school. After school, she walked 2 blocks south and 1 block west to the library. From the library, she traveled 1 block south and 2 blocks east to visit a friend. How far does Marta have to walk to return home? **5 blocks south and 4 blocks west**

46. **Aeronautics** One of the newer space shuttles is *Endeavour.* Its first flight left Earth on May 7, 1992. The oldest shuttle is *Columbia,* which flew its first flight on April 12, 1981. A space shuttle lands at an angle six times as great as that of the average commercial jet.

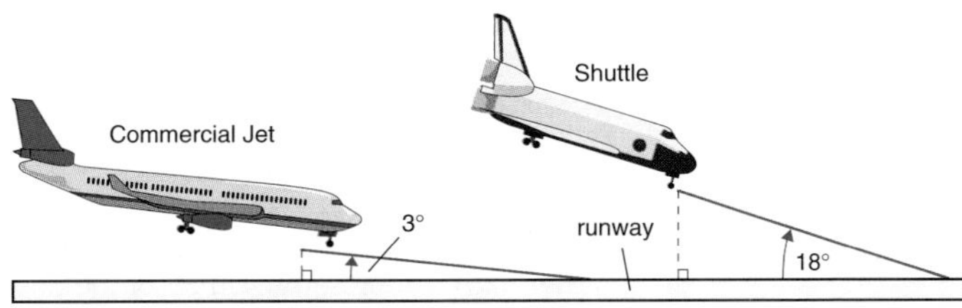

a. What is the measure of the angle the path of the jet makes with the vertical? **87°**

b. What is the measure of the angle the path of the shuttle makes with the vertical? **72°**

c. Which craft lands at a steeper angle? **shuttle**

47. **Mountain Climbing** *Rapelling* is a technique used by climbers to make a difficult descent. Climbers back to the edge and then spring off. To avoid slipping, the climber's legs should remain at a 90° angle with the side of the ledge. The ledge at the right is 50° off the vertical. What is the measure of the climber's angle with the vertical? **40°**

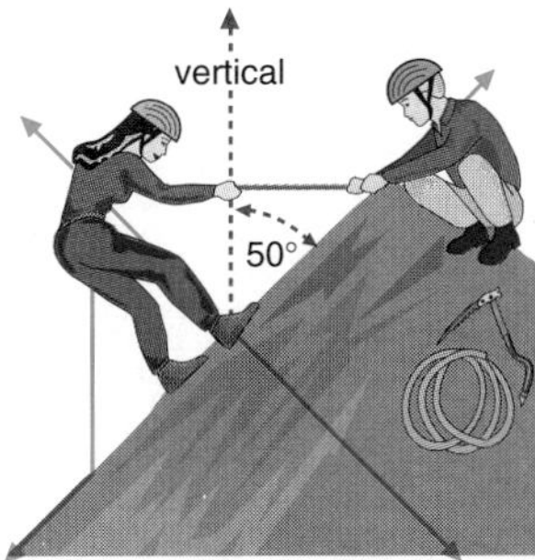

Mixed Review

48. Solve $4 + 7x = 39$. (Lesson 3–3) **5**

49. **Consumerism** When stores go out of business, there is generally a period when all of the items in the store can be purchased for a fraction of the original price. When Central Hardware went out of business, all of the items in the store began at $\frac{1}{4}$ off. (Lesson 3–2)

a. What was the original price of a circular saw if you save $20.00 during the sale? **$80**

b. How much would you save on patio furniture that originally cost $299.00? **$74.75**

50. Solve $h + (-13) = -5$. (Lesson 3–1) **8**

51. Simplify $5(3t - 2t) + 2(4t - 3t)$. (Lesson 2–6) **$7t$**

52. Find the sum $5y + (-12y) + (-21y)$. (Lesson 2–5) **$-28y$**

53. **American History** Each number below represents the age of a U.S. president on his first inauguration. (Lesson 2–2) **53a–b. See margin.**

57 61 57 57 58 57 61 54 68 51 49 64 50 48
65 52 56 46 54 49 50 47 55 55 54 42 51 56
55 51 54 51 60 62 43 55 56 61 62 69 64 46

a. Make a number line plot of the ages of the presidents.

b. Do any of the ages appear to be clustered? If so, which ones?

54. Name the property illustrated by $5a + 2b = 2b + 5a$. (Lesson 1–8)

55. **Statistics** Make a stem-and-leaf plot of the data in Exercise 53. (Lesson 1–4)

54. comm. (+)
55. See margin.

Extension

Reasoning Show that when the measure of a given angle is added to twice the measure of the complement of the angle, the result equals the measure of the supplement of the angle.

$$a + 2(90 - a) = a + 180 - 2a$$
$$= 180 - a$$

Tech Prep

Aeronautical Technician For students who are interested in aeronautics, you may wish to point out that an aeronautical technician is a vital part of the landing operations of an aircraft (see Exercise 46). For more information on tech prep, see the *Teacher's Handbook.*

4 ASSESS

Closing Activity

Modeling Have students use wooden craft sticks and glue to model angles of various measures. A third stick can be used to illustrate complementary and supplementary angles.

Chapter 3, Quiz B (Lessons 3-3 and 3-4), is available in the *Assessment and Evaluation Masters,* p. 72.

Mid-Chapter Test (Lessons 3-1 through 3-4) is available in the *Assessment and Evaluation Masters,* p. 71.

Additional Answers

53a.

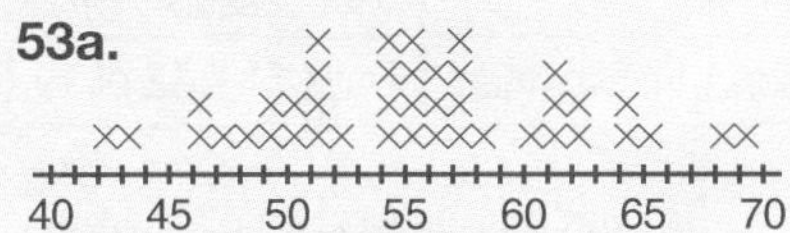

53b. yes; 51, 54, 55, and 57

55.

Stem	Leaf
4	2 3 6 6 7 8 9 9
5	0 0 1 1 1 1 2 4
	4 4 4 5 5 5 5 6
	6 6 7 7 7 7 8
6	0 1 1 1 2 2 4 4
	5 8 9

5|2 = 52

Enrichment Masters, p. 22

NAME______________ DATE______

Enrichment

Student Edition Pages 162–167

Angles of a Triangle

In geometry, many statements about physical space are proven to be true. Such statements are called **theorems.** Here are two examples of geometric theorems.

a. The sum of the measures of the angles of a triangle is 180°.

b. If two sides of a triangle have equal measure, then the two angles opposite those sides also have equal measure.

For each of the triangles, solve for x. (A tick mark on two or more sides of a triangle indicates that the sides have equal measure.)

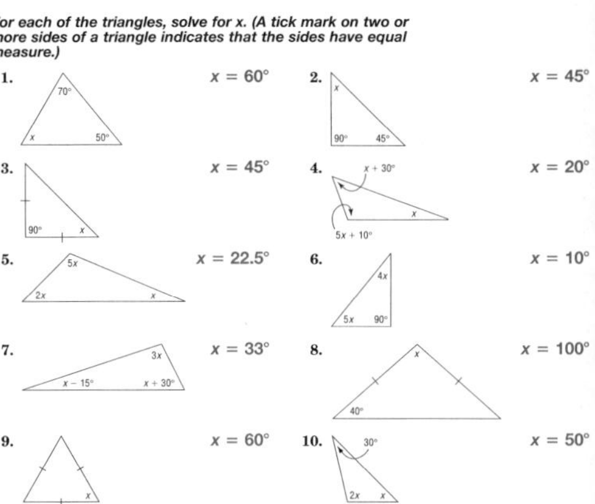

11. Two angles of a triangle have the same measure. The sum of the measures of these angles is one-half the measure of the third angle. Find the measures of the angles of the triangle.
30°, 30°, 120°

12. The measure of one angle of a triangle is twice the measure of a second angle. The measure of the third angle is 12 less than the sum of the other two. Find the measures of the angles of the triangle.
64°, 32°, 84°

3-5 LESSON NOTES

NCTM Standards: 1–5, 7

Instructional Resources

- Study Guide Master 3-5
- Practice Master 3-5
- Enrichment Master 3-5

Transparency 3-5A contains the 5-Minute Check for this lesson; **Transparency 3-5B** contains a teaching aid for this lesson.

Recommended Pacing	
Standard Pacing	Days 8 & 9 of 14
Honors Pacing	Days 7 & 8 of 13
Block Scheduling*	Day 4 of 6 (along with Lesson 3-4)
Alg. 1 in Two Years*	Days 11 & 12 of 19

*For more information on pacing and possible lesson plans, refer to the *Block Scheduling Booklet* and *Algebra 1 in Two Years.*

1 FOCUS

5-Minute Check
(over Lesson 3-4)

Find the measure of the third angle of each triangle in which the measure of two angles of the triangle are given.

1. 15°, 86° **79°**
2. 22°, 110° **48°**
3. 66°, 52° **62°**

Find the complement and supplement of each angle measure.

4. 24° **66°, 156°**
5. 67° **23°, 113°**
6. 80° **10°, 100°**

3–5 Solving Equations with the Variable on Both Sides

What YOU'LL LEARN

- To solve equations with the variable on both sides, and
- to solve equations containing grouping symbols.

Why IT'S IMPORTANT

You can use equations to solve problems involving track and field, business, and geometry.

APPLICATION

Track and Field

In the 1928 Olympics, the winner of the men's 800-meter run, Douglas Lowe of Great Britain, won the race in 1 minute 51.8 seconds, or 111.8 seconds. In that same year, the winner of the women's 800-meter run, Lina Radke of Germany, won the race in 2 minutes 16.8 seconds, or 136.8 seconds. Over the next 64 years, the men's winning times decreased an average of 0.127 seconds per year, and the women's winning times decreased an average of 0.332 seconds per year. Suppose the times continued to decrease at these rates. When will men and women have the same winning times in the 800-meter run?

fabulous FIRSTS

Wilma Rudolph (1940–1994)

Wilma Rudolph was the first woman to win three gold medals at a single Olympic games. She won the 100-meter sprint, the 200-meter dash, and the 4×100-meter relay at the 1960 Olympic Games.

After x years, the men's winning times can be represented by $111.8 - 0.127x$. After x years, the women's winning times can be represented by $136.8 - 0.332x$. The times would be the same when the two expressions are equal.

$$111.8 - 0.127x = 136.8 - 0.332x$$

Many equations contain variables on each side. To solve these types of equations, first use the addition or subtraction property of equality to write an equivalent equation that has all of the variables on one side. Then solve the equation.

Method 1

$$111.8 - 0.127x = 136.8 - 0.332x \quad \text{Use a calculator.}$$

$$111.8 - 0.127x + 0.332x = 136.8 - 0.332x + 0.332x \quad \text{Add } 0.332x \text{ to each side.}$$

$$111.8 + 0.205x = 136.8$$

$$111.8 - 111.8 + 0.205x = 136.8 - 111.8 \quad \text{Subtract 111.8 from each side.}$$

$$0.205x = 25$$

$$\frac{0.205x}{0.205} = \frac{25}{0.205} \quad \text{Divide each side by 0.205.}$$

$$x \approx 122$$

At these rates, the men's and women's winning times will be equal about 122 years after 1928, or in the year 2050.

fabulous FIRSTS

Wilma Rudolph overcame many obstacles on her way to becoming a champion. At 4, she was stricken with double pneumonia and scarlet fever. The illnesses crippled her left leg. When she was 8 she began walking with a corrective shoe. At 16, in 1956, she won a bronze medal as a member of the U.S. Olympic 4×100-meter relay team.

Method 2

Since two of the decimals involve thousandths, another way to solve the equation would be to multiply each side by 1000 to clear the decimals.

$$111.8 - 0.127x = 136.8 - 0.332x$$

Multiply each side by 1000. $$1000(111.8) - 1000(0.127x) = 1000(136.8) - 1000(0.332x)$$

$$111{,}800 - 127x = 136{,}800 - 332x$$

Add 332x to each side. $$111{,}800 + 205x = 136{,}800$$

Subtract 111,800 from each side. $$205x = 25{,}000$$

Divide each side by 205. $$x \approx 122$$

Example 1 **Solve $\frac{3}{8} - \frac{1}{4}x = \frac{1}{2}x - \frac{3}{4}$.**

Look back to Lesson 3–3 to see how you would solve an equation with the variable on both sides by using a graphing calculator.

$\frac{3}{8} - \frac{1}{4}x = \frac{1}{2}x - \frac{3}{4}$ *The least common denominator is 8.*

$8\left(\frac{3}{8} - \frac{1}{4}x\right) = 8\left(\frac{1}{2}x - \frac{3}{4}\right)$ *Multiply each side by 8.*

$8\left(\frac{3}{8}\right) - 8\left(\frac{1}{4}x\right) = 8\left(\frac{1}{2}x\right) - 8\left(\frac{3}{4}\right)$ *Use the distributive property.*

$3 - 2x = 4x - 6$ *The fractions are eliminated.*

$3 = 6x - 6$ *Add 2x to each side.*

$9 = 6x$ *Add 6 to each side.*

$\frac{3}{2} = x$ *Divide each side by 6.*

The solution is $\frac{3}{2}$. *Check this result.*

When solving equations that contain grouping symbols, first use the distributive property to remove the grouping symbols.

Example 2

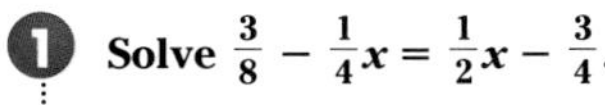

Geometry

One angle of a triangle measures 10° more than the second. The measure of the third angle is twice the sum of the first two angles. Find the measure of each angle.

Make a drawing. Let y = the measure of one angle. Let $y + 10$ = the measure of another angle. Then $2[y + (y + 10)]$ = the measure of the third angle. Recall that the sum of the measures of the angles in any triangle is 180°.

$2[y + (y + 10)]°$

$y°$

$(y + 10)°$

$y + (y + 10) + 2[y + (y + 10)] = 180$

$2y + 10 + 2(2y + 10) = 180$ *Simplify.*

$2y + 10 + 4y + 20 = 180$ *Use the distributive property.*

$6y + 30 = 180$ *Simplify.*

$6y + 30 - 30 = 180 - 30$ *Subtract 30 from each side.*

$6y = 150$

$y = 25$ *Divide each side by 6.*

The measures of the three angles are 25°, (25 + 10)° or 35°, and 2(25 + 35)° or 120°. *Check this result.*

Alternative Teaching Strategies

Student Diversity Review with students the different meanings of the equal sign. $3 = 1 + 2$ means that the left side is always equal to the right side. This is also true of $7x = 7x$. However, $2x = 8$ is not always true. For this to have a solution means that there is at least one value for x that makes the two sides equal.

Motivating the Lesson

Hands-On Activity Separate the class into two groups. Think of a trait, such as people wearing red sneakers, and count the number of students in each half of the class who exhibit the trait. Tell the class that you are thinking of a trait and that they need to identify it. Also tell them how many students in each half show the trait.

Students should examine each half of the class and use the clue to guess the trait. They can check the validity of their guess by having students move between the halves. Each time a student moves to a different half of the class, you must again announce how many students in each half show the trait. Continue in this manner until the trait has been identified.

Point out how the nature of the task changed as students moved between the halves. Show that the nature of an equation can also change if variables are not properly handled when solving equations.

2 TEACH

In-Class Examples

For Example 1

Solve.

a. $\frac{2}{5}n - 9 = 7 - \frac{3}{5}n$ **16**

b. $\frac{3}{7}t + 4 = \frac{1}{7}t - 6$ **−35**

c. $8 - \frac{1}{2}p = \frac{1}{4}p - 7$ **20**

For Example 2

The perimeter of a rectangle is 24 inches. Find its dimensions if its length is 3 inches greater than its width.
$4\frac{1}{2}$ in. by $7\frac{1}{2}$ in.

Teaching Tip Students usually have the least difficulty with equations similar to the one in Example 2 if they apply the distributive property first. Then they can finish solving the equation by applying properties of equality.

In-Class Examples

For Example 3
Solve.

a. $6k - 3 = 2k + 13$ **4**
b. $9t + 7 = 3t - 5$ **−2**
c. $3n + 1 = 7n - 5$ $1\frac{1}{2}$

For Example 4
Solve.

a. $8 - 2(t + 1) = -3t + 1$ **−5**
b. $5 + 2(k + 4) = 5(k - 3) + 10$ **6**
c. $8x - 5(2 + x) = 2(x + 1)$ **12**

Teaching Tip An identity is an equation that is true for all values of the variable except those for which the equation is not defined. At this point in the text, the equations are defined for all values of the variable. Later in the text, students will learn to solve identities such as $\frac{5}{x-1} = \frac{3}{x-1} + \frac{2}{x-1}$, which is not defined when x is 1.

3 PRACTICE/APPLY

Check for Understanding

Exercises 1–13 are designed to help you assess your students' understanding through reading, writing, speaking, and modeling. You should work through Exercises 1–4 with your students and then monitor their work on Exercises 5–13.

Study Guide Masters, p. 23

NAME________________ DATE ________

Study Guide

Student Edition
Pages 168–172

Solving Equations with the Variable on Both Sides

When an equation contains parentheses or other grouping symbols, first use the distributive property to remove the grouping symbols. If the equation has variables on each side, use addition and subtraction property of equality to write an equivalent equation that has all the variables on one side. Then solve the equation.

Example: Solve $4(2a - 1) = -10(a - 5)$.

$4(2a - 1) = -10(a - 5)$
$8a - 4 = -10a + 50$ **Use the distributive property.**
$8a + 10a - 4 = -10a + 10a + 50$ **Add 10a to each side.**
$18a - 4 = 50$
$18a - 4 + 4 = 50 + 4$ **Add 4 to each side.**
$18a = 54$
$\frac{18a}{18} = \frac{54}{18}$ **Divide each side by 18.**
$a = 3$

Check:
$4(2a - 1) = -10(a - 5)$
$4(2 \cdot 3 - 1) = -10(3 - 5)$
$4(6 - 1) = -10(-2)$
$4(5) = -10(-2)$
$20 = 20$ ✔

Some equations may have *no solution,* and some equations may have *every number* in their solution set. An equation that is true for every value of the variable is called an **identity.**

Solve each equation. Then check your solution.

1. $-3(x + 5) = 3(x - 1)$ −2
2. $6 - b = 5b + 30$ −4
3. $5y - 2y = 3y + 2$ no solution
4. $2(7 + 3t) = -t$ −2
5. $3(a + 1) - 5 = 3a - 2$ identity
6. $75 - 9g = 5(-4 + 2g)$ 5
7. $1.2x + 4.3 = 2.1 - x$ −1
8. $4.4s + 6.2 = 8.8s - 1.8$ $\frac{20}{11}$
9. $5(f + 2) = 2(3 - f)$ $-\frac{4}{7}$
10. $\frac{1}{2}b + 4 = \frac{1}{8}b + 88$ 224
11. $\frac{2}{5}w - w = -\frac{1}{5}(3w + 2)$ no solution
12. $5(p + 3) + 9 = 3(p - 2) + 6$ −12
13. $\frac{3}{4}k - 5 = \frac{1}{4}k - 1$ 8
14. $0.03g - (2g + 3) = 1.8$ $-\frac{480}{197}$
15. $-5(2r + 3) = 3(11 - 4r) - 58$ −5

Some equations with the variable on both sides may have no solution. That is, there is no value of the variable that will result in a true equation.

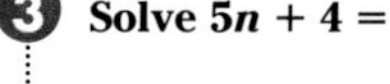

Example 3 **Solve $5n + 4 = 7(n + 1) - 2n$.**

$5n + 4 = 7(n + 1) - 2n$
$5n + 4 = 7n + 7 - 2n$ *Distributive property*
$5n + 4 = 5n + 7$
$4 = 7$ Since $4 = 7$ is a false statement, this equation has no solution.

An equation that is true for every value of the variable is called an **identity.**

Example 4 **Solve $7 + 2(x + 1) = 2x + 9$.**

$7 + 2(x + 1) = 2x + 9$
$7 + 2x + 2 = 2x + 9$ *Distributive property*
$2x + 9 = 2x + 9$ *Reflexive property of equality*

Since the expressions on each side of the equation are the same, this equation is an identity. The statement $7 + 2(x + 1) = 2x + 9$ is true for all values of x.

CHECK FOR UNDERSTANDING

Communicating Mathematics

1. See Solutions Manual.
5. $-\frac{1}{2}$ 6. $\frac{12}{11}$
12. $6x + (x - 3) + (3x + 7) = 180$; 105.6°, 14.6°, 59.8°
13. $\frac{1}{2}n + 16 = \frac{2}{3}n - 4$; 120

Study the lesson. Then complete the following. 2–3. See margin.

1. **a. Explain** why you think winning times in the 800-meter run for men and women are growing closer.
   b. Use a table to show how women's winning times in the 800-meter run could catch up to men's winning times in the same race.
   c. Do you think the times will ever be the same? Why or why not?
2. **Describe** the difference between an identity and an equation with no solution.
3. **You Decide** Lauren says that to solve the equation $2[x + 3(x - 1)] = 18$, the first step should be to multiply 2 by $[x + 3(x - 1)]$. Carmen says the first step should be to multiply 3 by $(x - 1)$. Which one is correct, and why?

4. **Assess Yourself** Describe your favorite sport and explain how math is used in it. See students' work.

Guided Practice

Explain how to solve each equation. Then solve each equation and check your solution. 5–10. See margin for explanations.

5. $3 - 4x = 10x + 10$
6. $8y - 10 = -3y + 2$
7. $\frac{3}{5}x + 3 = \frac{1}{5}x - 7$ −25
8. $5.4y + 8.2 = 9.8y - 2.8$ 2.5
9. $5x - 7 = 5(x - 2) + 3$ all numbers
10. $4(2x - 1) = -10(x - 5)$ 3

Define a variable, write an equation, and solve each problem. Then check your solution.

11. Twice the greater of two consecutive odd integers is 13 less than three times the lesser. Find the integers. $2(n + 2) = 3n - 13$; 17, 19
12. **Geometry** The measures of the angles of a triangle are given as $6x°$, $(x - 3)°$, and $(3x + 7)°$. What are the measures of each angle?
13. One half of a number increased by 16 is four less than two thirds of the number. Find the number.

Reteaching

Using Substeps List the first two steps you would take to solve each equation. Do *not* solve.

1. $6x + 1 = 15 - x$ **Add x to each side. Then subtract 1 from each side.**
2. $2x + 3 = 5x - 9$ **Subtract $2x$ from each side. Then add 9 to each side.**
3. $x + 36 = 1 - 4(x - 5)$ **Multiply 4 and $x - 5$. Then subtract x from each side.**
4. $3(2x + 4) = 5x + 12 + x$ **Multiply 3 and $2x + 4$. Then add $5x$ and x.**
5. $3(2x - 5) = 6x + 7$ **Multiply 3 and $2x - 5$. Then subtract $6x$ from each side.**

EXERCISES

Practice

Solve each equation. Then check your solution.

14. $6x + 7 = 8x - 13$ **10**
15. $17 + 2n = 21 + 2n$ **no solution**
16. $\frac{3n - 2}{5} = \frac{7}{10}$ **1.83 or $1\frac{5}{6}$**
17. $\frac{7 + 3t}{4} = -\frac{t}{8}$ **−2**
18. $13.7b - 6.5 = -2.3b + 8.3$ **0.925**
19. $18 - 3.8x = 7.36 - 1.9x$ **5.6**
20. $\frac{3}{2}y - y = 4 + \frac{1}{2}y$ **no solution**
21. $\frac{3}{4}n + 16 = 2 - \frac{1}{8}n$ **−16**
22. $-7(x - 3) = -4$
23. $4(x - 2) = 4x$ **no solution**
24. $28 - 2.2y = 11.6y + 262.6$ **−17**
25. $1.03x - 4 = -2.15x + 8.72$ **4**
26. $7 - 3x = x - 4(2 + x)$ **no solution**
27. $6 = 3 + 5(y - 2)$
28. $6(y + 2) - 4 = -10$ **−3**
29. $5 - \frac{1}{2}(b - 6) = 4$ **8**

B

30. $-8(4 + 9x) = 7(-2 - 11x)$
31. $2(x - 3) + 5 = 3(x - 1)$ **2**
32. $4(2a - 8) = \frac{1}{7}(49a + 70)$ **42**
33. $-3(2n - 5) = \frac{1}{2}(-12n + 30)$

22. 3.57 or $3\frac{4}{7}$

27. 2.6 or $2\frac{3}{5}$

30. 3.6 or $3\frac{3}{5}$

33. all numbers

Define a variable, write an equation, and solve each problem. Then check your solution.

34. Three times the greatest of three consecutive even integers exceeds twice the least by 38. Find the integers. **$3(n + 4) = 2n + 38$; 26, 28, 30**
35. One fifth of a number plus five times that number is equal to seven times the number less 18. Find the number.
36. The difference of two numbers is 12. Two fifths of the greater number is six more than one third of the lesser number. Find both numbers. **$\frac{2}{5}(n + 12) = 6 + \frac{1}{3}(n)$; 18, 30**

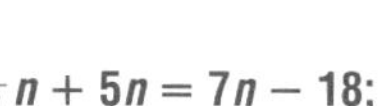

35. $\frac{1}{5}n + 5n = 7n - 18$; 10

INTEGRATION
Geometry

Make a diagram to represent each situation. Then define a variable, write an equation, and solve each problem. Check your solution.

37. The measures of the angles of a certain triangle are consecutive even integers. Find their measures. **$n + (n + 2) + (n + 4) = 180$; 58°, 60°, 62°**

C

38. One angle of a triangle measures 30° more than another angle. The measure of the third angle is three times the sum of the first two angles. Find the measure of each angle.

38. $x + (x + 30) + 3[x + (x + 30)] = 180$; 7.5°, 37.5°, 135°

Programming

39. The graphing calculator program at the right can help you solve equations of the form $ax + b = cx + d$.

```
PROGRAM:SOLVE
: Disp "ENTER A, B, C, D"
: Input A: Input B: Input C:
  Input D
: If A - C ≠ 0
: Goto 2
: If D - B ≠ 0
: Goto 1
: Disp "THIS IS AN"
: Disp "IDENTITY"
: Goto 3
: Lbl 1
: Disp "NO SOLUTION"
: Goto 3
: Lbl 2
: Disp "X = ", (D - B)/(A - C)
: Lbl 3
```

Write each equation in the form $ax + b = cx + d$. Then run the program to find the solution.

a. $2(2x + 3) = 4x + 6$
b. $5x - 7 = x + 3$
c. $6 - 3x = 3x - 6$
d. $6.8 + 5.4x = 4.6x + 2.8$
e. $5x - 8 - 3x = 2(x - 3)$

39a. $4x + 6 = 4x + 6$; identity

39b. $5x - 7 = x + 3$; 2.5

39c. $-3x + 6 = 3x - 6$; 2

39d. $5.4x + 6.8 = 4.6x + 2.8$; −5

39e. $2x - 8 = 2x - 6$; no solution

Assignment Guide

Core: 15–39 odd, 40, 41, 43–49
Enriched: 14–38 even, 40–49

For **Extra Practice**, see p. 763.

The red A, B, and C flags, printed only in the Teacher's Wraparound Edition, indicate the level of difficulty of the exercises.

Teaching Tip You may wish to review the geometry formulas that are needed to solve some of the problems in the exercise set.

Using the Programming Exercises The program given in Exercise 39 is for use with a TI-82 graphing calculator. For other programmable calculators, have students consult their owner's manuals for commands similar to those presented here.

Additional Answers

2. An identity is true for *all* values of the variable. An equation with no solution is true for *no* values of the variable.
3. Carmen; you begin with the numerals in the innermost grouping symbols—in this case, $(x - 1)$.

Practice Masters, p. 23

3-5 NAME__________ DATE ________

Practice

Student Edition Pages 168–172

Solving Equations with the Variable on Both Sides

Solve each equation. Then check your solution.

1. $4x - 9 = 7x + 12$ −7
2. $6y - 3 = 6y + 8$ no solution
3. $8m + 13 = 13 + 8m$ identity
4. $8n - 13 = 13 - 8n$ $\frac{13}{8}$
5. $\frac{x + 3}{2} = 15$ 27
6. $\frac{2r - 3}{3} = \frac{3}{5}$ $\frac{12}{5}$
7. $1.4f + 1.1 = 8.3 - f$ 3
8. $0.4r - 1.2 = 0.3r + 0.6$ 18
9. $\frac{1}{2}d + \frac{3}{8} = -2d$ $-\frac{3}{20}$
10. $\frac{3}{5}x - 2 = 6 + \frac{1}{4}x$ $\frac{160}{7}$
11. $\frac{5}{2}t - t = 3 + \frac{3}{2}t$ no solution
12. $4.2z = -4(0.6z - 1.2)$ $\frac{8}{11}$
13. $-3(b - 8) - 5 = 9(b + 2) + 1$ 0
14. $8p - 5(p + 3) = (7p - 1)3$ $-\frac{2}{3}$
15. $-4(2 - 3x) = 7 - 2(x - 3)$ $\frac{3}{2}$
16. $2(a - 8) + 7 = 5(a + 2) - 3a - 19$ identity

Write an equation and solve. Then check your solution.

18. Twice a number increased by 12 is equal to 31 less than three times the number. Find the number. 43
19. Eight minus two times a number is equal to the number plus 17. Find the number. −3
20. Twice the greater of two consecutive odd integers is 13 less than three times the lesser. Find the integers. 17, 19
21. The perimeter of a rectangle is 24 inches. Find the dimensions if its length is 3 inches greater than its width. $4\frac{1}{2}$ in. by $7\frac{1}{2}$ in.

Additional Answers

5. Add $4x$ to each side, subtract 10 from each side, then divide each side by 14.
6. Add $3y$ to each side, add 10 to each side, then divide each side by 11.
7. Subtract $\frac{1}{5}x$ from each side, subtract 3 from each side, then multiply by $\frac{5}{2}$.
8. Subtract $5.4y$ from each side, add 2.8 to each side, then divide each side by 4.4.
10. Multiply 4 by $(2x - 1)$ and −10 by $(x - 5)$. Subtract $8x$ from each side, and subtract 50 from each side. Then divide each side by −18.

4 ASSESS

Closing Activity

Speaking Ask students to explain the difference between an equation whose solution set is the empty set and an equation that is an identity.

Critical Thinking

40. **Mathematics History** Diophantus of Alexandria was one of the greatest mathematicians of the Greek civilization. Little is known of his life except for the following problem, which was first printed in a work called the *Greek Anthology.* Although the problem was not written by Diophantus, it is believed to accurately describe his life.

    Diophantus passed one sixth of his life in childhood, one twelfth in youth, and one seventh more as a bachelor. Five years after his marriage, there was born a son who died four years before his father, at half his father's (final) age. How old was Diophantus when he died? **84 years old**

Applications and Problem Solving

For Exercises 41–42, assume that the rates of change continue indefinitely.

41. **Sales** According to the Association of Home Appliance Manufacturers, sales of room air conditioners were 4.6 million units in 1988, and sales of window fans were 0.975 million units in 1988. Since 1988, sales of room air conditioners have decreased about 0.425 million units per year, and sales of window fans have increased about 0.106 million units per year.
    a. If the trend continues, after how many years will sales of room air conditioners and window fans be equal? **6.827 years**
    b. How would you explain these trends? **Air conditioner sales are decreasing while fan sales are increasing.**

42. **Pet Ownership** According to the American Veterinary Medical Association, in 1987, 34.7 million households owned a dog, and 27.7 million households owned a cat. Since 1987, dog ownership has decreased by about 0.025 million households per year, and cat ownership has increased by about 0.375 million households per year. If the trend continues, after how many years will the number of households that own dogs and cats be equal? **17.5 years**

Mixed Review

43. Find the supplement of a 32° angle. (Lesson 3–4) **148°**

44. **Work Backward** A number is decreased by 35, then multiplied by 6, then added to 87, then divided by 3. The result is 67. What is the number? (Lesson 3–3) **54**

45. Solve $x + 4.2 = 1.5$. Check your solution. (Lesson 3–1) **−2.7**

46. Define a variable, then write an equation for the following problem. (Lesson 2–9)

    Karen is 10, and she has always admired her 15-year-old sister Kristy. Karen wants to be exactly like Kristy. She knows that she will never catch up with Kristy in age, but she has heard that when one person is 0.9 times as old as another, you can't really tell them apart. How long will it be before Karen is 0.9 times as old as Kristy? **$10 + x = 0.9(15 + x)$, 35 years**

47. Which is greater, $\frac{11}{9}$ or $\frac{12}{10}$? (Lesson 2–4) **$\frac{11}{9}$**

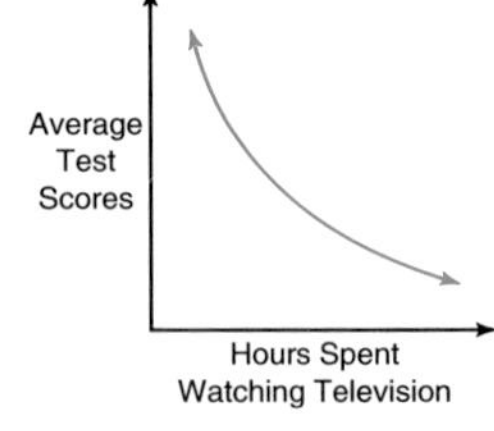

48. Explain what the graph at the right tells you about test scores as a function of hours spent watching television. (Lesson 1–9)

    48. As students spend more time watching TV, their test scores go down.

49. Solve $14.8 - 3.75 = t$. (Lesson 1–5) **11.05**

Tech Prep

Veterinary Technician Students who are interested in veterinary medicine may wish to do further research on the statistics given in Exercise 42 and explore the potential growth of this career. For more information on tech prep, see the *Teacher's Handbook.*

Extension

Connections The lengths of the sides of a seven-sided figure are consecutive odd integers. The sum of the shortest and longest sides is 34 cm. Find the perimeter of the figure. **119 cm**

Enrichment Masters, p. 23

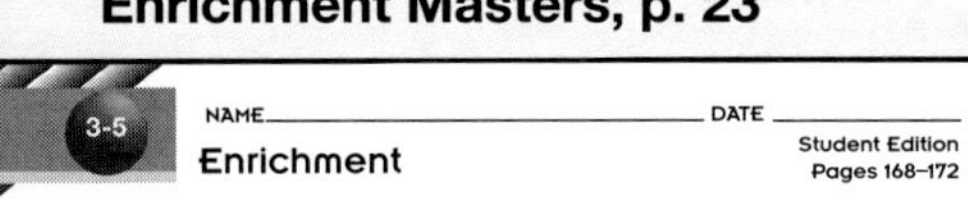

3-5 NAME__________ DATE ________

Enrichment

Student Edition Pages 168–172

Using Equations

Use what you have learned about equations to solve each problem.

1. In many universities, grade point averages are figured out using a table like this.

grade	A	B	C	D	F
point value	4	3	2	1	0

Different classes often meet for a different number of hours each week. If you had the following schedule and grades, your average would be 2.45.

Course	Hours	Grade
Chemistry	3	C
Chemistry Lab	2	F
Calculus	5	A
Phys. Ed.	1	C
Total	11	

$$\frac{3(2) + 2(0) + 5(4) + 1(2)}{11} = \frac{28}{11} = 2.54$$

Suppose you are taking the following courses and you know all your grades but your math grade. What is the lowest math grade you could receive and still have a grade point average of over 3.00? **B**

Course	Hours	Grade
Biology	5	A
Spanish	3	B
Math	4	?
English	4	B
Phys. Ed.	1	D
Total	17	

2. Read the information below the soccer standings at the right to understand how a team's points are determined. A shootout is a form of tie-breaker used when the score is still tied after an overtime period.
   a. A team had 5 wins, 2 shootout victories and 23 bonus points. Show that the number of points must be 61.
      $6(5) + 4(2) + 23 = 61$
   b. To find the number of shootout victories for Golden Bay, let x = number of shootout victories. Then, $5 - x$ = number of regulation or overtime wins.
      $6(5 - x) + 4x + 32 = 60$
      Solve this equation.
      $30 - 6x + 4x + 32 = 60$
      $-2x = -2$
      $x = 1$
   c. Find the number of shootout victories for Chicago.
      $6(10 - x) + 4x + 42 = 94$
      $60 - 6x + 4x + 42 = 94$
      $-2x = -8$
      $x = 4$

Soccer

NASL

Eastern Division	W	L	GF	GA	BP	Pts.
Chicago	10	9	42	40	42	94
Cosmos	10	6	31	27	29	89
Toronto	8	8	33	22	29	71
Tampa Bay	7	12	31	49	27	67
Western Division	**W**	**L**	**GF**	**GA**	**BP**	**Pts.**
Vancouver	11	6	39	33	32	96
Minnesota	10	6	30	30	26	82
San Diego	9	9	33	38	27	75
Tulsa	7	10	31	32	29	71
Golden Bay	5	11	43	42	32	60

Six points are awarded for a regulation or overtime win, four points for a shootout victory, and one bonus point for every goal scored with a maximum of three per game. No bonus point is awarded for overtime or shootout goals.

3–6 Solving Equations and Formulas

What YOU'LL LEARN
- To solve equations and formulas for a specified variable.

Why IT'S IMPORTANT
You can use equations and formulas to solve problems involving physics and geometry.

INTEGRATION Geometry

Some equations contain more than one variable. At times you will need to solve these equations for one of the variables. For example, suppose the variables x, y, and z were all in the same equation. To *solve for x* would mean to get x by itself on one side of the equation, with no x's on the other side. In the same way, to *solve for y* would mean to get y by itself on one side of the equation, with no y's on the other side.

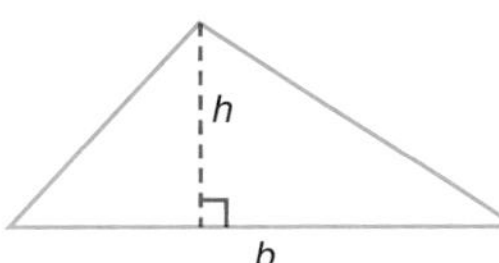

The formula for the area of a triangle is $A = \frac{1}{2}bh$ where b represents the length of the base and h represents the height of the triangle. Suppose you know the areas and the lengths of the bases of several triangles and you want to find the height of each triangle. Rather than solve the formula over and over for different values of A and b, it would be easier to solve the formula for h before substituting the values for the other variables.

$$A = \frac{1}{2}bh$$

$$2A = 2\left[\frac{1}{2}bh\right] \quad \textit{Multiply each side by 2.}$$

$$2A = bh$$

$$\frac{2A}{b} = \frac{bh}{b} \quad \textit{Divide each side by b.}$$

$$\frac{2A}{b} = h$$

When you divide by a variable in an equation, remember that division by 0 is undefined. For example, in the formula above, b cannot equal zero.

LOOK BACK
You can refer to Lesson 2-9 for information on writing equations and formulas.

Example 1 **Solve the equation $-5x + y = -56$**
a. for y, and
b. for x.

a.

$$-5x + y = -56$$

$$-5x + 5x + y = -56 + 5x \quad \textit{Add 5x to each side.}$$

$$y = -56 + 5x$$

b.

$$-5x + y = -56$$

$$-5x + y - y = -56 - y \quad \textit{Subtract y from each side.}$$

$$-5x = -56 - y$$

$$\frac{-5x}{-5} = \frac{-56 - y}{-5} \quad \textit{Divide each side by –5.}$$

$$x = \frac{-56 - y}{-5} \text{ or } \frac{56 + y}{5}$$

3-6 LESSON NOTES

NCTM Standards: 1–5, 7

Instructional Resources

- Study Guide Master 3-6
- Practice Master 3-6
- Enrichment Master 3-6
- Assessment and Evaluation Masters, p. 73

Transparency 3-6A contains the 5-Minute Check for this lesson; **Transparency 3-6B** contains a teaching aid for this lesson.

Recommended Pacing	
Standard Pacing	Days 10 & 11 of 14
Honors Pacing	Days 9 & 10 of 13
Block Scheduling*	Day 5 of 6 (along with Lesson 3-7)
Alg. 1 in Two Years*	Days 13 & 14 of 19

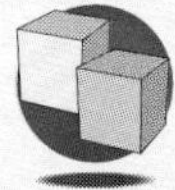

*For more information on pacing and possible lesson plans, refer to the *Block Scheduling Booklet* and *Algebra 1 in Two Years.*

1 FOCUS

5-Minute Check
(over Lesson 3-5)

Solve.
1. $8m + 3 = 5m + 15$ **4**
2. $\frac{2}{3}t - 7 = 5 - \frac{2}{3}t$ **9**
3. $\frac{3}{4}x + \frac{1}{2} = \frac{1}{3}x - \frac{1}{4}$ $\mathbf{-\frac{9}{5}}$
4. $4(x + 3) - 5 = 7(x - 1) + 9$ $\mathbf{\frac{5}{3}}$
5. $6(4 + y) - 3 = 4(y - 3) + 2y$ **no solution**

Motivating the Lesson

Questioning Begin the lesson by having students draw a number of geometric figures such as a parallelogram, trapezoid, and triangle. Then they should write the formulas used to calculate the perimeter and area of each figure.

2 TEACH

Teaching Tip Point out that solving equations for one of two variables will help students in their future work. For example, when writing linear equations in slope-intercept form, students will have to solve the equations for y.

In-Class Examples

For Example 1
Solve $2x - 4y = 7$ for x. Then solve $2x - 4y = 7$ for y.
$x = \frac{4y + 7}{2}$; $y = \frac{2x - 7}{4}$

For Example 2
Solve for *x*.

a. $3ax - b = d - 4cx$ $\frac{b + d}{3a + 4c}$

b. $4x - 3m = 2mx - 5$ $\frac{3m - 5}{4 - 2m}$

c. $x + 3ab = ax + n$ $\frac{n - 3ab}{1 - a}$

For Example 3
Solve for w in the formula for the perimeter of a rectangle, $P = 2\ell + 2w$. $\frac{P - 2\ell}{2}$

Teaching Tip Emphasize that although there are more variables in the equation in Example 2, the process of solving for a single variable does not change. Operations being performed on that variable still must be undone.

CAREER CHOICES

Computer programming is only one career choice in a wide and varied field. Other career options for people with degrees in computer science include systems analysis, management of information systems (MIS), data processing, and local area network (LAN) administration.

Example  **Solve for y in $3y + z = am - 4y$.**

$$3y + z = am - 4y$$
$$3y + z - z = am - 4y - z \quad \textit{Subtract } z \textit{ from each side.}$$
$$3y = am - 4y - z$$
$$3y + 4y = am + 4y - 4y - z \quad \textit{Add } 4y \textit{ to each side.}$$
$$3y + 4y = am - z$$
$$(3 + 4)y = am - z \quad \textit{Use the distributive property.}$$
$$7y = am - z$$
$$\frac{7y}{7} = \frac{am - z}{7} \quad \textit{Divide each side by 7.}$$
$$y = \frac{am - z}{7}$$

CAREER CHOICES

A **computer programmer** writes a detailed plan for solving a problem with an ordered sequence of instructions, often including equations or formulas.

Programming requires a degree in computer science with a working knowledge of computer languages and mathematics.

For more information, contact:

Association for Computing Machinery
11 W. 42nd St.
3rd Floor
New York, NY 10036

Many real-world problems require the use of formulas. When using formulas, you may need to use **dimensional analysis.** This is the process of carrying units throughout a computation. You may also be asked to solve the formula for a specific variable. This may make it easier to use certain formulas.

Example 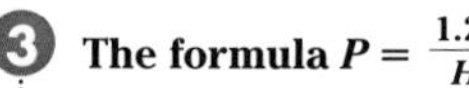 **The formula $P = \frac{1.2W}{H^2}$ represents the amount of pressure exerted on the floor by the heel of a shoe. In this formula, P represents the pressure in pounds per square inch (lb/in^2), W represents the weight of the person wearing the shoe in pounds, and H is the width of the heel of the shoe in inches.**

a. Find the amount of pressure exerted if a 130-pound person wore shoes with heels $\frac{1}{2}$ inch wide.

b. Solve the formula for W.

c. Find the weight of the person if the heel is 3 inches wide and the pressure exerted is 40 lb/in^2.

a. $P = \frac{1.2W}{H^2}$

$= \frac{1.2(130 \text{ lb})}{\left(\frac{1}{2} \text{ in}\right)^2}$ *$W = 130$ lb, $H = \frac{1}{2}$ in.*

$= \frac{156 \text{ lb}}{\frac{1}{4} \text{ in}^2}$

$= 624 \text{ lb/in}^2$

624 lb/in^2 of pressure are exerted.

b. $P = \frac{1.2W}{H^2}$

$H^2P = H^2\left(\frac{1.2W}{H^2}\right)$ *Multiply each side by H^2.*

$H^2P = 1.2W$

$\frac{H^2P}{1.2} = \frac{1.2W}{1.2}$ *Divide each side by 1.2.*

$\frac{H^2P}{1.2} = W$

GLENCOE Technology

CD-ROM Interaction

A multimedia simulation connects tables, graphs, and equations to business decisions in a fireworks factory. A blackline master activity with teacher's notes provides a follow-up to the CD-ROM simulation.

For Windows & Macintosh

c. $\frac{H^2P}{1.2} = W$

$\frac{(3 \text{ in.})^2(40 \text{ lb/in}^2)}{1.2} = W$ *Estimate: $[(3)^2 - 40] \div 1 = 360$ Will the actual answer be greater or less? Why?*

$\frac{360 \text{ lb}}{1.2} = W$ *$(9 \text{ in}^2)(40 \text{ lb/in}^2) = 360$ lb*

$300 \text{ lb} = W$ The person weighs 300 pounds.

CHECK FOR UNDERSTANDING

Communicating Mathematics

Study the lesson. Then complete the following.

1. **Show** how you would solve the equation in Example 2 for m. $\frac{7y + z}{a} = m$
2. **Compare and contrast** the amount of pressure being exerted in the case of the two people described in Example 3. **See margin.**

3. Find a formula in a newspaper or magazine. Define all of the variables and explain what the purpose of the formula is. **See students' work.**

Guided Practice

4. $\frac{7 + 4y}{3}$ 5. $\frac{3x - 7}{4}$ 6. $\frac{b}{a} - 1$

Solve each equation or formula for the variable specified.

4. $3x - 4y = 7$, for x
5. $3x - 4y = 7$, for y
6. $a(y + 1) = b$, for y
7. $4x + b = 2x + c$, for x $\frac{c - b}{2}$
8. $F = G\left(\frac{Mm}{d^2}\right)$, for M $\frac{Fd^2}{Gm}$
9. $S = \frac{n}{2}(A + t)$, for A $\frac{2S - nt}{n}$

Write an equation and solve for the variable specified.

10. Twice a number x and 12 is 31 less than three times another number y. Solve for y. $2x + 12 = 3y - 31$; $\frac{2x + 43}{3}$

EXERCISES

Practice

Solve each equation or formula for the variable specified.

11. $-3x + b = 6x$, for x $\frac{b}{9}$
12. $ex - 2y = 3z$, for x $\frac{3z + 2y}{e}$
13. $\frac{y + a}{3} = c$, for y $3c - a$
14. $\frac{3}{5}y + a = b$, for y $\frac{5(b - a)}{3}$
15. $v = r + at$, for a $\frac{v - r}{t}$
16. $y = mx + b$, for m $\frac{y - b}{x}$
17. $I = prt$, for r $\frac{I}{pt}$
18. $\frac{by + 2}{3} = c$, for y $\frac{3c - 2}{b}$

19. $H = (0.24)I^2Rt$, for R $\frac{H}{0.24I^2t}$
20. $P = \frac{E^2}{R}$, for R $\frac{E^2}{P}$

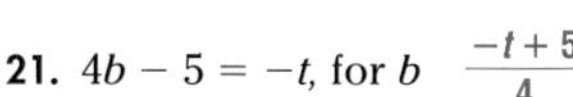

21. $4b - 5 = -t$, for b $\frac{-t + 5}{4}$
22. $km + 5x = 6y$, for m $\frac{6y - 5x}{k}$
23. $c = \frac{3}{4}y + b$, for y $\frac{4}{3}(c - b)$
24. $p(t + 1) = -2$, for t $\frac{-2 - p}{p}$
25. $\frac{5x + y}{a} = 2$, for a $\frac{5x + y}{2}$
26. $\frac{3ax - n}{5} = -4$, for x $\frac{-20 + n}{3a}$

Reteaching

Checking Solutions To check the solution to solving an equation for a specific variable, substitute a set of values for all other variables in both the original equation and its equivalent that has been solved for a given variable. Find the value of the given variable in each and compare.

3 PRACTICE/APPLY

Check for Understanding

Exercises 1–10 are designed to help you assess your students' understanding through reading, writing, speaking, and modeling. You should work through Exercises 1–3 with your students and then monitor their work on Exercises 4–10.

Assignment Guide

Core: 11–33 odd, 35–41
Enriched: 12–30 even, 31–41

For **Extra Practice,** see p. 763.

The red A, B, and C flags, printed only in the Teacher's Wraparound Edition, indicate the level of difficulty of the exercises.

Study Guide Masters, p. 24

3-6 NAME________ DATE________

Study Guide Student Edition Pages 173–177

Solving Equations and Formulas

If an equation that contains more than one variable is to be solved for a specific variable, use the properties of equality to isolate the specified variable on one side of the equation.

Example: Solve $ax - b = c$ for x.

$ax - b = c$

$ax - b + b = c + b$ Add b to each side.

$ax = c + b$

$\frac{ax}{a} = \frac{c + b}{a}$ Divide each side by a.

$x = \frac{c + b}{a}$ $a \neq 0$

Solve for x.

1. $15x + 1 = y$ $x = \frac{y - 1}{15}$
2. $x + 45z = 90$ $x = 90 - 45z$
3. $(x + f) + 2 = j$ $x = j - 2 - f$
4. $xy + z = 9$ $x = \frac{9 - z}{y}, y \neq 0$
5. $x(4 - k) = p$ $x = \frac{p}{4 - k}, k \neq 4$
6. $7x + 3y = m$ $x = \frac{m - 3y}{7}$
7. $2x + b = c$ $x = \frac{c - b}{2}$
8. $x(1 + y) = z$ $x = \frac{z}{1 + y}, y \neq -1$
9. $16z + 4x = y$ $x = \frac{y - 16z}{4}$
10. **Health** The formula $H = \frac{34 - A}{2}$ is sometimes used to relate a person's age, A, to the number of hours of sleep they need every day, H.
    a. Does this formula work for you? If not, why not? probably
    b. For what ages does the formula seem to work best? 0–20 years
    c. Solve the formula for A. $A = 34 - 2H$
    d. How old is a person who is getting his or her optimal amount of sleep, 8 hours per day? 13

Write an equation and solve for the variable specified.

27. Twice a number x increased by 12 is 31 less than three times another number y. Solve for x. $2x + 12 = 3y - 31$; $\frac{3y - 43}{2}$

28. Five eighths of a number x is three more than one half of another number y. Solve for y. $\frac{5}{8}x = \frac{1}{2}y + 3$; $\frac{5}{4}x - 6$

29. Five more than two thirds of a number x is the same as three less than one half of another number y. Solve for x. $\frac{2}{3}x + 5 = \frac{1}{2}y - 3$; $\frac{3}{4}y - 12$

Critical Thinking

30. **Weather** Here's a conversation between a mother and her teenage daughter, adapted from the pages of *Games* magazine.

> Mrs. Weatherby: "This thermometer's no good. It's in Celsius, and I want to know the temperature in Fahrenheit."
>
> Mercuria: "You're so old-fashioned, Mother. Just use the formula $\frac{9}{5}C + 32 = F$, where C is the temperature in Celsius and F is the temperature in Fahrenheit. If that's too hard for you, then just double the number on the thermometer and add 30. Of course, you won't get a precise answer."

The two women each computed the temperature in degrees Fahrenheit, Mercuria with her formula and Mrs. Weatherby by the approximation. Surprisingly, they got exactly the same answer. Just how hot was it? Justify your reasoning.

30. $\frac{9}{5}C + 32 = 2C + 30$; 10°C or 50°F

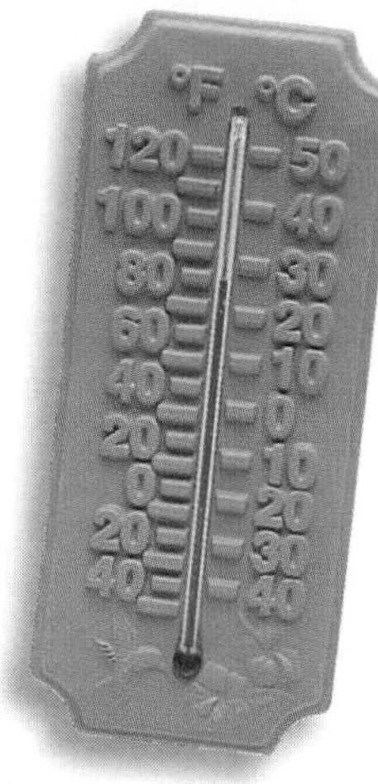

Applications and Problem Solving

31. **Business** The formula in Exercise 17 is the formula for computing simple interest, where I is the interest, p is the principal or amount invested, r is the interest rate, and t is the time in years. Find the amount of interest earned if you were to invest \$5000 at 6% interest (use 0.06), for 3 years. **\$900**

32. **Physics** The formula $d = vt + \frac{1}{2}at^2$ is the formula for distance d, given the initial velocity v, the time t, and the acceleration a. Suppose a biker going 4.5 meters per second (m/s) passes a lamppost at the top of a hill and then accelerates down the hill at a constant rate of 0.4 m/s^2 for 12 seconds. How far does she go down the hill during this time? **82.8 meters**

33. **Work** The formula $s = \frac{w - 10e}{m}$ is often used by placement services to find keyboarding speeds. In the formula, w represents the number of words typed, e represents the number of errors, m represents the number of minutes typed, and s represents the speed in words per minute.
    a. On a 10-minute keyboarding test, Sally typed 420 words with 6 errors. Find Sally's speed in words per minute. **36 words per minute**
    b. Who has the greater speed on a 5-minute test: Isabel, who types 500 words with 14 errors, or Clarence, who types 410 words with 4 errors?

33b. Clarence, with 74 words per minute

Practice Masters, p. 24

3-6 NAME________________ DATE ________

Practice — Student Edition Pages 173–177

Solving Equations and Formulas

Solve for x.

1. $7x = t$ $x = \frac{t}{7}$
2. $\frac{x - c}{2} = d$ $x = c + 2d$
3. $ax - c = b$ $x = \frac{b + c}{a}$
4. $fx + 3y = 2z$ $x = \frac{2z - 3y}{f}$
5. $e = rx$ $x = \frac{e}{r}$
6. $2p = kx - q$ $x = \frac{2p + q}{k}$

Solve for y.

7. $ry + s = tx - m$ $y = \frac{tx - m - s}{r}$
8. $x + 3y = 1$ $y = \frac{1 - x}{3}$
9. $x - 2y = 1$ $y = \frac{x - 1}{2}$
10. $\frac{2}{3}y + k = j$ $y = \frac{3j - 3k}{2}$
11. $5(2a + y) = 3b$ $y = \frac{3b - 10a}{5}$
12. $\frac{2}{3}y + a = a + b$ $\frac{3}{2}b$

Write an equation and solve for the variable specified.

13. Five more than a number x is six less than twice a number y. Solve for x.
$x + 5 = 2y - 6$; $x = 2y - 11$

14. One fourth of a number x is three more than three times a number y. Solve for x.
$\frac{1}{4}x = 3y + 3$; $x = 12y + 12$

15. Five times a number x minus 9 is two thirds of a number y. Solve for y.
$5x - 9 = \frac{2}{3}y$; $y = \frac{15x - 27}{2}$

Extension

Problem Solving Experiment with a sheet of $8\frac{1}{2} \times 11$-inch paper to make a figure with the greatest number of sides possible. Then use another sheet of paper to create a figure with the least number of sides possible.

Mixed Review

34. Solve $2(x - 2) = 3x - (4x - 5)$. (Lesson 3–5) **3**

35. Find the complement of an 85° angle. (Lesson 3–4) **5°**

36. **Work Backward** Greta Moore's plane is scheduled to leave for Dallas at 9:30 A.M. She is to pick up her tickets at the boarding gate thirty minutes before departure. It takes Ms. Moore fifty minutes to drive to the airport, ten minutes to park her car in the overnight lot, and ten more minutes to get to the gate. On her way to the airport, Ms. Moore needs to pick up some materials from her office. This detour will take about fifteen minutes. If it takes her an hour to get dressed and out the door, for what time should Ms. Moore set her alarm? (Lesson 3–3) **6:35 A.M.**

37. Write an inequality for the graph shown below. (Lesson 2–4) $x \neq 2$

38. Use a number line to find the sum of 9 and −5. (Lesson 2–1) **4**

39. Find the solution set for $x - 3 > \frac{x+1}{2}$ if the replacement set is{4, 5, 6, 7, 8}. (Lesson 1–5) **{8}**

40. **Health** Your optimum exercise heart rate per minute is given by the expression $0.7(220 - a)$, where a represents your age. Find your optimum exercise heart rate. (Lesson 1–1) **See students' work.**

WORKING ON THE

In·ves·ti·ga·tion

Refer to the Investigation on pages 68–69.

the Greenhouse Effect

Two scales are used to measure temperature. The Celsius scale, part of the metric system, was invented in 1741 by Swedish astronomer Anders Celsius (1701–1744). The Celsius scale has the freezing point of water at 0°C and the boiling point of water at 100° C. The scale most frequently used in the United States is the Fahrenheit scale, invented in 1714 by German physicist Gabriel Daniel Fahrenheit (1686–1736). On the Fahrenheit scale, the boiling point of water is 212°F, and the freezing point is 32°F.

To convert Celsius temperatures to Fahrenheit, use the formula $F = \frac{9}{5}C + 32$.

1 Using the formula, convert the average temperatures of Mars, Earth, and Venus from Celsius to Fahrenheit. How do the temperatures compare?

2 The greenhouse effect may raise the average temperature of Earth 8°F by the year 2050. What will be the average temperature of Earth in degrees Celsius at that time?

3 Change the formula for converting Celsius to Fahrenheit to a formula for converting Fahrenheit to Celsius. **1–3. See margin.**

Add the results of your work to your Investigation Folder.

4 ASSESS

Closing Activity

Modeling Have students write each letter from *a* to *d* on an index card. Then have them write the numbers 1–5, two multiplication signs , and an equal sign on eight other index cards. Have a volunteer arrange these cards to make an equation of the form $ax + b = cx + d$. Then have the class rearrange the cards to solve for the particular variable asked for by the volunteer who made the equation.

Chapter 3, Quiz C (Lessons 3-5 and 3-6), is available in the *Assessment and Evaluation Masters*, p. 73.

Enrichment Masters, p. 24

3-6 NAME________________ DATE________

Enrichment

Student Edition Pages 173–177

Diophantine Equations

The first great algebraist, Diophantus of Alexandria (about A.D. 300), devoted much of his work to the solving of indeterminate equations. An indeterminate equation has more than one variable and an unlimited number of solutions. An example is $x + 2y = 4$.

When the coefficients of an indeterminate equation are integers and you are asked to find solutions that must be integers, the equation is called *diophantine*. Such equations can be quite difficult to solve, often involving trial and error—and some luck!

Solve each diophantine equation by finding at least one pair of positive integers that makes the equation true. Some hints are given to help you.

1. $2x + 5y = 32$
   a. First solve the equation for x. $x = 16 - \frac{5y}{2}$
   b. Why must y be an even number? If y is odd, then x won't be an integer.
   c. Find at least one solution. Any of these: (11, 2), (6, 4), (1, 6)

2. $5x + 2y = 42$
   a. First solve the equation for x. $x = \frac{42 - 2y}{5}$
   b. Rewrite your answer in the form $x = 8 +$ some expression. $x = 8 + \frac{2 - 2y}{5}$
   c. Why must $(2 - 2y)$ be a multiple of 5? Otherwise, x won't be an integer.
   d. Find at least one solution. Any of these: (8, 1), (6, 6), (4, 11), (2, 16)

3. $2x + 7y = 29$
   (11, 1) or (4, 3)

4. $7x + 5y = 118$
   Any of these: (14, 4), (9, 11), (4, 18)

5. $8x - 13y = 100$
   (19, 4), (32, 12) or any pair when $y = 4n$ and n is a positive odd number

6. $3x + 4y = 22$
   (6, 1) or (2, 4)

7. $5x - 14y = 11$
   (5, 1), (19, 6) or any pair when $y = 5m - 4$ and m is a positive number

8. $7x + 3y = 40$
   (4, 4) or (1, 11)

In·ves·ti·ga·tion

Working on the Investigation

The Investigation on pages 68–69 is designed to be a long-term project that is completed over several days. Encourage students to keep their materials in their Investigation Folder as they work on the Investigation.

Sample Answers

Planet	Earth	Mars	Venus
Temp. with CO_2	15°C 59°F	1°C 33.8°F	454°C 849.2°F
Temp. without CO_2	−17°C 1.4°F	−31°C −23.8°F	422°C 791.6°F

−13.3°C change in temperature; $\frac{5}{9}(F - 32) = C$

3-7 LESSON NOTES

NCTM Standards: 1–5, 10

Instructional Resources

- Study Guide Master 3-7
- Practice Master 3-7
- Enrichment Master 3-7
- Assessment and Evaluation Masters, p. 73
- Real-World Applications, 9

Transparency 3-7A contains the 5-Minute Check for this lesson; **Transparency 3-7B** contains a teaching aid for this lesson.

Recommended Pacing	
Standard Pacing	Day 12 of 14
Honors Pacing	Day 11 of 13
Block Scheduling*	Day 5 of 6 (along with Lesson 3-6)
Alg. 1 in Two Years*	Days 15 & 16 of 19

*For more information on pacing and possible lesson plans, refer to the *Block Scheduling Booklet* and *Algebra 1 in Two Years.*

1 FOCUS

5-Minute Check
(over Lesson 3-6)

Solve for *n*.

1. $\frac{1}{2}n + b = n$ $2b$
2. $ab - bn = 1$ $a - \frac{1}{b}$
3. $cn - 7 = 4$ $\frac{11}{c}$
4. $\frac{a}{c} = n + bn$ $\frac{a}{c + bc}$
5. $an - bn + cn = 5$ $\frac{5}{a - b + c}$

Motivating the Lesson

Situational Problem Reports in newspapers and on TV often refer to the "average." The "average" American earns about $25,000 per year, the "average" cost of housing is about $10,000 per year, and so on. Ask students how averages are determined. Are they determined in the same manner for all types of situations?

Teaching Tip You may wish to suggest that students use a calculator to aid in computation.

3-7 Integration: Statistics
Measures of Central Tendency

What YOU'LL LEARN

- To find and interpret the mean, median, and mode of a set of data.

Why IT'S IMPORTANT

Measures of central tendency can help you easily describe a set of data.

Charities

The table at the right shows the top 10 charities in amount of contributions for 1993.

When analyzing data, it is helpful to have one number that describes the entire set of data. Numbers known as **measures of central tendency** are often used to describe sets of data because they represent a centralized, or *middle,* value. Three of the most commonly used measures of central tendency are the **mean, median,** and **mode.**

Charity	Millions of Dollars
Salvation Army	$726
Catholic Charities USA	411
United Jewish Appeal	407
Second Harvest	407
American Red Cross	395
American Cancer Society	355
YMCA	317
American Heart Association	235
YWCA	218
Boy Scouts of America	211

Source: *The Chronicle of Philanthropy,* Nov., 1993

Definition of Mean	**The mean of a set of data is the sum of the numbers in the set divided by the number of numbers in the set.**

FYI

Until 1994, the Cardwell sisters of Sweetwater, Texas, were the oldest living set of triplets. However, on October 2, 1994, the firstborn triplet, Faith, died at the age of 95. She was survived by her sisters Hope and Charity.

To find the mean of the charity data, find the sum of the dollar values and divide by 10, the number of numbers in the set. ***Estimate:*** *Is an answer of $500 reasonable?*

$$\text{mean} = \frac{726 + 411 + 407 + 407 + 395 + 355 + 317 + 235 + 218 + 211}{10}$$
$$= \frac{3682}{10}$$
$$= 368.2$$

The mean is $368.2 million. Notice that the amount collected by the Salvation Army, $726 million, is far greater than the amounts collected by the other charities. Because the mean is an average of several numbers, a single number that is so much greater than the others can affect the mean a great deal. In extreme cases, the mean becomes less representative of the values in a set of data.

FYI

When the Cardwell sisters were born, their doctor suggested they be named Lillie, Lola, and Lula. Their parents did not like those names or any others, so for six months, the sisters went unnamed. Finally, Frances Cleveland, the wife of then-President Grover Cleveland, suggested that the babies be named Faith, Hope, and Charity, and thus, they were.

The median is another measure of central tendency.

Definition of Median

The median of a set of data is the middle number when the numbers in the set are arranged in numerical order.

To find the median of the charity data, arrange the dollar values in order.

726 411 407 407 395 355 317 235 218 211

If there were an odd number of dollar values, the middle one would be the median. However, since there is an even number of dollar values, the median is the average of the two middle values, 395 and 355.

$$\text{median} = \frac{395 + 355}{2}$$
$$= \frac{750}{2}$$
$$= 375$$

The median is $375 million. Notice that the number of values that are greater than the median is the same as the number of values that are less than the median.

A third measure of central tendency is the mode.

Definition of Mode

The mode of a set of data is the number that occurs most often in the set.

To find the mode of the charity data, look for the number that occurs most often.

726 411 407 407 395 355 317 235 218 211

In this set, 407 appears twice. Thus, $407 million is the mode of the data.

It is possible for a set of data to have more than one mode. For example, the set of data {2, 3, 3, 4, 6, 6} has two modes, 3 and 6.

Based on our results, the charity data has mean $368.2 million, median $375 million, and mode $407 million. As you can see, the mean, median, and mode are rarely the same value.

Example 1

History

When the Declaration of Independence was signed in 1776, George III was king of Great Britain. After George III, Great Britain had seven monarchs before Queen Elizabeth II was crowned queen in 1952. The table at the right shows the number of years that each monarch reigned. Find the mean, median, and mode of the data.

Monarch	Reign (years)
George III	59
George IV	10
William IV	7
Victoria	63
Edward VII	9
George V	25
Edward VIII	1
George VI	15

$$\textbf{mean} = \frac{59 + 10 + 7 + 63 + 9 + 25 + 1 + 15}{8}$$
$$= \frac{189}{8}$$
$$= 23.625$$ The mean is 23.625 years.

(continued on the next page)

2 TEACH

In-Class Example

For Example 1
The heights of players on South High School's basketball team are 72", 74", 70", 77", 75", and 70". Find the mean, median, and mode of the heights.
Mean is 73", median is 73", and mode is 70".

GLENCOE Technology

Interactive Mathematics Tools Software

This multimedia software provides an interactive lesson that uses histograms to observe the changes in the relationships between the mean, mode, and median as the data change. A **Computer Journal** gives students an opportunity to write about what they have learned.

For Windows & Macintosh

In-Class Example

For Example 2

The stem-and-leaf plot shown below represents the scores on an algebra test. Find the mean, median, and mode of the scores.

Stem	Leaf
3	2 4
4	0 2
5	1 1 3 7
6	2 3 4 4 6 8
7	1 3 5 6 7 9
8	2 5 6 8 8
9	3 6

7 | 1 = 71

Mean is 67.3; median is 68; there are three modes: 51, 64, and 88.

Teaching Tip Emphasize that in order to find the median, the data must be arranged in numerical order, either from greatest to least value or from least to greatest value.

Teaching Tip Point out that when there are an even number of elements in a set of data, the median is found by determining the mean of the two middle elements.

Teaching Tip Point out that in some cases, the median and mode can be more representative of a set of data than the mean.

EXPLORATION

If students forget one or both of the braces, they will get an error message. They can use the REPLAY function to correct their error. To use it, press 2nd ENTER. The expression they entered will appear on the screen, and they can use their arrow keys to correct it.

median 1 7 9 10 15 25 59 63

The median is $\frac{10 + 15}{2}$ or $12\frac{1}{2}$ years. Why do you think the mean and median are so different? Which value, mean or median, do you think is more representative of the data?

mode Since each number appears exactly once, there is no mode.

You can also determine the mean, median, and mode of a set of data by examining stem-and-leaf plots.

Example 2

Since it would be inconvenient to list every stem, we have used dots to indicate that some stems are missing.

The stem-and-leaf plot at the right shows population data for the 20 largest cities in the United States to the nearest ten thousand. Find the mean, median, and mode of the data.

Stem	Leaf
73	2
⋮	⋮
34	9
⋮	⋮
27	8
⋮	⋮
16	9 3
⋮	⋮
10	3 1
⋮	⋮
8	8
7	9 9 4 3
6	8 4 4 3 3 3 1
5	7

8 | 8 = 880,000 people

mean Add the 20 values and then divide by 20. Since the sum of the 20 values is 2791, the mean is $\frac{2791}{20}$ or 139.55. This represents a mean population of 1,395,500.

median Since there are 20 values, the median is the average of the 10th and 11th values. Counting from the top down, you will find that the 10th and the 11th values are 79 and 74.

The average of 79 and 74 is $\frac{79 + 74}{2}$ or 76.5. Thus, the median population is 765,000.

mode There are three entries for 63. Thus, the mode is 630,000.

EXPLORATION GRAPHING CALCULATORS

You can use a graphing calculator and the MEAN and MEDIAN functions to find the mean and median of a list of numbers. The format is as follows.

mean(*{a,b,c,d,...z}*) median(*{a,b,c,d,...z}*)

To find the mean of the data in Example 2, access the LIST MATH function by pressing 2nd LIST ▶. Then choose 3, for mean or 4, for median. Enter the list of numbers, beginning with a left brace (2nd {) and ending with a right brace (2nd }) and a right parenthesis. When you press ENTER, the mean or median appears on the screen.

Your Turn

Take time to work through the examples in this lesson using a graphing calculator.

Cooperative Learning

Brainstorming You may wish to have students collect data from which they could calculate the mean, median, and mode. Have students work in groups of four to brainstorm survey topics, methods, and how the data will be displayed. For more information on the brainstorming strategy, see *Cooperative Learning in the Mathematics Classroom,* one of the titles in the Glencoe Mathematics Professional Series, page 30.

CHECK FOR UNDERSTANDING

Communicating Mathematics

Study the lesson. Then complete the following.

1. **Explain** why you think the mean, median, and mode are rarely the same value. **See margin.**
2. **Discuss** which of the measures of central tendency you would use to describe the data in Example 2. Explain your reasoning. **See margin.**
3. **Describe** the steps you would take to find the mean, median, and mode of the data represented by the line plot shown at the right. **See margin.**

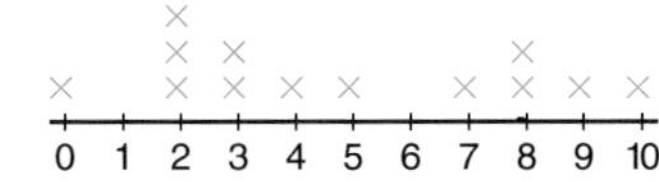

4. Suppose you conducted a survey in which you asked your classmates, "What's your favorite rock group?" Which measure of central tendency could you use to analyze the data, and why? **See margin.**

5. Find some data that interests you in a newspaper or magazine. Find the mean, median, and mode and discuss which of these measures you think best describes the data. **See students' work.**

Complete.

6. Measures of central tendency represent __?__ values of a set of data. **middle**
7. If the numbers in a set of data are arranged in numerical order, then the __?__ of the set is the middle number. **median**
8. Extremely high or low values affect the __?__ of a set of data. **mean**
9. If all the numbers in a set of data occur the same number of times, then the set has no __?__ . **mode**

Guided Practice

Find the mean, median, and mode for each set of data.

10. 4, 6, 12, 5, 8 **7; 6; none**
11. 8, 8, 8, 8, 9 **8.2; 8; 8**
12.

Stem	Leaf
7	3 5
8	2 2 4
9	0 4 7 9
10	5 8
11	4 6

9|4 = 94

12. 93.77; 94; 82

13–14. Sample answers are given.

List six numbers that satisfy each set of conditions.

13. The mean is 50, the median is 40, and the mode is 20. **20, 20, 30, 50, 90, 90**
14. The mean is 70, the median is 70, and the modes are 65 and 70. **65, 65, 70, 70, 73, 77**
15. **Basketball** The table at the right shows the names of players most often named most valuable player of the NBA.
    a. Find the mean, median, and mode of the data. **3.857; 3; 3**
    b. Which measure of central tendency best represents these data, and why? **mean**

Player	MVP Wins
Kareem Abdul-Jabbar	6
Bill Russell	5
Wilt Chamberlain	4
Larry Bird	3
Magic Johnson	3
Moses Malone	3
Michael Jordan	3

Source: *World Almanac, 1995*

3 PRACTICE/APPLY

Check for Understanding

Exercises 1–15 are designed to help you assess your students' understanding through reading, writing, speaking, and modeling. You should work through Exercises 1–9 with your students and then monitor their work on Exercises 10–15.

Additional Answers

1. **The mean is affected by extreme values, where median is not. Mode is affected by the amount of repetition of values.**
2. **The mean, since it finds the average amount of the contributions. The mean is affected by extreme values.**
3. **Count the total number of x's. To find the mean, multiply the number by the number of x's in that column. Then divide by the total number of x's (4.8). To find the median, divide the total number of x's by 2 (13 ÷ 2 or ≈ 7) and count from left to right. The seventh x represents the median (4). The mode is the number with the greatest number of x's in its column (2).**

Study Guide Masters, p. 25

3-7 NAME ______ DATE ______

Study Guide — Student Edition Pages 178–183

Integration: Statistics
Measures of Central Tendency

In working with statistical data, it is often useful to have one value represent the complete set of data. For example, **measures of central tendency** represent centralized values of the data. Three measures of central tendency are the **mean, median,** and **mode.**

	Definitions	Examples
Mean	Sum of the elements in the set divided by tha number of elements in the set.	**Data:** 24, 36, 21, 30, 21, 30 $\frac{24+36+21+30+21+30}{6}=27$
Median	The middle of a set of data when the numbers are arranged in numerical order. In an even number of elements, the median is halfway between the two middle elements.	**Data:** 21, 21, 25, 30, 31, 42 $\frac{25+30}{2}=27.5$
Mode	The number that occurs most often in a set of data.	**Data:** 21, 21, 24, 30, 30, 36 There are two modes, 21 and 30.

Find the mean, median, and mode for each set of data.

1.

Month	Days above 90°F
May	4
June	7
July	14
August	12
September	8

9; 8; none

2. (line plot, 0 to 7) 3.7; 4; 1 and 5

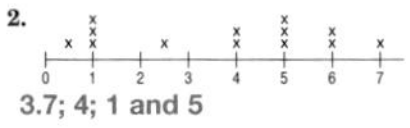

3. 3, 6, 6, 3, 6, 6, 3, 3 — 4.5, 4.5, 3 and 6
4. 19, 3, 0, 1 — 5.75; 2; none
5. $\frac{1}{4}, \frac{2}{5}, \frac{2}{8}, \frac{1}{3}$ — $\frac{37}{120}; \frac{7}{24}; \frac{1}{4}$
6. $1, \frac{1}{2}, 2, \frac{1}{3}, 3, \frac{1}{4}, 4, \frac{1}{8}$ — $\frac{269}{192}; \frac{3}{4}$; none

Find the median and mode(s) of the data shown in each stem-and-leaf plot.

7.

Stem	Leaf
2	4 7 7
3	1 2 6 6 6 9
4	0
5	8 8 9

3|2 = 32
36; 36

8.

Stem	Leaf
9	0 0 1 3 9
10	2 2 5
11	
12	0 3 3 8 8 9

10|5 = 105
103.5; 102, 123, 128

Reteaching

Using Data Give students a list of the scores from their last test. Have them find the mean, median, and mode of these scores and then discuss their results.

Additional Answer

4. **Mode; since the data are nonnumeric, the mean and median would have no meaning in this context.**

Assignment Guide

Core: 17–31 odd, 32–37
Enriched: 16–26 even, 27–37

For **Extra Practice,** see p. 764.

The red A, B, and C flags, printed only in the Teacher's Wraparound Edition, indicate the level of difficulty of the exercises.

Additional Answers

27b. Yes, because the article indicates that the median value is in the middle of the data set.

27c. No, he should have used the term *mean* since the mean can be affected by extremely high values, causing the majority to be below the mean value.

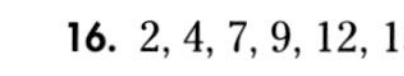

EXERCISES

Practice

Find the mean, median, and mode for each set of data.

A

16. 2, 4, 7, 9, 12, 15 — $8\frac{1}{6}$; 8; none
17. 300, 34, 40, 50, 60 — 96.8; 50; none
18. 23, 23, 23, 12, 12, 12 — 17.5; 17.5; 23, 12
19. 10, 3, 17, 1, 8, 6, 12, 15 — 9; 9; none
20. 7, 19, 9, 4, 7, 2 — 8; 7; 7
21. 2.1, 7.4, 13.9, 1.6, 5.21, 3.901 — 5.69; 4.56; none

B

22.

Stem	Leaf
5	3 6 8
6	5 8
7	0 3 7 7 9
8	1 4 8 8 9
9	9

6|8 = 68

75.3; 77; 77, 88

23.

Stem	Leaf
19	3 5 5
20	2 2 5 8
21	5 8 8 9 9 9
22	0 1 7 8 9

21|5 = 215

212.94; 218; 219

C

24. List ten numbers that satisfy each set of conditions.
    a. The mean is less than all but one of the numbers.
    b. The mean is greater than all but one of the numbers.
    c. The mean is greater than all of the numbers. not possible
    d. Would you be able to complete parts a, b, and c if the word *mean* were replaced by the word *median*? Why or why not?

24. Sample answers are given.
a. 6, 6, 6, 6, 6, 6, 6, 6, 6, 0
b. 5, 5, 5, 5, 5, 5, 5, 5, 5, 25.
d. No, because it is always in the middle of the set of data.

25. The mean of a set of ten numbers is 5. When the greatest number in the set is eliminated, the mean of the new set of numbers is 4. What number was eliminated from the original set of numbers? 14

26. The first of three consecutive odd integers is $2n + 1$.
    a. What is the mean of the three integers? $\frac{6n + 9}{3}$ or $2n + 3$
    b. What is the median? $2n + 3$

Critical Thinking

27. This paragraph appeared in the *San Diego Union-Tribune* on August 8, 1993.

> Lawyers at the top of their profession earn more than $1 million a year, while the typical attorney makes almost $67,000—far more than the average American. . . . According to census figures, lawyers and judges earned a median pay of $66,784 in 1991, which means that half of them made more than that amount and half made less. . . . However R. Wilson Montjoy II of Jackson, Mississippi, said, "We have lots more people at or below the median than above it."

a. To which measure of central tendency do you think the word *average* refers?
b. Has *median* been defined correctly in this passage? Why or why not?
c. In the statement by R. Wilson Montjoy II, did he use the term *median* correctly? If not, which term should he have used? Justify your answer.

27a. mean

27b–c. See margin.

Applications and Problem Solving

28. **Demographics** The table at the right shows population data for 10 South American countries, according to the 1990 Census.
    a. Find the mean, median, and mode of the data.
    b. If you were reporting these data in a newspaper article, which measure would you use, and why?

Country	Population (millions)
Argentina	32
Bolivia	7
Brazil	150
Chile	13
Colombia	33
Ecuador	10
Paraguay	5
Peru	22
Uruguay	3
Venezuela	19

28a. 29.4; 16; none

28b. The mean is most representative.

Practice Masters, p. 25

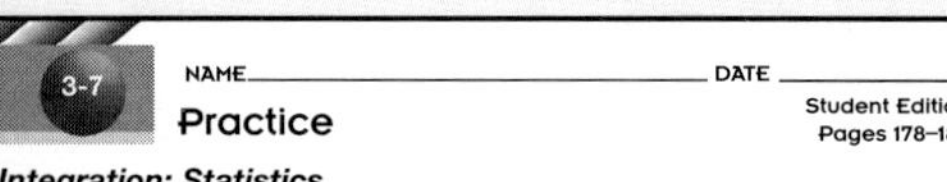

3-7 NAME______________________ DATE __________

Practice — Student Edition Pages 178–183

Integration: Statistics
Measures of Central Tendency

Find the mean, median, and mode for each set of data.

1. 2, 5, 11, 3, 9 — 6; 5; no mode
2. 6, 3, 3, 12, 14, 17 — 9.2; 9; 3
3. 1, 6, 7, 6, 8, 10, 10 — 6.9; 7; 6 and 10
4. 5, 0, 11, 1, 14, 3, 4, 7, 11 — 6.2; 5; 11

Find the median and mode of the data shown in each line plot.

5. 5; 4
6. 50; 59

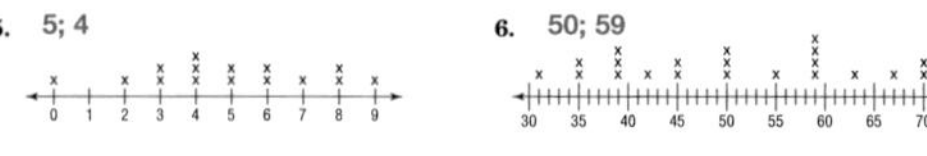

Find the median and mode of the data shown in each stem-and-leaf plot.

7.

Stem	Leaf
2	1 3 7 7 8
3	0 1 9
4	2 4 4 4 6 8

2|3 = 23

35; 44

8.

Stem	Leaf
5	5 7 8
6	1 3 7 9
7	0 4 6 6

5|7 = 57

67; 76

9.

Stem	Leaf
5	2 4 6
6	4 5
7	0 2 6 6 9
8	1 5 7 7 7
9	8

5|2 = 5.2

7.6; 8.7

10. The prices of a sweater in six different stores are $31.95, $26.30, $29.00, $32.50, $25.25, and $38.00. Find the mean and median prices of the sweater. $30.50; $30.48

11. Jack's scores on 20-point quizzes are 16, 18, 10, 15, 19, 16, 15, 19, 20, 16, 12, 17, 16, 15, 18, 14, and 16. Find the mean, median, and mode of his quiz scores. 16, 16, 16

29. **Geography** The areas of each of the South American countries listed in the table on the previous page are as follows (in millions of square miles): 1085, 139, 7778, 670, 3166, 1068, 177, 1135, 154, and 1234, respectively. Find the mean, median, and mode area of the countries. **$1660\frac{3}{5}$; $1076\frac{1}{2}$; none**

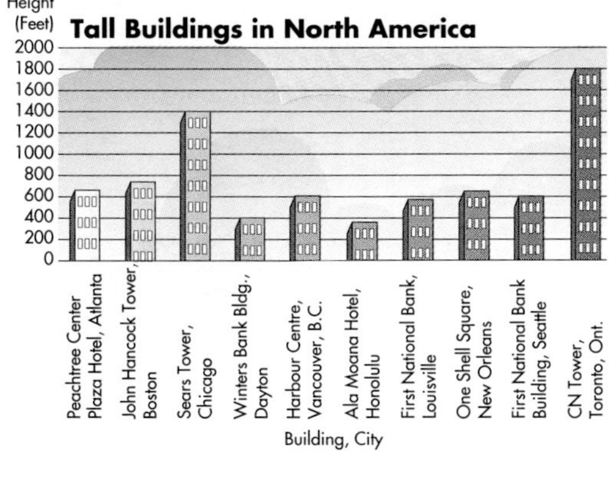

30. **American Landmarks** The graph at the left shows the height in feet of several tall buildings in North America. Find the mean and median height of the buildings. **about 798; about 625**

31. **Commuting** On her way home from work, Luisa drives 10 miles through the city at about 30 mph and then 10 miles on the highway at about 50 mph. What is her average speed? Use $d = rt$. (*Hint:* The answer is *not* 40 mph.) **37.5 mph**

Mixed Review

32. Solve $\frac{a+5}{3} = 7x$ for x. (Lesson 3–6) $\frac{a+5}{21}$

33. Solve $\frac{3}{4}n - 3 = 9$. Then check your solution. (Lesson 3–3) **16**

34. Evaluate $|k| + |m|$ if $k = 3$ and $m = -6$. (Lesson 2–3) **9**

35. Use a number line to add $-6 + (-14)$. (Lesson 2–1) **−20**

36. Evaluate $\left(13 + \frac{2}{5} \cdot 5\right)(3^2 - 2^3)$. Indicate the property used in each step. (Lesson 1–8) **15**

37. Simplify $5 + 7(ac + 2b) + 2ac$. (Lesson 1–7) **$5 + 9ac + 14b$**

Mathematics and SOCIETY

Teen Talk Barbies

The article below appeared in *Antique Week* on August 30, 1993.

Some time ago, Mattel recalled any Teen Talk Barbies programmed to say "Math class is tough" because some women's groups complained the comment perpetuated the view that girls are bad at mathematics. According to a news piece in a December issue of *The Atlanta Journal and Constitution,* only five Barbie buyers in the United States turned in their dolls because of Mattel's offer. While some talking Barbies may have found math tough, the consumers/collectors who own the doll won't have trouble adding up the profits.... Savvy collectors were tearing open Teen Talk boxes trying to find a Barbie who talked about math. Mike Huen, a professional toy collector, estimates the doll, which sold for $35, will be worth several hundred dollars now. ■

1. Why do you think some women's groups objected to what the dolls say? Do you think that what the dolls say has any influence on the girls who play with them? Why or why not? **1–3. See students' work.**
2. Why do you think only five Barbie buyers have returned their dolls? Explain.
3. Do you think it is right for people to make money on this doll? Justify your reasoning.

4 ASSESS

Closing Activity

Speaking Have students explain the difference between the mean, median, and mode of a set of data.

Chapter 3, Quiz D (Lesson 3-7), is available in the *Assessment and Evaluation Masters,* p. 73.

Enrichment Masters, p. 25

NAME______ DATE______
Enrichment Student Edition Pages 178–184

Other Kinds of Means

There are many different kinds of means besides the arithmetic mean. A mean for a set of numbers has these two properties:

a. It typifies or represents the set.

b. It is not less than the least number and it is not greater than the greatest number.

Here are the formulas for the arithmetic mean and three other means.

Arithmetic Mean Add the numbers in the set. Then divide the sum by n, the number of elements in the set.

$$\frac{x_1 + x_2 + x_3 + \cdots + x_n}{n}$$

Geometric Mean Multiply all the numbers in the set. Then find the nth root of their product.

$$\sqrt[n]{x_1 \cdot x_2 \cdot x_3 \cdot \cdots \cdot x_n}$$

Harmonic Mean Divide the number of elements in the set by the sum of the reciprocals of the numbers.

$$\frac{n}{\frac{1}{x_1} + \frac{1}{x_2} + \frac{1}{x_3} + \cdots + \frac{1}{x_n}}$$

Quadratic Mean Add the squares of the numbers. Divide their sum by the number in the set. Then, take the square root.

$$\sqrt{\frac{x_1^2 + x_2^2 + x_3^2 + \cdots + x_n^2}{n}}$$

Find the four different means for each set of numbers.

1. 10, 100
A = 55 G = 31.62
H = 18.18 Q = 71.06

2. 50, 60
A = 55 G = 54.77
H = 54.55 Q = 55.23

3. 1, 2, 3, 4, 5,
A = 3 G = 2.61
H = 2.19 Q = 3.32

4. 2, 2, 4, 4
A = 3 G = 2.83
H = 2.67 Q = 3.16

5. Use the results from Exercises 1 to 4 to compare the relative sizes of the four types of means.
From least to greatest, the means are the harmonic, geometric, arithmetic, and quadratic means.

Extension

Problem Solving Find the mean, median, and mode of the hourly wages of 20 workers. Ten workers earn $4.75 per hour, one earns $5.50 per hour, one earns $6.75 per hour, two earn $4.80 per hour, and six earn $5.25 per hour. **mean = $5.04, median = $4.78, mode = $4.75**

Mathematics and SOCIETY

Point out that there are articles and information that relate to mathematics in many popular magazines and journals. For example, this article appeared in *Antique Week.*

Discuss the feelings and opinions of those who wanted the dolls recalled and those who opposed the recall.

In·ves·ti·ga·tion

TEACHER NOTES

Closing the Investigation

This activity provides students an opportunity to bring their work on the Investigation to a close. For each Investigation, students should present their findings to the class. Here are some ways students can display their work.

- Conduct and report on an interview or survey.
- Write a letter, proposal, or report.
- Write an article for the school or local paper.
- Make a display, including graphs and/or charts.
- Plan an activity.

Assessment

To assess students' understanding of the concepts and topics explored in the Investigation and its follow-up activities, you may wish to examine students' Investigation Folders.

The scoring guide provided in the *Investigations and Projects Masters,* p. 3, provides a means for you to score students' work on the Investigation.

Investigations and Project Masters, p. 3

Scoring Guide
Chapters 2 and 3
Investigation

Level	Specific Criteria
3 Superior	• Shows thorough understanding of the concepts of *the greenhouse effect, global warming, data analysis, subtraction of rational numbers, equations,* and *formulas.* • Uses appropriate strategies to solve problems. • Computations are correct. • Written explanations are exemplary. • Charts, graph, and letter are appropriate and sensible. • Goes beyond requirements of all or some problems.
2 Satisfactory, with Minor Flaws	• Shows understanding of the concepts of *the greenhouse effect, global warming, data analysis, subtraction of rational numbers, equations,* and *formulas.* • Uses appropriate strategies to solve problems. • Computations are mostly correct. • Written explanations are effective. • Charts, graph, and letter are appropriate and sensible. • Satisfies all requirements of problems.
1 Nearly Satisfactory, with Obvious Flaws	• Shows understanding of most of the concepts of *the greenhouse effect, global warming, data analysis, subtraction of rational numbers, equations,* and *formulas.* • May not use appropriate strategies to solve problems. • Computations are mostly correct. • Written explanations are satisfactory. • Charts, graph, and letter are appropriate and sensible. • Satisfies most requirements of problems.
0 Unsatisfactory	• Shows little or no understanding of the concepts of *the greenhouse effect, global warming, data analysis, subtraction of rational numbers, equations,* and *formulas.* • Does not use appropriate strategies to solve problems. • Computations are incorrect. • Written explanations are not satisfactory. • Charts, graph, and letter are not appropriate or sensible. • Does not satisfy requirements of problems.

CLOSING THE

In·ves·ti·ga·tion

the Greenhouse Effect

Refer to the Investigation on pages 68–69.

For more than a decade, scientists have warned that cars and factories spew so many gases into the atmosphere that Earth could soon be affected by disastrous climatic changes. The loss of rain forests is reducing the number of trees to offset the large increases in carbon dioxide in our atmosphere. The possible consequences are so frightening that it makes sense to slow the buildup of CO_2 through preventive measures, such as encouraging energy conservation, developing alternatives to fossil fuels, and preventing the destruction of the rain forests.

Analyze

You have conducted experiments and done research on this important problem. It is now time to analyze your findings and state your conclusions.

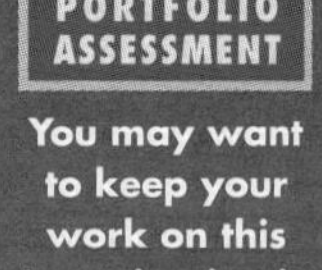

PORTFOLIO ASSESSMENT

You may want to keep your work on this Investigation in your portfolio.

1 Create a graph of the data from your experiments. Let the vertical axis represent the changes in temperature. Let the horizontal axis represent the distances between the lamp and the thermometer. Use different colors to indicate the control experiment and the greenhouse experiment.

2 Use the data from your experiments to draw a conclusion about the relationship between the temperature and the distance between the thermometer and the lamp. Describe this relationship as a function, and justify your reasoning.

3 Use the data from Working on the Investigation in Lesson 2–6 to make a chart. Use the headings *Planet, Temperature Without CO_2, Temperature With CO_2,* and *Distance from the Sun.*

4 Draw a conclusion about the relationship between the distance from the sun and the temperature of Venus and Earth with CO_2. Describe this relationship as a function, and justify your reasoning.

5 Draw a conclusion about the relationship between the distance from the sun and the temperature of Mars and Earth with CO_2. Describe this relationship as a function, and justify your reasoning.

6 How does your experiment explore the temperatures of the planets? How are the relationships similar and different?

7 Predict what the temperature of Mars would be if it had the same amount of CO_2 that Earth has. Explain your calculations.

Write

Some scientists predict that by the year 2050, Earth's atmosphere will contain between 500 and 700 ppm of CO_2. Write a one-page paper explaining the situation to a group of concerned citizens. Consider some of the following questions to address in your paper.

8 What factors will cause this increase in CO_2?

9 How different is this estimate from the amount of CO_2 calculated earlier?

10 In your opinion, what will the approximate average temperature on Earth be at that time?

11 What might occur if the temperature reaches that point? What are some of the dangers involved?

Sample Answers

Planet	Earth	Mars	Venus
Temp. with CO_2	15°C	1°C	454°C
Temp. without CO_2	−17°C	−31°C	422°C
Distance from the sun	150 million km	228 million km	108 million km

Answers will vary based on the data collected. However, general observations should be that the farther the object is from the light source, the colder the object becomes. The plastic bag over the thermometer holds in the heat in much the same way that CO_2 does for a planet.

CHAPTER 3 HIGHLIGHTS

VOCABULARY

After completing this chapter, you should be able to define each term, property, or phrase and give an example or two of each.

Algebra

addition property of equality (p. 144)
consecutive integers (p. 158)
dimensional analysis (p. 174)
division property of equality (p. 151)
equivalent equation (p. 144)
identity (p. 170)
multi-step equations (p. 157)
multiplication property of equality (p. 150)
number theory (p. 158)
solve an equation (p. 145)
subtraction property of equality (p. 146)

Statistics

mean (p. 178)
measures of central tendency (p. 178)
median (p. 178)
mode (p. 178)

Geometry

acute triangle (p. 165)
complementary angles (p. 163)
congruent (p. 164)
equilateral triangle (p. 164)
isosceles triangle (p. 164)
obtuse triangle (p. 165)
right triangle (p. 165)
supplementary angles (p. 162)
triangle (p. 163)

Problem Solving

work backward (p. 156)

UNDERSTANDING AND USING THE VOCABULARY

Choose the letter of the term that best matches each statement or phrase.

1. If $a = b$, then $a + c = b + c$. b
2. If $a = b$, then $a - c = b - c$. l
3. If $a = b$, then $ac = bc$. i
4. If $a = b$ and $c \neq 0$, then $\frac{a}{c} = \frac{b}{c}$. c
5. a triangle in which each angle has the same measure d
6. a triangle in which two angles have the same measure e
7. a triangle in which one angle measures 90° k
8. a triangle in which one angle measures more than 90° j
9. a triangle in which all of the angles measure less than 90° a
10. the sum of the numbers divided by the number of numbers f
11. the middle number when data is in numerical order g
12. the number that occurs most often in a set of data h

a. acute triangle
b. addition property of equality
c. division property of equality
d. equilateral triangle
e. isosceles triangle
f. mean
g. median
h. mode
i. multiplication property of equality
j. obtuse triangle
k. right triangle
l. subtraction property of equality

Using the CHAPTER HIGHLIGHTS

The Chapter Highlights begins with a listing of the new terms, properties, and phrases that were introduced in this chapter. Have students define each term and provide an example or two of it, if appropriate.

Assessment and Evaluation Masters, pp. 59–60

NAME__________ DATE ______

Chapter 3 Test, Form 1B

Write the letter for the correct answer in the blank at the right of each problem.

1. What is the solution of $m - (-4) = 7$? 1. A
   A. 3 B. 23 C. 11 D. 211
2. What is the solution of $-11 = n - (-11)$? 2. C
   A. 0 B. 22 C. −22 D. none of these
3. The sum of two integers is −46. The greater integer is 13. What is the lesser integer? 3. B
   A. 59 B. −59 C. −33 D. 33
4. What is the solution of $-26y = 884$? 4. D
   A. 910 B. 858 C. 34 D. −34
5. What is the solution of $\frac{3}{5}x = 15$? 5. C
   A. 45 B. 5 C. 25 D. 75
6. What is the solution of $5x + 3 = 23$? 6. A
   A. 4 B. −4 C. 5 D. −5
7. What is the solution of $\frac{n}{3} - 8 = -2$? 7. D
   A. −30 B. 30 C. −18 D. 18
8. What is the solution of $-14 = \frac{c + 12}{-6}$? 8. B
   A. −72 B. 72 C. 96 D. −96
9. What is the supplement of an angle whose measure is $(x - 28)°$? 9. C
   A. $(62 - x)°$ B. $(118 - x)°$ C. $(208 - x)°$ D. $(152 - x)°$
10. What is the measure of an angle that is twice its complement? 10. B
    A. 30° B. 60° C. 90° D. 120°
11. The measures of two angles of a triangle are 44° and 32°. What is the measure of the third angle? 11. D
    A. 76° B. 14° C. 44° D. 104°

NAME__________ DATE ______

Chapter 3 Test, Form 1B (continued)

12. What is the solution of $2x + 7 = 5x + 16$? 12. A
    A. −3 B. $\frac{2}{3}$ C. $-\frac{23}{3}$ D. 3
13. What is the solution of $\frac{2}{3}(6x - 3) = x + 5(x + 8) - 2x$? 13. B
    A. identity B. no solution C. 6 D. 0
14. What is the solution of $-3(h - 6) = 5(2h + 3)$? 14. B
    A. $-\frac{3}{13}$ B. $\frac{3}{13}$ C. $-\frac{9}{13}$ D. $\frac{9}{13}$
15. Let x represent the lesser of two consecutive even integers. Which equation would you use to find the two integers if their sum is 126? 15. A
    A. $x + (x + 2) = 126$ B. $x + 2x = 126$
    C. $(x + 2) + (x + 4) = 126$ D. $x(x + 2) = 126$
16. A number is added to 9. The result is then multiplied by 4 to give a new result of 120. What is the number? 16. D
    A. $27\frac{3}{4}$ B. 489 C. 39 D. 21
17. What is the solution for x of $2x - y = y$? 17. C
    A. $2y - 2$ B. $y - 2$ C. y D. 0
18. Find the mean of the set of data 4, 6, 7, 7, 9, 11, 17, 21, 25 rounded to the nearest tenth. 18. C
    A. 7 B. 9 C. 11.9 D. 11.5
19. Find the median of the set of data 5.11, 7.07, 8.95, 6.89, 8.63, 8.95. 19. A
    A. 7.85 B. 8.95 C. 7.60 D. no median
20. Which is affected by extremely high or low values in a set of data? 20. C
    A. median B. mode C. mean D. All are affected.

Bonus **Bonus**

A high school team had just won a conference mathematics contest. A reporter from the local paper asked the team captain, "How many gold, silver, and bronze medals did the team win?" The captain replied, "The product of the three numbers is 72 and their sum is the same as today's date." The reporter thought for a while and said, "I know the date but I still can't find the answer." The captain exclaimed, "I'm sorry. I forgot to tell you that the smallest number of medals was of bronze and the greatest number of medals was gold." "Thank you," replied the reporter, who then left to write the story. How many medals of each type were won?

9 gol
8 silve
1 bronz

Instructional Resources

Three multiple-choice tests and three free-response tests are provided in the *Assessment and Evaluation Masters*. Forms 1A and 2A are for honors pacing, and Forms 1B, 1C, 2B, and 2C are for average pacing. Chapter 3 Test, Form 1B is shown at the right. Chapter 3 Test, Form 2B is shown on the next page.

Applications and Problem Solving Encourage students to work through the exercises in the Applications and Problem Solving section to strengthen their problem-solving skills.

OBJECTIVES AND EXAMPLES

- find and interpret the mean, median, and mode of a set of data (Lesson 3–7)

Find the mean, median, and mode of 1.5, 3.4, 5.4, 5.6, 5.7, 6.2, 6.8, 7.1, 7.1, 8.4, and 9.9.

mean: The sum of the 11 values is 67.1.

$\frac{67.1}{11} = 6.1$

median: The 6th value is 6.2.

mode: 7.1

REVIEW EXERCISES

59. **School** Marisa's scores on the 25-point quizzes in her English class are 20, 21, 18, 21, 22, 22, 24, 21, 20, 19, and 23. Find the mean, median, and mode of her scores. **21; 21; 21**

60. **Business** Of the 42 employees at Pirate Printing, sixteen make \$4.75 an hour, four make \$5.50 an hour, three make \$6.85 an hour, six make \$4.85 an hour, and thirteen make \$5.25 an hour. Find the mean, median, and mode of the hourly wages. **\$5.14; \$4.85; \$4.75**

APPLICATIONS AND PROBLEM SOLVING

61. **Food and Nutrition** According to the National Eating Trends Service, in 1984, the average person ate 71 meals in a restaurant and purchased 20 meals to go. The number of meals eaten in a restaurant decreased 0.7 meals per year, and the number of meals purchased to go increased 1.3 meals per year. (Lesson 3–5)
    a. After how many years will people eat the same number of meals in restaurants as they order to go and eat at home? **25.5 years**
    b. How would you explain these trends? **Answers will vary.**

62. **Health** You can estimate your ideal weight with the following formulas.

Male	Female
$w = 100 + 6(h - 60)$	$w = 100 + 5(h - 60)$

In each formula, w is your ideal weight in pounds and h represents your height in inches. (Lesson 3–6)

a. Find your ideal weight. **See students' work.**
b. Solve each formula for h.
c. How tall is a man that is at his ideal weight of 170 pounds?

b. $\frac{w + 260}{6}$; $\frac{w}{5} + 40$ **c.** $71\frac{2}{3}$ **in.**

63. **Home Economics** According to the *Dallas Morning News,* the average refrigerator lasts $2\frac{1}{4}$ years less than twice as long as the average color television. The average color television lasts for 8 years. How long does the average refrigerator last? (Lesson 3–3)

$13\frac{3}{4}$ years

64. **Postal Service** The graph below shows the cost of mailing a first-class letter in several countries in 1994. (Lesson 3–7)
    a. Find the mean, median, and mode cost.
    b. Round each cost to the nearest cent. Then find the mean, median, and mode cost and compare your two answers.

a. 40.4¢; 32.2¢; 32¢ b. 40.6¢; 32¢; 32¢

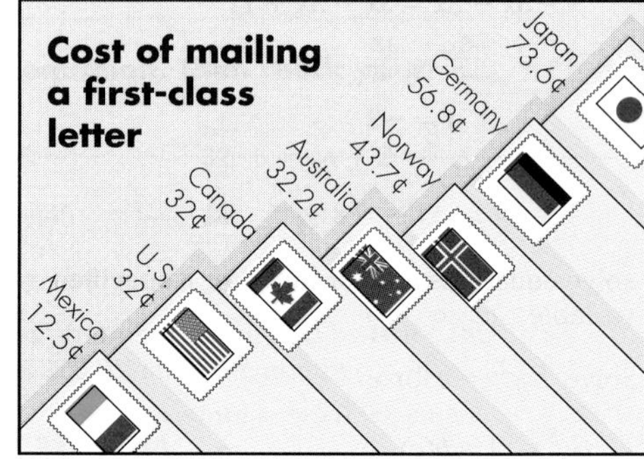

Source: *USA TODAY* research

A practice test for Chapter 3 is provided on page 789.

ALTERNATIVE ASSESSMENT

COOPERATIVE LEARNING PROJECT

Balances and Solving Equations In this chapter, you learned the addition, subtraction, multiplication, and division properties of equality. Each definition begins with an equation in *balance*, $a = b$, and states that if something is done to one side of the equation, the equation can be kept in balance by doing the same thing to the other side.

A basic principle is that two quantities in balance will remain in balance until something is done to only one of the quantities. To bring the quantities back into balance, you must do the same thing to the other quantity or undo what was done to the first quantity.

In this project, you will construct a device to demonstrate the principle of balance and how it relates to solving equations. All balances have certain features in common. A horizontal bar is balanced on a vertical support. Pans are suspended from the ends of the bar. A pointer attached to the bar moves along a scale on the support, indicating when the pans are balanced.

Follow these steps to design and build your balance.

- Outline a plan you can follow.
- Use materials you can easily find. The precision of your balance depends more on the care and accuracy with which you build it than it does on the materials you use.
- Carry out your plan.
- Determine how you might use your balance to represent equations. Include ways to solve equations by using the balance.
- Write several paragraphs describing how you can solve equations using the balance. Be sure to give examples of equations and their solutions.

THINKING CRITICALLY

- Write an equation that has no solution. Explain why there is no solution.
- Write an equation that has an infinite number of solutions. Explain why.

PORTFOLIO

Select one of the assignments from this chapter that you found especially challenging. Revise your work if necessary and place it in your portfolio. Explain why you found it to be a challenge.

SELF EVALUATION

One characteristic of a good problem solver is persistence. If you don't succeed the first time you attempt something, you should learn from your mistakes and try again.

Assess yourself. How persistent are you? How do you react when you fail at something? List two or three ways that you can make a conscious effort to be more persistent in solving problems both in your mathematics studies and in your daily life.

Alternative Assessment

The Alternative Assessment section provides students with the opportunity to assess their own work by thinking critically, working with others, keeping a portfolio, and honestly evaluating their own progress. For more information on alternative forms of assessment, see *Alternative Assessment in the Mathematics Classroom,* one of the titles in the Glencoe Mathematics Professional Series.

Performance Assessment

Performance Assessment tasks for this chapter are included in the *Assessment and Evaluation Masters.* A scoring guide is also provided.

Assessment and Evaluation Masters, pp. 70, 81

NAME ______________________ DATE ____________

Chapter 3 Performance Assessment

Instructions: ***Demonstrate your knowledge by giving a clear, concise solution to each problem. Be sure to include all relevant drawings and justify your answers. You may show your solution in more than one way or investigate beyond the requirements of the problem.***

1. Use the equation $10x + 18 = 48$ for parts a–d.
   - **a.** Add 4 to each side of the equation. Tell, in your own words, how you know that the two sides of the equation are still equal.
   - **b.** How are the solutions of the original equation and the equation in part a related? What does the term *equivalent equations* mean?
   - **c.** Use subtraction, multiplication, and division to form three equations equivalent to the original equation.
   - **d.** Solve the original equation. Explain each step in finding the solution.
2. Use the equation $\frac{1}{2}(x - 3) = \frac{1}{3}x + 1$ for parts a–c.
   - **a.** How can you eliminate the fraction from each side of the equation to make finding the solution easier?
   - **b.** Write a word problem for the equation.
   - **c.** Solve the equation. Explain each step and give the meaning of the answer.
3. **a.** Solve $ry + s = tx - m$ for y. Explain each step in your solution.
   - **b.** Would there be any limitations for the value of each variable? If so, explain the limitation.
4. The opening day scores in a local golf tournament are 78, 83, 70, 84, 89, 67, 84, 92, 78, 91, 85, 77, 68, 80, 71, 78, 99, 81, 75, 88, 90, 71, 73. Find the mean, median, and mode for the set of scores. Give a reason for finding each of these values.

Scoring Guide
Chapter 3
Performance Assessment

Level	Specific Criteria
3 Superior	• Shows thorough understanding of the concepts of *solving equations, solving formulas for a given variable,* and *finding measures of central tendency.* • Uses appropriate strategies to solve problems. • Computations are correct. • Written explanations are exemplary. • Word problem concerning equation is appropriate and makes sense. • Goes beyond requirements of all or some problems.
2 Satisfactory, with Minor Flaws	• Shows understanding of the concepts of *solving equations, solving formulas for a given variable,* and *finding measures of central tendency.* • Uses appropriate strategies to solve problems. • Computations are mostly correct. • Written explanations are effective. • Word problem concerning equation is appropriate and makes sense. • Satisfies all requirements of problems.
1 Nearly Satisfactory, with Serious Flaws	• Shows understanding of most of the concepts of *solving equations, solving formulas for a given variable,* and *finding measures of central tendency.* • May not use appropriate strategies to solve problems. • Computations are mostly correct. • Written explanations are satisfactory. • Word problem concerning equation is mostly appropriate and sensible. • Satisfies most requirements of problems.
0 Unsatisfactory	• Shows little or no understanding of the concepts of *solving equations, solving formulas for a given variable,* and *finding measures of central tendency.* • May not use appropriate strategies to solve problems. • Computations are incorrect. • Written explanations are not satisfactory. • Word problem concerning equation is not appropriate or sensible. • Does not satisfy requirements of problems.

In•ves•ti•ga•tion

TEACHER NOTES

NCTM Standards: 1–5

This Investigation is designed to be completed over several days or weeks. It may be considered optional. You may want to assign the Investigation and the follow-up activities to be completed at the same time.

Objective

Use the capture-recapture method and proportions to estimate the number of fish in a large lake.

Mathematical Overview

This Investigation will use the following mathematical skills and concepts from Chapters 4 and 5.

- Solve proportions.
- Find the probability of a simple event.
- Show relations.
- Determine and interpret the range, quartiles, and interquartile range of a set of data.

Recommended Time		
Part	Pages	Time
Investigation	190–191	1 class period
Working on the Investigation	200, 227, 269, 302	20 minutes each
Closing the Investigation	314	1 class period

Instructional Resources

Investigations and Projects Masters, pp. 5–8

A recording sheet, teacher notes, and scoring guide are provided for each investigation in the *Investigations and Projects Masters*.

1 MOTIVATION

This Investigation uses common household items to investigate fish population in a lake. Ask students if they are familiar with the capture-recapture method used to study animal populations. Discuss reasons why it is important to estimate populations of animals and ways to do this without harming the animals.

LONG-TERM PROJECT

In•ves•ti•ga•tion

MATERIALS NEEDED

- paper lunch bag
- 2 bags of dry beans, each a different color
- 5-oz. paper cup

Imagine that you are asked to determine the number of fish in a nearby pond. To count the fish one by one, you could remove the fish from the pond and stack them to one side, or mark each fish so you would not count them over and over again. Counting like this could be hazardous to a fish's health!

To determine the number of animals in a population, scientists often use the *capture-recapture* method. A number of animals are captured, carefully tagged, and returned to their native habitat. Then a second group of animals is captured and counted, and the number of tagged animals is noted. Scientists then use proportions to estimate the number in the entire population.

In this Investigation, you will work in pairs to model the process used by scientists to estimate the number of fish in a large lake. The lunch bag will represent the lake, the beans will represent fish, and the paper cup will represent a net.

Make an Investigation Folder in which you can store all of your work on this Investigation for future use.

Cooperative Learning

This Investigation offers an excellent opportunity for using cooperative learning groups. For more information on cooperative learning strategies and group management, see *Cooperative Learning in the Mathematics Classroom*, one of the titles in the Glencoe Mathematics Professional Series.

CASTING	A Total Number of Tagged Fish	B Total Number of Fish in Sample	C Number of Tagged Fish in Sample	D Estimate
1				
2				
3				
4				
5				

CAPTURE

1 Begin by copying the chart above onto a sheet of paper.

2 Write your name on the paper bag. Empty one bag of beans into the bag.

3 Use your net to remove a sample of fish. Count the number of fish you netted. Since this number will remain constant for all casts, you can record this number in each row of column A.

4 Replace all of the beans you counted with beans of a different color. Put these "tagged fish" back into the lake and gently shake your bag to mix the fish.

RECAPTURE

5 For your first casting, use your net to remove a sample of fish. Count the total number of fish in your sample and record this number in column B. Then count the number of tagged fish in this sample and record this number in column C of your chart. Return these fish to the lake and gently shake your bag to mix the fish.

6 Cast your net a second time and record your findings. Continue casting and recording until you have counted five samples.

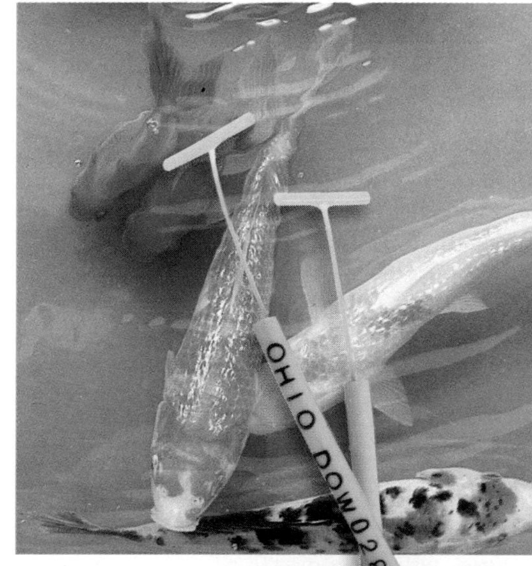

You will continue working on this Investigation throughout Chapters 4 and 5.

Be sure to keep your chart and materials in your Investigation Folder.

Go Fish! Investigation

Working on the Investigation
Lesson 4–1, p. 200

Working on the Investigation
Lesson 4–5, p. 227

Working on the Investigation
Lesson 5–2, p. 269

Working on the Investigation
Lesson 5–6, p. 302

Closing the Investigation
End of Chapter 5, p.314

2 SETUP

You may wish to have a student read the first two paragraphs of the Investigation to provide information about how we can study animal populations without threatening the animals' health. You may then wish to read the next two paragraphs, which introduce the activity. Discuss the activity with your students. Then separate the class into pairs.

3 MANAGEMENT

Each group member should be responsible for a specific task.

Recorder Collects data.
Measurer Puts beans into bag. Uses net to remove sample fish from bag.

At the end of the activity, each member should turn in his or her respective equipment.

Sample Answers

Answers will vary as they are based on the number of fish in the sample.

Investigations and Projects Masters, p. 8

4, 5 NAME______________ DATE______

Investigation, Chapters 4 and 5 — Student Edition Pages 190–191, 200, 227, 269, 302, 314

Go Fish!

Work with your group to add to the list of questions to be considered.

1. Find the population of fish in a pond given the following information:
   Number of fish tagged: 35
   Number of tagged fish from cast sample 1: 7
   Number of untagged fish from cast sample 2: 48
   Number of tagged fish from cast sample 2: 9
   Number of untagged fish from cast sample 2: 65

2. If you knew that 20% of a population of fish were tagged, could you determine how many total tagged fish there were? Explain how you could find that number or, if you couldn't, what else you would need to know.

3. Why is it important for scientists to keep track of animal population data?

Use this chart to record your calculations from page 227 of the Investigation.

Actual Population	
Estimated Population	
Percent in Lake Tagged	
Percent in Sample Tagged	
Probability of Catching a Tagged Fish	
Reasoning	

Use this diagram for the mapping on page 269.

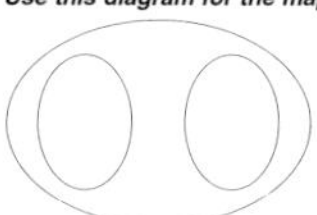

CHAPTER

4

Using Proportional Reasoning

PREVIEWING THE CHAPTER

In this chapter, students apply the process of mathematical modeling to real-world problem situations, making the connection with the equation-solving skills and concepts that they mastered in the previous chapter. Students solve proportions in the first lesson. Then, they study relationships between similar triangles and use similarity properties to find the measures of corresponding parts of similar triangles. This background leads to the study of the sine, cosine, and tangent ratios associated with an acute angle in a right triangle. Next, students solve problems involving percent, simple interest, percent of increase or decrease, discount, and sales tax. Students learn to find the probability and the odds of a simple event. The chapter concludes with lessons involving mixture problems, uniform motion problems, and direct and inverse variation problems.

Lesson (Pages)	Lesson Objectives	NCTM Standards	State/Local Objectives
4-1A (194)	Collect data to determine ratios.	1–5	
4-1 (195–200)	Solve proportions.	1–5	
4-2 (201–205)	Find the unknown measures of the sides of two similar triangles.	1–5, 7	
4-3 (206–214)	Use trigonometric ratios to solve right triangles.	1–5, 9	
4-4 (215–221)	Solve percent problems. Solve problems involving simple interest.	1–5	
4-5 (222–227)	Solve problems involving percent of increase or decrease. Solve problems involving discounts or sales tax.	1–5	
4-6 (228–232)	Find the probability of a simple event. Find the odds of a simple event.	1–5, 11	
4-7 (233–238)	Solve mixture problems. Solve problems involving uniform motion.	1–5	
4-8 (239–244)	Solve problems involving direct and inverse variation.	1–5	

A complete, 1-page lesson plan is provided for each lesson in the ***Lesson Planning Guide***. Answer keys for each lesson are available in the ***Answer Key Masters.***

ORGANIZING THE CHAPTER

You may want to refer to the **Course Planning Calendar** on page T12 for detailed information on pacing.
PACING: Standard—15 days; **Honors**—15 days; **Block**—7 days; **Two Years**—20 days

LESSON PLANNING CHART

Lesson (Pages)	Materials/ Manipulatives	Extra Practice (Student Edition)	Blackline Masters									Real-World Applications	Interactive Mathematics Tools Software	Teaching Transparencies
			Study Guide	Practice	Enrichment	Assessment and Evaluation	Modeling Mathematics	Multicultural Activity	Tech Prep Applications	Graphing Calculator	Science and Math Lab Manual			
4-1A (194)	tape measure*						p. 25							
4-1 (195–200)	calculator	p. 764	p. 26	p. 26	p. 26								4-1	4-1A 4-1B
4-2 (201–205)		p. 764	p. 27	p. 27	p. 27	p. 100						10		4-2A 4-2B
4-3 (206–214)	calculator protractor* straws tape paper clips graphing calculator string	p. 765	p. 28	p. 28	p. 28								4-3.1 4-3.2 4-3.3	4-3A 4-3B
4-4 (215–221)	calculator compass* protractor*	p. 765	p. 29	p. 29	p. 29	pp. 99, 100		p. 7	p. 7					4-4A 4-4B
4-5 (222–227)	graphing calculator	p. 765	p. 30	p. 30	p. 30		p. 75					11		4-5A 4-5B
4-6 (228–232)	graph paper	p. 766	p. 31	p. 31	p. 31	p. 101							4-6	4-6A 4-6B
4-7 (233–238)		p. 766	p. 32	p. 32	p. 32		pp. 45–47					12		4-7A 4-7B
4-8 (239–244)		p. 766	p. 33	p. 33	p. 33	p. 101		p. 8	p. 8	p. 4	pp. 13–16	13	4-8	4-8A 4-8B
Study Guide/ Assessment (245–249)						pp. 85–98, 102–104								

*Included in Glencoe's Student Manipulative Kit and Overhead Manipulative Resources.

ORGANIZING THE CHAPTER

Other Chapter Resources

Student Edition
Investigation, pp. 190–191
Chapter Opener, pp. 192–193
Mathematics and Society, p. 214
Working on the Investigation, pp. 200, 227

Teacher's Classroom Resources
Investigations and Projects Masters, pp. 37–40
Algebra and Geometry Overhead Manipulative Resources, pp. 13–16

Technology
Test and Review Software (IBM and Macintosh)
CD-ROM Interactions (Windows and Macintosh)

Professional Publications
Block Scheduling Booklet
Glencoe Mathematics Professional Series

Outside Resources

Books/Periodicals
Chazan, Daniel, *Similarity,* Sunburst
Lovell, Robert, *Probability Activities,* Dale Seymour Publications

Software
A Chance Look, Sunburst
Taking Chances, Sunburst

Videos/CD-ROMs
Mathmedia, ITSCO
Mathwise, ITSCO

Assessment Resources

Student Edition
Math Journal, pp. 224, 236
Mixed Review, pp. 200, 205, 214, 220, 227, 232, 238, 244
Self Test, p. 221
Chapter Highlights, p. 245
Chapter Study Guide and Assessment, pp. 246–248
Alternative Assessment, p. 249
Portfolio, p. 249
Cumulative Review, pp. 250–251

Teacher's Wraparound Edition
5-Minute Check, pp. 195, 201, 206, 215, 222, 228, 233, 239
Check for Understanding, pp. 198, 203, 210, 218, 224, 230, 236, 242
Closing Activity, pp. 200, 205, 214, 220, 227, 232, 238, 244
Cooperative Learning, pp. 197, 217

Assessment and Evaluation Masters
Multiple-Choice Tests, Forms 1A (Honors), 1B (Average), 1C (Basic), pp. 85–90
Free-Response Tests, Forms 2A (Honors), 2B (Average), 2C (Basic), pp. 91–96
Calculator-Based Test, p. 97
Performance Assessment, p. 98
Mid-Chapter Test, p. 99
Quizzes A–D, pp. 100–101
Standardized Test Practice, p. 102
Cumulative Review, pp. 103–104

ENHANCING THE CHAPTER

Examples of some of the materials for enhancing Chapter 4 are shown below.

Multicultural Activity Masters, pp. 7, 8

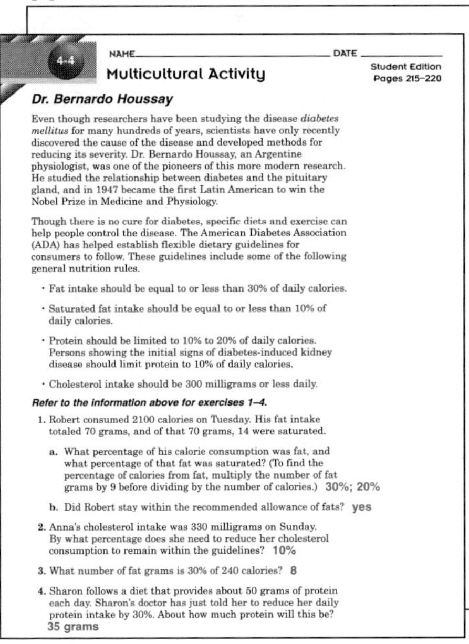

NAME ________ DATE ________

4-4 **Multicultural Activity** — Student Edition Pages 215–220

Dr. Bernardo Houssay

Even though researchers have been studying the disease *diabetes mellitus* for many hundreds of years, scientists have only recently discovered the cause of the disease and developed methods for reducing its severity. Dr. Bernardo Houssay, an Argentine physiologist, was one of the pioneers of this more modern research. He studied the relationship between diabetes and the pituitary gland, and in 1947 became the first Latin American to win the Nobel Prize in Medicine and Physiology.

Though there is no cure for diabetes, specific diets and exercise can help people control the disease. The American Diabetes Association (ADA) has helped establish flexible dietary guidelines for consumers to follow. These guidelines include some of the following general nutrition rules.

- Fat intake should be equal to or less than 30% of daily calories.
- Saturated fat intake should be equal to or less than 10% of daily calories.
- Protein should be limited to 10% to 20% of daily calories. Persons showing the initial signs of diabetes-induced kidney disease should limit protein to 10% of daily calories.
- Cholesterol intake should be 300 milligrams or less daily.

Refer to the information above for exercises 1–4.

1. Robert consumed 2100 calories on Tuesday. His fat intake totaled 70 grams, and of that 70 grams, 14 were saturated.
   a. What percentage of his calorie consumption was fat, and what percentage of that fat was saturated? (To find the percentage of calories from fat, multiply the number of fat grams by 9 before dividing by the number of calories.) 30%; 20%
   b. Did Robert stay within the recommended allowance of fats? yes
2. Anna's cholesterol intake was 330 milligrams on Sunday. By what percentage does she need to reduce her cholesterol consumption to remain within the guidelines? 10%
3. What number of fat grams is 30% of 240 calories? 8
4. Sharon follows a diet that provides about 50 grams of protein each day. Sharon's doctor has just told her to reduce her daily protein intake by 30%. About how much protein will this be? 35 grams

Real-World Applications, 10, 11, 12, 13

TECHNOLOGY

Graphing Calculator Masters, p. 4

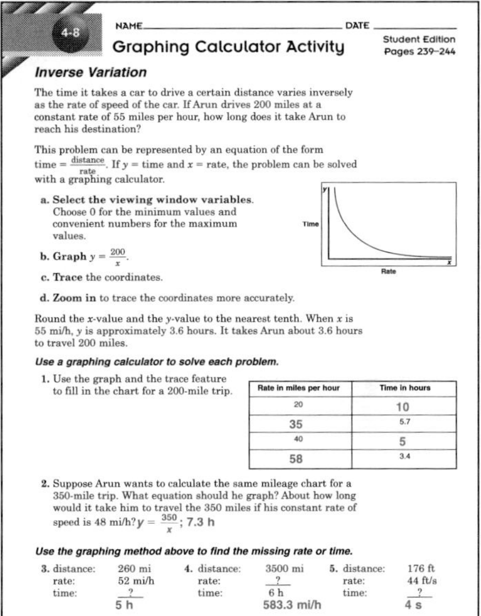

NAME ________ DATE ________

4-8 **Graphing Calculator Activity** — Student Edition Pages 239–244

Inverse Variation

The time it takes a car to drive a certain distance varies inversely as the rate of speed of the car. If Arun drives 200 miles at a constant rate of 55 miles per hour, how long does it take Arun to reach his destination?

This problem can be represented by an equation of the form time $= \frac{\text{distance}}{\text{rate}}$. If y = time and x = rate, the problem can be solved with a graphing calculator.

a. **Select the viewing window variables.** Choose 0 for the minimum values and convenient numbers for the maximum values.

b. **Graph** $y = \frac{200}{x}$.

c. **Trace** the coordinates.

d. **Zoom in** to trace the coordinates more accurately.

y Time x Rate

Round the x-value and the y-value to the nearest tenth. When x is 55 mi/h, y is approximately 3.6 hours. It takes Arun about 3.6 hours to travel 200 miles.

Use a graphing calculator to solve each problem.

1. Use the graph and the trace feature to fill in the chart for a 200-mile trip.

Rate in miles per hour	Time in hours
20	10
35	5.7
40	5
58	3.4

2. Suppose Arun wants to calculate the same mileage chart for a 350-mile trip. What equation should he graph? About how long would it take him to travel the 350 miles if his constant rate of speed is 48 mi/h? $y = \frac{350}{x}$; 7.3 h

Use the graphing method above to find the missing rate or time.

3. distance: 260 mi; rate: 52 mi/h; time: ? — 5 h
4. distance: 3500 mi; rate: ?; time: 6 h — 583.3 mi/h
5. distance: 176 ft; rate: 44 ft/s; time: ? — 4 s

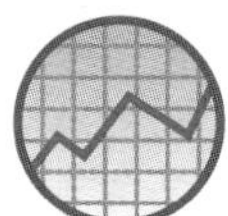

TECH PREP

Tech Prep Applications Masters, pp. 7, 8

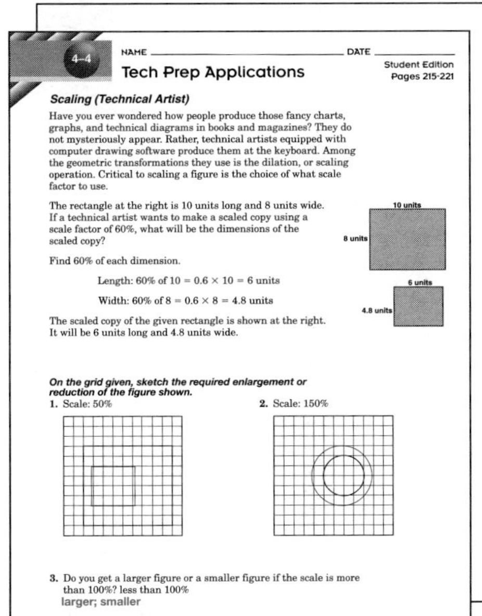

NAME ________ DATE ________

4–4 **Tech Prep Applications** — Student Edition Pages 215-221

Scaling (Technical Artist)

Have you ever wondered how people produce those fancy charts, graphs, and technical diagrams in books and magazines? They do not mysteriously appear. Rather, technical artists equipped with computer drawing software produce them at the keyboard. Among the geometric transformations they use is the dilation, or scaling operation. Critical to scaling a figure is the choice of what scale factor to use.

The rectangle at the right is 10 units long and 8 units wide. If a technical artist wants to make a scaled copy using a scale factor of 60%, what will be the dimensions of the scaled copy?

10 units 8 units 6 units 4.8 units

Find 60% of each dimension.

Length: 60% of 10 = 0.6 × 10 = 6 units

Width: 60% of 8 = 0.6 × 8 = 4.8 units

The scaled copy of the given rectangle is shown at the right. It will be 6 units long and 4.8 units wide.

On the grid given, sketch the required enlargement or reduction of the figure shown.

1. Scale: 50%
2. Scale: 150%
3. Do you get a larger figure or a smaller figure if the scale is more than 100%? less than 100% larger; smaller

Science and Math Lab Manual, pp. 13–16

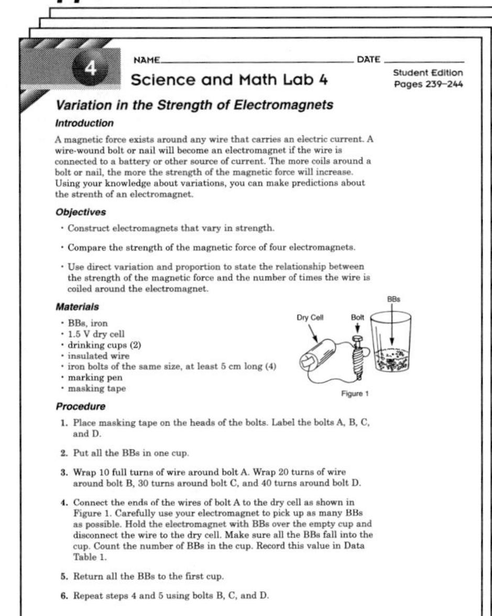

NAME ________ DATE ________

4 **Science and Math Lab 4** — Student Edition Pages 239–244

Variation in the Strength of Electromagnets

Introduction

A magnetic force exists around any wire that carries an electric current. A wire-wound bolt or nail will become an electromagnet if the wire is connected to a battery or other source of current. The more coils around a bolt or nail, the more the strength of the magnetic force will increase. Using your knowledge about variations, you can make predictions about the strenth of an electromagnet.

Objectives

- Construct electromagnets that vary in strength.
- Compare the strength of the magnetic force of four electromagnets.
- Use direct variation and proportion to state the relationship between the strength of the magnetic force and the number of times the wire is coiled around the electromagnet.

Materials

- BBs, iron
- 1.5 V dry cell
- drinking cups (2)
- insulated wire
- iron bolts of the same size, at least 5 cm long (4)
- marking pen
- masking tape

Dry Cell Bolt BBs

Figure 1

Procedure

1. Place masking tape on the heads of the bolts. Label the bolts A, B, C, and D.
2. Put all the BBs in one cup.
3. Wrap 10 full turns of wire around bolt A. Wrap 20 turns of wire around bolt B, 30 turns around bolt C, and 40 turns around bolt D.
4. Connect the ends of the wires of bolt A to the dry cell as shown in Figure 1. Carefully use your electromagnet to pick up as many BBs as possible. Hold the electromagnet with BBs over the empty cup and disconnect the wire to the dry cell. Make sure all the BBs fall into the cup. Count the number of BBs in the cup. Record this value in Data Table 1.
5. Return all the BBs to the first cup.
6. Repeat steps 4 and 5 using bolts B, C, and D.

Problem of the Week Cards, 10, 11, 12

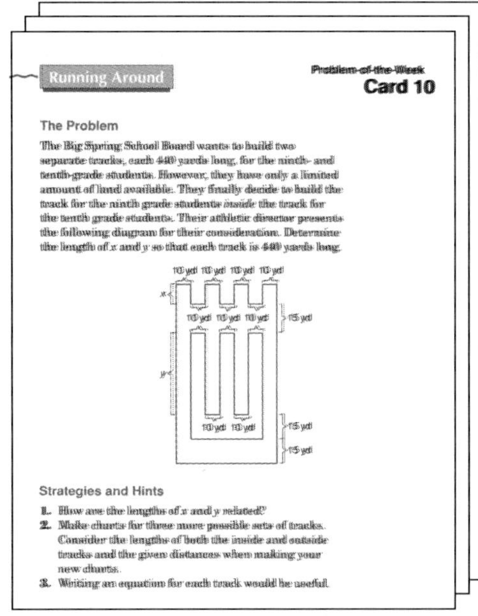

Running Around — Problem-of-the-Week **Card 10**

The Problem

The Big Spring School Board wants to build two separate tracks, each 440 yards long, for the ninth- and tenth-grade students. However, they have only a limited amount of land available. They finally decide to build the track for the ninth grade students *inside* the track for the tenth grade students. Their athletic director presents the following diagram for their consideration. Determine the length of x and y so that each track is 440 yards long.

10 yd 10 yd 10 yd 10 yd x 10 yd 10 yd 10 yd 15 yd y 10 yd 10 yd 15 yd 15 yd

Strategies and Hints

1. How are the lengths of x and y related?
2. Make charts for three more possible sets of tracks. Consider the lengths of both the inside and outside tracks and the given distances when making your new charts.
3. Writing an equation for each track would be useful.

MAKING MATHEMATICS RELEVANT

This two-page introduction to the chapter provides students with an opportunity to explore contemporary topics and their applications to mathematics.

Background Information

Incoming Immigrants Immigrants continue to come to America from diverse backgrounds, from every continent. Contrary to many people's opinions, immigrants have about the same level of education, employment, and income as natural citizens.

CHAPTER 4

Using Proportional Reasoning

Objectives

In this chapter, you will:

- solve proportions,
- find the unknown measures of the sides of two similar triangles,
- use trigonometric ratios to solve right triangles,
- solve percent problems,
- find the probability and odds of a simple event, and
- solve problems involving direct and inverse variation.

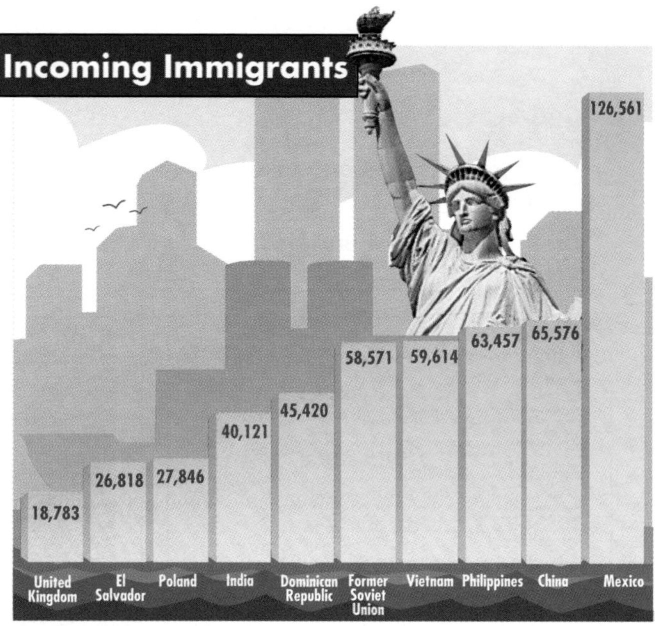

Source: U.S. Immigration and Naturalization Service

Some people blame immigrants, both legal and illegal, for many of the economic and social problems in the United States. Others believe immigrants bring vitality, diversity, and economic strength to the United States. What effect does immigration have on American society? Do you have any personal experiences with new immigrants?

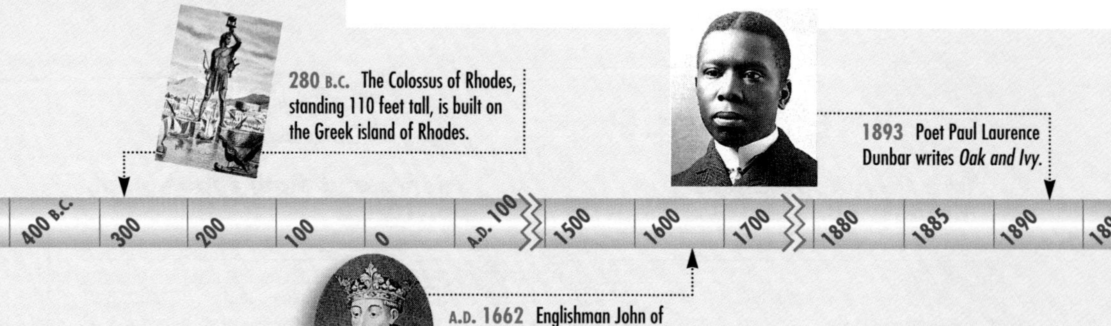

280 B.C. The Colossus of Rhodes, standing 110 feet tall, is built on the Greek island of Rhodes.

1893 Poet Paul Laurence Dunbar writes *Oak and Ivy*.

A.D. 1662 Englishman John of Gaunt publishes the first book of statistics.

Have students pick a country and research one famous person from that country. How did that person contribute to our society in the United States?

interNET CONNECTION

Explore the Immigration Issues Menu for articles on the cost of immigration, the quota system, and legal and illegal immigration.

World Wide Web
http://www.fairus/org/issues/04001604.htm

PEOPLE IN THE NEWS

Ayrris Layug Aunario won the grand prize in *Filipinas* magazine's 1994 National Essay Contest with an essay describing what it means to be Filipino-American. The 16-year-old from Waukegan, Illinois, is a student at the Illinois Mathematics and Science Academy, a public school for gifted students in science and mathematics.

Ayrris' essay explains what it was like to be a 10-year-old immigrant to the U.S., arriving in Chicago and learning to live with people of many different cultures and beliefs. He had to master English quickly and soon learned the American virtues of independence, self-reliance, and assertiveness. Ayrris plans to attend college in the field of physics.

Chapter Project

Work in cooperative groups to take a heritage survey of students in your school.

- Determine the geographic birthplace of each student and the birthplace of their parents or guardians. If possible, gather information about the birthplace of each student's grandparents and great-grandparents.
- Use a world map to pinpoint each geographic location. Connect the generations of each student using colored yarn.
- Draw some conclusions about the birthplaces of the students you surveyed. Share your findings with the class.

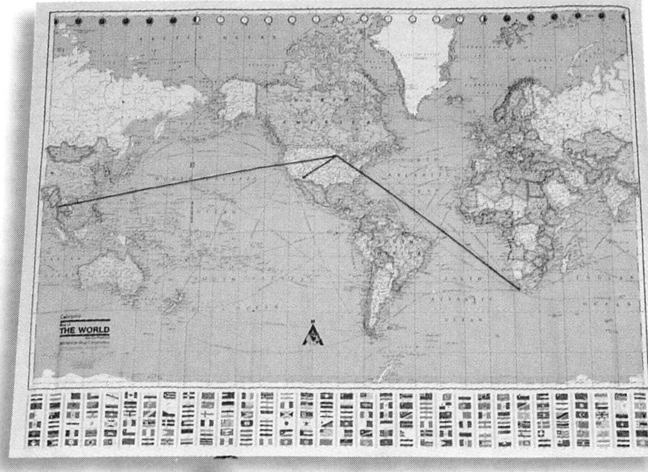

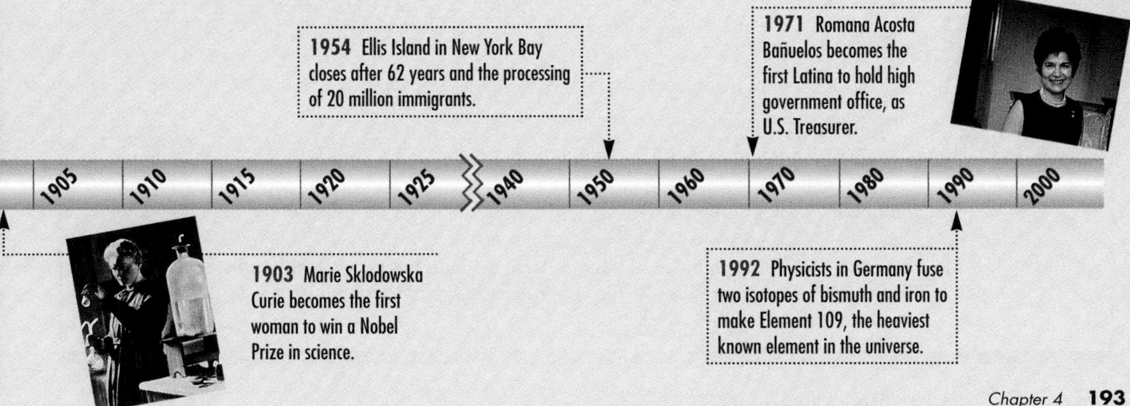

1954 Ellis Island in New York Bay closes after 62 years and the processing of 20 million immigrants.
1971 Romana Acosta Bañuelos becomes the first Latina to hold high government office, as U.S. Treasurer.
1905
1910
1915
1920
1925
1940
1950
1960
1970
1980
1990
2000
1903 Marie Sklodowska Curie becomes the first woman to win a Nobel Prize in science.
1992 Physicists in Germany fuse two isotopes of bismuth and iron to make Element 109, the heaviest known element in the universe.

Alternative Chapter Projects

Two other chapter projects are included in the *Investigations and Projects Masters*. In Chapter 4 Project A, pp. 37–38, students extend the topic in the chapter opener. In Chapter 4 Project B, pp. 39–40, students investigate sales and discounts.

PEOPLE IN THE NEWS

Once, Ayrris was competing in a spelling bee. He was asked to spell the word *illusion.* He spelled it correctly, but the pronouncer insisted it was spelled "illusian." He respectfully sat down, but checked a dictionary and showed it to the judges. He rejoined the competition and won first place.

Chapter Project

Cooperative Learning Students might also want to do library research on the immigration profile for the general population of the United States. Compare the findings for the school with the general population of the U.S.

Investigations and Projects Masters, p. 37

NAME______________ DATE ______

Chapter 4 Project A

Student Editi
Pages 194–2

A World of Names

1. For this project, you will work in a small group to research names and naming customs of different ethnic groups. In some groups, for example, the family name comes before the other names. Sometimes, a person's full family name is made up of both parents' names. Use reference books to find out more about such customs. You can also ask friends from different ethnic backgrounds about customs connected with their names. Find out what some common family names mean.
2. Do research to find out the most common family names of students at your school. You may be able to obtain lists of students' last names from classroom teachers or from the principal's office.
3. Make a list of the five most common names at your school. Determine what percent of the students at your school has each of these names.
4. Do research to find out the most common family names in your community. Use one or more telephone directories and the list of the most common family names at your school. Make a list of the five most common family names in your community and determine what percent of the population has each name.
5. If possible, interview an expert in the field of demographics, the study of human populations. Ask how the results of your research would have been different 25, 50, and 100 years ago. Also, ask the interviewee what a list of common last names might look like 100 years from now.
6. Write an article for your school newspaper about names and naming customs and the results of your investigation.

4-1A LESSON NOTES

NCTM Standards: 1–5

Objective
Collect data to determine ratios of the length of parts of the human body to the length of the head.

Recommended Time
Demonstration and discussion: 15 minutes; Exercises: 30 minutes

Instructional Resources
For each student or group of students
Student Manipulative Kit
• tape measure
Modeling Mathematics Masters
• p. 25 (worksheet)
For teacher demonstration
Algebra and Geometry Overhead Manipulative Resources

1 FOCUS

Motivating the Lesson
This modeling activity presents the idea that there are certain ratios that exist in nature. Such constants are fundamental to science. Students will also see that, in the case of the human body, nothing is exact.

2 TEACH

Teaching Tip Make it clear to students that the unit of measure is not the inch, but their heads.

3 PRACTICE/APPLY

Assignment Guide

Core: 1–2
Enriched: 1–2

A Preview of Lesson 4–1

4–1A Ratios

Materials: tape measure

You can collect data to determine whether there is a relationship between the length of an individual's head and an individual's body.

Activity 1 Use a tape measure to find the length of your head.

Step 1 Measure your head from the top of your skull to the bottom of your chin. Since the top of your head is round, hold something flat, such as a piece of cardboard, across the top of your head to get an accurate measurement.

Step 2 Make a measuring "stick" with the length of your head as one unit to find the following measurements.
- total height
- chin to waist
- waist to hip
- knee to ankle
- ankle to bottom of bare heel
- underarm to elbow
- elbow to wrist
- wrist to tip of finger
- shoulder to tip of finger

Step 3 Make a table to record the relationship, or *ratio,* of each measurement to the measurement of your head.

Body Proportions of an Average Adult

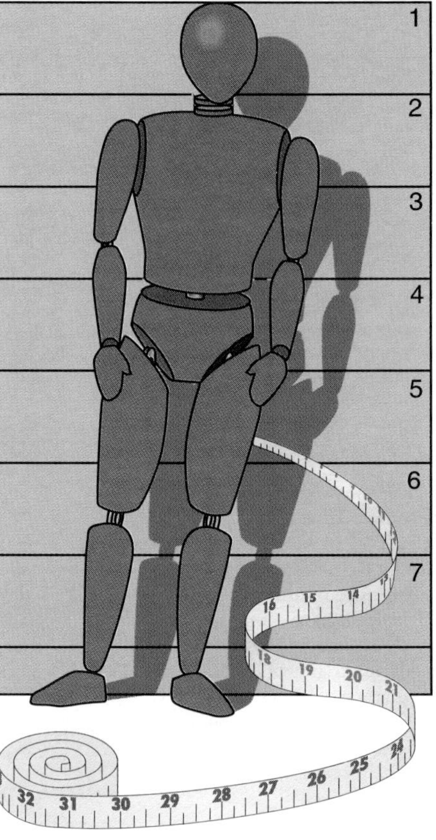

Draw 1. The average adult is seven and one-half heads tall. Use the information in your table to draw an outline of the proportions of your body. How does your drawing compare to the one above? **See students' work.**

Write 2. Compare your ratios to those of your classmates. Find the number of heads tall of a random student in your class. Write a paragraph explaining how you compare to this average. **See students' work.**

4 ASSESS

Observing students working in cooperative groups is an excellent method of assessment.

Cooperative Learning

This lesson offers an excellent opportunity for using cooperative learning groups. For more information on cooperative learning strategies and group management, see *Cooperative Learning in the Mathematics Classroom*, one of the titles in the Glencoe Mathematics Professional Series.

4–1 Ratios and Proportions

What YOU'LL LEARN
- To solve proportions.

Why IT'S IMPORTANT

You can use proportions to solve problems involving food and entertainment.

APPLICATION
Food

How would you like to eat chocolate for a living? That's exactly what Carl Wong does every day as associate director of product development for Hershey Foods. His job is to create new candy bars as well as taste test existing candy bars in order to improve them.

Hershey Foods keeps their candy recipes confidential, but the basic ingredients found in a batch of Hershey's chocolate bars are sugar, cocoa beans, milk, and flavorings. The ingredients for a batch of candy are shown in the table below.

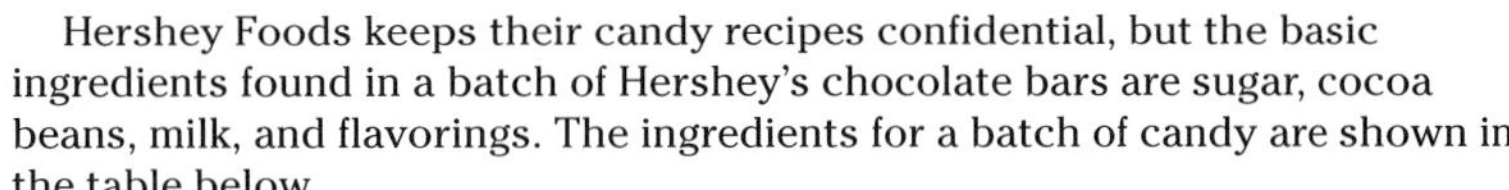

Ingredient	Parts Per Batch
Sugar	10
Cocoa Beans	5
Milk	4
Flavorings	1

A **ratio** is a comparison of two numbers by division. The ratio of x to y can be expressed in the following ways.

x to y $\qquad x{:}y \qquad \frac{x}{y}$

Ratios are often expressed as fractions in simplest form. A ratio that is equivalent to a whole number is written with a denominator of 1.

In the application at the beginning of the lesson, the table shows that for every 10 parts sugar in a batch of chocolate bars, there are 5 parts cocoa beans. The ratio of sugar to cocoa beans is $\frac{10}{5}$. Suppose Hershey uses 30 pounds of sugar and 15 pounds of cocoa beans in a batch of chocolate bars. This ratio is $\frac{30}{15}$. Is this ratio different than the first ratio of $\frac{10}{5}$? When simplified, both ratios are equivalent to $\frac{2}{1}$.

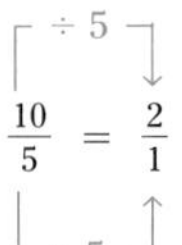

$$\frac{10}{5} = \frac{2}{1} \quad (\div 5) \qquad \frac{30}{15} = \frac{2}{1} \quad (\div 15)$$

FYI

Almost 500 new candy and chocolate products were introduced in the first four months of 1995.

An equation stating that two ratios are equal is called a **proportion**. So, $\frac{10}{5} = \frac{30}{15}$ is a proportion.

4-1 LESSON NOTES

NCTM Standards: 1–5

Instructional Resources

- Study Guide Master 4-1
- Practice Master 4-1
- Enrichment Master 4-1

Transparency 4-1A contains the 5-Minute Check for this lesson; **Transparency 4-1B** contains a teaching aid for this lesson.

Recommended Pacing	
Standard Pacing	Day 2 of 15
Honors Pacing	Day 2 of 15
Block Scheduling*	Day 1 of 7
Alg. 1 in Two Years*	Days 2 & 3 of 20

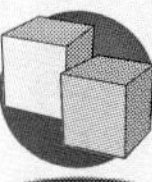

*For more information on pacing and possible lesson plans, refer to the *Block Scheduling Booklet* and *Algebra 1 in Two Years.*

1 FOCUS

5-Minute Check
(over Chapter 3)

Solve.

1. $\frac{1}{2}x + 4 = \frac{1}{3}x$ **−24**
2. $\frac{5-y}{6} = \frac{1}{6}y$ **$2\frac{1}{2}$**
3. $\frac{x+3}{5} = \frac{2}{5}x$ **3**

Solve for x.

4. $x + p = 2b$ **$2b - p$**
5. $ax - b = c$ **$\frac{c+b}{a}$**

Motivating the Lesson

Questioning Use equivalent fractions to introduce proportions through the following activity. Write $\frac{3}{4}$ on the chalkboard. Have students identify a number of equivalent fractions. Continue the activity by having one student name a fraction, and then have other students call out an equivalent fraction. A new fraction can be named when a student cannot think of an equivalent fraction. Continue in this way until all students have had a turn naming a fraction.

FYI

Chocolate is made from cocoa beans. It also contains small amounts of two stimulants—theobromine and caffeine.

GLENCOE Technology

CD-ROM Interaction

A multimedia simulation links the use of ratios, graphs, and estimation to bicycling. A blackline master activity with teacher's notes provides a follow-up to the CD-ROM simulation.

For Windows & Macintosh

2 TEACH

In-Class Examples

For Example 1
Use cross products to determine whether each pair of ratios forms a proportion.

a. $\frac{4}{6}, \frac{12}{16}$ **64 ≠ 72; This is not a proportion.**

b. $\frac{3}{5}, \frac{6}{10}$ **30 = 30; This is a proportion.**

For Example 2
Sixteen fish are captured from Spring Lake, tagged, and returned to the lake. Later, 13 fish are captured. Of these, 4 have tags. Estimate the number of fish in the lake. **52 fish**

Teaching Tip For Example 2, you may wish to explain that this type of problem can be solved using the following method.

$$\frac{5}{4} = \frac{75}{m}$$
$$m\left(\frac{5}{4}\right) = m\left(\frac{75}{m}\right)$$
$$\frac{5}{4}m = 75$$
$$\frac{4}{5}\left(\frac{5m}{4}\right) = \frac{4}{5}(75)$$
$$m = 60$$

Teaching Tip The proportion $\frac{a}{b} = \frac{c}{d}$ is generally read "a is to b as c is to d."

Teaching Tip A proportion may also be written as $a{:}b = c{:}d$. In this form, it is easy to see that a and d are the extremes, since they are at the extreme ends. The means are the middle terms.

Cocoa production began in Central and South America, but now is worldwide within 20° of the Equator. Cocoa is mainly grown on small peasant farms, two to five acres in size.

To solve a proportion, cross multiply.

One way to determine if two ratios form a proportion is to check their cross products. In the proportion at the right, the cross products are 10 · 15 and 5 · 30. In this proportion, 10 and 15 are called the **extremes,** and 5 and 30 are called the **means.**

$$\frac{10}{5} = \frac{30}{15}$$
$$10(15) = 5(30)$$
$$\text{extremes} = \text{means}$$
$$150 = 150$$

The cross products of a proportion are equal.

Means-Extremes Property of Proportions	In a proportion, the product of the extremes is equal to the product of the means. If $\frac{a}{b} = \frac{c}{d}$, then $ad = bc$.

Example  **Use cross products to determine whether each pair of ratios forms a proportion.**

Chocolate is made from the seeds, or cacao beans, of a tree called the *cacao*. The word cacao comes from two Maya Indian words meaning bitter juice. Because of a mistake in spelling by English importers, these beans became known as cocoa beans.

a. $\frac{2}{3}, \frac{12}{18}$

$2 \cdot 18 \stackrel{?}{=} 3 \cdot 12$

$36 = 36$

So, $\frac{2}{3} = \frac{12}{18}$.

This is a proportion.

b. $\frac{2.5}{6}, \frac{3.4}{5.2}$

Enter: 2.5 [×] 5.2 [=] *13*

Enter: 6 [×] 3.4 [=] *20.4*

Since $13 \neq 20.4$, $\frac{2.5}{6} \neq \frac{3.4}{5.2}$.

This is not a proportion.

You can write proportions that involve a variable and then use cross products to solve the proportion.

Example **2** **Refer to the application at the beginning of the lesson. Suppose Hershey makes a batch of chocolate with 75 pounds of cocoa beans. How many gallons of milk will they use?**

Food

You know that the ratio of cocoa beans to milk is 5:4. Let m represent the gallons of milk.

$$\frac{5 \text{ parts cocoa beans}}{4 \text{ parts milk}} = \frac{75 \text{ pounds of cocoa beans}}{m \text{ gallons of milk}}$$

$$\frac{5}{4} = \frac{75}{m}$$

$5m = 4(75)$ *Find the cross products.*

$5m = 300$

$m = 60$

In a batch of chocolate with 75 pounds of cocoa beans, Hershey needs to use 60 gallons of milk.

Alternative Learning Styles

Kinesthetic Have students bring in several cases of gears. Ask students to study the moving gears and then to determine the relationship between each gear pair. Challenge students to write a ratio to represent each relationship.

GLENCOE Technology

Interactive Mathematics Tools Software

This multimedia software provides an interactive lesson that uses counters to determine ratios. A **Computer Journal** gives students an opportunity to write about what they have learned.

For Windows & Macintosh

A ratio called a **scale** is used when making a model to represent something that is too large or too small to be conveniently drawn at actual size. The scale compares the size of the model to the actual size of the object being modeled.

Example 3

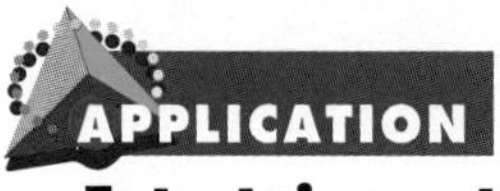

APPLICATION Entertainment

In the movie *Jurassic Park*, the dinosaurs were scale models and so was the sport utility vehicle that the T-Rex overturned. The vehicle was made to the scale of 1 inch to 8 inches. The actual vehicle was about 14 feet long. What was the length of the model sport utility vehicle?

First, change 14 feet to 168 inches. Then let ℓ represent the length of the model vehicle.

scale → actual → $\frac{1}{8} = \frac{\ell}{168}$

$168 = 8\ell$ *Find the cross products.*

$21 = \ell$

The model sport utility vehicle was 21 inches long.

You can solve proportions by using a calculator.

Example 4 **Solve each proportion.**

a. $\frac{5}{4.25} = \frac{11.32}{m}$

Method 1
Use paper and a pencil.

$\frac{5}{4.25} = \frac{11.32}{m}$

$5m = 4.25(11.32)$

$5m = 48.11$

$m = 9.622$

Method 2
Use a calculator.

By the means-extremes property, $5m = 4.25(11.32)$. Multiply the means. Then divide by 5.

Enter: 4.25 [×] 11.32 [÷] 5 [=] 9.622

Rounded to the nearest hundredth, the solution is 9.62.

b. $\frac{x}{3} = \frac{x + 5}{15}$

$\frac{x}{3} = \frac{x + 5}{15}$

$15x = 3(x + 5)$ *Means-extremes property*

$15x = 3x + 15$ *Distributive property*

$12x = 15$

$x = \frac{5}{4}$ *Check this result.*

The solution is $\frac{5}{4}$.

The ratio of two measurements having different units of measure is called a **rate.** For example, 30 miles per gallon is a rate. Proportions are often used to solve problems involving rates.

In-Class Examples

For Example 3
A model airplane is made to the scale 1 inch to 24 inches. If the wing of the actual plane is 18 feet long, how long will the model wing be? **9 inches**

For Example 4
Solve each proportion.

a. $\frac{7.8}{a} = \frac{4}{3.82}$ **7.449**

b. $\frac{t}{9} = \frac{13.4}{2.71}$ **44.502**

c. $\frac{3}{7} = \frac{x + 2}{x}$ $-\frac{7}{2}$

d. $\frac{3}{5} = \frac{x}{x + 6}$ **9**

Cooperative Learning

Group Discussion Separate students into groups of three to solve this problem.
In the proportion $\frac{e}{f} = \frac{g}{h}$, e is less than g, and e, f, g, and h are nonzero. What can be said about the relationship between f and h? $h > f$

For more information on the group discussion strategy, see *Cooperative Learning in the Mathematics Classroom,* one of the titles in the Glencoe Mathematics Professional Series, page 31.

In-Class Example

For Example 5
Ted travels 110 miles in 2 hours. At that rate, how long will it take Ted to travel 385 miles?
7 hours

3 PRACTICE/APPLY

Check for Understanding

Exercises 1–13 are designed to help you assess your students' understanding through reading, writing, speaking, and modeling. You should work through Exercises 1–4 with your students and then monitor their work on Exercises 5–13.

Additional Answers

1. **If the cross products of two ratios in a proportion are equal, then the two ratios are equivalent.**
2. **Sample answer: In the proportion $\frac{x}{2.5} = \frac{3}{4}$, multiply 2.5 and 3, then divide by 4.**

Study Guide Masters, p. 26

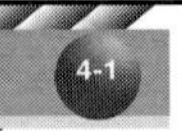

4-1 NAME__________ DATE ______

Study Guide

Student Edition
Pages 195–200

Ratios and Proportions

In mathematics, a **ratio** compares two numbers by division. A ratio that compares a number a to a number b can be written in the following ways.

a to b $\qquad a{:}b \qquad \frac{a}{b}$

When a ratio compares two quantities with different units of measure, that ratio is called a **rate.** For example, a 5°C rise in temperature per hour is a rate and can be expressed as $\frac{5 \text{ degrees}}{1 \text{ hour}}$, or 5 degrees per hour.

Proportions are often used to solve problems involving ratios. You can use the means-extremes property of proportions to solve equations that have the form of a proportion.

Definition of Proportion	An equation of the form $\frac{a}{b} = \frac{c}{d}$ stating that two ratios are equal is called a proportion.
Means-Extremes Property of Proportions	In a proportion, the product of the extremes is equal to the product of the means. If $\frac{a}{b} = \frac{c}{d}$, then $ad = bc$.

Example: Solve $\frac{x}{5} = \frac{10}{13}$.

$$\frac{x}{5} = \frac{10}{13}$$
$$13x = 50$$
$$x = 3\frac{11}{13}$$

The solution is $3\frac{11}{13}$.

Solve each proportion.

1. $\frac{0.1}{2} = \frac{0.5}{x}$ 10
2. $\frac{x+1}{4} = \frac{3}{4}$ 2
3. $\frac{4}{6} = \frac{8}{x}$ 12
4. $\frac{x}{21} = \frac{3}{63}$ 1
5. $\frac{9}{y+1} = \frac{18}{54}$ 26
6. $\frac{3-x}{4+x} = \frac{8}{48}$ 2
7. $\frac{4x}{25} = \frac{85-x}{100}$ 5
8. $\frac{x+8}{-3} = \frac{17-x}{-2}$ 7

Use a proportion to solve each problem.

9. To make a model of the Guadelupe River bed, Hermie used 1 inch of clay for 5 miles of the actual river's length. His model river was 50 inches long. How long is the Guadelupe River? 250 miles
10. Josh finished 24 math problems in one hour. At that rate, how many hours will it take him to complete 72 problems? 3 hours

Example 5

State Fair

In the first 30 minutes of the opening day of the Texas State Fair, 1252 people entered the gates. If this attendance rate continued, how many people visited the fair during the operating hours of 8:00 A.M. and 12:00 midnight the first day?

Explore Let p represent the number of people attending the fair on opening day.

Plan Write a proportion for the problem.

$\frac{1252}{0.5} = \frac{p}{16}$ *Notice that both ratios compare the number of people per hour.*

Solve

$$\frac{1252}{0.5} = \frac{p}{16}$$
$$1252(16) = 0.5p$$
$$40{,}064 = p$$

If the attendance rate continued, 40,064 people visited the fair on opening day.

Examine Use estimation to check your answer. About 1250 people entered the fair every 30 minutes. This means that about 2500 people entered every hour. The fair was open for 16 hours each day. Therefore, about 16 × 2.5 thousand or 40,000 people entered, and the answer is reasonable.

CHECK FOR UNDERSTANDING

Communicating Mathematics

Study the lesson. Then complete the following. **1–2. See margin.**

1. **Explain** how to determine whether two ratios are equivalent.
2. **Explain** how to use a calculator to solve a proportion.
3. **Find** three examples of ratios in a newspaper or magazine. **3–4. See students' work.**

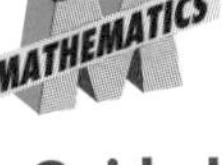

4. **Draw** an outline of an infant's body if the average infant is three heads long.

Guided Practice

Use cross products to determine whether each pair of ratios forms a proportion.

5. $\frac{3}{2}, \frac{21}{14}$ $=$
6. $\frac{2.3}{3.4}, \frac{0.3}{3.6}$ $\neq$

Solve each proportion.

7. $\frac{2}{3} = \frac{8}{x}$ **12**
8. $\frac{4}{w} = \frac{2}{10}$ **20**
9. $\frac{3}{15} = \frac{1}{y}$ **5**
10. $\frac{5.22}{13.92} = \frac{b}{48}$ **18**
11. $\frac{1.1}{0.6} = \frac{8.47}{n}$ **4.62**
12. $\frac{x}{1.5} = \frac{2.4}{1.6}$ **2.25**
13. **Travel** A 96-mile trip requires 6 gallons of gasoline. At that rate, how many gallons would be required for a 152-mile trip? **9.5 gallons**

Reteaching

Using Alternative Methods Solve these proportions by multiplying each side by the LCD to clear fractions. First tell by what number each side can be multiplied to clear fractions. Then solve.

1. $\frac{x}{5} = \frac{12}{15}$ **15; $x = 4$**
2. $\frac{6}{9} = \frac{10}{x}$ **9x; $x = 15$**

EXERCISES

Practice

Use cross products to determine whether each pair of ratios forms a proportion.

14. $\frac{6}{8}, \frac{22}{28}$ $\neq$ 15. $\frac{4}{5}, \frac{16}{20}$ $=$ 16. $\frac{4}{11}, \frac{12}{33}$ $=$

17. $\frac{8}{9}, \frac{16}{17}$ $\neq$ 18. $\frac{2.1}{3.6}, \frac{5}{7}$ $\neq$ 19. $\frac{0.4}{0.8}, \frac{0.7}{1.4}$ $=$

Solve each proportion.

22. $-\frac{63}{16}$; -3.938

27. $-\frac{149}{6}$; -24.8

28. $\frac{3}{5}$; 0.6

34. 63.368

20. $\frac{3}{4} = \frac{x}{8}$ 6
21. $\frac{a}{45} = \frac{3}{15}$ 9
22. $\frac{y}{9} = \frac{-7}{16}$
23. $\frac{3}{5} = \frac{x+2}{6}$ $\frac{8}{5}$; 1.6
24. $\frac{w+2}{5} = \frac{7}{5}$ 5
25. $\frac{x}{8} = \frac{0.21}{2}$ 0.84

C

26. $\frac{5+y}{y-3} = \frac{14}{10}$ 23
27. $\frac{m+9}{5} = \frac{m-10}{11}$
28. $\frac{r+7}{-4} = \frac{r-12}{6}$
29. $\frac{85.8}{t} = \frac{70.2}{9}$ 11
30. $\frac{z}{33} = \frac{11.75}{35.25}$ 11
31. $\frac{0.19}{2} = \frac{0.5x}{12}$ 2.28
32. $\frac{2.405}{3.67} = \frac{g}{1.88}$ 1.232
33. $\frac{x}{4.085} = \frac{5}{16.33}$ 1.251
34. $\frac{3t}{9.65} = \frac{21}{1.066}$

Critical Thinking

35. Mariah is exactly eight years older than her cousin Louis.
    a. Copy and complete the table below.

Louis' age	1	2	3	6	10	20	30
Mariah's age	9	10	11	14	18	28	38

35c. $r = \frac{y+8}{y}$

35e. No; if the ratio equaled 1, Mariah and Louis would be the same age.

    b. Find the ratio in decimal form of Mariah's age to Louis' age for each pair of ages listed in the table. 9, 5, $3.\overline{6}$, $2.\overline{3}$, 1.8, 1.4, $1.\overline{26}$
    c. Write a formula for the ratio of Mariah's age to Louis' age when Louis is y years old.
    d. As Mariah and Louis grow older, explain what happens to the ratio of their ages. The ratio gets smaller.
    e. Explain whether the ratio of their ages will ever equal 1.

Applications and Problem Solving

36. **Biology** A flea, usually less than an eighth of an inch long, can long jump about thirteen inches and high jump about eight inches. If humans could jump in proportion to the flea, it would take only nine leaps to go one mile! At this rate, how many leaps would it take a human to travel 693 miles from Louisville, Kentucky to Norfolk, Virginia? 6237 leaps

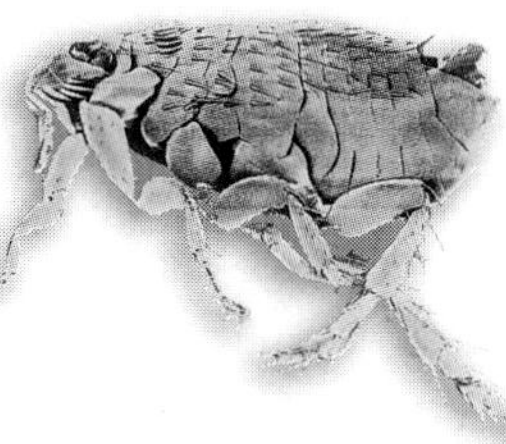

38. 50 pounds

37. **Movies** When rating movies in 1994, critic Gene Siskel gave four thumbs up to every five thumbs up given by his partner Roger Ebert. If Mr. Siskel gave thumbs up to 68 movies, how many movies did Mr. Ebert rate favorably? 85 movies

38. **Recycling** When a pair of blue jeans is made, the leftover denim scraps can be recycled to make stationery, pencils, and more denim. One pound of denim is left after making every five pairs of jeans. How many pounds of denim would be left from 250 pairs of jeans?

Assignment Guide

Core: 15–37 odd, 39–46
Enriched: 14–34 even, 35–46

For **Extra Practice,** see p. 764.

The red A, B, and C flags, printed only in the Teacher's Wraparound Edition, indicate the level of difficulty of the exercises.

Practice Masters, p. 26

4-1 NAME________ DATE________

Practice

Student Edition Pages 195–200

Ratios and Proportions

Solve each proportion.

1. $\frac{x}{4} = \frac{3}{2}$ 6
2. $\frac{3}{a} = \frac{1}{6}$ 18
3. $\frac{-3}{1} = \frac{12}{y}$ −4
4. $\frac{2}{7} = \frac{x}{-42}$ −12
5. $\frac{x}{-3} = \frac{3}{-2}$ $\frac{9}{2}$
6. $\frac{4}{5} = \frac{3}{w}$ $\frac{15}{4}$
7. $\frac{-5}{c} = \frac{1}{3}$ −15
8. $\frac{k}{8} = \frac{-2}{5}$ $-\frac{16}{5}$
9. $\frac{6}{z} = \frac{-3}{5}$ −10
10. $\frac{5}{12} = \frac{x+1}{4}$ $\frac{2}{3}$
11. $\frac{r+2}{7} = \frac{5}{7}$ 3
12. $\frac{-6}{13} = \frac{8}{x+1}$ $-\frac{55}{3}$
13. $\frac{x-1}{-4} = \frac{2}{3}$ $-\frac{5}{3}$
14. $\frac{3}{7} = \frac{x-2}{6}$ $\frac{32}{7}$
15. $\frac{7}{5} = \frac{1}{x+3}$ $-\frac{16}{7}$
16. $\frac{x-1}{3} = \frac{-1}{5}$ $\frac{2}{5}$
17. $\frac{1}{3} = \frac{2}{x+2}$ 4
18. $\frac{x+2}{-3} = \frac{9}{-7}$ $\frac{13}{7}$
19. $\frac{x+2}{3} = \frac{x-1}{6}$ −5
20. $\frac{z-1}{7} = \frac{z-2}{3}$ $\frac{11}{4}$
21. $\frac{n+8}{-5} = \frac{n-3}{-1}$ $\frac{23}{4}$
22. $\frac{x+5}{3} = \frac{x-2}{-4}$ −2
23. $\frac{x+1}{x-1} = \frac{2}{3}$ −5
24. $\frac{5-t}{3+t} = \frac{13}{12}$ $\frac{21}{25}$

4 ASSESS

Closing Activity

Speaking Discuss with students the similarities and differences among proportions, fractions, and ratios.

Enrichment Masters, p. 26

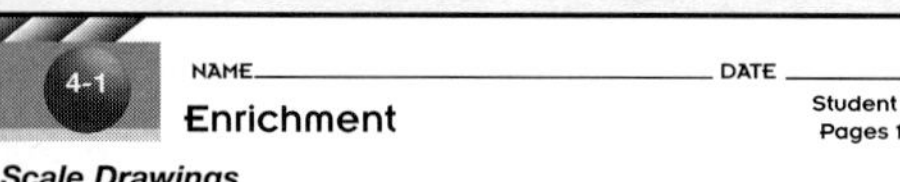

NAME ______ DATE ______

4-1 Enrichment

Student Edition Pages 195–200

Scale Drawings

The map at the left below shows building lots for sale. The scale ratio is 1:2400. At the right below is the floor plan for a two-bedroom apartment. The length of the living room is 6 m. On the plan the living room is 6 cm long.

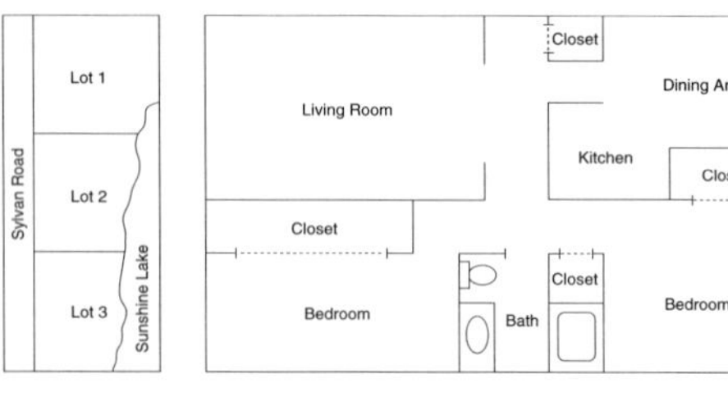

Answer each question.

1. On the map, how many feet are represented by an inch? 200 ft
2. On the map, measure the frontage of Lot 2 on Sylvan Road in inches. What is the actual frontage in feet? 200 ft
3. What is the scale ratio represented on the floor plan? 1:100
4. On the floor plan, measure the width of the living room in centimeters. What is the actual width in meters? 4 m
5. About how many square meters of carpeting would be needed to carpet the living room? 24 m²
6. Make a scale drawing of your classroom using an appropriate scale. Answers will vary.
7. On the scale for a map of Lancaster, Pennsylvania, 2.5 cm equals 3 km. Find the scale ratio. 1:120,000

Mixed Review

39. **Statistics** Find the mean, median, and mode for the following set of data. (Lesson 3–7) **23; 19; 18**

    19, 21, 18, 22, 46, 18, 17

40. Solve $a = \frac{v}{t}$ for t. (Lesson 3–6) $t = \frac{v}{a}$

41. Define a variable, write an equation, and then solve the following problem. (Lesson 3–2) $4x - 2x = 100$; **50**

    Four times a number decreased by twice the number is 100. What is the number?

42. Solve $-15 + d = 13$. (Lesson 3–1) **28**

43. **Sales** Tanya is a sales representative for Incredible Universe. A CD and tape player manufacturer is having a sales competition. Anyone who sells more than 40 CD players in one day wins a trip to Hawaii. Tanya sold 93 CD and tape players in one day for a total of \$9695. The CD players sold for \$135 and the tape players for \$80. (Lesson 2–9) **43c. more than 40**
    a. How many CD and tape players were sold? **93**
    b. What was the total amount of Tanya's CD and tape player sales? **\$9695**
    c. How many CD players does Tanya need to sell for her trip award?
    d. If p represents the number of CD players sold, what was the number of tape players sold? $93 - p$
    e. Will Tanya win the trip to Hawaii? **yes**

44. **Golf** In four rounds of a recent golf tournament, Heather shot 3 under par, 2 over par, 4 under par, and 1 under par. What was her score for the tournament? (Lesson 2–5) **6 under par, or −6**

45. Evaluate $|m - 4|$ if $m = -6$. (Lesson 2–3) **10**

46. Simplify $x^2 + \frac{7}{8}x - \frac{x}{8}$. (Lesson 1–7) $x^2 + \frac{3}{4}x$

Refer to the Investigation on pages 190–191.

Since it is impossible to count all the fish in the lake, you can use your samples to estimate the population in the lake.

1. Write a proportion that relates the numbers in your chart and the estimated number of fish in the lake.
2. Imagine that the only information you had was the data you recorded after your first casting. What would have been your estimate of the number of fish in the lake? Justify your reasoning. Record your estimate in the last column of your chart.
3. Make estimates for each of your castings and record them in the last column of your chart.
4. Taking all of your castings into consideration, if you had to give an official estimate of the number of fish in the lake, what would it be? Justify your reasoning.

Add the results of your work to your Investigation Folder.

Extension

Reasoning Solve each proportion for x.

1. $\frac{x + a}{x} = \frac{b + a}{2b}$ $\quad \frac{2ab}{a - b}$
2. $\frac{3 - x}{x} = \frac{4 - b}{2a}$ $\quad \frac{6a}{4 - b + 2a}$

In·ves·ti·ga·tion

Working on the Investigation

The Investigation on pages 190–191 is designed to be a long-term project that is completed over several days or weeks. Encourage students to keep their materials in their Investigation Folder as they work on the Investigation.

4-2 Integration: Geometry Similar Triangles

- To find the unknown measures of the sides of two similar triangles.

Why IT'S IMPORTANT

You can use similar triangles to solve problems involving tennis, architecture, and billiards.

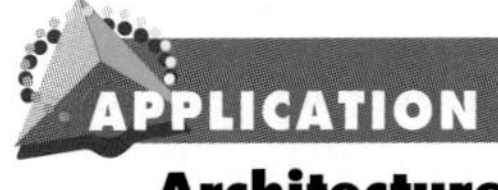

Architecture

"Rock n' roll is here to stay," is what you might hear after visiting the Rock and Roll Hall of Fame and Museum in Cleveland, Ohio. Memorabilia from musicians such as the Beatles, U2, and Carlos Santana cover the walls inside. The automatic teller machines in the building are even shaped like jukeboxes.

The building itself, designed by architect I.M. Pei, is made up of several geometric forms—a triangle, a cylinder, a rectangle, and a trapezoid. The triangular section of the building is similar to the one Pei designed for the Louvre in Paris, France. Both are made up of similar triangles.

fabulous FIRSTS

Maya Ying Lin (1959–)

Maya Lin became the first woman and the first student to design a major Washington, D.C., monument. In 1981, as a 21-year-old Yale architecture student, Lin won the design competition for the Vietnam Veterans Memorial. The black granite wall rising from Earth is the most visited site in the capital.

Two figures are **similar** if they have the same shape, but not necessarily the same size.

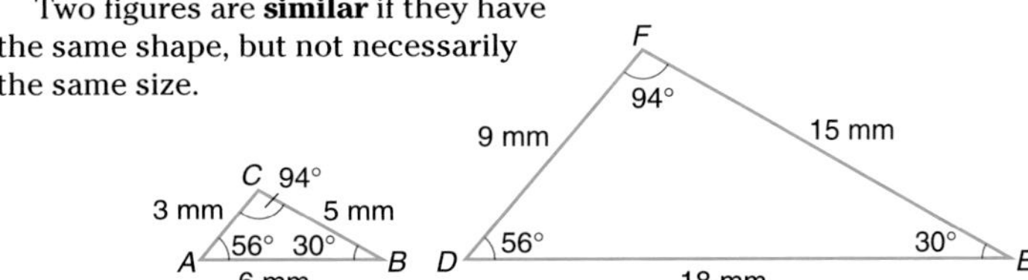

If **corresponding angles** of two triangles have equal measures, the triangles are similar. The two triangles above are similar. We write this as $\Delta ABC \sim \Delta DEF$. The order of the letters indicates the angles that correspond.

corresponding angles	corresponding sides
$\angle A$ and $\angle D$	$\overline{AB}$ and $\overline{DE}$
$\angle B$ and $\angle E$	$\overline{BC}$ and $\overline{EF}$
$\angle C$ and $\angle F$	$\overline{AC}$ and $\overline{DF}$

The sides opposite corresponding angles are called **corresponding sides**. Compare the measures of the corresponding sides. Note that AC means the measure of $\overline{AC}$, DF means the measure of $\overline{DF}$ and so on.

$$\frac{AB}{DE} = \frac{6}{18} = \frac{1}{3} \qquad \frac{BC}{EF} = \frac{5}{15} = \frac{1}{3} \qquad \frac{AC}{DF} = \frac{3}{9} = \frac{1}{3}$$

When the measures of the corresponding sides form equal ratios, the measures are said to be *proportional.*

Similar Triangles	**If two triangles are similar, the measures of their corresponding sides are proportional, and the measures of their corresponding angles are equal.**

fabulous FIRSTS

Maya Lin designed a permanent yet everchanging artwork at The Wexner Center for the Arts at The Ohio State University. It is made of mounds of crushed safety glass located in various areas outside the building. It was placed using a cement mixer, crane, and giant funnel.

4-2 LESSON NOTES

NCTM Standards: 1–5, 7

Instructional Resources

- Study Guide Master 4-2
- Practice Master 4-2
- Enrichment Master 4-2
- Assessment and Evaluation Masters, p. 100
- Real-World Applications, 10

Transparency 4-2A contains the 5-Minute Check for this lesson; **Transparency 4-2B** contains a teaching aid for this lesson.

Recommended Pacing	
Standard Pacing	Day 3 of 15
Honors Pacing	Day 3 of 15
Block Scheduling*	Day 2 of 7 (along with Lesson 4-3)
Alg. 1 in Two Years*	Days 4 & 5 of 20

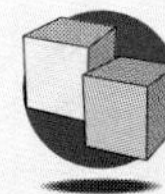

*For more information on pacing and possible lesson plans, refer to the *Block Scheduling Booklet* and *Algebra 1 in Two Years.*

1 FOCUS

5-Minute Check
(over Lesson 4-1)

Use cross products to determine whether each pair of ratios forms a proportion.

1. $\frac{9}{19}, \frac{8}{18}$ $\neq$
2. $\frac{16}{24}, \frac{10}{15}$ $=$

Solve each proportion.

3. $\frac{x}{20} = \frac{3}{5}$ 12
4. $\frac{8}{y} = \frac{4}{7}$ 14
5. $\frac{z-5}{8} = \frac{3}{4}$ 11

Motivating the Lesson

Hands-On Activity Before class, draw a triangle with side lengths 4 cm, 10 cm, and 8 cm on an overhead transparency. Project the triangle on the chalkboard and determine how far from the chalkboard the projector must be to obtain a triangle with sides 8, 20, and 16 cm. Then move the projector to the right and back until the side lengths are 12, 30, and 24 cm and again note the position of the projector. In class, move the projector to the predetermined locations, trace the triangles with chalk, and have students give you the measures of the angles and sides of the two triangles. Show that the corresponding angles are congruent and the corresponding sides are proportional.

2 TEACH

In-Class Examples

For Example 1

A tree 9 feet tall casts a shadow 5 feet long. How high is a flagpole that casts a shadow 15 feet long? **27 ft**

For Example 2

a. Find the distance, *UV*, across the pond shown below. ΔSTU is similar to ΔWVU. **The pond is 105 meters across.**

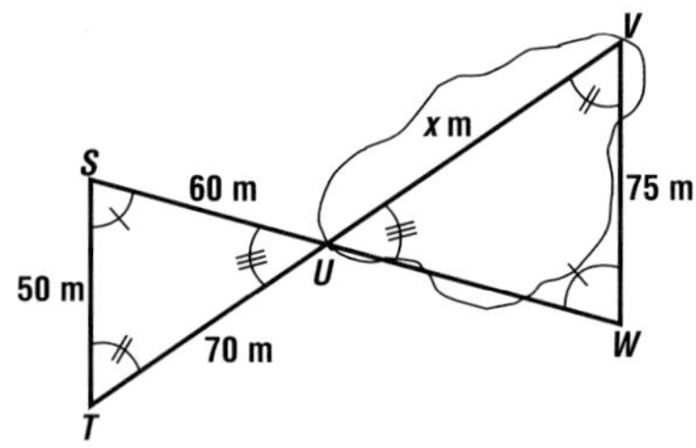

b. ΔABC is similar to ΔADE. Find *x*. **5**

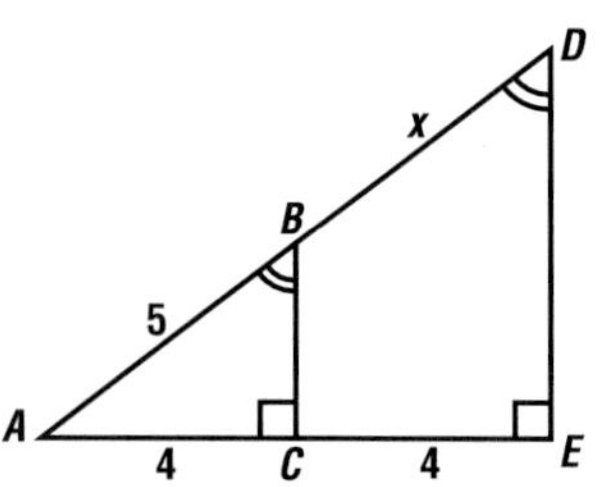

Proportions can be used to find the measures of the sides of similar triangles when some measurements are known.

Example 1 **Surveyors normally use instruments to measure objects that are too large or too far away to measure by hand. They also can use the shadows that objects cast to find the height of the objects without measuring them directly. How can a surveyor use a telephone pole that is 25 feet tall and casts a shadow 20 feet long to find the height of a building that casts a shadow 52 feet long?**

ΔNOP is similar to ΔMRQ.

$\frac{RQ}{OP} = \frac{QM}{PN}$ *Corresponding sides of similar triangles are proportional.*

$\frac{25}{x} = \frac{20}{52}$

$20x = 25(52)$ *Find the cross products.*

$20x = 1300$

$x = 65$

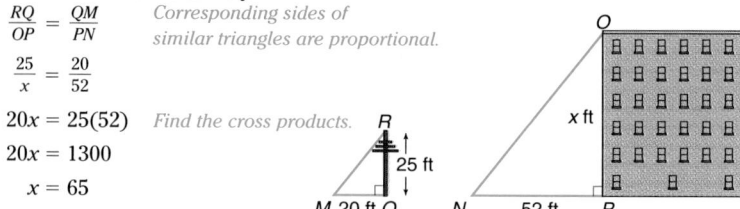

The building is 65 feet high.

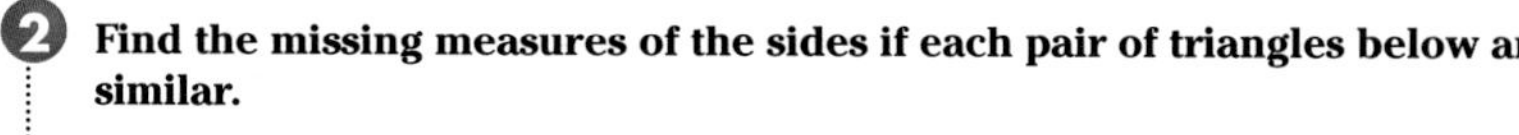

Similar triangles may be positioned so that the corresponding parts are not obvious.

Example 2 **Find the missing measures of the sides if each pair of triangles below are similar.**

Remember that the sum of the measures of the angles in a triangle is 180°.

a. The measure of $\angle Q$ is $180° - (117° + 27°)$ or 36°.

The measure of $\angle T$ is $180° - (117° + 36°)$ or 27°.

Since the corresponding angles have equal measures, $\Delta QRS \sim \Delta VUT$. This means that the lengths of the corresponding sides are proportional.

$\frac{SQ}{TV} = \frac{QR}{VU}$	$\frac{SQ}{TV} = \frac{SR}{TU}$
$\frac{33}{12} = \frac{x}{6}$	$\frac{33}{12} = \frac{y}{8}$
$12x = 33(6)$	$12y = 33(8)$
$12x = 198$	$12y = 264$
$x = 16.5$	$y = 22$

The missing measures are 16.5 and 22.

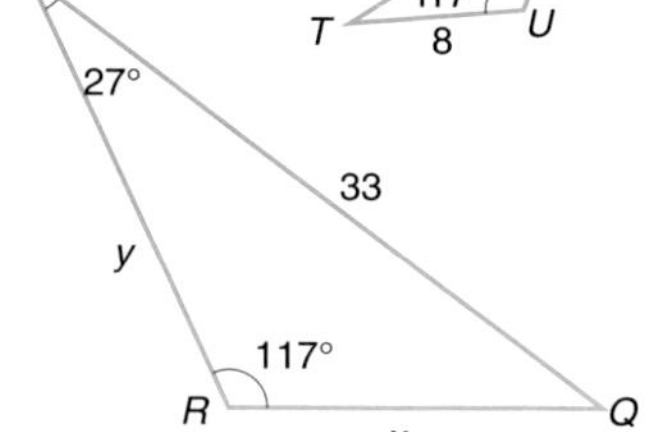

b. ΔNRT is similar to ΔMST.

$\frac{RN}{SM} = \frac{NT}{MT}$

$\frac{9}{x} = \frac{12}{28}$

$12x = 9(28)$ *Find the cross products.*

$12x = 252$

$x = 21$

The measure of $\overline{SM}$ is 21 meters.

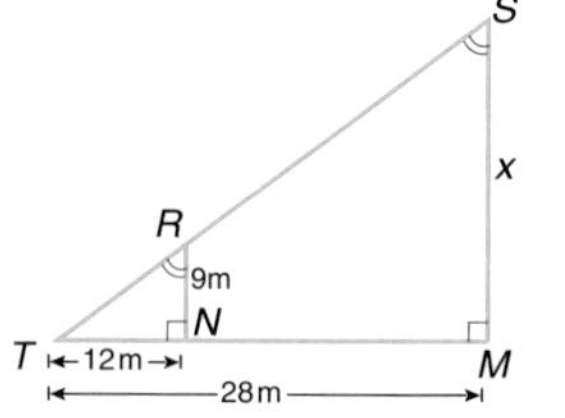

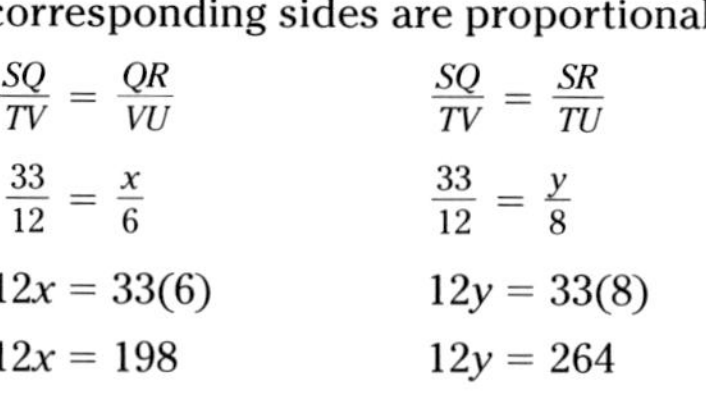

Land Surveyors use technical expertise to determine official land, space, and water boundaries. They plan field work, research legal records, record survey results, and prepare maps and reports. High school students should take courses in algebra, geometry, trigonometry, and drafting. Many colleges offer 4-year programs leading to a B.S. degree in surveying.

For more information, contact:

American Congress on Surveying and Mapping
5410 Grosvenor Lane
Bethesda, MD 20814–2122

Alternative Teaching Strategies

Student Diversity Have students draw a variety of triangles on plain pieces of paper. Use a copy machine to enlarge and reduce the triangles. Then have students cut them out and match similar triangles.

CAREER CHOICES

There is evidence that property boundaries were surveyed in Egypt as early as 1400 B.C. The Romans gained surveying instruments when they occupied Egypt.

CHECK FOR UNDERSTANDING

Communicating Mathematics

Study the lesson. Then complete the following. 2–3. See margin.

1. **Name** the pairs of corresponding angles in Example 2a. $\angle R$ and $\angle U$, $\angle S$ and $\angle T$, $\angle Q$ and $\angle V$
2. **List** what you know about the angles and sides of two triangles that are similar.
3. **List** the corresponding angles and corresponding sides if $\Delta RED \sim \Delta SOX$. Then write three proportions that correspond to these triangles.

4. **Model** a triangle using toothpicks or straws. Then make a similar triangle using more toothpicks or straws. Write a proportion to show that the two triangles are similar. See students' work.

Guided Practice

5. For the pair of similar triangles at the right, name the triangle that is similar to ΔXYZ. Make sure you have the letters in the correct order. ΔDEF

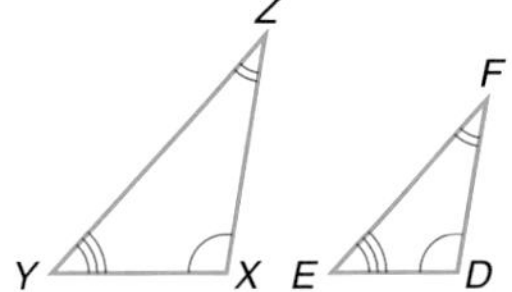

Determine whether each pair of triangles is similar. Justify your answer.

6. no

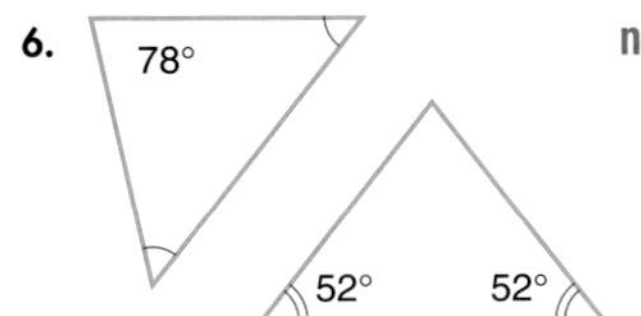

7. yes

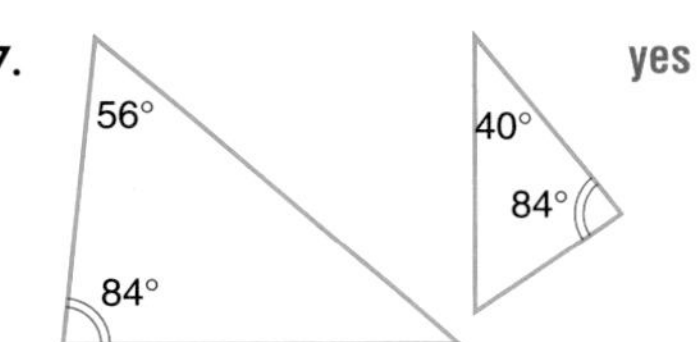

ΔKLM and ΔNOP are similar. For each set of measures given, find the measures of the remaining sides.

8. $o = 20$, $p = 10$ — **8.** $k = 24$, $\ell = 30$, $m = 15$, $n = 16$
9. $\ell = 12$, $m = 6$ — **9.** $k = 9$, $n = 6$, $o = 8$, $p = 4$
10. $k = 3$, $o = 8$ — **10.** $n = 6$, $p = 2.5$, $\ell = 4$, $m = 1.25$

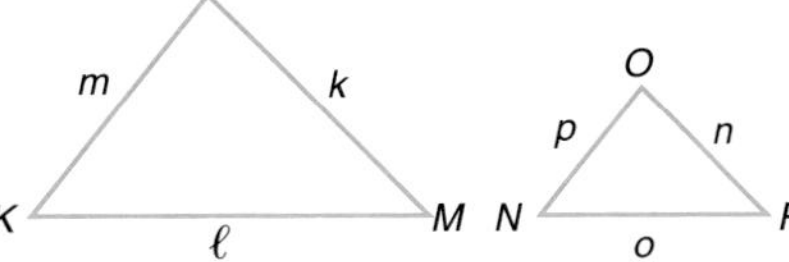

11. If a tree 6 feet tall casts a shadow 4 feet long, how high is a flagpole that casts a shadow 18 feet long at the same time of day? 27 feet high

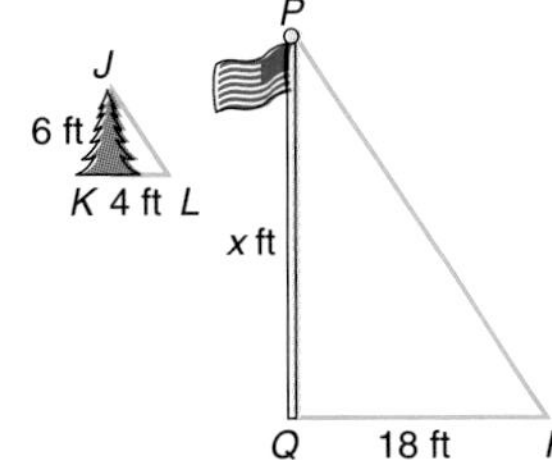

Reteaching

Using Models Have students move a mirror along the ground between an observer and an object until the top of the observer's line of sight is reflected in the mirror. Then use similar triangles to find the height of the object.

3 PRACTICE/APPLY

Check for Understanding

Exercises 1–11 are designed to help you assess your students' understanding through reading, writing, speaking, and modeling. You should work through Exercises 1–4 with your students and then monitor their work on Exercises 5–11.

Error Analysis

In identifying similar triangles, students are sometimes careless in making sure corresponding vertices are given in the same order. To check, make sure the angles at vertices given in corresponding order are congruent.

Additional Answers

2. Sample answer: The measures of their corresponding sides are proportional, and the measures of their corresponding angles are equal.
3. $\angle R$ and $\angle S$, $\angle E$ and $\angle O$, $\angle D$ and $\angle X$

$$\frac{RE}{SO} = \frac{ED}{OX}, \frac{RE}{SO} = \frac{RD}{SX}, \frac{ED}{OX} = \frac{RD}{SX}$$

Study Guide Masters, p. 27

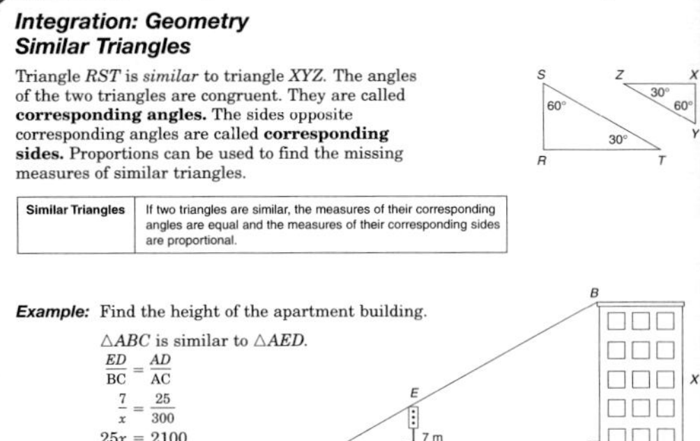

4-2 NAME________ DATE________

Study Guide — Student Edition Pages 201–205

Integration: Geometry
Similar Triangles

Triangle RST is *similar* to triangle XYZ. The angles of the two triangles are congruent. They are called **corresponding angles.** The sides opposite corresponding angles are called **corresponding sides.** Proportions can be used to find the missing measures of similar triangles.

Similar Triangles	If two triangles are similar, the measures of their corresponding angles are equal and the measures of their corresponding sides are proportional.

Example: Find the height of the apartment building.

$\triangle ABC$ is similar to $\triangle AED$.

$$\frac{ED}{BC} = \frac{AD}{AC}$$
$$\frac{7}{x} = \frac{25}{300}$$
$$25x = 2100$$
$$x = 84$$

The apartment building is 84 meters high.

Refer to the triangles at the right to answer the questions.

1. Which triangles are similar? $\triangle ABC$ and $\triangle HIG$
2. Name corresponding angles of the similar triangles. $\angle A$ and $\angle H$; $\angle B$ and $\angle I$; $\angle C$ and $\angle G$
3. Name corresponding sides of the similar triangles. $\overline{AB}$ and $\overline{HI}$; $\overline{AC}$ and $\overline{HG}$; $\overline{BC}$ and $\overline{IG}$

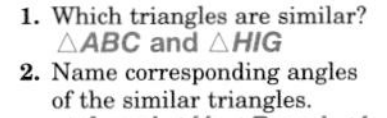

Solve.

4. Bruce likes to amuse his brother by shining a flashlight on his hand and making a shadow on the wall. How far is it from the flashlight to the wall? 4.3 ft

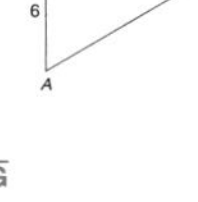

Assignment Guide

Core: 13–33 odd, 35–41
Enriched: 12–28 even, 30–41

For **Extra Practice,** see p. 764.

The red A, B, and C flags, printed only in the Teacher's Wraparound Edition, indicate the level of difficulty of the exercises.

Teaching Tip For Exercise 33, some students may feel these answers are not realistic. They are, though, for a very powerful serve. No player who is only 5 feet tall can ever match the 125 mph serves of Boris Becker (6 feet 3 inches), not because of the strength required, but because no serve that fast from a player that short could ever land in the service court.

Additional Answer

31. Sample answer:

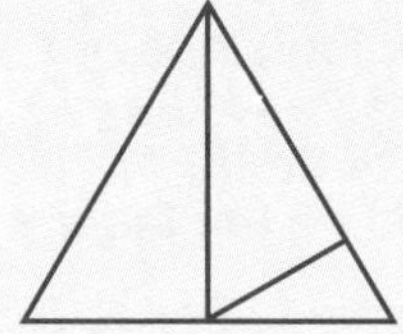

Practice Masters, p. 27

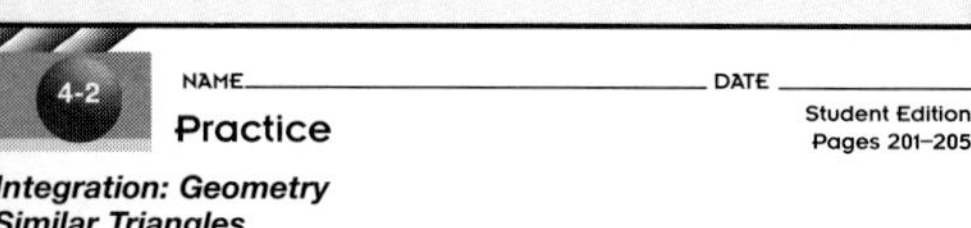

NAME__________________ DATE __________

Practice

Student Edition
Pages 201–205

Integration: Geometry
Similar Triangles

Determine whether each pair of triangles is similar. Justify your answer.

1. 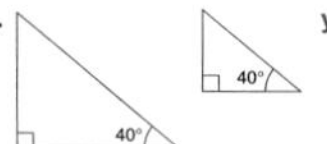

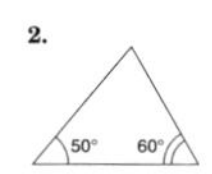

yes

2. 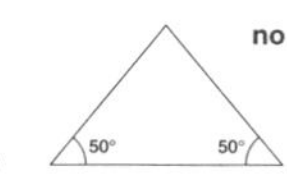

no

△ABC and △DEF are similar. For each set of measures given, find the measures of the remaining sides.

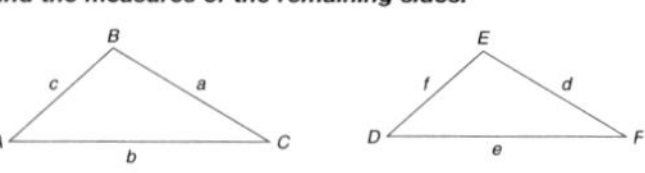

3. $a = 10, b = 12, c = 6, d = 6$
$e = 7.2, f = 3.6$

4. $a = 16, b = 22, c = 12, f = 8$
$d = 10\frac{2}{3}, e = 14\frac{2}{3}$

5. $a = 4, d = 6, e = 4, f = 3$
$c = 2, b = 2\frac{2}{3}$

6. $b = 15, d = 16, e = 20, f = 10$
$c = 7\frac{1}{2}, a = 12$

7. $c = 4, d = 12, e = 16, f = 8$
$a = 6, b = 8$

8. $e = 20, a = 24, b = 30, c = 15$
$f = 10, d = 16$

9. $a = 2.5, b = 3, f = 5.5, e = 7$
$c = 2\frac{5}{14}, d = 5\frac{5}{6}$

10. $c = 1.25, d = 6, e = 3.75, f = 2.50$
$a = 3, b = 1.875$

EXERCISES

Practice

For each pair of similar triangles, name the triangle that is similar to Δ*WXY*.

12. Δ*PNY*
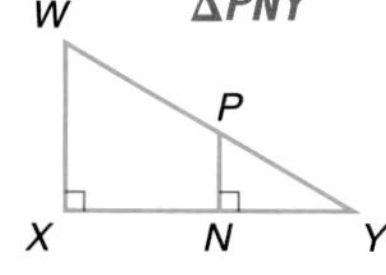

13. Δ*DFE*
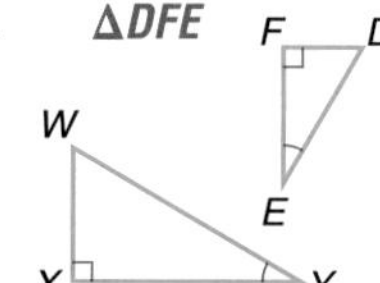

14. Δ*RQY*
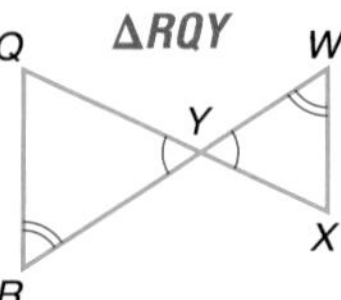

Determine whether each pair of triangles is similar. Justify your answer.

15. no
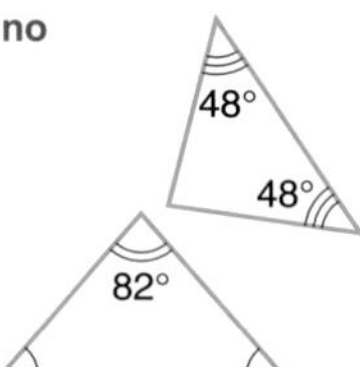

16. yes
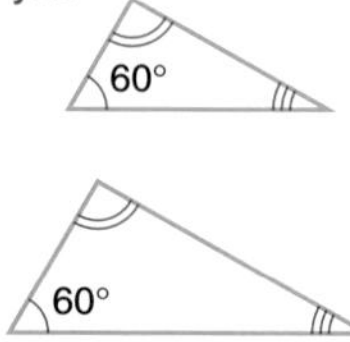

17. yes
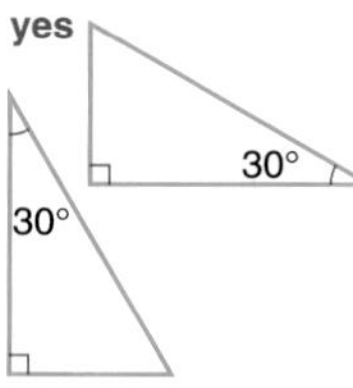

18. no
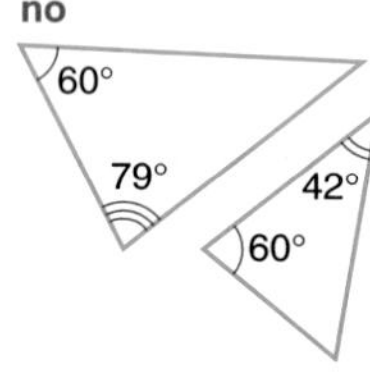

19. no
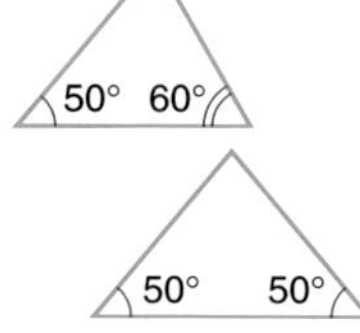

20. no
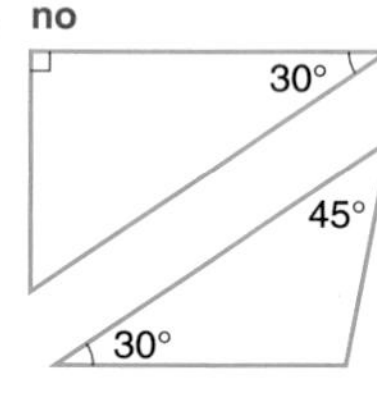

Δ*ABC* and Δ*DEF* are similar. For each set of measures given, find the measures of the remaining sides.

C

21. $c = 11, f = 6, d = 5, e = 4$

22. $a = 5, d = 7, f = 6, e = 5$

23. $a = 17, b = 15, c = 10, f = 6$

24. $a = 16, e = 7, b = 13, c = 12$

25. $d = 2.1, b = 4.5, f = 3.2, e = 3.4$ **$a = 2.78, c = 4.24$**

26. $f = 12, d = 18, c = 18, e = 16$ **$a = 27, b = 24$**

27. $c = 5, a = 12.6, e = 8.1, f = 2.5$ **$b = 16.2, d = 6.3$**

28. $f = 5, a = 10.5, b = 15, c = 7.5$ **$d = 7, e = 10$**

29. $a = 4\frac{1}{4}, b = 5\frac{1}{2}, e = 2\frac{3}{4}, f = 1\frac{3}{4}$ **$c = \frac{7}{2}, d = \frac{17}{8}$**

B c a A C b E f d D F e

21. $a = \frac{55}{6}, b = \frac{22}{3}$

22. $b = \frac{25}{7}, c = \frac{30}{7}$

23. $d = \frac{51}{5}, c = 9$

24. $d = \frac{112}{13}, f = \frac{84}{13}$

Critical Thinking

30. Δ*ABC* and Δ*DEF* are similar. If the ratio of *AC* to *DF* is 2 to 3, what is the ratio of the area of Δ*ABC* to the area of Δ*DEF*? **4:9**

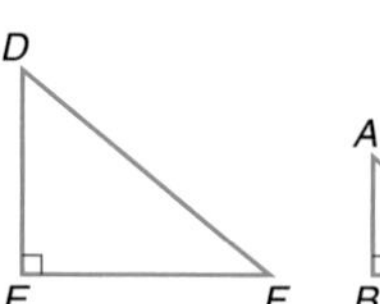

31. Draw an equilateral triangle. Try to cut it into three similar parts with no two parts that are congruent. Describe the process you used. **31. See margin.**

Applications and Problem Solving

32. **Billiards** Meda is playing billiards on a table like the one show at the right. If she can make her next shot, she figures she will have no trouble winning. She wants to strike the cue ball at D, bank it at C, and hit another ball at the mouth of pocket A. Use similar triangles to find where Meda's cue ball should strike the rail. **24 inches from pocket B**

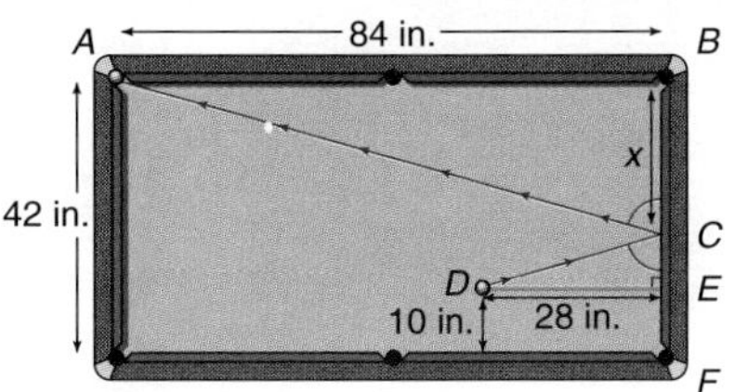

33. **Tennis** When serving in tennis, a player hits the ball standing at point E as shown in the figure at the right. To be a good serve, the ball must go over the net and land in the shaded area. Assume that all serves travel in a straight line and that all serves just clear the net. The figure below shows the view from the sideline.

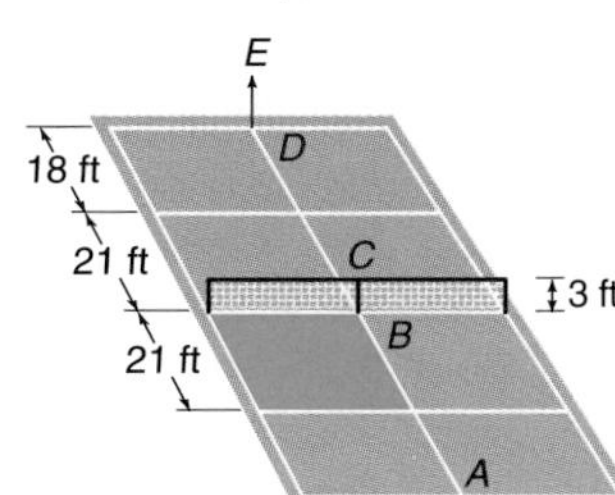

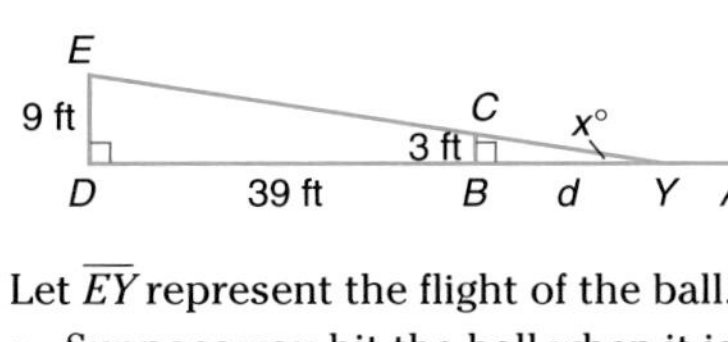

Let $\overline{EY}$ represent the flight of the ball. **a.** $\frac{3}{9} = \frac{d}{d + 39}$; **$d = 19.5$ ft; yes**

a. Suppose you hit the ball when it is 9 feet above the ground. Write a proportion and solve for d to determine whether the serve is good.

b. Repeat the calculation of the distance d if the ball is hit from 8 feet above the ground and from 10 feet above the ground. Use the results to help you determine which players have an easier time hitting a good serve, tall players or short players. Explain. **See margin.**

34. **Bridges** Haylee is making a model of the Mark Clark Expressway, a continuous truss bridge that spans the Cooper River in Charleston, South Carolina. Parts of the trusses are made up of triangles. Haylee plans to make the model bridge to the following scale: 1 inch to 12 feet. If the height of the triangle on the actual bridge is 40 feet, what will the height be on the model? **$3\frac{1}{3}$ inches**

Mark Clark Expressway

Mixed Review

35. **Drafting** The scale on the blueprint for a house is 1 inch to 3 feet. If the living room on the blueprint is $5\frac{1}{2}$ inches by 7 inches, what are the dimensions of the actual room? (Lesson 4–1) **$16\frac{1}{2}$ feet by 21 feet**

36. Solve $7 = 3 - \frac{n}{3}$. (Lesson 3–3) **−12**

37. **Table Tennis** Each Ping-Pong™ ball weighs about $\frac{1}{10}$ of an ounce. How many Ping-Pong balls weigh 1 pound altogether? (Lesson 3–2) **160**

38. Solve $x + (-8) = -31$. (Lesson 3–1) **−23**

39. 60 million; 420 million

39. **Ecology** Americans use about 2.5 million plastic bottles every hour. How many do they use in one day? How many in one week? (Lesson 2–6)

40. Replace _?_ with <, >, or = to make $\frac{8}{15}$ _?_ $\frac{9}{16}$ true. (Lesson 2–4) **<**

41. Evaluate $(19 - 12) \div 7 \cdot 23$. Name the property used in each step. (Lesson 1–6) **23; See margin.**

Extension

Problem Solving Suppose a man is 180 centimeters tall. His image on the film of a camera is 1.5 centimeters tall. If the film is 2 centimeters from the lens of the camera, how far is the man from the camera? **240 cm**

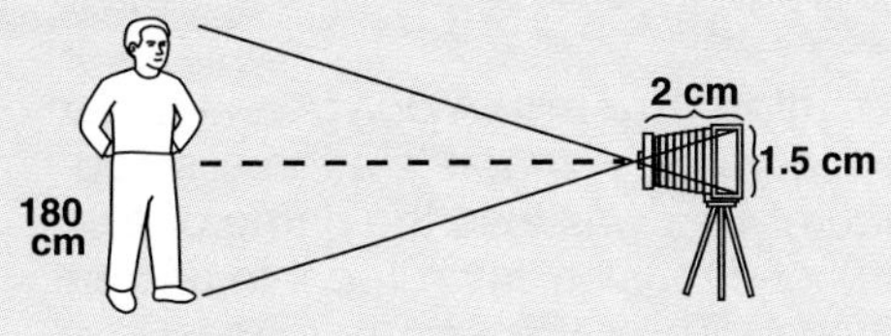

4 ASSESS

Closing Activity

Speaking Have students explain how they can use proportions to find the missing measures of similar triangles.

Chapter 4, Quiz A (Lessons 4-1 and 4-2), is available in the *Assessment and Evaluation Masters*, p. 100.

Additional Answers

33b. When the ball is served from 8 feet, $d = 23.4$ ft, and the serve is a fault. When the ball is served from 10 feet, $d = 16.7$ ft. Therefore, tall players have an easier time serving.

41. $(19 - 12) \div 7 \times 23$
$= 7 \div 7 \times 23$ Substitution property of equality
$= 1 \times 23$ Substitution property of equality
$= 23$ Multiplicative identity property

Enrichment Masters, p. 27

NAME______ DATE______ Student Edition Pages 201–205

Enrichment

A Curious Construction

Many mathematicians have been interested in ways to construct the number π. Here is one such geometric construction.

In the drawing, triangles ABC and ADE are right triangles. The length of $\overline{AD}$ equals the length of $\overline{AC}$ and $\overline{FB}$ is parallel to $\overline{EG}$.

The length of $\overline{BG}$ gives a decimal approximation of the fractional part of π to six decimal places.

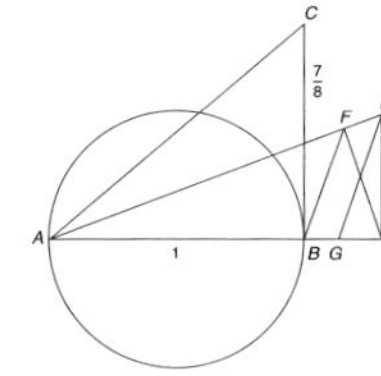

Follow the steps to find the length of $\overline{BG}$.

1. Use the length of $\overline{BC}$ and the Pythagorean theorem to find the length of $\overline{AC}$.
$AC = \sqrt{1^2 + \left(\frac{7}{8}\right)^2} = 1.3287682$
2. Find the length of $\overline{AD}$.
$AD = AC = 1.3287682$
3. Use the length of $\overline{AD}$ and the Pythagorean theorem to find the length of $\overline{AE}$.
$AE = \sqrt{(AD)^2 + \left(\frac{1}{2}\right)^2} = 1.4197271$
4. The sides of the similar triangles FED and DEA are in proportion. So, $\frac{FE}{0.5} = \frac{0.5}{AE}$. Find the length of $\overline{FE}$.
$FE = \frac{1}{4(AE)}$ $FE = 0.1760902$
5. Find the length of $\overline{AF}$.
$AF = AE - FE = 1.2436369$
6. The sides of the similar triangles AFB and AEG are in proportion. So, $\frac{AF}{AE} = \frac{AB}{AG}$. Find the length of $\overline{AG}$.
$AG = \frac{AB \cdot AE}{AF} = \frac{AE}{AF} = 1.1415929$
7. Now, find the length of $\overline{BG}$.
$BG = AG - AB = AG - 1 = 0.1415929$
8. The value of π to seven decimal places is 3.1415927. Compare the fractional part of π with the length of $\overline{AG}$.
$0.1415929 - 0.1415927 = 0.0000002$, an error of less than 1 part in a million

4-3 LESSON NOTES

NCTM Standards: 1–5, 9

Instructional Resources

- Study Guide Master 4-3
- Practice Master 4-3
- Enrichment Master 4-3

Transparency 4-3A contains the 5-Minute Check for this lesson; **Transparency 4-3B** contains a teaching aid for this lesson.

Recommended Pacing	
Standard Pacing	Days 4 & 5 of 15
Honors Pacing	Days 4 & 5 of 15
Block Scheduling*	Day 2 of 7 (along with Lesson 4-2)
Alg. 1 in Two Years*	Days 6 & 7 of 20

*For more information on pacing and possible lesson plans, refer to the *Block Scheduling Booklet* and *Algebra 1 in Two Years.*

1 FOCUS

5-Minute Check
(over Lesson 4-2)

1. If $\Delta NBC \sim \Delta FOX$, then $\frac{BC}{?} = \frac{CN}{?}$. ***OX; XF***
2. Name the triangle that is similar to ΔSIM. ΔLPM

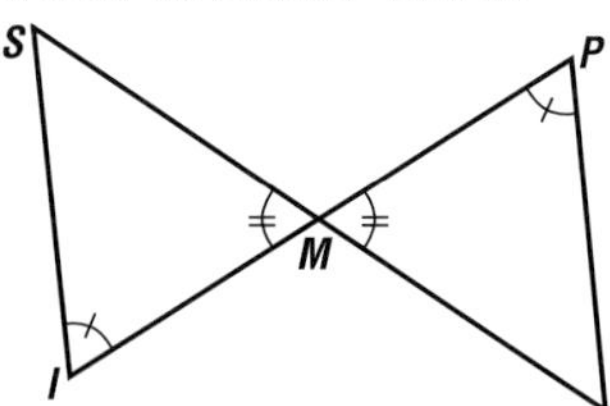

3. What is the value of x? **35**

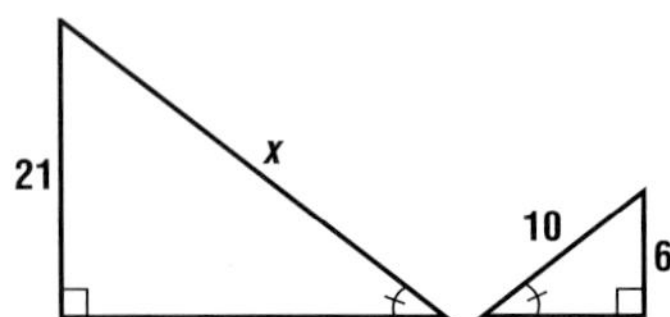

4. A tree five feet tall casts a shadow two feet long. At the same time, a flagpole casts a shadow nine feet long. How tall is the flagpole? **22.5 ft tall**

4-3 Integration: Trigonometry Trigonometric Ratios

What YOU'LL LEARN
- To use trigonometric ratios to solve right triangles.

Why IT'S IMPORTANT

You can use the trigonometric ratios to solve problems involving architecture and archaeology.

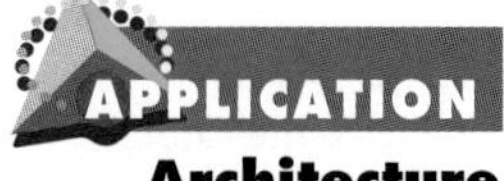

Architecture

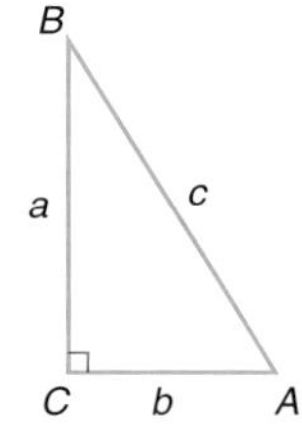

The Leaning Tower of Pisa, a bell tower in Pisa, Italy, currently leans 16.5 feet off center. The tower was closed in 1990 in order for workers to attempt to stabilize the tower's foundation to prevent it from eventually collapsing. By 1994, the tower's lean had decreased about $\frac{2}{5}$ of an inch. No date has been scheduled for reopening the tower.

If you were to draw a line from the top of the tower to the ground, a right triangle would be formed with the ground. If enough is known about a right triangle, certain ratios can be used to find the measures of the remaining parts of the triangle. These ratios are called **trigonometric ratios.** *Trigonometry is an area of mathematics that studies angles and triangles.*

A typical right triangle is shown at the right.

a is the measure of the side *opposite* $\angle A$, $\overline{BC}$.

b is the measure of the side *opposite* $\angle B$, $\overline{AC}$.

c is the measure of the side *opposite* $\angle C$, $\overline{AB}$.

a is *adjacent* to $\angle B$ and $\angle C$.

b is *adjacent* to $\angle A$ and $\angle C$.

c is *adjacent* to $\angle A$ and $\angle B$.

The side opposite $\angle C$, the right angle, is called the **hypotenuse.** The other two sides are called **legs.**

Three common trigonometric ratios are defined as follows.

Definition of Trigonometric Ratios	sine of $\angle A = \frac{\text{measure of leg opposite } \angle A}{\text{measure of hypotenuse}}$ $\sin A = \frac{a}{c}$ cosine of $\angle A = \frac{\text{measure of leg adjacent to } \angle A}{\text{measure of hypotenuse}}$ $\cos A = \frac{b}{c}$ tangent of $\angle A = \frac{\text{measure of leg opposite } \angle A}{\text{measure of leg adjacent to } \angle A}$ $\tan A = \frac{a}{b}$

Notice that sine, cosine, and tangent are abbreviated as sin, cos, and tan, respectively.

Alternative Teaching Strategies

Reading Algebra A *polygon* is a many-sided figure. A *trigon* is a three-sided figure. This was also the name of a stringed musical instrument in ancient Greece. The suffix *-metry* means "measure." So, *trigonometry* means "the study of finding measures of three-sided figures, or triangles."

Example 1 **Find the sine, cosine, and tangent of each acute angle. Round your answers to the nearest thousandth.**

$\sin J = \frac{\text{opposite leg}}{\text{hypotenuse}}$ $\quad$ $\sin L = \frac{\text{opposite leg}}{\text{hypotenuse}}$

$= \frac{7}{25}$ or 0.280 $\quad$ $= \frac{24}{25}$ or 0.960

$\cos J = \frac{\text{adjacent leg}}{\text{hypotenuse}}$ $\quad$ $\cos L = \frac{\text{adjacent leg}}{\text{hypotenuse}}$

$= \frac{24}{25}$ or 0.960 $\quad$ $= \frac{7}{25}$ or 0.280

$\tan J = \frac{\text{opposite leg}}{\text{adjacent leg}}$ $\quad$ $\tan L = \frac{\text{opposite leg}}{\text{adjacent leg}}$

$= \frac{7}{24}$ or about 0.292 $\quad$ $= \frac{24}{7}$ or about 3.429

J, K, L; 24, 25, 7

Consider triangles QTV and MSO. Since corresponding angles have the same measure, $\Delta QTV \sim \Delta MSO$. Recall that in similar triangles, the corresponding sides are proportional.

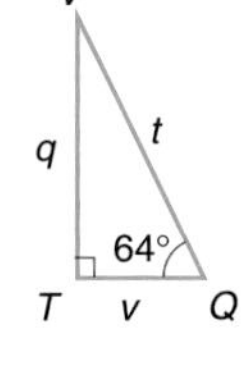

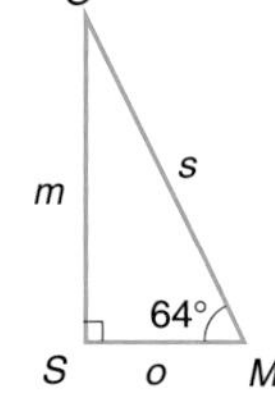

$$\frac{m}{q} = \frac{s}{t}$$

$$\frac{q}{s} \cdot \frac{m}{q} = \frac{q}{s} \cdot \frac{s}{t} \quad \textit{Multiply each side by } \frac{q}{s}.$$

$$\frac{m}{s} = \frac{q}{t}$$

$\sin M = \sin Q$

In general, the sine of a 64° angle of a right triangle will be the same number no matter how large or small the triangle is. A similar result holds for cosine and tangent.

You can use a calculator to find the values of trigonometric functions or to find the measure of an angle.

Example 2 **Find the value of sin 64° to the nearest ten thousandth.**

Use a scientific calculator.

Enter: 64 [SIN] 0.898794046 *The calculator must be in degree mode.*

Rounded to the nearest ten thousandth, $\sin 64° \approx 0.8988$.

Example 3 **Find the measure of ∠P to the nearest degree.**

Since the length of the opposite and adjacent sides are known, use the tangent ratio.

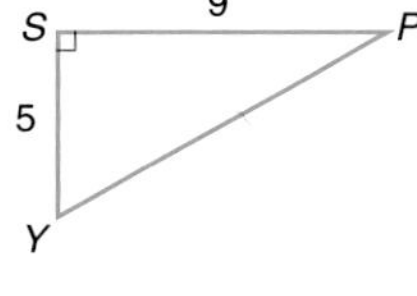

$\tan P = \frac{\text{opposite leg}}{\text{adjacent leg}}$

$= \frac{5}{9}$

Use a scientific calculator to find the angle measure with a tangent of $\frac{5}{9}$.

Enter: 5 [÷] 9 [=] [2nd] [TAN⁻¹] 29.0546041 *The TAN^{-1} key "undoes" the original function.*

To the nearest degree, the measure of ∠P is 29°.

Motivating the Lesson

Situational Problem The steepest streets in the U.S. are Filbert Street, 22nd Street, and Russian Hill in San Francisco. They rise 1 foot for every 3.17 feet. Have students use trigonometry to find the angle these streets form with the horizontal. **approximately 17.5°**

2 TEACH

Teaching Tip SOH-CAH-TOA is a helpful mnemonic device for remembering these ratios.

$\sin = \frac{\text{opposite}}{\text{hypotenuse}}$ or $s = \frac{o}{h}$

$\cos = \frac{\text{adjacent}}{\text{hypotenuse}}$ or $c = \frac{a}{h}$

$\tan = \frac{\text{opposite}}{\text{adjacent}}$ or $t = \frac{o}{a}$

In-Class Examples

For Example 1

Find the sine, cosine, and tangent of each acute angle to the nearest thousandth.

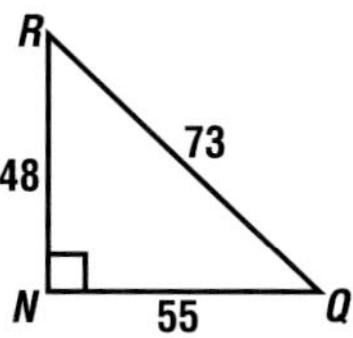

sin Q = 0.658; cos Q = 0.753; tan Q = 0.873; sin R = 0.753; cos R = 0.658; tan R = 1.146

For Example 2

Find the value of cos 72° to the nearest ten thousandth. **0.3090**

For Example 3

Find the measure of ∠A to the nearest degree. **18°**

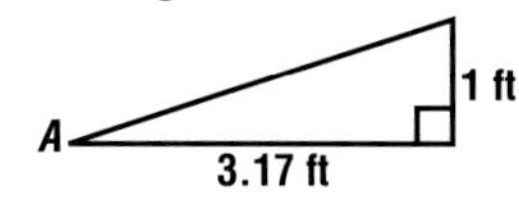

Alternative Learning Styles

Visual Given a trigonometric ratio of an angle, find the other two ratios for the angle by using a drawing of the relative lengths of a representative right triangle. For example, given $\sin A = \frac{3}{5}$, then the triangle at the right would be a representative right triangle.

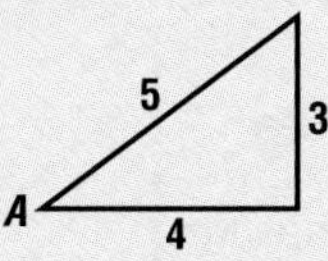

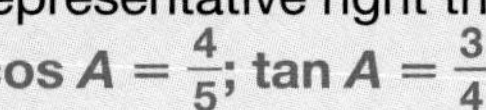

$\cos A = \frac{4}{5}$; $\tan A = \frac{3}{4}$

In-Class Examples

For Example 4

Guy wires are attached to a radio tower. They are 52 yards long and are anchored 30 yards from the base of the tower. Find the angle the wires make with the base of the tower to the nearest degree. **about 55°**

For Example 5

Solve ΔABC.

$x \approx 24.2, y \approx 16.2$

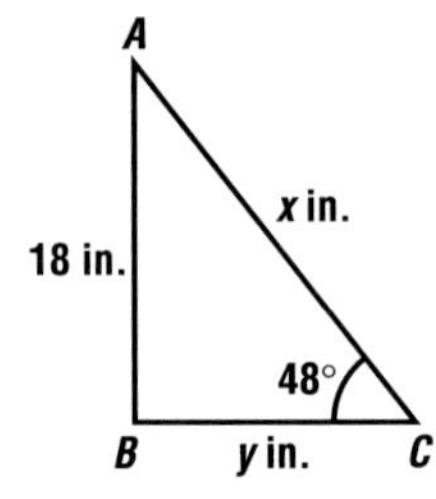

Many real-world applications involve trigonometry.

Example 4

Architecture

Refer to the application at the beginning of the lesson. Engineers working to correct the lean of the Leaning Tower of Pisa could check the angle the Tower makes with the ground to find any increase in the lean. Use the measurements to find the angle to the nearest degree that the Tower made with the ground.

Since the length of adjacent side and hypotenuse are known, use the cosine.

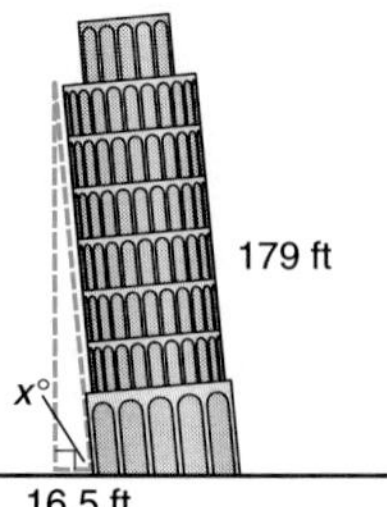

$\cos x = \frac{\text{adjacent leg}}{\text{hypotenuse}}$

$\cos x = \frac{16.5}{179}$

$\cos x = 0.09217877$ *Use a calculator to find x.*

$x \approx 85°$

To the nearest degree, the measure of the angle is 85°.

You can find the missing measures of a right triangle if you know the measure of two sides of the triangle or the measure of one side and one acute angle. Finding all of the measures of the sides and the angles in a right triangle is called **solving the triangle**.

Example 5 **Solve ΔABC.**

The measure of $\angle C$ is 90° − 42° or 48°.

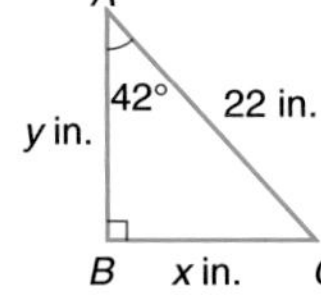

$\sin 42° = \frac{x}{22}$

$0.6691 \approx \frac{x}{22}$

$14.7 \approx x$

Thus, $\overline{BC}$ is about 14.7 inches long.

$\cos 42° = \frac{y}{22}$

$0.7431 \approx \frac{y}{22}$

$16.3 \approx y$

Thus, $\overline{AB}$ is about 16.3 inches long.

In the real world, trigonometric ratios are often used to find distances or lengths that cannot be measured directly. In these applications, you will sometimes use an angle of elevation or an angle of depression. An **angle of elevation** is formed by a horizontal line of sight and a line of sight above it. An **angle of depression** is formed by a horizontal line of sight and a line of sight below it.

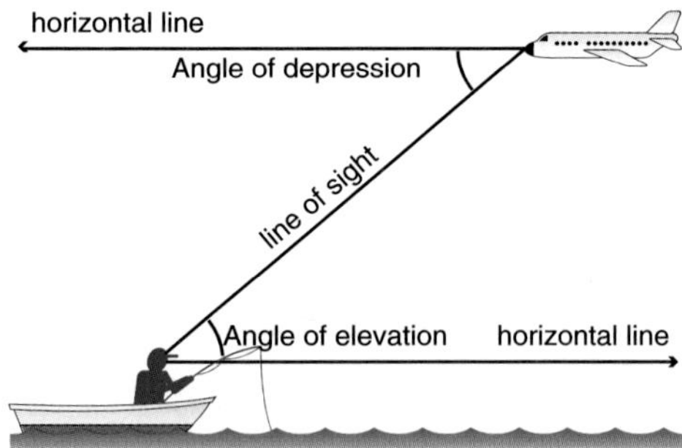

GLENCOE Technology

Interactive Mathematics Tools Software

This multimedia software provides interactive lessons by helping students explore the sine, cosine, and tangent ratios of a right triangle by manipulating the lengths of the sides of the triangle. **Computer Journals** give students an opportunity to write about what they have learned.

For Windows & Macintosh

Make a Hypsometer

Materials:  protractor  straw  tape  paper clip string

You can make a hypsometer to find the angle of elevation of an object that is too tall to measure.

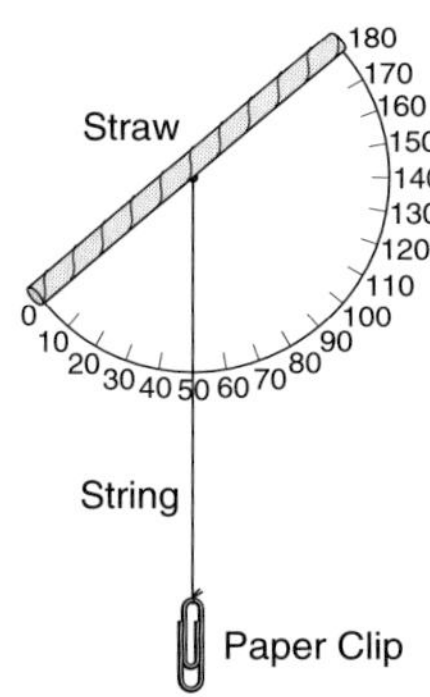

Your Turn

a. Tie one end of a piece of string to the middle of a straw. Tie the other end of string to a paper clip.

b. Tape a protractor to the side of the straw. Make sure that the string hangs freely to create a vertical or plumb line.

c. Find an object outside that is too tall to measure directly, such as a basketball hoop, a flagpole, or the school building.

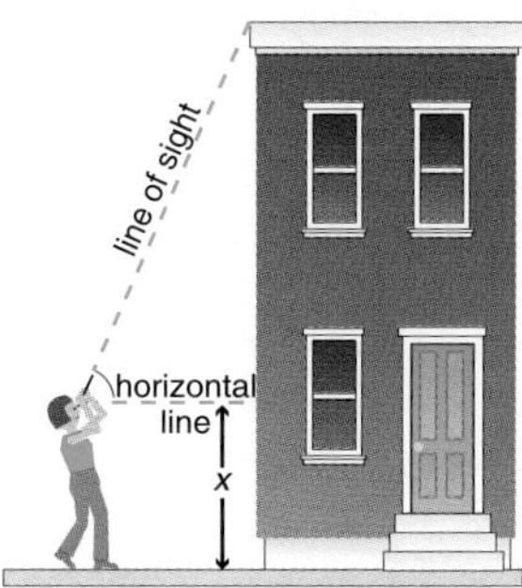

d. Look through the straw to the top of the object you are measuring. Find a horizontal line of sight as shown above. Find the measurement where the string and protractor intersect. Determine the angle of elevation by subtracting this measurement from 90°.

e. Use the equation tan (angle sighted) = $\frac{\text{height of object} - x}{\text{distance of object}}$, where x represents the distance from the ground to your eye level, to find the height of the object.

f. Compare your answer with someone who measured the same object. Did your heights agree? Why or why not? **See students' work.**

Example 6

Balloons

What would the Macy's Thanksgiving Day Parade be without balloons? In 1994, Dr. Suess' The Cat in the Hat made its first appearance. It took 36 people to maneuver the cat along the parade route. Suppose you were watching the parade. You noticed that two of the handling lines formed a right triangle with the ground. You estimated the angle of elevation to the bottom of the balloon to be 60°, and the handlers to be about 35 feet away from each other. Find the altitude of the bottom of the balloon to the nearest foot.

Explore You know the distance from one handler to another and the approximate angle of elevation. You need to find the height from the handlers to the bottom of the balloon.

(continued on the next page)

MODELING MATHEMATICS In this modeling activity, students build a simple version of a sextant. Since the instrument in the activity uses a counterweight, it can measure only vertical angles.

In-Class Example

For Example 6

If the angle of elevation to the top of the balloon is about 65° and the handler is about 45 feet away from the bottom of the balloon, what is the height of the balloon to the nearest foot? **97 ft**

3 PRACTICE/APPLY

Check for Understanding

Exercises 1–20 are designed to help you assess your students' understanding through reading, writing, speaking, and modeling. You should work through Exercises 1–6 with your students and then monitor their work on Exercises 7–20.

Error Analysis
Students sometimes take the terms *opposite leg* and *adjacent leg* to be absolute rather than relative. In a right triangle with acute angles A and B, the opposite leg of A is the adjacent leg of B. These labels are relative to the acute angle named.

Additional Answer

3. **Use the sine when the opposite side and hypotenuse are known, use the cosine when the adjacent side and hypotenuse are known, and use the tangent when the opposite side and adjacent side are known.**

Plan Let h represent the distance to the bottom of the balloon in feet. Use the tangent ratio to solve this problem.

Solve

$$\tan 60^\circ = \frac{\text{opposite leg}}{\text{adjacent}}$$

$$\tan 60^\circ = \frac{h}{35}$$

$$1.7321 \approx \frac{h}{35}$$

$$1.7321(35) \approx h$$

$$60.6235 \approx h$$

The altitude of the bottom of the balloon is about 61 feet above the handlers.

Examine You can examine the solution by determining the angle of elevation in the equation $\tan x = \frac{61}{35}$. Since $x = 60.15^\circ$, the solution is reasonable.

CHECK FOR UNDERSTANDING

Communicating Mathematics

Study the lesson. Then complete the following.

1. **Use** the figure at the right to answer the following questions.
   a. What is the measure of the leg adjacent to $\angle C$? **4**
   b. What is the measure of the leg opposite $\angle C$? **9**

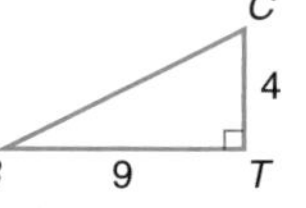

2. **List** the trigonometric ratio or ratios that involve the measure of the hypotenuse. **sine and cosine**

3. **Explain** how to determine which trigonometric ratio to use when solving for an unknown measure of a right triangle. **See margin.**

4. **Name** the angles of elevation and depression in the drawing at the right. **∠QRP; ∠SPR**

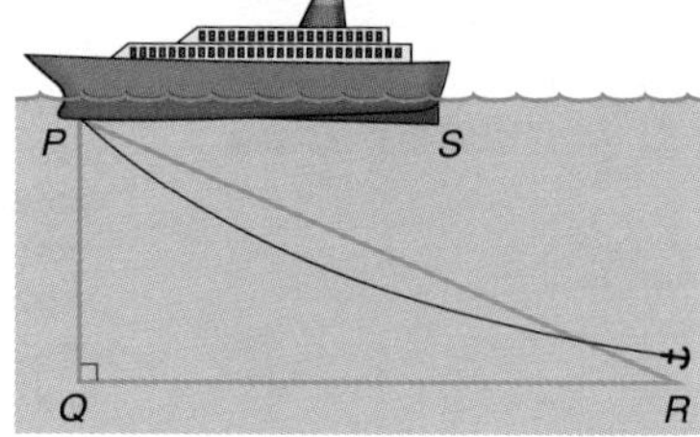

5. State which trigonometric ratio you could use to find x.

a.
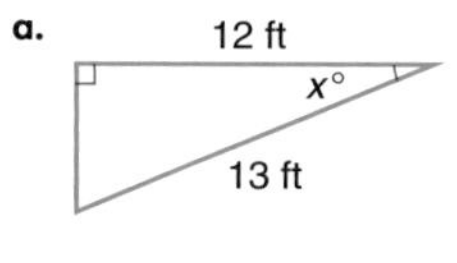

cosine

b. 39°, x m, 12 m
tangent

c.
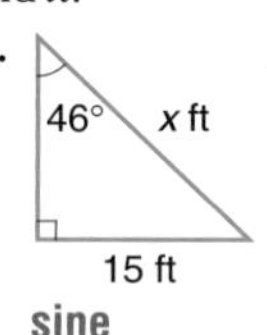

sine

6. **Use** a hypsometer to find the height of a tree in your community. Then write a paragraph explaining how you found the height. **See students' work.**

Reteaching

Using Alternative Methods Have students make flash cards with the name of a trigonometric ratio on one side and its definition on the other side. Encourage students to practice with these until they know them well.

GLENCOE Technology

Interactive Mathematics Tools Software

A multimedia software provides an interactive lesson by helping students explore the relationship of the sine ratio to the length of the hypotenuse of the right triangle. A **Computer Journal** gives students an opportunity to write about what they have learned.

For Windows & Macintosh

Guided Practice

For each triangle, find sin *Y*, cos *Y*, and tan *Y* to the nearest thousandth. 7–8. See margin.

7.
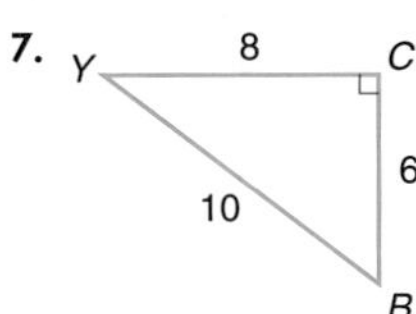

8.
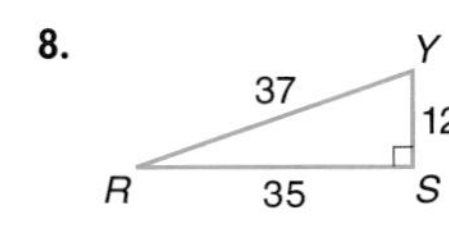

Use a calculator to find the value of each trigonometric ratio to the nearest ten thousandth.

9. cos 35° **0.8192** 10. sin 63° **0.8910** 11. tan 7° **0.1228**

Use a calculator to find the measure of each angle to the nearest degree.

12. cos $C = 0.9613$ **16°** 13. sin $X = 0.7193$ **46°** 14. tan $W = 2.4752$ **68°**

For each triangle, find the measure of the marked acute angle to the nearest degree.

15. **36°**
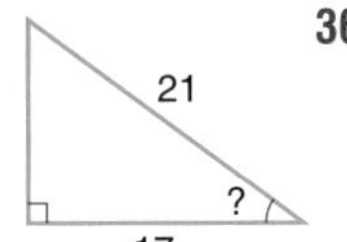

16. **10°**
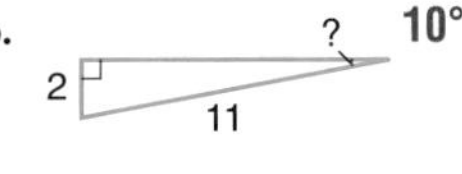

Solve each right triangle. State the side lengths to the nearest tenth and the angle measures to the nearest degree. 17–19. See margin.

17.
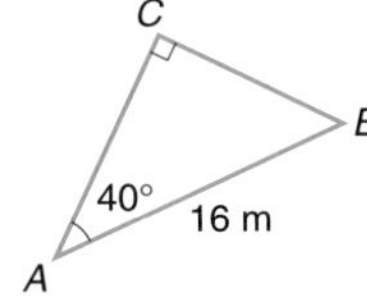

18.
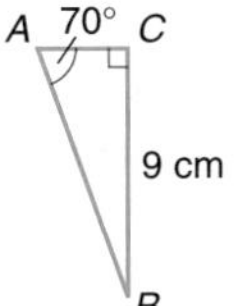

19.
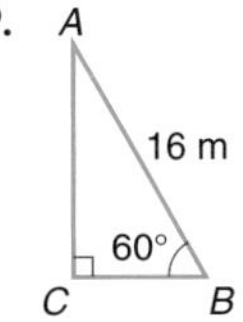

20. **Transportation** Find the angle of elevation of a train track if a train in the mountains rises 15 feet for every 250 feet it moves along the track. **about 3.4°**

EXERCISES

Practice

For each triangle, find sin *G*, cos *G*, and tan *G* to the nearest thousandth. 21–23. See margin.

21.
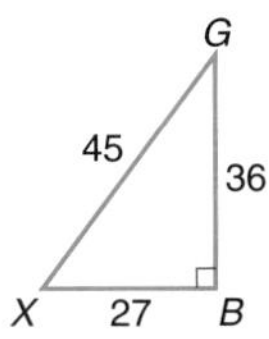

22.
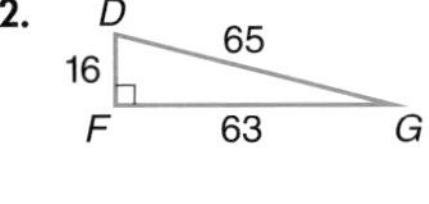

23.
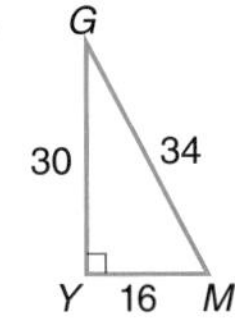

Assignment Guide

Core: 21–57 odd, 58–69
Enriched: 22–54 even, 55–69

For **Extra Practice,** see p. 765.

The red A, B, and C flags, printed only in the Teacher's Wraparound Edition, indicate the level of difficulty of the exercises.

Additional Answers

7. sin $Y = 0.600$, cos $Y = 0.800$, tan $Y = 0.750$
8. sin $Y \approx 0.946$, cos $Y \approx 0.324$, tan $Y \approx 2.917$
17. $\angle B = 50°$, $AC \approx 12.3$ m, $BC \approx 10.3$ m
18. $\angle B = 20°$, $AB \approx 9.6$ cm, $AC \approx 3.3$ cm
19. $\angle A = 30°$, $AC \approx 13.9$ m, $BC = 8$ m
21. sin $G = 0.600$, cos $G = 0.800$, tan $G = 0.750$
22. sin $G \approx 0.246$, cos $G \approx 0.969$, tan $G \approx 0.254$
23. sin $G \approx 0.471$, cos $G \approx 0.882$, tan $G \approx 0.533$

For each triangle, find sin *G*, cos *G*, and tan *G* to the nearest thousandth. **24–26. See margin.**

24. 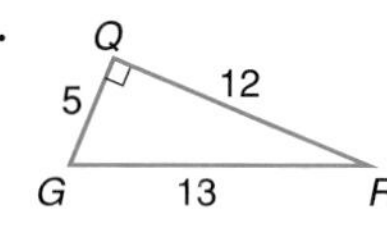

25. 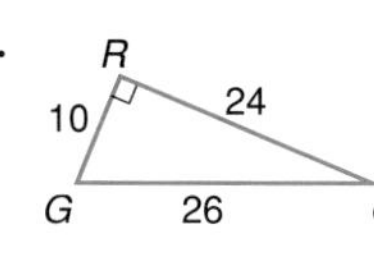

26. 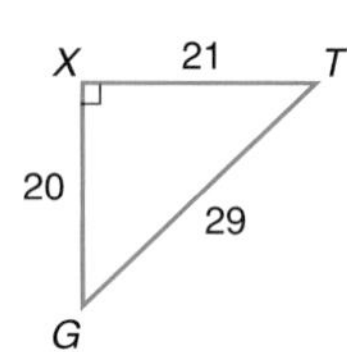

Use a calculator to find the value of each trigonometric ratio to the nearest ten thousandth.

27. sin 21° **0.3584** 28. cos 15° **0.9659** 29. sin 76° **0.9703**
30. tan 56° **1.4826** 31. cos 68° **0.3746** 32. tan 30° **0.5774**

Use a calculator to find the measure of each angle to the nearest degree.

33. $\tan A = 0.6473$ **33°** 34. $\cos B = 0.7658$ **40°** 35. $\sin F = 0.3823$ **22°**
36. $\cos Q = 0.2993$ **73°** 37. $\sin R = 0.8827$ **62°** 38. $\tan J = 8.8988$ **84°**

For each triangle, find the measure of the marked acute angle to the nearest degree.

39. **77°** 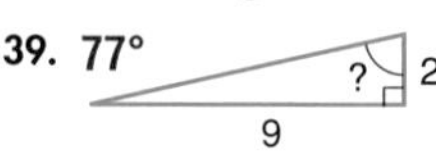

40. **33°** 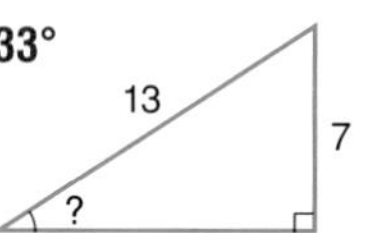

41. **58°** 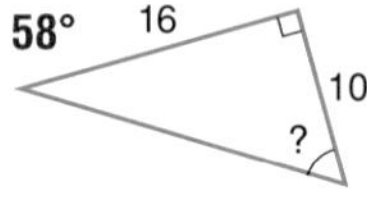

42. 9.3, ?, 9.7 **17°**

43. 12, 37, 20, ? **18°**

44. 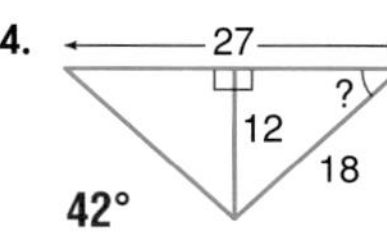

42°

Solve each right triangle. State the side lengths to the nearest tenth and the angle measures to the nearest degree. **45–53. See margin.**

45. 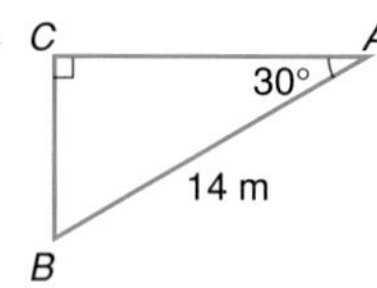

46. 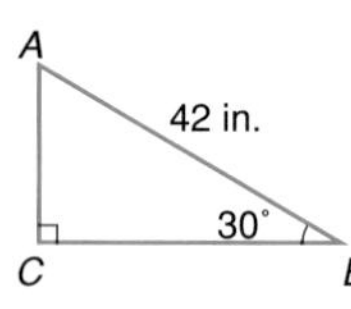

47. 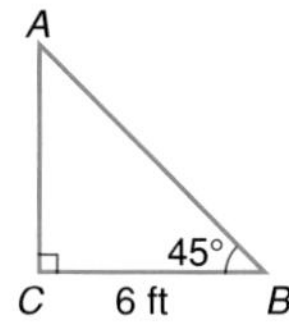

48. 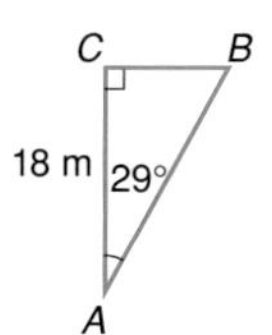

49. 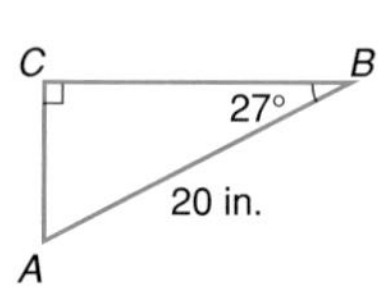

50. 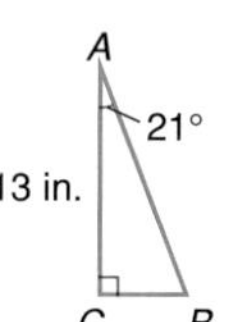

51. 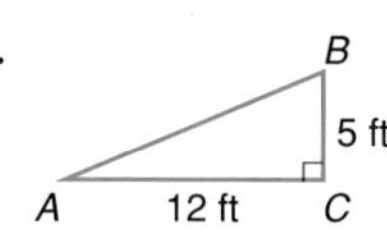

52. 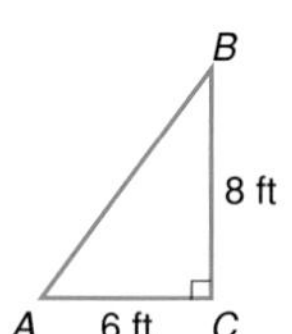

53. 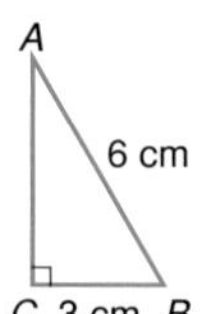

Additional Answers

24. $\sin G = 0.923$, $\cos G = 0.385$, $\tan G = 2.4$
25. $\sin G = 0.923$, $\cos G = 0.385$, $\tan G = 2.4$
26. $\sin G = 0.724$, $\cos G = 0.690$, $\tan G = 1.050$
45. $\angle B = 60°$, $AC = 12.1$ m, $BC = 7$ m
46. $\angle A = 60°$, $AC = 21$ in., $BC = 36.4$ in.
47. $\angle A = 45°$, $AC = 6$ ft, $AB = 8.5$ ft
48. $\angle B = 61°$, $AB = 20.6$ m, $BC = 10$ m
49. $\angle A = 63°$, $AC = 9.1$ in., $BC = 17.8$ in.
50. $\angle B = 69°$, $AB = 13.9$ in., $BC = 5.0$ in.
51. $\angle A = 23°$, $\angle B = 67°$, $AB \approx 12.8$ ft
52. $\angle A = 53°$, $\angle B = 37°$, $AB = 10$ ft
53. $\angle A = 30°$, $\angle B = 60°$, $AC = 5.2$ cm

Study Guide Masters, p. 28

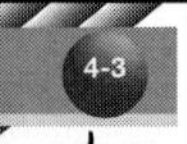

NAME________________ DATE________

Study Guide Student Edition Pages 206–214

Integration: Trigonometry Trigonometric Ratios

For each angle of every right triangle, certain ratios can be set up. These ratios are called **trigonometric ratios.** Study the triangle and then the chart below.

For angle *A*:
$\overline{BC}$ is *opposite* angle *A*.
$\overline{AC}$ is *adjacent* to angle *A*.
$\overline{AB}$ is the *hypotenuse* and is opposite the right angle *C*.

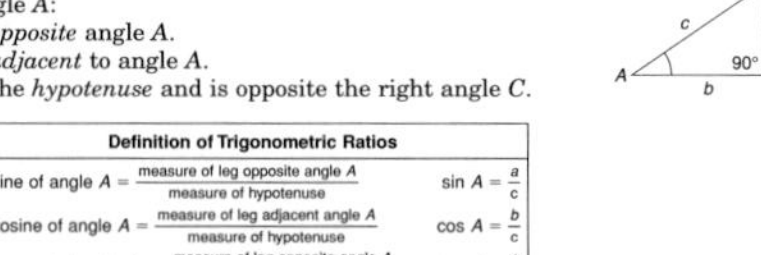

Definition of Trigonometric Ratios	
sine of angle $A = \frac{\text{measure of leg opposite angle } A}{\text{measure of hypotenuse}}$	$\sin A = \frac{a}{c}$
cosine of angle $A = \frac{\text{measure of leg adjacent angle } A}{\text{measure of hypotenuse}}$	$\cos A = \frac{b}{c}$
tangent of angle $A = \frac{\text{measure of leg opposite angle } A}{\text{measure of leg adjacent to angle } A}$	$\tan A = \frac{a}{b}$

Example 1: Find sin *A* to the nearest thousandth.
$\sin A = \frac{a}{c} = \frac{5}{10} = 0.500$

Example 2: Use a calculator to find the measure of angle *A* if $\sin A = 0.500$.
ENTER: 0.5 [INV] [SIN] Result: 30
The measure of angle *A* is 30°.

Find the sine, cosine, and tangent of each acute angle. Round your answers to the nearest thousandth.

1. 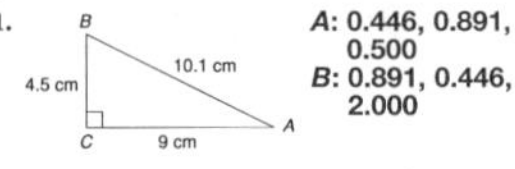

A: 0.446, 0.891, 0.500
B: 0.891, 0.446, 2.000

Use a calculator to find the value of each trigonometric ratio to the nearest ten thousandth.

2. sin 45° 0.7071 3. cos 45° 0.7071 4. tan 45° 1.0000 5. sin 47° 0.7314 6. cos 48° 0.6691

Use a calculator to find the measure of each angle to the nearest degree.

7. $\sin A = 0.7547$ 49° 8. $\tan C = 1.1106$ 48° 9. $\cos B = 0.6947$ 46°

Programming

To run the program repeatedly, press ENTER *after each run is completed and the prompt will appear again for new values of a and b.*

54. Suppose $\angle ABC$ has a right angle C, and the measures of its legs are a and b. The graphing calculator program at the right asks you to enter the lengths of the legs of the triangle. It then calculates the measure of the hypotenuse c, the measures of the acute angles, and the trigonometric ratios for each acute angle. The PAUSE commands in the program stop the calculator's computations and allow you to record the information shown on the screen. Press ENTER to continue the program. Make sure your calculator is in degree mode before running the program. **a–d. See margin.**

```
PROGRAM: LEGS
: Disp "ENTER LENGTHS OF",
  "LEGS A AND B"
: Prompt A, B
: √(A² + B²)→C
: Disp "SIDE C =",C
: Disp "ANGLE A =", tan⁻¹ (A/B)
: Disp "ANGLE B =", tan⁻¹ (B/A)
: Pause
: Disp "sin A=", A/C
: Disp "sin B=", B/C
: Disp "cos A=", B/C
: Pause
: Disp "cos B=", A/C
: Disp "tan A=", A/B
: Disp "tan B=", B/A
```

Use the program for each pair of values for *a* and *b*.

a. 3, 4
b. 5, 12
c. 20, 40
d. 125, 100

Critical Thinking

55. Use the triangle below to determine which of the following statements are true. Justify your answers. **a, d, e; See margin.**

a. $\sin C = \cos B$

b. $\cos B = \dfrac{1}{\sin B}$

c. $\tan C = \dfrac{\cos C}{\sin C}$

d. $\tan C = \dfrac{\sin C}{\cos C}$

e. $\sin B = (\tan B)(\cos B)$

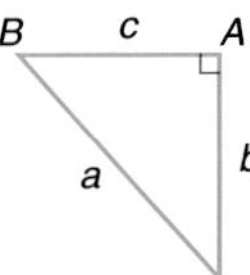

Applications and Problem Solving

56. Archaeology The largest of the pyramids of Egypt has a square base with sides 755 feet long. $\angle PQR$ has a measure of 52°. The top of the pyramid is no longer there. What was the pyramid's original height ($\overline{RP}$) to the nearest foot? **483 ft**

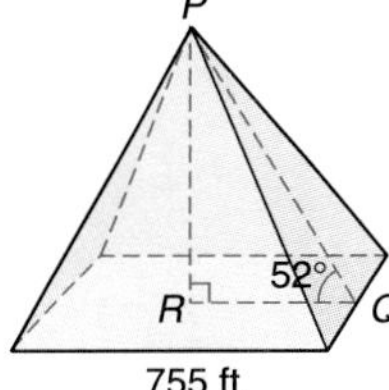

57. Make a Drawing On September 30, 1995, the Ohio State Buckeyes defeated the Notre Dame Fighting Irish in football as the Goodyear blimp relayed aerial views from above. Suppose the blimp was directly over the 50-yard line. A photographer on the ground estimated the angle of elevation to be 85° when she was 65 yards away from the point on the ground directly under the blimp. How high is the blimp? Round your answer to the nearest foot. **2229 ft**

Using the Programming Exercises The program given in Exercise 54 is for use with a TI-82 graphing calculator. For other programmable calculators, have students consult their owner's manual for commands similar to those presented here.

Additional Answers

54a. 5; 36.9°, 53.1°; $\sin A = 0.600$, $\sin B = 0.800$, $\cos A = 0.800$, $\cos B = 0.600$, $\tan A = 0.750$, $\tan B = 1.333$

54b. 13; 22.6°, 67.4°; $\sin A = 0.385$, $\sin B = 0.923$; $\cos A = 0.923$, $\cos B = 0.385$, $\tan A = 0.417$, $\tan B = 2.400$

54c. 44.7; 26.6°, 63.4°; $\sin A = 0.447$, $\sin B = 0.894$, $\cos A = 0.894$, $\cos B = 0.447$, $\tan A = 0.500$, $\tan B = 2.000$

54d. 160.1; 51.3°, 38.7°; $\sin A = 0.781$, $\sin B = 0.625$, $\cos A = 0.625$, $\cos B = 0.781$, $\tan A = 1.250$, $\tan B = 0.800$

55a. $\sin C = \dfrac{c}{a}$, $\cos B = \dfrac{c}{a}$

55b. $\cos B = \dfrac{c}{a}$, $\dfrac{1}{\sin B} = \dfrac{a}{b}$

55c. $\tan C = \dfrac{c}{b}$, $\dfrac{\cos C}{\sin C} = \dfrac{b}{a} \div \dfrac{c}{a} = \dfrac{b}{c}$

Practice Masters, p. 28

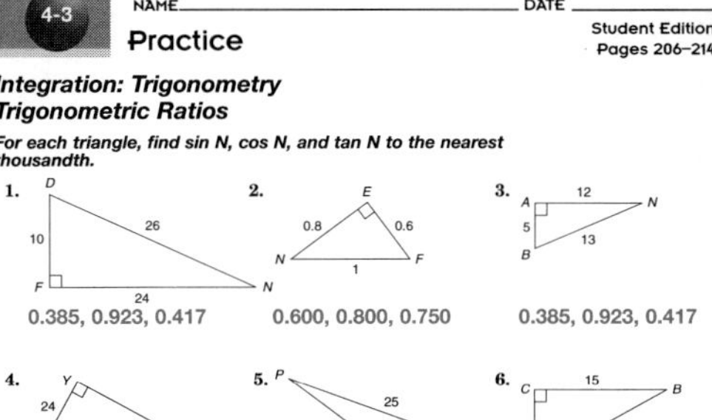

4-3 NAME_____ DATE_____
Practice
Student Edition Pages 206–214

Integration: Trigonometry Trigonometric Ratios

For each triangle, find sin N, cos N, and tan N to the nearest thousandth.

1. 0.385, 0.923, 0.417
2. 0.600, 0.800, 0.750
3. 0.385, 0.923, 0.417
4. 0.324, 0.946, 0.343
5. 0.960, 0.280, 3.429
6. 0.882, 0.471, 1.875

Use a calculator to find the measure of each angle to the nearest degree.

7. $\sin A = 0.3779$ 22°
8. $\cos B = 0.8931$ 27°
9. $\cos G = 0.4545$ 63°

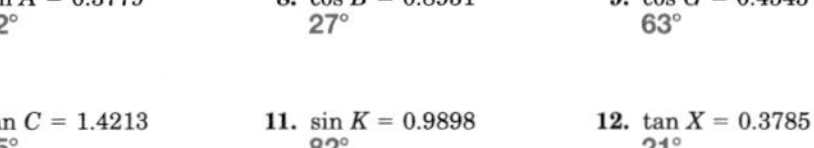

10. $\tan C = 1.4213$ 55°
11. $\sin K = 0.9898$ 82°
12. $\tan X = 0.3785$ 21°

For each triangle, find the measure of the marked acute angle to the nearest degree.

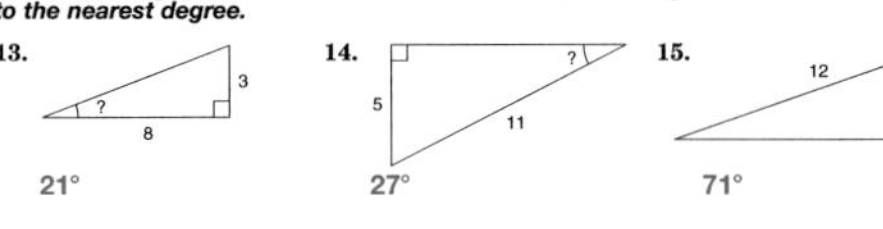

13. 21°
14. 27°
15. 71°

Extension

Connections The length in miles ℓ along the surface of Earth of one degree longitude is given by the formula $\ell = 2\pi \dfrac{(4000 \cos x)}{360}$, where x is the number of degrees latitude. Find the length to the nearest mile of one degree longitude at each latitude.

a. 30° latitude **60 miles**
b. 60° latitude **35 miles**
c. 90° latitude **0 miles**

Additional Answers

55d. $\tan C = \dfrac{c}{b}$, $\dfrac{\sin C}{\cos C} = \dfrac{c}{a} \div \dfrac{b}{a} = \dfrac{c}{b}$

55e. $\sin B = \dfrac{b}{a}$, $(\tan B)(\cos B) = \dfrac{b}{c} \cdot \dfrac{c}{a} = \dfrac{b}{a}$

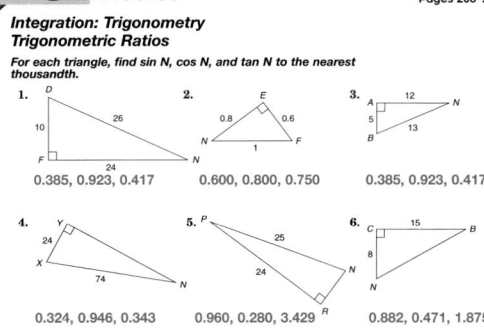

4 ASSESS

Closing Activity

Speaking Ask students to give the definition of trigonometry ratios in their own words without using terms such as *hypotenuse.* If students have to use simpler definitions, students are more likely to remember them.

Enrichment Masters, p. 28

NAME__________ DATE ______

Enrichment

Student Edition Pages 206–214

Modern Art

The painting below, aptly titled, "Right Triangles," was painted by that well-known artist Two-loose La-Rectangle. Using the information below, find the dimensions of this masterpiece. (*Hint:* The triangle that includes $\angle F$ is isosceles.)

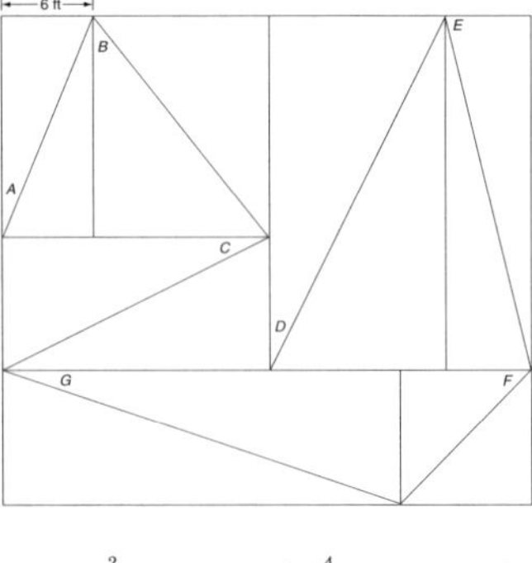

$\tan A = \frac{2}{5}$ $\tan B = \frac{4}{5}$ $\tan C = \frac{1}{2}$

$\tan D = \frac{1}{2}$ $\tan E = 4$ $\tan F = 1$

$\tan G = \frac{1}{3}$

1. What is the length of the painting? 33 ft
2. What is the width of the painting? 36 ft

Mixed Review

58. 121m

58. **Civil Engineering** The city of Mansfield plans to build a bridge across Spring Lake. Use the information in the diagram at the right to find the distance across Spring Lake. (Lesson 4–2)

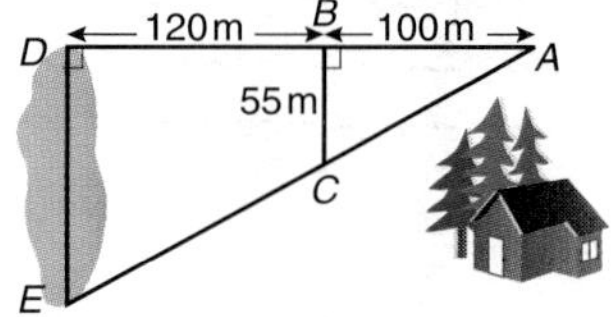

59. Solve $\frac{x}{5} = \frac{x+3}{10}$. (Lesson 4–1) **3**
60. **Baseball** The batting averages for 10 players on a baseball team are 0.234, 0.253, 0.312, 0.333, 0.286, 0.240, 0.183, 0.222, 0.297, and 0.275. Find the median batting average for these players. (Lesson 3–7) **0.264**
61. Solve $\frac{2}{5}y + \frac{y}{2} = 9$. (Lesson 3–3) **10**
62. Write an equation to represent the situation below. Then solve the problem. (Lesson 3–2) **$5x = \$47.50$; $9.50**

    Joyce Conners paid $47.50 for five football tickets. What was the cost per ticket?
63. Name the sets of numbers to which $\sqrt{36}$ belongs. Use N for natural numbers, W for whole numbers, Z for integers, Q for rational numbers, and I for irrational numbers. (Lesson 2–8) **N, W, Z, Q**
64. Simplify $\frac{3a+9}{3}$. (Lesson 2–7) **$a + 3$**
65. Find $\left(-\frac{2}{3}\right)\left(-\frac{1}{5}\right)$. (Lesson 2–6) **$\frac{2}{15}$**
66. Find $-13 + (-8)$. (Lesson 2–3) **−21**
67. Simplify $8x + 2y + x$. (Lesson 1–8) **$9x + 2y$**
68. Evaluate $(9 - 2 \cdot 3)^3 - 27 + 9 \cdot 2$. (Lesson 1–6) **18**
69. Evaluate 5^3. (Lesson 1–1) **125**

Morgan's Conjecture

The excerpt below appeared in an article in *The Columbus Dispatch* on December 21, 1994.

RYAN MORGAN WOULD HAVE GOTTEN an "A" in geometry even if he hadn't unearthed a mathematical treasure. But the persistent Baltimore high school sophomore pushed a hunch into a theory. He calls it "Morgan's Conjecture" and is hoping it will soon be "Morgan's Theorem." In geometric circles, developing a theorem is a big deal—especially if you're only 15....What did Ryan see?...When the sides of a triangle are divided by an odd number larger than one, and when lines are drawn from the division points on the sides of the triangle to the vertices, there always will be a hexagon in the interior of the triangle. And the area of that hexagon always will be a predictable fraction of the area of the larger triangle. The fraction is determined by a complex formula that Ryan worked out. ■

1. In talking about his work, the student said, "It's mostly just luck in playing around with it." What other factors besides luck might have been involved? **See margin.**
2. Are you surprised to learn that students can discover theorems? Why or why not? **See students' work.**
3. Is there an area of mathematics that interests you that you would like to explore further? How would you do it? **See students' work.**

Mathematics and SOCIETY

Mathematics has always been a field for young prodigies. The great mathematicians have usually been most productive in their twenties and thirties. One reason may be that success in mathematics does not depend on knowledge and experience of the real world, but on intellectual insight.

Answer for Mathematics and Society

1. **Answers will vary. Sample answers: curiosity, determination, a systematic approach, willingness to try new methods, looking for patterns, learning from mistakes, and keeping good records.**

4-4 Percents

What YOU'LL LEARN

- To solve percent problems, and
- to solve problems involving simple interest.

Why IT'S IMPORTANT

You can use percents to solve problems involving finance, nutrition, and statistics.

Cosmetics

According to *Good Housekeeping*, 3 out of 5 teenage girls wear mascara. What is the rate of teenage girls wearing mascara per 100 girls?

You can solve this problem by using a proportion. The ratio of teenage girls who wear mascara to the total number of teenage girls is $\frac{3}{5}$. Write a proportion that sets $\frac{3}{5}$ equal to a ratio with a denominator of 100.

$$\frac{3}{5} = \frac{n}{100}$$

$$60 = n$$

Teenage girls wear mascara at a rate of 60 per 100, or 60 percent. A **percent** is a ratio that compares a number to 100. Percent also means *per hundred*, or *hundredths*. Percents can be expressed with a percent symbol (%), as fractions, or as decimals.

$$60\% = \frac{60}{100} = 0.60$$

Example  **Write $\frac{3}{4}$ as a percent.**

Method 1

Use a proportion.

$\frac{3}{4} = \frac{n}{100}$

$300 = 4n$

$75 = n$

Thus, $\frac{3}{4}$ is equal to $\frac{75}{100}$ or 75%.

Method 2

Use a calculator.

$\frac{3}{4} = \frac{n}{100}$

$100\left(\frac{3}{4}\right) = n$

Enter: 100 [×] 3 [÷] 4 [=] 75

FYI

Sixty percent of teenage girls wear lipstick and lip gloss, 72% wear nail polish, 55% wear eye shadow, and 50% wear blush.

Proportions are often used to solve percent problems. One of the ratios in these proportions is always a comparison of two numbers called the **percentage** and the **base**. The other ratio, called the **rate**, is a fraction with a denominator of 100.

$$\begin{matrix} \text{percentage} \rightarrow \\ \text{base} \rightarrow \end{matrix} \frac{3}{4} = \frac{75}{100} \leftarrow \text{Rate}$$

Percent Proportion	$\frac{\text{Percentage}}{\text{Base}} = \text{Rate}$ or $\frac{\text{Percentage}}{\text{Base}} = \frac{r}{100}$

The $\frac{\text{percentage}}{\text{base}}$ represents $\frac{\text{part}}{\text{whole}}$.

FYI

The female teenage population in 1995 is estimated to be about 15 million. Based on this number, 9 million wear lipstick and gloss, 10.8 million wear nail polish, 8.25 million wear eye shadow, and 7.5 million wear blush.

4-4 LESSON NOTES

NCTM Standards: 1–5

Instructional Resources

- Study Guide Master 4-4
- Practice Master 4-4
- Enrichment Master 4-4
- Assessment and Evaluation Masters, pp. 99–100
- Multicultural Activity Masters, p. 7
- Tech Prep Applications Masters, p. 7

Transparency 4-4A contains the 5-Minute Check for this lesson; **Transparency 4-4B** contains a teaching aid for this lesson.

Recommended Pacing	
Standard Pacing	Days 6 & 7 of 15
Honors Pacing	Days 6 & 7 of 15
Block Scheduling*	Day 3 of 7 (along with Lesson 4-5)
Alg. 1 in Two Years*	Days 8 & 9 of 20

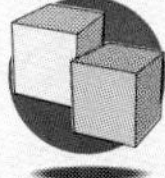

*For more information on pacing and possible lesson plans, refer to the *Block Scheduling Booklet* and *Algebra 1 in Two Years.*

1 FOCUS

5-Minute Check
(over Lesson 4-3)

For the triangle below, express each trigonometric ratio as a fraction in simplest form.

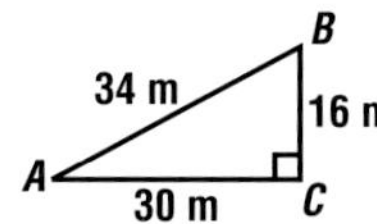

1. sin A $\frac{8}{17}$
2. sin B $\frac{15}{17}$
3. cos B $\frac{8}{17}$
4. cos A $\frac{15}{17}$
5. tan A $\frac{8}{15}$

Motivating the Lesson

Questioning Ask students to state one quick way to find the following percents of any number.

1. 50% **Take one-half of the number.**
2. 25% **Take one-half of the number twice or take one-fourth of the number.**
3. 10% **Move the decimal point one place to the left.**
4. 1% **Move the decimal point two places to the left.**

2 TEACH

Teaching Tip Example 1 can also be solved by finding a ratio that is equivalent to $\frac{3}{4}$ and has a denominator of 100. This ratio is $\frac{3.35}{4.25}$ or $\frac{75}{100}$.

In-Class Examples

For Example 1
Write each ratio as a percent.

a. $\frac{9}{10}$ **90%**
b. $\frac{4}{5}$ **80%**

For Example 2
Solve.

a. 30 is what percent of 70? **43%**
b. 46 is what percent of 80? **57.5%**

For Example 3
Solve.

a. 70% of what number is 56? **80**
b. 45% of what number is 270? **600**
c. Find 17% of 32.6. **5.542**
d. Find 42% of 180. **75.6**

Teaching Tip The formula $\frac{P}{B} = R$ can easily be rewritten as $P = R \cdot B$. This form is used if the rate and base are known and the percentage is to be found. If the rate is given as a percent, you can write it as a decimal.

Teaching Tip Some students confuse the terms *percentage* and *percent*. If you prefer, you could use the word *part* instead of *percentage*. Then the percent proportion would be $\frac{\text{part}}{\text{base}} = \frac{r}{100}$.

Example 

a. 30 is what percent of 50?

Use the percent proportion.

$\frac{\text{Percentage}}{\text{Base}} = \frac{r}{100}$

The percentage is 30. The base is 50. $\frac{30}{50} = \frac{r}{100}$

$3000 = 50r$

$60 = r$

Thus, 30 is 60% of 50.

b. 20 is what percent of 30?

Use a calculator.

$\frac{\text{Percentage}}{\text{Base}} = \frac{r}{100}$

$\frac{20}{30} = \frac{r}{100}$

Enter: 20 [÷] 30 [×] 100 [=] 66.66666667

Thus, 20 is $66.\overline{6}$% of 30. This can also be written as $66\frac{2}{3}$%.

You can also write equations to solve problems with percents.

Example 

a. 60% of what number is 54?

60% of what number is 54?

$\frac{60}{100} \cdot x = 54$

$0.6x = 54$

$x = 90$

Thus, 60% of 90 is 54.

b. What number is 40% of 37.5?

What number is 40% of 37.5?

$x = \frac{40}{100} \cdot 37.5$

$x = 15$

Thus, 15 is 40% of 37.5.

TECHNOLOGY Tip

Most calculators have a key labeled [%]. To find 28% of 58.4 on a scientific calculator, enter:

28 [2nd] [%] [×]
58.4 [=] 16.352.

MODELING MATHEMATICS 

Making Circle Graphs

Materials:  compass  protractor

You can use a circle graph to compare parts of a whole. Follow the steps to display the data below in a circle graph. **a–e. See margin.**

Where's the Remote?	
Number of Times TV Remote Misplaced Per Week	**Number of People Responding**
Never	220
1–5	190
5 or more	85
Don't Know	5

Your Turn

a. Find the total number of people surveyed.

b. Find the ratio that compares the number of people that responded for each category to the total number of people that responded.

c. Since there are 360° in a circle, multiply each ratio by 360 to find the number of degrees for each section of the graph. Round to the nearest degree.

d. Use a compass to draw a circle and a radius as shown at the right.

e. Use a protractor to draw angles. Start with the least number of degrees. Repeat for the remaining sections. Label each section and give the graph a title.

f. What is the sum of the percents in your graph? **f. 100%**

Answers for Modeling Mathematics

a. 500
b. $\frac{220}{500}, \frac{190}{500}, \frac{85}{500}, \frac{5}{500}$
c. 158°, 137°, 61°, 4°
d. See students' work.

e. 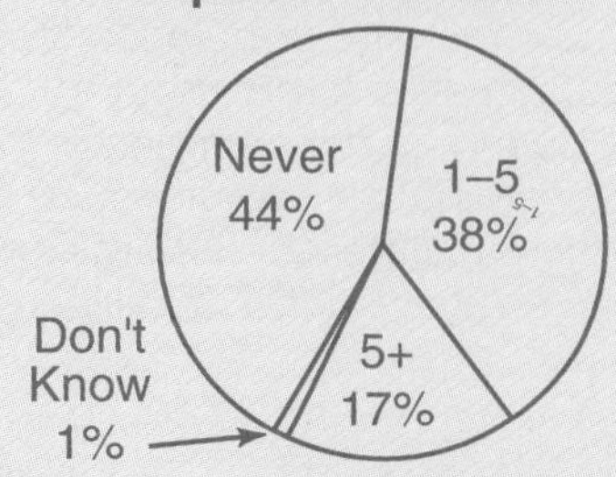

Percents are also used in simple interest problems. **Simple interest** is the amount paid or earned for the use of money. The formula $I = prt$ is used to solve problems involving simple interest. In the formula, I represents the interest, p represents the amount of money invested, called the *principal*, r represents the annual interest rate, and t represents time in years.

Example 4

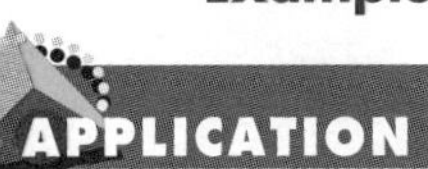

APPLICATION Finance

a. Luis Hernandez has some money he has saved over the summer from mowing lawns. He wants to put some of it in a 6-month certificate of deposit (CD) account at BancOne that would pay 6% annual interest. He doesn't want to put all of his money in the CD because he wants some spending money. He is hoping to earn \$45 in interest to buy a new video game. How much money should Luis put in the CD?

Explore Let p = the amount of money Luis should deposit.

Plan

$I = prt$

$45 = p(0.06)(0.5)$ *Write 6% as 0.06 and 6 months as 0.5 years.*

Solve

$45 = 0.03p$ *Multiply 0.06 by 0.5.*

$1500 = p$ *Divide each side by 0.03.*

Examine When p is 1500, $I = (1500)(0.06)(0.5)$ or \$45. Luis should deposit \$1500.

b. Whitney Williamson has \$30,000 she would like to invest. She has a choice of two interest-paying bonds, one offering a 6% annual interest and the other paying 7.5% annual interest. She would like to earn \$2100 in interest in one year. If she earns any more than \$2100, she would need to pay a higher rate of income tax. How much money should she invest in each bond?

Explore Let n = the amount of money invested at 6%. Then $30{,}000 - n$ = the amount of money invested at 7.5%.

Plan

$\underbrace{n(0.06)(1) \text{ or } 0.06n}_{\text{interest on 6\% investment}}$

$\underbrace{(30{,}000 - n)(0.075)(1) \text{ or } 2250 - 0.075n}_{\text{interest on 7.5\% investment}}$

Solve

$$\underbrace{0.06n}_{\text{interest at 6\%}} + \underbrace{(2250 - 0.075n)}_{\text{interest at 7.5\%}} = \underbrace{2100}_{\text{total interest}}$$

$$-0.015n + 2250 = 2100 \quad \text{Add } 0.06n \text{ and } -0.075n.$$

$$-0.015n = -150$$

$$n = 10{,}000$$

Whitney should invest \$10,000 at 6% and \$30,000 − \$10,000 or \$20,000 at 7.5%.

Examine $(10{,}000)(0.06)(1) + (20{,}000)(0.075)(1) = 2100$

The answer checks.

Teaching Tip Emphasize that when a problem asks for the percent, the students must find the value of r.

MODELING MATHEMATICS This modeling activity shows students how the circle graph displays parts of a whole. Each part is a percent of the total area of the circle.

In-Class Example

For Example 4

a. Luis sees that another bank has a 3-month certificate of deposit that would pay 8% annual interest. Which would be better for Luis? **BancOne; his \$1500 would earn only \$30 at the other bank.**

b. If Whitney had invested part of her money at 5% and the rest at 8%, how much had she invested at each rate if the interest after 1 year was \$1995? **\$13,500 at 5%; \$16,500 at 8%**

Teaching Tip Remind students that interest rate is annual. Therefore, time should be expressed in years when working with the simple interest formula.

Cooperative Learning

Categorizing Separate students into groups of five for this library research project. Have students list five occupations in which salary is dependent upon commission. Have students research these occupations to find the usual rate of commission. Challenge students to analyze their lists to find a relationship between the dollar value of the item sold and the size of the commission awarded an employee. For more information on the categorizing strategy, see *Cooperative Learning in the Mathematics Classroom*, one of the titles in the Glencoe Mathematics Professional Series, pages 31–32.

3 PRACTICE/APPLY

Check for Understanding

Exercises 1–15 are designed to help you assess your students' understanding through reading, writing, speaking, and modeling. You should work through Exercises 1–5 with your students and then monitor their work on Exercises 6–15.

Additional Answer

2. Enter: 57 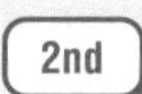 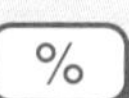  42 (=) 23.94

CHECK FOR UNDERSTANDING

Communicating Mathematics

Study the lesson. Then complete the following. **2. See margin.**

1. **Compare** $\frac{r}{100}$ in the percent proportion and r in the simple interest formula. How are they the same? How are they different? **1. They are the same.**
2. **Explain** how to find 57% of 42 using a scientific calculator.
3. **Explain** why an investor might invest a greater amount of money in a lower rate of interest instead of a higher rate. **3. The investment at the higher rate may be too risky.**
4. **Read** the following excerpt from an article about the America's Cup yacht race in *Motor Boating & Sailing* magazine. Explain what is incorrect in this excerpt and why. **4. The sum of the percents is greater than 100.**

> With a Ph.D. from MIT, (Dr. Koch) takes a scientific view of things, and quantifies the effort involved in winning the Cup as 55 percent boat speed, 20 percent tactics, 20 percent crew work, 5 percent luck, and less than 5 percent physical strength.

MODELING MATHEMATICS

5. **Draw** a circle graph to represent the number of students in your school by grade level. **See students' work.**

Guided Practice

Write each ratio as a percent and then as a decimal.

6. $\frac{3}{4}$ **75%, 0.75**
7. $\frac{43}{100}$ **43%, 0.43**
8. $\frac{2}{25}$ **8%, 0.08**

Use a proportion to answer each question.

9. Eleven is what percent of 20? **55%**
10. What percent of 80 is 45? **10. $56\frac{1}{4}$%; 56.25%**

Use an equation to answer each question.

11. What number is 30% of 50? **15**
12. What percent of 75 is 16? **12. $21\frac{1}{3}$%; $21.\overline{3}$%**

Use $I = prt$ to find the missing quantity.

13. Find I if p = \$1500, r = 7%, and t = 6 months. **\$52.50**
14. Find p if I = \$196, r = 10%, and t = 7 years. **\$280**
15. **Finance** Kaylee Richardson invested \$7200 for one year, part at 10% annual interest and the rest at 14% annual interest. Her total interest for the year was \$960. How much money did she invest at each rate? **\$1200 at 10%, \$6000 at 14%**

EXERCISES

Practice 

Write each ratio as a percent and then as a decimal.

16. $\frac{67}{100}$ **67%, 0.67**
17. $\frac{6}{20}$ **30%, 0.30**
18. $\frac{5}{8}$ **$62\frac{1}{2}$%, 0.625**
19. $\frac{7}{10}$ **70%, 0.70**
20. $\frac{5}{6}$ **$83\frac{1}{3}$%, $0.8\overline{3}$**
21. $\frac{9}{5}$ **180%, 1.8**
22. $\frac{2}{3}$ **$66\frac{2}{3}$%, $0.\overline{6}$**
23. $\frac{25}{40}$ **$62\frac{1}{2}$%, 0.625**
24. $\frac{20}{8}$ **250%, 2.5**

Reteaching

Using Models Have students go through the business sections of newspapers. They can compare different terms and rates of certificates of deposit for different banks. They can also compare interest rates on mortgages offered through different companies.

25. 50%
27. 
$12\frac{1}{2}\%$; 12.5%

Use a proportion to answer each question.

25. Thirty-five is what percent of 70?
26. What percent of 60 is 18? **30%**
27. Eight is what percent of 64?
28. Six is what percent of 15? **40%**
29. What percent of 2 is 8? **400%**
30. What percent of 14 is 4.34? **31%**

Use an equation to answer each question.

33. 72.3
34. 40%
35. 702.4

B
31. What percent of 160 is 4? **2.5%**
32. What number is 25% of 56? **14**
33. Twelve is 16.6% of what number?
34. Thirty-two is what percent of 80?
35. 17.56 is 2.5% of what number?
36. What percent of 75 is 30? **40%**

Solve.

37. \$64.93 is what percent of \$231.90? **28%**

C
38. Find 112% of \$500. **\$560**
39. Find 81% of 32. **25.92**

Use $I = prt$ to find the missing quantity.

40. Find r if I = \$5920, p = \$4000, and t = 3 years. **49.3%**
41. Find r if I = \$780, p = \$6500, and t = 1 year. **12%**
42. Find I if p = \$3200, r = 9%, and t = 18 months. **\$432**
43. Find I if p = \$5000, $r = 12\frac{1}{2}\%$, and t = 5 years. **\$3125**

44. $4\frac{1}{2}$ years

44. Find t if I = \$2160, p = \$6000, and r = 8%.
45. Find p if I = \$756, r = 9%, and $t = 3\frac{1}{2}$ years. **\$2400**

Critical Thinking

46. If x is 225% of y, then y is what percent of x? **$44.\overline{4}\%$**

Applications and Problem Solving

47. **Health** Physicians in the United States treated 4.1 million broken bones in 1992. The chart at the right shows the fractures by age. Find the percentage of fractures by age. **39%, 34%, 15%, 12%**

Age	Fractures
Under 18	1.6 million
18–44	1.4 million
45–64	600,000
65 and older	500,000

Source: American Academy of Orthopedic Surgeons

48. **Finance** Melanie Morgan invested \$5000 for one year, part at 9% annual interest and the rest at 12% annual interest. The interest from the investment at 9% was \$198 more than the interest from the investment at 12%. How much money did she invest at 9%? **\$3800**

49. **Finance** Juan and Rosita Diaz have invested \$2500 at 10% annual interest. They have \$6000 more to invest. At what rate must they invest the \$6000 to have a total of \$9440 at the end of the year? **11.5%**

50. **Statistics** According to The Alden Group, desserts in the home are not always eaten in the kitchen or dining room. In fact, only 30% of desserts are eaten in the kitchen and 14% in the dining room. Eighteen percent are most often eaten in the living room, 10% in the den, and an amazing 28% are eaten in the bedroom. Make a circle graph showing these data. **See margin.**

Assignment Guide

Core: 17–45 odd, 46, 47–53 odd, 55–64
Enriched: 16–44 even, 46–64
All: Self Test, 1–10

For **Extra Practice,** see p. 765.

The red A, B, and C flags, printed only in the Teacher's Wraparound Edition, indicate the level of difficulty of the exercises.

Additional Answer

50. Desserts Eaten in the Home

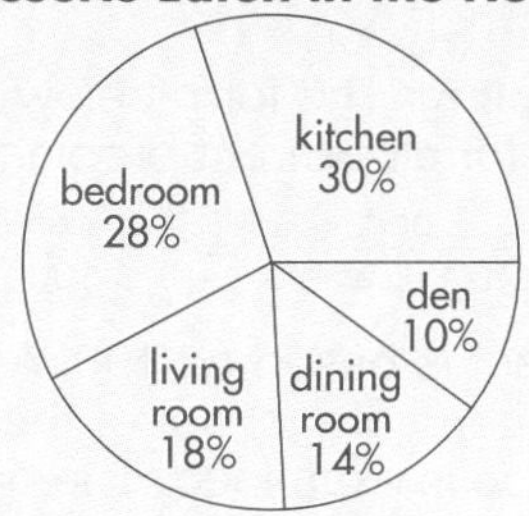

Study Guide Masters, p. 29

Percents

A percent problem may be easier to solve if a proportion is used.

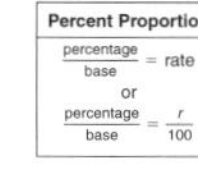

Example 1: 25 is what percent of 30?

$\frac{25}{30} = \frac{r}{100}$ (percentage → 25, base → 30, ← rate)

$2500 = 30r$

$83\frac{1}{3} = r$

25 is $83\frac{1}{3}\%$ of 30.

Example 2: What number is 24% of 200?

$\frac{n}{200} = \frac{24}{100}$ (percentage → n, base → 200, ← rate)

$n = \frac{24}{100}(200)$

$= 48$

48 is 24% of 200.

Use a proportion to answer each question.

1. Eight is what percent of 20? 40%
2. Thirty is what percent of 50? 60%
3. What is 75% of 24? 18
4. Find 60% of 90. 54
5. Twelve is 20% of what number? 60
6. 19.3 is 25% of what number? 77.2
7. On Wednesday Jean's Nursery received a shipment of 60 flowering crabapple trees. Jean had ordered 80 trees. What percent of her order arrived on Wednesday? 75%
8. Phil received a commission of 5% on the sale of a house. If the amount of his commission was \$4780, what was the selling price of the house? \$95,600

Classroom Vignette

"We looked at the Nutritional Information on food boxes and used proportions to find the percentage of fat calories in one serving. For example, a muffin mix has 2 grams of fat and 140 calories in one serving. One gram of fat is about 9 calories.

Fat calories → $\underline{18}$ = $\underline{\%}$ ← percentage of fat
Total calories → 140 = 100 ← total percentage

So, one serving of the muffin mix is about 13% fat calories. I encourage students to find foods that have less than 25% fat calories."

Lara Dillinger
Rock Creek High School
St. George, Kansas

Additional Answer

54. **U.S. Coins Minted in 1994**

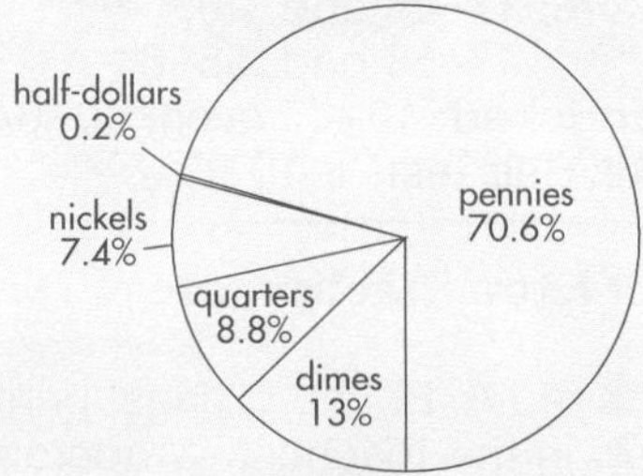

4 ASSESS

Closing Activity

Modeling The form below can be used for all percent problems.

$$\frac{\text{part}}{\text{base or whole}} = \frac{\text{\# of percent}}{100}$$

Percent is based on 100. A certain part (percentage) out of the whole or base is examined. The base follows the word *of.* Percent problems give two of the three quantities: part, base, and number of percent. So, there are three types of percent problems:

1. find the part (percentage),
2. find the base, or
3. find the number of percent.

Go through the exercises to determine the type of each problem.

Practice Masters, p. 29

4-4 NAME________________ DATE________

Practice — Student Edition Pages 215–221

Percents

Use a proportion to answer each question.

1. Eight is what percent of 64? $12\frac{1}{2}\%$
2. What percent of 40 is 25? $62\frac{1}{2}\%$
3. What percent of 50 is 80? 160%
4. One is what percent of 6? $16\frac{2}{3}\%$
5. Six is what percent of 5? 120%
6. What percent of 125 is 2? 1.6%

Use an equation to answer each question.

7. What number is 60% of 95? 57
8. 14 is 28% of what number? 50
9. Fifteen is what percent of 40? $37\frac{1}{2}\%$
10. What percent of 52 is 39? 75%
11. Twenty is what percent of 30? $66\frac{2}{3}\%$
12. What number is 85% of 40? 34

Solve.

13. Thirty is 20% of what number? 150
14. Find 8% of 75. 6
15. Find 6% of $7500. $450
16. Forty is what percent of 90? 44.4%
17. Fifty is 15% of what number? $333\frac{1}{3}$
18. Sixty is what percent of 40? 150%
19. Find 9% of $10,000. $900
20. Fifteen is what percent of 120? $12\frac{1}{2}\%$
21. Thirty-three is what percent of 100? 33%
22. Find 2.5% of 90. 2.25

51. **Consumerism** Luke's father used his credit card to purchase a $330 television. The bank that issued the credit card charges an annual interest rate of 19.8%. Each month they charge interest on the unpaid balance of the account. If Luke's father does not make any more purchases and makes no payments, how much interest will he be charged next month? **$5.45**

52. **Nutrition** The table at the right lists the number of calories and grams of fat in eight popular pizza toppings. The estimates are based on two slices of a 14-inch pizza. To find the percentage of calories from fat, multiply the fat grams by 9 and divide the result by the number of calories. Which of the toppings gets the highest percentage of its calories from fat? **black olives**

Topping	Calories	Fat Grams
Extra Cheese	168	8
Sausage	97	8
Pepperoni	80	7
Black Olives	56	5
Ham	41	2
Onion	11	<1
Green Pepper	5	0
Mushrooms	5	0

53. **Data Analysis** A survey was taken recently to determine what Americans would serve President Clinton and his family for dinner. The survey results are shown at the right.
   a. Suppose 258 people were surveyed. How many people said they would serve chicken marsala? How many said pizza?
   b. The sum of the percentages in the survey is not 100%. Why do you think this is the case? **Respondents could choose more than one entree.**

53a. about 54 people; about 23 people

What Would You Serve?	
steak and potatoes	39%
lasagna	25%
chicken marsala	21%
poached salmon	18%
hamburgers or hot dogs	15%
spaghetti and meatballs	14%
pasta primavera	13%
chef salad	10%
pizza	9%
spaghetti with tomato sauce	8%
soup and sandwiches	5%

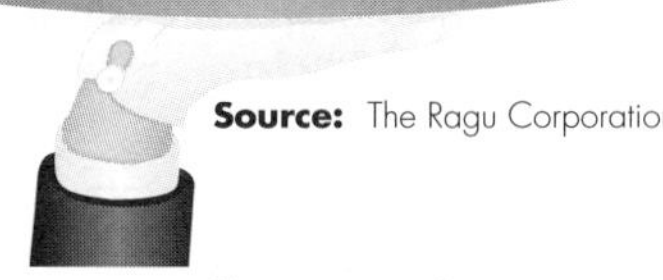

Source: The Ragu Corporation

54. **Money** The number of coins minted in 1994 is shown in the table at the right. Make a circle graph of the data. **See margin.**

U.S. Coins Minted in 1994	
Pennies	14,120,000,000
Nickels	1,480,000,000
Dimes	2,600,000,000
Quarters	1,760,000,000
Half-Dollars	40,000,000

Mixed Review

General Sherman

55. **Trees** While visiting the Sequoia National Park in California, Tyler stops near a tree called the "General Sherman." The guide tells him it is the world's largest tree according to its volume of wood. Tyler is standing about 160 feet from the tree. If he is looking at the top of the tree at a 60° angle, about how tall is the tree? (Lesson 4–3) **about 277 feet**

56. **Geometry** ΔABC and ΔRQP are similar. If $a = 10$, $r = 5$, $q = 3$, and $p = 4$, find the measures of the remaining sides. (Lesson 4–2) $b = 6$; $c = 8$

C, A, B, a, b, c; P, R, Q, r, q, p

57. Solve $\frac{9}{x-8} = \frac{4}{5}$. (Lesson 4–1) $\frac{77}{4}$; **19.25**

Extension

Problem Solving At Joe's Clothing, the manager increased the price of a coat *x* percent and then reduced the price *x* percent. At the local department store, the same coat that originally sold for the same price was reduced *x* percent and then the price was increased by *x* percent. If you purchased the coat now, which store would have the cheaper price? **Both are the same.**

58. **Statistics** In a girl's basketball game between East High School and Monroe High School, the East players' individual scores were 12, 4, 5, 3, 11, 23, 4, 6, 7, and 8. Find the mean, median, and mode of the individual points. (Lesson 3–7) **8.3, 6.5, 4**

59. Solve $y + (-7.5) = -12.2$. (Lesson 3-1) **−4.7**

60. **Diving** Greg Louganis of the United States has won a record of five world titles in diving. He won the highboard title in 1978, the highboard and springboard in 1982 and 1986, and four Olympic gold medals in 1984 and 1988. The total score for a dive is the sum of the scores of each dive times the degree of difficulty of the dive. His final dive in 1988 had a degree of difficulty of 3.4. If his scores were 9, 8, and 8.5, find his total score for the dive. (Lesson 2–6) **86.7**

61. Find $5y + (-12y) + (-21y)$. (Lesson 2–5) **−28y**

62. **Statistics** Mr. Crable asked his students how many hours of sleep they got on the average each night for a week. He labeled the data with G for girls and B for boys. Use Gs and Bs to make a line plot of these data. (Lesson 2–2) **See margin.**

10-G	7.5-B	9-B	9-G	8-G	8.5-G	9.5-B	8.5-B	8.5-G
9.5-B	8-G	8-B	10-B	10-G	9.5-G	7.5-G	9-G	8.5-B

63. Solve $\frac{21-3}{12-3} = x$. (Lesson 1–5) **2**

64. **Patterns** Find the next three terms in the sequence 3, 8.5, 14, 19.5.... (Lesson 1–2) **25, 30.5, 36**

SELF TEST

Solve each proportion. (Lesson 4–1)

1. $\frac{2}{10} = \frac{1}{a}$ **5**
2. $\frac{3}{5} = \frac{24}{x}$ **40**
3. $\frac{y}{4} = \frac{y+5}{8}$ **5**

4. ΔWYQ and ΔMVT shown at the right are similar. Find the measures of the missing sides. (Lesson 4–2) **x = 15, y = 25**

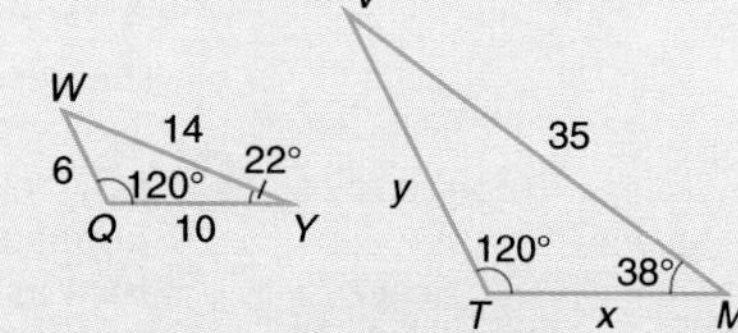

Solve each right triangle. State the side lengths to the nearest tenth and the angle measures to the nearest degree. (Lesson 4–3)

5.
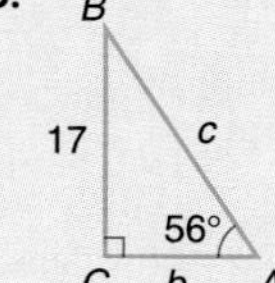

∠B = 34°, b = 11.5, c = 20.5

6.
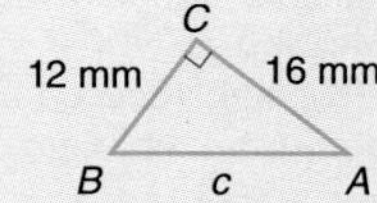

∠A = 37°, ∠B = 53°, c = 20 mm

7. Refer to the triangles in Exercises 5 and 6. Determine whether the triangles are similar. (Lesson 4–2) **7. no**
8. Thirty-six is 45% of what number? (Lesson 4–4) **80**
9. Fifty-five is what percent of 88? (Lesson 4–4) **$62\frac{1}{2}$%, 62.5%**
10. **Research** Suppose 6% of 8000 people polled regarding an election expressed no opinion. How many people had an opinion? (Lesson 4–4) **7520 people**

Chapter 4, Quiz B (Lessons 4-3 and 4-4), is available in the *Assessment and Evaluation Masters,* p. 100.

Mid-Chapter Test (Lessons 4-1 through 4-4) is available in the *Assessment and Evaluation Masters,* p. 99.

Additional Answer

62.

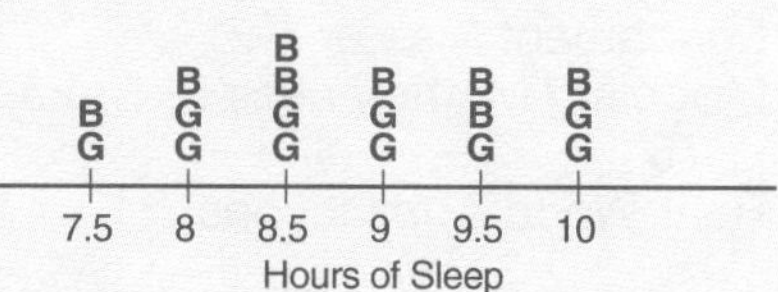

SELF TEST

The Self Test provides students with a brief review of the concepts and skills in Lessons 4-1 through 4-4. Lesson numbers are given to the right of exercises or instruction lines so students can review concepts not yet mastered.

Enrichment Masters, p. 29

NAME ____________ DATE ____________

Enrichment — Student Edition Pages 215–220

Compound Interest

In most banks, interest on savings accounts is compounded at set time periods such as three or six months. At the end of each period, the bank adds the interest earned to the account. During the next period, the bank pays interest on all the money in the bank, including interest. In this way, the account earns interest on interest.

Suppose Ms. Tanner has $1000 in an account that is compounded quarterly at 5%. Find the balance after the first two quarters.

Use $I = prt$ to find the interest earned in the first quarter if $p = 1000$ and $r = 5\%$. Why is t equal to $\frac{1}{4}$?

First quarter: $I = 1000 \times 0.05 \times \frac{1}{4}$
$I = 12.50$

The interest, $12.50, earned in the first quarter is added to $1000. The principal becomes $1012.50.

Second quarter: $I = 1012.50 \times 0.05 \times \frac{1}{4}$
$I = 12.65625$ The interest in the second quarter is $12.66.

The balance after two quarters is $1012.50 + 12.66 or $1025.16.

Answer each of the following questions.

1. How much interest is earned in the third quarter of Ms. Tanner's account? **I = $12.81**
2. What is the balance in her account after three quarters? **$1037.97**
3. What is the balance in her account after one year? **$1050.94**
4. Suppose Ms. Tanner's account is compounded semiannually. What is the balance at the end of six months? **$1025.00**
5. What is the balance after one year if her account is compounded semiannually? **$1050.63**

4-5 LESSON NOTES

NCTM Standards: 1–5

Instructional Resources

- Study Guide Master 4-5
- Practice Master 4-5
- Enrichment Master 4-5
- Modeling Mathematics Masters, p. 75
- Real-World Applications, 11

Transparency 4-5A contains the 5-Minute Check for this lesson; **Transparency 4-5B** contains a teaching aid for this lesson.

Recommended Pacing	
Standard Pacing	Day 8 of 15
Honors Pacing	Day 8 of 15
Block Scheduling*	Day 3 of 7 (along with Lesson 4-4)
Alg. 1 in Two Years*	Days 10 & 11 of 20

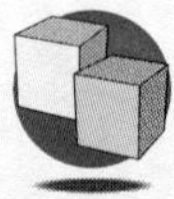

*For more information on pacing and possible lesson plans, refer to the *Block Scheduling Booklet* and *Algebra 1 in Two Years.*

1 FOCUS

5-Minute Check
(over Lesson 4-4)

Solve.

1. What number increased by 40% equals 14? **10**
2. What number increased by $33\frac{1}{3}$% equals 52? **39**
3. Twenty is 20% less than what number? **25**
4. Which earns more interest, \$1000 at 8% interest for one year or \$2000 at 2% interest for two years? **They both earn the same amount.**
5. Which earns more interest, \$1500 at 10% interest for one year or \$500 at 4% interest for ten years? **\$500 at 4%**

4–5 Percent of Change

What YOU'LL LEARN

- To solve problems involving percent of increase or decrease, and
- to solve problems involving discounts or sales tax.

Why IT'S IMPORTANT

You can use percents to solve problems involving sales, health, and shopping.

APPLICATION
Food

mallow plant

If you've ever bought a package of marshmallows, you've mostly bought air. That's right, marshmallows are 80% air! Today, marshmallows come in a variety of flavors such as vanilla, chocolate, toasted coconut, and fruit flavors like laser lime and astroberry. The marshmallow was invented in ancient Egypt and made with honey and the sap of the mallow plant that grows in swamps, or marshes. Since then, the marshmallow has gone through many changes.

Today, marshmallows are mass produced and made from cornstarch and gelatin. In 1955, the height of marshmallow making in the United States, there were 30 marshmallow companies. Today there are only a handful.

Let's say there are only six marshmallow companies today. You can write a ratio that compares the amount of decrease in marshmallow companies to the original number of companies in 1955. This ratio can be written as a percent. First, subtract to find the amount of change: $30 - 6 = 24$. Then divide by the original number of companies.

$$\frac{\text{amount of decrease} \rightarrow}{\text{original amount} \rightarrow} = \frac{r}{100}$$

$24 \cdot 100 = 30 \cdot r$ *Find the cross products.*

$80 = r$ *Solve for r.*

The amount of decrease is $\frac{80}{100}$ or 80% of the original number of companies. So, we can say that the **percent of decrease** is 80%.

The **percent of increase** can be found in a similar way.

Example 1 **CONNECTION** **Biology**

In 1982, the California condor was on the verge of extinction. The population had dwindled to a total of 21, making the condor the rarest bird in North America. According the Refuge Department of the U.S. Fish and Wildlife Service, however, there were 64 condors in existence in 1992 and 103 as of October 1995. Find the percent of increase in the population of California condors from 1992 to 1995.

Method 1
First, subtract to find the amount of change.

$103 - 64 = 39$

$$\frac{\text{amount of increase}}{\text{original number}} = \frac{39}{64}$$

$$\frac{39}{64} = \frac{r}{100}$$

$60.9375 = r$

The percent of increase is about 61%.

Method 2
Divide the new amount by the original amount.

$103 \div 64 = 1.609375$

Subtract 1 from the result and write the decimal as a percent.

$1.609375 - 1 = 0.609375$ or 60.9%

The percent of increase is about 61%.

FYI

In 1982, the California Condor Recovery Program, made up of a team of specialists and zoos, was formed to assist in the capturing, breeding, and freeing of the condors into the wild.

Sometimes an increase or decrease is given as a percent, rather than an amount. Two applications of percent of change are discounts and sales tax.

FYI

The condor range includes both California and Oregon.

Alternative Learning Styles

Auditory Have students conduct a telephone survey aimed at gathering information about how merchants determine selling prices of items. Challenge the class to review the results of the survey to determine general markup percentages used in various industries.

Example 2

Sales

A discount of 20% means that the price is decreased by 20%.

Ayita received a coupon in the mail offering her a 20% discount on the price of a pair of jeans at The Limited. Ayita visited the store and found a pair of jeans she wants to buy that cost $48. What will be the discounted price?

Explore The original price is $48.00, and the discount is 20%.

Plan You want to find the amount of discount, then subtract that amount from $48.00. The result is the discounted price.

Estimate: *20% of $48 → 2(10% of $50) = 2($5) or $10.*

Solve 20% of $48.00 = 0.20(48.00) *Note that* $20\% = \frac{20}{100} = 0.20$.
= 9.60

Subtract this amount from the original price.

$48.00 − $9.60 = $38.40

The discounted price will be $38.40.

Examine Solve the problem another way. The discount was 20%, so the discounted price will be 80% of the original price. *100% − 20% = 80%*

Find 80% of $48.00. 0.80(48.00) = 38.40

This method produces the same discounted price, $38.40.
How does this compare to the estimate?

Example 3

Sales

Sales tax of $5\frac{3}{4}\%$ *means that the price is increased by* $5\frac{3}{4}\%$.

Karen Danko, a sophomore at Westerville North High School, is ordering a class ring. The ring she has chosen costs $169.99. She also needs to pay a sales tax of $5\frac{3}{4}\%$**. What is the total price?**

Explore The price is $169.99 and the tax rate is $5\frac{3}{4}\%$.

Plan First find $5\frac{3}{4}\%$ of $169.99. Then add the result to $169.99.

Estimate: $5\frac{3}{4}\%$ *of $169.99 →* $\frac{1}{2}$*(10% of 170) =* $\frac{1}{2}$ *($17) or $8.50. The total will be about $170 + $8.50 or $178.50.*

Solve $5\frac{3}{4}\%$ of $169.99 = 0.0575(169.99) *Note that* $5\frac{3}{4}\% = 0.0575$.
= 9.774425 *Round 9.774425 to 9.78, since tax is always rounded up.*

Add this amount to the original price.

$169.99 + $9.78 = $179.77

The total price is $179.77.

Examine The tax rate is $5\frac{3}{4}\%$, so the total price was 100% + $5\frac{3}{5}\%$ or 105.75% of the purchase price. Find 105.75% of $169.99.

(1.0575)(169.99) = 179.76442

Thus, the total price of $179.77 is correct.
How does this compare to the estimate?

Motivating the Lesson

Situational Problem Have students consider the following. A pair of sneakers is reduced 10%. The sales tax on the sneakers is 6%. Is the final price of the sneakers equal to the original price less 4%? **No, because the sales tax is calculated on the sale price, not on the original price of the sneakers.**

2 TEACH

Teaching Tip Point out that the difference is always compared to the original price.

In-Class Examples

For Example 1
A retailer increased the price of ball-point pens from $0.40 to $0.45 each. What was the percent of increase? **12.5%**

For Example 2
Mieko received a 5% discount on a calculator. The original price was $15.00. What was the sale price? **$14.25**

For Example 3
Tim bought a pair of running shoes for $35. There was a 4% sales tax. Find the tax and the total price Tim paid.
tax = $1.40, total = 36.40

Teaching Tip In Example 2, you may wish to explain that since there is no monetary unit that represents $\frac{1}{10}$ of a cent, amounts of money are generally rounded to the nearest whole cent. If the original price of the jeans was $48.99, the discount would be $9.798, which would be rounded to $9.80.

Classroom Vignette

"I have students make a sign or display to advertise a percent-off sale in their imaginary store. We put the displays up in the classroom and then, as a review each day, we make an imaginary purchase at each store, calculating how much is saved. We used play money to illustrate the purchase and how much change is received."

Emma Lou Crawford
Roanoke High School
Robersonville, North Carolina

Emma Lou Crawford

EXPLORATION

This Exploration demonstrates the compounding effect of successive discounts. Caution students that the second discount is taken on the already discounted price, not the original price.

3 PRACTICE/APPLY

Check for Understanding

Exercises 1–11 are designed to help you assess your students' understanding through reading, writing, speaking, and modeling. You should work through Exercises 1–4 with your students and then monitor their work on Exercises 5–11.

Additional Answers

1. **Subtract old from new. Divide the result by old. Move decimal point 2 places to the right.**
2. **Multiply tax rate by original price.**
3. $\frac{1773}{17{,}972} = \frac{r}{100}$

EXPLORATION — PROGRAMMING

Suppose a newspaper ad states that a discount store's microwave ovens are discounted 20% from the manufacturer's suggested retail price (MSRP). The store is running a special sale that says all kitchen appliances have been discounted an additional 15%. What is the final sale price of a microwave oven whose MSRP is $250?

There are two ways you might interpret the ad in the paper.

- The discounts can be successive. That is, 20% is taken off the MSRP and then 15% is taken off the resulting price.
- The discounts can be combined. So the price is 35% off the MSRP.

The graphing calculator program at the right finds the cost of an item for each interpretation of the ad. You must enter the original price and each discount amount (as a decimal) when prompted.

```
PROGRAM:DISCOUNT
: Disp "ORIG. PRICE"
: Input P
: Disp "1ST DISCOUNT?"
: Input A
: Disp "2ND DISCOUNT?"
: Input B
: P(1-A)(1-B)→C
: P(1-(A+B))→D
: Disp "SUCCESSIVE",
  "DISCOUNT",C
: Disp "COMBINED",
  "DISCOUNT",D
```

e. The combined discount always yields a lower price. Stores often use successive discounts.

Your Turn

Copy the following table and use the program to complete it.

	Price	First Discount	Second Discount	Sale Price Successive Discount	Sale Price Combined Discount
a.	$49.00	20%	10%	**$35.28**	**$34.30**
b.	$185.00	25%	10%	**$124.88**	**$120.25**
c.	$12.50	30%	12.5%	**$7.66**	**$7.19**
d.	$156.95	30%	15%	**$93.39**	**$86.32**

e. What is the relationship between the sale price using successive discounts and the sale price using combined discounts? Which one is usually used in stores?

CHECK FOR UNDERSTANDING

Communicating Mathematics

Study the lesson. Then complete the following. **1–3. See margin.**

1. **Explain** how to find a percent of increase.
2. **Explain** how to find the sales tax if you know the rate of tax.
3. **Write** the equation you would use to find the percent of increase on a car that cost $17,972 in 1995 and $19,705 in 1996.

4. **Assess Yourself** Find an advertisement or article in the newspaper that shows a percent of change. Determine whether it is a percent of increase or decrease. Explain your reasoning. **See students' work.**

Reteaching

Using Substeps Make certain that students realize that two steps are required to solve problems involving percent of change. First, they must subtract to find the amount of change. Then, they must compare that amount, increasing or decreasing, to the *original*. In each exercise at the right, require students to show both steps to find the percent of increase or decrease.

1. old: $74.00
   new: $79.18 **$5.18, 7%**
2. old: $62.00
   new: $65.72 **$3.72, 6%**
3. old: $139.40
   new: $164.49 **$25.09, 18%**
4. old: $65.48
   new: $60.24 **$5.24, 8%**

Guided Practice

7. D; 50%

State whether each percent of change is a percent of increase or a percent of decrease. Then find the percent of increase or decrease.

5. original: $50 new: $70 I; 40%
6. original: $200 new: $172 D; 14%
7. original: 72 ounces new: 36 ounces

Find the final price of each item. When there is a discount and sales tax, first compute the discount price and then compute the sales tax and final price.

8. CD player: $149 discount: 15% $126.65
9. athletic shoes: $89.99 discount: 10% sales tax: 6% $85.85
10. sweater: $45 sales tax: 6.5% $47.93
11. **Sales** Kevin Mason paid $205.80 for his senior class photographs. This included 5% sales tax. What was the cost of the pictures before taxes? $196

EXERCISES

Practice

State whether each percent of change is a percent of increase or a percent of decrease. Then find the percent of increase or decrease. Round to the nearest whole percent. 14. I; 162%

12. original: $100 new: $59 D; 41%
13. original: 324 people new: 549 people I; 69%
14. original: 58 homes new: 152 homes
15. original: 66 dimes new: 30 dimes D; 55%
16. original: $53 new: $75 I; 42%
17. original: 15.6 liters new: 11.4 liters D; 27%

B

18. original: $3.78 new: $2.50 D; 34%
19. original: 231.2 mph new: 236.4 mph I; 2%
20. original: 124 tons new: 137 tons I; 10%

Find the final price of each item. When there is a discount and sales tax, first compute the discount price and then compute the sales tax and final price.

23. $19.90

21. VCR: $219 sales tax: 6.5% $233.24
22. jeans: $39.99 discount: 15% sales tax: 4% $35.35
23. book: $19.95 discount: 5% sales tax: 5%

26. $56.02

C

24. concert tickets: $52.50 sales tax: 7% $56.18
25. in-line skates: $99.99 discount: 20% sales tax: 6.75% $85.39
26. hiking boots: $59 discount: 10% sales tax: 5.5%

29. $30.09

27. compact disc: $15.88 sales tax: 4.5% $16.59
28. amusement park tickets: $37.50 sales tax: 6% $39.75
29. software: $29.99 discount: 6% sales tax: 6.75%

Graphing Calculator

30. Use a graphing calculator to find the sale price of each item using successive discounts and then one combined discount.
    a. price, $89; discounts, 25% and 10% $60.08; $57.85
    b. price, $254; discounts 30% and 14.5% $152.02; $140.97

Critical Thinking

31. An amount is increased by 15%. The result is decreased by 15%. Is the final result equal to the original amount? Explain your reasoning with an example. No; see students' work.

Assignment Guide

Core: 13–39 odd, 40–46
Enriched: 12–30 even, 31–46

For **Extra Practice,** see p. 765.

The red A, B, and C flags, printed only in the Teacher's Wraparound Edition, indicate the level of difficulty of the exercises.

Teaching Tip Before students work on Exercises 21–29, you may wish to discuss why it is necessary to deduct an amount of discount from an item before computing the sales tax on the item.

Study Guide Masters, p. 30

NAME____________ DATE ________

Study Guide

Student Edition Pages 222–227

Percent of Change

Some percent problems involve finding a percent of increase or decrease.

Percent of Increase	Percent of Decrease
A coat that cost $50 last year costs $55 this year. The price increased by $5 since last year.	Slacks that originally cost $30 are now on sale for $22. Find the percent of decrease.
$\frac{\text{amount of increase}}{\text{original price}} \rightarrow \frac{5}{50} = \frac{r}{100}$ $500 = 50r$ $10 = r$, or $r = 10$	$\frac{\text{amount of decrease}}{\text{original price}} \rightarrow \frac{8}{30} = \frac{r}{100}$ $800 = 30r$ $26\frac{2}{3} = r$, or $r = 26\frac{2}{3}$
The percent of increase is 10%.	The percent of decrease is $26\frac{2}{3}$%, or about 27%.

The sales tax on a purchase is a percent of the purchase price. To find the total price, you must calculate the amount of sales tax and add it to the purchase price.

Find the final price of each item. When there is a discount and sales tax, compute the discount price first.

1. Compact Disc: $16.00 Discount: 15% $13.60
2. Two concert tickets: $28.00 Student discount: 28% $20.16
3. Airline Ticket: $248.00 Superair discount: 33% $166.16
4. Celebrity Photo Calendar: $10.95 Sales tax: 7.5% $11.77
5. Class Ring: $89.00 Group discount: 17% Sales tax: 5% $77.56
6. Computer Software: $44.00 Discount: 21% Sales tax: 6% $36.85

Solve each problem.

7. The original selling price of a new sports video was $65.00. Due to demand the price was increased to $87.75. What was the percent of increase over the original price? 35%
8. A high school paper increased its sales by 75% when it ran an issue featuring a contest to win a class party. Before the contest issue, 10% of the school's 800 students bought the paper. How many students bought the contest issue? 140

32. **Sales** The advertisement below was used by a local optician's office in Walkersville, Maryland.

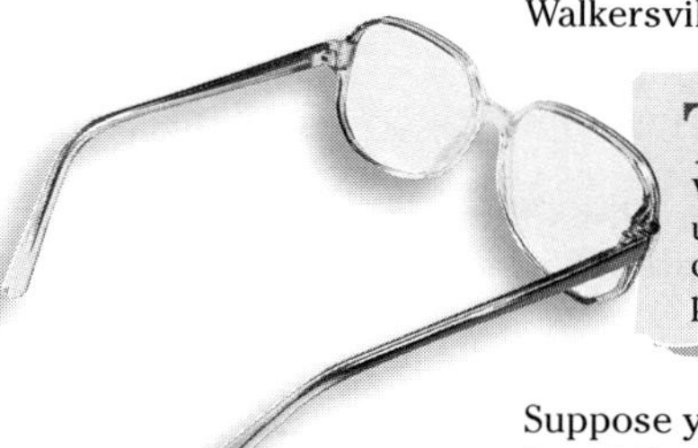

The 100% Price Pledge!
We guarantee that we will simply not be undersold. If you find the same eyewear as ours at a lower price within 30 days of your purchase, we'll pay you **110%** of the difference.

Suppose you bought a pair of glasses from this optician for \$150. You then found the same glasses from a competitor for \$125. What should the optician do to honor the pledge? **pay the customer \$27.50**

Applications and Problem Solving

33. 133%

35. 8.6%

33. **Predictions** The World Future Society predicts that by the year 2020, airplanes will be able to carry 1400 passengers. Today's biggest jets can carry 600 people. What will be the percent of increase of airplane passengers?

34. **Computers** In 1995, America Online had about 3,000,000 users. Over the next decade, users are expected to increase from a few million to the tens of millions. Suppose the number of users increases by 150% by the year 2000. How many users will there be in the year 2000? **7,500,000 users**

35. **Taxes** As the saying goes, time is money. The graph at the right shows a function of how much time it takes out of each eight-hour day to earn enough money to pay a day's worth of taxes. What was the percent of increase from 1970 to 1994?

Tax Bite in an Eight-Hour Day

Hours: 3:00, 2:45, 2:30, 2:15, 2:00, 1:45, 1:30, 1:15, 1:00, 0:45, 0:30, 0:15, 0:00

1929: 0:52; 1940: 1:29; 1950: 2:02; 1960: 2:20; 1970: 2:32; 1980: 2:40; 1994: 2:45

Source: *The Universal Almanac*, 1995

36. **Sales** Music Systems, Inc. allows a 10% discount if a purchase is paid for within 30 days. An additional 5% discount is given if the purchase is paid for within 15 days. Brent Goodson buys a sound system that originally cost \$360. If he pays the entire amount at the time of purchase, how much does he pay for his system after the successive discounts? **\$307.80**

37. **Health** The excerpt below is from the March 1994 issue of *Runner's World* magazine. **an increase of more than 22%**

When you first get up in the morning, your muscles and soft tissues are tight. In fact, at that time your muscles are generally about 10 percent shorter than their normal resting length. As you move around, they stretch to their normal length. Then when you start to exercise, your muscles stretch even more, to about 10 percent longer than resting length. This means you have a 20 percent change in muscle length from the time you get out of bed until your muscles are well warmed up.

According to the basic laws of physics, muscles work more efficiently when they are longer; they can exert more force with less effort. This means, too, that longer muscles are much less prone to injury.

What is the percent change in muscle length from the time a runner gets up in the morning until after he or she has exercised for a while?

Practice Masters, p. 30

4-5 NAME________ DATE________

Practice — Student Edition Pages 222–227

Percent of Change

Solve each problem.

1. What number increased by 30% equals 260? 200
2. What number decreased by $33\frac{1}{3}$% is 30? 45
3. Sixteen is 25% more than what number? 12.8
4. Forty is 40% less than what number? $66.\overline{6}$
5. What is 50% more than 50? 75
6. What is $66\frac{2}{3}$% less than 45? 15
7. A price decreased from \$25 to \$10. Find the percent of decrease. 60%
8. A price increased from \$30 to \$50. Find the percent of increase. $66\frac{2}{3}$%
9. A price plus 6% tax is equal to \$4.24. Find the original price. \$4.00
10. An item sells for \$45 after a 20% discount. Find the original price. \$56.25
11. An item sells for \$75 after a 50% discount. Find the original price. \$150.00

Find the final price of each item. When there is a discount and sales tax, compute the discount price and then compute the sales tax and final price.

12. sweater: \$55.00 discount: 15% \$46.75
13. compact disc player: \$269.00 discount: 20% \$215.20
14. television: \$375.00 discount: 25% \$281.25
15. coat: \$195.00 discount: 40% sales tax: 5% \$122.85

Tech Prep

Health and Fitness Expert Students who are interested in health and fitness may wish to do further research on the data given in Exercise 37 and explore the potential growth of this career. For more information on tech prep, see the *Teacher's Handbook*.

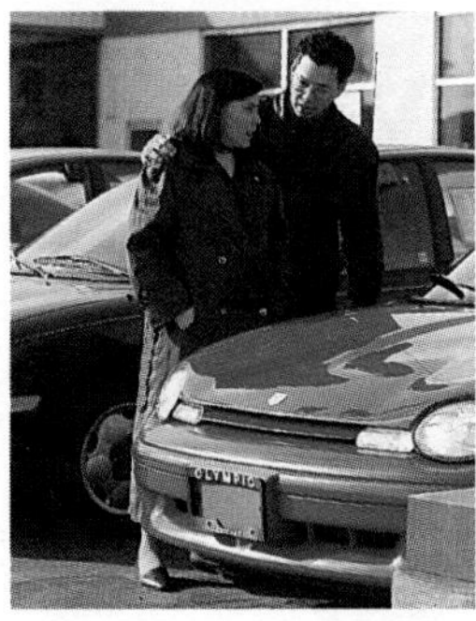

38. **Consumer Awareness** A windsuit that costs \$75 is discounted 25%. The sales tax is 6%.
    a. Would it be better for the customer to have the discount taken off before adding the sales tax or after the sales tax has been added? Explain your reasoning. **no difference**
    b. Do you think stores have a choice in which order to calculate discounts and sales tax? Explain your reasoning. **See students' work.**
39. **Automobiles** As soon as a new car is purchased and driven away from the dealership, it begins to lose its value, or depreciate. Alonso bought a 1994 Plymouth Neon for \$9559. One year later, the value of the car was \$8500. What is the percent of decrease of the value of the car? **11%**

Mixed Review

40. **Entertainment** A theater was filled to 75% capacity. How many of the 720 seats were filled? (Lesson 4–4) **540 seats**
41. **Geometry** A chimney casts a shadow 75 feet long when the angle of elevation of the sun is 41°. How tall is the chimney? (Lesson 4–3) **65 ft**
42. **Construction** To paint his house, Lonnie needs to purchase an extension ladder that reaches at least 24 feet off the ground. Ladder manufacturers recommend the angle formed by the ladder and the ground be no more than 75°. What is the shortest ladder he could buy to reach 24 feet safely? (Lesson 4–3) **at least 25 feet long**
43. Solve $0.2x + 1.7 = 3.9$. (Lesson 3–6) **11**
44. **Geometry** The measures of two angles of a triangle are 38° and 41°. Find the measure of the third angle. (Lesson 3–4) **101°**
45. **Dietetics** A hospitalized diabetic is on a carefully controlled 2000-calorie diet distributed over five feedings a day. There are two heavy feedings consisting each of $\frac{2}{7}$ of the total calories. There are also three light feedings consisting each of $\frac{1}{7}$ of the total calories. Determine the number of calories for a heavy feeding and for a light feeding. (Lesson 2–6)
46. **Statistics** State the scale you would use to make a line plot for the following data. Then draw the line plot. 4.2, 5.3, 7.6, 9.6, 7.3, 6.7 (Lesson 2–2) **See margin.**

45. heavy, 571; light, 286

WORKING ON THE In•ves•ti•ga•tion

Refer to the Investigation on pages 190–191.

1 Count all of the fish in your lake. Record the actual population and your estimates on a class chart.

2 How close were your estimates? How would you account for the difference between your estimates and the actual number of fish in the lake?

3 What percent of the fish in the lake was tagged? What percent of the fish in your samples was tagged?

4 Suppose you went fishing at that lake and caught one fish. What is the probability, or chance, that the fish you caught would be tagged? Justify your reasoning.

5 Write a paragraph or two that relates proportions, percents, and probability in this situation.

You will learn more about probability in the next lesson.

Add the results of your work to your Investigation Folder.

4 ASSESS

Closing Activity

Writing Ask students to write two different equations that can be used to solve the following problem. Fred bought a VCR that originally sold for \$245. He received a 20% discount. How much did he pay for the VCR?

Additional Answer

46. From 4 to 10, intervals of 0.2

Extension

Reasoning A \$75 blazer is marked down 25%. The buyer has to pay sales tax on the regular price. How much could have been saved if the sale price was taxed instead? **\$1.12**

In•ves•ti•ga•tion

Working on the Investigation

The Investigation on pages 190–191 is designed to be a long-term project that is completed over several days or weeks. Encourage students to keep their materials in their Investigation Folder as they work on the Investigation.

Enrichment Masters, p. 30

NAME________ DATE ________

Enrichment

Student Edition Pages 222–227

Using Percent

Use what you have learned about percent to solve each problem.

A TV movie had a "rating" of 15 and a 25 "share." The rating of 15 means that 15% of the nation's total TV households were tuned in to this show. The 25 share means that 25% of the homes with TVs turned on were tuned to the movie. How many TV households had their TVs turned off at this time?

To find out, let T = the number of TV households
and x = the number of TV households with the TV off.
Then $T - x$ = the number of TV households with the TV on.

Since $0.15T$ and $0.25(T - x)$ both represent the number of households tuned to the movie,

$$0.15T = 0.25(T - x)$$
$$0.15T = 0.25T - 0.25x.$$

Solve for x.

$$0.25x = 0.10T$$
$$x = \frac{0.10T}{0.25} = 0.40T$$

Forty percent of the TV households had their TVs off when the movie was aired.

Answer each question.

1. During that same week, a sports broadcast had a rating of 22.1 and a 43 share. Show that the percent of TV households with their TVs off was about 48.6%. $0.221T = 0.43T - 0.43x$; $x = \frac{0.221T - 0.43T}{-0.43} = 0.486T$
2. Find the percent of TV households with their TVs turned off during a show with a rating of 18.9 and a 29 share. 34.8%
3. Show that if T is the number of TV households, r is the rating, and s is the share, then the number of TV households with the TV off is $\frac{(s - r)T}{s}$. Solve $rT = s(T - x)$ for x.
4. If the fraction of TV households with no TV on is $\frac{s - r}{s}$ then show that the fraction of TV households with TVs on is $\frac{r}{s}$. $1 - \frac{s - r}{s} = \frac{r}{s}$
5. Find the percent of TV households with TVs on during the most watched serial program in history: the last episode of *M*A*S*H*, which had a 60.3 rating and a 77 share. $\frac{60.3}{77} = 78.3\%$
6. A local station now has a 2 share. Each share is worth \$50,000 in advertising revenue per month. The station is thinking of going commercial free for the three months of summer to gain more listeners. What would its new share have to be for the last 4 months of the year to make more money for the year than it would have made had it not gone commercial free? greater than 3.5

4-6 LESSON NOTES

NCTM Standards: 1–5, 11

Instructional Resources

- Study Guide Master 4-6
- Practice Master 4-6
- Enrichment Master 4-6
- Assessment and Evaluation Masters, p. 101

Transparency 4-6A contains the 5-Minute Check for this lesson; **Transparency 4-6B** contains a teaching aid for this lesson.

Recommended Pacing	
Standard Pacing	Days 9 & 10 of 15
Honors Pacing	Days 9 & 10 of 15
Block Scheduling*	Day 4 of 7
Alg. 1 in Two Years*	Days 12 & 13 of 20

*For more information on pacing and possible lesson plans, refer to the *Block Scheduling Booklet* and *Algebra 1 in Two Years.*

5-Minute Check
(over Lesson 4-5)

Solve.

1. What is 30% more than 30? 39
2. What is 15% less than 40? 34
3. A price decreased from \$36 to \$27. What is the percent of decrease? 25%
4. A price increased from \$68 to \$85. What is the percent of increase? 25%
5. An item sells for \$48 after a 20% discount. What was the original price of the item? \$60

Motivating the Lesson

Hands-On Activity Select a student randomly from the class. Ask the student, "What is the probability that you would have been selected?" Separate the class in half. Select a student from one half. Ask this student the same question.

4-6 Integration: Probability
Probability and Odds

What YOU'LL LEARN
- To find the probability of a simple event, and
- to find the odds of a simple event.

Why IT'S IMPORTANT

You can use probability to calculate the chances that something will happen. This is especially helpful in sports.

Contests

Have you ever heard anyone say that one man's trash is another man's treasure? For garbage truck driver Craig Randall of East Bridgewater, Massachusetts, the treasure was a Wendy's soft-drink cup. By peeling off a contest sticker on the discarded cup, Randall won the grand prize of \$200,000 toward the purchase of a home! Since there was only one grand prize available and 24,059,900 cups were distributed, the probability of winning the grand prize was $\frac{1}{24,059,900}$.

You can calculate the chance, or **probability,** that a certain event, such as winning a particular prize, will happen. The **probability of an event** is a ratio of the number of ways a certain event can occur to the number of possible outcomes. The numerator is the number of favorable outcomes, and the denominator is the total number of possible outcomes. *The probability of an event may be written as a percent, a fraction, or a decimal.*

For example, suppose you want to know the probability of getting a 2 on one roll of a die. When you roll a die, there are six possible outcomes. Of these outcomes, only one is favorable, a 2. Therefore, the probability is $\frac{1}{6}$. We write $P(2)$ to represent *the probability of getting a 2 on one roll of a die.*

Definition of Probability $P(\text{event}) = \frac{\text{number of favorable outcomes}}{\text{total number of possible outcomes}}$

Example 1

Games

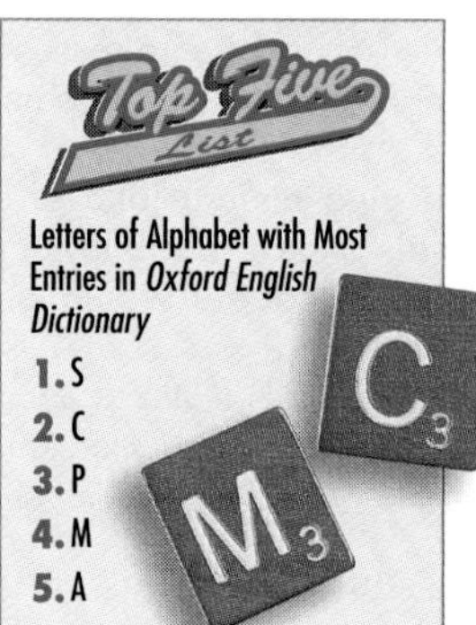

Top Five List

Letters of Alphabet with Most Entries in Oxford English Dictionary
1. S
2. C
3. P
4. M
5. A

Lauren Slocum is representing her sophomore class in a *Scrabble* tournament. She is the first player to select her seven tiles out of the 100 available tiles. The distribution of tiles is shown at the right. Find the probability of each selection below.

Letters	Number of Tiles
J, K, Q, X, Z	1
B, C, F, H, M, P, V, W, Y, blank	2
G	3
D, L, S, U	4
N, R, T	6
O	8
A, I	9
E	12

a. an O on the very first selection

There are eight O tiles and 100 tiles in all. $P(\text{selecting an O}) = \frac{8}{100}$ or $\frac{2}{25}$

The probability that Lauren selects an O is $\frac{2}{25}$ or 8%.

b. an E if 5 tiles have been selected and 2 of them were Es

There are 12 E tiles. If 2 have been chosen, 10 remain. There are 100 − 5 or 95 tiles from which to select.

$P(\text{choosing an E}) = \frac{10}{95}$ or $\frac{2}{19}$

The probability of selecting an E is $\frac{2}{19}$ or about 10%.

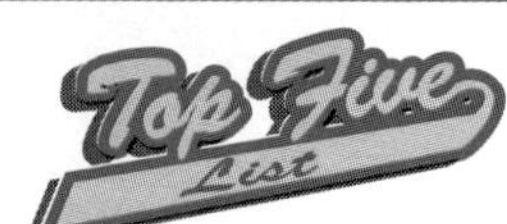

Top Five Languages Spoken in the U.S.

	Language	Speakers
1.	English	224,900,000
2.	Spanish	19,640,000
3.	French	1,930,000
4.	German	1,750,000
5.	Italian	1,480,000

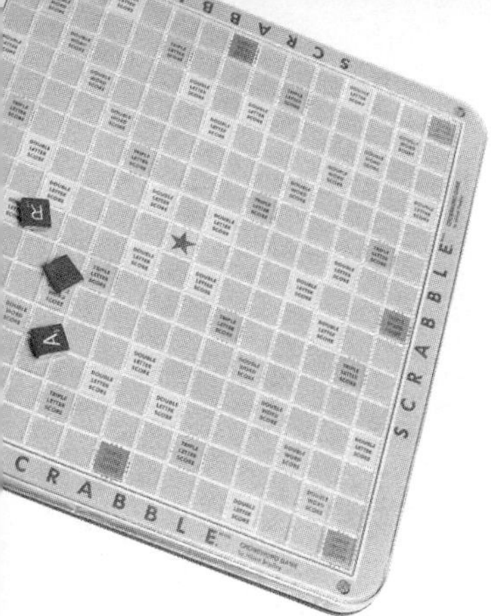

The probability that an event will occur is somewhere between 0 and 1.

- A probability of 0 means that it is impossible for an event to occur.
- A probability of 1 means that an event is certain to occur.
- A probability between 0 and 1 means that an event is neither impossible nor certain.

Based on the situations above, the probability of any event can be expressed as $0 \leq P(\text{event}) \leq 1$.

Some outcomes have an equal chance of occurring. We say that such outcomes are **equally likely.** When an outcome is chosen without any preference, we say that the outcome occurs at **random.**

MODELING MATHEMATICS **Probability**

Materials:  grid paper

What are the chances that two of the Oscar-winning actors or actresses from 1983 to 1993 share the same birthday? Considering that there are only 24 people in this group, you would probably say the chances are slim.

You may be surprised that in a group of only 24 people, the likelihood that two of them have the same birthday is just about 50%, or one out of two.

For example, in this group of Academy Award winners, Holly Hunter and William Hurt were both born on March 20.

Your Turn **a–d. See students' work.**

a. Survey a group of 24 people in your school. Tally the results.

b. Does your group contain a shared birthday? Display your findings in a chart or graph.

c. Combine your results with those of your classmates. Does your class graph differ from your individual graph? Explain.

d. Use the graph below to find the likelihood of birthdays shared by the Presidents of the United States. Research to find any shared Presidential birthdays. Compare your findings. Do you agree or disagree with the percentages? Explain your reasoning.

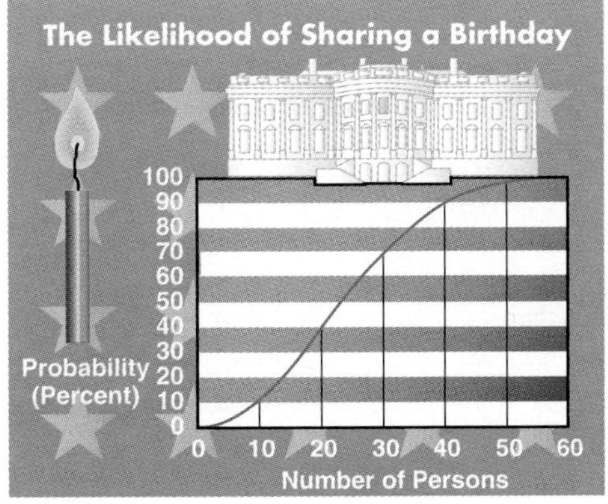

Another way to express the chance of an event's occurring is with **odds.** The odds of an event is the ratio that compares the number of ways an event can occur to the number of ways it *cannot* occur.

Definition of Odds	**The odds of an event occurring is the ratio of the number of ways the event can occur (successes) to the number of ways the event cannot occur (failures).**

Odds = number of successes: number of failures

2 TEACH

In-Class Example

For Example 1
Lynn collects stamps from different countries. He has five from Canada, two from France, one from Russia, four from Great Britain, and one from Germany. If he accidentally loses one stamp, what is the possibility that it is the stamp from Russia? $\frac{1}{13}$

MODELING MATHEMATICS Encourage students to follow this graphing method with other probability exercises. Survey and graph other characteristics such as height, eye color, or foot size. What is the likelihood of each characteristic?

Teaching Tip Develop student understanding of probability and odds by asking the following questions.

1. What is the meaning of the word *random?*
2. What are examples of situations that have a probability of 0? of 1?

GLENCOE Technology

 **Interactive Mathematics Tools Software**

This multimedia software provides an interactive lesson that uses histograms to observe the results when considering various chance outcome activities. Students will compare the results of the mean, mode, and median of these activities. A **Computer Journal** gives students an opportunity to write about what they have learned.

For Windows & Macintosh

Example 2 **Jersey Mike's Submarine Shop has a business card drawing for a free lunch every Tuesday. Four coworkers from Invo Accounting put their business cards in the bowl for the drawing. If 80 cards were in the bowl, what are the odds that one of the coworkers will win a free lunch?**

The coworkers have 4 of the 80 business cards. Thus, there are 80 − 4 or 76 business cards that will not be winning cards for the coworkers.

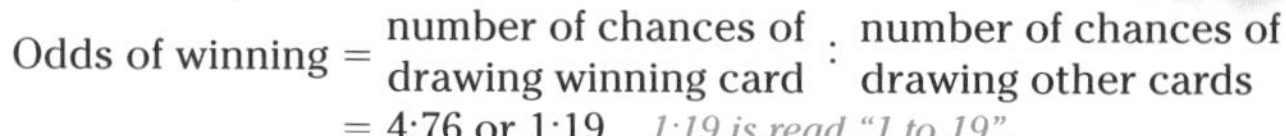

$$\text{Odds of winning} = \text{number of chances of drawing winning card} : \text{number of chances of drawing other cards}$$
$$= 4:76 \text{ or } 1:19$$ *1:19 is read "1 to 19".*

Example 3 **Refer to the application at the beginning of the lesson. The probability of winning any prize or food discount in Wendy's peel-off sticker contest was 20% or $\frac{1}{5}$. Find the odds of winning a prize or discount.**

Do you have a better chance of winning a prize or not winning a prize?

If the probability of winning a prize or discount is $\frac{1}{5}$, then the number of successes (prize) is 1, while the total number of outcomes is 5. This means that the number of failures (no prize) must be 5 − 1 or 4.

Odds of prize = number of successes : number of failures
= 1:4 *This is read "1 to 4."*

CHECK FOR UNDERSTANDING

Communicating Mathematics

Study the lesson. Then complete the following. **4–5. See students' work.**

1. **Refer** to Example 1. Find the probability of selecting a Z if 25 tiles have been selected and one of them was a Z. **0**
2. **Give** examples of an impossible event and a certain event when rolling a die.
3. **Tell** what the odds would be for an event not occurring if the odds for the event occurring are 3:5. **5:3**
4. **Write** a problem in which the answer will be a probability of $\frac{3}{4}$.

2. 7; a natural number less than 7

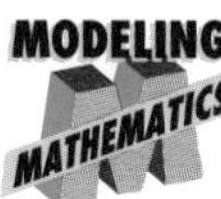

5. Choose another group of 24 people to survey. Record their birthdays. Does your group contain a shared birthday?

Guided Practice

Determine the probability of each event.

6. This is an algebra book. **1**
7. A coin will land tails up. $\mathbf{\frac{1}{2}}$

Find the probability of each outcome if a die is rolled.

8. a 6 $\mathbf{\frac{1}{6}}$
9. a number greater than 2 $\mathbf{\frac{2}{3}}$

Find the odds of each outcome if a die is rolled.

10. a number greater than 2 **2:1**
11. not a 3 **5:1**
12. If the probability that an event will occur is $\frac{3}{7}$, what are the odds that the event will *not* occur? **4:3**

In-Class Examples

For Example 2

The door prize at a party with 25 people is given by writing numbers 1 through 25 on the bottom of the paper plates used.

a. What is the probability that an individual had the winning plate? **1:25**
b. What are the odds of winning the door prize? **1:24**
c. What are the odds of not winning the door prize? **24:1**

For Example 3

The probability that Jim Jackson will make his next free throw is 75% or $\frac{3}{4}$. Find the odds that he will not make his next free throw. **1:3**

3 PRACTICE/APPLY

Check for Understanding

Exercises 1–12 are designed to help you assess your students' understanding through reading, writing, speaking, and modeling. You should work through Exercises 1–5 with your students and then monitor their work on Exercises 6–12.

Study Guide Masters, p. 31

NAME ______________ DATE ______

Study Guide — Student Edition Pages 228–232

Integration: Probability
Probability and Odds

The **probability** of an event is a ratio that tells how likely it is that the event will take place.

Definition of Probability
$P(\text{event}) = \frac{\text{number of favorable outcomes}}{\text{number of possible outcomes}}$

Example 1: Mr. Babcock picks 5 of the 25 students in his algebra class at random for a special project. What is the probability of being picked?

$$P(\text{being picked}) = \frac{\text{number of students picked}}{\text{total number of students}}$$

The probability of being picked is $\frac{5}{25}$ or $\frac{1}{5}$.

The probability of any event has a value from 0 to 1. If the probability of an event is 0, it is impossible for the event to occur. An event that is certain to occur has a probability of 1. This can be expressed as $0 \le P(\text{event}) \le 1$.

The odds of an event occurring is the ratio of the number of ways an event can occur (successes) to the number of ways the event cannot occur (failures).

Definition of Odds
$\text{Odds} = \frac{\text{number of successes}}{\text{number of failures}}$

Example 2: Find the odds that a member of Mr. Babcock's class will be picked for the special project.

Number of successes: 5 Number of failures: 20
Odds of being picked = number of successes : number of failures
= 5:20, or 1:4

Solve each problem.

1. There were 2 colas, 1 ginger ale, 5 cherry sodas, and 4 root beers in the cooler. What is the probability of pulling out a cola? $\mathbf{\frac{1}{6}}$
2. It will rain 8 times in November and snow 3 times. The other days it will be sunny. What is the probability of sun? the odds of sun? $\mathbf{\frac{19}{30}}$**; 19:11**

There is a bowl of money at the carnival. The bowl contains 50 quarters, 75 dimes, 100 nickels, and 125 pennies.

3. If one coin is chosen, what is the probability that a quarter will be chosen? $\mathbf{\frac{1}{7}}$
4. What would be the odds of choosing a quarter if all the dimes were removed first? **2:9**
5. What are the odds that a dime will not be chosen? **11:3**
6. What are the odds of choosing a penny? **5:9**

Reteaching

Using Probability Have students work in pairs. Have one student present an experiment and ask the probability of a certain event happening. Have the other student find the probability and then ask the first student the probability of another event happening in that same experiment, and so on. Have students record their questions and answers.

EXERCISES

Practice

Determine the probability of each event.

13. A coin will land heads up. $\frac{1}{2}$
14. There is a January 1 this year. 1
15. A baby will be a boy. $\frac{1}{2}$
16. Pigs will fly. 0

17. $\frac{1}{2}$ 18. $\frac{1}{12}$

17. You will roll a multiple of 2 on a die.
18. A person's birthday is in June.

Find the probability of each outcome if a computer randomly chooses a letter in the word "probability."

19. the letter b $\frac{2}{11}$
20. $P(not\ i)$ $\frac{9}{11}$
21. the letter e 0
22. $P(\text{vowel})$ $\frac{4}{11}$
23. *not b* or *y* $\frac{8}{11}$
24. the letters *a* or *t* $\frac{2}{11}$

Find the odds of each outcome if you randomly select a coin from a jar containing 80 pennies, 100 nickels, 70 dimes, and 50 quarters.

25. selecting a quarter 1:5
26. selecting a nickel 1:2
27. selecting a penny 4:11
28. *not* selecting a nickel 2:1
29. *not* selecting a dime 23:7
30. selecting a penny or dime 1:1

A card is selected at random from a standard deck of 52 cards.

31. What is the probability of selecting a red card? $\frac{1}{2}$
32. What is the probability of selecting a queen? $\frac{1}{13}$
33. What are the odds of selecting a heart? 1:3
34. What are the odds of *not* selecting a black 6? 25:1

35. $\frac{8}{13}$

35. If the odds that an event will occur are 8:5, what is the probability that the event will occur?
36. If the probability that an event will occur is $\frac{2}{3}$, what are the odds that the event will occur? 2:1

Critical Thinking

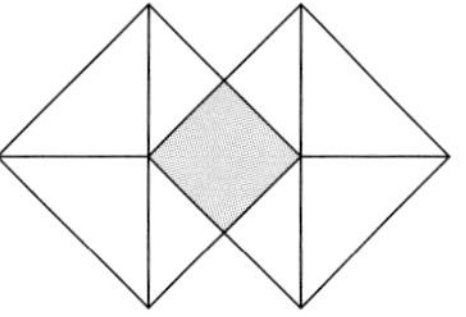

37. **Geometry** If a point inside the figure at the right is chosen at random, what is the probability that it will be in the shaded region? $\frac{1}{7}$

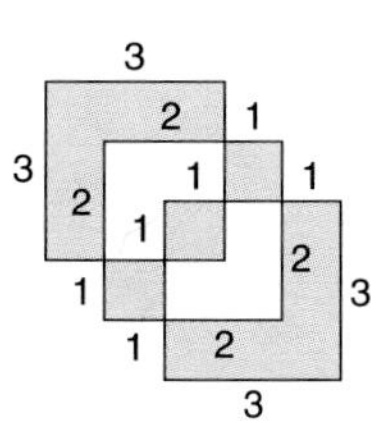

38. **Geometry** If a point inside the figure at the right is chosen at random, what is the probability that it will *not* be in the shaded region? $\frac{6}{19}$

Assignment Guide

Core: 13–41 odd, 42–49
Enriched: 14–36 even, 37–49

For **Extra Practice,** see p. 766.

The red A, B, and C flags, printed only in the Teacher's Wraparound Edition, indicate the level of difficulty of the exercises.

Practice Masters, p. 31

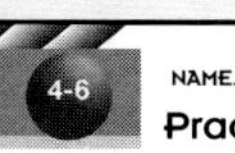

NAME__________ DATE________

Practice — Student Edition Pages 228–232

Integration: Probability
Probability and Odds

Find the probability of each outcome if a die is rolled.

1. a 1 $\frac{1}{6}$
2. a number less than 3 $\frac{1}{3}$
3. an odd number $\frac{1}{2}$
4. a number greater than 6 0
5. an even number less than 6 $\frac{1}{3}$
6. a number greater than 0 1

Find the odds of each outcome if a die is rolled.

7. a multiple of 2 1:1
8. a number less than 2 1:5
9. a factor of 6 2:1
10. a 5 1:5
11. not a 1 5:1
12. a number greater than 5 1:5

A card is selected at random from a standard deck of 52 cards.

13. What is the probability of selecting a red card? $\frac{1}{2}$
14. What is the probability of selecting an ace? $\frac{1}{13}$
15. What are the odds of selecting a diamond? 1:3
16. What are the odds of not selecting a black 2? 25:1

The number of male and female doctors in Camron are listed by age in the table at the right. Use this data to answer the following questions.

Camron Doctors

Age	Male	Female
Under 35	29	18
35–44	36	8
45–54	19	3
55–64	17	2

17. What is the probability that a doctor chosen is in the 45-54 age group? $\frac{1}{6}$
18. What is the probability that a doctor chosen is a female? $\frac{31}{132}$
19. What are the odds that a doctor chosen is a female under 35 years? 3:19
20. What are the odds that a doctor chosen is a male between the ages of 35 and 54 years? 5:7

4 ASSESS

Closing Activity

Speaking Suppose you have seven notebooks, three of which are blue, one red, two green, and one yellow. If you select one at random, do you have a greater chance of choosing any one color? What two colors do you have an equal chance of selecting? **yes, blue; red and yellow**

Chapter 4, Quiz C (Lessons 4-5 and 4-6), is available in the *Assessment and Evaluation Masters*, p. 101.

Additional Answer

48a.

Stem	Leaf
1	6 7 7 8 8 8 8 8 9 9 9 9 9 9
2	0 0 2 3 4 4 5 5 6 7 7
3	0 2 3 3 5 6
4	5 8
5	5

$3|5 = 35$

Enrichment Masters, p. 31

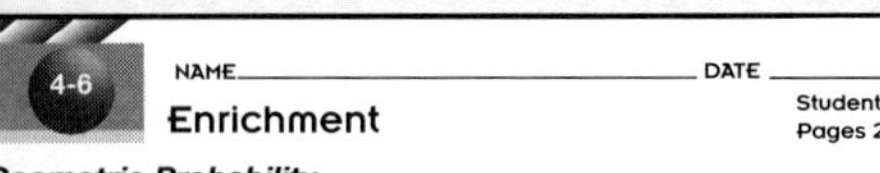
4-6 NAME ____________ DATE ______

Enrichment

Student Edition Pages 228–232

Geometric Probability

If a dart, thrown at random, hits a triangular board shown at the right, what is the probability that it will hit the shaded region? This probability can be determined by comparing the area of the shaded region to the area of the board. This ratio indicates what fraction of the tosses should hit in the shaded region.

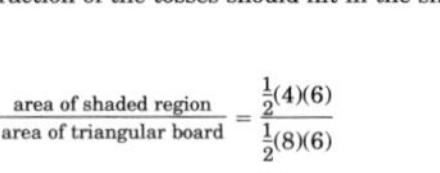

$$\frac{\text{area of shaded region}}{\text{area of triangular board}} = \frac{\frac{1}{2}(4)(6)}{\frac{1}{2}(8)(6)} = \frac{12}{24} \text{ or } \frac{1}{2}$$

In general, if S is a subregion of some region R, then the probability, $P(S)$, that a point, chosen at random, belongs to subregion S is given by the following.

$$P(S) = \frac{\text{area of subregion } S}{\text{area of region } R}$$

Find the probability that a point, chosen at random, belongs to the shaded subregions of the following regions.

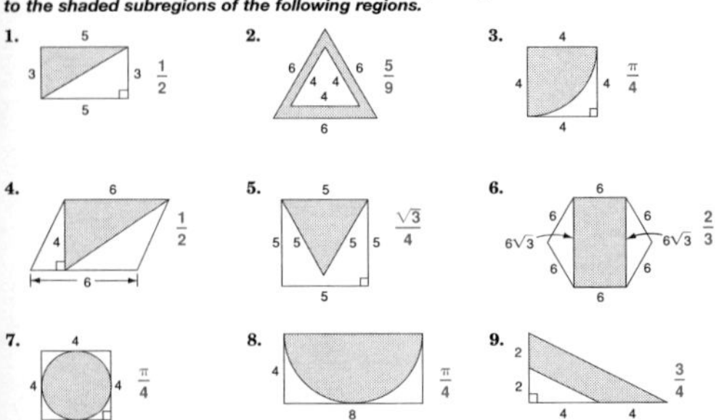
1. $\frac{1}{2}$ 2. $\frac{5}{9}$ 3. $\frac{\pi}{4}$ 4. $\frac{1}{2}$ 5. $\frac{\sqrt{3}}{4}$ 6. $\frac{2}{3}$ 7. $\frac{\pi}{4}$ 8. $\frac{\pi}{4}$ 9. $\frac{3}{4}$

Applications and Problem Solving

39a. 1:14 **39b.** $\frac{1}{27}$

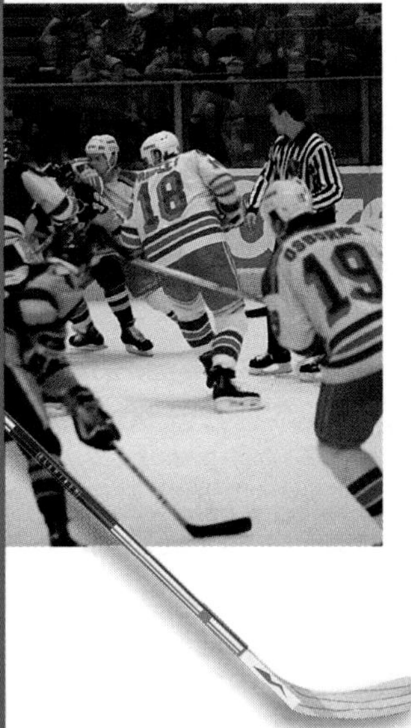

39. Games The game board on *Jeopardy!* is divided into 30 squares. There are six categories with five answers in each category. In the Double Jeopardy! round, 2 Daily Double squares are hidden among the 30 squares.

a. What are the odds of choosing a Daily Double square on the first selection?

b. If you are making the fourth selection of the Double Jeopardy! round, what is the probability of choosing a Daily Double square if one has been chosen?

40. School Have you ever had to make an oral presentation before your classmates? Did the teacher tell everyone to be prepared and then choose the first day's presenters at random?

a. If there are 35 students in your class, and four students will be chosen to give a presentation each day until everyone has made a presentation, what is the probability that you will be chosen to make your presentation the first day?

b. If you have not been chosen by the fourth day, what is the probability that you will be chosen on the fourth day?

41. Statistics The stem-and-leaf plot at the right shows the total points earned by each of the 26 teams in the National Hockey League for the 1993–1994 season.

Stem	Leaf
11	2
10	0 1 6
9	1 5 6 7 7 7 8
8	0 2 3 4 5 7 8
7	1 1 6
6	3 4 6
5	7
4	
3	7 $9|1 = 91$

a. What is the probability that a team earned 97 points?

b. What is the probability that a team earned less than 80 points?

c. What are the odds in favor of a team scoring more than 100 points? **3:23**

40a. $\frac{4}{35}$ **40b.** $\frac{4}{23}$

41a. $\frac{3}{26}$ **41b.** $\frac{4}{13}$

Mixed Review

42. Chocolate Illinois is the number one candy-producing state. The Little Chocolatier shop in Sterling, Illinois, is known for its pecan dumplings. The owners used to make nine pounds of pecan dumplings at one time. Now they make 300 pounds in the same amount of time! Find the percent of change. (Lesson 4–5) **3233%**

43. Geometry In a parking garage, there are 20 feet between each level. Each ramp to a level is 130 feet long. Find the measure of the angle of elevation of each ramp. (Lesson 4–3) **9°**

44. Solve $\frac{5}{8}x + \frac{3}{5} = x$. (Lesson 3–5) $\frac{8}{5}$; **1.6**

45. Work Backward Kristen spent one fifth of her money for gasoline. Then she spent half of what was left for a haircut. She bought lunch for $7. When she got home, she had $13 left. How much did Kristen have originally? (Lesson 3–3) **$50**

46. Simplify $\pm\sqrt{1764}$. (Lesson 2–8) **±42**

47. Find $\frac{17}{21} + \left(-\frac{13}{21}\right)$. (Lesson 2–5) $\frac{4}{21}$

48. School Each number below represents the age of a student in Ms. Wallace's evening calculus class at DeSantis Community College. (Lesson 1–4)

22 17 25 24 19 27 33 16 35 26 20 18 24 33 18 19 48
36 19 23 55 18 18 19 27 18 19 25 17 32 19 45 19 20 30

a. Make a stem-and-leaf plot of the data. **See margin.**

b. How many people attend Ms. Wallace's class? **35 people**

c. What is the difference in ages between the oldest and youngest person in class? **39 years**

d. What is the most common age for a student in the class? **19 years**

e. Which age group is most widely represented in the class? **teens**

49. Evaluate $8(a - c)^2 + 3$ if $a = 6$, $b = 4$, and $c = 3$. (Lesson 1–3) **75**

Extension

Connections Twenty-one Little League baseball teams each played seven games. The stem-and-leaf plot shows the sum of the runs scored by each team in the seven games. Use this to answer each question.

Stem	Leaf
6	0 2 3
7	1 5 8 9
8	2 4 5 7 9
9	5 7 8 8 8
10	1 4 6 7

$7|5 = 75$

1. What is the probability that a team scored 98 runs? $\frac{1}{7}$
2. What is the probability that a team scored fewer than 80 runs? $\frac{1}{3}$
3. What are the odds in favor of a team scoring more than 100 runs? **4:17**

4-7 Weighted Averages

What YOU'LL LEARN

- To solve mixture problems, and
- to solve problems involving uniform motion.

Why IT'S IMPORTANT

You can use weighted averages to solve problems involving travel, transportation, and chemistry.

School

Semester Grades	
Item	Score
Unit Exam A	79%
Unit Exam B	83
Unit Exam C	96
Unit Exam D	91
Unit Exam E	89
Semester Exam	90
Project	95

In Mr. Calloway's American History class, semester grades are based on five unit exams, one semester exam, and a long-term project. Each unit exam is worth 15% of the grade, the semester exam is worth 20% of the grade, and the project is worth 5% of the grade.

Parker's scores for the semester are given in the table at the right. If 100% is a perfect score for each item, find Parker's average for the semester.

In order to find Parker's average for the semester, you may want to add the percentages and then divide by the number of items, 7. But this assumes that each has the same weight. So, you need to find the **weighted average** of the scores.

Definition of Weighted Average	The weighted average M of a set of data is the sum of the product of each number in the set and its weight divided by the sum of all the weights.

You can find the weighted average for the application above by multiplying each score by the percentage of the semester grade that it represents and then dividing by the sum of the weights, or 100. $5(15) + 1(20) + 1(5) = 100$

$$M = \frac{15(79 + 83 + 96 + 91 + 89) + 20(90) + 5(95)}{100}$$

$$= \frac{8845}{100} \text{ or } 88.45$$

So, Parker's average for the semester is about 88%.

Mixture problems involve weighted averages. In a mixture problem, the weight is usually a price or a percentage of something.

Example 1

PROBLEM SOLVING
Make a Chart

Suppose the Central Perk coffee shop sells a cup of espresso for \$2.00 and a cup of cappuccino for \$2.50. On Friday, Rachel sold 30 more cups of cappuccino than espresso, and she sold \$178.50 worth of espresso and cappuccino. How many cups of each were sold?

Explore Let e represent the number of cups of espresso sold. Then $e + 30$ represents the number of cups of cappuccino sold.

(continued on the next page)

4-7 LESSON NOTES

NCTM Standards: 1–5

Instructional Resources

- Study Guide Master 4-7
- Practice Master 4-7
- Enrichment Master 4-7
- Modeling Mathematics Masters, pp. 45–47
- Real-World Applications, 12

Transparency 4-7A contains the 5-Minute Check for this lesson; **Transparency 4-7B** contains a teaching aid for this lesson.

Recommended Pacing	
Standard Pacing	Days 11 & 12 of 15
Honors Pacing	Days 11 & 12 of 15
Block Scheduling*	Day 5 of 7
Alg. 1 in Two Years*	Days 14 & 15 of 20

*For more information on pacing and possible lesson plans, refer to the *Block Scheduling Booklet* and *Algebra 1 in Two Years*.

5-Minute Check
(over Lesson 4-6)

Find the probability of each event when a single die is rolled.

1. 3 $\frac{1}{6}$
2. even number $\frac{1}{2}$
3. less than 3 $\frac{1}{3}$

Find the probability of each event when 5 coins are tossed.

4. 5 heads $\frac{1}{32}$
5. 1 head $\frac{5}{32}$
6. The odds of an event not occurring are 7:3. What is the probability that the event will occur? $\frac{3}{10}$

Motivating the Lesson

Questioning Open the lesson by asking what units of measure should be used for each rate described below.

1. the speed of a person running
2. the speed of a caterpillar
3. the rate of growth of a child
4. the speed of light

2 TEACH

In-Class Examples

For Example 1
Barrett's bookstore sells pencils for $0.10 each and erasers for $0.15 each. Last Tuesday, the store sold 17 more pencils than erasers for a total of $23.45. How many of each item were sold? **104 pencils, 87 erasers**

For Example 2
Zina has 48 mL of a solution that is 25% acid. How much water should she add to obtain a solution that is 15% acid? **32 mL of water**

Teaching Tip For Examples 1 and 2, emphasize to students that in mixture problems they are trying to find the perfect blend or balance of several parts to achieve the desired result.

Plan Make a chart of the information.

	Number of Cups	Price Per Cup	Total Price
espresso	e	$2.00	$2e$
cappuccino	$e + 30$	$2.50	$2.5(e + 30)$

Solve *total sales of espresso* + *total sales of cappuccino* = *total sales*

$$2e + 2.5(e + 30) = 178.50$$
$$2e + 2.5e + 75 = 178.50$$
$$4.5e + 75 = 178.50$$
$$4.5e = 103.50$$
$$e = 23$$

There were 23 cups of espresso sold. There were 23 + 30, or 53 cups of cappuccino sold.

Examine If 23 cups of espresso were sold, the total sales of those cups of coffee would be 23($2.00) or $46.00. If 53 cups of cappuccino were sold, the total sales of those cups of coffee would be 53($2.50) or $132.50.

Since $46.00 + $132.50 = $178.50, the solution is correct.

Sometimes mixture problems are expressed in terms of percents.

Example 2

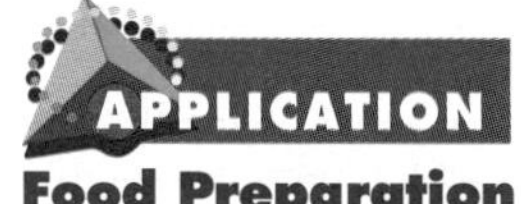

APPLICATION
Food Preparation

An advertisement for an orange drink claims that the drink contains 10% orange juice. Jamel needs 6 quarts of the drink to serve at a party and he wants the drink to contain 40% orange juice. How much of the 10% drink and pure orange juice should Jamel mix to obtain 6 quarts of a mixture that contains 40% orange juice?

Explore Let p represent the amount of pure juice to be added.

Plan Make a chart of the information.

	Quarts	Amount of Orange Juice
10% juice	$6 - p$	$0.10(6 - p)$
Pure juice	p	$1.00p$
40% juice	6	$0.40(6)$

Solve *amount of orange juice in 10% juice* + *amount of orange juice in pure juice* = *amount of orange juice in 40% juice*

$$0.10(6 - p) + 1.00p = 0.40(6)$$
$$0.6 - 0.1p + 1.00p = 2.4$$
$$0.9p = 1.8$$
$$p = 2$$

Jamel needs to combine 2 quarts of pure orange juice with 6 − 2 or 4 quarts of 10% juice to obtain a 6-quart mixture that is 40% orange juice. *Examine this solution.*

Motion problems are another application of weighted averages. When an object moves at a constant speed, or rate, it is said to be in **uniform motion**. The formula $d = rt$ is used to solve uniform motion problems. In the formula, d represents distance, r represents rate, and t represents time. You can also use equations and charts when solving motion problems.

Example 3

Travel

On Friday, Shenae and her brother Rafiel went to visit their grandparents for the weekend. Luckily, traffic was light and they were able to make the 50-mile trip in exactly one hour. On Sunday, they weren't so lucky. The trip home took exactly two hours. What was their average speed for the round trip?

To find the average speed for each leg of the trip, rewrite $d = rt$ as $r = \frac{d}{t}$.

Going

$r = \frac{d}{t}$

$= \frac{50 \text{ miles}}{1 \text{ hour}}$ or 50 miles per hour

Returning

$r = \frac{d}{t}$

$= \frac{50 \text{ miles}}{2 \text{ hours}}$ or 25 miles per hour

You may think that the average speed of the trip would be $\frac{50 + 25}{2}$ or 37.5 miles per hour. However, Shenae did not drive at these speeds for equal amounts of time. You can find the weighted average for their trip.

Round Trip

$M = \frac{50(1) + 25(2)}{3}$

$= \frac{100}{3}$ or $33\frac{1}{3}$

Their average speed was $33\frac{1}{3}$ miles per hour.

Example 4

Transportation

Two city buses leave their station at the same time, one heading east and the other heading west. The eastbound bus travels at 35 miles per hour, and the westbound bus travels at 45 miles per hour. In how many hours will they be 60 miles apart?

Explore Draw a diagram to help analyze the problem.

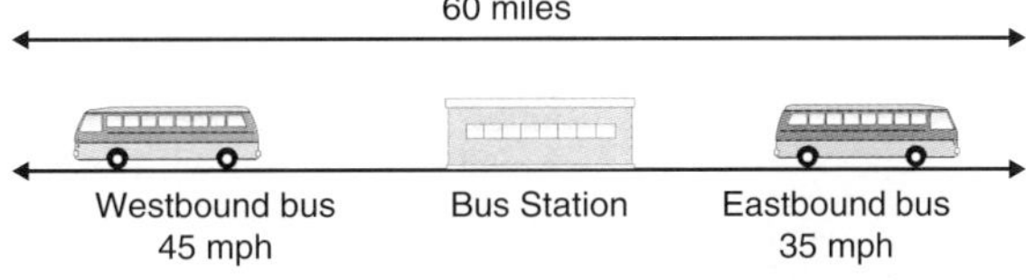

Plan Organize the information in a chart. Let t represent the number of hours until they are 60 miles apart. Remember that $rt = d$.

Bus	r	t	d
Eastbound	35	t	$35t$
Westbound	45	t	$45t$

The eastbound bus travels 35t miles.
The westbound bus travels 45t miles.

$$\underbrace{35t}_{\text{eastbound distance}} + \underbrace{45t}_{\text{westbound distance}} = \underbrace{60}_{\text{total distance}}$$

(continued on the next page)

In-Class Examples

For Example 3

Lisa and Jamaal traveled 20 miles in $\frac{1}{2}$ hour. What was their average speed? **40 mph**

For Example 4

Suppose John and Gina leave at the same time traveling in the same direction. Gina drives at a rate of 85 km/h and John at 70 km/h. How long until they are 90 km apart? **6 hours**

Teaching Tip For Examples 3 and 4, remind students that the actual solution of a uniform motion problem is a number with an appropriate unit of measure, not the equation containing the isolated variable.

Classroom Vignette

"For Example 4, have two students stand back to back, headed away from each other. Start them walking at the same time. Clap loudly when they are "60 miles" apart. They will quickly come up with the fact that the time is the same and that the *eastbound distance* + *westbound distance* = *total distance*."

William A. Brinkman

William Brinkman
Paynesville High School
Paynesville, Minnesota

3 PRACTICE/APPLY

Check for Understanding

Exercises 1–7 are designed to help you assess your students' understanding through reading, writing, speaking, and modeling. You should work through Exercises 1–3 with your students and then monitor their work on Exercises 4–7.

Additional Answers

1. d = distance, r = rate, t = time
2. to organize information and model the situation

Solve

$$35t + 45t = 60$$
$$80t = 60$$
$$t = \frac{3}{4}$$

In $\frac{3}{4}$ hour, or 45 minutes, the buses will be 60 miles apart.

Examine To check the answer, find the distance each bus could travel in $\frac{3}{4}$ hour and see if it totals 60 miles.

$$\left(\frac{3}{4}\text{h}\right)(35 \text{ mph}) + \left(\frac{3}{4}\text{h}\right)(45 \text{ mph}) \stackrel{?}{=} 60 \text{ mi}$$
$$26\tfrac{1}{4} \text{ mi} + 33\tfrac{3}{4} \text{ mi} \stackrel{?}{=} 60 \text{ mi}$$
$$60 \text{ mi} = 60 \text{ mi} \checkmark$$

CHECK FOR UNDERSTANDING

Communicating Mathematics

Study the lesson. Then complete the following. 1–2. See margin.

1. **Tell** what the d, r, and t represent in the formula $d = rt$.
2. **Explain** why it is sometimes helpful to use charts and diagrams.

3. **Describe** your favorite mode of transportation. Then write a problem using the formula $d = rt$ and your favorite mode. **See students' work.**

Guided Practice

4. **Chemistry** Joshua is doing a chemistry experiment that calls for a 30% solution of copper sulfate. He has 40 mL of 25% solution. How many milliliters of 60% solution should Joshua add to obtain the required 30% solution? **about 6.7 mL**

	Amount of Solution (mL)	Amount of Copper Sulfate
25% solution	40	0.25(40)
60% solution	x	$0.60x$
30% solution	$40 + x$	$0.30(40 + x)$

5. **Sales** The Cookie Crumbles Company sells two kinds of cookies daily: peanut butter at $6.50 per dozen and chocolate chip at $9.00 per dozen. Yesterday, Cookie Crumbles sold 85 dozen more peanut butter than chocolate chip cookies. The total sales for both were $4055.50. How many dozen of each were sold?

5. 226 dozen chocolate chip; 311 dozen peanut butter

6. **Air Travel** An airplane flies 1000 miles due east in 2 hours and 1000 miles due south in 3 hours. What is the average speed of the airplane? **400 mph**
7. **Boating** *The Yankee Clipper* leaves the pier at 9:00 A.M. at 8 knots (nautical miles per hour). A half hour later, *The Riverboat Rover* leaves the same pier in the same direction traveling at 10 knots. At what time will *The Riverboat Rover* overtake *The Yankee Clipper*? **11:30 A.M.**

Study Guide Masters, p. 32

NAME________________ DATE ________

Study Guide

Student Edition Pages 233–238

Weighted Averages

You can use charts to solve mixture problems.

Example 1: Alice Gleason invested a portion of $32,000 at 9% interest and the balance at 11% interest. How much did she invest at each rate if her total income from both investments was $3200.

Amount Invested	Rate	Annual Income
x	0.09	$0.09x$
$32{,}000 - x$	0.11	$0.11(32{,}000 - x)$

income from 9% investment + income from 11% investment = total income

$$0.09x + 0.11(32{,}000 - x) = 3200$$
$$0.09x + 3520 - 0.11x = 3200$$
$$-0.02x = -320$$
$$x = 16{,}000$$
$$32{,}000 - x = 16{,}000$$

Ms. Gleason invested $16,000 at 9% and $16,000 at 11%.

When an object moves without changing its speed, it is said to be in **uniform motion.** The formula $d = rt$ is used to solve uniform motion problems.

Example 2: Bill Gutierrez left home driving at a speed of 54 miles per hour. How many hours did it take him to reach his destination 243 miles away?

$$d = rt$$
$$243 = 54t$$
$$4\tfrac{1}{2} = t$$

It will take Bill $4\frac{1}{2}$ hours to drive 243 miles.

Solve.

1. How many grams of sugar must be added to 60 grams of a 32% solution to obtain a 50% solution? 21.6 grams
2. Mr. Anders and Ms. Rich each drove home from a business meeting. Mr. Anders traveled east at 100 kilometers per hour and Ms. Rich traveled west at 80 kilometers per hour. In how many hours were they 100 kilometers apart? $5\frac{5}{9}$ hours
3. The Quick Mart has two kinds of nuts. Pecans sell for $1.55 per pound and walnuts sell for $1.95 per pound. How many pounds of walnuts must be added to 15 pounds of pecans to make a mixture that sells for $1.75 per pound? 15 pounds
4. Justin left home at 8:00 A.M., riding his bike at 6 miles per hour. His brother Joshua left 2 hours later, riding his bike at 8 miles per hour. At what time will Joshua catch up to Justin if Joshua is delayed 15 minutes by a freight train? 5 P.M.

EXERCISES

Applications and Problem Solving

A

8. **Sales** The Madison Local High School marching band sold gift wrap to earn money for a band trip to Orlando, Florida. The gift wrap in solid colors sold for $4.00 per roll, and the print gift wrap sold for $6.00 per roll. The total number of rolls sold was 480, and the total amount of money collected was $2340. How many rolls of each kind of gift wrap were sold? **270 solid; 210 print**

	Number of Rolls	Price Per Roll	Total Price
Solid	r	$4	$4r$
Print	$480 - r$	$6	$6(480 - r)$

B

9. **Money** Rochelle has $2.55 in dimes and quarters. She has eight more dimes than quarters. How many quarters does she have? **5 quarters**

10. **Sales** The Nut House sells walnuts for $4.00 a pound and cashews for $7.00 a pound. How many pounds of cashews should be mixed with 10 pounds of walnuts to obtain a mixture that sells for $5.50 a pound? **10 pounds**

11. **Travel** Ryan and Jessica Wilson leave their home at the same time, traveling in opposite directions. Ryan travels at 57 miles per hour and Jessica travels at 65 miles per hour. In how many hours will they be 366 miles apart? **3 hours**

12. **Travel** At 7:00 A.M., Brooke leaves home to go on a business trip driving 35 miles per hour. Fifteen minutes later, Bart discovers that Brooke forgot her presentation materials. He drives 50 miles per hour to catch up with her. If Bart is delayed 30 minutes with a flat tire, when will he catch up with Brooke? **9:30 A.M.**

13. **Cycling** Two cyclists begin traveling in the same direction on the same bike path. One travels at 20 miles per hour, and the other at 14 miles per hour. After how many hours will they be 15 miles apart? **2.5 hours**

14. 5 adult tickets, 3 student tickets

C

14. **Entertainment** A local radio station is having a drawing to win tickets to see En Vogue in concert. The winners will be driven to the concert in a limousine that holds eight passengers. Concert tickets cost $22 for adults and $19 for students. After renting the limousine, the radio station has $167 left in their giveaway budget to buy the eight tickets. How many adult tickets and student tickets can the station buy?

15. **Travel** Pablo is driving 40 miles per hour. After he has driven 30 miles, his brother Ricardo starts driving in the same direction. At what rate must Ricardo drive to catch up with Pablo in 5 hours? **46 mph**

16. **Automotives** A car radiator has a capacity of 16 quarts and is filled with a 25% antifreeze solution. How much must be drained off and replaced with pure antifreeze to obtain a 40% antifreeze solution? **3.2 quarts**

17. **Law Enforcement** A state patrol officer is chasing a car he believes is speeding. The officer is traveling 70 miles per hour to try to catch up with the car. But he can't say later that the other car was going 70 mph because the driver could ask, "If we were going the same speed, how did you catch up? Obviously I was going slower than you. I wasn't speeding." The officer sees the driver pass a mile marker, one-fourth of a mile ahead. From the marker, it takes the officer five minutes $\left(\frac{1}{12}\text{ of an hour}\right)$ to catch up. How fast was the driver going? **67 mph**

Assignment Guide

Core: 9–21 odd, 23–29
Enriched: 8–20 even, 21–29

For **Extra Practice,** see p. 766.

The red A, B, and C flags, printed only in the Teacher's Wraparound Edition, indicate the level of difficulty of the exercises.

Practice Masters, p. 32

NAME__________ DATE______

Practice

Student Edition Pages 233–238

Weighted Averages

Solve.

1. **Entertainment** At a spring concert, tickets for adults cost $4.00 and tickets for students cost $2.50. How many of each kind of tickets were purchased if 125 tickets were bought for $413?
   67 adult tickets; 58 student tickets

2. Two trains leave Raleigh at the same time, one traveling north, the other south. The first train travels at 50 miles per hour and second at 60 miles per hour. In how many hours will the trains be 275 miles apart?
   2.5 h

3. A train leaves a city heading west and travels at 50 miles per hour. Three hours later, a second train leaves from the same place and travels in the same direction at 65 miles per hour. How long will it take for the second train to overtake the first train?
   10 h

4. **Sales** Susan wants to mix 10 pounds of Virginia peanuts that cost $3.50 a pound with Spanish peanuts that cost $3.00 a pound to obtain a mixture that costs $3.40 a pound. How many pounds of Spanish peanuts should she use?
   2.5 lb

5. A ship travels from New York to Southampton, England at 33 miles per hour. A plane traveling 605 miles per hour takes 104 hours less time to make the same trip. How far apart are the two cities?
   3630 mi

6. **Advertising** An advertisement for a pineapple drink claims that the drink contains 15% pineapple juice. How much pure pineapple juice would have to be added to 8 quarts of the drink to obtain a mixture containing 50% pineapple juice?
   5.6 qt

Reteaching

Using Discussion Have students discuss each problem in groups of two or three. Have them talk through what information is given and what is needed. Have each group develop a plan of attack; then have students solve the exercises individually.

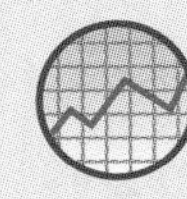

Tech Prep

Law Enforcement Students who are interested in law enforcement may wish to do further research on the data provided in Exercise 17 and explore the potential growth of this career. For more information on tech prep, see the *Teacher's Handbook*.

4 ASSESS

Closing Activity

Modeling Have each student create a diagram that presents a motion problem. An example might be drawing two cars moving in opposite directions with labels showing the starting time of each as well as their rates. Tell students to switch drawings and then create and solve a motion problem from their classmate's diagram.

Additional Answer

21. Sample answer: How much pure antifreeze must be mixed with a 20% solution to produce 40 quarts of a 28% solution?

Enrichment Masters, p. 32

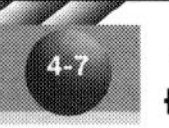

NAME____________________ DATE __________

Enrichment — Student Edition Pages 233–238

Partnerships

Two or more people who agree to combine their money, goods, or skill in some business often do so by forming a *partnership*. Gains and losses in a partnership are usually divided in proportion to the amount of money each person has invested.

Sharon and Tina start a business to print restaurant menus using Tina's computer. For supplies and software, Sharon invests \$1200 and Tina, \$800. If their profit after the first year is \$500, what is each person's share?

The entire investment is \$1200 plus \$800, or \$2000.
Sharon's share is $\frac{\$1200}{\$2238}$, or $\frac{3}{5}$. Tina's share is $\frac{\$800}{\$2238}$, or $\frac{2}{5}$.

Answer each question.

1. In the problem above, what is Sharon's share of the profit? \$300
2. What is Tina's share? \$200
3. Sal and Mario were partners in a construction business. Sal put in \$5238 and Mario, \$4238. Their profit in 3 years was \$4500. What was each partner's share? Sal: \$2500; Mario: \$2238
4. Tim, Bob, and Alice entered into a business partnership for 2 years. Tim put in \$3600, Bob put in \$2400, and Alice put in \$2238. If their profit was \$3238, what was each person's share? Tim: \$1350; Bob: \$900; Alice: \$750
5. Fred, Vito, and Hal began with a capital of \$28,238. Fred furnished \$7238; Vito, \$6238; and Hal, the remainder. If they gained 14% on their investment, what was each person's share? Fred: \$980; Vito: \$840; Hal: \$2100
6. A storeroom belonging to Taylor, Bartinelli, & Wong was entirely destroyed by fire. They received \$9675 insurance. What was each person's share, if Taylor owned 15%, Bartinelli, 40%, and Wong the remainder? Taylor: \$1451.25; Bartinelli: \$3870; Wong: \$4353.75
7. Divide a profit of \$7200 between two partners so that the ratio of the shares is 4 to 5. \$3200, \$4238
8. Divide a profit of \$18,238 among three partners so that the ratio of the shares is 1:2:3. \$3238, \$6238, \$9238

18. **Travel** An express train travels 80 kilometers per hour from Ironton to Wildwood. A local train, traveling at 48 kilometers per hour, takes 2 hours longer for the same trip. How far apart are Ironton and Wildwood? **240 km**

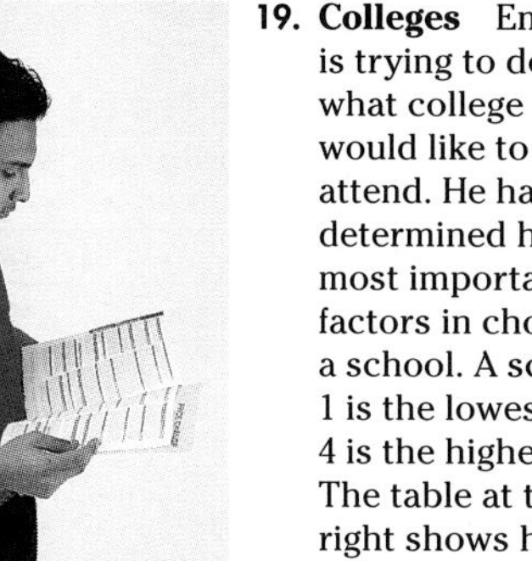

19. **Colleges** Emilio is trying to decide what college he would like to attend. He has determined his five most important factors in choosing a school. A score of 1 is the lowest and 4 is the highest. The table at the right shows his ratings for one school he is considering.

	Importance Factor (1 to 4)	Rating (1 to 4)	Total
Distance from home	4	4	16
Academic reputation	3	2	6
Social climate	3	1	3
Quality of dorms	1	4	4
Community	2	3	6

a. Find the weighted average of his ratings for this school. **7**
b. What is the highest rating that one of Emilio's schools can receive? **16**

20. **School** Many schools base a student's grade point average, or GPA, on the grade a student receives as well as on the number of credits received for the class. Mercedes' grades for this semester are listed in the table at the right. Find Mercedes' GPA if a grade of A equals 4 and a B equals 3. **3.56**

Grade Card

Class	Credits Rating	Grade
Honors Algebra	1	A
Biology	1	B
English 1	1	A
Spanish 2	1	B
Phys. Ed.	$\frac{1}{2}$	A

Critical Thinking

21. See margin.

21. Write a mixture problem for the equation $1.00x + 0.20(40 - x) = 0.28(40)$.
22. Monica Morrison drove to work on Monday at 40 miles per hour and arrived one minute late. She left at the same time on Tuesday, drove at 45 miles per hour, and arrived one minute early. How far does Monica drive to work? **12 miles**

Mixed Review

24. 12,000 bats
26. 58 feet
29. associative prop. (+)

23. **Probability** Find the probability of getting a number less than 1 if a die is rolled. (Lesson 4–6) **0**
24. **Baseball** In 1994, DeMarini Sports Inc. sold 7500 aluminum softball bats to softball enthusiasts. In 1995, the company expected to increase sales by 60%. How many bats did the company expect to sell in 1995? (Lesson 4–5)
25. **Finance** Selena Cruz wants to invest a portion of her \$16,000 savings at 8% annual interest and the balance at a safer 5% annual interest. She hopes to earn \$1130 in the next year to pay for a Caribbean cruise. How much money should she invest at 8%? (Lesson 4–4) **\$11,000**
26. **Model Trains** Model trains are scaled-down replicas of real trains that come in a variety of sizes, called scales. One of the most popular-sized models is called the HO. Every dimension of the HO model measures $\frac{1}{87}$ that of a real engine. The HO model of a modern diesel locomotive is about 8 inches long. About how many feet long is the real locomotive? (Lesson 4–1)
27. Solve $5.3 - 0.3x = -9.4$. (Lesson 3–3) **49**
28. Solve $24 = -2a$. (Lesson 3–2) **−12**
29. Name the property illustrated in $(a + 3b) + 2c = a + (3b + 2c)$. (Lesson 1–8)

Extension

Connections Ben, who is bicycling at the rate of x mph, has a head start of 2 miles on Rachel, who is bicycling at the rate of y mph ($y > x$). In how many hours will Rachel overtake Ben?
$\frac{2}{y - x}$ hours

4-8 Direct and Inverse Variation

What YOU'LL LEARN

- To solve problems involving direct and inverse variation.

Why IT'S IMPORTANT

You can use direct and inverse variation to solve problems involving work and music.

APPLICATION Water Usage

Do you find yourself singing in the shower? If you are like the average American, you have enough time to sing several songs. The national average for time spent in the shower is 12.2 minutes. A standard shower head uses about 6 gallons of water per minute. That means, you use 73.2 gallons of water for each shower you take. Taking one shower per day for one year would use 26,718 gallons of water!

FYI

A low-flow shower head uses only 2.5 gallons of water per minute.

The number of gallons of water used depends *directly* on the amount of time spent in the shower. The table below shows the number of gallons of water used y as a function of time in the shower x.

x (minutes)	3	6	9	12	15
y (gallons)	18	36	54	72	90

The relationship between the number of minutes in the shower and the gallons of water used is shown by the equation $y = 6x$. This type of equation is called a **direct variation.** We say that *y varies directly as x* or *y is directly proportional to x.* This means that as x increases in value, y increases in value, or as x decreases in value, y decreases in value.

Definition of Direct Variation	A direct variation is described by an equation of the form $y = kx$, where $k \neq 0$.

In the equation $y = kx$, k is called the **constant of variation**. To find the constant of variation, divide each side by x.

$$\frac{y}{x} = k$$

Example 1

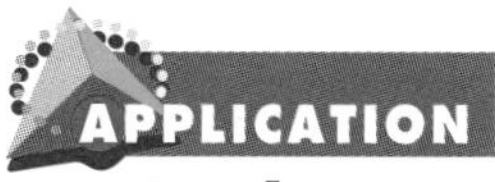

APPLICATION Employment

In this situation, Julio's pay is the depedent quantity and the number of hours Julio works is the independent quantity.

Julio's wages vary directly as the number of hours that he works. If his wages for 5 hours are $29.75, how much will they be for 30 hours?

First, find Julio's hourly pay. Let x = number of hours Julio works, and let y = Julio's pay. Find the value of k in the equation $y = kx$. The value of k is the amount of money Julio is paid per hour.

$k = \frac{y}{x}$

$k = \frac{29.75}{5}$

$k = 5.95$

Julio is paid $5.95 per hour.

(continued on the next page)

FYI

This means that a 5-minute shower uses 12.5 gallons, while a longer 20-minute shower uses 50 gallons of water.

4-8 LESSON NOTES

NCTM Standards: 1–5

Instructional Resources

- Study Guide Master 4-8
- Practice Master 4-8
- Enrichment Master 4-8
- Assessment and Evaluation Masters, p. 101
- Graphing Calculator Masters, p. 4
- Multicultural Activity Masters, p. 8
- Real-World Applications, 13
- Science and Math Lab Manual, pp. 13–16
- Tech Prep Applications Masters, p. 8

Transparency 4-8A contains the 5-Minute Check for this lesson; **Transparency 4-8B** contains a teaching aid for this lesson.

Recommended Pacing	
Standard Pacing	Day 13 of 15
Honors Pacing	Day 13 of 15
Block Scheduling*	Day 6 of 7
Alg. 1 in Two Years*	Days 16 & 17 of 20

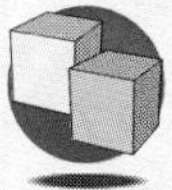

*For more information on pacing and possible lesson plans, refer to the *Block Scheduling Booklet* and *Algebra 1 in Two Years.*

1 FOCUS

5-Minute Check
(over Lesson 4-7)

Solve.

1. Lois has 27 coins in nickels and dimes. In all she as $1.90. How many of each coin does she have? **16 nickels, 11 dimes**
2. How much pure copper must be added to 50.255 kg of an alloy containing 12% copper to raise the copper content to 21%? **5.725 kg**
3. Juan traveled 270 miles. What is his rate if he made the trip in 9 hours? **30 mph**
4. David rode his bicycle 36 kilometers. What is his rate if he began riding 4 hours ago? **9 km/h**

Motivating the Lesson

Hands-On Activity Put 400 mL of water in a 500-mL beaker. Then add an object such as a spoon or glass stirring rod to the beaker. Ask students to describe what happens. **The water level rises.** Add additional spoons or rods to the beaker and observe. Challenge students to identify the relationship between the water level and the objects. **The rise of water level is directly proportional to the number of spoons or rods added to the beaker.**

2 TEACH

In-Class Examples

For Example 1

A car uses 8 gallons of gasoline to travel 290 miles. How much gasoline will the car use to travel 400 miles? **11.034 gallons**

For Example 2

Solve each problem. Assume that *y* varies directly as *x*.

a. If $y = 6$ when $x = 8$, find y when $x = 12$. **9**
b. If $y = 8$ when $x = 20$, find x when $y = 6$. **15**
c. If $y = 3$ when $x = 9$, find y when $x = 40.5$. **13.5**
d. If $x = -5$ when $y = 6$, find y when $x = -8$. **9.6**

Teaching Tip For Example 2, point out that the phrase *y varies directly as x* can also be phrased *y is directly proportional to x*. Both phrases mean that as *y* increases or decreases in value, *x* increases or decreases in value, respectively.

Next find out how much Julio's wages will be for 30 hours.

$y = kx$
$y = 5.95(30)$
$y = 178.50$

Thus, Julio's wages will be \$178.50 for 30 hours of work.

Using the table in the application at the beginning of the lesson, many proportions can be formed. Two examples are shown below.

number of minutes
$$\frac{3}{18} = \frac{9}{54}$$
gallons of water

number of minutes $\frac{3}{9} = \frac{18}{54}$ *gallons of water*

x_1 is read "x sub 1."

Two general forms for proportions like these can be derived from the equation $y = kx$. Let (x_1, y_1) be a solution for $y = kx$. Let a second solution be (x_2, y_2). Then $y_1 = kx_1$ and $y_2 = kx_2$.

$y_1 = kx_1$ *This equation describes a direct variation.*

$\frac{y_1}{y_2} = \frac{kx_1}{kx_2}$ *Use the division property of equality. Since y_2 and kx_2 are equivalent, you can divide the left side by y_2 and the right side by kx_2.*

$\frac{y_1}{y_2} = \frac{x_1}{x_2}$ *Simplify.*

Another proportion can be derived from this proportion.

$x_2y_1 = x_1y_2$ *Find the cross products of the proportion above.*

$\frac{x_2y_1}{y_1y_2} = \frac{x_1y_2}{y_1y_2}$ *Divide each side by y_1y_2.*

$\frac{x_2}{y_2} = \frac{x_1}{y_1}$ *Simplify.*

You can use either of these forms to solve problems involving direct proportion.

Example 2 **If *y* varies directly as *x*, and *y* = 28 when *x* = 7, find *x* when *y* = 52.**

Use $\frac{y_1}{y_2} = \frac{x_1}{x_2}$ to solve the problem.

$\frac{28}{52} = \frac{7}{x_2}$ *Let $y_1 = 28$, $x_1 = 7$, and $y_2 = 52$.*

$28x_2 = 52(7)$ *Find the cross products.*
$28x_2 = 364$
$x_2 = 13$

Thus, $x = 13$ when $y = 52$.

GLENCOE Technology

Interactive Mathematics Tools Software

This multimedia software provides an interactive lesson that uses a manipulative resistor to determine the relationship between current and resistance on a circuit. A **Computer Journal** gives students an opportunity to write about what they have learned.

For Windows & Macintosh

The reverse of direct variation is **inverse variation**. We say that y *varies inversely as* x. This means that as x increases in value, y decreases in value, or as y decreases in value, x increases in value. For example, the more miles you drive, the less gasoline you have in the tank.

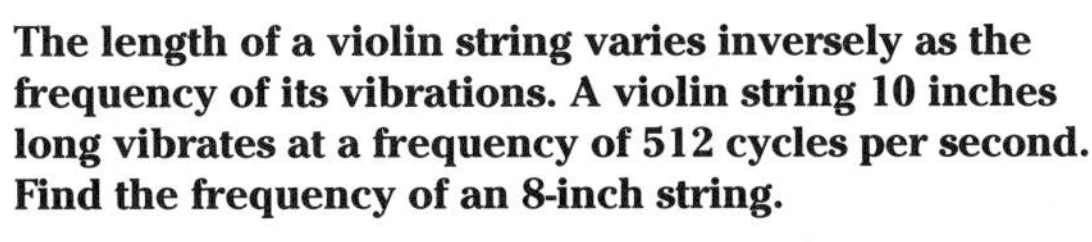

An inverse variation is described by an equation of the form xy = k, where k ≠ 0.

Sometimes inverse variations are written in the form $y = \frac{k}{x}$.

Example 3 APPLICATION **Music**

The length of a violin string varies inversely as the frequency of its vibrations. A violin string 10 inches long vibrates at a frequency of 512 cycles per second. Find the frequency of an 8-inch string.

Let ℓ represent the length in inches and f represent the frequency in cycles per second. Find the value of k.

$\ell f = k$

$(10)(512) = k$ *Substitute 10 for* ℓ *and 512 for* f.

$5120 = k$ *The constant of variation is 5120.*

Next, find the frequency, in cycles per second, of the 8-inch string.

$\ell f = k$ *Use the same inverse variation equation.*

$8 \cdot f = 5120$ *Replace* ℓ *with 8 and* k *with 5120.*

$f = \frac{5120}{8}$ *Divide each side by 8.*

$f = 640$

The frequency of an 8-inch string is 640 cycles per second.

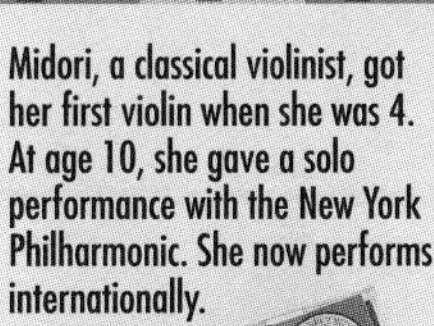

Midori, a classical violinist, got her first violin when she was 4. At age 10, she gave a solo performance with the New York Philharmonic. She now performs internationally.

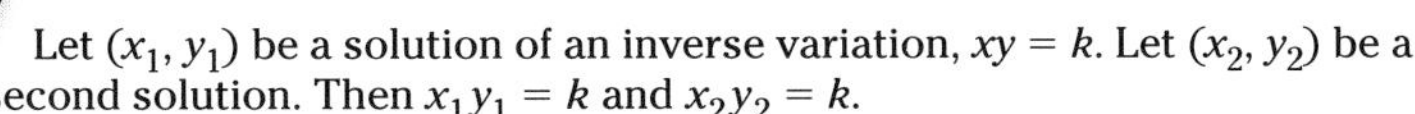

Let (x_1, y_1) be a solution of an inverse variation, $xy = k$. Let (x_2, y_2) be a second solution. Then $x_1y_1 = k$ and $x_2y_2 = k$.

$x_1y_1 = k$

$x_1y_1 = x_2y_2$ *You can substitute* x_2y_2 *for* k *because* $x_2y_2 = k$.

The equation $x_1y_1 = x_2y_2$ is called the product rule for inverse variations. Study how it can be used to form a proportion.

$x_1y_1 = x_2y_2$

$\frac{x_1y_1}{x_2y_1} = \frac{x_2y_2}{x_2y_1}$ *Divide each side by* x_2y_1.

$\frac{x_1}{x_2} = \frac{y_2}{y_1}$ *Notice that this proportion is different from the proportion for direct variation.*

In-Class Example

For Example 3

If 4 pounds of peanuts cost \$7.50, how much will 2.5 pounds cost? **\$4.69**

Teaching Tip Emphasize the difference between the proportions below.

Direct Proportion	Inverse Proportion
$\frac{x_1}{x_2} = \frac{y_1}{y_2}$	$\frac{x_1}{x_2} = \frac{y_2}{y_1}$

FYI

Midori Goto was born in Osaka, Japan, on October 25, 1971. She attended the Julliard School of Music in New York City.

Classroom Vignette

"I actually bring a 12-foot board to the classroom. I allow students to find the center of the board and then move on the board until they are balanced. The students measure the distances and weights to find the constant of variation."

John Cline
Greenfield Central High School
Greenfield, Indiana

In-Class Examples

For Example 4
Solve each problem. Assume that y varies inversely as x.

a. If $y = 6$ when $x = 2$, find x when $y = 4$. **3**

b. If $y = 15$ when $x = 21$, find x when $y = 27$. $\mathbf{\frac{35}{3}}$

For Example 5
A 200-pound weight is located 5 feet from the fulcrum. How far from the fulcrum should 125 pounds be placed to balance the lever? **8 ft**

3 PRACTICE/APPLY

Check for Understanding

Exercises 1–11 are designed to help you assess your students' understanding through reading, writing, speaking, and modeling. You should work through Exercises 1–4 with your students and then monitor their work on Exercises 5–11.

Study Guide Masters, p. 33

NAME______________________ DATE ____________

Study Guide

Student Edition
Pages 239–244

Direct and Inverse Variation

If two variables x and y are related by the equation $y = kx$, where k is a nonzero constant, then the equation is called a **direct variation,** and k is called the **constant of variation.** If two variables x and y are related by the equation $xy = k$, where $k \neq 0$, then the equation is called an **inverse variation.**

Example 1: If y varies directly as x, and $y = 12$ when $x = 4$, find y when $x = 7$.

$y = kx$ | $y = kx$
$12 = 4k$ | $y = 3(7)$
$\frac{12}{4} = \frac{4k}{4}$ | $y = 21$
$3 = k$

Example 2: If y varies inversely as x, and $y = 3$ when $x = 12$, find x when $y = 4$.

$xy = k$ | $xy = k$
$12(3) = k$ | $4x = 36$
$36 = k$ | $\frac{4x}{4} = \frac{36}{4}$
$x = 9$

Solve. Assume that y varies directly as x.

1. If $y = 4$ when $x = 2$, find y when $x = 16$.
32

2. If $y = 9$ when $x = -3$, find x when $y = 6$.
−2

3. If $y = \frac{2}{5}$ when $x = \frac{1}{3}$, find y when $x = \frac{1}{4}$.
$\frac{3}{10}$

4. If $y = \frac{1}{4}$ when $x = \frac{1}{8}$, find x when $y = \frac{3}{16}$.
$\frac{3}{32}$

Solve. Assume that y varies inversely as x.

5. If $y = 9$ when $x = 7$, find y when $x = 2$.
$\frac{63}{2}$

6. If $y = 4.3$ when $x = 12.9$, find y when $x = 15.8$.
$\frac{5547}{1580}$

7. If $x = \frac{1}{2}$ when $y = \frac{1}{3}$, find y when $x = \frac{1}{4}$.
$\frac{2}{3}$

8. If $y = -6$ when $x = \frac{1}{4}$, find y when $x = \frac{3}{16}$.
−8

You can use either the product rule or the proportion rule to solve problems involving inverse variation.

Example 4 **If y varies inversely as x, and $y = 5$ when $x = 15$, find x when $y = 3$.**

Let $x_1 = 15$, $y_1 = 5$, and $y_2 = 3$. Solve for x_2.

Method 1

Use the product rule.

$$x_1y_1 = x_2y_2$$
$$15 \cdot 5 = x_2 \cdot 3$$
$$\frac{75}{3} = x_2$$
$$25 = x_2$$

Method 2

Use the proportion.

$$\frac{x_1}{x_2} = \frac{y_2}{y_1}$$
$$\frac{15}{x_2} = \frac{3}{5}$$
$$75 = 3x_2$$
$$25 = x_2$$

Thus, $x = 25$ when $y = 3$.

If you have observed people on a seesaw, you may have noticed that the heavier person must sit closer to the fulcrum (pivot point) to balance the seesaw. A seesaw is a type of *lever,* and all lever problems involve inverse variation.

Suppose weights w_1 and w_2 are placed on a lever at distances d_1 and d_2, respectively, from the fulcrum. The lever is balanced when $w_1d_1 = w_2d_2$. This property of levers is illustrated at the right.

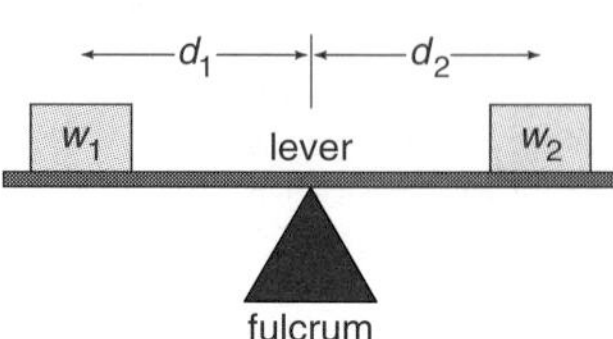

Example 5 **The fulcrum is placed in the middle of a 20-foot seesaw. Cholena, who weighs 120 pounds, is seated 9 feet from the fulcrum. How far from the fulcrum should Antonio sit if he weighs 135 pounds?**

Physics

Let $w_1 = 120$, $d_1 = 9$, and $w_2 = 135$. Solve for d_2.

$$w_1d_1 = w_2d_2$$
$$120 \cdot 9 = 135 \cdot d_2$$
$$1080 = 135d_2$$
$$d_2 = 8$$

Antonio should sit 8 feet from the fulcrum.

CHECK FOR UNDERSTANDING

Communicating Mathematics

Study the lesson. Then complete the following.

1. **Tell** whether an equation of the form $xy = k$, where $k \neq 0$, represents a direct or inverse variation. **inverse**
2. Refer to the application at the beginning of the lesson. Describe a reasonable domain and range.
3. **Explain** how you find the constant of variation in a direct variation.
4. **You Decide** Morgan says that a person's height varies directly as his or her age. Do you agree? Explain. **No, because growth rates vary widely.**

2. The domain is the minutes in the shower; the range is the gallons of water.

3. Divide each side of $y = kx$ by x.

Reteaching

Using Manipulatives Have a small group of students use a triple-beam balance and 4 stacks of uniform washers. Each stack should contain a different number of washers tied together with strong thread so students cannot weigh just one washer at a time.

# of Washers (n)	Weight (W)

Count and record the number of washers in each stack. Weigh the first stack. Use this result to predict the weights of each of the three remaining stacks. Then weigh these stacks. Write and solve a proportion for the first stack to find the *constant of variation k* in $W = kn$. What does it represent as a rate? **weight of 1 washer**

Guided Practice

Determine which equations represent inverse variations and which represent direct variations. Then find the constant of variation.

5. $mn = 5$ **I, 5**
6. $a = -3b$ **D, −3**

Solve. Assume that *y* varies directly as *x*.

7. If $y = 27$, when $x = 6$, find x when $y = 45$. **10**
8. If $y = -7$ when $x = -14$, find y when $x = 20$. **10**

Solve. Assume that *y* varies inversely as *x*.

10. $\frac{12}{5}$ or 2.4

9. If $y = 99$ when $x = 11$, find x when $y = 11$. **99**
10. If $y = -6$ when $x = -2$, find y when $x = 5$.
11. **Physics** An 8-ounce weight is placed at one end of a yardstick. A 10-ounce weight is placed at the other end. Where should the fulcrum be placed to balance the yardstick? **20 inches from the 8-ounce weight, 16 inches from the 10-ounce weight**

EXERCISES

Practice

Determine which equations represent inverse variations and which represent direct variations. Then find the constant of variation.

12. $c = 3.14d$ **D, 3.14**
13. $15 = rs$ **I, 15**
14. $\frac{35}{p} = q$ **I, 35**
15. $s = \frac{9}{t}$ **I, 9**
16. $\frac{1}{3}x = z$ **D, $\frac{1}{3}$**
17. $4a = b$ **D, 4**

Solve. Assume that *y* varies directly as *x*.

18. $2\frac{1}{4}$ 19. $26\frac{1}{4}$

18. If $y = -8$ when $x = -3$, find x when $y = 6$.
19. If $y = 12$ when $x = 15$, find x when $y = 21$.

20. If $y = 2.5$ when $x = 0.5$, find y when $x = 20$. **100**
21. If $y = 4$ when $x = 12$, find y when $x = -24$. **−8**
22. If $y = -6$ when $x = 9$, find y when $x = 6$. **−4**
23. If $y = 2\frac{2}{3}$ when $x = \frac{1}{4}$, find y when $x = 1\frac{1}{8}$. **12**

Solve. Assume that *y* varies inversely as *x*.

24. If $y = 9$ when $x = 8$, find y when $x = 6$. **12**
25. If $x = 2.7$ when $y = 8.1$, find y when $x = 3.6$. **6.075**
26. If $y = 24$ when $x = -8$, find y when $x = 4$. **−48**
27. If $x = 6.1$ when $y = 4.4$, find x when $y = 3.2$ **8.3875**

28. $\frac{2}{3}$ 29. $\frac{1}{4}$

28. If $y = 7$ when $x = \frac{2}{3}$, find y when $x = 7$.
29. If $x = \frac{1}{2}$ when $y = 16$, find x when $y = 32$.

Critical Thinking

30. Assume that *y* varies inversely as *x*. **30a. It is halved. 30b. It is divided by 3.**
    a. If the value of *x* is doubled, what happens to the value of *y*?
    b. If the value of *y* is tripled, what happens to the value of *x*?

Classroom Vignette

"After the initial introduction, examples, and discussion, I have students work in groups of four to solve the problems in this section. Each student completes the assignment, and one paper is collected at random from each group."

Barbara Ferguson
Enumclaw High School
Enumclaw, Washington

Barbara K. Ferguson

Assignment Guide

Core: 13–29 odd, 30, 31, 33–40
Enriched: 12–28 even, 30–40

For **Extra Practice,** see p. 766.

The red A, B, and C flags, printed only in the Teacher's Wraparound Edition, indicate the level of difficulty of the exercises.

Practice Masters, p. 33

NAME__________ DATE __________

Practice

Student Edition Pages 239–244

Direct and Inverse Variation

Solve. Assume that y varies directly as x.

1. If $y = -4$ when $x = 2$, find y when $x = -6$. **12**
2. If $y = 16$ when $x = 4$, find y when $x = 6$. **24**
3. If $y = -5$ when $x = 12.5$, find x when $y = 15$. **−37.5**
4. If $y = 7$ when $x = 4$, find y when $x = 12$. **21**
5. If $y = 80$ when $x = 32$, find x when $y = 100$. **40**
6. If $y = 198$ when $x = 22$, find y when $x = 182$. **1638**
7. If $y = 28$ when $x = 168$, find y when $x = 108$. **18**
8. If $y = 24$ when $x = 6$, find y when $x = -4$. **−16**

Solve. Assume that y varies inversely as x.

9. If $y = -4$ when $x = 2$, find y when $x = -6$. $\frac{4}{3}$
10. If $y = 16$ when $x = 4$, find y when $x = 6$. $\frac{32}{3}$
11. If $y = -5$ when $x = 12.5$, find x when $y = 15$. $-4.1\overline{6}$
12. If $y = 7$ when $x = 4$, find y when $x = 12$. $\frac{7}{3}$
13. If $y = 27$ when $x = 12$, find x when $y = -12$. **−27**
14. If $y = 6$ when $x = -4$, find x when $y = \frac{12}{5}$. **−10**
15. If $y = 60$ when $x = 80$, find x when $y = -20$. **−240**
16. If $y = 40$ when $x = 16$, find y when $x = 10$. **64**

4 ASSESS

Closing Activity

Modeling Have students create a diagram that illustrates how a lever works.

Chapter 4, Quiz D (Lessons 4-7 and 4-8), is available in the *Assessment and Evaluation Masters*, p. 101.

Enrichment Masters, p. 33

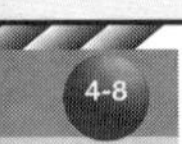

NAME________________ DATE ________

Enrichment Student Edition Pages 239–244

nth Power Variation

An equation of the form $y = kx^n$, where $k \neq 0$, describes an nth power variation. The variable n can be replaced by 2 to indicate the second power of x (the square of x) or by 3 to indicate the third power of x (the cube of x).

Assume that the weight of a person of average build varies directly as the cube of that person's height. The equation of variation has the form $w = kh^3$.

The weight that a person's legs will support is proportional to the cross-sectional area of the leg bones. This area varies directly as the square of the person's height. The equation of variation has the form $s = kh^2$.

Answer each question.

1. For a person 6 feet tall who weighs 200 pounds, find a value for k in the equation $w = kh^3$. **$k = 0.93$**
2. Use your answer from Exercise 1 to predict the weight of a person who is 5 feet tall. **about 116 pounds**
3. Find the value for k in the equation $w = kh^3$ for a baby who is 20 inches long and weighs 6 pounds. **$k = 1.296$ for $h = \frac{5}{3}$ ft**
4. How does your answer to Exercise 3 demonstrate that a baby is significantly fatter in proportion to its height than an adult? **k has a greater value.**
5. For a person 6 feet tall who weighs 200 pounds, find a value for k in the equation $s = kh^2$. **$k = 5.55$**
6. For a baby who is 20 inches long and weighs 6 pounds, find an "infant value" for k in the equation $s = kh^2$. **$k = 2.16$ for $h = \frac{5}{3}$ ft**
7. According to the adult equation you found (Exercise 1), how much would an imaginary giant 20 feet tall weigh? **7440 pounds**
8. According to the adult equation for weight supported (Exercise 5), how much weight could a 20-foot tall giant's legs actually support? **only 2222 pounds**
9. What can you conclude from Exercises 7 and 8? **Answers will vary. For example, bone strength limits the size humans can attain.**

Applications and Problem Solving

31. 18 pounds

31. **Space** The weight of an object on the moon varies directly as its weight on Earth. With all of his gear on, Neil Armstrong weighed 360 pounds on Earth. When he became the first person to step on the moon on July 20, 1969, he weighed 60 pounds. Tara weighs 108 pounds on Earth. What would she weigh on the moon?

32. **Physics** Pam and Adam are seated on the same side of a seesaw. Pam is 6 feet from the fulcrum and weighs 115 pounds. Adam is 8 feet from the fulcrum and weighs 120 pounds. Kam is seated on the other side of the seesaw, 10 feet from the fulcrum. If the seesaw is balanced, how much does Kam weigh? **165 pounds**

33. **Music** The pitch of a musical tone varies inversely as its wavelength. If one tone has a pitch of 440 vibrations per second and a wavelength of 2.4 feet, find the wavelength of a tone that has a pitch of 660 vibrations per second. **1.6 feet**

Neil Armstrong on the moon

Mixed Review

34. **Travel** At 8:00 A.M., Alma drove west at 35 miles per hour. At 9:00 A.M., Reiko drove east from the same point at 42 miles per hour. At what time will they be 266 miles apart? (Lesson 4–7) **12:00 noon**

35. **Probability** If the probability that an event will occur is $\frac{2}{3}$, what are the odds that the event will occur? (Lesson 4–6) **2:1**

36. **Retail Sales** A department store buys clothing at wholesale prices and then marks the clothing up 25% to sell at retail price to customers. If the retail price of a jacket is \$79, what was the wholesale price? (Lesson 4–5) **\$63.20**

37. **Finance** Hiroko invested \$11,700, part at 5% interest and the balance at 7% interest. If her annual earnings from both investments is \$733, how much is invested at each rate? (Lesson 4–4) **\$4300 at 5%, \$7400 at 7%**

38. Solve $\frac{2x}{5} + \frac{x}{4} = \frac{26}{5}$ for x. (Lesson 3–6) **8**

39. Solve $\frac{x}{4} + 9 = 6$. (Lesson 3–3) **−12**

40. Simplify $\frac{4a + 32}{4}$. (Lesson 2–7) **$a + 8$**

Extension

Problem Solving Six feet of steel wire weigh 0.7 kilogram. How much do 100 feet of steel wire weigh? **$11.\overline{6}$ kg**

CHAPTER 4 HIGHLIGHTS

VOCABULARY

After completing this chapter, you should be able to define each term, property, or phrase and give an example or two of each.

Algebra

base (p. 215)
constant of variation (p. 239)
direct variation (p. 239)
extremes (p. 196)
inverse variation (p. 241)
means (p. 196)
odds (p. 229)
percent (p. 215)
percentage (p. 215)
percent of decrease (p. 222)
percent of increase (p. 222)
percent proportion (p. 215)
proportion (p. 195)
rate (pp. 197, 215)
ratio (p. 195)
scale (p. 197)
simple interest (p. 217)
uniform motion (p. 235)
weighted average (p. 233)

Geometry

angle of depression (p. 208)
angle of elevation (p. 208)
corresponding angles (p. 201)
corresponding sides (p. 201)
hypotenuse (p. 206)
legs (p. 206)
similar (p. 201)
similar triangles (p. 201)
trigonometric ratios (p. 206)

Probability

equally likely (p. 229)
probability (p. 228)
probability of an event (p. 228)
random (p. 229)

Trigonometry

cosine (p. 206)
sine (p. 206)
solving the triangle (p. 208)
tangent (p. 206)

UNDERSTANDING AND USING THE VOCABULARY

Choose the correct term to complete each sentence.

1. The angle formed by a person in a radio tower looking up at an airplane in flight is called the (*angle of depression, angle of elevation*).
2. The equation $\frac{3}{5} = \frac{9}{15}$ is a (*proportion, ratio*).
3. If $\Delta ABC \sim \Delta DEF$, then their corresponding angles are (*equal, proportional*), and their corresponding sides are (*equal, proportional*).
4. A ratio (*can, cannot*) be expressed in the following ways: $\frac{4}{5}$, 4:5, and 4 to 5.
5. In a right triangle, the sides that form the right angle are called the (*hypotenuse, legs*) .
6. In $\frac{2}{x} = \frac{3}{9}$, 3 and x are called the (*extremes, means*).
7. The (*odds, probability*) that a certain event will occur is the ratio of the number of ways it can occur to the number of ways it cannot occur.
8. In the ratio $\frac{5}{28}$, 5 is the (*base, percentage*), and 28 is the (*base, percentage*).
9. A(n) (*direct variation, inverse variation*) is described by an equation of the form $xy = k$, where $k \neq 0$.
10. The tangent of an angle is defined as the measure of the (*adjacent, opposite*) leg divided by the measure of the (*adjacent leg, hypotenuse*).

Instructional Resources

Three multiple-choice tests and three free-response tests are provided in the *Assessment and Evaluation Masters*. Forms 1A and 2A are for honors pacing, and Forms 1B, 1C, 2B, and 2C are for average pacing. Chapter 4 Test, Form 1B is shown at the right. Chapter 4 Test, Form 2B is shown on the next page.

Using the CHAPTER HIGHLIGHTS

The Chapter Highlights begins with a listing of the new terms, properties, and phrases that were introduced in this chapter. Have students define each term and provide an example or two of it, if appropriate.

Assessment and Evaluation Masters, pp. 87–88

4 NAME_______ DATE_______

Chapter 4 Test, Form 1B

Write the letter for the correct answer in the blank at the right of each problem.

1. Solve the proportion $\frac{5}{4c} = \frac{1}{3}$. — 1. C
   A. $\frac{4}{15}$ B. $\frac{1}{15}$ C. $3\frac{3}{4}$ D. $\frac{4}{3}$
2. A 12-ounce can of cola contains 150 calories. How many ounces should you drink if your diet limits you to 100 calories? — 2. D
   A. 12 B. 50 C. 10 D. 8
3. By federal law, the ratio of the width to the length of the U. S. flag is 10 to 19. If you want to make a flag with an 8-foot width, what should be its length? — 3. B
   A. 23.75 ft B. 15.2 ft C. 4.21 ft D. 152 ft

Triangles ABC and DEF are similar. Use the information for exercises 4 and 5.

4. If $a = 10$, $b = 12$, and $d = 6$, find e. — 4. A
   A. 7.2 B. 3.6 C. 6 D. not enough information
5. If $a = 4$, $d = 6$, $e = 4$, and $f = 3$, find b and c. — 5. C
   A. $b = 6, c = \frac{9}{2}$ B. $b = \frac{9}{2}, c = 6$ C. $b = \frac{8}{3}, c = 2$ D. $b = 2, c = \frac{8}{3}$

Use the triangle at the right for exercises 6 and 7.

6. What is sin X? — 6. C
   A. $\frac{13}{12}$ B. $\frac{5}{12}$ C. $\frac{12}{13}$ D. $\frac{12}{5}$
7. What is the measure of angle Z to the nearest degree? — 7. B
   A. 43° B. 23° C. 67° D. 5°
8. An inclined plane links Johnstown, Pennsylvania, to suburban Westmont 504 feet above. Cable cars move up and down an incline 895 feet long. Find the measure of the angle of the incline to the nearest degree. — 8. A
   A. 34° B. 29° C. 56° D. 61°
9. What is 14% of 32? — 9. B
   A. 4480 B. 4.48 C. 228 D. 43.75
10. Ninety is what percent of 200? — 10. C
    A. 55% B. 40.5% C. 45% D. 20%
11. Lonnie earns 6% of the value of each house he sells. His commission on a sale was \$8700. What was the value of the house? — 11. D
    A. \$52,200 B. \$152,000 C. \$450,000 D. \$145,000

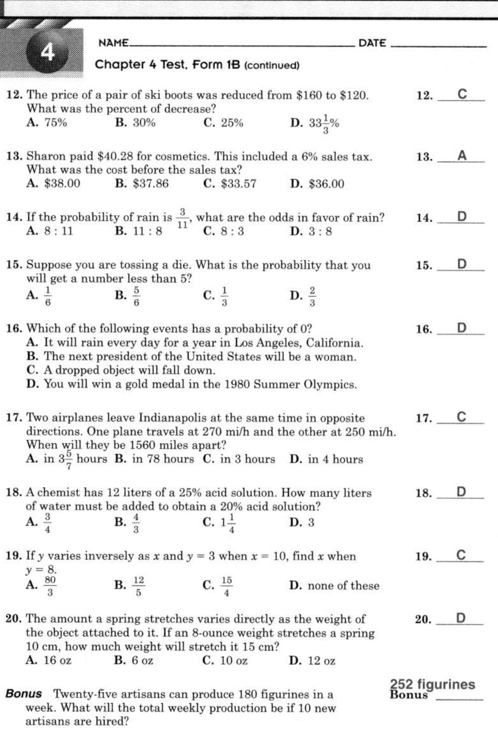

4 NAME_______ DATE_______

Chapter 4 Test, Form 1B (continued)

12. The price of a pair of ski boots was reduced from \$160 to \$120. What was the percent of decrease? — 12. C
    A. 75% B. 30% C. 25% D. $33\frac{1}{3}$%
13. Sharon paid \$40.28 for cosmetics. This included a 6% sales tax. What was the cost before the sales tax? — 13. A
    A. \$38.00 B. \$37.86 C. \$33.57 D. \$36.00
14. If the probability of rain is $\frac{3}{11}$, what are the odds in favor of rain? — 14. D
    A. 8 : 11 B. 11 : 8 C. 8 : 3 D. 3 : 8
15. Suppose you are tossing a die. What is the probability that you will get a number less than 5? — 15. D
    A. $\frac{1}{6}$ B. $\frac{5}{6}$ C. $\frac{1}{3}$ D. $\frac{2}{3}$
16. Which of the following events has a probability of 0? — 16. D
    A. It will rain every day for a year in Los Angeles, California.
    B. The next president of the United States will be a woman.
    C. A dropped object will fall down.
    D. You will win a gold medal in the 1980 Summer Olympics.
17. Two airplanes leave Indianapolis at the same time in opposite directions. One plane travels at 270 mi/h and the other at 250 mi/h. When will they be 1560 miles apart? — 17. C
    A. in $3\frac{5}{7}$ hours B. in 78 hours C. in 3 hours D. in 4 hours
18. A chemist has 12 liters of a 25% acid solution. How many liters of water must be added to obtain a 20% acid solution? — 18. D
    A. $\frac{3}{4}$ B. $\frac{4}{3}$ C. $1\frac{1}{4}$ D. 3
19. If y varies inversely as x and $y = 3$ when $x = 10$, find x when $y = 8$. — 19. C
    A. $\frac{80}{3}$ B. $\frac{12}{5}$ C. $\frac{15}{4}$ D. none of these
20. The amount a spring stretches varies directly as the weight of the object attached to it. If an 8-ounce weight stretches a spring 10 cm, how much weight will stretch it 15 cm? — 20. D
    A. 16 oz B. 6 oz C. 10 oz D. 12 oz

Bonus Twenty-five artisans can produce 180 figurines in a week. What will the total weekly production be if 10 new artisans are hired? — Bonus 252 figurines

Using the STUDY GUIDE AND ASSESSMENT

Skills and Concepts Encourage students to refer to the objectives and examples on the left as they complete the review exercises on the right.

Assessment and Evaluation Masters, pp. 93–94

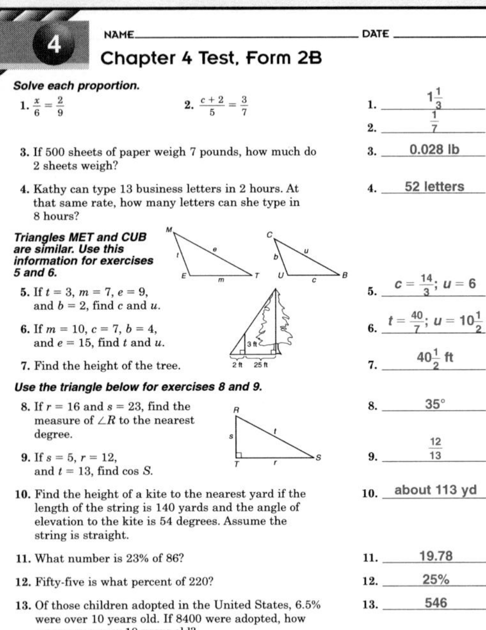

4 NAME ______ DATE ______

Chapter 4 Test, Form 2B

Solve each proportion.

1. $\frac{x}{6} = \frac{2}{9}$ 2. $\frac{c+2}{5} = \frac{3}{7}$

1. $1\frac{1}{3}$
2. $\frac{1}{7}$

3. If 500 sheets of paper weigh 7 pounds, how much do 2 sheets weigh? 3. 0.028 lb

4. Kathy can type 13 business letters in 2 hours. At that same rate, how many letters can she type in 8 hours? 4. 52 letters

Triangles MET and CUB are similar. Use this information for exercises 5 and 6.

5. If $t = 3$, $m = 7$, $e = 9$, and $b = 2$, find c and u. 5. $c = \frac{14}{3}$; $u = 6$

6. If $m = 10$, $c = 7$, $b = 4$, and $e = 15$, find t and u. 6. $t = \frac{40}{7}$; $u = 10\frac{1}{2}$

7. Find the height of the tree. 7. $40\frac{1}{2}$ ft

Use the triangle below for exercises 8 and 9.

8. If $r = 16$ and $s = 23$, find the measure of $\angle R$ to the nearest degree. 8. 35°

9. If $s = 5$, $r = 12$, and $t = 13$, find $\cos S$. 9. $\frac{12}{13}$

10. Find the height of a kite to the nearest yard if the length of the string is 140 yards and the angle of elevation to the kite is 54 degrees. Assume the string is straight. 10. about 113 yd

11. What number is 23% of 86? 11. 19.78

12. Fifty-five is what percent of 220? 12. 25%

13. Of those children adopted in the United States, 6.5% were over 10 years old. If 8400 were adopted, how many were over 10 years old? 13. 546

14. Johnny paid $24.36 including 5% sales tax for a pair of pants. What was the price of the pants without sales tax? 14. $23.20

15. A price increased from $145 to $156.60. What was the percent of increase? 15. 8%

4 NAME ______ DATE ______

Chapter 4 Test, Form 2B (continued)

16. A television normally selling for $423.00 is discounted 22%. Find the sale price. 16. $329.94

17. If the probability in favor of an event is $\frac{5}{12}$, what are the odds of that event occurring? 17. 5:7

18. A computer randomly selects a letter of the alphabet. What is the probability that the letter is one of the letters in the word "probability"? 18. $\frac{9}{26}$

19. The results of a survey of 284 randomly chosen students is shown below. What is the probability that a student chosen at random favors strawberry? 19. $\frac{121}{284}$

Preference for Frozen Yogurt Flavors			
Students	Strawberry	Chocolate	No Preference
Male	25	87	20
Female	96	34	22

20. Joe and Janna leave home at the same time, traveling in opposite directions. Joe drives 45 mi/h and Janna drives 40 mi/h. In how many hours will they be 510 miles apart? 20. 6 h

21. A factory has an order for 23,500 paper clips. Machine A can make 1500 paper clips per hour while Machine B can make 2500 paper clips per hour. Machine A starts at 8:00 A.M. and Machine B starts at 1:00 P.M. At what time will the two machines complete the job? 21. 5 P.M.

22. A chemist has 20 liters of a 10% acid solution. How many liters of pure acid must be added to produce a 25% acid solution? 22. 4 liters

23. If y varies directly as x and $y = 25$ when $x = 10$, find y when $x = 4$. 23. 10

24. If y varies inversely as x and $y = 9$ when $x = 5$, find y when $x = 6$. 24. $\frac{15}{2}$, or $7\frac{1}{2}$

25. Suppose that in a certain restaurant the number of hamburgers sold varies inversely as the price. If 70 burgers are sold at $1.50 each, how many burgers would be sold at $1.25 each? 25. 84 burgers

Bonus The intensity of illumination, I, on a movie screen varies inversely as the square of its distance from the projector. If the distance between the projector and the screen is changed from 200 feet to 100 feet, by what factor is the intensity of illumination changed? Bonus 4

CHAPTER 4 STUDY GUIDE AND ASSESSMENT

SKILLS AND CONCEPTS

OBJECTIVES AND EXAMPLES

Upon completing this chapter, you should be able to:

- solve proportions (Lesson 4–1)

$$\frac{x}{3} = \frac{x+1}{2}$$
$$2(x) = 3(x+1) \quad \text{Find the cross products.}$$
$$2x = 3x + 3$$
$$-3 = x$$

REVIEW EXERCISES

Use these exercises to review and prepare for the chapter test.

Use cross products to determine whether each pair of ratios forms a proportion.

11. $\frac{10}{3}, \frac{150}{45}$ $=$
12. $\frac{8}{7}, \frac{2}{1.75}$ $=$
13. $\frac{30}{12}, \frac{5}{4}$ $\neq$
14. $\frac{2.7}{3.1}, \frac{8.1}{9.3}$ $=$

Solve each proportion.

15. $\frac{6}{15} = \frac{n}{45}$ **18**
16. $\frac{x}{11} = \frac{35}{55}$ **7**
17. $\frac{y+4}{y-1} = \frac{4}{3}$ **16**
18. $\frac{z-7}{6} = \frac{z+3}{7}$ **67**

- find the unknown measures of the sides of two similar triangles (Lesson 4–2)

$$\frac{10}{5} = \frac{6}{a}$$
$$10a = 5(6)$$
$$10a = 30$$
$$a = 3$$

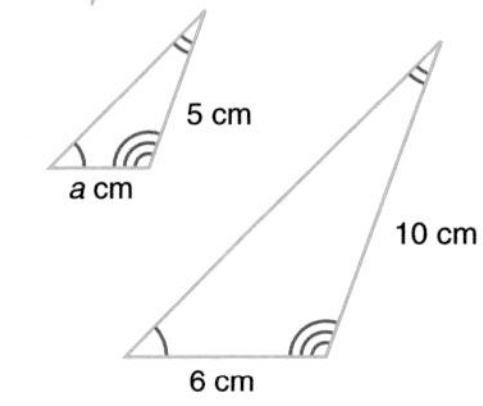

19. $d = \frac{45}{8}$, $e = \frac{27}{4}$ **20.** $d = \frac{48}{5}$, $e = \frac{36}{5}$

21. $b = \frac{44}{3}$, $d = 6$ **22.** $a = \frac{35}{2}$, $e = 8$

ΔABC and ΔDEF are similar. For each set of measures given, find the measures of the remaining sides.

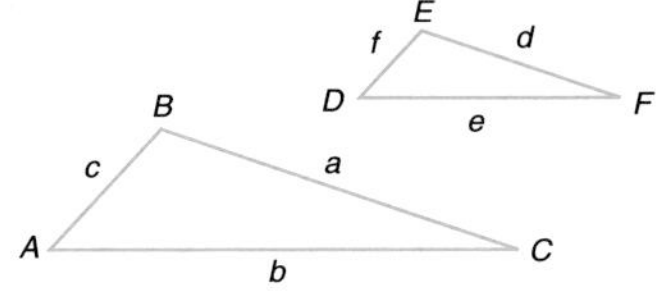

19. $c = 16$, $b = 12$, $a = 10$, $f = 9$
20. $a = 8$, $c = 10$, $b = 6$, $f = 12$
21. $c = 12$, $f = 9$, $a = 8$, $e = 11$
22. $b = 20$, $d = 7$, $f = 6$, $c = 15$

- use trigonometric ratios to solve right triangles (Lesson 4–3)

$$\sin A = \frac{\text{measure of leg opposite } \angle A}{\text{measure of hypotenuse}}$$

$$\cos A = \frac{\text{measure of leg adjacent } \angle A}{\text{measure of hypotenuse}}$$

$$\tan A = \frac{\text{measure of leg opposite } \angle A}{\text{measure of leg adjacent } \angle A}$$

For ΔABC, find each value to the nearest thousandth.

23. $\cos B$ **0.528**
24. $\tan A$ **0.622**
25. $\sin B$ **0.849**
26. $\cos A$ **0.849**
27. $\tan B$ **1.607**
28. $\sin A$ **0.528**

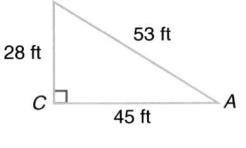

- use trigonometric ratios to solve right triangles (Lesson 4–3)

$\cos M = 0.3245$

Enter: 0.3245 [2nd] [COS⁻¹] 71.0647115

The measure of $\angle M$ is about 71°.

Use a calculator to find the measure of each angle to the nearest degree.

29. $\tan M = 0.8043$ **39°**
30. $\sin T = 0.1212$ **7°**
31. $\tan Q = 5.9080$ **80°**
32. $\cos F = 0.7443$ **42°**

GLENCOE Technology

Test and Review Software

You may use this software, a combination of an item generator and item bank, to create your own tests or worksheets. Types of items include free response, multiple choice, short answer, and open ended.

For IBM & Macintosh

OBJECTIVES AND EXAMPLES	REVIEW EXERCISES

- use trigonometric ratios to solve right triangles (Lesson 4–3) **33. $m\angle B = 45°$, $AB = 8.5$, $BC = 6$**

Solve right ΔABC *if* $m\angle B = 40°$ and $c = 6$.

$m\angle A = 180° - (90° + 40°)$ or $50°$

$\cos 40° = \frac{a}{6}$ $\quad$ $\sin 40° = \frac{b}{6}$

$0.7660 \approx \frac{a}{6}$ $\quad$ $0.6428 \approx \frac{b}{6}$

$0.7660(6) \approx a$ $\quad$ $0.6428(6) \approx b$

$4.596 \approx a$ $\quad$ $3.857 \approx b$

34. $m\angle B = 12°$, $AB = 134.7$, $BC = 131.7$

35. $m\angle A = 67°$, $AB = 9.8$, $m\angle B = 23°$

36. $m\angle B = 60°$, $AC = 22.5$, $BC = 13°$

Solve each right triangle. State the side lengths to the nearest tenth and the angle measures to the nearest degree.

33. 
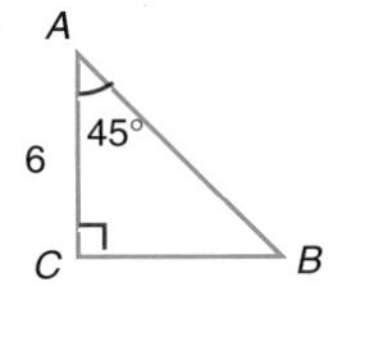

34. 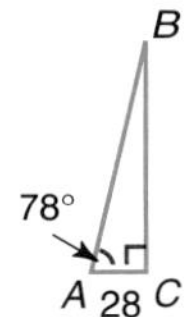

35.

36. 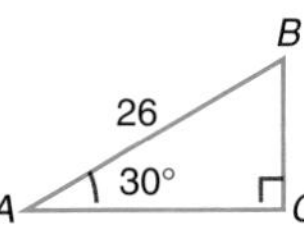

- solve percent problems (Lesson 4–4)

Nine is what percent of 15?

$\frac{9}{15} = \frac{r}{100}$ *Use a proportion.*

$9(100) = 15r$ *Find the cross products.*

$900 = 15r$

$60 = r$

Solve.

37. What number is 60% of 80? **48**
38. Twenty-one is 35% of what number? **60**
39. Eighty-four is what percent of 96? **87.5%**
40. What percent of 17 is 34? **200%**
41. What number is 0.3% of 62.7? **0.1881**
42. Find 0.12% of $5200. **6.24**

- solve problems involving percent of increase or decrease (Lesson 4–5)

original: $120 $\quad$ new: $114

amount of decrease: $120 − $114 = $6

$\frac{\text{amount of decrease}}{\text{original}} = \frac{r}{100}$

$\frac{6}{120} = \frac{r}{100}$

$5 = r$

The percent of decrease is 5%.

43. decrease, 13%
44. increase, 19%
45. increase, 6%
46. increase, 8%
47. decrease, 10%
48. decrease, 6%

State whether each percent of change is a percent of increase or a percent of decrease. Then find the percent of increase or decrease. Round to the nearest whole percent.

43. original: $40
new: $35

44. original: 97 cases
new: 115 cases

45. original: $35
new: $37.10

46. original: $50
new: $88

47. original: 1500 employees
new: 1350 employees

48. original: 12,500 students
new: 11,800 students

- solve problems involving discounts or sales tax (Lesson 4–5)

running shoes: $74; discount: 15%; sales tax: 6%

15% of $74 = (0.15)(74) or $11.10

$74 − $11.10 = $62.90 ← *sale price*

6% of $62.90 = (0.06)(62.90) or $3.77

$62.90 + $3.77 = $66.67 ← *price with tax*

The total price is $66.67 for the shoes.

Find the final price of each item. When there is a discount and sales tax, compute the discount price first.

49. calculator: $81
sales tax: 5.75%
$85.66

50. gasoline: $21.50
discount: 2%
$21.07

51. used car: $8,690
sales tax: 6.7%
$9272.23

52. dress: $89
discount: 25%
sales tax: 7% **$71.42**

Applications and Problem Solving Encourage students to work through the exercises in the Applications and Problem Solving section to strengthen their problem-solving skills.

OBJECTIVES AND EXAMPLES

- find the probability of a simple event and find the odds of a simple event (Lesson 4–6)

Find the probability of a computer randomly choosing the letter I in the word MISSISSIPPI.

$$\frac{\text{number of favorable outcomes}}{\text{number of possible outcomes}} = \frac{4}{11}$$

Find the odds that you will randomly *not* select the letter S in the word MISSISSIPPI.

number of successes:number of failures = 7:4

REVIEW EXERCISES

Find the probability of each outcome if a computer randomly chooses a letter in the word REPRESENTING.

53. the letter S **$\frac{1}{12}$**

54. the letter E **$\frac{1}{4}$**

55. P(not N) **$\frac{5}{6}$**

56. the letters R or P **$\frac{1}{4}$**

Find the odds of each outcome if you randomly select a coin from a jar containing 90 pennies, 75 nickels, 50 dimes, and 30 quarters.

57. selecting a dime **10:39**

58. selecting a penny **18:31**

59. *not* selecting a nickel **34:15**

60. selecting a nickel or a dime **25:24**

OBJECTIVES AND EXAMPLES

- solve problems involving direct and inverse variation (Lesson 4–8)

If x varies directly as y and $x = 15$ when $y = 1.5$, find x when $y = 9$.

$$\frac{1.5}{9} = \frac{15}{x}$$

$1.5x = 9(15)$

$1.5x = 135$

$x = 90$

If y varies inversely as x and $y = 24$ when $x = 30$, find x when $y = 10$.

$$\frac{30}{x} = \frac{10}{24}$$

$(30)(24) = 10x$

$720 = 10x$

$72 = x$

REVIEW EXERCISES

Solve. Assume that y varies directly as x.

61. If $y = 15$ when $x = 5$, find y when $x = 7$. **21**

62. If $y = 35$ when $x = 175$, find y when $x = 75$. **15**

63. If $y = 10$ when $x = 0.75$, find x when $y = 80$. **6**

64. If $y = 3$ when $x = 99.9$, find y when $x = 522.81$. **15.7**

Solve. Assume that y varies inversely as x.

65. If $y = 28$ when $x = 42$, find y when $x = 56$. **21**

66. If $y = 15$ when $x = 5$, find y when $x = 3$. **25**

67. If $y = 18$ when $x = 8$, find x when $y = 3$. **48**

68. If $y = 35$ when $x = 175$, find y when $x = 75$. **$81.\overline{6}$**

APPLICATIONS AND PROBLEM SOLVING

69. Travel Two airplanes leave Dallas at the same time and fly in opposite directions. One airplane travels 80 miles per hour faster than the other. After three hours, they are 2940 miles apart. What is the rate of each airplane? (Lesson 4–7) **450 mph and 530 mph**

70. Coffee Anne Leibowitz owns "The Coffee Pot," a specialty coffee store. She wants to create a special mix using two coffees, one priced at $8.40 per pound and the other at $7.28 per pound. How many pounds of the $7.28 coffee should she mix with 9 pounds of the $8.40 coffee to sell the mixture for $7.95 per pound? (Lesson 4–7) **6 pounds**

A practice test for Chapter 4 is provided on page 790.

ALTERNATIVE ASSESSMENT

COOPERATIVE LEARNING PROJECT

Letter Occurrence In this project, you will determine percentages, percent of change, and probability and odds.

There is an uneven distribution of letters which occur in words. Christofer Sholes, the inventor of the typewriter, may not have been aware of this unevenness. He gave the left hand 56 percent of all the strokes, and the two most agile fingers on the right hand hit two of the least often used letters of the alphabet, j and k.

The table below shows the percentages of the occurrence of letters in large samples.

E	12.3%	R	6.0%	F	2.3%	K	1.5%
T	9.6%	H	5.1%	M	2.2%	Q	0.2%
A	8.1%	L	4.0%	W	2.0%	X	0.2%
O	7.9%	D	3.7%	Y	1.9%	J	0.1%
N	7.2%	C	3.2%	B	1.6%	Z	0.1%
I	7.2%	U	3.1%	G	1.6%		
S	6.6%	P	2.3%	V	0.9%		

Use the second paragraph on this page to count the number of times that e, s, c, w, and k occur and compute their percentages. Compare your answers with the table. What is the probability that a letter picked at random from the paragraph is a vowel? Are the odds better that a vowel or a consonant was used in the above paragraph? What are these odds?

Write a paragraph describing similar triangles. Then determine the percentages of the occurrence of the above mentioned letters in your paragraph. Do they coincide with the percentages in the table? Determine the percent of change for the above letters in the paragraph above and the paragraph you wrote.

Follow these guidelines.

- Determine what type of data you will need in order to calculate your answers.
- Devise a chart that will be helpful in organizing the data you will need.
- Write a paragraph describing what your answers mean.
- Using the percentages from the above table, draw your idea of where the letters should be on a typewriter and explain why.

THINKING CRITICALLY

- Use an example to describe the difference between odds and probability. Explain why some events are described in terms of odds and others in terms of probability.
- Use the definitions of the trigonometric ratios to explain why, in any right ΔABC with right angle C, $\sin A = \cos B$, and $\cos A = \sin B$.

PORTFOLIO

Even after working through a chapter of material, you can get to the end and still feel uncertain about a concept or a certain type of problem. Find a problem that you still cannot solve or that you are worried you might not be able to solve on a test. Write out the question and as much of the solution as you can until you get to the hard part. Then explain what it is that keeps you from solving the problem. Be clear and precise. Place this in your portfolio.

SELF EVALUATION

Theme learning occurs when the subject matter being studied is all related to the same theme, such as water, animals, and so on. When the course material is presented, it is all applied to that theme.

Assess yourself. Are you a person who would enjoy theme learning? Would you like to relate all your course work to one area or do you prefer having a variety of application themes? Describe a theme that would be of interest to you and tell how you would use that theme to introduce and apply the contents of a specific math subject.

Alternative Assessment

The Alternative Assessment section provides students with the opportunity to assess their own work by thinking critically, working with others, keeping a portfolio, and honestly evaluating their own progress. For more information on alternative forms of assessment, see *Alternative Assessment in the Mathematics Classroom,* one of the titles in the Glencoe Mathematics Professional Series.

Performance Assessment

Performance Assessment tasks for this chapter are included in the *Assessment and Evaluation Masters.* A scoring guide is also provided.

Assessment and Evaluation Masters, pp. 98, 109

NAME____________ DATE ________

Chapter 4 Performance Assessment

Instructions: ***Demonstrate your knowledge by giving a clear, concise solution to each problem. Be sure to include all relevant drawings and justify your answers. You may show your solution in more than one way or investigate beyond the requirements of the problem.***

1. The following questions are related to the formula $d = rt$, where d represents distance, r represents rate, and t represents time.

a. Write what is meant by direct variation.

b. Write what is meant by inverse variation.

c. Every day at lunchtime Kelli walks 30 minutes before returning to work. On cooler days she walks more briskly than on warm days. Does the distance she walks vary directly or inversely with her rate of walk? Justify your answer.

d. At the Indianapolis 500 auto race, each car finishing the race travels 500 miles. Does the average speed of the cars finishing the race vary directly or inversely with the time taken to complete the race? Justify your answer.

2. A landscaper wishes to make a scale drawing of a rectangular garden. She makes the measurements indicated.

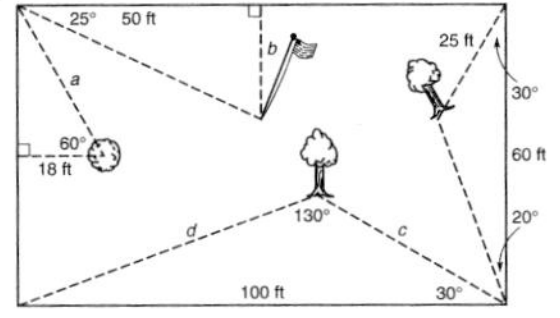

a. Find the measurements labeled a, b, c, and d. Show your work and explain each step.

b. Make a scale drawing based on the length of the garden being represented by 10 inches on your drawing. Show your work in determining each length in your drawing.

3. Explain how to find the sale price when you know the original price and the percent of discount.

4. Explain how to find the odds that an event will occur if you know the probability that it will occur.

Scoring Guide
Chapter 4
Performance Assessment

Level	Specific Criteria
3 Superior	• Shows thorough understanding of the concepts of *proportion; direct and inverse variation; similar triangles; sine, cosine, and tangent*; *percent of change;* and *probability.* • Uses appropriate strategies to solve problems. • Computations are correct. • Written explanations are exemplary. • Diagrams are accurate and appropriate. • Goes beyond requirements of some or all problems.
2 Satisfactory, with Minor Flaws	• Shows understanding of the concepts of *proportion; direct and inverse variation; similar triangles; sine, cosine, and tangent*; *percent of change;* and *probability.* • Uses appropriate strategies to solve problems. • Computations are mostly correct. • Written explanations are effective. • Diagrams are mostly accurate and appropriate. • Satisfies all requirements of problems.
1 Nearly Satisfactory, with Serious Flaws	• Shows understanding of most of the concepts of *proportion; direct and inverse variation; similar triangles; sine, cosine, and tangent*; *percent of change;* and *probability.* • May not use appropriate strategies to solve problems. • Computations are mostly correct. • Written explanations are satisfactory. • Diagrams are mostly accurate and appropriate. • Satisfies most requirements of problems.
0 Unsatisfactory	• Shows little or no understanding of the concepts of *proportion; direct and inverse variation; similar triangles; sine, cosine, and tangent*; *percent of change;* and *probability.* • May not use appropriate strategies to solve problems. • Computations are incorrect. • Written explanations are not satisfactory. • Diagrams are not accurate or appropriate. • Does not satisfy requirements of problems.

Using the CUMULATIVE REVIEW

These two pages review the skills and concepts presented in Chapters 1–4. This review is formatted to reflect new trends in standardized testing.

A more traditional cumulative review is provided in the *Assessment and Evaluation Masters,* pp. 103–104.

Assessment and Evaluation Masters, pp. 103–104

4 NAME__________ DATE ________

Chapter 4 Cumulative Review

1. Write a mathematical expression for twice the cube of a number. (Lesson1-1) — 1. $2n^3$
2. Write the expression $x \cdot y \cdot x \cdot x \cdot y$ using exponents. (Lesson 1-1) — 2. x^3y^2
3. Evaluate $5ab^2$ if $a = 3$ and $b = 4$. (Lesson 1-3) — 3. 240
4. Solve $6(2 + 6) = k$.(Lesson 1-5) — 4. 48
5. What number is the multiplicative identity element? (Lesson 1-6) — 5. 1
6. Simplify $5m + 8n - 3m + n$. (Lesson 1-8) — 6. $2m + 9n$

Find each sum or difference. (Lessons 2-3 and 2-5)

7. $-43 + 26$ — 7. -17
8. $-3.4 + 10$ — 8. 6.6
9. $|4 - (-7)|$ — 9. 11
10. $\begin{bmatrix} -1 & 0 \\ 2 & 3 \end{bmatrix} + \begin{bmatrix} -2 & 5 \\ 4 & -1 \end{bmatrix}$ — 10. $\begin{bmatrix} -3 & 5 \\ 6 & 2 \end{bmatrix}$

Solve. (Lesson 3-1)

11. $m + 5 = -23$ — 11. -28
12. $-4 = 8 - k$ — 12. 12
13. Define a variable, write an equation, and solve. (Lesson 3-1) The temperature at noon was 16°C. By dusk, the temperature had dropped 20°. What was the temperature at dusk? — 13. x = temp at dusk; $16 - 20 = x$; -4°C

Simplify. (Lessons 2-6 and 2-7)

14. $-9(4 - 7)$ — 14. 27
15. $\frac{16b}{4}$ — 15. 46

Find the measure of the third angle of each triangle in which the measures of two angles of the triangle are given. (Lesson 3-4)

16. 38°, 75° — 16. 67°
17. 114°, 27° — 17. 39°

4 NAME__________ DATE ________

Chapter 4 Cumulative Review (continued)

Find each square root. Round to the nearest hundredth if the result is not a whole number. (Lesson 2-8)

18. $\sqrt{576}$ — 18. 24
19. $-\sqrt{60}$ — 19. -7.75

Solve each question. (Lessons 3-2, 3-3, and 3-5)

20. $\frac{5}{6}b = -5$ — 20. -18
21. $-7x + 23 = 37$ — 21. -2
22. $8(x - 5) = 48x$ — 22. -1
23. Find two consecutive odd integers whose sum is 56. (Lesson 3-3) — 23. 27, 29
24. Solve the proportion $\frac{n}{500} = \frac{2}{40}$. (Lesson 4-1) — 24. 25
25. 93 is what percent of 300? (Lesson 4-4) — 25. 31%
26. If y varies directly as x and $y = 8$ when $x = 3$, find y when $x = 5$. (Lesson 4-8) — 26. $13\frac{1}{3}$
27. If y varies inversely as x and $y = 15$ when $x = 3$, find y when $x = 5$. (Lesson 4-8) — 27. 9
28. How heavy is an 800-foot reel of cable if a piece 6 feet long weighs 5 pounds? (Lesson 4-8) — 28. $666\frac{2}{3}$ pounds
29. A price of \$180 was increased to \$207. Find the percent of increase. (Lesson 4-5) — 29. 15%
30. A stereo normally selling for \$320 is discounted 18%. Find the sale price. (Lesson 4-5) — 30. \$262.40

A single die is rolled. (Lesson 4-6)

31. Find P (at least 5). — 31. $\frac{1}{3}$
32. What are the odds of rolling a number less than 5? — 32. 2:1
33. A five-foot-tall person casts a shadow 18 feet long. At the same time, a tree casts a shadow 45 feet long. How tall is the tree? (Lesson 4-2) — 33. 12.5 feet

CUMULATIVE REVIEW

CHAPTERS 1–4

SECTION ONE: MULTIPLE CHOICE

There are nine multiple-choice questions in this section. After working each problem, write the letter of the correct answer on your paper.

1. Solve the equation $7(3b + x) = 5a$ for x. **D**

   A. $5a - 21b = x$ B. $\frac{5a}{7} - 21b = x$

   C. $\frac{5a - 3b}{7} = x$ D. $\frac{5a - 21b}{7} = x$

2. **Sales** A Tampa Bay Buccaneers sweatshirt decreased in price from \$40 to \$35 during a clearance sale. Find the percent of decrease. **B**

   A. 5% B. 12.5%

   C. 14.3% D. 114.3%

3. A caterpillar was trying to crawl out of a bucket into which it had fallen. It crawled up $1\frac{1}{4}$ feet before sliding down $\frac{3}{8}$ feet. It then crawled up $\frac{5}{6}$ feet before resting. How much farther must the caterpillar crawl to reach the top of the 2-foot bucket? **A**

   A. $\frac{7}{24}$ feet B. $\frac{7}{8}$ feet

   C. $1\frac{1}{8}$ feet D. $1\frac{17}{24}$ feet

4. **Geometry** Choose the correct proportion to find the distance across the lake (AB) given ΔABC is similar to ΔEDC. **C**

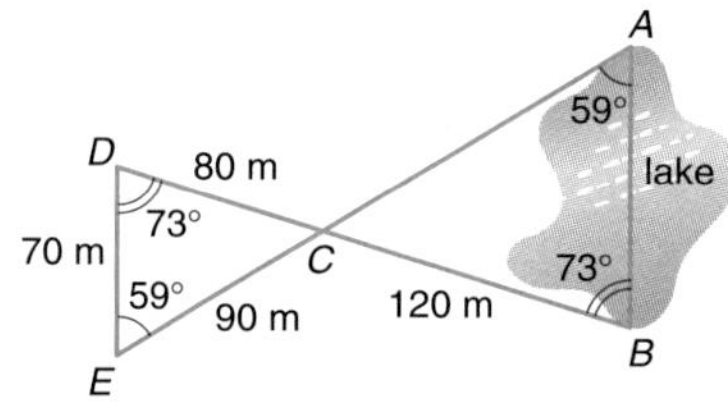

   A. $\frac{70}{AB} = \frac{90}{120}$ B. $\frac{80}{70} = \frac{AB}{120}$

   C. $\frac{70}{AB} = \frac{80}{120}$ D. $\frac{80}{90} = \frac{120}{AB}$

5. Kassim read 36 pages of a novel in 2 hours. At the same rate, find the number of hours it will take him to read the remaining 135 pages. **B**

   A. $3\frac{3}{4}$ B. $7\frac{1}{2}$

   C. $9\frac{1}{2}$ D. $10\frac{1}{2}$

6. Choose an equivalent equation to represent "h is the product of g and the difference of the square of b and t". **D**

   A. $h - t = gb^2$

   B. $h = g(b^2 - t)$

   C. $h = g(b - t)^2$

   D. $h + t = gb^2$

7. **Probability** The freshman class of Perry High School is planning an end-of-the-year dance. The dance committee decides to award door prizes. They want the odds of winning a prize to be 1:5. If they plan for 180 tickets to be sold for the dance, how many prizes will be needed? **C**

   A. 6 prizes B. 36 prizes

   C. 30 prizes D. 150 prizes

8. **Sales** Raven sold her concert ticket to Sandra for 75% of the original ticket price. If Sandra paid \$26.25 for the ticket, choose the proportion that could be used to find the ticket's original price. **A**

   A. $\frac{75}{100} = \frac{26.25}{x}$ B. $\frac{x}{26.25} = \frac{75}{100}$

   C. $\frac{x}{100} = \frac{75}{26.25}$ D. $\frac{x}{75} = \frac{26.25}{100}$

9. **Statistics** Carita's bowling scores for the first four games of a five-game series are $b + 2$, $b + 3$, $b - 2$, and $b - 1$. What must her score be for the last game to have an average of $b + 2$? **D**

   A. $b - 2$ B. $b + 2$

   C. $b - 8$ D. $b + 8$

Standardized Test Practice Questions are also provided in the *Assessment and Evaluation Masters,* p. 102.

SECTION TWO: FREE RESPONSE

This section contains nine questions for which you will provide short answers. Write your answer on your paper.

10. Describe the steps required to solve the equation $4t - 5 = 7t - 23$. **See students' work.**

11. **Geometry** Find the dimensions of a rectangle whose width is 5 inches less than its length and has a perimeter of 70 inches. **15 in. × 20 in.**

12. **Finance** In the beginning of the month, the balance in Marisa Fuente's checking account was $428.79. After writing checks totaling $1097.31, depositing 2 checks of $691.53 each, and withdrawing $100 from a bank machine, what was the new balance in Marisa's account? **$614.54**

13. **Finance** Maria Cruz wants to invest part of her $8000 savings in a bond paying 12% annual interest, and the rest of it in a savings account paying 8%. If she wants to earn a total of $760, how much money should Maria invest at each rate? **$5000 at 8%, $3000 at 12%**

14. **Weather** The time t in hours, that a storm will last is given by the formula $t = \sqrt{\frac{d^3}{216}}$, where d is the diameter of the storm in miles. Suppose the umpires in a baseball game declared rain delay at 4:00 P.M. The storm causing the rain delay has a diameter of 12 miles. After the rain has stopped, can the game be restarted before 6:00 P.M.?
No, since the rain will end about 6:50 P.M.

15. **Statistics** Find the mean, median, and mode of the data represented in the line plot below to the nearest tenth. **47.5, 46, 35 and 63**

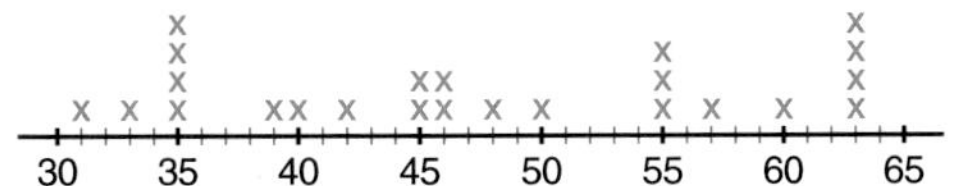

16. **Running** Raul ran a distance of 385 yards in $1\frac{3}{4}$ minutes. His younger sister, Luisa, estimated it would take her $3\frac{1}{2}$ minutes to run the same distance. If Luisa ran at a rate of 94 yards per minute, was her estimation correct? Verify your answer. **See students' work.**

17. **Government** The term of office for a United States Senator is 150% as long as the term of office for the President. How long is the term of office for a United States Senator? **6 years**

18. **Physics** Shannon weighs 126 pounds and Minal weighs 154 pounds. They are seated at opposite ends of a seesaw. Shannon and Minal are 16 feet apart, and the seesaw is balanced. How far is Shannon from the fulcrum? **8.8 ft**

SECTION THREE: OPEN-ENDED

This section contains two open-ended problems. Demonstrate your knowledge by giving a clear, concise solution to each problem. Your score on these problems will depend on how well you do the following.

- Explain your reasoning.
- Show your understanding of the mathematics in an organized manner.
- Use charts, graphs, and diagrams in your explanation.
- Show the solution in more than one way or relate it to other situations.
- Investigate beyond the requirements of the problem.

19. **Geometry** One angle of a triangle measures 15° more than the second. The measure of the third angle is the sum of the first two angles tripled. Find the measure of each angle.
15°, 30°, 135°

20. **Biology** An adult African elephant can weigh up to 12,000 pounds. An adult blue whale can weigh as much as 37 adult African elephants. A newborn blue whale weighs one-third of an adult African elephant's weight. The newborn whale will gain 200 pounds per day until it reaches its adult weight. How many days will it take for the newborn to reach adult weight? **2200 days**

CHAPTER

5 Graphing Relations and Functions

PREVIEWING THE CHAPTER

This chapter makes the connection between algebraic equations and their geometric models. Students identify coordinates and locate points on the coordinate plane, identify the domain, range, and inverse of a relation, and solve linear equations for a specific variable or a given domain. They graph relations from a table of ordered pairs and determine whether the relation is a function. Graphing technology is integrated throughout the chapter. Finally, students examine measures of variation and their connection to the distribution of data.

Lesson (Pages)	Lesson Objectives	NCTM Standards	State/Local Objectives
5-1 (254–259)	Graph ordered pairs on a coordinate plane. Solve problems by making a table.	1–5, 8	
5-2A (260–261)	Use a graphing calculator to graph relations.	1–5, 8	
5-2 (262–269)	Identify the domain, range, and inverse of a relation. Show relations as sets of ordered pairs, tables, mappings, and graphs.	1–5	
5-3A (270)	Use a graphing calculator to investigate relations and determine the ranges when the domains are given.	1–5	
5-3 (271–277)	Determine the range for a given domain. Graph the solution set for the given domain.	1–5	
5-4A (278–279)	Use a graphing calculator to graph linear relations and functions.	1–5, 8	
5-4 (280–286)	Graph linear equations.	1–5, 8	
5-5 (287–294)	Determine whether a given relation is a function. Find the value of a function for a given element of the domain.	1–6	
5-6 (295–302)	Write equations to represent relations, given some of the solutions for the equation.	1–6	
5-7A (303–304)	Use a graphing calculator to calculate measures of variation.	1–4, 10	
5-7 (305–313)	Calculate and interpret the range, quartiles, and interquartile range of sets of data.	1–4, 10	

A complete, 1-page lesson plan is provided for each lesson in the ***Lesson Planning Guide***. Answer keys for each lesson are available in the ***Answer Key Masters.***

ORGANIZING THE CHAPTER

You may want to refer to the **Course Planning Calendar** on page T12 for detailed information on pacing.
PACING: Standard—14 days; **Honors**—13 days; **Block**—7 days; **Two Years**—21 days

LESSON PLANNING CHART

Lesson (Pages)	Materials/ Manipulatives	Extra Practice (Student Edition)	Blackline Masters									Real-World Applications	Interactive Mathematics Tools Software	Teaching Transparencies
			Study Guide	Practice	Enrichment	Assessment and Evaluation	Modeling Mathematics	Multicultural Activity	Tech Prep Applications	Graphing Calculator	Science and Math Lab Manual			
5-1 (254–259)	geoboards graphing calculator	p. 767	p. 34	p. 34	p. 34			p. 9				14	5-1.1 5-1.2	5-1A 5-1B
5-2A (260–261)	graphing calculator									pp. 14, 15				
5-2 (262–269)	grid paper straightedge colored pencils	p. 767	p. 35	p. 35	p. 35	p. 128			p. 9		pp. 17–22		5-2	5-2A 5-2B
5-3A (270)	graphing calculator									pp. 16, 17				
5-3 (271–277)	graphing calculator	p. 767	p. 36	p. 36	p. 36			p. 10						5-3A 5-3B
5-4A (278–279)	graphing calculator									pp. 18, 19				
5-4 (280–286)	graphing calculator	p. 768	p. 37	p. 37	p. 37	pp. 127, 128				p. 5				5-4A 5-4B
5-5 (287–294)		p. 768	p. 38	p. 38	p. 38		pp. 48–50, 76		p. 10			15	5-5	5-5A 5-5B
5-6 (295–302)		p. 768	p. 39	p. 39	p. 39	p. 129								5-6A 5-6B
5-7A (303–304)	graphing calculator									pp. 20, 21				
5-7 (305–313)		p. 769	p. 40	p. 40	p. 40	p. 129						16		5-7A 5-7B
Study Guide/ Assessment (315–319)						pp. 113–126, 130–132								

ORGANIZING THE CHAPTER

OTHER CHAPTER RESOURCES

Student Edition
Chapter Opener, pp. 252–253
Mathematics and Society, p. 313
Working on the Investigation, pp. 269, 302
Closing the Investigation, p. 314

Teacher's Classroom Resources
Investigations and Projects Masters, pp. 41–44
Algebra and Geometry Overhead Manipulative Resources, pp. 16–18

Technology
Test and Review Software (IBM and Macintosh)
CD-ROM Interactions (Windows and Macintosh)

Professional Publications
Block Scheduling Booklet
Glencoe Mathematics Professional Series

OUTSIDE RESOURCES

Books/Periodicals
Algebra Experiments I: Exploring Linear Functions, Addison-Wesley
Cartesian Cartoons, Dale Seymour Publications

Software
Coordinate Math, MECC
Visualizing Algebra: The Function Analyzer, Sunburst

Videos/CD-ROMs
Solving Equations 1, Coronet, MTI

ASSESSMENT RESOURCES

Student Edition
Math Journal, pp. 274, 283, 291
Mixed Review, pp. 259, 269, 277, 285, 294, 302, 312
Self Test, p. 286
Chapter Highlights, p. 315
Chapter Study Guide and Assessment, pp. 316–318
Alternative Assessment, p. 319
Portfolio, p. 319

Teacher's Wraparound Edition
5-Minute Check, pp. 254, 262, 271, 280, 287, 295, 305
Check for Understanding, pp. 257, 266, 274, 283, 291, 298, 309
Closing Activity, pp. 259, 269, 276, 285, 294, 302, 312
Cooperative Learning, pp. 281, 307

Assessment and Evaluation Masters
Multiple-Choice Tests, Forms 1A (Honors), 1B (Average), 1C (Basic), pp. 113–118
Free-Response Tests, Forms 2A (Honors), 2B (Average), 2C (Basic), pp. 119–124
Calculator-Based Test, p. 125
Performance Assessment, p. 126
Mid-Chapter Test, p. 127
Quizzes A–D, pp. 128–129
Standardized Test Practice, p. 130
Cumulative Review, pp. 131–132

ENHANCING THE CHAPTER

Examples of some of the materials for enhancing Chapter 5 are shown below.

DIVERSITY

Multicultural Activity Masters, pp. 9, 10

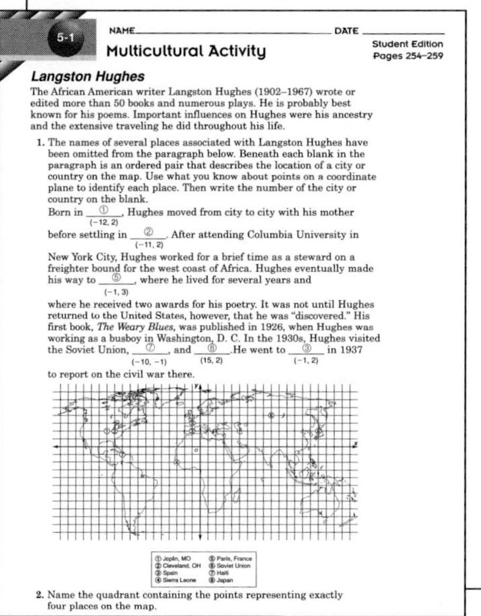

5-1 NAME ______ DATE ______

Multicultural Activity

Student Edition Pages 254–259

Langston Hughes

The African American writer Langston Hughes (1902–1967) wrote or edited more than 50 books and numerous plays. He is probably best known for his poems. Important influences on Hughes were his ancestry and the extensive traveling he did throughout his life.

1. The names of several places associated with Langston Hughes have been omitted from the paragraph below. Beneath each blank in the paragraph is an ordered pair that describes the location of a city or country on the map. Use what you know about points on a coordinate plane to identify each place. Then write the number of the city or country on the blank.

   Born in __①__ (−12, 2), Hughes moved from city to city with his mother before settling in __②__ (−11, 2). After attending Columbia University in New York City, Hughes worked for a brief time as a steward on a freighter bound for the west coast of Africa. Hughes eventually made his way to __⑤__ (−1, 3), where he lived for several years and where he received two awards for his poetry. It was not until Hughes returned to the United States, however, that he was "discovered." His first book, *The Weary Blues,* was published in 1926, when Hughes was working as a busboy in Washington, D. C. In the 1930s, Hughes visited the Soviet Union, __⑦__ (−10, −1), and __⑧__ (15, 2). He went to __③__ (−1, 2) in 1937 to report on the civil war there.

① Joplin, MO
② Cleveland, OH
③ Spain
④ Sierra Leone
⑤ Paris, France
⑥ Soviet Union
⑦ Haiti
⑧ Japan

2. Name the quadrant containing the points representing exactly four places on the map.

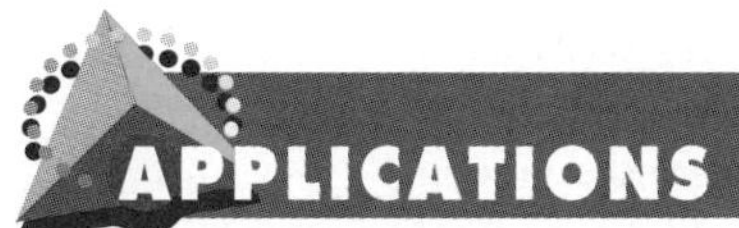

APPLICATIONS

Real-World Applications, 14, 15, 16

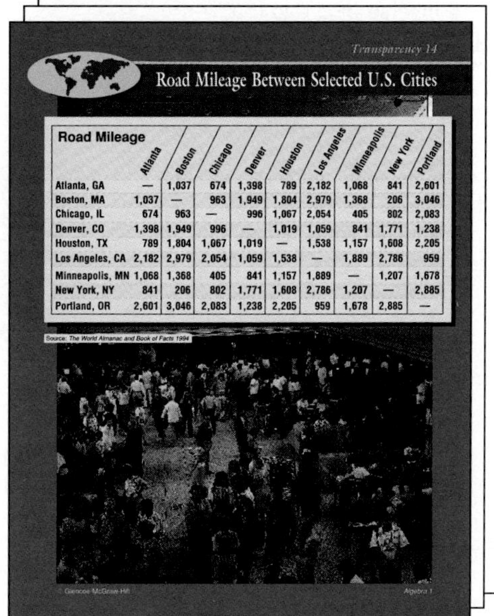

Transparency 14

Road Mileage Between Selected U.S. Cities

Road Mileage	Atlanta	Boston	Chicago	Denver	Houston	Los Angeles	Minneapolis	New York	Portland
Atlanta, GA	—	1,037	674	1,398	789	2,182	1,068	841	2,601
Boston, MA	1,037	—	963	1,949	1,804	2,979	1,368	206	3,046
Chicago, IL	674	963	—	996	1,067	2,054	405	802	2,083
Denver, CO	1,398	1,949	996	—	1,019	1,059	841	1,771	1,238
Houston, TX	789	1,804	1,067	1,019	—	1,538	1,157	1,608	2,205
Los Angeles, CA	2,182	2,979	2,054	1,059	1,538	—	1,889	2,786	959
Minneapolis, MN	1,068	1,368	405	841	1,157	1,889	—	1,207	1,678
New York, NY	841	206	802	1,771	1,608	2,786	1,207	—	2,885
Portland, OR	2,601	3,046	2,083	1,238	2,205	959	1,678	2,885	—

Source: *The World Almanac and Book of Facts 1994*

© Glencoe McGraw-Hill   Algebra 1

TECHNOLOGY

Graphing Calculator Masters, p. 5

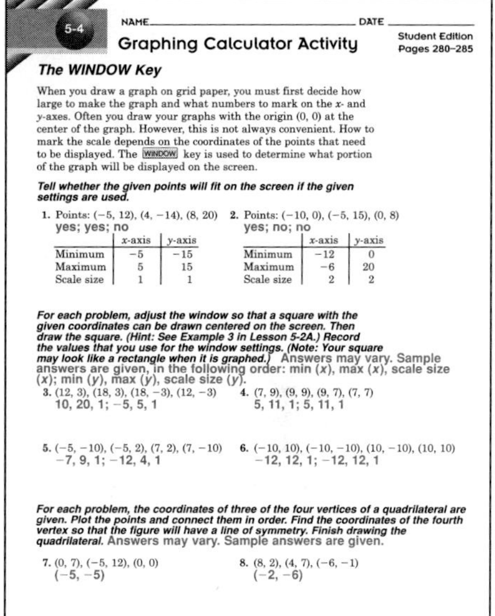

5-4 NAME ______ DATE ______

Graphing Calculator Activity

Student Edition Pages 280–285

The WINDOW Key

When you draw a graph on grid paper, you must first decide how large to make the graph and what numbers to mark on the x- and y-axes. Often you draw your graphs with the origin (0, 0) at the center of the graph. However, this is not always convenient. How to mark the scale depends on the coordinates of the points that need to be displayed. The WINDOW key is used to determine what portion of the graph will be displayed on the screen.

Tell whether the given points will fit on the screen if the given settings are used.

1. Points: (−5, 12), (4, −14), (8, 20)
   yes; yes; no

	x-axis	y-axis
Minimum	−5	−15
Maximum	5	15
Scale size	1	1

2. Points: (−10, 0), (−5, 15), (0, 8)
   yes; no; no

	x-axis	y-axis
Minimum	−12	0
Maximum	−6	20
Scale size	2	2

For each problem, adjust the window so that a square with the given coordinates can be drawn centered on the screen. Then draw the square. (Hint: See Example 3 in Lesson 5-2A.) Record the values that you use for the window settings. (Note: Your square may look like a rectangle when it is graphed.) Answers may vary. Sample answers are given, in the following order: min (x), max (x), scale size (x); min (y), max (y), scale size (y).

3. (12, 3), (18, 3), (18, −3), (12, −3)
   10, 20, 1; −5, 5, 1
4. (7, 9), (9, 9), (9, 7), (7, 7)
   5, 11, 1; 5, 11, 1
5. (−5, −10), (−5, 2), (7, 2), (7, −10)
   −7, 9, 1; −12, 4, 1
6. (−10, 10), (−10, −10), (10, −10), (10, 10)
   −12, 12, 1; −12, 12, 1

For each problem, the coordinates of three of the four vertices of a quadrilateral are given. Plot the points and connect them in order. Find the coordinates of the fourth vertex so that the figure will have a line of symmetry. Finish drawing the quadrilateral. Answers may vary. Sample answers are given.

7. (0, 7), (−5, 12), (0, 0)
   (−5, −5)
8. (8, 2), (4, 7), (−6, −1)
   (−2, −6)

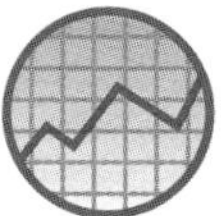

TECH PREP

Tech Prep Applications Masters, pp. 9, 10

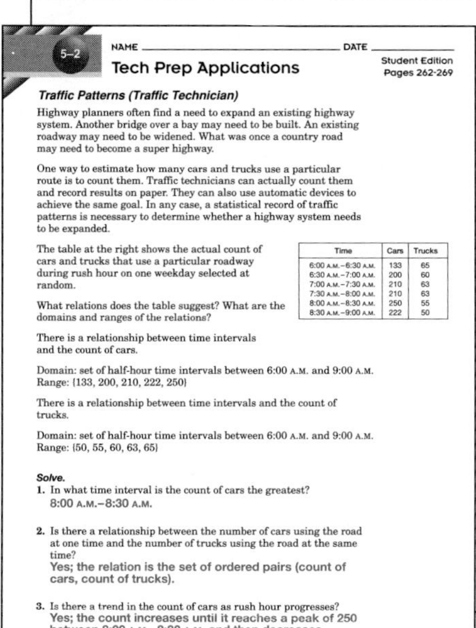

5–2 NAME ______ DATE ______

Tech Prep Applications

Student Edition Pages 262-269

Traffic Patterns (Traffic Technician)

Highway planners often find a need to expand an existing highway system. Another bridge over a bay may need to be built. An existing roadway may need to be widened. What was once a country road may need to become a super highway.

One way to estimate how many cars and trucks use a particular route is to count them. Traffic technicians can actually count them and record results on paper. They can also use automatic devices to achieve the same goal. In any case, a statistical record of traffic patterns is necessary to determine whether a highway system needs to be expanded.

The table at the right shows the actual count of cars and trucks that use a particular roadway during rush hour on one weekday selected at random.

Time	Cars	Trucks
6:00 A.M.–6:30 A.M.	133	65
6:30 A.M.–7:00 A.M.	200	60
7:00 A.M.–7:30 A.M.	210	63
7:30 A.M.–8:00 A.M.	210	63
8:00 A.M.–8:30 A.M.	250	55
8:30 A.M.–9:00 A.M.	222	50

What relations does the table suggest? What are the domains and ranges of the relations?

There is a relationship between time intervals and the count of cars.

Domain: set of half-hour time intervals between 6:00 A.M. and 9:00 A.M.
Range: {133, 200, 210, 222, 250}

There is a relationship between time intervals and the count of trucks.

Domain: set of half-hour time intervals between 6:00 A.M. and 9:00 A.M.
Range: {50, 55, 60, 63, 65}

Solve.

1. In what time interval is the count of cars the greatest?
   8:00 A.M.–8:30 A.M.
2. Is there a relationship between the number of cars using the road at one time and the number of trucks using the road at the same time?
   Yes; the relation is the set of ordered pairs (count of cars, count of trucks).
3. Is there a trend in the count of cars as rush hour progresses?
   Yes; the count increases until it reaches a peak of 250 between 8:00 A.M.–8:30 A.M. and then decreases.

CONNECTIONS

Science and Math Lab Manual, pp. 17–22

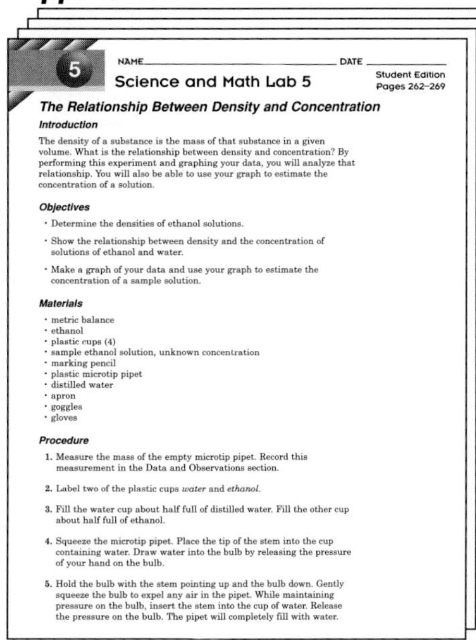

5 NAME ______ DATE ______

Science and Math Lab 5

Student Edition Pages 262–269

The Relationship Between Density and Concentration

Introduction

The density of a substance is the mass of that substance in a given volume. What is the relationship between density and concentration? By performing this experiment and graphing your data, you will analyze that relationship. You will also be able to use your graph to estimate the concentration of a solution.

Objectives

- Determine the densities of ethanol solutions.
- Show the relationship between density and the concentration of solutions of ethanol and water.
- Make a graph of your data and use your graph to estimate the concentration of a sample solution.

Materials

- metric balance
- ethanol
- plastic cups (4)
- sample ethanol solution, unknown concentration
- marking pencil
- plastic microtip pipet
- distilled water
- apron
- goggles
- gloves

Procedure

1. Measure the mass of the empty microtip pipet. Record this measurement in the Data and Observations section.
2. Label two of the plastic cups *water* and *ethanol.*
3. Fill the water cup about half full of distilled water. Fill the other cup about half full of ethanol.
4. Squeeze the microtip pipet. Place the tip of the stem into the cup containing water. Draw water into the bulb by releasing the pressure of your hand on the bulb.
5. Hold the bulb with the stem pointing up and the bulb down. Gently squeeze the bulb to expel any air in the pipet. While maintaining pressure on the bulb, insert the stem into the cup of water. Release the pressure on the bulb. The pipet will completely fill with water.

PROBLEM SOLVING

Problem of the Week Cards, 13, 14, 15

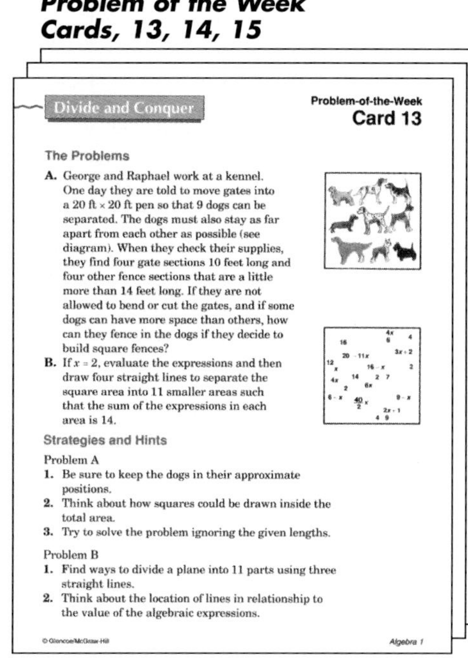

Divide and Conquer

Problem-of-the-Week **Card 13**

The Problems

A. George and Raphael work at a kennel. One day they are told to move gates into a 20 ft × 20 ft pen so that 9 dogs can be separated. The dogs must also stay as far apart from each other as possible (see diagram). When they check their supplies, they find four gate sections 10 feet long and four other fence sections that are a little more than 14 feet long. If they are not allowed to bend or cut the gates, and if some dogs can have more space than others, how can they fence in the dogs if they decide to build square fences?

B. If $x = 2$, evaluate the expressions and then draw four straight lines to separate the square area into 11 smaller areas such that the sum of the expressions in each area is 14.

Strategies and Hints

Problem A
1. Be sure to keep the dogs in their approximate positions.
2. Think about how squares could be drawn inside the total area.
3. Try to solve the problem ignoring the given lengths.

Problem B
1. Find ways to divide a plane into 11 parts using three straight lines.
2. Think about the location of lines in relationship to the value of the algebraic expressions.

© Glencoe/McGraw-Hill   Algebra 1

MAKING MATHEMATICS RELEVANT

This two-page introduction to the chapter provides students with an opportunity to explore contemporary topics and their applications to mathematics.

Background Information

What Kind of Music Do You Like? Have students review the following table of data collected by a survey. Which types of music cross generations, educational level, and income level? Which do not?

Country/western music tends to cut across age, education, and income levels. Rock music is much more popular with younger as compared to older listeners. Jazz music tends to be more popular among listeners with more education.

	Country/ Western	Rock	Jazz
TOTAL	51.7%	43.5%	34.2%
AGE			
18 to 24	39	70	30
25 to 34	50	59	41
35 to 44	53	57	39
45 to 54	61	39	33
55 to 64	58	14	30
65 to 74	54	9	27
75 to 96	46	7	21
EDUCATION			
Grade school	48	12	10
Some high school	59	27	15
High school graduate	57	42	28
Some college	50	54	42
College graduate	42	54	50
Graduate school	46	53	54
FAMILY INCOME			
Under $5,000	43	36	27
$5,000 to $9,999	52	32	21
$10,000 to $14,999	55	33	25
$15,000 to $24,999	57	39	29
$25,000 to $49,999	54	50	36
$50,000 or more	48	55	47
Not ascertained	42	35	35

Graphing Relations and Functions

Objectives

In this chapter, you will:

- graph ordered pairs, relations, and equations,
- solve problems by making a table,
- identify the domain, range, and inverse of a relation,
- determine if a relation is a function,
- write an equation to represent a relation, and
- calculate and interpret the range, quartiles, and interquartile range of a set of data.

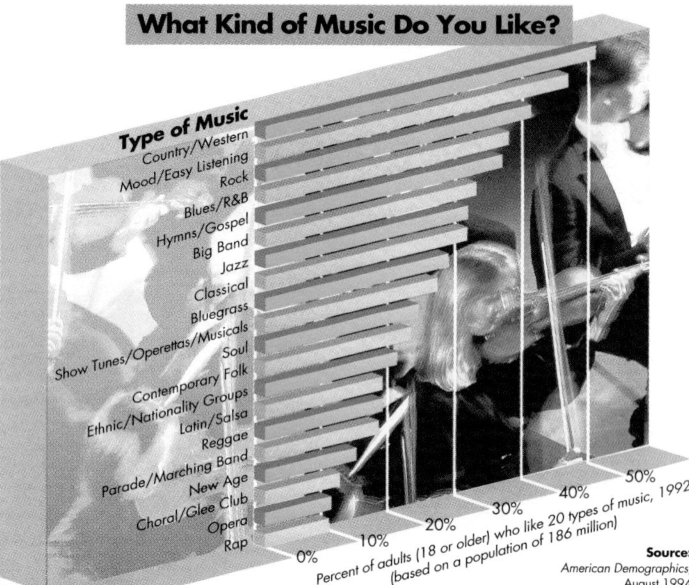

What kinds of music do you listen to most often? Most teenagers have predictable tastes in music—listening to and buying popular rock and country/western favorites. As you become an adult, you may learn to enjoy listening to many other types of music. Who knows, you may become a fan of 12th century Gregorian chants, or perhaps even Zydeco!

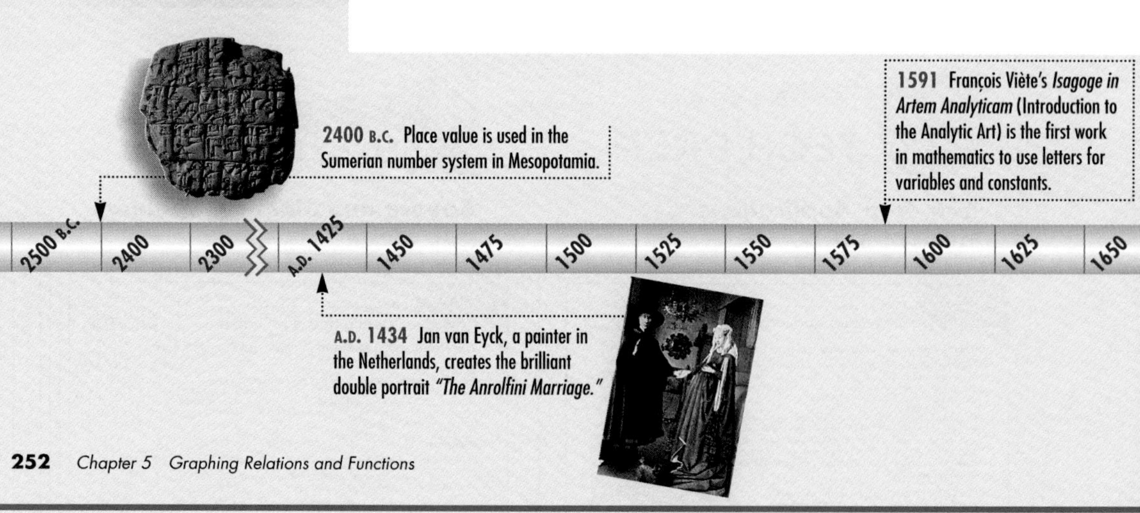

2400 B.C. Place value is used in the Sumerian number system in Mesopotamia.
1591 François Viète's *Isagoge in Artem Analyticam* (Introduction to the Analytic Art) is the first work in mathematics to use letters for variables and constants.
2500 B.C.
2400
2300
A.D. 1425
1450
1475
1500
1525
1550
1575
1600
1625
1650
A.D. 1434 Jan van Eyck, a painter in the Netherlands, creates the brilliant double portrait *"The Anrolfini Marriage."*

The Sumerian number system, which is based on 60, still survives today. Our hour is divided into 60 minutes, and our minute is divided into 60 seconds. Students might want to look further into the mathematics and applications of the base 60 system.

interNET CONNECTION

The Jazz Fan Attic site contains information on blues, ragtime, Dixieland, contemporary jazz, and general interest.

World Wide Web
http://www2.magmacom.com/~rbour/

PEOPLE IN THE NEWS

By the young age of 10, **Jamie Knight** was already an accomplished jazz singer. She has even performed at New York City's Carnegie Hall, following in the footsteps of Ella Fitzgerald and Mel Torme. Jamie is from Philadelphia, Pennsylvania, and has been performing since the age of 6. She has appeared at jazz festivals and on *Good Morning America.* She also has her own CD.

Chapter Project

Work in groups of three to conduct your own music survey.

- Each person in the group should survey 20 people of varying ages to determine their musical tastes. Use the graph on the previous page for ideas.
- Combine your results and compare them to the graph. Does your survey match the national results?
- Create a bar graph to show your results for each type of music. Also create bar graphs for the five most popular types of music you surveyed showing the age distribution for each type of music. Then create a line graph comparing male and female musical tastes in relation to age groups.
- Compare your graphs, make some conclusions, and report your findings.
- Do you think certain types of music are associated with different cultures? You may want to include your thoughts on this in your report.

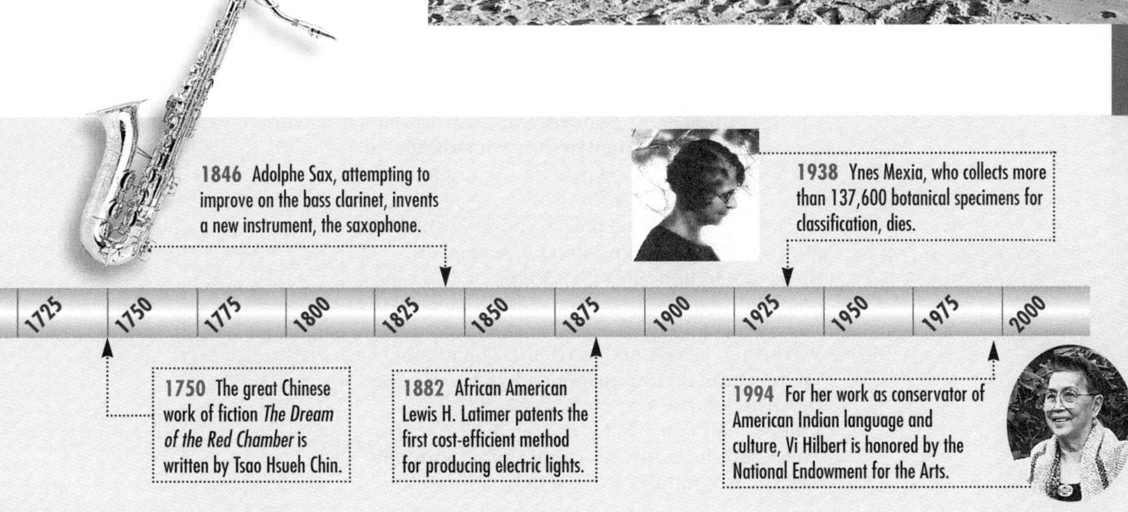

1846 Adolphe Sax, attempting to improve on the bass clarinet, invents a new instrument, the saxophone.
1938 Ynes Mexia, who collects more than 137,600 botanical specimens for classification, dies.
1725
1750
1775
1800
1825
1850
1875
1900
1925
1950
1975
2000
1750 The great Chinese work of fiction *The Dream of the Red Chamber* is written by Tsao Hsueh Chin.
1882 African American Lewis H. Latimer patents the first cost-efficient method for producing electric lights.
1994 For her work as conservator of American Indian language and culture, Vi Hilbert is honored by the National Endowment for the Arts.

Alternative Chapter Projects

Two other chapter projects are included in the *Investigations and Projects Masters.* In Chapter 5 Project A, pp. 41–42, students extend the topic in the chapter opener. In Chapter 5 Project B, pp. 43–44, students investigate fundraising.

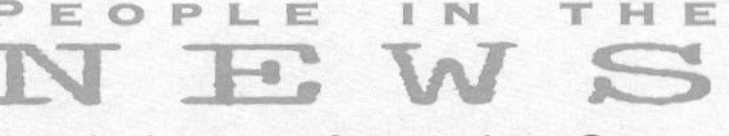

Jamie has performed at Carnegie Hall, with her own rendition of "It Don't Mean a Thing If It Ain't Got That Swing." While she likes all kinds of music, Jamie says, "Jazz is different. It has its own flavor."

Chapter Project

Cooperative Learning Students might want to organize into groups of four and then work in pairs within their groups. They can compile the results from the two teams to create a larger sample for their data.

Investigations and Projects Masters, p. 41

NAME__________ DATE________

Chapter 5 Project A

Student Edition Pages 254–314

Venues and Revenues

1. Work with a partner to make a list of places in your area that feature live performances by musicians. You may need to look in a newspaper for ideas.
2. Do research to find out the seating capacity of each place, or venue, you listed in exercise 1. Also, find out what kind(s) of music is (are) usually performed in each venue. Present your findings in the form of a graph.
3. Write an equation that can be used to estimate the amount of money that would be generated by a sold-out show at each venue.
4. Find the average ticket price for upcoming concerts at the venues you researched. You may be able to do this by looking at newspaper ads.
5. Use the information about ticket prices from exercise 4 and the equation from exercise 3 to determine how much money different concerts would make at each venue. Then find out how the money from ticket sales is divided among a venue's owners, the musicians, the promoter, the company selling the tickets, and others.
6. Use what you've learned to write a newspaper article about the business of performing live music. Be sure to include any graphs you've drawn.

5-1 LESSON NOTES

NCTM Standards: 1–5, 8

Instructional Resources

- Study Guide Master 5-1
- Practice Master 5-1
- Enrichment Master 5-1
- Multicultural Activity Masters, p. 9
- Real-World Applications, 14

Transparency 5-1A contains the 5-Minute Check for this lesson; **Transparency 5-1B** contains a teaching aid for this lesson.

Recommended Pacing	
Standard Pacing	Day 1 of 14
Honors Pacing	Day 1 of 13
Block Scheduling*	Day 1 of 7
Alg. 1 in Two Years*	Days 1 & 2 of 21

*For more information on pacing and possible lesson plans, refer to the *Block Scheduling Booklet* and *Algebra 1 in Two Years.*

1 FOCUS

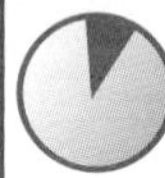

5-Minute Check
(over Chapter 4)

Solve.

1. $\frac{6}{15} = \frac{x}{45}$ **18**
2. What number is 75% of 200? **150**
3. Bill is driving at 50 kilometers per hour. How far will he travel in 2 hours? In h hours? **100 km; 50*h* km**
4. If $y = -13$ when $x = -6$, find y when $x = 3$. **6.5**
5. If $\Delta NBC \sim \Delta FOX$, then $\frac{BC}{?} = \frac{CN}{?}$. ***OX; XF***

5-1 The Coordinate Plane

What YOU'LL LEARN

- To graph ordered pairs on a coordinate plane, and
- to solve problems by making a table.

Why IT'S IMPORTANT

You can use graphing to solve problems involving geography and cartography.

APPLICATION
Landmarks

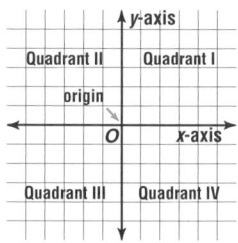

In 1988, as part of Cincinnati's 200th birthday celebration, the city's Bicentennial Commission created parks along the Ohio River. They sold over 33,800 personalized bricks that were placed in the pathways in the parks. However, people had difficulty finding their bricks. So, in 1989, Mr. Frank Albi of Business Information Storage (BIS) developed a master directory, in which each patron's name was listed along with the section of the park and the column in which his or her brick was laid. BIS then added locator bricks to identify the columns in each section.

Perpendicular lines are lines that meet to form 90° angles.

Unless otherwise designated, you can assume that each division on the axes represents 1 unit.

The system used in Cincinnati helped to locate the specific site of each brick. In mathematics, points are located in reference to two perpendicular number lines called **axes.** The axes intersect at their zero points, a point called the **origin.** The horizontal number line, called the ***x*-axis,** and the vertical number line, called the ***y*-axis,** divide the plane into four **quadrants.** The quadrants are numbered as shown at the right. Notice that neither the axes nor any point on the axes are located in any quadrant. The plane containing the x- and y-axes is called the **coordinate plane.**

Points in the coordinate plane are named by *ordered pairs* of the form (x, y). The first number, or ***x*-coordinate,** corresponds to the numbers on the x-axis. The second number, or ***y*-coordinate,** corresponds to the numbers on the y-axis. The ordered pair for the origin is (0, 0).

A point can be referred to by just a capital letter. Thus, Z can be used to mean point Z.

To find the ordered pair for point Z, shown at the right, first follow along a vertical line through Z to find the x-coordinate on the x-axis and along a horizontal line through Z to find the y-coordinate on the y-axis. The number on the x-axis that corresponds to Z is 4. The number on the y-axis that corresponds to Z is -6. Thus, the ordered pair for point Z is $(4, -6)$. This ordered pair can also be written as $Z(4, -6)$. The x-coordinate of Z is 4, and the y-coordinate of Z is -6. Point Z is located in Quadrant IV.

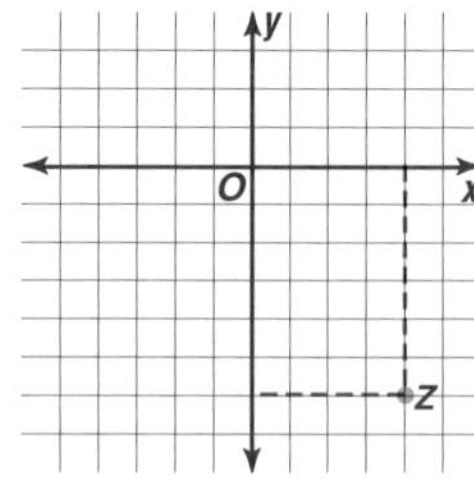

Classroom Vignette

"I give each student a slip of paper as he or she enters the room. The paper has a coordinate pair written on it. The abscissa is a letter and the ordinate is a whole number. The room is laid out like a coordinate grid. They use the coordinates to locate their new seats."

Stan Musgrave
Rochester High School
Rochester, Indiana

A table is often used when working with ordered pairs. It is a helpful organizational tool in problem solving.

Example 1 **Write ordered pairs for points *E, F, G,* and *H* shown at the right. Name the quadrant in which each point is located.**

PROBLEM SOLVING
Use a Table

Use a table to help find the coordinates of each point.

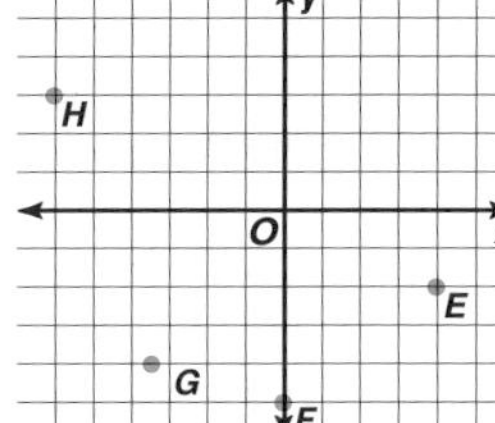

LOOK BACK
You can review more about using tables in Lesson 4-7.

Point	*x*-coordinate	*y*-coordinate	Ordered Pair	Quadrant
E	4	–2	(4, –2)	IV
F	0	–5	(0, –5)	none
G	–3.5	–4	(–3.5, –4)	III
H	–6	3	(–6, 3)	II

To **graph** an ordered pair means to draw a dot at the point on the coordinate plane that corresponds to the ordered pair. This is sometimes called *plotting a point.* When graphing an ordered pair, start at the origin. The *x*-coordinate indicates how many units to move right (positive) or left (negative). The *y*-coordinate indicates how many units to move up (positive) or down (negative).

Example 2 **Plot the following points on a coordinate plane.**

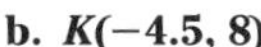

a. $N(-3, -5)$

Start at the origin, *O.* Move left 3 units since the *x*-coordinate is -3. Then move down 5 units since the *y*-coordinate is -5. Draw a dot and label it *N.*

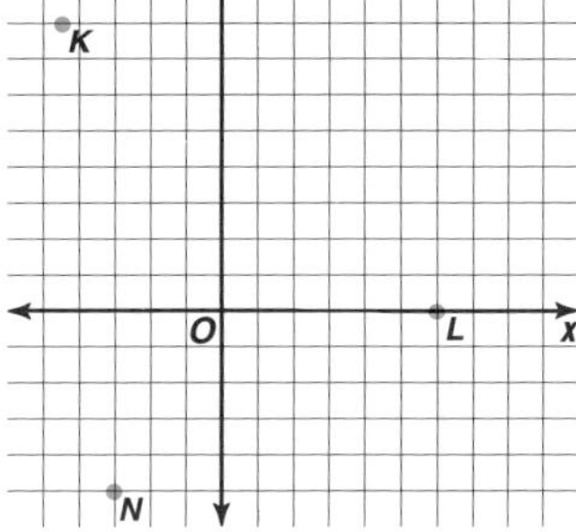

b. $K(-4.5, 8)$

Start at the origin, *O.* Move left 4.5 units and up 8 units. Draw a dot and label it *K.*

c. $L(6, 0)$

Start at the origin, *O.* Move right 6 units. Since the *y*-coordinate is 0, the point will be located on the *x*-axis. Draw the dot and label it *L.*

Motivating the Lesson

Questioning Identify the student sitting in the second seat of the third row of desks in your classroom. Tell students that this student will now be referred to as (3, 2). Identify the student sitting in the third seat of the second row. Tell students that this person will now be referred to as (2, 3). Challenge students to identify the procedure you are using to name the students. Then have students name their new identities based on their seat assignments.

2 TEACH

Teaching Tip Another name for the first coordinate of an ordered pair is the *abscissa.* The second coordinate is also known as the *ordinate.*

In-Class Examples

For Example 1
Write ordered pairs for points *M, N,* and *R.* Name the quadrant in which each point is located.

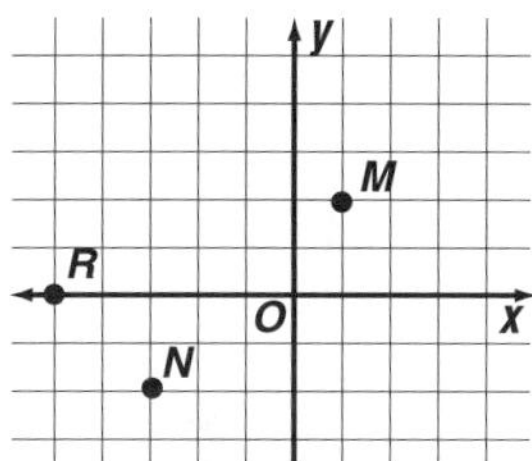

***M*(1, 2), I; *N*(−3, −2), III; *R*(−5, 0), none**

For Example 2
Plot *A*(4, −2) and *B*(0, 5) on a coordinate plane.

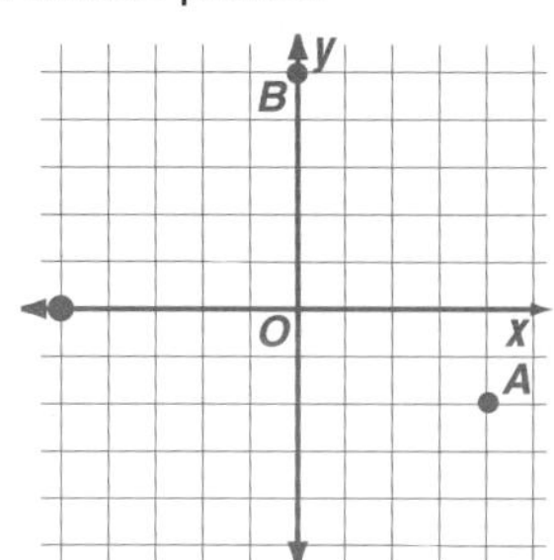

Teaching Tip In Example 2, help students to see an ordered pair as a set of directions when plotting the point. Discuss why the moves for positive numbers are to the right or up, and the moves for negative numbers are to the left or down.

Alternative Learning Styles

Kinesthetic Have students use geoboards to create triangles, listing the vertices and the points inside the triangle as ordered pairs. For example, a triangle with vertices at (0, 0), (6, 0), and (0, 4) would contain the points (1, 1), (1, 2), (1, 3), (2, 1), (2, 2), (3, 1), and (4, 1).

GLENCOE Technology

Interactive Mathematics Tools Software

This multimedia software provides an interactive lesson that uses the graph of the relationship of speed and the distance it takes to stop a car, at various blood-alcohol levels. A **Computer Journal** gives students an opportunity to write about what they have learned.

For Windows & Macintosh

Teaching Tip Students often confuse horizontal and vertical. Emphasize that a horizontal line runs in the same direction as the *horizon.* A vertical line runs up and down, or perpendicular to the horizontal line.

In-Class Example

For Example 3
In which sector is each landmark located?

a. San Antonio International Airport **(3, *A*)**
b. Stinson Airport **(3, *D*)**
c. St. Mary's University **(2, *B*)**
d. The University of Texas at San Antonio **(1, *A*)**

Cassini published his major work on cartography, *Geometric Description of the Earth,* in 1775.

3 PRACTICE/APPLY

Check for Understanding

Exercises 1–12 are designed to help you assess your students' understanding through reading, writing, speaking, and modeling. You should work through Exercises 1–3 with your students and then monitor their work on Exercises 4–12.

Error Analysis
A common mistake that students make is to plot the first coordinate of an ordered pair on the vertical axis and the second coordinate on the horizontal axis. Picture graphing as in Exercises 37 and 38 is a fun, effective way to practice graphing points and provides a self-check for students.

Each point in the coordinate plane can be named by exactly one ordered pair, and each ordered pair names exactly one point in the plane. Thus, there is a one-to-one correspondence between points and ordered pairs.

Completeness Property for Points in the Plane	**1. Exactly one point in the plane is named by a given ordered pair of numbers.** **2. Exactly one ordered pair of numbers names a given point in the plane.**

On city and state maps, letters and numbers are usually used to locate landmarks on the map. The number representing the horizontal coordinate is usually listed first followed by the letter representing the vertical coordinate. However, unlike the coordinate plane where a specific point is denoted by the ordered pair, the coordinates on a map define a region or sector in which the landmark is located.

Example 3 **Use ordered pairs to name all of the sectors that Interstate Highway 35 passes through on the map of San Antonio, Texas, shown below.**

FYI

César-François Cassini, who was born in Thury, France, in 1714, was the first cartographer to create a map according to modern principles. In 1744, he directed the first National Geographic survey, the triangulation of France.

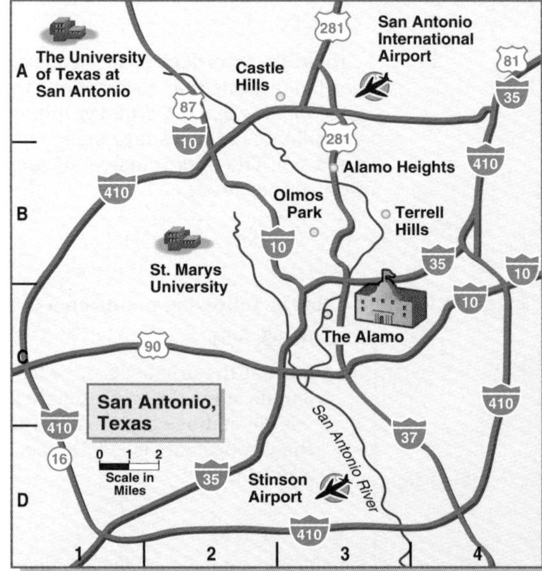

I-35 begins in the southwest corner and extends through San Antonio to the northeast corner. Part of it shares the same path as I-410.

I-35 goes through (1, *D*), continuing through (2, *D*), across the southeast corner of (2, *C*), turning upward through (3, *C*), across the southern portion of (3, *B*), sharing the path with I-410 through (4, *B*), and exiting the map at the eastern edge of (4, *A*).

GLENCOE Technology

Interactive Mathematics Tools Software

This multimedia software provides an interactive lesson by graphing ordered pairs on a coordinate plane and observing what occurs when the *x* and *y* values are exchanged. A **Computer Journal** gives students an opportunity to write about what they have learned.

For Windows & Macintosh

Alternative Teaching Strategies

Student Diversity

- Review lesson vocabulary words. Make sure students can identify the origin, *x*-axis, *y*-axis, quadrants, and the *x*- and *y*-coordinates within an ordered pair.
- Use a variety of maps with different scales (city, state, etc.) to help students better understand the concept of using coordinates.

CHECK FOR UNDERSTANDING

Communicating Mathematics

Study the lesson. Then complete the following.

1. **Draw** a coordinate plane. Label the origin, x-axis, y-axis, and the quadrants. See coordinate plane on page 254.
2. **Explain** how you can tell which quadrant a point is in by just looking at the signs of its coordinates. (+, +), I; (−, +), II; (−, −), III; (+, −), IV

3. **Copy or sketch** a map of the major thoroughfares in your community. Using a different color, draw a coordinate system on the map, making your home the origin. Find the coordinates of your school, a library, a shopping center or department store, and your favorite restaurant. See students' work.

Guided Practice

5. (−3, −1); III
7. (0, −2); none

Write the ordered pair for each point shown at the right. Name the quadrant in which the point is located.

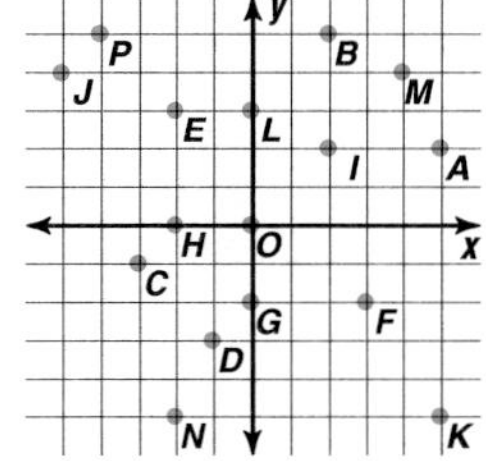

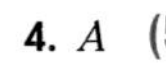
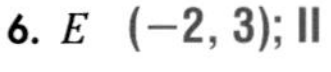
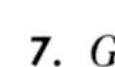

4. *A* (5, 2); I
5. *C*
6. *E* (−2, 3); II
7. *G*

Graph each point. 8–11. See margin.

8. $W(-5, 0)$
9. $X(-3, -4)$
10. $Y(3, -5)$
11. $Z(0, 4)$
12. Write the ordered pair that describes a point 15 units up from and 13 units to the left of the origin. (−13, 15)

EXERCISES

Practice

Refer to the coordinate plane for Exercises 4–7. Write the ordered pair for each point. Name the quadrant in which the point is located.

13. *D* (−1, −3); III
14. *H* (−2, 0); none
15. *L* (0, 3); none
16. *P* (−4, 5); II
17. *O* (0, 0); none
18. *K* (5, −5); IV
19. *F* (3, −2); IV
20. *B* (2, 5); I
21. *I* (2, 2); I
22. *M* (4, 4); I
23. *J* (−5, 4); II
24. *N* (−2, −5); III

Graph each point. 25–36. See margin.

25. $A(2, -1)$
26. $B(-3, -3)$
27. $C(0, 3.5)$
28. $D(5, -2)$
29. $E(-3, 0)$
30. $F(3, 5)$
31. $G(-3, -4)$
32. $H(-5, -2)$
33. $I(0, -2)$
34. $J(-\frac{1}{2}, 1)$
35. $K(4, 0)$
36. $L(4, 2)$

Graph each point. Then connect the points in alphabetical order and identify the figure. 37–38. See margin for graphs.

37. $A(2, 0)$, $B(2, 3)$, $C(1, 3)$, $D(-0.5, -1)$, $E(-2, 3)$, $F(-3, 3)$, $G(-3, -3)$, $H(-2, -3)$, $I(-2, 0)$, $J(-0.5, -3)$, $K(1, 0)$, $L(1, -3)$, $M(2, -3)$, $N(2, 0)$ M

38. $A(9, 0.5)$, $B(5, 1)$, $C(2, 1)$, $D(-1, 5)$, $E(-2, 5)$, $F(-1, 1)$, $G(-4, 1)$, $H(-5, 3)$, $I(-6, 3)$, $J(-5.5, 1)$, $K(-5.5, 0)$, $L(-6, -2)$, $M(-5, -2)$, $N(-4, 0)$, $P(-1, 0)$, $Q(-2, -4)$, $R(-1, -4)$, $S(2, 0)$, $T(5, 0)$, $U(9, 0.5)$ airplane

Reteaching

Using Visual Models To practice reading coordinates of points, have students make a picture using straight lines on a coordinate grid. List the coordinates of the points in order. Exchange ordered lists of points, and plot and connect the points in the order given as in dot-to-dot puzzles.

Additional Answer

38.

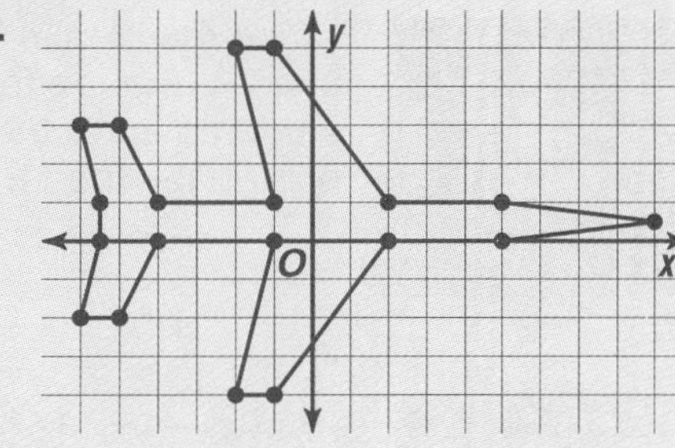

Additional Answers

8–11.

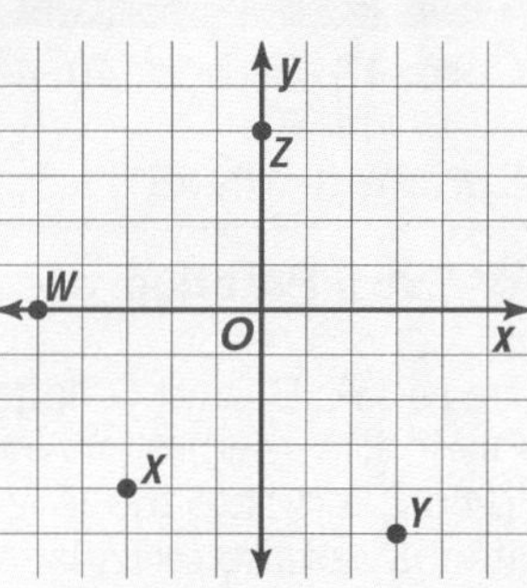

25–36.

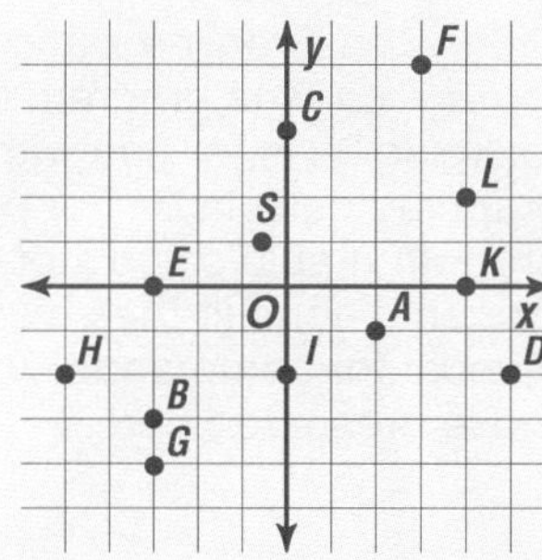

37.

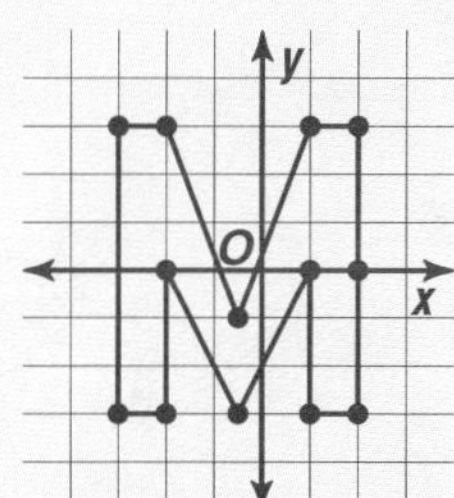

Study Guide Masters, p. 34

NAME__________ DATE______

Student Edition Pages 254–259

Study Guide

The Coordinate Plane

In the diagram at the right, the two perpendicular lines, called the x-axis and the y-axis, divide the coordinate plane into Quadrants I, II, III, and IV. The point where the two axes intersect is called the origin. The origin is represented by the ordered pair (0, 0).

Every other point in the coordinate plane is also represented by an ordered pair of numbers. The ordered pair for point Q is (5, −4). We say that 5 is the x-coordinate of Q and −4 is the y-coordinate of Q.

Example: Write the ordered pair for the point R above.

The x-coordinate is 0 and the y-coordinate is 4. Thus, the ordered pair for R is (0, 4).

To graph any ordered pair (x, y), begin at the origin. Move left or right x units. From there, move up or down y units. Draw a dot at that point.

Graph each point on the coordinate plane at the right.

1. A (0, 0)
2. B (5, 0)
3. C (−3, 4)
4. D (4, −5)
5. E (−2, −3)
6. F (2, −1)

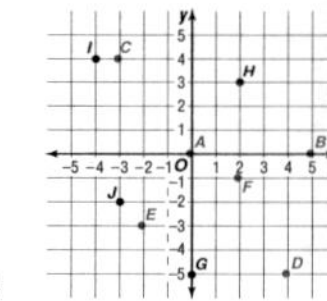

Write the ordered pair for each point shown at the right. Name the quadrant in which the point is located.

7. G (0, −5); none
8. H (2, 3); I
9. I (−4, 4); II
10. J (−3, −2); III

Assignment Guide

Core: 13–39 odd, 40, 41, 43, 44–51
Enriched: 14–38 even, 40–51

For **Extra Practice,** see p. 767.

The red A, B, and C flags, printed only in the Teacher's Wraparound Edition, indicate the level of difficulty of the exercises.

Using the Programming Exercises The program given in Exercise 39 is for use with a TI-82 graphing calculator. For other programmable calculators, have students consult their owner's manual for commands similar to those presented here.

Practice Masters, p. 34

5-1 NAME________________ DATE________
Practice
Student Edition Pages 254–259

The Coordinate Plane

Write the ordered pair for each point shown at the right. Name the quadrant in which the point is located.

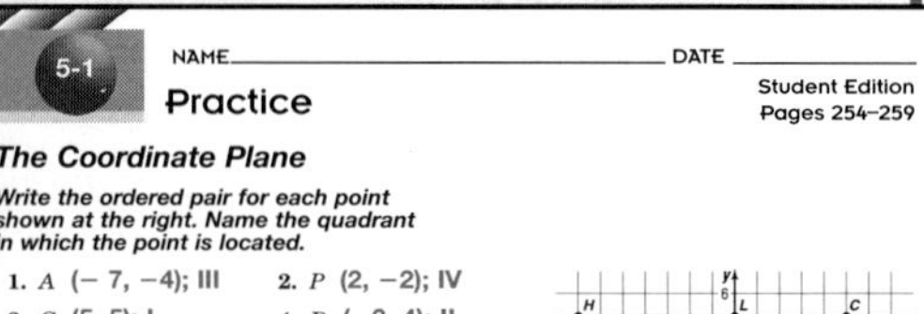

1. A (−7, −4); III
2. P (2, −2); IV
3. C (5, 5); I
4. D (−3, 4); II
5. E (0, 0); none
6. F (1, 2); I
7. G (−3, −3); III
8. Q (0, −6); none
9. B (5, −4); IV
10. H (−7, 5); III
11. K (−2, 0); none
12. L (0, 5); none

Graph each point.

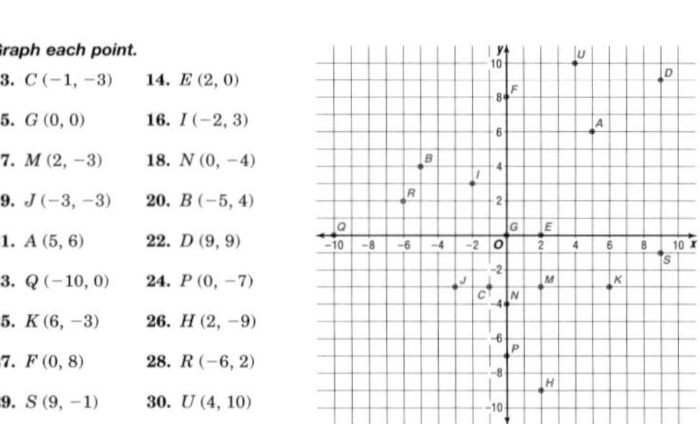

13. C (−1, −3)
14. E (2, 0)
15. G (0, 0)
16. I (−2, 3)
17. M (2, −3)
18. N (0, −4)
19. J (−3, −3)
20. B (−5, 4)
21. A (5, 6)
22. D (9, 9)
23. Q (−10, 0)
24. P (0, −7)
25. K (6, −3)
26. H (2, −9)
27. F (0, 8)
28. R (−6, 2)
29. S (9, −1)
30. U (4, 10)

Programming

39. The graphing calculator program at the right will plot a group of points on the same coordinate plane. You will need to set an appropriate window so that you will be able to see all of the points. After each point is plotted, press ENTER to plot another point. To end the program, press 2nd QUIT while you are at the home screen. Use this program to check the points you plotted in Exercises 37 and 38. **See students' work.**

```
PROGRAM:PLOTPTS
:ClrDraw
:Lbl 1
:Prompt X,Y
:Pt-On(X,Y)
:Pause
:Goto 1
```

Critical Thinking

40. Describe the possible locations, in terms of quadrants or axes, for the graph of (x, y) if x and y satisfy the given conditions.
  a. $xy > 0$ **I or III**
  b. $xy < 0$ **II or IV**
  c. $xy = 0$ **x- or y-axis**

Applications and Problem Solving

41. **Geography** On the map of the United States shown below, latitude and longitude can be used to form ordered pairs that name the sectors on the map. The longitude is represented along the horizontal axis and should be the first coordinate of the ordered pair. The latitude is represented along the vertical axis and should be the second coordinate.

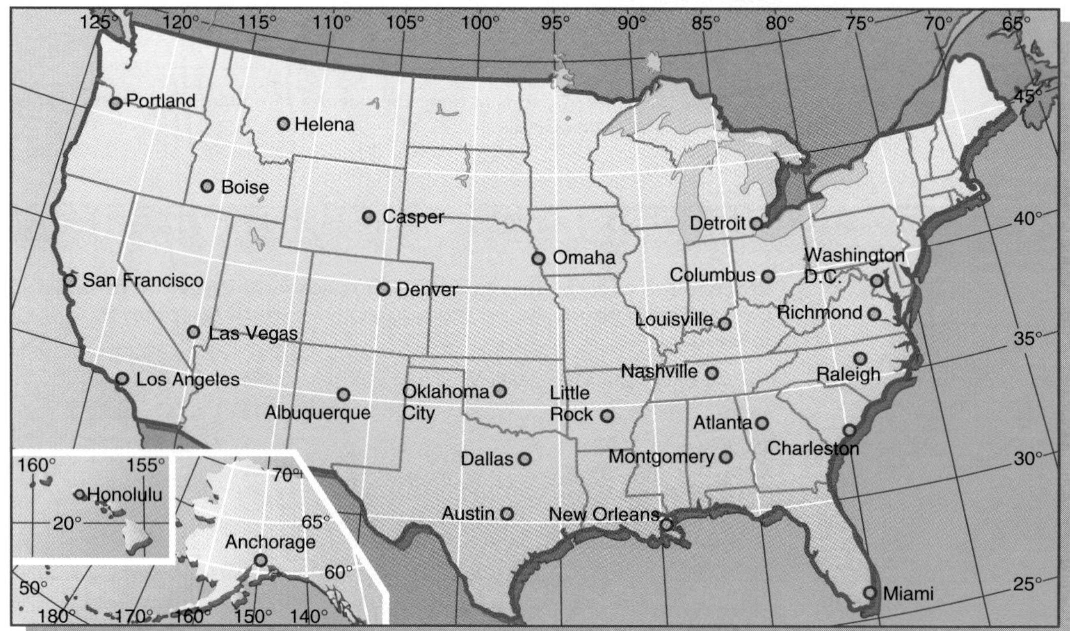

  a. Name the city at (90°, 30°). **New Orleans**
  b. Name the state in which (120°, 45°) is located. **Oregon**
  c. Estimate the latitude and longitude of our nation's capital to the nearest 5°. **Answers may vary. Sample answer: (75°, 40°)**
  d. What state capital has its location at (157°, 21°)? **Honolulu, Hawaii**
  e. What is the approximate latitude and longitude of your community?
  f. Look at the longitude and latitude lines on a globe and compare them to the ones shown on this map. What do you find?

41e. See students' work.
41f. Sample answer: The longitude lines are not the same distance apart; they meet at the poles.

Classroom Vignette

"I use the idea in Exercise 43 every fall to make up my seating chart. I arrange the desks in rows and columns, and as the students enter the class, I hand them a card with an ordered pair on it—then they find their seats!"

Deborah Phillips
St. Vincent–St. Mary High School
Akron, Ohio

Deborah Patonai Phillips

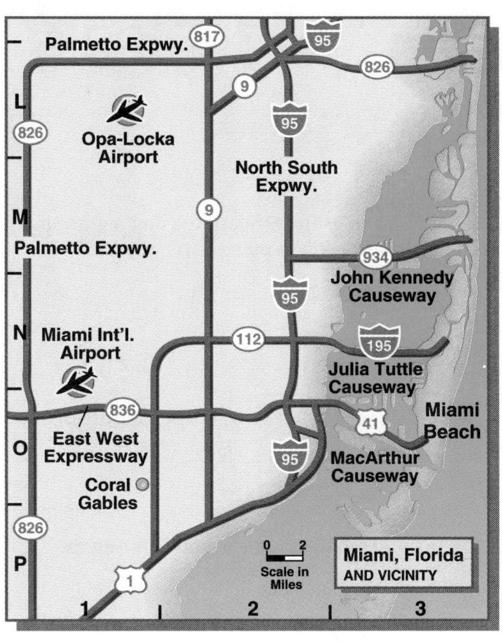

42. **Tourism** Diana works for a travel agency and has been asked to highlight key landmarks on a map of the Miami area for a customer. Use the map of Miami at the left to answer the following questions.
    a. What causeway is in sector (3, M)?
    b. In what sector is the city of Coral Gables? **(1, 0)**
    c. Which interstate goes from sector (2, L) to (2, O)? **I-95**
    d. Name all of the sectors through which Highway 826 passes.

43. **Draw a Diagram** Make a seating chart of the desks in your mathematics classroom. Create a coordinate plane on your diagram so that each desk corresponds to an ordered pair in the plane. **See students' work.**
    a. What are the coordinates of your desk in the room?
    b. Name another student in your class. What is the coordinate of this person's desk?

42a. John Kennedy 42d. (1, P), (1, O), (1, N), (1, M), (1, L), (2, L), (3, L)

Mixed Review

44. If y varies directly as x, and $y = 3$ when $x = 15$, find y when $x = -25$. (Lesson 4–8) **−5**

45. Two trains leave York at the same time, one traveling north, the other south. The first train travels at 40 miles per hour and the second at 30 miles per hour. In how many hours will the trains be 245 miles apart? (Lesson 4–7) **$3\frac{1}{2}$ hours**

46. Use a calculator to find the measure of $\angle A$ to the nearest degree if $\sin A = 0.2756$. (Lesson 4–3) **16°**

47. Find the mean, median, and mode for {5, 9, 1, 2, 3}. (Lesson 3–7) **4; 3; none**

Simplify each expression.

48. $\sqrt{\frac{16}{25}}$ (Lesson 2–8) **$\frac{4}{5}$**

49. $\frac{7a + 35}{-7}$ (Lesson 2–7) **$-a - 5$**

50. **Agriculture** Agribusiness technicians assist in the organization and management of farms. They may prepare graphs of weather data for a given region. For example, the graph at the right displays normal maximum, minimum, and mean temperatures for the region around Des Moines, Iowa. (Lesson 1–9)

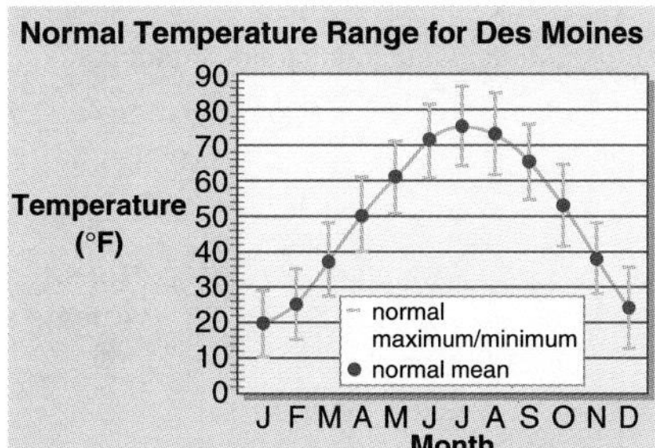

    a. What is the normal maximum temperature for October?
    b. What is the normal range of temperatures for May?

50a. 65°
50b. 51° to 72°

51. Simplify $7(5a + 3b) - 4a$. (Lesson 1–8) **$31a + 21b$**

4 ASSESS

Closing Activity

Speaking Ask students to explain why the x-axis and the y-axis are not contained in any of the quadrants.

Enrichment Masters, p. 34

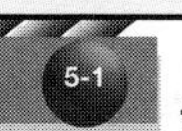

NAME _______________ DATE _______

Enrichment

Student Edition Pages 254–259

Coordinate Geometry and Area

How would you find the area of a triangle whose vertices have the coordinates $A(-1, 2)$, $B(1, 4)$, and $C(3, 0)$?

When a figure has no sides parallel to either axis, the height and base are difficult to find.

One method of finding the area is to enclose the figure in a rectangle and subtract the area of the surrounding triangles from the area of the rectangle.

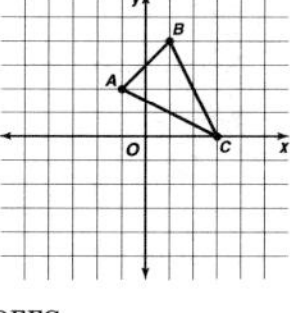

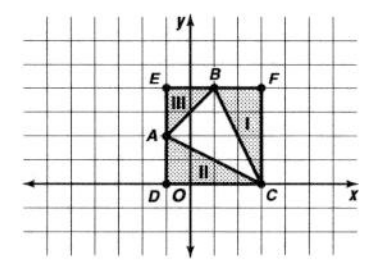

Area of rectangle $DEFC$
$= 4 \times 4$
$= 16$ square units

Area of triangle I $= \frac{1}{2}(2)(4) = 4$

Area of triangle II $= \frac{1}{2}(2)(4) = 4$

Area of triangle III $= \frac{1}{2}(2)(2) = 2$

Total $= 10$ square units

Area of triangle $ABC = 16 - 10$, or 6 square units

Find the areas of the figures with the following vertices.

1. $A(-4, -6)$, $B(0, 4)$, $C(4, 2)$ **24 square units**
2. $A(6, -2)$, $B(8, -10)$, $C(12, -6)$ **20 square units**
3. $A(0, 2)$, $B(2, 7)$, $C(6, 10)$, $D(9, -2)$ **55 square units**

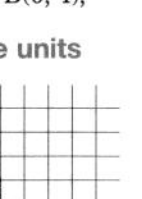

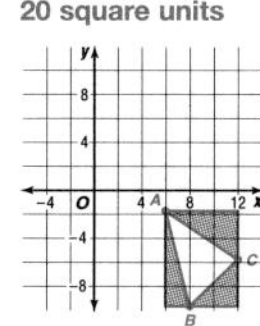

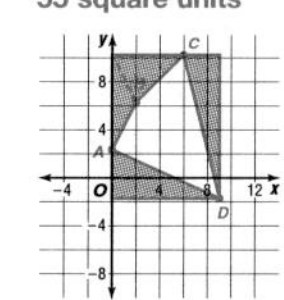

Extension

Connections A parallelogram has three of its vertices at (0, 0), (2, 2), and (7, 1). Find the possible ordered pairs that could correspond to its fourth vertex. **(9, 3) and (5, −1)**

Tech Prep

Travel Agent Students who are interested in a career in tourism may wish to do further research on the use of maps by travel agents, as in Exercise 42. For more information on tech prep, see the *Teacher's Handbook.*

5-2A LESSON NOTES

NCTM Standards: 1–5, 8

Objective

Use a graphing calculator to graph relations.

Recommended Time

25 minutes

Instructional Resources

Graphing Calculator Masters, pp. 14 and 15

These masters provide keystroking instruction for this lesson for the TI-81 and Casio graphing calculators.

1 FOCUS

Motivating the Lesson

Review with students and have them practice methods for setting standard windows and customizing a window by specifying Xmin, Xmax, Ymin, and Ymax.

2 TEACH

Teaching Tip When plotting a relation on the graphing calculator, the window must be chosen carefully. One way to do this is to set Xmin to 1 less than the smallest x value for all the pairs given and Xmax at 1 greater than the largest x value. Then set Ymin and Ymax in the same way.

5–2A Graphing Technology Relations

A Preview of Lesson 5–2

A **relation** is a set of ordered pairs. You can graph a relation using a graphing calculator. Points can be plotted on the coordinate plane using the DRAW menu.

If you press 2nd DRAW ▶ *1 before pressing* GRAPH*, the prompt Pt-On(appears. To plot a point from this prompt, enter the coordinates of the point, a closing parenthesis, and* ENTER*. Then proceed with the cursor for the other points.*

Example 1 **Plot the following points on a coordinate plane: $M(4, 7)$, $A(-10, 25)$, $T(-17, -17)$, and $H(23, -11)$.**

Begin with a viewing window of $[-47, 47]$ by $[-31, 31]$ using a scale of 10 for each axis. This can be done quickly by entering ZOOM 6 ZOOM 8 ENTER. *This sets the scale to integer values.*

Clear the coordinate plane by turning off all stat plots and clearing the Y= list.

- To plot the first point, enter GRAPH 2nd DRAW ▶ 1. *The cursor appears at the origin and X = 0 and Y = 0 appears at the bottom of the screen.*
- Now use the arrow keys to move the cursor to the coordinates of the point you wish to graph. *The coordinates of each position of the cursor appear at the bottom of the viewing screen.*

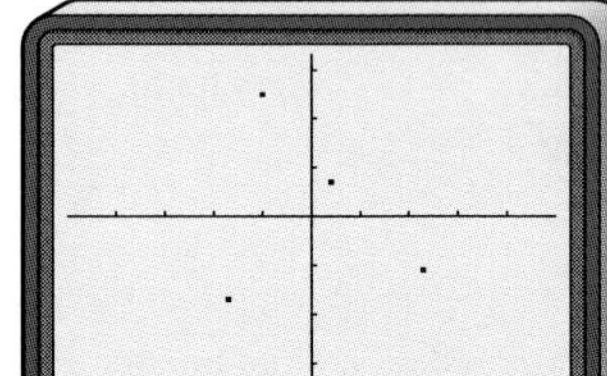

- Press ENTER to put a dot at that point.
- Continue to move the cursor and press ENTER to graph the other three points.
- When finished, you can clear the screen by pressing 2nd DRAW 1.

If you do not need a range as great as the one you get by pressing ZOOM 8, you can set the WINDOW values in the usual manner or press ZOOM 6. Once you set your new window and try the method shown in Example 1, you will find that you cannot plot points at exact integer values by using the cursor. However, there is another way to plot exact points.

Example 2 **Plot $G(-4, 5)$, $R(1, 6)$, $A(-3, -6)$, $P(5, -8)$, and $H(0, 7)$ in the standard viewing window.**

- Set the standard viewing window. **Enter:** ZOOM 6
- Return to the home screen by pressing 2nd QUIT.
- You can graph all of the points by linking commands into a series using a colon.

Enter: 2nd DRAW ▶ 1 *Pt-On(appears on the screen.*

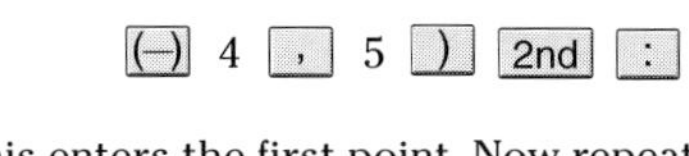

This enters the first point. Now repeat the steps without pressing ENTER to enter all the other points. Finally, press ENTER to graph the points.

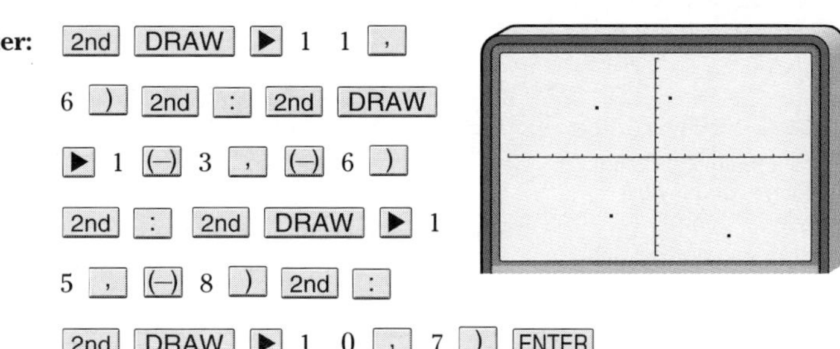

You can also use the Line(feature on the DRAW menu to connect the points. This can help you analyze the points in a relation. You must first clear the screen by pressing 2nd DRAW 1 ENTER before beginning a new graph.

Example 3 **Determine if the graphs of (3, 5), (−3, −5), and (0, 7) are on the same line.**

If you connect the points and only a segment appears, the three points are probably on the same line.

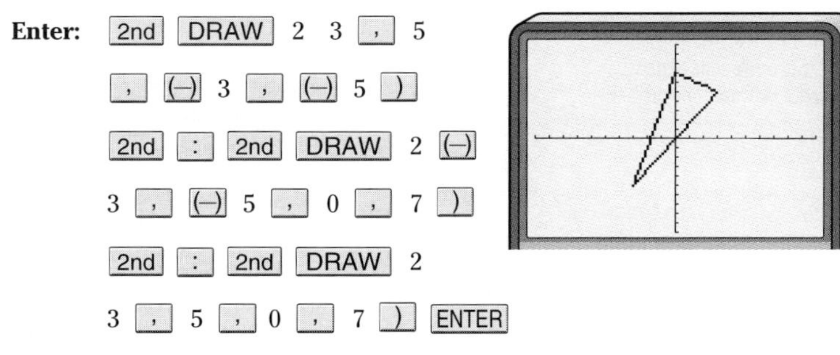

Because the resulting figure is a triangle, the three points are not on the same line.

EXERCISES

1–6. See margin.

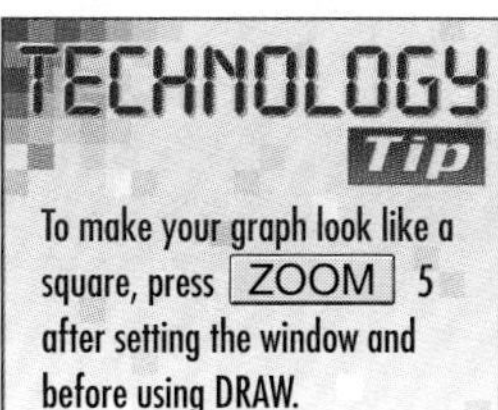

Use a graphing calculator to plot each point in the viewing window [−47, 47] by [−31, 31].

1. $X(10, 10)$
2. $Y(0, -6)$
3. $Z(-26, 11)$
4. $Q(-17, -19)$
5. $T(-21, 4)$
6. $P(-20, 0)$
7. Explain why, when you graph the points in Exercises 2 and 6, you do not see them on the screen. **They are points on the axes.**
8. a. Plot $H(9, 9)$ and $J(29, 9)$. **8a–c. See margin.**
   b. Find two additional points E and F so that these four points form the vertices of a square. Are there other possible coordinates of E and F?
   c. Plot E and F to complete the square.

3 PRACTICE/APPLY

Assignment Guide

Core: 1–8
Enriched: 1–8

4 ASSESS

Observing students working with technology is an excellent method of assessment.

Additional Answers

1–6.

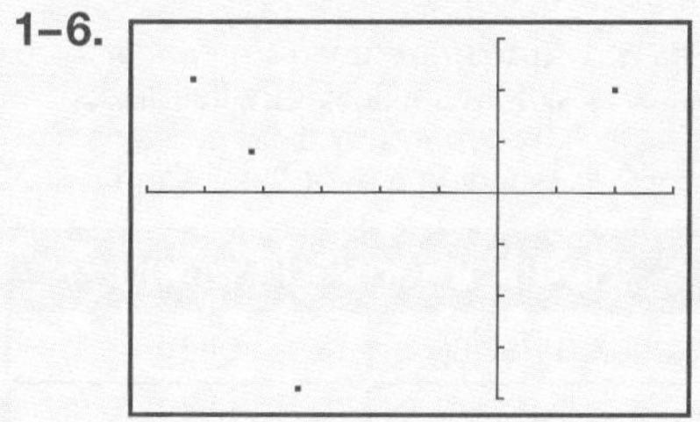

8a.

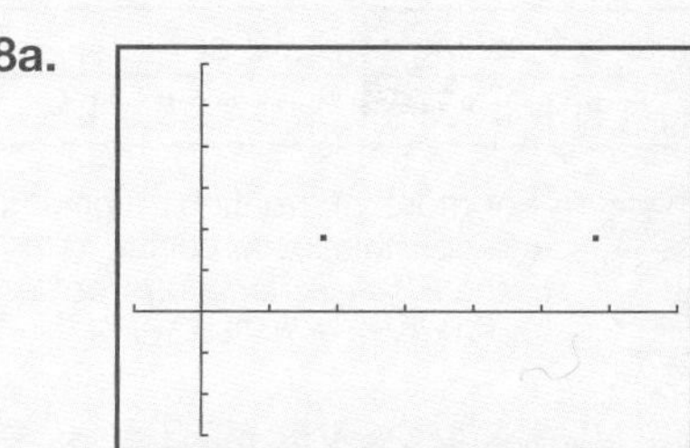

8b. 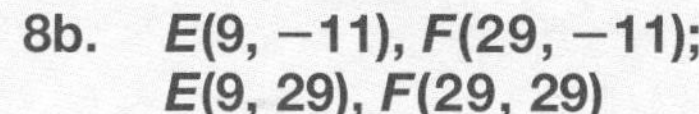
E(9, −11), *F*(29, −11); *E*(9, 29), *F*(29, 29)

8c.

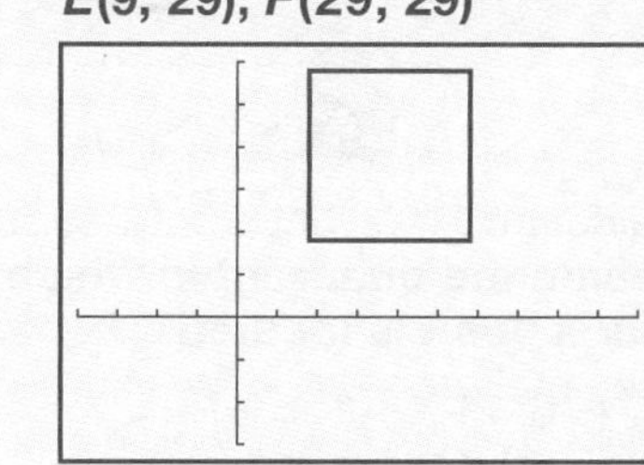

Using Technology

This lesson offers an excellent opportunity for using technology in your algebra classroom. For more information on using technology, see *Graphing Calculators in the Mathematics Classroom,* one of the titles in the Glencoe Mathematics Professional Series.

5-2 LESSON NOTES

NCTM Standards: 1–5

Instructional Resources

- Study Guide Master 5-2
- Practice Master 5-2
- Enrichment Master 5-2
- Assessment and Evaluation Masters, p. 128
- Science and Math Lab Manual, pp. 17–22
- Tech Prep Applications Masters, p. 9

Transparency 5-2A contains the 5-Minute Check for this lesson; **Transparency 5-2B** contains a teaching aid for this lesson.

Recommended Pacing	
Standard Pacing	Day 3 of 14
Honors Pacing	Day 3 of 13
Block Scheduling*	Day 2 of 7
Alg. 1 in Two Years*	Days 4 & 5 of 21

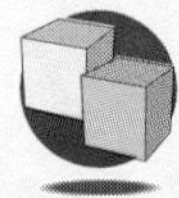

*For more information on pacing and possible lesson plans, refer to the *Block Scheduling Booklet* and *Algebra 1 in Two Years*.

1 FOCUS

5-Minute Check
(over Lesson 5-1)

Name the quadrant in which each point is located.

1. (−3, 2) **Quadrant II**
2. (6, 2) **Quadrant I**
3. (−8, −1) **Quadrant III**
4. (4, −7) **Quadrant IV**
5. (9, 0) **It lies on the x-axis.**

Motivating the Lesson

Questioning Tell students you are going to say two pairs of numbers. The numbers in each pair are related in the same way. The students need to identify the relationship and then name an additional pair that is related in the same way. For example, you might say "7 and 2, 4 and −1." The students might respond with "difference of the numbers in each pair is 5." Continue the activity with different number pairs.

5-2 Relations

What YOU'LL LEARN

- To identify the domain, range, and inverse of a relation, and
- to show relations as sets of ordered pairs, tables, mappings, and graphs.

Why IT'S IMPORTANT

You can use relations to solve problems involving economics and probability.

FYI

The U.S. Congress passed the Rare and Endangered Species Act in 1966, setting up rules by which the Fish and Wildlife Service can list and protect species in danger of extinction.

CONNECTION
Biology

There are 1.4 million classified species of microorganisms, invertebrates, plants, fish, birds, reptiles, amphibians, and mammals. Currently, 930 species worldwide have been listed as *endangered.* A species becomes endangered when its numbers are so low that the species is in danger of extinction.

In the United States, the manatee, an aquatic mammal, is considered to be endangered. Although manatees were once common throughout the coastal areas of North America, they now live mainly in Florida. In the table below, you will find the number of manatees that have been found dead since 1976.

What is the trend over time?

Manatee Mortality In Florida									
Year	1976	1977	1978	1979	1980	1981	1982	1983	1984
Number of Manatees	62	114	84	77	63	116	114	81	128
Year	1985	1986	1987	1988	1989	1990	1991	1992	1993
Number of Manatees	119	122	114	133	168	206	174	163	145

Source: Florida Marine Research Institute

The manatee data could also be represented by a set of ordered pairs, as shown in the list below. Each first coordinate is the year, and the second coordinate is the number of manatees. Each ordered pair can then be graphed.

{(1976, 62), (1977, 114), (1978, 84), (1979, 77), (1980, 63), (1981, 116), (1982, 114), (1983, 81), (1984, 128), (1985, 119), (1986, 122), (1987, 114), (1988, 133), (1989, 168), (1990, 206), (1991, 174), (1992, 163), (1993, 145)}

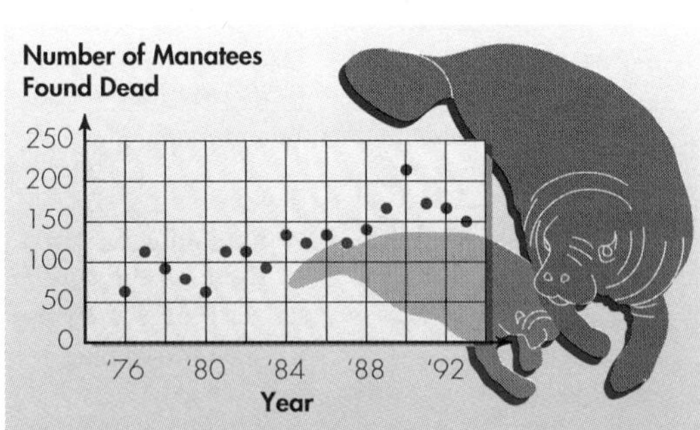

FYI

As of June 1993, the following animals were also on the list of endangered species in the United States: American alligator, American crocodile, brown bear, ocelot, Carolina northern flying squirrel.

Recall that a *relation* is a set of ordered pairs, like the one shown on the previous page. The set of first coordinates of the ordered pairs is called the *domain* of the relation. The domain usually contains the x-coordinates. The set of second coordinates is called the *range* of the relation. It usually contains the y-coordinates.

Definition of the Domain and Range of a Relation	**The domain of a relation is the set of all first coordinates from the ordered pairs in the relation. The range of the relation is the set of all second coordinates from the ordered pairs.**

For the relation representing the manatee data, the domain and range are as follows.

Domain = {1976, 1977, 1978, 1979, 1980, 1981, 1982, 1983, 1984, 1985, 1986, 1987, 1988, 1989, 1990, 1991, 1992, 1993}

Range = {62, 114, 84, 77, 63, 116, 81, 128, 119, 122, 133, 168, 206, 174, 163, 145}

An element is a member of a set.

In addition to ordered pairs, tables, and graphs, a relation can be represented by a **mapping.** A mapping illustrates how each element of the domain is paired with an element in the range. For example, the relation {(3, 3), (−1, 4), (0, −4)} can be modeled in each of the following ways.

Ordered Pairs

(3, 3)
(−1, 4)
(0, −4)

Table

x	y
3	3
−1	4
0	−4

Graph

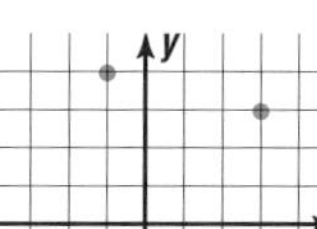

Mapping

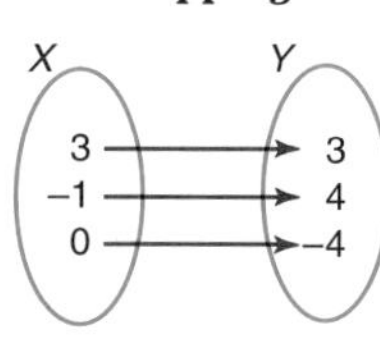

Example 1 **Represent the relation shown in the graph at the right as**

a. a set of ordered pairs,
b. a table, and
c. a mapping.
d. Then determine the domain and range.

a. The set of ordered pairs for the relation is {(−3, 3), (−1, 2), (1, 1), (1, 3), (3, −2), (4, −2)}.

b.

x	y
−3	3
−1	2
1	1
1	3
3	−2
4	−2

c.

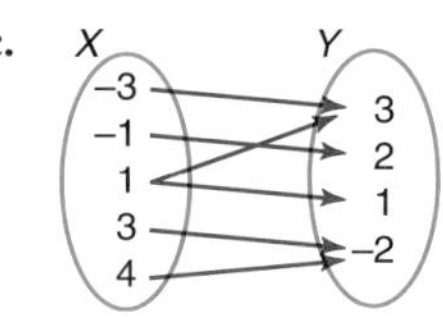

When giving the domain and range, if a value is repeated, you list it only once.

d. The domain for this relation is {−3, −1, 1, 3, 4}, and the range is {−2, 1, 2, 3}.

Alternative Learning Styles

Visual Graph the relation {(2, 4), (−1, 3), (0, 2), (1, −2)} and indicate the domain and range by marking those values on the x-axis and the y-axis.

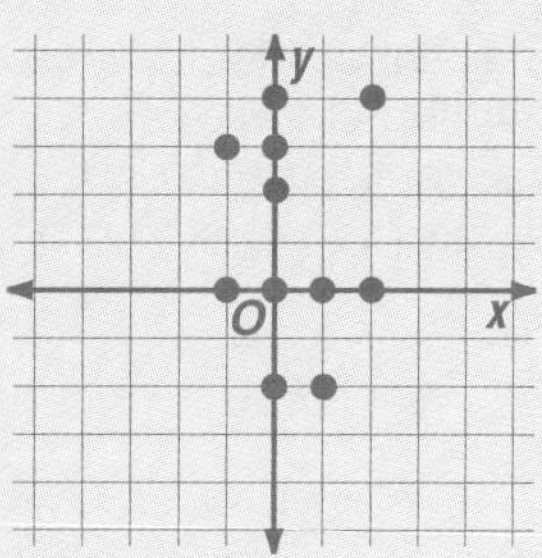

2 TEACH

In-Class Example

For Example 1
Represent the relation shown in the graph below as

a. a set of ordered pairs,
b. a table, and
c. a mapping.

Then determine the domain and range.

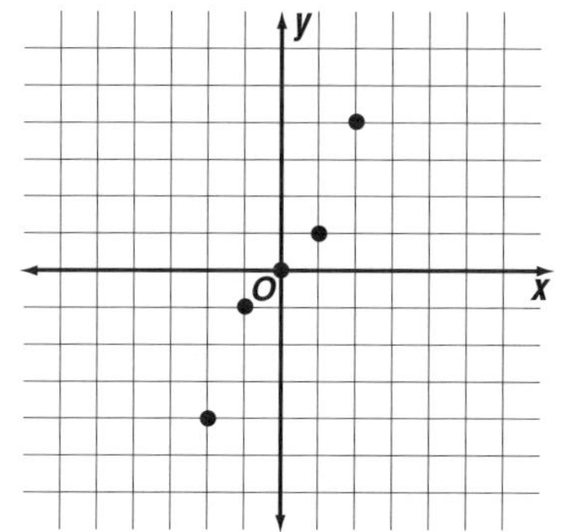

a. (−2, −4), (−1, −1), (0, 0), (1, 1), (2, 4)

b.

x	y
−2	−4
−1	−1
0	0
1	1
2	4

c.

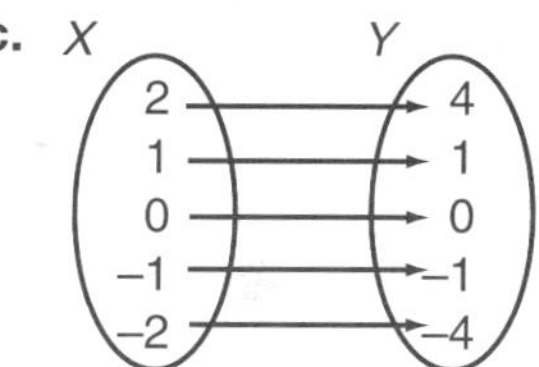

The domain is {−2, −1, 0, 1, 2} and the range is {−4, −1, 0, 1, 4}.

Teaching Tip The word *relation* simply indicates a set of ordered pairs. There may or may not be a reason for the pairing. Stress that a relation can be shown in many different ways. Remind students that the domain is a set of numbers, not a set of ordered pairs. The same is true of the range.

In-Class Example

For Example 2
The following table lists the high temperatures for certain days in June.

Date (June)	1	6	11	16	21	26
Temperature (°F)	70	85	82	90	82	87

a. Determine the domain and range of the relation. **domain = {1, 6, 11, 16, 21, 26}, range = {70, 82, 85, 87, 90}**

b. Graph the data.

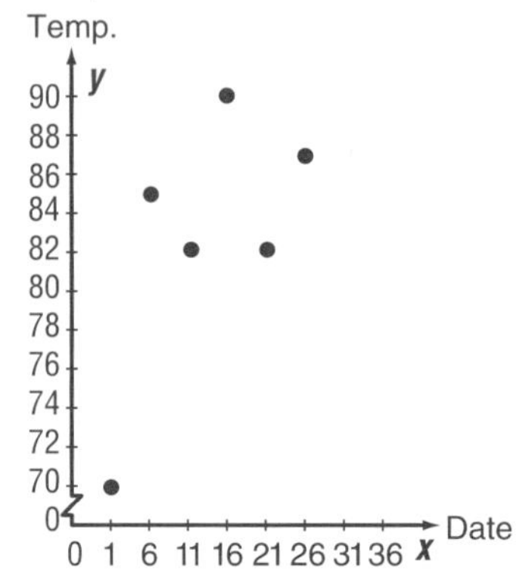

CAREER CHOICES

Median annual earnings for jobs in the biological and life sciences were about $34,500 in 1992. The middle 50 percent earned between $26,000 and $46,800. Ten percent earned less than $20,400 and ten percent earned over $56,900.

In real-life situations, you may need to select a range of values for the x- or y-axis that does not begin with 0 and does not have units of 1.

Example 2

Biology

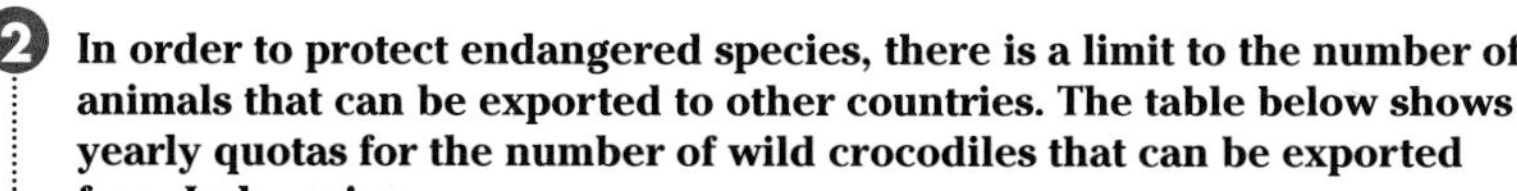

In order to protect endangered species, there is a limit to the number of animals that can be exported to other countries. The table below shows yearly quotas for the number of wild crocodiles that can be exported from Indonesia.

a. Determine the domain and range of the relation.
b. Graph the data.
c. What conclusions might you make from the graph of the data?

CAREER CHOICES

A **zoologist** studies animals, their origin, life process, behavior, and diseases. They are usually identified by the animal group in which they specialize, for example herpetology (reptiles), ornithology (birds), and ichthyology (fish).

Zoology requires a minimum of a bachelor's degree. Many college positions in this field require a Ph.D. degree.

For more information, contact:

Occupational Outlook Handbook, 1994–95, U.S. Department of Labor.

Maximum Exportation of Crocodiles from Indonesia

Year	1986	1987	1988	1989	1990	1991	1992	1993	1994
Number of Crocodiles	2000	2000	4000	4000	3000	3000	2700	1500	1500

Source: *Traffic USA*, World Wildlife Fund, August 1992

a. The domain is {1986, 1987, 1988, 1989, 1990, 1991, 1992, 1993, 1994}. The range is {1500, 2000, 2700, 3000, 4000}.

Values are usually listed in numerical order.

b. The values for the x-axis need to go from 1986 to 1994. It is not efficient to begin the scale at 0. The values on the y-axis need to include values from 1500 to 4000. You can include 0 and use units of 500.

A jagged line is used to represent the values prior to 1986 that have been omitted.

c. While the number of crocodiles that could be exported rose in 1988, it has decreased since 1991. This may indicate that crocodiles in Indonesia are even more endangered.

The **inverse** of any relation is obtained by switching the coordinates in each ordered pair.

Relation	Inverse
(1, 4)	(4, 1)
(−3, 2)	(2, −3)
(7, −9)	(−9, 7)

Notice that the domain of the relation becomes the range of the inverse. Similarly, the range of the relation becomes the domain of the inverse.

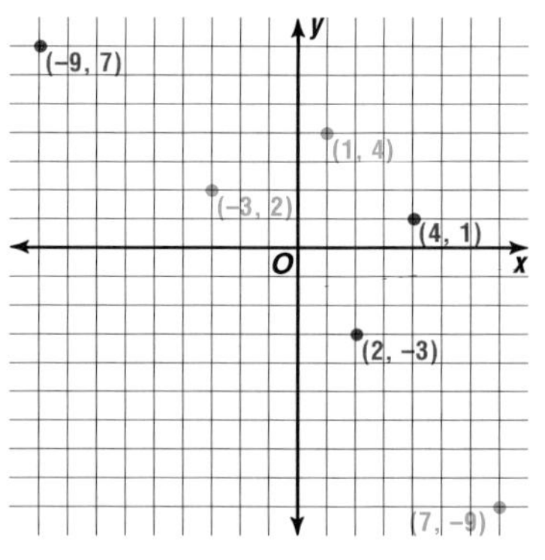

Alternative Teaching Strategies

Reading Algebra Have students look up the various definitions of *inverse*. Use these to help students develop a better understanding of the meaning of *inverse relation*.

Definition of the Inverse of a Relation	Relation Q is the inverse of relation S if and only if for every ordered pair (a, b) in S, there is an ordered pair (b, a) in Q.

Example 3 **Express the relation shown in the mapping as a set of ordered pairs. Write the inverse of the relation and draw a mapping to model the inverse.**

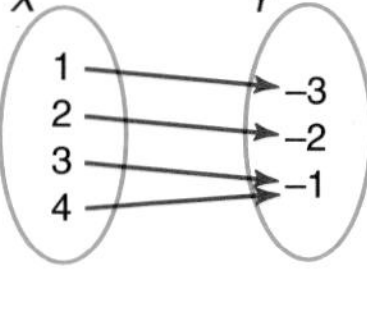

The mapping shows the relation {(1, −3), (2, −2), (3, −1), (4, −1)}.

The inverse of the relation is {(−3, 1), (−2, 2), (−1, 3), (−1, 4)} .

The mapping at the right shows the inverse.

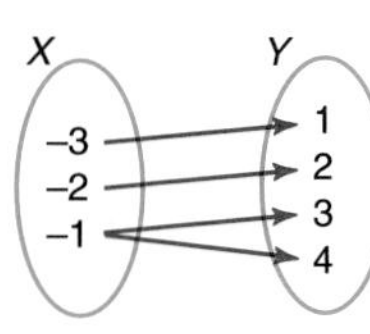

Relations and their inverses display some interesting characteristics, as you will discover below.

MODELING MATHEMATICS — Relations and Inverses

Materials:  grid paper  straightedge  colored pencils

There is a special relationship between a relation and its inverse, which you can discover by paper folding. **c. Sample ordered pairs: (−2, −2), (−1, −1), (0, 0), (1, 1). For each (x, y), $x = y$.**

Step 1 Graph the relation {(4, 5), (−3, 6), (−5, −3), (4, −7), (5, 0), (0, −3)} on a coordinate plane using one color of pencil.

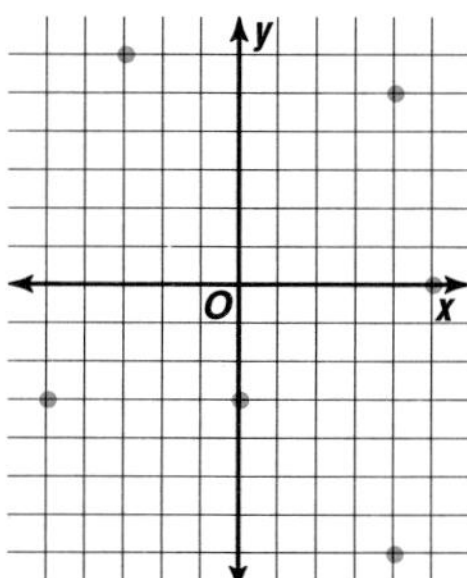

Connect the points in order using the same color pencil.

Step 2 Use a different color pencil to graph the inverse of the relation, connecting its points in order.

Step 3 Fold the graph paper through the origin so that the positive y-axis lies on top of the positive x-axis. Hold the paper up to a light so that you can see all of the points you graphed.

Your Turn

a. What do you notice about the location of the points you graphed when you looked at the folded paper? **See margin.**

b. Unfold the paper. Describe the pattern formed by the lines connecting the points in the relation and its inverse. **They are mirror images of each other.**

c. What do you think are the ordered pairs that represent the points on the fold line? Describe these in terms of x and y.

d. How could you graph the inverse of a function without writing ordered pairs first? **See margin.**

In-Class Example

For Example 3

Express the relation shown in the mapping below as a set of ordered pairs. Then write the inverse of this relation.

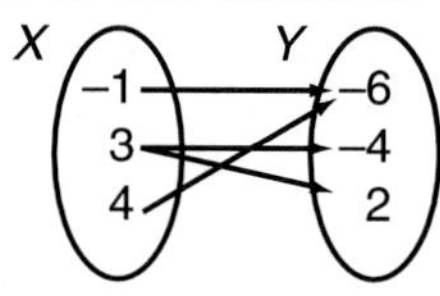

Relation = {(3, 2), (4, −6), (3, −4), (−1, −6)}
Inverse = {(2, 3), (−6, 4), (−4, 3), (−6, −1)}

MODELING MATHEMATICS This modeling activity introduces the idea that a relation and its inverse are symmetric with respect to each other about the line $y = x$, the diagonal.

Answers for Modeling Mathematics

a. The inverse of each point matches the point.

d. Draw the line through the ordered pairs (x, y) where $x = y$. Count the squares up/down until you meet the line. From that point, count the same number of squares right. Plot the point.

GLENCOE Technology

Interactive Mathematics Tools Software

This multimedia software provides an interactive lesson for determining a relation, given the domain and range of the relation. A **Computer Journal** gives students an opportunity to write about what they have learned.

For Windows & Macintosh

3 PRACTICE/APPLY

Check for Understanding

Exercises 1–15 are designed to help you assess your students' understanding through reading, writing, speaking, and modeling. You should work through Exercises 1–4 with your students and then monitor their work on Exercises 5–15.

Additional Answers

1. The domain values are the x values and the range values are the y values. If you do not know which is which, your plots will be incorrect.
2. After a high in 1990, the number of deaths seems to be decreasing.
4.

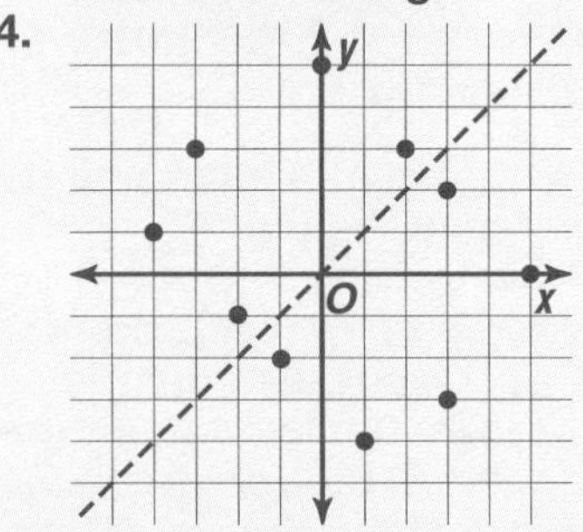

7. {(1, 3), (2, 4), (3, 5), (5, 7)}; D = {1, 2, 3, 5}; R = {3, 4, 5, 7}; Inv = {(3, 1), (4, 2), (5, 3), (7, 5)}
8. {1, 4), (3, −2), (4, 4), (6,−2)}; D = {1, 3, 4, 6}; R = {−2 , 4}; Inv = {(4, 1), (−2, 3), (4, 4), (−2, 6)}
9. {(1, 3), (2, 2), (4, 9), (6, 5)}; D = {1, 2, 4, 6}; R = {2, 3, 5, 9}; Inv = {(3, 1), (2, 2), (9, 4), (5, 6)}
10. {(−3, −2), (−2, −1), (0, 0), (1, 1)}; D = {−3, −2, 0, 1}; R = {−2, −1, 0, 1}; Inv = {(−2, −3), (−1, −2), (0, 0), (1, 1)}
11. {(−2, 2), (−1, 1), (0, 1), (1, 1), (1, −1), (2, −1), (3, 1)}; D = {−2, −1, 0, 1, 2, 3}; R = {−1, 1, 2}; Inv = {(2, −2), (1, −1), (1, 0), (1, 1), (−1, 1), (−1, 2), (1, 3)}

15c. Except for the first 6 months of 1992, the unemployment rate seems to be decreasing.

CHECK FOR UNDERSTANDING

Communicating Mathematics

3. The domain of the relation becomes the range of the inverse. The range becomes the domain of the inverse.

4. See margin.

Study the lesson. Then complete the following. 1–2. See margin.

1. **Explain** why it is important to identify the domain and range values of a relation when graphing the relation.
2. Refer to the application at the beginning of the lesson. Explain what the manatee graph suggests about patterns in the rate of manatee deaths.
3. **State** the relationship between the domain and range of a relation and the domain and range of its inverse.
4. Graph the relation {(0, 5), (2, 3), (1, −4), (−3, 3), (−1, −2)}. Draw a line that goes through the points (−3, −3) and (3, 3). Use this line to graph the inverse of the relation without writing the ordered pairs of the inverse.

Guided Practice

State the domain and range of each relation.

5. {(0, 2), (1, −2), (2, 4)}
   D = {0, 1, 2}; R = {2, −2, 4}
6. {(−4, 2), (−2, 0), (0, 2), (2, 4)}
   D = {−4, −2, 0, 2}; R = {2, 0, 4}

Express the relation shown in each table, mapping, or graph as a set of ordered pairs. Then state the domain, range, and inverse of the relation. 7–11. See margin.

7.

x	y
1	3
2	4
3	5
5	7

8.

x	y
1	4
3	−2
4	4
6	−2

9.

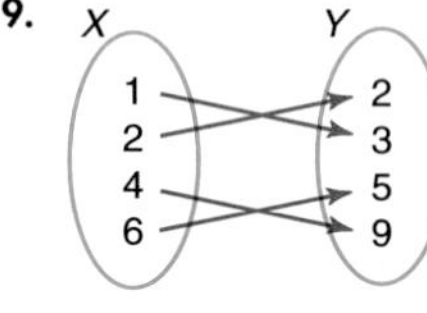

10.

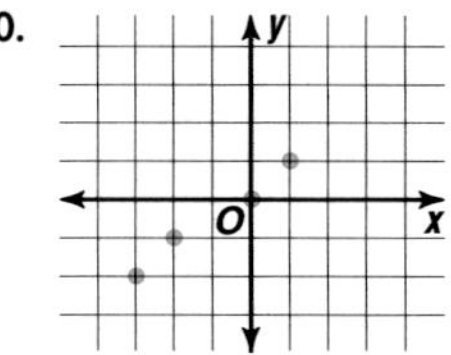

11.

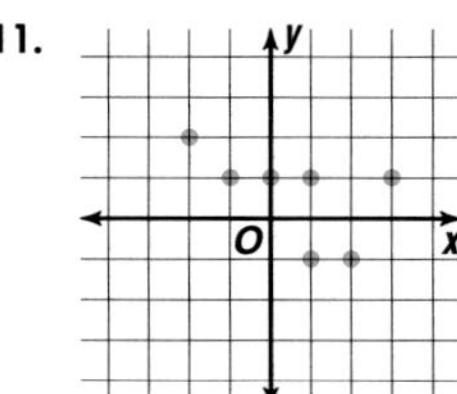

12–14. Graph the relations shown in Exercises 7–9. See Solutions Manual.

15. **Economics** The graph at the right shows the unemployment rate in the United States from January 1992 to December 1994.
    a. Name three ordered pairs shown by the graph.
    b. Estimate the least value in the range. 5.6%
    c. What conclusions might you make from the graph? See margin.

    15a. See students' work.

U.S. Unemployment Rate 1992–1994

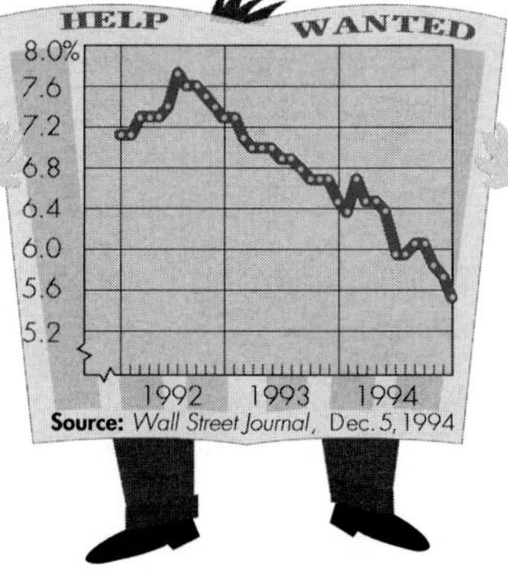

Reteaching

Using Real-World Examples

Relations in which elements of a first set are linked to elements of a second set are common in the real world. For example, names of students → homeroom, names of U.S. citizens → Social Security number. Ask students to name other relations and identify the domain and range of each.

EXERCISES

Practice 

State the domain and range of each relation.

16. $\{(1, 3), (2, 5), (1, -7), (2, 9), (3, 3)\}$ **D = {1, 2, 3}; R = {−7, 3, 5, 9}**
17. $\{(1, 7), (-2, 7), (3, 7), (-5, 7)\}$ **D = {−5, −2, 1, 3}, R = {7}**
18. $\{3.1, -1), (-4.7, 3.9), (2.4, -3.6), (-9, 2)\}$
19. $\left\{\left(\frac{1}{2}, \frac{1}{4}\right), \left(1\frac{1}{2}, -\frac{2}{3}\right), \left(-3, \frac{2}{5}\right), \left(-5\frac{1}{4}, -6\frac{2}{7}\right)\right\}$

18. $D = \{-9, -4.7, 2.4, 3.1\}$; $R = \{-3.6, -1, 2, 3.9\}$

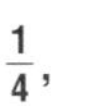

19. $D = \left\{-5\frac{1}{4}, -3, \frac{1}{2}, 1\frac{1}{2}\right\}$; $R = \left\{-6\frac{2}{7}, -\frac{2}{3}, \frac{1}{4}, \frac{2}{5}\right\}$

Express the relation shown in each table, mapping, or graph as a set of ordered pairs. Then state the domain, range, and inverse of the relation. **20–28. See margin.**

20.

x	y
0	4
1	5
2	6
3	6

21.

x	y
6	4
4	−2
3	4
1	−2

22.

x	y
−4	2
−2	0
0	2
2	4

23. 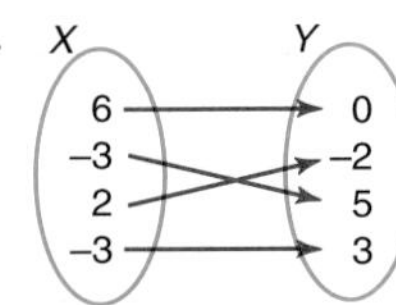

24.

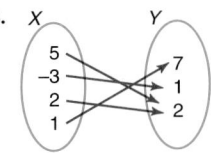

25. 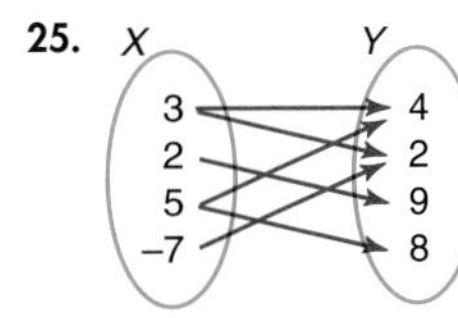

26. temperature of boiled water as it cools

time (minutes)	0	5	10	15	20	25	30
temperature (°C)	100	90	81	73	66	60	55

27. cost of car repairs

time (hours)	0	1	2	3
cost (dollars)	25	50	75	100

28. distance traveled at a rate of 55 mph

time (hours)	1.25	3.75	4.5	5.5	6
distance (miles)	68.75	206.25	247.5	302.5	330

29–34. See Solutions Manual.

29. 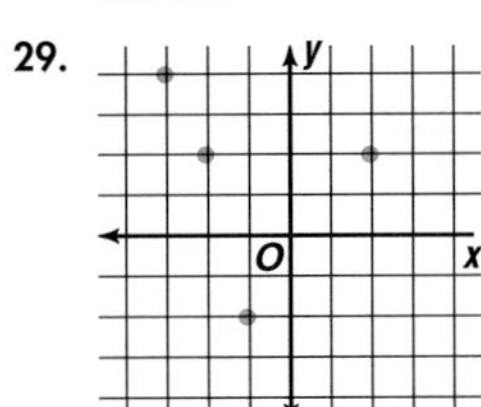

30. 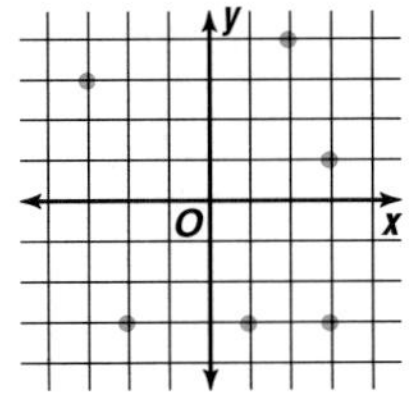

31. 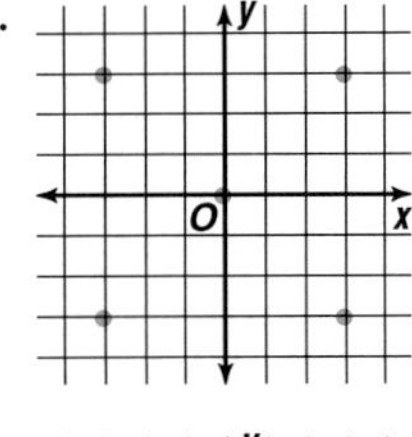

32. 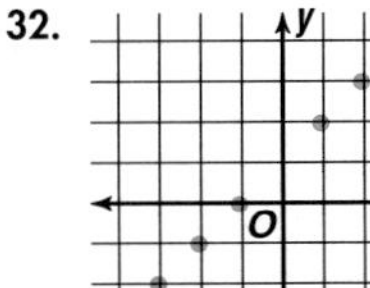

33. 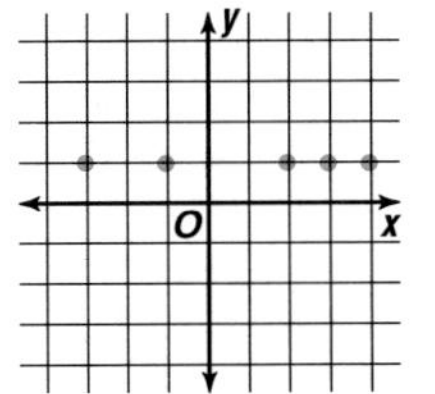

34. 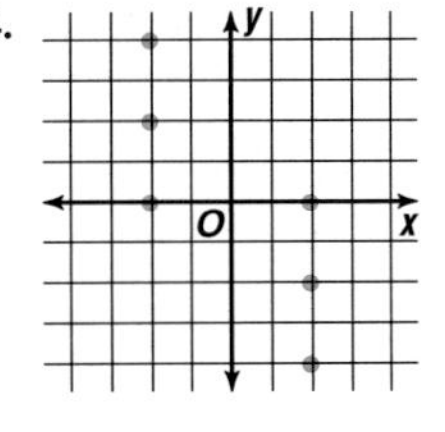

Assignment Guide

Core: 17–41 odd, 42, 43, 45, 46–53
Enriched: 16–40 even, 42–53

For **Extra Practice,** see p. 767.

The red A, B, and C flags, printed only in the Teacher's Wraparound Edition, indicate the level of difficulty of the exercises.

Additional Answers

20. {(0, 4), (1, 5), (2, 6), (3, 6)}; D = {0, 1, 2, 3}; R = {4, 5, 6}; Inv = {(4, 0), (5, 1), (6, 2), (6, 3)}
21. {(6, 4), (4, −2), (3, 4), (1, −2)}; D = {1, 3, 4, 6}; R = {−2, 4}; Inv = {(4, 6), (−2, 4), (4, 3), (−2, 1)}
22. {(−4, 2), (−2, 0), (0, 2), (2, 4)}; D = {−4, −2, 0, 2}; R = {0, 2, 4}; Inv = {(2, −4), (0, −2), (2, 0), (4, 2)}
23. {(6, 0), (−3, 5), (2, −2), (−3, 3)}; D = {−3, 2, 6}; R = {−2, 0, 3, 5}; Inv = {(0, 6), (5, −3), (−2, 2), (3, −3)}
24. {(5, 2), (−3, 1), (2, 2), (1, 7)}; D = {−3, 1, 2, 5}; R = {1, 2, 7}; Inv = {(2, 5), (1, −3), (2, 2), (7, 1)}
25. {(3, 4), (3, 2), (2, 9), (5, 4), (5, 8), (−7, 2)}; D = {−7, 2, 3, 5 }; R = { 2, 4, 8, 9}; Inv = {(4, 3), (2, 3), (9, 2), (4, 5), (8, 5), (2, −7)}

Study Guide Masters, p. 35

NAME__________________ DATE ________

Study Guide

Student Edition Pages 262–269

Relations

A **relation** is a set of ordered pairs. The **domain** of a relation is the set of all first coordinates of the ordered pairs, and the **range** is the set of all second coordinates.

Example 1: State the domain and range of each relation.
1. {(3, 3), (3, 4), (3, 5)} Domain = {3}; Range = {3, 4, 5}
2. {(1, 2), (2, 1), (3, 2)} Domain = {1, 2, 3}; Range = {1, 2}

Relations may be expressed in the form of ordered pairs, tables, graphs, and mappings.

Example 2: The relation {(1, 1), (0, 2), (3, −2)} can be expressed in each of the following ways.

Ordered pairs
(1, 1)
(0, 2)
(3, −2)

Table

x	y
1	1
0	2
3	−2

Graph

Mapping

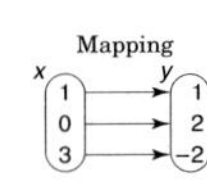

The **inverse** of any relation is obtained by switching the coordinates in each ordered pair.

State the domain and range of each relation.

1. {(−6, 5), (−3, 8), (−6, 9), (3, 11)} D = {−6, −3, 3}; R = {5, 8, 9, 11}
2. {(0.8, −0.8), (1.2, 0), (3.5, 4)} D = {0.8, 1.2, 3.5}; R = {−0.8, 0, 4}
3. $\left\{\left(\frac{1}{2}, \frac{1}{4}\right), \left(1\frac{1}{2}, 1\frac{1}{4}\right), \left(3\frac{1}{2}, 2\right)\right\}$ $D = \left\{\frac{1}{2}, 1\frac{1}{2}, 3\frac{1}{2}\right\}$; $R = \left\{\frac{1}{4}, 1\frac{1}{4}, 2\right\}$

Express the relations shown in each table, mapping, or graph as a set of ordered pairs. Then state the domain, range, and inverse of the relation.

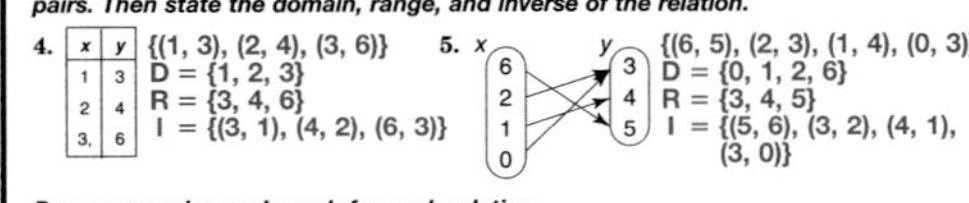

4.

x	y
1	3
2	4
3	6

{(1, 3), (2, 4), (3, 6)}
D = {1, 2, 3}
R = {3, 4, 6}
I = {(3, 1), (4, 2), (6, 3)}

5. {(6, 5), (2, 3), (1, 4), (0, 3)}
D = {0, 1, 2, 6}
R = {3, 4, 5}
I = {(5, 6), (3, 2), (4, 1), (3, 0)}

Draw a mapping and graph for each relation.

6. {(−2, −1), (3, 3), (4, 3)}

7. {(0, 0), (1, 1), (2, 2)}

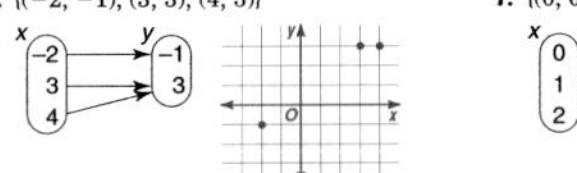

Additional Answers

26. {(0, 100), (5, 90), (10, 81), (15, 73), (20, 66), (25, 60), (30, 55)}; D = {0, 5, 10, 15, 20, 25, 30}; R = {55, 60, 66, 73, 81, 90, 100}; Inv = {(100, 0), (90, 5), (81, 10), (73, 15), (66, 20), (60, 25), (55, 30)}
27. {(0, 25), (1, 50), (2, 75), (3, 100)}; D = {0, 1, 2, 3}; R = {25, 50, 75, 100}; Inv = {(25, 0), (50, 1), (75, 2), (100, 3)}
28. {(1.25, 68.75), (3.75, 206.25), (4.5, 247.5), (5.5, 302.5), (6, 330)}; D = {1.25, 3.75, 4.5, 5.5, 6}; R = {68.75, 206.25, 247.5, 302.5, 330}; Inv = {(68.75, 1.25), (206.25, 3.75), (247.5, 4.5), (302.5, 5.5), (330, 6)}

Additional Answers

45a.

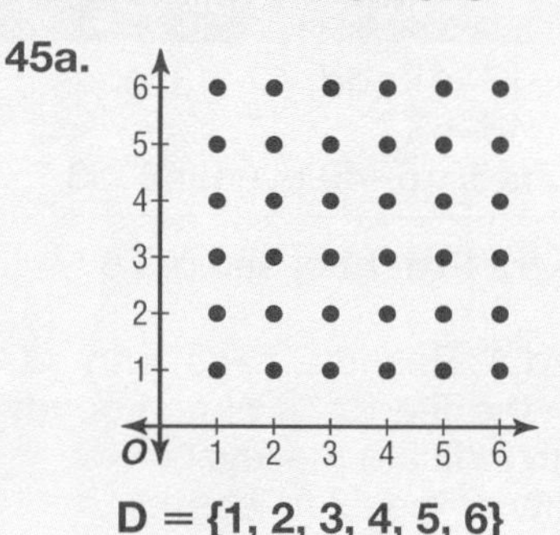

D = {1, 2, 3, 4, 5, 6}
R = {1, 2, 3, 4, 5, 6}

45c. 11 possible sums

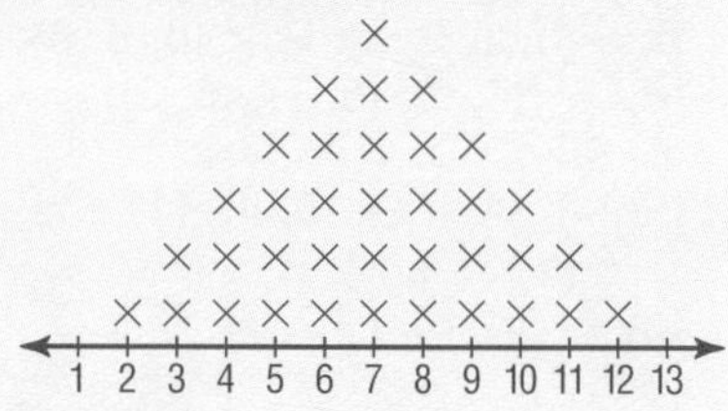

45d.

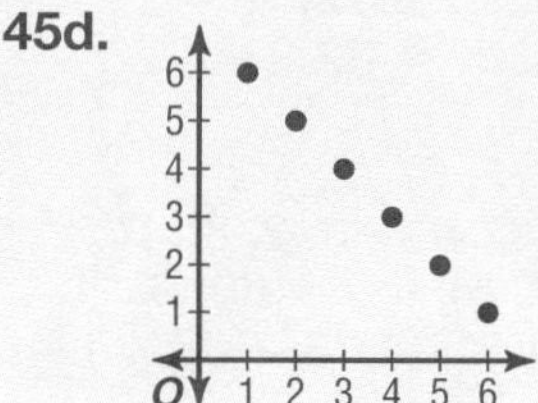

Practice Masters, p. 35

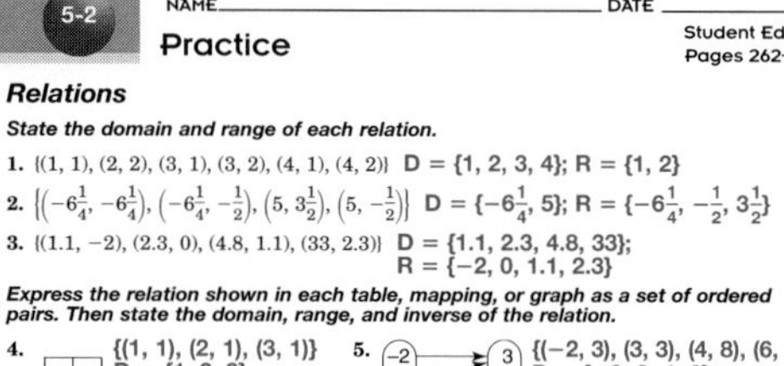

5-2 NAME________________ DATE________

Practice — Student Edition Pages 262–269

Relations

State the domain and range of each relation.

1. $\{(1, 1), (2, 2), (3, 1), (3, 2), (4, 1), (4, 2)\}$ D = {1, 2, 3, 4}; R = {1, 2}
2. $\left\{\left(-6\frac{1}{4}, -6\frac{1}{4}\right), \left(-6\frac{1}{4}, -\frac{1}{2}\right), \left(5, 3\frac{1}{2}\right), \left(5, -\frac{1}{2}\right)\right\}$ D = $\{-6\frac{1}{4}, 5\}$; R = $\{-6\frac{1}{4}, -\frac{1}{2}, 3\frac{1}{2}\}$
3. $\{(1.1, -2), (2.3, 0), (4.8, 1.1), (33, 2.3)\}$ D = {1.1, 2.3, 4.8, 33}; R = {−2, 0, 1.1, 2.3}

Express the relation shown in each table, mapping, or graph as a set of ordered pairs. Then state the domain, range, and inverse of the relation.

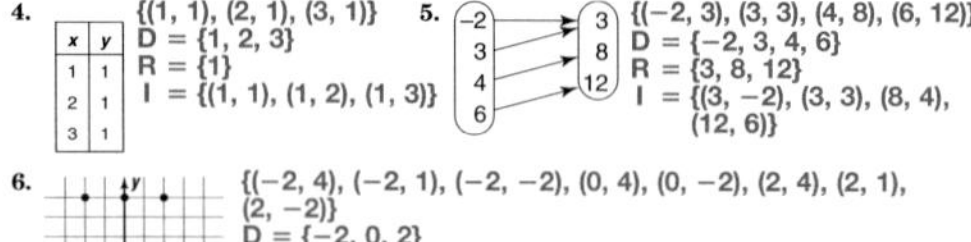

4.

x	y
1	1
2	1
3	1

{(1, 1), (2, 1), (3, 1)}
D = {1, 2, 3}
R = {1}
I = {(1, 1), (1, 2), (1, 3)}

5. {(−2, 3), (3, 3), (4, 8), (6, 12)}
D = {−2, 3, 4, 6}
R = {3, 8, 12}
I = {(3, −2), (3, 3), (8, 4), (12, 6)}

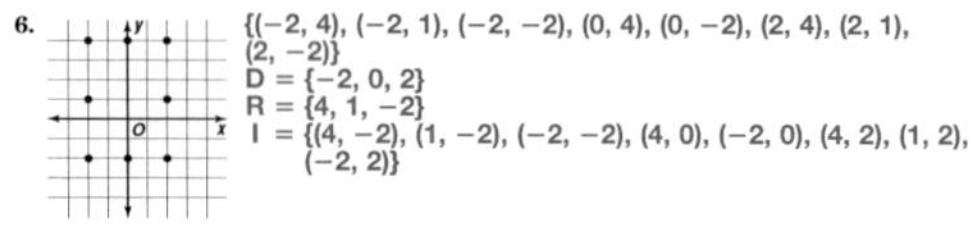

6. {(−2, 4), (−2, 1), (−2, −2), (0, 4), (0, −2), (2, 4), (2, 1), (2, −2)}
D = {−2, 0, 2}
R = {4, 1, −2}
I = {(4, −2), (1, −2), (−2, −2), (4, 0), (−2, 0), (4, 2), (1, 2), (−2, 2)}

7. Cost of Admission to Water World

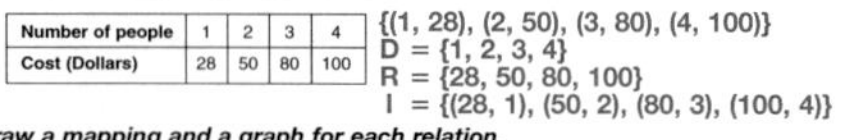

Number of people	1	2	3	4
Cost (Dollars)	28	50	80	100

{(1, 28), (2, 50), (3, 80), (4, 100)}
D = {1, 2, 3, 4}
R = {28, 50, 80, 100}
I = {(28, 1), (50, 2), (80, 3), (100, 4)}

Draw a mapping and a graph for each relation.

8. $\left\{\left(\frac{1}{2}, \frac{1}{4}\right), \left(0, \frac{1}{2}\right), \left(2\frac{1}{2}, 3\right)\right\}$

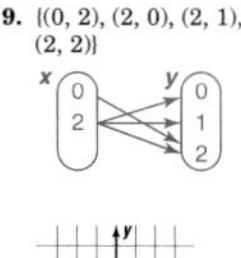

9. $\{(0, 2), (2, 0), (2, 1), (2, 2)\}$

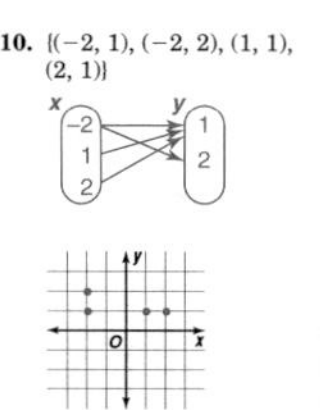

10. $\{(-2, 1), (-2, 2), (1, 1), (2, 1)\}$

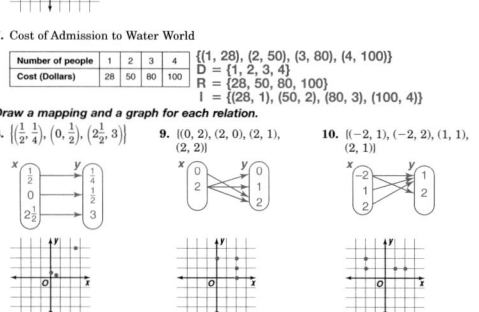

Draw a mapping and a graph for each relation.

35–37. See Solutions Manual.

35. $\{(8, 2), (4, 2), (8, -9), (7, 5), (-3, 2)\}$
36. $\{(3, 3), (2, 7), (-3, 3), (1, 3), (4, 1)\}$
37. $\{(6, 0), (6, -4), (4, -3), (5, -3)\}$

Graphing Calculator

Graph each relation on a graphing calculator.

a. **State the WINDOW settings that you used.**
b. **Write the coordinates of the inverse. Then graph the inverse.**
c. **Name the quadrants in which each point of the relation and its inverse lies.**

38–41. See Solutions Manual.

38.

x	y
1992	77
1993	200
1994	550
1995	880

39.

x	y
0	10
2	−8
6	6
9	−4

40.

x	y
−1	18
−2	23
−3	28
−4	33

41. Review your answers to part c of Exercises 38–40. What conclusions can you draw about the quadrant in which the inverse of a point will lie?

Critical Thinking

42. Write a relation with five ordered pairs such that the relation is its own inverse. Then describe the graph of the relation and its inverse. **Relations will vary. Sample answer: {(5, 3), (3, 5), (2, 2), (−4, −2) (−2, −4)}; the graphs are the same.**

Applications and Problem Solving

43. **Retail Sales** The graph at the right shows U.S. retail sales in billions of dollars. The amounts for each month have been adjusted to account for seasonal increases in sales.
a. Approximate the least and greatest values in the range.
b. In your own words, describe what the pattern formed by the points represents.
c. Compare this graph with the one in Exercise 15. How do you think unemployment rates might be compared with retail spending trends? Explain.

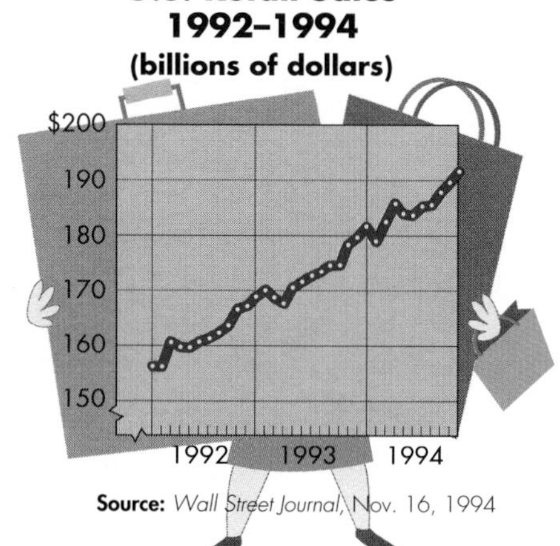

43. Sample answers:

43a. 157 billion, 191 billion

43b. Retail sales have increased from 1992 to 1994.

43c. As unemployment decreases, retail sales increase because people have more money to spend.

44. **Make a Table** Mandy Chin is saving money to pay for the first year of her car insurance, which will cost about \$1200. She already has \$500 in her savings account. She gets a job working at a fast-food restaurant and plans to save \$45 per week.
a. Make a table to show the number of weeks and how much money Mandy will have in her account after each week. For example, on Week 1, she will have \$545. Graph the results. **See Solutions Manual.**
b. When will she have enough money to pay her insurance premium?

44b. after 16 weeks

Extension

Reasoning A relation has domain $A = \{2, 4, 6\}$ and range $B = \{1, 3\}$. The Cartesian product $A \times B$ is defined as the relation consisting of all ordered pairs (a, b), where a is a member of A and b is a member of B.

1. What is the relation $A \times B$? **{(2, 1), (2, 3), (4, 1), (4, 3), (6, 1), (6, 3)}**
2. How many elements are contained in $A \times B$? **6**
3. Suppose the domain A contains 213 elements and range B contains 432 elements. How many elements are contained in $A \times B$? **213 × 432 = 92,016 ordered pairs**

45. **Probability** There are 36 different possible outcomes when you roll two dice. The outcomes can be expressed as ordered pairs, such as (4, 3).
   a. Graph all 36 possible outcomes. What is the domain and range? **See margin.**
   b. What is the domain and range of the inverse? What do you notice about the relation and its inverse?
   c. How many different sums are possible when two dice are rolled? Draw a line plot showing how many ways each sum can be rolled. Describe your line plot. **See margin.**
   d. Graph the outcomes that have a sum of 7. **See margin.**
   e. What is the probability of rolling a sum of 7? Explain.

45b. D = R = {1, 2, 3, 4, 5, 6}; relation = inverse

45e. $\frac{6}{36}$; There are 6 out of 36 ways to roll a sum of 7.

Mixed Review

46. Graph $A(6, 2)$, $B(-3, 6)$, and $C(-5, -4)$ on the same coordinate plane. (Lesson 5–1) **See margin.**
47. If y varies inversely as x, and $y = 8$ when $x = 24$, find y when $x = 6$. (Lesson 4–8) **32**

48. 4 adult, 12 children

48. **Entertainment** At Backintime Cinema, tickets for adults cost \$5.75 and tickets for children cost \$3.75. How many of each kind of ticket were purchased if 16 tickets were bought for \$68? (Lesson 4–7)
49. Use $I = prt$ to find I if $p = \$8000$, $r = 6\%$, and $t = 1$ year. (Lesson 4–4) **\$480**
50. Solve $E = mc^2$ for m. (Lesson 3–6) $m = \frac{E}{c^2}$
51. **Space Science** Halley's Comet flashes through the sky every 76.3 years. It last appeared in 1986. In what year of the 23rd century is Halley's Comet expected to reappear? (Lesson 2–6) **2214, 2290**
52. **Agriculture** Refer to the graph in Exercise 50 on page 259. (Lesson 1–9)
   a. What is the normal minimum temperature for March? **27°F**
   b. What happens to the temperature after July? **It continually decreases.**
53. Simplify $4[1 + 4(5x + 2y)]$. (Lesson 1–8) $4 + 80x + 32y$

WORKING ON THE

In•ves•ti•ga•tion

Refer to the Investigation on pages 190–191.

In tracking populations of fish, the Department of Natural Resources and Wildlife often chart their findings in a graph. They use these graphs to monitor the populations of the different types of fish. In this way they track which species are on the increase and which are in decline. This can be especially important in monitoring species that are rare or in danger of extinction.

1. Use the information in your chart to make a table of ordered pairs (x, y) such that x represents the number of the casting and y represents your estimate for the number of fish in the lake.
2. Draw a mapping for this relation.
3. Graph the ordered pairs on a coordinate grid. Describe the graph.

Add the results of your work to your Investigation Folder.

In•ves•ti•ga•tion

Working on the Investigation

The Investigation on pages 190–191 is designed to be a long-term project that is completed over several days or weeks. Encourage students to keep their materials in their Investigation Folder as they work on the Investigation.

4 ASSESS

Closing Activity

Speaking Ask students to describe the terms *relation*, *domain*, *range*, and *inverse* in their own words. Then have them identify any three ordered pairs and apply each term.

Chapter 5, Quiz A (Lessons 5-1 and 5-2), is available in the *Assessment and Evaluation Masters*, p. 128.

Additional Answers

46.

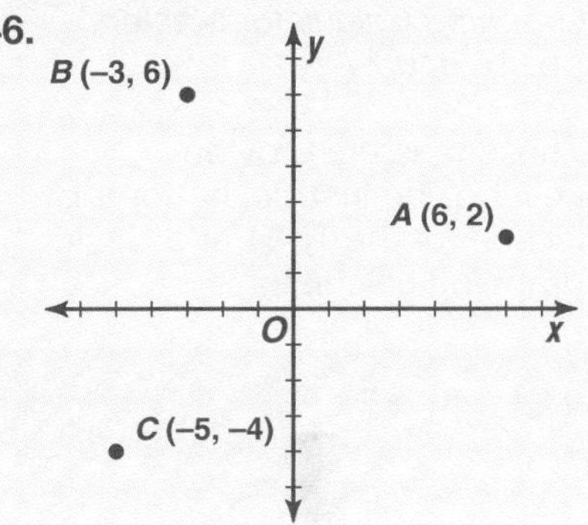

Enrichment Masters, p. 35

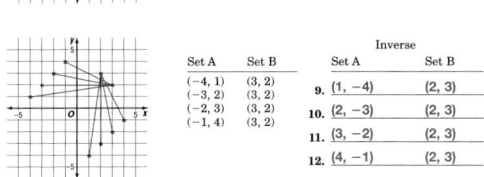

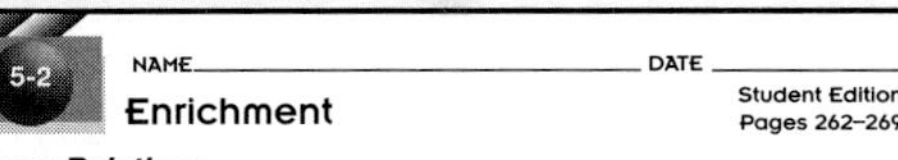

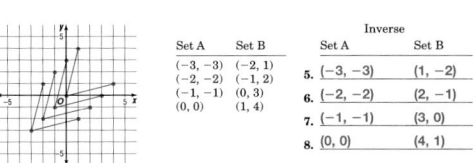

5-2 NAME ____________ DATE ________

Enrichment

Student Edition Pages 262–269

Inverse Relations

On each grid below, plot the points in Sets A and B. Then connect the points in Set A with the corresponding points in Set B. Then find the inverses of Set A and Set B, plot the two sets, and connect those points.

Set A	Set B		Inverse Set A	Inverse Set B
(−4, 0)	(0, 1)	1.	(0, −4)	(1, 0)
(−3, 0)	(0, 2)	2.	(0, −3)	(2, 0)
(−2, 0)	(0, 3)	3.	(0, −2)	(3, 0)
(−1, 0)	(0, 4)	4.	(0, −1)	(4, 0)

Set A	Set B		Inverse Set A	Inverse Set B
(−3, −3)	(−2, 1)	5.	(−3, −3)	(1, −2)
(−2, −2)	(−1, 2)	6.	(−2, −2)	(2, −1)
(−1, −1)	(0, 3)	7.	(−1, −1)	(3, 0)
(0, 0)	(1, 4)	8.	(0, 0)	(4, 1)

Set A	Set B		Inverse Set A	Inverse Set B
(−4, 1)	(3, 2)	9.	(1, −4)	(2, 3)
(−3, 2)	(3, 2)	10.	(2, −3)	(2, 3)
(−2, 3)	(3, 2)	11.	(3, −2)	(2, 3)
(−1, 4)	(3, 2)	12.	(4, −1)	(2, 3)

13. What is the graphical relationship between the line segments you drew connecting points in Sets A and B and the line segments connecting points in the inverses of those two sets? **Answers will vary. A possible answer is that the graphs are reflected across the line $x = y$ by their inverses.**

5-3A LESSON NOTES

NCTM Standards: 1–5

Objective
Use a graphing calculator to investigate relations and determine the ranges when the domains are given.

Recommended Time
15 minutes

Instructional Resources
Graphing Calculator Masters, pp. 16 and 17

These masters provide keystroking instruction for this lesson for the TI-81 and Casio graphing calculators.

1 FOCUS

Motivating the Lesson
You may want to have students review the use of the GRAPH feature on the calculator.

2 TEACH

Teaching Tip Encourage students to check their solutions by replacing the variable with the *x* values and computing the *y* values by hand.

3 PRACTICE/APPLY

Assignment Guide

Core: 1–4
Enriched: 1–4

4 ASSESS

Observing students working with technology is an excellent method of assessment.

5–3A Graphing Technology Equations

A Preview of Lesson 5–3

Ordered pairs can be used to represent solutions to equations in two variables. Thus, the solution set of an equation for a defined domain is a relation. We can use a graphing calculator to investigate the relation and determine the range when the domain is given.

Example **Solve $y = 2.5x + 4$ if the domain is {−8, −6, −4, −2, 0, 2, 4, 6, 8}.**

Method 1

Store the elements of the domain as a list of values in L1.

Enter: 2nd { (−) 8 , (−) 6 , (−) 4 , (−) 2 , 0 , 2 , 4 , 6 , 8 2nd } STO▸ 2nd L1 ENTER {-8 -6 -4 -2 0...

To calculate the range values, compute the right-hand side of the equation, substituting L1 for *x*.

Enter: 2.5 2nd L1 + 4 ENTER {-16 -11 -6 -1...

Press the arrow keys to scroll right or left through the display to see all the values in the range.

The solution set is {(−8, −16), (−6, −11), (−4, −6), (−2, −1), (0, 4), (2, 9), (4, 14), (6, 19), (8, 24)}.

Method 2

Store the equation in the Y= menu of the calculator and view the table of values it creates.

Enter: Y= 2.5 X,T,θ + 4 2nd TblSet (−) 8 ENTER 2 2nd TABLE

ΔTbl means the increments in which the x values will appear in the table.

A table with columns X and Y_1 appears on the screen. The values in the Y_1 column represent the corresponding range values for each *x* value in the domain. This table helps you find the ordered pairs that are solutions for the equation. The solution set is {(−8, −16), (−6, −11), (−4, −6), (−2, −1), (0, 4), (2, 9), (4, 14), (6, 19), (8, 24)}, which agrees with the list in Method 1.

EXERCISES

Use a graphing calculator to solve each equation if the domain is {−3, −2, −1, 0, 1, 2, 3}. **1–4. See margin.**

1. $y = 4x - 7$
2. $y = x^2 + 11$
3. $1.2x - y = 6.8$
4. $6x + 2y = 12$

Additional Answers

1. {(−3, −19), (−2, −15), (−1, −11), (0, −7), (1, −3), (2, −1), (3, 5)}
2. {(−3, 20), (−2, 15), (−1, 12), (0, 11), (1, 12), (2, 15), (3, 20)}
3. {(−3, −10.4), (−2, −9.2), (−1, −8), (0, −6.8), (1, −5.6), (2, −4.4), (3, −3.2)}
4. {(−3, 15), (−2, 12), (−1, 9), (0, 6), (1, 3), (2, 0), (3, −3)}

Using Technology
This lesson offers an excellent opportunity for using technology in your algebra classroom. For more information on using technology, see *Graphing Calculators in the Mathematics Classroom,* one of the titles in the Glencoe Mathematics Professional Series.

5-3 Equations as Relations

What YOU'LL LEARN
- To determine the range for a given domain, and
- to graph the solution set for the given domain.

Why IT'S IMPORTANT

You can use equations to explore the relations in physical science and anatomy.

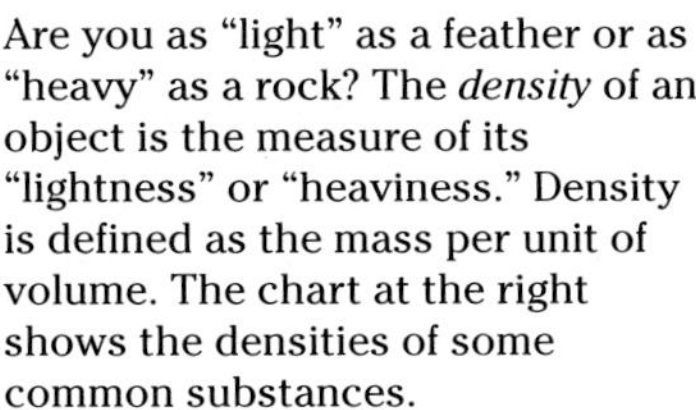

CONNECTION Physical Science

Are you as "light" as a feather or as "heavy" as a rock? The *density* of an object is the measure of its "lightness" or "heaviness." Density is defined as the mass per unit of volume. The chart at the right shows the densities of some common substances.

Substance	Density (g/cm^3)
air	0.0013
aluminum	2.7
gasoline	0.7
gold	19.3
lead	11.3
mercury	13.5
silver	10.5
steel	7.8
water	1.0

FYI Water has a density of 1.0. Any substance with a density less than 1.0 will float in water. Any substance with a density greater than 1.0 will sink in water.

An equation relating density, mass, and volume is $m = DV$, where m is the mass of a substance, D is the density, and V is the volume. For example, the equation for gold is $m = 19.3V$ because the density of gold is 19.3 g/cm^3. This equation is called an **equation in two variables,** m and V.

No one knows exactly when gold was first discovered. However, there have been gold artifacts dug up in Ur in Mesopotamia (modern day Iraq) that date as early as 3500 B.C. Suppose we wanted to know the mass of four gold artifacts with volumes of 10 cm^3, 50 cm^3, 100 cm^3, and 150 cm^3. We can use a table to find ordered pairs (V, m) that satisfy the equation.

V	$D \cdot V$	m	Ordered Pair
10	19.3(10)	193	(10, 193)
50	19.3(50)	965	(50, 965)
100	19.3(100)	1930	(100, 1930)
150	19.3(150)	2895	(150, 2895)

These four ordered pairs are graphed at the right. Since each ordered pair satisfies the equation, each ordered pair is a solution of the equation.

Definition of the Solution of an Equation in Two Variables	If a true statement results when the numbers in an ordered pair are substituted into an equation in two variables, then the ordered pair is a solution of the equation.

Since the solutions of an equation in two variables are ordered pairs, such an equation describes a relation. In an equation involving x and y, the set of x values is the domain of the relation. The set of corresponding y values is the range of the relation.

FYI

The densities of some other common materials include the following.

cork	0.24
wood (elm)	0.60
ice	0.92
human body	1.07

Which of the substances will float in water?

5-3 LESSON NOTES

NCTM Standards: 1–5

Instructional Resources

- Study Guide Master 5-3
- Practice Master 5-3
- Enrichment Master 5-3
- Multicultural Activity Masters, p. 10

Transparency 5-3A contains the 5-Minute Check for this lesson; **Transparency 5-3B** contains a teaching aid for this lesson.

Recommended Pacing	
Standard Pacing	Day 5 of 14
Honors Pacing	Day 5 of 13
Block Scheduling*	Day 3 of 7
Alg. 1 in Two Years*	Days 7 & 8 of 21

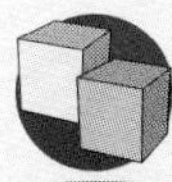

*For more information on pacing and possible lesson plans, refer to the *Block Scheduling Booklet* and *Algebra 1 in Two Years.*

1 FOCUS

5-Minute Check *(over Lesson 5-2)*

State the domain and range of each relation.

1. {(0, −3), (1, 4), (1, −3)} **D = {0, 1}, R = {−3, 4}**
2. {(1, 5), (2, 5), (3, 5), (4, 5)} **D = {1, 2, 3, 4}, R = {5}**
3. Show the inverse of the relation in Exercise 1 by using a table.

x	y
−3	0
4	1
−3	1

4. A relation contains five ordered pairs. If its domain is {2, 4, 6, 8, 10} and each y-coordinate is three more than the x-coordinate, then what are the ordered pairs in the relation? **{(2, 5), (4, 7), (6, 9), (8, 11), (10, 13)}**

Motivating the Lesson

Hands-On Activity Have students work in pairs. Provide each pair with an ample amount of play money. One member of each pair should give a random amount of money to the other member. The second student computes 15% of the amount and returns the original amount along with the interest to the first student. Ask students to think about the equation involved in the transaction.

2 TEACH

In-Class Examples

For Example 1

a. Solve $y = -\frac{1}{3}x$ if the domain is $\{-3, 0, 2, 6, 8\}$. **The solution set is $\left\{(-3, 1), (0, 0), \left(2, -\frac{2}{3}\right), (6, -2), \left(8, -\frac{8}{3}\right)\right\}$.**

b. Solve $y = -2x + 1$ if the domain is $\{-4, -2, 0, 2, 4\}$. **The solution set is $\{(-4, 9), (-2, 5), (0, 1), (2, -3), (4, -7)\}$.**

For Example 2

Solve $2x - 3y = 12$ if the domain is $\{-9, -6, -3, 0, 3, 6, 9\}$. **The solution set is $\{(-9, -10), (-6, -8), (-3, -6), (0, -4), (3, -2), (6, 0), (9, 2)\}$.**

Teaching Tip Transforming equations, as in Example 2, provides good practice in applying the properties of solving equations.

Teaching Tip Make sure students realize that the domain and range are not *x*'s and *y*'s. The domain and range are sets of numbers from which values for a variable are chosen. Letters other than *x* and *y* can be used for the variables.

Example 1 **Solve each equation for the given domain values. Graph the solution set.**

a. **$y = 4x$ if the domain is $\{-3, -2, 0, 1, 2\}$**

Make a table. The values of *x* come from the domain. Substitute each value of *x* into the equation to determine the corresponding values of *y*.

Domain x	$4x$	Range y	Ordered Pair (x, y)
-3	$4(-3)$	-12	$(-3, -12)$
-2	$4(-2)$	-8	$(-2, -8)$
0	$4(0)$	0	$(0, 0)$
1	$4(1)$	4	$(1, 4)$
2	$4(2)$	8	$(2, 8)$

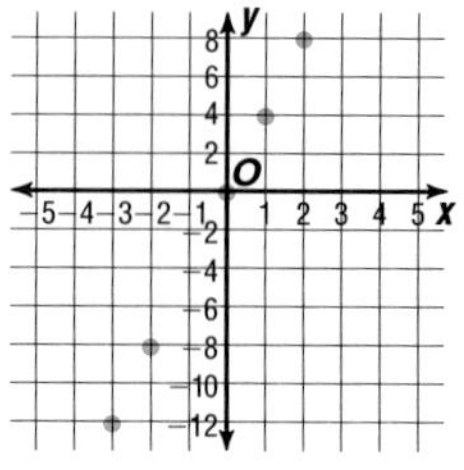

Then graph the solution set $\{(-3, -12), (-2, -8), (0, 0), (1, 4), (2, 8)\}$.

b. **$y = x + 6$ if the domain is $\{-4, -3, -1, 2, 4\}$**

Make a table. Find the values of *y* in the range by substituting the values of *x* from the domain into the equation.

x	$x + 6$	y	(x, y)
-4	$-4 + 6$	2	$(-4, 2)$
-3	$-3 + 6$	3	$(-3, 3)$
-1	$-1 + 6$	5	$(-1, 5)$
2	$2 + 6$	8	$(2, 8)$
4	$4 + 6$	10	$(4, 10)$

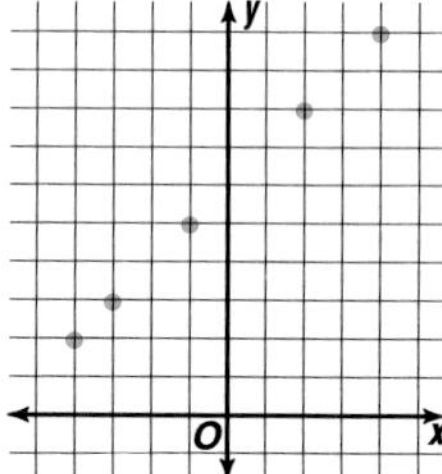

Then graph the solution set $\{(-4, 2), (-3, 3), (-1, 5), (2, 8), (4, 10)\}$.

It is often helpful to solve an equation for *y* before substituting each domain value into the equation. This makes creating a table of values easier.

Example 2 **Solve $8x + 4y = 24$ if the domain is $\{-2, 0, 5, 8\}$.**

First solve the equation for *y* in terms of *x*.

$$8x + 4y = 24$$
$$4y = 24 - 8x \quad \textit{Subtract 8x from each side.}$$
$$y = 6 - 2x \quad \textit{Divide each side by 4.}$$

Now substitute each value of *x* from the domain to determine the corresponding values of *y* in the range.

x	$6 - 2x$	y	(x, y)
-2	$6 - 2(-2)$	10	$(-2, 10)$
0	$6 - 2(0)$	6	$(0, 6)$
5	$6 - 2(5)$	-4	$(5, -4)$
8	$6 - 2(8)$	-10	$(8, -10)$

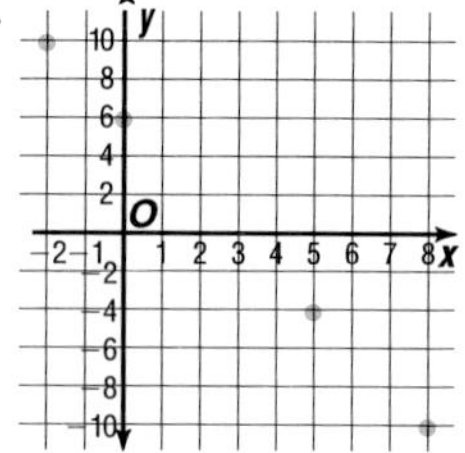

Then graph the solution set $\{(-2, 10), (0, 6), (5, -4), (8, -10)\}$.

Classroom Vignette

"When I teach my students to solve for *y*, I like the *x* value to be first, in slope-intercept form. So, in Example 2, when we solve for *y*, we would write $y = -2x + 6$. That way, when we get to that, the students do not have to relearn how to solve for *y*. It's already in the correct form."

Deborah Phillips
St. Vincent–St. Mary High School
Akron, Ohio

Deborah Patonai Phillips

EXPLORATION — GRAPHING CALCULATORS

You can enter selected x values in the TABLE feature of the graphing calculator, and it will calculate the corresponding y values for a given equation.

- Solve the equation for y and enter the equation into the Y= list.
- Press 2nd TblSet and use the arrow keys to go to Indpnt:. Highlight Ask.
- Press 2nd TABLE . Scroll up to the top of the X list. Enter the first value of the domain and press ENTER . The corresponding Y value appears in the second column. Enter the other values of the domain and record the range values.

Your Turn a. $\{(-4, 32), (-2, 22), (0, 12), (1, 7), (3, -3), (5, -13)\}$

a. Use a graphing calculator to find the corresponding y values for $y = -5x + 12$ if $x = \{-4, -2, 0, 1, 3, 5\}$. Write the solutions as ordered pairs.

b. Why do you think the Ask feature on the TABLE menu is helpful in finding the ordered pairs that are solutions for equations? **See margin.**

EXPLORATION

The object of this Exploration is to introduce students to the GRAPH and ASK features on the TABLE menu.

Answer for the Exploration

b. You can use it to find y values for specific x values that may not appear in the general table setup. It also saves time from scrolling through the TABLE list of values.

In-Class Example

For Example 3
Suppose a rectangle has a perimeter of 42 mm.

a. Solve the perimeter formula for w. $w = 21 - \ell$

b. State the independent and dependent variables. **w is the dependent variable, and ℓ is the independent variable.**

c. Choose five values for ℓ and find the corresponding values of w. **Answers will vary.**

LOOK BACK

You can look back to Lesson 1-9 to review independent and dependent variables.

Variables other than x and y are often used in equations representing real situations. The domain contains values represented by the *independent variable*. Graph the domain values on the horizontal axis. The range contains the corresponding values represented by the *dependent variable*, determined by the given equation. Graph the range values on the vertical axis.

When you solve an equation for a given variable, that variable becomes the dependent variable. That is, its value depends upon the domain values chosen for the other variable.

Example 3

INTEGRATION Geometry

The equation for the perimeter of a rectangle is $2w + 2\ell = P$. Suppose the perimeter of a rectangle is 24 centimeters.

a. Solve the equation for ℓ.

b. State the independent and dependent variables and determine the domain and range values for which this equation makes sense.

c. Choose five values for w and find the corresponding values of ℓ.

a. First substitute 24 for P in the equation. Then solve for ℓ.

$$2w + 2\ell = P$$
$$2w + 2\ell = 24$$
$$2\ell = 24 - 2w \quad \text{Subtract } 2w \text{ from each side.}$$
$$\frac{2\ell}{2} = \frac{24 - 2w}{2} \quad \text{Divide each side by 2.}$$
$$\ell = 12 - w$$

b. Since the value of ℓ depends on the value of w, ℓ is the dependent variable and w is the independent variable. Since distance can only be positive, the values in the domain and range must both be greater than zero.

(continued on the next page)

3 PRACTICE/APPLY

Check for Understanding

Exercises 1–12 are designed to help you assess your students' understanding through reading, writing, speaking, and modeling. You should work through Exercises 1–6 with your students and then monitor their work on Exercises 7–12.

Additional Answers

6. **The independent variable's values are chosen or assigned. The dependent variable's value varies with the value of the independent variable. The independent is graphed along the horizontal axis and the dependent along the vertical axis.**

11.

19. {(−3, −12), (−2, −8), (0, 0), (3, 12), (6, 24)}
20. {(−3, −10), (−2, −7), (0, −1), (3, 8), (6, 17)}
21. {(−3, 10), (−2, 9), (0, 7), (3, 4), (6, 1)}
22. {(−3, −7), (−2, −6), (0, −4), (3, −1), (6, 2)}
23. {(−3, 11), (−2, 9.5), (0, 6.5), (3, 2), (6, −2.5)}
24. {(−3, 19), (−2, 14), (0, 4), (3, −11), (6, −26)}
25. {(−3, −12), (−2, −7), (0, 3), (3, 18), (6, 33)}
26. {(−3, −5.5), (−2, −5), (0, −4), (3, −2.5), (6, −1)}
27. {(−3, 1.8), (−2, 1.4), (0, 0.6), (3, −0.6), (6, −1.8)}

c. When choosing values for w, you can choose any number greater than zero. Suppose we choose the domain {1, 2, 3, 4, 5}. Make a table of values.

w	$12 - w$	ℓ	(ℓ, w)
1	12 − 1	11	(1, 11)
2	12 − 2	10	(2, 10)
3	12 − 3	9	(3, 9)
4	12 − 4	8	(4, 8)
5	12 − 5	7	(5, 7)

The solution set for the domain we chose is {(1, 11), (2, 10), (3, 9), (4, 8), (5, 7)}. *Check to see if each solution satisfies the equation $2w + 2\ell = 24$.*

CHECK FOR UNDERSTANDING

Communicating Mathematics

Study the lesson. Then complete the following.

1. Refer to the connection at the beginning of the lesson.
   a. Which of the substances listed will float on water? on mercury?
   b. State the domain and range of the graphed relation.

1a. air, gasoline; air, aluminum, gasoline, lead, silver, steel, water
1b. $D = \{10, 50, 100, 150\}$; $R = \{193, 965, 1930, 2895\}$

2. **Show** why (1, −1) is or is not a solution of $x + 2y = 3$. **not; $1 + 2(-1) \neq 3$**
3. **Determine** which of the points shown in the graph at the right represent ordered pairs that are solutions for $y = 2x + 1$. ***A, C, F, G***

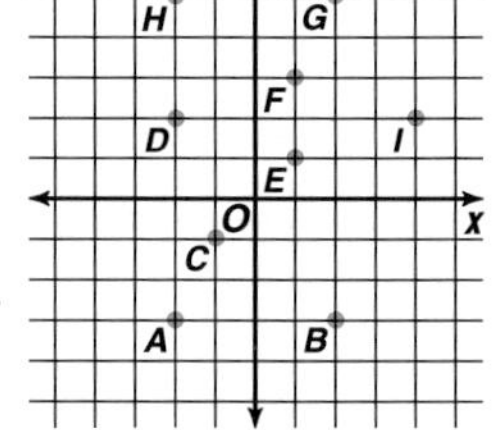

4. **Explain** why, in Example 3, you only consider values greater than zero.
   4. Length of a side cannot be zero or a negative number.
5. Why should you solve for y before finding ordered pairs that are solutions for an equation? **It makes calculating the y value easier.**

6. In your own words, explain the difference between an independent variable and a dependent variable and how you know on which axis to graph each of them. **See margin.**

Guided Practice

Solve each equation for the domain given in the table.

7. $y = 2x + 3$

x	y	(x, y)
−2	**−1**	**(−2, −1)**
−1	**1**	**(−1, 1)**
0	**3**	**(0, 3)**
1	**5**	**(1, 5)**
2	**7**	**(2, 7)**
3	**9**	**(3, 9)**

8. $a = \frac{3b - 5}{2}$

b	a	(b, a)
−5	**−10**	**(−5, −10)**
−2	$-\frac{11}{2}$	$\left(-2, -\frac{11}{2}\right)$
0	$-\frac{5}{2}$	$\left(0, -\frac{5}{2}\right)$
2	$\frac{1}{2}$	$\left(2, \frac{1}{2}\right)$
5	**5**	**(5, 5)**

Reteaching

Using Different Approaches

Demonstrate two approaches to solving an equation given its domain. First, substitute each value of the domain for x and solve for the corresponding y value. Second, solve for y in general. Then substitute each value of the domain for x into the phrase found equal to y.

9. a,d
10. b,c
11. {(−2, 8), (−1, 7.5), (0, 7), (1, 6.5), (2, 6)}; See margin for graph.

Which ordered pairs are solutions of each equation?

9. $1 + 5y = 2x$	**a.** $(-7, -3)$	**b.** $(7, 3)$	**c.** $(2, 1)$	**d.** $(-2, -1)$
10. $11 - 2y = 3x$	**a.** $(1, 3)$	**b.** $(3, 1)$	**c.** $(5, -2)$	**d.** $(-1, 4)$

11. Find the solution set for $x + 2y = 14$ if the domain is $\{-2, -1, 0, 1, 2\}$. Then graph the solution set.

12. Geometry The formula for the area of a rectangle is $A = \ell w$. Suppose the area of a rectangle is 36 square meters.

a. Solve the equation for ℓ. $\ell = \frac{A}{w}$ or $\ell = \frac{36}{w}$

b. Choose five values for w and find the corresponding values of ℓ. **Sample answer: (1, 36), (2, 18), (3, 12), (4, 9), (6, 6)**

EXERCISES

Practice

Which ordered pairs are solutions of each equation?

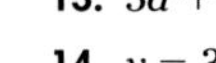

A

14. b,c
18. a,b,c,d

13. $3a + b = 8$	**a.** $(4, -4)$	**b.** $(8, 0)$	**c.** $(2, 2)$	**d.** $(3, 1)$	a,c
14. $y = 3x$	**a.** $(6, 2)$	**b.** $(-2, -6)$	**c.** $(0, 0)$	**d.** $(-15, -5)$	
15. $3a - 8b = -4$	**a.** $(0, 0.5)$	**b.** $(4, 2)$	**c.** $(2, 0.75)$	**d.** $(2, 4)$	a,b
16. $x = 3y - 7$	**a.** $(2, -1)$	**b.** $(2, 4)$	**c.** $(-1, 2)$	**d.** $(2, 3)$	c,d
17. $2y + 4x = 8$	**a.** $(0, 2)$	**b.** $(-3, 0.5)$	**c.** $(2, 0)$	**d.** $(-2, 1)$	c
18. $3x + 3y = 0$	**a.** $(1, -1)$	**b.** $(2, -2)$	**c.** $(-1, 1)$	**d.** $(-2, 2)$	

Solve each equation if the domain is {−3, −2, 0, 3, 6}.

19–27. See margin.

19. $y = 4x$	**20.** $y = 3x - 1$	**21.** $2x + 2y = 14$
22. $x = 4 + y$	**23.** $3x = 13 - 2y$	**24.** $y = 4 - 5x$
25. $5x + 3 = y$	**26.** $5x - 10y = 40$	**27.** $2x + 5y = 3$

B

Make a table and graph the solution set for each equation and domain.

28. $y = 3x$ for $x = \{-3, -2, -1, 0, 1, 2, 3\}$ **28–31. See Solutions Manual.**

29. $y = 2x + 1$ for $x = \{-5, -3, 0, 1, 3\ 6\}$

30. $3x - 2y = 5$ for $x = \{-3, -1, 2, 4, 7\}$

31. $5x = 8 - 4y$ for $x = \{-2, -1, 0, 1, 3, 4, 5\}$

32. Geometry A regular pentagon has five angles that have equal measures. Suppose the measure of one angle is $(a + b)°$. The sum of the measures of the angles of any pentagon is 540°.

a. Write an equation for the sum of the angle measures of a regular pentagon. $5a + 5b = 540$

b. Solve the equation for a. $a = 108 - b$

c. Choose five values for b and find the corresponding values of a.

32c. Sample answer: (1, 107), (2, 106), (3, 105), (4, 104), (5, 103)

33. Geometry Two angles that are supplementary have measures whose sum is 180°. Suppose the measures of two supplementary angles are $(x + y)°$ and $(2x + 3y)°$. **a.** $3x + 4y = 180$

a. Write an equation for the sum of the measures of these two angles.

b. Solve the equation for y.

c. Choose five values for y and find the corresponding values of x.

33b. $y = 45 - \frac{3}{4}x$

33c. Sample answer: (1, 44.25), (2, 43.5), (3, 42.75), (4, 42), (5, 41.25)

Find the domain for each equation if the range is {−2, −1, 0, 2, 3}.

34. $y = x + 7$ $\{-9, -8, -7, -5, -4\}$ **35.** $y = 3x$ $\left\{-\frac{2}{3}, -\frac{1}{3}, 0, \frac{2}{3}, 1\right\}$

36. $6x - y = -3$ **37.** $5y = 8 - 4x$

36. $\left\{-\frac{5}{6}, -\frac{2}{3}, -\frac{1}{2}, -\frac{1}{6}, 0\right\}$

37. $\left\{-\frac{7}{4}, -\frac{1}{2}, 2, \frac{13}{4}, \frac{9}{2}\right\}$

Assignment Guide

Core: 13–49 odd, 51–59
Enriched: 14–44 even, 45–59

For **Extra Practice**, see p. 767.

The red A, B, and C flags, printed only in the Teacher's Wraparound Edition, indicate the level of difficulty of the exercises.

Study Guide Masters, p. 36

NAME ____________ DATE ________

Study Guide

Student Edition Pages 271–277

Equations as Relations

An equation in two variables describes a relation. It is often easier to determine the solution of such an equation by solving for one of the variables.

Example: Solve $3y + 2x = 10$ if the domain is $\{-7, -1, 8\}$.
First solve for y in terms of x.

$$3y + 2x = 10$$
$$3y = 10 - 2x$$
$$y = \frac{10 - 2x}{3}$$

Then substitute values of x.

x	$\frac{10 - 2x}{3}$	y	(x, y)
−7	$\frac{10 - 2(-7)}{3}$	8	(−7, 8)
−1	$\frac{10 - 2(-1)}{3}$	4	(−1, 4)
8	$\frac{10 - 2(8)}{3}$	−2	(8, −2)

Which ordered pairs are solutions of each equation?

1. $y = 3x + 1$ **a.** $(0, 1)$ **b.** $\left(\frac{1}{3}, 2\right)$ **c.** $\left(-1, -\frac{2}{3}\right)$ **d.** $(-1, -2)$
a, b, d

2. $2a = 5 - b$ **a.** $(5, 0)$ **b.** $(5, -5)$ **c.** $\left(\frac{5}{2}, 0\right)$ **d.** $(1, -3)$
b, c

Solve each equation if the domain is {−4, −2, 0, 2, 4}.

3. $x + y = 4$
{(−4, 8), (−2, 6), (0, 4), (2, 2), (4, 0)}

4. $y = -4x - 6$
{(−4, 10), (2, 2), (0, −6), (2, −14}, (4, −22)}

5. $5a - 3b = 15$
$\left\{\left(-4, -11\frac{2}{3}\right), \left(-2, -8\frac{1}{3}\right), (0, -5), \left(2, -1\frac{2}{3}\right), \left(4, 1\frac{2}{3}\right)\right\}$

6. $3x - 5y = 8$
$\left\{(-4, -4), \left(-2, -\frac{14}{5}\right), \left(0, -\frac{8}{5}\right), \left(2, -\frac{2}{5}\right), \left(4, -\frac{4}{5}\right)\right\}$

7. $6x + 3y = 18$
{(−4, 14), (−2, 10), (0, 6), (2, 2), (4, −2)}

8. $4x + 8 = 6y$
$\left\{\left(-4, -\frac{4}{3}\right), (-2, 0), \left(0, \frac{4}{3}\right), \left(2, \frac{8}{3}\right), (4, 4)\right\}$

4 ASSESS

Closing Activity

Writing For each equation, find the range for the given domain.

1. $5y - 3x = 8$, where the domain is $\{-2, 0, 5\}$. $R = \left\{\frac{2}{5}, 1\frac{3}{5}, 4\frac{3}{5}\right\}$
2. $3z - 6w = 1$, where the domain is $\{-1, 0, 2\}$. $R = \left\{-1\frac{2}{3}, \frac{1}{3}, 4\frac{1}{3}\right\}$

Additional Answers

41. $\{(-2.5, -4.26), (-1.75, -3.21), (0, -0.76), (1.25, 0.99), (3.33, 3.902)\}$
42. $\{(-125, -425.5), (-37, -117.5), (-6, -9), (12, 54), (57, 211.5), (150, 537)\}$
43. $\{(-100, 350), (-30, 116.\overline{6}), (0, 16.\overline{6}), (120, -383.\overline{3}), (360, -1183.\overline{3}), (720, -2383.\overline{3})\}$
44. $\{(-10, 4.\overline{6}), (-5, 3), (0, 1.\overline{3}), (5, -0.\overline{3}), (10, -2), (15, -3.\overline{6})\}$

46a. The other graphs have points that seem to lie in a straight line. These points do not.

46b. $y = x^2 - 3x - 10$ is U-shaped; $y = x^3$ is both an upward and downward curve; $y = 3^x$ is J-shaped.

48b. North America, 19; South America, 33; Europe, 54; Asia, 7; Africa, 5

Practice Masters, p. 36

NAME__________ DATE __________

Practice

Student Edition Pages 271–277

Equations as Relations

Which ordered pairs are solutions of each equation?

1. $3x = 2y - 1$ a. $(1, -2)$ (b.) $(-1, -1)$ (c.) $\left(-2, -\frac{5}{2}\right)$ d. $\left(0, -\frac{1}{2}\right)$
2. $2y = x + 3$ a. $(-1, -1)$ b. $(1, -2)$ (c.) $(-3, 0)$ d. $\left(0, -\frac{3}{2}\right)$
3. $5x = 2 - y$ a. $(3, 12)$ b. $(-3, -17)$ (c.) $(2, -8)$ (d.) $(-1, 7)$

Solve each equation if the domain is {−2, −1, 0, 2, 5}.

4. $y = 3x + 2$ $\{(-2, -4), (-1, -1), (0, 2), (2, 8), (5, 17)\}$
5. $x - y = 5$ $\{(-2, -7), (-1, -6), (0, -5), (2, -3), (5, 0)\}$
6. $y = 2x - 7$ $\{(-2, -11), (-1, -9), (0, -7), (2, -3), (5, 3)\}$
7. $2y = 8 - 4x$ $\{(-2, 8), (-1, 6), (0, 4), (2, 0), (5, -6)\}$

Make a table and graph the solution set for each equation and domain.

8. $2x + y = 8$ for $x = \{-3, -1, 1, 3, 5, 7\}$

x	y
−3	14
−1	10
1	6
3	2
5	−2
7	−6

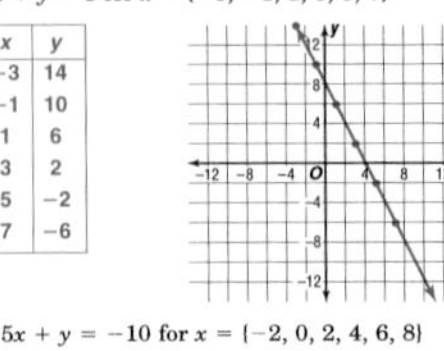

9. $-5x + y = -10$ for $x = \{-2, 0, 2, 4, 6, 8\}$

x	y
−2	−20
0	−10
2	0
4	10
6	20
8	30

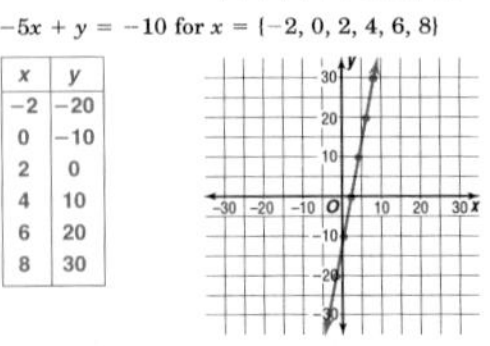

Make a table and graph the solution set for each equation.

38. $y = x^2 - 3x - 10$ if the domain is $\{-3, -1, 0, 1, 3, 5\}$
39. $y = x^3$ if the domain is $\{-2, -1, 0, 1, 2\}$
40. $y = 3^x$ if the domain is $\{1, 2, 3, 4\}$

38–40. See Solutions Manual.

Graphing Calculator

Use a graphing calculator to find the solution set for each domain.

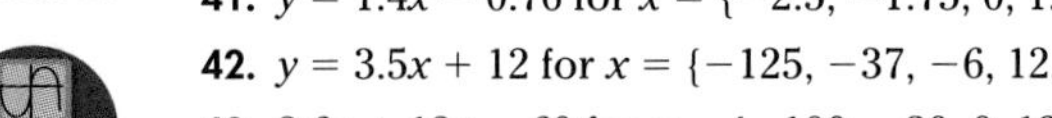

41. $y = 1.4x - 0.76$ for $x = \{-2.5, -1.75, 0, 1.25, 3.33\}$
42. $y = 3.5x + 12$ for $x = \{-125, -37, -6, 12, 57, 150\}$
43. $3.6y + 12x = 60$ for $x = \{-100, -30, 0, 120, 360, 720\}$
44. $75y + 25x = 100$ for $x = \{-10, -5, 0, 5, 10, 15\}$

41–44. See margin.

Critical Thinking

45. Find the domain values of each relation if the range is $\{0, 16, 36\}$.
    a. $y = x^2$ b. $y = |4x| - 16$ c. $y = |4x - 16|$
46. Compare the graphs you drew in Exercises 38–40 with the other graphs in this lesson. See margin.
    a. What is different about the pattern of the points for these relations?
    b. Study the equations associated with the graphs. How do you think you could predict the pattern of the points by looking at the equation?

45a. $\{-6, -4, 0, 4, 6\}$

45b. $\{-13, -8, -4, 4, 8, 13\}$

45c. $\{-5, 0, 4, 8, 13\}$

Applications and Problem Solving

47. **Physical Science** Refer to the connection at the beginning of the lesson. Suppose you have an unknown substance with a mass of 378 grams and a volume of 36 cm^3 and a second unknown substance with a mass of 87.5 grams and a volume of 125 cm^3.
    a. Solve the formula $m = DV$ for D. $D = \frac{m}{V}$
    b. Identify the unknown substances. silver and gasoline
48. **Demographics** The time T (in years) needed for a population to double is calculated using the formula $T = \frac{70}{R}$, where R represents the percent growth rate of the population.
    a. Determine the domain and range values for which this equation makes sense. D: $R > 0$; R: $T > 0$
    b. Use the formula to approximate the doubling time to the nearest year for the continents shown in the graph at the left.
    c. The population of the United States in 1990 was 248,709,873. If the U.S. maintains the same growth rate as North America, what will be the population of the U.S. in the year 2000? in 2010? in 2020?

48b. See margin.

48c. 354,234,275; 504,531,323; 718,597,477

Growth Rates of Continents

North America 3.6
South America 2.1
Europe 1.3
Asia 9.7
Africa 15.1

0 4 8 12 16
Percent

49. **Anatomy** The formula for relating the shoe size S and the length of a woman's foot in inches L is $S = 3L - 22$. The formula for relating a man's shoe size with his foot length is $S = 3L - 26$. Copy and complete each table below to determine shoe sizes given lengths of feet.

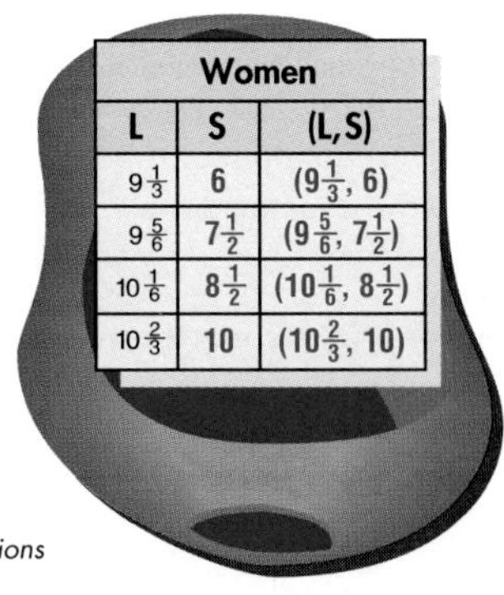

Women		
L	S	(L, S)
$9\frac{1}{3}$	6	$(9\frac{1}{3}, 6)$
$9\frac{5}{6}$	$7\frac{1}{2}$	$(9\frac{5}{6}, 7\frac{1}{2})$
$10\frac{1}{6}$	$8\frac{1}{2}$	$(10\frac{1}{6}, 8\frac{1}{2})$
$10\frac{2}{3}$	10	$(10\frac{2}{3}, 10)$

Men		
L	S	(L, S)
$11\frac{1}{3}$	8	$(11\frac{1}{3}, 8)$
$11\frac{5}{6}$	$9\frac{1}{2}$	$(11\frac{5}{6}, 9\frac{1}{2})$
$12\frac{1}{3}$	11	$(12\frac{1}{3}, 11)$
$12\frac{5}{6}$	$12\frac{1}{2}$	$(12\frac{5}{6}, 12\frac{1}{2})$

50. **Make a Table** Winona Brownsman is doing small jobs for her rich aunt. Her aunt has asked Winona how she wishes to be paid for her services. Winona has two choices.

- *Plan A:* Winona can be paid one dollar the first day of the month and an additional dollar for each day for a month. For example, she will be paid \$1 on Day 1, \$2 on Day 2, \$3 on Day 3, and so on.
- *Plan B:* Winona will be paid one cent on the first day and for each succeeding day, double the amount of the previous day. For example she will be paid 1¢ on Day 1, 2¢ on Day 2, 4¢ on Day 3, 8¢ on Day 4, and so on. **a–b. See margin.**

a. Use two tables to determine which plan would be more profitable.

b. On which day will the better plan become greater than the other?

Mixed Review

51. **Basketball** The table at the right shows the number of years that specific NBA basketball players have been playing and the average number of points each player made during the 1993–1994 season. (Lesson 5–2)

Player	Years	Points
Charles Barkley	10	21.6
Glen Rice	5	21.1
Chris Mullin	9	16.8
Jamal Mashburn	1	19.2
Scottie Pippen	7	22.0
Dominique Wilkins	12	26.0

Source: *Hawes Fantasy Basketball Guide, 1994–1995*

a. Plot the points and describe the graph. **a–b. See margin.**

b. What do you think the graph suggests about the relationship of the number of years played and the point average?

52. Graph each point below. Then connect the points in alphabetical order and identify the figure. (Lesson 5–1) **star**

$A(0, 5), B(4, -3), C(-5, 2), D(5, 2), E(-4, -3), F(0, 5)$

53. **Employment** Hugo's wages vary directly with the time he works. If his wages for 4 days are \$110, how much will they be for 17 days? (Lesson 4–8) **\$467.50**

54. What number decreased by 80% is 14? (Lesson 4–5) **70**

55. **\$18,000**

55. **Finance** If Li Fong had earned one fourth of a percent more in annual interest on an investment, the interest for one year would have been \$45 greater. How much did he invest at the beginning of the year? (Lesson 4–4)

56. Solve $\frac{1}{3}a - 2b = -9c$ for a. (Lesson 3–6) $a = -27c + 6b$

57. **Geometry** The perimeter of a rectangle is 148 inches.

a. Write an equation to represent this situation. $2w + 2\ell = 148$

b. Find its dimensions if the length is 17 inches greater than three times the width. (Lesson 3–3) **w, 14.25 in.; ℓ, 59.75 in.**

58. Find a number between $-\frac{8}{17}$ and $\frac{1}{9}$. (Lesson 2–4) **Sample answer: 0**

59. Evaluate $3x^3 - 2y$ if $x = 0.2$ and $y = 4$. (Lesson 1–3) **−7.976**

Additional Answers

50a.

Day	Plan A	Plan B
1	\$1	\$0.01
2	\$2	\$0.02
3	\$3	\$0.04
4	\$4	\$0.08
5	\$5	\$0.16
6	\$6	\$0.32
7	\$7	\$0.64
8	\$8	\$1.28
9	\$9	\$2.56
10	\$10	\$5.12
11	\$11	\$10.24
12	\$12	\$20.48
13	\$13	\$40.96
14	\$14	\$81.92
15	\$15	\$163.84
16	\$16	\$327.68
17	\$17	\$655.36
18	\$18	\$1310.72
19	\$19	\$2621.44
20	\$20	\$5242.88
21	\$21	\$10,485.76
22	\$22	\$20,971.52
23	\$23	\$41,943.04
24	\$24	\$83,886.08
25	\$25	\$167,772.16
26	\$26	\$335,544.32
27	\$27	\$671,088.64
28	\$28	\$1,342,177.28
29	\$29	\$2,684,354.56
30	\$30	\$5,368,709.12

50b. On day 12, Plan B exceeds Plan A.

Enrichment Masters, p. 36

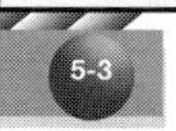

NAME______ DATE______

Enrichment

Student Edition Pages 271–277

Dissection Puzzles: Make the Square

In a dissection puzzle you are to cut apart one figure and then rearrange the pieces to make a new figure. Only straight cuts are allowed. Usually the puzzle-solver must figure out where to make a given number of cuts. However, for these puzzles, the cut lines are shown. You must find out how to rearrange the pieces.

Cut apart the figure shown. Then rearrange the pieces to form a square. Record your solution in the square at the right.

1.

2.

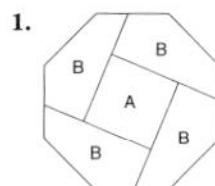

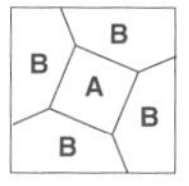

3.

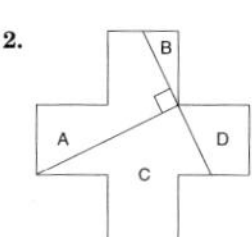

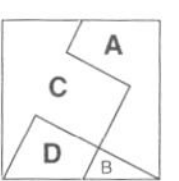

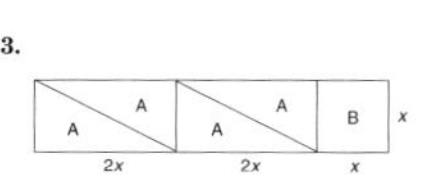

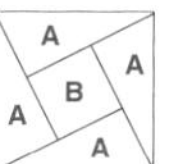

4. For this dissection, you must cut one of the triangles into two pieces to make the square.

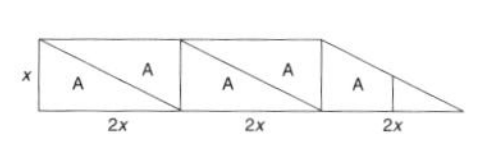

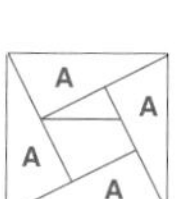

Extension

Reasoning A relation is called transitive if whenever (a, b) and (b, c) are in the relation, then so is the ordered pair (a, c).

a. Is "is taller than" a transitive relation? **yes**

b. Is "is the father of" a transitive relation? **no**

Additional Answers

51a.

Point Avg. (0, 16–28) vs. Years (0–12)

51b. **The more years played, the higher the point-per-game average of the player.**

5-4 LESSON NOTES

NCTM Standards: 1–5, 8

Instructional Resources

- Study Guide Master 5-4
- Practice Master 5-4
- Enrichment Master 5-4
- Assessment and Evaluation Masters, pp. 127–128
- Graphing Calculator Masters, p. 5

Transparency 5-4A contains the 5-Minute Check for this lesson; **Transparency 5-4B** contains a teaching aid for this lesson.

Recommended Pacing	
Standard Pacing	Day 7 of 14
Honors Pacing	Day 7 of 13
Block Scheduling*	Day 4 of 7
Alg. 1 in 2 Years*	Days 10 & 11 of 21

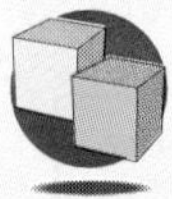

*For more information on pacing and possible lesson plans, refer to the *Block Scheduling Booklet* and *Algebra 1 in Two Years*.

1 FOCUS

5-Minute Check
(over Lesson 5-3)

Which ordered pairs are solutions of each equation?

1. $2m - 5n = 1$
  a. $(-2, -1)$ **b.** $(2, 1)$
  c. $(7, 3)$ **d.** $(-7, -3)$ **a, d**

2. $2x + 3y = 11$
  a. $(3, 1)$ **b.** $(1, 3)$
  c. $(-2, 5)$ **d.** $(4, -1)$ **b, c**

3. $4m + n = 7$
  a. $(1, 3)$ **b.** $(2, 1)$
  c. $(-1, 11)$ **d.** $(4, 7)$ **a, c**

Solve each equation if the domain is $\{-2, -1, 0, 1, 2\}$.

4. $y = 4x - 1$ $\{(-2, -9), (-1, -5), (0, -1), (1, 3), (2, 7)\}$

5. $3r + 2s = 4$ $\left\{(-2, 5), \left(-1, \frac{7}{2}\right), (0, 2), \left(1, \frac{1}{2}\right), (2, -1)\right\}$

5-4 Graphing Linear Equations

What YOU'LL LEARN
- To graph linear equations.

Why IT'S IMPORTANT

You can graph equations to solve problems involving health and physical science.

FYI

In science, a calorie is defined as the quantity of heat needed to raise the temperature of 1 gram of water 1°C. The word *calorie* that we use when referring to food actually refers to kilocalories (1000 calories). These are referred to as Calories (C) to differentiate them from the thermal unit.

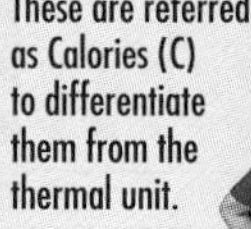

CONNECTION
Health

The manner in which you burn Calories depends on your weight and the activity you are doing. The chart at the right shows the number of Calories burned per minute per kilogram (C/min/kg).

Activity	C/min/kg
Basketball	0.138
Cycling (leisure)	0.064
Cycling (racing)	0.169
Dancing (aerobic)	0.135
Dancing (normal)	0.075
Drawing	0.036
Eating	0.023
Football	0.132
Free weights	0.086
Golf	0.085
Gymnastics	0.066
Jumping rope	0.162
Playing drums	0.066
Playing flute	0.035
Playing horn	0.029
Playing piano	0.040
Playing trumpet	0.031
Nautilus® training	0.092
Running (7.2 min/km)	0.135
Running (5.0 min/km)	0.208
Running (3.7 min/km)	0.252
Sitting Quietly	0.021
Swimming	0.156
Walking	0.080
Writing	0.029
Word processing	0.027

Manuel weighs 70 kilograms and wants to know how many Calories he burns playing football. The formula $C = wtr$, where w is his weight in kilograms, t is the time in minutes, and r is C/min/kg, can be used to determine how many Calories he burns.

$$C = wtr$$
$$C = 70 \cdot t \cdot 0.132$$
$$C = 9.24t$$

The number of Calories burned is dependent upon how long he exercises. So, t is the independent variable and C is the dependent variable.

t	$9.24t$	C	(t, C)
0	9.24(0)	0	(0, 0)
10	9.24(10)	92.4	(10, 92.4)
20	9.24(20)	184.8	(20, 184.8)
30	9.24(30)	277.2	(30, 277.2)
40	9.24(40)	369.6	(40, 369.6)

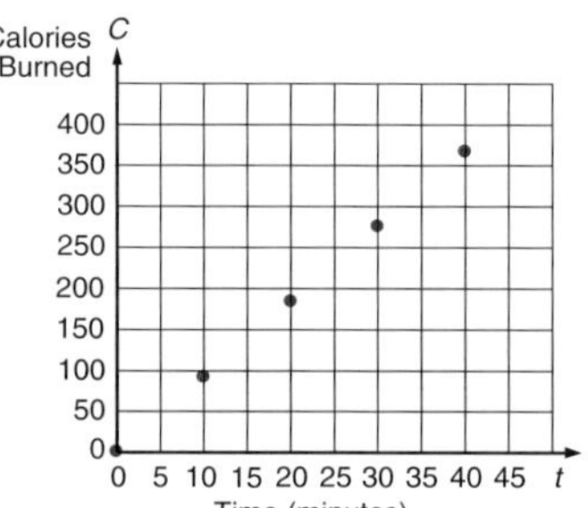

When you graph the ordered pairs, a pattern begins to form. The points seem to lie in a line.

What is the range for this relation?

Suppose the domain of $C = 9.24t$ is the set of positive real numbers. There would be an infinite number of ordered pairs that are solutions for the equation. If you graphed all the solutions, they would form a line. The line shown in the graph at the right represents all the solutions for $C = 9.24t$.

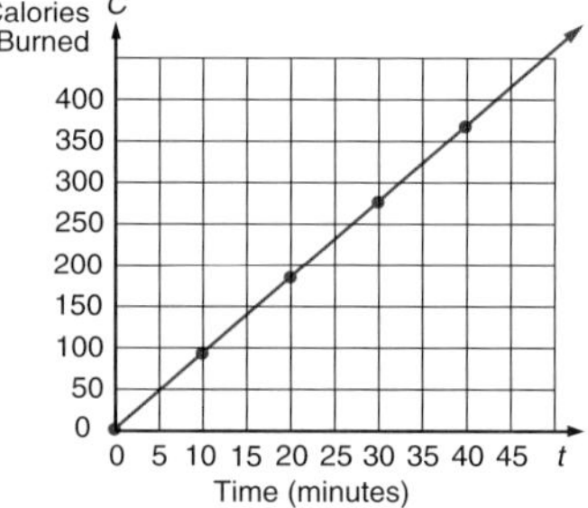

Since the graph of $C = 9.24t$ is a line, $C = 9.24t$ is called a **linear equation.**

FYI

Listed below are the caloric contents of some common foods.

1 apple	66 C
4 oz ground beef	326 C
1 oz American cheese	105 C
8 oz whole milk	151 C
1 tomato	33 C

Linear equations may contain one or two variables with no variable having an exponent other than 1.

Definition of a Linear Equation in Standard Form	A linear equation is an equation that can be written in the form $Ax + By = C$, where A, B, and C are any real numbers, and A and B are not both zero.

Example  **Determine whether each equation is a linear equation. If so, identify *A*, *B*, and *C*.**

a. $4x = 7 + 2y$

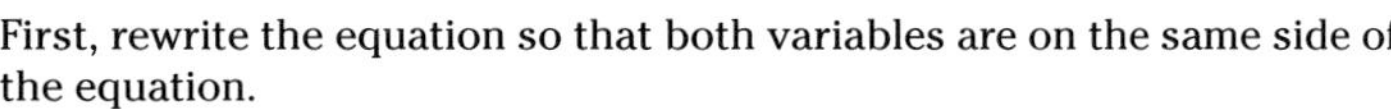

First, rewrite the equation so that both variables are on the same side of the equation.

$$4x = 7 + 2y$$

$$4x - 2y = 7 \quad \textit{Subtract 2y from each side.}$$

The equation is now in the form $Ax + By = C$, where $A = 4$, $B = -2$, and $C = 7$. This is a linear equation.

b. $2x^2 - y = 7$

The exponents of the variables in a linear equation must be 1. Since the exponent of x is 2, this is not a linear equation.

c. $x = 12$

This equation can be rewritten as $x + 0y = 12$. Therefore, it is a linear equation in the form $Ax + By = C$, where $A = 1$, $B = 0$, and $C = 12$.

To graph a linear equation, it is often helpful to make a table of ordered pairs that satisfy the equation. Then graph the ordered pairs and connect the points with a line.

Example  **Graph each equation.**

a. $y = 8x - 4$

Select five values for the domain and make a table. Then graph the ordered pairs and connect them to draw the line.

x	8x – 4	y	(x, y)
–2	8(–2) – 4	–20	(–2, –20)
–1	8(–1) – 4	–12	(–1, –12)
0	8(0) – 4	–4	(0, –4)
1	8(1) – 4	4	(1, 4)
2	8(2) – 4	12	(2, 12)

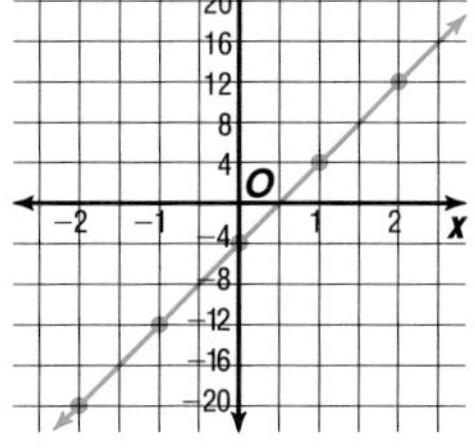

From the graph, you can find other ordered pairs that are solutions; for example, (2.5, 16).

The range of this relation is the set of real numbers.

(continued on the next page)

Cooperative Learning

Think-Pair-Share Have students work in pairs to solve Example 2b. One student in each pair should follow the procedure in the text. The other student should graph the equation on a graphing calculator, and then use the TRACE feature to verify the values in the table. Have pairs discuss their findings with the class. For more information on the think-pair-share strategy, see *Cooperative Learning in the Mathematics Classroom,* one of the titles in the Glencoe Mathematics Professional Series, pages 24–25.

Motivating the Lesson

Questioning Ask students to identify applications from previous lessons in which there appears to be a linear relationship between the two variables in a relation. Some examples might include a salary that is dependent upon the number of hours worked and commission based on the amount of sales.

2 TEACH

Teaching Tip Make sure students understand the definition of a linear equation. The phrase "can be written" does not mean the equation has to appear in that form. However, the equation must be equivalent to an equation in the form $Ax + By = C$.

In-Class Examples

For Example 1

Determine whether each equation is a linear equation. If so, identify *A*, *B*, and *C*.

a. $9y = x + 16$ **linear; $A = 1$, $B = -9$, $C = -16$**

b. $2x^2 + 3y = 4$ **not linear, has x^2 term**

c. $x = -5$ **linear; $A = 1$, $B = 0$, $C = -5$**

For Example 2

Graph each equation.

a. $y = 3x - 4$

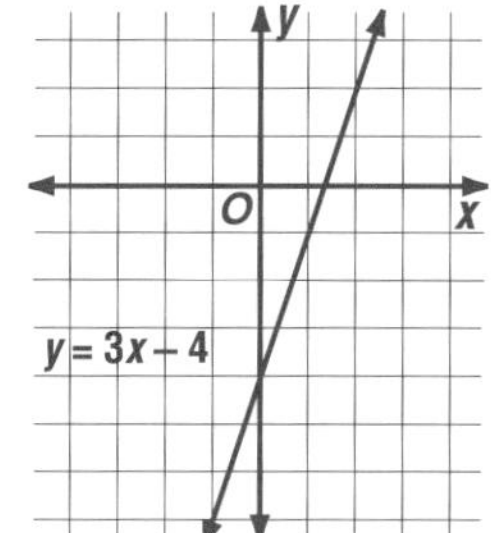

b. $2x + 5y = 8$

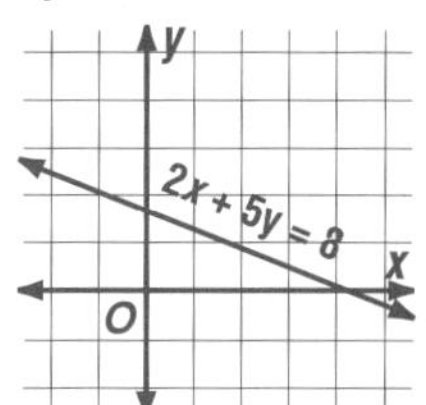

Teaching Tip In Example 2a, students should be aware that any value can be chosen for x.

Teaching Tip In Example 2b, when $2x + 5y = 10$ has been expressed as y in terms of x, the result is a fraction. In order to avoid fractional components in the ordered pairs, students could choose values for x that make the numerator an even integer. The result is an integral value for y.

In-Class Example

For Example 3
Yukari averages 40 miles per hour when she drives from Los Angeles to San Francisco.

a. What equation relates the distance traveled to the number of hours traveled? ***$d = 40t$***

b. Graph the relation by letting the horizontal axis represent the time and the vertical axis represent the distance.

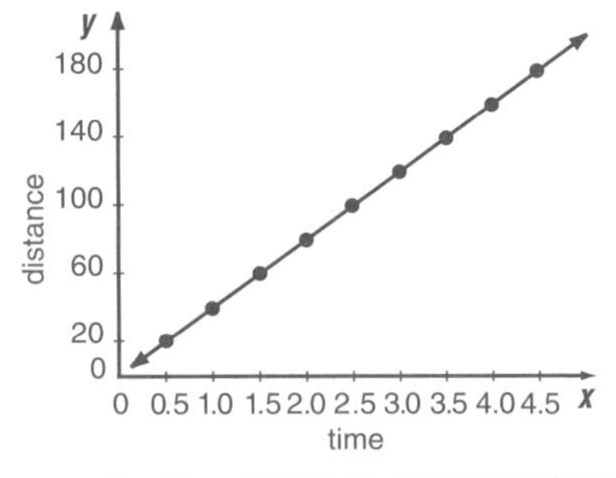

3 PRACTICE/APPLY

Check for Understanding

Exercises 1–15 are designed to help you assess your students' understanding through reading, writing, speaking, and modeling. You should work through Exercises 1–5 with your students and then monitor their work on Exercises 6–15.

Error Analysis
Students often confuse the graphs of $y = a$ and $x = b$. Invite students to reconstruct which line, horizontal or vertical, corresponds to which graph by making (x, y) tables that fit the equations. To differentiate these similar concepts, focus on just one of the ideas.

b. $2x + 5y = 10$

In order to find values for y more easily, solve the equation for y.

$2x + 5y = 10$

$5y = 10 - 2x$ *Subtract 2x from each side.*

$y = \frac{10 - 2x}{5}$ *Divide each side by 5.*

Now make a table and draw the graph.

x	$\frac{10 - 2x}{5}$	y	(x, y)
−10	$\frac{10 - 2(-10)}{5}$	6	(−10, 6)
−5	$\frac{10 - 2(-5)}{5}$	4	(−5, 4)
0	$\frac{10 - 2(0)}{5}$	2	(0, 2)
5	$\frac{10 - 2(5)}{5}$	0	(5, 0)
10	$\frac{10 - 2(10)}{5}$	−2	(10, −2)

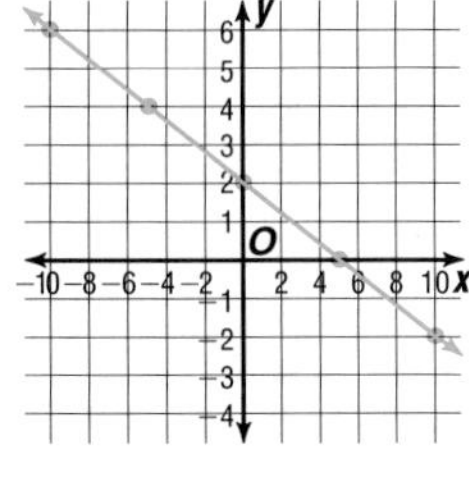

Using graphs is a good way to make comparisons in real-life situations.

Example 3

Health

Refer to the connection at the beginning of the lesson. Carmen Delgado is running in a cross-country meet at the rate of 5 min/km. Her weight is 50 kg. How does the number of Calories she is burning compare to those that she would be burning if she were at home word processing her research paper or if she were playing golf with her dad?

Explore Look up the Calories burned for each of the three types of activities in the chart on page 280.

Plan Write the equation for each activity using the formula $C = wtr$ and 50 for w.

Running	$C = 50(0.208)t$ or $C = 10.4t$
Word processing	$C = 50(0.027)t$ or $C = 1.35t$
Golfing	$C = 50(0.085)t$ or $C = 4.25t$

Solve Use a graphing calculator to graph all three equations. Enter the equations as Y_1, Y_2, and Y_3. Let x represent the t in each equation. Make a sketch of each graph and label each with its activity.

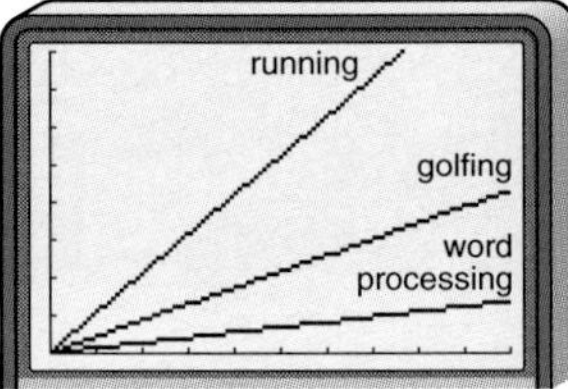

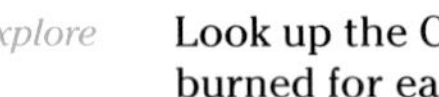

Graphing calculators are helpful when comparing equations that are not easily graphed with pencil and paper.

Reteaching

Using Examples To convince students that different samples from a domain of real numbers will produce the same line for a linear relation, use two or three different sets of numbers for x.

For each value of t, Carmen burns many more Calories running than she does golfing or word processing. Carmen would burn more Calories golfing than she would word processing.

Examine Think about the physical movement required for each activity and how much energy that would take. Does the result make sense?

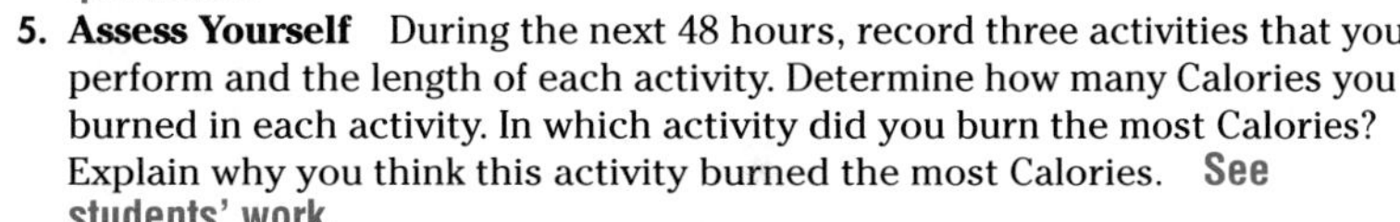

CHECK FOR UNDERSTANDING

Communicating Mathematics

Study the lesson. Then complete the following. 1–2. See margin.

1. **Explain** why the ordered pair $(-1, 3)$ is a solution of $y = -2x + 1$.
2. **Describe** the graph of a linear equation in the form $Ax + By = C$ for which
   a. $A = 0$ b. $B = 0$ c. $C = 0$
3. **Show** how the graph of $y = 2x + 1$ for the domain $\{-1, 0, 2, 3\}$ differs from the graph of $y = 2x + 1$ for the domain of all real numbers.
4. **Explain** why the x and y values used to set the viewing window in Example 3 do not include negative numbers. Time and Calories are nonnegative quantities.
5. **Assess Yourself** During the next 48 hours, record three activities that you perform and the length of each activity. Determine how many Calories you burned in each activity. In which activity did you burn the most Calories? Explain why you think this activity burned the most Calories. See students' work.

3. The first graph is a set of points; the second is a line; see margin for graphs.

Guided Practice

Determine whether each equation is a linear equation. If an equation is linear, rewrite it in the form $Ax + By = C$.

6. $3x - 5y = 0$ yes; $3x - 5y = 0$
7. $2x = 6 - y$ yes; $2x + y = 6$
8. $3x^2 + 3y = 4$ no
9. $3x = 7 - 2y$ yes; $3x + 2y = 7$

Graph each equation. 10–15. See margin.

10. $3x + y = 4$
11. $4x + 3y = 12$
12. $\frac{1}{2}x = 8 - y$
13. $x = 6$
14. $y = -5$
15. $x - y = 0$

EXERCISES

Practice

Determine whether each equation is a linear equation. If an equation is linear, rewrite it in the form $Ax + By = C$.

A

16. $\frac{3}{x} + \frac{4}{y} = 2$ no
17. $\frac{3}{5}x - \frac{2}{3}y = 5$ yes
18. $x + y^2 = 25$ no
19. $x + \frac{1}{y} = 7$ no
20. $3y + 2 = 0$ yes
21. $2y = 3x - 8$ yes
22. $5x - 7 = 0$ yes
23. $2x + 5x = 7y$ yes
24. $4x^2 - 3x = y$ no
25. $3m = 2n$ yes
26. $\frac{x}{2} = 10 + \frac{2y}{3}$ yes
27. $8a - 7b = 2a - 5$ yes

17. $\frac{3}{5}x - \frac{2}{3}y = 5$
20. $3y = -2$
21. $3x - 2y = 8$
22. $5x = 7$
23. $7x - 7y = 0$
25. $3m - 2n = 0$
26. $\frac{1}{2}x - \frac{2}{3}y = 10$
27. $6a - 7b = -5$

Additional Answers

1. Because $3 = -2(-1) + 1$ is a true sentence.
2a. horizontal line
2b. vertical line
2c. slanted line
3. $y = 2x + 1$ for $x = \{-1, 0, 2, 3\}$

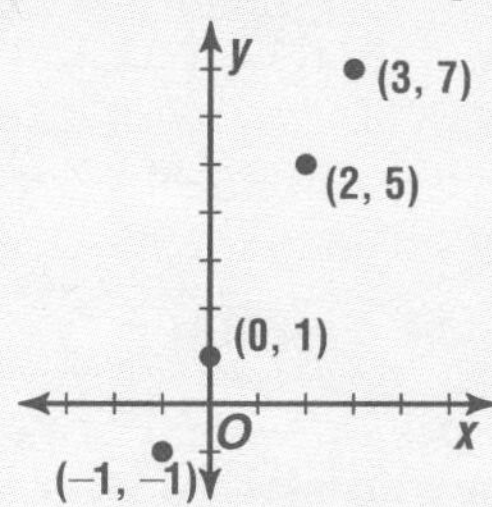

$y = 2x + 1$ for $x \in$ {real numbers}

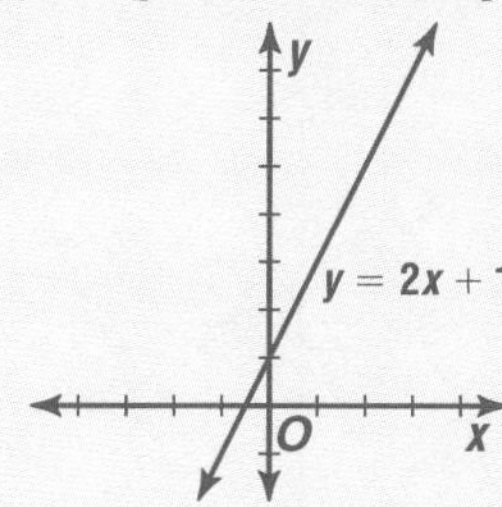

10.

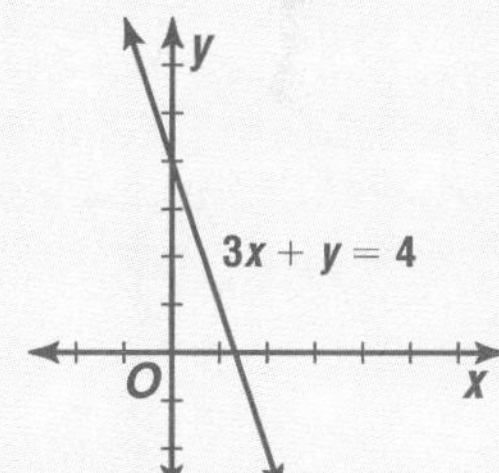

11.

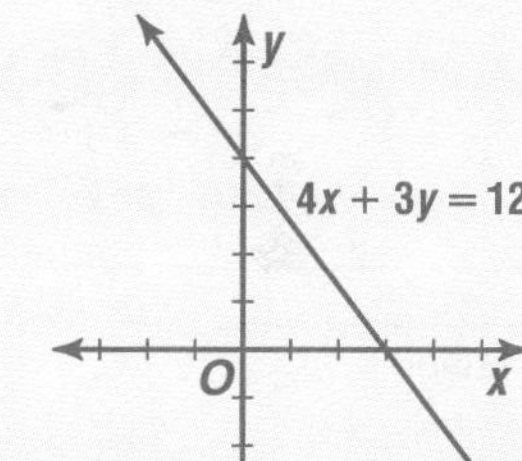

12.

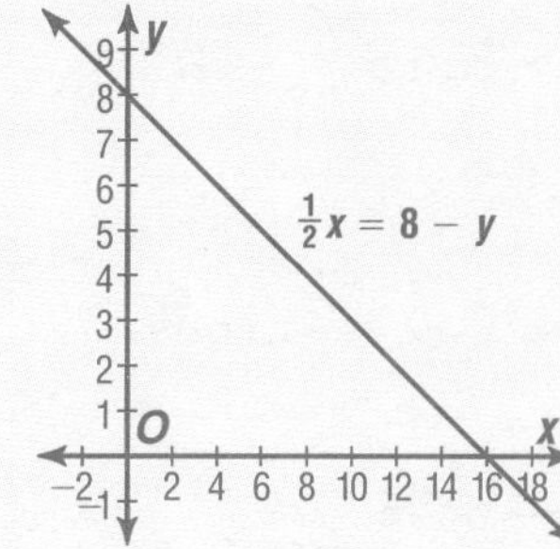

13.

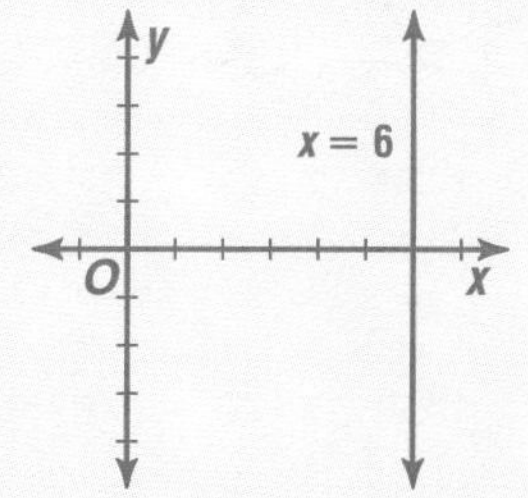

14.

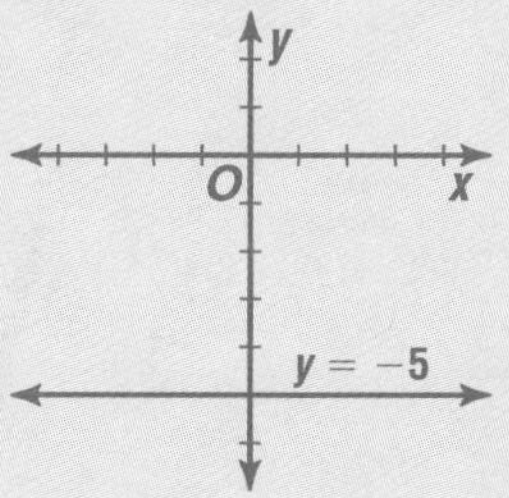

15.

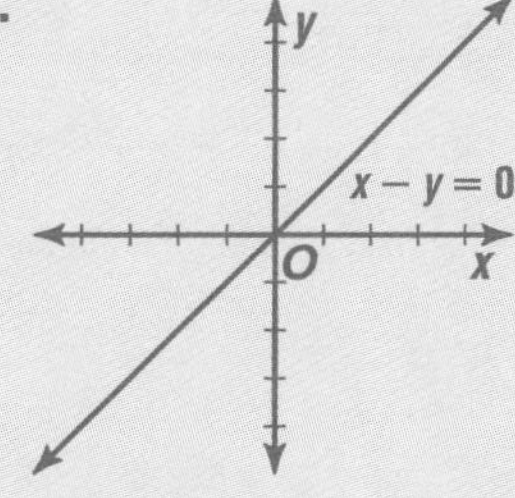

Assignment Guide

Core: 17–59 odd, 61–70
Enriched: 16–56 even, 57–70
All: Self Test, 1–10

For **Extra Practice,** see p. 768.

The red A, B, and C flags, printed only in the Teacher's Wraparound Edition, indicate the level of difficulty of the exercises.

Additional Answers

57a. Sample answer: Parallel lines that slant upward and intersect the *x*-axis at −7, −2.5, 0, and 4.5.

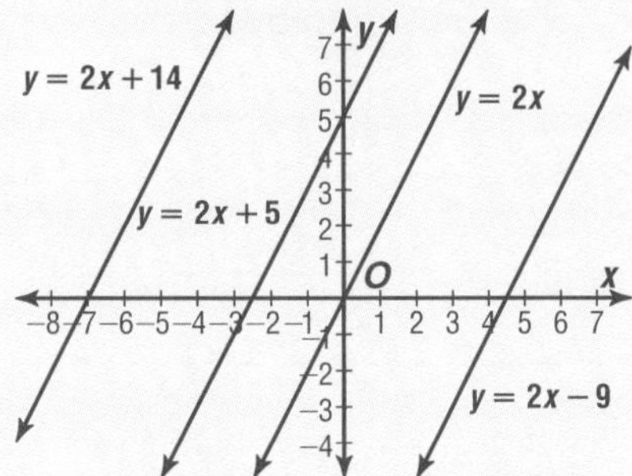

57b. Sample answer: Parallel lines that slant downward and intersect the *x*-axis at 0, $1\frac{1}{3}$, $2\frac{1}{3}$, and $-3\frac{1}{3}$.

Study Guide Masters, p. 37

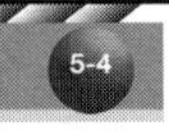

NAME__________ DATE________

Study Guide Student Edition Pages 280–286

Graphing Linear Equations

An equation whose graph is a straight line is called a **linear equation.**

Definition of Linear Equation

A **linear equation** is an equation that can be written in the form $Ax + By = C$, where *A*, *B*, and *C* are any numbers and *A* and *B* are not both 0.

Drawing the Graph of a Linear Equation

1. Solve the equation for one variable.
2. Set up a table of values for the variable.
3. Graph the ordered pairs and connect them with a line.

Example: Draw the graph of $y - 3x = 1$.
$y - 3x = 1$
$y = 3x + 1$

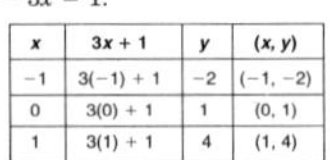

x	3x + 1	y	(x, y)
−1	3(−1) + 1	−2	(−1, −2)
0	3(0) + 1	1	(0, 1)
1	3(1) + 1	4	(1, 4)

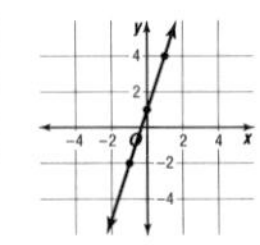

Determine whether each equation is a linear equation. If an equation is linear, rewrite it in the form $Ax + By = C$.

1. $4x - 2y = -1$ 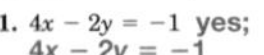 yes; $4x - 2y = -1$
2. $\frac{x}{3} = 5$ 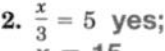 yes; $x = 15$
3. $y = x^2 + 7$ no

Graph each equation.

4. $3x + 2y = 6$
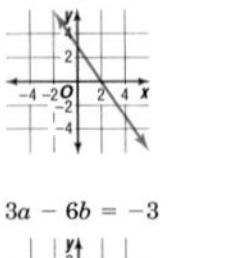
5. $m + 2n = 4$
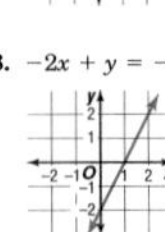
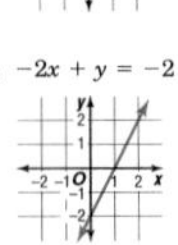
6. $3p + q = -2$
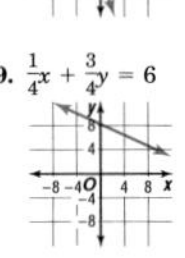
7. $3a - 6b = -3$
8. $-2x + y = -2$
9. $\frac{1}{4}x + \frac{3}{4}y = 6$

Graph each equation. 28–45. See Solutions Manual.

28. $x + 6 = -5$
29. $y = 3x + 1$
30. $6x + 7 = -14y$
31. $2x + 7y = 9$
32. $y + 3 = 4$
33. $x - 6 = -\frac{1}{3}y$
34. $8x - y = 16$
35. $3x + 3y = 12$
36. $6x = 24 - 6y$
37. $x - \frac{7}{2} = 0$
38. $3x - 4y = 60$
39. $4x - \frac{3}{8}y = 1$
40. $2.5x + 5y = 7.5$
41. $x + 5y = 16$
42. $y + 0.25 = 2$
43. $\frac{4x}{3} = \frac{3y}{4} + 1$
44. $y + \frac{1}{3} = \frac{1}{4}x - 3$
45. $\frac{3x}{4} + \frac{y}{2} = 6$

Each table below represents points on a linear graph. Copy and complete each table.

46.

x	y	
0	?	4
1	5	
2	6	
3	7	
4	8	
5	?	9

47.

x	y	
10	?	−5
5	−2.5	
0	0	
−5	2.5	
−10	5	
−15	?	7.5

48.

x	y	
0	0	
3	6	
6	12	
9	?	18
12	?	24
15	?	30

49.

x	y	
−6	5	
−4	?	6
−2	7	
0	8	
2	?	9
4	?	10

Graphing Calculator

Graph each equation using a graphing calculator. Determine a WINDOW setting to use so that a complete graph is shown. Make a sketch of each graph, noting the scale used on each axis.

50. $y = 2x + 4$
51. $4x - 9y = 45$
52. $27x + 75y = 100$
53. $17y = 22$
54. $0.2x - 9.7y = 8.9$
55. $\frac{1}{2}x - \frac{2}{3}y = 10$

50–55. See Solutions Manual.

56. Clear the screen. Press 2nd DRAW 4 3 ENTER. Describe the graph. Explain how you could use this approach to graph $x = -25$. Graph of $x = 3$; use the same procedure but change 3 to −25.

Critical Thinking

57a–b. See margin.

57. The graphs of each group of equations form a family of graphs. Graph each family on the same coordinate plane. Then write an explanation of the similarities and differences that exist in the graphs of each family.
    a. $y = 2x$ $y = 2x + 5$ $y = 2x - 9$ $y = 2x + 14$
    b. $y = -3x$ $y = -3x + 4$ $y = -3x - 10$ $y = -3x + 7$

Applications and Problem Solving

58. **Science** As a thunderstorm approaches, you see lightning as it occurs, but you hear the accompanying sound of the thunder a short time afterward. The distance *y*, in miles, that sound travels in *t* seconds is given by the equation $y = 0.21t$.
    a. Determine the domain and range values for which this equation makes sense.
    b. Graph the equation.
    c. Use the graph to estimate how long it will take you to hear the thunder from a storm that is 3 miles away.

58a. D: $t \geq 0$; R: $y \geq 0$
58b. See margin.
58c. about 14 seconds

Additional Answers

57b. cont.

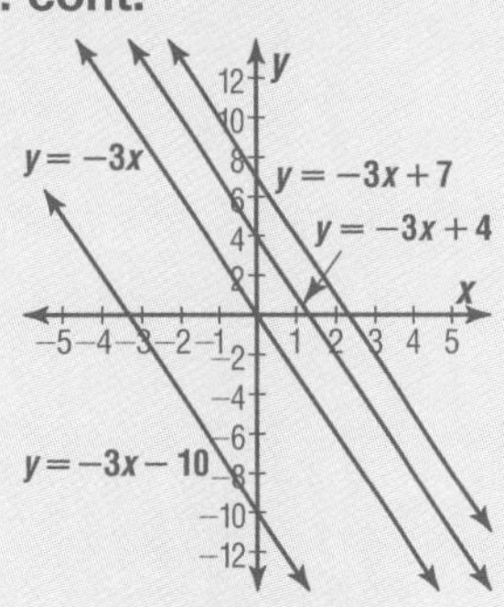

58b.

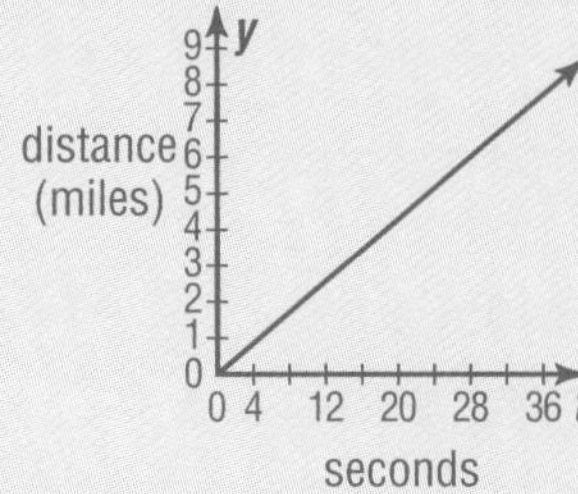

59. **Employment** Elva Duran works as a sales representative for Quasar Electronics. She receives a salary of $1800 per month plus a 6% commission on monthly sales over her target. Her sales in July will be $800, $1300, or $2000 over target, depending on when her orders are filled by the company's distribution center.

a. Graph ordered pairs that represent each of her possible incomes y for the three sales figures x. **See margin.**

b. Will her total income for July be more than $1850 if the equation that represents her income is $y = 1800 + 0.06x$? Explain.

59b. yes, but only if her sales are $1300 or $2000 over target

60. **Weight Training** Hector Farentez, who weighs 68.2 kg, wanted to increase his body size by beginning a weight-training program. A local fitness club suggests that he begin his program using Nautilus® equipment before he attempts the free weights. The Nautilus® equipment includes various machines, each of which exercises a certain group of muscles. After several weeks of training, Hector has expanded his workout time from 30 minutes to 1 hour. **a. $C = 0.092(68.2)t$; See Solutions Manual for graph.**

a. Use the information in the table at the beginning of the lesson to determine a formula for determining how many Calories Hector is burning. Graph the equation showing each 5-minute interval.

b. Later in his training at a weight of 72.3kg, Hector plans to spend one-half hour on Nautilus® and one-half hour on free weights. Will he use more or fewer Calories in this type of workout than an hour of Nautilus®? Explain.

He will burn more calories in the new routine because $0.092(30)(72.3) + 0.132(30)(72.3) > 0.092(60)(72.3)$.

Mixed Review

61. Solve $8x + 2y = 6$ for y. (Lesson 5–3) $y = 3 - 4x$

62. Draw a mapping for the relation {(2, 7), (−3, 7), (2, 4)}. (Lesson 5–2)

62. See Solutions Manual.

63. **Geometry** (Lesson 5–1)

a. Graph (3, 2), (6, 2), and (6, 5). **See Solutions Manual.**

b. If these three points represent the vertices of a square, what would be the coordinates of the fourth point? **(3, 5)**

c. Assume that each unit on each axis represents one inch. What is the perimeter of the square? **12 inches**

64. **Probability** A gumball machine has 24 cherry gumballs, 5 apple gumballs, 18 grape gumballs, 14 orange gumballs, and 9 licorice gumballs. What is the probability of getting an orange gumball? (Lesson 4–6) $\frac{1}{5}$

65. Jimmy got a discount of $4.50 on a new radio. The discounted price was $24.65. What was the percent of discount to the nearest percent? (Lesson 4–5) **15%**

Extension

Reasoning Write a linear equation for y in terms of x for the following ordered pairs: (1, 4), (2, 9), (3, 14), (5, 24), (8, 39). $y = 5x - 1$

4 ASSESS

Closing Activity

Modeling Rest a ladder against a wall. Have students measure the height above the ground and the distance from the wall at each step. Then have them determine the linear equation. **Answers will vary.**

Chapter 5, Quiz B (Lessons 5-3 and 5-4), is available in the *Assessment and Evaluation Masters*, p. 128.

Mid-Chapter Test (Lessons 5-1 through 5-4) is available in the *Assessment and Evaluation Masters*, p. 127.

Additional Answer

59a.

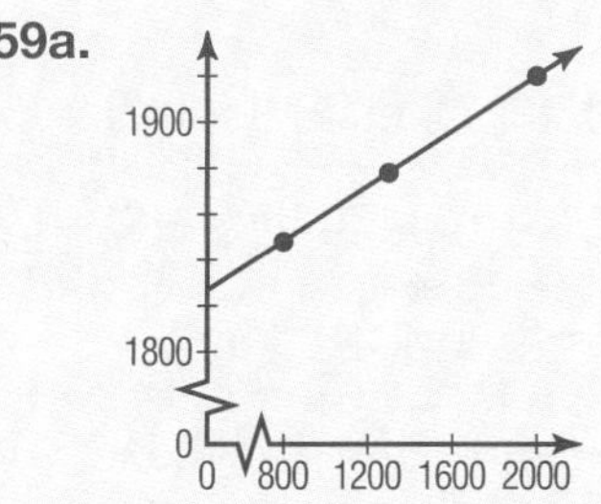

Practice Masters, p. 37

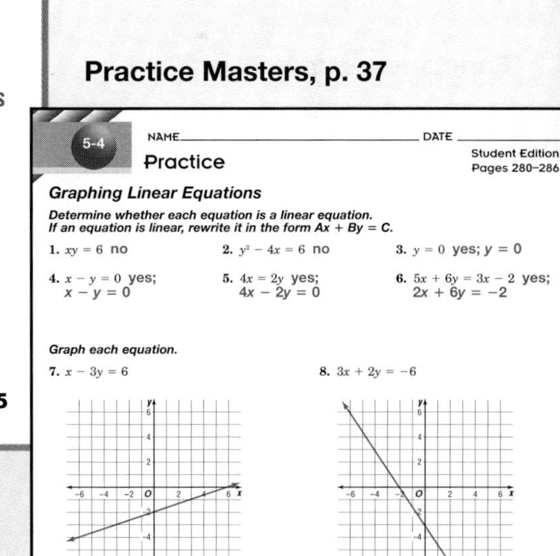

5-4 NAME ____ DATE ____
Practice
Student Edition Pages 280–286

Graphing Linear Equations

Determine whether each equation is a linear equation. If an equation is linear, rewrite it in the form Ax + By = C.

1. $xy = 6$ no
2. $y^2 - 4x = 6$ no
3. $y = 0$ yes; $y = 0$
4. $x - y = 0$ yes; $x - y = 0$
5. $4x = 2y$ yes; $4x - 2y = 0$
6. $5x + 6y = 3x - 2$ yes; $2x + 6y = -2$

Graph each equation.

7. $x - 3y = 6$

8. $3x + 2y = -6$

9. $\frac{1}{3}x - \frac{1}{6}y = 0$

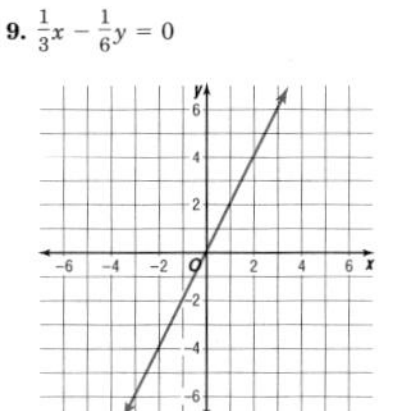

10. $\frac{1}{2}x + \frac{3}{4}y = 1$

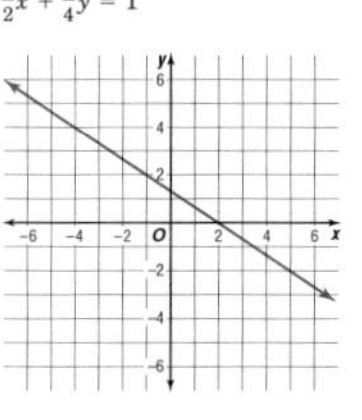

SELF TEST

The Self Test provides students with a brief review of the concepts and skills in Lessons 5-1 through 5-4. Lesson numbers are given to the right of exercises or instruction lines so students can review concepts not yet mastered.

Answers for the Self Test

1.

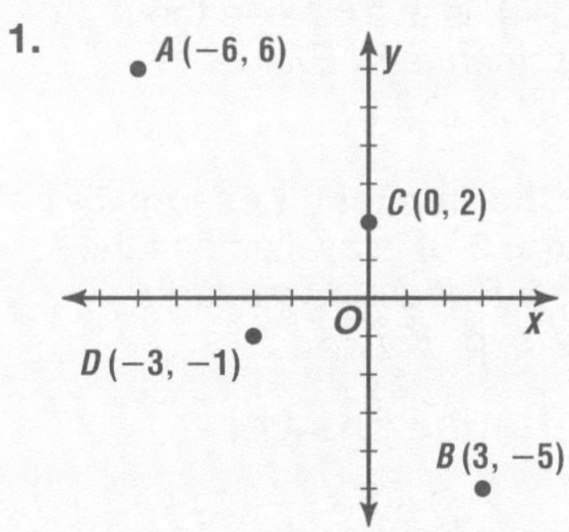

2. {(1, 3), (2, 7), (4, 1), (−3, 3), (−3, −3)}; D = {−3, 1, 2, 4}; R = {−3, 1, 3, 7); Inv = {(3, 1), (7, 2), (1, 4), (3, −3), (−3, −3)}
3. {(−5, −3), (−1, 4), (4, 4), (4, −3)}; D = {−5, −1, 4}; R = {−3, 4}; Inv = {(−3, −5), (4, −1), (4, 4), (−3, 4)}

Enrichment Masters, p. 37

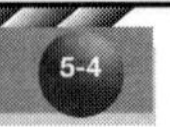

NAME________________________ DATE __________

Enrichment Student Edition Pages 280–285

Taxicab Graphs

You have used a rectangular coordinate system to graph equations such as $y = x - 1$ on a coordinate plane. In a coordinate plane, the numbers in an ordered pair (x, y) can be any two real numbers.

A **taxicab plane** is different from the usual coordinate plane. The only points allowed are those that exist along the horizontal and vertical grid lines. You may think of the points as taxicabs that must stay on the streets.

The taxicab graph shows the equations $y = -2$ and $y = x - 1$. Notice that one of the graphs is no longer a straight line. It is now a collection of separate points.

Taxicab Graph of $y = x - 1$

Graph these equations on the taxicab plane at the right.

1. $y = x + 1$
2. $y = -2x + 3$
3. $y = 2.5$
4. $x = -4$

Use your graphs for these problems.

5. Which of the equations would have the same graph in both the usual coordinate plane and the taxicab plane? $x = -4$

6. Describe the form of equations that have the same graph in both the usual coordinate plane and the taxicab plane. $x = A$ and $y = B$, where A and B are integers

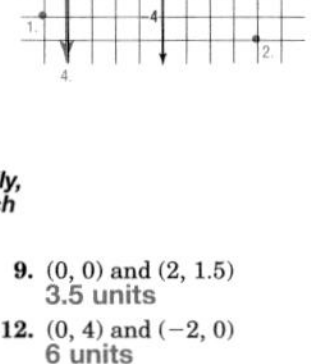

In the taxicab plane, distances are not measured diagonally, but along the streets. Write the taxi-distance between each pair of points.

7. (0, 0) and (5, 2) 7 units
8. (0, 0) and (−3, 2) 5 units
9. (0, 0) and (2, 1.5) 3.5 units
10. (1, 2) and (4, 3) 4 units
11. (2, 4) and (−1, 3) 4 units
12. (0, 4) and (−2, 0) 6 units

Draw these graphs on the taxicab grid at the right.

13. The set of points whose taxi-distance from (0, 0) is 2 units. indicated by crosses

14. The set of points whose taxi-distance from (2, 1) is 3 units. indicated by dots

66. **Running** One member of a cross-country team placed fourth in a meet. The next four finishers on the team placed in consecutive order, but farther behind. The team score was 70. In what places did the other members finish? (Lesson 3–5) **15th, 16th, 17th, 18th**

67. **Miniature Golf** In miniature golf, balls bounce off the walls of the course at the same angle at which they hit. If a ball strikes the wall at 30°, what is the measure of the angle between the two paths of the ball? (Lesson 3–4) **120°**

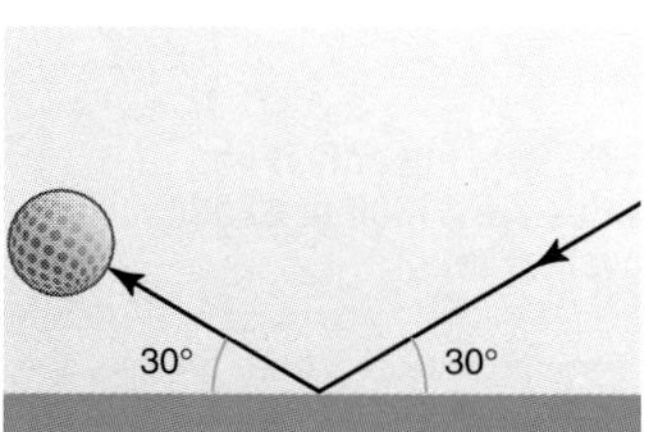

68. Two meters of copper tubing weigh 0.25 kilograms. How much do 50 meters of the same tubing weigh? (Lesson 3–3) **6.25 kg**

69. Simplify $\frac{\frac{-3}{4}}{-36}$. (Lesson 2–7) $\frac{1}{48}$

70. Find $-21 + 52$. (Lesson 2–3) **31**

SELF TEST

1. Graph $A(-6, 6)$, $B(3, -5)$, $C(0, 2)$, and $D(-3, -1)$. (Lesson 5–1) **1–3. See margin.**

Express each relation as a set of ordered pairs. Then state the domain, the range, and the inverse of the relation. (Lesson 5–2)

2.

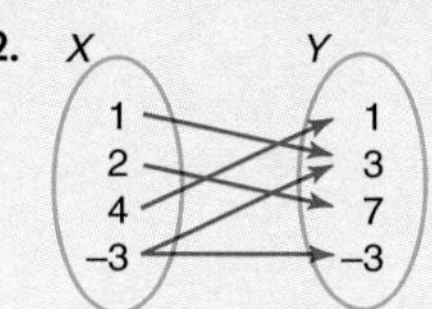

3.

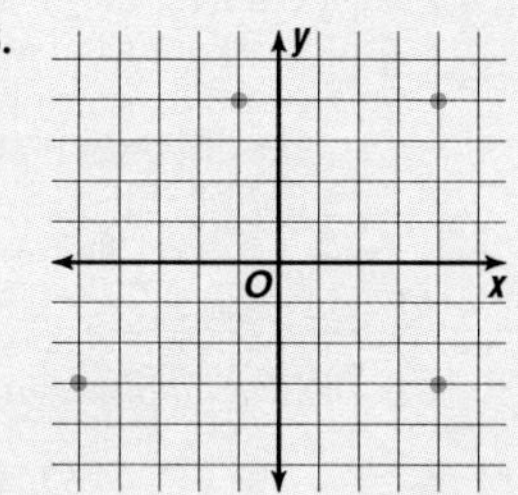

Solve each equation if the domain is {−2, −1, 0, 1, 3}. Then graph the solution set. (Lesson 5–3) **4–6. See Solutions Manual for graphs.**

4. $3y + 6x = 12$ for y $\mathbf{y = 4 - 2x}$
5. $2a + 3b = 9$ for b $\mathbf{b = 3 - \frac{2}{3}a}$
6. $5r = 8 - 4s$ for s $\mathbf{s = \frac{8 - 5r}{4}}$

Graph each equation. (Lesson 5–4) **7–9. See Solutions Manual.**

7. $y = x - 1$
8. $y = 2x - 1$
9. $3x + 2y = 4$

10. **Geometry** The formula for the area of a trapezoid is $A = \frac{1}{2}h(b_1 + b_2)$, where b_1 and b_2 are the lengths of the bases and h is the height. Find the height of the trapezoid shown at the right if its area is 40 m². (Lesson 5–3) **about 5.33 m**

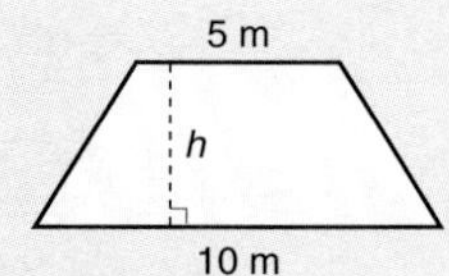

5-5 Functions

What **YOU'LL LEARN**

- To determine whether a given relation is a function, and
- to find the value of a function for a given element of the domain.

Why **IT'S IMPORTANT**

You can use functions to solve problems involving accounting, Earth science, and health.

Air Travel

During certain times of the year, airlines offer lower rates to fly to selected cities. The advertisement below shows discount ticket fares. The mileage between the cities is also listed for those people who are enrolled in frequent flier programs.

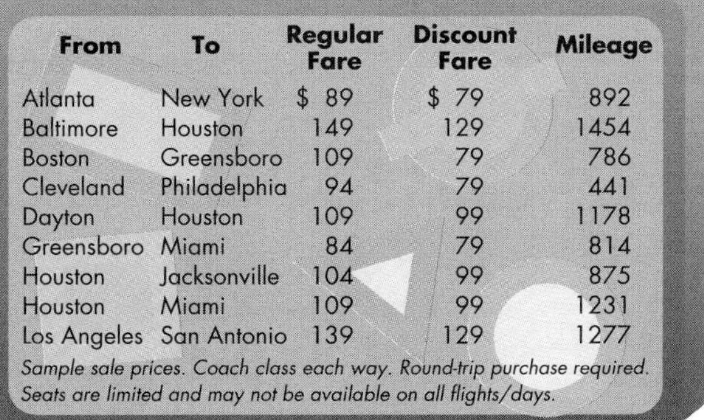

Plan Your Vacation Now!

From	To	Regular Fare	Discount Fare	Mileage
Atlanta	New York	$ 89	$ 79	892
Baltimore	Houston	149	129	1454
Boston	Greensboro	109	79	786
Cleveland	Philadelphia	94	79	441
Dayton	Houston	109	99	1178
Greensboro	Miami	84	79	814
Houston	Jacksonville	104	99	875
Houston	Miami	109	99	1231
Los Angeles	San Antonio	139	129	1277

Sample sale prices. Coach class each way. Round-trip purchase required. Seats are limited and may not be available on all flights/days.

Suppose we let r represent the regular fare, d the discount fair, and m the mileage. The relation whose ordered pairs are of the form (r, d) is graphed below on the left. The relation whose ordered pairs are of the form (m, d) is graphed below on the right.

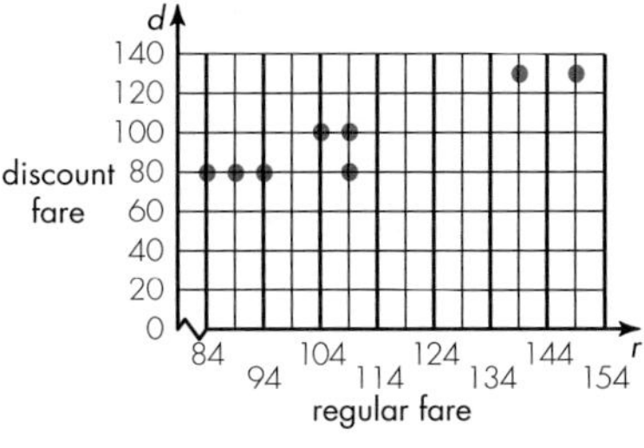

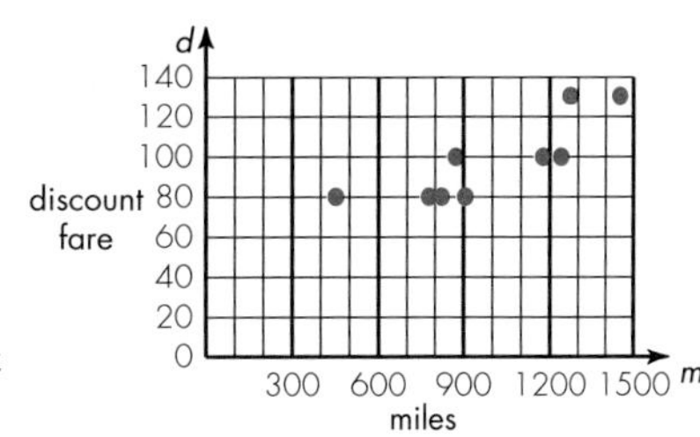

Notice that in the graph on the left, when $r = 109$, there is more than one value of d, 79 and 99. However, in the other graph, there is exactly one value of d for each value of m. Relations with this characteristic are called **functions.**

Definition of Function	**A function is a relation in which each element of the domain is paired with *exactly* one element of the range.**

5-5 LESSON NOTES

NCTM Standards: 1–6

Instructional Resources

- Study Guide Master 5-5
- Practice Master 5-5
- Enrichment Master 5-5
- Modeling Mathematics Masters, pp. 48–50, 76
- Real-World Applications, 15
- Tech Prep Applications Masters, p. 10

Transparency 5-5A contains the 5-Minute Check for this lesson; **Transparency 5-5B** contains a teaching aid for this lesson.

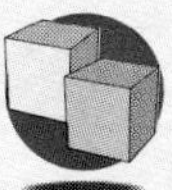

Recommended Pacing	
Standard Pacing	Days 8 & 9 of 14
Honors Pacing	Day 8 of 13
Block Scheduling*	Day 5 of 7 (along with Lesson 5-6)
Alg. 1 in Two Years*	Days 12 & 13 of 21

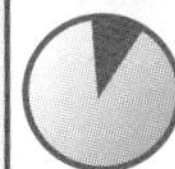

*For more information on pacing and possible lesson plans, refer to the *Block Scheduling Booklet* and *Algebra 1 in Two Years*.

1 FOCUS

5-Minute Check
(over Lesson 5-4)

Determine whether each equation is a linear equation. If so, identify *A, B,* and *C*.

1. $2a - 3b = 7$ **linear; $A = 2$, $B = -3$, $C = 7$**
2. $y = \frac{1}{x}$ **not linear**
3. $3x = 2y - 14$ **linear; $A = 3$, $B = -2$, $C = -14$**
4. Graph $2x - y = 8$.

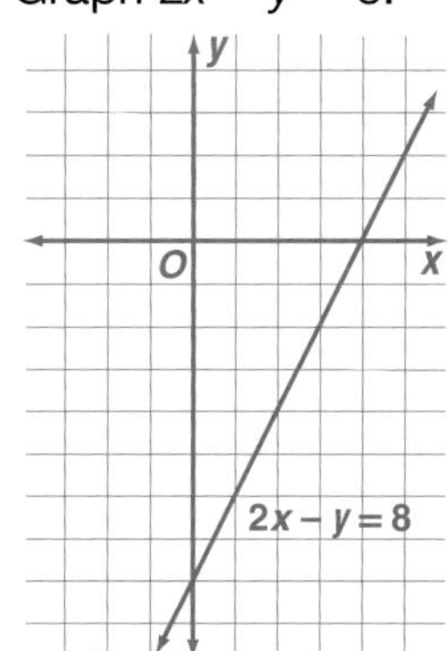

Motivating the Lesson

Questioning Ask students to consider the statement, "Your shoe size is a function of the size of your foot." Develop the idea that "is a function of" actually means "depends on." Have students give examples of additional situations in which the value of one variable results in only one true value for a second variable. Then have students express the definition of a function in their own words.

2 TEACH

In-Class Examples

For Example 1
Determine whether each relation is a function. Explain your answer.

a. {(3, 2), (4, 5), (6, 8), (7, 1)} **Yes; each element of the domain is paired with exactly one element of the range.**

b.

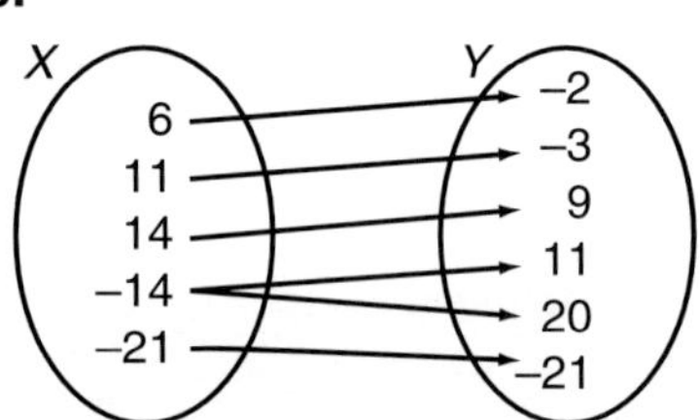

No; the element −14 in the domain is paired with both 11 and 20 in the range.

c.

x	y
0	5
4	3
6	5
−7	−4

Yes; for each element in the domain there is only one corresponding element in the range.

For Example 2
Determine whether each relation is a function.

a. $y = 2x + 1$ **yes**
b. $y = x^2 + 1$ **yes**
c. $y^2 = x$ **no**
d. $y = 4x + 5$ **yes**

Example 1 **Determine whether each relation is a function. Explain your answer.**

a. {(2, 3), (3, 0), (5, 2), (−1, −2), (4, 1)}

Since each element of the domain is paired with exactly one element of the range, this relation is a function.

b.

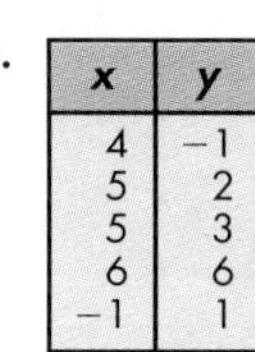

This mapping represents a function since, for each element of the domain, there is *only one* corresponding element in the range.
It does not matter if two elements of the domain are paired with the same element in the range.

c.

x	y
4	−1
5	2
5	3
6	6
−1	1

This table represents a relation that is not a function. The element 5, in the domain, is paired with both 2 and 3 in the range.

There are several ways to determine whether an equation represents a function.

Example 2 **Determine whether $x - 4y = 12$ is a function.**

Method 1: Make a table of solutions.

First solve for y.

$x - 4y = 12$

$-4y = -x + 12$ *Subtract x from each side.*

$y = \frac{1}{4}x - 3$ *Divide each side by −4.*

Next make a table of values like the one at the right.

x	y
−8	−5
−4	−4
−2	−3.5
0	−3
2	−2.5
4	−2
8	−1

It appears that for any given value of x, there is only one value for y that will satisfy the equation. Therefore, the equation $x - 4y = 12$ is a function.

Method 2: Graph the equation.

Since the equation is in the form $Ax + By = C$, the graph of the equation will be a line. Graph the ordered pairs from Method 1 and connect them with a line.

Now place your pencil at the left of the graph to represent a vertical line. Slowly move the pencil to the right across the graph.

For each value of x, this vertical line passes through no more than one point on the graph. Thus, the line represents a function.

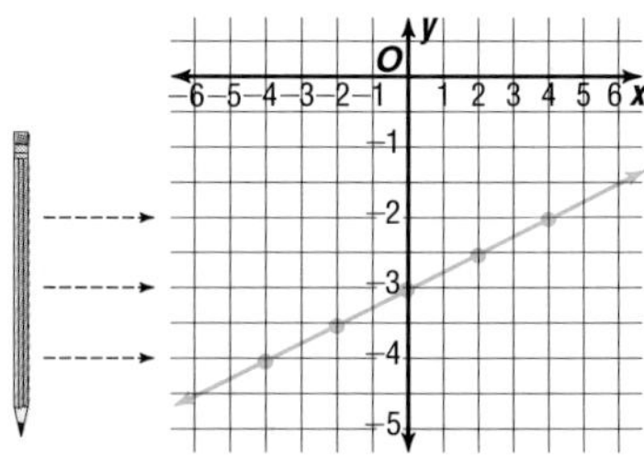

Alternative Learning Styles

Auditory Lead a class discussion about the differences between a relation and a function. Make sure students can clearly explain how the two differ.

Using a pencil to see if a graph represents a function is one way to perform the **vertical line test.**

Vertical Line Test for a Function	**If any vertical line passes through no more than one point of the graph of a relation, then the relation is a function.**

Example  **Use the vertical line test to determine if each relation is a function.**

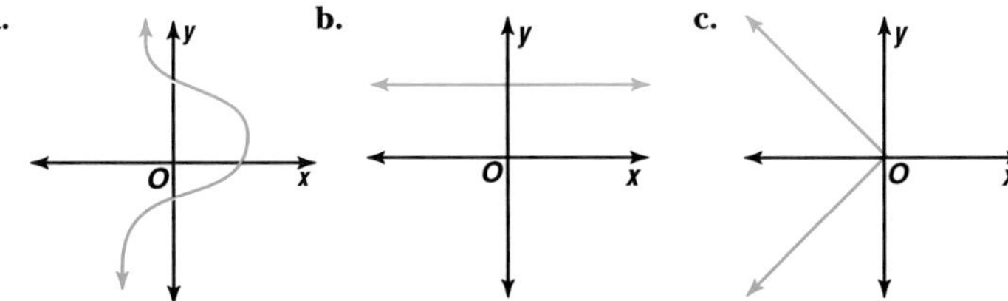

Graph b is the only relation to pass the vertical line test. Thus, it is the only function. With graphs a and c, a vertical line intersects the graph in more than one point. Thus, they are *not* functions.

Letters other than f are also used for names of functions. For example, g(x) and h(x) are sometimes used.

The ordered pair (4, f(4)) is a solution of the function f.

Equations that are functions can be written in a form called **functional notation.** For example, consider the equation $y = 3x - 7$.

equation	**functional notation**
$y = 3x - 7$	$f(x) = 3x - 7$
	The symbol f(x) is read "f of x."

In a function, x represents the elements of the domain and $f(x)$ represents the elements of the range. Suppose you want to find the value in the range that corresponds to the element 4 in the domain. This is written $f(4)$ and is read "f of 4." The value of $f(4)$ is found by substituting 4 for x in the equation. So, $f(4) = 3(4) - 7$ or 5.

Example  **If $f(x) = 2x - 9$, find each value.**

a. $f(6)$

$$\begin{aligned} f(6) &= 2(6) - 9 \\ &= 12 - 9 \\ &= 3 \end{aligned}$$

b. $f(-2)$

$$\begin{aligned} f(-2) &= 2(-2) - 9 \\ &= -4 - 9 \\ &= -13 \end{aligned}$$

c. $f(k + 1)$

$$\begin{aligned} f(k + 1) &= 2(k + 1) - 9 \\ &= 2k + 2 - 9 \\ &= 2k - 7 \end{aligned}$$

The functions we have studied thus far have been linear functions. Many functions are not linear. However, you can find values of functions in the same way.

Example  **If $h(z) = z^2 - 4z + 9$, find each value.**

a. $h(-3)$

$$\begin{aligned} h(-3) &= (-3)^2 - 4(-3) + 9 \\ &= 9 + 12 + 9 \text{ or } 30 \end{aligned}$$

(continued on the next page)

Teaching Tip Have students perform the vertical line test with the graphs in the application at the beginning of the lesson.

Teaching Tip Have students look at the graphs in Lesson 5-2 and determine which of the graphs are functions.

In-Class Examples

For Example 3
Use the vertical line test to determine if the relation is a function. **It is a function.**

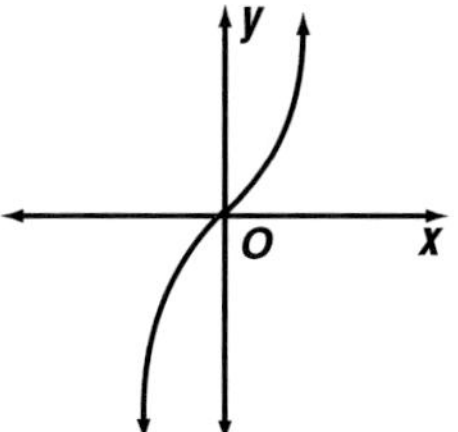

For Example 4
Find each value.

a. $f(1)$ if $f(x) = 3x - 2$ **1**
b. $f(2)$ if $f(x) = 7 - 4x$ **−1**

Teaching Tip It may be helpful for some students to think of $f(x)$ as "f at x" since it represents the value of f at x.

In-Class Examples

For Example 5
Find each value.

a. $g(3)$ if $g(x) = x^2 + 3x + 3$ **21**
b. $g(4a)$ if $g(x) = x^3 - 2x - 3$ $\mathbf{64a^3 - 8a - 3}$
c. $3[f(2)]$ if $f(t) = 8t + 3t^2$ **84**
d. $2[f(-2)]$ if $f(t) = 4t^2 - 6t$ **56**

For Example 6
For the equation $T = 3b + 100$:

a. Write the equation in functional notation. $\mathbf{T(b) = 3b + 100}$
b. Find $T(1)$, $T(2)$, $T(5)$, and $T(10)$. **103, 106, 115, 130**
c. Graph the function.

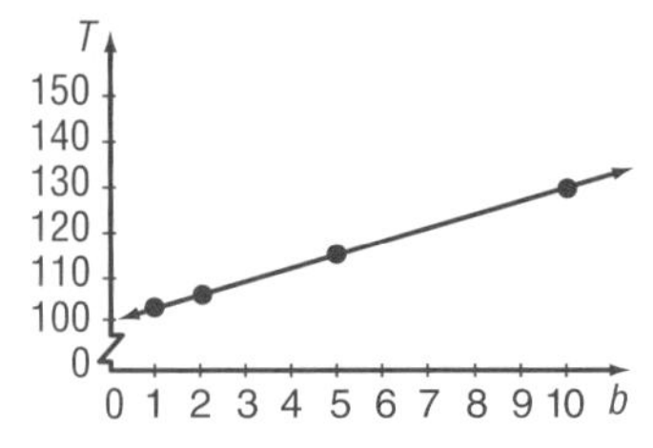

Listed below are the world's top drug companies and their profits in 1994.

1. Merck, \$1.5 billion
2. American Home Products, \$1.1 billion
3. Smith Kline Beecham, \$780 million
4. Squibb, \$747 million
5. Pfizer, \$681 million

b. $h(5c)$

$h(5c) = (5c)^2 - 4(5c) + 9$ *Substitute 5c for z.*
$= 5 \cdot 5 \cdot c \cdot c - 4 \cdot 5 \cdot c + 9$
$= 25c^2 - 20c + 9$

c. $5[h(c)]$

$5[h(c)] = 5[(c)^2 - 4(c) + 9]$ *5[h(c)] means 5 times the value of h(c).*
$= 5 \cdot c^2 - 5 \cdot 4c + 5 \cdot 9$ *Distributive property*
$= 5c^2 - 20c + 45$

Notice that when comparing parts b and c, $5[h(c)] \neq h(5c)$.

Functions are often used in solving real-life problems.

Example 6

Health

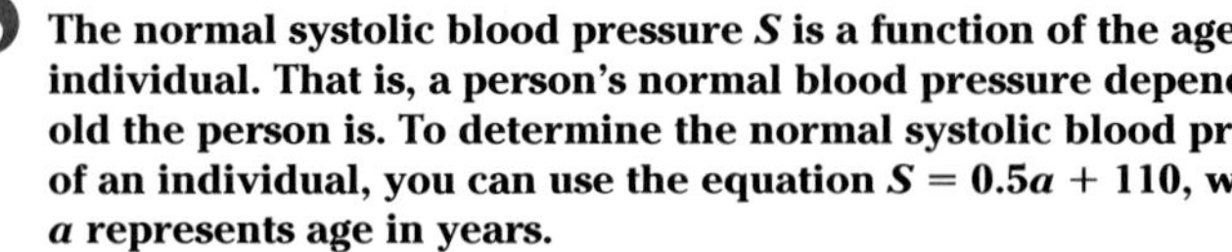

The normal systolic blood pressure S is a function of the age a of the individual. That is, a person's normal blood pressure depends on how old the person is. To determine the normal systolic blood pressure of an individual, you can use the equation $S = 0.5a + 110$, where a represents age in years.

a. Write the equation in functional notation.
b. Find $S(10)$, $S(30)$, $S(50)$, and $S(70)$.
c. Graph the function. Name the independent and dependent quantities.
d. Use the graph of the function to estimate whether blood pressure increases or decreases with age. Then estimate the blood pressure of an 80-year-old person.

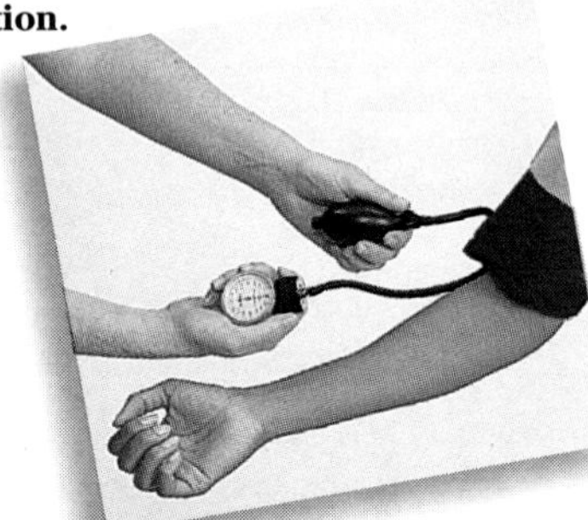

Top Five List

Best Selling Prescription Drugs in the World

Brand Name	Used for
1. Zantac	ulcers
2. Vasotec	hypertension*
3. Capoten	hypertension*
4. Voltaren	arthritis
5. Tenormin	hypertension*

*high blood pressure

a. Let $S(a)$ represent the function. The equation becomes $S(a) = 0.5a + 110$.

b. Make a table with 10, 30, 50, and 70 as values of a.

a	$S(a) = 0.5a + 110$	$S(a)$	$(a, S(a))$
10	$S(10) = 0.5(10) + 110$	115	(10, 115)
30	$S(30) = 0.5(30) + 110$	125	(30, 125)
50	$S(50) = 0.5(50) + 110$	135	(50, 135)
70	$S(70) = 0.5(70) + 110$	145	(70, 145)

c. Use the ordered pairs from the table to graph the function. Age is the independent quantity, and systolic blood pressure is the dependent quantity.

d. The graph indicates that as you age, your blood pressure is expected to rise. An 80-year-old person would expect to have a systolic pressure of about 150.

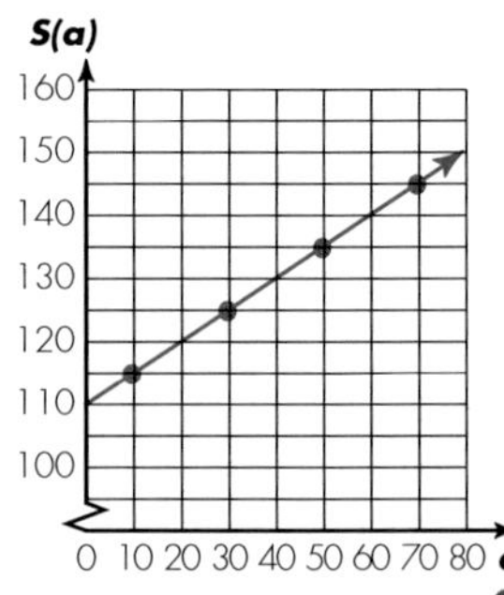

GLENCOE Technology

Interactive Mathematics Tools Software

In this interactive lesson, students use the function for height h of a rocket after t seconds to find the maximum height of the rocket on a graph. A **Computer Journal** gives students an opportunity to write about what they have learned.

For Windows & Macintosh

CHECK FOR UNDERSTANDING

Communicating Mathematics

Study the lesson. Then complete the following. 1–2. See margin.

1. **Explain** any differences between a relation and a function.
2. **Describe** how the phrase "is a function of" relates to the mathematical definition of function. For example, the diameter of a tree is a function of its age.
3. **Write** how you would find $g(1)$ if $g(x) = 3x + 12$.
4. **You Decide** Is the statement "All linear equations are functions" true? Support your answer with examples or counterexamples.

3. Substitute 1 for x in the equation and evaluate; $g(1) = 15$.
4. False; $x = 4$ is not a function.

5. In your own words, explain how and why the vertical line test works. See margin.

Guided Practice

Determine whether each relation is a function.

6. 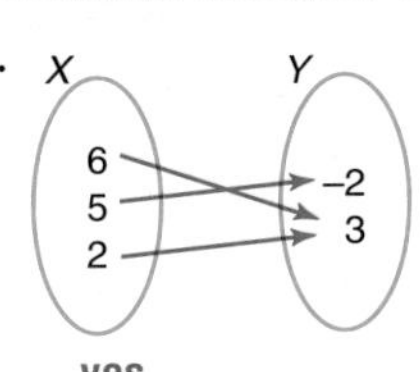

yes

7.

x	y
−2	3
5	3
5	4
4	0

no

8. 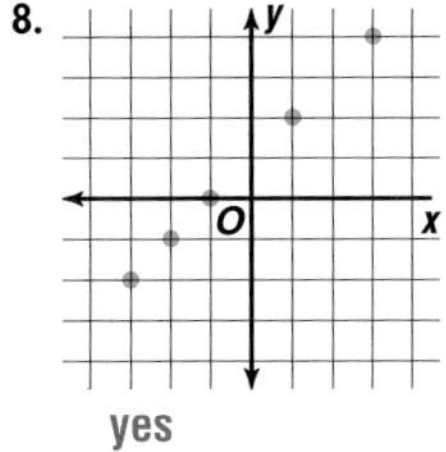

yes

9. $\{(-3, 1), (-1, 3), (1, -2), (3, 2)\}$ yes
10. $\{(3, 1), (-2, 2), (1, -1), (1, 2), (-3, 1), (-2, 5)\}$ no
11. $y + 5 = 7x$ yes
12. $2x^2 + 3y^2 = 36$ no
13. The graph of $y^2 = x + 4$ is shown at the right. Use the vertical line test to determine if this relation is a function. no

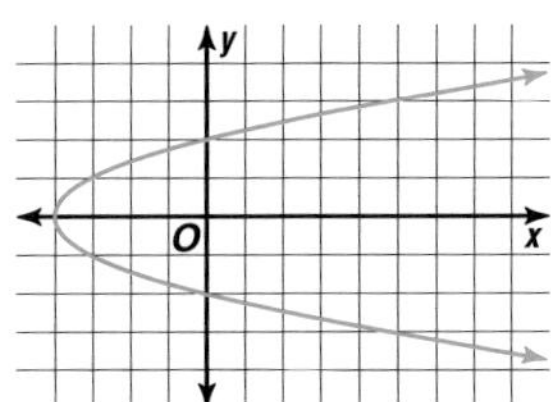

14. Which of the following graphs represent a function? a, c

a. $y = |x - 1|$

b. $x = |y| + 1$

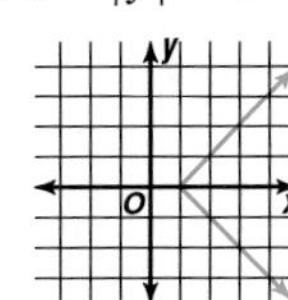

c. $y = |x| + 1$

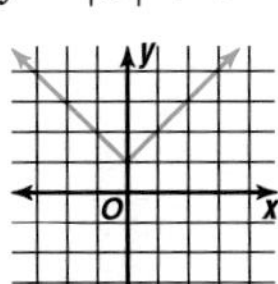

d. $x = 1 - |y|$

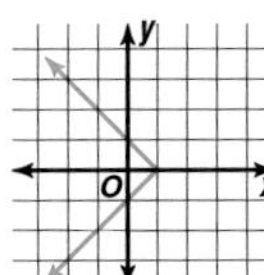

LOOK BACK

You can refer to Lesson 2-3 for more information on absolute value. $|x|$.

If $h(x) = 3x + 2$, find each value.

15. $h(-4)$ -10 16. $h(2)$ 8 17. $h(w)$ $3w + 2$ 18. $h(r - 6)$ $3r - 16$

3 PRACTICE/APPLY

Check for Understanding

Exercises 1–18 are designed to help you assess your students' understanding through reading, writing, speaking, and modeling. You should work through Exercises 1–5 with your students and then monitor their work on Exercises 6–18.

Additional Answers

1. A relation is a set of ordered pairs. A function is a relation in which each member of the domain is paired with only one member of the range.
2. The value of one variable is dependent upon the value of the other variable. That is, if w is a function of f, the value of w is dependent upon the value of f in the expression defining w.
5. Sample answer: If you can draw a vertical line through the graph at any point and it intersects the graph more than once, the graph is not the graph of a function.

Reteaching

Using Alternatives Demonstrate how functional notation $f(x)$ is easier to use than y.

$y = 2x - 3$	vs. $f(x) = 2x - 3$
Find y when x is 4.	Find $f(4)$.
$y = 2(4) - 3$	$f(4) = 2(4) - 3$
$y = 5$ when $x = 4$	$f(4) = 5$

Assignment Guide

Core: 19–53 odd, 54, 55, 57, 58–67
Enriched: 20–52 even, 54–67

For **Extra Practice,** see p. 768.

The red A, B, and C flags, printed only in the Teacher's Wraparound Edition, indicate the level of difficulty of the exercises.

EXERCISES

Practice

Determine whether each relation is a function.

19.
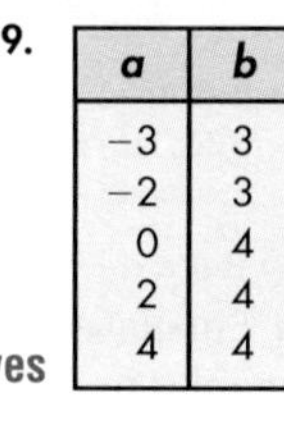

a	b
−3	3
−2	3
0	4
2	4
4	4

yes

20.
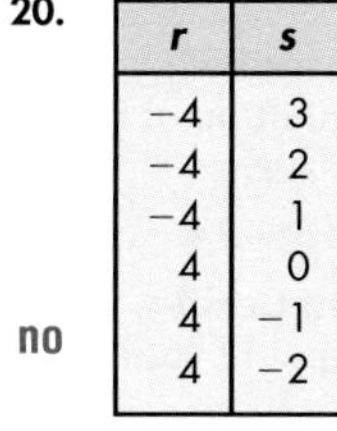

r	s
−4	3
−4	2
−4	1
4	0
4	−1
4	−2

no

21.
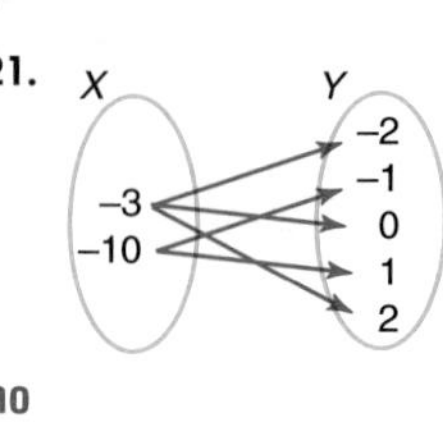

no

22.
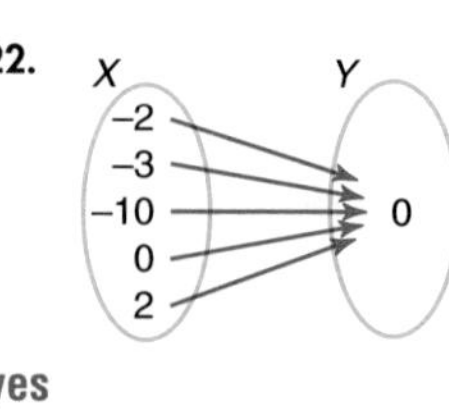

yes

23.
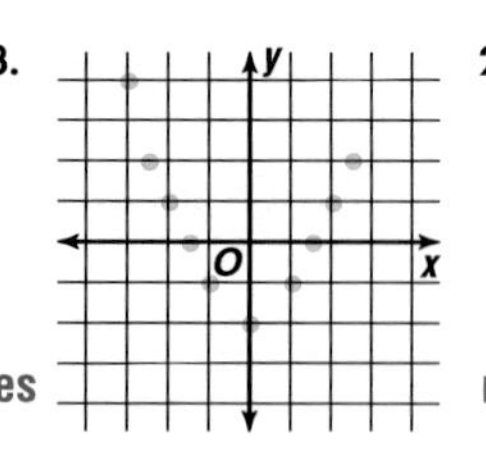

yes

24.
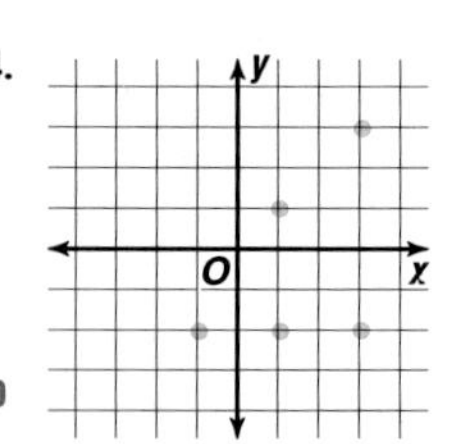

no

25. yes 26. no
27. yes 28. yes

25. $\{(6, 3), (5, -2), (2, 3), (12, -12)\}$
26. $\{(4, 5), (3, -2), (-2, 5), (4, 7)\}$
27. $\{(5, -1), (6, -1), (-8, -1), (0, -1)\}$
28. $\{(4, -2), (-4, -2), (9, -2), (0, -2)\}$

B

29. $y = -15$ yes
30. $x = 13$ no
31. $y = 3x - y$ yes
32. $y = |x|$ yes
33. $x = |y|$ no
34. $yx = 36$ yes

41. −18
42. 3.84
43. $9b^2 - 6b$
46. $12w^2 - 12w$
47. $5a^4 - 10a^2$
48. $4c + 14$
49. $24p - 36$
50. $-60w - 30$

If $f(x) = 4x + 2$ and $g(x) = x^2 - 2x$, find each value.

35. $f(-4)$ −14
36. $g(4)$ 8
37. $g\left(\frac{1}{5}\right)$ $-\frac{9}{25}$
38. $f\left(\frac{3}{4}\right)$ 5
39. $g(3.5)$ 5.25
40. $f(6.2)$ 26.8
41. $3[f(-2)]$
42. $-6[g(0.4)]$
43. $g(3b)$
44. $f(2y)$ $8y + 2$
45. $-3[g(1)]$ 3
46. $3[g(2w)]$

C

47. $5[g(a^2)]$
48. $f(c + 3)$
49. $6[f(p - 2)]$
50. $-3[f(5w + 2)]$

51. a. Create and graph a relation that is a function and has an inverse that is also a function. Answers will vary; a sample answer is $f(x) = x$.

b. Create and graph a relation that is a function and has an inverse that is not a function. Answers will vary; a sample answer is $f(x) = x^2$.

Choose the graph that best represents the information given. Explain your choice. Determine whether the graph represents a function.

52. **Communication** A call to London, England costs $1.78 for the first minute and $1.00 for each additional minute. Fractions of a minute are charged as an entire minute. b; Each step represents a minute or any part thereof.

a.
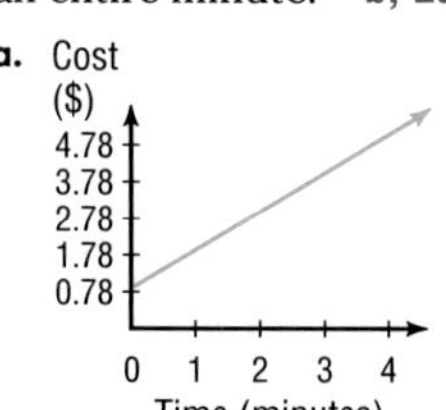

b.
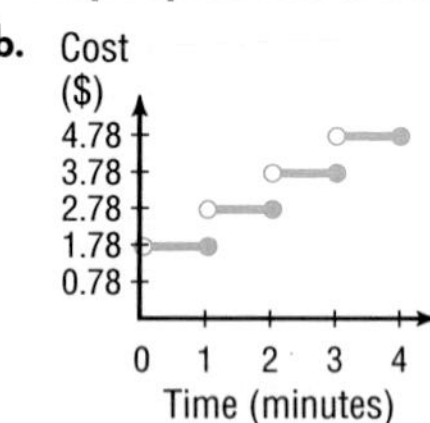

c.
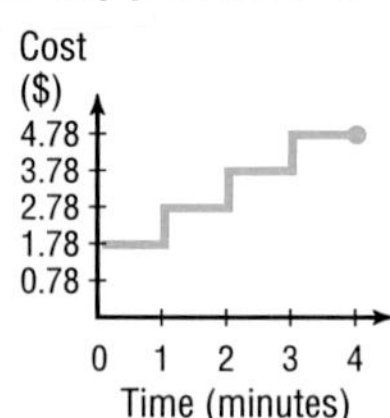

Study Guide Masters, p. 38

NAME______________________ DATE ____________

Study Guide

Student Edition Pages 287–294

Functions

A special type of relation is called a **function.**

Definition of Function
A **function** is a relation in which each element of the domain is paired with *exactly* one element of the range.

Example 1: Is $\{(6, -3), (4, 1), (7, -2), (-3, 1)\}$ a function? Is the inverse a function?

Since each element of the domain is paired with exactly one element of the range, the relation *is* a function. The inverse is not a function because 1 is paired with more than one element of the range.

The equation $y = 2x + 1$ can be written as $f(x) = 2x + 1$. If $x = 3$, then $f(3) = 2(3) + 1$, or 7. Thus, $f(3)$, which is read "f of 3" is a way of referring to the value of y that corresponds to $x = 3$.

Example: If $f(x) = 3x - 4$, find $f(3)$ and $f(-2)$.

$f(3) = 3(3) - 4 = 9 - 4 = 5$

$f(-2) = 3(-2) - 4 = -6 - 4 = -10$

Determine whether each relation is a function.

1.
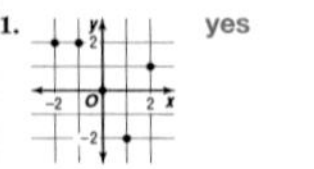

yes

2.
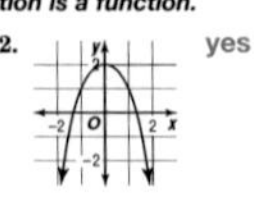

yes

3.
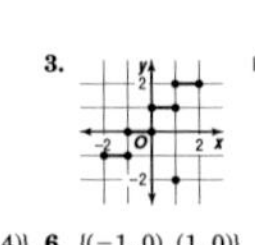

no

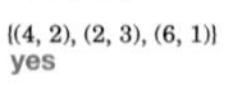

4. $\{(4, 2), (2, 3), (6, 1)\}$ yes

5. $\{(-3, -3), (-3, 4), (-2, 4)\}$ no

6. $\{(-1, 0), (1, 0)\}$ yes

7. $-2x + 4y = 0$ yes

8. $x^2 + y^2 = 8$ no

9. $-\frac{1}{4}x = -\frac{1}{4}y - 2$ yes

Given $f(x) = 2x - 4$ and $g(x) = x^2 - 4x$, find each value.

10. $f(4)$ 4
11. $g(2)$ −4
12. $f(-5)$ −14
13. $g(-3)$ 21
14. $f\left(\frac{1}{4}\right)$ $-3\frac{1}{2}$
15. $g\left(\frac{1}{4}\right)$ $-\frac{15}{16}$

53. **Manufacturing** A company charges \$9.95 for each custom-printed T-shirt. If you order 8 or more, but less than 15, the charge is \$8.25 each. If you order 15 or more, the charge is \$6.50 each. **See margin.**

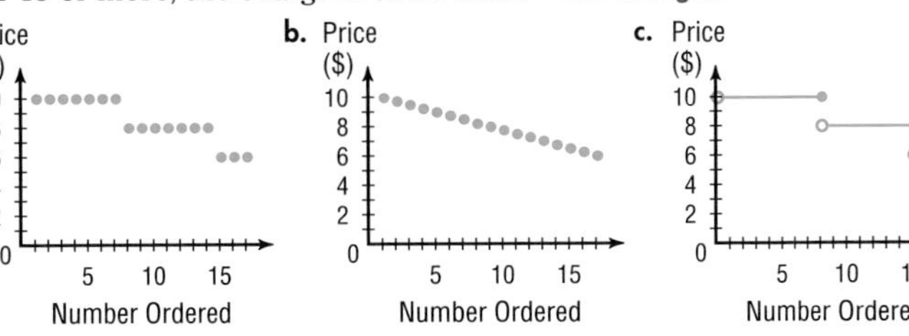

Critical Thinking

54c. The inverse is a reflection of the function with $y = x$ acting as the mirror.

54. Let the domain of $f(x) = x^2 + 3x + 2$ be $\{-5, -4, -3, -2, -1, 0, 1, 2\}$.
   a. Graph the function, the ordered pairs that make up the inverse of the function, and the line $y = x$ on the same coordinate plane. **See margin.**
   b. Is the inverse of this function a function? **no**
   c. How are the graphs of the function and its inverse related to the line $y = x$?

Applications and Problem Solving

55. **Metallurgy** Since pure gold is very soft, other metals are often added to gold to make an alloy that is stronger and more durable. The relative amount of gold in a piece of jewelry is measured in karats. The formula for the relationship is $g = \frac{25k}{6}$, where k represents the number of karats and g is the percent of gold in the piece.
   a. Determine the domain and range values for which this function makes sense.
   b. Graph the function and describe the graph. **See margin.**
   c. How many karats are in a ring made of pure gold? **24 karats**

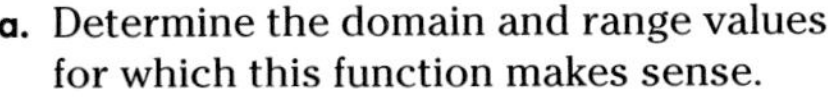

55a. D: $0 \leq k \leq 24$; R: $0 \leq g \leq 100$

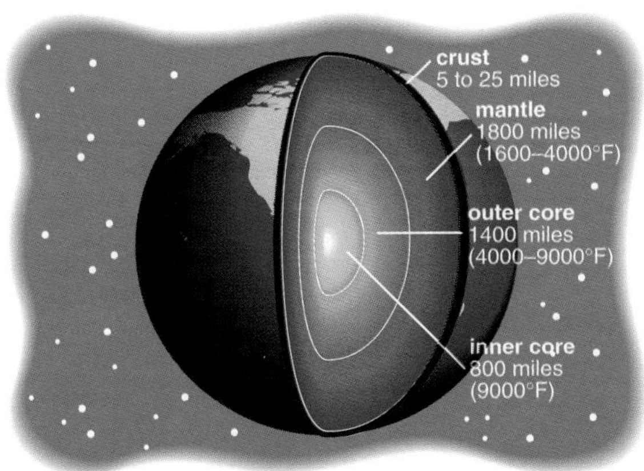

56. **Earth Science** Earth's interior is composed of four levels: the crust, the mantle, the outer core, and the inner core. The inner core is believed to be solid while the outer core is molten. The temperature in the first 62 miles below Earth's surface can be approximated with the formula $T = 35d + S$, where T is the temperature in degrees Fahrenheit, d is the distance in miles, and S is the temperature (°F) on the surface.
   a. Suppose the temperature at the surface is 75°F. Find the temperature at the lower edge of the crust if the crust is 30 miles deep at that point. **1125°F**
   b. Russian scientists are credited with drilling the deepest hole ever on the Kola Peninsula, which is located near the Arctic Circle in northwest Russia about 250 miles from Finland. Drilling began in 1970, but is only 7 miles deep thus far. What is the temperature at the bottom of this hole if the outside temperature is 0°F? **245°F**
   c. Why do you think that there is a range of values given for the depth of the crust?

56c. Because the crust is thinnest when measured from the bottom of the ocean and thicker from the tops of mountains.

Extension

Problem Solving If $h(x) = 4x^3 + 3x^2 + x - 12$, find $3[h(2)]$. **102**

Additional Answers

53. a; It is a step function using individual points, since you cannot have a fraction of a T-shirt.

54a.

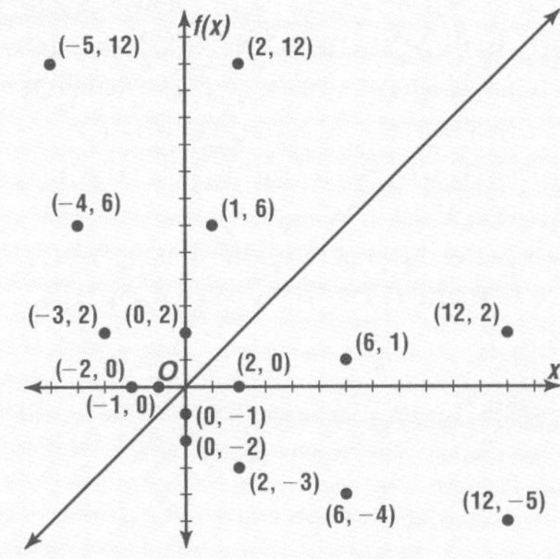

55b.

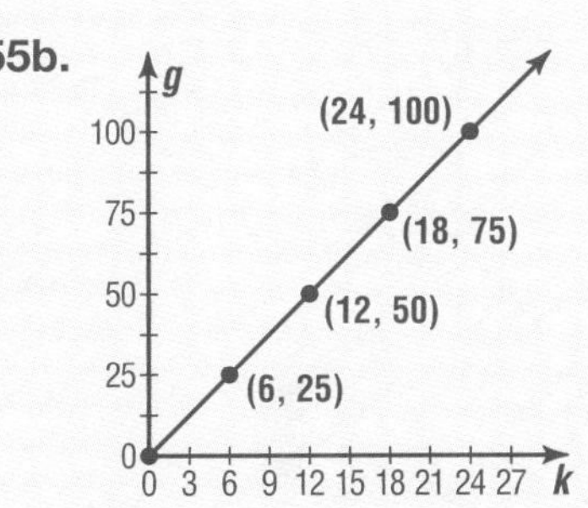

It is a line.

Practice Masters, p. 38

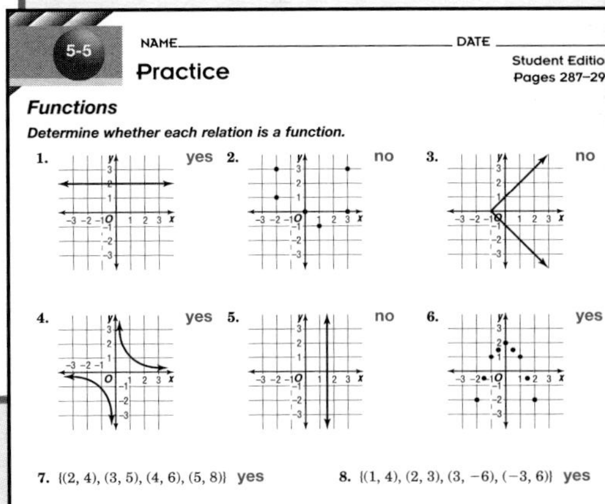

5-5 Practice

NAME ______ DATE ______

Student Edition Pages 287–294

Functions

Determine whether each relation is a function.

1. yes 2. no 3. no

4. yes 5. no 6. yes

7. {(2, 4), (3, 5), (4, 6), (5, 8)} yes
8. {(1, 4), (2, 3), (3, −6), (−3, 6)} yes
9. {(1, 5), (1, 7), (1, −5), (1, −7)} no
10. {(0, 4), (1, 4), (2, 4), (3, 4)} yes
11. $2x - 3y = 4$ yes
12. $xy = 1$ yes
13. $y = -2$ yes
14. $x^2 = y + 1$ yes
15. $x = 3$ no
16. $x^2 + y^2 = 4$ no

If $f(x) = 2x - 6$ and $g(x) = x - 2x^2$, find each value.

17. $f(2)$ −2
18. $f\left(-\frac{1}{2}\right)$ −7
19. $f(-2.5)$ −11
20. $g(-1)$ −3
21. $g(3)$ −15
22. $g\left(-\frac{1}{3}\right)$ $-\frac{5}{9}$

4 ASSESS

Closing Activity

Modeling Using materials on hand, construct a simple fulcrum-and-lever device. With objects of different weights, demonstrate for students how varying amounts of pressure are required to lift the different objects using the lever. Explain that the amount of pressure applied to lift the object is a function of the weight of the object.

Additional Answer

58.

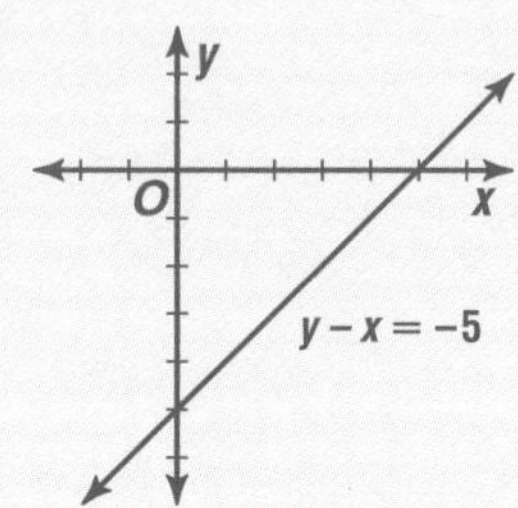

Enrichment Masters, p. 38

NAME______________________ DATE __________

Enrichment

Student Edition Pages 287–294

Composite Functions

Three things are needed to have a function—a set called the *domain*, a set called the *range*, and a *rule* that matches each element in the domain with only one element in the range. Here is an example.

Rule: $f(x) = 2x + 1$

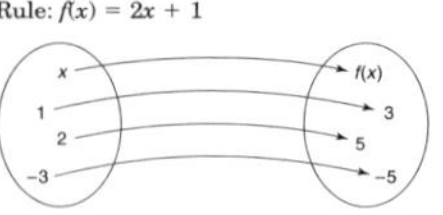

$f(x) = 2x + 1$
$f(1) = 2(1) + 1 = 2 + 1 = 3$
$f(2) = 2(2) + 1 = 4 + 1 = 5$
$f(-3) = 2(-3) + 1 = -6 + 1 = -5$

Suppose we have three sets A, B, and C and two functions described as shown below.

Rule: $f(x) = 2x + 1$ Rule: $g(y) = 3y - 4$

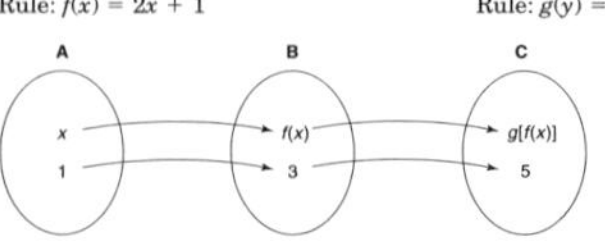

$g(y) = 3y - 4$
$g(3) = 3(3) - 4 = 5$

Let's find a rule that will match elements of set A with elements of set C without finding any elements in set B. In other words, let's find a rule for the **composite function** $g[f(x)]$.

Since $f(x) = 2x + 1$, $g[f(x)] = g(2x + 1)$.
Since $g(y) = 3y - 4$, $g(2x + 1) = 3(2x + 1) - 4$, or $6x - 1$.
Therefore, $g[f(x)] = 6x - 1$.

Find a rule for the composite function g[f(x)].

1. $f(x) = 3x$ and $g(y) = 2y + 1$ **$g[f(x)] = 6x + 1$**
2. $f(x) = x^2 + 1$ and $g(y) = 4y$ **$g[f(x)] = 4x^2 + 4$**
3. $f(x) = -2x$ and $g(y) = y^2 - 3y$ **$g[f(x)] = 4x^2 + 6x$**
4. $f(x) = \frac{1}{x - 3}$ and $g(y) = y^{-1}$ **$g[f(x)] = x - 3$**
5. Is it always the case that $g[f(x)] = f[g(x)]$? Justify your answer. **No. For example, in Exercise 1, $f[g(x)] = f(2x + 1) = 3(2x + 1) = 6x + 3$, not $6x + 1$.**

57a. $260.87

57b. Equations may vary; Sample equation: $B = \frac{3}{23}P$

57. Accounting The following is adapted from an article by Gail A. Eisner, CPA (certified public accountant), in *The Mathematics Teacher.*

> Early in my accounting career, the partner who was directly supervising me, a very bright, young CPA, called me into his office and asked how a person could possibly calculate a bonus if the company's formula required that the bonus be 15 percent of the net profits *after* the bonus has been deducted? I showed him the linear equation $B = 0.15(P - B)$, where B is the bonus, and P (a known quantity) represents the profits before subtracting the bonus.

a. Suppose the profits are $2000. Determine the amount of the bonus.
b. Solve the equation for B and graph the resulting equation.

Mixed Review

59. $\{-16, -13, 5, 11\}$
60. $\{(1, -1), (9, -5), (6, 4)\}$
61. $\frac{3}{13}$

58. Graph $y - x = -5$. (Lesson 5–4) **See margin.**

59. Solve $3a - b = 7$ if the domain is $\{-3, -2, 4, 6\}$. (Lesson 5–3)

60. State the inverse of the relation $\{(-1, 1), (-5, 9), (4, 6)\}$. (Lesson 5–2)

61. Probability A card is selected at random from a deck of 52 cards. What is the probability of selecting a jack, queen, or king? (Lesson 4–6)

62. 44 is what percent of 89? (Lesson 4–4) **49.4%**

63. For what angle are the sine and cosine equal? (Lesson 4–3) **45°**

64. Trigonometry Find the measure of the third angle of a triangle if the other angles have measures 167° and 4°. (Lesson 3–4) **9°**

65. Solve $\frac{z}{-4} - 9 = 3$. (Lesson 3–3) **−48**

66. Budgeting The total of Jon Young's gas and electric bills was $210.87. His electric bill was $95.25. How much was his gas bill? (Lesson 3–2) **$115.62**

67. Sports The players with the most runs batted in (RBI) for the National League are listed below. (Lesson 2–2)

Year	Name	RBI	Year	Name	RBI
1971	Joe Torre	137	1983	Dale Murphy	121
1972	Johnny Bench	125	1984	Mike Schmidt	106
1973	Willie Stargell	119		Gary Carter	
1974	Johnny Bench	129	1985	Dave Parker	125
1975	Greg Luzinski	120	1986	Mike Schmidt	119
1976	George Foster	121	1987	Andre Dawson	137
1977	George Foster	149	1988	Will Clark	109
1978	George Foster	120	1989	Kevin Mitchell	125
1979	Dave Winfield	118	1990	Matt Williams	122
1980	Mike Schmidt	121	1991	Howard Johnson	117
1981	Mike Schmidt	91	1992	Darren Daulton	109
1982	Dale Murphy	109	1993	Barry Bonds	123
	Al Oliver		1994	Jeff Bagwell	116

Source: *The World Almanac, 1995*

a. Make a line plot of the data. **See margin.**
b. What was the greatest number of RBIs during one season? **149**
c. What was the least number of RBIs during one season? **91**
d. What was the most frequently-occurring number of RBIs during one season? **121 and 125**
e. How many of the players had from 119 to 129 RBIs? **13 players**

Additional Answer

67a.

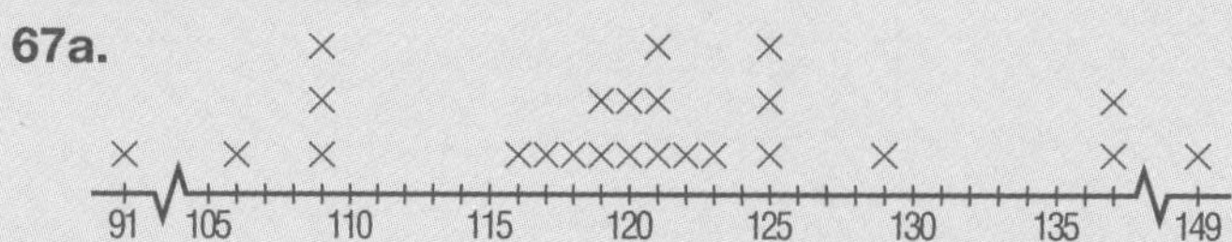

5-6 Writing Equations from Patterns

What YOU'LL LEARN

- To write equations to represent relations, given some of the solutions for the equations.

Why IT'S IMPORTANT

You can write equations to represent relations in scuba diving and aquatics.

APPLICATION

Scuba Diving

As scuba divers descend, the pressure of the water increases. Scuba divers can determine their depth by the pressure. Pressure can be expressed in atmospheres. An atmosphere is equivalent to 14.7 psi (pounds per square inch) of pressure. The table below shows the relationship between atmospheres of pressure and ocean depth.

Pressure (atmospheres)	1	2	3	4	5
Depth of Ocean (feet)	0	33	66	99	132

Suppose we let p represent the atmospheres of pressure, and let d represent the ocean depth. The relation shown in the table can also be represented by a graph. Graph the points and observe the pattern.

Notice that when the ordered pairs are graphed, they form a linear pattern. You can use a straightedge to draw the line through these points. Points that lie in a linear pattern can be described by an equation. Look at the relationship between the domain and range to find a pattern that can be described by an equation.

FYI

In 1943, Jacques-Yves Cousteau and E. Gagnon developed the Aqua-Lung™, to free divers from hoses connected to air sources on the surface. It is called *scuba* gear, which is an acronym for ***s**elf-**c**ontained **u**nderwater **b**reathing **a**pparatus.*

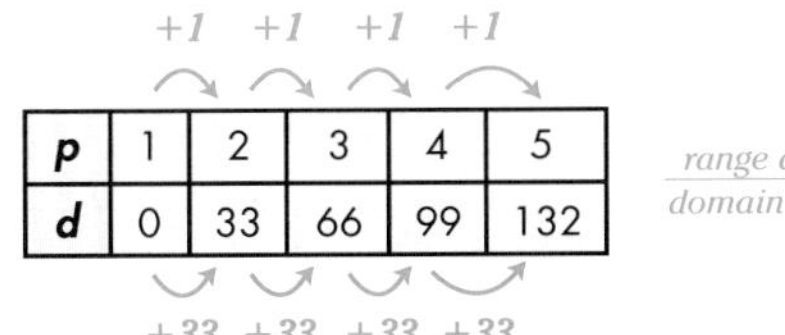

+1 +1 +1 +1

p	1	2	3	4	5
d	0	33	66	99	132

+33 +33 +33 +33

$\frac{\text{range differences}}{\text{domain differences}} = \frac{33}{1}$

Notice that the differences in the d values are 33 times the differences of the corresponding p values. It seems that this relationship can be described by the equation $d = 33p$. Check to see if this equation is correct by substituting values of p into the equation.

p	1	2	3	4	5
$33p$	33	66	99	132	165

These values do not match those in the first chart. However, note that each corresponding value for d using the equation is 33 more than the d in the relation. Thus, we need to adjust our equation to compensate for this difference. The equation that describes this relation is $d = 33p - 33$. Since this relation is also a function, we can write this equation as $d(p) = 33p - 33$.

FYI

Jacques-Yves Cousteau is known to many for his adventures aboard his research vessel, *The Calypso*, and for his three widely read books, *The Silent World* (1953), *The Living Sea* (1962), and *World Without Sun* (1964).

5-6 LESSON NOTES

NCTM Standards: 1–6

Instructional Resources

- Study Guide Master 5-6
- Practice Master 5-6
- Enrichment Master 5-6
- Assessment and Evaluation Masters, p. 129

Transparency 5-6A contains the 5-Minute Check for this lesson; **Transparency 5-6B** contains a teaching aid for this lesson.

Recommended Pacing	
Standard Pacing	Day 10 of 14
Honors Pacing	Day 9 of 13
Block Scheduling*	Day 5 of 7 (along with Lesson 5-5)
Alg. 1 in Two Years*	Days 14 & 15 of 21

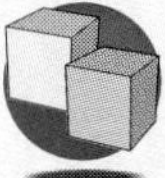

*For more information on pacing and possible lesson plans, refer to the *Block Scheduling Booklet* and *Algebra 1 in Two Years*.

1 FOCUS

5-Minute Check
(over Lesson 5-5)

1. Is the relation {(1, 3), (2, 4), (3, 5), (−1, 3), (−2, 4), (−3, 5)} a function? **yes**
2. Is the inverse of the relation shown in Exercise 1 a function? **no**
3. Is the relation described by $y = -2$ a function? **yes**

If $f(x) = 3x^2 - 4$, find each value.

4. $f(-2)$ **8**
5. $f(0)$ **−4**

Motivating the Lesson

Situational Problem Present the following number sequences to students. Ask them to determine what number is added to each term as well as what the missing terms are.

1. 3, 7, 11, 15, __, __
   Add 4; 19, 23
2. 6, 4, 2, 0, −2, __, __
   Add −2; −4, −6
3. 3, −1, −5, −9, __, __
   Add −4; −13, −17

2 TEACH

Teaching Tip In Example 1, it may help students to consider the question, "What must I do to x to get y?"

In-Class Examples

For Example 1
Write an equation in functional notation for the relation given in the chart.

x	1	2	3	4	5
y	6	12	18	24	30

$f(x) = 6x$

For Example 2
Write an equation in functional notation for the relation given in the chart.

x	1	2	3	4	5	6
y	−1	1	3	5	7	9

$f(x) = 2x - 3$

Example 1 **Plot the points in the relation shown in the table. Then write an equation in functional notation for the relation.**

x	−4	−8	−12	−16	−20
y	1	2	3	4	5

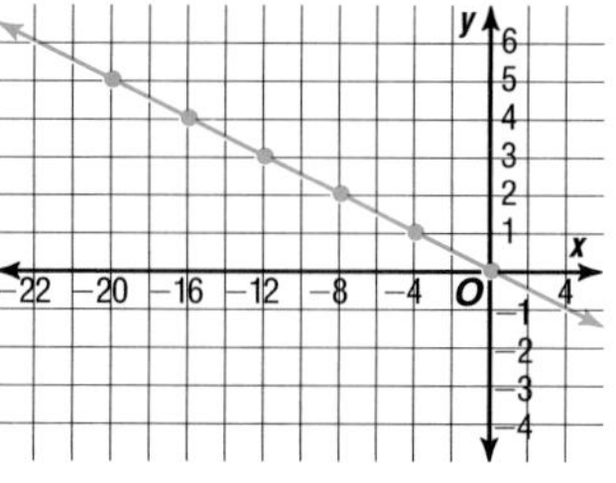

Since the points form a linear pattern, we know that there is a linear equation that describes this relation.

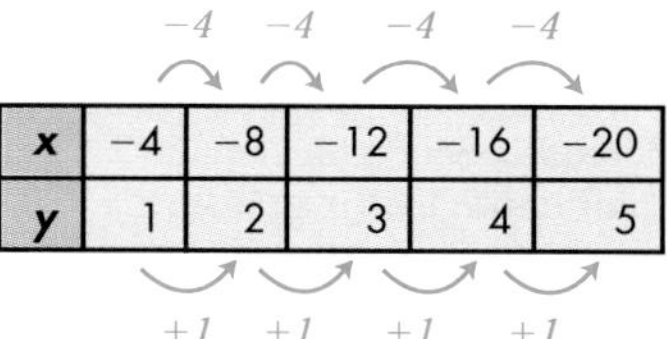

x	−4	−8	−12	−16	−20
y	1	2	3	4	5

$\frac{\text{range differences}}{\text{domain differences}} = \frac{1}{-4}$

The differences in the y values are one-fourth the differences in the x values. This pattern implies that the equation $y = -\frac{1}{4}x$ may describe the relation. Check the equation for the domain values in the relation.

Check: If $x = -4$, then $y = -\frac{1}{4}(-4)$ or 1. ✓

If $x = -8$, then $y = -\frac{1}{4}(-8)$ or 2. ✓

The equation that represents the relation is $y = -\frac{1}{4}x$. Since this relation is also a function, we can write this equation in functional notation as $f(x) = -\frac{1}{4}x$.

Sometimes you may need to find the equation for a graph.

Example 2 **Write an equation in functional notation for the relation graphed at the right.**

First make a table of ordered pairs for several points on the graph.

k	1	2	4	5	6	9
m	8	6	2	0	−2	−8

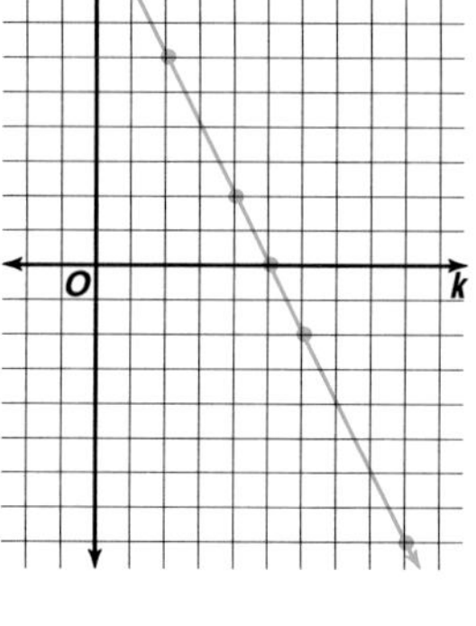

Now find the common differences of the domain and range.

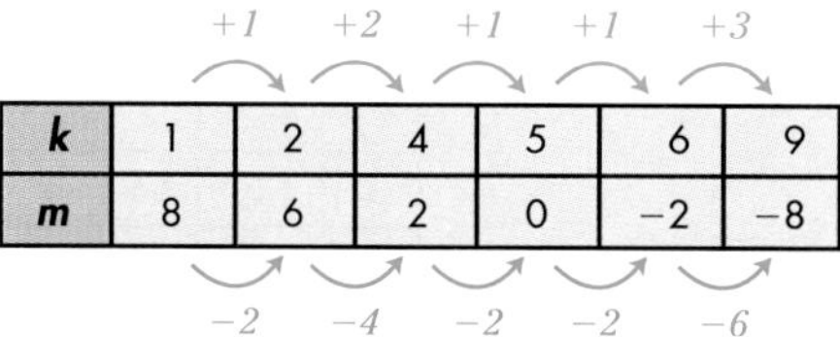

k	1	2	4	5	6	9
m	8	6	2	0	−2	−8

Notice that there is not a consistent change in the domain and range values. However, you can still find the ratio of the range difference to the domain difference for each ordered pair. There are three possible ratios.

GLENCOE Technology

CD-ROM Interaction

A multimedia simulation allows students to use data, graphs, and equations to analyze ocean depth and pressure. A blackline master activity with teacher's notes provides a follow-up to the CD-ROM simulation.

For Windows & Macintosh

$\frac{\text{range differences}}{\text{domain differences}}$ $\frac{-2}{1} = -2$ $\frac{-4}{2} = -2$ $\frac{-6}{3} = -2$

Since all of the ratios are the same, it appears that the differences in the m values are -2 times that of the k values. This pattern suggests that $m = -2k$. Check the equation to see if it is correct.

Check: If $k = 2$, then $m = -2(2)$ or -4. But the range value for $k = 2$ is 6, a difference of 10. Try some other values in the domain to see if the same difference occurs.

k	1	4	5	6	9
$-2k$	-2	-8	-10	-12	-18
m	8	2	0	-2	-8

m is 10 more than −2k.

This pattern suggests that 10 should be added to one side of the equation in order to correctly describe the relation. Thus, the equation for this relation is $m = -2k + 10$. Check this equation.

Check: If $k = 2$, then $m = -2(2) + 10$ or 6. ✓
If $k = 9$, then $m = -2(9) + 10$ or -8. ✓

Therefore, $m = -2k + 10$ describes this relation. Since this relation is also a function, we can write this equation in functional notation as $f(k) = -2k + 10$.

Many times you can generalize the relationships among data you have collected with an equation.

EXPLORATION — SPREADSHEETS

In the relation $\{(-2, 2), (-1, 5), (0, 8), (1, 11), (2, 14)\}$, the range difference is 3 when the domain difference is 1. You might suggest that the equation of the relation is $y = 3x$. Let's use spreadsheet software to check this equation.

- Each cell of a spreadsheet is named by a letter and number. The letter refers to the column and the number of the row. Enter each x value into cells A1 through A5. Enter the y values into cells B1 through B5.
- In cell C1, enter the formula A1∗3. This formula means take the value in cell A1 and multiply it by 3. Copy this formula to cells C2 through C5. The spreadsheet will automatically change the formula so that the appropriate cell in column A will be used.
- In cell D1, enter the formula B1 − C1. This formula will subtract the range value of the equation from the range value of the relation. Copy this formula to cells D2 through D5.

Column D tells what number to add to the equation so that it is correct. In this case, the entries in column D are 8s. The correct equation should be $y = 3x + 8$.

Your Turn **a. A1 * −2; See students' work.**

a. What formula would you use in cell C1 to test the first formula found in Example 2? Use a spreadsheet to check all the values in the relation.

b. What number would you expect to appear in column D if the equation you found is correct for a given relation? **0**

EXPLORATION

Students might check their conjecture by testing it with paper and pencil on the first few ordered pairs.

In-Class Example

For Example 3
Akbar notices that the counter on his VCR advances by 12 units every minute. Letting c = counter reading and t = time in minutes, find the equation for the counter as a function of time if Akbar begins the tape at $c = 0$ and $t = 0$.
$c(t) = 12t$

The penny terminology has been largely replaced today by more precise designations. A 6-common or a 6-finish nail gives information about length and usage.

3 PRACTICE/APPLY

Check for Understanding

Exercises 1–10 are designed to help you assess your students' understanding through reading, writing, speaking, and modeling. You should work through Exercises 1–4 with your students and then monitor their work on Exercises 5–10.

Additional Answers

2. Test the values of the domain in the equation. If the resulting values match the range, the equation is correct.
3. Yes; she can use points on the line since the line represents all solutions to the equation.

Example 3

Hardware

FYI

This pattern for nails does not hold for all penny sizes. For example, a 60-penny nail is only 6 inches long.

***Pennies* are units for measuring the lengths of nails. Nails measured in pennies can range from 2-penny nails to 60-penny nails. The graph at the right illustrates the lengths of several penny nails.**

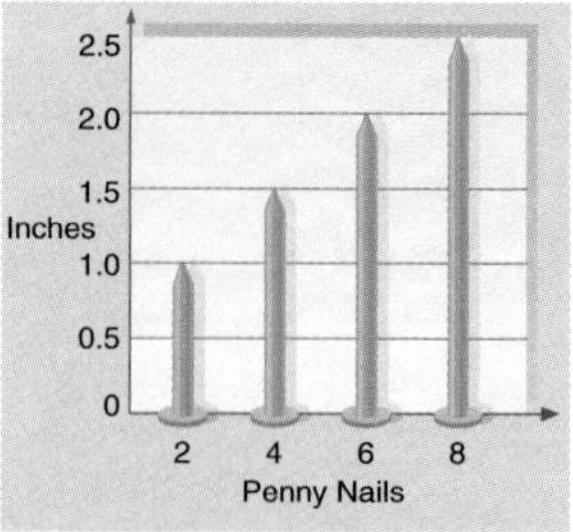

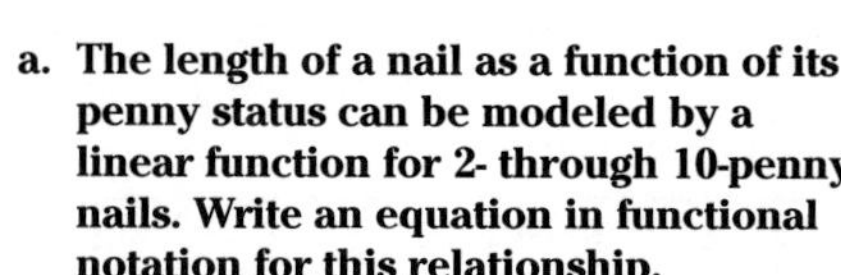

a. The length of a nail as a function of its penny status can be modeled by a linear function for 2- through 10-penny nails. Write an equation in functional notation for this relationship.

b. Find the lengths of a 3-penny nail, a 5-penny nail, and a 10-penny nail.

a. We can see from the graph that the points form a linear pattern. Thus, there is a linear equation that describes the relation for 2- through 10-penny nails.

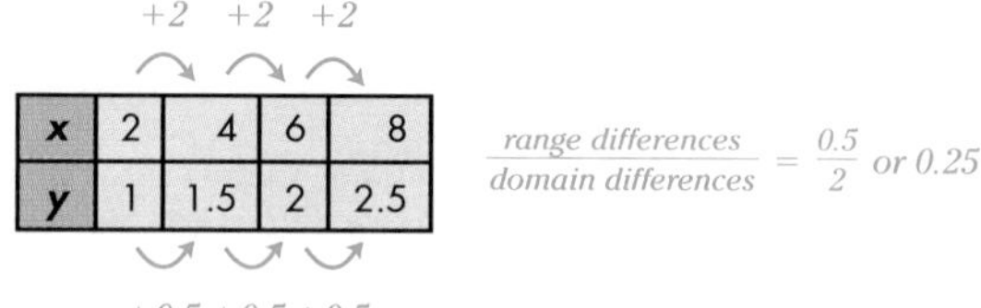

+2 +2 +2

x	2	4	6	8
y	1	1.5	2	2.5

+0.5 +0.5 +0.5

$\frac{\text{range differences}}{\text{domain differences}} = \frac{0.5}{2}$ or 0.25

This pattern suggests that the equation $f(x) = 0.25x$ might model this situation. However, if you check the domain values with this equation, you will find that the values of $f(x)$ are off by 0.5. The correct equation is $f(x) = 0.25x + 0.5$.

b. *3-penny nail*
$f(3) = 0.25(3) + 0.5$
$= 1.25$

5-penny nail
$f(5) = 0.25(5) + 0.5$
$= 1.75$

10-penny nail
$f(10) = 0.25(10) + 0.5$
$= 3$

A 3-penny nail is 1.25 inches long, a 5-penny nail is 1.75 inches long, and a 10-penny nail is 3 inches long.

CHECK FOR UNDERSTANDING

Communicating Mathematics

Study the lesson. Then complete the following.

1. **Analyze** the graphs in Examples 1 and 2 along with the ratios of the range differences to the domain differences. Then complete these sentences.
   a. If the ratio is positive, then the line slants __?__. **upward**
   b. If the ratio is negative, then the line slants __?__. **downward**
2. **Explain** how you can determine if an equation correctly represents a relation given as a table. **See margin.**
3. **You Decide** When the teacher was explaining Example 2, Narissa said she had another way to find the equation without calculating three ratios. She said she found other points that lie on the line so the differences were common. Is Narissa's method valid? Explain. **See margin.**
4. **Write** a linear equation in functional notation that has both (1, 1) and (0, 3) as solutions. Is this the only linear equation that has these two solutions? Explain.

4. Sample answer: $2x + y = 3$; No, $4x + 2y = 6$ and other equivalent equations also have these solutions.

Reteaching

Using Review Games For a fun way to practice finding equations from relations, play the function rule game. A person thinks of an equation like $y = 10 - 2x$. Students give input values and are given the functional output values. For example, $1 \rightarrow 8$, $3 \rightarrow 4$. The goal is to guess the function rule.

Guided Practice

Write an equation in functional notation for each relation.

5.

x	3	4	5	6	7
$f(x)$	12	14	16	18	20

$f(x) = 2x + 6$

6.

x	2	4	6	8	10
$f(x)$	−4	−3	−2	−1	0

$f(x) = \frac{1}{2}x - 5$

Write an equation for each graphed relation.

7. 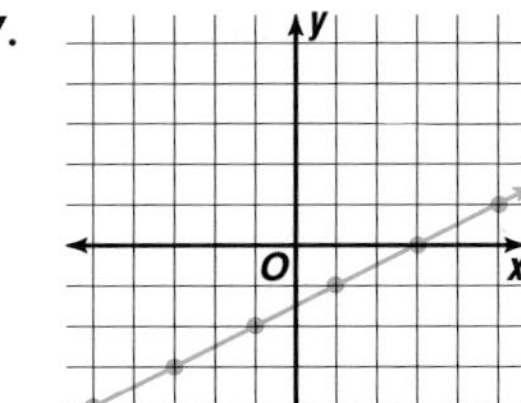

8.

> **TECHNOLOGY Tip**
> You can use a spreadsheet to check all of your equations in this exercise set.

7. $y = \frac{1}{2}x - \frac{3}{2}$

8. $y = x$

9. The table at the right represents values for the function $N(m)$. Copy and complete the table at the right. Explain how you determined the missing value. **−6; See students' work.**

m	8	4	2	0
$N(m)$	−10	−8	−7	?

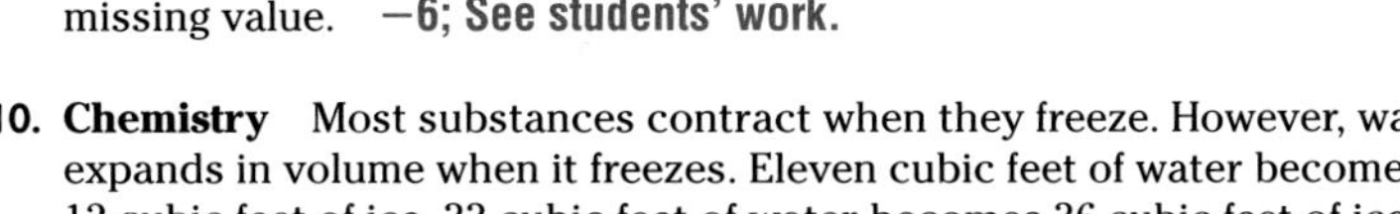

10. **Chemistry** Most substances contract when they freeze. However, water expands in volume when it freezes. Eleven cubic feet of water becomes 12 cubic feet of ice, 33 cubic feet of water becomes 36 cubic feet of ice, and 66 cubic feet of water becomes 72 cubic feet of ice.
    a. Make a graph of these data. **See margin.**
    b. Write a functional equation for the relationship between the volume of water and the corresponding volume of ice.

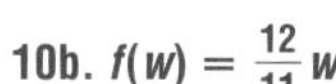

10b. $f(w) = \frac{12}{11}w$

Calvin and Hobbes by Bill Watterson

EXERCISES

Practice

Copy and complete the table for each function.

11.

x	1	2	3	4	5
$f(x)$	12	24	36	**48**	**60**

12.

x	−4	−2	0	2	4
$g(x)$	−2	−1	**0**	**1**	2

13.

x	−3	−1	1	2	4
$h(x)$	18	**8**	**−2**	−7	−17

14.

x	−2	0	2	4	6
$p(x)$	**0**	**1**	2	3	4

Technology Tip
Remind students that spreadsheets will verify a finite number of solutions. A linear equation has an infinite number of solutions.

Assignment Guide

Core: 11–35 odd, 36–42
Enriched: 12–30 even, 31–42

For **Extra Practice,** see p. 768.

The red A, B, and C flags, printed only in the Teacher's Wraparound Edition, indicate the level of difficulty of the exercises.

Additional Answer

10a.

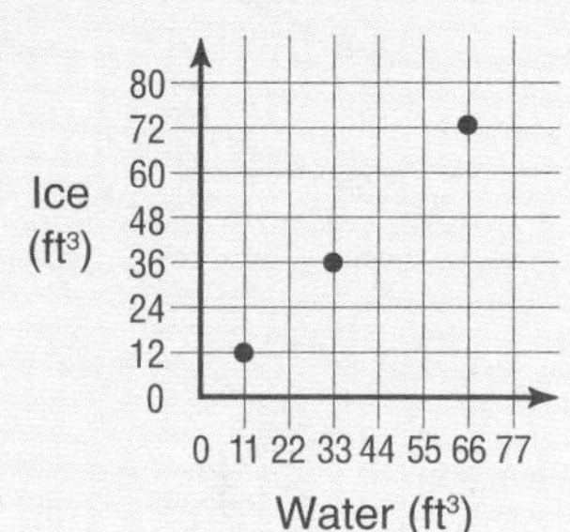

Write an equation for each relation.

15.

x	1	2	3	4	5
$f(x)$	5	10	15	20	25

$f(x) = 5x$

16.

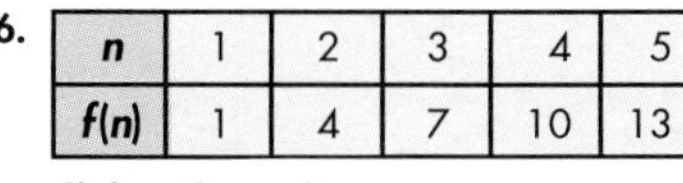

n	1	2	3	4	5
$f(n)$	1	4	7	10	13

$f(n) = 3n - 2$

17.

x	−2	−1	1	2	4
$g(x)$	13	12	10	9	7

$g(x) = 11 - x$

18.

n	−4	−2	0	2	4
$m(n)$	−11	−3	5	13	21

$m(n) = 4n + 5$

B

19.

x	0	6	12	18	24
$h(x)$	−2	0	2	4	6

$h(x) = \frac{1}{3}x - 2$

20.

n	−4	0	4	6	8
$m(n)$	26	18	10	6	2

$m(n) = 18 - 2n$

21. $y = -3x$

22.

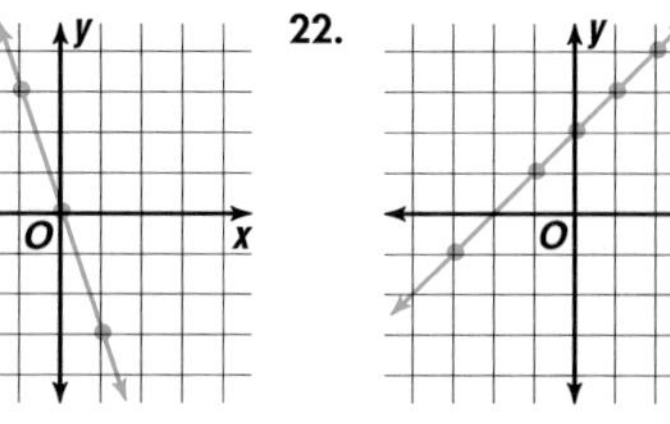

$y = x + 2$

23.

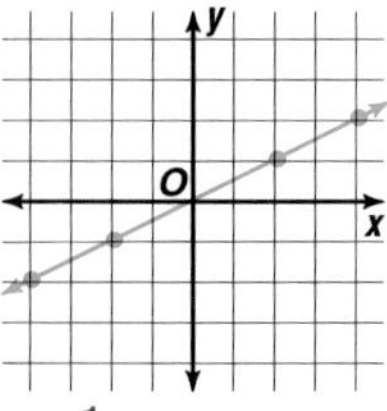

$y = \frac{1}{2}x$

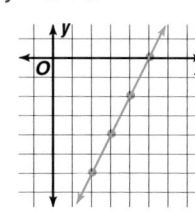

24. $y = 6 - x$

25. $y = 2x - 10$

26.

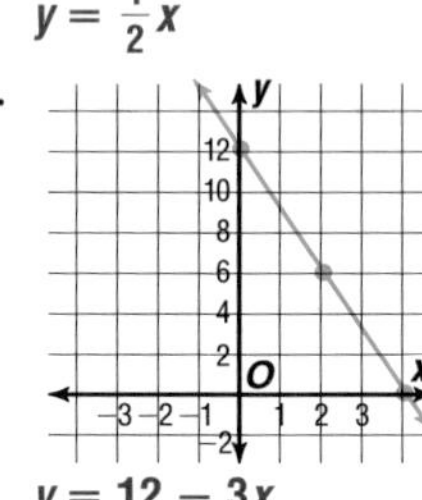

$y = 12 - 3x$

C

27. $\{(6, -4), (2, -12), (1, -24), (-3, 8), (-6, 4)\}$ $xy = -24$

28. $\{(-3, 10), (-2, 5), (-1, 2), (0, 1), (1, 2), (2, 5), (3, 10)$ $y = x^2 + 1$

29. $\{(-3, -27), (-1, -1), (2, 8), (3, 27), (10, 1000)\}$ $y = x^3$

30. $\{2, 12), (1, 48), (-1, 48), (-2, 12), (-4, 3)\}$ $y = \frac{48}{x^2}$

Critical Thinking

31. The y- and x-intercepts are those points at which a graph intersects the y- and x-axes, respectively. Use functional notation to describe these points. **y-intercept: $f(0)$, x-intercept: $f(x) = 0$**

32. See margin.

32. Suppose the ratio of range differences to domain differences in a linear function $g(x)$ is $\frac{2}{3}$ and the graph of $g(x)$ goes through the point $(-3, 2)$. Draw the graph of $g(x)$ and explain why you drew it the way you did.

Applications and Problem Solving

33. **Aquatics** The table below illustrates the relationship between atmospheres of pressure and the depth of fresh water.

Pressure (atmospheres)	1	2	3	4	5
Depth of Fresh Water (ft)	0	34	68	102	136

a. Write an equation in functional notation for the relation.

b. Refer to the application at the beginning of the lesson. How does the pressure in fresh water compare to the pressure in ocean water?

33a. $f(x) = 34x - 34$

33b. You must go deeper in fresh water to get the same pressure as in ocean water.

Additional Answers

32.

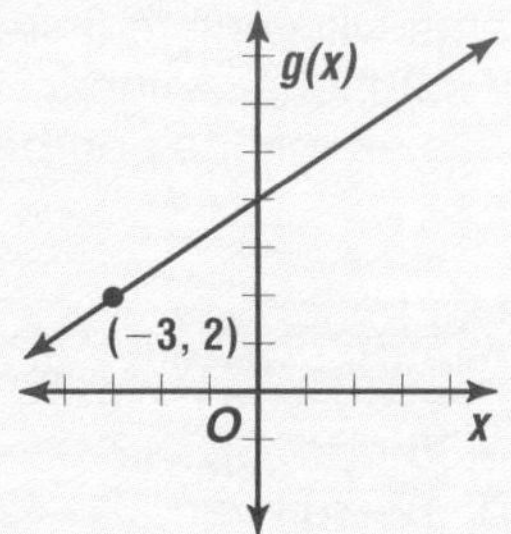

34a. Sample answer:

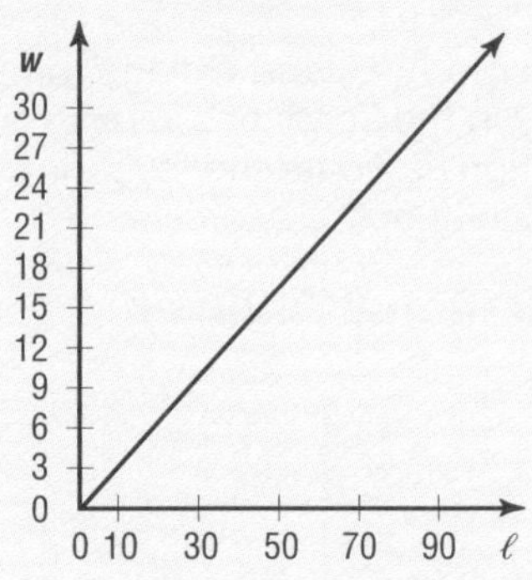

Study Guide Masters, p. 39

NAME ______ DATE ______

Study Guide

Student Edition Pages 295–302

Writing Equations From Patterns

You can find equations from relations. Suppose you purchased a number of packages of blank cassette tapes. If each package contained three tapes, you could make a chart to show the relationship between the number of packages of blank cassette tapes and the number of tapes purchased. Use x for the number of packages and y for the number of tapes.

x	1	2	3	4	5	6
y	3	6	9	12	15	18

This relationship can also be shown as an equation. Since y is always three times x, the equation is $y = 3x$. Another way to discover this relationship is to study the difference between successive values of x and y.

+1 +1 +1 +1 +1

x	1	2	3	4	5	6
y	3	6	9	12	15	18

+3 +3 +3 +3 +3

This suggests the relation $y = 3x$.

Write an equation for each relation. Then complete each chart.

1.

x	−1	0	1	2	3	4
y	−2	2	6	10	14	18

$y = 4x + 2$

2.

x	−2	−1	0	1	2	3
y	10	7	4	1	−2	−5

$y = -3x + 4$

3.

x	−4	−3	−2	−1	0	1
y	$\frac{5}{2}$	$\frac{9}{4}$	2	$\frac{7}{4}$	$\frac{3}{2}$	$\frac{5}{4}$

$y = -\frac{1}{4}x + 1\frac{1}{2}$

4.

x	0	1	2	3	4	5
y	3	$\frac{12}{5}$	$\frac{9}{5}$	$\frac{6}{5}$	$\frac{3}{5}$	0

$y = -\frac{3}{5}x + 3$

5. $\{(-10, -5), (-4, -2), (0, 0), (2, 1), (5, \frac{5}{2})\}$ $y = \frac{1}{2}x$

6. $\{(-3, -10), (-1, -4), (0, -1), (2, 5), (4, 11)\}$ $y = 3x - 1$

34a. See margin.
34c. $W(\ell) = \frac{23}{70}\ell$

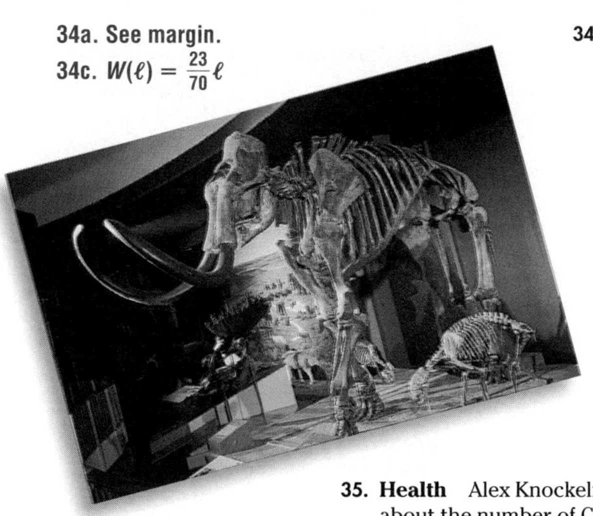

34. Archaeology Archaeologists sometimes use lengths of heavy-duty wire to hold together the bones that compose the spinal columns of various vertebrates. Suppose a 70-foot length of wire weighs 23 pounds.

a. If this sample of wire is representative of all lengths of this wire, draw a graph showing the weight of the wire as a function of its length.
b. Use your graph to estimate the weight of 50 feet of wire to the nearest pound. **about 16 lb**
c. Write an equation to describe the relationship between the length of a wire and its weight.
d. Evaluate the equation in part c for $\ell = 50$. How does this result compare with your estimate in part b? **16.4 lb, slightly more than estimate**

35. Health Alex Knockelmann is concerned about the number of Calories that she needs to intake during the basketball season so she will not lose weight. She wants to eat as many Calories before practice as she will use during practice. The table below shows some fast foods from restaurants near her school and their Calories.

Fast Food	Calories
Arby's® Ham and Cheese	380
Arby's® Roast Beef	350
Burger King's® Whopper™ with Cheese	760
Burger King's® Regular Fries	240
Burger King's® Chocolate Milk Shake	380
Wendy's® Cheeseburger	577
Wendy's® Double Cheeseburger	797
Wendy's® Fries	327
Wendy's® Frosty™	390
Pepsi® (20 oz)	241
7-Up® (20 oz)	219

a. Before practice, Alex went to Wendy's. She ate a double cheeseburger, fries, and a Frosty™. What was her Calorie intake for this meal? **1514 C**
b. In Lesson 5–4, you learned that playing basketball burns 0.138 C/min/kg. At Alex's weight of 80 kg, determine how many Calories she would burn each minute. **11.04 C/min**
c. Make a table to show the number of Calories that Alex burns each minute for the first 10 minutes of practice. Graph the points. **See margin.**
d. Let C represent the number of Calories burned and t represent the number of minutes. Write a functional equation to represent the relation. Is it a linear function? **$C(t) = 11.04t$; yes**
e. Use the graph and your equation to predict the number of Calories Alex would use during two hours of practice. Did she eat enough food so that she had Calories remaining at the end of practice?

35e. 1324.8 C burned; yes, 189.2 C remaining

Additional Answer

35c.

Minute	Calories
1	11.04
2	22.08
3	33.12
4	44.16
5	55.2
6	66.24
7	77.28
8	88.32
9	99.36
10	110.4

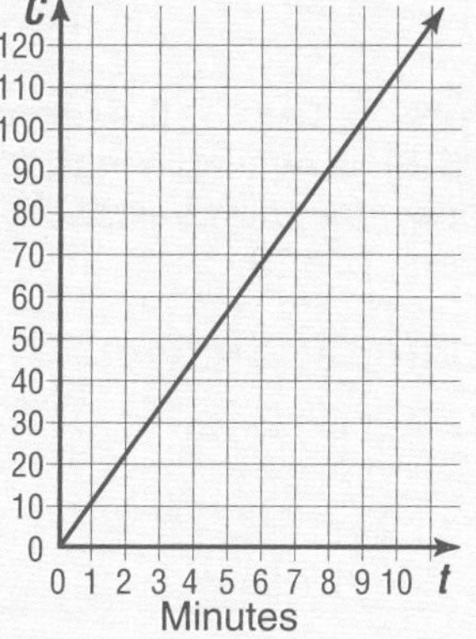

Extension

Problem Solving Tables are given at the right for two linear functions. If the domain of each of these functions is the set of all numbers, find the value(s) of x that make $f(x) = g(x)$. $x = 3$

x	2	4	6	8
$f(x)$	5	9	13	17

$y = 2x + 1$

x	2	4	6	8
$g(x)$	4	10	16	22

$y = 3x - 2$

Practice Masters, p. 39

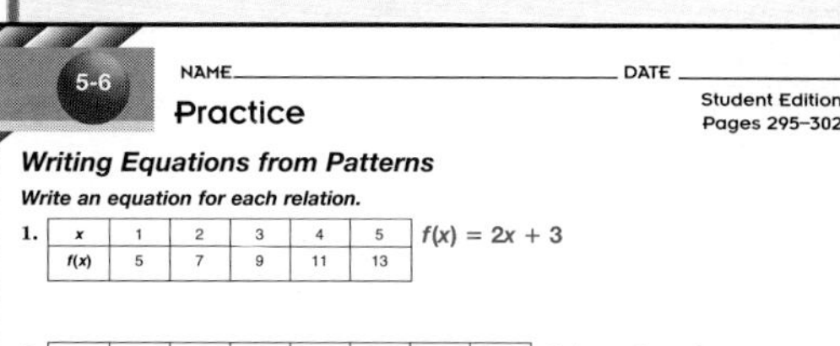

5-6 NAME______________ DATE__________

Practice

Student Edition Pages 295–302

Writing Equations from Patterns

Write an equation for each relation.

1.

x	1	2	3	4	5
$f(x)$	5	7	9	11	13

$f(x) = 2x + 3$

2.

x	−3	−2	−1	0	1	2	3
$f(x)$	14	10	6	2	−2	−6	−10

$f(x) = -4x + 2$

3.

x	−3	−2	−1	0	1	2	3
$f(x)$	4	4	4	4	4	4	4

$f(x) = 4$

4.

x	−3	−2	−1	0	1	2	3
$f(x)$	9	6	3	0	−3	−6	−9

$f(x) = -3x$

5.

x	−3	−2	−1	0	1	2	3
$f(x)$	3	2	1	0	−1	−2	−3

$f(x) = -x$

6.

x	−6	−4	−2	0	2	4	6
$f(x)$	−5	−4	−3	−2	−1	0	1

$f(x) = \frac{1}{2}x - 2$

7. $\{(-3, 3), (-1, 1), (0, 0), (2, 2), (4, 4)\}$ $y = |x|$

8. $\{(1, 2), (2, 4), (3, 8), (4, 16), (5, 32)\}$ $y = 2^x$

4 ASSESS

Closing Activity

Writing The table below shows a linear relation between x and y. Find the values of a, b, c, and d. Then write an equation that relates x and y.

x	8	9	10	d
y	a	5	b	c

$a = 3, b = 7, c = 9, d = 11$; $y = 2x - 13$

Chapter 5, Quiz C (Lessons 5-5 and 5-6), is available in the *Assessment and Evaluation Masters*, p. 129.

Additional Answers

41. 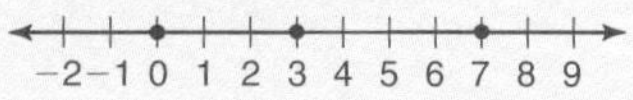 (number line from −2 to 9 with points at 0, 3, and 7)

42.

Stem	Leaf
2	3
3	2
4	3 5 7
5	3 7 8 9 9 9
6	0 0 2 2 4 4 5
•	5 6 6 6 6 6 6
•	6 7 8 8 8 8 9
7	0 2 3 4 4 4 4
•	4 4 4 4 5 5 5
•	5 5 5 5 5 5 5

$2|3 = 23¢$

Enrichment Masters, p. 39

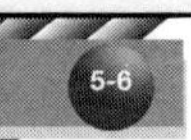

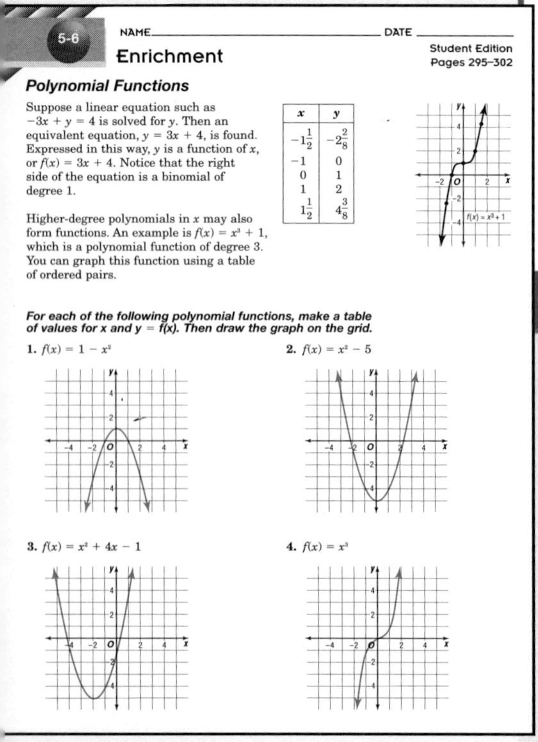

NAME ______________ DATE ______

Enrichment Student Edition Pages 295–302

Polynomial Functions

Suppose a linear equation such as $-3x + y = 4$ is solved for y. Then an equivalent equation, $y = 3x + 4$, is found. Expressed in this way, y is a function of x, or $f(x) = 3x + 4$. Notice that the right side of the equation is a binomial of degree 1.

Higher-degree polynomials in x may also form functions. An example is $f(x) = x^3 + 1$, which is a polynomial function of degree 3. You can graph this function using a table of ordered pairs.

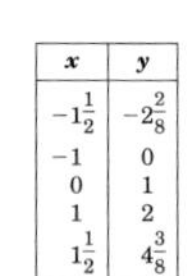

x	y
$-1\frac{1}{2}$	$-2\frac{2}{8}$
-1	0
0	1
1	2
$1\frac{1}{2}$	$4\frac{3}{8}$

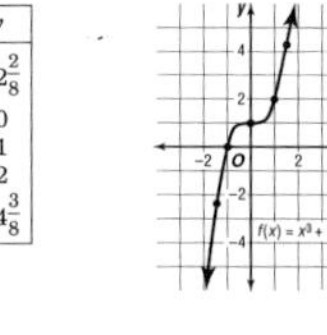

For each of the following polynomial functions, make a table of values for x and y = f(x). Then draw the graph on the grid.

1. $f(x) = 1 - x^2$

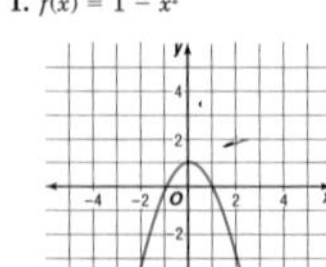

2. $f(x) = x^2 - 5$

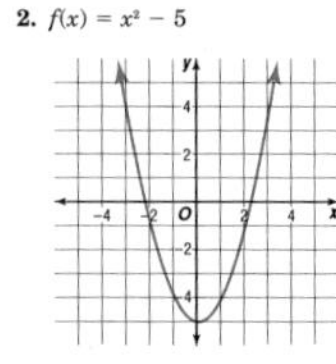

3. $f(x) = x^2 + 4x - 1$

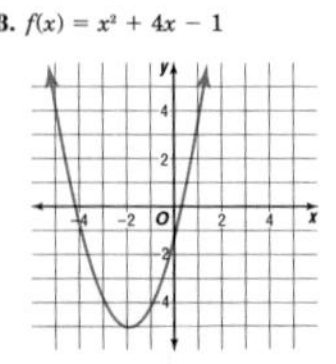

4. $f(x) = x^3$

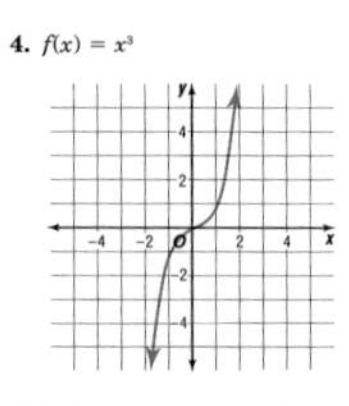

Mixed Review

36. Biology The function $f(t) = \frac{t}{0.2} - 32$ shows that the number of times a cricket chirps in an hour is a function of the temperature in degrees Celsius. (Lesson 5–5) **36a. 28 times 36b. 10°C**

a. If it is 12°C outside, how many times will a cricket chirp in an hour?

b. If a cricket chirps 54 times in 3 hours, what is the outside temperature?

37. State the domain and range of {(1, 6), (3, 4), (5, 2)}. (Lesson 5–2) **37. D = {1, 3, 5}; R = {2, 4, 6}**

38. What number is 16% of 50? (Lesson 4–4) **8**

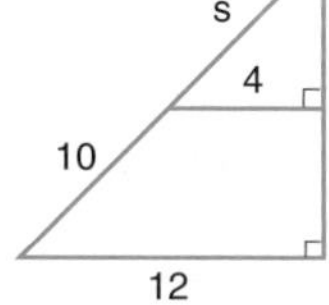

39. Trigonometry Given the similar triangles at the right, find s. (Lesson 4–2) **5**

40. Solve $x + (-7) = 36$. (Lesson 3–1) **43**

41. Graph {0, 3, 7} on a number line. (Lesson 2–1) **41. See margin.**

42. Personal Finance Many credit cards offer a protection plan so you can pay off your balance if you lose your job or have an accident and cannot work. The table below shows the monthly premium (in cents per $100 outstanding balance) for the plan on a Discover® card in the United States and its territories. Round each amount to the nearest cent and make a stem-and-leaf plot of these fees. (Lesson 1–4) **See margin.**

Area	Fee	Area	Fee	Area	Fee	Area	Fee	Area	Fee	Area	Fee
AK	68.2	FL	75	LA	74	NC	64.6	OK	75	UT	68.9
AL	66	GA	75	MA	63.5	ND	64.8	OR	69.8	VA	59.3
AR	66	HI	57.5	MD	75	NE	68	PA	22.5	VI	66
AZ	73.7	IA	58.8	ME	62	NH	46.9	PR	74	VT	56.7
CA	75	ID	60	MI	66	NJ	64.2	RI	67	WA	60
CO	68.1	IL	75	MN	32.3	NM	74	SC	71.9	WI	59
CT	53.2	IN	62.1	MO	75	NV	74	SD	66	WV	74
DC	66	KS	68.3	MS	74	NY	45.3	TN	75	WY	73.1
DE	66	KY	75	MT	74	OH	75	TX	42.7		

WORKING ON THE

In•ves•ti•ga•tion

Refer to the Investigation on pages 190–191.

Many of the prcedures from one area of science can be applied to other areas of science. Estimating populations is not reserved just for fish. Conservationists also keep track of species of birds, insects, plants, and other wildlife in a similar way.

1 Design a different experiment that illustrates the capture-recapture method. Complete your experiment a few times to see if it is as accurate as the one you have completed.

2 What was the range of your estimates using the method described in the Investigation? What was the range of your estimates using your method? Which do you think is the most reliable method?

Add the results of your work to your Investigation Folder.

In•ves•ti•ga•tion

Working on the Investigation

The Investigation on pages 190–191 is designed to be a long-term project that is completed over several days or weeks. Encourage students to keep their materials in their Investigation. Folder as they work on the Investigation.

5–7A Graphing Technology Measures of Variation

A Preview of Lesson 5–7

The graphing calculator is a powerful tool for observing patterns and analyzing data that come from research or experimentation. Many of the measures of variation you will learn about in Lesson 5–7 can be calculated by using a graphing calculator.

Before analyzing any data, you must know how to enter it into a list.

LOOK BACK You can review another method for using the graphing calculator to find the median of a set of data in Lesson 3-7.

Example 1 **The Glencoe Publishing Golf Tournament is held every September to raise money for a scholarship fund. The list below includes the scores for the employees of the manufacturing department. Enter the data and find the range of these scores.**

139, 99, 105, 115, 88, 91, 105, 80, 102, 101, 103, 95, 99, 77, 112

First enter the data.

Method 1 Assign the data to List 1 (L1) from the home screen.

Enter: [2nd] [{] 139 [,] 99 [,] 105 [,] 115 [,] 88 [,] 91 [,] 105 [,] 80 [,] 102 [,] 101 [,] 103 [,] 95 [,] 99 [,] 77 [,] 112 [2nd] [}] [STO▸] [2nd] [L1] [ENTER]

You can use the calculator to sort the data from least to greatest by pressing [STAT] 2 [2nd] [L1] [ENTER]. *You can view the list by pressing* [STAT] [ENTER].

Method 2 Enter the data directly into the list display.

Before entering your data, you must clear the L1 list of previously entered data.

Enter: [STAT] 4 [2nd] [L1] [ENTER]

Enter each score into the L1 list.

Enter: [STAT] 1

The cursor appears in the L1 list. Enter the data one by one pressing [ENTER] after each entry, including the last one.

Then find the range.

The range of the scores is the difference of the maximum and minimum values in the list.

Enter: [2nd] [QUIT] [2nd] [LIST] [▶] 2 [2nd] [L1] [)] [–] [2nd] [LIST] [▶] 1 [2nd] [L1] [)] [ENTER] *62*

If you do not use [2nd] [QUIT] *to leave the LIST menu, the range will be added to L1, which alters your data.*

The range of the data is 62.

5-7A LESSON NOTES

NCTM Standards: 1–4, 10

Objective

Use a graphing calculator to determine measures of variation.

Recommended Time

25 minutes

Instructional Resources

Graphing Calculator Masters, pp. 20 and 21

These masters provide keystroking instruction for this lesson for the TI-81 and Casio graphing calculators.

1 FOCUS

Motivating the Lesson

Ask students what the first step would be in finding the median for a long list of values such as a basketball player's points per game for an entire season. Students should suggest that arranging the values in increasing order would have to be done first. Explain that on a calculator, the data could be input in any order to find the median.

2 TEACH

Teaching Tip Review with students the difference between median and mean.

3 PRACTICE/APPLY

Assignment Guide

Core: 1–8
Enriched: 1–8

4 ASSESS

Observing students working with technology is an excellent method of assessment.

The data can be arranged into four groups that have approximately the same number of data in each group. The points separating these groups are called **quartiles** and are symbolized by Q_1, Q_2, and Q_3. The median of the set is Q_2. The **interquartile range** is $Q_3 - Q_1$. Half of the data lie in this range. You can use a graphing calculator to help you find each of these measures.

Example  **Find the quartiles and the interquartile range of the golf data in Example 1.**

Before finding the measures you want, you must first make sure the calculator is set to refer to the correct set of data. Our data is in L_1.

Enter: STAT ▶ 3 and highlight L_1 under 1-Var Stats. Make sure that the frequency is 1.

Enter: STAT  1 

1–Var Stats
↑n=15
minX=77
Q_1=91
Med=101
Q_3=105
maxX=139

Use the down arrow key to scroll to the end of the statistics. Your screen should look like the one at the right. This tells you there are 15 items of data, the least value (minX) is 77, the greatest value (maxX) is 139, the lower quartile (Q_1) is 91, the median is 101, and the upper quartile (Q_3) is 105.

Use Q_3 and Q_1 to find the interquartile range: $Q_3 - Q_1 = 105 - 91$ or 14.

EXERCISES

Remember to change the Setup when calculating the measures for each set of data.

Enter each set of data into the list given. Then find the lower quartile, median, upper quartile, range, and interquartile range of the data.

1. 12, 17, 16, 23, 18 in L_1 **Q_1, 14; Med, 17; Q_3, 20.5; R, 11; IQR, 6.5**
2. 56, 45, 37, 43, 10, 34 in L_2 **Q_1, 34; Med, 40; Q_3, 45; R, 46; IQR, 11**
3. 77, 78, 68, 96, 99, 84, 65 in L_3 **Q_1, 68; Med, 78; Q_3, 96; R, 34; IQR, 28**
4. 30, 90, 40, 70, 50, 100, 80, 60 in L_4 **Q_1, 45, Med, 65; Q_3, 85; R, 70; IQR, 40**
5. 3, 3.2, 6, 45, 7, 26, 1, 3.4, 4, 5.3, 5, 78, 8, 21, 5 in L_5
6. 85, 77, 58, 69, 62, 73, 55, 82, 67, 77, 59, 92, 75, 69, 76 in L_6
7. Look at the range and the interquartile range of each set of data. What do you think it means if the range is great but the interquartile range is small?
8. Suppose you entered a set of data and the interquartile range was 0. What does that mean? **Each item of data from Q_1 to Q_3 is the same number.**

5. Q_1, 3.4; Med, 5.3; Q_3, 21; R, 77; IQR, 17.6

6. Q_1, 62; Med, 73; Q_3, 77; R, 37; IQR, 15

7. At least 50% of the data is clustered around the median.

Using Technology

This lesson offers an excellent opportunity for using technology in your algebra classroom. For more information on using technology, see *Graphing Calculators in the Mathematics Classroom*, one of the titles in the Glencoe Mathematics Professional Series.

5-7 Integration: Statistics
Measures of Variation

What YOU'LL LEARN

- To calculate and interpret the range, quartiles, and interquartile range of sets of data.

Why IT'S IMPORTANT

Measures of variation can help you easily describe the spread of a set of data.

Climate

Justin Williams has been promoted and has the option of relocating either to Columbus, Ohio, or San Francisco, California. The Williams family lives in Tampa, Florida, where the average temperature is about 72°F and it is usually warm all year long. His family's wardrobe includes basically all lightweight apparel. He wondered if moving would mean they would have to get lots of new clothes. He looked up the average high temperature for the two cities and found the following information in a table of average high temperatures of U.S. cities.

	Columbus	San Francisco
Jan.	35°	56°
Feb.	38°	59°
Mar.	49°	60°
Apr.	62°	61°
May	73°	63°
June	81°	64°
July	84°	64°
Aug.	83°	65°
Sept.	77°	69°
Oct.	65°	68°
Nov.	51°	63°
Dec.	39°	57°

Source: U.S. National Oceanic and Atmospheric Administration

LOOK BACK

You can review finding the mean and median of a set of data in Lesson 3-7.

Mr. Williams calculated the mean and median of the temperatures of each city. The mean high temperature for Columbus was 63.5°, and for San Francisco, it was 62.4°. The median high temperature for Columbus was 63.5°, and for San Francisco, it was 63°. He figured that meant the climate in the two cities was just about the same. But then he decided to take another look at the temperatures and find another way to analyze them.

The months and temperatures for each city form a relation. Mr. Williams decided to graph the relation for each city. He used different colors for each city.

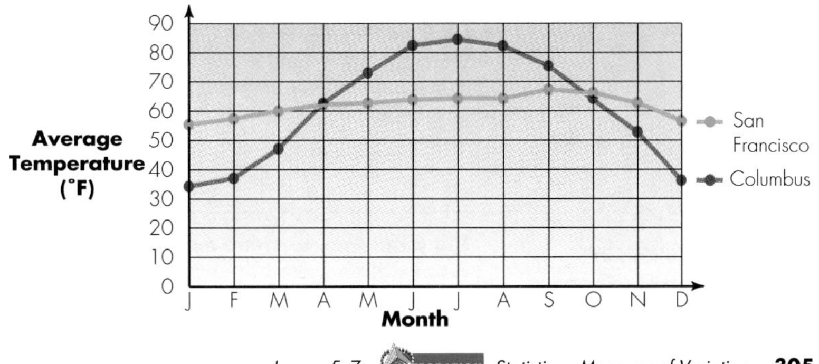

5-7 LESSON NOTES

NCTM Standards: 1–4, 10

Instructional Resources

- Study Guide Master 5-7
- Practice Master 5-7
- Enrichment Master 5-7
- Assessment and Evaluation Masters, p. 129
- Real-World Applications, 16

Transparency 5-7A contains the 5-Minute Check for this lesson; **Transparency 5-7B** contains a teaching aid for this lesson.

Recommended Pacing	
Standard Pacing	Day 12 of 14
Honors Pacing	Day 11 of 13
Block Scheduling*	Day 6 of 7
Alg. 1 in Two Years*	Days 17 & 18 of 21

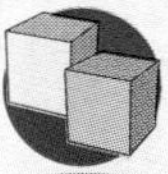

*For more information on pacing and possible lesson plans, refer to the *Block Scheduling Booklet* and *Algebra 1 in Two Years.*

1 FOCUS

5-Minute Check
(over Lesson 5-6)

Complete the table for the function given.

1.

x	1	3	5	7	9
$f(x)$	4	12	20	28	36

Write an equation for each relation.

2.

x	3	6	9	12	15
$f(x)$	1	2	3	4	5

$f(x) = \frac{1}{3}x$

3.

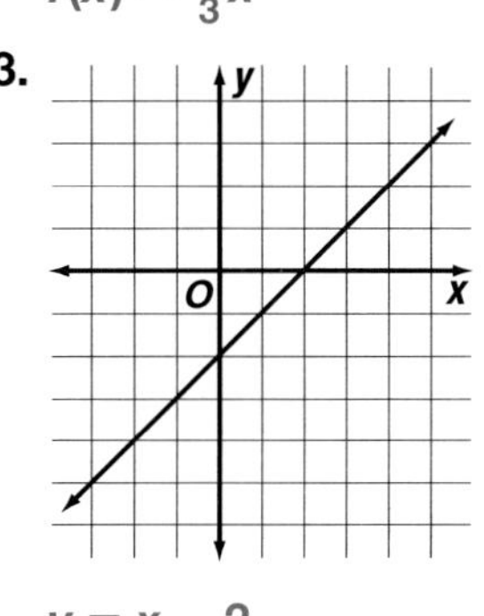

$y = x - 2$

Motivating the Lesson

Situational Problem Have students find the mean, median, and mode of the following temperatures that occurred in Phoenix, Arizona, over a 7-day period: 83°, 84°, 95°, 85°, 85°, 80°, and 48°. Ask students to identify a temperature that deviates greatly from these measures of central tendency. Introduce the idea that, in this lesson, students will learn about measures that can be used to determine variations that occur in a given set of data. **mean = 80°, median = 84°, mode = 85°; 48°**

2 TEACH

The climate of a region is due in part to its latitude. But many other factors may influence climate, including proximity to a large body of water, direction of prevailing winds, and altitude.

Teaching Tip Emphasize the division shown by the quartiles. The median divides the data in half. The upper and lower quartiles divide each half into two parts.

He noticed that the pattern in the temperatures was very different. This case shows that measures of central tendency may not give an accurate enough description of the data. Often **measures of variation** are also used to describe the distribution of the data.

One of the most common measures of variation is the **range.** Unlike the other definition of range in this chapter, the range of a set of data is a measure of the spread of the data.

Definition of Range	**The range of a set of data is the difference between the greatest and the least values of the set.**

If you lived in Managua, Nicaragua, you would experience hot, humid weather averaging 80°F all year. In Largeau, Chad, daytime temperatures can reach 115°F. In Punta Arenas, Chile, it is constantly cold and windy because it is so close to Antarctica. Life in Reykjavik, Iceland, is not all snow and ice. It has long mild winters averaging 32°F and cool, short summers.

Let's find the range for each set of temperatures. Make a table to help organize the data.

City	Greatest Temperature	Least Temperature	Range
Columbus	84°	35°	84° − 35° or 49°
San Francisco	69°	56°	69° − 56° or 13°

The range of temperatures for Columbus is greater than that for San Francisco. This means that temperatures in Columbus vary more than temperatures in San Francisco. If his family moves to Columbus, he will definitely need some new clothes for their colder winters.

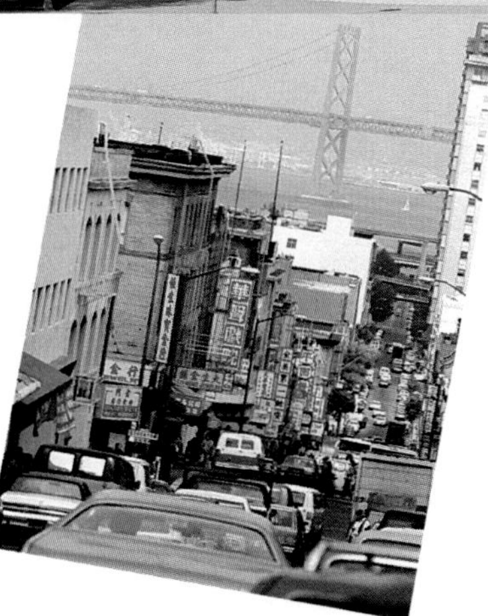

Another commonly used measure of variation is called the **interquartile range.** In a set of data, the **quartiles** are values that divide the data into four equal parts. Statisticians often use Q_1, Q_2, and Q_3 to represent the three quartiles. Remember that the median separates the data into two equal parts. Q_2 is the median. Q_1 is the **lower quartile.** It divides the lower half of the data into two equal parts. Likewise, Q_3 is the **upper quartile.** It divides the upper half of the data into two equal parts. The difference between the upper and lower quartiles is the interquartile range (IQR).

The abbreviations LQ and UQ are often used to represent the lower quartile and upper quartile.

Definition of Interquartile Range	**The difference between the upper quartile and the lower quartile of a set of data is called the interquartile range. It represents the middle half, or 50%, of the data in the set.**

Example 1

APPLICATION

Football

The table below shows the heights and weights of 13 veteran players on the 1994 Dallas Cowboys football team. Find the median, upper and lower quartiles, and the interquartile range for the weights of the players.

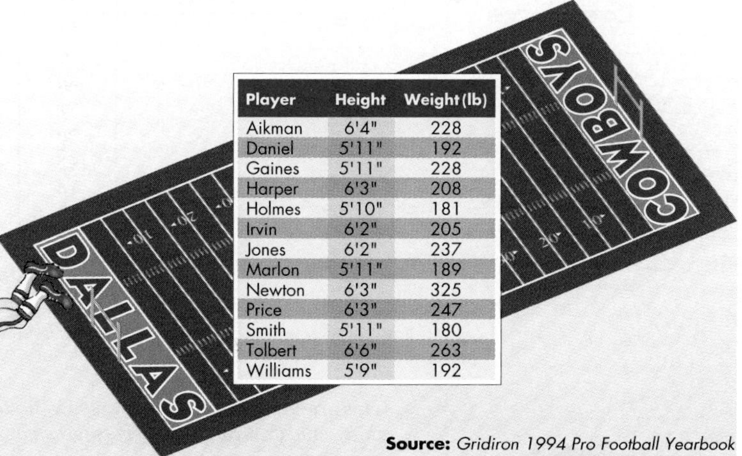

Player	Height	Weight (lb)
Aikman	6'4"	228
Daniel	5'11"	192
Gaines	5'11"	228
Harper	6'3"	208
Holmes	5'10"	181
Irvin	6'2"	205
Jones	6'2"	237
Marlon	5'11"	189
Newton	6'3"	325
Price	6'3"	247
Smith	5'11"	180
Tolbert	6'6"	263
Williams	5'9"	192

Source: *Gridiron 1994 Pro Football Yearbook*

Order the 13 weights from least to greatest. Then find the median.

180 181 189 192 192 205 208 228 228 237 247 263 325

↑ median (208)

Remember that the median is the middle number when the data are arranged in numerical order.

The lower quartile is the median of the lower half of the data, and the upper quartile is the median of the upper half of the data. If the median is an item in the set of data, it is not included in either half.

180 181 189 ↑ 192 192 205 208 228 228 237 ↑ 247 263 325

$Q_1 = 190.5$ $Q_2 =$ median $Q_3 = 242$

The interquartile range is 242 − 190.5 or 51.5. Therefore, the middle half, or 50% of the weights of the football players, varies within 51.5 pounds.

In Example 1, one weight, 325 pounds, is much greater than the others. In a set of data, a value that is much greater or much less than the rest of the data can be called an **outlier.** An outlier is defined as any element of a set of data that is at least 1.5 interquartile ranges greater than the upper quartile or less than the lower quartile.

An outlier will not affect the quartiles, but it will affect the mean.

The interquartile range of the football data is 51.5. So 1.5(51.5) is 77.25.

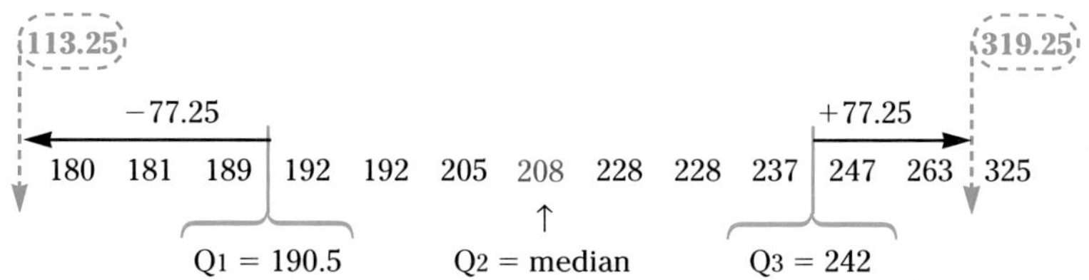

From the diagram above, you can see that the only item of data occurring beyond the dashed lines is 325. So, this is the only outlier.

In-Class Example

For Example 1

Mrs. Kollar gave a quiz in her statistics class. The scores were 23, 30, 22, 20, 20, 14, 15, 19, 19, 20, 23, 20, 22, and 24. Find the median, the upper and lower quartiles, the interquartile range, and any outliers. **median = 20, UQ = 23, LQ = 19, interquartile range = 4, outlier = 30**

Cooperative Learning

Trade-A-Problem Separate the class into two groups. Have group members measure and record one another's heights in inches or centimeters and then work together to determine the median, upper and lower quartiles, interquartile range, and any outliers for the values. Then have groups trade sets of values and determine the same information for the other group's heights. Have the whole class discuss their findings and methods. For more information on the trade-a-problem strategy, see *Cooperative Learning in the Mathematics Classroom,* one of the titles in the Glencoe Mathematics Professional Series, pages 25–26.

In-Class Example

For Example 2
The double stem-and-leaf plot below represents the number of points scored by the varsity and junior varsity basketball teams in 15 games last season.

JV	Stem	Varsity
	9	8
	8	2
6 3	7	1 4 5
2	6	0 1 1 3 6 8
9 3 1 0	5	7 9
8 5 3 2 0	4	6 8
7 3 1	3	

Find the median, quartiles, interquartile ranges, and any outliers for both sets of data.
JV: median–48, Q_3–59, Q_1–40, IQR–19, outliers–none; Varsity: median–63, Q_3–74, Q_1–59, IQR–15, outliers–98

Example 2

Commerce

The double stem-and-leaf plot below represents the tonnage of domestic and foreign imports being delivered at the top 19 busiest ports in the United States during 1992.

Domestic	Stem	Foreign
[5	10	
	9	[5
	8	
5	7	3
5	6	
	5	
8 0	4	1 4
8 1	3	6 6
8 5] △3 [1	2	0 0 3 4 5 5] △5 [5 9 9
9 9 7 5 2 0	1	8 8
5|2 = 25,000,000 tons 9 8]	0	8] 7|3 = 73,000,000 tons

Source: Army Corps of Engineers, U.S. Dept. of Defense

LOOK BACK

You can review stem-and-leaf plots in Lesson 1-4.

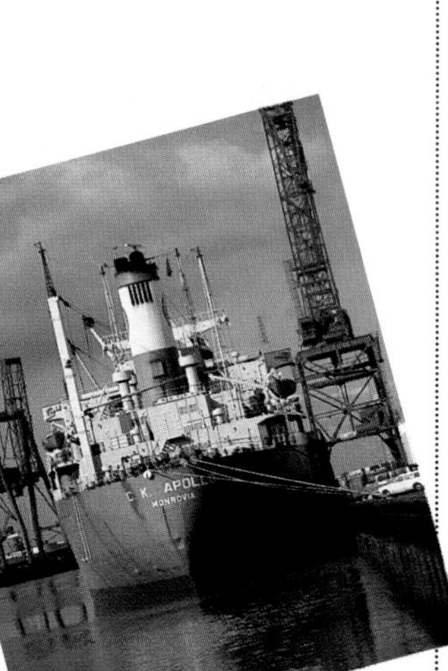

a. The brackets group the values in the lower half and the values in the upper half. What do the triangles contain? What do the boxes contain?

b. Find the interquartile ranges of the domestic and the foreign tonnage.

c. Which type of imports had a more consistent tonnage—the domestic or the foreign?

d. Find any outliers.

a. The triangles contain the medians.
The boxes contain the lower and upper quartiles.

b. Domestic

Q_1 is 15 and Q_3 is 40.

$\text{IQR} = Q_3 - Q_1$

$= 40 - 15$ or 25

Foreign

Q_1 is 20 and Q_3 is 36.

$\text{IQR} = Q_3 - Q_1$

$= 36 - 20$ or 16

c. The foreign imports had more consistent tonnage because it had the smaller interquartile range.

d. Outliers are those values that lie 1.5 IQRs above Q_3 or below Q_1.

Domestic

$1.5(\text{IQR}) + Q_3 = 1.5(25) + 40$ or 77.5

$Q_1 - 1.5\,(\text{IQR}) = 15 - 1.5(25)$ or -22.5

Since $105 > 77.5$, 105 is an outlier. There are no outliers below -22.5.

Foreign

$1.5(\text{IQR}) + Q_3 = (1.5)(16) + 36$ or 60

$Q_1 - 1.5\,(\text{IQR}) = 20 - (1.5)(16)$ or -4

Since $95 > 60$ and $73 > 60$, 95 and 73 are outliers.

There are no outliers below -4.

CHECK FOR UNDERSTANDING

Communicating Mathematics

1. San Francisco, 5; Columbus, 35
2. Outliers may make the mean much higher or lower than the mean of the data excluding the outliers.

Study the lesson. Then complete the following.

1. Refer to the application at the beginning of the lesson. Compare the interquartile ranges of the temperatures in San Francisco and Columbus.
2. **Describe** how the mean is affected by an outlier.
3. **Predict** the following. Sample answers are given.
   a. If you were to measure the height of all of the students in your school, in which group of students might you find outliers? the basketball teams
   b. If you were to measure the weight of all of the students in your school, in which group of students might you find outliers? the football team
4. **Explain** the difference between range as defined in Lesson 5–2 and range as defined in this lesson. The range in Lesson 5–2 is a set of values that correspond to a set of domain values. Range in this lesson describes the spread of data.

5. Along with your classmates, line up according to your heights.
   a. "Fold" the line in half by having the shortest person meet the tallest person. What is the median height? See students' work.
   b. Do the same with each half of the line. What is lower quartile and upper quartile of the heights of the students in your classroom? See students' work.

Guided Practice

Find the range, median, upper quartile, lower quartile, and interquartile range of each set of data. Identify any outliers.

6. 16, 24, 11, 17, 19 13, 17, 21.5, 13.5, 8
7. 43, 45, 56, 37, 11, 34

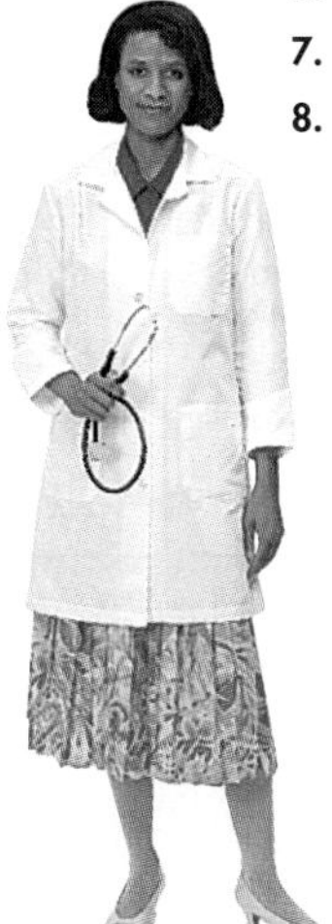

8.

Top 10 Jobs of the Future

Job	Average Salary
Civil engineer	$55,800
Computer system analyst	42,700
Electrical engineer	59,100
Geologist	50,800
High school teacher	32,500
Pharmacist	47,500
Physical therapist	37,200
Physician	148,000
Psychologist	53,000
School principal	57,300

Source: *American Careers,* Fall 1994

7. 45, 40, 45, 34, 11; 11
8. 115,500; 51,900; 57,300; 42,700; 14,600; $148,000

9.

Stem	Leaf
0	0 2 3
1	1 7 9
2	2 3 5 6
3	3 4 4 5 9
4	0 7 8 8

2|2 = 22 48, 26, 39, 17, 22

10. med, 6′2″; Q3, 6′3.5″; Q1, 5′11″; IQR, 4.5″

10. **Football** Refer to the data in Example 1. Find the median, upper and lower quartiles, and the interquartile range for the heights of the players.

3 PRACTICE/APPLY

Check for Understanding

Exercises 1–10 are designed to help you assess your students' understanding through reading, writing, speaking, and modeling. You should work through Exercises 1–5 with your students and then monitor their work on Exercises 6–10.

Error Analysis
In order to find the mean and mode, the data do not necessarily have to be in order. However, watch that the students do order the data to find the median, upper and lower quartiles, and the interquartile range.

Reteaching

Using Alternative Methods Have students work the processes they have just learned in reverse. Provide students with a median and a range and have them develop sets of data to satisfy these measures. You might also want to specify the number of data points within the set.

4 ASSESS

Closing Activity

Writing Have students briefly define each of the following terms: range, quartile, upper quartile, lower quartile, interquartile range, and outlier.

Chapter 5, Quiz D (Lesson 5-7), is available in the *Assessment and Evaluation Masters*, p. 129.

Practice Masters, p. 40

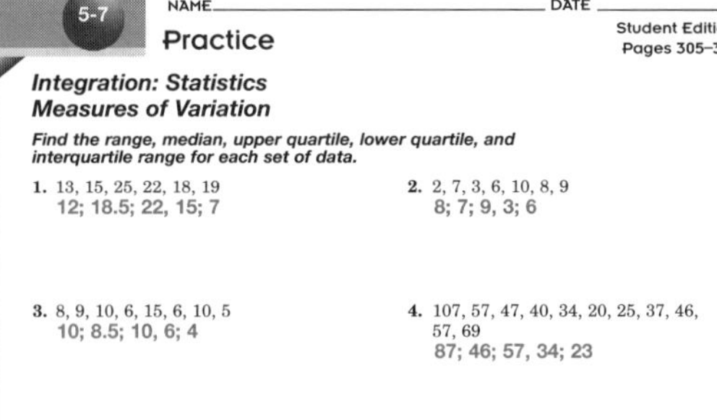

5-7 NAME ________ DATE ________

Practice

Student Edition Pages 305–313

Integration: Statistics
Measures of Variation

Find the range, median, upper quartile, lower quartile, and interquartile range for each set of data.

1. 13, 15, 25, 22, 18, 19
   12; 18.5; 22, 15; 7
2. 2, 7, 3, 6, 10, 8, 9
   8; 7; 9, 3; 6
3. 8, 9, 10, 6, 15, 6, 10, 5
   10; 8.5; 10, 6; 4
4. 107, 57, 47, 40, 34, 20, 25, 37, 46, 57, 69
   87; 46; 57, 34; 23
5. 200, 250, 230, 180, 160, 140
   110; 190; 230, 160; 70
6. 2.1, 3.2, 4.6, 7.1, 2.9
   5; 3.2; 5.85; 2.5; 3.35

7.

Stem	Leaf
2	0 1
3	2 4 6
4	3 3 8 9
5	2 2 4
6	3 6

$4|3 = 43$

46; 45.5; 52; 34; 18

8.

Stem	Leaf
1	0 0 1
2	4 6
3	2 3 3
4	7 8 9
5	1 2 2 3

$2|6 = 26$

43; 33; 51; 24; 27

9.

Stem	Leaf
7	0 2 5
8	0 0
9	3 5 9
10	1 1 4 6
11	6 7 7 8

$10|1 = 101$

48; 100; 111; 80; 31

10.

Stem	Leaf
8	1 9
9	2 3 6
10	1 9
11	2 4 8 9
12	3 7

$10|1 = 101$

46; 109; 118.5, 92.5; 26

27. **Sports** The double stem-and-leaf plot below represents the ages of the top 20 male and female golfers for the 1994 season, according to *Golfer's Almanac*, 1995.

Male	Stem	Female
8	1	
9 9 8 8 5	2	3 4 5 5 9 9
8 8 7 6 6 5	3	1 2 2 2 4 4 5 8 9 9 9
6 2 0 0	4	0 4 6
4|5 = 54 years old 4 3 1 0	5	*4|6 = 46 years old*

a. Find the ranges, quartiles, and interquartile ranges of the average golfer's age for male and female golfers.
b. Identify any outliers. **There are no outliers.**
c. Compare the ranges and interquartile ranges for male and female golfers. What can you conclude from these statistics?

27a. males: 36, 29, 37.5, 44, 15; females: 23, 29, 33, 39, 10
27c. The ages of the top female golfers are less varied than those of the top male golfers.

28. **Entertainment** The total sales of movie rentals as of April 1994 for selected Academy Award winners are listed below.

Title	Total Rental Sales ($ million)	Title	Total Rental Sales ($ million)
Schindler's List	38.7	*Ordinary People*	23.1
Silence of the Lambs	59.8	*Kramer vs. Kramer*	60.0
Dances with Wolves	81.5	*The Deer Hunter*	27.5
Driving Miss Daisy	50.5	*Annie Hall*	19.0
Rain Man	86.8	*Rocky*	56.5
Platoon	70.0	*One Flew Over the*	
Out of Africa	43.5	*Cuckoo's Nest*	60.0
Amadeus	23.0	*The Godfather, Part II*	30.6
Terms of Endearment	50.25	*The Sting*	78.2
Ghandi	25.0	*The Godfather*	86.3
Chariots of Fire	30.6	*The Sound of Music*	80.0

Source: *Variety*, May 1994

a. Find the range, median, upper and lower quartiles, and interquartile range for the total rental sales.
b. Identify any outliers. **There are none.**
c. The least value on the list of all-time top 20 movie rentals as of April, 1994, is $96.3 million. What can you conclude about Academy Award winners in relation to this data? **None of these are on the top 20 list.**

28a. $67.8 million; $50.5 million; $74.1 million; $29.05 million; $45.05 million

29. a. List the grades you have made during this grading period in mathematics. Determine the range, the quartiles, and the interquartile range of the scores. **See students' work.**
b. List the grades you have made during this grading period in English. Determine the range, the quartiles, and the interquartile range of the scores. **See students' work.**
c. In which of the two courses are your grades more consistent? Why do you think this is the case? **See students' work.**

Mixed Review

30. **Geology** The underground temperature of rocks varies with their depth below the surface. The temperature at the surface is about 20°C. At a depth of 2 km, the temperature is about 90°C, and at a depth of 10 km, the temperature is about 370°C. (Lesson 5–6)
a. Write an equation to describe this relationship. $y = 35x + 20$
b. Use the equation to predict the temperature at a depth of 13 km. **475°C**

Extension

Reasoning A set of data contains 7 elements. The median is 8900, the interquartile range is 1200, and there is one outlier above the upper quartile. Find a set of data that satisfies all the conditions. Compile the student responses to investigate how different the data might be. **Answers will vary.**

31. Which ordered pairs are solutions to $5 - 1.5x = 2y$? (Lesson 5–3) **c, d**
a. (0, 1) b. (8, 2) c. $\left(4, -\frac{1}{2}\right)$ d. (2, 1)

32. **Ecology** For every 20 grams of smog produced by a typical automobile in an hour, gasoline-powered lawn mowers produce 50 grams. If it takes Jodi 1.5 hours to mow her lawn, how long would she be able to drive her car and produce the same amount of smog? (Lesson 4–1) **3.75 hours**

33. **Sports** The table at the right shows how many millions of people participated in in-line skating and mountain biking in 1991 and 1992. Write an equation to represent each situation. Then solve. (Lesson 3–1)

Year	1991	1992
In-Line Skating	6.2	9.4
Mountain Biking	6.0	6.9

Source: *Men's Fitness*

a. How many more people were participating in in-line skating in 1992 than in 1991?
b. How many more people were mountain biking in 1992 than in 1991? **$6.0 + p = 6.9$; about 0.9 million people**

33a. $6.2 + p = 9.4$; about 3.2 million people

34. Simplify $3(-4) + 2(-7)$. (Lesson 2–6) **−26**

Mathematics and SOCIETY

Meteorology makes extensive use of computer/mathematical models to represent the changes in the atmosphere. Dynamic meteorology deals with air flow in the atmosphere. Mathematics' chaos theory developed from research carried out in dynamic meteorology.

Mathematics and SOCIETY

Record Cold Temperatures

The excerpt below appeared in an article in *Mother Earth News* in January, 1995. All of the temperatures are given in °F.

How cold does it get in Alaska? The all-time low temperature is −80 degrees at Prospect Creek Camp on January 23, 1971. Think of it: that temperature is to 0 degrees what 0 degrees is to a typical summer afternoon. Still, you might say, Alaska is up there near the North Pole. Surely, down here in the "lower 48," we never approach such levels of incredible frigidity. Yes we do. On January 20, 1954, the temperature in Rogers Pass, Montana, nudged −70 degrees.... Is there anywhere that temperatures can get colder than in Alaska? Siberia can get a little colder. But there is one place on Earth that is much more frigid: Antarctica. Earth's all-time record low was set at Vostok Station in Antarctica in 1983: −128 degrees. ■

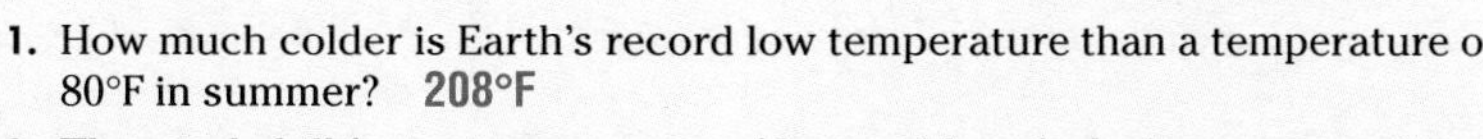

Antarctica

1. How much colder is Earth's record low temperature than a temperature of 80°F in summer? **208°F**
2. The wind-chill factor is a measure of how cold we feel when we experience wind along with a low temperature. Do you think that the wind-chill factor also applies to nonliving objects such as bicycles and cars? Why or why not?
3. The hottest temperature recorded on Earth was 136°F at Azizia, Tripolitaina, in northern Africa, on September 13, 1922. If every country's record high and low temperatures for the 20th century were entered in a computer, what would be the range of the data? **264°F**
4. If the interquartile range of temperatures in a country is very small, what does that tell you about the climate of that country? **The temperature stays the same about half of the time.**

2. No; the wind-chill factor applies to people and animals, but not to inanimate objects.

Enrichment Masters, p. 40

5-7 NAME __________ DATE ________

Enrichment — Student Edition Pages 305–314

Standard Deviation

The most commonly used measure of variation is called the **standard deviation.** It shows how far the data are from their mean. You can find the standard deviation using the steps given below.

a. Find the mean of the data.
b. Find the difference between each value and the mean.
c. Square each difference.
d. Find the mean of the squared differences.
e. Find the square root of the mean found in Step **d.** The result is the standard deviation.

Example: Calculate the standard deviation of the test scores 82, 71, 63, 78, and 66.

mean of the data $(m) = \frac{82 + 71 + 63 + 78 + 66}{5} = \frac{360}{5} = 72$

x	$x - m$	$(x - m)^2$
82	$82 - 72 = 10$	$10^2 = 100$
71	$71 - 72 = -1$	$(-1)^2 = 1$
63	$63 - 72 = -9$	$(-9)^2 = 81$
78	$78 - 72 = 6$	$6^2 = 36$
66	$66 - 72 = -6$	$(-6)^2 = 36$

mean of the squared differences $= \frac{100 + 1 + 81 + 36 + 36}{5} = \frac{254}{5} = 50.8$

standard deviation $= \sqrt{50.8} \approx 7.13$

Use the test scores 94, 48, 83, 61, and 74 to complete Exercises 1-3.

1. Find the mean of the scores. 72
2. Show that the standard deviation of the scores is about 16.2.

x	$x - m$	$(x - m)^2$
94	$94 - 72 = 22$	484
48	$48 - 72 = -24$	576
83	$83 - 72 = 11$	121
61	$61 - 72 = -11$	121
74	$74 - 72 = 2$	4

mean of $(x - m)^2 = 261.2$
S.D. $= \sqrt{261.2} \approx 16.2$

3. Which had less variations, the test scores listed above or the test scores in the example? the test scores in the example

Closing the Investigation

This activity provides students an opportunity to bring their work on the Investigation to a close. For each Investigation, students should present their findings to the class. Here are some ways students can display their work.

- Conduct and report on an interview or survey.
- Write a letter, proposal, or report.
- Write an article for the school or local paper.
- Make a display, including graphs and/or charts.
- Plan an activity.

Assessment

To assess students' understanding of the concepts and topics explored in the Investigation and its follow-up activities, you may wish to examine students' Investigation Folders.

The scoring guide provided in the *Investigations and Projects Masters,* p. 7, provides a means for you to score students' work on the Investigation.

Investigations and Projects Masters, p. 7

Scoring Guide
Chapters 4 and 5
Investigation

Level	Specific Criteria
3 Superior	• Shows thorough understanding of the concepts of *the capture-recapture method, data collection and analysis, proportions, percents,* and *probability.* • Uses appropriate strategies to solve problems. • Computations are correct. • Written explanations are exemplary. • Charts, graphs, and speech are appropriate and sensible. • Goes beyond the requirements of some or all problems.
2 Satisfactory, with Minor Flaws	• Shows understanding of the concepts of *the capture-recapture method, data collection and analysis, proportions, percents,* and *probability.* • Uses appropriate strategies to solve problems. • Computations are mostly correct. • Written explanations are effective. • Charts, graphs, and speech are appropriate and sensible. • Satisfies the requirements of problems.
1 Nearly Satisfactory, with Obvious Flaws	• Shows understanding of most of the concepts of *the capture-recapture method, data collection and analysis, proportions, percents,* and *probability.* • May not use appropriate strategies to solve problems. • Computations are mostly correct. • Written explanations are satisfactory. • Charts, graphs, and speech are appropriate and sensible. • Satisfies most requirements of problems.
0 Unsatisfactory	• Shows little or no understanding of the concepts of *the capture-recapture method, data collection and analysis, proportions, percents,* and *probability.* • Does not use appropriate strategies to solve problems. • Computations are incorrect. • Written explanations are not satisfactory. • Charts, graphs, and speech are not appropriate or sensible. • Does not satisfy the requirements of problems.

Refer to the Investigation on pages 190–191.

Sampling is a research method frequently used in many fields. Manufacturers use sampling to determine the quality of their products. Surveys are a way of sampling the opinions of the public about items they encounter in everyday life such as what television programs they watch or what food is their favorite. Capture-recapture is only one of many methods used in sampling a population.

Analyze

You have conducted experiments and organized your data in various ways. It is now time to analyze your findings and state your conclusions.

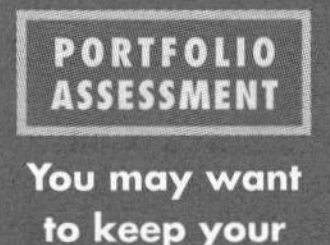

PORTFOLIO ASSESSMENT

You may want to keep your work on this Investigation in your portfolio.

1 What are the strengths and weaknesses of the capture-recapture method of sampling?
2 How many samples do you think give the best results? Why?
3 Write a one-page report on how the capture-recapture method is used in real life. Be sure to include references for your report.

Write

Imagine that the fish population in Moonlit Lake had been estimated each year for nine years and that these estimates are reflected in the following graph.

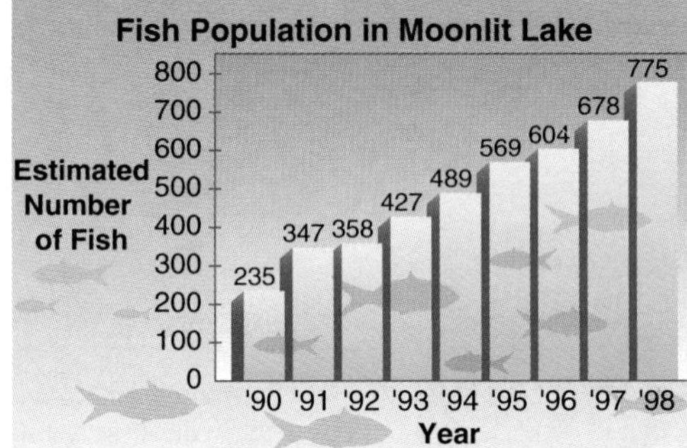

Notice that the fish population is increasing over time. Suppose city council member Aubrey Howard wanted to show other members of the city council that the growth in the fish population has been rapid. Mr. Howard might choose to find a best-fit line for the graph.

4 Plot the information from the graph as ordered pairs on a coordinate plane.
5 Use two vertical lines to separate the data into three sets of equal size.
6 Locate a point in each set whose x- and y-coordinates represent the medians of the x- and y-coordinates of all the points in that set. In other words, locate the middle point of the three sets.
7 Now take a straightedge and align it with the outer two median points. Then slide it one third of the distance to the median point of the center set of points. Then draw the line. This is a *best-fit line.*
8 What might Mr. Howard's argument be? Write a one-page speech for the council member, using the graph as evidence for your opinion.

CHAPTER 5 HIGHLIGHTS

VOCABULARY

After completing this chapter, you should be able to define each term, property, or phrase and give an example or two of each.

Algebra

axes (p. 254)
complete graph (p. 278)
completeness property for points in the plane (p. 256)
coordinate plane (p. 254)
domain (p. 263)
equation in two variables (p. 271)
function (p. 287)
functional notation (p. 289)
graph (p. 255)
inverse of a relation (p. 264)
linear equation (p. 280)
linear equation in standard form (p. 281)
linear function (p. 278)
mapping (p. 263)
origin (p. 254)
quadrants (p. 254)
range (p. 263)
relation (pp. 260, 263)
solution of an equation in two variables (p. 271)
vertical line test (p. 289)
x-axis (p. 254)
x-coordinate (p. 254)
y-axis (p. 254)
y-coordinate (p. 254)

Statistics

interquartile range (pp. 304, 306)
lower quartile (p. 306)
measures of variation (p. 306)
outlier (p. 307)
quartiles (pp. 304, 306)
range (p. 306)
upper quartile (p. 306)

Problem Solving

use a table (p. 255)

UNDERSTANDING AND USING THE VOCABULARY

Choose the letter of the term that best matches each statement or phrase.

1. In the coordinate plane, the axes intersect at the __?__. e
2. A __?__ is a set of ordered pairs. g
3. __?__ are graphed on a coordinate plane. d
4. In a coordinate system, the __?__ is a horizontal line. h
5. In the ordered pair, A(2, 7), 7 is the __?__. j
6. The coordinate axes separate a plane into four __?__. f
7. An equation whose graph is a non-vertical straight line is called a __?__. c
8. In the relation $\{(4, -2), (0, 5), (6, 2), (-1, 8)\}$, the __?__ is the set $\{-1, 0, 4, 6\}$. a
9. The domain contains values represented by the __?__. b

a. domain
b. independent variable
c. linear function
d. ordered pairs
e. origin
f. quadrants
g. relation
h. *x*-axis
i. *y*-axis
j. *y*-coordinate

Instructional Resources

Three multiple-choice tests and three free-response tests are provided in the *Assessment and Evaluation Masters*. Forms 1A and 2A are for honors pacing, and Forms 1B, 1C, 2B, and 2C are for average pacing. Chapter 5 Test Form 1B is shown at the right. Chapter 5 Test Form 2B is shown on the next page.

Using the CHAPTER HIGHLIGHTS

The Chapter Highlights begins with a listing of the new terms, properties, and phrases that were introduced in this chapter. Have students define each term and provide an example or two of it, if appropriate.

Assessment and Evaluation Masters, pp. 115–116

5 NAME ______ DATE ______

Chapter 5 Test, Form 1B

Write the letter for the correct answer in the blank at the right of each problem.

Use the graph to answer questions 1–3.

1. What is the ordered pair for point L? 1. C
   A. $(1, -4)$ B. $(-4, 1)$
   C. $(-1, -4)$ D. $(-4, -1)$
2. Name the quadrant in which the point E is located. 2. C
   A. I B. II
   C. III D. IV
3. What is the range of the relation graphed? 3. A
   A. $\{-4, -1, 0, 2, 3, 4\}$ B. $\{-4, -2, -1, 0, 1, 2, 3, 4\}$
   C. $\{-4, -2, -1, 0, 1, 4\}$ D. {all the integers}
4. Which relation is the inverse of $\{(-2, -1), (2, 1), (-2, 4)\}$? 4. D
   A. $\{(-1, 2), (-2, -1), (-2, -4)\}$ B. $\{(2, 1), (-2, -1), (2, -4)\}$
   C. $\{(-2, 4,), (2, 1), (-2, -1)\}$ D. $\{(-1, -2), (1, 2), (4, -2)\}$
5. What are the domain and range of the relation shown in the table? 5. D
   A. domain: $\{-3, 4\}$; range: $\{0, 6\}$
   B. domain: $\{4, 6\}$; range: $\{-3, 0\}$
   C. domain: $\{0, 6\}$; range: $\{-3, 4\}$
   D. domain: $\{-3, 0\}$; range: $(4, 6\}$

x	y
−3	4
0	6

6. The y-coordinate of each point on the x-axis is 6. C
   A. the same as the x-coordinate. B. always positive.
   C. always zero. D. any number.
7. To plot the point $(-5, 2)$, you start at the origin and go 7. B
   A. down 5 and right 2. B. up 2 and left 5.
   C. left 5 and down 2. D. right 2 and up 5.
8. When $3x - 5y = 12$ is solved for y, which equation results? 8. C
   A. $y = 7 - 3x$ B. $-5y = 12 - 3x$
   C. $y = \frac{3x - 12}{5}$ D. $y = \frac{12 - 3x}{5}$
9. Which set of ordered pairs shows the relation in the mapping? 9. C
   A. $\{(-3, 5), (1, -1), (1, 6), (8, 2)\}$
   B. $\{(5, 3), (2, 1), (-1, 1), (6, 8)\}$
   C. $\{(5, 3), (2, 8), (-1, 1), (6, 1)\}$
   D. $\{(3, 5), (1, 2), (1, -1), (8, 6)\}$
10. What is the solution set of the equation $2r + s = 8$ if the domain is $\{-2, 0, 4\}$? 10. A
    A. $\{(-2, 12), (0, 8), (4, 0)\}$ B. $\{(12, -2), (8, 0), (0, 4)\}$
    C. $\{(-2, 6), (0, 4), (4, 2)\}$ D. $\{(6, -2), (4, 0), (2, 4)\}$

5 NAME ______ DATE ______

Chapter 5 Test, Form 1B (continued)

11. Which equation is a linear equation? 11. B
    A. $2x^2 + 5y = 3$ B. $y = -10$ C. $5 = 3xy$ D. $y = \frac{1}{x} + 4$
12. The graph of $\frac{1}{2}x + \frac{2}{3}y = 1$ is which line? 12. D
    A. r B. s
    C. t D. v
13. The graph of $x = 0$ is 13. C
    A. the origin.
    B. the x-axis.
    C. the y-axis.
    D. not possible to draw.
14. Which equation has a graph that is a horizontal line? 14. C
    A. $x - 7 = 0$ B. $x = y$ C. $2y + 3 = 4$ D. $x + y = 0$
15. Determine which relation is a function. 15. D
    A. $\{(2, 8), (-1, 3), (2, -2)\}$ B. $y^2 = 3x + 2$
    C. $3x - 11 = 0$ D. $y = x^2 + 5x + 1$
16. Which is the graph of a function? 16. C
    A. B. C. D. none of these
17. If $f(x) = 7 - x$, which of the following is true? 17. D
    A. $f(-1) = (-1, 8)$ B. $2[f(3)] = f(6)$
    C. $f(0) = 0$ D. $f(7 - w) = w$
18. When an equation for the relation shown in the chart is solved for y, the constant term is 18. A
    A. −16. B. −4. C. 0. D. 4.

x	3	4	5
y	−10	−8	−6

19. If the lower quartile = 38 and the upper quartile = 54 for a set of data, find the least value above 54 that could be an outlier. 19. D
    A. 55.5 B. 70 C. 109 D. 78
20. Find the upper quartile for the data represented in the stem-and-leaf plot at the right. 9|7 represents 97. 20. D
    A. 102 B. 86
    C. 23 D. 103

Stem	Leaf
8	4 5 6 7
9	7 9
10	2 2 3 4 9

Bonus Graph all ordered pairs (x, y) that make $2x^2 - 5x - 3 = 0$ true. **Bonus**

Using the STUDY GUIDE AND ASSESSMENT

Skills and Concepts Encourage students to refer to the objectives and examples on the left as they complete the review exercises on the right.

Assessment and Evaluation Masters, pp. 121–122

5 NAME________ DATE________

Chapter 5 Test, Form 2B

Use the table to answer questions 1 and 2.

x	y
6	−2
−3	4
0	5

1. Write the relation as a set of ordered pairs. 1. {(6, −2), (−3, 4), (0, 5)}
2. State the domain, range, and inverse of the relation. 2. D: {6, −3, 0}; R: {−2, 4, 5}; I: {(−2, 6), (4, −3), (5, 0)}
3. Graph these points on the coordinate plane provided. $A(3, 2)$ $B(-3, 0)$ $C(0, -2)$ $D(3, -4)$ 3.
4. Name the quadrant in which each point is located. $A(-5, 7)$ $B(2, -4)$ $C(6, 0)$ 4. II; IV; none
5. Draw a mapping for the relation {(1, 4), (−5, 0), (7, 4)}. 5.

Use the graph to answer questions 6 and 7.

6. Write the relation in the graph as a set of ordered pairs. 6. {(−3, 2), (−3, −1), (0, 1), (4, 3)}
7. State the domain, range, and inverse of the relation. 7. D: {−3, 0, 4}; R: {−1, 1, 2, 3}; I: {(−1, −3), (1, 0), (2, −3), (3, 4)}
8. Solve $3x + y = 6$ for y. 8. $y = 6 - 3x$
9. Solve the equation $m - 2n = 9$ if the domain is {−1, 0, 5}. 9. $\{(-1, -5), (0, -\frac{9}{2}), (5, -2)\}$
10. Which of the following equations are linear equations? Write the letter for each linear equation in the answer blank. 10. b, c, d
    a. $xy = 6$ b. $y = 0$ c. $4x = 2y$
    d. $x - y = 0$ e. $y^2 - 4x = 0$
11. Graph $x - 4y = 2$. Use the coordinate plane provided. 11.

5 NAME________ DATE________

Chapter 5 Test, Form 2B (continued)

12. Graph $2x = -3y$. Use the coordinate plane provided. 12.
13. Determine whether each relation is a function. Then write the letter for each function in the answer blank. 13. a, d, e
    a. {(5, 3), (2, −1), (−1, 3)}
    b. {(2, 1), (2, −4), (−3, 0)}
    c. $x - 8 = 0$
    d. $4y = 9x$
    e.
14. Given $f(x) = x^2 + 3x - 2$, find $f(2)$. 14. 8
15. Given $h(x) = 4x + 8$, find $4[h(-\frac{3}{4})]$. 15. 20

Write an equation for the relation given in each chart.

16.

x	1	2	3	4	5
y	5	7	9	11	13

16. $y = 2x + 3$

17.

x	−2	0	2	4
y	10	2	−6	−14

17. $y = -4x + 2$

18. Li Teng is ordering graduation announcements for the senior class. If 100 announcements are ordered, the cost is \$225. The cost for 175 announcements is \$375, and 200 announcements cost \$425. Write an equation to describe this relationship and use it to find the cost of 250 announcements. 18. $c = 25 + 2a$; \$525
19. Find the median and upper and lower quartiles for this set of data: 200, 250, 230, 180, 160, 140. 19. median: 190; UQ: 230; LQ: 160
20. Find the interquartile range for the data in question 19. 20. interq. range: 70

Bonus Graph $x = 3$, $y = -1$, and $2x - 2y = 0$ on the coordinate plane provided. Name the geometric figure formed. Then state the coordinates of each vertex. Bonus triangle; (3, 3), (3, −1), (−1, −1)

SKILLS AND CONCEPTS

OBJECTIVES AND EXAMPLES

Upon completing this chapter, you should be able to:

- graph ordered pairs on a coordinate plane (Lesson 5–1)

Graph $T(3, -2)$ and state the quadrant in which the point is located.

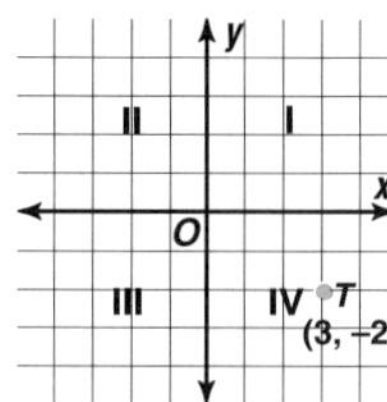

$T(3, -2)$ is located in Quadrant IV.

REVIEW EXERCISES

Use these exercises to review and prepare for the chapter test. **10–13. See Solutions Manual.**

Graph each point.

10. $A(4, 2)$
11. $B(-1, 3)$
12. $C(0, -5)$
13. $D(-3, -2)$

Write the ordered pair for each point shown at the right. State the quadrant in which the point is located.

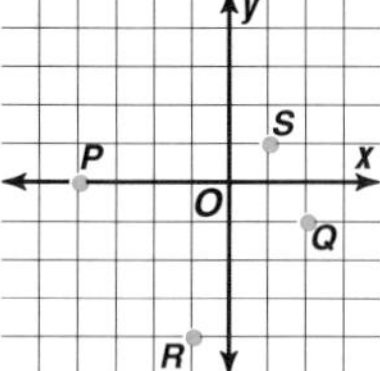

14. P **(−4, 0); none**
15. Q **(2, −1); IV**
16. R **(−1, −4); III**
17. S **(1, 1); I**

- identify the domain, range, and inverse of a relation (Lesson 5–2)

Determine the domain, range, and inverse of {(6, 6), (4, −3), (6, 0)}.

The domain is {4, 6}.

The range is {−3, 0, 6}.

The inverse is {(6, 6), (−3, 4), (0, 6)}.

19. D = {−3, 4}, R = {5, 6}
20. D = {−7, −5, −2}, R = {1}
21. D = {−3, −2, −1, 0}, R = {0, 1, 2}
22. {(4, 4), (5, −3), (−1, 4), (3, 0)}
23. {(2, 0), (−1, 3), (2, 2), (−1, −2)}

State the domain and range of each relation.

18. {(4, 1), (4, 6), (4, −1)} **D = {4}, R = {−1, 1, 6}**
19. {(−3, 5), (−3, 6), (4, 5), (4, 6)}
20. {(−2, 1), (−5, 1), (−7, 1)}
21. {(−3, 1), (−2, 0), (−1, 1), (0, 2)}

22–23. See Solutions Manual for mappings and graphs.

Draw a mapping and a graph for each relation. State the inverse of each relation.

22. {(4, 4), (−3, 5), (4, −1), (0, 3)}
23. {(0, 2), (3, −1), (2, 2), (−2, −1)}

- determine the range for a given domain (Lesson 5–3)

Solve $2x + y = 8$ if the domain is {3, 2, 1}.

Solve for y.

$2x + y = 8$
$y = 8 - 2x$

x	8 − 2x	y	(x, y)
3	8 − 2(3)	2	(3, 2)
2	8 − 2(2)	4	(2, 4)
1	8 − 2(1)	6	(1, 6)

Solve each equation if the domain is {−4, −2, 0, 2, 4}. **24–27. See margin.**

24. $y = 4x + 5$
25. $x - y = 9$
26. $3x + 2y = 9$
27. $4x - 3y = 0$

28–29. See Solutions Manual.

Make a table and graph the solution set for each equation and domain.

28. $y = 7 - 3x$ for $x = \{-3, -2, -1, 0, 1, 2, 3\}$
29. $5x - y = -3$ for $x = \{-2, 0, 2, 4, 6\}$

GLENCOE Technology

Test and Review Software

You may use this software, a combination of an item generator and item bank, to create your own tests or worksheets. Types of items include free response, multiple choice, short answer, and open ended.

For IBM & Macintosh

Additional Answers

24. {(−4, −11), (−2, −3), (0, 5), (2, 13), (4, 21)}
25. {(−4, −13), (−2, −11), (0, −9), (2, −7), (4, −5)}
26. $\left\{\left(-4, 10\frac{1}{2}\right), \left(-2, 7\frac{1}{2}\right), \left(0, 4\frac{1}{2}\right), \left(2, 1\frac{1}{2}\right), \left(4, -1\frac{1}{2}\right)\right\}$
27. $\left\{\left(-4, -5\frac{1}{3}\right), \left(-2, -2\frac{2}{3}\right), (0, 0), \left(2, 2\frac{2}{3}\right), \left(4, 5\frac{1}{3}\right)\right\}$

OBJECTIVES AND EXAMPLES	REVIEW EXERCISES

- graph linear equations (Lesson 5–4)

Graph $y = 3x - 4$.

x	3x − 4	y	(x, y)
0	3(0) − 4	−4	(0, −4)
1	3(1) − 4	−1	(1, −1)
2	3(2) − 4	2	(2, 2)

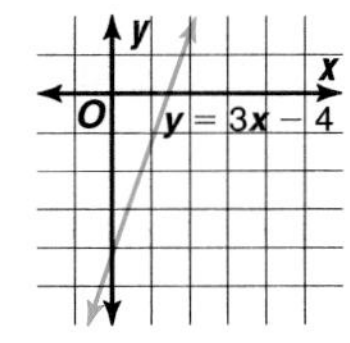

Graph each equation. 30–35. See margin.

30. $y = -x + 2$
31. $x + 5y = 4$
32. $2x - 3y = 6$
33. $5x + 2y = 10$
34. $\frac{1}{2}x + \frac{1}{3}y = 3$
35. $y - \frac{1}{3} = \frac{1}{3}x + \frac{2}{3}$

- determine whether a given relation is a function (Lesson 5–5)

Is {(3, 2),(5, 3), (4, 3), (5, 2)} a function?

Because there are two values of y for one value of x, 5, the relation is *not* a function.

Determine whether each relation is a function.

36.

a	b
−2	6
3	−2
3	0
4	6

no

37.

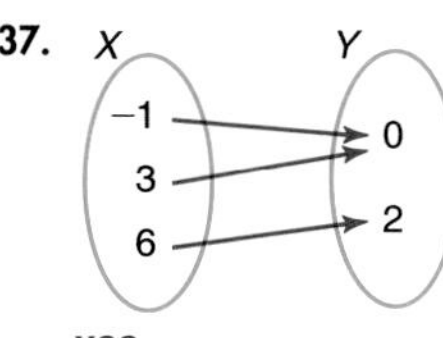

yes

38. {(3, 8), (9, 3), (−3, 8), (5, 3)} **yes**
39. $x - y^2 = 4$ **no**
40. $xy = 6$ **yes**
41. $3x - 4y = 7$ **yes**

- find the value of a function for a given element of the domain (Lesson 5–5)

Given $g(x) = 2x - 1$, find $g(-6)$.

$$g(-6) = 2(-6) - 1$$
$$= -12 - 1$$
$$= -13$$

If $g(x) = x^2 - x + 1$, find each value.

42. $g(2)$ **3**
43. $g(-1)$ **3**
44. $g\left(\frac{1}{2}\right)$ $\frac{3}{4}$
45. $g(a + 1)$ $a^2 + a + 1$
46. $g(-2a)$ $4a^2 + 2a + 1$
47. $2g(a - 3)$ $2a^2 - 14a + 26$

- write an equation to represent a relation, given some of the solutions for the equation (Lesson 5–6)

Write an equation for the relation given in the chart below.

a	−2	−1	0	1	2
b	20	17	14	11	8

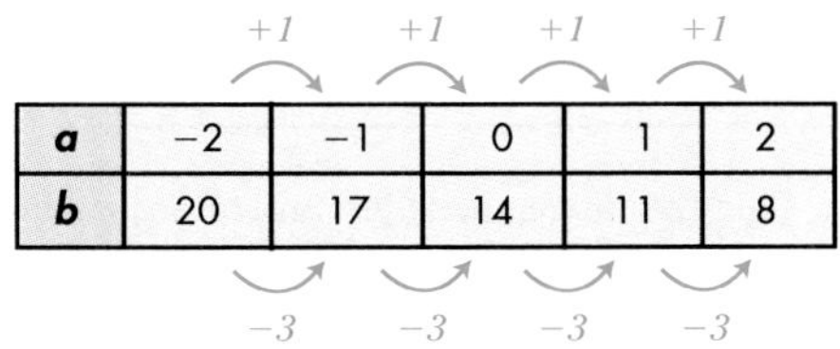

The equation is $b = 14 - 3a$.

Write an equation for each relation.

48.

x	0	1	2	3	4
y	5	8	11	14	17

49.

x	2	4	5	7	10
y	−2	0	1	3	6

50.

x	3	6	9	12	15
y	−1	−3	−5	−7	−9

48. $y = 3x + 5$ 49. $y = x - 4$ 50. $y = -\frac{2}{3}x + 1$

30.
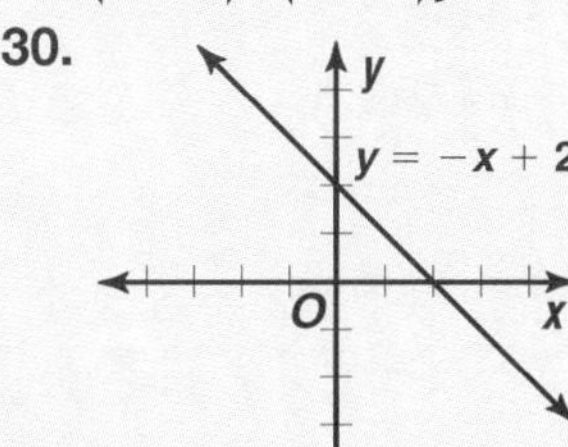

31.
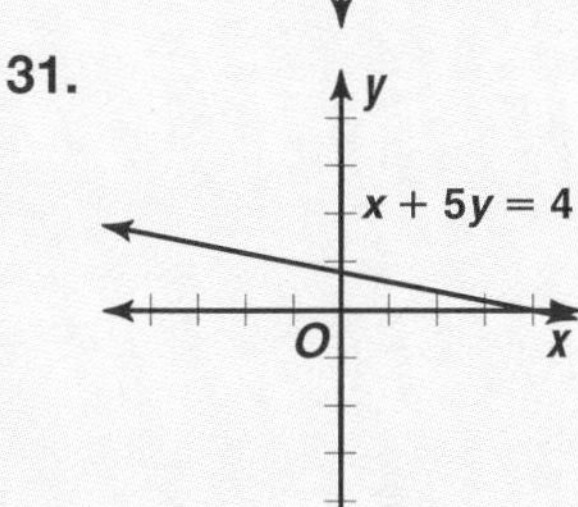

32.
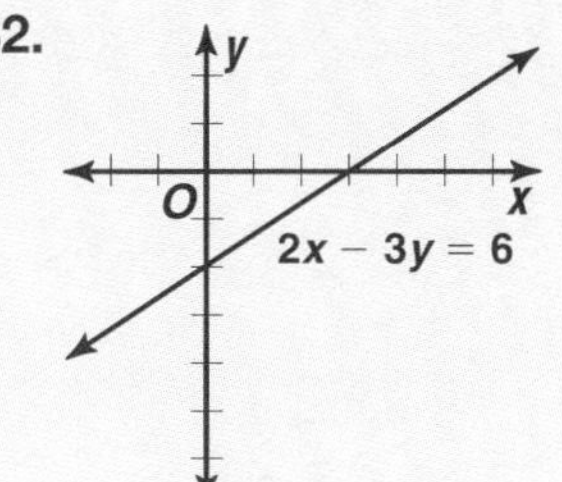

33.
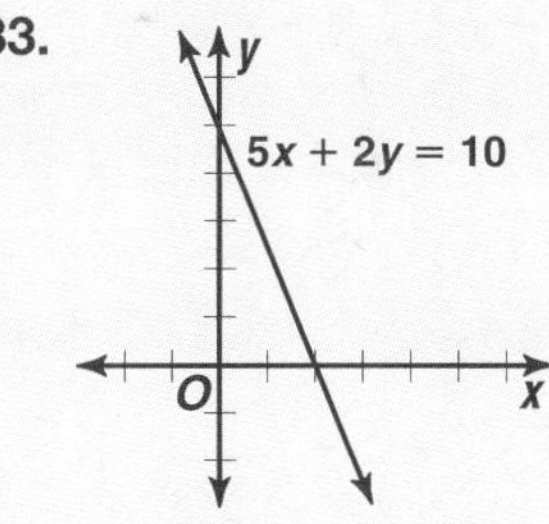

34.
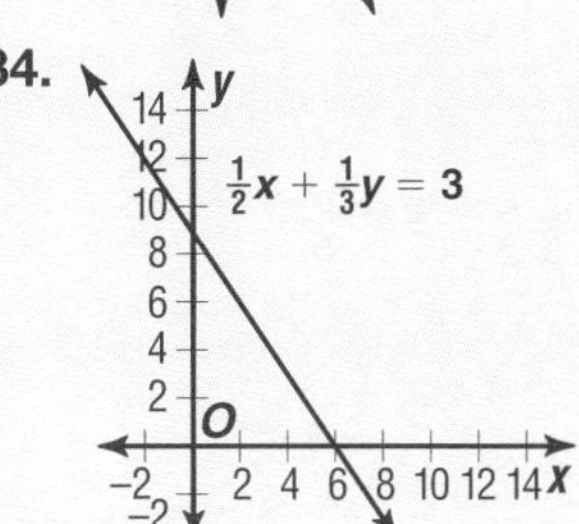

35.
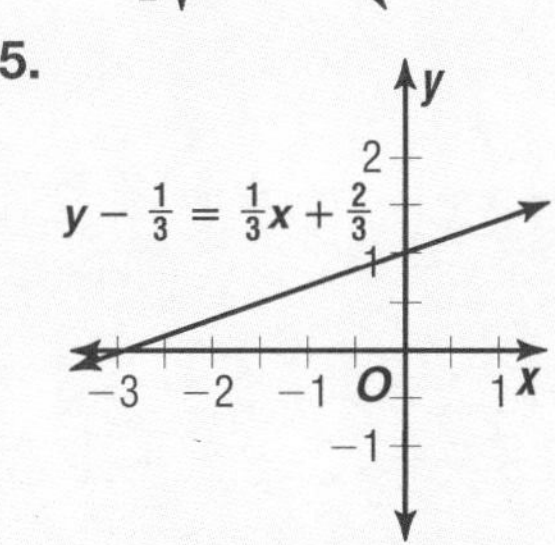

Applications and Problem Solving Encourage students to work through the exercises in the Applications and Problem Solving section to strengthen their problem-solving skills.

Additional Answer

55a.

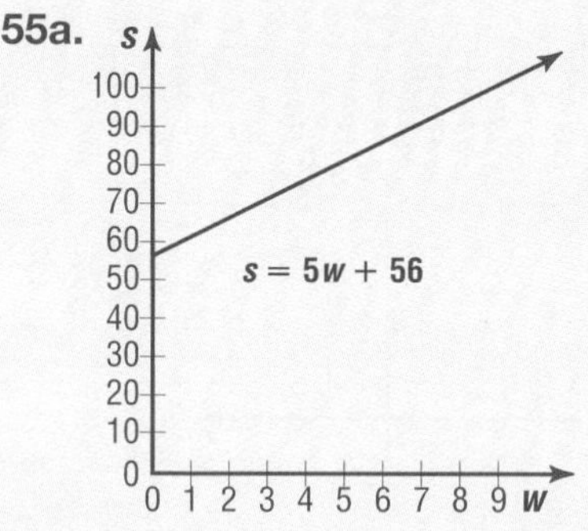

OBJECTIVES AND EXAMPLES

- calculate and interpret the range, quartiles, and interquartile range of sets of data (Lesson 5–7)

Find the range, median, upper quartile, lower quartile, and interquartile range for the set of data below.

25, 20, 30, 24, 22, 26, 28, 29, 19

Order the set of data from least to greatest.

19 20 22 24 25 26 28 29 30

The range is $30 - 19 = 11$.

The median is the middle number, 25.

The lower quartile is $\frac{20 + 22}{2}$ or 21.

The upper quartile is $\frac{28 + 29}{2}$ or 28.5.

The interquartile range is $28.5 - 21$ or 7.5.

REVIEW EXERCISES

Find the range, median, upper quartile, lower quartile, and interquartile range for each set of data.

51. 30, 90, 40, 70, 50, 100, 80, 60 **70, 65, 85, 45, 40**

52. 3, 3.2, 45, 7, 2, 1, 3.4, 4, 5.3, 5, 78, 8, 21, 5

53. 85, 77, 58, 69, 62, 73, 55, 82, 67, 77, 59, 92, 75, 69, 76

54. The average annual snowfall, in inches, for each of 12 northeastern cities is listed below.

111.5 70.7 59.8 68.6 63.8 254.8
64.3 82.3 91.7 88.9 110.5 77.1

52. 77, 5, 8, 3.2, 4.8

53. 37, 73, 77, 62, 15

54. 195, 79.7, 101.1, 66.45, 34.65

APPLICATIONS AND PROBLEM SOLVING

55. Finance To save for a new bicycle, Ralph begins a savings plan. Ralph's savings are described by the equation $s = 5w + 56$, where s represents his total savings in dollars and w represents the number of weeks since the start of the savings plan. (Lesson 5–4)

a. Draw a graph of this equation. **See margin.**

b. Use the graph to determine how much money Ralph had already saved when the plan started. **$56**

56. Car Rental The cost of a one-day car rental from A-1 Car Rental is $41 if you drive 100 miles, $51.80 if you drive 160 miles, and $63.50 if you drive 225 miles. (Lesson 5–6)

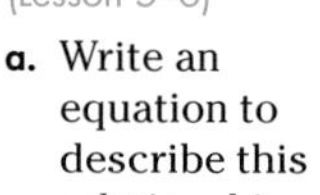

a. Write an equation to describe this relationship. $y = 0.18x + 23$

b. Use the equation to determine the per-mile charge. **$0.18 per mile**

57. Entertainment The ratings of the top 20 favorite syndicated television programs for 1993–94 are listed below. (Lesson 5–7)

Program	Rating
Action Pack Network	6.1
Baywatch	6.5
Cops	5.7
Current Affair	6.6
Donahue	5.1
Entertainment Tonight	8.4
Family Matters	5.9
Hard Copy	6.7
Inside Edition	7.3
Jeopardy!	12.6
Married With Children	6.9
National Geographic on Assignment	7.5
Oprah Winfrey	9.7
Roseanne	7.9
Sally Jessy Raphael	5.2
Star Trek	10.8
Star Trek: Deep Space Nine	8.2
Wheel of Fortune	14.7
Wheel of Fortune-Weekend	7.2
World Wrestling Federation	5.7

a. Find the range, quartiles, and interquartile range for the ratings. **9.6, 6, 7.05, 8.3, 2.3**

b. Identify any outliers. **12.6, 14.7**

A practice test for Chapter 5 is provided on page 791.

ALTERNATIVE ASSESSMENT

COOPERATIVE LEARNING PROJECT

Graphs and Business In this chapter, you learned how to graph relations and determine whether the relation is a linear equation. Once a relation has been graphed, the process of analyzing it to determine more applicable information is necessary in order to completely understand the problem or situation.

In this project, imagine that your family owns a catering business and you want to look at the amount that you charge for banquets. You have been asked to present a report to the other members of your family that shows recommended charges and how they will affect your profit. After some research, you find that the initial basic charge for a banquet should be $250 plus $4 per person. Your food costs per person are $2.50, and your other costs such as labor, time, etc. total $1.00 per person.

Graph the various charges determined by the number of persons. Graph the various costs determined by the number of persons.

Follow these steps to organize your presentation.

- Determine the charges for various banquet sizes.
- Determine the costs for various banquet sizes.
- Determine how to plot and display your graphs appropriately.
- Write an algebraic model that describes the total charge for a banquet.
- Write an algebraic model that describes the total cost for a banquet.
- Write several paragraphs for your presentation and incorporate your algebraic models and graphs to help your presentation.

THINKING CRITICALLY

- Use one set of data to construct two graphs in which each represents the data, but portrays different meanings.
- Create a relation that is a function and one that is not a function. Compare and contrast the two relations.

PORTFOLIO

You have been asked to show another student how to graph an equation, but it must be completely written and nonverbal. Select one of the graphing assignments from this chapter and list the steps involved in graphing this equation. Be sure to include your steps in the appropriate order.

When you have completed this paper, give it to another student in your class and have him or her read it and follow the written steps to graph a different equation. Collect his or her graph and check for accuracy. Place both of these in your portfolio.

SELF EVALUATION

Graphs can be used to analyze functions or data. To *analyze* means to separate or distinguish all the parts of something in order to discover more information about the complex whole element.

Assess yourself. How analytical are you? Do you delve deeper into a problem and look for smaller components that will help you understand the larger problem, or do you look at the big picture and attempt to solve the problem within the larger view of things? Describe two or three ways in which you could divide a complex problem or situation in mathematics and/or in your daily life in order to analyze it.

Alternative Assessment

The Alternative Assessment section provides students with the opportunity to assess their own work by thinking critically, working with others, keeping a portfolio, and honestly evaluating their own progress. For more information on alternative forms of assessment, see *Alternative Assessment in the Mathematics Classroom,* one of the titles in the Glencoe Mathematics Professional Series.

Performance Assessment

Performance Assessment tasks for this chapter are included in the *Assessment and Evaluation Masters.* A scoring guide is also provided.

Assessment and Evaluation Masters, pp. 126, 137

5 NAME______ DATE ______

Chapter 5 Performance Assessment

Instructions: ***Demonstrate your knowledge by giving a clear, concise solution to each problem. Be sure to include all relevant drawings and justify your answers. You may show your solution in more than one way or investigate beyond the requirements of the problem.***

1. The members of sets X and Y are shown at the right.

X	Y
−1	2
3	0
4	3
−3	

   a. Draw a mapping of X to Y that shows a function. Tell how you know it is a function.

   b. Draw a mapping of X to Y that represents a relation but not a function. Tell how you know it is a relation. Tell how you know it is not a function.

   c. Graph the relation $\{(-1, 3), (3, 5), (3, -2), (4, 0), (-3, 2)\}$. Use the vertical line test to determine if the relation is a function. Write a sentence to explain your reasoning.

   d. Express the relation shown in the mapping for part a as a set of ordered pairs. Then write the inverse of the relation. Is the inverse a function? Tell why or why not.

2. a. Complete the pattern in the chart.

x	−2	−1	0	1	2	3		
y	−9	−5	−1	3	7			

   b. Write an equation for the relation shown in the chart.

   c. Is the equation in part b a linear equation? Explain your answer.

   d. Graph the equation in part b.

   e. Is the inverse of the relation a linear function? Explain your answer.

3. The opening day scores in a local golf tournament are 78, 83, 70, 84, 89, 67, 84, 92, 78, 91, 85, 77, 68, 80, 71, 78, 99, 81, 75, 88, 90, 71, 73.

   a. Find the range and interquartile range of the scores. Tell in your own words the purpose of finding the range or interquartile range.

   b. Is the range or the interquartile range more affected by outliers?

   c. Display the results in a stem-and-leaf plot.

Scoring Guide
Chapter 5
Performance Assessment

Level	Specific Criteria
3 Superior	• Shows thorough understanding of the concepts of *relation, function, inverse function, domain, range, linear equation,* and *measures of variation.* • Computations are correct. • Written explanations are exemplary. • Graphs and diagrams are accurate and appropriate. • Goes beyond requirements of some or all problems.
2 Satisfactory, with Minor Flaws	• Shows understanding of the concepts of *relation, function, inverse function, domain, range, linear equation,* and *measures of variation.* • Computations are mostly correct. • Written explanations are effective. • Graphs and diagrams are accurate and appropriate. • Satisfies all requirements of problems.
1 Nearly Satisfactory, with Serious Flaws	• Shows understanding of most of the concepts of *relation, function, inverse function, domain, range, linear equation,* and *measures of variation.* • Computations are mostly correct. • Written explanations are satisfactory. • Graphs and diagrams are mostly accurate and appropriate. • Satisfies most requirements of problems.
0 Unsatisfactory	• Shows little or no understanding of the concepts of *relation, function, inverse function, domain, range, linear equation,* and *measures of variation.* • May not use appropriate strategies to solve problems. • Computations are incorrect. • Written explanations are not satisfactory. • Graphs and diagrams are not accurate or appropriate. • Does not satisfy requirements of problems.

In·ves·ti·ga·tion

TEACHER NOTES

NCTM Standards: 1–5

This Investigation is designed to be completed over several days or weeks. It may be considered optional. You may want to assign the Investigation and the follow-up activities to be completed at the same time.

Objective

Use mathematics to collect and analyze data regarding the effects of secondhand smoke.

Mathematical Overview

This Investigation will use the following mathematical skills and concepts from Chapters 6 and 7.

- making charts and graphs
- using the formulas for radius and volume

Recommended Time		
Part	**Pages**	**Time**
Investigation	320–321	1 class period
Working on the Investigation	338, 361, 398, 426	20 minutes each
Closing the Investigation	442	1 class period

Instructional Resources

Investigations and Projects Masters, pp. 9–12

A recording sheet, teacher notes, and scoring guide are provided for each Investigation in the *Investigations and Projects Masters.*

1 MOTIVATION

In this Investigation, students use simple materials and background information to investigate the effects of secondhand smoke. Ask students to discuss why this problem might be harmful for our society.

LONG-TERM PROJECT

In·ves·ti·ga·tion

Smoke Gets In Your Eyes

MATERIALS NEEDED

- stopwatch
- grid paper
- balloons
- tape measure
- string

In the 1950s and '60s, scientists piled up a mountain of evidence on the life-threatening health consequences of smoking. Recently there have been studies done on the effects of secondhand smoke. Secondhand smoke is the smoke others breathe when they are close to a person who is smoking. In 1993, the Environmental Protection Agency (EPA) estimated that secondhand smoke is responsible for several thousands of cases of lung cancer among nonsmokers each year in the U.S. Passive smoke joins a select company of only about a dozen other environmental pollutants in this risk category.

As an aware citizen and health-conscious teenager, you understand this problem and want to start a campaign to limit the effects of secondhand smoke. In this Investigation, you will gather evidence about how much air is inhaled when you breathe and compare that to smoke in the room. You will also examine the financial and health aspects of this problem. Work in groups of three. Make an Investigation Folder in which you can store all of your work on this Investigation for future use.

Adult Cigarette Use in the United States

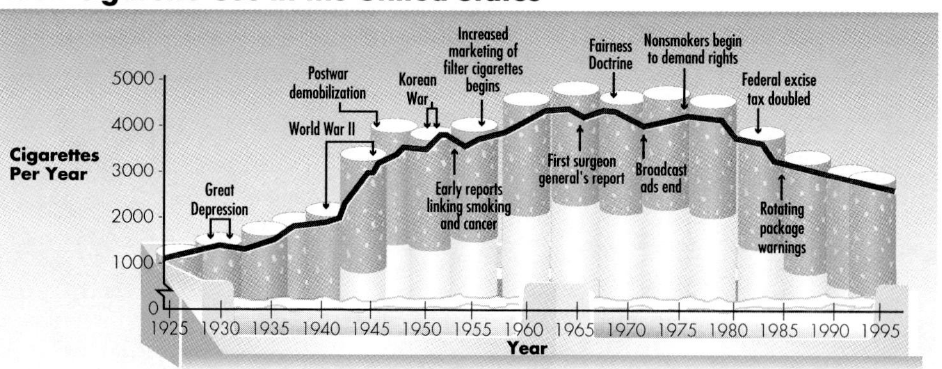

Source: U.S. Department of Health and Human Resources

Cooperative Learning

This Investigation offers an excellent opportunity for using cooperative learning groups. For more information on cooperative learning strategies and group management, see *Cooperative Learning in the Mathematics Classroom,* one of the titles in the Glencoe Mathematics Professional Series.

GROUP MEMBER	1	2	3
Number of breaths			
Circumference of balloon			
Radius of balloon			
Volume of balloon			
Volume of each breath			

COLLECT THE DATA

1. Begin by copying the chart above onto a sheet of paper.
2. Stretch each balloon. Have each member of the group blow up at least three round balloons to full size and let the air out.
3. Give each team member time to rest.
4. Cut a piece of string that is approximately 30 inches long. Tie the string so that it can form a circle that is approximately 8 inches in diameter when laid on a flat surface.
5. One member of the group should hold the string hoop around the balloon and another member should blow the balloon up again. The remaining group member should count and record the number of puffs it takes to blow up each balloon to the size of the hoop. Each group member should blow up one balloon.
6. Measure each inflated balloon to estimate its circumference at the fullest part. Find the approximate radius of the balloon by using the formula $r = \frac{C}{2\pi}$. Record the radius.

ANALYZE THE DATA

7. Display your data on a graph.
8. Calculate the average number of breaths it takes to fill a balloon.
9. The volume of the spherical balloon can be estimated by the formula $V = \frac{4}{3}\pi r^3$. Compute the volume of each balloon. Then estimate the volume of each breath.

You will continue working on this Investigation throughout Chapters 6 and 7.

Be sure to keep your chart and materials in your Investigation Folder.

Smoke Gets In Your Eyes Investigation

2 SETUP

You may wish to have a student read the first paragraph of the Investigation to provide information about why secondhand smoke is an important issue. You may then wish to read the last paragraph, which introduces the activity. Discuss the activity with students. Then separate the class into groups of three.

3 MANAGEMENT

Each group member should be responsible for a specific task.

Recorder Collects data.
Measurer Sets, measures, and ties the string and holds the string hoop around the balloon.
Timer Counts and records number of puffs that it takes to blow up each balloon to the size of the hoop.

At the end of the activity, each member should turn in his or her respective equipment.

Sample Answers

Answers will vary based on the efforts of each team member.

Investigations and Projects Masters, p. 12

6, 7 NAME________ DATE ______

Investigation, Chapters 6 and 7 — Student Edition Pages 320–321, 338, 361, 398, 426, 442

Smoke Gets In Your Eyes. . .

Work with your group to answer the following questions and add others to be considered.

- If you were in a closed room with someone who smoked two packs of cigarettes (there are 20 cigarettes in a pack), what estimated volume of smoke would you breathe? How many cigarettes does that equal?
- Based on what you have found in the Investigation, would you support or oppose laws that restrict people from smoking in public buildings and enclosed spaces? Explain why you think so.

Reread the information about secondhand smoke on pages 320 and 442. Then do research to find out more about this topic. Record your findings in the chart below.

Topic	Cancer	Heart Disease	Other Respiratory Diseases
Risks to children in smokers' households			
Risks to nonsmoking adults in smokers' households			
Risks to general population encountering secondhand smoke in public places			

CHAPTER 6

Analyzing Linear Equations

PREVIEWING THE CHAPTER

In this chapter, students begin by finding the slope of a line given the coordinates of two points on the line. Then, students study the point-slope, standard, and slope-intercept forms of linear equations. The concepts of scatter plots and best-fit lines are integrated into the chapter to provide students with another tool for data representation and problem solving. Next, students use a graphing calculator and the concepts of slopes and intercepts to graph equations. Students apply the slope-intercept form to write the equation of a line and then learn how to write equations for parallel or perpendicular lines. The chapter concludes with a lesson in which students find the coordinates of the midpoint of a line given the coordinates of the endpoints.

Lesson (Pages)	Lesson Objectives	NCTM Standards	State/Local Objectives
6-1A (324)	Use a geoboard to model a line segment and calculate its slope.	1–5, 8	
6-1 (325–331)	Find the slope of a line, given the coordinates of two points on the line.	1–5, 8	
6-2 (332–338)	Write linear equations in point-slope form. Write linear equations in standard form.	1–5, 8	
6-3 (339–345)	Graph and interpret points on a scatter plot. Draw and write equations for best-fit lines, and make predictions by using those equations. Solve problems by using models.	1–5, 10	
6-4 (346–353)	Determine the *x*- and *y*-intercepts of linear graphs from their equations. Write equations in slope-intercept form. Write and solve direct variation equations.	1–5	
6-5A (354–355)	Use a graphing calculator to determine whether a group of graphs forms a family.	1–5	
6-5 (356–361)	Graph a line given any linear equation.	1–5	
6-6 (362–368)	Determine if two lines are parallel or perpendicular by their slopes. Write equations of lines that pass through a given point, parallel or perpendicular to the graph of a given equation.	1–5, 8	
6-7 (369–374)	Find the coordinates of the midpoint of a line segment in the coordinate plane.	1–5, 8	

A complete, 1-page lesson plan is provided for each lesson in the ***Lesson Planning Guide***. Answer keys for each lesson are available in the ***Answer Key Masters.***

ORGANIZING THE CHAPTER

You may want to refer to the **Course Planning Calendar** on page T12 for detailed information on pacing.
PACING: Standard—13 days; **Honors**—12 days; **Block**—7 days; **Two Years**—18 days

LESSON PLANNING CHART

			Blackline Masters											
Lesson (Pages)	**Materials/ Manipulatives**	**Extra Practice (Student Edition)**	**Study Guide**	**Practice**	**Enrichment**	**Assessment and Evaluation**	**Modeling Mathematics**	**Multicultural Activity**	**Tech Prep Applications**	**Graphing Calculator**	**Science and Math Lab Manual**	**Real-World Applications**	**Interactive Mathematics Tools Software**	**Teaching Transparencies**
6-1A (324)	geoboard* rubber bands*						p. 26						6-1A	
6-1 (325–331)	spreadsheet	p. 769	p. 41	p. 41	p. 41			p. 11				17		6-1A 6-1B
6-2 (332–338)		p. 769	p. 42	p. 42	p. 42	p. 156								6-2A 6-2B
6-3 (339–345)	graphing calculator	p. 770	p. 43	p. 43	p. 43						pp. 23–28		6-3	6-3A 6-3B
6-4 (346–353)		p. 770	p. 44	p. 44	p. 44	pp. 155, 156		p. 12	p. 11			18		6-4A 6-4B
6-5A (354–355)	graphing calculator									pp. 22, 23				
6-5 (356–361)	graphing calculator	p. 770	p. 45	p. 45	p. 45				p. 12	p. 6			6-5.1 6-5.2	6-5A 6-5B
6-6 (362–368)	grid paper scissors*	p. 771	p. 46	p. 46	p. 46	p. 157	pp. 51–53, 77						6-6	6-6A 6-6B
6-7 (369–374)	grid paper ruler* colored pencils	p. 771	p. 47	p. 47	p. 47	p. 157								6-7A 6-7B
Study Guide/ Assessment (375–379)						pp. 141–154, 158–160								

*Included in Glencoe's Student Manipulative Kit and Overhead Manipulative Resources.

ORGANIZING THE CHAPTER

Other Chapter Resources

Student Edition
Investigation, pp. 320–321
Chapter Opener, pp. 322–323
Mathematics and Society, p. 374
Working on the Investigation, pp. 338, 361

Teacher's Classroom Resources
Investigations and Projects Masters, pp. 45–48
Algebra and Geometry Overhead Manipulative Resources, pp. 18–22

Technology
Test and Review Software (IBM and Macintosh)
CD-ROM Interactions (Windows and Macintosh)

Professional Publications
Block Scheduling Booklet
Glencoe Mathematics Professional Series

Outside Resources

Books/Periodicals
Taylor, Harold, *Developing Skills in Algebra One,* Dale Seymour Publications
Winter, Mary Jean and Ronald J. Carlson, *Algebra Experiments I,* Dale Seymour Publications

Software
Algebra Concepts, Vol. I, Ventura Educational Systems
Topics in Discrete Mathematics: Computer-Supported Problem Solving, Dale Seymour Publications

Videos/CD-ROMs
Algebra for Everyone, Videotape and Discussion Guide, NCTM

Assessment Resources

Student Edition
Math Journal, pp. 343, 350, 359
Mixed Review, pp. 331, 337, 345, 352, 361, 368, 374
Self Test, p. 353
Chapter Highlights, p. 375
Chapter Study Guide and Assessment, pp. 376–378
Alternative Assessment, p. 379
Portfolio, p. 379
Cumulative Review, pp. 380–381

Teacher's Wraparound Edition
5-Minute Check, pp. 325, 332, 339, 346, 356, 362, 369
Check for Understanding, pp. 328, 335, 343, 349, 359, 366, 371
Closing Activity, pp. 331, 338, 345, 352, 361, 368, 373
Cooperative Learning, pp. 340, 347

Assessment and Evaluation Masters
Multiple-Choice Tests, Forms 1A (Honors), 1B (Average), 1C (Basic), pp. 141–146
Free-Response Tests, Forms 2A (Honors), 2B (Average), 2C (Basic), pp. 147–152
Calculator-Based Test, p. 153
Performance Assessment, p. 154
Mid-Chapter Test, p. 155
Quizzes A–D, pp. 156–157
Standardized Test Practice, p. 158
Cumulative Review, pp. 159–160

ENHANCING THE CHAPTER

Examples of some of the materials for enhancing Chapter 6 are shown below.

Multicultural Activity Masters, pp. 11, 12

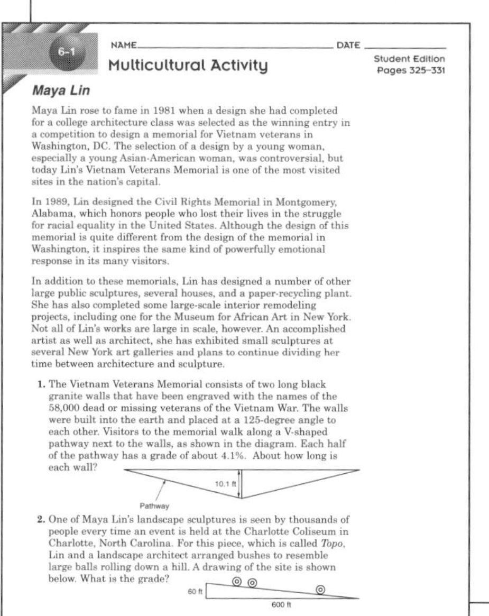

6-1 NAME________ DATE________

Multicultural Activity — Student Edition Pages 325–331

Maya Lin

Maya Lin rose to fame in 1981 when a design she had completed for a college architecture class was selected as the winning entry in a competition to design a memorial for Vietnam veterans in Washington, DC. The selection of a design by a young woman, especially a young Asian-American woman, was controversial, but today Lin's Vietnam Veterans Memorial is one of the most visited sites in the nation's capital.

In 1989, Lin designed the Civil Rights Memorial in Montgomery, Alabama, which honors people who lost their lives in the struggle for racial equality in the United States. Although the design of this memorial is quite different from the design of the memorial in Washington, it inspires the same kind of powerfully emotional response in its many visitors.

In addition to these memorials, Lin has designed a number of other large public sculptures, several houses, and a paper-recycling plant. She has also completed some large-scale interior remodeling projects, including one for the Museum for African Art in New York. Not all of Lin's works are large in scale, however. An accomplished artist as well as architect, she has exhibited small sculptures at several New York art galleries and plans to continue dividing her time between architecture and sculpture.

1. The Vietnam Veterans Memorial consists of two long black granite walls that have been engraved with the names of the 58,000 dead or missing veterans of the Vietnam War. The walls were built into the earth and placed at a 125-degree angle to each other. Visitors to the memorial walk along a V-shaped pathway next to the walls, as shown in the diagram. Each half of the pathway has a grade of about 4.1%. About how long is each wall?

2. One of Maya Lin's landscape sculptures is seen by thousands of people every time an event is held at the Charlotte Coliseum in Charlotte, North Carolina. For this piece, which is called *Topo*, Lin and a landscape architect arranged bushes to resemble large balls rolling down a hill. A drawing of the site is shown below. What is the grade?

Real-World Applications, 17, 18

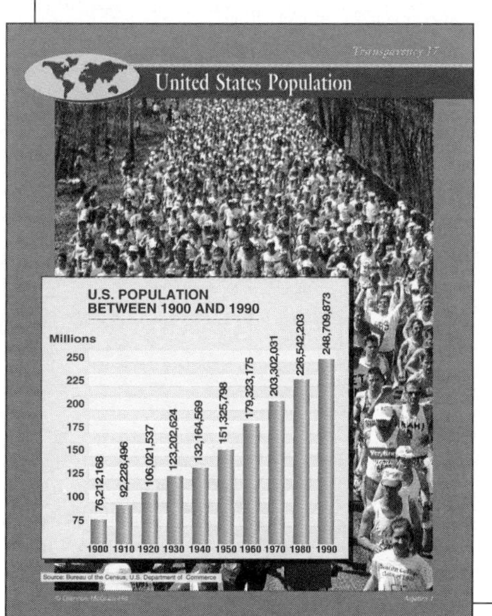

TECHNOLOGY

Graphing Calculator Masters, p. 6

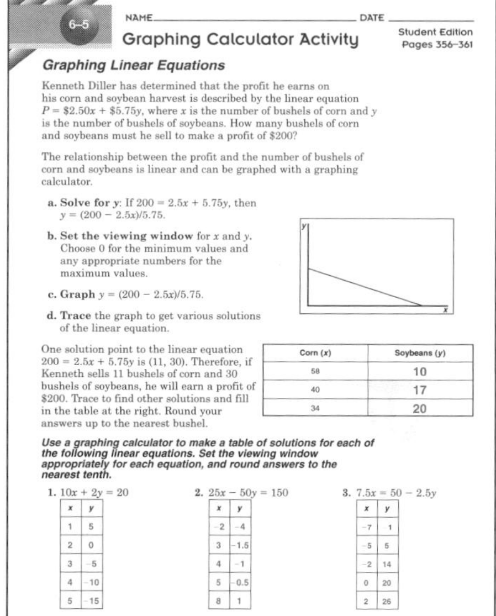

6-5 NAME________ DATE________

Graphing Calculator Activity — Student Edition Pages 356–361

Graphing Linear Equations

Kenneth Diller has determined that the profit he earns on his corn and soybean harvest is described by the linear equation $P = \$2.50x + \$5.75y$, where x is the number of bushels of corn and y is the number of bushels of soybeans. How many bushels of corn and soybeans must he sell to make a profit of $200?

The relationship between the profit and the number of bushels of corn and soybeans is linear and can be graphed with a graphing calculator.

a. **Solve for** y. If $200 = 2.5x + 5.75y$, then $y = (200 - 2.5x)/5.75$.

b. **Set the viewing window** for x and y. Choose 0 for the minimum values and any appropriate numbers for the maximum values.

c. **Graph** $y = (200 - 2.5x)/5.75$.

d. **Trace** the graph to get various solutions of the linear equation.

One solution point to the linear equation $200 = 2.5x + 5.75y$ is (11, 30). Therefore, if Kenneth sells 11 bushels of corn and 30 bushels of soybeans, he will earn a profit of $200. Trace to find other solutions and fill in the table at the right. Round your answers up to the nearest bushel.

Corn (x)	Soybeans (y)
58	10
40	17
34	20

Use a graphing calculator to make a table of solutions for each of the following linear equations. Set the viewing window appropriately for each equation, and round answers to the nearest tenth.

1. $10x + 2y = 20$

x	y
1	5
2	0
3	-5
4	-10
5	-15

2. $25x - 50y = 150$

x	y
-2	-4
3	-1.5
4	-1
5	-0.5
8	1

3. $7.5x = 50 - 2.5y$

x	y
-7	1
-5	5
-2	14
0	20
2	26

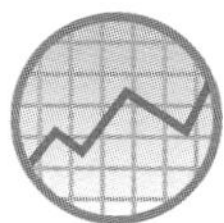

TECH PREP

Tech Prep Applications Masters, pp. 11, 12

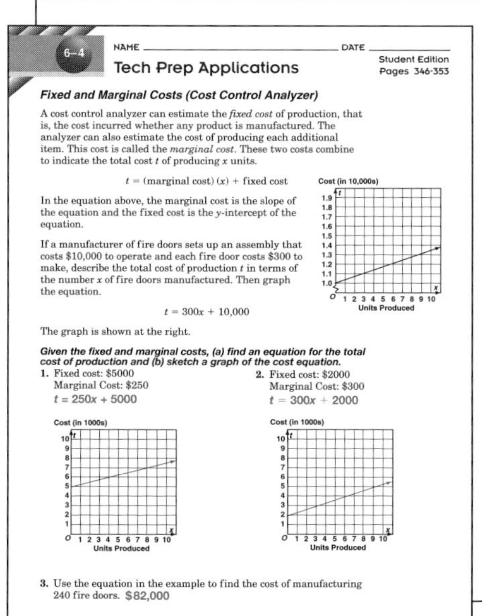

6-4 NAME________ DATE________

Tech Prep Applications — Student Edition Pages 346–353

Fixed and Marginal Costs (Cost Control Analyzer)

A cost control analyzer can estimate the *fixed cost* of production, that is, the cost incurred whether any product is manufactured. The analyzer can also estimate the cost of producing each additional item. This cost is called the *marginal cost*. These two costs combine to indicate the total cost t of producing x units.

$$t = (\text{marginal cost})(x) + \text{fixed cost}$$

In the equation above, the marginal cost is the slope of the equation and the fixed cost is the y-intercept of the equation.

If a manufacturer of fire doors sets up an assembly that costs $10,000 to operate and each fire door costs $300 to make, describe the total cost of production t in terms of the number x of fire doors manufactured. Then graph the equation.

$$t = 300x + 10{,}000$$

The graph is shown at the right.

(graph labels: Cost (in 10,000s); Units Produced)

Given the fixed and marginal costs, (a) find an equation for the total cost of production and (b) sketch a graph of the cost equation.

1. Fixed cost: $5000
   Marginal Cost: $250
   $t = 250x + 5000$

2. Fixed cost: $2000
   Marginal Cost: $300
   $t = 300x + 2000$

(graph labels: Cost (in 1000s); Units Produced)

3. Use the equation in the example to find the cost of manufacturing 240 fire doors. $82,000

Science and Math Lab Manual, pp. 23–28

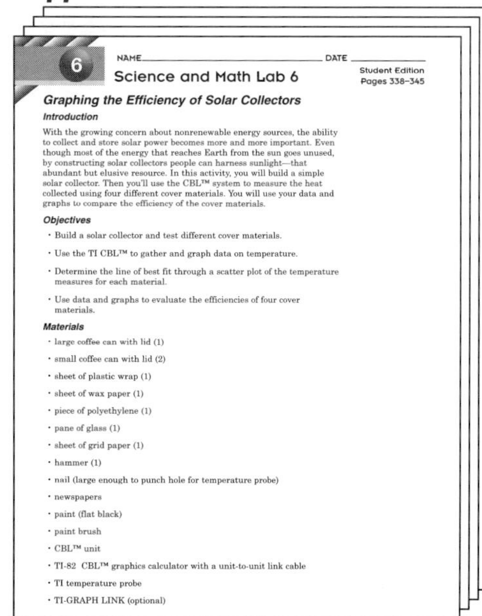

6 NAME________ DATE________

Science and Math Lab 6 — Student Edition Pages 338–345

Graphing the Efficiency of Solar Collectors

Introduction

With the growing concern about nonrenewable energy sources, the ability to collect and store solar power becomes more and more important. Even though most of the energy that reaches Earth from the sun goes unused, by constructing solar collectors people can harness sunlight—that abundant but elusive resource. In this activity, you will build a simple solar collector. Then you'll use the CBL™ system to measure the heat collected using four different cover materials. You will use your data and graphs to compare the efficiency of the cover materials.

Objectives

- Build a solar collector and test different cover materials.
- Use the TI CBL™ to gather and graph data on temperature.
- Determine the line of best fit through a scatter plot of the temperature measures for each material.
- Use data and graphs to evaluate the efficiencies of four cover materials.

Materials

- large coffee can with lid (1)
- small coffee can with lid (2)
- sheet of plastic wrap (1)
- sheet of wax paper (1)
- piece of polyethylene (1)
- pane of glass (1)
- sheet of grid paper (1)
- hammer (1)
- nail (large enough to punch hole for temperature probe)
- newspapers
- paint (flat black)
- paint brush
- CBL™ unit
- TI-82 CBL™ graphics calculator with a unit-to-unit link cable
- TI temperature probe
- TI-GRAPH LINK (optional)

Problem of the Week Cards, 16, 17, 18

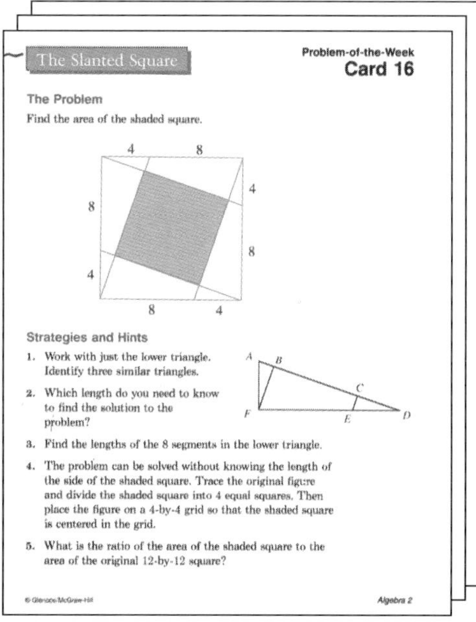

The Slanted Square — Problem-of-the-Week Card 16

The Problem

Find the area of the shaded square.

Strategies and Hints

1. Work with just the lower triangle. Identify three similar triangles.
2. Which length do you need to know to find the solution to the problem?
3. Find the lengths of the 8 segments in the lower triangle.
4. The problem can be solved without knowing the length of the side of the shaded square. Trace the original figure and divide the shaded square into 4 equal squares. Then place the figure on a 4-by-4 grid so that the shaded square is centered in the grid.
5. What is the ratio of the area of the shaded square to the area of the original 12-by-12 square?

© Glencoe/McGraw-Hill — Algebra 2

MAKING MATHEMATICS RELEVANT

This two-page introduction to the chapter provides students with an opportunity to explore contemporary topics and their applications to mathematics.

Background Information

Immigration on the Rise Many people believe that America is experiencing a flood of immigrants as never before. But the graph shows that the rate of immigration was much higher early in this century. The map shows that non-English speaking Americans are concentrated in the Southwest. In Los Angeles, 42.7% of students speak Spanish as their native language.

CHAPTER
6

Analyzing Linear Equations

Objectives

In this chapter, you will:

- find the slope of a line, given the coordinates of two of its points,
- write linear equations in point-slope, standard, and slope-intercept forms,
- draw a scatter plot and find the equation of a best-fit line for the data,
- solve problems by using models,
- graph linear equations, and
- use slope to determine if two lines are parallel or perpendicular.

Immigration on the Rise

Foreign-born residents as share of U.S. population
16%
12%
8%
4%
0%
1900
1910
1920
1930
1940
1950
1960
1970
1980
1990
1994

Source: *U.S. News and World Report*, September 25, 1995

TIME Line

Almost $20 billion a year is spent on bilingual programs in U.S. schools. These programs were launched in 1968 to help immigrant children learn English. The percentage of foreign-born Americans is rising. More than 22 states have enacted laws proclaiming English as the official language. What are your views on making English the official language of the U.S.?

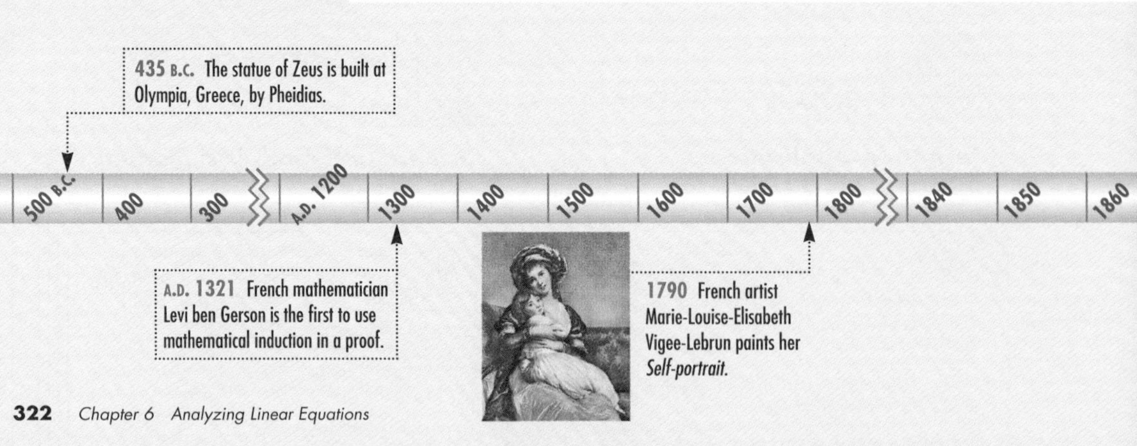
435 B.C. The statue of Zeus is built at Olympia, Greece, by Pheidias.
500 B.C.
400
300
A.D. 1200
1300
1400
1500
1600
1700
1800
1840
1850
1860
A.D. 1321 French mathematician Levi ben Gerson is the first to use mathematical induction in a proof.
1790 French artist Marie-Louise-Elisabeth Vigee-Lebrun paints her *Self-portrait*.

TIME Line

Students should investigate the Sieve of Eratosthenes, developed by the Libyan mathematician in 230 B.C. During their investigation, students should notice that the method does not generate prime numbers, but tests each integer to determine if it is prime.

interNET CONNECTION

Learn about languages of the world through The Human-Languages Page.

World Wide Web
http://www.willamette.edu/~tjones/Language-Page.html

PEOPLE IN THE NEWS

One student from Webster, Texas, **Jorge Arturo Pineda Aguilar,** is a member of the new immigrant minority. He and his family are from Monterrey, Mexico, and recently moved to Texas. At age 13, Jorge is one of the recent winners of the 1995 Youth Honor Awards presented by *Skipping Stones,* a multicultural children's magazine. His 1995 winning essay, written in both Spanish and English, deals with racial and ethnic prejudice.

Jorge regrets that so many people have ill-feelings about someone who doesn't speak English. He writes, "All of us are human beings and the color or nationality of a person should not be an issue. What should matter are the feelings of each person. We have to be together—work together for a common well being."

Chapter Project

The table below shows the per capita income and the total non-English speaking population (ages 5 years or older) in 15 selected states.

State	1993 Per Capita Income ($)	Non-English Speaking Population
AL	17,234	107,866
AR	16,143	60,781
FL	20,857	2,098,315
IL	22,582	1,499,112
IN	19,203	245,826
KY	17,173	86,482
LA	16,667	391,994
NC	17,488	240,866
NM	16,297	493,999
NY	24,623	3,908,720
OK	17,020	145,798
TN	17,666	131,550
TX	19,189	3,970,304
VA	21,634	418,521
WV	16,209	44,203

The population is given according to the 1990 U.S. Census.

Round the data to the nearest thousand. Graph the data for each state by plotting the per capita income data along the horizontal axis and the non-English-speaking population data along the vertical axis.

- What pattern, if any, do you notice?
- Can a best-fit line be drawn? If so, what might be the equation of this line?
- Can a valid relationship exist between these data? Why or why not?

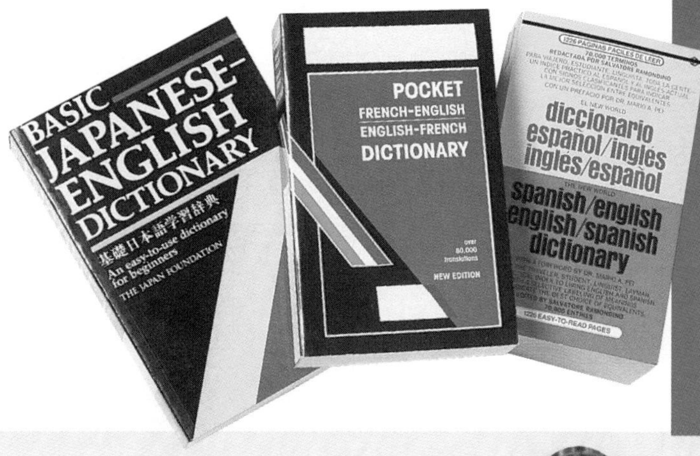

1968 The U.S. Postal Service issues a stamp honoring the life of Chief Joseph of the Nez Perce Nation.

1973 Marian Wright Edelman founds the Children's Defense Fund.

1890 1900 1910 1920 1930 1940 1950 1960 1970 1980 1990 2000

1926 Mexican artist Diego Rivera paints *The Tortilla Maker.*

1995 Twins Chris and Courtney Salthouse from Chamblee, Georgia *both* score a perfect 1600 on their SATs.

Alternative Chapter Projects

Two other chapter projects are included in the *Investigations and Projects Masters*. In Chapter 6 Project A, pp. 45–46, students extend the topic in the chapter opener. In Chapter 6 Project B, pp. 47–48, students investigate the growth of plants under various conditions.

PEOPLE IN THE NEWS

A quote from Jorge's essay follows.

"All of us are human beings and the color or nationality of a person should not be an issue. What should matter are the feelings of each person. . . . We should always find a place for peace in our hearts."

Chapter Project

Cooperative Learning In setting up cooperative teams, try to include in each group at least one student from a bilingual family. Students might also want to compare the results of their community with a profile for the entire country.

Investigations and Projects Masters, p. 45

NAME__________ DATE ________

Chapter 6 Project A

Student Edition Pages 324–374

Analyzing Immigration

1. Today, many people consider illegal immigration a significant problem for our nation. For this project you will work in a small group to research illegal immigration to the United States. Use reference works to find out more about this subject, including answers to the following questions.
   - How many illegal immigrants are estimated to live in the United States at this time?
   - How many people have entered the United States illegally in each of the last ten years?
   - What draws people to the United States and why are many willing to enter illegally?
   - What do politicians and public figures cite as positive and negative effects of illegal immigration?
2. Research immigration laws in the United States, particularly recent ones. What effects have these laws had on the number of persons entering the United States illegally?
3. Decide whether your group would be in favor of or opposed to spending large amounts of government money on keeping illegal immigrants out of the United States. Prepare an oral presentation of your group's position, in which each group member will participate. Use a scatter plot to show whether the correlation between the number of illegal immigrants and some other variable, such as a state's spending on welfare, prisons, or schools, is strong or weak.
4. Pair up with a group that takes the position opposite to your own and make your presentations to the rest of the class.

6-1A LESSON NOTES

NCTM Standards: 1–5, 8

Objective

Use a geoboard to model a line segment and calculate its slope.

Recommended Time

Demonstration and discussion: 15 minutes; Exercises: 30 minutes

Instructional Resources

For each student or group of students
Student Manipulative Kit
- geoboard
- red rubber band, green rubber band, blue or yellow rubber band

Modeling Mathematics Masters
- p. 6 (geoboard pattern)
- p. 26 (worksheet)

For teacher demonstration
Algebra and Geometry Overhead Manipulative Resources

1 FOCUS

Motivating the Lesson

Make certain that the desks in the room are arranged in rows and columns. Select two seated students. Count the number of rows one must walk, then the number of columns one must walk, to get from one student to the other. Do this for several pairs of students.

2 TEACH

Teaching Tip Point out to students that they must be careful to divide the vertical distance by the horizontal distance. Doing the opposite gives an incorrect answer.

3 PRACTICE/APPLY

Assignment Guide

Core: 1–5
Enriched: 1–5

MODELING MATHEMATICS

A Preview of Lesson 6–1

6–1A Slope

Materials:  geoboard  rubber bands

The **slope** of a line segment is the ratio of the change in the vertical distance between the endpoints of the segment to the change in the horizontal distance. You can use a geoboard to model a line segment and calculate its slope.

Activity **Model line segment *AB* whose endpoints are *A*(1, 2) and *B*(3, 5). Then find the slope of segment *AB*.**

Step 1 Suppose the pegs of the geoboard represent ordered pairs on a coordinate plane. Let's define the lower left peg as the point representing the ordered pair (1, 1). Locate the pegs that represent (1, 2) and (3, 5). Place a rubber band around these pegs to model segment *AB*.

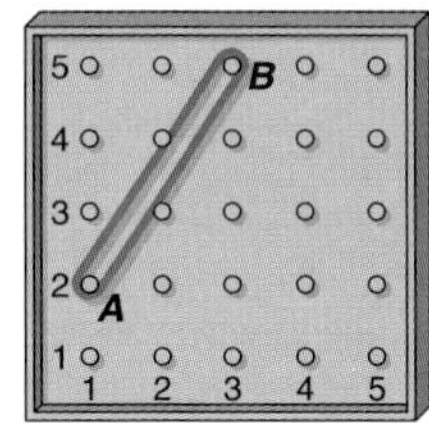

Step 2 Use a different color rubber band to show the horizontal distance from the *x* value of *A* to the *x* value of *B*. Use another color rubber band to show the vertical distance from the *y* value of *A* to the *y* value of *B*.

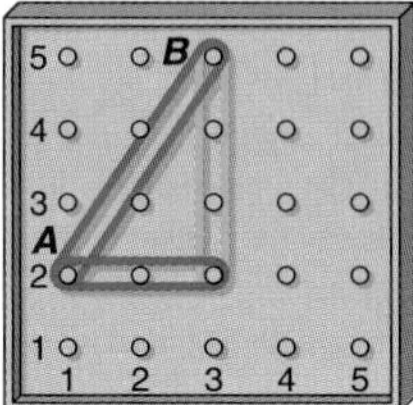

Step 3 As you move from point *A* to point *B*, you can go right and then up. These are positive directions on the coordinate plane. So, the red rubber band represents 2 units. The green rubber band represents 3 units. The slope of segment *AB* is the ratio of the *y* distance to the *x* distance or $\frac{3}{2}$.

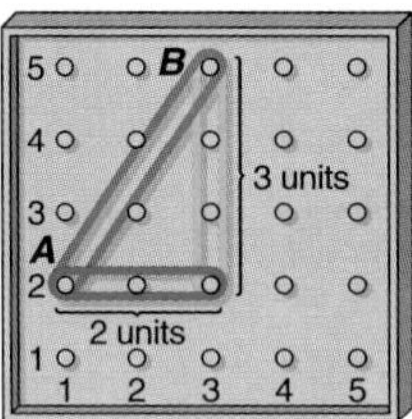

1. $-\frac{1}{2}$

2. $\frac{2-5}{1-3} = \frac{3}{2}$; $\frac{4-2}{1-5} = \frac{2}{-4}$ or $-\frac{1}{2}$; **It is the same ratio as the slope of the segments.**

Model

1. Use a geoboard to find the slope of a segment whose endpoints are at (5, 2) and (1, 4).

Write

2. For the activity above and Exercise 1, write ratios that compare the differences in the *y* values of the ordered pairs to the differences in the *x* values of the ordered pairs. How do these compare with the slopes of the respective segments?
3. What value could you assign to the lower left peg if you were modeling segment *CD* with endpoints *C*(3, 5) and *D*(6, 7)? **Sample answer: (3, 3)** **4. See Solutions Manual.**
4. Write a rule for finding the slope of any line segment shown on a coordinate plane.
5. Do you think the rule applies to ordered pairs that involve negative numbers? Explain your answer and give examples. **See Solutions Manual.**

4 ASSESS

Observing students working in cooperative groups is an excellent method of assessment.

GLENCOE Technology

Interactive Mathematics Tools Software

This multimedia software provides an interactive lesson by students observing the effect of the changes of the coefficients of *x*-terms on the slope of the line as well as when the lines have the same slope. A **Computer Journal** gives students an opportunity to write about what they have learned.

For Windows & Macintosh

6–1 Slope

What YOU'LL LEARN

- To find the slope of a line, given the coordinates of two points on the line.

Why IT'S IMPORTANT

You can use slope to describe lines and solve problems involving construction and architecture.

CAREER CHOICES

Construction **contractors** plan, budget, and direct residential and commercial construction projects. They use computers to graphically implement a building plan. A bachelors degree in construction science is helpful.

For more information, contact:

American Institute of Constructors
9887 N. Gandy Blvd.
Suite 104
St. Petersburg, FL 33702

APPLICATION

Construction

Alan and Mabel Wong bought a lot in San Francisco on which to build a house. The lot is located on a street having an 11% grade. The length of the sidewalk along the lot is 43 feet. City ordinances state that there must be a 5-foot clearance on each side of the house to the property line. The house they would like to build is 32 feet wide. Will they be able to build their house on the lot they bought? *This problem will be solved in Example 4.*

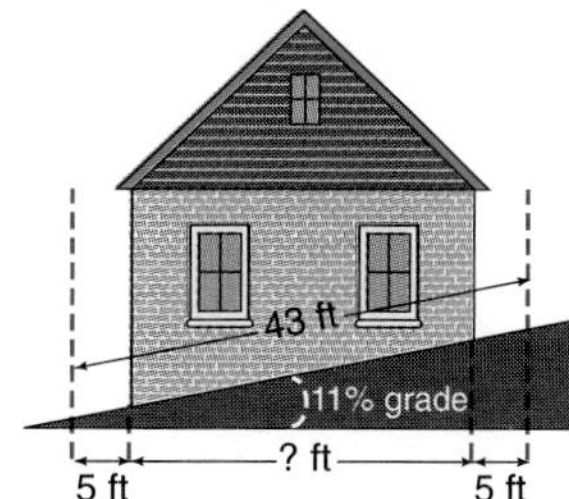

What do you think of when you hear the word *slope*? You might think of how something slants either uphill or downhill. Suppose you placed a ladder against a wall. Then you moved the base of the ladder out about a foot. What happens to the slant of the ladder? Suppose you move it again. What happens? Observe the pattern in the diagram below.

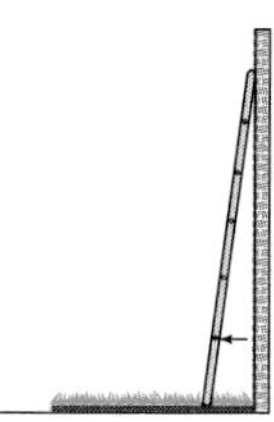
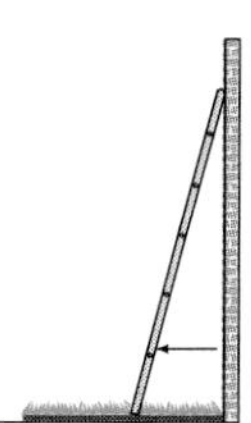
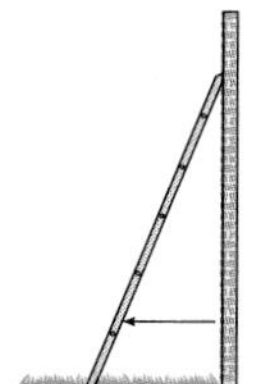
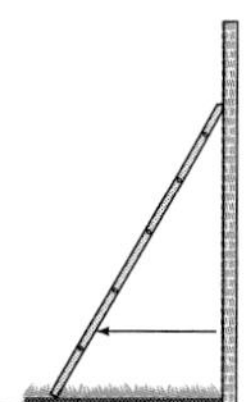

Notice that the top of the ladder moves down the wall when the bottom of the ladder moves out. As the top moves down the wall, the slant, or slope, becomes less steep. If you continued to move the ladder out, eventually it would lie flat on the ground, and there would be no slant at all.

The steepness of the line representing the ladder is called the **slope** of the line. It is defined as the ratio of the **rise,** or vertical change, to the **run,** or horizontal change, as you move from one point on the line to another. The graph at the right shows a line that passes through the origin and the point at (5, 4).

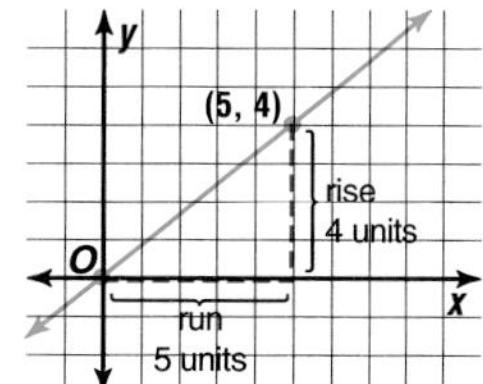
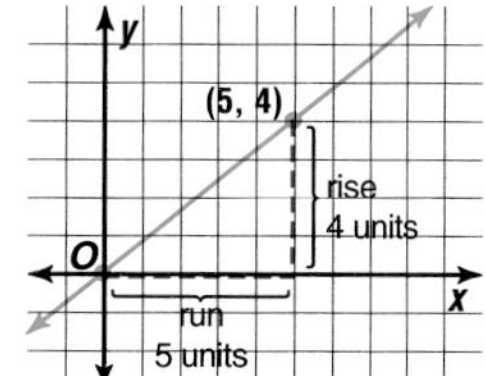

$$\text{slope} = \frac{\text{change in } y \text{ (rise)}}{\text{change in } x \text{ (run)}}$$

$$= \frac{4}{5}$$

So, the slope of this line is $\frac{4}{5}$.

CAREER CHOICES

There are many special trades within the construction industry. These include bricklayers, stone and concrete masons, carpenters, carpet installers, insulation and drywall workers, electricians, glaziers, painters, paper hangers, plasterers, plumbers, roofers, roustabouts, sheet metal workers, and tilesetters.

6-1 LESSON NOTES

NCTM Standards: 1–5, 8

Instructional Resources

- Study Guide Master 6-1
- Practice Master 6-1
- Enrichment Master 6-1
- Multicultural Activity Masters, p. 11
- Real-World Applications, 17

Transparency 6-1A contains the 5-Minute Check for this lesson; **Transparency 6-1B** contains a teaching aid for this lesson.

Recommended Pacing	
Standard Pacing	Day 2 of 13
Honors Pacing	Day 2 of 12
Block Scheduling*	Day 1 of 7
Alg. 1 in Two Years*	Day 2 of 18

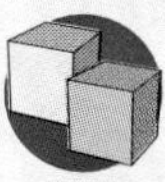

*For more information on pacing and possible lesson plans, refer to the *Block Scheduling Booklet* and *Algebra 1 in Two Years.*

1 FOCUS

5-Minute Check
(over Chapter 5)

State the domain and range of each relation.

1. {(5, 2), (0, 0), (−9, −1)} **Domain = {5, 0, −9}; Range = {2, 0, −1}**
2. {(1, 1), (2, 4), (−2, 4), (3, 9)} **Domain = {1, 2, −2, 3}; Range = {1, 4, 9}**

Determine whether each relation is a function. Then state the inverse relation and determine whether it is a function.

3. {(3, 1), (5, 1), (7, 1)} **yes; {(1, 3), (1, 5), (1, 7)}; no**
4. {(5, 4), (−6, 5), (4, 5), (0, 4)} **yes; {(4, 5), (5, −6), (5, 4), (4, 0)}; no**
5. If $f(x) = 3x - 5$, determine the value of $f(2)$. **1**

Motivating the Lesson

Questioning Introduce this lesson on slope by discussing the pitch of a roof. Ask why the pitch varies in different parts of the country. What determines the pitch of a roof when an architect draws up building plans? Develop the idea that a relationship exists between the amount of precipitation an area generally receives and the roof structures used in the region.

2 TEACH

Teaching Tip In Example 1, point out that a line with a slope of 0 is horizontal (parallel to the x-axis) and a line with an undefined slope is vertical (parallel to the y-axis).

In-Class Examples

For Example 1
Determine the slope of the line through the origin and the point named.

A	3
B	2
C	1
D	$\frac{1}{2}$
E	$\frac{1}{3}$
F	$-\frac{1}{3}$
G	$-\frac{1}{2}$
H	-1
I	-2
J	-3

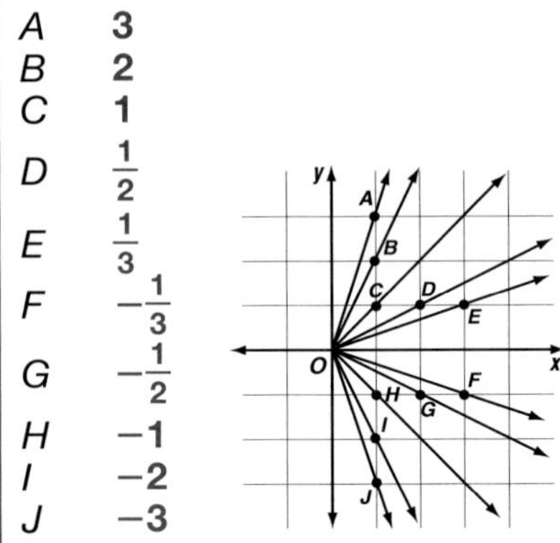

For Example 2
Determine the slope of the line that passes through each pair of points.

a. $(-5, 3), (2, 1)$ $-\frac{2}{7}$
b. $(4, 3), (1, -1)$ $\frac{4}{3}$

Teaching Tip In Example 2, students should be aware that the y-coordinates can be subtracted in any order as long as the corresponding x-coordinates are subtracted in that same order. $\frac{y_1 - y_2}{x_1 - x_2}$ or $\frac{y_2 - y_1}{x_2 - x_1}$, not $\frac{y_2 - y_1}{x_1 - x_2}$.

Definition of Slope 
The slope m of a line is the ratio of the change in the y-coordinates to the corresponding change in the x-coordinates.

For simplicity, we will refer to the change in the y-coordinates as the y change and the corresponding change in the x-coordinates as the x change.

Example 1 **Determine the slope of each line.**

a.
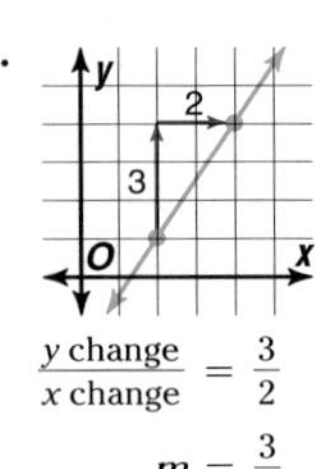

$\frac{y \text{ change}}{x \text{ change}} = \frac{3}{2}$

$m = \frac{3}{2}$

b.
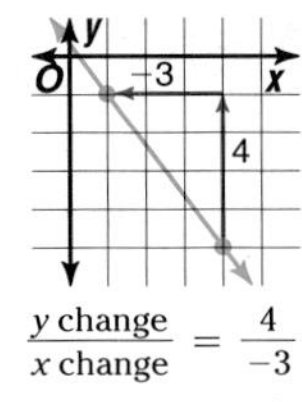

$\frac{y \text{ change}}{x \text{ change}} = \frac{4}{-3}$

$m = \frac{4}{-3}$

c.
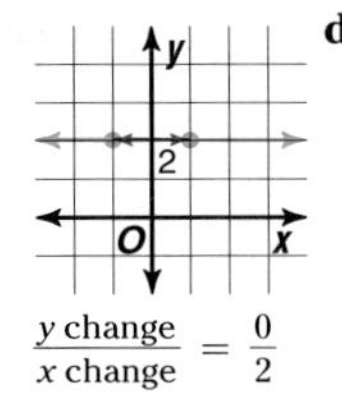

$\frac{y \text{ change}}{x \text{ change}} = \frac{0}{2}$

$m = 0$

d.
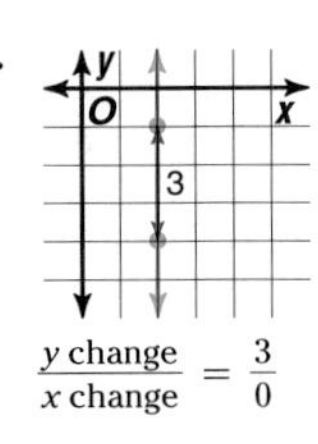

$\frac{y \text{ change}}{x \text{ change}} = \frac{3}{0}$

Since division by 0 is undefined, the slope is undefined.

Let's analyze the slopes of the graphs in Example 1.

- Graph a slopes upward from left to right and has a positive slope.
- Graph b slopes downward from left to right and has a negative slope.
- Graph c is a horizontal line and has a slope of 0.
- Graph d is a vertical line and its slope is undefined.

These observations are true of other lines that have the same characteristics.

Since a line is made up of an infinite number of points, you can use any two points on that line to find the slope of the line. So, we can generalize the definition of slope for any two points on the line.

Determining Slope Given Two Points 
Given the coordinates of two points, (x_1, y_1) and (x_2, y_2), on a line, the slope m can be found as follows:

$$m = \frac{y_2 - y_1}{x_2 - x_1}, \text{ where } x_1 \neq x_2.$$

y_2 is read "y sub 2." The 2 is called a subscript.

Example 2 **Determine the slope of the line that passes through $(2, -5)$ and $(7, -10)$.**

Method 1
Let $(2, -5) = (x_1, y_1)$ and $(7, -10) = (x_2, y_2)$.

$$m = \frac{y_2 - y_1}{x_2 - x_1} = \frac{-10 - (-5)}{7 - 2} = \frac{-5}{5} \text{ or } -1$$

Method 2
Let $(7, -10) = (x_1, y_1)$ and $(2, -5) = (x_2, y_2)$.

$$m = \frac{y_2 - y_1}{x_2 - x_1} = \frac{-5 - (-10)}{2 - 7} = \frac{5}{-5} \text{ or } -1$$

The slope of the line is -1.

As you can see in Example 2, it does not matter which ordered pair is selected as (x_1, y_1).

Alternative Learning Styles

Kinesthetic Have students determine the slope of their driveway or curb ramp. Ask them to describe the procedure they used in calculating the value of this slope.

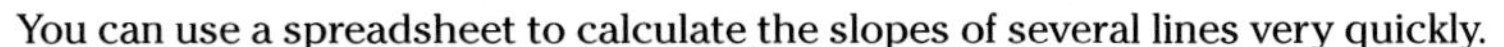

You can use a spreadsheet to calculate the slopes of several lines very quickly.

EXPLORATION — SPREADSHEETS

A spreadsheet is a table of cells that can contain text (labels) or numbers and formulas. Each cell is named by the column and row in which it is located. In the spreadsheet below, cells A1, B1, C1, D1, E1, and F1 contain labels.

a. Enter the formula (E2−C2)/(D2−B2) into cell F2.

b. Copy the formula to all of the cells in the F column. When you do, the spreadsheet automatically changes the formula to correspond to the values in that row. That is, for F3 the formula becomes (E3−C3)/(D3−B3), for F4 it becomes (E4−C4)/(D4−B4), and so on.

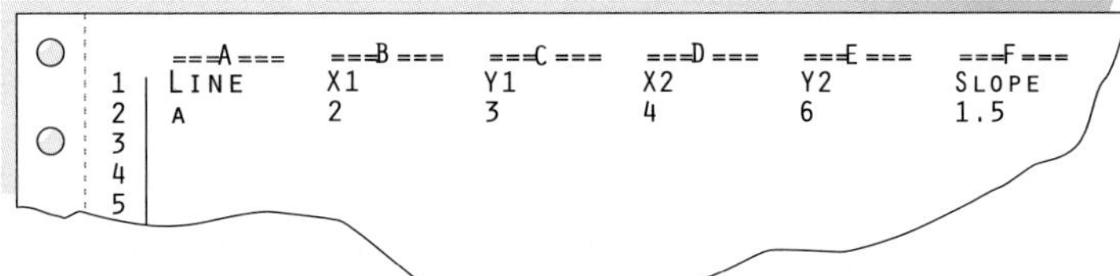

	A	B	C	D	E	F
1	LINE	X1	Y1	X2	Y2	SLOPE
2	A	2	3	4	6	1.5
3						
4						
5						

Your Turn **b. Slopes are given for each line in part a.**

a. Enter the following ordered pairs from each line in the spreadsheet.

line a	(5, 5), (11, 11)	**1**	line e	(−1, 5), (6, 3)	**−0.28571**
line b	(4, −4), (3, 5)	**−9**	line f	(5, 3), (4, 0)	**3**
line c	(6, −1), (4, −1)	**0**	line g	(2, 3), (11, 14)	**1.22222**
line d	(11, 3), (1, 1)	**0.2**	line h	(10, −5), (10, 3)	**undefined**

b. Use the calculate command to have the spreadsheet compute the slope.

c. Use the ordered pairs to graph each line on grid paper. Label each line.

d. From the results of your spreadsheet, make lists of lines with positive slope, negative slope, and 0 slope. **c–e. See margin.**

e. Compare the lists with the graphs of the lines. Write a statement to summarize your observations.

If you know the slope of a line and the coordinates of one of the points on a line, you can find the coordinates of other points on that line.

Example 3 **Determine the value of r so the line through $(r, 6)$ and $(10, -3)$ has a slope of $-\frac{3}{2}$.**

$m = \frac{y_2 - y_1}{x_2 - x_1}$

$-\frac{3}{2} = \frac{-3 - 6}{10 - r}$ $(x_1, y_1) = (r, 6)$ $(x_2, y_2) = (10, -3)$

$-\frac{3}{2} = \frac{-9}{10 - r}$

$-3(10 - r) = -9(2)$ *Find the cross products.*

$-30 + 3r = -18$ *Solve for r.*

$3r = 12$

$r = 4$

EXPLORATION

This Exploration illustrates the use of spreadsheet software to calculate a large number of slopes. Point out to students that since a graph is created, input errors are harder to detect.

Answers for the Exploration

c.

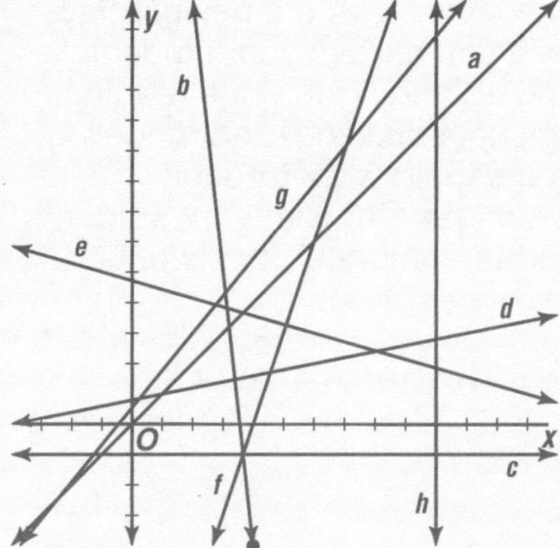

d. positive slopes: *a, d, f,* and *g;* negative slopes: *b* and *e;* zero slope: *c;* undefined slope: *h*

e. Sample answer: Positive slope means that the graphs slant upward to the right. Negative slope means that the graphs slant downward to the right. Zero slope means the lines are horizontal.

In-Class Example

For Example 3
Determine the value of r so the line through the two given points has the indicated slope.

a. $(-2, 4)$, $(r, 5)$, $m = \frac{1}{5}$ 3

b. $(3, 4)$, $(-1, r)$, $m = -\frac{3}{4}$ 7

Teaching Tip Have students graph the line through (9, −2) with slope $-\frac{3}{2}$ to verify that the point (5, 4) is on this line. They can think of slope as directions from one point to another on a line. That is, from (9, −2), go 2 units right and 3 units down to locate another point on the line.

In-Class Example

For Example 4
How many feet does a road with a 4% grade drop in 5 miles? **1056 feet**

3 PRACTICE/APPLY

Check for Understanding

Exercises 1–14 are designed to help you assess your students' understanding through reading, writing, speaking, and modeling. You should work through Exercises 1–6 with your students and then monitor their work on Exercises 7–14.

Additional Answers

Sample answers:

4a.

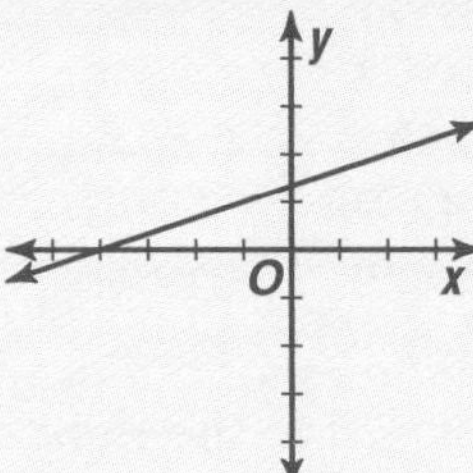

4b.

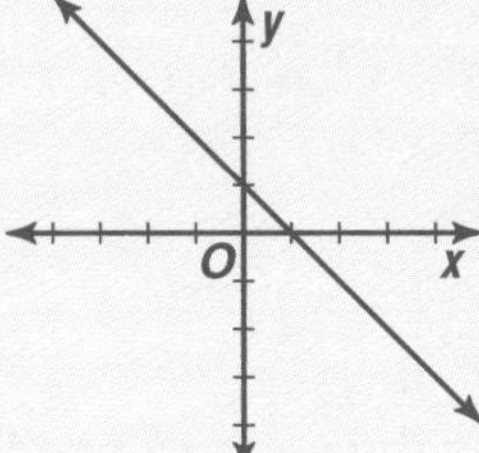

4c.

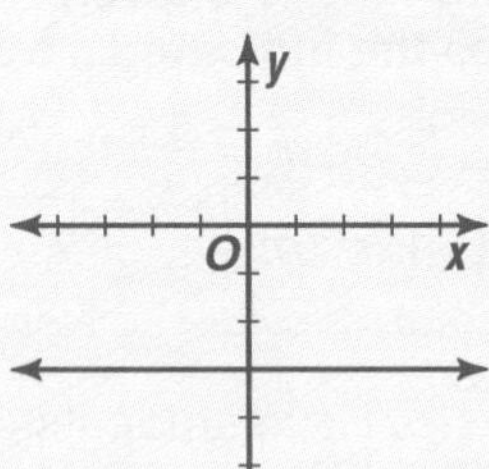

4d.

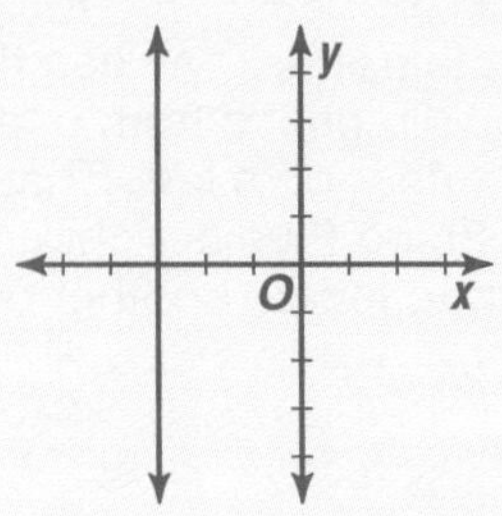

You can use slope with other mathematical skills to solve real-world problems like the one presented at the beginning of this lesson.

Example 4 **Refer to the application at the beginning of the lesson. Will the Wongs be able to build their house on their lot?**

APPLICATION **Construction**

The width of the house and the two 5-foot clearances equal $32 + 2(5)$ or 42 feet. Thus, the horizontal measure between the property lines must be at least 42 feet.

LOOK BACK
You can refer to Lesson 4-3 for more information on trigonometric functions.

We can use trigonometric ratios to determine if there is adequate room for the house. A grade of 11 percent means that the rise is 11 feet while the run is 100 feet. We can use this information to find the measure of $\angle A$.

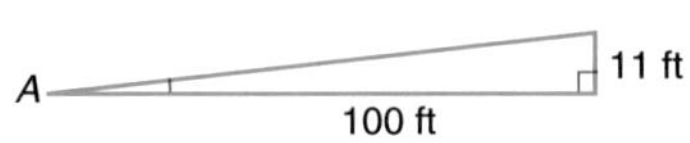

$\tan A = \frac{11}{100}$ ← *opposite* ← *adjacent*

$\tan A = 0.11$

$A \approx 6.3°$ *Use a calculator.*

We can use the measure of $\angle A$ to find the horizontal measure x of the property.

43 ft
A 6.3° x ft

$\cos A = \frac{x}{43}$ ← *adjacent* ← *hypotenuse*

$43(\cos 6.3°) = x$ *Multiply each side by 43.*

$42.74 = x$ *Use a calculator.*

The lot is 42.74 feet wide. The house and clearances will fit the 43-foot lot. So, the Wongs can build their house on the lot they bought.

CHECK FOR UNDERSTANDING

Communicating Mathematics

Study the lesson. Then complete the following.

1. **Explain** how you would find the slope of the line in the graph at the right. **From one point to another, go down 3 units and right 5 units. The slope is $-\frac{3}{5}$.**
2. **Describe** what it means if a road has an 8% grade. **It means that as you travel 100 ft horizontally, your altitude increases 8 ft.**
3. Would it matter if you subtracted the x-coordinates in the opposite order from the way you subtracted the y-coordinates? **Yes, it would affect the sign of the slope.**
4. **Draw** the graph of a line having each slope.
   a. a positive slope
   b. a negative slope
   c. a slope of 0
   d. undefined slope **a–d. See margin.**
5. **Explain** why the formula for determining the slope using two points does not apply to vertical lines. **The difference in the x values is always 0, and division by 0 is undefined.**

6. Use a geoboard to model line segments with each slope. Make a sketch of your model. **a–e. See margin.**
   a. 4
   b. $\frac{3}{4}$
   c. 0
   d. undefined
   e. $-\frac{3}{2}$

Reteaching

Using Alternative Methods Review the concept of slope by looking at four lines that pass through the origin: (1) a horizontal line, (2) a line that goes upward from left to right, (3) a vertical line, and (4) a line that goes downward from left to right. Emphasize the signs of the slopes.

Guided Practice

Determine the slope of each line.

7. 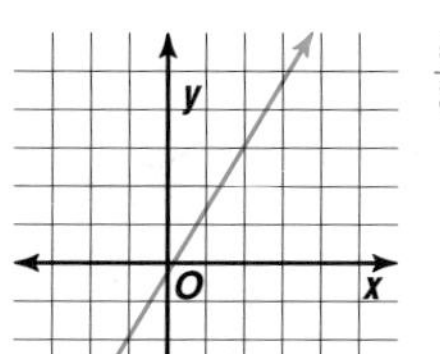  $\frac{5}{3}$

8. 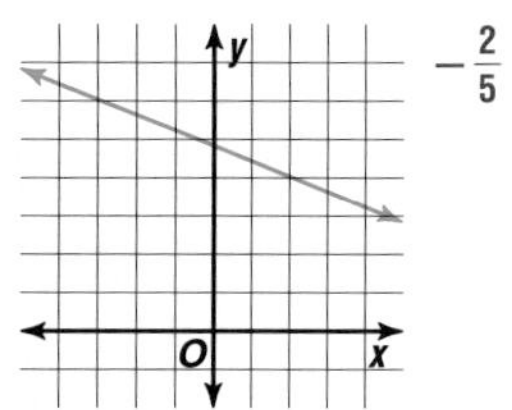 $-\frac{2}{5}$

Determine the slope of the line that passes through each pair of points.

9. $(7, -4), (9, -1)$ $\frac{3}{2}$
10. $(5, 7), (-2, -3)$ $\frac{10}{7}$
11. $(0.75, 1), (0.75, -1)$ undefined

Determine the value of r so the line that passes through each pair of points has the given slope.

12. $(6, -2), (r, -6), m = -4$ 7
13. $(9, r), (6, 3), m = -\frac{1}{3}$ 2

14. **Architecture** The slope of the roof line of a building is often referred to as the *pitch* of the roof.
    a. Use a millimeter ruler to find the rise and run of the roof of the tower on the horse barn in Versailles, Kentucky, shown at the right.
    b. Then write the pitch as a decimal. based on sample answer, about 1.5

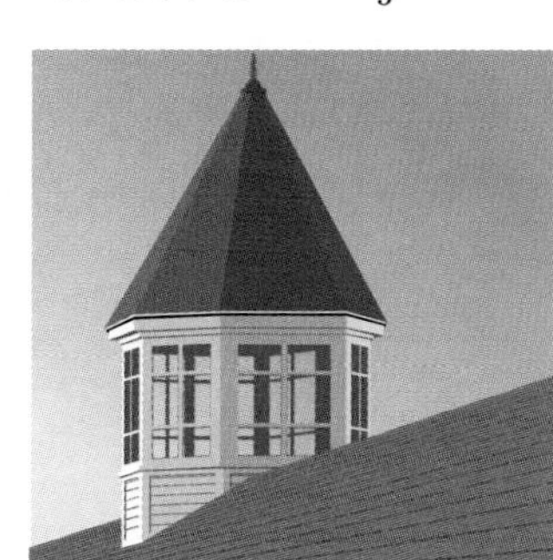

14a. Sample answer: rise, 23 mm; run, 15 mm

EXERCISES

Practice

Determine the slope of each line.

15. a
16. b
17. c 0
18. d undefined
19. e 1
20. f -6

15. $-\frac{1}{5}$
16. $\frac{1}{4}$

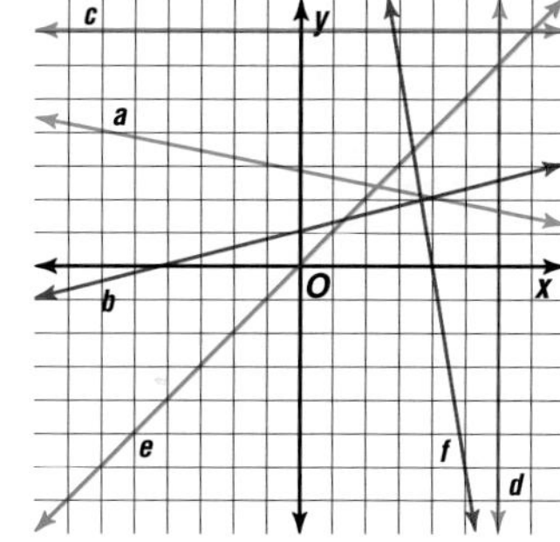

Determine the slope of the line that passes through each pair of points.

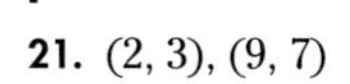

21. $(2, 3), (9, 7)$ $\frac{4}{7}$
22. $(-3, -4), (5, -1)$ $\frac{3}{8}$
23. $(2, -1), (5, -3)$ $-\frac{2}{3}$
24. $(2, 6), (-1, 3)$ 1
25. $(-5, 4), (-5, -1)$
26. $(-2, 3), (8, 3)$ 0
27. $(4, -5), (4, 2)$
28. $\left(2\frac{1}{2}, -1\frac{1}{2}\right), \left(-\frac{1}{2}, \frac{1}{2}\right)$
29. $\left(\frac{3}{4}, 1\frac{1}{4}\right), \left(-\frac{1}{2}, -1\right)$ $\frac{9}{5}$

25. undefined
27. undefined
28. $-\frac{2}{3}$

Additional Answers

6c. 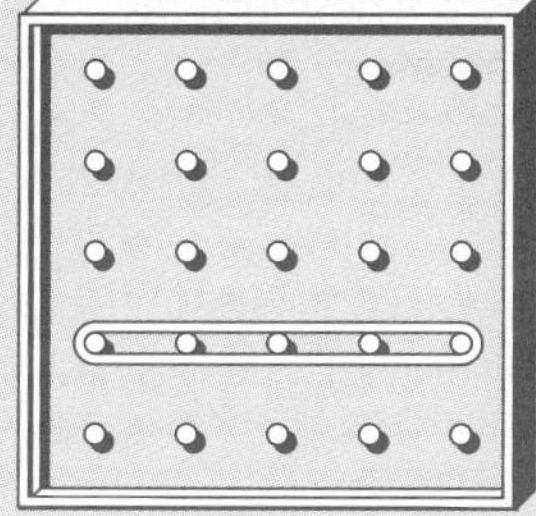

6d. 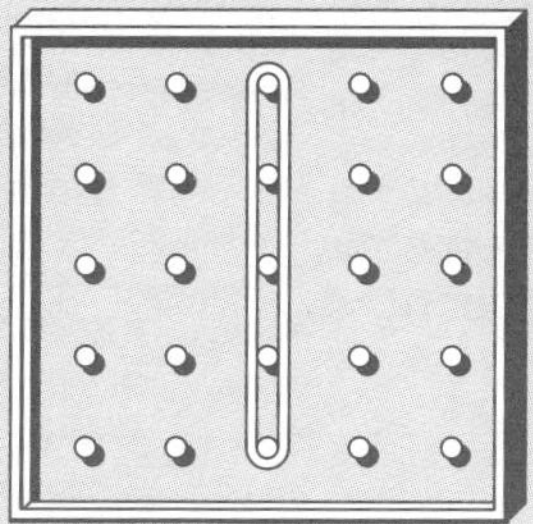

6e. 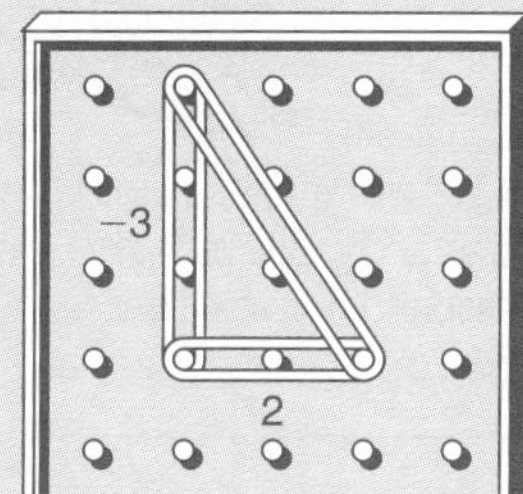

Assignment Guide

Core: 15–39 odd, 40, 41, 43, 45–52
Enriched: 16–38 even, 40–52

For **Extra Practice,** see p. 769.

The red A, B, and C flags, printed only in the Teacher's Wraparound Edition, indicate the level of difficulty of the exercises.

Additional Answers

6a. 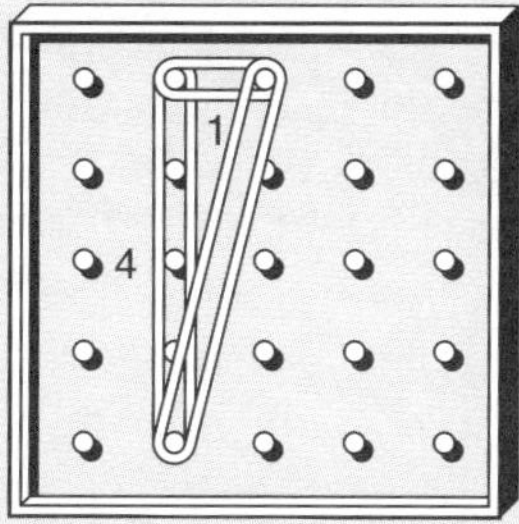

6b. 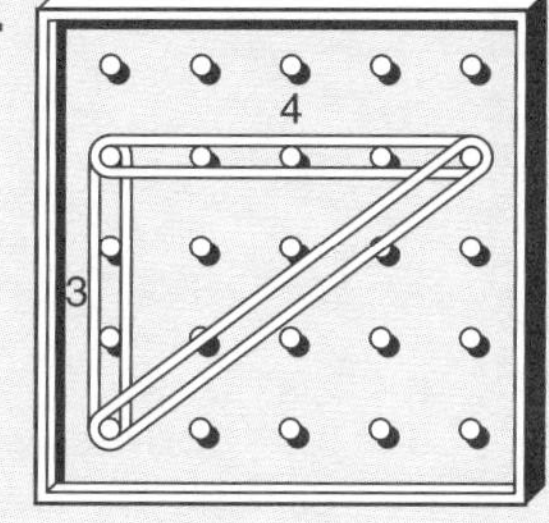

Study Guide Masters, p. 41

NAME ______ DATE ______

Study Guide — Student Edition Pages 325–331

Slope

The ratio of *rise* to *run* is called **slope.** The slope of a line describes its steepness, or rate of change.

On a coordinate plane, a line extending from lower left to upper right has a positive slope. A line extending from upper left to lower right has a negative slope. The slope of a horizontal line is zero. A vertical line has *no slope.*

The slope of a nonvertical line can be determined from the coordinates of any two points on the line.

Definition of Slope

The slope m of a line is the ratio of the change in the y-coordinates to the corresponding change in the x-coordinates.

$$\text{Slope} = \frac{\text{change in } y}{\text{change in } x} \text{ or } m = \frac{\text{change in } y}{\text{change in } x}$$

Determining Slope Given Two Points

Given the coordinates of two points, (x_1, y_1) and (x_2, y_2), on a line, the slope m can be found as follows:

$$m = \frac{y_2 - y_1}{x_2 - x_1}, \text{ where } x_2 \neq x_1$$

Example: Determine the slope of the line that passes through $(-1, 5)$ and $(4, -2)$.

$$m = \frac{y_2 - y_1}{x_2 - x_1} = \frac{-2 - 5}{4 - (-1)} = \frac{-7}{5} = -\frac{7}{5}$$

Determine the slope of the line that passes through each pair of points.

1. $(2, 1), (8, 9)$ $\frac{4}{3}$
2. $(4, 9), (1, 6)$ 1
3. $(7, -8), (14, -6)$ $\frac{2}{7}$
4. $(-10, 7), (-20, 8)$ $-\frac{1}{10}$
5. $(3, 11), (-12, 18)$ $-\frac{7}{15}$
6. $(-4, -1), (-2, -5)$ -2

Determine the value of r so the line that passes through each pair of points has the given slope.

7. $(10, r), (3, 4), m = -\frac{2}{7}$ 2
8. $(-1, -3), (7, r), m = \frac{3}{4}$ 3
9. $(-2, r), (10, 4), m = -\frac{1}{2}$ 10
10. $(12, r), (r, 6), m = 2$ 10
11. $(6, 8), (r, -2), m = -3$ $\frac{28}{3}$
12. $(r, 9), (7, 5), m = 6$ $\frac{23}{3}$

Additional Answers

36.

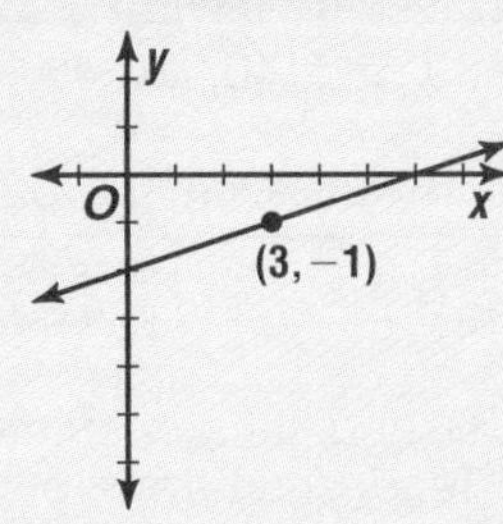

37.

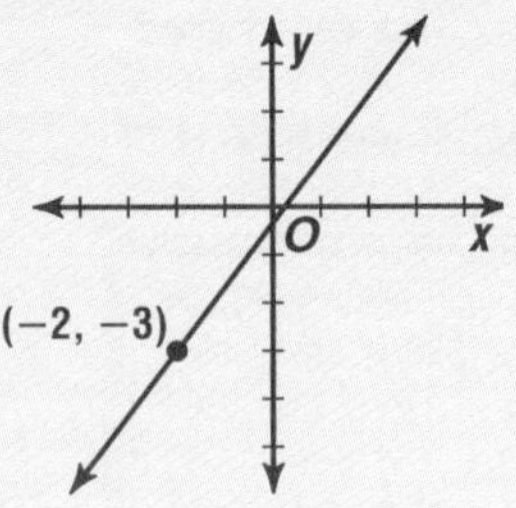

38.

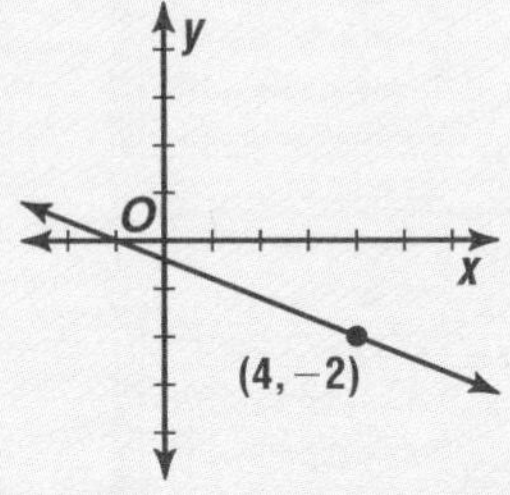

39. Sample answer: (7, 6). The slope of the line containing *A* and *B* is –2. Use the slope to go 2 units down and 1 unit to the right of either point.

Practice Masters, p. 41

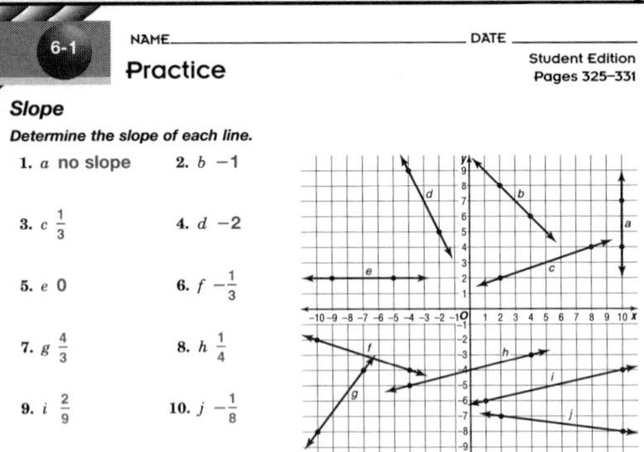

NAME________ DATE________

6-1 Practice — Student Edition Pages 325–331

Slope

Determine the slope of each line.

1. a no slope
2. b -1
3. c $\frac{1}{3}$
4. d -2
5. e 0
6. f $-\frac{1}{3}$
7. g $\frac{4}{3}$
8. h $\frac{1}{4}$
9. i $\frac{2}{9}$
10. j $-\frac{1}{8}$

Determine the slope of the line that passes through each pair of points.

11. (2, 5), (3, 6) 1
12. (4, 1), (–4, 1) 0
13. (4, 1), (–6, –4) $\frac{1}{2}$
14. (–2, 4), (10, 0) $-\frac{1}{3}$
15. (3, 2), (3, –2) no slope
16. (2, 7), (–2, –3) $\frac{5}{2}$

Determine the value of r so the line that passes through each pair of points has the given slope.

17. (6, 3), (r, 2), $m = \frac{1}{2}$ $r = 4$
18. (r, 3), (–4, 5), $m = -\frac{2}{5}$ $r = 1$

Determine the value of r so the line that passes through each pair of points has the given slope.

30. $(5, r), (2, -3), m = \frac{4}{3}$ **1**
31. $(-2, 7), (r, 3), m = \frac{4}{3}$ **–5**
32. $(4, -5), (3, r), m = 8$ **–13**
33. $(6, 2), (9, r), m = -1$ **–1**
34. $(4, r), (r, 2), m = -\frac{5}{3}$ **7**
35. $(r, 5), (-2, r), m = -\frac{2}{9}$ **7**

Draw a line through the given point that has the given slope.

36–38. See margin. 

36. $(3, -1), m = \frac{1}{3}$
37. $(-2, -3), m = \frac{4}{3}$
38. $(4, -2), m = -\frac{2}{5}$

39. The points $A(12, -4)$ and $B(6, 8)$ lie on a line. Find the coordinates of a third point that lies on line AB. Describe how you determined the coordinates of this point. **See margin.**

Critical Thinking

40. **Carpentry** A carpenter was a member of a crew building a roof that is 30 feet long at the base. The roof had a pitch of 4 inches for every foot of length along the base. His task on the crew was to put in vertical supports every 16 inches along the base. He would climb up the ladder, measure 16 inches horizontally, and then measure the vertical height to the roof. He then climbed down the ladder, sawed the piece he needed, and went back up the ladder to put it in place. He wondered if there wasn't some way he could figure out how long each support would be ahead of time so he wouldn't have to climb the ladder so many times. Explain how you can use what you have learned in this chapter to help him out. **See margin.**

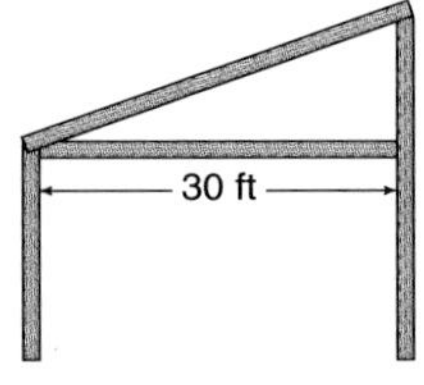

Applications and Problem Solving

41. about 1477 feet

41. **Road Construction** The Castaic Grade in Southern California has a grade of 5%. The length of the roadway is 5.6 miles. What is the change in elevation from the top of the grade to the bottom of the grade in feet?

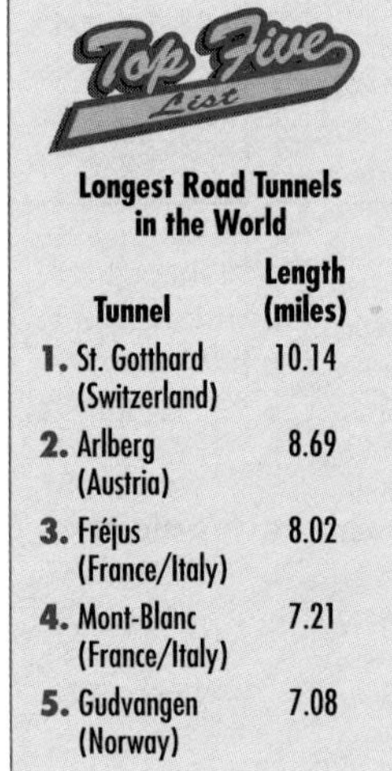

Top Five List

Longest Road Tunnels in the World

Tunnel	Length (miles)
1. St. Gotthard (Switzerland)	10.14
2. Arlberg (Austria)	8.69
3. Fréjus (France/Italy)	8.02
4. Mont-Blanc (France/Italy)	7.21
5. Gudvangen (Norway)	7.08

42. **Carpentry** Julio Mendez is a carpenter. He is building a staircase between the first and second floors of a house, a height of 9 feet. The *tread*, or depth of each step, must be 10 inches. The slope of the staircase cannot exceed $\frac{3}{4}$.

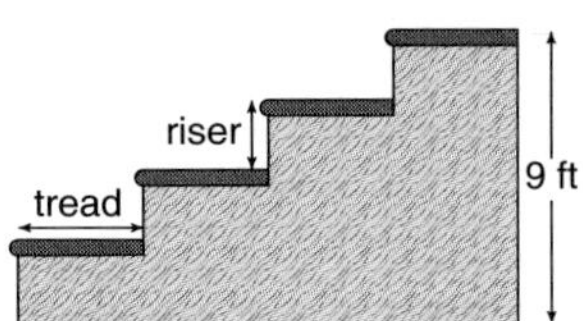

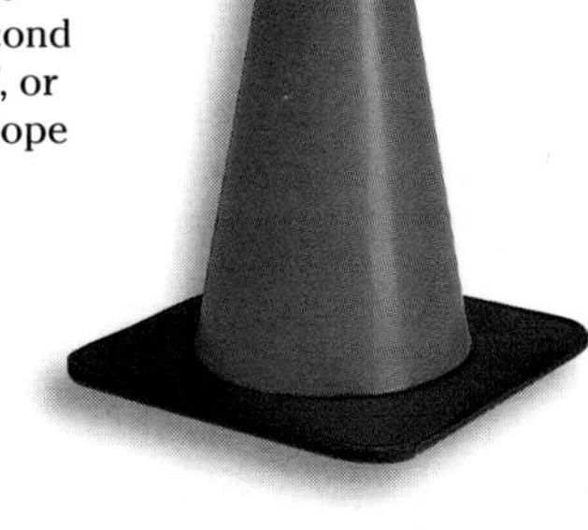

a. How many steps should he plan? **15 steps**

b. What is the measure of the riser, or height of each step? **7.2 in.**

43. **Driving** The Eisenhower Tunnel in Colorado was completed in 1973 and is the fourth longest road tunnel in the United States. The eastern entrance of the Eisenhower Tunnel is at an elevation of 11,080 feet. The tunnel is 8941 feet long and has an upgrade of 0.895% toward the western end. What is the elevation of the western end of the tunnel? **11,160 feet**

Top Five List

The very first tunnels were built by cave dwellers, who cut through clay and soft rock to connect adjacent caves.

Extension

Problem Solving The highway approaching Castaic Lake comes down a 5% grade for a distance of 5.2 miles. If the elevation of the lake is 1440 feet, calculate the drop from the top of the grade to the lake. What is the elevation of the top of the grade? **The drop is about 1373 feet. The top of the grade is about 2813 feet.**

44. **Architecture** Use a millimeter ruler to estimate the pitch, or slope, of each object.

a. 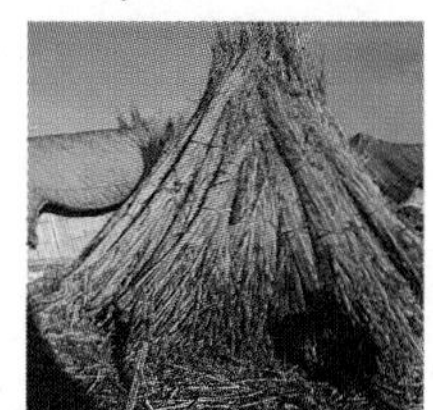
Uro Indian huts of Peru
1

b. 
S. Miniato, Florence, Italy
$\frac{3}{2}$

c. 
Great Pyramid at Giza, Eqypt
1

Mixed Review

45. **Statistics** Find the range, median, upper and lower quartiles, and interquartile range for the set of data. (Lesson 5–7)
3, 3.2, 6, 45, 7, 26, 2, 3.4, 4, 5.3, 5, 78, 8, 1, 5 **77; 5; 8; 3.2; 4.8**

46. **Patterns** Copy and complete the table below. (Lesson 5–5)

n	1	2	3	4	5
$f(n)$	14	13	12	11	10

47. Use the formula $I = prt$ to find r if $I = \$2430$, $p = \$9000$, and $t = 2$ years 3 months. (Lesson 4–4) **12%**

48. Solve $\frac{6}{14} = \frac{7}{x-3}$. (Lesson 4–1) $\frac{58}{3}$

Solve each equation. Check your solution.

49. $\frac{2}{3}x + 5 = \frac{1}{2}x + 4$ (Lesson 3–5) **−6**

50. $3x - 8 = 22$ (Lesson 3–3) **10**

51. **Animals** The average lifespans of 20 different animals are listed below. (Lesson 2–2)

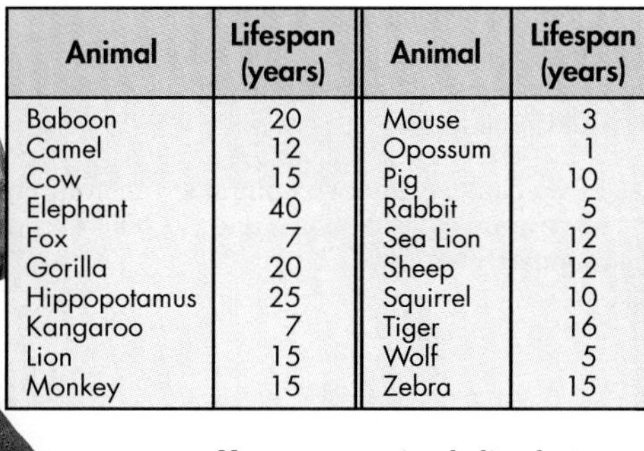

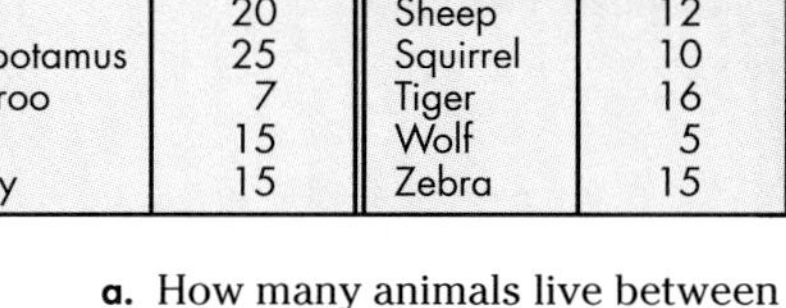

Animal	Lifespan (years)	Animal	Lifespan (years)
Baboon	20	Mouse	3
Camel	12	Opossum	1
Cow	15	Pig	10
Elephant	40	Rabbit	5
Fox	7	Sea Lion	12
Gorilla	20	Sheep	12
Hippopotamus	25	Squirrel	10
Kangaroo	7	Tiger	16
Lion	15	Wolf	5
Monkey	15	Zebra	15

a. How many animals live between 7 and 16 years, inclusive?
b. Make a line plot of the average lifespans of the animals from part a. **See margin.**
c. Which number occurred most frequently? **15 years**
d. How many animals have an average lifespan of at least 20 years?

51a. 12 animals
51d. 4 animals

52. Evaluate $29 - 3(9 - 4)$. (Lesson 1–3) **14**

Additional Answer

51b.

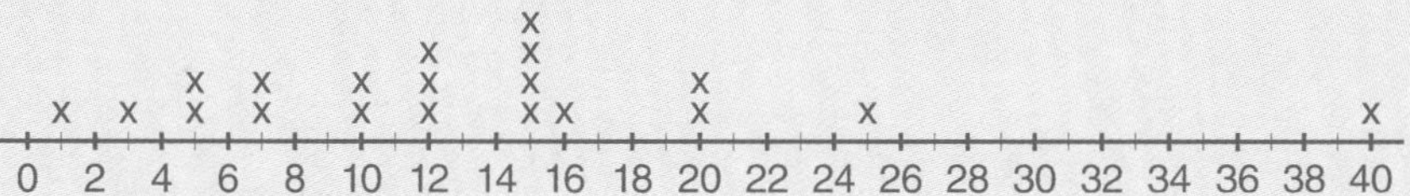

4 ASSESS

Closing Activity

Speaking Ask students to explain how they would determine the slope of a line containing two given points.

Additional Answer

40. The slope is $\frac{120 \text{ in.}}{30 \text{ ft}} = \frac{10 \text{ ft}}{30 \text{ ft}}$ or $\frac{1}{3}$. Use the slope to figure the length of the support.

distance (in.)	16	32	64	96
length (in.)	$\frac{16}{3}$	$\frac{32}{3}$	$\frac{64}{3}$	$\frac{96}{3}$

Enrichment Masters, p. 41

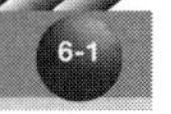

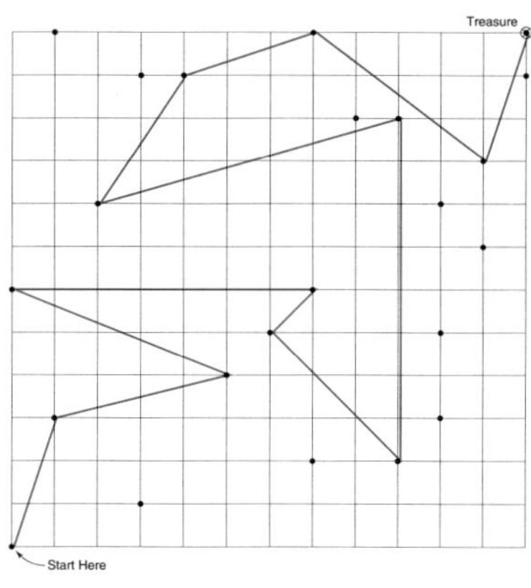

NAME ______________________ DATE ____________

6-1 Enrichment

Student Edition Pages 325–331

Treasure Hunt with Slopes

Using the definition of slope, draw lines with the slopes listed below. A correct solution will trace the route to the treasure.

1. 3 2. $\frac{1}{4}$ 3. $-\frac{2}{5}$ 4. 0 5. 1 6. −1
7. no slope 8. $\frac{2}{7}$ 9. $\frac{3}{2}$ 10. $\frac{1}{3}$ 11. $-\frac{3}{4}$ 12. 3

6-2 LESSON NOTES

NCTM Standards: 1–5, 8

Instructional Resources

- Study Guide Master 6-2
- Practice Master 6-2
- Enrichment Master 6-2
- Assessment and Evaluation Masters, p. 156

Transparency 6-2A contains the 5-Minute Check for this lesson; **Transparency 6-2B** contains a teaching aid for this lesson.

Recommended Pacing	
Standard Pacing	Days 3 & 4 of 13
Honors Pacing	Day 3 of 12
Block Scheduling*	Day 2 of 7
Alg. 1 in Two Years*	Days 3 & 4 of 18

*For more information on pacing and possible lesson plans, refer to the *Block Scheduling Booklet* and *Algebra 1 in Two Years.*

1 FOCUS

5-Minute Check
(over Lesson 6-1)

Determine the slope of the line that passes through each pair of points.

1. (5, −2), (−2, 1) $-\frac{3}{7}$

2. (−3, −8), (4, −2) $\frac{6}{7}$

3. (8, 3), (−1, 3) **0**

Find *r* so the given line through each pair of points has the given slope.

4. (5, *r*), (−2, 4), *m* = −1 **−3**

5. (2, 6), (*r*, −3), $m = \frac{3}{4}$ **−10**

6-2 Writing Linear Equations in Point-Slope and Standard Forms

What **YOU'LL LEARN**

- To write linear equations in point-slope form, and
- to write linear equations in standard form.

Why **IT'S IMPORTANT**

You can write linear equations to solve problems involving geometry.

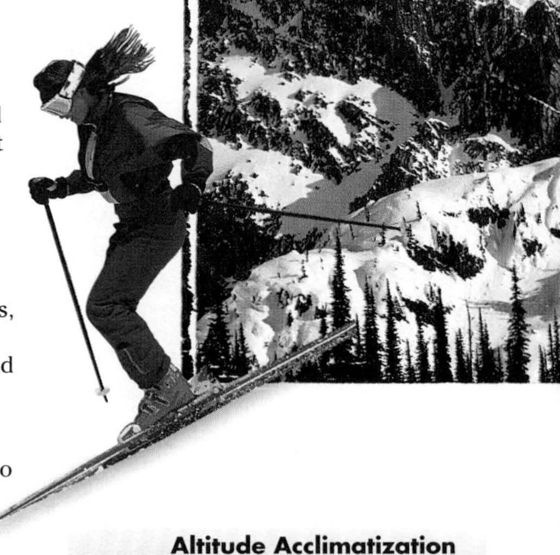

CONNECTION
Geography

If you lived in Miami, Florida, and moved to Denver, Colorado, what adjustments do you think you might have to make? Obviously, the weather is different, but did you know you would also have to adjust to living at a higher altitude? As the altitude increases, the oxygen in the air decreases. This can affect your breathing and cause dizziness, headache, insomnia, and loss of appetite.

Over time, people who move to higher elevations experience acclimatization, or the process of getting accustomed to the new climate. Their bodies develop more red blood cells to carry oxygen to the muscles. Generally, long-term acclimatization to an altitude of 7000 feet takes about 2 weeks. After that, it takes 1 week for each additional 2000 feet of altitude. The graph at the right shows the acclimatization for altitudes greater than 7000 feet.

What are the independent and dependent variables of this relation?

Altitude Acclimatization

Weeks (0–9) vs. Altitude (thousands of feet) (7–14); points (9000, 3) and (11,000, 4)

We can use any two points on the graph to find the slope of the line. For example, let (9000, 3) and (11,000, 4) represent (x_1, y_1) and (x_2, y_2), respectively.

$$m = \frac{y_2 - y_1}{x_2 - x_1}$$
$$= \frac{4 - 3}{11{,}000 - 9000} \text{ or } \frac{1}{2000}$$

Suppose we let (x, y) represent any other point on the line. We can use the slope and one of the given ordered pairs to write an equation for the line.

$$m = \frac{y_2 - y_1}{x_2 - x_1}$$

$$\frac{1}{2000} = \frac{y - 3}{x - 9000} \quad m = \frac{1}{2000}, (x_1, y_1) = (9000, 3), (x_2, y_2) = (x, y)$$

$$\frac{1}{2000}(x - 9000) = y - 3 \quad \textit{Multiply each side by } (x - 9000).$$

$$y - 3 = \frac{1}{2000}(x - 9000) \quad \textit{Reflexive property } (=)$$

Since this form of the equation was generated using the coordinates of a known point and the slope of the line, it is called the **point-slope form.**

Point-Slope Form of a Linear Equation	**For a given point (x_1, y_1) on a nonvertical line having slope *m*, the point-slope form of a linear equation is as follows.** $y - y_1 = m(x - x_1)$

You can write an equation in point-slope form for the graph of any nonvertical line if you know the slope of the line and the coordinates of one point on that line.

Example 1 **Write the point-slope form of an equation for each line.**

a. a line that passes through (−3, 5) and has a slope of $-\frac{3}{4}$

$y - y_1 = m(x - x_1)$

$y - 5 = -\frac{3}{4}[x - (-3)]$ *Replace x_1 with −3, y_1 with 5, and m with $-\frac{3}{4}$.*

$y - 5 = -\frac{3}{4}(x + 3)$

An equation of the line is $y - 5 = -\frac{3}{4}(x + 3)$.

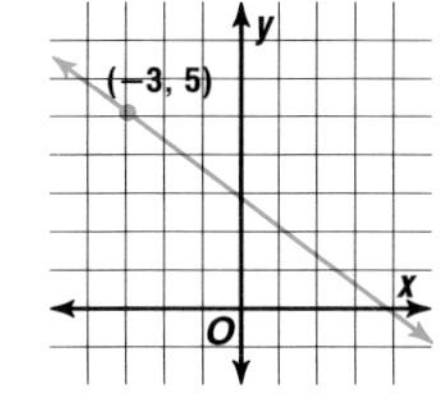

b. a horizontal line that passes through (−6, 2)

Horizontal lines have a slope of 0.

$y - y_1 = m(x - x_1)$

$y - 2 = 0[x - (-6)]$

$y - 2 = 0$

$y = 2$

An equation of the line is $y = 2$.

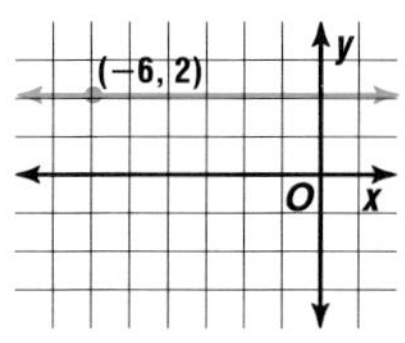

FYI

Tibet has one of the highest altitudes in the world. Nomads in Tibet tend their yaks for 5 months at altitudes of 14,000–17,000 feet above sea level. Then they return to their permanent homes, which are located 12,000 to 14,000 feet above sea level.

You can also write the equations of horizontal lines by using the y-coordinate as in Example 1b.

A vertical line has a slope that is undefined, so you cannot use the point-slope form of an equation. However, you can write the equation of a vertical line by using the coordinates of the points through which it passes. Suppose the line passes through (3, 5) and (3, −2). The equation of the line is $x = 3$, since the x-coordinate of every point on the line is 3.

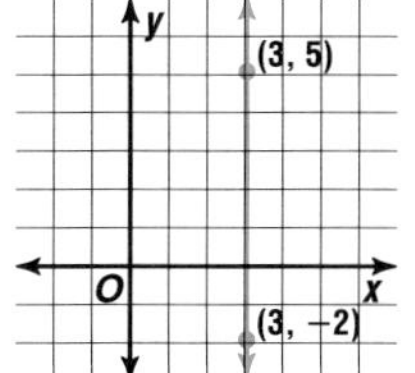

Any linear equation can also be expressed in the form $Ax + By = C$, where A, B, and C are integers and A and B are not both zero. This is called the **standard form** of a linear equation. *Usually, $A > 0$.*

Standard Form of a Linear Equation	**The standard form of a linear equation is $Ax + By = C$, where A, B, and C are integers, $A \geq 0$, and A and B are not both zero.**

Motivating the Lesson

Hands-On Activity Students need to develop an understanding of approximate slopes of lines. Have them make a paper cut-out of a straight line. Draw a grid on paper and place the cut-out in various positions. Students should estimate a reasonable slope for each position.

2 TEACH

In-Class Example

For Example 1
Write the point-slope form of an equation of the line that passes through the given point and has the given slope.

a. (3, 5), $\frac{4}{3}$ $y - 5 = \frac{4}{3}(x - 3)$

b. (−2, 0), $-\frac{3}{2}$ $y = -\frac{3}{2}(x + 2)$

c. (−3, 2), $-\frac{1}{2}$ $y - 2 = -\frac{1}{2}(x + 3)$

d. (0, 5), −3 $y - 5 = -3x$

Teaching Tip The point-slope form of a linear equation is not unique. This form is used as an intermediate step when finding the standard form or slope-intercept form. Ask students why the definition of the point-slope form includes the phrase "a nonvertical line."

Before the 1950s, Tibet was largely isolated from the rest of the world. In 1950, China occupied Tibet. There was an uprising in 1959, but Tibet still remains under Chinese control.

Alternative Teaching Strategies

Student Diversity Have students make a chart with the following headings: *Standard Form*, *Point-Slope Form*, *Slope*, *x-Intercept*, *y-Intercept*, *Points*. On the first line, have them write $3x - 2y = 12$ under *Standard Form.* On the second line, have them write 4 under *x-Intercept* and −1 under *y-Intercept*. On the third line, write (4, 1) and (−3, 5) under *Points*. Then have students complete the chart.

Assignment Guide

Core: 19–47 odd, 48–55
Enriched: 18–44 even, 45–55

For **Extra Practice,** see p. 769.

The red A, B, and C flags, printed only in the Teacher's Wraparound Edition, indicate the level of difficulty of the exercises.

Additional Answers

33. $y - 2 = -\frac{1}{3}(x + 5)$ or $y + 1 = -\frac{1}{3}(x - 4)$
34. $y - 1 = -5(x - 6)$ or $y + 4 = -5(x - 7)$
35. $y + 1 = \frac{3}{7}(x + 8)$ or $y - 5 = \frac{3}{7}(x - 6)$
36. $y - 3 = -\frac{2}{3}(x - 2)$ or $y - 1 = -\frac{2}{3}(x - 5)$
37. $y + 2 = 0(x - 4)$ or $y + 2 = 0(x - 8)$
38. $y - 3 = 2.5(x - 2.5)$ or $y + 4.5 = 2.5(x + 0.5)$

Study Guide Masters, p. 42

NAME______________________ DATE ____________

Study Guide — Student Edition Pages 332–338

Writing Linear Equations in Point-Slope and Standard Forms

If you know the slope of a line and the coordinates of one point on the line, you can write an equation of the line by using the **point-slope form.** For a given point (x_1, y_1) on a nonvertical line with slope m, the point-slope form of a linear equation is $y - y_1 = m(x - x_1)$.

> Any linear equation can be expressed in the form $Ax + By = C$ where A, B, and C are integers and A and B are not both zero. This is called the **standard form.** An equation that is written in point-slope form can be changed to standard form.

Example 1: Write the point-slope form of an equation of the line that passes through (6, 1) and has a slope of $-\frac{5}{2}$.

$y - y_1 = m(x - x_1)$
$y - 1 = -\frac{5}{2}(x - 6)$

Example 2: Write $y + 5 = 3(x - 4)$ in standard form.

$y + 5 = 3(x - 4)$
$y + 5 = 3x - 12$
$-3x + y = -17$
$3x - y = 17$

You can also find an equation of a line if you know the coordinates of two points on the line. First, find the slope of the line. Then write an equation of the line by using the point-slope form or the standard form.

Write the standard form of an equation of the line that passes through the given point and has the given slope.

1. (2, 1), 4 — $4x - y = 7$
2. (−7, 2), 6 — $6x - y = -44$
3. $\left(\frac{1}{2}, 3\right)$, 5 — $10x - 2y = -1$
4. (4, 9), $\frac{3}{4}$ — $3x - 4y = -24$
5. (−6, 7), 0 — $y = 7$
6. (8, 3), 1 — $x - y = 5$

Write the point-slope form of an equation of the line that passes through each pair of points.

7. (6, 3), (−8, 5) — $y - 3 = -\frac{1}{7}(x - 6)$
8. (−1, 9), (10, 7) — $y - 7 = -\frac{2}{11}(x - 10)$
9. (8, 5), (0, −4) — $y - 5 = \frac{9}{8}(x - 8)$
10. (−3, −4), (5, −6) — $y + 6 = -\frac{1}{4}(x - 5)$
11. (2, 9), (9, 2) — $y - 2 = -(x - 9)$
12. (−1, −4), (−6, −10) — $y + 10 = \frac{6}{5}(x + 6)$

Write the point-slope form of an equation of the line that passes through the given point and has the given slope.

8. (9, 1), $m = \frac{2}{3}$ $\quad y - 1 = \frac{2}{3}(x - 9)$
9. (−2, 4), $m = -3$ $\quad y - 4 = -3(x + 2)$
10. (−3, 6), $m = 0$ $\quad y = 6$

Write each equation in standard form.

11. $y + 3 = -\frac{3}{4}(x - 1)$ $\quad 3x + 4y = -9$
12. $y - \frac{1}{2} = \frac{5}{6}(x + 2)$ $\quad 5x - 6y = -13$
13. $y - 3 = 2(x + 1.5)$ $\quad 2x - y = -6$

Write the point-slope form and the standard form of an equation of the line that passes through each pair of points.

14. (−6, 1), (−8, 2)
15. (−1, −2), (−8, 2)
16. (4, 8), (−2.5, 8)

14. $y - 1 = -\frac{1}{2}(x + 6)$ or $y - 2 = -\frac{1}{2}(x + 8)$; $x + 2y = -4$

17. **Geometry** Write the standard form of the equation of the line containing the hypotenuse of the right triangle shown at the right. $3x - 5y = -2$

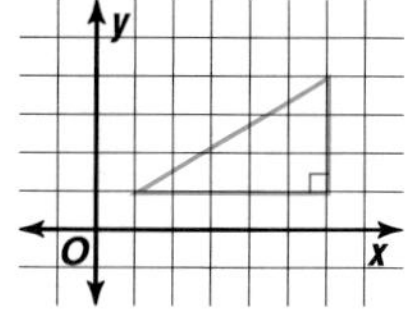

15. $y + 2 = -\frac{4}{7}(x + 1)$ or $y - 2 = -\frac{4}{7}(x + 8)$; $4x + 7y = -18$

16. $y - 8 = 0$; $y = 8$

EXERCISES

Practice

Write the point-slope form of an equation of the line that passes through the given point and has the given slope.

A

18. (3, 8), $m = 2$
19. (4, 5), $m = 3$
20. (−4, −3), $m = 1$
21. (−6, 1), $m = -4$
22. (0, 5), $m = 0$
23. (1, 3), $m = -2$
24. (3, 5), $m = \frac{2}{3}$
25. (8, −3), $m = \frac{3}{4}$
26. (−6, 3), $m = -\frac{2}{3}$

18. $y - 8 = 2(x - 3)$
19. $y - 5 = 3(x - 4)$
20. $y + 3 = x + 4$
21. $y - 1 = -4(x + 6)$
22. $y - 5 = 0$
23. $y - 3 = -2(x - 1)$
24. $y - 5 = \frac{2}{3}(x - 3)$
25. $y + 3 = \frac{3}{4}(x - 8)$
26. $y - 3 = -\frac{2}{3}(x + 6)$

Write the standard form of an equation of the line that passes through the given point and has the given slope.

27. (2, 13), $m = 4$ $\quad 4x - y = -5$
28. (−5, −3), $m = 4$ $\quad 4x - y = -17$
29. (−4, 6), $m = \frac{3}{2}$ $\quad 3x - 2y = -24$
30. (−2, −7), $m = 0$ $\quad y = -7$

B

31. (8, 2), $m = -\frac{2}{5}$ $\quad 2x + 5y = 26$
32. (−5, 5), m = undefined $\quad x = -5$

Write the point-slope form of an equation of the line that passes through each pair of points. 33–38. See margin.

33. (−5, 2), (4, −1)
34. (6, 1), (7, −4)
35. (−8, −1), (6, 5)
36. (2, 3), (5, 1)

C

37. (4, −2), (8, −2)
38. (2.5, 3), (−0.5, −4.5)

Write the standard form of an equation of the line that passes through each pair of points.

39. $(6, 5), (12, -3)$ **$4x + 3y = 39$**

40. $(-2, -7), (1, 2)$ **$3x - y = 1$**

41. $(-5, 9), (3, -2)$ **$11x + 8y = 17$**

42. $(0.7, -1.3), (-0.4, 3.1)$

43. $\left(-\frac{2}{3}, -\frac{5}{6}\right), \left(\frac{3}{4}, -\frac{1}{3}\right)$

44. $\left(-2, 7\frac{2}{3}\right), \left(-2, \frac{16}{5}\right)$ **$x = -2$**

42. $4x + y = 1.5$

43. $36x - 102y = 61$

Critical Thinking

45. A line contains the points at (9, 1) and (5, 5). Write a convincing argument that the same line intersects the x-axis at (10, 0). **See margin.**

Applications and Problem Solving

46. **Geometry** Three lines intersect to form triangle ABC, as shown at the right.
    a. Write the equation for each line in point-slope form.
    b. Write the equation for each line in standard form.

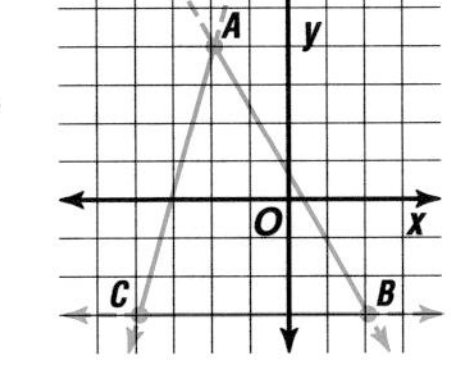

46a. AC: $y - 4 = \frac{7}{2}(x + 2)$ or $y + 3 = \frac{7}{2}(x + 4)$
AB: $y - 4 = -\frac{7}{4}(x + 2)$ or $y + 3 = -\frac{7}{4}(x - 2)$
BC: $y + 3 = 0$

46b. $7x - 2y = -22$; $7x + 4y = 2$; $y = -3$

47. **Draw a Diagram** The Americans with Disabilities Act (ADA) of 1990 states that ramps should have at least a 12-inch run for each rise of 1 inch, with a maximum rise of 30 inches. The post office in Meyersville is changing one of its entrances to accommodate a wheelchair ramp. The distance from the street to the building is 18 feet. The current sidewalk has steps up to the entrance. The entranceway is 30 inches above ground level.

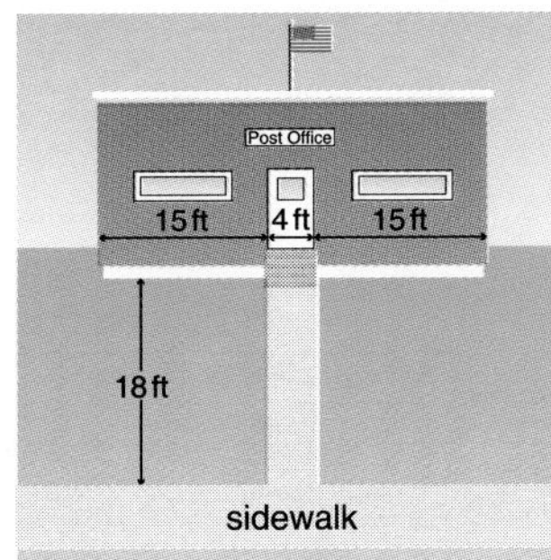

a. Determine if a wheelchair ramp that meets ADA standards can be placed over the existing sidewalk. Explain why or why not.
b. Draw a diagram of an alternative plan for the ramp. **See margin.**

47a. No, for a rise of 30 inches the ramp must be 30 feet long, but there is only 18 feet available.

Mixed Review

48. **Aviation** An airplane flying over Albuquerque at an elevation of 33,000 feet begins its descent to land at Santa Fe, 50 miles away. If the elevation of Santa Fe is 7000 feet, what should be the approximate slope of descent, expressed as a percent? (Lesson 6–1) **9.85%**

49. **Statistics** The stem-and-leaf plot at the right represents the cost per cup of various brands of coffee. (Lesson 5–7)
    a. Find the range and interquartile range for the costs. **\$0.26, \$0.09**
    b. Identify any outliers. **\$0.32**

Stem	Leaf
0	6 6 6 6 8 9 9 9 9
1	0 2 3 4 5 7 7 8 8
2	4 8 9
3	0 2 $1 \mid 2 = \$0.12$

50. Draw a mapping for $\{(-6, 0), (-1, 2), (-3, 4)\}$. (Lesson 5–2) **See margin.**

51. If $y = 12$ when $x = 3$, find y when $x = 7$. Assume that y varies directly as x. (Lesson 4–8) **28**

Extension

Connections The vertices of a triangle are $A(2, 1)$, $B(-3, 4)$, and $C(-5, -2)$. Write the standard form of the equations of the lines that form the sides of the triangle.

side $\overline{AB}$: $3x + 5y = 11$
side $\overline{AC}$: $3x - 7y = -1$
side $\overline{BC}$: $3x - y = -13$

Additional Answers

45. $\frac{5 - 1}{5 - 9}$ or -1
An equation of the line is $(y - 1) = -1(x - 9)$. Let $y = 0$ in the equation and see if $x = 10$.
$0 - 1 = -x + 9$
$-10 = -x$
$10 = x$
(10, 0) lies on the line. Since (10, 0) is a point on the x-axis, the line intersects the x-axis at (10, 0).

47b. Sample answer:

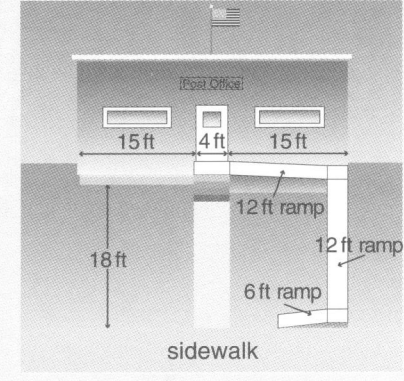

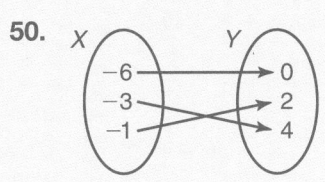

Post Office
15 ft | 4 ft | 15 ft
12 ft ramp
12 ft ramp
6 ft ramp
18 ft
sidewalk

50. X → Y
$-6 \to 0$
$-3 \to 4$
$-1 \to 2$

Practice Masters, p. 42

6-2 NAME ______ DATE ______
Practice
Student Edition Pages 332–338

Writing Linear Equations in Point-Slope and Standard Forms

Write the standard form of an equation of the line that passes through the given point and has the given slope.

1. $(1, 1), \frac{1}{4}$ $x - 4y = -3$
2. $(6, 0), -\frac{1}{2}$ $x + 2y = 6$
3. $(-2, 1), 1$ $x - y = -3$
4. $(-6, -2), -\frac{1}{3}$ $x + 3y = -12$
5. $(3, -4), 0$ $y = -4$
6. $(-4, 1), \frac{3}{2}$ $3x - 2y = -14$
7. $(0, 0), -3$ $3x + y = 0$
8. $(5, -3)$, none $x = 5$

Write the point-slope form of an equation of the line that passes through each pair of points.

9. $(-1, -7), (1, 3)$ $y - 3 = 5(x - 1)$ or $y + 7 = 5(x + 1)$
10. $(5, 3), (-4, 3)$ $y - 3 = 0$
11. $(-4, 6), (-2, 5)$ $y - 5 = -\frac{1}{2}(x + 2)$ or $y - 6 = -\frac{1}{2}(x + 4)$
12. $(2, -6), (2, 5)$ $x = 2$
13. $(-3, -2), (4, 5)$ $y - 5 = x - 4$ or $y + 2 = x + 3$
14. $(-5, 1), (0, -2)$ $y + 2 = -\frac{3}{5}x$ or $y - 1 = -\frac{3}{5}(x + 5)$

4 ASSESS

Closing Activity

Writing Place several lines on a coordinate grid and have the students estimate their slopes and *y*-intercepts. Have them write a linear equation in point-slope form for each line.

Chapter 6, Quiz A (Lessons 6-1 and 6-2), is available in the *Assessment and Evaluation Masters*, p. 156.

Enrichment Masters, p. 42

NAME ________________ DATE ________

Enrichment

Student Edition Pages 332–338

Equations of Lines and Planes in Intercept Form

One form that a linear equation may take is **intercept form.** The constants a and b are the x- and y-intercepts of the graph.

$$\frac{x}{a} + \frac{y}{b} = 1$$

In three-dimensional space, the equation of a plane takes a similar form.

$$\frac{x}{a} + \frac{y}{b} + \frac{z}{c} = 1$$

Here, the constants a, b, and c are the points where the plane meets the x, y, and z-axes.

$\frac{x}{8} + \frac{y}{7} + \frac{z}{6} = 1$

Solve each problem.

1. Graph the equation $\frac{x}{3} + \frac{y}{2} + \frac{z}{1} = 1$.

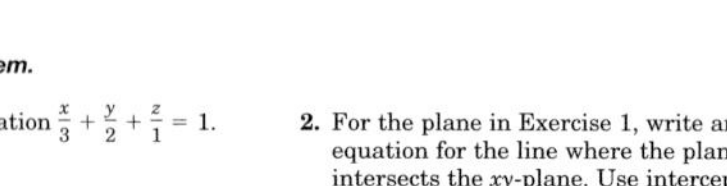

2. For the plane in Exercise 1, write an equation for the line where the plane intersects the xy-plane. Use intercept forms. $\frac{x}{3} + \frac{y}{2} = 1$
3. Write an equation for the line where the plane intersects the xz-plane. $\frac{x}{3} + \frac{z}{1} = 1$
4. Write an equation for the line where the plane intersects the yz-plane. $\frac{y}{2} + \frac{z}{1} = 1$
5. Graph the equation $\frac{x}{1} + \frac{y}{4} + \frac{z}{2} = 1$.

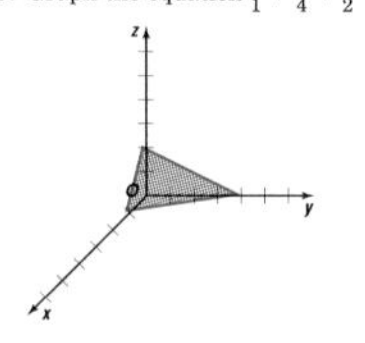

6. Write an equation for the xy-plane. $z = 0$
7. Write an equation for the yz-plane. $x = 0$
8. Write an equation for a plane parallel to the xy-plane with a z-intercept of 2. $z = 2$
9. Write an equation for a plane parallel to the yz-plane with an x-intercept of −3. $x = -3$

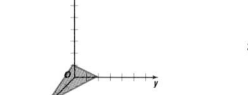

52. **Motion** At 1:30 P.M., an airplane leaves Tucson for Baltimore, a distance of 2240 miles. The plane flies at 280 miles per hour. A second airplane leaves Tucson at 2:15 P.M., and is scheduled to land in Baltimore 15 minutes before the first airplane. At what rate must the second airplane travel to arrive on schedule? (Lesson 4–7) **320 mph**

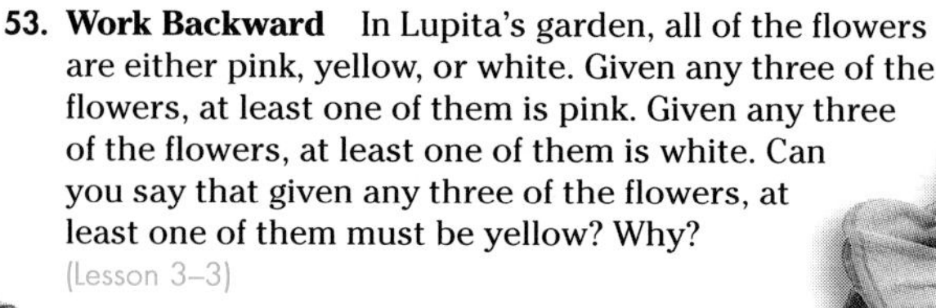

53. **Work Backward** In Lupita's garden, all of the flowers are either pink, yellow, or white. Given any three of the flowers, at least one of them is pink. Given any three of the flowers, at least one of them is white. Can you say that given any three of the flowers, at least one of them must be yellow? Why? (Lesson 3–3)

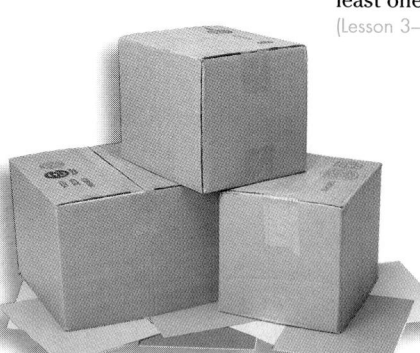

53. No; because there could be 2 pink and 1 white or 1 pink and 2 white.

54. **Printing** A customer orders a job requiring 2500 sheets of paper. He has a choice of ordering a full carton of 3000 sheets at \$27.50 per thousand or "breaking" a carton (ordering exactly the number of sheets required) and paying \$38.40 per thousand. Which would be least expensive in this case? (Lesson 2–6) **full; \$82.50 vs. \$96**
55. Name the property illustrated by $(a + 3b) + 2c = a + (3b + 2c)$. (Lesson 1–8) **associative (+)**

WORKING ON THE

Refer to the Investigation on pages 320–321.

Smoke Gets In Your Eyes

The amount of smoke exhaled into a room on each puff of a cigarette is comparable to the amount of air you exhale in a deep breath. The average smoker takes 10 puffs per cigarette.

1. Suppose that each breath into a balloon approximates the smoke from each exhale during smoking. How many cigarettes are represented by the air in the balloon?
2. Make a graph showing the number of breaths and the cumulative smoke in the air from smoking a cigarette. Describe the graph, and write a function that fits the data graphed.
3. Graph the relationship between the radius of the balloon and the circumference of the balloon. Write an equation to represent this relationship.
4. Find the volume of each balloon in milliliters if 1 $in^3 \approx 16.4$ milliliters. The average at-rest exhale is approximately 500 milliliters. How does the volume of the balloon compare with the volume of air you would expect from the number of breaths it took to inflate the balloon? What accounts for the difference?

Add the results of your work to your Investigation Folder.

In•ves•ti•ga•tion

Working on the Investigation

The Investigation on pages 320–321 is designed to be a long-term project that is completed over several days or weeks. Encourage students to keep their materials in their Investigation Folder as they work on the Investigation.

6-3

Integration: Statistics
Scatter Plots and Best-Fit Lines

What YOU'LL LEARN
- To graph and interpret points on a scatter plot,
- to draw and write equations for best-fit lines and make predictions by using those equations, and
- to solve problems by using models.

Why IT'S IMPORTANT

You can use scatter plots to display data, examine trends, and make predictions.

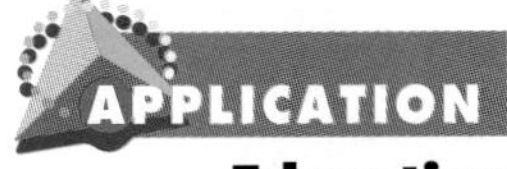

Education

If you get a good score on your SAT does that mean you have a good chance of graduating from college? The table below shows the average SAT (Scholastic Assessment Test) scores for freshmen at selected universities and the graduation rate of those students.

School (State)	Average SAT	Graduation Rate
Baylor University (TX)	1045	69%
Brandeis University (MA)	1215	81%
Case Western Reserve University (OH)	1235	65%
College of William and Mary (VA)	1240	90%
Colorado School of Mines (CO)	1200	78%
Georgia Institute of Technology (GA)	1240	68%
Lehigh University (PA)	1140	88%
New York University (NY)	1145	69%
Penn State University, Main Campus (PA)	1096	61%
Pepperdine University (CA)	1070	64%
Rensselaer Polytechnic Institute (NY)	1190	68%
Rutgers at New Brunswick (NJ)	1110	74%
Tulane University (LA)	1168	71%
University of Florida (FL)	1135	64%
University of North Carolina at Chapel Hill (NC)	1045	81%
University of Texas at Austin (TX)	1135	62%
Wake Forest University (NC)	1250	86%

Source: *America's Best Colleges*, June, 1995

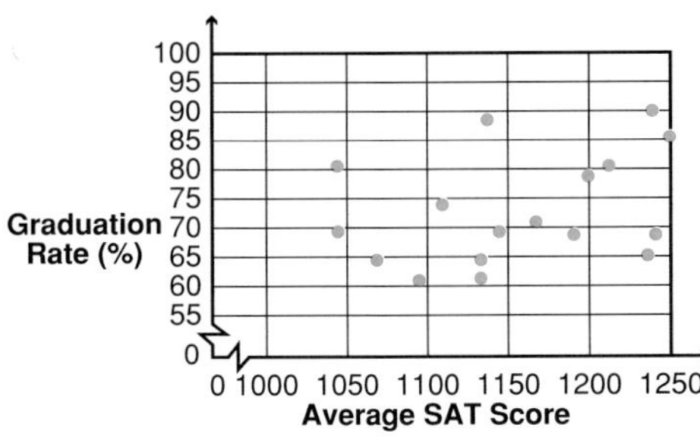

To determine if there is a relationship between SAT scores and graduation rates, we can display the data points in a graph called a **scatter plot.** In a scatter plot, the two sets of data are plotted as ordered pairs in the coordinate plane. In this example, the independent variable is the average SAT score, and the dependent variable is the graduation rate. The scatter plot is shown at the left. The graph indicates that higher SAT scores do not necessarily result in higher graduation rates.

One way to solve real-world problems is to **use a model.** Scatter plots are an excellent way to model real-life data to observe patterns and trends.

6-3 LESSON NOTES

NCTM Standards: 1–5, 10

Instructional Resources
- Study Guide Master 6-3
- Practice Master 6-3
- Enrichment Master 6-3
- Science and Math Lab Manual, pp. 23–28

Transparency 6-3A contains the 5-Minute Check for this lesson; **Transparency 6-3B** contains a teaching aid for this lesson.

Recommended Pacing	
Standard Pacing	Day 5 of 13
Honors Pacing	Day 4 of 12
Block Scheduling*	Day 3 of 7
Alg. 1 in Two Years*	Days 5 & 6 of 18

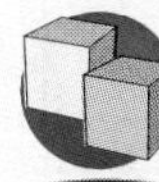

*For more information on pacing and possible lesson plans, refer to the *Block Scheduling Booklet* and *Algebra 1 in Two Years.*

1 FOCUS

5-Minute Check
(over Lesson 6-2)

Write the standard form of an equation of the line that passes through the given point and has the given slope.

1. $(-2, 6), 0$ $y = 6$
2. $(4, -3), 2$ $2x - y = 11$
3. $(5, 7), 0$ $y = 7$

Write the standard form of an equation of the line that passes through each pair of points.

4. $(5, 3), (-6, 3)$ $y = 3$
5. $(9, 1), (8, 2)$ $x + y = 10$

Motivating the Lesson

Situational Problem A local football team has gained the following yardage in its first seven games. Have students analyze the data in the table on the left to identify a pattern. Develop the idea that neither a stem-and-leaf plot nor a box-and-whisker plot would show this pattern.

Game	Yardage
1	348
2	350
3	300
4	295
5	280
6	250
7	225

The yardage per game is decreasing.

2 TEACH

In-Class Example

For Example 1

The table below lists the number of shots attempted and the number of shots made by each of eleven members of a basketball team. Make a scatter plot of the data, and answer the questions below.

Player	Shots Attempted	Shots Made
A	240	160
B	200	120
C	135	50
D	80	50
E	75	70
F	35	20
G	20	10
H	12	5
I	9	7
J	5	5
K	2	0

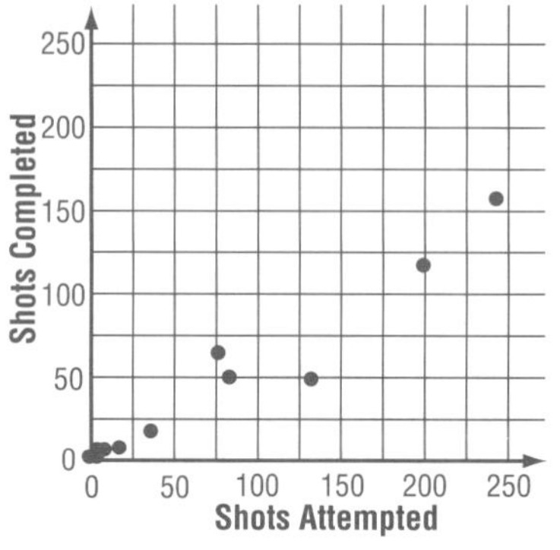

a. Do the points graphed seem to fall on a line? **Yes, with the exception of a few, the points appear to form a line slanting upward.**

b. Is there a negative, a positive, or no correlation between the variables? **positive**

Teaching Tip Point out that with real-world data, rarely will all the data points fall on the line.

Look for a relationship between x and y in the graphs below.

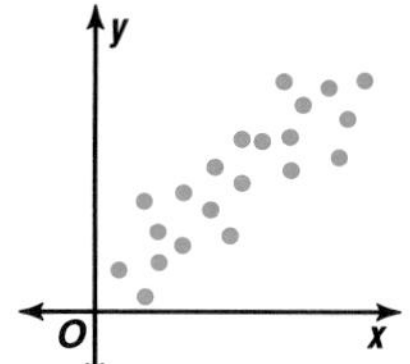

In this graph, x and y have a **positive correlation**. That is, the values are related in the same way. As x increases, y increases.

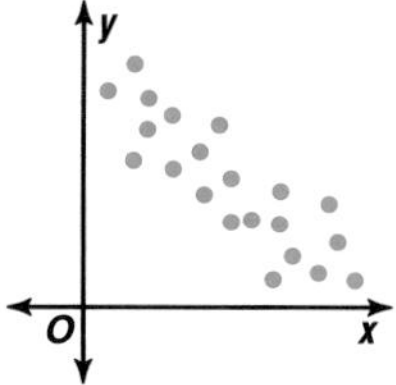

In this graph, x and y have a **negative correlation**. That is, the values are related in opposite ways. As x increases, y decreases.

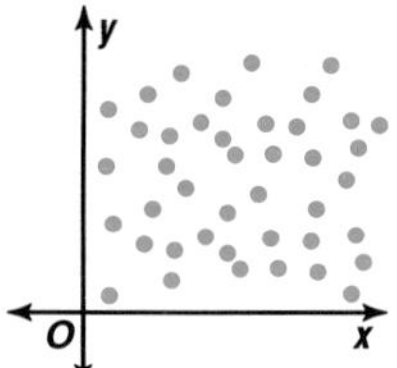

In this graph, x and y have *no correlation*. In this case, x and y are not related and are said to be *independent.*

Example 1 **The table below shows 13 of the fastest-growing cities in the United States and their latitude and longitude.**

PROBLEM SOLVING
Use a Model

Fastest-Growing Cities	Ranking	North Latitude	West Longitude
Austin, TX	9	30°	98°
Bakersfield, CA	1	35°	119°
Colorado Springs, CO	13	39°	105°
Durham, NC	8	36°	79°
Fresno, CA	2	37°	120°
Laredo, TX	10	28°	100°
Las Vegas, NV	3	36°	115°
Raleigh, NC	6	36°	79°
Reno, NV	12	40°	120°
Sacramento, CA	11	39°	121°
San Bernardino, CA	7	34°	117°
Stockton, CA	5	38°	121°
Tallahassee, FL	4	30°	84°

Source: Bureau of the Census, National Oceanic and Atmospheric Administration

a. **Draw one scatter plot to represent the correlation between the rankings and each city's latitude, and another to represent the correlation between the rankings and each city's longitude.**

b. **Is there a correlation between the cities' locations and their popularity?**

a. The scatter plots are shown below.

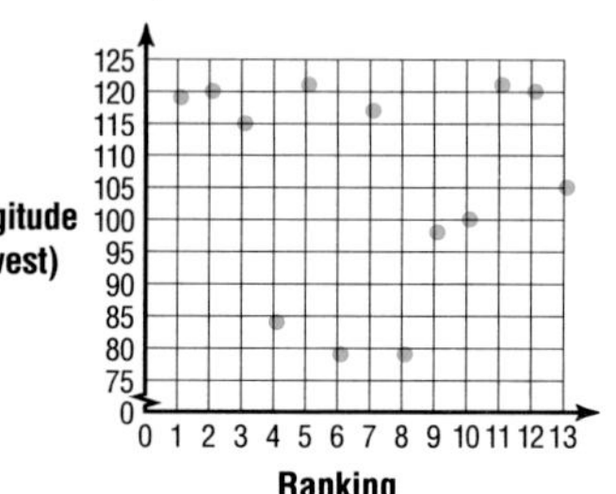

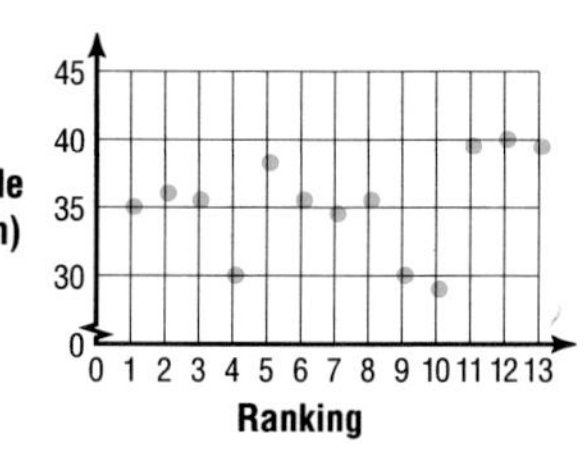

b. Both of the correlations are fairly weak. However, the correlation between the rankings and latitude is stronger than the correlation between the rankings and longitude because the data approximate a line a little more. It is impossible to tell whether either correlation is positive or negative. Thus, we could say that there is a weak correlation between the cities' location and popularity.

Cooperative Learning

Pairs Check Have students work in pairs for this activity. One student is to come up with several relationships of the following type: (1) miles per gallon and speed (2) hourly wage and weekly pay. Then the other student should determine which value is determined by the other value. For more information on the pairs check strategy, see *Cooperative Learning in the Mathematics Classroom*, one of the titles in the Glencoe Mathematics Professional Series, page 12.

The best-fit line may not pass through any of the data points.

You can use a scatter plot to make predictions. To help you do this, you can draw a line, called a **best-fit line**, that passes close to most of the data points. Use a ruler to draw a line that is close to most or all of the points. Then use the ordered pairs representing points on the line to make predictions.

A best-fit line shows if the correlation between two variables is *strong* or *weak*. The correlation is strong if the data points come close to, or lie on, the best-fit line. The correlation is weak if the data points do not come close to the line.

Example 2

Animals

Dogs age differently than humans do. You may have heard someone say that a dog ages 1 year for every 7 human years. However, that is not the case. The table at the right shows the relationship between dog years and human years.

Dog Years	Human Years
1	15
2	24
3	28
4	32
5	37
6	42
7	47

Source: *National Geographic World*, January, 1995

a. Draw a scatter plot to model the data and determine what relationship, if any, exists in the data.

b. Draw a best-fit line for the scatter plot.

c. Find an equation for the best-fit line.

d. Use the equation to determine how many human years are comparable to 13 dog years.

FYI

Almost 40% of U.S. households have at least one dog kept as a pet, accounting for over 50 million dogs. According to the American Kennel Club, the most popular dogs in America are Labrador retrievers.

a. Let the independent variable d be dog years, and let the dependent variable h be human years. The scatter plot is shown below.

The plot seems to indicate that there is nearly a linear relationship between dog years and human years.

There also appears to be a positive correlation between the two variables. As x increases, y increases.

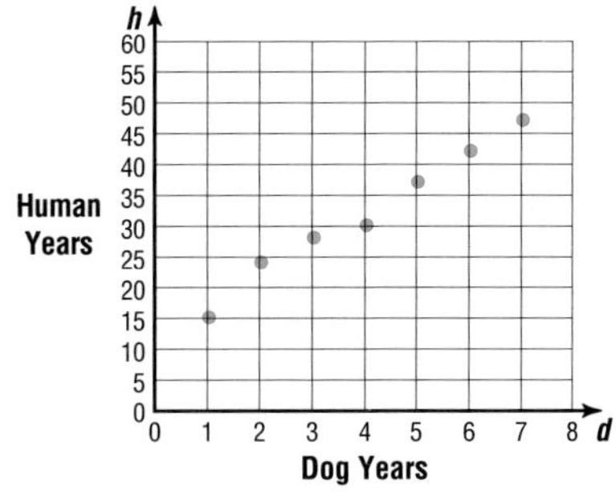

b. No line will pass through all of the data points. However, the best-fit line describes the trend of the data. Draw a line that seems to represent the data.

The slope of the best-fit line is positive. Thus, the correlation between the two variables is also positive.

The best-fit line drawn in this example is arbitrary. You may draw another best-fit line that is equally as valid.

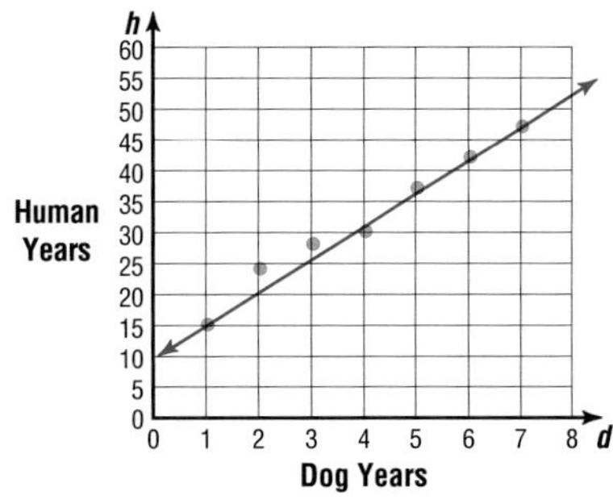

c. As you can see, the best-fit line we drew passes through three of the data points: (4, 32), (5, 37), and (6, 42). Use two of these points to write the equation of the line.

(continued on the next page)

In-Class Example

For Example 2
Use the scatter plot from In-Class Example 1 and draw a best-fit line for it. Then answer the following questions.

a. Does there appear to be any correlation between the number of shots attempted and the number of shots completed?
Yes; the plot seems to indicate that as the number of shots attempted increases, so does the number of shots completed.

b. Find an equation for the best-fit line.

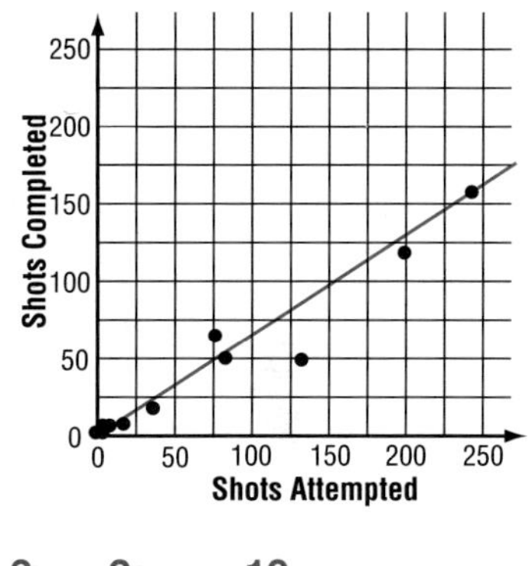

$3y - 2x = -10$

Stanley Coven, professor of psychology at the University of British Columbia, has developed a ranking of breeds by intelligence. The top 5 are:

1. Border Collie
2. Poodle
3. German Shepherd
4. Golden Retriever
5. Doberman Pinscher

Alternative Teaching Strategies

Reading Algebra A dictionary defines *scatter* using the words, *separate, irregular, widely, diffusely,* and *randomly*. This suggests that the points on a scatter plot are unrelated. This may be true where there is no correlation. Remind students that the point of correlation and regression is to discover some relationship among the data.

GLENCOE Technology

Interactive Mathematics Tools Software

This multimedia software provides an interactive lesson by helping students observe positive and negative relationships of points on a scatter plot. A **Computer Journal** gives students an opportunity to write about what they have learned.

For Windows & Macintosh

EXPLORATION

This Exploration provides students with the experience of working with real-world data. They will begin to see the relationship between real-world data and fitting a line to these data.

First, find the slope.

$\frac{y_2 - y_1}{x_2 - x_1} = \frac{37 - 32}{5 - 4}$ or 5 $(x_1, y_1) = (4, 32)$ *and* $(x_2, y_2) = (5, 37)$

Then, use the point-slope form.

$h - y_1 = m(d - x_1)$

$h - 32 = 5(d - 4)$ *Let (4, 32) represent* (x_1, y_1).

$h - 32 = 5d - 20$ *Distributive property*

$5d - h = -12$

d. To determine how many human years are comparable to 13 dog years, let $d = 13$ and solve for h.

$5d - h = -12$

$5(13) - h = -12$

$h = 77$

A dog that is 13 years old is comparable to a human that is 77 years old.

A **regression line** is the most accurate best-fit line for a set of data, and can be determined with a graphing calculator or computer. A graphing calculator assigns each regression line an r value. This value $(-1 \leq r \leq 1)$ measures how closely the data are related. -1 indicates a strong negative correlation and 1 indicates a strong positive correlation.

EXPLORATION GRAPHING CALCULATORS

The table at the right shows the years of rate increases and the fares charged to ride the subway in New York City.

Year	Fare
1953	$0.15
1966	0.20
1970	0.30
1972	0.35
1975	0.50
1981	0.75
1984	0.90
1986	1.00
1990	1.15
1992	1.25

Source: *New York Times*, July 25, 1993

a. Use the Edit option on the STAT menu to enter the year data in L1 and the fare data into L2.

b. Use the window [1950, 1995] with a scale factor of 5 and [0, 1.5] with a scale factor of 0.25. Make sure any equations are deleted from the Y= list. Then press [2nd] [STAT PLOT] 1. Make sure the following items are highlighted: On, the first type of graph (scatter plot), L1 as the Xlist, and L2 as the Ylist. Press [GRAPH], and describe the graph.

c. Then find an equation for the regression line.

Enter: [STAT] [▶] 5 [2nd] [L1] [,] [2nd] [L2] [ENTER].

The screen displays the equation $y = ax + b$ and gives values for a, b, and r. Record these values on your paper.

d. Now graph the best-fit line.

Enter: [Y=] [VARS] 5 [▶] [▶] 7 [GRAPH]

LOOK BACK

Refer to Lesson 5-7A for more information on entering data into lists on a graphing calculator.

Your Turn

a. How well do you feel the graph of the equation fits the data? Justify your answer. **See students' work.**

b. Do any of the data points lie on the best-fit line? If so, name them. **no**

c. Remove the ordered pair (1953, $0.15) from the set of data and repeat the process. What impact has removing this ordered pair had on the fit of the line to the data? (*Hint:* Compare the r values.)

c. The data are more closely related.

d. Use the equation and graph to predict the fares for the year 2000. **$1.40**

CHECK FOR UNDERSTANDING

Communicating Mathematics

1. If the slope is positive (or negative), the correlation is positive (or negative).

2–3. See Solutions Manual.

Study the lesson. Then complete the following.

1. **Explain** how to determine whether a scatter plot has a positive or negative correlation.
2. **Draw** sketches of scatter plots that have each type of correlation.
   a. very strong positive b. very strong negative c. no correlation
3. **Describe** a situation that illustrates each type of plot in Exercise 2.
4. What does the r value on a graphing calculator tell you? Give examples.
5. What situations in your own life would present a negative correlation? Sample answer: distance driven vs. gallons of gasoline in tank

4. how strongly the data are correlated; -1 is very negative; 1 is very positive

Guided Practice

9b. yes; negative

9c. (7, 18) and (8, 16)

Explain whether a scatter plot for each pair of variables would probably show a *positive, negative,* or *no* correlation between the variables.

6. money earned by a restaurant worker and time spent on the job **positive**
7. heights of fathers and sons **positive**
8. grades in school and how many times you miss class **negative**
9. **Archery** The graph at the right shows the results when three student archers shot at a target nine times each.
   a. As the archers take more and more shots, what happens to the accuracy of the shots? **It gets better.**
   b. Is there a correlation between the variables? Is it positive or negative?
   c. Which data points seem to be extremes from the rest of the data?

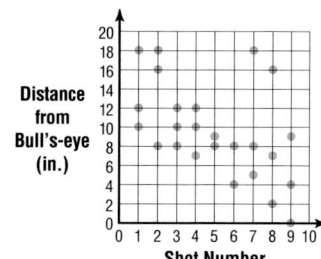

Distance from Bull's-eye (in.)

Shot Number

10. **Biology** The table below shows the average body temperature of 14 insects at a given air temperature.

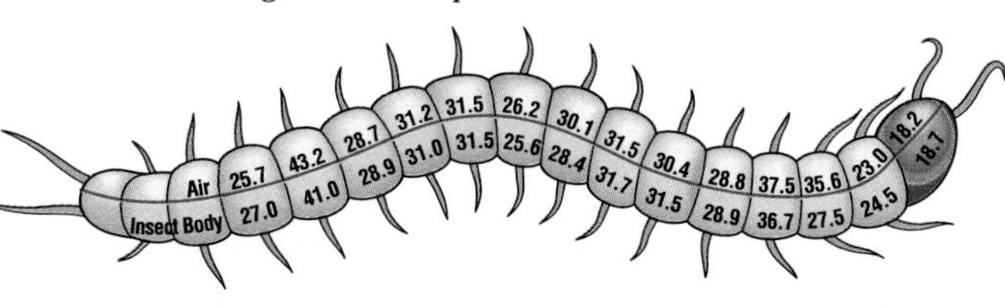

10c. An insect's body temperature closely approximates that of the air around it, and thus, is cold-blooded.

   a. Draw a scatter plot and a best-fit line for these data. **See margin.**
   b. Write an equation for the best-fit line. **Sample answer: $y = x$**
   c. What conclusion might you draw from these data?

EXERCISES

Practice

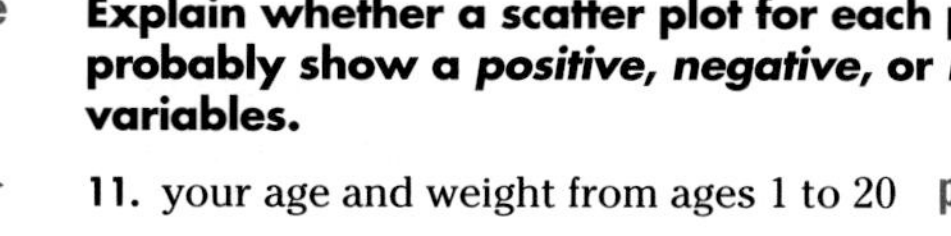

Explain whether a scatter plot for each pair of variables would probably show a *positive, negative,* or *no* correlation between the variables.

A

11. your age and weight from ages 1 to 20 **positive**
12. temperature of a cup of coffee and the time it sits on a table **negative**
13. the amount a mail carrier earns each day and the weight of the mail delivered each day **no**

3 PRACTICE/APPLY

Check for Understanding

Exercises 1–10 are designed to help you assess your students' understanding through reading, writing, speaking, and modeling. You should work through Exercises 1–5 with your students and then monitor their work on Exercises 6–10.

Error Analysis

It does not matter which data are placed on which axis, but because the data are in ordered pairs, the order in which they are graphed is important. Have students write the data in ordered pair form before graphing.

Assignment Guide

Core: 11–29 odd, 31–35
Enriched: 12–24 even, 25–35

For **Extra Practice,** see p. 770.

The red A, B, and C flags, printed only in the Teacher's Wraparound Edition, indicate the level of difficulty of the exercises.

Study Guide Masters, p. 43

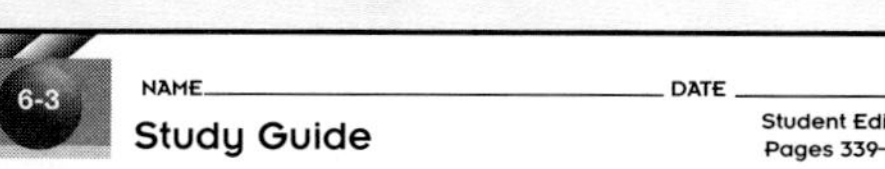

6-3 NAME_____ DATE_____
Study Guide
Student Edition Pages 339–345

Integration: Statistics
Scatter Plots and Best-Fit Lines

A **scatter plot** is a graph that shows the relationship between paired data. The scatter plot may reveal a pattern, or association, between the paired data. This association can be negative or positive. The association is said to be positive when a line suggested by the points slants upward.

The scatter plot at the right represents the relationship between the amount of money Carmen earned each week and the amount she deposited to her savings account. Since the points suggest a line that slants upward, there seems to be a positive relationship between the paired data. In general, the scatter plot seems to show that the more Carmen earned, the more she saved.

Solve each problem.

1. The table below shows the number of bull's-eyes attempted and the number of bull's-eyes made during a few dart games.
   a. Draw a scatter plot at the right from the data in the table.

Name	Bull's-eyes Attempted	Bull's-eyes Made
Darlene	5	4
Chris	7	7
Mark	5	1
Kathy	6	2

Bull's-eyes Made / Bull's-eyes Attempted

   b. What are the paired data? the number of bull's-eyes attempted and the number of bull's-eyes made
   c. Is there a relationship between attempts and successes? There is a weak suggestion of a positive correlation.

2. Anna's running speed after 5 minutes was 10 mi/h; at 10 minutes, 8 mi/h; at 15 minutes, 5 mi/h; and at 20 minutes, 4 mi/h.
   a. Make a scatter plot pairing time run with running speed.
   b. How is the data related, positively, negatively, or not at all? negatively

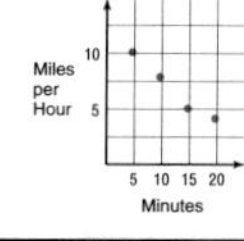

Reteaching

Using Modeling Separate the class into groups and give each group a different age level, but give all groups the same survey question. Have them collect data on the question and make a scatter plot of the data. Show the different age level scatter plots to the whole class to see what happens to each plot as age is varied. Discuss their findings.

Additional Answer

10a.

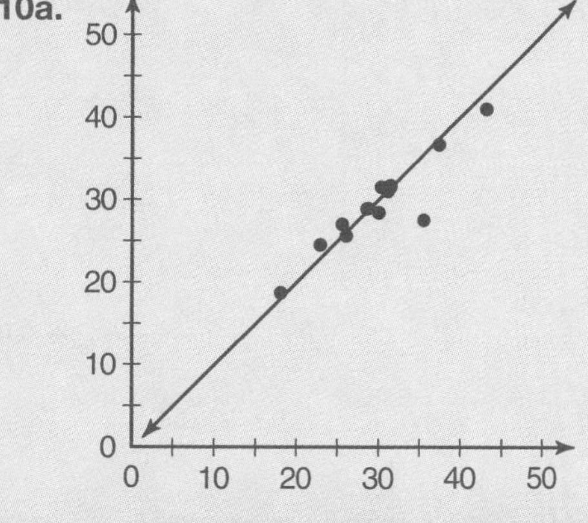

Additional Answers

24a.

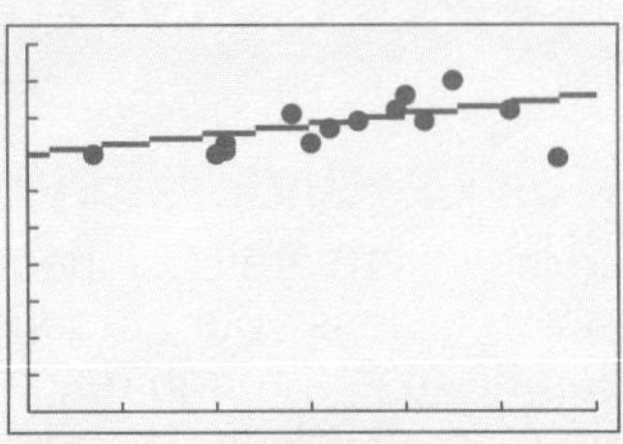

24b. The points seem to be positively correlated.

24c. $y = 2.48x + 68.94$, $r = 0.501$

24d. Sample answer: As the January temperature increases, so does the July temperature.

Sample answers:

25a. The more taxation increases, the more in debt the government becomes.

25b. You work harder, and your grades go up.

25c. As more money is spent on research, fewer people die of cancer.

25d. comparing the number of professional golfers with the number of holes-in-one

27a. The correlation shown by the graph shows a slightly positive correlation between SAT scores and graduation rate.

27b. See students' graphs. Sample equation: $y = 0.89x + 1142.17$

Practice Masters, p. 43

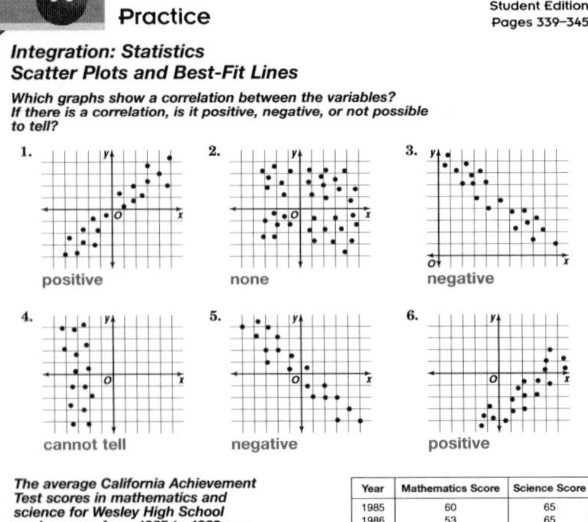

6-3 NAME ______ DATE ______

Practice

Student Edition Pages 339–345

Integration: Statistics
Scatter Plots and Best-Fit Lines

Which graphs show a correlation between the variables? If there is a correlation, is it positive, negative, or not possible to tell?

1. positive
2. none
3. negative
4. cannot tell
5. negative
6. positive

The average California Achievement Test scores in mathematics and science for Wesley High School sophomores from 1985 to 1989 are given in the table at the right.

Year	Mathematics Score	Science Score
1985	60	65
1986	53	65
1987	44	57
1988	61	61
1989	70	67

7. Make a scatter plot of the data. Use the coordinate plane provided.
8. Does the scatter plot show a positive or negative correlation? positive
9. Predict the mathematics score corresponding to a science score of 50. about 35
10. Predict the science score corresponding to a mathematics score of 65. about 65

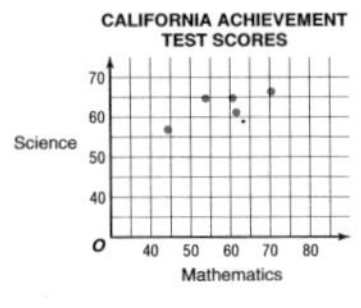

Explain whether a scatter plot for each pair of variables would probably show a *positive, negative,* or *no* correlation between the variables. 16. negative 17. positive

14. the distance traveled and the time driving positive
15. a person's height and their birth month no
16. the amount of snow on the ground and the daily temperature
17. the amount of time you exercise and the amount of calories you burn
18. the time water boils and the amount of water in the pot positive
19. the number of files stored on a disk and the amount of memory left on the disk negative

20. Sample reason: Dots lie in a horizontal pattern.

21. Sample reason: Dots are grouped in an upward diagonal pattern.

22. Sample reason: Dots are everywhere, no linear pattern exists.

Determine whether a best-fit line should be drawn for each set of data graphed below. Explain your reasoning.

20. yes 21. yes 22. no

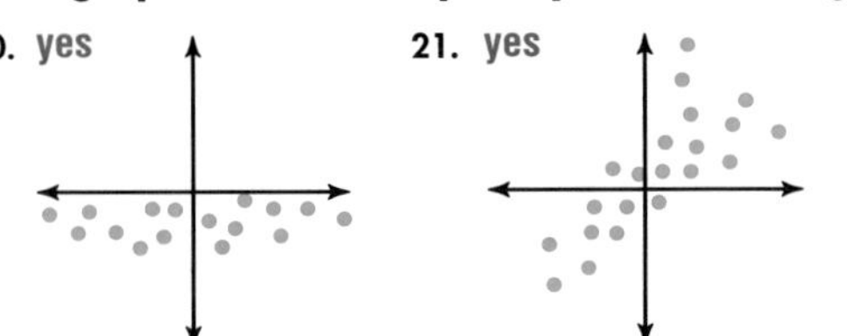

23. c; if 1 is correct then 19 are wrong, if 2 are correct, then 18 are wrong, and so on. Graph c shows these pairs of numbers.

23. A test contains 20 true-false questions. Which scatter plot below best illustrates all the possible combinations of the numbers of correct and incorrect answers if the horizontal axis represents the number of responses that are correct and the vertical axis represents the number that are incorrect? Explain your choice.

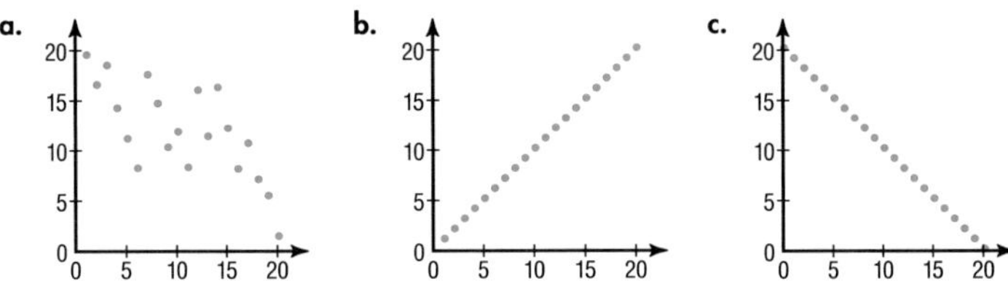

Graphing Calculator

24. The average January and July temperatures (°F) for 14 cities in various parts of the United States are listed below.

Jan.	21	35	42	32	51	7	30	21	20	40	28	45	56	39
July	71	79	79	77	82	70	73	73	70	86	81	90	69	82

a. Use a graphing calculator to enter the January temperatures as L1 and the July temperatures as L2 and create a scatter plot. a–d. See margin.
b. Describe the pattern of the points in the scatter plot.
c. Record the equation of the regression line and the r value.
d. Write a sentence that generally describes the relationship between the January temperature in a city and its July temperature.

Critical Thinking

25a–d. See margin for sample answers.

25. Different correlation values are acceptable for different situations. Give an example for each situation. Explain your reasoning.
a. When is a strongly positive correlation a bad thing?
b. When is a mildly positive correlation an acceptable outcome?
c. When would a strongly negative correlation be the desired outcome?
d. When would no correlation be acceptable?

Additional Answer

28a.

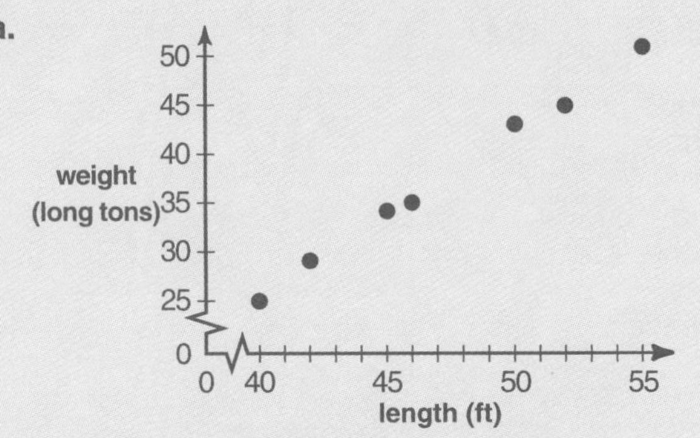

There seems to be a positive correlation between the length and weight of humpback whales.

26. Many verbal expressions can be translated into algebraic expressions or equations. How do you think you could illustrate the expression "The bigger they are, the harder they fall" as a scatter plot? **See students' work.**

Applications and Problem Solving

27. **Education** Refer to the application at the beginning of the lesson.
   a. Describe the correlation of the SAT scores to the graduation rate.
   b. Copy the graph and draw a best-fit line. What is the equation of the best-fit line you drew? **a–b. See margin for sample answers.**

28. **Biology** Scientists use samples to generalize relationships they observe in nature. The table at the right shows the length and weight of several humpback whales. A long ton is about 2240 pounds.

Length (feet)	Weight (long tons)
40	25
42	29
45	34
46	35
50	43
52	45
55	51

   a. Make a scatter plot of these data. Does a relationship exist? If so, what is the relationship? **See margin.**
   b. Use your scatter plot as a model to predict what you think a humpback whale 51 feet long might weigh. **about 43–45 long tons**
   c. How long might a whale that weighs between 30 long tons and 35 long tons be? **43–46 feet**
   d. A 46-foot long humpback whale weighing 36 long tons was observed. Does this alter your conclusion about the correlation of the data? Explain.

28d. Yes; sample answer: the point (46, 36) falls in the same general area as the other data.

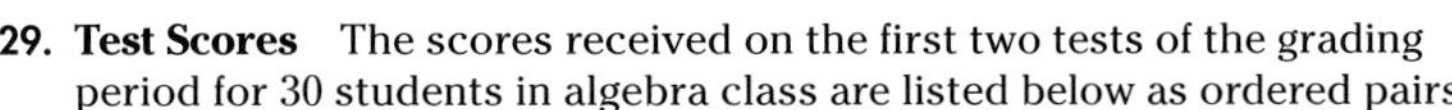

29. **Test Scores** The scores received on the first two tests of the grading period for 30 students in algebra class are listed below as ordered pairs.

30, 43	58, 57	55, 61	4, 71	32, 27	68, 59
54, 47	38, 27	56, 47	72, 63	50, 23	70, 67
60, 53	73, 79	55, 68	19, 58	38, 41	71, 59
68, 75	74, 83	58, 67	66, 73	72, 67	42, 57
71, 69	70, 89	94, 59	8, 84	84, 71	82, 73

   a. Make a scatter plot of the data. **See margin.**
   b. What correlation, if any, do you observe in the data? **positive**
   c. Draw a best-fit line and write the equation of the line.

29c. Sample answer: $y = \frac{2}{3}x + 22$

30. **Use a Model** Collect 10 round objects around your home. Treat each of them as a circle and measure the diameter and circumference in millimeters.
   a. Use a scatter plot as a model to relate the diameter and circumference of each object. **a–c. See students' work.**
   b. Write an equation for the best-fit line.
   c. The actual formula that relates the circumference and diameter of a circle is $C = \pi d$. How does this formula compare with your equation?

Mixed Review

31. **Air Travel** A Boeing 747 takes off from the runway, and 20 minutes after liftoff, the plane is 30,000 feet from the ground. At 22 minutes after liftoff, the plane is 45,000 feet from the ground. (Lesson 6–2)
   a. Write an equation in standard form to represent the flight of the plane during this time, with x = time in minutes, and y = altitude in feet.
   b. Use the equation to find the plane's altitude 18 minutes after takeoff.
   c. Can this equation be used to determine the height of the plane at any time during the flight? If so, explain how.

32. Solve $6b - a = 32$ if the domain is $\{-2, -1, 0, 2, 5\}$. (Lesson 5–3)

33. **Trigonometry** Use a calculator to solve $\cos W = 0.2598$. (Lesson 4–3) **75°**

34. Simplify $1.2(4x - 5y) - 0.2(-1.5x + 8y)$. (Lesson 2–6) **$5.1x - 7.6y$**

35. Solve $8.2 - 6.75 = m$. (Lesson 1–5) **1.45**

31a. $7500x - y = 120{,}000$
31b. 15,000 feet
31c. No, it only describes the plane's path in that part of the flight.
32. $\left\{(-2, 5), \left(-1, 5\frac{1}{6}\right), \left(0, 5\frac{1}{3}\right), \left(2, 5\frac{2}{3}\right), \left(5, 6\frac{1}{6}\right)\right\}$

Extension

Reasoning The daily high temperatures in Tampa were recorded for the month of November 1994. The data are shown in the form of a scatter plot, with the horizontal scale representing the day of the month and the vertical scale representing the temperature.

1. What information can be determined from the plot? **weather trends that occurred during the month**
2. What information cannot be determined from the plot? **how the daily temperatures in 1994 compared with average temperatures over an extended period of time**

4 ASSESS

Closing Activity

Writing Have students describe a situation in which a scatter plot would be a better representation of a set of data than a box-and-whisker plot.

Additional Answers

29a and 29c.

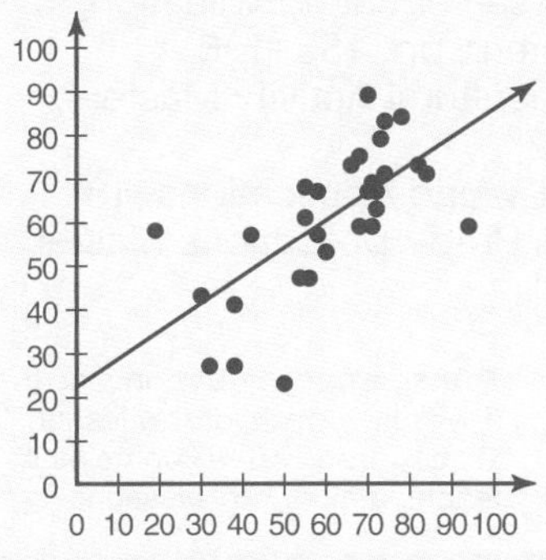

Enrichment Masters, p. 43

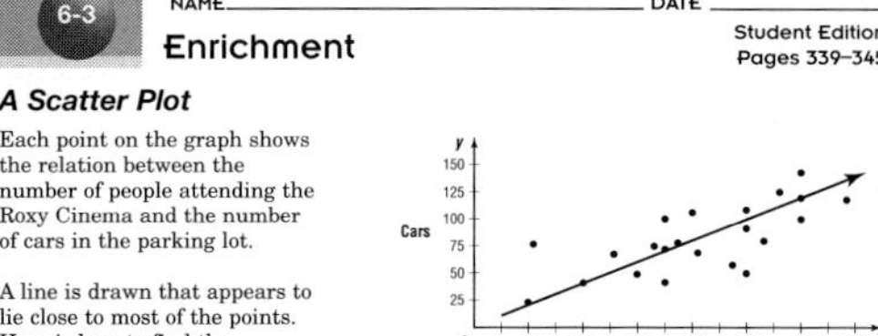
6-3 NAME ______ DATE ______

Enrichment

Student Edition Pages 339–345

A Scatter Plot

Each point on the graph shows the relation between the number of people attending the Roxy Cinema and the number of cars in the parking lot.

A line is drawn that appears to lie close to most of the points. Here is how to find the equation of this line.

The line passes through (100, 40) and (300, 120). Use the slope-intercept form.

$m = \frac{120 - 40}{300 - 100} = \frac{80}{200} = \frac{2}{5}$

$y = mx + b$
$40 = \frac{2}{5}(100) + b$
$0 = b$

An equation for the line is $y = \frac{2}{5}x$.

Solve each problem.

1. Suppose the owner of the Roxy decides to increase the seating capacity of the theater to 1000. How many cars should the parking lot be prepared to accommodate? 400 cars

2. The points (240, 60) and (340, 120) lie on the scatter plot. Write an equation for the line through these points. $y = \frac{3}{5}x - 84$

3. Do you think the equation in Exercise 2 is a good representation of the relationship in this problem? No. The graph of the equation does not appear to go through the center of the data points.

4. Suppose the equation for the relationship between attendance at the theater and cars in the parking lot is $y = 2x + 20$. What might you suspect about the users of the parking lot? Many people are parking in the lot who are not going to the Roxy.

6-4 LESSON NOTES

NCTM Standards: 1–5

Instructional Resources

- Study Guide Master 6-4
- Practice Master 6-4
- Enrichment Master 6-4
- Assessment and Evaluation Masters, pp. 155–156
- Multicultural Activity Masters, p. 12
- Real-World Applications, 18
- Tech Prep Applications Masters, p. 11

Transparency 6-4A contains the 5-Minute Check for this lesson; **Transparency 6-4B** contains a teaching aid for this lesson.

Recommended Pacing	
Standard Pacing	Day 6 of 13
Honors Pacing	Day 5 of 12
Block Scheduling*	Day 4 of 7
Alg. 1 in Two Years*	Days 7 & 8 of 18

*For more information on pacing and possible lesson plans, refer to the *Block Scheduling Booklet* and *Algebra 1 in Two Years.*

1 FOCUS

5-Minute Check
(over Lesson 6-3)

Use the scatter plot shown below to answer the questions.

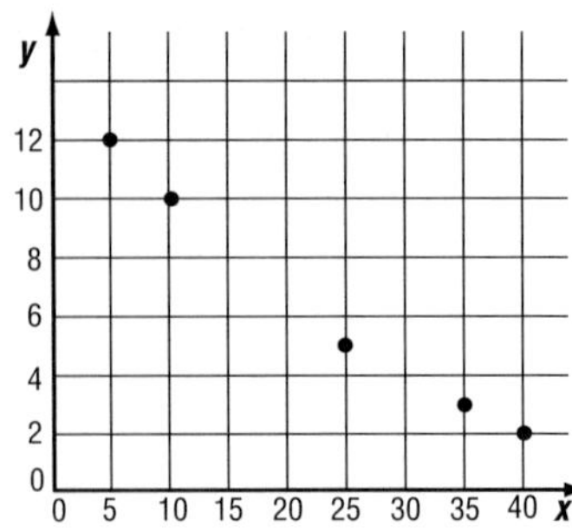

1. What ordered pairs are represented? {(5, 12), (10, 10), (25, 5), (35, 3), (40, 2)}
2. Does the plot indicate a negative, a positive, or no correlation? negative

6–4 Writing Linear Equations in Slope-Intercept Form

What **YOU'LL LEARN**

- To determine the *x*- and *y*-intercepts of linear graphs from their equations,
- to write equations in slope-intercept form, and
- to write and solve direct variation equations.

Why **IT'S IMPORTANT**

You can write linear equations to solve problems involving swimming and banking.

APPLICATION
Banking

Have you seen advertisements like the one at the right in the newspaper? The bank is offering free checking. However, in the fine print of this ad, you will find that the free checking only occurs if you maintain a minimum daily balance of at least \$2000.

Suppose this bank charges a monthly service fee of \$3 plus 10¢ for each check or withdrawal transaction if you fail to maintain a minimum balance of \$2000. The graph at the right shows the line that represents this situation. *You will write an equation for this line in Exercise 57.*

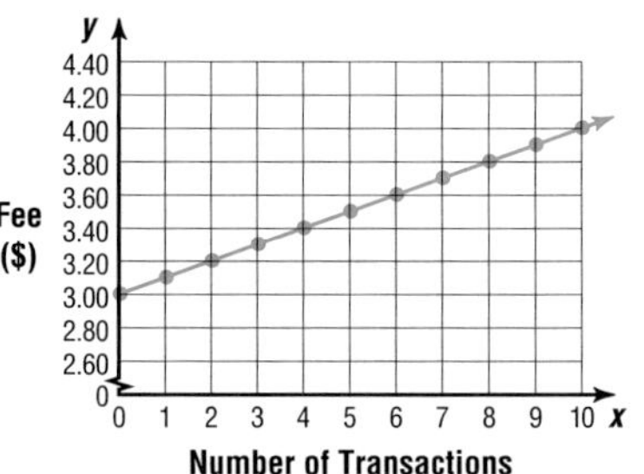

The coordinates at which a graph intersects the axes are known as the **y-intercept** and the **x-intercept**. Since this graph intersects the *y*-axis at (0, 3), the *y*-intercept is 3. If you were to extend the graph to the left, you would find that it crosses the *x*-axis at (−30, 0), so the *x*-intercept is −30.

Notice that the x-coordinate of the ordered pair for the y-intercept and the y-coordinate of the ordered pair for the x-intercept are both 0.

Example 1 **Find the *x*- and *y*-intercepts of the graph of $3x + 4y = 6$.**

Remember that the *x*-intercept is the point at which $y = 0$.

$3x + 4y = 6$
$3x + 4(0) = 6$ *Let y = 0.*
$3x = 6$
$x = \frac{6}{3}$ or 2 *Divide each side by 3.*

Now let $x = 0$ to find the *y*-intercept.

$3x + 4y = 6$
$3(0) + 4y = 6$ *Let x = 0.*
$4y = 6$
$y = \frac{6}{4}$ or $\frac{3}{2}$ *Divide each side by 4.*

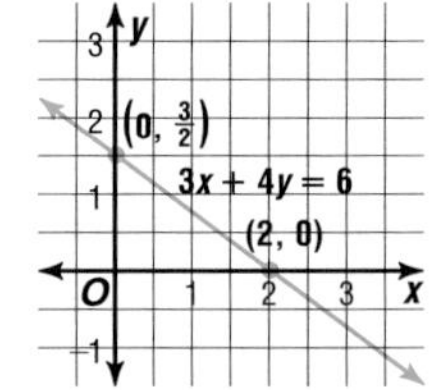

The *x*-intercept is 2, and the *y*-intercept is $\frac{3}{2}$. This means that the graph crosses the *x*-axis at (2, 0) and the *y*-axis at $\left(0, \frac{3}{2}\right)$. You can use these points to graph the equation $3x + 4y = 6$.

Classroom Vignette

"I usually give students real-world examples to help them understand this concept.
Example: You have 4 gallons of gas in the tank of your car. You pump gas at the rate of 5 gallons/minute.
Students make a table showing time (minutes) and gallons pumped for each minute. They then graph these ordered pairs and draw the graph to determine its *y*-intercept and equation."

Eva Gates
Independent Mathematics Consultant
Pearland, Texas

Consider the graph at the right. The line crosses the y-axis at $(0, b)$, so its y-intercept is b. Write the point-slope form of an equation for this line using $(0, b)$ as (x_1, y_1).

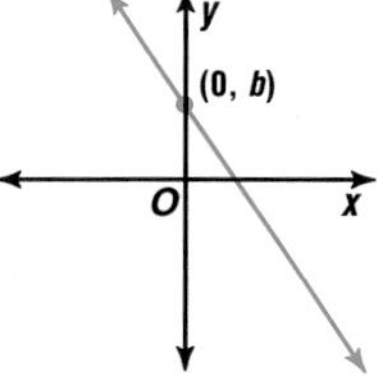

$y - y_1 = m(x - x_1)$ *Point-slope form*

$y - b = m(x - 0)$ *Replace (x_1, y_1) with $(0, b)$.*

$y - b = mx$

$y = mx + b$ *Add b to each side.*

This form of an equation is called the **slope-intercept form** of a linear equation, because in this form the slope and y-intercept are easily identified.

$y = mx + b$

↑ slope ↖ *y-intercept*

Slope-Intercept Form of a Linear Equation	**Given the slope m and the y-intercept b of a line, the slope-intercept form of an equation of the line is $y = mx + b$.**

If you know the slope and y-intercept of a line, you can write an equation of the line in slope-intercept form. This equation can also be written in standard form.

Example 2 **Write an equation of a line in slope-intercept form if the line has a slope of $\frac{2}{3}$ and a y-intercept of 6. Then write the equation in standard form.**

$y = mx + b$

$y = \frac{2}{3}x + 6$ *$b = 6$, $m = \frac{2}{3}$*

Now rewrite the equation in standard form.

$y = \frac{2}{3}x + 6$

$3y = 2x + 18$ *Multiply by 3 to eliminate the fraction.*

$-2x + 3y = 18$ *Subtract 2x from each side.*

or

$2x - 3y = -18$ *Multiply by −1 so that A is positive.*

Let's compare the slope-intercept form with the standard form of a linear equation. First solve $Ax + By = C$ for y.

$Ax + By = C$

$Ax - Ax + By = C - Ax$ *Subtract Ax from each side.*

$By = -Ax + C$ *Commutative property of addition*

$y = -\frac{A}{B}x + \frac{C}{B}$ *Divide each side by B.*

↑ slope ↑ *y-intercept*

Thus, you can identify the slope and y-intercept from an equation written in standard form by using $m = -\frac{A}{B}$ and $b = \frac{C}{B}$.

Cooperative Learning

Group Discussion Separate students into small groups. Have students measure three sets of stairs in the school. Tell them to determine the slope of the stairs. Ask them to make some generalizations about the run (tread) and rise of the steps. Have them discuss how the slope of each set of stairs is determined. For more information on the group discussion strategy, see *Cooperative Learning in the Mathematics Classroom*, one of the titles in the Glencoe Mathematics Professional Series, page 31.

Motivating the Lesson

Questioning Graph $y = 2x - 1$.

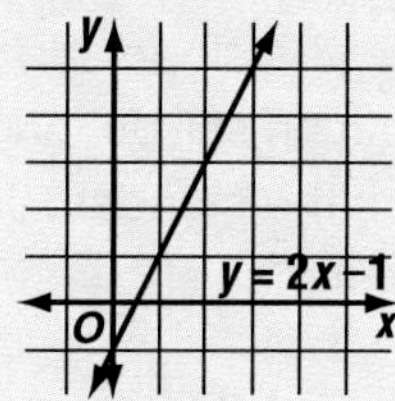

Ask students what the y-intercept and slope of the line are. Then ask them if they can find these numbers in the equation $y = 2x - 1$. You may have to graph several linear equations to convince students of the relationship.

2 TEACH

Teaching Tip You may wish to have advanced students discover the formula for the x-intercept on their own.

In-Class Examples

For Example 1

Find the x- and y-intercepts of the graph of each equation.

a. $x - 2y = 12$ **12, −6**

b. $3x - 5y = 9$ **$3, -\frac{9}{5}$**

c. $2x + 7y = 10$ **$5, \frac{10}{7}$**

d. $y = 7$ **none, 7**

For Example 2

Write an equation in slope-intercept form if the line has a slope of 3 and a y-intercept of 1. **$y = 3x + 1$**

Teaching Tip For Example 2, make students aware that each time they are given an intercept, they know a point on the line. For example, for a y-intercept of 3, the point (0, 3) is on the line. For an x-intercept of −2, the point (−2, 0) is on the line.

Teaching Tip For Example 3, explain that if students forget $-\frac{A}{B}$ and $\frac{C}{B}$, the slope and y-intercept can still be found. Have them solve the standard-form equation for y as shown below.

$$3x + 2y = 6$$
$$2y = -3x + 6$$
$$y = -\frac{3}{2}x + 3$$

slope y - intercept

In-Class Examples

For Example 3
Find the slope and y-intercept of the graph of each equation.

a. $y = 3x - 7$ **3, −7**
b. $y = \frac{2}{3}x$ **$\frac{2}{3}$, 0**
c. $y = 5$ **0, 5**
d. $y = -\frac{3}{4}x + 11$ **$-\frac{3}{4}$, 11**

For Example 4
Write the slope-intercept and standard forms of the equation for a line that passes through $(-2, -1)$ and $(4, 2)$.
$y = \frac{1}{2}x$; $x - 2y = 0$

Teaching Tip After completing Example 4, point out that vertical lines have no y-intercept, while horizontal lines have no x-intercept.

Example 3 **Find the slope and y-intercept of the graph of $5x - 3y = 6$.**

Method 1
In $5x - 3y = 6$, which is in standard form, $A = 5$, $B = -3$, and $C = 6$.
Find the slope.
$$m = -\frac{A}{B}$$
$$= -\frac{5}{-3} \text{ or } \frac{5}{3}$$
Find the y-intercept.
$$b = \frac{C}{B}$$
$$= \frac{6}{-3} \text{ or } -2$$

Method 2
Solve for y to find the slope-intercept form.
$$5x - 3y = 6$$
$$5x - 5x - 3y = 6 - 5x$$
$$-3y = -5x + 6$$
$$\frac{-3y}{-3} = \frac{-5x}{-3} + \frac{6}{-3}$$
$$y = \frac{5}{3}x + (-2)$$
$$m = \frac{5}{3},\ b = -2$$

In Lesson 6–2, you learned how to use the point-slope form to find an equation of a line passing through two given points. Now you have another tool that you can use for the same situation.

Example 4 **Write the slope-intercept and standard forms of the equation for a line that passes through $(-3, -1)$ and $(6, -4)$.**

First find the slope.
$$m = \frac{y_2 - y_1}{x_2 - x_1}$$
$$= \frac{-4 - (-1)}{6 - (-3)} \quad (x_1, y_1) = (-3, -1) \text{ and } (x_2, y_2) = (6, -4)$$
$$= \frac{-3}{9} \text{ or } -\frac{1}{3}$$

Method 1
In $Ax + By = C$, the slope is $-\frac{A}{B}$.
If $-\frac{A}{B} = -\frac{1}{3}$, then $A = 1$ and $B = 3$.
So $Ax + By = C$ becomes $1x + 3y = C$.
To find the value of C, substitute either ordered pair into the equation. For example, choose $(6, -4)$.
$$x + 3y = C$$
$$6 + 3(-4) = C \quad (x, y) = (6, -4)$$
$$-6 = C$$
An equation in standard form for the line is $x + 3y = -6$.

Solve the equation for y to find the slope-intercept form.
$$x + 3y = -6$$
$$3y = -x - 6$$
$$y = -\frac{1}{3}x - \frac{6}{3}$$
$$y = -\frac{1}{3}x - 2$$

Method 2
Use one of the points in the slope-intercept form to find b.
$$y = mx + b$$
$$-4 = -\frac{1}{3}(6) + b \quad (x, y) = (6, -4)$$
$$-4 = -2 + b$$
$$-2 = b$$
An equation in slope-intercept form for the line is $y = -\frac{1}{3}x - 2$.

Now write the equation in standard form.
$$y = -\frac{1}{3}x - 2$$
$$3y = 3\left(-\frac{1}{3}x - 2\right)$$
$$3y = -x - 6$$
$$x + 3y = -6$$

Check: Make sure that both points satisfy the equation.
$$x + 3y = -6$$
$$-3 + 3(-1) \stackrel{?}{=} -6 \quad (x, y) = (-3, -1)$$
$$-6 = -6 \ \checkmark$$
$$x + 3y = -6$$
$$6 + 3(-4) \stackrel{?}{=} -6 \quad (x, y) = (6, -4)$$
$$-6 = -6 \ \checkmark$$

Alternative Learning Styles

Auditory Discuss the following questions. Is it possible to write an equation of a line given its graph? How? **Yes; find the coordinates of two points on the line.**

A special case of the slope-intercept form occurs when the y-intercept is 0. When $m = k$ and $b = 0$, $y = mx + b$ becomes $y = kx$. You may recognize this as the equation for *direct variation*. Another way to write this equation is $\frac{y}{x} = k$.

Example 5

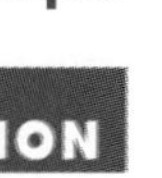

APPLICATION
Swimming

Penny Dean holds the record for the fastest time swimming the English Channel. The 23-year-old Californian swam the 21-mile distance in $7\frac{2}{3}$ hours on July 29, 1978. At the same rate, how long would it take her to swim the Catalina Channel, a 26-mile distance near Los Angeles?

Explore Make a table to relate the information given in the problem.

Channel	Distance	Time
English	21 miles	$7\frac{2}{3}$ hours
Catalina	26 miles	? hours

Plan The formula that relates time t and distance d is $d = rt$, where r is the rate at which she travels. This formula is an example of direct variation. Use the information from the English Channel swim to find her rate. Then apply that rate to the Catalina Channel swim.

Solve For her English Channel swim, $d = 21$ miles, and $t = 7\frac{2}{3}$ or $\frac{23}{3}$ hours. Find the rate r.

$$d = rt$$
$$21 = r\left(\frac{23}{3}\right) \quad d = 21,\ r = \frac{23}{3}$$
$$\frac{3}{23}(21) = \left(\frac{23}{3}\right)\left(\frac{3}{23}\right)r$$
$$\frac{63}{23} = r$$

Now use the same formula and the rate r to find the time for the Catalina Channel swim.

$$d = rt$$
$$26 = \frac{63}{23}t$$
$$\frac{23}{63}(26) = \frac{23}{63}\left(\frac{63}{23}\right)t$$
$$9.5 \approx t \quad \text{Use a calculator.}$$

At the same rate that she swam the English Channel, Ms. Dean could swim the Catalina Channel in 9.5 hours.

Examine You can also use a proportion to solve this problem since the rate r will be the same in each situation. If $d = rt$, then $r = \frac{d}{t}$, a form of the direct variation equation.

$$\frac{d_1}{d_2} = \frac{t_1}{t_2}$$
$$\frac{21}{26} = \frac{\frac{23}{3}}{t}$$
$$21t = \frac{23}{3}(26) \quad \text{Find the cross products.}$$
$$t = \frac{23 \cdot 26}{3 \cdot 21} \quad \text{Divide each side by 21.}$$
$$t \approx 9.5 \quad \text{Use a calculator.}$$

The answer checks.

***fabulous* FIRSTS**

Gertrude Ederle (1907–)

The first woman to swim the English Channel was an American, Gertrude Ederle. She was only 19 years old and made the swim in 14 hours, 39 minutes on August 6, 1926.

CHECK FOR UNDERSTANDING

Communicating Mathematics

Study the lesson. Then complete the following. **1–3. See margin.**

1. **Explain** how to find the intercepts of a line given its equation. Then find the intercepts of the standard form $Ax + By = C$.
2. **Write** a sentence to explain how direct variation and the slope-intercept form of a linear equation are related.
3. **Describe** two ways in which a direct variation problem may be solved.

In-Class Example

For Example 5
At the same rate, how long would it take Penny Dean to swim a distance of 40 miles?
14.6 hours

3 PRACTICE/APPLY

Check for Understanding

Exercises 1–18 are designed to help you assess your students' understanding through reading, writing, speaking, and modeling. You should work through Exercises 1–6 with your students and then monitor their work on Exercises 7–18.

Additional Answers

1. **Let $x = 0$ to find the y-intercept; let $y = 0$ to find the x-intercept; x: $\frac{C}{A}$; y: $\frac{C}{B}$.**
2. **Sample answer: Direct variation is the slope-intercept form when $b = 0$.**
3. **Direct variation may be solved by using the formulas $y = kx$ or $\frac{y}{x} = k$, where $k = m$, which represents slope.**

Reteaching

Checking Data Have students find the x- and y-intercepts of a linear equation and graph them. Then write the equation in slope-intercept form. Using the graph, find another point on the line and substitute the coordinates of that point into the original equation to check for equality. Also, find the slope from the graph and check with the slope-intercept form equation.

***fabulous* FIRSTS**

Gertrude Ederle's success in swimming led to financial success. She earned $2000 a week traveling the country giving swimming exhibitions. In 1953 she received the Helms Foundation Hall of Fame Award, and in 1965 she was inducted into the International Swimming Hall of Fame.

Assignment Guide

Core: 19–53 odd, 54, 55, 57–67
Enriched: 20–52 even, 53–67
All: Self Test, 1–10

For **Extra Practice,** see p. 770.

The red A, B, and C flags, printed only in the Teacher's Wraparound Edition, indicate the level of difficulty of the exercises.

Additional Answer

6. Sample answer:
   (1) Use the y-intercept (3) and slope, $-\frac{2}{5}$.
   $y = mx + b$
   $y = -\frac{2}{5}x + 3$
   (2) Use a point (5, 2) and the slope, $-\frac{2}{5}$.
   $y - y_1 = m(x - x_1)$
   $y - 2 = -\frac{2}{5}(x - 5)$
   (3) Use two points, (1, −3) and (4, 6).
   $m = \frac{6 - (-3)}{4 - 1}$ or 3
   $y - (-3) = 3(x - 1)$
   $y + 3 = 3x - 3$
   $y = 3x - 6$

4. Because its slope is undefined.

11. $y = \frac{2}{3}x - 10$, $2x - 3y = 30$

18a. $y = \frac{1}{4}x + 12$

4. Why can't you write the equation of a vertical line in slope-intercept form?
5. **You Decide** Taka wrote $y = 3.5x - 8$ as an equation of a line having a slope of 3.5 and a y-intercept of -8. Chuma wrote $7x - 2y = 16$ as an equation of the same line. Who is correct, and why? **Both; Chuma's equation is the standard form of Taka's slope-intercept form.**

6. List all the types of information you can use to write an equation of a line. Be sure to include examples of each. **See margin for sample answer.**

Guided Practice

7. For the line shown in the graph at the right:
   a. state the slope, **2**
   b. state the x- and y-intercepts, and **2, −4**
   c. write an equation in slope-intercept form. **$y = 2x - 4$**

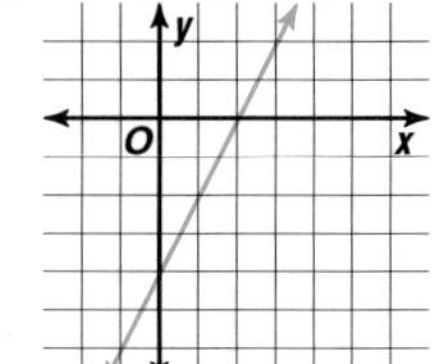

Find the x- and y-intercepts of the graph of each equation.

8. $3x + 4y = 24$ **8, 6**
9. $\frac{3}{4}x - 2y + 7 = 0$ **$-\frac{28}{3}, \frac{7}{2}$**

Write an equation in slope-intercept form of a line with the given slope and y-intercept. Then write the equation in standard form.

10. $m = -4, b = 5$ **$y = -4x + 5$, $4x + y = 5$**
11. $m = \frac{2}{3}, b = -10$

Find the slope and y-intercept of the graph of each equation.

12. $y = -3x + 7$ **−3, 7**
13. $2x + y = -4$ **−2, −4**
14. $x = 3y - 2$ **$\frac{1}{3}, \frac{2}{3}$**

Write an equation in standard form for a line that passes through each pair of points.

15. (8, 1), (−2, 3) **$x + 5y = 13$**
16. (5, 7), (1, 9) **$x + 2y = 19$**
17. Write a direct variation equation that has (3, 11) as a solution. Then find y when $x = 12$. **$y = \frac{11}{3}x$; 44**
18. **Construction** The roof line of a house starts 12 feet above the ground and has a slope of $\frac{1}{4}$.
    a. If the outer wall of the house represents the y-axis and the floor represents the x-axis, find the equation, in slope-intercept form, of the line that represents the roof.
    b. If you are standing inside the house 6 feet from the wall, how high is the roof at that point? **13.5 feet**

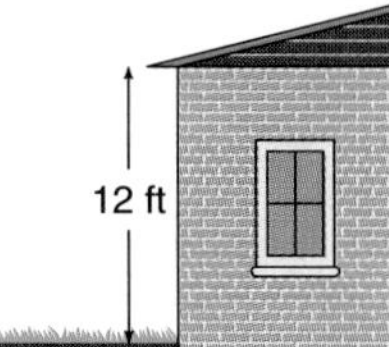

EXERCISES

Practice

For each line graphed at the right:
a. state the slope,
b. state the x- and y-intercepts, and
c. write an equation in slope-intercept form.

19. p **$-\frac{3}{2}$; 0, 0; $y = -\frac{3}{2}x$**
20. q **−3; 3, 9; $y = -3x + 9$**
21. r **$\frac{2}{5}$; 5, −2, $y = \frac{2}{5}x - 2$**

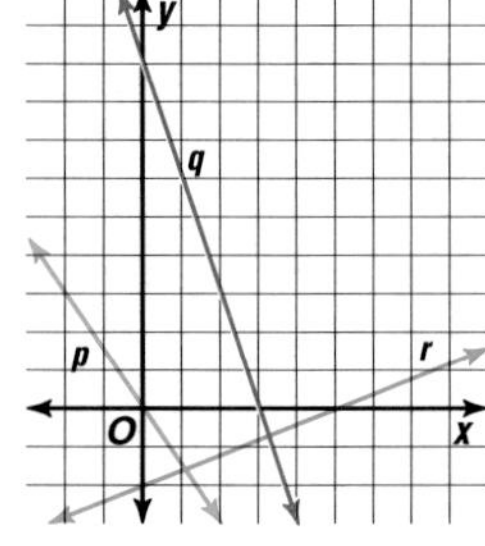

22. $-\frac{12}{5}, 4$
25. $-\frac{5}{6}, 5$
26. $-3, \frac{3}{4}$
28. $y = 3x + 5$, $3x - y = -5$
29. $y = 7x - 2$, $7x - y = 2$
30. $y = -6x$, $6x + y = 0$
31. $y = -1.5x + 3.75$, $6x + 4y = 15$
32. $y = \frac{1}{4}x - 10$, $x - 4y = 40$
33. $y = -7$, $y = -7$
34. $\frac{2}{3}, -3$ 35. $-\frac{5}{4}, \frac{5}{2}$
44. $5x + 12y = 83$
46. $y = -\frac{8}{5}x, -\frac{55}{8}$
47. $y = \frac{11}{24}x, \frac{33}{2}$
48. $y = -\frac{3}{17}x, -85$
49. $y = \frac{2}{3}x - \frac{8}{3}$
50. $\left(-\frac{5}{2}, -\frac{15}{2}\right)$

Find the *x*- and *y*-intercepts of the graph of each equation.

22. $5x - 3y = -12$
23. $4x + 7y = 8$ $2, \frac{8}{7}$
24. $5y - 2 = 2x$ $-1, \frac{2}{5}$
25. $y - 6x = 5$
26. $4y - x = 3$
27. $3y = 18$ none, 6

Write an equation in slope-intercept form of a line with the given slope and *y*-intercept. Then write the equation in standard form.

B

28. $m = 3, b = 5$
29. $m = 7, b = -2$
30. $m = -6, b = 0$
31. $m = -1.5, b = 3.75$
32. $m = \frac{1}{4}, b = -10$
33. $m = 0, b = -7$

Find the slope and *y*-intercept of the graph of each equation.

34. $3y = 2x - 9$
35. $5x + 4y = 10$
36. $4x - \frac{1}{3}y = -2$ 12, 6
37. $\frac{2}{3}x + \frac{1}{6}y = 2$ $-4, 12$
38. $5(x - 3y) = 2(x + 3)$ $\frac{1}{5}, -\frac{2}{5}$
39. $4(3x + 9) - 3(5y + 7) = 11$ $\frac{4}{5}, \frac{4}{15}$

Write an equation in standard form for a line that passes through each pair of points. 40. $10x - 7y = 5$ 41. $y = -2$ 42. $3x + 10y = -8$

40. $(-3, -5), (4, 5)$
41. $(7, -2), (-4, -2)$
42. $(-6, 1), (4, -2)$
43. $(3, 5), (3, -6)$ $x = 3$
44. $(7, 4), (-5, 9)$
45. $(2, 9), (-5, 9)$ $y = 9$

Write a direct variation equation that has each ordered pair as a solution. Then find the missing value.

46. $(-5, 8), (?, 11)$
47. $(24, 11), (36, ?)$
48. $(-17, 3), (?, 15)$

49. Write an equation in slope-intercept form of a line with slope $\frac{2}{3}$ and an x-intercept of 4.
50. Find the coordinates of a point on the graph of $5x - 3y = 10$ in which the x-coordinate is 5 greater than the y-coordinate.
51. Find the coordinates of a point on the graph of $3x + 7y = -16$ in which the x-coordinate is 3 times the y-coordinate. $(-3, -1)$
52. Write the equation in standard form of the line with an x-intercept of 7 and a y-intercept of -2. $2x - 7y = 14$

Programming

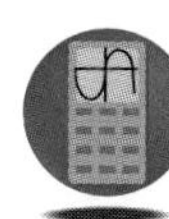

53. The program below finds the slope of a line, given the coordinates of two of the points through which it passes. Then it displays the graph of the line and gives the equation of the line in slope-intercept form.

```
PROGRAM:SLOPE
: ClrDraw
: Input "X1=",Q
: Input "Y1=",R
: Input "X2=",S
: Input "Y2=",T
: If Q-S≠0
: Then
: (R-T)/(Q-S)→A
: R-AQ→B
: DrawF AX+B
: Text(2,5, "EQUATION IS"
: Text(8,5, "Y=",A,"X+",B)
: Else
: ClrHome
: Text(2,5,"SLOPE UNDEFINED")
: Text(8,5,"EQUATION OF LINE")
: Text(15,5,"X=",Q)
: Vertical Q
```

a. Use the program to write an equation of a line that passes through (8.57, −3.82) and (11.09, 1.31). Round numbers to the nearest hundredth. $y = 2.04x - 21.32$
b. Use the program to check your equations in Exercises 40–45.

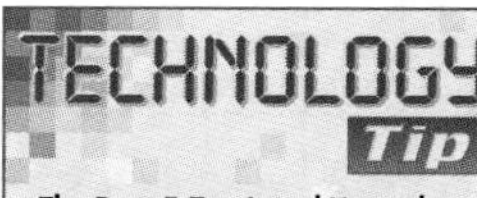

The DrawF, Text(, and Vertical commands are located in the DRAW menu.

Using the Programming Exercises The program given in Exercise 53 is for use with a TI-82 graphing calculator. For other programmable calculators, have students consult their owner's manual for commands similar to those presented here.

Technology Tip
Remind students to use the Clr Home command to clear the screen and return to home position if earlier work is still on the screen.

Study Guide Masters, p. 44

NAME__________ DATE__________

Study Guide

Student Edition Pages 346–353

Writing Linear Equations in Slope-Intercept Form

The x-coordinate of the point where a line crosses the x-axis is called the **x-intercept.** Similarly, the y-coordinate of the point where the line crosses the y-axis is called the **y-intercept.**

Slope-Intercept Form of a Linear Equation
Given the slope m and the y-intercept b of a line, the slope-intercept form of an equation of the line is $y = mx + b$.

If an equation is given in standard form $Ax + By = C$ and B is not zero, the slope of the line is $-\frac{A}{B}$ and the y-intercept is $\frac{C}{B}$. The x-intercept is $\frac{C}{A}$ where $A \neq 0$.

Example: Find the x- and y-intercepts of the graph of $5x - 2y = 10$. Then write the equation in slope-intercept form.

Since $A = 5$, $B = -2$, and $C = 10$,

$\frac{C}{A} = \frac{10}{5} = 2$ $\frac{C}{B} = \frac{10}{-2} = -5$ $m = -\frac{A}{B} = \frac{5}{2}$

Thus, the x-intercept is 2, and the y-intercept is -5. The equation of the line in slope-intercept form is $y = \frac{5}{2}x - 5$.

Find the x- and y-intercepts of the graph of each equation.

1. $5x + 4y = 20$ 4, 5
2. $2x - 5y = -7$ $-\frac{7}{2}, \frac{7}{5}$
3. $4x - 8y = 10$ $\frac{5}{2}, -\frac{5}{4}$
4. $9x + y = -1$ $-\frac{1}{9}, -1$

Write an equation in slope-intercept form of a line with the given slope and y-intercept. Then write the equation in standard form.

5. $m = 6, b = 10$ $y = 6x + 10$; $6x - y = -10$
6. $m = 4, b = 0$ $y = 4x$; $4x - y = 0$
7. $m = -1, b = 3$ $y = -x + 3$; $x + y = 3$
8. $m = 2, b = -3$ $y = 2x - 3$; $2x - y = 3$

Find the slope and y-intercept of the graph of each equation. Then write each equation in slope-intercept form.

9. $0.2x + 0.5y = 1.6$ $-\frac{2}{5}; \frac{16}{5}; y = -\frac{2}{5}x + \frac{16}{5}$
10. $3x + 7y = 10$ $-\frac{3}{7}; \frac{10}{7}; y = -\frac{3}{7}x + \frac{10}{7}$
11. $6x - y = 9$ $6; -9; y = 6x - 9$
12. $14x - 21y = 7$ $\frac{2}{3}; -\frac{1}{3}; y = \frac{2}{3}x - \frac{1}{3}$

4 ASSESS

Closing Activity

Speaking Ask students to explain the procedure they would use to determine the slope and y-intercept of the equation $3x + 4y = 12$. Then have them use the procedure described to find each. $-\frac{3}{4}$, 3

Chapter 6, Quiz B (Lessons 6-3 and 6-4), is available in the *Assessment and Evaluation Masters,* p. 156.

Mid-Chapter Test (Lessons 6-1 through 6-4) is available in the *Assessment and Evaluation Masters,* p. 155.

Additional Answers

56c–56d.

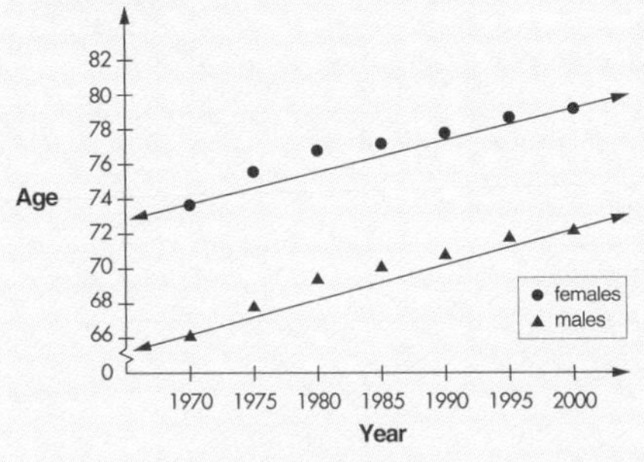

Practice Masters, p. 44

6-4

NAME____________ DATE ________

Practice

Student Edition Pages 346–353

Writing Linear Equations in Slope-Intercept Form

For each line graphed at the right:
a. state the slope,
b. state the x- and y-intercepts, and
c. write an equation in slope-intercept form.

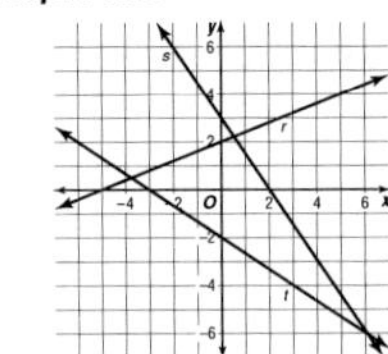

1. r $\frac{2}{5}$; -5, 2; $y = \frac{2}{5}x + 2$

2. s $-\frac{3}{2}$; 2, 3; $y = -\frac{3}{2}x + 3$

3. t $-\frac{2}{3}$; -3, -2; $y = -\frac{2}{3}x - 2$

Find the x- and y-intercepts of the graph of each equation.

4. $4x - 3y = 12$ 3, -4
5. $2x - 5y = 20$ 10, -4
6. $x + 5y = 8$ 8, $\frac{8}{5}$
7. $3x - y = -7$ $-\frac{7}{3}$, 7

Find the slope and y-intercept of the graph of each equation.

8. $5x + 2y = 7$ $-\frac{5}{2}$, $\frac{7}{2}$
9. $-2x + 3y = 8$ $\frac{2}{3}$, $\frac{8}{3}$
10. $8x - y = 12$ 8, -12
11. $7x - 2y = 9$ $\frac{7}{2}$, $-\frac{9}{2}$

Write an equation in standard form for a line that passes through each pair of points.

12. $(3, 4), (-1, -4)$ $2x - y = 2$
13. $(2, 5), (0, 3)$ $x - y = -3$
14. $(-3, 1), (-1, -3)$ $2x + y = -5$
15. $(-5, -2), (1, -6)$ $2x + 3y = -16$

Critical Thinking

54. The x-intercept of a line is p, and the y-intercept is q. Write an equation of the line. $y = -\frac{qx}{p} + q$

Applications and Problem Solving

55. **Chemistry** Charles' Law states that when the pressure is constant, the volume of a gas is directly proportional to the temperature on the Kelvin scale. Write an equation for each situation and solve. **55a. 42.24 ft^3**
   a. If the volume of a gas is 35 ft^3 at 290 K, what is the volume at 350 K?
   b. If the volume of a gas is 200 ft^3 at 300 K, at what temperature is the volume 180 ft^3? **270 K**

56. **Life Expectancy** The life expectancies at birth of the average male and female born between 1970 and 2000 (estimated) are shown in the table at the right.

Life Expectancy		
Year Born	Males (years)	Females (years)
1970	67.1	74.7
1975	68.8	76.6
1980	70.4	77.8
1985	71.1	78.2
1990	71.8	78.8
1995	72.8	79.7
2000 (est.)	73.2	80.2

   a. Write an equation for the line that passes through (1970, 67.1) and (2000, 73.2).
   56a. Sample answer: $y = 0.20x - 326.9$
   b. Write an equation for the line that passes through (1970, 74.7) and (2000, 80.2).
   56b. Sample answer: $y = 0.18x - 279.9$
   c. Graph the ordered pairs for male life expectancy and the equation from part a. **c–e. See margin.**
   d. On the same graph, use a different color pencil to graph the ordered pairs for female life expectancy and the equation from part b.
   e. Write a paragraph to describe the relationship between the points and lines you graphed.
   f. Use the data to predict the life expectancy for males and females in the year 2100. **based on sample equations: males, 93.2; females, 98.2**

57. **Banking** Refer to the application at the beginning of the lesson.
   a. Write an equation to represent the line that shows the service fees charged by the bank for an account that maintains a daily balance that is less than \$2000. **$y = 0.1x + 3$**
   b. Suppose you are not able to maintain \$2000 in an account and you write about 25 checks each month. Use the equation from part a to determine whether you should open an account at this bank or go to the bank across the street that charges a flat \$5 each month. **Go to the other bank, since this one would charge you \$5.50.**

Mixed Review

58. **Statistics** The table below shows the keyboarding speeds of 12 students in words per minute (wpm) and their weeks of experience. (Lesson 6–3)

Experience (weeks)	4	7	8	1	6	3	5	2	9	6	7	10
Keyboarding Speed (wpm)	33	45	46	20	40	30	38	22	52	44	42	55

   a. Make a scatter plot of these data. **a–b. See Solutions Manual.**
   b. Draw a best-fit line for the data. Find the equation of the line.
   c. Use the equation to predict the keyboarding speed of a student after a 12-week course. **about 65 wpm**
   d. Why can't this equation be used to predict the speed for any number of weeks of experience? **There's a limit as to how fast one can type.**

Extension

Problem Solving Write an equation in slope-intercept form of the line with y-intercept 6 that passes through the point (3, 4). $y = -\frac{2}{3}x + 6$

Additional Answer

56e. The life expectancy of females is higher than the life expectancy of males and, as time passes, the life expectancy of both increases. According to the graph, at approximately the year 2350, men and women both will have a life expectancy of about 136 years.

59. Determine the slope of the line that passes through (14, 3) and (−11, 3). (Lesson 6–1) **0**

60. Write an equation for the function shown in the chart below. (Lesson 5–6)

m	1	2	4	5	6	9
n	9	6	0	−3	−6	−15

$n = 12 - 3m$

62. $\frac{1}{2}$

63. 13 ft 11 in.

61. Graph $A(5, -2)$ on a coordinate plane. (Lesson 5–1) **See margin.**

62. **Probability** A card is selected at random from a deck of 52 cards. What is the probability of selecting a black card? (Lesson 4–6)

63. **Construction** When building a roof, a 5-foot support is to be placed at point B as shown on the diagram at the right. Find the length of the support that is to be placed at point A. (Lesson 4–2)

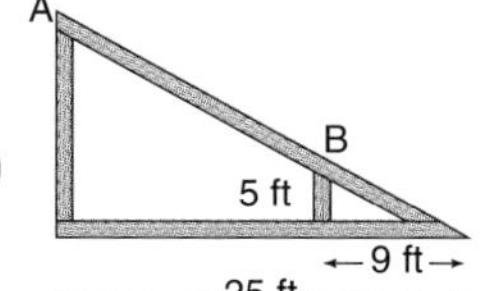

64. Solve $8x + 2y = 6$ for y. (Lesson 3–6) **$y = 3 - 4x$**

65. Solve $-36 = 4z$. (Lesson 3–2) **−9**

66. Evaluate $|a + k|$ if $a = -5$ and $k = 3$. (Lesson 2–3) **2**

67. Simplify $4(3x + 2) + 2(x + 3)$. (Lesson 1–7) **$14x + 14$**

SELF TEST

Determine the slope of the line that passes through each pair of points. (Lesson 6–1)

1. (−7, 10) and (−2, 5) **−1**
2. (−6, 3) and (−12, 3) **0**
3. (−5, 7) and (−5, −15) **3. undefined**
4. Determine the value of r so the line through $(r, 3)$ and $(6, -2)$ has a slope of $-\frac{5}{2}$. (Lesson 6–1) **4**

Write the point-slope form of an equation of the line that passes through the given point and has the given slope. (Lesson 6–2)

5. $(-6, 4), m = \frac{1}{2}$ **$y - 4 = \frac{1}{2}(x + 6)$**
6. $(-12, 12), m = 0$ **$y - 12 = 0$**
7. Write the standard form of an equation of the line that passes through (−3, 4) and (2, 3). (Lesson 6–2) **$x + 5y = 17$**
8. **Computers** A newspaper article stated that with advancements in technology, computers would become cheaper and cheaper. Is this true? The table at the right shows the average cost of a computer system for each year from 1991 to 1994. (Lesson 6–3) **a–b. See margin.**
   a. Make a scatter plot of these data. **d–e. See margin.**
   b. Draw a best-fit line and write an equation that describes it.
   c. Does the scatter plot show a positive, negative, or no correlation between the variables? **positive**
   d. Write what this correlation means in words.
   e. With each year, computer systems become faster at handling more and more data. How can you use this fact to substantiate the newspaper article claim?

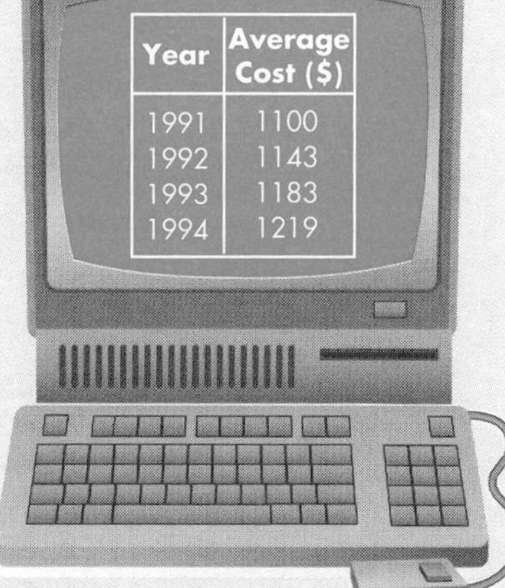

Year	Average Cost ($)
1991	1100
1992	1143
1993	1183
1994	1219

Source: *Vitality*, 1995

9. Find the x- and y-intercepts of the graph of $7x + 3y = -42$. (Lesson 6–4) **−6, −14**
10. Write the slope-intercept form of an equation for the line that passes through (0, −3) and (6, 0). (Lesson 6–4) **$y = \frac{1}{2}x - 3$**

Additional Answer

61.

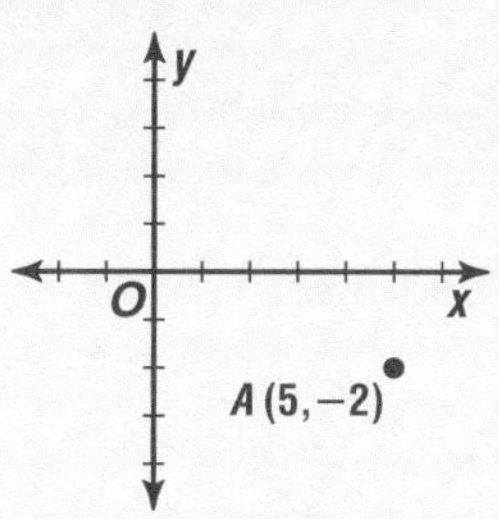

Answers for the Self Test

8a, b.

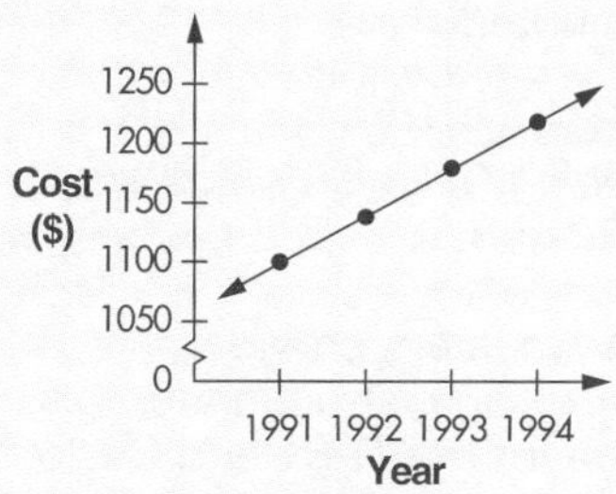

8b. Sample equation: $y = 40x - 78{,}537$

8d. Each year the cost of a computer increases.

8e. The number of features on a computer increases more quickly than the cost. Thus, for specific features on a computer, it costs less to buy those features now than it did a year ago.

Tech Prep

Roofing Contractor Students who are interested in construction may wish to do further research on the data presented in Exercise 63 and explore the potential growth of this career. For more information on tech prep, see the *Teacher's Handbook*.

SELF TEST

The Self Test provides students with a brief review of the concepts and skills in Lessons 6-1 through 6-4. Lesson numbers are given to the right of exercises or instruction lines so students can review concepts not yet mastered.

Enrichment Masters, p. 44

NAME______________________ DATE __________

Enrichment — Student Edition Pages 346–353

Analyzing Data

Fill in each table below. Then write inversely, or directly to complete each conclusion.

1.

l	2	4	8	16	32
w	4	4	4	4	4
A	8	16	32	64	128

For a set of rectangles with a width of 4, the area varies **directly** as the length.

2.

Hours	2	4	5	6
Speed	55	55	55	55
Distance	165	220	275	330

For a car traveling at 55 mi/h, the distance covered varies **directly** as the hours driven.

3.

Oat bran	$\frac{1}{3}$ cup	$\frac{2}{3}$ cup	1 cup
Water	1 cup	2 cup	3 cup
Servings	1	2	3

The number of servings of oat bran varies **directly** as the number of cups of oat bran.

4.

Hours of Work	128	128	128
People Working	2	4	8
Hours per Person	64	32	16

A job requires 128 hours of work. The number of hours each person works varies **inversely** as the number of people working.

5.

Miles	100	100	100	100
Rate	20	25	50	100
Hours	5	4	2	1

For a 100-mile car trip, the time the trip takes varies **inversely** as the average rate of speed the car travels.

6.

b	3	4	5	6
h	10	10	10	10
A	15	20	25	30

For a set of right triangles with a height of 10, the area varies **directly** as the base.

Use the table at the right.

7. x varies **directly** as y.
8. z varies **inversely** as y.
9. x varies **inversely** as z.

x	1	1.5	2	2.5	3
y	2	3	4	5	6
z	60	40	30	24	20

6-5A LESSON NOTES

NCTM Standards: 1–5

Objective
Use a graphing calculator to determine if a group of graphs forms a family.

Recommended Time
25 minutes

Instructional Resources
Graphing Calculator Masters, pp. 22 and 23

These masters provide keystroking instruction for this lesson for the TI-81 and Casio graphing calculators.

1 FOCUS

Motivating the Lesson
Have students graph, with paper and pencil, the equations $y = \frac{x}{2}$, $y = x$, $y = 2x$, and $y = 5x$. This lesson shows students one of the benefits of the graphing calculator.

2 TEACH

Teaching Tip Emphasize that *same slope* means "parallel" and *same intercept* means "spokes of a wheel."

3 PRACTICE/APPLY

Assignment Guide
Core: 1–10 **Enriched:** 1–10

4 ASSESS

Observing students working with technology is an excellent method of assessment.

6–5A Graphing Technology Parent and Family Graphs

A Preview of Lesson 6–5

What is a family? Generally, a family is a group of people that are related either by birth, marriage, or adoption. Graphs can also form families. A **family of graphs** includes graphs and equations of graphs that have at least one characteristic in common. That characteristic differentiates the group of graphs from other groups.

Families of linear graphs often fall into two categories—those with the same slope or those with the same intercept. The **parent graph** is the simplest of the graphs in a family. For many linear functions, the parent graph is of the form $y = mx$, where m is any number.

A graphing calculator is a useful tool in studying a group of graphs to determine if they form a family.

Example 1 **Graph $y = x$, $y = 2x$, and $y = 4x$ in the standard viewing window. Describe any similarities and differences among the graphs. Write a description of the family.**

Clear the Y= list of all other equations.

Enter: [Y=] [X,T,θ] [ENTER] 2 [X,T,θ] [ENTER] 4 [X,T,θ] [ENTER] [ZOOM] 6

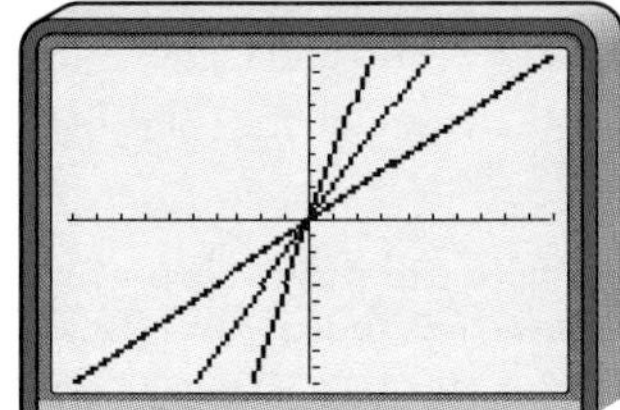

These three graphs form a family in which the slope of each graph is positive and each graph goes through the origin. However, each graph has a different slope. The parent graph is the graph of $y = x$. The graph of $y = 4x$ is the steepest, and the graph of $y = x$ is the least steep.

You might describe this family of graphs as lines that pass through the origin. Another description might be lines that have 0 as their y-intercept.

You can use braces to enter equations that have a common characteristic. Since all of the equations in Example 1 are of the form $y = mx$, you can use braces to enter the different values of m using one step as follows.

Enter: [Y=] [2nd] [{] 1 [,] 2 [,] 4 [2nd] [}] [X,T,θ] [GRAPH]

This tells the calculator to graphs equations for which the coefficients of x are 1, 2, and 4 and the y-intercept is 0.

Example 2 **Graph $y = x$, $y = x + 3$, and $y = x - 3$ in the standard viewing window. Describe any similarities and differences among the graphs. Write a description of the family.**

You can enter each of the graphs into the Y= list as you did in Example 1 or use braces to enter the equations in one step. Notice that all three equations can be written in the form $y = x + b$. The braces can be used to enter the different values of b.

Enter: [Y=] [X,T,θ] [+] [2nd] [{] 0 [,] 3 [,] [(–)] 3 [2nd] [}] 6 [ZOOM]

The parent graph in this example is the graph of $y = x$.

The slope of each graph is positive. However, each graph has a different y-intercept.

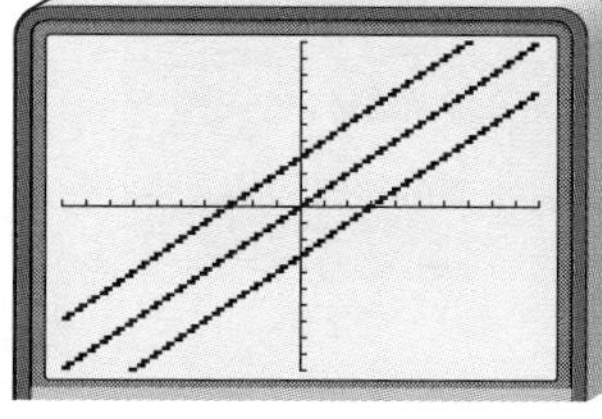

The graph of $y = x$ has a y-intercept of 0.
The graph of $y = x + 3$ has a y-intercept of 3.
The graph of $y = x - 3$ has a y-intercept of -3.

Since the slope of each line is 1, we can describe this family as linear graphs whose slope is 1.

A function that is closely related to linear functions is the **absolute value function.**

Example 3 **Graph $y = |x|$, $y = |x + 2|$, and $y = |x| + 4$ on the same screen. Describe any similarities and differences among the graphs.**

Enter: Y= 2nd ABS X,T,θ ENTER 2nd ABS (X,T,θ + 2) ENTER 2nd ABS X,T,θ + 4 ZOOM 6

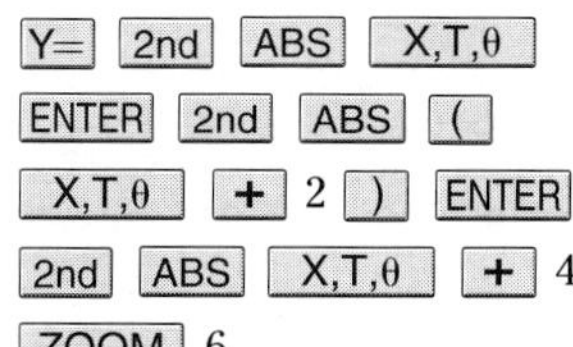

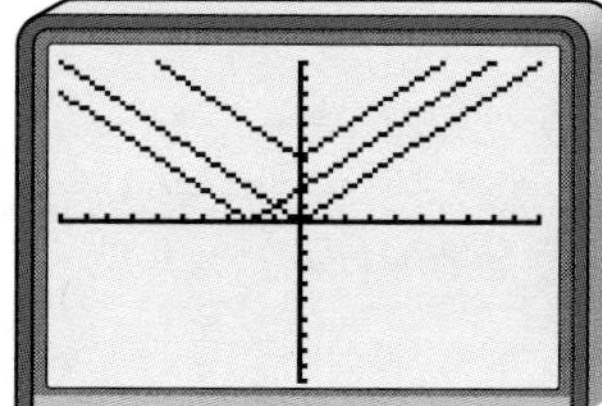

The parent graph in this example is the graph of $y = |x|$.

Each graph is shaped like the letter v. The graph of $y = |x + 2|$ is shaped like the graph of $y = |x|$, but is shifted 2 units to the left. The graph of $y = |x| + 4$ is shaped like the graph of $y = |x|$, but is shifted 4 units up.

EXERCISES

Graph each set of equations on the same screen. Describe any similarities or differences among the graphs. State what the graphs have in common. 1–4. See margin.

1. $y = -x$
   $y = -2x$
   $y = -4x$
2. $y = |x|$
   $y = 2|x|$
   $y = 0.5|x|$
3. $y = -x$
   $y = -x + 2$
   $y = -x - 3$
4. Write a sentence comparing the graphs of equations with a positive coefficient of x and graphs with a negative coefficient of x.

Sketch a graph on paper to represent how you think the graph of each equation will appear. Describe any similarities to the graphs in Examples 1 and 2 and Exercises 1–3. Then use a graphing calculator to confirm your prediction and describe how accurate your sketch was.

5–8. See Solutions Manual.

5. $y = x - 5$
6. $y = -|x| + 6$
7. $y = 0.1x$
8. $y = \frac{1}{3}x + 4$
9. Write an equation of a line whose graph lies between the graphs of $y = -x + 2$ and $y = -x + 3$. Sample answer: $y = -x + 2.5$
10. Write a paragraph explaining how the values of m and b in the slope-intercept form affect the graph of the equation. Include several graphs with your paragraph. See margin.

Additional Answers

1. All graphs are of the family $y = -ax + 0$, where a represents different negative slopes.

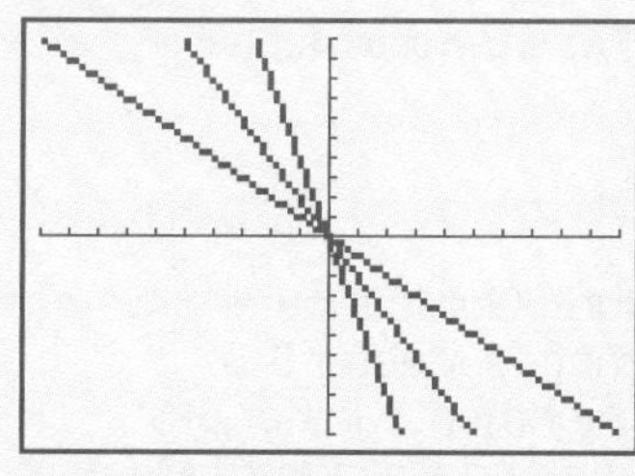

2. All graphs are of the family $y = ax + 0$, where a represents different positive slopes.

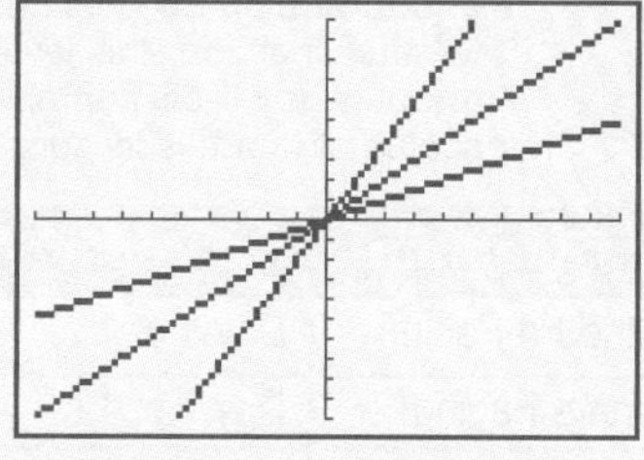

3. All graphs have the same slope, -1, but have different y-intercepts.

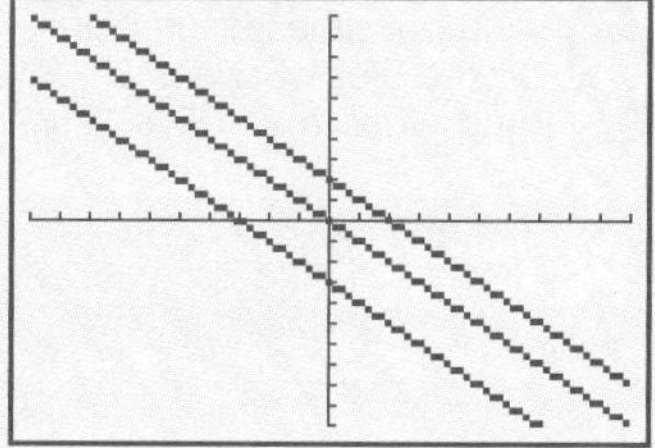

4. Graphs with positive x-coefficients slant upward from left to right. Graphs with negative x-coefficients slant downward from left to right.

10. Sample answer: The value of m tells how the line will slant—upward or downward—and how steep it will be. The value of b tells where the line will intersect the y-axis. See students' work for graphs.

Using Technology

This lesson offers an excellent opportunity for using technology in your algebra classroom. For more information on using technology, see *Graphing Calculators in the Mathematics Classroom,* one of the titles in the Glencoe Mathematics Professional Series.

6-5 LESSON NOTES

NCTM Standards: 1–5

Instructional Resources

- Study Guide Master 6-5
- Practice Master 6-5
- Enrichment Master 6-5
- Graphing Calculator Masters, p. 6
- Tech Prep Applications Masters, p. 12

Transparency 6-5A contains the 5-Minute Check for this lesson; **Transparency 6-5B** contains a teaching aid for this lesson.

Recommended Pacing	
Standard Pacing	Day 8 of 13
Honors Pacing	Day 7 of 12
Block Scheduling*	Day 5 of 7
Alg. 1 in Two Years*	Days 10 & 11 of 18

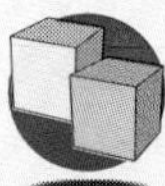

*For more information on pacing and possible lesson plans, refer to the *Block Scheduling Booklet* and *Algebra 1 in Two Years.*

1 FOCUS

5-Minute Check
(over Lesson 6-4)

Find the slope, x-intercept, and y-intercept for the graph of each equation.

1. $4x - y = 16$ **4, 4, −16**

2. $2x + 3y = 8$ $-\frac{2}{3}, 4, \frac{8}{3}$

3. $y = 2x + 7$ $2, -\frac{7}{2}, 7$

4. $3x = y - 1$ $3, -\frac{1}{3}, 1$

5. $y = \frac{3}{4}x - 7$ $\frac{3}{4}, \frac{28}{3}, -7$

Motivating the Lesson

Questioning Introduce this lesson with a review of graphing a linear equation using two ordered pairs that satisfy the equation. Ask students to describe the procedure they would use to graph $x + 3y = 3$. Then have them graph the equation.

6–5 Graphing Linear Equations

What YOU'LL LEARN

- To graph a line given any linear equation.

Why IT'S IMPORTANT

You can graph linear equations to show trends in fields like health and physical science.

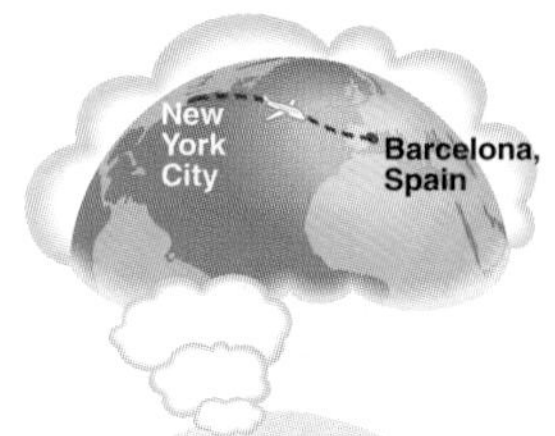

CONNECTION
Earth Science

Solome was flying from New York to Spain to be an exchange student for the spring quarter. After climbing to cruising level, the pilot announced that they were at an altitude of 13,700 meters and the outside temperature was about −76°C. You may think that since you get closer to the sun as you get farther from Earth, you might get warmer as altitude increases. However, as altitude increases, the air gets thinner and colder.

The air temperature above Earth on a day when the temperature at ground level is 15°C can be calculated using the formula $a + 150t = 2250$, where a is the altitude in meters and t is the temperature in degrees Celsius.

Suppose you wanted to graph $a + 150t = 2250$. In Chapter 5, you learned that you could graph a linear equation by finding and then graphing several ordered pairs that satisfy the equation.

Using the x- and y-intercepts is a convenient way to find ordered pairs to graph a linear equation. In $a + 150t = 2250$, these can be called the a-intercept and the t-intercept.

Find the a-intercept.

$$a + 150t = 2250$$
$$a + 150(0) = 2250 \quad \textit{Let } t = 0.$$
$$a = 2250$$

Find the t-intercept.

$$a + 150t = 2250$$
$$0 + 150t = 2250 \quad \textit{Let } a = 0.$$
$$t = 15$$

Now graph (2250, 0) and (0, 15), the ordered pairs for the intercepts. Draw the line representing the equation by connecting the points. Thus, you can graph a line if you know two points on that line.

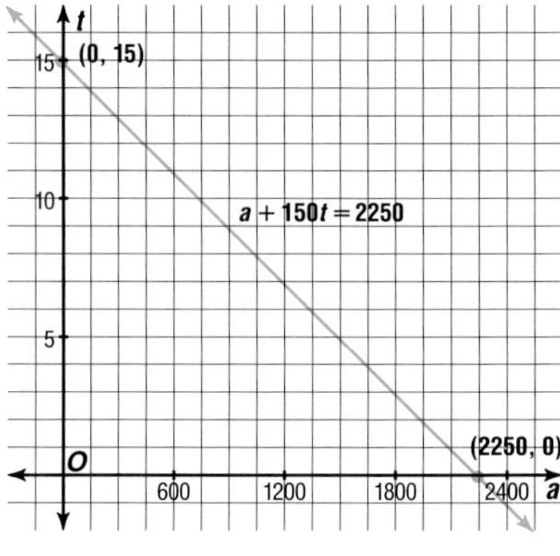

Use the equation to confirm the pilot's announcement that the temperature was −76°C at 13,700 meters altitude.

$$a + 150t = 2250$$
$$13{,}700 + 150t = 2250$$
$$150t = -11{,}450$$
$$t \approx -76.33$$

So, the pilot was correct in saying that the outside temperature was about −76°C.

You can also graph a line if you know its slope and a point on the line. The forms of linear equations you have learned in this chapter can help you find the slope and a point on the line.

GLENCOE Technology

CD-ROM Interaction

A multimedia simulation connects slopes and equations with family vacations. A blackline master activity with teacher's notes provides a follow-up to the CD-ROM simulation.

For Windows & Macintosh

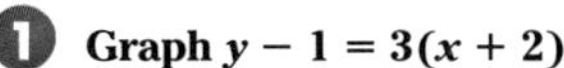

Example 1 **Graph $y - 1 = 3(x + 2)$.**

This equation is in point-slope form. The slope is 3, and a point on the line is $(-2, 1)$.

Graph the point $(-2, 1)$. Remember that the slope represents the change in y and x.

$$\frac{3}{1} = \frac{\text{change in } y}{\text{change in } x}$$

Thus, from the point $(-2, 1)$, you can move up 3 and right 1. Draw a dot. You can repeat this process to find another point on the graph. Draw a line connecting the points. This line is the graph of $y - 1 = 3(x + 2)$.

Check: Check $(-1, 4)$ to make sure that it satisfies the equation.

$$y - 1 = 3(x + 2)$$
$$4 - 1 \stackrel{?}{=} 3(-1 + 2)$$
$$3 = 3 \checkmark$$

You can also use the slope-intercept form to help you graph an equation.

Example 2 **Graph $\frac{3}{4}x + \frac{1}{2}y = 4$.**

Solve the equation for y to find the slope-intercept form.

$\frac{3}{4}x + \frac{1}{2}y = 4$

$\frac{1}{2}y = -\frac{3}{4}x + 4$ *Subtract $\frac{3}{4}x$ from each side.*

$y = -\frac{3}{2}x + 8$ *Multiply each side by 2.*

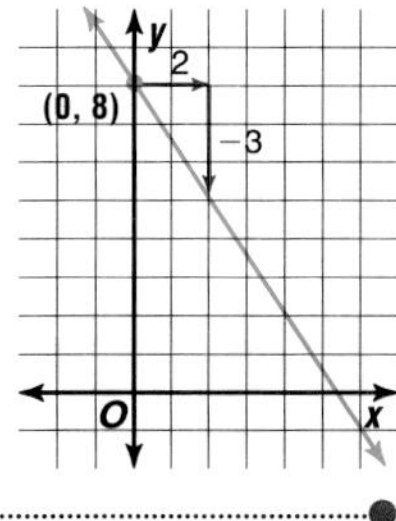

The slope-intercept form of this equation tells us that the slope is $-\frac{3}{2}$ and the y-intercept is 8. Graph the y-intercept, and then use the slope to find another point on the line. Draw the line.

Be sure to check your second point to make sure it satisfies the *original* equation.

You frequently need to rewrite equations before you can determine the best way to graph them.

Example 3 **Graph $\frac{4}{5}(2x - y) = 6x + \frac{2}{5}y - 10$.**

First, simplify the equation.

$\frac{4}{5}(2x - y) = 6x + \frac{2}{5}y - 10$

$4(2x - y) = 30x + 2y - 50$ *Multiply each side by 5 to eliminate the fractions.*

$8x - 4y = 30x + 2y - 50$

$0 = 22x + 6y - 50$ *Add $-8x$ and $4y$ to each side.*

$50 = 22x + 6y$ *Add 50 to each side.*

$25 = 11x + 3y$ *Divide each side by 2.*

$11x + 3y = 25$ *Standard form*

(continued on the next page)

2 TEACH

Teaching Tip The method shown in Example 1 for checking graphs should be used only when a point on the graph appears to have integral coordinates. Point out that if the coordinates are not integers, the equation probably will not check. The best way to check such graphs is to find another ordered pair that satisfies the equation and then determine whether the point for the ordered pair lies on the line.

In-Class Examples

For Example 1

Graph $2x + 5y = 20$ using the x- and y-intercepts.

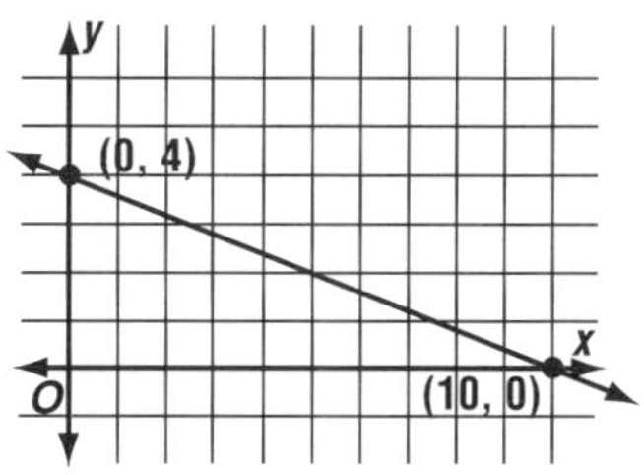

For Example 2

Graph $y = 4x - 1$ using the slope and y-intercept.

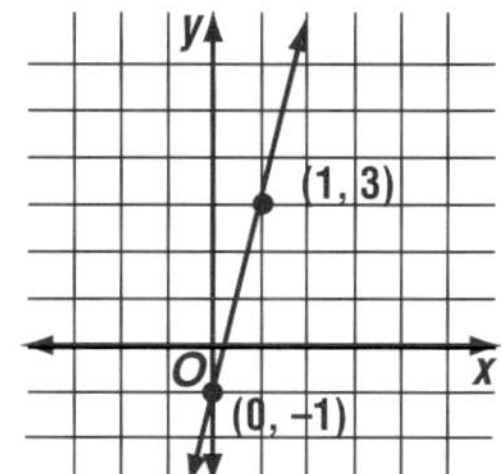

For Example 3

Graph $\frac{2}{3}(6x - 9y) = -\frac{8}{3}x - 4y - 4$.

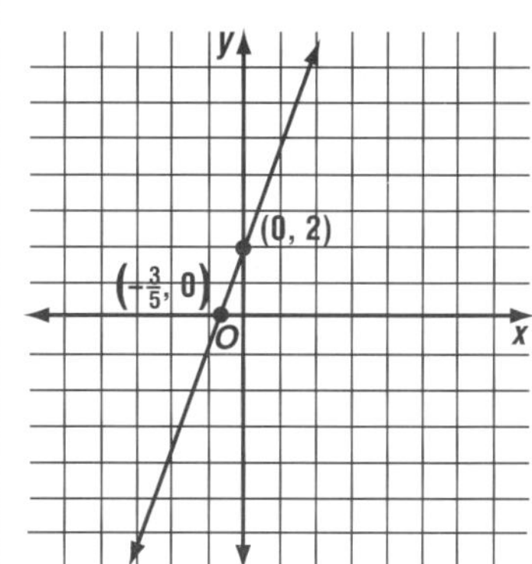

GLENCOE Technology

Interactive Mathematics Tools Software

This multimedia software provides an interactive lesson that uses linear coefficients and constants to find the equations of a line, given the graph of the line. A **Computer Journal** gives students an opportunity to write about what they have learned.

For Windows & Macintosh

In-Class Examples

For Example 3

Find the slope of each line. Are the lines parallel?

slope of $a = \frac{3}{2}$, slope of $b = 2$. They are not parallel.

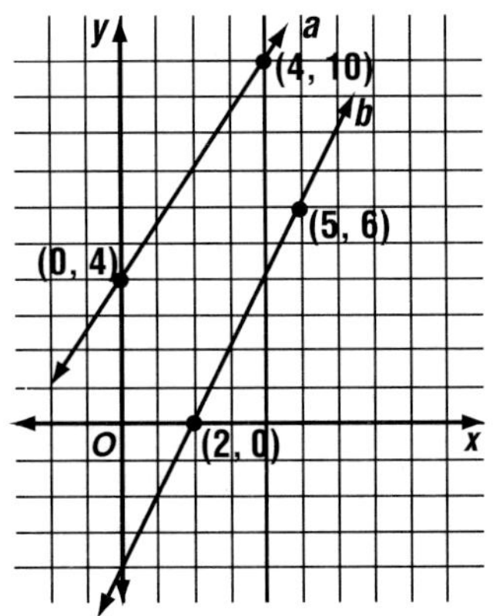

For Example 4

Write the equations of the lines in slope-intercept form.

a: $y = \frac{1}{2}x + 2$

b: $y = -2x - 2$

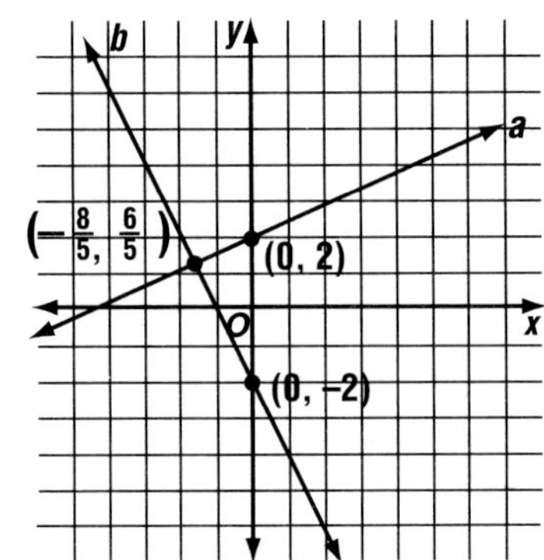

Sometimes graphs can be drawn in a way that leads you to the wrong conclusion. Mathematics can help you determine if a graph is misleading.

Example 3

Health

The American Heart Association recommends that men keep their weight within a certain "healthy weight band" as shown in the graph at the right. Are the lines parallel as they appear in the graph?

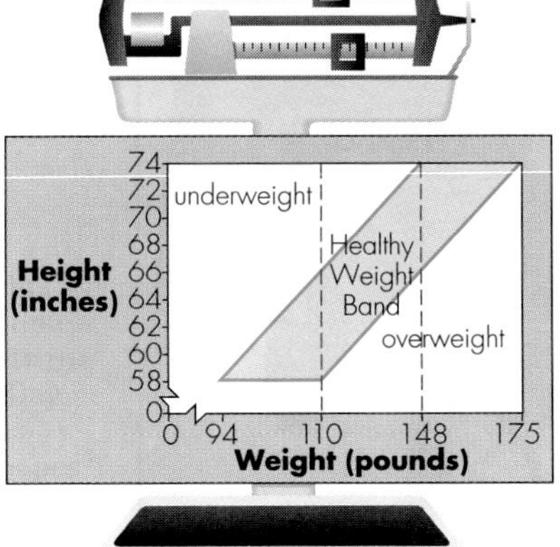

From the graph, we can determine that the endpoints of the minimum edge are (94, 58) and (148, 74). The endpoints of the maximum edge are (110, 58) and (175, 74). Find the slopes of each segment.

$$m_1 = \frac{74 - 58}{148 - 94} = \frac{16}{54} \text{ or about } 0.30$$

$$m_2 = \frac{74 - 58}{175 - 110} = \frac{16}{65} \text{ or about } 0.25$$

The segments do not have the same slope. Therefore, they are not parallel as they appear in this graph.

We have seen that the slopes of parallel lines are equal. What is the relationship of the slopes of perpendicular lines?

Example 4

Kite-Making

Refer to the application at the beginning of the lesson.

a. Write equations of the lines containing the slats of the kite in slope-intercept form.

b. Determine the relationship between the slopes of perpendicular lines.

a. The endpoints of the longer slat are (0, 0) and (8, 4). The endpoints of the shorter slat are (5, 5) and (7, 1). Find the slope of each slat and then use the point-slope form to find an equation for each slat.

long slat

$m = \frac{4 - 0}{8 - 0} = \frac{4}{8} \text{ or } \frac{1}{2}$

Let $(x_1, y_1) = (0, 0)$.

$y - y_1 = m(x - x_1)$

$y - 0 = \frac{1}{2}(x - 0)$

$y = \frac{1}{2}x$

short slat

$m = \frac{1 - 5}{7 - 5} = \frac{-4}{2} \text{ or } -2$

Let $(x_1, y_1) = (5, 5)$.

$y - y_1 = m(x - x_1)$

$y - 5 = -2(x - 5)$

$y - 5 = -2x + 10$

$y = -2x + 15$

Equations of the slats are $y = \frac{1}{2}x$ and $y = -2x + 15$.

b. The slats of the kite are perpendicular. The slopes of the slats are $\frac{1}{2}$ and -2. The two slopes are negative reciprocals of each other. Note that their product is $\frac{1}{2}(-2)$ or -1.

GLENCOE Technology

Interactive Mathematics Tools Software

This multimedia software provides an interactive lesson by having students graph and compare the slopes of parallel and perpendicular lines. A **Computer Journal** gives students an opportunity to write about what they have learned.

For Windows & Macintosh

These results suggest the following definition.

Definition of Perpendicular Lines in a Coordinate Plane	If the product of the slopes of two lines is -1, then the lines are perpendicular. In a plane, vertical lines and horizontal lines are perpendicular.

The slopes of perpendicular lines are negative reciprocals of each other.

You can use a right triangle and a coordinate grid to model the slopes of perpendicular lines.

MODELING MATHEMATICS Perpendicular Lines on a Coordinate Plane

Materials:  grid paper  scissors

a. A scalene triangle is one in which no two sides are equal. Cut out a scalene right triangle *ABC* so that $\angle C$ is the right angle. Label the vertices and the sides as shown at the right.

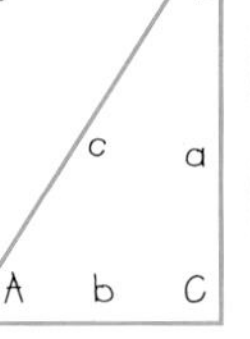

b. Draw a coordinate plane on the grid paper. Place ΔABC on the coordinate plane so that *A* is at the origin and side *b* lies along the positive *x*-axis.

c. Name the coordinates of *B*. **(b, a)**

d. What is the slope of side *c*? $\frac{a}{b}$

Your Turn

a. Rotate the triangle 90° counterclockwise so that *A* is still at the origin and side *b* is along the positive *y*-axis.

b. Name the coordinates of *B*. **$(-a, b)$**

c. What is the slope of *c*? $-\frac{b}{a}$

d. What is the relationship between the first position of *c* and the second? **perpendicular lines**

e. What is the relationship between the slopes of *c* in each position? **negative reciprocals**

You can use your knowledge of perpendicular lines to write equations of lines perpendicular to a given line.

Example  **Write the slope-intercept form of an equation that passes through $(8, -2)$ and is perpendicular to the graph of $5x - 3y = 7$.**

First find the slope of the given line.

$$5x - 3y = 7$$
$$-3y = -5x + 7$$
$$y = \frac{5}{3}x - \frac{7}{3}$$

The slope of the line is $\frac{5}{3}$. The slope of the line perpendicular to this line is the negative reciprocal of $\frac{5}{3}$, or $-\frac{3}{5}$.

Use the point-slope form to find the equation.

$$y - y_1 = m(x - x_1)$$
$$y - (-2) = -\frac{3}{5}(x - 8) \quad m = -\frac{3}{5}, (x_1, y_1) = (8, -2)$$
$$y + 2 = -\frac{3}{5}x + \frac{24}{5} \quad \text{Distributive property}$$
$$y = -\frac{3}{5}x + \frac{14}{5} \quad \text{Subtract 2 or } \frac{10}{5} \text{ from each side.}$$

An equation for the line is $y = -\frac{3}{5}x + \frac{14}{5}$.

MODELING MATHEMATICS This hands-on exercise uses the hypotenuse of a scalene right triangle to generate perpendicular lines. Using the lengths of the adjacent sides enables students to visualize slopes.

In-Class Example

For Example 5

Write the slope-intercept form of an equation that passes through each point and is perpendicular to the graph of each equation.

a. $7x - 2y = 3$, $(4, -1)$
$y = -\frac{2}{7}x + \frac{1}{7}$

b. $2x + 5y = -3$, $(2, -3)$
$y = \frac{5}{2}x - 8$

3 PRACTICE/APPLY

Check for Understanding

Exercises 1–17 are designed to help you assess your students' understanding through reading, writing, speaking, and modeling. You should work through Exercises 1–5 with your students and then monitor their work on Exercises 6–17.

Additional Answers

3. Example 2

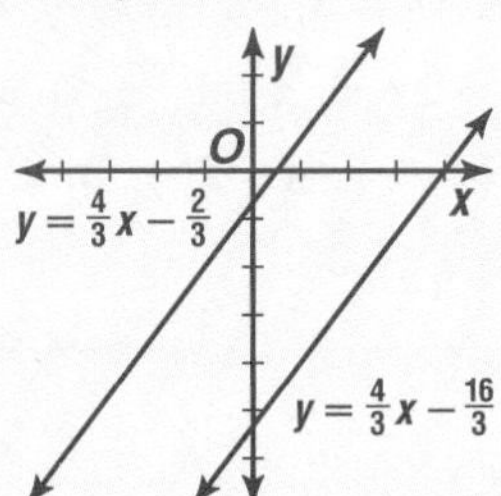

Example 5

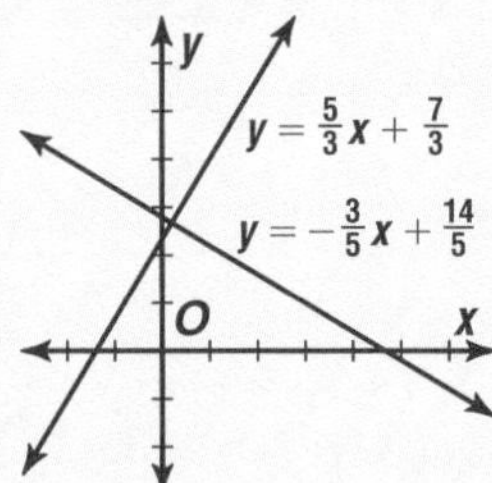

5b. The relationship would not work because there would be no perpendicular lines.

Study Guide Masters, p. 46

NAME________________ DATE ________

Study Guide

Student Edition Pages 362–368

Integration: Geometry
Parallel and Perpendicular Lines

When you graph two lines, you may encounter the two special types of graphs described at the right.

Parallel Lines and Perpendicular Lines
If two nonvertical lines have the same slope, then they are **parallel.** All vertical lines are parallel.
If the product of the slope of two lines is −1, then the lines are **perpendicular.** In a plane, vertical lines and horizontal lines are perpendicular.

Example: Write an equation in slope-intercept form of the line that passes through (0, 6) and is parallel to the graph of $2x - y = -12$.

The slope of the graph is 2. $-\frac{A}{B} = -\frac{2}{-1}$, or 2

The slope-intercept form of an equation whose graph is parallel to the original graph is $y = 2x + b$. Substitute (0, 6) into the equation and solve for b.

$6 = 0x + b$
$b = 6$ The y-intercept is b.

The equation of the line is $y = 2x + 6$.

Since $2 \cdot \left(-\frac{1}{2}\right) = -1$, any line that is perpendicular to the line in the Example has an equation of the form $y = -\frac{1}{2}x + b$. If the line includes the point (−4, 3), then $3 = -\frac{1}{2}(-4) + b$ and thus $b = 1$. The equation of the line is $y = -\frac{1}{2}x + 1$.

Write an equation in slope-intercept form of the line that passes through the given point and is parallel to the graph of each equation.

1. $y = -\frac{2}{3}x + 4$; (1, −3) $y = -\frac{2}{3}x - \frac{7}{3}$
2. $y = \frac{1}{2}x + 1$; (4, 2) $y = \frac{1}{2}x$
3. $2x + y = 5$; (4, −6) $y = -2x + 2$
4. $3x - y = 0$; (7, 2) $y = 3x - 19$
5. $x - 8y = 10$; (0, 5) $y = \frac{1}{8}x + 5$
6. $7x + 2y = 3$; (1, 6) $y = -\frac{7}{2}x + \frac{19}{2}$

Write an equation in slope-intercept form of the line that passes through the given point and is perpendicular to the graph of each equation.

7. $4x - 3y = 7$; (5, −2) $y = -\frac{3}{4}x + \frac{7}{4}$
8. $6x + 16y = 8$; (0, 4) $y = \frac{8}{3}x + 4$
9. $y = 7x + 1$; (6, 4) $y = -\frac{1}{7}x + \frac{34}{7}$

CHECK FOR UNDERSTANDING

Communicating Mathematics

Study the lesson. Then complete the following.

1. **Describe** the relationship between the slopes of each of the following.
   a. two parallel lines **The slopes are equal.**
   b. two perpendicular lines **The slopes are negative reciprocals.**
2. **Explain** what negative reciprocals are. Give an example.
   2. Negative reciprocals are two numbers whose product is −1; −2 and $\frac{1}{2}$.
3. **Check** the results of Examples 2 and 5 by graphing the original equation and the answer equation for each example on the same coordinate plane. Describe what happens. **See margin.**

4. Refer to the Modeling Mathematics activity. Why do you think the right triangle should be scalene? **If it were not, *a* and *b* would be equal.**
5. Repeat the activity with a different size right triangle.
   a. Are the results different? **Slopes are different; relationship is not.**
   b. What do you think would happen if you used a triangle that was not a right triangle? **See margin for sample answer.**

Guided Practice

State the slopes of the lines parallel to and perpendicular to the graph of each equation.

6. $6x - 5y = 11$ $\frac{6}{5}, -\frac{5}{6}$
7. $y = \frac{2}{3}x - \frac{4}{5}$ $\frac{2}{3}, -\frac{3}{2}$
8. $3x = 10y - 3$ $\frac{3}{10}, -\frac{10}{3}$

Determine whether the graphs of each pair of equations are *parallel, perpendicular,* or *neither.*

9. $3x - 7y = 1$, $7x + 3y = 4$ **perpendicular**
10. $5x - 2y = 6$, $4y - 10x = -48$ **parallel**

Write an equation in slope-intercept form of the line that passes through the given point and is parallel to the graph of each equation.

11. (9, −3), $5x - 6y = 2$ $y = \frac{5}{6}x - \frac{21}{2}$
12. (0, 4), $2y = 5x - 7$ $y = \frac{5}{2}x + 4$
13. (7, −2), $x - y = 0$ $y = x - 9$

Write an equation in slope-intercept form of the line that passes through the given point and is perpendicular to the graph of each equation.

14. (8, 5), $7x + 4y = 23$ $y = \frac{4}{7}x + \frac{3}{7}$
15. (0, 0), $9y = 3 - 5x$ $y = \frac{9}{5}x$
16. (−2, 7), $2x - 5y = 3$ $y = -\frac{5}{2}x + 2$
17. **Geometry** The diagonals of a square are segments that connect the opposite vertices. Use what you've learned in this lesson to determine the relationship between the diagonals $\overline{AC}$ and $\overline{BD}$ of the square graphed at the right. **perpendicular**

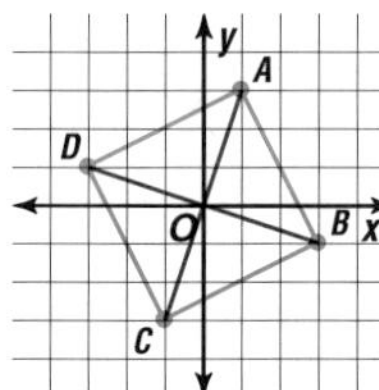

Reteaching

Using Alternative Methods Put a graph of a line on an overhead and ask each row of students to pick a point not on the line. Students should then find the line that is perpendicular to the original line and that passes through the point they chose. Then they should find the line that is parallel to the original line and that passes through the point they chose. Use slope-intercept form.

EXERCISES

Practice

Determine whether the graphs of each pair of equations are *parallel, perpendicular, or neither.*

18. $y = -2x + 11$
    $y + 2x = 23$
    parallel
19. $3y = 2x + 14$
    $2x - 3y = 2$
    parallel
20. $y = -5x$
    $y = 5x - 18$
    neither
21. $y = 0.6x + 7$
    $3y = -5x + 30$
    perpendicular
22. $y = 3x + 5$
    $y = 5x + 3$
    neither
23. $y + 6 = -5$
    $y + x = y + 7$
    perpendicular

Write an equation in slope-intercept form of the line that passes through the given point and is parallel to the graph of each equation.

26. $y = -\frac{2}{3}x - \frac{22}{3}$
27. $y = \frac{8}{7}x + \frac{2}{7}$

24. $(2, -7), y = x - 2$ **$y = x - 9$**
25. $(2, 3), y = x + 5$ **$y = x + 1$**
26. $(-5, -4), 2x + 3y = -1$
27. $(5, 6), 8x - 7y = 23$
28. $(1, -2), y = -2x + 7$ **$y = -2x$**
29. $(0, -5), 5x - 2y = -7$ **$y = 2.5x - 5$**
30. $(5, -4), x - 3y = 8$ **$y = \frac{1}{3}x - \frac{17}{3}$**
31. $(-3, 2), 2x - 3y = 6$ **$y = \frac{2}{3}x + 4$**
32. $(2, -1), y = -0.5x + 2$ **$y = -0.5x$**

Write an equation in slope-intercept form of the line that passes through the given point and is perpendicular to the graph of each equation.

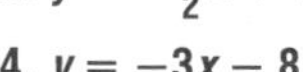

33. $y = -\frac{9}{2}x + 14$
34. $y = -3x - 8$
35. $y = 3x - 19$
36. $y = -\frac{5}{3}x + 8$
37. $y = -\frac{1}{5}x - 1$
38. $y = -\frac{3}{5}x + \frac{14}{5}$

33. $(6, -13), 2x - 9y = 5$
34. $(-3, 1), y = \frac{1}{3}x + 2$
35. $(6, -1), 3y + x = 3$
36. $(6, -2), y = \frac{3}{5}x - 4$
37. $(0, -1), 5x - y = 3$
38. $(8, -2), 5x - 7 = 3y$
39. $(4, -3), 2x - 7y = 12$ **$y = -\frac{7}{2}x + 11$**
40. $(3, 7), y = \frac{3}{4}x - 1$ **$y = -\frac{4}{3}x + 11$**
41. $(3, -3), 3x + 7 = 2x$ **$y = -3$**

Write an equation of the line having the following properties.

42. passes through $(-5, 3)$ and is perpendicular to the x-axis **$x = -5$**
43. is parallel to the graph of $y = \frac{5}{4}x - 3$ and passes through the origin.
44. has an x-intercept of 3 and is perpendicular to the graph of $5x - 3y = 2$
45. has a y-intercept of -6 and is parallel to the graph of $x - 3y = 8$
    $y = \frac{1}{3}x - 6$

43. $y = \frac{5}{4}x$
44. $y = -\frac{3}{5}x + \frac{9}{5}$

Critical Thinking

46. Lines a, b, and c lie on the same coordinate plane. Line a is perpendicular to line b and intersects it at $(3, 6)$. Line b is perpendicular to line c, which is the graph of $y = 2x + 3$.
    a. Write the equation of lines a and b. **$y = 2x$, $y = -\frac{1}{2}x + \frac{15}{2}$**
    b. Graph the three lines. **See margin.**
    c. What is the relationship of lines a and c? **parallel**

47. No, because the slope of $\overline{AC}$ is $\frac{6}{7}$ and the slope of $\overline{BD}$ is $-\frac{2}{3}$. These are not negative reciprocals, so the diagonals are not perpendicular.

Applications and Problem Solving

47. **Geometry** A rhombus is a parallelogram that has perpendicular diagonals. Determine if quadrilateral $ABCD$ with $A(-2, 1)$, $B(3, 3)$, $C(5, 7)$, and $D(0, 5)$ is a rhombus. Explain.

Error Analysis

Watch that students do not just use the coefficient of x for the slope. They must solve for y first.
Example: $-x - 4y = 2$
The slope is not -1. Solve for y.

$$-4y = x + 2$$
$$y = -\frac{1}{4}x - \frac{1}{2}$$

The slope is $-\frac{1}{4}$.

Assignment Guide

Core: 19–45 odd, 46, 47, 49–58
Enriched: 18–44 even, 46–58

For **Extra Practice,** see p. 771.

The red A, B, and C flags, printed only in the Teacher's Wraparound Edition, indicate the level of difficulty of the exercises.

Additional Answer

46b.

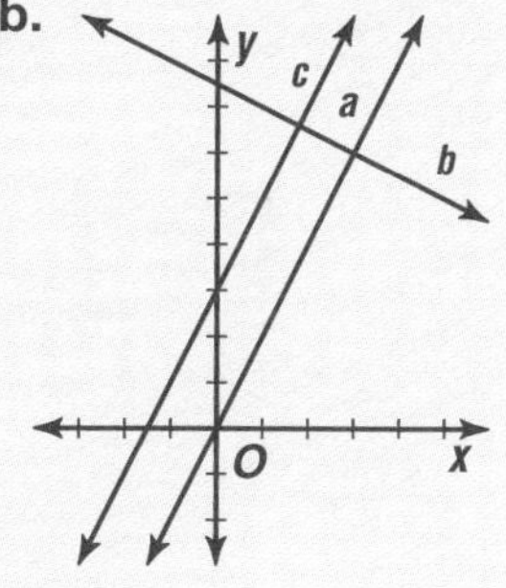

Practice Masters, p. 46

NAME ________ DATE ________

Practice

Student Edition Pages 362–368

Integration: Geometry
Parallel and Perpendicular Lines

Write an equation in slope-intercept form of the line that passes through the given point and is parallel to the graph of each equation.

1. $2x + y = 5; (3, 1)$ $y = -2x + 7$
2. $3x - y = 5; (-1, -2)$ $y = 3x + 1$
3. $5x - 4y = 1; (-8, 2)$ $y = \frac{5}{4}x + 12$
4. $9x + 3y = 8; (-1, -4)$ $y = -3x - 7$
5. $y = \frac{4}{3}x + 5; (12, 3)$ $y = \frac{4}{3}x - 13$
6. $y = -\frac{3}{4}x + \frac{1}{4}; (4, -2)$ $y = -\frac{3}{4}x + 1$

Write an equation in slope-intercept form of the line that passes through the given point and is perpendicular to the graph of each equation.

7. $x - 6y = 2; (2, 4)$ $y = -6x + 16$
8. $3x + 2y = -7; (1, 1)$ $y = \frac{2}{3}x + \frac{1}{3}$
9. $5x + 4y = 8; (10, 5)$ $y = \frac{4}{5}x - 3$
10. $4x + 3y = -6; (2, 1)$ $y = \frac{3}{4}x - \frac{1}{2}$
11. $y = \frac{1}{4}x - 4; (-2, 3)$ $y = -4x - 5$
12. $2x + 10y = 3; (2, 3)$ $y = 5x - 7$
13. $x = 2y - 1; (0, 0)$ $y = -2x$
14. $4x + 7y = 6; (-4, 1)$ $y = \frac{7}{4}x + 8$

4 ASSESS

Closing Activity

Speaking Ask students to explain the relationship of the line represented by $ax + by = c$ and all the lines parallel and perpendicular to it.

Chapter 6, Quiz C (Lessons 6-5 and 6-6), is available in the *Assessment and Evaluation Masters*, p. 157.

Additional Answers

50.

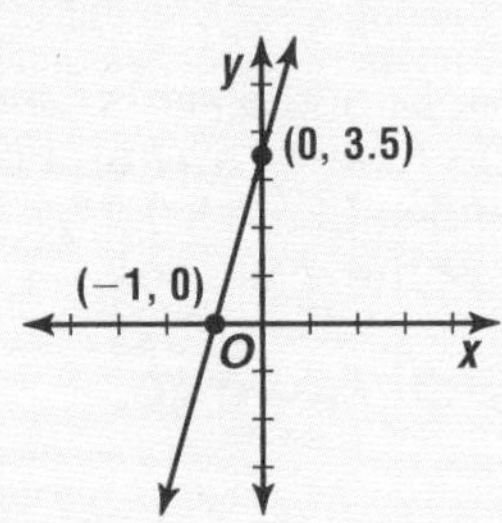

51.

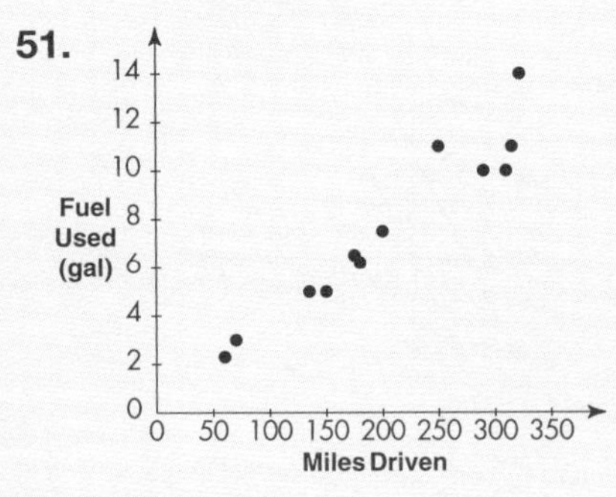

Enrichment Masters, p. 46

NAME ____ DATE ____

Enrichment

Student Edition Pages 362–368

Pencils of Lines

All of the lines that pass through a single point in the same plane are called a **pencil of lines.**

All lines with the same slope, but different intercepts, are also called a "pencil," a **pencil of parallel lines.**

Graph some of the lines in each pencil.

1. A pencil of lines through the point (1, 3)

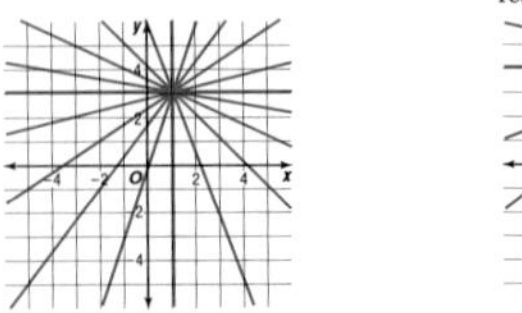

2. A pencil of lines described by $y - 4 = m(x - 2)$, where m is any real number

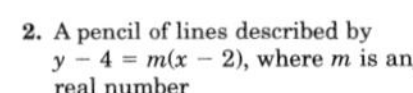

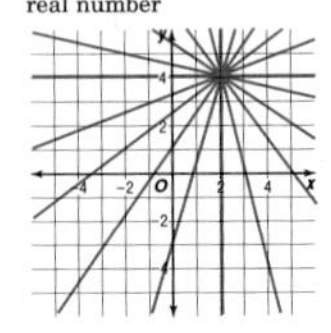

3. A pencil of lines parallel to the line $x - 2y = 7$

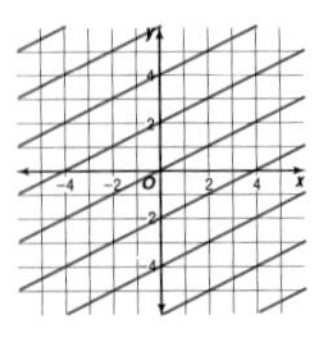

4. A pencil of lines described by $y = mx + 3m - 2$

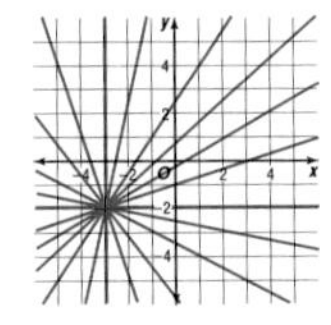

48b. $y = -\frac{x}{10} + \frac{39}{10}$, $y = 10x + 14$

48. **Construction** A large roller is being used to flatten new asphalt on a hill that has a slope of $-\frac{1}{10}$. The steering mechanism for the front roller is perpendicular to the hill.
    a. What is the slope of the steering mechanism? **10**
    b. If the bottom of the hill has coordinates (9, 3) and the point directly below the steering mechanism is at (−1, 4), write the equations of the lines representing the hill and the steering mechanism.

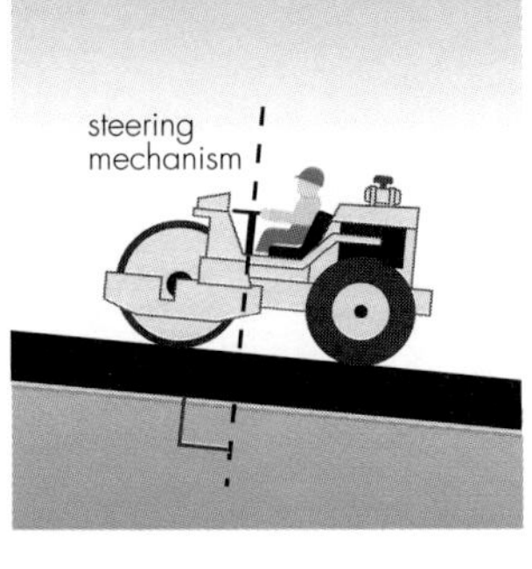

49a. $y = \frac{1}{2}x + \frac{7}{2}$, $y = -2x + 11$

49. **Geometry** An angle inscribed in a circle is one in which the vertex of the angle lies on the circle and the two sides intersect the circle, as shown at the right.
    a. Find the equations of the lines that contain the sides of $\angle Z$.
    b. What type of angle is $\angle Z$? **right or 90° angle**

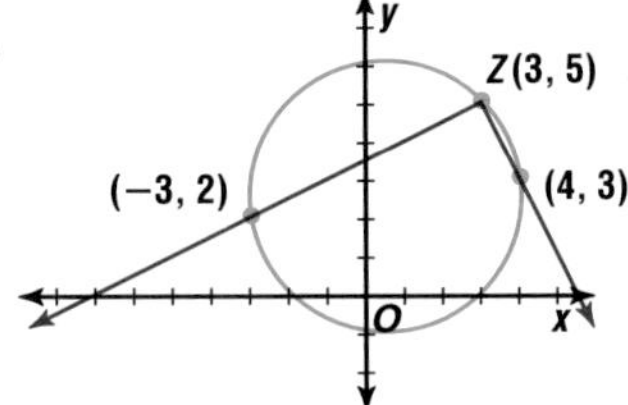

Mixed Review

50. See margin.

50. Graph $7x - 2y = -7$ by using the x- and y-intercepts. (Lesson 6–5)

51. **Statistics** Make a scatter plot of the data below. (Lesson 6–3) **See margin.**

Miles Driven	200	322	250	290	310	135	60	150	180	70	315	175
Fuel Used (gallons)	7.5	14	11	10	10	5	2.3	5	6.2	3	11	6.5

52. **Patterns** Copy and complete the table below. (Lesson 5–4)

x	−1	2	5	8	11	14
f(x)	5	−1	−7	−13	−19	−25

55a. $\frac{5}{12}$; $\frac{7}{12}$

53. Find 4% of $6070. (Lesson 4–4) **$242.80**

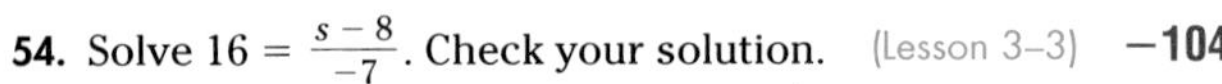

54. Solve $16 = \frac{s - 8}{-7}$. Check your solution. (Lesson 3–3) **−104**

55. **Metallurgy** The gold content of jewelry is given in karats. For example, 24-karat gold is pure gold, and 18-karat gold is $\frac{18}{24}$ or 0.75 gold. (Lesson 2–7)
    a. What fraction of 10-karat gold is pure gold? What fraction is *not* gold?
    b. If a piece of jewelry is $\frac{2}{3}$ gold, how would you describe it using karats? **16-karat gold**

56. Find $-0.0005 + (-0.3)$. (Lesson 2–5) **−0.3005**

57. Name the property illustrated by $1(87) = 87$. (Lesson 1–6) **identity (×)**

58. Solve $x = 6 + 0.28$. (Lesson 1–5) **6.28**

Extension

Communication Have students explain the method for finding an equation in standard form when they know one point and a line to which it is parallel.

Tech Prep

Metallurgist Students who are interested in metallurgy may wish to do further research on the data provided in Exercise 55 and explore the potential growth of this career. For more information on tech prep, see the *Teacher's Handbook*.

6–7

Integration: Geometry Midpoint of a Line Segment

What YOU'LL LEARN

- To find the coordinates of the midpoint of a line segment in the coordinate plane.

Why IT'S IMPORTANT

You can find the midpoint of a line segment to solve problems involving geometry.

APPLICATION

Woodworking

Maria Hernandez made a frame for one of her watercolor paintings in a class offered at the community college. Because her canvas was heavy and the wooden frame was light, her instructor suggested that she stabilize her frame by adding crossbars connecting the consecutive sides. One of the books she used as a reference showed the crossbars inserted so that the bars are placed on the sides of the frame at the **midpoint** of each side. The midpoint of a line segment is the point that is halfway between the endpoints of the segment.

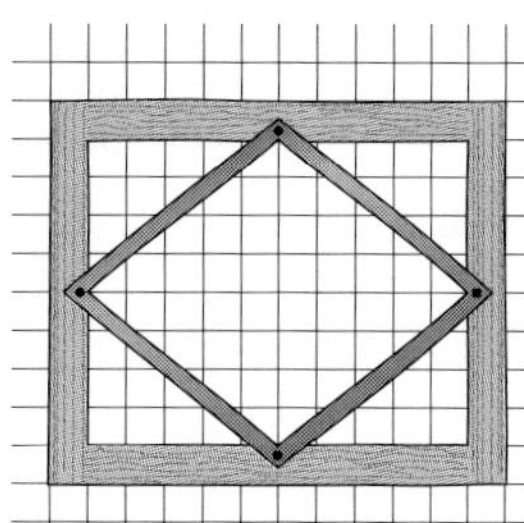

Maria decided to use some grid paper to model her frame and the crossbars she is going to insert. Each square of the paper represented 2 inches of her frame. To find the midpoint of the interior vertical edge, she counted how many units it was from one corner to the next and divided by 2. The inside of the frame is 16 inches high, so the midpoint is 8 inches from the corner. She counted 4 units on the grid and placed a bullet on her drawing. She used a similar method to find the midpoint of the interior horizontal edge. The inside is 20 inches wide, so the midpoint is 10 inches from the end.

Suppose we place Maria's frame drawing on a coordinate plane. Let's list the coordinates of the corners of the frame and the midpoints of the sides. What pattern do you notice?

$A(2, 2)$
$B(22, 2)$
$C(22, 18)$
$D(2, 18)$

Midpoint of $\overline{AB}$: (12, 2)
Midpoint of $\overline{BC}$: (22, 10)
Midpoint of $\overline{CD}$: (12, 18)
Midpoint of $\overline{AD}$: (2, 10)

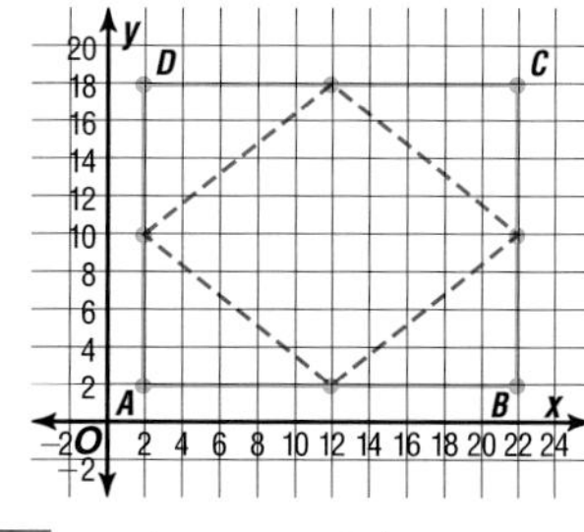

Notice that the coordinates of the midpoint of $\overline{AB}$ are the mean of the corresponding coordinates of A and B.

Midpoint of $\overline{AB}$: (12, 2)

mean of x-coordinates of A(2, 2) and B(22, 2) $\frac{2+22}{2} = 12$

mean of y-coordinates of A(2, 2) and B(22, 2) $\frac{2+2}{2} = 2$

The woodworking example uses the midpoints of horizontal and vertical line segments. The Modeling Mathematics activity on the next page applies to any line segment.

6-7 LESSON NOTES

NCTM Standards: 1–5, 8

Instructional Resources

- Study Guide Master 6-7
- Practice Master 6-7
- Enrichment Master 6-7
- Assessment and Evaluation Masters, p. 157

Transparency 6-7A contains the 5-Minute Check for this lesson; **Transparency 6-7B** contains a teaching aid for this lesson.

Recommended Pacing	
Standard Pacing	Day 11 of 13
Honors Pacing	Day 10 of 12
Block Scheduling*	Day 6 of 7 (along with Lesson 6-6)
Alg. 1 in Two Years*	Days 14 & 15 of 18

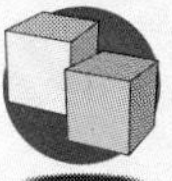

*For more information on pacing and possible lesson plans, refer to the *Block Scheduling Booklet* and *Algebra 1 in Two Years.*

1 FOCUS

5-Minute Check
(over Lesson 6-6)

Write an equation in slope-intercept form of the lines that pass through the given point and are parallel and perpendicular to the graph of each equation.

1. $5x - 3y = 8, (4, -2)$
   $\mathbf{5x - 3y = 26, 3x + 5y = 2}$
2. $7x + 2y = 4, (-6, 1)$
   $\mathbf{7x + 2y = -40,}$
   $\mathbf{2x - 7y = -19}$
3. $x = 7, (4, -3)$
   $\mathbf{x = 4, y = -3}$
4. $x - y = 3, (4, 6)$
   $\mathbf{x - y = -2, x + y = 10}$

Motivating the Lesson

Hands-On Activity Students are to hang a framed painting. The hook needs to be placed at a point one third the distance from the top to the bottom and at the midpoint of the horizontal dimension. They can use only a ruler to locate the point.

 This hands-on modeling activity uses color to help students visualize the concept of midpoint.

2 TEACH

Teaching Tip Point out that the midpoint is computed using the average of the x-coordinates and the y-coordinates.

In-Class Example

For Example 1
If *P* is the midpoint of line segment *AB*, find the coordinates of the missing point *A* or *B*.

a. $A(3, 2)$, $P(6, 3)$ **$B(9, 4)$**
b. $A(6, 3)$, $P(3, 2)$ **$B(0, 1)$**
c. $A(3, 2)$, $P(1, 5)$ **$B(-1, 8)$**
d. $P(6, 3)$, $B(1, 5)$ **$A(11, 1)$**
e. $P(1, 5)$, $B(6, 3)$ **$A(-4, 7)$**

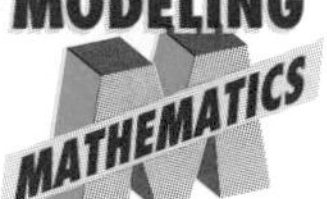

MODELING MATHEMATICS — Midpoint of a Line Segment

Materials:  grid paper  ruler colored pencils

- Create a coordinate plane on your grid paper.
- Use a ruler to draw any line segment on the coordinate plane. The segment should not be vertical or horizontal. Label the endpoints of the segment *A* and *B* with their coordinates.
- Hold the paper up to the light and fold it so that *A* and *B* coincide. Crease the paper at the fold.
- Unfold the paper. Label the point *M* at which the crease meets the segment and write the coordinates of *M*.
- Use a different color pencil to draw a vertical line through the lower endpoint of your segment. Then draw a horizontal line through the upper endpoint of your segment. Label the point of intersection *P*.
- Write the coordinates of the midpoints of $\overline{AP}$ and $\overline{BP}$.
- How do the *x*-coordinate of the vertical segment and the *y*-coordinate of the horizontal segment compare with the coordinates of *M*?

Your Turn **a–c. See student's work.**

a. Create another coordinate plane and draw a segment with a different slope from the first one you drew.

b. Repeat the activity with this segment. What are your results?

c. Write a general rule for finding the midpoint of any segment.

This activity suggests the following rule for finding the midpoint of any segment, given the coordinates of its endpoints.

Midpoint of a Line Segment on a Coordinate Plane	**The coordinates of the midpoint of a line segment whose endpoints are at (x_1, y_1) and (x_2, y_2) are given by $\left(\frac{x_1 + x_2}{2}, \frac{y_1 + y_2}{2}\right)$.**

Example  **If the vertices of parallelogram *WXYZ* are $W(3, 0)$, $X(9, 3)$, $Y(7, 10)$, and $Z(1, 7)$, prove that the diagonals bisect each other. That is, prove that they intersect at their midpoints.**

Explore Graph the vertices and draw the parallelogram and its diagonals.

Plan Find the coordinates of the midpoints of the diagonals to see if they are the same point.

Solve Find the midpoints of $\overline{WY}$ and $\overline{XZ}$.

$W(x_1, y_1) = W(3, 0)$ $\quad X(x_1, y_1) = X(9, 3)$
$Y(x_2, y_2) = Y(7, 10)$ $\quad Z(x_2, y_2) = Z(1, 7)$

midpoint of $\overline{WY} = \left(\frac{3+7}{2}, \frac{0+10}{2}\right) = \left(\frac{10}{2}, \frac{10}{2}\right) = (5, 5)$

midpoint of $\overline{XZ} = \left(\frac{9+1}{2}, \frac{3+7}{2}\right) = \left(\frac{10}{2}, \frac{10}{2}\right) = (5, 5)$

Examine Since the midpoint of $\overline{XZ}$ has the same coordinates as the midpoint of $\overline{WY}$, the diagonals bisect each other.

Alternative Teaching Strategies

Reading Algebra In non-mathematical uses, the word *middle* and the prefix *mid-* both mean "between." But in mathematics, they mean "exactly halfway between."

If you know the midpoint and one endpoint of a segment, you can find the other endpoint.

Example  **The center of a circle is $M(0, 2)$, and the endpoint of one of its radii is $A(-6, -4)$. If $\overline{AB}$ is a diameter of the circle, what are the coordinates of B?**

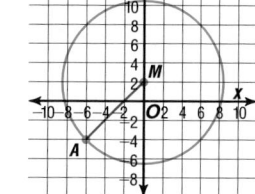

Use the midpoint formula and substitute the values you know.

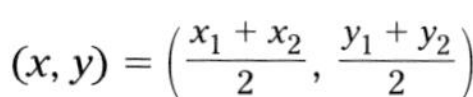

$$(x, y) = \left(\frac{x_1 + x_2}{2}, \frac{y_1 + y_2}{2}\right)$$

$$(0, 2) = \left(\frac{-6 + x_2}{2}, \frac{-4 + y_2}{2}\right) \quad (x, y) = (0, 2), (x_1, y_1) = (-6, -4)$$

The x- and y-coordinates of the ordered pairs are equal since the ordered pairs are equal. So, two equations can be formed.

Find the x-coordinate.

$$0 = \frac{-6 + x_2}{2}$$
$$0 = -6 + x_2$$
$$6 = x_2$$

Find the y-coordinate.

$$2 = \frac{-4 + y_2}{2}$$
$$4 = -4 + y_2$$
$$8 = y_2$$

The coordinates of endpoint $B(x_2, y_2)$ are (6, 8).

Check: Copy the graph of the circle and points A and M. Extend the radius through M to intersect the circle. This segment is the diameter of the circle. The intersection point is at (6, 8), which are the coordinates of point B.

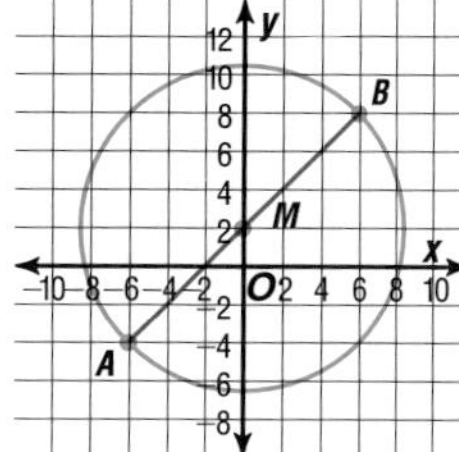

CHECK FOR UNDERSTANDING

Communicating Mathematics

Study the lesson. Then complete the following.

1. **Explain** how you can use the midpoint formula to find the midpoint of each of the following. **a–b. See margin.**
   a. a vertical line segment
   b. a horizontal line segment
2. **Demonstrate** how to find the coordinates of vertex C of rectangle $ABCD$ if its two diagonals meet at $M(-3, 4)$ and the coordinates of A are $(0, -1)$. **$C(-6, 9)$**

MODELING MATHEMATICS

3. Draw a rectangle on a coordinate plane. **a–c. See students' work.**
   a. Find the midpoint of each side of the rectangle.
   b. Find the midpoint of the diagonals.
   c. How do the coordinates of the midpoint of the diagonal relate to the coordinates of the midpoints of the sides?

In-Class Example

For Example 2
The endpoints of a diameter of a circle are at (4, 2) and (8, 26). Find the coordinates of the center of the circle. **(6, 22)**

3 PRACTICE/APPLY

Check for Understanding

Exercises 1–12 are designed to help you assess your students' understanding through reading, writing, speaking, and modeling. You should work through Exercises 1–3 with your students and then monitor their work on Exercises 4–12.

Additional Answers

1a. The x-coordinates will be the same, so calculate the average of the y-coordinates.

1b. The y-coordinates will be the same, so calculate the average of the x-coordinates.

Reteaching

Using Problem Solving Have students find the midpoint of a given segment, then the midpoint of half of the original segment, and so on. For example, given points $A(-4, -5)$ and $B(12, 3)$, find the coordinates for X, Y, Z if X is the midpoint of $\overline{AB}$, Y is the midpoint of $\overline{AX}$, and Z is the midpoint of $\overline{AY}$.

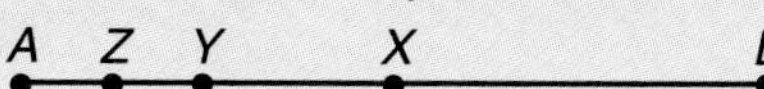

Assignment Guide

Core: 13–39 odd, 40, 41, 43–51
Enriched: 14–38 even, 40–51

For **Extra Practice,** see p. 771.

The red A, B, and C flags, printed only in the Teacher's Wraparound Edition, indicate the level of difficulty of the exercises.

Guided Practice

Find the coordinates of the midpoint of a segment with each pair of endpoints.

4. $A(5, -2), B(5, 8)$ **(5, 3)**
5. $C(-5, 6), D(8, 6)$ **(1.5, 6)**
6. $E(6, 5), F(14, 7)$ **(10, 6)**
7. $G(-9, -6), H(-3, 8)$ **(−6, 1)**

Find the coordinates of the other endpoint of a segment given one endpoint and the midpoint *M*.

8. $A(3, 6), M(-1.5, 5)$ **(−6, 4)**
9. $B(-3, 7), M(-7, 7)$ **(−11, 7)**
10. $C(8, 11), M\left(5, \frac{17}{2}\right)$ **(2, 6)**
11. $D(8, 4.5), M(8, 7.15)$ **(8, 9.8)**

12. **Archery** The target for archery competition is composed of 10 rings and 5 colored circles. Hitting the bull's-eye is worth ten points, and hitting the outer ring is worth one point. A target is shown on a coordinate plane at the right. The two labeled points are the endpoints of a diameter of the target. Find the coordinates of the center of the bull's-eye. **(−3, 5)**

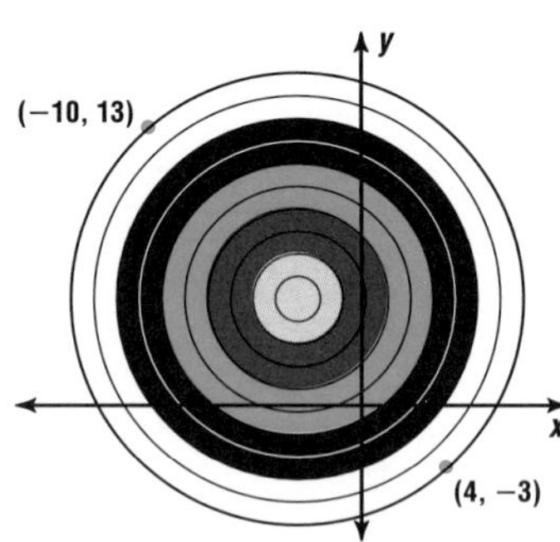

EXERCISES

Practice

Find the coordinates of the midpoint of a segment with each pair of endpoints.

13. $J(6, 6), K(19, 6)$ **(12.5, 6)**
14. $L(5, 9), M(-7, 3)$ **(−1, 6)**
15. $P(-5, 9), Q(7, 1)$ **(1, 5)**
16. $R(7, 4), S(11, -10)$ **(9, −3)**
17. $T(8, -7), U(-2, 11)$ **(3, 2)**
18. $V(-8, 1.2), W(-8, 7.4)$ **(−8, 4.3)**
19. $X(-3, 9), Y(4, -7)$
20. $B(9, -4), C(2, 7)$
21. $D(4.7, -2.9), E(-3.1, 8.3)$
22. $F(a, b), G(c, d)$
23. $Y(6x, 14y), Z(2x, 4y)$ **(4x, 9y)**
24. $M(-2w, -7v), N(6w, 2v)$ $\left(2w, -\frac{5}{2}v\right)$

19. $\left(\frac{1}{2}, 1\right)$
20. $\left(\frac{11}{2}, \frac{3}{2}\right)$
21. **(0.8, 2.7)**
22. $\left(\frac{a+c}{2}, \frac{b+d}{2}\right)$

Find the coordinates of the other endpoint of a segment given one endpoint and the midpoint *M*.

25. $E(-7, 8), M(-7, 4)$ **(−7, 0)**
26. $F(4, 2), M(2, 1)$ **(0, 0)**
27. $G(3, -6), M(12, -6)$ **(21, −6)**
28. $H(5, 3), M(6, 4)$ **(7, 5)**
29. $L(-8, 4), M\left(\frac{1}{2}, 7\right)$ **(9, 10)**
30. $N(5, -9), M(8, -7.5)$ **(11, −6)**
31. $R\left(\frac{1}{6}, \frac{1}{3}\right), M\left(\frac{1}{2}, \frac{1}{3}\right)$ $\left(\frac{5}{6}, \frac{1}{3}\right)$
32. $S(a, b), M\left(\frac{x+a}{2}, \frac{y+b}{2}\right)$ **(x, y)**

If *P* is the midpoint of segment *AB*, find the coordinates of the missing point.

33. $A(6.5, -8.2), P(4.4, -0.7)$
34. $P(1.2, 4.5), B(5.3, 1.9)$
35. $A(9.7, -5.4), B(3.6, 1.7)$
36. $A(-5.9, 7.2), P(-1.05, 5.85)$

33. **B(2.3, 6.8)**
34. **A(−2.9, 7.1)**
35. **P(6.65, −1.85)**
36. **B(3.8, 4.5)**

Study Guide Masters, p. 47

NAME______________________ DATE __________

Study Guide

Student Edition Pages 369–373

Integration: Geometry
Midpoint of a Line Segment

The **midpoint** of a line segment is the point that is halfway between the endpoints of the segment.

Midpoint of a Line Segment
The coordinates of the midpoint of a line segment whose endpoints are at (x_1, y_1) and (x_2, y_2) are given by $\left(\frac{x_1 + x_2}{2}, \frac{y_1 + y_2}{2}\right)$.

Example: Find the coordinates of the midpoint of a segment whose endpoints are $F(-4, 5)$ and $G(6, -1)$.

$$(x, y) = \left(\frac{x_1 + x_2}{2}, \frac{y_1 + y_2}{2}\right)$$
$$= \left(\frac{-4 + 6}{2}, \frac{5 + (-1)}{2}\right)$$
$$= (1, 2)$$

The coordinates of the midpoint of $\overline{FG}$ are (1, 2).

Find the coordinates of the midpoint of a segment with each pair of endpoints.

1. $C(6, 2), D(4, 1)$ $\left(5, \frac{3}{2}\right)$
2. $M(5, 7), N(3, 9)$ (4, 8)
3. $I(0, -2), J\left(\frac{3}{4}, 5\right)$ $\left(\frac{3}{8}, \frac{3}{2}\right)$
4. $K\left(\frac{1}{4}, 1\right), L\left(\frac{1}{8}, 0\right)$ $\left(\frac{3}{16}, \frac{1}{2}\right)$
5. $G(10, -8), H(4, -3)$ $\left(7, -\frac{11}{2}\right)$
6. $A(6, 5), B(3, -1)$ $\left(\frac{9}{2}, 2\right)$
7. $E(2, 7), F(8, 4)$ $\left(5, \frac{11}{2}\right)$
8. $R(13, 10), S(7, 27)$ $\left(10, \frac{37}{2}\right)$
9. $P(x, y), Q(3x, 5y)$ (2x, 3y)

If P is the midpoint of segment AB, find the coordinates of the missing point.

10. $A(4, 1), B(6, 3)$ P(5, 2)
11. $B(-14, 24), P(-2, 7)$ A(10, −10)
12. $B(-8, -10), P(-9, 0)$ A(−10, 10)
13. $A(11, 12), P(2, 17)$ B(−7, 22)
14. $A(9, -18), B(5, -16)$ P(7, −17)
15. $B(0, 12), P(-5, -1)$ A(−10, −14)

37. The coordinates of the endpoints of a diameter of a circle are (5, 8) and (−7, 2). Find the coordinates of the center. **(−1, 5)**

38. Find the coordinates of the point on $\overline{AB}$ that is one-fourth the distance from A to B for $A(8, 3)$ and $B(10, -5)$. **(8.5, 1)**

39. The endpoints of $\overline{AB}$ are $A(-2, 7)$ and $B(6, -5)$. Find the coordinates of P if P lies on $\overline{AB}$ and is $\frac{3}{8}$ the distance from A to B. $\left(1, \frac{5}{2}\right)$

Critical Thinking

40. Points W, X, Y, and Z are the midpoints of the sides of quadrilateral $QUAD$. Write a convincing argument that quadrilateral $WXYZ$ is a parallelogram.
The vertices are $W(2, -3)$, $X(6, 2)$, $Y(2, 4)$, and $Z(-2, -1)$. The slope of $\overline{XW}$ = slope of $\overline{YZ} = \frac{5}{4}$. The slope of $\overline{WZ}$ = slope of $\overline{XY} = -\frac{1}{2}$. The opposite sides are parallel, so $WXYZ$ is a parallelogram.

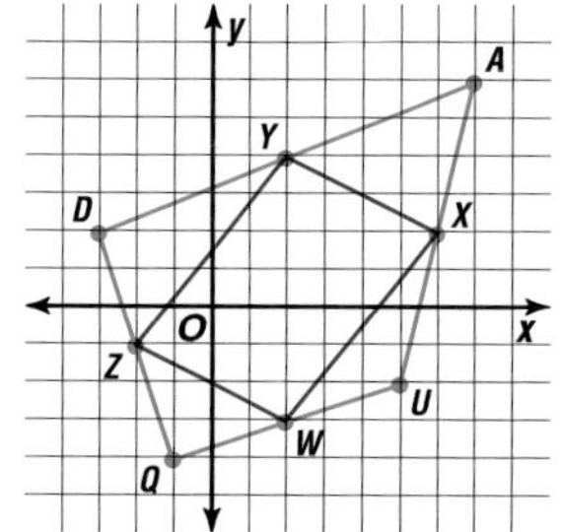

Applications and Problem Solving

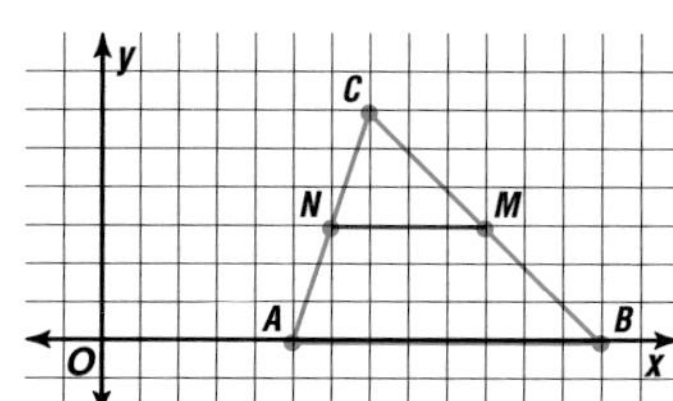

41. **Geometry** N is the midpoint of $\overline{AC}$ in the triangle shown at the left. M is the midpoint of $\overline{BC}$. **a. $N(6, 3)$, $M(10, 3)$**
a. Write the coordinates of M and N.
b. Compare $\overline{MN}$ and $\overline{AB}$. Verify your findings. **parallel, $MN = \frac{1}{2}AB$**

42. **Technology** The screen of a TI-82 graphing calculator is composed of tiny dots called *pixels* that are turned on or off during a display. Each pixel is named by the row and column in which it lies.

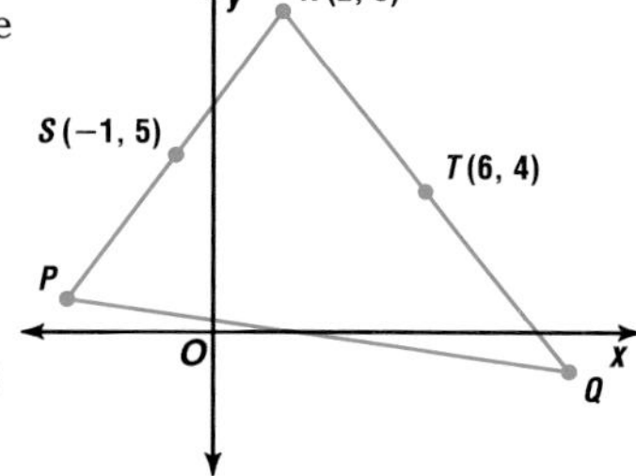

a. If a segment drawn on the screen has one endpoint at pixel (20, 43) and its midpoint is at pixel (30, 60), what is the pixel name for the other endpoint of the segment? **(40, 77)**
b. What is the pixel name of the origin? **(31, 47)**
c. How does this coordinate system differ from the coordinate system we usually use for graphing? **Ordered pairs are (y, x) and the origin is in a different place.**

43. **Geometry** S and T are the midpoints of their respective sides in the triangle shown at the right.
a. Find the coordinates of P and Q.
$P(-4, 1)$, $Q(10, -1)$
b. If the area of triangle RST is 15.5 square units, what is the area of triangle RPQ? Explain how you arrived at your answer. **62 square units; See margin for sample explanation.**

Extension

Connections The diameter of a circle has its endpoints at (8, 6) and (2, −2). Find the center of the circle and at least two other points on the circle. **Center is at (5, 2); answers may vary for other points on the circle. Possible answers include (2, 6), (8, −2), (9, 5), (9, −1), (1, 5), (1, −1), (10, 2), and (0, 2).**

4 ASSESS

Closing Activity

Speaking Ask students to explain the procedure they would use to determine the midpoint of a line segment.

Chapter 6, Quiz D (Lesson 7), is available in the *Assessment and Evaluation Masters*, p. 157.

Additional Answer

43b. Sample answer: The area of the smaller triangle is $\frac{1}{2}bh$. Since the base of the larger triangle is twice that of the smaller one and the height is also twice the length of the small one, the area of the larger is $\frac{1}{2}(2b)(2h)$, or $2bh$. This is 4 times the area of the small one.

Practice Masters, p. 47

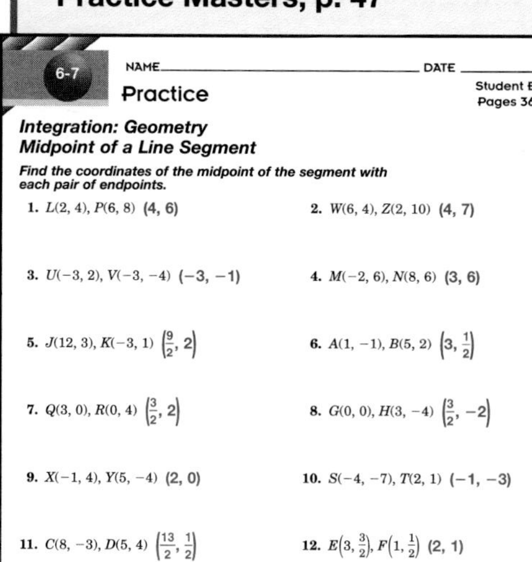

6-7 NAME______ DATE______

Practice — Student Edition Pages 369–374

Integration: Geometry
Midpoint of a Line Segment

Find the coordinates of the midpoint of the segment with each pair of endpoints.

1. $L(2, 4)$, $P(6, 8)$ (4, 6)
2. $W(6, 4)$, $Z(2, 10)$ (4, 7)
3. $U(-3, 2)$, $V(-3, -4)$ (−3, −1)
4. $M(-2, 6)$, $N(8, 6)$ (3, 6)
5. $J(12, 3)$, $K(-3, 1)$ $\left(\frac{9}{2}, 2\right)$
6. $A(1, -1)$, $B(5, 2)$ $\left(3, \frac{1}{2}\right)$
7. $Q(3, 0)$, $R(0, 4)$ $\left(\frac{3}{2}, 2\right)$
8. $G(0, 0)$, $H(3, -4)$ $\left(\frac{3}{2}, -2\right)$
9. $X(-1, 4)$, $Y(5, -4)$ (2, 0)
10. $S(-4, -7)$, $T(2, 1)$ (−1, −3)
11. $C(8, -3)$, $D(5, 4)$ $\left(\frac{13}{2}, \frac{1}{2}\right)$
12. $E\left(3, \frac{3}{2}\right)$, $F\left(1, \frac{1}{2}\right)$ (2, 1)

If P is the midpoint of line segment AB, find the coordinates of the missing point.

13. $A(6, 3)$, $P\left(1, -\frac{1}{2}\right)$ $B(-4, -4)$
14. $B\left(\frac{2}{5}, \frac{1}{10}\right)$, $P\left(\frac{3}{5}, \frac{9}{10}\right)$ $A\left(\frac{4}{5}, \frac{17}{10}\right)$
15. $A(1, 5)$, $P(3, 7)$ $B(5, 9)$
16. $B(2, 11)$, $P(5, 7)$ $A(8, 3)$
17. $A(4, -1)$, $B(7, -3)$ $P\left(\frac{11}{2}, -2\right)$
18. $P(-2, -1)$, $A(-3, -5)$ $B(-1, 3)$

Mathematics and SOCIETY

This excerpt introduces students to a relationship that is not linear. Students might like to collect population data for the planet or a particular country over many decades and plot the data. They will see that it is not a linear graph.

Answers for Mathematics and Society

1. **Sample answer: The more people there are on Earth, the more need for the natural resources and the less resources to share. Without conservation of resources, there will not be enough to go around.**
2. **Sample answers: Less space; less water, power, gas, and other natural resources; smaller yards due to more homes being built; and so on.**

Enrichment Masters, p. 47

NAME ______________________ DATE ____________

Enrichment

Student Edition Pages 369–374

Celsius and Kelvin Temperatures

If you blow up a balloon and put it in the refrigerator, the balloon will shrink as the temperature of the air in the balloon decreases.

The volume of a certain gas is measured at 30° Celsius. The temperature is decreased and the volume is measured again.

Temperature (t)	Volume (v)
30°C	202 mL
21°C	196 mL
0°C	182 mL
−12°C	174 mL
−27°C	164 mL

1. Graph this table on the coordinate plane provided below.

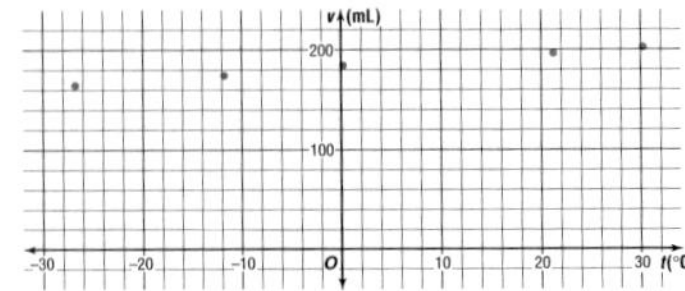

2. Find the equation of the line that passes through the points you graphed in Exercise 1. $y = \frac{2}{3}x + 182$ or $v = \frac{2}{3}t + 182$

3. Use the equation you found in Exercise 2 to find the temperature that would give a volume of zero. This temperature is the lowest one possible and is called "absolute zero." −273°C

4. In 1848 Lord Kelvin proposed a new temperature scale with 0 being assigned to absolute zero. The size of the degree chosen was the same size as the Celsius degree. Change each of the Celsius temperatures in the table above to degrees Kelvin. 303°, 294°, 273°, 261°, 246°

Mixed Review

44. $y = -\frac{3}{5}x + \frac{14}{5}$

45. $y = -\frac{7}{9}x - \frac{8}{3}$

44. Write an equation in slope-intercept form of the line that passes through $(8, -2)$ and is perpendicular to the graph of $5x - 3y = 7$. (Lesson 6–6)

45. Write an equation in slope-intercept form of the line that passes through $(-6, 2)$ and $(3, -5)$. (Lesson 6–4)

46. Given $g(x) = x^2 - x$, find $g(4b)$. (Lesson 5–5) **$16b^2 - 4b$**

47. Determine whether $9x + (-7) = 6y$ is a linear equation. If it is, rewrite it in standard form. (Lesson 5–4) **yes; $9x - 6y = 7$**

48. Trigonometry If $\sin N = 0.6124$, use a calculator to find the measure of angle N to the nearest degree. (Lesson 4–3) **38°**

49. Solve $-27 - b = -7$. (Lesson 3–1) **−20**

50. In Mr. Tucker's algebra class, students can get extra points for finding the correct solution to the "Riddle of the Week." One week, Mr. Tucker posed this riddle. *Josh is 10 years older than his brother. Next year, he will be three times as old as his brother.* Answer the following questions. (Lesson 2–9)

a. How old was Josh when his brother was born? **10 years old**

b. If his brother's age is represented by a, what is Josh's age? **$a + 10$**

c. India says that Josh is 12 years old and his brother is 4. Does India get the extra points for her answer? Why or why not? **no; $4 + 10 \neq 12$**

51. Simplify $0.2(3x + 0.2) + 0.5(5x + 3)$. (Lesson 1–8) **$3.1x + 1.54$**

3b. Sample answer: No, this does not take into account deaths and births that may occur during the immigration.

Mathematics and SOCIETY

People Around the World

This excerpt is from an article in *The Amicus Journal*, Winter 1994.

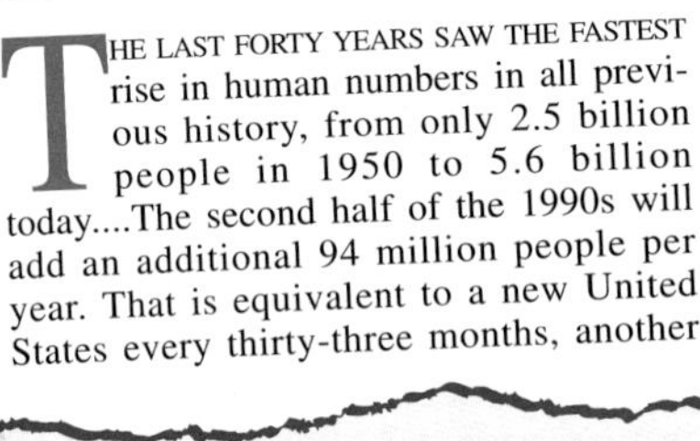

THE LAST FORTY YEARS SAW THE FASTEST rise in human numbers in all previous history, from only 2.5 billion people in 1950 to 5.6 billion today....The second half of the 1990s will add an additional 94 million people per year. That is equivalent to a new United States every thirty-three months, another Britain every seven months, a Washington (DC) every six days. A whole Earth of 1800 was added in just one decade, according to United Nations Population Division statistics. After 2000, annual additions will slow, but by 2050 the United Nations expects the human race to total just over 10 billion—an extra Earth of 1980 on top of today's. ■

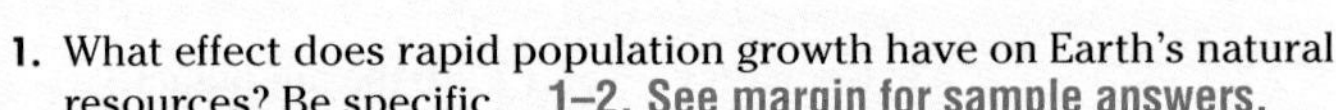

1. What effect does rapid population growth have on Earth's natural resources? Be specific. **1–2. See margin for sample answers.**
2. What changes would you expect to see if the community in which you live doubles in population during your lifetime?
3. A village is deserted, but its buildings are intact and there are ample resources. Suppose 1000 refugees per day flock to this village.
   a. At this rate, how long would it take to reach the midpoint to a population of five million? **about 6.85 years**
   b. Is the proposition of a village growing at this rate realistic? Why or why not?

CHAPTER 6 HIGHLIGHTS

VOCABULARY

After completing this chapter, you should be able to define each term, property, or phrase and give an example or two of each.

Algebra

family of graphs (p. 354)
parent graph (p. 354)
point-slope form (p. 333)
rise (p. 325)
run (p. 325)
slope (pp. 324, 325)
slope-intercept form (p. 347)
standard form (p. 333)
x-intercept (p. 346)
y-intercept (p. 346)

Problem Solving

use a model (p. 339)

Geometry

midpoint (p. 369)
parallel lines (p. 362)
parallelogram (p. 362)
perpendicular lines (p. 362)

Statistics

best-fit line (p. 341)
negative correlation (p. 340)
positive correlation (p. 340)
regression line (p. 342)
scatter plot (p. 339)

UNDERSTANDING AND USING THE VOCABULARY

Choose the correct term to complete each sentence.

1. The lines with equations $y = -2x + 7$ and $y = -2x - 6$ are (*parallel, perpendicular*) lines.
2. The equation $y - 2 = -3(x - 1)$ is written in (*point-slope, slope-intercept*) form.
3. If the endpoints of $\overline{AB}$ are $A(6, -3)$ and $B(2, 4)$, then the (*midpoint, slope*) of $\overline{AB}$ is at $\left(4, \frac{1}{2}\right)$.
4. The (*x-intercept, y-intercept*) of the equation $2x + 3y = -1$ is $-\frac{1}{2}$.
5. The lines with equations $y = \frac{1}{3}x + 1$ and $y = -3x - 5$ are (*parallel, perpendicular*) lines.
6. The slope of a line is defined as the ratio of the (*rise, run*), or vertical change, to the (*rise, run*), or horizontal change, as you move from one point on the line to another.
7. The equation $y = -\frac{1}{2}x + 3$ is written in (*slope-intercept, standard*) form.
8. The (*x-intercept, y-intercept*) of the equation $-x - 4y = 2$ is $-\frac{1}{2}$.
9. The (*midpoint, slope*) of the line with equation $-3y = 2x + 3$ is $-\frac{2}{3}$.
10. The equation $3x - 4y = -5$ is written in (*point-slope, standard*) form.

Instructional Resources

Three multiple-choice tests and three free-response tests are provided in the *Assessment and Evaluation Masters*. Forms 1A and 2A are for honors pacing, and Forms 1B, 1C, 2B, and 2C are for average pacing. Chapter 6 Test, Form 1B is shown at the right. Chapter 6 Test, Form 2B is shown on the next page.

Using the CHAPTER HIGHLIGHTS

The Chapter Highlights begins with a listing of the new terms, properties, and phrases that were introduced in this chapter. Have students define each term and provide an example or two of it, if appropriate.

Assessment and Evaluation Masters, pp. 143–144

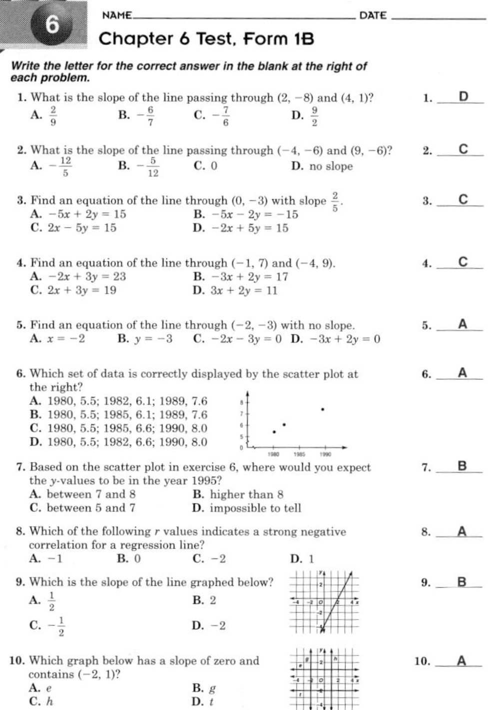

6 NAME________ DATE ________

Chapter 6 Test, Form 1B

Write the letter for the correct answer in the blank at the right of each problem.

1. What is the slope of the line passing through (2, −8) and (4, 1)? **1.** D
   A. $\frac{2}{9}$ B. $-\frac{6}{7}$ C. $-\frac{7}{6}$ D. $\frac{9}{2}$
2. What is the slope of the line passing through (−4, −6) and (9, −6)? **2.** C
   A. $-\frac{12}{5}$ B. $-\frac{5}{12}$ C. 0 D. no slope
3. Find an equation of the line through (0, −3) with slope $\frac{2}{5}$. **3.** C
   A. $-5x + 2y = 15$ B. $-5x - 2y = -15$
   C. $2x - 5y = 15$ D. $-2x + 5y = 15$
4. Find an equation of the line through (−1, 7) and (−4, 9). **4.** C
   A. $-2x + 3y = 23$ B. $-3x + 2y = 17$
   C. $2x + 3y = 19$ D. $3x + 2y = 11$
5. Find an equation of the line through (−2, −3) with no slope. **5.** A
   A. $x = -2$ B. $y = -3$ C. $-2x - 3y = 0$ D. $-3x + 2y = 0$
6. Which set of data is correctly displayed by the scatter plot at the right? **6.** A
   A. 1980, 5.5; 1982, 6.1; 1989, 7.6
   B. 1980, 5.5; 1985, 6.1; 1989, 7.6
   C. 1980, 5.5; 1985, 6.6; 1990, 8.0
   D. 1980, 5.5; 1982, 6.6; 1990, 8.0
7. Based on the scatter plot in exercise 6, where would you expect the *y*-values to be in the year 1995? **7.** B
   A. between 7 and 8 B. higher than 8
   C. between 5 and 7 D. impossible to tell
8. Which of the following *r* values indicates a strong negative correlation for a regression line? **8.** A
   A. −1 B. 0 C. −2 D. 1
9. Which is the slope of the line graphed below? **9.** B
   A. $\frac{1}{2}$ B. 2
   C. $-\frac{1}{2}$ D. −2
10. Which graph below has a slope of zero and contains (−2, 1)? **10.** A
    A. *e* B. *g*
    C. *h* D. *t*

6 NAME________ DATE ________

Chapter 6 Test, Form 1B (continued)

11. What is the equation in slope-intercept form of the line whose graph has a slope of 5 and *y*-intercept of −8? **11.** D
    A. $y = -8x + 5$ B. $y = 8x - 5$ C. $5x - y = -8$ D. $y = 5x - 8$
12. Find an equation in slope-intercept form of the line that passes through (−4, −9) and the origin. **12.** A
    A. $y = \frac{9}{4}x$ B. $9x + 4y = 0$ C. $y = -9$ D. $y + 9 = x$
13. A baby blue whale weighs 3 tons at birth. In 10 days, it weighs 4 tons. Assume that the growth can be represented by a straight line. Which equation shows the weight, *w*, when the whale is *d* days old? **13.** C
    A. $w = 10d + 3$ B. $w = 10d + 4$
    C. $w = \frac{1}{10}d + 3$ D. $w = d + 10$
14. What is the slope of the graph of $6x - 2y = 24$? **14.** B
    A. −12 B. 3 C. −6 D. 6
15. What is the *x*-intercept of the graph of $9x - y = 18$? **15.** A
    A. 2 B. −2 C. 18 D. −18
16. The lines $2x + y = 7$ and $2y = 9 - 4x$ are **16.** A
    A. parallel. B. perpendicular.
    C. the same line. D. none of these.
17. Find the equation of the line parallel to the graph of $12x - 3y = 10$ and passing through (−5, 3). **17.** D
    A. $y = -4x - 17$ B. $y = 4x - 13$
    C. $y = -4x + 13$ D. $y = 4x + 23$
18. Find an equation of the line perpendicular to the graph of $y = 2x - 3$ and passing through (2, −5). **18.** A
    A. $x + 2y = -8$ B. $x - 2y = 12$
    C. $2x - y = 9$ D. $2x - y = -12$
19. What are the coordinates of the midpoint of the segment whose endpoints are (8, −5) and (−6, 11)? **19.** C
    A. (3, 1) B. (7, 8) C. (1, 3) D. (7, −8)
20. *P* is the midpoint of line segment *AB*. If *A* has coordinates (7, 1) and *P* has coordinates (−3, −9), what are the coordinates of *B*? **20.** B
    A. (2, −4) B. (−13, −19) C. (17, 11) D. none of these

Bonus Find the equation in slope-intercept form of the line with *x*-intercept −5 and *y*-intercept 8. **Bonus** $y = \frac{8}{5}x + 8$

Using the STUDY GUIDE AND ASSESSMENT

Skills and Concepts Encourage students to refer to the objectives and examples on the left as they complete the review exercises on the right.

Assessment and Evaluation Masters, pp. 149–150

6 NAME__________ DATE __________

Chapter 6 Test, Form 2B

Determine the slope of the line passing through each pair of points. If the slope is undefined, write "no slope."

1. $(6, -4)$ and $(-3, 7)$ — 1. $-\frac{11}{9}$
2. $(-2, -6)$ and $(7, -6)$ — 2. 0
3. Determine the value of r so that the line through $(-4, 8)$ and $(r, -6)$ has a slope of $\frac{2}{3}$. — 3. -25

Write an equation in standard form of the line satisfying the given conditions.

4. has no slope and passes through $(-6, 4)$ — 4. $x = -6$
5. passes through $(1, -4)$ and $(-6, 8)$ — 5. $12x + 7y = -16$
6. passes through $(7, -3)$ and has a y-intercept of 2 — 6. $5x + 7y = 14$

The average SAT math and verbal scores for selected years from 1967 to 1985 are listed in the table below. Use the data for problems 7–9.

Year	Verbal Score	Math Score
1967	466	492
1970	460	488
1975	434	472
1980	424	466
1985	431	475

7. Make a scatter plot of the data with the verbal score on the horizontal axis and the math score on the vertical axis. — 7.
8. Is the correlation positive or negative? — 8. positive
9. Predict the mathematics score corresponding to a verbal score of 445. — 9. about 478
10. Graph the line with x-intercept -3 and y-intercept 2. Use the coordinate plane at the right. — 10.

6 NAME__________ DATE __________

Chapter 6 Test, Form 2B (continued)

11. Graph the line with y-intercept 3 and slope $-\frac{3}{4}$. Use the coordinate plane at the right. — 11.

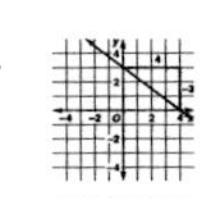

12. Graph the equation $4x - 3y = -9$. Use the coordinate plane at the right. — 12.

Write an equation in slope-intercept form of the line satisfying the given conditions.

13. passes through $(5, 4)$ and $(6, -1)$ — 13. $y = -5x + 29$
14. slope $\frac{1}{3}$ and passes through $(-2, 8)$ — 14. $y = \frac{1}{3}x + \frac{26}{3}$
15. passes through $(10, -3)$ and has a y-intercept of 2 — 15. $y = -\frac{1}{2}x + 2$
16. parallel to the graph of $9x + 3y = 6$ and passes through $(5, 3)$ — 16. $y = -3x + 18$
17. perpendicular to the graph of $4x - y = 12$ and passes through $(8, 2)$ — 17. $y = -\frac{1}{4}x + 4$
18. parallel to the x-axis and passes through $(4, 2)$ — 18. $y = 2$
19. Find the coordinates of the midpoint of the segment whose endpoints are $(-2, 14)$ and $(3, -9)$. — 19. $\left(\frac{1}{2}, 2\frac{1}{2}\right)$
20. The center of a circle is $M(-3, 0)$. If $\overline{AB}$ is a diameter, and $A(-7, 3)$ is one endpoint on the circle, find the coordinates of B. — 20. $(1, -3)$

Bonus In a certain lake, a 1-year-old bluegill fish is 3 inches long, while a 4-year-old bluegill is 6.6 inches long. Assuming the growth rate can be approximated by a linear equation, write an equation in slope-intercept form for the length of a bluegill fish in inches (y) after x years. Use this equation to estimate the length of a 10-year-old bluegill. About how old is a 9-inch-long bluegill? — Bonus $y = 1.2x + 1.8$; 13.8 in.; 6 yr

CHAPTER 6 STUDY GUIDE AND ASSESSMENT

SKILLS AND CONCEPTS

OBJECTIVES AND EXAMPLES

Upon completing this chapter, you should be able to:

- find the slope of a line, given the coordinates of two points on the line (Lesson 6–1)

Determine the slope of the line that passes through $(-6, 5)$ and $(3, -2)$.

$$m = \frac{y_2 - y_1}{x_2 - x_1} = \frac{-2 - 5}{3 - (-6)} = -\frac{7}{9}$$

The slope is $-\frac{7}{9}$.

REVIEW EXERCISES

Use these exercises to review and prepare for the chapter test.

Determine the slope of the line that passes through each pair of points. 12. undefined

11. $(8, 3), (2, 5)$ $-\frac{1}{3}$
12. $(-2, 5), (-2, 9)$
13. $(-3, 6), (-8, 4)$ $\frac{2}{5}$
14. $(4, 3), (-5, 3)$ 0

Determine the value of r so the line that passes through each pair of points has the given slope.

15. $(r, 4), (7, 3), m = \frac{3}{4}$ $\frac{25}{3}$
16. $(4, -7), (-2, r), m = \frac{8}{3}$ -23

17. $y + 3 = -2(x - 4)$ 18. $y - 5 = 5(x - 8)$ 19. $y - 7 = 0$ 20. $y - 2 = \frac{1}{2}(x - 6)$ 21. $y - 3 = \frac{3}{5}x$

- write linear equations in point-slope form (Lesson 6–2)

Write the point-slope form of an equation of the line that passes through the points at $(-4, 7)$ and $(-2, 3)$.

$m = \frac{3 - 7}{-2 - (-4)}$ or -2

$y - y_1 = m(x - x_1)$

$y - 7 = -2[x - (-4)]$

$y - 7 = -2(x + 4)$

An equation of the line is $y - 7 = -2(x + 4)$.

Write the point-slope form of an equation of the line that passes through the given point and has the given slope.

17. $(4, -3), m = -2$
18. $(8, 5), m = 5$
19. $(-5, 7), m = 0$
20. $(6, 2), m = \frac{1}{2}$

Write the point-slope form of an equation of the line that passes through each pair of points.

21. $(0, 3), (-5, 0)$
22. $(5, 4), (6, 3)$
23. $(4, 1), (-3, 7)$
24. $(2, -5), (0, 4)$

22. $y - 3 = -(x - 6)$ 23. $y - 1 = -\frac{6}{7}(x - 4)$ 24. $y + 5 = -\frac{9}{2}(x - 2)$

- write linear equations in standard form (Lesson 6–2)

Write the standard form of an equation of the line that passes through the points at $(6, -4)$ and $(-1, 5)$.

$m = \frac{5 - (-4)}{-1 - 6}$ or $-\frac{9}{7}$

$y - y_1 = m(x - x_1)$

$y - (-4) = -\frac{9}{7}(x - 6)$

$y + 4 = -\frac{9}{7}(x - 6)$

$y + 4 = -\frac{9}{7}(x - 6)$

$7(y + 4) = 7 \cdot -\frac{9}{7}(x - 6)$

$7y + 28 = -9(x - 6)$

$7y + 28 = -9x + 54$

$9x + 7y = 26$

An equation of the line is $9x + 7y = 26$.

Write the standard form of an equation of the line that passes through the given point and has the given slope. 25–28. See margin.

25. $(4, -6), m = 3$
26. $(1, 5), m = 0$
27. $(6, -1), m = \frac{3}{4}$
28. $(8, 3), m =$ undefined

Write the standard form of an equation of the line that passes through each pair of points. 29–32. See margin.

29. $(-2, 5), (9, 5)$
30. $(0, 5), (-2, 0)$
31. $\left(-2, \frac{2}{3}\right), \left(-2, \frac{2}{7}\right)$
32. $(-5, 7), \left(0, \frac{1}{2}\right)$

GLENCOE Technology

Test and Review Software

You may use this software, a combination of an item generator and item bank, to create your own tests or worksheets. Types of items include free response, multiple choice, short answer, and open ended.

For IBM & Macintosh

Additional Answers

25. $3x - y = 18$
26. $y = 5$
27. $3x - 4y = 22$
28. $x = 8$
29. $y = 5$
30. $5x - 2y = -10$
31. $x = -2$
32. $13x + 10y = 5$

Additional Answers

OBJECTIVES AND EXAMPLES

- graph and interpret points on a scatter plot; draw and write equations for best-fit lines and make predictions by using those equations (Lesson 6–3)

The scatter plot below shows a negative relationship since the line suggested by the points would have a negative slope.

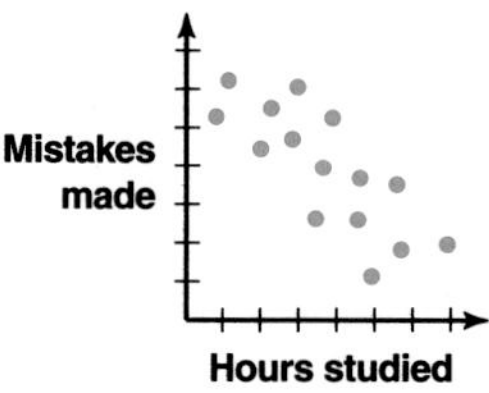

33b. Sample answer: 35 stories

33c. See margin for graph.

REVIEW EXERCISES

33. Draw a scatter plot of the data below. Let the independent variable be height.

Buildings in Oklahoma City	Height (feet)	Number of Stories
Liberty Tower	500	36
First National Center	493	28
City Place	440	33
First Oklahoma Tower	425	31
Kerr-McGee Center	393	30
Mid-America Tower	362	29

a. Does a relationship exist? If so, what is the relationship? Yes; it is positive.

b. Use your scatter plot as a model to predict how many stories you might think a building 475 feet high would be.

c. Draw a best-fit line and write the equation of the line. $y = 0.03x + 21$

- determine the x- and y-intercepts of linear graphs from their equations and write equations in slope-intercept form (Lesson 6–4)

Determine the slope and x- and y-intercepts of the graph of $3x - 2y = 7$.

Solve for y to find the slope and y-intercept.	Let $y = 0$ to find the x-intercept.
$3x - 2y = 7$	$3x - 2(0) = 7$
$-2y = -3x + 7$	$3x = 7$
$y = \frac{3}{2}x - \frac{7}{2}$	$x = \frac{7}{3}$

The slope is $\frac{3}{2}$, the y-intercept is $-\frac{7}{2}$, and the x-intercept is $\frac{7}{3}$.

Write an equation in slope-intercept form of a line with the given slope and *y*-intercept.

34. $m = 2, b = 4$ $y = 2x + 4$

35. $m = -3, b = 0$ $y = -3x$

36. $m = -\frac{1}{2}, b = -9$ $y = -\frac{1}{2}x - 9$

37. $m = 0, b = 5.5$ $y = 5.5$

Find the slope and *y*-intercept of the graph of each equation.

38. $x = 2y - 7$ $\frac{1}{2}, \frac{7}{2}$

39. $8x + y = 4$ $-8, 4$

40. $y = \frac{1}{4}x + 3$ $\frac{1}{4}, 3$

41. $\frac{1}{2}x + \frac{1}{4}y = 3$ $-2, 12$

- graph a line given any linear equation (Lesson 6–5)

Graph $-2x - y = 5$.

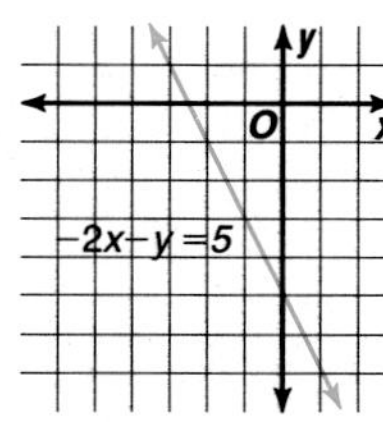

Graph each equation.

42. $3x - y = 9$

43. $5x + 2y = 12$

44. $y = \frac{2}{3}x + 4$

45. $y = -\frac{3}{2}x - 6$

46. $3x + 4y = 6$

47. $5x - \frac{1}{2}y = 2$

48. $y - 4 = -2(x + 1)$

49. $y + 5 = -\frac{3}{4}(x - 6)$

42–49. See margin for graphs.

Additional Answers

33c.

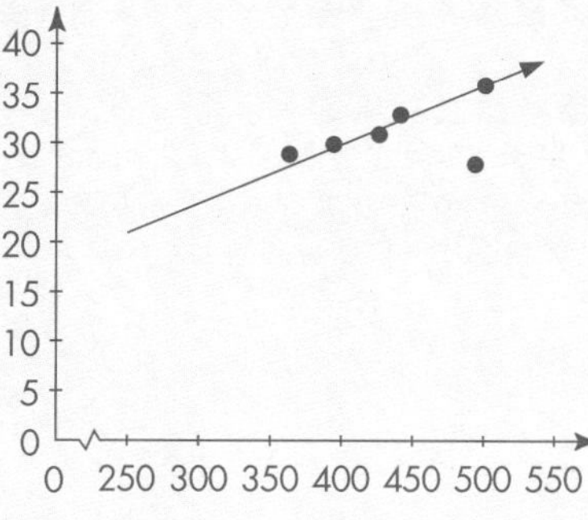

42.

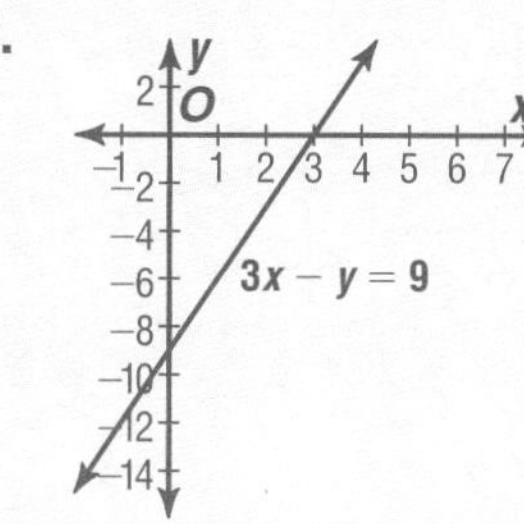

43.

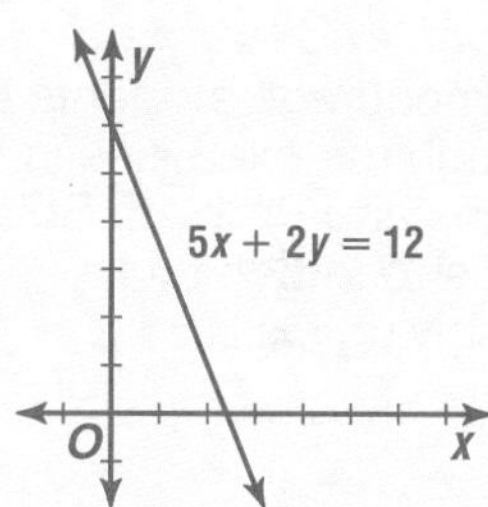

44.

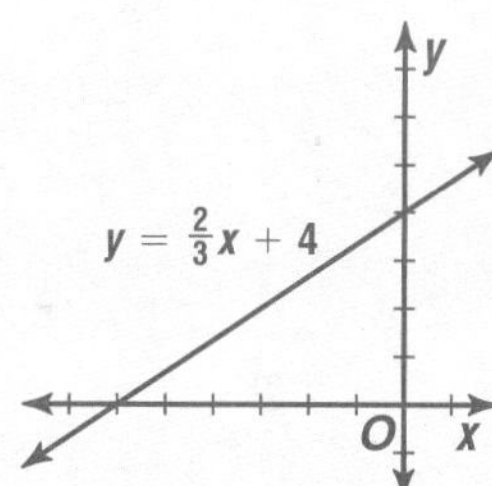

45.

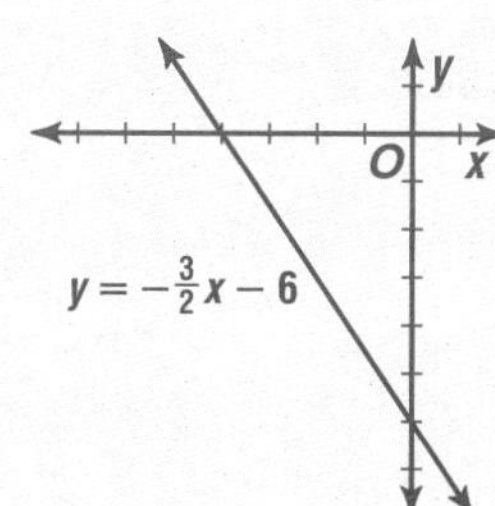

46.

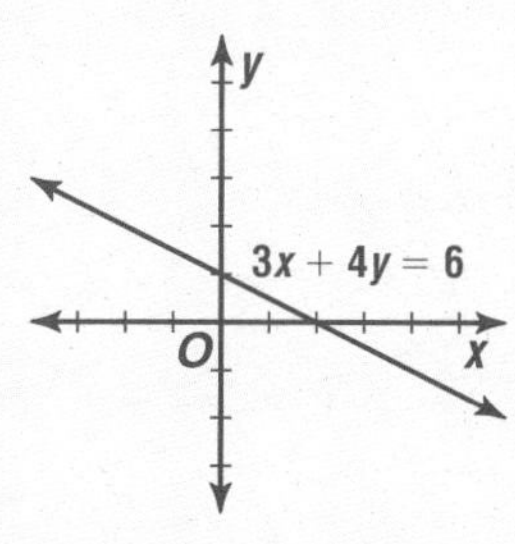

47.

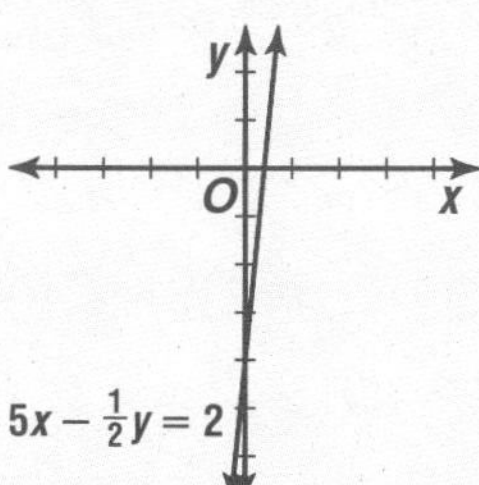

48.

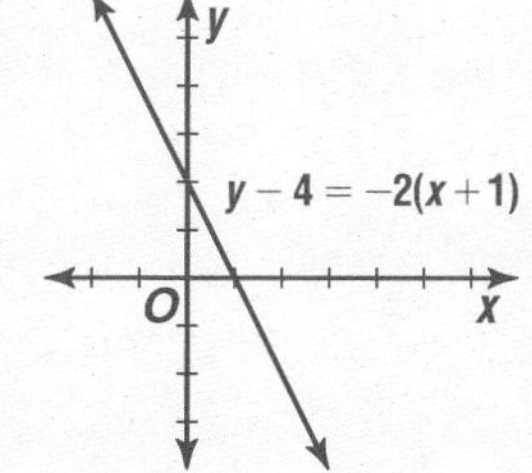

49.

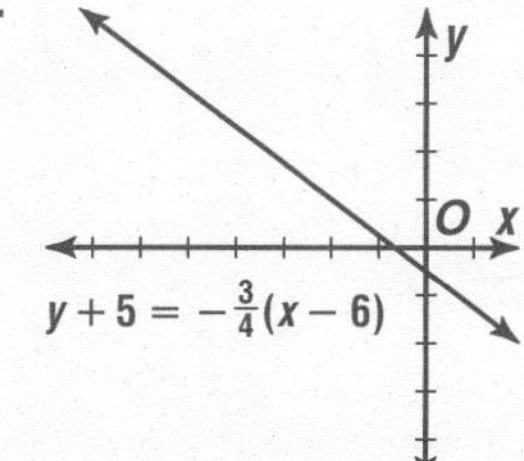

Additional Answers

50. $y = 4x - 9$
51. $y = -\frac{7}{2}x - 14$
52. $y = -\frac{1}{3}x + 1$
53. $y = -\frac{3}{8}x + \frac{13}{2}$
54. $y = 2x - 16$
55. $y = 5x - 15$
56. (6, 1)
57. $\left(1, -\frac{5}{2}\right)$
58. (8, 2)
59. $\left(5, \frac{11}{2}\right)$
60. (3, 2)
61. $\left(\frac{7}{2}, -\frac{3}{2}\right)$
62. (19, 9)
63. (13, 11)
64. (6, −7)
65. (−5, 20)

Applications and Problem Solving Encourage students to work through the exercises in the Applications and Problem Solving section to strengthen their problem-solving skills.

OBJECTIVES AND EXAMPLES

- write equations of lines that pass through a given point, parallel or perpendicular to the graph of a given equation (Lesson 6–6)

Write an equation of the line that is perpendicular to the graph of $2x + y = 6$ and passes through (2, 3).

$2x + y = 6$

$y = -2x + 6$

The slope of this line is −2. Therefore, the slope of the line perpendicular to it is $\frac{1}{2}$.

$y - 3 = \frac{1}{2}(x - 2)$ $m = \frac{1}{2}, (x_1, y_1) = (2, 3)$

$y = \frac{1}{2}x + 2$

REVIEW EXERCISES

Write an equation of the line having the following properties 50–55. See margin.

50. parallel to $4x - y = 7$ and passes through (2, −1)
51. perpendicular to $2x - 7y = 1$ and passes through (−4, 0)
52. parallel to $3x + 9y = 1$ and passes through (3, 0)
53. perpendicular to $8x - 3y = 7$ and passes through (4, 5)
54. parallel to $-y = -2x + 4$ and passes through (5, −6)
55. perpendicular to $5y = -x + 1$ and passes through (2, −5)

- find the coordinates of the midpoint of a line segment in the coordinate plane (Lesson 6–7)

The endpoints of a segment are at (11, 4) and (9, 2). Find its midpoint.

$(x, y) = \left(\frac{x_1 + x_2}{2}, \frac{y_1 + y_2}{2}\right)$

$= \left(\frac{11+9}{2}, \frac{4+2}{2}\right)$

$= (10, 3)$

56–65. See margin.

Find the coordinates of the midpoint of a segment with each pair of endpoints.

56. $A(3, 5), B(9, -3)$
57. $A(-6, 6), B(8, -11)$
58. $A(14, 4), B(2, 0)$
59. $A(2, 7), B(8, 4)$
60. $A(2, 5), B(4, -1)$
61. $A(10, 4), B(-3, -7)$

Find the coordinates of the other endpoint of a segment given one endpoint and the midpoint *M*.

62. $A(3, 5), M(11, 7)$
63. $A(5, 3), M(9, 7)$
64. $A(4, -11), M(5, -9)$
65. $A(11, -4), M(3, 8)$

APPLICATIONS AND PROBLEM SOLVING

66. **Skiing** A course for cross-country skiing is regulated so that the slope of any hill cannot be greater than 0.33. Suppose a hill rises 60 meters over a horizontal distance of 250 meters.
    a. What is the slope of the hill? $\frac{6}{25}$
    b. Does the hill meet the requirements? (Lesson 6–1) yes, $\frac{6}{25} < \frac{1}{3}$

67. **Travel** Jon Erlanger is taking a long trip. In the first two hours, he drives 80 miles. After that, he averages 45 miles per hour. Write an equation in slope-intercept form relating distance traveled and time. (Lesson 6–4) $d = 45t - 10$

A practice test for Chapter 6 is provided on page 792.

68. **Entertainment** Carolyn Parks owns stock in Star Gazer Motion Picture Company. Every other week, she graphs the closing value of a share of the stock. What is the midpoint between the highest and lowest values of the stock? (Lesson 6–7) (7, $5.50)

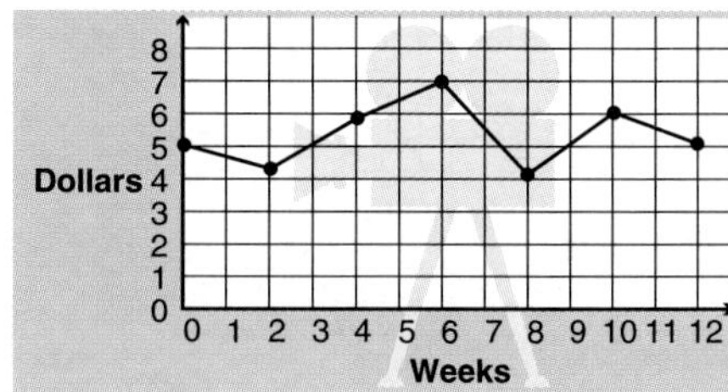

ALTERNATIVE ASSESSMENT

COOPERATIVE LEARNING PROJECT

Video Rental In this chapter, you investigated graphing linear equations. You used slope and data points to graph equations. These graphs represented information that could be analyzed and extended in order to interpret further information than what was given.

In this project, you will analyze two video-store rental policies. There are two stores in Joe's neighborhood where he can rent videos: Video World and Mega Video. At one of the two stores, you have to buy a membership card first. The other one does not have these cards; you simply pay per video. Both have a daily charge (rate per day). Mega Video has a membership card that costs \$15 per year and then each video that is rented costs \$2 per day. Video World simply rents each video for \$3.25 per day.

Which is the most economical video store for Joe to use? If he rarely rents videos, which store should he use? If he often rents videos, which store should he use?

Follow these steps to determine the store he should use.

- Determine a way to graph this information.
- Develop a linear equation for each store that describes the amount of money it costs to rent a video.
- Determine whether the graphs should be drawn separately or on the same grid.
- Investigate the graphs and determine what each graph represents.
- What factors other than price might Joe want to consider before he chooses a store?
- Write a paragraph describing various situations for Joe and the appropriate solutions.

THINKING CRITICALLY

- Give an example of an equation of a line that has the same x- and y-intercepts. What is the slope of the line?
- Give an example of an equation of a line whose x- and y-intercepts are opposites. What is the slope of the line?
- Make a conjecture about the slopes of lines whose x- and y-intercepts are equal or opposites.
- The x-intercept of a line is s, and the y-intercept is t. Write the equation of the line.

PORTFOLIO

In mathematics, there is often more than one way to solve a problem. In order to graph a line for example, various information can be used. Sometimes you use two points, sometimes you use a point and a slope, and sometimes you use x- and y-intercepts. Find an equation from your work in this chapter and describe at least three ways in which to graph it. Place this in your portfolio.

SELF EVALUATION

Checking your answer is the final step in solving a problem. This step is the most critical, but it is also the step that is most often forgotten. When an answer is finally reached, it should be checked to see whether it makes sense and also whether it can be verified.

Assess yourself. Do you use all four steps when you solve a problem? Do you reach an answer, assume that it is the correct answer, and then move on without checking it? Do you ask yourself questions about your answer to verify that it is correct and that it makes sense? Can you think of an example of a problem in your daily life in which you didn't ask yourself questions about your solution to the problem and later you discovered that the solution didn't work?

Assessment and Evaluation Masters, pp. 154, 165

NAME________________ DATE ________

Chapter 6 Performance Assessment

Instructions: ***Demonstrate your knowledge by giving a clear, concise solution to each problem. Be sure to include all relevant drawings and justify your answers. You may show your solution in more than one way or investigate beyond the requirements of the problem.***

1. a. Write a linear equation in standard form. Tell how you know the equation is linear.

b. Use the coefficients and constant term to name the slope and y-intercept of the graph. Then write the equation in slope-intercept form.

c. Use the y-intercept and slope to graph the equation. Explain how you graphed the equation.

d. Write an equation whose graph is parallel to the graph in part c. Tell how you know they are parallel.

e. Write an equation of a line perpendicular to the graph in part c. Tell how you know they are perpendicular.

2. a. Describe the pattern of the points in the scatter plot at the right.

b. Give at least two examples of real-life situations that, if graphed, would result in a correlation like the one shown in this scatter plot.

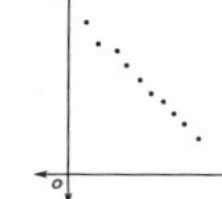

3. The graphs of each group of equations have at least one characteristic in common. Name the characteristic(s) and then graph each group of equations on the same axes to verify your answer.

a. $y = x, y = 3x, y = 5x$

b. $y = 2x, y = 2x + 1, y = 2x - 3$

c. $y = \frac{1}{2}x, y = \frac{1}{3}x, y = \frac{2}{5}x$

d. $2y = 6x, 4y - 12x = 0, 3x - y = 0$

e. $2x - 4y = 12, 3y + 2x = -2, x + y = 0$

Scoring Guide
Chapter 6
Performance Assessment

Level	Specific Criteria
3 Superior	• Shows thorough understanding of the concepts of *standard form, linear equation, slope, y-intercept, slope-intercept form, parallel and perpendicular lines, scatter plot,* and *correlation.* • Uses appropriate strategies to solve problems. • Computations are correct. • Written explanations are exemplary. • Word problem concerning a linear function is appropriate and makes sense. • Graphs are accurate and appropriate. • Goes beyond requirements of some or all problems.
2 Satisfactory, with Minor Flaws	• Shows understanding of the concepts of *standard form, linear equation, slope, y-intercept, slope-intercept form, parallel and perpendicular lines, scatter plot,* and *correlation.* • Uses appropriate strategies to solve problems. • Computations are mostly correct. • Written explanations are effective. • Word problem concerning a linear function is appropriate and makes sense. • Graphs are mostly accurate and appropriate. • Satisfies all requirements of problems.
1 Nearly Satisfactory, with Serious Flaws	• Shows understanding of most of the concepts of *standard form, linear equation, slope, y-intercept, slope-intercept form, parallel and perpendicular lines, scatter plot,* and *correlation.* • May not use appropriate strategies to solve problems. • Computations are mostly correct. • Written explanations are satisfactory. • Word problem concerning a linear function is appropriate and sensible. • Graphs are mostly accurate and appropriate. • Satisfies most requirements of problems.
0 Unsatisfactory	• Shows little or no understanding of the concepts of *standard form, linear equation, slope, y-intercept, slope-intercept form, parallel and perpendicular lines, scatter plot,* and *correlation.* • May not use appropriate strategies to solve problems. • Computations are incorrect. • Written explanations are not satisfactory. • Word problem concerning a linear function is not appropriate or sensible. • Graphs are not accurate or appropriate. • Does not satisfy requirements of problems.

Alternative Assessment

The Alternative Assessment section provides students with the opportunity to assess their own work by thinking critically, working with others, keeping a portfolio, and honestly evaluating their own progress. For more information on alternative forms of assessment, see *Alternative Assessment in the Mathematics Classroom,* one of the titles in the Glencoe Mathematics Professional Series.

Performance Assessment

Performance Assessment tasks for this chapter are included in the *Assessment and Evaluation Masters.* A scoring guide is also provided.

Using the CUMULATIVE REVIEW

These two pages review the skills and concepts presented in Chapters 1–6. This review is formatted to reflect new trends in standardized testing.

A more traditional cumulative review, shown below, is provided in the *Assessment and Evaluation Masters,* pp. 159–160.

Assessment and Evaluation Masters, pp. 159–160

6 NAME ______ DATE ______

Chapter 6 Cumulative Review

1. Write the expression $2 \cdot r \cdot r \cdot s \cdot s$ using exponents. (Lesson 1-1) — 1. $2r^2s^2$
2. Evaluate $2xy - y^2$ if $x = 6$ and $y = 12$. (Lesson 1-3) — 2. 0

Simplify. (Lessons 2-5, 2-6, and 2-7)

3. $4\frac{7}{8} - 2\frac{5}{8}$ — 3. $2\frac{1}{4}$
4. $\frac{3}{8} \times 2\frac{7}{18}$ — 4. $\frac{43}{48}$
5. $-\frac{2}{3} + \frac{3}{4}$ — 5. $\frac{1}{12}$
6. $-3.9 + (-2.5) + (-8.7)$ — 6. -15.1
7. $4(2y + y) - 6(4y + 3y)$ — 7. $-30y$
8. $\frac{12a - 18b}{-6}$ — 8. $-2a + 3b$

Solve each equation. (Lessons 3-1 and 3-2)

9. $13 - m = 21$ — 9. -8
10. $\frac{3}{4}x = \frac{2}{3}$ — 10. $\frac{8}{9}$
11. Find three consecutive even integers whose sum is 132. (Lesson 3-3) — 11. 42, 44, 46
12. Which is a better buy: a liter of milk for 59¢ or 1.5 liters for 81¢? — 12. 1.5 liters for 81¢
13. Six is what percent of 80? — 13. 7.5%

Find the final price of each item. When there is a discount and sales tax, compute the discount price first.

14. calculator: \$90
    sales tax: 8% — 14. \$97.20
15. magazine: \$3.95
    discount: 10%
    sales tax: 6.5% — 15. \$3.79

6 NAME ______ DATE ______

Chapter 6 Cumulative Review (continued)

16. A can contains two different kinds of nuts, weighs 5 pounds, and cost \$16. One type of nut costs \$3.50 a pound. The other type costs \$2.75 a pound. How many pounds of \$2.75 nuts are there? (Lesson 4-1) — 16. 2 pounds at \$2.75
17. Solve $x - 2y = 12$ if the domain is $\{-3, -1, 0, 2, 5\}$. (Lesson 5-3) — 17. $\{(-3, -7.5), (-1, -6.5), (0, -6), (2, -5), (5, -3, 5)\}$
18. Graph $3x - y = 1$. Use the coordinate plane provided. (Lesson 5-4) — 18.
19. State whether the following relation is a function. (Lesson 5-5) $\{(1, 4), (2, 6), (3, 7), (4, 4)\}$ — 19. yes
20. Write an equation for the relationship between the variables in the chart at the right. (Lesson 5-6) — 20. $y = \frac{3}{2}x + 2$

x	0	2	4	6
y	2	5	8	11

21. Determine the slope of the line passing through (2, 7) and (−5, 2). (Lesson 6-1) — 21. $\frac{5}{7}$
22. Write an equation in standard form for the line passing through (2, 6) and having a slope of −3. (Lesson 6-2) — 22. $3x + y = 12$
23. Write an equation in slope-intercept form for the line in exercise 22. (Lesson 6-4) — 23. $y = -3x + 12$
24. Write an equation for the line that is parallel to the graph of $5x - 3y = 1$ and passes through (0, −4). (Lesson 6-6) — 24. $y = \frac{5}{3}x - 4$
25. Graph $y = 2x - 3$. Use the coordinate plane at the right. (Lesson 6-5) — 25.

CUMULATIVE REVIEW

CHAPTERS 1–6

SECTION ONE: MULTIPLE CHOICE

There are eight multiple-choice questions in this section. After working each problem, write the letter of the correct answer on your paper.

1. Name the quadrant in which point $P(x, y)$ is located if it satisfies the condition $x < 0$ and $y = -3$. C
   - A. I
   - B. II
   - C. III
   - D. IV

2. Raini made a map of his neighborhood using a coordinate grid. His school has coordinates (3, 8), where 3 is the number of blocks east he must walk and 8 is the number of blocks north. From school, he walks 4 blocks east and 2 blocks south to get to his friend's house. Find the coordinates of the midpoint of the line segment that could be drawn to connect his school and his friend's house. A
   - A. (5, 7)
   - B. (8, −2)
   - C. (7, 6)
   - D. (5, −6)

3. Determine the slope of the line that passes through the points at (−3, 6) and (−5, 9). C
   - A. $\frac{3}{2}$
   - B. $\frac{2}{3}$
   - C. $-\frac{3}{2}$
   - D. none of these

4. **Probability** A shipment of 100 CDs was just received at Tunes R Us. There is a 4% probability that one of the CDs was damaged during shipment, even though the package may not be cracked. If Craig buys a CD from this shipment, what are the odds that he is buying a damaged CD? D
   - A. 1:25
   - B. $\frac{4}{100}$
   - C. 96:4
   - D. 1:24

5. Choose the graph that represents a function. B

A. 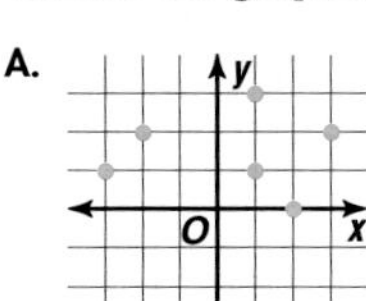

B. 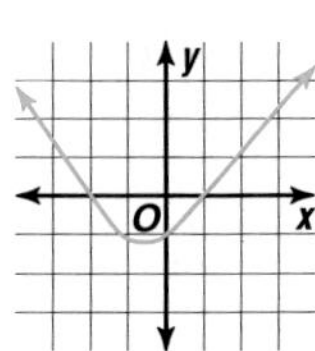

C. 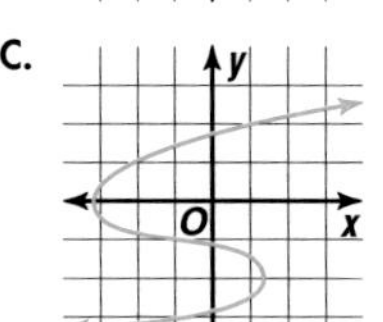

D. 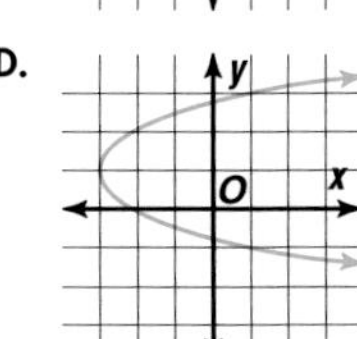

6. Choose the property illustrated by the statement *The quantity g plus h times b equals g times b plus h times b.* A
   - A. distributive property
   - B. commutative property of multiplication
   - C. multiplicative identity property
   - D. associative property of addition

7. Choose the line that represents the graph of the equation $y = -\frac{2}{3}x + 2$. D

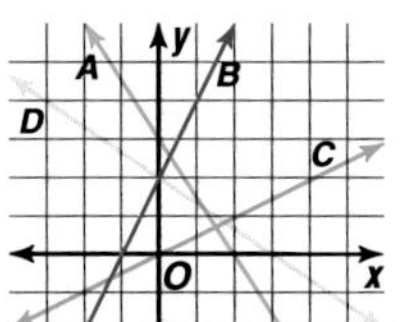

   - A. line A
   - B. line B
   - C. line C
   - D. line D

8. **Geometry** The formula Area $= \frac{1}{2}bc \sin A$ can be used to find the area of any triangle ABC. Find the area of the triangle shown below to the nearest tenth of a square centimeter if $b = 16$, $c = 9$, and $\angle A$ is a 36° angle. B

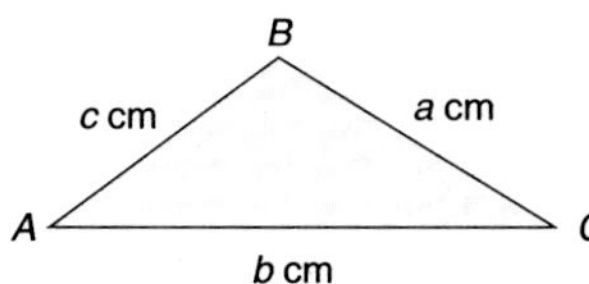

   - A. 72 cm²
   - B. 42.3 cm²
   - C. 84.6 cm²
   - D. 144 cm²

Standardized Test Practice Questions are also provided in the *Assessment and Evaluation Masters,* p. 158.

SECTION TWO: SHORT ANSWER

This section contains seven questions for which you will provide short answers. Write your answer on your paper.

9. Express the relation shown in the graph as a set of ordered pairs. Then draw a mapping of the relation. **See margin.**

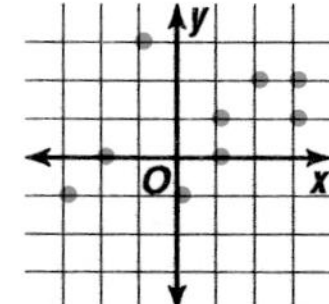

10. Write an equation of the line that passes through the point at $(2, -1)$ and is perpendicular to $4x - y = 7$. $y = -\frac{1}{4}x - \frac{1}{2}$

11. The average person can live for eleven days without water, assuming a mean temperature of 60° F. Determine if survival is possible given daily high temperatures of 59°, 70°, 49°, 62°, 46°, 63°, 71°, 64°, 55°, 68°, and 54° F. Verify your answer. **yes**

12. Write the solution set for $5x + y = 4$, if the domain is $\{-2, -1, 0, 2, 5\}$. **See margin.**

13. Peanuts sell for \$3.00 per pound. Cashews sell for \$6.00 per pound. How many pounds of cashews should be mixed with 12 pounds of peanuts to obtain a mixture that sells for \$4.20 per pound? **8 pounds**

14. At Videoville, Eshe earns a weekly salary of \$150 plus \$0.30 for each video over 100 that she sells each week. If Eshe sells v videos in a week, then her total weekly salary is $C(v) = 150 + 0.30(v - 100)$ for $v > 100$. If she wants to earn \$225 each week to save up for a stereo system, how many videos must she sell each week? **350 videos**

15. Graph the solution set {integers greater than or equal to -4} on a number line. **See margin.**

SECTION THREE: OPEN ENDED

This section contains two open-ended problems. Demonstrate your knowledge by giving a clear, concise solution to each problem. Your score on these problems will depend on how well you do the following.

- Explain your reasoning.
- Show your understanding of the mathematics in an organized manner.
- Use charts, graphs, and diagrams in your explanation.
- Show the solution in more than one way or relate it to other situations.
- Investigate beyond the requirements of the problem.

16. For long-distance phone calls, a telephone company charges \$1.72 for a 4-minute call, \$2.40 for a 6-minute call, and \$5.46 for a 15-minute call. Determine the charge for a 1-minute call. **\$0.70**

17. The table below shows the annual income and the number of years of college education for eleven people. **a–b. See margin.**

Income (thousands)	College Education (years)
\$23	3.0
20	2.0
25	4.0
47	6.0
19	2.5
48	7.5
35	6.5
10	1.0
39	5.5
26	4.5
36	4.0

a. Make a scatter plot of the data.

b. Based on this plot, how does the number of years of college affect income?

Additional Answers

9.

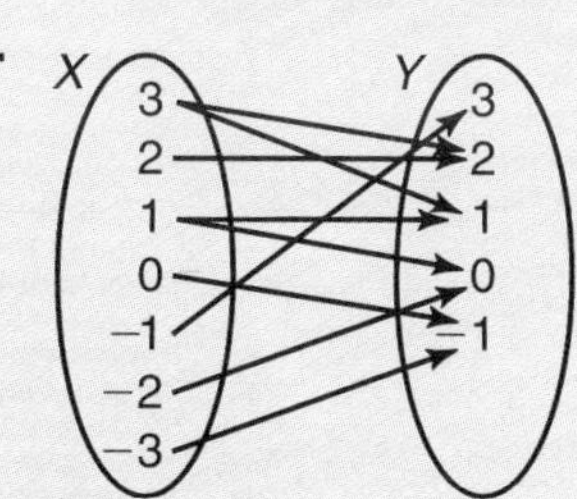

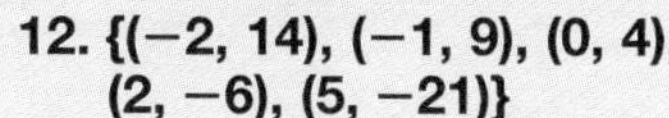

12. {(−2, 14), (−1, 9), (0, 4), (2, −6), (5, −21)}

15.

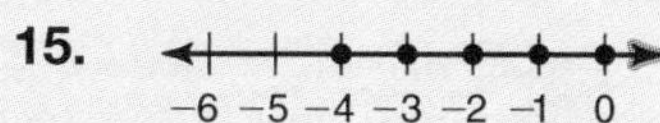

17a.

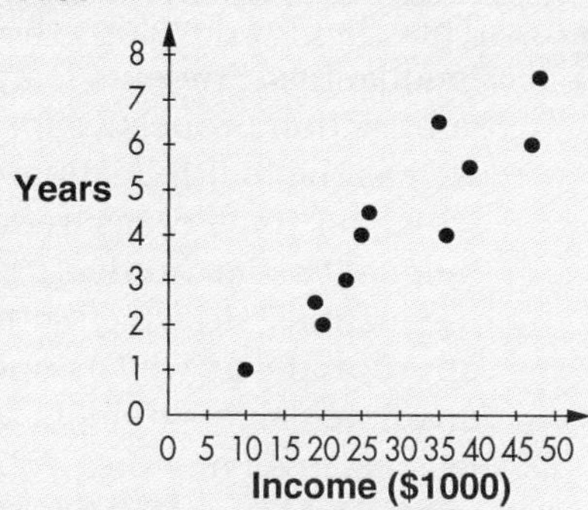

17b. The more years of education you have, the higher your income.

CHAPTER

7

Solving Linear Inequalities

PREVIEWING THE CHAPTER

This chapter develops the properties for solving inequalities. Students learn the addition and subtraction properties for solving inequalities, as well as the multiplication and division properties which, in some cases, require the reversal of the inequality sign. Students solve compound inequalities, drawing pictures or diagrams when appropriate, as well as open sentences involving absolute value. Students apply their knowledge in lessons involving the probability of compound events and the statistical tool known as a box-and-whisker plot. The chapter concludes with a lesson on graphing inequalities in two variables.

Lesson (Pages)	Lesson Objectives	NCTM Standards	State/Local Objectives
7-1 (384–390)	Solve inequalities by using addition and subtraction.	1–5	
7-2A (391)	Use cups and counters to solve inequalities.	1–5	
7-2 (392–398)	Solve inequalities by using multiplication and division.	1–5	
7-3 (399–404)	Solve linear inequalities involving more than one operation. Find the solution set for a linear inequality when replacement values are given for the variables.	1–5	
7-4 (405–412)	Solve problems by making a diagram. Solve compound inequalities and graph their solution sets. Solve problems that involve compound inequalities.	1–5	
7-5 (413–419)	Find the probability of a compound event.	1–4, 11	
7-6 (420–426)	Solve open sentences involving absolute value and graph the solutions.	1–5	
7-7 (427–432)	Display and interpret data on box-and-whisker plots.	1–4, 10	
7-7B (433–434)	Use a graphing calculator to compare two sets of data by using a double box-and-whisker plot.	1–5, 10	
7-8A (435)	Use a graphing calculator to graph inequalities in two variables.	1–6	
7-8 (436–441)	Graph inequalities in the coordinate plane.	1–5	

A complete, 1-page lesson plan is provided for each lesson in the ***Lesson Planning Guide***. Answer keys for each lesson are available in the ***Answer Key Masters.***

ORGANIZING THE CHAPTER

You may want to refer to the **Course Planning Calendar** on page T12 for detailed information on pacing.
PACING: Standard—14 days; **Honors**—13 days; **Block**—7 days; **Two Years**—21 days

LESSON PLANNING CHART

			Blackline Masters											
Lesson (Pages)	**Materials/ Manipulatives**	**Extra Practice (Student Edition)**	**Study Guide**	**Practice**	**Enrichment**	**Assessment and Evaluation**	**Modeling Mathematics**	**Multicultural Activity**	**Tech Prep Applications**	**Graphing Calculator**	**Science and Math Lab Manual**	**Real-World Applications**	**Interactive Mathematics Tools Software**	**Teaching Transparencies**
7-1 (384–390)	graphing calculator	p. 771	p. 48	p. 48	p. 48		p. 78		p. 13				7-1	7-1A 7-1B
7-2A (391)	equation mat* cups and counters* self-adhesive note						p. 27						7-2A.1 7-2A.2	
7-2 (392–398)		p. 772	p. 49	p. 49	p. 49	p. 184			p. 14					7-2A 7-2B
7-3 (399–404)	graphing calculator	p. 772	p. 50	p. 50	p. 50			p. 13				19		7-3A 7-3B
7-4 (405–412)	graphing calculator	p. 772	p. 51	p. 51	p. 51	pp. 183, 184	pp. 54–56			p. 7		20		7-4A 7-4B
7-5 (413–419)		p. 773	p. 52	p. 52	p. 52									7-5A 7-5B
7-6 (420–426)		p. 773	p. 53	p. 53	p. 53	p. 185								7-6A 7-6B
7-7 (427–432)		p. 773	p. 54	p. 54	p. 54						pp. 29–34		7-7	7-7A 7-7B
7-7B (433–434)	graphing calculator									pp. 24, 25				
7-8A (435)	graphing calculator									pp. 26, 27				
7-8 (436–441)	graphing calculator	p. 774	p. 55	p. 55	p. 55	p. 185		p. 14						7-8A 7-8B
Study Guide/ Assessment (443–447)						pp. 169–182, 186–188								

*Included in Glencoe's Student Manipulative Kit and Overhead Manipulative Resources.

ORGANIZING THE CHAPTER

Other Chapter Resources

Student Edition
Chapter Opener, pp. 382–383
Mathematics and Society, p. 419
Working on the Investigation, pp. 398, 426
Closing the Investigation, p. 442

Teacher's Classroom Resources
Investigations and Projects Masters, pp. 49–52
Algebra and Geometry Overhead Manipulative Resources, pp. 22–25

Technology
Test and Review Software (IBM and Macintosh)
CD-ROM (Windows and Macintosh)

Professional Publications
Block Scheduling Booklet
Glencoe Mathematics Professional Series

Outside Resources

Books/Periodicals
Algebra Experiments I: Exploring Linear Functions, Dale Seymour Publications
Developing Skills in Algebra 1: Book C, Tricon Publishing

Software
Math Connections: Algebra I, Sunburst

Videos/CD-ROMs
The Laws of Algebra/Solving Equations I, Coronet/MTI

Assessment Resources

Student Edition
Math Journal, pp. 388, 415, 439
Mixed Review, pp. 390, 397, 404, 412, 418, 425, 432, 441
Self Test, p. 412
Chapter Highlights, p. 443
Chapter Study Guide and Assessment, pp. 444–446
Alternative Assessment, p. 447
Portfolio, p. 447

Teacher's Wraparound Edition
5-Minute Check, pp. 384, 392, 399, 405, 413, 420, 427, 436
Check for Understanding, pp. 387, 395, 402, 409, 415, 423, 429, 439
Closing Activity, pp. 390, 398, 404, 412, 419, 426, 432, 441
Cooperative Learning, pp. 408, 425

Assessment and Evaluation Masters
Multiple-Choice Tests, Forms 1A (Honors), 1B (Average), 1C (Basic), pp. 169–174
Free-Response Tests, Forms 2A (Honors), 2B (Average), 2C (Basic), pp. 175–180
Calculator-Based Test, p. 181
Performance Assessment, p. 182
Mid-Chapter Test, p. 183
Quizzes A–D, pp. 184–185
Standardized Test Practice, p. 186
Cumulative Review, pp. 187–188

ENHANCING THE CHAPTER

Examples of some of the materials for enhancing Chapter 7 are shown below.

Multicultural Activity Masters, pp. 13, 14

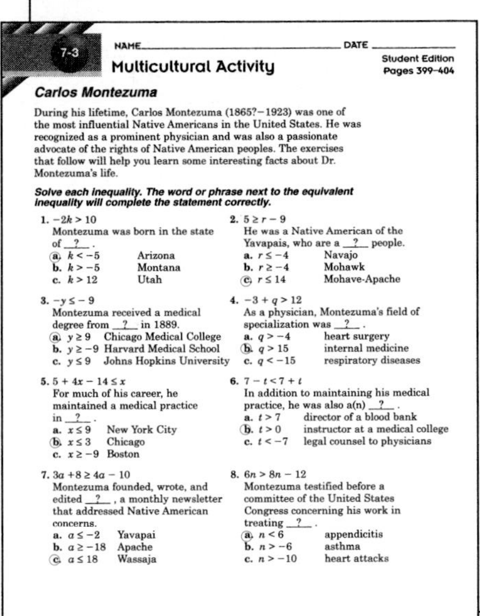

NAME ________________ DATE ________

7-3 Multicultural Activity

Student Edition Pages 399–404

Carlos Montezuma

During his lifetime, Carlos Montezuma (1865?–1923) was one of the most influential Native Americans in the United States. He was recognized as a prominent physician and was also a passionate advocate of the rights of Native American peoples. The exercises that follow will help you learn some interesting facts about Dr. Montezuma's life.

Solve each inequality. The word or phrase next to the equivalent inequality will complete the statement correctly.

1. $-2k > 10$
   Montezuma was born in the state of __?__ .
   (a.) $k < -5$ Arizona
   b. $k > -5$ Montana
   c. $k > 12$ Utah

2. $5 \geq r - 9$
   He was a Native American of the Yavapais, who are a __?__ people.
   a. $r \leq -4$ Navajo
   b. $r \geq -4$ Mohawk
   (c.) $r \leq 14$ Mohave-Apache

3. $-y \leq -9$
   Montezuma received a medical degree from __?__ in 1889.
   (a.) $y \geq 9$ Chicago Medical College
   b. $y \geq -9$ Harvard Medical School
   c. $y \leq 9$ Johns Hopkins University

4. $-3 + q > 12$
   As a physician, Montezuma's field of specialization was __?__ .
   a. $q > -4$ heart surgery
   (b.) $q > 15$ internal medicine
   c. $q < -15$ respiratory diseases

5. $5 + 4x - 14 \leq x$
   For much of his career, he maintained a medical practice in __?__ .
   a. $x \leq 9$ New York City
   (b.) $x \leq 3$ Chicago
   c. $x \geq -9$ Boston

6. $7 - t < 7 + t$
   In addition to maintaining his medical practice, he was also a(n) __?__ .
   a. $t > 7$ director of a blood bank
   (b.) $t > 0$ instructor at a medical college
   c. $t < -7$ legal counsel to physicians

7. $3a + 8 \geq 4a - 10$
   Montezuma founded, wrote, and edited __?__ , a monthly newsletter that addressed Native American concerns.
   a. $a \leq -2$ Yavapai
   b. $a \geq -18$ Apache
   (c.) $a \leq 18$ Wassaja

8. $6n > 8n - 12$
   Montezuma testified before a committee of the United States Congress concerning his work in treating __?__ .
   (a.) $n < 6$ appendicitis
   b. $n > -6$ asthma
   c. $n > -10$ heart attacks

Real-World Applications, 19, 20

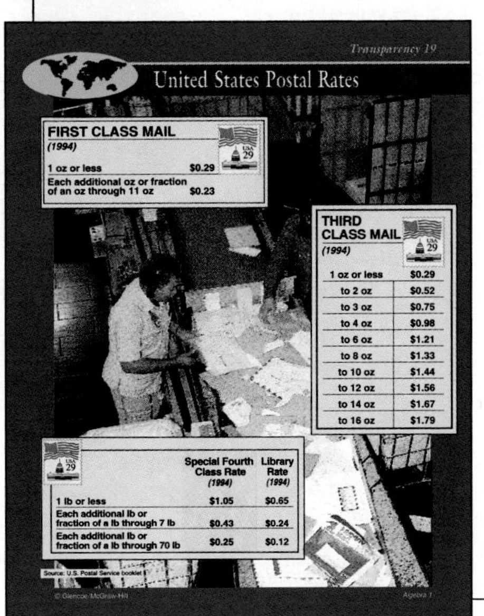

Transparency 19

United States Postal Rates

FIRST CLASS MAIL *(1994)*

1 oz or less	\$0.29
Each additional oz or fraction of an oz through 11 oz	\$0.23

THIRD CLASS MAIL *(1994)*

1 oz or less	\$0.29
to 2 oz	\$0.52
to 3 oz	\$0.75
to 4 oz	\$0.98
to 6 oz	\$1.21
to 8 oz	\$1.33
to 10 oz	\$1.44
to 12 oz	\$1.56
to 14 oz	\$1.67
to 16 oz	\$1.79

	Special Fourth Class Rate *(1994)*	Library Rate *(1994)*
1 lb or less	\$1.05	\$0.65
Each additional lb or fraction of a lb through 7 lb	\$0.43	\$0.24
Each additional lb or fraction of a lb through 70 lb	\$0.25	\$0.12

Source: U.S. Postal Service booklet

© Glencoe/McGraw-Hill Algebra 1

TECHNOLOGY

Graphing Calculator Masters, p. 7

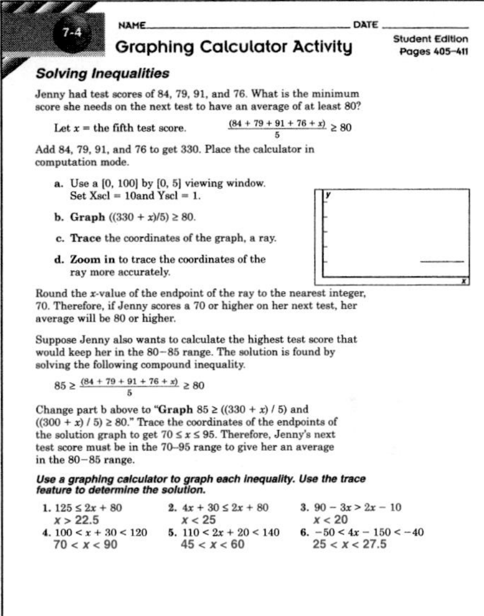

NAME ________________ DATE ________

7-4 Graphing Calculator Activity

Student Edition Pages 405–411

Solving Inequalities

Jenny had test scores of 84, 79, 91, and 76. What is the minimum score she needs on the next test to have an average of at least 80?

Let x = the fifth test score. $\frac{(84 + 79 + 91 + 76 + x)}{5} \geq 80$

Add 84, 79, 91, and 76 to get 330. Place the calculator in computation mode.

a. Use a [0, 100] by [0, 5] viewing window. Set Xscl = 10and Yscl = 1.

b. **Graph** $((330 + x)/5) \geq 80$.

c. **Trace** the coordinates of the graph, a ray.

d. **Zoom in** to trace the coordinates of the ray more accurately.

Round the x-value of the endpoint of the ray to the nearest integer, 70. Therefore, if Jenny scores a 70 or higher on her next test, her average will be 80 or higher.

Suppose Jenny also wants to calculate the highest test score that would keep her in the 80–85 range. The solution is found by solving the following compound inequality.

$85 \geq \frac{(84 + 79 + 91 + 76 + x)}{5} \geq 80$

Change part b above to "**Graph** $85 \geq ((330 + x) / 5)$ and $((300 + x) / 5) \geq 80$." Trace the coordinates of the endpoints of the solution graph to get $70 \leq x \leq 95$. Therefore, Jenny's next test score must be in the 70–95 range to give her an average in the 80–85 range.

Use a graphing calculator to graph each inequality. Use the trace feature to determine the solution.

1. $125 \leq 2x + 80$ — $x > 22.5$
2. $4x + 30 \leq 2x + 80$ — $x < 25$
3. $90 - 3x > 2x - 10$ — $x < 20$
4. $100 < x + 30 < 120$ — $70 < x < 90$
5. $110 < 2x + 20 < 140$ — $45 < x < 60$
6. $-50 < 4x - 150 < -40$ — $25 < x < 27.5$

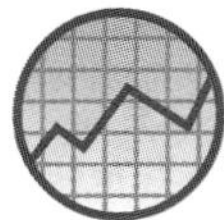

TECH PREP

Tech Prep Applications Masters, pp. 13, 14

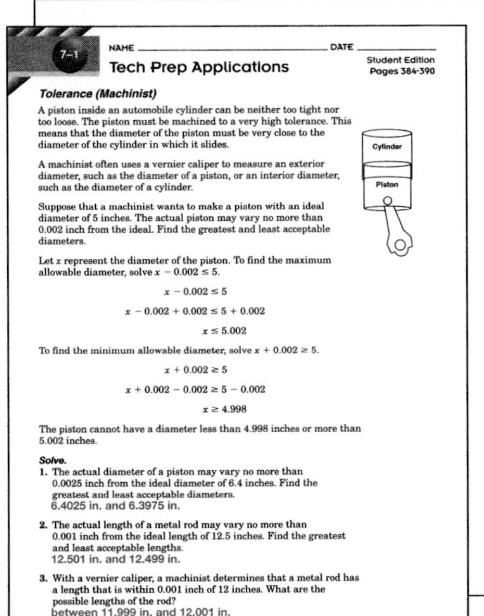

NAME ________________ DATE ________

7–1 Tech Prep Applications

Student Edition Pages 384-390

Tolerance (Machinist)

A piston inside an automobile cylinder can be neither too tight nor too loose. The piston must be machined to a very high tolerance. This means that the diameter of the piston must be very close to the diameter of the cylinder in which it slides.

Cylinder
Piston

A machinist often uses a vernier caliper to measure an exterior diameter, such as the diameter of a piston, or an interior diameter, such as the diameter of a cylinder.

Suppose that a machinist wants to make a piston with an ideal diameter of 5 inches. The actual piston may vary no more than 0.002 inch from the ideal. Find the greatest and least acceptable diameters.

Let x represent the diameter of the piston. To find the maximum allowable diameter, solve $x - 0.002 \leq 5$.

$$x - 0.002 \leq 5$$
$$x - 0.002 + 0.002 \leq 5 + 0.002$$
$$x \leq 5.002$$

To find the minimum allowable diameter, solve $x + 0.002 \geq 5$.

$$x + 0.002 \geq 5$$
$$x + 0.002 - 0.002 \geq 5 - 0.002$$
$$x \geq 4.998$$

The piston cannot have a diameter less than 4.998 inches or more than 5.002 inches.

Solve.

1. The actual diameter of a piston may vary no more than 0.0025 inch from the ideal diameter of 6.4 inches. Find the greatest and least acceptable diameters.
   6.4025 in. and 6.3975 in.
2. The actual length of a metal rod may vary no more than 0.001 inch from the ideal length of 12.5 inches. Find the greatest and least acceptable lengths.
   12.501 in. and 12.499 in.
3. With a vernier caliper, a machinist determines that a metal rod has a length that is within 0.001 inch of 12 inches. What are the possible lengths of the rod?
   between 11.999 in. and 12.001 in.

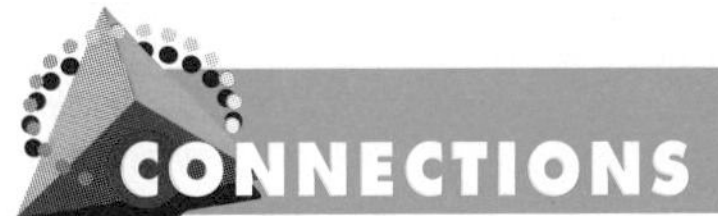

Science and Math Lab Manual, pp. 29–34

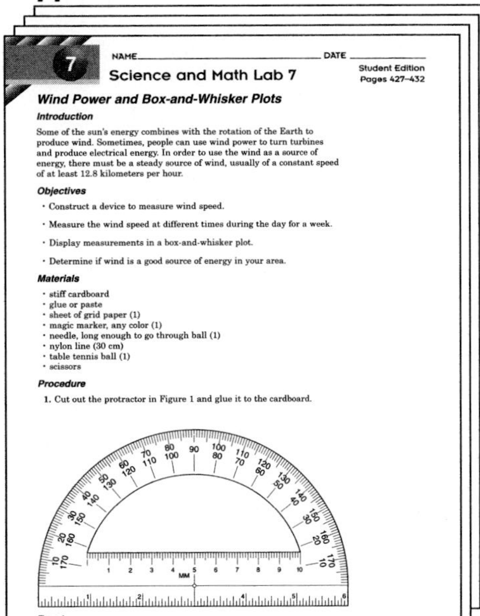

NAME ________________ DATE ________

7 Science and Math Lab 7

Student Edition Pages 427–432

Wind Power and Box-and-Whisker Plots

Introduction

Some of the sun's energy combines with the rotation of the Earth to produce wind. Sometimes, people can use wind power to turn turbines and produce electrical energy. In order to use the wind as a source of energy, there must be a steady source of wind, usually of a constant speed of at least 12.8 kilometers per hour.

Objectives

- Construct a device to measure wind speed.
- Measure the wind speed at different times during the day for a week.
- Display measurements in a box-and-whisker plot.
- Determine if wind is a good source of energy in your area.

Materials

- stiff cardboard
- glue or paste
- sheet of grid paper (1)
- magic marker, any color (1)
- needle, long enough to go through ball (1)
- nylon line (30 cm)
- table tennis ball (1)
- scissors

Procedure

1. Cut out the protractor in Figure 1 and glue it to the cardboard.

Figure 1

Problem of the Week Cards, 19, 20, 21

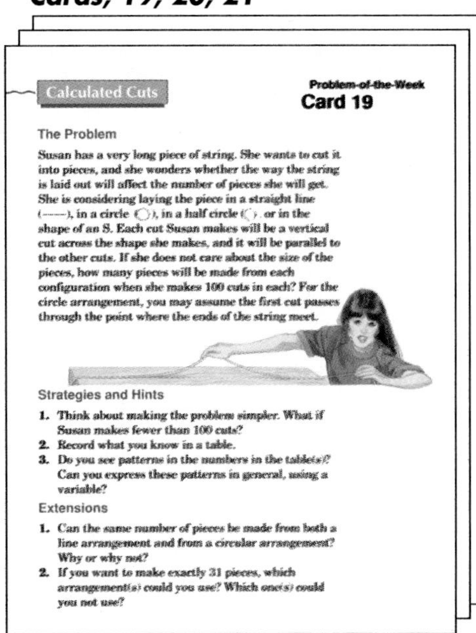

Calculated Cuts

Problem-of-the-Week Card 19

The Problem

Susan has a very long piece of string. She wants to cut it into pieces, and she wonders whether the way the string is laid out will affect the number of pieces she will get. She is considering laying the piece in a straight line (——), in a circle ( ), in a half circle ( ), or in the shape of an S. Each cut Susan makes will be a vertical cut across the shape she makes, and it will be parallel to the other cuts. If she does not care about the size of the pieces, how many pieces will be made from each configuration when she makes 100 cuts in each? For the circle arrangement, you may assume the first cut passes through the point where the ends of the string meet.

Strategies and Hints

1. Think about making the problem simpler. What if Susan makes fewer than 100 cuts?
2. Record what you know in a table.
3. Do you see patterns in the numbers in the table(s)? Can you express these patterns in general, using a variable?

Extensions

1. Can the same number of pieces be made from both a line arrangement and from a circular arrangement? Why or why not?
2. If you want to make exactly 31 pieces, which arrangement(s) could you use? Which one(s) could you not use?

MAKING MATHEMATICS RELEVANT

This two-page introduction to the chapter provides students with an opportunity to explore contemporary topics and their applications to mathematics.

Background Information

Increase Your Earnings Potential The percentage of Americans who have completed college has been increasing steadily. But so has the cost of a college education. Average yearly costs at a public university in 1995 were slightly over $10,000. Still, the data presented here show that college tuition is a wise investment. While a bachelor's degree at a state university averages $44,000, the return on that investment is a $600,000 increase in earnings over a lifetime.

CHAPTER 7

Solving Linear Inequalities

Objectives

In this chapter, you will:

- solve inequalities,
- graph solutions of inequalities,
- graph solutions of open sentences that involve absolute value,
- solve problems by drawing a diagram, and
- use box-and-whisker plots to display and analyze data.

Increase Your Earnings Potential

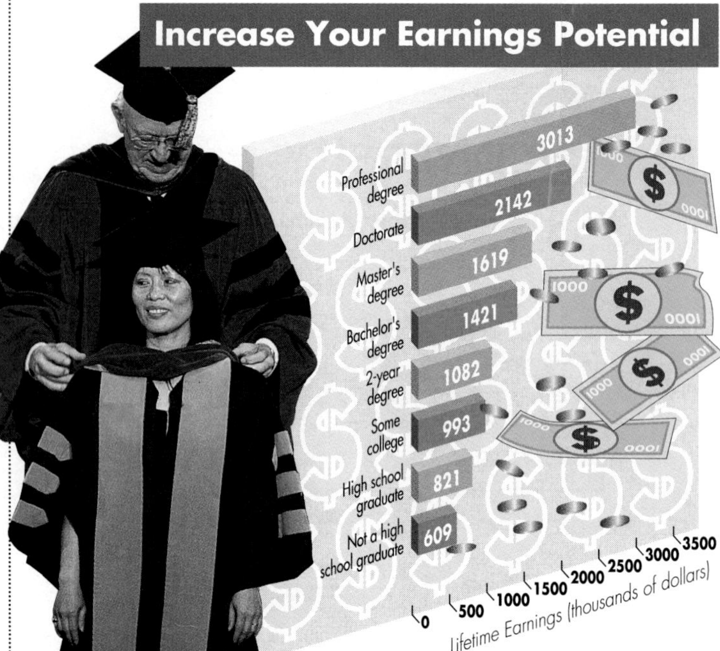

Source: *Chicago Tribune*, 1995

There is a direct relationship between lifetime earnings and educational attainment. College graduation and advanced degrees really do make a difference. Set your sights on a career and aim to be the best you can be.

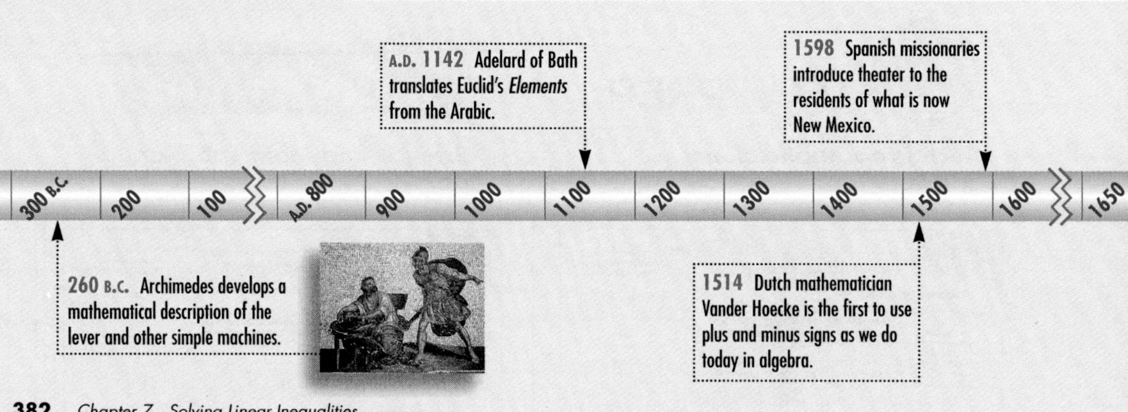

A.D. 1142 Adelard of Bath translates Euclid's *Elements* from the Arabic.
1598 Spanish missionaries introduce theater to the residents of what is now New Mexico.
300 B.C.
200
100
A.D. 800
900
1000
1100
1200
1300
1400
1500
1600
1650
260 B.C. Archimedes develops a mathematical description of the lever and other simple machines.
1514 Dutch mathematician Vander Hoecke is the first to use plus and minus signs as we do today in algebra.

382 *Chapter 7 Solving Linear Inequalities*

Students might find it interesting to write a short report on Archimedes. He contributed greatly both to the sciences and to mathematics.

interNET CONNECTION

You can learn about medical, microbiological, and molecular biological sciences and use links to other sites of such material on the Internet.

World Wide Web
http://www.qmw.ac.uk/~rhbm001/

PEOPLE IN THE NEWS

Chapter Project

Shakema Hodge is aiming for an advanced degree in microbiology. The young scientist from St. Thomas, Virgin Islands, earned her bachelor's degree in biology at the University of the Virgin Islands. Now she is enrolled in a 5-year Ph.D. program in molecular microbiology at the University of Rochester in New York. She would love to teach high school students, believing that "to capture a young person's mind, you have to expose them to the sciences at an early age."

Kimana plans to study either botany, genetics, or microbiology at a nearby university. She hasn't ruled out teaching as a possible career choice, but is interested in finding other businesses in which she can apply the science she will be studying.

- Visit the public library or local college to investigate careers that relate to Kimana's interests.
- Which careers might offer Kimana an opportunity to earn an amount greater than that presented in the graph for a bachelor's degree?
- Which careers require a degree for which the years studied are greater than those for a bachelor's degree? Choose a career. Write an inequality that might express the time needed to study and train for that career.

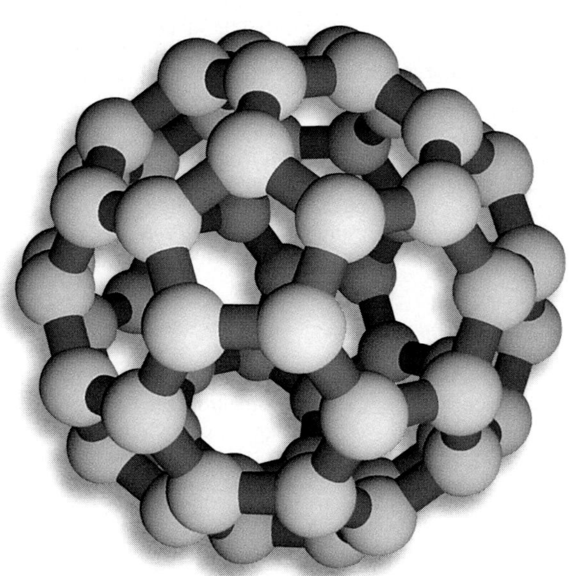

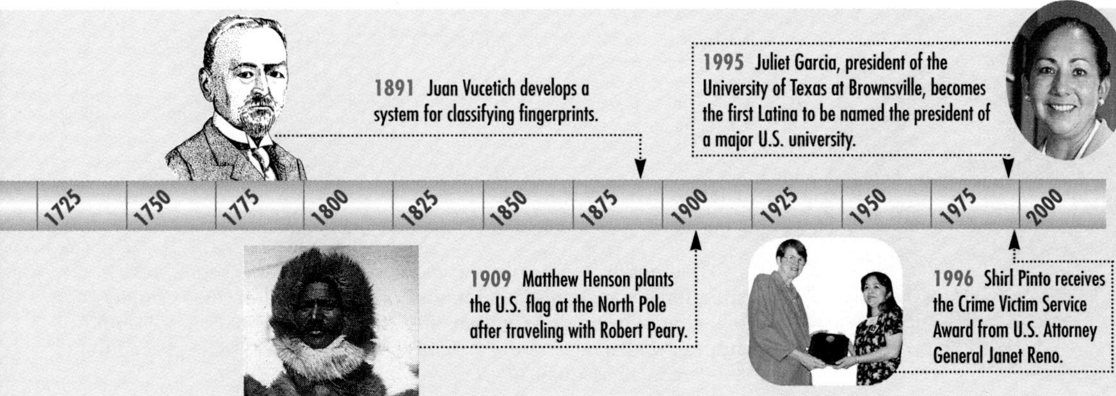

1891 Juan Vucetich develops a system for classifying fingerprints.
1995 Juliet Garcia, president of the University of Texas at Brownsville, becomes the first Latina to be named the president of a major U.S. university.
1725
1750
1775
1800
1825
1850
1875
1900
1925
1950
1975
2000
1909 Matthew Henson plants the U.S. flag at the North Pole after traveling with Robert Peary.
1996 Shirl Pinto receives the Crime Victim Service Award from U.S. Attorney General Janet Reno.

PEOPLE IN THE NEWS

Even as a child in St. Thomas, Shakema was interested in "weird" science. She spent much of her play time in creative activities, such as making perfume out of flowers and leaves. Shakema was heavily recruited by a number of American graduate schools, but chose Rochester because of its research in the area of molecular biology.

Chapter Project

Cooperative Learning In groups of two or three, have students investigate careers. Students should prepare a 10-minute class presentation concerning the career, education required, and earning potential.

Alternative Chapter Projects

Two other chapter projects are included in the *Investigations and Projects Masters.* In Chapter 7 Project A, pp. 49–50, students extend the topic in the chapter opener. In Chapter 7 Project B, pp. 51–52, students compare the heights, heart rates, and temperatures of their classmates with the national norm.

Investigations and Projects Masters, p. 49

NAME__________ DATE________

Chapter 7 Project A

Student Edition Pages 384–442

Choose a Career

1. What career do you want to pursue after you graduate from high school? You may be considering several different careers; if so, which of these are you most likely to choose? Do research to find out more about this career. Are there any courses you can take in high school that will help you prepare for your career?
2. What are the requirements for entering your chosen field? Do research to find out about where you can get the necessary education and/or training. Be sure to find out about the money and time you will need to invest to receive such education and/or training.
3. Think about how you will pay for your education and/or training. Write a plan for saving the money required. Consider the answers to the following questions as you work on your savings plan.
   - Can you start saving money now?
   - Are your parents or guardians saving money for you?
   - Are there scholarships for which you might be eligible?

   Write an inequality that expresses the minimum amount you plan to save from all possible sources.
4. Share your plan with your classmates. Ask them if they think the plan is realistic and if they can offer any suggestions for improving it.

7-1 LESSON NOTES

NCTM Standards: 1–5

Instructional Resources

- Study Guide Master 7-1
- Practice Master 7-1
- Enrichment Master 7-1
- Modeling Mathematics Masters, p. 78
- Tech Prep Applications Masters, p. 13

Transparency 7-1A contains the 5-Minute Check for this lesson; **Transparency 7-1B** contains a teaching aid for this lesson.

Recommended Pacing	
Standard Pacing	Day 1 of 14
Honors Pacing	Day 1 of 13
Block Scheduling*	Day 1 of 7
Alg. 1 in Two Years*	Days 1 & 2 of 21

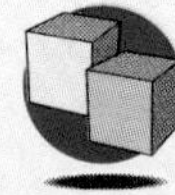

*For more information on pacing and possible lesson plans, refer to the *Block Scheduling Booklet* and *Algebra 1 in Two Years.*

1 FOCUS

5-Minute Check
(over Chapter 6)

State the slopes of the lines parallel and perpendicular to the graph of each equation.

1. $4r = 2p - 7$ $\frac{1}{2}, -2$

2. $t = 3m - 7$ $3, -\frac{1}{3}$

3. $5x - y = 7$ $5, -\frac{1}{5}$

Find the coordinates of the midpoint of the line segment whose endpoints are given.

4. (9, 5), (17, 3) (13, 4)

5. (11, 4), (9, 2) (10, 3)

7-1 Solving Inequalities by Using Addition and Subtraction

What YOU'LL LEARN

- To solve inequalities by using addition and subtraction.

Why IT'S IMPORTANT

You can use inequalities to solve problems involving nutrition and personal finance.

CAREER CHOICES

Nutritionists plan dietary programs and supervise the preparation and serving of meals for hospitals, nursing homes, health maintenance organizations, and individuals. A bachelor's degree in dietetics, foods, or nutrition is required that includes studies in biology, physiology, mathematics, and chemistry.

For more information, contact:

The American Dietetic Association
216 West Jackson Blvd.
Chicago, IL 60606

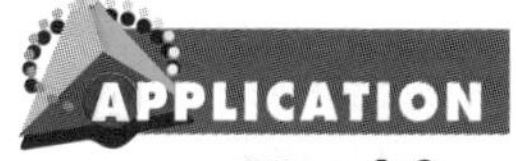

Nutrition

In 1990, the U.S. Department of Agriculture issued new dietary guidelines. These guidelines recommend that people greatly reduce their fat intake. Your recommended calorie intake depends on your height, desired weight, and physical activity. The average 14-year-old is 5 feet 2 inches tall and weighs 107 pounds. Boys should consume about 2434 calories per day and girls 2208 calories per day to maintain this weight.

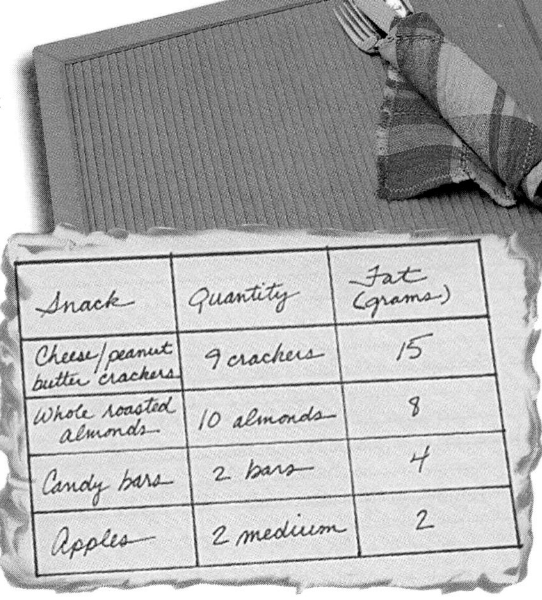

Snack	Quantity	Fat (grams)
Cheese/peanut butter crackers	9 crackers	15
Whole roasted almonds	10 almonds	8
Candy bars	2 bars	4
Apples	2 medium	2

Oliana learned in health class that no more than 30% of her calorie intake should come from fat. For her 2030-calorie-a-day diet, that means no more than 68 grams of fat. She keeps track of her snacks for one day and records their fat content, as shown in the table above. How many grams of fat can Oliana have in the other foods she eats that day and stay within the guidelines?

Let's write an inequality to represent the problem. Let g represent the remaining grams of fat that Oliana can eat that day.

grams of fat in snacks	*plus*	*grams of fat remaining*	*is less than or equal to*	*total grams*
$15 + 8 + 4 + 2$	$+$	g	$\leq$	68

That is, $29 + g \leq 68$.

The symbol $\leq$ indicates *less than or equal to*. It is used in this situation because the total number of grams of fat in Oliana's daily diet should be no greater than 68. If this were an equation, we would subtract 29 from (or add -29 to) each side. Can the same procedure be used in an inequality? *This problem will be solved in Example 1.*

CAREER CHOICES

Nutritionists with less than 5 years experience earned a median annual salary of $28,500 in 1991. Those with 6–10 years experience earned a median salary of $32,900; 11–15 years, $36,000; 16–20 years, $38,400; and more than 20 years, $40,000.

Let's explore what happens if inequalities are solved in the same manner as equations. We know that $7 > 2$. What happens when you add or subtract the same quantity to each side of the inequality? We can use number lines to model the situation.

Add 3 to each side.

$7 > 2$

$7 + 3 \overset{?}{>} 2 + 3$

$10 > 5$

Subtract 4 from each side.

$7 > 2$

$7 - 4 \overset{?}{>} 2 - 4$

$3 > -2$

-2 -1 0 1 2 3 4 5 6 7 8 9

-4 -4

-4 -3 -2 -1 0 1 2 3 4 5 6 7

In each case, the inequality holds true. These examples illustrate two properties of inequalities.

Addition and Subtraction Properties for Inequalities	For all numbers a, b, and c, the following are true. 1. If $a > b$, then $a + c > b + c$ and $a - c > b - c$. 2. If $a < b$, then $a + c < b + c$ and $a - c < b - c$.

These properties are also true when $>$ and $<$ are replaced by $\geq$ and $\leq$. So, we can use these properties to obtain a solution to the application at the beginning of the lesson.

Example 1

Nutrition

Refer to the application at the beginning of the lesson. Solve $29 + g \leq 68$.

$$29 + g \leq 68$$

$$29 - 29 + g \leq 68 - 29 \quad \textit{Subtract 29 from each side.}$$

$$g \leq 39 \quad \textit{This means all numbers less than or equal to 39.}$$

The solution set can be written as {all numbers less than or equal to 39}.

Check: To check this solution, substitute 39, a number less than 39, and a number greater than 39 into the inequality.

Let $g = 39$.	Let $g = 20$.	Let $g = 40$.
$29 + g \leq 68$	$29 + g \leq 68$	$29 + g \leq 68$
$29 + 39 \overset{?}{\leq} 68$	$29 + 20 \overset{?}{\leq} 68$	$29 + 40 \overset{?}{\leq} 68$
$68 \leq 68$ true	$49 \leq 68$ true	$69 \leq 68$ false

So, Oliana can have 39 or fewer grams of fat in other foods that day and stay within the dietary guidelines.

Top Five List

Preferred snack foods and % of calories from fat

1. potato chips	58%
2. tortilla chips	47%
3. popcorn	45%
4. pretzels	8%
5. mixed nuts	80%

The solution to the inequality in Example 1 was expressed as a set. A more concise way of writing a solution set is to use **set-builder notation**. The solution in set-builder notation is $\{g \mid g \leq 39\}$. This is read *the set of all numbers g such that g is less than or equal to 39.*

Motivating the Lesson

Questioning Introduce the lesson by having students name some values for a and b that make all three of the following inequalities true.

$a < 0, b < 0, a - b < 0$

Then ask them to describe the relationship between a and b for the above inequalities to be true. $a < b$

2 TEACH

In-Class Example

For Example 1
Solve each inequality.

a. $x + 14 < 16$ $\{x \mid x < 2\}$

b. $y + (-21) > 7$ $\{y \mid y > 28\}$

c. $(-11) + t > 5$ $\{t \mid t > 16\}$

Teaching Tip Emphasize that as in Example 1, the solutions of open sentences are numbers. The set of all solutions for a sentence is the solution set.

Suggest that students go to a grocery store and examine the labels on five different items that come in both a regular version and a "lite" version. Have them compare the grams of fat in each version and write down the results. Discuss their findings in class.

Alternative Learning Styles

Kinesthetic Use the concept of an unbalanced scale to solve inequalities. Use a ruler, algebra tiles or counters, and a triangular piece of cardboard. Place the ruler on the triangle so that the greater side of the inequality is up. The representation of $x - 5 > -3$ is shown at the right.

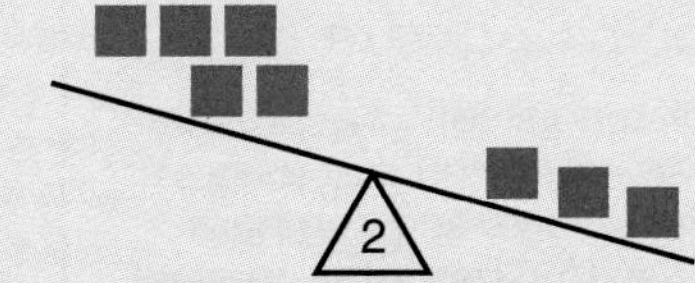

To solve, place 5 positive tiles on each side of the scale. Thus, the solution is $x > 2$.

Teaching Tip In Example 2, make sure students understand that $52 < z$ and $z > 52$ are equivalent inequalities. Students may find it easier to determine the solution set by writing the inequality with the variable on the left side.

In-Class Examples

For Example 2
Solve each inequality. Then graph the solution.

a. $3r - 17 > 2r + 14$ $\{r \mid r > 31\}$

29 30 31 32 33 34 35

b. $9t + 5 < 10t - 3$ $\{t \mid t > 8\}$

5 6 7 8 9 10 11 12 13 14

For Example 3
A stove and a freezer together weigh at least 260 kg. The stove weighs 115 kg. What is the weight of the freezer? **at least 145 kg**

LOOK BACK
You can refer to Lesson 1-5 for information on solution sets.

In Lesson 2–4, you learned that you can show the solution to an inequality on a graph. The solution to Example 1 is shown on the number line below.

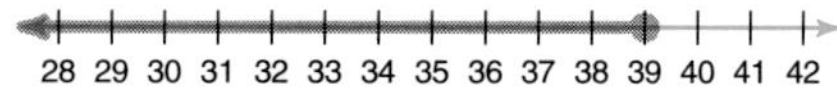

28 29 30 31 32 33 34 35 36 37 38 39 40 41 42

The closed circle at 39 tells us that 39 is included in the inequality. The heavy arrow pointing to the left shows that it also includes all numbers less than 39. *If the inequality was <, the circle would be open.*

Example 2 **Solve $13 + 2z < 3z - 39$. Then graph the solution.**

$13 + 2z < 3z - 39$

$13 + 2z - 2z < 3z - 2z - 39$ *Subtract 2z from each side.*

$13 < z - 39$

$13 + 39 < z - 39 + 39$ *Add 39 to each side.*

$52 < z$

Since $52 < z$ is the same as $z > 52$, the solution set is $\{z \mid z > 52\}$.

The graph of the solution contains an open circle at 52 since 52 is not included in the solution, and the arrow points to the right.

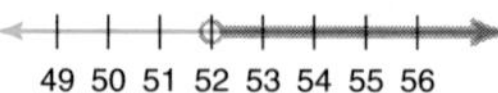

49 50 51 52 53 54 55 56

Verbal problems containing phrases like *greater than* or *less than* can often be solved by using inequalities. The following chart shows some other phrases that indicate inequalities.

Inequalities			
<	>	≤	≥
• less than • fewer than	• greater than • more than	• at most • no more than • less than or equal to	• at least • no less than • greater than or equal to

Example 3

Budgeting

Alvaro, Chip, and Solomon have earned $500 to buy equipment for their band. They have already spent $275 on a used guitar and a drum set. They are now considering buying a $125 amplifier. What is the most they can spend on promotional materials and T-shirts for the band if they buy the amplifier?

Explore *At most* means that they cannot go over what is left of their budget of $500. They have spent $275, so they have $225 left. Let m = the amount of money for promotional materials and T-shirts.

GLENCOE Technology

Interactive Mathematics Tools Software

This multimedia software provides an interactive lesson that uses counters to solve and compare inequalities. A **Computer Journal** gives students an opportunity to write about what they have learned.

For Windows & Macintosh

Alternative Teaching Strategies

Student Diversity To help students relate the language used in verbal problems to algebraic symbols, read the phrases on page 386 that indicate inequalities. Have students write the symbol that is used in each expression.

Plan: *Total to spend* / *is at most* / *$225.*

$125 + m \quad \leq \quad 225$

Solve

$$125 + m \leq 225$$

$$125 - 125 + m \leq 225 - 125 \quad \textit{Subtract 125 from each side.}$$

$$m \leq 100$$

The members of the band can spend $100 or less on promotional materials and T-shirts.

Examine: Since $275 + $125 + $100 = $500, Alvaro, Chip, and Solomon can spend $100 or less on promotional materials and T-shirts.

When solving problems involving equations, it is often necessary to write an equation that represents the words in the problem. This is also true of inequalities.

Example 4 **Write an inequality for the sentence below. Then solve the inequality and check the solution.**

Three times a number is more than the difference of twice that number and three.

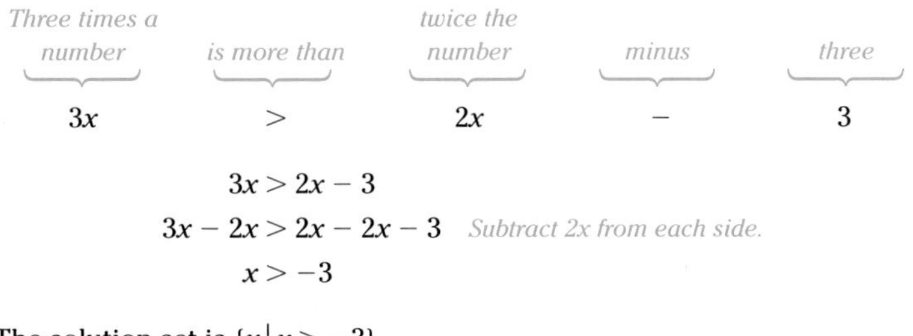

Three times a number
is more than
twice the number
minus
three
3x
>
2x
−
3

$$3x > 2x - 3$$

$$3x - 2x > 2x - 2x - 3 \quad \textit{Subtract 2x from each side.}$$

$$x > -3$$

The solution set is $\{x \mid x > -3\}$.

CHECK FOR UNDERSTANDING

Communicating Mathematics

Study the lesson. Then complete the following.

1. **Write** three inequalities that are equivalent to $x < -10$.
2. **Explain** what $\{w \mid w > -3\}$ means.
3. **Explain** the difference between the solution sets for $x + 24 < 17$ and $x + 24 \leq 17$. **One (≤) includes −7; the other does not.**
4. **Describe** how you would graph the solution to an inequality. Include examples and graphs in your explanation. **See margin.**
5. Is it possible for the solution set of an inequality to be the empty set? If so, give an example. **Yes, sample answer: $x + 6 > x + 8$.**

1. Sample answer: $x + 5 < -5$, $x - 5 < -15$, $2x < x - 10$

2. The set of all numbers w such that w is greater than −3.

In-Class Example

For Example 4
Write an inequality for the sentence below. Then solve the inequality.
One-fourth a number is less than five minus the number.
$\frac{1}{4}x < 5 - x$; $\{x \mid x < 4\}$

3 PRACTICE/APPLY

Check for Understanding

Exercises 1–16 are designed to help you assess your students' understanding through reading, writing, speaking, and modeling. You should work through Exercises 1–6 with your students and then monitor their work on Exercises 7–16.

Error Analysis
Students may assume that they have solved an inequality correctly by unknowingly substituting incorrect representative values into the original inequality. For example, suppose they solve $x + 3 < 7$ incorrectly as $x < 1$. Substituting 3 into the original inequality makes it true. However, 3 is not a solution of the inequality.

Additional Answer

4. **Answers and graphs will vary. Sample answer: The circle on the graph will be filled if the inequality involves ≤ or ≥. The circle will be open if it involves > or <. The arrow goes to the left if it involves < or ≤. It goes to the right for > or ≥.**

Reteaching

Using Diagrams Use a two-pan balance to demonstrate how to solve inequalities. If no balance is handy, draw a large balance on the chalkboard. Boxes have identical weights. What can be concluded about the weight of one box? Explain.

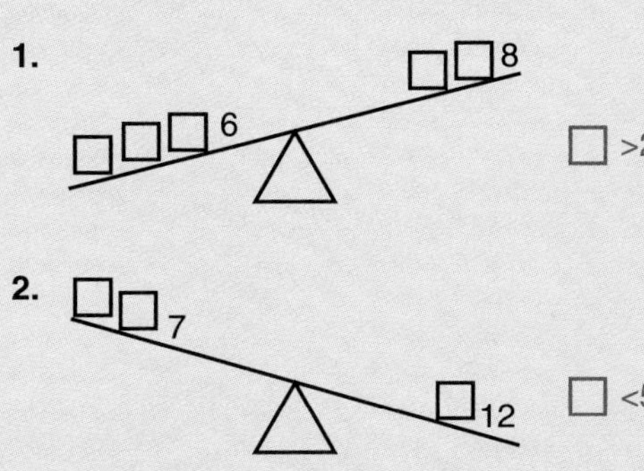

Assignment Guide

Core: 17–49 odd, 50–58
Enriched: 18–44 even, 45–58

For **Extra Practice,** see p. 771.

The red A, B, and C flags, printed only in the Teacher's Wraparound Edition, indicate the level of difficulty of the exercises.

Additional Answers

17.

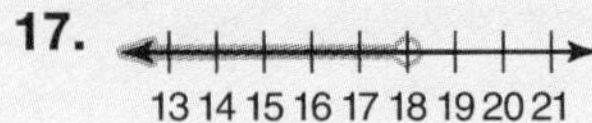

18. −19 −18 −17 −16 −15 −14 −13 −12

19. −4 −3 −2 −1 0 1 2 3 4

20. 13 14 15 16 17 18 19 20 21

21. 1 2 3 $3\frac{11}{3}$ 5 6 7 8 9

22. −3 −2 −0.15 1 2 3 4 5

23. −1 0 1 2 3 4 5 6 7

24. −4 −3 −2 −1 0 1 2 3 4

Study Guide Masters, p. 48

NAME________________ DATE ________

Study Guide — Student Edition Pages 384–390

Solving Inequalities by Using Addition and Subtraction

You can use the addition and subtraction properties for inequalities when solving problems involving inequalities.

Addition and Subtraction Properties for Inequalities
For all numbers a, b, and c, 1. if $a > b$, then $a + c > b + c$ and $a - c > b - c$; 2. if $a < b$, then $a + c < b + c$ and $a - c < b - c$.

Example: Solve $3a + 5 > 4 + 2a$.

$$3a + 5 > 4 + 2a$$
$$3a - 2a + 5 > 4 + 2a - 2a$$
$$a + 5 > 4$$
$$a + 5 - 5 > 4 - 5$$
$$a > -1$$

The solution set is {all numbers greater than −1}.

To check your solution, choose two numbers, one greater than −1, and one less than −1. Substitute both numbers in the original inequality. Only those numbers greater than −1 should yield a true statement.

The solution set in the above example, written in set-builder notation is $\{a \mid a > -1\}$. This is read "The set of all numbers a such that a is greater than −1."

Solve each inequality. Then check your solution.

1. $a + 4 < 14$ $\{a \mid a < 10\}$
2. $9k - 12 > 80 + 8k$ $\{k \mid k > 92\}$
3. $-19 + x < 2x - 33$ $\{x \mid x > 14\}$
4. $6y > 14 - 2 + 7y$ $\{y \mid y < -12\}$
5. $3n + 17 < 4n - 6$ $\{n \mid n > 23\}$
6. $\frac{3}{2}q - \frac{25}{5} \geq \frac{2q}{4}$ $\{q \mid q \geq 5\}$
7. $h + \frac{2}{3} \leq 2 - \frac{2}{3}$ $\left\{h \mid h \leq \frac{2}{3}\right\}$
8. $4p - 3.2 \geq 3p + 0.7$ $\{p \mid p \geq 3.9\}$
9. $-2\frac{1}{2}z \leq 3\frac{1}{3} + 2\frac{1}{3} - 3\frac{1}{2}z$ $\left\{z \mid z \leq 5\frac{2}{3}\right\}$
10. $\frac{1}{b} + 4 \leq 10 + \frac{1}{b}, b \neq 0$ all numbers
11. $6r > 10r - r - 3r$ no solution
12. $3.2x < 2x - (9 - 1.2x)$ no solution

MATH JOURNAL

6. Sometimes statements we make can be translated into inequalities. For example, *In some states, you have to be at least 16 years old to have a driver's license* can be expressed as $a \geq 16$, and *Tomás cannot lift more than 72 pounds* can be translated into $w \leq 72$. Following these examples, write three statements that deal with your everyday life. Then translate each into a corresponding inequality. **See students' work.**

Guided Practice

Match each inequality with the graph of its solution.

7. $b - 18 > -3$ **c**
8. $10 \geq -3 + x$ **a**
9. $x + 11 < 6$ **d**
10. $4c - 3 \leq 5c$ **b**

a. 10 11 12 13 14 15 16 17
b. −6 −5 −4 −3 −2 −1 0 1
c. 10 11 12 13 14 15 16 17
d. −7 −6 −5 −4 −3 −2 −1 0

Solve each inequality. Then check your solution.

11. $x + 7 > 2$ $\{x \mid x > -5\}$
12. $10 \geq x + 8$ $\{x \mid x \leq 2\}$
13. $y - 7 < -12$ $\{y \mid y < -5\}$
14. $-81 + q > 16 + 2q$ $\{q \mid q < -97\}$

Define a variable, write an inequality, and solve each problem. Then check your solution.

15. A number decreased by 17 is less than −13. $x - 17 < -13, \{x \mid x < 4\}$
16. A number increased by 4 is at least 3. $x + 4 \geq 3, \{x \mid x \geq -1\}$

EXERCISES

Practice

Solve each inequality. Then check your solution, and graph it on a number line. **17–24. See margin for graphs.**

17. $a - 12 < 6$ $\{a \mid a < 18\}$
18. $m - 3 < -17$ $\{m \mid m < -14\}$
19. $2x \leq x + 1$ $\{x \mid x \leq 1\}$
20. $-9 + d > 9$ $\{d \mid d > 18\}$
21. $x + \frac{1}{3} > 4$ $\left\{x \mid x > \frac{11}{3}\right\}$
22. $-0.11 \leq n - (-0.04)$ $\{n \mid n \geq -0.15\}$
23. $2x + 3 > x + 5$ $\{x \mid x > 2\}$
24. $7h - 1 \leq 6h$ $\{h \mid h \leq 1\}$

Solve each inequality. Then check your solution.

25. $x + \frac{1}{8} < \frac{1}{2}$ $\left\{x \mid x < \frac{3}{8}\right\}$
26. $3x + \frac{4}{5} \leq 4x + \frac{3}{5}$ $\left\{x \mid x \geq \frac{1}{5}\right\}$
27. $3x - 9 \leq 2x + 6$ $\{x \mid x \leq 15\}$
28. $6w + 4 \geq 5w + 4$ $\{w \mid w \geq 0\}$
29. $-0.17x - 0.23 < 0.75 - 1.17x$
30. $0.8x + 5 \geq 6 - 0.2x$ $\{x \mid x \geq 1\}$
31. $3(r - 2) < 2r + 4$ $\{r \mid r < 10\}$
32. $-x - 11 \geq 23$ $\{x \mid x \leq -34\}$

29. $\{x \mid x < 0.98\}$

Define a variable, write an inequality, and solve each problem. Then check your solution.

33. A number decreased by -4 is at least 9. $x-(-4) \geq 9, \{x \mid x \geq 5\}$
34. The sum of a number and 5 is at least 17. $x+5 \geq 17, \{x \mid x \geq 12\}$

35. $3x < 2x + 8$

35. Three times a number is less than twice the number added to 8. $\{x \mid x < 8\}$

36. $21 \geq x + (-2)$

36. Twenty-one is no less than the sum of a number and -2. $\{x \mid x \leq 23\}$
37. The sum of two numbers is less than 53. One number is 20. What is the other number? $20 + x < 53, \{x \mid x < 33\}$

C

38. The sum of four times a number and 7 is less than 3 times that number.

38. $4x + 7 < 3x$, $\{x \mid x < -7\}$

39. Twice a number is more than the difference of that number and 6.

39. $2x > x - 6$, $\{x \mid x > -6\}$

40. The sum of two numbers is 100. One number is at least 16 more than the other number. What are the two numbers? $100 - x \geq x + 16, \{x \mid x \leq 42\}$

If $3x \geq 2x + 5$, then complete each inequality.

41. $3x + 7 \geq 2x +$ _?_ 12
42. $3x - 10 \geq 2x -$ _?_ 5
43. $3x +$ _?_ $\geq 2x + 3$ -2
44. _?_ $\leq x$ 5

Programming

45. **Geometry** For three line segments to form a triangle, the sum of the lengths of any two sides must exceed the length of the third side. Let the lengths of the possible sides be a, b, and c. Then these three inequalities must be true: $a + b > c$, $a + c > b$, and $b + c > a$. The graphing calculator program at the right uses these inequalities to determine if the three lengths can be measures of the sides of a triangle.

```
PROGRAM: TRIANGLE
: Disp "ENTER THREE LENGTHS"
: Prompt A, B, C
: If C≥A+B
: Then
: Goto 1
: End
: If B≥A+C
: Then
: Goto 1
: End
: If A≥B+C
: Then
: Goto 1
: End
: Disp "THIS IS A TRIANGLE."
: Stop
: Lbl 1
: Disp "NOT A TRIANGLE"
```

Use the program to determine whether segments with the given lengths can form a triangle.

Tip: To run the program again after trying one set of numbers, press ENTER.

a. 10 in., 12 in., 27 in. no
b. 3 ft, 4 ft, 5 ft yes
c. 125 cm, 140 cm, 150 cm yes
d. 1.5 m, 2.0 m, 2.5 m yes

47. The value of x falls between -2.4 and 3.6.

Critical Thinking

46. Using an example, show that even though $x > y$ and $t > w$, $x - t > y - w$ may be false. Answers will vary. Sample answer: $x = 4$, $y = 2$, $t = 3$, $w = 0$
47. What does the sentence $-2.4 < x < 3.6$ mean?

Using the Programming Exercises The program given in Exercise 45 is for use with a TI-82 graphing calculator. For other programmable calculators, have students consult their owner's manual for commands similar to those presented here.

Practice Masters, p. 48

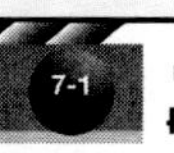

NAME__________ DATE ________

Practice

Student Edition Pages 384–390

Solving Inequalities by Using Addition and Subtraction

Solve each inequality. Then check your solution.

1. $n + 5 \geq 32$ $n \geq 27$
2. $v - 8 \leq 35$ $v \leq 43$
3. $r - 6 < -15$ $r < -9$
4. $-81 > 16 + q$ $q < -97$
5. $-51 \leq x - (-38)$ $x \geq -89$
6. $m - (-3.4) \geq 12.7$ $m \geq 9.3$
7. $4.2 > -11 + t$ $t < 15.2$
8. $2\ell < \ell - 6$ $\ell < -6$
9. $16w \geq 15w - 8$ $w \geq -8$
10. $6p > 5p + 19$ $p > 19$
11. $u - 12 < 2u$ $u > -12$
12. $2y - 17 \geq y - 6$ $y \geq 11$
13. $3e - 0.2 \geq 4e + 0.5$ $e \leq -0.7$
14. $-3 + 12a < -4 + 11a$ $a < -1$
15. $-11.4 + s > 2s - 6.5$ $s < -4.9$
16. $2.1k - 4 \geq 3.1k + 4$ $k \leq -8$
17. $-\frac{3}{10} + d < \frac{9}{10}$ $d < \frac{6}{5}$
18. $\frac{3}{4}c \leq \frac{7}{4}c - 1$ $c \geq 1$

Define a variable, write an inequality, and solve each problem. Then check your solution.

19. A number decreased by 10 is greater than -5. $n - 10 > -5$; $n > 5$
20. A number increased by 2 is at most 6. $n + 2 \leq 6$; $n \leq 4$
21. A number increased by -1 is less than 10. $n - 1 < 10$; $n < 11$
22. A number decreased by -4 is at least 9. $n - (-4) \geq 9$; $n \geq 5$

4 ASSESS

Closing Activity

Writing Without solving, rewrite each inequality in a form that has the same meaning.

1. $17 < x + 4$ $x + 4 > 17$
2. $-3 + x > 2x - 4$ $2x - 4 < -3 + x$
3. $4 + x < 8$ $8 > 4 + x$

Additional Answer

48b. 

Enrichment Masters, p. 48

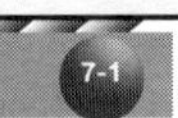

NAME__________ DATE ______

Enrichment

Student Edition Pages 384–390

Inequalities with Triangles

Recall that a line segment can be named by the letters of its endpoints. Line segment AB (written as "$\overline{AB}$") has points A and B for endpoints. The *length* of AB is written without the bar as AB.

$AB < BC$ $\quad \angle A < \angle B$

The statement on the left above shows that $\overline{AB}$ is shorter than $\overline{BC}$. The statement on the right above shows that the measure of angle A is less than that of angle B.

These three inequalities are true for any triangle ABC, no matter how long the sides are.

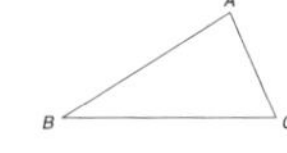

a. $AB + BC > AC$
b. If $AB > AC$, then $\angle C > \angle B$.
c. If $\angle C > \angle B$, then $AB > AC$.

Use the three triangle inequalities for these problems.

1. List the sides of triangle DEF in order of increasing length. $\overline{DF}, \overline{DE}, \overline{EF}$

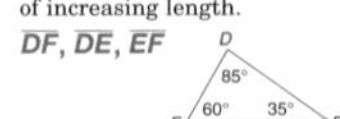
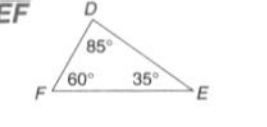

2. In the figure below, which line segment is the shortest? $\overline{LM}$

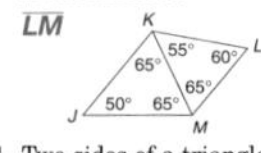

3. Explain why the lengths 5 cm, 10 cm, and 20 cm could not be used to make a triangle. 5 + 10 is not greater than 20.

4. Two sides of a triangle measure 3 in. and 7 in. Between which two values must the third side be? 4 in. and 10 in.

5. In triangle XYZ, $XY = 15$, $YZ = 12$, and $XZ = 9$. Which is the greatest angle? Which is the least? $\angle Z$; $\angle Y$

6. List the angles $\angle A$, $\angle C$, $\angle ABC$, and $\angle ABD$, in order of increasing size. $\angle ABD$, $\angle A$, $\angle ABC$, $\angle C$

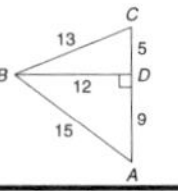

Applications and Problem Solving

Define a variable, write an inequality, and solve each problem.

48. **Academics** Josie must have at least 320 points in her math class to get a B. She needs a B or better to maintain her grade-point average so she can play on the basketball team. The grade is based on four 50-point tests, three 20-point quizzes, two 20-point projects, and a final exam worth 100 points. Josie's record of her grades are shown in the table below.

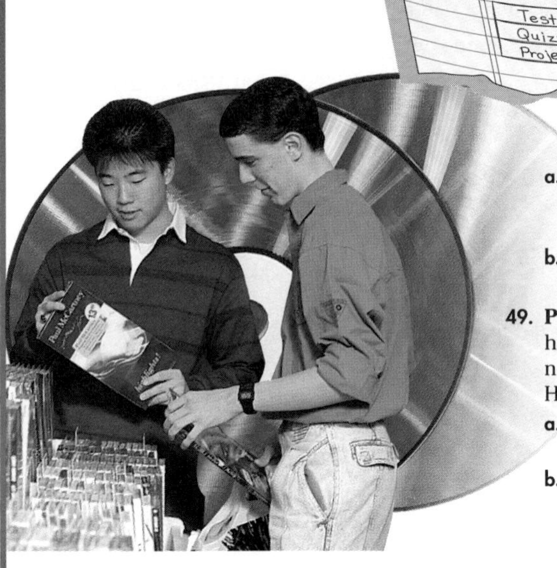

Points
Total Points
Tests
40, 42, 41, 45
168
Quizzes
15, 12, 19
46
Projects
15, 18
33

a. Write an inequality to represent the range of scores on the final exam needed in order for Josie to make a B. $247 + x \geq 320$

b. Solve the inequality and graph its solution. $x \geq 73$; See margin for graph.

49. **Personal Finances** Tanaka had \$75 to buy presents for his family. He bought a \$21.95 shirt for his dad, a \$23.42 necklace for his mother, and a \$16.75 CD for his sister. He still has to buy a present for his brother.

a. How much can he spend on his brother's present? $x \leq \$12.88$

b. What factors not stated in the problem may affect how much money he can spend? Sample answer: There may be sales tax on his purchases.

Mixed Review

50. Find the midpoint of the line segment whose endpoints are at $(-1, 9)$ and $(-5, 5)$. (Lesson 6–7) $(-3, 7)$

Write an equation in slope-intercept form for a line that satisfies each condition. (Lesson 6–6)

51. perpendicular to $x + 7 = 3y$ that passes through $(1, 0)$ $y = -3x + 3$

52. parallel to $\frac{1}{5}y - 3x = 2$ that passes through $(0, -3)$ $y = 15x - 3$

53. **Statistics** Find the range, quartiles, and interquartile range for the set of data at the right. (Lesson 5–7) 42, 131, 145, 159, 28

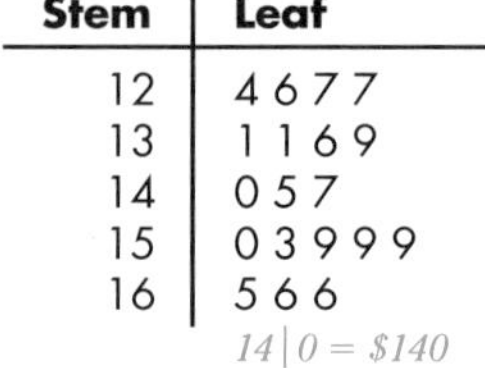

Stem	Leaf
12	4 6 7 7
13	1 1 6 9
14	0 5 7
15	0 3 9 9 9
16	5 6 6

$14|0 = \$140$

54. Solve $4x + 3y = 16$ if the domain is $\{-2, -1, 0, 2, 5\}$. (Lesson 5–3)

54. $\left\{(-2, 8), \left(-1, \frac{20}{3}\right), \left(0, \frac{16}{3}\right), \left(2, \frac{8}{3}\right), \left(5, -\frac{4}{3}\right)\right\}$

55. If y varies inversely as x and $y = 32$ when $x = 3$, find y when $x = 8$. (Lesson 4–8) 12

56. Solve $y - \frac{7}{16} = -\frac{5}{8}$ (Lesson 3–1) $-\frac{3}{16}$

57. Replace the _?_ with $<$, $>$, or $=$ to make $\frac{6}{13}$ _?_ $\frac{1}{2}$ true. (Lesson 2–4) $<$

58. **Look for a Pattern** How many triangles are shown at the right? Count only the triangles pointing upward. (Lesson 1–2) 120 triangles

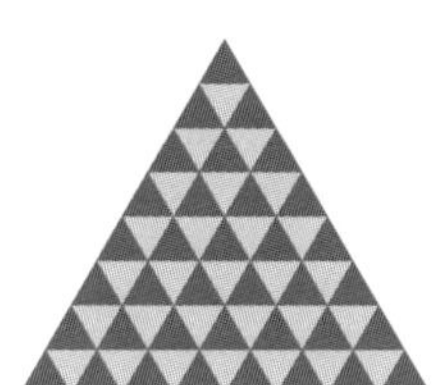

Extension

Problem Solving Find the least prime number greater than 80. 83

7–2A Solving Inequalities

Materials: 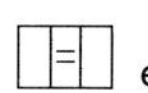 equation mat cups and counters

 self-adhesive note

A Preview of Lesson 7–2

You can use an equation model to solve inequalities.

Activity Model the solution for $-2x < 4$.

Step 1 Use the note to cover the equals sign on the equation mat. Then write a < symbol on the note. Label 2 cups with a negative sign, and place them on the left side. Place 4 positive counters on the right.

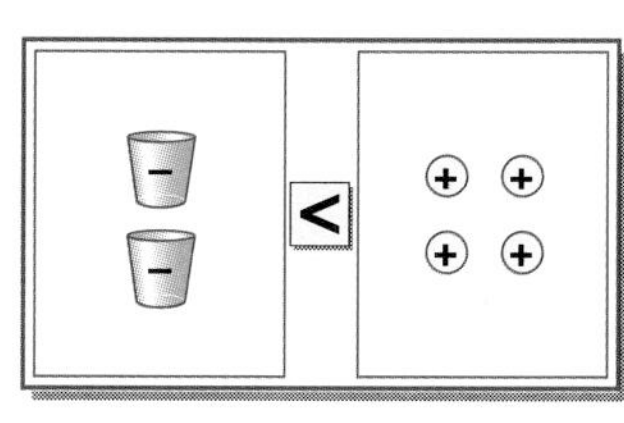

$-2x < 4$

Step 2 Since we cannot solve for a negative cup, we must eliminate the negative cups by adding 2 positive cups to each side. Remove the zero pairs.

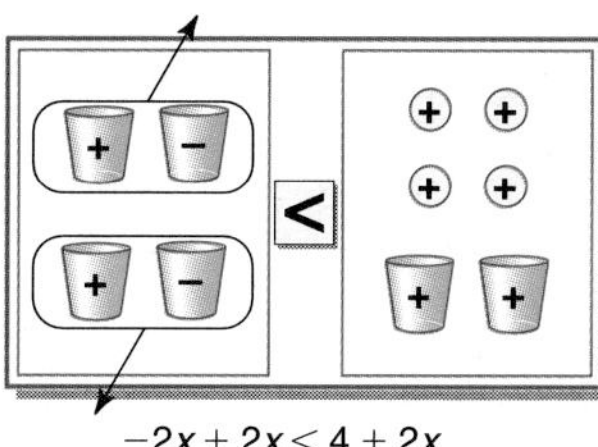

$-2x + 2x < 4 + 2x$

Step 3 Add 4 negative counters to each side to isolate the cups. Remove the zero pairs.

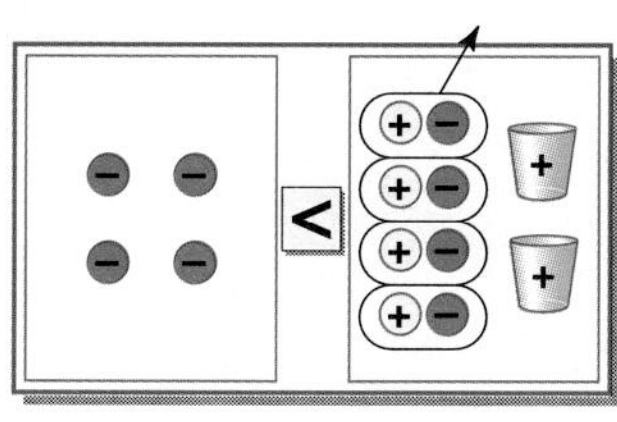

$-4 < 2x$

Step 4 Separate the counters into 2 groups.

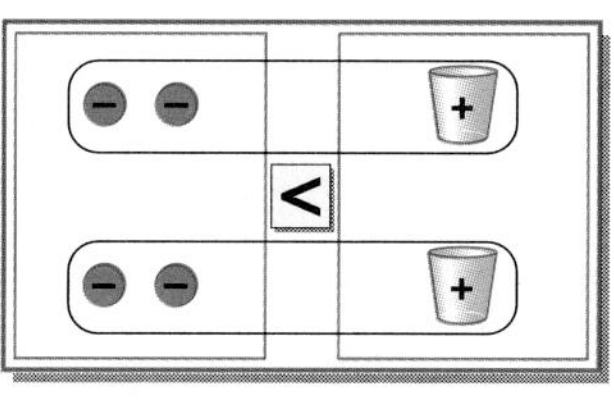

$-2 < x$ or $x > -2$

Model

1. Compare the symbol and location of the variable in the original problem with those in the solution. What do you find? **1–4. See Solutions Manual.**
2. Model the solution for $3x > 12$. What do you find? How is this different from solving $-2x < 4$?

Write

3. Write a rule for solving inequalities involving multiplication.
4. Do you think the rule applies to inequalities involving division?
Remember that dividing by a number is the same as multiplying by its reciprocal.

7-2A LESSON NOTES

NCTM Standards: 1–5

Objective

Use cups and counters to solve inequalities.

Recommended Time

Demonstration and discussion: 15 minutes; Exercises: 30 minutes

Instructional Resources

For each student or group of students
Student Manipulative Kit
- 20 red/yellow 2-sided counters
- 5 cups

Modeling Mathematics Masters
- pp. 2–3 (cups and counters)
- p. 13 (equation mat)
- p. 27 (worksheet)

For teacher demonstration
Algebra and Geometry Overhead Manipulative Resources

1 FOCUS

Motivating the Lesson

Write the inequality $-5 < -2$ on the board. Ask students if the inequality would still be true if the negative signs were removed. Ask what would have to be done with the inequality symbol to make the sentence true again.

2 TEACH

Teaching Tip Point out that there is an alternative method for the Activity. After Step 1, they could replace every cup and every counter with its opposite and reverse the direction of the inequality symbol.

3 PRACTICE/APPLY

Assignment Guide

Core: 1–4
Enriched: 1–4

4 ASSESS

Observing students working in cooperative groups is an excellent method of assessment.

GLENCOE Technology

Interactive Mathematics Tools Software

This multimedia software provides interactive lessons that allow students to use counters and number lines to explore inequalities. A **Computer Journal** gives students an opportunity to write about what they have learned.

For Windows & Macintosh

7-2 LESSON NOTES

NCTM Standards: 1–5

Instructional Resources

- Study Guide Master 7-2
- Practice Master 7-2
- Enrichment Master 7-2
- Assessment and Evaluation Masters, p. 184
- Tech Prep Applications Masters, p. 14

Transparency 7-2A contains the 5-Minute Check for this lesson; **Transparency 7-2B** contains a teaching aid for this lesson.

Recommended Pacing	
Standard Pacing	Day 3 of 14
Honors Pacing	Day 3 of 13
Block Scheduling*	Day 2 of 7
Alg. 1 in Two Years*	Day 4 of 21

*For more information on pacing and possible lesson plans, refer to the *Block Scheduling Booklet* and *Algebra 1 in Two Years*.

1 FOCUS

5-Minute Check *(over Lesson 7-1)*

Solve each inequality.

1. $y + 3 > 11$ $\{y \mid y > 8\}$
2. $t - 9 < 6$ $\{t \mid t < 15\}$
3. $3t + 11 < 4$ $\left\{t \mid t < -\frac{7}{3}\right\}$
4. $2y - 10 < y + 6$ $\{y \mid y < 16\}$
5. $0.9 - p < -3.4$ $\{p \mid p > 4.3\}$

7-2 Solving Inequalities by Using Multiplication and Division

What YOU'LL LEARN

- To solve inequalities by using multiplication and division.

Why IT'S IMPORTANT

You can use inequalities to solve problems involving the physical and political sciences.

GLOBAL CONNECTIONS

About 15,000 years ago, an important application of the lever appeared as hunters used an *atlatl*, the Aztec word for spear-thrower. This simple device was a handle that provided extra mechanical advantage by adding length to the hunter's arm. This enabled the hunter to use less motion to get greater results when hunting.

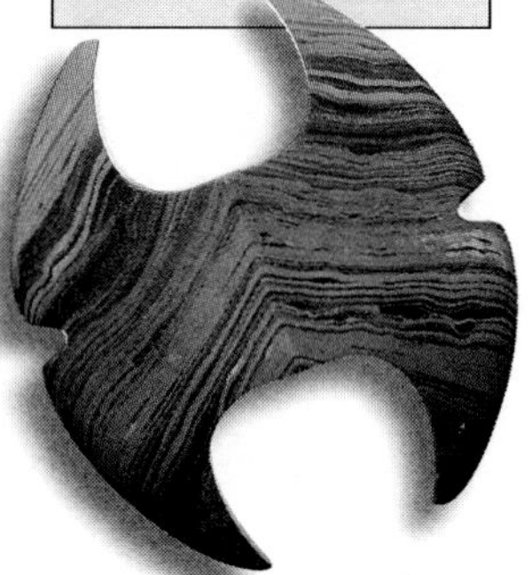

Physical Science

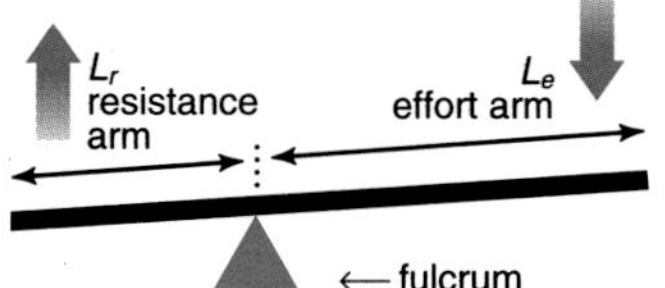

A lever can be used to multiply the effort force you exert when trying to move something. The fixed point or *fulcrum* of a lever separates the length of the lever into two sections—the *effort arm* on which the effort force is applied and the *resistance arm* that exerts the resistance force. The *mechanical advantage* of a lever is the number of times a lever multiplies that effort force.

The formula for determining the mechanical advantage *MA* of a lever can be expressed as $MA = \frac{L_e}{L_r}$, where L_e represents the length of the effort arm and L_r represents the length of the resistance arm.

Suppose a group of volunteers is clearing hiking trails at Yosemite National Park. They need to position a lever so that a mechanical advantage of at least 7 is achieved in order to remove a boulder blocking the trail. The volunteers place the lever on a rock so they can use the rock as a fulcrum. They will need the resistance arm to be 1.5 ft long so that it is long enough to get under the boulder. What should be the length of the lever in order to move the boulder?

We need to find the length of the effort arm to find the total length. Let L_e represent the length of the effort arm. We know that 1.5 feet is the length of the resistance arm L_r. Since the mechanical advantage must be at least 7, we can write an inequality using the formula.

$MA \geq 7$

$\frac{L_e}{L_r} \geq 7$ *Replace MA with $\frac{L_e}{L_r}$.*

$\frac{L_e}{1.5} \geq 7$ *Replace L_r with 1.5.* *You will solve this problem in Example 1.*

GLOBAL CONNECTIONS

The atlatl was probably developed by hunters in the open, treeless landscape of the Siberian tundra. From there, its use spread around the world. Often made from bone, antler, or wood, atlatls provided increased accuracy and power.

If you were solving the equation $\frac{L_e}{1.5} = 7$, you would multiply each side by 1.5. Will this method work when solving inequalities? Before answering this question, let's explore how multiplying (or dividing) an inequality by a positive or negative number affects the inequality. Consider the inequality $10 < 15$, which we know is true.

Multiply by 2.

$10 < 15$

$10(2) < 15(2)$

$20 < 30$ true

Multiply by −2.

$10 < 15$

$10(-2) < 15(-2)$ false

$-20 < -30$ false

$-20 > -30$ true

Divide by 5.

$10 < 15$

$\frac{10}{5} < \frac{15}{5}$

$2 < 3$ true

Divide by −5.

$10 < 15$

$\frac{10}{-5} < \frac{15}{-5}$ false

$-2 < -3$ false

$-2 > -3$ true

These results suggest the following.

- If each side of a true inequality is multiplied or divided by the same positive number, the resulting inequality is also true.
- If each side of a true inequality is multiplied or divided by the same negative number, the direction of the inequality symbol must be *reversed* so that the resulting inequality is also true.

Multiplication and Division Properties for Inequalities

For all numbers, *a*, *b*, and *c*, the following are true.

1. **If *c* is positive and $a < b$, then $ac < bc$ and $\frac{a}{c} < \frac{b}{c}$, $c \neq 0$, and if *c* is positive and $a > b$, then $ac > bc$ and $\frac{a}{c} > \frac{b}{c}$, $c \neq 0$.**
2. **If *c* is negative and $a < b$, then $ac > bc$ and $\frac{a}{c} > \frac{b}{c}$, $c \neq 0$, and if *c* is negative and $a > b$, then $ac < bc$ and $\frac{a}{c} < \frac{b}{c}$, $c \neq 0$.**

These properties also hold for inequalities involving ≤ and ≥.

Example 1

Physical Science

Refer to the connection at the beginning of the lesson. What should the minimum length of the lever be?

$\frac{L_e}{1.5} \geq 7$

$1.5 \cdot \frac{L_e}{1.5} \geq 1.5(7)$ *Multiply each side by 1.5.*

$L_e \geq 10.5$

The effort arm must be at least 10.5 feet long.

In order to find the length of the lever, add the lengths of the effort arm and the resistance arm. The lever should be at least 10.5 + 1.5 or 12 feet long.

Motivating the Lesson

Hands-On Activity Have students draw a number line and graph the following points: $A = 0.5$, $B = -1.5$. Then have students consider the following.

1. If the coordinates of *A* and *B* are multiplied by 2, where are the corresponding points located on the number line? $A = 1, B = -3$
2. If the coordinates of *A* and *B* are multiplied by −2, where are the corresponding points located on the number line? $A = -1, B = 3$

2 TEACH

In-Class Example

For Example 1
Suppose a mechanical advantage of 10 is desired and the resistance arm is 20 centimeters long. What should be the minimum length of the lever?
at least 220 centimeters

Teaching Tip In Example 2, remind students that $\frac{x}{12}$ is equal to $\frac{1}{12}x$. The reciprocal of $\frac{1}{12}$ is $\frac{12}{1}$, or 12.

Teaching Tip In Example 3, emphasize that when multiplying or dividing an inequality by a negative number, the inequality sign is reversed when the negative factor first appears.

In-Class Examples

For Example 2
Solve each inequality.

a. $\frac{k}{-4} > 13$ **$k < -52$**
b. $\frac{x}{3} < 10$ **$x < 30$**
c. $\frac{x}{28} \leq \frac{6}{7}$ **$x < 24$**

For Example 3
Solve each inequality.

a. $-8p \geq -96$ **$p \leq 12$**
b. $-7v \leq -105$ **$v \geq 15$**
c. $-14c \geq 70$ **$c \leq -5$**

For Example 4
Mr. Samuels works for a real estate office that pays its agents 7% of their sales. How much real estate will Mr. Samuels have to sell to earn a minimum of $42,000? **at least $600,000**

Example  **Solve $\frac{x}{12} \leq \frac{3}{2}$.**

$\frac{x}{12} \leq \frac{3}{2}$

$12 \cdot \frac{x}{12} \leq 12 \cdot \frac{3}{2}$ *Multiply each side by 12.*

$x \leq 18$ *Since we multiplied by a positive number, the inequality symbol stays the same.*

The solution set is $\{x \mid x \leq 18\}$.

Since dividing is the same as multiplying by the reciprocal, there can be two methods to solve an inequality that involves multiplication.

Example  **Solve $-3w > 27$.**

Method 1

$-3w > 27$

$\frac{-3w}{-3} < \frac{27}{-3}$ *Divide each side by −3 and change > to <.*

$w < -9$

Method 2

$-3w > 27$

$\left(-\frac{1}{3}\right)(-3w) < \left(-\frac{1}{3}\right)(27)$ *Multiply each side by $-\frac{1}{3}$ and change > to <.*

$w < -9$

Check: Let w be any number less than −9.

$-3w > 27$

$-3(-10) \overset{?}{>} 27$ *Suppose we select −10.*

$30 > 27$ true

Numbers less than −9 compose the solution set. The solution set is $\{w \mid w < -9\}$.

Example 4 **Angelica Moreno is a sales representative for an appliance distributor. She needs at least $5000 in weekly sales of a particular TV model to qualify for a sales competition to win a trip to the Bahamas. If the TVs sell for $250 each, how many TVs will Ms. Moreno have to sell to qualify?**

Business

Explore Let t represent the number of TVs to be sold. At least $5000 means greater than or equal to $5000.

Plan The price of one TV times the number of sets sold must be greater than or equal to the total amount of sales needed.

The price of one TV	*times*	*the number of TVs sold*	*is at least*	*$5000.*
$250	×	t	≥	$5000

Solve $250t \geq 5000$

$\frac{250t}{250} \geq \frac{5000}{250}$ *Divide each side by 250.*

$t \geq 20$

Examine Ms. Moreno must sell a minimum of 20 TVs to qualify for the contest. The amount of money from the sale of 20 TVs is $250(20) or $5000.

Tech Prep

Real Estate Associate Students who are interested in a career in real estate, as mentioned in Example 4, may wish to do further research into the use of math in that occupation. For more information on tech prep, see the *Teacher's Handbook.*

Example 5

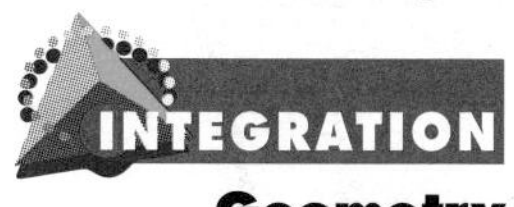

Geometry

Triangle XYZ is not an acute triangle. The greatest angle in the triangle has a measure of $(6d)°$. What are the possible values of d?

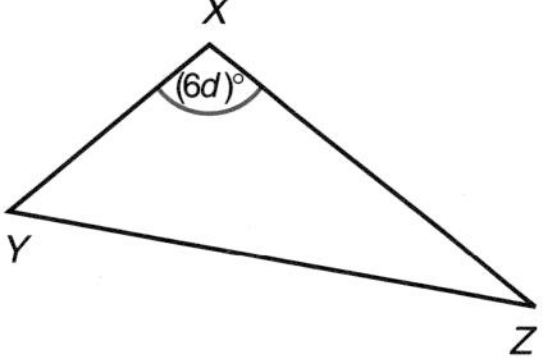

Since ΔXYZ is not acute, the measure of the greatest angle must be 90° or larger, but less than 180°. Thus, $6d \geq 90$ and $6d < 180$.

$6d \geq 90$	$6d < 180$
$\frac{6d}{6} \geq \frac{90}{6}$ *Divide each side by 6.*	$\frac{6d}{6} < \frac{180}{6}$ *Divide each side by 6.*
$d \geq 15$	$d < 30$

The value of d must be greater than or equal to 15 but less than 30.

In-Class Example

For Example 5

The acute angle shown below has a measure greater than 0° and less than 90°. What is the set of all possible values for r?

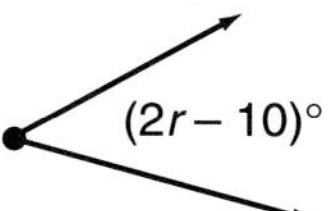

$r > 5$ and $r < 50$

CHECK FOR UNDERSTANDING

Communicating Mathematics

Study the lesson. Then complete the following.

1. **Classify** each statement as *true* or *false*. If false, explain how to change the inequality to make it true.
   a. If $x > 9$, then $-3x > -27$. **False; change > to < in the second inequality.**
   b. If $x < 4$, then $3x < 12$. **true**
2. **Complete** each statement.
   a. If each side of an inequality is multiplied by the same __?__ number, the direction of the inequality symbol must be reversed so that the resulting inequality is true. **negative**
   b. Multiplying by __?__ is the same as dividing by -6. $-\frac{1}{6}$
   c. An acute triangle has angles whose measures are all less than __?__. **90°**
3. **You Decide** Utina and Paige are discussing the rules for changes in the direction of the inequality symbol when solving inequalities. Paige says the rule is "Whenever you have a negative sign in the problem, the direction of the inequality symbol will change." Utina says that is not always true. Decide who is correct and give examples to support your answer. **See margin.**

MODELING MATHEMATICS

Use models to solve each inequality. Write the answer in set-builder notation.

4. $3x < 15$ $\{x \mid x < 5\}$
5. $-6x < 18$ $\{x \mid x > -3\}$
6. $2x + 6 > x - 7$ $\{x \mid x > -13\}$
7. $-4x + 8 \geq 14$ $\{x \mid x \leq -1.5\}$
8. Use models to determine the appropriate symbol (>, <, or =) to complete each comparison.
   a. $x + 5$ __?__ $x - 7$ $>$
   b. $x + 3 + (x - 6)$ __?__ $(2x + 7) - 4$ $<$
   c. $2x + 8$ __?__ $2x + 6 - x$ **cannot be determined**

3 PRACTICE/APPLY

Check for Understanding

Exercises 1–19 are designed to help you assess your students' understanding through reading, writing, speaking, and modeling. You should work through Exercises 1–8 with your students and then monitor their work on Exercises 9–19.

Error Analysis

Some students may add an extra step when solving inequalities that involve dividing or multiplying. Example:

$-3x \leq 12$
$x \leq -4$ *Divide by −3.*
$x \geq -4$ *Reverse the sign.*

Remind students that the sign is reversed in the same step as the multiplying or dividing.

Reteaching

Using Manipulatives Use a two-pan balance to demonstrate how to solve these inequalities.

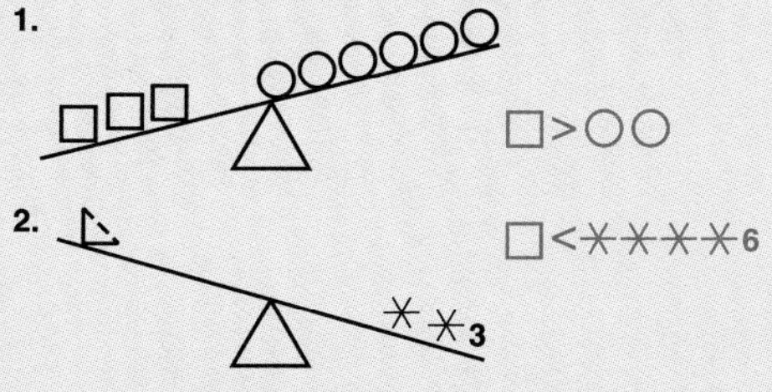

Additional Answer

3. **Utina is correct because the only time the inequality symbol changes in solving an inequality is when you must multiply or divide by a negative to solve for the variable. For example, in $-3x > 12$, the inequality symbol changes. However, in $3x > -12$, the inequality symbol does not change.**

Assignment Guide

Core: 21–49 odd, 50, 51, 53–62
Enriched: 20–48 even, 50–62

For **Extra Practice,** see p. 772.

The red A, B, and C flags, printed only in the Teacher's Wraparound Edition, indicate the level of difficulty of the exercises.

Guided Practice

State the number by which you multiply or divide to solve each inequality. Indicate whether the direction of the inequality symbol reverses. Then solve.

9. $-6y \geq -24$ $\times -\frac{1}{6}$, $\div -6$; yes; $\{y \mid y \leq 4\}$
10. $10x > 20$ $\times \frac{1}{10}$, $\div 10$; no; $\{x \mid x > 2\}$
11. $\frac{x}{4} < -5$ $\times 4$; no; $\{x \mid x < -20\}$
12. $-\frac{2}{7}z \geq -12$ $\times -\frac{7}{2}$; yes; $\{z \mid z \leq 42\}$

Solve each inequality. Then check your solution.

13. $\frac{4}{5}x < 24$ $\{x \mid x < 30\}$
14. $-\frac{v}{3} \geq 4$ $\{v \mid v \leq -12\}$
15. $-0.1t \geq 3$ $\{t \mid t \leq -30\}$
16. $5y > -25$ $\{y \mid (y > -5\}$

Define a variable, write an inequality, and solve each problem. Then check your solution.

17. One fifth of a number is at most 4.025. $\frac{1}{5}x \leq 4.025$; $\{x \mid x \leq 20.125\}$
18. The opposite of six times a number is less than 216. $-6x < 216$; $\{x \mid x > -36\}$
19. **Geometry** Determine the value of s so that the area of the square is at least 144 square feet. $s \geq 12$

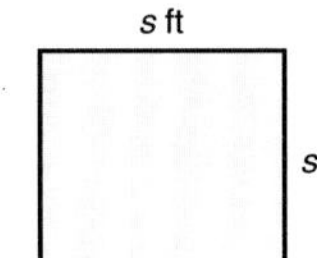

EXERCISES

Practice

Solve each inequality. Then check your solution.

20. $7a \leq 49$ $\{a \mid a \leq 7\}$
21. $12b > -144$ $\{b \mid b > -12\}$
22. $-5w > -125$ $\{w \mid w < 25\}$
23. $-x \leq 44$ $\{x \mid x \geq -44\}$
24. $4 < -x$ $\{x \mid x < -4\}$
25. $-102 > 17r$ $\{r \mid r < -6\}$
26. $\frac{b}{-12} \leq 3$ $\{b \mid b \geq -36\}$
27. $\frac{t}{13} < 13$ $\{t \mid t < 169\}$
28. $\frac{2}{3}w > -22$ $\{w \mid w > -33\}$

B

29. $6 \leq 0.8g$ $\{g \mid g \geq 7.5\}$
30. $-15b < -28$ $\left\{b \mid b > \frac{28}{15}\right\}$
31. $-0.049 \leq 0.07x$ $\{x \mid x \geq -0.7\}$
32. $\frac{3}{7}h < \frac{3}{49}$ $\left\{h \mid h < \frac{1}{7}\right\}$
33. $\frac{12r}{-4} > \frac{3}{20}$ $\left\{r \mid r < -\frac{1}{20}\right\}$
34. $\frac{3b}{4} \leq \frac{2}{3}$ $\left\{b \mid b \leq \frac{8}{9}\right\}$
35. $-\frac{1}{3}x > 9$ $\{x \mid x < -27\}$
36. $\frac{y}{6} \geq \frac{1}{2}$ $\{y \mid y \geq 3\}$
37. $\frac{-3m}{4} \leq 18$ $\{m \mid m \geq -24\}$

Define a variable, write an inequality, and solve each problem. Then check your solution.

38. Four times a number is at most 36. $4x \leq 36$; $\{x \mid x \leq 9\}$
39. Thirty-six is at least one half of a number. $36 \geq 0.5x$; $\{x \mid x \leq 72\}$
40. The opposite of three times a number is more than 48. $-3x > 48$; $\{x \mid x < -16\}$
41. Three fourths of a number is at most -24. $\frac{3}{4}x \leq -24$; $\{x \mid x \leq -32\}$
42. Eighty percent of a number is less than 24. $0.80x < 24$; $\{x \mid x < 30\}$

43. The product of two numbers is no greater than 144. One of the numbers is -8. What is the other number? $-8x \leq 144$; -18 or greater

Study Guide Masters, p. 49

NAME________________ DATE ________

Study Guide   Student Edition Pages 392–398

Solving Inequalities by Using Multiplication and Division

You can solve inequalities by using the same methods you have already used to solve equations. However, when solving inequalities, if you multiply or divide each side by the same negative number, you must *reverse* the direction of the inequality symbol. The following chart shows the multiplication and division properties for solving inequalities.

Multiplication Property for Inequalities	Division Property for Inequalities
For all numbers a, b, and c, 1. if c is positive and $a < b$, then $ac < bc$;	For all numbers a, b, and c, 1. if c is positive and $a < b$, then $\frac{a}{c} < \frac{b}{c}$;
if c is positive and $a > b$, then $ac > bc$;	if c is positive and $a > b$, then $\frac{a}{c} > \frac{b}{c}$;
2. if c is negative and $a < b$, then $ac > bc$;	2. if c is negative and $a < b$, then $\frac{a}{c} > \frac{b}{c}$;
if c is negative and $a > b$, then $ac < bc$.	if c is negative and $a > b$, then $\frac{a}{c} < \frac{b}{c}$.

State the number by which you multiply each side to solve each inequality. Indicate whether the direction of the inequality symbol reverses.

1. $4s < 24$ $\frac{1}{4}$; no
2. $-9k > 2$ $-\frac{1}{9}$; yes
3. $-5\frac{1}{3} > 2n$ $\frac{1}{2}$; no
4. $\frac{7}{3} < -\frac{1}{9}n$ -9; yes

Solve each inequality. Then check your solution.

5. $77 < 12r$ $\left\{r \mid r > 6\frac{5}{12}\right\}$
6. $-5c > 2$ $\left\{c \mid c < -\frac{2}{5}\right\}$
7. $25g \geq -100$ $\{g \mid g \geq -4\}$
8. $\frac{n}{-50} > 22$ $\{n \mid n < -1100\}$
9. $0.24 < 0.6w$ $\{w \mid w > 0.4\}$
10. $-2.51 \leq \frac{2h}{-4}$ $\{h \mid h \leq 5.02\}$
11. $\frac{2a}{7} \geq -6$ $\{a \mid a \geq -21\}$
12. $-\frac{1}{3} < \frac{2p}{9}$ $\left\{p \mid p > -\frac{3}{2}\right\}$

Define a variable, write an inequality, and solve each problem. Then check your solution.

13. Four times a number is no more than 108. $4n \leq 108$; $\{n \mid n \leq 27\}$
14. A number divided by 5 is at least -10. $\frac{n}{5} \geq -10$; $\{n \mid n \geq -50\}$

44. **Geometry** Determine the value of x so that the area of the rectangle at the right is at least 918 square feet. $x \geq 8.5$ **feet**

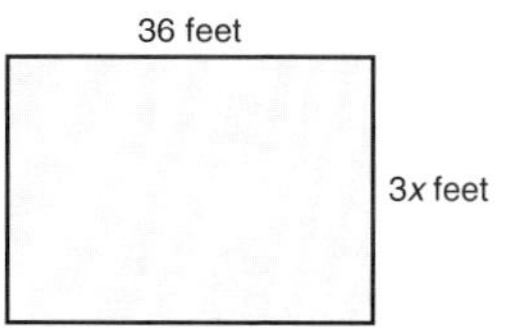

45. **Geometry** Determine the value of y so that the perimeter of the triangle at the right is less than 100 meters. $y < 7.14$ **meters**

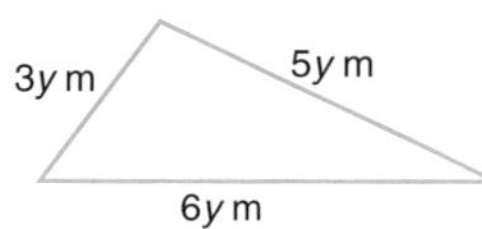

Complete.

46. If $24m \geq 16$, then __?__ ≥ 12. **18m**

47. If $-9 \leq 15b$, then $25b$ __?__ -15. $\geq$

48. If $5y < -12$, then $20y <$ __?__. **−48**

49. If $-10a > 21$, then $30a$ __?__ -63. $<$

Critical Thinking

50. Use an example to show that if $x > y$, then $x^2 > y^2$ is not necessarily true. **Answers will vary. Sample answer:** $x = -1$, $y = -2$

Applications and Problem Solving

Define a variable, write an inequality, and solve each problem.

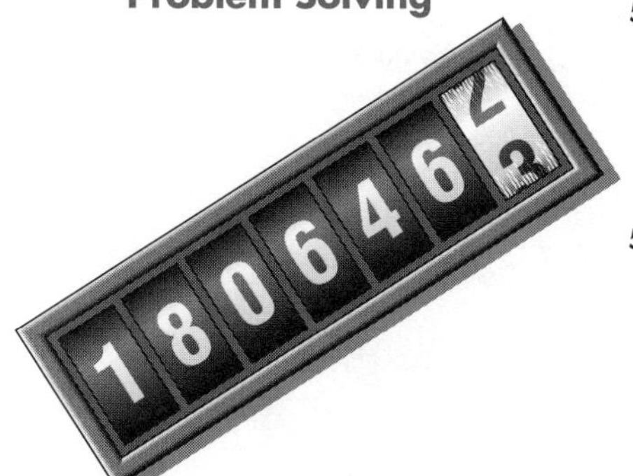

51. **Travel** The charge per mile for a compact rental car at 4-D Rentals is \$0.12. Mrs. Rodriguez is on a business trip and must rent a car to attend various meetings. She has a budget of \$50 per rental for mileage charges. What is the greatest number of miles Mrs. Rodriguez can travel without going over her budget? **up to 416 miles**

52. **Physics** Refer to the application at the beginning of the lesson. A city worker needs to raise a utility access cover. She has an iron bar to use as a lever. When using a lever, the resistance force is equal to the effort force multiplied by the mechanical advantage, or $F_r = MA \cdot F_e$. The worker weighs 120 pounds, so she can supply that much effort force. All utility covers in the city weigh at least 360 pounds. What mechanical advantage does she need to lift the cover? **at least 3**

fabulous **FIRSTS**

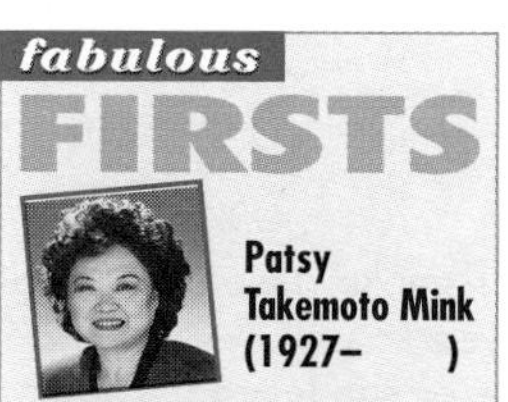

Patsy Takemoto Mink (1927–)

In 1964, Patsy Takemoto Mink, a Democrat from Hawaii, became the first Asian-American woman elected to Congress. She served until 1977 and then was reelected in 1990.

53. **Political Science** A candidate needs 5000 signatures on a petition before she can run for a township office. Experience shows that 15% of the signatures on petitions are not valid. What is the smallest number of signatures the candidate should get to end up with 5000 valid signatures? **at least 5883 signatures**

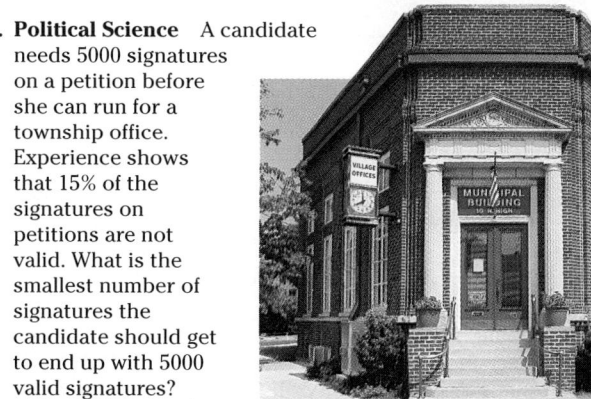

Mixed Review

54. Define a variable, write an inequality, and solve the following problem. The difference of five times a number less four times that number plus seven is at most 34. (Lesson 7–1) $5a - 4a + 7 \leq 34$; $a \leq 27$

Extension

Problem Solving A plumber charges \$36 for the first half hour of work and \$16 for every half hour or any part of a half hour thereafter. Find the longest amount of time this plumber can work without having the bill exceed \$150. **4 hours**

fabulous **FIRSTS**

In 1974, Congresswoman Mink sponsored the Women's Educational Equity Act, which provides funding for the development of nonsexist teaching materials and model programs that encourage full educational opportunities for women, without stereotypes.

Practice Masters, p. 49

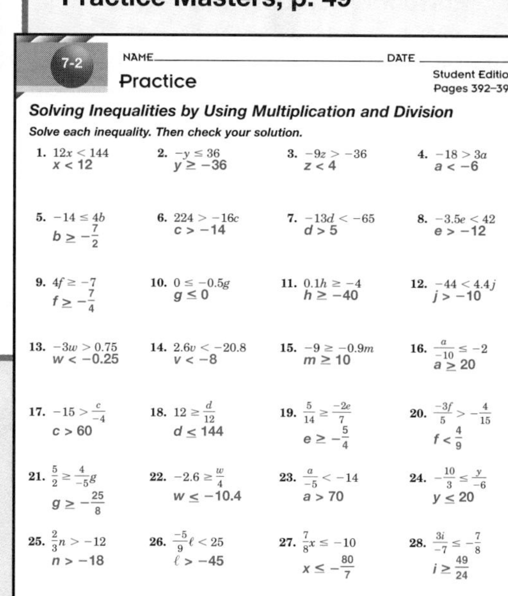

7-2 NAME ____________ DATE ________

Practice — Student Edition Pages 392–398

Solving Inequalities by Using Multiplication and Division

Solve each inequality. Then check your solution.

1. $12x < 144$ — $x < 12$
2. $-y \leq 36$ — $y \geq -36$
3. $-9z > -36$ — $z < 4$
4. $-18 > 3a$ — $a < -6$
5. $-14 \leq 4b$ — $b \geq -\frac{7}{2}$
6. $224 > -16c$ — $c > -14$
7. $-13d < -65$ — $d > 5$
8. $-3.5e < 42$ — $e > -12$
9. $4f \geq -7$ — $f \geq -\frac{7}{4}$
10. $0 \leq -0.5g$ — $g \leq 0$
11. $0.1h \geq -4$ — $h \geq -40$
12. $-44 < 4.4j$ — $j > -10$
13. $-3w > 0.75$ — $w < -0.25$
14. $2.6v < -20.8$ — $v < -8$
15. $-9 \geq -0.9m$ — $m \geq 10$
16. $\frac{a}{-10} \leq -2$ — $a \geq 20$
17. $-15 > \frac{c}{-4}$ — $c > 60$
18. $12 \geq \frac{d}{12}$ — $d \leq 144$
19. $\frac{5}{14} \geq \frac{-2e}{7}$ — $e \geq -\frac{5}{4}$
20. $\frac{-3f}{5} > -\frac{4}{15}$ — $f < \frac{4}{9}$
21. $\frac{5}{2} \geq \frac{4}{-5}g$ — $g \geq -\frac{25}{8}$
22. $-2.6 \geq \frac{w}{4}$ — $w \leq -10.4$
23. $\frac{a}{-5} < -14$ — $a > 70$
24. $-\frac{10}{3} \leq \frac{y}{-6}$ — $y \leq 20$
25. $\frac{2}{3}n > -12$ — $n > -18$
26. $\frac{-5}{9}\ell < 25$ — $\ell > -45$
27. $\frac{7}{8}x \leq -10$ — $x \leq -\frac{80}{7}$
28. $\frac{3i}{-7} \leq -\frac{7}{8}$ — $i \geq \frac{49}{24}$

Define a variable, write an inequality, and solve each problem. Then check your solution.

29. Four times a number is greater than −48. $4n > -48$; $n > -12$

30. Negative three times a number is at least 57. $-3n \geq 57$; $n \leq -19$

31. A number divided by negative six is no less than 5. $\frac{n}{-6} \geq 5$; $n \leq -30$

32. Two thirds of a number is no more than −10. $\frac{2}{3}n \leq -10$; $n \leq -15$

4 ASSESS

Closing Activity

Speaking Discuss the following. If $-2a < b$, is it possible that

1. *a* and *b* are both positive? **yes**
2. *a* is positive and *b* is negative? **yes**
3. *a* is negative and *b* is positive? **yes**
4. *a* and *b* are both negative? **no**

Chapter 7, Quiz A (Lessons 7-1 and 7-2), is available in the *Assessment and Evaluation Masters*, p. 184.

Enrichment Masters, p. 49

7-2 NAME______________________ DATE ________

Enrichment — Student Edition Pages 392–398

Traceable Figures

Try to trace over each of the figures below without tracing the same segment twice.

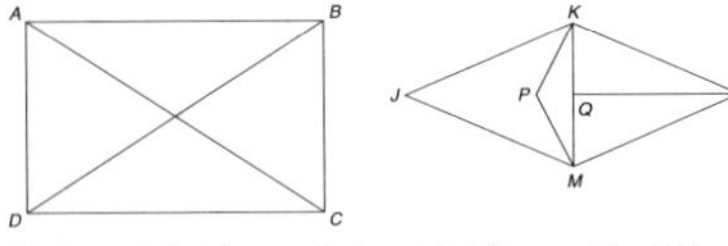

The figure at the left cannot be traced, but the one at the right can. The rule is that a figure is traceable if it has no points, or exactly two points where an odd number of segments meet. The figure at the left has three segments meeting at each of the four corners. However, the figure at the right has exactly two points, *L* and *Q*, where an odd number of segments meet.

Determine whether each figure can be traced. If it can, then name the starting point and number the sides in the order in which they should be traced.

1. Yes *E*
2. No

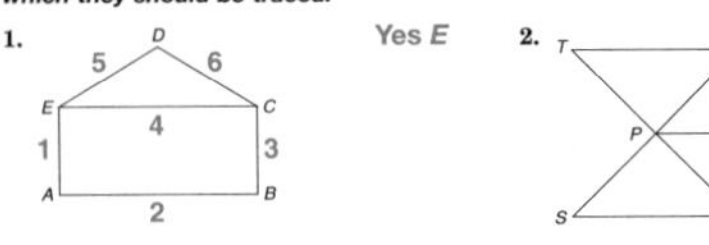

3. Yes *X*
4. Yes *D*

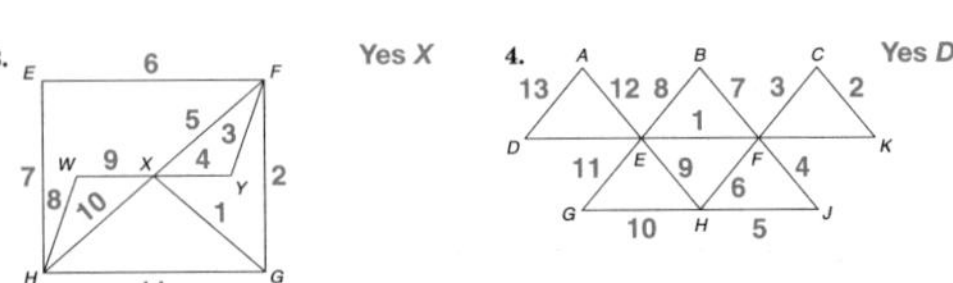

55. **(−1, 1)**

56. **-3; $\frac{1}{3}$**

55. **Geometry** Three vertices of a square are at $(-5, -3)$, $(-5, 5)$, and $(3, -3)$. Suppose you were to inscribe a circle in this square. What would the coordinates of the center of the circle be? (Lesson 6–7)
56. **Geometry** Determine the slopes of the lines parallel and perpendicular to the graph of $3x - 6 = -y$. (Lesson 6–6)
57. Determine the slope of the line that passes through $(4, -9)$ and $(-2, 3)$. (Lesson 6–1) **−2**
58. Write an equation in functional notation for the relation at the right. (Lesson 5–6) **$b = a - 3$**

a	−2	0	2	4	6	8	10
b	−5	−3	−1	1	3	5	7

59. **Budgeting** JoAnne Paulsen's take-home pay is \$1782 per month. She spends \$325 on rent, \$120 on groceries, and \$40 on gas. She allows herself 12% of the remaining amount for entertainment. How much can she spend on entertainment each month? (Lesson 4–4) **\$155.64**
60. What number is 47% of 27? (Lesson 4–4) **12.69**

61. **Soccer** A soccer field is 75 yards shorter than 3 times its width. Its perimeter is 370 yards. Find its dimensions. (Lesson 3–5) **65 yd by 120 yd**
62. Evaluate $5y + 3$ if $y = 1.3$. (Lesson 1–3) **9.5**

WORKING ON THE

In•ves•ti•ga•tion

Refer to the Investigation on pages 320–321.

Smoke Gets In Your Eyes

Currently, about 50 million Americans smoke, and each smoker averages about 30 cigarettes a day. Consider the amount of smoke produced by a single cigarette. Make some projections on the volume of cigarette smoke generated by all 50 million smoking Americans.

1. Find the volume of smoke produced by the average smoker.
2. Project the amount of smoke generated by 50 million smokers in a year.
3. Estimate the volume of air in your room at home. Write an inequality that would estimate the number of puffs *p* it would take to fill your room. Solve the inequality.
4. Suppose a smoker was locked in your room with several cartons of cigarettes. Would it be possible for the smoker to fill the room completely with smoke in the same concentration as the smoke exhaled? Explain your answer.

Add the results of your work to your Investigation Folder.

In•ves•ti•ga•tion

Working on the Investigation

The Investigation on pages 320–321 is designed to be a long-term project that is completed over several days or weeks. Encourage students to keep their materials in their Investigation Folder as they work on the Investigation.

7-3 Solving Multi-Step Inequalities

What YOU'LL LEARN
- To solve linear inequalities involving more than one operation, and
- to find the solution set for a linear inequality when replacement values are given for the variables.

Why IT'S IMPORTANT
You can use inequalities to solve problems involving engineering and real estate.

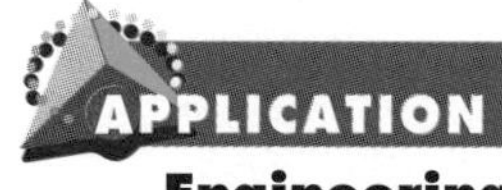

APPLICATION
Engineering

Rosa Whitehair is a partner in an engineering consulting firm. Her fee for consulting on large construction projects is \$1000 plus 10% of the design fee that the company charges its clients. Ms. Whitehair is considering two construction projects: a 50-story office building and the design of an airport terminal. She is interested in both projects, but decides to choose the one for which her fee is higher. The company that is designing the office building has agreed to pay Ms. Whitehair a flat fee of \$5000. How much does the design company's fee need to be for her to choose the terminal?

Let x represent the design fee for the airport terminal.

The flat fee	plus	10% of the design fee	is more than	the flat fee for the office building.
1000	+	$0.10x$	>	5000

LOOK BACK
You can review solving multi-step equations in Lesson 3-3.

This inequality involves more than one operation. It can be solved by undoing the operations in reverse of the order of operations in the same way you would solve an equation with more than one operation.

$$1000 + 0.10x > 5000$$
$$1000 - 1000 + 0.10x > 5000 - 1000 \quad \text{Subtract 1000 from each side.}$$
$$0.10x > 4000$$
$$\frac{0.10x}{0.10} > \frac{4000}{0.10} \quad \text{Divide each side by 0.10.}$$
$$x > 40{,}000$$

If the company charges design fees higher than \$40,000 for the terminal, Ms. Whitehair will choose them since her consulting fee is higher.

GLENCOE Technology

CD-ROM Interaction

A multimedia simulation links graphing inequalities to determining safety constraints for a fireworks factory. A blackline master activity with teacher's notes provides a follow-up to the CD-ROM simulation.

For Windows & Macintosh

7-3 LESSON NOTES

NCTM Standards: 1–5

Instructional Resources

- Study Guide Master 7-3
- Practice Master 7-3
- Enrichment Master 7-3
- Multicultural Activity Masters, p. 13
- Real-World Applications, 19

Transparency 7–3A contains the 5-Minute Check for this lesson; **Transparency 7–3B** contains a teaching aid for this lesson.

Recommended Pacing	
Standard Pacing	Day 4 of 14
Honors Pacing	Day 4 of 13
Block Scheduling*	Day 3 of 7 (along with Lesson 7-4)
Alg. 1 in Two Years*	Days 5 & 6 of 21

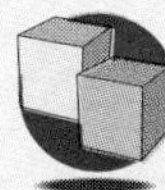

*For more information on pacing and possible lesson plans, refer to the *Block Scheduling Booklet* and *Algebra 1 in Two Years.*

1 FOCUS

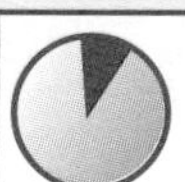

5-Minute Check
(over Lesson 7-2)

Solve each inequality.

1. $\frac{a}{4} > 16$ $\{a \mid a > 64\}$
2. $15 < \frac{x}{3}$ $\{x \mid x > 45\}$
3. $\frac{5}{7}p > -20$ $\{p \mid p > -28\}$
4. $-9v \leq -108$ $\{v \mid v \geq 12\}$
5. $-13x \geq 208$ $\{x \mid x \leq -16\}$

Motivating the Lesson

Questioning Write the following equations on the chalkboard.
$-3x + 5 = -10$
$-3(a + 2) = 8(a - 4)$
Ask students to identify the steps needed to solve each equation. Then replace each = with >. Challenge students to compare the steps needed to solve the inequalities with those for equations.

2 TEACH

In-Class Examples

For Example 1
Solve each inequality.

a. $5m - 8 > 12$ **$m > 4$**
b. $12 + 3a > 18$ **$a > 2$**
c. $20 - 6c < 44$ **$c > -4$**

For Example 2
Solve each inequality.

a. $5m - 4 < 2m + 11$ **$m < 5$**
b. $2r - 18 < 5r + 3$ **$r > -7$**
c. $26p - 20 > 14p + 64$ **$p > 7$**

For Example 3
Determine the solution set for $2x + 5 < 17$ if the replacement set for x is $\{-10, -6, -2, 2, 6, 10\}$. **$\{-10, -6, -2, 2\}$**

Teaching Tip In Example 2, suggest to students that they check several numbers in the solution set to determine whether true statements result.

Example  **Determine the value of x so that $\angle A$ is acute.**

Assume that x is positive.

For $\angle A$ to be acute, its measure must be less than 90°.

Thus, $3x - 15 < 90$.

$3x - 15 < 90$
$3x - 15 + 15 < 90 + 15$ *Add 15 to each side.*
$3x < 105$
$\frac{3x}{3} < \frac{105}{3}$ *Divide each side by 3.*
$x < 35$

For $\angle A$ to be acute, x must be less than 35.

A (3x − 15)°

Sometimes inequalities, like equations, involve variables on each side of the inequality.

Example  **Solve $-4w + 9 \le w - 21$.**

$-4w + 9 \le w - 21$
$-w - 4w + 9 \le w - 21 - w$ *Subtract w from each side.*
$-5w + 9 \le -21$
$-5w + 9 - 9 \le -21 - 9$ *Subtract 9 from each side.*
$-5w \le -30$
$\frac{-5w}{-5} \ge \frac{-30}{-5}$ *Divide each side by −5 and change ≤ to ≥.*
$w \ge 6$

The solution set is $\{w \mid w \ge 6\}$.

When we solve an inequality, the solution set usually includes all numbers for a certain criteria, such as $\{x \mid x > 4\}$. Sometimes a replacement set is given from which the solution set can be chosen.

Example  **Determine the solution set for $3x + 6 > 12$ if the replacement set for x is $\{-2, -1, 0, 1, 2, 3, 4, 5\}$.**

Method 1

Substitute values into the inequality to find the values that satisfy the inequality. Try −2.

$3x + 6 > 12$
$3(-2) + 6 > 12$
$0 > 12$ false

From this trial, we can estimate that the value of x must be much greater than −2 to make the inequality true. Try 2.

$3x + 6 > 12$
$3(2) + 6 > 12$
$12 > 12$ false

Alternative Learning Styles

Visual Have students find solutions to the inequality $2x - 3 \le 5$ using trial-and-error substitution. Each time the substituted value works, have them plot the value on a number line. After students have tried about ten values, they should guess at the solution.

From this trial, we see that values greater than 2 must be in the solution set. Try 3.

$3x + 6 > 12$
$3(3) + 6 > 12$
$15 > 12$ true The solution set is {3, 4, 5}.

Method 2

Solve the inequality for all values of x. Then determine which values from the replacement set belong to the solution set.

$$3x + 6 > 12$$
$$3x + 6 - 6 > 12 - 6$$
$$3x > 6$$
$$\frac{3x}{3} > \frac{6}{3}$$
$$x > 2$$

The solution set is those numbers from the replacement set that are greater than 2. Thus, the solution set is {3, 4, 5}.

EXPLORATION GRAPHING CALCULATORS

You can use the inequality symbols in the TEST menu on the TI-82 graphing calculator to find the solution to an inequality in one variable.

Your Turn a. You see part of the graph of $y = 1$.

a. Clear the  list. Enter $3x + 6 > 4x + 9$ as Y1. (The symbol > is item 3 on the TEST menu.) Press GRAPH. Describe what you see.

b. Use the TRACE function to scan the values along the graph. What do you notice about the values of y on the graph?

c. Solve the inequality algebraically. How does your solution compare to the pattern you noticed in part **b**?

b. $y = 1$ until $x = -3$. Then $y = 0$.

c. $x < -3$; $y = 1$ if the inequality is true for those values of x; $y = 0$ for those values of x for which the inequality is not true.

When solving some inequalities that contain grouping symbols, remember to first use the distributive property to remove the grouping symbols.

Example 4 **Solve $5(k + 4) - 2(k + 6) \geq 5(k + 1) - 1$. Then graph the solution.**

$5(k + 4) - 2(k + 6) \geq 5(k + 1) - 1$
$5k + 20 - 2k - 12 \geq 5k + 5 - 1$ *Distributive property.*
$3k + 8 \geq 5k + 4$ *Combine like terms.*
$3k - 5k + 8 \geq 5k - 5k + 4$ *Subtract 5k from each side.*
$-2k + 8 \geq 4$
$-2k + 8 - 8 \geq 4 - 8$ *Subtract 8 from each side.*
$-2k \geq -4$
$\frac{-2k}{-2} \leq \frac{-4}{-2}$ *Divide each side by −2 and change ≥ to ≤.*
$k \leq 2$

The solution set is $\{k \mid k \leq 2\}$.
The graph of $k \leq 2$ is shown at the right.

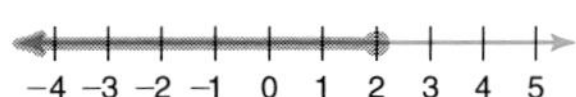

EXPLORATION

The basis for the circuitry of a computer is the bit, with a value of 0 for false and 1 for true. In the TEST menu, the calculator is assigning a 1 to any value of x that makes the inequality true, and a 0 to any value of x that makes it false.

In-Class Example

For Example 4
Solve each inequality. Then graph the solution.

a. $\frac{1}{3}(3t + 6) < \frac{1}{3}(6t - 9)$ $t > 5$

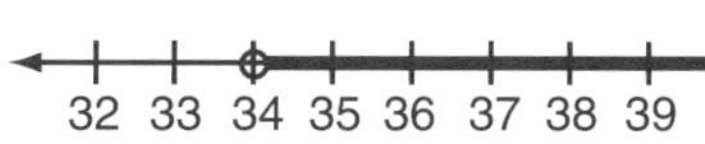

b. $0.5(n - 2) > 0.4(n + 6)$
$n > 34$

32 33 34 35 36 37 38 39

3 PRACTICE/APPLY

Check for Understanding

Exercises 1–15 are designed to help you assess your students' understanding through reading, writing, speaking, and modeling. You should work through Exercises 1–6 with your students and then monitor their work on Exercises 7–15.

Additional Answers

1. Sample answer:
$$\begin{aligned} -3x + 7 &< 4x - 5 \\ 3x - 3x + 7 &< 4x - 5 + 3x \\ 7 &< 7x - 5 \\ 7 + 5 &< 7x - 5 + 5 \\ 12 &< 7x \\ \frac{12}{7} &< \frac{7x}{7} \\ \frac{12}{7} &< x \text{ or } x > \frac{12}{7} \end{aligned}$$
4. Add $5w$ to each side. Subtract 29 from each side. Divide each side by 5. The solution is $\left\{w \mid w < -\frac{13}{5}\right\}$.

Study Guide Masters, p. 50

NAME__________ DATE ______

Study Guide — Student Edition Pages 399–404

Solving Multi-Step Inequalities

Solving an inequality may require more than one operation. Use the same procedure you used for solving equations to solve inequalities.

Procedure For Solving Inequalities
1. Use the distributive property to remove any grouping symbols. 2. Simplify each side of the inequality. 3. Undo any indicated additions and subtractions. 4. Undo any indicated multiplications and divisions involving the variable.

Example: Solve $21 > -7(m + 2)$.

$$\begin{aligned} 21 &> -7(m + 2) \\ 21 &> -7m - 14 && \text{Use the distributive property.} \\ 21 + 14 &> -7m - 14 + 14 && \text{Subtraction is indicated; use addition.} \\ 35 &> -7m \\ \frac{35}{-7} &< \frac{-7m}{-7} && \text{Multiplication is indicated; use division. Reverse the inequality symbol.} \\ -5 &< m \end{aligned}$$

The solution set is $\{m \mid -5 < m\}$, or $\{m \mid m > -5\}$.

Solve each inequality. Then check your solution.

1. $11y + 13 \geq -1$ $\left\{y \mid y \geq -1\frac{3}{11}\right\}$
2. $-3v + 3 \leq -12$ $\{v \mid v \geq 5\}$
3. $\frac{q}{7} + 1 > -5$ $\{q \mid q > -42\}$
4. $-1 - \frac{m}{4} \leq 5$ $\{m \mid m \geq -24\}$
5. $\frac{3x}{7} - 2 < -3$ $\left\{x \mid x < -2\frac{1}{3}\right\}$
6. $\frac{4x - 2}{5} \geq -4$ $\left\{x \mid x \geq -4\frac{1}{2}\right\}$
7. $9n - 24n + 42 > 0$ $\left\{n \mid n < 2\frac{4}{5}\right\}$
8. $4.6(x - 3.4) \geq 5.1x$ $\{x \mid x \leq -31.28\}$
9. $7.3y - 3.02 > 4.9y$ $\left\{y \mid y > 1\frac{31}{120}\right\}$
10. $6y + 10 > 8 - (y + 14)$ $\left\{y \mid y > -2\frac{2}{7}\right\}$
11. $m + 17 \leq -(4m - 13)$ $\left\{m \mid m \leq -\frac{4}{5}\right\}$
12. $-5x - (2x + 3) \geq 1$ $\left\{x \mid x \leq -\frac{4}{7}\right\}$

CHECK FOR UNDERSTANDING

Communicating Mathematics

Study the lesson. Then complete the following.

1. **Explain** each step in solving $-3x + 7 < 4x - 5$. **See margin.**
2. **Write** an inequality that expresses the fact that when you add 3 feet to the perimeter of a square of sides with length s, the resulting perimeter does not exceed 50 feet. $4s + 3 \leq 50$
3. **Write** an inequality that has no solution.
4. **Describe** how you would solve $16 - 5w > 29$ without dividing by -5 or multiplying by $-\frac{1}{5}$. **See margin.**
5. Refer to Example 3. Which of the two methods seems more efficient in solving the inequality for the given replacement set? Explain your selection.

3. Answers will vary. Sample answer: $x + 1 < x - 1$.

5. Sample answer: Method 2, because it is shorter.

6. Use the method shown in Lesson 7–2A to model the solutions for each inequality.
   a. $3 - 4x \geq 15$ $x \leq -3$
   b. $6x - 1 < 5 + 3x$ $x < 2$

Guided Practice

Choose the correct solution for each inequality.

7. Solve $2m + 5 \leq 4m - 1$. **c**
   a. $m > -3$ b. $m < 3$ c. $m \geq 3$ d. $m \leq -3$
8. Solve $13r - 11 \geq 7r + 37$. **b**
   a. $r < 8$ b. $r \geq 8$ c. $r \leq -8$ d. $r > -8$

Solve each inequality. Then check your solution.

9. $9x + 2 > 20$ $\{x \mid x > 2\}$
10. $-4h + 7 > 15$ $\{h \mid h < -2\}$
11. $-2 - \frac{d}{5} < 23$ $\{d \mid d > -125\}$
12. $6a + 9 < -4a + 29$ $\{a \mid a < 2\}$

Find the solution set of each inequality given the replacement set.

13. $3x - 1 > 4$, $\{-1, 0, 1, 2, 3\}$ $\{2, 3\}$
14. $-7a + 6 \leq 48$, $\{-10, -9, -8, -7, -6, -5, -4, -3\}$ $\{-6, -5, -4, -3\}$
15. **Number Theory** Consider the sentence *The sum of two consecutive even integers is greater than 75.*
    a. Write an inequality for this statement. $x + (x + 2) > 75$
    b. Solve the inequality. $x > 36.5$
    c. Name two consecutive even integers that meet the requirements of the statement. **Answers will vary; sample answer: 38 and 40.**

EXERCISES

Practice

18. $\{-10, -9, -8\ldots, 8, 9, 10\}$

19. $\{-10, -9, \ldots, -5, -4\}$

Find the solution set of each inequality if the replacement set for each variable is $\{-10, -9, -8, \ldots, 8, 9, 10\}$.

A

16. $n - 3 \geq \frac{n + 1}{2}$ $\{7, 8, 9, 10\}$
17. $\frac{2(x + 2)}{3} < 4$ $\{-10, -9, \ldots, 2, 3\}$
18. $1.3y - 12 < 0.9y + 4$
19. $-20 \geq 8 + 7k$

Reteaching

Using Manipulatives Use a two-pan balance to demonstrate how to solve these inequalities. □ □ 3 represents 2 identical objects of unknown weight and a weight of 3 units.

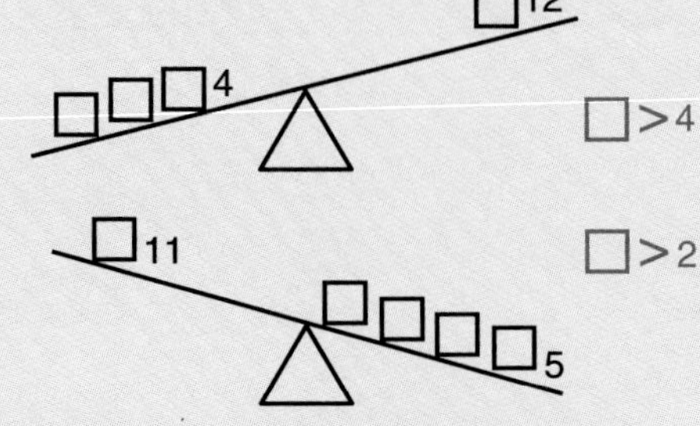

Solve each inequality. Then check your solution. 27. $\{m \mid m < 15\}$

20. $2m + 7 > 17$ $\{m \mid m > 5\}$
21. $-3 > -3t + 6$ $\{t \mid t > 3\}$
22. $-2 - 3x \geq 2$ $\left\{x \mid x \leq -\frac{4}{3}\right\}$
23. $\frac{2}{3}w - 3 \leq 7$ $\{w \mid w \leq 15\}$
24. $7x - 1 < 29 - 2x$ $\left\{x \mid x < \frac{10}{3}\right\}$
25. $8n + 2 - 10n < 20$ $\{n \mid n > -9\}$
26. $2x + 5 < 3x - 7$ $\{x \mid x > 12\}$
27. $5 - 4m + 8 + 2m > -17$

B

28. $\frac{2x - 3}{5} < 7$ $\{x \mid x < 19\}$
29. $x < \frac{2x - 15}{3}$ $\{x \mid x < -15\}$
30. $9r + 15 \geq 24 + 10r$ $\{r \mid r \leq -9\}$
31. $6p - 2 \leq 3p + 12$ $\left\{p \mid p \leq \frac{14}{3}\right\}$
32. $4y + 2 < 8y - (6y - 10)$ $\{y \mid y < 4\}$
33. $3(x - 2) - 8x < 44$ $\{x \mid x > -10\}$
34. $3.1q - 1.4 > 1.3q + 6.7$ $\{q \mid q > 4.5\}$
35. $-5(k + 4) \geq 3(k - 4)$ $\{k \mid k \leq -1\}$
36. $5(2h - 6) - 7(h + 7) > 4h$
37. $7 + 3y > 2(y + 3) - 2(-1 - y)$

36. $\{h \mid h < -79\}$
37. $\{y \mid y < -1\}$

Define a variable, write an inequality, and solve each problem. Then check your solution.

38. $\frac{2}{3}x - 27 \geq 9$; $\{x \mid x \geq 54\}$

38. Two thirds of a number decreased by 27 is at least 9.
39. Three times the sum of a number and 7 is greater than 5 times the number less 13. $3(x + 7) > 5x - 13$; $\{x \mid x < 17\}$

C **INTEGRATION** **Number Theory**

40. The sum of two consecutive odd integers is at most 123. Find the pair with the greatest sum. $2x + 2 \leq 123$; 59 and 61
41. Find all sets of two consecutive positive odd integers whose sum is no greater than 18. $2x + 2 \leq 18$ for $x > 0$; 7 and 9; 5 and 7; 3 and 5; 1 and 3
42. Find all sets of three consecutive positive even integers whose sum is less than 40. $3x + 6 < 40$; 2, 4, 6; 4, 6, 8; 6, 8, 10; 8, 10, 12; 10, 12, 14

Solve each inequality. Then check your solution.

43. $3x + 4 > 2(x + 3) + x$ no solution ($\varnothing$)
44. $3 - 3(y - 2) < 13 - 3(y - 6)$ $\{y \mid y$ is a real number.$\}$

Graphing Calculator

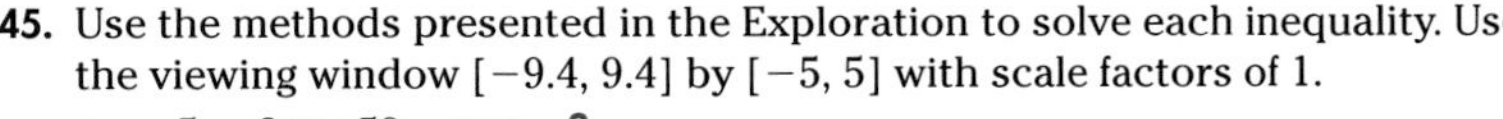

45. Use the methods presented in the Exploration to solve each inequality. Use the viewing window $[-9.4, 9.4]$ by $[-5, 5]$ with scale factors of 1.
    a. $-5 - 8x \geq 59$ $x \leq -8$
    b. $13x - 11 > 7x + 37$ $x > 8$
    c. $8x - (x - 5) > x + 17$ $x > 2$
    d. $-5(x + 4) \geq 3(x - 4)$ $x \leq -1$

Critical Thinking

46. We can write the expression *the numbers between −3 and 4* as $-3 < x < 4$. What does the algebraic sentence $-3 < x + 2 < 4$ represent? the numbers between −5 and 2 or $\{x \mid -5 < x < 2\}$

Applications and Problem Solving

Define a variable, write an inequality, and solve each problem.

47. **Personal Finances** A couple does not want to have a charge of more than \$50 for dinner at a restaurant. A sales tax of 4% is added to the bill, and they plan to tip 15% after the tax has been added. What can the couple spend on the meal? $x + 0.04x + 0.15(x + 0.04x) \leq \$50$, $x \leq \$41.80$
48. **Recreation** The admission fee to a video game arcade is \$1.25 per person, and it costs \$0.50 for each game played. Latoya and Donnetta have a total of \$10.00 to spend. What is the greatest number of games they will be able to play? 15 games

Assignment Guide

Core: 17–45 odd, 46, 47, 49, 51–60
Enriched: 16–44 even, 45–60

For **Extra Practice,** see p. 772.

The red A, B, and C flags, printed only in the Teacher's Wraparound Edition, indicate the level of difficulty of the exercises.

Practice Masters, p. 50

7-3 NAME________________ DATE________
Practice Student Edition Pages 399–404

Solving Multi-Step Inequalities

Solve each inequality. Then check your solution.

1. $7a - 5 < 9$ $a < 2$
2. $6 - 11b \leq -3$ $b \geq \frac{9}{11}$
3. $15 < 5 - 8c$ $c < -\frac{5}{4}$
4. $\frac{d}{6} - 16 < -9$ $d < 42$
5. $\frac{e}{-4} + 8 \leq 1$ $e \geq 28$
6. $-12 \geq \frac{r}{8} - 5$ $r \leq -56$
7. $\frac{5s}{6} + 7 > -3$ $s > -12$
8. $5 \geq \frac{3x}{4} + 12$ $x \leq -\frac{28}{3}$
9. $-3 - \frac{k}{5} \geq -10$ $k \leq 35$
10. $15 > \frac{2m}{3} - 1$ $m < 24$
11. $\frac{t + 3}{2} < -8$ $t < -19$
12. $\frac{3x - 1}{6} < 2$ $x < \frac{13}{3}$
13. $4n - 6 > 6n - 20$ $n < 7$
14. $2.4x + 13 \leq 5x$ $x \geq 5$
15. $5p - 3(p - 6) \leq 0$ $p \leq -9$

Define a variable, write an inequality, and solve each problem. Then check your solution.

16. The sum of a number and 81 is greater than the product of −3 and that number. $n + 81 > -3n$; $n > -\frac{81}{4}$
17. Four more than the quotient of a number and 3 is at least that number. $\frac{n}{3} + 4 \geq n$; $n \leq 6$

4 ASSESS

Closing Activity

Writing Have students supply the property or operation applied in each step of the solution to the problem below.

$3(x - 7) - 14 > 2(x + 5) - 5$

$3x - 21 - 14 > 2x + 10 - 5$

distributive property

$3x - 35 > 2x + 5$

substitution property

$3x > 2x + 40$

addition property for inequalities

$x > 40$

subtraction property for inequalities

Enrichment Masters, p. 50

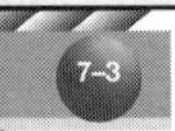

7-3 NAME ____________ DATE ______

Enrichment

Student Edition Pages 399–404

Some Properties of Inequalities

The two expressions on either side of an inequality symbol are sometimes called the *first* and *second* members of the inequality.

If the inequality symbols of two inequalities point in the same direction, the inequalities have the same sense. For example, $a < b$ and $c < d$ have the same sense; $a < b$ and $c > d$ have opposite senses.

In the problems on this page, you will explore some properties of inequalities.

Three of the four statements below are true for all numbers a and b (or a, b, c, and d). Write each statement in algebraic form. If the statement is true for all numbers, prove it. If it is not true, give an example to show that it is false.

1. Given an inequality, a new and equivalent inequality can be created by interchanging the members and reversing the sense.
   If $a > b$, then $b < a$.
   $a > b$, $a - b > 0$, $-b > -a$, $(-1)(-b) < (-1)(-a)$, $b < a$
2. Given an inequality, a new and equivalent inequality can be created by changing the signs of both terms and reversing the sense.
   If $a > b$, then $-a < -b$.
   $a > b$, $a - b > 0$, $-b > -a$, $-a < -b$
3. Given two inequalities with the same sense, the sum of the corresponding members are members of an equivalent inequality with the same sense.
   If $a > b$ and $c > d$, then $a + c > b + d$.
   $a > b$ and $c > d$, so $(a - b)$ and $(c - d)$ are positive numbers, so the sum $(a - b) + (c - d)$ is also positive.
   $a - b + c - d > 0$, so $a + c > b + d$.
4. Given two inequalities with the same sense, the difference of the corresponding members are members of an equivalent inequality with the same sense.
   If $a > b$ and $c > d$, then $a - c > b - d$. The statement is false. $5 > 4$ and $3 > 2$, but $5 - 3 \not> 4 - 2$.

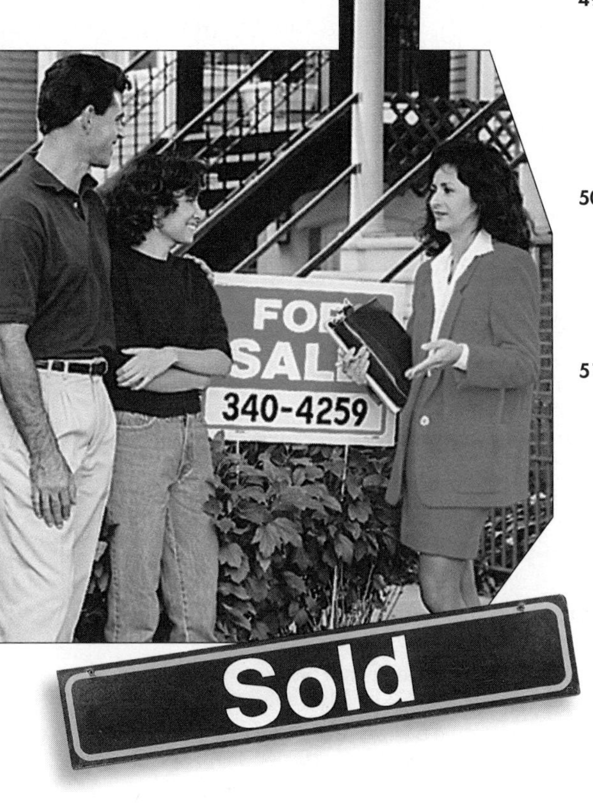

49. **Fund-raising** A university is running a drive to raise money. A corporation has promised to match 40% of whatever the university can raise from other sources. How much must the school raise from other sources to have a total of at least \$800,000 after the corporation's donation? **at least \$571,428.57**

50. **Real Estate** A homeowner is selling her house. She must pay 7% of the selling price to her real estate agent after the house is sold. To the nearest dollar, what must be the selling price of her house to have at least \$90,000 after the agent is paid? **at least \$96,774**

51. **Labor** A union worker is currently making \$400 per week. His union is seeking a one-year contract. If there is a strike, the new contract will provide for a 6% raise; if there is no strike, the worker expects no increase in salary.
   a. Assuming the worker will not be paid during the strike, how many weeks could he go on strike and still make at least as much for the year as he would have made without a strike?
   b. How would your answer change if the worker was currently making \$575 per week?
   c. How would your answer to part a change if the worker's union provided him with \$120 per week during the strike? **at most 4.1 weeks**

   51a. at most 2.9 weeks **51b. no change**

Mixed Review

Solve each inequality. **52. $\{r \mid r < -6.6\}$ 53. $\{y \mid y > 10\}$**

52. $2r - 2.1 < -8.7 + r$ (Lesson 7–2) 53. $7 - 2y < -y - 3$ (Lesson 7–1)

54. Write the slope-intercept form of an equation for the line that passes through $(-12, 12)$ and $(-2, 7)$. (Lesson 6–4) **54. $y = -\frac{1}{2}x + 6$**

55. Write the standard form of an equation of the line that passes through $(2, 4)$ with slope $-\frac{3}{2}$. (Lesson 6–2) **55. $3x + 2y = 14$**

56. **Business** The owner of No Spots City Car Wash found that if c cars were washed in a day, the average daily profit $P(c)$ was given by the formula $P(c) = -0.027c^2 + 8c - 280$. Find the values of $P(c)$ for various values of c to determine the least number of cars that must be washed each day for No Spots City to make a profit. (Lesson 5–5) **41 cars**

57. Find the domain for $3y - 2 = x$ if the range is $\left\{-1, 6, 0, -\frac{1}{3}, 2\right\}$. (Lesson 5–3) **57. $\{-5, -3, -2, 4, 16\}$**

58. **Advertising** For many years, the slogan of Crest® toothpaste has been "Four out of five dentists recommend that their patients use Crest." What are the odds that your dentist will *not* recommend you use Crest? (Lesson 4–6) **1:4**

59. Jason scored the following points for his basketball team during the past ten games: 18, 32, 20, 21, 34, 9, 33, 37, 22, 25. Find the mean, median, and mode of his total points. (Lesson 3–7) **25.1; 23.5; no mode**

60. Find $18 - (-34)$. (Lesson 2–3) **52**

Extension

Communication Give an example of an inequality involving more than one operation for which the solution set is the empty set. **Possible answers include $3x - 6 > 3(x + 2)$.**

7–4 Solving Compound Inequalities

YOU'LL LEARN

- To solve problems by making a diagram,
- to solve compound inequalities and graph their solution sets, and
- to solve problems that involve compound inequalities.

IT'S IMPORTANT

You can use inequalities to solve problems involving chemistry and physics.

Chemistry

The largest fish that spends its whole life in fresh water is the rare Pla beuk, found in the Mekong River in China, Laos, Cambodia, and Thailand. The largest specimen was reportedly 9 feet $10\frac{1}{4}$ inches long and weighed 533.5 pounds.

Such rare fish are sometimes displayed in aquariums. Aquariums can house freshwater or marine life and must be closely monitored to maintain the correct temperature and pH for the animals to survive. pH is a measure of acidity. To determine pH, a scale with values from 0 to 14 is used. One such scale is shown in the diagram below.

pH SCALE

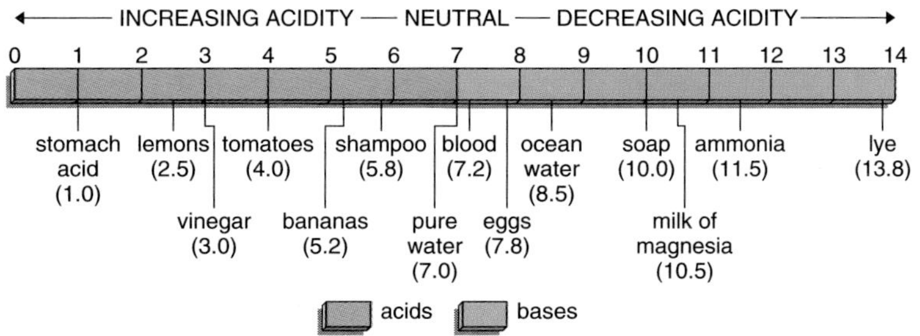

FYI

The largest fish ever caught is a whale shark. Caught near Karachi, Pakistan in 1949, it was 41.5 feet long and weighed 16.5 tons. The smallest fish is a dwarf goby found in the Indian Ocean. Males of this species are only 0.34 inches long.

whale shark

dwarf goby

If we let p represent the value of the pH scale, we can express the different pH levels by using inequalities. For example, an acid solution will have a pH level of $p \geq 0$ and $p < 7$. When considered together, these two inequalities form a **compound inequality.** This compound inequality can also be written without using *and* in two ways.

$$0 \leq p < 7 \quad \text{or} \quad 7 > p \geq 0$$

The statement $0 \leq p < 7$ can be read *0 is less than or equal to p, which is less than 7.* The statement $7 > p \geq 0$ can be read *7 is greater than p, which is greater than or equal to 0.*

The pH levels of bases could be written as follows.

$p > 7$ and $p \leq 14$ or $7 < p \leq 14$ or $14 \geq p > 7$

You will graph this inequality in Exercise 6.

FYI

There are over 350 species of sharks. While there are 50 to 75 shark attacks on humans recorded annually, only 30 species are responsible for unprovoked attacks, and most can be traced to three species: the great white shark, the tiger shark, and the bull shark.

7-4 LESSON NOTES

NCTM Standards: 1–5

Instructional Resources

- Study Guide Master 7-4
- Practice Master 7-4
- Enrichment Master 7-4
- Assessment and Evaluation Masters, pp. 183–184
- Graphing Calculator Masters, p. 7
- Modeling Mathematics Masters, pp. 54–56
- Real-World Applications, 20

Transparency 7-4A contains the 5-Minute Check for this lesson; **Transparency 7-4B** contains a teaching aid for this lesson.

Recommended Pacing	
Standard Pacing	Days 5 & 6 of 14
Honors Pacing	Day 5 of 13
Block Scheduling*	Day 3 of 7 (along with Lesson 7-3)
Alg. 1 in Two Years*	Days 7 & 8 of 21

*For more information on pacing and possible lesson plans, refer to the *Block Scheduling Booklet* and *Algebra 1 in Two Years.*

1 FOCUS

5-Minute Check
(over Lesson 7-3)

Solve each inequality.

1. $3x - 15 < 45$ $\{x \mid x < 20\}$
2. $2p - 22 \geq 4p + 14$ $\{p \mid p \leq -18\}$
3. $7y + 12 > 46 - 10y$ $\{y \mid y > 2\}$
4. $3(m + 2) \leq 6(2m - 5)$ $\{m \mid m \geq 4\}$
5. $5(r + 3) > 2(4r - 6)$ $\{r \mid r < 9\}$

Motivating the Lesson

Questioning Ask students to list any integers that are less than 7 *and* greater than 2. **3, 4, 5, 6** Then have them list any integers that are greater than 2 *or* less than 7. **all integers** Challenge students to identify how the use of the words *and* or *or* changes the way a solution set is determined. **and requires the intersection of sets; *or* requires the union**

2 TEACH

Teaching Tip In Example 1, challenge students to identify other types of diagrams that could be used to solve the problem.

In-Class Examples

For Example 1
Five people come together for a meeting. Each person shakes each other person's hand once. Draw a diagram to find the total number of handshakes that will occur. **10**

For Example 2
Graph the solution set of $x < 10$ and $x < 5$.

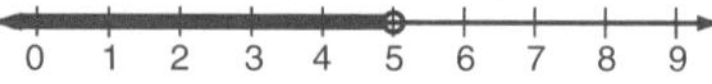

Teaching Tip Before introducing Example 2, discuss possible solutions to the compound sentence in the example. Have students choose a number and substitute it for x. Then have them determine whether the number is a solution to the compound sentence. For example, try 6. Is 6 greater than or equal to -2 and less than 5? Note that 6 is greater than -2, but it is not less than 5. Therefore, the number 6 is not a solution to the compound sentence. Other numbers that students could try are -3, 0, 4, and 8.

You can **draw a diagram** to help solve many problems. Sometimes a picture will help you decide how to work the problem. Other times the picture will show you the answer to the problem.

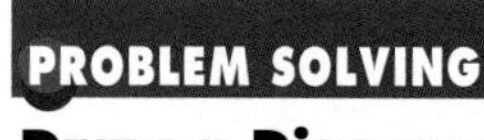

Draw a Diagram

Example 1 **On May 6, 1994, President Francois Mitterrand of France and Queen Elizabeth II of England officially opened the Channel Tunnel connecting England and France. After the ceremonies, a group of 36 English and French government officials had dinner at a restaurant in Calais, France, to celebrate the occasion. Suppose the restaurant staff used small tables that seat four people each, placed end to end, to form one long table. How many tables were needed to seat everyone?**

Draw a diagram to represent the tables placed end to end. Use Xs to indicate where the people are sitting. Let's start with a guess, say 10 tables.

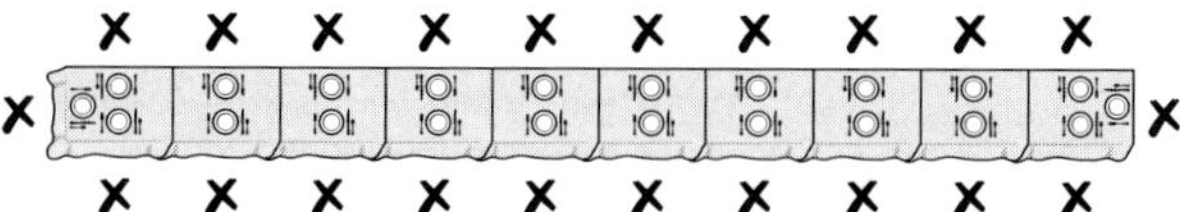

Ten tables will seat 22 people. If we use an extra table, we can seat 2 more people. Now, let's look for a pattern.

Number of tables	10	11	12	13	14	15	16	17
Number of people seated	22	24	26	28	30	32	34	36

This pattern shows that the restaurant needed 17 tables to seat all 36 officials.

The logic symbol for and *is $\wedge$. You can write $x > 7$ and $x < 10$ as $(x > 7) \wedge (x < 10)$.*

A compound inequality containing *and* is true only if *both* inequalities are true. Thus, the graphs of a compound inequality containing *and* is the **intersection** of the graphs of the two inequalities. The intersection can be found by graphing the two inequalities and then determining where these graphs overlap. In other words, draw a diagram to solve the inequality.

Example  **Graph the solution set of $x \geq -2$ and $x < 5$.**

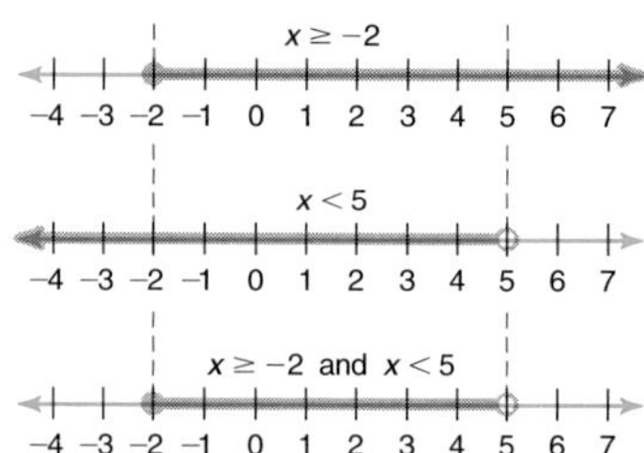

The solution set, shown in the bottom graph, is $\{x \mid -2 \leq x < 5\}$. Note that the graph of $x \geq -2$ includes the point -2. The graph of $x < 5$ does not include 5.

Example 3 **Solve $-1 < x + 3 < 5$. Then graph the solution set.**

First express $-1 < x + 3 < 5$ using *and.* Then solve each inequality.

$$-1 < x + 3 \quad \text{and} \quad x + 3 < 5$$
$$-1 - 3 < x + 3 - 3 \qquad x + 3 - 3 < 5 - 3$$
$$-4 < x \qquad x < 2$$

Now graph each solution and find the intersection of the solutions.

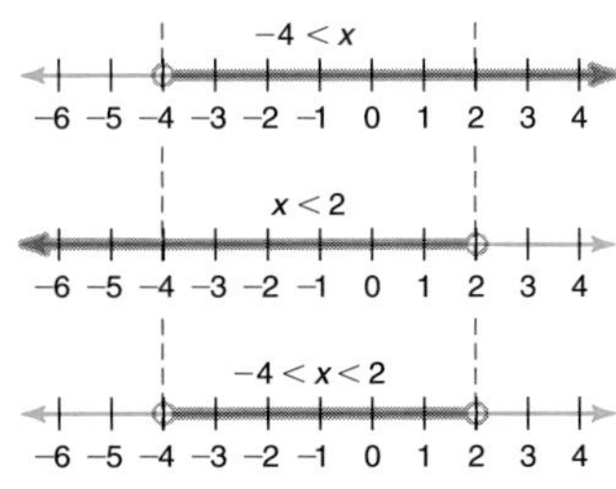

The solution set is $\{x \mid -4 < x < 2\}$.

The following example shows how you can solve a problem by using geometry, a diagram, and a compound inequality.

Example 4

PROBLEM SOLVING

Draw a Diagram

Mai and Luis hope someday to compete in the Olympics in pairs ice skating. Each day they travel from their homes to an ice rink to practice before going to school. Luis lives 17 miles from the rink, and Mai lives 20 miles from it. If this were all the information you were given, determine how far apart Mai and Luis live.

Explore Mai lives 20 miles from the rink and Luis lives 17 miles from the rink. We do not know the relative positions of Mai's and Luis's homes in relation to the rink.

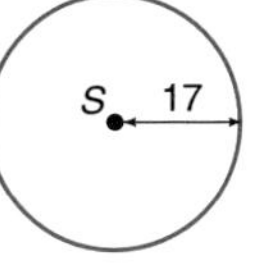

Plan Draw a diagram of the situation. Let S be the location of the skating rink. Since Luis lives 17 miles from the rink, we can draw a circle with a radius of 17. Luis will live somewhere on that circle. Likewise, we can draw another circle with radius of 20 for the location of Mai's home.

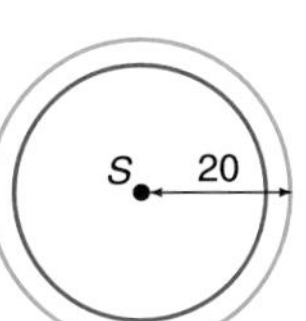

(continued on the next page)

In-Class Examples

For Example 3
Solve $-3 < x - 2 \leq 0$.
Then graph the solution set.
$\{x \mid -1 < x \leq 2\}$

For Example 4
Suppose that Luis' family moves and now lives 24 miles from the rink. Determine how far apart Mai and Luis now live.
$4 \leq d \leq 44$

Teaching Tip In Example 4, point out to students that the triangle inequality is needed only to establish an inequality. It is not needed to compute exact distances.

In-Class Example

For Example 5

Graph the solution set of $x \geq 3$ or $x < 2$.

Solve Let's examine three possibilities for the locations of Mai's and Luis's homes.

(1) Mai and Luis live along the same radius from the skating rink.
(2) Mai and Luis live on opposite radii from the skating rink.
(3) Mai and Luis live somewhere other than the locations described in (1) and (2).

Let L and M represent where Luis and Mai live, respectively.

(1) the same radius	(2) opposite radii	(3) somewhere other than (1) or (2)
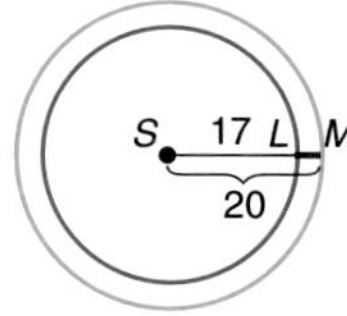	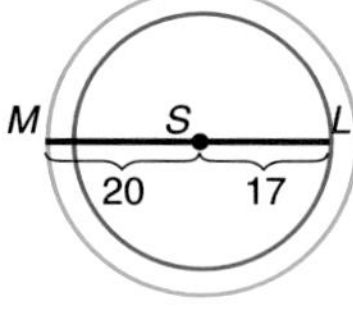	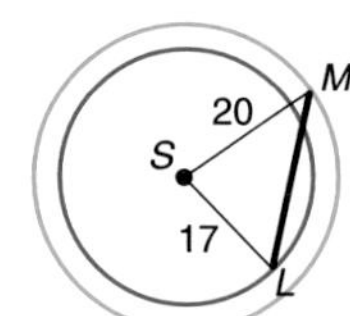
The distance is $20 - 17$ or 3 miles.	The distance is $20 + 17$ or 37 miles.	By the triangle inequality theorem, the distance must be less than 37 miles and greater than 3 miles.

LOOK BACK

You can find more information about the triangle inequality theorem in Exercise 45 (Programming) of Lesson 7–1.

The distance d can be described by the inequality $3 \leq d \leq 37$.

Examine Diagrams (1) and (2) show the least and greatest possiblities. To convince yourself that the statement with diagram (3) is always true, you may want to sketch a different triangle.

Another type of compound inequality contains the word *or* instead of *and.* A compound inequality containing *or* is true if one or more of the inequalities is true. The graph of a compound inequality containing *or* is the **union** of the graphs of the two inequalities.

Example  **Graph the solution set of $x \geq -1$ or $x < -4$.**

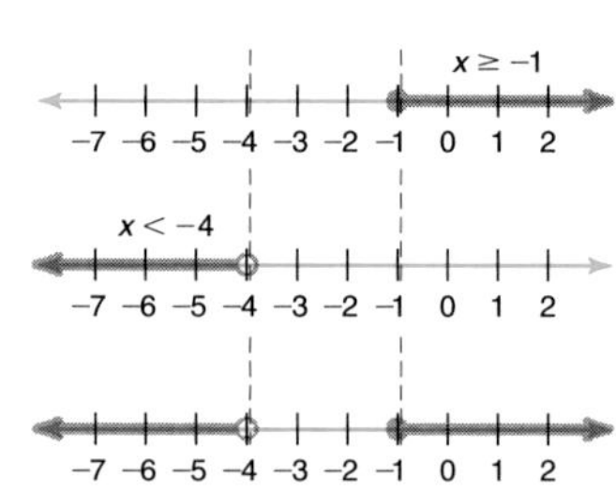

The last graph shows the solution set, $\{x \mid x \geq -1 \text{ or } x < -4\}$.

Cooperative Learning

Numbered Heads Together Have students work in groups of four to solve the following problem. Make sure that each person in the group understands the solution and is able to explain it to the class.

When changing a dollar bill, you can give one coin (silver dollar), two coins (2 half-dollars), three coins (2 quarters and 1 half-dollar), and so forth. What is the least number of coins that is impossible to give as change for a dollar bill? **77 coins** For more information on the numbered heads together strategy, see *Cooperative Learning in the Mathematics Classroom,* one of the titles in the Glencoe Mathematics Professional Series, pages 9–12.

Example 6 **Solve $3w + 8 < 2$ or $w + 12 > 2 - w$. Graph the solution set.**

$$3w + 8 < 2 \qquad \text{or} \qquad w + 12 > 2 - w$$
$$3w + 8 - 8 < 2 - 8 \qquad w + w + 12 > 2 - w + w$$
$$3w < -6 \qquad 2w + 12 > 2$$
$$\frac{3w}{3} < \frac{-6}{3} \qquad 2w + 12 - 12 > 2 - 12$$
$$w < -2 \qquad 2w > -10$$
$$\frac{2w}{2} > \frac{-10}{2}$$
$$w > -5$$

Now graph each solution and find the union of the two.

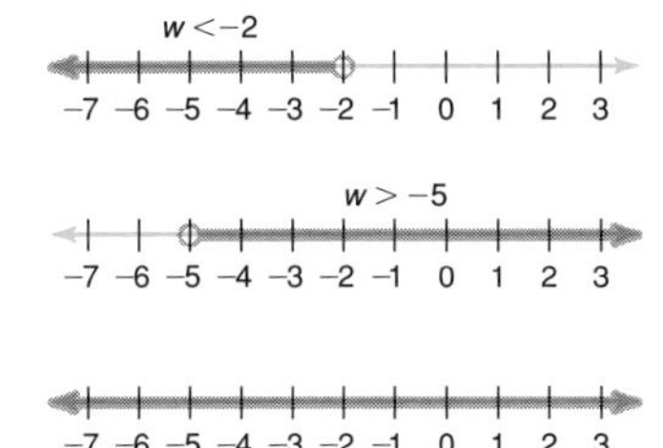

−7 −6 −5 −4 −3 −2 −1 0 1 2 3

Find the union.

The last graph shows the solution set, $\{x \mid x \text{ is a real number}\}$.

CHECK FOR UNDERSTANDING

Communicating Mathematics

Study the lesson. Then complete the following.

1. **Write** a statement that could be represented by $\$7.50 < p \leq \$18.50$.
2. **Write** a compound inequality that describes the graph below.

   −6 −5 −4 −3 −2 −1 0 1 2 3 **$x < -4$ or $x \geq 1$**

3. **List** two problem-solving strategies used in Example 1. Describe how each strategy helped in the solution of the problem. **See margin.**
4. **Describe** the difference between a compound inequality containing *and* and a compound inequality containing *or*. **See margin.**
5. **Name** two ways in which a picture or diagram can help you solve a problem more easily. **Helps you plan the solution; provides the solution.**
6. **Refer** to the connection at the beginning of the lesson. Graph the inequality that represents the pH levels of bases. **See margin.**

1. Answers will vary. Sample answer: The price of an item is at most \$18.50 but more than \$7.50.

Guided Practice

Write each compound inequality without using *and*. Then graph the solution set. **7–8. See margin for graphs.**

7. $x < 9$ and $0 \leq x$ **$0 \leq x < 9$**
8. $x > -2$ and $x < 3$ **$-2 < x < 3$**

Write a compound inequality for each solution set shown below.

9. 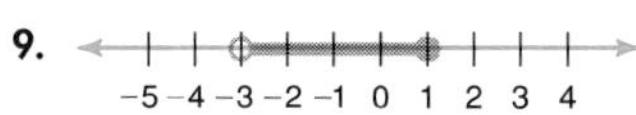

10. 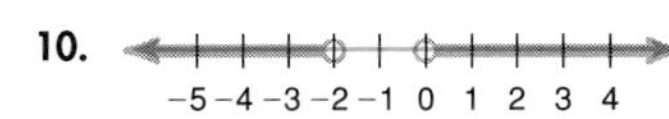

9. $-3 < x \leq 1$
10. $x < -2$ or $x > 0$

In-Class Example

For Example 6
Solve $3m > m + 4$ or $-2m > m - 6$. Graph the solution set.
$\{m \mid m > 2$ or $m < 2\}$

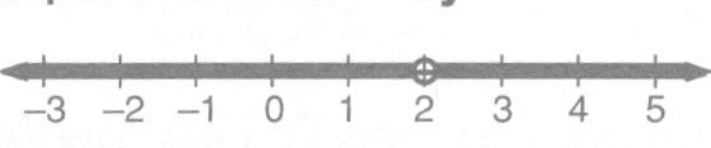

3 PRACTICE/APPLY

Check for Understanding

Exercises 1–17 are designed to help you assess your students' understanding through reading, writing, speaking, and modeling. You should work through Exercises 1–6 with your students and then monitor their work on Exercises 7–17.

Additional Answers

3. Sample answers: Draw a diagram (helps you to see the solution), make a table (helps you to see pattern), look for a pattern (determines how the tables relate to the number of people seated).
4. The one containing *and* is true when both inequalities are true; the one containing *or* is true if either inequality is true.
6. 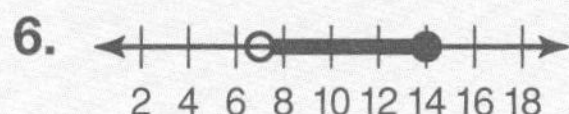

7. 

8. 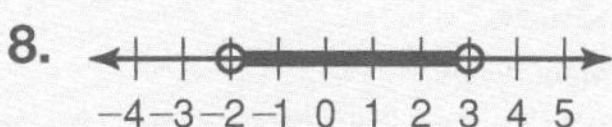

Reteaching

Using Graphing Graph sets A, B, and C.
$A = \{x \mid x > 3\}$ $B = \{x \mid x \leq 1\}$
$C = \{x \mid x < 6\}$
Use these graphs to graph the intersections and unions.

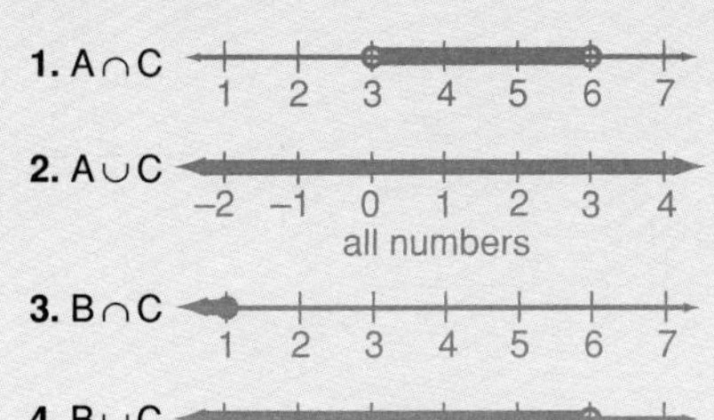

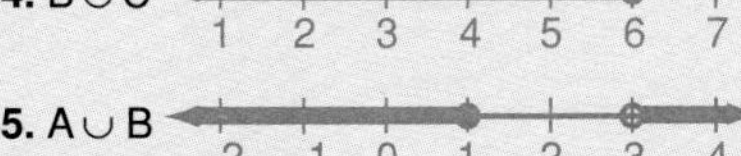

Assignment Guide

Core: 19–57 odd, 59–68
Enriched: 18–52 even, 53–68
All: Self Test, 1–10

For **Extra Practice,** see p. 772.

The red A, B, and C flags, printed only in the Teacher's Wraparound Edition, indicate the level of difficulty of the exercises.

Additional Answers

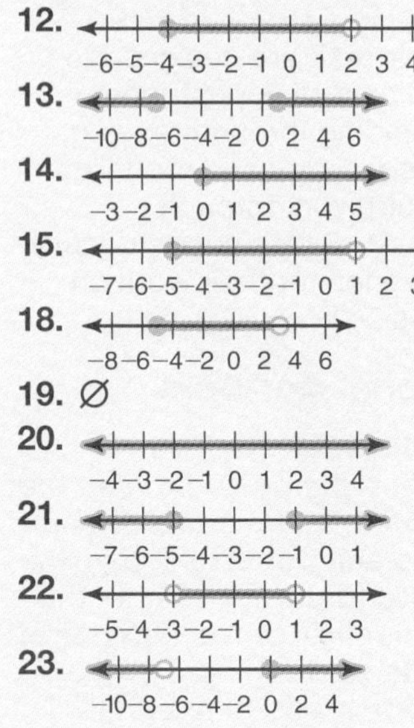

Study Guide Masters, p. 51

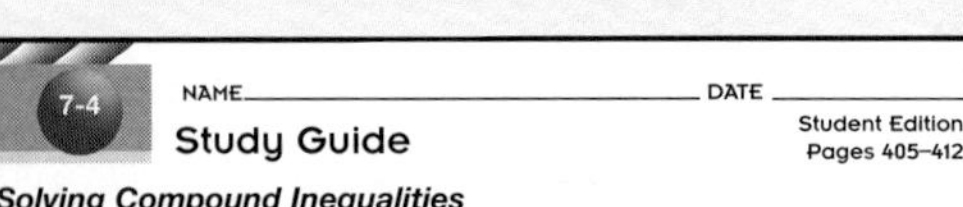

Solving Compound Inequalities

A **compound inequality** consists of two inequalities that are connected by the words *and* or *or*. A compound inequality containing *and* is true only if *both* inequalities are true. Its graph is the **intersection** of the graphs of the two inequalities. A compound inequality containing *or* is true if one or more of the inequalities is true. Its graph is the **union** of the graphs of the two inequalities.

Example 1: $x > -3$ and $x \le 4$

Example 2: $t \ge 8$ or $t < 5$

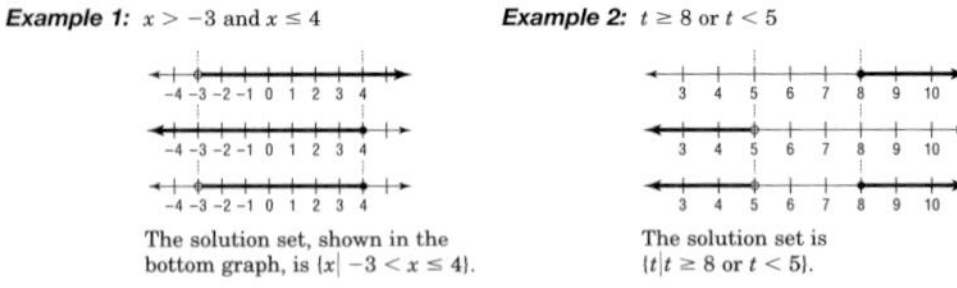

The solution set, shown in the bottom graph, is $\{x \mid -3 < x \le 4\}$.

The solution set is $\{t \mid t \ge 8 \text{ or } t < 5\}$.

Sometimes it is better to first solve each inequality and then graph the solution. Study the examples below.

Example 3: $-3 \le p - 5 < 2$

$-3 \le p - 5$ and $p - 5 < 2$
$-3 + 5 \le p - 5 + 5$ $\quad p - 5 + 5 < 2 + 5$
$2 \le p$ $\quad p < 7$

Example 4: $2a + 1 < 11$ or $a > 3a + 2$

$2a + 1 < 11$ or $a > 3a + 2$
$2a + 1 - 1 < 11 - 1$ $\quad a - 3a > 3a - 3a + 2$
$2a < 10$ $\quad -2a > 2$
$\frac{2a}{2} < \frac{10}{2}$ $\quad \frac{-2a}{-2} > \frac{2}{-2}$
$a < 5$ $\quad a < -1$

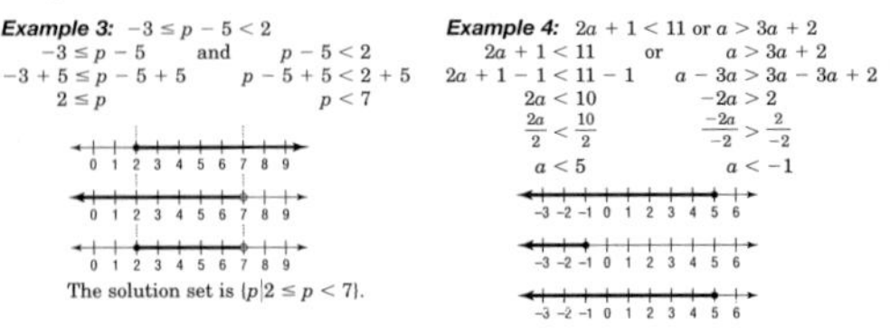

The solution set is $\{p \mid 2 \le p < 7\}$.

The solution set is $\{a \mid a \le 5\}$.

Graph the solution set of each compound inequality.

1. $b > -1$ and $b \le 3$
2. $y \le -4$ or $y > 0$
3. $2 \ge q \ge -5$

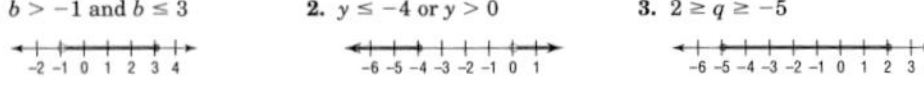

Solve each compound inequality. Then graph the solution set.

4. $2x + 4 \le 6$ or $x \ge 2x - 4$
5. $d - 3 < 6d + 12 < 2d + 32$

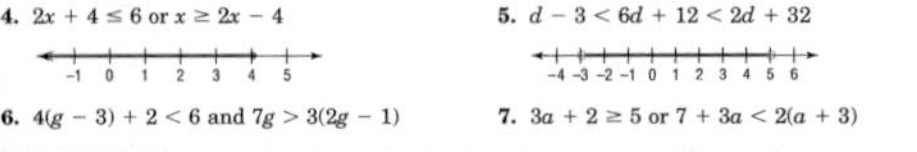

6. $4(g - 3) + 2 < 6$ and $7g > 3(2g - 1)$
7. $3a + 2 \ge 5$ or $7 + 3a < 2(a + 3)$

11. Graph the solution set of $y > 5$ and $y < -3$. Describe what the solution set means. **The solution is the empty set. There are no numbers greater than 5 but less than −3.**

Solve each compound inequality. Then graph the solution set.

12–15. See margin for graphs.
12. $\{y \mid -4 \le y < 2\}$
13. $\{h \mid h \le -7 \text{ or } h \ge 1\}$
15. $\{w \mid 1 > w \ge -5\}$

12. $2 \le y + 6$ and $y + 6 < 8$
13. $4 + h \le -3$ or $4 + h \ge 5$
14. $b + 5 > 10$ or $b \ge 0$ **$\{b \mid b \ge 0\}$**
15. $2 + w > 2w + 1 \ge -4 + w$

16. Write a compound inequality without using *and* for the following situation. Solve the inequality and then check the solution.
It costs the same to register mail containing articles with values from \$0 to \$100. **$\$0 \le c \le \$100$**

17. **Draw a Diagram** You can cut a pizza into seven pieces with only three straight cuts as shown at the right. Draw a diagram to show the greatest number of pieces you can make with five straight cuts. **Drawings will vary; 16 pieces.**

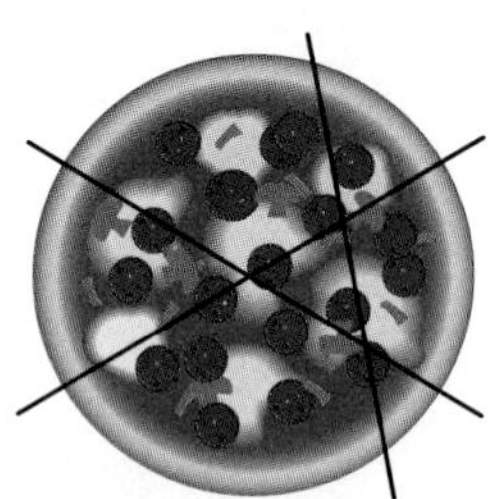

EXERCISES

Practice

Graph the solution set of each compound inequality.

18. $m \ge -5$ and $m < 3$
19. $p < -8$ and $p > 4$
20. $s < 3$ or $s \ge 1$
21. $n \le -5$ or $n \ge -1$
22. $w > -3$ and $w < 1$
23. $x < -7$ or $x \ge 0$

18–23. See margin.

Write a compound inequality for each solution set shown below.

24. 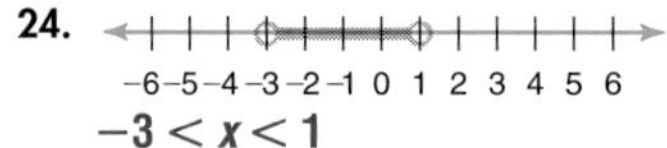
$-3 < x < 1$

25. (–6 –5 –4 –3 –2 –1 0 1 2 3 4 5 6)
$-4 \le x \le 5$

26. (–6 –5 –4 –3 –2 –1 0 1 2 3 4 5 6)
$x < 0$ or $x > 2$

27. (–6 –5 –4 –3 –2 –1 0 1 2 3 4 5 6)
$x \le -2$ or $x > 1$

Solve each compound inequality. Then graph the solution set.

28–43. See margin for graphs.
28. $\{m \mid m > 3 \text{ or } m < -1\}$
29. $\{x \mid -1 < x < 5\}$
30. $\{y \mid -7 < y < 6\}$
31. $\{x \mid x < -2 \text{ or } x > 3\}$
32. $\{p \mid p < -1\}$
33. $\{c \mid c < 7\}$
36. $\{q \mid -1 < q < 6\}$
37. $\{x \mid x \text{ is a real number}\}$
38. $\{n \mid n \le 4\}$
39. $\{y \mid y > 3 \text{ and } y \ne 6\}$
40. $\left\{z \mid z \ge -\frac{2}{3}\right\}$
41. $\{x \mid x \text{ is a real number.}\}$
42. $\left\{y \mid y < \frac{3}{2}\right\}$
44. Sample answer: $x > -2$ or $x < 4$
45. Sample answer: $x > 5$ and $x < -4$

28. $4m - 5 > 7$ or $4m - 5 < -9$
29. $x - 4 < 1$ and $x + 2 > 1$
30. $y + 6 > -1$ and $y - 2 < 4$
31. $x + 4 < 2$ or $x - 2 > 1$
32. $10 - 2p > 12$ and $7p < 4p + 9$
33. $6 - c > c$ or $3c - 1 < c + 13$
34. $4 < 2x - 2 < 10$ **$\{x \mid 3 < x < 6\}$**
35. $14 < 3h + 2 < 2$ **∅**
36. $8 > 5 - 3q$ and $5 - 3q > -13$
37. $-1 + x \le 3$ or $-x \le -4$
38. $3n + 11 \le 13$ or $2n \ge 5n - 12$
39. $3y + 1 > 10$ and $y \ne 6$
40. $4z + 8 \ge z + 6$ or $7z - 14 \ge 2z - 4$
41. $5x + 7 > 2x + 4$ or $3x + 3 < 24 - 4x$
42. $2 - 5(2y - 3) > 2$ or $3y < 2(y - 8)$
43. $5w > 4(2w - 3)$ and $5(w - 3) + 2 < 7$ **$\{w \mid w < 4\}$**

Write a compound inequality for each solution set shown below.

44. 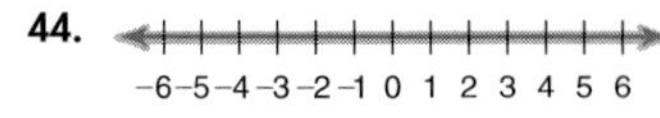

45. 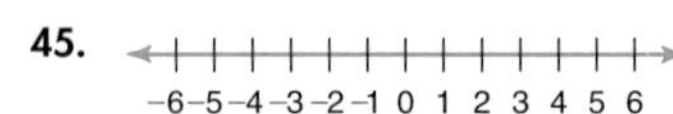

Additional Answers

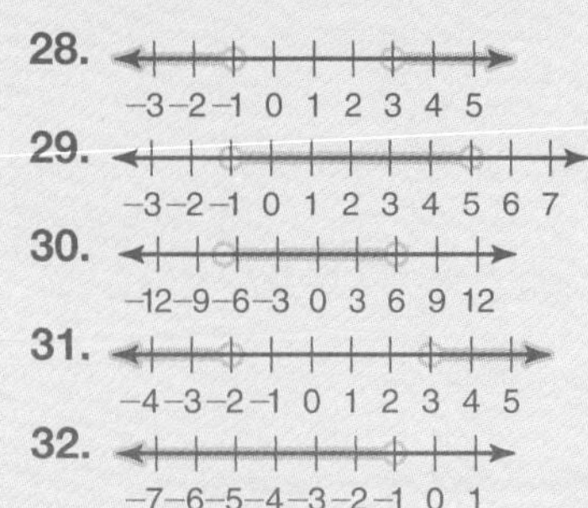

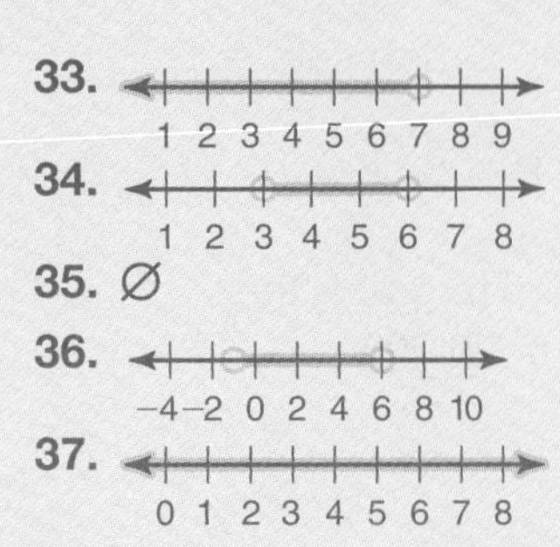

Define a variable, write a compound inequality, and solve each problem. Then check your solution.

46. $50 < 3d + 5 < 89$; $\{d \mid 15 < d < 28\}$

47. $n + 2 \leq 6$ or $n + 2 \geq 10$; $\{n \mid n \leq 4$ or $n \geq 8\}$

48. $7 < 2n + 5 < 11$; $\{n \mid 1 < n < 3\}$

46. When three times the distance to the finish line is increased by 5 km, the total kilometers covered in the race will be between 50 and 89 km.

47. The sum of a number and 2 is no more than 6 or no less than 10.

48. The sum of twice a number and 5 lies between 7 and 11.

49. Five less than 6 times a number is at most 37 and at least 31. $31 \leq 6n - 5 \leq 37$; $\{n \mid 6 \leq n \leq 7\}$

Solve each inequality. Then graph the solution set.

50–52. See margin for graphs.

50. $3 + y > 2y > -3 - y$ $\{y \mid -1 < y < 3\}$ **51.** $m > 2m - 1 > m - 5$ $\{m \mid -4 < m < 1\}$ **52.** $\frac{5}{x} + 3 > 0$ $\{x \mid x < -\frac{5}{3}$ or $x > 0\}$

Graphing Calculator

53. In Lesson 7–3, you learned how to use a graphing calculator to find graphically which values of x make a given inequality true. You can also use this method to test compound inequalities. The words *and* and *or* can be found in the LOGIC submenu of the TEST menu. Use this method to solve each of the following using your graphing calculator.

a. $3 + x < -4$ or $3 + x > 4$ $\{x \mid x < -7$ or $x > 1\}$

b. $-2 \leq x + 3$ and $x + 3 < 4$ $\{x \mid -5 \leq x < 1\}$

Critical Thinking

54. For what values of a does the compound inequality $-a < x < a$ have no solution? $a \leq 0$

55. Write a compound inequality for the solution set shown below.

−6 −5 −4 −3 −2 −1 0 1 2 3 4 5 6

$-4 \leq x \leq -1.5$ or $x \geq 2$

Applications and Problem Solving

56. Draw a Diagram There are eight houses on McArthur Street, all in a row. These houses are numbered from 1 to 8. Allison, whose house number is greater than 2, lives next door to her best friend, Adrienne. Belinda, whose house number is greater than 5, lives two doors away from her boyfriend, Benito. Cheri, whose house number is greater than Benito's, lives three doors away from her piano teacher, Mr. Crawford. Darryl, whose house number is less than 4, lives four doors away from his teammate, Don. Who lives in each house? 1-Darryl, 2-Adrienne, 3-Allison, 4-Mr. Crawford, 5-Don, 6-Benito, 7-Cheri, 8-Belinda

Define a variable, write a compound inequality, and solve each problem.

57. Physics According to Hooke's Law, the force F (in pounds) required to stretch a certain spring x inches beyond its natural length is given by $F = 4.5x$. If forces between 20 and 30 pounds, inclusive, are applied to the spring, what will be the range of the increased lengths of the stretched spring? $4.4 < x < 6.7$

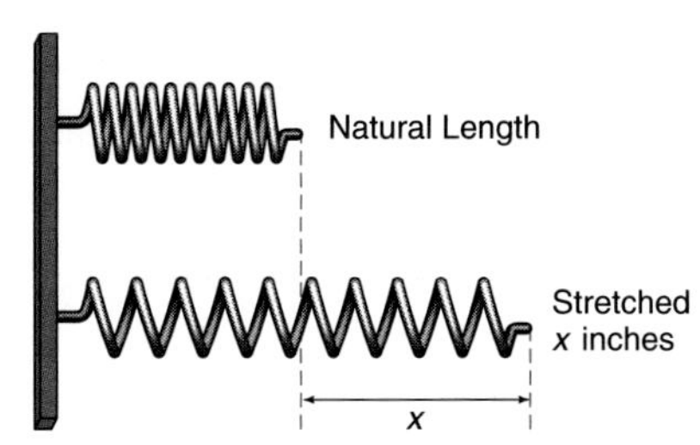

58. Statistics Clarissa must have an average of 92, 93, or 94 points to receive a grade of A– in social studies. She earned 92, 96, and 88 on the first three tests of the grading period. What range of scores on the fourth test will give her an A–? $92 \leq s \leq 100$

Additional Answers

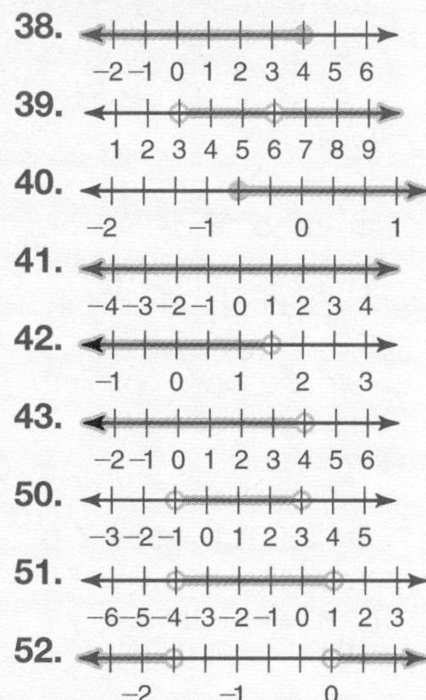

Practice Masters, p. 51

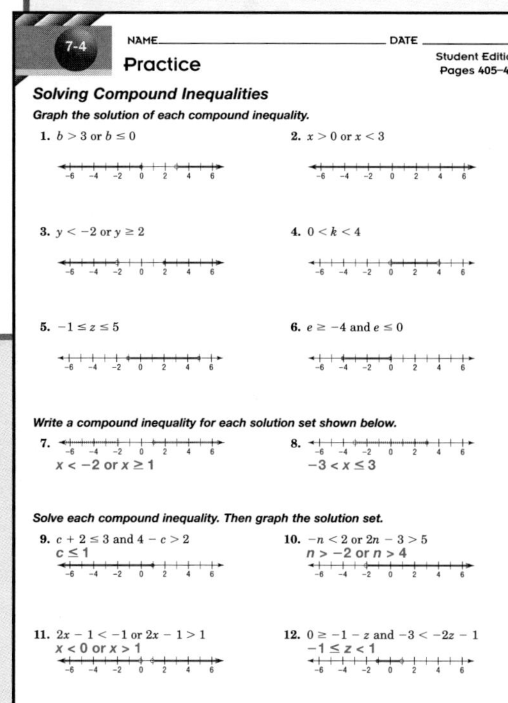

7-4 NAME ______ DATE ______

Practice

Student Edition Pages 405–412

Solving Compound Inequalities

Graph the solution of each compound inequality.

1. $b > 3$ or $b \leq 0$
2. $x > 0$ or $x < 3$
3. $y < -2$ or $y \geq 2$
4. $0 < k < 4$
5. $-1 \leq z \leq 5$
6. $e \geq -4$ and $e \leq 0$

Write a compound inequality for each solution set shown below.

7. $x < -2$ or $x \geq 1$
8. $-3 < x \leq 3$

Solve each compound inequality. Then graph the solution set.

9. $c + 2 \leq 3$ and $4 - c > 2$ $c \leq 1$
10. $-n < 2$ or $2n - 3 > 5$ $n > -2$ or $n > 4$
11. $2x - 1 < -1$ or $2x - 1 > 1$ $x < 0$ or $x > 1$
12. $0 \geq -1 - z$ and $-3 < -2z - 1$ $-1 \leq z < 1$

Extension

Reasoning Points A, B, C, and D are located on the same number line. If A is to the left of B, A is to the right of C, and C is to the right of D, what is the order of the points from left to right? ***D, C, A, B***

4 ASSESS

Closing Activity

Modeling Have each student create a diagram to model a word problem. Models can resemble those used in Examples 1 and 4. Then have students switch models and write a word problem that could be solved using their classmate's model.

Chapter 7, Quiz B (Lessons 7-3 and 7-4), is available in the *Assessment and Evaluation Masters,* p. 184.

Mid-Chapter Test (Lessons 7-1 through 7-4) is available in the *Assessment and Evaluation Masters,* p. 183.

Additional Answer

63.

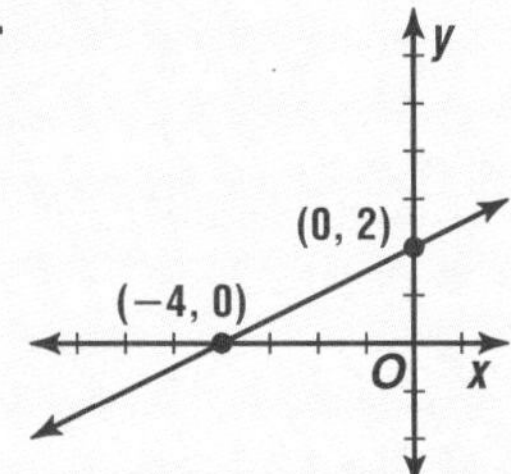

Enrichment Masters, p. 51

NAME__________ DATE________

Enrichment

Student Edition Pages 405–411

Precision of Measurement

The precision of a measurement depends both on your accuracy in measuring and the number of divisions on the ruler you use. Suppose you measured a length of wood to the nearest one-eighth of an inch and got a length of $6\frac{5}{8}$ in.

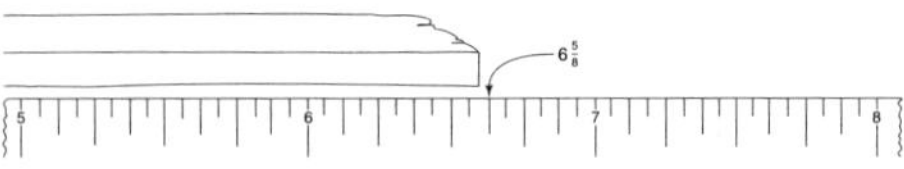

The drawing shows that the actual measurement lies somewhere between $6\frac{9}{16}$ in. and $6\frac{11}{16}$ in. This measurement can be written using the symbol ±, which is read "plus or minus." It can also be written as a compound inequality.

$6\frac{5}{8} \pm \frac{1}{16}$ in. $\qquad 6\frac{9}{16}$ in. $\le m \le 6\frac{11}{16}$ in.

In this example, $\frac{1}{16}$ in. is the absolute error. The absolute error is one-half the smallest unit used in a measurement.

Write each measurement as a compound inequality. Use the variable m.

1. $3\frac{1}{2} \pm \frac{1}{4}$ in. **$3\frac{1}{4}$ in. $\le m \le 3\frac{3}{4}$ in.**
2. 9.78 ± 0.005 cm **9.775 cm $\le m \le$ 9.785 cm**
3. 2.4 ± 0.05 g **2.35 g $\le m \le$ 2.45 g**
4. $28 \pm \frac{1}{2}$ ft **$27\frac{1}{2}$ ft $\le m \le 28\frac{1}{2}$ ft**
5. 15 ± 0.5 cm **14.5 cm $\le m \le$ 15.5 cm**
6. $\frac{11}{16} \pm \frac{1}{64}$ in. **$\frac{43}{64}$ in. $\le m \le \frac{45}{64}$ in.**

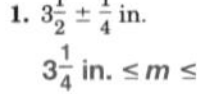

For each measurement, give the smallest unit used and the absolute error.

7. 12.5 cm $\le m \le 13.5$ cm **1 cm, 0.5 cm**
8. $12\frac{1}{8}$ in. $\le m \le 12\frac{3}{8}$ in. **$\frac{1}{4}$ in., $\frac{1}{8}$ in.**
9. $56\frac{1}{2}$ in. $\le m \le 57\frac{1}{2}$ in. **1 in., $\frac{1}{2}$ in.**
10. 23.05 mm $\le m \le 23.15$ mm **0.1 mm, 0.05 mm**

Mixed Review

Define a variable, write an inequality, and solve each problem.

59. Three fourths of a number decreased by 8 is at least 3. (Lesson 7–3) **$\{m \mid m \ge \frac{44}{3}\}$**

60. A number added to 23 is at most 5. (Lesson 7–1) **$\{x \mid x \le -18\}$**

61. If $7x - 2 \le 9x + 3$, then $2x \ge$ _?_ . (Lesson 7–1) **−5**

62. **Geometry** Line a is perpendicular to a line that is perpendicular to a line whose equation is $3x - 7y = -3$. If all lines are in the same plane, what is the slope of line a? (Lesson 6–6) **$\frac{3}{7}$**

63. Graph $x = 2y - 4$ using the x- and y-intercepts. (Lesson 6–5) **See margin.**

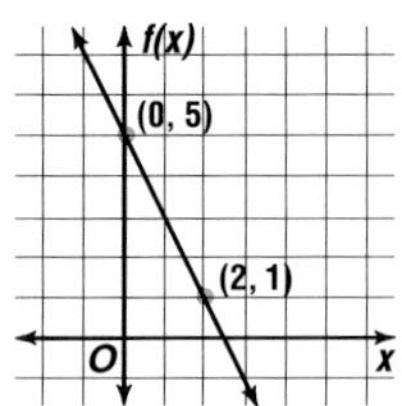

64. Write an equation in functional notation for the relation graphed at the right. (Lesson 5–6) **$f(x) = -2x + 5$**

65. Sarah is visiting a friend about two miles away from her house. She sees a hot-air balloon directly above her house. Sara estimates that the angle of elevation formed by the hot-air balloon is about 15°. About how high is the hot-air balloon? (Lesson 4–3) **a little more than half a mile**

66. Solve $\frac{4m - 3}{-2} = 12$. (Lesson 3–3) **$-\frac{21}{4}$**

67. Simplify $-17px + 22bg + 35px + (-37bg)$. (Lesson 2–5) **$18px - 15bg$**

68. Evaluate $3ab - c^2$ if $a = 6$, $b = 4$, and $c = 3$. (Lesson 1–3) **63**

SELF TEST

Solve each inequality. Then check your solution. (Lessons 7–1, 7–2, and 7–3)

1. $y + 15 \ge -2$ **$\{y \mid y \ge -17\}$**
2. $-102 > 17r$ **$\{r \mid r < -6\}$**
3. $5 - 6n > -19$ **$\{n \mid n < 4\}$**
4. $\frac{11 - 6w}{5} > 10$ **$\{w \mid w < -6.5\}$**
5. $7(g + 8) < 3(g + 12)$ **$\{g \mid g < -5\}$**
6. $0.1y - 2 \le 0.3y - 5$ **$\{y \mid y \ge 15\}$**

7. Choose an equivalent statement for $4 \ge x - 1 \ge -3$. (Lesson 7–4) **c**
   a. $5 \ge x \ge -4$ b. $3 \ge x \ge -4$ c. $5 \ge x \ge -2$ d. $3 \ge x \ge -2$

8. Solve $8 + 3t < 2$ or $-12 < 11t - 1$. Then graph the solution set. (Lesson 7–4) **$\{t \mid t < -2$ or $t > -1\}$; See margin for graph.**

9. **Sports** Jennifer has scored 18, 15, and 30 points in her last three starts on the junior varsity girls basketball team. How many points must she score in her next start so that her four-game average is greater than 20 points? (Lesson 7–3) **more than 17 points**

10. **Draw a Diagram** Suppose you roll a die two times. (Lesson 7–4)
   a. Draw a diagram to show the possibilities of rolling an even number the first time and an odd number the second time. **See margin for the nine results.**
   b. How many of the possibilities have a sum of 7? **3**

SELF TEST

The Self Test provides students with a brief review of the concepts and skills in Lessons 7-1 through 7-4. Lesson numbers are given to the right of exercises or instruction lines so students can review concepts not yet mastered.

Answers for Self Test

8. −5 −4 −3 −2 −1 0 1 2

10a. (2, 1), (2, 3), (2, 5), (4, 1), (4, 3), (4, 5), (6, 1), (6, 3), (6, 5)

7-5 Integration: Probability Compound Events

What YOU'LL LEARN

- To find the probability of a compound event.

Why IT'S IMPORTANT

You can use tree diagrams to solve problems involving business, sports, and games.

APPLICATION **Archaeology**

A group of college students are planning a trip to visit Virgin Islands National Park, where they will study the Carib Indian relics and the remnants of the forts built by the Danes. The Carib Indians were the original occupants of the islands, but had died or left by the early 1600s. The Danes formally claimed the islands in 1666, and they remained under Danish control until 1917.

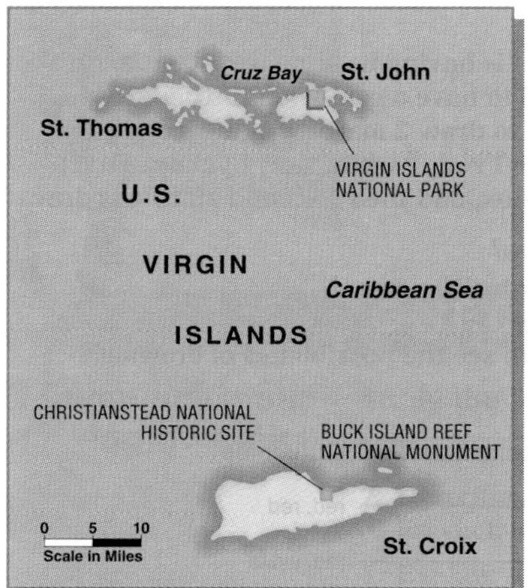

The group's advisor must plan how the group will get to the Virgin Islands. From their college, they will travel to Miami, Florida, by car, bus, train, or plane. Then to travel to St. Thomas in the Virgin Islands, they could take a plane or a ship. Suppose the advisor picks a mode of transportation at random. What is the probability that they will travel by car first and then fly?

To calculate this probability, you need to know all of the possible ways to get to St. Thomas. One method for finding this out is to draw a **tree diagram.** The tree diagram below shows how to get from the college to Miami and then to St. Thomas. The last column details all the possible combinations or **outcomes** of transportation.

Note that the number of outcomes is the product of the number of ways to Miami (4) and the number of ways to the Virgin Islands (2).

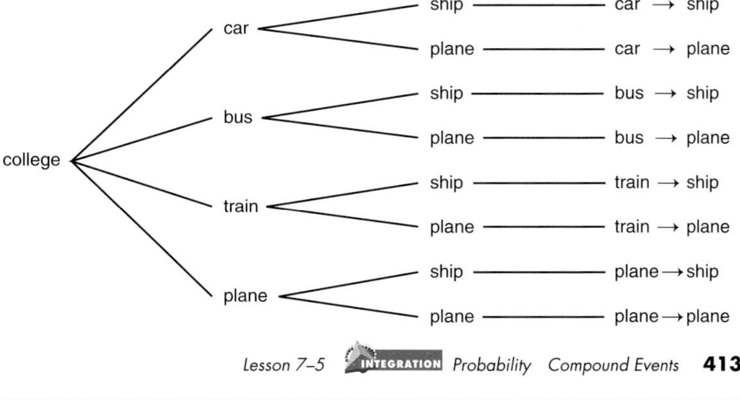

Alternative Learning Styles

Auditory Have students close their books. Read aloud the archaeology application at the beginning of the lesson. Have students construct the tree diagram as you read the passage.

7-5 LESSON NOTES

NCTM Standards: 1–4, 11

Instructional Resources

- Study Guide Master 7-5
- Practice Master 7-5
- Enrichment Master 7-5

Transparency 7-5A contains the 5-Minute Check for this lesson; **Transparency 7-5B** contains a teaching aid for this lesson.

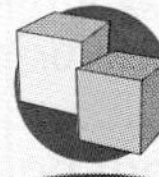

Recommended Pacing	
Standard Pacing	Day 7 of 14
Honors Pacing	Day 6 of 13
Block Scheduling*	Day 4 of 7 (along with Lesson 7-6)
Alg. 1 in Two Years*	Days 9 & 10 of 21

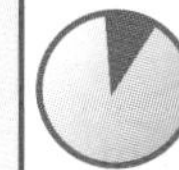

*For more information on pacing and possible lesson plans, refer to the *Block Scheduling Booklet* and *Algebra 1 in Two Years.*

1 FOCUS

5-Minute Check (over Lesson 7-4)

Write the compound inequality whose solution set is graphed.

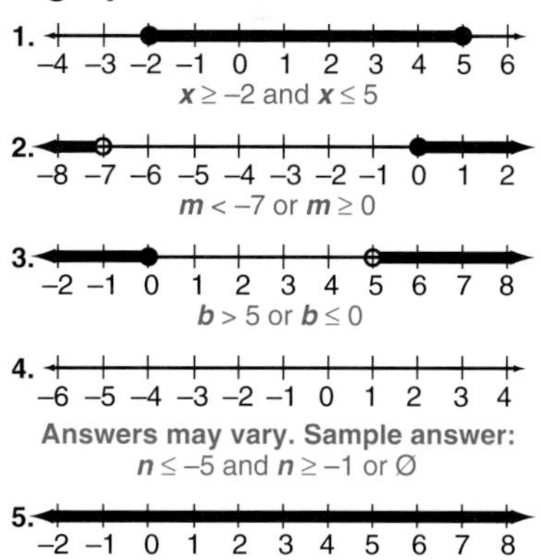

1. $x \geq -2$ and $x \leq 5$
2. $m < -7$ or $m \geq 0$
3. $b > 5$ or $b \leq 0$
4. Answers may vary. Sample answer: $n \leq -5$ and $n \geq -1$ or Ø
5. Answers may vary. Sample answer: $t \leq 8$ or $t \geq 3$

Assignment Guide

Core: 9–17 odd, 19–23
Enriched: 8–16 even, 17–23

For **Extra Practice,** see p. 773.

The red A, B, and C flags, printed only in the Teacher's Wraparound Edition, indicate the level of difficulty of the exercises.

Additional Answers

7a. See Solutions Manual for tree diagram. Possible combinations are: burger, soup, lemonade; burger, soup, soft drink; burger, salad, lemonade; burger, salad, soft drink; burger, French fries, lemonade; burger, French fries, soft drink; sandwich, soup, lemonade; sandwich, soup, soft drink; sandwich, salad, lemonade; sandwich, salad, soft drink; sandwich, French fries, lemonade; sandwich, French fries, soft drink; taco, soup, lemonade; taco, soup, soft drink; taco, salad, lemonade; taco, salad, soft drink; taco, French fries, lemonade; taco, French fries, soft drink; pizza, soup, lemonade; pizza, soup, soft drink; pizza, salad, lemonade; pizza, salad, soft drink; pizza, French fries, lemonade; pizza, French fries, soft drink.

8a. Let 1–5 represent the roads, and let A and B represent the doors.

1 — A — 1A
1 — B — 1B
2 — A — 2A
2 — B — 2B
3 — A — 3A
3 — B — 3B
4 — A — 4A
4 — B — 4B
5 — A — 5A
5 — B — 5B

7. Dining For lunch at the 66 Diner, you can select one item from each of the following categories for express service guaranteed in 15 minutes.

Entree	Side Dish	Beverage
Burger	Soup	Lemonade
Sandwich	Salad	Soft Drink
Taco	French Fries	
Pizza		

a. Draw a tree diagram showing all of the meal combinations.
b. What is the probability that a customer will have soup with the meal?
c. What is the probability of selecting a burger with French fries?
d. What is the probability of having pizza with a salad and a soft drink?

7a. See margin.
7b. $\frac{1}{3}$ or $0.\overline{3}$
7c. $\frac{1}{12}$ or $0.08\overline{3}$
7d. $\frac{1}{24}$ or $0.041\overline{6}$

EXERCISES

Applications and Problem Solving

8. Video Games In a computer video game, you have a choice of five roads to collect an important clue to solve a mystery. At the end of each road, there are two doors. The clue can only be found behind one door.

a. Draw a tree diagram to show the possibilities you have to find the clue.
b. What is the probability that you will find the clue? $\frac{1}{10}$ or 0.1

8a. See margin.

9. Travel Three different airlines fly from Bowling Green to Lexington. Those same three airlines and two other airlines fly from Lexington to Louisville. There are no direct flights from Bowling Green to Louisville.

a. How many ways can a traveler book flights from Bowling Green to Louisville? 15
b. What is the probability that flights booked at random from Bowling Green to Louisville use the same airline?

9b. $\frac{1}{5}$ or 0.2

10. Business On Wednesday, Ralph and Linda were talking about what they were going to wear on Friday since their company had declared it T-Shirt Day. Ralph said he only had three T-shirts he could wear to work. One was a solid color, one was an Earth Day shirt, and the other was a shirt with the company name and logo on it. He also said he would either wear tan pants or jeans. Linda is going to wear jeans and her company T-shirt. If Ralph selects a pair of pants and a T-shirt at random, what is the probability that he, too, will be wearing jeans and the company T-shirt?

10. $\frac{1}{6}$ or $0.1\overline{6}$

Alternative Teaching Strategies

Reading Algebra Students must be cautious not to read too much into the phrase, "The probability of event E." The fact that event E is more probable than an alternative event E_2 does not mean that it will happen. In fact, unless the probability is 1, it may not happen.

11. $\frac{1}{3}$ or $0.\overline{3}$

11. Food Kita and Jason are working on the school newspaper's final layout for October. They decide that after they finish, they will go get pizza. They have a coupon for a large pizza with three toppings for $8.99. In discussing what the three toppings should be, they find they have four favorite toppings, and choosing three that both agreed upon was going to be difficult. After some discussion, they decided that pepperoni was the one topping they both agreed upon and they would put the names of the other three toppings in a bag and choose two at random. If the other three toppings are mushrooms, olives, and sausage, what is the probability they will get a pizza with mushrooms?

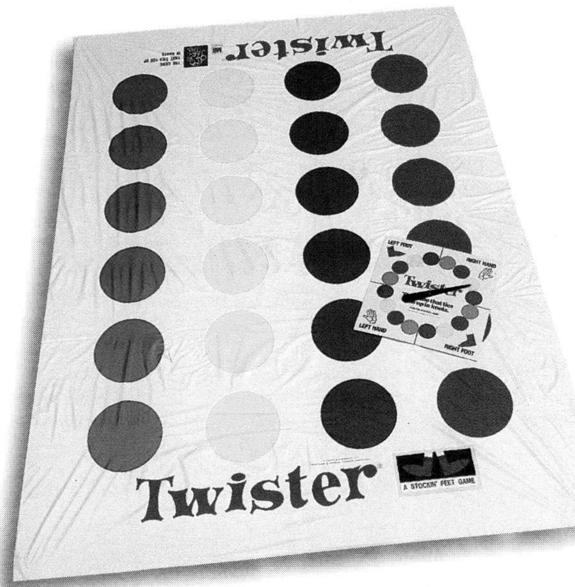

12. Games Twister® is a game composed of a mat with four rows, each having six circles of the same color. One person uses a spinner and calls out *hand* or *foot, right* or *left,* and a color (*blue, red, green,* or *yellow*). Each player must then place that particular body part on a circle of that color. The caller continues to call out body parts and colors with the players moving in the appropriate manner. If a player falls down in the process, he or she is disqualified. The winner is the last player still in position.

a. Draw a tree diagram to show the possibilities for body parts and colors. **See margin.**

b. What is the probability that the caller will spin and say *"right hand yellow?"* $\frac{1}{16}$ **or 0.0625**

13. Playing Cards The deck of cards for Rook® is composed of 57 cards. There are four suits (red, yellow, black, and green), each containing the numbers 1 to 14, and a rook card. Suppose there are two piles of five cards each with the cards turned face down. This first pile contains a red 3, a black 3, a red 5, a red 14, and a yellow 10. The second pile contains a green 5, a red 10, a black 10, a green 1, and a yellow 14. You are asked to draw a card from each pile.

a. List all the possibilities of drawing two cards. **See margin.**

13b. $\frac{3}{25}$ or 0.12

b. What is the probability that both cards are red?

13c. $\frac{2}{25}$ or 0.08

c. What is the probability that both cards are 10s?

d. What is the probability that both cards are green? **0**

13e. $\frac{14}{25}$ or 0.56

e. What is the probability that the sum of the cards is at least 15?

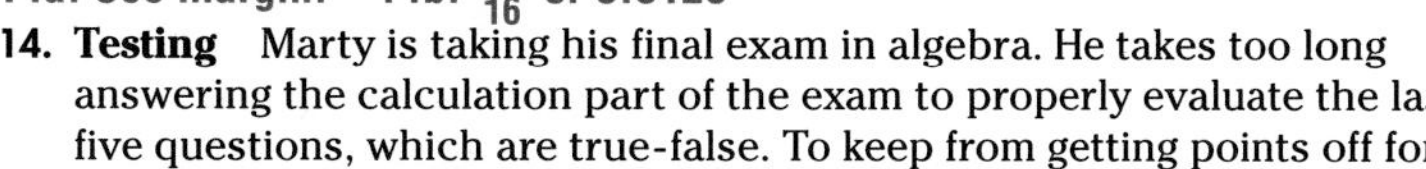

14a. See margin. 14b. $\frac{13}{16}$ or 0.8125

14. Testing Marty is taking his final exam in algebra. He takes too long answering the calculation part of the exam to properly evaluate the last five questions, which are true-false. To keep from getting points off for not answering these questions at all, he guesses at them just as the bell rings.

a. Draw a tree diagram to show the possible answer outcomes.

b. What is the probability that he got at least two of the questions correct if the answers were T, F, F, T, F?

c. Suppose Marty answered all the questions as true. Is this a good strategy to try to get the most correct without reading the question? Explain. **Answers will vary. See students' work.**

Additional Answers

12a. See Solutions Manual for tree diagram. Possible combinations are: blue, right, hand; blue, right, foot; blue, left, hand; blue, left, foot; red, right, hand; red, right, foot; red, left, hand; red, left, foot; green, right, hand; green, right, foot; green, left, hand; green, left, foot; yellow, right, hand; yellow, right, foot; yellow, left, hand; and yellow, left, foot.

13a. R3-G5, R3-R10, R3-B10, R3-G1, R3-Y14, B3-G5, B3-R10, B3-B10, B3-G1, B3-Y14, R5-G5, R5-R10, R5-B10, R5-G1, R5-Y14, R14-G5, R14-R10, R14-B10, R14-G1, R14-Y14, Y10-G5, Y10-R10, Y10-B10, Y10-G1, Y10-Y14

14a. See Solutions Manual for tree diagram. Possible combinations are: TTTTT, TTTTF, TTTFT, TTTFF, TTFTT, TTFTF, TTFFT, TTFFF, TFTTT, TFTTF, TFTFT, TFTFF, TFFTT, TFFTF, TFFFT, TFFFF, FTTTT, FTTTF, FTTFT, FTTFF, FTFTT, FTFTF, FTFFT, FTFFF, FFTTT, FFTTF, FFTFT, FFTFF, FFFTT, FFFTF, FFFFT, and FFFFF.

Study Guide Masters, p. 52

7-5 NAME______________________ DATE__________

Study Guide — Student Edition Pages 413–419

Integration: Probability Compound Events

A **compound event** consists of two or more simple events. The probability of a compound event can be calculated if its outcomes are equally likely. To calculate probability, determine the possible outcomes. A tree diagram is a good way of doing this.

Example: A die is rolled and a coin is tossed. What is the probability of getting a three on the die and heads on the coin?

To calculate the probability, you need to know how many outcomes are possible. There are 12 possible outcomes. The probability of getting a three on the die and a heads on the coin is $\frac{1}{12}$.

Solve each problem.

1. Max makes three piles of cards, each consisting of only a Queen and a Jack. Lisa draws one card from each pile. What is the probability that Lisa draws two Queens and a Jack in any order? Make a tree diagram of all the choices. $\frac{3}{8}$

2. Phyllis drops three pennies in a wishing pond. What is the probability that only one lands with tails showing? Make a tree diagram of all the choices. $\frac{3}{8}$

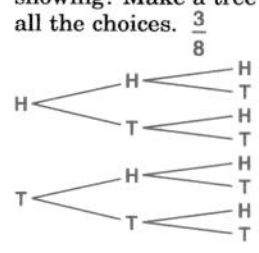

3. George has 1 pair of red socks and 1 pair of white socks in a drawer. What is the probability of drawing 2 red socks? $\frac{1}{6}$

4. In Exercise 3, what is the probability of picking a red sock and a white sock in that order. $\frac{1}{3}$

tree for Exercises 3 and 4: R W W R W W R R W R R W

15. **Babies** According to the National Center for Health Statistics, each time a baby is born, the probability that it is a boy is always 51.3% and a girl is 48.7%, regardless of the sex of any previous births in the family. Suppose a couple has four children. **15b. about 26.3%**
   a. What is the probability they are all girls? **about 5.6%**
   b. What is the probability there are one girl and three boys?
   c. How many children are there in your family? Find the probability of your family situation. **See students' work.**

16. **Computers** The results of a survey in which students were asked how they used their computers are shown in the Venn diagram at the right. Based on these findings, what is the probability that a student, chosen at random, will use a computer only for word processing, and a second student will use it only for games? **14%**

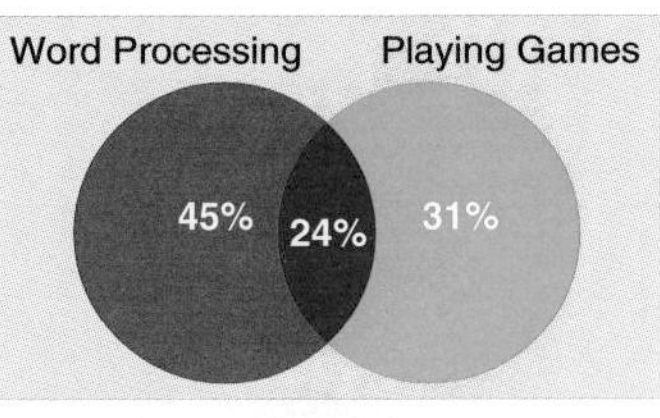

Critical Thinking

17. Every working day Adrianna either bicycles or drives to work. If she oversleeps, she drives to work 80% of the time. If she does not oversleep, she bicycles 80% of the time. She oversleeps 20% of the time. What is the probability that she will drive to work on any particular day? **32%**

18. A bag of marbles contains 2 yellow marbles, 1 blue marble, and 3 red marbles. Suppose you select one marble at random and *do not* put it back in the bag. Then you select another marble.
   a. Draw a tree diagram to model this situation. How many possible outcomes are there?
   b. What is the probability of selecting a red marble first and then a yellow one?

18a. See students' work; 30 outcomes.

18b. $\frac{1}{5}$

Mixed Review

19. **Statistics** Amy's scores on the first three of four 100-point biology tests were 88, 90, and 91. To get an A− in the class, her average must be between 88 and 92, inclusive, on all tests. What score must she receive on the fourth test to get an A− in biology? (Lesson 7–4) **between 83 and 99, inclusive**

20. Define a variable, write an inequality, and solve the following. Two thirds of a number is more than 99. (Lesson 7–2)

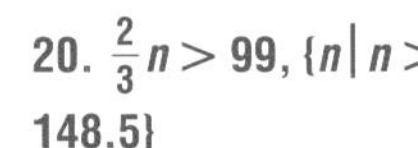

20. $\frac{2}{3}n > 99$, $\{n \mid n > 148.5\}$

21. Determine the slope and y-intercept of the graph of $3x - y = 9$. (Lesson 6–4)

21. 3; −9

22. Solve $8r - 7t + 2 = 5(r + 2t) - 9$ for r. (Lesson 5–3) $r = \frac{17t - 11}{3}$

Extension

Connections A four-sided (tetrahedral) die has faces numbered 1, 2, 5, 6. If the die is tossed twice, what is the probability that the sum of the numbers facing down is 7? $\frac{1}{4}$

Practice Masters, p. 52

7-5 NAME________ DATE________

Practice — Student Edition Pages 413–419

Integration: Probability Compound Events

1. Draw a tree diagram to show the possibilities for boys and girls in a family of 3 children. Assume that the probabilities for girls and boys being born are the same.

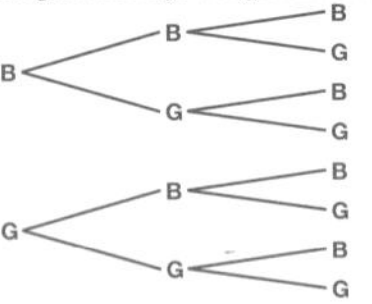

a. What is the probability that the family has exactly 3 boys? $\frac{1}{8}$
b. What is the probability that the family has more boys than girls? $\frac{1}{2}$
c. What is the probability that the family has at most two girls? $\frac{7}{8}$
d. What is the probability that the family has at least two girls? $\frac{1}{2}$

2. Use a tree diagram to find the probability of getting at most two heads when three fair coins are tossed. $\frac{7}{8}$

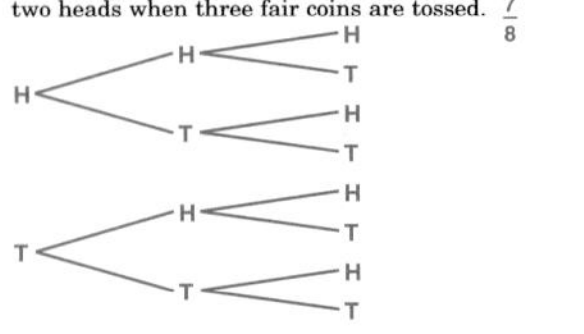

3. Compare the tree diagram used for Exercise 2 with the one used for Exercise 1.
Both tree diagrams show 3 simple events and 8 outcomes.

4. Box A contains one blue and two green marbles. Box B contains one green marble and two red marbles. Box C contains one green marble and two white marbles. A marble is drawn at random from each box.
   a. What is the probability that all of the marbles are green? $\frac{2}{27} \approx 0.074$
   b. What is the probability that at least one marble is not green? $\frac{25}{27} \approx 0.926$

23. **Optics** Peripheral vision is the ability to see around you when you keep your head and eyes pointed straight ahead. The average person with good peripheral vision can see about 180° when they are still. For a person in a moving vehicle, peripheral vision decreases as speed increases, as shown in the illustration below. (Lesson 3–4)

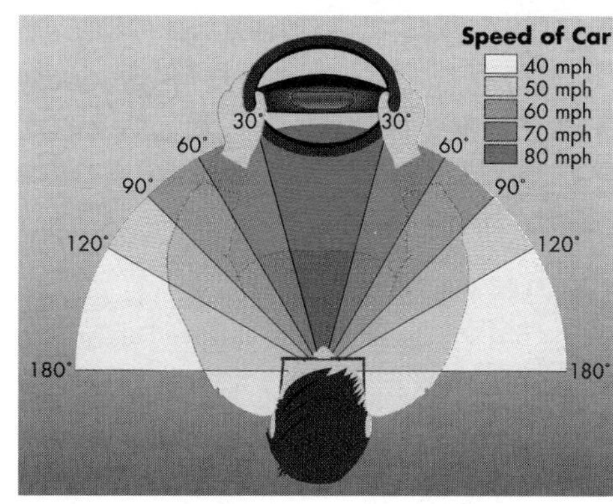

a. If Seth Lytle is driving at 65 mph, approximately how large is his angle of vision? **50°**

b. About how large an angle is he not seeing compared to his stationary angle of vision? **130°**

c. If there were a deer standing 50° to the right of Seth's center view and he was driving at 45 mph, would he be able to see the deer? **yes**

Computers and Chess

The excerpt below appeared in an article in the January 1995 issue of *Discover* magazine.

HOW SMART HAVE COMPUTERS BECOME? Very smart, at least by one measure: this past August, a computer defeated the human world champion at chess for the first time. Russian grand master Garry Kasparov was outwitted by Genius 2, a computer program designed by an English physicist, Richard Lang, who calls himself a mediocre chess player.... Chess programs like Lang's work by focusing on what computers do best: long and intricate calculations. Before deciding on a move, the program examines each of the roughly 36 possibilities in an average game situation. For each of these 36 "branches," it looks 16 moves ahead, determining its opponent's possible countermoves, its own possible responses to each countermove, and so on. Sixteen moves ahead, the number of possible board configurations is huge. The genius of Genius 2 lies in the way it prunes the possibility tree, ruling out bad moves from the start. But the software also had help from the hardware; Lang's computer uses a speedy Intel Pentium microprocessor that can execute 166 million instructions per second. "We wouldn't have won (the computer world championship) had we not had this processor," Lang concedes. ■

1–2. **See margin.**

1. Many chess-playing programs are written by people who are not top-level chess players. How then can these programs defeat very good players?
2. If you want to improve your chess game by practicing against a computer, would you choose a program that is below your playing level, at your level, a little above your level, or far above your level? Why?
3. Do you think it's fair to match human chess players against computers? Why or why not? **See students' work.**
4. How might a computer use probability in determining the proper move to make? **See margin.**

Answers for Mathematics and Society

1. The computer's speed can evaluate several options for this and future moves better than the best of players.
2. First choose a level a little above your level to stimulate your playing skills. Next choice would be at your level to improve your technique and speed. Playing at a level far above your level might be too frustrating.
4. The computer tests all combinations of possible moves now and in the future. It is programmed to take into consideration the likelihood of certain combinations of moves. Those combinations that are least likely can probably be ignored as possibilities for your opponent's next move.

4 ASSESS

Closing Activity

Modeling Place one red and one green marble in one jar, one blue and one white marble in another jar, and one black, one pink, and one yellow marble in a third jar. Have students reenact each of the twelve possible outcomes of the experiment of selecting one marble from each jar.

Mathematics and SOCIETY

Students might find it helpful to begin writing a very simple game tree for the game tic-tac-toe. Play the game for a short while, such as five moves. Now, from this configuration, construct a tree for all possible games from this point. Students will quickly see how large a tree would be for a complete tree for all possible tic-tac-toe games.

Enrichment Masters, p. 52

NAME ______ DATE ______

Enrichment — Student Edition Pages 413–419

Conditional Probability

The probability of an event given the occurrence of another event is called **conditional probability.** The conditional probability of event A given event B is denoted $P(A|B)$.

Example: Suppose a pair of number cubes is rolled. It is known that the sum is greater than seven. Find the probability that the number cubes match.

There are 15 sums greater than seven and there are 36 possible pairs altogether.

$P(B) = \frac{15}{36}$

There are three matching pairs greater than seven, (4, 4), (5, 5), and (6, 6).

$P(A \text{ and } B) = \frac{3}{36}$

$$P(A|B) = \frac{P(A \text{ and } B)}{P(B)} = \frac{\frac{3}{36}}{\frac{15}{36}} \text{ or } \frac{1}{5}$$

The conditional probability is $\frac{1}{5}$.

A card is drawn from a standard deck of 52 cards and is found to be red. Given that event, find each of the following probabilities.

1. $P(\text{heart})$ $\frac{1}{2}$
2. $P(\text{ace})$ $\frac{1}{13}$
3. $P(\text{face card})$ $\frac{3}{13}$
4. $P(\text{jack or ten})$ $\frac{2}{13}$
5. $P(\text{six of spades})$ 0
6. $P(\text{six of hearts})$ $\frac{1}{26}$

A sports survey taken at Stirers High School shows that 48% of the respondents liked soccer, 66% liked basketball, and 38% liked hockey. Also, 30% liked soccer and basketball, 22% liked basketball and hockey and 28% liked soccer and hockey. Finally, 12% liked all three sports.

7. Find the probability that Meg likes soccer if she likes basketball. $\frac{5}{11}$
8. Find the probability that Juan likes basketball if he likes soccer. $\frac{5}{8}$
9. Find the probability that Mieko likes hockey if she likes basketball. $\frac{1}{3}$
10. Find the probability that Greg likes hockey if he likes soccer. $\frac{7}{12}$

7-6 LESSON NOTES

NCTM Standards: 1–5

Instructional Resources

- Study Guide Master 7-6
- Practice Master 7-6
- Enrichment Master 7-6
- Assessment and Evaluation Masters, p. 185

Transparency 7-6A contains the 5-Minute Check for this lesson; **Transparency 7-6B** contains a teaching aid for this lesson.

Recommended Pacing	
Standard Pacing	Day 8 of 14
Honors Pacing	Day 7 of 13
Block Scheduling*	Day 4 of 7 (along with Lesson 7-5)
Alg. 1 in Two Years*	Days 11 & 12 of 21

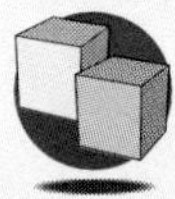

*For more information on pacing and possible lesson plans, refer to the *Block Scheduling Booklet* and *Algebra 1 in Two Years.*

1 FOCUS

5-Minute Check
(over Lesson 7-5)

Two red marbles and four green marbles are placed in a jar. Two marbles are to be selected from the jar at random. A marble is drawn, put back in the box, and then a second marble is drawn.

1. What is the probability that both are red? $\frac{1}{9}$
2. What is the probability that both are green? $\frac{4}{9}$
3. What is the probability that at least one is green? $\frac{8}{9}$
4. What is the probability that one is green and one is red? $\frac{4}{9}$
5. Which is more likely, at least one is red or at least one is green? **at least one is green**

7-6 Solving Open Sentences Involving Absolute Value

What YOU'LL LEARN

- To solve open sentences involving absolute value and graph the solutions.

Why IT'S IMPORTANT

You can use open sentences to solve problems involving space exploration, law enforcement, and entertainment.

Space Exploration

On Tuesday, February 7, 1995, the space shuttle *Discovery* maneuvered within 37 feet of the Russian space station *Mir,* 245 miles above Earth. To accomplish this feat, *Discovery* had to be launched within 2.5 minutes of a designated time (12:45 A.M.). This time period is known as the *launch window.* If t represents the time elapsed since the launch countdown began and there are 300 minutes scheduled from the beginning of the countdown to blast-off, you can write the following inequality to represent the launch window.

$|300 - t| \leq 2.5$ *The difference between 300 minutes and the actual time elapsed since the countdown began must be less than or equal to 2.5 minutes.*

We use absolute value because $300 - t$ cannot be negative.

fabulous FIRSTS

Sally Ride (1951–)

Sally Ride was the first American woman in space. She was a member of the *Challenger STS-7* crew that was launched June 18, 1983. She has a Ph.D. in physics from Stanford University.

There are three types of open sentences that can involve absolute value. They are as follows, when n is nonnegative.

$|x| = n$ $\qquad |x| < n$ $\qquad |x| > n$

First let's consider the case of $|x| = n$.

If $|x| = 4$, this means that the difference from 0 to x is 4 units.

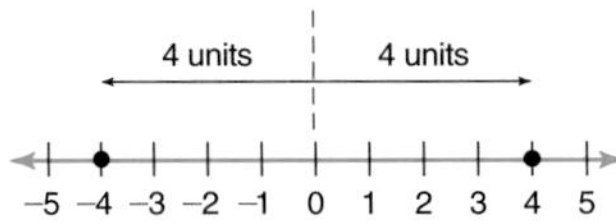

The solution set can also be written as $\{x \mid x = -4 \text{ or } x = 4\}$.

Therefore, if $|x| = 4$, then $x = -4$ or $x = 4$. The solution set is $\{-4, 4\}$. So, if $|x| = n$, then $x = -n$ or $x = n$.

Equations involving absolute value can be solved by graphing them on a number line or by writing them as a compound sentence and solving it.

Example  **Solve $|x - 3| = 5$.**

Method 1: Graphing

$|x - 3| = 5$ means the distance between x and 3 is 5 units. To find x on the number line, start at 3 and move 5 units in either direction.

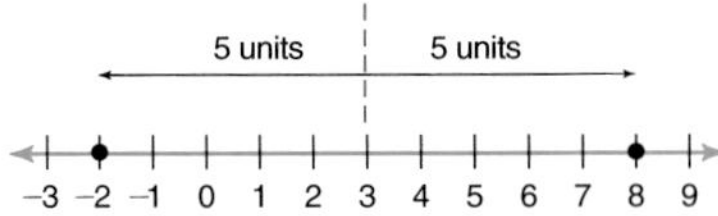

The solution set is $\{-2, 8\}$.

fabulous FIRSTS

On her first shuttle flight, Ride helped launch communications satellites, conducted experiments on the production of pharmaceuticals, and tested the space shuttle's remote manipulator arm. Ride flew a second shuttle mission in October 1984.

Method 2: Compound Sentence

$|x - 3| = 5$ also means $x - 3 = 5$ or $-(x - 3) = 5$.

$$x - 3 = 5 \quad \text{or} \quad -(x - 3) = 5$$
$$x - 3 + 3 = 5 + 3 \qquad x - 3 = -5 \quad \textit{Multiply each side by } -1.$$
$$x = 8 \qquad x - 3 + 3 = -5 + 3$$
$$x = -2$$

This verifies the solution set.

Now let's consider the case of $|x| < n$. Inequalities involving absolute value can also be represented on a number line or as compound inequalities. Let's examine $|x| < 4$.

$|x| < 4$ means that the distance from 0 to x is less than 4 units.

4 units 4 units

−5 −4 −3 −2 −1 0 1 2 3 4 5

When the inequality is < or ≤, the compound sentence uses *and.*

Therefore, $x > -4$ and $x < 4$. The solution set is $\{x \mid -4 < x < 4\}$. So, if $|x| < n$, then $x > -n$ and $x < n$.

Example 2 **Solve $|3 + 2x| < 11$ and graph the solution set.**

$|3 + 2x| < 11$ means $3 + 2x < 11$ and $3 + 2x > -11$.

$$3 + 2x < 11 \quad \text{and} \quad 3 + 2x > -11$$
$$3 - 3 + 2x < 11 - 3 \qquad 3 - 3 + 2x > -11 - 3$$
$$2x < 8 \qquad 2x > -14$$
$$\frac{2x}{2} < \frac{8}{2} \qquad \frac{2x}{2} > \frac{-14}{2}$$
$$x < 4 \qquad x > -7$$

The solution set is $\{x \mid x > -7 \text{ and } x < 4\}$, which can be written as $\{x \mid -7 < x < 4\}$.

Now graph the solution set.

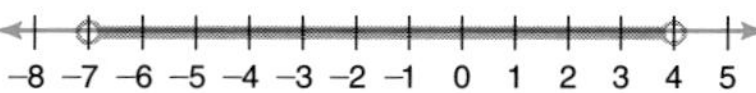

Finally let's examine $|x| > 4$. This means that the distance from 0 to x is greater than 4 units.

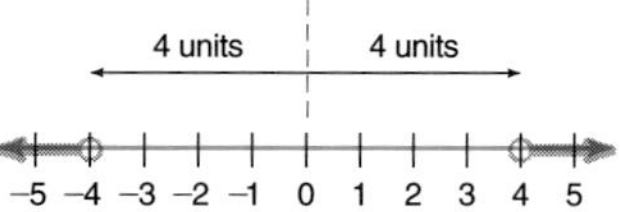

When the inequality is > or ≥, the compound sentence uses *or.*

Therefore, $x < -4$ or $x > 4$. The solution set is $\{x \mid x < -4 \text{ or } x > 4\}$. So, if $|x| > n$, then $x < -n$ or $x > n$.

Motivating the Lesson

Situational Problem Jim's average is less than 5 points from Sam's. Sam's average is 74. What conclusion can be drawn about Jim's average? **His average is less than 79 (74 + 5) and greater than 69 (74 − 5).**

2 TEACH

Teaching Tip When presenting Example 1, remind students that $|x| = 5$ means $x = 5$ or $-x = 5$. Therefore, $x = 5$ or $x = -5$. $|x - 3| = 5$ means $x - 3 = 5$ or $-(x - 3) = 5$. Therefore, $x - 3 = 5$ or $x - 3 = -5$.

In-Class Examples

For Example 1
Solve each equation.

a. $|4p - 3| = 7$
$\left\{p \mid p = \frac{5}{2} \text{ or } p = -1\right\}$

b. $|4k + 6| = 5$
$\left\{k \mid k = -\frac{1}{4} \text{ or } k = -\frac{11}{4}\right\}$

For Example 2
Solve $|2x + 1| < 8$ and graph the solution set.
$\left\{x \mid x > -\frac{9}{2} \text{ and } x < \frac{7}{2}\right\}$

−5 −4 −3 −2 −1 0 1 2 3 4

Teaching Tip When presenting Example 2, point out to students that the open sentence $|3 + 2x| < 11$ says that the distance from 0 to $3 + 2x$ is less than 11 units. This means $3 + 2x < 11$ and $-(3 + 2x) < 11$, or $3 + 2x > -11$.

Classroom Vignette

"I compare absolute value inequalities to the amount of gas in the tank of a student's truck and the direction the student can travel (left or right) out of the driveway. For example, if they have $5 worth of gas, they can drive a certain distance either way until the gas is gone (that is, $|x| = 5$). If they have more than $5 worth, they can drive farther either way ($|x| > 5$)."

Linda Weissinger

Linda Weissinger
Rochester High School
Rochester, Indiana

In-Class Examples

For Example 3
Solve $|4x - 5| \geq 3$ and graph the solution set.
$\left\{x \mid x \leq \frac{1}{2} \text{ or } x \geq 2\right\}$

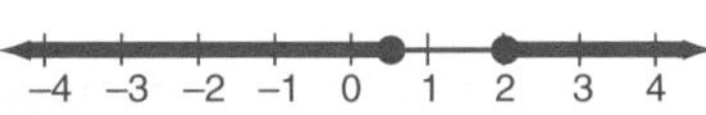

For Example 4
If the price of a stock varies more than 3.4 points from its opening price, Mr. Witter buys ten more shares of the stock. If the opening price is 15.8, for what range of values will he buy stock? $\{x \mid x > 19.2 \text{ or } x < 12.4\}$

Teaching Tip If students are having difficulty determining when they should use *and* or *or* in an inequality involving an absolute value, explain that an alternate approach is available.
Solve $|3 + 2x| < 11$.
Solve the inequality as if it were an equation.

$$3 + 2x = -11 \quad 3 + 2x = 11$$
$$2x = -14 \quad 2x = 8$$
$$x = -7 \quad x = 4$$

Plot these points with open circles on the number line.
The two points separate the number line into three regions, $x < -7$, $-7 < x < 4$, and $x > 4$. Have students test a point from each region in the original inequality, for example, -10, 0, and 5. The region that contains 0 is the region that is the solution. This region is $-7 < x < 4$. This represents a compound sentence involving *and.* Draw a line between the two open circles to represent the answer.

Example 3 Solve $|5 + 2y| \geq 3$ and graph the solution set.

$|5 + 2y| \geq 3$ means $5 + 2y \leq -3$ or $5 + 2y \geq 3$.

	or	
$5 + 2y \leq -3$		$5 + 2y \geq 3$
$5 - 5 + 2y \leq -3 - 5$		$5 - 5 + 2y \geq 3 - 5$
$2y \leq -8$		$2y \geq -2$
$\frac{2y}{2} \leq \frac{-8}{2}$		$\frac{2y}{2} \geq \frac{-2}{2}$
$y \leq -4$		$y \geq -1$

The solution set is $\{y \mid y \leq -4 \text{ or } y \geq -1\}$.

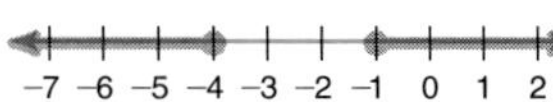

Organizations such as OSHA set standards for buildings to meet the needs of those using the building. Building code standards are often written as maximums or minimums that must be met. These standards can be written as inequalities.

Example 4

There are a number of specifications in the building industry that address the needs of physically-challenged persons. For example, hallways in hospitals must have handrails. The handrails must be placed within a range of 2 inches from a height of 36 inches.

a. Write an open sentence that involves absolute value to represent the range of acceptable heights for hallway handrails.

b. Find and graph the corresponding compound sentence.

a. Let h be an acceptable height for the handrail. Then h can differ from 36 inches by no more than 2 inches. Write an open sentence that represents the range of acceptable heights.

h differs from 36	*by less than or equal to*	*2.*
$\lvert h - 36\rvert$	$\leq$	2

Now, solve $|h - 36| \leq 2$ to find the compound sentence.

	and	
$h - 36 \geq -2$		$h - 36 \leq 2$
$h - 36 + 36 \geq -2 + 36$		$h - 36 + 36 \leq 2 + 36$
$h \geq 34$		$h \leq 38$

The compound sentence is $34 \leq h \leq 38$.

Construction Manager Students who are interested in a career in construction may wish to do further research into the use of math as it relates to building codes, as mentioned in Example 4. For more information on tech prep, see the *Teacher's Handbook*.

b. The graph of this sentence is shown below.

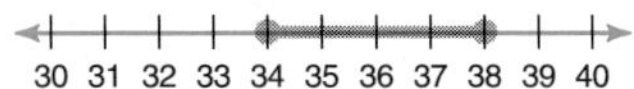

So, the handrail must be placed between 34 and 38, inclusive, inches from the floor.

CHECK FOR UNDERSTANDING

Communicating Mathematics

Study the lesson. Then complete the following. 1–2. See margin.

1. **Describe** two ways you can solve an open sentence involving an absolute value.
2. **Explain** the difference in the solution for $|x + 7| > 4$ and the solution for $|x + 7| < 4$.
3. If $x < 0$ and $|x| = n$, describe n in terms of x. $n = -x$
4. **You Decide** Jamila's teacher said they should work out each problem and test points to determine whether their solutions are correct. Jamila thinks that she knows the solution for problems like $|w + 5| < 0$, where the absolute value quantity is always less than 0, without testing points. Is she correct? Why or why not? **Yes, it will always be ∅.**

Guided Practice

Choose the replacement set that makes each open sentence true.

5. $|x - 7| = 2$ **c**
   a. {9, −5} b. {5, −9} c. {5, 9} d. ∅
6. $|x - 2| > 4$ **d**
   a. $\{x \mid -2 < x < 6\}$ b. $\{x \mid x = 6 \text{ or } x = 2\}$
   c. $\{x \mid x > 6 \text{ or } x > -2\}$ d. $\{x \mid x < -2 \text{ or } x > 6\}$

State which graph below matches each open sentence in Exercises 7–10.

7. $|y| = 3$ **c**
8. $|y| > 3$ **a**
9. $|y| < 3$ **d**
10. $|y| \leq 3$ **b**

a. −5 −4 −3 −2 −1 0 1 2 3 4 5
b. −5 −4 −3 −2 −1 0 1 2 3 4 5
c. −5 −4 −3 −2 −1 0 1 2 3 4 5
d. −5 −4 −3 −2 −1 0 1 2 3 4 5

Solve each open sentence. Then graph the solution set.

11–14. See margin for graphs.

11. $|m| \geq 5$ $\{m \mid m \leq -5 \text{ or } m \geq 5\}$
12. $|n| < 6$ $\{n \mid -6 < n < 6\}$
13. $|r + 3| < 6$ $\{r \mid -9 < r < 3\}$
14. $|8 - t| \geq 3$ $\{t \mid t \leq 5 \text{ or } t \geq 11\}$

For each graph, write an open sentence involving absolute value.

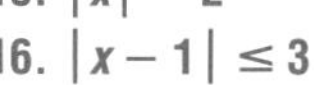

15. $|x| = 2$
16. $|x - 1| \leq 3$

15. 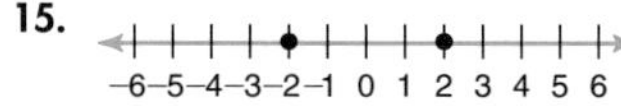

16. 

3 PRACTICE/APPLY

Check for Understanding

Exercises 1–16 are designed to help you assess your students' understanding through reading, writing, speaking, and modeling. You should work through Exercises 1–4 with your students and then monitor their work on Exercises 5–16.

Error Analysis
In interpreting $|x + 2|$ as a distance, students sometimes overlook the sign. Students must think of this as $|x - (-2)|$ since distance is the *difference* of the coordinates. Also, inequalities of the form $|x - a| < 0$ have an empty set solution since distances cannot be negative.

Additional Answers

1. You can graph the meaning of the absolute value inequality on a number line, or you can solve the compound inequality it represents algebraically.
2. $|x + 7| > 4$ is solved as $x + 7 > 4$ or $x + 7 > -4$; $|x + 7| < 4$ is solved as $x + 7 < 4$ and $x + 7 > -4$.
11. −8 −6 −4 −2 0 2 4 6 8
12. −10 −8 −6 −4 −2 0 2 4 6 8 10
13. −12 −10 −8 −6 −4 −2 0 2 4 6
14. 3 4 5 6 7 8 9 10 11 12 13

Reteaching

Using Graphs Graph each open sentence.

1. $|x - 2| = 4$
   −3 −2 −1 0 1 2 3 4 5 6
2. $|x - 2| < 4$
   −3 −2 −1 0 1 2 3 4 5 6
3. $|x - 2| > 4$

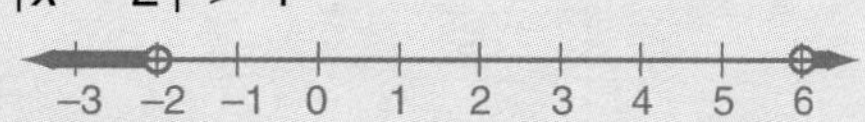

Have students compare and contrast the three graphs and equations.

Assignment Guide

Core: 17–55 odd, 57–66
Enriched: 18–46 even, 48–66

For **Extra Practice,** see p. 773.

The red A, B, and C flags, printed only in the Teacher's Wraparound Edition, indicate the level of difficulty of the exercises.

Additional Answers

17. 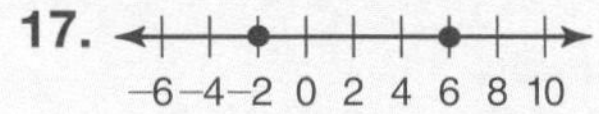
−6 −4 −2 0 2 4 6 8 10

18. −3 −2 −1 0 1 2 3 4 5

19. ∅

20. −12 −11 −10 −9 −8 −7 −6 −5 −4

21. −2 −1 0 1 2 3 4 5 6

22. −9 −8 −7 −6 −5 −4 −3 −2 −1 0 1

23. ∅

24. −5 −4 −3 −2 −1 0

25. 1 2 3 4

26. −6 −5 −4 −3 −2 −1 0 1 2 3 4

27. −4 −3 −2 −1 0 1 2 3 4

Study Guide Masters, p. 53

NAME ______________ DATE ______

Study Guide Student Edition Pages 420–426

Solving Open Sentences Involving Absolute Value

An open sentence involving absolute value should be interpreted, solved, and graphed as a compound sentence. Study the examples that follow.

Example 1:
$|5y - 2| = 7$ means
$5y - 2 = 7$ or $5y - 2 = -7$.
$5y = 9$ $5y = -5$
$y = 1\frac{4}{5}$ $y = 1$
Graph: −4 −2 0 2 4

Example 2:
$|3a + 4| < 10$ means
$3a + 4 < 10$ and $3a + 4 > -10$.
$3a < 6$ $3a > -14$
$a < 2$ $a > -4\frac{2}{3}$
Graph: −6 −4 −2 0 2

Example 3: $|4z + 1| \geq 5$ means
$4z + 1 \leq -5$ or $4z + 1 \geq 5$.
$4z \leq -6$ $4z \geq 4$
$z \leq -1\frac{1}{2}$ $z \geq 1$
Graph: −4 −2 0 2 4

Solve each open sentence. Then graph the solution set on the number line provided.

1. $|c - 2| > 6$ −6 −4 −2 0 2 4 6 8
{c | c < −4 or c > 8}

2. $|x - 9| < 0$ −3 −2 −1 0 1 2 3
no solution

3. $|3f + 10| \leq 4$ −6 −5 −4 −3 −2 −1 0 1
$\left\{f \mid -4\frac{2}{3} \leq f \leq -2\right\}$

Express each statement in terms of an inequality involving absolute value. Do not try to solve.

4. The price was within \$10.00 of his \$100.00 limit. $|p - 100| < 10$

5. Her score was more than 20 points from the 53 point record. $|s - 53| > 20$

6. The poll showed the incumbent mayor had the support of 12% of the voters plus or minus 3 percent points. $|v - 0.12| \leq 0.03$

7. Mike swam his daily laps no more than 5 seconds off his previous record of 40 seconds. $|x - 40| \leq 5$

For each graph, write an open sentence involving absolute value.

8. −4 −3 −2 −1 0 1 2 3 4

9. −4 −3 −2 −1 0 1 2 3 4

EXERCISES

Practice

Solve each open sentence. Then graph the solution set.

17–34. See margin for graphs.

20. $\{w \mid w \leq -9 \text{ or } w \geq -7\}$

17. $|y - 2| = 4$ **{−2, 6}**
18. $|3 - 3x| = 0$ **{1}**
19. $|7x + 2| = -2$ ∅
20. $|w + 8| \geq 1$
21. $|2 - y| \leq 1$ **$\{y \mid 1 \leq y \leq 3\}$**
22. $|t + 4| \geq 3$ **$\{t \mid t \leq -7 \text{ or } t \geq -1\}$**
23. $|4y - 8| < 0$ ∅
24. $|2x + 5| < 4$

25. $|3e - 7| < 2$
26. $|3x + 4| < 8$
27. $|1 - 3y| > -2$ **{all numbers}**
28. $3 + |x| > 3$ **$\{x \mid x \neq 0\}$**
29. $|8 - (w - 1)| \leq 9$
30. $|6 - (11 - b)| = -3$ ∅
31. $|2.2y - 1.1| = 5.5$ **{−2, 3}**
32. $|3r - 0.5| \geq 5.5$
33. $\left|\frac{2 - 3x}{5}\right| \geq 2$
34. $\left|\frac{1}{2} - 3p\right| < \frac{7}{2}$

24. $\left\{x \mid -\frac{9}{2} < x < -\frac{1}{2}\right\}$
25. $\left\{e \mid \frac{5}{3} < e < 3\right\}$
26. $\left\{x \mid -4 < x < \frac{4}{3}\right\}$
29. $\{w \mid 0 \leq w \leq 18\}$
32. $\{r \mid r \leq -1.\overline{6} \text{ or } r \geq 2\}$
33. $\left\{x \mid x \leq -\frac{8}{3} \text{ or } x \geq 4\right\}$
34. $\left\{p \mid -1 < p < \frac{4}{3}\right\}$

Express each statement in terms of an inequality involving absolute value. Do *not* try to solve.

35. The diameter of the lead in a pencil p must be within 0.01 millimeters of 1 millimeter. **$|p - 1| \leq 0.01$**
36. The cruise control of a car set at 55 mph should keep the speed s within 3 mph of 55 mph. **$|s - 55| \leq 3$**
37. A liquid at 50°C will change to a gas or liquid if the temperature t increases or decreases more than 50°C. **$|t - 50| > 50$**

For each graph, write an open sentence involving absolute value.

38. −3 −2 −1 0 1 2 3 4 5 6 7
39. −6 −5 −4 −3 −2 −1 0 1 2 3 4

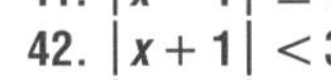

40. −5 −4 −3 −2 −1 0 1 2 3 4 5
41. −4 −3 −2 −1 0 1 2 3 4 5
42. −6 −5 −4 −3 −2 −1 0 1 2 3 4
43. 3 4 5 6 7 8 9 10 11 12 13

38. $|x - 2| = 4$
39. $|x + 1| = 3$
40. $|x| \geq 2$
41. $|x - 1| \leq 1$
42. $|x + 1| < 3$
43. $|x - 8| \geq 3$

44. Find all integer solutions of $|x| < 4$. **{−3, −2, −1, 0, 1, 2, 3}**
45. Find all integer solutions of $|x| \leq 2$. **{−2, −1, 0, 1, 2}**
46. If $a > 0$, how many integer solutions exist for $|x| < a$? **$2a - 1$**
47. If $a > 0$, how many integer solutions exist for $|x| \leq a$? **$2a + 1$**

Critical Thinking

48. Solve $|y - 3| = |2 + y|$. **$\left\{\frac{1}{2}\right\}$**
49. Under what conditions is $-|a|$ negative? positive? **$a \neq 0$; never**
50. Suppose $8 \leq x \leq 12$. Write an absolute value inequality that is equivalent to this compound inequality. **$|x - 10| \leq 2$**

Applications and Problem Solving

51. **Probability** Suppose $|x| \leq 6$ and x is an integer. Find the probability of $|x|$ being a factor of 18. **$\frac{8}{13}$ or 0.61**
52. **Chemistry** For hydrogen to be a liquid, its temperature must be within 2°C of −257°C. What is the range of temperatures for this substance to remain a liquid? **−259°C < t < 255°C**

Additional Answers

28. 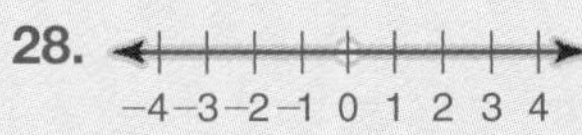
−4 −3 −2 −1 0 1 2 3 4

29. 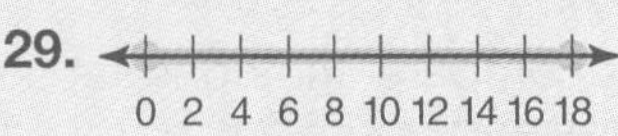
0 2 4 6 8 10 12 14 16 18

30. ∅

31. −4 −3 −2 −1 0 1 2 3 4

32. −4 −3 −2 −1 0 1 2 3 4

33. −4 −3 −2 −1 0 1 2 3 4 5 6

34. 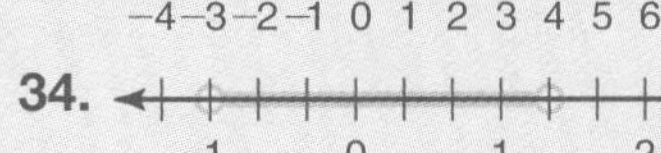
−1 0 1 2

53. **Law Enforcement** A radar gun used to determine the speed of passing cars must be within 7 mph of the actual speed of a selected car. If a highway patrol officer reads a speed of 59 mph for a car, does he have irrefutable evidence that the car was speeding in a 55 mph zone? Explain your reasoning and include a graph. **no; $52 \leq s \leq 66$**

54. **Space Exploration** Refer to the application at the beginning of the lesson. Find how much time can elapse from the beginning of the countdown to remain within the launch window.

55. **Entertainment** Luis Gomez is a contestant on the *Price is Right*. He must guess within $1500 of the actual price of a Jeep Cherokee without going over in order to win the vehicle. The actual price of the Jeep is $18,000. What is the range of guesses in which Luis can win the vehicle? **$\$16,500 \leq p \leq \$18,000$**

54. $297.5 \leq t \leq 302.5$ minutes

56. **Spending** The graph below shows the spending power of kids aged 3 to 17.

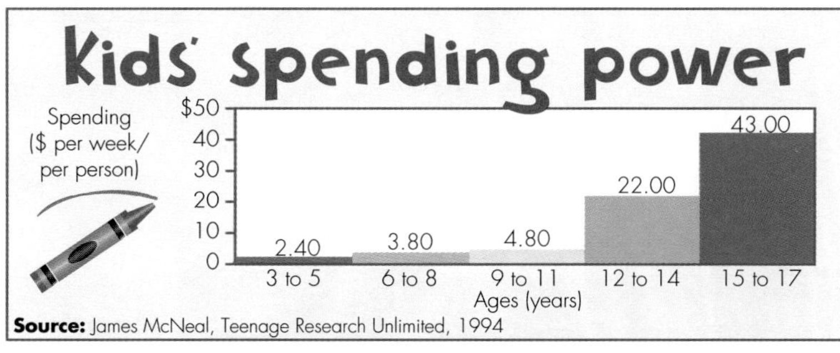

a. Write an inequality that represents the spending power of 3- to 17-year olds. Then write an absolute value inequality that describes their spending. **$2.40 \leq s \leq 43.00$; $|s - 22.70| \leq 20.30$**

b. Write an inequality that represents the spending power of 12- to 17-year olds. Then write an absolute value inequality that describes their spending. **$22 \leq s \leq 43$; $|s - 32.50| \leq 10.50$**

c. Keep a record of how much money you spend in a week. How does your spending compare with the data in this graph? **See students' work.**

Mixed Review

57. **Draw a Diagram** The Sanchez family acts as a host family for foreign exchange students during each quarter of the year. Suppose it is equally likely that they get a boy or a girl each quarter. (Lesson 7–5)

a. Draw a tree diagram to represent the possible orders of boys (B) and girls (G) during the four quarters of the year. List the possible outcomes. **See margin.**

b. From the tree diagram, what is the probability that all of the students will be girls? boys?

c. From the tree diagram, what is the probability that they will host two boys and two girls?

57b. $\frac{1}{16}$ or 0.0625, regardless of gender

57c. $\frac{3}{8}$ or 0.375

Additional Answer

57a. See student's work for diagram. List of outcomes: BBBB, BBBG, BBGB, BBGG, BGBB, BGBG, BGGB,BGGG, GBBB, GBBG, GBGB, GBGG, GGBB, GGBG, GGGB, and GGGG.

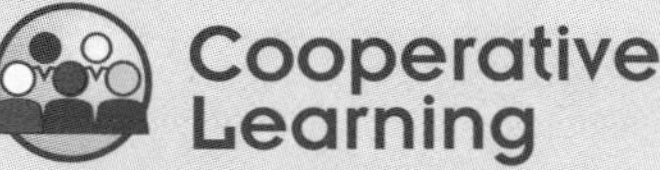

Cooperative Learning

Send-A-Problem Separate the class into four or five small groups. Have each group member create an inequality and then graph its solution set. Have groups switch graphs and identify the inequality that each solution set solves. Have groups return papers and verify each other's answers. For more information on the send-a-problem strategy, see *Cooperative Learning in the Mathematics Classroom*, one of the titles in the Glencoe Mathematics Professional Series, pages 23–24.

Practice Masters, p. 53

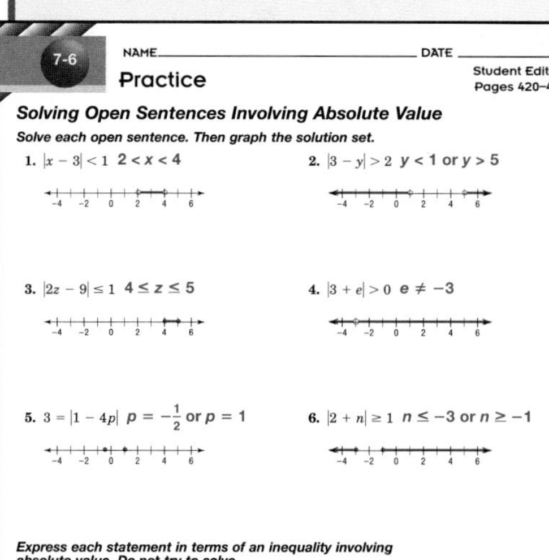

7-6 NAME______ DATE______ Student Edition Pages 420–426

Practice

Solving Open Sentences Involving Absolute Value

Solve each open sentence. Then graph the solution set.

1. $|x - 3| < 1$ $2 < x < 4$
2. $|3 - y| > 2$ $y < 1$ or $y > 5$
3. $|2z - 9| \leq 1$ $4 \leq z \leq 5$
4. $|3 + e| > 0$ $e \neq -3$
5. $3 = |1 - 4p|$ $p = -\frac{1}{2}$ or $p = 1$
6. $|2 + n| \geq 1$ $n \leq -3$ or $n \geq -1$

Express each statement in terms of an inequality involving absolute value. Do not try to solve.

7. Mary's bowling score was within 10 pins of her average score of 105. $|x - 105| < 10$
8. The temperature was at least 7 Fahrenheit degrees from 0°F. $|t| \geq 7$

For each graph, write an open sentence involving absolute value.

9. $|x| = 3$
10. $|x + 2| \geq 1$

4 ASSESS

Closing Activity

Speaking Have students explain what must be true of $|m|$ if

a. $m > 5$. $|m| > 5$

b. $m < -5$. $|m| > 5$

Chapter 7, Quiz C (Lessons 7-5 and 7-6), is available in the *Assessment and Evaluation Masters*, p. 185.

Additional Answer

63.

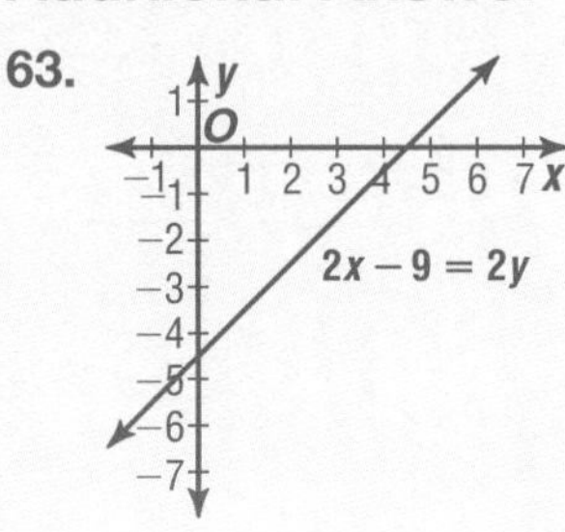

Enrichment Masters, p. 53

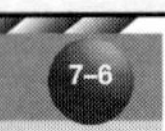

NAME ____________ DATE ________

Enrichment — Student Edition Pages 420–426

Consecutive Integer Problems

Many types of problems and puzzles involve the idea of consecutive integers. Here is an example.

Find four consecutive odd integers whose sum is −80.

An odd integer can be written as $2n + 1$, where n is any of the numbers 0, 1, 2, 3, and so on. Then, the equation for the problem is as follows.

$$(2n + 1) + (2n + 3) + (2n + 5) + (2n + 7) = -80$$

Solve these problems. Write an equation or inequality for each.

1. Complete the solution to the problem in the example. $n = -12$; The integers are −23, −21, −19, −17.
2. Find three consecutive even integers whose sum is 132. $2n + (2n + 2) + (2n + 4) = 132$; $n = 21$; Integers are 42, 44, 46.
3. Find the two least consecutive integers whose sum is greater than 20. $n + (n + 1) > 20$; $n > 9.5$; Integers are 10 and 11.
4. Find the two greatest consecutive integers whose sum is less than 100. $n + (n + 1) < 100$; $n < 49.5$; Integers are 49 and 50.
5. The lesser of two consecutive even integers is 10 more than one-half the greater. Find the integers. $2n = 10 + \frac{1}{2}(2n + 2)$; $n = 11$; Integers are 22 and 24.
6. The greater of two consecutive even integers is 6 less than three times the lesser. Find the integers. $2n + 2 = 3(2n) - 6$; $n = 2$; Integers are 4 and 6.
7. Find four consecutive integers such that twice the sum of the two greater integers exceeds three times the first by 91. $2[(n + 2) + (n + 3)] = 3n + 91$; $n = 81$; Integers are 81, 82, 83, 84.
8. Find all sets of four consecutive positive integers such that the greatest integer in the set is greater than twice the least integer in the set. $n + 3 > 2n$; $n < 3$; The two sets are {1, 2, 3, 4} and {2, 3, 4, 5}.

58. $\$1932 < p < \$2500$

60. $\{t \mid t > -36\}$

61. $\{k \mid k \geq -15\}$

58. Peter wants to buy Crystal an engagement ring. He wants to spend between \$1700 and \$2200. If he goes shopping during a 12%-off sale to celebrate the store's 12th anniversary, what would his price range be? (Lesson 7–4)

59. Solve $10x - 2 \geq 4(x - 2)$. (Lesson 7–3) $\{x \mid x \geq -1\}$

Solve each inequality.

60. $396 > -11t$ (Lesson 7–2)

61. $-11 \leq k - (-4)$ (Lesson 7–1)

62. Find the coordinates of the midpoint of the line segment whose endpoints are at (−4, 1) and (10, 3). (Lesson 6–7) **(3, 2)**

63. Graph $2x - 9 = 2y$. (Lesson 5–4) **See margin.**

64. Architecture Julie is making a model of a house for her drafting class. One part of the house has a triangle like the one below. She wants the base of the triangle to measure 2 inches. What will be the measures of the other two sides? (Lesson 4–2) **1.2 in.; 1.6 in.**

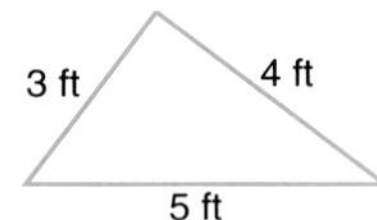

65. Solve $2m + \frac{3}{4}n = \frac{1}{2}m - 9$ for m. (Lesson 3–6) $m = -6 - \frac{n}{2}$

66. Simplify $\frac{-42r + 18}{3}$. (Lesson 2–7) $-14r + 6$

WORKING ON THE In•ves•ti•ga•tion

Refer to the Investigation on pages 320–321.

Smoke Gets In Your Eyes

Smoking increases the chance of health problems. Constant exposure to secondhand smoke also puts the nonsmoker at risk. According to a statement from a tobacco company, a nonsmoker working among smoking coworkers inhales the smoke of 1.25 cigarettes per month. A restaurant worker in the smoking section breathes just 2 cigarettes per month. The Environmental Protection Agency (EPA) disagrees with these data.

1. Write an equation to express the equivalent number of cigarettes one would breathe if he/she worked in the smoking section of a restaurant for one year.
2. A lawsuit involving the tobacco industry and the EPA states that the EPA did not use the standard significance level of 5% in evaluating their data from various studies. A 5% level means that their numerical conclusions have a 5% margin of error. How does the margin of error relate to absolute value?
3. In January, 1995, the average price of a pack of cigarettes was \$2.06, of which 56¢ is federal and state excise tax. The price of cigarettes is on the rise. Write an inequality to represent how much is spent by the average smoker in a year. Solve the inequality.
4. How much tax revenue is generated by the sale of cigarettes to the estimated number of smokers in the United States?

Add the results of your work to your Investigation Folder.

Extension

Problem Solving Find all integer values of x for which $|x - 2| < x$. $x > 1$

In•ves•ti•ga•tion

Working on the Investigation

The Investigation on pages 320–321 is designed to be a long-term project that is completed over several days or weeks. Encourage students to keep their materials in their Investigation Folder as they work on the Investigation.

7-7 Integration: Statistics Box-and-Whisker Plots

What YOU'LL LEARN

- To display and interpret data on box-and-whisker plots.

Why IT'S IMPORTANT

Box-and-whisker plots are a useful way to display data. They allow you to see important characteristics of the data at a glance.

APPLICATION

Olympics

In 1992, the site of the Summer Games of the XXV Olympiad was Barcelona, Spain. More than 14,000 athletes from 172 nations competed for medals in 257 events. The table at the right shows the number of gold medals won by the top 16 medal-winning teams.

1992 SUMMER OLYMPICS Top 16 Medal Winners

Team	Number of Medals Won: Gold	Total
Unified Team (formerly USSR)	45	112
USA	37	108
Germany	33	82
China	16	54
Cuba	14	31
Hungary	11	30
South Korea	12	29
France	8	29
Australia	7	27
Spain	13	22
Japan	3	22
Britain	5	20
Italy	6	19
Poland	3	19
Canada	6	18
Romania	4	18

Source: *The World Almanac, 1995*

We can describe these data using the mean, median, and mode. We can also use the median, along with the quartiles and interquartile range, to obtain a graphic representation of the data. A type of diagram, or graph, that shows quartiles and extreme values of data is called a **box-and-whisker plot.**

Box-and-whisker plots are sometimes called box plots.

Suppose we wanted to make a box-and-whisker plot of the numbers of gold medals won by each of the nations in the table. First, arrange the data in numerical order. Next, compute the median and quartiles. Also, identify the extreme values.

3 3 4 5 6 6 7 (8 ▲ 11) 12 13 14 16 33 37 45

median (Q2)

LOOK BACK

You can refer to Lesson 3-7 for information on measures of central tendency and Lesson 5-7 for measures of variation.

The median for this set of data is the average of the eighth and ninth values.

$$\text{median} = \frac{8 + 11}{2} \text{ or } 9.5$$

Recall that the *lower quartile* (Q1) is the median of the lower half of the distribution of values. The *upper quartile* (Q3) is the median of the upper half of the data.

The median is not included in either half of the data.

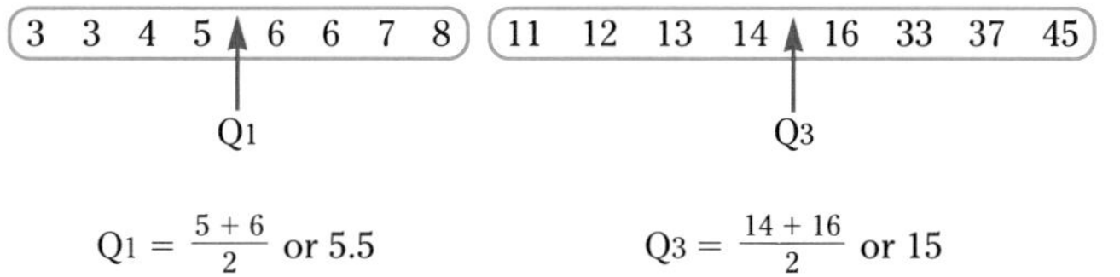

$$Q1 = \frac{5 + 6}{2} \text{ or } 5.5 \qquad Q3 = \frac{14 + 16}{2} \text{ or } 15$$

The **extreme values** are the least value (LV), 3, and the greatest value (GV), 45.

7-7 LESSON NOTES

NCTM Standards: 1–4, 10

Instructional Resources

- Study Guide Master 7-7
- Practice Master 7-7
- Enrichment Master 7-7
- Science and Math Lab Manual, pp. 29–34

Transparency 7-7A contains the 5-Minute Check for this lesson; **Transparency 7-7B** contains a teaching aid for this lesson.

Recommended Pacing	
Standard Pacing	Day 9 of 14
Honors Pacing	Day 8 of 13
Block Scheduling*	Day 5 of 7
Alg. 1 in Two Years*	Days 13 & 14 of 21

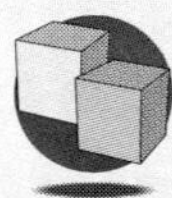

*For more information on pacing and possible lesson plans, refer to the *Block Scheduling Booklet* and *Algebra 1 in Two Years.*

1 FOCUS

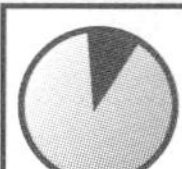

5-Minute Check *(over Lesson 7-6)*

Solve each open sentence.

1. $|e - 5| = -3$ **∅**
2. $|a - 1| < 4$ **$\{a \mid -3 < a < 5\}$**
3. $|b + 8| \geq 1$ **$\{b \mid b \leq -9 \text{ or } b \geq -7\}$**
4. $|9 - c| \geq 13$ **$\{c \mid c \geq 22 \text{ or } c \leq -4\}$**
5. $|3d - 12| < 12$ **$\{d \mid 0 < d < 8\}$**

Motivating the Lesson

Hands-On Activity Reinforce student understanding of the terms *median, upper quartile,* and *lower quartile* with the following activity. Ask students to arrange themselves in a row in ascending height. Have them determine which class member has the median height. Then ask students to identify the class member who represents the upper quartile and the class member who represents the lower quartile.

2 TEACH

Teaching Tip Remind students that items of data that occur more than once must appear the same number of times when arranging the data in ascending order.

Technology Tip Point out that graphing calculators do not identify outliers.

Now we have the information we need to draw a box-and-whisker plot.

Step 1 Draw a number line. Assign a scale to the number line that includes the extreme values. Plot dots to represent the extreme values (LV and GV), the upper and lower quartile points (Q_3 and Q_1), and the median (Q_2).

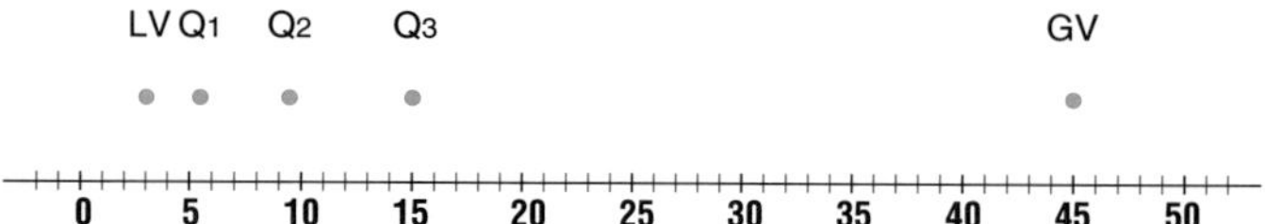

The median line will not always divide the box into equal parts.

Step 2 Draw a box to designate the data falling between the upper and lower quartiles. Draw a vertical line through the point representing the median. Draw a segment from the lower quartile to the least value and one from the upper quartile to the greatest value. These segments are the **whiskers** of the plot.

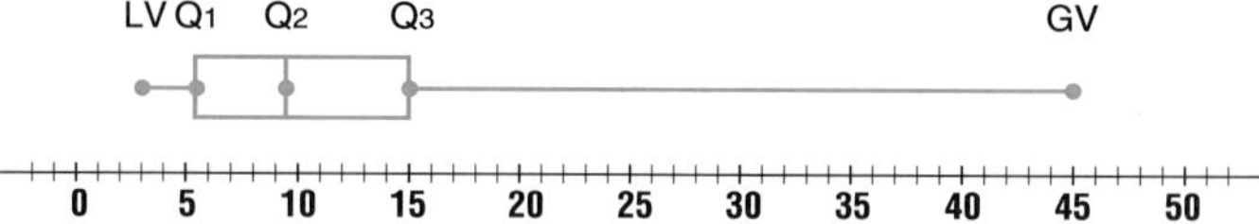

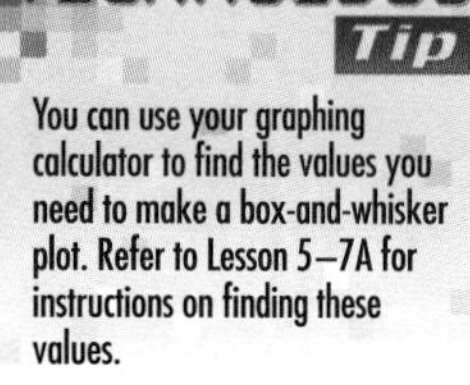

You can use your graphing calculator to find the values you need to make a box-and-whisker plot. Refer to Lesson 5–7A for instructions on finding these values.

Even though the whiskers are different lengths, each whisker contains at least one fourth of the data while the box contains one half of the data. Compound inequalities can be used to describe the data in each fourth. Assume that the replacement set for x is the set of data.

1st fourth	$\{x \mid x < 5.5\}$
2nd fourth	$\{x \mid 5.5 < x < 9.5\}$
3rd fourth	$\{x \mid 9.5 < x < 15\}$
4th fourth	$\{x \mid x > 15\}$

Step 3 Before finishing the box-and-whisker plot, check for outliers. In Lesson 5–7, you learned that an outlier is any element of the set of data that is at least 1.5 interquartile ranges above the upper quartile or below the lower quartile. Recall that the *interquartile range* (IQR) is the difference between the upper and lower quartiles, or in this case, $15 - 5.5$ or 9.5.

$x \geq Q_3 + 1.5(\text{IQR})$	or	$x \leq Q_1 - 1.5(\text{IQR})$
$x \geq 15 + 1.5(9.5)$		$x \leq 5.5 - 1.5(9.5)$
$x \geq 15 + 14.25$		$x \leq 5.5 - 14.25$
$x \geq 29.25$		$x \leq -8.75$

Step 4 If x is an outlier in this set of data, then the outliers can be described as $\{x \mid x \leq -8.75 \text{ or } x \geq 29.5\}$. In this case, there are no data less than -8.75. However, 45, 37, and 33 are greater than 29.25, so they are outliers. We now need to revise the box-and-whisker plot. Outliers are plotted as isolated points, and the right whisker is shortened to stop at 16.

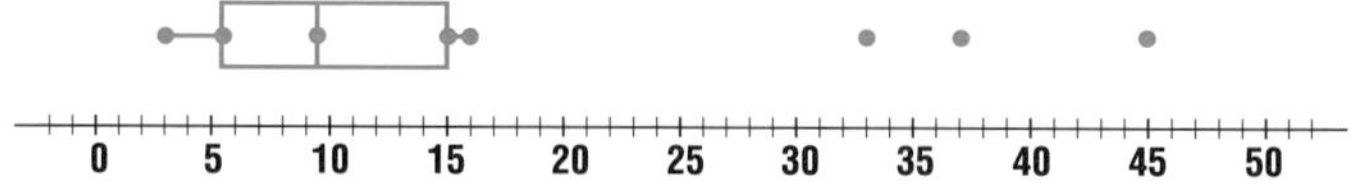

GLENCOE Technology

Interactive Mathematics Tools Software

This multimedia software provides an interactive lesson by helping students observe the effect of changing data on box-and-whisker plots. A **Computer Journal** gives students an opportunity to write about what they have learned.

For Windows & Macintosh

Example

APPLICATION
Sports

Refer to the application at the beginning of the lesson. Use the box-and-whisker plot for the gold medals to answer each question.

a. What percent of the teams won between 6 and 15 gold medals?

b. What does the box-and-whisker plot tell us about the upper half of the data compared to the lower half?

a. The box in the plot indicates 50% of the values in the distribution. Since the box goes from 5.5 to 15, we know that 50% of the teams won between 6 and 15 gold medals.

b. The upper half of the data is spread out while the lower half is fairly clustered together.

CHECK FOR UNDERSTANDING

Communicating Mathematics

Study the lesson. Then complete the following.

1. **Explain** how to determine the scale of the number line in a box-and-whisker plot. **Scale must be large enough to include the least and greatest values.**
2. **Describe** which two points the two whiskers of a box-and-whisker plot connect. **LV and Q_1; Q_3 and GV**
3. What does Q_2 represent? **the median**
4. Refer to the box-and-whisker plot at the right. Assume that LV, Q_1, Q_2, Q_3, and GV are whole numbers.
   a. What percent of the data is between 85 and 90? **25%**
   b. Between what two values does the middle 50% of the data lie? **90–120**
   c. What outliers are represented in the box-and-whisker plot? **none**
5. **Describe** what characteristics of a set of data you can gather from its box-and-whisker plot. What are some things you cannot gather from a box-and-whisker plot?

5. Sample answer: quartiles, interquartile range, outliers, whether the data are clustered or diverse; individual point of data, number of data

Guided Practice

6. The table at the right shows the birthrates for 15 selected countries.
   a. Find the median, upper quartile, lower quartile, and interquartile range for each year's data.
   b. Are there any outliers? If so, name them. **21.5, 23.3**
   c. Draw a box-and-whisker plot for each set of data on the same number line. **See margin.**
   d. Which set of data seems to be more clustered? Why? **1992; because the plot is not as wide.**

6a. 1985: 14.0, 18.0, 11.9, 6.1; 1992: 13.1, 15.7, 11.9, 3.8

Birth Rates for Selected Countries (per 1000 population)

Country	1985	1992
Australia	15.7	15.1
Cuba	18.0	14.5
Denmark	10.6	13.1
France	13.9	12.9
Hong Kong	14.0	11.9
Israel	23.5	21.5
Italy	10.1	9.9
Japan	11.9	9.7
Netherlands	12.3	13.0
Panama	26.6	23.3
Poland	18.2	13.4
Portugal	12.8	11.4
Singapore	16.6	17.7
Switzerland	11.6	12.6
United States	15.7	15.7

Source: United Nations, *Monthly Bulletin of Statistics*, May 1994

In-Class Example

For the Example
Identify the extreme values and quartiles and draw a box-and-whisker plot for the following data: 26, 42, 57, 12, 38, 63, 45, 18.
LV = 12, GV = 63, median = 40, Q_3 = 51, Q_1 = 22

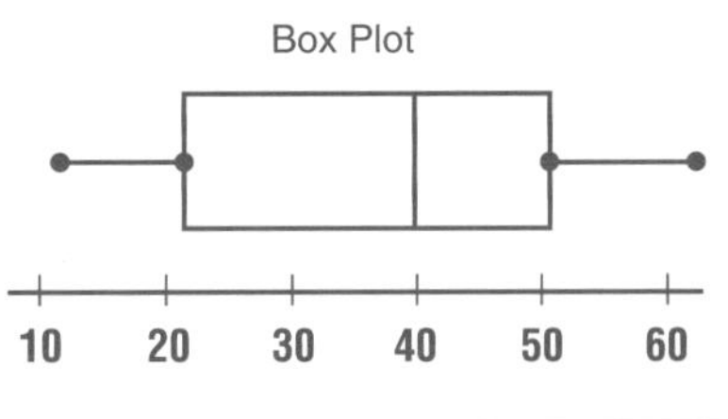

3 PRACTICE/APPLY

Check for Understanding

Exercises 1–7 are designed to help you assess your students' understanding through reading, writing, speaking, and modeling. You should work through Exercises 1–5 with your students and then monitor their work on Exercises 6–7.

Additional Answer

6c.

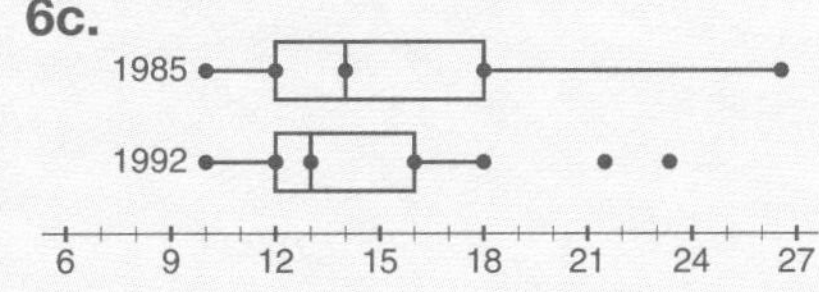

Reteaching

Using Class Discussion Give each row of students a different set of data for a box-and-whisker plot. Have each student draw his or her own plot on an overhead transparency. Put the plots on the overhead so that the class can see them. Ask questions about the plots to see if they were drawn correctly. Discuss problems and good points of the plots.

Assignment Guide

Core: 9–15 odd, 16–20
Enriched: 8–20

For **Extra Practice,** see p. 773.

The red A, B, and C flags, printed only in the Teacher's Wraparound Edition, indicate the level of difficulty of the exercises.

Additional Answers

9c.

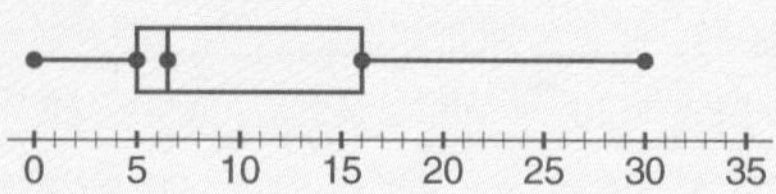

10.

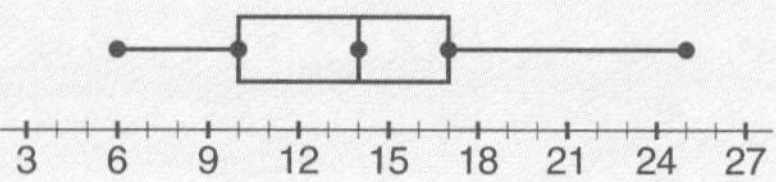

Study Guide Masters, p. 54

NAME____________ DATE______

Study Guide

Student Edition Pages 427–432

Integration: Statistics Box-and-Whisker Plots

A box-and-whisker plot displays the extremes, the quartiles, and the median for a set of data. It may be drawn horizontally or vertically.

Study the box at the right. Note the dots marking the extremes, the quartiles, and the median.

Data: 3, 7, 9, 14, 16, 19, 19, 25

Stem	Leaf
0	3 7 9
1	4 6 9 9
2	5

1|6 = 16
Interquartile range is 19 – 8, or 11.

Box-and-Whisker Plot:
Lower extreme is 3.
Lower quartile is 8.
Median is 15.
Upper quartile is 19.
Upper extreme is 25.

Answer each question for the indicated box-and-whisker plot.

1. What are the upper and lower extremes? 4; 16
2. What is the median? 12
3. What are the upper and lower quartiles? 15; 10
4. What is the interquartile range? 5

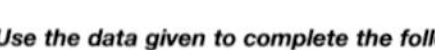

Use the data given to complete the following. Data: 19, 49, 73, 30, 32, 46, 51, 30

5. Graph the data on a stem-and-leaf plot.

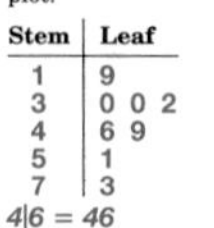

Stem	Leaf
1	9
3	0 0 2
4	6 9
5	1
7	3

4|6 = 46

6. Transfer the data to a vertical box-and-whisker plot.

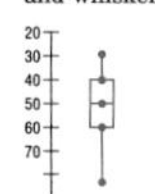

7. What are the extremes? 19; 73
8. What is the interquartile range? 20
9. Why are the whiskers unequal? The difference between the LQ and the LV is less than the difference between the GV and the UQ.
10. How is an outlier determined? Are there any outlier values? Multiply 1.5 times the interquartile range, 20. Then add this product to the UQ, 50, and subtract it from the LQ, 30. There are no outliers.

7a. A: 25, 65, 30, 60, 40; B: 20, 70, 40, 60, 45

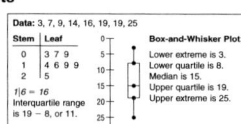

7. Refer to box-and-whisker plots A and B.

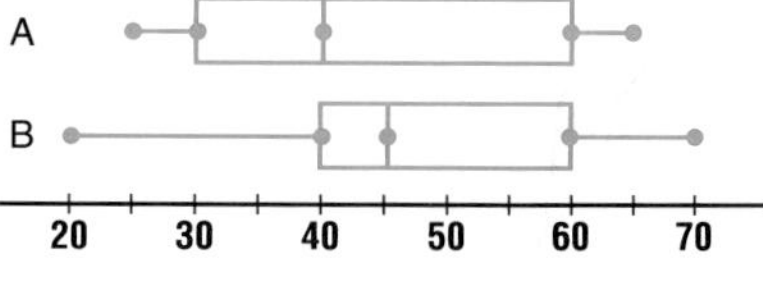

a. Estimate the least value, greatest value, lower quartile, upper quartile, and median for each plot. Assume that these values are whole numbers.
b. Which set of data contains the least value? **B**
c. Which plot has the greatest interquartile range? **A**
d. Which plot has the greatest range? **B**

EXERCISES

Applications and Problem Solving

8. **Manufacturing** The box-and-whisker plots at the right show the results of testing the useful life of 10 light bulbs each from two manufacturers.

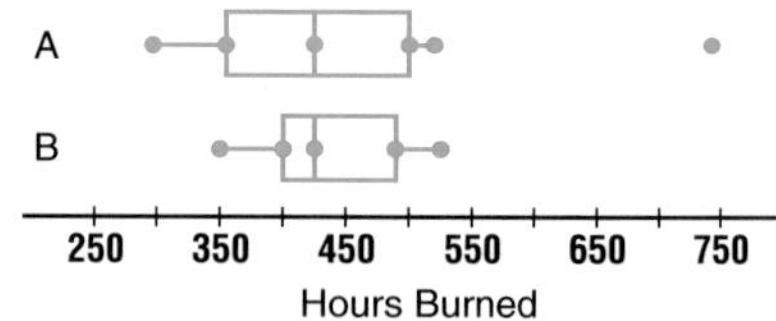

a. Which test had the most varied results? **A**

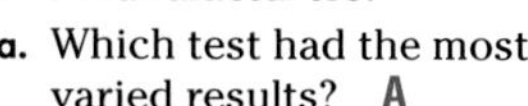

b. Were there any outliers? If so, for which brand? **yes; A**
c. How would you compare the medians of the tests? **about the same**
d. Based on this plot, from which manufacturer would you buy your light bulbs? Why? **B; The lives of the bulbs were more consistent.**

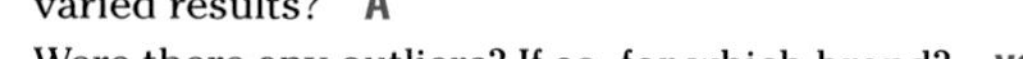

9. **Meteorology** Meteorologists keep track of temperatures for four 90-day periods during the year to predict the trends in weather for future years. The following low temperatures were recorded during a 2-week cold snap in Indianapolis during 1993. **9a. $Q_2 = 6.5$, $Q_3 = 16$, $Q_1 = 5$, IQR = 11**

30°, 20°, 2°, 12°, 5°, 4°, 17°, 7°, 6°, 16°, 5°, 0°, 5°, 16°

a. Find the median, upper quartile, lower quartile, and interquartile range.
b. Are there any outliers? If so, name them. **no**
c. Draw a box-and-whisker plot of the data. **See margin.**

10. **Football** The table below shows the American Football Conference's leading quarterbacks in touchdowns for the 1993 season. Make a box-and-whisker plot for the data. **See margin.**

1993 AFC Individual Leaders in Passing

Player	Team	Touchdowns
Steve DeBerg	Miami	7
John Elway	Denver	25
Boomer Esiason	N.Y. Jets	16
John Friesz	San Diego	6
Jeff George	Indianapolis	8
Jeff Hostetler	L.A. Raiders	14
Jim Kelly	Buffalo	18
Scott Mitchell	Miami	12
Joe Montana	Kansas City	13
Warren Moon	Houston	21
Neil O'Donnell	Pittsburgh	14
Vinny Testaverde	Cleveland	14

Source: *The World Almanac*, 1995

B **11. Demographics** According to the 1990 Census, the American Indian population in the United States is 1.959 million. Many American Indian people live on reservations or trust lands. The stem-and-leaf plot shows the number of reservations in the 34 states that have them.

Stem	Leaf
0	1 1 1 1 1 1 1 1 1 1 1 1 2
•	3 3 3 3 3 4 4 4 4 7 7 8 8 9
1	1 4 9
2	3 5 7
9	6 9|6 = 96

11c. There are four western states that have more American Indian people than other states.

a. Make a box-and-whisker plot of these data. **See margin.**

b. Describe the distribution of the data. **clustered with lots of outliers**

c. Why do you think there are so many outliers?

d. The mean of these data is about 8.8. How does this compare with the median? **It is greater than the median.**

12. Environment The table below shows the number of hazardous waste sites in 25 states.

Hazardous Waste Sites in the United States (selected states)									
State	No.	State	No.	State	No.	State	No.	State	No.
AL	13	FL	57	LA	13	OH	38	TN	17
AR	12	GA	13	MD	13	OK	11	TX	30
CA	96	IL	37	MI	77	OR	12	VA	25
CO	18	IN	33	NY	85	PA	101	WA	56
CT	16	KY	20	NC	22	SC	24	WV	6

Source: Environmental Protection Agency, May 1994

a. Make a box-and-whisker plot of the data. **See margin.**

b. What is the median number of waste sites for the states listed? **22**

c. Which states, if any, are outliers? **PA**

d. Which half of the data is more widely dispersed? **the upper half**

13. Education The graph below shows the average American College Testing (ACT) Program mathematics scores for students from 1985–1993.

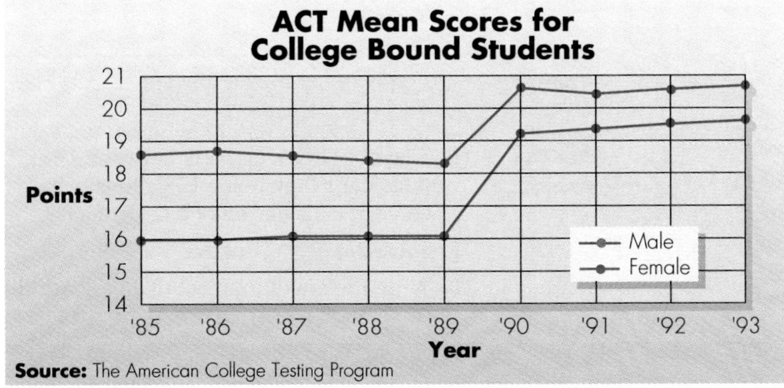

Source: The American College Testing Program

(continued on the next page)

Additional Answers

11a.

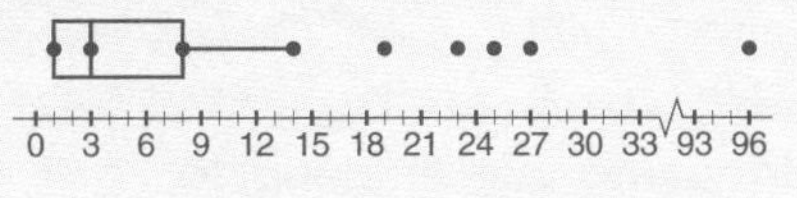

12a.

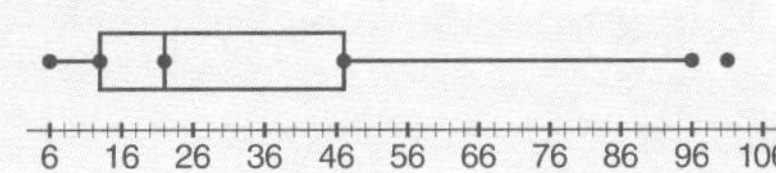

Practice Masters, p. 54

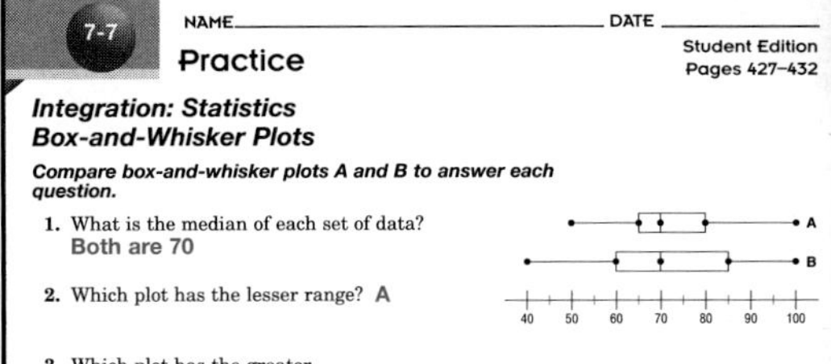

7-7 NAME________ DATE ________

Practice — Student Edition Pages 427–432

Integration: Statistics Box-and-Whisker Plots

Compare box-and-whisker plots A and B to answer each question.

1. What is the median of each set of data? Both are 70
2. Which plot has the lesser range? A
3. Which plot has the greater interquartile range? B
4. What is the upper quartile of each set of data? A: 80; B: 85
5. What is the lower quartile of each set of data? A: 65; B: 60
6. What is the least value in plot A? 50
7. What is the greatest value in plot B? 100
8. Which plot illustrates the larger range of data? B
9. What percent of the data in plot B is between 60 and 85? 50%
10. What percent of the data in plot A is greater than 80? 25%
11. What percent of the data in plot A is less than 65? 25%
12. The numbers below represent forty eighteen-hole scores of a golfer at Crystal Springs Golf Club. Make a box-and-whisker plot of this data in the space provided.

88	80	80	77	84	74	80	79	83	77
75	75	78	79	77	81	72	85	75	76
78	76	77	72	75	70	78	77	72	79
78	75	79	77	77	73	77	78	82	86

70 75 80 85 90

13. What is the median score? 77
14. What percent of the scores were higher than 75? 75%

4 ASSESS

Closing Activity

Speaking Ask students to describe the steps to follow when making a box-and-whisker plot.

Additional Answers

13a.

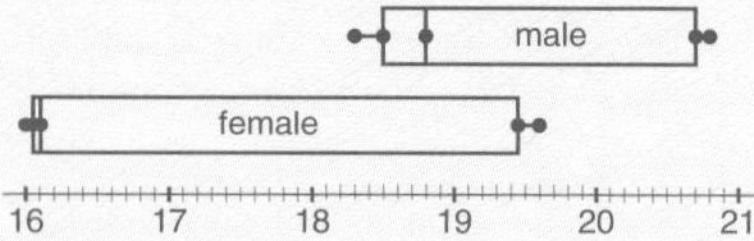

Male data are more condensed and generally higher than female data.

14a–b.

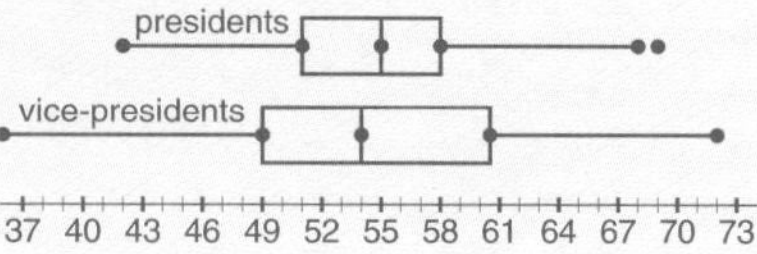

19.

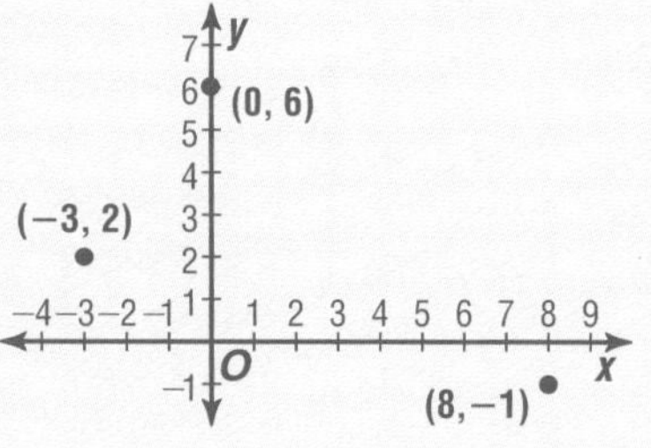

Enrichment Masters, p. 54

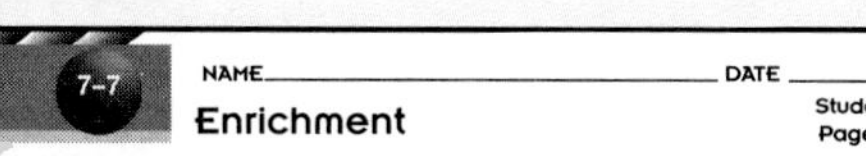

Curious Circles

Two circles can be arranged in four ways: one circle can be inside the other, they can be separate, they can overlap, or they can coincide.

In how many ways can a given number of circles be either separate or inside each other? (The situations in which the circles overlap or coincide are not counted here.)

Here is the answer for 3 circles. There are 4 different possibilities.

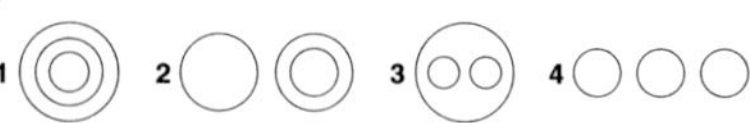

Solve each problem. Make drawings to show your answers.

1. Show the different ways in which 2 circles can be separate or inside each other. How many ways are there? two ways

2. Show the different ways for 4 circles. How many ways are there? nine ways

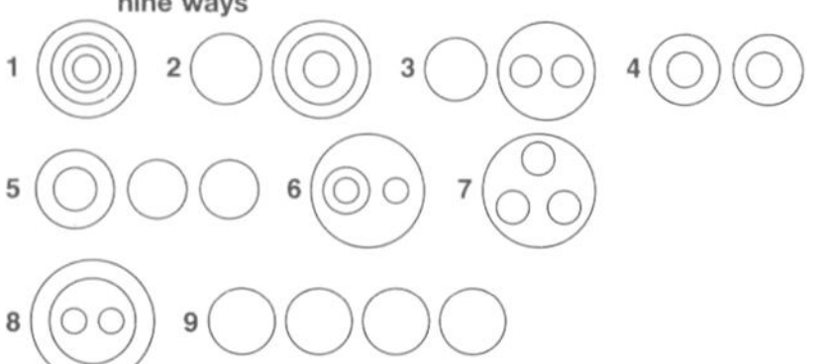

3. Use your answer for Exercise 2 to show that the number of ways for 5 circles is at least 18. see below

4. Find the number of ways for 5 circles. Show your drawings on a separate sheet of paper. 20 ways

3. First, draw an extra circle next to each of the ways for 4 circles. Then draw a circle around each of the ways for 4 circles.

a. Make a box-and-whisker plot of the scores for the male students and another for the female students using the same scale. Compare the plots. **See margin.**

b. In a particular year, an entirely new ACT Assessment was given that emphasized rhetorical skills, advanced mathematics items, and a new reading test. From the data on the graph, in which year do you think this occurred? Why? **1990; See students' work.**

14. History Did you know that there have been more U.S. vice-presidents than presidents? As of 1995, there have been 41 presidents and 45 vice-presidents. Some presidents had more than one vice-president, and some had none. The line plot below shows the ages of vice-presidents on their inauguration days.

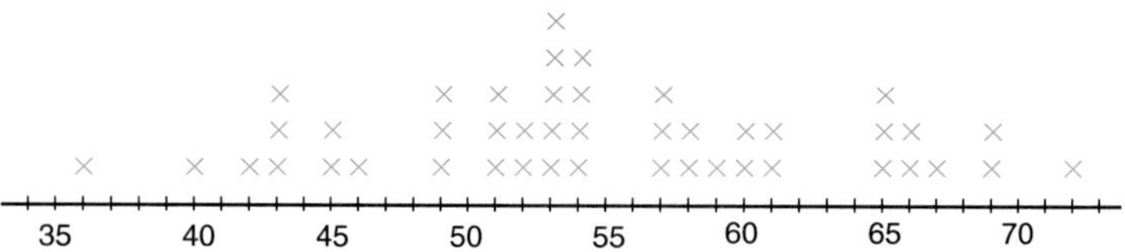

Source: *The World Almanac, 1995*

14a–b. See margin.

a. Make a box-and-whisker plot of these data.

b. The ages for presidents on their inauguration days have the following statistics: LV = 42, Q1 = 51, Q2 = 55, Q3 = 58, GV = 68, and 69 is an outlier. Make a box-and-whisker plot to represent these ages.

c. Which set of data is more clustered? Explain your answer.

14c. The presidents seem more clustered because the box portion, representing 50% of the data, is narrower.

d. Fourteen vice-presidents went on to become presidents. Which ages of vice-presidents were definitely not those who went on to become president? **ages 66–72, assuming no vice-president age 66 or 67 succeeded a president during the president's term**

Critical Thinking

15. The box-and-whisker plots shown below picture the distribution of test scores in two algebra classes taught by two different teachers. If you could select one of the classes to be in, which one would it be, and why?

15. Sample answer: Class A appears to be a more difficult class than B because the students don't do as well.

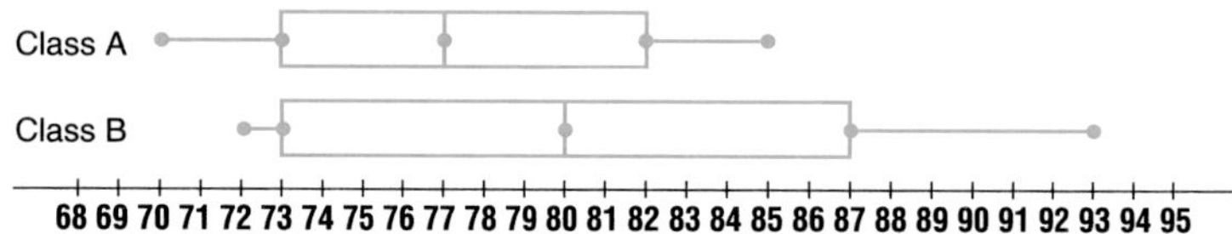

Mixed Review

16. Travel Greg's car gets between 18 and 21 miles per gallon of gasoline. If his car's tank holds 15 gallons, what is the range of distances that Greg can drive his car on one tank of gas? (Lesson 7–6) **270 to 315 miles**

17. Solve $2m - 3 > 7$ or $2m + 7 > 9$. (Lesson 7–4) $\{m \mid m > 1\}$

18. Write the standard form of an equation of the line that passes through (4, 7) and (1, −2). (Lesson 6–2) $3x - y = 5$

19. Graph (0, 6), (8, −1), and (−3, 2). (Lesson 5–1) **See margin.**

20. Solve $\frac{6}{x-3} = \frac{3}{4}$. (Lesson 4–1) **11**

Extension

Connections Have students research the median ages of men and women at the time of their first marriage for each decade from 1900 to the present time. Have students determine the mean, mode, quartiles, range, interquartile range, and any outliers for each set of data, and display the data using box-and-whisker plots. Then compare the plots.

7–7B Graphing Technology Box-and-Whisker Plots

An Extension of Lesson 7–7

You can use a graphing calculator to compare two sets of data by using a double box-and-whisker plot. The plots that the calculator draws, however, do not account for outliers. If you want to use a graphing calculator to help you sketch a plot, it will be necessary for you to check for outliers and adjust the graph as needed.

In an experiment, the ability of boys and girls to identify objects held in their left hands versus those held in their right hands was tested. The left side of the body is controlled by the right side of the brain and vice versa. The results of the experiment found that the boys did not identify objects with their right hand as well as with their left. The girls could identify objects equally as well with either hand.

Texas Instruments decided to test this premise by conducting the same tests with 12 male and 10 female employees, chosen at random. Thirty small objects were selected and separated into two groups—one for the right hand and one for the left hand. Blindfolded employees felt each of the objects with the prescribed hand and tried to identify them. The results are presented in the chart below.

LOOK BACK

For more information on entering data into lists on the graphing calculator, see Lesson 5-7A.

Each employee had a score for left and right hands.

Correct Responses			
Female Left Hand	Female Right Hand	Male Left Hand	Male Right Hand
8	4	7	12
9	3	8	6
12	7	7	12
11	12	5	12
10	11	7	7
8	11	8	11
12	13	11	12
7	12	4	8
9	11	10	12
11	12	14	11
		13	9
		5	9

Source: *TI-82 Graphics Calculator Guidebook*

7-7B LESSON NOTES

NCTM Standards: 1–5, 10

Objective
Use a graphing calculator to compare two sets of data by using a double box-and-whisker plot.

Recommended Time
25 minutes

Instructional Resources
Graphing Calculator Masters, pp. 24 and 25

These masters provide keystroking instruction for this lesson for the TI-81 and Casio graphing calculators.

1 FOCUS

Motivating the Lesson
Have students pair off. Select several objects that can be held in one hand. One student is to be blindfolded. The other student will place the objects in the blindfolded student's hand, one object at a time, alternating hands. The blindfolded student will try to guess what each object is. As a class, discuss whether there were any marked differences for the number of correct responses for each hand or for the results between boys and girls. Ask students what a box-and-whisker plot comparing left-hand and right-hand results might look like.

2 TEACH

Teaching Tip Point out to students that for such an experiment to be valid, many factors must be controlled—for example, the number of left-handed individuals, objects that are equally difficult to recognize, and so on.

3 PRACTICE/APPLY

Assignment Guide

Core: 1–6
Enriched: 1–6

4 ASSESS

Observing students working with technology is an excellent method of assessment.

Make a double box-and-whisker plot of the data to compare the results of Female Left Hand with Female Right Hand using a graphing calculator.

Step 1 Clear lists L1, L2, L3, and L4. Enter the data from each column of the table into lists L1, L2, L3, and L4, respectively.

Step 2 Select the box-and-whisker plot and define which list will be used.

Enter: [2nd] [STAT PLOT] 1 [ENTER] *Turns plot on.*

[▼] [▶] [▶] [ENTER] *Selects box-and-whisker plot.*

If L1 is not highlighted in the Xlist, use the down arrow and [ENTER] to highlight it and make sure the frequency is set for 1.

Repeat the process to assign Plot2 as a box-and-whisker plot using L2.

Step 3 Clear the Y= list. Set the WINDOW settings for Xscl = 1, Ymin = 0, and Yscl = 0. Ignore the other settings. Press [ZOOM] 9 to select ZoomStat. This sets the other settings and displays the box-and-whisker plots. It will only display those plots that you have turned on.

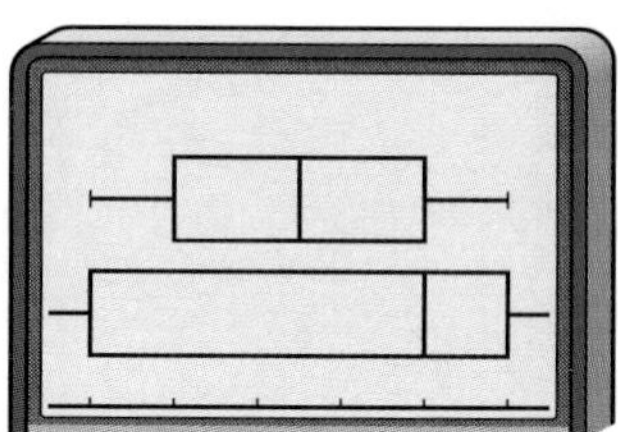

Step 4 Use [TRACE] to examine the minX (least value), Q1 (lower quartile), Med (median), Q3 (upper quartile), and maxX (greatest value).

EXERCISES

1. Which set of data does the upper plot represent? **1. women identifying objects with left hand**
2. Observe the two graphs. Does it appear that the females guessed correctly more often with the left hand or the right? How do you know? **2. With the left hand; the data are clustered.**
3. Reset your calculator to define Plot1 as L3 and Plot 2 as L4 to examine the males' data. What do you observe in these plots? **3. The left hand data are more clustered.**
4. Reset the calculator to compare the left-hand results of males and females. Were the males or females better at guessing with their left hands? **females**
5. Reset the calculator to compare the right-hand results of males and females. Which group seemed more adept at identifying objects with their right hands? **males**
6. How do the results of this experiment compare with the study of boys and girls mentioned at the beginning of this lesson? What reasons may account for any discrepancies? **6. It does not agree. Sample reason: Adults may have better recognition skills or there may have been a large number of left-handed people in sample.**

Using Technology

This lesson offers an excellent opportunity for using technology in your algebra classroom. For more information on using technology, see *Graphing Calculators in the Mathematics Classroom*, one of the titles in the Glencoe Mathematics Professional Series.

7–8A Graphing Technology Graphing Inequalities

A Preview of Lesson 7–8

Inequalities in two variables can be graphed on a graphing calculator using the "Shade(" command, which is option 7 on the DRAW menu. You must enter *two* functions to activate the shading since the calculator always shades between two specified functions. The first function entered defines the lower boundary of the region to be shaded. The second function defines the upper boundary of the region. The calculator graphs both functions and shades between the two.

Before using the "Shade(" option, be sure to clear any equations stored in the  *list, and press* ZOOM *6 for the standard viewing window.*

When finished, press  DRAW *1 to clear the screen.*

Example 1 **Graph $y \geq 2x - 3$ in the standard viewing window.**

The inequality refers to points at which y is *greater than or equal to* $2x - 3$. This means we want to shade above the graph of $y = 2x - 3$. Since the calculator screen shows only part of the coordinate plane, we can use the top of the screen, Ymax or 10, as the upper boundary and $2x - 3$ as the lower boundary.

Enter:  DRAW 7 2 

 3 , 10  

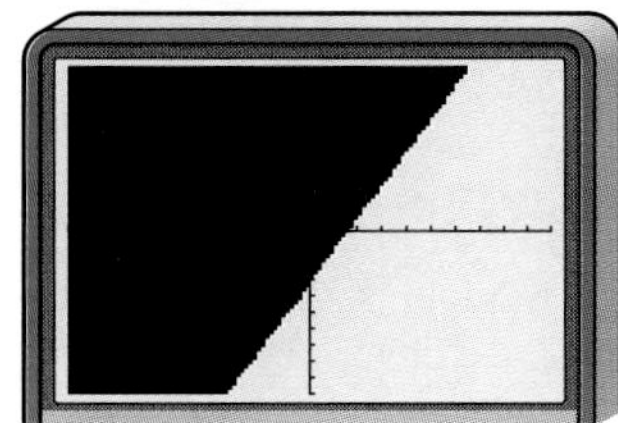

Since both the x- and y-intercepts of the graph and the origin are within the current viewing window, the graph of the inequality is complete.

Example 2 **Graph $y - x \leq 1$ in the standard viewing window.**

First solve the inequality for y: $y \leq x + 1$. This inequality refers to points where y is *less than or equal to* $x + 1$. This means we want to shade below the graph of $y = x + 1$. We can use the bottom of the screen, Ymin or -10, as the lower boundary and $x + 1$ as the upper boundary.

Enter:  DRAW 7 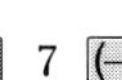 10 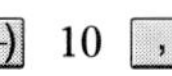

 + 1  ENTER

Don't forget to clear the screen when finished.

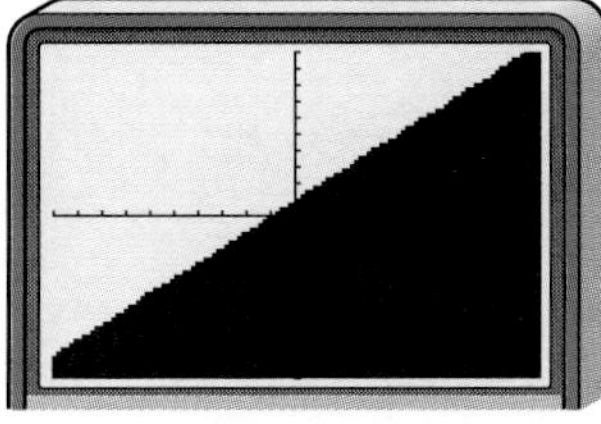

EXERCISES

Use a graphing calculator to graph each inequality. Sketch each graph on a sheet of paper. 1–9. See Solutions Manual.

1. $y \geq x + 2$	2. $y \leq -2x - 4$	3. $y + 1 \leq 0.5x$
4. $y \geq 4x$	5. $x + y \leq 0$	6. $2y + x \geq 4$
7. $3x + y \leq 18$	8. $y \geq 3$	9. $0.2x + 0.1y \leq 1$

7-8A LESSON NOTES

NCTM Standards: 1–6

Objective
Use a graphing calculator to graph inequalities in two variables.

Recommended Time
15 minutes

Instructional Resources
Graphing Calculator Masters, pp. 26 and 27

These masters provide keystroking instruction for this lesson for the TI-81 and Casio graphing calculators.

1 FOCUS

Motivating the Lesson
Which of the following ordered pairs, (3, 1), (5, 3), (−2, −6), (−1, −8), satisfy the inequality $2x - y \geq 5$?
(3, 1), (5, 3), (−1, −8)

2 TEACH

Teaching Tip Remind students to carefully enter the function at the appropriate time. To shade above a graph, the function must be entered as the first function. To shade below a graph, the function must be entered as the second function.

3 PRACTICE/APPLY

Assignment Guide

Core: 1–9
Enriched: 1–9

Using Technology
This lesson offers an excellent opportunity for using technology in your algebra classroom. for more information on using technology, see *Graphing Calculators in the Mathematics Classroom,* one of the titles in the Glencoe Mathematics Professional Series.

4 ASSESS

Observing students working with technology is an excellent method of assessment.

7-8 LESSON NOTES

NCTM Standards: 1–5

Instructional Resources

- Study Guide Master 7-8
- Practice Master 7-8
- Enrichment Master 7-8
- Assessment and Evaluation Masters, p. 185
- Multicultural Activity Masters, p. 14

Transparency 7-8A contains the 5-Minute Check for this lesson; **Transparency 7-8B** contains a teaching aid for this lesson.

Recommended Pacing	
Standard Pacing	Day 12 of 14
Honors Pacing	Day 11 of 13
Block Scheduling*	Day 6 of 7
Alg. 1 in Two Years*	Days 17 & 18 of 21

*For more information on pacing and possible lesson plans, refer to the *Block Scheduling Booklet* and *Algebra 1 in Two Years.*

1 FOCUS

5-Minute Check
(over Lesson 7-7)

Use the box-and-whisker plot shown below to answer the following questions.

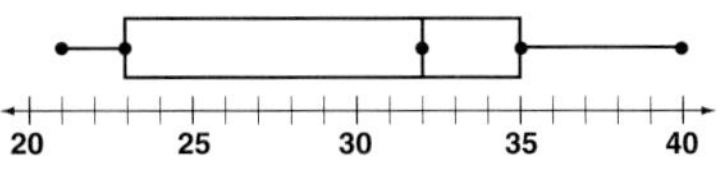

1. What is the lower extreme? **21**
2. What is the range? **19**
3. What is the median? **32**
4. What is the upper quartile? **35**
5. What is the interquartile range? **12**

7–8 Graphing Inequalities in Two Variables

What YOU'LL LEARN

- To graph inequalities in the coordinate plane.

Why IT'S IMPORTANT

You can graph inequalities to solve problems involving manufacturing and health.

Manufacturing

Rapid Cycle, Inc. is a manufacturer and distributor of racing bicycles. It takes 3 hours to assemble a bicycle and 1 hour to road test a bicycle. Each technician in the company works no more than 45 hours a week. How many racing bikes can one technician assemble, and how many can he or she road test in one week?

Let x represent the number of bikes that are assembled in a week, and let y represent the number of bikes that are road-tested. Then the following inequality can be used to represent the solution.

Total time to assemble x bikes	*plus*	*Total time to road test y bikes*	*is no more than*	*45 hours.*
$3x$	$+$	y	$\leq$	45

There are an infinite number of ordered pairs that are solutions to this inequality. The easiest way to show all of these solutions is to draw a *graph* of the inequality. Before doing this, let's consider some simpler inequalities.

This problem will be solved in Example 3.

Example  **From the set {(3, 4), (0, 1), (1, 4), (1, 1)}, which ordered pairs are part of the solution set for $4x + 2y < 8$?**

Let's use a table to substitute the x and y values of each ordered pair into the inequality.

x	y	$4x + 2y < 8$	True or False?
3	4	$4(3) + 2(4) < 8$ $20 < 8$	false
0	1	$4(0) + 2(1) < 8$ $2 < 8$	true
1	4	$4(1) + 2(4) < 8$ $12 < 8$	false
1	1	$4(1) + 2(1) < 8$ $6 < 8$	true

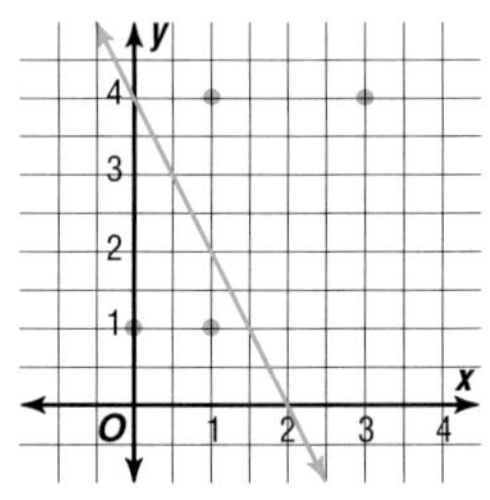

The ordered pairs {(0, 1), (1, 1)} are part of the solution set of $4x + 2y < 8$. The graph above shows the four ordered pairs of the replacement set and the equation $4x + 2y = 8$. Notice the location of the two ordered pairs that are solutions for $4x + 2y < 8$ in relation to the graph of the line.

EXPLORATION PROGRAMMING

You can use the following graphing calculator program to find out if a given ordered pair (x, y) is a solution for the inequality $5x - 3y \geq 15$.

```
PROGRAM: XYTEST
: Disp "IS (X, Y) A ","SOLUTION?"
: Prompt X,Y
: If 5X−3Y ≥ 15
: Then
: Disp "YES"
: Else
: Disp "NO"
```

To run the program for other ordered pairs, simply press ENTER and the program will begin again.

c. {(−1, −2), (0, −2), (0, −1), (0, 0), (1, −2), (1, −1), (1, 0) (1, 1)}

Your Turn **a. See students' work.**

a. Try the program for ten ordered pairs (x, y). Keep a list of which ordered pairs you tried and which ones were solutions.

b. How do you think you could change this program to test the inequality $2x + y \geq 2y$? **Change the sentence in the If statement.**

c. Use your changed program to find the solution set if $x = \{-1, 0, 1\}$ and $y = \{-2, -1, 0, 1\}$.

The solution set for an inequality contains many ordered pairs when the domain and range are the set of real numbers. The graphs of all of these ordered pairs fill an area on the coordinate plane called a **half-plane.** An equation defines the **boundary** or edge for each half-plane. For example, suppose you wanted to graph the inequality $y > 5$ on the coordinate plane.

First determine the boundary by graphing $y = 5$.

Since the inequality involves only $>$, the line should be dashed. The boundary divides the coordinate plane into two half-planes.

If the inequality contains $\geq$ or $\leq$, the graph of the boundary equation would be drawn as a solid line.

To determine which half-plane contains the solution, choose a point from each half-plane and test it in the inequality.

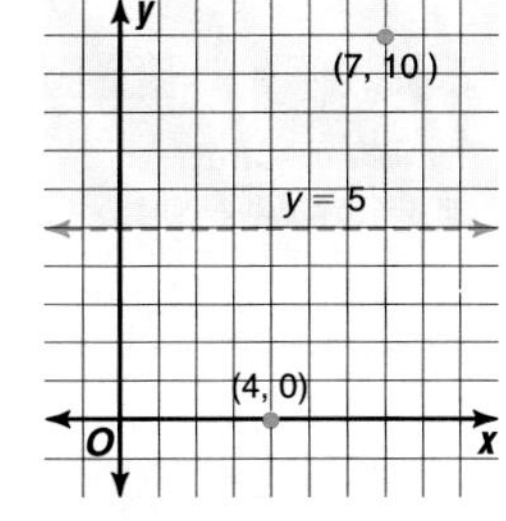

This graph is called an open half-plane because the boundary is not part of the graph.

Try (7, 10).		Try (4, 0).	
$y > 5$	$y = 10$	$y > 5$	$y = 0$
$10 > 5$	true	$0 > 5$	false

The half-plane that contains (7, 10) contains the solution. Shade that half-plane.

Example 2 **Graph $y + 2x \leq 3$.**

First solve for y in terms of x.

$$y + 2x \leq 3$$
$$y + 2x - 2x \leq 3 - 2x \quad \text{Subtract } 2x \text{ from each side.}$$
$$y \leq 3 - 2x$$

(continued on the next page)

Motivating the Lesson

Situational Problem Have students consider each of the following situations.

1. Your height must be less than 4 feet to go on a kiddy ride at the amusement park.
2. The center must score at least 24 points to set a new record.
3. The cost of five pencils and three pens is at most $4.25.

Ask students to show a relation to represent each of the above sentences.

2 TEACH

In-Class Example

For Example 1
From the set {(1, 1), (2, 8), (5, 0), (8, 1)}, which ordered pairs are part of the solution set for $2x - 3y < 10$? **(1, 1), (2, 8)**

EXPLORATION

This program has a typical if-then-else structure. The *else* case handles the negation of $5x - 3y \geq 15$; that is, $5x - 3y < 15$.

Teaching Tip Point out that the graphs shown are graphs of relations but not of functions. Every x value is paired with many y values. Have students consider the vertical line test to verify that these relations are not functions.

Teaching Tip Relate the open and closed half-planes used when graphing inequalities in two variables to the open and closed endpoints used in graphing inequalities in one variable.

In-Class Examples

For Example 2
Graph $3y - 2x \leq 6$.

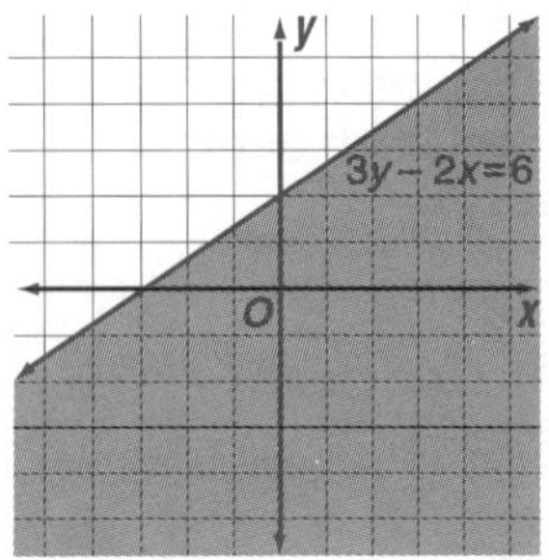

For Example 3
The drama club would like to attend a play at the local playhouse. Tickets for the play cost \$15 and \$20. If the club can spend no more than \$240 on tickets, how many of each ticket can the club purchase? Show the inequality used to represent the problem, and graph the answer.
$\mathbf{15x + 20y \leq 240}$

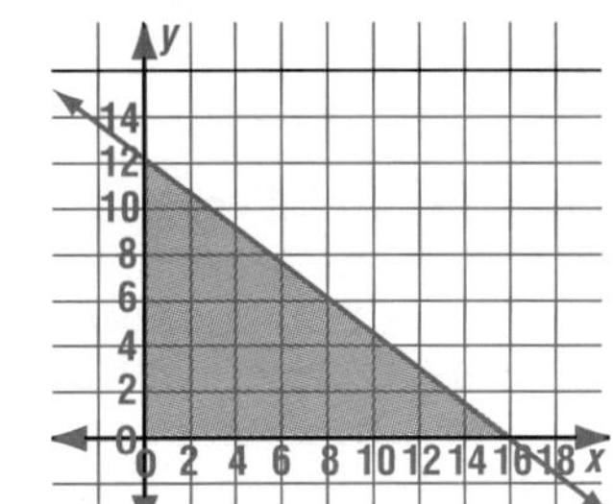

Teaching Tip For Example 3, make sure students understand that the shaded area represents only whole number and rational solutions. The area appears totally shaded because many solutions are graphed in a small space. Point out that ordered pairs like $\left(\sqrt{50}, \sqrt{425}\right)$ are located in the shaded region but are not solutions of the real-world problem.

Graph $y = 3 - 2x$. Since $y \leq 3 - 2x$ means $y < 3 - 2x$ or $y = 3 - 2x$, the boundary is included in the graph and should be drawn as a solid line.

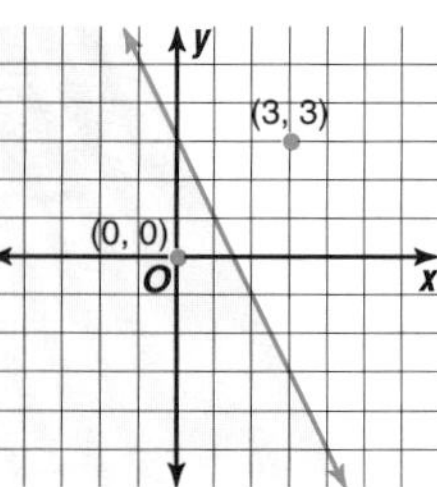

This graph is called a closed half-plane because the boundary line is included.

Select a point in one of the half-planes and test it. For example, use the origin (0, 0).

The origin is often used as a test point because the values are easy to substitute into the inequality.

The half-plane that contains the origin should be shaded.

Check: Test a point in the other half-plane, for example (3, 3).

Since the statement is false, the half-plane containing (3, 3) is not part of the solution.

When solving real-life inequalities, the domain and range of the inequality are often restricted to nonnegative numbers or whole numbers.

Example 3 **APPLICATION Manufacturing**

Refer to the application at the beginning of the lesson. How many racing bikes will the technician be able to assemble and road test?

First, solve for y in terms of x.

$3x + y \leq 45$

$3x - 3x + y \leq 45 - 3x$

$y \leq 45 - 3x$

Since the open sentence includes the equation, graph $y = 45 - 3x$ as a solid line. Test a point in one of the half-planes, for example (0, 0). The half-plane containing (0, 0) represents the solution since $3(0) + 0 \leq 45$ is true.

FYI

The world speed record for bicycles over a 200-meter course using a flying start is 65.484 mph by Fred Markham at Mono Lake, California, set in 1986.

Let's examine what the solution means. The technician cannot complete negative numbers of racing bikes. So, any point in the half-plane whose x- and y-coordinates are whole numbers is a possible solution. That is, only the portion of the shading in the first quadrant is the solution for this problem.
One solution is (5, 30).
This represents 5 bicycles assembled and 30 road tested by the technician in a 45-hour week.

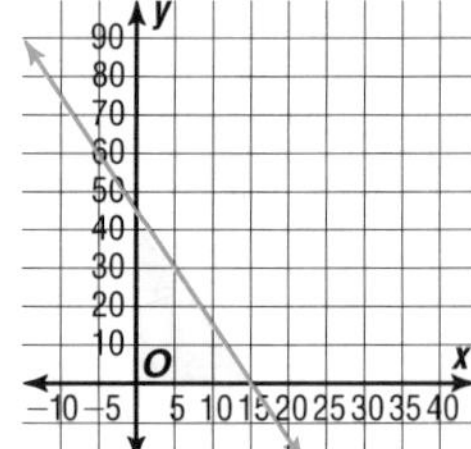

FYI

The fastest speed ever achieved on a bicycle was 152.284 mph, by John Howard at Bonneville Salt Flats, Utah, on July 20, 1985. The bicycle followed a car that was pulling a wind barrier, thus reducing air resistance.

CHECK FOR UNDERSTANDING

Communicating Mathematics

Study the lesson. Then complete the following.

1. a. **Graph** $y \geq x + 1$. **See margin.**
   b. **Identify** the boundary and indicate whether it is included or not.
   c. **Identify** the half-plane that is part of the graph.
   d. **Write** the coordinates of a point not on the boundary that satisfy the inequality.
2. **Explain** how you would check whether a point is part of the graph of an inequality. **See margin.**

3. **Assess Yourself** What do you think was the most challenging concept you learned in this chapter? Give an example of that concept and tell why you thought it was challenging. **See students' work.**

Guided Practice

Match each inequality with its graph.

4. $y \geq \frac{1}{2}x - 2$ **c**
5. $y \leq 0.5x - 2$ **a**
6. $y \geq \frac{2}{3}x + 2$ **d**
7. $y \leq \frac{2}{3}x + 2$ **b**

a.

b. 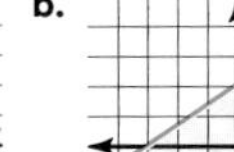

c. 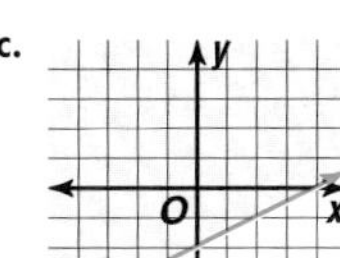

d. 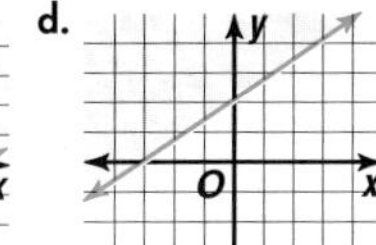

Determine which ordered pairs are solutions to the inequality. State whether the boundary is included in the graph. **8. b, c; yes**

8. $y \leq x$ a. $(-3, 2)$ b. $(1, -2)$ c. $(0, -1)$
9. $y > x - 1$ a. $(0, 0)$ b. $(2, 0)$ c. $(1, 3)$ **a, c; no**
10. Find which ordered pairs from the set $\{(-2, 2), (-2, 3), (2, 2), (2, 3)\}$ are part of the solution set for $a + b < 1$. **$\{(-2, 2)\}$**

Graph each inequality. **11–14. See Solutions Manual.**

11. $y > 3$
12. $x + y > 1$
13. $2x + 3y \geq -2$
14. $-x < -y$

EXERCISES

Practice

Copy each graph. Shade the appropriate half-plane to complete the graph of the inequality.

15. $x > 4$
16. $3y < x$
17. $3x + y > 4$
18. $2x - y \leq -2$

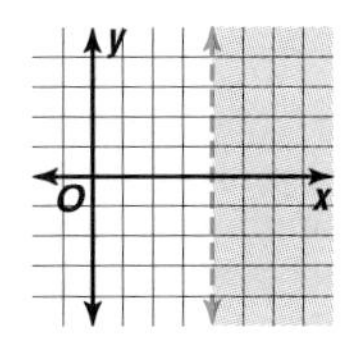

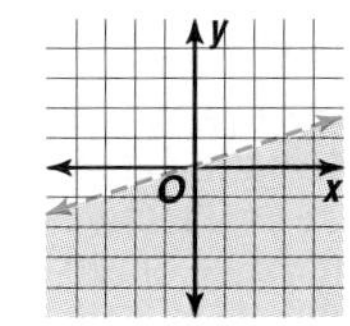

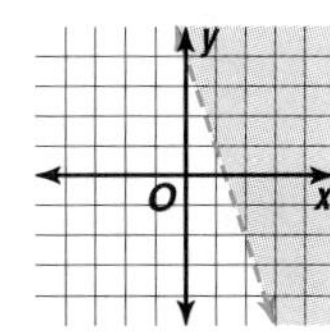

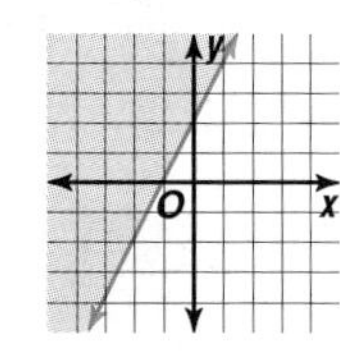

Reteaching

Using Alternative Methods An alternate strategy in determining which half-plane contains the solution to an inequality in two variables starts by transforming the inequality so y is the left number. Shade the lower half-plane for the "$y <$" form or shade the upper half-plane for the "$y >$" form.

3 PRACTICE/APPLY

Check for Understanding

Exercises 1–14 are designed to help you assess your students' understanding through reading, writing, speaking, and modeling. You should work through Exercises 1–3 with your students and then monitor their work on Exercises 4–14.

Additional Answers

1a.

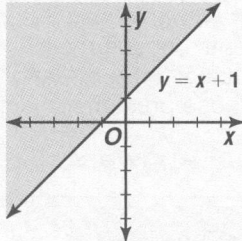

1b. The boundary is the graph of $y = x + 1$ and it is included.
1c. The half-plane above the line is shaded.
1d. Sample answer: $(-3, 3)$
2. Replace the x and y in the inequality with the values of the ordered pair. If the inequality holds true, then the ordered pair is part of the solution set.

Study Guide Masters, p. 55

7-8 NAME ______________ DATE ________

Study Guide — Student Edition Pages 436–441

Graphing Inequalities in Two Variables

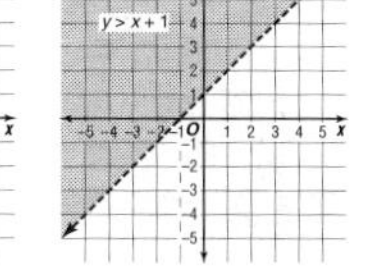

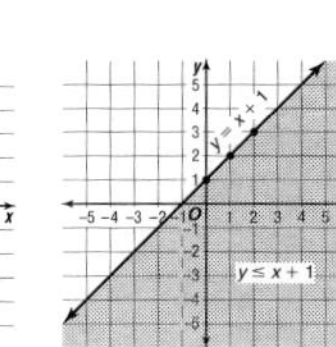

The graph of the equation $y = x + 1$ is a line that separates the coordinate plane into two regions. Each region is called a **half-plane.** The line for $y = x + 1$ is called the **boundary** for each half-plane.

The boundary line in both regions is the line for $y = x + 1$. In $y > x + 1$, the boundary is *not* part of the graph. The boundary is shown as a dashed line. All points above the line are part of the graph. This graph is called an **open half-plane.** In $y \leq x + 1$, the boundary *is* part of the graph and is shown as a solid line. The graph also contains all points below the line. This graph is called a **closed half-plane.**

Graph each inequality.

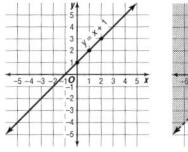

1. $y < 4$ 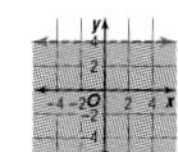
2. $3x < y$ 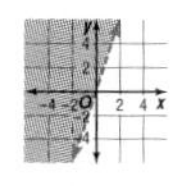
3. $2x - 3y \leq 6$ 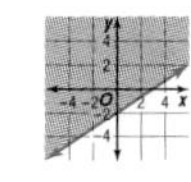
4. $-5x + 2 \geq y$ 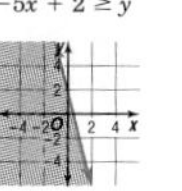
5. $x - y \geq 1$
6. $-x > y$

Assignment Guide

Core: 15–45 odd, 46, 47, 49–56
Enriched: 16–46 even, 47–56

For **Extra Practice,** see p. 774.

The red A, B, and C flags, printed only in the Teacher's Wraparound Edition, indicate the level of difficulty of the exercises.

Using the Programming Exercises The program given in Exercises 43–45 is for use with a TI-82 graphing calculator. For other programmable calculators, have students consult their owner's manual for commands similar to those presented here.

Additional Answers

40.

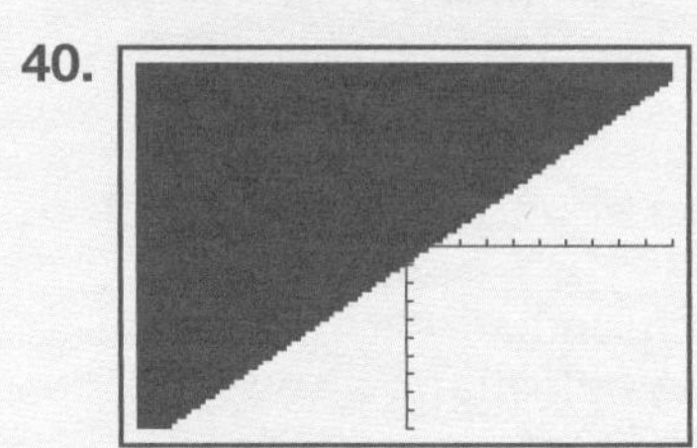

41.

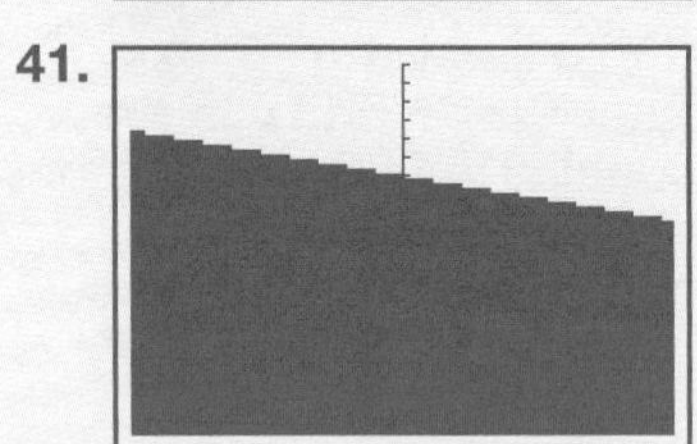

Practice Masters, p. 55

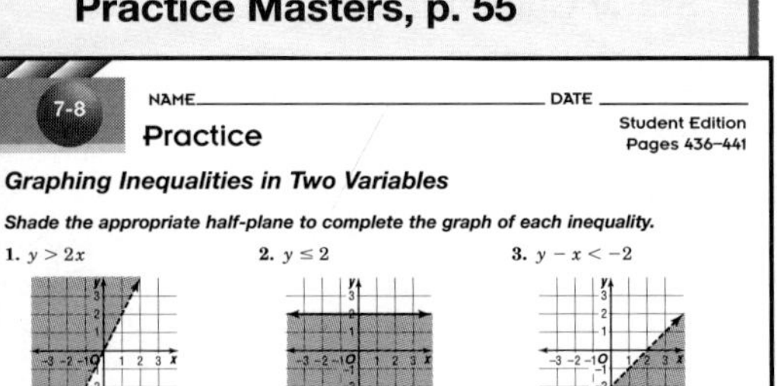

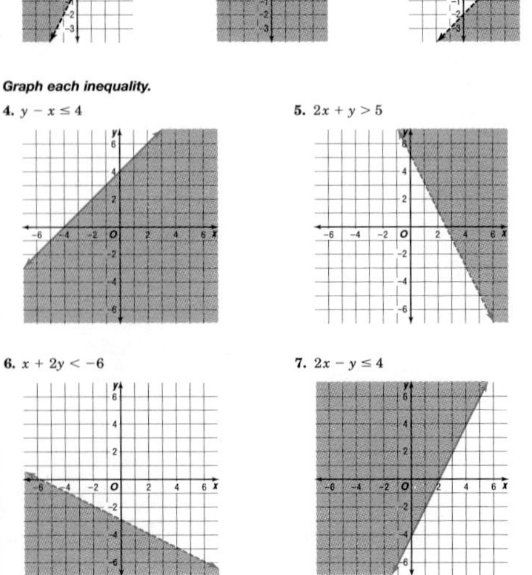

7-8 NAME______ DATE______

Practice — Student Edition Pages 436–441

Graphing Inequalities in Two Variables

Shade the appropriate half-plane to complete the graph of each inequality.

1. $y > 2x$ 2. $y \leq 2$ 3. $y - x < -2$

Graph each inequality.

4. $y - x \leq 4$ 5. $2x + y > 5$

6. $x + 2y < -6$ 7. $2x - y \leq 4$

Find which ordered pairs from the given set are part of the solution set for each inequality.

19. $y < 3x$, $\{(-3, 1), (-3, 2), (1, 1), (1, 2)\}$ $\{(1, 1), (1, 2)\}$
20. $y - x > 0$, $\{(1, 1), (1, 2), (4, 1), (4, 2)\}$ $\{(1, 2)\}$
B 21. $2y + x \geq 4$, $\{(-1, -3), (-1, 0), (-2, -3), (-2, 0)\}$ $\varnothing$

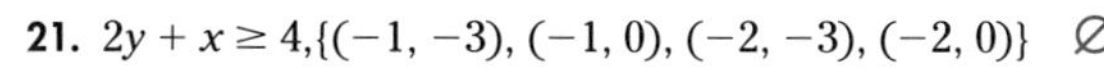

Graph each inequality. 22–39. See Solutions Manual.

22. $x > -5$	23. $y < -3$	24. $3y + 6 > 0$
25. $4x + 8 < 0$	26. $y \leq x + 1$	27. $x + y > 2$
28. $x + y < -4$	29. $3x - 1 \geq y$	30. $3x + y < 1$
31. $x - y \geq -1$	32. $x < y$	33. $-y > x$
C 34. $2x - 5y \leq -10$	35. $8y + 3x < 16$	36. $\lvert y \rvert \geq 2$
37. $y > \lvert x + 2 \rvert$	38. $y > 2$ and $x < 3$	39. $y \leq -x$ and $x \geq -3$

Graphing Calculator

Use a graphing calculator to graph each inequality. Make a sketch of the graph. 40–42. See margin.

40. $y > x - 1$ 41. $4y + x < 16$ 42. $x - 2y < 4$

Programming

Use the program in the Exploration to determine which pairs are solutions for each inequality.

43. $x + 2y \geq 3$ a. $(-2, 2)$ b. $(4, -1)$ c. $(3, 1)$ d. $(0, 0)$ c
44. $2x - 3y \leq 1$ a. $(2, 1)$ b. $(5, -1)$ c. $(1, 1)$ d. $(0, 0)$
45. $-2x < 8 - y$ a. $(5, 10)$ b. $(3, 6)$ c. $(-4, 0)$ d. $(0, 0)$
a, b, d 44. a, c, d

Critical Thinking

46. What compound inequality is described by the graph at the right? Find a simple inequality that also describes this graph. $\{x \geq 0$ and $y \geq 0\}$ or $\{x \leq 0$ and $y \leq 0\}$; $xy \geq 0$

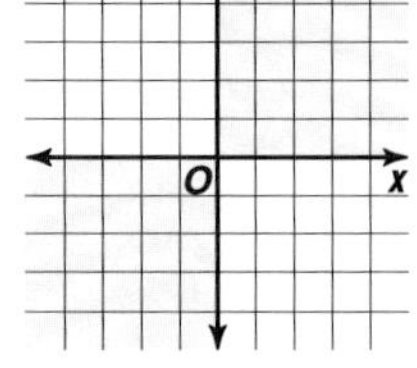

Applications and Problem Solving

47. **Health** The graph below shows the effective heart rate ranges for each type of exercise goal.

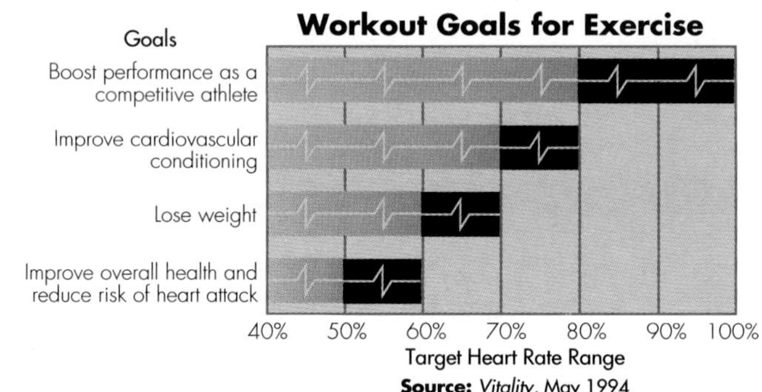

Source: *Vitality*, May 1994

Arrio is 35 years old and just beginning a bench-stepping aerobics class. In the orientation at the beginning of the first class, he learned that during exercise an effective minimum heart rate (beats/minute) should be 70% of the difference of 220 and his age. A maximum heart rate should be 80% of the difference of 220 and his age.

Additional Answer

42.

a. Write a compound inequality that expresses the effective rate zone for a person a years of age. **$0.7(220 - a) \leq z \leq 0.8(220 - a)$**

b. In class, the participants take a break and count their heart beats for 15 seconds. What should be Arrio's effective heart rate zone for that 15-second count? **$32 \leq z \leq 37$**

c. According to the graph and the heart rate range given, what is the goal of the bench-stepping class? **improve cardiovascular conditioning**

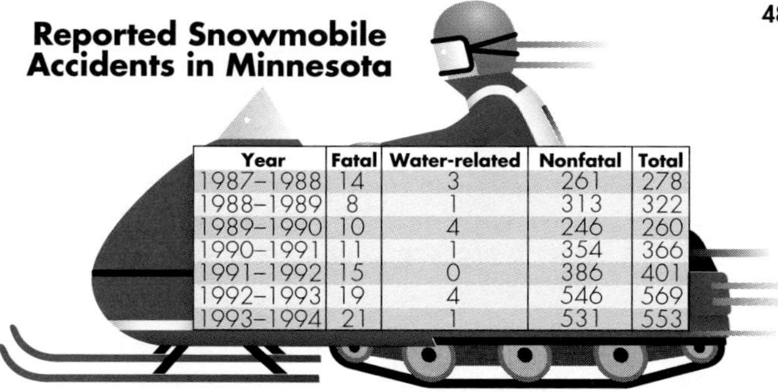

Reported Snowmobile Accidents in Minnesota

Year	Fatal	Water-related	Nonfatal	Total
1987–1988	14	3	261	278
1988–1989	8	1	313	322
1989–1990	10	4	246	260
1990–1991	11	1	354	366
1991–1992	15	0	386	401
1992–1993	19	4	546	569
1993–1994	21	1	531	553

Source: Minnesota Department of Natural Resources

48. **Snowmobiling** Although snowmobiling is an exhilarating sport, it can be very dangerous. The chart below shows the reported accidents in Minnesota involving snowmobiles.

a. A study suggests that the average number of nonfatal snowmobile accidents per year nationwide is about 350. Suppose x represents the total number of fatal snowmobile accidents in Minnesota and y represents the total number of snowmobile accidents in Minnesota. In what years is $x + 350 > y$? **1987–88, 1988–89, 1989–90**

b. Graph $x + 350 > y$. **See margin.**

c. The snowmobiling season in Minnesota goes from November to March. The lack of snow during the 1994–1995 season caused many to resort to riding their snowmobiles on icy lakes instead of on land. As of December 1, there had been 7 fatal accidents, 1 water-related fatal accident, and 81 nonfatal accidents. Do you think that the 1994–1995 season's accident count was greater or less than the 1993–1994 season? Explain your answer. **See students' work.**

Mixed Review

49. See margin.

49. **Statistics** Make a box-and-whisker plot of the total number of snowmobiling accidents shown in the chart for Exercise 48. (Lesson 7–7)

50. Solve $5 - |2x - 7| > 2$. (Lesson 7–6) **$\{x \mid 2 < x < 5\}$**

51. Write an equation in slope-intercept of the line that is parallel to the graph of $8x - 2y = 7$ and whose y-intercept is the same as the line whose equation is $2x - 9y = 18$. (Lesson 6–4) **$y = 4x - 2$**

52. $-\frac{12}{5}$

52. Determine the value of r so that the line passing through $(r, 4)$ and $(-4, r)$ has a slope of 4. (Lesson 6–1)

53. Graph $-y + \frac{2}{7}x = 1$. (Lesson 5–4) **See margin.**

54. State the inverse of the relation $\{(4, -1), (3, 2), (-4, 0), (17, 9)\}$. (Lesson 5–2) **$\{(-1, 4), (2, 3), (0, -4), (9, 17)\}$**

55. What is 98.5% of \$140.32? (Lesson 4–4) **\$138.22**

56. Find three consecutive integers whose sum is 87. (Lesson 3–3) **28, 29, 30**

Extension

Connections Find the area of the triangular region determined as the intersection of the closed half-planes $x \leq 4$, $y \geq -\frac{1}{2}x + 3$, and $y \leq \frac{1}{2}x + 5$. **18 square units**

Additional Answer

53.

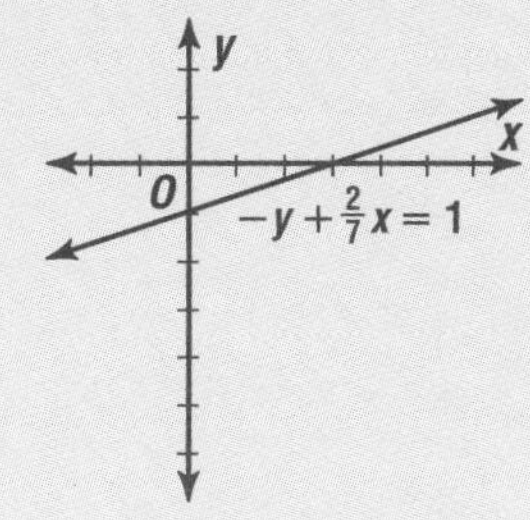

4 ASSESS

Closing Activity

Speaking Ask students to explain the steps they would use to graph $3x + 4y \leq 12$.

Chapter 7, Quiz D (Lessons 7-7 and 7-8), is available in the *Assessment and Evaluation Masters,* p. 185.

Additional Answers

48b.

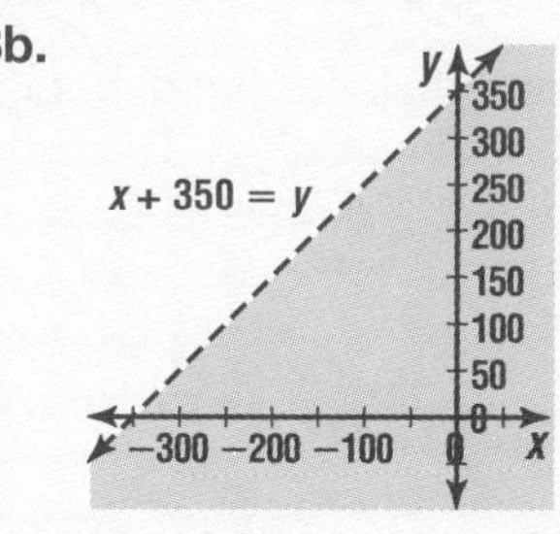

49.

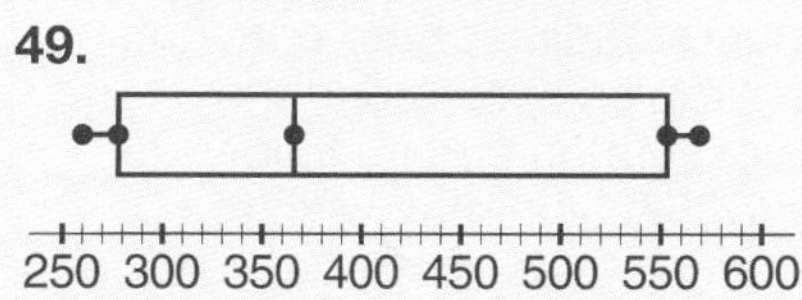

Enrichment Masters, p. 55

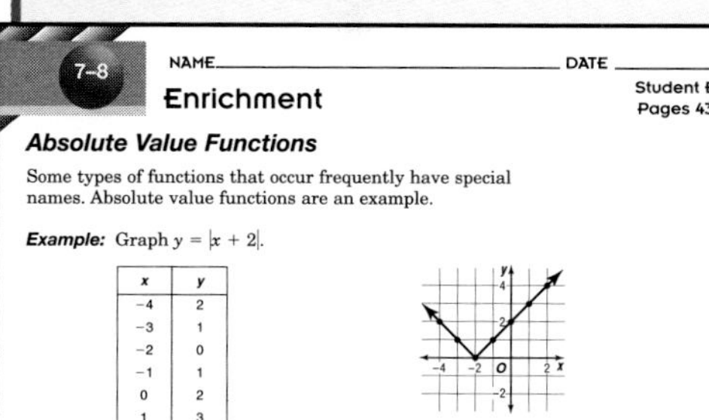

7-8 NAME ______ DATE ______

Enrichment

Student Edition Pages 436–442

Absolute Value Functions

Some types of functions that occur frequently have special names. Absolute value functions are an example.

Example: Graph $y = |x + 2|$.

x	y
−4	2
−3	1
−2	0
−1	1
0	2
1	3
2	4

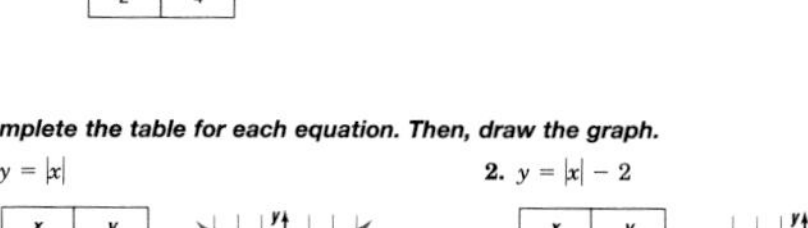

Complete the table for each equation. Then, draw the graph.

1. $y = |x|$

x	y
−3	3
−2	2
−1	1
0	0
1	1
2	2
3	3

2. $y = |x| - 2$

x	y
−3	1
−2	0
−1	−1
0	−2
1	−1
2	0
3	1

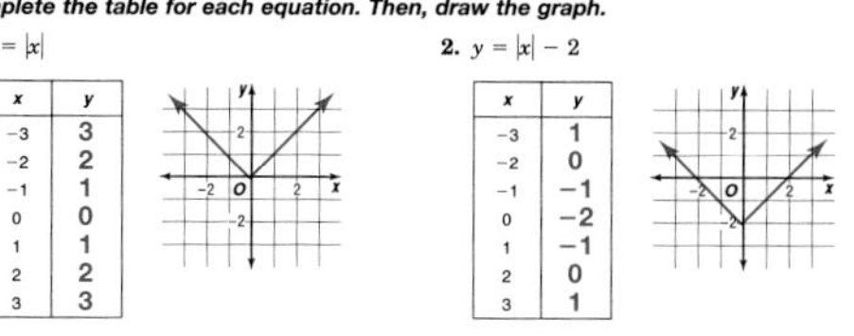

3. $y = |x - 1|$

x	y
−2	3
−1	2
0	1
1	0
2	1
3	2
4	3

4. $y = |2 - x|$

x	y
−2	4
−1	3
0	2
1	1
2	0
3	1
4	2

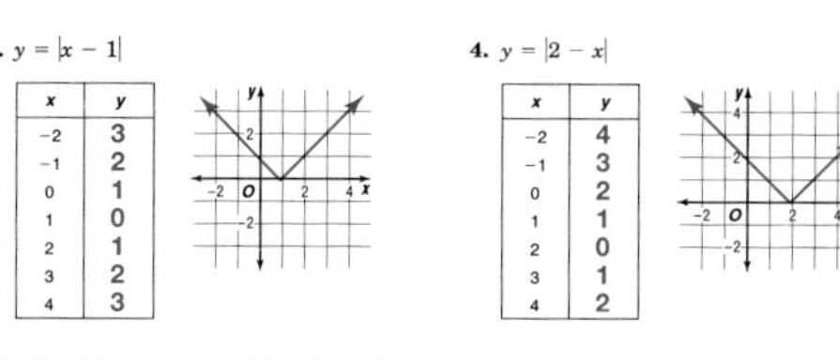

Closing the Investigation

This activity provides students an opportunity to bring their work on the Investigation to a close. For each Investigation, students should present their findings to the class. Here are some ways students can display their work.

- Conduct and report on an interview or survey.
- Write a letter, proposal, or report.
- Write an article for the school or local paper.
- Make a display, including graphs and/or charts.
- Plan an activity.

Assessment

To assess students' understanding of the concepts and topics explored in the Investigation and its follow-up activities, you may wish to examine students' Investigation Folders.

The scoring guide provided in the *Investigations and Projects Masters,* p. 11, provides a means for you to score students' work on the Investigation.

Investigations and Projects Masters, p. 11

Scoring Guide
Chapters 6 and 7
Investigation

Level	Specific Criteria
3 Superior	• Shows thorough understanding of the concepts of *data collection and analysis, using formulas to calculate radius and volume of a sphere, making graphs,* and *writing equations and inequalities.* • Uses appropriate strategies to solve problems. • Computations are correct. • Written explanations are exemplary. • Charts, graphs, and report are appropriate and sensible. • Goes beyond the requirements of some or all problems.
2 Satisfactory, with Minor Flaws	• Shows understanding of the concepts of *data collection and analysis, using formulas to calculate radius and volume of a sphere, making graphs,* and *writing equations and inequalities.* • Uses appropriate strategies to solve problems. • Computations are mostly correct. • Written explanations are effective. • Charts, graphs, and report are appropriate and sensible. • Satisfies the requirements of problems.
1 Nearly Satisfactory, with Obvious Flaws	• Shows understanding of most of the concepts of *data collection and analysis, using formulas to calculate radius and volume of a sphere, making graphs,* and *writing equations and inequalities.* • May not use appropriate strategies to solve problems. • Computations are mostly correct. • Written explanations are satisfactory. • Charts, graphs, and report are appropriate and sensible. • Satisfies the requirements of problems.
0 Unsatisfactory	• Shows little or no understanding of the concepts of *data collection and analysis, using formulas to calculate radius and volume of a sphere, making graphs,* and *writing equations and inequalities.* • Does not use appropriate strategies to solve problems. • Computations are incorrect. • Written explanations are not satisfactory. • Charts, graphs, and report are not appropriate or sensible. • Does not satisfy the requirements of problems.

CLOSING THE In·ves·ti·ga·tion

Smoke Gets In Your Eyes

Refer to the Investigation on pages 320–321. Add the results of your work below to your Investigation Folder.

The EPA's most recent long-term study on the effects of secondhand smoke shows that non-smokers married to smokers have a 19% increased risk of having lung cancer. Lung cancer is not the only danger of secondhand smoke. Twelve studies show that heart disease is another danger. Nonsmokers who are exposed to their spouses' smoke have a 30% increased chance of death from heart disease than do other nonsmokers. After reviewing a number of studies, the EPA's risk analysis has also concluded that secondhand smoke causes an extra 150,000 to 300,000 respiratory infections a year among the nations 5.5 million children under the age of 18 months.

Katharine Hammond, an environmental-health expert at the University of California, Berkeley, has also conducted a study on the *carcinogenic* components of secondhand smoke. The carcinogenic components are the parts of smoke that are known to cause cancer in humans. She found that "in the same room, at the same time, the nonsmoker is getting as much benzene (a chemical that is known to cause cancer in humans) as a smoker gets smoking six cigarettes."

James Repace and Alfred Lowery, two statistical researchers who study the effects of secondhand smoke, have concluded that a lifetime increase in lung-cancer risk of 1 in 1000 could be caused by long-term exposure to air containing more than 6.8 micrograms of nicotine per cubic meter of air.

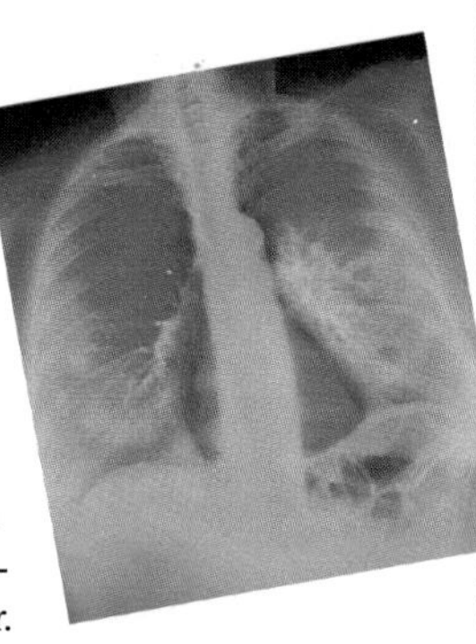

Analyze

You have conducted experiments and organized your data in various ways. It is now time to analyze your findings and state your conclusions.

PORTFOLIO ASSESSMENT

You may want to keep your work on this Investigation in your portfolio.

1. True secondhand smoke consists mostly of sidestream smoke. This is the smoke that comes from the smoldering cigarette. This smoke is much more toxic than inhaled smoke. How does this information affect your conclusions about the amount of smoke inhaled by nonsmokers in a room?
2. If you are a nonsmoker and live with a smoker, what are the cost factors involved? Explain your calculations.
3. Describe your personal experience with secondhand smoke.

Write

You want to inform people of the effects of secondhand smoke. You decide to write a letter to the editor of a local paper describing your investigation on the effects of secondhand smoke.

4. Use the information above and the results of your experiments and explorations to write a paper regarding the health risks of secondhand smoke.
5. You may want to do further research. The American Cancer Society and other agencies have information regarding the health risks of smoking and secondhand smoke.
6. Use data, charts, and graphs to justify your position. Use mathematics to help convince your readers of the conclusions you drew from this Investigation.

CHAPTER 7 HIGHLIGHTS

VOCABULARY

After completing this chapter, you should be able to define each term, property, or phrase and give an example or two of each.

Algebra

addition property for inequality (p. 385)
boundary (p. 437)
compound inequality (p. 405)
division property for inequality (p. 393)
half-plane (p. 437)
intersection (p. 406)
multiplication property for inequality (p. 393)
set-builder notation (p. 385)
subtraction property for inequality (p. 385)
union (p. 408)

Statistics

box-and-whisker plot (p. 427)
extreme values (p. 427)
whiskers (p. 428)

Probability

compound event (p. 414)
outcomes (p. 413)
simple events (p. 414)
tree diagram (p. 413)

Problem Solving

draw a diagram (p. 406)

UNDERSTANDING AND USING THE VOCABULARY

Choose the letter of the term that best matches each statement, algebraic expression, or algebraic sentence.

1. If $\frac{1}{2}x \leq -5$, then $x \leq -10$. **h**
2. If $8 > 4$, then $8 + 5 > 4 + 5$. **b**
3. $\{h \mid h > 43\}$ **i**
4. $x \geq -3$ or $x < -10$ **c, k**
5. $x \geq -4$ and $x < 2$ **c, f**
6. If $4x - 1 < 7$, then $4x - 4 < 4$. **j**
7. If $-3x < 9$, then $x > -3$. **d**
8. $>$ **e**
9. $<$ **g**
10. $7 > x > 1$ **c, f**
11. $|x + 6| > 12$ means $x + 6 > 12$ or $-(x + 6) > 12$. **a**

a. absolute value inequality
b. addition property for inequality
c. compound inequality
d. division property for inequality
e. greater than
f. intersection
g. less than
h. multiplication property for inequality
i. set builder notation
j. subtraction property for inequality
k. union

Using the CHAPTER HIGHLIGHTS

The Chapter Highlights begins with a listing of the new terms, properties, and phrases that were introduced in this chapter. Have students define each term and provide an example or two of it, if appropriate.

Instructional Resources

Three multiple-choice tests and three free-response tests are provided in the *Assessment and Evaluation Masters*. Forms 1A and 2A are for honors pacing, and Forms 1B, 1C, 2B, and 2C are for average pacing. Chapter 7 Test, Form 1B is shown at the right. Chapter 7 Test, Form 2B is shown on the next page.

Assessment and Evaluation Masters, pp. 171–172

NAME ____________ DATE ________

Chapter 7 Test, Form 1B

Write the letter for the correct answer in the blank at the right of each problem.

1. Solve $2x - 7 \geq 3x$. 1. **B**
   A. $\{x \mid x \leq \frac{5}{7}\}$ B. $\{x \mid x \leq -7\}$ C. $\{x \mid x \geq 7\}$ D. $\{x \mid x \geq -7\}$
2. Solve $-13 > w - (-12)$. 2. **A**
   A. $\{w \mid w < -25\}$ B. $\{w \mid w > -25\}$ C. $\{w \mid w > -1\}$ D. $\{w \mid w < -1\}$
3. Solve $\frac{m}{5} < -3$. 3. **B**
   A. $\{m \mid m > -15\}$ B. $\{m \mid m < -15\}$ C. $\{m \mid m < 15\}$ D. $\{m \mid m > 15\}$
4. Solve $-\frac{2}{3}s > 6$. 4. **D**
   A. $\{s \mid s > -9\}$ B. $\{s \mid s > 9\}$ C. $\{s \mid s < 9\}$ D. $\{s \mid s < -9\}$
5. Solve $-1.1t \leq 4.62$. 5. **D**
   A. $\{t \mid t \leq -5.06\}$ B. $\{t \mid t \geq -5.06\}$ C. $\{t \mid t \leq -4.2\}$ D. $\{t \mid t \geq -4.2\}$
6. Solve $6d + 10 < 46$. 6. **A**
   A. $\{d \mid d < 6\}$ B. $\{d \mid d > 6\}$ C. $\{d \mid d < -6\}$ D. $\{d \mid d > -6\}$
7. Solve $5z - 4 > 2z + 8$. 7. **A**
   A. $\{z \mid z > 4\}$ B. $\{z \mid z < 1\}$ C. $\{z \mid z < 4\}$ D. $\{z \mid z > 1\}$
8. Solve $5w - (w - 8) > 9 + 3(2w - 3)$. 8. **D**
   A. $\{w \mid w < \frac{11}{5}\}$ B. $\{w \mid w < -\frac{11}{5}\}$ C. $\{w \mid w < -4\}$ D. $\{w \mid w < 4\}$
9. Solve $(0.5)(r + 1) \leq (0.6)(r - 2)$. 9. **B**
   A. $\{r \mid r \leq 30\}$ B. $\{r \mid r \geq 17\}$ C. $\{r \mid r \geq 30\}$ D. $\{r \mid r \leq 17\}$
10. The sum of two consecutive positive integers is at most 3. What is the greater integer? 10. **D**
    A. 5 B. 1 C. 3 D. 2
11. If one coin and one die are tossed, what is *P*(one head and one 5)? 11. **C**
    A. $\frac{1}{24}$ B. $\frac{1}{2}$ C. $\frac{1}{12}$ D. $\frac{1}{6}$
12. Renata's World Wide Web home page has 9 possible background colors and 3 possible text colors. How many combinations of background color and text color does the page have? 12. **A**
    A. 27 B. 12 C. 6 D. 30
13. If 2 six-sided dice are tossed, what is the probability of getting two different numbers? 13. **D**
    A. $\frac{1}{12}$ B. $\frac{1}{6}$ C. $\frac{1}{2}$ D. $\frac{5}{6}$

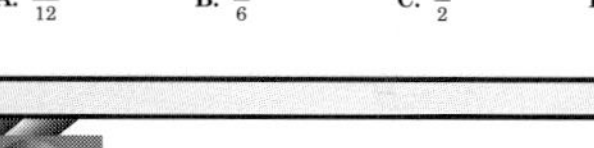

NAME ____________ DATE ________

Chapter 7 Test, Form 1B (continued)

14. What percent of the data shown in the box-and-whisker plot at right is located between 30 and 70? 14. **D**
    A. 25 B. 40 C. 50 D. 75
15. What compound sentence is graphed below? 15. **D**
    A. $-2 < y < 3$ B. $-2 < y \leq 3$ C. $y \geq -2$ or $y < 3$ D. $-2 \leq y < 3$
16. Which of the following is a graph of the solution set of the compound sentence $x > 0$ or $x < -4$? 16. **B**
    A. B. C. D.
17. Which of the following is the graph of the solution set of $-3 < 2x + 7 \leq 13$? 17. **A**
    A. B. C. D.
18. Which of the following is the solution set of $|x - 3| = 6$? 18. **C**
    A. $\{9\}$ B. $\{-3\}$ C. $\{-3, 9\}$ D. $\{-9, 3\}$
19. Which inequality is graphed at the right? 19. **A**
    A. $2x - 4y \leq -6$
    B. $2x - 4y < -6$
    C. $2x - 4y \geq -6$
    D. $2x - 4y > -6$

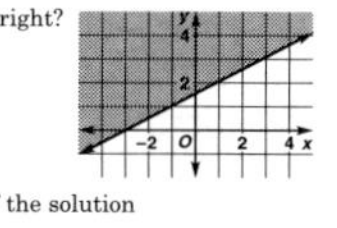

20. Which of the following is a graph of the solution set of $|4x - 8| \leq 16$? 20. **C**
    A. B. C. D.

Bonus

Bonus Graph $y = \begin{cases} 3 \text{ if } x \geq 1 \\ 2 \text{ if } x < 1 \end{cases}$. Use the coordinate plane provided.

Using the STUDY GUIDE AND ASSESSMENT

Skills and Concepts Encourage students to refer to the objectives and examples on the left as they complete the review exercises on the right.

CHAPTER 7 STUDY GUIDE AND ASSESSMENT

SKILLS AND CONCEPTS

OBJECTIVES AND EXAMPLES	REVIEW EXERCISES
Upon completing this chapter, you should be able to:	Use these exercises to review and prepare for the chapter test.

OBJECTIVES AND EXAMPLES

- solve inequalities by using addition and subtraction (Lesson 7–1)

$56 > m + 16$

$56 - 16 > m + 16 - 16$

$40 > m$

$\{m \mid m < 40\}$

12. $\{r \mid r > -12\}$ **13.** $\{n \mid n > -35\}$
14. $\{t \mid t \leq 6\}$ **15.** $\{p \mid p \leq -18\}$
16. $n - 3 \geq 2, \{n \mid n \geq 5\}$
17. $3n > 4n - 8, \{n \mid n < 8\}$

REVIEW EXERCISES

Solve each inequality. Then check your solution.

12. $r + 7 > -5$ **13.** $-35 + 6n < 7n$

14. $2t - 0.3 \leq 5.7 + t$ **15.** $-14 + p \geq 4 - (-2p)$

Define a variable, write an inequality, and solve each problem. Then check your solution.

16. The difference of a number and 3 is at least 2.

17. Three times a number is greater than four times the number less eight.

OBJECTIVES AND EXAMPLES

- solve inequalities by using multiplication and division (Lesson 7–2)

$\frac{-5}{6}m > 25$

$\frac{-6}{5}\left(\frac{-5}{6}m\right) > \frac{-6}{5}(25)$

$m < -30$

$\{m \mid m < -30\}$

20. $\left\{k \mid k \geq \frac{1}{5}\right\}$ **21.** $\{x \mid x > 32\}$
22. $6n \leq 32.4, \{n \mid n \leq 5.4\}$
23. $-\frac{3}{4}n \leq 30, \{n \mid n \geq -40\}$

REVIEW EXERCISES

Solve each inequality. Then check your solution. **18.** $\{x \mid x \geq -8\}$ **19.** $\{w \mid w \leq -15\}$

18. $7x \geq -56$ **19.** $90 \leq -6w$

20. $\frac{2}{3}k \geq \frac{2}{15}$ **21.** $9.6 < 0.3x$

Define a variable, write an inequality, and solve each problem. Then check your solution.

22. Six times a number is at most 32.4.

23. Negative three fourths of a number is no more than 30.

OBJECTIVES AND EXAMPLES

- solve linear inequalities involving more than one operation (Lesson 7–3)

$15b - 12 > 7b + 60$

$15b - 7b - 12 > 7b - 7b + 60$

$8b - 12 > 60$

$8b - 12 + 12 > 60 + 12$

$8b > 72$

$\frac{8b}{8} > \frac{72}{8}$

$b > 9$

$\{b \mid b > 9\}$

26. $\{r \mid r > 1.8\}$ **27.** $\left\{y \mid y \leq -\frac{9}{2}\right\}$

REVIEW EXERCISES

Find the solution set of each inequality if the replacement set for each variable is {−5, −4, −3, . . . 3, 4, 5}.

24. $\frac{x-5}{3} > -3$ **{−3, −2, . . . 4, 5}**

25. $3 \leq -4x + 7$ **{−5, −4, . . . 0, 1}**

Solve each inequality. Then check your solution.

26. $2r - 3.1 > 0.5$ **27.** $4y - 11 \geq 8y + 7$

28. $-3(m - 2) > 12$ $\{m \mid m < -2\}$

29. $-5x + 3 < 3x + 23$ $\left\{x \mid x > -\frac{5}{2}\right\}$

30. $4(n - 1) < 7n + 8$ $\{n \mid n > -4\}$

31. $0.3(z - 4) \leq 0.8(0.2z + 2)$ $\{z \mid z \leq 20\}$

Assessment and Evaluation Masters, pp. 177–178

NAME__________ DATE ________

Chapter 7 Test, Form 2B

Solve each inequality.

1. $-14 \leq n + 5$ — 1. $\{n \mid n \geq -19\}$
2. $3a < 6 + 4a$ — 2. $\{a \mid a > -6\}$
3. $\frac{3y}{8} > -\frac{2}{5}$ — 3. $\left\{y \mid y > \frac{16}{15}\right\}$
4. $-\frac{t}{6} \geq 14$ — 4. $\{t \mid t \leq -84\}$
5. $-19.8 \geq 3.6y$ — 5. $\{y \mid y \leq -5.5\}$
6. $4x - 5 < 2x + 11$ — 6. $\{x \mid x < 8\}$
7. $1.3(c - 4) \leq 2.6 + 0.7c$ — 7. $\{c \mid c \leq 13\}$
8. Write an inequality and solve: Forty less three times a number is no less than the number increased by 15. — 8. $40 - 3n \geq n + 15;\ n \leq 6\frac{1}{4}$

Define a variable, write an inequality, and solve.

9. Ray had scores of 75, 82, 94, and 77 on his first four science tests. What must he score on the next test to have an average of at least 85? — 9. s = score needed; $\frac{75 + 82 + 94 + 77 + s}{5} \geq 85$; 97 or greater
10. Felicita's bank charges $2.25 a month plus $0.10 per check. How many checks does she write if her bank charges are always between $3.50 and $5.00? — 10. c = no. of checks; $3.50 < 0.10c + 2.25 < 5.00$; between 13 and 27

Solve each open sentence and graph the solution set on the number lines provided.

11. $3w < 6$ and $-5 < w$ — 11. $-5 < w < 2$
12. $-4 \leq n$ or $3n + 1 < -2$ — 12. {all numbers}
13. $-2 \leq x + 1 < 4$ — 13. $-3 \leq x < 3$
14. $|1 - y| = 2$ — 14. $\{-1, 3\}$
15. $|3 - 2x| \geq 1$ — 15. $x \leq 1$ or $x \geq 2$
16. $|3w + 1| > -8$ — 16. {all numbers}

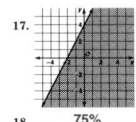

NAME__________ DATE ________

Chapter 7 Test, Form 2B (continued)

17. Graph $2y - 4x < 8$. Use the coordinate plane provided. — 17.
18. What percent of the data shown in the box-and-whisker plot below are between 20 and 70? — 18. 75%

Bob, Ted, and Al each have to make a presentation in Mrs. Small's speech class. Mrs. Small chooses students for the three presentations at random. Each student speaks exactly once.

19. Draw a tree diagram to show the possible outcomes. — 19. Bob < Ted — Al, Al — Ted; Al < Bob — Ted, Ted — Bob; Ted < Al — Bob, Bob — Al
20. What is the probability that Al speaks third? — 20. $\frac{1}{3}$
21. One spinner is divided into 4 sections labeled 0–3. A second spinner is divided into 3 sections labeled 0, 1, and 5. Customers at a department store spin both spinners to determine the amount they will save on large purchases; the first spinner gives the tens digit and the second spinner gives the ones digit of the discount. What is the probability that a customer will save at least $10? — 21. $\frac{3}{4}$

Use the following math quiz scores for exercises 22 and 23.

3, 3, 4, 5, 5, 6, 6, 6, 6, 7, 7, 7, 8, 8, 9, 10

22. Find the median. — 22. 6
23. Find the upper quartile. — 23. 7.5

Which ordered pairs from the given set are part of the solution set for each inequality?

24. $y > \frac{x}{2}$ {(−4, −2), (−6, −2), (0, 2), (0, 0), (6, 4), (6, 3)} — 24. {(−6, −2), (0, 2), (6, 4)}
25. $2y - x > 5$ {(−1, 2), (−2, 4), (0, 2), (0, 3), (1, 3), (1, 4)} — 25. {(−2, 4), (0, 3), (1, 4)}

Bonus Graph the solution set of the compound inequality $\frac{n + 4}{5} < 3$ and $\frac{n}{5} + 4 > 3$. **Bonus** $\left\{\frac{n}{-5} < n < 11\right\}$

GLENCOE Technology

Test and Review Software

You may use this software, a combination of an item generator and item bank, to create your own tests or worksheets. Types of items include free response, multiple choice, short answer, and open ended.

For IBM & Macintosh

OBJECTIVES AND EXAMPLES

- solve compound inequalities and graph their solution sets (Lesson 7–4)

$$2a > a - 3 \quad \text{and} \quad 3a < a + 6$$
$$2a - a > a - a - 3 \qquad 3a - a < a - a + 6$$
$$a > -3 \qquad 2a < 6$$
$$\frac{2a}{2} < \frac{6}{2}$$
$$a < 3$$
$$\{a \mid -3 < a < 3\}$$

−4 −3 −2 −1 0 1 2 3 4

REVIEW EXERCISES

Solve each compound inequality. Then graph the solution set. 33–36. See margin for graphs.

32. $x - 5 < -2$ and $x - 5 > 2$ $\varnothing$
33. $2a + 5 \leq 7$ or $2a \geq a - 3$ $\{a \mid a \text{ is a real number}\}$
34. $4r \geq 3r + 7$ and $3r + 7 < r + 29$ $\{r \mid 7 \leq r < 11\}$
35. $-2b - 4 \geq 7$ or $-5 + 3b \leq 10$ $\{b \mid b \leq 5\}$
36. $a \neq 6$ and $3a + 1 > 10$ $\{a \mid a > 3 \text{ and } a \neq 6\}$

- find the probability of a compound event (Lesson 7–5)

Draw a tree diagram to show the possibilities for boys and girls in a family of 3 children. Assume that the probabilities for girls and boys being born are the same.

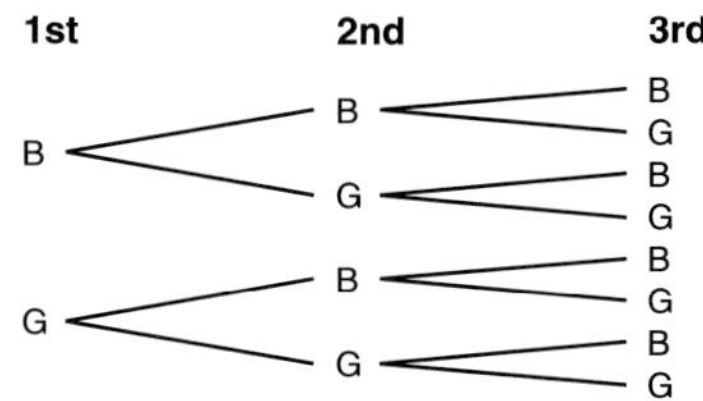

The probability that the family has exactly 3 girls is $\frac{1}{8}$ or 0.125, because there is 1 way out of 8 for this to happen. The probability that the family has exactly 2 boys and 1 girl is $\frac{3}{8}$ or 0.375, because there are 3 ways out of 8 for this to happen.

37. With each shrimp, salmon, or crab dinner at the Seafood Palace, you may have soup or salad. With shrimp, you may have broccoli or a baked potato. With salmon, you may have rice or broccoli. With crab, you may have rice, broccoli, or a potato. If all combinations are equally likely, find the probability of an order containing each item.
    a. salmon
    b. soup
    c. rice
    d. shrimp and rice
    e. salad and broccoli
    f. crab, soup, and rice
38. Matthew has 2 brown and 4 black socks in his dresser. While dressing one morning, he pulled out 2 socks without looking. What is the probability that he chose a matching pair? $\frac{7}{15} \approx 0.4\overline{6}$

37a. $\frac{2}{7}$ b. $\frac{1}{2}$ c. $\frac{2}{7}$ d. 0 e. $\frac{3}{14}$ f. $\frac{1}{14}$

- solve open sentences involving absolute value and graph the solutions (Lesson 7–6)

$$|2x + 1| > 1$$
$$2x + 1 > 1 \quad \text{or} \quad 2x + 1 < -1$$
$$2x + 1 - 1 > 1 - 1 \qquad 2x + 1 - 1 < -1 - 1$$
$$\frac{2x}{2} > \frac{0}{2} \qquad \frac{2x}{2} < \frac{-2}{2}$$
$$x > 0 \qquad x < -1$$
$$\{x \mid x > 0 \text{ or } x < -1\}$$

−4 −3 −2 −1 0 1 2 3

Solve each open sentence. Then graph the solution set. 39–44. See margin for graphs.

39. $|y + 5| > 0$ $\{y \mid y > -5 \text{ or } y < -5\}$
40. $|1 - n| \leq 5$ $\{n \mid 6 \geq n \geq -4\}$
41. $|4k + 2| \leq 14$ $\{k \mid 3 \geq k \geq -4\}$
42. $|3x - 12| < 12$ $\{x \mid 0 < x < 8\}$
43. $|13 - 5y| \geq 8$ $\left\{y \mid y \geq \frac{21}{5} \text{ or } y \leq 1\right\}$
44. $\left|2p - \frac{1}{2}\right| > \frac{9}{2}$ $\left\{p \mid p > \frac{5}{2} \text{ or } p < -2\right\}$

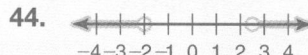

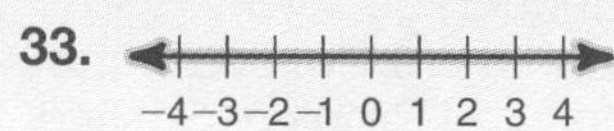

Additional Answers

33. −4 −3 −2 −1 0 1 2 3 4
34. 5 6 7 8 9 10 11 12 13
35. −1 0 1 2 3 4 5 6 7
36. 1 2 3 4 5 6 7 8 9
39. −9 −8 −7 −6 −5 −4 −3 −2 −1
40. −8 −6 −4 −2 0 2 4 6 8 10
41. −5 −4 −3 −2 −1 0 1 2 3 4 5
42. −4 −2 0 2 4 6 8 10 12
43. −2 −1 0 1 2 3 4 5 6
44. −4 −3 −2 −1 0 1 2 3 4

Applications and Problem Solving Encourage students to work through the exercises in the Applications and Problem Solving section to strengthen their problem-solving skills.

Additional Answers

45. 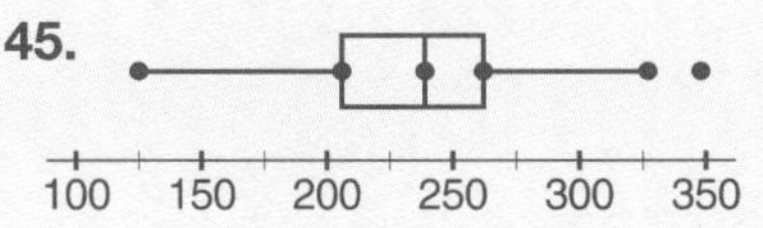

50. 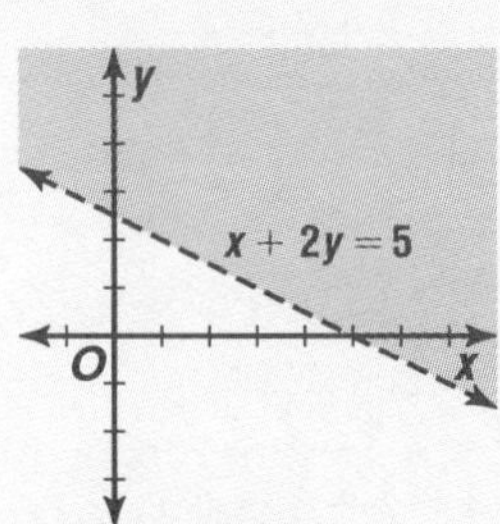

51. 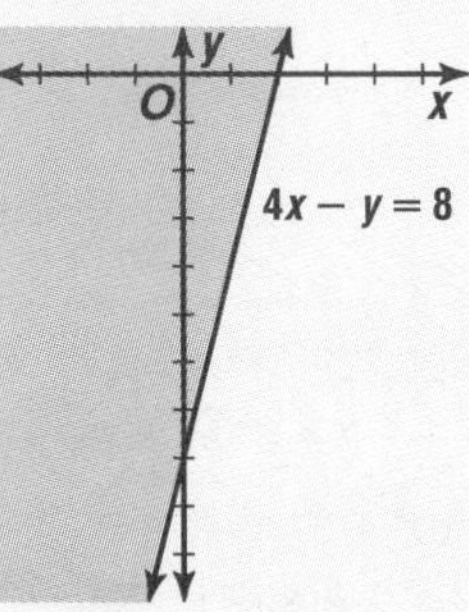

52. 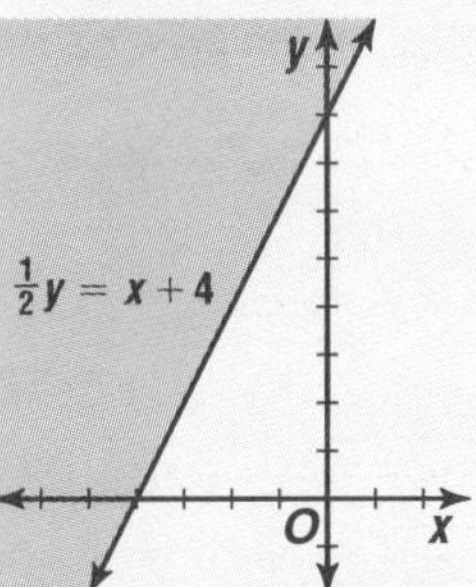

53. 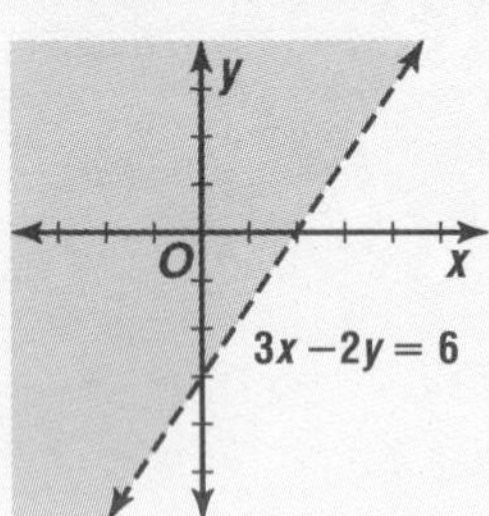

CHAPTER 7 STUDY GUIDE AND ASSESSMENT

OBJECTIVES AND EXAMPLES	REVIEW EXERCISES
• display and interpret data on box-and-whisker plots (Lesson 7–7) The following high temperatures were recorded during a two-week cold spell in St. Louis. Make a box-and-whisker plot of the temperatures. 20° 2° 12° 5° 4° 16° 17° 7° 6° 16° 5° 0° 5° 30° 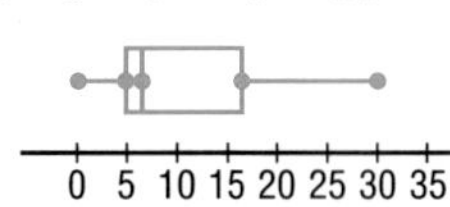	**The number of calories in a serving of french fries at 13 restaurants are 250, 240, 220, 348, 199, 200, 125, 230, 274, 239, 212, 240, and 327.** 45. See margin. **45.** Make a box-and-whisker plot of these data. **46.** Are there any outliers? If so, name them. yes, 348
• graph inequalities in the coordinate plane (Lesson 7–8) Graph $2x + 3y < 9$. 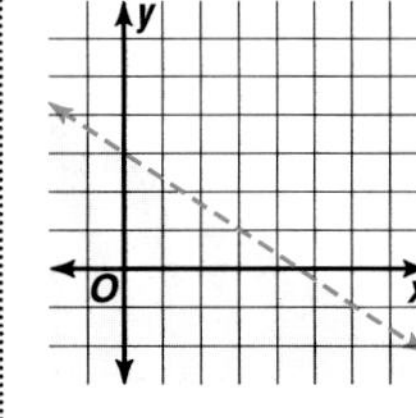 50–53. See margin for graphs.	**Find which ordered pairs from the given set are part of the solution set for each inequality.** **47.** $3x + 4y < 7$, $\{(1, 1), (2, -1), (-1, 1), (-2, 4)\}$ **48.** $4y - 8 \geq 0$, $\{(5, -1), (0, 2), (2, 5), (-2, 0)\}$ **49.** $-2x < 8 - y$, $\{(5, 10), (3, 6), (-4, 0), (-3, 6)\}$ 47. {(2, −1), (−1, 1)} 48. {(0, 2), (2, 5)} 49. {(5, 10), (3, 6)} **Graph each inequality.** **50.** $x + 2y > 5$ **51.** $4x - y \leq 8$ **52.** $\frac{1}{2}y \geq x + 4$ **53.** $3x - 2y < 6$

APPLICATIONS AND PROBLEM SOLVING

54. Number Theory The sum of three consecutive integers is less than 100. Find the three integers with the greatest sum. (Lesson 7–3) 32, 33, 34

56a. $\frac{1}{2}$ or 0.5 56b. $\frac{1}{4}$ or 0.25 56c. $\frac{1}{8}$ or 0.125

55. Shipping An empty book crate weighs 30 pounds. The weight of a book is 1.5 pounds. For shipping, the crate must weigh at least 55 pounds and no more than 60 pounds. What is the acceptable number of books that can be packed in the crate? (Lesson 7–4) 17 to 20 books

56. Automobiles An automobile dealer has cars available painted red or blue, with 4-cylinder or 6-cylinder engines, and with manual or automatic transmissions. (Lesson 7–5)

a. What is the probability of selecting a car with manual transmission?

b. What is the probability of selecting a car with a 4-cylinder engine and a manual transmission?

c. What is the probability of selecting a blue car with a 6-cylinder engine and an automatic transmission?

A practice test for Chapter 7 is provided on page 793.

ALTERNATIVE ASSESSMENT

COOPERATIVE LEARNING PROJECT

Statistics In this chapter, you learned how to make and interpret a box-and-whisker plot. Making a box-and-whisker plot and interpreting the data from a box-and-whisker plot are two different skills, however. One can go through the routine of drawing the plot but not be able to use the data portrayed by the plot to answer appropriate questions.

For this project, suppose there are two companies that manufacture window glass. They have each submitted a bid to a contractor who is building a library. Since glass that varies in thickness can cause distortions, the contractor has decided to measure the thickness of panes of glass from each factory, at several locations on each pane. The table below shows measurements for the two panes, one from each manufacturer.

Glass Thickness (mm)	
Company A	**Company B**
10.2	9.4
12.0	13.0
11.6	8.2
10.1	14.9
11.2	12.6
9.7	7.7
10.7	13.2
11.6	12.2
10.4	10.2
9.8	9.5
10.6	9.9
10.3	9.7
8.5	11.5
10.2	11.5
9.7	10.5
9.2	10.6
8.6	6.4
11.3	13.5

Prepare a stem-and-leaf plot and a box-and-whisker plot to organize and compare the two sets of data.

Follow these steps to organize your data.

- Determine how to set up the stem-and-leaf plot using decimals.
- Determine if using a common stem would be useful in comparing the two sets of data.
- Compare the ranges of the two sets of data.
- For which company is there "bunching" or "spreading out evenly" of the data?
- Compare the stem-and-leaf plot shape with the box-and-whisker plot shape.
- Compare the middle half of the data for each company.
- Write a comparative description of the sets of data and determine, with support, which company's glass should have less distortion.

THINKING CRITICALLY

- Why are multiplication and division the only two out of the four operations for which it is necessary to distinguish between positive and negative numbers when solving linear inequalities?
- Under what conditions will the compound sentence $x < a$ and $-a < x$ have no solutions?

PORTFOLIO

Select one of the assignments from this chapter for which you felt organization of the problem and reevaluation of the answer were important in order to get an accurate answer. Revise your work as necessary and place it in your portfolio. Explain why organization and reevaluation were important.

SELF EVALUATION

Do you look beyond the obvious in your math answers? Many times math students will work through a math problem rather routinely and not evaluate or check their answer. An answer must make sense and be accurate.

Assess yourself. Do you take the obvious solution as the whole answer or do you evaluate your answers for accuracy and rationalness? List two problems in mathematics and/or your daily life whereby the obvious answer was incorrect, so you needed to evaluate your solution for accuracy.

Assessment and Evaluation Masters, pp. 182, 193

NAME_______________ DATE _______

Chapter 7 Performance Assessment

Instructions: ***Demonstrate your knowledge by giving a clear, concise solution to each problem. Be sure to include all relevant drawings and justify your answers. You may show your solution in more than one way or investigate beyond the requirements of the problem.***

1. An architect is designing a house for the Frazier family. In the design he must consider the desires of the family and the local building codes. The rectangular lot on which the house will be built has 91 feet of frontage on a lake and is 158 feet deep.

158 ft
91 ft

a. The building codes state that one can build no closer than 10 feet to the lot line. Write an inequality and solve to see how long the front of the house facing the lake may be.

b. The Fraziers requested that the house contain no less than 2800 ft^2 and no more than 3200 ft^2 of floor space. Write an inequality to represent the range of permissible widths for the house.

c. The Fraziers have asked that the cost of the house be about $175,000 and are willing to deviate from this price no more than $20,000. Write an open sentence involving an absolute value and solve. Give the meaning of the answer.

2. a. Write a word problem involving an inequality with more than one operation.

b. Solve and give the meaning of the answer.

3. Students on a band trip are given a soft drink and a sandwich chosen at random for lunch. They may then trade with other band members if they wish to do so. Students have an equal chance of getting root beer or orange drink and peanut butter and jelly, bologna, or ham sandwich. Use a tree diagram to find the probability of getting an orange drink and meat sandwich. Give your reasoning.

4. Describe the data set used to make this box-and-whisker plot. What can't you tell about the data by looking at the box-and-whisker plot?

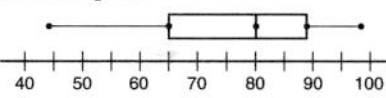

Scoring Guide
Chapter 7
Performance Assessment

Level	Specific Criteria
3 Superior	• Shows thorough understanding of the concepts of *inequality, compound inequality, absolute value, solving inequalities, compound events,* and *box-and-whisker plot.* • Uses appropriate strategies to solve problems. • Computations are correct. • Written explanations are exemplary. • Word problem concerning inequality with more than one operation is appropriate and makes sense. • Goes beyond requirements of some or all problems.
2 Satisfactory, with Minor Flaws	• Shows understanding of the concepts of *inequality, compound inequality, absolute value, solving inequalities, compound events,* and *box-and-whisker plot.* • Uses appropriate strategies to solve problems. • Computations are mostly correct. • Written explanations are effective. • Word problem concerning inequality with more than one operation is appropriate and makes sense. • Satisfies all requirements of problems.
1 Nearly Satisfactory, with Serious Flaws	• Shows understanding of most of the concepts of *inequality, compound inequality, absolute value, solving inequalities, compound events,* and *box-and-whisker plot.* • May not use appropriate strategies to solve problems. • Computations are mostly correct. • Written explanations are satisfactory. • Word problem concerning inequality with more than one operation is mostly appropriate and sensible. • Satisfies most requirements of problems.
0 Unsatisfactory	• Shows little or no understanding of the concepts of *inequality, compound inequality, absolute value, solving inequalities, compound events,* and *box-and-whisker plot.* • May not use appropriate strategies to solve problems. • Computations are incorrect. • Written explanations are not satisfactory. • Word problem concerning inequality with more than one operation is not appropriate or sensible. • Does not satisfy requirements of problems.

Alternative Assessment

The Alternative Assessment section provides students with the opportunity to assess their own work by thinking critically, working with others, keeping a portfolio, and honestly evaluating their own progress. For more information on alternative forms of assessment, see *Alternative Assessment in the Mathematics Classroom,* one of the titles in the Glencoe Mathematics Professional Series.

Performance Assessment

Performance Assessment tasks for this chapter are included in the *Assessment and Evaluation Masters.* A scoring guide is also provided.

In•ves•ti•ga•tion

TEACHER NOTES

NCTM Standards: 1–6, 9

This Investigation is designed to be completed over several days or weeks. It may be considered optional. You may want to assign the Investigation and the follow-up activities to be completed at the same time.

Objective

Use mathematics to examine the relationship between the speed of descent and the size of a hang glider.

Mathematical Overview

This Investigation will use the following mathematical skills and concepts from Chapters 8 and 9.

- making charts and graphs
- solving systems of equations and inequalities
- adding and subtracting polynomials
- using the looking for a pattern strategy
- checking for hidden assumptions

Recommended Time

Part	Pages	Time
Investigation	448–449	1 class period
Working on the Investigation	461, 519, 541	20 minutes each
Closing the Investigation	548	1 class period

Instructional Resources

Investigations and Projects Masters, pp. 13–16

A recording sheet, teacher notes, and scoring guide are provided for each investigation in the *Investigations and Projects Masters.*

1 MOTIVATION

This Investigation uses simple items to investigate hang gliders. Ask students if any of them have ever seen a hang glider on TV or in person. Discuss the importance of math in the construction of a safe hang glider.

LONG-TERM PROJECT

In•ves•ti•ga•tion

MATERIALS NEEDED

- construction paper
- metric ruler
- paper clips
- scissors
- stopwatch
- tape
- tissue paper

- washers
- wire

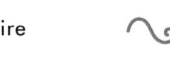

Hang gliding became popular in the United States in the early 1970s. In most states, a hang gliding certification is required before you are allowed to participate in the sport. The U.S. Hang Gliding Association is located in Los Angeles and certifies instructors and safety officers to train would-be hang gliders.

A hang glider looks like a manned kite. It consists of a triangular sail of synthetic fabric attached to an aluminum frame. The pilot hangs from a harness and steers the glider with a control bar that adjusts as the pilot shifts his or her body weight.

Hang gliders can be launched in several ways. The pilot can hold the glider and run down a hill until the glider is airborne. In areas with high cliffs, the pilot can run and jump from the cliff's edge, using the air currents below to fly. In flatter landscapes, the glider is often launched by towing it with a rope from a truck or boat and releasing it at an altitude of 400–500 feet.

Imagine that you are an engineer for an aeronautical engineering firm. A group of people who are interested in hang gliding have asked your firm to design a hang glider that can be used for recreational purposes. Your task is to design a hang glider that is as compact as possible, yet is safe for flight and landings. You have no previous experience designing hang gliders. You don't know what size hang glider is needed or whether or not the size of a hang glider depends on the size of its load. (The people range in size.)

With so many unknowns, you decide to conduct some tests to understand the principles involved. In this Investigation, you will use mathematics to examine the relationship between the speed of descent and the size of a hang glider. As part of a three-member research team, you will use tissue-paper triangles to study hang gliders.

Make an Investigation Folder in which you can store all of your work on this Investigation for future use.

Cooperative Learning

This Investigation offers an excellent opportunity for using cooperative learning groups. For more information on cooperative learning strategies and group management, see *Cooperative Learning in the Mathematics Classroom,* one of the titles in the Glencoe Mathematics Professional Series.

TRIANGLE TEST				
Test	5 cm	10 cm	20 cm	35 cm
perimeter				
surface area				
1				
2				
3				
4				
5				

THE EXPERIMENT

1. Begin by copying the chart above.
2. Cut out four equilateral triangles from tissue paper. The sides of the triangles should be 5, 10, 20, and 35 centimeters long, respectively. These triangles will serve as models of hang gliders. Find the perimeter and surface area of each of these triangles and record them in your chart.
3. Measure a height of five feet on a wall. Mark this height with a piece of masking tape.
4. Hold the smallest triangle parallel to the ground at a height of 5 feet. Have a second person ready to use a stopwatch to time how long it takes the triangle to reach the floor. A third person should give the verbal command, "Ready, set, drop." At the drop command, the person holding the tissue paper glider should let go of the paper. The timer starts the stopwatch at the verbal command and stops it when the glider hits the ground. Repeat this process until five drops have been made, recording your data after each drop.
5. Repeat Step 4 for the other three gliders, recording the data for each drop.
6. Review the data that you collected. What observations can you make? Are there any relationships that you can see from the data?

Do the perimeter and surface area have a relationship with the time of the drop? Explain.

You will continue working on this Investigation throughout Chapters 8 and 9.

Be sure to keep your triangle models, charts, and other materials in your Investigation Folder.

Ready, Set, Drop! Investigation

Working on the Investigation
Lesson 8–1, p. 461

Working on the Investigation
Lesson 9–4, p. 519

Working on the Investigation
Lesson 9–7, p. 541

Closing the Investigation
End of Chapter 9, p. 548

2 SETUP

You may wish to have a student read the first three paragraphs of the Investigation to provide information about hang gliders. Then read the next two paragraphs to introduce the activity. Discuss the activity with your students. Then separate the class into groups of four.

3 MANAGEMENT

Each group member should be responsible for a specific task.

Recorder Collects data.
Measurer Measures sides of triangles and height on the wall.
Triangles Builder Makes triangles and drops them from indicated height.
Timer Times how long it takes each triangle to reach the floor.

At the end of the activity, each member should turn in his or her respective equipment.

Sample Answers

Answers will vary as they are based on the size of the hang glider.

Investigations and Projects Masters, p. 16

NAME________ DATE ________

Investigation, Chapters 8 and 9

Student Edition Pages 448–449, 461, 519, 541, 548

Ready, Set, Drop!

Use this chart to record your data from Working on the Investigation, page 461.

Launch Method		
Launch Trial	Launch Height	Horizontal Distance
1		
2		
3		
4		
5		
6		
7		
8		
9		
10		

Linear Equation:

Slope:

Use this chart to record your data from Working on the Investigation, page 541.

Triangle	Average glide time
1	
2	
3	
4	

Work with your group to answer the following questions and add others to be considered.

- You have investigated the effect of forward motion on the hang gliders. Using that information, what recommendation would you make to a hang glider in regards to their starting motions?
- What predictions about glide speed would you make for a hang glider that is larger than 32 feet wide?

CHAPTER 8

Solving Systems of Linear Equations and Inequalities

PREVIEWING THE CHAPTER

This chapter introduces students to systems of linear equations by having them solve systems by graphing and classify systems as consistent or inconsistent and independent or dependent. Students learn to use algebraic methods, including the substitution method and the elimination method with addition or subtraction and with multiplication and addition. They also learn to assess systems of equations to decide which solution method to employ. Graphing calculator technology is integrated throughout to help students graph systems of equations. The chapter concludes with students solving systems of inequalities by graphing.

Lesson (Pages)	Lesson Objectives	NCTM Standards	State/Local Objectives
8-1A (452–453)	Use a graphing calculator to solve systems of equations.	1–5, 8	
8-1 (454–461)	Solve systems of equations by graphing. Determine by graphing whether a system of equations has one solution, no solutions, or infinitely many solutions.	1–5, 8	
8-2 (462–468)	Solve systems of equations by using the substitution method. Organize data to solve problems.	1–5	
8-3 (469–474)	Solve systems of equations by using the elimination method with addition or subtraction.	1–5	
8-4 (475–481)	Solve systems of equations by using the elimination method with multiplication and addition. Determine the best method for solving systems of equations.	1–5	
8-5 (482–486)	Solve systems of inequalities by graphing.	1–5, 8	

A complete, 1-page lesson plan is provided for each lesson in the ***Lesson Planning Guide***. Answer keys for each lesson are available in the ***Answer Key Masters.***

ORGANIZING THE CHAPTER

You may want to refer to the **Course Planning Calendar** on page T12 for detailed information on pacing.
PACING: Standard—9 days; **Honors**—8 days; **Block**—5 days; **Two Years**—14 days

LESSON PLANNING CHART

			Blackline Masters											
Lesson (Pages)	**Materials/ Manipulatives**	**Extra Practice (Student Edition)**	**Study Guide**	**Practice**	**Enrichment**	**Assessment and Evaluation**	**Modeling Mathematics**	**Multicultural Activity**	**Tech Prep Applications**	**Graphing Calculator**	**Science and Math Lab Manual**	**Real-World Applications**	**Interactive Mathematics Tools Software**	**Teaching Transparencies**
8-1A (452–453)	graphing calculator									pp. 28, 29				
8-1 (454–461)	geoboard rubber bands graphing calculator	p. 774	p. 56	p. 56	p. 56		p. 79		p. 15	p. 8		21	8-1	8-1A 8-1B
8-2 (462–468)	equation mat* cups and counters*	p. 774	p. 57	p. 57	p. 57	p. 212					pp. 35–38			8-2A 8-2B
8-3 (469–474)	graphing calculator	p. 775	p. 58	p. 58	p. 58	pp. 211, 212								8-3A 8-3B
8-4 (475–481)	graphing calculator	p. 775	p. 59	p. 59	p. 59	p. 213		p. 15	p. 16			22		8-4A 8-4B
8-5 (482–486)		p. 775	p. 60	p. 60	p. 60	p. 213	pp. 57–59	p. 16						8-5A 8-5B
Study Guide/ Assessment (487–491)						pp. 197–210, 214–216								

*Included in Glencoe's Student Manipulative Kit and Overhead Manipulative Resources.

ORGANIZING THE CHAPTER

Other Chapter Resources

Student Edition
Investigation, pp. 448–449
Mathematics and Society, p. 481
Working on the Investigation, p. 461

Teacher's Classroom Resources
Investigations and Projects Masters, pp. 53–56
Algebra and Geometry Overhead Manipulative Resources, pp. 25–26

Technology
Test and Review Software (IBM and Macintosh)
CD-ROM Interactions (Windows and Macintosh)

Professional Publications
Block Scheduling Booklet
Glencoe Mathematics Professional Series

Outside Resources

Books/Periodicals
Developing Skills in Algebra One, Book C, Dale Seymour Publications
Problem-Solving Experiences in Algebra, Addison-Wesley

Software
Alge-Blaster Plus, Davidson
Tools of Mathematics: Algebra, William K. Bradford Publishing Company

Videos/CD-ROMs
Solving Equations 2, The Wisconsin Foundation for Vocational, Technical, and Adult Education
Solving Simultaneous Equations and Inequalities Algebraically and Geometrically, Nasco

Assessment Resources

Student Edition
Math Journal, pp. 478, 484
Mixed Review, pp. 461, 468, 474, 481, 486
Self Test, p. 474
Chapter Highlights, p. 487
Chapter Study Guide and Assessment, pp. 488–490
Alternative Assessment, p. 491
Portfolio, p.491
Cumulative Review, pp. 492–493

Teacher's Wraparound Edition
5-Minute Check, pp. 454, 462, 469, 475, 482
Check for Understanding, pp. 458, 466, 472, 478, 484
Closing Activity, pp. 461, 468, 473, 481, 486
Cooperative Learning, pp. 471, 477

Assessment and Evaluation Masters
Multiple-Choice Tests, Forms 1A (Honors), 1B (Average), 1C (Basic), pp. 197–202
Free-Response Tests, Forms 2A (Honors), 2B (Average), 2C (Basic), pp. 203–208
Calculator-Based Test, p. 209
Performance Assessment, p. 210
Mid-Chapter Test, p. 211
Quizzes A–D, pp. 212–213
Standardized Test Practice, p. 214
Cumulative Review, pp. 215–216

ENHANCING THE CHAPTER

Examples of some of the materials for enhancing Chapter 8 are shown below.

Multicultural Activity Masters, pp. 15, 16

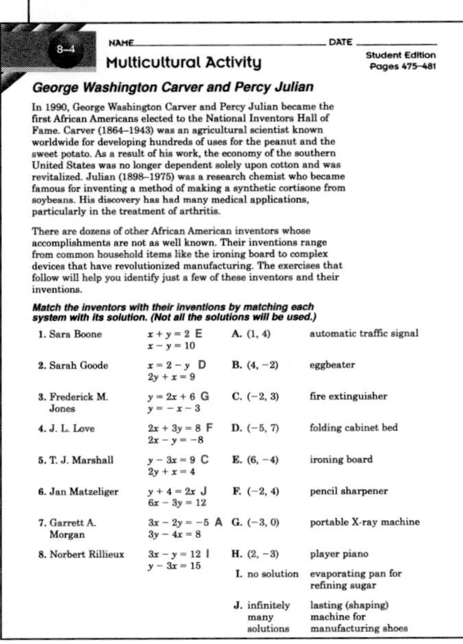

8–4 NAME ______ DATE ______

Multicultural Activity — Student Edition Pages 475–481

George Washington Carver and Percy Julian

In 1990, George Washington Carver and Percy Julian became the first African Americans elected to the National Inventors Hall of Fame. Carver (1864–1943) was an agricultural scientist known worldwide for developing hundreds of uses for the peanut and the sweet potato. As a result of his work, the economy of the southern United States was no longer dependent solely upon cotton and was revitalized. Julian (1898–1975) was a research chemist who became famous for inventing a method of making a synthetic cortisone from soybeans. His discovery has had many medical applications, particularly in the treatment of arthritis.

There are dozens of other African American inventors whose accomplishments are not as well known. Their inventions range from common household items like the ironing board to complex devices that have revolutionized manufacturing. The exercises that follow will help you identify just a few of these inventors and their inventions.

Match the inventors with their inventions by matching each system with its solution. (Not all the solutions will be used.)

Inventor	System	Answer	Solution	Invention
1. Sara Boone	$x + y = 2$ $x - y = 10$	E	A. (1, 4)	automatic traffic signal
2. Sarah Goode	$x = 2 - y$ $2y + x = 9$	D	B. (4, −2)	eggbeater
3. Frederick M. Jones	$y = 2x + 6$ $y = -x - 3$	G	C. (−2, 3)	fire extinguisher
4. J. L. Love	$2x + 3y = 8$ $2x - y = -8$	F	D. (−5, 7)	folding cabinet bed
5. T. J. Marshall	$y - 3x = 9$ $2y + x = 4$	C	E. (6, −4)	ironing board
6. Jan Matzeliger	$y + 4 = 2x$ $6x - 3y = 12$	J	F. (−2, 4)	pencil sharpener
7. Garrett A. Morgan	$3x - 2y = -5$ $3y - 4x = 8$	A	G. (−3, 0)	portable X-ray machine
8. Norbert Rillieux	$3x - y = 12$ $y - 3x = 15$	I	H. (2, −3)	player piano
			I. no solution	evaporating pan for refining sugar
			J. infinitely many solutions	lasting (shaping) machine for manufacturing shoes

Real-World Applications, 21, 22

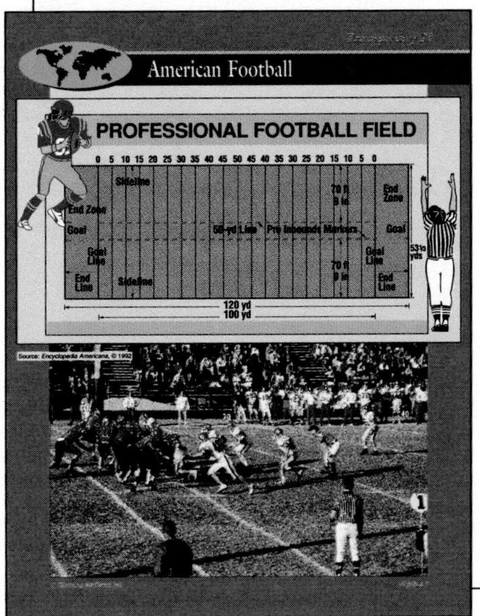

TECHNOLOGY

Graphing Calculator Masters, p. 8

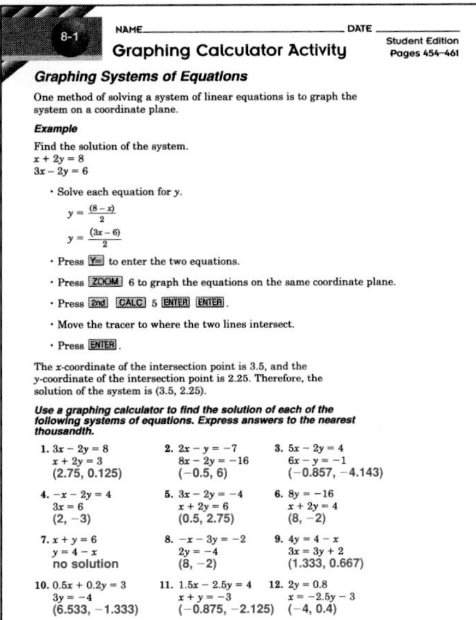

8-1 NAME ______ DATE ______

Graphing Calculator Activity — Student Edition Pages 454–461

Graphing Systems of Equations

One method of solving a system of linear equations is to graph the system on a coordinate plane.

Example

Find the solution of the system.

$x + 2y = 8$
$3x - 2y = 6$

- Solve each equation for y.

  $y = \frac{(8 - x)}{2}$

  $y = \frac{(3x - 6)}{2}$

- Press [Y=] to enter the two equations.
- Press [ZOOM] 6 to graph the equations on the same coordinate plane.
- Press [2nd] [CALC] 5 [ENTER] [ENTER].
- Move the tracer to where the two lines intersect.
- Press [ENTER].

The x-coordinate of the intersection point is 3.5, and the y-coordinate of the intersection point is 2.25. Therefore, the solution of the system is (3.5, 2.25).

Use a graphing calculator to find the solution of each of the following systems of equations. Express answers to the nearest thousandth.

1. $3x - 2y = 8$, $x + 2y = 3$ (2.75, 0.125)
2. $2x - y = -7$, $8x - 2y = -16$ (−0.5, 6)
3. $5x - 2y = 4$, $6x - y = -1$ (−0.857, −4.143)
4. $-x - 2y = 4$, $3x = 6$ (2, −3)
5. $3x - 2y = -4$, $x + 2y = 6$ (0.5, 2.75)
6. $8y = -16$, $x + 2y = 4$ (8, −2)
7. $x + y = 6$, $y = 4 - x$ no solution
8. $-x - 3y = -2$, $2y = -4$ (8, −2)
9. $4y = 4 - x$, $3x = 3y + 2$ (1.333, 0.667)
10. $0.5x + 0.2y = 3$, $3y = -4$ (6.533, −1.333)
11. $1.5x - 2.5y = 4$, $x + y = -3$ (−0.875, −2.125)
12. $2y = 0.8$, $x = -2.5y - 3$ (−4, 0.4)

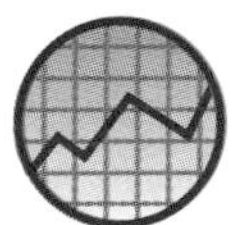

TECH PREP

Tech Prep Applications Masters, pp. 15, 16

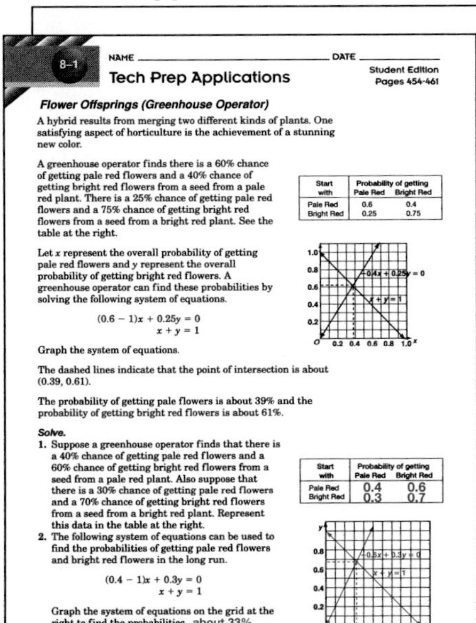

8–1 NAME ______ DATE ______

Tech Prep Applications — Student Edition Pages 454-461

Flower Offsprings (Greenhouse Operator)

A hybrid results from merging two different kinds of plants. One satisfying aspect of horticulture is the achievement of a stunning new color.

A greenhouse operator finds there is a 60% chance of getting pale red flowers and a 40% chance of getting bright red flowers from a seed from a pale red plant. There is a 25% chance of getting pale red flowers and a 75% chance of getting bright red flowers from a seed from a bright red plant. See the table at the right.

Start with	Probability of getting Pale Red	Bright Red
Pale Red	0.6	0.4
Bright Red	0.25	0.75

Let x represent the overall probability of getting pale red flowers and y represent the overall probability of getting bright red flowers. A greenhouse operator can find these probabilities by solving the following system of equations.

$$(0.6 - 1)x + 0.25y = 0$$
$$x + y = 1$$

Graph the system of equations.

The dashed lines indicate that the point of intersection is about (0.39, 0.61).

The probability of getting pale flowers is about 39% and the probability of getting bright red flowers is about 61%.

Solve.

1. Suppose a greenhouse operator finds that there is a 40% chance of getting pale red flowers and a 60% chance of getting bright red flowers from a seed from a pale red plant. Also suppose that there is a 30% chance of getting pale red flowers and a 70% chance of getting bright red flowers from a seed from a bright red plant. Represent this data in the table at the right.

Start with	Probability of getting Pale Red	Bright Red
Pale Red	0.4	0.6
Bright Red	0.3	0.7

2. The following system of equations can be used to find the probabilities of getting pale red flowers and bright red flowers in the long run.

$$(0.4 - 1)x + 0.3y = 0$$
$$x + y = 1$$

Graph the system of equations on the grid at the right to find the probabilities. about 33%, about 67%

Science and Math Lab Manual, pp. 35–38

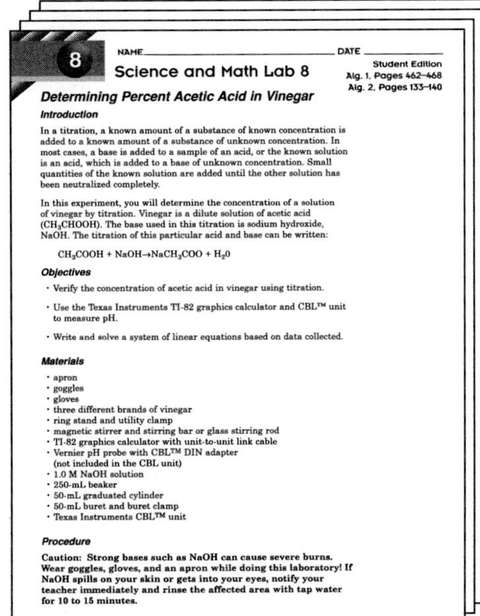

8 NAME ______ DATE ______

Science and Math Lab 8 — Student Edition Alg. 1, Pages 462–468; Alg. 2, Pages 133–140

Determining Percent Acetic Acid in Vinegar

Introduction

In a titration, a known amount of a substance of known concentration is added to a known amount of a substance of unknown concentration. In most cases, a base is added to a sample of an acid, or the known solution is an acid, which is added to a base of unknown concentration. Small quantities of the known solution are added until the other solution has been neutralized completely.

In this experiment, you will determine the concentration of a solution of vinegar by titration. Vinegar is a dilute solution of acetic acid (CH_3CHOOH). The base used in this titration is sodium hydroxide, NaOH. The titration of this particular acid and base can be written:

$$CH_3COOH + NaOH \rightarrow NaCH_3COO + H_20$$

Objectives

- Verify the concentration of acetic acid in vinegar using titration.
- Use the Texas Instruments TI-82 graphics calculator and CBL™ unit to measure pH.
- Write and solve a system of linear equations based on data collected.

Materials

- apron
- goggles
- gloves
- three different brands of vinegar
- ring stand and utility clamp
- magnetic stirrer and stirring bar or glass stirring rod
- TI-82 graphics calculator with unit-to-unit link cable
- Vernier pH probe with CBL™ DIN adapter (not included in the CBL unit)
- 1.0 M NaOH solution
- 250-mL beaker
- 50-mL graduated cylinder
- 50-mL buret and buret clamp
- Texas Instruments CBL™ unit

Procedure

Caution: Strong bases such as NaOH can cause severe burns. Wear goggles, gloves, and an apron while doing this laboratory! If NaOH spills on your skin or gets into your eyes, notify your teacher immediately and rinse the affected area with tap water for 10 to 15 minutes.

Problem of the Week Cards, 22, 23, 24

Troublesome Toothpicks — Problem-of-the-Week **Card 22**

The Problem

For this problem you actually will be solving several problems by rearranging toothpicks into different configurations with given properties. Good Luck!

1. In the figure shown, 13 toothpicks have made a shape with 6 equal areas. Use 12 toothpicks to create a shape that also has 6 regions of equal area.
   a. Using the shape you formed above, remove 4 toothpicks to make 3 triangles.
2. With 9 toothpicks, make the figure at the right.
   a. Remove 3 toothpicks to make 1 triangle.
   b. Remove 6 toothpicks to make 1 triangle.
   c. Remove 2 toothpicks to make 3 triangles.
   d. Remove 3 toothpicks to make 2 triangles.
   e. Remove 2 toothpicks to make 2 triangles.
   f. Remove 4 toothpicks to make 2 triangles.
3. With 12 toothpicks, make 5 squares in such a way that you can do the following.
   a. Remove 4 toothpicks to make 1 square.
   b. Remove 4 toothpicks to make 2 squares.
   c. Remove 2 toothpicks to make 3 squares.
   d. Remove 1 toothpick to make 3 squares and 1 rectangle.
   e. Remove 2 toothpicks to make 1 square and 2 rectangles.
4. With 6 toothpicks, make 4 equilateral triangles.

Strategies and Hints

1. Use actual toothpicks to solve the problems.
2. Find other 5-square arrangements.
3. Consider different-sized squares.

MAKING MATHEMATICS RELEVANT

This two-page introduction to the chapter provides students with an opportunity to explore contemporary topics and their applications to mathematics.

Background Information

Games People Play A great deal of information can be contained in a table. Often the information can be displayed differently to create a different emphasis. For example, the table here could be written with the fastest-growing sport, in-line roller skating, at the top decreasing to the fastest-declining sport, volleyball. Students could also use the data to compute the percentage of participants in each sport in 1993.

CHAPTER 8

Solving Systems of Linear Equations and Inequalities

Objectives

In this chapter, you will:

- graph systems of equations,
- solve systems of equations using various methods,
- organize data to solve problems, and
- solve systems of inequalities by graphing.

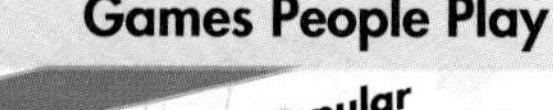

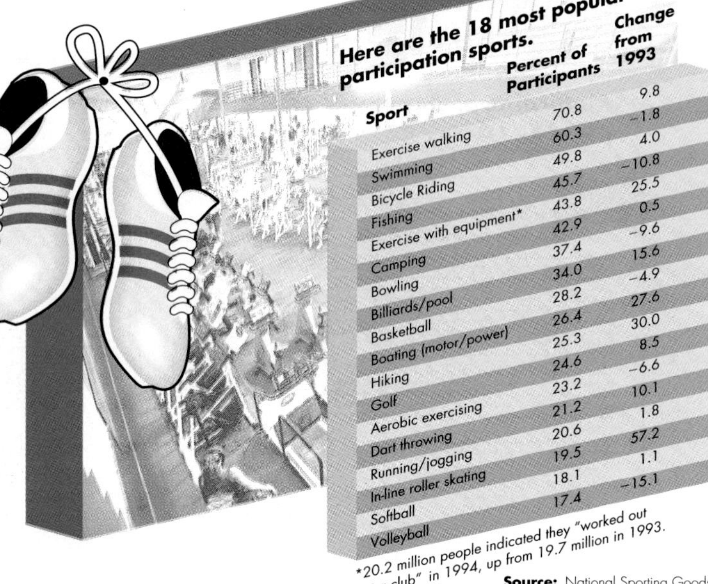

Here are the 18 most popular participation sports.

Sport	Percent of Participants	Change from 1993
Exercise walking	70.8	9.8
Swimming	60.3	–1.8
Bicycle Riding	49.8	4.0
Fishing	45.7	–10.8
Exercise with equipment*	43.8	25.5
Camping	42.9	0.5
Bowling	37.4	–9.6
Billiards/pool	34.0	15.6
Basketball	28.2	–4.9
Boating (motor/power)	26.4	27.6
Hiking	25.3	30.0
Golf	24.6	8.5
Aerobic exercising	23.2	–6.6
Dart throwing	21.2	10.1
Running/jogging	20.6	1.8
In-line roller skating	19.5	57.2
Softball	18.1	1.1
Volleyball	17.4	–15.1

*20.2 million people indicated they "worked out at a club" in 1994, up from 19.7 million in 1993.

Source: National Sporting Goods Association, 1994

Do you dream of starring in the NBA or playing for the New York Yankees? Do you devote all your leisure time to mainly one sport? Maybe you should try something new in the world of sports. Lacrosse, in-line skating, or judo, anyone?

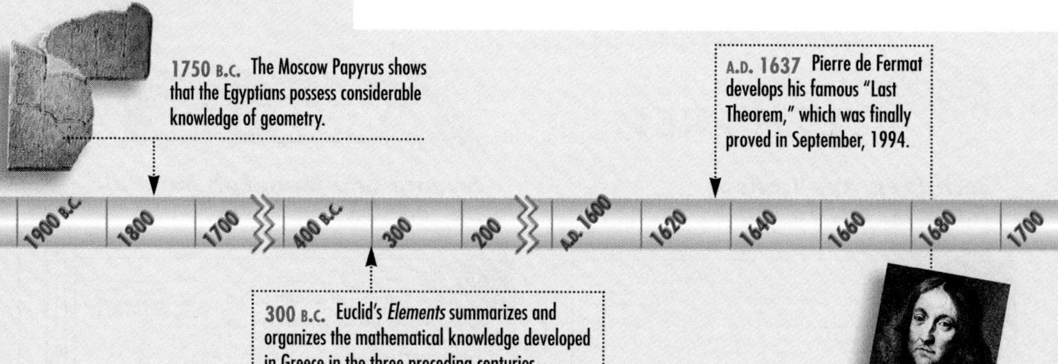

1750 B.C. The Moscow Papyrus shows that the Egyptians possess considerable knowledge of geometry.

A.D. 1637 Pierre de Fermat develops his famous "Last Theorem," which was finally proved in September, 1994.

300 B.C. Euclid's *Elements* summarizes and organizes the mathematical knowledge developed in Greece in the three preceding centuries.

The Egyptians practiced an applied approach to mathematics. The Greeks employed what is known as pure mathematics. Have students research each approach and report on the influence of Egyptian and Greek mathematicians.

interNET CONNECTION

Use the Affirmative Fitness page on nutritional services, personal training, and other fitness links to learn more about physical fitness.

World Wide Web
http://student-www.uchicago.edu/users/tkhafen/affirmative.html

PEOPLE IN THE NEWS

Short-track speed skaters **Julie Goskowicz**, 15, and **Tony Goskowicz**, 18, are a brother-and-sister team aiming for the 1998 Olympics. They started skating eight years ago in their hometown of New Berlin, Wisconsin, when their father gave them each a pair of skates. Although both finished last in their first race, they enjoyed the sport and continued to train. Hard work has earned them a place at the U.S. Olympic Education Center in Marquette, Michigan, where they study and train while participating in the racing circuit.

Chapter Project

In 1994, five-time Boston Marathon winner Jim Knaub tested his wheelchair's aerodynamics in the same wind tunnel that the Chrysler Corporation uses to test its car and truck designs. Knaub gained invaluable information concerning racing posture as well as helmet, wheel, and seat design. Earlier in the year, Knaub's Boston-Marathon-winning streak ended when he had to pull over twice during the race to make repairs. The winner was Heinz Frei of Switzerland.

- Suppose during a race, Frei's speed is 45 mph and Knaub is 264 feet ahead of him, racing at 36 mph.
- Write a system of equations to represent this situation. (*Hint:* Convert units from miles per hour to feet per second.) $y = 66x$, $y = 264 + 52.8x$
- If their speeds remained constant, when would Frei catch up with Knaub? Explain how you know using graphing. **After 20 seconds; the graphs intersect at (20, 1320); that is, after 20 seconds, Frei catches Knaub.**

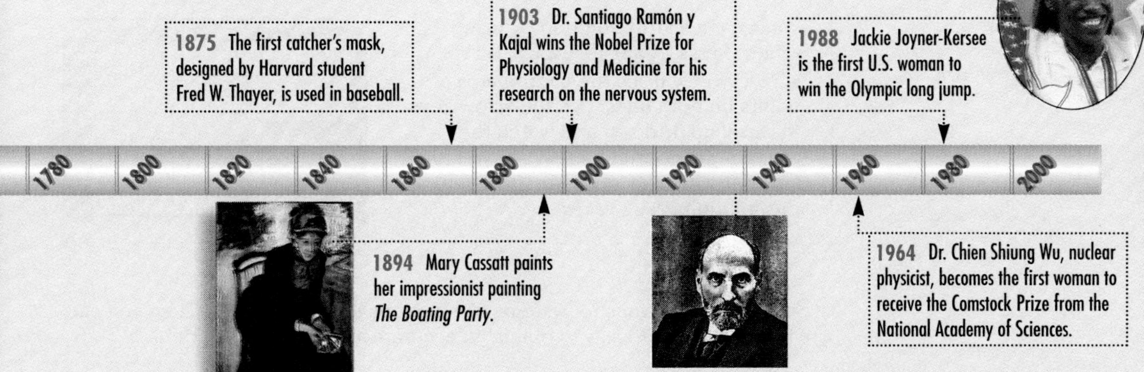

1875 The first catcher's mask, designed by Harvard student Fred W. Thayer, is used in baseball.
1903 Dr. Santiago Ramón y Kajal wins the Nobel Prize for Physiology and Medicine for his research on the nervous system.
1988 Jackie Joyner-Kersee is the first U.S. woman to win the Olympic long jump.
1780
1800
1820
1840
1860
1880
1900
1920
1940
1960
1980
2000
1894 Mary Cassatt paints her impressionist painting *The Boating Party*.
1964 Dr. Chien Shiung Wu, nuclear physicist, becomes the first woman to receive the Comstock Prize from the National Academy of Sciences.

PEOPLE IN THE NEWS

Like most people, the Goskowiczes did not begin a new activity with instant success. In their first race, both Tony and Julie fell and finished last. "But we thought it was fun," Julie says, so they persevered and eventually mastered the sport.

Chapter Project

Cooperative Learning You may choose to have students pair off or arrange them in cooperative groups to discuss possible solutions for this activity. This project challenges students to examine the concepts of rate, time, and distance as they apply to a race situation and to represent that situation as a system of equations.

Alternative Chapter Projects

Two other chapter projects are included in the *Investigations and Projects Masters.* In Chapter 8 Project A, pp. 53–54, students extend the topic in the chapter opener. In Chapter 8 Project B, pp. 55–56, students explore and research linear programming.

Investigations and Projects Masters, p. 53

NAME________________ DATE________

Chapter 8 Project A

Student Edition Pages 450–484

Games People Play

1. Work with a partner. Select 50–100 people over the age of 7 in your neighborhood or family for a survey.
2. Ask the people you have selected to name any sports they have participated in at least once during the last year. (People who name swimming, exercise walking, exercising with equipment, running/jogging, or aerobics can be counted only if they have done one of these particular activities at least 6 times during the year.)
3. Determine the number of people who participated in each sport. What percent of the people you surveyed participated in each sport?
4. List the sports in order from greatest to least participation.
5. Compare your results with the table in the chapter opener. Are your results similar? If not, how do the results differ? Are you surprised at the results? Explain.
6. Complete one of the following activities based on your data:
   - Put your data together with that of another group or two. Find the percent of participation for each sport and make a new list of sports, ranked from greatest to least participation. Share your findings with the class by making a poster or giving an oral report.
   - Write an article for the school newspaper.
   - Make a pictograph for the school newspaper.
   - Do some library research and discover the participation in sports of past generations. What sporting pastimes were the most popular in 1850? 1900? 1950? What do you think will be the most popular sports in 2000? Write a report or make an oral presentation of your findings.

8-1A LESSON NOTES

NCTM Standards: 1–5, 8

Objective
Use a graphing calculator to solve systems of equations.

Recommended Time
25 minutes

Instructional Resources
Graphing Calculator Masters, pp. 28 and 29

These masters provide keystroking instruction for this lesson for the TI-81 and Casio graphing calculators.

1 FOCUS

Motivating the Lesson
Ask students how many points of intersection are possible when two straight lines are drawn in the same plane. **0, 1, or infinitely many** Remind students that such pairs of lines are called *parallel, intersecting,* and *coincident,* respectively.

8–1A Graphing Technology Systems of Equations

A Preview of Lesson 8–1

When solving systems of linear equations graphically, each equation is graphed on the same coordinate plane. The coordinates of the point at which the graphs intersect is the solution of the system. The graphing calculator permits us to graph several equations on the same coordinate plane and approximate the coordinates of the intersection point.

Example 1 **Use a graphing calculator to solve the system of equations.**

$x + y = 9$
$2x - y = 15$

Begin by rewriting each equation in an equivalent form by solving for y.

$x + y = 9$ $2x - y = 15$
$y = -x + 9$ $2x - 15 = y$

Graph each equation in the integer window $[-47, 47]$ by $[-31, 31]$. Recall that the integer window can be obtained by entering ZOOM 6 ZOOM 8 ENTER.

Enter: Y= (–) X,T,θ + 9 ENTER 2 X,T,θ – 15 ZOOM 6 ZOOM 8 ENTER

The graphs intersect in one point. The coordinates of this point are the solution to the system of equations. Press the TRACE key and use the arrow keys to move the cursor to the point of intersection. The coordinates of the point are (8, 1). Thus, the solution is (8, 1).

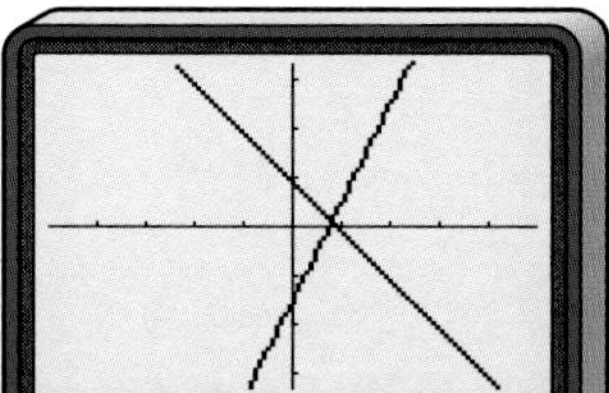

We can check this solution by using tables. Press 2nd TABLE. On the screen you will see the coordinates of points on both lines. Use the arrow keys to scroll up or down and watch the trend of the coordinates. When you find a row at which $Y_1 = Y_2$, you have found the solution. *The solution checks.*

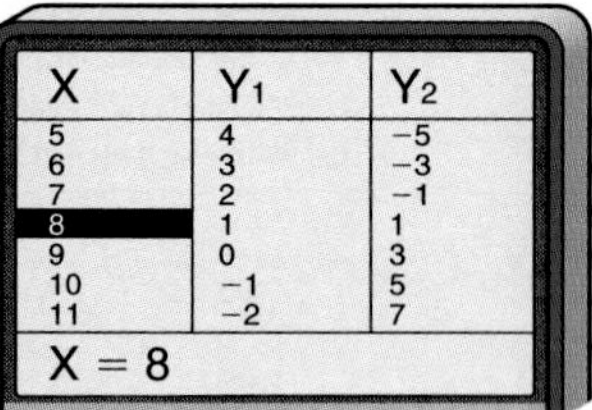

X	Y_1	Y_2
5	4	−5
6	3	−3
7	2	−1
8	1	1
9	0	3
10	−1	5
11	−2	7

X = 8

Sometimes solutions to systems of equations are not integers. Then you can use the ZOOM IN process to obtain an accurate approximate solution.

Example 2 **Use a graphing calculator to solve this system of equations to the nearest hundredth.**

$$y = 0.35x - 1.12$$
$$y = -2.25x - 4.05$$

Begin by graphing the equations in the standard viewing window.

Enter: [Y=] .35 [X,T,θ] [–] 1.12

[ENTER] [(–)] 2.25 [X,T,θ]

[–] 4.05 [ZOOM] 6

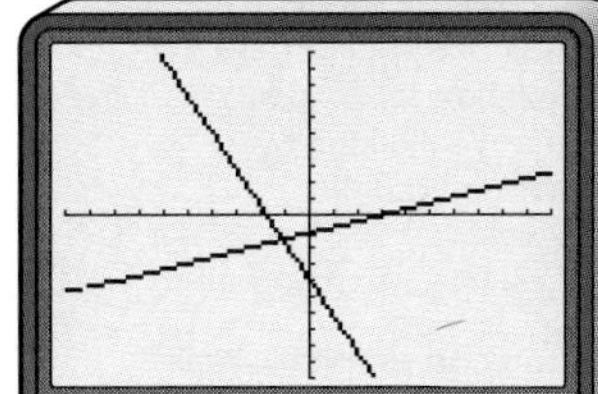

The graphs intersect at a point in the third quadrant. Use the TRACE function and the arrow keys to determine an approximation for the coordinates of the point of intersection. The ZOOM IN feature of the calculator is very useful for determining the coordinates of the intersection point with greater accuracy. Begin by placing the cursor on the intersection point and observing the coordinates, then press [ZOOM] 2 [ENTER]. Repeat this process as many times as necessary to get a more accurate answer.

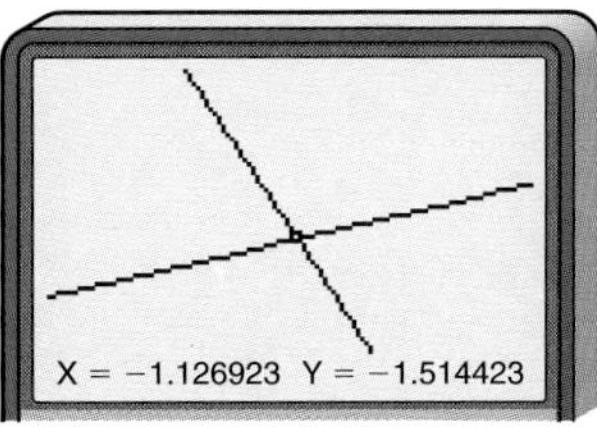

You may want to use the INTERSECT feature to find the coordinates of the point of intersection. Press [2nd] [CALC] 5 [ENTER] [ENTER] [ENTER].

The solution is $(-1.13, -1.51)$.

EXERCISES

Use a graphing calculator to solve each system of equations. State each decimal solution to the nearest hundredth.

1. $y = x + 7$, $y = -x + 9$ **(1, 8)**
2. $x + y = 27$, $3x - y = 41$ **(17, 10)**
3. $y = 3x - 4$, $y = -0.5x + 6$ **(2.86, 4.57)**
4. $x - y = 6$, $y = 9$ **(15, 9)**
5. $x + y = 5.35$, $3x - y = 3.75$ **(2.28, 3.08)**
6. $5x - 4y = 26$, $4x + 2y = 53.3$ **(10.2, 6.25)**
7. $2x + 3y = 11$, $4x + y = -6$ **(−2.9, 5.6)**
8. $2.93x + y = 6.08$, $8.32x - y = 4.11$ **(0.91, 3.43)**
9. $125x - 200y = 800$, $65x - 20y = 140$ **(1.14, −3.29)**
10. $0.22x + 0.15y = 0.30$, $-0.33x + y = 6.22$ **(−2.35, 5.44)**

2 TEACH

Teaching Tip In both examples, the initial viewing window was specified. Sometimes, however, a point of intersection may lie outside the window. In this case, students would have to increase the window to locate the intersection.

3 PRACTICE/APPLY

Assignment Guide

Core: 1–10
Enriched: 1–10

4 ASSESS

Observing students working with technology is an excellent method of assessment.

Using Technology

This lesson offers an excellent opportunity for using technology in your algebra classroom. For more information on using technology, see *Graphing Calculators in the Mathematics Classroom,* one of the titles in the Glencoe Mathematics Professional Series.

8-1 LESSON NOTES

NCTM Standards: 1–5, 8

Instructional Resources

- Study Guide Master 8-1
- Practice Master 8-1
- Enrichment Master 8-1
- Graphing Calculator Masters, p. 8
- Modeling Mathematics Masters, p. 79
- Real-World Applications, 21
- Tech Prep Applications Masters, p. 15

Transparency 8-1A contains the 5-Minute Check for this lesson; **Transparency 8-1B** contains a teaching aid for this lesson.

Recommended Pacing	
Standard Pacing	Days 2 & 3 of 9
Honors Pacing	Day 2 of 8
Block Scheduling*	Day 1 of 5
Alg. 1 in Two Years*	Days 2 & 3 of 14

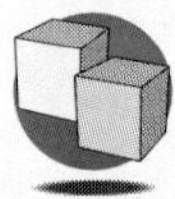

*For more information on pacing and possible lesson plans, refer to the *Block Scheduling Booklet* and *Algebra 1 in Two Years.*

1 FOCUS

5-Minute Check
(over Chapter 7)

Solve.

1. $6 + n > 40$ $n > 34$
2. $14p < 84$ $p < 6$
3. $12 + 4x < 20$ $x < 2$
4. $13r - 11 > 7r + 37$ $r > 8$
5. $3 + x < -4$ *or* $3 + x > 4$ $x < -7$ or $x > 1$

Motivating the Lesson

Hands-On Activity Take two rulers and move them around in relation to each other. It should quickly become clear that they can intersect in, at most, one point. Then take two pieces of string and move them in relation to each other. Students should see that the strings can intersect many times.

8–1 Graphing Systems of Equations

What YOU'LL LEARN

- To solve systems of equations by graphing, and
- to determine whether a system of equations has one solution, no solution, or infinitely many solutions by graphing.

Why IT'S IMPORTANT

You can graph systems of equations to solve problems involving business and geometry.

APPLICATION
World Records

Cape Verde is a group of islands located off the westernmost point of Africa. In December, 1994, Frenchman Guy Delage set off from these islands for a 2400-mile swim across the Atlantic Ocean. He arrived at Barbados in the West Indies eight weeks later. Every day he would swim a while and then rest while floating with the current on a huge raft equipped with a fax machine, a computer, and a two-way radio.

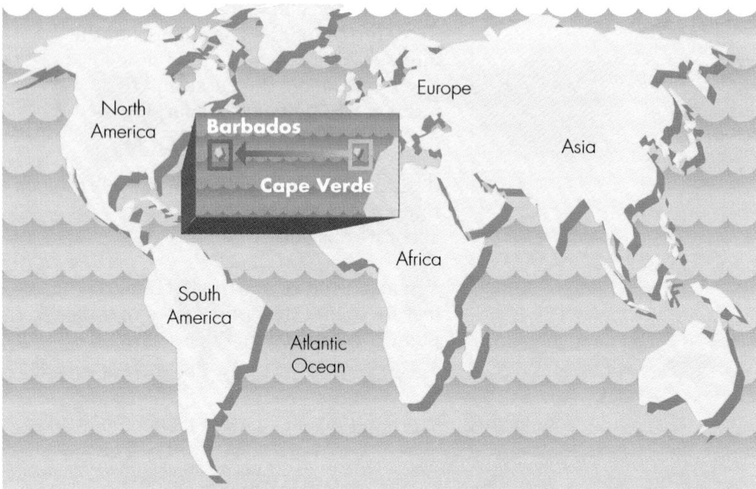

People are already considering trying to break Guy's record, but before a challenger makes the attempt, he or she should know what is required. Guy traveled approximately 44 miles per day. A good swimmer like Guy can swim about 3 miles per hour for an extended period, and the Atlantic currents will float a raft about 1 mile per hour. To match Guy's record, how many hours per day would one have to swim? How many hours would one be able to spend floating on the raft?

To solve this problem, let s represent the number of hours Guy swam, and let f represent the number of hours he floated. Then $3s$ represents the number of miles he traveled while swimming and $1f$ represents the number of miles he traveled while floating. You can write two equations to represent this situation.

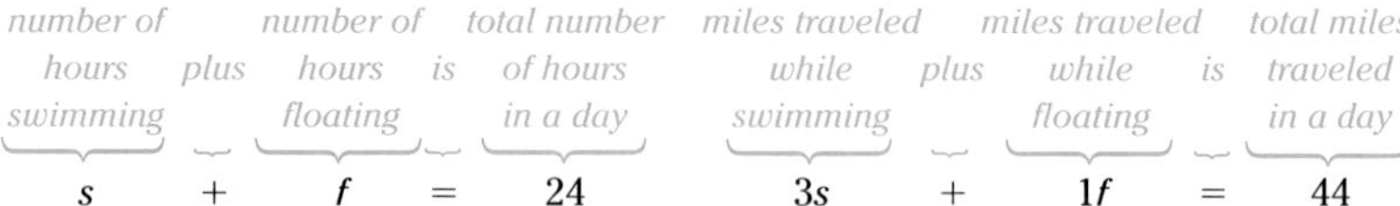

number of hours swimming plus number of hours floating is total number of hours in a day
s + f = 24
miles traveled while swimming plus miles traveled while floating is total miles traveled in a day
3s + 1f = 44

Top Five List

Fastest swimmers to cross the English Channel (hr:min)

1. Penny Lee Dean, 7:40
2. Philip Rush, 7:55
3. Richard Davey, 8:05
4. Irene van der Laan, 8:06
5. Paul Asmuth, 8:12

While the fastest human swimmer, the world freestyle champion, has a top speed of 5.37 miles per hour (mph), many animals are much faster. The fastest fish, the sailfish, can swim at 68 mph. The fastest mammal, the killer whale, can reach a speed of 34.5 mph.

The equations $s + f = 24$ and $3s + f = 44$ together are called a **system of equations**. The solution to this problem is the ordered pair of numbers that satisfies both of these equations.

One method for solving a system of equations is to carefully graph the equations on the same coordinate plane. The coordinates of the point at which the graphs intersect is the solution of the system.

Guy Delage's raft

With most graphs of systems of equations, we can only estimate the solution. In this case, the graphs of $s + f = 24$ and $3s + f = 44$ appear to intersect at the point with coordinates (10, 14).

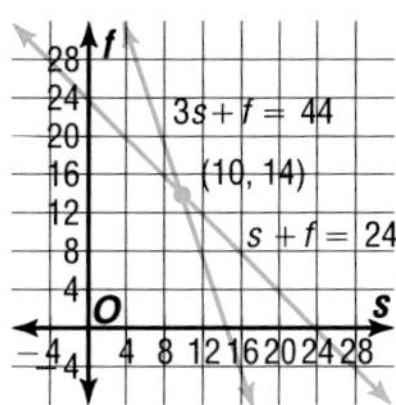

Check: In each equation, replace s with 10 and f with 14.

$$s + f = 24 \qquad 3s + f = 44$$
$$10 + 14 \stackrel{?}{=} 24 \qquad 3(10) + 14 \stackrel{?}{=} 44$$
$$24 = 24 \checkmark \qquad 44 = 44 \checkmark$$

The solution of the system of equations $s + f = 24$ and $3s + f = 44$ is (10, 14). The ordered pair (10, 14) means that a person trying to match Guy Delage's record would have to spend approximately 10 hours a day swimming and 14 hours floating.

Example 1 **Graph the system of equations to find the solution.**

$$x + 2y = 1$$
$$2x + y = 5$$

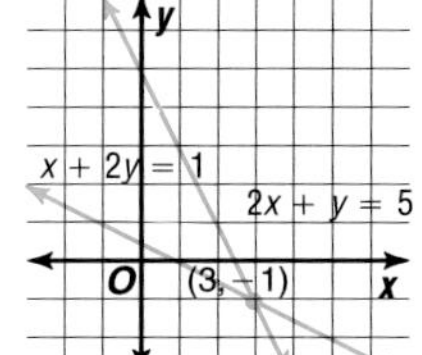

The graphs appear to intersect at the point with coordinates (3, −1). Check this estimate by replacing x with 3 and y with −1 in each equation.

Check:
$$x + 2y = 1 \qquad 2x + y = 5$$
$$3 + 2(-1) \stackrel{?}{=} 1 \qquad 2(3) + (-1) \stackrel{?}{=} 5$$
$$1 = 1 \checkmark \qquad 5 = 5 \checkmark$$

The solution is (3, −1).

CAREER CHOICES

Graphing systems of equations often arises in the study of populations of groups, and is used by people with careers in biological science, such as **ecologists**. An ecologist uses graphs of systems to study populations of organisms and how they relate to their environment.

A career as an ecologist usually requires a Ph.D. in a biological science and several years of laboratory work.

For more information, contact:

Ecological Society of America
2010 Massachusetts Ave.
Suite 400
Washington, D.C. 20036

A system of two linear equations has exactly one ordered pair as its solution when the graphs of the equations intersect at exactly one point. If the graphs coincide, they are the same line and have infinitely many points in common. In either case, the system of equations is said to be **consistent**. That is, it has *at least* one ordered pair that satisfies both equations.

2 TEACH

In-Class Example

For Example 1
Graph the system of equations to find the solution.
$y = -3x + 1$
$x + \frac{1}{2}y = -\frac{1}{2}$
(2, −5)

Teaching Tip If necessary, review graphing linear equations using x- and y-intercepts or slope-intercept form before beginning this lesson.

Teaching Tip Point out that since there are two variables in Example 1, the solution, if it exists, will be an ordered pair or ordered pairs.

CAREER CHOICES

In the science of ecology, predator-prey relationships are often studied. If there are too few predators, the population of prey will increase. If there are too many predators, the prey population can vanish. In general, nature attempts to maintain a balance in the populations of these different groups. Systems of equations can be used to model these various relationships.

GLENCOE Technology

Interactive Mathematics Tools Software

This multimedia software provides an interactive lesson that changes the speeds of two cyclists as they manipulate the distance and time. A **Computer Journal** gives students an opportunity to write about what they have learned.

For Windows & Macintosh

In-Class Example

For Example 2
Graph each pair of equations. Then determine the number of solutions for the system of equations.

a. $x - 2y = 4$
$x - 2y = -2$

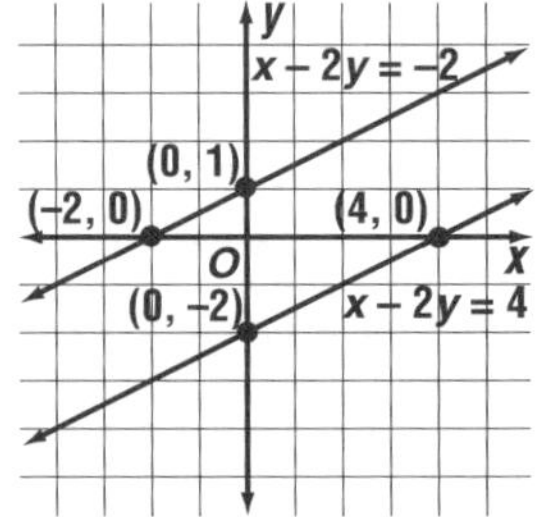

no solution

b. $3x - y = 2$
$12x - 4y = 8$

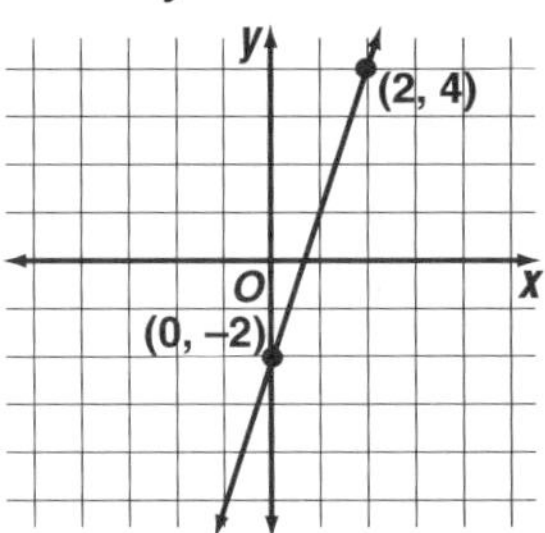

infinitely many solutions

Teaching Tip In Example 2b, explain that the graphs of these equations coincide.

It is also possible for the two graphs to be *parallel*. In this case, the system of equations is **inconsistent** because there is *no* ordered pair that satisfies both equations.

Another way to classify a system is by the number of solutions it has.

- If a system has exactly one solution, it is **independent**.
- If a system has an infinite number of solutions, it is **dependent**.

Thus, the system in Example 1 is said to be *consistent and independent*.

The chart below summarizes the possible solutions to systems of linear equations.

Graphs of Equations	Number of Solutions	Terminology
intersecting lines	exactly one	consistent and independent
same line	infinitely many	consistent and dependent
parallel lines	none	inconsistent

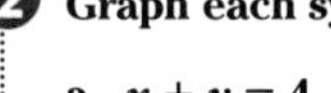

Example 2 **Graph each system of equations to determine the number of solutions.**

a. $x + y = 4$
$x + y = 1$

The graphs of the equations are parallel lines. Since they do not intersect, there is no solution to this system of equations. Notice that the two lines have the same slope but different y-intercepts.

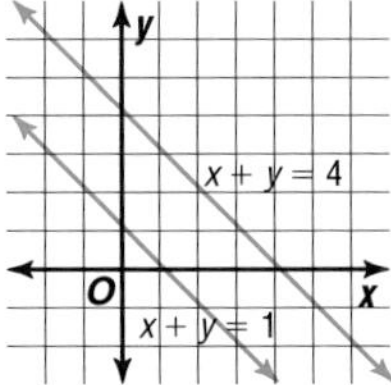

Recall that a system of equations that has no solution is said to be inconsistent.

b. $x - y = 3$
$2x - 2y = 6$

Each equation has the same graph. Any ordered pair on the graph will satisfy both equations. Therefore, there are infinitely many solutions of this system of equations. Notice that the graphs have the same slope and intercepts.

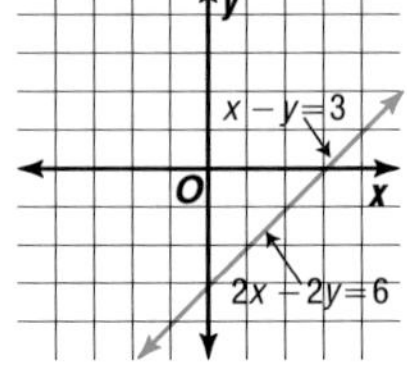

Recall that a system of equations that has infinitely many solutions is said to be consistent and dependent.

Check: Verify that the point at (4, 1) lies on both lines.

$$x - y = 3 \qquad 2x - 2y = 6$$
$$4 - 1 \stackrel{?}{=} 3 \qquad 2(4) - 2(1) \stackrel{?}{=} 6$$
$$3 = 3 \checkmark \qquad 6 = 6 \checkmark$$

Alternative Learning Styles

Auditory Read to students the World Records application at the beginning of the lesson. At each stage, encourage students to think about how each part of the problem can be represented algebraically.

The methods you use to solve algebra problems are often useful in solving problems involving geometry.

Example 3

Geometry

The points $A(-1, 6)$, $B(4, 8)$, $C(8, 3)$ and $D(-2, -1)$ are vertices of a quadrilateral.

a. Use a graph to determine the point of intersection of the diagonals of quadrilateral $ABCD$.

b. Find the equations of the lines containing the diagonals to verify the solution.

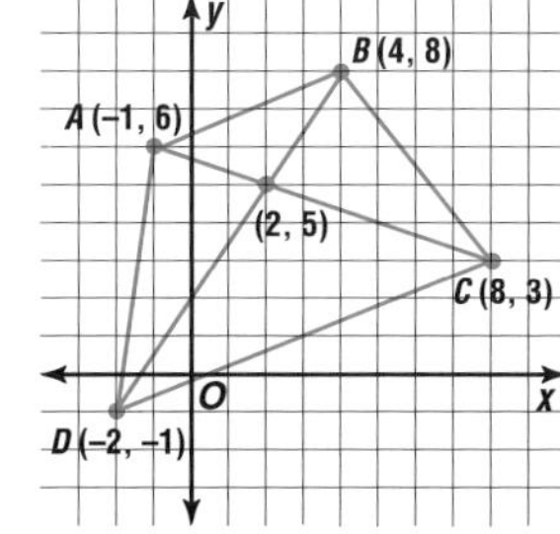

a. Draw quadrilateral $ABCD$ with diagonals $\overline{AC}$ and $\overline{BD}$. The diagonals appear to intersect at the point (2, 5).

b. To check the solution, find the equations of lines AC and BD and then verify that (2, 5) is a solution of both equations. First, find the slope of each line using $m = \frac{y_2 - y_1}{x_2 - x_1}$.

Slope of $\overleftrightarrow{AC}$

$$m = \frac{3 - 6}{8 - (-1)}$$
$$= \frac{-3}{9} \text{ or } -\frac{1}{3}$$

Slope of $\overleftrightarrow{BD}$

$$m = \frac{-1 - 8}{-2 - 4}$$
$$= \frac{-9}{-6} \text{ or } \frac{3}{2}$$

Then use the slope-intercept form, $y = mx + b$, to determine the equations.

Equation for $\overleftrightarrow{AC}$

$y = mx + b$

$6 = -\frac{1}{3}(-1) + b$ *Replace m with $-\frac{1}{3}$ and (x, y) with (−1, 6).*

$\frac{17}{3} = b$ The equation for $\overleftrightarrow{AC}$ is $y = -\frac{1}{3}x + \frac{17}{3}$.

Equation for $\overleftrightarrow{BD}$

$y = mx + b$

$8 = \frac{3}{2}(4) + b$ *Replace m with $\frac{3}{2}$ and (x, y) with (4, 8).*

$2 = b$ The equation for $\overleftrightarrow{BD}$ is $y = \frac{3}{2}x + 2$.

Check that (2, 5) is a solution to both equations.

$y = -\frac{1}{3}x + \frac{17}{3}$

$5 \stackrel{?}{=} -\frac{1}{3}(2) + \frac{17}{3}$ *(x, y) = (2, 5)*

$5 = 5$ ✔

$y = \frac{3}{2}x + 2$

$5 \stackrel{?}{=} \frac{3}{2}(2) + 2$ *(x, y) = (2, 5)*

$5 = 5$ ✔

The solution checks.

In-Class Example

For Example 3

The graphs of the equations $2x + y = 8$, $-x + y = 6$, and $7x + y = 3$ contain the sides of a triangle. Find the coordinates of the vertices of the triangle. **$\left(\frac{2}{3}, 6\frac{2}{3}\right)$, $\left(-\frac{3}{8}, 5\frac{5}{8}\right)$, (−1, 10)**

3 PRACTICE/APPLY

Check for Understanding

Exercises 1–20 are designed to help you assess your students' understanding through reading, writing, speaking, and modeling. You should work through Exercises 1–7 with your students and then monitor their work on Exercises 8–20.

Error Analysis
Stress that students should check their answers in the original equations. Make sure it is in the original equation, too, because they could have made a mistake along the way.

Additional Answers

6.

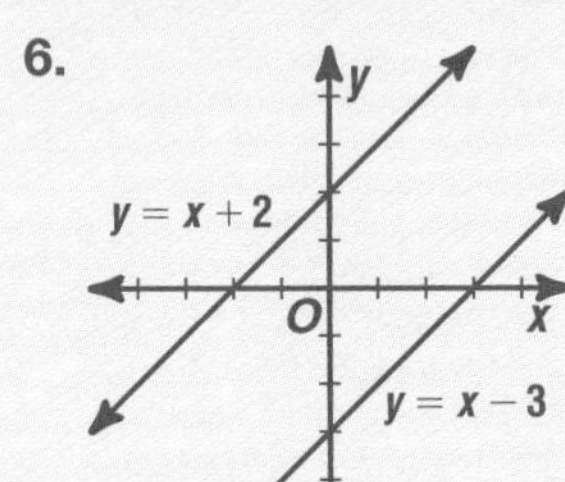

14.

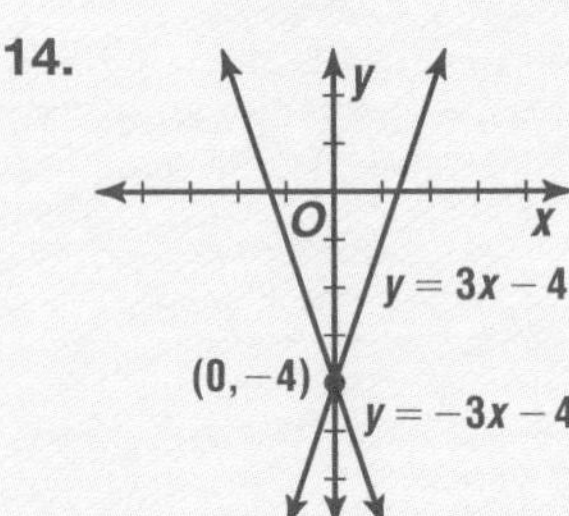

15.

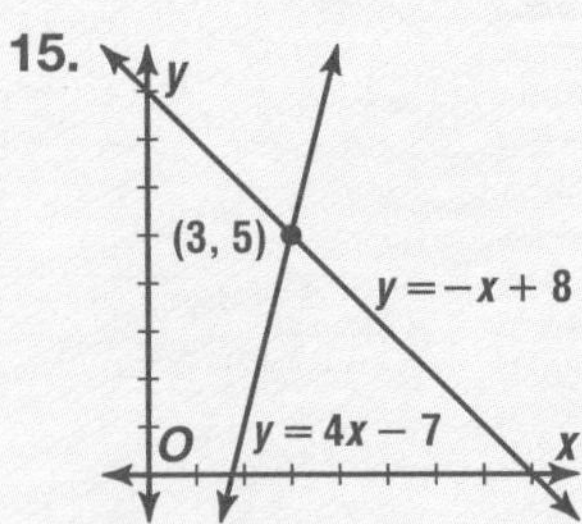

16.

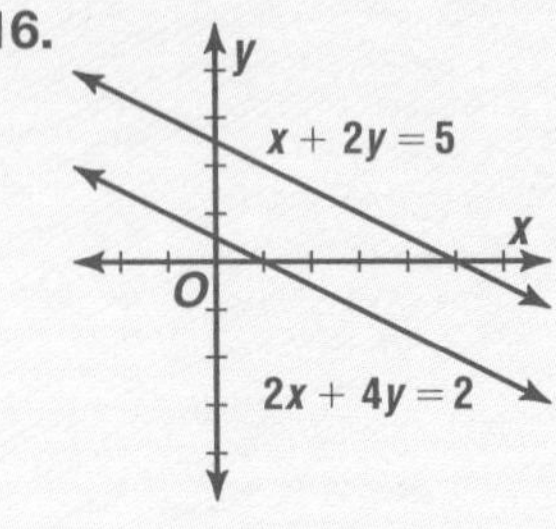

CHECK FOR UNDERSTANDING

Communicating Mathematics

Study the lesson. Then complete the following.

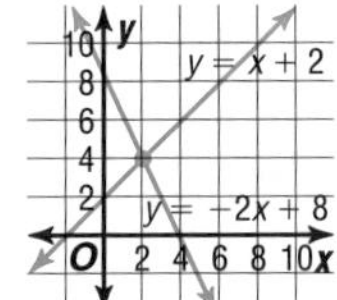

1. **State** the solution of the system of equations shown in the graph at the right. Justify your answer.
2. **Explain** what it means to *solve* a system of linear equations.
3. **Describe** the graph of a linear system that has infinitely many solutions.
4. **Name** two of the solutions for the system of equations in Example 2b. Verify your answers algebraically. **Sample answer: (3, 0), (4, 1)**
5. **Write** a system of linear equations that has $(-3, 5)$ as its only solution.
6. **Sketch** the graph of a linear system that has *no* solution. **Graphs will vary, but lines must be parallel; see margin for sample graph.**

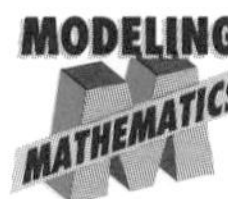

7. Use a geoboard and rubber bands to model a system of two equations that has the solution (3, 2). Let the lower left point on the geoboard represent the origin. **See students' work.**

1. (2, 4); The graphs intersect at the point (2, 4) and (2, 4) is a solution of both equations.
2. Find an ordered pair that satisifies both equations.
3. The graphs are the same line.
5. Sample answer: $x + y = 2$, $x - y = -8$

Guided Practice

Use the graphs at the right to determine whether each system has *one* solution, *no* solution, or *infinitely many* solutions. If the system has one solution, name it.

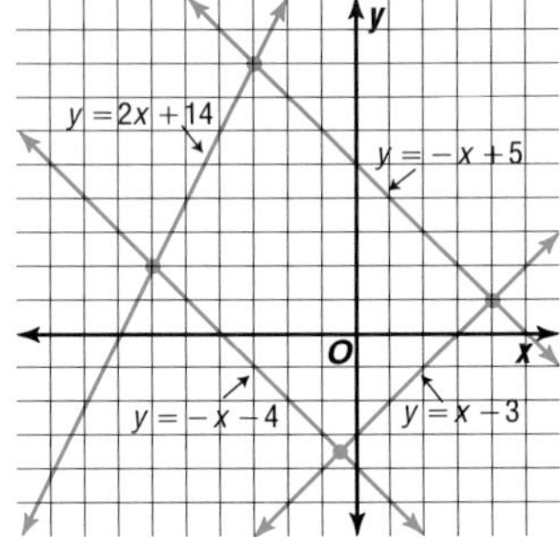

8. $y = -x + 5$, $y = x - 3$ **one; (4, 1)**
9. $y = -x - 4$, $y = -x + 5$
10. $y = 2x + 14$, $y = -x + 5$ **one; (−3, 8)**
11. $y = -x - 4$, $y = 2x + 14$

9. no solution
11. one; (−6, 2)

State whether the given ordered pair is a solution to each system. Write *yes* or *no*.

12. $x - y = 6$, $2x + y = 0$ $(-2, -4)$ **no**
13. $2x - y = 4$, $3x + y = 1$ $(1, -2)$ **yes**

Graph each system of equations. Then determine whether the system has *one* solution, *no* solution, or *infinitely many* solutions. If the system has one solution, name it. **14–19. See margin for graphs.**

14. $y = 3x - 4$, $y = -3x - 4$ **(0, −4)**
15. $y = -x + 8$, $y = 4x - 7$ **(3, 5)**
16. $x + 2y = 5$, $2x + 4y = 2$ **no solution**
17. $y = -6$, $4x + y = 2$ **(2, −6)**
18. $2x + 3y = 4$, $-4x - 6y = -8$ **infinitely many**
19. $2x + y = -4$, $5x + 3y = -6$ **(−6, 8)**
20. a. Graph the line $y - x = 6$. **a–b. See margin.**
    b. Slide the entire line four units to the right and down one unit. Draw the new line.
    c. Describe this system of equations.

20c. The graphs of the system are parallel lines. The system has no solution and is inconsistent.

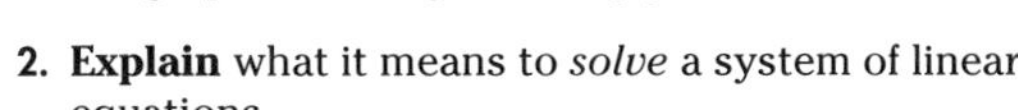

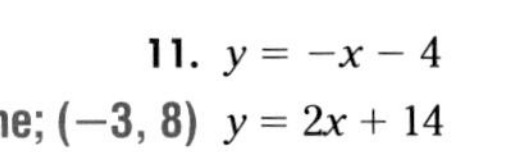

Reteaching

Using Alternative Methods Have half of the class determine the number of solutions for a system of equations by graphing. The other half can determine the number of solutions using the following method. They should write the equations in slope-intercept form. If the slopes and the *y*-intercepts are the same, the lines are identical. If the slopes are equal but the *y*-intercepts differ, the lines are parallel. If the equations have different slopes, the lines intersect in exactly one point. Have students discuss and compare the methods.

EXERCISES

Practice

Use the graphs below to determine whether each system has *one* solution, *no* solution, or *infinitely many* solutions. If the system has one solution, name it.

21. one, (3, −1)

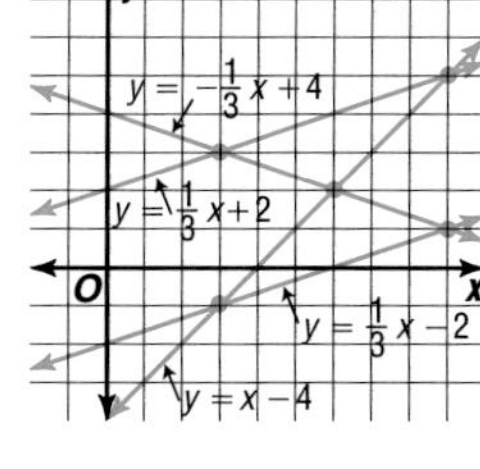

21. $y = x - 4$
$y = \frac{1}{3}x - 2$

22. $y = x - 4$ one, (6, 2)
$y = -\frac{1}{3}x + 4$

23. no solution

23. $y = \frac{1}{3}x + 2$
$y = \frac{1}{3}x - 2$

24. $y = x - 4$ one, (9, 5)
$y = \frac{1}{3}x + 2$

25. one, (3, 3)
26. one, (9, 1)

25. $y = -\frac{1}{3}x + 4$
$y = \frac{1}{3}x + 2$

26. $y = \frac{1}{3}x - 2$
$y = -\frac{1}{3}x + 4$

Graph each system of equations. Then determine whether the system has *one* solution, *no* solution, or *infinitely many* solutions. If the system has one solution, name it.

27–41. See Solutions Manual for graphs.

27. $y = -x$ (2, −2)
$y = 2x - 6$

28. $y = 2x + 6$ (−3, 0)
$y = -x - 3$

29. $x + y = 2$ (−1, 3)
$y = 4x + 7$

32. no solution

30. $2x + y = 10$ (4, 2)
$y = \frac{1}{2}x$

31. $x + y = 2$ (−2, 4)
$2y - x = 10$

32. $3x + 2y = 12$
$3x + 2y = 6$

33. $x - 2y = 2$ (2, 0)
$3x + y = 6$

34. $x - y = 2$ (3, 1)
$3y + 2x = 9$

35. $3x + y = 3$ infinitely many
$2y = -6x + 6$

36. (−1, −5)

36. $2x + 3y = -17$
$y = x - 4$

37. $y = \frac{2}{3}x - 5$ no solution
$3y = 2x$

38. $4x + 3y = 24$ (3, 4)
$5x - 8y = -17$

40. infinitely many

39. $\frac{1}{2}x + \frac{1}{3}y = 6$ (8, 6)
$y = \frac{1}{2}x + 2$

40. $6 - \frac{3}{8}y = x$
$\frac{2}{3}x + \frac{1}{4}y = 4$

41. $2x + 4y = 2$ infinitely many
$3x + 6y = 3$

Geometry

42. (2, 4), (0, 3), (−1, 10)

42. The graphs of the equations $-x + 2y = 6$, $7x + y = 3$, and $2x + y = 8$ contain the sides of a triangle. Find the coordinates of the vertices of the triangle.

43. Graph the system of equations below. Then find the area of the geometric figure.
$2x - 4 = 0$
$y = 8$
$x = 5$
$3y - 9 = 0$ See Solutions Manual for graph; 15 square units.

Graphing Calculator

Use a graphing calculator to solve each system of equations. Approximate the coordinates of the point of intersection to the nearest hundredth.

44. $y = x + 2$ (−1.50, 0.50)
$y = -x - 1$

45. $y = \frac{1}{4}x - 3$ (1.71, −2.57)
$y = -\frac{1}{3}x - 2$

46. $6x + y = 5$ (−0.44, 7.67)
$y = 9 + 3x$

47. $3 + y = x$ (−0.25, −3.25)
$2 + y = 5x$

Assignment Guide

Core: 21–47 odd, 48, 49, 51, 52–60
Enriched: 22–46 even, 48–60

For **Extra Practice,** see p. 774.

The red A, B, and C flags, printed only in the Teacher's Wraparound Edition, indicate the level of difficulty of the exercises.

Additional Answers

20a & 20b.

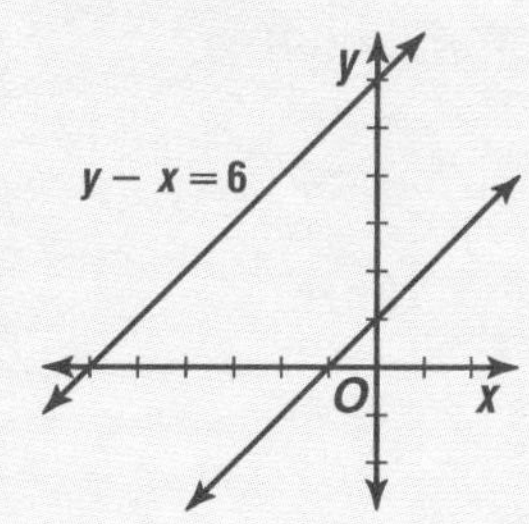

Study Guide Masters, p. 56

NAME______ DATE ______

Student Edition Pages 454–461

Study Guide

Graphing Systems of Equations

Two or more linear equations involving the same variables form a **system of equations.** The solution set for the system is the set of ordered pairs that satisfy both equations. One method for solving a system of equations is to graph the equations on the same coordinate plane.

Examples: Solve each system of equations by graphing.

$x + y = 2$
$x - y = 4$

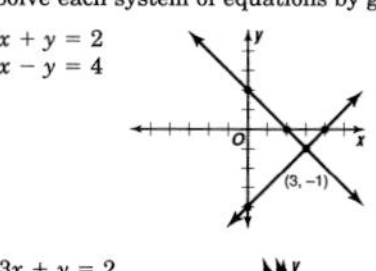

The point (3, −1) lies on both lines, thus (3, −1) is the solution set for the system of equations.

$3x + y = 2$
$3x + y = 4$
no solution

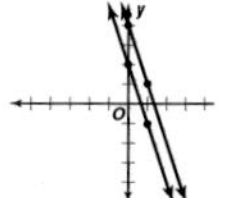

$y = 2x + 1$
$2y = 4x + 2$
infinitely many solutions

Use the graphs below to determine whether each system has one solution, no solution, or infinitely many solutions. If the system has one solution, name it.

1. $x - y = 1$, $x + y = -1$ — one solution (0, −1)

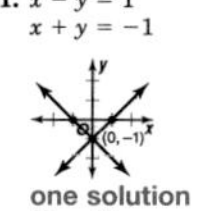

2. $x + 3y = 3$, $2x + y = 2$ — one solution $\left(\frac{3}{5}, \frac{4}{5}\right)$
3. $2x + 3y = 6$, $2x + 3y = -6$ — no solution
4. $x - y = 2$, $2x - 2y = 4$ — infinitely many solutions

Graph each system of equations on a separate sheet of paper. Then determine whether the system has one solution, no solution, or infinitely many solutions. If the system has one solution, name it.

5. $4x - 2y = 4$, $4x - 2y = 0$ — no solution
6. $y = x + 2$, $y = 2x - 1$ — one solution (3, 5)
7. $x + y = 1$, $3x + 3y = 3$ — infinitely many solutions
8. $2x - 2y = 2$, $y = x$ — no solution

Additional Answers

17.

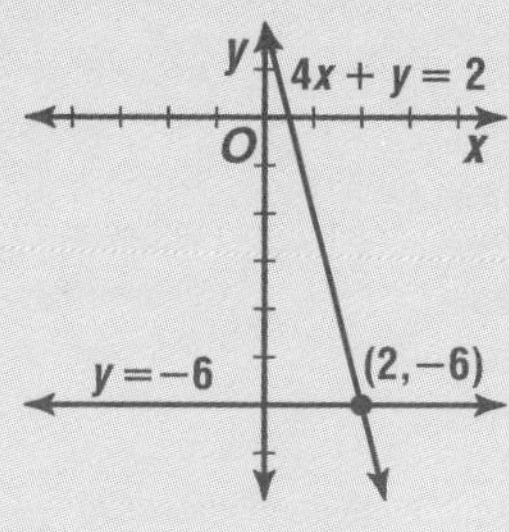

18.

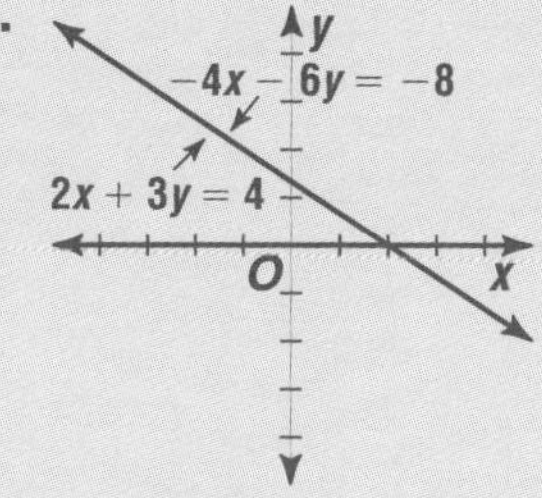

19.

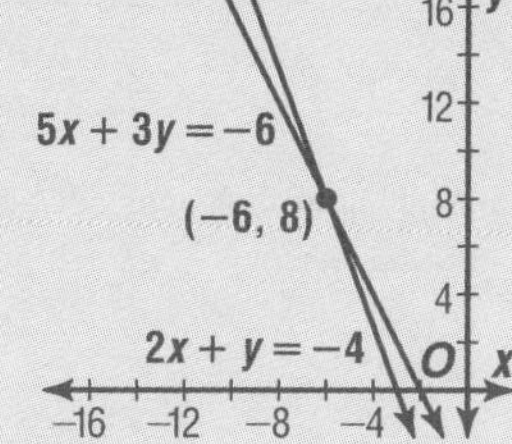

Additional Answers

50a. The number of toys sold for which expenses equal income.

50b. After selling more than 1000 toys; the line representing income is above the line representing expenses, so income is greater than expenses.

50c. When selling fewer than 1000 toys; the line representing expenses is above the line representing income, so expenses are greater than income.

51.

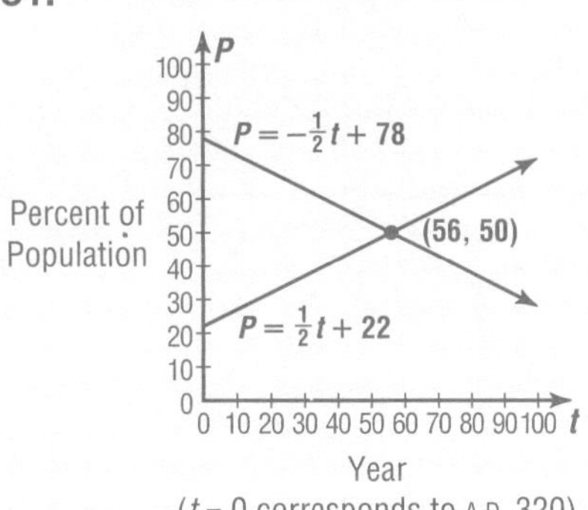

The Gupta Dynasty ruled from A.D. 320 until the late sixth century A.D. The founder was Candragupta.

Practice Masters, p. 56

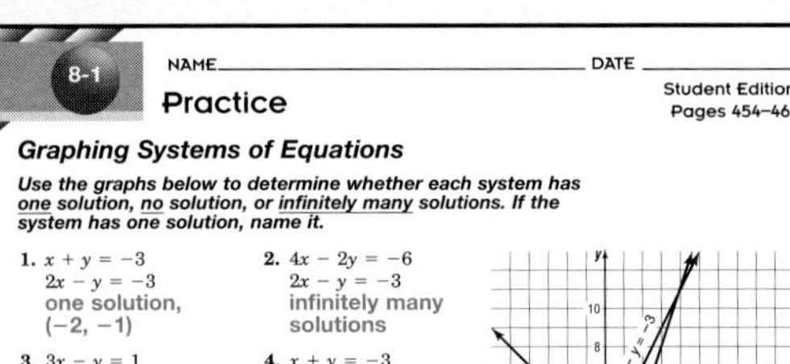

8-1 NAME______ DATE______ Student Edition Pages 454–461

Practice

Graphing Systems of Equations

Use the graphs below to determine whether each system has one solution, no solution, or infinitely many solutions. If the system has one solution, name it.

1. $x + y = -3$, $2x - y = -3$ — one solution, (−2, −1)
2. $4x - 2y = -6$, $2x - y = -3$ — infinitely many solutions
3. $3x - y = 1$, $x + y = 3$ — one solution, (1, 2)
4. $x + y = -3$, $x + y = 3$ — no solution
5. $x + y = 3$, $2x - y = -3$ — one solution, (0, 3)
6. $2x - y = -3$, $3x - y = 1$ — one solution, (4, 11)

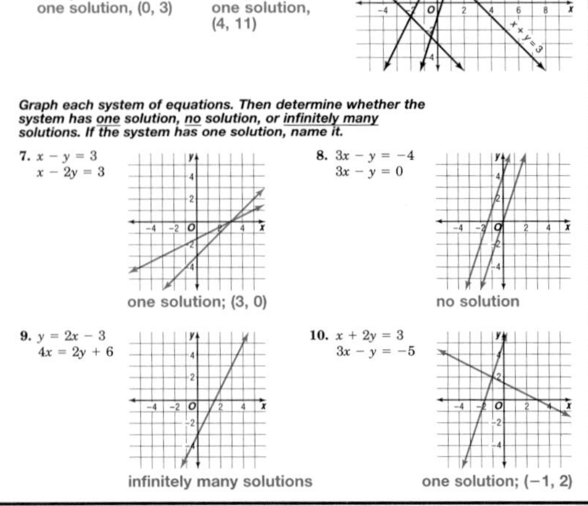

Graph each system of equations. Then determine whether the system has one solution, no solution, or infinitely many solutions. If the system has one solution, name it.

7. $x - y = 3$, $x - 2y = 3$ — one solution; (3, 0)
8. $3x - y = -4$, $3x - y = 0$ — no solution
9. $y = 2x - 3$, $4x = 2y + 6$ — infinitely many solutions
10. $x + 2y = 3$, $3x - y = -5$ — one solution; (−1, 2)

Critical Thinking

48. If (0, 0) and (2, 2) are known to be solutions of a system of two linear equations, does the system have any other solutions? Justify your answers. **Yes, the lines are coincident.**

49. The solution to the system of equations $Ax + y = 5$ and $Ax + By = 7$ is $(-1, 2)$. What are the values of A and B? $A = -3$, $B = 2$

Applications and Problem Solving

50a. (1000, 6000); See margin.

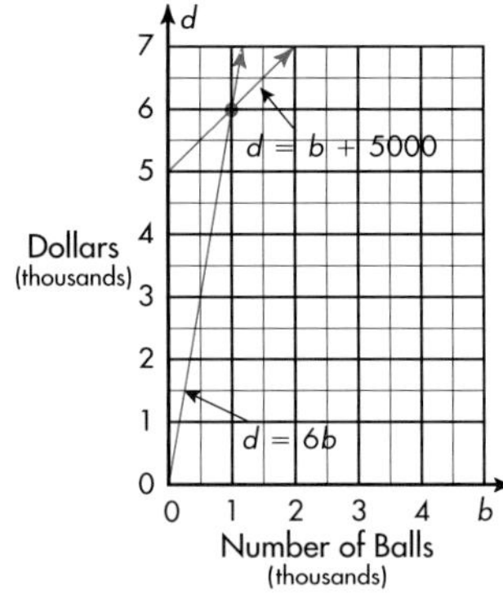

50. Business Mary Rodas is an 18-year-old toy specialist who tests and evaluates products at Catco, Inc., a company in New York City. She also helps design new toys, such as the Balzac Balloon Balls. Suppose the income from Balzac Balloon Balls is represented by the equation $d = 6b$ and the expenses are represented by the equation $d = b + 5000$. In both equations, b is the number of balls, and d is the number of dollars. Use the graph at the left to answer the following questions.

a. Find the solution to this system of equations. This solution is called the *break-even point.* What does this point represent? **a–c. See margin.**

b. A profit is made if income is greater than expenses. When is a profit made from the toys? How can you tell this from the graph?

c. Money is lost if expenses are greater than income. When is money lost from the Balzac Balloon Balls? How can you tell this from the graph?

FYI

Gupta doctors developed plastic surgery. Metallurgists made iron columns that are still free from rust after more than 1500 years. Gupta mathematicians developed a system of numbers that was later adopted by the Arabs.

51. World Cultures The Golden Age of India was during the expansion of the Gupta Empire, beginning in A.D. 320. India became a center of art, medicine, science, and mathematics. Suppose $P = \frac{1}{2}t + 22$ represents the percent of Indian people in the Gupta Empire, at time t. Let $P = -\frac{1}{2}t + 78$ represent the percent of Indian people that were not Guptas. Graph the system of equations and estimate the year in which the percent of Guptas equaled the percent of Indians that were not Gupta. (*Hint:* Let $t = 0$ correspond to A.D. 320.) **A.D. 376; See margin for graph.**

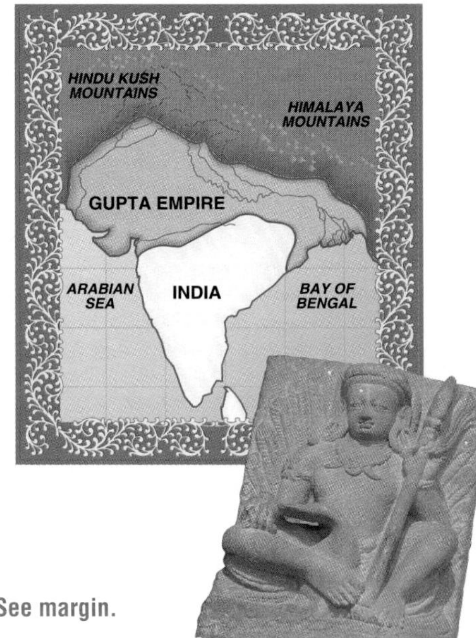

Mixed Review

52. Graph $y - 7 > 3x$. (Lesson 7–8) **See margin.**

53. Solve $|2m + 15| = 12$. (Lesson 7–6) $-1.5, -13.5$

54. Solve $10p - 14 < 8p - 17$. (Lesson 7–3) $\left\{p \mid p < -\frac{3}{2}\right\}$

55. Write an equation for the line that passes through the point at $(2, -2)$ and is parallel to $y = -2x + 21$. (Lesson 6–6) $y = -2x + 2$

Additional Answer

52.

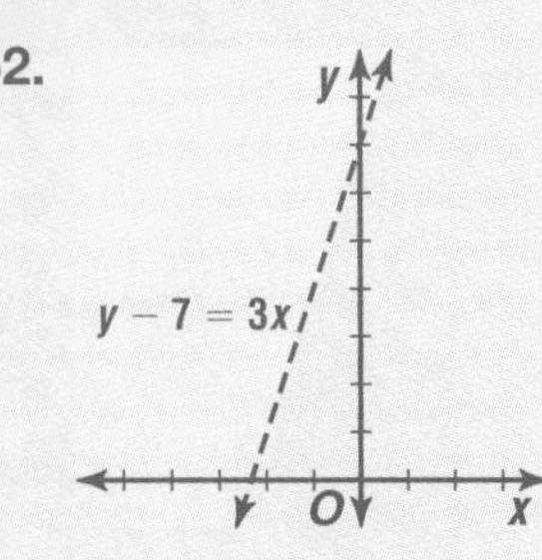

56. **Statistics** Find the range, median, upper and lower quartiles, and interquartile range of the data in the stem-and-leaf plot at the right. (Lesson 5–7) **34; 446.5; 457; 439; 18**

Stem	Leaf
43	3 5 6 6 9
44	1 4 4 4 9 9
45	0 2 7 7 8
46	5 7 $44 \mid 9 = 449$

57. **Finance** Patricia invested \$5000 for one year. Martin also invested \$5000 for one year. Martin's account earned interest at a rate of 10% per year. At the end of the year, Martin's account had earned \$125 more than Patricia's account. What was the annual interest rate on Patricia's account? (Lesson 4–4) $7\frac{1}{2}\%$

58. Solve $\frac{a-x}{-3} = \frac{-2}{b}$ for x. (Lesson 3–6) $-\frac{6}{b} + a$

59a. how many three- and four-bedroom homes will be built

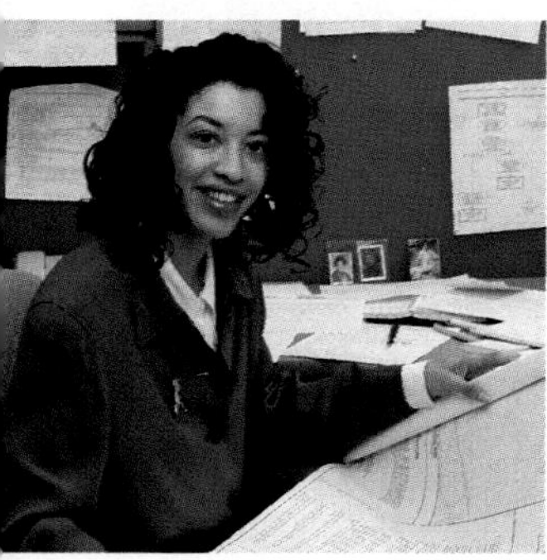

59. **Architecture** Answer the related questions for the verbal problem below. A developer is designing a housing development. She proposes to have four times as many three-bedroom homes as four-bedroom homes. If the development is planned for 100 homes, how many three- and four-bedroom homes will be built? (Lesson 2–9)
    a. What does the problem ask?
    b. If h represents the number of four-bedroom homes that are planned, how many three-bedroom homes are planned? **$100 - h$ or $4h$**
    c. If 20 four-bedroom homes are planned, how many three-bedroom homes should be built? **80 homes**

60. Evaluate $\frac{6ab}{3x + 2y}$ if $a = 6$, $b = 4$, $x = 0.2$, and $y = 1.3$. (Lesson 1–3) **45**

WORKING ON THE

In•ves•ti•ga•tion

Refer to the Investigation on pages 448–449.

Your research team determines that hang gliders are not just dropped from a point as you did with your tissue paper triangles. They are always launched into forward motion before gliding. The team decides that scale models are needed in order to get a feel for the launching and landing aspects of a real hang glider.

1 Each team in your class will construct a hang glider model using tissue paper and wire. A table top will act as the top of the cliff from which the glider is to be launched.

2 Each team should discuss different types of methods for launching their hang glider models from the table top. They should present their ideas to the class, and the class should agree upon which method they prefer to use. Then each team tests their glider using the method that the class has chosen.

3 Launch the glider 10 times. For each trial, measure the horizontal distance (along the floor) from the table to the spot at which the glider lands. Record this measurement and the height of the launch site.

4 Use these data to write a linear equation that describes the path of your glider. What is the slope of the path for your glider?

5 Using the linear equations from each of the other teams' data, would your glider collide with any of the other teams' gliders if they were launched at the same time from cliffs that are opposite each other? Write a detailed report on your conclusions.

Add the results of your work to your Investigation Folder.

4 ASSESS

Closing Activity

Writing At the end of class, have students write down the key points to consider when solving a system of equations by graphing.

Extension

Connections The graphs of the equations $y = 2$, $3x + 2y = 1$, and $3x - 4y = -29$ contain the sides of a triangle. Find the measure of the area of this triangle. (*Hint:* Use the formula $A = \frac{1}{2}bh$.) **9**

In•ves•ti•ga•tion

Working on the Investigation

The Investigation on pages 448–449 is designed to be a long-term project that is completed over several days or weeks. Encourage students to keep their materials in their Investigation Folder as they work on the Investigation.

Enrichment Masters, p. 56

NAME__________ DATE ________

Enrichment

Student Edition Pages 454–461

Graphing a Trip

The distance formula, $d = rt$, is used to solve many types of problems. If you graph an equation such as $d = 50t$, the graph is a model for a car going at 50 mi/h. The time the car travels is t; the distance in miles the car covers is d. The slope of the line is the speed.

Suppose you drive to a nearby town and return. You average 50 mi/h on the trip out but only 25 mi/h on the trip home. The round trip takes 5 hours. How far away is the town?

The graph at the right represents your trip. Notice that the return trip is shown with a negative slope because you are driving in the opposite direction.

Solve each problem.

1. Estimate the answer to the problem in the above example. About how far away is the town? **about 80 miles**
2. Graph this trip and solve the problem. An airplane has enough fuel for 3 hours of safe flying. On the trip out the pilot averages 200 mi/h flying against a headwind. On the trip back, the pilot averages 250 mi/h. How long a trip out can the pilot make? **about $1\frac{2}{3}$ hours and 330 miles**
3. Graph this trip and solve the problem. You drive to a town 100 miles away. On the trip out you average 25 mi/h. On the trip back you average 50 mi/h. How many hours do you spend driving? **6 hours**

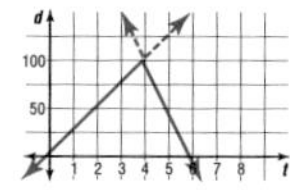

4. Graph this trip and solve the problem. You drive at an average speed of 50 mi/h to a discount shopping plaza, spend 2 hours shopping, and then return at an average speed of 25 mi/h. The entire trip takes 8 hours. How far away is the shopping plaza? **100 miles**

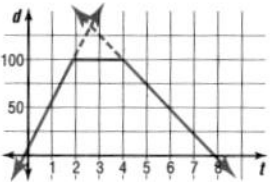

8-2 LESSON NOTES

NCTM Standards: 1–5

Instructional Resources

- Study Guide Master 8-2
- Practice Master 8-2
- Enrichment Master 8-2
- Assessment and Evaluation Masters, p. 212
- Science and Math Lab Manual, pp. 35–38

Transparency 8-2A contains the 5-Minute Check for this lesson; **Transparency 8-2B** contains a teaching aid for this lesson.

Recommended Pacing	
Standard Pacing	Day 4 of 9
Honors Pacing	Day 3 of 8
Block Scheduling*	Day 2 of 5 (along with Lesson 8-3)
Alg. 1 in Two Years*	Days 4 & 5 of 14

*For more information on pacing and possible lesson plans, refer to the *Block Scheduling Booklet* and *Algebra 1 in Two Years.*

1 FOCUS

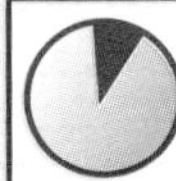

5-Minute Check
(over Lesson 8-1)

Graph each system of equations and state its solution.

1. $x + y = 3$
   $y = x + 3$ **(0, 3)**
2. $y = x$
   $x + y = 4$ **(2, 2)**
3. $y = -x$
   $y = 2x$ **(0, 0)**
4. If two distinct lines intersect, how many solutions exist? **one**
5. If a system of two linear equations has no solution, what is true of their graphs? **They are parallel lines.**

8-2 Substitution

What YOU'LL LEARN

- To solve systems of equations by using the substitution method, and
- to organize data to solve problems.

Why IT'S IMPORTANT

You can use systems of equations to solve problems involving geography and accounting.

FYI

California has 12% of the entire U.S. population.

Alaska, the largest state in the U.S., has the second smallest population.

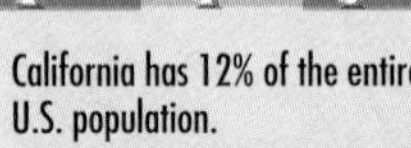

CONNECTION
Geography

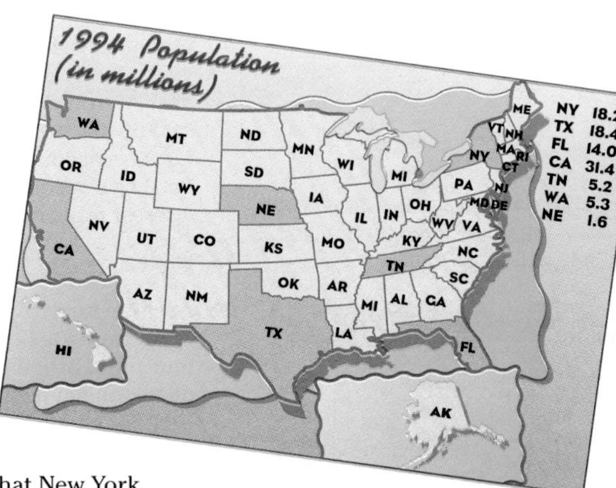

A recent article in *USA Today* reported that New York lost its position as the second most populous state when Texas slipped into the No. 2 spot at the end of 1994. Census Bureau projections show that New York will likely be pushed even further down the population ladder when Florida catches up early in the 21st century.

If New York's population grows at a constant rate of 0.02 million people per year and Florida's population grows at a constant rate of 0.26 million people per year, when would Florida catch up to New York in population? What would their populations be then?

Let P represent the population in millions, and let t represent the amount of time in years. The information above can be described by the following system of equations.

$$P = 18.2 + 0.02t$$
$$P = 14 + 0.26t$$

You could try to solve this system of equations by graphing, as shown at the right. Notice that the *exact* coordinates of the point where the lines intersect cannot be easily determined from this graph. An estimate is (18, 18).

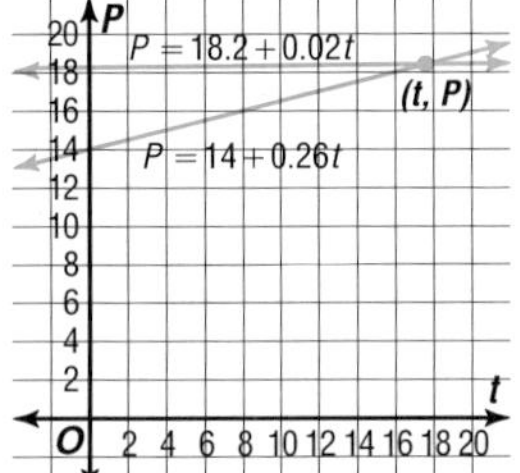

The exact solution of this system of equations can be found by using algebraic methods. One such method is called **substitution**.

From the first equation in the system, $P = 18.2 + 0.02t$, you know that P is equal to $18.2 + 0.02t$. Since P must have the same value in *both* equations, you can substitute $18.2 + 0.02t$ for P in the second equation $P = 14 + 0.26t$.

$$\begin{aligned} P &= 14 + 0.26t \\ 18.2 + 0.02t &= 14 + 0.26t \quad \text{Substitute } 18.2 + 0.02t \text{ for } P \text{ so the equation will have only one variable.} \\ 0.02t &= -4.2 + 0.26t \\ -0.24t &= -4.2 \\ t &= 17.5 \end{aligned}$$

FYI

The population of the United States is not evenly distributed over its land area. The east and west coasts have the largest population centers. Current trends show the largest population growth occurring in the southern states, where the climate is warmer.

Now find the value of P by substituting 17.5 for t in either equation.

$$P = 18.2 + 0.02t$$
$$= 18.2 + 0.02(17.5)$$ *You could also substitute 17.5 for t in P = 14 + 0.26t.*
$$= 18.55$$

Check: In each equation, replace t with 17.5 and P with 18.55.

$P = 18.2 + 0.02t$	$P = 14 + 0.26t$
$18.55 \stackrel{?}{=} 18.2 + 0.02(17.5)$	$18.55 \stackrel{?}{=} 14 + 0.26(17.5)$
$18.55 = 18.55$ ✓	$18.55 = 18.55$ ✓

The solution of the system of equations is (17.5, 18.55). Therefore after 17.5 years, or in 2011, the populations of New York and Florida would both be 18.55 million. *Compare this result to the estimate we obtained from the graph.*

You can use substitution to solve systems of equations even when the equations are more complex.

Example 1 **Use substitution to solve each system of equations.**

a. $x + 4y = 1$
$2x - 3y = -9$

Solve the first equation for x since the coefficient of x is 1.

$$x + 4y = 1$$
$$x = 1 - 4y$$

Next, find the value of y by substituting $1 - 4y$ for x in the second equation.

$$2x - 3y = -9$$
$$2(1 - 4y) - 3y = -9$$
$$2 - 8y - 3y = -9$$
$$-11y = -11$$
$$y = 1$$

Then substitute 1 for y in either of the original equations and find the value of x. *Choose the equation that is easier for you to solve.*

$$x + 4y = 1$$
$$x + 4(1) = 1$$
$$x + 4 = 1$$
$$x = -3$$

The solution of this system is $(-3, 1)$. *Use the graph at the left to verify this result.*

b. $\frac{5}{2}x + y = 4$
$5x + 2y = 8$

Solve the first equation for y since the coefficient of y is 1.

$$\frac{5}{2}x + y = 4$$
$$y = 4 - \frac{5}{2}x$$

Next, find the value of x by substituting $4 - \frac{5}{2}x$ for y in the second equation.

$$5x + 2y = 8$$
$$5x + 2\left(4 - \frac{5}{2}x\right) = 8$$
$$5x + 8 - 5x = 8$$
$$8 = 8$$

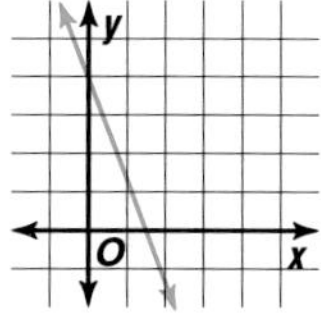

The statement $8 = 8$ is true. This means that there are infinitely many solutions to the system of equations. This is true because the slope intercept form of both equations is $y = 4 - \frac{5}{2}x$. That is, the equations are equivalent, and both have the same graph.

Motivating the Lesson

Situational Problem Review solving equations with the following activity. Let $y = 3$. Find the value of x in each equation by substituting 3 for y and solving for x.

1. $y - x = 6$ **−3**
2. $4x - y = 9$ **3**
3. $2y - 3x = 6$ **0**

2 TEACH

In-Class Example

For Example 1
Use substitution to solve each system of equations.

a. $y = 2x$
$2x + 5y = -12$ **(−1, −2)**

b. $2y = -3x$
$4x + y = 5$ **(2, −3)**

c. $x + y = 6$
$3x + y = 15$ $\left(\frac{9}{2}, \frac{3}{2}\right)$

d. $\frac{5}{2}x + 3y = 7$
$5x + 6y = 9$ **no solution**

e. $3x + 4y = 7$
$\frac{3}{2}x + 2y = 11$ **no solution**

Teaching Tip When an equation contains fractions, as in Example 1b, some students may find it easier to eliminate the fractions first. This can be done by multiplying each side of the equation by the LCM of the denominators.

The process students follow in this modeling activity mirrors the process of substitution. The arrangement of cups and counters on the mat represents the result of substitution: $4x + 3(x - 2) = 8$.

In-Class Example

For Example 2
A metal alloy is 25% copper. Another metal alloy is 50% copper. How much of each alloy should be used to make 1000 grams of a metal alloy that is 45% copper?
200 grams of the 25% copper alloy and 800 grams of the 50% copper alloy

In general, if you solve a system of linear equations and the result is a true statement (an identity such as $8 = 8$), the system has an infinite number of solutions; if the result is a false statement (for example, $8 = 12$), the system has no solution.

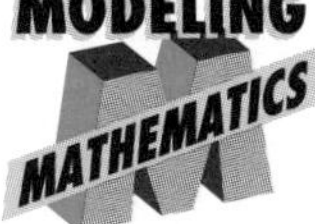

Systems of Equations

Materials: cups and counters  equation mat

Use a model to solve the system of equations.

$4x + 3y = 8$
$y = x - 2$

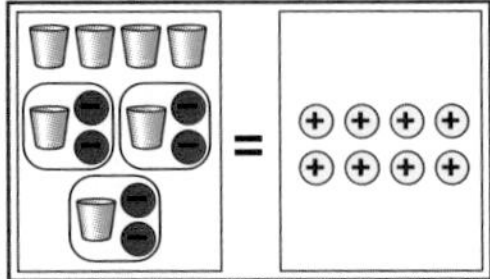

Your Turn **a. a cup and 2 negative counters**

a. Let a cup represent the unknown value x. If $y = x - 2$, how can you represent y?

b. Represent $4x + 3y = 8$ on the equation mat. On one side of the mat, place four cups to represent $4x$ and three representations of y from step a. On the other side of the mat, place eight positive counters.

c. Use what you know about equation mats and zero pairs to solve the equation. What value of x is the solution of the system of equations? **2**

d. Use the value of x from step c and the equation $y = x - 2$ to find the value of y. **0** **e. (2, 0)**

e. What is the solution of the system of equations?

Sometimes it is helpful to **organize data** before solving a problem. Some ways to organize data are to use tables, charts, different types of graphs, or diagrams.

Example 2

Organize Data

EJH Labs needs to make 1000 gallons of a 34% acid solution. The only solutions available are 25% acid and 50% acid. How many gallons of each solution should be mixed to make the 34% solution?

Explore Let a represent the number of gallons of 25% acid. Let b represent the number of gallons of 50% acid.

Make a table to organize the information in the problem.

	25% Acid	50% Acid	34% Acid
Total Gallons	a	b	1000
Gallons of Acid	$0.25a$	$0.50b$	0.34(1000)

Plan The system of equations is $a + b = 1000$ and $0.25a + 0.50b = 0.34(1000)$. Use substitution to solve this system.

Solve Since $a + b = 1000$, $a = 1000 - b$.

$0.25a + 0.50b = 0.34(1000)$
$0.25(1000 - b) + 0.50b = 340$ *Substitute 1000 − b for a.*
$250 - 0.25b + 0.50b = 340$ *Solve for b.*
$0.25b = 90$
$b = 360$

$a + b = 1000$
$a + 360 = 1000$ *Substitute 360 for b.*
$a = 640$ *Solve for a.*

Thus, 640 gallons of the 25% acid solution and 360 gallons of the 50% acid solution should be used.

Examine The 34% acid solution contains $0.25(640) + 0.50(360) = 160 + 180$ or 340 gallons of acid. Since $0.34(1000) = 340$, the answer checks.

Systems of equations can be useful in representing real-life situations and solving real-life problems.

Example 3

Entertainment

The Williams family is going to the Johnstown Summer Carnival. They have two ticket options, as shown in the table below.

Ticket Option	Admission Price	Price Per Ride
A	\$5	30¢
B	\$3	80¢

a. Write an equation that represents the cost per person for each option.

b. Graph the equations and estimate a solution. Explain what the solution means.

c. Solve the system using substitution.

d. Write a short paragraph advising the Williams family which option to choose.

a. Let r represent the number of rides. The total cost C for each person will be the cost of admission plus the cost of the rides.

Option A: $C = 5 + 0.30r$ *The cost of the rides is the price per*

Option B: $C = 3 + 0.80r$ *ride × number of rides, r.*

b. We can estimate from the graph that the solution is about (4, 6). This means that when the number of rides equals 4, both ticket options cost about \$6 per person.

C, $C = 3 + 0.80r$, $C = 5 + 0.30r$, O, r

c. Use substitution to solve this system.

$C = 5 + 0.30r$

$3 + 0.80r = 5 + 0.30r$ *Replace C with 3 + 0.80r.*

$0.50r = 2$ *Solve for r.*

$r = 4$

$C = 5 + 0.30r$

$= 5 + 0.30(4)$ *Replace r with 4.*

$= 6.2$ *Solve for C.*

The solution is (4, 6.2). This means that if a person rides 4 rides, both options cost the same, \$6.20. From the graph, you can see that Option A tickets will cost less if a person rides more than 4 rides. Option B tickets will cost less if a person rides less than 4 rides.

d. You should advise the Williams family to purchase Option A tickets for those who plan to ride more than 4 rides and purchase Option B tickets for the rest of the family.

In-Class Example

For Example 3

Use the two ticket options below for Example 3 and advise when each option is appropriate.

A	\$8	10
B	\$1	1

For more than 8 rides, Option A is less expensive; for fewer than 8 rides, Option B is less expensive.

Classroom Vignette

"An alternative problem-solving strategy for part a of Example 3 is to have students observe a pattern like the one below and then determine the equation based on the pattern."

Admission Price	Number of Rides @ 30¢	Cost
\$5	1	5 + 0.30(1)
5	2	5 + 0.30(2)
•	3	5 + 0.30(3)
•	•	•
•	•	•
•	•	•
5	*r*	5 + 0.30(*r*)

Cindy Boyd
Abilene High School
Abilene, Texas

Cindy Boyd

3 PRACTICE/APPLY

Check for Understanding

Exercises 1–16 are designed to help you assess your students' understanding through reading, writing, speaking, and modeling. You should work through Exercises 1–6 with your students and then monitor their work on Exercises 7–16.

Additional Answers

1. From the first equation, y is equal to $2x - 4$, and y must have the same value in both equations.

5a. Yolanda is walking faster than Adele and catches up after about 5 seconds; 1 solution.

5b. Yolanda never catches up with Adele because they both are walking at the same rate, and Adele has a 30-ft headstart; no solution.

5c. Yolanda never catches up with Adele because Adele is walking at a faster rate and has a 30-ft headstart.

7. $x = 8 - 4y;\ y = 2 - \frac{1}{4}x$

8. $x = 4 + \frac{5}{3}y;\ y = -\frac{12}{5} + \frac{3}{5}x$

9. $x = -\frac{0.75}{0.8}y - 7.5;$
$y = -\frac{0.8}{0.75}x - 8$

Study Guide Masters, p. 57

8-2 NAME________________ DATE ________

Study Guide

Student Edition Pages 462–468

Substitution

One method of solving systems of equations is by algebraic **substituton.**

Example: Solve $x + 3y = 7$ and $2x - 4y = 6$.

Solve the first equation for x.

$$x + 3y = 7$$
$$x = 7 - 3y$$

Substitute $7 - 3y$ for x in the second equation. Solve for y.

$$2(7 - 3y) - 4y = -6$$
$$14 - 6y - 4y = -6$$
$$-10y = -20$$
$$y = 2$$

Substitute 2 for y in either one of the two original equations to find the value of x.

$$x + 3(2) = 7$$
$$x + 6 = 7$$
$$x = 1$$

The solution of this system is (1, 2).

Use substitution to solve each system of equations. If the system does not have exactly one solution, state whether it has no solution or infinitely many solutions.

1. $x = 3$, $2y + x = 3$ (3, 0)
2. $y = 2$, $2x - 4y = 1$ $\left(\frac{9}{2}, 2\right)$
3. $y = 3x - 7$, $3x - y = 7$ infinitely many
4. $y = -x + 3$, $2y + 2x = 4$ no solution
5. $x + y = 16$, $2y = -2x + 2$ no solution
6. $x = 2y$, $0.25x + 0.5y = 10$ (20, 10)

Use a system of equations and substitution to solve each problem.

7. How much of a 10% saline solution should be mixed with a 20% saline solution to obtain 1000 milliliters of a 12% saline solution? 800 mL of 10% solution and 200 mL of 20% solution
8. The tens digit of a two-digit number is 3 greater than the units digit. Eight times the sum of the digits is 1 less than the number. Find the number. 41

CHECK FOR UNDERSTANDING

Communicating Mathematics

Study the lesson. Then complete the following.

1. **Explain** why, when solving the system $y = 2x - 4$ and $4x - 2y = 0$, you can substitute $2x - 4$ for y in the second equation. **See margin.**
2. **State** what you would conclude if the solution to a system of linear equations yields the equation $8 = 0$. **There is no solution.**
3. **Describe** how you can tell just by looking at the equations $y = 9x + 2$ and $y = 9x - 5$ whether or not the system has a solution.
   3. The graphs of both equations have the same slope, 9, but different y-intercepts, so the lines are parallel and the system has no solution.
4. **Explain** why graphing a system of equations may not give you an exact solution. **Answers will vary.**
5. Yolanda is walking across campus when she sees Adele walking about 30 feet ahead of her. In each graph, t represents time in seconds and d represents distance in feet. Describe what happens in each case and how it relates to the solution. **See margin.**

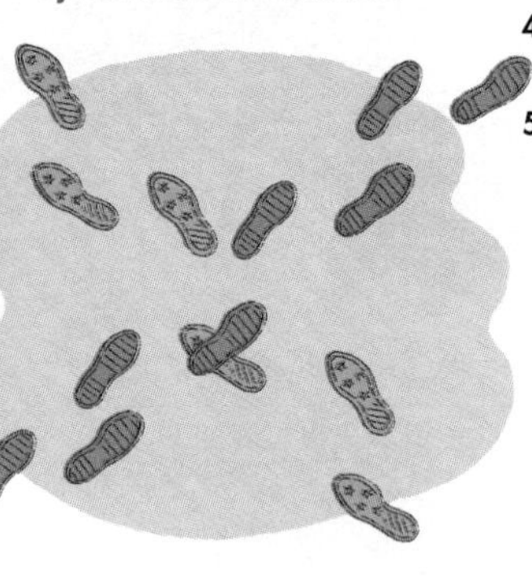

a.

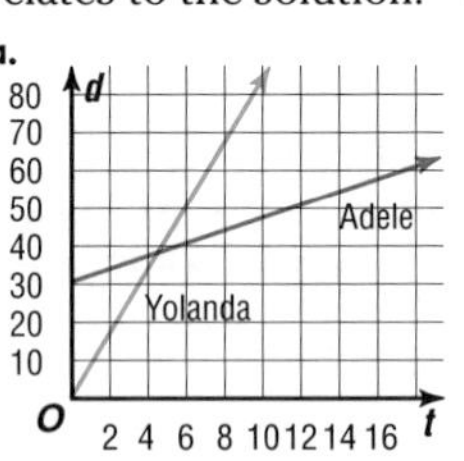

b.

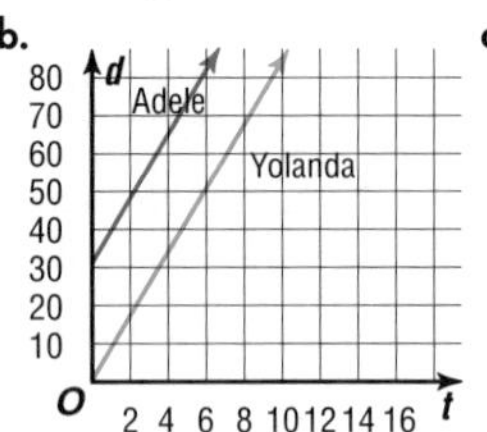

c.

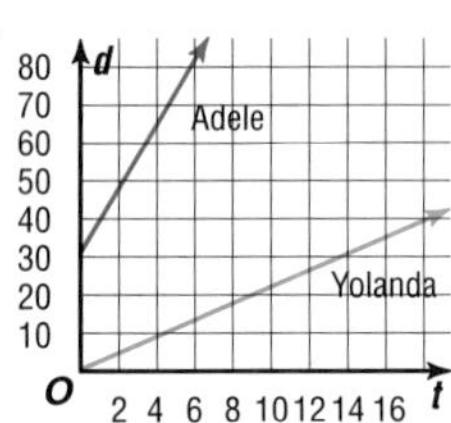

MODELING MATHEMATICS

6. Use cups and counters to model and solve the system of equations. **(3, 0)**
   $y = 2x - 6$
   $3x + 2y = 9$

Guided Practice

Solve each equation for x. Then, solve each equation for y.

7. $x + 4y = 8$
8. $3x - 5y = 12$
9. $0.8x + 6 = -0.75y$

7–9. See margin.

Use substitution to solve each system of equations. If the system *does not* have exactly one solution, state whether it has *no* solution or *infinitely many* solutions.

10. $y = 3x$, $x + 2y = -21$ **(−3, −9)**
11. $x = 2y$, $4x + 2y = 15$ **$\left(3, \frac{3}{2}\right)$**
12. $x + 5y = -3$, $3x - 2y = 8$ **(2, −1)**
13. $8x + 2y = 13$, $4x + y = 11$ **no solution**
14. $2x - y = -4$, $-3x + y = -9$ **(13, 30)**
15. $6x - 2y = -4$, $y = 3x + 2$ **infinitely many**

16. **Sales** Maria spent a long day working the cash register at Musicville during a sale on CDs. For this sale, all CDs in the store were marked either \$12 or \$10. Just when she thought she could go home, the store manager gave Maria the job of figuring out how many CDs they had sold at each price, so they could write the total in the store records. Maria doesn't want to sort through hundreds of sales slips, so she decided on an easier way. The counter at the exit of the store says that 500 people left with CDs (limit one per customer) during the sale, and the cash register contains \$5750 from the day's sales. Maria wrote a system of equations for the number of \$10 CDs and the number of \$12 CDs.
    a. What was the system of equations? **$x + y = 500$, $12x + 10y = 5750$**
    b. How many CDs were sold at each price? **\$12: 375; \$10: 125**

466 *Chapter 8 Solving Systems of Linear Equations and Inequalities*

Reteaching

Using Comparisons Give the class the following system of equations and have half the class use substitution by solving the first equation for y and the other half of the class use substitution by solving the second equation for x.
$3x - 2y = 2$
$x + 4y = 10$ **(2, 2)**

Discuss whether it matters which equation is substituted into which equation and how to look for the "best" way.

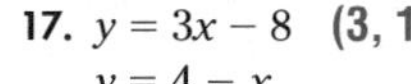

EXERCISES

Practice **Use substitution to solve each system of equations. If the system *does not* have exactly one solution, state whether it has *no* solution or *infinitely many* solutions.**

17. $y = 3x - 8$, $y = 4 - x$ **(3, 1)**
18. $2x + 7y = 3$, $x = 1 - 4y$ **(5, −1)**
19. $x + y = 0$, $3x + y = -8$ **(−4, 4)**
20. $4c = 3d + 3$, $c = d - 1$ **(6, 7)**
21. $4x + 5y = 11$, $y = 3x - 13$ **(4, −1)**
22. $3x - 5y = 11$, $x - 3y = 1$ **(7, 2)**
23. $c - 5d = 2$, $2c + d = 4$ **(2, 0)**
24. $3x - 2y = 12$, $x + 2y = 6$
25. $x + 3y = 12$, $x - y = 8$ **(9, 1)**

B

26. $x - 3y = 0$, $3x + y = 7$
27. $5r - s = 5$, $-4r + 5s = 17$ **(2, 5)**
28. $2x + 3y = 1$, $-3x + y = 15$ **(−4, 3)**
29. $8x + 6y = 44$, $x - 8y = -12$ **(4, 2)**
30. $0.5x - 2y = 17$, $2x + y = 104$ **(50, 4)**
31. $-0.3x + y = 0.5$, $0.5x - 0.3y = 1.9$
32. $x = \frac{1}{2}y + 3$, $2x - y = 6$ **infinitely many**
33. $y = \frac{1}{2}x + 3$, $y = 2x - 1$
34. $y = \frac{3}{5}x$, $3x - 5y = 15$ **no solution**

24. $\left(\frac{9}{2}, \frac{3}{4}\right)$
26. $\left(\frac{21}{10}, \frac{7}{10}\right)$
31. (5, 2)
33. $\left(\frac{8}{3}, \frac{13}{3}\right)$

Use substitution to solve each system of equations. Write each solution as an ordered triple of the form (*x*, *y*, *z*).

35. $x + y + z = -54$, $x = -6y$, $z = 14y$ **(36, −6, −84)**
36. $2x + 3y - z = 17$, $y = -3z - 7$, $2x = z + 2$ **(−1, 5, −4)**
37. $12x - y + 7z = 99$, $x + 2z = 2$, $y + 3z = 9$

37. (14, 27, −6)

Critical Thinking

38. **Number Theory** If 36 is subtracted from certain two-digit positive integers, their digits are reversed. Find all integers for which this is true. **40, 51, 62, 73, 84, 95**

Applications and Problem Solving

39. **Entertainment** American songwriter Cole Porter completed his first professional score in 1916 at age 23. At Harding High, this year's spring musical is *Anything Goes,* which Porter completed in 1934. The production is going to be part of a dinner theater; each ticket includes dinner and the show. The total cost of producing the show (stage, costumes, and so on) is $1000, and each dinner costs $5 to prepare. The drama club is going to sell tickets for $13 each.
   a. Write a system of equations to represent the cost of and the income from the production. $y = 1000 + 5x$, $y = 13x$
   b. How many tickets do they need to sell to break even? **125 tickets**
40. **Humor** Refer to the cartoon below. Solve the problem that is sending Peppermint Patty into a frenzy. Find how much cream and milk must be mixed together to obtain 50 gallons of cream containing $12\frac{1}{2}$% butterfat. **20.93 gal of cream, 29.07 gal of milk**

Peanuts®

Assignment Guide

Core: 17–37 odd, 38, 39, 41, 43, 44–51
Enriched: 18–36 even, 38–51

For **Extra Practice,** see p. 774.

The red A, B, and C flags, printed only in the Teacher's Wraparound Edition, indicate the level of difficulty of the exercises.

Practice Masters, p. 57

NAME ______________ DATE ______

Practice

Student Edition Pages 462–468

Substitution

Use substitution to solve each system of equations. If the system does not have exactly one solution, state whether it has no solution or infinitely many solutions.

1. $y = 4x$, $x + y = 5$ **(1, 4)**
2. $x = -4y$, $3x + 2y = 20$ **(8, −2)**
3. $y = x - 1$, $x + y = 3$ **(2, 1)**
4. $3x - y = 4$, $2x - 3y = -9$ **(3, 5)**
5. $x + 5y = 4$, $3x + 15y = -1$ **no solution**
6. $x - 5y = 10$, $2x - 10y = 20$ **infinitely many**
7. $x + 4y = 8$, $2x - 5y = 29$ **(12, −1)**
8. $4x + y = 0$, $x + 2y = -7$ **(1, −4)**
9. $2x - 3y = -24$, $x + 6y = 18$ **(−6, 4)**
10. $x + 14y = 84$, $2x - 7y = -7$ **(14, 5)**
11. $0.3x - 0.2y = 0.5$, $x + 2y = 15$ **(5, 5)**
12. $x - 3y = -4$, $2x + 6y = 5$ $\left(-\frac{3}{4}, \frac{13}{12}\right)$
13. $3x - 2y = 11$, $x - \frac{y}{2} = 4$ **(5, 2)**
14. $\frac{1}{2}x + 2y = 12$, $x - 2y = 6$ **(12, 3)**
15. $\frac{x}{3} - y = 3$, $2x + y = 25$ **(12, 1)**

4 ASSESS

Closing Activity

Modeling Have students create a flowchart diagram to model the steps used in solving a system of equations using the substitution method.

Chapter 8, Quiz A (Lessons 8-1 and 8-2), is available in the *Assessment and Evaluation Masters*, p. 212.

Additional Answers

43a.

	75% Gold (18-carat)	50% Gold (12-carat)	58% Gold (14-carat)
Total Grams	x	y	300
Grams of Pure Gold	$0.75x$	$0.50y$	0.58(300)

43b. $x + y = 300$
$0.75x + 0.50y = 0.58(300)$

44.

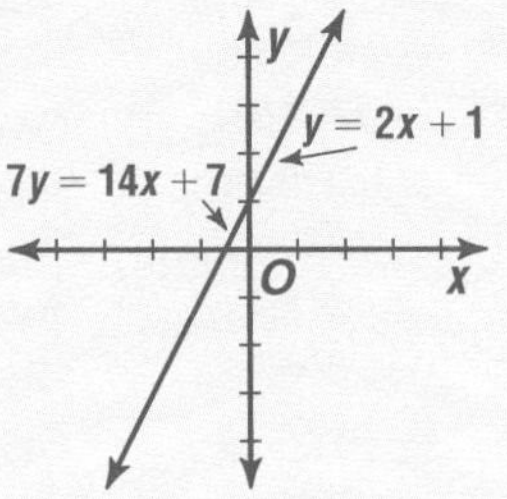

Enrichment Masters, p. 57

8-2 NAME______ DATE______

Enrichment Student Edition Pages 462–468

Using Equations: Ideal Weight

You can find your ideal weight as follows.
A woman should weigh 100 pounds for the first 5 feet of height and 5 additional pounds for each inch over 5 feet (5 feet = 60 inches). A man should weigh 106 pounds for the first 5 feet of height and 6 additional pounds for each inch over 5 feet. These formulas apply to people with normal bone structures.

To determine your bone structure, wrap your thumb and index finger around the wrist of your other hand. If the thumb and finger just touch, you have normal bone structure. If they overlap, you are small-boned. If they don't overlap, you are large-boned. Small-boned people should decrease their calculated ideal weight by 10%. Large-boned people should increase the value by 10%.

Calculate the ideal weights of these people.

1. woman, 5 ft 4 in., normal-boned 120 lb
2. man, 5 ft 11 in., large-boned 189.2 lb
3. man, 6 ft 5 in., small-boned 187.2 lb
4. you, if you are at least 5 ft tall Answers will vary.

Suppose a normal-boned man is x inches tall. If he is at least 5 feet tall, then x − 60 represents the number of inches this man is over 5 feet tall. For each of these inches, his ideal weight is increased by 6 pounds. Thus, his proper weight (y) is given by the formula y = 6(x − 60) + 106 or y = 6x − 254. If the man is large-boned, the formula becomes y = 6x − 254 + 0.10(6x − 254).

5. Write the formula for the weight of a large-boned man in slope-intercept form. $y = 6.6x - 279.4$
6. Derive the formula for the ideal weight (y) of a normal-boned female with height x inches. Write the formula in slope-intercept form. $y = 5x - 200$
7. Derive the formula in slope-intercept form for the ideal weight (y) of a large-boned female with height x inches. $y = 5.5x - 220$
8. Derive the formula in slope-intercept form for the ideal weight (y) of a small-boned male with height x inches. $y = 5.4x - 228.6$
9. Find the heights at which normal-boned males and large-boned females would weigh the same. 68 in., or 5 ft 8 in.

41. **Athletes** According to *Health* magazine, top women athletes are narrowing the gap between their performances and those of their male counterparts. Speed skater Bonnie Blair's fastest time in the 500-meter would have won an Olympic gold medal in every men's 500-meter competition through 1976. The women's record time for the 500-meter in speed skating is 39.1 seconds, and the men's is 36.45 seconds. Suppose the women's record time decreases at an average rate of 0.20 second per year and the men's record time decreases at an average rate of 0.10 second per year. **c. See students' work.**
    a. When would the women's record time equal the men's? **26.5 years**
    b. What would the time be? **33.8 seconds**
    c. Do you think this could actually happen? Why or why not?

42. **Accounting** Sometimes accountants must figure out how many stock shares to transfer from one person to another to reach a certain proportion of ownership. Suppose Rebeca Avila owns \$3000 worth of stock in a new company that has no other stockholders. For tax purposes, the company is going to issue new stock to Muriel Eppick so that Ms. Avila owns 80%, rather than 100% of the total stock. Let S represent the new total value of company stock and let x represent the value of stock that Ms. Eppick is to receive. Use the equations below to find the value of stock to be issued to Ms. Eppick. **\$750**
    $S = 3000 + x$ *New total stock = Ms. Avila's share + Ms. Eppick's share.*
    $3000 = 0.80S$ *Ms. Avila's share is 80% of new total stock.*
    $x = 0.20S$ *Ms. Eppick's share is 20% of new total stock.*

43. **Organize Data** For thousands of years, gold has been considered one of Earth's most precious metals. When archaeologist Howard Carter discovered King Tutankhamun's tomb in 1922, he exclaimed that the tomb was filled with "strange animals, statues, and gold—everywhere the glint of gold." One hundred percent pure gold is 24-carat gold. If 18-carat gold is 75% gold and 12-carat gold is 50% gold, how much of each would be used to make a 14-carat gold bracelet weighing 300 grams? (*Hint:* 14-carat gold is about 58% gold.)
    a. Make a table to organize the data. **See margin.**
    b. Write a system of equations that represents this problem. **See margin.**
    c. How much 18-carat gold and 12-carat gold would it take to make a 14-carat gold bracelet weighing 300 grams? **96 grams of 18-carat gold, 204 grams of 12-carat gold**

Mixed Review

44. Graph the system of equations below. Determine whether the system has *one* solution, *no* solutions, or *infinitely many* solutions. If the system has one solution, name it. (Lesson 8–1)
    $y = 2x + 1$
    $7y = 14x + 7$ **Infinitely many; see margin for graph.**

45. **Finance** Michael uses at most 60% of his annual FlynnCo stock dividend to purchase more shares of FlynnCo stock. If his dividend last year was \$885 and FlynnCo stock is selling for \$14 per share, what is the greatest number of shares that he can purchase? (Lesson 7–2) **37 shares**

46. Graph $y = \frac{1}{5}x - 3$ using the slope and y-intercept. (Lesson 6–5) **See margin.**

47. Solve $3a - 4 = b$ if the domain is $\{-1, 4, 7, 13\}$. (Lesson 5–3)

48. What is 25% less than 94? (Lesson 4–5) **70.5**

49. Solve $-8 - 12x = 28$. (Lesson 3–3) **−3**

50. Graph the solution set of $n \leq -2$ on a number line. (Lesson 2–8)

51. Write an algebraic expression for *twelve less than m*. (Lesson 1–1) **m − 12**

47. $\{(-1, -7), (4, 8), (7, 17), (13, 35)\}$

50. See margin.

Extension

Problem Solving The sum of the digits of a three-digit number is 15. The tens digit exceeds the units digit by the same amount that the hundreds digit exceeds the tens digit. If the digits are reversed, the new number is 76 times the original hundreds digit. Find the original number. **654**

Additional Answers

46.

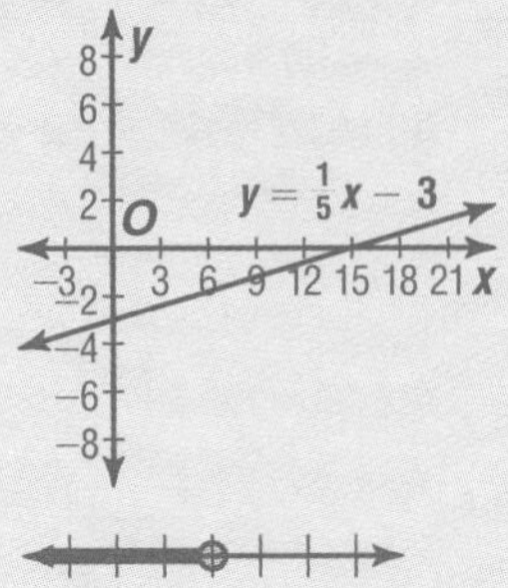

50.

8-3 Elimination Using Addition and Subtraction

What YOU'LL LEARN
- To solve systems of equations by using the elimination method with addition or subtraction.

Why IT'S IMPORTANT

You can use systems of equations to solve problems involving entertainment and testing.

APPLICATION
Entertainment

Disney cartoons are animated using an expensive computer process that makes the action flow smoothly and seem lifelike. In 1994, Disney's animated feature *The Lion King* was the top-grossing film of the year, making an estimated \$300.4 million at the box office.

On a Saturday afternoon, the Johnson and Olivera families decided to go see *The Lion King* together. The Johnson family, two adults and four children, can afford to spend \$30 from their entertainment budget this weekend for the movie tickets, while the Olivera family, two adults and two children, can afford to spend \$21.50. Different theaters around town charge different amounts for adult and child tickets. What price can the Johnsons and Oliveras afford to pay for each adult and each child?

Let a represent the ticket price for one adult, and let c represent the ticket price for one child. Then the information in this problem can be represented by the following system of equations.

$$2a + 4c = 30$$
$$2a + 2c = 21.5$$

From the graph at the right, an estimated solution is (\$7, \$4). To get an exact solution, solve algebraically. You could solve this system by first solving either of the equations for a or c and then using substitution.

$2a + 2c = 21.5$ $2a + 4c = 30$ c a O

However, a simpler method of solution is to subtract one equation from the other since the coefficients of the variable a are the same. This method is called **elimination** because the subtraction eliminates one of the variables. First, write the equations in column form and subtract.

Recall that subtraction is the same as adding the opposite.

$$\begin{array}{r} 2a + 4c = 30 \\ (-)\ 2a + 2c = 21.5 \\ \hline \end{array} \quad \xrightarrow{\text{Multiply by } -1.} \quad \begin{array}{r} 2a + 4c = 30 \\ (+)\ -2a - 2c = -21.5 \\ \hline 2c = 8.5 \\ c = 4.25 \end{array}$$

Then, substitute 4.25 for c in either equation and find the value of a.

$$\begin{aligned} 2a + 2c &= 21.5 \\ 2a + 2(4.25) &= 21.5 \quad \textit{Substitute 4.25 for c.} \\ 2a + 8.5 &= 21.5 \\ 2a &= 13 \\ a &= 6.5 \end{aligned}$$

Is (6.5, 4.25) a solution of the system?

fabulous FIRSTS

Elmer Simms Campbell (1906–1971)

Elmer Simms Campbell was the first African-American cartoonist to work for national publications. He contributed cartoons and other art work to *Esquire*, *Cosmopolitan*, and *Redbook*, as well as syndicated features in 145 newspapers.

fabulous FIRSTS

The first African-American woman filmmaker with a feature film in theatrical release was Julie Dash. Her film, *Daughters of the Dust*, opened in 1992 and told the story of a day in the lives of an African-American family early in this century.

8-3 LESSON NOTES

NCTM Standards: 1–5

Instructional Resources

- Study Guide Master 8-3
- Practice Master 8-3
- Enrichment Master 8-3
- Assessment and Evaluation Masters, pp. 211–212

Transparency 8-3A contains the 5-Minute Check for this lesson; **Transparency 8-3B** contains a teaching aid for this lesson.

Recommended Pacing	
Standard Pacing	Day 5 of 9
Honors Pacing	Day 4 of 8
Block Scheduling*	Day 2 of 5 (along with Lesson 8-2)
Alg. 1 in Two Years*	Days 6 & 7 of 14

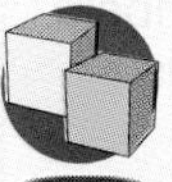

*For more information on pacing and possible lesson plans, refer to the *Block Scheduling Booklet* and *Algebra 1 in Two Years.*

1 FOCUS

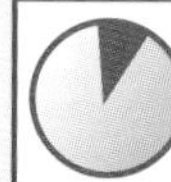

5-Minute Check
(over Lesson 8-2)

Solve each equation for *x*. Then solve each equation for *y*.

1. $y + 1 = x$ **$x = y + 1$; $y = x - 1$**
2. $2x + 3y = 6$ **$x = \frac{1}{2}(6 - 3y)$; $y = \frac{1}{3}(6 - 2x)$**
3. $y + 2x = 3$ **$x = \frac{1}{2}(3 - y)$; $y = 3 - 2x$**

Motivating the Lesson

Situational Problem Tell students that you are thinking of two numbers whose sum is 30 and whose difference is 10. Ask them to identify the numbers. **20, 10** Have student volunteers make up additional problems like this one for the class to solve.

3 PRACTICE/APPLY

Check for Understanding

Exercises 1–11 are designed to help you assess your students' understanding through reading, writing, speaking, and modeling. You should work through Exercises 1–3 with your students and then monitor their work on Exercises 4–11.

Error Analysis

When solving a system of equations by elimination using subtraction, have students change every sign of the equation they are subtracting in order to alleviate the problem of only subtracting the column of like variables and not the other columns too.

Additional Answers

1a. When the coefficients of one of the variables are the same.

1b. When the coefficients of one of the variables are additive inverses of each other.

Study Guide Masters, p. 58

NAME______________________ DATE ________

Study Guide — Student Edition Pages 469–474

Elimination Using Addition and Subtraction

In systems of equations where the coefficient of the x or y terms are additive inverses, solve the system by adding the equations. Because one of the variables is eliminated, this method is called **elimination**.

Example: Use elimination to solve the system of equations $x - 3y = 7$ and $3x + 3y = 9$.

Add the two equations.
$$\begin{aligned} x - 3y &= 7 \\ 3x + 3y &= 9 \\ \hline 4x &= 16 \\ x &= 4 \end{aligned}$$

Substitute 4 for x in either original equation and solve for y.
$$\begin{aligned} 4 - 3y &= 7 \\ -3y &= 7 - 4 \\ -3y &= 3 \\ y &= -1 \end{aligned}$$

The solution of the system is $(4, -1)$.

Use elimination to solve each system of equations.

1. $2x + 2y = -2$, $3x - 2y = 12$ $(2, -3)$
2. $4x - 2y = -1$, $-4x + 4y = -2$ $\left(-1, -\frac{3}{2}\right)$
3. $x - y = 2$, $x + y = -3$ $\left(-\frac{1}{2}, -\frac{5}{2}\right)$
4. $6x + 5y = 4$, $6x - 7y = -20$ $(-1, 2)$
5. $2x - 3y = 12$, $4x + 3y = 24$ $(6, 0)$
6. $0.1x + 0.3y = 0.9$, $0.1x = 0.3y + 0.2$ $\left(\frac{11}{2}, \frac{7}{6}\right)$

Use a system of equations and elimination to solve each problem.

7. Two angles are supplementary. The measure of one angle is 10 more than three times the other. Find the measure of each angle. 42.5°, 137.5°
8. Rema is older than Ken. The difference of their ages is 12 and the sum of their ages is 50. Find the age of each. Rema is 31 and Ken is 19
9. The sum of two numbers is 70 and their difference is 24. Find the two numbers. 23 and 47
10. The sum of the digits of a two-digit number is 12. The difference of the digits is 2. Find the number if the units digit is larger than the tens digit. 57

CHECK FOR UNDERSTANDING

Communicating Mathematics

Study the lesson. Then complete the following. 1–3. See margin.

1. **Explain** when it is easier to solve a system of equations in each way.
   a. by elimination using subtraction
   b. by elimination using addition
2. a. **State** the result when you add $3x - 8y = 29$ and $-3x + 8y = 16$. What does this result tell you about the system of equations?
   b. What does this result tell you about the graph of the system?
3. **You Decide** Maribela says that a system of equations has no solution if both variables are eliminated by addition or subtraction. Devin argues that there may be an infinite number of solutions. Who is correct? Explain your answer.

Guided Practice

4. Refer to the graph at the right.
   a. Estimate the solution of the system. Sample answer: $(-3, 4)$
   b. Use elimination to find the exact solution. $(-3.4, 4.2)$

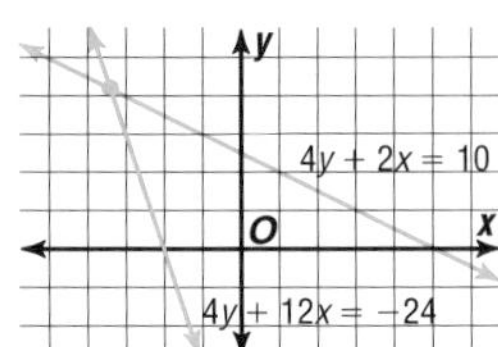

State whether addition, subtraction, or substitution would be most convenient to solve each system of equations. Then solve the system.

5. $3x - 5y = 3$, $4x + 5y = 4$
6. $3x + 2y = 7$, $y = 4x - 2$
7. $-4m + 2n = 6$, $-4m + n = 8$
8. $8a + b = 1$, $8a - 3b = 3$
9. $3x + y = 7$, $2x + 5y = 22$ substitution, $(1, 4)$
10. $2b + 4c = 8$, $c - 2 = b$ substitution, $(0, 2)$
11. **Statistics** The mean of two numbers is 28. Find the numbers if three times one of the numbers equals half the other number. 8, 48

5. addition, $(1, 0)$
6. substitution, $(1, 2)$
7. subtraction, $\left(-\frac{5}{2}, -2\right)$
8. subtraction, $\left(\frac{3}{16}, -\frac{1}{2}\right)$

EXERCISES

Practice

For Exercises 12–14,
a. estimate the solution of each system of linear equations, and
b. use elimination to find the exact solution of each system.

12. $y + 3x = -5$, $2y - 3x = 9$ — $(-2, 1); \left(-\frac{19}{9}, \frac{4}{3}\right)$

13. $1.6y + 8.9x = 5$, $1.6y + 2.7x = -3$ — $(1, -4); (1.29, -4.05)$

14.

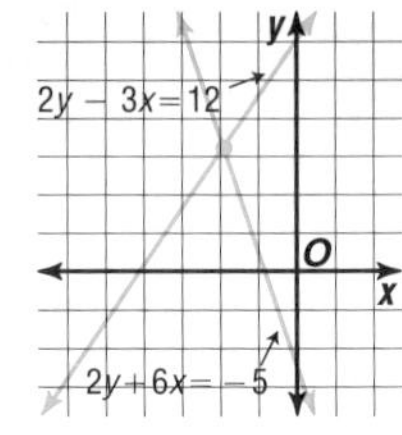

14. $(-2, 3); \left(-\frac{17}{9}, \frac{19}{6}\right)$

Reteaching

Using Models Let paper rectangles represent the coefficients of x, circles represent the coefficients of y, and squares represent constants. Use red for negatives and black for positives. Solve the system $x - 2y = 5$ and $2x + 2y = 7$. Since there are 3 rectangles after adding, separate the squares into groups of 3. Since there are 4 groups, $x = 4$. Substitute 4 for x in either equation and solve for y. $\left(4, -\frac{1}{2}\right)$

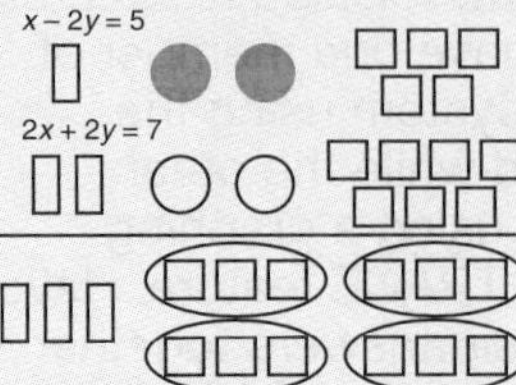

18. −; (−5, 20)
19. −; (4, −7)
20. + or −; $\left(\frac{9}{2}, 2\right)$

22. −; (7, 4)
23. sub; infinitely many
24. −; (−5, 4)
25. −; (−2, 3)
26. −; $\left(2, -\frac{3}{2}\right)$
27. −; $\left(\frac{1}{2}, 1\right)$

29. +; (10, −15)

State whether addition, subtraction, or substitution would be most convenient to solve each system of equations. Then solve the system.

15. $x + y = 8$, $x - y = 4$ +; (6, 2)
16. $2r + s = 5$, $r - s = 1$ +; (2, 1)
17. $x - 3y = 7$, $x + 2y = 2$ −; (4, −1)
18. $3x + y = 5$, $2x + y = 10$
19. $5s + 2t = 6$, $9s + 2t = 22$
20. $4x - 3y = 12$, $4x + 3y = 24$

B

21. $2x + 3y = 13$, $x - 3y = 2$ +; (5, 1)
22. $2m - 5n = -6$, $2m - 7n = -14$
23. $x - 2y = 7$, $-3x + 6y = -21$
24. $3r - 5s = -35$, $2r - 5s = -30$
25. $13a + 5b = -11$, $13a + 11b = 7$
26. $a - 2b - 5 = 0$, $3a - 2b - 9 = 0$
27. $4x = 7 - 5y$, $8x = 9 - 5y$
28. $\frac{2}{3}x + y = 7$, $\frac{10}{3}x + 5y = 11$ sub; ∅
29. $\frac{3}{5}c - \frac{1}{5}d = 9$, $\frac{7}{5}c + \frac{1}{5}d = 11$
30. $0.6m - 0.2n = 0.9$, $0.3m = 0.45 - 0.1n$ sub; (1.5, 0)
31. $1.44x - 3.24y = -5.58$, $1.08x + 3.24y = 9.99$ +; (1.75, 2.5)
32. $7.2m + 4.5n = 129.06$, $7.2m + 6.7n = 136.54$ −; (15.8, 3.4)

Number Theory

Use a system of equations and elimination to solve each problem.

33. Find two numbers whose sum is 64 and whose difference is 42. 11, 53
34. Find two numbers whose sum is 18 and whose difference is 22. 20, −2
35. Twice one number added to another number is 18. Four times the first number minus the other number is 12. Find the numbers. 5, 8
36. If $x + y = 11$ and $x - y = 5$, what does xy equal? 24

Use elimination twice to solve each system of equations. Write the solution as an ordered triple of the form (*x*, *y*, *z*).

37. $x + y = 5$, $y + z = 10$, $x + z = 9$ (2, 3, 7)
38. $2x + y + z = 13$, $x - y + 2z = 8$, $4x - 3z = 7$ (4, 2, 3)
39. $x + 2z = 2$, $y + 3z = 9$, $12x - y + 7z = 99$ (14, 27, −6)

Critical Thinking

40. The graphs of $Ax + By = 7$ and $Ax - By = 9$ intersect at (4, −1). Find A and B. (2, 1)

Applications and Problem Solving

41. Ling, 1.45 hours or 1 hour, 27 minutes; José, 1.15 hours, or 1 hour, 9 minutes

41. **On-Line Entertainment** On June 27, 1994, Aerosmith became the first major rock band to release a song distributed exclusively in the U.S. through a computer on-line service. Users of the commercial service, CompuServe, were able to download the Aerosmith song *Head First* for free. However, it took a long time to download the song, which itself lasted only 3 minutes, 14 seconds, because of the high-fidelity sound. José and Ling share a personal computer, and one evening they each downloaded the song without realizing that the other had done it. Ling also wasted 18 minutes because he typed the wrong word and had to start over. At the end of the month, the bill from CompuServe said they had used a total of 2.6 hours of time that evening. How long did it take to download the song each time? (*Hint:* 18 minutes = 0.3 hour.)

Assignment Guide

Core: 13–39 odd, 40, 41, 43–49
Enriched: 12–38 even, 40–49
All: Self Test, 1–10

For **Extra Practice,** see p. 775.

The red A, B, and C flags, printed only in the Teacher's Wraparound Edition, indicate the level of difficulty of the exercises.

4 ASSESS

Closing Activity

Writing Have students write in outline form the steps to follow when solving a system of equations by the elimination method.

Chapter 8, Quiz B (Lesson 8-3), is available in the *Assessment and Evaluation Masters*, p. 212.

Mid-Chapter Test (Lessons 8-1 through 8-3) is available in the *Assessment and Evaluation Masters*, p. 211.

Practice Masters, p. 58

NAME ______ DATE ______

Practice — Student Edition Pages 469–473

Elimination Using Addition and Subtraction

State whether addition, subtraction, or substitution would be most convenient to solve each system of equations. Then solve the system.

1. $x - y = 1$, $x + y = 3$ addition; (2, 1)
2. $-x + y = 1$, $x + y = 11$ addition; (5, 6)
3. $x + 4y = 11$, $x - 6y = 11$ subtraction; (11, 0)
4. $-x + 3y = 6$, $x + 3y = 18$ addition; (6, 4)
5. $3x + 4y = 19$, $3x + 6y = 33$ subtraction; (−3, 7)
6. $x + 4y = -8$, $x - 4y = -8$ addition; (−8, 0)
7. $3a + 4b = 2$, $4a - 4b = 12$ addition; (2, −1)
8. $3x - y = -1$, $-3x - y = 5$ addition; (−1, −2)
9. $2x - 3y = 9$, $-5x - 3y = 30$ subtraction; (−3, −5)
10. $x - y = 4$, $2x + y = -4$ addition; (0, −4)
11. $3y - x = 2$, $-2y - x = -18$ subtraction; (10, 4)
12. $5x - y = -6$, $-x + y = 2$ addition; (−1, 1)
13. $6r - 3t = 6$, $6r + 8t = -16$ subtraction; (0, −2)
14. $-3x + y = 3$, $3x + 2y = -12$ substitution; (−2, −3)
15. $-\frac{1}{3}x - \frac{4}{3}y = -2$, $\frac{1}{3}x - \frac{2}{3}y = 4$ addition; (10, −1)

Use a system of equations and elimination to solve each problem.

16. The sum of two numbers is 28. Their difference is 4. What are the two numbers? 12, 16
17. A two-digit number is 11 times its units digit. The sum of the digits is 12. Find the number. 66

Additional Answers

2a. The result is 0 = 45, which is false. Thus, the system has no solution.
2b. The graph is two parallel lines.
3. Both are correct. If the resulting statement is false, there is no solution. If the resulting statement is true, there is an infinite number of solutions.

King Khufu is also sometimes known by his Greek name, Cheops. He was the builder of the Great Pyramid at Giza, the largest single building in its time.

Additional Answer

46.

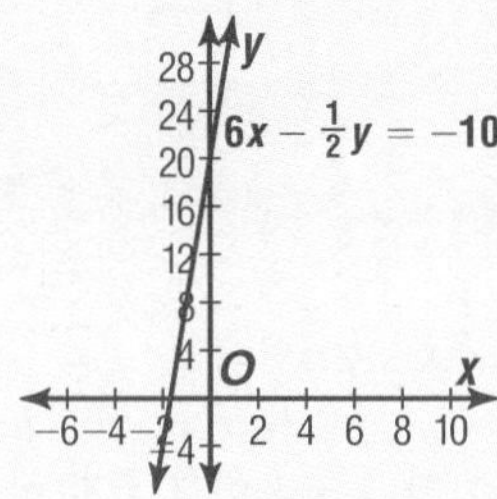

F Y I

The Great Pyramid of King Khufu, built around 2600 B.C., has only one entrance, which is nearly impossible to find. Centuries later, however, robbers stole everything from the pyramid, including the mummy of the king.

42. **World Cultures** The ancient Egyptians believed that the pharaohs lived forever after death in their houses of eternity, the pyramids. Suppose the side of the pyramid containing the entrance is represented by the line $13x + 10y = 9600$, the opposite side of the pyramid, by the line $13x - 10y = 0$, and the descending corridor leading to the entrance, by the line $3x - 10y = 1500$, where x is the distance in feet and y is the height in feet.
   a. Find the coordinates of the entrance. **(693.75, 58.125)**
   b. Find the height of the pyramid. **480 ft**

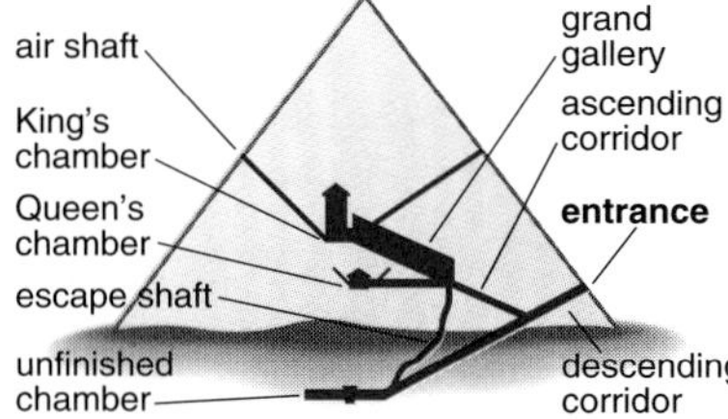

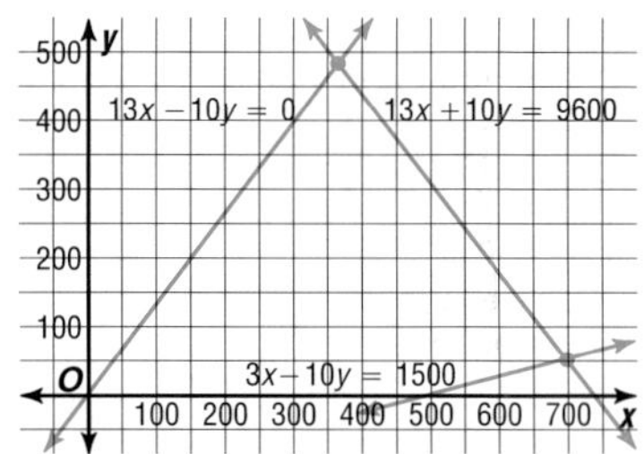

Mixed Review

43. **Chemistry** MX Labs needs to make 500 gallons of a 34% acid solution. The only solutions available are 25% acid and 50% acid. How many gallons of each solution should be mixed to make the 34% solution? Write and solve a system of equations by using substitution. (Lesson 8–2) **320 gal of 25% and 180 gal of 50%**

44. $\left\{h \mid h \geq -\frac{1}{2}\right\}$

44. Solve $5 - 8h \leq 9$. (Lesson 7–3)

45. Determine the slope of the line that passes through the points at $(2, -9)$ and $(-1, 0)$. (Lesson 6–1) **−3**

46. Graph $6x - \frac{1}{2}y = -10$. (Lesson 5–4) **See margin.**

47. $-\frac{7}{6}$

48. 3.87

47. Solve $-2(3t + 1) = 5$. (Lesson 3–3)

48. Find an approximation to the nearest hundredth for $\sqrt{15}$. (Lesson 2–8)

49. Name the property illustrated by the following statement. (Lesson 1–6)
If $6 = 2a$ *and* $a = 3$, *then* $6 = 2 \cdot 3$. **substitution =**

SELF TEST

Graph each system of equations. Then determine if the system has *one* solution, *no* solution, or *infinitely many* solutions. If the system has one solution, name it. (Lesson 8–1)

1. $x - y = 3$
   $3x + y = 1$ **(1, −2)**
2. $2x - 3y = 7$
   $3y = 7 + 2x$ **no solution**
3. $4x + y = 12$
   $x = 3 - \frac{1}{4}y$ **infinitely many**

1–3. See Solutions Manual for graphs.

Use substitution to solve each system of equations. (Lesson 8–2)

4. $y = 5x$
   $x + 2y = 22$ **(2, 10)**
5. $2y - x = -5$
   $y - 3x = 20$ **(−9, −7)**
6. $3x + 2y = 18$
   $x + \frac{8}{3}y = 12$ **(4, 3)**

Use elimination to solve each system of equations. (Lesson 8–3)

7. $x - y = -5$
   $x + y = 25$ **(10, 15)**
8. $3x + 5y = 14$
   $2x - 5y = 1$ **(3, 1)**
9. $5x + 4y = 12$
   $3x + 4y = 4$ **(4, −2)**
10. **Recreation** At a recreation and sports facility, 3 members and 3 nonmembers pay a total of \$180 to take an aerobics class. A group of 5 members and 3 nonmembers pay \$210 to take the same class. How much does it cost members and nonmembers to take an aerobics class? (Lesson 8–3) **\$15 for members, \$45 for nonmembers**

Enrichment Masters, p. 58

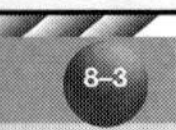

NAME______ DATE______

8–3 Enrichment — Student Edition Pages 469–473

Arithmetic Series

An **arithmetic series** is a series in which each term after the first may be found by adding the same number to the preceding term. Let S stand for the following series in which each term is 3 more than the preceding one.

$S = 2 + 5 + 8 + 11 + 14 + 17 + 20$

The series remains the same if we reverse the order of all the terms. So let us reverse the order of the terms and add one series to the other, term by term. This is shown at the right.

$S = 2 + 5 + 8 + 11 + 14 + 17 + 20$
$S = 20 + 17 + 14 + 11 + 8 + 5 + 2$
$2S = 22 + 22 + 22 + 22 + 22 + 22 + 22$
$2S = 7(22)$
$S = \frac{7(22)}{2} = 7(11) = 77$

Let a represent the first term of the series.
Let l represent the last term of the series.
Let n represent the number of terms in the series.
In the preceding example, $a = 2$, $l = 20$, and $n = 7$. Notice that when you add the two series, term by term, the sum of each pair of terms is 22. That sum can be found by adding the first and last terms, $2 + 20$ or $a + l$. Notice also that there are 7, or n, such sums. Therefore, the value of $2S$ is 7(22), or $n(a + l)$ in the general case. Since this is twice the sum of the series, you can use the following formula to find the sum of any arithmetic series.

$$S = \frac{n(a + l)}{2}$$

Example 1: Find the sum: $1 + 2 + 3 + 4 + 5 + 6 + 7 + 8 + 9$
$a = 1, l = 9, n = 9$, so $S = \frac{9(1 + 9)}{2} = \frac{9 \cdot 10}{2} = 45$

Example 2: Find the sum: $-9 + (-5) + (-1) + 3 + 7 + 11 + 15$
$a = -9, l = 15, n = 7$, so $S = \frac{7(-9 + 15)}{2} = \frac{7 \cdot 6}{2} = 21$

Find the sum of each arithmetic series.

1. $3 + 6 + 9 + 12 + 15 + 18 + 21 + 24$ **108**
2. $10 + 15 + 20 + 25 + 30 + 35 + 40 + 45 + 50$ **270**
3. $-21 + (-16) + (-11) + (-6) + (-1) + 4 + 9 + 14$ **−28**
4. even whole numbers from 2 through 100 **2550**
5. odd whole numbers between 0 and 100 **2500**

Extension

Connections The graphs of the equations $x - y = 1$, $x + y = 3$, and $-x + 5y = 19$ contain the sides of a triangle. Write the equation of the line that passes through the midpoint of the shortest side of the triangle and the vertex opposite the shortest side.
$x - 2y = -4$

SELF TEST

The Self Test provides students with a brief review of the concepts and skills in Lessons 8-1 through 8-3. Lesson numbers are given to the right of exercises or instruction lines so students can review concepts not yet mastered.

8–4 Elimination Using Multiplication

What YOU'LL LEARN
- To solve systems of equations by using the elimination method with multiplication and addition, and
- to determine the best method for solving systems of equations.

Why IT'S IMPORTANT

You can use systems of equations to solve problems involving telecommunications and geography.

FYI

On January 14, 1876, Alexander Graham Bell beat Elisha Gray by only a few hours when filing a patent application for his telephone. Both men had invented workable prototypes simultaneously.

Telecommunications

GBT Mobilnet provides monthly plans for cellular phone customers. Carla Ramos and Robert Johnson both selected Plan B for which monthly charges are based on per-minute rates of calls during peak and nonpeak hours. In one month, Carla made 75 minutes of peak calls and 30 minutes of nonpeak calls. Her bill was \$40.05. During the same period, Robert made 50 minutes of peak calls and 60 minutes of nonpeak calls. His bill was \$35.10. What is GBT Mobilnet's charge per minute for peak and nonpeak calls on Plan B?

Let p represent the rate per minute for peak calls, and let n represent the rate per minute for nonpeak calls. Then the information in this problem can be represented by the following system of equations.

$$75p + 30n = 40.05$$
$$50p + 60n = 35.10$$

So far, you have learned four methods for solving a system of two linear equations.

Method	The Best Time to Use
Graphing	if you want to estimate the solution, since graphing usually does not give an exact solution
Substitution	if one of the variables in either equation has a coefficient of 1 or −1
Addition	if one of the variables has opposite coefficients in the two equations
Subtraction	if one of the variables has the same coefficient in the two equations

The system above is not easily solved using any of these methods. However, there is an extension of the elimination method that can be used. Multiply one of the equations by some number so that adding or subtracting eliminates one of the variables.

For this system, multiply the first equation by −2 and add. Then the coefficient of n in both equations will be 60 or −60.

$$75p + 30n = 40.05 \quad \xrightarrow{\text{Multiply by } -2.} \quad -150p - 60n = -80.10$$
$$50p + 60n = 35.10 \qquad\qquad (+)\ 50p + 60n = 35.10$$
$$-100p = -45$$
$$p = 0.45$$

Now, solve for n by replacing p with 0.45.

$$75p + 30n = 40.05$$
$$75(0.45) + 30n = 40.05 \quad \textit{Substitute 0.45 for p.}$$
$$33.75 + 30n = 40.05 \quad \textit{Solve for n.}$$
$$30n = 6.3$$
$$n = 0.21$$

Is (0.45, 0.21) a solution?

FYI

For years, both Canada and the United States claimed that Bell invented the telephone in their country. The courts recently decided in favor of Canada. Born in Scotland, Bell spent most of his life in Canada, tested the telephone in the United States, and lived his final years in the United States.

8-4 LESSON NOTES

NCTM Standards: 1–5

Instructional Resources

- Study Guide Master 8-4
- Practice Master 8-4
- Enrichment Master 8-4
- Assessment and Evaluation Masters, p. 213
- Multicultural Activity Masters, p. 15
- Real-World Applications, 22
- Tech Prep Applications Masters, p. 16

Transparency 8-4A contains the 5-Minute Check for this lesson; **Transparency 8-4B** contains a teaching aid for this lesson.

Recommended Pacing	
Standard Pacing	Day 6 of 9
Honors Pacing	Day 5 of 8
Block Scheduling*	Day 3 of 5 (along with Lesson 8-3)
Alg. 1 in Two Years*	Days 8 & 9 of 14

*For more information on pacing and possible lesson plans, refer to the *Block Scheduling Booklet* and *Algebra 1 in Two Years.*

1 FOCUS

5-Minute Check
(over Lesson 8-3)

Use elimination to solve each system of equations.

1. $2x + y = 8$
   $x - y = 3$ $\left(\frac{11}{3}, \frac{2}{3}\right)$
2. $y + 5x = 9$
   $y - 5x = 7$ $\left(\frac{1}{5}, 8\right)$
3. $y + 3x = 12$
   $3y - 3x = 6$ $\left(\frac{5}{2}, \frac{9}{2}\right)$
4. The sum of two numbers is 20. Their difference is 4. Find the numbers. **12, 8**
5. Half the perimeter of a rectangle is 56 feet. The width of the rectangle is 8 feet less than the length. Find the dimensions of the rectangle. **24 ft by 32 ft**

Motivating the Lesson

Questioning Ask students if multiplying each side of an equation, such as $2x = 4$, by the same number will affect the value of the variable. For example, if each side of $2x = 4$ were multiplied by 2, the equation would read $4x = 8$. Students should recognize that the value of the variable is unaffected. Ask students if the same would hold true for an equation such as $2x + 3y = 8$. Explain that when simple addition or subtraction cannot be used to eliminate a variable in a system of equations, multiplication may be used in the elimination process.

2 TEACH

In-Class Example

For Example 1
Use elimination to solve each system of equations.

a. $3x + 5y = 11$
$2x + 3y = 7$ **(2, 1)**

b. $4x + 3y = 8$
$3x - 5y = -23$ **(−1, 4)**

c. $2x - 3y = 8$
$-5x + 2y = 13$ **(−5, −6)**

Teaching Tip In Example 1, emphasize that there are many other combinations of multipliers that can be used. You may want to have students suggest other multipliers.

Check:

$$75p + 30n = 40.05 \qquad 50p + 60n = 35.10$$
$$75(0.45) + 30(0.21) \stackrel{?}{=} 40.05 \qquad 50(0.45) + 60(0.21) \stackrel{?}{=} 35.10$$
$$40.05 = 40.05 \ \checkmark \qquad 35.10 = 35.10 \ \checkmark$$

The solution of this system is (0.45, 0.21). Thus, the per-minute rate for peak-hour calls is 45¢, and the per-minute rate for nonpeak calls is 21¢ on this plan.

For some systems of equations, it is necessary to multiply *each* equation by a different number in order to solve the system by elimination. You can choose to eliminate either variable.

Example  **Use elimination to solve the system of equations in two different ways.**

$$\mathbf{2x + 3y = 5}$$
$$\mathbf{5x + 4y = 16}$$

Method 1
You can eliminate the variable x by multiplying the first equation by 5 and the second equation by −2 and then adding the resulting equations.

$2x + 3y = 5$ → Multiply by 5. → $10x + 15y = 25$

$5x + 4y = 16$ → Multiply by −2. → $(+)\ -10x - 8y = -32$

$$7y = -7$$
$$y = -1$$

Now find x using one of the original equations.

$2x + 3y = 5$
$2x + 3(-1) = 5$ *Substitute −1 for y.*
$2x - 3 = 5$ *Solve for x.*
$2x = 8$
$x = 4$

The solution of the system is (4, −1).

Method 2
You can also solve this system by eliminating the variable y. Multiply the first equation by −4 and the second equation by 3. Then add.

$2x + 3y = 5$ → Multiply by −4. → $-8x - 12y = -20$

$5x + 4y = 16$ → Multiply by 3. → $(+)\ 15x + 12y = 48$

$$7x \qquad = 28$$
$$x = 4$$

Now find y.

$2x + 3y = 5$
$2(4) + 3y = 5$ *Substitute 4 for x.*
$8 + 3y = 5$ *Solve for y.*
$3y = -3$
$y = -1$

The solution is (4, −1), which matches the result obtained with Method 1.

Alternative Teaching Strategies

Reading Algebra The word *system* has many meanings. Have students research these meanings and determine which applies in the phrase "system of linear equations."

Example 2

Testing

Luis Diaz discovered while entering test scores into his computer that he had accidentally reversed the digits of a test and shorted a student 36 points. Mr. Diaz told the student that the sum of the digits was 14 and agreed to give the student his correct score plus extra credit if he could determine his actual score without looking at his test. What was his actual score on the test?

Explore Let t represent the tens digit of the score.
Let u represent the units digit.

The actual score on the test can be represented by $10t + u$. The amount entered in the computer can be represented by $10u + t$. *Why?*

Plan Since the sum of the digits is 14, one equation is $t + u = 14$. Since the teacher accidentally shorted the student by 36 points, another equation is $(10t + u) - (10u + t) = 36$ or $9t - 9u = 36$.

Solve

$$t + u = 14 \quad \xrightarrow{\text{Multiply by 9.}} \quad 9t + 9u = 126$$
$$9t - 9u = 36 \qquad\qquad\qquad (+)\ 9t - 9u = 36$$
$$18t \quad = 162$$
$$t = 9$$

Now find u using one of the original equations.

$t + u = 14$

$9 + u = 14$ *Substitute 9 for t.*

$u = 5$ *Solve for u.*

The solution is (9, 5), which means that the student's actual test score was $10(9) + 5$, or 95 points.

Examine The sum of the digits, $9 + 5$, is 14 and $95 - 59$ is 36.

You can use systems of equations to solve problems involving the distance formula, $rt = d$.

Example 3

Uniform Motion

FYI

The world's largest riverboat is the 418-ft *American Queen.* Passengers paid up tp $9400 to take its first cruise in June, 1995.

A riverboat on the Mississippi River travels 48 miles upstream in 4 hours. The return trip takes the riverboat only 3 hours. Find the rate of the current.

Explore Let r represent the rate of the riverboat in still water. Let c represent the rate of the current.

Then $r + c$ represents the rate of the riverboat traveling downstream *with* the current and $r - c$ represents the rate of the riverboat traveling upstream *against* the current.

(continued on the next page)

In-Class Examples

For Example 2
A bank teller reversed the digits in the amount of a check and overpaid a customer by $9. The sum of the digits in the two-digit amount was 9. Find the amount of the check. **$45**

For Example 3
A boat is rowed 10 miles downstream in 2 hours, and then rowed the same distance upstream in $3\frac{1}{3}$ hours. Find the rate of the boat in still water and the rate of the current. **4 mph, 1 mph**

Teaching Tip Discuss the chart in Example 3. What is the rate of the boat when it travels upstream? Why? ***r − c*; Because of the current, the boat travels upstream slower than in still water.** What is the rate of the boat when it travels downstream? Why? ***r + c*; Because of the current, the boat travels downstream faster than in still water.**

The first commercially successful steamship, *The Clermont,* was created by Robert Fulton in 1807.

Cooperative Learning

Trade-A-Problem Separate the class into groups of four. One student selects a pair of equations for the other group members to solve. One group member should attempt to solve the equations by using graphing, another by substitution, another by elimination using addition and subtraction, and the last group member by elimination using multiplication. Group members should then discuss which methods worked the best. Have group members switch roles. For more information on the trade-a-problem strategy, see *Cooperative Learning in the Mathematics Classroom,* one of the titles in the Glencoe Mathematics Professional Series, pages 25–26.

3 PRACTICE/APPLY

Check for Understanding

Exercises 1–14 are designed to help you assess your students' understanding through reading, writing, speaking, and modeling. You should work through Exercises 1–4 with your students and then monitor their work on Exercises 5–14.

Error Analysis
When working a "digit problem," carefully explain the difference between the terms *digit* and *number*. Have students read the problem word for word and translate into mathematics.
Example: tens digit = T
units digit = U
the number = $10T + U$
A digit has no value, but a number has a value.

Additional Answers

5. Multiply the first equation by −3, then add.
6. Multiply the first equation by 3, multiply the second equation by −2, then add.
7. Multiply the second equation by 5, then add.
8. Multiply the first equation by 5, multiply the second equation by −7, then add.
9. Multiply the first equation by 5, multiply the second equation by −8, then add.
10. Multiply the first equation by 4, multiply the second equation by 3, then add.

Plan Use the formula rate × time = distance, or $rt = d$, to write a system of equations. Then solve the system to find the value of c.

	r	t	d	$rt = d$
Downstream	$r + c$	3	48	$3r + 3c = 48$
Upstream	$r - c$	4	48	$4r - 4c = 48$

Solve $3r + 3c = 48$ → Multiply by 4. → $12r + 12c = 192$

$4r - 4c = 48$ → Multiply by −3. → $(+)\ -12r + 12c = -144$

$24c = 48$

$c = 2$

The rate of the current is 2 miles per hour.

Examine Find the value of r for this system and then check the solution.

CHECK FOR UNDERSTANDING

Communicating Mathematics

Study the lesson. Then complete the following.

1. **Write** a problem about a real-life situation in which only an estimate of the solution is needed rather than the exact solution. The problem should involve a system of equations. **See students' work.**
2. **Explain** why you might need to multiply each equation by a different number when using elimination to solve a system of equations.
3. **Write** a system of equations that could best be solved by using multiplication and then elimination using addition or subtraction. **Sample answer: $3x + 5y = 7$, $4x - 10y = 1$**
4. **Assess Yourself** Describe the method you like to use best when solving systems of linear equations. Explain your reasons. **See students' work.**

2. To make either the x-term or y-term coefficients additive inverses.

Guided Practice

Explain the steps you would follow to eliminate the variable x in each system of equations. Then solve the system. **5–7. See margin for steps.**

5. $x + 5y = 4$, $3x - 7y = -10$ **(−1, 1)**
6. $2x - y = 6$, $3x + 4y = -2$ **(2, −2)**
7. $-5x + 3y = 6$, $x - y = 4$ **(−9, −13)**

Explain the steps you would follow to eliminate the variable y in each system of equations. Then solve the system. **8–10. See margin for steps.**

8. $4x + 7y = 6$, $6x + 5y = 20$ **(5, −2)**
9. $3x - 8y = 13$, $4x - 5y = 6$ **(−1, −2)**
10. $2x - 3y = 2$, $5x + 4y = 28$ **(4, 2)**

Match each system of equations with the method that could be most efficiently used to solve it. Then solve the system.

11. $3x - 7y = 6$, $2x + 7y = 4$ **b; (2, 0)**
12. $y = 4x + 11$, $3x - 2y = -7$ **a; (−3, −1)**
13. $4x + 3y = 19$, $3x - 4y = 8$ **c; (4, 1)**

a. substitution
b. elimination using addition or subtraction
c. elimination using multiplication

Reteaching

Using Decision Making Have students go through a list of systems of equations and write which process would be best to use in order to solve (substitution, elimination using addition or subtraction, or elimination using multiplication). If elimination using multiplication is the most efficient process, have students use it to solve the problem.

14. **Uniform Motion** A riverboat travels 36 miles downstream in 2 hours. The return trip takes 3 hours.

a. Find the rate of the riverboat in still water. **15 mph**

b. Find the rate of the current. **3 mph**

EXERCISES

Practice

Use elimination to solve each system of equations. 23. $(-1, -2)$

15. $2x + y = 5$; $3x - 2y = 4$ **(2, 1)**
16. $4x - 3y = 12$; $x + 2y = 14$ **(6, 4)**
17. $3x - 2y = 19$; $5x + 4y = 17$ **(5, −2)**
18. $9x = 5y - 2$; $3x = 2y - 2$ **(2, 4)**
19. $7x + 3y = -1$; $4x + y = 3$ **(2, −5)**
20. $6x - 5y = 27$; $3x + 10y = -24$ **(2, −3)**
21. $8x - 3y = -11$; $2x - 5y = 27$ **(−4, −7)**
22. $11x - 5y = 80$; $9x - 15y = 120$ **(5, −5)**
23. $4x - 7y = 10$; $3x + 2y = -7$

B

24. $3x - \frac{1}{2}y = 10$; $5x + \frac{1}{4}y = 8$ **(2, −8)**
25. $2x + \frac{2}{3}y = 4$; $x - \frac{1}{2}y = 7$ **(4, −6)**
26. $\frac{2x + y}{3} = 15$; $\frac{3x - y}{5} = 1$ **(10, 25)**
27. $7x + 2y = 3(x + 16)$; $x + 16 = 5y + 3x$ **(13, −2)**
28. $0.4x + 0.5y = 2.5$; $1.2x - 3.5y = 2.5$ **(5, 1)**
29. $1.8x - 0.3y = 14.4$; $x - 0.6y = 2.8$ **(10, 12)**

Number Theory

Use a system of equations and elimination to solve each problem.

30. The sum of the digits of a two-digit number is 14. If the digits are reversed, the new number is 18 less than the original number. Find the original number. **86**
31. Three times one number equals twice a second number. Twice the first number is 3 more than the second number. Find the numbers. **6, 9**
32. The ratio of the tens digit to the units digit of a two-digit number is 1:4. If the digits are reversed, the sum of the new number and the original number is 110. Find the original number. **28**

Determine the best method to solve each system of equations. Then solve the system.

33. elim, addn; $\left(2, \frac{1}{8}\right)$

35. substitution or elim, mult.; infinitely many

33. $9x - 8y = 17$; $4x + 8y = 9$
34. $3x - 4y = -10$; $5x + 8y = -2$ **elim, mult (−2, 1)**
35. $x + 2y = -1$; $2x + 4y = -2$
36. $5x + 3y = 12$; $4x - 5y = 17$ **elim, mult (3, −1)**
37. $\frac{2}{3}x - \frac{1}{2}y = 14$; $\frac{5}{6}x - \frac{1}{2}y = 18$ **elim, subtr (24, 4)**
38. $\frac{1}{2}x - \frac{2}{3}y = \frac{7}{3}$; $\frac{3}{2}x + 2y = -25$ **elim, mult (−6, −8)**

Use elimination to solve each system of equations.

39. (11, 12)

39. $\frac{1}{x-5} - \frac{3}{y+6} = 0$; $\frac{2}{x+7} - \frac{1}{y-3} = 0$
40. $\frac{2}{x} + \frac{3}{y} = 16$; $\frac{1}{x} + \frac{1}{y} = 7$ $\left(\frac{1}{5}, \frac{1}{2}\right)$
41. $\frac{1}{x-y} = \frac{1}{y}$; $\frac{1}{x+y} = 2$ $\left(\frac{1}{3}, \frac{1}{6}\right)$

Assignment Guide

Core: 15–45 odd, 47–54
Enriched: 16–42 even, 43–54

For **Extra Practice,** see p. 775.

The red A, B, and C flags, printed only in the Teacher's Wraparound Edition, indicate the level of difficulty of the exercises.

Study Guide Masters, p. 59

NAME______________________ DATE __________

Study Guide

Student Edition Pages 475–481

Elimination Using Multiplication

Some systems of equations cannot be solved simply by adding or subtracting the equations. One or both equations must first be multiplied by a number before the system can be solved by elimination. Consider the following example.

Example: Use elimination to solve the system of equations $x + 10y = 3$ and $4x + 5y = 5$.

$x + 10y = 3$
$4x + 5y = 5$

Multiply $x + 10y = 3$ by -4. Then add the equations.

$$\begin{aligned} -4x - 40y &= -12 \\ 4x + 5y &= 5 \\ \hline -35y &= -7 \\ y &= \tfrac{1}{5} \end{aligned}$$

Substitute $\frac{1}{5}$ for y into either original equation and solve for x.

$$\begin{aligned} x + 10\left(\tfrac{1}{5}\right) &= 3 \\ x + 2 &= 3 \\ x &= 1 \end{aligned}$$

The solution of the system is $\left(1, \frac{1}{5}\right)$.

Use elimination to solve each system of equations.

1. $3x + 2y = 0$; $x - 5y = 17$ (2, −3)
2. $2x + 3y = 6$; $x + 2y = 5$ (−3, 4)
3. $3x - y = 2$; $x + 2y = 3$ (1, 1)
4. $4x + 5y = 6$; $6x - 7y = -20$ (−1, 2)

Use a system of equations and elimination to solve each problem.

5. The length of Sally's garden is 4 meters greater than 3 times the width. The perimeter of her garden is 72 meters. What are the dimensions of Sally's garden? 28 meters by 8 meters
6. Anita is $4\frac{1}{2}$ years older than Basilio. Three times Anita's age added to six times Basilio's age is 36. How old are Anita and Basilio? Basilio is $2\frac{1}{2}$ yr and Anita is 7 yr.

Using the Programming Exercises The program given in Exercise 42 is for use with a TI-82 graphing calculator. For other programmable calculators, have students consult their owner's manual for commands similar to those presented here.

Practice Masters, p. 59

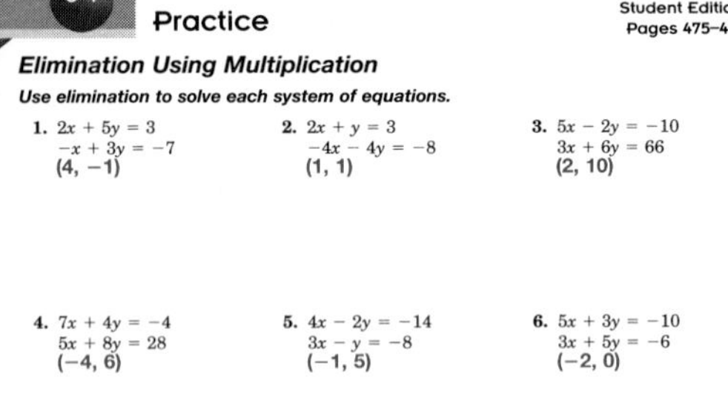

8-4 NAME ______ DATE ______

Practice — Student Edition Pages 475–481

Elimination Using Multiplication

Use elimination to solve each system of equations.

1. $2x + 5y = 3$, $-x + 3y = -7$ (4, −1)
2. $2x + y = 3$, $-4x - 4y = -8$ (1, 1)
3. $5x - 2y = -10$, $3x + 6y = 66$ (2, 10)
4. $7x + 4y = -4$, $5x + 8y = 28$ (−4, 6)
5. $4x - 2y = -14$, $3x - y = -8$ (−1, 5)
6. $5x + 3y = -10$, $3x + 5y = -6$ (−2, 0)
7. $2x + y = 0$, $5x + 3y = 2$ (−2, 4)
8. $9x - 6y = -12$, $x + 2y = 0$ $\left(-1, \frac{1}{2}\right)$
9. $0.5x + 0.5y = -2$, $x - 0.25y = 6$ (4, −8)

Use a system of equations and elimination to solve each problem.

10. The sum of the digits of a two-digit number is 11. If 45 is added to the number, the result is the number with the digits reversed. Find the number. 38
11. Suppose you invested $10,000, part at 6% annual interest and the rest at 9% annual interest. If you received $684 in interest after one year, how much did you invest at each rate? $7200 at 6%, $2800 at 9%

Programming

42. The graphing calculator program at the right finds the solution of two linear equations written in standard form.

$ax + by = c$

$dx + ey = f$

The formulas for the solution of this system are as follows.

$$x = \frac{ce - bf}{ae - bd}, \quad y = \frac{af - cd}{ae - bd}$$

```
PROGRAM:SOLVE
: Disp "ENTER COEFFICIENTS"
: Prompt A, B, C, D, E, F
: If AE–BD = 0
: Then
: Goto 1
: End
: (CE–BF)/(AE-BD) → X
: (AF–CD)/(AE-BD) → Y
: Disp "THE SOLUTION IS"
: Disp "X= ", X
: Disp "Y= ", Y
: Stop
: Lbl 1
: If CE–BF=0 or AF–CD=0
: Then
: Disp "INFINITELY", "MANY"
: Else
: Disp "NO SOLUTION"
```

Use the program to solve each system.

a. $8x + 2y = 0$, $12x + 3y = 0$ **infinitely many**

b. $x - 2y = 5$, $3x - 5y = 8$ **(−9, −7)**

c. $5x + 5y = 16$, $2x + 2y = 5$ **no solution**

d. $7x - 3y = 5$, $14x - 6y = 10$ **infinitely many**

Critical Thinking

43. The graphs of the equations $5x + 4y = 18$, $2x + 9y = 59$, and $3x - 5y = -4$ contain the sides of a triangle. Determine the coordinates of the vertices of the triangle. **(−2, 7), (2, 2), (7, 5)**

Applications and Problem Solving

44. **Geography** Benjamin Banneker, a self-taught mathematician and astronomer, was the first African-American to publish an almanac. He is most noted for being the assistant surveyor on the team that designed the ten-mile square of Washington, D.C. The White House is located in the center of the square, at the intersection of Pennsylvania Avenue and New York Avenue. Let $-5x + 7y = 0$ represent New York Avenue and let $3x + 8y = 305$ represent Pennsylvania Avenue. Find the coordinates for the White House. **(35, 25)**

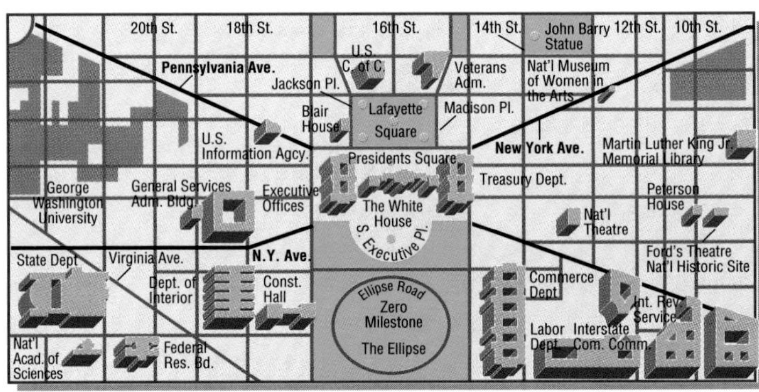

45. **Organize Data** At the new Cozy Inn Restaurant, which is still under construction, the owners have hired enough waiters and waitresses to handle 17 tables of customers. The fire marshall has looked at the plans for the restaurant and says he will approve it for a limit of 56 customers. The restaurant owners are now deciding how many two-seat tables and how many four-seat tables to buy for the restaurant. How many of each kind should they buy? **6 2-seat tables, 11 4-seat tables**

Extension

Connections The graphs of the equations $5x + 4y = 18$, $2x + 9y = 59$, and $3x - 5y = -4$ contain the sides of a triangle. Determine the vertices of the triangle. **(−2, 7), (2, 2), (7, 5)**

Tech Prep

Surveyor Students who are interested in surveying may wish to do further research on the role of Benjamin Banneker in the building of Washington, D.C., as mentioned in Exercise 44, and on the role of surveyors in the construction of planned communities today. For more information on tech prep, see the *Teacher's Handbook*.

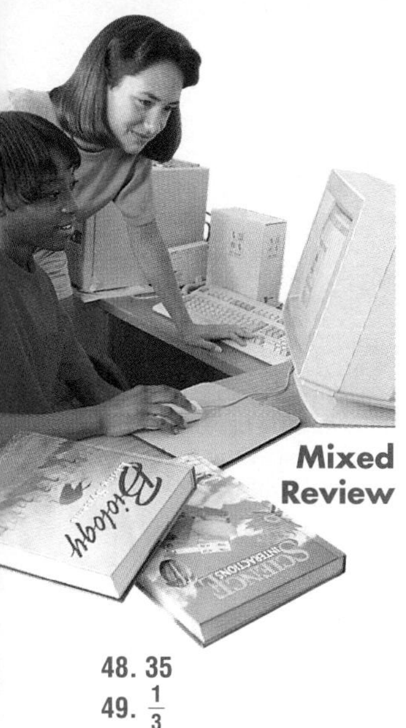

46. **Information Highway** The Mercury Center provides a reference and research service for on-line computer users based on peak and nonpeak usage. Miriam and Lesharo are both subscribers. The chart below displays the number of peak and nonpeak minutes each of them spent on-line in one month and how much it cost. Use the information to find the Mercury Center's rate per minute for its peak and nonpeak on-line research service. **\$0.45 and \$0.15**

User	Number of Peak Minutes	Number of Nonpeak Minutes	Cost
Lesharo	45	50	\$27.75
Miriam	70	30	\$36

Mixed Review

47. Use elimination to solve the system of equations. (Lesson 8–3) **(3, −4)**
$2x - y = 10$
$5x + 3y = 3$

48. **Statistics** What is the outlier in the box-and-whisker plot? (Lesson 7–7)

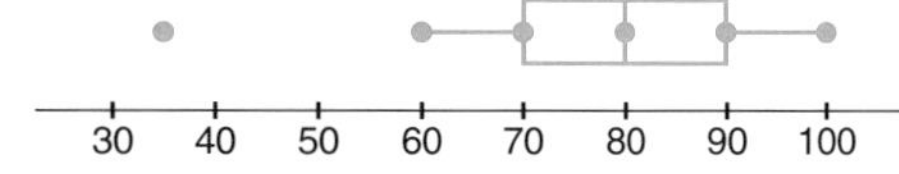

48. 35

49. $\frac{1}{3}$

49. **Probability** If Bill, Raul, and Kenyatta each have an equal chance of winning a bicycle race, find the probability that Raul finishes last. (Lesson 7–5)

50. Find the coordinates of the midpoint of the line segment whose endpoints are at (1, 6) and (−3, 4). (Lesson 6–7) **(−1, 5)**

51. **Track** Alfonso runs a 440-yard race in 55 seconds, and Marcus runs it in 88 seconds. To have Alfonso and Marcus finish at the same time, how much of a head start should Alfonso give Marcus? (Lesson 4–7) **165 yd**

52. If $12m = 4$, then $3m =$ __?__. (Lesson 3–2) **1**

53. 2 cups

53. **Cooking** If there are four sticks in a pound of butter and each stick is $\frac{1}{2}$ cup, how many cups of butter are in a pound of butter? (Lesson 2–6)

54. Evaluate $288 \div [3(9 + 3)]$. (Lesson 1–3) **8**

Mathematics and SOCIETY

High-Tech Checkout Lanes **1–2. See margin.**

The article below appeared in *Progressive Grocer* in February, 1994.

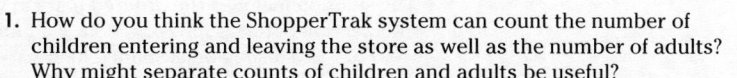

> K MART...HAS INSTALLED A NEW technology in 48 stores that helps ensure that enough checklanes are open to serve customers in a store. The system counts the number of adults and children who are in a store at any given moment. The system, called ShopperTrak...uses infrared technology on door-mounted units to give a continuous count of shoppers entering and exiting a store...The data is channeled to software in a PC called FastLane, which uses it to calculate how many checklanes should be open during the next 20 minutes so that no more than two to three people are in line at each lane. Managers read the data at monitors stationed at the checkout area. ■

1. How do you think the ShopperTrak system can count the number of children entering and leaving the store as well as the number of adults? Why might separate counts of children and adults be useful?
2. Do you think the average shopping times would differ between men and women, boys and girls, or senior citizens and young people? Why do you think the ShopperTrak system doesn't consider these factors?

4 ASSESS

Closing Activity

Speaking Ask students to describe how to solve a system of equations by the elimination method using multiplication and addition.

Chapter 8, Quiz C (Lesson 8-4), is available in the *Assessment and Evaluation Masters,* p. 213.

Mathematics and SOCIETY

Although the actual system is very complex, have students discuss its operation in simple terms. For example, suggest that the system keeps a count of the number of shoppers who have entered the store, *N,* and a count of the number of shoppers who have exited, *X,* and stores the count as $N - X$. Whenever $N - X$ increases by 10, a lane is opened. Each time it decreases by 10, a lane is closed. Have students suggest other methods that the system might employ.

Enrichment Masters, p. 59

NAME ______ DATE ______

Enrichment Student Edition Pages 475–481

Geometric Vanishing Acts

Puzzles of this type use a "trick" drawing. It appears that rearranging the pieces of each figure causes one or more squares to disappear.

Make figures of your own on graph paper. Then explain the "trick" in each puzzle.

1. The rectangle has an area of 65 square units, but the square has an area of only 64 square units.

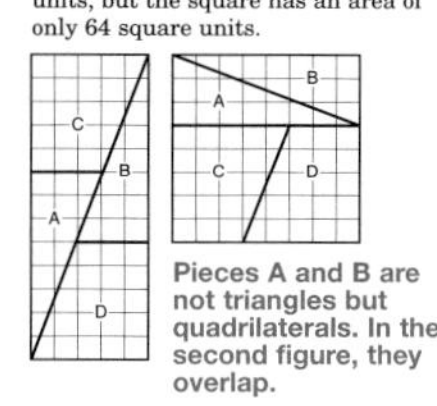

Pieces A and B are not triangles but quadrilaterals. In the second figure, they overlap.

2. The square has an area of 64 square units, but the rectangle has an area of only 63 square units.

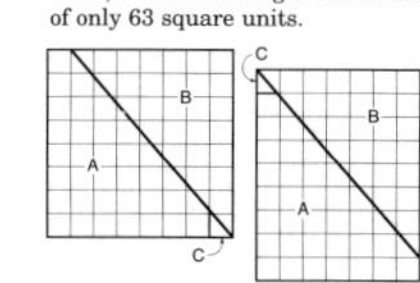

The triangle C actually has a height of $1\frac{1}{7}$ units, so the rectangle is really $9\frac{1}{7}$ units high.

3. The square has an area of 64 units, but the new figure has an area of only 63 units.

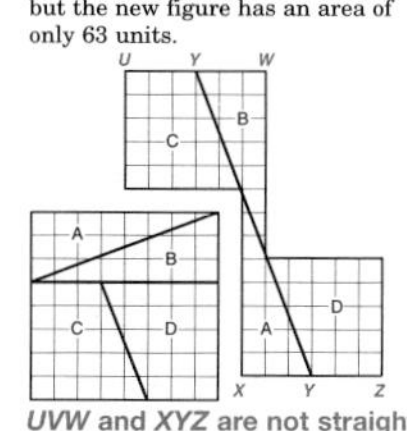

UVW and *XYZ* are not straight line segments. Thus, some "unit squares" of the second figure are not squares at all.

4. Rearranging the square on the left causes a 2-unit "hole" to appear.

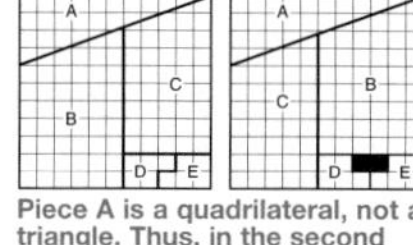

Piece A is a quadrilateral, not a triangle. Thus, in the second figure, the tops of pieces B and C overlap portions of piece A.

Answers for Mathematics and Society

1. The infrared detection system can measure the heights of persons entering and leaving the store, and uses the height measurements to attempt to differentiate between adults and children. The higher the ratio of adults to children, the more checkout clerks would be needed.
2. Sample answer: The age and gender factors may not, by themselves, yield any useful data. Even if there were measurable differences, it may be too complicated and costly to include them in a system. For example, the device can count people in the ShopperTrak system, but it can't detect age or gender.

8-5 LESSON NOTES

Standards: 1–5, 8

Instructional Resources

- Study Guide Master 8-5
- Practice Master 8-5
- Enrichment Master 8-5
- Assessment and Evaluation Masters, p. 213
- Modeling Mathematics Masters, pp. 57–59
- Multicultural Activity Masters, p. 16

Transparency 8-5A contains the 5-Minute Check for this lesson; **Transparency 8-5B** contains a teaching aid for this lesson.

Recommended Pacing	
Standard Pacing	Day 7 of 9
Honors Pacing	Day 6 of 8
Block Scheduling*	Day 4 of 5
Alg. 1 in Two Years*	Days 10 & 11 of 14

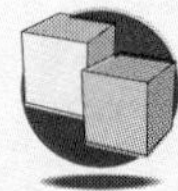

*For more information on pacing and possible lesson plans, refer to the *Block Scheduling Booklet* and *Algebra 1 in Two Years.*

1 FOCUS

5-Minute Check
(over Lesson 8-4)

Use elimination to solve each system of equations.

1. $\frac{1}{3}x - \frac{5}{6}y = -6$
   $4x + 7y = 30$ **(−3, 6)**
2. $2a + b = 19$
   $3a - 2b = -3$ **(5, 9)**
3. $x - y = 6$
   $x - 3y = 7$ $\left(\frac{11}{2}, -\frac{1}{2}\right)$
4. $x + y = 8$
   $2x - y = -6$ $\left(\frac{2}{3}, \frac{22}{3}\right)$
5. $y = 2x$
   $2x + y = 10$ $\left(\frac{5}{2}, 5\right)$

8–5 Graphing Systems of Inequalities

What YOU'LL LEARN
- To solve systems of inequalities by graphing.

Why IT'S IMPORTANT
You can use systems of inequalities to solve problems involving travel and nutrition.

APPLICATION
Employment

Unita likes her job as a baby-sitter, but it pays only \$3 per hour. She has been offered a job as a tutor that pays \$6 per hour. Because of school, her parents only allow her to work a maximum of 15 hours per week. How many hours can Unita tutor *and* baby-sit and still make at least \$65 per week?

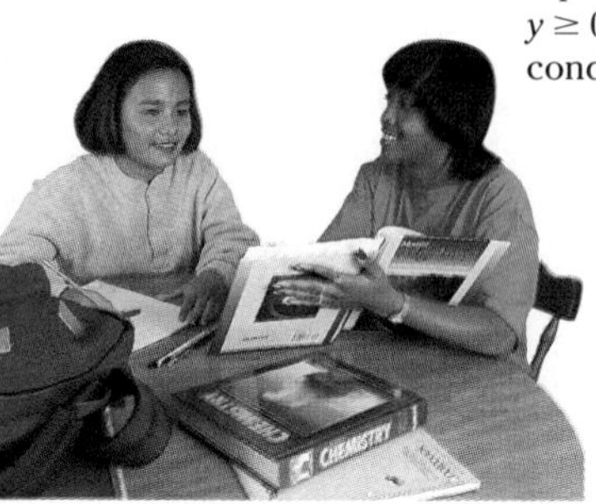

Let x represent the number of hours Unita can baby-sit each week. Let y represent the number of hours she can tutor each week. Since both x and y represent a number of hours, neither can be a negative number. Thus, $x \geq 0$ and $y \geq 0$. Then the following **system of inequalities** can be used to represent the conditions of this problem.

$x \geq 0$
$y \geq 0$
$3x + 6y \geq 65$ *She wants to earn at least \$65.*
$x + y \leq 15$ *She can work up to 15 hours.*

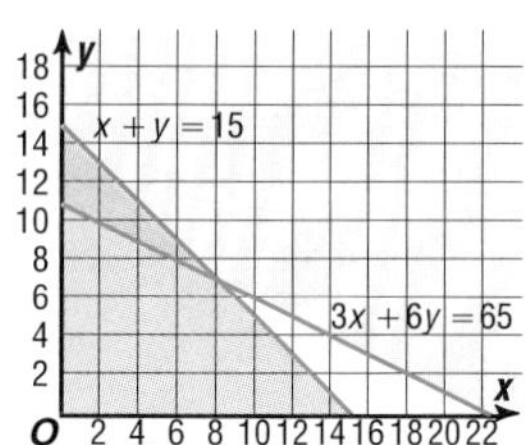

The solution of this system is the set of all ordered pairs that satisfies both inequalities and lies in the first quadrant. The solution can be determined by graphing each inequality on the same coordinate plane.

Recall that the graph of each inequality is called a *half-plane*. The intersection of the two half-planes represents the solution to the system of inequalities. This solution is a region that contains the graphs of an infinite number of ordered pairs. The boundary line of the half-plane is solid and is included in the graph if the inequality is $\leq$ or $\geq$. The boundary line of the half-plane is dashed and is not included in the graph if the inequality is $<$ or $>$.

LOOK BACK
You can refer to Lesson 7-8 for information on graphing inequalities in two variables.

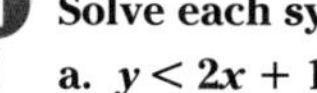

The graphs of $3x + 6y = 65$ and $x + y = 15$ are the boundaries of the region and are included in the graph of this system. This region is shown in green above. Only the portion in the first quadrant is shaded since $x \geq 0$ and $y \geq 0$. Every point in this region is a possible solution to the system. For example, since the graph of (5, 9) is a point in the region, Unita could baby-sit for 5 hours and tutor for 9 hours. In this case, she would make \$3(5) + \$6(9) or \$69. *Does this meet her requirements of time and earnings?*

Example  **Solve each system of inequalities by graphing.**

a. $y < 2x + 1$
$y \geq -x + 3$

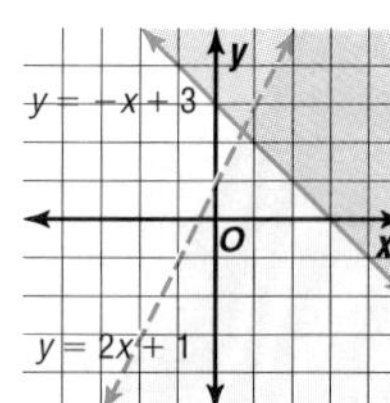

The solution includes the ordered pairs in the intersection of the graphs of $y < 2x + 1$ and $y \geq -x + 3$. This region is shaded in green at the right. The graphs of $y = 2x + 1$ and $y = -x + 3$ are the boundaries of this region. The graph of $y = 2x + 1$ is dashed and is *not* included in the graph of $y < 2x + 1$. The graph of $y = -x + 3$ is included in the graph of $y \geq -x + 3$.

GLOBAL CONNECTIONS

Europeans first learned of the Hawaiian Islands in 1778 when Captain James Cook travelled to the islands and named them the Sandwich Islands.

GLENCOE Technology

CD-ROM Interaction

A multimedia simulation allows students to use graphing to determine which cellular phone service is the most cost effective. A blackline master activity with teacher's notes provides a follow-up to the CD-ROM simulation.

For Windows & Macintosh

b. $2x + y \geq 4$
$y \leq -2x - 1$

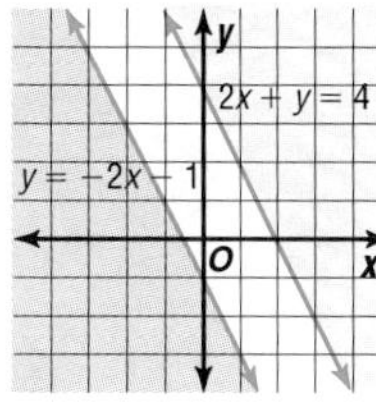

The graphs of $2x + y = 4$ and $y = -2x - 1$ are parallel lines. Because the two regions have no points in common, the system of inequalities has no solution.

Sometimes in real-life problems involving systems, only whole-number solutions make sense.

Example 2

Vacations

Elena Ayala wants to spend no more than \$700 for hotels while vacationing in Hawaii. She wants to stay at the Hyatt Resort at least one night and at the Coral Reef Hotel for the remainder of her stay. The Hyatt Resort costs \$130 per night, and the Coral Reef Hotel costs \$85 per night.

a. If she wants to stay in Hawaii at least 6 nights, how many nights could she spend at each hotel and still stay within her budget?

b. What advice might you give Elena concerning her options?

GLOBAL CONNECTIONS

Polynesians from the Marquesas Islands settled in Hawaii in about A.D. 400. A second wave of immigration arrived from Tahiti approximately 400 to 500 years later.

a. Let c represent the number of nights she will stay at the Coral Reef Hotel. Let h represent the number of nights she will stay at the Hyatt Resort.

Then the following system of inequalities can be used to represent the conditions of this problem.

$h + c \geq 6$	*Elena wants to stay at least 6 nights.*
$h \geq 1$	*She wants to stay at least 1 night at the Hyatt Resort.*
$130h + 85c \leq 700$	*She wants to spend no more than \$700.*

The solution is the set of all ordered pairs whose graphs are in the intersection of the graphs of these inequalities. This region is shown in brown at the right.

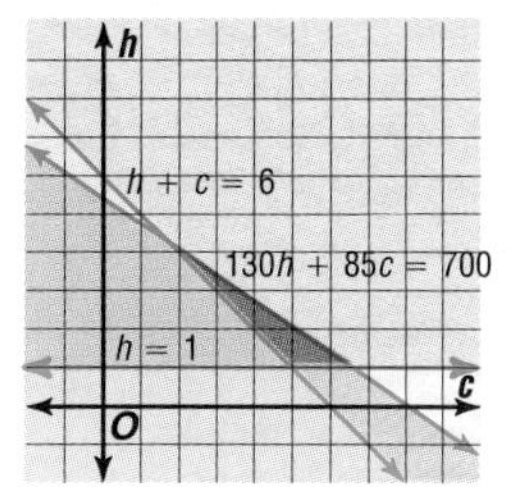

Any point in this region is a possible solution; however, only whole-number solutions make sense in this problem. *Why?* For example, since (3, 3) is a point in the region, Elena could stay 3 nights at each hotel. In this case, she would spend 3(\$130) or \$390 at the Hyatt Resort and 3(\$85) or \$255 at the Coral Reef Hotel for a total of \$645. The other solutions are (5, 1), (6, 1), (4, 2), (5, 2) and (2, 4). *Check this result.*

b. You could advise Elena that she could stay in Hawaii a maximum of 7 nights if she stayed at the Hyatt Resort only 1 or 2 nights and stayed at the Coral Reef Hotel for the remainder of her vacation.

Motivating the Lesson

Questioning Review the methods for graphing a linear equation and a linear inequality. Examine the method for testing an ordered pair on one side of the boundary line of the equation to determine which side of the graph to shade.

2 TEACH

In-Class Examples

For Example 1
Solve each system of inequalities by graphing.

a. $y < -2x + 4$ and $y > 3x - 4$

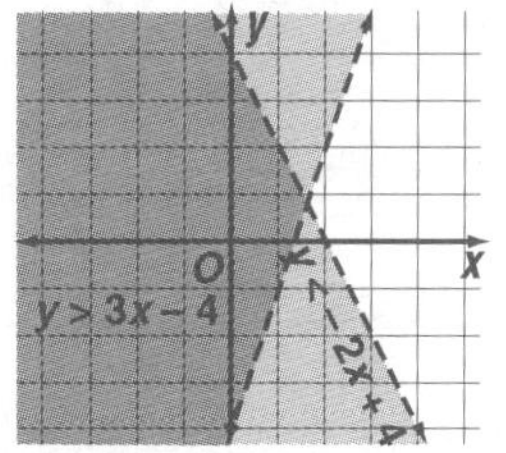

The solution is the intersection of the shaded regions.

b. $x + y \geq 5$ and $x + y \leq 1$

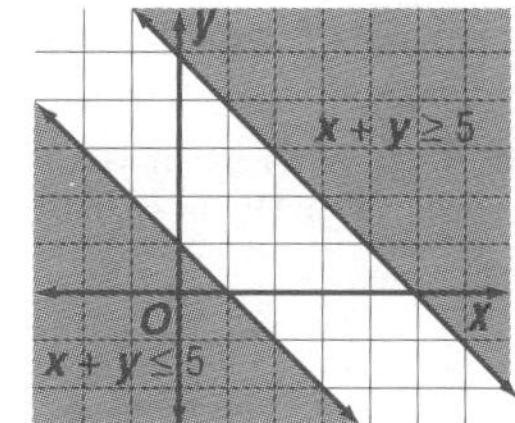

The system has no solution.

For Example 2
Solve the system of inequalities by graphing.

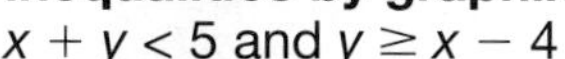

$x + y < 5$ and $y \geq x - 4$

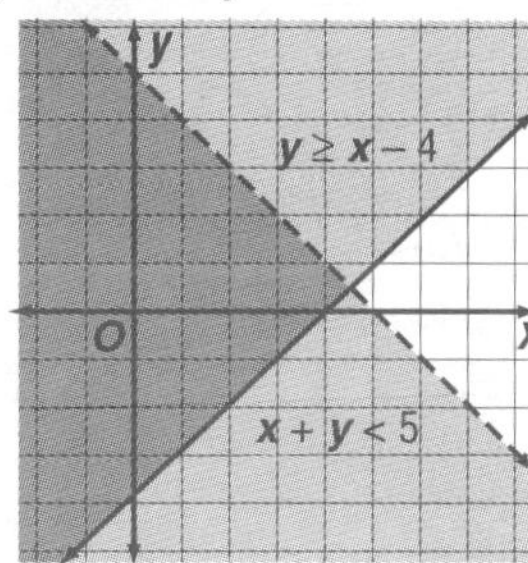

The solution is the intersection of the shaded regions.

Alternative Learning Styles

Kinesthetic Arrange desks in 6 rows and 6 columns. Explain that each row represents 2 months of the year: row 1 = January, February; row 2 = March, April; and so on. Each column will represent a range of heights: column 1 = less than 5'; column 2 = 5' to 5'2"; column 3 = 5'2" to 5'4"; and so on, with column 6 = more than 5'10". Have students arrange themselves on this grid according to their birth month and height. If b = number of birth month (1–12) and h = students' height (expressed as a decimal), have students identify those classmates who satisfy the inequality $12 \leq b + h$, and note the pattern created on the "graph."

Teaching Tip Remind students that when solving the exercises in this lesson, the word *and* in the question tells them to graph the intersection of two regions.

EXPLORATION

It is essential that students enter the equation that forms the lower boundary first. The shade feature shades the region above the first equation that is below the second equation.

Answer for the Exploration

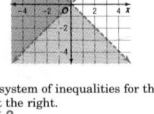

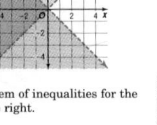
a.

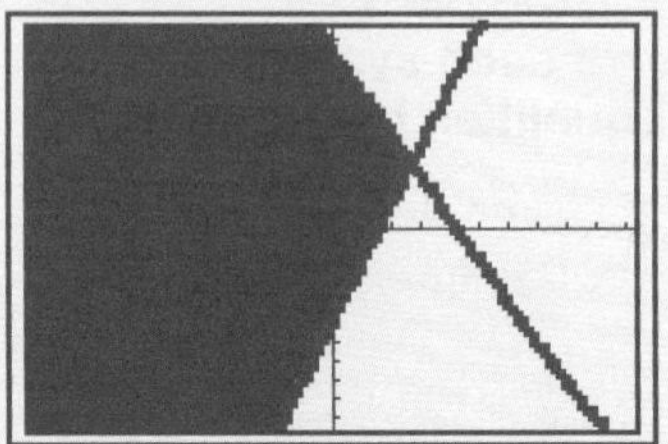

3 PRACTICE/APPLY

Check for Understanding

Exercises 1–14 are designed to help you assess your students' understanding through reading, writing, speaking, and modeling. You should work through Exercises 1–5 with your students and then monitor their work on Exercises 6–14.

Study Guide Masters, p. 60

8-5

NAME__________ DATE________

Study Guide

Student Edition Pages 482–486

Graphing Systems of Inequalities

The solution of a system of inequalities is the set of all ordered pairs that satisfy both inequalities. To find the solution of the system

$y > x + 2$
$y \le -2x - 1,$

graph each inequality. The graph of each inequality is called a **half-plane.** The intersection of the half-planes represents the solution of the system. The graphs of $y = x + 2$ and $y = -2x - 1$ are the boundaries of the region.

An inequality containing an absolute value expression can be graphed by graphing an equivalent system of two inequalities.

Solve each system of inequalities by graphing.

1. $y \ge 2x$, $y \ge -1$
2. $5x - 2y < 6$, $y > -x + 1$
3. $|y| > x$
4. $-x + y \le 6$, $x + y \le 2$

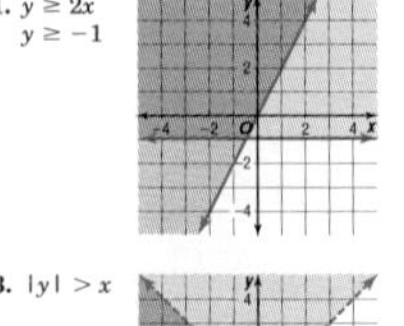

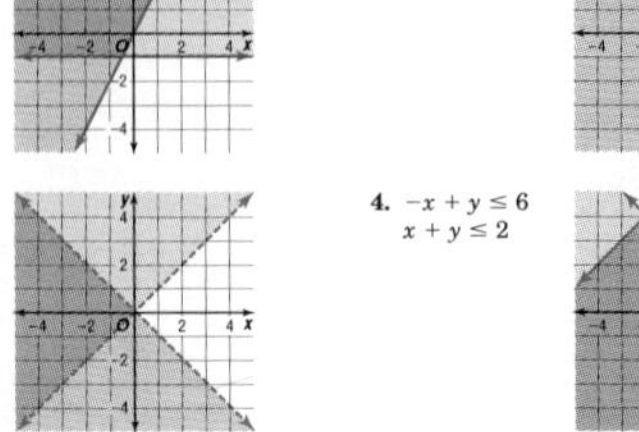

5. Write a system of inequalities for the graph at the right. $x + y \le 2$, $x > 1$

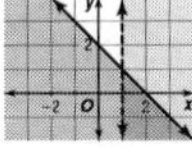

A graphing calculator is a useful tool for graphing systems of inequalities. It is important to enter the functions in the correct order, since this determines the shading.

EXPLORATION — GRAPHING CALCULATORS

You can use a graphing calculator to solve systems of inequalities. The TI-82 graphs functions and shades above the first function entered and below the second function entered. Select 7 on the DRAW menu to choose the SHADE feature. First, enter the function that is the lower boundary of the region to be shaded. (Note that inequalities that have $>$ or $\ge$ are lower boundaries and inequalities that have $<$ or $\le$ are upper boundaries.) Press [,]. Then enter the function that is the upper boundary of the region. Press [ )] [ENTER].

Your Turn **a. See margin.** **b. See students' work.**

a. Use a graphing calculator to graph the system of inequalities.
$y \ge 4x - 3$
$y \le -2x + 9$

b. Use a graphing calculator to work through the examples in this lesson. List and explain any disadvantages that you discovered when using the graphing calculator to graph systems of inequalities.

c. Describe the process of using a graphing calculator to solve systems of linear inequalities in your own words. **See students' work.**

CHECK FOR UNDERSTANDING

Communicating Mathematics

Study the lesson. Then complete the following.

1. **Explain** how to determine whether boundary lines should be included in the graph of a system of inequalities.
   1. A boundary line is included in the graph if the inequality is $\le$ or $\ge$ and not included if the inequality is $<$ or $>$.
2. **You Decide** Joshua says that the intersection point of the boundary lines is always a solution of a system of inequalities. Rolanda says the point of intersection may not be part of the solution set. Explain who is correct and give an example to support your answer. **See margin.**
3. **Write** a system of inequalities that has no solutions. Describe the graph of your system.
   3. Sample answer: $y < x + 1$ and $y > x + 3$. There is no intersection.
4. **State** which points are solutions to the system of inequalities graphed at the right. Explain how you know.
   a. (0, 0) **yes** b. (−1, 4) **yes**
   c. (2, 5) **no** d. (0.5, −1.7) **no**

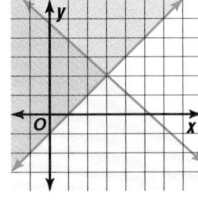

5. Describe a real-life situation that you can model using a system of linear inequalities. **Answers will vary.**

Reteaching

Using Graphs Have students use the graph at the right to determine whether each point is a solution of the system of inequalities.

1. (3, 0) **no**
2. (−3, 0) **yes**
3. (−1, −5) **yes**
4. (1, −2) **no**
5. (−4, 3) **no**
6. (−3, −3) **yes**

Write a system of inequalities for the graph and check the answers by substituting into the inequalities.

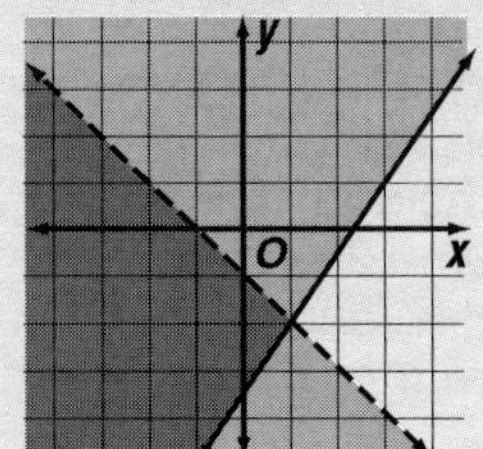

$y < -x - 1,$
$y \le \frac{3}{2}x - \frac{7}{2}$

Guided Practice

6–11. See Solutions Manual.

Solve each system of inequalities by graphing.

6. $x < 1$
 $x > -4$
7. $y \geq -2$
 $y - x < 1$
8. $y \geq 2x + 1$
 $y \leq -x + 1$
9. $y \geq 3x$
 $3y < 5x$
10. $y - x < 1$
 $y - x > 3$
11. $2x + y \leq 4$
 $3x - y \geq 6$

Write a system of inequalities for each graph.

12. 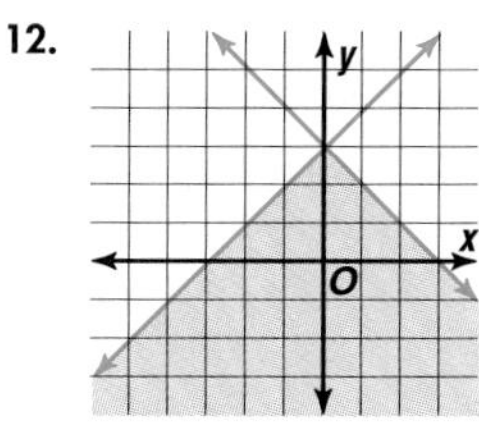

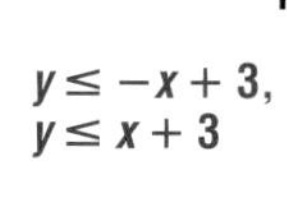

$y \leq -x + 3$, $y \leq x + 3$

13. 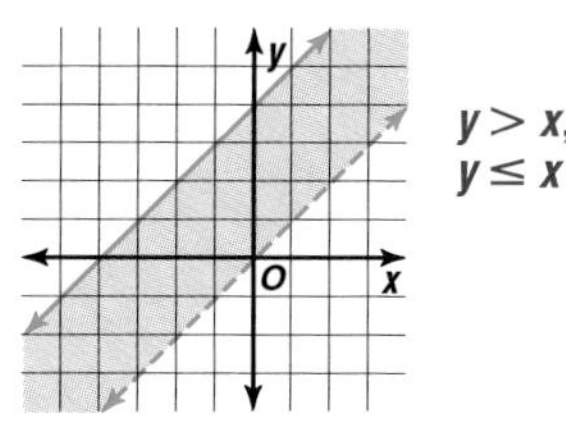

$y > x$, $y \leq x + 4$

14. **Sales** Ms. Johnson's homeroom class can order up to \$90 of free pizzas from Angelino's Pizza as a reward for selling the most magazines during the magazine drive. They need to order at least 6 large pizzas in order to serve the entire class. If a pepperoni pizza costs \$9.95 and a supreme pizza costs \$12.95, how many of each type can they order? List three possible solutions. See margin.

EXERCISES

Practice

Solve each system of inequalities by graphing.

15–32. See Solutions Manual.

15. $x > 5$
 $y \leq 4$
16. $y < 0$
 $x \geq 0$
17. $y > 3$
 $y > -x + 4$
18. $x \leq 2$
 $y - 4 \geq 5$
19. $x \geq 2$
 $y + x \leq 5$
20. $y < -3$
 $x - y > 1$

B

21. $y \leq 2x + 3$
 $y < -x + 1$
22. $y - x < 3$
 $y - x \geq 2$
23. $y \geq 3x$
 $7y < 2x$
24. $x - y < -1$
 $x - y > 3$
25. $2y + x < 6$
 $3x - y > 4$
26. $3x - 4y < 1$
 $x + 2y \leq 7$
27. $y - 4 > x$
 $y + x < 4$
28. $5y \geq 3x + 10$
 $2y \leq 4x - 10$
29. $y + 2 \leq x$
 $2y - 3 > 2x$
30. $2x + y \geq -4$
 $-5x + 2y < 1$
31. $x + y > 4$
 $-2x + 3y < -12$
32. $-4x + 5y \leq 41$
 $x + y > -1$

Write a system of inequalities for each graph.

33. $y > -1$, $x \geq -2$
34. $3x - 5y \geq -25$, $y \geq 0$, $x \leq 0$
35. $y \leq x$, $y > x - 3$

33. 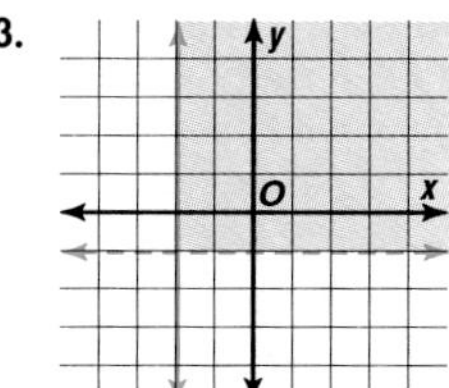

34. 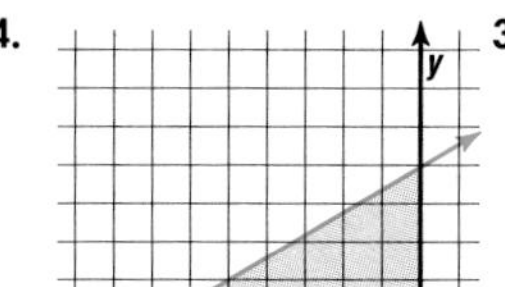

35. 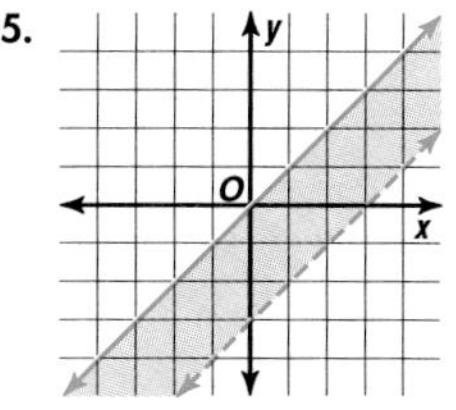

Alternative Teaching Strategies

Student Diversity When students are deciding which area in a graph to shade, have them think it through logically. Emphasize using a solution that makes sense rather than guessing.

Assignment Guide

Core: 15–47 odd, 48–56
Enriched: 16–44 even, 45–56

For **Extra Practice,** see p. 775.

The red A, B, and C flags, printed only in the Teacher's Wraparound Edition, indicate the level of difficulty of the exercises.

Additional Answers

2. Rolanda is correct; Sample answer: $y < 2x + 4$ and $y > x - 1$; the solution of $y = 2x + 4$ and $y = x - 1$, $(-5, -6)$, is not a solution to the system of inequalities.

14. 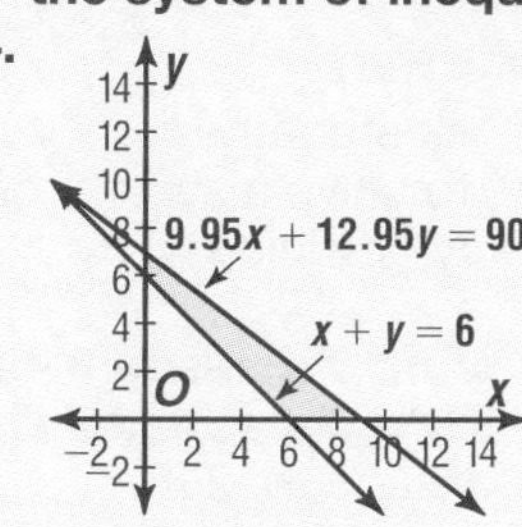

Sample answer:
7 pepperoni, 1 supreme;
5 pepperoni, 3 supreme;
2 pepperoni, 5 supreme

Practice Masters, p. 60

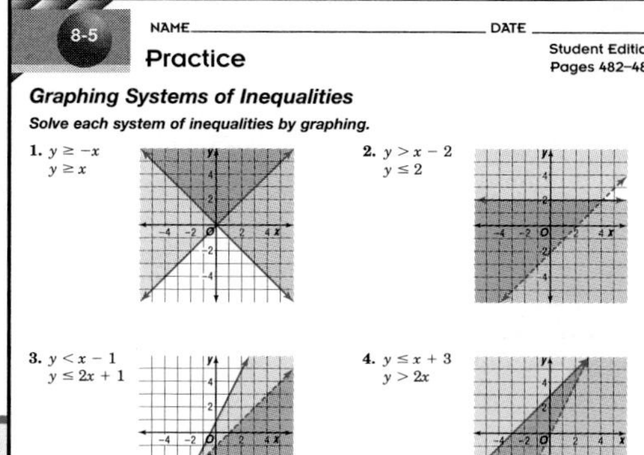

8-5 NAME______ DATE______

Practice — Student Edition Pages 482–486

Graphing Systems of Inequalities

Solve each system of inequalities by graphing.

1. $y \geq -x$
 $y \geq x$
2. $y > x - 2$
 $y \leq 2$
3. $y < x - 1$
 $y \leq 2x + 1$
4. $y \leq x + 3$
 $y > 2x$
5. $y < 1 - x$
 $y \leq 1 - 3x$

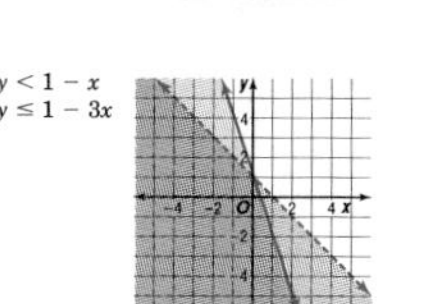

6. $x < 2$
 $y - x \leq 2$

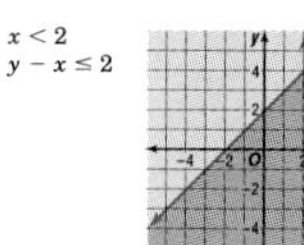

7. $y < x$
 $2x - y \geq 3$

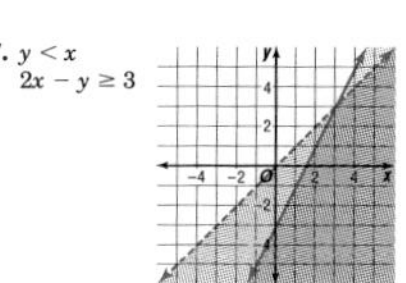

8. $|x + 1| \leq y$

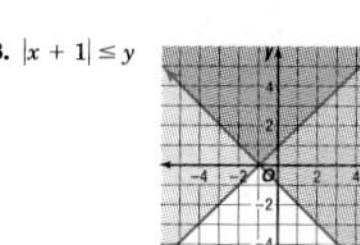

Additional Answers

46. Sample answers: 3 ounces of mozzarella, 4 ounces of Swiss; 4 ounces of mozzarella, 3 ounces of Swiss; 5 ounces of mozzarella, 3 ounces of Swiss

47. Sample answer: walk, 15 min, jog, 15 min; walk 10 min, jog 20 min; walk 5 min, jog 25 min

50. 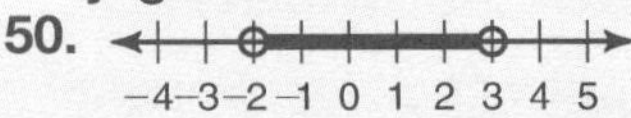

4 ASSESS

Closing Activity

Speaking Ask students to explain how to solve a system of inequalities by graphing.

Chapter 8, Quiz D (Lesson 8-5), is available in the *Assessment and Evaluation Masters*, p. 213.

Enrichment Masters, p. 60

NAME________________ DATE ________

Enrichment

Student Edition Pages 482–486

Describing Regions

The shaded region inside the triangle can be described with a system of three inequalities.

$y < -x + 1$
$y > \frac{1}{3}x - 3$
$y > -9x - 31$

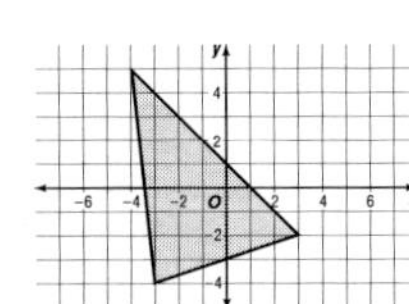

Write systems of inequalities to describe each region. You may first need to divide a region into triangles or quadrilaterals.

1\. 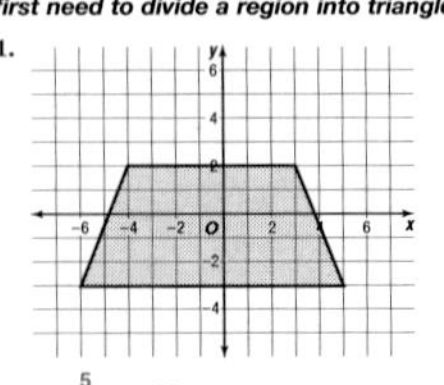

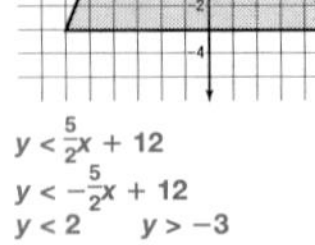

$y < \frac{5}{2}x + 12$
$y < -\frac{5}{2}x + 12$
$y < 2$ $\quad y > -3$

2\. 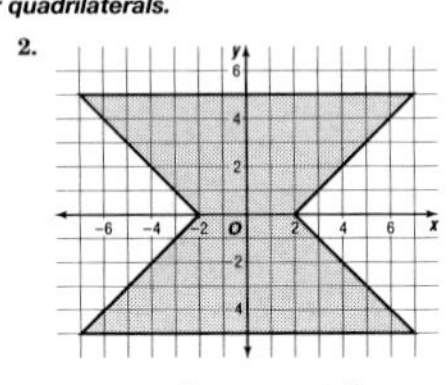

$y > -x - 2$ $\quad y < x + 2$
$y > x - 2$ $\quad y < -x + 2$
$y < 5$ $\quad y > -5$

3\. 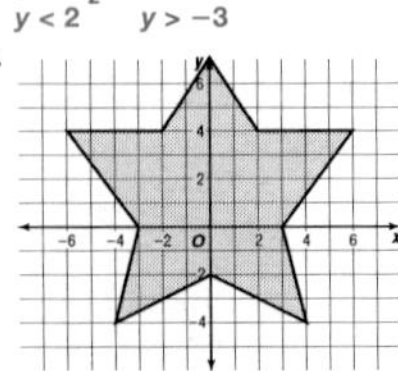

top: $y < \frac{3}{2}x + 7, y < -\frac{3}{2}x + 7, y > 4$

middle: $y < 4, y > 0, y > -\frac{4}{3}x - 4, y > \frac{4}{3}x - 4$

bottom left: $y < 4x + 12, y > \frac{1}{2}x - 2, y < 0, x < 0$

bottom right: $y < -4x + 12, y > -\frac{1}{2}x - 2, y < 0, x < 0$

Write a system of inequalities for each graph.

36. 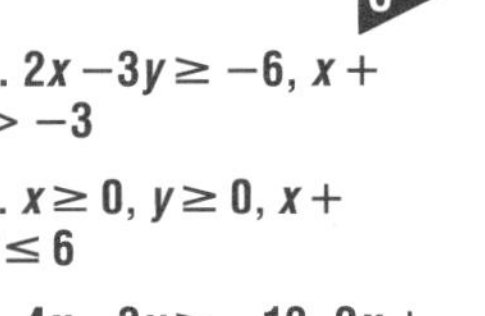

37. 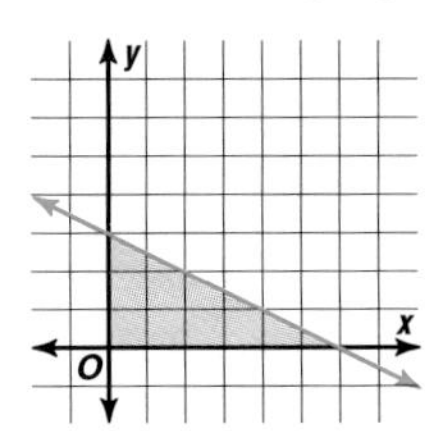

38. 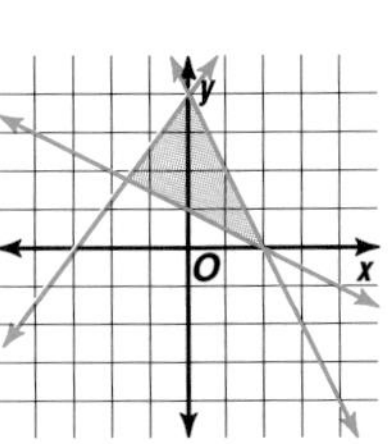

36. $2x - 3y \geq -6$, $x + y > -3$

37. $x \geq 0$, $y \geq 0$, $x + 2y \leq 6$

38. $4x - 3y \geq -12$, $2x + y \leq 4$, $x + 2y \geq 2$

Solve each system of inequalities by graphing.

39. $x - 2y \leq 2$
$3x + 4y \leq 12$
$x \geq 0$

40. $x - y \leq 5$
$5x + 3y \geq -6$
$y \leq 3$

41. $x < 2$
$4y > x$
$2x - y < -9$
$x + 3y < 9$

39–41. See Solutions Manual.

Graphing Calculator

Use a graphing calculator to solve each system of inequalities.

42. $y \geq 3x - 6$
$y \leq x + 1$

43. $y \leq x + 9$
$y > -x - 4$

44. $y < 2x + 10$
$y \geq 7x + 15$

42–44. See Solutions Manual.

Critical Thinking

45. Solve the inequality $|y| \leq 3$ by graphing. (*Hint:* Graph as a system of inequalities.) **See Solutions Manual.**

Applications and Problem Solving

Graph a system of inequalities to solve each problem.

46. Nutrition Young people between the ages of 11 and 18 should get at least 1200 milligrams of calcium each day. One ounce of mozzarella cheese has 147 milligrams of calcium, and one ounce of Swiss cheese has 219 milligrams. If you wanted to eat no more than 8 ounces of cheese, how much of each type could you eat and still get your daily requirement of calcium? List three possible solutions. **See margin.**

47. Organize Data Kenny Choung likes to exercise every day by walking and jogging at least 3 miles. Kenny walks at a rate of 4 mph and jogs at a rate of 8 mph. If he has only a half hour to exercise, how much time can he spend walking and jogging and cover at least 3 miles? List 3 possible solutions.

47. See margin.

Mixed Review

48. Number Theory If the digits of a two-digit positive integer are reversed, the result is 6 less than twice the original number. Find all such integers for which this is true. (Lesson 8–4) **24**

49. Organize Data When Roberta cashed her check for \$180, the bank teller gave her 12 bills, each one worth either \$5 or \$20. How many of each bill did she receive? (Lesson 8–2) **4 \$5 bills, 8 \$20 bills**

50. Solve $4 > 4a + 12 > 24$ and graph the solution set. (Lesson 7–4)

50. $\{a \mid -2 > a > 3\}$; See margin for graph.

51. Write an equation in slope-intercept form of a line that passes through the points at (3, 3) and (−1, 5). (Lesson 6–2)

51. $y = -\frac{1}{2}x + \frac{9}{2}$

52. Solve $y = -\frac{1}{2}x + 3$ if the domain is {2, 4, 6}. (Lesson 5–3) **{2, 1, 0}**

53. What number increased by 40% equals 14? (Lesson 4–5) **10**

54. Travel Paloma Rey drove to work on Wednesday at 40 miles per hour and arrived one minute late. She left home at the same time on Thursday, drove 45 miles per hour, and arrived one minute early. How far does Ms. Rey drive to work? (*Hint:* Convert hours to minutes.) (Lesson 3–5) **12 miles**

55. Define a variable, then write an equation for the following problem. Diego gained 134 yards running. This was 17 yards more than in the previous game. How many yards did he gain in both games? (Lesson 2–9)

55. Let y = the number of yards gained in both games; $y = 134 + (134 - 17)$

56. Name the property illustrated by $(3 \cdot x) \cdot y = 3 \cdot (x \cdot y)$. (Lesson 1–8)

56. assoc. (×)

Extension

Reasoning Solve the system of inequalities: $x + 4y \leq 13$, $x - 5y \leq 16$, $x + y \leq 4$, $x + 2y \geq -5$, and $2x - y \geq -10$.

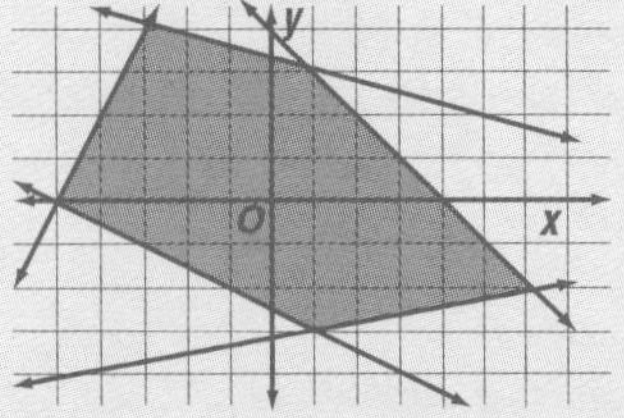

Tech Prep

Nutritionist Nutritionists often plan meals to provide certain vitamins and minerals, such as the calcium mentioned in Exercise 46. Students who are interested in nutrition may wish to further explore the role of mathematics in this career. For more information on tech prep, see the *Teacher's Handbook*.

VOCABULARY

After completing this chapter, you should be able to define each term, property, or phrase and give an example or two of each.

Algebra
consistent (p. 455)
dependent (p. 456)
elimination (p. 469)
independent (p. 456)
inconsistent (p. 456)
substitution (p. 462)
system of equations (p. 455)
system of inequalities (p. 482)

Problem Solving
organize data (p. 464)

UNDERSTANDING AND USING THE VOCABULARY

Choose the correct term to complete each statement.

1. The method used in solving the following system of equations is (*elimination, substitution*). **substitution**

$$\left.\begin{array}{l} x = 4y + 1 \\ x + y = 6 \end{array}\right\} \rightarrow \quad \begin{aligned} (4y + 1) + y &= 6 \\ 5y + 1 &= 6 \\ 5y &= 5 \\ y &= 1 \end{aligned} \qquad \begin{aligned} x &= 4(1) + 1 \\ x &= 4 + 1 \\ x &= 5 \\ \text{solution: } & (5, 1) \end{aligned}$$

2. If a system of equations has exactly one solution, it is (*dependent, independent*). **independent**
3. If the graph of a system of equations is parallel lines, the system of equations is said to be (*consistent, inconsistent*). **inconsistent**
4. A system of equations that has infinitely many solutions is (*dependent, independent*). **dependent**
5. The method used in solving the following system of equations is (*elimination, substitution*). **elimination**

$$\left.\begin{array}{l} -2c + b = 3 \\ -b - c = -6 \end{array}\right\} \rightarrow \quad \begin{aligned} b - 2c &= 3 \\ (+)\ -b - c &= -6 \\ \hline -3c &= -3 \\ c &= 1 \end{aligned} \qquad \begin{aligned} -b - (1) &= -6 \\ -b - 1 &= -6 \\ -b &= -5 \\ b &= 5 \end{aligned} \qquad \text{solution: } (5, 1)$$

6. If a system of equations has the same slope and different intercepts, the graph of the system is (*intersecting lines, parallel lines*). **parallel lines**
7. If a system of equations has the same slope and intercepts, the system has (*exactly one, infinitely many*) solutions. **infinitely many**
8. The solution to a system of equations is $(3, -5)$; therefore, this system is (*consistent, inconsistent*). **consistent**
9. The graph of a system of equations is shown at the right. This system has (*infinitely many, no*) solution. **no**
10. The solution to a system of inequalities is the (*intersection, union*) of two half-planes. **intersection**
11. A system of inequalities that includes $x < 0$ and $y > 0$ is in the (*second, fourth*) quadrant. **second**

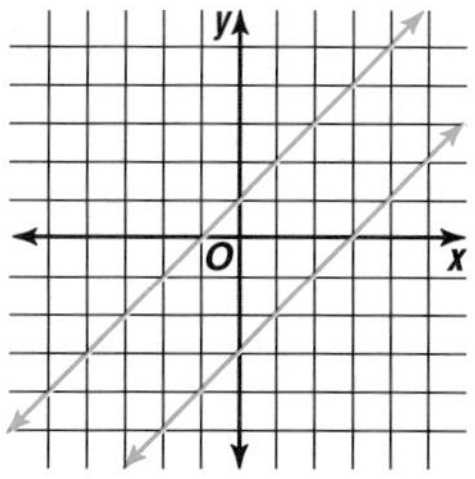

Using the CHAPTER HIGHLIGHTS

The Chapter Highlights begins with a listing of the new terms, properties, and phrases that were introduced in this chapter. Have students define each term and provide an example or two of it, if appropriate.

Instructional Resources

Three multiple-choice tests and three free-response tests are provided in the *Assessment and Evaluation Masters*. Forms 1A and 2A are for honors pacing, and Forms 1B, 1C, 2B, and 2C are for average pacing. Chapter 8 Test, Form 1B is shown at the right. Chapter 8 Test, Form 2B is shown on the next page.

Assessment and Evaluation Masters, pp. 199–200

8 NAME__________ DATE ________

Chapter 8 Test, Form 1B

Write the letter for the correct answer in the blank at the right of each problem. Use the graph to answer questions 1–4.

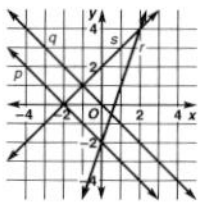

1. Which ordered pair is the solution of the system of equations represented by lines q and s? 1. **D**
   A. $(0, -2)$ B. $(2, 4)$ C. $(1, -1)$ D. $(-1, 1)$
2. Which pair of lines is the graph of the system of equations $\begin{array}{l} x - y = -2 \\ 3x - y = 2 \end{array}$? 2. **D**
   A. p and q B. q and s
   C. r and p D. r and s
3. How many solutions are there of the system of equations represented by the line q and the equation $y = -x$? 3. **D**
   A. exactly one B. exactly two C. no solutions D. infinitely many
4. How many solutions are there of the system of equations represented by lines p and q? 4. **C**
   A. exactly one B. exactly two C. no solutions D. infinitely many
5. If the system of equations $\begin{array}{l} 3x + y = 14 \\ x + 4y = 3 \end{array}$ is to be solved by substitution, which expression can be substituted for y in the second equation? 5. **B**
   A. $3 - 4y$ B. $14 - 3x$ C. $\frac{14 - y}{3}$ D. $\frac{3 - x}{4}$
6. Solve the system of equations $\begin{array}{l} x = 5y \\ 2x + 5y = 15 \end{array}$ by substitution. 6. **C**
   A. $\{(1, 5)\}$ B. $\{(0, 3)\}$ C. $\{(5, 1)\}$ D. none of these
7. Use substitution to solve the system of equations $\begin{array}{l} x + 6y = 12 \\ 2x + 12y = 1 \end{array}$. 7. **D**
   A. $\{(0, 2)\}$ B. $\{(12, 0)\}$
   C. There are infinitely many solutions. D. There are no solutions.
8. Identify the best estimate of the solution of this system. 8. **D**
   A. $\{(0, 0)\}$ B. $\{(0, -3)\}$
   C. $\{(-1, 1)\}$ D. $\{(1, -1)\}$

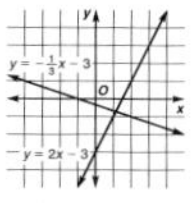

9. The sum of the digits of a 2-digit number is 6. The number is 6 times the units digit. If t = the tens digit, and u = the units digit, which system must be true? 9. **B**
   A. $\begin{array}{l} t = 6 - u \\ 10t + 6u = 6 \end{array}$ B. $\begin{array}{l} t + u = 6 \\ 10t + u = 6u \end{array}$ C. $\begin{array}{l} t + u = 6u \\ 10t + u = 6u \end{array}$ D. $\begin{array}{l} t + u = 6 \\ 10t = 6u \end{array}$
10. If the system of equations $\begin{array}{l} 3x + 5y = 16 \\ 8x - 5y = 28 \end{array}$ is solved by elimination, what is the value of x? 10. **A**
   A. 4 B. 5 C. −6 D. 7

8 NAME__________ DATE ________

Chapter 8 Test, Form 1B (continued)

11. What is the value of y if the system of equations $\begin{array}{l} x + y = 5 \\ x + 2y = 3 \end{array}$ is solved by elimination? 11. **A**
   A. −2 B. 3 C. $\frac{8}{3}$ D. $-\frac{8}{3}$
12. The system of equations $\begin{array}{l} 4x + 7y = 8 \\ -3x - 2y = 5 \end{array}$ is to be solved by elimination of x. If the first equation is multiplied by 3, then by which number should the second equation be multiplied? 12. **D**
   A. 3 B. − 3 C. 7 D. 4
13. What is the value of x if the system of equations $\begin{array}{l} 4x + 7y = -4 \\ 8x + 5y = 28 \end{array}$ is solved by elimination? 13. **B**
   A. 2 B. 6 C. 4 D. −4
14. Solve the system of equations $\begin{array}{l} 5x + 4y = -10 \\ 3x + 6y = -6 \end{array}$ by elimination. 14. **C**
   A. $\{(0, -2)\}$ B. $\{(2, -5)\}$ C. $\{(-2, 0)\}$ D. none of these
15. Which system of inequalities is graphed? 15. **C**
   A. $\begin{array}{l} y > 2 \\ y > x - 2 \end{array}$
   B. $\begin{array}{l} y < 2 \\ y < x - 2 \end{array}$
   C. $\begin{array}{l} y < 2 \\ x < y + 2 \end{array}$
   D. $\begin{array}{l} y > 2 \\ x > y - 2 \end{array}$

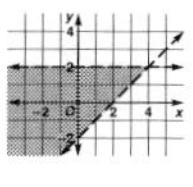

16. State which region in the graph is the solution of the system $|x + 1| \le y$. 16. **B**

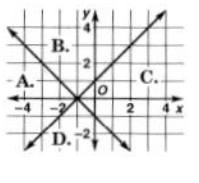

17. A lab technician has a 15% alcohol solution and a 35% alcohol solution. She wants to make 100 gallons of a 29% alcohol solution. How much of the 15% solution should she use? 17. **B**
   A. 70 gal B. 30 gal C. 40 gal D. 55 gal
18. How many pounds of cashew nuts that sell for \$2 a pound should be mixed with peanuts, which sell for 80¢ a pound, to make a 10-pound mixture that sells for \$1.28 a pound? 18. **A**
   A. 4 pounds B. 6 pounds C. 8 pounds D. 10 pounds
19. An airplane travels 1560 miles in 3 hours flying with the wind. The return trip takes 4 hours. Find the rate of the plane in still air. 19. **C**
   A. 65 mi/h B. 195 mi/h C. 455 mi/h D. 260 mi/h
20. Roger reversed the digits in the amount of a check and overpaid a customer by \$9. The sum of the digits in the two-digit amount was 7. Find the amount of the check. 20. **B**
   A. \$43 B. \$34 C. \$63 D. \$54

Bonus Find the value of a and b so that the graph of $y = ax^2 + b$ will pass through the points $(3, 1)$ and $(-2, -1)$. Bonus $a = \frac{2}{5}$; $b = -\frac{13}{5}$

Using the STUDY GUIDE AND ASSESSMENT

Skills and Concepts Encourage students to refer to the objectives and examples on the left as they complete the review exercises on the right.

Assessment and Evaluation Masters, pp. 205–206

8 NAME________________ DATE ________

Chapter 8 Test, Form 2B

Graph each system of equations using the coordinate plane provided. Then determine whether the system has one solution, no solution, or infinitely many solutions. If the system has one solution, name it.

1. $x + y = 4$
   $x - y = 4$ — 1. (4, 0)

2. $2x - y = -3$
   $6x - 3y = -9$ — 2. infinitely many

3. $x + y = -2$
   $x + y = 3$ — 3. no solution

Solve each system of equations.

4. $y = 3x$, $x + y = 4$ — 4. (1, 3)
5. $5x - y = 10$, $7x - 2y = 11$ — 5. (3, 5)
6. $x - 6y = 4$, $3x - 18y = 4$ — 6. no solution
7. $x - 5y = 10$, $2x - 10y = 20$ — 7. infinitely many
8. $x + 4y = -8$, $x - 4y = -8$ — 8. (−8, 0)
9. $2x + 5y = 3$, $-x + 3y = -7$ — 9. (4, −1)
10. $2x - 5y = -16$, $3y = 2x + 12$ — 10. (−3, 2)
11. $2x - 3y = 1$, $5x + 4y = 14$ — 11. (2, 1)

Solve a system of equations to find the two numbers described.

12. The sum of two numbers is 17 and their difference is 29. — 12. 23, −6
13. The sum of a number and twice a greater number is 8. The sum of the greater number and twice the lesser number is −6. — 13. $-6\frac{2}{3}, 7\frac{1}{3}$

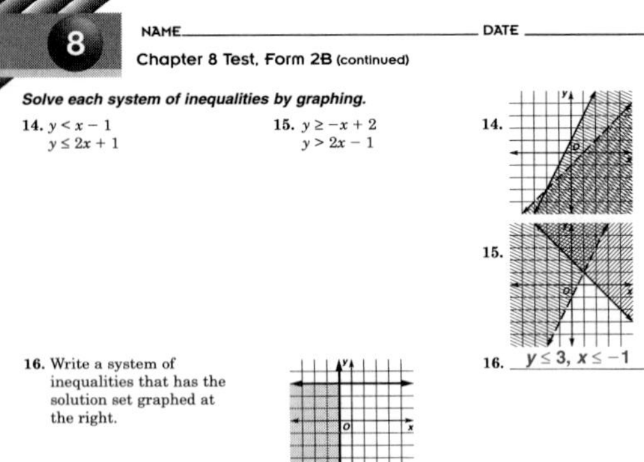

8 NAME________________ DATE ________

Chapter 8 Test, Form 2B (continued)

Solve each system of inequalities by graphing.

14. $y < x - 1$
    $y \le 2x + 1$

15. $y \ge -x + 2$
    $y > 2x - 1$

16. Write a system of inequalities that has the solution set graphed at the right. — 16. $y \le 3, x \le -1$

Use a system of equations to solve each problem.

17. Adult tickets for the school musical sold for \$3.50 and student tickets sold for \$2.50. Three hundred twenty-one tickets were sold altogether for \$937.50. How many of each kind of ticket were sold? — 17. 135 adult; 186 student

18. Trinidad has \$2.35 in nickels and dimes. If she has 33 coins in all, find the number of nickels and dimes. — 18. 14 dimes; 19 nickels

19. Amal reversed the digits in the amount of a check and underpaid a customer by \$36. The sum of the digits in the 2-digit amount was 8. Find the amount of the check. — 19. \$62

20. A boat travels 12 miles downstream in $1\frac{1}{2}$ hours. On the return trip the boat travels the same distance upstream in 2 hours. Find the rate of the boat in still water and the rate of the current. — 20. 7 mi/h; 1 mi/h

Bonus Graph the solution set of $-3 \le 4x + y < 1$.

CHAPTER 8 STUDY GUIDE AND ASSESSMENT

SKILLS AND CONCEPTS

OBJECTIVES AND EXAMPLES

Upon completing this chapter, you should be able to:

- solve systems of equations by graphing (Lesson 8–1)

Graph $x + y = 6$ and $x - y = 2$. Then find the solution.

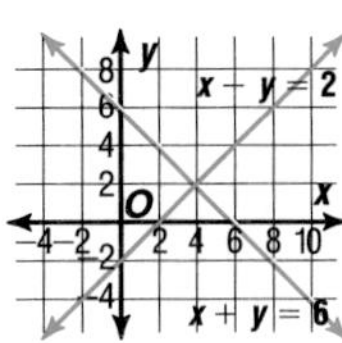

The solution is (4, 2).

REVIEW EXERCISES

Use these exercises to review and prepare for the chapter test. **12–15. See margin for graphs.**

Graph each system of equations to find the solution.

12. $y = 2x - 7$, $x + y = 11$ **(6, 5)**
13. $x + 2y = 6$, $2y - 8 = -x$ **13. no solution**
14. $3x + y = -8$, $x + 6y = 3$ **(−3, 1)**
15. $5x - 3y = 11$, $2x + 3y = -25$ **(−2, −7)**

- determine whether a system of equations has one solution, no solution, or infinitely many solutions by graphing (Lesson 8–1)

Graph $3x + y = -4$ and $6x + 2y = -8$. Then determine the number of solutions.

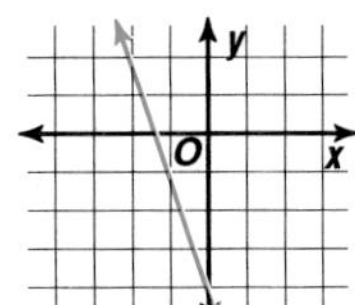

There are infinitely many solutions.

Graph each system of equations. Then determine whether the system of equations has *one* solution, *no* solution, or *infinitely many* solutions. If the system has one solution, name it.

16. $x - y = 9$, $x + y = 11$
17. $9x + 2 = 3y$, $y - 3x = 8$
18. $2x - 3y = 4$, $6y = 4x - 8$
19. $3x - y = 8$, $3x = 4 - y$

16. one, (10, 1) **17. no solution**
18. infinitely many **19. one, (2, −2)**
16–19. See margin for graphs.

- solve systems of equations by using the substitution method (Lesson 8–2)

Use substitution to solve the system of equations.

$y = x - 1$
$4x - y = 19$

$4x - y = 19$ $4x - (x - 1) = 19$ $4x - x + 1 = 19$ $3x + 1 = 19$ $3x = 18$ $x = 6$	$y = x - 1$ $y = 6 - 1$ $y = 5$

The solution is (6, 5).

Use substitution to solve each system of equations. If the system *does not* have exactly one solution, state whether it has *no* solution or *infinitely many* solutions.

20. $2m + n = 1$, $m - n = 8$ **(3, −5)**
21. $3a - 2b = -4$, $3a + b = 2$ **(0, 2)**
22. $x = 3 - 2y$, $2x + 4y = 6$ **infinitely many**
23. $3x - y = 1$, $2x + 4y = 3$ $\left(\frac{1}{2}, \frac{1}{2}\right)$

GLENCOE Technology

Test and Review Software

You may use this software, a combination of an item generator and item bank, to create your own tests or worksheets. Types of items include free response, multiple choice, short answer, and open ended.

For IBM & Macintosh

OBJECTIVES AND EXAMPLES	REVIEW EXERCISES

- solve systems of equations by using the elimination method with addition or subtraction (Lesson 8–3)

Use elimination to solve the system of equations.

$2m - n = 4$
$m + n = 2$

$$\begin{array}{r} 2m - n = 4 \\ (+)\ m + n = 2 \\ \hline 3m = 6 \\ m = 2 \end{array} \quad \Bigg| \quad \begin{array}{r} m + n = 2 \\ 2 + n = 2 \\ n = 0 \end{array}$$

The solution is (2, 0).

Use elimination to solve each system of equations.

24. $x + 2y = 6$ (2, 2)
$x - 3y = -4$

25. $2m - n = 5$ (2, −1)
$2m + n = 3$

26. $3x - y = 11$ (4, 1)
$x + y = 5$

27. $3s + 6r = 33$ (5, 1)
$6r - 9s = 21$

28. $3x + 1 = -7y$
$6x + 7y = 0$
$\left(\frac{1}{3}, -\frac{2}{7}\right)$

29. $12x - 9y = 114$
$7y + 12x = 82$
(8, −2)

- solve systems of equations by using the elimination method with multiplication and addition (Lesson 8–4)

Use elimination to solve the system of equations.

$3x - 4y = 7$
$2x + y = 1$

$3x - 4y = 7$ → $3x - 4y = 7$

$2x + y = 1$ **Multiply by 4.** → $(+)\ 8x + 4y = 4$

$11x = 11$

$x = 1$

$2x + y = 1$
$2(1) + y = 1$
$y = -1$

The solution is (1, −1).

Use elimination to solve each system of equations.

30. $x - 5y = 0$ (5, 1)
$2x - 3y = 7$

31. $x - 2y = 5$
$3x - 5y = 8$

32. $2x + 3y = 8$
$x - y = 2$

33. $-5x + 8y = 21$
$10x + 3y = 15$

34. $5m + 2n = -8$
$4m + 3n = 2$

35. $6x + 7y = 5$ (2, −1)
$2x - 3y = 7$

31. (−9, −7)
32. $\left(\frac{14}{5}, \frac{4}{5}\right)$
33. $\left(\frac{3}{5}, 3\right)$
34. (−4, 6)

- determine the best method for solving systems of equations (Lesson 8–4)

Use the best method to solve the system of equations.

$x + 2y = 8$
$3x + 2y = 6$

$$\begin{array}{r} 3x + 2y = 6 \\ (-)\ x + 2y = 8 \\ \hline 2x = -2 \\ x = -1 \end{array} \quad \Bigg| \quad \begin{array}{r} x + 2y = 8 \\ -1 + 2y = 8 \\ 2y = 9 \\ y = \frac{9}{2} \end{array}$$

The solution is $\left(-1, \frac{9}{2}\right)$.

Determine the best method to solve each system of equations. Then solve the system.

36. $y = 2x$
$x + 2y = 8$

37. $9x + 8y = 7$
$18x - 15y = 14$

38. $2x - y = 36$
$3x - 0.5y = 26$

39. $3x + 5y = 2x$ (0, 0)
$x + 3y = y$

40. $5x - 2y = 23$
$5x + 2y = 17$

41. $2x + y = 3x - 15$
$x + 5 = 4y + 2x$

36. $\left(\frac{8}{5}, \frac{16}{5}\right)$
37. $\left(\frac{7}{9}, 0\right)$
38. (4, −28)
40. $\left(4, -\frac{3}{2}\right)$
41. (13, −2)

Additional Answers

12.
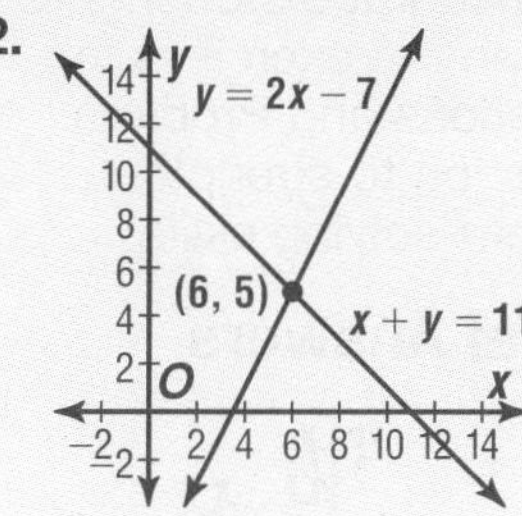

13.
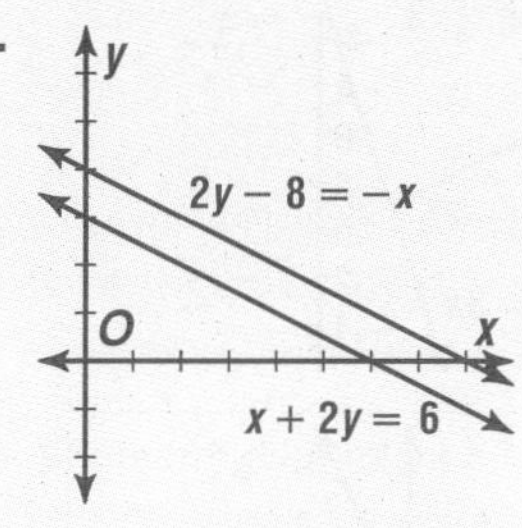

14.
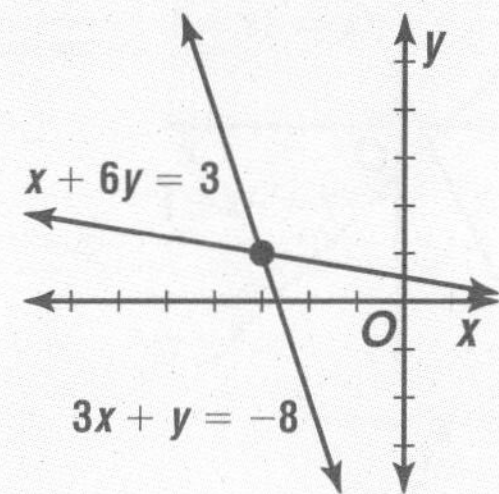

15.
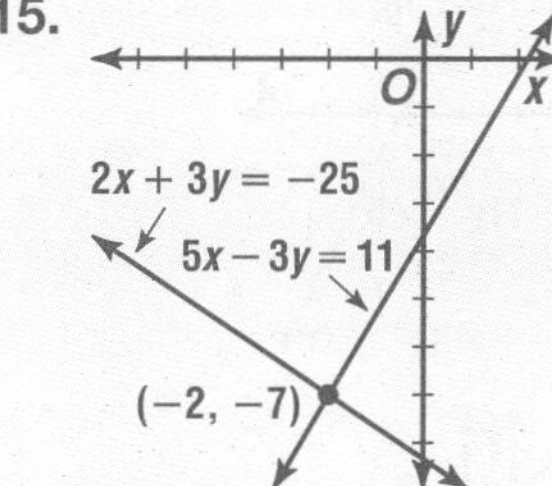

16.
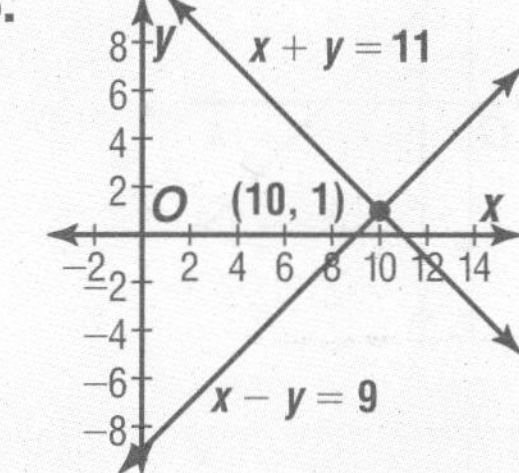

17.
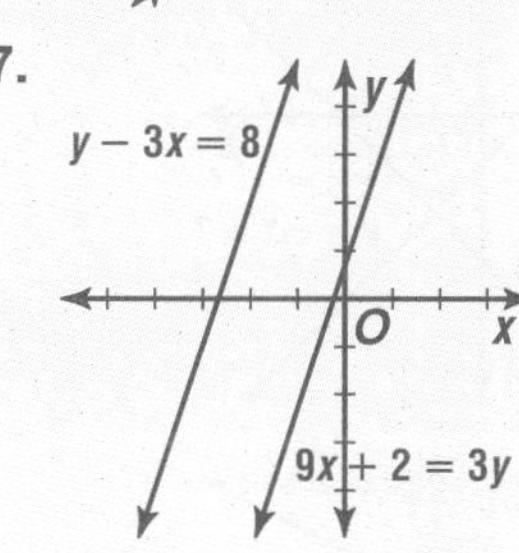

Additional Answers

18.
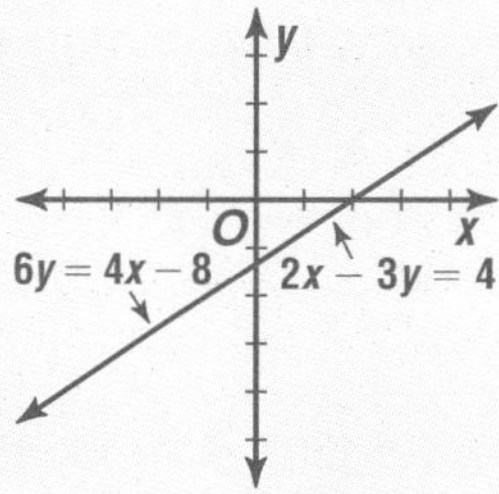

19.
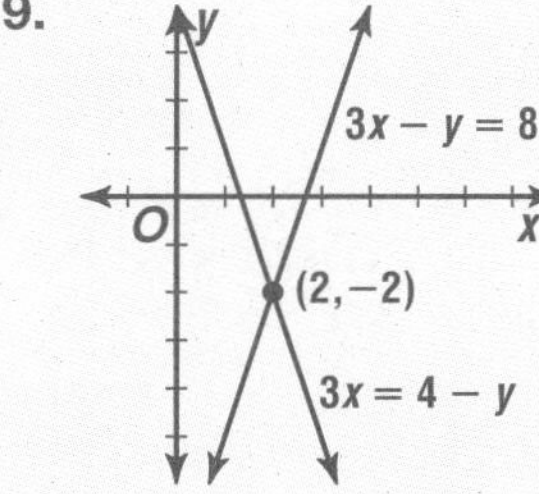

Using the CUMULATIVE REVIEW

These two pages review the skills and concepts presented in Chapters 1–8. This review is formatted to reflect new trends in standardized testing.

A more traditional cumulative review is provided in the *Assessment and Evaluation Masters,* pp. 215–216.

Assessment and Evaluation Masters, pp. 215–216

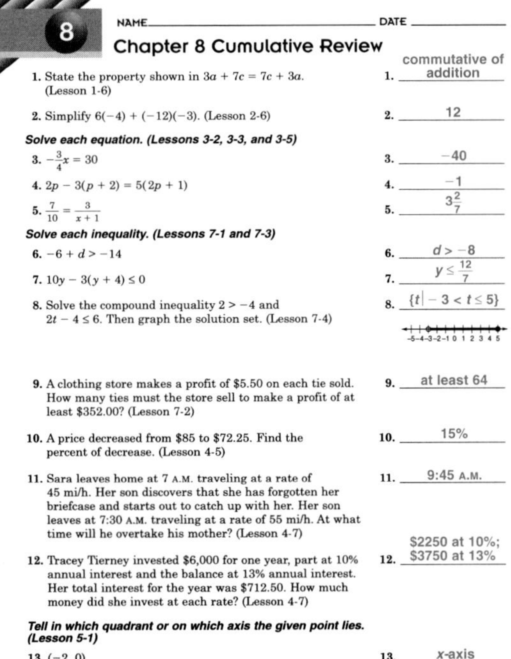

8 NAME______ DATE______

Chapter 8 Cumulative Review

1. State the property shown in $3a + 7c = 7c + 3a$. (Lesson 1-6) — 1. commutative of addition
2. Simplify $6(-4) + (-12)(-3)$. (Lesson 2-6) — 2. 12

Solve each equation. (Lessons 3-2, 3-3, and 3-5)

3. $-\frac{3}{4}x = 30$ — 3. -40
4. $2p - 3(p + 2) = 5(2p + 1)$ — 4. -1
5. $\frac{7}{10} = \frac{3}{x+1}$ — 5. $3\frac{2}{7}$

Solve each inequality. (Lessons 7-1 and 7-3)

6. $-6 + d > -14$ — 6. $d > -8$
7. $10y - 3(y + 4) \le 0$ — 7. $y \le \frac{12}{7}$
8. Solve the compound inequality $2 > -4$ and $2t - 4 \le 6$. Then graph the solution set. (Lesson 7-4) — 8. $\{t \mid -3 < t \le 5\}$
9. A clothing store makes a profit of \$5.50 on each tie sold. How many ties must the store sell to make a profit of at least \$352.00? (Lesson 7-2) — 9. at least 64
10. A price decreased from \$85 to \$72.25. Find the percent of decrease. (Lesson 4-5) — 10. 15%
11. Sara leaves home at 7 A.M. traveling at a rate of 45 mi/h. Her son discovers that she has forgotten her briefcase and starts out to catch up with her. Her son leaves at 7:30 A.M. traveling at a rate of 55 mi/h. At what time will he overtake his mother? (Lesson 4-7) — 11. 9:45 A.M.
12. Tracey Tierney invested \$6,000 for one year, part at 10% annual interest and the balance at 13% annual interest. Her total interest for the year was \$712.50. How much money did she invest at each rate? (Lesson 4-7) — 12. \$2250 at 10%; \$3750 at 13%

Tell in which quadrant or on which axis the given point lies. (Lesson 5-1)

13. $(-2, 0)$ — 13. x-axis
14. $(5, -3)$ — 14. Quadrant IV

8 NAME______ DATE______

Chapter 8 Cumulative Review (continued)

For exercises 15 and 16, refer to the triangle shown below. (Lesson 4-3)

15. Find b to the nearest tenth. — 15. 13.5
16. Find c to the nearest tenth. — 16. 3.7
17. Determine the slope of the line passing through $(-2, 1)$ and $(-6, 5)$. (Lesson 6-1) — 17. -1
18. Find the x- and y-intercepts for the graph of $x - 4y = 4$. (Lesson 6-4) — 18. 4; -1
19. Write an equation in slope-intercept form for the line that is perpendicular to the graph of $y - 3x = -4$ and passes through $(1, 2)$. (Lesson 6-6) — 19. $y = -\frac{1}{3}x + \frac{7}{3}$
20. Find the coordinates of the midpoint of the line segment whose endpoints have the coordinates $(-1, 4)$ and $(2, 7)$. (Lesson 6-7) — 20. $\left(\frac{1}{2}, 5\frac{1}{2}\right)$
21. Use the coordinate plane provided to graph the equations $x + y = 4$ and $y = x$. Then state the solution set. (Lesson 8-1) — 21. $\{(2, 2)\}$
22. Use substitution to solve the system of equations $-x - 5y = 7$ and $x + y = 1$. (Lesson 8-2) — 22. $(3, -2)$

Use elimination to solve each system of equations. (Lessons 8-3 and 8-4)

23. $2x + 4y = 1$, $x - 4y = 5$ — 23. $\left(2, -\frac{3}{4}\right)$
24. $2x - 2y = 6$, $x + y = 3$ — 24. $(3, 0)$
25. With the wind, an airplane travels 2200 miles in 4 hours. Against the wind, it takes 5 hours to travel 2350 miles. Find the rate of the wind and the rate of the plane in still air. (Lesson 8-4) — 25. 40 mi/h; 510 mi/h

CUMULATIVE REVIEW

CHAPTERS 1–8

SECTION ONE: MULTIPLE CHOICE

There are eight multiple-choice questions in this section. After working each problem, write the letter of the correct answer on your paper.

1. Choose the statement that is true for a system of two linear equations. **D**
   - A. There are no solutions when the graphs of the equations are perpendicular lines.
   - B. There is exactly one solution when the graphs of the equations are one line.
   - C. A system can only be solved by graphing the equations.
   - D. There are infinitely many solutions when the graphs of the equations have the same slope and intercepts.

2. The scale on a map is 2 centimeters to 5 kilometers. Doe Creek and Kent are 15.75 kilometers apart. How far apart are they on the map? **C**
   - A. 7.88 cm
   - B. 39.38 cm
   - C. 6.3 cm
   - D. 31.5 cm

3. Choose the graph that represents $2x - y < 6$. **C**

A. 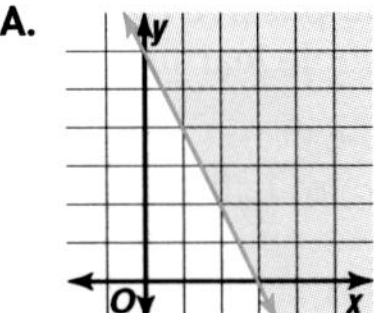

B. 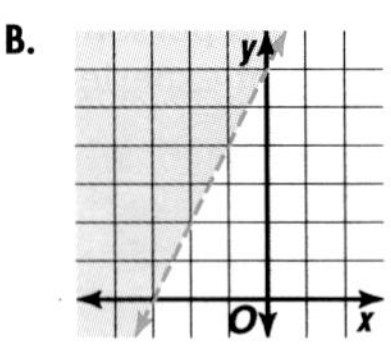

C. 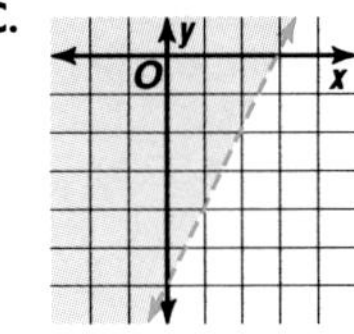

D. 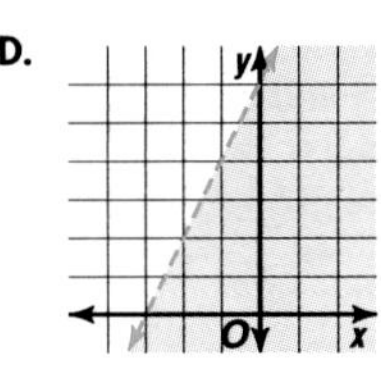

4. The units digit of a two-digit number exceeds twice the tens digit by 1. Find the number if the sum of its digits is 7. **A**
   - A. 25
   - B. 16
   - C. 34
   - D. 61

5. State which region in the graph shown below is the solution of the system. **B**

$y \ge 2x + 2$

$y \le -x - 1$

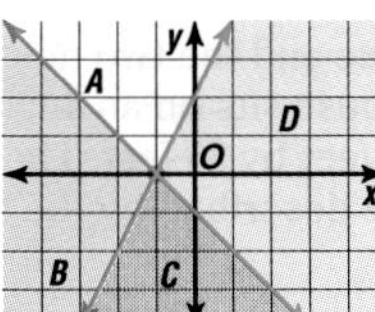

   - A. Region A
   - B. Region B
   - C. Region C
   - D. Region D

6. Abeytu's scores on the first four of five 100-point tests were 85, 89, 90, and 81. What score must she receive on the fifth test to have an average of at least 87 points for all of the tests? **B**
   - A. at least 86 points
   - B. at least 90 points
   - C. at least 69 points
   - D. at least 87 points

7. The frequency of a vibrating violin string is inversely proportional to its length. If a 10-inch violin string vibrates at a frequency of 512 cycles per second, find the frequency of an 8-inch violin string. **D**
   - A. 284.4 cycles per second
   - B. 409.6 cycles per second
   - C. 514 cycles per second
   - D. 640 cycles per second

8. Find the mean of $3\frac{1}{2}$, 5, $4\frac{1}{8}$, $7\frac{3}{4}$, 4, and $6\frac{5}{8}$. **A**
   - A. $5\frac{1}{6}$
   - B. 6
   - C. $6\frac{1}{5}$
   - D. 31

Standardized Test Practice Questions are also provided in the *Assessment and Evaluation Masters*, p. 214.

SECTION TWO: SHORT ANSWER

This section contains seven questions for which you will provide short answers. Write your answer on your paper.

9. Eric is preparing to run in the Bay Marathon. One day, he ran and walked a total of 16 miles. He ran the first mile, and after that walked one mile for every two miles he ran. How many miles did he run, and how many did he walk? **11 miles, 5 miles**

10. The number of Calories in a serving of French fries at 13 restaurants are 250, 240, 220, 348, 199, 200, 125, 230, 274, 239, 212, 240, and 327. Make a box-and-whisker plot of these data. **See Solutions Manual.**

11. Find three consecutive odd integers whose sum is 81. **25, 27, 29**

12. The concession stand sells hot dogs and soda during Beck High School football games. John bought 6 hot dogs and 4 sodas and paid $6.70. Jessica bought 4 hot dogs and 3 sodas and paid $4.65. At what prices are the hot dogs and sodas sold? **hot dogs: $0.75, soda: $0.55**

13. Draw a tree diagram to show the possible meals that could be created from the following choices.

    Meat: chicken, steak

    Vegetable: broccoli, baked potato, tossed salad, carrots

    Drink: milk, cola, juice **See Solutions Manual.**

14. Patricia is going to purchase graduation gifts for her friends, Sarah and Isabel. She wants to spend at least $5 more on Isabel's gift than on Sarah's. She can afford to spend at most a total of $56. Draw a graph showing the possible amounts she can spend on each gift. **See Solutions manual.**

15. The cost of a one-day car rental from Rossi Rentals is given by the formula $C(m) = 31 + 0.13m$, where m is the number of miles that the car is driven, $0.13 is the cost per mile driven, and $C(m)$ is the total cost. If Sheila drove a distance of 110 miles and back in one day, what is the cost of the car rental? **$45.30**

SECTION THREE: OPEN-ENDED

This section contains two open-ended problems. Demonstrate your knowledge by giving a clear, concise solution to each problem. Your score on these problems will depend on how well you do the following.

- Explain your reasoning.
- Show your understanding of the mathematics in an organized manner.
- Use charts, graphs, and diagrams in your explanation.
- Show the solution in more than one way or relate it to other situations.
- Investigate beyond the requirements of the problem.

16. Curtis has a coupon for 33% off any purchase of $50 or more at First Place Sports Shop. His grandmother has given him $75 for his birthday. Write a problem about how Curtis spends his money if he wants to buy several items, including a baseball hat for $18.75, a pair of cleats for $53.95, and a $26.95 sweatshirt discounted 15%. Then solve the problem. **See students' work.**

17. The graphs of the equations $y = x + 3$, $2x - 7y = 4$, and $2y + 3x = 6$ contain the sides of a triangle. Write a problem about these graphs that requires using systems of equations to solve. Then solve the problem. **See students' work.**

CHAPTER

9 Exploring Polynomials

PREVIEWING THE CHAPTER

This chapter helps students master the skills necessary to perform operations with monomials and polynomials. Multiplication and division of monomials are developed first as students connect their knowledge of the powers of whole numbers with the concept of monomials containing variables and exponents. Students also learn to look for a pattern, an important problem-solving strategy. Polynomials are then defined, and students learn to add, subtract, and multiply polynomials, modeling the various operations with algebra tiles. The chapter concludes with a look at some special polynomial products.

Lesson (Pages)	Lesson Objectives	NCTM Standards	State/Local Objectives
9-1 (496–500)	Multiply monomials. Simplify expressions involving powers of monomials. Solve problems by looking for a pattern.	1–5	
9-2 (501–505)	Simplify expressions involving quotients of monomials. Simplify expressions containing negative exponents.	1–5	
9-3 (506–512)	Express numbers in scientific and standard notation. Find products and quotients of numbers expressed in scientific notation.	1–5	
9-4A (513)	Use algebra tiles to model polynomials.	1–5	
9-4 (514–519)	Find the degree of a polynomial. Arrange the terms of a polynomial so that the powers of a variable are in ascending or descending order.	1–5	
9-5A (520–521)	Use algebra tiles to add and subtract polynomials.	1–5	
9-5 (522–527)	Add and subtract polynomials.	1–5	
9-6A (528)	Use algebra tiles to model the product of simple polynomials.	1–5	
9-6 (529–533)	Multiply a polynomial by a monomial. Simplify expressions involving polynomials.	1–5	
9-7A (534–535)	Use algebra tiles to find the product of binomials.	1–5	
9-7 (536–541)	Use the FOIL method to multiply two binomials. Multiply any two polynomials by using the distributive property.	1–5	
9-8 (542–547)	Use patterns to find $(a + b)^2$, $(a - b)^2$, and $(a + b)(a - b)$.	1–5	

A complete, 1-page lesson plan is provided for each lesson in the ***Lesson Planning Guide***. Answer keys for each lesson are available in the ***Answer Key Masters.***

ORGANIZING THE CHAPTER

You may want to refer to the **Course Planning Calendar** on page T12 for detailed information on pacing.
PACING: Standard—15 days; **Honors**—14 days; **Block**—8 days; **Two Years**—26 days

LESSON PLANNING CHART

			BLACKLINE MASTERS											
Lesson (Pages)	**Materials/ Manipulatives**	**Extra Practice (Student Edition)**	**Study Guide**	**Practice**	**Enrichment**	**Assessment and Evaluation**	**Modeling Mathematics**	**Multicultural Activity**	**Tech Prep Applications**	**Graphing Calculator**	**Science and Math Lab Manual**	**Real-World Applications**	**Interactive Mathematics Tools Software**	**Teaching Transparencies**
9-1 (496–500)	calculator	p. 776	p. 61	p. 61	p. 61					p. 9		23	9-1	9-1A 9-1B
9-2 (501–505)	scientific calculator	p. 776	p. 62	p. 62	p. 62	p. 240		p. 17						9-2A 9-2B
9-3 (506–512)	graphing calculator	p. 776	p. 63	p. 63	p. 63				p. 17		pp. 39–42	24		9-3A 9-3B
9-4A (513)	algebra tiles*						p. 28						9-4A	
9-4 (514–519)		p. 777	p. 64	p. 64	p. 64	pp. 239, 240		p. 18						9-4A 9-4B
9-5A (520–521)	algebra tiles*						p. 29						9-5A.1 9-5A.2	
9-5 (522–527)		p. 777	p. 65	p. 65	p. 65									9-5A 9-5B
9-6A (528)	algebra tiles* product mat*						p. 30							
9-6 (529–533)		p. 777	p. 66	p. 66	p. 66	p. 241	p. 80		p. 18					9-6A 9-6B
9-7A (534–535)	algebra tiles* product mat*						p. 31						9-7A	
9-7 (536–541)		p. 778	p. 67	p. 67	p. 67		pp. 60–62							9-7A 9-7B
9-8 (542–547)	algebra tiles* product mat*	p. 778	p. 68	p. 68	p. 68	p. 241								9-8A 9-8B
Study Guide/ Assessment (549–553)						pp. 225–238, 242–244								

*Included in Glencoe's Student Manipulative Kit and Overhead Manipulative Resources.

ORGANIZING THE CHAPTER

Other Chapter Resources

Student Edition
Chapter Opener, pp. 494–495
Mathematics and Society, p. 512
Working on the Investigation, pp. 519, 541
Closing the Investigation, p. 548

Teacher's Classroom Resources
Investigations and Projects Masters, pp. 57–60
Algebra and Geometry Overhead Manipulative Resources, pp. 26–31

Technology
Test and Review Software (IBM and Macintosh)
CD-ROM Interactions (Windows and Macintosh)

Professional Publications
Block Scheduling Booklet
Glencoe Mathematics Professional Series

Outside Resources

Books/Periodicals
Developing Skills in Algebra One: Book B, Dale Seymour Publications

Software
Tools of Mathematics: Algebra, William K. Bradford
Visualizing Algebra: The Function Analyzer, Sunburst

Videos/CD-ROMs
Adding and Subtracting Algebraic Expressions/Multiplying Polynomials, Coronet/MTI
Polynomials, Dale Seymour Publications

Assessment Resources

Student Edition
Math Journal, pp. 499, 504
Mixed Review, pp. 500, 505, 511, 518, 526, 533, 540, 547
Self Test, p. 527
Chapter Highlights, p. 549
Chapter Study Guide and Assessment, pp. 550–552
Alternative Assessment, p. 553
 Portfolio, p. 553

Teacher's Wraparound Edition
5-Minute Check, pp. 496, 501, 506, 514, 522, 529, 536, 542
Check for Understanding, pp. 499, 503, 509, 516, 524, 531, 538, 546
Closing Activity, pp. 500, 505, 512, 519, 527, 533, 541, 547
Cooperative Learning, pp. 498, 523

Assessment and Evaluation Masters
Multiple-Choice Tests, Forms 1A (Honors), 1B (Average), 1C (Basic), pp. 225–230
Free-Response Tests, Forms 2A (Honors), 2B (Average), 2C (Basic), pp. 231–236
Calculator-Based Test, p. 237
Performance Assessment, p. 238
Mid-Chapter Test, p. 239
Quizzes A–D, pp. 240–241
Standardized Test Practice, p. 242
Cumulative Review, pp. 243–244

ENHANCING THE CHAPTER

Examples of some of the materials for enhancing Chapter 9 are shown below.

DIVERSITY

Multicultural Activity Masters, pp. 17, 18

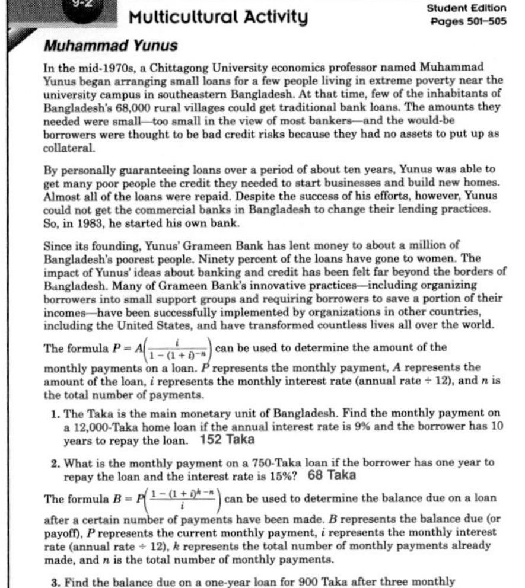

9-2 NAME ______________________ DATE __________

Multicultural Activity Student Edition Pages 501–505

Muhammad Yunus

In the mid-1970s, a Chittagong University economics professor named Muhammad Yunus began arranging small loans for a few people living in extreme poverty near the university campus in southeastern Bangladesh. At that time, few of the inhabitants of Bangladesh's 68,000 rural villages could get traditional bank loans. The amounts they needed were small—too small in the view of most bankers—and the would-be borrowers were thought to be bad credit risks because they had no assets to put up as collateral.

By personally guaranteeing loans over a period of about ten years, Yunus was able to get many poor people the credit they needed to start businesses and build new homes. Almost all of the loans were repaid. Despite the success of his efforts, however, Yunus could not get the commercial banks in Bangladesh to change their lending practices. So, in 1983, he started his own bank.

Since its founding, Yunus' Grameen Bank has lent money to about a million of Bangladesh's poorest people. Ninety percent of the loans have gone to women. The impact of Yunus' ideas about banking and credit has been felt far beyond the borders of Bangladesh. Many of Grameen Bank's innovative practices—including organizing borrowers into small support groups and requiring borrowers to save a portion of their incomes—have been successfully implemented by organizations in other countries, including the United States, and have transformed countless lives all over the world.

The formula $P = A\left(\frac{i}{1-(1+i)^{-n}}\right)$ can be used to determine the amount of the monthly payments on a loan. P represents the monthly payment, A represents the amount of the loan, i represents the monthly interest rate (annual rate ÷ 12), and n is the total number of payments.

1. The Taka is the main monetary unit of Bangladesh. Find the monthly payment on a 12,000-Taka home loan if the annual interest rate is 9% and the borrower has 10 years to repay the loan. 152 Taka
2. What is the monthly payment on a 750-Taka loan if the borrower has one year to repay the loan and the interest rate is 15%? 68 Taka

The formula $B = P\left(\frac{1-(1+i)^{k-n}}{i}\right)$ can be used to determine the balance due on a loan after a certain number of payments have been made. B represents the balance due (or payoff), P represents the current monthly payment, i represents the monthly interest rate (annual rate ÷ 12), k represents the total number of monthly payments already made, and n is the total number of monthly payments.

3. Find the balance due on a one-year loan for 900 Taka after three monthly payments of 80 Taka have been made at an annual interest rate of 12%. 685 Taka
4. Find the balance due on a 10-year loan for 6500 Taka after 60 monthly payments of 83 Taka have been made at an annual interest rate of 9%. 3998 Taka

APPLICATIONS

Real-World Applications, 23, 24

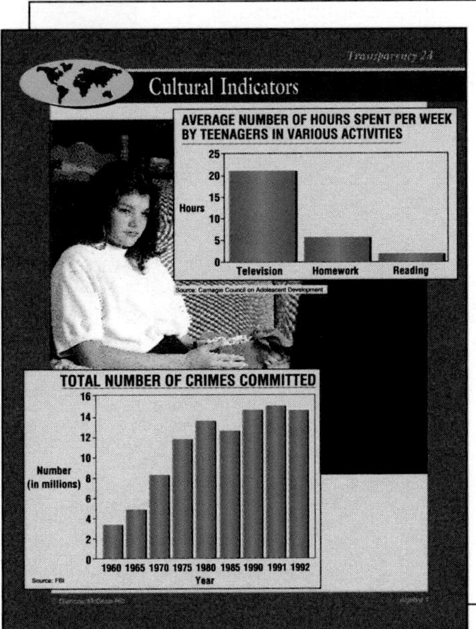

TECHNOLOGY

Graphing Calculator Masters, p. 9

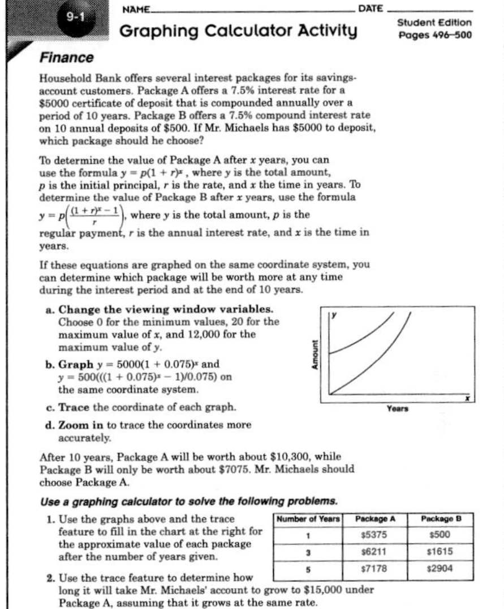

9-1 NAME ______________________ DATE __________

Graphing Calculator Activity Student Edition Pages 496–500

Finance

Household Bank offers several interest packages for its savings-account customers. Package A offers a 7.5% interest rate for a $5000 certificate of deposit that is compounded annually over a period of 10 years. Package B offers a 7.5% compound interest rate on 10 annual deposits of $500. If Mr. Michaels has $5000 to deposit, which package should he choose?

To determine the value of Package A after x years, you can use the formula $y = p(1 + r)^x$, where y is the total amount, p is the initial principal, r is the rate, and x the time in years. To determine the value of Package B after x years, use the formula $y = p\left(\frac{(1+r)^x - 1}{r}\right)$, where y is the total amount, p is the regular payment, r is the annual interest rate, and x is the time in years.

If these equations are graphed on the same coordinate system, you can determine which package will be worth more at any time during the interest period and at the end of 10 years.

a. **Change the viewing window variables.** Choose 0 for the minimum values, 20 for the maximum value of x, and 12,000 for the maximum value of y.

b. **Graph** $y = 5000(1 + 0.075)^x$ and $y = 500(((1 + 0.075)^x - 1)/0.075)$ on the same coordinate system.

c. **Trace** the coordinate of each graph.

d. **Zoom in** to trace the coordinates more accurately.

After 10 years, Package A will be worth about $10,300, while Package B will only be worth about $7075. Mr. Michaels should choose Package A.

Use a graphing calculator to solve the following problems.

1. Use the graphs above and the trace feature to fill in the chart at the right for the approximate value of each package after the number of years given.

Number of Years	Package A	Package B
1	$5375	$500
3	$6211	$1615
5	$7178	$2904

2. Use the trace feature to determine how long it will take Mr. Michaels' account to grow to $15,000 under Package A, assuming that it grows at the same rate.
about $15\frac{1}{4}$ yr

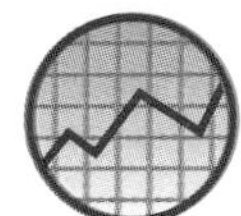

TECH PREP

Tech Prep Applications Masters, pp. 17, 18

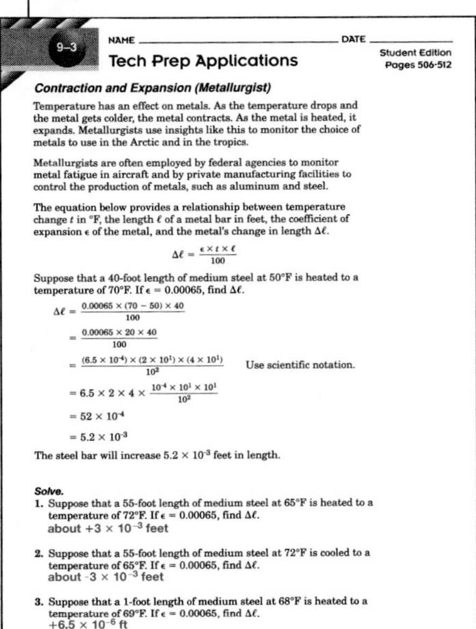

9–3 NAME ______________________ DATE __________

Tech Prep Applications Student Edition Pages 506-512

Contraction and Expansion (Metallurgist)

Temperature has an effect on metals. As the temperature drops and the metal gets colder, the metal contracts. As the metal is heated, it expands. Metallurgists use insights like this to monitor the choice of metals to use in the Arctic and in the tropics.

Metallurgists are often employed by federal agencies to monitor metal fatigue in aircraft and by private manufacturing facilities to control the production of metals, such as aluminum and steel.

The equation below provides a relationship between temperature change t in °F, the length ℓ of a metal bar in feet, the coefficient of expansion ϵ of the metal, and the metal's change in length $\Delta\ell$.

$$\Delta\ell = \frac{\epsilon \times t \times \ell}{100}$$

Suppose that a 40-foot length of medium steel at 50°F is heated to a temperature of 70°F. If $\epsilon = 0.00065$, find $\Delta\ell$.

$$\begin{aligned}\Delta\ell &= \frac{0.00065 \times (70 - 50) \times 40}{100}\\ &= \frac{0.00065 \times 20 \times 40}{100}\\ &= \frac{(6.5 \times 10^{-4}) \times (2 \times 10^1) \times (4 \times 10^1)}{10^2} \quad \text{Use scientific notation.}\\ &= 6.5 \times 2 \times 4 \times \frac{10^{-4} \times 10^1 \times 10^1}{10^2}\\ &= 52 \times 10^{-4}\\ &= 5.2 \times 10^{-3}\end{aligned}$$

The steel bar will increase 5.2×10^{-3} feet in length.

Solve.

1. Suppose that a 55-foot length of medium steel at 65°F is heated to a temperature of 72°F. If $\epsilon = 0.00065$, find $\Delta\ell$.
about $+3 \times 10^{-3}$ feet
2. Suppose that a 55-foot length of medium steel at 72°F is cooled to a temperature of 65°F. If $\epsilon = 0.00065$, find $\Delta\ell$.
about -3×10^{-3} feet
3. Suppose that a 1-foot length of medium steel at 68°F is heated to a temperature of 69°F. If $\epsilon = 0.00065$, find $\Delta\ell$.
$+6.5 \times 10^{-6}$ ft

CONNECTIONS

Science and Math Lab Manual, pp. 39–42

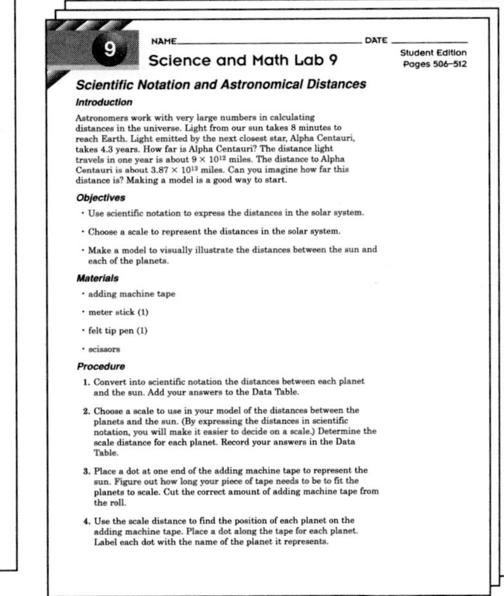

9 NAME ______________________ DATE __________

Science and Math Lab 9 Student Edition Pages 506–512

Scientific Notation and Astronomical Distances

Introduction

Astronomers work with very large numbers in calculating distances in the universe. Light from our sun takes 8 minutes to reach Earth. Light emitted by the next closest star, Alpha Centauri, takes 4.3 years. How far is Alpha Centauri? The distance light travels in one year is about 9×10^{12} miles. The distance to Alpha Centauri is about 3.87×10^{13} miles. Can you imagine how far this distance is? Making a model is a good way to start.

Objectives

- Use scientific notation to express the distances in the solar system.
- Choose a scale to represent the distances in the solar system.
- Make a model to visually illustrate the distances between the sun and each of the planets.

Materials

- adding machine tape
- meter stick (1)
- felt tip pen (1)
- scissors

Procedure

1. Convert into scientific notation the distances between each planet and the sun. Add your answers to the Data Table.
2. Choose a scale to use in your model of the distances between the planets and the sun. (By expressing the distances in scientific notation, you will make it easier to decide on a scale.) Determine the scale distance for each planet. Record your answers in the Data Table.
3. Place a dot at one end of the adding machine tape to represent the sun. Figure out how long your piece of tape needs to be to fit the planets to scale. Cut the correct amount of adding machine tape from the roll.
4. Use the scale distance to find the position of each planet on the adding machine tape. Place a dot along the tape for each planet. Label each dot with the name of the planet it represents.

PROBLEM SOLVING

Problem of the Week Cards, 25, 26, 27

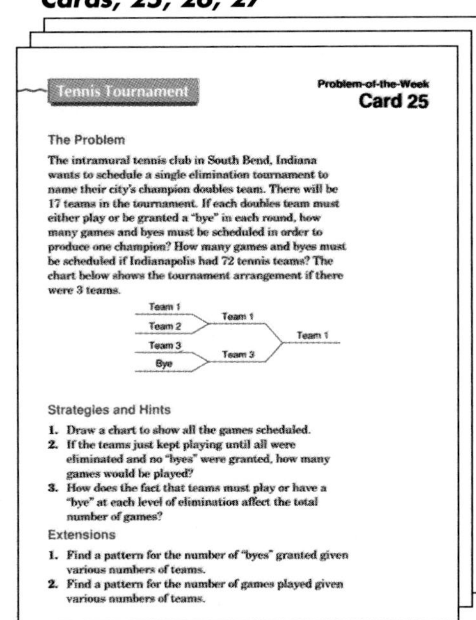

Tennis Tournament Problem-of-the-Week **Card 25**

The Problem

The intramural tennis club in South Bend, Indiana wants to schedule a single elimination tournament to name their city's champion doubles team. There will be 17 teams in the tournament. If each doubles team must either play or be granted a "bye" in each round, how many games and byes must be scheduled in order to produce one champion? How many games and byes must be scheduled if Indianapolis had 72 tennis teams? The chart below shows the tournament arrangement if there were 3 teams.

Strategies and Hints

1. Draw a chart to show all the games scheduled.
2. If the teams just kept playing until all were eliminated and no "byes" were granted, how many games would be played?
3. How does the fact that teams must play or have a "bye" at each level of elimination affect the total number of games?

Extensions

1. Find a pattern for the number of "byes" granted given various numbers of teams.
2. Find a pattern for the number of games played given various numbers of teams.

MAKING MATHEMATICS RELEVANT

This two-page introduction to the chapter provides students with an opportunity to explore contemporary topics and their applications to mathematics.

Background Information

Keeping Up with Technology Mechanical calculating machines were first developed in Europe in the 17th century. Blaise Pascal created the first one in 1642, and his design was improved upon by Gottfried Wilhelm Leibniz in 1673. The first automated computer was designed in 1830 by Charles Babbage, but it remained only an idea on paper. It is generally accepted that the first modern computer was created by John V. Atanasoff and Clifford Berry of the University of Iowa in 1942. Improved models quickly followed. These early modern computers were huge machines, containing thousands of vacuum tubes and filling several rooms. The invention of the transistor in 1948 paved the way for the much smaller computers we use today.

CHAPTER 9

Exploring Polynomials

Objectives

In this chapter, you will:

- solve problems by looking for a pattern,
- multiply and divide monomials,
- express numbers in scientific notation, and
- add, subtract, and multiply polynomials.

Keeping Up With Technology

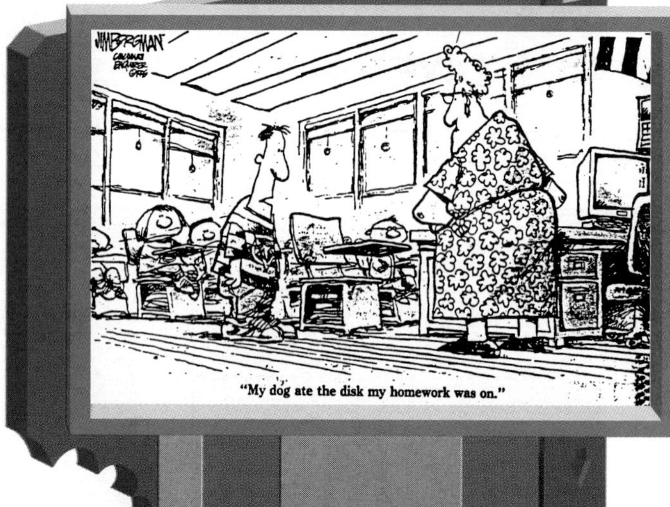
"My dog ate the disk my homework was on."

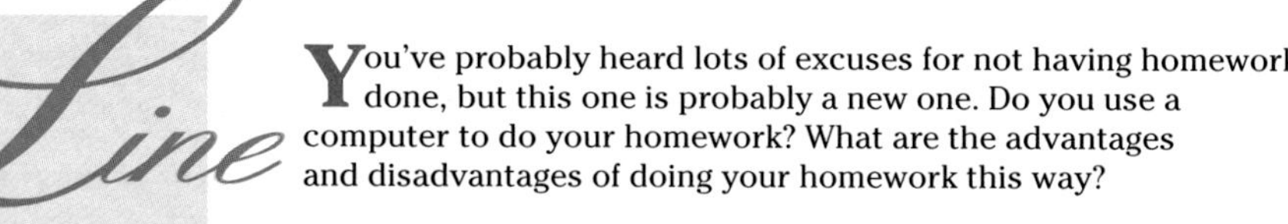

You've probably heard lots of excuses for not having homework done, but this one is probably a new one. Do you use a computer to do your homework? What are the advantages and disadvantages of doing your homework this way?

TIMELine

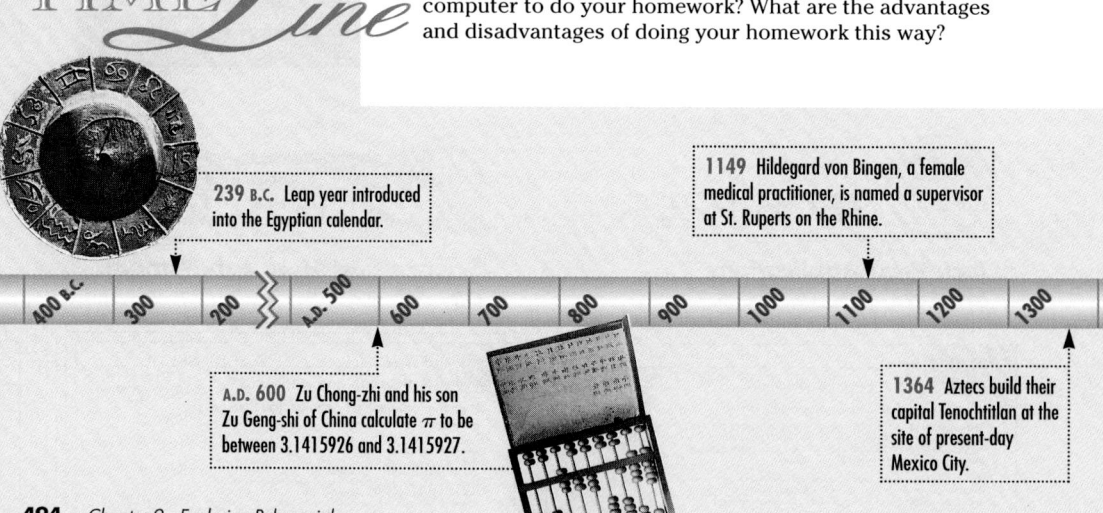

239 B.C. Leap year introduced into the Egyptian calendar.

1149 Hildegard von Bingen, a female medical practitioner, is named a supervisor at St. Ruperts on the Rhine.

400 B.C. | 300 | 200 | A.D. 500 | 600 | 700 | 800 | 900 | 1000 | 1100 | 1200 | 1300 | 1400

A.D. 600 Zu Chong-zhi and his son Zu Geng-shi of China calculate π to be between 3.1415926 and 3.1415927.

1364 Aztecs build their capital Tenochtitlan at the site of present-day Mexico City.

Earth revolves around the sun once every 365.25 days. Because a year must have a whole number of days, the accepted solution is to have three years of 365 days followed by a year with 366 days—a leap year.

interNET CONNECTION

Explore technology and school reform, link to PBS's home page, and learn about child safety on-line.

World Wide Web
http://edweb.cnidr.org:90/resource.cntnts.html

PEOPLE IN THE NEWS

Jenny Slabaugh and **Amy Gusfa** help produce a weekly video show created by students at Dearborn High School in Michigan. When they were only 15 years old, their expertise in electronics and video won each a $1000 scholarship to the Sony Institute of Technology in Hollywood. They were the first females to receive a scholarship from the Sony Institute, and they spent a week there studying theoretical and applied electronics.

Their interest in video production came from Dearborn High's nationally honored video program. They are combining their interests in science, math, electronics, music, and art to create video programs. They both think that more girls should investigate video technology, and they see themselves contributing to this field in the future.

Chapter Project

A byte is a single unit of information such as a number or a letter processed by a computer. A megabyte is 1.048576×10^6 bytes, which is just over 1 million bytes.

- Research three personal computers and three laptop computers. Find out the amount of RAM and hard drive memory in megabytes for each computer. Write each number in standard notation and in scientific notation.
- Use scientific notation to write the amount of memory in bytes for each computer.

- Find the mean of the memory in megabytes for the three personal computers. Find the mean of the memory in megabytes for the three laptop computers. Write the answers in scientific notation.
- Write the ratio of the mean memory for the personal computers to the mean memory for the laptop computers. Write the ratio as a decimal. Discuss the significance of this ratio.
- Investigate the size of the memory of computers that are being developed for the future. How many bytes are in a gigabyte? How many times larger is a gigabyte than a megabyte?

1738 First cuckoo clocks are produced in the Black Forest district.

1881 Booker T. Washington establishes Tuskegee Institute.

1830 Latina Eulalia Elias runs the first major cattle ranch in Arizona.

1995 Microsoft spends billions of dollars to launch Windows '95.

1725 1750 1775 1800 1825 1850 1875 1900 1925 1950 1975 2000

Alternative Chapter Projects

Two other chapter projects are included in the *Investigations and Projects Masters.* In Chapter 9 Project A, pp. 57–58, students extend the topic in the chapter opener. In Chapter 9 Project B, pp. 59–60, students investigate animal populations and endangered species.

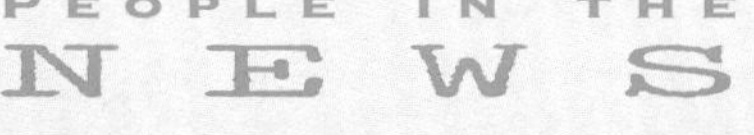

Both Jenny and Amy feel that they were the first girls to receive the Sony Institute scholarship because many girls are intimidated by technology. Encourage a class discussion on this issue.

Chapter Project

Cooperative Learning You may choose to have students work in pairs or in cooperative groups to work on this project. One student, or several group members, can research personal computers. The other student or students can research laptop computers.

Investigations and Projects Masters, p. 57

NAME________ DATE________

9 Chapter 9 Project A

Student Edition Pages 496–548

Make History

1. Working in a small group, interview several adults to find out how their lives have been affected by the development of faster and more powerful computers over the years. If possible, make an audiotape or videotape recording of each interview.
2. Research important events in the history of computers—for example, the development of the semiconductor and the Internet. Be sure to find out how scientists and other researchers have achieved dramatic increases in the speed of computers (as measured by the number of calculations that can be performed in a certain amount of time) and in the amount of information that computers can store.
3. Decide how the group will present the information group members have gathered about the history of computers. You might consider doing one of the following.
   - Make an illustrated timeline.
   - Create an interactive museum exhibit.
   - Conduct additional interviews. Then write a newspaper article or put together a radio or TV documentary about people's attitudes toward computers.
4. Make your presentation to the rest of the class at a time designated by the teacher.

9-1 LESSON NOTES

NCTM Standards: 1–5

Instructional Resources

- Study Guide Master 9-1
- Practice Master 9-1
- Enrichment Master 9-1
- Graphing Calculator Masters, p. 9
- Real-World Applications, 23

Transparency 9-1A contains the 5-Minute Check for this lesson; **Transparency 9-1B** contains a teaching aid for this lesson.

Recommended Pacing	
Standard Pacing	Days 1 & 2 of 15
Honors Pacing	Day 1 of 14
Block Scheduling*	Day 1 of 8 (along with Lesson 9-2)
Alg. 1 in Two Years*	Days 1, 2, & 3 of 26

*For more information on pacing and possible lesson plans, refer to the *Block Scheduling Booklet* and *Algebra 1 in Two Years.*

1 FOCUS

5-Minute Check
(over Chapter 8)

Solve each system of equations.

1. $3x + 5y = 11$
   $2x + 3y = 7$ **(2, 1)**
2. $x + y = 8$
   $2x - 3y = -9$ **(3, 5)**
3. $3x + 3y = 6$
   $2x - y = 1$ **(1, 1)**
4. $r - s = -5$
   $r + s = 25$ **(10, 15)**
5. $-3x + 2y = 16$
   $x - 2y = -12$ **(−2, 5)**

Motivating the Lesson

Questioning A certain number is multiplied by itself four times. The sum of the digits in the product is equal to the number. What is the number? **7**

Teaching Tip Point out that *monomial* is pronounced mah-NOH-mee-ul.

9–1 Multiplying Monomials

What YOU'LL LEARN

- To multiply monomials,
- to simplify expressions involving powers of monomials, and
- to solve problems by looking for a pattern.

Why IT'S IMPORTANT

You can use monomials to solve problems involving finance and geometry.

Muriel Siebert (1932–)

The first woman to hold a seat on the New York Stock Exchange was Muriel ("Mickey") Siebert. She paid $445,000 and was admitted as a full member on December 28, 1967.

Finance

Since 1983, the Beardstown Business and Professional Women's Investment Club ("Beardstown Ladies") has been able to earn enough of a profit in the stock market to make any market expert envious. The club was started when each of 16 women from a small town in Illinois contributed $100 to start an investment fund. Dividends and monthly dues of $25 are also invested. The Beardstown Ladies have earned an average annual profit of 23% on their investments. How has each woman's initial investment of $100 increased over the years?

At the end of the first year (1984), each initial investment would be worth $\$100(1 + 0.23)$, or $123. By the end of the second year, the initial investment would have grown to $\$100(1 + 0.23)(1 + 0.23)$, which is the same as $\$100(1 + 0.23)^2$ or $151.29. The table below shows the value of an initial investment of $100 for each of the first 13 years.

Year	Yearly Calculation	Value
1984	$100(1 + 0.23)$	$123.00
1985	$100(1 + 0.23)^2$	$151.29
1986	$100(1 + 0.23)^3$	$186.09
1987	$100(1 + 0.23)^4$	$228.89
1988	$100(1 + 0.23)^5$	$281.53
1989	$100(1 + 0.23)^6$	$346.28
1990	$100(1 + 0.23)^7$	$425.93
1991	$100(1 + 0.23)^8$	$523.89
1992	$100(1 + 0.23)^9$	$644.39
1993	$100(1 + 0.23)^{10}$	$792.59
1994	$100(1 + 0.23)^{11}$	$974.89
1995	$100(1 + 0.23)^{12}$	$1199.12
1996	$100(1 + 0.23)^{13}$	$1474.91

After 13 years, the initial $100 investment was worth $1474.91! If we let x equal the factor $(1 + 0.23)$, then the value of the initial investment after 13 years can be represented by $100x^{13}$.

An expression like $100x^{13}$ is called a **monomial.** A monomial is a number, a variable, or a product of a number and one or more variables. Monomials that are real numbers are called **constants.**

Monomials	Not Monomials
12	$a + b$
q	$\frac{a}{b}$
$4x^3$	$5 - 7d$
$11ab$	$\frac{5}{a^2}$
$\frac{1}{3}xyz^{12}$	$\frac{5a}{7b}$

fabulous FIRSTS

Full membership for Muriel Siebert brought with it the privilege of handling customers' orders on the floor.

Recall that an expression of the form x^n is a *power*. The base is x, and the exponent is n. A table of powers of 2 is shown below.

2^1	2^2	2^3	2^4	2^5	2^6	2^7	2^8	2^9	2^{10}
2	4	8	16	32	64	128	256	512	1024

In the following products, each number can be expressed as a power of 2. Study the pattern of the exponents.

Number	8(32) = 256	8(64) = 512	4(16) = 64	16(32) = 512
Power	$2^3(2^5) = 2^8$	$2^3(2^6) = 2^9$	$2^2(2^4) = 2^6$	$2^4(2^5) = 2^9$
Pattern of Exponents	3 + 5 = 8	3 + 6 = 9	2 + 4 = 6	4 + 5 = 9

These examples suggest that you can multiply powers with the same base by adding exponents.

Product of Powers

For any number a, and all integers m and n,
$$a^m \cdot a^n = a^{m+n}.$$

Example  **Simplify each expression.**

a. $(3a^6)(a^8)$

$(3a^6)(a^8) = 3a^{6+8}$
$= 3a^{14}$

b. $(8y^3)(-3x^2y^2)\left(\frac{3}{8}xy^4\right)$

$(8y^3)(-3x^2y^2)\left(\frac{3}{8}xy^4\right)$
$= \left(8 \cdot (-3) \cdot \frac{3}{8}\right)(x^2 \cdot x)(y^3 \cdot y^2 \cdot y^4)$
$= -9x^{2+1}y^{3+2+4}$
$= -9x^3y^9$

We looked for a pattern to discover the product of powers property. **Look for a pattern** is an important strategy in problem solving.

Example  **Solve by extending the pattern.**

PROBLEM SOLVING
Look for a Pattern

4 × 6 = 24
14 × 16 = 224
24 × 26 = 624
34 × 36 = 1224
124 × 126 = ?

Explore Look at the problem. You need to find a pattern to determine the product of 124 and 126.

Plan The last two digits of the product are always 24. To find the first digit(s) of the product, look at the tens place of each pair of factors. Notice that 0 × 1 = 0, 1 × 2 = 2, 2 × 3 = 6, and 3 × 4 = 12. Extend this pattern to find the product.

Solve 12 × 13 = 156

Therefore, 124 × 126 = 15,624

Examine Use a calculator to verify that the product is 15,624. The pattern remains true and the product is correct.

2 TEACH

Teaching Tip Emphasize the difference between a term and a monomial. A term may represent division of variables, but a monomial may not. The expressions in the text are not monomials for these reasons.
$a + b$ shows addition, not multiplication.
$\frac{a}{b}$, $\frac{5}{a^2}$, and $\frac{5a}{7b}$ show division, not multiplication. $5 - 7d$ shows subtraction, not multiplication.

Teaching Tip Point out that the product of powers rule applies to all integers m and n, but only positive integers will be treated in this lesson.

In-Class Examples

For Example 1
Simplify each expression.

a. $(21c^6)(c^7)$ $\mathbf{21c^{13}}$
b. $(8x^4)(3x)$ $\mathbf{24x^5}$
c. $(2a^4)(2a^3b^2)(-3ab^3)$ $\mathbf{-12a^8b^5}$

For Example 2
Solve by extending the pattern.

2 × 3 = 6
12 × 13 = 156
22 × 23 = 506
32 × 33 = 1056
122 × 123 = ? **15,006**

GLENCOE Technology

Interactive Mathematics Tools Software

This multimedia software provides an interactive lesson that uses real-life problems and patterns to use and find the Fibonacci sequence. A **Computer Journal** gives students an opportunity to write about what they have learned.

For Windows & Macintosh

Alternative Teaching Strategies

Student Diversity Review the definitions and properties discussed in the lesson. As you read a definition or property aloud, have students provide a number of examples using real numbers and variables, and write the examples on the board.

Teaching Tip Ask students to identify the base in $(5^2)^4$ for the exponent 4 and for the exponent 2. **5^2, 5** Emphasize the need to determine the base for each exponent. In $(5^2)^4$, for the exponent 4, the base 5^2 is multiplied by itself four times.

In-Class Example

For Example 3
Simplify.

a. $(6b^4y)^2\,[(-y)^2]^4$ **$36b^8y^{10}$**

b. $(x^4y^3z^6)^2\,[(2x^2y^2)^2]^4$ **$256x^{24}y^{22}z^{12}$**

Study the examples below.

$$(8^3)^5 = (8^3)(8^3)(8^3)(8^3)(8^3) \qquad (y^7)^3 = (y^7)(y^7)(y^7)$$
$$= 8^{3+3+3+3+3} \longleftarrow \text{Product of powers} \longrightarrow = y^{7+7+7}$$
$$= 8^{15} \qquad = y^{21}$$

Therefore, $(8^3)^5 = 8^{15}$ and $(y^7)^3 = y^{21}$. These examples suggest that you can find the power of a power by multiplying the exponents.

Power of a Power	**For any number *a*, and all integers *m* and *n*, $(a^m)^n = a^{mn}$.**

Look for a pattern in the examples below.

$$(ab)^4 = (ab)(ab)(ab)(ab)$$
$$= (a \cdot a \cdot a \cdot a)(b \cdot b \cdot b \cdot b)$$
$$= a^4b^4$$
$$(5pq)^5 = (5pq)(5pq)(5pq)(5pq)(5pq)$$
$$= (5 \cdot 5 \cdot 5 \cdot 5 \cdot 5)(p \cdot p \cdot p \cdot p \cdot p)(q \cdot q \cdot q \cdot q \cdot q)$$
$$= 5^5p^5q^5 \text{ or } 3125p^5q^5$$

LOOK BACK

You can refer to Lesson 1-1 for information on using a calculator to find a power of a number.

These examples suggest that the power of a product is the product of the powers.

Power of a Product	**For all numbers *a* and *b*, and any integer *m*, $(ab)^m = a^mb^m$.**

The power of a power property and the power of a product property can be combined into the following property.

Power of a Monomial	**For all numbers *a* and *b*, and all integers *m*, *n*, and *p*, $(a^mb^n)^p = a^{mp}b^{np}$.**

Example 3 **Simplify $(2a^4b)^3[(-2b)^3]^2$.**

$(2a^4b)^3[(-2b)^3]^2 = (2a^4b)^3(-2b)^6$ *Power of a power property*

$= 2^3(a^4)^3b^3(-2)^6b^6$ *Power of a product property*

$= 8a^{12}b^3(64)b^6$ *Power of a power property*

$= 512a^{12}b^9$ *Product of powers property*

To simplify an expression involving monomials, write an equivalent expression in which:

- there are no powers of powers,
- each base appears exactly once, and
- all fractions are in simplest form.

Study Guide Masters, p. 61

NAME____________ DATE ________

Study Guide — Student Edition Pages 496–500

Multiplying Monomials

When you multiply monomials, you use the following rules for all numbers a and b and any integers m, n, and p.

	Rule	Example
Product of Powers	For any number a, and all integers m and n, $a^m \cdot a^n = a^{m+n}$.	$a^2 \cdot a^6 = a^{2+6} = a^8$
Power of a Power	For any number a, and all integers m and n, $(a^m)^n = a^{mn}$.	$(x^2)^6 = x^{2 \cdot 6} = x^{12}$
Power of a Product	For all numbers a and b, and any integer m, $(ab)^m = a^mb^m$.	$(pq)^4 = p^4q^4$
Power of a Monomial	For all numbers a and b, and all integers m, n, and p, $(a^mb^n)^p = a^{mp}b^{np}$.	$(s^4t)^3 = (s^4t^1)^3 = s^{4 \cdot 3}t^{1 \cdot 3} = s^{12}t^3$

Simplify.

1. $[n^5(n^2)]$ n^7
2. $b(b^4)$ b^5
3. $(-7x^2)(x^4)$ $-7x^6$
4. $(2a^2)(8a)$ $16a^3$
5. $(rs)(rs^3)(s^2)$ r^2s^6
6. $(x^2y)(4xy^3)$ $4x^3y^4$
7. $\frac{1}{3}(2a^3b)(6b^3)$ $4a^3b^4$
8. $(-5nx)(4x^2)(n^4)$ $-20n^5x^3$
9. $(n^3)^5$ n^{15}
10. $(a^4)^6$ a^{24}
11. $-3(ab^4)^3$ $-3a^3b^{12}$
12. $(-3ab^4)^3$ $-27a^3b^{12}$
13. $(4x^2b)^3$ $64x^6b^3$
14. $(4x)^2(b^3)$ $16x^2b^3$
15. $(-2m^5n^6)^2$ $4m^{10}n^{12}$
16. $-2m^5(n^6)^2$ $-2m^5n^{12}$
17. $2(3x)^3$ $54x^3$
18. $-3(2x)^5$ $-96x^5$
19. $(-2n^6y^5)(-6n^3y^2)(ny)^3$ $12n^{12}y^{10}$
20. $(-3a^3n^4)(-3a^3n)^4$ $-243a^{15}n^8$

Cooperative Learning

Trade-A-Problem Separate the class into small groups. Have each group draw a square and a rectangular solid on a piece of grid paper and label the measure of one side of the square and three edges of the rectangular solid using monomials. Have groups switch papers and find the area of each square and volume of each rectangular solid. Group members should all agree on their answers before checking them with other groups. For more information on the trade-a-problem strategy, see *Cooperative Learning in the Mathematics Classroom,* one of the titles in the Glencoe Mathematics Professional Series, pages 25–26.

CHECK FOR UNDERSTANDING

Communicating Mathematics

Study the lesson. Then complete the following.

1. **Write** in your own words. See margin.
   a. the product of powers property
   b. the power of a power property
   c. the power of a product property
2. **Explain** why the product of powers property does not apply when the bases are different. See Solutions Manual.
3. **You Decide** Luisa says $10^4 \times 10^5 = 100^9$, but Taryn says that $10^4 \times 10^5 = 10^9$. Who is correct? Explain your answer. See Solutions Manual.

4. **Write** 64 in six different ways using exponents; for example, $64 = (2^2)^3$. Answers will vary. Sample answers: 2^6, $(2^3)^2$, $2^2 \times 2^4$, 2×2^5, $2^3 \times 2^3$, $(2^2)^2 \times 2^2$

Guided Practice

Determine whether each pair of monomials is equivalent. Write *yes* or *no*.

5. $2d^3$ and $(2d)^3$ no
6. $(xy)^2$ and x^2y^2 yes
7. $-x^2$ and $(-x)^2$ no
8. $5(y^2)^2$ and $25y^4$ no

Simplify.

9. $a^4(a^7)(a)$ a^{12}
10. $(xy^4)(x^2y^3)$ x^3y^7
11. $[(3^2)^4]^2$ 3^{16} or 43,046,721
12. $(2a^2b)^2$ $4a^4b^2$
13. $(-27ay^3)\left(-\frac{1}{3}ay^3\right)$ $9a^2y^6$
14. $(2x^2)^2\left(\frac{1}{2}y^2\right)^2$ x^4y^4
15. **Geometry** Find the measure of the area of the rectangle at the right. $15a^4b^3$

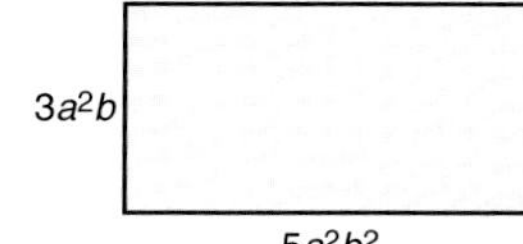

EXERCISES

Practice

Simplify.

16. $b^3(b)(b^5)$ b^9
17. $(m^3n)(mn^2)$ m^4n^3
18. $(a^2b)(a^5b^4)$ a^7b^5
19. $[(2^3)^2]^2$ 2^{12} or 4096
20. $(3x^4y^3)(4x^4y)$ $12x^8y^4$
21. $(a^3x^2)^4$ $a^{12}x^8$
22. $m^7(m^3b^2)$ $m^{10}b^2$
23. $(3x^2y^2z)(2x^2y^2z^3)$ $6x^4y^4z^4$

24. $(0.6d)^3$ $0.216d^3$
25. $(ab)(ac)(bc)$ $a^2b^2c^2$
26. $-\frac{5}{6}c(12a^3)$ $-10a^3c$
27. $\left(\frac{2}{5}d\right)^2$ $\frac{4}{25}d^2$
28. $-3(ax^3y)^2$ $-3a^2x^6y^2$
29. $(0.3x^3y^2)^2$ $0.09x^6y^4$
30. $(-3ab)^3(2b^3)$ $-54a^3b^6$
31. $\left(\frac{3}{10}y^2\right)^2(10y^2)^3$ $90y^{10}$
32. $(3x^2)^2\left(\frac{1}{3}y^2\right)^2$ x^4y^4
33. $\left(\frac{2}{5}a\right)^2(25a)(13b)\left(\frac{1}{13}b^4\right)$ $4a^3b^5$

34. $(3a^2)^3 + 2(a^3)^2$ $29a^6$
35. $(-2x^3)^3 - (2x)^9$ $-520x^9$

3 PRACTICE/APPLY

Check for Understanding

Exercises 1–15 are designed to help you assess your students' understanding through reading, writing, speaking, and modeling. You should work through Exercises 1–4 with your students and then monitor their work on Exercises 5–15.

Assignment Guide

Core: 17–39 odd, 40–47
Enriched: 16–34 even, 36–47

For **Extra Practice,** see p. 776.

The red A, B, and C flags, printed only in the Teacher's Wraparound Edition, indicate the level of difficulty of the exercises.

Additional Answers

1a. When numbers with the same base are multiplied, the exponents are added.
1b. When a number is raised to a power and then raised to another power, the two exponents are multiplied.
1c. When two numbers are multiplied and the product is raised to a power, each number can be raised to the power before multiplying.

Practice Masters, p. 61

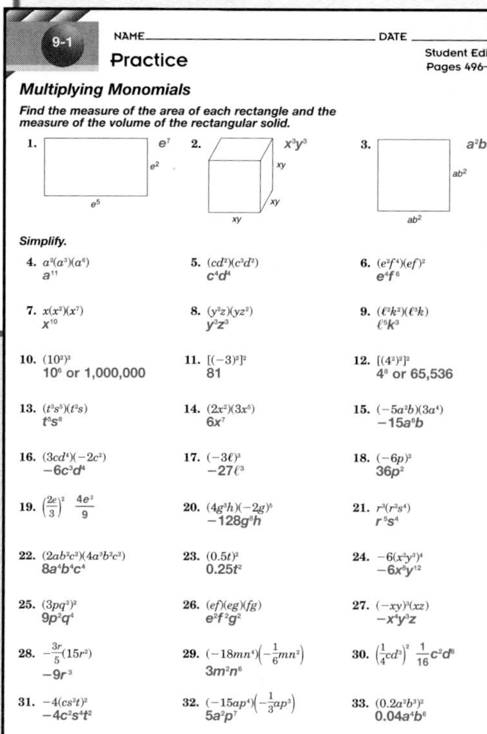
9-1 NAME_____ DATE_____

Practice

Student Edition Pages 496–500

Multiplying Monomials

Find the measure of the area of each rectangle and the measure of the volume of the rectangular solid.

1. (rectangle: e^5 by e^2) e^7
2. (solid: xy, xy, xy) x^3y^3
3. (square: ab^2 by ab^2) a^2b^4

Simplify.

4. $a^2(a^3)(a^6)$ a^{11}
5. $(cd^2)(c^3d^2)$ c^4d^4
6. $(e^2f^4)(ef)^2$ e^4f^6
7. $x(x^2)(x^7)$ x^{10}
8. $(y^2z)(yz^2)$ y^3z^3
9. $(\ell^2k^2)(\ell^3k)$ ℓ^5k^3
10. $(10^2)^3$ 10^6 or 1,000,000
11. $[(-3)^2]^2$ 81
12. $[(4^2)^2]^2$ 4^8 or 65,536
13. $(t^3s^5)(t^2s)$ t^5s^6
14. $(2x^2)(3x^5)$ $6x^7$
15. $(-5a^2b)(3a^4)$ $-15a^6b$
16. $(3cd^4)(-2c^2)$ $-6c^3d^4$
17. $(-3\ell)^3$ $-27\ell^3$
18. $(-6p)^2$ $36p^2$
19. $\left(\frac{2e}{3}\right)^2$ $\frac{4e^2}{9}$
20. $(4g^3h)(-2g)^5$ $-128g^8h$
21. $r^3(r^2s^4)$ r^5s^4
22. $(2ab^2c^2)(4a^3b^2c^2)$ $8a^4b^4c^4$
23. $(0.5t)^2$ $0.25t^2$
24. $-6(x^2y^3)^4$ $-6x^8y^{12}$
25. $(3pq^2)^2$ $9p^2q^4$
26. $(ef)(eg)(fg)$ $e^2f^2g^2$
27. $(-xy)^3(xz)$ $-x^4y^3z$
28. $-\frac{3r}{5}(15r^2)$ $-9r^3$
29. $(-18mn^4)\left(-\frac{1}{6}mn^2\right)$ $3m^2n^6$
30. $\left(\frac{1}{4}cd^3\right)^2$ $\frac{1}{16}c^2d^6$
31. $-4(cs^2t)^2$ $-4c^2s^4t^2$
32. $(-15ap^4)\left(-\frac{1}{3}ap^3\right)$ $5a^2p^7$
33. $(0.2a^2b^3)^2$ $0.04a^4b^6$

Reteaching

Using Models Use circular markers or chips to illustrate the difference between the product of powers and a power of a power. For example, consider $2^3 \times 2^2$ and $(2^3)^2$. Let each marker represent the base 2. For the first example, lay out a group of 3 markers and a group of 2 markers. This is a total of 5 markers. Thus, $2^3 \times 2^2 = 2^5$. For the second example, lay out 2 groups of 3 markers. This is a total of 6. Thus, $(2^3)^2 = 2^6$. This model can be used with any examples in which the bases are the same.

4 ASSESS

Closing Activity

Speaking Discuss each property of monomials described in this lesson. Have students provide examples to illustrate each property.

Additional Answers

36. For example, $(x + y)^2 = (x + y)(x + y)$ and not $x^2 + y^2$.

37. -2^4 equals $-(2)(2)(2)(2)$ or -16 and $(-2)^4$ equals $(-2)(-2)(-2)(-2)$ or 16.

41.

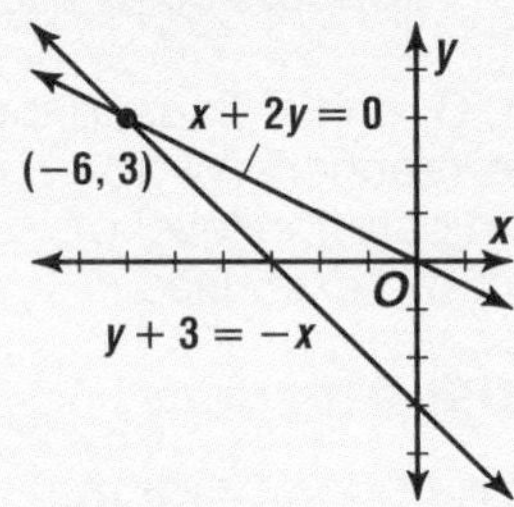

Enrichment Masters, p. 61

9-1 NAME_______ DATE_______

Enrichment — Student Edition Pages 496–500

Circular Areas and Volumes

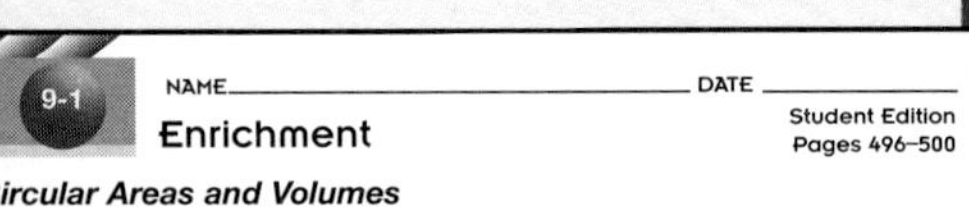

Write an algebraic expression for each shaded area. (Recall that the diameter of a circle is twice its radius.)

1.
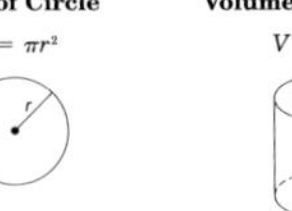

2.
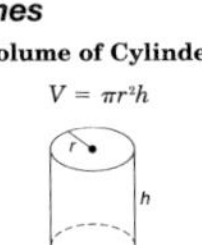

3.
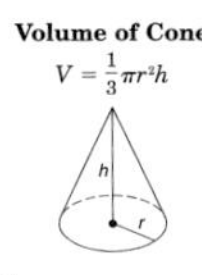

$\pi x^2 - \pi\left(\frac{x}{2}\right)^2 = \frac{3}{4}\pi x^2$ $\quad \frac{\pi}{2}(y^2 + 2xy)$ $\quad \frac{23\pi}{4}x^2$

Write an algebraic expression of the total volume of each figure.

4.
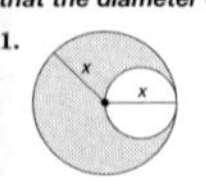
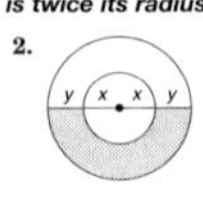

5.
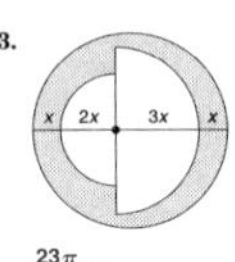
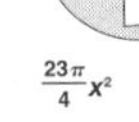

$5\frac{2}{3}\pi x^3$ $\quad \frac{\pi}{12}[13x^3 + (4a + 9b)x^2]$

Each figure has a cylindrical hole with a radius of 2 inches and a height of 5 inches. Find each volume.

6.
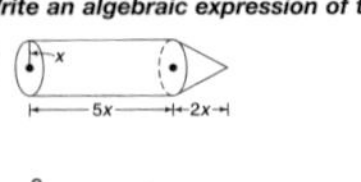

7.
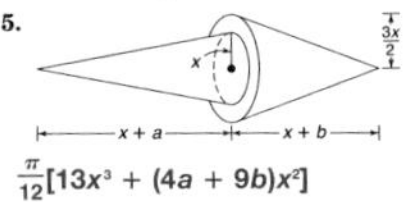

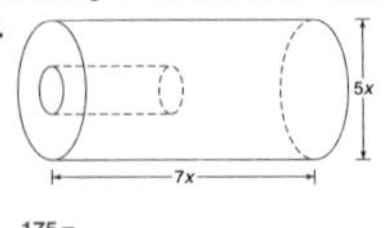
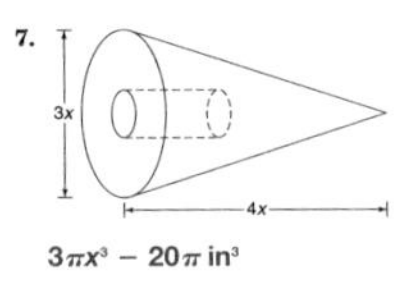

$\frac{175\pi}{4}x^3 - 20\pi$ in³ $\quad 3\pi x^3 - 20\pi$ in³

Critical Thinking

36. Explain why $(x + y)^z$ does not equal $x^z + y^z$. **See margin.**

37. Explain why -2^4 does not equal $(-2)^4$. **See margin.**

Applications and Problem Solving

38. **Investments** Refer to the application at the beginning of the lesson. Each of the Beardstown Ladies added \$25 to the investments each month. This amounts to \$300 a year. Assume that each member invested the \$300 at the beginning of each year starting in 1984. You can use the formula $T = p\left[\frac{(1 + r)^t - 1}{r}\right]$ to determine how each member's money grew. T represents the total amount, p represents the regular payment, r represents the annual interest rate, and t represents the time in years.
   a. How much money did each member make from their additional investments from 1984 to 1996? **\$14,336.30**
   b. What was the total value of each member's investment in 1996? **\$15,811.21**

39. **Look for a Pattern** The symbol used for a U.S. dollar is a capital S with a vertical line through it. The line separates the S into 4 parts, as shown at the right. How many parts would there be if the S had 100 vertical lines through it? **301 parts**

Mixed Review

40. Write a system of inequalities for the graph at the right. (Lesson 8–5) $x \leq 2,\ y \geq 2$

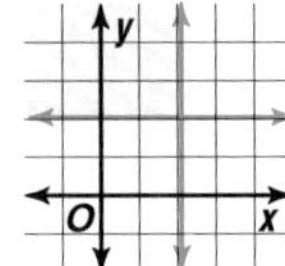

41. Graph the system of equations. Determine whether the system has *one* solution, *no* solution, or *infinitely many* solutions. If the system has one solution, name it. (Lesson 8–1) **See margin for graph; one; (−6, 3).**
   $x + 2y = 0$
   $y + 3 = -x$

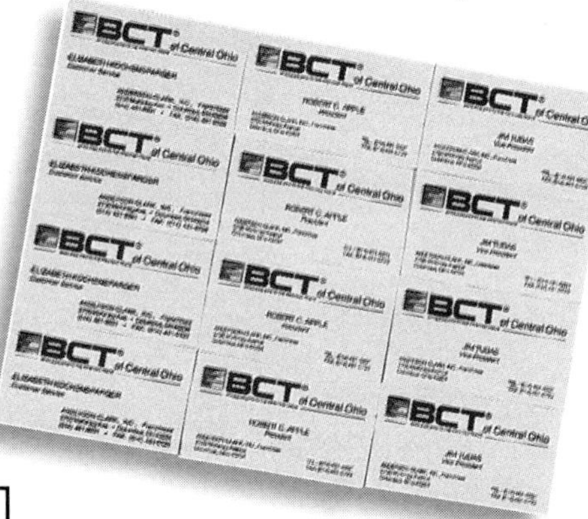

42. **Business** Jorge Martinez has budgeted \$150 to have business cards printed. A card printer charges \$11 to set up each job and an additional \$6 per box of 100 cards printed. What is the greatest number of cards Mr. Martinez can have printed? (Lesson 7–3) **2300 cards**

43. Write an equation from the relation shown in the chart below. Then copy and complete the chart. (Lesson 5–6) $n = 2m + 1$

m	−3	−2	−1	0	1
n	−5	−3	−1	**1**	**3**

44. **Travel** Tiffany wants to reach Dallas at 10 A.M. If she drives at 36 miles per hour, she would reach Dallas at 11 A.M. But if she drives 54 miles per hour, she would arrive at 9 A.M. At what average speed should she drive to reach Dallas exactly at 10 A.M.? (Lesson 4–7) **43.2 mph**

45. **Geometry** Find the supplement of 44°. (Lesson 3–4) **136°**

46. Simplify $-16 \div 8$. (Lesson 2–7) **−2**

47. Simplify $0.3(0.2 + 3y) + 0.21y$. (Lesson 1–8) $\mathbf{1.11y + 0.06}$

Extension

Reasoning Find a if $(3^{a+2})^2 \times 3^{3a-10} = 81$. (*Hint:* Write 81 as a power of 3.) $a = 2$

9-2 Dividing by Monomials

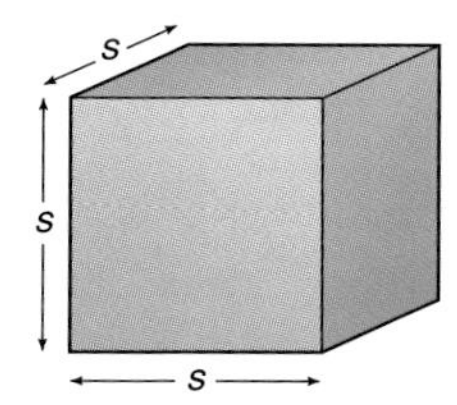

What YOU'LL LEARN

- To simplify expressions involving quotients of monomials, and
- to simplify expressions containing negative exponents.

Why IT'S IMPORTANT

You can use monomials to solve problems involving finance and geometry.

INTEGRATION Geometry

The volume of a cube with each side s units long is s^3 cubic units. So, the ratio of the measure of the volume of a cube to the measure of the length of each side is $\frac{s^3}{s}$. How can you express this ratio in simplest form?

$V = s \cdot s \cdot s$ or s^3

Just as we used a pattern to discover the product of powers property, we can use a pattern to discover a property for a quotient of powers such as $\frac{s^3}{s}$. In the following quotients, each number can be expressed as a power of 2. Study the pattern of exponents.

Number	$\frac{64}{32} = 2$	$\frac{32}{8} = 4$	$\frac{64}{8} = 8$	$\frac{32}{2} = 16$
Power	$\frac{2^6}{2^5} = 2^1$	$\frac{2^5}{2^3} = 2^2$	$\frac{2^6}{2^3} = 2^3$	$\frac{2^5}{2^1} = 2^4$
Pattern of Exponents	$6 - 5 = 1$	$5 - 3 = 2$	$6 - 3 = 3$	$5 - 1 = 4$

These examples suggest that you can divide powers with the same base by subtracting exponents.

Quotient of Powers

For all integers *m* and *n*, and any nonzero number *a*,
$$\frac{a^m}{a^n} = a^{m-n}.$$

To write $\frac{s^3}{s}$ in simplest form, subtract the exponents.

$\frac{s^3}{s} = s^{3-1}$ *Recall that $s = s^1$, and apply the quotient of powers property.*

$= s^2$

In simplest form, $\frac{s^3}{s} = s^2$.

Example 1 **Simplify $\frac{y^4z^3}{y^2z^2}$.**

$\frac{y^4z^3}{y^2z^2} = \left(\frac{y^4}{y^2}\right)\left(\frac{z^3}{z^2}\right)$ *Group the powers with the same base, y^4 with y^2 and z^3 with z^2.*

$= (y^{4-2})(z^{3-2})$ *Quotient of powers property.*

$= y^2z$

9-2 LESSON NOTES

NCTM Standards: 1–5

Instructional Resources

- Study Guide Master 9-2
- Practice Master 9-2
- Enrichment Master 9-2
- Assessment and Evaluation Masters, p. 240
- Multicultural Activity Masters, p. 17

Transparency 9-2A contains the 5-Minute Check for this lesson; **Transparency 9-2B** contains a teaching aid for this lesson.

Recommended Pacing	
Standard Pacing	Day 3 of 15
Honors Pacing	Day 2 of 14
Block Scheduling*	Day 1 of 8 (along with Lesson 9-1)
Alg. 1 in Two Years*	Days 4, 5, & 6 of 26

*For more information on pacing and possible lesson plans, refer to the *Block Scheduling Booklet* and *Algebra 1 in Two Years.*

1 FOCUS

5-Minute Check
(over Lesson 9-1)

Simplify.

1. $(a^4)(a^7)$ a^{11}
2. $(2p^3)(5p)$ $10p^4$
3. $(x^4y^5)^2$ x^8y^{10}
4. $(2x^3y^4)^2$ $4x^6y^8$
5. $(-5mn)^2[(2mn^4)^2]^2$ $400m^6n^{18}$

Motivating the Lesson

Questioning A number multiplied by itself five times divided by the number multiplied by itself twice is the number multiplied by itself three times. What is the number? any positive whole integer

Assignment Guide

Core: 15–41 odd, 43–50
Enriched: 14–34 even, 35–50

For **Extra Practice,** see p. 776.

The red A, B, and C flags, printed only in the Teacher's Wraparound Edition, indicate the level of difficulty of the exercises.

5. You Decide Taigi and Isabel each simplified $\left(\frac{x^{-2}y^3}{x}\right)^{-2}$ correctly as shown below.

Taigi	Isabel
$\left(\frac{x^{-2}y^3}{x}\right)^{-2} = \frac{x^4y^{-6}}{x^{-2}}$	$\left(\frac{x^{-2}y^3}{x}\right)^{-2} = (x^{-2-1}y^3)^{-2}$
$= x^{4-(-2)}y^{-6}$	$= (x^{-3}y^3)^{-2}$
$= x^6y^{-6}$	$= x^6y^{-6}$
$= \frac{x^6}{y^6}$	$= \frac{x^6}{y^6}$

Whose method do you prefer? Why? See students' work.

6. Assess Yourself Which properties of monomials do you find easy to understand? Which properties do you need to study more? See students' work.

Guided Practice

Simplify. Assume that no denominator is equal to zero.

7. 11^{-2} $\frac{1}{121}$
8. $(6^{-2})^2$ $\frac{1}{1296}$
9. $\left(\frac{1}{4}\cdot\frac{2}{3}\right)^{-2}$ 36
10. $a^4(a^{-7})(a^0)$ $\frac{1}{a^3}$
11. $\frac{6r^3}{r^7}$ $\frac{6}{r^4}$
12. $\frac{(a^7b^2)^2}{(a^{-2}b)^{-2}}$ $a^{10}b^6$

13. Geometry Write the ratio of the area of the circle to the area of the square in simplest form.
$\frac{\pi}{4}$

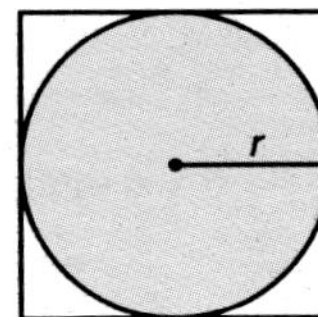

EXERCISES

Practice

Simplify. Assume that no denominator is equal to zero.

14. $a^0b^{-2}c^{-1}$ $\frac{1}{b^2c}$
15. $\frac{a^0}{a^{-2}}$ a^2
16. $\frac{5n^5}{n^8}$ $\frac{5}{n^3}$
17. $\frac{m^2}{m^{-4}}$ m^6
18. $\frac{b^5d^2}{b^3d^8}$ $\frac{b^2}{d^6}$
19. $\frac{10m^4}{30m}$ $\frac{m^3}{3}$

B

20. $\frac{(-y)^5m^8}{y^3m^{-7}}$ $-y^2m^{15}$
21. $\frac{b^6c^5}{b^{14}c^2}$ $\frac{c^3}{b^8}$
22. $\frac{22a^2b^5c^7}{-11abc^2}$ $-2ab^4c^5$
23. $\frac{(a^{-2}b^3)^2}{(a^2b)^{-2}}$ b^8
24. $\frac{7x^3z^5}{4z^{15}}$ $\frac{7x^3}{4z^{10}}$
25. $\frac{(-r)^5s^8}{r^5s^2}$ $-s^6$
26. $\frac{(r^{-4}k^2)^2}{(5k^2)^2}$ $\frac{1}{25r^8}$
27. $\frac{16b^4}{-4bc^3}$ $-\frac{4b^3}{c^3}$
28. $\frac{27a^4b^6c^9}{15a^3c^{15}}$ $\frac{9ab^6}{5c^6}$
29. $\frac{(4a^{-1})^{-2}}{(2a^4)^2}$ $\frac{1}{64a^6}$
30. $\left(\frac{3m^2n^2}{6m^{-1}k}\right)^0$ 1
31. $\frac{r^{-5}s^{-2}}{(r^2s^5)^{-1}}$ $\frac{s^3}{r^3}$

32. $\left(\frac{7m^{-1}n^3}{n^2r^{-1}}\right)^{-1}$ $\frac{m}{7rn}$
33. $\frac{(-b^{-1}c)^0}{4a^{-1}c^2}$ $\frac{a}{4c^2}$
34. $\left(\frac{3xy^{-2}z}{4x^{-2}y}\right)^{-2}$ $\frac{16y^6}{9x^6z^2}$

Extension

Problem Solving Simplify $\frac{a^0(a)(a^2)^2(a^3)^3}{(a^{-1})^{-1}(a^{-2})^{-2}(a^{-3})^{-3}}$. 1

Practice Masters, p. 62

9-2 NAME ________ DATE ________

Practice — Student Edition Pages 501–505

Dividing by Monomials

Simplify. Assume that no denominator is equal to zero.

1. $\frac{c^0}{c^{-3}}$ c^3
2. $\frac{1}{x^{-4}}$ x^4
3. $\frac{m}{m^3}$ $\frac{1}{m^2}$
4. $\frac{a^2b^3}{a^3b^2}$ $\frac{b}{a}$
5. $\frac{(-c)^2d^7}{c^2d^6}$ d
6. $\frac{(-e)^6f}{e^6f^4}$ $\frac{1}{f^3}$
7. $\frac{9d^7}{3d^6}$ $3d$
8. $\frac{12n^5}{36n}$ $\frac{n^4}{3}$
9. $\frac{r^3s^2}{r^3s^4}$ $\frac{1}{s^2}$
10. $\frac{w^4u^3}{w^4u}$ u^2
11. $\frac{c^6(-d)^7}{c^{10}d^5}$ $\frac{-d^2}{c^4}$
12. $\frac{a^5b^3}{a^7b^2}$ $\frac{b}{a^2}$
13. $\frac{15x^3y^5}{-3x^5y^2}$ $-\frac{5y^3}{x^2}$
14. $\frac{-24kd^8}{8k^2d^7}$ $-\frac{3d}{k}$
15. $\frac{16x^2y^4}{8x^2(-y)^3}$ $-2y$
16. $\frac{12x^7}{4x^5}$ $3x^2$
17. $\frac{3y^6}{y^6}$ 3
18. $\frac{10a^{10}b^5}{2a^{11}b^5}$ $\frac{5}{a}$
19. $\frac{48r^{12}s^6}{16r^{14}s^4}$ $\frac{3s^2}{r^2}$
20. $\frac{16x^4y^2}{2x^2y^3}$ $\frac{8x^2}{y}$
21. $\frac{-27x^5y^4}{9x^3y^4}$ $-3x^2$
22. $\frac{32x^3y^2z^5}{-8xyz^2}$ $-4x^2yz^3$
23. $\frac{8a^{11}b^7}{-2a^5b^9}$ $\frac{-4a^6}{b^2}$
24. $\frac{48x^6y^7z^5}{-6xy^5z^6}$ $-\frac{8x^5y^2}{z}$
25. $\frac{54x^7y^4z}{27x^6y}$ $2xy^3z$
26. $\frac{-21w^5u^2}{7w^4u^5}$ $-\frac{3w}{u^3}$
27. $\frac{12d^2ef^2}{2de^2f^2}$ $\frac{6d}{e}$

Critical Thinking

Simplify. Assume that no denominator is equal to zero.

35. $m^3(m^n)$ m^{3+n}
36. $y^{2c}(y^{5c})$ y^{7c}
37. $(3^{2x+1})(3^{2x-7})$ 3^{4x-6}
38. $\frac{r^{y-2}}{r^{y+3}}$ $\frac{1}{r^5}$
39. $\frac{(q^{y-7})^2}{(q^{y+2})^2}$ $\frac{1}{q^{18}}$
40. $\frac{y^x}{y^{a-x}}$ y^{2x-a}

Applications and Problem Solving

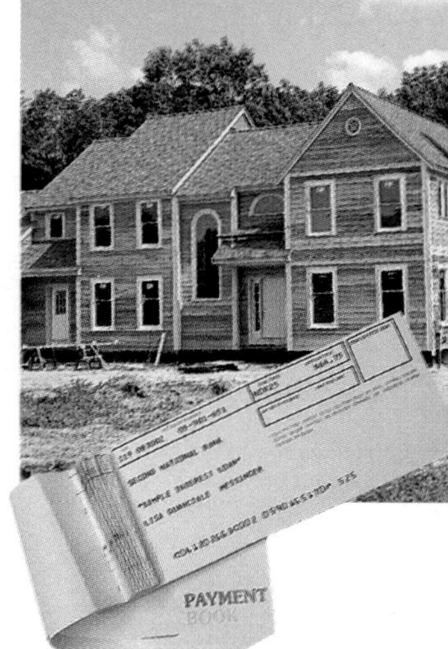

41. **Finance** You can use the formula $P = A\left[\frac{i}{1-(1+i)^{-n}}\right]$ to determine the monthly payment on a home. P represents the monthly payment, A represents the price of the home less the down payment, i represents the *monthly* interest rate (annual rate ÷ 12), and n is the total number of monthly payments. Find the monthly payment on a \$180,000 home with 10% down and an *annual* interest rate of 8.6% over 30 years. **\$1257.14**

42. **Finance** You can use the formula $B = P\left[\frac{1-(1+i)^{k-n}}{i}\right]$ to calculate the balance due on a car loan after a certain number of payments have been made. B represents the balance due (or payoff), P represents the current monthly payment, i represents the *monthly* interest rate (annual rate ÷ 12), k represents the total number of monthly payments already made, and n is the total number of monthly payments. Find the balance due on a 48-month loan for \$10,562 after 20 monthly payments of \$265.86 have been made at an *annual* interest rate of 9.6%. **\$6645.57**

Mixed Review

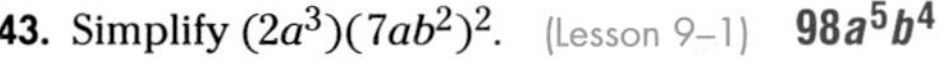

43. Simplify $(2a^3)(7ab^2)^2$. (Lesson 9–1) $98a^5b^4$

44. **Aviation** Flying with the wind, a plane travels 300 miles in 40 minutes. Flying against the wind, it travels 300 miles in 45 minutes. Find the air speed of the plane. (Lesson 8–4) **425 mph**

45. Write an open sentence involving the absolute value for the graph at the right. (Lesson 7–6) $|x + 1| < 3$

−5 −4 −3 −2 −1 0 1 2 3

46. Solve $-\frac{2}{5} > \frac{4z}{7}$. (Lesson 7–2) $-\frac{7}{10} > z$

47. **Geometry** Find the coordinates of the midpoint of the line segment whose endpoints are (5, −3) and (1, −7). (Lesson 6–7) **(3, −5)**

48. **Sales** Latoya bought a new dress for \$32.86. This included 6% sales tax. What was the cost of the dress before tax? (Lesson 4–5) **\$31.00**

49. Solve $\frac{4-x}{3+x} = \frac{16}{25}$. (Lesson 4–1) $\frac{52}{41}$

50. Simplify $41y - (-41y)$. (Lesson 2–3) $82y$

Classroom Vignette

"To extend Exercise 42, I have students select a car they would like to buy and use the graphing calculator to determine the interest they would pay, the total cost of the loan, and the monthly payments. They also work with the remaining balance formula to discover how different numbers of payments affect the interest paid. This is a real eye-opener for students."

Harvey A. Johnecheck
Gaylord High School
Gaylord, Michigan

4 ASSESS

Closing Activity

Writing Have students write down the key points to consider when dividing monomials using negative and zero exponents.

Chapter 9, Quiz A (Lessons 9-1 and 9-2), is available in the *Assessment and Evaluation Masters,* p. 240.

Enrichment Masters, p. 62

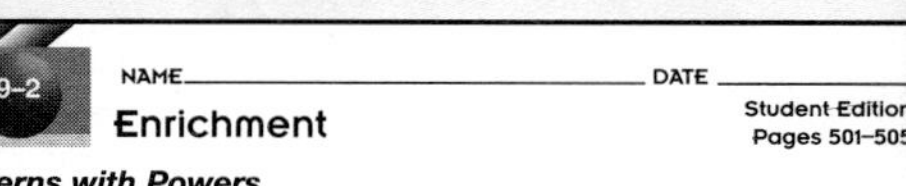

9–2 NAME ______ DATE ______

Enrichment

Student Edition Pages 501–505

Patterns with Powers

Use your calculator, if necessary, to complete each pattern.

a.		b.		c.	
2^{10} =	1024	5^{10} =	9,765,625	4^{10} =	1,048,576
2^9 =	512	5^9 =	1,953,125	4^9 =	262,144
2^8 =	256	5^8 =	390,625	4^8 =	65,536
2^7 =	128	5^7 =	78,125	4^7 =	16,384
2^6 =	64	5^6 =	15,625	4^6 =	4096
2^5 =	32	5^5 =	3125	4^5 =	1024
2^4 =	16	5^4 =	625	4^4 =	256
2^3 =	8	5^3 =	125	4^3 =	64
2^2 =	4	5^2 =	25	4^2 =	16
2^1 =	2	5^1 =	5	4^1 =	4

Study the patterns for a, b, and c above. Then answer the questions.

1. Describe the pattern of the exponents from the top of each column to the bottom. **The exponents decrease by one from each row to the one below.**
2. Describe the pattern of the powers from the top of the column to the bottom. **To get each power, divide the power on the row above by the base (2, 5, or 4).**
3. What would you expect the following powers to be?
   2^0 **1** 5^0 **1** 4^0 **1**
4. Write a rule. Test it on patterns that you obtain using −2, −5, and −4 as bases. **Any nonzero number to the zero power equals one.**

Study the pattern below. Then answer the questions.

$0^3 = 0$ $0^2 = 0$ $0^1 = 0$ $0^0 = \underline{\ ?\ }$ 0^{-1} does not exist. 0^{-2} does not exist. 0^{-3} does not exist.

5. Why do 0^{-1}, 0^{-2}, and 0^{-3} not exist? **Negative exponents are not defined unless the base is nonzero.**
6. Based upon the pattern, can you determine whether 0^0 exists? **No, since the pattern $0^n = 0$ breaks down for $n < 1$.**
7. The symbol 0^0 is called an **indeterminate,** which means that it has no unique value. Thus it does not exist as a unique real number. Why do you think that 0^0 cannot equal 1? **Answers will vary. One answer is that if $0^0 = 1$, then $1 = \frac{1}{1} = \frac{1^0}{0^0} = \left(\frac{1}{0}\right)^0$, which is a false result, since division by zero is not allowed. Thus, 0^0 cannot equal 1.**

9-3 LESSON NOTES

NCTM Standards: 1–5

Instructional Resources

- Study Guide Master 9-3
- Practice Master 9-3
- Enrichment Master 9-3
- Real-World Applications, 24
- Science and Math Lab Manual, pp. 39–42
- Tech Prep Applications Masters, p. 17

Transparency 9-3A contains the 5-Minute Check for this lesson; **Transparency 9-3B** contains a teaching aid for this lesson.

Recommended Pacing	
Standard Pacing	Day 4 of 15
Honors Pacing	Day 3 of 14
Block Scheduling*	Day 2 of 8
Alg. 1 in Two Years*	Days 7 & 8 of 26

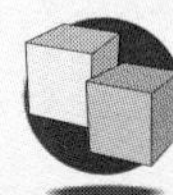

*For more information on pacing and possible lesson plans, refer to the *Block Scheduling Booklet* and *Algebra 1 in Two Years.*

1 FOCUS

5-Minute Check
(over Lesson 9-2)

Simplify.

1. $\frac{b^5}{b^7}$ b^{-2}
2. $\frac{18m^8n^3}{36m^8n}$ $\frac{n^2}{2}$
3. $\frac{x^{-5}y^{-6}}{(9x^2y^6)^{-2}}$ $\frac{81y^6}{x}$
4. $\frac{36x^4y^2}{12x^6y}$ $\frac{3y}{x^2}$
5. $\frac{10t^6r}{40t^3r^5}$ $\frac{t^3}{4r^4}$

Motivating the Lesson

Situational Problem Begin the lesson by having students examine large numbers, such as the distance between Earth and the sun or the population of China. Ask students to describe problems that could arise when working with such large numbers.

9-3 Scientific Notation

What YOU'LL LEARN

- To express numbers in scientific and standard notation, and
- to find products and quotients of numbers expressed in scientific notation.

Why IT'S IMPORTANT

You can use scientific notation to express the solutions to many problems.

APPLICATION
Transportation

The numbers of passengers arriving at and departing from some major U. S. airports for 1993 are listed in the table below.

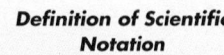

Airport	City	Number of Passengers Arriving and Departing (nearest million)
O'Hare International	Chicago	65,000,000
Dallas/Ft. Worth International	Dallas/Ft. Worth	50,000,000
Los Angeles International	Los Angeles	48,000,000
Hartsfield Atlanta International	Atlanta	48,000,000
San Francisco International	San Francisco	32,000,000
Miami International	Miami	29,000,000
J. F. Kennedy International	New York	27,000,000
Newark International	Newark	26,000,000
Detroit Metropolitan Wayne County	Detroit	24,000,000
Logan International	Boston	24,000,000

Source: Air Transport Association of America

When dealing with very large numbers, keeping track of place value can be difficult. For this reason, it is not always desirable to express numbers in standard notation as shown in the chart. Large numbers such as these may be expressed in **scientific notation.**

Definition of Scientific Notation	**A number is expressed in scientific notation when it is in the form $a \times 10^n$, where $1 \leq a < 10$ and n is an integer.**

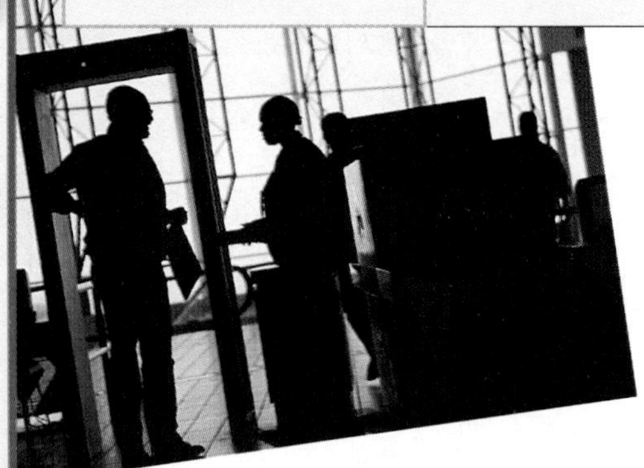

For example, the number of passengers arriving at and departing from Miami International was about 29,000,000. To write this number in scientific notation, express it as a product of a number greater than or equal to 1, but less than 10, and a power of 10.

$$29{,}000{,}000 = 2.9 \times 10{,}000{,}000$$
$$= 2.9 \times 10^7$$

Scientific notation is also used to express very small numbers. When numbers between zero and one are written in scientific notation, the exponent of 10 is negative.

Example 1 **Express each number in scientific notation.**

a. 98,700,000,000

$$98{,}700{,}000{,}000 = 9.87 \times 10{,}000{,}000{,}000 = 9.87 \times 10^{10}$$

b. 0.0000056

$$0.0000056 = 5.6 \times 0.000001 = 5.6 \times \frac{1}{1{,}000{,}000} = 5.6 \times \frac{1}{10^6} = 5.6 \times 10^{-6}$$

Example 2 **Express each number in standard notation.**

a. 3.45×10^5

$$3.45 \times 10^5 = 3.45 \times 100{,}000 = 345{,}000$$

b. 9.72×10^{-4}

$$9.72 \times 10^{-4} = 9.72 \times \frac{1}{10^4} = 9.72 \times \frac{1}{10{,}000} = 9.72 \times 0.0001 = 0.000972$$

CAREER CHOICES

A **pilot** is a highly trained professional who flies planes and helicopters for the transportation of people and cargo, law enforcement, or emergency rescues.

Pilots are required to have a license, which is earned through extensive flight training and written examinations. A college degree is recommended for pilot candidates.

For more information, contact:

Future Aviation Professionals of America
4291J Memorial Dr.
Atlanta, GA 30032

You can use scientific notation to simplify computation with very large numbers and/or very small numbers.

EXPLORATION GRAPHING CALCULATORS

You can use a graphing calculator to solve problems using scientific notation. First, put your calculator in scientific mode. To enter 3.5×10^9, enter 3.5 [X] 10 [^] 9.

Your Turn

a. Use your calculator to find $(3.5 \times 10^9)(2.36 \times 10^{-3})$. **$8.26 \times 10^6$**

b. Explain how the calculator calculated the product in part a. **See margin.**

c. Write the product for part a in standard notation. **8,260,000**

d. Use your calculator to find $(5.544 \times 10^3) \div (1.54 \times 10^7)$. **$3.6 \times 10^{-4}$**

e. Explain how the calculator calculated the quotient in part d. **See margin.**

f. Write the quotient for part d in standard notation. **0.00036**

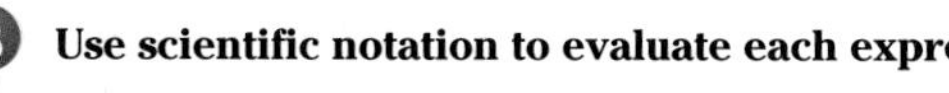

Example 3 **Use scientific notation to evaluate each expression.**

a. (610)(2,500,000,000) *Estimate: 2.5 billion × 600 = 1.5 trillion*

$$(610)(2{,}500{,}000{,}000) = (6.1 \times 10^2)(2.5 \times 10^9)$$
$$= (6.1 \times 2.5)(10^2 \times 10^9) \quad \textit{Associative property}$$
$$= 15.25 \times 10^{11}$$
$$= 1.525 \times 10^{12} \text{ or } 1{,}525{,}000{,}000{,}000$$

b. (0.000009)(3700) *Estimate: 0.00001 × 3700 = 0.037*

$$(0.000009)(3700) = (9 \times 10^{-6})(3.7 \times 10^3)$$
$$= (9 \times 3.7)(10^{-6} \times 10^3) \quad \textit{Associative property}$$
$$= 33.3 \times 10^{-3}$$
$$= 3.33 \times 10^{-2} \text{ or } 0.0333$$

2 TEACH

Teaching Tip Emphasize the difference between scientific notation and standard notation. Students should feel comfortable with the terminology.

Teaching Tip When expressing numbers in scientific notation, such as those in Example 1, help students derive a method to recognize when base 10 will have a negative or a positive exponent.

In-Class Examples

For Example 1
Express each number in scientific notation.

a. 635,000,000 **6.35×10^8**
b. 4,000,000,000,000 **4×10^{12}**
c. 0.00234 **2.34×10^{-3}**
d. 0.000000028 **2.8×10^{-8}**

For Example 2
Express each number in standard notation.

a. 2.36×10^2 **236**
b. 8.04×10^{-7} **0.000000804**

For Example 3
Use scientific notation to evaluate each expression.

a. (170) (475,000,000) **8.075×10^{10}**
b. (3,600,000,000) (0.0023) **8.28×10^6**
c. $\frac{4.473 \times 10^{12}}{4.26 \times 10^5}$ **1.05×10^7**

EXPLORATION

It might be helpful to remind students that since multiplication is commutative, the expression in part a can be written as follows.
$(3.5 \times 2.36) \times (10^9 \times 10^{-3})$

CAREER CHOICES

In 1992, the average salary for pilots was $80,000. Flight engineers averaged $42,000, copilots averaged $65,000, and captains averaged $107,000.

Answers for the Exploration

b. Multiply 3.5 and 3.36 and then 10^9 and 10^{-3}.

e. Divide 5.544 by 1.54 and then 10^3 by 10^7.

In-Class Example

For Example 4
What would be the Schwarzschild radius of a star with a mass of 6.1×10^{31} kilograms?
9.0822×10^4 meters

Scientists believe that when a large star runs out of nuclear fuel, it begins to break down under the force of its own gravity. The gravity eventually becomes so intense that even light cannot escape, making the star look black.

c. $\dfrac{2.0286 \times 10^8}{3.15 \times 10^3}$ *Estimate:* $\dfrac{210{,}000{,}000}{3000} = 70{,}000$

$$\frac{2.0286 \times 10^8}{3.15 \times 10^3} = \left(\frac{2.0286}{3.15}\right)\left(\frac{10^8}{10^3}\right)$$
$$= 0.644 \times 10^5$$
$$= 6.44 \times 10^4 \text{ or } 64{,}400$$

Scientific notation is extensively used by scientists in fields such as physics and astronomy.

Example 4

Astronomy

A black hole is a region in space where matter seems to disappear. A star becomes a black hole when the radius of the star reaches a certain critical value called the *Schwarzschild radius.* The value is given by the equation $R_s = \frac{2GM}{c^2}$, where R_s is the Schwarzschild radius in meters, G is the gravitational constant (6.7×10^{-11}), M is the mass in kilograms, and c is the speed of light (3×10^8 meters per second).

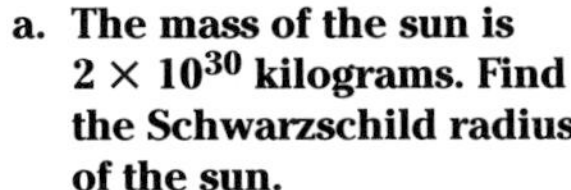

a. The mass of the sun is 2×10^{30} kilograms. Find the Schwarzschild radius of the sun.

b. The actual radius of the sun is 700,000 kilometers. Is it in danger of becoming a black hole in the near future?

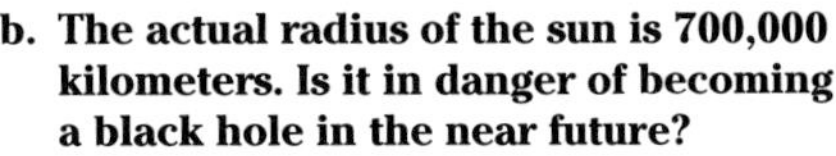

Stephen W. Hawking is a theoretical physicist from Great Britain. Dr. Hawking has made many contributions to the field of science including extensive work dealing with black holes.

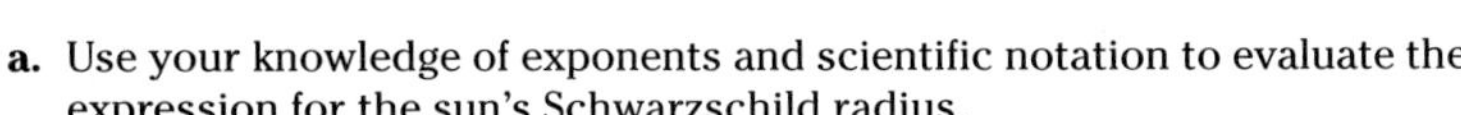

a. Use your knowledge of exponents and scientific notation to evaluate the expression for the sun's Schwarzschild radius.

$$R_s = \frac{2GM}{c^2}$$
$$= \frac{2(6.7 \times 10^{-11})(2 \times 10^{30})}{(3 \times 10^8)^2}$$
$$= \frac{2(6.7 \times 10^{-11})(2 \times 10^{30})}{3^2 \times 10^{16}} \quad \textit{Power of a monomial property}$$
$$= \left(\frac{2(6.7)(2)}{3^2}\right)\left(\frac{10^{-11}(10^{30})}{10^{16}}\right)$$
$$= \left(\frac{26.8}{9}\right)10^{-11+30-16} \quad \textit{Product and quotient of powers}$$
$$\approx 2.98 \times 10^3 \text{ or } 2980$$

The Schwarzschild radius for the sun is about 2980 meters.

b. Since the actual radius of the sun is 700,000 kilometers, it does not seem to be in danger of becoming a black hole in the near future.

Classroom Vignette

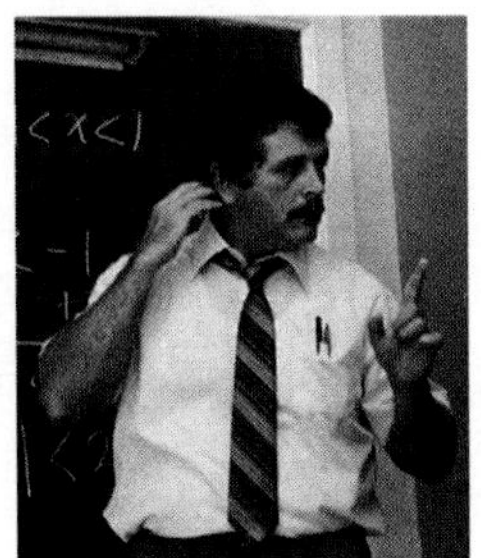

"My students use the TI-82 graphing calculator throughout the year. I have the students write their earth science problems in standard scientific notation and then in TI-82 notation and compare the answers."

Frank Arcuri
Croton Harmon High School
Croton-on-Hudson, New York

CHECK FOR UNDERSTANDING

Communicating Mathematics

Study the lesson. Then complete the following. **1–4. See margin.**

1. When do you use positive exponents in scientific notation?
2. When do you use negative exponents in scientific notation?
3. **Explain** how you can find the product of (1.2×10^5) and (4×10^8) without using pencil and paper or a calculator.
4. **Explain** how you can find the quotient of (4.4×10^4) and (4×10^7) without using pencil and paper or a calculator.

Guided Practice

Express each number in the second column in standard notation. Express each number in the third column in scientific notation.

	Planet	Maximum Distance from Sun (miles)	Radius (miles)
5.	Mercury	4.34×10^7	1515
6.	Earth	9.46×10^7	3963
7.	Jupiter	5.07×10^8	44,419
8.	Uranus	1.8597×10^9	15,881
9.	Pluto	4.5514×10^9	714

5–9. See margin.

Express each number in scientific notation.

10. **Chemistry** The wavelength of cadmium's green line is 0.0000509 centimeters.
11. **Physics** The mass of a proton is 0.000000000000000000001672 milligrams.
12. **Health** The length of the AIDS virus is 0.00011 millimeters. **1.1×10^{-4} mm**
13. **Biology** The diameter of an organism called the *Mycoplasma laidlawii* is 0.000004 inch. **4×10^{-6} in.**

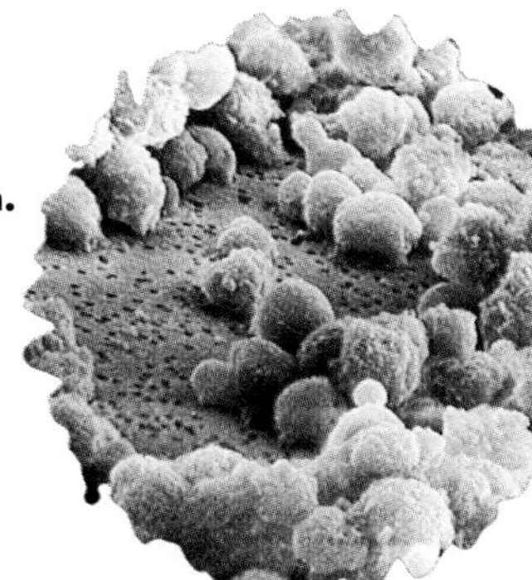
AIDS virus

Evaluate. Express each result in scientific and standard notation.

14. $(3.24 \times 10^3)(6.7 \times 10^4)$
15. $(0.2 \times 10^{-3})(31 \times 10^{-4})$
16. $\frac{8.1 \times 10^2}{2.7 \times 10^{-3}}$ **3×10^5; 300,000**
17. $\frac{52{,}440{,}000{,}000}{(2.3 \times 10^6)(38 \times 10^{-5})}$ **6×10^7; 60,000,000**

10. 5.09×10^{-5} cm
11. 1.672×10^{-21} mg

14. 2.1708×10^8; 217,080,000
15. 6.2×10^{-7}; 0.00000062

EXERCISES

Practice

Express each number in scientific notation.

18. 9500 **9.5×10^3**
19. 0.0095 **9.5×10^{-3}**
20. 56.9 **5.69×10**
21. 87,600,000,000 **8.76×10^{10}**
22. 0.000000000761 **7.61×10^{-10}**
23. 312,720,000 **3.1272×10^8**
24. 0.00000008 **8×10^{-8}**
25. 0.090909 **9.0909×10^{-2}**

B

26. 355×10^7 **3.55×10^9**
27. 78.6×10^3 **7.86×10^4**
28. 112×10^{-8} **1.12×10^{-6}**
29. 0.007×10^{-7} **7×10^{-10}**
30. 7830×10^{-2} **7.83×10**
31. 0.99×10^{-5} **9.9×10^{-6}**

Reteaching

Using Cooperative Learning Have students work in groups of three. Student A will write a number in standard notation and give it to Student B. Student B will convert it to scientific notation and give it to Student C. Student C will convert it to standard notation and give it back to Student A, who will verify that it is the original number that he or she wrote.

3 PRACTICE/APPLY

Check for Understanding

Exercises 1–17 are designed to help you assess your students' understanding through reading, writing, speaking, and modeling. You should work through Exercises 1–4 with your students and then monitor their work on Exercises 5–17.

Assignment Guide

Core: 19–55 odd, 56–67
Enriched: 18–48 even, 49–67

For **Extra Practice,** see p. 776.

The red A, B, and C flags, printed only in the Teacher's Wraparound Edition, indicate the level of difficulty of the exercises.

Additional Answers

1. when the number is greater than or equal to 10
2. when the number is less than 1
3. In your head, first multiply 1.2 and 4 and then multiply 10^5 and 10^8. The answer is 4.8×10^{13}.
4. In your head, first divide 4.4 by 4 and then divide 10^4 by 10^7. The answer is 1.1×10^{-3}.
5. 43,400,000; 1.515×10^3
6. 94,600,000; 3.963×10^3
7. 507,000,000; 4.4419×10^4
8. 1,859,700,000; 1.5881×10^4
9. 4,551,400,000; 7.14×10^2

Using the Programming Exercises The program given in Exercise 48 is for use with a TI-82 graphing calculator. For other programmable calculators, have students consult their owner's manual for commands similar to those presented here.

Additional Answer

49b. Multiply 3.7 and 5.6 and multiply 10^{112} and 10^{10}. Then write the product in scientific notation.

Answers: 32. 4.48×10^6; 4,480,000 33. 6×10^{-3}; 0.006 34. 2.088×10^{10}; 20,880,000,000 35. 8.992×10^{-7}; 0.0000008992 36. 4×10^7; 40,000,000 37. 4×10^{-2}; 0.04 38. 2.3×10^{-6}; 0.0000023 39. 6.5×10^{-6}; 0.0000065 40. 3×10^{-5}; 0.00003 41. 6.6×10^{-6}; 0.0000066

Evaluate. Express each result in scientific and standard notation.

32. $(6.4 \times 10^3)(7 \times 10^2)$ **33.** $(4 \times 10^2)(15 \times 10^{-6})$

34. $360(5.8 \times 10^7)$ **35.** $(5.62 \times 10^{-3})(16 \times 10^{-5})$

36. $\frac{6.4 \times 10^9}{1.6 \times 10^2}$ **37.** $\frac{9.2 \times 10^3}{2.3 \times 10^5}$ **38.** $\frac{1.035 \times 10^{-3}}{4.5 \times 10^2}$

39. $\frac{2.795 \times 10^{-7}}{4.3 \times 10^{-2}}$ **40.** $\frac{3.6 \times 10^2}{1.2 \times 10^7}$ **41.** $\frac{5.412 \times 10^{-2}}{8.2 \times 10^3}$

C ▶ **42.** $\frac{(35{,}921{,}000)(62 \times 10^3)}{3.1 \times 10^5}$ 7.1842×10^6; 7,184,200 **43.** $\frac{1.6464 \times 10^5}{(98{,}000)(14 \times 10^3)}$ 1.2×10^{-4}; 0.00012

Graphing Calculator

Use a graphing calculator to evaluate each expression. Express each answer in scientific notation. 44. 2.7504×10^9 45. 2.4336×10^{-1}

44. $(4.8 \times 10^6)(5.73 \times 10^2)$ **45.** $(5.07 \times 10^{-4})(4.8 \times 10^2)$

46. $(9.1 \times 10^6) \div (2.6 \times 10^{10})$ 3.5×10^{-4} **47.** $(9.66 \times 10^3) \div (3.45 \times 10^{-2})$ 2.8×10^5

Programming

48. The graphing calculator program at the right evaluates the expression $(2ab^2)^3$ for the values you input. It expresses the result in scientific notation. If $a = 9$ and $b = 10$, the result is 5.832E9.

```
PROGRAM:SCINOT
: Sci
: Prompt A, B, C
: Disp "(2AB^2)^3 =",
  (2AB^2)^3
```

Edit the program to evaluate each expression. Then evaluate for $a = 4$, $b = 6$, and $c = 8$.

a. $a^2b^3c^4$ 1.416E7
b. $(-2a)^2(4b)^3$ 8.84736E5
c. $(4a^2b^4)^3$ 5.706E14
d. $(ac)^3 + (3b)^2$ 3.3092E4

Critical Thinking

49a. Sample answer: overflow
49b. See margin.

49. Use a calculator to multiply 3.7×10^{112} and 5.6×10^{10}.

a. Describe what happens when you multiply these values.
b. Describe how you could find the product.
c. Write the product in scientific notation. 2.072×10^{123}

Applications and Problem Solving

50. about 6.57×10^6 times greater

50. Biology Seeds come in all sizes. The largest seed is the seed from the double coconut tree, which can have a mass as great as 23 kilograms. In contrast, the mass of the seed of an orchid is about 3.5×10^{-6} grams. Use a calculator to find how many times greater the mass of the seed of the double coconut tree is than the mass of the seed of the orchid. Express your answer in scientific notation.

Germination of wheat seedling

51. Movies In the movie *I.Q.*, Albert Einstein, played by Walter Matthau, tries to start a romance between his niece Catherine Boyd, played by Meg Ryan, and an auto mechanic named Ed Walters, played by Tim Robbins. When Ed asks Catherine to estimate the number of stars in the sky, she answers "$10^{12} + 1$." Write this number in standard notation. 1,000,000,000,001

Study Guide Masters, p. 63

9-3 NAME ____________ DATE ________

Study Guide — Student Edition Pages 506–512

Scientific Notation

Scientists use scientific notation to express very large or very small numbers. A number is expressed in scientific notation when it is expressed as the product of a number between one and ten, and a power of ten.

Example 1: $37{,}000 = 3.7 \times 10^4$ ***Example 2:*** $0.00391 = 3.91 \times 10^{-3}$

You can find products or quotients of numbers that are expressed in scientific notation by following the rules for multiplying and dividing monomials and powers.

Example 3: $\frac{8.4 \times 10^6}{4.2 \times 10^3} = \frac{8.4}{4.2} \times \frac{10^6}{10^3} = 2 \times 10^3$, or 2000

Example 4: $(6700)(0.00002) = (6.7 \times 10^3)(2 \times 10^{-5}) = (6.7 \times 2)(10^3 \times 10^{-5}) = 13.4 \times 10^{-2} = 1.34 \times 10^{-1}$, or 0.134

Express each number in scientific notation.

1. 0.0000456 4.56×10^{-4}
2. 0.01 10^{-2}
3. 590,000,000 5.9×10^8
4. 640×10^5 6.4×10^7

Evaluate. Express each result in scientific and standard notation.

5. $\frac{1.4 \times 10^4}{0.2 \times 10^2}$ 7×10^2; 700
6. $\frac{3.3 \times 10^{-12}}{1.1 \times 10^{-14}}$ 3×10^2; 300
7. $\frac{0.000042}{600}$ 7×10^{-8}; 0.00000007
8. $(2 \times 10^4)(4 \times 10^{-5})$ 8×10^{-1}; 0.8
9. $(77 \times 10^4)(0.02 \times 10^3)$ 1.54×10^7; 15,400,000
10. $(3.2 \times 10^{-2})(2.0 \times 10^2)$ 6.4×10^0; 6.4
11. $250{,}000 \div 0.0005$ 5×10^8; 500,000,000
12. $(0.00004)(20{,}000)$ 8×10^{-1}; 0.8
13. $(1.5 \times 10^5)(3.3 \times 10^{-4})$ 4.95×10^1; 49.5

54b. The smallest bacteria would probably not be seen with such a microscope.

52. **National Parks** Each year, millions of people visit our national parks and recreational areas. The five most popular locations are listed at the right. Write the number of visitors to each location in scientific notation. **See margin.**

Location	State	Visitors in 1994 (nearest thousand)
Golden Gate National Recreation Area	California	15,309,000
Great Smokey Mountains National Park	Tennessee	9,227,000
Lake Mead National Recreational Area	Nevada	9,022,000
Gulf Islands National Seashore	Florida	5,460,000
Cape Cod National Seashore	Massachusetts	5,153,000

Source: *Good Housekeeping*, Sept., 1994

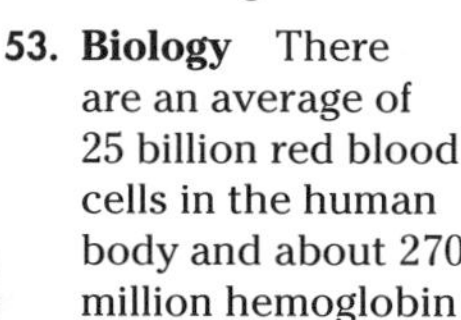

53. **Biology** There are an average of 25 billion red blood cells in the human body and about 270 million hemoglobin molecules in each red blood cell. Use a calculator to find the average number of hemoglobin molecules in the human body. Write your answer in scientific notation. **6.75×10^{18} molecules**

54. **Health** Laboratory technicians look at bacteria through microscopes. A microscope set on 1000× makes an organism appear to be 1000 times larger than its actual size. Most bacteria are between 3×10^{-4} and 2×10^{-3} millimeters in diameter.
    a. How large would bacteria appear under a microscope set on 1000×? **0.3 mm to 2 mm**
    b. Do you think a microscope set on 1000× would allow the technician to see all the bacteria? Explain your answer.

55. **Economics** Suppose you try to feed all of the people on Earth using the 4.325×10^{11} kilograms of food produced each year in the United States and Canada. You need to divide this food among the population of the world so that each person receives the same amount of food each day. The population of Earth is about 4.8×10^{9}. **55a. about 0.25 kg**
    a. How much food will each person have each day?
    b. Do you think the amount in part a is enough to live on? Justify your answer. **Answers may vary.**

Mixed Review

56. Simplify $\frac{24x^2y^7z^3}{-6x^2y^3z}$. (Lesson 9–2) $-4y^4z^2$

57. Graph the system of inequalities. (Lesson 8–5) **See margin.**
    $y \leq -x$
    $x \geq -3$

58. Use substitution to solve the system of equations. (Lesson 8–2) **(-52, -20)**
    $x = 2y - 12$
    $x - 3y = 8$

59. **Statistics** What percent of the data represented at the right is between 15 and 25? (Lesson 7–7) **25%**

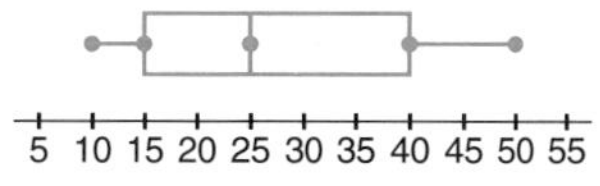

Additional Answers

52. 1.5309×10^7;
9.227×10^6;
9.022×10^6;
5.460×10^6;
5.153×10^6

57.

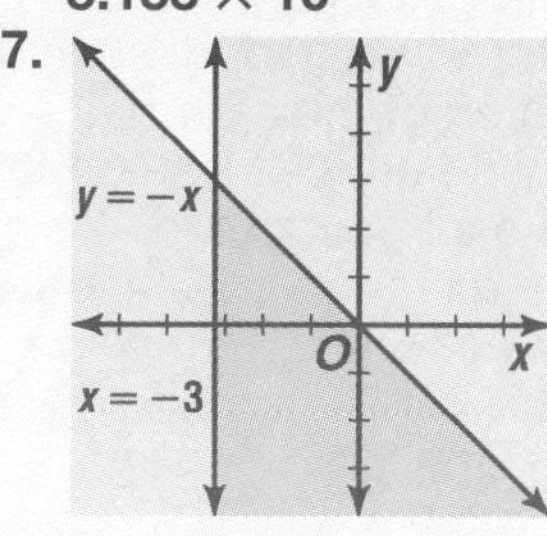

Tech Prep

Park Ranger Students who are interested in a career as a park ranger may wish to do further research on the statistics given in Exercise 52 and explore the potential growth of this career. For more information on tech prep, see the *Teacher's Handbook.*

Extension

Reasoning Evaluate. Express the result in scientific notation.

$$\frac{(3.266 \times 10^3)^{-2}(0.815 \times 10^{-4})^{-2}}{(1.96 \times 10^{-5})^{-2}}$$

5.422×10^{-9}

Practice Masters, p. 63

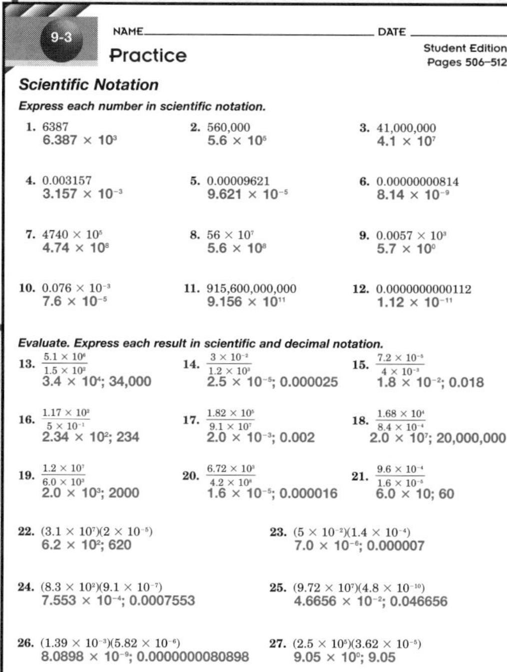

9-3 NAME________ DATE________

Practice

Student Edition Pages 506–512

Scientific Notation

Express each number in scientific notation.

1. 6387 — 6.387×10^3
2. 560,000 — 5.6×10^5
3. 41,000,000 — 4.1×10^7
4. 0.003157 — 3.157×10^{-3}
5. 0.00009621 — 9.621×10^{-5}
6. 0.00000000814 — 8.14×10^{-9}
7. 4740×10^5 — 4.74×10^8
8. 56×10^7 — 5.6×10^8
9. 0.0057×10^3 — 5.7×10^0
10. 0.076×10^{-3} — 7.6×10^{-5}
11. 915,600,000,000 — 9.156×10^{11}
12. 0.0000000000112 — 1.12×10^{-11}

Evaluate. Express each result in scientific and decimal notation.

13. $\frac{5.1 \times 10^6}{1.5 \times 10^2}$ — 3.4×10^4; 34,000
14. $\frac{3 \times 10^{-2}}{1.2 \times 10^3}$ — 2.5×10^{-5}; 0.000025
15. $\frac{7.2 \times 10^{-5}}{4 \times 10^{-3}}$ — 1.8×10^{-2}; 0.018
16. $\frac{1.17 \times 10^2}{5 \times 10^{-1}}$ — 2.34×10^2; 234
17. $\frac{1.82 \times 10^5}{9.1 \times 10^7}$ — 2.0×10^{-3}; 0.002
18. $\frac{1.68 \times 10^4}{8.4 \times 10^{-4}}$ — 2.0×10^7; 20,000,000
19. $\frac{1.2 \times 10^7}{6.0 \times 10^3}$ — 2.0×10^3; 2000
20. $\frac{6.72 \times 10^3}{4.2 \times 10^8}$ — 1.6×10^{-5}; 0.000016
21. $\frac{9.6 \times 10^{-4}}{1.6 \times 10^{-5}}$ — 6.0×10; 60
22. $(3.1 \times 10^7)(2 \times 10^{-5})$ — 6.2×10^2; 620
23. $(5 \times 10^{-2})(1.4 \times 10^{-4})$ — 7.0×10^{-6}; 0.000007
24. $(8.3 \times 10^2)(9.1 \times 10^{-7})$ — 7.553×10^{-4}; 0.0007553
25. $(9.72 \times 10^7)(4.8 \times 10^{-10})$ — 4.6656×10^{-2}; 0.046656
26. $(1.39 \times 10^{-3})(5.82 \times 10^{-6})$ — 8.0898×10^{-9}; 0.0000000080898
27. $(2.5 \times 10^5)(3.62 \times 10^{-5})$ — 9.05×10^0; 9.05
28. $(9.8 \times 10^6)(1.12 \times 10^2)$ — 1.0976×10^9; 1,097,600,000
29. $(7.8 \times 10^5)(3.45 \times 10^{-6})$ — 2.691×10^0; 2.691

4 ASSESS

Closing Activity

Modeling Have students identify three very large or very small numbers in the real world, such as the population of India or the mass of an element. Tell students to express each number in scientific and standard notation. Extend the activity by compiling a class list that ranks numbers from greatest to smallest.

Additional Answer

60. 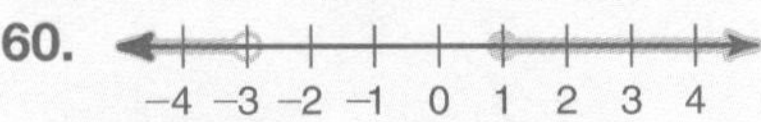

Enrichment Masters, p. 63

NAME ______________________ DATE ________

Enrichment — Student Edition Pages 506–512

Converting Metric Units

Scientific notation is convenient to use for unit conversions in the metric system.

Example 1: How many kilometers are there in 4,300,000 meters?

Divide the measure by the number of meters (1000) in one kilometer. Express both numbers in scientific notation.

$\frac{4.3 \times 10^6}{1 \times 10^3} = 4.3 \times 10^3$ The answer is 4.3×10^3 km.

Example 2: Convert 3700 grams into milligrams.

Multiply by the number of milligrams (1000) in 1 gram.

$(3.7 \times 10^3)(1 \times 10^3) = 3.7 \times 10^6$ There are 3.7×10^6 mg in 3700 g.

Complete the following. Express each answer in scientific notation.

1. 250,000 m = 2.5×10^2 km
2. 375 km = 3.75×10^5 m
3. 247 m = 2.47×10^4 cm
4. 5000 m = 5.0×10^6 mm
5. 0.0004 km = 4.0×10^{-1} m
6. 0.01 mm = 1.0×10^{-5} m
7. 6000 m = 6×10^{-6} mm
8. 340 cm = 3.4×10^{-3} km
9. 52,000 mg = 5.2×10^1 g
10. 420 kL = 4.2×10^5 L

Solve.

11. The planet Mars has a diameter of 6.76×10^3 km. What is the diameter of Mars in meters? Express the answer in both scientific and decimal notation.
6,760,000 m; 6.76×10^6 m
12. The distance of the earth from the sun is 149,590,000 km. Light travels 3.0×10^8 meters per second. How long does it take light from the sun to reach the earth in minutes?
8.31 min
13. A light-year is the distance that light travels in one year. (See Exercise 12.) How far is a light year in kilometers? Express your answer in scientific notation.
9.46×10^{12} km

60. Graph the solution set of $x < -3$ or $x \geq 1$. (Lesson 7–4) **See margin.**
61. **Geometry** Write an equation of the line that is perpendicular to the line $5x + 5y = 35$ and passes through the point at $(-3, 2)$. (Lesson 6–6) $y = x + 5$
62. **Aviation** An airplane passing over Sacramento at an elevation of 37,000 feet begins its descent to land at Reno, 140 miles away. If the elevation of Reno is 4500 feet, what should be the approximate slope of descent? (Lesson 6–1) **−232 ft/mi or about −0.044**

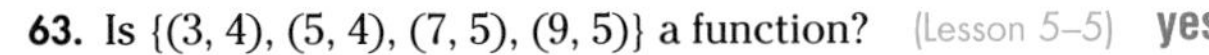

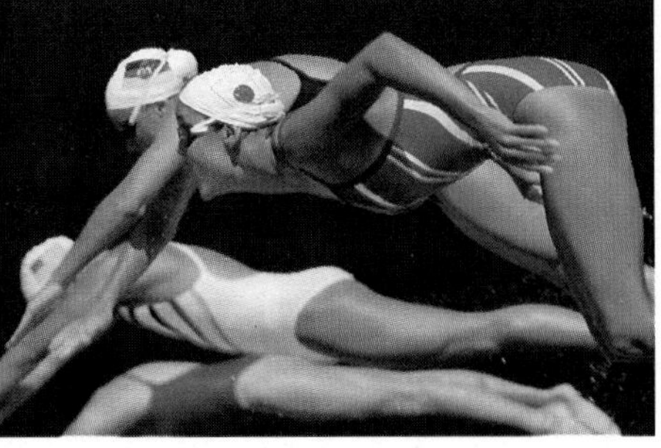

63. Is $\{(3, 4), (5, 4), (7, 5), (9, 5)\}$ a function? (Lesson 5–5) **yes**
64. State the domain of $\{(4, 4), (0, 1), (21, 5), (13, 0), (3, 9)\}$. (Lesson 5–2) **{4, 0, 21, 13, 3}**
65. **Geometry** Suppose $\sin K = 0.4563$. Find the measure of $\angle K$ to the nearest degree. (Lesson 4–3) **27º**
66. **Swimming** Rosalinda swims the 50-yard freestyle for the Wachung High School swim team. Her times in the last six meets were 26.89 seconds, 26.27 seconds, 25.18 seconds, 25.63 seconds, 27.16 seconds, and 27.18 seconds. Find the mean and median of her times. (Lesson 3–7) **26.385, 26.58**
67. Solve $x - 44 = -207$. (Lesson 3–1) **−163**

Mathematics and SOCIETY

Measurements Great and Small

The excerpt below appeared in an article in the *New York Times* on September 12, 1993.

ONCE UPON A TIME, WHEN THE WORLD was simpler, a foot was really as long as someone's foot, and a cubit the distance from someone's elbow to the end of the middle finger. Now we measure things that are much smaller than shoes and larger than arms, but there is still a desire to put them into a human context. Take the micron, for example, a unit so small—one millionth of a meter—that it seems hard to cast in a human context. But that doesn't stop newspapers from trying; they inevitably hitch it to something else: a human hair. Earlier this year, describing the one-micron width of a computer circuit, one newspaper article said a hair, by comparison, is 100 microns across. But another article, on cancer-causing soot particles smaller than 10 microns, said a human hair was 75 microns in diameter. Last year, in an article on fiber-optic beams, a human hair was 70 microns; in 1982, in one on coatings for cutting tools, the hair was down to 25 microns. . . . It can be hard to grope with the big, too. *Strategically Speaking*, a marketing newsletter, said recently that 1.8 billion slices of frozen pizza are sold each year—enough to cover 511,366 square miles. How big is that? The newsletter had the sense not to give the answer in microns: enough to cover New York, California, Texas, Maine, Delaware, and Rhode Island combined. ■

1–2. See margin.

1. Based on the data above, what is the average size of one slice of frozen pizza? (State the answer in square feet.) Does this size seem reasonable to you? Explain.
2. Did you have any difficulty performing the calculations for Exercise 1? How is scientific notation useful in calculations with very large numbers?
3. When you can, do you check the numbers you see presented in newspapers or magazines to see if they are reasonable? Why or why not?

3. See students' work.

Mathematics and SOCIETY

Provide students with the area of your state in square miles. Have them determine how many of a certain item of their choice, for example, algebra textbooks or pizza boxes, are necessary to cover the state. Have students present their findings in scientific notation.

Answers for Mathematics and Society

1. 7920 ft^2 per slice; this is clearly not a reasonable answer. One or more of the figures stated in the article must be incorrect.
2. See students' work; the problem can be made more manageable by using scientific notation to arrive at the solution.

A Preview of Lesson 9–4

9–4A Polynomials

Materials: algebra tiles

Algebra tiles can be used as a model for polynomials. A **polynomial** is a monomial or the sum of monomials. The diagram below shows the models.

Polynomial Models	
Polynomials are modeled using three types of tiles.	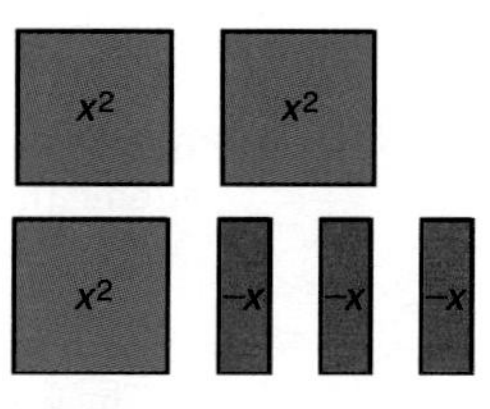
Each tile has an opposite.	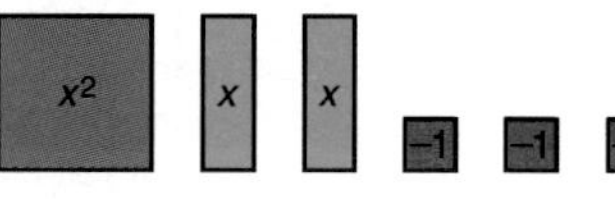

Activity **Use algebra tiles to model each polynomial.**

a. $2x^2$

To model this monomial, you will need 2 blue x^2-tiles.

b. $x^2 - 3x$

To model this polynomial, you will need 1 blue x^2-tile and 3 red x-tiles.

c. $x^2 + 2x - 3$

To model this polynomial, you will need 1 blue x^2-tile, 2 green x-tiles, and 3 red 1-tiles.

Model **Use algebra tiles to model each monomial or polynomial. Then draw a diagram of your model.** 1–4. See Solutions Manual.

1. $-3x^2$
2. $2x^2 - 3x + 5$
3. $2x^2 - 7$
4. $6x - 4$

Write **Write each model as an algebraic expression.**

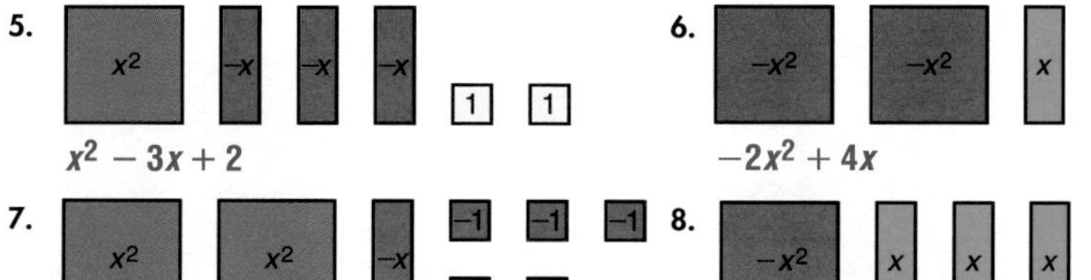

5. $x^2 - 3x + 2$
6. $-2x^2 + 4x$
7. $2x^2 - x - 5$
8. $-x^2 + 3x - 1$
9. Write a few sentences giving reasons why algebra tiles are sometimes called *area tiles.* x^2, x, and 1 represent the areas of the tiles.

9-4A LESSON NOTES

NCTM Standards: 1–5

Objective

Use algebra tiles to model polynomials.

Recommended Time

Demonstration and discussion: 15 minutes; Exercises: 30 minutes

Instructional Resources

For each student or group of students
Student Manipulative Kit
- algebra tiles

Modeling Mathematics Masters
- p. 1 (algebra tiles)
- p. 28 (worksheet)

For teacher demonstration
Algebra and Geometry Overhead Manipulative Resources

1 FOCUS

Motivating the Lesson

Ask students the following questions.

- Does an x-tile need to be any particular length? Does its length matter? **No; because it represents a variable, its length does not matter.**
- Do the sides of an x^2-tile need to be a particular length? **No**

2 TEACH

Teaching Tip Be sure that students lay down all the x^2-tiles first, then the x-tiles, and finally the constant tiles.

3 PRACTICE/APPLY

Assignment Guide
Core: 1–9 **Enriched:** 1–9

4 ASSESS

Observing students working in cooperative groups is an excellent method of assessment.

GLENCOE Technology

Interactive Mathematics Tools Software

This multimedia software provides an interactive lesson that uses a manipulation of the variables in the formula of compound interest on the return. A **Computer Journal** gives students an opportunity to write about what they have learned.

For Windows & Macintosh

9-4 LESSON NOTES

NCTM Standards: 1–5

Instructional Resources

- Study Guide Master 9-4
- Practice Master 9-4
- Enrichment Master 9-4
- Assessment and Evaluation Masters, pp. 239–240
- Multicultural Activity Masters, p. 18

Transparency 9-4A contains the 5-Minute Check for this lesson; **Transparency 9-4B** contains a teaching aid for this lesson.

Recommended Pacing	
Standard Pacing	Day 6 of 15
Honors Pacing	Day 5 of 14
Block Scheduling*	Day 3 of 8
Alg. 1 in Two Years*	Days 10 & 11 of 26

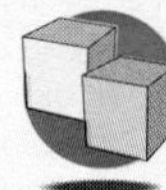

*For more information on pacing and possible lesson plans, refer to the *Block Scheduling Booklet* and *Algebra 1 in Two Years.*

1 FOCUS

5-Minute Check
(over Lesson 9-3)

Express in scientific notation.

1. 42,345 **4.2345×10^4**
2. 0.000567 **5.67×10^{-4}**

Evaluate. Express each result in scientific and standard notation.

3. $(1.3 \times 10^5)(3 \times 10^4)$ **3.9×10^9; 3,900,000,000**
4. $(8 \times 10^6)(2.6 \times 10^{-5})$ **2.08×10^2; 208**
5. $\dfrac{2.416 \times 10^{-4}}{3.02 \times 10^5}$ **8×10^{-10}; 0.0000000008**

9-4 Polynomials

What YOU'LL LEARN

- To find the degree of a polynomial, and
- to arrange the terms of a polynomial so that the powers of a variable are in ascending or descending order.

Why IT'S IMPORTANT

You can use polynomials to solve problems involving agriculture and biology.

Art

The picture at the right shows the painting *Composition with Red, Yellow, and Blue* by the Dutch painter Piet Mondrian. Mondrian preferred abstraction and simplification in his art. He liked to limit his palette to the primary colors and to use straight lines and right angles. The style of painting that Mondrian developed is called *neoplasticism,* and he is considered one of the most influential painters of the 20th century.

Top Five List

Selling Prices of Paintings

1. *Portrait du Dr. Gachet* by van Gogh, $75,000,000
2. *Au Moulin de la Galette* by Renoir, $71,000,000
3. *Les Noces de Pierrette* by Picasso, $51,700,000
4. *Irises* by van Gogh, $49,000,000
5. *Yo Picasso* by Picasso, $43,500,000

Consider a portion of the painting *Composition with Red, Yellow, and Blue.* The length of the sides and the area of each section are given.

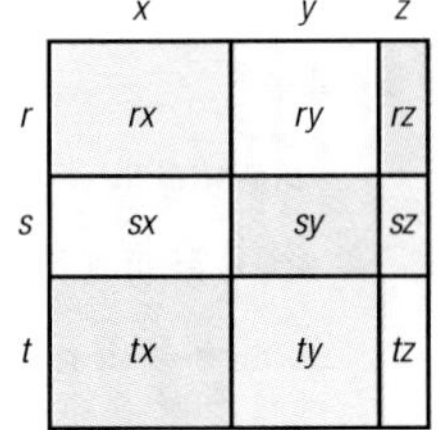

We can find the area of this portion of the painting by adding the areas of each section.

$$rx + ry + rz + sx + sy + sz + tx + ty + tz$$

The expression representing the area of this portion of the painting is called a **polynomial**. A polynomial is a monomial or a sum of monomials. Recall that a monomial is a number, a variable, or a product of numbers and variables. The exponents of the variables of a monomial must be positive. A **binomial** is the sum of two monomials. A **trinomial** is the sum of three monomials. Here are some examples of each. Polynomials with more than three terms have no special names.

Monomial	Binomial	Trinomial
$3y^2$	$4x - 7$	$a + 2b + 4c$
$2abc^2$	$2x + 9y$	$x^2 + 8x + 9$
-9	$3x^2 - 11xy$	$x^2 + 2xy + y^2$
$14m$	$2 + 13x$	$3a - 7b^2 - 4c$

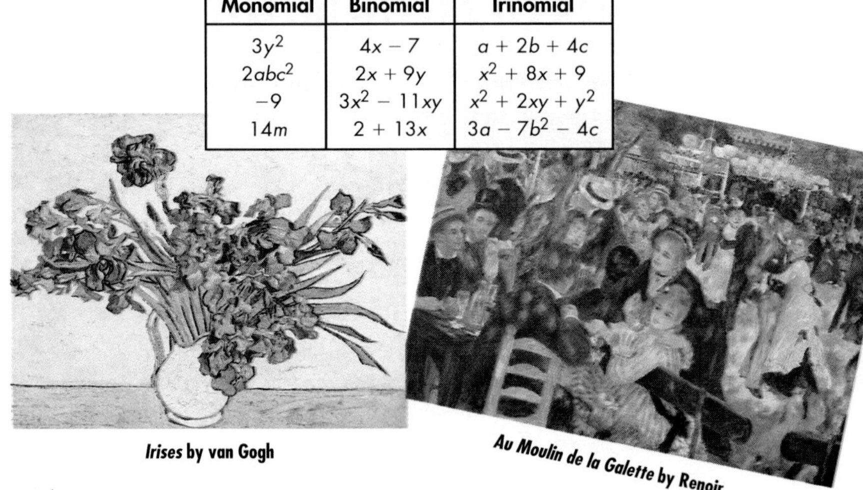

Irises by van Gogh

Au Moulin de la Galette by Renoir

During his lifetime (1853–1890), Vincent van Gogh was always desperately poor. Few exhibitions of his work were mounted while he was alive. Van Gogh's fame grew in the twentieth century.

Alternative Teaching Strategies

Reading Algebra Students often confuse the degree of a polynomial with the number of terms in the expression. Read the following polynomials to them and have them identify the degree and the number of terms.

$3x^3 + 2x^2 + 4$ **3, 3**
$2x^3 + 5x^2 - x + 6$ **3, 4**
$x^3 - 4x$ **3, 2**
$3x^2 + x - 2$ **2, 3**

Example 1 **State whether each expression is a polynomial. If it is a polynomial, identify it as either a *monomial, binomial,* or *trinomial.***

a. $3a - 7bc$

The expression $3a - 7bc$ can be written as $3a + (-7bc)$. Therefore, it is a polynomial. Since $3a - 7bc$ can be written as a sum of two monomials, $3a$ and $-7bc$, it is a binomial.

LOOK BACK

You can refer to Lesson 1-7 for information on simplifying expressions.

b. $3x^2 + 7a - 2 + a$

The expression $3x^2 + 7a - 2 + a$ can be written as $3x^2 + 8a + (-2)$. Therefore, it is a polynomial. Since it can be written as the sum of three monomials, $3x^2$, $8a$, and -2, it is a trinomial.

c. $\frac{7}{2r^2} + 6$

The expression $\frac{7}{2r^2} + 6$ is not a polynomial because $\frac{7}{2r^2}$ is not a monomial.

Polynomials can be used to represent savings accumulated over several years.

Example 2

APPLICATION

Finance

Each of the three summers before Li Chiang attends college, she plans to work as a lifeguard and save \$2000 towards her college expenses. She plans to invest the money in a savings account at her local bank. The value of each year's savings can be found by the expression px^t where p represents the amount invested, x represents the sum of 1 and the annual interest rate, and t represents the time in years.

a. Write a polynomial that represents the total amount of money Li will have when she starts college.

b. Find the amount of money Li will have if the bank is paying 6% annually.

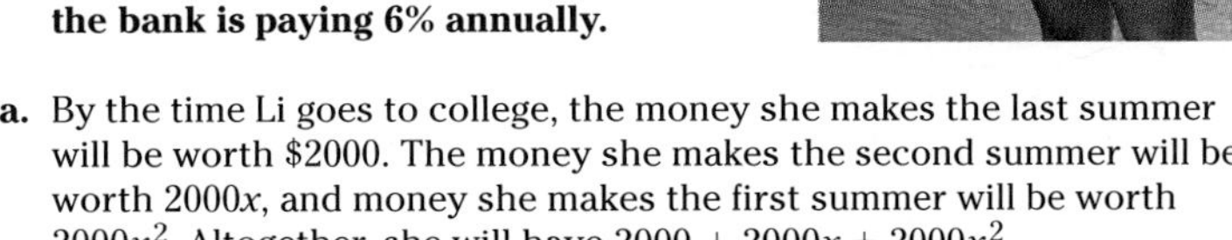

a. By the time Li goes to college, the money she makes the last summer will be worth \$2000. The money she makes the second summer will be worth $2000x$, and money she makes the first summer will be worth $2000x^2$. Altogether, she will have $2000 + 2000x + 2000x^2$.

b. Replace x with 1.06 and calculate the total amount of money Li will have.

$2000 + 2000(1.06) + 2000(1.06)^2 = 6367.20$ *$x = 1 + 6\%$ or 1.06*

Li will have \$6367.20 for college.

The **degree** of a monomial is the sum of the exponents of its variables.

Monomial	Degree	
$8y^3$	3	
$4y^2ab$	$2 + 1 + 1 = 4$	*Remember that $a = a^1$ and $b = b^1$.*
-14	0	*Remember that $x^0 = 1$ and $-14 = -14x^0$.*
$42abc$	$1 + 1 + 1 = 3$	

Motivating the Lesson

Questioning Introduce the lesson with a discussion of the prefixes *mono-*, *bi-*, *tri-*, and *poly-*. Ask students to give examples of common terms that contain these prefixes. Develop the idea that *mono-* means "one," *bi-* means "two," *tri-* means "three," and *poly-* means "more than one."

2 TEACH

In-Class Examples

For Example 1
State whether each expression is a polynomial. If it is a polynomial, identify it as either a monomial, binomial, or trinomial.

a. $7y^3 - 4y^2 + 2$ **trinomial**
b. $10x^3y^2z$ **monomial**
c. $\frac{8}{r^6} + 11r$ **not a polynomial**

For Example 2
If Li's savings account pays 12% annually, how much will Li save for her first year of college? **\$6748.80**

Teaching Tip Explain to students that the degree of a constant, such as -9, is 0 since -9 can be expressed as $-9x^0$. Emphasize that only a nonzero constant has a degree of 0. The number 0 and expressions such as $0x^2$ and $0a^2b^2$, which represent 0, have no degree.

In-Class Example

For Example 3
Find the degree of each polynomial.

a. $3x^2 + 5$ **2**
b. $y^7 + y^6 + 3x^4m^4$ **8**
c. $p^5 + p^3m^3 + 4m$ **6**
d. $3a^2 + 2ab + 4b^2$ **2**

Teaching Tip Explain to students that when placing terms in ascending or descending order (in x), the degree of y has no bearing on the placement of terms unless the degree of x is the same in more than one term. For $x^5y^3 + x^5y^2$, the terms are in descending powers of x. Even though the degree of x is the same in both terms, the first term is of degree 8, and the second term is of degree 7.

3 PRACTICE/APPLY

Check for Understanding

Exercises 1–13 are designed to help you assess your students' understanding through reading, writing, speaking, and modeling. You should work through Exercises 1–4 with your students and then monitor their work on Exercises 5–13.

Additional Answer

3. Sample answers:

poly- more than one
polygon, a plane closed figure formed by several line segments; *polygraph,* an instrument used to measure several different pulsations at one time

mono- one
monologue, a conversation by one person; *monotone,* a series of sounds in the same key or pitch

bi- two
bicycle, a vehicle with two wheels; *bifocal,* eyeglasses with two focal lengths

tri- three
triangle, a polygon with three sides; *tricycle,* vehicle with three wheels

To find the degree of a polynomial, you must find the degree of each term. The greatest degree of any term is the degree of the polynomial.

Polynomial	Terms	Degree of the Terms	Degree of the Polynomial
$3x^2 + 8a^2b - 4$	$3x^2, 8a^2b, -4$	2, 3, 0	3
$7x^4 - 9x^2y^7 + 4x$	$7x^4, -9x^2y^7, 4x$	4, 9, 1	9

Example 3 **Find the degree of each polynomial.**

a. $9xy + 2$

The degree of $9xy$ is 2.
The degree of 2 is 0.
Thus, the degree of $9xy + 2$ is 2.

b. $18x^2 + 21xy^2 + 13x - 2abc$

The degree of $18x^2$ is 2.
The degree of $21xy^2$ is 3.
The degree of $13x$ is 1.
The degree of $2abc$ is 3.
Thus, the degree of $18x^2 + 21xy^2 + 13x - 2abc$ is 3.

The terms of a polynomial are usually arranged so that the powers of one variable are in ascending or descending order. Later in this chapter, you will learn to add, subtract, and multiply polynomials. These operations are easier to perform if the polynomials are arranged in one of these orders.

Ascending Order	Descending Order
$4 + 5a - 6a^2 + 2a^3$	$2a^3 - 6a^2 + 5a + 4$
$-5 - 2x + 4x^5$	$4x^5 - 2x - 5$
(in x) $8xy - 3x^2y + x^5 - 2x^7y$	(in x) $-2x^7y + x^5 - 3x^2y + 8xy$
(in y) $2x^4 - 3x^3y + 2x^2y^2 - y^{12}$	(in y) $-y^{12} + 2x^2y^2 - 3x^3y + 2x^4$

CHECK FOR UNDERSTANDING

Communicating Mathematics

Study the lesson. Then complete the following.

1. **Explain** why the degree of an integer like -27 is 0. $-27 = -27x^0$
2. **Explain** why $m + \frac{34}{n}$ is not a binomial. $\frac{34}{n}$ **is not a monomial.**
3. In this lesson, you were introduced to the words *polynomial, monomial, binomial,* and *trinomial.* These words begin with the prefixes poly-, mono-, bi-, and tri- respectively. Find the meaning of each prefix. List two other words that begin with each prefix, and define each of these words. **See margin.**

4. The model below represents a polynomial.

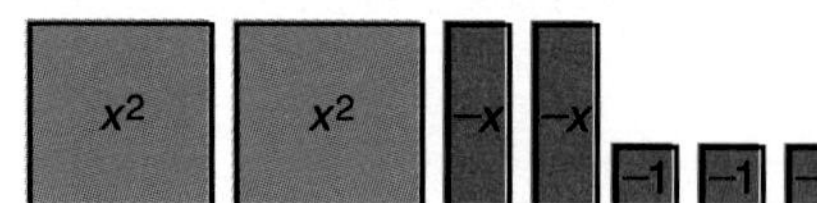

Write this polynomial in simplest form. $2x^2 - 2x - 3$

Reteaching

Using Questioning If the expression is a polynomial, give the degree. If not, tell why.

1. $-3x + 7$ **yes, 1**
2. $3x^{-2} + 4x + 1$ **no, not the sum of monomials**
3. $4x^2y - 3xy^2$ **yes, 3**
4. $6x + 9 - 5x^2$ **yes, 2**

Guided Practice

State whether each expression is a polynomial. If the expression is a polynomial, identify it as a *monomial*, a *binomial*, or a *trinomial*.

5. yes; trinomial
7. yes; binomial

5. $4x^3 - 11ab + 6$
6. $x^3 - \frac{7}{4}x + \frac{y}{x^2}$ no
7. $4c + ab - c$

Find the degree of each polynomial.

8. $11d$ 1
9. 10 0
10. $42x^{12}y^3 - 23x^8y^6$ 15

11. Arrange the terms of $-11x + 5x^3 - 12x^6 + x^8$ so that the powers of x are in descending order. $x^8 - 12x^6 + 5x^3 - 11x$

12. Arrange the terms of $y^4x + y^5x^3 - x^2 + yx^5$ so that the powers of x are in ascending order. $y^4x - x^2 + y^5x^3 + yx^5$

13. **Geometry** The area of a rectangle equals the length times the width. The area of a square equals the square of the length of a side. The area of a circle equals the number pi (π) times the square of the radius. The figure at the right consists of a rectangle, a circle, and a square.

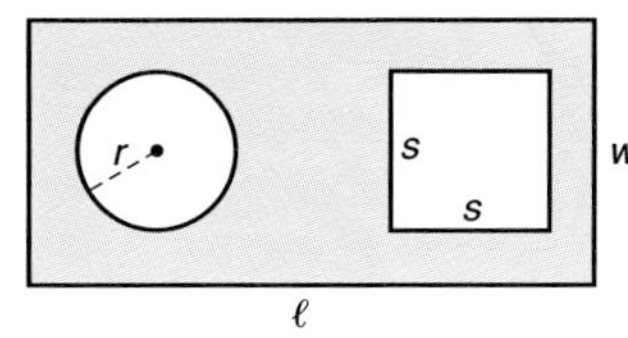

a. Write a polynomial expression that represents the area of the shaded region at the right. $\ell w - \pi r^2 - s^2$

b. Find the area of the shaded region if $w = 8$, $\ell = 15$, $r = 2$, and $s = 4$. about 91.43 square units

EXERCISES

Practice

State whether each expression is a polynomial. If the expression is a polynomial, identify it as a *monomial*, a *binomial*, or a *trinomial*.

15. yes; trinomial
17. yes; binomial
19. yes; trinomial

14. $\frac{r^5}{26}$ yes; monomial
15. $x^2 - \frac{1}{3}x + \frac{y}{234}$
16. $\frac{y^3}{12x}$ no
17. $5a - 6b - 3a$
18. $\frac{7}{t} + t^2$ no
19. $9ag^2 + 1.5g^2 - 0.7ag$

Find the degree of each polynomial.

20. $6a^2$ 2
21. $15t^3y^2$ 5
22. 24 0
23. $m^2 + n^3$ 3
24. $x^2y^3z - 4x^3z$ 6
25. $3x^2y^3z^4 - 18a^5f^3$ 9
26. $8r - 7y + 5d - 6h$ 1
27. $9 + t^2 - s^2t^2 + rs^2t$ 4
28. $-4yzw^4 + 10x^4z^2w$ 7

Arrange the terms of each polynomial so that the powers of x are in descending order.

30. $-2x^5 - 9x^2y + 8x + 5$

29. $5 + x^5 + 3x^3$ $x^5 + 3x^3 + 5$
30. $8x - 9x^2y + 5 - 2x^5$
31. $abx^2 - bcx + 34 - x^7$ $-x^7 + abx^2 - bcx + 34$
32. $7a^3x + 9ax^2 - 14x^7 + \frac{12}{19}x^{12}$ $\frac{12}{19}x^{12} - 14x^7 + 9ax^2 + 7a^3x$

Arrange the terms of each polynomial so that the powers of x are in ascending order. 33. $1 + x^2 + x^3 + x^5$ 34. $y^4 + 3xy^4 - x^2y^3 + 4x^3y$

35. $7a^3x + \frac{2}{3}x^2 - 8a^3x^3 + \frac{1}{5}x^5$
36. $4 + \frac{2}{3}x - x^2 + \frac{3}{4}x^3y$

33. $1 + x^3 + x^5 + x^2$
34. $4x^3y + 3xy^4 - x^2y^3 + y^4$
35. $7a^3x - 8a^3x^3 + \frac{1}{5}x^5 + \frac{2}{3}x^2$
36. $\frac{3}{4}x^3y - x^2 + 4 + \frac{2}{3}x$

Assignment Guide

Core: 15–43 odd, 45–52
Enriched: 14–40 even, 41–52

For **Extra Practice,** see p. 777.

The red A, B, and C flags, printed only in the Teacher's Wraparound Edition, indicate the level of difficulty of the exercises.

Study Guide Masters, p. 64

NAME__________ DATE ______

Study Guide

Student Edition
Pages 514–519

Polynomials

A **polynomial** is a monomial or a sum of monomials. A **binomial** is the sum of two monomials, and a **trinomial** is the sum of three monomials.

Examples of each type of polynomial are given in the following chart.

Monomial	Binomial	Trinomial
$5x^2$ $4abc$	$3x + 2$ $4x + 5y$	$5x^2 - 2x + 7$ $a^2 + 2ab + b^2$

The **degree** of a monomial is the sum of the exponents of its variables.

Monomial	Degree
$5x^2$ $4ab^3c^4$	2 $1 + 3 + 4 = 8$

To find the degree of a polynomial, first find the degree of each of its terms. The degree of the polynomial is the greatest of the degrees of its terms. The terms of a polynomial are usually arranged so that the powers of one variable are in either ascending or descending order.

Ascending Order: $3 + 5a - 8a^2 + a^3$
Descending Order: (in x) $x^5y^2 - x^4 + x^3y^2 + 5xy$

Find the degree of each polynomial.

1. $4x^2y^3z$ 6
2. $-2abc$ 3
3. $15m$ 1
4. $s + 5t$ 1
5. 22 0
6. $18x^2y + 4yz - 10y$ 3
7. $x^4 - 6x^2 - 2x^3 - 10$ 4
8. $2x^3y^2 - 4xy^3$ 5
9. $-2r^8s^4 + 7r^2s - 4r^7s^6$ 13

Arrange the terms of each polynomial so that the powers of x are in descending order.

10. $24x^2y - 12x^3y^2 + 6x^4$ $6x^4 - 12x^3y^2 + 24x^2y$
11. $20x - 10x^2 + 5x^3$ $5x^3 - 10x^2 + 20x$
12. $9bx + 3bx^2 - 6x^3$ $-6x^3 + 3bx^2 + 9bx$
13. $-15x^3 + 10x^4y^2 + 7xy^2$ $10x^4y^2 - 15x^3 + 7xy^2$
14. $ax^2 + 8a^2x^5 - 4$ $8a^2x^5 + ax^2 - 4$
15. $x^5 + x^2 - x^3$ $x^5 - x^3 + x^2$

Arrange the terms of each polynomial so that the powers of x are in ascending order.

16. $x^4 + x^3 + x^2$ $x^2 + x^3 + x^4$
17. $2x^3 - x + 3x^7$ $-x + 2x^3 + 3x^7$
18. $-5cx + 10c^2x^3 + 15cx^2$ $-5cx + 15cx^2 + 10c^2x^3$
19. $3 + 9x^4 + 9x^3$ $3 + 9x^3 + 9x^4$
20. $-4nx - 5n^3x^3 + 5$ $5 - 4nx - 5n^3x^3$
21. $4xy + 2y + 5x^2$ $2y + 4xy + 5x^2$

9-5A LESSON NOTES

NCTM Standards: 1–5

Objective

Use algebra tiles to add and subtract polynomials.

Recommended Time

Demonstration and discussion: 30 minutes; Exercises: 30 minutes

Instructional Resources

For each student or group of students
Student Manipulative Kit
- algebra tiles

Modeling Mathematics Masters
- p. 1 (algebra tiles)
- p. 29 (worksheet)

For teacher demonstration
Algebra and Geometry Overhead Manipulative Resources

1 FOCUS

Motivating the Lesson

Review the concept of additive inverses with students. Present several numbers and have students name the inverses. Explain that students will use opposite, or inverse, tiles in some of the activities.

2 TEACH

Teaching Tip In Activity 1, stress that adding polynomials is like adding integers.

Teaching Tip Watch for students who have difficulty modeling the opposite of a polynomial. Point out that if they combine the original model with its opposite, they get all zero pairs.

Teaching Tip In Activity 3, another form of subtraction of polynomials is presented. This form is closer to the actual algebraic manipulation used when simplifying a difference.

9–5A Adding and Subtracting Polynomials

A Preview of Lesson 9–5

Materials: algebra tiles

Monomials such as $4x^2$ and $-7x^2$ are called *like terms* because they have the same variable to the same power. When you use algebra tiles, you can recognize like terms because the individual tiles have the same size and shape.

Polynomial Models

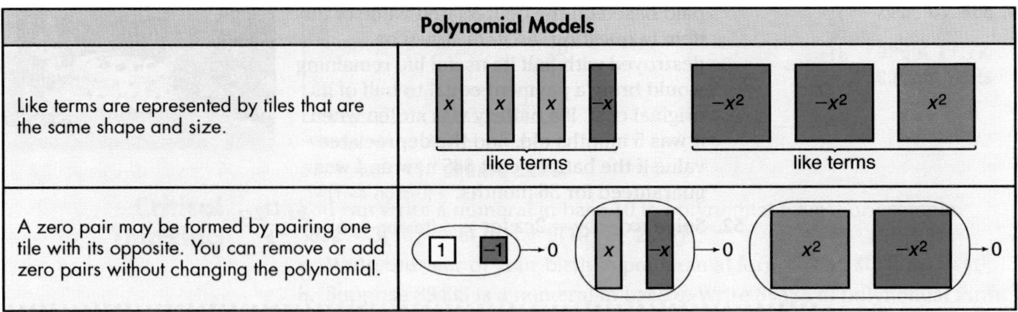

Like terms are represented by tiles that are the same shape and size.
x
x
x
−x
like terms
−x²
−x²
x²
like terms
A zero pair may be formed by pairing one tile with its opposite. You can remove or add zero pairs without changing the polynomial.
1
−1
0
x
−x
0
x²
−x²
0

Activity 1 **Use algebra tiles to find $(2x^2 + 3x + 2) + (x^2 - 5x - 5)$.**

Step 1 Model each polynomial. You may want to arrange like terms in columns for convenience.

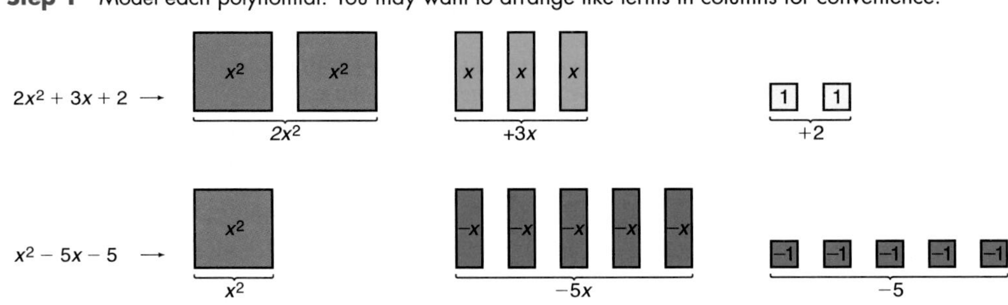

Step 2 Combine like terms and remove all zero pairs.

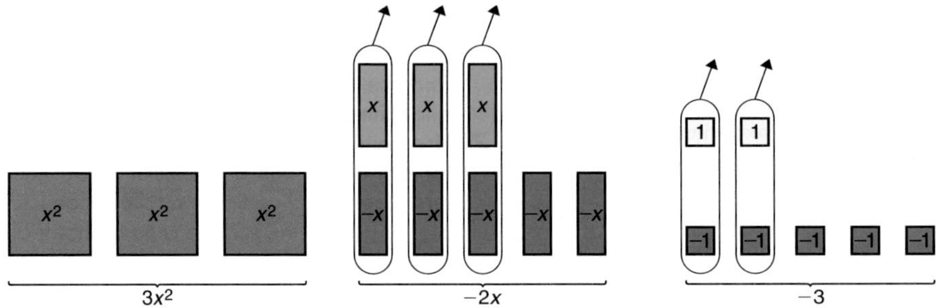

Step 3 Write the polynomial for the tiles that remain.
$(2x^2 + 3x + 2) + (x^2 - 5x - 5) = 3x^2 - 2x - 3$

Activity 2 **Use algebra tiles to find (2x + 5) − (−3x + 2).**

Step 1 Model the polynomial $2x + 5$.

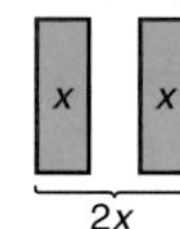

2x +5

Step 2 To subtract $-3x + 2$, you must remove 3 red x-tiles and 2 yellow 1-tiles. You can remove the yellow 1-tiles, but there are no red x-tiles. Add 3 zero pairs of x-tiles. Then remove the 3 red x-tiles.

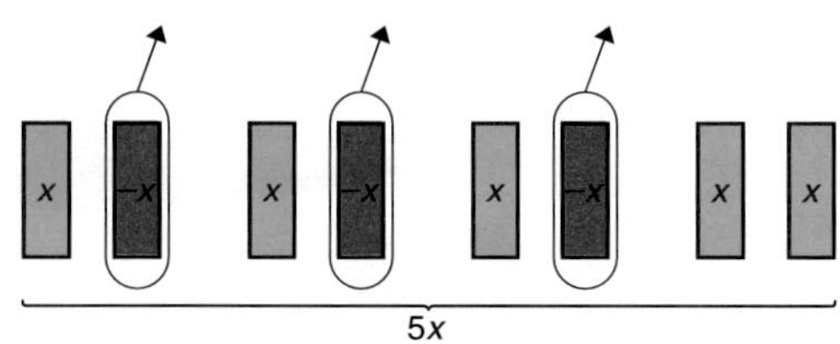

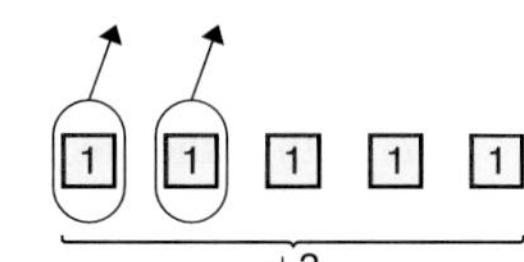

5x +3

Remember that the value of a zero pair is 0.

Step 3 Write the polynomial for the tiles that remain.
$(2x + 5) - (-3x + 2) = 5x + 3$

Recall that you can subtract a number by adding its additive inverse or opposite. Similarly, you can subtract a polynomial by adding its opposite.

Activity 3 **Use algebra tiles and the additive inverse, or opposite, to find (2x + 5) − (−3x + 2).**

Step 1 To find the difference of $2x + 5$ and $-3x + 2$, add $2x + 5$ and the opposite of $-3x + 2$.

$2x + 5 \rightarrow$

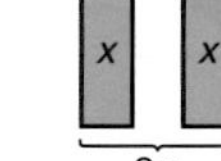

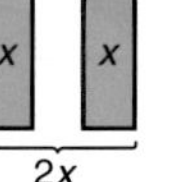

2x

The opposite of $-3x + 2$ is $3x - 2$. $\rightarrow$

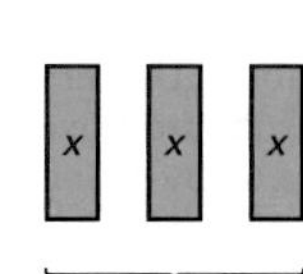

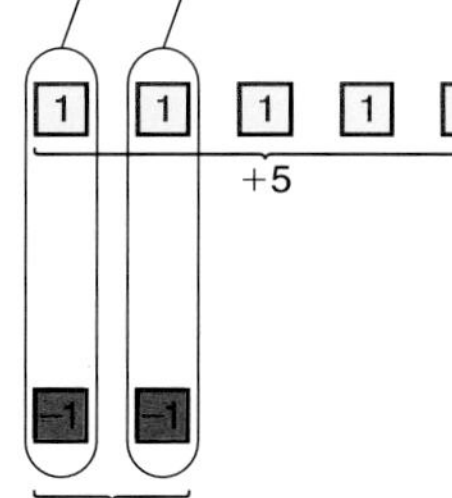

+5 3x −2

Step 2 Write the polynomial for the tiles that remain.
$(2x + 5) - (-3x + 2) = 5x + 3$ *This is the same answer as in Activity 2.*

Model **Use algebra tiles to find each sum or difference.** **1.** $-x^2 + 6$ **2.** $-2x^2 - 4x - 2$

1. $(2x^2 - 7x + 6) + (-3x^2 + 7x)$
2. $(-2x^2 + 3x) + (-7x - 2)$
3. $(x^2 - 4x) - (3x^2 + 2x)$
4. $(3x^2 - 5x - 2) - (x^2 - x + 1)$
5. $(x^2 + 2x) + (2x^2 - 3x + 4)$
6. $(2x^2 + 3x - 4) - (3x^2 - 4x + 1)$

3. $-2x^2 - 6x$ **4.** $2x^2 - 4x - 3$ **5.** $3x^2 - x + 4$ **6.** $-x^2 + 7x - 5$

Draw **Is each statement *true* or *false*? Justify your answer with a drawing.**

7–8. See margin for drawings.

7. $(3x^2 + 2x - 4) + (-x^2 + 2x - 3) = 2x^2 + 4x - 7$ **true**
8. $(x^2 - 2x) - (-3x^2 + 4x - 3) = -2x^2 - 6x - 3$ **false**

Write

9. Find $(x^2 - 2x + 4) - (4x + 3)$ using each method from Activity 2 and Activity 3. Illustrate with drawings and explain in writing how zero pairs are used in each case. **See margin.**

GLENCOE Technology

Interactive Mathematics Tools Software

This multimedia software provides an interactive lesson that uses algebra tiles to add and subtract like terms in polynomials. A **Computer Journal** gives students an opportunity to write about what they have learned.

For Windows & Macintosh

3 PRACTICE/APPLY

Assignment Guide

Core: 1–9
Enriched: 1–9

Additional Answers

7.

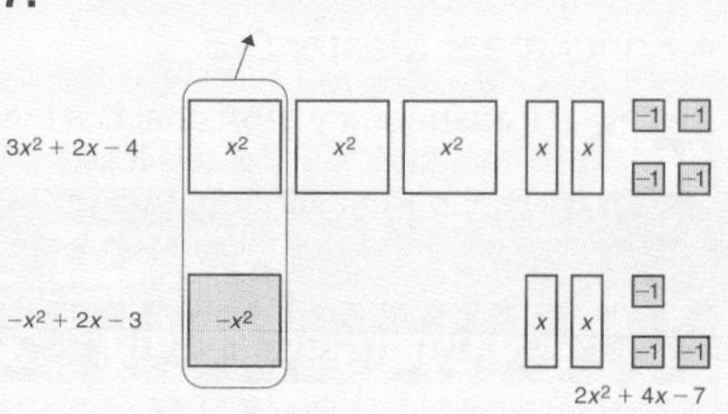

8.

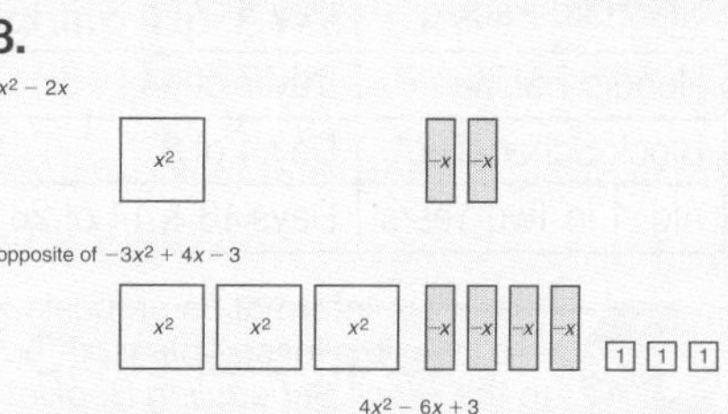

9. Method from Activity 2:

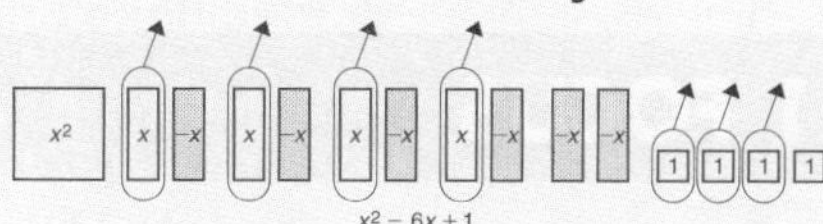

You need to add zero pairs so that you can remove 4 green x-tiles.

Method from Activity 3:

$x^2 - 2x + 4$

opposite of $4x + 3$

You remove all zero pairs to find the difference in simplest form.

4 ASSESS

Observing students working in cooperative groups is an excellent method of assessment.

9-6A LESSON NOTES

NCTM Standards: 1–5

Objective
Use algebra tiles to model the product of simple polynomials.

Recommended Time
Demonstration and discussion: 15 minutes; Exercises: 30 minutes

Instructional Resources
For each student or group of students
Student Manipulative Kit
- algebra tiles

Modeling Mathematics Masters
- p. 1 (algebra tiles)
- p. 15 (product mat)
- p. 30 (worksheet)

For teacher demonstration
Algebra and Geometry Overhead Manipulative Resources

1 FOCUS

Motivating the Lesson
Encourage students to develop a method for using algebra tiles to model the multiplication of a polynomial by a monomial.

2 TEACH

Teaching Tip Students should recognize that this is an application of the distributive property. Emphasize that the top and side of the rectangle define the algebra tile to be used. That is, x by x requires an $x \times x$ or x^2-tile.

Teaching Tip Students may apply their integer knowledge to determine the appropriate tile to use. In Activity 1, $(-1)(1x) = -x$, which indicates that a red x-tile should be used.

3 PRACTICE/APPLY

Assignment Guide

Core: 1–9
Enriched: 1–9

A Preview of Lesson 9–6

9–6A Multiplying a Polynomial by a Monomial

Materials: algebra tiles, product mat

You have used rectangles to model multiplication. In this activity, you will use algebra tiles to model the product of simple polynomials. The width and length of a rectangle will represent a monomial and a polynomial, respectively. The area of the rectangle will represent the product of the monomial and the polynomial.

Activity 1 **Use algebra tiles to find $x(x - 4)$.**

The rectangle will have a width of x units and a length of $(x - 4)$ units. Use your algebra tiles to mark off the dimensions on a product mat. Then make the rectangle with algebra tiles.

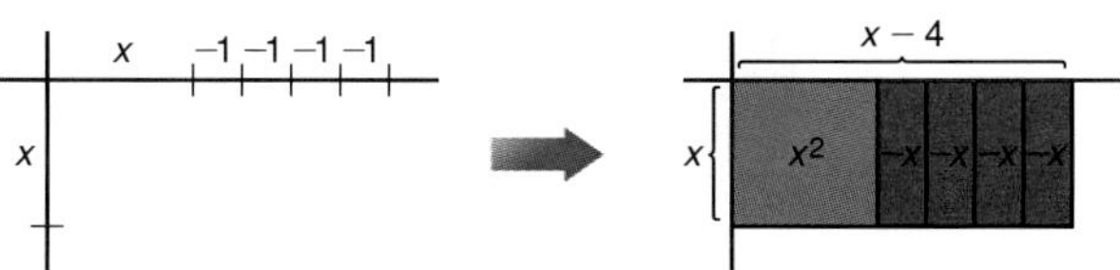

The rectangle consists of 1 blue x^2-tile and 4 red x-tiles.
The area of the rectangle is $x^2 - 4x$. Therefore, $x(x - 4) = x^2 - 4x$.

Activity 2 **Use algebra tiles to find $2x(x + 2)$.**

The rectangle will have a width of $2x$ units and a length of $(x + 2)$ units. Make the rectangle with algebra tiles.

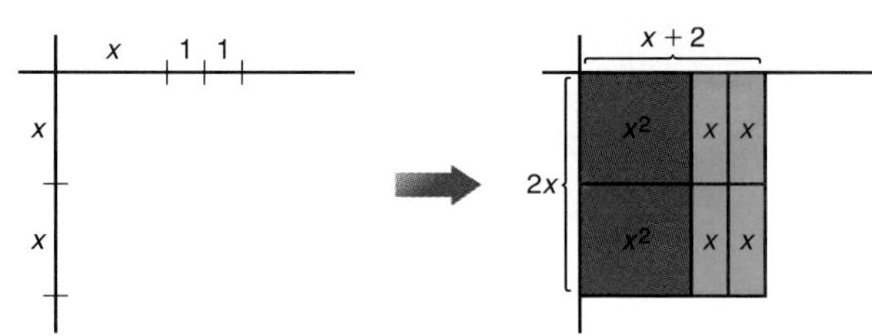

The rectangle consists of 2 blue x^2-tiles and 4 green x-tiles.
The area of the rectangle is $2x^2 + 4x$. Therefore, $2x(x + 2) = 2x^2 + 4x$.

Model **Use algebra tiles to find each product.**

1. $x(x + 2)$ **$x^2 + 2x$**
2. $x(x - 3)$ **$x^2 - 3x$**
3. $2x(x + 1)$ **$2x^2 + 2x$**
4. $2x(x - 3)$ **$2x^2 - 6x$**
5. $x(2x + 1)$ **$2x^2 + x$**
6. $3x(2x - 1)$ **$6x^2 - 3x$**

7–8. See margin for drawings.

Draw **Is each statement *true* or *false*? Justify your answer with a drawing.**

7. $x(2x + 4) = 2x^2 + 4x$ **true**
8. $2x(3x - 4) = 6x^2 - 8$ **false**

Write

9. Suppose you have a square storage building that measures x feet on a side. You triple the length of the building and increase the width by 15 feet.
   a. What will be the dimensions of the new building? **$3x$ and $x + 15$**
   b. What is the area of the new building? Write your solution in paragraph form, complete with drawings. **$(3x^2 + 45x)$ square feet; See Solutions Manual for drawing.**

4 ASSESS

Observing students working in cooperative groups is an excellent method of assessment. You may wish to ask a student from each group to explain how to model and solve a problem. Also watch for and acknowledge students who are helping others understand the concept being taught.

Additional Answers

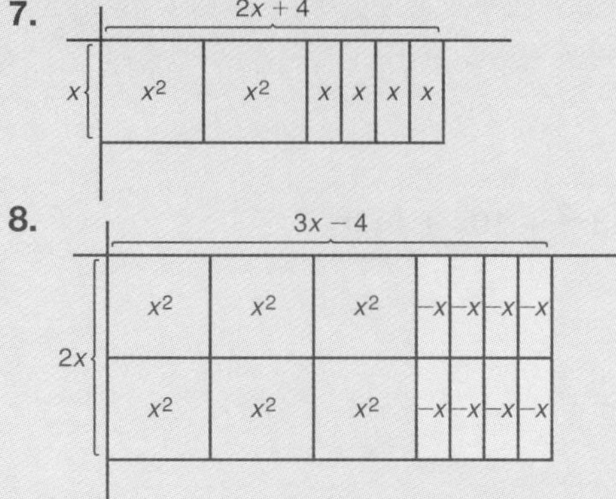

9–6 Multiplying a Polynomial by a Monomial

What YOU'LL LEARN

- To multiply a polynomial by a monomial, and
- to simplify expressions involving polynomials.

Why IT'S IMPORTANT

You can use monomials and polynomials to solve problems involving travel and recreation.

Recreation

Have you ever played the game of hopscotch? Children from all over the world like to play some form of this game. The following diagrams show versions from different countries.

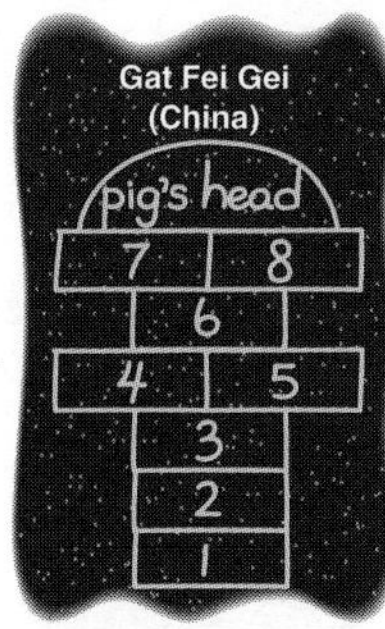

FYI

Tourists can still see the hopscotch pattern carved into the floor of the ancient Forum in Rome.

On the Caribbean island of Trinidad, the children play Jumby. The pattern for this game is shown at the right. Suppose the dimensions of each rectangle are x and $x + 14$. To find the area of each rectangle, you must multiply its length by its width.

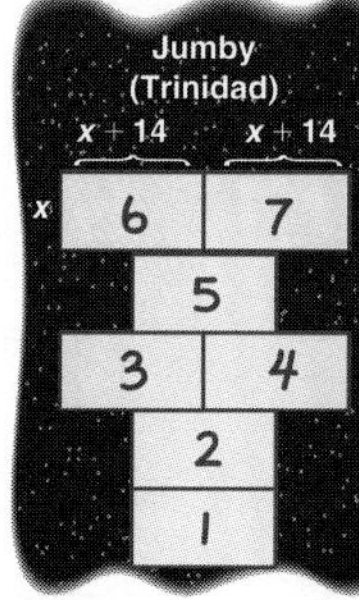

This diagram of one of the rectangles shows that the area is $x(x + 14)$.

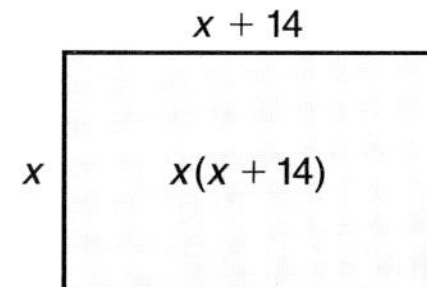

This diagram of the same rectangle shows that the area is $x^2 + 14x$.

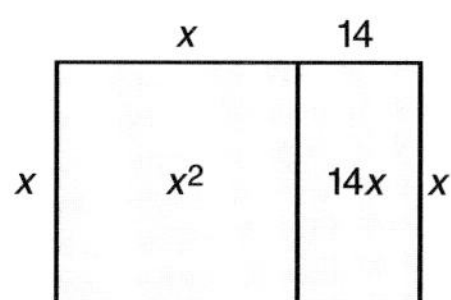

Since the areas are equal, $x(x + 14) = x^2 + 14x$.

The application above shows how the distributive property can be used to multiply a polynomial by a monomial.

The “scotch” refers to the line drawn on the ground. The name therefore refers to the fact that the player must “hop” over the “scotch.”

9-6 LESSON NOTES

NCTM Standards: 1–5

Instructional Resources

- Study Guide Master 9-6
- Practice Master 9-6
- Enrichment Master 9-6
- Assessment and Evaluation Masters, p. 241
- Modeling Mathematics Masters, p. 80
- Tech Prep Applications Masters, p. 18

Transparency 9-6A contains the 5-Minute Check for this lesson; **Transparency 9-6B** contains a teaching aid for this lesson.

Recommended Pacing	
Standard Pacing	Day 10 of 15
Honors Pacing	Day 9 of 14
Block Scheduling*	Day 5 of 8
Alg. 1 in Two Years*	Days 16 & 17 of 26

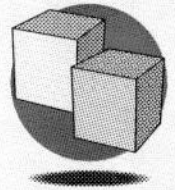

*For more information on pacing and possible lesson plans, refer to the *Block Scheduling Booklet* and *Algebra 1 in Two Years.*

1 FOCUS

5-Minute Check
(over Lesson 9-5)

Find each sum or difference.

1. $(6a + 7b) + (11a + 4b)$ $17a + 11b$
2. $(5x^2 - 7x + 3) + (16x + 4x^2 - 9)$ $9x^2 + 9x - 6$
3. $(3a^2 + 7ab - 9b^2) + (4a^2 + ab - 11b^2)$ $7a^2 + 8ab - 20b^2$
4. $(8d^2 - 3d - 7) - (d^2 - 13)$ $7d^2 - 3d + 6$
5. $(6x^2 + 2x - 9) - (3x^2 - 8x + 11)$ $3x^2 + 10x - 20$

Motivating the Lesson

Hands-On Activity Have groups of students draw hopscotch patterns on the pavement outside or on large sheets of paper. Have the groups measure all the dimensions and compute the total area for their hopscotch patterns.

2 TEACH

In-Class Examples

For Example 1
Find each product.

a. $5n(7n - 2)$ $\mathbf{35n^2 - 10n}$

b. $\frac{3}{4}a(8a^2 + 12a + 24)$
$\mathbf{6a^3 + 9a^2 + 18a}$

For Example 2
Find each product.

a. $6rs(r^2s - 3r + 2r^2s)$
$\mathbf{18r^3s^2 - 18r^2s}$

b. $-4s^2t^2(5s^2t + 3s + 4t^2 - 6s^2t - st^2)$
$\mathbf{4s^4t^3 + 4s^3t^4 - 12s^3t^2 - 16s^2t^4}$

For Example 3
If each lane is about 4 feet wide, how far apart should officials start the runners in the two inside lanes?
about 12.6 feet apart

Teaching Tip In Example 2, make sure that students are careful when distributing the negative term.

Example 1 **Find each product.**

a. $\mathbf{7b(4b^2 - 18)}$

You can multiply horizontally or vertically.

Method 1: Horizontal

$$7b(4b^2 - 18) = 7b(4b^2) - 7b(18)$$
$$= 28b^3 - 126b$$

Method 2: Vertical

$$\begin{array}{r} 4b^2 - 18 \\ (\times) \quad 7b \\ \hline 28b^3 - 126b \end{array}$$

b. $\mathbf{-3y^2(6y^2 - 8y + 12)}$

$$-3y^2(6y^2 - 8y + 12) = -3y^2(6y^2) - (-3y^2)(8y) + (-3y^2)(12)$$
$$= -18y^4 + 24y^3 - 36y^2$$

Some expressions may contain like terms. In these cases, you will need to simplify by combining like terms.

Example 2 **Find** $\mathbf{-3pq(p^2q + 2p - 3p^2q)}$.

Method 1

Multiply first and then simplify by combining like terms.

$$-3pq(p^2q + 2p - 3p^2q) = -3pq(p^2q) + (-3pq)(2p) - (-3pq)(3p^2q)$$
$$= -3p^3q^2 - 6p^2q + 9p^3q^2$$
$$= 6p^3q^2 - 6p^2q$$

Method 2

Simplify by combining like terms and then multiply.

$$-3pq(p^2q + 2p - 3p^2q) = -3pq(-2p^2q + 2p)$$
$$= -3pq(-2p^2q) + (-3pq)(2p)$$
$$= 6p^3q^2 - 6p^2q$$

p^2q and $-3p^2q$ are like terms.

Example 3

Track

The runners in a 200-meter dash race around the curved part of a track. If the runners start and finish at the same line, the runner on the outside lane would run farther than the other runners. To compensate for this situation, the starting points of the runners are staggered. If the radius of the inside lane is x and each lane is 2.5 feet wide, how far apart should the officials start the runners in the two inside lanes?

The formula for the circumference C of a circle is $C = 2\pi r$, where r is the radius of the circle. The distance around half of a circle is πr. Use this information to find the distance around the curve for the two inside lanes and subtract the quantities to find the stagger distance.

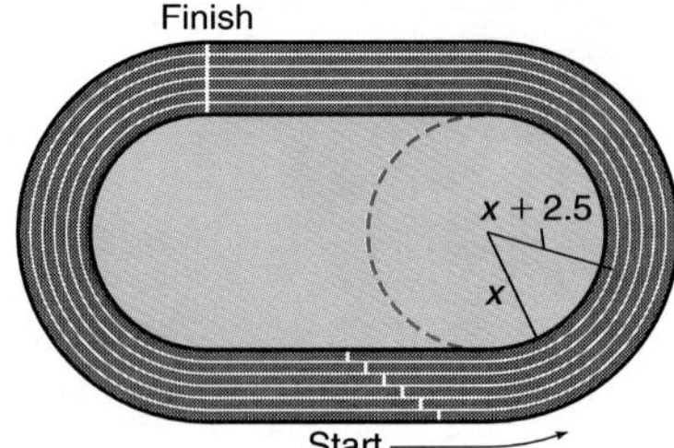

Alternative Learning Styles

Kinesthetic Have students design a hopscotch pattern with a total area of 100 square feet. The pattern may contain only squares and rectangles. Once the pattern has been designed, students may draw it on the ground and measure it to verify the total area.

outside semicircle, inside semicircle

$\pi(x + 2.5) - \pi x = \pi x + 2.5\pi - \pi x$ *Distributive property*

$= 2.5\pi$ *Combine like terms.*

The two runners should start 2.5π, or about 7.9, feet apart.

Many equations contain polynomials that must be added, subtracted, or multiplied before the equation can be solved. The distributive property is frequently used as at least one of the steps in solving equations.

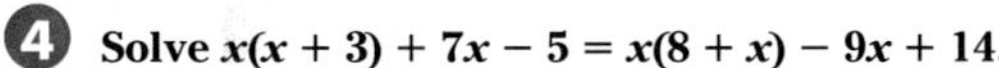

Example 4 **Solve $x(x + 3) + 7x - 5 = x(8 + x) - 9x + 14$.**

$x(x + 3) + 7x - 5 = x(8 + x) - 9x + 14$

$x^2 + 3x + 7x - 5 = 8x + x^2 - 9x + 14$ *Distributive property*

$x^2 + 10x - 5 = x^2 - x + 14$ *Combine like terms.*

$10x - 5 = -x + 14$ *Subtract x^2 from each side.*

$11x - 5 = 14$ *Add x to each side.*

$11x = 19$ *Add 5 to each side.*

$x = \frac{19}{11}$ *Divide each side by 11.*

The solution is $\frac{19}{11}$.

CHECK FOR UNDERSTANDING

Communicating Mathematics

Study the lesson. Then complete the following. 1. distributive property

1. **Name** the property used to simplify $3a(5a^2 + 2b - 3c^2)$.
2. Refer to the application at the beginning of the lesson.
   a. **Describe** how you could find the area of the entire pattern used to play Jumby. **Multiply 7 times $x^2 + 14x$.** **2b. $7x^2 + 98x$**
   b. **Write** an expression in simplest form for the area of this pattern.
   c. If x represents 8 inches, find the area of the pattern. **1232 in²**
3. Refer to Example 4. **3a. See margin.**
   a. **Explain** how you could check the solution to the equation.
   b. **Check** the solution. **See Solutions Manual.**

4. A rectangular garden is $2x + 3$ units long and $3x$ units wide.
   a. Draw a model of the garden.
   b. Find the area of the garden. **$6x^2 + 9x$**

4a. See margin.

Guided Practice

Find each product. **6. $-32a^5c + 4a^2c - 44a^2$**

5. $-7b(9b^3c + 1)$ **$-63b^4c - 7b$**
6. $4a^2(-8a^3c + c - 11)$
7. $5y - 13$ $(\times)\ 2y$ **$10y^2 - 26y$**
8. $2ab - 5a$ $(\times)\ 11ab$ **$22a^2b^2 - 55a^2b$**

Reteaching

Using Diagrams Have students examine the diagram below and answer the questions that follow.

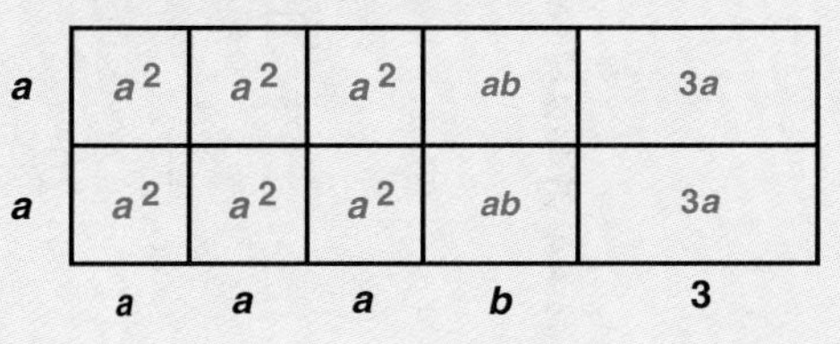

1. What is the width of the large rectangle? **2a**
2. What is the length? **3a + b + 3**
3. Find the area. **$2a(3a + b + 3) = 6a^2 + 2ab + 6a$**
4. Inside each square, write its area. **See diagram.**
5. Add the areas in Exercise 4. Compare with your answer in Exercise 3. **same**

In-Class Example

For Example 4

Solve $t(t - 5) + 2t - 1 = 7t + 3 + t(8 + t)$. $-\frac{2}{9}$

3 PRACTICE/APPLY

Check for Understanding

Exercises 1–13 are designed to help you assess your students' understanding through reading, writing, speaking, and modeling. You should work through Exercises 1–4 with your students and then monitor their work on Exercises 5–13.

Additional Answers

3a. Replace x with $\frac{19}{11}$ and check to see if both sides of the equation represent the same number.

4a.

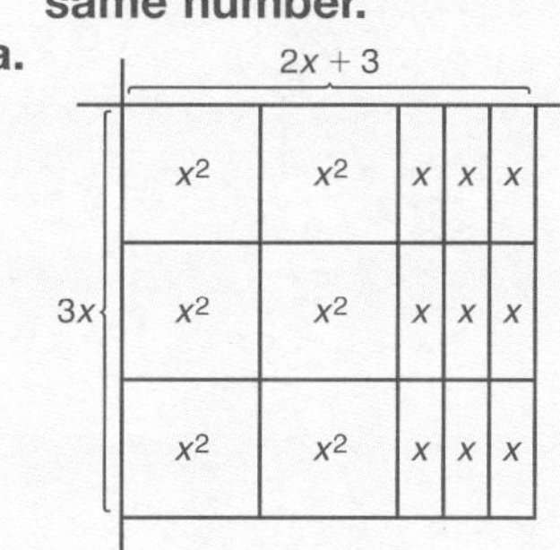

Study Guide Masters, p. 66

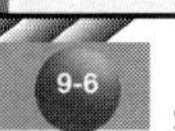

NAME____________ DATE ________

Study Guide Student Edition Pages 529–533

Multiplying a Polynomial by a Monomial

The example below shows how the distributive property can be used to multiply a polynomial by a monomial.

Example 1: $6a(a^2 + 5) = 6a(2a^2) + 6a(5)$
$= 12a^3 + 30a$

Many equations contain polynomials that must be added, subtracted, or multiplied before the equation can be solved.

Example 2:
Solve $-5 - (10 - 3p) = 12$.

$-5 - (10 - 3p) = 12$
$-5 - 10 + 3p = 12$
$-15 + 3p = 12$
$-15 + 15 + 3p = 12 + 15$
$3p = 27$
$\frac{3p}{3} = \frac{27}{3}$
$p = 9$

Example 3:
Solve $4(n - 2) + 5n = 6(3 - n) + 18$.

$4(n - 2) + 5n = 6(3 - n) + 18$
$4n - 8 + 5n = 18 - 6n + 18$
$9n - 8 = 36 - 6n$
$9n + 6n - 8 + 8 = 36 + 8 - 6n + 6n$
$15n = 44$
$\frac{15n}{15} = \frac{44}{15}$
$n = 2\frac{14}{15}$

Find each product.

1. $5(2a + 3b)$ $10a + 15b$
2. $3x(4x - 2y)$ $12x^2 - 6xy$
3. $-4(a^2 - b^2)$ $-4a^2 + 4b^2$
4. $8b\left(b^2 - \frac{1}{2}b\right)$ $8b^4 - 4b^2$
5. $-5t^2\left(\frac{1}{25}t^2 - \frac{1}{25}t + \frac{1}{5}\right)$ $-\frac{1}{5}t^4 + \frac{1}{5}t^3 - t^2$
6. $\left(\frac{1}{2}x + y\right)\left(\frac{1}{2}y\right)$ $\frac{1}{4}xy + \frac{1}{2}y^2$

Simplify.

7. $4r(2r^2 - 3r + 5) + 6r(4r^2 + 2r + 8)$ $32r^3 + 68r$
8. $2b(b^2 + 4b + 8) - 3b(3b^2 + 9b - 18)$ $-7b^3 - 19b^2 + 70b$

Solve each equation.

9. $3(x + 5) - 6 = 18$ 3
10. $3x(x - 5) - 3x^2 = -30$ 2
11. $(4y - 3) - 8y + 6 = 19$ −4
12. Ria's corn field has a perimeter of 18,000 meters. The length of one side is 3 meters less than twice the width. What are the dimensions of the field? 5999 m × 3001 m
13. Three sides of a triangle have measures that are consecutive even integers. What are the lengths of the sides if the perimeter is 114 meters? 36 m, 38 m, 40 m

Assignment Guide

Core: 15–47 odd, 48–55
Enriched: 14–42 even, 43–55

For **Extra Practice,** see p. 777.

The red A, B, and C flags, printed only in the Teacher's Wraparound Edition, indicate the level of difficulty of the exercises.

Additional Answers

43. Sample answer:
$1(8a^2b + 18ab)$
$a(8ab + 18b)$
$b(8a^2 + 18a)$
$2(4a^2b + 9ab)$
$(2a)(4ab + 9b)$
$(2b)(4a^2 + 9a)$
$(ab)(8a + 18)$
$(2ab)(4a + 9)$

47a. See page 533.

52.

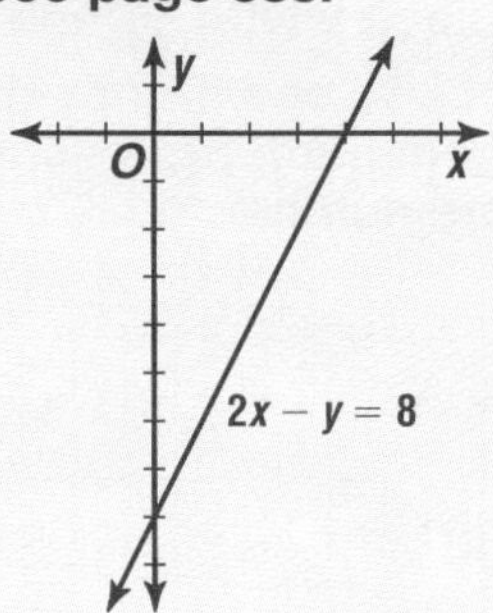

Practice Masters, p. 66

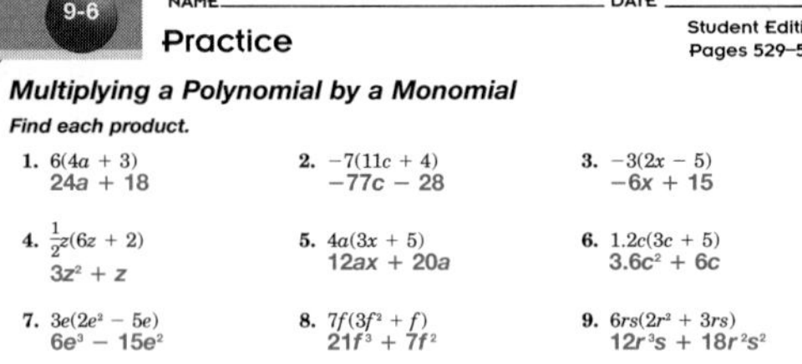

9-6 NAME ____________ DATE ________
Practice
Student Edition Pages 529–533

Multiplying a Polynomial by a Monomial

Find each product.

1. $6(4a + 3)$ $24a + 18$
2. $-7(11c + 4)$ $-77c - 28$
3. $-3(2x - 5)$ $-6x + 15$
4. $\frac{1}{2}z(6z + 2)$ $3z^2 + z$
5. $4a(3x + 5)$ $12ax + 20a$
6. $1.2c(3c + 5)$ $3.6c^2 + 6c$
7. $3e(2e^2 - 5e)$ $6e^3 - 15e^2$
8. $7f(3f^2 + f)$ $21f^3 + 7f^2$
9. $6rs(2r^2 + 3rs)$ $12r^3s + 18r^2s^2$
10. $2c^2d(4cd - d^2)$ $8c^3d^2 - 2c^2d^3$
11. $5m(11m^2 - 4n)$ $55m^3 - 20mn$
12. $-\frac{1}{3}y(9y + 3yz)$ $-3y^2 - y^2z$
13. $-4ay(ay - ab)$ $-4a^2y^2 + 4a^2by$
14. $2.1p(3p - q)$ $6.3p^2 - 2.1pq$
15. $-3mn^2(2m - 3n)$ $-6m^2n^2 + 9mn^3$
16. $-9w^2(7 - w)$ $-63w^2 + 9w^3$
17. $\frac{3}{4}k^2(8k - 12k^2)$ $6k^3 - 9k^4$
18. $-4.5cd(d^2 - 2c^2)$ $-4.5cd^3 + 9c^3d$
19. $3x(5x^2 - x + 4)$ $15x^3 - 3x^2 + 12x$
20. $7c(5 - 2c^2 + c^3)$ $35c - 14c^3 + 7c^4$
21. $2m(7m^3 + 5m^2 - m)$ $14m^4 + 10m^3 - 2m^2$
22. $-3u(8u^2 - 3u + 1)$ $-24u^3 + 9u^2 - 3u$
23. $4r^2s(2r^2 - 5rs + s^2)$ $8r^4s - 20r^3s^2 + 4r^2s^3$
24. $6i^2(5i^2x + ix - 3x^2)$ $30i^4x + 6i^3x - 18i^2x^2$
25. $-10pq(3pq + 4p - 5q^2)$ $-30p^2q^2 - 40p^2q + 50pq^3$
26. $-\frac{1}{4}h(8h^2 + h - 7)$ $-2h^3 - \frac{1}{4}h^2 + \frac{7}{4}h$

Simplify.

27. $5w(w^2 - 7w + 3) + 2w(2 + 19w - 2w^2)$ $w^3 + 3w^2 + 19w$
28. $6t(2t - 3) - 5(2t^2 + 9t - 3)$ $2t^2 - 63t + 15$
29. $8b(b + 2b^3c + b^4) + 4b^2(1 + 6b^2c + b^3)$ $12b^2 + 40b^4c + 12b^5$
30. $4x^2(x - 8x^2 - 3x^3) - 7x^3(8xy - 1)$ $11x^3 - 32x^4 - 56x^4y - 12x^5$

Simplify.

9. $w(3w - 5) + 3w$ $3w^2 - 2w$
10. $4y(2y^3 - 8y^2 + 2y + 9) - 3(y^2 + 8y)$ $8y^4 - 32y^3 + 5y^2 + 12y$

Solve each equation.

11. $12(b + 14) - 20b = 11b + 65$ $\frac{103}{19}$
12. $x(x - 4) + 2x = x(x + 12) - 7$ $\frac{1}{2}$
13. **Number Theory** Suppose a is an even integer. **13a.** $a^2 + a$
    a. Write the product, in simplest form, of a and the next integer after it.
    b. Write the product, in simplest form, of a and the next even integer after it. $a^2 + 2a$

EXERCISES

Practice

A

Find each product.

14. $-7(2x + 9)$ $-14x - 63$
15. $\frac{1}{3}x(x - 27)$ $\frac{1}{3}x^2 - 9x$
16. $3st(5s^2 + 2st)$ $15s^3t + 6s^2t^2$
17. $-4m^3(5m^2 + 2m)$ $-20m^5 - 8m^4$
18. $3d(4d^2 - 8d - 15)$
19. $5m^3(6m^2 - 8mn + 12n^3)$
20. $7x^2y(5x^2 - 3xy + y)$
21. $-4d(7d^2 - 4d + 3)$
22. $2m^2(5m^2 - 7m + 8)$
23. $-8rs(4rs + 7r - 14s^2)$
24. $-\frac{3}{4}ab^2\left(\frac{1}{3}abc + \frac{4}{9}a - 6\right)$
25. $\frac{4}{5}x^2\left(9xy + \frac{5}{4}x - 30y\right)$

18. $12d^3 - 24d^2 - 45d$
19. $30m^5 - 40m^4n + 60m^3n^3$
20. $35x^4y - 21x^3y^2 + 7x^2y^2$
21. $-28d^3 + 16d^2 - 12d$
22. $10m^4 - 14m^3 + 16m^2$
23. $-32r^2s^2 - 56r^2s + 112rs^3$
24. $-\frac{1}{4}a^2b^3c -$

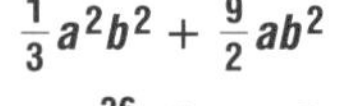

$\frac{1}{3}a^2b^2 + \frac{9}{2}ab^2$
25. $\frac{36}{5}x^3y + x^3 - 24x^2y$

B

Simplify.

26. $b(4b - 1) + 10b$ $4b^2 + 9b$
27. $3t(2t - 4) + 6(5t^2 + 2t - 7)$ $36t^2 - 42$
28. $8m(-9m^2 + 2m - 6) + 11(2m^3 - 4m + 12)$ $-50m^3 + 16m^2 - 92m + 132$
29. $8y(11y^2 - 2y + 13) - 9(3y^3 - 7y + 2)$ $61y^3 - 16y^2 + 167y - 18$
30. $\frac{3}{4}t(8t^3 + 12t - 4) + \frac{3}{2}(8t^2 - 9t)$ $6t^4 + 21t^2 - \frac{33}{2}t$
31. $6a^2(3a - 4) + 5a(7a^2 - 6a + 5) - 3(a^2 + 6a)$ $53a^3 - 57a^2 + 7a$

Solve each equation.

32. $2(5w - 12) = 6(-2w + 3) + 2$ 2
33. $7(x - 12) = 13 + 5(3x - 4)$ $-\frac{77}{8}$
34. $\frac{1}{2}(2d - 34) = \frac{2}{3}(6d - 27)$ $\frac{1}{3}$
35. $p(p + 2) + 3p = p(p - 3)$ 0
36. $y(y + 12) - 8y = 14 + y(y - 4)$
37. $x(x - 3) - x(x + 4) = 17x - 23$
38. $a(a + 8) - a(a + 3) - 23 = 3a + 11$ 17
39. $t(t - 12) + t(t + 2) + 25 = 2t(t + 5) - 15$ 2

36. $\frac{7}{4}$
37. $\frac{23}{24}$

INTEGRATION Geometry

Find the measure of the area of each shaded region in simplest terms. **40.** $x^2 + 6x$ **41.** $15p^2 + 32p$ **42.** $12a^3 - 17a^2b + 15a^2 + 10b + 10$

C

40. 3x; 2x; x + 2; x

41.

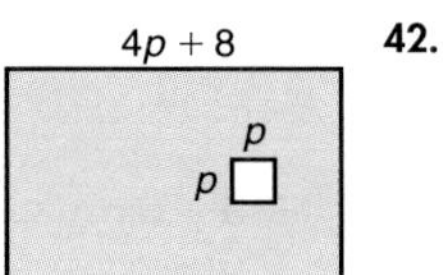

42. 4a + b + 5; 3a²; 2a²b − b − 1; 10

Critical Thinking

43. Write eight multiplication problems whose product is $8a^2b + 18ab$. **See margin.**

Applications and Problem Solving

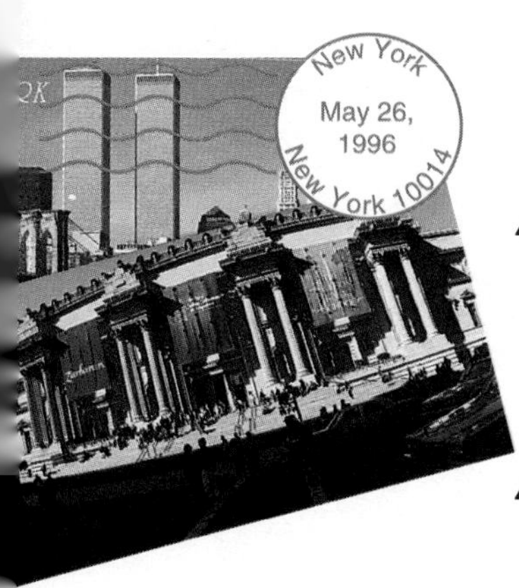

44. **Recreation** In Honduras, children play a form of hopscotch called La Rayuela. The pattern for this game is shown at the right. Suppose that each rectangle is $2y + 1$ units long and y units wide.

La Rayuela (Honduras)
cabeza
brazo casa brazo
tercera
segunda
primera

   a. Write an expression in simplest form for the area of the pattern. **$14y^2 + 7y$**
   b. If y represents 9 inches, find the area of the pattern. **1197 in^2**

45. **Travel** The Drama Club of Lincoln High School is visiting New York City. They plan to take taxis from the World Trade Center to the Metropolitan Museum of Art. The fare for a taxi is \$2.75 for the first mile and \$1.25 for each additional mile. Suppose the distance between the two locations is m miles and t taxis are needed to transport the entire group. Write an expression in simplest form for the cost to transport the group to the Metropolitan Museum of Art excluding the tip. **$1.50t + 1.25mt$**

46. **Construction** A landscaper is designing a rectangular garden for an office complex. There will be a concrete walkway on three sides of the garden, as shown at the right. The width of the garden will be 24 feet, and the length will be 42 feet. The width of the longer portion of the walkway will be 3 feet. The concrete will cost \$20 per square yard, and the builders have told the landscaper that she can spend \$820 on the concrete. How wide should the two remaining sides be? **4.5 ft**

3 feet
42 feet
x
x
24 feet

47. **Geometry** The number of diagonals that can be drawn for a polygon with n sides is represented by the expression $\frac{1}{2}n(n - 3)$.
   a. Draw polygons with 3, 4, 5, and 6 sides. Show that this expression is true for these polygons. **See margin.**
   b. Find the product of this expression.
   c. How many diagonals can be drawn for a polygon with 15 sides? **90 diagonals**

47b. $\frac{1}{2}n^2 - \frac{3}{2}n$

Mixed Review

49. 75 gal of 50%, 25 gal of 30%

50. $\frac{5}{4}$

51. $a^2 - 1$

52. See margin.

48. Find $(3a - 4ab + 7b) - (7a - 3b)$. (Lesson 9–5) **$-4a - 4ab + 10b$**

49. **Chemistry** One solution is 50% glycol, and another is 30% glycol. How much of each solution should be mixed to make a 100-gallon solution that is 45% glycol? (Lesson 8–2)

50. Determine the slope of the line that passes through $A(1, 5)$ and $B(-3, 0)$. (Lesson 6–1)

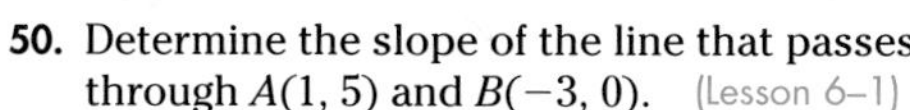

51. If $g(x) = x^2 + 2x$, find $g(a - 1)$. (Lesson 5–5)

52. Draw the graph of $2x - y = 8$. (Lesson 5–4)

53. **Finance** Antonio earned \$340 in 4 days by mowing lawns and doing yard work. At this rate, how long will it take him to earn \$935? (Lesson 4–8) **11 days**

54. **Geometry** Find the measure of an angle that is 44° less than its complement. (Lesson 3–4) **23°**

55. Evaluate $(15x)^3 - y$ if $x = 0.2$ and $y = 1.3$. (Lesson 1–3) **25.7**

Extension

Connections The perimeter of a square can be found by using the formula $P = 4s$, where s represents the length of a side. Find the perimeter of a square if its sides are each $(3x^2 + 2)$ cm long. **$(12x^2 + 8)$ cm**

Additional Answer (continued)

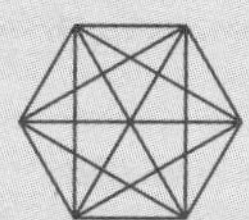

9 diagonals

$\frac{1}{2}(6)(6 - 3) = \frac{1}{2}(6)(3)$
$= 9$

4 ASSESS

Closing Activity

Writing Have students write a paragraph summarizing the process of multiplying a polynomial by a monomial.

Chapter 9, Quiz C (Lessons 9-5 and 9-6), is available in the *Assessment and Evaluation Masters,* p. 241.

Additional Answer

47a.

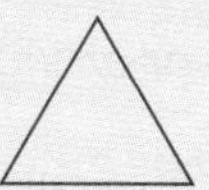

0 diagonals

$\frac{1}{2}(3)(3 - 3) = \frac{1}{2}(3)(0)$
$= 0$

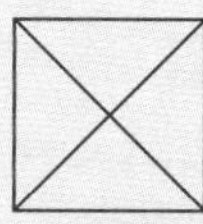

2 diagonals

$\frac{1}{2}(4)(4 - 3) = \frac{1}{2}(4)(1)$
$= 2$

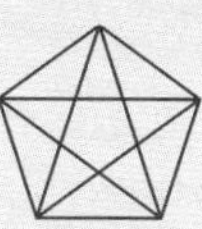

5 diagonals

$\frac{1}{2}(5)(5 - 3) = \frac{1}{2}(5)(2)$
$= 5$

Enrichment Masters, p. 66

NAME ______ DATE ______

Enrichment

Student Edition Pages 529–533

Geometric Series

The terms of this polynomial form a geometric series.

$$a + ar + ar^2 + ar^3 + ar^4$$

The first term is the constant a. Then each term after that is found by multiplying by a constant multiplier r.

Use the equation $S = a + ar + ar^2 + ar^3 + ar^4$ for Exercises 1–3.

1. Multiply each side of the equation by r.
   $rS = ar + ar^2 + ar^3 + ar^4 + ar^5$
2. Subtract the original equation from your result in Exercise 1.
   $rS - S = ar^5 - a$
3. Solve the result from Exercise 2 for the variable S. $S = \frac{a(r^5 - 1)}{r - 1}$

Use the polynomial $a + ar + ar^2 + ar^3 + ar^4 + \cdots + ar^{n-1}$ for Exercises 4–8.

4. Write the 10th term of the polynomial. ar^9
5. If $a = 5$ and $r = 2$, what is the 8th term? $ar^7 = 640$
6. Follow the steps in Exercises 1–3 to write a formula for the sum of this polynomial. $S = \frac{a(r^n - 1)}{r - 1}$
7. If the 3rd term is 20 and the 6th term is 160, solve for r^3 and then find r. Then solve $ar^2 = 20$ for a and find the value of the first six terms of the polynomial.
   $\frac{ar^5}{ar^2} = \frac{160}{20}$, $r^3 = 8$, $r = 2$; $ar^2 = 20$; $a = 5$; 5, 10, 20, 40, 80, 160
8. Find the sum of the first six terms of the geometric series that begins 3, 6, 12, 24, …. First write the values for a and r.
   $a = 3$, $r = 2$
   $S = \frac{3(2^6 - 1)}{2 - 1} = 3 \times 63 = 189$

9-7A LESSON NOTES

NCTM Standards: 1–5

Objective

Use algebra tiles to find the product of binomials.

Recommended Time

Demonstration and discussion: 30 minutes; Exercises: 30 minutes

Instructional Resources

For each student or group of students
Student Manipulative Kit
- algebra tiles

Modeling Mathematics Masters
- p. 1 (algebra tiles)
- p. 15 (product mat)
- p. 31 (worksheet)

For teacher demonstration
Algebra and Geometry Overhead Manipulative Resources

1 FOCUS

Motivating the Lesson

Now that students have seen how the multiplication of a polynomial by a monomial is modeled by algebra tiles, encourage students to develop a method for using tiles to model the multiplication of a polynomial by another polynomial.

2 TEACH

Teaching Tip Stress that each multiplication is an extension of the distributive property.

Teaching Tip Some students may not require rearranging the tiles to determine the simplest form of the polynomial. You may want to encourage them to do so as a check of their mental math.

Teaching Tip After they have completed the activities, ask students to look for visual patterns in the models. When is a checkerboard pattern visible? When are exactly two different color rectangles visible?

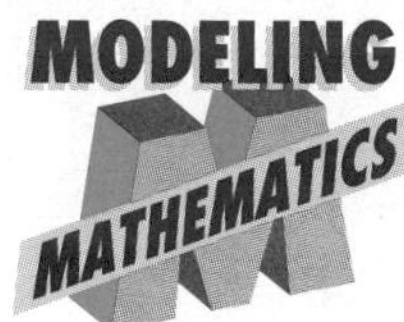

A Preview of Lesson 9–7

9–7A Multiplying Polynomials

Materials: algebra tiles   product mat

You can find the product of binomials by using algebra tiles.

Activity 1 **Use algebra tiles to find $(x + 2)(x + 3)$.**

The rectangle will have a width of $x + 2$ and a length of $x + 3$. Use your algebra tiles to mark off the dimensions on a product mat. Then make the rectangle with algebra tiles.

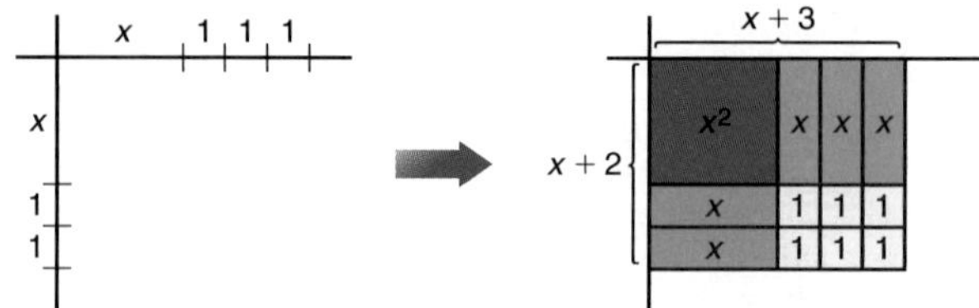

The rectangle consists of 1 blue x^2-tile, 5 green x-tiles, and 6 yellow 1-tiles.
The area of the rectangle is $x^2 + 5x + 6$. Therefore, $(x + 2)(x + 3) = x^2 + 5x + 6$.

Activity 2 **Use algebra tiles to find $(x - 1)(x - 3)$.**

Step 1 The rectangle will have a width of $(x - 1)$ units and a length of $(x - 3)$ units. Use your algebra tiles to mark off the dimensions on a product mat. Then begin to make the rectangle with algebra tiles.

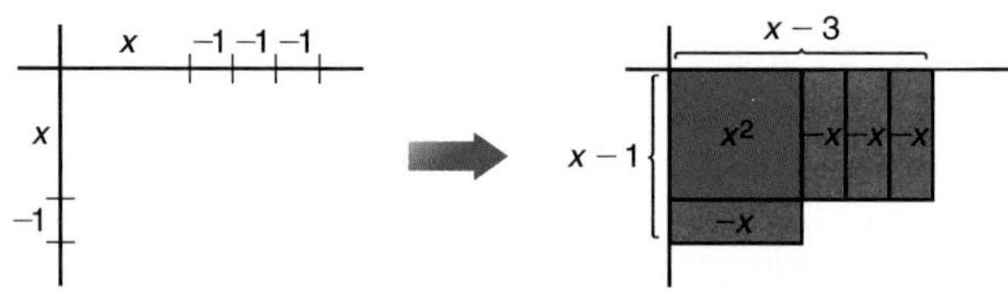

Step 2 Determine whether to use 3 yellow 1-tiles or 3 red 1-tiles to complete the rectangle. Remember that the numbers at the top and side give the dimensions of the tile needed. The area of each tile is the product of -1 and -1. This is represented by a yellow 1-tile. Fill in the space with 3 yellow 1-tiles to complete the rectangle.

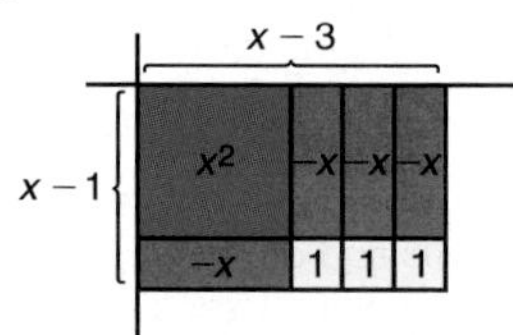

The rectangle consists of 1 blue x^2-tile, 4 red x-tiles, and 3 yellow 1-tiles.
The area of the rectangle is $x^2 - 4x + 3$. Therefore, $(x - 1)(x - 3) = x^2 - 4x + 3$.

GLENCOE Technology

Interactive Mathematics Tools Software

This multimedia software provides an interactive lesson that uses tiles to multiply like terms in polynomials. A **Computer Journal** gives students an opportunity to write about what they have learned.

For Windows & Macintosh

Activity 3 **Use algebra tiles to find $(x + 1)(2x - 1)$.**

Step 1 The rectangle will have a width of $(x + 1)$ units and a length of $(2x - 1)$. Use your algebra tiles to mark off the dimensions on a product mat. Then begin to make the rectangle with algebra tiles.

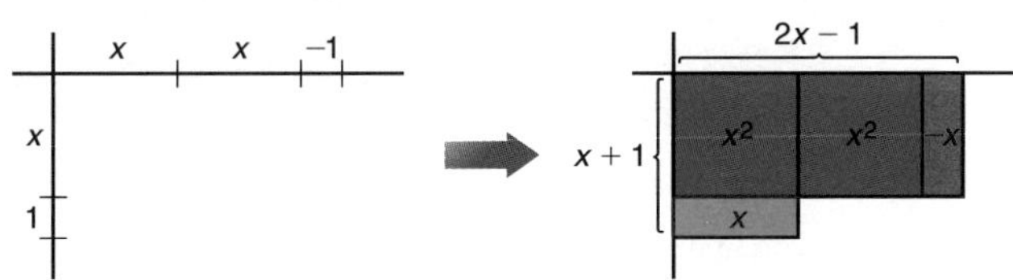

Step 2 Determine what color x-tile and what color 1-tile to use to complete the rectangle. The area of the x-tile is the product of x and 1. This is represented by a green x-tile. The area of the 1-tile is represented by the product of -1 and 1. This is represented by a red 1-tile.

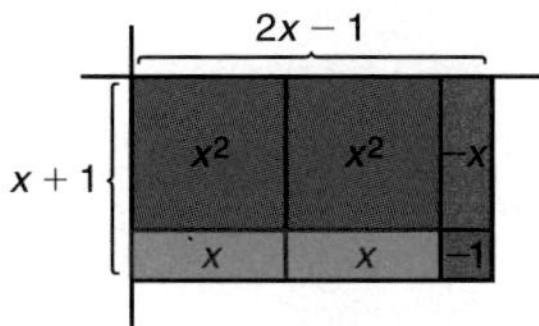

Step 3 Rearrange the tiles to simplify the polynomial you have formed. Notice that a zero pair is formed by the x-tiles.

There are 2 blue x^2-tiles, 1 green x-tile, and 1 red 1-tile left. In simplest form, $(x + 1)(2x - 1) = 2x^2 + x - 1$.

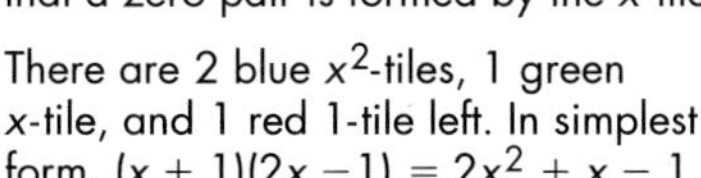

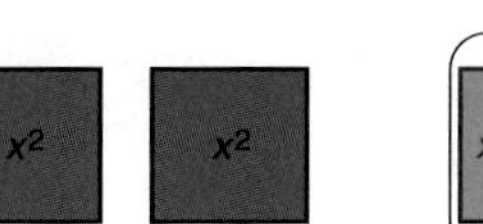

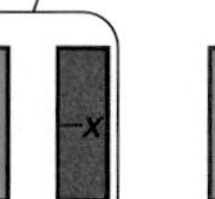

Model **Use algebra tiles to find each product.**

1. $x^2 + 3x + 2$
2. $x^2 - 2x - 3$
3. $x^2 - 6x + 8$

1. $(x + 1)(x + 2)$
2. $(x + 1)(x - 3)$
3. $(x - 2)(x - 4)$
4. $(x + 1)(2x + 2)$ **$2x^2 + 4x + 2$**
5. $(x - 1)(2x + 2)$ **$2x^2 - 2$**
6. $(x - 3)(2x - 1)$ **$2x^2 - 7x + 3$**

Draw ***Is each statement true or false? Justify your answer with a drawing.***

7. $(x + 4)(x + 6) = x^2 + 24$ **false**
8. $(x + 3)(x - 2) = x^2 + x - 6$ **true**
9. $(x - 1)(x + 5) = x^2 - 4x - 5$ **false**
10. $(x - 2)(x - 3) = x^2 - 5x + 6$ **true**

7–10. See margin for drawings.

Write

11. You can also use the distributive property to find the product of two binomials. The figure at the right shows the model for $(x + 3)(x + 2)$ separated into four parts. Write a paragraph explaining how this model shows the use of the distributive property. **By the distributive property, $(x + 3)(x + 2) = x(x + 2) + 3(x + 2)$. The top row represents $x(x + 2)$ or $x^2 + 2x$. The bottom row represents $3(x + 2)$ or $3x + 6$.**

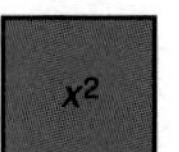

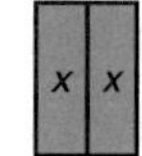

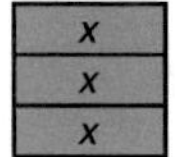

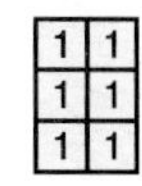

Cooperative Learning

This lesson offers an excellent opportunity for using cooperative learning groups. For more information on cooperative learning strategies and group management, see *Cooperative Learning in the Mathematics Classroom,* one of the titles in the Glencoe Mathematics Professional Series.

3 PRACTICE/APPLY

Assignment Guide

Core: 1–11
Enriched: 1–11

Additional Answers

7.

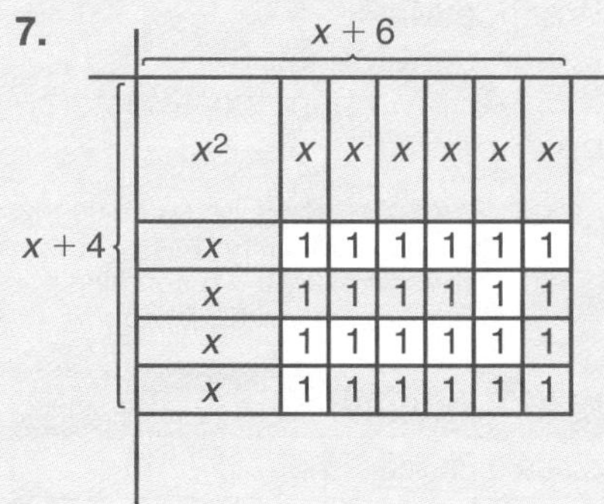

8.

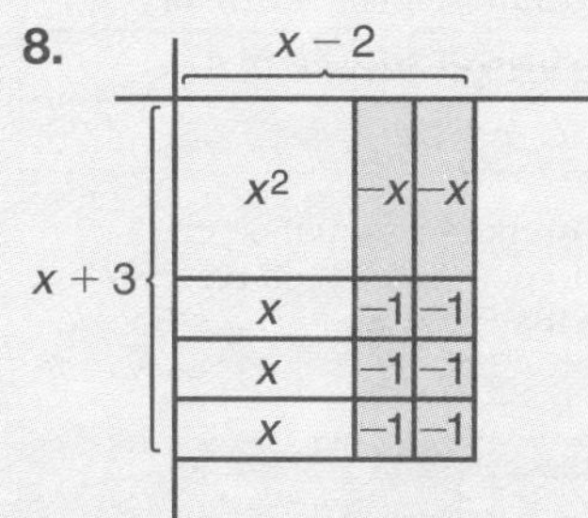

9.

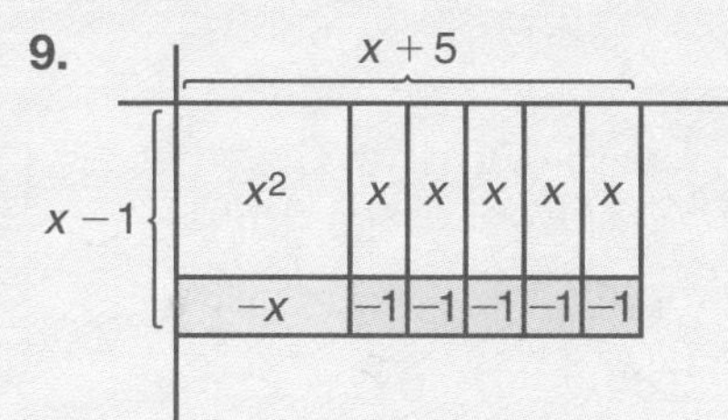

10.

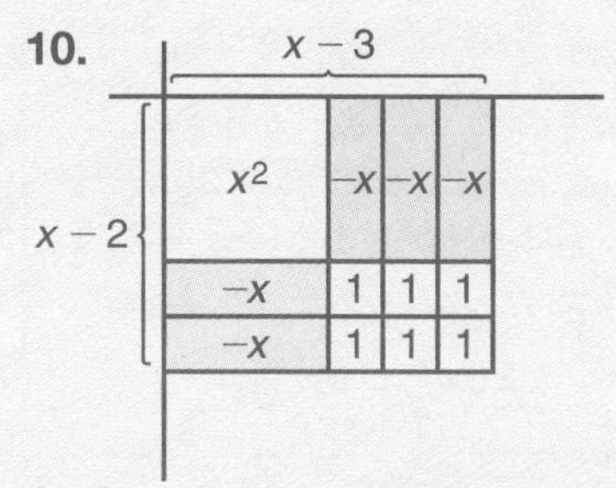

4 ASSESS

Observing students working in cooperative groups is an excellent method of assessment.

9-7 LESSON NOTES

NCTM Standards: 1–5

Instructional Resources

- Study Guide Master 9-7
- Practice Master 9-7
- Enrichment Master 9-7
- Modeling Mathematics Masters, pp. 60–62

Transparency 9-7A contains the 5-Minute Check for this lesson; **Transparency 9-7B** contains a teaching aid for this lesson.

Recommended Pacing	
Standard Pacing	Day 12 of 15
Honors Pacing	Day 11 of 14
Block Scheduling*	Day 6 of 8
Alg. 1 in Two Years*	Days 19 & 20 of 26

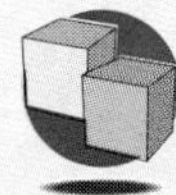

*For more information on pacing and possible lesson plans, refer to the *Block Scheduling Booklet* and *Algebra 1 in Two Years.*

1 FOCUS

5-Minute Check
(over Lesson 9-6)

Simplify.

1. $3x(2x^2 - 7)$ **$6x^3 - 21x$**
2. $3c^2(4c^2 + 3c - 9)$ **$12c^4 + 9c^3 - 27c^2$**
3. $5ab(4a^2b - 7ab^2 - 6b^3)$ **$20a^3b^2 - 35a^2b^3 - 30ab^4$**
4. $2x(3x^2 - 4x + 7) - 5(4x^2 + 3x - 9)$ **$6x^3 - 28x^2 - x + 45$**

Solve.

5. $3a(a + 2b) + 7b(a + 2b)$ **$3a^2 + 13ab + 14b^2$**

Motivating the Lesson

Questioning Introduce the lesson by having students describe possible ways of finding the product of $(2a + y)$ and $(2a + 5t - 6y + 5r)$.

9–7 Multiplying Polynomials

What YOU'LL LEARN

- To use the FOIL method to multiply two binomials, and
- to multiply any two polynomials by using the distributive property.

Why IT'S IMPORTANT

You can use polynomials to solve problems involving art and business.

Art

Have you ever flown over farm land and looked down? You probably saw various fields that gave the appearance of a patchwork quilt. However, if you had flown over a field designed by Stan Herd, you may have seen some sunflowers in a vase or a picture of Will Rogers. Since 1981, Stan Herd has been combining his interests in art and agriculture to form crop art. Most of Herd's work is harvested, and therefore is only visible for a short time.

In 1991, however, Herd created the picture above using native perennials. This picture is called *Little Girl in the Wind.* It depicts a Kickapoo Indian girl by the name of Carole Cadue. If you fly near Salina, Kansas, you may see this work of art.

GLOBAL CONNECTIONS

There is an historic work of art in Ohio that is best experienced from a bird's eye view. It is a mound built by the Adena Indians perhaps 30 centuries ago. The mound is in the shape of a snake and is about one-quarter mile long.

Suppose the measure of the length of the field used for *Little Girl in the Wind* can be represented by the polynomial $7x + 2$ units and the width can be represented by $5x + 1$. You know that the area of a rectangle is the product of its length and width. You can multiply $7x + 2$ and $5x + 1$ to find the area of the rectangle.

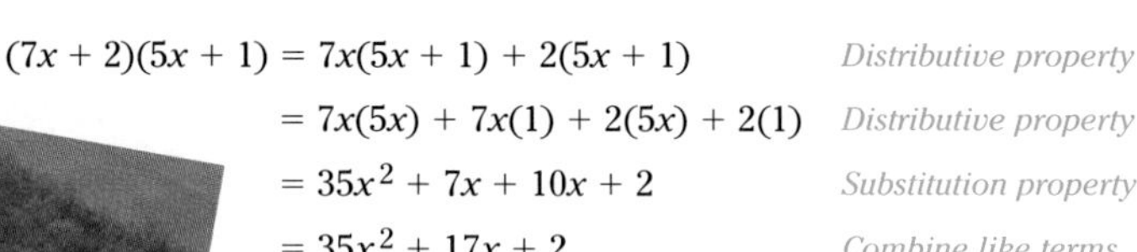

$(7x + 2)(5x + 1) = 7x(5x + 1) + 2(5x + 1)$	*Distributive property*
$= 7x(5x) + 7x(1) + 2(5x) + 2(1)$	*Distributive property*
$= 35x^2 + 7x + 10x + 2$	*Substitution property*
$= 35x^2 + 17x + 2$	*Combine like terms.*

The area can also be determined by finding the sum of the areas of four smaller rectangles.

	$5x$	1
$7x$	$7x \cdot 5x$	$7x \cdot 1$
2	$2 \cdot 5x$	$2 \cdot 1$

$(7x + 2)(5x + 1) = 7x \cdot 5x + 7x \cdot 1 + 2 \cdot 5x + 2 \cdot 1$	*Find the sum of the four areas.*
$= 35x^2 + 7x + 10x + 2$	*Substitution property*
$= 35x^2 + 17x + 2$	*Combine like terms.*

The serpent mound is twenty feet wide and averages five feet in height.

GLENCOE Technology

CD-ROM Interaction

A multimedia simulation connects polynomial functions with the construction of a swimming pool. A blackline master activity with teacher's notes provides a follow-up to the CD-ROM simulation.

For Windows & Macintosh

This example illustrates a shortcut of the distributive property called the **FOIL method**. You can use the FOIL method to multiply two binomials.

$$(7x + 2)(5x + 1) = \underset{F}{(7x)(5x)} + \underset{O}{(7x)(1)} + \underset{I}{(2)(5x)} + \underset{L}{(2)(1)}$$

$$= 35x^2 + 7x + 10x + 2$$

$$= 35x^2 + 17x + 2$$

FOIL Method for Multiplying Two Binomials	To multiply two binomials, find the sum of the products of **F** the *First* terms, **O** the *Outer* terms, **I** the *Inner* terms, and **L** the *Last* terms.

Example 1 **Find each product.**

a. $(x - 4)(x + 9)$

$$(x - 4)(x + 9) = \overset{F}{(x)(x)} + \overset{O}{(x)(9)} + \overset{I}{(-4)(x)} + \overset{L}{(-4)(9)}$$

$$= x^2 + 9x - 4x - 36$$

$$= x^2 + 5x - 36 \quad \text{Combine like terms.}$$

b. $(4x + 7)(3x - 8)$

$$(4x + 7)(3x - 8) = \overset{F}{(4x)(3x)} + \overset{O}{(4x)(-8)} + \overset{I}{(7)(3x)} + \overset{L}{(7)(-8)}$$

$$= 12x^2 - 32x + 21x - 56$$

$$= 12x^2 - 11x - 56 \quad \text{Combine like terms.}$$

The distributive property can be used to multiply any two polynomials.

Example 2 **Find each product.**

a. $(2y + 5)(3y^2 - 8y + 7)$

$$(2y + 5)(3y^2 - 8y + 7)$$

$$= 2y(3y^2 - 8y + 7) + 5(3y^2 - 8y + 7) \quad \text{Distributive property}$$

$$= (6y^3 - 16y^2 + 14y) + (15y^2 - 40y + 35) \quad \text{Distributive property}$$

$$= 6y^3 - 16y^2 + 14y + 15y^2 - 40y + 35$$

$$= 6y^3 - y^2 - 26y + 35 \quad \text{Combine like terms.}$$

b. $(x^2 + 4x - 5)(3x^2 - 7x + 2)$

$$(x^2 + 4x - 5)(3x^2 - 7x + 2)$$

$$= x^2(3x^2 - 7x + 2) + 4x(3x^2 - 7x + 2) - 5(3x^2 - 7x + 2)$$

$$= (3x^4 - 7x^3 + 2x^2) + (12x^3 - 28x^2 + 8x) - (15x^2 - 35x + 10)$$

$$= 3x^4 - 7x^3 + 2x^2 + 12x^3 - 28x^2 + 8x - 15x^2 + 35x - 10$$

$$= 3x^4 + 5x^3 - 41x^2 + 43x - 10 \quad \text{Combine like terms.}$$

2 TEACH

Teaching Tip Remind students that FOIL is a memory device, not a mathematical theorem or rule. Stress that this works only with two binomials.

In-Class Examples

For Example 1
Find each product.

a. $(m - 3)(m + 4)$
$\mathbf{m^2 + m - 12}$

b. $(3a + 11)(5a - 2)$
$\mathbf{15a^2 + 49a - 22}$

For Example 2
Find each product.

a. $(2x + 3)(x^2 + 3x + 8)$
$\mathbf{2x^3 + 9x^2 + 25x + 24}$

b. $(4x^2 - 10x - 3)(2x^2 + 6x - 9)$
$\mathbf{8x^4 + 4x^3 - 102x^2 + 72x + 27}$

Teaching Tip In Example 2, emphasize to students that they must multiply each term in the first polynomial by each term in the second polynomial.

Classroom Vignette

"I cut out colored transparencies of the rectangles in the diagram on page 536 and put them on the overhead. I ask the students to describe two different ways to find the total area. By doing this activity, they discover the 'FOIL' method on their own. Then I explain the two different methods to the entire class."

Linda Glover

Linda Glover
Conway Junior High School
Conway, Arkansas

In-Class Examples

For Example 3
Find $(4x^3 - 2x^2 + 5)(5x - 6)$ using column form. $\mathbf{20x^4 - 34x^3 + 12x^2 + 25x - 30}$

For Example 4
Write a polynomial expression that represents the volume of the prism shown here. Then find the volume if $y = 3$.

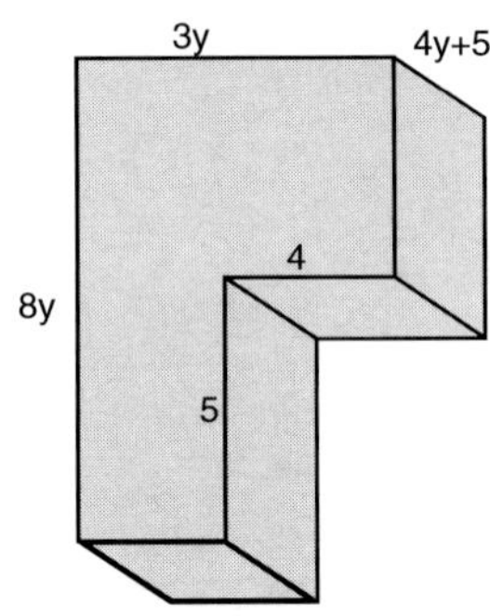

$(96y^3 + 120y^2 - 80y - 100)$ cubic units; 3332 cubic units

3 PRACTICE/APPLY

Check for Understanding

Exercises 1–13 are designed to help you assess your students' understanding through reading, writing, speaking, and modeling. You should work through Exercises 1–4 with your students and then monitor their work on Exercises 5–13.

Additional Answers

1a. Sample answer:

$(42)(27)$
$= (40 + 2)(20 + 7)$
F O I L
$= 40(20) + 40(7) + 2(20) + 2(7)$
$= 800 + 280 + 40 + 14$
$= 1134$

1b. Sample answer:

$4\frac{1}{2} \cdot 6\frac{3}{4}$
$= \left(4 + \frac{1}{2}\right)\left(6 + \frac{3}{4}\right)$
F O I L
$= 4(6) + 4\left(\frac{3}{4}\right) + \frac{1}{2}(6) + \frac{1}{2}\left(\frac{3}{4}\right)$
$= 24 + 3 + 3 + \frac{3}{8}$
$= 30\frac{3}{8}$

Polynomials can also be multiplied in column form. Be careful to align the like terms.

Example 3 **Find $(x^3 - 8x^2 + 9)(3x + 4)$ using column form.**

Since there is no x term in $x^3 - 8x^2 + 9$, $0x$ is used as a placeholder.

$$\begin{array}{rrrrrl}
 & x^3 & -\ 8x^2 & +\ 0x & +\ 9 & \\
(\times) & & & 3x & +\ 4 & \\
\hline
 & 4x^3 & -\ 32x^2 & +\ 0x & +\ 36 & \leftarrow \textit{product of } x^3 - 8x^2 + 0x + 9 \textit{ and } 4 \\
3x^4 & -\ 24x^3 & +\ 0x^2 & +\ 27x & & \leftarrow \textit{product of } x^3 - 8x^2 + 0x + 9 \textit{ and } 3x \\
\hline
3x^4 & -\ 20x^3 & -\ 32x^2 & +\ 27x & +\ 36 & \leftarrow \textit{sum of the partial products}
\end{array}$$

Example 4

INTEGRATION
Geometry

The volume V of a prism equals the area of the base B times the height h.

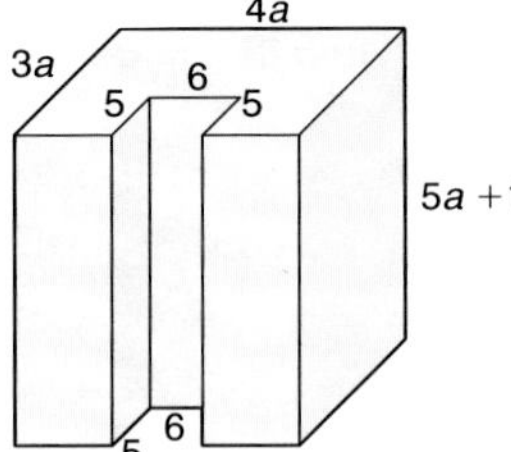

a. Write a polynomial expression that represents the volume of the prism shown at the right.

b. Find the volume if $a = 5$.

a. A diagram of the base is shown below.

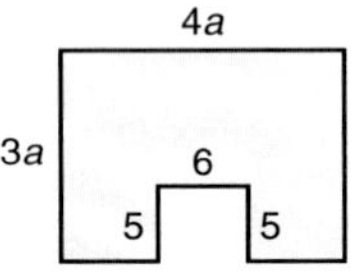

To find the area of the base, first find the area of a rectangle that is $3a$ by $4a$. Then subtract the area of a rectangle that is 6 by 5.

$B = 3a(4a) - 6(5)$
$= 12a^2 - 30$

The volume of this prism equals the product of the base and the height, $12a^2 - 30$ and $5a + 1$, respectively. Use FOIL to find the product.

$$(12a^2 - 30)(5a + 1) = \overset{F}{(12a^2)(5a)} + \overset{O}{(12a^2)(1)} + \overset{I}{(-30)(5a)} + \overset{L}{(-30)(1)}$$
$$= 60a^3 + 12a^2 - 150a - 30$$

The volume of the prism is $(60a^3 + 12a^2 - 150a - 30)$ cubic units.

b. Substitute 5 for a and evaluate the expression.

$60(5^3) + 12(5^2) - 150(5) - 30 = 7500 + 300 - 750 - 30$
$= 7020$

If $a = 5$, the volume of the prism is 7020 cubic units.

CHECK FOR UNDERSTANDING

Communicating Mathematics

Study the lesson. Then complete the following.

1. Use the FOIL method to evaluate each product. **a–b. See margin.**
   a. $42(27)$ (*Hint:* Rewrite as $(40 + 2)(20 + 7)$ or $(40 + 2)(30 - 3)$.)
   b. $4\frac{1}{2} \cdot 6\frac{3}{4}$

Alternative Learning Styles

Visual Have students draw a diagram of the garden in the following problem to illustrate its dimensions.
Pete has a rectangular garden 15 feet longer than it is wide. A cement divider that is 4 feet wide surrounds the garden. The total area of the divider is 632 square feet. What are the dimensions of the garden?
28 feet by 43 feet

2. **You Decide** Adita and Delbert used the following methods to find the product of $(t^3 - t^2 + 5t)$ and $(6t^2 + 8t - 7)$.

Adita:

$$\begin{aligned}(t^3 - t^2 + 5t)(6t^2 + 8t - 7) &= t^3(6t^2 + 8t - 7) - t^2(6t^2 + 8t - 7) + 5t(6t^2 + 8t - 7)\\ &= 6t^5 + 8t^4 - 7t^3 - 6t^4 - 8t^3 + 7t^2 + 30t^3 + 40t^2 - 35t\\ &= 6t^5 + 2t^4 + 15t^3 + 47t^2 - 35t\end{aligned}$$

Delbert:

$$\begin{array}{r} t^3 - t^2 + 5t\\ (\times)\ 6t^2 + 8t - 7\\ \hline -7t^3 + 7t^2 - 35t\\ 8t^4 - 8t^3 + 40t^2\qquad\\ 6t^5 - 6t^4 + 30t^3\qquad\qquad\\ \hline 6t^5 + 2t^4 + 15t^3 + 47t^2 - 35t \end{array}$$

Which method do you prefer? Why? **See students' work.**

3. **Draw** a diagram to show how you would use algebra tiles to find the product of $(2x - 3)$ and $(x + 2)$. **See margin.**

MODELING MATHEMATICS

4. **Write** two binomials whose product is represented at the right.
Sample answer: $(a + x)(2x + 3)$

2ax	3a
$2x^2$	3x

Guided Practice

Find each product.

5. $(d + 2)(d + 8)$
6. $(r - 5)(r - 11)$
7. $(y + 3)(y - 7)$
8. $(3p - 5)(5p + 2)$
9. $(2x - 1)(x + 5)$
10. $(2m + 5)(3m - 8)$
11. $(2a + 3b)(5a - 2b)$
12. $(2x - 5)(3x^2 - 5x + 4)$

13. a. **Number Theory** Find the product of three consecutive integers if the least integer is a. $a^3 + 3a^2 + 2a$
b. Choose an integer as the first of three consecutive integers. Find their product. **See students' work.**
c. Evaluate the polynomial in part a for these integers. Describe the result. **See students' work. The result is the same as the product in part b.**

Answers: 5. $d^2 + 10d + 16$ 6. $r^2 - 16r + 55$ 7. $y^2 - 4y - 21$ 8. $15p^2 - 19p - 10$ 9. $2x^2 + 9x - 5$ 10. $6m^2 - m - 40$ 11. $10a^2 + 11ab - 6b^2$ 12. $6x^3 - 25x^2 + 33x - 20$

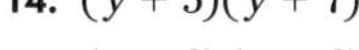

EXERCISES

Practice

Find each product. **23–33. See margin.**

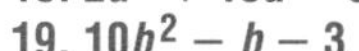

A

14. $(y + 5)(y + 7)$
15. $(c - 3)(c - 7)$
16. $(x + 4)(x - 8)$
17. $(w + 3)(w - 9)$
18. $(2a - 1)(a + 8)$
19. $(5b - 3)(2b + 1)$
20. $(11y + 9)(12y + 6)$
21. $(13x - 3)(13x + 3)$
22. $(8x + 9y)(3x + 7y)$

B

23. $(0.3v - 7)(0.5v + 2)$
24. $\left(3x + \frac{1}{3}\right)\left(2x - \frac{1}{9}\right)$
25. $\left(a - \frac{2}{3}b\right)\left(\frac{2}{3}a + \frac{1}{2}b\right)$
26. $(2r + 0.1)(5r - 0.3)$
27. $(0.7p + 2q)(0.9p + 3q)$
28. $(x + 7)(x^2 + 5x - 9)$
29. $(3x - 5)(2x^2 + 7x - 11)$
30. $\begin{array}{r} a^2 - 3a + 11\\ (\times)\ 5a + 2\\ \hline \end{array}$
31. $\begin{array}{r} 3x^2 - 7x + 2\\ (\times)\ 3x - 8\\ \hline \end{array}$
32. $\begin{array}{r} 5x^2 + 8x - 11\\ (\times)\ x^2 - 2x - 1\\ \hline \end{array}$
33. $\begin{array}{r} 5d^2 - 6d + 9\\ (\times)\ 4d^2 + 3d + 11\\ \hline \end{array}$

Answers: 14. $y^2 + 12y + 35$ 15. $c^2 - 10c + 21$ 16. $x^2 - 4x - 32$ 17. $w^2 - 6w - 27$ 18. $2a^2 + 15a - 8$ 19. $10b^2 - b - 3$ 20. $132y^2 + 174y + 54$ 21. $169x^2 - 9$ 22. $24x^2 + 83xy + 63y^2$

Assignment Guide

Core: 15–47 odd, 48–56
Enriched: 14–40 even, 42–56

For **Extra Practice,** see p. 778.

The red A, B, and C flags, printed only in the Teacher's Wraparound Edition, indicate the level of difficulty of the exercises.

Additional Answers

3.

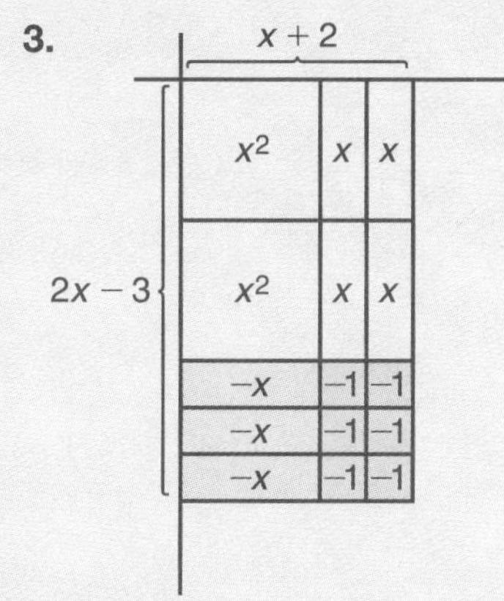

23. $0.15v^2 - 2.9v - 14$
24. $6x^2 + \frac{1}{3}x - \frac{1}{27}$
25. $\frac{2}{3}a^2 + \frac{1}{18}ab - \frac{1}{3}b^2$
26. $10r^2 - 0.1r - 0.03$
27. $0.63p^2 + 3.9pq + 6q^2$
28. $x^2 + 12x^2 + 26x - 63$
29. $6x^3 + 11x^2 - 68x + 55$
30. $5a^3 - 13a^2 + 49a + 22$

(continued on p. 540)

Study Guide Masters, p. 67

9-7 NAME ____________ DATE ________

Study Guide Student Edition Pages 536–541

Multiplying Polynomials

The following example shows how the distributive property can be used to multiply any two polynomials.

Example 1: Find $(2x - 6)(3x + 1)$.

$$\begin{aligned}(2x - 6)(3x + 1) &= \overset{F}{2x \cdot 3x} + \overset{O}{2x \cdot 1} + \overset{I}{(-6) \cdot 3x} + \overset{L}{(-6) \cdot 1}\\ &= 6x^2 + 2x - 18x - 6\\ &= 6x^2 - 16x - 6\end{aligned}$$

You can also multiply polynomials vertically.

Example 2: Find $(3x^2 - x + 1)(5x + 2)$.

$$\begin{array}{rl} 3x^2 - x + 1 & \\ \times\quad 5x + 2 & \\ \hline 6x^2 - 2x + 2 & \text{Multiply } 3x^2 - x + 1 \text{ by 2.}\\ 15x^3 - 5x^2 + 5x\qquad & \text{Multiply } 3x^2 - x + 1 \text{ by } 5x.\\ \hline 15x^3 + x^2 + 3x + 2 & \text{Combine like terms.} \end{array}$$

Find each product.

1. $(5t + 4)(2t - 6)$ $10t^2 - 22t - 24$
2. $(5m - 3n)(4m - 2n)$ $20m^2 - 22mn + 6n^2$
3. $(a - 3b)(2a - 5b)$ $2a^2 - 11ab + 15b^2$
4. $(3x - 0.1)(x + 0.1)$ $3x^2 + 0.2x - 0.01$
5. $(8x + 5)(8x - 5)$ $64x^2 - 25$
6. $(x + 5)(x + 2)$ $x^2 + 7x + 10$
7. $(2x - 4)(2x + 5)$ $4x^2 + 2x - 20$
8. $\begin{array}{r} y^2 - 5y + 3\\ \times\ 2y^2 + 7y - 4\\ \hline \end{array}$ $2y^4 - 3y^3 - 33y^2 + 41y - 12$
9. $\begin{array}{r} 3b^3 - 2b^2 + b\\ \times\quad 2b - 3\\ \hline \end{array}$ $6b^4 - 13b^3 + 8b^2 - 3b$

Reteaching

Using Diagrams Have students use the diagram below to answer the following questions.

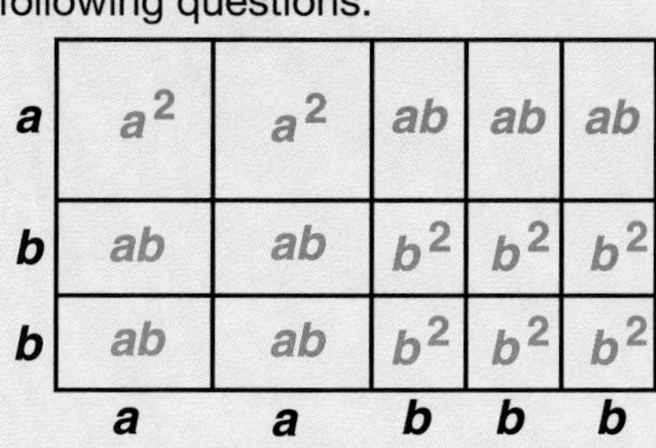

1. What is the width of the large rectangle? $a + 2b$
2. What is the length? $2a + 3b$
3. What is the area? $(a + 2b)(2a + 3b)$ or $2a^2 + 7ab + 6b^2$
4. Inside each small square or rectangle, write its area. See diagram.
5. Add the areas in Exercise 4. Compare with your answer in Exercise 3. same

Additional Answers

31. $9x^3 - 45x^2 + 62x - 16$
32. $5x^4 - 2x^3 - 32x^2 + 14x + 11$
33. $20d^4 - 9d^3 + 73d^2 - 39d + 99$

34. $2x^4 - 20x^3 + 39x^2 - 68x - 9$

Geometry

35. $10x^4 + 3x^3 + 51x^2 - 16x - 48$

36. $-35b^5 + 14b^4 - 18b^3 - 19b^2 + 14b - 12$

Find each product.

34. $(x^2 - 8x - 1)(2x^2 - 4x + 9)$
35. $(5x^2 - x - 4)(2x^2 + x + 12)$
36. $(-7b^3 + 2b - 3)(5b^2 - 2b + 4)$
37. $(a^2 + 2a + 5)(a^2 - 3a - 7)$ $a^4 - a^3 - 8a^2 - 29a - 35$

Find the measure of the volume of each prism.

38.

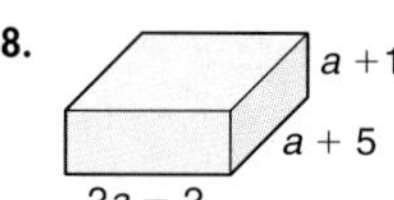

39.

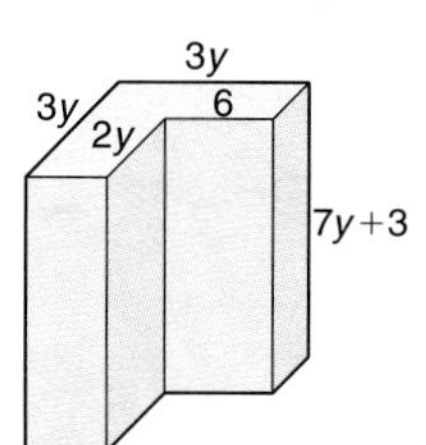

40.

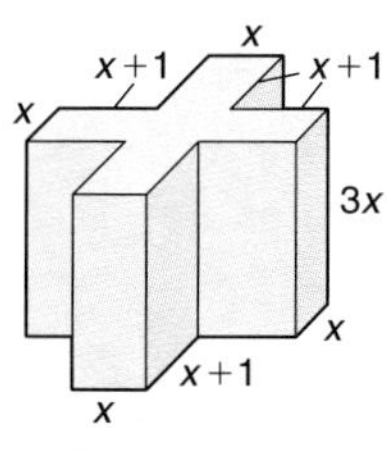

38. $2a^3 + 10a^2 - 2a - 10$ 39. $63y^3 - 57y^2 - 36y$ 40. $15x^3 + 12x^2$

41. **Geometry** Refer to the prism in Exercise 38. Suppose a represents 15 centimeters.
   a. Find the length, width, and height of the prism. **28 cm, 20 cm, 16 cm**
   b. Use the values in part a to find the volume of the prism. **8960 cm³**
   c. Evaluate your answer for Exercise 38 if $a = 15$. **8960**
   d. How do your answers for parts b and c compare? **They are the same measure.**

Critical Thinking

If $A = 3x + 4$, $B = x^2 + 2$, and $C = x^2 + 3x - 2$, find each of the following. 44. $3x^5 + 13x^4 + 12x^3 + 18x^2 + 12x - 16$

43. $-8x^4 - 6x^3 + 24x^2 - 12x + 80$

42. $AC + B$ $3x^3 + 14x^2 + 6x - 6$
43. $2B(3A - 4C)$
44. ABC
45. $(A + B)(B - C)$ $-3x^3 - 5x^2 - 6x + 24$

Applications and Problem Solving

46. **Construction** A homeowner is considering installing a swimming pool in his backyard. He wants its length to be 5 yards longer than its width, to make room for a diving area at one end. Then he wants to surround it with a concrete walkway 4 yards wide. After finding out the price of concrete, he decides that he can afford 424 square yards of it for the walkway. What should the dimensions of the pool be? **20 yd by 25 yd**

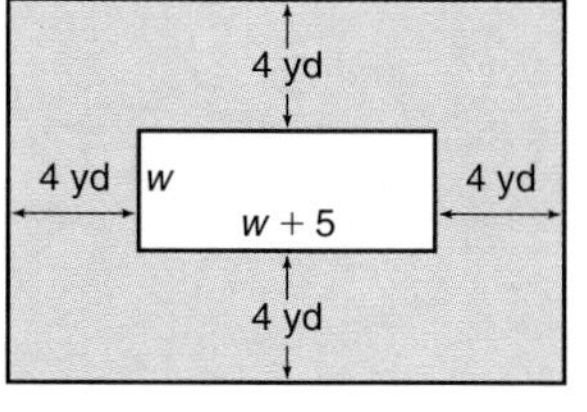

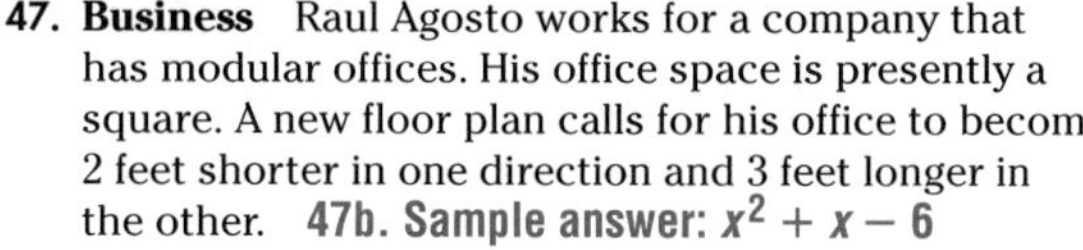

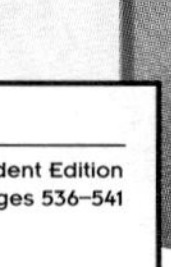

47a. Sample answer: $x - 2$, $x + 3$

47. **Business** Raul Agosto works for a company that has modular offices. His office space is presently a square. A new floor plan calls for his office to become 2 feet shorter in one direction and 3 feet longer in the other. 47b. Sample answer: $x^2 + x - 6$
   a. Write expressions that represent the new dimensions of Mr. Agosto's office.
   b. Find the area of his new office.
   c. Suppose his office is presently 8 feet by 8 feet. Will his new office be bigger or smaller than this office? by how much? **larger, 2 sq ft**

Mixed Review

48. Find $\frac{3}{4}a(6a + 12)$. (Lesson 9–6) $\frac{9}{2}a^2 + 9a$
49. Solve $6 - 9y < -10y$. (Lesson 7–3) $\{y \mid y < -6\}$

Practice Masters, p. 67

9-7 NAME________ DATE ________

Practice Student Edition Pages 536–541

Multiplying Polynomials

Find each product.

1. $(x + 2)(x + 7)$ $x^2 + 9x + 14$
2. $(y + 6)(y + 3)$ $y^2 + 9y + 18$
3. $(a - 3)(a - 8)$ $a^2 - 11a + 24$
4. $(r - 5)(r - 1)$ $r^2 - 6r + 5$
5. $(x - 10)(x + 2)$ $x^2 - 8x - 20$
6. $(3t + 2q)(7t - 9q)$ $21t^2 - 13qt - 18q^2$
7. $(7y - 4t)(2y + 5t)$ $14y^2 + 27ty - 20t^2$
8. $(5w - 3y)(w - 2y)$ $5w^2 - 13yw + 6y^2$
9. $(a + b)(2a - 3b)$ $2a^2 - ab - 3b^2$
10. $(0.3n + 5)(0.4n - 11)$ $0.12n^2 - 1.3n - 55$
11. $(1.3g + 3)(0.4g + 5)$ $0.52g^2 + 7.7g + 15$
12. $(x + y)(x + y)$ $x^2 + 2xy + y^2$
13. $(z - 2x)(z - 2x)$ $z^2 - 4xz + 4x^2$
14. $(3x + 2y)(2y + 3x)$ $4y^2 + 12xy + 9x^2$
15. $(y^2 + 3y - 6) \times (2y + 3)$ $2y^3 + 9y^2 - 3y - 18$
16. $(a^2 + 2a - 9) \times (a + 3)$ $a^3 + 5a^2 - 3a - 27$
17. $(e^2 - 2ef + f^2) \times (e - f)$ $e^3 - 3e^2f + 3ef^2 - f^3$
18. $(9m^2 - 12m + 4) \times (3m + 2)$ $27m^3 - 18m^2 - 12m + 8$
19. $(2a - 1)(a^2 + 3a + 5)$ $2a^3 + 5a^2 + 7a - 5$
20. $(2 + 3c)(4 + 6c + 9c^2)$ $8 + 24c + 36c^2 + 27c^3$

50. **National Landmarks** At the Royal Gorge in Colorado, an inclined railway takes visitors down to the Arkansas River. Suppose the slope is 50% and the vertical drop is 1015 feet. What is the horizontal change of the railway? (Lesson 6–1) **2030 ft**

51. Write an equation to represent the relation. (Lesson 5–6) **$y = 2x + 1$**
$\{(-1, -1), (0, 1), (1, 3), (2, 5), (3, 7)\}$

52. Determine the domain, range, and inverse of the relation. (Lesson 5–2)
$\{(8, 1), (4, 2), (6, -4), (5, -3), (6, 0)\}$ **See margin.**

53. **Geometry** ΔABC and ΔXYZ are similar. Find the values of a and y. (Lesson 4–2) **$a = 4$, $y = 9$**

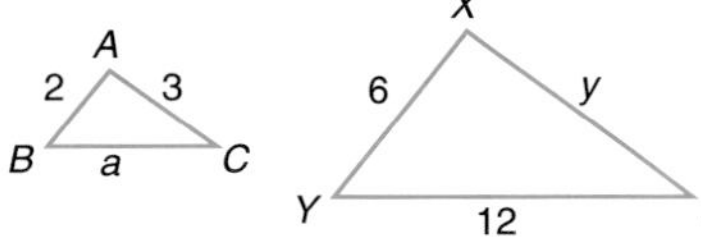

54. **Consumerism** The prices of six different models of printers in a computer store are $299, $369, $359, $228, $525, and $398. Find the mean and median prices for the printers. (Lesson 3–7) **$363; $364**

55. **Temperature** The formula for finding the Celsius temperature C when you know the Fahrenheit temperature F is $C = \frac{5}{9}(F - 32)$. Find the Celsius temperature when the Fahrenheit temperature is 59°. (Lesson 2–9) **15°C**

56. Replace the variable to make the sentence $\frac{3}{4}s = 6$ true. (Lesson 1–5) **8**

Refer to the Investigation on pages 448–449.

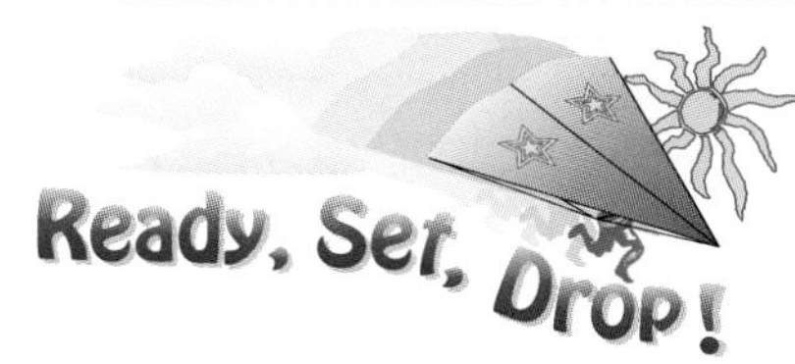

You have experimented with various sizes of gliders to explore their flying abilities. You now need to investigate how the size of the glider and the weight of the load are related.

1 Cut out two equilateral triangles from construction paper so that the side of one is 6 centimeters long, and the side of the other is 12 centimeters long. What is the surface area and perimeter of each triangle?

2 Straighten two paper clips and bend them into the shape shown at the right. Punch the bent end of the paper clip through the center of the triangle, and tape it into place so that the hook-end hangs down under the other side of the triangle.

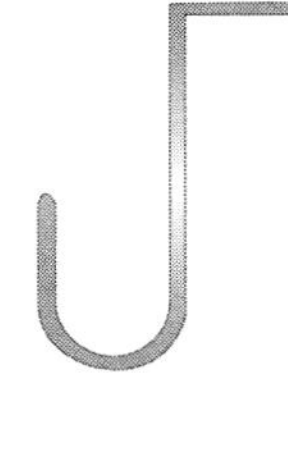

3 One at a time, drop each hang glider from the top of the bleachers or out of a second-floor window. (In order to get more accurate data, a greater height is needed than those used for previous experiments.) Record the glide time of each hang glider.

4 Add one washer onto the hook of each glider. Repeat the dropping procedure and record the times. How do those times compare with the first drop? Continue to add washers, one at a time, to each glider and record the glide times in a chart that compares the glide times with the number of washers carried by each glider.

5 Graph the relationship between the weight and the glide times. Let the independent variable be the weight, and let the dependent variable be the glide time. Analyze your findings.

6 Look at the weight, glide time, surface area, and perimeter in your data. Write a polynomial expression to relate some, if not all, of these measures.

Add the results of your work to your Investigation Folder.

4 ASSESS

Closing Activity

Modeling The following diagram illustrates all the multiplications that must be performed when multiplying two polynomials.

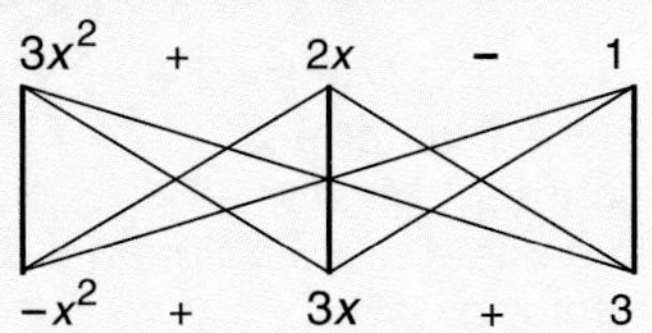

Have students draw such diagrams for several problems.

Additional Answer

52. $\{8, 4, 6, 5\}$; $\{1, 2, -4, -3, 0\}$; $\{(1, 8), (2, 4), (-4, 6), (-3, 5), (0, 6)\}$

Enrichment Masters, p. 67

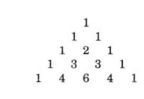

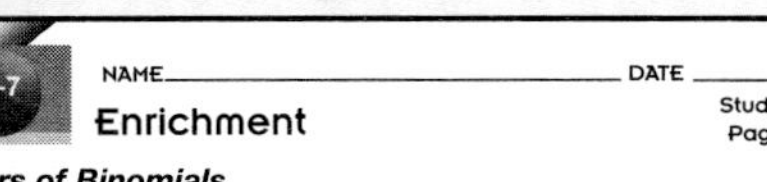

9-7 NAME______ DATE______

Enrichment

Student Edition Pages 536–541

Powers of Binomials

This arrangement of numbers is called Pascal's Triangle. It was first published in 1665, but was known hundreds of years earlier.

1
1 1
1 2 1
1 3 3 1
1 4 6 4 1

1. Each number in the triangle is found by adding two numbers. What two numbers were added to get the 6 in the 5th row?
3 and 3

2. Describe how to create the 6th row of Pascal's Triangle.
The first and last numbers are 1. Evaluate 1 + 4, 4 + 6, 6 + 4, and 4 + 1 to find the other numbers.

3. Write the numbers for rows 6 through 10 of the triangle.

Row 6: 1 5 10 10 5 1
Row 7: 1 6 15 20 15 6 1
Row 8: 1 7 21 35 35 21 7 1
Row 9: 1 8 28 56 70 56 28 8 1
Row 10: 1 9 37 84 126 126 84 37 9 1

Multiply to find the expanded form of each product.

4. $(a + b)^2$ $a^2 + 2ab + b^2$

5. $(a + b)^3$ $a^3 + 3a^2b + 3ab^2 + b^3$

6. $(a + b)^4$ $a^4 + 4a^3b + 6a^2b^2 + 4ab^3 + b^4$

Now compare the coefficients of the three products in Exercises 4–6 with Pascal's Triangle.

7. Describe the relationship between the expanded form of $(a + b)^n$ and Pascal's Triangle. The coefficients of the expanded form are found in row $n + 1$ of Pascal's Triangle.

8. Use Pascal's Triangle to write the expanded form of $(a + b)^6$.
$a^6 + 6a^5b + 15a^4b^2 + 20a^3b^3 + 15a^2b^4 + 6ab^5 + b^6$

Extension

Communication Have students find the product of $(4m^2 - m + 8)$ and $(m^3 + 2m^2 + 3m + 4)$ and then describe the process they used to solve the problem. **$4m^5 + 7m^4 + 18m^3 + 29m^2 + 20m + 32$**

In·ves·ti·ga·tion

Working on the Investigation

The Investigation on pages 448–449 is designed to be a long-term project that is completed over several days or weeks. Encourage students to keep their materials in their Investigation Folder as they work on the Investigation.

9-8 LESSON NOTES

NCTM Standards: 1–5

Instructional Resources

- Study Guide Master 9-8
- Practice Master 9-8
- Enrichment Master 9-8
- Assessment and Evaluation Masters, p. 241

Transparency 9-8A contains the 5-Minute Check for this lesson; **Transparency 9-8B** contains a teaching aid for this lesson.

Recommended Pacing	
Standard Pacing	Day 13 of 15
Honors Pacing	Day 12 of 14
Block Scheduling*	Day 7 of 8
Alg. 1 in Two Years*	Days 21, 22, & 23 of 26

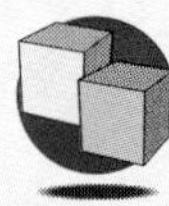

*For more information on pacing and possible lesson plans, refer to the *Block Scheduling Booklet* and *Algebra 1 in Two Years.*

1 FOCUS

5-Minute Check
(over Lesson 9-7)

Find each product.

1. $(3w + 7)(2w + 5)$
   $6w^2 + 29w + 35$
2. $(a + 6)(a - 3)$
   $a^2 + 3a - 18$
3. $(3m - 4)(5m + 6)$
   $15m^2 - 2m - 24$
4. $(2b^2 - 5b - 3)(5b^2 + 3b - 2)$
   $10b^4 - 19b^3 - 34b^2 + b + 6$
5. $(8j^2 - 2j + 1)(2j^2 + 8j + 1)$
   $16j^4 + 60j^3 - 6j^2 + 6j + 1$

Motivating the Lesson

Hands-On Activity Have students use algebra tiles to represent the following perfect squares: 1, 4, 9, 16, 25, 36, 49, 64, and 81. Using the tile representations, students should determine the square root of each number.
1, 2, 3, 4, 5, 6, 7, 8, 9

9–8 Special Products

What YOU'LL LEARN
- To use patterns to find $(a + b)^2$, $(a - b)^2$, and $(a + b)(a - b)$.

Why IT'S IMPORTANT
You can use polynomials to solve problems involving biology and history.

CONNECTION
Biology

Punnett squares are diagrams that are used to show the possible ways that genes can combine at fertilization. In a Punnett square, *dominant* genes are shown with capital letters. Recessive genes are shown with lowercase letters. Letters representing the parents' genes are placed on two of the outer sides of the Punnett square. Letters inside the boxes of the square show the possible gene combinations of their offspring.

FYI
Punnett squares are named after the English zoologist and geneticist Reginald Crundall Punnett (1875–1967). He held the first chair in genetics at Cambridge University and helped build the science of genetics in the 20th century.

The Punnett square below represents a cross between tall pea plants and short pea plants. Let T represent the dominant gene for tallness. Let t represent the recessive gene for shortness. The parents are called *hybrids,* since they have one of each kind of gene.

Hybrid tall × Hybrid tall

Tall = T

Short = t

Offspring

$\frac{1}{4}$ or 25% pure tall (TT)

$\frac{2}{4}$ or 50% hybrid tall (Tt)

$\frac{1}{4}$ or 25% pure short (tt)

hybrid tall	T	t
T	TT	Tt
t	Tt	tt

Because the parent plants have both a dominant tall gene and a recessive short gene, biologists know that their offspring can be predicted by squaring the binomial $(0.5T + 0.5t)^2$. Therefore, the following must be true.

$$\begin{aligned}(0.5T + 0.5t)^2 &= (0.5T + 0.5t)(0.5T + 0.5t)\\ &= 0.5T(0.5T) + 0.5T(0.5t) + 0.5t(0.5T) + 0.5t(0.5t)\\ &= 0.25T^2 + 0.25Tt + 0.25Tt + 0.25t^2\\ &= 0.25T^2 + 0.50Tt + 0.25t^2\end{aligned}$$

T^2 and t^2 represent TT and tt, respectively.

FYI
Squares very similar to Punnett squares are also used in logic, probability, and the design of digital computer circuits.

You can use the diagram below to derive a general form for the expression $(a + b)^2$.

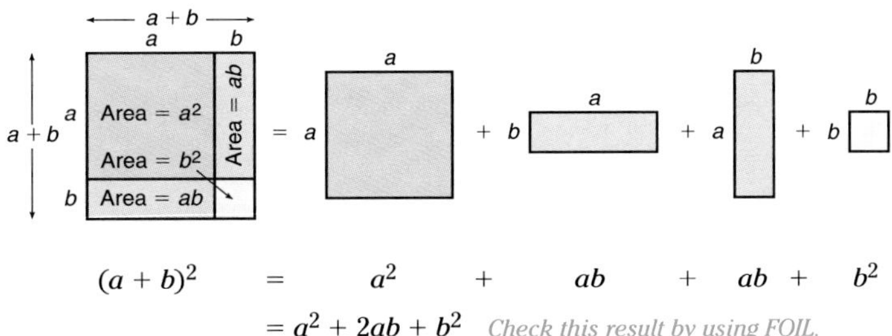

$(a + b)^2 \quad = \quad a^2 \quad + \quad ab \quad + \quad ab \quad + \quad b^2$

$= a^2 + 2ab + b^2$ *Check this result by using FOIL.*

In general, the square of a binomial that is a sum can be found by using the following rule.

Square of a Sum	$(a + b)^2 = (a + b)(a + b)$ $= a^2 + 2ab + b^2$

Example 1 **Find each product.**

a. $(y + 7)^2$

Method 1

Use the square of a sum rule.

$(a + b)^2 = a^2 + 2ab + b^2$

$(y + 7)^2 = y^2 + 2(y)(7) + 7^2$

$= y^2 + 14y + 49$

Method 2

Use FOIL.

$(y + 7)^2 = (y + 7)(y + 7)$

$= y^2 + 7y + 7y + 49$

$= y^2 + 14y + 49$

b. $(6p + 11q)^2$

$(a + b)^2 = a^2 + 2ab + b^2$

$(6p + 11q)^2 = (6p)^2 + 2(6p)(11q) + (11q)^2$ *$a = 6p$ and $b = 11q$*

$= 36p^2 + 132pq + 121q^2$

The square of a sum rule can be used with other rules to simplify products of polynomials.

Example 2

History

Tourists to the southern part of England can visit the historic Gwennap Pit. In the 16th century, the pit of a tin mine was converted into an amphitheater. During the 18th century, John Wesley spoke to overflow crowds in this amphitheater. Gwennap Pit consists of a circular stage surrounded by circular levels used for seating. Each seating level is 1 meter wide. Suppose the radius of the stage is *s* meters. Find the area of the third seating level.

(continued on the next page)

2 TEACH

In-Class Examples

For Example 1
Find each product.

a. $(3a + 2)^2$ $\mathbf{9a^2 + 12a + 4}$

b. $(5b + 7c)^2$
$\mathbf{25b^2 + 70bc + 49c^2}$

For Example 2
What would be the area of the third seating level if each seating level were 2 meters wide?
12.6*s* + 62.8 square meters

In-Class Example

For Example 3
Find each product.

a. $(y - 13)^2$ $y^2 - 26y + 169$
b. $(6a^2 - 10x)^2$
$36a^4 - 120a^2x + 100x^2$

Teaching Tip Emphasize to students that the square of a difference is another case of multiplying binomials by FOIL.

$$(a - b)^2 = (a - b)(a - b)$$
F O I L
$$= a^2 - ab - ba + b^2$$
$$= a^2 - 2ab + b^2$$

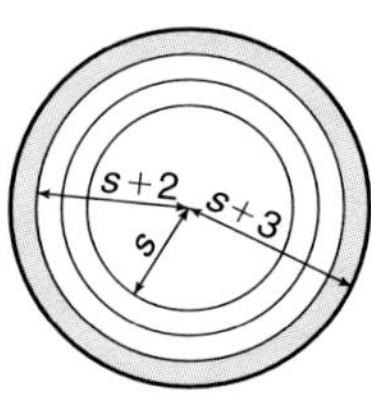

The area of a circle equals πr^2. The radius of the second seating level is $s + 2$ meters, and the radius of the third seating level is $s + 3$. The area of the third seating level can be found by subtracting the areas of two circles.

$$A = \underbrace{\pi(s + 3)^2}_{\text{area of third level}} - \underbrace{\pi(s + 2)^2}_{\text{area of second level}}$$
$$= \pi(s^2 + 6s + 9) - \pi(s^2 + 4s + 4) \quad \textit{Square of a sum rule}$$
$$= (\pi s^2 + 6\pi s + 9\pi) - (\pi s^2 + 4\pi s + 4\pi) \quad \textit{Distributive property}$$
$$= \pi s^2 + 6\pi s + 9\pi - \pi s^2 - 4\pi s - 4\pi$$
$$= 2\pi s + 5\pi \quad \textit{Combine like terms.}$$

The area of the third seating level is $2\pi s + 5\pi$, or about $6.3s + 15.7$ square meters.

To find $(a - b)^2$, write $(a - b)$ as $[a + (-b)]$ and square it.

$$(a - b)^2 = [a + (-b)]^2$$
$$= a^2 + 2(a)(-b) + (-b)^2$$
$$= a^2 - 2ab + b^2$$

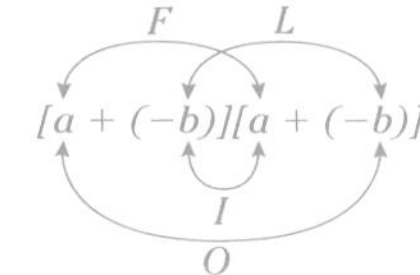

In general, the square of a binomial that is a difference can be found by using the following rule.

Square of a Difference	$(a - b)^2 = (a - b)(a - b)$ $= a^2 - 2ab + b^2$

Example 3 **Find each product.**

a. $(r - 6)^2$

Method 1

Use the square of a difference rule.

$$(a - b)^2 = a^2 - 2ab + b^2$$
$$(r - 6)^2 = r^2 - 2(r)(6) + 6^2$$
$$= r^2 - 12r + 36$$

Method 2

Use FOIL.

$$(r - 6)^2 = (r - 6)(r - 6)$$
$$= r^2 - 6r - 6r + 36$$
$$= r^2 - 12r + 36$$

b. $(4x^2 - 7t)^2$

$$(a - b)^2 = a^2 - 2ab + b^2$$
$$(4x^2 - 7t)^2 = (4x^2)^2 - 2(4x^2)(7t) + (7t)^2 \quad a = 4x^2 \text{ and } b = 7t$$
$$= 16x^4 - 56x^2t + 49t^2$$

Product of a Sum and a Difference

Materials:  algebra tiles  product mat

You have learned how to use algebra tiles to find the product of two binomials. In this activity, you will use algebra tiles to study a special situation.

Your Turn

a. Use algebra tiles to find each product.

$(x + 3)(x - 3)$ $\mathbf{x^2 - 9}$ $(x + 5)(x - 5)$ $\mathbf{x^2 - 25}$

$(x + 1)(x - 1)$ $\mathbf{x^2 - 1}$ $(x + 2)(x - 2)$ $\mathbf{x^2 - 4}$

$(x + 6)(x - 6)$ $\mathbf{x^2 - 36}$ $(x + 4)(x - 4)$ $\mathbf{x^2 - 16}$

b. What do you notice about the binomials used as factors in part a?

b. They are the sum and difference of the same values.

c. What pattern do the products in part a have?

c. They are the differences of the square of the first number and the square of the second number.

You can use the FOIL method to find the product of a sum and a difference of the same two numbers.

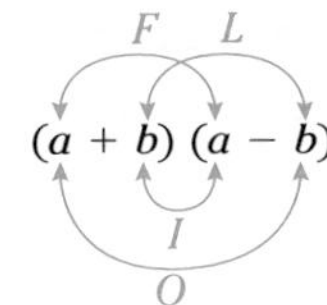

$$\begin{aligned}(a + b)(a - b) &= a(a) + a(-b) + b(a) + b(-b)\\ &= a^2 - ab + ab - b^2\\ &= a^2 - b^2\end{aligned}$$

The resulting product, $a^2 - b^2$, has a special name. It is called a **difference of squares.**

Difference of Squares	$(a + b)(a - b) = (a - b)(a + b)$ $= a^2 - b^2$

Example  **Find each product.**

a. $(m - 2n)(m + 2n)$

$$(a - b)(a + b) = a^2 - b^2$$

$$(m - 2n)(m + 2n) = m^2 - (2n)^2 \quad a = m \text{ and } b = 2n$$

$$= m^2 - 4n^2$$

b. $(0.3t + 0.25w^2)(0.3t - 0.25w^2)$

$$(a + b)(a - b) = a^2 - b^2$$

$$(0.3t + 0.25w^2)(0.3t - 0.25w^2) = (0.3t)^2 - (0.25w^2)^2 \quad a = 0.3t \text{ and } b = 0.25w^2$$

$$= 0.09t^2 - 0.0625w^4$$

Reteaching

Using Reasoning Use the pattern of the square of a sum, $(a + b)^2 = a^2 + 2ab + b^2$, to square numbers. For example, $(27)^2 = (20 + 7)^2 = 20^2 + 2 \cdot 140 + 49$ or 729.

MODELING MATHEMATICS Students must be careful not to confuse algebraic products with perfect squares in number theory: $9 = 3 \times 3$, not $9 = 3 \times (-3)$; $x^2 - 9 = (x + 3)(x - 3)$, not $x^2 - 9 = (x + 3)(x + 3)$.

Teaching Tip Have students compare these expressions.

square of a difference
$(a - b)^2$
and
difference of two squares
$a^2 - b^2$

Point out that the difference of two squares is the only time the product of two binomials is a binomial.

In-Class Example

For Example 4
Find each product.

a. $(6x - 20y)(6x + 20y)$
$\mathbf{36x^2 - 400y^2}$

b. $(14k + 9t^2)(14k - 9t^2)$
$\mathbf{196k^2 - 81t^4}$

Study Guide Masters, p. 68

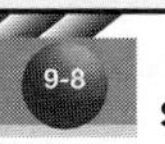

NAME________________ DATE ________

Study Guide

Student Edition Pages 542–547

Special Products

You can use the FOIL method to find some special products.

Square of a Sum	$(a + b)^2 = (a + b)(a + b) = a^2 + 2ab + b^2$
Square of a Difference	$(a - b)^2 = (a - b)(a - b) = a^2 - 2ab + b^2$
Difference of Squares	$(a + b)(a - b) = (a - b)(a + b) = a^2 - b^2$

Study the examples below.

Binomials	Product
$(3n + 4)^2$	$9n^2 + 24n + 16$
$(2z - 9)^2$	$4z^2 - 36z + 81$
$(5x - 3y)(5x + 3y)$	$25x^2 - 9y^2$

Find each product.

1. $(x - 6)^2$ $x^2 - 12x + 36$
2. $(3p + 4)^2$ $9p^2 + 24p + 16$
3. $(x + 11)(x - 11)$ $x^2 - 121$
4. $(2x + 3)(2x - 3)$ $4x^2 - 9$
5. $(4x - 5)^2$ $16x^2 - 40x + 25$
6. $(9x - y)(9x + y)$ $81x^2 - y^2$
7. $(m + 5)^2$ $m^2 + 10m + 25$
8. $(8a - 7b)(8a + 7b)$ $64a^2 - 49b^2$
9. $(4a - 3b)^2$ $16a^2 - 24ab + 9b^2$
10. $(3 - 5q)(3 + 5q)$ $9 - 25q^2$
11. $(x^2 - 2)^2$ $x^4 - 4x^2 + 4$
12. $(2.5 + q)^2$ $6.25 + 5q + q^2$
13. $\left(\frac{3}{4}x + 1\right)\left(\frac{3}{4}x - 1\right)$ $\frac{9}{16}x^2 - 1$
14. $(0.3p - 2q)^2$ $0.09p^2 - 1.2pq + 4q^2$
15. $\left(\frac{1}{2}y + z\right)^2$ $\frac{1}{4}y^2 + yz + z^2$
16. $(8 + x)^2$ $64 + 16x + x^2$
17. $(6c - 10)(6c + 10)$ $36c^2 - 100$
18. $(x^3 - 1)^2$ $x^6 - 2x^3 + 1$

3 PRACTICE/APPLY

Check for Understanding

Exercises 1–12 are designed to help you assess your students' understanding through reading, writing, speaking, and modeling. You should work through Exercises 1–5 with your students and then monitor their work on Exercises 6–12.

Error Analysis

It is very common for students to assume that $(a + b)^2 = a^2 + b^2$. They reason that to square a quantity, it is acceptable to square its parts. Remind students to write out the multiplication. $(a + b)^2$ means $(a + b)(a + b)$ or $a^2 + 2ab + b^2$.

Assignment Guide
Core: 13–35 odd, 37–43
Enriched: 14–32 even, 33–43

For **Extra Practice,** see p. 778.

The red A, B, and C flags, printed only in the Teacher's Wraparound Edition, indicate the level of difficulty of the exercises.

Practice Masters, p. 68

NAME______________________ DATE ____________

Practice

Student Edition Pages 542–547

Special Products

Find each product.

1. $(z + 4)^2$ — $z^2 + 8z + 16$
2. $(3p + 9)^2$ — $9p^2 + 54p + 81$
3. $(4u - 5)^2$ — $16u^2 - 40u + 25$
4. $(5v - 8)^2$ — $25v^2 - 80v + 64$
5. $(10r - 3s)^2$ — $100r^2 - 60rs + 9s^2$
6. $(6 - p)^2$ — $36 - 12p + p^2$
7. $(3k + 2h)^2$ — $9k^2 + 12kh + 4h^2$
8. $(y - 2z)^2$ — $y^2 - 4yz + 4z^2$
9. $(11a^4 - p^3)^2$ — $121a^8 - 22a^4p^3 + p^6$
10. $(3.1y + 0.4w)^2$ — $9.61y^2 + 2.48wy + 0.16w^2$
11. $(0.4x - 3.2)^2$ — $0.16x^2 - 2.56x + 10.24$
12. $(2.2c + d)^2$ — $4.84c^2 + 4.4cd + d^2$
13. $(t^3 - 3r^2)^2$ — $t^6 - 6r^2t^3 + 9r^4$
14. $(x^2 - 5y^2)^2$ — $x^4 - 10x^2y^2 + 25y^4$
15. $(2x - 3)(2x + 3)$ — $4x^2 - 9$
16. $(c - 8)(c + 8)$ — $c^2 - 64$
17. $(2u + 1)(2u - 1)$ — $4u^2 - 1$
18. $(7n - 9)(7n + 9)$ — $49n^2 - 81$
19. $(3d + 8x)(3d - 8x)$ — $9d^2 - 64x^2$
20. $(2x - 3y)(2x + 3y)$ — $4x^2 - 9y^2$
21. $\left(\frac{4}{5}y + m\right)\left(\frac{4}{5}y - m\right)$ — $\frac{16}{25}y^2 - m^2$
22. $\left(\frac{1}{3}x - \frac{2}{3}y\right)\left(\frac{1}{3}x + \frac{2}{3}y\right)$ — $\frac{1}{9}x^2 - \frac{4}{9}y^2$
23. $(6x^2 - 5y^3)(6x^2 + 5y^3)$ — $36x^4 - 25y^6$
24. $(2y^3 - 7z^2)(2y^3 + 7z^2)$ — $4y^6 - 49z^4$
25. $(8 - n^3)(8 + n^3)$ — $64 - n^6$
26. $(2a^2 + 3b^2)(2a^2 + 3b^2)$ — $4a^4 + 12a^2b^2 + 9b^4$
27. $(x + 1)(x + 2)(x - 1)$ — $x^3 + 2x^2 - x - 2$
28. $(2p - 1)(2p + 1)(2p - 1)$ — $8p^3 - 4p^2 - 2p + 1$

CHECK FOR UNDERSTANDING

Communicating Mathematics

Study the lesson. Then complete the following.

1. **Explain** how the square of a difference and the square of a sum are different. **The middle terms have different signs.**
2. **Compare and contrast** the square of a difference and the difference of two squares. **See margin.**
3. **Explain** how you could mentally multiply 29×31. (*Hint:* $29 = 30 - 1$ and $31 = 30 + 1$) $(30 - 1)(30 + 1) = 900 - 1$ **or 899**

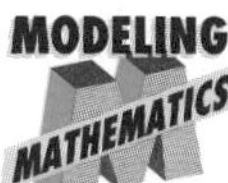

4. Draw a diagram to represent each of the following. **a–b. See margin.**
   a. $(x + y)^2$ b. $(x - y)^2$
5. What does the diagram at the right represent if the shading represents regions to be removed or subtracted?
   $(a - b)^2 = a^2 - 2ab + b^2$

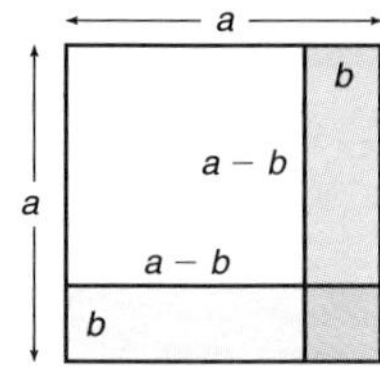

Guided Practice

Find each product.

6. $(2x + 3y)^2$ $4x^2 + 12xy + 9y^2$
7. $(m - 3n)^2$ $m^2 - 6mn + 9n^2$
8. $(2a + 3)(2a - 3)$ $4a^2 - 9$
9. $(m^2 + 4n)^2$ $m^4 + 8m^2n + 16n^2$
10. $(4y + 2z)(4y - 2z)$ $16y^2 - 4z^2$
11. $(5 - x)^2$ $25 - 10x + x^2$
12. **Recreation** In India, children play a form of hopscotch called Chilly. One of the three possible patterns for this game is shown at the right. Suppose each side of the small squares is $2x + 5$ units long. Find the area of this Chilly pattern. $16x^2 + 80x + 100$ **square units**

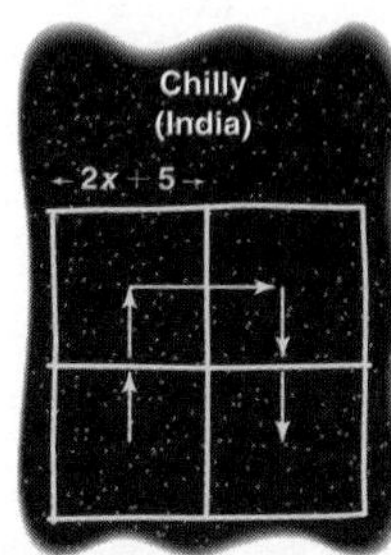

EXERCISES

Practice

Find each product.

13. $(x + 4y)^2$ $x^2 + 8xy + 16y^2$
14. $(m - 2n)^2$ $m^2 - 4mn + 4n^2$
15. $(3b - a)^2$ $9b^2 - 6ab + a^2$
16. $(3x + 5)(3x - 5)$ $9x^2 - 25$
17. $(9p - 2q)(9p + 2q)$ $81p^2 - 4q^2$
18. $(5s + 6t)^2$ $25s^2 + 60st + 36t^2$

B

19. $(5b - 12a)^2$ $25b^2 - 120ab + 144a^2$
20. $(2a + 0.5y)^2$ $4a^2 + 2ay + 0.25y^2$
21. $(x^3 + a^2)^2$ $x^6 + 2x^3a^2 + a^4$
22. $\left(\frac{1}{2}b^2 - a^2\right)^2$ $\frac{1}{4}b^4 - a^2b^2 + a^4$
23. $(8x^2 - 3y)(8x^2 + 3y)$ $64x^4 - 9y^2$
24. $(7c^2 + d^3)(7c^2 - d^3)$ $49c^4 - d^6$
25. $(1.1g + h^5)^2$ $1.21g^2 + 2.2gh^5 + h^{10}$
26. $(9 - z^9)(9 + z^9)$ $81 - z^{18}$
27. $\left(\frac{4}{3}x^2 - y\right)\left(\frac{4}{3}x^2 + y\right)$ $\frac{16}{9}x^4 - y^2$
28. $\left(\frac{1}{3}v^2 - \frac{1}{2}w^3\right)^2$

C

29. $(3x + 1)(3x - 1)(x - 5)$
30. $(x - 2)(x + 5)(x + 2)(x - 5)$
31. $(a + 3b)^3$ $a^3 + 9a^2b + 27ab^2 + 27b^3$
32. $(2m - n)^4$ $16m^4 - 32m^3n + 24m^2n^2 - 8mn^3 + n^4$

Critical Thinking

33. Find $(x + y + z)^2$. Draw a diagram to show each term of the polynomial.

28. $\frac{1}{9}v^4 - \frac{1}{3}v^2w^3 + \frac{1}{4}w^6$
29. $9x^3 - 45x^2 - x + 5$
30. $x^4 - 29x^2 + 100$
33. $x^2 + y^2 + z^2 + 2xy + 2yz + 2xz$; See margin for diagram.

Additional Answers

2. The square of a difference is $(a - b)^2$, which equals $a^2 - 2ab + b^2$. The difference of two squares is the product $(a + b)(a - b)$ or $a^2 - b^2$.

4a.

	x	y
x	x^2	xy
y	xy	y^2

4b.

	x	−y
x	x^2	$-xy$
−y	$-xy$	y^2

33.

	x	y	z
x	x^2	xy	xz
y	xy	y^2	yz
z	xz	yz	z^2

Applications and Problem Solving

34. **Biology** Refer to the application at the beginning of the lesson.
    a. Make a Punnett square for pea plants if one parent is pure short (tt) and the other parent is hybrid tall (Tt). **See margin.**
    b. What percent of the offspring will be pure short? **50%**
    c. What percent of the offspring will be hybrid tall? **50%**
    d. What percent of the offspring will be pure tall? **0%**

35. **History** Refer to Example 2.
    a. Write an expression for the area of the fourth seating level in the Gwennap Pit. **$2\pi s + 7\pi$ square meters**
    b. The radius of the stage level of the Gwennap Pit is 3 meters. Find the area of the stage. **about 28.27 square meters**
    c. Find the area of the fourth seating level in the Gwennap Pit. **about 40.84 square meters**

36. **Photography** Lenora cut off a 0.75-inch strip all around a square photograph so it would fit in an envelope she was mailing to her aunt. She decided to have the photo lab make a copy of the photo from the negative, but she forgot to measure how large the original photo was. All she had was the strips she cut off, whose area was 33.75 square inches. What were the original dimensions of the photograph? **12 in. by 12 in.**

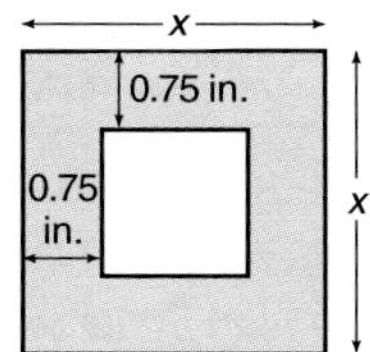

Mixed Review

37. Find $(3t - 3)(2t + 1)$. (Lesson 9–7) **$6t^2 - 3t - 3$**

38. Solve $-13z > -1.04$. (Lesson 7–2) **$\{z \mid z < 0.08\}$**

39. **Statistics** The table below shows the heights and weights of each of 12 players on a pro basketball team. (Lesson 6–3)

Height (in.)	75	82	75	74	80	80	75	79	80	78	76	81
Weight (lb)	180	235	184	185	230	205	185	230	221	195	205	215

    a. Make a scatter plot of these data. **See margin.**
    b. Describe the correlation between height and weight.

39b. the taller the player, the greater the weight

40. Write the standard form of the line that passes through (3, 1) and has a slope of $\frac{2}{7}$. (Lesson 6–2) **$2x - 7y = -1$**

41. Graph the points $A(4, 2)$, $B(-3, 1)$, and $C(-2, -3)$. (Lesson 5–1) **See margin.**

42. **Electricity** The resistance R of a power circuit is 4.5 ohms. How much current I, in amperes, can the circuit generate if it can produce at most 1500 watts of power? Use $I^2R = P$. (Lesson 2–8)

42. about 18.26 amperes

43. Evaluate $5(9 \div 3^2)$. (Lesson 1–6) **5**

Extension

Connections The formula for the volume of a cube is $V = e^3$, where e represents the length of an edge. Find the volume of a cube if the length of its edge is $a + b$.
$a^3 + 3a^2b + 3ab^2 + b^3$

4 ASSESS

Closing Activity

Writing Have students write a paragraph summarizing the steps to follow when squaring binomials.

Chapter 9, Quiz D (Lessons 9-7 and 9-8), is available in the *Assessment and Evaluation Masters*, p. 241.

Additional Answers

34a.

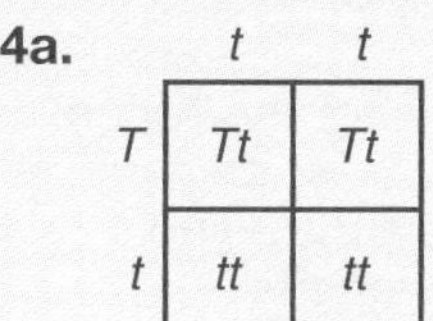

39a.

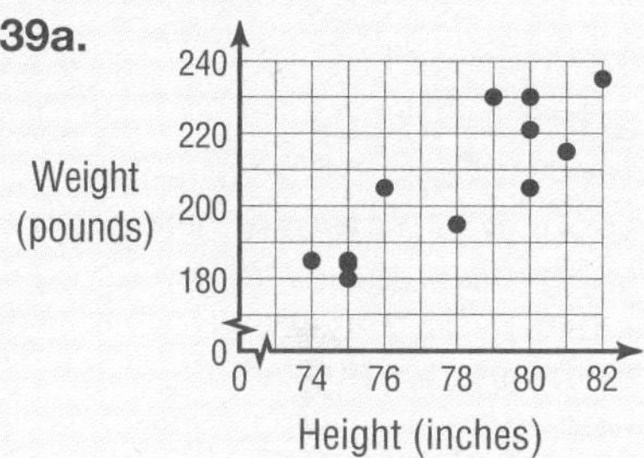

41.

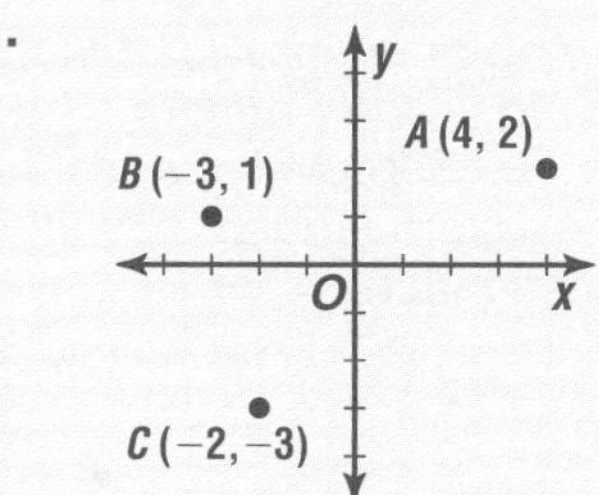

Enrichment Masters, p. 68

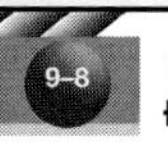

NAME ____________ DATE ________

Enrichment

Student Edition Pages 542–548

Special Polynomial Products

Sometimes the product of two polynomials can be found readily with the use of one of the special products of binomials.

For example, you can find the square of a trinomial by recalling the square of a binomial.

Example 1: Find $(x + y + z)^2$.

$$(a + b)^2 = a^2 + 2 \cdot a \cdot b + b^2$$
$$[(x + y) + z]^2 = (x + y)^2 + 2(x + y)z + z^2$$
$$= x^2 + 2xy + y^2 + 2xz + 2yz + z^2$$

Example 2: Find $(3t + x + 1)(3t - x - 1)$.

(*Hint:* $(3t + x + 1)(3t - x - 1)$ is the product of a sum $3t + (x + 1)$ and a difference $3t - (x + 1)$.)

$$(3t + x + 1)(3t - x - 1) = [3t + (x + 1)][3t - (x + 1)]$$
$$= 9t^2 - (x + 1)^2$$
$$= 9t^2 - x^2 - 2x - 1$$

Use a special product of binomials to find each product.

1. $(x + y - z)^2$ $x^2 + 2xy - 2xz + y^2 - 2yz + z^2$
2. $(r + s + 5)^2$ $r^2 + 2rs + s^2 + 10r + 10s + 25$
3. $(b - 3 + d)^2$ $b^2 - 6b + 2bd + 9 - 6d + d^2$
4. $(k - m - 2)^2$ $k^2 - 2km + m^2 - 4k + 4m + 4$
5. $(x + 1 + 2b)(x + 1 - 2b)$ $x^2 + 2x + 1 - 4b^2$
6. $(y - 2 + x)(y - 2 - x)$ $y^2 - 4y + 4 - x^2$
7. $(5 + b - x)(5 + b + x)$ $25 + 10b + b^2 - x^2$
8. $(j - 5 - f)(j + 5 + f)$ $j^2 - f^2 - 10f - 25$
9. $[(x + y) + (z + w)][(x + y) - (z + w)]$ $x^2 + 2xy + y^2 - z^2 - 2zw - w^2$
10. $(2a + 1 + 3b - c)(2a + 1 - 3b + c)$ $4a^2 + 4a + 1 - 9b^2 + 6bc - c^2$

In•ves•ti•ga•tion

TEACHER NOTES

Closing the Investigation

This activity provides students an opportunity to bring their work on the Investigation to a close. For each Investigation, students should present their findings to the class. Here are some ways students can display their work.

- Conduct and report on an interview or survey.
- Write a letter, proposal, or report.
- Write an article for the school or local paper.
- Make a display, including graphs and/or charts.
- Plan an activity.

Assessment

To assess students' understanding of the concepts and topics explored in this Investigation and its follow-up activities, you may wish to examine students' Investigation Folders.

The scoring guide provided in the *Investigations and Projects Masters,* p. 15, provides a means for you to score students' work on this Investigation.

Investigations and Projects Masters, p. 15

Scoring Guide
Chapters 8 and 9
Investigation

Level	Specific Criteria
3 Superior	• Shows thorough understanding of the concepts of *surface area, perimeter, linear equation, polynomial expression, scatter plot, slope, best-fit line,* and *using mean, median, or mode to determine an average.* • Uses appropriate strategies to solve problems. • Computations are correct. • Written explanations are exemplary. • Charts, graphs, and report are appropriate and sensible. • Goes beyond the requirements of some or all problems.
2 Satisfactory, with Minor Flaws	• Shows understanding of the concepts of *surface area, perimeter, linear equation, polynomial expression, scatter plot, slope, best-fit line,* and *using mean, median, or mode to determine an average.* • Uses appropriate strategies to solve problems. • Computations are mostly correct. • Written explanations are effective. • Charts, graphs, and report are appropriate and sensible. • Satisfies the requirements of problems.
1 Nearly Satisfactory, with Obvious Flaws	• Shows understanding of most of the concepts of *surface area, perimeter, linear equation, polynomial expression, scatter plot, slope, best-fit line,* and *using mean, median, or mode to determine an average.* • May not use appropriate strategies to solve problems. • Computations are mostly correct. • Written explanations are satisfactory. • Charts, graphs, and report are appropriate and sensible. • Satisfies the requirements of problems.
0 Unsatisfactory	• Shows little or no understanding of the concepts of *surface area, perimeter, linear equation, polynomial expression, scatter plot, slope, best-fit line,* and *using mean, median, or mode to determine an average.* • Does not use appropriate strategies to solve problems. • Computations are incorrect. • Written explanations are not satisfactory. • Charts, graphs, and report are not appropriate or sensible. • Does not satisfy the requirements of problems.

CLOSING THE

In•ves•ti•ga•tion

Refer to the Investigation on pages 448–449.

Analyze

You have conducted several experiments and organized your data in various ways. It is now time to analyze your findings and state your conclusions.

PORTFOLIO ASSESSMENT

You may want to keep your work on this Investigation in your portfolio.

1. Look over your data and organize it in such a way that the various relationships are obvious.
2. Describe the relationships in the data. What does weight have to do with glide time? Does perimeter or surface area have an effect on glide time? What other factors need to be considered?

Write

The report to the people interested in hang gliding should explain your process for investigating these hang glider models and what you found from your investigations.

3. Begin the report by stating the process you used to investigate the matter. Explain all the experiments conducted. State the purpose and findings for each.
4. Show the data you collected in tables, charts, and graphs. Explain your analysis of the data and conclusions you found.
5. Make a recommendation to the group about the size of hang glider(s) that is most suited for them. Include the weight of the object that the hang glider(s) should carry to be most efficient.

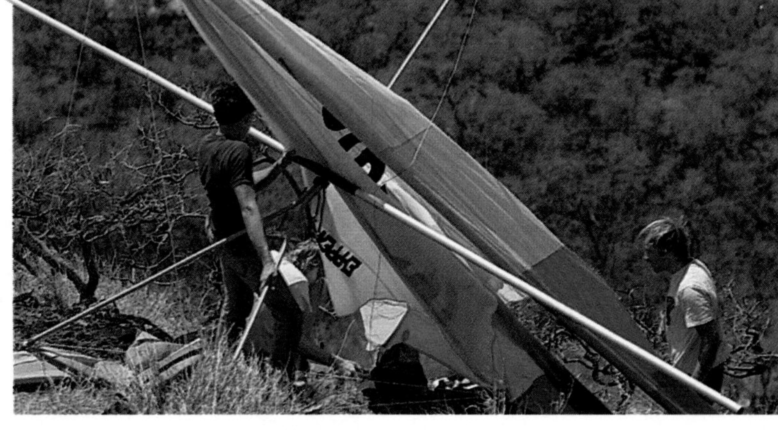

6. While you were conducting experiments, one of your team members found that the frames for most hang gliders are 32 feet wide. Explain how this information affects your generalizations regarding the weight and size of hang gliders.
7. Summarize your findings in a concluding statement to the group.

HIGHLIGHTS

VOCABULARY

After completing this chapter, you should be able to define each term, property, or phrase and give an example or two of each.

Algebra

binomial (p. 514)
constants (p. 496)
degree of monomial (p. 515)
degree of polynomial (p. 516)
difference of squares (p. 545)
FOIL method (p. 537)
monomial (p. 496)
negative exponent (p. 503)
polynomial (pp. 513, 514)
power of a monomial (p. 498)
power of a power (p. 498)
power of a product (p. 498)
product of powers (p. 497)
quotient of powers (p. 501)
scientific notation (p. 506)
square of a difference (p. 544)
square of a sum (p. 543)
trinomial (p. 514)
zero exponent (p. 502)

Problem Solving

look for a pattern (p. 497)

UNDERSTANDING AND USING THE VOCABULARY

Choose the letter of the term that best matches each example.

1. $4^{-3} = \frac{1}{4^3}$ or $\frac{1}{64}$ e
2. $(x + 2y)(x - 2y) = x^2 - 4y^2$ b
3. $\frac{4x^2y}{8xy^3} = \frac{x}{2y^2}$ h
4. $4x^2$ d
5. $x^2 - 3x + 1$ i
6. $2^0 = 1$ j
7. $x^4 - 3x^3 + 2x^2 - 1$ f
8. $(x + 3)(x - 4) = x^2 - 4x + 3x - 12$ c
9. $x^2 + 2$ a
10. $(a^3b)(2ab^2) = 2a^4b^3$ g

a. binomial
b. difference of squares
c. FOIL method
d. monomial
e. negative exponent
f. polynomial
g. product of powers
h. quotient of powers
i. trinomial
j. zero exponent

Instructional Resources

Three multiple-choice tests and three free-response tests are provided in the *Assessment and Evaluation Masters.* Forms 1A and 2A are for honors pacing, and Forms 1B, 1C, 2B, and 2C are for average pacing. Chapter 9 Test, Form 1B is shown at the right. Chapter 9 Test, Form 2B is shown on the next page.

Using the CHAPTER HIGHLIGHTS

The Chapter Highlights begins with a listing of the new terms, properties, and phrases that were introduced in this chapter. Have students define each term and provide an example or two of it, if appropriate.

Assessment and Evaluation Masters, pp. 227–228

NAME ______________ DATE ______

Chapter 9 Test, Form 1B

Write the letter for the correct answer in the blank at the right of each problem.

1. What is the simplest form of $m^4 \cdot m^2$? 1. C
A. m^2 B. m^8 C. m^6 D. $2m^8$

2. What is the simplest form of $(3w^2v)(-2w^5v^2)(4w^6v^5)$? 2. A
A. $-24w^{13}v^8$ B. $5w^{60}v^{10}$ C. $-24w^{60}v^{10}$ D. $5w^{13}v^8$

3. What is the simplest form of $(y^2n^3)^4 + (3y^4n^6)^2$? 3. A
A. $10y^8n^{12}$ B. $9y^{16}n^{27}$ C. $10y^8 + 2n^{12}$ D. $y^6n^7 + 9y^6n^8$

4. What is the simplest form of $\frac{m^6n^3}{m^2n^6}$? Assume the denominator is not equal to zero. 4. A
A. $\frac{m^4}{n^3}$ B. $-\frac{m^4}{n^3}$ C. $-\frac{m^8}{n^3}$ D. $\frac{m^8}{n^3}$

5. What is the simplest form of $\frac{6n^{-3}yw^{-5}}{2n^{-1}y^{-3}w^2}$? Assume the denominator is not equal to zero. 5. B
A. $\frac{4y^3}{n^2w^3}$ B. $\frac{3y^4}{n^2w^7}$ C. $\frac{3}{n^4w^3y^2}$ D. $\frac{3n^2}{y^4w^7}$

6. What is the simplest form of $\frac{(a^{-2}b^4)^{-6}}{(a^4b^{-8})^3}$? Assume the denominator is not equal to zero. 6. B
A. ab^3 B. 1 C. $\frac{a^{24}}{b^{48}}$ D. $\frac{b^{48}}{a^{24}}$

7. What is 0.000543 expressed in scientific notation? 7. D
A. 5.43×10^4 B. 5.43×10^{-3}
C. 543×10^6 D. 5.43×10^{-4}

8. What is the value of $\frac{2.88 \times 10^3}{2.4 \times 10^{-7}}$? 8. C
A. 0.48×10^{-4} B. 1.2×10^{-4} C. 1.2×10^{10} D. 0.48×10^{10}

9. What is the degree of $2x^3y - 4xy^2 + 9x^3y^2$? 9. D
A. 4 B. 3 C. 12 D. 5

10. Which of the following shows the terms of $x^2y^3 + 4xy^2 - 3x^3y + 6$ arranged so that the powers of x are in ascending order? 10. A
A. $6 + 4xy^2 + x^2y^3 - 3x^3y$ B. $x^2y^3 - 3x^3y + 4xy^2 + 6$
C. $6 + 4xy^2 - 3x^3y + x^2y^3$ D. $6 - 3x^3y + 4xy^2 + x^2y^3$

11. What is the simplest form of $(3c^2 - 8c + 5) + (c^2 - 8c - 6)$? 11. C
A. $3c^2 - 1$ B. $4c^2 + 11$
C. $4c^2 - 16c - 1$ D. $2c^2 - 16c - 1$

NAME ______________ DATE ______

Chapter 9 Test, Form 1B (continued)

12. What is the simplest form of $(6w^2 - 5wz - 3z^2) - (4w^2 + 6wz - 8z^2)$? 12. B
A. $10w^2 + wz - 11z^2$ B. $2w^2 - 11wz + 5z^2$
C. $2w^2 + wz - 11z^2$ D. $2w^2 - 11wz - 5z^2$

13. What is the simplest form of $8a^2b(3a + 2ab^3)$? 13. C
A. $24a^2b + 16a^2b^3$ B. $24a^3b + 2ab^3$
C. $24a^3b + 16a^3b^4$ D. $11a^2b + 10a^2b^3$

14. What is the simplest form of $(x + 2)(x + 4)$? 14. D
A. $x^2 + 8$ B. $2x + 6$ C. $2x + 8$ D. $x^2 + 6x + 8$

15. What is the simplest form of $(2n - 6)(4n + 2)$? 15. C
A. $6n^2 + 28n - 12$ B. $8n^2 - 12$
C. $8n^2 - 20n - 12$ D. $8n^2 + 20n - 12$

16. What is the simplest form of $(x + 6)^2$? 16. B
A. $x^2 + 36$ B. $x^2 + 12x + 36$
C. $x^2 + 12$ D. $2x + 12$

17. What is the simplest form of $(3y + 4z)(3y - 4z)$? 17. A
A. $9y^2 - 16z^2$ B. $9y^2 - 24yz - 16z^2$
C. $9y^2 + 16z^2$ D. $9y^2 - 24yz + 16z^2$

18. Find the solution of $-4(5 - 2n) = 8(-6 - 5n)$. 18. D
A. $-\frac{1}{9}$ B. $-\frac{28}{3}$ C. $-\frac{7}{8}$ D. $-\frac{7}{12}$

19. Find the solution of $x(x + 3) - 2 = 2 + x(x + 1)$. 19. A
A. 2 B. −2 C. 1 D. 0

20. A picture is 4 inches longer than it is wide. It is surrounded by a mat 2 inches wide. The total area of the mat is 112 square inches. If w is the width of the picture, which equation is true? 20. C
A. $(w + 4)(w + 8) + w(w + 4) = 112$
B. $(w + 2)(w + 6) - w(w + 4) = 112$
C. $(w + 4)(w + 8) - w(w + 4) = 112$
D. $(w + 2)(w + 6) + w(w + 4) = 112$

Bonus Solve $3^{2n-1} \cdot 3^{5n} = (3^4)^{n+2}$ for n. Bonus 3

Applications and Problem Solving Encourage students to work through the exercises in the Applications and Problem Solving section to strengthen their problem-solving skills.

CHAPTER 9 STUDY GUIDE AND ASSESSMENT

OBJECTIVES AND EXAMPLES

- use the FOIL method to multiply two binomials and multiply any two polynomials by using the distributive property (Lesson 9–7)

$$\begin{aligned}(3x+2)(x-2) &= \overset{F}{(3x)(x)} + \overset{O}{(3x)(-2)} + \overset{I}{(2)(x)} + \overset{L}{(2)(-2)}\\ &= 3x^2 - 6x + 2x - 4\\ &= 3x^2 - 4x - 4\end{aligned}$$

$$\begin{aligned}(4x-3)(3x^2-x+2)\\ &= 4x(3x^2-x+2) - 3(3x^2-x+2)\\ &= (12x^3-4x^2+8x) - (9x^2-3x+6)\\ &= 12x^3-4x^2+8x-9x^2+3x-6\\ &= 12x^3-13x^2+11x-6\end{aligned}$$

REVIEW EXERCISES

Find each product.

57. $(r-3)(r+7)$ **$r^2+4r-21$**
58. $(x+5)(3x-2)$ **$3x^2+13x-10$**
59. $(4x-3)(x+4)$ **$4x^2+13x-12$**
60. $(2x+5y)(3x-y)$ **$6x^2+13xy-5y^2$**
61. $(3x+0.25)(6x-0.5)$ **$18x^2-0.125$**
62. $(5r-7s)(4r+3s)$ **$20r^2-13rs-21s^2$**
63. x^2+7x-9
 $(\times)\ 2x+1$
64. $a^2-17ab-3b^2$
 $(\times)\ 2a+b$

63. **$2x^3+15x^2-11x-9$**
64. **$2a^3-33a^2b-23ab^2-3b^3$**

- use patterns to find $(a+b)^2$, $(a-b)^2$, and $(a+b)(a-b)$ (Lesson 9–8)

$$\begin{aligned}(x+4)^2 &= x^2+2(4x)+4^2\\ &= x^2+8x+16\end{aligned}$$

$$\begin{aligned}(r-5)^2 &= r^2-2(5r)+5^2\\ &= r^2-10r+25\end{aligned}$$

$$\begin{aligned}(b+9)(b-9) &= b^2-9^2\\ &= b^2-81\end{aligned}$$

Find each product.

65. $(x-6)(x+6)$ **x^2-36**
66. $(7-2x)(7+2x)$ **$49-4x^2$**
67. $(4x+7)^2$ **$16x^2+56x+49$**
68. $(8x-5)^2$ **$64x^2-80x+25$**
69. $(5x-3y)(5x+3y)$ **$25x^2-9y^2$**
70. $(a^2+b)^2$ **$a^4+2a^2b+b^2$**
71. $(6a-5b)^2$ **$36a^2-60ab+25b^2$**
72. $(3m+4n)^2$ **$9m^2+24mn+16n^2$**

APPLICATIONS AND PROBLEM SOLVING

73. **Finance** Find the current monthly payment on a 36-month car loan for $18,543. Twenty-five monthly payments have already been made at an annual interest rate of 8.7%. There is a balance due of $3216.27 at this time. Use the formula $B = P\left[\frac{1-(1+i)^{k-n}}{i}\right]$, where B represents the balance, P represents the current monthly payment, i represents the *monthly* interest rate (annual rate ÷ 12), k represents the total number of monthly payments already made, and n is the total number of monthly payments. (Lesson 9–2) **$305.26**

74. **Health** A radio station advertised the Columbus Marathon by saying that about 19,500,000 Calories would be burned in one day. If there were 6500 runners, about how many Calories did each runner burn? (Use scientific notation to solve.) (Lesson 9–3) **about 3×10^3 Calories**

75. **Finance** Upon his graduation from college, Mark Price received $10,000 in a trust fund from his grandparents. If he invests this money in an account with an annual interest rate of 6% and adds $1000 of his own money to the account at the end of each year, will his money have doubled after 5 years? If not, when? (Lesson 9–4) **no; after 6 years**

A practice test for Chapter 9 is provided on page 795.

ALTERNATIVE ASSESSMENT

COOPERATIVE LEARNING PROJECT

Saving for College In this chapter, you developed the concept of polynomials. You performed operations on polynomials, simplified polynomials, and solved polynomial equations. They were helpful in setting up a general formula to be used for inputting various data.

In this project, you will forecast a friend's finances. Jane has received \$75 from her grandparents on every birthday since she was one year old. She has been saving the money in an account that pays 5% interest. She is saving her money to help pay for her college education, which she will start this fall after her 18th birthday. She also has been receiving birthday checks from her other relatives, but these didn't start until she was 12 years old. The amounts of these checks from her 12th birthday until her 18th birthday are \$45, \$45, \$55, \$50, \$55, \$60, and \$65.

How much money will she have saved just from her birthdays by the time she starts college? Is this a reasonable amount to pay for a used car during her junior year in college? If she had invested her money in a different account that had earned 7% interest, how much more money would she have saved?

Follow these steps to accomplish your task.

- Construct a pattern for this situation.
- Develop a polynomial model to describe the amount of money she has each year.
- Determine the amount of money she received on birthdays 12 through 18.
- Determine what needs to be changed in your model when changing the interest rate.
- Write a paragraph describing the problem and your solution.

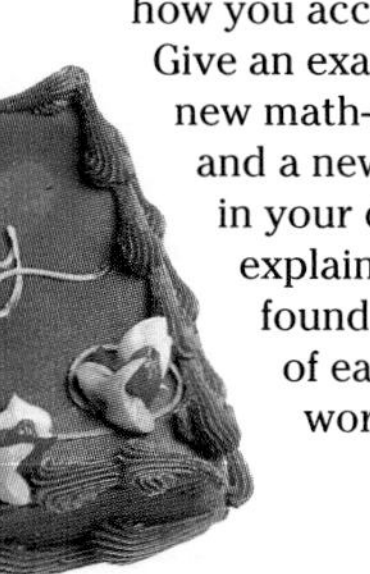

THINKING CRITICALLY

- Can $(-b)^2$ ever equal $-b^2$? Explain and give an example to support your answer.
- For all numbers a and b and any integer m, is $(a+b)^m = a^m + b^m$ a true sentence? Explain and give examples.

PORTFOLIO

Error analysis shows common mistakes that happen when performing an operation. Here is an example of an error when multiplying like bases.

$$4^3 \cdot 4^4 = 16^7$$

Actually, $4^3 \cdot 4^4 = 4^7$. The error of multiplying the bases while adding the exponents was incorrect. The base should stay the same while adding the exponent.

From the material in this chapter, find a problem that occurs often and write an error analysis for it. Describe the situation, give an example of the incorrect method, give the correct method for that example, and write a paragraph about it. Place this in your portfolio.

SELF EVALUATION

In this chapter, there are several words that have prefixes or suffixes that can be analyzed to determine what the word means. Do you break down words to find their meanings or do you just skip over those words and look for the meaning in the context of the sentence or paragraph? Maybe you go straight to the dictionary to get the meaning.

Assess yourself. How do you best learn new vocabulary words? After learning the meaning of the new word, do you then try to use that new word in your speaking and/or writing? Describe the plan that you use when learning a new word and how you accomplish it. Give an example of a new math-related word and a new word used in your daily life and explain how you found the meaning of each of these words.

Alternative Assessment

The Alternative Assessment section provides students with the opportunity to assess their own work by thinking critically, working with others, keeping a portfolio, and honestly evaluating their own progress. For more information on alternative forms of assessment, see *Alternative Assessment in the Mathematics Classroom*, one of the titles in the Glencoe Mathematics Professional Series.

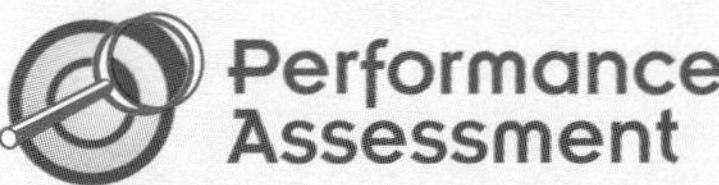

Performance Assessment

Performance Assessment tasks for this chapter are included in the *Assessment and Evaluation Masters*. A scoring guide is also provided.

Assessment and Evaluation Masters, pp. 238, 249

9 NAME ______ DATE ______

Chapter 9 Performance Assessment

Instructions: ***Demonstrate your knowledge by giving a clear, concise solution to each problem. Be sure to include all relevant drawings and justify your answers. You may show your solution in more than one way or investigate beyond the requirements of the problem.***

1. Rectangular areas can be used to represent products of binomials.
   a. Find the product $(2x+3)(x+1)$. Tell how the product and the area at the right are related.

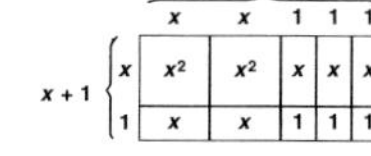

   b. Draw an area model demonstrating the product $(3x+1)(x+2)$. Find the product algebraically to verify your model.
   c. Write the two binomials whose product is demonstrated by the area model at the right. Find the product and show how it relates to the model.
   d. If the shaded regions represent a negative area and can be used to subtract or remove area from positive areas, tell how $x(2x-1) = 2x^2 - x$ relates to the model.
   e. Draw an area model demonstrating the product $x(3x-2)$. Find the product and tell how it relates to the model.
   f. The product of two binomials is $x^2 + 3x + 2$. Use the area models at the right to form a rectangle and find the two binomials. Check your answer.
2. Simplify $\frac{(-3a^3b^2)(4ab)}{2a^7b}$ in at least two ways. Explain each step.
3. Write a product of two numbers in scientific notation. Find the product and explain each step.

Scoring Guide
Chapter 9
Performance Assessment

Level	Specific Criteria
3 Superior	• Shows thorough understanding of the concepts of *multiplication of binomials, multiplication and division of monomials,* and *scientific notation.* • Computations are correct. • Written explanations are exemplary. • Diagrams are accurate and appropriate. • Goes beyond requirements of some or all problems.
2 Satisfactory, with Minor Flaws	• Shows understanding of the concepts of *multiplication of binomials, multiplication and division of monomials,* and *scientific notation.* • Computations are mostly correct. • Written explanations are effective. • Diagrams are mostly accurate and appropriate. • Satisfies all requirements of problems.
1 Nearly Satisfactory, with Serious Flaws	• Shows understanding of most of the concepts of *multiplication of binomials, multiplication and division of monomials,* and *scientific notation.* • Computations are mostly correct. • Written explanations are satisfactory. • Diagrams are mostly accurate and appropriate. • Satisfies most requirements of problems.
0 Unsatisfactory	• Shows little or no understanding of the concepts of *multiplication of binomials, multiplication and division of monomials,* and *scientific notation.* • Computations are incorrect. • Written explanations are not satisfactory. • Diagrams are not accurate or appropriate. • Does not satisfy requirements of problems.

Investigation

TEACHER NOTES

NCTM Standards: 1–5, 7, 9

This Investigation is designed to be completed over several days or weeks. It may be considered optional. You may want to assign the Investigation and the follow-up activities to be completed at the same time.

Objective

Design brick rectangular patios using small square bricks, large square bricks, and rectangular bricks.

Mathematical Overview

This Investigation will use the following mathematical skills and concepts from Chapters 10 and 11.

- making charts and tables
- identifying and factoring binomials that are the differences of squares
- using the zero product property to solve equations
- estimating roots of quadratic equations
- graphing quadratic functions

Recommended Time

Part	Pages	Time
Investigation	554–555	1 class period
Working on the Investigation	586, 600, 617, 627	20 minutes each
Closing the Investigation	650	1 class period

Instructional Resources

Investigations and Projects Masters, pp. 17–20

A recording sheet, teacher notes, and scoring guide are provided for each investigation in the *Investigations and Projects Masters.*

1 MOTIVATION

This Investigation uses common materials to investigate the specifications given to design rectangular brick patios. Ask students if they have seen different designs of brick patios. Discuss the importance of having the correct measurements and the correct materials for each design.

LONG-TERM PROJECT

Investigation

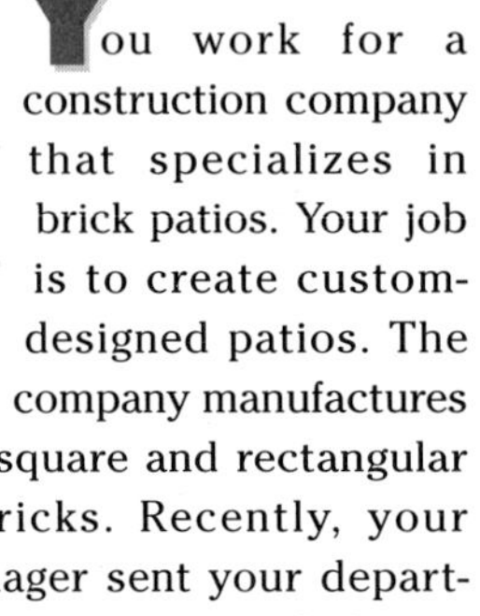

MATERIALS NEEDED

construction paper

scissors

ruler

You work for a construction company that specializes in brick patios. Your job is to create custom-designed patios. The company manufactures square and rectangular bricks. Recently, your manager sent your department the memo shown below.

In this Investigation, you must design brick patios that fit the specifications given in the memo. The design plans must be explicit and detailed, so that the construction crew can build them accordingly. Your design team consists of three people.

Make an Investigation Folder in which you can store all of your work on this Investigation for future use.

MEMO

To: Custom Design Department
From: Joanna Brown, Manager *JB*

We have a problem, and I need your help in solving it. We have an excess inventory of three types of bricks:

- small square bricks,
- large square bricks, and
- rectangular bricks that are as long as the large square brick and as wide as the small square brick.

We need to move this inventory, so I am asking you to investigate the possible patio design patterns using these three types of bricks.

I don't know if this is helpful, but the length of the larger brick is the same length as the diagonal of the smaller square brick.

Our other custom designs include using triangles, rectangles, and hexagons exclusively to create repeating patterns. However, for these surplus bricks, we need to concentrate on repeating rectangular patterns.

Please create several patio designs that will utilize these bricks. Submit at least three different plans, explaining the materials required for each patio. I am anxious to see the different ways in which these bricks can be arranged to form rectangular patios. Is there a general formula or pattern we can use to design these in the future? I look forward to your report on helping us solve our inventory problem.

Cooperative Learning

This Investigation offers an excellent opportunity for using cooperative learning groups. For more information on cooperative learning strategies and group management, see *Cooperative Learning in the Mathematics Classroom,* one of the titles in the Glencoe Mathematics Professional Series.

PATIO SKETCH #____	DIMENSIONS			MATERIALS USED	
	small square: ___ × ___	bricks	number	total area
	large square: ___ × ___	sm. squares		
	rectangle: ___ × ___	lg. squares		
	design size: ___ × ___	rectangles		
		TOTAL		

CREATE MODELS

1 Copy the table above. You will use it, along with other tables, to record the data as you explore the designs that are possible.

2 Using the measurement requirements from Ms. Brown's memo, create a model of each of the three sizes of bricks.

3 After determining that your three models comply with the given specifications, use construction paper to make several copies of each model.

ANALYZE THE MODELS

4 Share the dimensions of your models with the other design teams in your class. Obviously the models of each design team will not be the same size, but each set of models must comply with the requirements Ms. Brown wrote in her memo.

5 Explain how the three sizes relate to each other. Make a chart listing the dimensions of each of the models from the other design teams. Do each set of dimensions relate in the same way that your dimensions do? Should they? Explain.

6 How do the lengths of the two square bricks relate? If you were to make one row of small squares and below it make one row of large squares, how many small squares would it take to match the exact length of the row of large squares? Explain your answer mathematically.

You will continue working on this Investigation throughout Chapters 10 and 11.

Be sure to keep your individual brick models, chart, and other materials in your Investigation Folder.

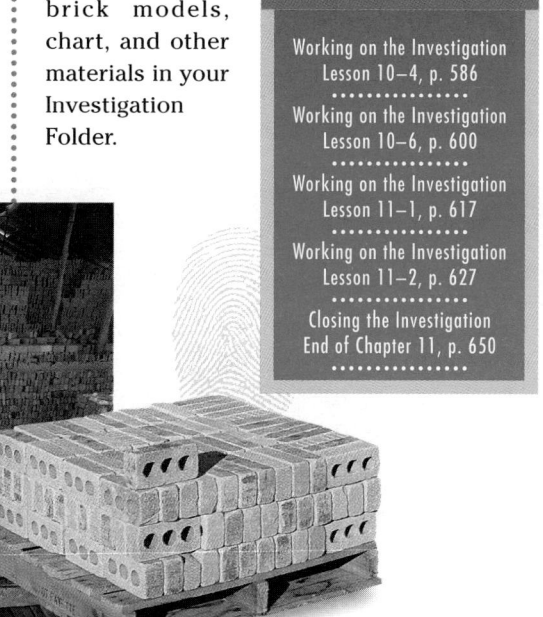

The Brickyard Investigation

2 SETUP

You may wish to have a student read the first two paragraphs of the Investigation to provide background information about the patio designs and specifications. You may then wish to read the memo to introduce the activity. Discuss the activity with your students. Then separate the class into groups of three.

3 MANAGEMENT

Each group member should be responsible for a specific task.

Recorder Collects data. Calculates dimensions.
Measurer Obtains measurement requirements from memo. Gathers materials.
Designer Creates models.

At the end of the activity, each member should turn in his or her respective equipment.

Sample Answers

Answers will vary as they are based on each design.

Investigations and Projects Masters, p. 20

10, 11 NAME________ DATE______

Investigation, Chapters 10 and 11 Student Edition Pages 554–555, 586, 600, 617, 627, 650

The Brickyard

Use this chart to record the data that you found from Working on the Investigation, page 627.

Situation 1

Dimensions			
Perimeter			
Area			

Situation 2

Dimensions			
Perimeter			
Area			

Situation 3

Dimensions			
Perimeter			
Area			

Work with your group to answer the following questions and add others to be considered.

- If you had two large square bricks, five small square bricks, and five rectangular bricks, what size rectangular patterns could you make?
- For each of the patterns that you designed, express the dimensions, perimeter, and area of each in terms of x and y.
- Which of the patterns would use up the inventory more quickly?

CHAPTER

10 Using Factoring

PREVIEWING THE CHAPTER

This chapter covers factors and factoring. Students are introduced to factors and the greatest common factor. They then learn to factor using the distributive property. Students learn to factor trinomials and differences of squares. Finally, they learn to factor perfect square trinomials and apply all of their skills to solve quadratic equations.

Lesson (Pages)	Lesson Objectives	NCTM Standards	State/Local Objectives
10-1 (558–563)	Find prime factorizations of integers. Find greatest common factors (GCF) for sets of monomials.	1–5	
10-2A (564)	Use algebra tiles to factor binomials.	1–5	
10-2 (565–571)	Use the GCF and the distributive property to factor polynomials. Use grouping techniques to factor polynomials with four or more terms.	1–5	
10-3A (572–573)	Use algebra tiles to factor simple trinomials.	1–5	
10-3 (574–580)	Solve problems by using guess and check. Factor quadratic trinomials.	1–5	
10-4 (581–586)	Identify and factor binomials that are the differences of squares.	1–5	
10-5 (587–593)	Identify and factor perfect square trinomials.	1–5	
10-6 (594–600)	Use the zero product property to solve equations.	1–5	

A complete, 1-page lesson plan is provided for each lesson in the ***Lesson Planning Guide***. Answer keys for each lesson are available in the ***Answer Key Masters***.

ORGANIZING THE CHAPTER

You may want to refer to the **Course Planning Calendar** on page T12 for detailed information on pacing.
PACING: Standard—12 days; **Honors**—11 days; **Block**—6 days; **Two Years**—22 days

LESSON PLANNING CHART

			Blackline Masters											
Lesson (Pages)	**Materials/ Manipulatives**	**Extra Practice (Student Edition)**	**Study Guide**	**Practice**	**Enrichment**	**Assessment and Evaluation**	**Modeling Mathematics**	**Multicultural Activity**	**Tech Prep Applications**	**Graphing Calculator**	**Science and Math Lab Manual**	**Real-World Applications**	**Interactive Mathematics Tools Software**	**Teaching Transparencies**
10-1 (558–563)		p. 778	p. 69	p. 69	p. 69			p. 19					10-1	10-1A 10-1B
10-2A (564)	algebra tiles* product mat*						p. 32							
10-2 (565–571)		p. 779	p. 70	p. 70	p. 70	p. 268			p. 19					10-2A 10-2B
10-3A (572–573)	algebra tiles* product mat*						p. 33						10-3A	
10-3 (574–580)	graphing calculator	p. 779	p. 71	p. 71	p. 71	pp. 267, 268	pp. 63–65							10-3A 10-3B
10-4 (581–586)	algebra tiles* product mat*	p. 779	p. 72	p. 72	p. 72							25		10-4A 10-4B
10-5 (587–593)		p. 780	p. 73	p. 73	p. 73	p. 269	pp. 66–68, 81					26		10-5A 10-5B
10-6 (594–600)		p. 780	p. 74	p. 74	p. 74	p. 269		p. 20	p. 20	p. 10	pp. 43–48		10-6	10-6A 10-6B
Study Guide/ Assessment (601–605)						pp. 253–266, 270–272								

*Included in Glencoe's Student Manipulative Kit and Overhead Manipulative Resources.

ORGANIZING THE CHAPTER

OTHER CHAPTER RESOURCES

Student Edition
Investigation, pp. 554–555
Chapter Opener, pp. 556–557
Mathematics and Society, p. 563
Working on the Investigation, pp. 586, 600

Teacher's Classroom Resources
Investigations and Projects Masters, pp. 61–64
Algebra and Geometry Overhead Manipulative Resources, pp. 31–36

Technology
Test and Review Software (IBM and Macintosh)
CD-ROM Interactions (Windows and Macintosh)

Professional Publications
Block Scheduling Booklet
Glencoe Mathematics Professional Series

OUTSIDE RESOURCES

Books/Periodicals
The Ideas of Algebra, K–12, NCTM
Olson, Melfried, "A Geometric Look at Greatest Common Divisor," *Mathematics Teacher*, March 1991

Software
Mathematics Curriculum and Teaching Program, NCTM

Videos/CD-ROMs
Mathematics Assessment: Alternative Approaches, NCTM

ASSESSMENT RESOURCES

Student Edition
Math Journal, pp. 561, 591
Mixed Review, pp. 563, 571, 580, 585, 593, 600
Self Test, p. 580
Chapter Highlights, p. 601
Chapter Study Guide and Assessment, pp. 602–604
Alternative Assessment, p. 605
Portfolio, p. 605
Cumulative Review, pp. 606–607

Teacher's Wraparound Edition
5-Minute Check, pp. 558, 565, 574, 581, 587, 594
Check for Understanding, pp. 561, 568, 578, 584, 591, 598
Closing Activity, pp. 562, 571, 580, 586, 593, 600
Cooperative Learning, pp. 560, 566

Assessment and Evaluation Masters
Multiple-Choice Tests, Forms 1A (Honors), 1B (Average), 1C (Basic), pp. 253–258
Free-Response Tests, Forms 2A (Honors), 2B (Average), 2C (Basic), pp. 259–264
Calculator-Based Test, p. 265
Performance Assessment, p. 266
Mid-Chapter Test, p. 267
Quizzes A–D, pp. 268–269
Standardized Test Practice, p. 270
Cumulative Review, pp. 271–272

ENHANCING THE CHAPTER

Examples of some of the materials for enhancing Chapter 10 are shown below.

Multicultural Activity Masters, pp. 19, 20

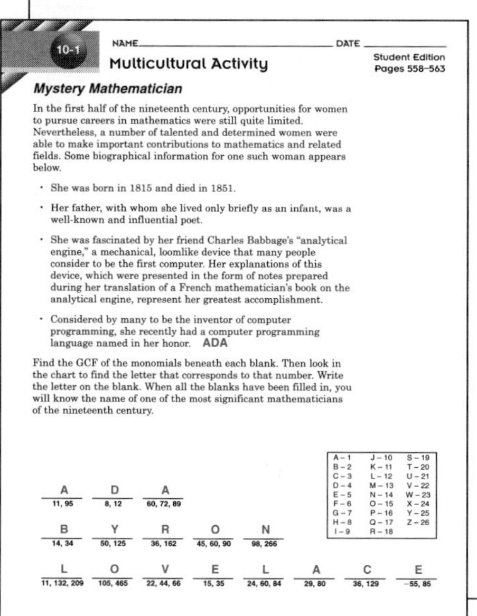
10-1 NAME ____________ DATE ________

Multicultural Activity

Student Edition Pages 558–563

Mystery Mathematician

In the first half of the nineteenth century, opportunities for women to pursue careers in mathematics were still quite limited. Nevertheless, a number of talented and determined women were able to make important contributions to mathematics and related fields. Some biographical information for one such woman appears below.

- She was born in 1815 and died in 1851.
- Her father, with whom she lived only briefly as an infant, was a well-known and influential poet.
- She was fascinated by her friend Charles Babbage's "analytical engine," a mechanical, loomlike device that many people consider to be the first computer. Her explanations of this device, which were presented in the form of notes prepared during her translation of a French mathematician's book on the analytical engine, represent her greatest accomplishment.
- Considered by many to be the inventor of computer programming, she recently had a computer programming language named in her honor. **ADA**

Find the GCF of the monomials beneath each blank. Then look in the chart to find the letter that corresponds to that number. Write the letter on the blank. When all the blanks have been filled in, you will know the name of one of the most significant mathematicians of the nineteenth century.

A – 1	J – 10	S – 19
B – 2	K – 11	T – 20
C – 3	L – 12	U – 21
D – 4	M – 13	V – 22
E – 5	N – 14	W – 23
F – 6	O – 15	X – 24
G – 7	P – 16	Y – 25
H – 8	Q – 17	Z – 26
I – 9	R – 18	

A	D	A
11, 95	8, 12	60, 72, 89

B	Y	R	O	N
14, 34	50, 125	36, 162	45, 60, 90	98, 266

L	O	V	E	L	A	C	E
11, 132, 209	105, 465	22, 44, 66	15, 35	24, 60, 84	29, 80	36, 129	–55, 85

Real-World Applications, 25, 26

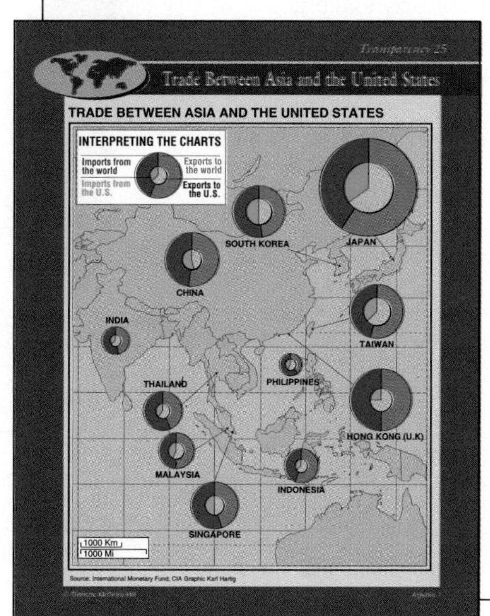

TECHNOLOGY

Graphing Calculator Masters, p. 10

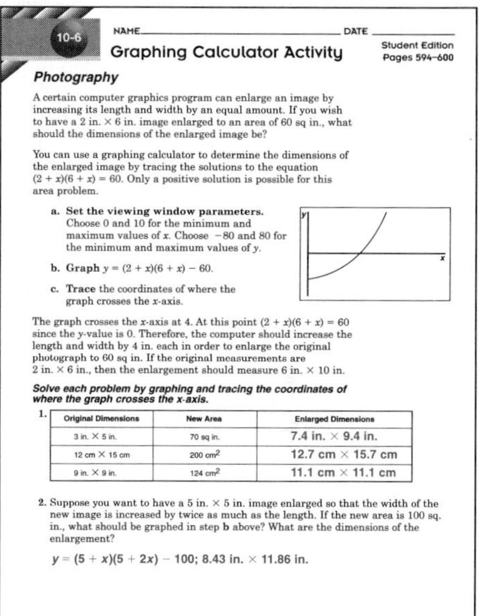
10-6 NAME ____________ DATE ________

Graphing Calculator Activity

Student Edition Pages 594–600

Photography

A certain computer graphics program can enlarge an image by increasing its length and width by an equal amount. If you wish to have a 2 in. × 6 in. image enlarged to an area of 60 sq in., what should the dimensions of the enlarged image be?

You can use a graphing calculator to determine the dimensions of the enlarged image by tracing the solutions to the equation $(2 + x)(6 + x) = 60$. Only a positive solution is possible for this area problem.

a. **Set the viewing window parameters.** Choose 0 and 10 for the minimum and maximum values of x. Choose -80 and 80 for the minimum and maximum values of y.

b. **Graph** $y = (2 + x)(6 + x) - 60$.

c. **Trace** the coordinates of where the graph crosses the x-axis.

The graph crosses the x-axis at 4. At this point $(2 + x)(6 + x) = 60$ since the y-value is 0. Therefore, the computer should increase the length and width by 4 in. each in order to enlarge the original photograph to 60 sq in. If the original measurements are 2 in. × 6 in., then the enlargement should measure 6 in. × 10 in.

Solve each problem by graphing and tracing the coordinates of where the graph crosses the x-axis.

1.

Original Dimensions	New Area	Enlarged Dimensions
3 in. × 5 in.	70 sq in.	7.4 in. × 9.4 in.
12 cm × 15 cm	200 cm²	12.7 cm × 15.7 cm
9 in. × 9 in.	124 cm²	11.1 cm × 11.1 cm

2. Suppose you want to have a 5 in. × 5 in. image enlarged so that the width of the new image is increased by twice as much as the length. If the new area is 100 sq. in., what should be graphed in step **b** above? What are the dimensions of the enlargement?

$y = (5 + x)(5 + 2x) - 100$; 8.43 in. × 11.86 in.

TECH PREP

Tech Prep Applications Masters, pp. 19, 20

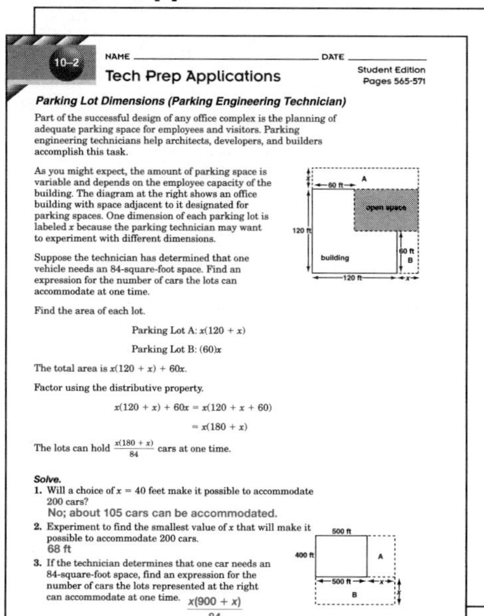
10–2 NAME ____________ DATE ________

Tech Prep Applications

Student Edition Pages 565-571

Parking Lot Dimensions (Parking Engineering Technician)

Part of the successful design of any office complex is the planning of adequate parking space for employees and visitors. Parking engineering technicians help architects, developers, and builders accomplish this task.

As you might expect, the amount of parking space is variable and depends on the employee capacity of the building. The diagram at the right shows an office building with space adjacent to it designated for parking spaces. One dimension of each parking lot is labeled x because the parking technician may want to experiment with different dimensions.

Suppose the technician has determined that one vehicle needs an 84-square-foot space. Find an expression for the number of cars the lots can accommodate at one time.

Find the area of each lot.

Parking Lot A: $x(120 + x)$

Parking Lot B: $(60)x$

The total area is $x(120 + x) + 60x$.

Factor using the distributive property.

$$x(120 + x) + 60x = x(120 + x + 60)$$
$$= x(180 + x)$$

The lots can hold $\frac{x(180 + x)}{84}$ cars at one time.

Solve.

1. Will a choice of $x = 40$ feet make it possible to accommodate 200 cars?
   No; about 105 cars can be accommodated.
2. Experiment to find the smallest value of x that will make it possible to accommodate 200 cars.
   68 ft
3. If the technician determines that one car needs an 84-square-foot space, find an expression for the number of cars the lots represented at the right can accommodate at one time. $\frac{x(900 + x)}{84}$

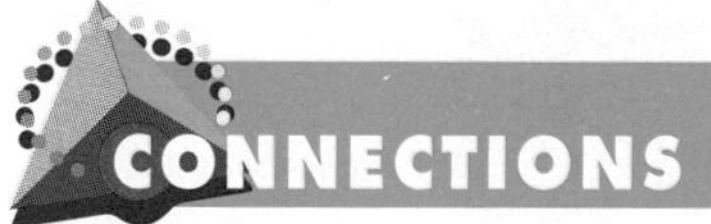

Science and Math Lab Manual, pp. 43–48

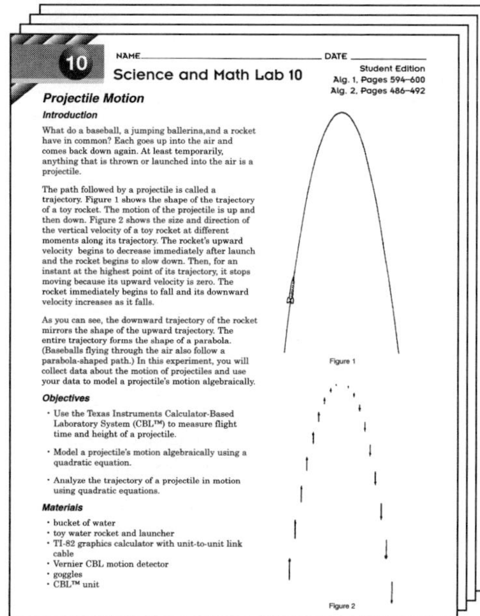
10 NAME ____________ DATE ________

Science and Math Lab 10

Student Edition
Alg. 1, Pages 594–600
Alg. 2, Pages 486–492

Projectile Motion

Introduction

What do a baseball, a jumping ballerina,and a rocket have in common? Each goes up into the air and comes back down again. At least temporarily, anything that is thrown or launched into the air is a projectile.

The path followed by a projectile is called a trajectory. Figure 1 shows the shape of the trajectory of a toy rocket. The motion of the projectile is up and then down. Figure 2 shows the size and direction of the vertical velocity of a toy rocket at different moments along its trajectory. The rocket's upward velocity begins to decrease immediately after launch and the rocket begins to slow down. Then, for an instant at the highest point of its trajectory, it stops moving because its upward velocity is zero. The rocket immediately begins to fall and its downward velocity increases as it falls.

As you can see, the downward trajectory of the rocket mirrors the shape of the upward trajectory. The entire trajectory forms the shape of a parabola. (Baseballs flying through the air also follow a parabola-shaped path.) In this experiment, you will collect data about the motion of projectiles and use your data to model a projectile's motion algebraically.

Figure 1

Figure 2

Objectives

- Use the Texas Instruments Calculator-Based Laboratory System (CBL™) to measure flight time and height of a projectile.
- Model a projectile's motion algebraically using a quadratic equation.
- Analyze the trajectory of a projectile in motion using quadratic equations.

Materials

- bucket of water
- toy water rocket and launcher
- TI-82 graphics calculator with unit-to-unit link cable
- Vernier CBL motion detector
- goggles
- CBL™ unit

Problem of the Week Cards, 28, 29

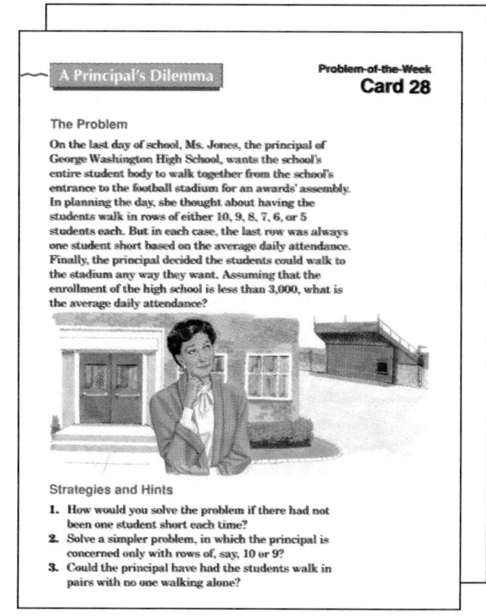
A Principal's Dilemma

Problem-of-the-Week
Card 28

The Problem

On the last day of school, Ms. Jones, the principal of George Washington High School, wants the school's entire student body to walk together from the school's entrance to the football stadium for an awards' assembly. In planning the day, she thought about having the students walk in rows of either 10, 9, 8, 7, 6, or 5 students each. But in each case, the last row was always one student short based on the average daily attendance. Finally, the principal decided the students could walk to the stadium any way they want. Assuming that the enrollment of the high school is less than 3,000, what is the average daily attendance?

Strategies and Hints

1. **How would you solve the problem if there had not been one student short each time?**
2. **Solve a simpler problem, in which the principal is concerned only with rows of, say, 10 or 9?**
3. **Could the principal have had the students walk in pairs with no one walking alone?**

MAKING MATHEMATICS RELEVANT

This two-page introduction to the chapter provides students with an opportunity to explore contemporary topics and their applications to mathematics.

Background Information

The Impact of the Media The table in the text reveals that most Americans are concerned about violence in movies, television, and popular music, and that most Americans do want restrictions. But what kind of restrictions do Americans favor? The table provides the answer.

CHAPTER 10

Using Factoring

The Impact of the Media

Objectives

In this chapter, you will:

- find the prime factorization of integers,
- find the greatest common factors (GCF) for sets of monomials,
- factor polynomials,
- solve problems by using guess and check, and
- use the zero product property to solve equations.

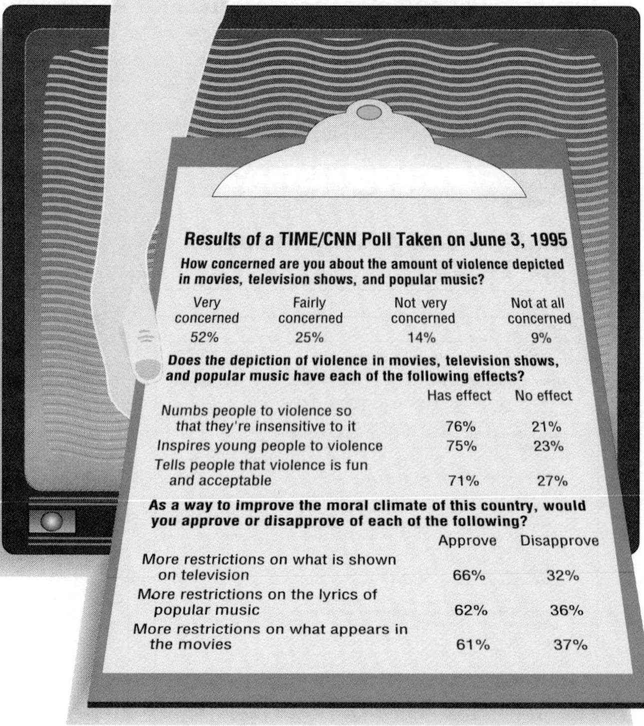

Results of a TIME/CNN Poll Taken on June 3, 1995

How concerned are you about the amount of violence depicted in movies, television shows, and popular music?

Very concerned	Fairly concerned	Not very concerned	Not at all concerned
52%	25%	14%	9%

Does the depiction of violence in movies, television shows, and popular music have each of the following effects?

	Has effect	No effect
Numbs people to violence so that they're insensitive to it	76%	21%
Inspires young people to violence	75%	23%
Tells people that violence is fun and acceptable	71%	27%

As a way to improve the moral climate of this country, would you approve or disapprove of each of the following?

	Approve	Disapprove
More restrictions on what is shown on television	66%	32%
More restrictions on the lyrics of popular music	62%	36%
More restrictions on what appears in the movies	61%	37%

Source: *Time Magazine,* June 12, 1995

Is American culture too violent? Do movies, television, magazines, and music reflect a true picture of America or do they contribute to the violence in our culture? What impact does the media have on American youth?

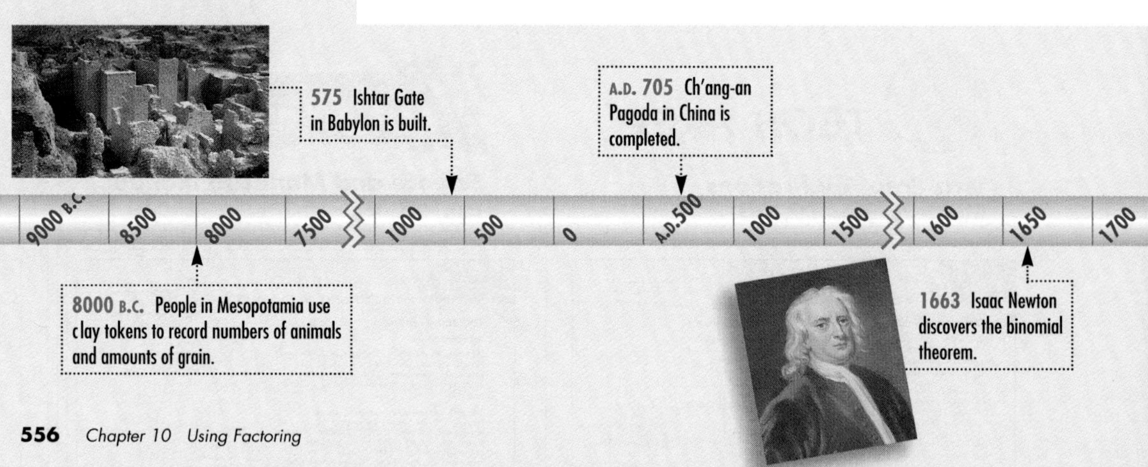
575 Ishtar Gate in Babylon is built.
A.D. 705 Ch'ang-an Pagoda in China is completed.
9000 B.C.
8500
8000
7500
1000
500
0
A.D. 500
1000
1500
1600
1650
1700
8000 B.C. People in Mesopotamia use clay tokens to record numbers of animals and amounts of grain.
1663 Isaac Newton discovers the binomial theorem.

TIME Line

Students might want to study the 1663 discovery of the binomial theorem by Isaac Newton. Like many mathematical discoveries, it was invented for just one purpose, but later many more applications were found.

interNET CONNECTION

Learn about such topics as "Caught in the Crossfire" and "Healthy Children, Healthy Learning" on PBS's quarterly Merrow Report.

World Wide Web
http://www.merrow.oa.net:8003/

PEOPLE IN THE NEWS

The title of **Robert Rodriquez's** new book is *Rebel Without a Crew: Or How a 23-Year-Old Filmmaker with $7,000 Became a Hollywood Player.* It is the story of how the University of Texas film student from Austin, Texas, made a feature film on a very small budget. He used friends as actors, wrote the script, directed the 14-day shoot, and handled the camera work as a one-man crew. The film, *El Mariachi*, went on to win the Audience Award at the Sundance Film Festival, was released by Columbia, and is now on video.

Robert's second film, a full budget production, was *Desperado*, released in 1995. He told Columbia that he would sign a contract if he could stay in Texas, near his family and his inspiration. His advice to future film makers, "Grab your camera and just do it."

Chapter Project

- Pick five of your favorite movies. List five hints for each movie that will help your classmates to guess the names of the movies.
- Exchange your list of hints with a classmate. Try to guess your classmate's favorite movies. Explain how each hint helped you to eliminate some movies and to concentrate on others. Did you guess your classmate's favorite movies correctly?
- Explain how guessing can help when factoring polynomials.
- List five polynomials for one of your classmates to factor. Make sure that one of your polynomials cannot be factored.
- Exchange your polynomials with a classmate and factor the polynomials on the list you receive.

1850 1900 1910 1920 1930 1940 1950 1960 1970 1980 1990 2000

1939 Marian Anderson gives a concert for 75,000 at the Lincoln Memorial.

1975 The VHS format (Video Home System) is launched by the Japanese company JVC.

1986 Franklin Chang-Diaz, plasmaphysicist, is a member of the crew of the space shuttle *Columbia.*

PEOPLE IN THE NEWS

Rodriguez lists Alfred Hitchcock, Sergio Leone, and Sam Raimi as his mentors. The plot of *El Mariachi* is that of a typical action film, but the camera work (Rodriguez is the cameraman) is innovative. Because of the success of his films, Hollywood is interested in him. But, as is often the case, studio executives wanted to change his movie ideas to resemble other Hollywood blockbuster movies. Rodriguez plans to remain true to his own cinematic vision.

Chapter Project

There is a difference between guessing and intelligent guessing. Intelligent guessing requires knowing in advance what counts as a good guess and a bad guess. Factoring polynomials involves intelligent guessing.

Alternative Chapter Projects

Two other chapter projects are included in the *Investigations and Projects Masters.* In Chapter 10 Project A, pp. 61–62, students extend the topic in the chapter opener. In Chapter 10 Project B, pp. 63–64, students research formulas and their use.

Investigations and Projects Masters, p. 61

NAME________________ DATE ________

Chapter 10 Project A

Student Edition Pages 558–600

Movie Numbers

1. Numbers often appear in the titles of movies. Examples include *101 Dalmatians, Apollo 13,* and *Twelve Monkeys.* For this project, you will work in a small group to make a list of at least 20 such movies. Use newspaper movie listings for current films; visit a video store or consult a video catalogue for older titles.
2. Determine which of the numbers in your titles are prime and which are composite.
3. Find the prime factorization for each composite number in your list.
4. Create a board game in which players advance their game pieces by guessing movie titles with numbers in them. Write clues for each movie title you listed in exercise 1. For example, the clues for *Apollo 13* might be "Greek god of the sun" and "$1 \cdot 13$," and those for *Twelve Monkeys* might be "$2^2 \cdot 3$" and "small, lively primates that live together in social groups."
5. Exchange games with other groups and play them.

10-1 LESSON NOTES

NCTM Standards: 1–5

Instructional Resources

- Study Guide Master 10-1
- Practice Master 10-1
- Enrichment Master 10-1
- Multicultural Activity Masters, p. 19

Transparency 10-1A contains the 5-Minute Check for this lesson; **Transparency 10-1B** contains a teaching aid for this lesson.

Recommended Pacing	
Standard Pacing	Day 1 of 12
Honors Pacing	Day 1 of 11
Block Scheduling*	Day 1 of 6
Alg. 1 in Two Years*	Days 1, 2, & 3 of 22

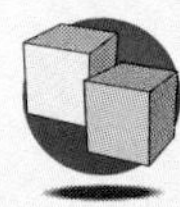

*For more information on pacing and possible lesson plans, refer to the *Block Scheduling Booklet* and *Algebra 1 in Two Years.*

1 FOCUS

5-Minute Check
(over Chapter 9)

Find each product.

1. $(a + 2b)^2$ $\mathbf{a^2 + 4ab + 4b^2}$
2. $(x - 3r)^2$ $\mathbf{x^2 - 6rx + 9r^2}$
3. Simplify $\frac{63x^5}{9x^2}$. $\mathbf{7x^3}$
4. Express $(2 \times 10^4)(1.5 \times 10^5)$ in scientific notation. $\mathbf{3 \times 10^9}$
5. Simplify $(8x^3 - 2x^2 + 7) - (3x^3 + 5x - 2)$. $\mathbf{5x^3 - 2x^2 - 5x + 9}$

Motivating the Lesson

Questioning Introduce the lesson by having students identify factors of numbers such as 12, 36, 64, and 100.

$12 = 2 \times 2 \times 3$
$36 = 2 \times 2 \times 3 \times 3$
$64 = 2 \times 2 \times 2 \times 2 \times 2 \times 2$
$100 = 2 \times 2 \times 5 \times 5$

10-1 Factors and Greatest Common Factors

What YOU'LL LEARN

- To find prime factorizations of integers, and
- to find greatest common factors (GCF) for sets of monomials.

Why IT'S IMPORTANT

You can use factors to solve problems involving packaging and gardening.

INTEGRATION Geometry

Suppose you were asked to use grid paper to draw all of the possible rectangles with whole number dimensions that have areas of 12 square units each. The figure at the right shows one possible drawing.

Rectangles A and B are both 3 by 4 and, therefore, can be considered the same. Likewise, Rectangles C and D and Rectangles E and F are considered the same.

Recall that when two or more numbers are multiplied to form a product, each number is a *factor* of the product. In the example above, 12 is expressed as the product of different pairs of whole numbers.

$$12 = 3 \times 4 \qquad 12 = 2 \times 6 \qquad 12 = 12 \times 1$$
$$12 = 4 \times 3 \qquad 12 = 6 \times 2 \qquad 12 = 1 \times 12$$

The whole numbers 1, 2, 3, 4, 6, and 12 are factors of 12.

Example  **Find the factors of 72.**

To find the factors of 72, list all the pairs of numbers whose product is 72.

$1 \times 72 \qquad 2 \times 36 \qquad 3 \times 24 \qquad 4 \times 18 \qquad 6 \times 12 \qquad 8 \times 9$

Therefore, the factors of 72, in increasing order, are 1, 2, 3, 4, 6, 8, 9, 12, 18, 24, 36, and 72.

Some whole numbers have exactly two factors, the number itself and 1. These numbers are called **prime numbers**. Whole numbers that have more than two factors are called **composite numbers**.

Definitions of Prime and Composite Numbers	A prime number is a whole number, greater than 1, whose only factors are 1 and itself. A composite number is a whole number, greater than 1, that is not prime.

0 and 1 are neither prime nor composite.

FYI

The greatest known prime number was discovered by computer scientists Paul Gage and David Slowinski in 1994, using a Cray supercomputer. The number has 258,716 digits and is expressed as $2^{859,433} - 1$.

The number 6 is a factor of 12, but not a *prime factor* of 12, since 6 is not a prime number. When a whole number is expressed as a product of factors that are all prime numbers, the expression is called the **prime factorization** of the number. Thus, the prime factorization of 12 is $2 \cdot 2 \cdot 3$ or $2^2 \cdot 3$.

The prime factorization of every number is unique except for the order in which the factors are written. For example, $2 \cdot 3 \cdot 2$ is also a prime factorization of 12, but it is the same as $2 \cdot 2 \cdot 3$. This property of numbers is called the **unique factorization theorem**.

FYI

Investigating prime numbers belongs to a branch of mathematics called number theory. This field of research began around 600 B.C. At that time it was associated with numerology—a field that investigated the mystical properties of numbers.

GLENCOE Technology

Interactive Mathematics Tools Software

This multimedia software provides an interactive lesson that uses the arrangement of cells to find common factors and greatest common factors of two numbers. A **Computer Journal** gives students an opportunity to write about what they have learned.

For Windows & Macintosh

Example 2 **Find the prime factorization of 140.**

Method 1

$140 = 2 \cdot 70$ *The least prime factor of 140 is 2.*

$\quad = 2 \cdot 2 \cdot 35$ *The least prime factor of 70 is 2.*

$\quad = 2 \cdot 2 \cdot 5 \cdot 7$ *The least prime factor of 35 is 5.*

All the factors in the last row are prime. Thus, the prime factorization of 140 is $2 \cdot 2 \cdot 5 \cdot 7$ or $2^2 \cdot 5 \cdot 7$.

Method 2

Use a factor tree.

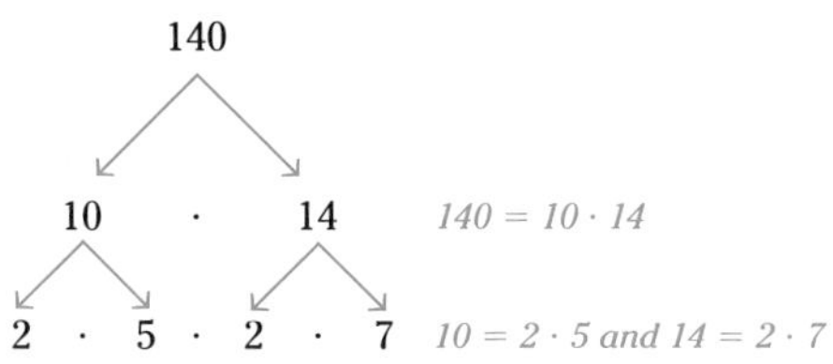

All of the factors in the last branch of the factor tree are prime. Thus, the prime factorization of 140 is $2 \cdot 2 \cdot 5 \cdot 7$ or $2^2 \cdot 5 \cdot 7$.

FYI

Computer scientists have been able to find the prime factors of a 155-digit number. This is particularly significant because some secret codes are based on the factors of large numbers.

A negative integer is factored completely when it is expressed as the product of -1 and prime numbers.

Example 3 **Factor -150 completely.**

$-150 = -1 \cdot 150$ *Express -150 as -1 times 150.*

$\quad = -1 \cdot 2 \cdot 75$ *Find the prime factors of 150.*

$\quad = -1 \cdot 2 \cdot 3 \cdot 25$

$\quad = -1 \cdot 2 \cdot 3 \cdot 5 \cdot 5$ or $-1 \cdot 2 \cdot 3 \cdot 5^2$

A monomial is in **factored form** when it is expressed as the product of prime numbers and variables and no variable has an exponent greater than 1.

Example 4 **Factor $45x^3y^2$.**

$45x^3y^2 = 3 \cdot 15 \cdot x \cdot x \cdot x \cdot y \cdot y$

$\quad = 3 \cdot 3 \cdot 5 \cdot x \cdot x \cdot x \cdot y \cdot y$

Two or more numbers may have some common factors. Consider the numbers 84 and 70, for example.

Factors of 84: 1, 2, 3, 4, 6, 7, 12, 14, 21, 28, 42, 84

Factors of 70: 1, 2, 5, 7, 10, 14, 35, 70

There are some factors that appear on both lists. The greatest of these numbers is 14, which is called the **greatest common factor (GCF)** of 84 and 70.

Definition of Greatest Common Factor	**The greatest common factor of two or more integers is the greatest number that is a factor of all of the integers.**

2 TEACH

In-Class Examples

For Example 1
Find the factors of 84.
1, 2, 3, 4, 6, 7, 12, 14, 21, 42, 84

For Example 2
Find the prime factorization of each number.

a. 200 **$2 \times 2 \times 2 \times 5 \times 5$**
b. 650 **$2 \times 5 \times 5 \times 13$**
c. 420 **$2 \times 2 \times 3 \times 5 \times 7$**
d. 168 **$2 \times 2 \times 2 \times 3 \times 7$**

For Example 3
Factor each number completely.

a. -210 **$-1 \times 2 \times 3 \times 5 \times 7$**
b. -448 **$-1 \times 2 \times 2 \times 2 \times 2 \times 2 \times 2 \times 7$**

For Example 4
Factor each monomial.

a. $45a^2b^2$ **$3 \times 3 \times 5 \times a \times a \times b \times b$**
b. $77ab^2$ **$7 \times 11 \times a \times b \times b$**

Teaching Tip The order of prime factors in a prime factorization is not critical. Usually the factors are ordered from the least prime factor to the greatest.

FYI

How big is a 155-digit number? The largest one is 155 nines. This is one less than 10^{155}.

Alternative Learning Styles

Kinesthetic Give students 72 marbles or counters. Their task is to group the marbles or counters into smaller groupings, such as 8 groups of 9 marbles or counters. Point out that 8 groups of 9 marbles represents 8×9. Therefore, 72 can be factored into 8×9. But 72 is not uniquely factorable. Have them explore other groupings.

Teaching Tip Emphasize that 7 is the greatest whole number that will divide 84 and 70 evenly.

In-Class Examples

For Example 5
Find the GCF of the following.

a. 40, 60, and 80 **20**
b. 64, 80, and 96 **16**

For Example 6
Suppose the bakery packed its cookies in boxes of 20 cookies and boxes of 32 cookies. How many cookies should be placed in each cellophane package? **4 cookies**

For Example 7
Find the GCF of the following.

a. $40a^2b$ and $48ab^4$ $8ab$
b. $18x^3$ and $6x$ $6x$

CAREER CHOICES

In 1992 there were 3.1 million people working as chefs and cooks.

Short-order & fast-food cooks	714,000
Restaurant cooks	602,000
Institutional cooks	406,000
Bread and pastry bakers	146,000
Other kitchen workers	1,233,000

There is an easier way to find the GCF of numbers without having to find all of their factors. Look at the prime factorizations of the numbers, and multiply all the prime factors they have in common. *If there are no common prime factors, the GCF is 1.*

$$84 = 2 \cdot 2 \cdot 3 \cdot 7 \qquad 70 = 2 \cdot 5 \cdot 7$$

The integers 84 and 70 have 2 and 7 as common prime factors. The product of these common prime factors is 14, the GCF of 84 and 70.

Example  **Find the GCF of 54, 63, and 180.**

$54 = 2 \cdot 3 \cdot 3 \cdot 3$ *Factor each number.*

$63 = 3 \cdot 3 \cdot 7$ *Circle the common factors.*

$180 = 2 \cdot 2 \cdot 3 \cdot 3 \cdot 5$

The GCF of 54, 63, and 180 is $3 \cdot 3$ or 9.

Example 6

APPLICATION
Packaging

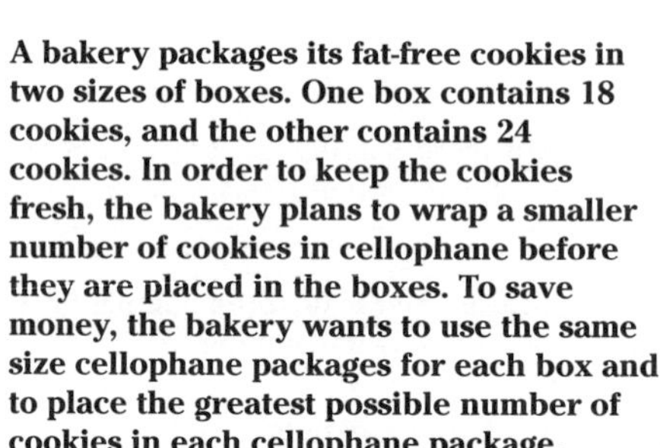

A bakery packages its fat-free cookies in two sizes of boxes. One box contains 18 cookies, and the other contains 24 cookies. In order to keep the cookies fresh, the bakery plans to wrap a smaller number of cookies in cellophane before they are placed in the boxes. To save money, the bakery wants to use the same size cellophane packages for each box and to place the greatest possible number of cookies in each cellophane package.

a. How many cookies should the bakery place in each cellophane package?

b. How many cellophane packages will go in each size of box?

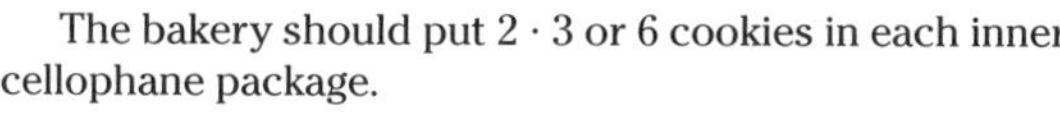

a. Find the GCF of 18 and 24.

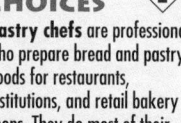

$18 = 2 \cdot 3 \cdot 3$

$24 = 2 \cdot 2 \cdot 2 \cdot 3$

The bakery should put $2 \cdot 3$ or 6 cookies in each inner cellophane package.

b. The box of 18 cookies will contain $18 \div 6$ or 3 cellophane packages. The box of 24 cookies will contain $24 \div 6$ or 4 cellophane packages.

CAREER CHOICES

Pastry chefs are professionals who prepare bread and pastry goods for restaurants, institutions, and retail bakery shops. They do most of their work by hand, taking pride in their creations.

Apprenticeship programs include studying food preparation, decoration techniques, purchasing, and business mathematics.

For more information, contact:

The Educational Foundation of the National Restaurant Association
250 South Wacker Dr.
Suite 1400
Chicago, IL 60606

The GCF of two or more monomials is the product of their common factors, when each monomial is expressed in factored form.

Example  **Find the GCF of $12a^2b$ and $90a^2b^2c$.**

$12a^2b = 2 \cdot 2 \cdot 3 \cdot a \cdot a \cdot b$ *Factor each monomial.*

$90a^2b^2c = 2 \cdot 3 \cdot 3 \cdot 5 \cdot a \cdot a \cdot b \cdot b \cdot c$ *Circle the common factors.*

The GCF of $12a^2b$ and $90a^2b^2c$ is $2 \cdot 3 \cdot a \cdot a \cdot b$ or $6a^2b$.

Cooperative Learning

Pairs Check Separate the class into pairs. Give each pair 20 index cards. Students should write two numbers on each of 10 cards and the GCF of the numbers on separate index cards. Then ask them to exchange cards. Tell students to shuffle the cards and place them face down on a flat surface. Have students take turns flipping over two cards to match numbers and their GCF. If a match is made, the student keeps the cards. The student holding the greatest number of cards once all the cards have been picked up is the winner. For more information on the pairs check strategy, see *Cooperative Learning in the Mathematics Classroom,* one of the titles in the Glencoe Mathematics Professional Series, pages 12–13.

CHECK FOR UNDERSTANDING

Communicating Mathematics

Study this lesson. Then complete the following.

1. **Draw** and label as many rectangles as possible with whole number dimensions that have an area of 48 square inches. See margin.
2. Is $2 \cdot 3^2 \cdot 4$ the prime factorization of 72? Why or why not? No; 4 is not prime.
3. If the GCF of two numbers is 1, must the numbers be prime? Explain.

4. How many prime numbers do you believe there are? Write a statement to support your opinion. See students' work.

3. No; the GCF of 4 and 9 is 1, but neither number is prime.

Guided Practice

Find the factors of each number.

5. 4 1, 2, 4
6. 56 1, 2, 4, 7, 8, 14, 28, 56

State whether each number is *prime* or *composite*. If the number is composite, find its prime factorization.

7. 89 prime
8. 39 composite; $3 \cdot 13$

Factor each expression completely. Do not use exponents.

9. -30 $-1 \cdot 2 \cdot 3 \cdot 5$
10. $22m^2n$ $2 \cdot 11 \cdot m \cdot m \cdot n$

Find the GCF of the given monomials. 16. $6a^2$

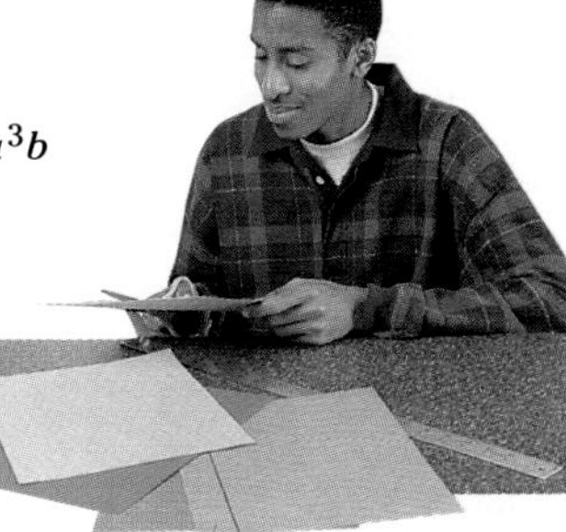

11. 4, 12 4
12. 10, 15 5
13. $24d^2, 30c^2d$ $6d$
14. 18, 35 1
15. $-20gh, 36g^2h^2$ $4gh$
16. $30a^2, 42a^3, 54a^3b$
17. **Geometry** Suppose Terrell cuts out a rectangle that has an area of 96 square inches. If the length and width are both whole numbers, what is the minimum perimeter of the rectangle? Explain how you found the answer. 40 in.

20. 1, 2, 3, 4, 6, 9, 12, 18, 36
21. 1, 2, 4, 5, 8, 10, 16, 20, 40, 80
22. 1, 2, 4, 5, 8, 10, 16, 20, 25, 40, 50, 80, 100, 200, 400

EXERCISES

Practice

Find the factors of each number.

18. 25 1, 5, 25
19. 67 1, 67
20. 36
21. 80
22. 400
23. 950

State whether each number is *prime* or *composite*. If the number is composite, find its prime factorization.

24. 17 prime
25. 63 composite; $3^2 \cdot 7$
26. 91 composite; $7 \cdot 13$
27. 97 prime
28. 304
29. 1540

Factor each expression completely. Do not use exponents.

30. -70 $-1 \cdot 2 \cdot 5 \cdot 7$
31. -117 $-1 \cdot 3 \cdot 3 \cdot 13$
32. $66z^2$ $2 \cdot 3 \cdot 11 \cdot z \cdot z$

33. $4b^3d^2$
34. $-102x^3y$
35. $-98a^2b$

23. 1, 5, 10, 19, 25, 38, 50, 95, 190, 950
28. composite; $2^4 \cdot 19$
29. composite; $2^2 \cdot 5 \cdot 7 \cdot 11$
33. $2 \cdot 2 \cdot b \cdot b \cdot b \cdot d \cdot d$
34. $-1 \cdot 2 \cdot 3 \cdot 17 \cdot x \cdot x \cdot x \cdot y$
35. $-1 \cdot 2 \cdot 7 \cdot 7 \cdot a \cdot a \cdot b$

3 PRACTICE/APPLY

Check for Understanding

Exercises 1–17 are designed to help you assess your students' understanding through reading, writing, speaking, and modeling. You should work through Exercises 1–4 with your students and then monitor their work on Exercises 5–17.

Assignment Guide

Core: 19–63 odd, 64–74
Enriched: 18–60 even, 61–74

For **Extra Practice,** see p. 778.

The red A, B, and C flags, printed only in the Teacher's Wraparound Edition, indicate the level of difficulty of the exercises.

Additional Answer

1.

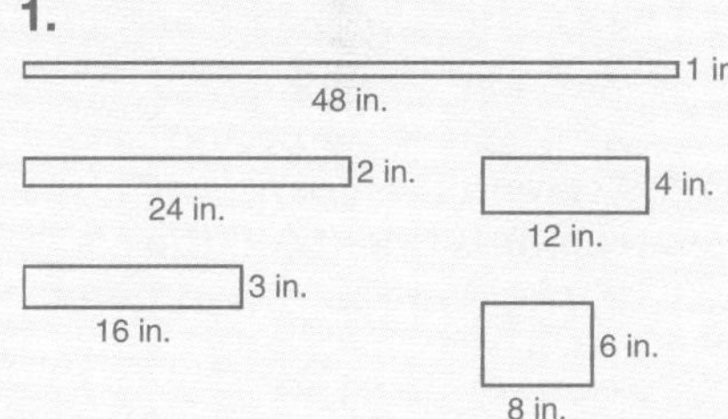

Study Guide Masters, p. 69

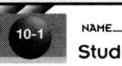

NAME__________ DATE ________

Study Guide

Student Edition Pages 558–563

Factors and Greatest Common Factor

If two or more numbers are multiplied, each number is a **factor** of the product.

	Definition	Example
Prime Number	A *prime number* is a whole number, greater than 1, whose only factors are 1 and itself.	5
Composite Number	A *composite number* is a whole number, greater than 1, that is not prime.	10
Prime Factorization	*Prime factorization* occurs when a whole number is expressed as a product of factors that are all prime.	$45 = 3^2 \cdot 5$
Greatest Common Factor (GCF)	The *greatest common factor* of two or more integers is the greatest number that is a factor of the integers.	The GCF of 12 and 30 is 6.

The GCF of two or more monomials is the product of their common factors, when each monomial is expressed in factored form.

Example: Find the GCF of $16xy^2z^2$ and $72xyz^3$.

$16xy^2z^2 = 2 \cdot 2 \cdot 2 \cdot 2 \cdot x \cdot y \cdot y \cdot z \cdot z$

$72xyz^3 = 2 \cdot 2 \cdot 2 \cdot 3 \cdot 3 \cdot x \cdot y \cdot z \cdot z \cdot z$

The GCF of $16xy^2z^2$ and $72xyz^3$ is $2 \cdot 2 \cdot 2 \cdot x \cdot y \cdot z \cdot z$, or $8xyz^2$.

State whether each number is prime or composite. If the number is composite, find its prime factorization.

1. 28 composite; $2^2 \cdot 7$
2. 61 prime
3. 112 composite; $2^4 \cdot 7$
4. 2865 composite; $3 \cdot 5 \cdot 191$

Factor each expression completely. Do not use exponents.

5. -34 $-1 \cdot 2 \cdot 17$
6. -150 $-1 \cdot 2 \cdot 3 \cdot 5 \cdot 5$
7. $56pq^3$ $2 \cdot 2 \cdot 2 \cdot 7 \cdot p \cdot q \cdot q \cdot q$
8. $-108(cd)^2$ $-1 \cdot 2 \cdot 2 \cdot 3 \cdot 3 \cdot 3 \cdot c \cdot c \cdot d \cdot d$

Find the GCF of the given monomials.

9. $-45, 15$ 15
10. $169, 13$ 13
11. $-20, 440$ 20
12. $49x, 343x^2$ $49x$
13. $4a^7b, 28ab$ $4ab$
14. $96y, 12x, -8y$ 8

Reteaching

Using Alternative Methods To find the prime factorization of an integer, use "factor stairs." Divide the integer by its least prime factor. Repeat for the quotient. When you reach 1, the factorization is complete. The answer is on the steps.
$252 = 2 \cdot 2 \cdot 3 \cdot 3 \cdot 7$

```
       1
    7⟌ 7
   3⟌ 21
   3⟌ 63
  2⟌ 126
  2⟌ 252
```

Alternative Teaching Strategies

Student Diversity Many students confuse the GCF with the LCM. It may help them if they remember that any factor of *A* is *less* than *A,* even the *greatest* common factor. Any multiple of *A* is *greater* than *A,* even the *least* common multiple.

4 ASSESS

Closing Activity

Speaking Have students explain how they would determine whether 1237 is a prime number.

Using the Programming Exercises The program given in Exercise 60 is for use with a TI-82 graphing calculator. For other programmable calculators, have students consult their owner's manual for commands similar to those presented here.

Additional Answers

61a. $2b^3 \times 1 \times 1$
$2b^2 \times b \times 1$
$2b \times b \times b$
$b^3 \times 2 \times 1$
$b^2 \times 2b \times 1$
$b^2 \times 2 \times b$

61c. $4b^3 + 1$ or 865
$2b^3 + 2b^2 + 1$ or 505
$5b^2$ or 180
$3b^3 + 2$ or 650
$2b^3 + b^2 + b$ or 474
$b^3 + 2b^2 + 2b$ or 300

61d. Though the volume remains constant, the surface areas vary greatly.

Practice Masters, p. 69

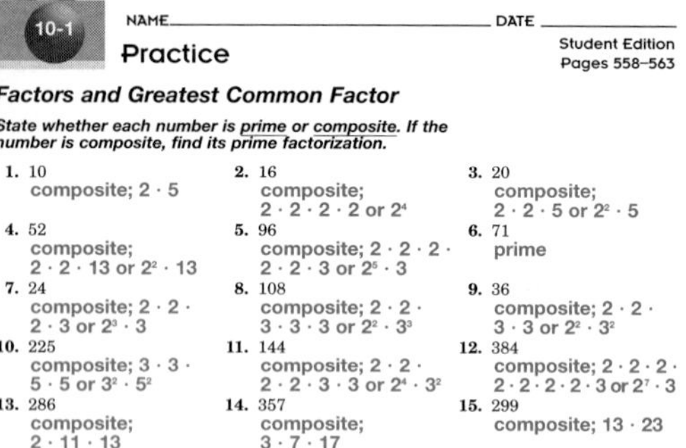

10-1 NAME ______ DATE ______

Practice — Student Edition Pages 558–563

Factors and Greatest Common Factor

State whether each number is prime or composite. If the number is composite, find its prime factorization.

1. 10 composite; $2 \cdot 5$
2. 16 composite; $2 \cdot 2 \cdot 2 \cdot 2$ or 2^4
3. 20 composite; $2 \cdot 2 \cdot 5$ or $2^2 \cdot 5$
4. 52 composite; $2 \cdot 2 \cdot 13$ or $2^2 \cdot 13$
5. 96 composite; $2 \cdot 2 \cdot 2 \cdot 2 \cdot 2 \cdot 3$ or $2^5 \cdot 3$
6. 71 prime
7. 24 composite; $2 \cdot 2 \cdot 2 \cdot 3$ or $2^3 \cdot 3$
8. 108 composite; $2 \cdot 2 \cdot 3 \cdot 3 \cdot 3$ or $2^2 \cdot 3^3$
9. 36 composite; $2 \cdot 2 \cdot 3 \cdot 3$ or $2^2 \cdot 3^2$
10. 225 composite; $3 \cdot 3 \cdot 5 \cdot 5$ or $3^2 \cdot 5^2$
11. 144 composite; $2 \cdot 2 \cdot 2 \cdot 2 \cdot 3 \cdot 3$ or $2^4 \cdot 3^2$
12. 384 composite; $2 \cdot 2 \cdot 2 \cdot 2 \cdot 2 \cdot 2 \cdot 2 \cdot 3$ or $2^7 \cdot 3$
13. 286 composite; $2 \cdot 11 \cdot 13$
14. 357 composite; $3 \cdot 7 \cdot 17$
15. 299 composite; $13 \cdot 23$

Factor each expression completely. Do not use exponents.

16. $36x^2y$ $2 \cdot 2 \cdot 3 \cdot 3 \cdot x \cdot x \cdot y$
17. $169x^3y^3$ $13 \cdot 13 \cdot x \cdot x \cdot x \cdot y \cdot y \cdot y$
18. $81b^2c^3$ $3 \cdot 3 \cdot 3 \cdot 3 \cdot b \cdot b \cdot c \cdot c \cdot c$
19. $-52ab^3$ $-2 \cdot 2 \cdot 13 \cdot a \cdot b \cdot b \cdot b$
20. $72a^4c^2$ $2 \cdot 2 \cdot 2 \cdot 3 \cdot 3 \cdot a \cdot a \cdot a \cdot a \cdot c \cdot c$
21. $-32abc^3$ $-2 \cdot 2 \cdot 2 \cdot 2 \cdot 2 \cdot a \cdot b \cdot c \cdot c \cdot c$

Find the GCF of the given monomials.

22. 42, 60 6
23. 36, 64 4
24. 80, 28 4
25. 88, 40 8
26. 24, 81 3
27. 700, 1000 100
28. 24, 336 24
29. 90, 480 30
30. 17, 85 17
31. $18abc$, $54ab^2$ $18ab$
32. $12x^3y^2$, $44xy^3$ $4xy^2$
33. $32xyz$, $48xy^4$ $16xy$
34. $18m^2n^2$, $66m^3$ $6m^2$
35. $72r^2s^2$, $36rs^3$ $36rs^2$
36. $18xp^2$, $48x^4$ $6x$
37. 16, 24, 48 8
38. 18, 45, 63 9
39. 12, 30, 114 6

Find the GCF of the given monomials.

36. 18, 36 **18**
37. 18, 45 **9**
38. 84, 96 **12**
39. 28, 75 **1**
40. $-34, 51$ **17**
41. $95, -304$ **19**
42. $17a$, $34a^2$ **$17a$**
43. $21p^2q$, $35pq^2$ **$7pq$**
44. $12an^2$, $40a^4$ **$4a$**
45. $-60r^2s^2t^2$, $45r^3t^3$ (C) — 45. $15r^2t^2$
46. 18, 30, 54 **6**
47. 24, 84, 168 **12**
48. $14a^2b^3$, $20a^3b^2c$, $35ab^3c^2$ **ab^2**
49. $18x^2$, $30x^3y^2$, $54y^3$ **6**
50. $14a^2b^2$, $18ab$, $2a^3b^3$ **$2ab$**
51. $32m^2n^3$, $8m^2n$, $56m^3n^2$ **$8m^2n$**

Find each missing factor.

52. $42a^2b^5c = 7a^2b^3(\underline{\ ?\ })$ **$6b^2c$**
53. $-48x^4y^2z^3 = 4xyz(\underline{\ ?\ })$ **$-12x^3yz^2$**
54. $48a^5b^5 = 2ab^2(4ab)(\underline{\ ?\ })$ **$6a^3b^2$**
55. $36m^5n^7 = 2m^3n(6n^5)(\underline{\ ?\ })$ **$3m^2n$**
56. **Geometry** The area of a rectangle is 116 square inches. What are its possible whole number dimensions? — 56. 1 in. by 116 in., 2 in. by 58 in., 4 in. by 29 in.
57. **Geometry** The area of a rectangle is 1363 square centimeters. If the measures of the length and width are both prime numbers, what are the dimensions of the rectangle? **29 cm by 47 cm**

58. **Number Theory** Check to see if your house number is a prime number and if the last four digits in your telephone number form a prime number. Explain how you decided. **See students' work.**
59. **Number Theory** *Twin primes* are two consecutive odd numbers that are prime, such as 11 and 13. List the twin primes where both primes are less than 100. — 59. 3, 5; 5, 7; 11, 13; 17, 19; 29, 31; 41, 43; 59, 61; 71, 73

Programming

60. Use the graphing calculator program below to find the GCF of two numbers.

```
PROGRAM:GCF
: Input "INTEGER",A          : Goto 4
: Input "INTEGER",B          : A-B→A
: A→E                        : Goto R
: B→F                        : Lbl 4
: Lbl R                      : B-A→B
: If A=B                     : Goto R
: Goto 5                     : Lbl 5
: If A<B                     : Disp "GCF IS", A
```

Use the program to find the GCF of each pair of numbers.

a. 896, 700 **28**
b. 1015, 3132 **29**
c. 567, 416 **1**
d. 486, 432 **54**
e. 891, 1701 **81**
f. 1105, 1445 **85**

Critical Thinking

61. **Geometry** Suppose the volume of a rectangular solid is $2b^3$ and the measure of each side is a monomial with integral coefficients.
    a. List the demensions of each such rectangular solid. (*Hint:* There are 6.) — 61a. See margin.
    b. Draw and label each solid. **See students' work.**
    c. Find the surface area of each solid if $b = 6$. **See margin.**
    d. What can you conclude about the surface areas of these solids, given that the volume remains constant? **See margin.**

Extension

Problem Solving Find the GCF of the given monomials so the second factor has integral coefficients.

$\left(\frac{3}{8}\right)a^3b^3c^7, \left(\frac{7}{20}\right)a^2b^2c^6, \left(\frac{1}{12}\right)ab^6c^4$

$\frac{1}{120ab^2c^4}$

Applications and Problem Solving

62. **Gardening** Marisela is planning to have 100 tomato plants in her garden. In what ways can she arrange them so that she has the same number of plants in each row, at least 5 rows of plants, and at least 5 plants in each row? **5 rows of 20 plants, 10 rows of 10 plants, 20 rows of 5 plants**

63. **Sports** A new athletic field is being sodded at Beck High School using 2-yard-by-2-yard squares of sod. If the length of the field is 70 yards longer than the width and its area is 6000 square yards, how many squares of sod will be needed? **1500 squares of sod**

Mixed Review

64. Find $(1.1x + y)^2$. (Lesson 9–8) $\mathbf{1.21x^2 + 2.2xy + y^2}$

65. Simplify $\frac{12b^5}{4b^4}$. (Lesson 9–2) $\mathbf{3b}$

66. Solve the system of equations by graphing. (Lesson 8–1) **(0, 0)**
$y = -x$
$y = 2x$

67. Graph the compound inequality $y > 2$ or $y < 1$. (Lesson 7–4) **See margin.**

68. Solve $16x < 96$. Check your solution. (Lesson 7–2) $\mathbf{x < 6}$

69. Write the standard form of an equation of the line that passes through $(4, -2)$ and $(4, 8)$. (Lesson 6–2) $\mathbf{x = 4}$

70. Graph $8x - y = 16$. (Lesson 5–4) **See margin.**

71. **Physics** Weights of 50 pounds and 75 pounds are placed on a lever. The two weights are 16 feet apart, and the lever is balanced. How far from the fulcrum is the 50-pound weight? (Lesson 4–8)

71. $\mathbf{9\frac{3}{5}}$ **feet**

72. **Waves** The highest wave ever sighted and recorded was 112 feet high. This wave was brought on by a wind of 74 mph. Using ratios, determine how high a wave brought on by a 25-mph wind could reach. (Lesson 4–1)

72. 37.8 feet

73. Solve $9 = x + 13$. (Lesson 3–1) $\mathbf{-4}$

74. Write a verbal expression for $z^7 + 2$. (Lesson 1–1)

74. the number *z* to the seventh power added to 2

Number Sieve

The article below appeared in *Science News* on October 1, 1994.

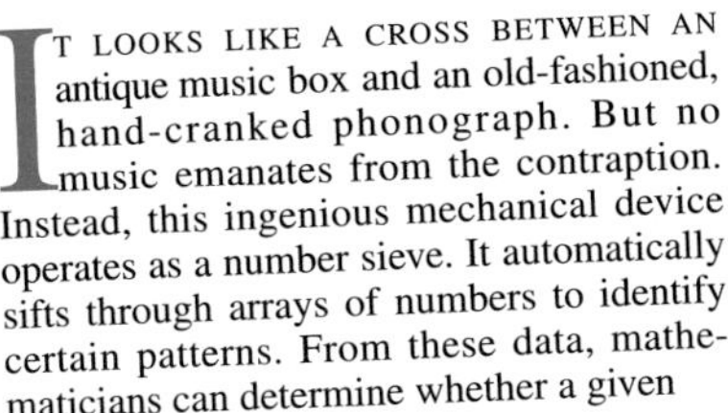

IT LOOKS LIKE A CROSS BETWEEN AN antique music box and an old-fashioned, hand-cranked phonograph. But no music emanates from the contraption. Instead, this ingenious mechanical device operates as a number sieve. It automatically sifts through arrays of numbers to identify certain patterns. From these data, mathematicians can determine whether a given number is a prime or the product of two or more primes multiplied together. Constructed 75 years ago, it also represents the first known, successful attempt to automate the factoring of whole numbers. Until three researchers tracked down the machine recently, few people knew of its existence. Now, this unique device can take its proper place in the history of computational number theory. ■

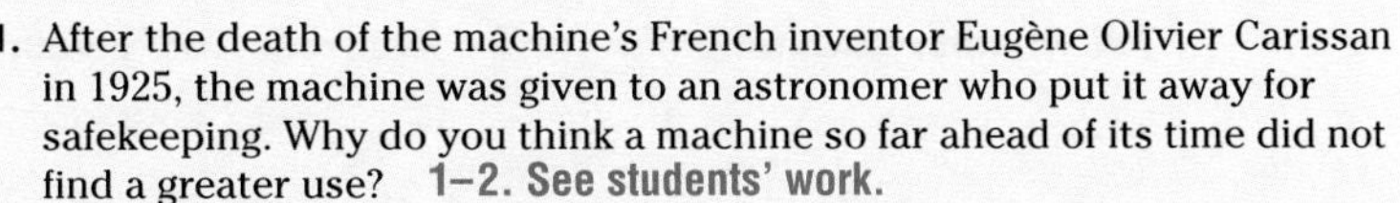

1. After the death of the machine's French inventor Eugène Olivier Carissan in 1925, the machine was given to an astronomer who put it away for safekeeping. Why do you think a machine so far ahead of its time did not find a greater use? **1–2. See students' work.**
2. One of the main uses of prime numbers today is in cryptography, the coding and decoding of data and messages. Why are more sophisticated codes needed today than they were in the 1920s?

Mathematics and SOCIETY

This machine, invented and built by Eugene Olivier Carissan, is about the size of a laptop computer. It consists of 14 brass gears of differing sizes. The machine works on a trial-and-error method, called the sieve method. It tries to divide the number by a smaller number to see whether it is or is not a factor.

Additional Answers

67. Number line from −1 to 5

70. Graph with axes labeled x and y; x-axis: −8, −4, O, 4, 8; y-axis: −4, −8, −12, −16

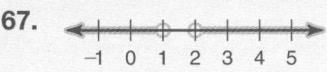

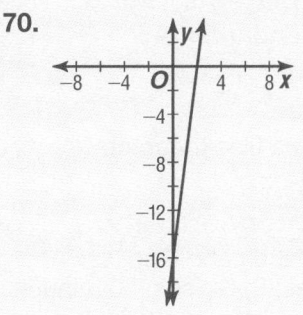

Enrichment Masters, p. 69

NAME ____________ DATE ________
10-1 **Enrichment** — Student Edition Pages 558–563

Finding the GCF by Euclid's Algorithm

Finding the greatest common factor of two large numbers can take a long time using prime factorizations. This method can be avoided by using *Euclid's Algorithm* as shown in the following example.

Example: Find the GCF of 209 and 532.

Divide the greater number, 532, by the lesser, 209.

209)532 = 2, 418; 114)209 = 1, 114; 95)114 = 1, 95; 19)95 = 5, 95, 0

Divide the remainder into the divisor above. Repeat this process until the remainder is zero. The last nonzero remainder is the GCF.

The divisor, 19, is the GCF of 209 and 532.

Suppose the GCF of two numbers is found to be 1. Then the numbers are said to be **relatively prime.**

Find the GCF of each group of numbers by using Euclid's Algorithm.

1. 187; 578 17
2. 1802; 106 106
3. 161; 943 23
4. 215; 1849 43
5. 1325; 3498 53
6. 3484; 5963 67
7. 33,583; 4257 473
8. 453; 484 1
9. 95; 209; 589 19
10. 518; 407; 851 37
11. $17a^2x^2z$; $1615axz^2$ $17axz$
12. $752cf^3$; $893c^3f^3$ $47cf^3$
13. $979r^2s^2$; $495rs^3$, $154r^3s^3$ $11rs^2$
14. $360x^5y^7$; $328xy$; $568x^3y^3$ $8xy$

10-2A LESSON NOTES

NCTM Standards: 1–5

Objective

Use algebra tiles to factor binomials.

Recommended Time

Demonstration and discussion: 15 minutes; Exercises: 30 minutes

Instructional Resources

For each student or group of students

Student Manipulative Kit
- algebra tiles
- product mat

Modeling Mathematics Masters
- p. 1 (algebra tiles)
- p. 15 (product mat)
- p. 32 (worksheet)

For teacher demonstration

Algebra and Geometry Overhead Manipulative Resources

1 FOCUS

Motivating the Lesson

The number 63 can be modeled by a rectangle of dimension 7 units by 9 units. Any product can be modeled by the area of a rectangle.

2 TEACH

Teaching Tip Students must form a rectangle with the given tiles. If a rectangle cannot be formed, the binomial is not factorable.

Teaching Tip Some students may want to add zero pairs to make the rectangle. Encourage this type of creative thinking.

3 PRACTICE/APPLY

Assignment Guide
Core: 1–9
Enriched: 1–9

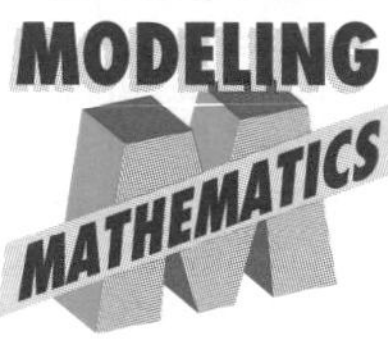

A Preview of Lesson 10–2

10–2A Factoring Using the Distributive Property

Materials: algebra tiles, product mat

When two or more numbers are multiplied, these numbers are factors of the product. Sometimes you know the product of binomials and are asked to find the factors. This is called **factoring.** You can use algebra tiles to factor binomials.

Activity 1 Use algebra tiles to factor $2x + 8$.

Step 1 Model the polynomial $2x + 8$.

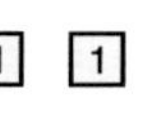

Step 2 Arrange the tiles into a rectangle. The total area of the tiles represents the product and its length and width represent the factors.

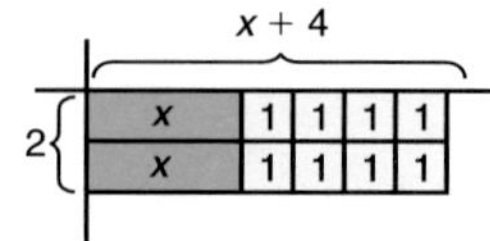

The rectangle has a width of 2 and a length of $x + 4$. Therefore, $2x + 8 = 2(x + 4)$.

Activity 2 Use algebra tiles to factor $x^2 - 3x$.

Step 1 Model the polynomial $x^2 - 3x$.

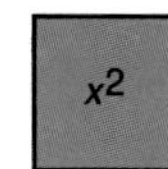
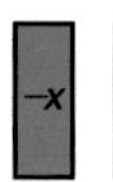
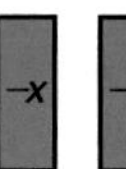

Step 2 Arrange the tiles into a rectangle.

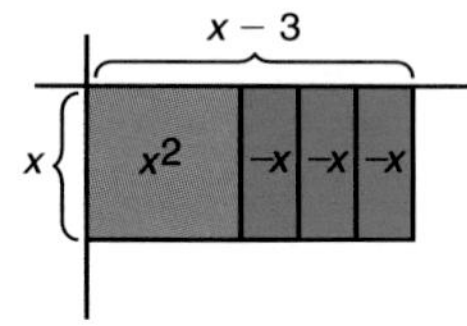

The rectangle has a width of x and a length of $x - 3$. Therefore, $x^2 - 3x = x(x - 3)$.

Model Use algebra tiles to factor each binomial.

1. $3x + 9$ $3(x + 3)$
2. $4x - 10$ $2(2x - 5)$
3. $3x^2 + 4x$ $x(3x + 4)$
4. $10 - 5x$ $5(2 - x)$

Draw Tell whether each binomial can be factored. Justify your answer with a drawing.

5. $2x + 3$ no
6. $3 - 9x$ yes
7. $x^2 - 5x$ yes
8. $3x^2 + 5$ no

5–8. See Solutions Manual for drawings.

Write

9. Write a paragraph that explains how you can determine whether a binomial can be factored. Include an example of one binomial that can be factored and one that cannot.

9. Binomials can be factored if they can be represented by a rectangle. Examples: $4x + 4$ can be factored and $4x + 3$ cannot be factored.

4 ASSESS

Observing students working in cooperative groups is an excellent method of assessment.

10-2 Factoring Using the Distributive Property

What YOU'LL LEARN

- To use the greatest common factor (GCF) and the distributive property to factor polynomials, and
- to use grouping techniques to factor polynomials with four or more terms.

Why IT'S IMPORTANT

You can use factoring to solve problems involving sports and construction.

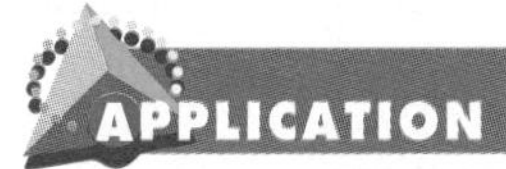

Sports

Rugby is a contact sport in which each of two teams tries to get an oval ball behind its opponent's goal line or kick it over its opponent's goal. It is similar to American football. However, the action in the game is almost nonstop, and the players wear little protective gear.

There are two versions of rugby—Rugby Union and Rugby League. Rugby Union is popular in Australia, Canada, England, France, Ireland, Japan, New Zealand, Scotland, South Africa, and Wales. It is played on a rectangular field.

LOOK BACK

You can refer to Lesson 1-7 for information on the distributive property.

If the width of this field is represented by x, its length can be represented by $x + 75$ and the area of the field by $x(x + 75)$, or $x^2 + 75x$. If $x(x + 75) = x^2 + 75x$, then $x^2 + 75x = x(x + 75)$. *Why?*

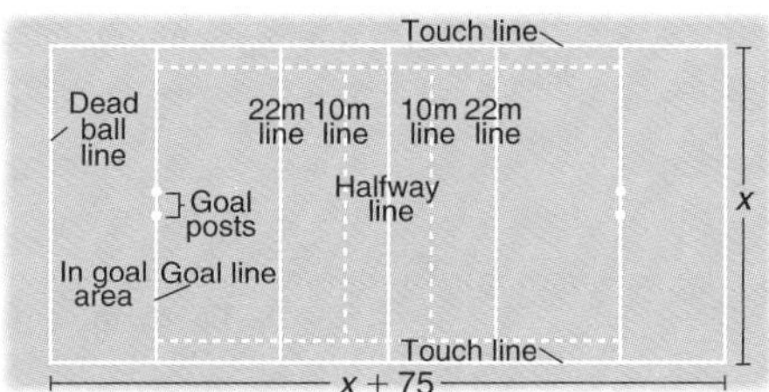

The expression $x(x + 75)$ is called the *factored form* of $x^2 + 75x$. A polynomial is in factored form, or **factored,** when it is expressed as the product of monomials and polynomials.

Rugby is named for Rugby School in England where the game was first played in 1823. The game has spread throughout the British Empire and into Asia. The Rugby World Cup, the World Championship played in 1995, was won by the South African team.

In Chapter 9, you multiplied a polynomial by a monomial using the distributive property. You can also reverse this process and express a polynomial in factored form by using the distributive property.

Model	Multiplying Polynomials	Factoring Polynomials
$2a$, b; 3; $6a$, $3b$	$3(2a + b) = 6a + 3b$	$6a + 3b = 3(a + 2b)$
$3x$, $-4y$; $5x$; $15x^2$, $-20xy$	$5x(3x - 4y) = 15x^2 - 20xy$	$15x^2 - 20xy = 5x(3x - 4y)$
x^2, $5x$; 3; $3x^2$, $15x$	$3(x^2 + 5x) = 3x^2 + 15x$	$3x^2 + 15x = 3(x^2 + 5x)$

GLOBAL CONNECTIONS

Rugby was excluded from the Football Association, established in 1863. In 1871, the Rugby Football Union was formed to standardize the rules and regulate the game.

10-2 LESSON NOTES

NCTM Standards: 1–5

Instructional Resources

- Study Guide Master 10-2
- Practice Master 10-2
- Enrichment Master 10-2
- Assessment and Evaluation Masters, p. 268
- Tech Prep Applications Masters, p. 19

Transparency 10-2A contains the 5-Minute Check for this lesson; **Transparency 10-2B** contains a teaching aid for this lesson.

Recommended Pacing	
Standard Pacing	Day 3 of 12
Honors Pacing	Day 3 of 11
Block Scheduling*	Day 2 of 6
Alg. 1 in Two Years*	Days 5 & 6 of 22

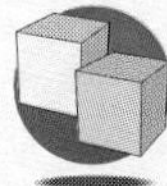

*For more information on pacing and possible lesson plans, refer to the *Block Scheduling Booklet* and *Algebra 1 in Two Years.*

1 FOCUS

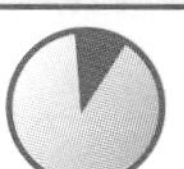

5-Minute Check
(over Lesson 10-1)

Find the prime factorization of each number.

1. 86 **2 × 43**
2. 168 **2 × 2 × 2 × 3 × 7**

Find the GCF.

3. 49 and 77 **7**
4. 24 and 104 **8**
5. $40xy^2$ and $64yz^2$ **8y**

Motivating the Lesson

Situational Problem Show that 2(3) can be found by using the distributive property and only 1s. For example:

$$\begin{aligned} 2(3) &= (1 + 1)(1 + 1 + 1) \\ &= (1 + 1)(1) + (1 + 1)(1) + (1 + 1)(1) \\ &= 1 + 1 + 1 + 1 + 1 + 1 \\ &= 6 \end{aligned}$$

Have students find 3(5) using only 1s and 2s.

2 TEACH

In-Class Examples

For Example 1
Use the distributive property to factor each polynomial.

a. $25a^4 + 15a^2$ $\mathbf{5a^2(5a^2 + 3)}$
b. $18x^2 - 12x^3$ $\mathbf{6x^2(3 - 2x)}$
c. $28a^2b + 56abc^2$ $\mathbf{28ab(a + 2c^2)}$
d. $20x^2 - 24xy^2$ $\mathbf{4x(5x - 6y^2)}$

For Example 2
The area of a rectangle is represented by $25a^2 - 30ab$. What are its length and width? $\mathbf{5a, 5a - 6b}$

Teaching Tip After the GCF is determined, the remaining factors may be found by dividing each term in the expression by the GCF. For example, in $10y^2 + 15y$, the GCF is $5y$. The remaining factors are found as follows.

$$\frac{10y^2}{5y} = 2y \text{ and } \frac{15y}{5y} = 3$$

$$10y^2 + 15y = 5y(2y) + 5y(3)$$
$$= 5y(2y + 3)$$

Teaching Tip Encourage students to check the factoring process by multiplying the factors.

Factoring a polynomial or finding the factored form of a polynomial means to find its *completely* factored form. The expression $3(x^2 + 5x)$ above is not considered completely factored since the polynomial $x^2 + 5x$ can be factored as $x(x + 5)$. The completely factored form of $3x^2 + 15x$ is $3x(x + 5)$.

Example  **Use the distributive property to factor each polynomial.**

a. $\mathbf{12mn^2 - 18m^2n^2}$

First find the GCF for $12mn^2$ and $18m^2n^2$.

$12mn^2 = 2 \cdot 2 \cdot 3 \cdot m \cdot n \cdot n$
$18m^2n^2 = 2 \cdot 3 \cdot 3 \cdot m \cdot m \cdot n \cdot n$ *The GCF is* $2 \cdot 3 \cdot m \cdot n \cdot n$ *or* $6mn^2$.

Notice that $12mn^2 = 6mn^2(2)$ and $18m^2n^2 = 6mn^2(3m)$. Then use the distributive property to express the polynomial as the product of the GCF and the remaining factor of each term.

$$12mn^2 - 18m^2n^2 = 6mn^2(2) - 6mn^2(3m)$$
$$= 6mn^2(2 - 3m) \quad \textit{Distributive property}$$

b. $\mathbf{20abc + 15a^2c - 5ac}$

$20abc = 2 \cdot 2 \cdot 5 \cdot a \cdot b \cdot c$
$15a^2c = 3 \cdot 5 \cdot a \cdot a \cdot c$
$5ac = 5 \cdot a \cdot c$ *The GCF is* $5ac$.

$$20abc + 15a^2c - 5ac = 5ac(4b) + 5ac(3a) - 5ac(1)$$
$$= 5ac(4b + 3a - 1)$$

Factoring a polynomial can simplify computations.

Example 

Construction

The Lopez family wants to build a swimming pool in the shape of the figure below. Although the family has not yet decided on the actual dimensions of the pool, they do know that they want to build a deck that is 4 feet wide around the pool.

a. Write an equation for the area of the deck.

b. If they decide to let *a* be 24 feet long, *b* be 6 feet long, and *c* be 10 feet, find the area of the deck.

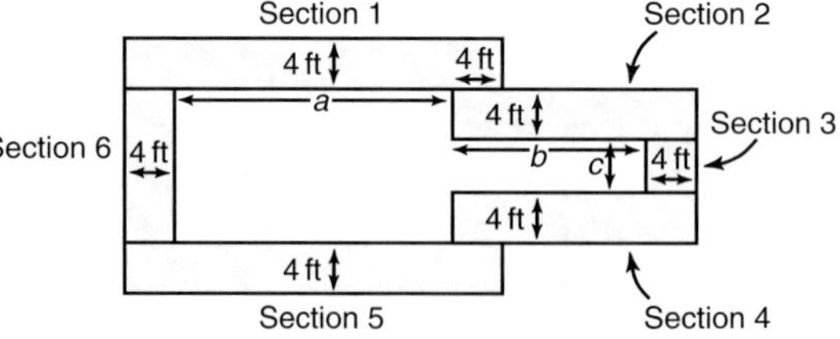

Cooperative Learning

Trade-A-Problem Factoring problems can be created by starting with the solution. For example, the solution $2x(a^2 + 4x)$ leads to the problem $2xa^2 + 8x^2$. Have students work in pairs. One student should create a factoring problem and give it to the other to solve. Then they should switch roles. For more information on the trade-a-problem strategy, see *Cooperative Learning in the Mathematics Classroom,* one of the titles in the Glencoe Mathematics Professional Series, pages 25–26.

a. You can find the area of the deck by finding the sum of the areas of the 6 rectangular sections shown in the figure. The resulting expression can be simplified by first using the distributive property and then factoring.

Section 1 Section 2 Section 3 Section 4 Section 5 Section 6

$A = 4(a + 4 + 4) + 4(b + 4) + 4c + 4(b + 4) + 4(a + 4 + 4) + 4(c + 4 + 4)$

$= 4a + 16 + 16 + 4b + 16 + 4c + 4b + 16 + 4a + 16 + 16 + 4c + 16 + 16$

$= 8a + 8b + 8c + 128$ *Combine like terms.*

$= 8(a + b + c + 16)$ *The GCF is 8.*

The area of the deck is $8(a + b + c + 16)$ square feet.
Would dividing the deck into different sections result in a different answer?

b. Replace a with 24, b with 6, and c with 10.

$A = 8(24 + 6 + 10 + 16)$

$= 8(56)$ or 448

The area of the deck would be 448 square feet.

Just as it is possible to use the distributive property to factor a polynomial into monomial and polynomial factors, it is possible to factor some polynomials containing four or more terms into the product of two polynomials. Consider $(3a + 2b)(4c + 7d) = 12ac + 21ad + 8bc + 14bd$. Here, the product of two binomials results in a polynomial with four terms. How can the process be reversed to factor the four-term polynomial into its two binomial factors?

Example 3 **Factor $12ac + 21ad + 8bc + 14bd$.**

$12ac + 21ad + 8bc + 14bd$

$= (12ac + 21ad) + (8bc + 14bd)$ *Apply the associative property, since a common factor of 3a appears in the first two terms and a common factor of 2b appears in the last two terms.*

$= 3a(4c + 7d) + 2b(4c + 7d)$ *Factor the first two terms and the last two terms.*

$= (3a + 2b)(4c + 7d)$ *4c + 7d is a common factor. Use the distributive property.*

LOOK BACK

You can refer to Lesson 9-6 for information on FOIL.

Check by using FOIL.

$(3a + 2b)(4c + 7d) = \overset{F}{(3a)(4c)} + \overset{O}{(3a)(7d)} + \overset{I}{(2b)(4c)} + \overset{L}{(2b)(7d)}$

$= 12ac + 21ad + 8bc + 14bd$ ✓

This method is called **factoring by grouping**. It is necessary to group the terms and factor each group separately so that the remaining polynomial factors of each group are the same. This allows the distributive property to be applied a second time with a polynomial as the common factor.

In-Class Example

For Example 3
Factor.

a. $6x^2y + 9xyz + 4xy^2 + 6y^2z$
$(3xy + 2y^2)(2x + 3z)$

b. $49m + 21n^2 + 35m^3n + 15m^2n^3$
$(7m + 3n^2)(5m^2n + 7)$

In-Class Example

For Example 4
Factor.

a. $20ab - 35b - 63 + 36a$
$(4a - 7)(5b + 9)$

b. $3x^3 + 2xy - 15x^2 - 10y$
$(3x^2 + 2y)(x - 5)$

c. $a^2 - ab - 7b + 7a$
$(a - b)(a + 7)$

d. $3a^2 - 2ab + 10b - 15a$
$(a - 5)(3a - 2b)$

3 PRACTICE/APPLY

Check for Understanding

Exercises 1–25 are designed to help you assess your students' understanding through reading, writing, speaking, and modeling. You should work through Exercises 1–6 with your students and then monitor their work on Exercises 7–25.

Error Analysis
Students may think they have factored by grouping when they factor problems like $3x + 3y + 4w + 4z$ as $3(x + y) + 4(x + z)$. Although some factoring was done in removing GCF from two binomials, the final form is a sum and not a product. For factoring by grouping to work, binomial factors must match.

Sometimes you can group the terms in more than one way when factoring a polynomial. For example, the polynomial in Example 3 could have been factored in the following way.

$$\begin{aligned} 12ac + 21ad + 8bc + 14bd &= (12ac + 8bc) + (21ad + 14bd) \\ &= 4c(3a + 2b) + 7d(3a + 2b) \\ &= (4c + 7d)(3a + 2b) \quad \textit{The result is the same as in Example 3.} \end{aligned}$$

Recognizing binomials that are additive inverses is often helpful in factoring. For example, the binomials $3 - a$ and $a - 3$ are additive inverses since the sum of $3 - a$ and $a - 3$ is 0. Thus, $3 - a$ and $-a + 3$ are equivalent. What is the additive inverse of $5 - y$?

$-1(a - 3) = -a + 3$
$= 3 - a$

Example 4 **Factor $15x - 3xy + 4y - 20$.**

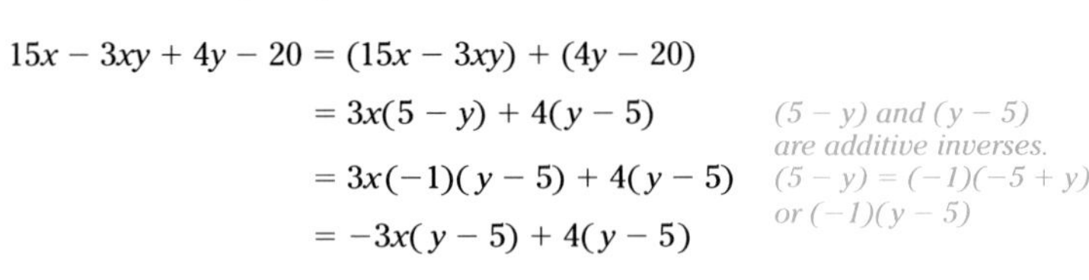

$$\begin{aligned} 15x - 3xy + 4y - 20 &= (15x - 3xy) + (4y - 20) \\ &= 3x(5 - y) + 4(y - 5) \\ &= 3x(-1)(y - 5) + 4(y - 5) \\ &= -3x(y - 5) + 4(y - 5) \\ &= (-3x + 4)(y - 5) \end{aligned}$$

$(5 - y)$ and $(y - 5)$ are additive inverses. $(5 - y) = (-1)(-5 + y)$ or $(-1)(y - 5)$

Check:
$$\begin{aligned} (-3x + 4)(y - 5) &= (-3x)(y) + (-3x)(-5) + 4(y) + 4(-5) \\ &= -3xy + 15x + 4y - 20 \\ &= 15x - 3xy + 4y - 20 \ \checkmark \end{aligned}$$

In summary, a polynomial can be factored by grouping if all of the following situations exist.

- There are four or more terms.
- Terms with common factors can be grouped together.
- The two common factors are identical or differ by a factor of -1.

CHECK FOR UNDERSTANDING

Communicating Mathematics

Study the lesson. Then complete the following.

1. **a. Express** $8d^2 - 14d$ as a product of factors in three different ways.
   b. Which of the three answers in part a is the completely factored form of $8d^2 - 14d$? Explain. **$2d(4d - 7)$; $2d$ is the GCF of $8d^2$ and $14d$.**
2. **a. Express** the area of the rectangle at the right by adding the areas of the smaller rectangles.
   b. Express the area as the product of the length and the width.
   c. What is the relationship between the expressions in parts a and b?

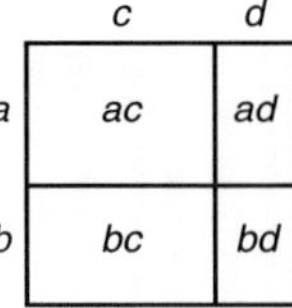

1a. $2(4d^2 - 7d)$, $d(8d - 14)$, $2d(4d - 7)$

2a. $ac + ad + bc + bd$

2b. $(a + b)(c + d)$

2c. They are equal.

Reteaching

Using Substeps The following format may be useful in helping students factor sums and differences of polynomials.
Factor $54x^2 + 72x$.

Terms	Prime Factors	GCF
$54x^2$ $72x$	$2 \cdot 3 \cdot 3 \cdot 3 \cdot x \cdot x$ $2 \cdot 2 \cdot 2 \cdot 3 \cdot 3 \cdot x$	$2 \cdot 3 \cdot 3 \cdot x$ or $18x$

So, $54x^2 + 72x = 18x(3x + 4)$.

3. distributive, associative, commutative

4. $(4gh + 8h) + (3g + 6)$, $(4gh + 3g) + (8h + 6)$

3. **List** the properties used to factor $4gh + 8h + 3g + 6$ by grouping.
4. **Group** the terms of $4gh + 8h + 3g + 6$ in pairs in two different ways so that the pairs of terms have a common monomial factor.
5. **Write** the additive inverse of $7p^2 - q$. $q - 7p^2$

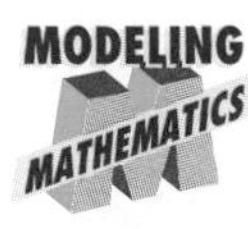

6. Use algebra tiles to factor $2x^2 - x$. $x(2x - 1)$

Guided Practice

Find the GCF of the terms in each expression.

7. $3y^2 + 12$ 3
8. $5n - n^2$ n
9. $5a + 3b$ 1
10. $6mn + 15m^2$ $3m$
11. $12x^2y^2 - 8xy^2$ $4xy^2$
12. $4x^2y - 6xy^2$ $2xy$

Express each polynomial in factored form. 13. $(a + b)(x + y)$

14. $(3m + 5n)(a - 2b)$

13. $a(x + y) + b(x + y)$
14. $3m(a - 2b) + 5n(a - 2b)$
15. $x(3a + 4b) - y(3a + 4b)$ $(x - y)(3a + 4b)$
16. $x^2(a^2 + b^2) + (a^2 + b^2)$ $(x^2 + 1)(a^2 + b^2)$

Complete. In Exercise 18, both blanks represent the same expression.

17. $20s + 12t = 4(5s + \underline{?})$ $3t$
18. $(6x^2 - 10xy) + (9x - 15y) = 2x(\underline{?}) + 3(\underline{?})$ $3x - 5y$

Factor each polynomial.

19. $29xy - 3x$ $x(29y - 3)$
20. $x^5y - x$ $x(x^4y - 1)$
21. $3c^2d - 6c^2d^2$ $3c^2d(1 - 2d)$
22. $ay - ab + cb - cy$ $(a - c)(y - b)$
23. $rx + 2ry + kx + 2ky$
24. $5a - 10a^2 + 2b - 4ab$
25. **Volleyball** Peta is scheduling the games for a volleyball league. To find the number of games she needs to schedule, she can use the equation $g = \frac{1}{2}n^2 - \frac{1}{2}n$, where g represents the number of games needed for each team to play each other team exactly once and n represents the number of teams.
    a. Write this equation in factored form.
    b. How many games are needed for 14 teams to play each other exactly once? 91 games
    c. How many games are needed for 7 teams to play each other exactly 3 times? 63 games

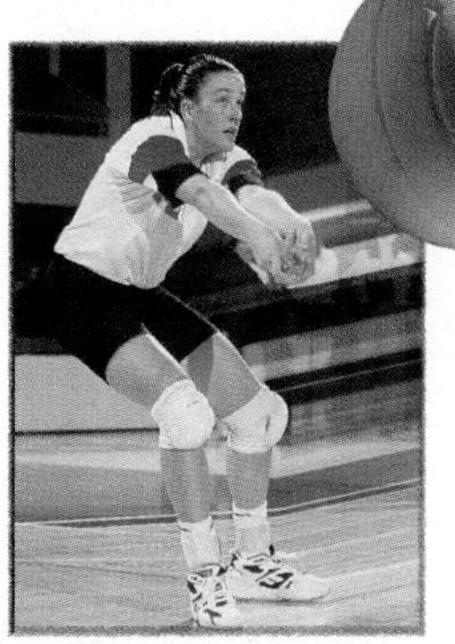

23. $(r + k)(x + 2y)$

24. $(5a + 2b)(1 - 2a)$

25a. $g = \frac{1}{2}n(n - 1)$

EXERCISES

Practice

Complete. In exercises with two blanks, both blanks represent the same expression.

26. $10g - 15h = 5(\underline{?} - 3h)$ $2g$
27. $8rst + 8rs^2 = \underline{?}\,(t + s)$ $8rs$
28. $11p - 55p^2q = \underline{?}\,(1 - 5pq)$ $11p$
29. $(6xy - 15x) + (-8y + 20) = 3x(\underline{?}) - 4(\underline{?})$ $2y - 5$
30. $(a^2 + 3ab) + (2ac + 6bc) = a(\underline{?}) + 2c(\underline{?})$ $a + 3b$
31. $(20k^2 - 28kp) + (7p^2 - 5kp) = 4k(\underline{?}) - p(\underline{?})$ $5k - 7p$

Assignment Guide

Core: 27–59 odd, 61–71
Enriched: 26–56 even, 57–71

For **Extra Practice,** see p. 779.

The red A, B, and C flags, printed only in the Teacher's Wraparound Edition, indicate the level of difficulty of the exercises.

Study Guide Masters, p. 70

NAME________________ DATE ________

Study Guide

Student Edition Pages 565–571

Factoring Using the Distributive Property

The distributive property has been used to multiply a polynomial by a monomial. It can also be used to express a polynomial in factored form. Compare the two columns in the table below.

Multiplying	Factoring
$3(a + b) = 3a + 3b$	$3a + 3b = 3(a + b)$
$x(y - z) = xy - xz$	$xy - xz = x(y - z)$
$6y(2x + 1) = 6y(2x) + 6y(1)$ $= 12xy + 6y$	$12xy + 6y = \mathbf{2} \cdot 2 \cdot \mathbf{3} \cdot x \cdot y + \mathbf{2} \cdot \mathbf{3} \cdot y$ $= 6y(2x) + 6y(1)$ $= 6y(2x + 1)$

Complete.

1. $9a + 18b = 9(\underline{\ a\ } + 2b)$
2. $12mn + 80m^2 = 4m(3n + \underline{20m})$
3. $7c^3 - 7c^4 = 7c^3(\underline{\ 1\ } - c)$
4. $4xy^3 + 16x^2y^2 = \underline{4xy^2}(y + 4x)$

Factor each polynomial.

5. $24x + 48y$ $24(x + 2y)$
6. $30mn^2 + m^2n - 6n$ $n(30mn + m^2 - 6)$
7. $q^4 - 18q^3 + 22q$ $q(q^3 - 18q^2 + 22)$
8. $a + 8a^2b - ab$ $a(1 + 8ab - b)$
9. $55p^2 - 11p^7 + 44p^5$ $11p^2(5 - p^5 + 4p^3)$
10. $14c^3 - 42c^5 - 49c^4$ $7c^3(2 - 6c^2 - 7c)$
11. $4m + 6n - 8mn$ $2(2m + 3n - 4mn)$
12. $14y^3 - 28y^2 + y$ $y(14y^2 - 28y + 1)$
13. $48w^2z + 18wz^2 - 36wz$ $6wz(8w + 3z - 6)$
14. $9x^2 - 3x$ $3x(3x - 1)$
15. $96ab + 12a^2b - 84ab^3$ $12ab(8 + a - 7b^2)$
16. $45s^3 - 15s^2$ $15s^2(3s - 1)$
17. $18b^2a - 4ba + 7ab^2$ $ab(25b - 4)$
18. $12p^3q^2 - 18p^2q^2 + 30p$ $6p(2p^2q^2 - 3pq^2 + 5)$
19. $-x^5 - 4x^4 + 23x^3 - x$ $x(-x^4 - 4x^3 + 23x^2 - 1)$

Factor each polynomial.

32. $9t^2 + 36t$ **$9t(t + 4)$**
33. $14xz - 18xz^2$ **$2xz(7 - 9z)$**
34. $15xy^3 + y^4$ **$y^3(15x + y)$**
35. $17a - 41a^2b$ **$a(17 - 41ab)$**

B

36. $2ax + 6xc + ba + 3bc$
37. $2my + 7x + 7m + 2xy$
38. $3m^2 - 5m^2p + 3p^2 - 5p^3$
39. $3x^3y - 9xy^2 + 36xy$
40. $5a^2 - 4ab + 12b^3 - 15ab^2$
41. $2x^3 - 5xy^2 - 2x^2y + 5y^3$
42. $12ax + 20bx + 32cx$
43. $4ax - 14bx + 35by - 10ay$

C

44. $3my - ab + am - 3by$
45. $28a^2b^2c^2 + 21a^2bc^2 - 14abc$
46. $6a^2 - 6ab + 3bc - 3ca$
47. $12mx - 8m + 6rx - 4r$
48. $2ax + bx - 6ay - 3by - bz - 2az$
49. $7ax + 7bx + 3at + 3bt - 4a - 4b$

36. $(2x + b)(a + 3c)$
37. $(m + x)(2y + 7)$
38. $(m^2 + p^2)(3 - 5p)$
39. $3xy(x^2 - 3y + 12)$

INTEGRATION
Geometry

Write an expression in factored form for the area of each shaded region.

50. 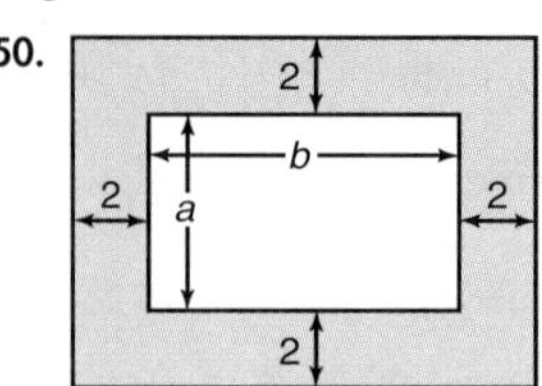

$4(a + b + 4)$

51. 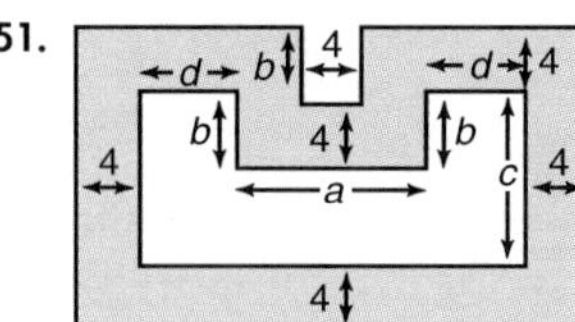

$8a - 4b + 8c + 16d + ab + 64$

52. 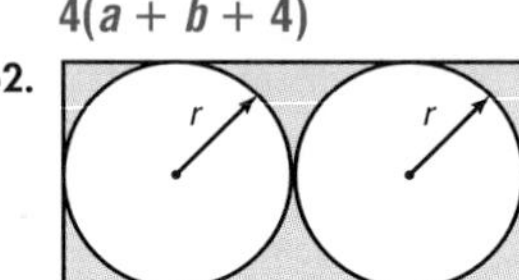

$2r^2(4 - \pi)$

53. 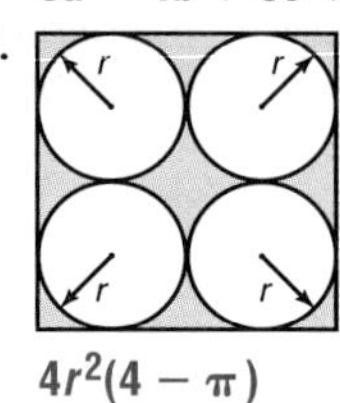

$4r^2(4 - \pi)$

40. $(a - 3b^2)(5a - 4b)$
41. $(2x^2 - 5y^2)(x - y)$
42. $4x(3a + 5b + 8c)$
43. $(2x - 5y)(2a - 7b)$
44. $(m - b)(3y + a)$
45. $7abc(4abc + 3ac - 2)$
46. $3(2a - c)(a - b)$
47. $2(2m + r)(3x - 2)$
48. $(x - 3y - z)(2a + b)$
49. $(7x + 3t - 4)(a + b)$

Find the dimensions of a rectangle having the given area if its dimensions can be represented by binomials with integral coefficients.

54. $(5xy + 15x - 6y - 18)$ cm^2 **$(5x - 6)$ cm by $(y + 3)$ cm**
55. $(4z^2 - 24z - 18m + 3mz)$ cm^2 **$(4z + 3m)$ cm by $(z - 6)$ cm**

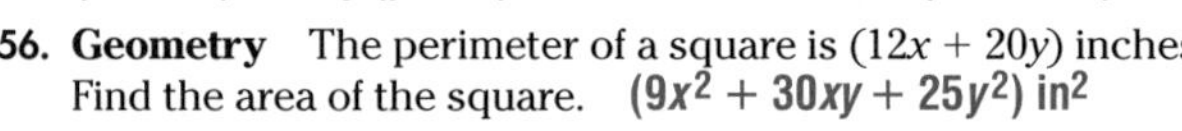

56. **Geometry** The perimeter of a square is $(12x + 20y)$ inches. Find the area of the square. **$(9x^2 + 30xy + 25y^2)$ in^2**

Critical Thinking

57. **Geometry** The perimeter of a rectangle is $(6a + 4b + 2ab + 12)$ centimeters. Find three possible expressions, in factored form, for the measure of its area. **Sample answer: $(3a + 2b)(ab + 6)$, $(3a + ab)(2b + 6)$, $(2b + ab)(3a + 6)$**

Applications and Problem Solving

58. **Gardening** The length of Eduardo's garden is 5 feet more than twice its width w. This year, Eduardo decided to make the garden 4 feet longer and double its width. How much additional area did Eduardo add to his garden?

58. $w(2w + 13)$ ft^2

Tech Prep

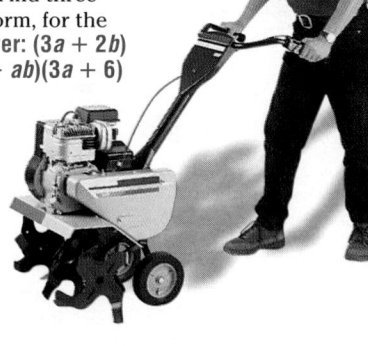

Landscape Designer Students interested in gardening should look into landscape design as in Exercise 58, or horticulture and the actual growing of plants. For more information on tech prep, see the *Teacher's Handbook.*

Practice Masters, p. 70

10-2 NAME__________ DATE ________

Practice

Student Edition Pages 565–571

Factoring Using the Distributive Property

Complete.

1. $8m - 6 = 2(4m - \underline{3})$
2. $36a^2 + 24b^2 = 12(\underline{3a^2} + 2b^2)$
3. $12x^3y - 15xy = \underline{3xy}(4x^2 - 5)$
4. $5a^2b - 10a^2b^2 = \underline{5a^2b}(1 - 2b)$

Factor each polynomial.

5. $5a^2 - 15$ — $5(a^2 - 3)$
6. $7x + 49$ — $7(x + 7)$
7. $2y + 6xy$ — $2y(1 + 3x)$
8. $8ax - 56a$ — $8a(x - 7)$
9. $36xy^2 - 48x^2y$ — $12xy(3y - 4x)$
10. $75b^2c^3 + 60bc^6$ — $15bc^3(5b + 4c^3)$
11. $64 - 40ab$ — $8(8 - 5ab)$
12. $81 - 36xy$ — $9(9 - 4xy)$
13. $t^2h + 3t$ — $t(th + 3)$
14. $6p - 72$ — $6(p - 12)$
15. $81r + 48rs$ — $3r(27 + 16s)$
16. $5c^3 - 2c^2$ — $c^2(5c - 2)$
17. $82e^3 - 122ef$ — $2e(41e^2 - 61f)$
18. $10q - 25q^2$ — $5q(2 - 5q)$
19. $xy^2 + xy$ — $xy(y + 1)$
20. $15cd + 30c^2d^2$ — $15cd(1 + 2cd)$
21. $a^2b^2 + a$ — $a(ab^2 + 1)$
22. $6r^2s - 3rs^2$ — $3rs(2r - s)$
23. $\ell^2 - 9\ell$ — $\ell(\ell - 9)$
24. $4d^2 + 16$ — $4(d^2 + 4)$
25. $6z^4 - 18z^3$ — $6z^3(z - 3)$
26. $20p^2 - 16p^2q^2$ — $4p^2(5 - 4q^2)$
27. $6m^4 - 60$ — $6(m^4 - 10)$
28. $7a^3 + 14a^2$ — $7a^2(a + 2)$
29. $16wv^4 + 12w^3v^2$ — $4wv^2(4v^2 + 3w^2)$
30. $9c^4d^3 - 6c^2d^4$ — $3c^2d^3(3c^2 - 2d)$
31. $6y + 15y^2$ — $3y(2 + 5y)$
32. $30x^3y + 35x^2y^2$ — $5x^2y(6x + 7y)$
33. $6e^3f - 11ef$ — $ef(6e^2 - 11)$
34. $20r^3s^2 + 25rs^3$ — $5rs^2(4r^2 + 5s)$
35. $34x^4y^3 - 17x^2y^5$ — $17x^2y^3(2x^2 - y^2)$
36. $35m^3n + 105m^2n^3$ — $35m^2n(m + 3n^2)$
37. $2d^2e^2 - 8d^6e^6$ — $2d^2e^2(1 - 4d^4e^4)$

59. **Construction** A 4-foot wide stone path is to be built along each of the longer sides of a rectangular flower garden. The length of the longer side of the garden is 3 feet less than twice the length of the shorter side s. Write an expression, in factored form, to represent the measure of the total area of the garden and path. **$(2s - 3)(s - 8)$**

60. **Rugby** Refer to the application at the beginning of the lesson.
    a. The width of a Rugby Union field is 69 meters. What is the length of the field? **144 m**
    b. What is the area of a Rugby Union field? **9936 m^2**
    c. The length of a Rugby League field is 52 meters longer than its width. Write an expression for the area of the field. **$[w(w + 52)]$ m^2**
    d. The width of a Rugby League field is 68 meters. Find the area of the field. **8160 m^2**
    e. Which type of rugby field has a greater area? **Rugby Union**

Mixed Review

61. A and C sharp

61. **Music** Two musical notes played at the same time produce harmony. The closest harmony is produced by frequencies with the greatest GCF. A, C, and C sharp have frequencies of 220, 264, and 275, respectively. Which pair of these notes produces the closest harmony? (Lesson 10–1)

62. **Government** In 1990, the population of the United States was 248,200,000. The area of the United States is 3,540,000 square miles. (Lesson 9–3)
    a. If the population were equally spaced over the land, how many people would there be for each square mile? **about 70 people**
    b. In 1990, the federal budget deficit was $220,000,000,000. How much would each American have had to pay in 1990 to erase the deficit? **$886.38**

63. **Geometry** The graphs of $3x + 2y = 1$, $y = 2$, and $3x - 4y = -29$ contain the sides of a triangle. Find the measure of the area of the triangle. (Hint: Use the formula $A = \frac{1}{2}bh$.) (Lesson 8–2) **9**

64. **Statistics** The ten highest-paying occupations in America are shown in the table at the right. (Lesson 7–7)
    a. Make a box-and-whisker plot of the data. **See margin.**
    b. Name any outliers. **$148,000**

Occupation	Median Salary
Physician	$148,000
Dentist	93,000
Lobbyist	91,300
Management Consultant	61,900
Lawyer	60,500
Electrical Engineer	59,100
School Principal	57,300
Aeronautical Engineer	56,700
Airline Pilot	56,500
Civil Engineer	55,800

Source: Bureau of Labor Statistics

65. Solve $17.42 - 7.029z \geq 15.766 - 8.029z$. (Lesson 7–1) **$\{z \mid z \geq -1.654\}$**

66. $\left(6, \frac{1}{2}\right)$

66. **Geometry** Find the coordinates of the midpoint of the line segment whose endpoints are $A(5, -2)$ and $B(7, 3)$. (Lesson 6–7)

67. $\{(-3, 10), \left(0, \frac{11}{2}\right), (1, 4), \left(2, \frac{5}{2}\right), (5, -2)\}$

69. $y = -\frac{4}{3}x + \frac{7}{3}$

67. Solve $3a + 2b = 11$ if the domain is $\{-3, 0, 1, 2, 5\}$. (Lesson 5–3)

68. Fourteen is 50% less than what number? (Lesson 4–5) **28**

69. Solve $4x + 3y = 7$ for y. (Lesson 3–6)

70. Solve $\frac{5}{2}x = -25$. (Lesson 3–2) **-10**

71. Find the value of $(2^5 - 5^2) + (4^2 - 2^4)$. (Lesson 1–6) **7**

fabulous **FIRSTS**

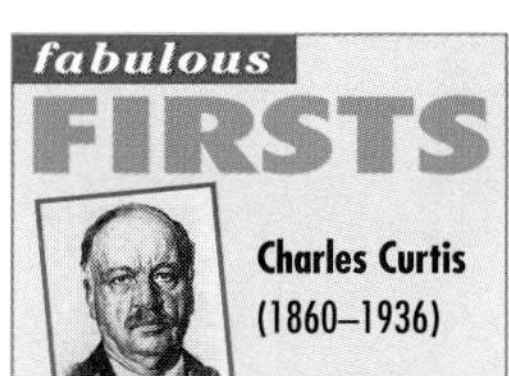

Charles Curtis (1860–1936)

In 1929, Charles Curtis became the first American Indian to serve as vice president of the United States, under Herbert Hoover. Curtis was a member of the U.S. House of Representatives from 1893–1907 and then won a seat in the U.S. Senate (1907–1913, 1915–1929) before becoming vice president.

Extension

Connections The area of a triangle is represented by $7a^3 - 28a^2b$. What could be the measures of its base and altitude?
$14a^2$, $a - 4b$

fabulous **FIRSTS**

Curtis' mother was one-quarter Kama Indian, and Charles spent his early youth growing up with the Kama tribe.

4 ASSESS

Closing Activity

Modeling The area of a rectangle is represented by $24x^3y^2 + 18x^2$. Ask students to identify possible lengths and widths of this rectangle and draw and label each rectangle.

Chapter 10, Quiz A (Lessons 10-1 and 10-2), is available in the *Assessment and Evaluation Masters,* p. 268.

Additional Answers

64a.

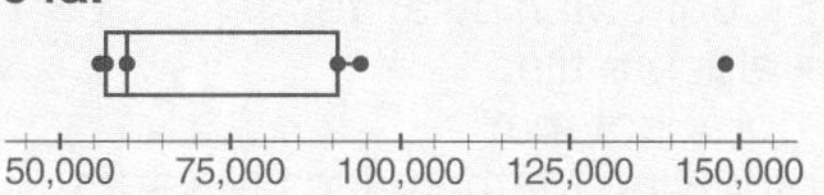

Enrichment Masters, p. 70

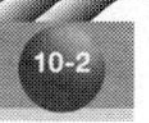

NAME________________ DATE ________

Enrichment

Student Edition Pages 565–571

Factoring Trinomials of Fourth Degree

Some trinomials of the form $a^4 + a^2b^2 + b^4$ can be written as the difference of two squares and then factored.

Example: Factor $4x^4 - 37x^2y^2 + 9y^4$.

Step 1: Find the square roots of the first and last terms.

$\sqrt{4x^4} = 2x^2$ $\quad$ $\sqrt{9y^4} = 3y^2$

Step 2: Find twice the product of the square roots.

$2(2x^2)(3y^2) = 12x^2y^2$

Step 3: Separate the middle term into two parts. One part is either your answer to Step 2 or its opposite. The other part should be the opposite of a perfect square.

$-37x^2y^2 = -12x^2y^2 - 25x^2y^2$

Step 4: Rewrite the trinomial as the difference of two squares and then factor.

$$\begin{aligned} 4x^4 - 37x^2y^2 + 9y^4 &= (4x^4 - 12x^2y^2 + 9y^4) - 25x^2y^2 \\ &= (2x^2 - 3y^2)^2 - 25x^2y^2 \\ &= [(2x^2 - 3y^2) + 5xy][(2x^2 - 3y^2) - 5xy] \\ &= (2x^2 + 5xy - 3y^2)(2x^2 - 5xy - 3y^2) \end{aligned}$$

Factor each trinomial.

1. $x^4 + x^2y^2 + y^4$
   $(x^2 + xy + y^2)(x^2 - xy + y^2)$

2. $x^4 + x^2 + 1$
   $(x^2 + x + 1)(x^2 - x + 1)$

3. $9a^4 - 15a^2 + 1$
   $(3a^2 + 3a - 1)(3a^2 - 3a - 1)$

4. $16a^4 - 17a^2 + 1$
   $(4a - 1)(a + 1)(4a + 1)(a - 1)$

5. $4a^4 - 13a^2 + 1$
   $(2a^2 + 3a - 1)(2a^2 - 3a - 1)$

6. $9a^4 + 26a^2b^2 + 25b^4$
   $(3a^2 + 2ab + 5b^2)(3a^2 - 2ab + 5b^2)$

7. $4x^4 - 21x^2y^2 + 9y^4$
   $(2x^2 + 3xy - 3y^2)(2x^2 - 3xy - 3y^2)$

8. $4a^4 - 29a^2c^2 + 25c^4$
   $(2a + 5c)(a - c)(2a - 5c)(a + c)$

10-3A LESSON NOTES

NCTM Standards: 1–5

Objective

Use algebra tiles to factor simple trinomials.

Recommended Time

Demonstration and discussion: 30 minutes; Exercises: 30 minutes

Instructional Resources

For each student or group of students

Student Manipulative Kit
- algebra tiles
- product mat

Modeling Mathematics Masters
- p. 1 (algebra tiles)
- p. 15 (product mat)
- p. 33 (worksheet)

For teacher demonstration

Algebra and Geometry Overhead Manipulative Resources

1 FOCUS

Motivating the Lesson

This modeling lesson extends the ideas of Lesson 10-2A to the case of trinomials. The strategy is to arrange the tiles in a rectangular shape.

2 TEACH

Teaching Tip In Activity 1, ask students to compare their models. Some may have factored $x^2 + 3x + 2$ as $(x + 2)(x + 1)$ instead of $(x + 1)(x + 2)$. Point out that these are both correct factorizations of the trinomial.

Teaching Tip Encourage students to experiment with different rectangles before they declare that a trinomial is not factorable. This is shown in Activity 2.

A Preview of Lesson 10–3

10–3A Factoring Trinomials

Materials: algebra tiles product mat

You can use algebra tiles to factor trinomials. If a rectangle cannot be formed to represent the trinomial, then the trinomial is not factorable.

Activity 1 Use algebra tiles to factor $x^2 + 4x + 3$.

Step 1 Model the polynomial $x^2 + 4x + 3$.

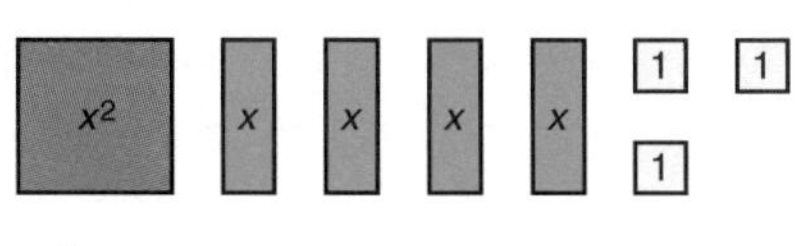

Step 2 Place the x^2-tile at the corner of the product mat. Arrange the 1-tiles into a 1-by-3 rectangular array as shown.

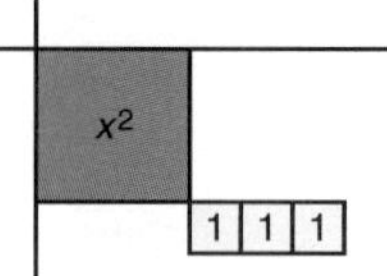

Step 3 Complete the rectangle with the x-tiles.

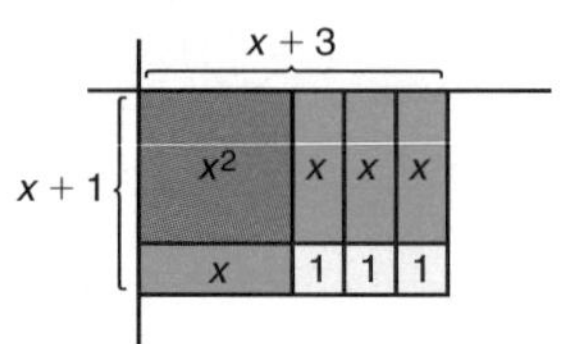

The rectangle has a width of $x + 1$ and a length of $x + 3$. Therefore, $x^2 + 4x + 3 = (x + 1)(x + 3)$.

You will need to use the guess-and-check strategy with many trinomials.

Activity 2 Use algebra tiles to factor $x^2 + 5x + 4$.

Step 1 Model the polynomial $x^2 + 5x + 4$.

Step 2 Place the x^2 tile at the corner of the product mat. Arrange the 1-tiles into a 2-by-2 rectangular array as shown. Try to complete the rectangle. Notice that there is an extra x-tile.

Step 3 Arrange the 1-tiles into a 1-by-4 rectangular array. This time you can complete the rectangle with the x-tiles.

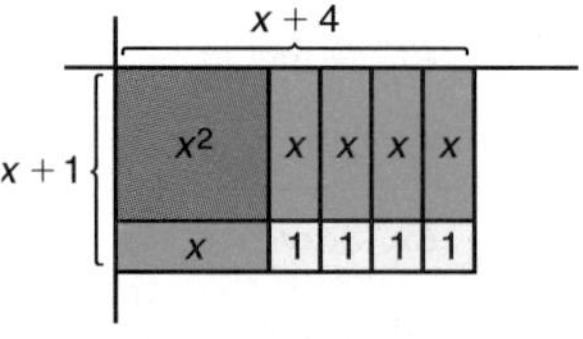

The rectangle has a width of $x + 1$ and a length of $x + 4$. Therefore, $x^2 + 5x + 4 = (x + 1)(x + 4)$.

GLENCOE Technology

Interactive Mathematics Tools Software

This multimedia software provides an interactive lesson that uses tiles to factor trinomials. A **Computer Journal** gives students an opportunity to write about what they have learned.

For Windows & Macintosh

Activity 3 Use algebra tiles to factor $x^2 - 4x + 4$.

Step 1 Model the polynomial $x^2 - 4x + 4$.

Step 2 Place the x^2-tile at the corner of the product mat. Arrange the 1-tiles into a 2-by-2 rectangular array as shown.

Step 3 Complete the rectangle with the x-tiles.

The rectangle has a width of $x - 2$ and a length of $x - 2$. Therefore, $x^2 - 4x + 4 = (x - 2)(x - 2)$.

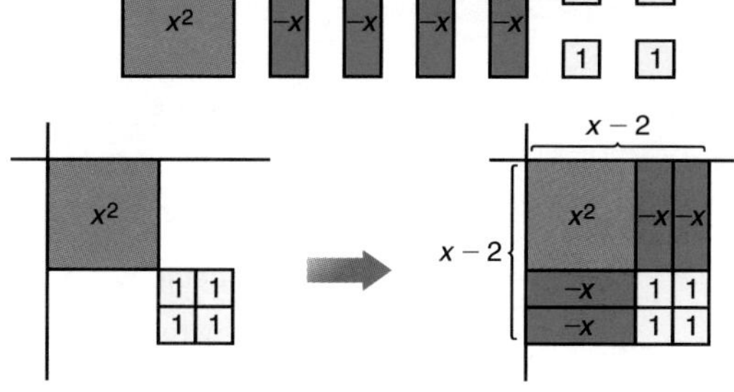

Activity 4 Use algebra tiles to factor $x^2 - x - 2$.

Step 1 Model the polynomial $x^2 - x - 2$.

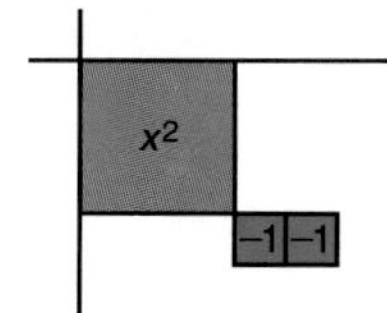

Step 2 Place the x^2-tile at the corner of the product mat. Arrange the 1-tiles into a 1-by-2 rectangular array as shown.

Step 3 Place the x-tile as shown. Recall that you can add zero-pairs without changing the value of the polynomial. In this case, add a zero pair of x-tiles.

The rectangle has a width of $x + 1$ and a length of $x - 2$. Therefore, $x^2 - x - 2 = (x + 1)(x - 2)$.

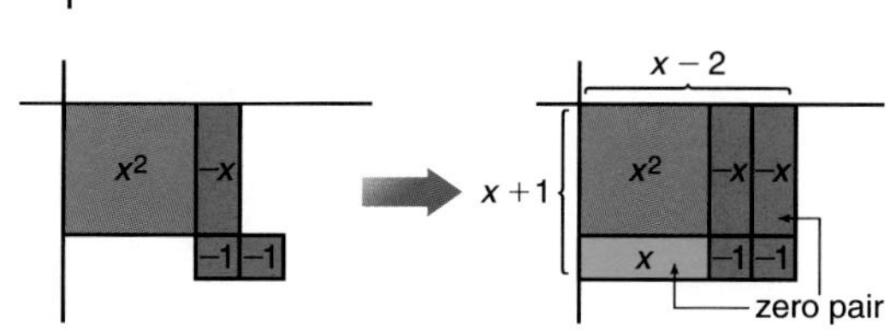

1. $(x+1)(x+5)$ **2.** $(x+2)(x+3)$ **3.** $(x+3)(x+4)$ **4.** $(x-3)(x-3)$
13. Trinomials can be factored if they can be represented by a rectangle. Sample answers: $x^2 + 4x + 4$ can be factored and $x^2 + 6x + 4$ cannot be factored.

Model **Use algebra tiles to factor each trinomial.**

1. $x^2 + 6x + 5$
2. $x^2 + 5x + 6$
3. $x^2 + 7x + 12$
4. $x^2 - 6x + 9$
5. $x^2 - 3x + 2$ $(x-1)(x-2)$
6. $x^2 - 6x + 8$ $(x-2)(x-4)$
7. $x^2 + 4x - 5$ $(x-1)(x+5)$
8. $x^2 - x - 6$ $(x+2)(x-3)$

Draw **Tell whether each trinomial can be factored. Justify your answer with a drawing.** 9–12. See margin for drawings.

9. $x^2 + 7x + 10$ yes
10. $x^2 - 4x + 5$ no
11. $x^2 + 5x - 4$ no
12. $x^2 + 2x + 6$ no

Write 13. Write a paragraph that explains how you can determine whether a trinomial can be factored. Include an example of one trinomial that can be factored and one that cannot.

Additional Answers

9.

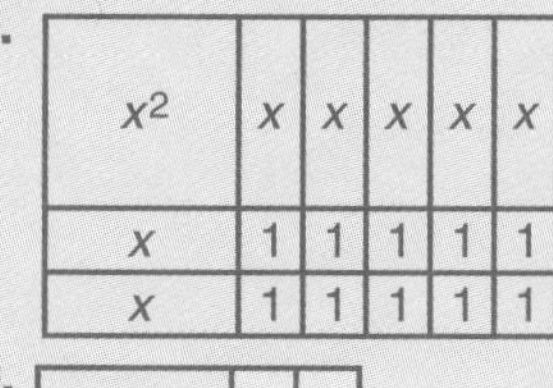

10.

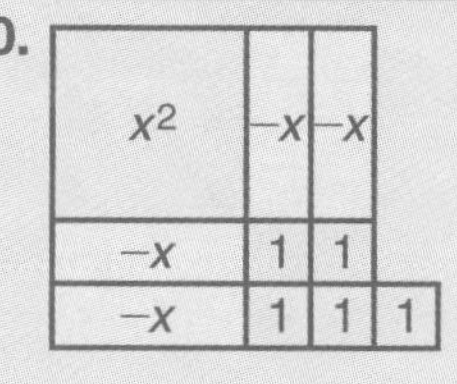

11.

12.

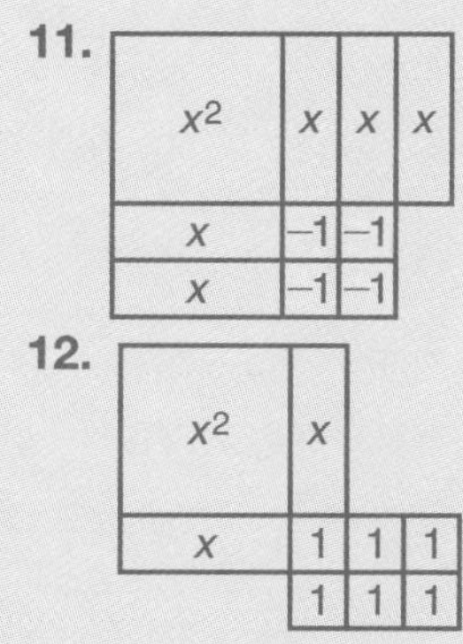

Teaching Tip Before proceeding with Activities 3 and 4, you may want students to review the visual patterns formed when multiplying binomials involving negatives.

Teaching Tip Before using Activity 4, remind students that after they try every arrangement they can think of, it may be necessary to add zero pairs to complete the rectangle. Remind them that adding zero pairs to any groups of tiles does not change the value of the groups.

3 PRACTICE/APPLY

Assignment Guide

Core: 1–13
Enriched: 1–13

4 ASSESS

Observing students working in cooperative groups is an excellent method of assessment.

10-3 LESSON NOTES

NCTM Standards: 1–5

Instructional Resources

- Study Guide Master 10-3
- Practice Master 10-3
- Enrichment Master 10-3
- Assessment and Evaluation Masters, pp. 267–268
- Modeling Mathematics Masters, pp. 63–65

Transparency 10-3A contains the 5-Minute Check for this lesson; **Transparency 10-3B** contains a teaching aid for this lesson.

Recommended Pacing	
Standard Pacing	Day 5 of 12
Honors Pacing	Day 5 of 11
Block Scheduling*	Day 3 of 6
Alg. 1 in Two Years*	Days 8, 9, & 10 of 22

*For more information on pacing and possible lesson plans, refer to the *Block Scheduling Booklet* and *Algebra 1 in Two Years.*

1 FOCUS

5-Minute Check
(over Lesson 10-2)

Factor by grouping.

1. $10xy + 25x - 14y - 35$ $(5x - 7)(2y + 5)$
2. $18x^2y^2 + 21x^3y - 30y - 35x$ $(3x^2y - 5)(6y + 7x)$
3. $49c - 7cd - 21 + 3d$ $(7c - 3)(7 - d)$

Motivating the Lesson

Questioning Have students use FOIL and write out all four terms for the following products.
$(3x + 2)(5x + 3)$
$(t + 7)(3t - 5)$
$(2a - 5)(3a - 4)$
In groups, ask students to list all the relationships they notice relative to the coefficients. This can be used as a springboard to the sum-product method of factoring the general trinomial.

10-3 Factoring Trinomials

What YOU'LL LEARN
- To solve problems by using guess and check, and
- to factor quadratic trinomials.

Why IT'S IMPORTANT
You can use factoring to solve problems involving shipping and geometry.

INTEGRATION Number Theory

The product of two consecutive odd integers is 3363. What are the numbers?

One way to find the two numbers is to use a problem-solving strategy called **guess and check**. To use this strategy, guess the answer to the problem, and then check whether the guess is correct. If the first guess is incorrect, guess and check again until you find the correct answer. Often, the results of one guess can help you make a better guess. Always keep an organized record of your guesses so you don't make the same guess twice.

Guess: Try 41 and 43.

Check: $41 \times 43 = 1763$
This product is considerably less than 3363.

Guess: Try two numbers greater than 41 and 43, such as 61 and 63.

Check: $61 \times 63 = 3843$
This product is greater than 3363.

Guess: Try 51 and 53.

Check: $51 \times 53 = 2703$
This product is less than 3363. Since the ones digit is not 0 or 5, do not use 55 as one of the numbers. *Why?*

Guess: Try 57 and 59.

Check: $57 \times 59 = 3363$

The two numbers are 57 and 59.

In Lesson 10–1, you learned that when two numbers are multiplied, each number is a factor of the product. Similarly, when two binomials are multiplied, each binomial is a factor of the product. You can use the guess-and-check strategy to find the factors of a trinomial. Consider the binomials $5x + 3$ and $2x + 7$. You can use the FOIL method to find their product.

$$\begin{aligned}(5x + 3)(2x + 7) &= \overset{F}{(5x)(2x)} + \overset{O}{(5x)(7)} + \overset{I}{(3)(2x)} + \overset{L}{(3)(7)} \\ &= 10x^2 + 35x + 6x + 21 \\ &= 10x^2 + (35 + 6)x + 21 \\ &= 10x^2 + 41x + 21\end{aligned}$$

Notice that $10 \cdot 21 = 210$ and $35 \cdot 6 = 210$.

The binomials $5x + 3$ and $2x + 7$ are factors of $10x^2 + 41x + 21$.

When using the FOIL method above, look at the product of the coefficients of the first and last terms, 10 and 21. Notice that this product, 210, is the same as the product of the coefficients of the two middle terms, 35 and 6. Their sum is the coefficient of the middle term of the final product.

You can use this pattern to factor quadratic trinomials, such as $3y^2 + 10y + 8$.

$3y^2 + 10y + 8$ The product of 3 and 8 is 24.

$3y^2 + (\underline{?} + \underline{?})y + 8$

You need to find two integers whose *product is 24* and whose *sum is 10.*

Use the guess-and-check strategy to find these numbers.

Factors of 24	Sum of Factors	
1, 24	$1 + 24 = 25$	*no*
2, 12	$2 + 12 = 14$	*no*
3, 8	$3 + 8 = 11$	*no*
4, 6	$4 + 6 = 10$	*yes*

$3y^2 + 10y + 8$

$= 3y^2 + (4 + 6)y + 8$ *Select the factors 4 and 6.*

$= 3y^2 + 4y + 6y + 8$

$= (3y^2 + 4y) + (6y + 8)$ *Group terms that have a common monomial factor.*

$= y(3y + 4) + 2(3y + 4)$ *Factor.*

$= (y + 2)(3y + 4)$ *Use the distributive property.*

Therefore, $3y^2 + 10y + 8 = (y + 2)(3y + 4)$. *Check by using FOIL.*

Example 1 **Factor $10x^2 - 27x + 18$.**

$10x^2 - 27x + 18$ The product of 10 and 18 is 180.

$10x^2 + (\underline{?} + \underline{?})x + 18$ Since the product is positive and the sum is negative, both factors of 180 must be negative. *Why?*

Factors of 180	Sum of Factors	
$-180, -1$	$-180 + (-1) = -181$	*no*
$-90, -2$	$-90 + (-2) = -92$	*no*
$-45, -4$	$-45 + (-4) = -49$	*no*
$-15, -12$	$-15 + (-12) = -27$	*yes* *You can stop listing factors when you find a pair that works.*

$10x^2 - 27x + 18$

$= 10x^2 + [-15 + (-12)]x + 18$

$= 10x^2 - 15x - 12x + 18$

$= (10x^2 - 15x) + (-12x + 18)$

$= 5x(2x - 3) + (-6)(2x - 3)$ *Factor the GCF from each group.*

$= (5x - 6)(2x - 3)$ *Use the distributive property.*

Therefore, $10x^2 - 27x + 18 = (5x - 6)(2x - 3)$. *Check by using FOIL.*

2 TEACH

In-Class Example

For Example 1
Factor.

a. $2x^2 + 9x + 10$
$(x + 2)(2x + 5)$

b. $3a^2 + 13a + 4$
$(3a + 1)(a + 4)$

c. $6t^2 + 25t + 14$
$(2t + 7)(3t + 2)$

d. $15y^2 + 34y + 15$
$(3y + 5)(5y + 3)$

Teaching Tip Point out that when the sum of factors is positive, only positive factors need to be considered.

Teaching Tip Remind students to check for a GCF first.

In-Class Example

For Example 2

a. The area of a rectangle is $(x^2 + 5x + 4)$ square inches. The area is decreased by 2 inches in both the length and the width. Find the area of the new rectangle. $\mathbf{x^2 + x - 2}$

b. The area of a rectangle is $(m^2 + 7m + 10)$ square inches. The area is changed by decreasing the width 1 inch and the length 2 inches. (The length of a rectangle is always the longer side.) Find the area of the new rectangle. $\mathbf{m^2 + 4m + 3}$

Example 2

Geometry

The area of a rectangle is $(a^2 - 3a - 18)$ square inches. This area is increased by adding 5 inches to both the length and the width. If the dimensions of the original rectangle are represented by binomials with integral coefficients, find the area of the new rectangle.

To determine the area of the new rectangle, you must first find the dimensions of the original rectangle by factoring $a^2 - 3a - 18$. The coefficient of a^2 is 1. Thus, you must find two numbers whose product is $1 \cdot (-18)$ or -18 and whose sum is -3.

Original Rectangle
Area = $(a^2 - 3a - 18)$ in^2
? in.
? in.

Factors of −18	Sum of Factors	
−18, 1	$-18 + 1 = -17$	*no*
−9, 2	$-9 + 2 = -7$	*no*
−6, 3	$-6 + 3 = -3$	*yes*

The factors of −18 should be chosen so that exactly one factor in each pair is negative and that factor has the greater absolute value. Why?

$$\begin{aligned} a^2 - 3a - 18 &= a^2 + [(-6) + 3]a - 18 \\ &= a^2 - 6a + 3a - 18 \\ &= (a^2 - 6a) + (3a - 18) \\ &= a(a - 6) + 3(a - 6) \\ &= (a + 3)(a - 6) \quad \textit{Check by using FOIL.} \end{aligned}$$

The dimensions of the original rectangle are $(a + 3)$ inches and $(a - 6)$ inches. Therefore, the dimensions of the new rectangle are $(a + 3) + 5$ or $(a + 8)$ inches and $(a - 6) + 5$ or $(a - 1)$ inches. Now, find an expression for the area of the new rectangle.

$$\begin{aligned} (a + 8)(a - 1) &= a^2 - a + 8a - 8 \\ &= a^2 + 7a - 8 \end{aligned}$$

The area of the new rectangle is $(a^2 + 7a - 8)$ square inches.

New Rectangle
Area = ? in^2
$(a - 6) + 5$ or $(a - 1)$ in.
$(a + 3) + 5$ or $(a + 8)$ in.

Let's study the factorization of $a^2 - 3a - 18$ from Example 2 more closely.

$$a^2 - 3a - 18 = (a + 3)(a - 6)$$

Notice that the sum of 3 and −6 is equal to −3, the coefficient of a in the trinomial. Also, the product of 3 and −6 is equal to −18, the constant term of the trinomial. This pattern holds for all trinomials whose quadratic term has a coefficient of 1.

Occasionally the terms of a trinomial will contain a common factor. In these cases, first use the distributive property to factor out the common factor, and then factor the trinomial.

Example 3 **Factor $14t - 36 + 2t^2$.**

First rewrite the trinomial so the terms are in descending order.

$$14t - 36 + 2t^2 = 2t^2 + 14t - 36$$
$$= 2(t^2 + 7t - 18) \quad \textit{The GCF of the terms is 2. Use the distributive property.}$$

Now factor $t^2 + 7t - 18$. Since the coefficient of t^2 is 1, we need to find two factors of -18 whose sum is 7.

Factors of -18	Sum of Factors	
18, -1	$18 + (-1) = 17$	*no*
9, -2	$9 + (-2) = 7$	*yes*

The desired factors are 9 and -2 and $t^2 + 7t - 18 = (t + 9)(t - 2)$.
Therefore, $2t^2 + 14t - 36 = 2(t + 9)(t - 2)$.

A polynomial that cannot be written as a product of two polynomials with integral coefficients is called a **prime polynomial**.

Example  **Factor $2a^2 - 11a + 7$.**

You must find two numbers whose product is $2 \cdot 7$ or 14 and whose sum is -11. Since the sum has to be negative, both factors of 14 have to be negative.

Factors of 14	Sum of Factors	
$-1, -14$	$-1 + (-14) = -15$	*no*
$-2, -7$	$-2 + (-7) = -9$	*no*

There are no factors of 14 whose sum is -11.

Therefore, $2a^2 - 11a + 7$ cannot be factored using integers.
Thus, $2a^2 - 11a + 7$ is a prime polynomial.

You can use your knowledge of factoring to write polynomials that can be factored using integers.

Example  **Find all values of k so the trinomial $3x^2 + kx - 4$ can be factored using integers.**

For $3x^2 + kx - 4$ to be factorable, k must equal the sum of the factors of $3(-4)$ or -12.

Factors of -12	Sum of Factors (k)
$-12, 1$	$-12 + 1 = -11$
$12, -1$	$12 + (-1) = 11$
$-6, 2$	$-6 + 2 = -4$
$6, -2$	$6 + (-2) = 4$
$-4, 3$	$-4 + 3 = -1$
$4, -3$	$4 + (-3) = 1$

Therefore, the values of k are -11, 11, -4, 4, -1, and 1.

In-Class Examples

For Example 3
Factor.

a. $6y^2 - 13y - 5$ **$(2y - 5)(3y + 1)$**
b. $8t^2 - 6t - 9$ **$(4t + 3)(2t - 3)$**
c. $20a^2 - 21a - 5$ **$(5a + 1)(4a - 5)$**
d. $12x^2 + 11x - 5$ **$(4x + 5)(3x - 1)$**

For Example 4
Factor.

a. $x^2 - 2x - 63$ **$(x + 7)(x - 9)$**
b. $a^2 - 9a + 20$ **$(a - 4)(a - 5)$**
c. $t^2 + 14t + 45$ **$(t + 9)(t + 5)$**
d. $r^2 + 4r - 21$ **$(r - 3)(r + 7)$**

For Example 5
Find all values of k so each trinomial can be factored using integers.

a. $x^2 + kx + 9$ **6, −6, 10, −10**
b. $y^2 + ky + 6$ **5, −5, 7, −7**
c. $b^2 + kb + 10$ **7, −7, 11, −11**
d. $z^2 + kz + 12$ **7, −7, 8, −8, 13, −13**

EXPLORATION

This Exploration provides a method of using a graphing calculator to check a factorization. Remind students that they can also verify a factorization by multiplying the factors.

3 PRACTICE/APPLY

Check for Understanding

Exercises 1–21 are designed to help you assess your students' understanding through reading, writing, speaking, and modeling. You should work through Exercises 1–4 with your students and then monitor their work on Exercises 5–21.

Additional Answers

1. So you do not waste time trying the same guess twice. It also helps you to make better guesses.

3a. It shows that the missing areas must have a sum of $8x$.

4.

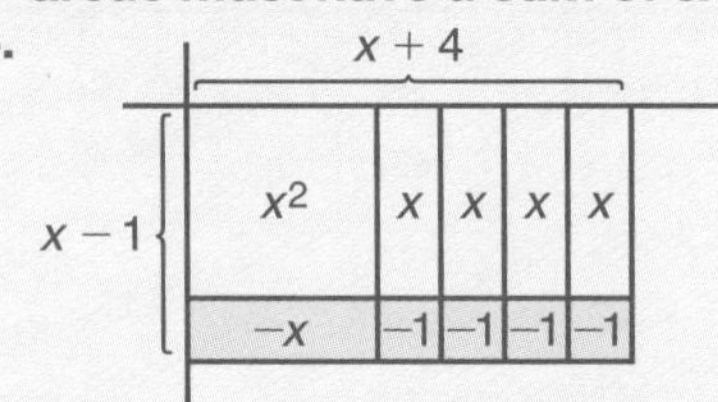

Study Guide Masters, p. 71

NAME______________________ DATE__________

Study Guide

Student Edition Pages 574–580

Factoring Trinomials

To factor a trinomial of the form $ax^2 + bx + c$, follow Example 1 below.

Example 1: Factor $2d^2 + 15d + 18$.
The product of 2 and 18 is 36. You need to find two integers whose *product* is 36 and whose *sum* is 15.

Factors of 36	Sum of Factors
1, 36	1 + 36 = 37
2, 18	2 + 18 = 20
3, 12	3 + 12 = 15

$$\begin{aligned} 2d^2 + 15d + 18 &= 2d^2 + (12 + 3)d + 18 \\ &= 2d^2 + 12d + 3d + 18 \\ &= (2d^2 + 12d) + (3d + 18) \\ &= 2d(d + 6) + 3(d + 6) && \text{Factor the GCF from each group.} \\ &= (2d + 3)(d + 6) && \text{Use the distributive property.} \end{aligned}$$

To factor a trinomial of the form given above when $a = 1$, you need to find only the factors of c whose sum is b.

Example 2: Factor $x^2 + 7x + 10$.

Since 2 and 5 are factors of 10 whose sum is 7, $x^2 + 7x + 10 = (x + 2)(x + 5)$.

The same pattern can be used to factor a trinomial $ax^2 + bx + c$ when $a = 1$ and c is negative. When this occurs, the factors of the trinomial are a *difference* and a *sum*.

Complete.

1. $x^2 - 5x - 14 = (x + \underline{2})(x - 7)$
2. $a^2 + 13a + 36 = (a + 9)(a \underline{+} 4)$
3. $p^2 - 25 = (p + 5)(p \underline{-} \underline{5})$
4. $x^2 - 6xy - 16y^2 = (x \underline{-} \underline{8y})(x + 2y)$
5. $49 - n^2 = (7 \underline{+} \underline{n})(\underline{7} \underline{-} n)$
6. $a^4 + 3xa^2 - 10x^2 = (a^2 - \underline{2x})(\underline{a^2} \underline{+} 5x)$

Factor each trinomial, if possible. If the trinomial cannot be factored using integers, write prime.

7. $x^2 + 12y + 32$ $(y + 4)(y + 8)$
8. $x^2 - x - 6$ $(x - 3)(x + 2)$
9. $x^2 - 4x - 21$ $(x - 7)(x + 3)$
10. $y^2 + 22y + 121$ $(y + 11)(y + 11)$
11. $9 - 7n + n^2$ prime
12. $a^2 - 16a + 64$ $(a - 8)(a - 8)$
13. $3x^2 + 2x - 8$ $(3x - 4)(x + 2)$
14. $18h^2 - 27h - 5$ $(3h - 5)(6h + 1)$
15. $28x^2 + 60x - 25$ $(2x + 5)(14x - 5)$
16. $48x^2 + 22x - 15$ $(6x + 5)(8x - 3)$
17. $-4y^2 + 19y - 21$ $(4y - 7)(3 - y)$
18. $6a^2 - 7a + 18$ prime

You can graph a polynomial and its factored form on the same axes to see if you have factored correctly. If the two graphs coincide, the factored form is probably correct.

EXPLORATION — GRAPHING CALCULATORS

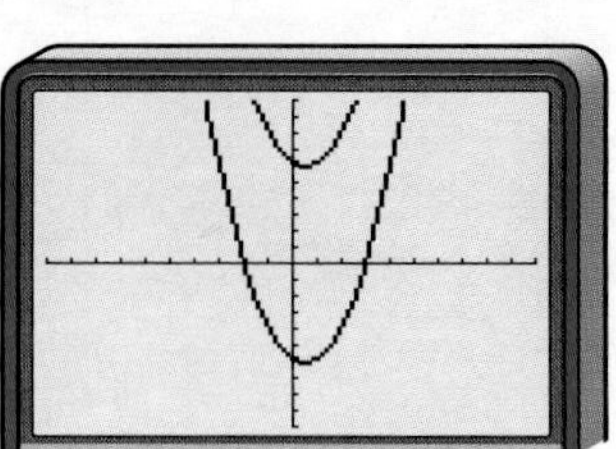

Suppose $x^2 - x + 6$ has been factored as $(x + 2)(x - 3)$.

a. Press [Y=]. Enter $x^2 - x + 6$ for Y1 and $(x + 2)(x - 3)$ for Y2.

b. Press [ZOOM] 6. Notice that two different graphs appear. Therefore, $x^2 - x + 6 \neq (x + 2)(x - 3)$.

Your Turn a. no; $(x + 1)(x - 2)$ d. no; $(2x - 1)(x + 2)$

Determine whether each equation is a true statement. If it is not correct, state the correct factorization of the trinomial.

a. $x^2 - x - 2 = (x - 1)(x + 2)$

b. $x^2 - 2x - 3 = (x - 3)(x + 1)$ yes

c. $x^2 - 5x + 4 = (x - 4)(x - 1)$ yes

d. $2x^2 + 3x - 2 = (2x + 2)(x - 1)$

CHECK FOR UNDERSTANDING

Communicating Mathematics

Study this lesson. Then complete the following. 2. See students' work.

1. **Explain** why you should keep a record of your guesses when you are using the guess-and-check strategy. See margin.
2. **Write** an example of a prime polynomial.

3. **Study** the model at the right.
   a. **Explain** how this model could help to factor $x^2 + 8x + 12$. See margin.
   b. **Factor** $x^2 + 8x + 12$. $(x + 6)(x + 2)$
4. Use algebra tiles to factor $x^2 + 3x - 4$. Make a drawing of the algebra tiles. $(x + 4)(x - 1)$; See margin for drawing.

Guided Practice

For each trinomial of the form $ax^2 + bx + c$, find two integers whose product is equal to ac and whose sum is equal to b.

5. $x^2 + 11x + 24$ 3, 8
6. $x^2 + 4x - 45$ −5, 9
7. $2x^2 + 13x + 20$ 8, 5
8. $3x^2 - 19x + 6$ −1, −18
9. $4x^2 - 8x + 3$ −2, −6
10. $5x^2 - 13x - 6$ −15, 2

Complete.

11. $r^2 - 5r - 14 = (r + 2)(r \underline{\ ?\ } 7)$ −
12. $2g^2 + 5g - 12 = (2g - 3)(g + \underline{\ ?\ })$ 4

Factor each trinomial, if possible. If the trinomial cannot be factored using integers, write *prime*.

13. $t^2 + 7t + 12$ — 13. $(t + 3)(t + 4)$
14. $c^2 - 13c + 36$ — 14. $(c - 4)(c - 9)$
15. $2y^2 - 2y - 12$ — 15. $2(y + 2)(y - 3)$
16. $3d^2 - 12d + 9$ $3(d - 3)(d - 1)$
17. $2x^2 + 5x - 2$ prime
18. $6p^2 + 15p - 9$ $3(2p - 1)(p + 3)$

Find all values of k so each trinomial can be factored using integers.

19. $x^2 + kx + 14$ 9, −9, 15, −15
20. $2b^2 + kb - 3$ 1, −1, 5, −5

Reteaching

Working Backward What values of b will make these trinomials factorable by unFOILing?

1. $x^2 + bx + 12$ {±7, ±8, ±13}
2. $x^2 + bx - 20$ {±19, ±8, ±1}
3. $x^2 + bx - 8$ {±7, ±2}
4. $2x^2 + bx + 15$ {±11, ±13, ±17, ±31}
5. $3x^2 + bx - 8$ {±2, ±5, ±10, ±23}
6. $6x^2 + bx + 7$ {±13, ±17, ±23, ±43}

21. **Geometry** The area of a rectangle is $(3x^2 + 14x + 15)$ square meters. This area is reduced by decreasing both the length and width by 3 meters. If the dimensions of the original rectangle are represented by binomials with integral coefficients, find the area of the new rectangle. **$(3x^2 + 2x)$ m^2**

EXERCISES

Practice

28. $(b+3)(b+4)$
29. $(m-4)(m-10)$
30. $(z-8)(z+3)$
32. $(s-12)(s+15)$
33. $(2x+7)(x-3)$
34. $(7a+1)(a+3)$
35. $(2x+3)(x-4)$
37. $(2n-7)(2n+5)$
38. $2(9-y)(4-y)$
39. $(2+3m)(5+2m)$
42. $(5x-y)(3x-2y)$
43. $2x(3x+8)(2x-5)$
47. 7, −7, 11, −11
48. 10, −10, 11, −11, 14, −14, 25, −25
49. 1, −1, 11, −11, 19, −19, 41, −41
50. 7, 12, 15, 16
51. 6, 4
53. r cm, $(5r+6)$ cm, $(3r-7)$ cm
57. no; $(x-3)(x-3)$

Complete.

22. $a^2 + a - 30 = (a - 5)(a \underline{\ ?\ } 6)$ **+**
23. $g^2 - 8g + 16 = (g - 4)(g \underline{\ ?\ } 4)$ **−**
24. $4y^2 - y - 3 = (\underline{\ ?\ } + 3)(y - 1)$ **4y**
25. $6t^2 - 23t + 20 = (3t - 4)(2t - \underline{\ ?\ })$ **5**
26. $4x^2 + 4x - 3 = (2x - 1)(\underline{\ ?\ } + 3)$ **2x**
27. $15g^2 + 34g + 15 = (5g + 3)(3g + \underline{\ ?\ })$ **5**

Factor each trinomial, if possible. If the trinomial cannot be factored using integers, write *prime*.

28. $b^2 + 7b + 12$
29. $m^2 - 14m + 40$
30. $z^2 - 5z - 24$
31. $t^2 - 2t + 35$ **prime**
32. $s^2 + 3s - 180$
33. $2x^2 + x - 21$
34. $7a^2 + 22a + 3$
35. $2x^2 - 5x - 12$
36. $3c^2 - 3c - 5$ **prime**
37. $4n^2 - 4n - 35$
38. $72 - 26y + 2y^2$
39. $10 + 19m + 6m^2$
40. $a^2 + 2ab - 3b^2$ **$(a + 3b)(a - b)$**
41. $12r^2 - 11r + 3$ **prime**
42. $15x^2 - 13xy + 2y^2$
43. $12x^3 + 2x^2 - 80x$
44. $5a^3b^2 + 11a^2b^2 - 36ab^2$ **$ab^2(5a - 9)(a + 4)$**
45. $20a^4b - 58a^3b^2 + 42a^2b^3$ **$2a^2b(5a - 7b)(2a - 3b)$**

Find all values of k so each trinomial can be factored using integers.

46. $r^2 + kr - 13$ **12, −12**
47. $x^2 + kx + 10$
48. $2c^2 + kc + 12$
49. $3s^2 + ks - 14$
50. $x^2 + 8x + k, k > 0$
51. $n^2 - 5n + k, k > 0$

52. **Geometry** The area of a rectangle is $(6x^2 - 31x + 35)$ square inches. If the dimensions of the rectangle are all whole numbers, what is the minimum possible area of the rectangle? **7 in^2**

53. **Geometry** The volume of a rectangular prism is $(15r^3 - 17r^2 - 42r)$ cubic centimeters. If the dimensions of the prism are represented by polynomials with integral coefficients, find the dimensions of the prism.

Graphing Calculator

Use a graphing calculator to determine whether each equation is a true statement. If it is not correct, state the correct factorization of the trinomial. **54. yes 55. no; $(2x+3)(x-1)$ 56. no; $(3x+2)(x-2)$**

54. $x^2 - 2x - 15 = (x - 5)(x + 3)$
55. $2x^2 + x - 3 = (2x - 1)(x + 3)$
56. $3x^2 - 4x - 4 = (3x - 2)(x + 2)$
57. $x^2 - 6x + 9 = (x + 3)(x - 3)$

Critical Thinking

58. Complete each polynomial in three different ways so that the resulting polynomial can be factored. Then factor each polynomial.
    a. $x^2 + 8x + \underline{\ ?\ }$
    b. $x^2 + \underline{\ ?\ }x - 10$ **See margin.**

Applications and Problem Solving

59. 27 ft^3

59. **Shipping** A shipping crate is to be built in the shape of a rectangular solid. The volume of the crate is $(45x^2 - 174x + 144)$ cubic feet where x is a positive integer. If the height of the crate is 3 feet, what is the minimum volume possible for this crate?

Additional Answers

58a. Sample answers:
$16, x^2 + 8x + 16 = (x + 4)(x + 4);$
$12, x^2 + 8x + 12 = (x + 6)(x + 2);$
$7, x^2 + 8x + 7 = (x + 7)(x + 1)$

58b. Sample answers:
$9, x^2 + 9x - 10 = (x + 10)(x - 1);$
$3, x^2 + 3x - 10 = (x + 5)(x - 2);$
$-3, x^2 + (-3)x - 10 = (x - 5)(x + 2)$

Assignment Guide

Core: 23–57 odd, 58, 59, 61–71
Enriched: 22–56 even, 58–71
All: Self Test, 1–10

For **Extra Practice,** see p. 779.

The red A, B, and C flags, printed only in the Teacher's Wraparound Edition, indicate the level of difficulty of the exercises.

Practice Masters, p. 71

NAME ______________________ DATE __________

Practice — Student Edition Pages 574–580

Factoring Trinomials

Factor each trinomial, if possible. If the trinomial cannot be factored using integers, write prime.

1. $t^2 + 8t + 12$ — $(t + 2)(t + 6)$
2. $w^2 + 24w + 144$ — $(w + 12)(w + 12)$
3. $m^2 - 7m + 12$ — $(m - 4)(m - 3)$
4. $n^2 + 3n - 18$ — $(n + 6)(n - 3)$
5. $v^2 - 18v + 80$ — $(v - 8)(v - 10)$
6. $p^2 - p - 56$ — $(p - 8)(p + 7)$
7. $b^2 + 8b - 22$ — prime
8. $x^2 + 7x - 44$ — $(x + 11)(x - 4)$
9. $y^2 - 5y - 84$ — $(y - 12)(y + 7)$
10. $32 + 18r + r^2$ — $(16 + r)(2 + r)$
11. $48 - 16e + e^2$ — $(12 - e)(4 - e)$
12. $s^2 + 17s + 52$ — $(s + 4)(s + 13)$
13. $102 - 23t + t^2$ — $(6 - t)(17 - t)$
14. $u^2 - 16u - 36$ — $(u - 18)(u + 2)$
15. $a^2b^2 + ab - 6$ — $(ab + 3)(ab - 2)$
16. $a^2b^2 + 5ab + 6$ — $(ab + 3)(ab + 2)$
17. $m^2 - mv - 56v^2$ — $(m - 8v)(m + 7v)$
18. $j^2 - 9jk - 10k^2$ — $(j - 10k)(j + k)$
19. $3h^2 + 2h - 16$ — $(3h + 8)(h - 2)$
20. $6c^2 + 7c + 2$ — $(3c + 2)(2c + 1)$
21. $5p^2 - 22p + 8$ — $(5p - 2)(p - 4)$
22. $8m^2 - 10m + 3$ — $(2m - 1)(4m - 3)$
23. $6z^2 - 5z - 4$ — $(3z - 4)(2z + 1)$
24. $15y^2 - y - 2$ — $(5y - 2)(3y + 1)$
25. $18x^2 + 9xz + z^2$ — $(6x + z)(3x + z)$
26. $20m^2 + 13mn + 2n^2$ — $(4m + n)(5m + 2n)$
27. $5\ell^2 - 26\ell x + 5x^2$ — $(5\ell - x)(\ell - 5x)$
28. $15s^2 - 16st + 4t^2$ — $(5s - 2t)(3s - 2t)$

4 ASSESS

Closing Activity

Speaking Have students explain the relationship between the sign of the last term in a quadratic trinomial and the signs of its binomial factors.

Chapter 10, Quiz B (Lesson 10-3), is available in the *Assessment and Evaluation Masters,* p. 268.

Mid-Chapter Test (Lessons 10-1 through 10-3) is available in the *Assessment and Evaluation Masters,* p. 267.

Enrichment Masters, p. 71

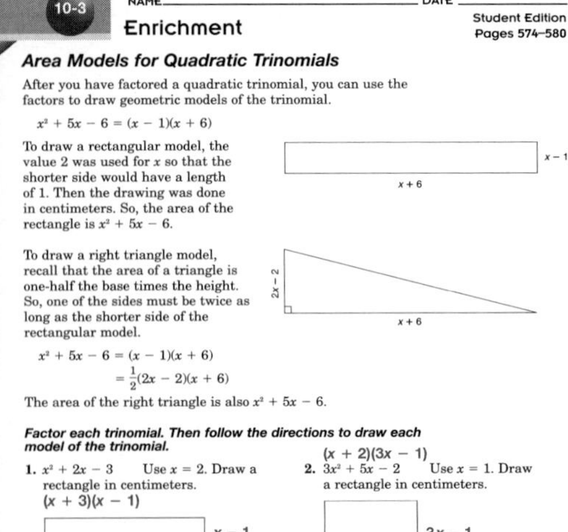

10-3 NAME ______ DATE ______

Enrichment — Student Edition Pages 574–580

Area Models for Quadratic Trinomials

After you have factored a quadratic trinomial, you can use the factors to draw geometric models of the trinomial.

$x^2 + 5x - 6 = (x - 1)(x + 6)$

To draw a rectangular model, the value 2 was used for x so that the shorter side would have a length of 1. Then the drawing was done in centimeters. So, the area of the rectangle is $x^2 + 5x - 6$.

To draw a right triangle model, recall that the area of a triangle is one-half the base times the height. So, one of the sides must be twice as long as the shorter side of the rectangular model.

$x^2 + 5x - 6 = (x - 1)(x + 6)$
$= \frac{1}{2}(2x - 2)(x + 6)$

The area of the right triangle is also $x^2 + 5x - 6$.

Factor each trinomial. Then follow the directions to draw each model of the trinomial.

1. $x^2 + 2x - 3$ Use $x = 2$. Draw a rectangle in centimeters. $(x + 3)(x - 1)$

2. $3x^2 + 5x - 2$ Use $x = 1$. Draw a rectangle in centimeters. $(x + 2)(3x - 1)$

3. $x^2 - 4x + 3$ Use $x = 4$. Draw two different right triangles in centimeters. $(x - 1)(x - 3)$

4. $9x^2 - 9x + 2$ Use $x = 2$. Draw two different right triangles. Use 0.5 centimeter for each unit. $(3x - 2)(3x - 1)$

60. **Guess and Check** Place the digits 1, 2, 3, 4, 5, 6, 8, 9, 10, 12 on the dots at the right so that the sum of the integers on any line equals the sum on any other line.

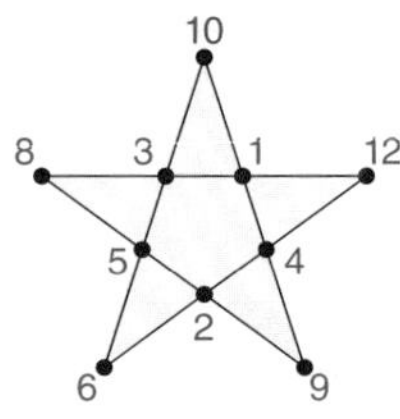

Mixed Review

61. **Finance** During the first hour of trading, John Sugarman sold x shares of stock that cost \$4 per share. During the next hour, he sold stock that cost \$8 per share. He sold 5 more shares during the first hour than the second hour. If he had sold only the stock that cost \$4 per share during the two hours, how many shares would he have needed to sell to have the same amount of total sales? (Lesson 10–2) **$(3x - 10)$ shares**

62. Find the degree of $7x^3 + 4xy + 3xz^3$. (Lesson 9–4) **4**

63. Use elimination to solve the system of equations. (Lesson 8–3) **(5, −2)**
$2x = 4 - 3y$
$3y - x = -11$

64. Solve $|2y - 7| \geq -6$. (Lesson 7–6) **{all numbers}**

65. Write an equation of the line that is parallel to the graph of $2x + 3y = 1$ and passes through (4, 2). (Lesson 6–6) **65.** $y = -\frac{2}{3}x + \frac{14}{3}$

66. Determine the slope of the line that passes through the points at (−3, 6) and (−5, 9). (Lesson 6–1) **66.** $-\frac{3}{2}$

67. State the domain and range of {(0, 2), (1, −2), (2, 4)}. (Lesson 5–2) **67. D = {0, 1, 2}; R = {2, −2, 4}**

68. **Geography** There is three times as much water as land on Earth's surface. What percent of Earth is covered by water? (Lesson 4–4) **75%**

69. Find the supplement of 90°. (Lesson 3–4) **90°**

70. **Time** When it is noon in Richmond, Virginia, it is 2:00 A.M. the following morning in Kanagawa, Japan. Joel, who teaches English in Kanagawa, would like to call his mother in Richmond at 7:30 on the morning of her birthday, October 26. On what day and at what time would Joel have to call her? (Lesson 2–3) **Oct. 26 at 9:30 P.M.**

71. Simplify $3(x + 2y) - 2y$. (Lesson 1–7) **$3x + 4y$**

SELF TEST

Find the GCF of the given monomials. (Lesson 10–1)

1. $50n^4, 40n^2p^2$ **$10n^2$**
2. $15abc, 35a^2c, 105a$ **$5a$**

Factor each polynomial, if possible. If the polynomial cannot be factored using integers, write *prime*. (Lessons 10–2 and 10–3) **4. $(2a - 1)(b + m)$** **8. $3mn(9mn - 25)$**

3. $18xy^2 - 24x^2y$ **$6xy(3y - 4x)$**
4. $2ab + 2am - b - m$
5. $2q^2 - 9q - 18$ **$(2q + 3)(q - 6)$**
6. $t^2 + 5t - 20$ **prime**
7. $3y^2 - 8y + 5$ **$(3y - 5)(y - 1)$**
8. $27m^2n^2 - 75mn$

9. **Guess and Check** Write an eight-digit number using the digits 1, 2, 3, and 4 each twice so that the 1s are separated by 1 digit, the 2s are separated by 2 digits, the 3s are separated by 3 digits, and the 4s are separated by 4 digits. (Lesson 10–3) **41,312,432 or 23,421,314**

10. **Geometry** The area of a rectangle is $(x^2 - x - 6)$ square meters. The length and width are each increased by 9 meters. If the dimensions of the original rectangle are binomials with integral coefficients, find the area of the new rectangle. (Lesson 10–3) **$(x^2 + 17x + 66)$ m²**

Extension

Problem Solving Factor $24x^6 - 114x^4 - 225x^2$.
$3x^2(2x - 5)(2x + 5)(2x^2 + 3)$

SELF TEST

The Self Test provides students with a brief review of the concepts and skills in Lessons 10-1 through 10-3. Lesson numbers are given to the right of exercises or instruction lines so students can review concepts not yet mastered.

10-4 Factoring Differences of Squares

What YOU'LL LEARN
- To identify and factor binomials that are the differences of squares.

INTEGRATION
Modeling

You have used algebra tiles to factor trinomials. You can also use algebra tiles to factor some binomials.

MODELING MATHEMATICS **Difference of Squares**

Materials: algebra tiles, product mat

Factor $x^2 - 9$.

Step 1
Model the polynomial $x^2 - 9$.

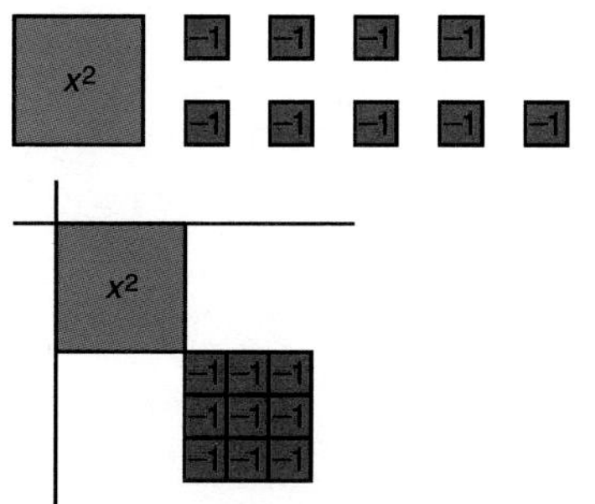

Step 2
Place the x^2-tile at the corner of the product mat. Arrange the 1-tiles into a 3-by-3 square.

Step 3
Complete the rectangle using 3 zero pairs as shown.

The rectangle has a width of $x - 3$ and a length of $x + 3$. Therefore, $x^2 - 9 = (x - 3)(x + 3)$.

Your Turn **b, c, e. See margin.**

a. Use algebra tiles to factor each binomial.
$x^2 - 16$ **$(x - 4)(x + 4)$** $x^2 - 4$ **$(x - 2)(x + 2)$** $x^2 - 1$ **$(x - 1)(x + 1)$**
$4x^2 - 9$ **$(2x - 3)(2x + 3)$** $9x^2 - 4$ **$(3x - 2)(3x + 2)$** $4x^2 - 1$ **$(2x - 1)(2x + 1)$**
b. Binomials such as those in part a are called the *difference of squares*. Explain why you think this term applies to these binomials.
c. Study the factors of the binomials in part a. What do you notice about the signs of the factors? about the terms of the factors?
d. Use the pattern that you observe to factor $x^2 - 100$. **$(x - 10)(x + 10)$**
e. Use FOIL to check your answer in part d. Was your answer correct?

Why IT'S IMPORTANT
You can use factoring to solve problems involving geometry and number theory.

Recall that the product of the sum and difference of two binomials such as $n + 8$ and $n - 8$ is called the *difference of squares.*

$(n + 8)(n - 8) = n^2 - 8n + 8n - 64$ *Use FOIL.*
$= n^2 - 64$ *Note that this is the difference of two squares, n^2 and 64.*

MODELING MATHEMATICS In this activity, students use factoring models developed earlier to study the problem of factoring a difference of squares. Emphasize to students that while in general the solution to a trinomial must be rectangular, the solution to a difference of squares must be square.

Answers for Modeling Mathematics

b. Each term is a square and the sign between them is always a subtraction sign.
c. One sign is minus and the other is plus. The terms are the same.
e. $(x - 10)(x + 10)$
$= (x)(x) + (-10)(x) + (10)(x) + (-10)(10)$
$= x^2 - 10x + 10x - 100$
$= x^2 - 100$

10-4 LESSON NOTES

NCTM Standards: 1–5

Instructional Resources

- Study Guide Master 10-4
- Practice Master 10-4
- Enrichment Master 10-4
- Real-World Applications, 25

Transparency 10-4A contains the 5-Minute Check for this lesson; **Transparency 10-4B** contains a teaching aid for this lesson.

Recommended Pacing	
Standard Pacing	Days 6 & 7 of 12
Honors Pacing	Day 6 of 11
Block Scheduling*	Day 4 of 6 (along with Lesson 10-5)
Alg. 1 in Two Years*	Days 11, 12, & 13 of 22

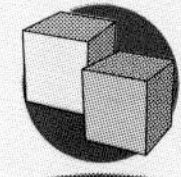

*For more information on pacing and possible lesson plans, refer to the *Block Scheduling Booklet* and *Algebra 1 in Two Years.*

1 FOCUS

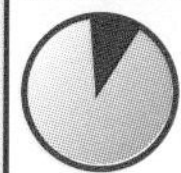

5-Minute Check
(over Lesson 10-3)

Factor each trinomial.

1. $10x^2 + 9x - 9$ **$(5x - 3)(2x + 3)$**
2. $y^2 - 15y + 56$ **$(y - 7)(y - 8)$**
3. $12t^2 + 35t + 8$ **$(4t + 1)(3t + 8)$**
4. $p^2 + 6p - 27$ **$(p + 9)(p - 3)$**
5. $6n^2 - 25n + 14$ **$(2n - 7)(3n - 2)$**

Motivating the Lesson

Hands-On Activity Cut out pieces of paper as shown in the lesson to show how they fit to form a rectangle after a square has been removed from a larger square. This can be illustrated very well on an overhead projector. Students may want to cut out a square of any size and discover that the remaining "L" can always be cut to form a rectangle.

2 TEACH

In-Class Examples

For Example 1
Factor each binomial.

a. $x^2 - 25$ $(x - 5)(x + 5)$
b. $a^2 - 121$ $(a - 11)(a + 11)$
c. $16x^4 - z^4$
$(4x^2 + z^2)(2x - z)(2x + z)$

For Example 2
Factor each polynomial.

a. $16x^2 - 25y^2$
$(4x - 5y)(4x + 5y)$
b. $81a^2 - 16y^2$
$(9a - 4y)(9a + 4y)$
c. $16x^2 - z^4$
$(4x + z^2)(4x - z^2)$
d. $a^4 - 81b^8$
$(a^2 + 9b^4)(a - 3b^2)(a + 3b^2)$

Teaching Tip Students should memorize the form of a difference of two squares.

Teaching Tip Review the FOIL method with students. For example, find $(x + 2)(x + 5)$.

$$\begin{aligned}(x + 2)(x + 5) &= \overset{F}{x^2} + \overset{O}{5x} + \overset{I}{2x} + \overset{L}{10}\\ &= x^2 + 7x + 10\end{aligned}$$

The Modeling Mathematics activity suggests the following rule for factoring the difference of squares.

Difference of Squares	$a^2 - b^2 = (a - b)(a + b) = (a + b)(a - b)$

You can use this rule to factor binomials that can be written in the form $a^2 - b^2$.

Example 1 **Factor each binomial.**

a. $m^2 - 81$

$$\begin{aligned}m^2 - 81 &= (m)^2 - (9)^2 && m \cdot m = m^2 \text{ and } 9 \cdot 9 = 81\\ &= (m - 9)(m + 9) && \text{Use the difference of squares.}\end{aligned}$$

b. $100s^2 - 25t^2$

$$\begin{aligned}100s^2 - 25t^2 &= 25(4s^2 - t^2) && \text{25 is the GCF.}\\ &= 25[(2s)^2 - t^2] && 2s \cdot 2s = 4s^2 \text{ and } t \cdot t = t^2\\ &= 25(2s - t)(2s + t) && \text{Use the difference of squares.}\end{aligned}$$

c. $\frac{1}{9}x^2 - \frac{4}{25}y^2$

$$\begin{aligned}\frac{1}{9}x^2 - \frac{4}{25}y^2 &= \left(\frac{1}{3}x\right)^2 - \left(\frac{2}{5}y\right)^2 && \text{Why?}\\ &= \left(\frac{1}{3}x - \frac{2}{5}y\right)\left(\frac{1}{3}x + \frac{2}{5}y\right) && \text{Check this result by using FOIL.}\end{aligned}$$

Sometimes the terms of a binomial have common factors. If so, the GCF should always be factored out first. Occasionally, the difference of squares needs to be applied more than once or along with grouping in order to completely factor a polynomial.

Example 2 **Factor each polynomial.**

a. $20cd^2 - 125c^5$

$$\begin{aligned}20cd^2 - 125c^5 &= 5c(4d^2 - 25c^4) && \text{The GCF of } 20cd^2 \text{ and } 125c^5 \text{ is } 5c.\\ &= 5c(2d - 5c^2)(2d + 5c^2) && 2d \cdot 2d = 4d^2 \text{ and } 5c^2 \cdot 5c^2 = 25c^4\end{aligned}$$

b. $3k^4 - 48$

$$\begin{aligned}3k^4 - 48 &= 3(k^4 - 16) && \text{Why?}\\ &= 3(k^2 - 4)(k^2 + 4) && k^2 \cdot k^2 = k^4\\ &= 3(k - 2)(k + 2)(k^2 + 4) && k^2 + 4 \text{ cannot be factored. Why not?}\end{aligned}$$

c. $9x^5 + 11x^3y^2 - 100xy^4$

To factor the trinomial, we need two numbers whose product is −900 and whose sum is 11.

$$\begin{aligned}9x^5 + 11x^3y^2 - 100xy^4 &= x(9x^4 + 11x^2y^2 - 100y^4)\\ &= x[(9x^4 - 25x^2y^2) + (36x^2y^2 - 100y^4)]\\ &= x[x^2(9x^2 - 25y^2) + 4y^2(9x^2 - 25y^2)]\\ &= x(x^2 + 4y^2)(9x^2 - 25y^2)\\ &= x(x^2 + 4y^2)(3x - 5y)(3x + 5y)\end{aligned}$$

Alternative Learning Styles

Visual The factorization $(a^2 - b^2) = (a - b)(a + b)$ can be modeled by drawing the rectangle for $(4^2 - 2^2) = (4 + 2)(4 - 2)$.

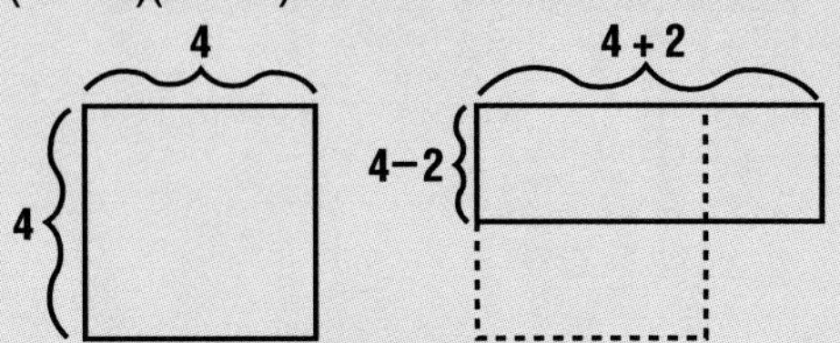

The difference of squares can be used to multiply numbers mentally.

Example 3 **Show a method for finding the product of 37 and 43 mentally.**

Since $37 = 40 - 3$ and $43 = 40 + 3$, the product of (37)(43) can be expressed as $(40 - 3)(40 + 3)$.

$$(43)(37) = (40 + 3)(40 - 3)$$
$$= 40^2 - 3^2$$
$$= 1600 - 9 \text{ or } 1591$$

Pythagoras

According to the Pythagorean theorem, the sum of the squares of the measures of the legs of a right triangle equals the square of the measure of the hypotenuse.

$$a^2 + b^2 = c^2$$

A **Pythagorean triple** is a group of three whole numbers that satisfy the equation $a^2 + b^2 = c^2$. For example, the numbers 3, 4, and 5 form a Pythagorean triple.

$$3^2 + 4^2 \stackrel{?}{=} 5^2$$
$$9 + 16 \stackrel{?}{=} 25$$
$$25 = 25 \checkmark$$

You can use the difference of squares to find Pythagorean triples.

Example 4 **Find a Pythagorean triple that includes 8 as one of its numbers.**

INTEGRATION
Number Theory

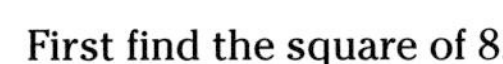

First find the square of 8. $8^2 = 64$

Factor 64 into two even factors or two odd factors. $64 = (2)(32)$

Find the mean of the two factors. $\frac{2 + 32}{2} = 17$

Complete the following statement.

$$(2)(32) = (17 - \underline{?})(17 + \underline{?})$$
$$= (17 - 15)(17 + 15)$$
$$= 17^2 - 15^2$$

Therefore, $8^2 = 17^2 - 15^2$ or $8^2 + 15^2 = 17^2$. The numbers 8, 15, and 17 form a Pythagorean triple.

FYI

Finding Pythagorean triples has interested mathematicians since ancient times. Some Pythagorean triples have been found on Babylonian cuneiform tablets.

Teaching Tip Point out that Example 3 shows a good method for doing multiplication without a calculator.

In-Class Examples

For Example 3
Find the product of 28 and 32 mentally.
$(28)(32) = (30 - 2)(30 + 2) = 30^2 - 2^2 = 900 - 4 = 896$

For Example 4
Find a Pythagorean triple that includes 6 as one of its numbers.
6, 8, 10

Teaching Tip A more in-depth discussion of the Pythagorean theorem is provided in Chapter 13.

FYI

Pythagoras was born on the island of Samos in the Aegean Sea in about 580 B.C. In addition to being a brilliant mathematician, he founded a mystical cult based on relationships between numbers.

3 PRACTICE/APPLY

Check for Understanding

Exercises 1–21 are designed to help you assess your students' understanding through reading, writing, speaking, and modeling. You should work through Exercises 1–6 with your students and then monitor their work on Exercises 7–21.

Error Analysis
Here are two common errors in factoring.

1. $5a^2 - 20 = 5(a^2 - 4)$
   $= (a + 2)(a - 2)$
   The factor of 5 was omitted.
2. $16x^2 - 36 = (4x + 6)(4x - 6)$
   This was not factored completely. The GCF should be removed first.

Assignment Guide

Core: 23–53 odd, 54, 55, 57, 59–68
Enriched: 22–52 even, 54–68

For **Extra Practice,** see p. 779.

The red A, B, and C flags, printed only in the Teacher's Wraparound Edition, indicate the level of difficulty of the exercises.

Study Guide Masters, p. 72

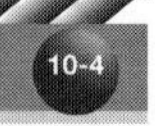

NAME________________ DATE ________

Study Guide — Student Edition Pages 581–586

Factoring Differences of Squares

Use the difference of squares to factor polynomials.

Difference of Squares	$a^2 - b^2 = (a - b)(a + b) = (a + b)(a - b)$

Example 1: Factor $4y^2 - 81z^2$.
$4y^2 - 81z^2 = (2y)^2 - (9z)^2$ — $2y \cdot 2y = 4y^2$ and $9z \cdot 9z = 81z^2$
$= (2y - 9z)(2y + 9z)$ — Use the difference of squares.

In some binomials you have to factor a GCF before you can factor the difference of squares.

Example 2: Factor $50a^2 - 72$.
$50a^2 - 72 = 2(25a^2 - 36)$ — The GCF is 2.
$= 2(5a - 6)(5a + 6)$ — Use the difference of squares.

State whether each binomial can be factored as a difference of squares.

1. $a^2 - b^2$ yes
2. $x^2 + y^2$ no
3. $a^2 - 36$ yes
4. $2p - \frac{1}{9}$ no
5. $\frac{1}{2}m^2 + \frac{1}{4}n^2$ no
6. $\frac{49}{289}x^2 - 1$ yes
7. $0.16m^2 + 0.25n^2$ no
8. $225b^2 - a^2$ yes
9. $a - 16$ no
10. $15x^2 + 5$ no
11. $9y^2 - 4x^2$ yes
12. $-p^2 + 9q^2$ yes

Factor each polynomial, if possible. If the polynomial cannot be factored, write prime.

13. $m^2 - 16n^2$ $(m - 4n)(m + 4n)$
14. $4a^2 - 9b^2$ $(2a + 3b)(2a - 3b)$
15. $x^2 - 64$ $(x - 8)(x + 8)$
16. $-81 + a^4$ $(a - 3)(a + 3)(a^2 + 9)$
17. $m^6 - 16n^4$ $(m^3 - 4n^2)(m^3 + 4n^2)$
18. $-2 + 2y^2$ $2(y - 1)(y + 1)$
19. $p^2q^2 - \frac{1}{16}$ $\left(pq - \frac{1}{4}\right)\left(pq + \frac{1}{4}\right)$
20. $\frac{1}{4}z^4 - 25$ $\left(\frac{1}{2}z^2 - 5\right)\left(\frac{1}{2}z^2 + 5\right)$
21. $\frac{2}{3}x^2 - 9$ prime
22. $12x^2 - 27y^2$ $3(2x - 3y)(2x + 3y)$
23. $6 - 54z^2$ $6(1 + 3z)(1 - 3z)$
24. $(x + y)^2 - w^2$ $(x + y - w)(x + y + w)$
25. $3x^4 - 75$ $3(x^2 - 5)(x^2 + 5)$
26. $(n + 7)^2 - 1$ $(n + 6)(n + 8)$
27. $2p^4 - 32q^4$ $2(p - 2q)(p + 2q)(p^2 + 4q^2)$

CHECK FOR UNDERSTANDING

Communicating Mathematics

Study this lesson. Then complete the following.

1. **Describe** a binomial that is the difference of two squares. **See margin.**
2. **Write** a polynomial that is the difference of two squares. Factor your polynomial. **Sample answer: $a^2 - 25 = (a - 5)(a + 5)$**
3. **Explain** how to factor a difference of squares by using the method for factoring trinomials presented in Lesson 10-3. **See margin.**
4. **You Decide** Patsy says that $28f^2 - 7g^2$ can be factored using the difference of squares. Sally says it cannot. Who is correct? Explain. **See margin.**
5. **Show** how to use the difference of squares to find $\frac{15}{16} \cdot \frac{17}{16}$.
6. Use algebra tiles to factor $4 - x^2$. **$(2 - x)(2 + x)$**

5. $\frac{15}{16} \cdot \frac{17}{16} = \frac{(16 - 1)(16 + 1)}{16^2} = \frac{16^2 - 1^2}{16^2} = \frac{255}{256}$

MODELING MATHEMATICS

Guided Practice

State whether each binomial can be factored as a difference of squares.

7. $p^2 - 49q^2$ **yes**
8. $25a^2 - 81b^4$ **yes**
9. $9x^2 + 16y^2$ **no**

Match each binomial with its factored form.

10. $4x^2 - 25$ **c**
11. $16x^2 - 4$ **d**
12. $25x^2 - 4$ **b**
13. $25x^2 - 25$ **a**

a. $25(x - 1)(x + 1)$
b. $(5x - 2)(5x + 2)$
c. $(2x - 5)(2x + 5)$
d. $4(2x - 1)(2x + 1)$

Factor each polynomial, if possible. If the polynomial cannot be factored, write *prime*. **14. $(t - 5)(t + 5)$ 15. $(1 - 4g)(1 + 4g)$**

14. $t^2 - 25$
15. $1 - 16g^2$
16. $2a^2 - 25$ **prime**
17. $20m^2 - 45n^2$
18. $(a + b)^2 - c^2$
19. $x^4 - y^4$
20. Find the product of 17 and 23 mentally using difference of squares. **391**
21. The difference of two numbers is 3. If the difference of their squares is 15, what is the sum of the numbers? **5**

17. $5(2m - 3n)(2m + 3n)$
18. $(a + b - c)(a + b + c)$
19. $(x - y)(x + y)(x^2 + y^2)$

EXERCISES

Practice

Factor each polynomial, if possible. If the polynomial cannot be factored, write *prime*. **22–45. See margin.**

22. $w^2 - 81$
23. $4 - v^2$
24. $4q^2 - 9$
25. $100d^2 - 1$
26. $16a^2 - 25b^2$
27. $2z^2 - 98$

B

28. $9g^2 - 75$
29. $4t^2 - 27$
30. $8x^2 - 18$
31. $17 - 68k^2$
32. $25y^2 - 49z^4$
33. $36x^2 - 125y^2$
34. $-16 + 49h^2$
35. $16b^2c^4 + 25d^8$
36. $-9r^2 + 81$
37. $a^2x^2 - 0.64y^2$
38. $\frac{1}{16}x^2 - 25z^2$
39. $\frac{9}{2}a^2 - \frac{49}{2}b^2$

C

40. $(4p - 9q)^2 - 1$
41. $(a + b)^2 - (c + d)^2$
42. $25x^2 - (2y - 7z)^2$
43. $x^8 - 16y^4$
44. $a^6 - a^2b^4$
45. $a^4 + a^2b^2 - 20b^4$

Find each product mentally by using differences of squares.

46. 29×31 **899**
47. 24×26 **624**
48. 94×106 **9964**

Reteaching

Using Alternative Methods The difference of two squares can be used to find the product of two numbers that are the same difference from a multiple of ten. For example:

$(17)(23) = (20 - 3)(20 + 3)$
$= 20^2 - 3^2$
$= 400 - 9$
$= 391$

Additional Answers

1. Each term of the binomial is a perfect square, and the binomial can be written as a difference of terms.
3. Write the binomial as a trinomial where the coefficient of the middle term is 0, and then factor this trinomial.
4. Patsy; if 7 is factored from each term, the binomial factor is the difference of squares $4f^2 - g^2$.

Geometry

Find the dimensions of a rectangle with the same area as the shaded region in each drawing. Assume that the dimensions of the rectangle must be represented by binomials with integral coefficients.

49. $(2a - b)$ in., $(2a + b)$ in.

50. $(\pi r - 5\pi)$ cm, $(r + 5)$ cm or $(r - 5)$ cm, $(\pi r + 5\pi)$ cm

51. $(x - 2)$ ft, $(x + 4)$ ft

49.
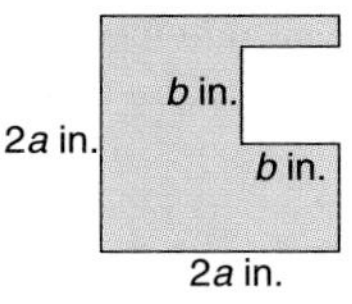

50.
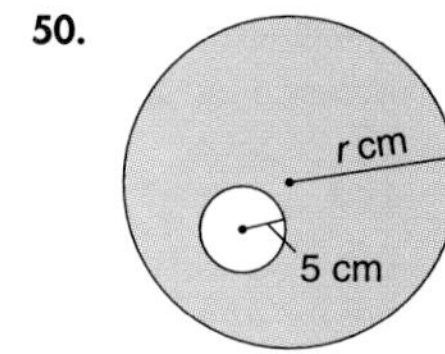

51.
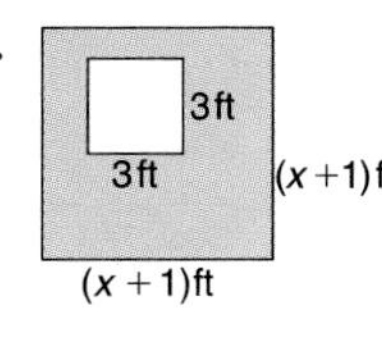

Find the dimensions of a rectangular solid having the given volume if each dimension can be written as a binomial with integral coefficients.

52. $(p - r)$ cm, $(p + r)$ cm, $(7m + 2n)$ cm

53. $(a - 5b)$ in., $(a + 5b)$ in., $(5a + 3b)$ in.

52. $(7mp^2 + 2np^2 - 7mr^2 - 2nr^2)$ cubic centimeters
53. $(5a^3 - 125ab^2 - 75b^3 + 3a^2b)$ cubic inches

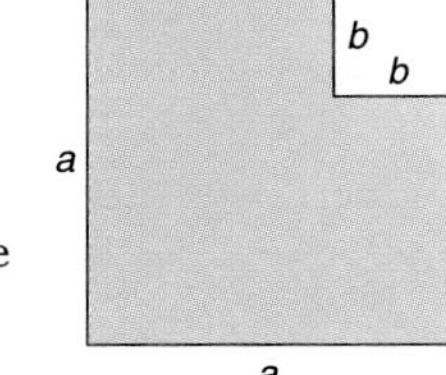

Critical Thinking

54. Show how to divide and rearrange the diagram at the right to show that $a^2 - b^2 = (a - b)(a + b)$. Make a diagram to show your reasoning. **See Solutions Manual.**

Applications and Problem Solving

55. **Geometry** The side of a square is x centimeters long. The length of a rectangle is 5 centimeters longer than a side of the square, and the width of the rectangle is 5 centimeters shorter than the side of the square.
  a. Which has the greater area, the square or the rectangle? **square**
  b. How much greater is that area? **25 cm^2**
56. **Number Theory** Find a Pythagorean triple that includes 7. **7, 24, 25**
57. **Number Theory** Find a Pythagorean triple that includes 9. **9, 12, 15**
58. **Geometry** Express the square of the length of the missing side of the triangle at the right as the product of two binomials. **$(2a + 1)(4a + 3)$**

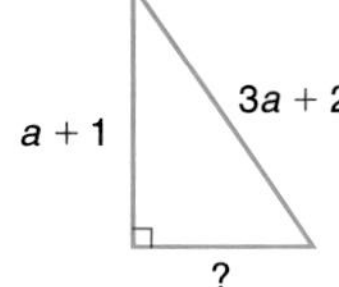

Mixed Review

59. **Guess and Check** Julie went to the corner store to buy four items. The clerk at the store had to use a calculator to add the four prices, since her cash register was broken. When the clerk figured Julie's bill, she mistakenly hit the multiplication key each time, instead of the plus key. Julie had already mentally computed her sum, so, realizing that the total was actually correct, she paid the clerk the amount of $7.11. How much was each item that Julie bought? (Lesson 10–3) **$3.16, $1.50, $1.25, $1.20**
60. Find $(n^2 + 5n + 3) + (2n^2 + 8n + 8)$. (Lesson 9–5) **$3n^2 + 13n + 11$**
61. Use elimination to solve the system of equations. (Lesson 8–4) **(4, 16)**
  $x + y = 20$
  $0.4x + 0.15y = 4$

62. $\frac{15}{16}$

62. **Probability** Use a tree diagram to find the probability of getting at least one tail when four fair coins are tossed. (Lesson 7–5)
63. Graph $6x - 3y = 6$. (Lesson 6–5) **See margin.**
64. Write an equation for the relation given in the chart below. (Lesson 5–6)

a	1	2	3	4	5	6
b	1	4	7	10	13	16

$b = 3a - 2$

Classroom Vignette

"I like to have my students be familiar with the squares and approximate square roots of the numbers 1 through 20. I let a student draw a question from a 'pot' and give the student five seconds to answer the question. You can give a score for the student's answer, for example 100, and reduce the score for each attempt the student must make until they get the correct answer."

Karen C. Painter
Smoky Mountain High School
Sylva, North Carolina

Additional Answers

22. $(w - 9)(w + 9)$
23. $(2 - v)(2 + v)$
24. $(2q - 3)(2q + 3)$
25. $(10d - 1)(10d + 1)$
26. $(4a - 5b)(4a + 5b)$
27. $2(z - 7)(z + 7)$
28. $3(3g^2 - 25)$
29. prime
30. $2(2x - 3)(2x + 3)$
31. $17(1 - 2k)(1 + 2k)$
32. $(5y - 7z^2)(5y + 7z^2)$
33. prime
34. $(7h - 4)(7h + 4)$
35. prime
36. $9(3 - r)(3 + r)$
37. $(ax + 0.8y)(ax - 0.8y)$
38. $\left(\frac{1}{4}x + 5z\right)\left(\frac{1}{4}x - 5z\right)$
39. $\frac{1}{2}(3a + 7b)(3a - 7b)$
40. $(4p - 9q + 1)(4p - 9q - 1)$
41. $(a + b + c + d)$ $(a + b - c - d)$
42. $(5x + 2y - 7z)(5x - 2y + 7z)$
43. $(x^2 - 2y)(x^2 + 2y)(x^4 + 4y^2)$
44. $a^2(a - b)(a + b)(a^2 + b^2)$
45. $(a^2 + 5b^2)(a - 2b)(a + 2b)$
63.

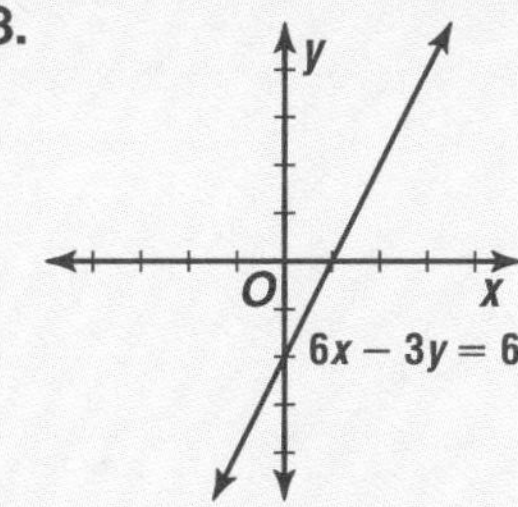

Practice Masters, p. 72

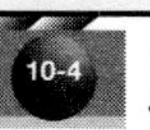

NAME ______________ DATE ______

Practice

Student Edition Pages 581–586

Factoring Differences of Squares

Factor each polynomial, if possible. If the polynomial cannot be factored, write prime.

1. $a^2 - 4$ $(a - 2)(a + 2)$
2. $y^2 - 1$ $(y - 1)(y + 1)$
3. $x^2 - 64$ $(x - 8)(x + 8)$
4. $1 - 49c^2$ $(1 - 7c)(1 + 7c)$
5. $-16 + p^2$ $(p - 4)(p + 4)$
6. $100r^2 - 9$ $(10r - 3)(10r + 3)$
7. $36 - n^2$ $(6 - n)(6 + n)$
8. $144 - 9f^2$ $(12 - 3f)(12 + 3f)$
9. $-r^2s^2 + 81$ $(9 - rs)(9 + rs)$
10. $5c^2 - 4d^2$ prime
11. $4g^2 - 81h^2$ $(2g - 9h)(2g + 9h)$
12. $36j^2 - 49m^2$ $(6j - 7m)(6j + 7m)$
13. $8n^2 - 72p^2$ $8(n - 3p)(n + 3p)$
14. $20q^2 - 5r^2$ $5(2q - r)(2q + r)$
15. $s^4t^2 - 4t^2$ $t^2(s^2 - 2)(s^2 + 2)$
16. $36n^2 - 25$ $(6n - 5)(6n + 5)$
17. $49 - 100k^2$ $(7 - 10k)(7 + 10k)$
18. $32 - 8n^2$ $8(2 - n)(2 + n)$
19. $t^2 - 64u^2$ $(t - 8u)(t + 8u)$
20. $121r^2 - 1$ $(11r - 1)(11r + 1)$
21. $2yz^4 - 50yz^2$ $2yz^2(z - 5)(z + 5)$
22. $25v^5x - 9v^3x$ $v^3x(5v - 3)(5v + 3)$
23. $4t^2 - s^4t^2$ $t^2(2 - s^2)(2 + s^2)$
24. $200y^2z^5 - 242y^4z^3$ $2y^2z^3(10z - 11y)(10z + 11y)$
25. $75x^2 - 147y^2$ $3(5x - 7y)(5x + 7y)$
26. $32h^2 - 18\ell^2$ $2(4h - 3\ell)(4h + 3\ell)$
27. $x^2 + y^2$ prime
28. $x^2y^2 - z^2$ $(xy - z)(xy + z)$
29. $-4c^2 + 25$ $(5 - 2c)(5 + 2c)$
30. $j^2 - 33k^2$ prime
31. $100b^4 - 169$ $(10b^2 - 13)(10b^2 + 13)$
32. $24e^2 - 54f^4$ $6(2e - 3f^2)(2e + 3f^2)$
33. $32a^2 - 50b^2$ $2(4a - 5b)(4a + 5b)$
34. $-98r^2 + 8t^2$ $2(2t - 7r)(2t + 7r)$
35. $x^{12} - 4x^2$ $x^2(x^5 - 2)(x^5 + 2)$
36. $3\ell^2 - \frac{1}{3}$ $3\left(\ell - \frac{1}{3}\right)\left(\ell + \frac{1}{3}\right)$
37. $\frac{1}{4}u^2 - \frac{9}{4}$ $\frac{1}{4}(u - 3)(u + 3)$
38. $9t^6m^4 - 196t^8m^4$ $t^6m^4(3 - 14t)(3 + 14t)$
39. $5v^2 - \frac{5}{4}$ $5\left(v - \frac{1}{2}\right)\left(v + \frac{1}{2}\right)$
40. $64v^7x^3 - 121vx^7$ $vx^3(8v^3 - 11x^2)(8v^3 + 11x^2)$
41. $2z^2 - 196c^2$ $2(z^2 - 98c^2)$
42. $85p^2 - 17q^2$ $17(5p^2 - q^2)$

4 ASSESS

Closing Activity

Modeling Have students explore $a^3 - b^3$ by removing a cube from the corner of a larger cube. This can be done by building a cube from centimeter blocks and removing a cube from the corner. Students should write expressions for each of the remaining sections to illustrate the factoring of $a^3 - b^3$.

Additional Answer

65.

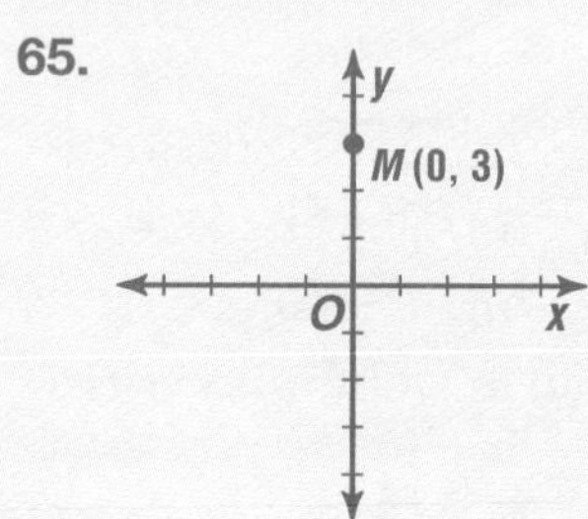

Enrichment Masters, p. 72

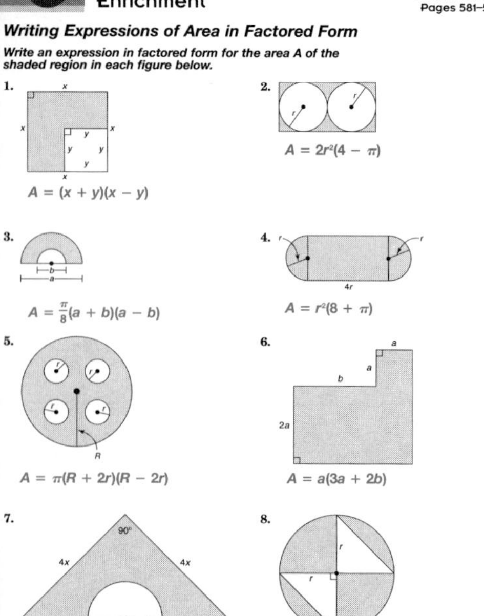

10-4 NAME ______ DATE ______

Enrichment — Student Edition Pages 581–586

Writing Expressions of Area in Factored Form

Write an expression in factored form for the area A of the shaded region in each figure below.

1. $A = (x + y)(x - y)$
2. $A = 2r^2(4 - \pi)$
3. $A = \frac{\pi}{8}(a + b)(a - b)$
4. $A = r^2(8 + \pi)$
5. $A = \pi(R + 2r)(R - 2r)$
6. $A = a(3a + 2b)$
7. $A = x^2\left(8 - \frac{\pi}{2}\right)$
8. $A = r^2(\pi - 1)$

65. Graph $M(0, 3)$. (Lesson 5–1) **See margin.**

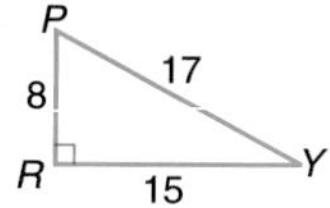

66. Trigonometry For the triangle at the right, find sin Y, cos Y, and tan Y to the nearest thousandth. (Lesson 4–3) **0.471; 0.882; 0.533**

67. Statistics The populations in millions of the 50 states in 1993 are shown below. Find the mean, median, and mode population. (Lesson 3–7)

1.2	1.1	0.6	6.0	1.0	3.3	18.2	7.9	12.0
11.1	5.7	11.7	9.5	5.0	4.5	2.8	5.2	0.6
0.7	1.6	2.5	0.7	5.0	6.5	1.8	6.9	3.6
6.9	13.7	3.8	5.1	4.2	2.6	2.4	4.3	3.2
18.0	0.8	1.1	0.5	3.6	1.6	3.9	1.9	1.4
5.3	3.0	31.2	0.6	1.2				

5.14; 3.6; 0.6

68. In Mr. Tucker's algebra class, students can get extra points for finding the correct solution to the "Riddle of the Week." One week, Mr. Tucker posed this riddle. (Lesson 2–9)

Luis is 10 years older than his brother. Next year, he will be three times as old as his brother. How old is Luis now?

Josh's answer is: *Luis is 12 years old and his brother is 4.*

a. If the brother's age is represented by a, what is Luis' age? $a + 10$

b. What will be the brother's age next year? $a + 1$

c. Does Josh get the extra points for his answer? Why or why not? **no; $12 - 4 \neq 10$**

Refer to the Investigation on pages 554–555.

the BRICKYARD

Suppose your manager, Ms. Brown, also gave you certain specifications as to which bricks to use and how many of each kind to use.

1 Using one of each type of brick, is there a rectangular pattern you can make with just three bricks? Justify your answer.

2 Select two of one type of brick and one each of the other two types. Is there a way to arrange these bricks into a rectangular pattern? If so, is there more than one pattern? Sketch a drawing of the rectangular pattern(s) you found and label the size of each brick.

3 If there is not a way to arrange the bricks into a rectangular pattern, explain why not. Is there more than one possible choice of four bricks that will make a rectangular pattern?

4 Now, using at least one brick of each type, find all the different patterns (if any) there are to arrange the bricks into a rectangular pattern of five bricks.

5 Now try designs using at least one of each type for a pattern of six bricks, seven bricks, eight bricks, nine bricks, and ten bricks.

6 Draw a diagram of all of the possible patterns. Label the size of each brick. Describe how you developed a process for finding the patterns. Are there generalizations you can make about the number of bricks and rectangular arrangements? Justify any generalizations. Explain how you know you have all of the possible patterns.

Add the results of your work to your Investigation Folder.

Extension

Problem Solving Factor $\frac{256}{6561}x^8 - 256$.

$\left(\frac{2}{3}x - 2\right)\left(\frac{2}{3}x + 2\right)\left(\frac{4}{9}x^2 + 4\right)\left(\frac{16}{81}x^4 + 16\right)$

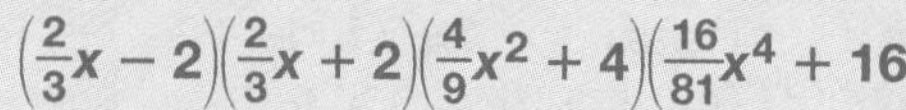

Working on the Investigation

The Investigation on pages 554–555 is designed to be a long-term project that is completed over several days or weeks. Encourage students to keep their materials in their Investigation Folder as they work on the Investigation.

10–5 Perfect Squares and Factoring

What YOU'LL LEARN

- To identify and factor perfect square trinomials.

Why IT'S IMPORTANT

You can use factoring to solve problems involving finance and construction.

INTEGRATION
Number Theory

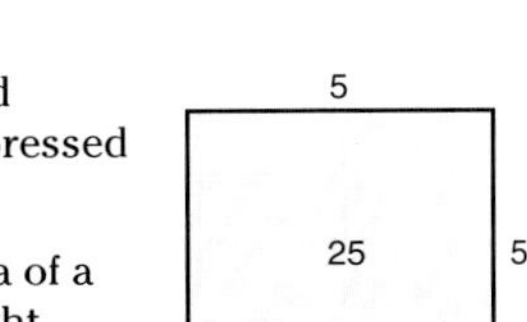

Recall that the numbers 1, 4, 9, 16, and 25 are called *perfect square* numbers, since they can each be expressed as the square of an integer.

The equation $5^2 = 25$ can be modeled as the area of a square having a side of length 5 as shown at the right.

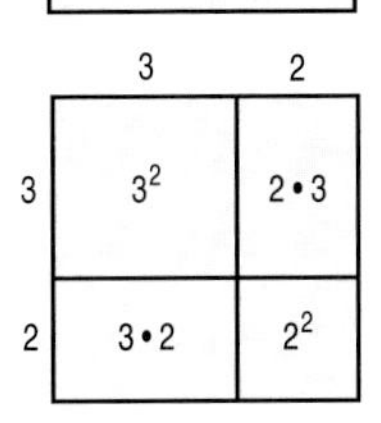

Suppose $(3 + 2)$ is substituted for 5. Then $(3 + 2)^2$ can be modeled by using a 5-by-5 square and divided it into four regions as shown at the right. The sum of the areas of the four regions equals the area of the square.

$$\begin{aligned}(3 + 2)^2 &= 3^2 + (2 \cdot 3) + (3 \cdot 2) + 2^2\\ &= 3^2 + (3 \cdot 2) + (3 \cdot 2) + 2^2\\ &= 3^2 + 2(3 \cdot 2) + 2^2\end{aligned}$$

The last line shows a very interesting relationship.

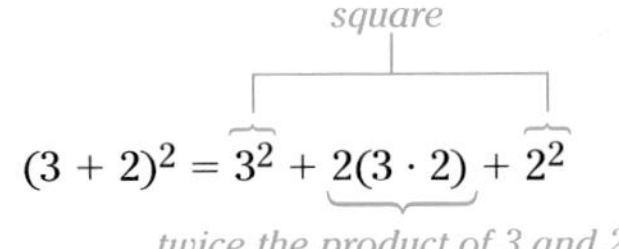

The square of this binomial is the sum of

- the square of the first term,
- twice the product of the first and second term, and
- the square of the second term.

This observation is generalized by the model at the right.

	a	b
a	a^2	ba
b	ab	b^2

$(a + b)^2 = a^2 + 2ab + b^2$

LOOK BACK

You can refer to Lesson 9-8 for information on squares of sums and squares of differences.

To square $(a - b)$, write the binomial as $[a + (-b)]$ and then square this binomial.

$$\begin{aligned}(a - b)^2 &= [a + (-b)]^2\\ &= a^2 + 2a(-b) + (-b)^2\\ &= a^2 - 2ab + b^2\end{aligned}$$

Products of the form $(a + b)^2$ and $(a - b)^2$ are called perfect squares, and their expansions are **perfect square trinomials.**

Perfect Square Trinomials	$(a + b)^2 = a^2 + 2ab + b^2$ $(a - b)^2 = a^2 - 2ab + b^2$

These patterns can be used to factor trinomials.

10-5 LESSON NOTES

NCTM Standards: 1–5

Instructional Resources

- Study Guide Master 10-5
- Practice Master 10-5
- Enrichment Master 10-5
- Assessment and Evaluation Masters, p. 269
- Modeling Mathematics Masters, pp. 66–68, 81
- Real-World Applications, 26

Transparency 10-5A contains the 5-Minute Check for this lesson; **Transparency 10-5B** contains a teaching aid for this lesson.

Recommended Pacing	
Standard Pacing	Day 8 of 12
Honors Pacing	Day 7 of 11
Block Scheduling*	Day 4 of 6 (along with Lesson 10-4)
Alg. 1 in Two Years*	Days 14, 15, & 16 of 22

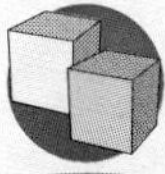

*For more information on pacing and possible lesson plans, refer to the *Block Scheduling Booklet* and *Algebra 1 in Two Years.*

1 FOCUS

5-Minute Check
(over Lesson 10-4)

Factor each polynomial.

1. $n^2 - 81$ **$(n - 9)(n + 9)$**

2. $64p^2 - 25g^2$
$(8p - 5g)(8p + 5g)$

3. $2a^5 - 162a$
$2a(a^2 + 9)(a - 3)(a + 3)$

4. $75a^4 - 12a^2$
$3a^2(5a - 2)(5a + 2)$

Motivating the Lesson

Situational Problem Winata said, "31^2 equals $30^2 + 1^2$." Do you agree or disagree? Emphasize the value of and need for the middle term when squaring a binomial.

2 TEACH

In-Class Example

For Example 1
Determine whether each trinomial is a perfect square trinomial. If so, factor it.

a. $x^2 - 12x + 36$ **$(x - 6)^2$**

b. $a^2 + 14a - 49$ **no**

Teaching Tip Caution students against confusing the difference of squares with the square of a difference.
$(a + b)(a - b) = a^2 - b^2$
$(a - b)^2 \neq (a^2 - b^2)$

Teaching Tip Point out that before factoring a trinomial, its terms should be arranged so that the powers of x are in descending or ascending order. For example, $8x + x^2 + 16$ should first be written as $x^2 + 8x + 16$ or $16 + 8x + x^2$.

Teaching Tip Answering the three questions gives students a system for identifying and factoring perfect square trinomials.

Model	Squaring a Binomial	Factoring a Perfect Square Trinomial
columns v, 3; rows v, 3: v^2, $3v$ / $3v$, 3^2	$(v + 3)^2 = v^2 + 2(v)(3) + 3^2$ $= v^2 + 6v + 9$	$v^2 + 6v + 9 = (v)^2 + 2(v)(3) + (3)^2$ $= (v + 3)^2$
columns $3p$, $-2q$; rows $3p$, $-2q$: $(3p)^2$, $(-2q)(3p)$ / $(3p)(-2q)$, $(-2q)^2$	$(3p - 2q)^2 = (3p)^2 + 2(3p)(-2q) + (-2q)^2$ $= 9p^2 - 12pq + 4q^2$	$9p^2 - 12pq + 4q^2 = (3p)^2 - 2(3p)(2q) + (2q)^2$ $= (3p - 2q)^2$

To determine whether a trinomial can be factored using these patterns, you must decide if it is a perfect square trinomial. In other words, you must determine if it can be written in the form $a^2 + 2ab + b^2$ or in the form $a^2 - 2ab + b^2$. For a trinomial to be in one of these forms, the following must be satisfied.

- The first term is a perfect square.
- The third term is a perfect square.
- The middle term is either 2 or -2 times the product of the square root of the first term and the square root of the last term.

Example 1 **Determine whether each trinomial is a perfect square trinomial. If so, factor it.**

a. $4y^2 + 36yz + 81z^2$

To determine whether $4y^2 - 36yz + 81z^2$ is a perfect square trinomial, answer each question.

- Is the first term a perfect square? $4y^2 \stackrel{?}{=} (2y)^2$ *yes*
- Is the last term a perfect square? $81z^2 \stackrel{?}{=} (9z)^2$ *yes*
- Is the middle term twice the product of $2y$ and $9z$? $36yz \stackrel{?}{=} 2(2y)(9z)$ *yes*

$4y^2 + 36yz + 81z^2$ is a perfect square trinomial.

$4y^2 + 36yz + 81z^2 = (2y)^2 + 2(2y)(9z) + (9z)^2$
$= (2y + 9z)^2$

b. $9n^2 + 49 - 21n$

First arrange the terms of $9n^2 + 49 - 21n$ so that the powers of n are in descending order.

$9n^2 + 49 - 21n = 9n^2 - 21n + 49$

- Is the first term a perfect square? $9n^2 \stackrel{?}{=} (3n)^2$ *yes*
- Is the last term a perfect square? $49 \stackrel{?}{=} (7)^2$ *yes*
- Is the middle term the product of -2, $3n$, and 7? $-21n \stackrel{?}{=} -2(3n)(7)$ *no*

$9n^2 - 21n + 49$ is not a perfect square trinomial.

Example 2 **Suppose the dimensions of a rectangle can be written as binomials with integral coefficients. Is the rectangle with the area of $(121x^2 - 198xy + 81y^2)$ square millimeters a square? If so, what is the measure of each side of the square?**

Geometry

Explore You know that the dimensions of the rectangle can be written as binomials with integral coefficients. The problem gives the area of a rectangle and asks whether it is a square. If it is a square, you need to find the dimension of each side.

Plan The rectangle is a square if $121x^2 - 198xy + 81y^2$ is a perfect square trinomial. You must answer three questions to determine if it is a perfect square trinomial. If it is a perfect square trinomial, you must factor it to find the measure of each side of the square.

Solve

- Is the first term a perfect square? $121x^2 \stackrel{?}{=} (11x)^2$ *yes*
- Is the last term a perfect square? $81y^2 \stackrel{?}{=} (9y)^2$ *yes*
- Is the middle term the product of -2, $11x$, and $9y$? $-198xy \stackrel{?}{=} -2(11x)(9y)$ *yes*

Since $121x^2 - 198xy + 81y^2$ is a perfect square trinomial, the rectangle is a square. To find the measure of each side, factor the trinomial.

$$\begin{aligned} 121x^2 - 198xy + 81y^2 &= (11x)^2 - 2(11x)(9y) + (9y)^2 \\ &= (11x - 9y)^2 \end{aligned}$$

The measure of each side is $(11x - 9y)$ millimeters.

Examine If each side of the square is $(11x - 9y)$ millimeters, then the area of the square is $(11x - 9y)^2$ square millimeters. Use FOIL to see if $(11x + 9y)^2$ equals $(121x^2 - 198xy + 81y^2)$.

$$\begin{aligned} (11x - 9y)^2 &= (11x - 9y)(11x - 9y) \\ &= (11x)(11x) + (11x)(-9y) + (-9y)(11x) + (-9y)(-9y) \\ &= 121x^2 - 99xy - 99xy + 81y^2 \\ &= 121x^2 - 198xy + 81y^2 \ \checkmark \end{aligned}$$

As you continue your study of mathematics, you will find that forming a perfect square trinomial can sometimes be a useful tool for solving problems.

Example 3 **Determine all values of k that make $25x^2 + kx + 49$ a perfect square trinomial.**

$25x^2 + kx + 49 = (5x)^2 + kx + (7)^2$

In order for this to be a perfect square trinomial, kx must equal either $2(5x)(7)$ or $-2(5x)(7)$. *Why?*

$kx = 2(5x)(7)$	or	$kx = -2(5x)(7)$
$kx = 70x$		$kx = -70x$
$k = 70$		$k = -70$

Check to see if $25x^2 + 70x + 49$ and $25x^2 - 70x + 49$ are perfect square trinomials.

In-Class Examples

For Example 2
Is it possible for $49a^2 - 42ab + 9b^2$ to be the measure of the area of a square? If so, what is the measure of each side of the square? **$(7a - 3b)$**

For Example 3
Determine all values of k that make each trinomial a perfect square trinomial.

a. $9x^2 + kx + 16$ **± 24**

b. $49x^2 + kx + 9$ **± 42**

Teaching Tip You may wish to suggest that students use FOIL to check their factorizations.

In-Class Example

For Example 4
Factor each polynomial.

a. $49k^2 - 81$ $\mathbf{(7k - 9)(7k + 9)}$

b. $10x^2 - x - 21$ $\mathbf{(2x - 3)(5x + 7)}$

c. $6p^4q + 3p^3q - 12p^2q^2 - 6pq^2$ $\mathbf{3pq(p^2 - 2q)(2p + 1)}$

Teaching Tip In Example 4c, remind students that factoring completely often takes many steps.

In this chapter, you have learned various methods to factor different types of polynomials. The following chart summarizes these methods and can help you decide when to use a specific method.

Check for:	Number of Terms		
	Two	Three	Four or More
greatest common factor	✓	✓	✓
difference of squares	✓		
perfect square trinomials		✓	
trinomial that has two binomial factors		✓	
pairs of terms that have a common monomial factor			✓

Whenever there is a GCF other than 1, always factor it out first. Then, check the appropriate factoring methods in the order shown in the table. Use these methods to factor until all of the factors are prime.

Example  **Factor each polynomial.**

a. $\mathbf{4k^2 - 100}$

First check for a GCF. Then, since the polynomial has two terms, check for the difference of squares.

$4k^2 - 100 = 4(k^2 - 25)$ *The GCF is 4.*

$= 4(k - 5)(k + 5)$ *$k^2 - 25$ is the difference of squares since $k \cdot k = k^2$ and $5 \cdot 5 = 25$.*

Therefore, $4k^2 - 25$ is completely factored as $4(k - 5)(k + 5)$.

b. $\mathbf{9x^2 - 3x - 20}$

The polynomial has three terms. The GCF is 1. $9x^2 = (3x)^2$, but -20 is not a perfect square. The trinomial is not a perfect square trinomial.

Are there two numbers whose product is $9(-20)$ or -180 and whose sum is -3? Yes, the product of -15 and 12 is -180, and their sum is -3.

$$\begin{aligned} 9x^2 - 3x - 20 &= 9x^2 - 15x + 12x - 20 \\ &= (9x^2 - 15x) + (12x - 20) \\ &= 3x(3x - 5) + 4(3x - 5) \\ &= (3x + 4)(3x - 5) \end{aligned}$$

Therefore, $9x^2 - 3x - 20$ is completely factored as $(3x + 4)(3x - 5)$.

c. $\mathbf{4m^4n + 6m^3n - 16m^2n^2 - 24mn^2}$

Since the polynomial has four terms, first check for the GCF and then check for pairs of terms that have a common factor.

$$\begin{aligned} 4m^4n + 6m^3n - 16m^2n^2 - 24mn^2 &= 2mn(2m^3 + 3m^2 - 8mn - 12n) \\ &= 2mn[(2m^3 + 3m^2) + (-8mn - 12n)] \\ &= 2mn[m^2(2m + 3) + (-4n)(2m + 3)] \\ &= 2mn(m^2 - 4n)(2m + 3) \end{aligned}$$

Therefore, $4m^4n + 6m^3n - 16m^2n^2 - 24mn^2$ is completely factored as $2mn(m^2 - 4n)(2m + 3)$.

CHECK FOR UNDERSTANDING

Communicating Mathematics

Study the lesson. Then complete the following.

1. a. **Draw** a rectangle to show how to factor $4x^2 + 12x + 9$. Label the dimensions and the area of the rectangle. **See margin.**
   b. **Explain** why the name *perfect square trinomial* is appropriate for this trinomial. **It is a trinomial that can be represented by a square.**

2a. See students' work.

2. a. **Write** a polynomial that is a perfect square trinomial.
   b. **Factor** your trinomial. **Sample answer: $x^2 + 2x + 1 = (x + 1)^2$**
3. a. **Describe** the first step in factoring any polynomial. **See margin.**
   b. **Explain** why this step is important. **See margin.**
4. **You Decide** Robert says that $12a^4 - 8a^2 - 4$ is completely factored as $4(3a^2 + 1)(a^2 - 1)$. Samuel says that he can factor it further. Who is correct? Explain your answer. **Samuel; $(a^2 - 1)$ can be factored as $(a - 1)(a + 1)$.**

5. **Assess Yourself** Describe the relationship between multiplying polynomials and factoring polynomials. Do you prefer to multiply polynomials or to factor polynomials? Explain. **Multiplying and factoring polynomials are inverse operations; see students' work.**

Guided Practice

Complete.

6. $b^2 + 10b + 25 = (b + \underline{?})^2$ **5**
7. $64a^2 - 16a + 1 = (\underline{?} - 1)^2$ **8a**
8. $81n^2 + 36n + 4 = (\underline{?} + 2)^2$ **9n**
9. $1 - 12c + 36c^2 = (1 - \underline{?})^2$ **6c**

Determine whether each trinomial is a perfect square trinomial. If so, factor it.

10. $t^2 + 18t + 81$ **yes; $(t + 9)^2$**
11. $4n^2 - 28n + 49$ **yes; $(2n - 7)^2$**
12. $9y^2 + 30y - 25$ **no**
13. $16b^2 - 56bc + 49c^2$ **yes; $(4b - 7c)^2$**

Factor each polynomial, if possible. If the polynomial cannot be factored, write *prime*.

14. $15g^2 + 25$ **$5(3g^2 + 5)$**
15. $4a^2 - 36b^2$ **$4(a - 3b)(a + 3b)$**
16. $x^2 + 6x - 9$ **prime**
17. $50g^2 + 40g + 8$ **$2(5g + 2)^2$**
18. $9t^3 + 66t^2 - 48t$ **$3t(3t - 2)(t + 8)$**
19. $20a^2x - 4a^2y - 45xb^2 + 9yb^2$

19. $(2a - 3b)(2a + 3b)(5x - y)$

20. a. Find the missing value that makes the following a perfect square trinomial.
   $9x^2 + 24x + \underline{?}$ **16**
   b. Copy and complete the model for this trinomial.

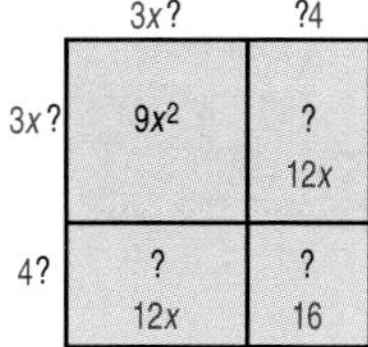

EXERCISES

Practice

Determine whether each trinomial is a perfect square trinomial. If so, factor it.

21. $r^2 - 8r + 16$ **yes; $(r - 4)^2$**
22. $d^2 + 50d + 225$ **no**
23. $49p^2 - 28p + 4$ **yes; $(7p - 2)^2$**
24. $4y^2 + 12yz + 9z^2$ **yes; $(2y + 3z)^2$**
25. $49s^2 - 42st + 36t^2$ **no**
26. $25y^2 + 20yz - 4z^2$ **no**
27. $4m^2 + 4mn + n^2$ **yes; $(2m + n)^2$**
28. $81t^2 - 180t + 100$ **yes; $(9t - 10)^2$**

Reteaching

Using Manipulatives Have students use algebra tiles to model the trinomials in the exercises. Remind students that if the tiles form a square, then the trinomial is a perfect square. The length of each side represents each factor of the trinomial.

3 PRACTICE/APPLY

Check for Understanding

Exercises 1–20 are designed to help you assess your students' understanding through reading, writing, speaking, and modeling. You should work through Exercises 1–5 with your students and then monitor their work on Exercises 6–20.

Error Analysis

Students may overlook the sign of the constant term of a trinomial when factoring. For example, students might factor $x^2 - 12x - 36$ into $(x - 6)(x - 6)$. You may wish to review the difference between -6^2 and $(-6)^2$.
$-6^2 = -36$ and $(-6)^2 = 36$

Additional Answers

1a.

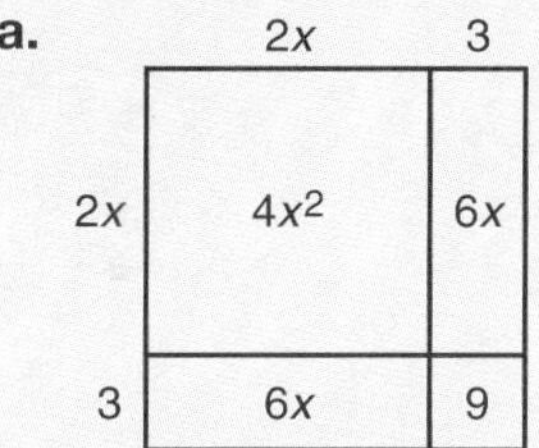

3a. Factor out the GCF of the terms.

3b. The other factoring patterns will be more apparent after the GCF has been factored out.

Study Guide Masters, p. 73

NAME ______ DATE ______

Study Guide — Student Edition Pages 587–593

Perfect Squares and Factoring

Perfect Square Trinomials	$(a + b)^2 = a^2 + 2ab + b^2$ $(a - b)^2 = a^2 - 2ab + b^2$

To factor a perfect square trinomial, the first and last terms must be perfect squares, and the middle term must be twice the product of the square roots of the first and last terms.

Example: Factor $m^2 + 26m + 169$.
$$m^2 + 26m + 169 = (m)^2 + 2(m)(13) + (13)^2 = (m + 13)^2$$

To determine whether a trinomial in mixed order is a perfect square, first arrange the terms in order of descending powers.

Determine whether each trinomial is a perfect square trinomial. If so, factor it.

1. $y^2 + 10y + 25$ yes; $(y + 5)^2$
2. $m^2 - 14mn + 49n$ yes; $(m - 7n)^2$
3. $p^2 + 8p + 64$ no

Factor each polynomial, if possible. If the polynomial cannot be factored, write prime.

4. $16x^2 + 48x + 36$ $4(2x + 3)^2$
5. $49a^4 - 112a^2b^2 + 64b^4$ $(7a^2 - 8b^2)^2$
6. $x^2y^2 - 6abxy + 9a^2b^2$ $(xy - 3ab)^2$
7. $81 + 18xy + x^2y^2$ $(9 + xy)^2$
8. $25x^2 - 10x - 1$ prime
9. $169 - 26r + r^2$ $(13 - r)^2$
10. $7x^2 - 9x + 2$ $(7x - 2)(x - 1)$
11. $a^2 + 22a + 121$ $(a + 11)^2$
12. $9x^2 - 12x + 4$ $(3x - 2)^2$
13. $x^2 - 9x - 81$ prime
14. $49b^2 - 126ab + 81a^2$ $(7b - 9a)^2$
15. $144 + 24x + x^2$ $(12 + x)^2$
16. $256 - 16b + b^2$ prime
17. $361 - 38x + x^2$ $(19 - x)^2$
18. $16a^2 - 40ab + 25b^2$ $(4a - 5b)^2$
19. $36x^2 - 12x + 1$ $(6x - 1)^2$
20. $9x^2 + 66x + 121y^2$ prime
21. $4a^2 - 20a + 25$ $(2a - 5)^2$

Assignment Guide

Core: 21–63 odd, 64–74
Enriched: 22–60 even, 61–74

For **Extra Practice,** see p. 780.

The red A, B, and C flags, printed only in the Teacher's Wraparound Edition, indicate the level of difficulty of the exercises.

Determine whether each trinomial is a perfect square trinomial. If so, factor it.

29. $2g^2 - 10g + 25$ **no**
30. $1 + 100h^2 + 20h$ **yes; $(10h + 1)^2$**
31. $64b^2 - 72b + 81$ **no**
32. $9a^2 - 24a + 16$ **yes; $(3a - 4)^2$**
33. $\frac{1}{4}a^2 + 3a + 9$ **yes; $\left(\frac{1}{2}a + 3\right)^2$**
34. $\frac{4}{9}x^2 - \frac{16}{3}x + 16$ **yes; $\left(\frac{2}{3}x - 4\right)^2$**

Factor each polynomial, if possible. If the polynomial cannot be factored, write *prime*.

B

35. $45a^2 - 32ab$ **$a(45a - 32b)$**
36. $c^2 - 5c + 6$ **$(c - 3)(c - 2)$**
37. $v^2 - 30v + 225$ **$(v - 15)^2$**
38. $m^2 - p^4$ **$(m - p^2)(m + p^2)$**
39. $9a^2 + 12a - 4$ **prime**
40. $3a^2b + 6ab + 9ab^2$
41. $3y^2 - 147$ **$3(y - 7)(y + 7)$**
42. $20n^2 + 34n + 6$ **$2(5n + 1)(2n + 3)$**
43. $18a^2 - 48a + 32$ **$2(3a - 4)^2$**
44. $3m^3 + 48m^2n + 192mn^2$
45. $x^2y^2 - y^2 - z^2 + x^2z^2$
46. $5a^2 + 7a + 6b^2 - 4b$ **prime**

C

47. $4a^3 + 3a^2b^2 + 8a + 6b^2$
48. $(x + y)^2 - (w - z)^2$
49. $0.7p^2 - 3.5pq + 4.2q^2$
50. $(x + 2y)^2 - 3(x + 2y) + 2$
51. $g^4 + 6g^3 + 9g^2 - 3g^2h - 18gh - 27h$ **$(g^2 - 3h)(g + 3)^2$**
52. $12mp^2 - 15np^2 - 16m + 20np - 16mp + 20n$ **$(4m - 5n)(3p + 2)(p - 2)$**

Determine all values of *k* that make each of the following a perfect square trinomial.

53. $25t^2 - kt + 121$ **−110, 110**
54. $64x^2 - 16xy + k$ **y^2**
55. $ka^2 - 72ab + 144b^2$ **9**
56. $169n^2 + knp + 100p^2$ **−260, 260**

Answers:
40. $3ab(a + 2 + 3b)$
44. $3m(m + 8n)^2$
45. $(y^2 + z^2)(x - 1)(x + 1)$
47. $(a^2 + 2)(4a + 3b^2)$
48. $(x + y - w + z)(x + y + w - z)$
49. $0.7(p - 3q)(p - 2q)$
50. $(x + 2y - 1)(x + 2y - 2)$
58. $(x - 3y)$ in., $(x + 3y)$ in., $(xy + 7)$ in.

INTEGRATION
Geometry

57. The area of a circle is $(9y^2 + 78y + 169)\pi$ square centimeters. What is the diameter of the circle? **$(6y + 26)$ cm**
58. The volume of a rectangular prism is $(x^3y - 63y^2 + 7x^2 - 9xy^3)$ cubic inches. Find the dimensions of the prism, if its dimensions can be represented by binomials with integral coefficients.
59. The length of a rectangle is 3 centimeters greater than the length of a side of a square. The width of the rectangle is one-half the length of the side of the square. If the area of the square is $(16x^2 - 56x + 49)$ square centimeters, what is the area of the rectangle? **$(8x^2 - 22x + 14)$ cm²**
60. The area of a square is $(81 - 90x + 25x^2)$ square meters. If x is a positive integer, what is the least possible perimeter measure for the square? **4 m**

LOOK BACK
You can refer to Lesson 2-8 to review square roots.

Critical Thinking

61. Consider the value of $\sqrt{a^2 - 2ab + b^2}$.
   a. Under what circumstances does the value equal $a - b$? **$a \geq b$**
   b. Under what circumstances does the value equal $b - a$? **$a \leq b$**
   c. Under what circumstances does the value equal $a - b$ and $b - a$? **$a = b$**

Applications and Problem Solving

62. **Construction** The builders of an office complex are looking for a square lot. They found a vacant lot that was long enough. However, its length was 60 yards more than its width w, so it was not a square. It also did not have enough area; they needed 900 additional square yards. So they are still looking for a lot. Write an expression for the length of a side of the square lot they should be looking for. **$(w + 30)$ yd**

Tech Prep

Construction Manager Students who are interested in construction management may wish to do further research on the data given in Exercise 62 and explore the potential growth of this career. For more information on tech prep, see the *Teacher's Handbook.*

Practice Masters, p. 73

10-5 NAME________________ DATE ________

Practice Student Edition Pages 587–593

Perfect Squares and Factoring

Determine whether each trinomial is a perfect square trinomial. If so, factor it.

1. $a^2 + 2a + 1$ $(a + 1)^2$
2. $2c^2 - 4c + 9$ no
3. $4d^2 - 4d + 1$ $(2d - 1)^2$
4. $r^2 + 4r + 4$ $(r + 2)^2$

Factor each polynomial, if possible. If the polynomial cannot be factored, write prime.

5. $x^2 - 6x + 9$ $(x - 3)^2$
6. $m^2 + 16m + 64$ $(m + 8)^2$
7. $s^2 - 14s + 49$ $(s - 7)^2$
8. $a^2 - 14 + 36$ prime
9. $c^2 + 24c + 144$ $(c + 12)^2$
10. $49z^2 - 56z + 16$ $(7z - 4)^2$
11. $25v^2 + 30v + 9$ $(5v + 3)^2$
12. $36s^2 - 24s + 4$ $4(3s - 1)^2$
13. $4 - 28t + 49t^2$ $(2 - 7t)^2$
14. $9p^2 + 12pq + 4q^2$ $(3p + 2q)^2$
15. $16m^2 - 24mn + 9n^2$ $(4m - 3n)^2$
16. $9r^2 + 48rt + 64t^2$ $(3r + 8t)^2$
17. $28m^2 - 28mp + 7p^2$ $7(2m - p)^2$
18. $16s^2 + 56s + 49$ $(4s + 7)^2$
19. $16c^2 + 72cd + 81d^2$ $(4c + 9d)^2$
20. $9k^2 - 30k - 25$ prime
21. $\frac{1}{9}h^2 - 4hj + 36j^2$ $\left(\frac{1}{3}h - 6j\right)^2$
22. $\frac{1}{4}x^2 - 5xz + 25z^2$ $\left(\frac{1}{2}x - 5z\right)^2$
23. $4e^2 - 44ef + 121f^2$ $(2e - 11f)^2$
24. $16a^4 - 40a^2b^3 + 25b^6$ $(4a^2 - 5b^3)^2$
25. $c^2d^2 - 2cde + e^2$ $(cd - e)^2$
26. $\frac{1}{16}m^2 - \frac{1}{4}mn + \frac{1}{4}n^2$ $\frac{1}{4}\left(\frac{1}{2}m - n\right)^2$

63. **Investments** Tamara plans to invest some money in a certificate of deposit. After 2 years, the value of the certificate will be $p + 2pr + pr^2$, where p represents the amount of money invested and r represents the annual interest rate.
   a. If Tamara invests \$1000 at an annual interest rate of 8%, find the value of the certificate after 2 years. **\$1166.40**
   b. Factor the expression that represents the value of the certificate after 2 years. **$p(1 + r)^2$**
   c. Suppose Tamara invests \$1000 at 7%. Use your expression in part b to find the value of the certificate after 2 years. **\$1144.90**
   d. Which form of the expression do you prefer to use to make your computations? Explain. **See students' work.**

Mixed Review

64. Factor $45x^2 - 20y^2z^2$. (Lesson 10–4) **$5(3x - 2yz)(3x + 2yz)$**
65. Simplify $2.5t(8t - 12) + 5.1(6t^2 + 10t - 20)$. (Lesson 9–6) **$50.6t^2 + 21t - 102$**
66. Simplify $(3a^2)(4a^3)$. (Lesson 9–1) **$12a^5$**
67. **Employment** Mike's parents allow him to work 30 hours a week. He would like to use this time to help out in his parents' hardware store, but it pays only \$5 per hour. He could mow lawns for \$7.50 per hour, but there is less than 20 hours of lawn work available. What is the maximum amount of time Mike can work in his parents' store and still make at least \$175 per week? (Lesson 8–5) **20 hours**

68. $95° \leq F \leq 104°$

68. **Physical Science** A European-made hot tub is advertised to have a temperature of 35°C to 40°C, inclusive. What is the temperature range for the hot tub in degrees Fahrenheit? (*Hint:* Use $F = \frac{9}{5}C + 32$.) (Lesson 7–4)

Year	Median Income
1970	\$ 8734
1975	11,800
1980	17,710
1981	19,074
1982	20,191
1983	21,018
1984	22,415
1985	23,618
1986	24,897
1987	26,061
1988	27,225
1989	28,905
1990	29,943
1991	30,126
1992	30,786

Source: U.S. Census Bureau

69. **Statistics** The median incomes of American families since 1970 are shown in the table at the left. (Lesson 6–3)
   a. Make a scatter plot of the data. **See margin.**
   b. Can the data be approximated by a straight line? If so, graph the line and write an equation of this line. **See margin.**
   c. Estimate the median family income for this year. **See students' work.**
70. Refer to Exercise 67 on page 586. Find the range and interquartile range of the state populations. (Lesson 5–7) **30.7, 4.6**
71. **Probability** If a card is selected at random from a deck of 52 cards, what are the odds of selecting a club? (Lesson 4–6) **1:3**
72. **Geometry** ΔABC and ΔDEF are similar. If $a = 5$, $d = 11$, $f = 6$, and $e = 14$, find the missing measures. (Lesson 4–2)

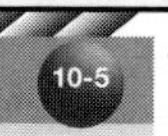

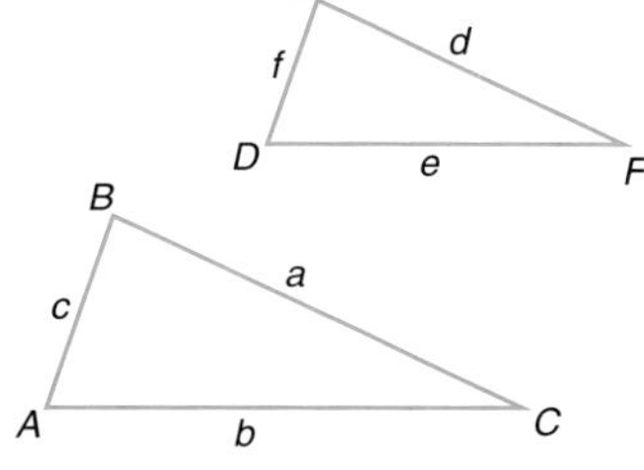

72. $b = \frac{70}{11}$, $c = \frac{30}{11}$

73. **National Landmarks** The Statue of Liberty and the pedestal on which it stands are 302 feet tall altogether. The pedestal is 2 feet shorter than the statue. How tall is the statue? (Lesson 3–3) **152 ft**
74. **Basketball** In 1962, Wilt Chamberlain set an NBA record by averaging 50.4 points per game for a few games. If he had been able to maintain this average over an 82-game season, how many total points would he have scored? Round to the nearest whole number. (Lesson 2–6) **4133 points**

Extension

Reasoning If $x^n - y^m$ is the difference of two squares, what are some possible values of n and m?
any two positive, even integers

4 ASSESS

Closing Activity

Writing Bernie said, "For a quadratic trinomial of the form $ax^2 + bx + c$, give me two of the constants a, b, or c and I can find the missing constant to make the trinomial a perfect square." Do you agree or disagree with him? Write a paragraph to support your opinion.

Chapter 10, Quiz C (Lessons 10-4 and 10-5), is available in the *Assessment and Evaluation Masters*, p. 269.

Additional Answers

69a.

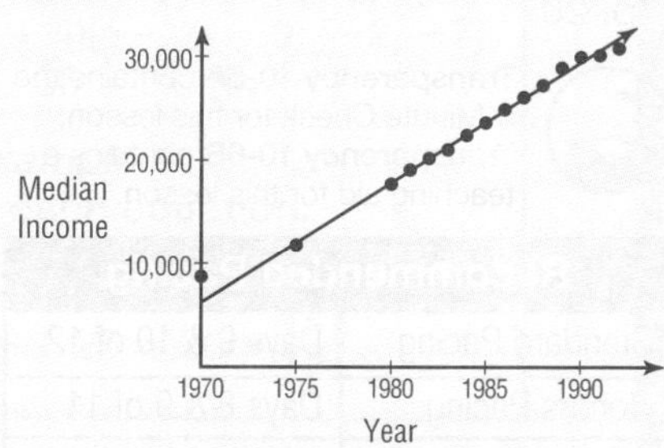

69b. Sample answer: Yes; $I = 1200y + 6000$, where I is the median income and y is the number of years since 1970.

Enrichment Masters, p. 73

10-5 NAME____________ DATE ________

Enrichment

Student Edition
Pages 587–591

Squaring Numbers: A Shortcut

A shortcut helps you to square a positive two-digit number ending in 5. The method is developed using the idea that a two-digit number may be expressed as $10t + u$. Suppose $u = 5$.

$$\begin{aligned}(10t + 5)^2 &= (10t + 5)(10t + 5)\\ &= 100t^2 + 50t + 50t + 25\\ &= 100t^2 + 100t + 25\\ (10t + 5)^2 &= 100t(t + 1) + 25\end{aligned}$$

In words, this formula says that the square of a two-digit number has $t(t + 1)$ in the hundreds place. Then 2 is the tens digit and 5 is the units digit.

Example: Using the formula for $(10t + 5)^2$, find 85^2.

$$\begin{aligned}85^2 &= 100 \cdot 8 \cdot (8 + 1) + 25\\ &= 7200 + 25\\ &= 7225\end{aligned}$$

Shortcut: First think $8 \cdot 9 = 72$. Then write 25.

Thus, to square a number, such as 85, you can write the product of the tens digit and the next consecutive integer $t + 1$. Then write 25.

Find each of the following using the shortcut.

1. 15^2 225
2. 25^2 625
3. 35^2 1225
4. 45^2 2025
5. 55^2 3025
6. 65^2 4225

Solve each problem.

7. What is the tens digit in the square of 95? 2
8. What are the first two digits in the square of 75? 56
9. Any three-digit number can be written as $100a + 10b + c$. Square this expression to show that if the last digit of a three-digit number is 5 then the last two digits of the square of the number are 2 and 5.
   $10{,}000a^2 + 2000ab + 200ac + 100b^2 + 20bc + c^2 =$
   $10{,}000a^2 + 2000ab + 1000a + 100b^2 + 100b + 25$
   The last two digits are not affected by the first five terms.

In-Class Examples

For Example 3
Solve each equation.

a. $x^3 + 2x^2 = 15x$ **$\{0, -5, 3\}$**
b. $a^3 - 13a^2 + 42a = 0$ **$\{0, 6, 7\}$**
c. $4r^3 - 9r = 0$ **$\left\{0, \frac{3}{2}, -\frac{3}{2}\right\}$**
d. $6t^3 + t^2 - 5t = 0$ **$\left\{0, \frac{5}{6}, -1\right\}$**

For Example 4
If the flare is launched with an initial upward velocity of 96 feet per second, how long will the flare stay aloft? What will be the maximum height attained by the flare? **6 seconds; 144 feet**

Teaching Tip Point out that the zero product property can be extended to more than two factors.

Check: $m^2 + 144 = 24m$

$$(12)^2 + 144 \stackrel{?}{=} 24(12)$$
$$144 + 144 \stackrel{?}{=} 288$$
$$288 = 288 \checkmark$$

The solution set is $\{12\}$.

You can apply the zero product property to an equation that is written as the product of any number of factors equal to zero.

Example 3 **Solve $5b^3 + 34b^2 = 7b$.**

$5b^3 + 34b^2 = 7b$

$5b^3 + 34b^2 - 7b = 0$ *Arrange the terms so the powers of b are in descending order.*

$b(5b^2 + 34b - 7) = 0$ *Factor the GCF, b.*

$b(5b - 1)(b + 7) = 0$ *Factor $5b^2 + 34b - 7$.*

$b = 0$ or $5b - 1 = 0$ or $b + 7 = 0$

$5b = 1$ $\quad b = -7$

$b = \frac{1}{5}$

The solution set is $\left\{0, \frac{1}{5}, -7\right\}$. *Check this result.*

If an object is launched from ground level, it reaches its maximum height in the air at the time halfway between the launch and impact times. Its height above the ground after t seconds is given by the formula $h = vt - 16t^2$. In this formula, h represents the height of the object in feet, and v represents the object's initial upward velocity in feet per second.

Example 4 **A flare is launched from a life raft with an initial upward velocity of 144 feet per second. How long will the flare stay aloft? What will be the maximum height attained by the flare?**

Rescue Missions

Explore You know the initial upward velocity of the flare is 144 feet per second. You need to determine the length of time the flare will be in the air and how high the flare will go.

Plan The flare will be in the air until the height is 0. Use the general formula $h = vt - 16t^2$ to determine how long the flare will be in the air. The flare will reach its maximum height halfway between the launch and impact times. Use the formula again to determine the height at the middle of its flight.

Solve

$h = vt - 16t^2$

$0 = 144t - 16t^2$ *Replace v with 144.*

$0 = 16t(9 - t)$

$16t = 0$ or $9 - t = 0$

$t = 0$ $\quad 9 = t$

Alternative Learning Styles

Auditory Ask students, "About how large would you expect the product of all integers −100 to 100, inclusive, to be? Take a few minutes to discuss it with your neighbor." The answer is zero since zero is one of the factors. This is a good way to introduce students to the zero product property.

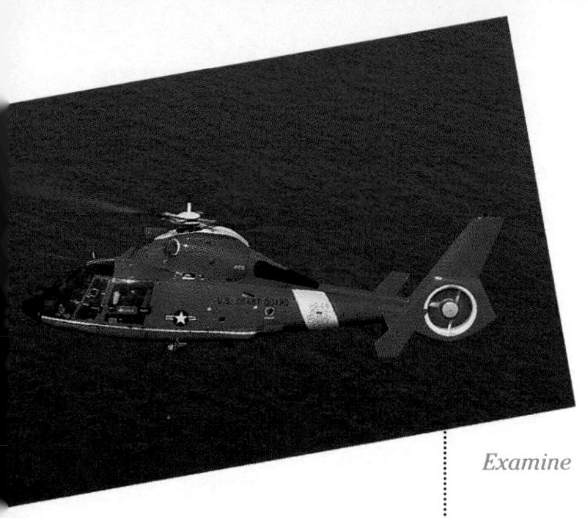

Since 0 seconds is the launch time, the landing time is 9 seconds. The flare will be aloft for 9 seconds. It will reach its maximum height halfway through its flight time at $\frac{1}{2}(9)$ or 4.5 seconds.

$h = 144t - 16t^2$

$= 144(4.5) - 16(4.5)^2$ *Replace t with 4.5.*

$= 648 - 324$

$= 324$

The flare will reach its maximum height of 324 feet after 4.5 seconds.

Examine Check to see if the flare will actually be at a height of 0 feet after 9 seconds.

$0 \stackrel{?}{=} 144(9) - 16(9)^2$

$0 \stackrel{?}{=} 1296 - 1296$

$0 = 0$ ✓

The flare will be aloft for 9 seconds and the maximum height will be reached after 4.5 seconds.

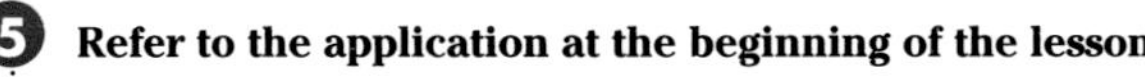

Example 5

Diving

Refer to the application at the beginning of the lesson.

a. What is the diver's maximum height?

b. When will the diver enter the water?

a. The diver will reach the maximum height after $\frac{1}{4}$ second.

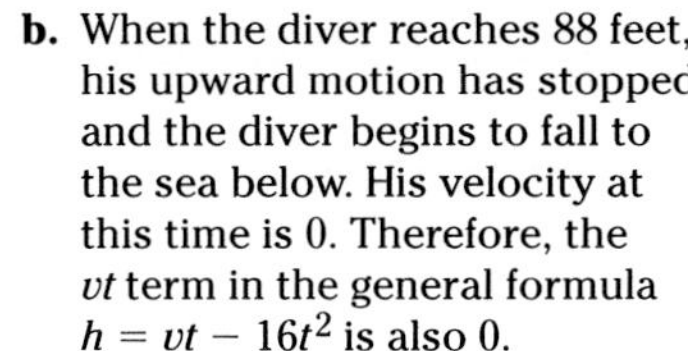

$h = 87 + 8t - 16t^2$

$= 87 + 8\left(\frac{1}{4}\right) - 16\left(\frac{1}{4}\right)^2$

$= 87 + 2 - 1$

$= 88$

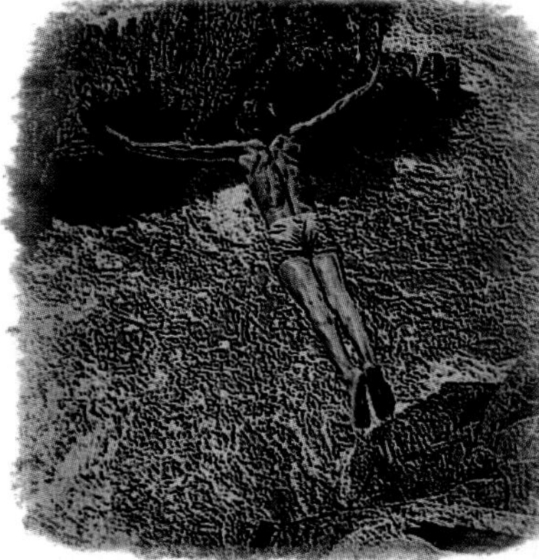

The diver's maximum height is 88 feet. *Does this seem reasonable?*

b. When the diver reaches 88 feet, his upward motion has stopped and the diver begins to fall to the sea below. His velocity at this time is 0. Therefore, the vt term in the general formula $h = vt - 16t^2$ is also 0.

$88 = 16t^2$ *Why does 88 equal $16t^2$ instead of $-16t^2$?*

$5.5 = t^2$

$2.35 \approx t$

It takes the diver $\frac{1}{4}$ or 0.25 second to reach the maximum height and about another 2.35 seconds to reach the water. The diver will reach the water about 0.25 + 2.35 or 2.60 seconds after he starts his dive.

In-Class Example

For Example 5

Solve for h if t is $\frac{1}{2}$.

$h = 43 + 8t - 16t^2$ **43**

Teaching Tip Explain to students that velocity is positive when an object is going up and negative when it is going down. Velocity is zero when it changes direction from up to down.

***GLENCOE** Technology*

Interactive Mathematics Tools Software

This multimedia software provides an interactive lesson for solving an equation by factoring using the formula for the height of a falling object, $D = -16t^2 + vt + h$. A **Computer Journal** gives students an opportunity to write about what they have learned.

For Windows & Macintosh

3 PRACTICE/APPLY

Check for Understanding

Exercises 1–12 are designed to help you assess your students' understanding through reading, writing, speaking, and modeling. You should work through Exercises 1–4 with your students and then monitor their work on Exercises 5–12.

Error Analysis

Emphasize that an expression must be set equal to zero before the zero product property can be applied. Students may attempt to solve an equation such as $x(x - 4) = 3$ by saying that either $x = 3$ or $x - 4 = 3$. Neither 3 nor 7 is a correct solution.

Assignment Guide

Core: 13–41 odd, 42–48
Enriched: 14–32 even, 34–48

For **Extra Practice,** see p. 780.

The red A, B, and C flags, printed only in the Teacher's Wraparound Edition, indicate the level of difficulty of the exercises.

Study Guide Masters, p. 74

NAME_______________ DATE _______

Study Guide

Student Edition Pages 594–600

Solving Equations by Factoring

Factoring can be used to solve many kinds of problems.

Example: A rocket is fired with an initial velocity of 2288 feet per second. How many seconds will it take for the rocket to hit the ground?

Explore Use the formula $h = vt - 16t^2$.

Plan Substitute the appropriate values into the formula.

Solve $0 = 2288t - 16t^2$
$16t = 0$ or $143 - t = 0$ Zero product property
$t = 0$ $143 = t$

Examine An answer of 0 seconds is not a reasonable answer, so use only the value 143. The rocket returns to the ground in 143 seconds.

Solve each equation. Check your solutions.

1. $n^2 - 16 = 0$ $\{-4, 4\}$
2. $x^2 + 10x + 25 = 0$ $\{-5\}$
3. $9x^2 + x = 0$ $\left\{-\frac{1}{9}, 0\right\}$
4. $x^2 = 24 - 10x$ $\{-12, 2\}$
5. $x^3 - 18x = 7x^2$ $\{-2, 0, 9\}$
6. $x^3 - 36x = 9x^2$ $\{-3, 0, 12\}$

For each problem below, define a variable. Then use an equation to solve the problem. Disregard any unreasonable solutions.

7. The difference of the squares of two consecutive odd integers is 24. Find the integers. −5, −7
8. The length of a Charlotte, North Carolina, conservatory garden is 20 yards greater than its width. The area is 300 square yards. What are the dimensions? 30 yards by 10 yards

Use the formula $h = vt - 16t^2$ to solve each problem.

9. A punter can kick a football with an initial velocity of 48 feet per second. How many seconds will it take the ball to return to the ground? 3 seconds
10. If a rocket is launched at Houston, Texas, with an initial velocity of 1600 feet per second, when will the rocket be 14,400 feet high? at 10 seconds and 90 seconds

CHECK FOR UNDERSTANDING

Communicating Mathematics

Study the lesson. Then complete the following.

1. If the product of two or more factors is zero, what must be true of the factors? **At least one of the factors equals 0.**
2. **Describe** the type of equation that can be solved by using the zero-product property. **equations that can be written as a product of factors that equal 0**
3. Can $(x + 3)(x - 5) = 0$ be solved by dividing each side of the equation by $x + 3$? Explain. **No; the division would eliminate −3 as a solution.**
4. **You Decide** Diana says that if $(x + 2)(x - 3) = 8$, then $x + 2 = 8$ or $x - 3 = 8$. Caitlin disagrees. Who is correct? **Caitlin; the zero product property only works for 0. If you multiply two numbers and get 8, that does not mean one of the numbers must have been 8.**

Guided Practice

Solve each equation. Check your solutions. **6. $\{-2, 4\}$** **7. $\left\{0, \frac{5}{3}\right\}$**

5. $g(g + 5) = 0$ **$\{0, -5\}$**
6. $(n - 4)(n + 2) = 0$
7. $5m = 3m^2$
8. $x^2 = 5x + 14$ **$\{-2, 7\}$**
9. $7r^2 = 70r - 175$ **$\{5\}$**
10. $a^3 - 29a^2 = -28a$ **$\{0, 1, 28\}$**
11. **Geometry** The dimensions of a rectangle are $(2x + 9)$ inches and $(2x - 1)$ inches. A square with side x inches is cut out of one of the corners. If the remaining area is 195 square inches, what is x? **6**

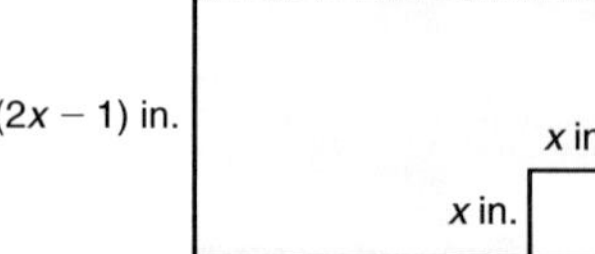

12. **Baseball** Nolan Ryan, the greatest strike-out pitcher in the history of baseball, had a fastball clocked at 103 miles per hour, which is 151 feet per second. **a. about 9.44 seconds**
    a. If he threw the ball directly upward with the same velocity, how many seconds would it take for the ball to return to his glove? (*Hint:* Use the general formula $h = vt - 16t^2$.)
    b. How high above his glove would the ball travel? **about 356 feet**

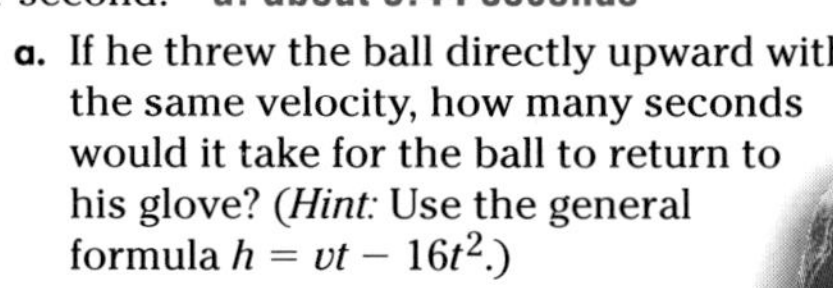

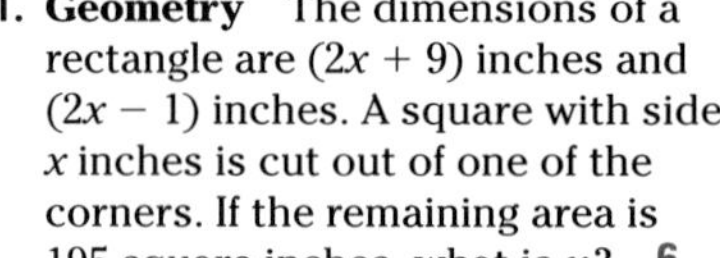

First professional pitchers to throw a perfect game
1. Lee Pichmond, June 12, 1880
2. Monte Ward, June 17, 1880
3. Cy Young, May 5, 1904
4. Adrian Joss, Oct. 2, 1908
5. Charlie Robertson, Apr. 30, 1922

EXERCISES

Practice

Solve each equation. Check your solutions.

15. $\left\{\frac{3}{2}, \frac{8}{3}\right\}$ **16.** $\left\{-\frac{5}{4}, \frac{7}{3}\right\}$ **20.** $\left\{0, \frac{5}{2}\right\}$ **23.** $\left\{-\frac{1}{3}, -\frac{5}{2}\right\}$ **24.** $\left\{-7, 0, \frac{1}{5}\right\}$ **30.** $\left\{-4, \frac{2}{3}\right\}$

A

13. $x(x - 24) = 0$ **$\{0, 24\}$**
14. $(q + 4)(3q - 15) = 0$ **$\{-4, 5\}$**
15. $(2x - 3)(3x - 8) = 0$
16. $(4a + 5)(3a - 7) = 0$

17. $a^2 + 13a + 36 = 0$ **$\{-9, -4\}$**
18. $x^2 - x - 56 = 0$ **$\{-7, 8\}$**
19. $y^2 - 64 = 0$ **$\{-8, 8\}$**
20. $5s - 2s^2 = 0$
21. $3z^2 = 12z$ **$\{0, 4\}$**
22. $m^2 - 24m = -144$ **$\{12\}$**
23. $6q^2 + 5 = -17q$
24. $5b^3 + 34b^2 = 7b$

C

25. $\frac{x^2}{12} - \frac{2x}{3} - 4 = 0$ **$\{12, -4\}$**
26. $t^2 - \frac{t}{6} = \frac{35}{6}$ **$\left\{-\frac{7}{3}, \frac{5}{2}\right\}$**
27. $n^3 - 81n = 0$ **$\{-9, 0, 9\}$**
28. $(x + 8)(x + 1) = -12$ **$\{-4, -5\}$**
29. $(r - 1)(r - 1) = 36$ **$\{-5, 7\}$**
30. $(3y + 2)(y + 3) = y + 14$

Reteaching

Logical Thinking Abe adds up the number of runs that the Reds score in each game for the season. Bob multiplies together the number of runs that the Mets score in each game. Who will probably have the larger number at the end of the season? Why? **Abe; think of the effects of a shutout.**

There have been a total of 13 pitchers who have pitched perfect games. The most recent is Kenny Rogers on July 28, 1994.

31. **Number Theory** Find two consecutive even integers whose product is 168. **−14 and −12 or 12 and 14**

32. **Number Theory** Find two consecutive odd integers whose product is 1023. **−33 and −31 or 31 and 33**

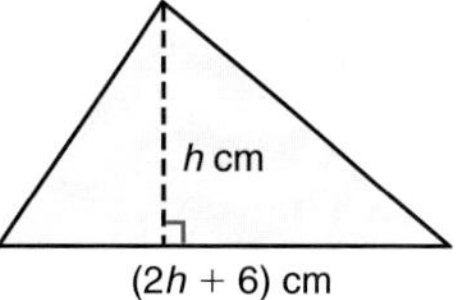

33. **Geometry** The triangle at the right has an area of 40 square centimeters. Find the height h of the triangle. **5 cm**

Critical Thinking

34. Write an equation with integral coefficients that has $\{-3, 0, 7\}$ as its solution set. $x^3 - 4x^2 - 21x = 0$

35. Consider the equations $a^2 + 5a = 6$ and $|2x + 5| = 7$.
    a. Solve each equation. **$\{-6, 1\}$; $\{-6, 1\}$**
    b. What is the relationship between the two equations? What does that mean? **They are equivalent; they have the same solutions.**

Applications and Problem Solving

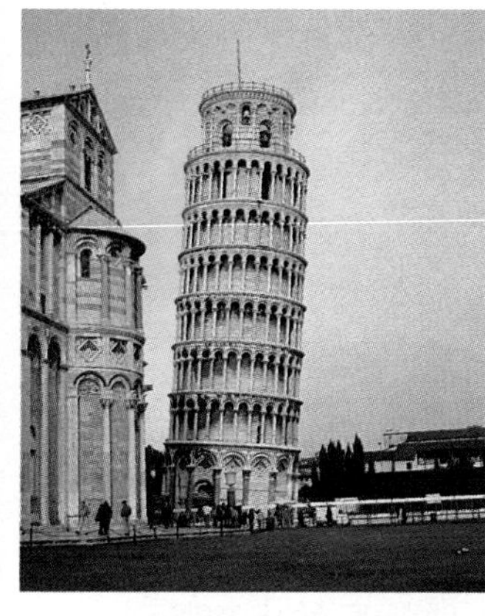

36. **Gardening** LaKeesha has enough bricks to make a 30-foot-long border around the rectangular vegetable garden she is planning. The booklet she got from the plant nursery when she bought the seeds says the plants will need space to grow, and it advises that the seeds should be planted in an area of 54 square feet. What should the dimensions of her garden be? **9 ft by 6 ft**

37. **History** During the late 16th century, Galileo was a professor at the University of Pisa, Italy. During this time, he demonstrated that objects of different weights fell at the same velocity by dropping two objects of different weights from the top of the Leaning Tower of Pisa.
    a. If he dropped the objects from a height of 180 feet, how long did it take them to hit the ground? **about 3.35 s**
    b. Research Galileo's life. Why was he criticized for experimenting with these weights? What other belief of his led to his trial by the Inquisition of 1633? **See margin.**

38. **Fountains** The tallest fountain in the world is at Fountain Hills, Arizona. When all three pumps are working, the nozzle speed of the water is 146.7 miles per hour (215.16 feet per second). It is claimed the water reaches a height of 625 feet. Do you agree or disagree with this claim? Explain. **See margin.**

FYI

Superman first appeared in a comic book in 1938 and in a newspaper strip in 1939. Jerry Seigel wrote these comics and Joe Shuster provided the artwork.

39. **Superheroes** A meteorite is headed towards Metropolis when Superman intercepts it. He takes it to the top of the *Daily Planet* (180 feet high) and tosses the meteorite into space with an upward velocity of 2400 feet per second. What height will the meteorite attain before returning to Earth? **about 90,180 ft or 17 mi**

40. **Bridges** The chart at the right compares the highest bridge in the world at Royal Gorge, Colorado with the highest railroad bridge in the world at Kolasin, Yugoslavia. If Elva accidentally drops her keys from the Royal Gorge Bridge, will the keys hit the Arkansas River below within 8 seconds? **yes**

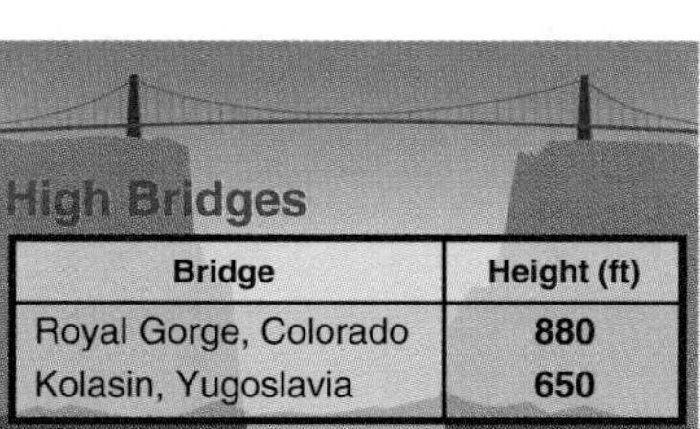

High Bridges

Bridge	Height (ft)
Royal Gorge, Colorado	880
Kolasin, Yugoslavia	650

FYI

The partnership between Siegel and Shuster began at Glenville High School in Cleveland, Ohio. The two teenagers used the high school's mimeograph machine to publish their first science fiction magazine.

Additional Answers

37b. His ideas about falling objects differed from what most people thought to be true. He believed that Earth is a moving planet and that the sun and planets did not revolve around Earth.

38. Sample answer: Agree, since a nozzle speed of 215.16 ft/s should launch an object to a height of about 723 ft in a vacuum. Because of wind resistance, it may only go to 625 feet.

Practice Masters, p. 74

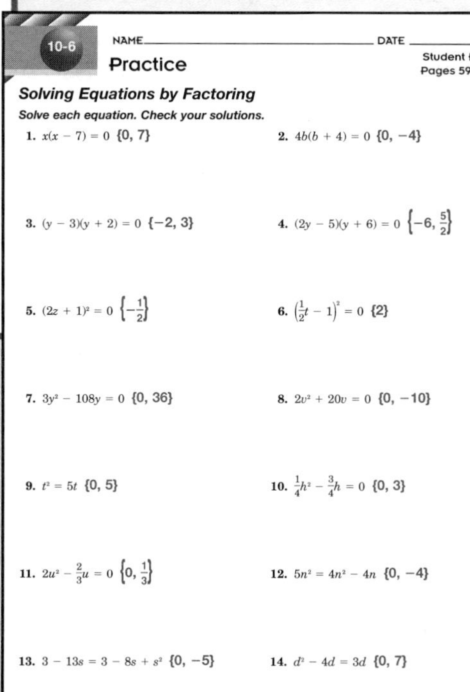

10-6 NAME______ DATE______

Practice — Student Edition Pages 594–600

Solving Equations by Factoring

Solve each equation. Check your solutions.

1. $x(x - 7) = 0$ $\{0, 7\}$
2. $4b(b + 4) = 0$ $\{0, -4\}$
3. $(y - 3)(y + 2) = 0$ $\{-2, 3\}$
4. $(2y - 5)(y + 6) = 0$ $\{-6, \frac{5}{2}\}$
5. $(2z + 1)^2 = 0$ $\{-\frac{1}{2}\}$
6. $(\frac{1}{2}t - 1)^2 = 0$ $\{2\}$
7. $3y^2 - 108y = 0$ $\{0, 36\}$
8. $2v^2 + 20v = 0$ $\{0, -10\}$
9. $t^2 = 5t$ $\{0, 5\}$
10. $\frac{1}{4}h^2 - \frac{3}{4}h = 0$ $\{0, 3\}$
11. $2u^2 - \frac{2}{3}u = 0$ $\{0, \frac{1}{3}\}$
12. $5n^2 = 4n^2 - 4n$ $\{0, -4\}$
13. $3 - 13s = 3 - 8s + s^2$ $\{0, -5\}$
14. $d^2 - 4d = 3d$ $\{0, 7\}$

4 ASSESS

Closing Activity

Writing Have students write about the advantage of factoring an equation in order to solve it.

Chapter 10, Quiz D (Lesson 10-6), is available in the *Assessment and Evaluation Masters,* p. 269.

41. **Oil Leases** A drilling company has the oil rights for a rectangular piece of land that is 6 kilometers by 5 kilometers. According to a lease agreement, the company can only dig their wells in the inner two-thirds of the land. A uniform strip around the edge of the land must remain untouched. What is the width of the strip of land that must remain untouched? **0.5 km**

Mixed Review

44. $\left\{x \mid x < \frac{3}{22}\right\}$
45. 14; −4

42. Factor $100x^2 + 20x + 1$. (Lesson 10–5) $(10x + 1)^2$
43. Find $(5q + 2r)(8q - 3r)$. (Lesson 9–7) $40q^2 + rq - 6r^2$
44. Solve $9x + 4 < 7 - 13x$. (Lesson 7–3)
45. Determine the x- and y-intercepts of the graph of $2x - 7y = 28$. (Lesson 6–4)
46. **Patterns** Copy and complete the table below. (Lesson 5–5)

s	4	2	0	−2	−4
$r(s)$	19	11	3	−5	−13

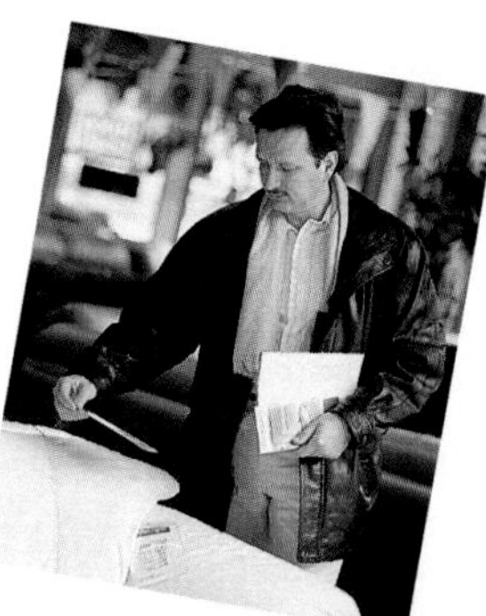

47. **World Records** The Huey P. Long Bridge in Metairie, Louisiana, is the longest railroad bridge in the world. If you were traveling on a train going 60 miles per hour across the 22,996-foot bridge, how long would it take you to cross it? (Lesson 4–7) **4.355 minutes or about 4 minutes 21 seconds**
48. **Consumerism** Julio is offered two payment plans when he buys a sofa. Under one plan, he pays $400 down and x dollars per month for 9 months. Under the other plan, he pays no money down and $x + 25$ dollars per month for 12 months. How much does the sofa cost? (Lesson 3–5) **$700**

WORKING ON THE

Refer to the Investigation on pages 554–555.

the BRICKYARD

The perimeter P of a square is given by the formula $P = 4s$, where s is the length of the side of the square. The formula for the area A of a square is $A = s^2$. The perimeter of a rectangle is found using $P = 2\ell + 2w$, where ℓ is the length of the rectangle and w is the width. The formula for the area of the rectangle is $A = \ell w$.

1. Suppose the large square brick has a side x units long and the small square brick has a side y units long. What are the dimensions of the rectangular brick? What is the area of each of the bricks? Draw a diagram to illustrate the dimensions of each brick.
2. Refer to the drawings of all the rectangular brick patterns you found using 4, 5, 6, 7, 8, 9, and 10 bricks. Calculate the perimeter and area in terms of x and y of all of the patterns you discovered.
3. Create a table to record the linear dimensions, the perimeter, and the area of the brick patterns you found. Use the table to record the dimensions, perimeter, and area in terms of x and y. The table should be set up like the one below.

Add the results of your work to your Investigation Folder.

# of bricks in pattern	length of pattern	width of pattern	perimeter of pattern	area of pattern
4				
5				
6				

Extension

Connections Two circles have radii that are consecutive numbers. Find their radii if their areas differ by 37π square feet. **18 and 19 feet**

In•ves•ti•ga•tion

Working on the Investigation

The Investigation on pages 554–555 is designed to be a long-term project that is completed over several days or weeks. Encourage students to keep their materials in their Investigation Folder as they work on the Investigation.

Enrichment Masters, p. 74

10-6 NAME ________________ DATE ________

Enrichment — Student Edition Pages 592–600

Using Factoring

The right side of each formula is a fraction. Factor the numerator and denominator of each fraction. Then reduce the fraction to simplest terms if possible. Use the result to answer the questions about each formula.

1. $t = \frac{\pi(d + r)^2w - \pi r^2w}{hws}$ where t = the approximate number of minutes of playing time remaining on a tape
   d = the depth of the tape in inches
   r = the radius of the core in inches
   w = the width of the tape in inches
   h = the thickness of the tape in inches
   and s = the speed at which the tape is played in inches per minute

   Find the approximate number of minutes left on a tape where $d = \frac{5}{8}$, $r = \frac{7}{16}$, $w = \frac{3}{32}$, $h = 0.0015$, $s = 100$, and $\pi = 3.14$.
   19.6 minutes

2. $R = \frac{M^2v^2 + 2eM^2v^2 + e^2M^2v^2}{2gM^2 + 4gMm + 2gm^2}$ where m = the mass of a golf ball in grams
   M = the mass of a golf-club head in grams
   v = the velocity of the club head in meters per second
   e = the coefficient of restitution between the ball and the club head
   g = the acceleration due to gravity (9.8 meters per second per second)
   and R = the maximum range of the golf ball in meters

   The value e, coefficient of restitution, is a measure of how elastic a golf ball is. As the temperature increases, the ball becomes more elastic. At 32°F, the value of e is about 0.64. At 80°F, the value of e is about 0.75. The mass of a golf ball is about 50 grams. Assume the mass of the club head is 250 grams and the velocity of the club head is 45 meters/second.

   a. Find the maximum range at 32°F in meters and convert it to yards (1 m = 1.093 yds). Round your answers to the nearest whole number. 211 yd
   b. Find the maximum range at 80°F in meters and yards to the nearest whole number. 219.72 m; 240 yd
   c. Why might a golfer want to buy a battery-heated glove to keep spare golf balls warm? Increasing the temperature of the ball appears to increase the distance the ball can travel.

CHAPTER 10 HIGHLIGHTS

VOCABULARY

After completing this chapter, you should be able to define each term, property, or phrase and give an example or two of each.

Algebra

composite numbers (p. 558)
difference of squares (p. 581)
factored form (p. 559)
factoring or factored (pp. 564, 565)
factoring by grouping (p. 567)
greatest common factor (GCF) (p. 559)
perfect square trinomials (p. 587)
prime factorization (p. 558)
prime numbers (p. 558)
prime polynomial (p. 577)
Pythagorean triple (p. 583)
unique factorization theorem (p. 558)
zero product property (p. 594)

Problem Solving

guess and check (p. 574)

UNDERSTANDING AND USING THE VOCABULARY

State whether each sentence is *true* or *false*. If false, replace the underlined word or number to make a true sentence.

1. The number 27 is an example of a <u>prime</u> number. **false, composite**
2. <u>$2x$</u> is the greatest common factor (GCF) of $12x^2$ and $14xy$. **true**
3. <u>66</u> is an example of a perfect square. **false, sample answer: 64**
4. 61 is a <u>factor</u> of 183. **true**
5. The prime factorization for 48 is <u>$3 \cdot 4^2$</u>. **false, $2^4 \cdot 3$**
6. $x^2 - 25$ is an example of a <u>perfect square trinomial</u>. **false, difference of squares**
7. The number 35 is an example of a <u>composite</u> number. **true**
8. <u>$x^2 - 3x - 70$</u> is an example of a prime polynomial. **false; sample answer: $x^2 + 2x + 2$**
9. The <u>unique factorization theorem</u> allows you to solve equations. **false, zero product property**
10. <u>$(b - 7)(b + 7)$</u> is the factorization of a difference of squares. **true**

Instructional Resources

Three multiple-choice tests and three free-response tests are provided in the *Assessment and Evaluation Masters.* Forms 1A and 2A are for honors pacing, and Forms 1B, 1C, 2B, and 2C are for average pacing. Chapter 10 Test, Form 1B is shown at the right. Chapter 10 Test, Form 2B is shown on the next page.

Using the CHAPTER HIGHLIGHTS

The Chapter Highlights begins with a listing of the new terms, properties, and phrases that were introduced in this chapter. Have students define each term and provide an example or two of it, if appropriate.

Assessment and Evaluation Masters, pp. 255–256

NAME ______ DATE ______

Chapter 10 Test, Form 1B

Write the letter for the correct answer in the blank at the right of each problem.

1. What is the prime factorization of 342? 1. C
   A. $2 \cdot 3 \cdot 57$ B. $1 \cdot 2 \cdot 3 \cdot 57$ C. $2 \cdot 3 \cdot 3 \cdot 19$ D. none of these
2. Find the GCF of $18abc^3$ and $54ab^2$. 2. C
   A. $18ab^2c^3$ B. $9ab$ C. $18ab$ D. $432a^2b^3c^3$
3. Find the GCF of $3ab$, $4bc$, and $5ac$. 3. B
   A. 0 B. 1 C. $2c$ D. $60a^2b^2c^2$
4. Factor $3x^2 + 75$. 4. B
   A. $3(x + 5)(x + 5)$ B. $3(x^2 + 25)$
   C. $3(x - 5)(x - 5)$ D. none of these
5. Factor $6y(y^2 - 6) + 5(6 - y^2)$. 5. A
   A. $(6y - 5)(y^2 - 6)$ B. $(6y + 5)(6 - y^2)$
   C. $(6y + 5)(y^2 - 6)$ D. none of these
6. If $x^3 - 7x^2 + 4x - 28$ is factored completely, one of the factors is 6. D
   A. $x + 2$. B. $x + 7$. C. $x^2 + 7$. D. $x^2 + 4$.
7. Factor $5x^2 - 13x + 6$. 7. D
   A. $(x + 3)(5x - 2)$ B. $(x - 3)(5x + 2)$
   C. $(x + 2)(5x + 3)$ D. $(x - 2)(5x - 3)$
8. If $14a^2 - 15a + 4$ is factored completely, one of the factors is 8. B
   A. $7a + 2$. B. $7a - 4$. C. $7a - 1$. D. $14a - 1$.
9. If $6x^2 + 11x - 2$ is factored completely, one of the factors is 9. B
   A. $3x - 1$. B. $6x - 1$. C. $2x + 1$. D. $3x - 2$.
10. Which polynomial is prime? 10. C
    A. $x^2 + 13x - 114$ B. $n^2 - 2ny - 195y^2$
    C. $x^2 + 2x - 18$ D. More than one of these is prime.
11. Factor $100v^2 - 64y^2$. 11. A
    A. $(10v - 8y)(10v + 8y)$ B. $(10v - 8y)(10v - 8y)$
    C. $(10v + 8y)(10v + 8y)$ D. none of these

NAME ______ DATE ______

Chapter 10 Test, Form 1B (continued)

12. Factor $n^4 - 5n^2 - 36$. 12. D
    A. $(n + 2)(n - 2)(n + 3)(n - 3)$ B. $(n^2 + 9)(n + 2)(n - 2)$
    C. $(n^2 - 4)(n^2 + 9)$ D. none of these
13. How much less than 1600 is the product of 38 and 42? 13. C
    A. 1 B. 2 C. 4 D. 16
14. Each of the following is a perfect square except 14. C
    A. $4x^2 + 4x + 1$. B. $x^2 - 6x + 9$.
    C. $x^2 + 10x - 25$. D. All are perfect squares.
15. If $6x^2 + 48x + 96$ is factored completely, one of the factors is 15. A
    A. $x + 4$. B. $3x + 8$. C. $3x + 12$. D. $6x + 16$.
16. Find the value of c that makes $25y^2 - 40y + c$ a perfect square trinomial. 16. D
    A. 64 B. 25 C. 20 D. 16
17. What is the solution set of the equation $(3w + 4)(2w - 7) = 0$? 17. C
    A. $\left\{-\frac{3}{4}, \frac{2}{7}\right\}$ B. $\left\{\frac{3}{4}, -\frac{2}{7}\right\}$ C. $\left\{-\frac{4}{3}, \frac{7}{2}\right\}$ D. $\left\{\frac{4}{3}, -\frac{7}{2}\right\}$
18. What is the solution set of the equation $4x^2 = 20x - 25$? 18. D
    A. $\left\{0, \frac{5}{4}\right\}$ B. $\left\{0, -\frac{2}{5}\right\}$ C. $\left\{-\frac{4}{5}\right\}$ D. $\left\{\frac{5}{2}\right\}$
19. What is the solution set of the equation $x^2 - 16x = 0$? 19. C
    A. $\{4, -4\}$ B. $\{0, 4, -4\}$ C. $\{0, 16\}$ D. $\{16\}$
20. The width of a rectangle is 6 centimeters less than the length. The area of the rectangle is 112 square centimeters. What is the length? 20. D
    A. 16 cm B. 31 cm C. 7 cm D. 14 cm

Bonus Insert parentheses so that a true equation results. Bonus $7 + 3 \cdot 5 - (6 - 2) \cdot 4 = 6$
$7 + 3 \cdot 5 - 6 - 2 \cdot 4 = 6$

Applications and Problem Solving Encourage students to work through the exercises in the Applications and Problem Solving section to strengthen their problem-solving skills.

OBJECTIVES AND EXAMPLES

- use the zero product property to solve equations (Lesson 10–6)

Solve $b^2 - b - 12 = 0$.

$b^2 - b - 12 = 0$

$(b - 4)(b + 3) = 0$

If $(b - 4)(b + 3) = 0$, then $(b - 4) = 0$ or $(b + 3) = 0$.

$b - 4 = 0$ or $b + 3 = 0$

$b = 4$ $b = -3$

The solution set is $\{4, -3\}$.

REVIEW EXERCISES

Solve each equation. Check your solution.

56. $y(y + 11) = 0$ $\{0, -11\}$

57. $(3x - 2)(4x + 7) = 0$ 57. $\left\{\frac{2}{3}, -\frac{7}{4}\right\}$

58. $2a^2 - 9a = 0$ 58. $\left\{0, \frac{9}{2}\right\}$

59. $n^2 = -17n$ $\{0, -17\}$

60. $\frac{3}{4}y = \frac{1}{2}y^2$ $\left\{0, \frac{3}{2}\right\}$

61. $y^2 + 13y + 40 = 0$ $\{-5, -8\}$

62. $2m^2 + 13m = 24$ 62. $\left\{-8, \frac{3}{2}\right\}$

63. $25r^2 + 4 = -20r$ 63. $\left\{-\frac{2}{5}\right\}$

APPLICATIONS AND PROBLEM SOLVING

64. Geometry The measure of the area of a rectangle is $4m^2 - 3mp + 3p - 4m$. If the dimensions of the rectangle are represented by polynomials with integral coefficients, find the dimensions of the rectangle. (Lesson 10–2) $(4m - 3p)$ by $(m - 1)$

65. Guess and Check Numero Uno says, "I am thinking of a three-digit number. If you multiply the digits together and then multiply the result by 4, the answer is the number I'm thinking of. What is my number?" (Lesson 10–3) 384

66. Photography To get a square photograph to fit into a rectangular frame, Li-Chih had to trim a 1-inch strip from one pair of opposite sides of the photo and a 2-inch strip from the other two sides. In all, he trimmed off 64 square inches. What were the original dimensions of the photograph? (Lesson 10–4) 12 in. by 12 in.

67. Guess and Check Fill in each box below with a digit from 1 to 6 to make this multiplication work. Use each digit exactly once. (Lesson 10–3)

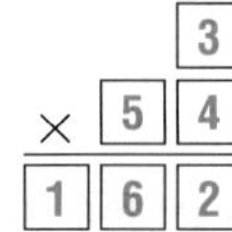

68. Geometry The measure of the area of a rectangle is $16x^2 - 9$. Find the measure of its perimeter. (Lesson 10–4) $16x$

69. Number Theory The product of two consecutive odd integers is 99. Find the integers. (Lesson 10–6) 9, 11; −11, −9

A practice test for Chapter 10 is provided on page 796.

ALTERNATIVE ASSESSMENT

COOPERATIVE LEARNING PROJECT

Art and Framing In this project, you will determine what size mat and frame will be the most visually stimulating for an abstract print that you just bought on your trip to The Art Institute of Chicago. The print is 9 inches by 12 inches and can be cropped an inch on either side and not spoil its effect.

You called a friend to get some advice on how to mat and frame your print. This friend majored in art and works in the art industry, but he also enjoys puzzling you. He gave you two problems and told you to solve them and decide which you think would be the most appropriate dimensions for your mat and frame. Here are the two problems.

- The inside rectangle has a width four inches less than its length. The length of the larger rectangle is twice its width. Find the dimensions of each rectangle if the matted area is twice as much as the inside rectangle area and the perimeter of the inside rectangle is 32 inches less than the perimeter of the larger rectangle.
- The inside rectangle has a width four inches less than its length. The length of the larger rectangle is twice its width. Find the dimensions of each rectangle if the matted area is seven times as much in square inches as the perimeter of the smaller rectangle is in inches, and the width of the larger rectangle is two inches more than the length of the smaller rectangle.

Which problem will give you the dimensions that will best suit your print? What are the dimensions? Will you have to crop your print? If so, how much?

Follow these steps to determine the appropriate dimensions for the mat and frame.

- Illustrate each of the problems.
- Label each of the drawings with as much information as possible.
- Develop an equation for each problem that describes the situation.
- Investigate your answers and determine the appropriate dimensions for the mat and frame.
- If possible, find a 9-inch by 12-inch photograph or advertisement from a magazine. Have two color photocopies made of it and frame each with a cardboard frame that meets the specifications.
- Write a paragraph describing the two problems and how you determined the solution.

THINKING CRITICALLY

- A *nasty* number is a positive integer with at least four different factors such that the difference between the numbers in one pair of factors equals the sum of the numbers in another pair. The first nasty number is 6 since $6 = 6 \cdot 1 = 2 \cdot 3$ and $6 - 1 = 2 + 3$. Find the next five nasty numbers. (*Hint*: They are all multiples of 6.)
- Write an equation with integral coefficients that has $\left\{\frac{2}{3}, -1\right\}$ as the solution.

PORTFOLIO

When using the guess-and-check strategy for solving a math problem, if a first guess does not work, you must make a second guess. How do you make a second guess when the first one is incorrect? In making this second guess, you must know how and why the first guess was incorrect. Find a problem from your work in this chapter in which you used the guess-and-check strategy. Write a step-by-step description of why you chose each successive guess after the previous guess for that problem. Place this in your portfolio.

SELF EVALUATION

Having a strategy before solving a problem is helpful. It helps you to focus on the problem and the solution and to be organized in the solution of your problem. Using strategies or check lists are fundamental skills used in critical thinking.

Assess yourself. Do you determine a strategy for solving a problem before you delve into it? Do you stay with the initial plan or do you reevaluate after a time period and try a new strategy? Do you keep a record of your strategy and attempts so that you can refer back to it to determine if you are still on track? Give an example of a math problem in which a strategy is beneficial in order to keep track of what has been attempted. Also, give an example of a daily life problem in which you used a strategy to help organize your plan for a solution.

Assessment and Evaluation Masters, pp. 266, 277

10 NAME________ DATE ________

Chapter 10 Performance Assessment

Instructions: ***Demonstrate your knowledge by giving a clear, concise solution to each problem. Be sure to include all relevant drawings and justify your answers. You may show your solution in more than one way or investigate beyond the requirements of the problem.***

1. One way to factor a trinomial such as $x^2 - 2x - 3$ is to assign a value such as 10 to x and evaluate the expression.

$$\begin{aligned} x^2 - 2x - 3 &= 10^2 - 2(10) - 3 \\ &= 100 - 20 - 3 \\ &= 77 \end{aligned}$$

A factorization of 77 is 7×11. Since $x = 10$, $7 = (x - 3)$ and $11 = (x + 1)$. Multiply to see if $(x - 3)(x + 1) = x^2 - 2x - 3$.

   a. Try the ab[illegible] method to factor $x^2 - 8x + 15$. Show your work and explain each step. Did the method give you the correct factors?

   b. Try the above method to factor $2x^2 - 13x - 24$. *Hint: The factors will be of the form* $(2x + a)(x + b)$. Show your work and explain each step.

   c. Try the above method to factor $x^2 - 2x - 8$. Why is it difficult to find the correct factors for this trinomial using the above method?

   d. Evaluate the expression in part c for $x = 7$. Then use the above method to find the factors. Explain each step. Remember in finding your factors that $x = 7$.

2. a. Tell how to factor a polynomial with four terms by grouping.

   b. Use grouping to factor $10ax - 5ay + 2bx - by$. Explain each step.

   c. Factor $6x^3 + 3x^2y - 2xy - y^2$ in two different ways. Show your work.

Scoring Guide
Chapter 10
Performance Assessment

Level	Specific Criteria
3 Superior	• Shows thorough understanding of the concepts of *prime factorization of integers, factorization of polynomials, evaluating polynomials,* and *the guess-and-check strategy for problem solving.* • Computations are correct. • Written explanations are exemplary. • Goes beyond requirements of some or all problems.
2 Satisfactory, with Minor Flaws	• Shows understanding of the concepts of *prime factorization of integers, factorization of polynomials, evaluating polynomials,* and *the guess-and-check strategy for problem solving.* • Computations are mostly correct. • Written explanations are effective. • Satisfies all requirements of problems.
1 Nearly Satisfactory, with Serious Flaws	• Shows understanding of most of the concepts of *prime factorization of integers, factorization of polynomials, evaluating polynomials,* and *the guess-and-check strategy for problem solving.* • Computations are mostly correct. • Written explanations are satisfactory. • Satisfies most requirements of problems.
0 Unsatisfactory	• Shows little or no understanding of the concepts of *prime factorization of integers, factorization of polynomials, evaluating polynomials,* and *the guess-and-check strategy for problem solving.* • Computations are incorrect. • Written explanations are not satisfactory. • Does not satisfy requirements of problems.

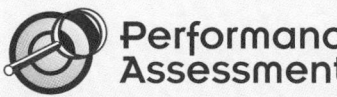

Alternative Assessment

The Alternative Assessment section provides students with the opportunity to assess their own work by thinking critically, working with others, keeping a portfolio, and honestly evaluating their own progress. For more information on alternative forms of assessment, see *Alternative Assessment in the Mathematics Classroom,* one of the titles in the Glencoe Mathematics Professional Series.

Performance Assessment

Performance Assessment tasks for this chapter are included in the *Assessment and Evaluation Masters.* A scoring guide is also provided.

Using the CUMULATIVE REVIEW

These two pages review the skills and concepts presented in Chapters 1–10. This review is formatted to reflect new trends in standardized testing.

A more traditional cumulative review, shown below, is provided in the *Assessment and Evaluation Masters,* pp. 271–272.

Assessment and Evaluation Masters, pp. 271–272

10 NAME________ DATE________

Chapter 10 Cumulative Review

1. Find $-\frac{3}{8} + \frac{5}{16}$. (Lesson 2-5) 1. $-\frac{1}{16}$
2. Simplify $\frac{2}{3}(6a - 9b) - \frac{1}{2}(2a - 12b)$. (Lesson 2-6) 2. $3a$
3. Solve $\frac{3y - 2}{5} = \frac{1}{10}y$. (Lesson 3-5) 3. $\frac{4}{5}$
4. Find two consecutive even integers whose sum is 74. (Lesson 3-3) 4. 36 and 38
5. Two airplanes leave Atlanta at the same time and fly in opposite directions. One plane travels 60 mi/h faster than the other. After $2\frac{1}{2}$ hours they are 1700 miles apart. What is the rate of each plane? (Lesson 4-7) 5. 310 mi/h; 370 mi/h

Find the probability of each outcome if two marbles are selected from a bag containing 3 red marbles, 4 blue marbles, and 2 green marbles. The first marble is not replaced before the second marble is selected. (Lesson 4-6)

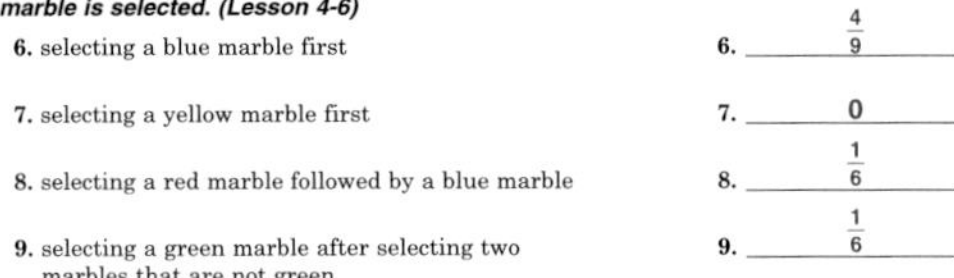

6. selecting a blue marble first 6. $\frac{4}{9}$
7. selecting a yellow marble first 7. 0
8. selecting a red marble followed by a blue marble 8. $\frac{1}{6}$
9. selecting a green marble after selecting two marbles that are not green 9. $\frac{1}{6}$

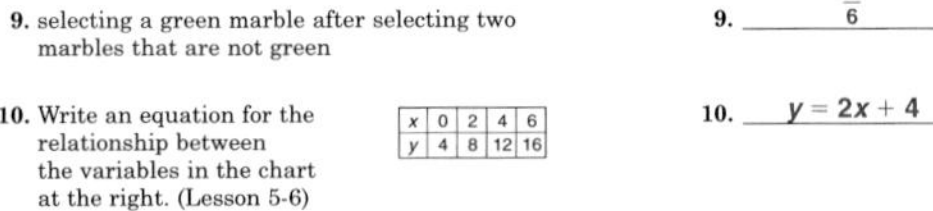

10. Write an equation for the relationship between the variables in the chart at the right. (Lesson 5-6) 10. $y = 2x + 4$

x	0	2	4	6
y	4	8	12	16

11. If $f(x) = 2x^2 - 5x + 4$, find $f(-3)$. (Lesson 5-5) 11. 37
12. Graph $4x + 2y = 1$ using the slope and y-intercept. (Lesson 6-5) 12.

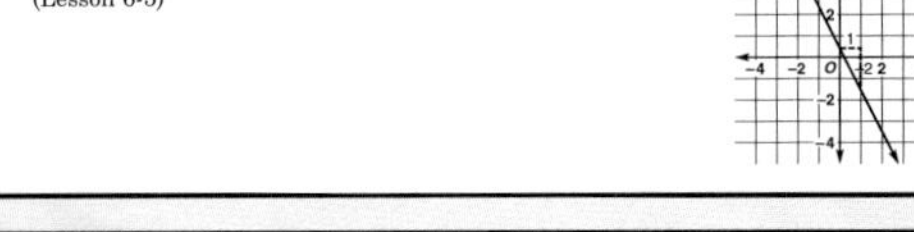

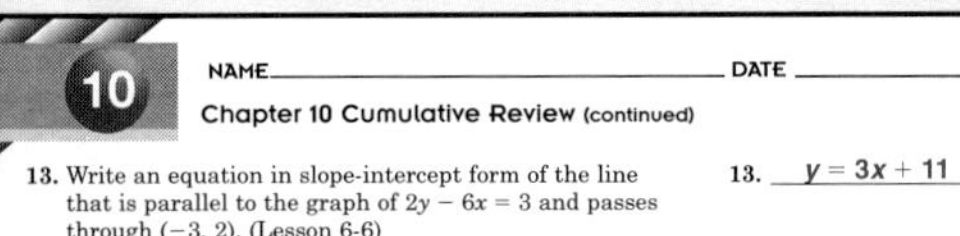

10 NAME________ DATE________

Chapter 10 Cumulative Review (continued)

13. Write an equation in slope-intercept form of the line that is parallel to the graph of $2y - 6x = 3$ and passes through $(-3, 2)$. (Lesson 6-6) 13. $y = 3x + 11$

14. Solve $\frac{2}{3}x > -10$. (Lesson 7-2) 14. $x > -15$
15. Graph the inequality $x + 2y < 4$. (Lesson 7-8) 15.

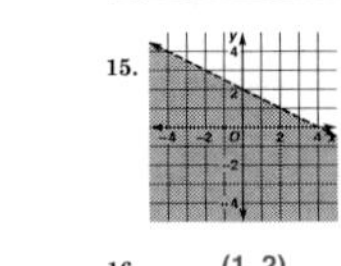

16. Solve the system of equations $3y - 2x = 4$ and $2y + 4x = 8$. (Lesson 8-4) 16. (1, 2)
17. Simplify $(-2x^2y^3)^4$. (Lesson 9-1) 17. $16x^8y^{12}$
18. Evaluate $\frac{3.6 \times 10^{-4}}{1.2 \times 10^{-8}}$. Express the result in scientific notation. (Lesson 9-3) 18. 3×10^4

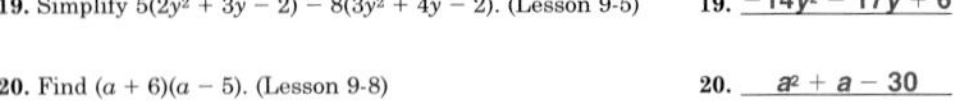

19. Simplify $5(2y^2 + 3y - 2) - 8(3y^2 + 4y - 2)$. (Lesson 9-5) 19. $-14y^2 - 17y + 6$
20. Find $(a + 6)(a - 5)$. (Lesson 9-8) 20. $a^2 + a - 30$

Factor each polynomial, if possible. (Lessons 10-2 and 10-3)

21. $m^2 + 12m + 36$ 21. $(m + 6)^2$
22. $2ar + 2rt - a - t$ 22. $(2r - 1)(a + t)$
23. $36c^3 + 6c^2 - 6c$ 23. $6c(3c - 1)(2c + 1)$
24. Solve $x(x + 12) = 0$. (Lesson 10-6) 24. 0, −12
25. Find two consecutive odd integers whose product is 255. (Lesson 10-6) 25. 15 and 17 or −17 and −15

CUMULATIVE REVIEW

CHAPTERS 1–10

SECTION ONE: MULTIPLE CHOICE

There are nine multiple-choice questions in this section. After working each problem, write the letter of the correct answer on your paper.

1. The square of a number subtracted from 8 times the number is equal to twice the number. Find the number. **A**
   - **A.** 0 or 6
   - **B.** 3
   - **C.** 2
   - **D.** 0 or 10

2. Choose the open sentence that represents the range of acceptable diameters for a lawn mower bolt that will work properly only if its diameter differs from 2 cm by no more than 0.04 cm. **C**
   - **A.** $|d - 0.04| \geq 2$
   - **B.** $|d| < 1.96$
   - **C.** $|d - 2| \leq 0.04$
   - **D.** $|d| > 2.04$

3. Choose an expression for the area of the shaded region shown below. **D**

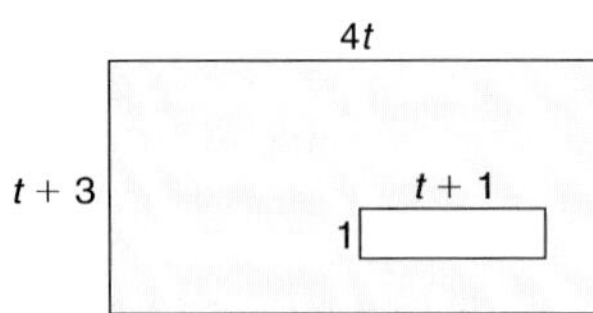

   - **A.** $(4t)(t + 3)$
   - **B.** $(4t - t)(t + 3) - (t + 1)$
   - **C.** $(t + 1)(t + 3) - (4t)(t)$
   - **D.** $(4t)(t + 3) - (t + 1)$

4. Choose the equivalent equation for $A = \frac{1}{2}h(a + b)$. **A**
   - **A.** $h = \frac{2A}{a + b}$
   - **B.** $h = 2A - (a + b)$
   - **C.** $h = \frac{\frac{1}{2}(a + b)}{A}$
   - **D.** $h = \frac{A - b}{2a}$

5. Bob and Vicki took a trip to Zuma Beach. On the way there, their average speed was 42 miles per hour. On the way home, their average speed was 56 miles per hour. If their total travel time was 7 hours, find the distance to the beach. **B**
   - **A.** 126 miles
   - **B.** 168 miles
   - **C.** 98 miles
   - **D.** 294 miles

6. Choose which statement is true. **B**
   - **A.** The range is the difference between the greatest value and the lower quartile in a set of data.
   - **B.** An outlier will only affect the mean of a set of data.
   - **C.** Quartiles are values that divide a set of data into equal halves.
   - **D.** The interquartile range is the sum of the upper and lower quartiles in a set of data.

7. Choose the prime polynomial. **C**
   - **A.** $y^2 + 12y + 27$
   - **B.** $6x^2 - 11x + 4$
   - **C.** $h^2 + 5h - 8$
   - **D.** $9k^2 + 30km + 25m^2$

8. Choose the equation of the line that passes through the points at (9, 5) and (−3, −4). **D**
   - **A.** $y = -\frac{1}{12}x + \frac{23}{4}$
   - **B.** $y = \frac{4}{3}x$
   - **C.** $y = -\frac{7}{2}x - \frac{3}{4}$
   - **D.** $y = \frac{3}{4}x - \frac{7}{4}$

9. Carrie's bowling scores for four games are $b + 2$, $b + 3$, $b - 2$, and $b - 1$. What must her score be on her fifth game to average $b + 2$? **A**
   - **A.** $b + 8$
   - **B.** b
   - **C.** $b - 2$
   - **D.** $b + 5$

Standardized Test Practice Questions are also provided in the *Assessment and Evaluation Masters,* p. 270.

SECTION TWO: SHORT ANSWER

This section contains ten questions for which you will provide short answers. Write your answer on your paper.

10. The measure of the perimeter of a square is $20m + 32p$. Find the measure of its area. **$25m^2 + 80mp + 64p^2$**

11. At a bake sale, cakes cost twice as much as pies. Pies were $4 more than triple the price of cookies. Darin bought a cake, three pies, and four cookies for $24.75. What is the price of each item? **cake: $9.50, pie: $4.75, cookie: $0.25**

12. Factor $3g^2 - 10gh - 8h^2$. **$(3g + 2h)(g - 4h)$**

13. On the first day of school, 264 school notebooks were sold. Some sold for 95¢ each, and the rest sold for $1.25 each. How many of each were sold if the total sales were $297? **110 at 95¢, 154 at $1.25**

14. A rectangular photograph is 8 centimeters wide and 12 centimeters long. The photograph is enlarged by increasing the length and width by an equal amount. If the area of the new photograph is 69 square centimeters greater than the area of the original photograph, what are the dimensions of the new photograph? **11 cm by 15 cm**

15. Linda plans to spend at most $50 on shorts and blouses. She bought 2 pairs of shorts for $14.20 each. How much can she spend on blouses? **$21.60 or less**

16. The difference of two numbers is 3. If the difference of their squares is 15, what is the sum of the numbers? **5**

17. Geometry The length of a rectangle is eight times its width. If the length was decreased by 10 meters and the width was decreased by 2 meters, the area would be decreased by 162 square meters. Find the original dimensions. **7 m by 56 m**

18. Ben's car gets between 18 and 21 miles per gallon of gasoline. If his car's tank holds 15 gallons, what is the range of distance that Ben can drive his car on one tank of gasoline? **270 miles to 315 miles**

19. Find the measure of the third side of a triangle if the perimeter of the triangle is $8x^2 + x + 15$ and the measures of two of its sides can be represented by the expressions $2x^2 - 5x + 7$ and $x^2 - x + 11$. **$5x^2 + 7x - 3$**

SECTION THREE: OPEN-ENDED

This section contains two open-ended problems. Demonstrate your knowledge by giving a clear, concise solution to each problem. Your score on these problems will depend on how well you do the following.

- Explain your reasoning.
- Show your understanding of the mathematics in an organized manner.
- Use charts, graphs, and diagrams in your explanation.
- Show the solution in more than one way or relate it to other situations.
- Investigate beyond the requirements of the problem.

20. A tinsmith is going to make a box out of tin by cutting a square from each corner and folding up the sides. The box needs to be 3 inches high (so he will be cutting out 3-inch squares), and it needs to be twice as long as it is wide, so that it can hold two smaller square cardboard boxes. Finally, its volume needs to be 1350 cubic inches. What should the dimensions be? **3 in. by 30 in. by 15 in.**

21. The length of a rectangle is five times its width. If the length was increased by 7 meters and the width was decreased by 4 meters, the area would be decreased by 132 square meters. Find the original dimensions. **8 m by 40 m**

CHAPTER

11 Exploring Quadratic and Exponential Functions

PREVIEWING THE CHAPTER

This chapter introduces students to quadratic equations by having them find the equation of the axis of symmetry and the coordinates of the vertex. Students graph quadratic functions, compare different parabolas, and approximate roots of quadratic functions. Next, students solve quadratic equations using the quadratic formula, focus on the discriminant, and study how it can be used to determine the number of real roots of an equation. Students graph exponential functions on the graphing calculator and on paper. Finally, students analyze exponential functions and use problem-solving techniques to solve equations that have real exponents.

Lesson (Pages)	Lesson Objectives	NCTM Standards	State/Local Objectives
11-1A (610)	Use a graphing calculator to graph a quadratic function and find the coordinates of its vertex.	1–6	
11-1 (611–617)	Find the equation of the axis of symmetry and the coordinates of the vertex of a parabola. Graph quadratic functions.	1–6, 13	
11-1B (618–619)	Use a graphing calculator to study the characteristics of families of parabolas.	1–6	
11-2 (620–627)	Use estimation to find roots of quadratic equations. Find roots of quadratic equations by graphing.	1–6, 13	
11-3 (628–633)	Solve quadratic equations by using the quadratic formula.	1–6	
11-4A (634)	Use a graphing calculator to graph exponential functions.	1–6, 13	
11-4 (635–642)	Graph exponential functions. Determine if a set of data displays exponential behavior. Solve exponential equations.	1–6, 13	
11-5 (643–649)	Solve problems involving growth and decay.	1–6	

A complete, 1-page lesson plan is provided for each lesson in the ***Lesson Planning Guide***. Answer keys for each lesson are available in the ***Answer Key Masters.***

ORGANIZING THE CHAPTER

You may want to refer to the **Course Planning Calendar** on page T12 for detailed information on pacing.
PACING: Standard—11 days; **Honors**—10 days; **Block**—5 days; **Two Years**—19 days

LESSON PLANNING CHART

Lesson (Pages)	Materials/ Manipulatives	Extra Practice (Student Edition)	Blackline Masters									Real-World Applications	Interactive Mathematics Tools Software	Teaching Transparencies
			Study Guide	Practice	Enrichment	Assessment and Evaluation	Modeling Mathematics	Multicultural Activity	Tech Prep Applications	Graphing Calculator	Science and Math Lab Manual			
11-1A (610)	graphing calculator									pp. 30, 31				
11-1 (611–617)	grid paper	p. 780	p. 75	p. 75	p. 75		p. 82			p. 11			11-1	11-1A 11-1B
11-1B (618–619)	graphing calculator									pp. 32, 33			11-1B	
11-2 (620–627)	graphing calculator	p. 781	p. 76	p. 76	p. 76	p. 296			p. 21					11-2A 11-2B
11-3 (628–633)	scientific calculator	p. 781	p. 77	p. 77	p. 77	pp. 295, 296		p. 21	p. 22			27	11-3.1 11-3.2	11-3A 11-3B
11-4A (634)	graphing calculator									pp. 34, 35				
11-4 (635–642)	large piece of paper graphing calculator	p. 781	p. 78	p. 78	p. 78	p. 297								11-4A 11-4B
11-5 (643–649)	spreadsheets	p. 782	p. 79	p. 79	p. 79	p. 297		p. 22			pp. 49–52	28		11-5A 11-5B
Study Guide/ Assessment (651–655)						pp. 281–294, 298–300								

ORGANIZING THE CHAPTER

Other Chapter Resources

Student Edition
Chapter Opener, pp. 608–609
Mathematics and Society, p. 649
Working on the Investigation, pp. 617, 627
Closing the Investigation, p. 650

Teacher's Classroom Resources
Investigations and Projects Masters, pp. 65–68

Technology
Test and Review Software (IBM and Macintosh)
CD-ROM Interactions (Windows and Macintosh)

Professional Publications
Block Scheduling Booklet
Glencoe Mathematics Professional Series

Outside Resources

Books/Periodicals
Isdell, Wendy, *A Gebra Named Al,* Free Spirit Publishing

Software
Algebra Xpresser, William K. Bradford

Videos/CD-ROMs
Algebra for Everyone, NCTM

Assessment Resources

Student Edition
Math Journal, pp. 624, 631
Mixed Review, pp. 616, 627, 633, 642, 649
Self Test, p. 633
Chapter Highlights, p. 651
Chapter Study Guide and Assessment, pp. 652–654
Alternative Assessment, p. 655
Portfolio, p. 655

Teacher's Wraparound Edition
5-Minute Check, pp. 611, 620, 628, 635, 643
Check for Understanding, pp. 614, 624, 631, 640, 646
Closing Activity, pp. 617, 627, 632, 641, 648
Cooperative Learning, pp. 616, 630

Assessment and Evaluation Masters
Multiple-Choice Tests, Forms 1A (Honors), 1B (Average), 1C (Basic), pp. 281–286
Free-Response Tests, Forms 2A (Honors), 2B (Average), 2C (Basic), pp. 287–292
Calculator-Based Test, p. 293
Performance Assessment, p. 294
Mid-Chapter Test, p. 295
Quizzes A–D, pp. 296–297
Standardized Test Practice, p. 298
Cumulative Review, pp. 299–300

ENHANCING THE CHAPTER

Examples of some of the materials for enhancing Chapter 11 are shown below.

DIVERSITY

Multicultural Activity Masters, pp. 21, 22

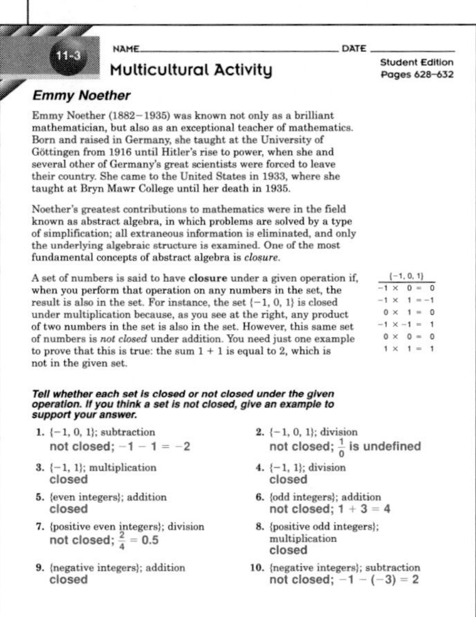

NAME ______________________ DATE __________

11-3 **Multicultural Activity** — Student Edition Pages 628–632

Emmy Noether

Emmy Noether (1882–1935) was known not only as a brilliant mathematician, but also as an exceptional teacher of mathematics. Born and raised in Germany, she taught at the University of Göttingen from 1916 until Hitler's rise to power, when she and several other of Germany's great scientists were forced to leave their country. She came to the United States in 1933, where she taught at Bryn Mawr College until her death in 1935.

Noether's greatest contributions to mathematics were in the field known as abstract algebra, in which problems are solved by a type of simplification; all extraneous information is eliminated, and only the underlying algebraic structure is examined. One of the most fundamental concepts of abstract algebra is *closure*.

A set of numbers is said to have **closure** under a given operation if, when you perform that operation on any numbers in the set, the result is also in the set. For instance, the set $\{-1, 0, 1\}$ is closed under multiplication because, as you see at the right, any product of two numbers in the set is also in the set. However, this same set of numbers is *not closed* under addition. You need just one example to prove that this is true: the sum $1 + 1$ is equal to 2, which is not in the given set.

$\{-1, 0, 1\}$
$-1 \times 0 = 0$
$-1 \times 1 = -1$
$0 \times 1 = 0$
$-1 \times -1 = 1$
$0 \times 0 = 0$
$1 \times 1 = 1$

Tell whether each set is closed or not closed under the given operation. If you think a set is not closed, give an example to support your answer.

1. $\{-1, 0, 1\}$; subtraction
   not closed; $-1 - 1 = -2$
2. $\{-1, 0, 1\}$; division
   not closed; $\frac{1}{0}$ is undefined
3. $\{-1, 1\}$; multiplication
   closed
4. $\{-1, 1\}$; division
   closed
5. {even integers}; addition
   closed
6. {odd integers}; addition
   not closed; $1 + 3 = 4$
7. {positive even integers}; division
   not closed; $\frac{2}{4} = 0.5$
8. {positive odd integers}; multiplication
   closed
9. {negative integers}; addition
   closed
10. {negative integers}; subtraction
    not closed; $-1 - (-3) = 2$

APPLICATIONS

Real-World Applications, 27, 28

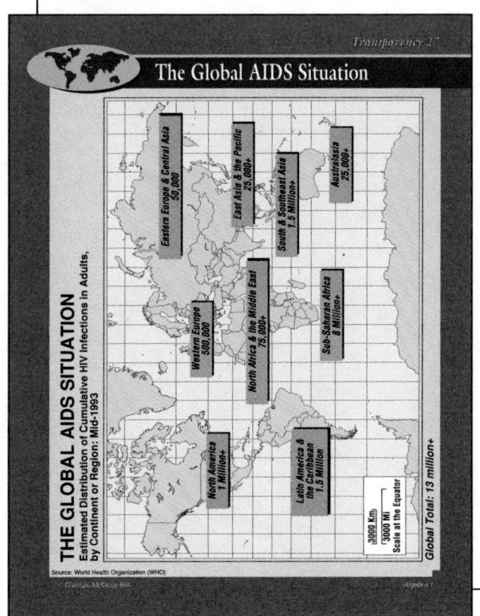

TECHNOLOGY

Graphing Calculator Masters, p. 11

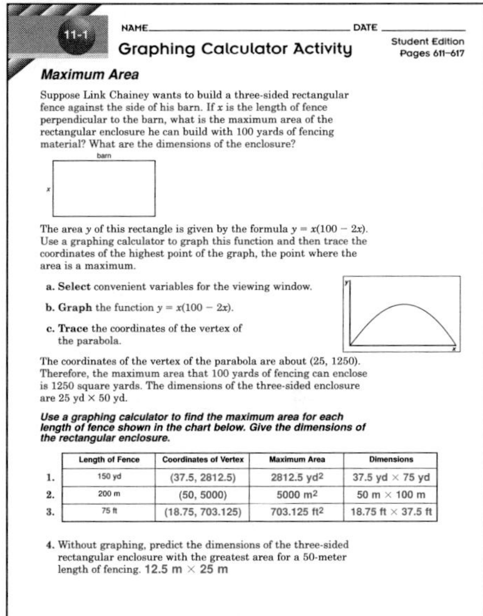

NAME ______________________ DATE __________

11-1 **Graphing Calculator Activity** — Student Edition Pages 611–617

Maximum Area

Suppose Link Chainey wants to build a three-sided rectangular fence against the side of his barn. If x is the length of fence perpendicular to the barn, what is the maximum area of the rectangular enclosure he can build with 100 yards of fencing material? What are the dimensions of the enclosure?

barn
x

The area y of this rectangle is given by the formula $y = x(100 - 2x)$. Use a graphing calculator to graph this function and then trace the coordinates of the highest point of the graph, the point where the area is a maximum.

a. **Select** convenient variables for the viewing window.

b. **Graph** the function $y = x(100 - 2x)$.

c. **Trace** the coordinates of the vertex of the parabola.

The coordinates of the vertex of the parabola are about (25, 1250). Therefore, the maximum area that 100 yards of fencing can enclose is 1250 square yards. The dimensions of the three-sided enclosure are 25 yd × 50 yd.

Use a graphing calculator to find the maximum area for each length of fence shown in the chart below. Give the dimensions of the rectangular enclosure.

	Length of Fence	Coordinates of Vertex	Maximum Area	Dimensions
1.	150 yd	(37.5, 2812.5)	2812.5 yd^2	37.5 yd × 75 yd
2.	200 m	(50, 5000)	5000 m^2	50 m × 100 m
3.	75 ft	(18.75, 703.125)	703.125 ft^2	18.75 ft × 37.5 ft

4. Without graphing, predict the dimensions of the three-sided rectangular enclosure with the greatest area for a 50-meter length of fencing. 12.5 m × 25 m

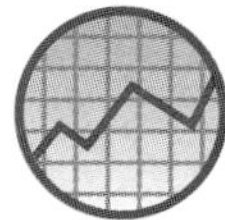

TECH PREP

Tech Prep Applications Masters, pp. 21, 22

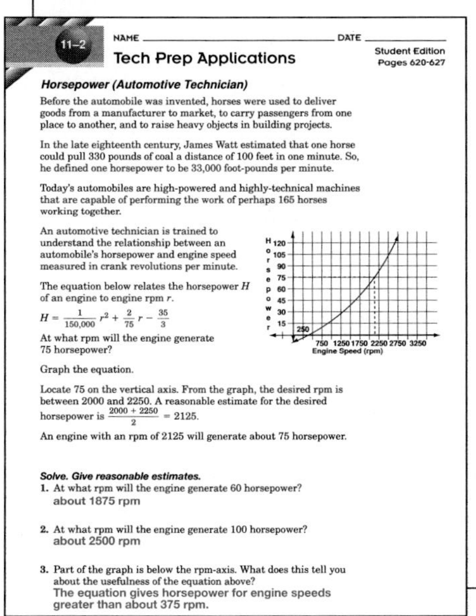

NAME ______________________ DATE __________

11–2 **Tech Prep Applications** — Student Edition Pages 620-627

Horsepower (Automotive Technician)

Before the automobile was invented, horses were used to deliver goods from a manufacturer to market, to carry passengers from one place to another, and to raise heavy objects in building projects.

In the late eighteenth century, James Watt estimated that one horse could pull 330 pounds of coal a distance of 100 feet in one minute. So, he defined one horsepower to be 33,000 foot-pounds per minute.

Today's automobiles are high-powered and highly-technical machines that are capable of performing the work of perhaps 165 horses working together.

An automotive technician is trained to understand the relationship between an automobile's horsepower and engine speed measured in crank revolutions per minute.

The equation below relates the horsepower H of an engine to engine rpm r.

$$H = \frac{1}{150{,}000}r^2 + \frac{2}{75}r - \frac{35}{3}$$

At what rpm will the engine generate 75 horsepower?

Horsepower: 120, 105, 90, 75, 60, 45, 30, 15
250, 750, 1250, 1750, 2250, 2750, 3250
Engine Speed (rpm)

Graph the equation.

Locate 75 on the vertical axis. From the graph, the desired rpm is between 2000 and 2250. A reasonable estimate for the desired horsepower is $\frac{2000 + 2250}{2} = 2125$.

An engine with an rpm of 2125 will generate about 75 horsepower.

Solve. Give reasonable estimates.

1. At what rpm will the engine generate 60 horsepower?
   about 1875 rpm
2. At what rpm will the engine generate 100 horsepower?
   about 2500 rpm
3. Part of the graph is below the rpm-axis. What does this tell you about the usefulness of the equation above?
   The equation gives horsepower for engine speeds greater than about 375 rpm.

CONNECTIONS

Science and Math Lab Manual, pp. 49–52

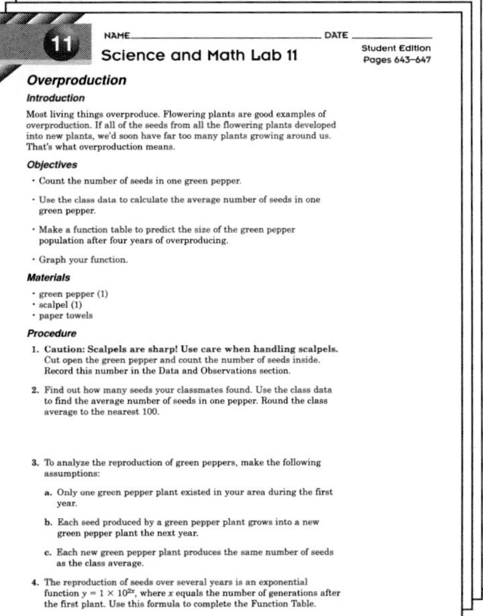

NAME ______________________ DATE __________

11 **Science and Math Lab 11** — Student Edition Pages 643–647

Overproduction

Introduction

Most living things overproduce. Flowering plants are good examples of overproduction. If all of the seeds from all the flowering plants developed into new plants, we'd soon have far too many plants growing around us. That's what overproduction means.

Objectives

- Count the number of seeds in one green pepper.
- Use the class data to calculate the average number of seeds in one green pepper.
- Make a function table to predict the size of the green pepper population after four years of overproducing.
- Graph your function.

Materials

- green pepper (1)
- scalpel (1)
- paper towels

Procedure

1. **Caution: Scalpels are sharp! Use care when handling scalpels.** Cut open the green pepper and count the number of seeds inside. Record this number in the Data and Observations section.
2. Find out how many seeds your classmates found. Use the class data to find the average number of seeds in one pepper. Round the class average to the nearest 100.
3. To analyze the reproduction of green peppers, make the following assumptions:
   a. Only one green pepper plant existed in your area during the first year.
   b. Each seed produced by a green pepper plant grows into a new green pepper plant the next year.
   c. Each new green pepper plant produces the same number of seeds as the class average.
4. The reproduction of seeds over several years is an exponential function $y = 1 \times 10^{2x}$, where x equals the number of generations after the first plant. Use this formula to complete the Function Table.

PROBLEM SOLVING

Problem of the Week Cards, 30, 31

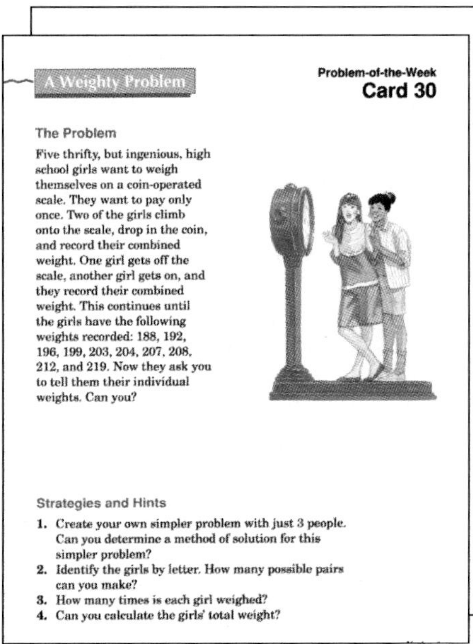

A Weighty Problem

Problem-of-the-Week **Card 30**

The Problem

Five thrifty, but ingenious, high school girls want to weigh themselves on a coin-operated scale. They want to pay only once. Two of the girls climb onto the scale, drop in the coin, and record their combined weight. One girl gets off the scale, another girl gets on, and they record their combined weight. This continues until the girls have the following weights recorded: 188, 192, 196, 199, 203, 204, 207, 208, 212, and 219. Now they ask you to tell them their individual weights. Can you?

Strategies and Hints

1. Create your own simpler problem with just 3 people. Can you determine a method of solution for this simpler problem?
2. Identify the girls by letter. How many possible pairs can you make?
3. How many times is each girl weighed?
4. Can you calculate the girls' total weight?

MAKING MATHEMATICS RELEVANT

This two-page introduction to the chapter provides students with an opportunity to explore contemporary topics and their applications to mathematics.

Background Information

Reaching Out to Help Others The tables in the text and the table below raise questions that can stimulate an interesting class discussion.

1. Why does the percent of volunteers increase as the level of education increases?
2. Why does the percent of volunteers increase as the level of income increases?

Education	Percent of Volunteers	Average Hours Volunteered per Week
Elementary school	25.0	NA
Some high school	22.1	5.1
High school graduate	44.7	4.1
Technical, trade, or business school	51.5	3.9
Some college	66.1	3.5
College graduate	76.6	4.6

CHAPTER 11

Exploring Quadratic and Exponential Functions

Objectives

In this chapter, you will:

- find the equation of the axis of symmetry and the coordinates of the vertex of a parabola,
- graph quadratic and exponential functions,
- use estimation to find roots of quadratic equations by graphing,
- find roots of quadratic equations by using the quadratic formula,
- solve problems by looking for and using a pattern, and
- solve problems involving growth and decay.

Reaching Out to Help Others

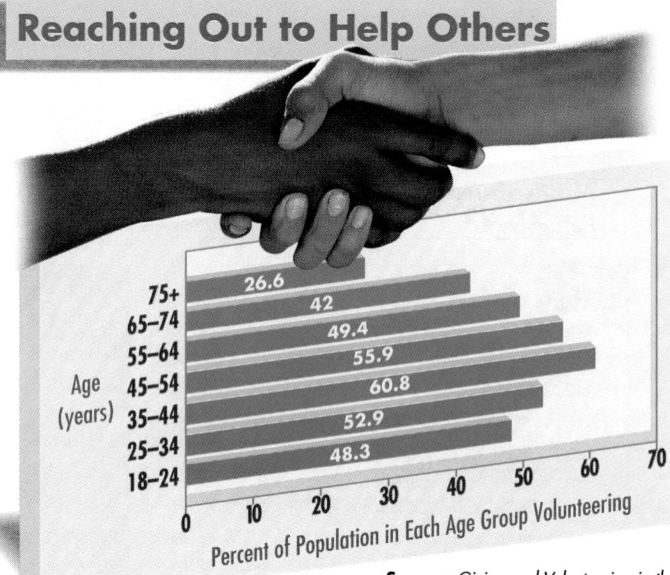

Source: *Giving and Volunteering in the United States*, 1992 Edition

Dr. Robert Coles, a psychiatrist and Harvard professor, calls it "the call of service," men and women who volunteer their time to help others. Many teens would like to get involved in volunteer service, but may not know where to start. Some opportunities may be available in your own community—helping senior citizens, working as a Big Brother or Big Sister, or volunteering at a food pantry. Volunteering will help others and may help you feel great, too.

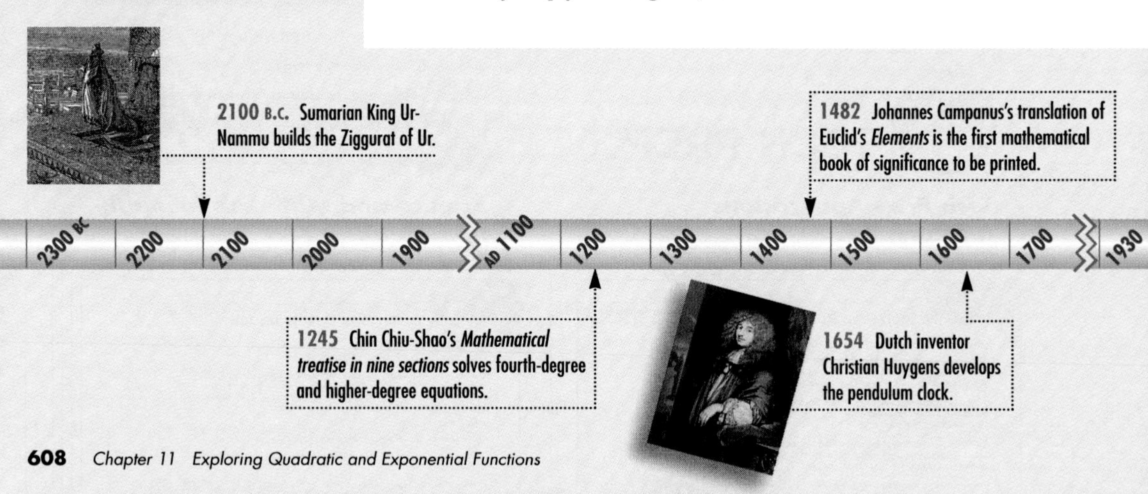

The pendulum is a phenomenon that illustrates fundamental principles of physical science. It also applies to these physical principles the mathematics of measurement of time. Students might find it interesting to report on Huygen's invention of 1654.

interNET CONNECTION

Read the Internet NonProfit Center page for information and ideas about nonprofit organizations.

World Wide Web
http://www.human.com/inc/

PEOPLE IN THE NEWS

A perfect example of a teenage volunteer is **Liz Alvarez** of St. Petersburg, Florida. She has been a community volunteer at St. Anthony's Hospital, giving more than 600 hours of her time to help the staff. She answers phones in the pastoral care department, does filing as an administrative assistant, and carries messages for staff members. She helps patients by delivering items to their rooms, works in the gift shop, and assists other volunteers.

She enjoys her time spent volunteering, is happy to help others, and believes that the experience is very rewarding. She encourages other teens to investigate the opportunities in their communities and join a volunteer team.

Chapter Project

Organizations in every community study the characteristics of those who volunteer their time. In this manner, they know what population to target when seeking volunteers for a particular project or event. Survey companies, such as the Gallup Organization, Inc., often provide statistics for these organizations.

One of the characteristics the Gallup Organization studied was the income of volunteers. They found the following information on income brackets and the percent of the people in each bracket who volunteer.

Income ($)	Percent of Population Volunteering
Under 10,000	31.6
10,000–19,999	37.9
20,000–29,999	51.3
30,000–39,999	56.4
40,000–49,999	67.4
50,000–59,999	67.7
60,000–74,999	55.0
75,000–99,999	62.8
100,000 +	73.7

Analyze these findings.

- Make a graph of the data.
- What type of behavior do these data present? **See margin.**
- Why do you think this behavior exists? **Answers will vary.**
- Research in an almanac or statistical abstract to find the average income for the age groups listed in the graph on the previous page.
- Use your research, the graph on the previous page, and the table to make a conjecture about the characteristics of the average volunteer.

1940 U.S. photographer Helen Levitt chronicles life in the streets of New York City with her print *Children.*

1981 The Peace Corps becomes an independent agency, sending volunteers to improve living conditions in developing countries.

1945 1950 1955 1960 1965 1970 1975 1980 1985 1990 1995 2000

1947 Dr. Bernardo A. Houssay receives the Nobel Prize for Medicine and Physiology for his research on the pituitary gland.

1991 Advances in music technology enable Natalie Cole to produce a duet album with her late father Nat "King" Cole.

PEOPLE IN THE NEWS

Encourage students to talk about what volunteer work they have done. Encourage them to explain why they do it and what they get out of it. Would their work be as helpful if they were being paid? Would they feel differently?

Chapter Project

Cooperative Learning Students should work in groups of three for this project. One will be responsible for drawing the graphs. Another will be responsible for library research. The third member will present the results to the class. All should participate in a group discussion to answer the questions posed in this project.

Alternative Chapter Projects

Two other chapter projects are included in the *Investigations and Projects Masters.* In Chapter 11 Project A, pp. 65–66, students extend the topic in the chapter opener. In Chapter 11 Project B, pp. 67–68, students investigate how engineers use parabolic curves in bridge designs.

Investigations and Projects Masters, p. 65

NAME__________ DATE ________

Chapter 11 Project A

Student Edition Pages 610–650

Volunteer Survey

1. Work with a partner to research and list organizations in your area that have volunteering opportunities. Possible places to check include health care organizations, schools, social service agencies, environmental programs, religious organizations, and museums.
2. Arrange the list you made in exercise 1 into different categories. Which categories have the most organizations listed? Which have the fewest?
3. Decide on a few organizations where you might be interested in volunteering. Interview a supervisor at one of these organizations to find out more about what the organization does. Find out how the organization uses its volunteers.
4. Find out how the organization uses the general equations for exponential growth and decay to make decisions about purchasing versus leasing, federal government funding, investments, population growth, and so on.
5. Write a newspaper article about volunteers and volunteering. At the end of your article, list different organizations that have volunteering opportunities and the phone numbers where people can get in touch with them.

11-1A LESSON NOTES

NCTM Standards: 1–6

Objective
Use a graphing calculator to graph a quadratic function and find the coordinates of its vertex.

Recommended Time
15 minutes

Instructional Resources
Graphing Calculator Masters, pp. 30 and 31

These masters provide keystroking instruction for this lesson for the TI-81 and Casio graphing calculators.

1 FOCUS

Motivating the Lesson
Plot the points given by the coordinates in the table below.

x	y
0	0
1	2
2	4
3	6
4	8
5	10

Note that each value of y is a constant, increased over the preceding value of y. Ask students what the graph would look like if every increase in y value were greater than the preceding increase.

2 TEACH

Teaching Tip For some quadratic functions, the standard viewing window may need to be adjusted until the vertex is visible.

3 PRACTICE/APPLY

Assignment Guide

Core: 1–6
Enriched: 1–6

11–1A Graphing Technology Quadratic Functions

A Preview of Lesson 11–1

Equations in the form $y = ax^2 + bx + c$ are called **quadratic functions,** and their graphs are called **parabolas.** A parabola is a U-shaped curve that can open upward or downward. The maximum or minimum point of a parabola is called its **vertex.** You can use a graphing calculator to graph a quadratic function and find the coordinates of its vertex.

Example **Graph $y = \frac{1}{4}x^2 - 4x - 2$ and locate its vertex. Use the integer window.**

A quadratic function is entered into the Y= list in the same way that you entered linear functions.

Sometimes the coordinates of the vertex may not be integers. Use the ZOOM feature several times until you get coordinates that are fairly consistent (within three decimal places).

Enter: [Y=] .25 [X,T,θ] [x^2] [−] 4 [X,T,θ] [−] 2 *Enters the function.*

[ZOOM] 6 [ZOOM] 8 [ENTER] *Selects the integer window.*

The parabola opens upward. The vertex is a minimum point.

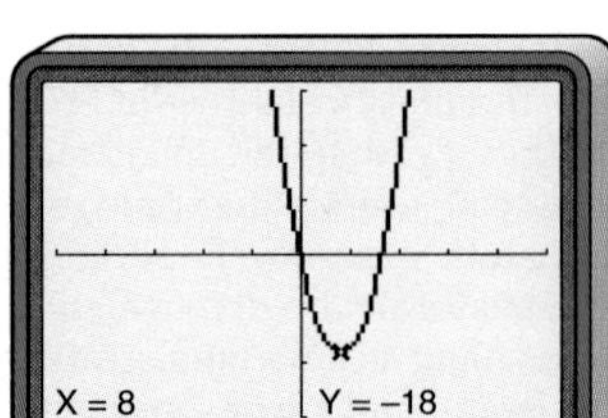

Method 1: Use TRACE.

Press [TRACE] and use the left and right arrow keys to move the cursor to the vertex. Watch the coordinates at the bottom of the screen as you move the cursor. The vertex will be the point of the least y value. *Why?*

Method 2: Use CALC.

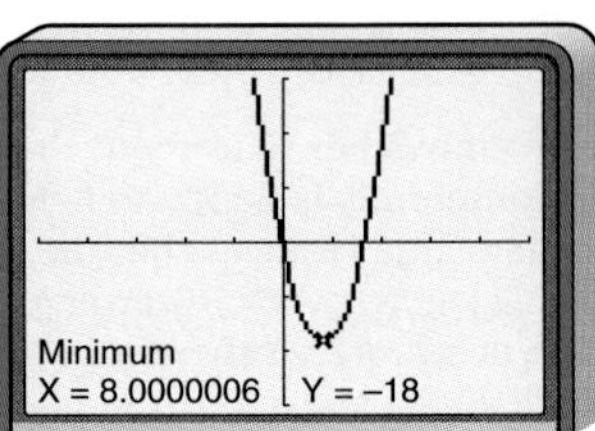

Since the parabola opens upward, the vertex is a minimum point. Press [2nd] [CALC] 3.

- A "Lower Bound?" prompt appears. Move the cursor to a location left of where you think the vertex is and press [ENTER].
- An "Upper Bound?" prompt appears. Move the cursor to a location right of where you think the vertex is and press [ENTER].
- When the "Guess?" prompt appears, move the cursor to your choice for the vertex and press [ENTER]. The screen will give you an approximation of the vertex coordinates.

You can also press [ENTER] at the Guess? prompt without entering a guess.

The coordinates of the vertex are (8, −18).

1–6. See Solutions Manual for graphs.

EXERCISES

Graph each function. Make a sketch of the graph and note the ordered pair representing the vertex on the graph.

1. $y = x^2 + 16x + 59$
2. $y = 12x^2 + 18x + 10$
3. $y = x^2 - 10x + 25$
4. $y = -2x^2 - 8x - 1$
5. $y = 2(x - 10)^2 + 14$
6. $y = -0.5x^2 - 2x + 3$

1. (−8, −5)
2. (−0.75, 3.25)
3. (5, 0) 4. (−2, 7)
5. (10, 14) 6. (−2, 5)

4 ASSESS

Observing students working with technology is an excellent method of assessment.

Using Technology
This lesson offers an excellent opportunity for using technology in your algebra classroom. For more information on using technology, see *Graphing Calculators in the Mathematics Classroom,* one of the titles in the Glencoe Mathematics Professional Series.

11-1 Graphing Quadratic Functions

What YOU'LL LEARN

- To find the equation of the axis of symmetry and the coordinates of the vertex of a parabola, and
- to graph quadratic functions.

Why IT'S IMPORTANT

You can graph quadratic functions to solve problems involving fireworks and football.

APPLICATION

Landmarks

The Gateway Arch of the Jefferson National Expansion Memorial in St. Louis, Missouri, is shaped like an upside-down U. This shape is actually a *catenary,* which resembles a geometric shape called a **parabola.** The shape of the arch can be approximated by the graph of the function $f(x) = -0.00635x^2 + 4.0005x - 0.07875$, where $f(x)$ is the height of the arch in feet and x is the horizontal distance from one base.

This type of function is an example of a **quadratic function.** A quadratic function can be written in the form $f(x) = ax^2 + bx + c$, where $a \neq 0$. Notice that this function has a degree of 2 and the exponents are positive.

Definition of a Quadratic Function	A quadratic function is a function that can be described by an equation of the form $y = ax^2 + bx + c$, where $a \neq 0$.

Since distance and height are involved, the domain and range must both be positive.

To graph a quadratic equation, you can use a table of values. The table at the right shows the distance (in 35-foot increments) from one base of the arch and the height of the arch at each increment (rounded to the nearest foot). Graph the ordered pairs and connect them with a smooth curve. The graph of $f(x) = -0.00635x^2 + 4.0005x - 0.07875$ is shown below.

x	f(x)
0	0
35	132
70	249
105	350
140	436
175	506
210	560
245	599
280	622
315	630
350	622
385	599
420	560
455	506
490	436
525	350
560	249
595	132
630	0

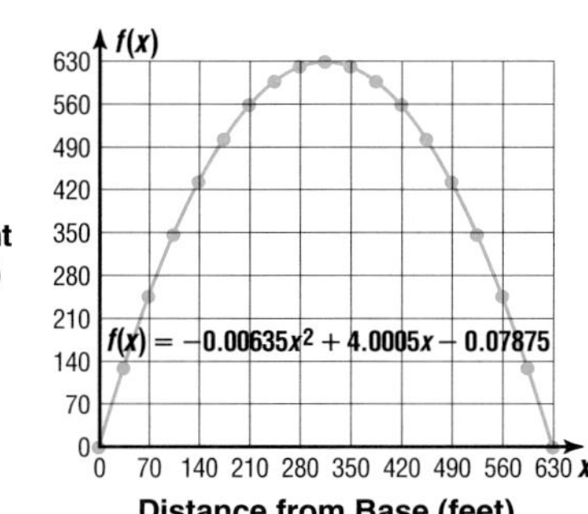

GLOBAL CONNECTIONS

The arch as an architectural form dates back to Egypt and Greece, but was first used extensively by the Romans in building bridges and aqueducts, like the Pont du Gard at Nimes, France. The architects in medieval times used majestic pointed arches in the construction of gothic cathedrals.

Notice that the value of a in this function is negative and the curve opens downward. The greatest value of $f(x)$ seems to be 630 feet, which occurs when the distance from the base is 315 feet. The point at (315, 630) would be the **vertex** of the parabola. For a parabola that opens downward, the vertex is a **maximum** point of the function. If the parabola opens upward, the vertex is a **minimum** point of the function.

GLOBAL CONNECTIONS

Many Roman architects did not use mortar. They relied on the stones fitting tightly for support. The Gateway Arch in St. Louis, Missouri, is made entirely of stainless steel.

11-1 LESSON NOTES

NCTM Standards: 1–6, 13

Instructional Resources

- Study Guide Master 11-1
- Practice Master 11-1
- Enrichment Master 11-1
- Graphing Calculator Masters, p. 11
- Modeling Mathematics Masters, p. 82

Transparency 11-1A contains the 5-Minute Check for this lesson; **Transparency 11-1B** contains a teaching aid for this lesson.

Recommended Pacing	
Standard Pacing	Day 2 of 11
Honors Pacing	Day 2 of 10
Block Scheduling*	Day 1 of 5
Alg. 1 in Two Years*	Days 2 & 3 of 19

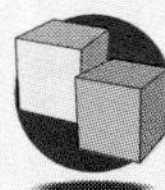

*For more information on pacing and possible lesson plans, refer to the *Block Scheduling Booklet* and *Algebra 1 in Two Years.*

1 FOCUS

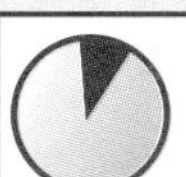

5-Minute Check
(over Chapter 10)

Solve.

1. $x^3 + 2x^2 = 15x$ **0, −5, 3**
2. $a^3 - 13a^2 + 42a = 0$ **0, 6, 7**
3. $4r^3 - 9r = 0$ **0, $\frac{3}{2}$, $-\frac{3}{2}$**
4. The sum of the squares of two consecutive odd integers is 802. Find the integers. **−21 and −19 or 19 and 21**
5. Find two consecutive odd integers such that the square of the greater decreased by the lesser is 274. **15 and 17**

Motivating the Lesson

Situational Problem Ask students to describe a parabola. Have a volunteer draw a parabola on the chalkboard. Ask students to identify common objects that have this shape. Possible examples include a satellite dish and an automobile headlight.

2 TEACH

This modeling activity illustrates that quadratic functions have a vertical axis of symmetry.

In-Class Example

For Example 1
Find the equation of the axis of symmetry and the coordinates of the vertex of the graph of each equation. Then graph each equation.

a. $y = x^2 - 4x - 5$
$x = 2$; $(2, -9)$

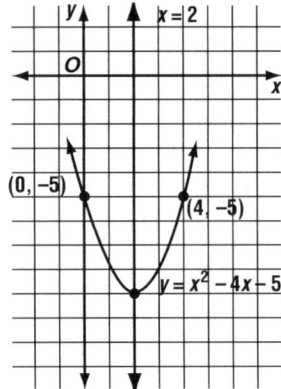

b. $y = -x^2 + 2x - 1$
$x = 1$; $(1, 0)$

Teaching Tip The axis of symmetry is helpful in Example 1 for determining which elements of the domain to choose when graphing a function.

Parabolas possess a geometric property called **symmetry.** Symmetrical figures are those in which the figure can be folded and each half matches the other exactly. The following activity explores the symmetry of a parabola.

MODELING MATHEMATICS — Symmetry of Parabolas

Materials:  grid paper

Your Turn

a. Graph $y = x^2 - 6x + 5$ on grid paper.

b. Hold your paper up to the light and fold the parabola in half so the two sides match exactly.

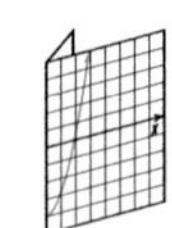

c. Unfold the paper. Which point on the parabola lies on the fold line? **$(3, -4)$**

d. Write an equation to describe the fold line **$x = 3$**

e. Write a few sentences to describe the symmetry of a parabola based on your findings in this activity. **See students' work.**

The fold line in the activity above is called the **axis of symmetry** for the parabola. Each point on the parabola that is on one side of the axis of symmetry has a corresponding point on the parabola on the other side of the axis. The vertex is the only point on the parabola that is on the axis of symmetry. The equation for the axis of symmetry can be determined from the equation of the parabola.

Equation of the Axis of Symmetry of a Parabola 	**The equation of the axis of symmetry for the graph of $y = ax^2 + bx + c$, where $a \neq 0$, is $x = -\frac{b}{2a}$.**

You can determine a lot of information about a parabola from its equation.

Example 1 **Given the equation $y = x^2 - 4x + 5$,**

a. find the equation of the axis of symmetry,

b. find the coordinates of the vertex of the parabola, and

c. graph the equation.

a. In the equation $y = x^2 - 4x + 5$, $a = 1$ and $b = -4$. Substitute these values into the equation of the axis of symmetry.

$$x = -\frac{b}{2a}$$
$$= -\frac{-4}{2(1)} \text{ or } 2$$

b. Since the equation for the axis of symmetry is $x = 2$ and the vertex lies on the axis, the x-coordinate for the vertex is 2.

$$y = x^2 - 4x + 5$$
$$= 2^2 - 4(2) + 5 \quad \textit{Replace x with 2.}$$
$$= 4 - 8 + 5 \text{ or } 1$$

The coordinates of the vertex are (2, 1).

GLENCOE Technology

Interactive Mathematics Tools Software

This multimedia software helps students observe changes in graphs when the quadratic coefficient, linear coefficient, and constant term are manipulated. A **Computer Journal** gives students an opportunity to write about what they have learned.

For Windows & Macintosh

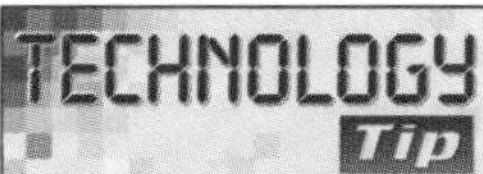

You can use the TABLE feature on a graphing calculator to help you find ordered pairs that satisfy an equation.

c. You can use the symmetry of the parabola to help you draw its graph. Draw a coordinate plane. Graph the vertex and axis of symmetry. Choose a value less than 2, say 0, and find the y-coordinate that satisfies the equation.

$y = x^2 - 4x + 5$
$= 0^2 - 4(0) + 5$ or 5

Graph (0, 5). Since the graph is symmetrical, you can find another point on the other side of the axis of symmetry. The point at (0, 5) is 2 units left of the axis. Go 2 units right of the axis and plot a point at (4, 5). Repeat this for several other points. Then sketch the parabola.

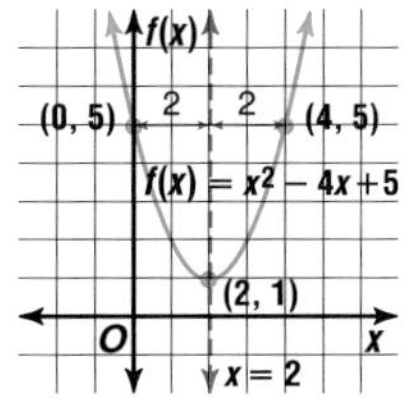

Check: Does (4, 5) satisfy the equation?

$y = x^2 - 4x + 5$
$5 \stackrel{?}{=} 4^2 - 4(4) + 5$
$5 \stackrel{?}{=} 16 - 16 + 5$
$5 = 5$ ✓

The point (4, 5) satisfies the equation $y = x^2 - 4x + 5$ and is part of the graph.

The graph of an equation describing the height of an object propelled into the air can be an example of a quadratic function. Because gravity overtakes the initial force of the object, the object falls back to Earth.

Example 2

APPLICATION

Pyrotechnics

The largest firework ever exploded was a 1543-pound shell called the Universe I Part II. It was used as part of the Lake Toya Festival in Hokkaido, Japan, on July 15, 1988, and had a burst 3937 feet in diameter.

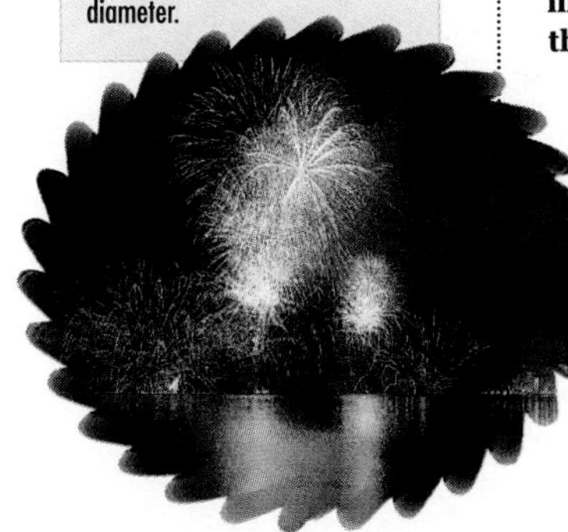

In Cincinnati, Ohio, radio station WEBN and Rozzi's Famous Fireworks, Inc. team together to develop a large fireworks display for Labor Day. The fireworks are coordinated with recorded music compiled by the radio station's DJs. Mr. Rozzi assumes that each rocket (firework) explodes at approximately its highest point after being propelled from a barge in the Ohio River. The formula that describes the height $H(t)$ of an object propelled into the air is $H(t) = v_0t - \frac{1}{2}gt^2 + h_0$, where v_0 represents the initial velocity in m/s, t represents time in seconds, g represents the acceleration of gravity (about 9.8 m/s^2), and h_0 is the initial height of the object at the time it is launched. A certain rocket has an initial velocity of 39.2 m/s and is launched 1.6 meters above the surface of the water.

a. Graph the equation representing the height of the rocket.

b. What is the maximum height that the rocket achieves?

c. One of these rockets is scheduled to explode 2 minutes and 28 seconds into the program. When should the rocket be fired from the barge?

(continued on the next page)

In-Class Example

For Example 2
Find the coordinates of the maximum or minimum point of the graph of each equation.

a. $y = 5x^2 - 10x - 15$ **(1, −20)**
b. $y = -2x^2 + 8x - 1$ **(2, 7)**

Teaching Tip In Example 2, point out that the maximum or minimum value of a quadratic function is always the y-coordinate of the point of intersection of the graph of the function and its axis of symmetry.

FYI

There are two main types of fireworks—force-and-spark and flame. Both use potassium nitrate. Colorful fireworks combine metal salts.

Alternative Teaching Strategies

Reading Algebra The prefix *quad-* usually means "four." Warn students not to be misled. Quadratic functions are of degree 2 and have at most 2 roots.

3 PRACTICE/APPLY

Check for Understanding

Exercises 1–14 are designed to help you assess your students' understanding through reading, writing, speaking, and modeling. You should work through Exercises 1–6 with your students and then monitor their work on Exercises 7–14.

Additional Answers

2. Sample answer: Once you determine the axis of symmetry, there is a matching point on the other side of the axis for each point you find on one side. Suppose the axis is $x = 4$ and one point is $(-2, 3)$. -2 is 6 units left of the axis; another point is 6 units right of the axis with the same y-coordinate, that is, $(10, 3)$.
4. Angie is correct. The graph of $y = (-x)^2$ is the same graph as $y = x^2$, which opens upward from the origin. The graph of $y = -x^2$ opens downward from the origin.

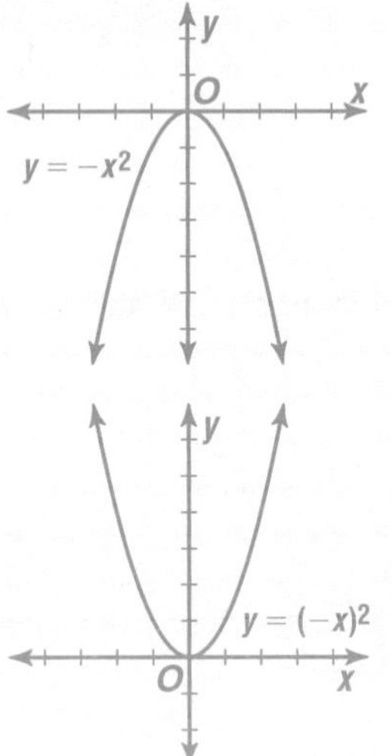

6.

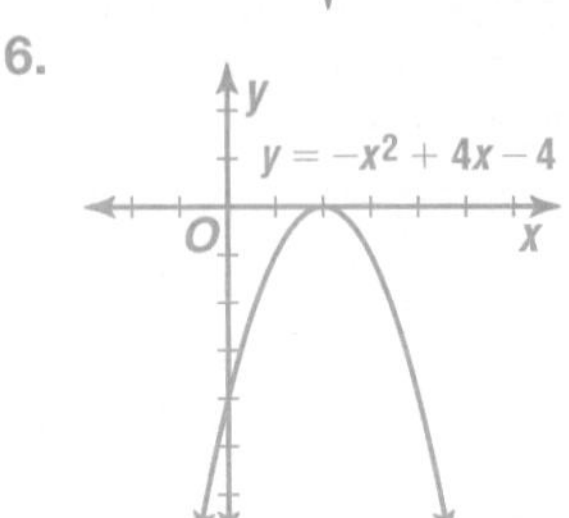

a. Use the given velocity and the force of gravity to determine the formula.

$H(t) = v_0t - \frac{1}{2}gt^2 + h_0$

$= 39.2t - \frac{1}{2}(9.8)t^2 + 1.6$ *Replace v_0, g, and h_0 with the given values.*

$= 39.2t - 4.9t^2 + 1.6$ *Simplify.*

The values of a and b are -4.9 and 39.2, respectively.

$H(t)$ corresponds to $f(x)$ and t corresponds to x.

Find the equation of the axis of symmetry to help you locate the vertex.

$t = -\frac{b}{2a}$

$= -\frac{39.2}{2(-4.9)}$ or 4 seconds

Use a calculator to find $H(t)$ when $t = 4$.

$H(4) = 39.2(4) - 4.9(4)^2 + 1.6$ or 80 meters

The vertex of the graph is at (4, 80).

We can use a table of values to find other points that satisfy the equation.

The graph shows the height of the rocket at any given time. It does not show the path that the rocket took.

t	$H(t)$
0	1.6
1	35.9
2	60.4
3	75.1
4	80.0
5	75.1
6	60.4
7	35.9
8	1.6
9	−42.5

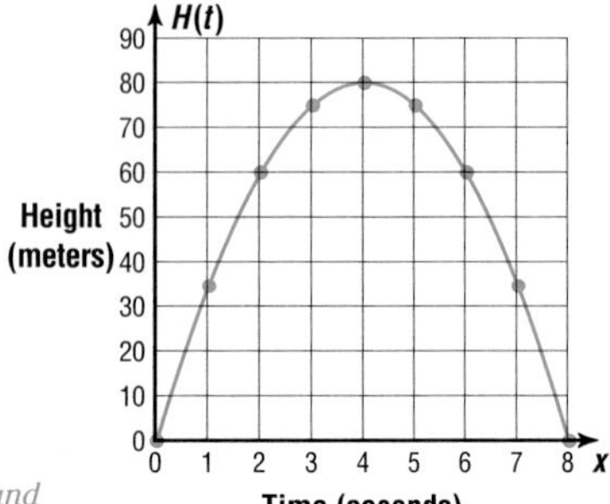

Since time is the independent variable and height is the dependent variable, the domain and range must both be positive.

b. The maximum height is at the vertex. This height is 80 meters.

c. Since the maximum height is achieved 4 seconds after launch, the rocket must be fired 4 seconds before its scheduled explosion at 2 minutes 28 seconds into the program. The rocket should be fired at 2 minutes 24 seconds into the program.

CHECK FOR UNDERSTANDING

Communicating Mathematics

Study the lesson. Then complete the following.

1. **Determine** the distance between the bases of the Gateway Arch. 630 ft
2. **Explain** how you can use symmetry to help you graph a parabola.
3. **Write** a sentence to explain how you can write the equation of the axis of symmetry if you know the coordinates of the vertex of a parabola.
4. **You Decide** Lisa says that $y = -x^2$ is the same as $y = (-x)^2$. Angie says that they are different. Who is correct? Explain and include graphs.
5. How can you determine if the vertex is a maximum or minimum without graphing the equation first?
6. Graph $y = -x^2 + 4x - 4$. How does this graph differ from the one in the activity on page 612? Use paper folding to determine the axis of symmetry.

2, 4. See margin.

3. If (p, q) is the vertex, $x = p$ is the equation of the axis of symmetry.

5. If the coefficient of x^2 is positive, it is a minimum. If the coefficient is negative, it is a maximum.

6. It opens downward; see margin for graph; $x = 2$.

Reteaching

Using Alternative Methods Have students graph the following equations on the same graph. Have them describe the changes that occur in the graph with each equation. Generalize the results and have students extend them to other equations.

1. $y = x^2$ **2.** $y = x^2 - 1$
3. $y = x^2 - 2x$ **4.** $y = x^2 - 2x - 1$

Guided Practice

7–12. See Solutions Manual for graphs.

9. $x = -2$, $(-2, -13)$, min.

10. $x = 7$, $(7, -36)$, min.

Write the equation of the axis of symmetry and find the coordinates of the vertex of the graph of each equation. State if the vertex is a maximum or minimum. Then graph the equation.

7. $y = x^2 + 2$ $x = 0$, $(0, 2)$, min.
8. $y = -2x^2$ $x = 0$, $(0, 0)$, max.
9. $y = x^2 + 4x - 9$
10. $y = x^2 - 14x + 13$
11. $y = -x^2 + 5x + 6$ $x = 2.5$, $(2.5, 12.25)$, max.
12. $y = -2x^2 + 4x + 6.5$ $x = 1$, $(1, 8.5)$, max.
13. Which equation describes the graph at the right? c
    a. $f(x) = x^2 - 6x + 9$
    b. $f(x) = -x^2 + 6x + 9$
    c. $f(x) = x^2 + 6x + 9$

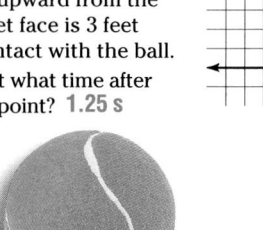

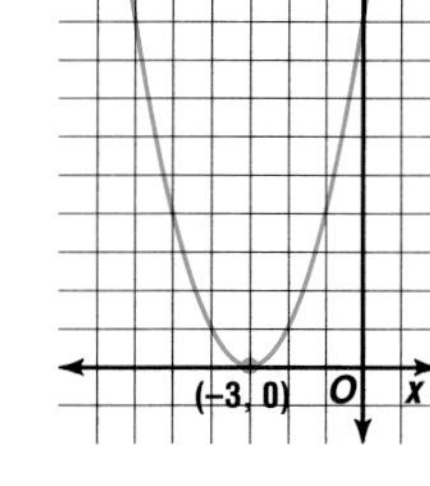

14. **Tennis** A tennis ball is propelled upward from the face of a racket at 40 ft/s. The racket face is 3 feet from the ground when it makes contact with the ball.
    a. If the force of gravity is 32 ft/s^2, at what time after hitting the ball is it at its highest point? 1.25 s
    b. How high does it go? 28 ft

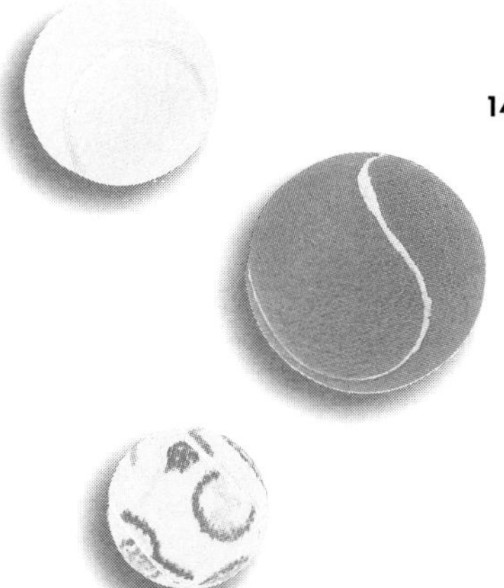
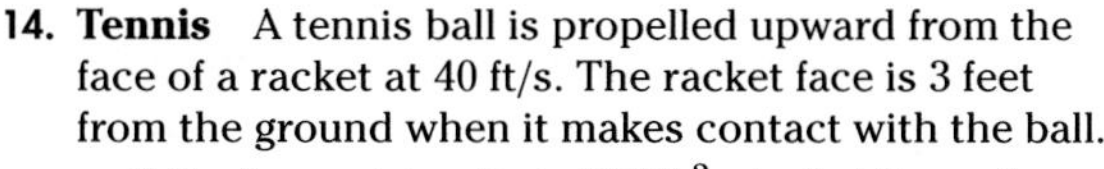

15–29. See Solutions Manual for graphs.

EXERCISES

Practice

15. $x = 0$, $(0, 0)$, min.
16. $x = 2$, $(2, 3)$, max.
17. $x = -1$, $(-1, 17)$, min.
18. $x = \frac{3}{2}$, $\left(\frac{3}{2}, -\frac{49}{4}\right)$, min.
19. $x = 0$, $(0, -5)$, min.
20. $x = 0$, $(0, 16)$, min.
21. $x = -3$, $(-3, -29)$, min.
22. $x = -4$, $(-4, 32)$, min.
23. $x = 0$, $(0, -25)$, min.
24. $x = -3$, $(-3, 24)$, max.
25. $x = -1$, $(-1, 7)$, max.
26. $x = 4$, $(4, 37)$, max.

Write the equation of the axis of symmetry and find the coordinates of the vertex of the graph of each equation. State if the vertex is a maximum or minimum. Then graph the equation.

A
15. $y = 4x^2$
16. $y = -x^2 + 4x - 1$
17. $y = x^2 + 2x + 18$
18. $y = x^2 - 3x - 10$
19. $y = x^2 - 5$
20. $y = 4x^2 + 16$
21. $y = 2x^2 + 12x - 11$
22. $y = 3x^2 + 24x + 80$
23. $y = x^2 - 25$
24. $y = 15 - 6x - x^2$
25. $y = -3x^2 - 6x + 4$
26. $y = 5 + 16x - 2x^2$

B
27. $y = 3(x + 1)^2 - 20$ $x = -1$, $(-1, -20)$, min.
28. $y = -(x - 2)^2 + 1$ $x = 2$, $(2, 1)$, max.
29. $y = \frac{2}{3}(x + 1)^2 - 1$ $x = -1$, $(-1, -1)$, min.

Match each equation with its graph.

30. $f(x) = \frac{1}{2}x^2 + 1$ b
31. $f(x) = -\frac{1}{2}x^2 + 1$ c
32. $f(x) = \frac{1}{2}x^2 - 1$ a

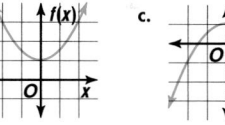

a. 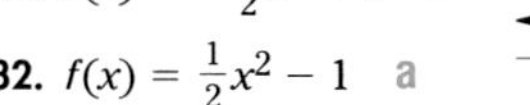
b. 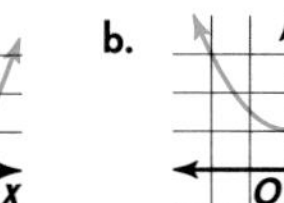
c. 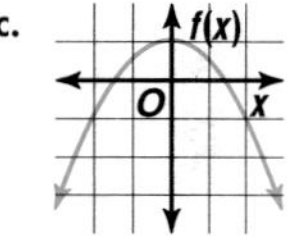

Graph each equation. 33–36. See margin.

C
33. $y + 2 = x^2 - 10x + 25$
34. $y + 1 = 3x^2 + 12x + 12$
35. $y + 3 = -2(x - 4)^2$
36. $y - 5 = \frac{1}{3}(x + 2)^2$
37. What is the equation of the axis of symmetry of a parabola if its x-intercepts are -6 and 4? $x = -1$

Additional Answers

35.

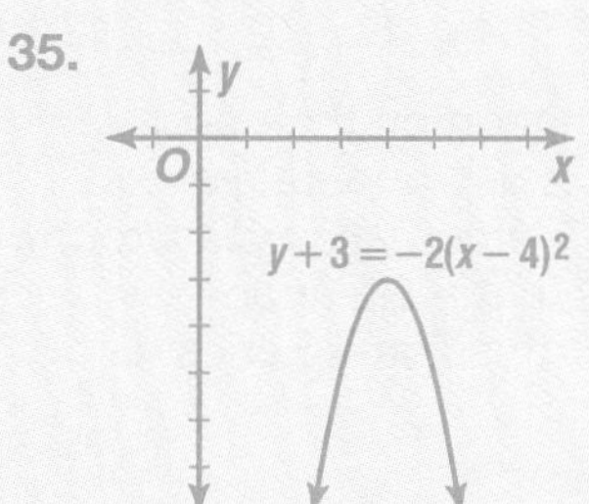

36.

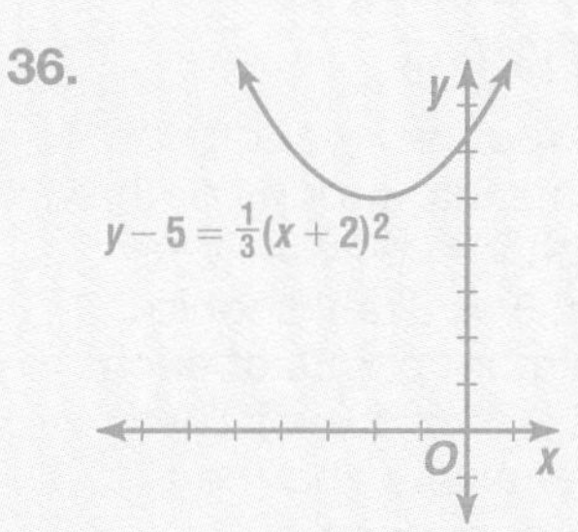

Assignment Guide

Core: 15–43 odd, 44, 45, 47–54
Enriched: 16–42 even, 44–54

For **Extra Practice,** see p. 780.

The red A, B, and C flags, printed only in the Teacher's Wraparound Edition, indicate the level of difficulty of the exercises.

Additional Answers

33.

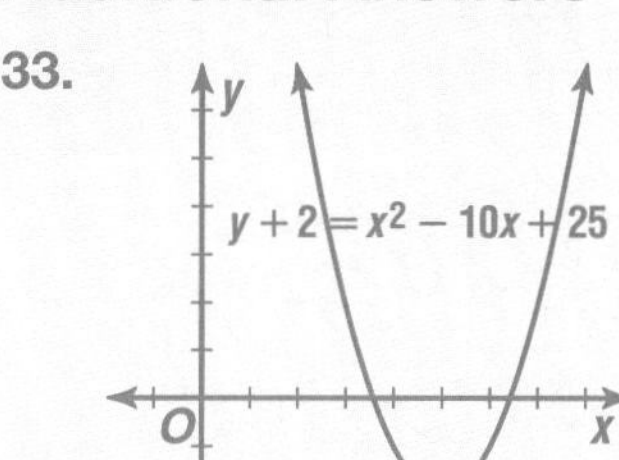

34.

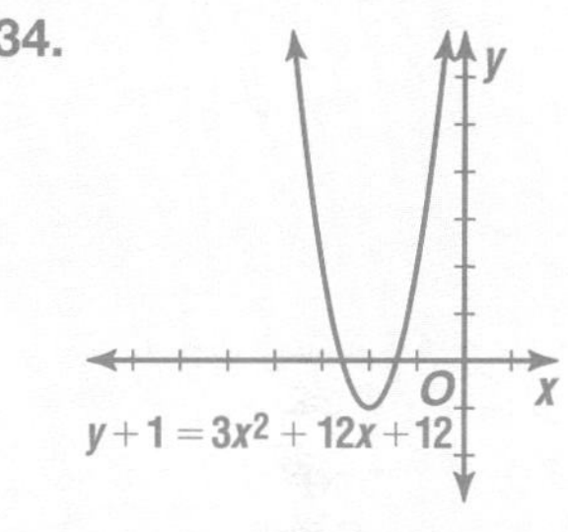

Study Guide Masters, p. 75

11-1 NAME ________ DATE ________

Study Guide

Student Edition Pages 611–617

Graphing Quadratic Functions

An equation such as $y = x^2 - 4x + 1$ describes a type of function known as a **quadratic function.**

Definition of Quadratic Function
A quadratic function is a function that can be described by an equation of the form $y = ax^2 + bx + c$, where $a \neq 0$.

Graphs of quadratic functions have certain common characteristics. They have a general shape called a **parabola.** They have a minimum or maximum point. In general, a parabola will open upward and have a minimum point when the coefficient of y and x^2 have the same sign. It will open downward and have a maximum point when the coefficient of y and x^2 are opposite in sign. The vertical line containing the minimum or maximum point is the **axis of symmetry.** The minimum or maximum point is called the **vertex** of the parabola.

Equation of Axis of Symmetry
The equation of the axis of symmetry of $y = ax^2 + bx + c$, where $a \neq 0$ is $x = -\frac{b}{2a}$.

Write the equation of the axis of symmetry and find the coordinates of the vertex of the graph of each equation. Then graph the equation.

1. $y = x^2 + 3x$
2. $y = 3x^2 + x + 1$
3. $y = -x^2 - 4$

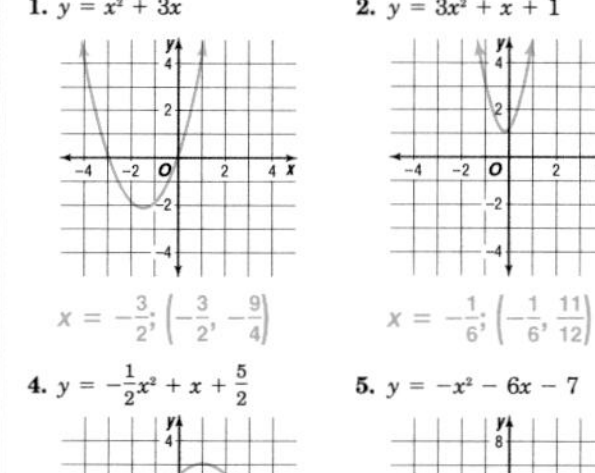
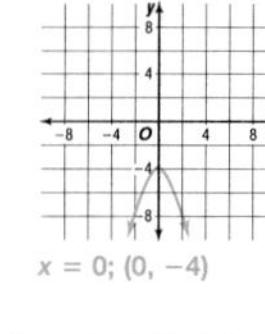

$x = -\frac{3}{2}; \left(-\frac{3}{2}, -\frac{9}{4}\right)$ $x = -\frac{1}{6}; \left(-\frac{1}{6}, \frac{11}{12}\right)$ $x = 0; (0, -4)$

4. $y = -\frac{1}{2}x^2 + x + \frac{5}{2}$
5. $y = -x^2 - 6x - 7$
6. $y = 2x^2 + 12x + 9$

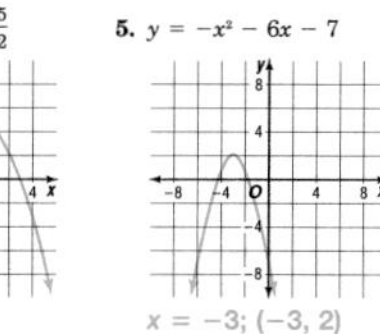
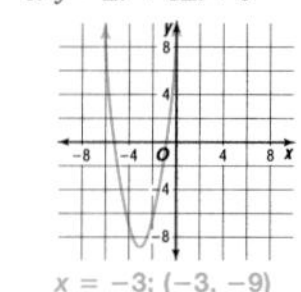

$x = 1; (1, 3)$ $x = -3; (-3, 2)$ $x = -3; (-3, -9)$

Additional Answers

40.
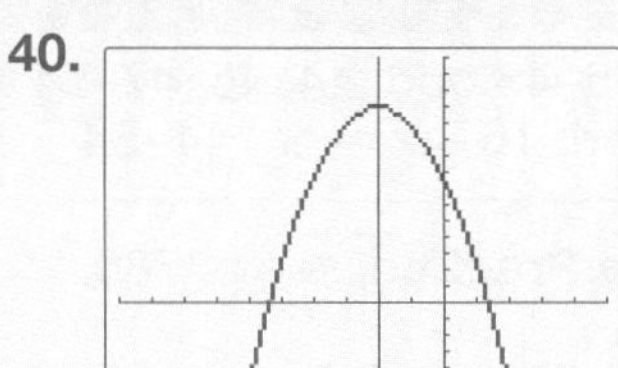

41.
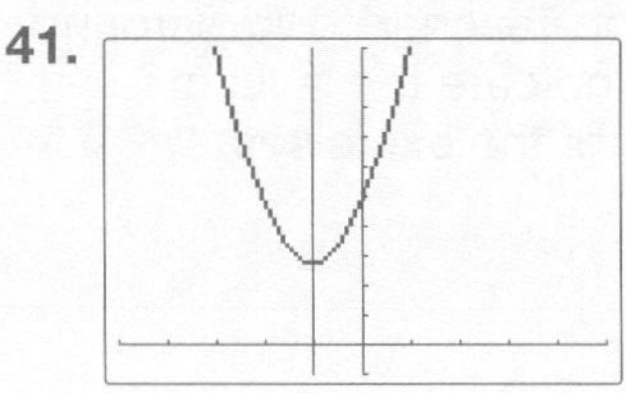

42.
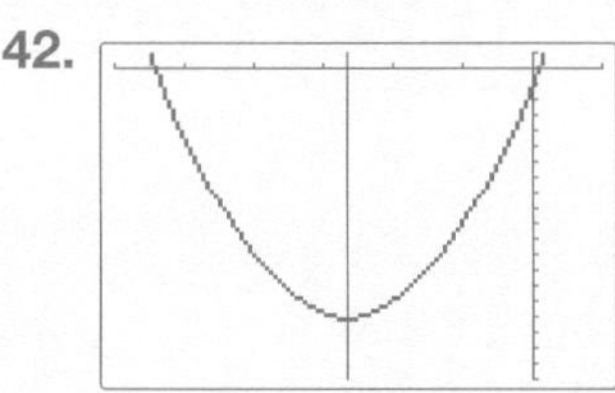

43.
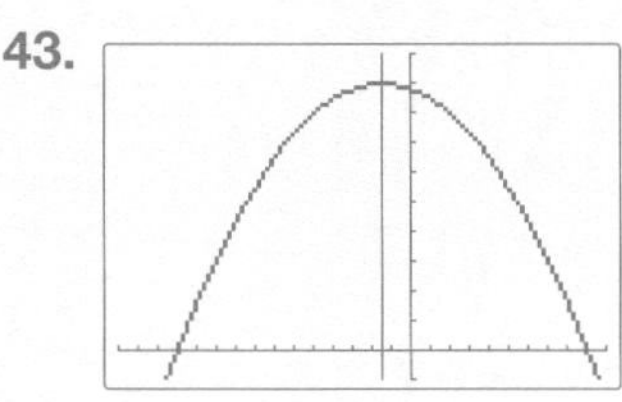

44.
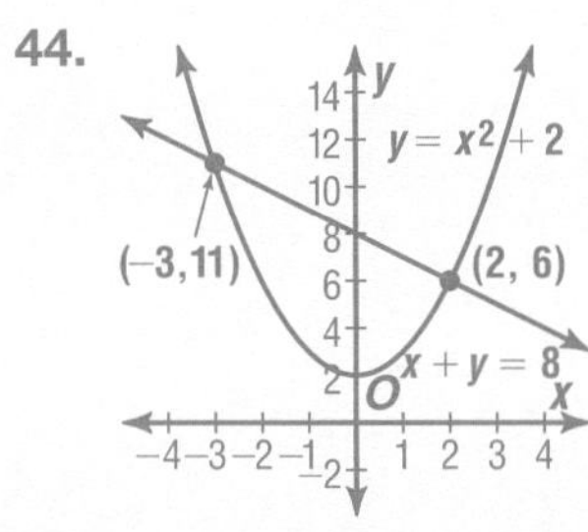

Practice Masters, p. 75

11-1 NAME ______ DATE ______

Practice

Student Edition Pages 611–617

Graphing Quadratic Functions

Write the equation of the axis of symmetry and find the coordinates of the vertex of the graph of each quadratic function.

1. $y = x^2 - 2x - 8$ — $x = 1$, $(1, -9)$
2. $y = -x^2 + x + 12$ — $x = \frac{1}{2}$, $\left(\frac{1}{2}, \frac{49}{4}\right)$
3. $y = 8x^2 + 12x$ — $x = -\frac{3}{4}$, $\left(-\frac{3}{4}, -\frac{9}{2}\right)$
4. $y = -2x^2 + x + 2$ — $x = \frac{1}{4}$, $\left(\frac{1}{4}, \frac{17}{8}\right)$

Write the equation of the axis of symmetry and find the coordinates of the vertex of the graph of each quadratic function. State if the vertex is a maximum or minimum. Then graph the equation.

5. $y = x^2 - 3x + 2$

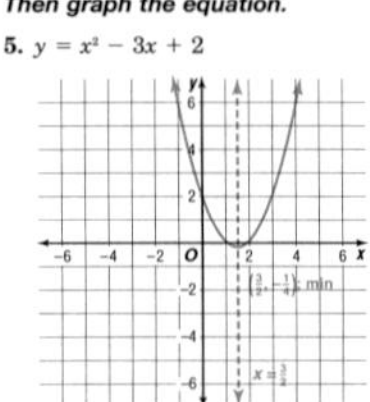

6. $y = -x^2 - 2x + 3$

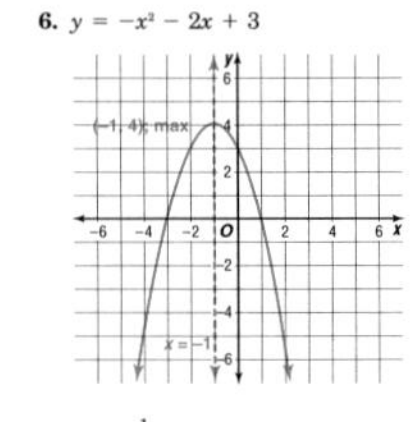

7. $y = x^2 - 2$ 8. $y = -\frac{1}{4}x^2$

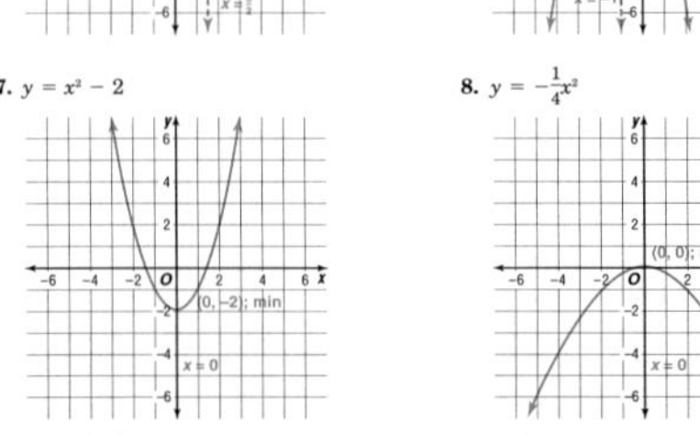

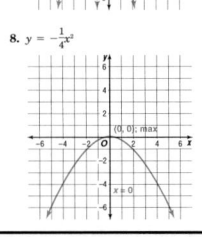

38. The vertex of a parabola is at $(-4, -3)$. If one x-intercept is -11, what is the other x-intercept? 3

39. Two points on a parabola are at $(-8, 7)$ and $(12, 7)$. What is the equation of the axis of symmetry? $x = 2$

Graphing Calculator

You can draw the axis of symmetry on a graphed parabola by using the VERTICAL command on the DRAW menu. Press [2nd] [DRAW] 4 and use the arrow keys to move the line into place. To graph the axis of symmetry before the graph is drawn, enter the equation in the Y= list and press [2nd] [DRAW] 4 (the *x* value through which the line will pass) [ENTER]. Graph each equation and its axis of symmetry. Make a sketch of each graph on your paper, labeling the vertex of the graph.

Sample answers:
40. (−2.00, 12)
41. (−1.10, 125.8)
42. (−268.04, −1719.47)
43. (−0.15, 90.14)
40–43. See margin for graphs.

40. $y = 8 - 4x - x^2$
41. $y = 20x^2 + 44x + 150$
42. $y = 0.023x^2 + 12.33x - 66.98$
43. $y = -78.23x^2 - 23.76x + 88.34$

Critical Thinking

44. Graph $y = x^2 + 2$ and $x + y = 8$ on the same coordinate plane. What are the coordinates of the points they have in common? Explain how you determined these points. See margin for graph; (−3, 11), (2, 6); see students' work.

Applications and Problem Solving

45b. D: $0 \leq t \leq 23$; R: $326 < U(t) < 1746$
45d. See students' work.
46a. $H(t) = 80t - 16t^2 + 2$
46b. 66 ft, 98 ft, 98 ft

45. **College** The cost of a college education is increasing every year. Many parents are starting college funds for their children before they are even born. The average tuition and fees for public college during the years 1970–1993 can be estimated using the function $U(t) = 2.97t^2 - 6.78t + 329.96$, where $U(t)$ represents tuition and fees (in dollars) for one year and t represents the number of years after 1970.

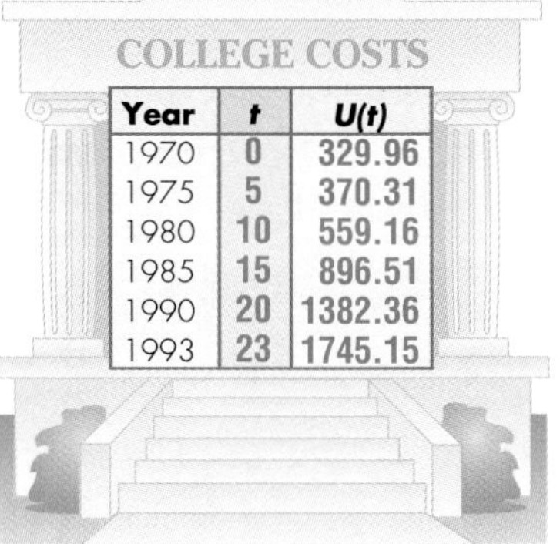
COLLEGE COSTS

Year	t	U(t)
1970	0	329.96
1975	5	370.31
1980	10	559.16
1985	15	896.51
1990	20	1382.36
1993	23	1745.15

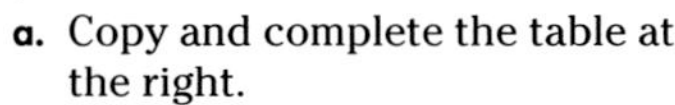

a. Copy and complete the table at the right.

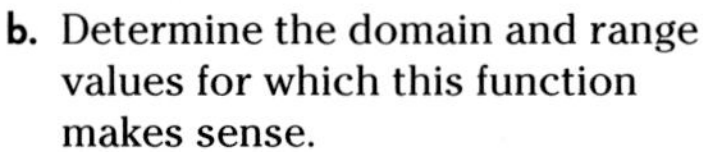

b. Determine the domain and range values for which this function makes sense.
c. Graph this function. See margin.
d. Assume that this function is a model for all years after 1993. How much will the tuition and fees for your first year of college be if you attend a public college?

46. **Football** When a football player punts a football, he hopes for a long "hang time," the total amount of time the ball stays in the air. Any hang time of more than about 4.5 seconds is usually good. Manuel is the punter for his high school team. He can kick the ball with an upward velocity of 80 ft/s, and his foot meets the ball 2 feet off the ground.
a. Write a quadratic equation to describe the height of the football at any given time t. Use 32 ft/s^2 as the acceleration of gravity.
b. How high is the ball after 1 second? 2 seconds? 3 seconds?
c. What is Manuel's hang time? about 5 seconds

Mixed Review

47. **Agriculture** A field is 1.2 kilometers long and 0.9 kilometers wide. A farmer begins plowing the field by starting at the outer edge and going all the way around the field. When he stops for lunch, a strip of uniform width has been plowed on all sides of the field and half the field is plowed. What is the width of the strip? (Lesson 10–7) 0.15 km

Cooperative Learning

Co-op Co-op You may want to separate students into groups of three. Have them graph $y = x^2$, $y = x^2 - 4$, and $y = (x + 2)^2 + 5$. They should identify the minimum points of each graph. (0, 0), (0, −4), (−2, 5) They should compare results and then predict the minimum point of $y = x^2 + 7$. (0, 7)

For more information on the co-op co-op strategy, see *Cooperative Learning in the Mathematics Classroom,* one of the titles in the Glencoe Mathematics Professional Series, page 31.

48. Find $(x - 4)(x - 8)$. (Lesson 9–7) $x^2 - 12x + 32$

49. **Astronomy** Mars, located 227,920,000 kilometers from the sun, has a diameter of 6.79×10^3 kilometers. (Lesson 9–3) **49b.** 2.2792×10^8
   a. Write the diameter of Mars in decimal notation. **6790**
   b. Write the distance Mars is from the sun in scientific notation.

50. Use elimination to solve the system of equations. (Lesson 8–4)

$3x + 4y = -25$

$2x - 3y = 6$ **(−3, −4)**

51. **Meteorology** The table below lists the record 24-hour precipitation for each state as of 1990. (Lesson 7–7)

State	Inches	State	Inches	State	Inches	State	Inches	State	Inches
AL	20.33	HI	38.00	MA	18.15	NM	11.28	SD	8.00
AK	15.20	ID	7.17	MI	9.78	NY	11.17	TN	11.00
AZ	11.40	IL	16.54	MN	10.84	NC	22.22	TX	43.00
AR	14.06	IN	10.50	MS	15.68	ND	8.10	UT	6.00
CA	26.12	IA	16.70	MO	18.18	OH	10.51	VT	8.77
CO	11.08	KS	12.59	MT	11.50	OK	15.50	VA	27.00
CT	12.77	KY	10.40	NE	13.15	OR	10.17	WA	12.00
DE	8.50	LA	22.00	NV	7.40	PA	34.50	WV	19.00
FL	38.70	ME	8.05	NH	10.38	RI	12.13	WI	11.72
GA	18.00	MD	14.75	NJ	14.81	SC	13.25	WY	6.06

Source: National Climatic Data Center

   a. Make a box-and-whisker plot of the data. **See margin.**
   b. Are there any outliers? If so, list them. **34.5, 38, 38.7, 43**

52. Graph $y = 3x + 4$. (Lesson 5–5) **See margin.**

53. Solve $3x = -15$. (Lesson 3–2) **−5**

54. **Patterns** Complete the pattern: 3, 6, 12, 24, __?__, __?__. (Lesson 1–2) **48, 96**

WORKING ON THE

In•ves•ti•ga•tion

Refer to the Investigation on pages 554–555.

Examine the table you created in Lesson 10–7 that included the length, width, perimeter, and area of the rectangular brick patterns. Measure each of your models in millimeters to determine the values of x and y and record the measures.

1 Graph the data, plotting the length of the pattern on the horizontal axis and the area of the pattern on the vertical axis. What kind of a graph is it? What relationship does it show?

2 Make another graph, plotting the perimeter of the pattern on the horizontal axis and the area of the pattern on the vertical axis. What kind of a graph is it? What kind of relationship does it show?

3 What is the relationship between the measures of the area, length, and width? Why are there some rectangular patterns that have the same area but different perimeters?

Add the results of your work to your Investigation Folder.

4 ASSESS

Closing Activity

Speaking Have students compare the graph of a quadratic function whose coefficients of y^2 and y have the same sign and the graph of a quadratic function whose coefficients have opposite signs.

Additional Answers

45c.

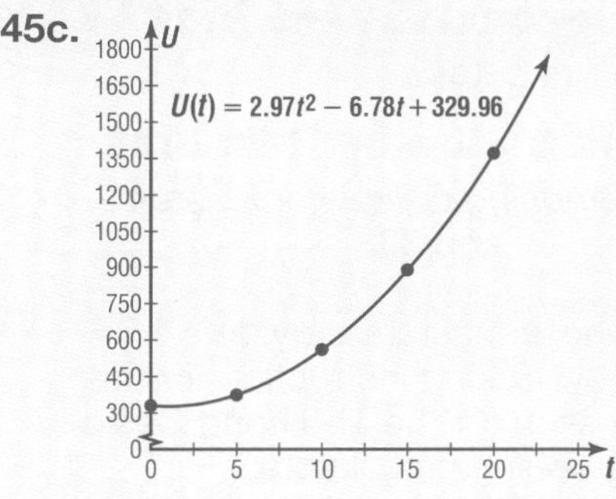

51a.

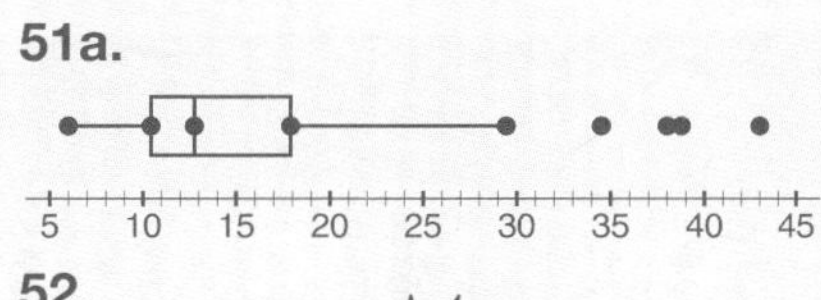

52.

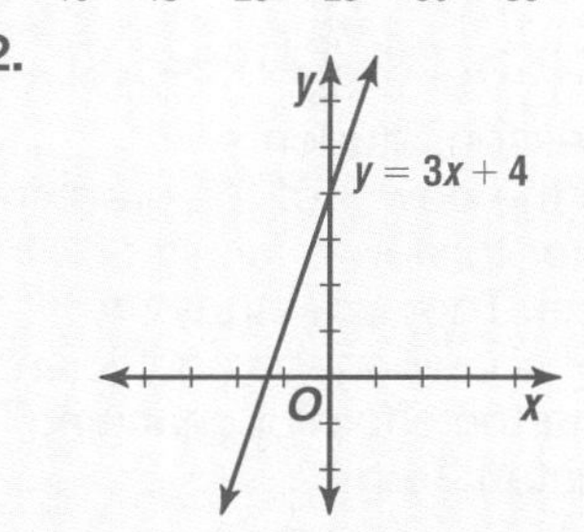

Enrichment Masters, p. 75

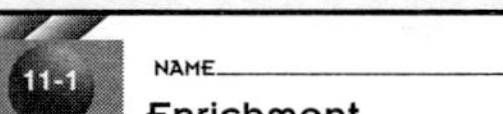

NAME ____ DATE ____

11-1 Enrichment

Student Edition Pages 611–617

Translating Quadratic Graphs

When a figure is moved to a new position without undergoing any rotation, then the figure is said to have been **translated** to the new position.

The graph of a quadratic equation in the form $y = (x - b)^2 + c$ is a translation of the graph of $y = x^2$.

Start with $y = x^2$.
Slide to the right 4 units.
$y = (x - 4)^2$
Then slide up 3 units.
$y = (x - 4)^2 + 3$

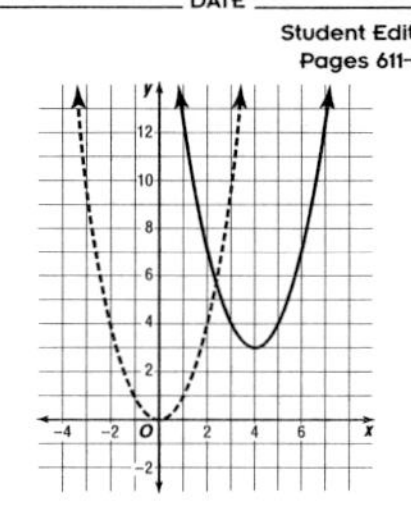

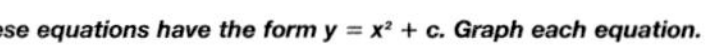

These equations have the form $y = x^2 + c$. Graph each equation.

1. $y = x^2 + 1$ 2. $y = x^2 + 2$ 3. $y = x^2 - 2$

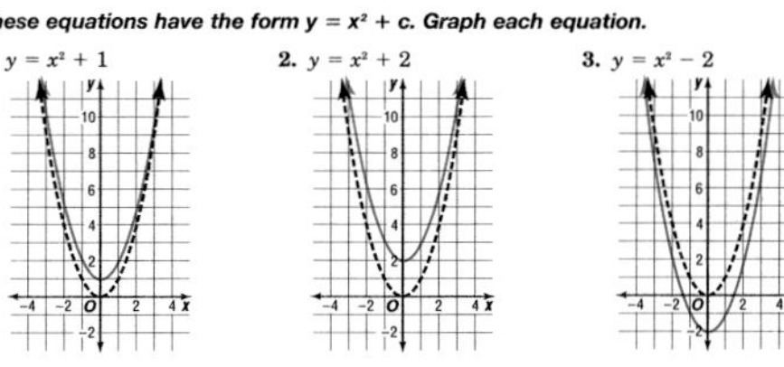

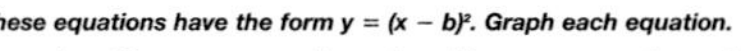

These equations have the form $y = (x - b)^2$. Graph each equation.

4. $y = (x - 1)^2$ 5. $y = (x - 3)^2$ 6. $y = (x + 2)^2$

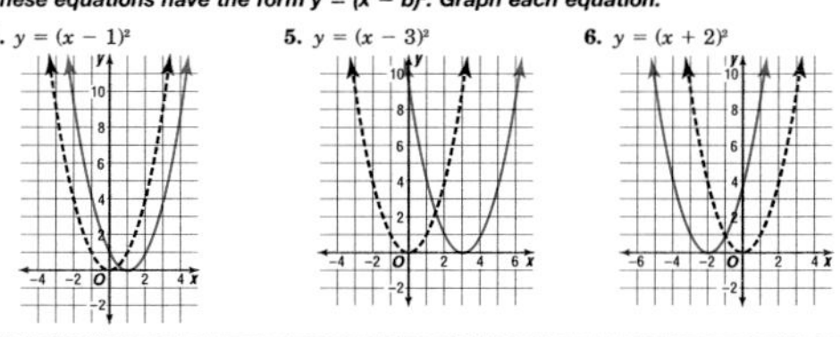

Extension

Problem Solving Graph the following pair of equations on the same coordinate plane. Then compare the graph of the second equation to that of the first equation.

$y = x^2$

$y = (x + 6)^2 - 8$

The second graph is 6 units to the left and 8 units below the first graph.

In•ves•ti•ga•tion

Working on the Investigation

The Investigation on pages 554–555 is designed to be a long-term project that is completed over several days or weeks. Encourage students to keep their materials in their Investigation Folder as they work on the Investigation.

NCTM Standards: 1–6

Objective
Use a graphing calculator to study the characteristics of families of parabolas.

Recommended Time
25 minutes

Instructional Resources
Graphing Calculator Masters, pp. 32 and 33

These masters provide keystroking instruction for this lesson for the TI-81 and Casio graphing calculators.

1 FOCUS

Motivating the Lesson
In this lesson, students will explore the power of the graphing calculator to view a family of functions. They will study the effect of changing the coefficient of x^2 and the effect of changing the constant term.

2 TEACH

Teaching Tip Once students have graphed each equation, ask them if a better viewing window than the standard one would give them a better look at the parabolas being graphed.

11–1B Graphing Technology Parent and Family Graphs

An Extension of Lesson 11–1

A family of graphs is a group of graphs that have at least one characteristic in common. In Lesson 6–5A, you learned about families of linear graphs that shared the same slope or y-intercept. Families of parabolas often fall into two categories—those that have the same vertex and those that have the same shape. Graphing calculators make it easy to study the characteristics of families of parabolas.

The parent function in each of these families is $y = x^2$.

LOOK BACK
Refer to Lesson 6-5A for an introduction to parent and family graphs.

Example 1 **Graph each group of equations on the same screen. Compare and contrast the graphs.**

a. $y = x^2$, $y = 2x^2$, $y = 3x^2$

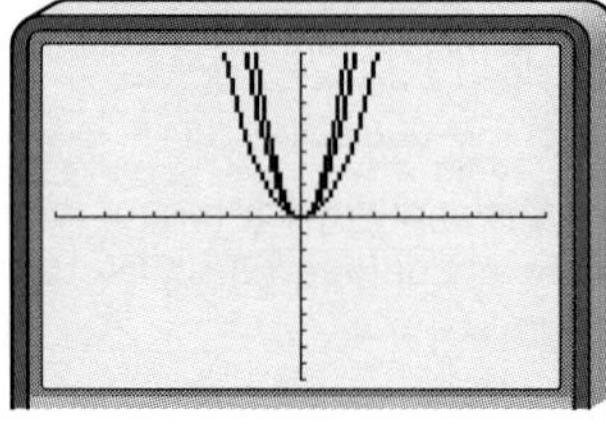

Each graph opens upward and has its vertex at the origin. The graphs of $y = 2x^2$ and $y = 3x^2$ are narrower than the graph of $y = x^2$.

b. $y = x^2$, $y = 0.5x^2$, $y = 0.3x^2$

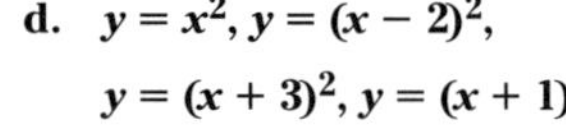

Each graph opens upward and has its vertex at the origin. The graphs of $y = 0.5x^2$ and $y = 0.3x^2$ are wider than the graph of $y = x^2$.

How does the value of a in $y = ax^2$ affect the shape of the graph?

c. $y = x^2$, $y = x^2 + 2$, $y = x^2 - 3$, $y = x^2 - 5$

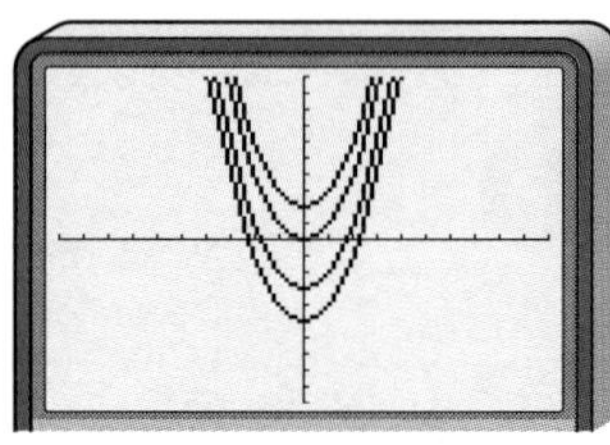

Each graph opens upward and has the same shape as $y = x^2$. However, each parabola has a different vertex, located along the y-axis. *How does the value of the constant affect the placement of the graph?*

d. $y = x^2$, $y = (x - 2)^2$, $y = (x + 3)^2$, $y = (x + 1)^2$

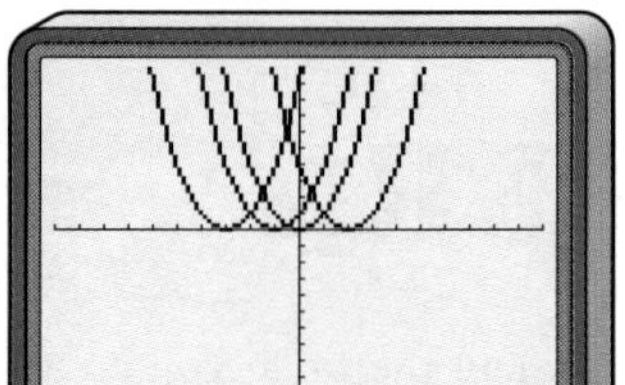

Each graph opens upward and has the same shape as $y = x^2$. However, each parabola has a different vertex, located along the x-axis. *How is the location of the vertex related to the equation of the graph?*

GLENCOE Technology

Interactive Mathematics Tools Software

This multimedia software provides an interactive lesson that uses a quadratic equation to find its axis of symmetry. A **Computer Journal** gives students an opportunity to write about what they have learned.

For Windows & Macintosh

When analyzing or comparing the shapes of various graphs on different screens, it is important to compare the graphs using the same parameters. That is, the window used to compare the graphs should be the same, with the same scale factor. Suppose we graph the same equation using a different window for each. How will this affect the appearance of the graph?

Example 2 **Graph $y = x^2 - 5$ in each viewing window. What conclusions can you draw about the appearance of a graph in the window used?** *The scale is 1 unless otherwise noted.*

a. standard viewing window

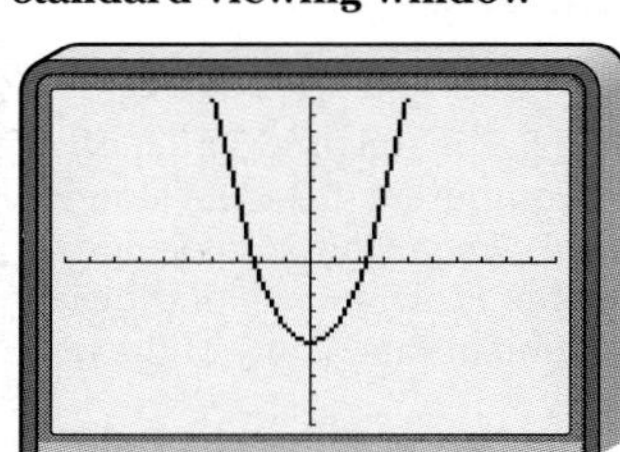

b. [−10, 10] by [−100, 100] Yscl: 20

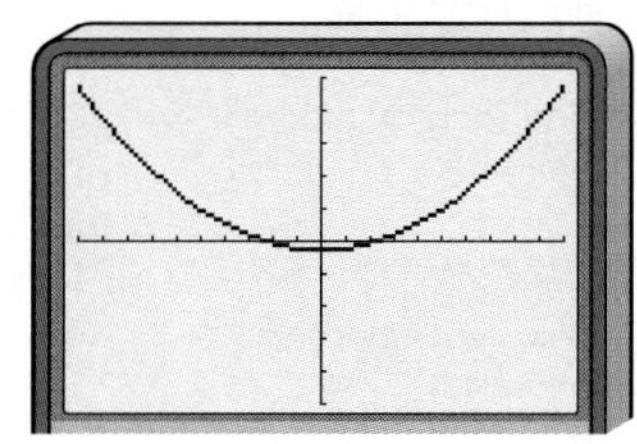

c. [−50, 50] Xscl: 5 by [−10, 10]

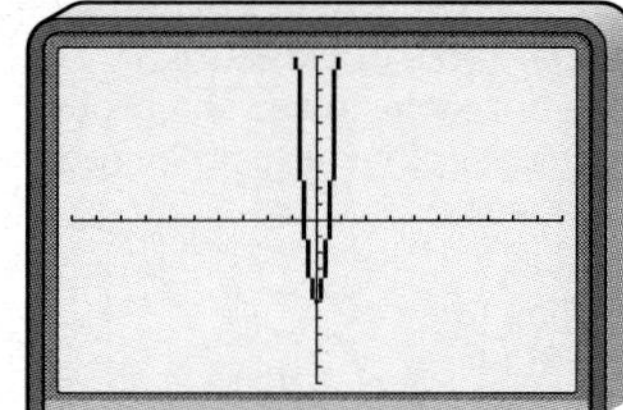

d. [−0.5, 0.5] Xscl: 0.1 by [−10, 10]

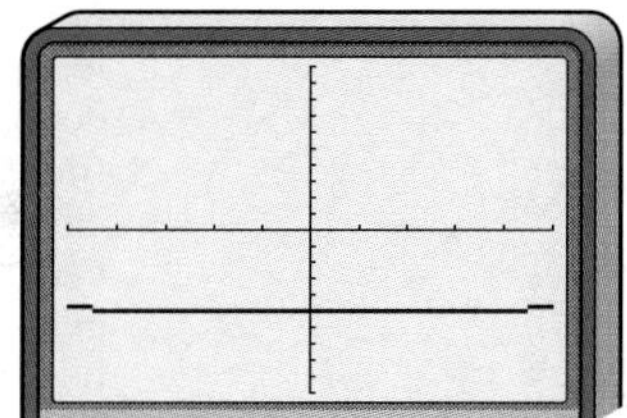

The window greatly affects the appearance of the parabola. Without knowing the window, graph b might be of the family $y = ax^2$, where $0 < a < 1$. Graph c makes the graph look like a member of $y = ax^2$, where $a > 1$. Graph d looks more like a line. However, all are graphs of the same equation.

EXERCISES

Graph each group of equations on the same screen. Make a sketch of the screen on grid paper, and compare and contrast the graphs.

1–4. See margin.

1. $y = -x^2$, $y = -2x^2$, $y = -5x^2$
2. $y = -x^2$, $y = -0.3x^2$, $y = -0.7x^2$
3. $y = -x^2$, $y = -(x + 4)^2$, $y = -(x - 8)^2$
4. $y = -x^2$, $y = -x^2 + 6$, $y = -x^2 - 4$

Use the families of graphs that have appeared in this lesson to predict the appearance of the graph of each equation. Then sketch the graph. 5–9. See Solutions Manual.

5. $y = 4x^2$
6. $y = x^2 - 6$
7. $y = -0.1x^2$
8. $y = (x + 1)^2$
9. Describe how each change in the equation of $y = x^2$ would affect the graph of $y = x^2$. Be sure to consider all values of a and b.
   a. $y = ax^2$
   b. $y = x^2 + a$
   c. $y = (x + a)^2$
   d. $y = (x + a)^2 + b$

Additional Answer

4. All open downward, have the same shape, and have vertices along the y-axis. However, each vertex is different.

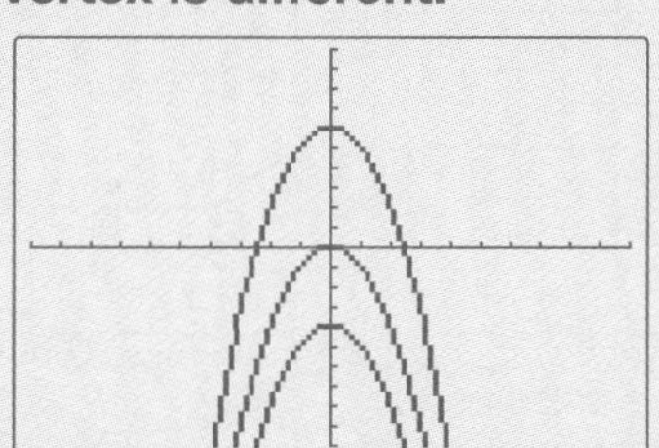

3 PRACTICE/APPLY

Assignment Guide

Core: 1–9
Enriched: 1–9

4 ASSESS

Observing students working with technology is an excellent method of assessment.

Additional Answers

1. All the graphs open downward from the origin. $y = -2x^2$ is narrower than $y = -x^2$, and $y = -5x^2$ is the narrowest.

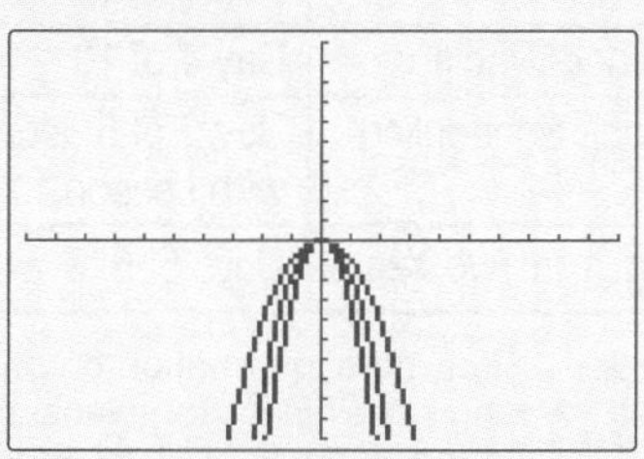

2. All the graphs open downward from the origin. $y = -0.3x^2$ is wider than $y = -x^2$, and $y = -0.7x^2$ is the widest.

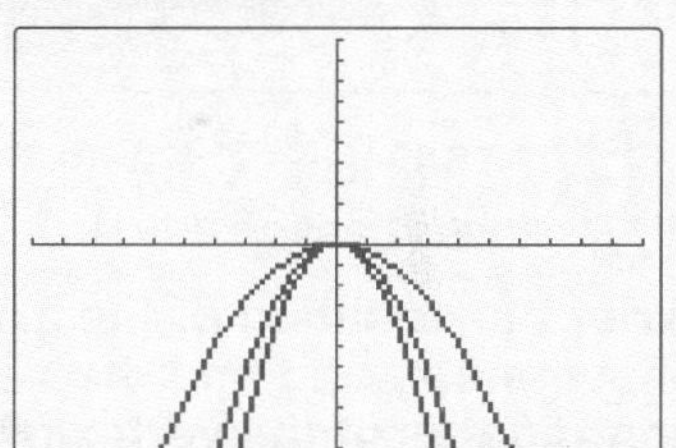

3. All open downward, have the same shape, and have vertices along the x-axis. However, each vertex is different.

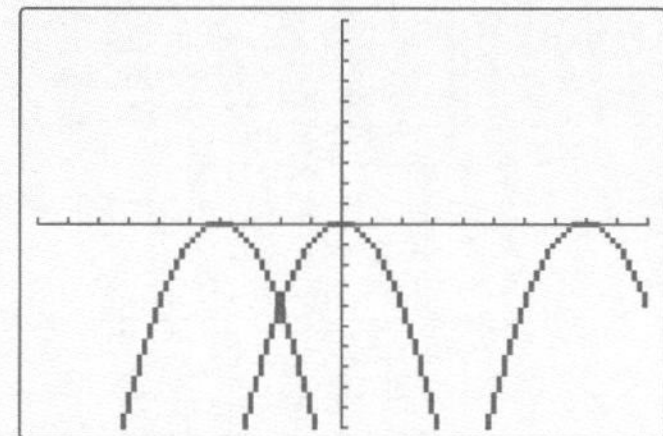

11-2 LESSON NOTES

NCTM Standards: 1–6, 13

Instructional Resources

- Study Guide Master 11-2
- Practice Master 11-2
- Enrichment Master 11-2
- Assessment and Evaluation Masters, p. 296
- Tech Prep Applications Masters, p. 21

Transparency 11-2A contains the 5-Minute Check for this lesson; **Transparency 11-2B** contains a teaching aid for this lesson.

Recommended Pacing	
Standard Pacing	Day 4 of 11
Honors Pacing	Day 4 of 10
Block Scheduling*	Day 2 of 5 (along with Lesson 11-3)
Alg. 1 in Two Years*	Days 5, 6, & 7 of 19

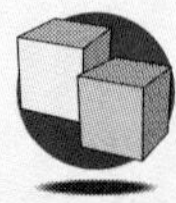

*For more information on pacing and possible lesson plans, refer to the *Block Scheduling Booklet* and *Algebra 1 in Two Years.*

1 FOCUS

5-Minute Check
(over Lesson 11-1)

Write the equation of the axis of symmetry and find the coordinates of the vertex of the graph of each equation. State if the vertex is a maximum or a minimum. Then graph the equation.

1. $y = x^2 - 4x + 7$
minimum, (2, 3), $x = 2$

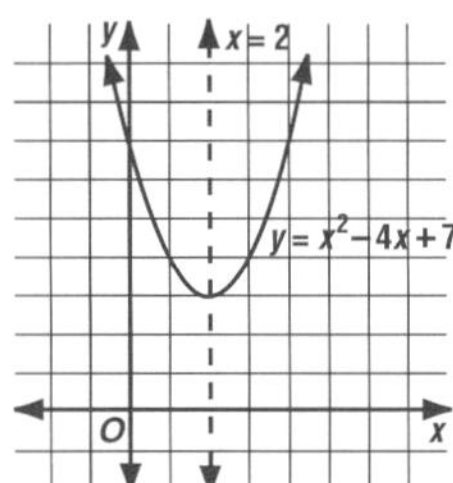

(continued at the right)

11–2 Solving Quadratic Equations by Graphing

***What* YOU'LL LEARN**

- To use estimation to find roots of quadratic equations, and
- to find roots of quadratic equations by graphing.

***Why* IT'S IMPORTANT**

You can use quadratic equations to solve problems involving architecture and number theory.

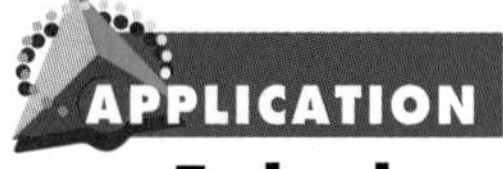

Technology

In the United States, one of the fastest growing industries is the production of CD-ROMs for computers. CD-ROM stands for *Compact Disc-Read Only Memory*. CD-ROMs can hold text, music, photographic images, or combinations of any of these.

Any company that produces a product to sell finds that the profit they can make depends on the number of employees they have (among other things). The relationship between profit and the number of employees looks like the parabola drawn at right. The company does not make much of a profit if there are too few employees; as they hire more employees, the work can be done more efficiently, leading to higher profits. But if the company hires too many employees, it may not have room for them or enough work for them to do, yet they still have to be paid, causing lower profits.

Since number of employees is the independent variable and profit is the dependent variable, the domain and range must both be positive.

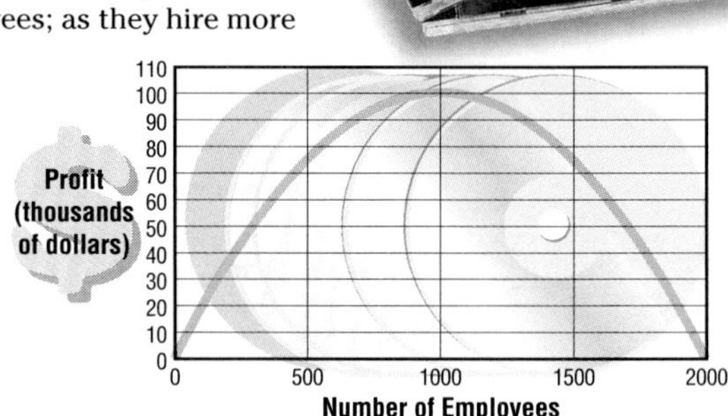

Suppose a company that produces CD-ROMs can express its profit as $P(x) = -0.1x^2 + 200x$, where x is the number of employees in the company. As the personnel manager of the company, it is your job to determine the least number of employees the company should have in order to reach its goal of having $75,000 in profit. *You will solve this problem in Exercise 4.*

A **quadratic equation** is an equation in which the value of the related quadratic function is 0. That is, for the quadratic equation $0 = x^2 + 6x - 7$, the related quadratic function is $f(x) = x^2 + 6x - 7$, where $f(x) = 0$. You have used factoring to solve equations like $x^2 + 6x - 7 = 0$. You can also use graphing to estimate the solutions to this equation.

The solutions of a quadratic equation are called the **roots** of the equation. The roots of a quadratic equation can be found by finding the x-intercepts or **zeros** of the related quadratic function.

2. $y = x^2 + 8x + 11$
minimum, (−4, −5), $x = -4$

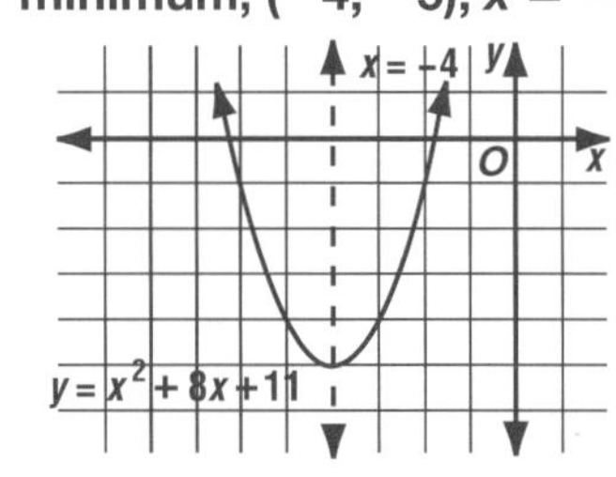

3. $y = -x^2 + 5$
maximum, (0, 5), $x = 0$

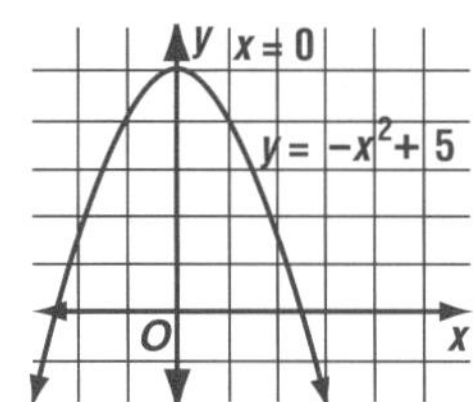

Example  **Solve $x^2 - 2x - 3 = 0$ by graphing. Check by factoring.**

Graph the related function $f(x) = x^2 - 2x - 3$. The equation of the axis of symmetry is $x = 1$, and the coordinates of the vertex are $(1, -4)$. Make a table of values to find other points to sketch the graph of $f(x) = x^2 - 2x - 3$.

x	$f(x)$
−1	0
0	−3
1	−4
2	−3
3	0

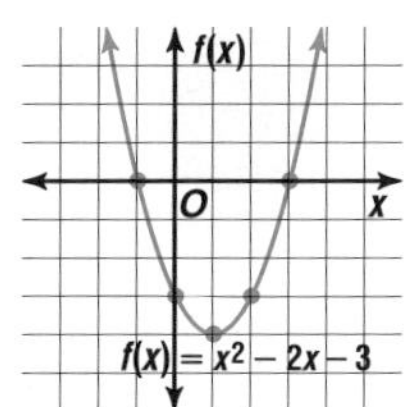

To solve the equation $x^2 - 2x - 3 = 0$, we need to know where the value of $f(x)$ is 0. On the graph, this occurs at the x-intercepts. The x-intercepts of the parabola appear to be −1 and 3.

LOOK BACK

You can refer to Lesson 10-6 for more information on the zero product property and solving equations by factoring.

Check: Solve by factoring.

$x^2 - 2x - 3 = 0$

$(x - 3)(x + 1) = 0$

Set each factor equal to 0.

$x - 3 = 0$ *Zero product property*

$x = 3$ *Add 3 to each side.*

$x + 1 = 0$ *Zero product property*

$x = -1$ *Add −1 to each side.*

The solutions of the equation are 3 and −1.

In Example 1, the zeros of the function were integers. Usually the zeros of a quadratic function are not integers. In these cases, use estimation to approximate the roots of the equation.

Example  **Solve $x^2 + 9x + 5 = 0$ by graphing. If integral roots cannot be found, estimate the roots by stating the consecutive integers between which the roots lie.**

Use a table of values to graph the related function $f(x) = x^2 + 9x + 5$.

Notice in the table of values that the value of the function changes from negative to positive between the x values of −9 and −8 and −1 and 0.

x	$f(x)$
−9	5
−8	−3
−7	−9
−6	−13
−5	−15
−4	−15
−3	−13
−2	−9
−1	−3
0	5

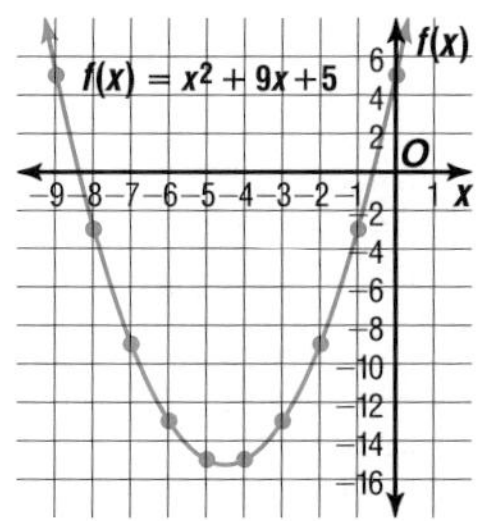

The x-intercepts of the graph are between −9 and −8 and between −1 and 0. So, one root of the equation is between −9 and −8, and the other root is between −1 and 0.

Motivating the Lesson

Hands-On Activity Toss a Ping-Pong™ ball and ask students to describe a curve that approximates the shape of its trajectory. Attach a Ping-Pong ball to each end of a piece of string and have the string hang in a parabolic shape. Remind students that the curve is not a parabola but a catenary, which is closely approximated by a parabola. Ask students to identify the axis of symmetry. Challenge them to determine if the axis of symmetry moves when the two Ping-Pong balls remain in the same place and a longer string is used. **Axis of symmetry does not move.**

2 TEACH

In-Class Examples

For Example 1

Solve $x^2 + 2x - 8 = 0$ by graphing. **$x = -4, 2$**

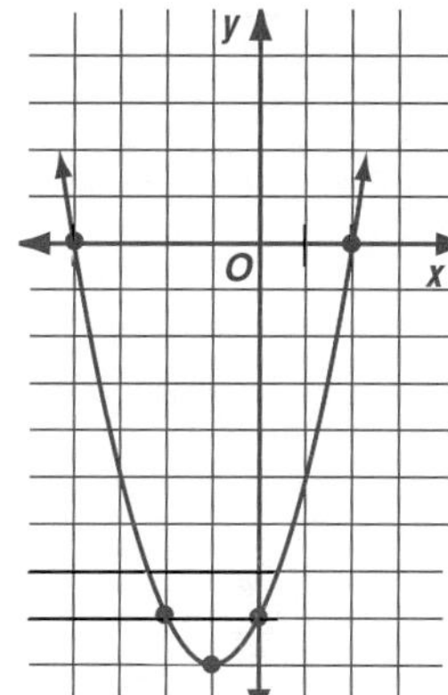

For Example 2

Solve $4x^2 - 4x - 3 = 0$ by graphing. If integral roots cannot be found, estimate the roots by stating the consecutive integers between which the roots lie. **One root is between 0 and −1. The other root is between 1 and 2.**

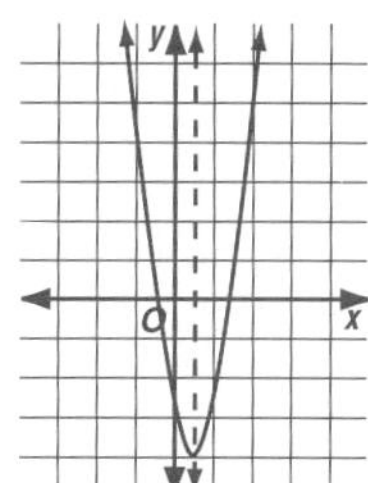

Teaching Tip Explain that the x-intercepts are the zeros of the function.

GLENCOE Technology

CD-ROM Interaction

A multimedia simulation allows students to explore the launch of fireworks by collecting data, graphing, and writing an equation. A blackline master activity with teacher's notes provides a follow-up to the CD-ROM simulation.

For Windows & Macintosh

EXPLORATION

Remind students that there are often two roots. If they see only one and it is not the vertex, they should zoom out to find the other root.

In-Class Example

For Example 3

Solve $x^2 + 4x + 4 = 0$ by graphing. **(−2, 0)**

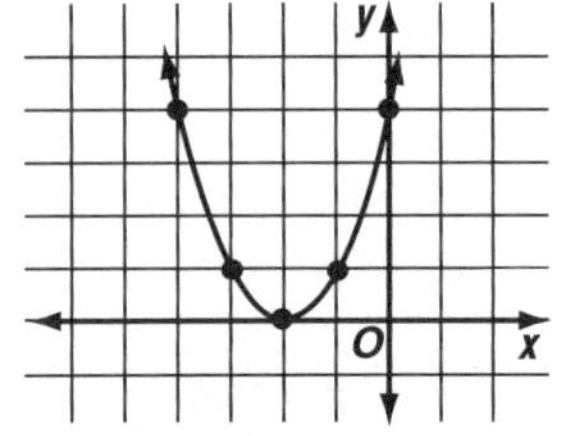

You can use a graphing calculator to find a more accurate estimate for the root of a quadratic equation than you can by using paper and pencil. One way to estimate the root is to ZOOM IN on the zero. Another method is to use the ROOT feature on the CALC menu.

EXPLORATION — GRAPHING CALCULATORS

Use a graphing calculator to solve $3x^2 - 6x - 2 = 0$ to the nearest hundredth.

Graph $y = 3x^2 - 6x - 2$ in the standard viewing window. To use the ROOT feature, you must use the cursor to define the interval in which the calculator will look. You define the interval in the same way the MAXIMUM and MINIMUM intervals are defined.

Enter: [2nd] [CALC] 2

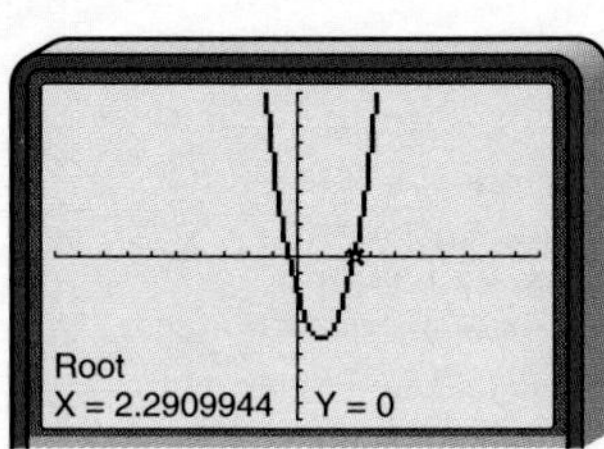

Now use the arrow keys to move the cursor to the left of one of the x-intercepts and press [ENTER] to define the lower bound. Then use the arrow keys to move the cursor to the right of that x-intercept and press [ENTER] to define the upper bound.

The y-coordinate value may appear as a decimal in scientific notation, such as −1E−12 rather than 0.

Press [ENTER] and the approximate coordinates of the root will appear. Repeat this process to find the coordinates of the other root.

Your Turn **a. −4.16, 2.16 b. −1.13, 1.13 c. no real roots**

Use a graphing calculator to estimate the roots of each equation.

a. $x^2 + 2x - 9 = 0$ **b.** $7.5x^2 - 9.5 = 0$ **c.** $6x^2 + 5x + 5 = 0$

d. Use a graphing calculator to solve $x^2 + 9x + 5 = 0$. Compare your results to those in Example 2. Which method of solution do you prefer? **See students' work.**

Quadratic equations always have two roots. However, these roots may not be two distinct numbers.

Example  **Solve $x^2 - 12x + 36 = 0$ by graphing.**

Graph the related function $f(x) = x^2 - 12x + 36$.

x	$f(x)$
4	4
5	1
6	0
7	1
8	4

The equation of the axis of symmetry is $x = 6$. The vertex of the parabola is at (6, 0).

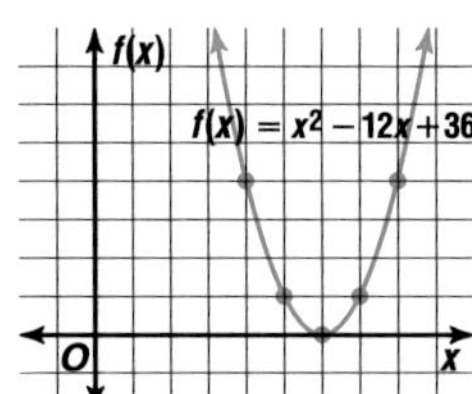

TECHNOLOGY Tip

If you graph a function with a graphing calculator and it appears that the vertex of a parabola is its x-intercept, use the ZOOM IN feature several times. You may find that the parabola actually does cross the x-axis twice.

Notice that the vertex of the parabola is the x-intercept. Thus, one solution is 6. What is the other solution?

Alternative Learning Styles

Kinesthetic Introduce students to the Jordan curve theorem with this activity. Draw a line on the floor. Have a student step over the line once. Then have a student step over the line twice and then three times. Ask the class on which side of the line a student would be standing if the student stepped over the line 17 times, 30 times, and so on.

Try solving by factoring.

$x^2 - 12x + 36 = 0$

$(x - 6)(x - 6) = 0$

Set each factor equal to 0.

$x - 6 = 0$ $\quad$ $x - 6 = 0$ $\quad$ *Zero product property*

$x = 6$ $\quad$ $x = 6$ $\quad$ *Add 6 to each side.*

There are two identical roots to this equation. So, there is only one distinct root. The solution for $x^2 - 12x + 36 = 0$ is 6.

Thus far, we have seen that quadratic equations can have two distinct real roots or one distinct real root. Is it possible that there may be no real roots?

Example  **Solve $x^2 + 2x + 5 = 0$ by graphing.**

Graph the function $f(x) = x^2 + 2x + 5$.

x	f(x)
−3	8
−2	5
−1	4
0	5
1	8

The equation of the axis of symmetry is $x = -1$. The vertex of the parabola is at $(-1, 4)$.

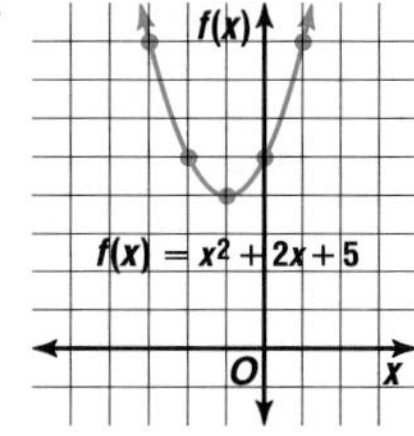

This graph has no x-intercept. Thus, there are no real number solutions for this equation.

The symbol ∅, indicating an empty set, is often used to represent no real solution.

Quadratic equations can be used to solve number problems.

Example 5 **Two numbers have a sum of 4. What are the numbers if their product is −12?**

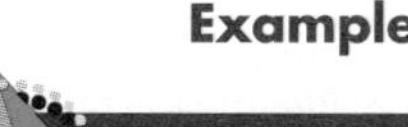

Number Theory

Explore Let n represent one of the numbers. Then the other number is $4 - n$.

Plan A function that describes the product of these two numbers is $f(n) = n(4 - n)$ or $f(n) = -n^2 + 4n$. Find the value of n if $f(n)$ equals −12.

Solve Solve $f(n) = -n^2 + 4n$ if $f(n) = -12$.

$f(n) = -n^2 + 4n$

$-12 = -n^2 + 4n$ $\quad$ *$f(n) = -12$*

$0 = -n^2 + 4n + 12$ $\quad$ *Rewrite the equation so one side is 0.*

Graph the related function $f(n) = -n^2 + 4n + 12$.

x	y
−3	−9
−2	0
−1	7
0	12
1	15
2	16
3	15
4	12
5	7
6	0
7	−9

The equation of the axis of symmetry is $n = 2$. The vertex of the parabola is at $(2, 16)$.

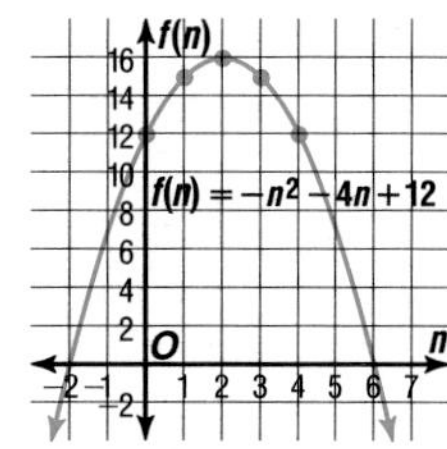

(continued on the next page)

In-Class Examples

For Example 4

Solve $5x^2 - 4x + 1 = 0$ by graphing. **There are no roots that are real numbers.**

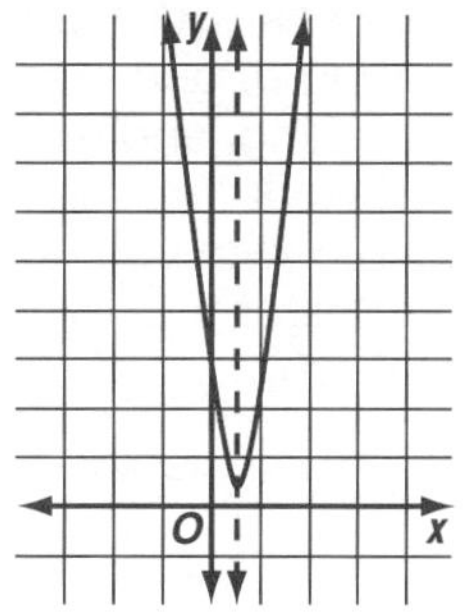

For Example 5

Two numbers have a sum of 11. What are these numbers if their product is 24? **3, 8**

3 PRACTICE/APPLY

Check for Understanding

Exercises 1–16 are designed to help you assess your students' understanding through reading, writing, speaking, and modeling. You should work through Exercises 1–5 with your students and then monitor their work on Exercises 6–16.

Additional Answers

10.

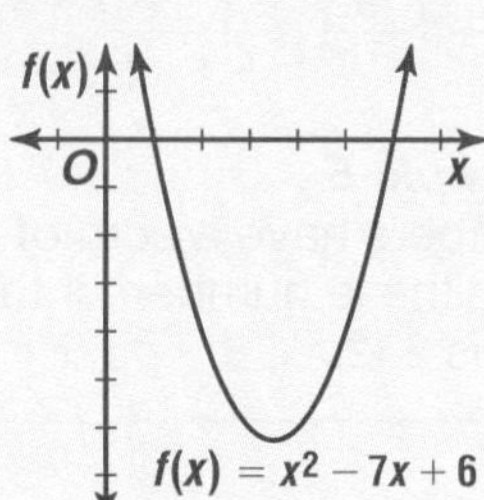

11.

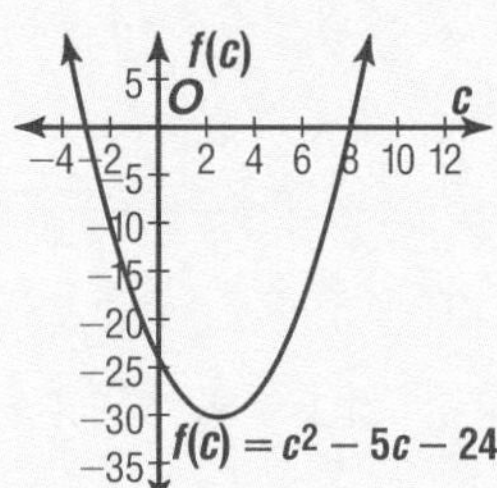

12.

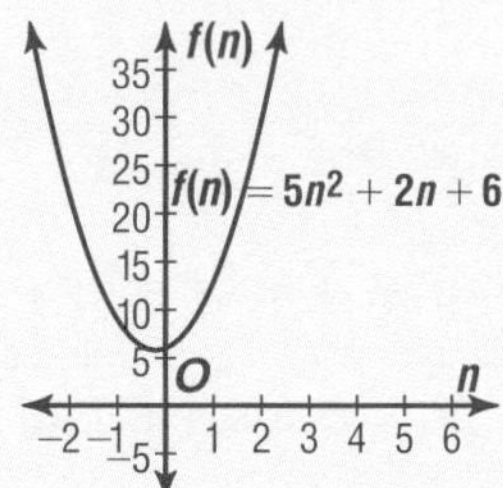

13.

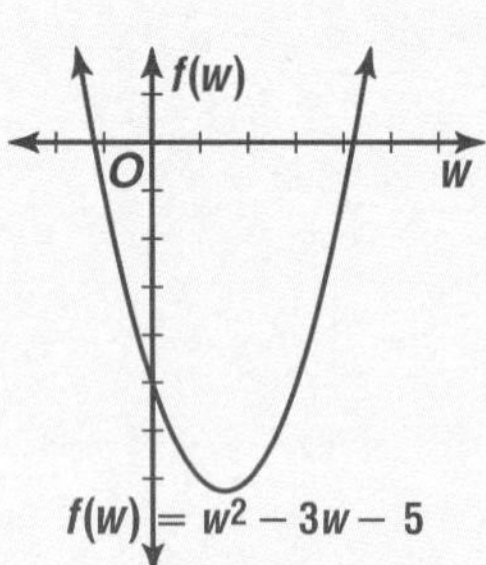

14.

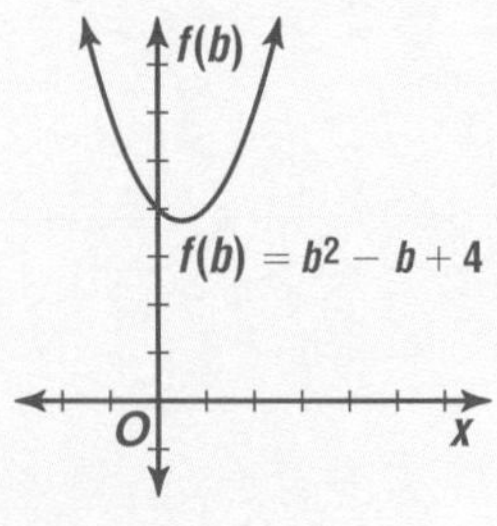

The n-intercepts of the graph are -2 and 6. Use these values of n to find the value of the other number $4 - n$.
If $n = -2$, then $4 - n = 4 - (-2)$ or 6.
If $n = 6$, then $4 - n = 4 - 6$ or -2.
So the numbers are 6 and -2.

Examine Test to see if the numbers satisfy the problem.

The sum of the numbers is 4.	The product of the numbers is -12.
$-2 + 6 \stackrel{?}{=} 4$	$-2(6) \stackrel{?}{=} -12$
$4 = 4$ ✓	$-12 = -12$ ✓

The two numbers are -2 and 6.

CHECK FOR UNDERSTANDING

Communicating Mathematics

Study the lesson. Then complete the following.

1. **Explain** why the x-intercepts of a quadratic function can be used to solve a quadratic equation. **The x-intercept is where the function equals 0.**
2. **You Decide** Joshua says he likes to solve a quadratic equation by factoring rather than graphing. Hanna says that she likes graphing because she can always get an answer. Who is correct? Give examples to support your answer.
3. What is the related function you would use to solve $x^2 + 9x + 2 = 3x - 4$ by graphing? **$f(x) = x^2 + 6x + 6$**
4. Refer to the application at the beginning of the lesson. What is the least number of employees that would help the company make its goal of $75,000 profit? **500 employees**

2. Hanna, because factoring only works when the polynomial making up the equation is factorable. Graphing will always give you an estimate of the solutions if they are real.

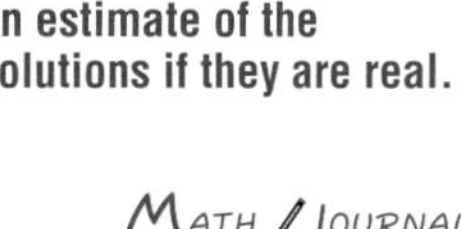

5. **Draw** an example of each type of situation that may occur when using graphing to find the solutions of a quadratic equation. Identify the number of real roots of the quadratic function in each situation. **See students' work.**

Guided Practice

Determine the number of real roots for each quadratic equation whose related function is graphed below.

6. **1 distinct root**

7.

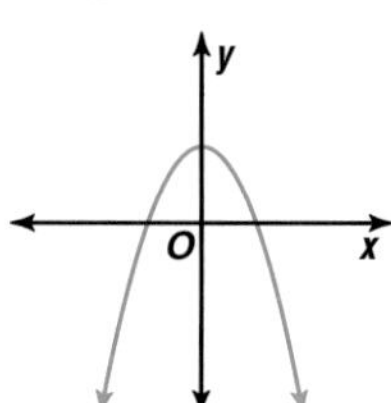

2 real roots

8.

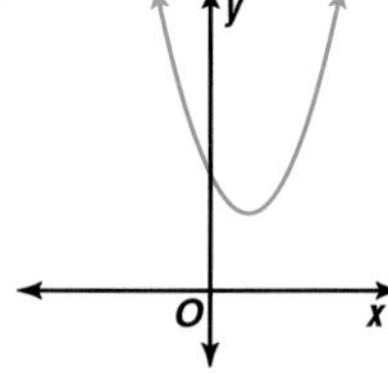

no real roots (∅)

9. State the real roots of the quadratic equation whose related function is graphed at the right. **−1, 2**

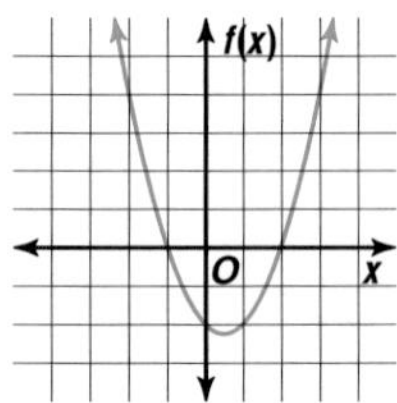

Reteaching

Decision Making Give students a list of maximum and minimum points of a parabola and the value of a in $y = ax^2 + bx + c$. Then have them describe the number of roots of each equation and how they determined that number.

Example: Minimum point is $(-1, -1)$ and $a = 2$. **There are two roots because the minimum point is below the x-axis and the parabola will open up, therefore crossing the x-axis twice.**

10–15. See margin for graphs.

10. 1, 6
11. −3, 8
12. ∅
13. $-2 < w < -1$, $4 < w < 5$

Solve each equation by graphing. If integral roots cannot be found, state the consecutive integers between which the roots lie.

10. $x^2 - 7x + 6 = 0$ **11.** $c^2 - 5c - 24 = 0$ **12.** $5n^2 + 2n + 6 = 0$

13. $w^2 - 3w = 5$ **14.** $b^2 - b + 4 = 0$ ∅ **15.** $a^2 - 10a = -25$ 5

16. Number Theory Use a quadratic equation to find two real numbers whose sum is 5 and whose product is −24. **8 and −3**

EXERCISES

Practice

State the real roots of each quadratic equation whose related function is graphed below.

17.

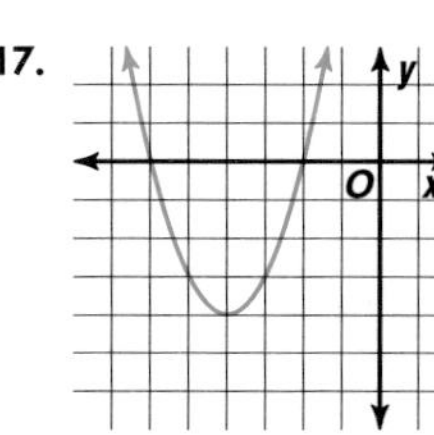

−2, −6

18.

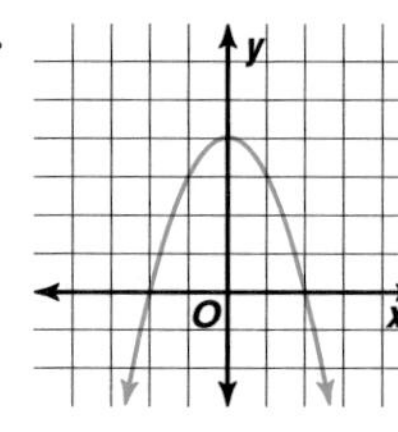

−2, 2

19.

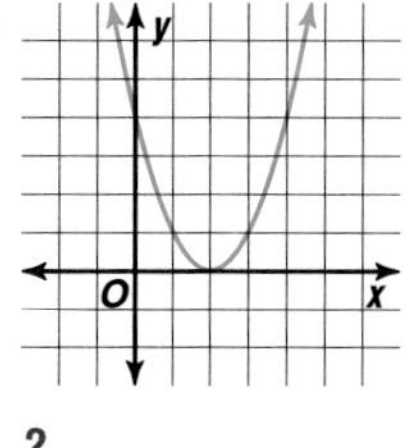

2

20–34. See Solutions Manual for graphs.

20. −3, −4
22. $-5 < a < -4$, $-2 < a < -1$
24. −6, 2
26. −8, −2
27. −4, 1
31. $8 < a < 9$, $-1 < a < 0$
33. 3, $-1 < n < 0$
34. $-4 < x < -3$, $2 < x < 3$

Solve each equation by graphing. If integral roots cannot be found, state the consecutive integers between which the roots lie.

20. $x^2 + 7x + 12 = 0$ **21.** $x^2 - 16 = 0$ **−4, 4** **22.** $a^2 + 6a + 7 = 0$

23. $x^2 + 6x + 9 = 0$ **−3** **24.** $r^2 + 4r - 12 = 0$ **25.** $c^2 + 3 = 0$ ∅

26. $2c^2 + 20c + 32 = 0$ **27.** $3x^2 + 9x - 12 = 0$ **28.** $2x^2 - 18 = 0$ **−3, 3**

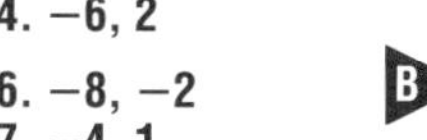

29. $p^2 + 16 = 8p$ **4** **30.** $w^2 - 10w = -21$ **3, 7** **31.** $a^2 - 8a = 4$

32. $m^2 - 2m = -2$ ∅ **33.** $12n^2 - 26n = 30$ **34.** $4x^2 - 35 = -4x$

35–38. See Solutions Manual.

The roots of a quadratic equation are given. Graph the related quadratic function if it has the indicated maximum or minimum point.

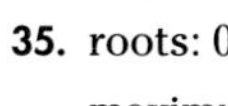

35. roots: 0, −6
maximum point: (−3, 4)

36. roots: −2, −6
minimum point : (−4, −2)

37. roots: no real roots
minimum point: (2, 5)

38. roots: $-4 < x < -3$, $1 < x < 2$
maximum point: (−1, 6)

INTEGRATION
Number Theory

Use a quadratic equation to determine the two numbers that satisfy each situation.

39. Their difference is 4 and their product is 32. **−4, −8 or 4, 8**

40. Their sum is 9 and their product is 20. **4, 5**

41. Their sum is 4 and their product is 5. **no solution**

42. They differ by 2. The sum of their squares is 130. **9 and 7 or −9 and −7**

43–45. See margin for graphs.
43. −2, −6
44. $0 < y < 1$, $3 < y < 4$
45. no y-intercepts

Estimate the y-intercepts of each quadratic equation by graphing.

43. $x = -0.75y^2 - 6y - 9$ **44.** $x = y^2 - 4y + 1$ **45.** $x = 3y^2 + 2y + 4$

Additional Answer

45.

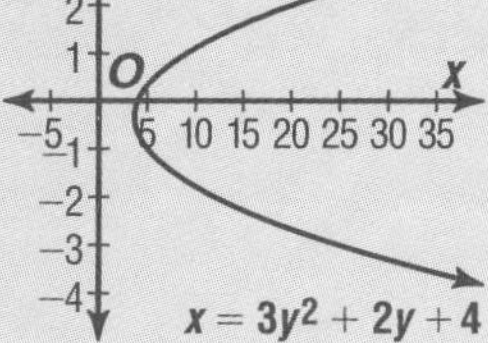

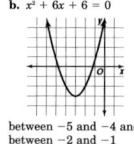

Assignment Guide

Core: 17–53 odd, 54, 55, 57–66
Enriched: 18–52 even, 54–66

For **Extra Practice,** see p. 781.

The red A, B, and C flags, printed only in the Teacher's Wraparound Edition, indicate the level of difficulty of the exercises.

Additional Answers

15.

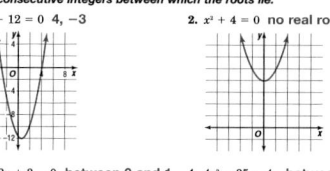

f(a)
$f(a) = a^2 - 10a + 25$
O
a

43.

y
O
x
$x = -0.75y^2 - 6y - 9$

44.

y
$x = y^2 - 4y + 1$
x
O

Study Guide Masters, p. 76

NAME______________________ DATE ______

Study Guide

Student Edition
Pages 620–627

Solving Quadratic Equations by Graphing

A **quadratic equation** is an equation in which the value of the related quadratic function is 0. The solutions of a quadratic equation are called the **roots** of the equation. The roots of a quadratic equation can be found by graphing the related quadratic function and finding the x-intercepts or **zeros** of the function.

Examples: Solve each equation by graphing. If exact roots cannot be found, estimate the roots by stating the consecutive integers between which the roots lie.

a. $x^2 + 7x + 12 = 0$

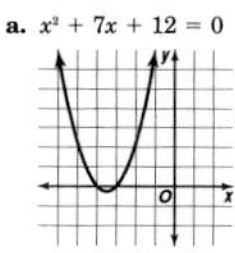

−3 and −4

b. $x^2 + 6x + 6 = 0$

between −5 and −4 and between −2 and −1

c. $x^2 - 4x + 5$

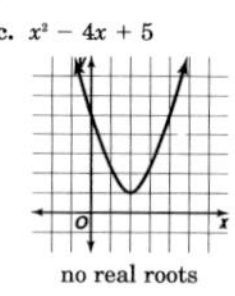

no real roots

Solve each equation by graphing. If exact roots cannot be found, state the consecutive integers between which the roots lie.

1. $x^2 - x - 12 = 0$ 4, −3

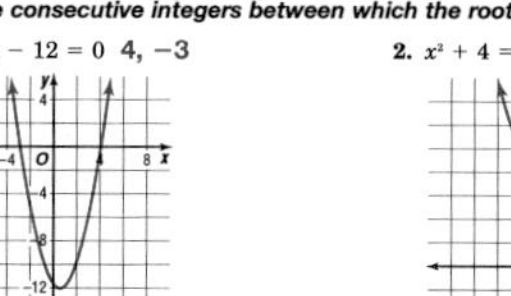

2. $x^2 + 4 = 0$ no real roots

3. $4x^2 - 12x + 3 = 0$ between 0 and 1 and 2 and 3

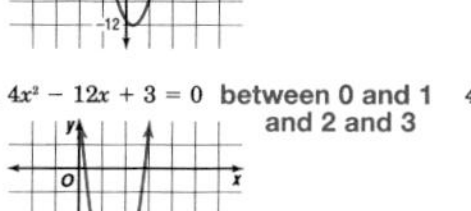

4. $4x^2 = 35 - 4x$ between −3 and −4 and 2 and 3

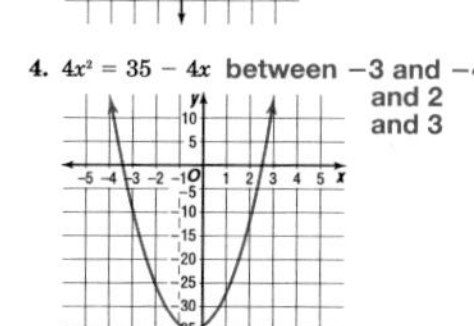

Additional Answers

54. The value of the function changes from negative to positive as it moves from $x = 10$ to $x = 11$. In order to do this, it must cross the x-axis between 10 and 11. Therefore, an x-intercept occurs between 10 and 11. Thus, the solution of the related equation lies between 10 and 11.

55a.

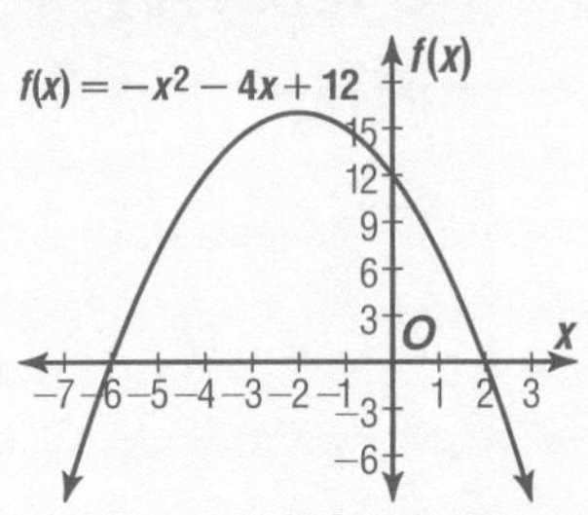

56a.

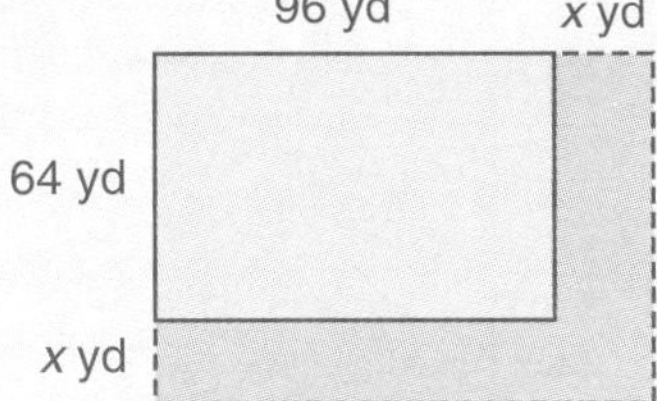

Practice Masters, p. 76

11-2 NAME ____ DATE ____

Practice — Student Edition Pages 620–627

Solving Quadratic Equations by Graphing

Solve each equation by graphing. If exact roots cannot be found, state the consecutive integers between which the roots are located.

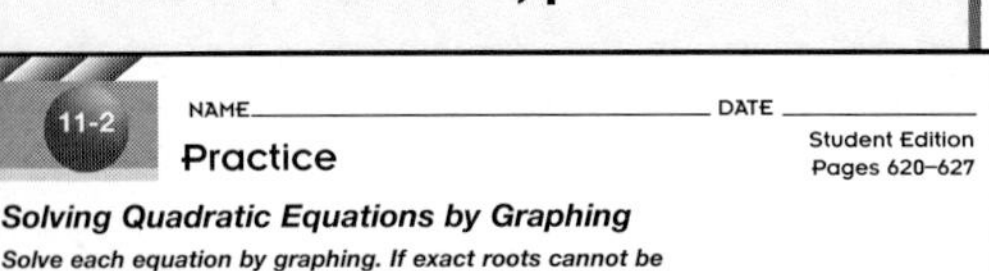

1. $x^2 - x - 6 = 0$

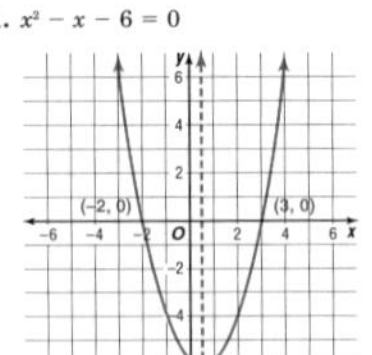

2. $-x^2 - x + 1 = 0$

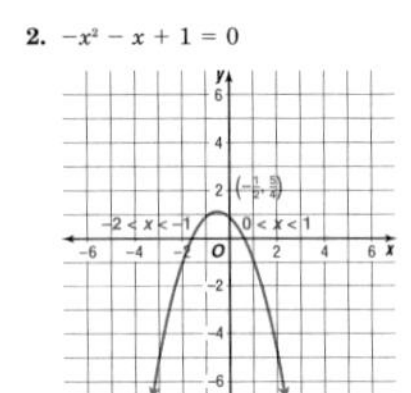

The roots of a quadratic equation are given. Graph the related quadratic function if it has the indicated maximum or minimum point.

3. roots: −2, 0; minimum point: (−1, −1)

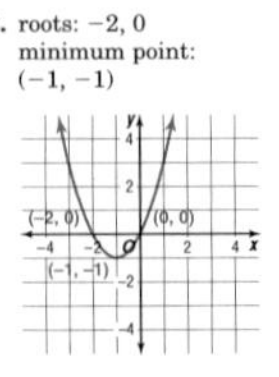

4. roots: −2, 2; maximum point: (0, 2)

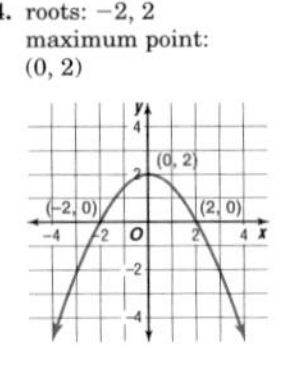

5. roots: −3, −1; maximum point: (−2, 1)

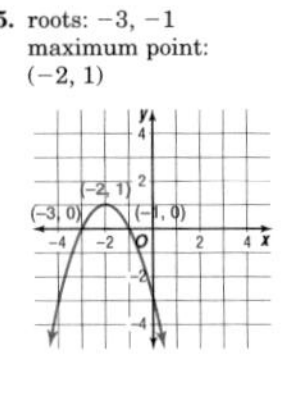

Graphing Calculator

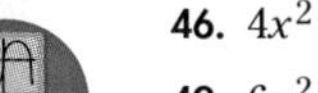

Use a graphing calculator to solve each equation to the nearest hundredth. 46. −1.66, 1.66 47. no real roots 48. −11

46. $4x^2 - 11 = 0$
47. $-2x^2 - x - 3 = 0$
48. $x^2 + 22x + 121 = 0$
49. $6x^2 - 12x + 3 = 0$ 0.29, 1.71
50. $5x^2 + 4x - 7 = 0$ −1.6, 0.85
51. $-4x^2 + 7x + 8 = 0$ −0.79, 2.54

Use a graphing calculator to find values for k for each equation so that it will have (a) one distinct root, (b) two real roots, and (c) no real roots.

52. $x^2 + 3x + k = 0$ 2.25, $k < 2.25$, $k > 2.25$
53. $kx^2 + 4x - 2 = 0$ −2, $k > -2$, $k < -2$

Critical Thinking

54. Suppose the value of a quadratic function is negative when $x = 10$ and positive when $x = 11$. Explain why it is reasonable to assume that the related equation has a solution between 10 and 11. See margin.

Applications and Problem Solving

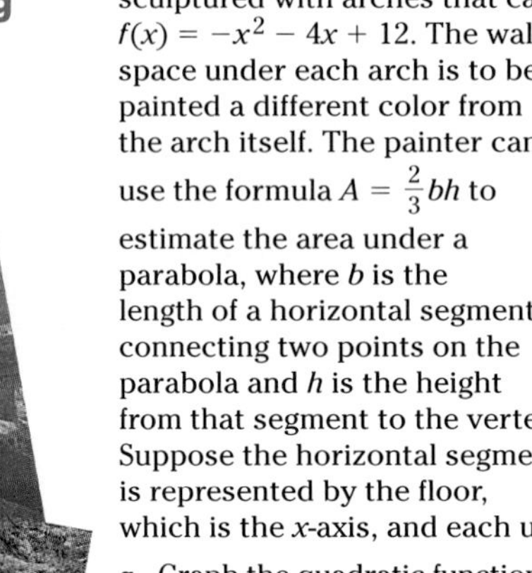

55. **Architecture** A painter is hired to paint an art gallery whose walls are sculptured with arches that can be represented by the quadratic function $f(x) = -x^2 - 4x + 12$. The wall space under each arch is to be painted a different color from the arch itself. The painter can use the formula $A = \frac{2}{3}bh$ to estimate the area under a parabola, where b is the length of a horizontal segment connecting two points on the parabola and h is the height from that segment to the vertex. Suppose the horizontal segment is represented by the floor, which is the x-axis, and each unit represents 1 foot.

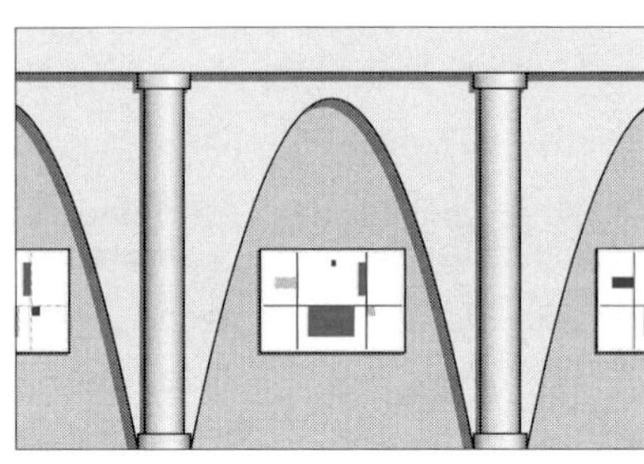

a. Graph the quadratic function and determine its x-intercepts. See margin.
b. What is the length of the segment along the floor? 8 feet
c. What is the height of the arch? 16 feet
d. How much wall space is there under each arch? $85\frac{1}{3}$ ft^2
e. How much would the paint cost to paint the walls under 12 arches if the paint is \$27/gallon, she applies two coats, and the manufacturer states that each gallon will cover 200 ft^2? *Remember you cannot buy part of a gallon.* \$297

Most-Visited U.S. National Parks in 1993

Park	Visitors
1. Great Smoky Mountains	9,283,848
2. Grand Canyon	4,575,602
3. Yosemite	3,839,645
4. Yellowstone	2,912,193
5. Rocky Mountains	2,780,342

56. **Make a Drawing** Banneker Park has set aside a section of the park as a nature preserve with an observation deck and telescopes so that people can observe the wildlife and plants up close. To accommodate the increased number of visitors they expect, the park commission has received funding to double the parking area. The current lot is 64 yards by 96 yards, and they are going to add strips of equal width to the end and side of the lot to create a larger rectangle that is twice the size of the original.

a. Make a drawing of the lot and the proposed additions. See margin.
b. Write a quadratic equation to find x, the width of the strips, so that the area of the parking lot is doubled. $x^2 + 160x - 6144 = 0$
c. How wide are the strips to be added? 32 yards
d. What are the dimensions of the new parking lot? 96 × 128 yards

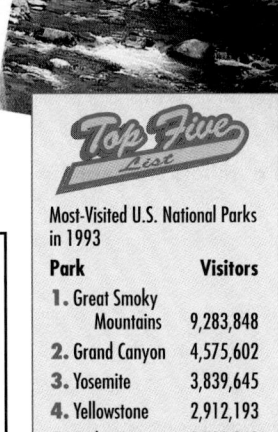

Top Five List

Listed below is the area in acres of each of the parks.

Park	Area (acres)
Great Smoky Mountains	520,004
Grand Canyon	1,218,375
Yosemite	748,542
Yellowstone	2,221,766
Rocky Mountains	265,668

Mixed Review

57. $x = 0$; (0, 4); See margin for graph.

58. $\left\{0, -\frac{2}{9}\right\}$

59. $(4x - 3)$ m by $(2x - 1)$ m

57. Find the equation of the axis of symmetry and the coordinates of the vertex of the graph of $y = -3x^2 + 4$. (Lesson 11–1)

58. Solve $81x^3 + 36x^2 = -4x$. (Lesson 10–6)

59. Geometry The area of a rectangle is $(8x^2 - 10x + 3)$ square meters. What are the dimensions of the rectangle? (Lesson 10–3)

60. Find $(x - 2y)^3$. (Lesson 9–8) $x^3 - 6x^2y + 12xy^2 - 8y^3$

61. Find the degree of $6x^2y + 5x^3y^2z - x + x^2y^2$. (Lesson 9–4) **6**

62. Use graphing to solve the system of equations. (Lesson 8–1)
$x + y = 3$
$x + y = 4$ **no solution**

63. Solve $|3x + 4| < 8$. (Lesson 7–6) $-4 < x < \frac{4}{3}$

64. Graph $y = -x + 6$. (Lesson 6–5) **See margin.**

65. Commerce Find the sale price of an item originally marked as \$33 with 25% off. (Lesson 4–4) **\$24.75**

66. Find $-4 + 6 + (-10) + 8$. (Lesson 2–3) **0**

WORKING ON THE

In•ves•ti•ga•tion

Refer to the Investigation on pages 554–555.

the BRICKYARD

In order to best use the excess inventory of bricks, you are going to investigate the possible combinations that will use up the inventory at a steady rate.

1. Suppose you have four large square bricks and three small square bricks. How many rectangular bricks would you need to create a rectangular pattern? Is there more than one pattern that can be made if you change the number of rectangular bricks?
   - What are the dimensions, perimeters, and areas of the rectangular patterns formed?
   - Use the variables x and y as the length of the large square and the length of the small square, respectively. List the dimensions, perimeter, and area of each pattern in terms of x and y.
   - Explain your findings. How do the perimeters and areas of these patterns relate in this matter?
2. If you had three large square bricks, five rectangular bricks, and two small square bricks, what size rectangular patterns can you make? Express the dimensions, perimeter, and area of each pattern in terms of x and y.
3. If you had two large bricks, six rectangular bricks, and four small bricks, what size rectangular patterns can you make? Express the dimensions, perimeter, and area of each pattern in terms of x and y.
4. Explain the method you used to solve these problems and any generalizations you found.

Add the results of your work to your Investigation Folder.

4 ASSESS

Closing Activity

Writing Have students list the steps to follow when locating the roots of a quadratic equation by graphing the related function.

Chapter 11, Quiz A (Lessons 11-1 and 11-2), is available in the *Assessment and Evaluation Masters*, p. 296.

Additional Answers

57.

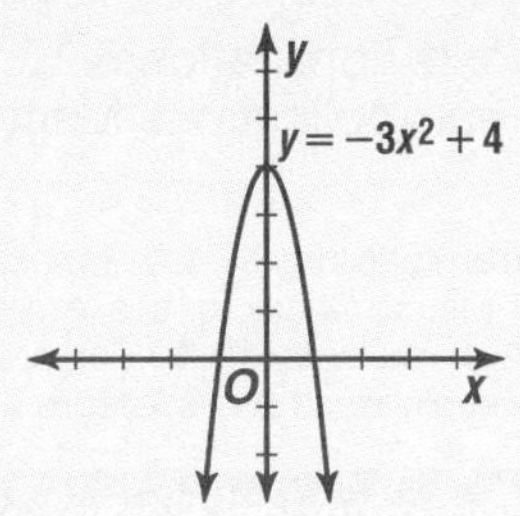

64.

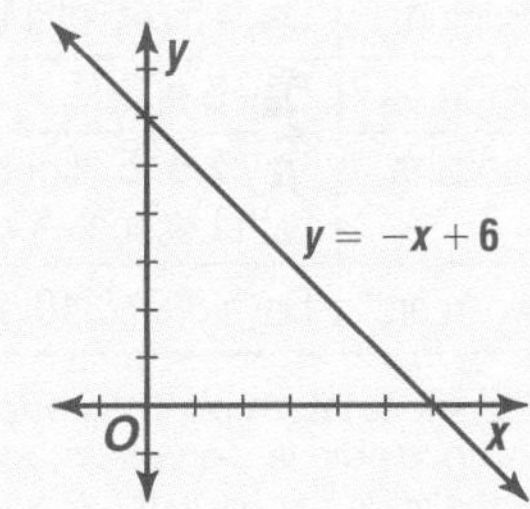

Enrichment Masters, p. 76

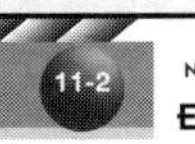

NAME ______ DATE ______

11-2 Enrichment

Student Edition Pages 620–627

Mechanical Constructions of Parabolas

A given line and a point determine a parabola. Here is one way to construct the curve.

Use a right triangle *ABC* (or a stiff piece of rectangular cardboard).

Place one leg of the triangle on the given line *d*. Fasten one end of a string with length *BC* at the given point *F* and the other end to the triangle at point *B*.

Put the tip of a pencil at point *P* and keep the string tight.

As you move the triangle along the line *d*, the point of your pencil will trace a parabola.

Draw the parabola determined by line d and point F.

1. 2.

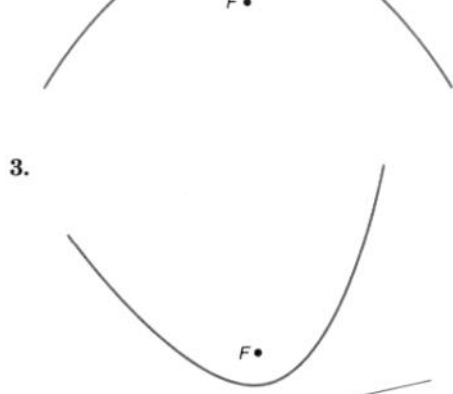

3. 4.

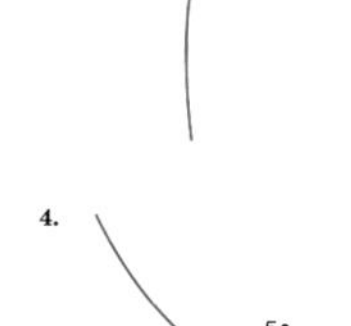

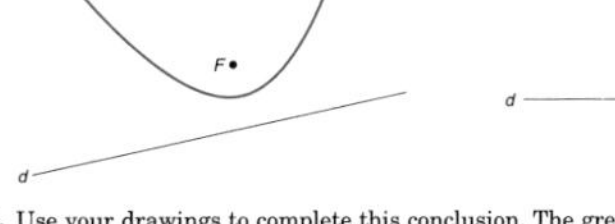

5. Use your drawings to complete this conclusion. The greater the distance of point *F* from line *d*, the wider the opening of the parabola.

Extension

Connections You can show students that the solution to a system of equations involving one or more quadratics is the point or points at which the graphs intersect.

1. $y = x + 2$, $y = x^2$ **(−1, 1), (2, 4)**
2. $y = -3x - 8$, $y = x^2 - 4$ **no real solution**
3. $y = -x^2 + 2x$, $y = x^2 - 2x$ **(0, 0), (2, 0)**

In•ves•ti•ga•tion

Working on the Investigation

The Investigation on pages 554–555 is designed to be a long-term project that is completed over several days or weeks. Encourage students to keep their materials in their Investigation Folder as they work on the Investigation.

11-3 LESSON NOTES

NCTM Standards: 1–6

Instructional Resources

- Study Guide Master 11-3
- Practice Master 11-3
- Enrichment Master 11-3
- Assessment and Evaluation Masters, pp. 295–296
- Multicultural Activity Masters, p. 21
- Real-World Applications, 27
- Tech Prep Applications Masters, p. 22

Transparency 11-3A contains the 5-Minute Check for this lesson; **Transparency 11-3B** contains a teaching aid for this lesson.

Recommended Pacing	
Standard Pacing	Days 5 & 6 of 11
Honors Pacing	Day 5 of 10
Block Scheduling*	Day 2 of 5 (along with Lesson 11-2)
Alg. 1 in Two Years*	Days 8, 9, & 10 of 19

*For more information on pacing and possible lesson plans, refer to the *Block Scheduling Booklet* and *Algebra 1 in Two Years.*

1 FOCUS

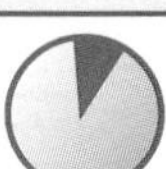

5-Minute Check
(over Lesson 11-2)

Solve each equation by graphing.

1. $0 = x^2 - 8x + 12$ **2, 6**
2. $x^2 + 6x + 9 = 0$ **−3**
3. $x^2 - 3x + 7 = 0$ **no real roots**
4. $x^2 - 4x + 5 = 0$ **no real roots**
5. $x^2 - 2x + 1 = 0$ **1**

11–3 Solving Quadratic Equations by Using the Quadratic Formula

What **YOU'LL LEARN**

- To solve quadratic equations by using the quadratic formula.

Why **IT'S IMPORTANT**

You can use quadratic equations to solve problems involving hydraulics and civics.

CONNECTION

Civics

The number of citizens (in millions) voting in each presidential election since 1824 can be approximated by the quadratic function $V(t) = 0.0046t^2 - 0.185t + 3.30$, where t represents the number of years since 1824. For her history project, Marcela Ruiz needed to determine in what year the number of voters in a presidential election was approximately 55 million. She used the function above, replacing $V(t)$ with 55.

$$V(t) = 0.0046t^2 - 0.185t + 3.30$$
$$55 = 0.0046t^2 - 0.185t + 3.30$$
$$0 = 0.0046t^2 - 0.185t - 51.7$$

Marcela knew that she could use her graphing calculator to estimate the solutions of this equation, but she wondered how she could get a good estimate of t if she didn't have a graphing calculator. *You will estimate these solutions in Example 4.*

You can use the **quadratic formula** to solve any quadratic equation.

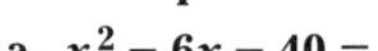

The Quadratic Formula — **The solutions of a quadratic equation in the form $ax^2 + bx + c = 0$, where $a \neq 0$, are given by the formula**

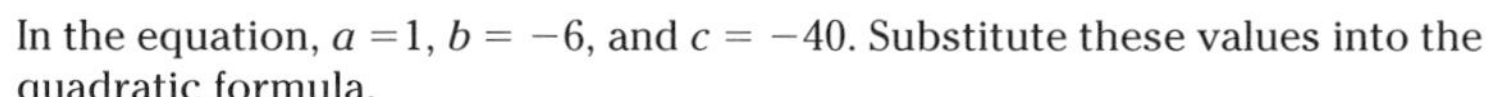

$$x = \frac{-b \pm \sqrt{b^2 - 4ac}}{2a}.$$

The quadratic formula can be used to solve any quadratic equation involving any variable.

Example 1 **Use the quadratic formula to solve each equation.**

a. $x^2 - 6x - 40 = 0$

In the equation, $a = 1$, $b = -6$, and $c = -40$. Substitute these values into the quadratic formula.

The symbol ± means to evaluate the expression first using + and then evaluate it again using −. This provides for the two solutions of the equation.

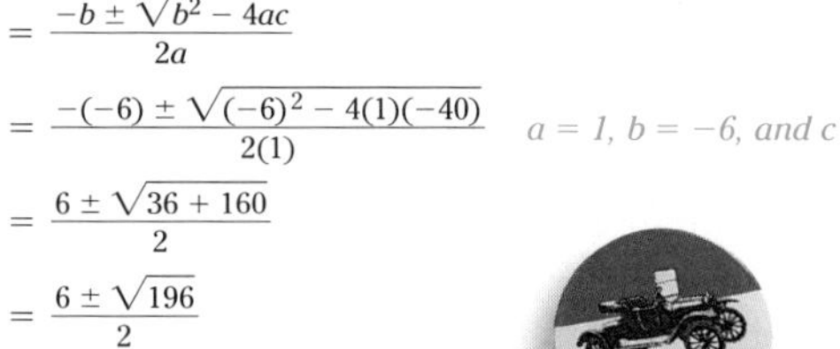

$$x = \frac{-b \pm \sqrt{b^2 - 4ac}}{2a}$$
$$= \frac{-(-6) \pm \sqrt{(-6)^2 - 4(1)(-40)}}{2(1)} \quad a = 1, b = -6, \text{ and } c = -40$$
$$= \frac{6 \pm \sqrt{36 + 160}}{2}$$
$$= \frac{6 \pm \sqrt{196}}{2}$$
$$= \frac{6 \pm 14}{2}$$

$x = \frac{6 + 14}{2}$ or $x = \frac{6 - 14}{2}$

$= \frac{20}{2}$ or 10 $= -\frac{8}{2}$ or -4

You can also check the solution to any equation by substituting each value into the original equation.

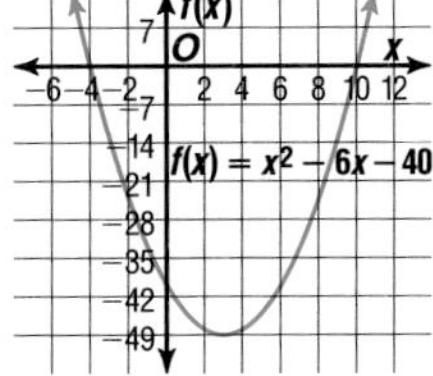

Check: Solve by graphing the related function $f(x) = x^2 - 6x - 40$. The x-intercepts appear to be 10 and -4. This agrees with the algebraic solution.

The solutions are 10 and -4.

b. $\mathbf{y^2 - 6y + 9 = 0}$

$y = \frac{-b \pm \sqrt{b^2 - 4ac}}{2a}$

$= \frac{-(-6) \pm \sqrt{(-6)^2 - 4(1)(9)}}{2(1)}$ *a = 1, b = −6, and c = 9*

$= \frac{6 \pm \sqrt{36 - 36}}{2}$

$= \frac{6 \pm \sqrt{0}}{2}$

$y = \frac{6 + 0}{2}$ or $y = \frac{6 - 0}{2}$

$= 3$ $= 3$

Check: Solve by factoring.

$y^2 - 6y + 9 = 0$

$(y - 3)(y - 3) = 0$

$y - 3 = 0$ $y - 3 = 0$ *Zero product property*

$y = 3$ $y = 3$ *Add 3 to each side.*

There is one distinct solution, 3.

LOOK BACK

You can refer to Lesson 2-8 for more information on irrational numbers.

Sometimes when you use the quadratic formula, you find the solutions are irrational numbers. It is helpful to use a calculator to estimate the values of the solutions in this case.

Example 2 **Use the quadratic formula to solve $2n^2 - 7n - 3 = 0$.**

$n = \frac{-b \pm \sqrt{b^2 - 4ac}}{2a}$

$= \frac{-(-7) \pm \sqrt{(-7)^2 - 4(2)(-3)}}{2(2)}$ *a = 2, b = −7, and c = −3*

$= \frac{7 \pm \sqrt{49 + 24}}{4}$

$= \frac{7 \pm \sqrt{73}}{4}$

$\sqrt{73}$ is an irrational number. We can approximate the solutions by using a calculator to find a decimal value for $\sqrt{73}$.

$n = \frac{7 + \sqrt{73}}{4} \approx 3.886$ $n = \frac{7 - \sqrt{73}}{4} \approx -0.386$

The two solutions are approximately -0.386 and 3.886.

Motivating the Lesson

Questioning Introduce the lesson with a review of the general form of a quadratic equation, $ax^2 + bx + c = 0$. Ask students to name a, b, and c for each quadratic equation.

1. $2x^2 - 4x + 3 = 0$ **2, −4, 3**
2. $x^2 + x - 5 = 0$ **1, 1, −5**
3. $7x^2 - 10x + 3 = 0$ **7, −10, 3**
4. $-3x^2 + 4x - 12 = 0$ **−3, 4, −12**

2 TEACH

In-Class Examples

For Example 1
Use the quadratic formula to solve each equation.

a. $x^2 + 6x + 8 = 0$ **−2, −4**
b. $x^2 - 2x - 15 = 0$ **−3, 5**
c. $2x^2 + 7x - 15 = 0$ $\mathbf{-5, \frac{3}{2}}$

For Example 2
Use the quadratic formula to solve $3n^2 - 8n - 2 = 0$. Use a calculator to approximate the roots.
$\mathbf{\frac{4 \pm \sqrt{22}}{3}}$ **or −0.230, 2.897**

Teaching Tip For Examples 1 and 2, encourage students to memorize the quadratic formula. Emphasize that it can be used to solve any quadratic equation.

Alternative Teaching Strategies

Student Diversity Discuss the fact that the quadratic formula can be seen as a "cheat-sheet" to help students solve equations more easily. Any equation that can be rearranged and simplified to the form of $ax^2 + bx + c = 0$ can be solved using the quadratic formula.

In-Class Examples

For Example 3
Use the quadratic formula to solve $3x^2 - x + 3 = 0$.
no real roots

For Example 4
Determine in what year the number of people voting in a presidential election was approximately 77 million. **1972**

Teaching Tip For Example 4, remind students that the square root of a negative number is not defined for the set of real numbers.

When we solved quadratic equations by graphing, we found that some quadratic equations have no real solutions. How does the quadratic formula work in this situation?

Example  **Use the quadratic formula to solve $z^2 - 5z + 12 = 0$.**

$$z = \frac{-b \pm \sqrt{b^2 - 4ac}}{2a}$$

$$= \frac{-(-5) \pm \sqrt{(-5)^2 - 4(1)(12)}}{2(1)} \quad a = 1,\ b = -5,\ \text{and } c = 12$$

$$= \frac{5 \pm \sqrt{25 - 48}}{2}$$

$$= \frac{5 \pm \sqrt{-23}}{2}$$

Since there is no real number that is the square root of a negative number, this equation has no real solutions.

It is often helpful to use a calculator when using the quadratic formula to solve real-world problems. If there are no real solutions for the equation, the calculator will give you an error message.

Example 4

Civics

Refer to the connection at the beginning of the lesson. Determine in what year the number of people voting in a presidential election was approximately 55 million.

Use a scientific calculator and the quadratic formula to find values for *t*. The values for *a*, *b*, and *c* are 0.0046, −0.185, and −51.7, respectively. Find the value of $\sqrt{b^2 - 4ac}$ and store it in the calculator's memory.

Enter: [( ] .185 [+/−] [x^2] [−] 4 [×] .0046 [×] 51.7 [+/−] [ )] [$\sqrt{x}$] [STO] 0.992726044

Now evaluate the quadratic formula.

Enter: [( ] [( ] .185 [+/−] [ )] [+/−] [+] [RCL] [ )] [÷] [( ] 2 [×] .0046 [ )] [=] 128.0137005

Enter: [( ] [( ] .185 [+/−] [ )] [+/−] [−] [RCL] [ )] [÷] [( ] 2 [×] .0046 [ )] [=] −87.79630922

The negative root has no meaning in this problem.

Since *t* represents the number of years since 1824, add 128 to 1824: 128 + 1824 = 1952.

So, 1952 was the year in which approximately 55 million people voted.

Cooperative Learning

Pairs Check Separate students into pairs. One student will use the quadratic formula and a scientific calculator. The other student will use the graphing calculator to find the roots of each equation to the nearest hundredth.

a. $5.97x^2 - 17.1x + 2.54 = 0$
2.71, 0.16

b. $1.9y^2 - 11.5y - 6.2 = 0$
6.55, −0.50

c. $9.35t^2 + 1.97t - 7.42 = 0$
0.79, −1.00

For more information on the pairs check strategy, see *Cooperative Learning in the Mathematics Classroom,* one of the titles in the Glencoe Mathematics Professional Series, pages 12–13.

CHECK FOR UNDERSTANDING

Communicating Mathematics

1. Evaluate the formula using + for one root; then reevaluate using − for the other.

Study the lesson. Then complete the following.

1. **Explain** how you get two solutions when using the quadratic formula.
2. **Explain** what happens in the quadratic formula when there are no real solutions for the equation. **You get the square root of a negative number.**
3. Refer to the connection at the beginning of the lesson.
   a. Predict how many will vote in the year 2000. **113.23 million**
   b. Describe the domain and range of this relation.
4. **Assess Yourself** You have learned to use graphing, factoring, and the quadratic formula to solve quadratic equations. Which method do you prefer and why? **See students' work.**

Guided Practice

3b. The domain is number of years since 1824 and the range is the number of citizens voting.

6. −1, −5, −2; −4.56, −0.44

State the values of *a*, *b*, and *c* for each quadratic equation. Then solve the equation by using the quadratic formula. Approximate irrational roots to the nearest hundredth.

5. $x^2 + 3x - 18 = 0$ **1, 3, −18; 3, −6**
6. $14 = 12 - 5x - x^2$
7. $4x^2 - 2x + 15 = 0$ **4, −2, 15; no real roots**
8. $x^2 = 25$ **1, 0, −25; ±5**

Solve each equation by using the quadratic formula. Approximate roots to the nearest hundredth if necessary.

9. $4x^2 + 2x - 17 = 0$ **1.83, −2.33**
10. $3b^2 + 5b + 11 = 0$ **∅**
11. $x^2 + 7x + 6 = 0$ **−1, −6**
12. $z^2 - 13z = 32$ **−2.12, 15.12**
13. **Hydraulics** Cox's formula for measuring the velocity of water escaping from a reservoir through a horizontal pipe is $4v^2 + 5v - 2 = \frac{1200HD}{L}$, where v represents the velocity of the water in feet per second, H the height of the reservoir in feet, D the diameter of the pipe in inches, and L the length of the pipe in feet. How fast is water flowing through a pipe 20 feet long with a diameter of 6 inches that is draining a swimming pool with a depth of 10 feet? Round your answer to the nearest tenth. **about 29.4 ft/s**

EXERCISES

Practice

14. −4, 6 16. −10, −2

A

17. $-\frac{4}{5}$, 1 19. $-\frac{5}{3}$, 4

21. −2.91, 2.41

25. $-\frac{3}{2}, -\frac{5}{3}$

27. −6, 6

29. $-\frac{1}{3}$, 1

B

30. $-\frac{3}{4}, \frac{5}{6}$ 31. −3, $-\frac{1}{2}$

35. 0.5, 2

36. 0.6, 1.6

37. −0.2, 1.4

C

Solve each equation by using the quadratic formula. Approximate irrational roots to the nearest hundredth. **15. −8.61, −1.39**

14. $x^2 - 2x - 24 = 0$
15. $a^2 + 10a + 12 = 0$
16. $c^2 + 12c + 20 = 0$
17. $5y^2 - y - 4 = 0$
18. $r^2 + 25 = 0$ **∅**
19. $3b^2 - 7b - 20 = 0$
20. $y^2 + 12y + 36 = 0$ **−6**
21. $2r^2 + r - 14 = 0$
22. $2x^2 + 4x = 30$ **−5, 3**
23. $2x^2 - 28x + 98 = 0$ **7**
24. $24x^2 - 14x = 6$ **−0.29, 0.87**
25. $6x^2 + 15 = -19x$
26. $12x^2 = 48$ **−2, 2**
27. $x^2 + 6x = 36 + 6x$
28. $1.34a^2 - 1.1a = -1.02$ **∅**
29. $3m^2 - 2m = 1$
30. $24a^2 - 2a = 15$
31. $2w^2 = -(7w + 3)$
32. $a^2 - \frac{3}{5}a + \frac{2}{25} = 0$ **$\frac{2}{5}, \frac{1}{5}$**
33. $-2x^2 + 0.7x = -0.3$ **0.60, −0.25**
34. $2y^2 - \frac{5}{4}y = \frac{1}{2}$ **−0.28, 0.90**

Without graphing, determine the *x*-intercepts of the graph of each function to the nearest tenth.

35. $f(x) = 2x^2 - 5x + 2$
36. $f(x) = 4x^2 - 9x + 4$
37. $f(x) = 13x^2 - 16x - 4$

Reteaching

Using Discussion Have half the class solve a quadratic equation by using the quadratic formula and the other half of the class solve by graphing. Compare answers and discuss which method was better to use for that problem. Given a new problem, exchange tasks for each half of the class. Solve and discuss.

GLENCOE Technology

Interactive Mathematics Tools Software

This multimedia software helps students observe the changes in maximum height of the graph of a projectile when the initial velocity and height are altered. A **Computer Journal** gives students an opportunity to write about what they have learned.

For Windows & Macintosh

3 PRACTICE/APPLY

Check for Understanding

Exercises 1–13 are designed to help you assess your students' understanding through reading, writing, speaking, and modeling. You should work through Exercises 1–4 with your students and then monitor their work on Exercises 5–13.

Error Analysis

Make sure students put quadratic equations in the form $ax^2 + bx + c = 0$. For Example, $2x^2 + 6x - 9 = (2)x^2 + (6)x + (-9)$ so $a = 2$, $b = 6$, and $c = -9$.

Assignment Guide

Core: 15–41 odd, 42, 43, 45–53
Enriched: 14–40 even, 41–53
All: Self Test, 1–10

For **Extra Practice,** see p. 781.

The red A, B, and C flags, printed only in the Teacher's Wraparound Edition, indicate the level of difficulty of the exercises.

Study Guide Masters, p. 77

NAME__________ DATE________

Study Guide

Student Edition
Pages 628–633

Solving Quadratic Equations by Using the Quadratic Formula

The method of completing the square can be used to develop a general formula called the *quadratic formula* that can be used to solve any quadratic equation.

The Quadratic Formula
The roots of a quadratic equation of the form $ax^2 + bx + c = 0$, where $a \neq 0$, are given by the formula $x = \frac{-b \pm \sqrt{b^2 - 4ac}}{2a}$.

In order to find a real value for $\sqrt{b^2 - 4ac}$, the value of $b^2 - 4ac$ must be nonnegative. If $b^2 - 4ac$ is negative, the equation has no real roots.

Example: Use the quadratic formula to solve $x^2 - 6x - 2 = 0$.

$$x = \frac{-b \pm \sqrt{b^2 - 4ac}}{2a}$$

$$= \frac{6 \pm \sqrt{(-6)^2 - 4(1)(-2)}}{2(1)} \quad a = 1, b = -6, \text{ and } c = -2$$

$$= \frac{6 \pm \sqrt{44}}{2}$$

$$x \approx 6.32 \text{ or } x \approx 1.78$$

The two roots are approximately 6.32 and 1.78.

Solve each equation by using the quadratic formula. Approximate irrational roots to the nearest hundredth.

1. $4t^2 = 144$ −6, 6
2. $2x^2 + 9x + 4 = 0$ −4, $-\frac{1}{2}$
3. $8x^2 + 17x + 2 = 0$ −2, $-\frac{1}{8}$
4. $3z^2 + 5z - 2 = 0$ −2, $\frac{1}{3}$
5. $-2m^2 + 8m + 4 = 0$ 4.45, −0.45
6. $x^2 + 3x - 2 = 0$ 0.56, −3.56
7. $2x^2 - 6x + 4 = 0$ 1, 2
8. $5x^2 + 10 = 15$ −1, 1
9. $2y^2 - 9y - 3 = 0$ 4.81, −0.31

Using the Programming Exercises The program given in Exercise 41 is for use with a TI-82 graphing calculator. For other programmable calculators, have students consult their owner's manual for commands similar to those presented here.

Additional Answers

43a. Discriminant is not a perfect square.

43b. Discriminant is a perfect square, but the expression is not an integer.

43c. Discriminant is a perfect square, and the equation can be factored.

43d. Discriminant is 0, and the equation is a perfect square.

Use the quadratic formula to determine values for *a*, *b*, and *c* if the given numbers are solutions of a quadratic equation. Then write the equation.

38–40. Sample answers are given.

38. $-1 \pm \sqrt{3}$ 38. 1, 2, −2; $x^2 + 2x - 2 = 0$

39. $\frac{-5 \pm \sqrt{2}}{2}$ 39. 4, 20, 23; $4x^2 + 20x + 23 = 0$

40. $\frac{4 \pm \sqrt{29}}{2}$ 40. 1, −4, $-\frac{13}{4}$; $x^2 - 4x - \frac{13}{4} = 0$

Programming

41. The graphing calculator program at the right determines what type of solutions a quadratic equation will have and then prints decimal approximations of the solutions if they exist.

```
PROGRAM:SOLUTIONS
: Prompt A, B, C
: B²−4AC→D
: If D<0
: Then
: Disp "NO REAL SOLUTIONS"
: Stop
: End
: If D=0
: Then
: Disp "1 DISTINCT SOLUTION:"
  −B/2A
: Else
: Disp "2 REAL SOLUTIONS",
  (−B+√ D)/2A, (−B−√ D)/2A
: End
```

Use the program to find the solutions of each equation.

a. $x^2 - 11x + 10 = 0$ 10, 1

b. $3x^2 - 2x + 1 = 0$ none

c. $4x^2 + 4x + 1 = 0$ −0.5

d. $7x^2 + 2x - 5 = 0$ −1, 0.7142857143

Critical Thinking

42. The expression $b^2 - 4ac$ is called the **discriminant** of a quadratic equation. The discriminant can help you determine what type of solutions to expect when you solve a quadratic equation. Copy and complete the table below.

Equation	$x^2 - 4x + 1 = 0$	$x^2 + 6x + 11 = 0$	$x^2 - 4x + 4 = 0$
Value of the Discriminant	12	−8	0
Graph of the Equation	See Solutions	Manual for	graphs.
Number of *x*-intercepts	2	none	1
Number of Real Solutions	2	none	1 distinct solution

34. See margin.

43. Use the results of Examples 1–3 and the table above to describe the discriminant of a quadratic equation for each type of solution.

a. two irrational solutions

b. two noninteger rational solutions

c. two integral solutions

d. 1 distinct integral solution

Applications and Problem Solving

44. Government Between 1980 and 1993, the income (billions of dollars) received by the federal government can be modeled by the quadratic function $I(t) = 0.26t^2 + 49.94t + 511.4$, where t represents the number of years since 1980.

a. Determine the domain and range values for which this function makes sense. D: $0 \leq t \leq 13$; R: $511.4 \leq I(t) < 1205$

b. Determine the income in 1993. $1204.6 billion

c. Assume that the pattern continues to hold. What is the projected federal income for the year 2000? $1614.2 billion

d. Determine in which year the federal income was $1000 billion or $1 trillion. 1989

Practice Masters, p. 77

11-3 NAME ________ DATE ________

Practice — Student Edition Pages 628–633

Solving Quadratic Equations by Using the Quadratic Formula

Solve each equation by using the quadratic formula. Approximate irrational roots to the nearest hundredth.

1. $4m^2 + 3m = 1$ $\frac{1}{4}, -1$

2. $3n^2 = 10n$ $0, \frac{10}{3}$

3. $p^2 + 4p + 1 = 0$ −0.27, −3.73

4. $q^2 - 5q + 2 = 0$ 4.56, 0.44

5. $r^2 - 7r + 3 = 0$ 6.54, 0.46

6. $a^2 + 6a + 3 = 0$ −0.55, −5.45

7. $3y^2 + 2y - 3 = 0$ 0.72, −1.39

8. $5z^2 - 3 = 0$ 0.77, −0.77

9. $x^2 + \frac{2}{3}x - \frac{15}{9} = 0$ $1, -\frac{5}{3}$

10. $f^2 + \frac{5}{12}f = \frac{1}{4}$ $\frac{1}{3}, -\frac{3}{4}$

GLENCOE Technology

Interactive Mathematics Tools Software

This multimedia software provides an interactive lesson that uses the graphs of two rockets' paths to determine when they will be at the same height. A **Computer Journal** gives students an opportunity to write about what they have learned.

For Windows & Macintosh

Tech Prep

Accountant Students who are interested in accounting may wish to do further research on the information given in Exercise 44 and explore the potential growth of this career. For more information on tech prep, see the *Teacher's Handbook.*

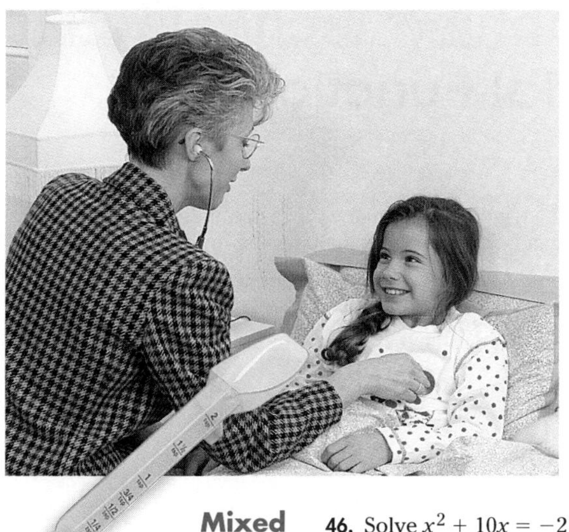

45. **Medicine** Two rules that govern the amount of medicine you should give a child if you know the adult dosage are Young's rule, $c = \frac{ad}{a + 12}$, and Cowling's rule, $c = \frac{(a + 1)d}{24}$. In both formulas, a represents the age of the child (in years), d represents the amount of the adult dosage, and c represents the amount of the child's dosage.
   a. The adult dosage for a drug is 30 mg/day. Calculate the dosage for a 6-year-old using each rule. **Y, 10 mg; C, 8.75 mg**
   b. Write a quadratic equation that represents the age(s) at which the two rules give the same dosage. Then solve the equation. **$0 = a^2 - 11a + 12$; 1.2 yr, 9.8 yr**

Mixed Review

48. $21a^3 - 6a^2 - 46a + 28$

46. Solve $x^2 + 10x = -21$ by graphing. (Lesson 11–2) **−7, −3**
47. Solve $4s^2 = -36s$. (Lesson 10–6) **{0, −9}**
48. Find $3a^2(a - 4) + 6a(3a^2 + a - 7) - 4(a - 7)$. (Lesson 9–6)
49. **Number Theory** The sum of two numbers is 42. Their difference is 6. Find the numbers. (Lesson 8–3) **24, 18**
50. Graph the solution set for $b > 5$ or $b \leq 0$. (Lesson 7–4) **See margin.**
51. Determine if the following relation is a function. (Lesson 5–5)
   {(3, 2), (−3, −2), (−4, −2), (4, −2)} **yes**
52. Solve $3x + 8 = 2x - 4$. (Lesson 3–5) **−12**
53. **Weather** The temperature at 8:00 A.M. was 36°F. A cold front went through that evening, and by 3:00 A.M. the next day, the temperature had fallen 40°. What was the temperature at 3:00 A.M.? (Lesson 2–1) **−4°F**

SELF TEST

Write the equation of the axis of symmetry and find the coordinates of the vertex of the graph of each equation. Then graph the equation. (Lesson 11–1) **1–3. See Solutions Manual for graphs.**

1. $y = x^2 - x - 6$ **$x = \frac{1}{2}, \left(\frac{1}{2}, -\frac{25}{4}\right)$**
2. $y = 2x^2 + 3$ **$x = 0$, (0, 3)**
3. $y = -x^2 + 7$ **$x = 0$, (0, 7)**

4. **Physics** The height h in feet of an experimental rocket t seconds after blast-off is given by the formula $h = -16t^2 + 2320t + 125$. (Lesson 11–1)
   a. Approximately how long after blast-off does the rocket reach a height of 84,225 feet? **72.5 s**
   b. How much longer from this height does it take the rocket to reach its maximum height? **0 seconds; this is the maximum height.**

Solve each equation by graphing. If integral roots cannot be found, state the consecutive integers between which the roots lie. (Lesson 11–2) **5–7. See Solutions Manual for graphs.**

5. $x^2 = 81$ **−9, 9**
6. $4x^2 = 35 - 4x$ **$2 < x < 3, -4 < x < -3$**
7. $6x^2 + 36 = 0$ **∅**

Solve each equation by using the quadratic formula. Approximate irrational roots to the nearest hundredth. (Lesson 11–3) **10. −2.12, 15.12**

8. $a^2 + 7a = -6$ **−6, −1**
9. $y^2 + 6y + 10 = 0$ **∅**
10. $z^2 - 13z - 32 = 0$

4 ASSESS

Closing Activity

Speaking Ask students to explain how to use the quadratic formula to solve any quadratic equation.

Chapter 11, Quiz B (Lesson 3), is available in the *Assessment and Evaluation Masters,* p. 296.

Mid-Chapter Test (Lessons 11-1 through 11-3) is available in the *Assessment and Evaluation Masters,* p. 295.

Additional Answer

50. 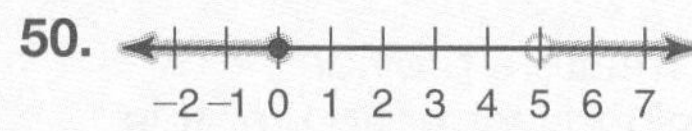

Extension

Problem Solving Solve each equation. Express irrational roots in simplest radical form and in decimal form to the nearest hundredth.

1. $(2n - 3)^2 - 7 = 6n + 1$
   $\frac{9 \pm \sqrt{77}}{4}$; 4.44, 0.06
2. $(3x + 2)^2 = (x - 1)^2 + 4$
   $\frac{-7 \pm \sqrt{57}}{8}$; 0.07, −1.82

SELF TEST

The Self Test provides students with a brief review of the concepts and skills in Lessons 11-1 through 11-3. Lesson numbers are given to the right of exercises or instruction lines so students can review concepts not yet mastered.

Enrichment Masters, p. 77

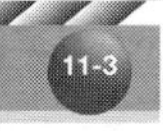

NAME________________ DATE________

Enrichment

Student Edition Pages 628–633

Odd Numbers and Parabolas

The solid parabola and the dashed stair-step graph are related. The parabola intersects the stair steps at their inside corners.

Use the figure for Exercises 1–3.

1. What is the equation of the parabola?
   $y = x^2$
2. Describe the horizontal sections of the stair-step graph.
   Each is 1 unit wide.
3. Describe the vertical sections of the stair-step graph.
   They form the sequence 1, 3, 5, 7.

Use the second figure for Exercises 4–6.

4. What is the equation of the parabola?
   $y = \frac{1}{2}x^2$
5. Describe the horizontal sections of the stair steps.
   Each is 1 unit wide.
6. Describe the vertical sections.
   They form the sequence $\frac{1}{2}, \frac{3}{2}, \frac{5}{2}, \frac{7}{2}, \frac{9}{2}$.

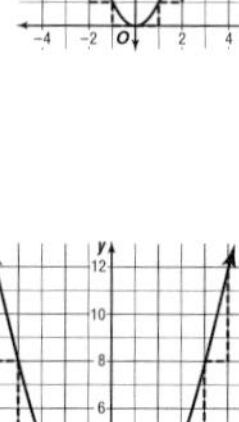

7. How does the graph of $y = \frac{1}{2}x^2$ relate to the sequence of numbers $\frac{1}{2}, \frac{3}{2}, \frac{5}{2}, \frac{7}{2}, \cdots$?
   If the x-values increase by 1, the y-values increase by the numbers in the sequence.
8. Complete this conclusion. To graph a parabola with the equation $y = ax^2$, start at the vertex. Then go over 1 and up a; over 1 and up $3a$;
   over 1 and up $5a$; over 1 and up $7a$; and so on. The coefficients of a are the odd numbers.

11-4A LESSON NOTES

NCTM Standards: 1–6, 13

Objective
Use a graphing calculator to graph exponential functions of the form $y = a^x$, where $a > 0$ and $a \neq 1$.

Recommended Time
15 minutes

Instructional Resources
Graphing Calculator Masters, pp. 34 and 35

These masters provide keystroking instruction for this lesson for the TI-81 and Casio graphing calculators.

1 FOCUS

Motivating the Lesson
This activity introduces students to the graphs of simple exponential functions. The symmetry of n^x and $\left(\frac{1}{n}\right)^x$ are illustrated.

2 TEACH

Teaching Tip Set the *y*-range at [0, 1] to see how the graph approaches but never touches the *x*-axis.

3 PRACTICE/APPLY

Assignment Guide

Core: 1–10
Enriched: 1–10

4 ASSESS

Observing students working with technology is an excellent method of assessment.

11–4A Graphing Technology Exponential Functions

A Preview of Lesson 11–4

Graphing calculators can be used to graph many types of functions easily so patterns in the functions can be studied. This includes **exponential functions** of the form $y = a^x$, where $a > 0$ and $a \neq 1$.

Example 1 **Graph each equation in the standard viewing window. Describe the graph.**

a. $y = 3^x$

Enter the equation in the Y= list.

Enter:  Y= 3 ^ X,T,θ ZOOM 6

Notice that the graph increases rapidly as *x* becomes greater. The graph passes through the point at (0, 1). The domain of the function is all real numbers, and the range is all positive real numbers.

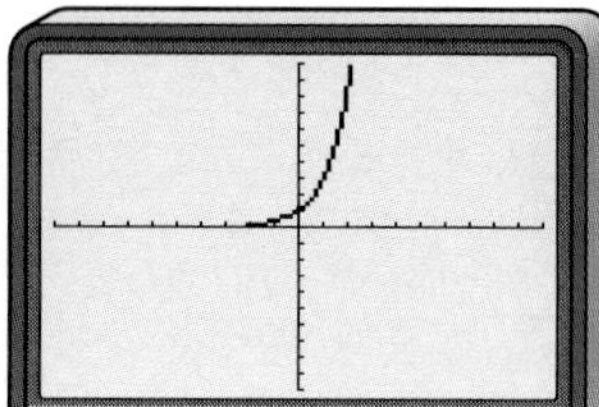

b. $y = \left(\frac{1}{3}\right)^x$

Enter: Y= (1 ÷ 3)  ^ X,T,θ GRAPH

The graph decreases as *x* increases. The graph passes through the point at (0, 1). The domain is all real numbers, and the range is all positive real numbers.

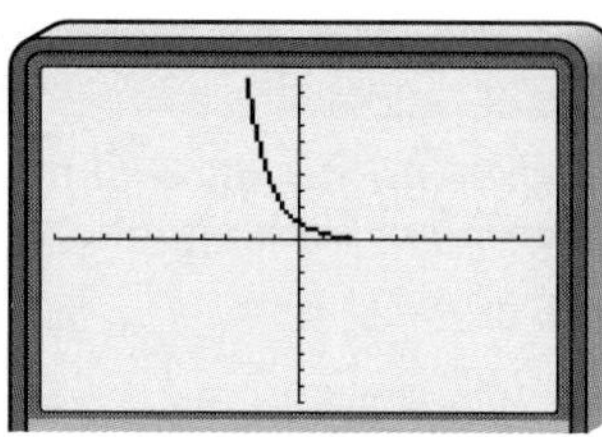

1–9. See Solutions Manual.

EXERCISES

Use a graphing calculator to graph each exponential equation. Sketch the graphs on a separate piece of paper.

1. $y = 2^x$
2. $y = 5^x$
3. $y = 0.1^x$
4. $y = \left(\frac{2}{3}\right)^x$
5. $y = 0.25^x$
6. $y = 1.6^x$
7. $y = 0.2^x$
8. $y = 0.5^{-x}$
9. $y = 10^x$
10. Solve $1.2^x = 10$ graphically. Explain how you solved the equation and write the solution accurately to the nearest hundredth. **12.63**

10. Sample answers: (1) Graph $y = 1.2^x$ and $y = 10$ and find their intersection, (2) Graph $y = 1.2^x - 10$ and find its zero, (3) Graph $y = 1.2^x$ and use trace to find where $y = 10$.

Using Technology
This lesson offers an excellent opportunity for using technology in your algebra classroom. For more information on using technology, see *Graphing Calculators in the Mathematics Classroom,* one of the titles in the Glencoe Mathematics Professional Series.

11–4 Exponential Functions

What YOU'LL LEARN

- To graph exponential functions,
- to determine if a set of data displays exponential behavior, and
- to solve exponential equations.

Why IT'S IMPORTANT

You can use exponential equations to solve problems involving biology and archaeology.

Folklore

A wise man asked his ruler to provide rice for feeding his people. Rather than receiving a constant daily supply of rice, the wise man asked the ruler to give him 2 grains of rice for the first square on a chessboard, 4 grains of rice for the second, 8 grains of rice for the third, 16 grains of rice for the fourth, and so on, doubling the amount of rice with each square of the board. How many grains of rice will he receive for the last (64th) square on the chessboard?

You could make a table and look for a pattern to determine how many grains of rice he received for each square of the chessboard.

Square	Grains	Pattern	Square	Grains	Pattern
1	2	2^1	19	524,288	2^{19}
2	4	2^2	20	1,048,576	2^{20}
3	8	2^3	21	2,097,152	2^{21}
4	16	2^4	22	4,194,304	2^{22}
5	32	2^5	23	8,388,608	2^{23}
6	64	2^6	24	16,777,216	2^{24}
7	128	2^7	25	33,554,432	2^{25}
8	256	2^8	26	67,108,864	2^{26}
9	512	2^9	27	134,217,728	2^{27}
10	1024	2^{10}	28	268,435,456	2^{28}
11	2048	2^{11}	29	536,870,912	2^{29}
12	4096	2^{12}	30	1,073,741,824	2^{30}
13	8192	2^{13}	31	2,147,483,648	2^{31}
14	16,384	2^{14}	32	4,294,967,296	2^{32}
15	32,768	2^{15}	33	8,589,934,592	2^{33}
16	65,536	2^{16}	34	17,179,869,184	2^{34}
17	131,072	2^{17}	35	34,359,738,368	2^{35}
18	262,144	2^{18}	36	68,719,476,736	2^{36}

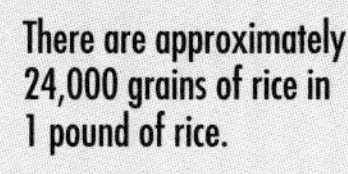

There are approximately 24,000 grains of rice in 1 pound of rice.

Notice that when only 36 of the 64 squares are calculated, there are over 68 billion grains of rice. How many grains would there be for square 64? *You will answer this question in Exercise 1.*

Study the pattern column. Notice that the exponent number matches the number of the square on the chessboard. So we can write an equation to describe y, the number of grains of rice for any given square x as $y = 2^x$. This type of function, in which the variable is the exponent, is called an **exponential function.**

Definition of Exponential Function

An exponential function is a function that can be described by an equation of the form $y = a^x$, where $a > 0$ and $a \neq 1$.

A simple computation on a scientific calculator reveals that the average grain of rice weighs 4.167×10^{-5} pounds.

11-4 LESSON NOTES

NCTM Standards: 1–6, 13

Instructional Resources

- Study Guide Master 11-4
- Practice Master 11-4
- Enrichment Master 11-4
- Assessment and Evaluation Masters, p. 297

Transparency 11-4A contains the 5-Minute Check for this lesson; **Transparency 11-4B** contains a teaching aid for this lesson.

Recommended Pacing	
Standard Pacing	Day 8 of 11
Honors Pacing	Day 7 of 10
Block Scheduling*	Day 3 of 5
Alg. 1 in Two Years*	Days 12 & 13 of 19

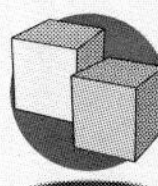

*For more information on pacing and possible lesson plans, refer to the *Block Scheduling Booklet* and *Algebra 1 in Two Years.*

1 FOCUS

5-Minute Check
(over Lesson 11-3)

Solve each equation by using the quadratic formula.

1. $y^2 + 8y + 15 = 0$ **−3, −5**
2. $2p^2 = 98$ **±7**
3. $-2x^2 + 8x = 3$ **0.42, 3.58**
4. $x^2 - 2x = 3$ **3, −1**
5. $-6x^2 + 15x = 6$ **$\frac{1}{2}$, 2**

Motivating the Lesson

Situational Problem Have students look up the U.S. population during this century. Plot the data on a graph and point out that it is not a linear graph.

2 TEACH

MODELING MATHEMATICS This hands-on activity illustrates how to generate the exponential function $y = 2^x$. Encourage students to figure out how to modify the activity to generate $y = 3^x$.

In-Class Example

For Example 1
Graph each function. State the *y*-intercept of each graph.

a. $y = 4^x$ ***y*-intercept = 1**

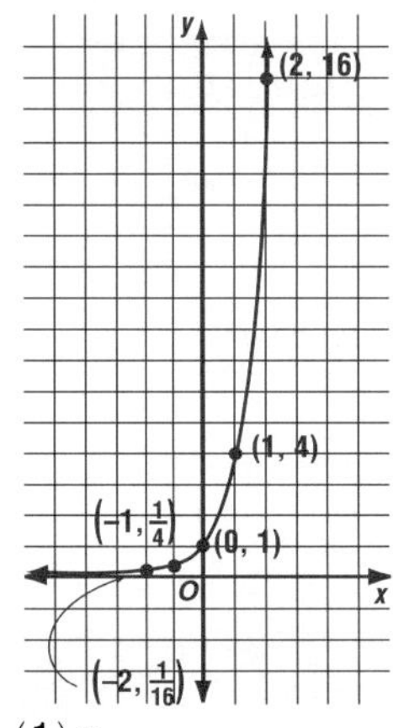

b. $y = \left(\frac{1}{4}\right)^x$ ***y*-intercept = 1**

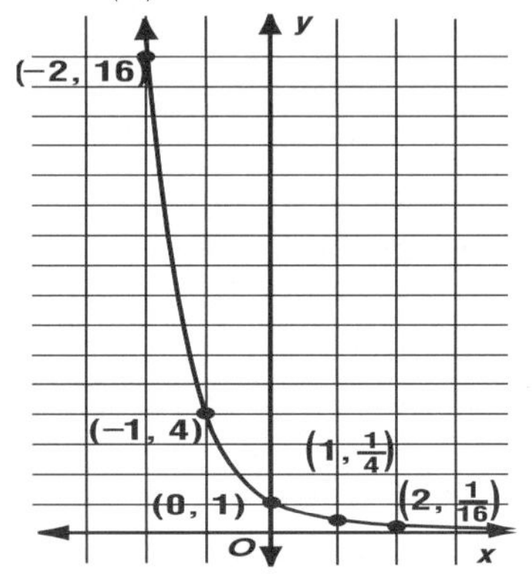

c. $y = 4^x - 8$ ***y*-intercept = −7**

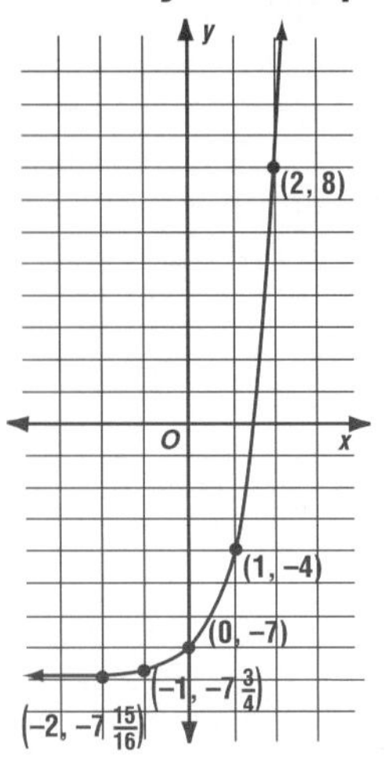

You can use paper-folding to illustrate an exponential function.

MODELING MATHEMATICS

Modeling an Exponential Function

Materials: large piece of paper

Your Turn

a. Fold a large rectangular piece of paper in half. Unfold it and record how many sections are formed by the creases. Refold the paper.

b. Fold the paper in half again. Record how many sections are formed by the creases. Refold the paper.

c. Continue folding in half and recording the number of sections until you can no longer fold the paper.

d. How many folds could you make? **about 6**

e. How many sections were formed? **64**

f. What exponential function is modeled by the folds and sections created? $y = 2^x$

As with other functions, you can use ordered pairs to graph an exponential function. Use a table of values and a calculator to find ordered pairs that satisfy $y = 2^x$. While the negative values of x have no meaning in the rice problem, they should be included in the graph of the function. Connect the points to form a smooth curve.

x	y
−5	0.03125
−4	0.0625
−3	0.125
−2	0.25
−1	0.5
0	1
1	2
2	4
3	8
4	16

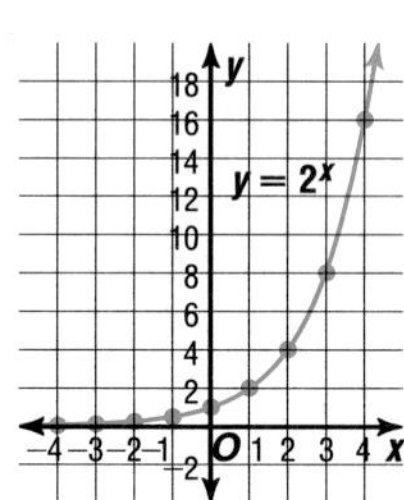

Notice that the graph has a *y*-intercept of 1. Does it have an *x*-intercept?
You will answer this question in Exercise 2.

The graph shown above represents all real values of x and their corresponding values of y for $y = 2^x$.

You can use a scientific calculator to help you find ordered pairs to graph other exponential functions. For example, suppose $y = 3^x$ and $x = -2$.

Enter: 3 [y^x] 2 [+/−] [=] *0.11111111*

Example 1 **Graph each function. State the *y*-intercept of each graph.**

a. $y = 3^x$

x	y
−3	0.037
−2	0.111
−1	0.333
0	1
1	3
2	9
3	27

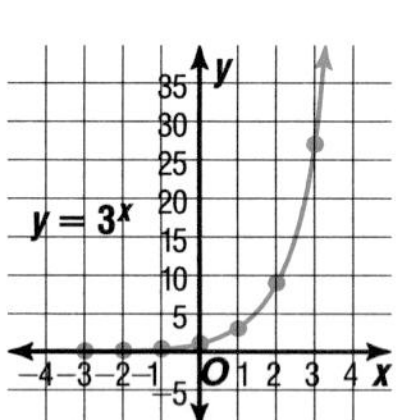

The *y*-intercept is 1.

b. $y = \left(\frac{1}{3}\right)^x$

x	y
−3	27
−2	9
−1	3
0	1
1	0.333
2	0.111
3	0.037
4	0.012

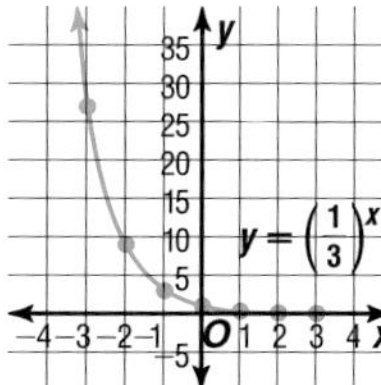

The y-intercept is 1.

Notice that the graph of $y = \left(\frac{1}{a}\right)^x$ decreases rapidly as x increases.

c. $y = 3^x - 7$

x	y
−3	−6.96
−2	−6.89
−1	−6.67
0	−6
1	−4
2	2
3	20
4	74

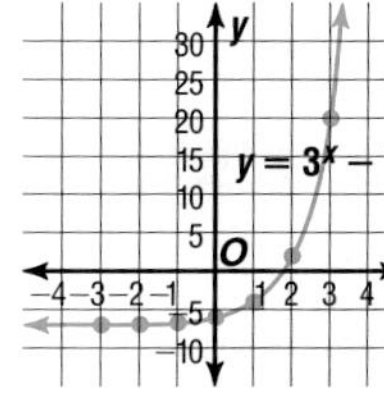

The y-intercept is −6.

LOOK BACK

You can refer to Lesson 10-1 for more information on factors.

How do you know if a set of data is exponential? One method is to observe the shape of the graph. But the graph of an exponential function may resemble part of the graph of a quadratic function. Another way is to **look for a pattern** in the data.

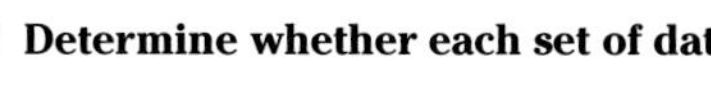

Example 2

PROBLEM SOLVING
Look for a Pattern

Determine whether each set of data displays exponential behavior.

a.

x	0	5	10	15	20	25
y	800	400	200	100	50	25

Method 1: Look for a Pattern
The domain values are at regular intervals of 5. Let's see if there is a common factor among the range values.

800 400 200 100 50 25
$\times\frac{1}{2}$ $\times\frac{1}{2}$ $\times\frac{1}{2}$ $\times\frac{1}{2}$ $\times\frac{1}{2}$

Since the domain values are at regular intervals and the range values have a common factor, the data are probably exponential. The equation for the data probably involves $\left(\frac{1}{2}\right)^x$.

Method 2: Graph the data.

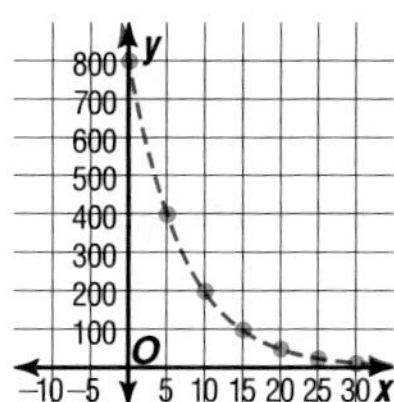

The graph shows a rapidly decreasing value of y as x increases. This is a characteristic of exponential behavior.

TECHNOLOGY Tip

You could also use a graphing calculator to make a scatter plot of the data to observe the patterns.

In-Class Example

For Example 2
Determine whether each set of data displays exponential behavior.

a.

x	0	1	2	3	4	5
y	3	9	27	81	243	729

yes

b.

x	0	1	2	3	4
y	4	8	12	16	20

no

Classroom Vignette

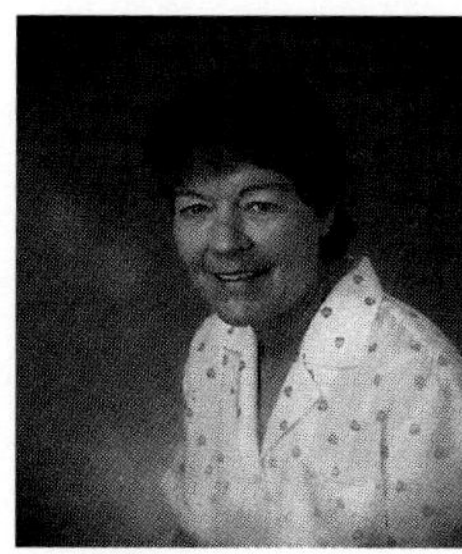

"Since students encountered scatter plots in Chapter 6, I would recommend using scatter plots as an alternative to the methods introduced. If students haven't been exposed to the graphing calculator option of graphing the exponential function over the scatter plot, this would be an appropriate place for it."

Joan M. Gell

Joan Gell
Palos Verdes High School
Palos Verdes Estates, California

3 PRACTICE/APPLY

Check for Understanding

Exercises 1–16 are designed to help you assess your students' understanding through reading, writing, speaking, and modeling. You should work through Exercises 1–6 with your students and then monitor their work on Exercises 7–16.

Error Analysis

When looking at the graph of an exponential function, students may assume that the function is equal to zero for all x values below a certain point, while in fact it only approaches zero.

Assignment Guide

Core: 17–47 odd, 48, 49, 51–59
Enriched: 18–46 even, 48–59

For **Extra Practice,** see p. 781.

The red A, B, and C flags, printed only in the Teacher's Wraparound Edition, indicate the level of difficulty of the exercises.

Study Guide Masters, p. 78

NAME__________ DATE ______

Study Guide

Student Edition Pages 635–642

Exponential Functions

An equation in the form $y = a^x$, where $a > 0$ and $a \neq 1$, is called an **exponential function.** You can use a calculator to evaluate exponential expressions.

Example: Use a calculator to approximate $0.5^{-3.5} + 2$ to the nearest hundredth.
ENTER: 0.5 [y^x] 3.5 [+/−] [=] [+] 2 [=] 13.3137085
$0.5^{-3.5} + 2 \approx 13.31$

You can use values of exponential expressions to find ordered pairs that satisfy an exponential function. Then you can use the ordered pairs to graph the function.

Example: Graph $y = \left(\frac{1}{2}\right)^x$.

x	y
−2	4
−1	2
0	1
1	0.5
2	0.25

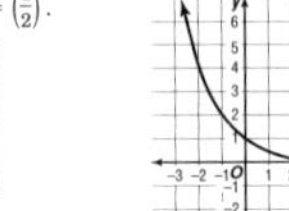

Graph each function.

1. $y = 0.3^x$

2. $y = 3^x + 1$

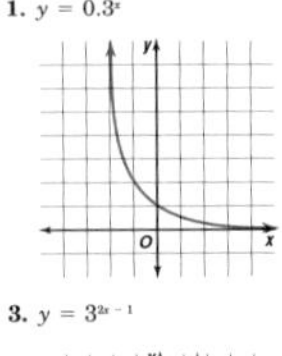

3. $y = 3^{2x-1}$

4. $y = \left(\frac{1}{2}\right)^x - 2$

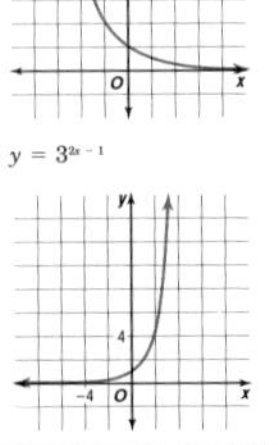

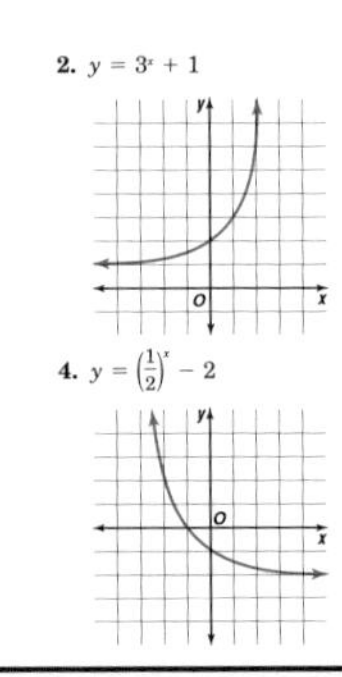

CHECK FOR UNDERSTANDING

Communicating Mathematics

Study the lesson. Then complete the following.

1. **Refer** to the application at the beginning of the lesson.
   a. Use a calculator to determine how many grains there would be for the 64th square of the chessboard. **$1.844674407 \times 10^{19}$**
   b. How many tons of rice is this? (*Hint:* Recall that 1 T = 2000 lb.)
2. a. **Determine** whether the graph of $y = 2^x$ has an x-intercept.
   b. **Describe** your method.
   c. Is this true of all exponential functions? **no; only those of form $y = a^x$**
3. a. **Determine** whether the graph of $y = 2^x$ has a vertex. **No, it does not.**
   b. **Describe** your method for answering part a. **See students' work.**
   c. Is this true for all exponential functions? **Yes.**
4. **Explain** why $a \neq 1$ in the definitions and properties involving exponential functions. **If $a = 1$, $a^x = 1$ for any value of x.**
5. **Write** a paragraph explaining why you think a graphing calculator might be a good tool to have when studying exponential functions. **See students' work.**
6. Refer to the Modeling Mathematics activity on page 636.
   a. The area of the large rectangle is 1. Find the area of each section after each set of folds. Record your findings in a table. **See margin.**
   b. Compare the number of folds to the area of each section. What pattern do you see? **See margin.**
   c. Write an exponential function relating the number of folds to the area of each section. **$y = 0.5^x$**

1b. over 3.8×10^{11} tons
2a. No, it does not.
2b. Sample answer: Use the graph of $y = 2^x$ and the TRACE feature on a graphing calculator to examine the y values.

MODELING MATHEMATICS

Guided Practice

Use a calculator to determine the approximate value of each expression to the nearest hundredth.

7. $3^{1.5}$ **5.20**
8. $3^{-0.9}$ **0.37**
9. $3^{2.3}$ **12.51**

Graph each function. State the y-intercept.

10. $y = 0.5^x$ **1**
11. $y = 2^x + 6$ **7**

10–11. See Solutions Manual for graphs.

12. Determine if the data in the table below displays exponential behavior. Describe the behavior.

x	0	1	2	3	4	5
y	1	6	36	216	1296	7776

Yes, increases by a factor of 6.

Solve each equation.

13. $5^{3y+4} = 5^y$ **−2**
14. $2^5 = 2^{2x-1}$ **3**
15. $3^x = 9^{x+1}$ **−2**
16. **Biology** Suppose $B = 100 \cdot 2^t$ represents the number of bacteria B in a petri dish after t hours if you began with 100 bacteria. How long would it take to obtain 1000 bacteria? **3.32 hours**

EXERCISES

Practice

Use a calculator to determine the approximate value of each expression to the nearest hundredth.

17. $4^{1.7}$ **10.56**
18. $10^{-0.5}$ **0.32**
19. $\left(\frac{2}{3}\right)^{-1.2}$ **1.63**
20. $\left(\frac{1}{3}\right)^{4.1}$ **0.01**
21. $50(3^{-0.6})$
22. $10(3^{-1.8})$ **1.38**
23. $0.4(3^{0.7})$ **0.86**
24. $20(0.25^{-2.7})$

21. 25.86 **24. 844.49**

Reteaching

Using Graphics Using a graphing calculator, graph a family of exponential functions $y = 2^x$, $y = 3^x$, $y = 5^x$, and $y = 11^x$. Note the effect of increasing the base.

Additional Answers

6a.

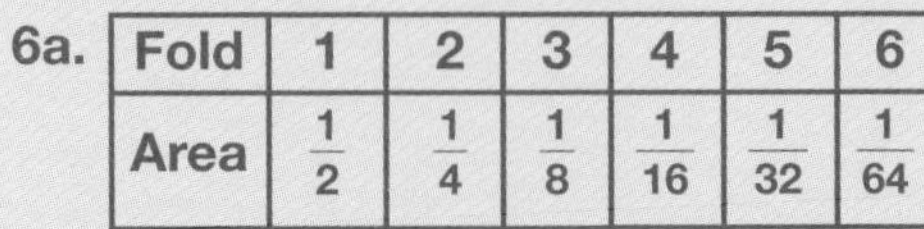

Fold	1	2	3	4	5	6
Area	$\frac{1}{2}$	$\frac{1}{4}$	$\frac{1}{8}$	$\frac{1}{16}$	$\frac{1}{32}$	$\frac{1}{64}$

6b. Each fold makes a rectangle whose area is $\frac{1}{2^x}$, where x is the number of folds.

25–30. See Solutions Manual for graphs.

Graph each function. State the y-intercept.

25. $y = 2^x + 4$ **5** 26. $y = 2^{x+4}$ **16** 27. $y = 3\left(\frac{1}{3}\right)^x$ **3**

28. $y = 2 \cdot 3^x$ **2** 29. $y = 4^x$ **1** 30. $y = \left(\frac{1}{4}\right)^x$ **1**

Determine if the data in each table display exponential behavior. Explain why or why not.

31.

x	y
−2	−5
−1	−2
0	1
1	4

no, linear

32.

x	y
0	1
1	0.5
2	0.25
3	0.125

yes, constant factor

33.

x	y
−1	−0.5
0	1.0
1	−2.0
2	4.0

no, no pattern

Solve each equation.

34. $5^{3x} = 5^{-3}$ **−1** 35. $2^{x+3} = 2^{-4}$ **−7** 36. $5^x = 5^{3x+1}$ **−0.5**

37. $10^x = 0.001$ **−3** 38. $2^{2x} = \frac{1}{8}$ $-\frac{3}{2}$ 39. $\left(\frac{1}{6}\right)^q = 6^{q-6}$ **3**

40. $16^{x-1} = 64^x$ **−2** 41. $81^x = 9^{x^2-3}$ **3, −1** 42. $4^{x^2-2x} = 8^{x^2+1}$ **−1, −3**

Graphing Calculator

As with linear graphs and quadratic graphs, exponential graphs can form families of graphs. Graph each set of equations on the same screen. Sketch the graphs and discuss any similarities or differences. **43–46. See margin.**

43. $y = 3^x$; $y = 3^{x+4}$; $y = 3^{x-2}$

44. $y = \left(\frac{1}{3}\right)^x$; $y = \left(\frac{1}{3}\right)^x + 5$; $y = \left(\frac{1}{3}\right)^x - 3$

45. $y = 2^x$; $y = 2^{x-7}$; $y = 2^{x-2}$

46. $y = 6^x$; $y = 6^{3x}$; $y = 6^{8x}$

47. a. Use a graphing calculator to solve $2.5^x = 10$. Write the solution to the nearest hundredth. **2.51**

b. Explain why you cannot solve this equation algebraically like you did the equation in Exercise 41. **You cannot write 10 as a power of 2.5.**

Critical Thinking

48. Refer to the equations in Example 1. Use a calculator to find additional values to complete the following.

a. For $y = 3^x$, as x decreases, the value of y approaches what number? **0**

b. For $y = \left(\frac{1}{3}\right)^x$, as x increases, the value of y approaches what number? **0**

48c. −7

c. For $y = 3^x - 7$, as x decreases, the value of y approaches what number?

d. For any equation $y = a^x + c$, where $a > 1$, what value does y approach as x decreases? **c**

e. For any equation $y = a^x + c$, where $0 < a < 1$, what value does y approach as x increases? **c**

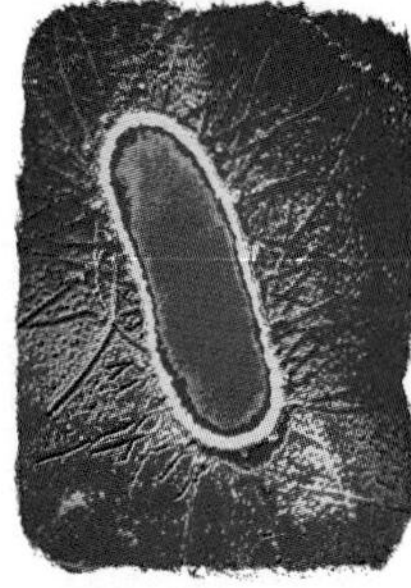

Applications and Problem Solving

49. **Biology** Mitosis is a process of cell reproduction in which one cell divides into two identical cells. *E. coli* is a fast-growing bacteria that is often responsible for food poisoning in uncooked meat. It can reproduce itself in 15 minutes. If you begin with 100 *e. coli* bacteria, how many bacteria will there be in 1 hour? **1600 bacteria**

Tech Prep

Medical Technician Students who are interested in medical laboratory assisting may wish to further investigate the data provided in Exercise 49 and explore the potential growth of this career. For more information on tech prep, see the *Teacher's Handbook.*

Additional Answer

46.

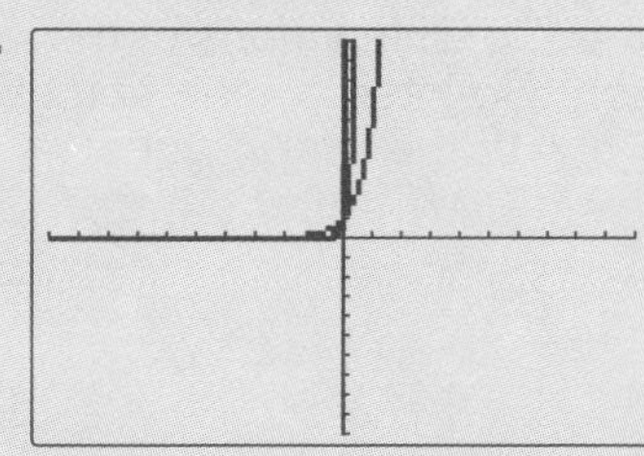

All have the same y-intercept, but each rises more steeply than the previous one.

Additional Answers

43.

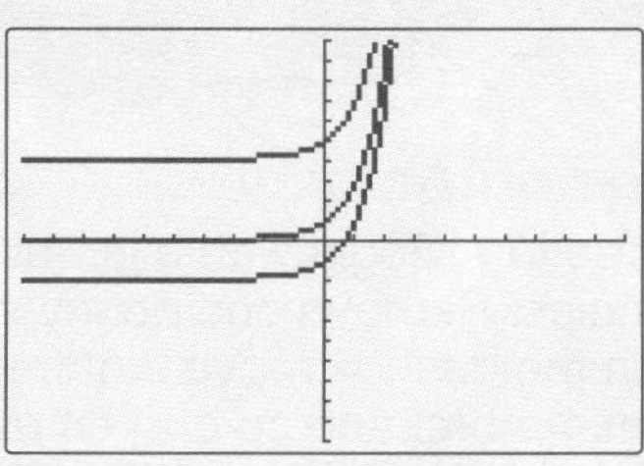

All have the same shape but different y-intercepts.

44.

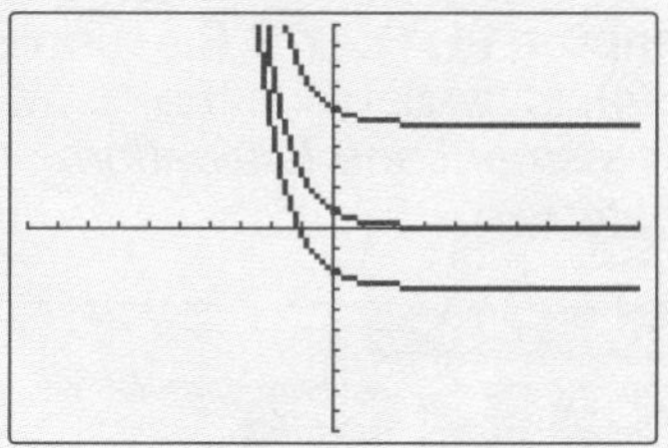

All have the same shape but different y-intercepts.

45.

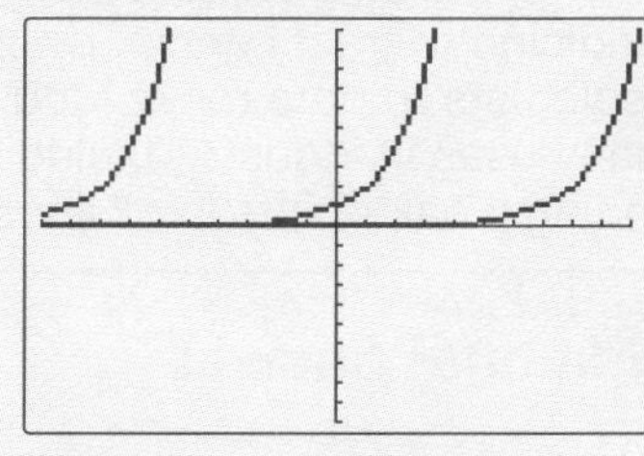

All have the same shape but are positioned at different places along the x-axis.

Practice Masters, p. 78

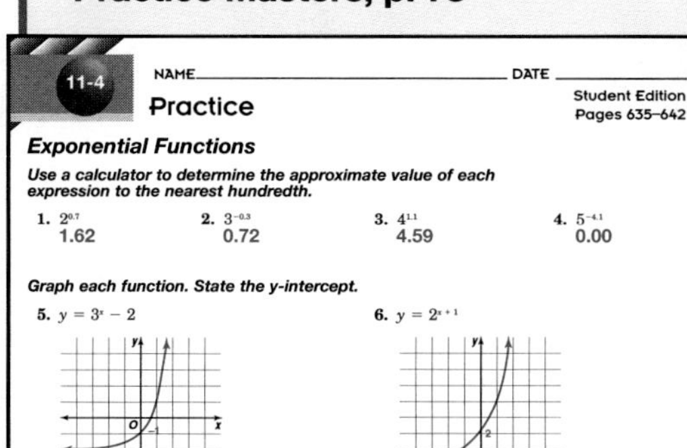

11-4 NAME__________ DATE ________

Practice — Student Edition Pages 635–642

Exponential Functions

Use a calculator to determine the approximate value of each expression to the nearest hundredth.

1. $2^{0.7}$ 1.62 2. $3^{-0.3}$ 0.72 3. $4^{1.1}$ 4.59 4. $5^{-4.1}$ 0.00

Graph each function. State the y-intercept.

5. $y = 3^x - 2$ 6. $y = 2^{x+1}$

7. $y = 3 \cdot 2^x$

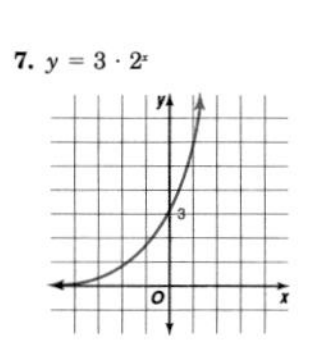

8. $y = \left(\frac{1}{2}\right)^x$

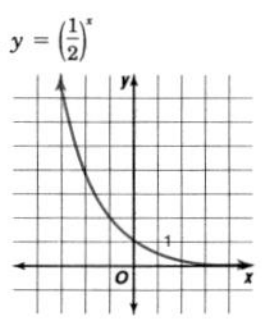

Solve each equation.

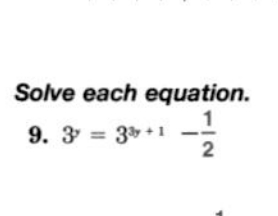

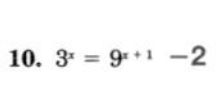

9. $3^y = 3^{3y+1}$ $-\frac{1}{2}$ 10. $3^x = 9^{x+1}$ −2

11. $8^{r-1} = 16^{3r}$ $-\frac{1}{3}$ 12. $9^{3y} = 27^{y+2}$ 2

13. $2^{2n-1} = 8^{n+7}$ −22 14. $4^{x-1} = 8^x$ −2

4 ASSESS

Closing Activity

Modeling Begin with a jar full of pennies. Remove one penny, put it on the desk, remove 2 more, then 4 more, and so on, removing twice the number each time. Note how fast the jar is emptied.

Chapter 11, Quiz C (Lesson 11-4), is available in the *Assessment and Evaluation Masters,* p. 297.

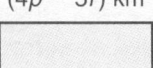

***fabulous* FIRSTS**

Radioisotopes are used as tracers in biological research, in medicine to treat cancer, by geologists to date rocks, and for many uses in industry, including finding weak points in oil lines.

Additional Answer

52b.

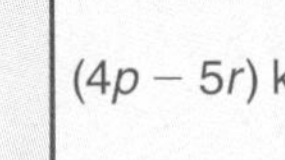

Enrichment Masters, p. 78

NAME ________________ DATE ________

Enrichment

Student Edition Pages 635–642

Parabolas Through Three Given Points

If you know two points on a straight line, you can find the equation of the line. To find the equation of a parabola, you need three points on the curve.

For example, here is how to approximate an equation of the parabola through the points (0, –2), (3, 0), and (5, 2).

Use the general equation $y = ax^2 + bx + c$. By substituting the given values for x and y, you get three equations.

$$(0, -2):\quad -2 = c$$
$$(3, 0):\quad 0 = 9a + 3b + c$$
$$(5, 2):\quad 2 = 25a + 5b + c$$

First, substitute –2 for c in the second and third equations. Then solve those two equations as you would any system of two equations. Multiply the second equation by 5 and the third equation by –3.

$$\begin{aligned} 0 &= 45a + 15b - 10 \\ -6 &= -75a - 15b + 6 \\ \hline -6 &= -30a - 4 \\ a &= \frac{1}{15} \end{aligned}$$

To find b, substitute $\frac{1}{15}$ for a in either the second or third equation.

$$\begin{aligned} 0 &= 9\left(\frac{1}{15}\right) + 3b - 2 \\ b &= \frac{7}{15} \end{aligned}$$

The equation of a parabola through the three points is $y = \frac{1}{15}x^2 + \frac{7}{15}x - 2$.

Find the equation of a parabola through each set of three points.

1. (1, 5), (0, 6), (2, 3) $y = -\frac{1}{2}x^2 - \frac{1}{2}x + 6$
2. (–5, 0), (0, 0), (8, 100) $y = \frac{25}{26}x^2 + \frac{125}{26}x$
3. (4, –4), (0, 1), (3, –2) $y = -\frac{1}{4}x^2 - \frac{1}{4}x + 1$
4. (1, 3), (6, 0), (0, 0) $y = -\frac{3}{5}x^2 + \frac{18}{5}x$
5. (2, 2), (5, –3), (0, –1) $y = -\frac{19}{30}x^2 + \frac{83}{30}x - 1$
6. (0, 4), (4, 0), (–4, 4) $y = \frac{1}{8}x^2 - \frac{1}{2}x + 4$

50. **Currency** In the United States between 1910 and 1994, the amount of currency in circulation $M(t)$ (billions of dollars) can be approximated by the function $M(t) = 2.08(1.06)^t$, where t represents the number of years since 1910.
    a. Determine $M(t)$ for the years 1920, 1950, 1980, and 2000.
    b. Determine the amount of currency in circulation in the years 1920, 1950, 1980, and 2000. **\$3.72 billion, \$21.39 billion, \$122.88 billion, \$394.09 billion**

50a. 10, 40, 70, 90

51. –3, $-\frac{1}{2}$

Mixed Review

51. Solve $2x^2 + 3 = -7x$ by using the quadratic formula. Check your solution by factoring. (Lesson 11–3)

52. **Geometry** A rectangle has an area of $(16p^2 - 40pr + 25r^2)$ square kilometers. (Lesson 10–5)
    a. Find the dimensions of the rectangle. **(4p – 5r) km by (4p – 5r) km**
    b. Sketch the rectangle, labeling its dimensions. **See margin.**

53. Factor $\frac{4}{5}a^2b - \frac{3}{5}ab^2 - \frac{1}{5}ab$. (Lesson 10–2) $\frac{1}{5}ab(4a - 3b - 1)$

54. **City Planning** A section of Lithopolis is shaped like a trapezoid with an area of 81 square miles. The distance between Union Street and Lee Street is 9 miles. The length of Union Street is 14 miles less than 3 times the length of Lee Street. Find the length of Lee Street. Use $A = \frac{h(a + b)}{2}$. (Lesson 9–8) **8 miles**

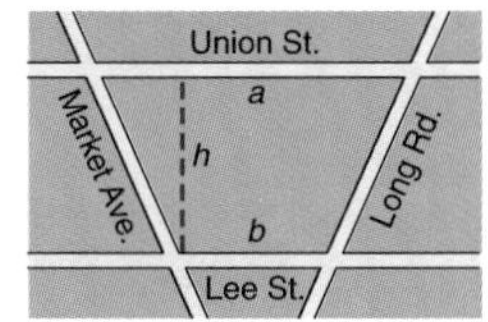

55. 184

55. **Basketball** On December 13, 1983, the Denver Nuggets and the Detroit Pistons broke the record for the highest score in a professional basketball game. The two teams scored a total of 370 points. If the Nuggets scored 2 points less than the Pistons, what was the Nuggets' final score? (Lesson 9–5)

56. Solve the system of inequalities by graphing. (Lesson 8–5)
    $y \le 2x + 2$
    $y \ge -x - 1$ **See margin.**

57a. See Solutions Manual.

57b. Sample answer: $y = 0.07x - 12$

57. **Astronomy** The table at the right shows the relationship between the distance from the sun, measured in millions of miles, and the time to complete an orbit, measured in Earth years. (Lesson 6–3)

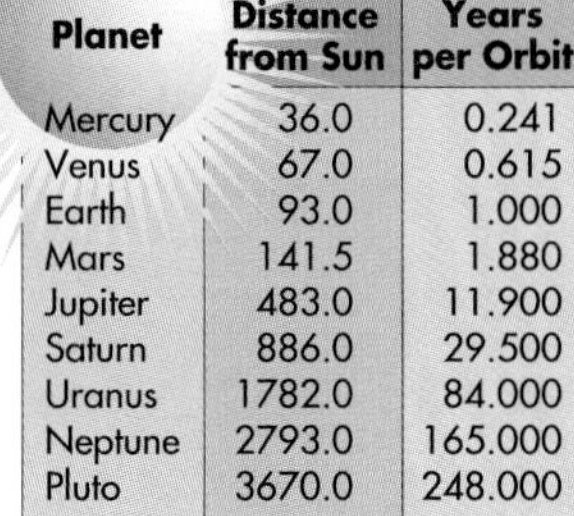

Planet	Distance from Sun	Years per Orbit
Mercury	36.0	0.241
Venus	67.0	0.615
Earth	93.0	1.000
Mars	141.5	1.880
Jupiter	483.0	11.900
Saturn	886.0	29.500
Uranus	1782.0	84.000
Neptune	2793.0	165.000
Pluto	3670.0	248.000

   a. Make a scatter plot of these data.
   b. Draw a best-fit line and write an equation for the line.
   c. Suppose a tenth planet was discovered at a distance of 4.1 billion miles from the sun. Use the scatter plot to estimate how long it would take it to orbit the sun. **275 years**

***fabulous* FIRSTS**

Clair C. Patterson (1922–1995)

Geochemist Clair Patterson, a professor at the California Institute of Technology, was the first scientist to establish Earth's age based on analysis of radioactive isotopes and their daughter products. In 1953, he established Earth's age to be 4.6 billion years.

58. **Geometry** Triangle ABC is similar to triangle ADE in the figure at the right. Find the value of s. (Lesson 4–2) **75 m**

59. Find $(-2)(3)(-10)$. (Lesson 2–6) **60**

Extension

Reasoning Do all functions of the form $y = (b \pm c)^x \pm d$ intersect the point (0, 1)? **If $d = 0$, the graph goes through (0, 1). If $d \neq 0$, the graph does not.**

Additional Answer

56.

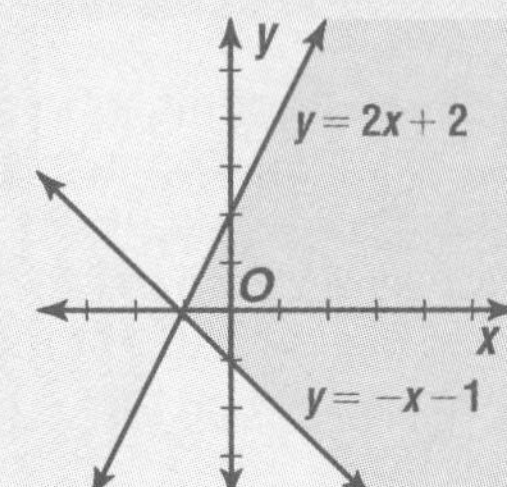

11-5 Growth and Decay

What YOU'LL LEARN

- To solve problems involving growth and decay.

Why IT'S IMPORTANT

You can solve growth and decay problems to learn more about demographics and energy.

CAREER CHOICES

Urban, or **city, planners** use statistics on demographics, traffic patterns, and economics to help officials make decisions on social, economic, and environmental issues in their communities. A bachelors degree in urban or regional planning is required, and a masters degree in civil engineering or planning is helpful for advancement.

For further information, contact:

American Planning Association
1776 Massachusetts Ave., NW
Washington, DC 20036

APPLICATION

Demographics

How quickly has the population of your state grown in the 20th century? Both California and Nebraska have grown in population during this century by a constant percent. The graph at the right shows the population of each state where t represents the number of years since 1900. Since 1900, the population in Nebraska has grown at a rate of 0.4% per year, while the population of California has grown at a rate of 3% per year.

The exponential functions that model the growth of each state are given below.

California: $y = 1.77(1.03)^t$ *The population in 1900 was 1.77 million.*

Nebraska: $y = 1.14(1.004)^t$ *The population in 1900 was 1.14 million.*

Which state exhibits the more rapid growth? What will the population of each state be in the year 2000? *You will answer these questions in Exercise 1.*

The equations for the two states' populations are variations of the equation $y = C(1 + r)^t$. This is the **general equation for exponential growth** in which the initial amount C increases by the same percent r over a given period of time t. This equation can be applied to many kinds of growth applications.

One of these applications is monetary growth. When solving problems involving compound interest, the growth equation becomes $A = P\left(1 + \frac{r}{n}\right)^{nt}$, where A is the amount of the investment over a period of time, P is the principal (initial amount of the investment), r is the annual rate of interest expressed as a decimal, n is the number of times that the interest is compounded each year, and t is the number of years that the money is invested.

Example 1

APPLICATION Finance

In the spring of 1994, Mr. and Mrs. Mitzu had $10,000 they wished to place in a bank certificate of deposit toward their retirement in the year 2004. The interest rate at that time was 2.5% compounded monthly. However, there were seven increases in the prime rate in a year so that in the spring of 1995, the interest rate had risen to 5.5%.

a. Determine the amount of their investment after 10 years if they invested the principal and let it remain at the 2.5% rate.

b. Determine the amount of the investment after 9 years if they waited and invested the principal at the 5.5% rate.

c. What are the best options for their investment?

(continued on the next page)

CAREER CHOICES

Urban planners develop a master plan of a city to show how land should be used in the future and how public facilities should be developed. Urban planners consult with many professionals with expertise in fields that include architecture, engineering, finance, air and water quality, law, economics, statistics, and sociology.

11-5 LESSON NOTES

NCTM Standards: 1–6

Instructional Resources

- Study Guide Master 11-5
- Practice Master 11-5
- Enrichment Master 11-5
- Assessment and Evaluation Masters, p. 297
- Multicultural Activity Masters, p. 22
- Real-World Applications, 28
- Science and Math Lab Manual, pp. 49–52

Transparency 11-5A contains the 5-Minute Check for this lesson; **Transparency 11-5B** contains a teaching aid for this lesson.

Recommended Pacing	
Standard Pacing	Day 9 of 11
Honors Pacing	Day 8 of 10
Block Scheduling*	Day 4 of 5
Alg. 1 in Two Years*	Days 14, 15, & 16 of 19

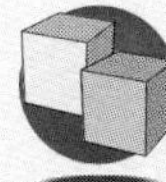

*For more information on pacing and possible lesson plans, refer to the *Block Scheduling Booklet* and *Algebra 1 in Two Years.*

1 FOCUS

5-Minute Check
(over Lesson 11-4)

Use a calculator to determine the approximate value of each expression to the nearest hundredth.

1. $3^{2.6}$ **17.40**
2. $8^{-0.5}$ **0.35**
3. Graph and state the y-intercept for $y = 6^x$. **y-intercept = 1**

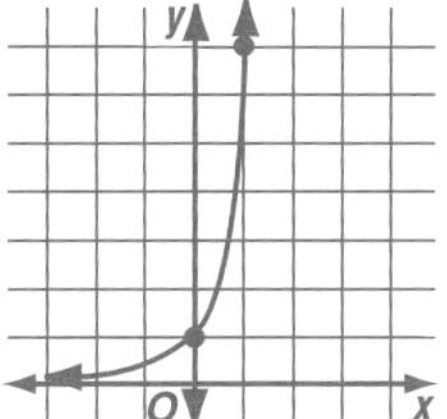

Solve each equation.

4. $2^{3x} = 2^{-3}$ **−1**
5. $7^{3x} = 7^{x-2}$ **−1**

Motivating the Lesson

Situational Problem In a small town, population 100, every married couple has two children. Everyone is married in this town. If this trend continues, what will the population be after 100 years? **100**

2 TEACH

In-Class Examples

For Example 1

a. \$5000 is invested at a rate of 4%. Determine the amount after five years. **\$6104.98**

b. \$5000 is invested at a rate of 8%. Determine the amount after five years. **\$7449.23**

For Example 2

If a city with a population of 50,000 is decreasing at a rate of 1.16%, what will its population be after 10 years? 50 years? 100 years? **44,493; 27,900; 15,568**

Teaching Tip Use numbers for your town or a local city to make exercises more realistic for students.

a. The interest rate r as a decimal is 0.025 and $n = 12$. *Why?*

$$A = P\left(1 + \frac{r}{n}\right)^{nt}$$

$$= 10{,}000\left(1 + \frac{0.025}{12}\right)^{12 \cdot 10} \quad P = 10{,}000,\ r = 0.025,\ n = 12, \text{ and } t = 10$$

Use a calculator.

Enter: 10000 [×] [(] 1 [+] [(] .025 [÷] 12 [)] [)] [y^x] 120 [=] 12836.91542

The amount of the account after 10 years at 2.5% is about \$12,836.92.

b. The interest rate as a decimal is 0.055, and t is 9 years.

$$A = P\left(1 + \frac{r}{n}\right)^{nt}$$

$$= 10{,}000\left(1 + \frac{0.055}{12}\right)^{12 \cdot 9} \quad P = 10{,}000,\ r = 0.055,\ n = 12, \text{ and } t = 9$$

$$\approx 16{,}386.44$$

c. If they waited until the rate went up, they would have had more money than leaving the money at the initial rate. However, if it is possible to reinvest the money each year, they could have invested the money for one year at 2.5% and then reinvested it the next year at 5.5%. That would yield even more money for their retirement in 2004.

A variation of the growth equation can be used as the **general equation for exponential decay.** In the formula $A = C(1 - r)^t$, A represents the final amount, C is the initial amount, r is the rate of decay, and t denotes time.

Example 2

Demographics

The cities listed at the right have experienced declining populations since 1970.

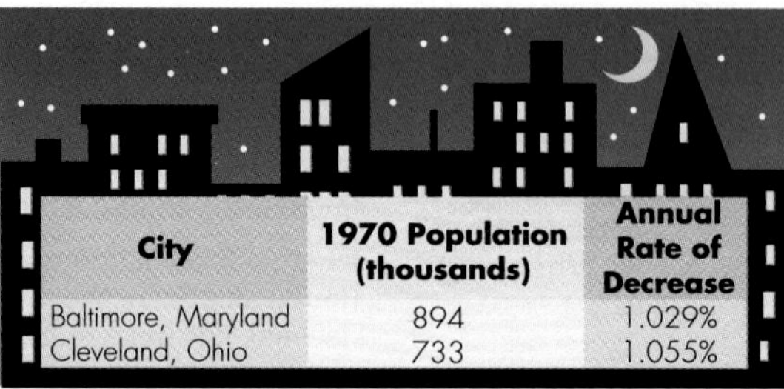

City	1970 Population (thousands)	Annual Rate of Decrease
Baltimore, Maryland	894	1.029%
Cleveland, Ohio	733	1.055%

a. If t represents the number of years since 1970 and C represents the 1970 population, write an exponential decay equation for each city.

b. Assume that each city maintains the same decrease rate into the next century. Calculate the population of each city in the year 2070.

c. How do the projected populations of the cities in 2070 compare to the actual populations in 1970?

a. Baltimore

$C = 894,\ r = 0.01029$

$A = C(1 - r)^t$

$A = 894(1 - 0.01029)^t$

$A = 894(0.98971)^t$

Cleveland

$C = 733,\ r = 0.01055$

$A = C(1 - r)^t$

$A = 733(1 - 0.01055)^t$

$A = 733(0.98945)^t$

Alternative Learning Styles

Auditory Initiate a class discussion about the increasing world population. What are the consequences of an increasing population and decreasing nonrenewable natural resources?

b. The year 2070 is 100 years later. Evaluate each decay equation for $t = 100$.

Baltimore	**Cleveland**
$A = 894(0.98971)^t$	$A = 733(0.98945)^t$
$A = 894(0.98971)^{100}$	$A = 733(0.98945)^{100}$
$A \approx 317.78$	$A \approx 253.80$

c. In 1970, the difference in the populations of Baltimore and Cleveland was 161,000. In 2070, the difference is only about 64,000. If the populations would continue to decrease at the same rates, they would grow even closer together.

Sometimes items decrease in value. For example, as equipment gets older, it *depreciates*. You can use the decay formula to determine the value of an item at a given time.

Example 3

Consumerism

Hogan Blackburn is considering the purchase of a new car. He is faced with the decision to lease or buy. If he leases the car, he pays \$369 a month for 2 years and then has the option to buy the car for \$13,642. The price of the car now is \$16,893.

a. If the car depreciates at 17% per year, how will the depreciated price compare with the buyout price of the lease?

b. At the end of the lease, the dealer offers a loan for 3 years at a rate of \$435.79 per month. If he buys the car now, he will pay \$464.56 monthly for 4 years. Which is the best way to go if he plans to keep the car for at least 5 years?

a. Use the decay formula.

$A = C(1 - r)^t$

$= 16{,}893(1 - 0.17)^2$ *$r = 0.17$, $C = 16{,}893$, and $t = 2$*

$= 11{,}637.59$

The depreciated value is \$2000 less than the buyout price.

b. Calculate the cost of each possibility over a 5-year period.

Lease and Buy: \$369(24 months) + \$435.79(36 months) = \$24,544.44

Buy: \$464.56(48 months) = \$22,298.88

The best way to go may depend on Hogan's financial status.

- Overall, the buy option costs less and lasts only 4 years while the lease and buy option costs more and lasts for 5 years.
- However, if Hogan really wants a new car and cannot afford the larger monthly payment now, the lease option may be best. However, he will eventually have to make the larger payment if he keeps the car.

You can use a spreadsheet to quickly evaluate different values for any exponential growth or decay formula.

In-Class Example

For Example 3

If the car depreciates at 15% per year, how will the depreciated price compare with the buyout price of the lease? **\$12,205.19; \$1436.81 less than the buyout price**

EXPLORATION

Spreadsheets have an advantage in that the values are exact. Graphing calculators have the advantage of allowing the student to visualize exponential growth and decay.

Answer for the Exploration

b. Change P to C. Delete the "n" column, which would make column C the "t" column and column D the "A" column. Change the formula to A2 * (1 – B2)^C2.

3 PRACTICE/APPLY

Check for Understanding

Exercises 1–8 are designed to help you assess your students' understanding through reading, writing, speaking, and modeling. You should work through Exercises 1–2 with your students and then monitor their work on Exercises 3–8.

Error Analysis

Students often assume that exponential functions must increase, however slightly. Make sure that students know how to distinguish between growth and decay.

EXPLORATION SPREADSHEETS

Recall that in a spreadsheet you can refer to each cell by its name (A1 means column A, row 1). Suppose we set up a spreadsheet to evaluate the formula for monetary exponential growth.

- You can enter labels into the first row to identify the data in each column. Let column A contain the values of P, column B the values of r, column C the values of n, column D the value of t, and column E the values of A.

LOOK BACK

Refer to the Exploration in Lesson 6-1 for more information about spreadsheets.

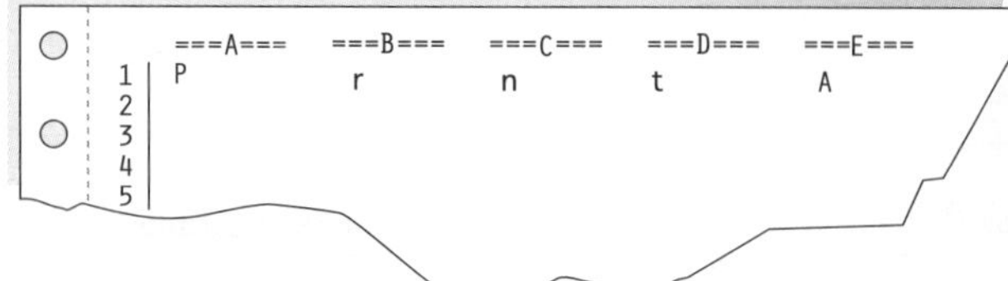

- You can enter a formula into cell E2 to evaluate the formula given the values entered into cells A2 through D2. The formula is A2 *(1 + (B2/C2))^(C2*D2).
- Copy this formula into the other cells in column E.

Your Turn

a. Use the spreadsheet to evaluate the data in Example 1.

b. How would you change the spreadsheet to evaluate the general formula for decay? **See margin.**

CHECK FOR UNDERSTANDING

Communicating Mathematics

Study the lesson. Then complete the following.

1. Refer to the application at the beginning of the lesson.
   a. Which state exhibits more rapid growth? **California**
   b. What will the population of each state be in the year 2000?

1b. CA: 34.02 million; NE: 1.70 million.

2. **Explain** how you could tell the difference between an exponential graph that shows 0.7% growth versus one that shows 7% growth. **The 7% graph would rise more steeply than the 0.7% graph when using the same scale.**

Guided Practice

3. **State** the value of n in the monetary growth formula for each period of compounded interest. **3b. 2**
   a. annually **1** b. semi-annually c. quarterly **4** d. daily **365**

Determine whether each exponential equation represents growth or decay.

4. $y = 10(1.03)^x$ **growth** 5. $y = 10(0.50)^x$ **decay** 6. $y = 10(0.75)^x$ **decay**

7. **History** In 1626, Peter Minuit, governor of the colony of New Netherland, bought the island of Manhattan from the Indians for beads, cloth, and trinkets worth 60 Dutch guilders ($24). If that $24 had been invested at 6% per year compounded annually, how much money would there be in the year 2000? **about $70,000,000,000**

Reteaching

Using Graphics Graph population growth and decay equations for several values of r. Display these on a single coordinate system to emphasize the effect of varying values of r.

8. **Farming** Many self-employed individuals such as farmers can depreciate the value of the machinery they buy as a part of their income tax returns. Suppose a tractor valued at \$50,000 depreciates 10% per year. Make a table to determine in how many years the value of the tractor would be less than \$25,000. **about 7 years**

EXERCISES

Applications and Problem Solving

Finance Determine the final amount from the investment for each situation.

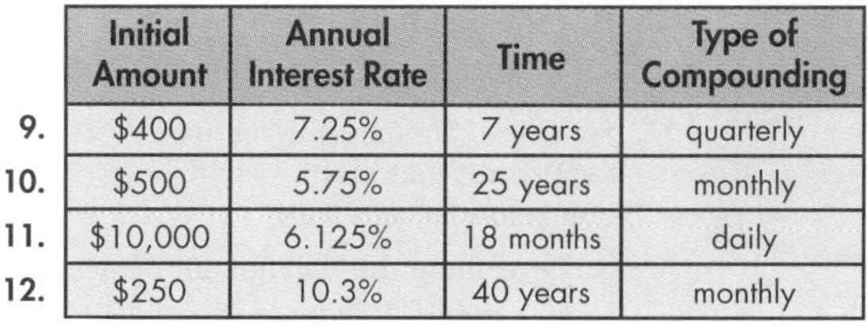

	Initial Amount	Annual Interest Rate	Time	Type of Compounding	
9.	\$400	7.25%	7 years	quarterly	**\$661.44**
10.	\$500	5.75%	25 years	monthly	**\$2097.86**
11.	\$10,000	6.125%	18 months	daily	**\$10,962.19**
12.	\$250	10.3%	40 years	monthly	**\$15,121.61**

13. **Demographics** The population in 1994 and the growth rate for four countries is given below.

Country	Continent	Growth Rate	1994 Population	
Ethiopia	Africa	2.5%	58.7 million	**b. 68.1 million**
India	Asia	1.9%	919.9 million	**1029.9 million**
Colombia	S. America	2.1%	35.6 million	**40.3 million**
Singapore	Asia	1.3%	2.9 million	**3.1 million**

a. Write an exponential equation for each country's growth.

b. Compute the estimated population for each country in the year 2000.

c. Exclude the data for India. Make a double-bar graph to show how the population in the other three countries in 1994 compares with the estimate for the year 2000. **See margin.**

13a. Each equation represents growth and t is the number of years since 1994.
$y = 58.7(1.025)^t$,
$y = 919.9(1.019)^t$,
$y = 35.6(1.021)^t$,
$y = 2.9(1.013)^t$

14. **Energy** The Environmental Protection Agency (EPA) has called for businesses to find cleaner sources of energy as we approach the 21st century. Coal is not considered to be a clean source of energy. In 1950, the use of coal by residential and commercial users was 114.6 million tons. Since then, the use of coal has decreased by 6.6% per year.

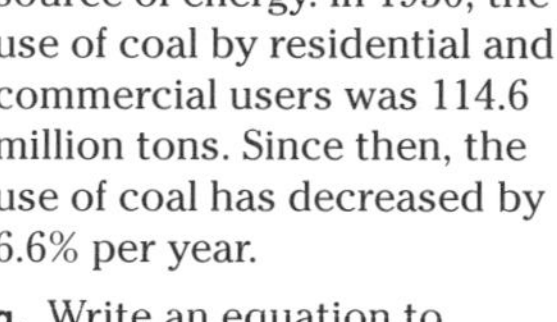

a. Write an equation to represent the use of coal since 1950.

b. Suppose the use of coal continues to decrease at the same rate. Use a calculator to estimate in what year the use of coal will end.

14a. $y = 114.6(0.934)^t$, t = years since 1950, decay

14b. It never reaches 0, but comes close in about 100 years.

Assignment Guide

Core: 9–21 odd, 23–27
Enriched: 10–20 even, 21–27

For **Extra Practice,** see p. 782.

The red A, B, and C flags, printed only in the Teacher's Wraparound Edition, indicate the level of difficulty of the exercises.

Additional Answer

13c.

70
65
60
55
50
45
40
35
30
25
20
15
10
5
0
1994
2000
Ethiopia
Columbia
Singapore

Study Guide Masters, p. 79

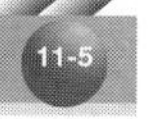

11-5 NAME__________ DATE ________

Study Guide

Student Edition Pages 643–649

Growth and Decay

Population increases and monetary investments are examples of **exponential growth.** This means that an initial amount increases at a steady rate over a period of time.

General Formula for Exponential Growth
$A = C(1 + r)^t$, where C is the initial amount, r is the rate of increase (a percent written as a decimal), t is time, and A is the final amount.

Radioactive decay and depreciation are examples of **exponential decay.** This means that an initial amount decreases at a steady rate over a period of time.

General Formula for Exponential Decay
$A = C(1 - r)^t$, where C is the initial amount, r is the rate of decrease (a percent written as a decimal), t is time, and A is the final amount.

Example 1: The population of Johnson City has grown at a rate of 3.2% each year for the last 10 years. If the population 10 years ago was 25,000, what is the population today?

$A = C(1 + r)^t$
$= 25{,}000(1 + 0.032)^{10}$ **3.2% = 0.032**
$\approx 34{,}256$ **Use a calculator.**

Example 2: A tractor has a depreciation rate of 19% per year. If the original price of the tractor was \$29,000, what is the value of the tractor 5 years later?

$A = C(1 - r)^t$
$= \$29{,}000(1 - 0.19)^5$ **19% = 0.19**
$\approx \$10{,}111.68$ **Use a calculator.**

Solve.

1. The formula for compound interest is $A = P\left(1 + \frac{r}{n}\right)^{nt}$, where P is the initial amount invested, r is the annual rate of interest, n is the number of times a year the interest is compounded, and t is the number of years. If \$100,000 is invested in an account that pays 5.2% compounded quarterly, how much money will be in the account after 12 years? **\$185,888.87**

2. Carl Gossell is a machinist. He bought some new machinery for \$125,000. He wants to calculate the value of the machinery over the next 10 years for tax purposes. If the machinery depreciates at 15% per year, what is the value of the machinery (to the nearest \$100) at the end of 10 years? **\$24,600**

4 ASSESS

Closing Activity

Writing Have students write a short essay explaining the difference between exponential growth and decay. Students need to make reference to the base of each function.

Chapter 11, Quiz D (Lesson 11-5), is available in the *Assessment and Evaluation Masters,* p. 297.

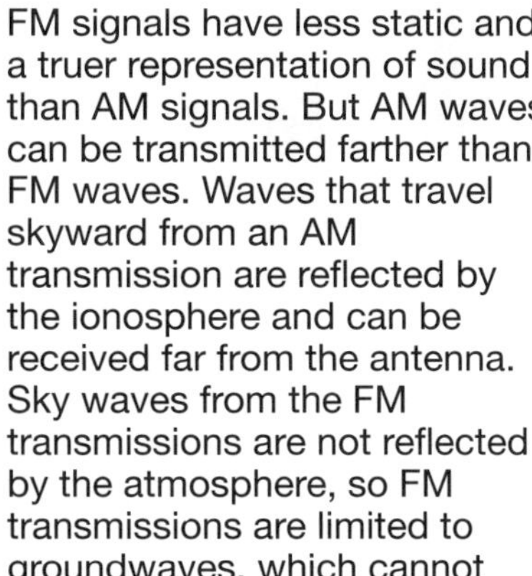

F Y I

FM signals have less static and a truer representation of sound than AM signals. But AM waves can be transmitted farther than FM waves. Waves that travel skyward from an AM transmission are reflected by the ionosphere and can be received far from the antenna. Sky waves from the FM transmissions are not reflected by the atmosphere, so FM transmissions are limited to groundwaves, which cannot travel beyond the curve of Earth.

Practice Masters, p. 79

11-5 NAME ________ DATE ________

Practice — Student Edition Pages 643–649

Growth and Decay

Solve.

1. Ten years ago, Mr. and Mrs. Boyce bought a house for $96,000. Their home is now worth $125,000. Assuming a steady rate of growth, what was the annual rate of appreciation? **about 3%**

2. The Greens bought a condominium for $83,000. Assuming that its value will appreciate 6% per year, how much will the condo be worth in five years when the Greens are ready to move? **$111,072.72**

3. Ten years ago, Tiffany's mother bought a new car for $9000. Tiffany is now going to buy the car for $900. Assuming a steady rate of depreciation, what was the approximate annual rate of depreciation? **21%**

4. A piece of machinery valued at $2500 depreciates at a steady rate of 10% yearly. The owner of the business plans to replace the equipment when its value has depreciated to $500. In how many years will the equipment be replaced? **$15\frac{1}{2}$ years**

5. Kyle has saved $500 of the money he earned working at Carousel Music. If he spends 10% of the money each week, after how many weeks will he have less than $1? **60 weeks**

6. Zeller Industries bought a piece of weaving equipment for $50,000. It is expected to depreciate at a steady rate of 10% per year. When will its value have depreciated to $25,000? **about 6.5 years**

7. The Fresh and Green Company has a savings plan for its employees. If the employee makes an initial contribution of $1000, the company pays 8% interest compounded quarterly.
   a. If an employee participating in the plan withdraws the balance of the account after five years, how much will the company have paid into the account? **$485.95**
   b. If an employee participating in the plan withdraws the balance of the account after thirty-five years, how much will the company have paid into the account? **$14,996.47**

16a. $y = 1.2(0.932)^t$, t = number of years since 1980, decay

16c. preservation efforts, conservation of land, domestic reproduction in zoos, poaching

F Y I

FM stations began surpassing AM stations in popularity during the 1970s. This may have been due to the FM waves producing a clearer signal that could be heard over a longer distance.

15. **Insurance** The total amount of life insurance sold in 1950 was $231.5 billion. Since then, the annual increase has been estimated at 9.63%.
    a. Write an equation to represent the amount of insurance sold annually since 1950. **$y = 231.5(1.0963)^t$, t = years since 1950, growth**
    b. Find the estimated amount of life insurance that will be sold in the year 2010. **$57.6 trillion**

16. **Wildlife** In 1980, there were 1.2 million elephants living in Africa. Because the natural grazing lands for the elephant are disappearing due to increased population and cultivation of the land, the number of elephants in Africa has decreased by about 6.8% per year.
    a. Write an equation to represent the population of elephants in Africa.
    b. In what year did the population of African elephants drop to less than half of the number in 1980? **1990**
    c. What factors might affect the rate of decline in elephant population?

17. **Savings** Sheena is investing her $5000 inheritance in a saving certificate that matures in 4 years. The interest rate is 8.25% compounded quarterly.
    a. Determine the balance in the account after 4 years. **$6931.53**
    b. Her friend, LaDonna, invests the same amount of money at the same interest rate but her bank compounds daily. Determine how much she will have after 4 years. **$6954.58**
    c. What is the difference in the amount Sheena and LaDonna have after 4 years? **$23.05**
    d. Which type of compounding appears to be more profitable? **daily**

18. **Radio** FM broadcast frequencies range from 88 to 108 MHz in tenths. Before digital displays existed, you turned a knob and a bar slide from across the display or *dial* to find the station you wanted to hear. The equation that relates the MHz reading to position of the slide is $f(d) = 88(1.0137)^d$, where d is the distance from the left of the dial. Suppose someone didn't know the number of their favorite station, but knew it was about halfway across the dial. If the dial is 15 cm long, what station is their favorite? **97.5 MHz**

19. **Population** Research to find the population in your community for the past 50 years. **See students' work.**
    a. Make a graph to show the population change.
    b. Does the population of your community show growth or decay?
    c. Estimate what the population of your community will be when you are 50 years old.

Graphing Calculator

20. **Radioactivity** A formula for examining the decay of radioactive materials is $y = Ne^{kt}$, where N is the beginning amount in grams, $e \approx 2.72$, k is a negative constant for the substance, and t is the number of years. Use a graphing calculator to estimate each of the following.
    a. How long will it take 250 grams of a radioactive substance to reduce to 50 grams if $k = -0.08042$? **20 years**
    b. In 10 years, 200 grams of a radioactive substance is reduced to 100 grams. Find an estimate for the constant k for this substance. **−0.0693**

Extension

Reasoning If interest were compounded daily instead of monthly, what would the formula for A look like?

$$A = P\left(1 + \frac{r}{365.25}\right)^{(365.25)t}$$

Critical Thinking

21. The general exponential formula for growth or decay is $y = Ca^x$. Determine what values of a describe growth or decay. What does x usually represent?

21. $a > 1$, growth; $0 < a < 1$, decay; x represents time

22. Consider the equation $y = C(1 + r)^x$. How do you determine the y-intercept of the graph of this equation without graphing or evaluating the function for values of x? **The y-intercept is C.**

Mixed Review

23. Solve $3^y = 3^{3y+1}$. (Lesson 11–4) $-\frac{1}{2}$

24. Find $(n^2 + 5n + 3) - (2n^2 + 8n + 8)$. (Lesson 9–5) $-n^2 - 3n - 5$

25. Simplify $\frac{-6r^3s^5}{18r^{-7}s^5t^{-2}}$. (Lesson 9–2) $-\frac{r^{10}t^2}{3}$

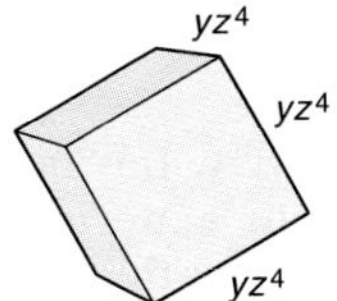

26. **Geometry** Find the volume of the cube shown at the right. (Lesson 9–2) y^3z^{12}

27. **Geometry** Write the equation of a line that passes through $(-2, 7)$ and is perpendicular to the line whose equation is $2x - 5y = 3$. Use slope-intercept form. (Lesson 6–6) $y = -2.5x + 2$

Minimizing Computers

The following excerpt appeared in an article in *The New York Times* on November 22, 1994.

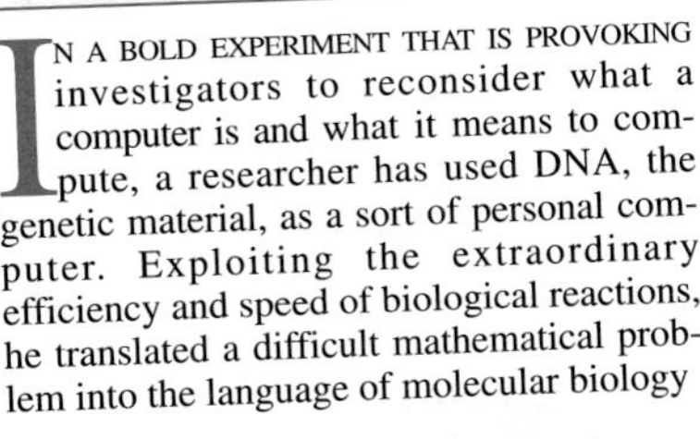

IN A BOLD EXPERIMENT THAT IS PROVOKING investigators to reconsider what a computer is and what it means to compute, a researcher has used DNA, the genetic material, as a sort of personal computer. Exploiting the extraordinary efficiency and speed of biological reactions, he translated a difficult mathematical problem into the language of molecular biology and solved it by carrying out a reaction in one-fiftieth of a teaspoon of solution in a test tube....Molecular computers can perform more than a trillion operations per second, which makes them a thousand times as fast as the fastest supercomputer. They are a billion times as energy efficient as conventional computers. And they can store information in a trillionth of the space required by ordinary computers. ■

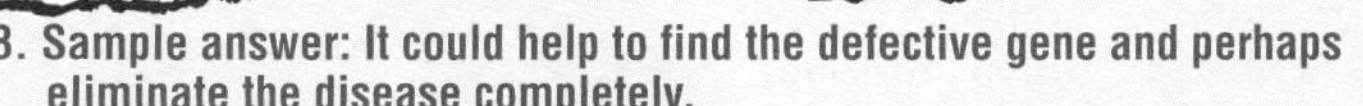

3. Sample answer: It could help to find the defective gene and perhaps eliminate the disease completely.

1. Does the idea of a molecular computer in a test tube surprise you? Why or why not? **See students' work.** **2. See margin for sample answer.**
2. If biological systems can have computational abilities, what effects might this have on computer scientists, programmers, and mathematicians?
3. One class of problems that molecular computers could help solve involves finding one desired solution or path out of a huge number of possibilities. How could this be used if the problem was a defective gene causing the rate of a deadly disease to grow exponentially?

Enrichment Masters, p. 79

11-5 NAME__________ DATE ______

Enrichment

Student Edition Pages 643–649

Perfect, Excessive, Defective, and Amicable Numbers

A **perfect number** is the sum of all of its factors except itself. Here is an example.

$28 = 1 + 2 + 4 + 7 + 14$

There are very few perfect numbers. Most numbers are either *defective* or *excessive.*

An **excessive number** is greater than the sum of all of its factors except itself.

A **defective number** is less than this sum.

Two numbers are **amicable** if the sum of the factors of the first number, except for the number itself, equals the second number, and vice versa.

Solve each problem.

1. Write the perfect numbers between 0 and 31. 6, 28
2. Write the excessive numbers between 0 and 31. 2, 3, 4, 5, 7, 8, 9, 10, 11, 13, 14, 15, 16, 17, 19, 21, 22, 23, 25, 26, 27, 29
3. Write the defective numbers between 0 and 31. 12, 18, 20, 24, 30
4. Show that 8128 is a perfect number. $8128 = 1 + 2 + 4 + 8 + 16 + 32 + 64 + 127 + 254 + 508 + 1016 + 2032 + 4064$
5. The sum of the reciprocals of all the factors of a perfect number (including the number itself) equals 2. Show that this is true for the first two perfect numbers. $\frac{1}{1} + \frac{1}{2} + \frac{1}{3} + \frac{1}{6} = \frac{12}{6} = 2$ $\quad \frac{1}{1} + \frac{1}{2} + \frac{1}{4} + \frac{1}{7} + \frac{1}{14} + \frac{1}{28} = \frac{56}{28} = 2$
6. More than 1000 pairs of amicable numbers have been found. One member of the first pair is 220. Find the other member. 284
7. One member of the second pair of amicable numbers is 2620. Find the other member. 2924
8. The Greek mathematician Euclid proved that the expression $2^{n-1}(2^n - 1)$ equals a perfect number if the expression inside the parentheses is prime. Use Euclid's expression with n equal to 19 to find the seventh perfect number. $2^{18}(2^{19} - 1) = 137{,}438{,}691{,}328$

Mathematics and SOCIETY

The actual circuitry of today's computers is referred to as *hardware.* The program run on the hardware is referred to as *software.* The new biocomputers have been referred to as *wetware.*

Answer for Mathematics and Society

2. Sample answer: The methods used for rapid calculation in biological computers might be applied to similar problems in other real-world applications.

Closing the Investigation

This activity provides students an opportunity to bring their work on the Investigation to a close. For each Investigation, students should present their findings to the class. Here are some ways students can display their work.

- Conduct and report on an interview or survey.
- Write a letter, proposal, or report.
- Write an article for the school or local paper.
- Make a display, including graphs and/or charts.
- Plan an activity.

Assessment

To assess students' understanding of the concepts and topics explored in the Investigation and its follow-up activities, you may wish to examine students' Investigation Folders.

The scoring guide in the *Investigations and Projects Masters,* p. 19, provides a means for you to score students' work on the Investigation.

Investigations and Projects Masters, p. 19

Scoring Guide
Chapters 10 and 11
Investigation

Level	Specific Criteria
3 Superior	• Shows thorough understanding of the concepts of *creating different brick pattern designs; data collection and analysis; determining the dimensions, perimeter, and area of rectangles and squares in terms of x and y;* and *making graphs.* • Uses appropriate strategies to solve problems. • Computations are correct. • Written explanations are exemplary. • Charts, graphs, and report are appropriate and sensible. • Goes beyond the requirements of some or all problems.
2 Satisfactory, with Minor Flaws	• Shows understanding of the concepts of *creating different brick pattern designs; data collection and analysis; determining the dimensions, perimeter, and area of rectangles and squares in terms of x and y;* and *making graphs.* • Uses appropriate strategies to solve problems. • Computations are mostly correct. • Written explanations are effective. • Charts, graphs, and report are appropriate and sensible. • Satisfies the requirements of problems.
1 Nearly Satisfactory, with Obvious Flaws	• Shows understanding of most of the concepts of *creating different brick pattern designs; data collection and analysis; determining the dimensions, perimeter, and area of rectangles and squares in terms of x and y;* and *making graphs.* • May not use appropriate strategies to solve problems. • Computations are mostly correct. • Written explanations are satisfactory. • Charts, graphs, and report are appropriate and sensible. • Satisfies the requirements of problems.
0 Unsatisfactory	• Shows little or no understanding of the concepts of *creating different brick pattern designs; data collection and analysis; determining the dimensions, perimeter, and area of rectangles and squares in terms of x and y;* and *making graphs.* • Does not use appropriate strategies to solve problems. • Computations are incorrect. • Written explanations are not satisfactory. • Charts, graphs, and report are not appropriate or sensible. • Does not satisfy the requirements of problems.

Refer to the Investigation on pages 554–555.

Review the knowledge you have gained from your experiments working with the bricks. Review the instructions given to you by your manager as you begin to close this Investigation.

> Please create several patio designs that will utilize these bricks. Submit at least three different plans, explaining the materials required for each patio. I am anxious to see the different ways in which these bricks can be arranged to form rectangular patios. Is there a general formula or pattern we can use to design these in the future? I look forward to your report on helping us solve our inventory problems.

Analyze

You have conducted experiments and organized your data in various ways. It is now time to analyze your findings and state your conclusions.

1 Look over your data and complete a chart like the one on page 555 for each design. Use actual measurements.

PORTFOLIO ASSESSMENT

You may want to keep your work on this Investigation in your portfolio.

2 What information does this chart reflect? Does it give information that can be used to generalize a method for forming future brick patio patterns? Explain.

Write

The report to your manager should explain your process for investigating these rectangular brick patterns and what you found from your investigations.

3 What size rectangular brick patterns are possible? Draw sketches of the possible patterns. Describe the numbers of bricks used and include the dimensions, perimeter, and area of each pattern in terms of x and y.

4 Write procedures or generalizations that may be followed to find rectangular patterns in the following situations.
- You have a certain number of large squares and small squares. How many rectangular tiles are necessary to create a rectangular pattern?
- How can you find the dimensions of a pattern given the type and number of bricks available?
- How can you find the different possible patterns for any set number of bricks?
- How do you know that no possible rectangular pattern can be made given a set of the three types of bricks?

5 Summarize your findings and give recommendations for possible brick patterns. Explain methods for exploring more patterns in the future.

CHAPTER 11 HIGHLIGHTS

VOCABULARY

After completing this chapter, you should be able to define each term, property, or phrase and give an example or two of each.

Algebra

axis of symmetry (p. 612)
discriminant (p. 632)
exponential function (pp. 634, 635)
general equation for exponential decay (p. 644)
general equation for exponential growth (p. 643)
half-life (p. 638)
maximum (p. 611)
minimum (p. 611)
parabola (pp. 610, 611)
quadratic equation (p. 620)
quadratic formula (p. 628)
quadratic function (pp. 610, 611)
roots (p. 620)
symmetry (p. 612)
vertex (pp. 610, 611)
zeros (p. 620)

Problem Solving

look for a pattern (p. 637)

UNDERSTANDING AND USING THE VOCABULARY

Choose the letter of the term that best matches each equation or phrase.

1. $y = C(1 + r)^t$ d
2. $f(x) = ax^2 + bx + c$ g
3. a geometric property of parabolas i
4. $x = \frac{-b}{2a}$ a
5. $y = a^x$ c
6. maximum or minimum point of a parabola j
7. $A = C(1 - r)^t$ b
8. solutions of a quadratic equation h
9. $x = \frac{-b \pm \sqrt{b^2 - 4ac}}{2a}$ f
10. the graph of a quadratic function e

a. equation of axis of symmetry
b. exponential decay formula
c. exponential function
d. exponential growth formula
e. parabola
f. quadratic formula
g. quadratic function
h. roots
i. symmetry
j. vertex

Instructional Resources

Three multiple-choice tests and three free-response tests are provided in the *Assessment and Evaluation Masters*. Forms 1A and 2A are for honors pacing, and Forms 1B, 1C, 2B, and 2C are for average pacing. Chapter 11 Test, Form 1B is shown at the right. Chapter 11 Test, Form 2B is shown on the next page.

Using the CHAPTER HIGHLIGHTS

The Chapter Highlights begins with a listing of the new terms, properties, and phrases that were introduced in this chapter. Have students define each term and provide an example or two of it, if appropriate.

Assessment and Evaluation Masters, pp. 283–284

11 NAME________ DATE ________

Chapter 11 Test, Form 1B

Write the letter for the correct answer in the blank at the right of each problem.

1. What are the equation of the axis of symmetry and the coordinates of the vertex of the graph of $y = -x^2 - 10x + 17$? 1. B
   A. $x = -5$; $(-5, -8)$ B. $x = -5$; $(-5, 42)$
   C. $x = 5$; $(5, 92)$ D. $x = 5$; $(5, 32)$
2. Find the equation of the axis of symmetry for the graph of $y = -2x^2 + x + 17$, and state whether the axis of symmetry contains the minimum or maximum point of the graph. 2. A
   A. $x = \frac{1}{4}$; maximum B. $x = -\frac{1}{4}$; maximum
   C. $x = \frac{1}{4}$; minimum D. $x = -\frac{1}{4}$; minimum
3. The equation of the axis of symmetry of a parabola is $x = -1$. If the point (0, 1) is on the parabola, which of the following points is also on the parabola? 3. C
   A. (−1, 0) B. (0, −2) C. (−2, 1) D. (−1, −2)
4. Which of the following equations describes the graph at the right? 4. A
   A. $y = x^2 - 1$ B. $y = x^2 + 1$
   C. $y = -x^2 - 1$ D. $y = x^2$
5. If a quadratic equation has exactly no real roots, then its graph intersects the x-axis at how many points? 5. C
   A. two B. one C. none D. It varies.
6. If the roots of $x^2 - 3x - 3 = 0$ are located by graphing the related function, between which pair of integers does a root of the equation lie? 6. A
   A. −1 and 0 B. 1 and 2 C. 2 and 3 D. −2 and −1
7. What must be the value of c in the function $y = x^2 + 3x + c$ if its graph has a minimum point at $\left(-\frac{3}{2}, 0\right)$? 7. C
   A. $-\frac{3}{2}$ B. $-\frac{9}{4}$ C. $\frac{9}{4}$ D. $\frac{3}{2}$
8. How many real roots does $x^2 - 2x - 3 = 0$ have? 8. C
   A. 0 B. 1 C. 2 D. cannot be determined
9. Solve $9x^2 + 16 = 24x$. 9. C
   A. 24 B. 0 C. $\frac{4}{3}$ D. $\frac{-8 \pm 2\sqrt{70}}{9}$
10. Solve $3x^2 + 5x - 16 = 0$. 10. D
    A. $\frac{-3 \pm 6\sqrt{3}}{5}$ B. $\frac{-5 \pm 6\sqrt{3}}{5}$ C. $\frac{-5 \pm \sqrt{167}}{6}$ D. $\frac{-5 \pm \sqrt{217}}{6}$
11. Solve $d^2 - 6d + 4 = 0$. 11. A
    A. $3 \pm \sqrt{5}$ B. $6 \pm 2\sqrt{5}$ C. $\frac{-6 \pm \sqrt{5}}{2}$ D. $\frac{3 \pm 2\sqrt{5}}{2}$

11 NAME________ DATE ________

Chapter 11 Test, Form 1B (continued)

12. If $3x^2 - \frac{2}{3}x = \frac{1}{2}$ is to be solved using the quadratic formula, which of these would be best to do first? 12. D
    A. Add $\frac{1}{9}$ to each side. B. Divide each side by 3.
    C. Multiply each side by 3. D. Multiply each side by 6.
13. Use a calculator to determine the value of $0.5(2.5^{-2.5})$ to the nearest hundredth. 13. C
    A. 0.10 B. −3.13 C. 0.05 D. 0.57
14. Which of the following is the graph of $y = 3^x + 1$? 14. C
    A. B. C. D. none of them
15. Which table of values displays exponential behavior? 15. B

A.

x	0	1	2	3
y	0	1	32	243

B.

x	0	1	2	3
y	1	5	25	125

C.

x	0	1	2	3
y	0	5	10	15

D. none of them

16. Solve $4^x = 2^{3x - 6}$. 16. D
    A. $1\frac{1}{5}$ B. −6 C. 3 D. 6
17. If \$1000 is invested at an annual interest rate of 6% compounded monthly, which equation can you solve to find the value of the investment at the end of 2 years? 17. A
    A. $A = 1000(1.005)^{24}$ B. $A = 1000(1.06)^2$
    C. $A = 1000(1.005)^2$ D. $A = 1000(1.06)^{12}$
18. Suppose a printing press valued at \$100,000 depreciates 10% per year. After how many years will the value of the printing press be less than \$50,000? 18. C
    A. 5 B. 2 C. 7 D. 20
19. In 1995, the annual rate of inflation for consumer goods was about 2.5%. Suppose the inflation rate remains constant until the year 2000. What would a bag of groceries that cost \$25 in 1995 cost in 2000? 19. A
    A. \$28.29 B. \$37.50 C. \$76.29 D. \$67.50
20. In 1980, the population of Honolulu, Hawaii, was about 763,000. Since then, the population has been increasing at an average rate of 1.5% per year. What will be the population of Honolulu in 2030 if it continues to increase at the same rate? 20. D
    A. about 1,335,000 B. about 826,831,000
    C. about 358,000 D. about 1,606,000

Bonus Solve $x + 7 + \frac{3}{x} = 0$. Bonus $\frac{-7 \pm \sqrt{37}}{2}$

Using the STUDY GUIDE AND ASSESSMENT

Skills and Concepts Encourage students to refer to the objectives and examples on the left as they complete the review exercises on the right.

Assessment and Evaluation Masters, pp. 289–290

11 Chapter 11 Test, Form 2B

NAME__________ DATE __________

1. Find the equation of the axis of symmetry and the coordinates of the vertex of the graph of $y = -2x^2 + 4x - 5$. 1. $x = 1$; $(1, -3)$
2. Find the equation of the axis of symmetry for the graph of $y = x^2 - 7x + 12$, and state whether the axis of symmetry contains a maximum point or a minimum point of the graph. 2. $x = \frac{7}{2}$; minimum
3. Find the coordinates of the vertex of the graph of $y = 6x^2 - 14x$. 3. $\left(\frac{7}{6}, -\frac{49}{6}\right)$
4. The equation of the axis of symmetry of a parabola is $x = 3$. The point (10, 0) is on the parabola. Name another point that is also on the parabola. 4. $(-4, 0)$
5. For the graph shown, is the value of the coefficient of x^2 in the related quadratic equation positive or negative? How many real roots does the equation have? 5. negative; none

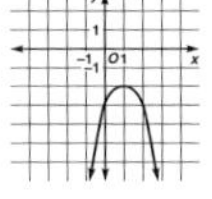

Solve each equation by graphing. If exact roots cannot be found, state the consecutive integers between which the roots lie.

6. $x^2 - 6x + 4 = 0$ 6. $0 < x < 1, 5 < x < 6$
7. $x^2 - 4x + 5 = 0$ 7. no real roots
8. $x^2 - 6x + 9 = 0$ 8. 3

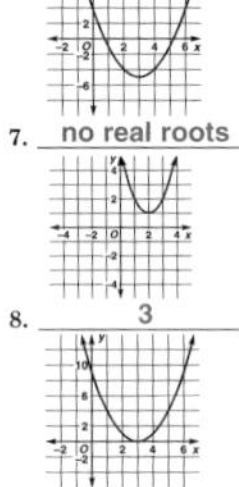

9. Suppose that the roots of a quadratic equation are given by $x = \frac{-5 \pm \sqrt{105}}{4}$. Approximate the roots of the equation to the nearest hundredth. 9. $-3.81, 1.31$

11 Chapter 11 Test, Form 2B (continued)

NAME__________ DATE __________

Solve using the quadratic formula.

10. $y^2 - 9y + 8 = 0$ 10. 1, 8
11. $3b^2 + 2b - 3 = 0$ 11. $\frac{-1 \pm \sqrt{10}}{3}$
12. $3x^2 = 10x$ 12. $0, \frac{10}{3}$
13. Use a calculator to determine the value of $10(0.4^{-1.5})$ to the nearest hundredth. 13. 39.53
14. What is the y-intercept of the graph of $y = 2.5 \cdot 10^x$? 14. 2.5
15. Graph $y = 0.4^x - 1$. 15.
16. Solve $3^{2x+5} = 9^{2x-5}$. 16. 7.5
17. Suppose \$500 is invested at an annual interest rate of 8% compounded semiannually. Write an equation for finding the value of the investment at the end of 3 years. 17. $A = 500(1.04)^6$
18. A new car costing \$16,500 depreciates 12% per year. What will be the value of the car after 10 years? 18. \$4595.27
19. An oil painting that originally cost \$2500 increases in value at an annual rate of 2%. What is the value of the painting after 10 years? 19. \$3047.49
20. In 1980, the population of Rochester, New York, was about 1,030,000. Since then, the population has been increasing at an average rate of 0.25% per year. At this rate of increase, what would be the approximate population of Rochester in 2030? 20. about 1,167,000

Bonus Graph the solution set of $y \geq x^2 + 6x + 9$. Bonus

CHAPTER 11 STUDY GUIDE AND ASSESSMENT

SKILLS AND CONCEPTS

OBJECTIVES AND EXAMPLES	REVIEW EXERCISES

Upon completing this chapter, you should be able to:

- find the equation of the axis of symmetry and the coordinates of the vertex of a parabola (Lesson 11–1)

In the equation $y = x^2 - 8x + 12$, $a = 1$ and $b = -8$.

The equation of the axis of symmetry is

$$x = -\frac{b}{2a} = -\frac{(-8)}{2(1)} \text{ or } 4.$$

Use the value $x = 4$ to find the coordinates of the vertex.

$$\begin{aligned} y &= x^2 - 8x + 12 \\ &= (4)^2 - 8(4) + 12 \\ &= 16 - 32 + 12 \text{ or } -4 \end{aligned}$$

The coordinates of the vertex are $(4, -4)$.

Use these exercises to review and prepare for the chapter test.

Write the equation of the axis of symmetry and find the coordinates of the vertex of the graph of each equation.

11. $y = -3x^2 + 4$ $x = 0$; $(0, 4)$
12. $y = x^2 - 3x - 4$ $x = \frac{3}{2}$; $\left(\frac{3}{2}, -\frac{25}{4}\right)$
13. $y = 3x^2 + 6x - 17$ $x = -1$; $(-1, -20)$
14. $y = 3(x + 1)^2 - 20$ $x = -1$; $(-1, -20)$
15. $y = x^2 + 2x$ $x = -1$; $(-1, -1)$

- graph quadratic functions (Lesson 11–1)

Graph $y = x^2 - 8x + 12$. Use the information above.

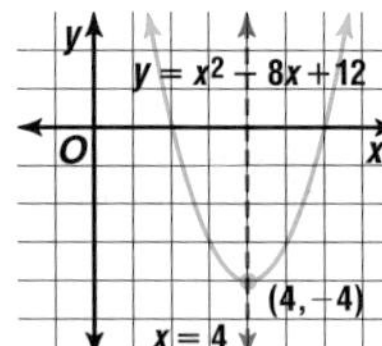

Using the results from Exercises 11–15, graph each equation. 16–20. See margin.

16. $y = -3x^2 + 4$
17. $y = x^2 - 3x - 4$
18. $y = 3x^2 + 6x - 17$
19. $y = 3(x + 1)^2 - 20$
20. $y = x^2 + 2x$

- find roots of quadratic equations by graphing (Lesson 11–2)

Based on the graph of $y = x^2 - 8x + 12$ shown above, the roots of the equation $x^2 - 8x + 12 = 0$ are 2 and 6. This is because $y = 0$ at the x-intercepts, which appear to be at 2 and 6.

Substitute these values into the original equation.

$x^2 - 8x + 12 = 0$	$x^2 - 8x + 12 = 0$
$(2)^2 - 8(2) + 12 = 0$	$(6)^2 - 8(6) + 12 = 0$
$4 - 16 + 12 = 0$	$36 - 48 + 12 = 0$
$0 = 0$ ✓	$0 = 0$ ✓

The solutions of the equation are 2 and 6.

Solve each equation by graphing. If integral roots cannot be found, state the consecutive integers between which the roots lie.

21. $x^2 - x - 12 = 0$ $-3, 4$
22. $x^2 + 6x + 9 = 0$ -3
23. $x^2 + 4x - 3 = 0$ $-5 < x < -4, 0 < x < 1$
24. $2x^2 - 5x + 4 = 0$ $\varnothing$
25. $x^2 - 10x = -21$ 3, 7
26. $6x^2 - 13x = 15$ $-1 < x < 0, 3$

21–26. See Solutions Manual for graphs.

GLENCOE Technology

Test and Review Software

You may use this software, a combination of an item generator and item bank, to create your own tests or worksheets. Types of items include free response, multiple choice, short answer, and open ended.

For IBM & Macintosh

OBJECTIVES AND EXAMPLES	REVIEW EXERCISES

- solve quadratic equations by using the quadratic formula (Lesson 11–3)

Solve $2x^2 + 7x - 15 = 0$.

In the equation, $a = 2$, $b = 7$, and $c = -15$. Substitute these values into the quadratic formula.

$$x = \frac{-(7) \pm \sqrt{(7)^2 - 4(2)(-15)}}{2(2)}$$

$$= \frac{-7 \pm \sqrt{169}}{4}$$

$$x = \frac{-7 + 13}{4} \quad \text{or} \quad x = \frac{-7 - 13}{4}$$

$$= \frac{3}{2} \qquad = -5$$

Solve each equation by using the quadratic formula. Approximate irrational roots to the nearest hundredth.

27. $x^2 - 8x = 20$ **10, −2**
28. $r^2 + 10r + 9 = 0$ **−1, −9**
29. $4p^2 + 4p = 15$ $\frac{3}{2}, -\frac{5}{2}$
30. $2y^2 + 3 = -8y$ **−0.42, −3.58**
31. $9k^2 - 13k + 4 = 0$ **1,** $\frac{4}{9}$
32. $9a^2 + 25 = 30a$ $\frac{5}{3}$
33. $-a^2 + 5a - 6 = 0$ **2, 3**
34. $-2d^2 + 8d + 3 = 3$ **0, 4**
35. $21a^2 + 5a - 7 = 0$ **0.47, −0.71**
36. $2m^2 = \frac{17}{6}m - 1$ $\frac{3}{4}, \frac{2}{3}$

- graph exponential functions (Lesson 11–4)

Graph $y = 2^x - 3$.

x	y
−3	−2.875
−2	−2.75
−1	−2.5
0	−2
1	−1
2	1
3	5

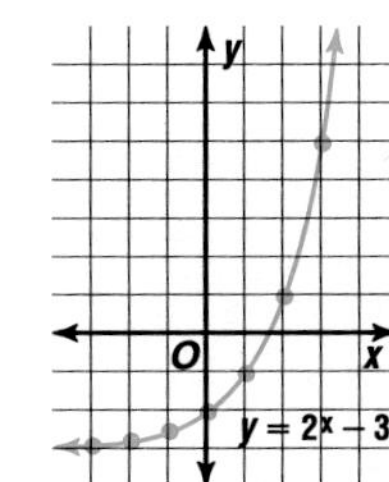

The y-intercept is −2.

Graph each function. State the y-intercept.

37. $y = 3^x + 6$ **7**
38. $y = 3^{x+2}$ **9**
39. $y = 2^x$ **1**
40. $y = 2\left(\frac{1}{2}\right)^x$ **2**

37–40. See margin for graphs.

- solve exponential equations (Lesson 11–4)

Solve $25^{b+4} = \left(\frac{1}{5}\right)^{2b}$.

$$25^{b+4} = \left(\frac{1}{5}\right)^{2b}$$
$$(5^2)^{b+4} = (5^{-1})^{2b}$$
$$5^{2b+8} = 5^{-2b}$$
$$2b + 8 = -2b$$
$$8 = -4b$$
$$-2 = b$$

The solution is −2.

Solve each equation.

41. $3^{4x} = 3^{-12}$ **−3**
42. $7^x = 7^{4x+9}$ **−3**
43. $\left(\frac{1}{3}\right)^t = 27^{t+8}$ **−6**
44. $0.01 = \left(\frac{1}{10}\right)^{4r}$ $\frac{1}{2}$
45. $64^{y-3} = \left(\frac{1}{16}\right)^{y^2}$ **−3,** $\frac{3}{2}$

Additional Answers

16.
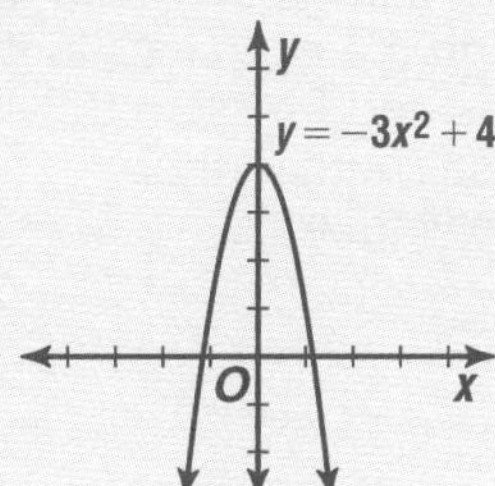

17.
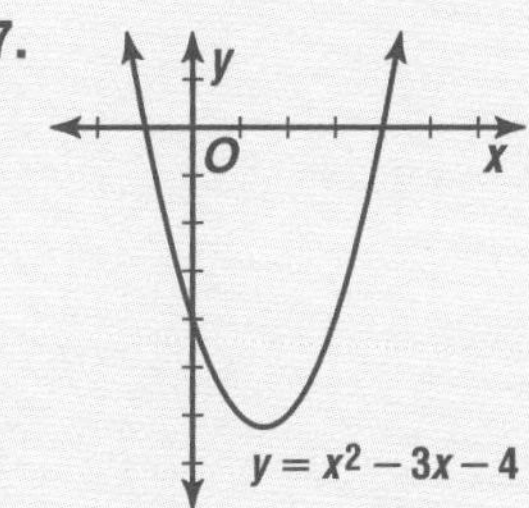

18.
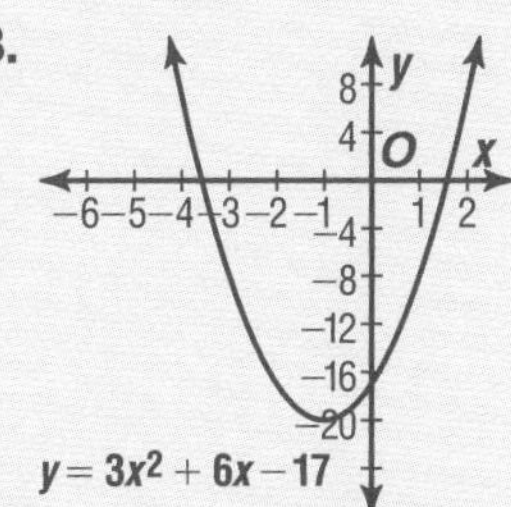

19.
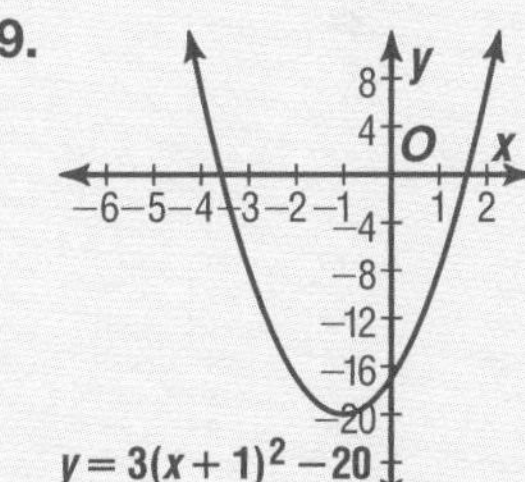

20.
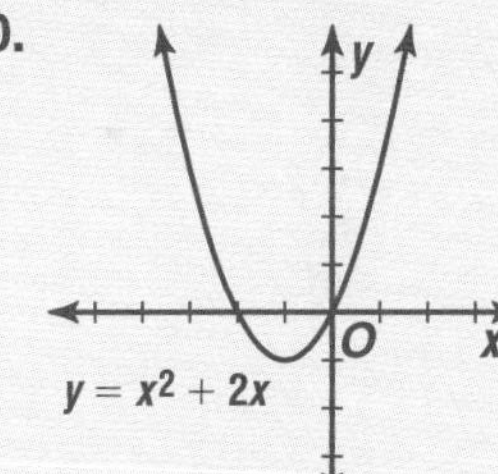

37.
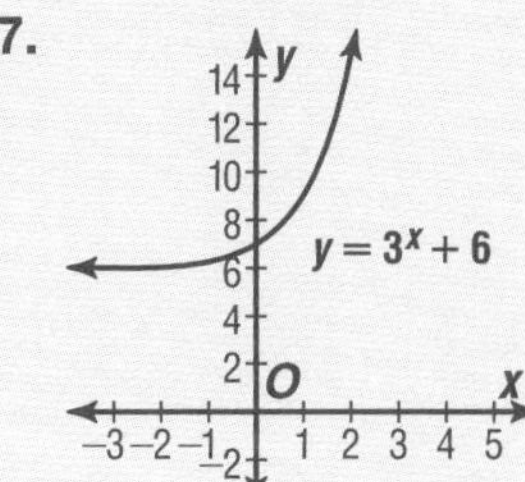

38.
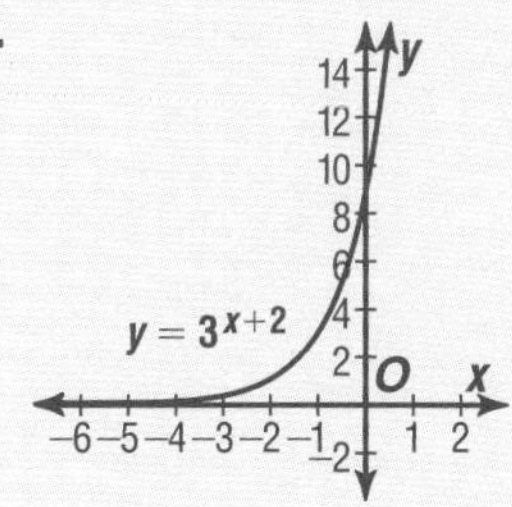

Additional Answers

39.
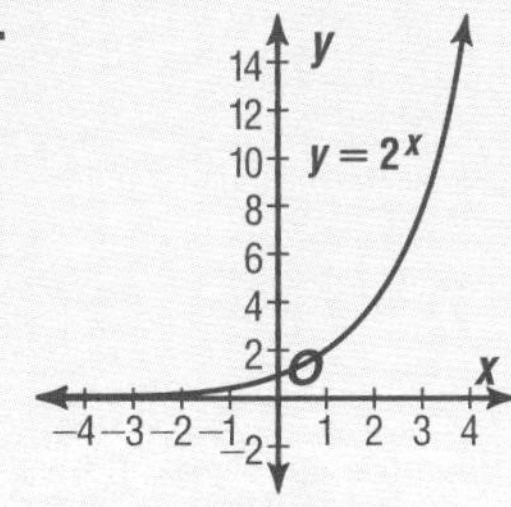

40.
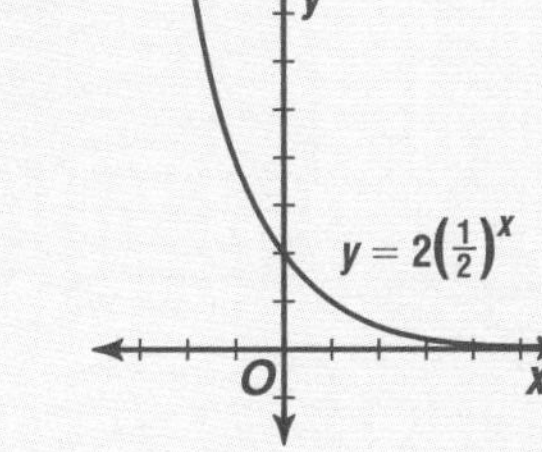

Applications and Problem Solving Encourage students to work through the exercises in the Applications and Problem Solving section to strengthen their problem-solving skills.

OBJECTIVES AND EXAMPLES

- solve problems involving growth and decay (Lesson 11–5)

Find the final amount from an investment of \$1500 invested at an interest rate of 7.5% compounded quarterly for 10 years.

$$A = P\left(1 + \frac{r}{n}\right)^{nt}$$

$$= 1500\left(1 + \frac{0.075}{4}\right)^{4 \cdot 10}$$

$$= 3153.523916$$

The amount of the account is about \$3153.52.

46. \$3769.08
47. \$12,067.68
48. \$97,243.21
49. \$24,688.37

REVIEW EXERCISES

Determine the final amount from the investment for each situation.

	Initial Amount	Annual Interest Rate	Time	Type of Compounding
46.	\$2000	8%	8 years	quarterly
47.	\$5500	5.25%	15 years	monthly
48.	\$15,000	7.5%	25 years	monthly
49.	\$500	9.75%	40 years	daily

APPLICATIONS AND PROBLEM SOLVING

50. **Archery** The height h, in feet, that a certain arrow will reach t seconds after being shot directly upward is given by the formula $h = 112t - 16t^2$. What is the maximum height for this arrow? (Lesson 11–1) **196 ft**

51. **Physics** A projectile is shot vertically up in the air. Its distance s, in feet, after t seconds is given by the equation $s = 96t - 16t^2$. Find the values of t when s is 96 feet. (Lesson 11–3) **1.3 second and 4.7 seconds**

52. **Finance** Kevin deposited \$1400 for 8 years at $6\frac{1}{2}$% interest compounded quarterly. How much will he have at the end of 8 years? (Lesson 11–5) **\$2345**

53. **Diving** Wyatt is diving from a 10-meter platform. His height h in meters above the water when he is x meters away from the platform is given by the formula $h = -x^2 + 2x + 10$. Approximately how far away from the platform is he when he enters the water? (Lesson 11–2) **between 4 meters and 5 meters**

54. **Number Theory** Find a number whose square is 168 greater than 2 times the number. (Lesson 11–3) **−12 or 14**

55. **Decision Making** Juanita wants to buy a new computer but she only has \$500. She decides to wait a year and invest her money. Should she put it in a 1-year CD with a rate of 8% compounded monthly or in a savings account with a rate of 6% compounded daily? Explain your answer. (Lesson 11–5) **CD, which yields \$541.50 vs. savings at \$530.92**

A practice test for Chapter 11 is provided on page 797.

ALTERNATIVE ASSESSMENT

COOPERATIVE LEARNING PROJECT

Sightseeing Tours In this project, you will model a business's profit. The Wash Student Tour Company offers one week tours of Washington, D.C. in small groups. While some of Wash's cost per person go down as the number of people on the tour increases, other costs go up because they must reserve rooms in another motel and rent extra vans. Wash has a function that enables them to predict their profit per student. If x is the number of students on the tour, and $f(x)$ is the profit (in dollars) per student, then $f(x) = -0.6x^2 + 18x - 45$.

Write a summary, using this model that describes in detail the profit structure for the Tour Company. Find the number of students that will give Wash the largest profit per student. What is the maximum profit? What does this represent? The company will offer tours as long as they do not lose money. What is the least or greatest number of students they should accept?

Follow these steps to accomplish your task.

- Substitute various numbers of students into the function to determine the profit for each.
- Substitute various amounts of profits into the function to determine the number of students that must be on the tour.
- Graph the function.
- Using the model, discuss terms such as profit, loss, break even, and maximum or minimum.
- Write a summary.

THINKING CRITICALLY

- If the value of a quadratic function is negative when $x = 1$ and positive when $x = 2$, explain what this means in terms of the roots and why.
- In the quadratic equation $ax^2 - bx + c = 0$, if $ac < 0$, what must be true about the nature of the roots of the equation?

PORTFOLIO

Use the quadratic formula and factoring to solve several quadratic equations. As you solve each quadratic equation in both ways, compare the similarities and the differences, if there are any. Write a description of how the quadratic formula can be used to determine whether or not a quadratic polynomial is factorable. Place this in your portfolio.

SELF EVALUATION

While the solution to a problem may be unforeseen, a good problem solver can use the details of the problem to plan a method of solution. Those details can often be part of the solution or preliminary steps needed to arrive at the solution.

Assess yourself. Do you pay attention to detail? Do you organize and evaluate as you go through a problem? Once you have a solution, do you analyze the solution by looking at all options or do you only look for the most obvious? Give an example of a math-related problem and a daily life problem in which detail was essential and explain how it helped.

Alternative Assessment

The Alternative Assessment section provides students with the opportunity to assess their own work by thinking critically, working with others, keeping a portfolio, and honestly evaluating their own progress. For more information on alternative forms of assessment, see *Alternative Assessment in the Mathematics Classroom,* one of the titles in the Glencoe Mathematics Professional Series.

Performance Assessment

Performance Assessment tasks for this chapter are included in the *Assessment and Evaluation Masters.* A scoring guide is also provided.

Assessment and Evaluation Masters, pp. 294, 305

NAME ______________ DATE ______

Chapter 11 Performance Assessment

***Instructions:** Demonstrate your knowledge by giving a clear, concise solution to each problem. Be sure to include all relevant drawings and justify your answers. You may show your solution in more than one way or investigate beyond the requirements of the problem.*

1. a. Write the equation of a quadratic function.

b. Determine what the axis of symmetry and the maximum or minimum point of the graph of the equation would be without graphing the equation.

c. Graph the function.

d. Tell how the graph indicates the nature of the function's roots.

2. a. Write a word problem involving finding two real numbers that could be solved with the equation $x(9 - x) = 14$.

b. Show how to solve the equation in part a by graphing.

3. Describe two real-life situations that could be represented by the graph shown below.

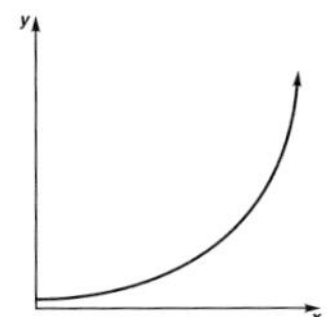

Scoring Guide
Chapter 11
Performance Assessment

Level	Specific Criteria
3 Superior	• Shows thorough understanding of the concepts of *quadratic functions, roots of quadratic equations,* and *exponential functions.* • Uses appropriate strategies to solve problems. • Computations are correct. • Written explanations are exemplary. • Word problem is appropriate and makes sense. • Graphs are accurate and appropriate. • Goes beyond requirements of some or all problems.
2 Satisfactory, with Minor Flaws	• Shows understanding of the concepts of *quadratic functions, roots of quadratic equations,* and *exponential functions.* • Uses appropriate strategies to solve problems. • Computations are mostly correct. • Written explanations are effective. • Word problem is appropriate and makes sense. • Graphs are mostly accurate and appropriate. • Satisfies all requirements of problems.
1 Nearly Satisfactory, with Serious Flaws	• Shows understanding of most of the concepts of *quadratic functions, roots of quadratic equations,* and *exponential functions.* • May not use appropriate strategies to solve problems. • Computations are mostly correct. • Written explanations are satisfactory. • Word problem is mostly appropriate and sensible. • Graphs are mostly accurate and appropriate. • Satisfies most requirements of problems.
0 Unsatisfactory	• Shows little or no understanding of the concepts of *quadratic functions, roots of quadratic equations,* and *exponential functions.* • May not use appropriate strategies to solve problems. • Computations are incorrect. • Written explanations are not satisfactory. • Word problem is not appropriate or sensible. • Graphs are not accurate or appropriate. • Does not satisfy requirements of problems.

In•ves•ti•ga•tion

TEACHER NOTES

NCTM Standards: 1–5, 7–8

This Investigation is designed to be completed over several days or weeks. It may be considered optional. You may want to assign the Investigation and the follow-up activities to be completed at the same time.

Objective

Create a model of a landscaping design and a bid for the Sanchez family.

Mathematical Overview

This Investigation will use the following mathematical skills and concepts from Chapters 12 and 13.

- simplifying rational expressions
- simplifying mixed expressions and complex functions
- using the Pythagorean theorem
- identifying subgoals
- completing squares

Recommended Time

Part	Pages	Time
Investigation	656–657	1 class period
Working on the Investigation	665, 695, 718, 741	20 minutes each
Closing the Investigation	748	1 class period

Instructional Resources

Investigations and Projects Masters, pp. 21–24

A recording sheet, teacher notes, and scoring guide are provided for each Investigation in the *Investigations and Projects Masters*.

1 MOTIVATION

This Investigation uses common materials to investigate the specifications, materials, and costs involved in creating a model and a bid for a landscaping design. Discuss the importance of requesting bids from different contractors before making the final decision.

LONG-TERM PROJECT

In•ves•ti•ga•tion

A Growing Concern

MATERIALS NEEDED

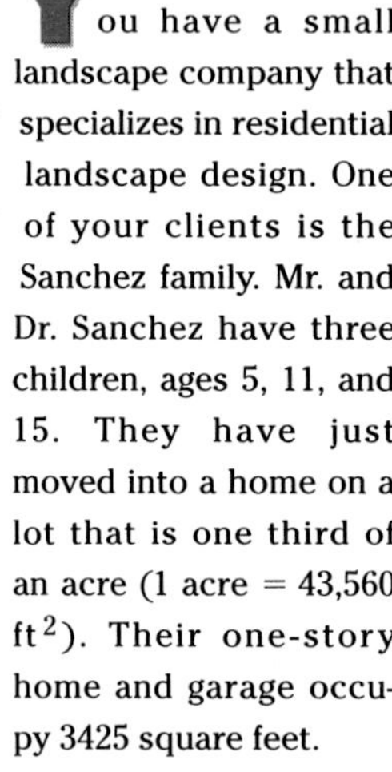

calculator

cardboard

construction paper

flashlight

glue

markers

modeling clay

paint

ruler

scissors

duct tape

You have a small landscape company that specializes in residential landscape design. One of your clients is the Sanchez family. Mr. and Dr. Sanchez have three children, ages 5, 11, and 15. They have just moved into a home on a lot that is one third of an acre (1 acre = 43,560 ft^2). Their one-story home and garage occupy 3425 square feet.

Dr. Sanchez loves to garden, and Mr. Sanchez likes to swim laps to keep in shape. So, the Sanchez family is interested in a backyard pool, hot tub, deck and/or patio, and a fairly good-sized lawn. They also want to leave room to later construct a play area for their youngest child and a garden for Dr. Sanchez. The front yard was fully landscaped by the construction company that built the house.

The family prefers a low-maintenance landscape, which consists mainly of lawn care. Because of the dry climate, daily watering must be done with a sprinkler system. They are also interested in concrete walkways, to match the driveway and the path to the front door.

For this job, the Sanchez family will request bids from several suppliers. They have asked your company to construct a model of your design, along with a bid, for them to view.

The Sanchezes are interested in a reasonably low price, but will choose a higher bid if they prefer the design and features. In either case, they plan to spend no more than $65,000.

When calculating bids, you must consider the cost of supplies, materials, labor, and a profit margin. Use the company's labor and material tables to estimate costs. Labor costs are determined either by the entire job or by the hour. This cost is dependent on the individual job. The profit margin is 20% of the total cost of materials, supplies, and labor.

Your design team consists of three people. Make an Investigation Folder in which you can store all of your work on this Investigation for future use.

Cooperative Learning

This Investigation offers an excellent opportunity for using cooperative learning groups. For more information on cooperative learning strategies and group management, see *Cooperative Learning in the Mathematics Classroom*, one of the titles in the Glencoe Mathematics Professional Series.

THE PROPERTY

Use the dimensions given in the diagram below and your knowledge of geometry to design and create a *rough draft* of a landscaping plan for the Sanchez family's backyard. Be sure to include a pool, hot tub, lawn area, flower beds, trees, deck and/or patio, and concrete walkways in your design. Think about the dimensions of all of the features in your design. Be sure to leave ample space for a play area and garden to be developed later.

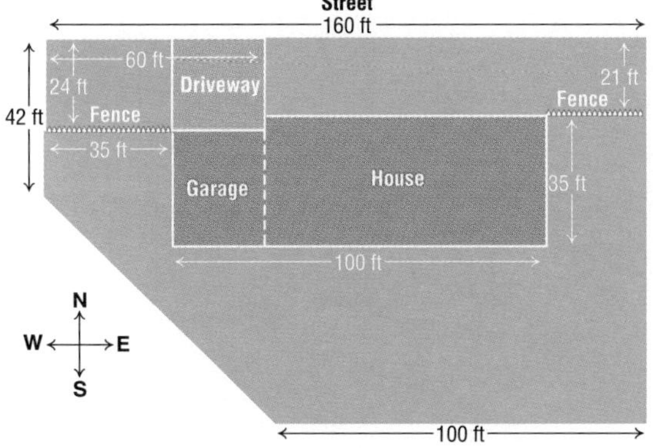

LABOR AND MATERIALS

POOL

Labor

digging: $1/cu ft

pool construction: $10/ft^2 of surface area

plumbing, filter, heating insulation: 8 h @ $35/h

Materials

pool construction: $18/ft^2 of surface area

plumbing and filter: $1250

heater: $1600

HOT TUB

Labor

digging: $1/cu ft

hot tub construction: $10/ft^2 of tub surface area

plumbing, filter, heating insulation: 8 h @ $35/h (3 hours if installed with a pool)

Materials

tub construction: $18/ft^2 of tub surface area

plumbing and filter: $1250

plumbing without filter: $250

heater: $1600 (pool and hot tub can share heater and filter)

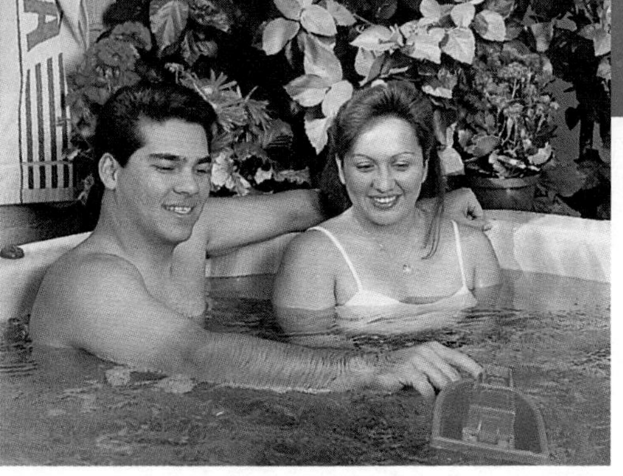

DECK

Labor

8 ft^2 of deck per hour @ $25/h

Materials

redwood deck materials (4 ft × 1 ft): $12

PLANTS and TREES

Labor

plants/trees: 4 per hour @ $20/h

grass seed: 125 ft^2 per hour @ $20/h

sprinkler installation: 20 ft per hour @ $20/h

Materials

1 plant: $6.25

1 tree: $22.50

soil preparation: $1.75 per ft^2 of seeded area

sprinkler: 14 ft of sprinkler for every 10 ft^2 of lawn @ $1.50/ft

PATIO and WALKWAYS

Labor

12 ft^2 of patio or walkway per hour @ $22/h

Materials

concrete: $16/10 ft^2

You will continue working on this Investigation throughout Chapters 12 and 13.

Be sure to keep your designs, models, charts, and other materials in your Investigation Folder.

A Growing Concern Investigation

Working on the Investigation Lesson 12–1, p. 665

Working on the Investigation Lesson 12–7, p. 695

Working on the Investigation Lesson 13–1, p. 718

Working on the Investigation Lesson 13–5, p. 741

Closing the Investigation End of Chapter 13, p. 748

2 SETUP

You may wish to have a student read the first six paragraphs of the Investigation to provide background information about creating the landscaping model and the bid. You may wish to read the next two paragraphs, which introduce the activity. Discuss the activity with your students. Then separate the class into groups of three.

3 MANAGEMENT

Each group member should be responsible for a specific task.

Recorder Collects data.
Measurer Sets dimensions of features for the model.
Contractor Calculates costs for the job.

At the end of the activity, each member should turn in his or her respective equipment.

Sample Answers

Answers will vary as they are based on the design chosen by each group.

Investigations and Projects Masters, p. 24

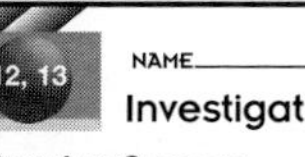

NAME__________ DATE ______

Investigation, Chapters 12 and 13

Student Edition Pages 656–657, 665, 695, 718, 741, 748

A Growing Concern

Use this chart to record your calculations for the cost of materials, labor, and profit margin for each item in your plan.

Item	Cost of Materials	Labor	Profit

Work with your group to consider the following questions.

- How did the movement of the sun affect your decisions about plant choices and placement?
- If the Sanchez family was able to finance another $10,000, thus making the total amount that they can spend $75,000, what changes or additions would you make in your plan?
- List the elements that your group proposed to include for an effective sales presentation.

CHAPTER

12

Exploring Rational Expressions and Equations

PREVIEWING THE CHAPTER

In this chapter, students build on their understanding of fractions to learn how to simplify, add, subtract, multiply, and divide rational expressions. The integration of graphing calculators helps students check the simplification of rational expressions. Students make a list of possible solutions to problems and determine which of the possibilities can be solutions. Students use their knowledge of mixed numbers and improper and complex fractions to study mixed expressions, rational expressions, and complex fractions containing variables. The chapter concludes with students applying their new knowledge to solve rational equations.

Lesson (Pages)	Lesson Objectives	NCTM Standards	State/Local Objectives
12-1 (660–665)	Simplify rational expressions. Identify values excluded from the domain of a rational expression.	1–5	
12-1B (666)	Use a graphing calculator to check the simplification of rational expressions.	1–6	
12-2 (667–670)	Multiply rational expressions.	1–5	
12-3 (671–674)	Divide rational expressions.	1–5	
12-4 (675–680)	Divide polynomials by monomials. Divide polynomials by binomials.	1–5	
12-5 (681–684)	Add and subtract rational expressions with like denominators.	1–5	
12-6 (685–689)	Add and subtract rational expressions with unlike denominators. Make an organized list of possibilities to solve problems.	1–5	
12-7 (690–695)	Simplify mixed expressions and complex fractions.	1–5	
12-8 (696–702)	Solve rational equations.	1–5	

A complete, 1-page lesson plan is provided for each lesson in the ***Lesson Planning Guide***. Answer keys for each lesson are available in the ***Answer Key Masters.***

ORGANIZING THE CHAPTER

You may want to refer to the **Course Planning Calendar** on page T12 for detailed information on pacing.
PACING: Standard—14 days; **Honors**—13 days; **Block**—6 days; **Two Years**—27 days

LESSON PLANNING CHART

			BLACKLINE MASTERS											
Lesson (Pages)	**Materials/ Manipulatives**	**Extra Practice (Student Edition)**	**Study Guide**	**Practice**	**Enrichment**	**Assessment and Evaluation**	**Modeling Mathematics**	**Multicultural Activity**	**Tech Prep Applications**	**Graphing Calculator**	**Science and Math Lab Manual**	**Real-World Applications**	**Interactive Mathematics Tools Software**	**Teaching Transparencies**
12-1 (660–665)	scientific calculator graphing calculator	p. 782	p. 80	p. 80	p. 80						pp. 53–56			12-1A 12-1B
12-1B (666)	graphing calculator									pp. 36, 37				
12-2 (667–670)		p. 782	p. 81	p. 81	p. 81	p. 324								12-2A 12-2B
12-3 (671–674)		p. 783	p. 82	p. 82	p. 82									12-3A 12-3B
12-4 (675–680)	algebra tiles* product mat* graphing calculator	p. 783	p. 83	p. 83	p. 83	pp. 323, 324								12-4A 12-4B
12-5 (681–684)		p. 783	p. 84	p. 84	p. 84									12-5A 12-5B
12-6 (685–689)		p. 784	p. 85	p. 85	p. 85	p. 325		p. 23	p. 23					12-6A 12-6B
12-7 (690–695)	graphing calculator	p. 784	p. 86	p. 86	p. 86			p. 24	p. 24			29		12-7A 12-7B
12-8 (696–702)		p. 784	p. 87	p. 87	p. 87	p. 325	p. 83			p. 12		30	12-8	12-8A 12-8B
Study Guide/ Assessment (703–707)						pp. 309–322, 326–328								

*Included in Glencoe's Student Manipulative Kit and Overhead Manipulative Resources.

ORGANIZING THE CHAPTER

Other Chapter Resources

Student Edition
Investigation, pp. 656–657
Chapter Opener, pp. 658–659
Mathematics and Society, p. 689
Working on the Investigation, pp. 665, 695

Teacher's Classroom Resources
Investigations and Projects Masters, pp. 69–72

Technology
Test and Review Software (IBM and Macintosh)
CD-ROM Interactions (Windows and Macintosh)

Professional Publications
Block Scheduling Booklet
Glencoe Mathematics Professional Series

Outside Resources

Books/Periodicals
Marcy, Steve, *Algebra with Pizzazz!*, Creative Publications
Pedersen, Katherine, *Trivia Math: Algebra*, Creative Publications

Software
Algebra Concepts, Vol. 1, Ventura Educational Systems

Videos/CD-ROMs
Algebra for Everyone, NCTM

Assessment Resources

Student Edition
Math Journal, pp. 668, 672, 699
Mixed Review, pp. 665, 670, 674, 680, 684, 689, 695, 702
Self Test, p. 680
Chapter Highlights, p. 703
Chapter Study Guide and Assessment, pp. 704–706
Alternative Assessment, p. 707
Portfolio, p. 707
Cumulative Review, pp. 708–709

Teacher's Wraparound Edition
5-Minute Check, pp. 660, 667, 671, 675, 681, 685, 690, 696
Check for Understanding, pp. 663, 669, 673, 678, 683, 687, 693, 699
Closing Activity, pp. 665, 670, 674, 680, 684, 689, 695, 702
Cooperative Learning, pp. 673, 697

Assessment and Evaluation Masters
Multiple-Choice Tests, Forms 1A (Honors), 1B (Average), 1C (Basic), pp. 309–314
Free-Response Tests, Forms 2A (Honors), 2B (Average), 2C (Basic), pp. 315–320
Calculator-Based Test, p. 321
Performance Assessment, p. 322
Mid-Chapter Test, p. 323
Quizzes A–D, pp. 324–325
Standardized Test Practice, p. 326
Cumulative Review, pp. 327–328

ENHANCING THE CHAPTER

Examples of some of the materials for enhancing Chapter 12 are shown below.

Multicultural Activity Masters, pp. 23, 24

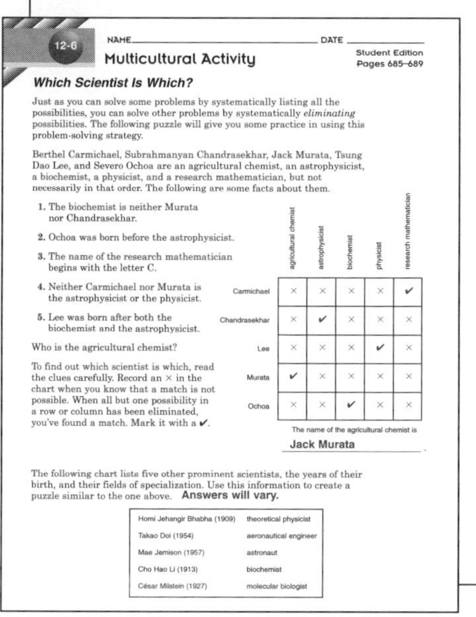

12-6 NAME ______ DATE ______

Multicultural Activity — Student Edition Pages 685–689

Which Scientist Is Which?

Just as you can solve some problems by systematically listing all the possibilities, you can solve other problems by systematically *eliminating* possibilities. The following puzzle will give you some practice in using this problem-solving strategy.

Berthel Carmichael, Subrahmanyan Chandrasekhar, Jack Murata, Tsung Dao Lee, and Severo Ochoa are an agricultural chemist, an astrophysicist, a biochemist, a physicist, and a research mathematician, but not necessarily in that order. The following are some facts about them.

1. The biochemist is neither Murata nor Chandrasekhar.
2. Ochoa was born before the astrophysicist.
3. The name of the research mathematician begins with the letter C.
4. Neither Carmichael nor Murata is the astrophysicist or the physicist.
5. Lee was born after both the biochemist and the astrophysicist.

Who is the agricultural chemist?

To find out which scientist is which, read the clues carefully. Record an × in the chart when you know that a match is not possible. When all but one possibility in a row or column has been eliminated, you've found a match. Mark it with a ✔.

	agricultural chemist	astrophysicist	biochemist	physicist	research mathematician
Carmichael	×	×	×	×	✔
Chandrasekhar	×	✔	×	×	×
Lee	×	×	×	✔	×
Murata	✔	×	×	×	×
Ochoa	×	×	✔	×	×

The name of the agricultural chemist is **Jack Murata**.

The following chart lists five other prominent scientists, the years of their birth, and their fields of specialization. Use this information to create a puzzle similar to the one above. **Answers will vary.**

Homi Jehangir Bhabha (1909)	theoretical physicist
Takao Doi (1954)	aeronautical engineer
Mae Jemison (1957)	astronaut
Cho Hao Li (1913)	biochemist
César Milstein (1927)	molecular biologist

Real-World Applications, 29, 30

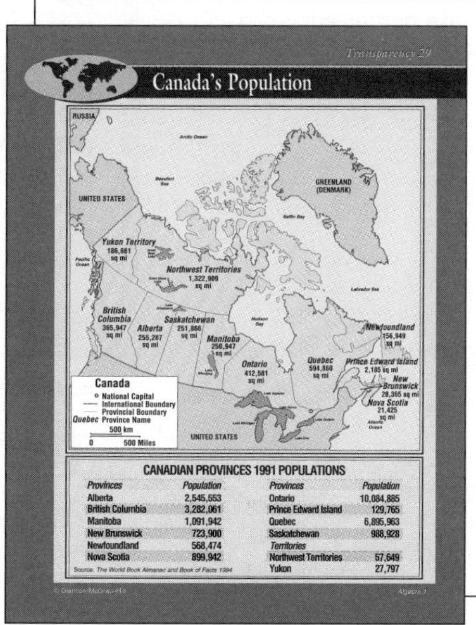

CANADIAN PROVINCES 1991 POPULATIONS

Provinces	*Population*	*Provinces*	*Population*
Alberta	2,545,553	Ontario	10,084,885
British Columbia	3,282,061	Prince Edward Island	129,765
Manitoba	1,091,942	Quebec	6,895,963
New Brunswick	723,900	Saskatchewan	988,928
Newfoundland	568,474	*Territories*	
Nova Scotia	899,942	Northwest Territories	57,649
		Yukon	27,797

Source: *The World Book Almanac and Book of Facts 1994*

TECHNOLOGY

Graphing Calculator Masters, p. 12

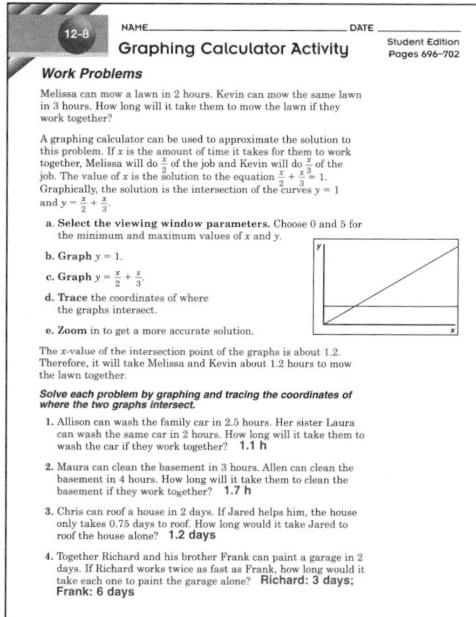

12-8 NAME ______ DATE ______

Graphing Calculator Activity — Student Edition Pages 696–702

Work Problems

Melissa can mow a lawn in 2 hours. Kevin can mow the same lawn in 3 hours. How long will it take them to mow the lawn if they work together?

A graphing calculator can be used to approximate the solution to this problem. If x is the amount of time it takes for them to work together, Melissa will do $\frac{x}{2}$ of the job and Kevin will do $\frac{x}{3}$ of the job. The value of x is the solution to the equation $\frac{x}{2} + \frac{x}{3} = 1$. Graphically, the solution is the intersection of the curves $y = 1$ and $y = \frac{x}{2} + \frac{x}{3}$.

a. **Select the viewing window parameters.** Choose 0 and 5 for the minimum and maximum values of x and y.
b. **Graph** $y = 1$.
c. **Graph** $y = \frac{x}{2} + \frac{x}{3}$.
d. **Trace** the coordinates of where the graphs intersect.
e. **Zoom** in to get a more accurate solution.

The x-value of the intersection point of the graphs is about 1.2. Therefore, it will take Melissa and Kevin about 1.2 hours to mow the lawn together.

Solve each problem by graphing and tracing the coordinates of where the two graphs intersect.

1. Allison can wash the family car in 2.5 hours. Her sister Laura can wash the same car in 2 hours. How long will it take them to wash the car if they work together? **1.1 h**
2. Maura can clean the basement in 3 hours. Allen can clean the basement in 4 hours. How long will it take them to clean the basement if they work together? **1.7 h**
3. Chris can roof a house in 2 days. If Jared helps him, the house only takes 0.75 days to roof. How long would it take Jared to roof the house alone? **1.2 days**
4. Together Richard and his brother Frank can paint a garage in 2 days. If Richard works twice as fast as Frank, how long would it take each one to paint the garage alone? **Richard: 3 days; Frank: 6 days**

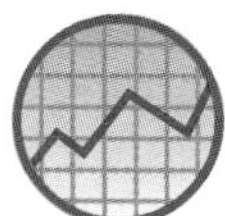

TECH PREP

Tech Prep Applications Masters, pp. 23, 24

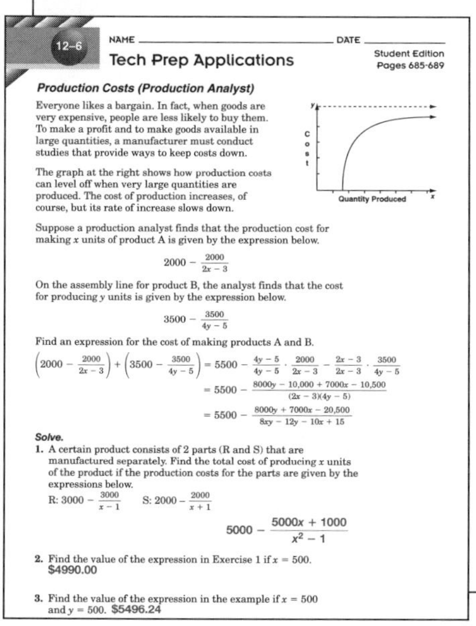

12-6 NAME ______ DATE ______

Tech Prep Applications — Student Edition Pages 685-689

Production Costs (Production Analyst)

Everyone likes a bargain. In fact, when goods are very expensive, people are less likely to buy them. To make a profit and to make goods available in large quantities, a manufacturer must conduct studies that provide ways to keep costs down.

The graph at the right shows how production costs can level off when very large quantities are produced. The cost of production increases, of course, but its rate of increase slows down.

Suppose a production analyst finds that the production cost for making x units of product A is given by the expression below.

$$2000 - \frac{2000}{2x-3}$$

On the assembly line for product B, the analyst finds that the cost for producing y units is given by the expression below.

$$3500 - \frac{3500}{4y-5}$$

Find an expression for the cost of making products A and B.

$$\left(2000 - \frac{2000}{2x-3}\right) + \left(3500 - \frac{3500}{4y-5}\right) = 5500 - \frac{4y-5}{4y-5}\cdot\frac{2000}{2x-3} - \frac{2x-3}{2x-3}\cdot\frac{3500}{4y-5}$$
$$= 5500 - \frac{8000y - 10{,}000 + 7000x - 10{,}500}{(2x-3)(4y-5)}$$
$$= 5500 - \frac{8000y + 7000x - 20{,}500}{8xy - 12y - 10x + 15}$$

Solve.

1. A certain product consists of 2 parts (R and S) that are manufactured separately. Find the total cost of producing x units of the product if the production costs for the parts are given by the expressions below.
   R: $3000 - \frac{3000}{x-1}$ S: $2000 - \frac{2000}{x+1}$
   $$5000 - \frac{5000x + 1000}{x^2 - 1}$$
2. Find the value of the expression in Exercise 1 if $x = 500$. **$4990.00**
3. Find the value of the expression in the example if $x = 500$ and $y = 500$. **$5496.24**

Science and Math Lab Manual, pp. 53–56

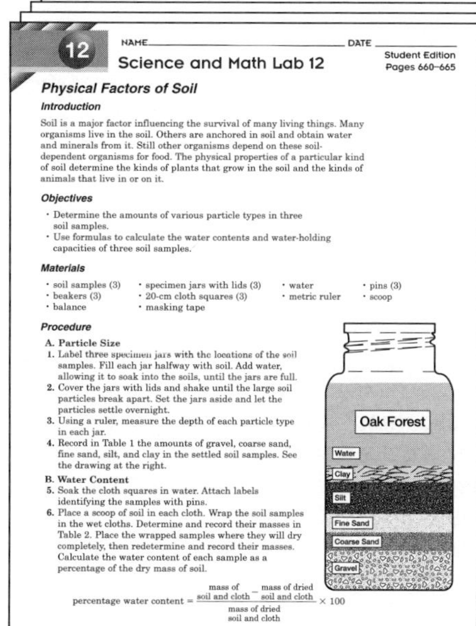

12 NAME ______ DATE ______

Science and Math Lab 12 — Student Edition Pages 660–665

Physical Factors of Soil

Introduction

Soil is a major factor influencing the survival of many living things. Many organisms live in the soil. Others are anchored in soil and obtain water and minerals from it. Still other organisms depend on these soil-dependent organisms for food. The physical properties of a particular kind of soil determine the kinds of plants that grow in the soil and the kinds of animals that live in or on it.

Objectives

- Determine the amounts of various particle types in three soil samples.
- Use formulas to calculate the water contents and water-holding capacities of three soil samples.

Materials

- soil samples (3)
- beakers (3)
- balance
- specimen jars with lids (3)
- 20-cm cloth squares (3)
- masking tape
- water
- metric ruler
- pins (3)
- scoop

Procedure

A. Particle Size
1. Label three specimen jars with the locations of the soil samples. Fill each jar halfway with soil. Add water, allowing it to soak into the soils, until the jars are full.
2. Cover the jars with lids and shake until the large soil particles break apart. Set the jars aside and let the particles settle overnight.
3. Using a ruler, measure the depth of each particle type in each jar.
4. Record in Table 1 the amounts of gravel, coarse sand, fine sand, silt, and clay in the settled soil samples. See the drawing at the right.

B. Water Content
5. Soak the cloth squares in water. Attach labels identifying the samples with pins.
6. Place a scoop of soil in each cloth. Wrap the soil samples in the wet cloths. Determine and record their masses in Table 2. Place the wrapped samples where they will dry completely, then redetermine and record their masses. Calculate the water content of each sample as a percentage of the dry mass of soil.

$$\text{percentage water content} = \frac{\text{mass of soil and cloth} - \text{mass of dried soil and cloth}}{\text{mass of dried soil and cloth}} \times 100$$

Problem of the Week Cards, 32, 33, 34

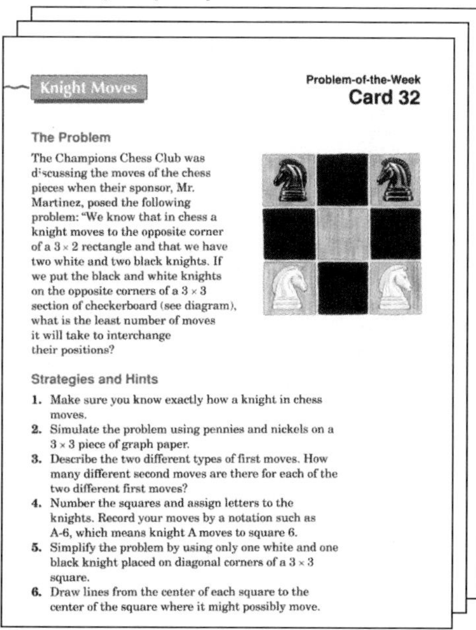

Knight Moves — Problem-of-the-Week **Card 32**

The Problem

The Champions Chess Club was d'scussing the moves of the chess pieces when their sponsor, Mr. Martinez, posed the following problem: "We know that in chess a knight moves to the opposite corner of a 3 × 2 rectangle and that we have two white and two black knights. If we put the black and white knights on the opposite corners of a 3 × 3 section of checkerboard (see diagram), what is the least number of moves it will take to interchange their positions?

Strategies and Hints

1. Make sure you know exactly how a knight in chess moves.
2. Simulate the problem using pennies and nickels on a 3 × 3 piece of graph paper.
3. Describe the two different types of first moves. How many different second moves are there for each of the two different first moves?
4. Number the squares and assign letters to the knights. Record your moves by a notation such as A-6, which means knight A moves to square 6.
5. Simplify the problem by using only one white and one black knight placed on diagonal corners of a 3 × 3 square.
6. Draw lines from the center of each square to the center of the square where it might possibly move.

MAKING MATHEMATICS RELEVANT

This two-page introduction to the chapter provides students with an opportunity to explore contemporary topics and their applications to mathematics.

Background Information

National Endowment for the Arts The table in the text indicates several trends that students should note. Among these trends is a steady decline in funds available for the arts since 1990. Encourage students to research other economic or social trends during these same years that may account for the fluctuations.

CHAPTER 12

Exploring Rational Expressions and Equations

Objectives

In this chapter, you will:

- simplify rational expressions,
- add, subtact, multiply, and divide rational expressions,
- divide polynomials, and
- make organized lists to solve problems.

National Endowment for the Arts

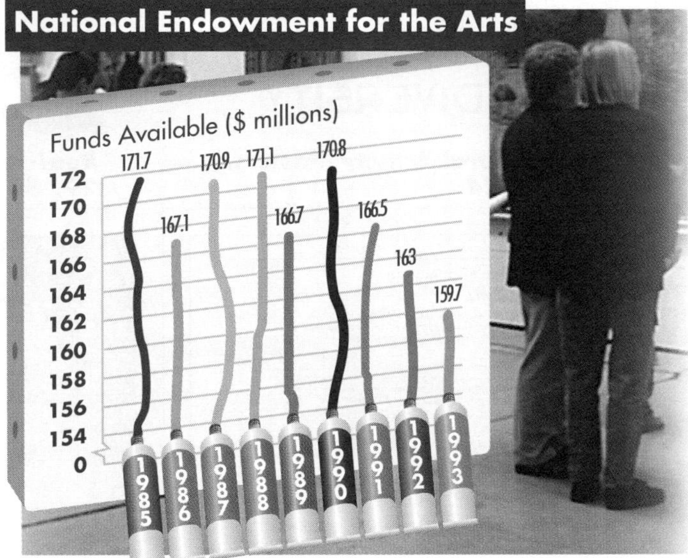

Source: U.S. National Endowment for the Arts, *Annual Report*

The term "starving artist" is not without merit. Many artists often have other jobs to survive while trying to pursue their careers as artists. The National Endowment for the Arts is authorized to assist individuals and nonprofit organizations financially in a wide range of artistic endeavors. Some of the artistic forms that are funded are music, museums, theater, dance, media arts, and visual arts.

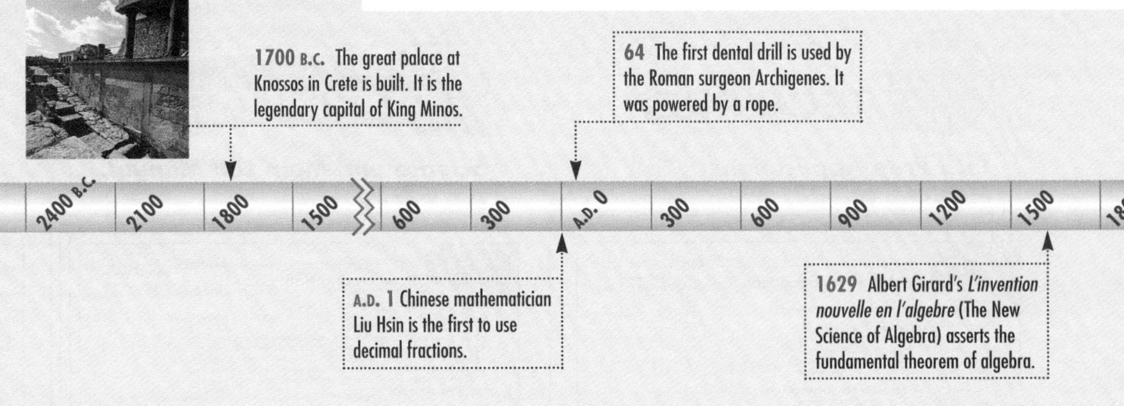

1700 B.C. The great palace at Knossos in Crete is built. It is the legendary capital of King Minos.

64 The first dental drill is used by the Roman surgeon Archigenes. It was powered by a rope.

A.D. 1 Chinese mathematician Liu Hsin is the first to use decimal fractions.

1629 Albert Girard's *L'invention nouvelle en l'algebre* (The New Science of Algebra) asserts the fundamental theorem of algebra.

The fundamental theorem of algebra was not proven until 170 years after its discovery. Students might find it interesting to write a report on why mathematicians prove theorems that they often "know" to be true.

interNET CONNECTION

The Smithsonian Institution's National Museum of the American Indian covers languages, literature, history, and arts with links to Native American sites.

World Wide Web
http://www.si.edu/organiza/museums/amerind/nmai/

PEOPLE IN THE NEWS

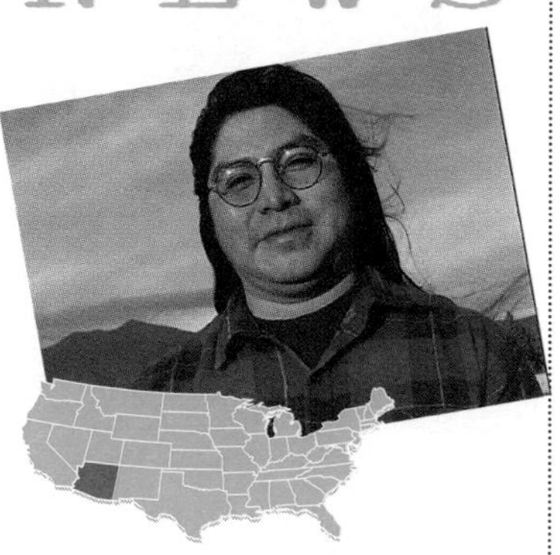

The expression of energy, vibrancy, and harmony are just some of the thoughts that flow through the creativity of **Joe Maktima** of Flagstaff, Arizona. The themes of his work in acrylic paints and mediums are deeply rooted in his pueblo culture. Joe began his artistic endeavors in high school as a hobby, but it was winning the Best of Show Award in a national competition of Native American high school art students that gave him the confidence to become a professional artist.

Chapter Project

Native American art also includes blanket weaving. Many of the designs include pictorial representations of everyday objects or characters from folklore about their ancestors.

- Design a 64″ × 80″ blanket. Choose either objects that represent you or some part of your personal history for the blanket's design.
- Make a drawing to represent your design. Include a scale expressed as a rational number to show the actual size of your blanket.
- Use yarn or construction paper to weave a portion of your design.
- Calculate how much yarn in each color you would need to create a real blanket.
- Estimate the cost of your blanket in materials and labor.

Suppose a blanket is priced at $375. How does this price compare with your estimate?

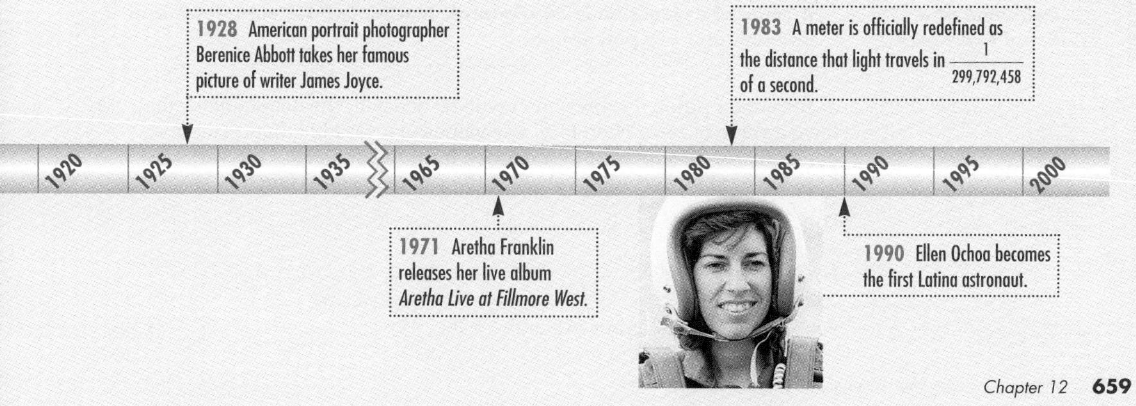

1928 American portrait photographer Berenice Abbott takes her famous picture of writer James Joyce.
1983 A meter is officially redefined as the distance that light travels in $\frac{1}{299,792,458}$ of a second.
1920 1925 1930 1935 1965 1970 1975 1980 1985 1990 1995 2000
1971 Aretha Franklin releases her live album *Aretha Live at Fillmore West.*
1990 Ellen Ochoa becomes the first Latina astronaut.

PEOPLE IN THE NEWS

Joe Maktima has works on display in galleries in eight states and in Germany, Japan, and England. Mr. Maktima studied art at the Institute of American Indian Arts.

Chapter Project

Cooperative Learning Students can work on this project in groups of three. As a group, they can discuss and decide on the objects that will be represented in their design. One student will then draw the design, another student will estimate the cost, while the third student will produce a sample of the design in yarn.

Alternative Chapter Projects

Two other chapter projects are included in the *Investigations and Projects Masters.* In Chapter 12 Project A, pp. 69–70, students extend the topic in the chapter opener. In Chapter 12 Project B, pp. 71–72, students investigate focal length in relationship to a lens.

Investigations and Projects Masters, p. 69

NAME______________________ DATE __________

Chapter 12 Project A

Student Edition Pages 660–702

Mayan Time

1. The Maya were a Native American people whose complex civilization reached a peak during the years from about 250 to 900. For this project, you will work with a small group to research Mayan civilization. Investigate such topics as art and architecture, mathematics and astronomy, religion, and way of life.

2. Using careful astronomical observations, the Maya developed a solar calendar. They also developed a sacred calendar that ran simultaneously with it. Research these two calendars to find out the answers to the following questions.
   - How many days made up a year in the solar calendar?
   - How were the days of the solar calendar divided into months?
   - How many days made up a year in the sacred calendar?
   - How were the days of the sacred calendar named?

3. Use the information you found about the number of days in each calendar to determine how many days would pass before a given day of the sacred calendar coincided with the same day of the solar calendar again. How many years on each calendar would pass before the calendars reached the same point again? Explain how these problems relate to the concept of the least common multiple. Make up some other problems using your knowledge of the Mayan calendars and exchange them with other groups to solve.

4. With your group, design a museum exhibit about the Mayan civilization. Include a timeline that would show how the two calendars correspond. Share your timelines with other groups.

12-1 LESSON NOTES

NCTM Standards: 1–5

Instructional Resources

- Study Guide Master 12-1
- Practice Master 12-1
- Enrichment Master 12-1
- Science and Math Lab Manual, pp. 53–56

Transparency 12-1A contains the 5-Minute Check for this lesson; **Transparency 12-1B** contains a teaching aid for this lesson.

Recommended Pacing	
Standard Pacing	Day 1 of 14
Honors Pacing	Day 1 of 13
Block Scheduling*	Day 1 of 6
Alg. 1 in Two Years*	Days 1 & 2 of 27

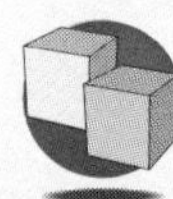

*For more information on pacing and possible lesson plans, refer to the *Block Scheduling Booklet* and *Algebra 1 in Two Years.*

1 FOCUS

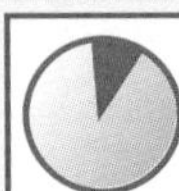

5-Minute Check
(over Chapter 11)

Solve each equation.

1. $n^2 + \frac{1}{6}n - \frac{1}{6} = 0$ $\left\{\frac{1}{3}, -\frac{1}{2}\right\}$
2. $t^2 - 14t + 49 = 0$ **{7}**
3. $y^2 + 5y = 0$ **{−5, 0}**
4. Find two numbers whose sum is 35 and whose product is 300. **15 and 20**
5. Find two numbers that differ by 7 and have a product of 144. **9 and 16, −9 and −16**

Motivating the Lesson

Hands-On Activity Set up a lever and fulcrum using a book and a ruler. One at a time, place objects of different weights and lift each by grasping the lever at different positions. Have students generalize their observations.

Teaching Tip In the examples at the bottom of the page, note that $x^2 - 5x + 6 = (x - 2)(x - 3)$. Use the zero product property to show that $x \neq 2$ and $x \neq 3$.

12–1 Simplifying Rational Expressions

What YOU'LL LEARN

- To simplify rational expressions, and
- to identify values excluded from the domain of a rational expression.

Why IT'S IMPORTANT

You can use rational expressions to solve problems involving physics and carpentry.

Physical Science

Many bicyclists carry small tool kits in case they need to make adjustments to their bicycles while on the road. A wrench, for example, might be needed to tighten the bolts that keep the seat aligned. The force applied to one end of the wrench is multiplied so that the bolt at the other end of the wrench is made very secure.

A wrench is an example of a simple machine called a *lever*. In Lesson 7–2, you learned that to calculate the *mechanical advantage* (MA) of a lever, you find the ratio of the length of the effort arm to the length of the resistance arm. MA is usually expressed as a decimal value.

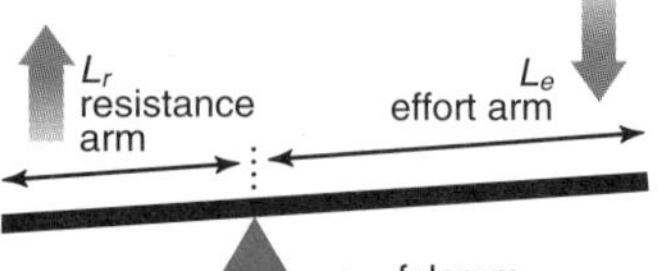

$$MA = \frac{\text{length of effort arm}}{\text{length of resistance arm}} = \frac{L_e}{L_r}$$

Suppose the total length of a lever is 6 feet. If the length of the resistance arm of the lever is x feet, then the length of the effort arm is $(6 - x)$ feet. The function $f(x) = \frac{6 - x}{x}$ represents the mechanical advantage of the lever. The table and graph at the right represent this function. They illustrate how varying the length of the resistance arm affects the mechanical advantage. Notice that, as x increases, the mechanical advantage $f(x)$ decreases. The expression that defines this function, $\frac{6-x}{x}$, is an example of a **rational expression.**

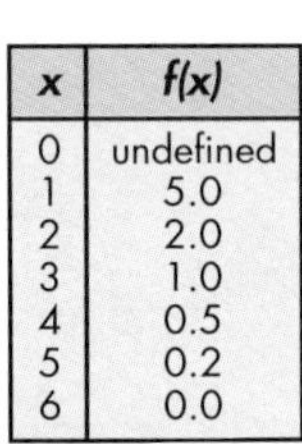

x	f(x)
0	undefined
1	5.0
2	2.0
3	1.0
4	0.5
5	0.2
6	0.0

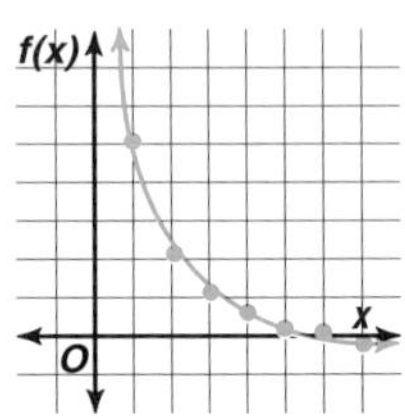

Definition of a Rational Expression	A rational expression is an algebraic fraction whose numerator and denominator are polynomials.

$f(x) = \frac{6 - x}{x}$ is called a rational function.

Because a rational expression involves division, the denominator may not have a value of zero. Therefore, any values of a variable that result in a denominator of zero must be excluded from the domain of the variable. These are called **excluded values** of the rational expression.

For $\frac{6 - x}{x}$, exclude $x = 0$.

For $\frac{5m + 3}{m + 6}$, exclude $m = -6$, since $-6 + 6 = 0$.

For $\frac{x^2 - 5}{x^2 - 5x + 6}$, exclude $x = 2$ and $x = 3$. *Why?*

Alternative Learning Styles

Visual The rectangle at the right has length $4a^2 - 8a + 4$ and area $3a^2 - 4a + 1$. Express its width in simplest form. $w = \frac{3a - 1}{4a - 4}$

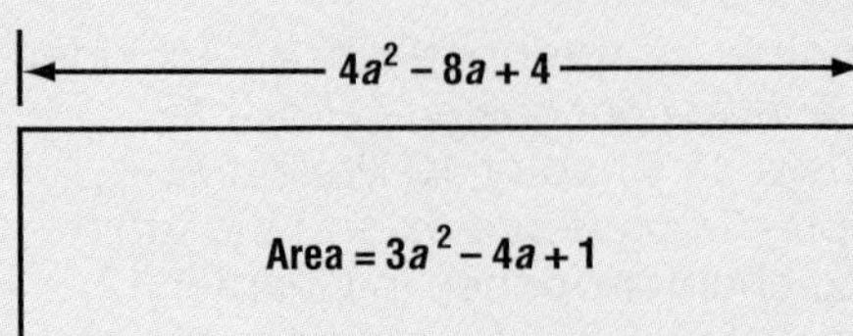

Example 1 **For each rational expression, state the values of the variable that must be excluded.**

a. $\dfrac{7b}{b+5}$

Exclude the values for which $b + 5 = 0$.

$b + 5 = 0$

$b = -5$

Therefore, b cannot equal -5.

b. $\dfrac{r^2 + 32}{r^2 + 9r + 8}$

Exclude the values for which $r^2 + 9r + 8 = 0$.

$r^2 + 9r + 8 = 0$

$(r + 1)(r + 8) = 0$

$r = -1$ or $r = -8$ *Zero product property*

Therefore, r cannot equal -1 or -8.

To simplify a rational expression, you must eliminate any common factors of the numerator and denominator. To do this, use their greatest common factor (GCF). Remember that $\frac{ab}{ac} = \frac{a}{a} \cdot \frac{b}{c}$ and $\frac{a}{a} = 1$. So, $\frac{ab}{ac} = 1 \cdot \frac{b}{c}$ or $\frac{b}{c}$.

Example 2 **Simplify $\frac{9x^2yz}{24xyz^2}$. State the excluded values of x, y, and z.**

$\dfrac{9x^2yz}{24xyz^2} = \dfrac{(3xyz)(3x)}{(3xyz)(8z)}$ *Factor; the GCF is 3xyz.*

$= \dfrac{\overset{1}{\cancel{(3xyz)}}(3x)}{\underset{1}{\cancel{(3xyz)}}(8z)}$ *Divide by the GCF.*

$= \dfrac{3x}{8z}$

Exclude the values for which $24xyz^2 = 0$: $x = 0$, $y = 0$, or $z = 0$. Therefore, neither x, y, nor z can equal 0.

You can use the same procedure to simplify a rational expression in which the numerator and denominator are polynomials.

Example 3 **Simplify $\frac{a+3}{a^2+4a+3}$. State the excluded values of a.**

$\dfrac{a+3}{a^2+4a+3} = \dfrac{a+3}{(a+1)(a+3)}$ *Factor the denominator.*

$= \dfrac{\overset{1}{\cancel{a+3}}}{(a+1)\underset{1}{\cancel{(a+3)}}}$ *The GCF is a + 3.*

$= \dfrac{1}{a+1}$

Exclude the values for which $a^2 + 4a + 3 = 0$.

$a^2 + 4a + 3 = 0$

$(a + 1)(a + 3) = 0$

$a = -1$ or $a = -3$

Therefore, a cannot equal -1 or -3.

2 TEACH

Teaching Tip For Example 1, it may be necessary to review factoring. You may want to draw a flowchart for factoring polynomials. Then have students test the flowchart with various examples.

In-Class Examples

For Example 1
For each rational expression, state the values of the variable that must be excluded.

a. $\dfrac{3r}{6-r}$ **6**

b. $\dfrac{m+5}{m^2-4m+3}$ **3, 1**

c. $\dfrac{15k}{k^2-64}$ **8, −8**

For Example 2
Simplify each rational expression. State the excluded values of the variables.

a. $\dfrac{3m^2-48}{3(m+4)}$

$\boldsymbol{m - 4, m \neq -4}$

b. $\dfrac{y^2-4}{y^2+y-2}$

$\boldsymbol{\dfrac{y-2}{y-1}, y \neq -2, 1}$

For Example 3
Simplify each rational expression. State the excluded values of the variables.

a. $\dfrac{b-2m}{12m^2-3b^2}$

$\boldsymbol{-\dfrac{1}{6m+3b}, b \neq \pm 2m}$

b. $\dfrac{2x^2-50}{5-x}$

$\boldsymbol{-2x + 10, x \neq 5}$

Teaching Tip For Example 3, remind students that they must first factor the numerator and denominator in order to find the GCF. Stress that factors, not terms, can be canceled.

Alternative Teaching Strategies

Student Diversity Ask students to compare and contrast the expressions $(x - 2)$ and $\frac{x^2-4}{x+2}$. Have them find the value for various values of x, including 2.

EXPLORATION

This Exploration has two goals.

1. Students learn to use the calculator to compute values of rational functions.
2. Students learn what it means for a rational function to be undefined for certain values.

In-Class Example

For Example 4

A screwdriver that is 26.5 cm long is used to pry the lid off a paint can. It is placed so that 0.5 cm of its length extends inward from the rim of the can. The same 5 pounds of force is applied. What is the force applied on the lid? **260 pounds**

EXPLORATION CALCULATORS

You can use a scientific or graphing calculator to evaluate rational expressions for given values of the variables. When you do this, you must be careful to use parentheses to group both the numerator and the denominator of the expression. For example, to evaluate $\frac{x^2 - 6x + 8}{x - 2}$ when $x = -3$, use a key sequence like this on a scientific calculator.

Enter: (3 +/− x^2 − 6 × 3 +/− + 8) ÷ (3 +/− − 2) = −7

The key sequence for a graphing calculator is very similar.

Enter: (((−) 3) x^2 − 6 × (−) 3 + 8) ÷ ((−) 3 − 2) ENTER −7

Your Turn

a. Copy and complete the table below.

x	−3	−2	−1	0	1	2	3
$\frac{x^2 - 6x + 8}{x - 2}$	−7	**−6**	**−5**	**−4**	**−3**	**undef.**	**−1**
$x - 4$	**−7**	**−6**	**−5**	**−4**	**−3**	**−2**	**−1**

b. For which value(s) of x is the value of $\frac{x^2 - 6x + 8}{x - 2}$ equal to the value of $x - 4$? Why does this result make sense? **all except $x = 2$**

c. For which value(s) of x is the value of $\frac{x^2 - 6x + 8}{x - 2}$ *not* equal to the value of $x - 4$? Why does this result make sense? **$x = 2$; It is undefined at $x = 2$.**

Example 4

Physical Science

To pry the lid off a paint can, a screwdriver that is 20.5 cm long is used as a lever. It is placed so that 0.5 cm of its length extends inward from the rim of the can. Then a force of 5 pounds is applied at the end of the screwdriver. What is the force placed on the lid?

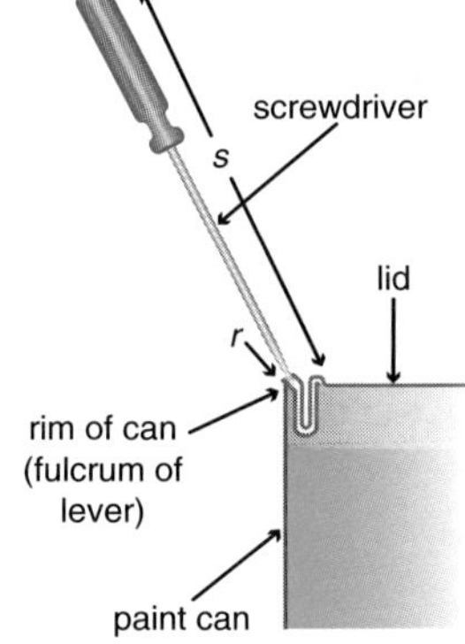

Let s represent the total length of the screwdriver.

Let r represent the length that extends inward from the rim. This is the length of the resistance arm of the lever.

Then $s - r$ represents the length that extends outward from the rim. This is the length of the effort arm of the lever.

Use the formula given in the connection at the beginning of the lesson to write an expression for mechanical advantage.

$$MA = \frac{\text{length of effort arm}}{\text{length of resistance arm}} \text{ or } \frac{s - r}{r}$$

Classroom Vignette

"When factoring to simplify expressions, I teach my students that in $\frac{x + 3}{3}$, the $x + 3$ is like a baseball player trying out for the team. The coach will cut only a whole player, not part of a player. So the $x + 3$ can be cut only with another $x + 3$, not part of a player like an x or a 3."

Joyce Anne MacKenzie

Joyce Anne MacKenzie
Camdenton R-III High School
Camdenton, Missouri

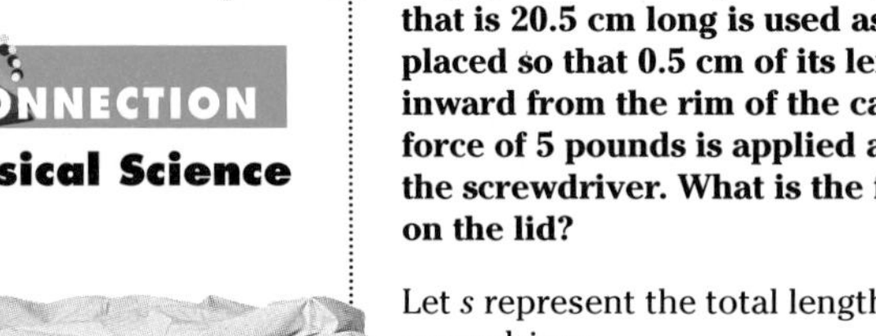

Now evaluate the expression for the given values.

$$\frac{s-r}{r} = \frac{20.5 - 0.5}{0.5} \quad s = 20.5 \text{ and } r = 0.5$$

$$= \frac{20}{0.5} \text{ or } 40 \quad \textit{Simplify.}$$

The mechanical advantage is 40. The force placed on the lid is the product of this number and the force applied at the end of the screwdriver.

$40 \cdot 5$ pounds $= 200$ pounds

The force placed on the lid is 200 pounds.

CHECK FOR UNDERSTANDING

Communicating Mathematics

Study the lesson. Then complete the following.

1. **Explain** why $x = 2$ is excluded from the domain of $f(x) = \frac{x^2 + 7x + 12}{x - 2}$.
2. **Estimate** the mechanical advantage of the lever shown on page 660 using the graph if x is 1.5. **about 3**
3. **Explain** how you would determine the values to be excluded from the domain of $f(x) = \frac{x + 5}{x^2 + 6x + 5}$. **See margin.**
4. **Write** a rational expression involving one variable for which the excluded values are -2 and 7.
5. **Write** the meaning of the term *mechanical advantage* in your own words. **The multiplication effect of using a tool or lever to carry out a task.**

1. When $x = 2$, the denominator becomes zero. Division by zero is undefined.

4. Sample answer: $\frac{x-2}{(x+2)(x-7)}$

Guided Practice

For each expression, find the GCF of the numerator and the denominator. Then simplify. State the excluded values of the variables.

6. $\frac{13a}{14ay}$ **a; $\frac{13}{14y}$; $a \neq 0$, $y \neq 0$**
7. $\frac{-7a^2b^3}{21a^5b}$ **$7a^2b$; $\frac{-b^2}{3a^3}$; $a \neq 0$, $b \neq 0$**
8. $\frac{a(m+3)}{a(m-2)}$ **a; $\frac{m+3}{m-2}$; $m \neq 2$, $a \neq 0$**
9. $\frac{3b}{b(b+5)}$ **b; $\frac{3}{b+5}$; $b \neq 0$, $b \neq -5$**
10. $\frac{(r+s)(r-s)}{(r-s)(r-s)}$
11. $\frac{m-3}{m^2-9}$ **$m - 3$; $\frac{1}{m+3}$; $m \neq \pm 3$**
12. Evaluate $\frac{x^2 + 7x + 12}{x + 3}$ if $x = 2$. **6**

10. $r - s$; $\frac{r+s}{r-s}$; $r - s \neq 0$

13. **Landscaping** To clear land for a garden, Chang needs to move some large rocks. He plans to use a 6-foot-long pinch bar as a lever. He positions it next to each rock as shown at the right.
    a. Calculate the mechanical advantage. **5**
    b. If Chang can apply a force of 150 pounds to the effort arm, what is the greatest weight he can lift? **750 lb**

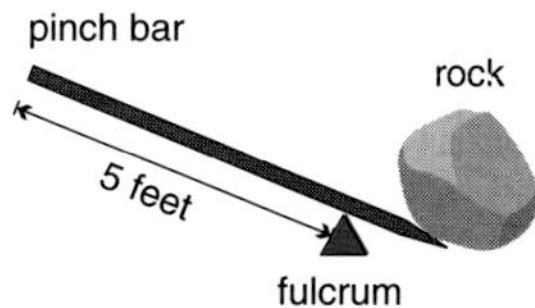

Reteaching

Using Alternative Methods Write a rational expression with one variable and simplify it. Choose values for the variable and complete a table with headings *Variable, Value of Rational Expression,* and *Value of Simplified Form.* Write the table on the chalkboard. This may help students see how the rational expression is equivalent to its simplified form.

3 PRACTICE/APPLY

Check for Understanding

Exercises 1–13 are designed to help you assess your students' understanding through reading, writing, speaking, and modeling. You should work through Exercises 1–5 with your students and then monitor their work on Exercises 6–13.

Error Analysis

Students sometimes simplify a rational expression and then find excluded values of the variable from this simplified form. This "loses" some excluded values of the variable. Excluded values of the variable must include all values that make any denominator zero at any step.

Additional Answer

3. Factor the denominator, set each of its factors equal to zero, and solve for x. This will tell you that the restricted values are -5 and -1.

Study Guide Masters, p. 80

12-1 NAME__________ DATE______

Study Guide

Student Edition Pages 660–665

Simplifying Rational Expressions

Factoring polynomials is a useful tool for simplifying fractions or **rational expressions.**

To simplify a rational number such as $\frac{9}{21}$, first factor the numerator and denominator. Then divide each by the greatest common factor.

$\frac{9}{21} = \frac{3 \cdot 3}{7 \cdot 3}$ **The greatest common factor is 3.**

$= \frac{3 \cdot \not 3}{7 \cdot \not 3}$ or $\frac{3}{7}$ **Notice that $\frac{3}{3} = 1$.**

The procedure can be used to simplify rational expressions.

Example 1: Simplify $\frac{54z^3}{24yz}$.

$\frac{54z^3}{24yz} = \frac{2 \cdot 3 \cdot 3 \cdot 3 \cdot z \cdot z \cdot z}{2 \cdot 2 \cdot 2 \cdot 3 \cdot y \cdot z}$

$= \frac{\not 2 \cdot \not 3 \cdot 3 \cdot 3 \cdot \not z \cdot z \cdot z}{\not 2 \cdot 2 \cdot 2 \cdot \not 3 \cdot y \cdot \not z} = \frac{9z^2}{4y}$

The excluded value of y and z is 0.

Example 2: Simplify $\frac{x+3}{x^2-9}$.

$\frac{x+3}{x^2-9} = \frac{(x+3) \cdot 1}{(x+3)(x-3)}$

$= \frac{\cancel{(x+3)} \cdot 1}{\cancel{(x+3)}(x-3)} = \frac{1}{x-3}$

The excluded values of x are 3 and -3.

Simplify each rational expression. State the excluded values of the variables.

1. $\frac{a^3}{4a^6}$ $\frac{1}{4a^3}$, $a \neq 0$
2. $\frac{7n^3}{21n^8}$ $\frac{1}{3n^5}$, $n \neq 0$
3. $\frac{6}{2x^2+4}$ $\frac{3}{x^2+2}$
4. $\frac{2b(x+1)}{4b(x-1)}$ $\frac{x+1}{2(x-1)}$, $b \neq 0, x \neq 1$
5. $\frac{12x}{8(x+1)}$ $\frac{3x}{2(x+1)}$, $x \neq -1$
6. $\frac{4a(x+1)}{2a}$ $2(x+1)$, $a \neq 0$
7. $\frac{x^2+5x+6}{x+3}$ $x + 2$, $x \neq -3$
8. $\frac{x^2-2x-3}{x^2-1}$ $\frac{x-3}{x-1}$, $x \neq -1, 1$
9. $\frac{x+8}{x^2+11x+24}$ $\frac{1}{x+3}$, $x = -8, -3$
10. $\frac{3a^5b^2}{18a^5b^3}$ $\frac{1}{6b}$, $a \neq 0, b \neq 0$
11. $\frac{6x^2+7x+2}{4x^2-1}$ $\frac{3x+2}{2x-1}$, $x \neq -\frac{1}{2}, \frac{1}{2}$
12. $\frac{4x-4}{7x-7}$ $\frac{4}{7}$, $x \neq 1$
13. $\frac{x+1}{x^3-x}$ $\frac{1}{x(x-1)}$, $x \neq -1, 0, 1$
14. $\frac{6x-30}{x^2-7x+10}$ $\frac{6}{x-2}$, $x \neq 2, 5$

Assignment Guide

Core: 15–35 odd, 36, 37, 39–46
Enriched: 14–34 even, 36–46

For **Extra Practice**, see p. 782.

The red A, B, and C flags, printed only in the Teacher's Wraparound Edition, indicate the level of difficulty of the exercises.

Additional Answers

14. $a \neq 0$
15. $y \neq 0, z \neq 0$
16. $a \neq 0, b \neq 0$
17. $x \neq 0, y \neq 0$
18. $a \neq -3$ or 0
19. $y \neq -\frac{1}{3}$
20. $x \neq -3$
21. $y \neq 7x$
22. $x \neq -5$ or 4
23. $a \neq 3$ or 4
24. $a \neq 2$ or 5
25. $x \neq -4$
26. $x \neq 4$ or -3
27. $a \neq -4$ or 2
28. $x \neq -6$ or 5
29. $x \neq 9$ or 4
30. $x \neq -1.5$ or -1
31. $x \neq -1$

Practice Masters, p. 80

12-1 NAME________________ DATE________

Practice — Student Edition Pages 660–665

Simplifying Rational Expressions

Simplify each rational expression. State the excluded values of the variables.

1. $\frac{12a}{48a^3}$ $\frac{1}{4a^2}$; $a \neq 0$
2. $\frac{4x^2y^2}{16yx^3}$ $\frac{y}{4x}$; $x \neq 0, y \neq 0$
3. $\frac{6abc^3}{3a^2b^2}$ $\frac{2c^3}{ab}$; $a \neq 0, b \neq 0$
4. $\frac{n+6}{3(n+6)}$ $\frac{1}{3}$; $n \neq -6$
5. $\frac{-2w^3}{w(w^2+4)}$ $\frac{-2w^2}{w^2+4}$; $w \neq 0$
6. $\frac{z+1}{(z+1)(z-1)}$ $\frac{1}{z-1}$; $z \neq 1, z \neq -1$
7. $\frac{(c+5)(c-5)}{(c-5)(c+2)}$ $\frac{c+5}{c+2}$; $c \neq 5, c \neq -2$
8. $\frac{-1(2r-3)}{(r+3)(2r-3)}$ $\frac{-1}{r+3}$; $r \neq -3, r \neq \frac{3}{2}$
9. $\frac{a^2-b^2}{a-b}$ $a+b$; $a \neq b$
10. $\frac{4x-4}{4x+4}$ $\frac{x-1}{x+1}$; $x \neq -1$
11. $\frac{3a+15}{a^2-25}$ $\frac{3}{a-5}$; $a \neq 5, a \neq -5$
12. $\frac{t^2+5t+6}{t^2+6t+8}$ $\frac{t+3}{t+4}$; $t \neq -2, t \neq -4$
13. $\frac{3s^2-27}{s^2+7s+12}$ $\frac{3(s-3)}{s+4}$; $s \neq -3, s \neq -4$
14. $\frac{7k^4}{5k^3-2k^2}$ $\frac{7k^2}{5k-2}$; $k \neq 0, k \neq \frac{2}{5}$
15. $\frac{m^2-4m-12}{m-6}$ $m+2$; $m \neq 6$
16. $\frac{v^2+3v-4}{v^2+2v-8}$ $\frac{v-1}{v-2}$; $v \neq 2, v \neq -4$
17. $\frac{5z^2+5z-30}{7z^2+7z-42}$ $\frac{5}{7}$; $z \neq -3, z \neq 2$
18. $\frac{p^2+p-2}{p^4-5p^2+4}$ $\frac{1}{(p-2)(p+1)}$; $p \neq 2, p \neq -2, p \neq 1, p \neq -1$

EXERCISES

Practice

14–31. See margin for excluded values.

Simplify each rational expression. State the excluded values of the variables.

14. $\frac{15a}{39a^2}$ $\frac{5}{13a}$
15. $\frac{35y^2z}{14yz^2}$ $\frac{5y}{2z}$
16. $\frac{28a^2}{49ab}$ $\frac{4a}{7b}$
17. $\frac{56x^2y}{70x^3y}$ $\frac{4}{5x}$
18. $\frac{4a}{3a+a^2}$ $\frac{4}{3+a}$
19. $\frac{y+3y^2}{3y+1}$ y
20. $\frac{x^2-9}{2x+6}$ $\frac{x-3}{2}$
21. $\frac{y^2-49x^2}{y-7x}$ $y+7x$
22. $\frac{x+5}{x^2+x-20}$ $\frac{1}{x-4}$
23. $\frac{a-3}{a^2-7a+12}$ $\frac{1}{a-4}$
24. $\frac{3x-15}{x^2-7x+10}$ $\frac{3}{x-2}$
25. $\frac{x+4}{x^2+8x+16}$ $\frac{1}{x+4}$
26. $\frac{x^2-2x-15}{x^2-x-12}$ $\frac{x-5}{x-4}$
27. $\frac{a^2+4a-12}{a^2+2a-8}$ $\frac{a+6}{a+4}$
28. $\frac{x^2-36}{x^2+x-30}$ $\frac{x-6}{x-5}$
29. $\frac{b^2-3b-4}{b^2-13b+36}$ $\frac{b+1}{b-9}$
30. $\frac{14x^2+35x+21}{12x^2+30x+18}$ $\frac{7}{6}$
31. $\frac{4x^2+8x+4}{5x^2+10x+5}$ $\frac{4}{5}$

Calculator

Use a calculator to evaluate each expression for the given values.

32. $\frac{x^2-x}{3x}, x=-3$ $-\frac{4}{3}$
33. $\frac{x^4-16}{x^4-8x^2+16}, x=-1$ $-\frac{5}{3}$
34. $\frac{x+y}{x^2+2xy+y^2}, x=3, y=2$ $\frac{1}{5}$
35. $\frac{x^3y^3+5x^3y^2+6x^3y}{xy^5+5xy^4+6xy^3}, x=1, y=-2$

35. not possible, $y \neq -2$

Critical Thinking

36. Explain why $\frac{m^2-16}{m+4}$ is not the same as $m-4$.

36. Because $m \neq -4$ in the original expression.

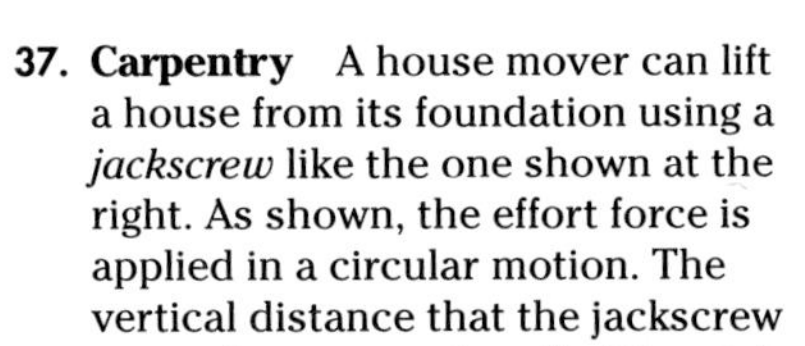

Applications and Problem Solving

37. **Carpentry** A house mover can lift a house from its foundation using a *jackscrew* like the one shown at the right. As shown, the effort force is applied in a circular motion. The vertical distance that the jackscrew moves in one turn is called its *pitch*. You can use the following formula to calculate the mechanical advantage of the jackscrew.

$$MA = \frac{\text{circumference of the circle}}{\text{pitch of the screw}}$$

Effort distance 48 in.
Effort force
Resistance distance, $\frac{1}{4}$ in.
Pitch

a. Calculate the mechanical advantage for the jackscrew shown above. 192

b. Suppose the effort force is 20 pounds. What is the greatest weight the jackscrew can lift? 3840 lb

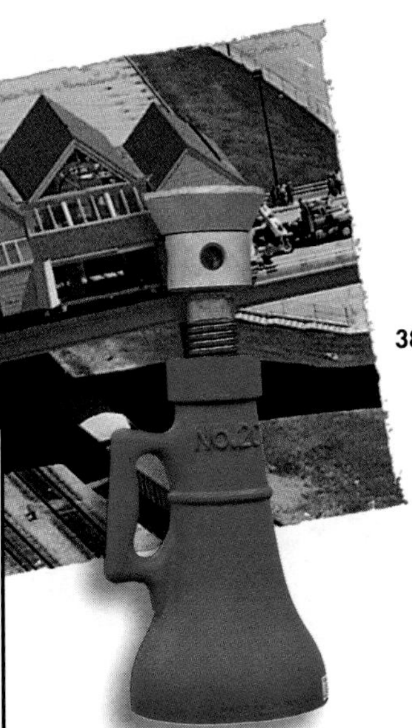

38. **Aeronautics** Aircraft engineers can use the following formula to calculate the atmospheric pressure P in pounds per square inch when an aircraft is flying at an altitude a in feet.

$$P = \frac{-9.05\left[\left(\frac{a}{1000}\right)^2 - \frac{65a}{1000}\right]}{\left(\frac{a}{1000}\right)^2 + 40\left(\frac{a}{1000}\right)}$$

a. Calculate the atmospheric pressure outside a plane flying at an altitude of 20,000 feet. 6.79 lb/in^2

b. Calculate the atmospheric pressure outside a plane flying at an altitude of 40,000 feet. 2.83 lb/in^2

c. Is your answer to part b twice that of part a? By how much do they differ? no, 3.96 lb/in^2

Extension

Connections The measure of the area of a rectangle is $6x^2 - 7x - 20$, and the measure of one side is $2x - 5$. Find the measure of the other side of the rectangle. $3x + 4$

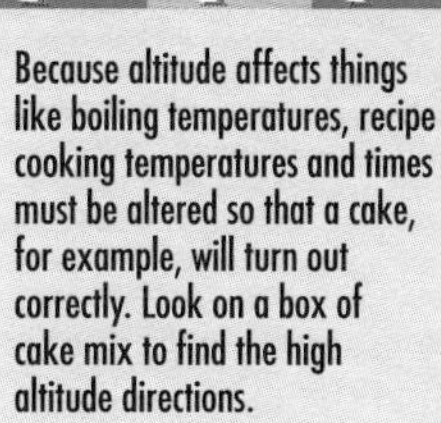

FYI

Because altitude affects things like boiling temperatures, recipe cooking temperatures and times must be altered so that a cake, for example, will turn out correctly. Look on a box of cake mix to find the high altitude directions.

39b. $-\frac{95}{10} = -\frac{19}{2}$

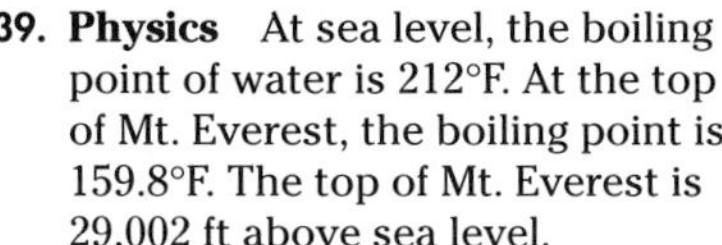

39. **Physics** At sea level, the boiling point of water is 212°F. At the top of Mt. Everest, the boiling point is 159.8°F. The top of Mt. Everest is 29,002 ft above sea level.

a. How many degrees does the boiling point drop for every mile up from sea level? Write the solution as a ratio involving units. **9.5°/mile**

b. Simplify the expression you wrote in part a.

c. Draw a graph that represents the boiling point in degrees Fahrenheit as a function of altitude in miles. **See margin.**

Mixed Review

40. **Finance** In the compound interest formula $A = P\left(1 + \frac{r}{n}\right)^{nt}$, A represents the value of the investment in the future, P is the amount of the original investment, r is the annual interest rate, t is the number of years of the investment, and n is the number of times the interest is compounded each year. Find the total amount after \$2500 is invested for 18 years at a rate of 6%, compounded quarterly. (Lesson 11–5) **\$7302.90**

41. **Number Theory** The square of a number decreased by 121 is 0. Find the number. (Lesson 10–7) **±11**

42. Find $3x(-5x^2 - 2x + 7)$. (Lesson 9–6) $-15x^3 - 6x^2 + 21x$

43. Graph the system of inequalities. (Lesson 8–5)
$y \geq x - 2$
$y \leq 2x - 1$ **See margin.**

44. Solve $4 + |x| = 12$. (Lesson 7–6) **{8, −8}**

45. Graph $y = 3x + 2$. (Lesson 5–4) **See margin.**

46. Solve $\frac{x}{4} + 7 = 6$. (Lesson 3–3) **−4**

WORKING ON THE In·ves·ti·ga·tion

Refer to the Investigation on pages 656–657.

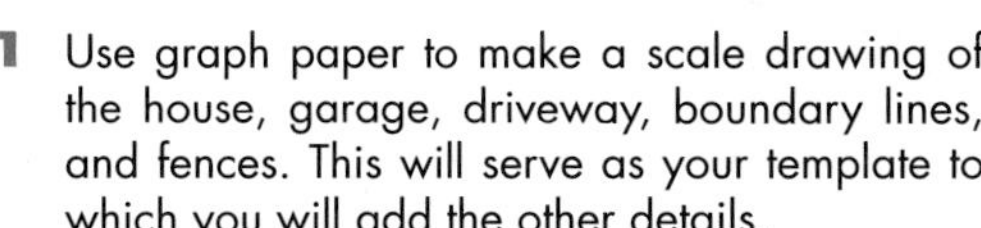

A Growing Concern

1 Use graph paper to make a scale drawing of the house, garage, driveway, boundary lines, and fences. This will serve as your template to which you will add the other details.

2 Develop a detailed plan for the pool and hot tub locations. Indicate the dimensions of both the pool and the hot tub. Explain why you chose the size, location, and orientation for the pool.

3 Figure the costs for materials, labor, and profit margin for construction of the pool and hot tub. Justify your costs and make a case for building the pool and hot tub as you have designed them.

4 Add the pool and hot tub to the scale drawing, indicating their measurements.

Add the results of your work to your Investigation Folder.

FYI

At higher altitudes water boils at lower temperatures because of the difference in pressure. For example, pasta may need to cook 7 to 10 minutes at high altitude to reach the same temperature it reaches in 6 to 8 minutes at low altitude.

In·ves·ti·ga·tion

Working on the Investigation

The Investigation on pages 656–657 is designed to be a long-term project that is completed over several days or weeks. Encourage students to keep their materials in their Investigation Folder as they work on the Investigation.

4 ASSESS

Closing Activity

Writing At the end of class, have students write down the key points to consider in simplifying rational expressions.

Additional Answers

39c.

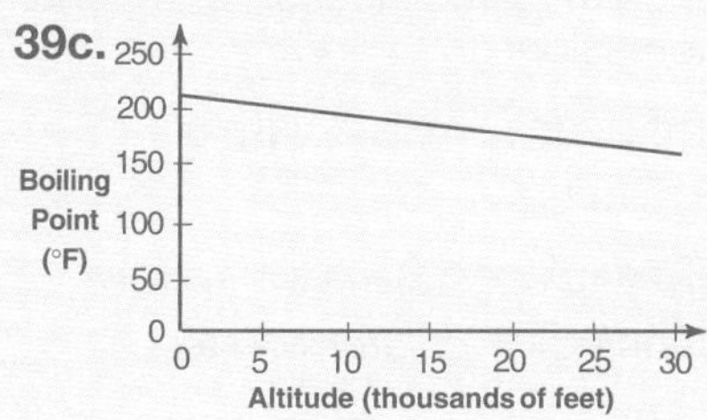

43.

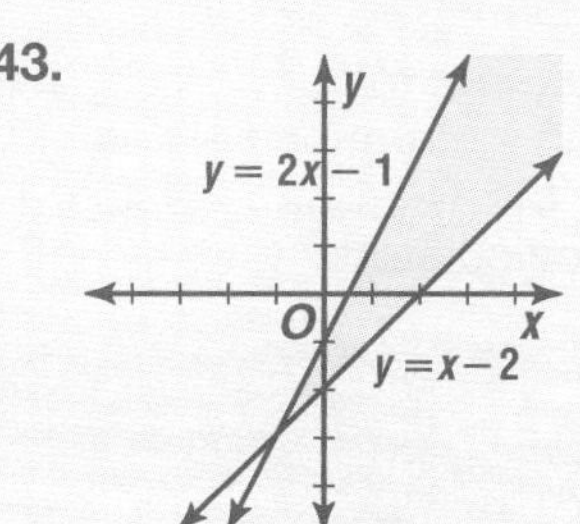

45.

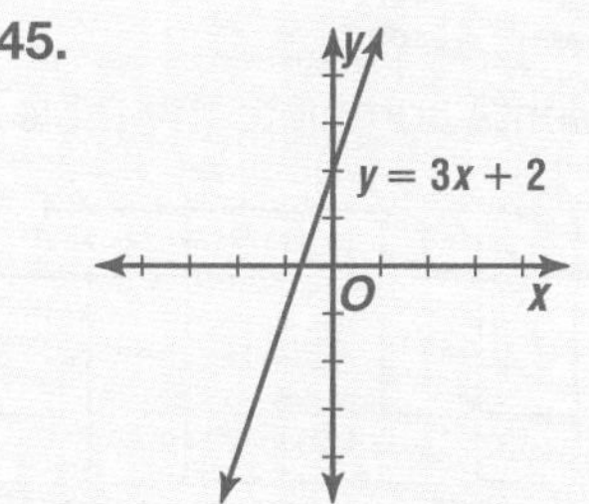

Enrichment Masters, p. 80

12-1 NAME ______ DATE ______

Enrichment Student Edition Pages 660–665

Continued Fractions

The following is an example of a continued fraction. By starting at the bottom you can simplify the expression to a rational number.

$$3 + \cfrac{4}{1 + \cfrac{6}{7}} = 3 + \cfrac{4}{\frac{13}{7}}$$
$$= 3 + \frac{28}{13} \text{ or } \frac{67}{13}$$

Example: Express $\frac{48}{19}$ as a continued fraction.

$$\frac{48}{19} = 2 + \frac{10}{19}$$

Notice that the numerator of the last fraction must be equal to 1 before the process stops.

$$= 2 + \cfrac{1}{\frac{19}{10}}$$
$$= 2 + \cfrac{1}{1 + \frac{9}{10}}$$
$$= 2 + \cfrac{1}{1 + \cfrac{1}{\frac{10}{9}}}$$
$$= 2 + \cfrac{1}{1 + \cfrac{1}{1 + \frac{1}{9}}}$$

Write each continued fraction as a rational number.

1. $6 + \cfrac{1}{1 + \cfrac{1}{3 + \frac{1}{3}}}$ $\frac{88}{13}$

2. $5 + \cfrac{7}{2 + \cfrac{3}{4 + \frac{2}{3}}}$ $\frac{283}{37}$

Write each rational number as a continued fraction.

3. $\frac{97}{17}$ $5 + \cfrac{1}{1 + \cfrac{1}{2 + \cfrac{1}{2 + \frac{1}{2}}}}$

4. $\frac{22}{65}$ $\cfrac{1}{2 + \cfrac{1}{1 + \frac{1}{21}}}$

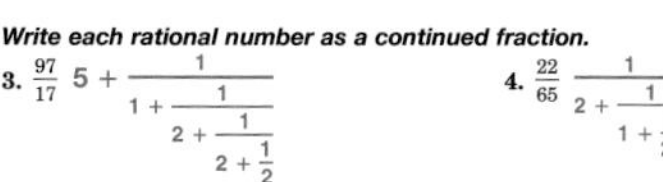

12-1B LESSON NOTES

NCTM Standards: 1–6

Objective
Use a graphing calculator to check the simplification of rational expressions.

Recommended Time
15 minutes

Instructional Resources
Graphing Calculator Masters, pp. 36 and 37

These masters provide keystroking instruction for this lesson for the TI-81 and Casio graphing calculators.

1 FOCUS

Motivating the Lesson
Have students complete the table below.

x	−1	0	1	2	3	4
$\frac{x^2-9}{x-3}$	2	3	4	5	–	7
$x+3$	2	3	4	5	6	7

2 TEACH

Teaching Tip Asymptotes are often difficult to detect on the graphing calculator. Reset the range of the viewing window and the x-scale to get a better view of a suspected asymptote.

3 PRACTICE/APPLY

Assignment Guide

Core: 1–9
Enriched: 1–9

4 ASSESS

Observing students working with technology is an excellent method of assessment.

12–1B Graphing Technology Rational Expressions

An Extension of Lesson 12–1

When simplifying rational expressions, you can use a graphing calculator to support your answer. You can also use a calculator to find the excluded values.

Example 1 **Simplify** $\frac{3x^2 - 8x + 5}{x^2 - 1}$.

$$\frac{3x^2 - 8x + 5}{x^2 - 1} = \frac{(3x-5)(x-1)}{(x+1)(x-1)} \quad \textit{Factor the numerator and denominator.}$$

$$= \frac{3x-5}{x+1} \quad \textit{Divide the common factor } (x-1).$$

When $x = -1$ or $x = 1$, $x^2 - 1 = 0$. Therefore, x cannot equal -1 or 1.

Now use a graphing calculator. Graph $y = \frac{3x^2 - 8x + 5}{x^2 - 1}$ using the window $[-4.7, 4.7]$ by $[-10, 10]$ with scale factors of 1.

Enter:

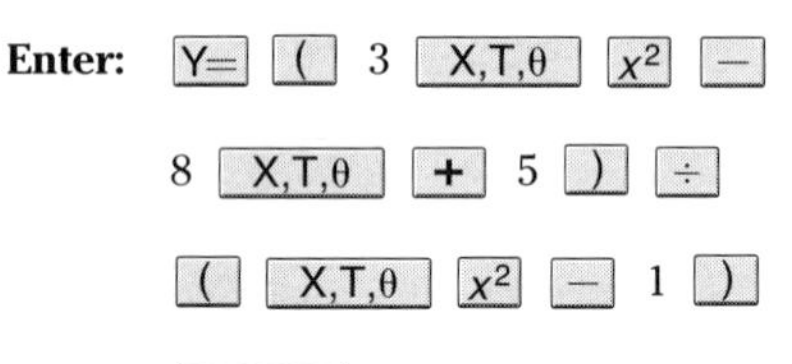

GRAPH

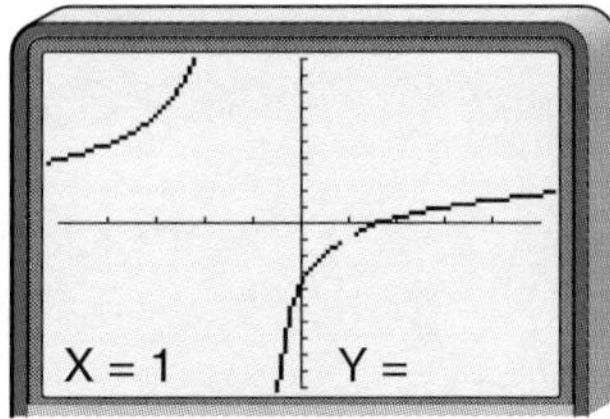

Use the TRACE key to observe the values of x and y. Notice that at $x = -1$ and $x = 1$, no value for y appears in the window. There are also "holes" in the graph at those points. These are the excluded values.

Now enter $y = \frac{3x-5}{x+1}$ as Y2 and observe the graphs.

Enter:

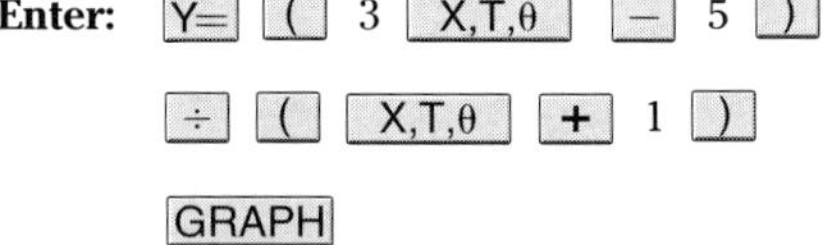

GRAPH

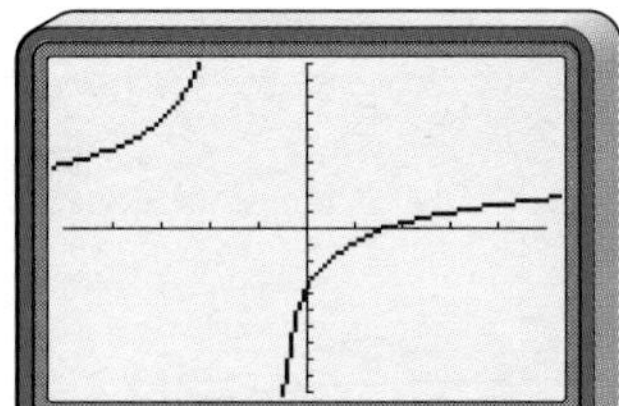

Notice that the two graphs appear to be identical. This means that the simplified expression is probably correct.

Excluded values:
1. −5
2. −2, −5
3. 0, 4.5
4. 2, −3
5. 8
6. 0, −16
7. 3
8. −1
9. −6, 5

EXERCISES

Simplify each expression. Then verify your answer graphically. Name the excluded values.

1. $\frac{x^2 - 25}{x^2 + 10x + 25}$ $\frac{x-5}{x+5}$
2. $\frac{3x + 6}{x^2 + 7x + 10}$ $\frac{3}{x+5}$
3. $\frac{2x - 9}{4x^2 - 18x}$ $\frac{1}{2x}$
4. $\frac{2x^2 - 4x}{x^2 + x - 6}$ $\frac{2x}{x+3}$
5. $\frac{x^2 - 9x + 8}{x^2 - 16x + 64}$ $\frac{x-1}{x-8}$
6. $\frac{3x^2}{12x^2 + 192x}$ $\frac{x}{4(x+16)}$
7. $\frac{-x^2 + 6x - 9}{x^2 - 6x + 9}$ -1
8. $\frac{5x^2 + 10x + 5}{3x^2 + 6x + 3}$ $\frac{5}{3}$
9. $\frac{25 - x^2}{x^2 + x - 30}$ $-\frac{x+5}{x+6}$

Using Technology
This lesson offers an excellent opportunity for using technology in your algebra classroom. For more information on using technology, see *Graphing Calculators in the Mathematics Classroom*, one of the titles in the Glencoe Mathematics Professional Series.

12-2 Multiplying Rational Expressions

What YOU'LL LEARN

- To multiply rational expressions.

Why IT'S IMPORTANT

You can use rational expressions to solve problems involving traffic and money.

APPLICATION

Traffic

On the Sunday after Thanksgiving in 1994, westbound traffic was backed up for 13 miles at one exit of the Massachusetts Turnpike. Assume that each vehicle occupied an average of 30 feet of space in a lane and that the turnpike has three lanes. You can use the expression below to estimate the number of vehicles involved in the backup.

$$3 \text{ lanes} \cdot \left(\frac{13 \text{ miles}}{\text{lane}}\right)\left(\frac{5280 \text{ feet}}{\text{mile}}\right)\left(\frac{1 \text{ vehicle}}{30 \text{ feet}}\right)$$

This expression will be simplified in Example 3.

LOOK BACK

You can refer to Lesson 2-6 for information about multiplying rational numbers.

This multiplication is similar to the multiplication of rational expressions. Recall that to multiply rational numbers that are expressed as fractions, you multiply numerators and multiply denominators. You can use this same method to multiply rational expressions. From this point on, you may assume that no denominator of a rational expression has a value of 0.

rational numbers

$$\frac{4}{5} \cdot \frac{3}{7} = \frac{4 \cdot 3}{5 \cdot 7} = \frac{12}{35}$$

rational expressions

$$\frac{3}{m} \cdot \frac{k}{4} = \frac{3 \cdot k}{m \cdot 4} = \frac{3k}{4m}$$

Example  **Find** $\frac{3x^2y}{2rs} \cdot \frac{24r^2s}{15xy^2}$.

Method 1: Divide by the common factors after multiplying.

$$\frac{3x^2y}{2rs} \cdot \frac{24r^2s}{15xy^2} = \frac{72r^2sx^2y}{30rsxy^2}$$

$$= \frac{\overset{1}{\cancel{6rsxy}}(12rx)}{\underset{1}{\cancel{6rsxy}}(5y)} \quad \text{The GCF is } 6rsxy.$$

$$= \frac{12rx}{5y}$$

Method 2: Divide by the common factors before multiplying.

$$\frac{3x^2y}{2rs} \cdot \frac{24r^2s}{15xy^2} = \frac{\overset{1\,x\,1}{\cancel{3}\cancel{x^2}\cancel{y}}}{\underset{1\,1\,1}{\cancel{2}\cancel{r}\cancel{s}}} \cdot \frac{\overset{12\,r\,1}{\cancel{24}\cancel{r^2}\cancel{s}}}{\underset{5\,1\,y}{\cancel{15}\cancel{x}\cancel{y^2}}}$$

$$= \frac{12rx}{5y}$$

Sometimes you must factor a quadratic expression before you can simplify a product of rational expressions.

12-2 LESSON NOTES

NCTM Standards: 1–5

Instructional Resources

- Study Guide Master 12-2
- Practice Master 12-2
- Enrichment Master 12-2
- Assessment and Evaluation Masters, p. 324

Transparency 12-2A contains the 5-Minute Check for this lesson; **Transparency 12-2B** contains a teaching aid for this lesson.

Recommended Pacing	
Standard Pacing	Day 3 of 14
Honors Pacing	Day 3 of 13
Block Scheduling*	Day 2 of 6 (along with Lesson 12-3)
Alg. 1 in Two Years*	Days 4, 5, & 6 of 27

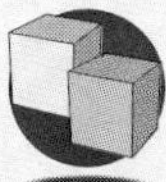

*For more information on pacing and possible lesson plans, refer to the *Block Scheduling Booklet* and *Algebra 1 in Two Years.*

1 FOCUS

5-Minute Check
(over Lesson 12-1)

Simplify each rational expression. State the excluded values of the variable.

1. $\frac{(y+4)}{(y-4)(y+4)}$

   $\frac{1}{y-4}$; **4, −4**

2. $\frac{4n^2-8}{4n-4}$

   $\frac{n^2-2}{n-1}$; **1**

3. $\frac{x^2-49}{x^2-2x-35}$

   $\frac{x+7}{x+5}$; **7, −5**

4. $\frac{m(m-1)}{m-1}$

   m; **1**

5. $\frac{y-3}{y^2-9}$

   $\frac{1}{y+3}$; **−3, 3**

Motivating the Lesson

Questioning Review the multiplication of arithmetic fractions. Have students find the following products expressed in simplest form.

1. $\frac{1}{2} \times \frac{4}{5}$ $\frac{2}{5}$
2. $\frac{2}{7} \times \frac{49}{50}$ $\frac{7}{25}$
3. $\frac{3}{8} \times \frac{20}{12}$ $\frac{5}{8}$
4. $\frac{12}{15} \times \frac{25}{16}$ $\frac{5}{4}$
5. $\frac{10}{18} \times \frac{27}{30}$ $\frac{1}{2}$

2 TEACH

In-Class Examples

For Example 1

Find $\frac{6rst^2}{5s^2t} \cdot \frac{4r^2s}{3st^2}$ $\frac{8r^3}{5st}$

For Example 2

Find each product.

a. $\frac{x+6}{2x^2} \cdot \frac{8x^4}{x^2-36}$ $\frac{4x^2}{x-6}$

b. $\frac{x^2-9}{3x-9} \cdot \frac{6x-12}{x+3}$
$2x - 4$

For Example 3

A star is measured to be 24 kiloparsecs from Earth. A kiloparsec is 1000 parsecs. A parsec is 3.26 light years. How many light years away is this star? **78,240 light years**

Study Guide Masters, p. 81

12-2 NAME__________ DATE ______

Study Guide

Student Edition Pages 667–670

Multiplying Rational Expressions

To multiply rational expressions, you multiply the numerators and multiply the denominators. Then simplify.

Multiplying Rational Expressions
For all rational numbers $\frac{a}{b}$ and $\frac{c}{d}$, where $b \neq 0$ and $d \neq 0$, $\frac{a}{b} \cdot \frac{c}{d} = \frac{ac}{bd}$.

Example 1: Find $\frac{2c^3d}{5ab^2} \cdot \frac{a^2b}{3cd}$. Assume that no denominator has a value of 0.

$$\frac{2c^3d}{5ab^2} \cdot \frac{a^2b}{3cd} = \frac{2c^3d \cdot a^2b}{5ab^2 \cdot 3cd} = \frac{2a^2bc^3d}{15ab^2cd} = \frac{2ac}{15b}$$

You can often find products of rational expressions by factoring.

Example 2: Find $\frac{y^3-4y}{y^2-4y+4} \cdot \frac{y^2+3y-10}{y}$. Assume that no denominator has a value of 0.

$$\frac{y^3-4y}{y^2-4y+4} \cdot \frac{y^2+3y-10}{y} = \frac{\overset{1}{\cancel{y}}\overset{1}{\cancel{(y-2)}}(y+2)}{\underset{1}{\cancel{(y-2)}}\underset{1}{\cancel{(y-2)}}} \cdot \frac{(y+5)\overset{1}{\cancel{(y-2)}}}{\underset{1}{\cancel{y}}} = (y+2)(y-5)$$

Find each product. Assume that no denominator has a value of 0.

1. $\frac{mn^2}{3} \cdot \frac{4}{mn}$ $\frac{4n}{3}$
2. $\frac{x+4}{x-4} \cdot \frac{1}{x+4}$ $\frac{1}{x-4}$
3. $\frac{m-5}{8} \cdot \frac{16}{m-5}$ 2
4. $\frac{5x+5y}{3x^2} \cdot \frac{xy^3}{15x+15y}$ $\frac{y^3}{9x}$
5. $\frac{8x^3y}{3x^2} \cdot \frac{9y}{4x^2}$ $\frac{6y^2}{x}$
6. $\frac{a-5}{a+2} \cdot \frac{a^2-4}{a-5}$ $a - 2$
7. $\frac{16x+8}{x^2-2x+1} \cdot \frac{x-1}{2x+2}$ $\frac{4(2x+1)}{x^2-1}$, or $\frac{8x+4}{x^2-1}$
8. $\frac{x^2-16}{2x+8} \cdot \frac{x+4}{x^2+8x+16}$ $\frac{x^2-16}{2(x+4)^2}$, or $\frac{x^2-16}{2x^2+16x+32}$
9. $\frac{v^2-4v-21}{3v^2+6v} \cdot \frac{v^2+8v}{v^2+11v+24}$ $\frac{v-7}{3(v+2)}$, or $\frac{v-7}{3v+6}$
10. $\frac{2b-2c}{10bc} \cdot \frac{5bc}{3b-3c}$ $\frac{1}{3}$
11. $\frac{3p-3q}{10pq} \cdot \frac{20p^2q^2}{p^2-q^2}$ $\frac{6pq}{p+q}$
12. $\frac{a^2+7a+12}{a^2+2a-8} \cdot \frac{a^2+3a-10}{a^2+2a-8}$ $\frac{(a+3)(a+5)}{(a-2)(a+4)}$, or $\frac{a^2+8a+15}{a^2+2a-8}$

Example 2 Find each product.

a. $\frac{m+3}{5m} \cdot \frac{20m^2}{m^2+8m+15}$

$$\frac{m+3}{5m} \cdot \frac{20m^2}{m^2+8m+15} = \frac{m+3}{5m} \cdot \frac{20m^2}{(m+3)(m+5)} \quad \textit{Factor the denominator.}$$

$$= \frac{\overset{4}{\cancel{20}}\overset{m}{\cancel{m^2}}\overset{1}{\cancel{(m+3)}}}{\underset{1}{\cancel{5}}\underset{1}{\cancel{m}}\underset{1}{\cancel{(m+3)}}(m+5)} \quad \textit{The GCF is } 5m(m+3).$$

$$= \frac{4m}{m+5}$$

b. $\frac{3(x+5)}{4(2x^2+11x+12)} \cdot (2x+3)$

$$\frac{3(x+5)}{4(2x^2+11x+12)} \cdot (2x+3) = \frac{3(x+5)(2x+3)}{4(2x^2+11x+12)}$$

$$= \frac{3(x+5)(2x+3)}{4(2x+3)(x+4)} \quad \textit{Factor the denominator.}$$

$$= \frac{3(x+5)\overset{1}{\cancel{(2x+3)}}}{4\underset{1}{\cancel{(2x+3)}}(x+4)} \quad \textit{The GCF is } (2x+3).$$

$$= \frac{3(x+5)}{4(x+4)}$$

$$= \frac{3x+15}{4x+16} \quad \textit{Simplify the numerator and the denominator.}$$

When you multiply fractions that involve units of measure, you can divide by the units in the same way that you divide by variables. Recall that this process is called *dimensional analysis*.

Traffic

Example 3 Refer to the application at the beginning of the lesson. About how many vehicles were involved in the backup?

$$3 \text{ lanes} \cdot \left(\frac{13 \text{ miles}}{\text{lane}}\right)\left(\frac{5280 \text{ feet}}{\text{mile}}\right)\left(\frac{1 \text{ vehicle}}{30 \text{ feet}}\right)$$

$$= \frac{\overset{1}{3 \cancel{\text{ lanes}}}}{1} \cdot \left(\frac{13 \cancel{\text{ miles}}}{\cancel{\text{lane}}}\right)\left(\frac{\overset{528}{\cancel{5280 \text{ feet}}}}{\cancel{\text{mile}}}\right)\left(\frac{1 \text{ vehicle}}{\underset{1}{\cancel{30 \text{ feet}}}}\right) \quad \textit{Divide by common units.}$$

$$= 6864 \text{ vehicles}$$

The backup involved about 6900 vehicles.

CHECK FOR UNDERSTANDING

Communicating Mathematics

Study the lesson. Then complete the following.

1. **Write** two rational expressions whose product is $\frac{xy}{x-4}$. $\frac{x}{x-4}$, y
2. **You Decide** The work of two students who simplified $\frac{4x+8}{4x-8} \cdot \frac{x^2+2x-8}{2x+8}$ is shown below. Which one is correct, and why? **2. Angie; she divided out the GCF.**

Matt: $\frac{\overset{1}{\cancel{4x}}+\overset{1}{\cancel{8}}}{\underset{1}{\cancel{4x}}-\underset{1}{\cancel{8}}} \cdot \frac{x^2+\overset{1}{\cancel{2x}}-\overset{1}{\cancel{8}}}{\underset{1}{\cancel{2x}}+\underset{1}{\cancel{8}}} = x^2$

Angie: $\frac{\overset{1}{\cancel{4}}(x+2)}{\underset{1}{\cancel{4}}\underset{1}{\cancel{(x-2)}}} \cdot \frac{\overset{1}{\cancel{(x-2)}}\overset{1}{\cancel{(x+4)}}}{2\underset{1}{\cancel{(x+4)}}} = \frac{x+2}{2}$

3. Write a paragraph explaining which is better—to simplify an expression first and then multiply, or to multiply and then simplify. Include examples to support your answers. **See students' work.**

Teaching Tip Stress that each polynomial must be factored and then simplified.

Teaching Tip Caution students to write the negative sign before the fraction bar and not with the numerator. For example,

$\frac{-x+4}{x+2}$ is not the same as $-\frac{x+4}{x+2}$.

Guided Practice

6. $\frac{r}{y+3}$

Find each product. Assume that no denominator has a value of 0.

4. $\frac{12m^2y}{5r^2} \cdot \frac{r^2}{my}$ $\frac{12m}{5}$

5. $\frac{16xy^2}{3m^2p} \cdot \frac{27m^3p}{32x^2y^3}$ $\frac{9m}{2xy}$

6. $\frac{x-5}{r} \cdot \frac{r^2}{(x-5)(y+3)}$

7. $\frac{x+4}{4y} \cdot \frac{16y}{x^2+7x+12}$ $\frac{4}{x+3}$

8. $\frac{a^2+7a+10}{a+1} \cdot \frac{3a+3}{a+2}$ $3(a+5)$

9. $\frac{4(x-7)}{3(x^2-10x+21)} \cdot (x-3)$ $\frac{4}{3}$

10. $3(x+6) \cdot \frac{x+3}{9(x^2+7x+6)}$ $\frac{x+3}{3(x+1)}$

11. a. Multiply $\frac{2.54 \text{ centimeters}}{1 \text{ inch}} \cdot \frac{12 \text{ inches}}{1 \text{ foot}} \cdot \frac{3 \text{ feet}}{1 \text{ yard}}$. Then simplify. **91.44 cm/yd**
   b. What does the simplified expression represent?

12. **Bicycling** Marisa says that bicycling at the rate of 20 miles per hour is equivalent to a rate of 1760 feet per minute. Is she correct? Write a product involving units of measure to justify your answer.

EXERCISES

Practice

Find each product. Assume that no denominator has a value of 0.

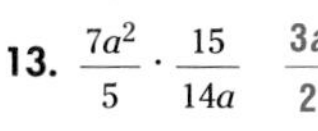

13. $\frac{7a^2}{5} \cdot \frac{15}{14a}$ $\frac{3a}{2}$

14. $\frac{3m^2}{2m} \cdot \frac{18m^2}{9m}$ $3m^2$

15. $\frac{10r^3}{6x^3} \cdot \frac{42x^2}{35r^3}$ $\frac{2}{x}$

16. $\frac{7ab^3}{11r^2} \cdot \frac{44r^3}{21a^2b}$ $\frac{4b^2r}{3a}$

17. $\frac{64y^2}{5y} \cdot \frac{5y}{8y}$ $8y$

18. $\frac{2a^2}{b} \cdot \frac{5bc}{6a}$ $\frac{5ac}{3}$

11b. changing centimeters to yards

19. $\frac{m+4}{3m} \cdot \frac{4m^2}{m^2+9m+20}$ $\frac{4m}{3(m+5)}$

20. $\frac{m^2+8m+15}{a+b} \cdot \frac{7a+14b}{m+3}$

12. yes; $\frac{20 \text{ mi}}{1 \text{ h}} \cdot \frac{5280 \text{ ft}}{1 \text{ mi}} \cdot \frac{1 \text{ h}}{60 \text{ min}}$

21. $\frac{5a+10}{10m^2} \cdot \frac{4m^3}{a^2+11a+18}$ $\frac{2m}{a+9}$

22. $\frac{6r+3}{r+6} \cdot \frac{r^2+9r+18}{2r+1}$ $3(r+3)$

23. $2(x+1) \cdot \frac{x+4}{x^2+5x+4}$ 2

24. $4(a+7) \cdot \frac{12}{3(a^2+8a+7)}$ $\frac{16}{a+1}$

20. $\frac{7(a+2b)(m+5)}{a+b}$

25. $\frac{x^2-y^2}{12} \cdot \frac{36}{x+y}$ $3(x-y)$

26. $\frac{3a+9}{a} \cdot \frac{a^2}{a^2-9}$ $\frac{3a}{a-3}$

27. $\frac{9}{3+2x} \cdot (12+8x)$ 36

28. $(3x+3) \cdot \frac{x+4}{x^2+5x+4}$ 3

30. $b+1$

29. $\frac{4x}{9x^2-25} \cdot (3x+5)$ $\frac{4x}{3x-5}$

30. $(b^2+12b+11) \cdot \frac{b+9}{b^2+20b+99}$

31. $\frac{4x+8}{x^2-25} \cdot \frac{x-5}{5x+10}$ $\frac{4}{5(x+5)}$

32. $\frac{a^2-a-6}{a^2-9} \cdot \frac{a^2+7a+12}{a^2+4a+4}$ $\frac{a+4}{a+2}$

Multiply. Explain what each expression represents.

33. $\frac{32 \text{ feet}}{1 \text{ second}} \cdot \frac{60 \text{ seconds}}{1 \text{ minute}} \cdot \frac{60 \text{ minutes}}{1 \text{ hour}} \cdot \frac{1 \text{ mile}}{5280 \text{ feet}}$ **21.8 mph; converts ft/s to mph**

34. $10 \text{ feet} \cdot 18 \text{ feet} \cdot 3 \text{ feet} \cdot \frac{1 \text{ yard}^3}{27 \text{ feet}^3}$ **20 yd³; converts ft³ to yd³**

Simplify each expression.

35. **Heat** 20 grams at 540 Calories per gram **10,800 Calories**

36. 40.25 grams

36. **Metal Refining** 4.025 grams per amp-hour at 2 amps for 5 hours

Critical Thinking

37. Find two different pairs of rational expressions whose product is $\frac{6x^2-6x-36}{x^2+3x-28}$. Sample answer: $\frac{3(x+2)}{x+7} \cdot \frac{2(x-3)}{x-4}$; $\frac{6}{x+7} \cdot \frac{x^2-x-1}{x-4}$

3 PRACTICE/APPLY

Check for Understanding

Exercises 1–12 are designed to help you assess your students' understanding through reading, writing, speaking, and modeling. You should work through Exercises 1–3 with your students and then monitor their work on Exercises 4–12.

Error Analysis

When all the factors in the numerator or denominator cancel, students may indicate the value of the expression as 0, rather than 1. For example,

$\frac{4x(x-3)}{8x^2(x-3)} = \frac{0}{2x}$ rather than $\frac{1}{2x}$.

Remind students to use small numbers when canceling to show that $\frac{a}{a} = 1$.

Assignment Guide

Core: 13–39 odd, 40–46
Enriched: 14–36 even, 37–46

For **Extra Practice**, see p. 782.

The red A, B, and C flags, printed only in the Teacher's Wraparound Edition, indicate the level of difficulty of the exercises.

Practice Masters, p. 81

NAME______________________ DATE ________

Practice

Student Edition Pages 667–670

Multiplying Rational Expressions

Find each product. Assume that no denominator has a value of 0.

1. $\frac{2a^2b}{b^2c} \cdot \frac{b}{a}$ $\frac{2a}{c}$

2. $\frac{18x^2}{10y^2} \cdot \frac{15y^3}{24x}$ $\frac{9xy}{8}$

3. $\frac{24st^2}{8s^4t^3} \cdot \frac{12s^3t^2}{36s^2t}$ $\frac{1}{s^2}$

4. $\frac{12a^2b}{4} \cdot \frac{4a+8b}{20a^2b^3}$ $\frac{3a+6b}{5b^2}$

5. $\frac{12m-18}{18n} \cdot \frac{9n^2}{8m-12}$ $\frac{3n}{4}$

6. $\frac{y^2-9}{4} \cdot \frac{8}{y+3}$ $2y-6$

7. $\frac{(x+2)^2}{8} \cdot \frac{72}{x^2-4}$ $\frac{9x+18}{x-2}$

8. $\frac{2a^2+a}{4a^2-1} \cdot \frac{6a-3}{4a}$ $\frac{3}{4}$

9. $\frac{5n+15}{8n+4} \cdot \frac{4n+2}{3n+9}$ $\frac{5}{6}$

10. $\frac{e^2-1}{2e-6} \cdot \frac{e^2-9}{3e-3}$ $\frac{e^2+4e+3}{6}$

11. $\frac{k^2-4}{8k^2+3k} \cdot \frac{16k+6}{k-2}$ $\frac{2k+4}{k}$

12. $\frac{4r+r^2}{8+2r} \cdot \frac{4}{2r^3}$ $\frac{1}{r^2}$

13. $\frac{25-c^2}{12} \cdot \frac{4}{5-c}$ $\frac{5+c}{3}$

14. $\frac{xy}{3x-3y} \cdot \frac{x^2-xy}{xy}$ $\frac{x}{3}$

15. $\frac{2c^2-5c-3}{c+d} \cdot \frac{c^2-d^2}{2c+1}$ $c^2-cd-3c+3d$

16. $\frac{t^2+6t+9}{t^2-10t+25} \cdot \frac{t^2-t-20}{t^2+7t+12}$ $\frac{t+3}{t-5}$

Reteaching

Using Models Have students choose values for a and b and complete the table below. The ratio of opposites simplifies to -1.

a	b	$a-b$	$b-a$	$\frac{a-b}{b-a}$
1	4	−3	3	−1
3	−2	5	−5	−1

Pick $a = b$ to see excluded values.

4 ASSESS

Closing Activity

Modeling Use the formula $V = \ell wh$ to represent the volume of the rectangular solid shown. Write the expression in simplest form.
$\frac{x^2 + 9x + 14}{2x^2 + x}$

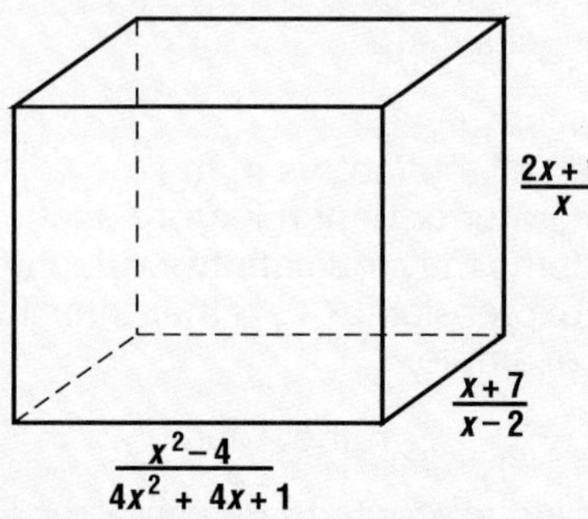

Chapter 12, Quiz A (Lessons 12-1 and 12-2), is available in the *Assessment and Evaluation Masters*, p. 324.

Additional Answer

38a. $\frac{24 \text{ sec}}{1 \text{ car}} \cdot \frac{6864 \text{ cars}}{8 \text{ collectors}} \cdot \frac{1 \text{ min}}{60 \text{ sec}} \cdot \frac{1 \text{h}}{60 \text{ min}}$

Enrichment Masters, p. 81

12-2 NAME__________ DATE______

Enrichment

Student Edition Pages 667–670

Surface Area of Solid Figures

Many solid objects are formed by rectangles and squares. A box is an example.

The dimensions of the box shown at the right are represented by letters. The length of the base is ℓ units, its width is w units, and the height of the box is h units.

Suppose the box is cut on the seams so that it can be spread out on a flattened surface as shown at the right. The area of this figure is the surface area of the box. Find a formula for the surface area of the box.

There are 6 rectangles in the figure. The surface area is the sum of the areas of the 6 rectangles.

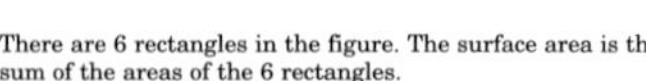

$$S = hw + h\ell + \ell w + h\ell + hw + \ell w$$
$$S = 2\ell w + 2h\ell + 2hw$$

Find the surface area of a box with the given dimensions.

1. $\ell = 14$ cm, $w = 8$ cm, $h = 2$ cm
312 cm^2

2. $\ell = 40$ cm, $w = 30$ cm, $h = 25$ cm
5900 cm^2

3. $\ell = x$ cm, $w = (x - 3)$ cm, $h = (x + 3)$ cm $6(x^2 - 3)$ cm^2

4. $\ell = (s + 9)$ cm, $w = (s - 9)$ cm, $h = (s + 9)$ cm
$(6s^2 + 36s - 162)$ cm^2

5. The surface area of a box is 142 cm^2. The length of the base is 2 cm longer than its width. The height of the box is 2 cm less than the width of the base. Find the dimensions of the box.
$h = x = 3$ cm; $w = x + 2 = 5$ cm; $\ell = x + 4 = 7$ cm

6. Write an expression that represents the surface area of the figure shown at the right. Include the surface area of the base. $22x^2$

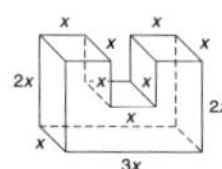

Applications and Problem Solving

38. **Traffic** Refer to Example 3. Suppose there are eight toll collectors at the exit and it takes each an average of 24 seconds to collect the toll from one vehicle. **a. See margin.**
  a. Write a product involving units to estimate the time it would take in hours to collect tolls from all the vehicles in the backup.
  b. Simplify the product.

38b. 5.72 hours

39. **Money** The chart at the right shows a recent foreign exchange rate table. It indicates how much of another country's currency you would receive in exchange for one American dollar at the time these rates were in effect.

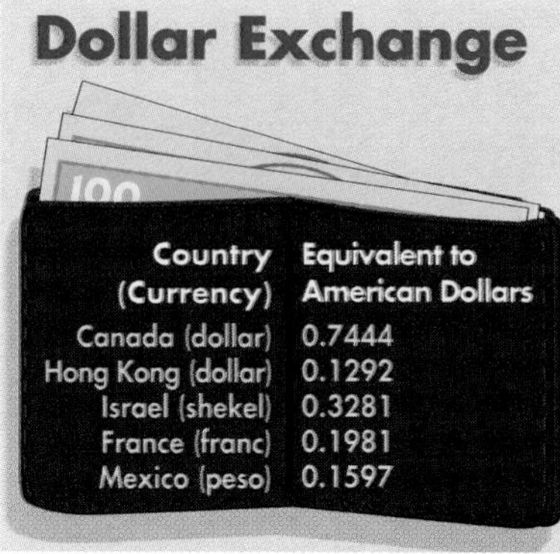

Country (Currency)	Equivalent to American Dollars
Canada (dollar)	0.7444
Hong Kong (dollar)	0.1292
Israel (shekel)	0.3281
France (franc)	0.1981
Mexico (peso)	0.1597

  a. Write a rational expression involving units of measure that indicates how many American dollars you would receive for one French franc. Write another expression that indicates how many French francs you would receive for one American dollar.
  b. Write a product involving units of measure that indicates how many French francs you would receive for 12,500 Mexican pesos. Then find the product. **about 10,077 francs**
  c. Explain how to determine an exchange rate between Canada's dollar and Hong Kong's dollar. **Convert to American dollars and then to Hong Kong's dollar.**

39a. $\frac{1 \text{ franc}}{0.1981 \text{ dollars}}$; $\frac{1 \text{ dollar}}{5.05 \text{ francs}}$

39b. 12,500 pesos $\cdot \frac{0.1597 \text{ dollars}}{1 \text{ peso}} \cdot \frac{1 \text{ franc}}{0.1981 \text{ dollars}}$

Mixed Review

40. **Cooking** The formula $t = \frac{40(25 + 1.85a)}{50 - 1.85a}$ relates the time t in minutes that it takes to bake an average-size potato in an oven that is at an altitude of a thousands of feet. (Lesson 12–1)
  a. What is the value of a for an altitude of 4500 feet? **4.5**
  b. Calculate the time is takes to bake a potato at an altitude of 3500 feet.
  c. Calculate the time it takes to bake a potato at an altitude of 7000 feet.
  d. The altitude in part c is twice that of part b. How do your baking times compare for those two altitudes?

40b. about 29 minutes

40c. about 41 minutes

40d. The times are not doubled; the difference is 12 minutes.

41. Find the value of y for $y = 3^{2x}$ if $x = 3$. (Lesson 11–4) **729**

42. Solve $5x^2 + 30x + 45 = 0$ by factoring. (Lesson 10–6) **−3**

43. Find the degree of $6x^2yz + 5xyz - x^3$. (Lesson 9–4) **4**

44. Solve the system of equations. (Lesson 8–3)
$5 = 2x - 3y$
$-1 = -4x + 3y$ **(−2, −3)**

45. Find the slope of the line that passes through (−3, 5) and (8, 5). (Lesson 5–1)

45. 0

46. What is 45% of \$1567 to the nearest dollar? (Lesson 4–4) **\$705**

Tech Prep

Traffic Technician Students who are interested in traffic control may wish to do further research on the data given in Exercise 38 and explore the potential growth of this career. For more information on tech prep, see the *Teacher's Handbook.*

Extension

Connections Latest estimates suggest that the known universe is about 14 billion light years across. One light year equals 5.9×10^{12} miles. The top speed of our present-day spacecrafts is 25,000 mph. With this spacecraft, how long would it take to travel across the universe?
376,000,000,000,000 or 3.76×10^{14} or 376 trillion years

12-3 Dividing Rational Expressions

What YOU'LL LEARN
- To divide rational expressions.

Why IT'S IMPORTANT

You can use rational expressions to solve problems involving construction and railroads.

APPLICATION
Parades

A crowd watching the Tournament of Roses Parade in Pasadena, California, fills the sidewalks along the route for about 5.5 miles on each side. Suppose these sidewalks are 10 feet wide and each person occupies an average of 4 square feet of space. To estimate the number of people along this part of the route, you can use dimensional analysis and the following expression.

$\left(5.5 \text{ miles} \cdot \frac{5280 \text{ ft}}{1 \text{ mile}} \cdot 10 \text{ feet}\right) \div \frac{4 \text{ ft}^2}{1 \text{ person}}$ *This expression will be simplified in Example 4.*

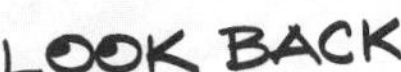

LOOK BACK

You can refer to Lesson 2-7 for information about dividing rational numbers.

This division is similar to the division of rational expressions. Recall that to divide rational numbers that are expressed as fractions you multiply by the reciprocal of the divisor. You can use this same method to divide algebraic rational expressions.

rational numbers

$\frac{3}{2} \div \frac{2}{5} = \frac{3}{2} \cdot \frac{5}{2}$ *The reciprocal of $\frac{2}{5}$ is $\frac{5}{2}$.*

$= \frac{15}{4}$

rational expressions

$\frac{a}{b} \div \frac{c}{d} = \frac{a}{b} \cdot \frac{d}{c}$ *The reciprocal of $\frac{c}{d}$ is $\frac{d}{c}$.*

$= \frac{ad}{bc}$

Example 1 Find $\frac{2a}{a+3} \div \frac{a+7}{a+3}$.

$\frac{2a}{a+3} \div \frac{a+7}{a+3} = \frac{2a}{a+3} \cdot \frac{a+3}{a+7}$ *The reciprocal of $\frac{a+7}{a+3}$ is $\frac{a+3}{a+7}$.*

$= \frac{2a}{\cancel{a+3}_1} \cdot \frac{\overset{1}{\cancel{a+3}}}{a+7}$ *Divide by the common factor $a + 3$.*

$= \frac{2a}{a+7}$

Sometimes a quotient of rational expressions involves a divisor that is a binomial.

Example 2 Find $\frac{2m+6}{m+5} \div (m+3)$.

$\frac{2m+6}{m+5} \div (m+3) = \frac{2m+6}{m+5} \cdot \frac{1}{m+3}$ *The reciprocal of $m + 3$ is $\frac{1}{m+3}$.*

$= \frac{2\overset{1}{\cancel{(m+3)}}}{(m+5)\underset{1}{\cancel{(m+3)}}}$ *Divide by the common factor $m + 3$.*

$= \frac{2}{m+5}$

Sometimes you must factor before you can simplify a quotient of rational expressions.

12-3 LESSON NOTES

NCTM Standards: 1–5

Instructional Resources

- Study Guide Master 12-3
- Practice Master 12-3
- Enrichment Master 12-3

Transparency 12-3A contains the 5-Minute Check for this lesson; **Transparency 12-3B** contains a teaching aid for this lesson.

Recommended Pacing	
Standard Pacing	Day 4 of 14
Honors Pacing	Day 4 of 13
Block Scheduling*	Day 2 of 6 (along with Lesson 12-2)
Alg. 1 in Two Years*	Days 7, 8, & 9 of 27

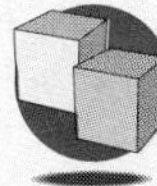

*For more information on pacing and possible lesson plans, refer to the *Block Scheduling Booklet* and *Algebra 1 in Two Years.*

1 FOCUS

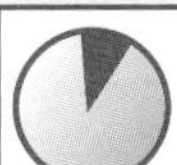

5-Minute Check
(over Lesson 12-2)

Find each product. Assume that no denominator has a value of 0.

1. $\frac{3}{5} \cdot \frac{35}{99}$ $\frac{7}{33}$
2. $\frac{7a}{9a} \cdot \frac{3a^2}{6b^2}$ $\frac{7a^2}{18b^2}$
3. $\frac{a^2b}{b^3c} \cdot \frac{ac}{a^3}$ $\frac{1}{b^2}$
4. $\frac{5x-5}{4} \cdot \frac{12}{x-1}$ 15
5. $\frac{w-5}{w^2-7w+10} \cdot \frac{w-2}{12}$ $\frac{1}{12}$

Motivating the Lesson

Questioning Review dividing fractions. Ask students to find each quotient in simplest form.

1. $\frac{2}{3} \div \frac{8}{9}$ $\frac{3}{4}$
2. $\frac{3}{4} \div \frac{5}{6}$ $\frac{9}{10}$
3. $\frac{10}{21} \div \frac{25}{42}$ $\frac{4}{5}$
4. $\frac{5}{8} \div \frac{1}{60}$ $\frac{75}{2}$

2 TEACH

In-Class Examples

For Example 1

Find $\frac{3x}{x+2} \div \frac{3x^2-3x}{x+2}$. $\frac{1}{x-1}$

For Example 2

Find $\frac{3a}{a+2} \div (a-2)$. $\frac{3a}{a^2-4}$

For Example 3

Find $\frac{y+5}{y^2-4y-5} \div \frac{y^2+4y-5}{y+1}$. $\frac{1}{y^2-6y+5}$

For Example 4

If the sidewalks are 18 feet wide and each person occupies an average of 3 square feet of space, estimate the number of people along this part of the parade route. **174,240 people**

Teaching Tip For Example 4, you may want to stress the idea of reciprocals and their importance in the division of rational expressions.

Study Guide Masters, p. 82

NAME________________ DATE________

Study Guide — Student Edition Pages 671–674

Dividing Rational Expressions

You should recall that two numbers whose product is 1 are called **multiplicative inverses** or **reciprocals.** To find the quotient of two rational expressions, you multiply the first rational expression by the reciprocal of the second rational expression.

Dividing Rational Expressions
For all rational numbers $\frac{a}{b}$ and $\frac{c}{d}$, where $b \neq 0$, $c \neq 0$ and $d \neq 0$, $\frac{a}{b} \div \frac{c}{d} = \frac{a}{b} \cdot \frac{d}{c}$.

Example 1: Find $\frac{2}{x} \div \frac{z}{y}$.

$$\frac{2}{x} \div \frac{z}{y} = \frac{2}{x} \cdot \frac{y}{z} = \frac{2y}{xz}$$

Example 2: Find $\frac{x^2+6x-27}{x^2+11x+18} \div \frac{x-3}{x^2+x-2}$.

$$\frac{x^2+6x-27}{x^2+11x+18} \div \frac{x-3}{x^2+x-2} = \frac{x^2+6x-27}{x^2+11x+18} \cdot \frac{x^2+x-2}{x-3}$$
$$= \frac{(x+9)(x-3)}{(x+9)(x+2)} \cdot \frac{(x+2)(x-1)}{x-3}$$
$$= x - 1$$

Find each quotient. Assume that no denominator has a value of 0.

1. $\frac{5}{6} \div \frac{1}{6}$ 5
2. $1 \div \frac{1}{3}$ 3
3. $\frac{n}{4} \div \frac{n}{m}$ $\frac{m}{4}$
4. $\frac{6}{11} \div \frac{12}{7x}$ $\frac{7x}{22}$
5. $\frac{x}{y} \div \frac{2}{y}$ $\frac{x}{2}$
6. $\frac{a+b}{a-b} \div (a^2+2ab+b^2)$ $\frac{1}{a^2-b^2}$
7. $\frac{3x^2y}{8} \div 6xy$ $\frac{x}{16}$
8. $\frac{y^2-36}{y^2-49} \div \frac{y+6}{y+7}$ $\frac{y-6}{y-7}$
9. $\frac{x^2-5x+6}{5} \div \frac{x-3}{15}$ $3(x-2)$
10. $\frac{49-m^2}{m} \div \frac{m^2-13m+42}{3m^2}$ $\frac{-3m(m+7)}{m-6}$, or $\frac{-3m^2-21m}{m-6}$
11. $\frac{3a^3-27a}{2a^2+13a-7} \div \frac{9a^2}{4a^2-1}$ $\frac{(a^2-9)(2a+1)}{3a(a+7)}$, or $\frac{2a^3+a^2-18a-9}{3a^2+21a}$
12. $\frac{p^2-5p+6}{p^2+3p} \div \frac{3-p}{4p+12}$ $\frac{4(2-p)}{p}$, or $\frac{8-4p}{p}$

Example 3 Find $\frac{x}{x+2} \div \frac{x^2}{x^2+5x+6}$.

$$\frac{x}{x+2} \div \frac{x^2}{x^2+5x+6} = \frac{x}{x+2} \cdot \frac{x^2+5x+6}{x^2}$$
$$= \frac{x}{x+2} \cdot \frac{(x+2)(x+3)}{x^2} \quad \text{Divide by common factors.}$$
$$= \frac{x+3}{x}$$

Example 4 **APPLICATION Parades**

Refer to the application at the beginning of the lesson. Estimate the number of people along the part of the parade route that is described.

$$\left(5.5 \text{ miles} \cdot \frac{5280 \text{ ft}}{1 \text{ mile}} \cdot 10 \text{ feet}\right) \div \frac{4 \text{ ft}^2}{1 \text{ person}}$$
$$= \left(5.5 \text{ miles} \cdot \frac{5280 \text{ ft}}{1 \text{ mile}} \cdot 10 \text{ feet}\right) \cdot \frac{1 \text{ person}}{4 \text{ ft}^2} \quad \text{The reciprocal of } \frac{4 \text{ ft}^2}{1 \text{ person}} \text{ is } \frac{1 \text{ person}}{4 \text{ ft}^2}.$$
$$= \frac{5.5 \text{ miles}}{1} \cdot \frac{5280 \text{ ft}}{1 \text{ mile}} \cdot \frac{10 \text{ ft}}{1} \cdot \frac{1 \text{ person}}{4 \text{ ft}^2} \quad \text{Divide by common units.}$$
$$= 72{,}600 \text{ people}$$

This means that there were about 72,600 people along *each side*. Since there were two sides, the total number of people along the route was about 2(72,600), or 145,200.

CHECK FOR UNDERSTANDING

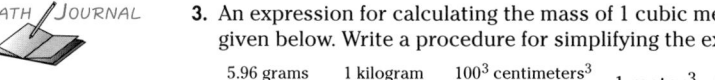

Communicating Mathematics

Study the lesson. Then complete the following.

1. **Analyze** the solution shown below.

$$\frac{a^2-9}{3a} \div \frac{a+3}{a-3} = \frac{3a}{a^2-9} \cdot \frac{a+3}{a-3}$$
$$= \frac{3a}{(a+3)(a-3)} \cdot \frac{(a+3)}{a-3}$$
$$= \frac{3a}{(a-3)^2}$$

What error was made? **The wrong reciprocal was used.**

2. **Write** two rational expressions whose quotient is $\frac{10r}{d^2}$. $\frac{10}{d}$, $\frac{d}{r}$

3. An expression for calculating the mass of 1 cubic meter of a substance is given below. Write a procedure for simplifying the expression.

$$\frac{5.96 \text{ grams}}{\text{centimeter}^3} \cdot \frac{1 \text{ kilogram}}{1000 \text{ grams}} \cdot \frac{100^3 \text{ centimeters}^3}{1 \text{ meter}^3} \cdot 1 \text{ meter}^3$$ **See students' work.**

Guided Practice

Find the reciprocal of each expression.

4. $\frac{m^2}{3}$ $\frac{3}{m^2}$
5. $\frac{x}{5}$ $\frac{5}{x}$
6. $\frac{-9}{4y}$ $\frac{4y}{-9}$
7. $\frac{x^2-9}{y+3}$ $\frac{y+3}{x^2-9}$
8. $m-3$ $\frac{1}{m-3}$
9. x^2+2x+5

9. $\frac{1}{x^2+2x+5}$

Reteaching

Logical Thinking Have each student create a division problem that has binomial factors that cancel. Take four arbitrary binomials (A, B, C, and D). The division problem is $\frac{E}{F} \div \frac{A}{C}$ where $AB = E$ and $CD = F$. The quotient is $\frac{B}{D}$. Have students exchange problems with a neighbor to check.

Find each quotient. Assume that no denominator has a value of 0.

10. $\frac{x}{x+7} \div \frac{x-5}{x+7}$ $\frac{x}{x-5}$

11. $\frac{m^2+3m+2}{4} \div \frac{m+1}{m+2}$ $\frac{(m+2)^2}{4}$

12. $\frac{5a+10}{a+5} \div (a+2)$ $\frac{5}{a+5}$

13. $\frac{x^2+7x+12}{x+6} \div (x+3)$ $\frac{x+4}{x+6}$

Simplify each dimensional expression. State what you think the expression represents.

14. $(8 \text{ feet} \cdot 3 \text{ feet} \cdot 12 \text{ feet}) \div \frac{27 \text{ feet}^3}{1 \text{ yard}^3}$ **10.7 cubic yards**

15. $(12 \text{ inches} \cdot 18 \text{ inches} \cdot 4 \text{ inches}) \div \frac{1728 \text{ inches}^3}{1 \text{ feet}^3}$ **0.5 cubic feet**

16. **Railroading** Jaheem is a railroad buff who likes to watch trains. At the railroad crossing near his house in Menominee Falls, Wisconsin, a Chicago & North Western freight train passes by at 40 miles per hour. Suppose each railroad car is 48 feet long.

a. Write an expression involving units that represents the number of cars that pass by him in one minute.

b. Find how many railroad cars would pass by him per minute. $73\frac{1}{3}$ **cars**

16a. $\frac{40 \text{ mi}}{1 \text{ h}} \cdot \frac{5280 \text{ ft}}{1 \text{ mi}} \cdot \frac{1 \text{ h}}{60 \text{ min}} \cdot \frac{1 \text{ car}}{48 \text{ ft}}$

EXERCISES

Practice

Find each quotient. Assume that no denominator has a value of 0.

17. $\frac{a}{a+3} \div \frac{a+11}{a+3}$ $\frac{a}{a+11}$

18. $\frac{m+7}{m} \div \frac{m+7}{m+3}$ $\frac{m+3}{m}$

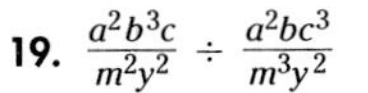

19. $\frac{a^2b^3c}{m^2y^2} \div \frac{a^2bc^3}{m^3y^2}$ $\frac{b^2m}{c^2}$

20. $\frac{5x^2}{7} \div \frac{10x^3}{21}$ $\frac{3}{2x}$

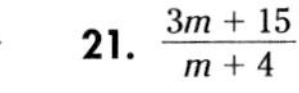

21. $\frac{3m+15}{m+4} \div \frac{3m}{m+4}$ $\frac{m+5}{m}$

22. $\frac{3x}{x+2} \div (x-1)$ $\frac{3x}{(x+2)(x-1)}$

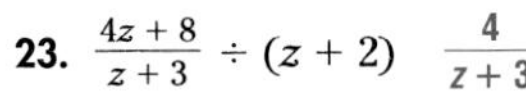

23. $\frac{4z+8}{z+3} \div (z+2)$ $\frac{4}{z+3}$

24. $\frac{x+3}{x+1} \div (x^2+5x+6)$ $\frac{1}{(x+2)(x+1)}$

25. $\frac{2x+4}{x^2+11x+18} \div \frac{x+1}{x^2+14x+45}$

26. $\frac{k+3}{m^2+4m+4} \div \frac{2k+6}{m+2}$ $\frac{1}{2(m+2)}$

25. $\frac{2(x+5)}{x+1}$

27. What is the quotient when $\frac{2x+6}{x+5}$ is divided by $\frac{2}{x+5}$? $x+3$

28. Find the quotient when $\frac{m-8}{m+7}$ is divided by m^2-7m-8. $\frac{1}{(m+7)(m+1)}$

Find each quotient. Assume that no denominator has a value of 0.

29. $\frac{x^2+5x+6}{x^2-x-12} \div \frac{x+2}{x^2+x-20}$ $x+5$

30. $\frac{m^2+m-6}{m^2+8m+15} \div \frac{m^2-m-2}{m^2+9m+20}$ $\frac{m+4}{m+1}$

31. $\frac{2x^2+7x-15}{x+2} \div \frac{2x-3}{x^2+5x+6}$ $(x+5)(x+3)$

32. $\frac{t^2-2t-8}{w-3} \div \frac{t-4}{w^2-7w+12}$ $(t+2)(w-4)$

Simplify each expression. State what you think the expression represents.

33. $\left(\frac{60 \text{ miles}}{1 \text{ hour}} \cdot \frac{5280 \text{ feet}}{1 \text{ mile}} \div \frac{60 \text{ minutes}}{1 \text{ hour}}\right) \div \frac{60 \text{ seconds}}{1 \text{ minute}}$ **88 ft/s**

34. $\frac{23.75 \text{ inches}}{1 \text{ revolution}} \cdot \frac{33\frac{1}{3} \text{ revolutions}}{1 \text{ minute}} \cdot 16.5 \text{ minutes}$ **13,062.5 in.**

35. $(5 \text{ feet} \cdot 16.5 \text{ feet} \cdot 9 \text{ feet}) \div \frac{27 \text{ feet}^3}{1 \text{ yard}^3}$ **27.5 yd³**

36. $\left[\left(\frac{60 \text{ kilometers}}{1 \text{ hour}} \cdot \frac{1000 \text{ meters}}{1 \text{ kilometer}}\right) \div \frac{60 \text{ minutes}}{1 \text{ hour}}\right] \div \frac{60 \text{ seconds}}{1 \text{ minute}}$ **16.7 meters/s**

Sample answers:
33. changes 60 mph to ft/s
34. calculates how far a needle travels on a record for a song 16.5 minutes long
35. changes cubic feet to cubic yards
36. changes 60 km/h to m/s

Cooperative Learning

Pairs Check Students can work in pairs on Exercises 17–26. One student can solve the even numbered problems while the other student solves the odd numbered problems. Each should verify the other student's work.

For more information on the pairs check strategy, see *Cooperative Learning in the Mathematics Classroom*, one of the titles in the Glencoe Mathematics Professional Series, pages 12–13.

3 PRACTICE/APPLY

Check for Understanding

Exercises 1–16 are designed to help you assess your students' understanding through reading, writing, speaking, and modeling. You should work through Exercises 1–3 with your students and then monitor their work on Exercises 4–16.

Assignment Guide

Core: 17–39 odd, 41–47
Enriched: 18–36 even, 37–47

For **Extra Practice**, see p. 783.

The red A, B, and C flags, printed only in the Teacher's Wraparound Edition, indicate the level of difficulty of the exercises.

Practice Masters, p. 82

NAME______________________ DATE ____________

Practice

Student Edition Pages 671–674

Dividing Rational Expressions

Find each quotient. Assume that no denominator has a value of 0.

1. $\frac{c^3}{d^3} \div \frac{d^3}{c^3}$ $\frac{c^6}{d^6}$
2. $\frac{x^3}{y^2} \div \frac{x^3}{y}$ $\frac{1}{y}$
3. $\frac{(-e)^3}{f^2} \div \frac{e^3}{f^2}$ -1
4. $\frac{2x}{x-1} \div (x+1)$ $\frac{2x}{x^2-1}$
5. $\frac{z^2-16}{3z} \div (z-4)$ $\frac{z+4}{3z}$
6. $\frac{a^2-a-12}{a^2} \div (a+3)$ $\frac{a-4}{a^2}$
7. $\frac{5m}{m+1} \div \frac{25m^2}{m^2+2m+1}$ $\frac{m+1}{5m}$
8. $\frac{\ell^2-49}{\ell^2-36} \div \frac{\ell+7}{\ell-6}$ $\frac{\ell-7}{\ell+6}$
9. $\frac{15b^2}{b^2-c^2} \div \frac{5b}{b^2-25}$ $\frac{3b^3-75b}{b^2-c^2}$
10. $(6x-24) \div \frac{x^2-16}{6x}$ $\frac{36x}{x+4}$
11. $\frac{z^2-1}{6z-10} \div \frac{2z-2}{9z^2-25}$ $\frac{3z^2+8z+5}{4}$
12. $\frac{x^2-x-6}{x^2+2x-15} \div \frac{x^2-4x-5}{x^2-25}$ $\frac{x+2}{x+1}$
13. $\frac{m^2+2m+1}{10m-10} \div \frac{m+1}{20}$ $\frac{2m+2}{m-1}$
14. $\frac{a^2+10a+25}{a^2-9} \div \frac{a+5}{a^2-3a}$ $\frac{a^2+5a}{a+3}$
15. $\frac{2y-6}{4y+28} \div \frac{y-3}{2y+14}$ 1
16. $\frac{x^2-1}{x^2-3x-10} \div \frac{x^2+3x+2}{x^2+4x+4}$ $\frac{x-1}{x-5}$
17. $\frac{6a}{a-4} \div \frac{a^2-7a}{a^2-11a+28}$ 6

4 ASSESS

Closing Activity

Writing Have students write key points to consider when dividing rational expressions.

Phonograph records were played at a speed of $33\frac{1}{3}$ RPMs, or revolutions per minute. This means that a typical 3-minute song required 100 revolutions.

Additional Answers

38a. $\dfrac{5\text{ ft }(18\text{ ft} + 15\text{ ft})}{2} \cdot 9\text{ ft} \div \dfrac{27\text{ ft}^3}{1\text{ yd}^3}$

40a. $\dfrac{2\pi\left(3\frac{3}{4}\text{ in.}\right)}{1\text{ revolution}} \cdot \dfrac{33\frac{1}{3}\text{ revolutions}}{1\text{ minute}} \cdot \dfrac{16.5\text{ minutes}}{1}$

Enrichment Masters, p. 82

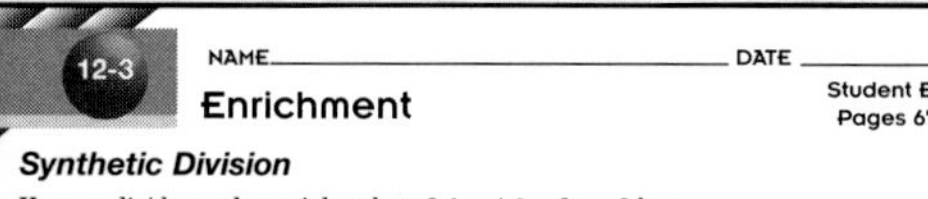

12-3 NAME__________ DATE ________

Enrichment

Student Edition Pages 671–674

Synthetic Division

You can divide a polynomial such as $3x^3 - 4x^2 - 3x - 2$ by a binomial such as $x - 3$ by a process called **synthetic division.** Compare the process with long division in the following explanation.

Example: Divide $(3x^3 - 4x^2 - 3x - 2)$ by $(x - 3)$ using synthetic division.

1. Show the coefficients of the terms in descending order.
2. The divisor is $x - 3$. Since 3 is to be subtracted, write 3 in the corner ⌋.
3. Bring down the first coefficient, 3.
4. Multiply. $3 \cdot 3 = 9$
5. Add. $-4 + 9 = 5$
6. Multiply. $3 \cdot 5 = 15$
7. Add. $-3 + 15 = 12$
8. Multiply. $3 \cdot 12 = 36$
9. Add. $-2 + 36 = 34$

$$\begin{array}{r|rrrr} & 3 & -4 & -3 & -2 \\ & & 9 & 15 & 36 \\ \hline 3 & 3 & 5 & 12 & 34 \end{array}$$

$3x^2 + 5x + 12$, remainder 34

Check: Use long division.

$$\begin{array}{r} 3x^2 + 5x + 12 \\ x - 3\overline{)3x^3 - 4x^2 - 3x - 2} \\ \underline{3x^3 - 9x^2} \qquad\qquad \\ 5x^2 - 3x \qquad \\ \underline{5x^2 - 15x} \qquad \\ 12x - 2 \\ \underline{12x - 36} \\ 34 \end{array}$$

The result is $3x^2 + 5x + 12 + \frac{34}{x - 3}$.

Divide by using synthetic division. Check your result using long division.

1. $(x^3 + 6x^2 + 3x + 1) \div (x - 2)$ $x^2 + 8x + 19 + \frac{39}{x - 2}$
2. $(x^3 - 3x^2 - 6x - 20) \div (x - 5)$ $x^2 + 2x + 4$
3. $(2x^3 - 5x + 1) \div (x + 1)$ $2x^2 - 2x - 3 + \frac{4}{x + 1}$
4. $(3x^3 - 7x^2 + 4) \div (x - 2)$ $3x^2 - x - 2$
5. $(x^3 + 2x^2 - x + 4) \div (x + 3)$ $x^2 - x + 2 - \frac{2}{x + 3}$
6. $(x^3 + 4x^2 - 3x - 11) \div (x - 4)$ $x^2 + 8x + 29 + \frac{105}{x - 4}$

Critical Thinking

37. **Geometry** The area of a rectangle is $\frac{x^2 - y^2}{2}$, and its length is $2x + 2y$. Find the width. $\frac{x - y}{4}$

Applications and Problem Solving

38. **Construction** A construction supervisor needs to determine how many truckloads of earth must be removed from a site before a foundation can be poured. The bed of the truck has the shape shown at the right.

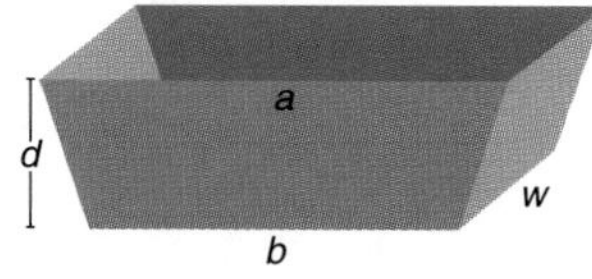

a. Use the formula $V = \frac{d(a + b)}{2} \cdot w$ to write an expression involving units that represents the volume of the truck bed in cubic yards if $a = 18$ ft, $b = 15$ ft, $w = 9$ ft, and $d = 5$ ft. **See margin.** **b. $727.\overline{27}$ truckloads**

b. There are 20,000 cubic yards of earth that must be removed from the excavation site. Write an expression involving units that represents the number of truckloads that will be required to remove all the earth.

39. **Railroads** The table at the right shows data about railroads in the United States. **a. 5500 miles**

Year	Miles of Track	Number of Freight Cars
1980	290,000	1,168,000
1985	257,000	867,000
1990	239,000	659,000
1992	227,000	605,000

a. If one freight car is 48 feet long on average, how many miles long would a train without a locomotive be if all the freight cars for 1992 were connected?

b. How many trains whose length is the answer to part a would it take to occupy all of the track in 1992? **41.3 trains**

c. Repeat parts a and b for the year 1980. **10,618.2 miles; 27.3 trains**

FYI

Phonograph records are played on a turntable that rotates at a constant speed. Each side of the record has a single groove, or *track*. As the record turns, the stylus of a *tone arm* vibrates in response to variations in the track. The vibration is converted to an electrical signal, which in turn is converted into sound.

40. **Music** Many old phonograph records turn at a rate of $33\frac{1}{3}$ revolutions per minute. Suppose a record of this type plays for 16.5 minutes and the average of the radii of the grooves on the record is $3\frac{3}{4}$ inches. **a. See margin.**

a. Write an expression involving units that represents how many inches the needle travels while playing the record.

b. Find the distance the needle travels. **about 1079.92 ft**

Mixed Review

41. Find $\frac{2m + 3}{4} \cdot \frac{32}{(2m + 3)(m - 5)}$. (Lesson 12–2) $\frac{8}{m - 5}$
42. Solve $x^2 + 6x + 8 = 0$ by graphing. (Lesson 11–2) **−4, −2**
43. Factor $16a^2 - 24ab^2 + 9b^4$. (Lesson 10–5) $(4a - 3b^2)^2$
44. **Health** If your heart beats once every second and you live to be 78 years old, your heart will have beat about 2,460,000,000 times. Write this number in scientific notation. (Lesson 9–3) 2.46×10^9
45. Solve $-3x + 6 > 12$. (Lesson 7–3) $x < -2$
46. Find $-3 + 4 - 10$. (Lesson 2–3) **−9**
47. Use the numbers 7 and 2 to write a mathematical sentence illustrating the commutative property of addition. (Lesson 1–8) $7 + 2 = 2 + 7$

Extension

Problem Solving Perform the indicated operations and simplify

$\dfrac{2y}{y^2 - 4} \div \dfrac{3}{y^2 - 4y + 4} \cdot \dfrac{y^2 + 4y + 4}{y^2 - y - 2}$. $\dfrac{2y^2 + 4y}{3y + 3}$

12–4 Dividing Polynomials

What YOU'LL LEARN

- To divide polynomials by monomials, and
- to divide polynomials by binomials.

Why IT'S IMPORTANT

You can use polynomials to solve problems involving science and interior design.

APPLICATION

Interior Design

Tomi wants to put a decorative border waist-high around his dining room. The perimeter of the room is 52 feet, and the widths of the two windows and two doorways total $12\frac{3}{4}$ feet. To find the number of yards of border needed, he can use the expression $\frac{52 \text{ ft} - 12\frac{3}{4} \text{ ft}}{3 \text{ ft/yd}}$. The expression $\frac{52 \text{ ft}}{3 \text{ ft/yd}} - \frac{12\frac{3}{4} \text{ ft}}{3 \text{ ft/yd}}$ could also be used. In each case, each term of the numerator was divided by the denominator. *This expression will be simplified in Example 4.*

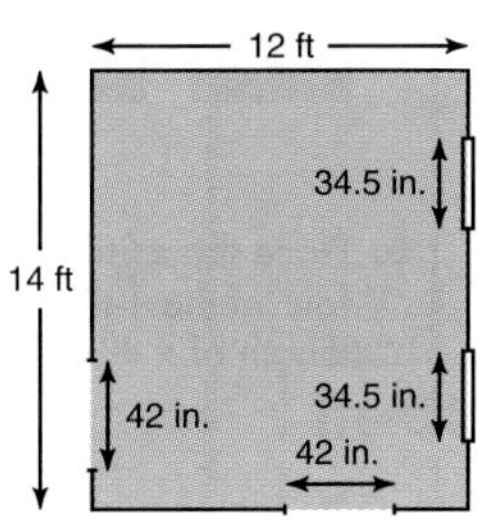

To divide a polynomial by a monomial, you divide each term of the polynomial by the monomial.

Example 1 **Find each quotient.**

a. $(3r^2 - 5) \div 12r$

$$(3r^2 - 5) \div 12r = \frac{3r^2 - 5}{12r} \quad \textit{Write as a rational expression.}$$
$$= \frac{3r^2}{12r} - \frac{5}{12r} \quad \textit{Divide each term by } 12r.$$
$$= \frac{\overset{1\ r}{\cancel{3}\cancel{r^2}}}{\underset{4\ 1}{\cancel{12}\cancel{r}}} - \frac{5}{12r} \quad \textit{Divide by the common factors.}$$
$$= \frac{r}{4} - \frac{5}{12r}$$

b. $(9n^2 - 15n + 24) \div 3n$

$$(9n^2 - 15n + 24) \div 3n = \frac{9n^2 - 15n + 24}{3n}$$
$$= \frac{9n^2}{3n} - \frac{15n}{3n} + \frac{24}{3n}$$
$$= \frac{\overset{3\ n}{\cancel{9}\cancel{n^2}}}{\underset{1}{\cancel{3n}}} - \frac{\overset{5}{\cancel{15n}}}{\underset{1}{\cancel{3n}}} + \frac{\overset{8}{\cancel{24}}}{\underset{1}{\cancel{3}n}} \quad \textit{Divide by the common factors.}$$
$$= 3n - 5 + \frac{8}{n}$$

Recall from Lesson 12–3 that, when you can factor, some divisions can be performed easily, as shown below.

$$(a^2 + 7a + 12) \div (a + 3) = \frac{a^2 + 7a + 12}{(a + 3)}$$
$$= \frac{\overset{1}{\cancel{(a+3)}}(a + 4)}{\underset{1}{\cancel{(a+3)}}} \quad \textit{Factor the dividend.}$$
$$= a + 4$$

12-4 LESSON NOTES

NCTM Standards: 1–5

Instructional Resources

- Study Guide Master 12-4
- Practice Master 12-4
- Enrichment Master 12-4
- Assessment and Evaluation Masters, pp. 323–324

Transparency 12-4A contains the 5-Minute Check for this lesson; **Transparency 12-4B** contains a teaching aid for this lesson.

Recommended Pacing	
Standard Pacing	Days 5 & 6 of 14
Honors Pacing	Day 5 of 13
Block Scheduling*	Day 3 of 6
Alg. 1 in Two Years*	Days 10, 11, & 12 of 27

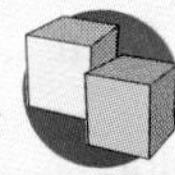

*For more information on pacing and possible lesson plans, refer to the *Block Scheduling Booklet* and *Algebra 1 in Two Years.*

1 FOCUS

5-Minute Check
(over Lesson 12-3)

Find each quotient. Assume that no denominator has a value of 0.

1. $\frac{5}{8} \div \frac{3}{4}$ $\frac{5}{6}$
2. $\frac{36}{64} \div \frac{27}{32}$ $\frac{2}{3}$
3. $\frac{5a}{6b} \div \frac{10d}{6c}$ $\frac{ac}{2bd}$
4. $\frac{a^2}{b} \div \frac{a^2}{b^2}$ b
5. $\frac{x^2 - 1}{x^2 + x - 2} \div \frac{x + 1}{x + 2}$ 1

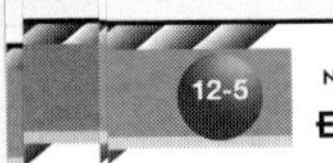

Motivating the Lesson

Hands-On Activity Turn a bicycle upside-down. Have one student spin the back tire at a rate of 1 revolution every 2 seconds. Have another student spin the front tire at a rate of 1 revolution every 3 seconds. Before spinning, put a chalk mark on the top of each tire. Record how long it takes for the two tires to return to this same arrangement. **6 seconds**

2 TEACH

In-Class Examples

For Example 1
Find the LCM for each group of rational expressions.

a. $6x^2y$, $9xy^2$ **18^2y^2**
b. $3a^2b$, $16ab^2$, $21ab$ **$336a^2b^2$**

For Example 2
Find the LCM for each pair of rational expressions.

a. $x^2 - 5x + 4$, $x^2 - 3x - 4$
$(x - 4)(x - 1)(x + 1)$
b. $x^2 - 9$, $x^2 + 6x + 9$
$(x - 3)(x + 3)^2$

For Example 3
Find each sum or difference.

a. $\frac{2}{3y^2} - \frac{1}{12y^3}$ **$\frac{8y - 1}{12y^3}$**
b. $\frac{4a}{mn} + \frac{2a}{np}$ **$\frac{4ap + 2am}{mnp}$**

For Example 4
Find each sum or difference.

a. $\frac{r}{r^2 - 16} - \frac{5}{r + 4}$
$\frac{-4r + 20}{r^2 - 16}$
b. $\frac{3}{y - 3} + \frac{2y}{y^2 - y - 6}$
$\frac{5y + 6}{y^2 - y - 6}$
c. $\frac{y}{(1 - y)(y + 1)} + \frac{y + 2}{(y + 1)^2}$
$\frac{-2}{(y - 1)(y + 1)^2}$
d. $\frac{a - 1}{(a + 3)^2} - \frac{a + 1}{(3 - a)(a + 3)}$
$\frac{2a^2 + 6}{(a - 3)(a + 3)^2}$

Teaching Tip Point out that the LCD of two rational expressions can be equal to the denominator of one of them if one denominator is a factor of the other.

Example 2 **Find the LCM of $x^2 - x - 6$ and $x^2 + 2x - 15$.**

$x^2 - x - 6 = (x - 3)(x + 2)$ *Factor each expression.*
$x^2 + 2x - 15 = (x + 5)(x - 3)$
$\text{LCM} = (x - 3)(x + 2)(x + 5)$

To add or subtract fractions with unlike denominators, first rename the fractions so the denominators are alike. Any common denominator could be used. However, the computation usually is easier if you use the **least common denominator (LCD)**. Recall that the least common denominator is the LCM of the denominators.

You usually will use the following steps to add or subtract rational expressions with unlike denominators.

1. Find the LCD.
2. Change each rational expression into an equivalent expression with the LCD as the denominator.
3. Add or subtract as with rational expressions with like denominators.
4. Simplify if necessary.

Example 3 **Find $\frac{7}{3m} + \frac{5}{6m^2}$.**

$3m = 3 \cdot m$ *Factor each denominator.*
$6m^2 = 2 \cdot 3 \cdot m \cdot m$
$\text{LCM} = 2 \cdot 3 \cdot m \cdot m$, or $6m^2$

Since the denominator of $\frac{5}{6m^2}$ is already $6m^2$, only $\frac{7}{3m}$ needs to be renamed.

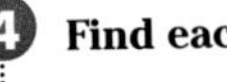

$$\frac{7}{3m} + \frac{5}{6m^2} = \frac{7(2m)}{3m(2m)} + \frac{5}{6m^2} \quad \textit{Multiply } \tfrac{7}{3m} \textit{ by } \tfrac{2m}{2m}\textit{. Why?}$$
$$= \frac{14m}{6m^2} + \frac{5}{6m^2}$$
$$= \frac{14m + 5}{6m^2} \quad \textit{Add the numerators.}$$

To combine rational expressions whose denominators are polynomials, follow the same procedure you use to combine expressions whose denominators are monomials.

Example 4 **Find each sum or difference.**

a. $\frac{s}{s + 3} + \frac{3}{s - 4}$

Since the denominators are $(s + 3)$ and $(s - 4)$, there are no common factors. The LCD is $(s + 3)(s - 4)$.

$$\frac{s}{s + 3} + \frac{3}{s - 4} = \frac{s}{s + 3} \cdot \frac{s - 4}{s - 4} + \frac{3}{s - 4} \cdot \frac{s + 3}{s + 3}$$
$$= \frac{s(s - 4)}{(s + 3)(s - 4)} + \frac{3(s + 3)}{(s + 3)(s - 4)}$$
$$= \frac{s^2 - 4s}{(s + 3)(s - 4)} + \frac{3s + 9}{(s + 3)(s - 4)}$$
$$= \frac{s^2 - 4s + 3s + 9}{(s + 3)(s - 4)}$$
$$= \frac{s^2 - s + 9}{(s + 3)(s - 4)}$$

Teaching Tip Students should be aware of how the LCD is determined. Remind them that each factor is used the greatest number of times it appears in either factorization.

b. $\dfrac{n-4}{(2-n)^2} - \dfrac{n-5}{n^2+n-6}$

$$\frac{n-4}{(2-n)^2} - \frac{n-5}{n^2+n-6}$$

$$= \frac{n-4}{(2-n)^2} - \frac{n-5}{(n-2)(n+3)} \quad \textit{Factor the denominators.}$$

$$= \frac{n-4}{(n-2)(n-2)} - \frac{n-5}{(n-2)(n+3)} \quad (2-n)^2 = (n-2)^2$$

$$= \frac{n-4}{(n-2)(n-2)} \cdot \frac{(n+3)}{(n+3)} - \frac{n-5}{(n-2)(n+3)} \cdot \frac{(n-2)}{(n-2)} \quad \textit{Rename each fraction using the LCD.}$$

$$= \frac{(n^2+3n-4n-12)-(n^2-2n-5n+10)}{(n-2)(n-2)(n+3)} \quad \textit{Subtract numerators}$$

$$= \frac{n^2-n-12-n^2+7n-10}{(n-2)^2(n+3)}$$

$$= \frac{6n-22}{(n-2)^2(n+3)}$$

CHECK FOR UNDERSTANDING

Communicating Mathematics

1. Yes; no; no common factors were divided.

6. $(a+6)(a+7)$
7. $(x-3)(x+3)$
8. $2(x-4)$

Study the lesson. Then complete the following.

1. **You Decide** Kaylee simplified $\frac{16}{64}$ as $\frac{1\not{6}}{\not{6}4} = \frac{1}{4}$. Is her answer correct? Was her method of simplifying correct? Explain.
2. **Describe** a situation in which the LCD of two or more rational expressions is equal to the denominator of one of the rational expressions. $\frac{3x}{4} + \frac{7}{8x}$; $8x$ is the LCD.

Guided Practice

Find the LCD for each pair of rational expressions.

3. $\frac{3}{x^2}, \frac{5}{x}$ x^2
4. $\frac{4}{a^2b}, \frac{3}{ab^2}$ a^2b^2
5. $\frac{4}{15m^2}, \frac{7}{18mb^2}$ $90m^2b^2$
6. $\frac{6}{a+6}, \frac{7}{a+7}$
7. $\frac{5}{x-3}, \frac{4}{x+3}$
8. $\frac{9}{2x-8}, \frac{10}{x-4}$

Find each sum or difference. 9–14. See margin.

9. $\frac{7}{15m^2} + \frac{3}{5m}$
10. $\frac{2}{x+3} + \frac{3}{x-2}$
11. $3x + 6 - \frac{9}{x+2}$
12. $\frac{m+2}{m^2+4m+3} - \frac{6}{m+3}$
13. $\frac{11}{3y^2} - \frac{7}{6y}$
14. $\frac{4}{2g-7} + \frac{5}{3g+1}$

15. **Parades** At the Veteran's Day parade, the local members of the Veterans of Foreign Wars (VFW) found that they could arrange themselves in rows of 6, 7, or 8, with no one left over. What was the least number of VFW members in the parade? 168 members

EXERCISES

Practice

Find each sum or difference.

16. $\frac{m}{4} + \frac{3m}{5}$ $\frac{17m}{20}$
17. $\frac{x}{7} - \frac{2x}{9}$ $\frac{-5x}{63}$
18. $\frac{m+1}{m} + \frac{m-3}{3m}$ $\frac{4}{3}$
19. $\frac{7}{x} + \frac{3}{xyz}$ $\frac{7yz+3}{xyz}$
20. $\frac{7}{6a^2} - \frac{5}{3a}$ $\frac{7-10a}{6a^2}$
21. $\frac{2}{st^2} - \frac{3}{s^2t}$ $\frac{2s-3t}{s^2t^2}$

3 PRACTICE/APPLY

Check for Understanding

Exercises 1–15 are designed to help you assess your students' understanding through reading, writing, speaking, and modeling. You should work through Exercises 1–2 with your students and then monitor their work on Exercises 3–15.

Assignment Guide

Core: 17–43 odd, 44–50
Enriched: 16–38 even, 39–50

For **Extra Practice**, see p. 784.

The red A, B, and C flags, printed only in the Teacher's Wraparound Edition, indicate the level of difficulty of the exercises.

Additional Answers

9. $\frac{9m+7}{15m^2}$
10. $\frac{5x+5}{(x+3)(x-2)}$
11. $\frac{3(x^2+4x+1)}{x+2}$
12. $\frac{-5m-4}{(m+3)(m+1)}$
13. $\frac{22-7y}{6y^2}$
14. $\frac{22g-31}{(2g-7)(3g+1)}$

Study Guide Masters, p. 85

NAME__________ DATE ________

Study Guide — Student Edition Pages 685–689

Rational Expressions with Unlike Denominators

When you add or subtract rational expressions with *unlike* denominators, you must first rename the rational expressions so the denominators are alike. Any common denominator could be used. However, the computation is easier if the **least common denominator** (LCD) is used. The least common denominator is the **least common multiple** (LCM) of the denominators.

Example: Find $\frac{1}{2x^2+6x} + \frac{3}{x^2}$.

$$\frac{1}{2x^2+6x} + \frac{3}{x^2} = \frac{1}{2x(x+3)} + \frac{3}{x^2}$$
$$= \frac{1}{2x(x+3)} \cdot \frac{x}{x} + \frac{3}{x^2} \cdot \frac{2(x+3)}{2(x+3)}$$
$$= \frac{x+3(2)(x+3)}{2x^2(x+3)}$$
$$= \frac{x+6x+18}{2x^2(x+3)}$$
$$= \frac{7x+18}{2x^2(x+3)}$$

Find each sum or difference.

1. $\frac{1}{x} + \frac{7}{3x}$ $\frac{10}{3x}$
2. $\frac{1}{6x} + \frac{3}{8}$ $\frac{4+9x}{24x}$
3. $\frac{x}{x-3} - \frac{3}{x+3}$ $\frac{x^2+9}{x^2-9}$
4. $\frac{4}{h+1} + \frac{2}{h+2}$ $\frac{6h+10}{(h+1)(h+2)}$, or $\frac{6h+10}{h^2+3h+2}$
5. $\frac{x}{6x+6} - \frac{1}{x+1}$ $\frac{x-6}{6(x+1)}$, or $\frac{x-6}{6x+6}$
6. $\frac{4x}{6x-2y} + \frac{3y}{9x-3y}$ $\frac{2x+y}{3x-y}$
7. $\frac{y+2}{y^2+5y+6} + \frac{2-y}{y^2+y-6}$ 0
8. $\frac{x}{x-7} - \frac{x+3}{x^2-4x-21}$ $\frac{x-1}{x-7}$
9. $\frac{a-6b}{2a^2-5ab+2b^2} - \frac{7}{a-2b}$ $\frac{-13a+b}{(2a-b)(a-2b)}$, or $\frac{-13a+b}{2a^2-5ab+2b^2}$
10. $\frac{m}{1-2m} + \frac{4}{2m-1}$ $\frac{4-m}{2m-1}$
11. $\frac{7d+4}{3d+9} - \frac{2d}{d+3}$ $\frac{d+4}{3(d+3)}$, or $\frac{d+4}{3d+9}$
12. $\frac{q}{q^2-16} + \frac{q+1}{q^2+5q+4}$ $\frac{2(q-2)}{(q-4)(q+4)}$, or $\frac{2q-4}{q^2-16}$

Reteaching

Using Problem Solving When adding rational expressions, a common denominator is needed. The LCD provides the shortest solution but any common denominator will do. Work the problem below twice, once using a common denominator of $(2x+6)(x+3)$ and once using the LCD.

$\frac{3x}{2x+6} - \frac{x-1}{x+3}$ $\frac{x+2}{2x+6}$

Additional Answer

38. $\frac{2(2a+5)}{(a-2)(a+1)}$; $\frac{-2(2a+5)}{(a-2)(a+1)}$; They are opposite.

Practice Masters, p. 85

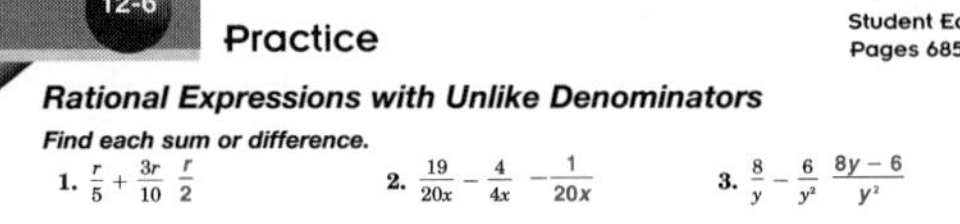

12-6 NAME ____________ DATE ________

Practice

Student Edition Pages 685–689

Rational Expressions with Unlike Denominators

Find each sum or difference.

1. $\frac{r}{5}+\frac{3r}{10}$ $\frac{r}{2}$
2. $\frac{19}{20x}-\frac{4}{4x}$ $-\frac{1}{20x}$
3. $\frac{8}{y}-\frac{6}{y^2}$ $\frac{8y-6}{y^2}$
4. $\frac{-3}{8a^2}+\frac{5}{2a}$ $\frac{20a-3}{8a^2}$
5. $\frac{7}{xy}-\frac{-9}{yz}$ $\frac{7z+9x}{xyz}$
6. $\frac{-7}{6x^2r}+\frac{-10}{3xr^2}$ $\frac{-7-20x}{6x^2r^2}$
7. $\frac{x+2}{y}+\frac{3}{x}$ $\frac{x^2+2x+3y}{xy}$
8. $\frac{k}{k-5}+\frac{k-1}{k+5}$ $\frac{2k^2-k+5}{k^2-25}$
9. $\frac{2a+6}{a^2-9}+\frac{1}{a+3}$ $\frac{3a+3}{a^2-9}$
10. $\frac{10}{a-b}-\frac{6b}{a^2-b^2}$ $\frac{10a+4b}{a^2-b^2}$
11. $\frac{3}{2h^2-2h}+\frac{5}{2h-2}$ $\frac{5h+3}{2h^2-2h}$
12. $\frac{2}{x+2}+\frac{3}{(x+2)^2}$ $\frac{2x+7}{(x+2)^2}$
13. $\frac{6}{x^2-4x+5}+\frac{1}{x+1}$ $\frac{1}{x-5}$
14. $\frac{6}{5y^2-45}-\frac{8}{3-y}$ $\frac{126+40y}{5y^2-45}$

23. $\frac{7z+8}{(z+5)(z-4)}$
24. $\frac{d^2+6d+12}{(d+4)(d+3)}$
25. $\frac{k^2+k-10}{(k+5)(k+3)}$
26. $\frac{-y^2+6y+12}{(y-3)(y+4)}$
27. $\frac{-17r-32}{(3r-2)(r-5)}$
28. $\frac{3a^2-2a+8}{3(a-2)(a+2)}$
29. $\frac{10+m}{2(2m-3)}$

Find each sum or difference.

22. $\frac{3}{7m}+\frac{4}{5m^2}$ $\frac{15m+28}{35m^2}$
23. $\frac{3}{z+5}+\frac{4}{z-4}$
24. $\frac{d}{d+4}+\frac{3}{d+3}$
25. $\frac{k}{k+5}-\frac{2}{k+3}$
26. $\frac{3}{y-3}-\frac{y}{y+4}$
27. $\frac{10}{3r-2}-\frac{9}{r-5}$

B

28. $\frac{4}{3a-6}+\frac{a}{2+a}$
29. $\frac{5}{2m-3}-\frac{m}{6-4m}$
30. $\frac{b}{3b+2}+\frac{2}{9b+6}$ $\frac{1}{3}$
31. $\frac{w}{5w+2}-\frac{4}{15w+6}$ $\frac{3w-4}{3(5w+2)}$
32. $\frac{h-2}{h^2+4h+4}+\frac{h-2}{h+2}$ $\frac{h^2+h-6}{(h+2)^2}$
33. $\frac{n+2}{n^2+4n+3}-\frac{6}{n+3}$ $\frac{-5n-4}{(n+3)(n+1)}$
34. $\frac{a}{5-a}-\frac{3}{a^2-25}$ $\frac{-a^2-5a-3}{(a+5)(a-5)}$
35. Find the difference when $\frac{2}{t+3}$ is subtracted from $\frac{3}{10t-9}$. $\frac{-17t+27}{(t+3)(10t-9)}$
36. Find the sum of $\frac{2y}{y^2+7y+12}$ and $\frac{y+2}{y+4}$. $\frac{y^2+7y+6}{(y+4)(y+3)}$

C

37. Find the sum of $\frac{2}{v+4}$, $\frac{v}{v-1}$, and $\frac{5v}{v^2+3v-4}$. $\frac{v^2+11v-2}{(v+4)(v-1)}$
38. Find the difference when $\frac{2}{a+1}$ is subtracted from $\frac{6}{a-2}$. Then find the difference when $\frac{6}{a-2}$ is subtracted from $\frac{2}{a+1}$. How are the differences related? See margin.

Critical Thinking

39. Copy and complete.
    a. $15 \cdot 24 =$ _?_ 360
       GCF of 15 and 24 = _?_ 3
       LCM of 15 and 24 = _?_ 120
       GCF · LCM = _?_ 360
    b. $18 \cdot 30 =$ _?_ 540
       GCF of 18 and 30 = _?_ 6
       LCM of 18 and 30 = _?_ 90
       GCF · LCM = _?_ 540
    c. Write a rule that describes the relationship between the GCF, the LCM, and the product of the two numbers.
    d. Use your rule to describe how to find the LCM of two numbers if you know their GCF. **Divide the GCF into the product of the two numbers to find the LCM.**

39c. The GCF times the LCM of two numbers is equal to the product of the two numbers.

Applications and Problem Solving

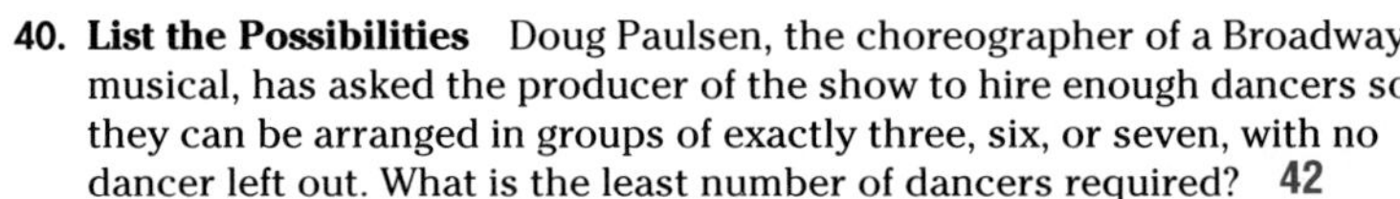

40. **List the Possibilities** Doug Paulsen, the choreographer of a Broadway musical, has asked the producer of the show to hire enough dancers so they can be arranged in groups of exactly three, six, or seven, with no dancer left out. What is the least number of dancers required? **42**

41. **Automobiles** Car owners need to follow a regular maintenance schedule to keep their cars running smoothly. The table below shows several of the checkups that should be performed on a regular basis, according to the Spring, 1995, issue of *Know-How* magazine.

Inspection or Service	Frequency
engine oil and oil filter change	every 3000 miles (about 3 months)
transmission fluid level check	every oil change
brake system inspection	every oil change
chassis lubrication	every other oil change
power steering pump fluid level check	twice a year
tire and wheel rotation and inspection	every 15,000 miles

If all these inspections and services are performed on April 20, 1996, and the owner follows the recommendations shown in the table, what is the next date on which they all should be performed again? **15 months later or July 20, 1997**

42. **Travel** Jaheed Toliver spent, in this order, a third of his life in the United States, a sixth of his life in Kenya, twelve years in Saudi Arabia, half the remainder in Australia, and as long in Canada as he spent in Hong Kong. How many years did Jaheed live if he spent his 45th birthday in Saudi Arabia and he spent a whole number of years in each country? **66 years**

Mixed Review

43. Find $\frac{8z+3}{3z+4} - \frac{2z-5}{3z+4}$. (Lesson 12–5) **2**

44. Simplify $\frac{x^2+7x+6}{3x^2+x-2}$. State the excluded values of x. (Lesson 12–1) **44.** $\frac{x+6}{3x-2}$; -1, $\frac{2}{3}$

45. Solve $2x^2 - 3x - 4 = 0$ by using the quadratic formula. (Lesson 11–3) **45.** $\frac{3 \pm \sqrt{41}}{4}$ **or about 2.35 and −0.85**

46. Factor $3x^2y + 6xy + 9y^2$. (Lesson 10–1) $3y(x^2 + 2x + 3y)$

47. Find $(3a^2b)(-5a^4b^2)$. (Lesson 9–1) $-15a^6b^3$

48. Graph the solution set of $3x > -15$ and $2x \le 6$. (Lesson 7–4) **See margin.**

49. Solve $3n - 12 = 5n - 20$. (Lesson 3–5) **4**

50. **Demographics** The populations of the capitals of some southern states are listed in the table below. Make a stem-and-leaf plot of the populations. (Lesson 1–4) **See margin.**

Capital	Population (thousands)	Capital	Population (thousands)
Atlanta, GA	394	Montgomery, AL	188
Austin, TX	466	Nashville, TN	488
Baton Rouge, LA	220	Oklahoma City, OK	445
Columbia, SC	98	Raleigh, NC	208
Frankfort, KY	26	Richmond, VA	203
Jackson, MS	197	Tallahassee, FL	125
Little Rock, AR	176		

Reading the Labels

The excerpt below appeared in an article in *Aging Magazine*, issue number 366, 1994.

> BY THIS SUMMER, ALMOST ALL PACKAGED foods found in your local supermarket included new, improved labels that make it easier to understand the nutritional information contained on them and to compare the nutritional value of different products. This revolution in food labeling is the result of years of work by consumer advocates who pressed for more complete, accurate, and pertinent information on food labels....The percent of daily fat consumption should be no more than 30 percent of calories....The labels also give percentages of daily allowances for sodium, sugar, fiber, and protein. ■

1. The food labels on items A, B, C, and D show fat contents of 25, 17, 4, and 33 grams per serving, respectively. If you are allowed a maximum daily allowance of 65 grams, which three-item combinations of these foods can you put together without exceeding the maximum allowance? **25, 17 and 4; 17, 4 and 33**
2. Numbers on the food labels are listed as "per serving" rather than per package, and the label also defines serving size. Do you think this is important? Explain your response. **Yes, because food intake is daily and per serving information better suits daily calculations.**
3. When you shop for food, do you use the nutritional data on the labels in deciding what to buy? Why or why not? **See students' work.**

Extension

Reasoning Find $\frac{2a^2+5a+3}{2a^2-a-6} - \frac{a^2-4}{a^2-a-2}$ and simplify. $\frac{2a+5}{a^2-a-2}$

Mathematics and SOCIETY

Even lower fat consumption has been shown to *reverse* heart disease. In 1990 Dr. Dean Ornish announced results of a study that showed that when fat consumption was kept to 10% of calories, plaque build-up around the heart had actually been reversed.

4 ASSESS

Closing Activity

Speaking Have students describe the similarities in the addition and subtraction of fractions with unlike denominators and in the addition and subtraction of rational expressions with unlike denominators.

Chapter 12, Quiz C (Lessons 12-5 and 12-6), is available in the *Assessment and Evaluation Masters,* p. 325.

Additional Answers

48. Number line graph from −6 to 4: −6 −5 −4 −3 −2 −1 0 1 2 3 4

50.

Stem	Leaf
0	3
1	0 3 8 9
2	0 0 1 2
3	9
4	5 7 9

0|3 = 30 thousand

Enrichment Masters, p. 85

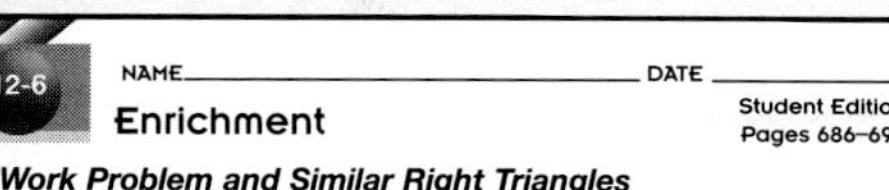

12-6 NAME ______ DATE ______

Enrichment — Student Edition Pages 686–690

The Work Problem and Similar Right Triangles

"The work problem" has been included in algebra textbooks for a very long time. In older books, the people in the problem always seemed to be digging ditches.

> If Olivia can dig a ditch in x hours and George can dig the same ditch in y hours, how long will it take them to dig the ditch if they work together?

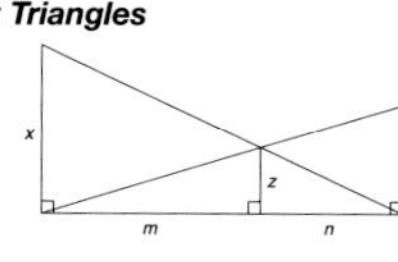

You have learned a way to solve this type of problem using rational equations. It can also be solved using a geometric model that uses two overlapping right triangles.

In the drawing, the length x is Olivia's time. The length y is George's time. The answer to the problem is the length of the segment z. The distance $m + n$ can be any convenient length.

Solve each problem. 1. z = 2 hours

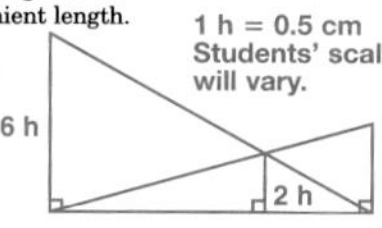

1. Solve the work problem for $x = 6$ and $y = 3$ by drawing a diagram and measuring.
2. Confirm your solution to problem 1 by writing and solving a rational equation.
3. On a separate sheet of paper, create a word problem to go with the values $x = 6$ and $y = 3$. Answers will vary.
4. On a separate sheet of paper, solve this problem with a diagram. Use centimeters and measure to the nearest tenth. Olivia can wash a car in 3 hours. George can wash a car in 4 hours. How long will it take them working together to wash one car? about 1.7 hours
5. Triangles that have the same shape are called *similar triangles.* You may have learned that corresponding sides of similar triangles form equal ratios. Using the drawing at the top of the page, you can thus conclude that Equations A and B below are true. Use the equations to prove the formula for the work problem.

2. $\frac{t}{6} + \frac{t}{3} = 1$ or $\frac{1}{6} + \frac{1}{3} = \frac{1}{t}$; $t = 2$

Equation A	Equation B	Work Formula
$\frac{z}{x} = \frac{n}{m+n}$	$\frac{z}{y} = \frac{m}{m+n}$	$\frac{1}{x} + \frac{1}{y} = \frac{1}{z}$

Adding equations A and B gives $\frac{z}{x} + \frac{z}{y} = \frac{m+n}{m+n}$. So, $\frac{z}{x} + \frac{z}{y} = 1$.
Dividing both sides by z gives $\frac{1}{x} + \frac{1}{y} = \frac{1}{z}$.

12-7 LESSON NOTES

NCTM Standards: 1–5

Instructional Resources

- Study Guide Master 12-7
- Practice Master 12-7
- Enrichment Master 12-7
- Multicultural Activity Masters, p. 24
- Real-World Applications, 29
- Tech Prep Applications Masters, p. 24

Transparency 12-7A contains the 5-Minute Check for this lesson; **Transparency 12-7B** contains a teaching aid for this lesson.

Recommended Pacing	
Standard Pacing	Day 10 of 14
Honors Pacing	Day 9 of 13
Block Scheduling*	Day 5 of 6 (along with Lesson 12-8)
Alg. 1 in Two Years*	Days 19, 20, & 21 of 27

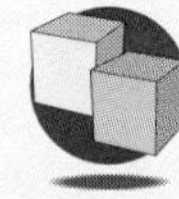

*For more information on pacing and possible lesson plans, refer to the *Block Scheduling Booklet* and *Algebra 1 in Two Years.*

1 FOCUS

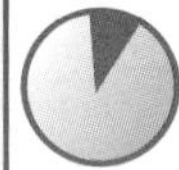

5-Minute Check
(over Lesson 12-6)

Find the LCD for each pair of rational expressions.

1. $\frac{7}{m^2n^3}, \frac{9}{mn^2}$ **m^2n^3**
2. $\frac{2x}{x+1}, \frac{5x}{x-1}$ **$(x+1)(x-1)$**

Find each sum or difference.

3. $\frac{5}{a} - \frac{3}{b}$ **$\frac{5b-3a}{ab}$**
4. $\frac{7}{5} + \frac{a}{5b}$ **$\frac{7b+a}{5b}$**
5. $\frac{6}{3x-9} + \frac{3}{x-3}$ **$\frac{5}{x-3}$**

Solve the problem by listing the possibilities.

6. There are 3 routes between Kelsey's house and the school. How many ways can Kelsey go from her house to the school and back home again? **9**

12–7 Mixed Expressions and Complex Fractions

What YOU'LL LEARN

- To simplify mixed expressions and complex fractions.

Why IT'S IMPORTANT

You can use rational expressions to solve problems involving acoustics and statistics.

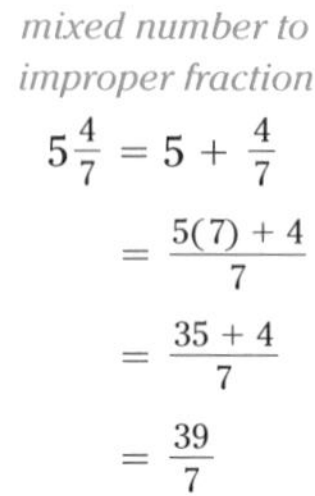

APPLICATION
Table Tennis

A Ping-Pong™ ball weighs about $\frac{1}{10}$ of an ounce. How many Ping-Pong balls weigh $1\frac{1}{2}$ pounds altogether? *This problem will be solved in Example 2.*

A number like $1\frac{1}{2}$ is a mixed number. Expressions like $a + \frac{b}{c}$ and $4 + \frac{x+y}{x-5}$ are **mixed expressions.** Changing mixed expressions to rational expressions is similar to changing mixed numbers to simple fractions (improper fractions).

mixed number to improper fraction

$$5\frac{4}{7} = 5 + \frac{4}{7} = \frac{5(7)+4}{7} = \frac{35+4}{7} = \frac{39}{7}$$

mixed expression to rational expression

$$4 + \frac{x+y}{x-5} = \frac{4(x-5)}{x-5} + \frac{x+y}{x-5} = \frac{4(x-5)+(x+y)}{(x-5)} = \frac{4x-20+x+y}{x-5} = \frac{5x+y-20}{x-5}$$

Example 1 **Simplify $7 + \frac{y-3}{y+4}$.**

$$7 + \frac{y-3}{y+4} = \frac{7(y+4)}{y+4} + \frac{y-3}{y+4} \quad \textit{The LCD is } y+4.$$

$$= \frac{7(y+4)+y-3}{y+4} \quad \textit{Add the numerators.}$$

$$= \frac{7y+28+y-3}{y+4}$$

$$= \frac{8y+25}{y+4} \quad \textit{Simplify.}$$

Now let's solve the Ping-Pong ball problem.

Example 2

APPLICATION
Table Tennis

Refer to the applications at the beginning of the lesson. How many Ping-Pong balls would weigh $1\frac{1}{2}$ pounds altogether?

$$\frac{1\frac{1}{2}\text{ pounds}}{\frac{1}{10}\text{ ounce}} = \frac{\frac{3}{2}\text{ pounds}}{\frac{1}{10}\text{ ounce}}$$

$$= \frac{\frac{3}{2}\text{ pounds}}{\frac{1}{10}\text{ ounce}} \cdot \frac{16\text{ ounces}}{1\text{ pound}} \quad \textit{Convert pounds to ounces. Divide by common units.}$$

$$= \frac{24}{\frac{1}{10}} \text{ or } 240$$

It would take 240 Ping-Pong balls to weigh $1\frac{1}{2}$ pounds.

Alternative Learning Styles

Auditory Have the first student in a row name a mixed number such as $3\frac{1}{2}$. The next student in the row must name an equivalent improper fraction. A third student then names another mixed number. Continue the activity until all class members have had a turn speaking aloud.

Recall that if a fraction has one or more fractions in the numerator or denominator, it is called a *complex fraction.* Some complex fractions are shown below.

$$\frac{5\frac{1}{2}}{3\frac{3}{4}} \qquad \frac{9}{\frac{x}{y}} \qquad \frac{\frac{x+y}{y}}{\frac{x-y}{x}} \qquad \frac{\frac{1}{a}-\frac{1}{b}}{\frac{1}{a}+\frac{1}{b}}$$

You simplify an algebraic complex fraction in the same way you simplify a numerical complex fraction.

numerical		*algebraic*	
$\frac{\frac{11}{2}}{\frac{15}{4}} = \frac{11}{2} \div \frac{15}{4}$		$\frac{\frac{a}{b}}{\frac{c}{d}} = \frac{a}{b} \div \frac{c}{d}$	
$= \frac{11}{2} \cdot \frac{4}{15}$	*The reciprocal of* $\frac{15}{4}$ *is* $\frac{4}{15}$.	$= \frac{a}{b} \cdot \frac{d}{c}$	*The reciprocal of* $\frac{c}{d}$ *is* $\frac{d}{c}$.
$= \frac{22}{15}$		$= \frac{ad}{bc}$	

Simplifying a Complex Fraction

Any complex fraction $\frac{\frac{a}{b}}{\frac{c}{d}}$, where $b \neq 0$, $c \neq 0$, and $d \neq 0$, can be expressed as $\frac{ad}{bc}$.

Example  **Simplify each rational expression.**

a. $\dfrac{1+\frac{4}{a}}{\frac{a}{6}+\frac{2}{3}}$

Simplify the numerator and denominator separately. Then divide.

$\dfrac{1+\frac{4}{a}}{\frac{a}{6}+\frac{2}{3}} = \dfrac{\frac{1}{1}\cdot\frac{a}{a}+\frac{4}{a}}{\frac{a}{6}+\frac{2}{3}\cdot\frac{2}{2}}$ *The LCD of the numerator is a. The LCD of the denominator is 6.*

$= \dfrac{\frac{a+4}{a}}{\frac{a+4}{6}}$ *Add to simplify both numerator and denominator.*

$= \frac{a+4}{a} \div \frac{a+4}{6}$ *Rewrite as a division sentence.*

$= \frac{\overset{1}{\cancel{a+4}}}{a} \cdot \frac{6}{\underset{1}{\cancel{a+4}}}$ *The reciprocal of* $\frac{a+4}{6}$ *is* $\frac{6}{a+4}$.

$= \frac{6}{a}$ *Divide by common factors.*

b. $\dfrac{m-\frac{m+5}{m-3}}{m+1}$

$\dfrac{m-\frac{m+5}{m-3}}{m+1} = \dfrac{\frac{m(m-3)}{(m-3)}-\frac{m+5}{m-3}}{m+1}$ *The LCD of the numerator is m − 3. The LCD of the denominator is m + 1.*

$= \dfrac{\frac{m^2-3m-m-5}{m-3}}{m+1}$ *Subtract to simplify the numerator.*

(continued on the next page)

Motivating the Lesson

Hands-On Activity Have students cut scrap paper to represent a mixed number. For example, cut three larger pieces and one half size to make $3\frac{1}{2}$. Have them cut the whole pieces to represent fractions such as $1 = \frac{2}{2}$. Then the students can name the improper fraction by counting the pieces.

2 TEACH

In-Class Examples

For Example 1
Simplify.

a. $4 + \frac{x-3}{x+7}$ $\frac{5x+25}{x+7}$

b. $3 + \frac{r-4}{r^2-4}$ $\frac{3r^2+r-16}{r^2-4}$

For Example 2
How many Ping-Pong balls will weigh $1\frac{7}{8}$ pounds altogether? **300**

For Example 3
Simplify each rational expression.

a. $\dfrac{\frac{1}{a}-\frac{1}{b}}{\frac{1}{a}+\frac{1}{b}}$ $\frac{b-a}{b+a}$

b. $\dfrac{r-\frac{r+5}{r-3}}{r+1}$ $\frac{r-5}{r-3}$

Teaching Tip Help students to see that they are changing numbers with unlike denominators to numbers with like or common denominators. For example, consider $4 + \frac{x+y}{x-5}$ as $\frac{4}{1} + \frac{x+y}{x-5}$. Changing the mixed expression to an algebraic fraction requires that $\frac{4}{1}$ be renamed with the LCD $x - 5$ as its denominator.

Alternative Teaching Strategies

Reading Algebra The term *improper fraction* can be confusing. There is nothing improper about it; it is just another way of writing a rational expression.

EXPLORATION

Have students graph $y = x - (x + 5) \div (x - 3) \div (x - 1)$ so that they can see how different the graph is when they leave out some of the parentheses. Emphasize the importance of using parentheses to group numbers and denominators.

Answers for the Exploration

a.

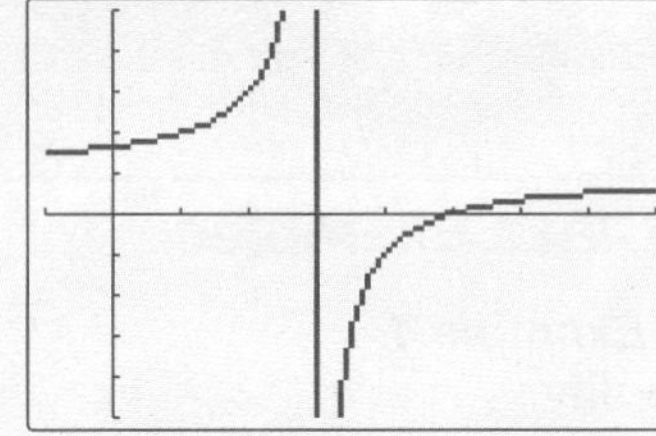

c.

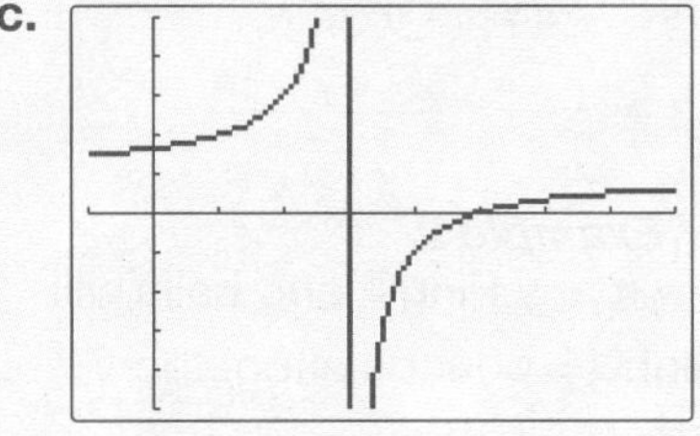

In-Class Example

For Example 4

Simplify $\dfrac{x - 2 + \frac{3}{x+2}}{x + 1 - \frac{10}{x+4}}$. $\dfrac{(x + 1)(x + 4)}{(x + 2)(x + 6)}$

$$= \frac{\frac{m^2 - 4m - 5}{m - 3}}{m + 1} \quad \textit{Simplify.}$$

$$= \frac{\frac{(m + 1)(m - 5)}{m - 3}}{m + 1} \quad \textit{Factor the numerator.}$$

$$= \frac{(m + 1)(m - 5)}{m - 3} \div (m + 1) \quad \textit{Rewrite as a division sentence.}$$

$$= \frac{(m + 1)(m - 5)}{m - 3} \cdot \frac{1}{(m + 1)} \quad \textit{The reciprocal of } m + 1 \textit{ is } \frac{1}{m+1}.$$

$$= \frac{\overset{1}{\cancel{(m + 1)}}(m - 5)}{m - 3} \cdot \frac{1}{\underset{1}{\cancel{(m + 1)}}} \quad \textit{Divide by the common factors.}$$

$$= \frac{m - 5}{m - 3}$$

You can use a graphing calculator to check the work you did in Example 3.

EXPLORATION — GRAPHING CALCULATORS

When you graph a function that involves rational expressions, remember to enclose every numerator and denominator in parentheses.

Your Turn b. $x \neq 3, x \neq -1$ a, c. See margin.

a. Graph $y = \dfrac{x - \frac{x+5}{x-3}}{x+1}$. Use the window [−1.4, 8] by [−5, 5].

b. What are the excluded values of x in the expression in part a? 3, −1

c. Graph $y = \dfrac{x-5}{x-3}$ on the same display used in part a.

d. What are the excluded values of x in the expression in part c? $x \neq 3$

e. Except for excluded values, do the graphs appear the same? yes

Example  **4** **Simplify** $\dfrac{a - 2 + \frac{3}{a+2}}{a + 1 - \frac{10}{a+4}}$.

$$\frac{a - 2 + \frac{3}{a+2}}{a + 1 - \frac{10}{a+4}} = \frac{\frac{(a-2)(a+2)}{a+2} + \frac{3}{a+2}}{\frac{(a+1)(a+4)}{(a+4)} - \frac{10}{a+4}} \quad \textit{The LCD of the numerator is } a + 2. \textit{ The LCD of the denominator is } a + 4.$$

$$= \frac{\frac{a^2 - 4 + 3}{a + 2}}{\frac{a^2 + 5a + 4 - 10}{(a + 4)}} \quad \textit{Add to simplify the numerator. Subtract to simplify the denominator.}$$

$$= \frac{\frac{a^2 - 1}{(a + 2)}}{\frac{a^2 + 5a - 6}{(a + 4)}} \quad \textit{Simplify.}$$

$$= \frac{\frac{(a + 1)(a - 1)}{(a + 2)}}{\frac{(a + 6)(a - 1)}{(a + 4)}} \quad \textit{Factor to simplify the numerator and denominator.}$$

$$= \frac{(a + 1)\overset{1}{\cancel{(a - 1)}}}{(a + 2)} \cdot \frac{(a + 4)}{(a + 6)\underset{1}{\cancel{(a - 1)}}} \quad \textit{Multiply by the reciprocal.}$$

$$= \frac{a^2 + 5a + 4}{a^2 + 8a + 12} \quad \textit{Multiply.}$$

CHECK FOR UNDERSTANDING

Communicating Mathematics

Study the lesson. Then complete the following.

1. **Determine** the simplified form of $\dfrac{\frac{3}{(x+1)(x+2)(x+3)}}{\frac{4}{(x+3)(x+2)(x+1)}}$ mentally. $\mathbf{\frac{3}{4}}$

2. a. What is the LCD for the expression $3 + \frac{x}{2} + \frac{4}{x} + \frac{x^2 + 3x + 15}{x - 2}$? $\mathbf{2x(x-2)}$

   b. Write the expression in simplified form. $\mathbf{\frac{3x^3 + 10x^2 + 26x - 16}{2x(x-2)}}$

5. $\mathbf{\frac{6m^2 + m + 1}{2m}}$

Guided Practice

Write each mixed expression as a rational expression.

3. $8 + \frac{3}{x}$ $\mathbf{\frac{8x+3}{x}}$ 4. $5 + \frac{8}{3m}$ $\mathbf{\frac{15m+8}{3m}}$ 5. $3m + \frac{m+1}{2m}$

Simplify.

6. $\dfrac{4\frac{1}{3}}{5\frac{4}{7}}$ $\mathbf{\frac{7}{9}}$ 7. $\dfrac{6\frac{2}{5}}{3\frac{5}{9}}$ $\mathbf{\frac{9}{5}}$ 8. $\dfrac{\frac{3}{x}}{\frac{x}{3}}$ $\mathbf{\frac{9}{x^2}}$

9. $\dfrac{\frac{5}{y}}{\frac{10}{y^2}}$ $\mathbf{\frac{y}{2}}$ 10. $\dfrac{\frac{x+4}{x-2}}{\frac{x+5}{x-2}}$ $\mathbf{\frac{x+4}{x+5}}$ 11. $\dfrac{\frac{a-b}{a+b}}{\frac{3}{a+b}}$ $\mathbf{\frac{a-b}{3}}$

12. Nakita performed an operation on $\frac{x+2}{3x-1}$ and $\frac{2x^2-8}{3x-1}$ and got $\frac{1}{2(x-2)}$. What operation was it? Explain. **Division; see margin for explanation.**

EXERCISES

Practice

Write each mixed expression as a rational expression.

13. $3 + \frac{6}{x+3}$ $\mathbf{\frac{3x+15}{x+3}}$ 14. $11 + \frac{a-b}{a+b}$ $\mathbf{\frac{12a+10b}{a+b}}$ 15. $3 - \frac{4}{2x+1}$ $\mathbf{\frac{6x-1}{2x+1}}$

17. $\mathbf{\frac{5r^2 + r - 48}{r^2 - 9}}$

16. $3 + \frac{x-4}{x+y}$ $\mathbf{\frac{4x+3y-4}{x+y}}$ 17. $5 + \frac{r-3}{r^2-9}$ 18. $3 + \frac{x^2+y^2}{x^2-y^2}$ $\mathbf{\frac{4x^2-2y^2}{x^2-y^2}}$

Simplify.

19. $\dfrac{7\frac{2}{3}}{5\frac{3}{4}}$ $\mathbf{\frac{4}{3}}$ 20. $\dfrac{6\frac{1}{7}}{8\frac{3}{5}}$ $\mathbf{\frac{5}{7}}$ 21. $\dfrac{\frac{a^3}{b}}{\frac{a^2}{b^2}}$ $\mathbf{ab}$

22. $\dfrac{\frac{x^2y^2}{a}}{\frac{x^2y}{a^3}}$ $\mathbf{a^2y}$ 23. $\dfrac{2 + \frac{5}{x}}{\frac{x}{3} + \frac{5}{6}}$ $\mathbf{\frac{6}{x}}$ 24. $\dfrac{4 + \frac{3}{y}}{\frac{3}{8} + \frac{y}{2}}$ $\mathbf{\frac{8}{y}}$

27. $\mathbf{\frac{1}{x+3}}$

25. $\dfrac{a - \frac{15}{a-2}}{a+3}$ $\mathbf{\frac{a-5}{a-2}}$ 26. $\dfrac{x + \frac{35}{x+12}}{x+7}$ $\mathbf{\frac{x+5}{x+12}}$ 27. $\dfrac{\frac{x^2-4}{x^2+5x+6}}{x-2}$

30. $\mathbf{\frac{(a+4)(a-1)(a+2)}{(a+5)(a^2+2)}}$

32. $\mathbf{\frac{(x+2)(x-1)(x+3)}{(x-2)(x^2+2x-5)}}$

33. $\mathbf{\frac{t^2+2t+2}{-t^2}}$

28. $\dfrac{m + \frac{3m+7}{m+5}}{m+1}$ $\mathbf{\frac{m+7}{m+5}}$ 2

$\dfrac{m + 5 + \frac{2}{m+2}}{m + 1 + \frac{6}{m+6}}$ $\mathbf{\frac{m+6}{m+2}}$ 30. $\dfrac{a + 1 + \frac{3}{a+5}}{a + 1 + \frac{3}{a-1}}$

31. $\dfrac{y + 6 + \frac{3}{y+2}}{y + 11 + \frac{48}{y-3}}$ $\mathbf{\frac{y-3}{y+2}}$ 3

$\dfrac{x + 3 + \frac{4}{x-2}}{x - 1 - \frac{2}{x+3}}$ 33. $\dfrac{t + 1 + \frac{1}{t+1}}{1 - t - \frac{1}{t+1}}$

3 PRACTICE/APPLY

Check for Understanding

Exercises 1–12 are designed to help you assess your students' understanding through reading, writing, speaking, and modeling. You should work through Exercises 1–2 with your students and then monitor their work on Exercises 3–12.

Error Analysis

Since division is not commutative, identifying the major division bar in a complex fraction is crucial.

$\dfrac{\frac{a}{b}}{c} = (a \div b) \div c$ or $a \div (b \div c)$?

The major division bar should be indicated by some emphasis (bolder, longer, or wider).

Assignment Guide

Core: 13–37 odd, 38, 39, 41–48
Enriched: 14–36 even, 37–48

For **Extra Practice**, see p. 784.

The red A, B, and C flags, printed only in the Teacher's Wraparound Edition, indicate the level of difficulty of the exercises.

Study Guide Masters, p. 86

NAME____________ DATE ________

Study Guide

Student Edition Pages 690–695

Mixed Expressions and Complex Fractions

Algebraic expressions such as $a + \frac{b}{c}$ and $5 + \frac{x+y}{x+3}$ are called **mixed expressions.** Changing mixed expressions to improper fractions is similar to changing mixed numbers to improper fractions.

If a fraction has one or more fractions in the numerator or denominator, it is called a **complex fraction.**

Simplifying Complex Fractions
Any complex fraction $\frac{\frac{a}{b}}{\frac{c}{d}}$, where $b \neq 0$, $c \neq 0$, and $d \neq 0$, may be expressed as $\frac{a}{b} \cdot \frac{d}{c}$, or $\frac{ad}{bc}$.

Example: Simplify $\dfrac{2 + \frac{4}{a}}{\frac{a+2}{3}}$.

$$\frac{2 + \frac{4}{a}}{\frac{a+2}{3}} = \frac{\frac{2a}{a} + \frac{4}{a}}{\frac{a+2}{3}} = \frac{\frac{2a+4}{a}}{\frac{a+2}{3}} = \frac{2a+4}{a} \cdot \frac{3}{a+2} = \frac{2(a+2)}{a} \cdot \frac{3}{a+2} = \frac{6}{a}$$

Simplify.

1. $10 + \frac{60}{x+5}$ $\frac{10x+110}{x+5}$ 2. $12 + \frac{x-y}{x+y}$ $\frac{13x+11y}{x+y}$ 3. $4 - \frac{4}{2x+1}$ $\frac{8x}{2x+1}$

4. $\dfrac{2\frac{2}{5}}{3\frac{3}{4}}$ $\frac{16}{25}$ 5. $\dfrac{\frac{3}{x}}{\frac{4}{y}}$ $\frac{3y}{4x}$ 6. $\dfrac{\frac{3}{y+2} - \frac{2}{y-2}}{\frac{1}{y+2} - \frac{2}{y-2}}$ $\frac{y-10}{-y-6}$

7. $\dfrac{1 - \frac{1}{x}}{1 - \frac{1}{x^2}}$ $\frac{x}{x+1}$ 8. $\dfrac{\frac{1}{x-3}}{\frac{2}{x^2-9}}$ $\frac{x+3}{2}$ 9. $\dfrac{\frac{x^2-25}{y}}{x^3-5x^2}$ $\frac{x+5}{x^2y}$

Reteaching

Using Models Have students complete the following table.

Single Fraction	Mixed Expression
$\frac{6x^2 + 4x + 1}{3}$	$2x^2 + \frac{4x+1}{3}$
$\frac{2x}{x+y}$	$1 + \frac{x-y}{x+y}$
$\frac{5x^2 + 5x + 6}{x+1}$	$5x + \frac{6}{x+1}$
$\frac{a+3b}{a+b}$	$2 - \frac{a-b}{a+b}$

Additional Answer

12. $\frac{x+2}{3x-1} \div \frac{2x^2-8}{3x-1} = \frac{x+2}{3x-1} \cdot \frac{3x-1}{2(x-2)(x+2)} = \frac{1}{2(x-2)}$

35. $\frac{32b^2}{35}$

34. What is the quotient when $b + \frac{1}{b}$ is divided by $a + \frac{1}{a}$? $\frac{a(b^2+1)}{b(a^2+1)}$

35. What is the product when $\frac{2b^2}{5c}$ is multiplied by the quotient of $\frac{4b^3}{2c}$ and $\frac{7b^3}{8c^2}$?

36. Write $1 + \cfrac{1}{1 + \cfrac{1}{1 + \cfrac{1}{1 + \cfrac{1}{x}}}}$ in simplest form. $\frac{5x+3}{3x+2}$

Graphing Calculator

37. a. Graph $y = \cfrac{\frac{x+3}{2x}}{\frac{3x+9}{4}}$. What are its excluded values? $x \neq 0, x \neq -3$

b. Graph $\frac{x+3}{2x} \cdot \frac{4}{3x+9}$ on the same display used in part a. What are its excluded values? $x \neq 0, x \neq -3$

c. Except for the excluded values, are the graphs the same? yes

Critical Thinking

38. Simplify $\cfrac{3}{1 - \cfrac{3}{3+y}} - \cfrac{3}{\cfrac{3}{3-y} - 1}$. 6

Applications and Problem Solving

39. **Acoustics** If a train is moving toward you at v miles per hour and blowing its whistle at a frequency of f, then you hear it as though it were blowing its whistle with a frequency h, which is defined as $h = \cfrac{f}{1 - \frac{v}{s}}$, where s is the speed of sound.

a. Simplify the right side of this formula. $\frac{fs}{s-v}$

b. Suppose a train whistle blows at 370 cycles per second (at the same frequency as the first F sharp (F#) above middle C on the piano). The train is moving toward you at a speed of 80 miles per hour. The speed of sound is 760 miles per hour. Find the frequency of the sound as you hear it. 413.5

c. The note F# and the notes to its right on the piano are listed with their frequencies in the table.

Note	F#	G	G#	A	A#	B	C	C#
Frequency	370.0	392.0	415.3	440.0	466.1	493.8	523.2	554.3

By approximately how many notes did the sound rise in part b? 2

d. Find the frequency of the same whistle as you would hear it from an approaching TGV, the French train that is the fastest in the world ($v = 236$ miles per hour). By approximately how many notes would it rise? 6

FYI

Have you noticed that the pitch (frequency) of a train whistle or a horn from a speeding car gets higher as it approaches you and lower as it moves away from you? This effect is known as the *Doppler effect*, named for the Austrian scientist Christian Doppler, who discovered it in 1842.

Practice Masters, p. 86

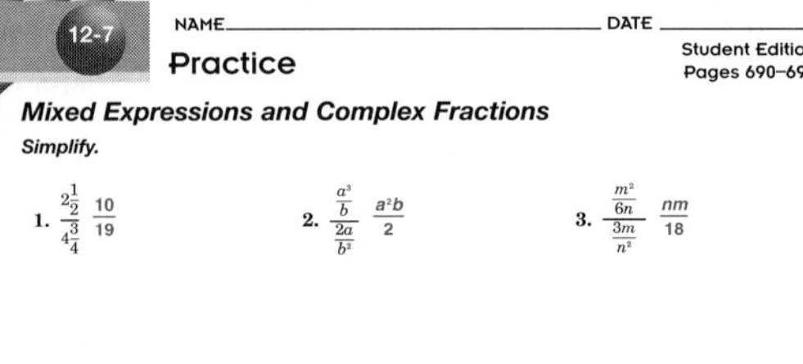

12-7 NAME ______ DATE ______

Practice

Student Edition Pages 690–695

Mixed Expressions and Complex Fractions

Simplify.

1. $\cfrac{2\frac{1}{2}}{4\frac{3}{4}}$ $\frac{10}{19}$

2. $\cfrac{\frac{a^3}{b}}{\frac{2a}{b^2}}$ $\frac{a^2b}{2}$

3. $\cfrac{\frac{m^3}{6n}}{\frac{3m}{n^2}}$ $\frac{nm}{18}$

4. $\cfrac{\frac{b-2}{b+3}}{\frac{b-2}{3}}$ $\frac{3}{b+3}$

5. $\cfrac{\frac{x^2-y^2}{x^2}}{\frac{x+y}{3x}}$ $\frac{3(x-y)}{x}$

6. $\cfrac{\frac{w+4}{w}}{\frac{w^2-16}{w}}$ $\frac{1}{w-4}$

7. $\cfrac{\frac{x^2-1}{x}}{\frac{x-1}{x^2}}$ $x(x+1)$

8. $\cfrac{\frac{a}{a-b}}{\frac{a^2}{a^2-b^2}}$ $\frac{a+b}{a}$

9. $\cfrac{\frac{r}{s} - \frac{s}{r}}{\frac{1}{s} + \frac{1}{r}}$ $r - s$

10. $\cfrac{\frac{4}{c^2-d^2}}{\frac{2}{c+d}}$ $\frac{2}{c-d}$

11. $\cfrac{\frac{m}{n}+1}{\frac{m}{n}-1}$ $\frac{m+n}{m-n}$

12. $\cfrac{\frac{3}{a}+\frac{4}{b}}{\frac{4}{a}-\frac{3}{b}}$ $\frac{3b+4a}{4b-3a}$

13. $\cfrac{\frac{x^2-y^2}{xy}}{\frac{x+y}{y}}$ $\frac{x-y}{x}$

14. $\cfrac{\frac{a^2+a-12}{a^2+3a-4}}{\frac{a-3}{a^2-a}}$ a

15. $\cfrac{x - \frac{6}{x+2}}{x + \frac{8}{x+5}}$ $\frac{x^3+7x^2+4x-30}{x^3+7x^2+18x+16}$

Extension

Reasoning

Simplify $\cfrac{\frac{a^2-a-1}{a-1}}{a - \frac{1}{a-1}}$. 1

FYI

The Doppler effect is used in astronomical measurements and in radar.

40. **Statistics** In 1993, New Jersey was the most densely populated state, and Alaska was the least densely populated. The population of New Jersey was 7,879,000, and the population of Alaska was 599,000. The land area of New Jersey is about 7419 square miles, and the land area of Alaska is about 570,374 square miles. How many more people were there per square mile in New Jersey than in Alaska in 1993? **approximately 1061**

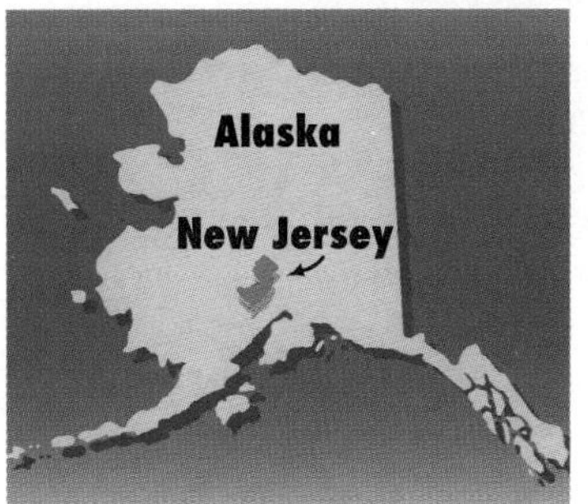

Mixed Review

41. Find $\frac{4x}{2x+6} + \frac{3}{x+3}$. (Lesson 12–6) $\frac{2x+3}{x+3}$
42. Find $\frac{b-5}{b^2-7b+10} \cdot \frac{b-2}{3}$. (Lesson 12–2) $\frac{1}{3}$
43. Factor $4x^2 - 1$. (Lesson 10–4) $(2x+1)(2x-1)$
44. Find $x^4y^5 \div xy^3$. (Lesson 9–2) x^3y^2
45. Solve the system of equations. (Lesson 8–4)
$3a + 4b = -25$
$2a - 3b = 6$ $(-3, -4)$
46. **Geometry** A quadrilateral has vertices $A(-1, -4)$, $B(2, -1)$, $C(5, -4)$, and $D(2, -7)$. (Lesson 6–6)
    a. Graph the quadrilateral and determine the relationship, if any, among its sides.
    b. What type of quadrilateral is $ABCD$? **square**
47. Solve $P = 2\ell + 2w$ for w. (Lesson 3–6)
48. Find $\sqrt{225}$. (Lesson 2–8) **15**

46a. See margin for graph; opposite sides are parallel, adjacent sides are perpendicular, perpendicular diagonals.

47. $w = \frac{P - 2\ell}{2}$

WORKING ON THE In•ves•ti•ga•tion

Refer to the Investigation on pages 656–657.

A Growing Concern

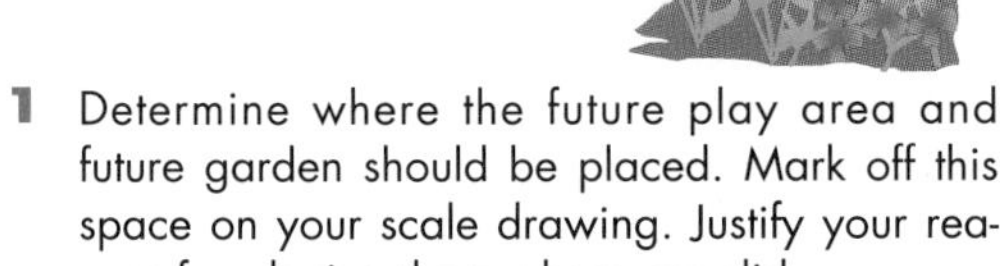

1. Determine where the future play area and future garden should be placed. Mark off this space on your scale drawing. Justify your reasons for placing them where you did.
2. Develop a detailed plan for the deck and/or patio and walkways for the Sanchez family. Indicate the dimensions. Explain why you chose the placement and the type of materials for each place.
3. Figure the cost of materials, labor, and profit margin of the construction of the deck and/or patio. Justify your costs and make a case for building a deck and/or patio and the walkways as you designed them.
4. Be sure to add the deck and/or patio and the walkways to your drawing.

Add the results of your work to your Investigation Folder.

Tech Prep

Statistician Students who are interested in statistics may wish to do further research on the data provided in Exercise 40 and explore the potential growth of this career. For more information on tech prep, see the *Teacher's Handbook*.

In•ves•ti•ga•tion

Working on the Investigation

The Investigation on pages 656–657 is designed to be a long-term project that is completed over several days or weeks. Encourage students to keep their materials in their Investigation Folder as they work on the Investigation.

4 ASSESS

Closing Activity

Speaking Have students describe the process used to simplify $\frac{\frac{5a^2-20}{2a+2}}{\frac{10a-20}{4a}} \cdot \frac{a^2+2a}{a+10}$

Additional Answer

46a.

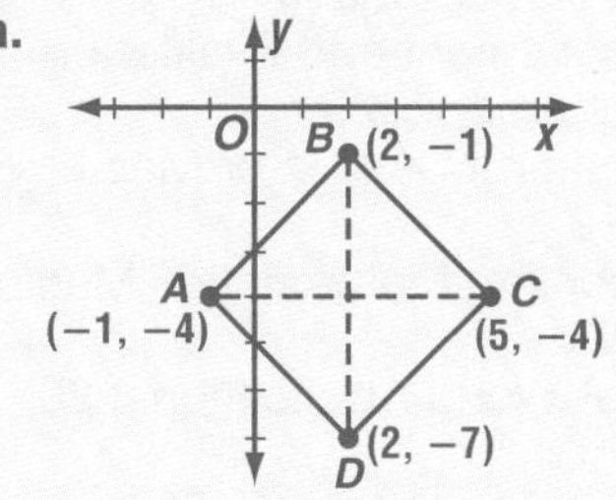

Enrichment Masters, p. 86

12-7 NAME ________ DATE ________

Enrichment

Student Editio
Pages 691–69

Complex Fractions

Complex fractions are really not complicated. Remember that a fraction can be interpreted as dividing the numerator by the denominator.

$$\frac{\frac{2}{3}}{\frac{5}{7}} = \frac{2}{3} \div \frac{5}{7} = \frac{2}{3} \cdot \frac{7}{5} = \frac{2(7)}{3(5)} = \frac{14}{15}$$

Let a, b, c, and d be numbers, with $b \neq 0$, $c \neq 0$, and $d \neq 0$.

$$\frac{\frac{a}{b}}{\frac{c}{d}} = \frac{a}{b} \div \frac{c}{d} = \frac{a}{b} \cdot \frac{d}{c} = \frac{ad}{bc}$$

Notice the pattern: *numerator of the answer (ad)* $\frac{\frac{a}{b}}{\frac{c}{d}}$ *denominator of the answer (bc)*

Example 1: Simplify $\frac{\frac{5x}{4}}{\frac{x+2}{3}}$.

$$\frac{\frac{5x}{4}}{\frac{x+2}{3}} = \frac{5x(3)}{4(x+2)} = \frac{15x}{4x+8}$$

Example 2: Simplify $\frac{\frac{x}{2}+4}{3x-2}$.

$$\frac{\frac{x}{2}+4}{3x-2} = \frac{\frac{x+8}{2}}{\frac{3x-2}{1}} = \frac{(x+8)(1)}{2(3x-2)} = \frac{x+8}{6x-4}$$

Simplify each complex fraction.

1. $\frac{\frac{2x}{5}}{\frac{y}{6}}$ $\frac{12x}{5y}$
2. $\frac{\frac{4}{5x}}{\frac{3}{x}}$ $\frac{4}{15}$
3. $\frac{x-3}{\frac{2x+1}{4}}$ $\frac{4x-12}{2x+1}$
4. $\frac{x^2+\frac{1}{3}}{4x+\frac{1}{3}}$ $\frac{3x^2+1}{12x+1}$
5. $\frac{1-x^{-1}}{\frac{2x}{5}-1}$ $\frac{5x-5}{2x^2-5x}$
6. $\frac{x+2x^{-2}}{2+\frac{x}{3}}$ $\frac{3(x^3+2)}{x^3+6x^2}$
7. $\frac{x}{x+\frac{1}{x+\frac{1}{x}}}$ $\frac{x^2+1}{x^2+2}$
8. $\frac{x+2}{x-2+\frac{1}{x+2+\frac{1}{x}}}$ $\frac{x^3+4x^2+5x+2}{x^3-2x-2}$

12-8 LESSON NOTES

NCTM Standards: 1–5

Instructional Resources

- Study Guide Master 12-8
- Practice Master 12-8
- Enrichment Master 12-8
- Assessment and Evaluation Masters, p. 325
- Graphing Calculator Masters, p. 12
- Modeling Mathematics Masters, p. 83
- Real-World Applications, 30

Transparency 12-8A contains the 5-Minute Check for this lesson; **Transparency 12-8B** contains a teaching aid for this lesson.

Recommended Pacing	
Standard Pacing	Days 11 & 12 of 14
Honors Pacing	Days 10 & 11 of 13
Block Scheduling*	Day 5 of 6 (along with Lesson 12-7)
Alg. 1 in Two Years*	Days 22, 23, & 24 of 27

*For more information on pacing and possible lesson plans, refer to the *Block Scheduling Booklet* and *Algebra 1 in Two Years*.

1 FOCUS

5-Minute Check
(over Lesson 12-7)

Simplify.

1. $8 + \frac{5}{3y}$ $\quad \frac{24y + 5}{3y}$

2. $3r^2 + \frac{4}{2r + 1}$ $\quad \frac{6r^3 + 3r^2 + 4}{2r + 1}$

3. $\frac{\frac{b^2}{t}}{\frac{t}{b^3}}$ $\quad \frac{b^5}{t^2}$

4. $\frac{\frac{j^3}{t^2}}{\frac{j+t}{j-t}}$ $\quad \frac{j^3(j - t)}{t^2(j + t)}$

Motivating the Lesson

Situational Problem If Gene addresses 200 envelopes in 8 hours and Marty addresses 200 envelopes in 8 hours, how long will it take them to address 200 envelopes working together? **4 hours**

12–8 Solving Rational Equations

What YOU'LL LEARN

- To solve rational equations.

Why IT'S IMPORTANT

You can use rational equations to solve problems involving work, psychology, and electronics.

Work

Tiko and Julio have decided to start a lawn care service in their neighborhood. In order to schedule their clients in an organized manner, they compared notes on how long it took each of them to mow the same yard. Tiko could mow and trim Mrs. Harris's lawn in 3 hours, while it took Julio 2 hours. How long would it take if they both worked together?

You can answer this question by solving a **rational equation**. A rational equation is an equation that contains rational expressions.

Explore Since it takes Tiko 3 hours to do the yard, he can finish $\frac{1}{3}$ of the yard in 1 hour. At his rate, Julio can finish $\frac{1}{2}$ of the yard in an hour. Use the following formula.

$$\underbrace{\text{rate of work}}_{r} \cdot \underbrace{\text{time}}_{t} = \underbrace{\text{work done}}_{w}$$

Plan In t hours, Tiko can do $t \cdot \frac{1}{3}$ or $\frac{t}{3}$ of the job and Julio can do $t \cdot \frac{1}{2}$ or $\frac{t}{2}$ of the job. Thus, $\frac{t}{3} + \frac{t}{2} = 1$, where 1 represents the finished job.

Solve

$$\frac{t}{3} + \frac{t}{2} = 1$$

$$6\left(\frac{t}{3} + \frac{t}{2}\right) = 6(1) \quad \textit{Multiply each side by the LCD, 3(2) or 6.}$$

$$2t + 3t = 6 \quad \textit{Use the distributive property.}$$

$$5t = 6$$

$$t = \frac{6}{5} \text{ or } 1\frac{1}{5} \quad \textit{Check this solution.}$$

Examine If Tiko and Julio work for $1\frac{1}{5}$ hours, will the whole yard get done?

$$\text{Tiko's } rt + \text{Julio's } rt \stackrel{?}{=} 1 \rightarrow \frac{1}{3} \cdot \frac{6}{5} + \frac{1}{2} \cdot \frac{6}{5} \stackrel{?}{=} 1$$

$$\frac{2}{5} + \frac{3}{5} \stackrel{?}{=} 1$$

$$1 = 1$$

Working together, Tiko and Julio can finish the yard in $1\frac{1}{5}$ hours.

Example 1 Solve $\frac{10}{3y} - \frac{5}{2y} = \frac{1}{4}$.

$$\frac{10}{3y} - \frac{5}{2y} = \frac{1}{4}$$

$$12y\left(\frac{10}{3y} - \frac{5}{2y}\right) = 12y\left(\frac{1}{4}\right) \quad \textit{Multiply each side by the LCD, 12y.}$$

$$12y \cdot \frac{10}{3y} - 12y \cdot \frac{5}{2y} = 12y \cdot \frac{1}{4} \quad \textit{Use the distributive property.}$$

$$40 - 30 = 3y$$

$$10 = 3y$$

$$y = \frac{10}{3} \text{ or } 3\frac{1}{3}$$

GLENCOE Technology

CD-ROM Interaction

A multimedia simulation connects rational equations with underwater sonar devices. A blackline master activity with teacher's notes provides a follow-up to the CD-ROM simulation.

For Windows & Macintosh

Check: $\frac{10}{3\left(3\frac{1}{3}\right)} - \frac{5}{2\left(3\frac{1}{3}\right)} = \frac{1}{4}$

$$\frac{10}{10} - \frac{15}{20} \stackrel{?}{=} \frac{1}{4}$$

$$\frac{20}{20} - \frac{15}{20} \stackrel{?}{=} \frac{1}{4}$$

$$\frac{1}{4} = \frac{1}{4} \checkmark$$

The solution is $3\frac{1}{3}$.

Multiplying each side of an equation by the LCD of two rational expressions can yield results that are not solutions to the original equation. Such solutions are called **extraneous solutions** or "false" solutions.

Example  **Solve $\frac{x}{x-1} + \frac{2x-3}{x-1} = 2$.**

$$\frac{x}{x-1} + \frac{2x-3}{x-1} = 2$$

$$(x-1)\left(\frac{x}{x-1} + \frac{2x-3}{x-1}\right) = (x-1)2 \quad \textit{Multiply each side by the LCD, } x-1.$$

$$(x-1)\left(\frac{x}{x-1}\right) + (x-1)\left(\frac{2x-3}{x-1}\right) = (x-1)2$$

$$x + 2x - 3 = 2x - 2$$

$$3x - 3 = 2x - 2$$

$$x = 1$$

The number 1 is not a solution, since 1 is an excluded value for x. Thus, the equation has no solution.

GLOBAL CONNECTIONS

Algonquin Indians named the Mississippi River the "Father of Waters." Their phrase *Misi Sipi* meant big water. The Mississippi River drains one eighth of the North American continent, discharging 350 trillion gallons of water a day.

You can use the distance formula to solve real-world problems.

Example 3

APPLICATION

Commerce

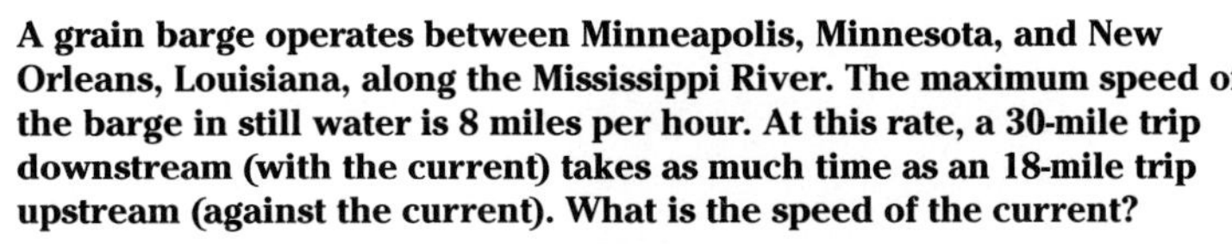

A grain barge operates between Minneapolis, Minnesota, and New Orleans, Louisiana, along the Mississippi River. The maximum speed of the barge in still water is 8 miles per hour. At this rate, a 30-mile trip downstream (with the current) takes as much time as an 18-mile trip upstream (against the current). What is the speed of the current?

Explore Let c = the speed of the current. The speed of the barge traveling downstream is 8 miles per hour plus the speed of the current, that is, $(8 + c)$ miles per hour. The speed of the barge traveling upstream is 8 miles per hour minus the speed of the current, that is, $(8 - c)$ miles per hour.

Plan To represent time t, solve $d = rt$ for t. Thus, $t = \frac{d}{r}$.

	d	r	$t = \frac{d}{r}$
downstream	30	$8 + c$	$\frac{30}{8+c}$
upstream	18	$8 - c$	$\frac{18}{8-c}$

(continued on the next page)

2 TEACH

Teaching Tip Emphasize that when solving rational equations, each term must be multiplied by the LCD in order to have a balanced equation. A common mistake is to multiply only those terms that are expressed in fractions.

In-Class Examples

For Example 1
Solve each equation.

a. $\frac{2-r}{4} + \frac{3r}{5} = \frac{r+2}{3}$ **10**

b. $\frac{4}{k+1} - \frac{5}{k} = \frac{3}{k}$ **−2**

For Example 2
Solve each equation.

a. $\frac{-a}{a+1} + \frac{a-1}{4-a} = \frac{-5}{a^2-3a-4}$ **3**

b. $\frac{1}{x-4} + \frac{2}{16-x^2} = \frac{3}{x+4}$ **7**

For Example 3
Use the formula $d = rt$ to solve the following problem.

An airplane can fly at a rate of 500 mph in calm air. It can fly 2575 miles with the wind in the same time it can fly 2425 miles against the wind. Find the speed of the wind.
$w = 15$ **mph**

Teaching Tip In Example 3, emphasize the difference between traveling downstream and upstream. Also emphasize the fact that the times are equal.

GLOBAL CONNECTIONS

The Mississippi River's source is in Minnesota. It then flows 2340 miles to the Gulf of Mexico. The Amazon River discharges ten times more water and the Congo River discharges twice as much water as the Mississippi.

Cooperative Learning

Trade-A-Problem Have students work in pairs. One student will add two fractions $\left(\text{for example, } \frac{2}{3} + \frac{4}{7} = \frac{26}{21}\right)$, then replace one number with a variable $\left(\frac{x}{3} + \frac{x+2}{7} = \frac{26}{21}\right)$, and then give it to the other student to solve. Then switch roles.

For more information on the trade-a-problem strategy, see *Cooperative Learning in the Mathematics Classroom*, one of the titles in the Glencoe Mathematics Professional Series, pages 25–26.

Longest Rivers in the World

River	Miles
1. Nile	4132
2. Amazon	4000
3. Yangtze	3915
4. Yenisey	3442
5. Huang Ho (Yellow)	3395

In-Class Example

For Example 4
R_1 and R_2 are connected in series. Find R_T.

a. $R_1 = 5$ ohms
$R_2 = 11$ ohms **16 ohms**
b. $R_1 = 8.2$ ohms
$R_2 = 4.6$ ohms **12.8 ohms**

Longest Rivers in North America

River	Miles
1. Mackenzie-Peace	2635
2. Missouri-Red Rock	2540
3. Mississippi	2348
4. Missouri	2315
5. Yukon	1979

Solve
$$\frac{30}{8+c} = \frac{18}{8-c}$$
$$30(8-c) = 18(8+c) \quad \textit{Find the cross products.}$$
$$240 - 30c = 144 + 18c$$
$$96 = 48c$$
$$c = 2$$

Examine Check the value to see if it makes sense.
The barge goes downstream at (8 + 2) or 10 mph.
A 30-mile trip would take 30 ÷ 10 or 3 hours.
The barge goes upstream at (8 − 2) or 6 mph.
An 18-mile trip takes 18 ÷ 6 or 3 hours.
Both trips take the same amount of time, so the speed of the current must be 2 miles per hour.

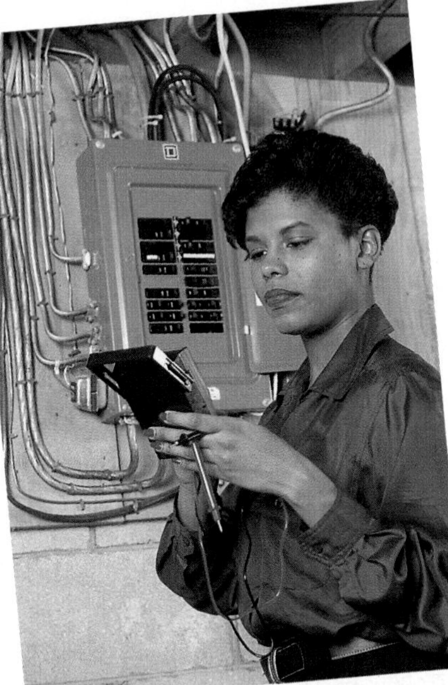

Electricity can be described as the flow of electrons through a conductor, such as a copper wire. Electricity flows more freely through some conductors than others. The force opposing the flow is called *resistance*. The unit of resistance commonly used is the *ohm*.

Resistances can occur one after another, that is, *in series*. Resistances can also occur in branches with the conductor going in the same direction, or *in parallel*.

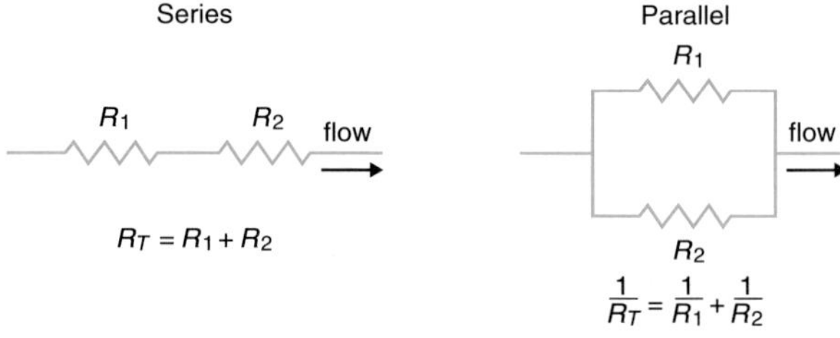

Example 4

Electronics

Assume that $R_1 = 4$ ohms and $R_2 = 3$ ohms. Compute the total resistance of the conductor when the resistances are in series and in parallel.

series
$$R_T = R_1 + R_2$$
$$= 4 + 3 \text{ or } 7$$

parallel
$$\frac{1}{R_T} = \frac{1}{R_1} + \frac{1}{R_2}$$
$$\frac{1}{R_T} = \frac{1}{4} + \frac{1}{3}$$
$$\frac{1}{R_T} = \frac{7}{12}$$
$$12 = 7R_T \quad \textit{Cross products}$$
$$R_T = \frac{12}{7} \text{ or } 1\frac{5}{7}$$

***GLENCOE* Technology**

Interactive Mathematics Tools Software

This multimedia software provides an interactive lesson that explores traveling upstream and downstream in a boat by manipulating the current and distance traveled. A **Computer Journal** gives students an opportunity to write about what they have learned.

For Windows & Macintosh

A circuit, or path for the flow of electrons, often has some resistances connected in series and others in parallel.

Example 5 

Electronics

A parallel circuit has one branch in series as shown at the right. Given that the total resistance is 2.25 ohms, R_1 = 3 ohms, and R_2 = 4 ohms, find R_3.

$\frac{1}{R_T} = \frac{1}{R_1} + \frac{1}{R_2 + R_3}$ *The total resistance of the branch in series is $R_2 + R_3$.*

$\frac{1}{2.25} = \frac{1}{3} + \frac{1}{4 + R_3}$ *$R_T = 2.25$, $R_1 = 3$, and $R_2 = 4$*

$\frac{1}{2.25} - \frac{1}{3} = \frac{1}{4 + R_3}$

$\frac{4}{9} - \frac{3}{9} = \frac{1}{4 + R_3}$

$\frac{1}{9} = \frac{1}{4 + R_3}$

$4 + R_3 = 9$ *Find the cross products.*

$R_3 = 5$

Thus, R_3 is 5 ohms.

R2 R3 flow R1

CHECK FOR UNDERSTANDING

Communicating Mathematics

Study the lesson. Then complete the following.

1. Refer to the application at the beginning of the lesson. How would the solution be different if Julio takes 6 hours to complete the lawn? **2 hours**
2. **You Decide** Antoinette solved $\frac{2m}{1-m} + \frac{m+3}{m^2-1} = 1$ and claimed that 1 and $-\frac{4}{3}$ were the solutions. Joel says that $-\frac{4}{3}$ is the only solution. Who is correct, and why? **2–3. See margin.**
3. **Define** a rational equation and distinguish it from a linear equation.

4. **Assess Yourself** **a–b. See students' work.**
   a. Describe an activity that you and a friend do together that each can complete separately. Estimate the time it would take for each of you to complete it working alone.
   b. Use the math you have learned to find out how long it would take to complete the activity if the two of you worked together.

Guided Practice

Solve each equation.

5. $\frac{1}{4} + \frac{4}{x} = \frac{1}{x}$ **−12**
6. $\frac{1}{5} + \frac{3}{2y} = \frac{3}{3y}$ **$-\frac{5}{2}$**
7. $\frac{4}{x+5} = \frac{4}{3(x+2)}$ **$-\frac{1}{2}$**
8. $\frac{x}{2} = \frac{3}{x+1}$ **−3, 2**
9. $\frac{a-1}{a+1} - \frac{2a}{a-1} = -1$ **0**
10. $\frac{w-2}{w} - \frac{w-3}{w-6} = \frac{1}{w}$ **3**
11. **Work** Olivia can wash and wax her car, vacuum the interior, and wash the insides of the windows in 5 hours. What part of the job can she do in
    a. 1 hour? **$\frac{1}{5}$** b. 3 hours? **$\frac{3}{5}$** c. x hours? **$\frac{x}{5}$**

In-Class Example

For Example 5
Solve each example for the circuit shown with Example 5.

a. Find R_3 if $R_T = \frac{14}{3}$ ohms, R_1 = 6 ohms, and R_2 = 9 ohms. **12 ohms**
b. Find R_1 if R_T = 10 ohms, R_2 = 12 ohms, and R_3 = 8 ohms. **20 ohms**

Teaching Tip For Example 5, remind students that when more than two resistances are connected in series, the total resistance is the sum of the individual resistances.
$R_T = R_1 + R_2 + R_3 + ...$
When more than two resistors are connected in parallel, the following formula is used for total resistance.
$\frac{1}{R_T} = \frac{1}{R_1} + \frac{1}{R_2} + \frac{1}{R_3} + ...$

3 PRACTICE/APPLY

Check for Understanding

Exercises 1–15 are designed to help you assess your students' understanding through reading, writing, speaking, and modeling. You should work through Exercises 1–4 with your students and then monitor their work on Exercises 5–15.

Additional Answers

2. She correctly solved the equation, but $m = 1$ makes a denominator zero and it must therefore be discarded as a solution.
3. A rational equation contains rational expressions whereas a linear equation has integer coefficients.

Reteaching

Using Guess and Check Give a set of work and uniform motion problems with a proposed answer to each student. Students are to determine if the answers are correct without solving the problems. Students will have to identify and examine the conditions in each problem.

Assignment Guide

Core: 17–49 odd, 50–58
Enriched: 16–40 even, 41–58

For **Extra Practice**, see p. 784.

The red A, B, and C flags, printed only in the Teacher's Wraparound Edition, indicate the level of difficulty of the exercises.

12. Recreation Sally and her brother rented a boat to go fishing in Jones Creek. The maximum speed of the boat in still water was 3 miles per hour. At this rate, a 9-mile trip downstream (with the current) took the same amount of time as a 3-mile trip upstream. Let c = the speed of the current. Copy and complete the table below.

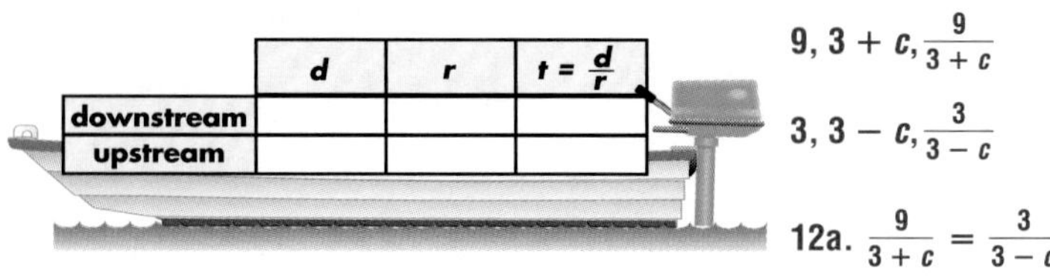

$9, 3 + c, \frac{9}{3 + c}$

$3, 3 - c, \frac{3}{3 - c}$

12a. $\frac{9}{3 + c} = \frac{3}{3 - c}$

a. Write an equation that represents the conditions in the problem.
b. Find the speed of the current. **1.5 mph**

Electronics: Exercises 13–15 refer to the diagram below.

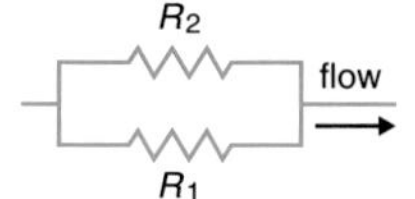

13. Find the total resistance, R_T, given that $R_1 = 8$ ohms and $R_2 = 6$ ohms. **3.429 ohms**
14. Find R_1, given that R_T is $2.\overline{2}$ ohms and $R_2 = 5$ ohms. **14. 4 ohms**
15. Find R_1 and R_2, given that the total resistance is $2.\overline{6}$ ohms and R_1 is twice as great as R_2. **8 ohms, 4 ohms**

EXERCISES

Practice **Solve each equation.**

16. $\frac{1}{4} + \frac{3}{x} = \frac{1}{x}$ **−8**
17. $\frac{1}{5} - \frac{4}{3m} = \frac{2}{m}$ $\mathbf{\frac{50}{3}}$
18. $x + 3 = -\frac{2}{x}$ **−1 or −2**
19. $\frac{m + 1}{m} + \frac{m + 4}{m} = 6$ $\mathbf{\frac{5}{4}}$
20. $\frac{x}{x + 1} + \frac{5}{x - 1} = 1$ $\mathbf{\frac{-3}{2}}$
21. $\frac{m - 1}{m + 1} - \frac{2m}{m - 1} = -1$ **0**
22. $\frac{-4}{a + 1} + \frac{3}{a} = 1$ **1 or −3**
23. $\frac{3x}{10} - \frac{1}{5x} = \frac{1}{2}$ **2 or $-\frac{1}{3}$**
24. $\frac{b}{4} + \frac{1}{b} = \frac{-5}{3}$ **−6 or $-\frac{2}{3}$**
25. $\frac{-4}{n} = 11 - 3n$ **4 or $-\frac{1}{3}$**
26. $\frac{x - 3}{x} = \frac{x - 3}{x - 6}$ **3**
27. $\frac{7}{a - 1} = \frac{5}{a + 3}$ **−13**
28. $\frac{3}{r + 4} - \frac{1}{r} = \frac{1}{r}$ **8**
29. $\frac{3}{x} + \frac{4x}{x - 3} = 4$ $\mathbf{\frac{3}{5}}$
30. $\frac{1}{4m} + \frac{2m}{m - 3} = 2$ $\mathbf{\frac{3}{25}}$
31. $\frac{a - 2}{a} - \frac{a - 3}{a - 6} = \frac{1}{a}$ **3**

B

32. $\frac{x + 3}{x + 5} + \frac{2}{x - 9} = \frac{5}{2x + 10}$ **3 or 1**
33. $\frac{-1}{w + 2} = \frac{w^2 - 7w - 8}{3w^2 + 2w - 8}$ **6**

Electronics: Refer to the diagram at the right.

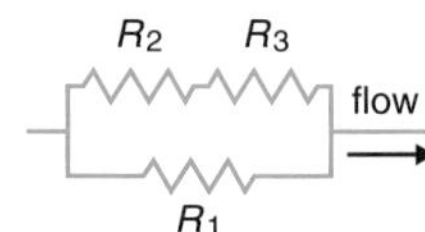

34. Find R_T, given that $R_1 = 5$ ohms, $R_2 = 4$ ohms, and $R_3 = 3$ ohms. **2.92 ohms**
35. Find R_1, given that $R_T = 2\frac{10}{13}$ ohms, $R_2 = 3$ ohms, and $R_3 = 6$ ohms. **4 ohms**

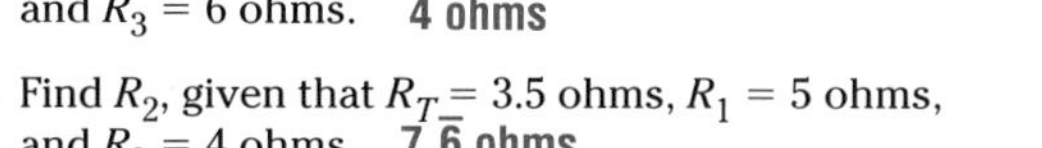

36. Find R_2, given that $R_T = 3.5$ ohms, $R_1 = 5$ ohms, and $R_3 = 4$ ohms. **7.6 ohms**

Study Guide Masters, p. 87

12-8 NAME ______ DATE ______
Study Guide Student Edition Pages 696–702

Solving Rational Equations

To solve equations containing fractions, eliminate the fractions by multiplying each side of the equation by a common denominator.

Example 1: Solve $\frac{x - 3}{3} + \frac{x}{2} = 4$.
$6\left(\frac{x - 3}{3} + \frac{x}{2}\right) = 6 \cdot 4$
$2(x - 3) + 3(x) = 24$
$2x - 6 + 3x = 24$
$5x = 30$
$x = 6$

Problems involving work can be solved by calculating the rate of work and the time spent working as shown below.

(rate of working) · (time) = (work done)
$r \cdot t = w$

Uniform motion problems can be solved by using a formula similar to the one used to solve work problems.

(rate) · (time) = (distance)
$r \cdot t = d$

Example 2: Marla can paint Percy's kitchen in 3 hours. Percy can paint it in 2 hours. Working together, how long will it take Marla and Percy to paint the kitchen?
$\frac{t}{3} + \frac{t}{2} = 1$
$2t + 3t = 6$
$5t = 6$
$t = \frac{6}{5}$ or $1\frac{1}{5}$ hours
Together Marla and Percy can paint the kitchen in $1\frac{1}{5}$ hours, or 1 hour and 12 minutes.

Solve each equation.

1. $\frac{x - 5}{5} + \frac{x}{4} = 8$ **20**
2. $\frac{4}{8y} - \frac{3y}{y - 1} = -2$ **−1 or $-\frac{1}{2}$**
3. $s - \frac{4}{s + 3} = s + 3$ **$-4\frac{1}{3}$**
4. $\frac{m + 4}{m} + \frac{m}{3} = \frac{m}{3}$ **−4**
5. $\frac{q + 4}{q - 1} + \frac{q}{q + 1} = 2$ **$-\frac{3}{2}$**
6. $\frac{5 - 2x}{2} - \frac{4x + 3}{6} = \frac{7x + 2}{6}$ **$\frac{10}{17}$**

7. A motorboat went upstream at 15 mi/h and returned downstream at 20 mi/h. How far did the boat travel one way if the round trip took 3.5 hours? **30 miles**

8. It takes Kenesha 45 minutes to prepare 20 greeting cards. It takes Paula 30 minutes to prepare the same cards. Working together at this rate, how long will it take them to prepare the cards? **18 minutes**

37. $R_1 = \dfrac{R_2R_T}{R_2 - R_T}$

Electronics: Solve each formula for the variable indicated.

C

37. $\frac{1}{R_T} = \frac{1}{R_1} + \frac{1}{R_2}$, for R_1

38. $I = \frac{E}{r + R}$, for R $R = \dfrac{E - Ir}{I}$

39. $I = \frac{nE}{nr + R}$, for n $n = \dfrac{IR}{E - Ir}$

40. $I = \frac{E}{\frac{r}{n} + R}$, for r $r = \dfrac{En - IRn}{I}$

Critical Thinking

41. What number would you add to both the numerator and denominator of $\frac{4}{11}$ to make a fraction equivalent to $\frac{2}{3}$? 10

42a. $R_T = R_1 + \dfrac{R_2R_3}{R_2 + R_3}$

42. Refer to the diagram at the right.
    a. Write an equation for the total resistance for the diagram.
    b. Find the total resistance, given that R_1 = 5 ohms, R_2 = 4 ohms, and R_3 = 6 ohms. 7.4 ohms

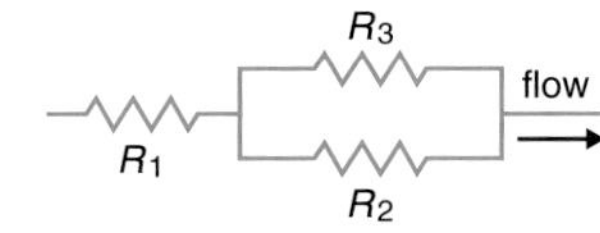

Applications and Problem Solving

44. 12.63 ohms

43. **Electricity** Eight lights on a decorated tree are connected in series. Each has a resistance of 12 ohms. What is the total resistance? 96 ohms

44. **Electricity** Three appliances are connected in parallel: a lamp with a resistance of 60 ohms, an iron with a resistance of 20 ohms, and a heating coil with a resistance of 80 ohms. Find the total resistance.

45. **Air Travel** The flying distance from Honolulu, Hawaii, to San Francisco, California, is approximately 2400 miles. The air speed of a 747-100 airplane is 520 miles per hour. A strong tailwind blows at 120 miles per hour. At 1000 miles into the trip (1400 miles from San Francisco), the aircraft loses power in one engine. As a navigator for the aircraft, you must answer the following questions.
    a. How long would it take to return to Honolulu? 2.5 hours
    b. How long would it take to continue on to San Francisco? 2.19 hours
    c. What is the point-of-no-return? That is, at what point into the flight (in miles) would it be quicker to continue the flight to San Francisco than return to Honolulu? after 923 miles

46a. $\frac{2}{3}$ h or 40 min, $10\frac{2}{3}$ miles from one end or $13\frac{1}{3}$ miles from the other

46b. 2 mph

46. **Bicycling** The Lake Pontchartrain Causeway in Louisiana is 24 miles long.

    a. Suppose Todd starts cycling at one end at 20 miles per hour and Kristie starts at the other end at 16 miles per hour. How long would it take them to meet? At what distance from either end would they meet?
    b. Todd's rate with no wind is 20 miles per hour, while Kristie's is 16 miles per hour. If Todd cycles against the wind, while Kristie cycles with the wind, and they meet at the midpoint of the bridge, what is the wind speed?
    c. The causeway consists of two parallel bridges. Assume Todd starts to bicycle from one end in one direction at 20 miles per hour. Kristie starts from the other end on the parallel bridge, bicycling at 16 miles per hour in the opposite direction. They pedal in a continuous loop on the two parallel causeways. How long does it take Todd to catch up with Kristie? (*Note:* She has a 24-mile head start.) 6 hours

Practice Masters, p. 87

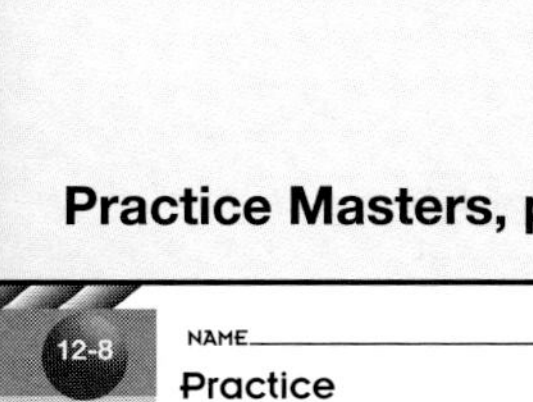

NAME ________ DATE ________

12-8 Practice

Student Edition Pages 696–702

Solving Rational Equations

Solve each equation.

1. $\frac{2a + 5}{6} - \frac{2a}{3} = -\frac{1}{2}$ $a = 4$

2. $\frac{z - 3}{10} + \frac{z - 5}{5} = \frac{1}{2}$ $z = 6$

3. $\frac{c + 2}{c} + \frac{c + 3}{c} = 7$ $c = 1$

4. $\frac{2d - 1}{6} - \frac{d}{3} = \frac{d + 4}{18}$ $d = -7$

5. $\frac{1}{m} + \frac{4m}{m + 1} = 4$ $m = \frac{1}{3}$

6. $\frac{3}{4r} - \frac{4r}{r - 3} = -4$ $r = -\frac{1}{5}$

7. $\frac{5}{p - 1} - \frac{3}{p + 2} = 0$ $p = -\frac{13}{2}$

8. $\frac{y - 2}{4} - \frac{y + 2}{5} = -1$ $y = -2$

9. $\frac{x + 2}{2} = 11 - \frac{3(28 - x)}{8}$ $x = -4$

10. $\frac{3}{4x} + \frac{4x}{x - 3} = 4$ $x = \frac{3}{17}$

11. $\frac{1}{z + 1} - \frac{6 - z}{6z} = 0$
    $z = -3$ or $z = 2$

12. $\frac{2}{m + 2} - \frac{m + 2}{m - 2} = \frac{7}{3}$
    $m = \frac{2}{5}$ or $m = -1$

13. $\frac{5e}{3e + 1} - \frac{1}{9e + 3} = \frac{7}{6}$ $e = 1$

14. $\frac{u}{u + 2} - \frac{2u}{2 - u} = \frac{u^2}{u^2 - 4}$
    $u = 0$ or $u = -1$

4 ASSESS

Closing Activity

Speaking Ask students to explain why 1 is not a solution to Example 4. **It makes the denominator 0.**

Chapter 12, Quiz D (Lessons 12-7 and 12-8), is available in the *Assessment and Evaluation Masters*, p. 325.

Additional Answer

53.

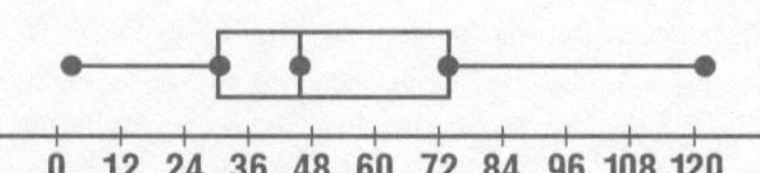

***fabulous* FIRSTS**

In the Olympics, Thorpe won the decathlon and the pentathlon. In 1913 he was deprived of his medals because he had played semiprofessional baseball in 1909 and 1910. His family regained his medals in 1983. Thorpe played baseball, football, basketball, lacrosse, and hockey, and he boxed and swam.

Enrichment Masters, p. 87

NAME ______________ DATE ______

Enrichment

Student Edition Pages 697–703

Using Rational Expressions and Equations

In 1985 Steve Cram set a world record for the mile run with a time of 3:46.31. In 1954, Roger Bannister ran the first mile under 4 minutes at 3:59.4. Had they run those times in the same race, how far in front of Bannister would Cram have been at the finish?

Use $\frac{d}{t} = r$. Since 3 min 46.31 s = 226.31 s, and 3 min 59.4 s = 239.4 s, Cram's rate was $\frac{5280 \text{ ft}}{226.31 \text{ s}}$ and Bannister's rate was $\frac{5280 \text{ ft}}{239.4 \text{ s}}$.

	r	t	d
Cram	$\frac{5280}{226.31}$	226.31	5280 feet
Bannister	$\frac{5280}{239.4}$	226.31	$\frac{5280}{239.4} \cdot 226.31$ or 4491.3 feet

Therefore, when Cram hit the tape, he would be 5280 − 4491.3, or 288.7 feet, ahead of Bannister. Let's see whether we can develop a formula for this type of problem.

Let D = the distance raced,
W = the winner's time,
and L = the loser's time.

	r	t	d
Winner	$\frac{D}{W}$	W	$\frac{D}{W} \cdot W = D$
Loser	$\frac{D}{L}$	W	$\frac{D}{L} \cdot W = \frac{DW}{L}$

Following the same pattern, you obtain the results shown in the table at the right.

The winning distance will be $D - \frac{DW}{L}$.

1. Show that the expression for the winning distance is equivalent to $\frac{D(L - W)}{L}$. $D - \frac{DW}{L} = \frac{DL}{L} - \frac{DW}{L}$ $\frac{DL - DW}{L}$ $\frac{D(L - W)}{L}$

Use the formula winning distance = $\frac{D(L - W)}{L}$ to find the winning distance for each of the following Olympic races.

2. women's 400 meter relay: Canada 48.4 s (1928); East Germany 41.6 s (1980) **56.2 meters**
3. men's 200 meter freestyle swimming: Mark Spitz 1 min 52.78 s (1972); Michael Gross 1 min 47.44 s (1984) **9.5 meters**
4. men's 50,000 meter walk: Thomas Green 4 h 50 min 10 s (1932); Hartwig Gauter 3 h 49 min 24 s (1980) **10,471 meters**
5. women's 400 meter freestyle relay: Great Britain 5 min 52.8 s (1912); East Germany 3 min 42.71 s (1980) **147.5 meters**

***fabulous* FIRSTS**

Jim Thorpe (1886–1953)

James Francis Thorpe was the first American-Indian sports hero, winner of Olympic Gold medals in 1912, a professional baseball and football player, and the first president of the American Professional Football Association, which later became the NFL. In 1950, he was selected by sportswriters and broadcasters as the greatest athlete of the first half of the 20th century.

47. **Anthropology** The formula $c = \frac{100w}{\ell}$ provides a measure called the cephalic index c. Anthropologists use this index to identify skulls by their ethnic characteristics. You calculate the cephalic index by using the width w of a person's head, ear to ear, and the length ℓ of the head from face to back.

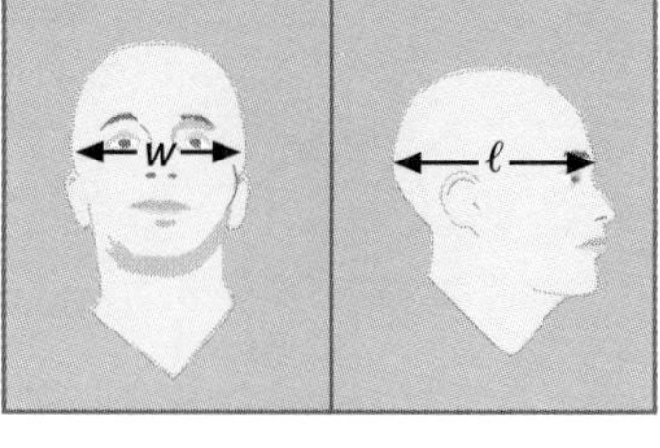

   a. Solve the formula for w in terms of c and ℓ.
   b. Solve the formula for ℓ in terms of c and w.

   47a. $w = \frac{c\ell}{100}$ b. $\ell = \frac{100w}{c}$

48. **Psychology** The formula $i = \frac{100m}{c}$ provides a measure i of intelligence, called the intelligence quotient, or I.Q. In the formula, m represents the person's mental age, and c represents the person's chronological age.
   a. Solve the formula for m in terms of i and c.
   b. Solve the formula for c in terms of i and m. **48a. $m = \frac{ic}{100}$ b. $c = \frac{100m}{i}$**

49. **Baseball** Chang has 32 hits in 128 times at bat. His current batting average is $\frac{32}{128} = 0.250$. How many consecutive hits must he get in his next x times at bat in order to get his average up to 0.300? **10**

Mixed Review

50. Simplify $\dfrac{\frac{x^2 - 5x}{x^2 + x - 30}}{\frac{x^2 + 2x}{x^2 + 9x + 18}}$. (Lesson 12–7) $\frac{x + 3}{x + 2}$

51. Simplify $\frac{a + 2}{b^2 + 4b + 4} \div \frac{4a + 8}{b + 4}$. (Lesson 12–3) $\frac{b + 4}{4(b + 2)^2}$

52. Find $(0.5a + 0.25b)^2$. (Lesson 9–8) $0.25a^2 + 0.25ab + 0.0625b^2$

53. **Sales** The amounts in the picture at the right are the cash register totals of 20 customers on the Wednesday before Thanksgiving. Make a box-and-whisker plot of these data. (Lesson 7–7) **See margin.**

$45.76 $46.03 $99.21 $35.43
$56.84 $35.35 $122.30 $56.90
$102.78 $1.99 $32.18 $37.50
$24.82 $60.82 $15.27 $88.12
$6.78 $29.15 $98.55 $52.98

54. **Statistics** Refer to the data in Exercise 53. Find the range, median, upper quartile, lower quartile, and interquartile range of the data. Identify any outliers. (Lesson 5–7) **120.31, 45.90, 74.47, 30.67, 43.80; no outliers**

55. Write 12 pounds to 100 ounces as a fraction in simplest form. (Lesson 4–1) $\frac{48}{25}$

56. Solve $3x = -15$. (Lesson 3–2) **−5**

57. Find $(-2)(3)(-3)$. (Lesson 2–6) **18**

58. Complete: $3(2 + x) = 6 + \underline{\ ?\ }$. (Lesson 1–7) **3x**

Extension

Connections A jet aircraft is flying from Honolulu to Los Angeles, a distance of 2570 miles. In still air, the plane flies at 600 mph. There is a 50 mph tailwind pushing the plane. At what point in the trip would it be quicker to go on to Los Angeles than to return to Honolulu? Give your answer in terms of distance from Honolulu. (*Hint:* Write expressions for equal times to continue or to return.) **1177.9 miles**

CHAPTER 12 HIGHLIGHTS

VOCABULARY

After completing this chapter, you should be able to define each term, property, or phrase and give an example or two of each.

Algebra

excluded values (p. 660)

extraneous solutions (p. 697)

least common denominator (LCD) (p. 686)

least common multiple (LCM) (p. 685)

mixed expressions (p. 690)

rational equation (p. 696)

rational expression (p. 660)

Problem Solving

make an organized list (p. 685)

UNDERSTANDING AND USING THE VOCABULARY

State whether each sentence is *true* or *false*. If false, replace the underlined word or number to make a true sentence.

1. A mixed expression is an algebraic fraction whose numerator and denominator are polynomials. **false, rational**
2. The complex fraction $\dfrac{\frac{4}{5}}{\frac{2}{3}}$ can be simplfied as $\underline{\frac{6}{5}}$. **true**
3. The equation $\dfrac{x}{x-1} + \dfrac{2x-3}{x-1} = 2$ has an extraneous solution of $\underline{1}$. **true**
4. The mixed expression $6 - \dfrac{a-2}{a+3}$ can be rewritten as the rational expression $\underline{\dfrac{5a+16}{a+3}}$. **false, $\dfrac{5a+20}{a+3}$**
5. The least common multiple for $(x^2 - 144)$ and $(x + 12)$ is $\underline{x + 12}$. **false, $x^2 - 144$**
6. The excluded values for $\dfrac{4x}{x^2 - x - 12}$ are $\underline{-3 \text{ and } 4}$. **true**
7. The least common denominator is the greatest common factor of the denominators. **false, least common multiple**

Using the CHAPTER HIGHLIGHTS

The Chapter Highlights begins with a listing of the new terms, properties, and phrases that were introduced in this chapter. Have students define each term and provide an example or two of it, if appropriate.

Instructional Resources

Three multiple-choice tests and three free-response tests are provided in the *Assessment and Evaluation Masters*. Forms 1A and 2A are for honors pacing, and Forms 1B, 1C, 2B, and 2C are for average pacing. Chapter 12 Test, Form 1B is shown at the right. Chapter 12 Test, Form 2B is shown on the next page.

Assessment and Evaluation Masters, pp. 311–312

12 NAME__________ DATE________

Chapter 12 Test, Form 1B

Write the letter for the correct answer in the blank at the right of each problem.

1. Simplify $\frac{4x-4}{4x+4}$. — 1. C
   A. $x-4$ B. $x+4$ C. $\frac{x-1}{x+1}$ D. -1
2. Simplify $\frac{k^2-8k+16}{k^2-16}$. — 2. C
   A. $k-4$ B. -1 C. $\frac{k-4}{k+4}$ D. $\frac{k+4}{k-4}$
3. What are the excluded values of x in $\frac{x-3}{x^2-9}$? — 3. D
   A. $x \neq 3$ B. $x \neq -3$ C. $x \neq 9$ D. $x \neq \pm 3$
4. What are the excluded values of n in $\frac{n^2-3n-28}{n^2+3n-4}$? — 4. D
   A. $n \neq -1, 4$ B. $n \neq 1$ C. $n \neq -1$ D. $n \neq 1, -4$
5. Find $\frac{3a^2b}{b^2c^3} \cdot \frac{2bc^2}{15a}$. — 5. C
   A. $\frac{ab}{3c}$ B. $\frac{2a^2b}{5c}$ C. $\frac{2a}{5c}$ D. $\frac{a^2}{3c^2}$
6. Find $\frac{(y+2)^2}{8} \cdot \frac{72}{y^2-4}$. — 6. A
   A. $\frac{9(y+2)}{y-2}$ B. $\frac{-9}{y+2}$ C. $\frac{9}{y-2}$ D. $\frac{-9}{8(y-2)}$
7. Find $\frac{3}{x+2} + \frac{2x}{x+2}$. — 7. B
   A. $\frac{5x}{x+2}$ B. $\frac{2x+3}{x+2}$ C. $\frac{2x+3}{2x+4}$ D. $\frac{5x}{2x+4}$
8. Find $\frac{u+t}{w-6} + \frac{u+t}{6-w}$. — 8. A
   A. 0 B. $\frac{2t}{6-w}$ C. $\frac{u+t}{w-6}$ D. $\frac{2u+2t}{w-6}$
9. Find $\frac{16a^2-9w^2}{4a^2-3aw} \cdot \frac{a^2}{a+w}$. — 9. B
   A. $\frac{4a-3}{a+w}$ B. $\frac{a(4a+3w)}{a+w}$
   C. $\frac{a^2(4a-3w)}{(a+w)(2a-3w)(2a+w)}$ D. a
10. Find $\frac{n^2+3n-10}{n^2+6n+8} \div \frac{n-2}{n^2+2n}$. — 10. A
    A. $\frac{n(n+5)}{n+4}$ B. $\frac{(n+5)(n-2)^2}{(n+3)^3}$
    C. $\frac{n+5}{n+2}$ D. $\frac{n(n-5)}{(n-2)^2}$
11. Use long division to find the quotient $x+3\overline{)5x^2-9x+20}$. — 11. C
    A. $5x+6+\frac{38}{x+3}$ B. $5x+6+\frac{2}{x+3}$
    C. $5x-24+\frac{92}{x+3}$ D. $5x-24+\frac{-52}{x+3}$

12 NAME__________ DATE________

Chapter 12 Test, Form 1B (continued)

12. What is the simplest form of $\dfrac{x-11+\frac{42}{x+2}}{x-5}$? — 12. D
    A. $\frac{2x+31}{(x-5)(x+2)}$ B. $\frac{x^2+20}{(x-5)(x+2)}$ C. $\frac{x^2-9x+20}{x-5}$ D. $\frac{x-4}{x+2}$
13. Solve $\frac{6}{y+2} + \frac{1}{y} = 3$. — 13. A
    A. $-\frac{2}{3}$ and 1 B. $\frac{2}{3}$ C. 1 D. $\frac{1}{7}$
14. Solve by making a list. In how many ways can the letters "C-A-T" be arranged to form a word? (A word is any arrangement of the letters.) — 14. C
    A. 3 B. 4 C. 6 D. 1
15. Find $\frac{2x+3}{x-4} + \frac{8}{x+1}$. — 15. D
    A. $\frac{2x+11}{x^2-3x-4}$ B. $\frac{2x^2+5x+12}{x^2-3x-4}$ C. $\frac{10x-9}{x+1}$ D. $\frac{2x^2+13x-29}{x^2-3x-4}$
16. Find $\frac{10}{a-b} - \frac{6b}{a^2-b^2}$. — 16. D
    A. $\frac{4}{a-b}$ B. $\frac{10a-6b}{a^2-b^2}$ C. $\frac{4b}{a-b}$ D. $\frac{10a+4b}{a^2-b^2}$
17. Solve $\frac{5x}{3x+1} - \frac{1}{9x+3} = \frac{7}{6}$. — 17. A
    A. 1 B. -1 C. $-\frac{9}{11}$ D. $\frac{9}{11}$
18. Solve $\frac{dn-n}{e} = f$ for n. — 18. B
    A. $\frac{f-e}{d}$ B. $\frac{fe}{d-1}$ C. $\frac{fe}{d}$ D. $\frac{f-e}{d-1}$
19. Simplify $\dfrac{1}{1+\frac{1}{1-\frac{1}{2}}}$. — 19. A
    A. $\frac{1}{3}$ B. $\frac{3}{5}$ C. $\frac{2}{3}$ D. $\frac{1}{2}$
20. The formula $\frac{1}{R_T} = \frac{1}{R_1} + \frac{1}{R_2} + \frac{1}{R_3}$ represents the total resistance of a circuit with 3 resistances connected in parallel. What is the resistance R_1 if $R_T = 2.5$ ohms, $R_2 = 10$ ohms, and $R_3 = 5$ ohms? — 20. C
    A. 17.5 B. 2.5 C. 10 D. 7

Bonus Find the value of k so that $x + 1$ is a factor of $3x^3 - 2x^2 + x + k$. — Bonus 6

Using the STUDY GUIDE AND ASSESSMENT

Skills and Concepts Encourage students to refer to the objectives and examples on the left as they complete the review exercises on the right.

Assessment and Evaluation Masters, pp. 317–318

12 NAME__________ DATE ________

Chapter 12 Test, Form 2B

Perform the indicated operations.

1. $\frac{a^2 - b^2}{a + b}$	2. $\frac{a^2 - 3a - 28}{a^2 + 3a - 4}$	1. $a - b$ 2. $\frac{a - 7}{a - 1}$
3. $\frac{y^2 - 9}{4} \cdot \frac{8}{y + 3}$	4. $\frac{s^2 + 2s - 8}{s^2 + 3s - 10} \cdot \frac{s + 5}{s^2 - 16}$	3. $2y - 6$ 4. $\frac{1}{s - 4}$
5. $\frac{10 - 5x}{6 + 3x} \div \frac{5}{12 + 6x}$	6. $\frac{5m}{m + 1} \div \frac{25m^2}{m^2 + 2m + 1}$	5. $4 - 2x$ 6. $\frac{m + 1}{5m}$
7. $\frac{r}{r - 5} - \frac{5}{r - 5}$	8. $\frac{r^2}{r + s} - \frac{s^2}{r + s}$	7. 1 8. $r - s$
9. $\frac{4n^2}{2n - 3} + \frac{9}{3 - 2n}$	10. $\frac{7}{6x^2r} + \frac{10}{3xr^2}$	9. $2n + 3$ 10. $\frac{7r + 20x}{6x^2r^2}$
11. $\frac{2a + 6}{a^2 + 6a + 9} + \frac{1}{a + 3}$	12. $\frac{8}{2a - 2b} - \frac{3b}{a^2 - b^2}$	11. $\frac{3}{a + 3}$ 12. $\frac{4a + b}{a^2 - b^2}$
13. $\dfrac{\frac{r^2 + r - 6}{r^2}}{\frac{r^2 + 3r}{r^2 + 2r}}$	14. $\dfrac{x - \frac{6}{x + 2}}{x + \frac{8}{x + 5}}$	13. $\frac{r^2 - 4}{r^2}$ 14. $\frac{(x + 5)(x^2 + 2x - 6)}{(x + 2)(x^2 + 5x + 8)}$

15. What are the excluded values of y in the expression $\frac{y^2 + y - 2}{y^2 - 2y - 15} \cdot \frac{y^2 - 5y + 5}{y^2 + 5y + 6}$? — 15. $y \neq 5, -3, -2$

16. Use long division to find the quotient $(t^2 + 4t - 6) \div (t + 5)$. — 16. $t - 1 - \frac{1}{t + 5}$

12 NAME__________ DATE ________

Chapter 12 Test, Form 2B (continued)

17. Solve by making a tree diagram. For dinner, you can choose spaghetti, tacos, or egg foo yung. For a side dish, you can choose between a salad or soup. Cake, pie, and ice cream are offered for dessert. How many possible meals can you arrange? — 17. 18 meals

Solve each equation.

18. $\frac{a - 3}{10} + \frac{a - 5}{5} = \frac{1}{2}$ — 18. 6

19. $\frac{2}{x + 2} = \frac{9}{8} - \frac{5x}{4x + 8}$ — 19. 2

20. $\frac{2}{n - 3} + \frac{2}{n + 5} = \frac{5n - 7}{n^2 + 2n - 15}$ — 20. 11

21. Solve $\frac{y}{c} + \frac{n}{d} = 5$ for n. — 21. $n = \frac{5cd - dy}{c}$

22. Simplify $\left[\left(\frac{45 \text{ miles}}{1 \text{ hour}} \cdot \frac{5280 \text{ feet}}{1 \text{ mile}}\right) \div \frac{60 \text{ minutes}}{1 \text{ hour}}\right] \div \frac{60 \text{ seconds}}{1 \text{ minute}}$. — 22. 66 ft/sec

23. The volume of a box is $2x^3 = 11x^2 + 17x + 6$. One dimension of the box is $x + 2$. What are the other two dimensions if they are polynomials in x with integer coefficients? — 23. $x + 3, 2x + 1$

24. An airplane can fly at a rate of 525 mi/h in still air. It can fly 2800 miles with the wind in the same time that it can fly 2450 miles against the wind. Find the rate of the wind. — 24. 35 mi/h

25. Given the formula $\frac{1}{R_T} = \frac{1}{R_1} + \frac{1}{R_2 + R_3}$ for the resistance of a circuit, determine the resistance R_2 if $R_T = 2$ ohms, $R_1 = 3$ ohms, and $R_3 = 2\frac{3}{5}$ ohms. — 25. $3\frac{2}{5}$ ohms

Bonus Cardiss can do a job in 4 hours. Rachel can do the same job in 6 hours. Suppose that Rachel works on the job for 2 hours and then is joined by Cardiss. Find the time it will take both working together to finish the job. — Bonus $1\frac{3}{5}$ h

CHAPTER 12 STUDY GUIDE AND ASSESSMENT

SKILLS AND CONCEPTS

OBJECTIVES AND EXAMPLES

Upon completing this chapter, you should be able to:

- simplify rational expressions (Lesson 12–1)

Simplify $\frac{x + y}{x^2 + 3xy + 2y^2}$.

$$\frac{x + y}{x^2 + 3xy + 2y^2} = \frac{\overset{1}{\cancel{x + y}}}{\underset{1}{\cancel{(x + y)}}(x + 2y)} = \frac{1}{x + 2y}$$

REVIEW EXERCISES

Use these exercises to review and prepare for the chapter test.

Simplify each rational expression. State the excluded values of the variables.

8. $\frac{3x^2y}{12xy^3z}$ $\quad \frac{x}{4y^2z}$, $x, y, z \neq 0$
9. $\frac{z^2 - 3z}{z - 3}$ $\quad z, z \neq 3$
10. $\frac{a^2 - 25}{a^2 + 3a - 10}$ $\quad \frac{a - 5}{a - 2}$, $a \neq -5, 2$
11. $\frac{3a^3}{3a^3 + 6a^2}$ $\quad \frac{a}{a + 2}$, $a \neq 0, -2$
12. $\frac{x^2 + 10x + 21}{x^3 + x^2 - 42x}$ $\quad \frac{x + 3}{x(x - 6)}$, $x \neq 0, -7, 6$
13. $\frac{b^2 - 5b + 6}{b^4 - 13b^2 + 36}$ $\quad \frac{1}{(b + 3)(b + 2)}$, $b \neq \pm 2, \pm 3$

- multiply rational expressions (Lesson 12–2)

$$\frac{1}{x^2 + x - 12} \cdot \frac{x - 3}{x + 5} = \frac{1}{(x + 4)\underset{1}{\cancel{(x - 3)}}} \cdot \frac{\overset{1}{\cancel{x - 3}}}{x + 5}$$
$$= \frac{1}{(x + 4)(x + 5)}$$
$$= \frac{1}{x^2 + 9x + 20}$$

Find each product. Assume that no denominator has a value of 0.

14. $\frac{7b^2}{9} \cdot \frac{6a^2}{b}$ $\quad \frac{14a^2b}{3}$
15. $\frac{5x^2y}{8ab} \cdot \frac{12a^2b}{25x}$ $\quad \frac{3axy}{10}$
16. $(3x + 30) \cdot \frac{10}{x^2 - 100}$ $\quad \frac{30}{x - 10}$
17. $\frac{3a - 6}{a^2 - 9} \cdot \frac{a + 3}{a^2 - 2a}$ $\quad \frac{3}{a^2 - 3a}$
18. $\frac{x^2 + x - 12}{x + 2} \cdot \frac{x + 4}{x^2 - x - 6}$ $\quad \frac{(x + 4)^2}{(x + 2)^2}$
19. $\frac{b^2 + 19b + 84}{b - 3} \cdot \frac{b^2 - 9}{b^2 + 15b + 36}$ $\quad b + 7$

- divide rational expressions (Lesson 12–3)

$$\frac{y^2 - 16}{y^2 - 64} \div \frac{y + 4}{y - 8} = \frac{y^2 - 16}{y^2 - 64} \cdot \frac{y - 8}{y + 4}$$
$$= \frac{(y - 4)\overset{1}{\cancel{(y + 4)}}}{\underset{1}{\cancel{(y - 8)}}(y + 8)} \cdot \frac{\overset{1}{\cancel{y - 8}}}{\underset{1}{\cancel{y + 4}}}$$
$$= \frac{y - 4}{y + 8}$$

Find each quotient. Assume that no denominator has a value of 0.

20. $\frac{p^3}{2q} \div \frac{p^2}{4q}$ $\quad 2p$
21. $\frac{y^2}{y + 4} \div \frac{3y}{y^2 - 16}$ $\quad \frac{y^2 - 4y}{3}$
22. $\frac{3y - 12}{y + 4} \div (y^2 - 6y + 8)$ $\quad \frac{3}{y^2 + 2y - 8}$
23. $\frac{2m^2 + 7m - 15}{m + 5} \div \frac{9m^2 - 4}{3m + 2}$ $\quad \frac{2m - 3}{3m - 2}$

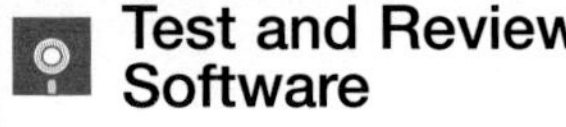

Test and Review Software

You may use this software, a combination of an item generator and item bank, to create your own tests or worksheets. Types of items include free response, multiple choice, short answer, and open ended.

For IBM & Macintosh

OBJECTIVES AND EXAMPLES	REVIEW EXERCISES

- divide a polynomial by a binomial. (Lesson 12–4)

$$\begin{array}{r} x^2 + x - 19 \\ x - 3\overline{)x^3 - 2x^2 - 22x + 21} \\ \underline{x^3 - 3x^2} \\ x^2 - 22x \\ \underline{x^2 - 3x} \\ -19x + 21 \\ \underline{-19x + 57} \\ -36 \end{array}$$

The quotient is $x^2 + x - 19 - \frac{36}{x-3}$.

Find each quotient. $24.\ 2ac^2 - 4a^2c + \frac{3c^2}{b}$

24. $(4a^2b^2c^2 - 8a^3b^2c + 6abc^2) \div (2ab^2)$
25. $(x^3 + 7x^2 + 10x - 6) \div (x + 3)$ $x^2 + 4x - 2$
26. $(x^3 - 7x + 6) \div (x - 2)$ $x^2 + 2x - 3$
27. $(x^4 + 3x^3 + 2x^2 - x + 6) \div (x - 2)$
28. $(48b^2 + 8b + 7) \div (12b - 1)$

27. $x^3 + 5x^2 + 12x + 23 + \frac{52}{x-2}$
28. $4b + 1 + \frac{8}{12b-1}$

- add and subtract rational expressions with like denominators. (Lesson 12–5)

$$\frac{m^2}{m+4} - \frac{16}{m+4} = \frac{m^2-16}{m+4}$$
$$= \frac{\overset{1}{\cancel{(m-4)(m+4)}}}{\underset{1}{\cancel{m+4}}}$$
$$= m - 4$$

Find each sum or difference. Express in simplest form.

29. $\frac{7a}{m^2} - \frac{5a}{m^2}$ $\frac{2a}{m^2}$
30. $\frac{2x}{x-3} - \frac{6}{x-3}$ 2
31. $\frac{m+4}{5} + \frac{m-1}{5}$ $\frac{2m+3}{5}$
32. $\frac{-5}{2n-5} + \frac{2n}{2n-5}$ 1
33. $\frac{a^2}{a-b} + \frac{-b^2}{a-b}$ $a + b$
34. $\frac{m^2}{m-n} - \frac{2mn-n^2}{m-n}$ $m - n$

- add and subtract rational expressions with unlike denominators (Lesson 12–6)

$$\frac{x}{x+3} - \frac{5}{x-2} = \frac{x}{x+3}\cdot\frac{x-2}{x-2} - \frac{5}{x-2}\cdot\frac{x+3}{x+3}$$
$$= \frac{x(x-2)}{(x+3)(x-2)} - \frac{5(x+3)}{(x+3)(x-2)}$$
$$= \frac{x^2-2x}{(x+3)(x-2)} - \frac{5x+15}{(x+3)(x-2)}$$
$$= \frac{x^2-2x-5x-15}{(x+3)(x-2)}$$
$$= \frac{x^2-7x-15}{x^2+x-6}$$

Find each sum or difference.

35. $\frac{7n}{3} - \frac{9n}{7}$ $\frac{22n}{21}$
36. $\frac{7}{3a} - \frac{3}{6a^2}$ $\frac{14a-3}{6a^2}$
37. $\frac{2c}{3d^2} + \frac{3}{2cd}$ $\frac{4c^2+9d}{6cd^2}$
38. $\frac{2a}{2a+8} - \frac{4}{5a+20}$ $\frac{5a-4}{5a+20}$
39. $\frac{r^2+21r}{r^2-9} + \frac{3r}{r+3}$ $\frac{4r}{r-3}$
40. $\frac{3a}{a-2} + \frac{5a}{a+1}$ $\frac{8a^2-7a}{(a-2)(a+1)}$

Using the CUMULATIVE REVIEW

These two pages review the skills and concepts presented in Chapters 1–12. This review is formatted to reflect new trends in standardized testing.

A more traditional cumulative review, is provided in the *Assessment and Evaluation Masters,* pp. 327–328.

Assessment and Evaluation Masters, pp. 327–328

NAME________________ DATE ________

12 Chapter 12 Cumulative Review

1. State the property shown in $3 + 2 = 3 + 2$. (Lesson 1-6) — 1. reflexive
2. Simplify $6x^2 + 9x^2 + 4x + 6x$. (Lesson 1-1) — 2. $15x^2 + 10x$
3. Find $-2.349 + 1.456$. (Lesson 2-5) — 3. -0.893
4. Use substitution to solve the system of equations $x + y = 2$ and $2x - y = 1$. (Lesson 8-2) — 4. (1, 1)
5. Use elimination to solve the system of equations $2x + 2y = -4$ and $7x + 4 = 1$. (Lesson 8-4) — 5. (3, −5)

Use the data 50, 46, 68, 62, 59, 74, 60, 70, and 64 for exercises 6–10. (Lessons 3-7, 5-7, and 7-7)

6. Find the mean to the nearest tenth. — 6. 61.4
7. Find the median. — 7. 62
8. Find the lower quartile. — 8. 54.5
9. Find the upper quartile. — 9. 69
10. Make a box-and-whisker plot for the data. — 10.

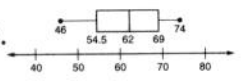

Simplify. (Lessons 9-1 and 9-2)

11. $(-4x^5y^3)(2xy^2)$ — 11. $-8x^6y^5$
12. $\left(\frac{2}{3}c^2d^3\right)^4$ — 12. $\frac{16}{81}c^8d^{12}$
13. $\frac{21hk^3j}{14h^5kj^8}$ — 13. $-\frac{3k^2}{2h^4j^7}$
14. Find $(2x^2y - 3xy + 4y^3) - (4xy - 3y^3 + x^2y)$. (Lesson 9-5) — 14. $x^2y - 7xy + 7y^3$
15. Find $(3m - 2n)(3m + 2n)$. (Lesson 9-8) — 15. $9m^2 - 4n^2$

Factor each polynomial, if possible. (Lessons 10-2 and 10-3)

16. $12c - 8d - 15rc + 10rd$ — 16. $(4 - 5r)(3c - 2d)$
17. $3a^2 - 13a + 12$ — 17. $(3a - 4)(a - 3)$

NAME________________ DATE ________

12 Chapter 12 Cumulative Review (continued)

Solve each equation. (Lessons 10-6 and 11-4)

18. $(x + 3)(x - 4) = 0$ — 18. $-3, 4$
19. $4r^2 - 7r + 3 = 0$ — 19. $\frac{3}{4}, 1$
20. $3^{5y+4} = 3^y$ — 20. -1
21. $4^{x-1} = 2^x$ — 21. 2
22. Simplify $\frac{2x^2 + 3x + 1}{4x^2 + 4x + 1}$. (Lesson 12-1) — 22. $\frac{x + 1}{2x + 1}$
23. Simplify $\frac{4x^2 - 8x - 5}{2x^2 + x - 15}$. (Lesson 12-1) — 23. $\frac{2x + 1}{x + 3}$
24. Find $\frac{a^3}{a - b} \cdot \frac{a^2 - b^2}{a}$. (Lesson 12-2) — 24. $a^2(a + b)$
25. Find $(2x^2 + 3x - 5) \div (x - 3)$. (Lesson 12-4) — 25. $2x + 9 + \frac{22}{x - 3}$
26. Find $\frac{2}{3k - 1} + \frac{k}{k + 3}$. (Lesson 12-6) — 26. $\frac{3k^2 + k + 6}{(3k - 1)(k + 3)}$
27. Find $\frac{c^2}{c^2 - 4} - \frac{5c + 14}{c^2 - 4}$. (Lesson 12-5) — 27. $\frac{c - 7}{c - 2}$
28. Simplify $\frac{z - \frac{3}{z - 2}}{z - \frac{6}{z - 1}}$. (Lesson 12-7) — 28. $\frac{(z + 1)(z - 1)}{(z + 2)(z - 2)}$
29. Solve $wx - w = \frac{x}{y}$ for w. (Lesson 12-8) — 29. $w = \frac{x}{y(x - 1)}$
30. Find three consecutive even integers such that the difference of the squares of the least and the greatest is 96. (Lesson 10-4) — 30. 10, 12, and 14
31. The length of a rectangular lot is 10 feet less than 3 times its width. The perimeter of the lot is 620 feet. Find the dimensions of the lot. (Lesson 3-3) — 31. 80 ft by 230 ft
32. Carlos paid $37.98 for some compact discs. This included $5\frac{1}{2}$% sales tax. What did the CDs cost without the tax? (Lesson 4-5) — 32. $36
33. Find the dimensions of the rectangle whose width is 8 cm less than its length and whose area is 20 cm². (Lesson 11-3) — 33. 2 cm and 10 cm

CUMULATIVE REVIEW

CHAPTERS 1–12

SECTION ONE: MULTIPLE CHOICE

There are eight multiple-choice questions in this section. After working each problem, write the letter of the correct answer on your paper.

1. **Geometry** The measure of the area of a square is 129 square inches. What is the perimeter of this square, rounded to the nearest hundredth? **C**

   A. 11.36 in. B. 32.25 in.
   C. 45.43 in. D. 1040.06 in.

2. Express the relation shown as a set of ordered pairs. **D**

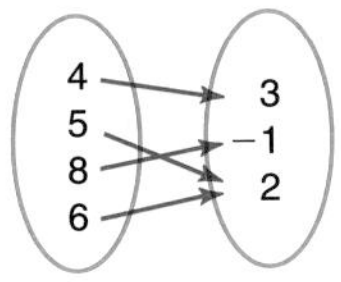

   A. {(3, 4), (−1, 8), (2, 5), (2, −6)}
   B. {(4, 3), (5, 2), (−1, 8), (−6, 2)}
   C. {4, 5, 8, −6, 3, −1, 2}
   D. {(4, 3), (5, 2), (8, −1), (−6, 2)}

3. **Geometry** Find the measure of the area of the rectangle shown below in simplest form. **B**

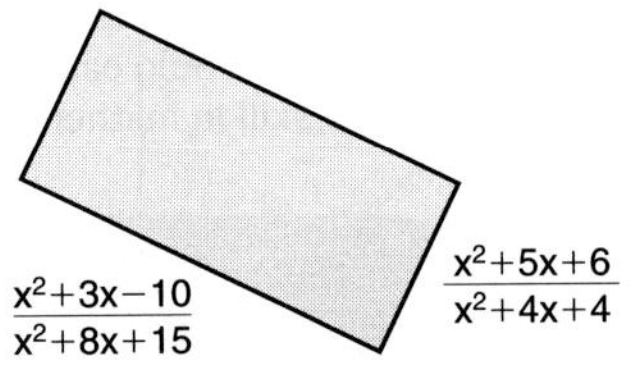

   A. $\frac{15}{32}x^2 - 1$ B. $\frac{x - 2}{x + 2}$
   C. $\frac{15x^2 - 32x - 60}{32x^2 + 92x + 60}$ D. $\frac{9x^2 - 42x + 49}{16x^2}$

4. **Probability** If a card is selected at random from a deck of 52 cards, what is the probability that it is not a face card? **D**

   A. $\frac{3}{13}$ B. 3:10
   C. $\frac{13}{25}$ D. $\frac{10}{13}$

5. **Physics** Rafael tossed a rock off the edge of a 10-meter-high cliff with an initial velocity of 15 meters per second. To the nearest tenth of a second, determine when the rock will hit the ground by using the formula $H = -4.9t^2 + vt + h$. **A**

   A. 3.6 s B. 3.1 s
   C. 4 s D. 3.9 s

6. Simplify $\frac{\frac{x^2 + 8x + 15}{x^2 + x - 6}}{\frac{x^2 + 2x - 15}{x^2 - 2x - 3}}$. **D**

   A. $\frac{23}{30}$ B. $\frac{x^2 + 4x + 3}{x^2 + x - 6}$
   C. $\frac{x^2 + 10x + 25}{x^2 - x - 2}$ D. $\frac{x + 1}{x - 2}$

7. **Geometry** What is the value of x if the perimeter of a square is 60 cm and its area is $4x^2 - 28x + 49$ cm²? **B**

   A. 4 only B. 11 only
   C. 4 and 11 D. −4 and 11

8. **Physics** The distance a force can move an object is $\frac{2a}{6a^2 - 17a - 3}$ yards. The distance a second force can move the same object is $\frac{a + 2}{a^2 - 9}$ yards. How much farther did the object move when the second force was applied than when the first force was applied? **A**

   A. $\frac{4a^2 + 7a + 2}{(a - 3)(a + 3)(6a + 1)}$ B. $\frac{a + 2}{5a^2 + 17a - 6}$
   C. $\frac{4a^2 + 17a - 2}{(a - 3)(a + 3)(6a - 1)}$ D. $\frac{10a^2 + 11a + 1}{(3a - 2)(2a + 1)}$

Standardized Test Practice Questions are also provided in the *Assessment and Evaluation Masters*, p. 326.

SECTION TWO: SHORT ANSWER

This section contains nine questions for which you will provide short answers. Write your answer on your paper.

9. Evaluate
$42 \div 7 - 1 - 5 + 8 \cdot 2 + 14 \div 2 - 8$. **15**

10. The rectangular penguin pond at the Bay Park Zoo is 12 meters long by 8 meters wide. The zoo wants to double the area of the pond by increasing the length and width by the same amount. By how much should the length and width be increased? **4 meters**

11. Determine the slope, y-intercept, and x-intercept of the graph of $2x - 3y = 13$. $\frac{2}{3}; -\frac{13}{3}; \frac{13}{2}$

12. Find the value of k if the remainder is 15 when $x^3 - 7x^2 + 4x + k$ is divided by $(x - 2)$. **27**

13. The tens digit of a two-digit number exceeds twice its units digit by 1. If the digits are reversed, the number is 4 more than 3 times the sum of the digits. Find the number. **52**

14. A long-distance cyclist pedaling at a steady rate travels 30 miles with the wind. He can travel only 18 miles against the wind in the same amount of time. If the rate of the wind is 3 miles per hour, what is the cyclist's rate without the wind? **12 mph**

15. Solve $-2 \leq 2x + 4 < 6$ and graph the solution set. $\{x \mid -3 \leq x < 1\}$

16. **Biology** The 2-inch long hummingbird flaps its wings about forty to fifty times each second. At this rate, how many times does it flap its wings in half of an hour? Express your answer in scientific notation. 7.2×10^4 **to** 9×10^4

17. **Construction** Muturi has 120 meters of fence to make a rectangular pen for his rabbits. If a shed is used as one side of the pen, what would be the maximum area of the pen? **1800 m**^{2}

SECTION THREE: OPEN-ENDED

This section contains three open-ended problems. Demonstrate your knowledge by giving a clear, concise solution to each problem. Your score on these problems will depend on how well you do the following.

- Explain your reasoning.
- Show your understanding of the mathematics in an organized manner.
- Use charts, graphs, and diagrams in your explanation.
- Show the solution in more than one way or relate it to other situations.
- Investigate beyond the requirements of the problem.

18. Mrs. Bloom bought some impatiens and petunias for her landscaping business for \$111.25. How many flats of each did she buy if each flat of impatiens was \$10.00, each flat of petunias was \$8.75, and if two fewer flats of impatiens than petunias were bought? **5 flats of impatiens; 7 flats of petunias**

19. Graph the quadratic function $y = -x^2 + 6x + 16$. Include the equation of the axis of symmetry, the coordinates of the vertex, and the roots of the related quadratic equation. **See margin.**

20. Solve the system of equations below by graphing. Explain how you determined the solution, and name at least three ordered pairs that satisfy the system.

$y \leq x + 3$

$2x - 2y < 8$

$2y + 3x > 4$ **See margin.**

Additional Answer

20.

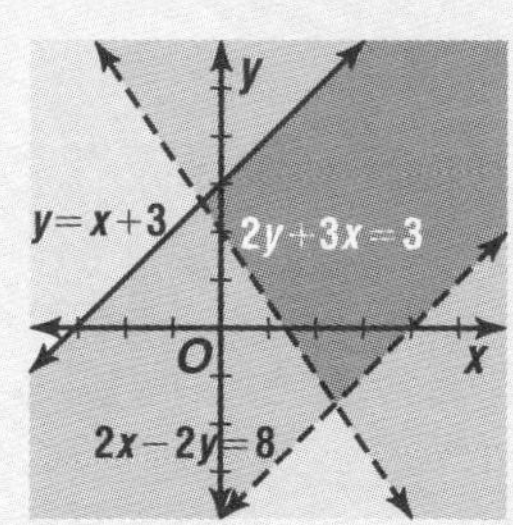

CHAPTER

13 Exploring Radical Expressions and Equations

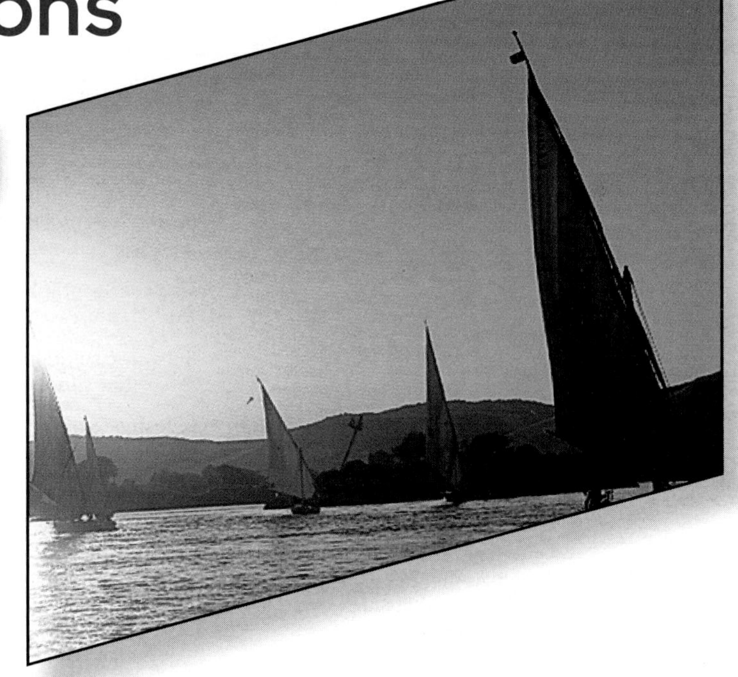

PREVIEWING THE CHAPTER

In this chapter, students apply the Pythagorean theorem to solve problems involving geometry. They simplify radical expressions with and without graphing technology. Students add and subtract radical expressions and apply this skill to solving radical equations. The distance formula is introduced as an integration with geometry. Students model completing the square and then apply this skill to solving equations.

Lesson (Pages)	Lesson Objectives	NCTM Standards	State/Local Objectives
13-1A (712)	Use a geoboard to find areas by using the Pythagorean theorem.	1–5, 7	
13-1 (713–718)	Use the Pythagorean theorem to solve problems.	1–5, 7	
13-2 (719–725)	Simplify square roots. Simplify radical expressions.	1–5	
13-2B (726)	Use a graphing calculator to simplify and approximate values of expressions containing radicals.	1–5	
13-3 (727–731)	Simplify radical expressions involving addition, subtraction, and multiplication.	1–5	
13-4 (732–736)	Solve radical equations.	1–5	
13-5 (737–741)	Find the distance between two points in the coordinate plane.	1–5, 8	
13-6A (742)	Use algebra tiles as a model for completing the square.	1–5	
13-6 (743–747)	Solve problems by identifying subgoals. Solve quadratic equations by completing the square.	1–5	

A complete, 1-page lesson plan is provided for each lesson in the ***Lesson Planning Guide***. Answer keys for each lesson are available in the ***Answer Key Masters.***

ORGANIZING THE CHAPTER

You may want to refer to the **Course Planning Calendar** on page T12 for detailed information on pacing.
PACING: Honors—11 days; **Block**—7 days; **Two Years**—22 days

LESSON PLANNING CHART

			BLACKLINE MASTERS											
Lesson (Pages)	**Materials/ Manipulatives**	**Extra Practice (Student Edition)**	**Study Guide**	**Practice**	**Enrichment**	**Assessment and Evaluation**	**Modeling Mathematics**	**Multicultural Activity**	**Tech Prep Applications**	**Graphing Calculator**	**Science and Math Lab Manual**	**Real-World Applications**	**Interactive Mathematics Tools Software**	**Teaching Transparencies**
13-1A (712)	geoboard* dot paper						p. 34						13-1A	
13-1 (713–718)		p. 785	p. 88	p. 88	p. 88		pp. 69–71	p. 25	p. 25			31		13-1A 13-1B
13-2 (719–725)	graphing calculator	p. 785	p. 89	p. 89	p. 89	p. 352			p. 26				13-2	13-2A 13-2B
13-2B (726)	graphing calculator									pp. 38, 39				
13-3 (727–731)	calculator	p. 785	p. 90	p. 90	p. 90	pp. 351, 352	p. 84							13-3A 13-3B
13-4 (732–736)	graphing software	p. 786	p. 91	p. 91	p. 91									13-4A 13-4B
13-5 (737–741)		p. 786	p. 92	p. 92	p. 92	p. 353		p. 26		p. 13	pp. 57–62		13-5.1 13-5.2	13-5A 13-5B
13-6A (742)	algebra tiles* equation mat*						p. 35							
13-6 (743–747)	algebra tiles*	p. 786	p. 93	p. 93	p. 93	p. 353						32		13-6A 13-6B
Study Guide/ Assessment (749–753)						pp. 337–350, 354–356								

*Included in Glencoe's Student Manipulative Kit and Overhead Manipulative Resources.

ORGANIZING THE CHAPTER

Other Chapter Resources

Student Edition
Chapter Opener, pp. 710–711
Mathematics and Society, p. 736
Working on the Investigation, pp. 718, 741
Closing the Investigation, p. 748

Teacher's Classroom Resources
Investigations and Projects Masters, pp. 73–76
Algebra and Geometry Overhead Manipulative Resources, pp. 36–38

Technology
Test and Review Software (IBM and Macintosh)
CD-ROM Interactions (Windows and Macintosh)

Professional Publications
Block Scheduling Booklet
Glencoe Mathematics Professional Series

Outside Resources

Books/Periodicals
Downing, Douglas, *Algebra the Easy Way,* Dale Seymour Publications

Software
Connecting to Algebra, Prometheus Software, Macintosh

Videos/CD-ROMs
The Theorem of Pythagorus, California Institute of Technology

 Assessment Resources

Student Edition
Math Journal, pp. 723, 729, 734
Mixed Review, pp. 718, 725, 731, 736, 741, 747
Self Test, p. 731
Chapter Highlights, p. 749
Chapter Study Guide and Assessment, pp. 750–752
Alternative Assessment, p. 753
Portfolio, p. 753

Teacher's Wraparound Edition
5-Minute Check, pp. 713, 719, 727, 732, 737, 743
Check for Understanding, pp. 715, 723, 729, 734, 738, 745
Closing Activity, pp. 718, 725, 731, 736, 741, 747
Cooperative Learning, pp. 733, 744

Assessment and Evaluation Masters
Multiple-Choice Tests, Forms 1A (Honors), 1B (Average), 1C (Basic), pp. 337–342
Free-Response Tests, Forms 2A (Honors), 2B (Average), 2C (Basic), pp. 343–348
Calculator-Based Test, p. 349
Performance Assessment, p. 350
Mid-Chapter Test, p. 351
Quizzes A–D, pp. 352–353
Standardized Test Practice, p. 354
Cumulative Review, pp. 355–356

ENHANCING THE CHAPTER

Examples of some of the materials for enhancing Chapter 13 are shown below.

Multicultural Activity Masters, pp. 25, 26

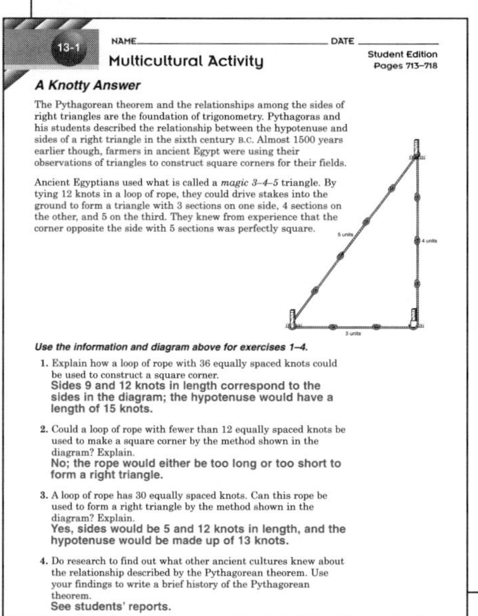

13-1 NAME ______ DATE ______

Multicultural Activity — Student Edition Pages 713–718

A Knotty Answer

The Pythagorean theorem and the relationships among the sides of right triangles are the foundation of trigonometry. Pythagoras and his students described the relationship between the hypotenuse and sides of a right triangle in the sixth century B.C. Almost 1500 years earlier though, farmers in ancient Egypt were using their observations of triangles to construct square corners for their fields.

Ancient Egyptians used what is called a *magic 3–4–5* triangle. By tying 12 knots in a loop of rope, they could drive stakes into the ground to form a triangle with 3 sections on one side, 4 sections on the other, and 5 on the third. They knew from experience that the corner opposite the side with 5 sections was perfectly square.

Use the information and diagram above for exercises 1–4.

1. Explain how a loop of rope with 36 equally spaced knots could be used to construct a square corner.
   Sides 9 and 12 knots in length correspond to the sides in the diagram; the hypotenuse would have a length of 15 knots.
2. Could a loop of rope with fewer than 12 equally spaced knots be used to make a square corner by the method shown in the diagram? Explain.
   No; the rope would either be too long or too short to form a right triangle.
3. A loop of rope has 30 equally spaced knots. Can this rope be used to form a right triangle by the method shown in the diagram? Explain.
   Yes, sides would be 5 and 12 knots in length, and the hypotenuse would be made up of 13 knots.
4. Do research to find out what other ancient cultures knew about the relationship described by the Pythagorean theorem. Use your findings to write a brief history of the Pythagorean theorem.
   See students' reports.

Real-World Applications, 31, 32

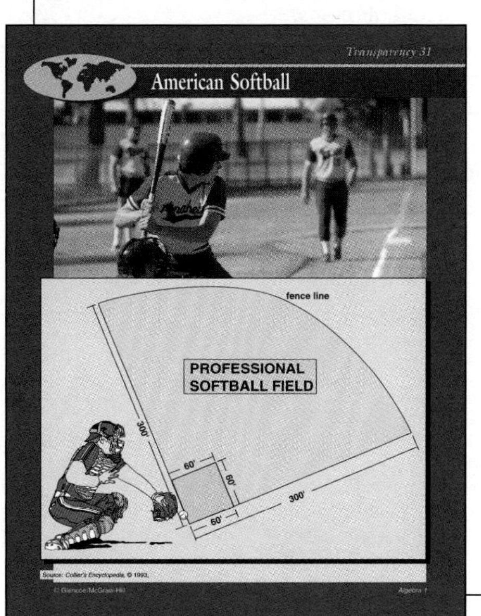

TECHNOLOGY

Graphing Calculator Masters, p. 13

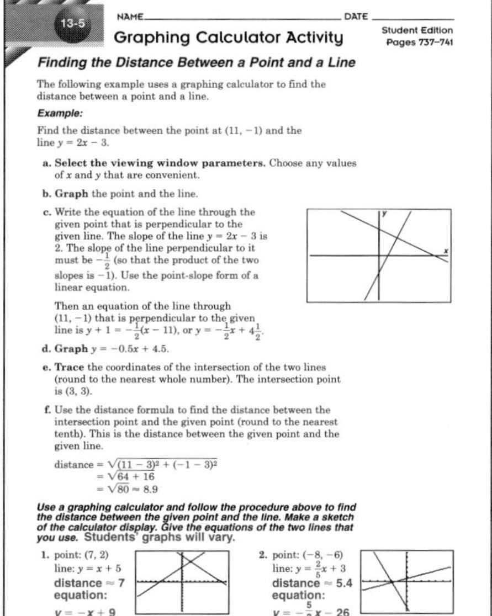

13-5 NAME ______ DATE ______

Graphing Calculator Activity — Student Edition Pages 737–741

Finding the Distance Between a Point and a Line

The following example uses a graphing calculator to find the distance between a point and a line.

Example:

Find the distance between the point at $(11, -1)$ and the line $y = 2x - 3$.

a. **Select the viewing window parameters.** Choose any values of x and y that are convenient.

b. **Graph** the point and the line.

c. Write the equation of the line through the given point that is perpendicular to the given line. The slope of the line $y = 2x - 3$ is 2. The slope of the line perpendicular to it must be $-\frac{1}{2}$ (so that the product of the two slopes is -1). Use the point-slope form of a linear equation.

   Then an equation of the line through $(11, -1)$ that is perpendicular to the given line is $y + 1 = -\frac{1}{2}(x - 11)$, or $y = -\frac{1}{2}x + 4\frac{1}{2}$.

d. **Graph** $y = -0.5x + 4.5$.

e. **Trace** the coordinates of the intersection of the two lines (round to the nearest whole number). The intersection point is (3, 3).

f. Use the distance formula to find the distance between the intersection point and the given point (round to the nearest tenth). This is the distance between the given point and the given line.

$$\begin{aligned} \text{distance} &= \sqrt{(11-3)^2 + (-1-3)^2} \\ &= \sqrt{64 + 16} \\ &= \sqrt{80} \approx 8.9 \end{aligned}$$

Use a graphing calculator and follow the procedure above to find the distance between the given point and the line. Make a sketch of the calculator display. Give the equations of the two lines that you use. **Students' graphs will vary.**

1. point: (7, 2)
   line: $y = x + 5$
   distance ≈ 7
   equation:
   $y = -x + 9$
2. point: $(-8, -6)$
   line: $y = \frac{2}{5}x + 3$
   distance ≈ 5.4
   equation:
   $y = -\frac{5}{2}x - 26$

TECH PREP

Tech Prep Applications Masters, pp. 25, 26

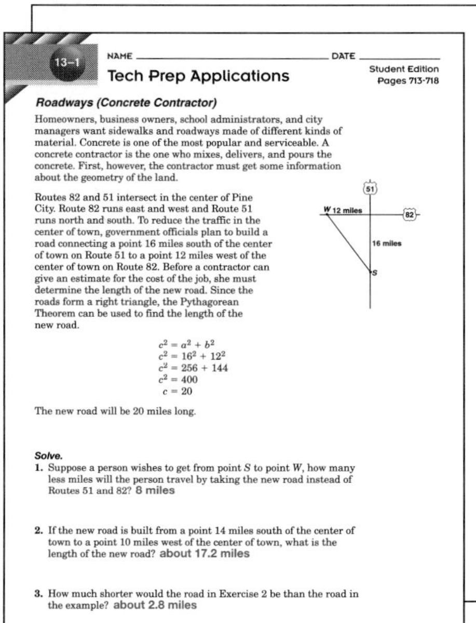

13–1 NAME ______ DATE ______

Tech Prep Applications — Student Edition Pages 713-718

Roadways (Concrete Contractor)

Homeowners, business owners, school administrators, and city managers want sidewalks and roadways made of different kinds of material. Concrete is one of the most popular and serviceable. A concrete contractor is the one who mixes, delivers, and pours the concrete. First, however, the contractor must get some information about the geometry of the land.

Routes 82 and 51 intersect in the center of Pine City. Route 82 runs east and west and Route 51 runs north and south. To reduce the traffic in the center of town, government officials plan to build a road connecting a point 16 miles south of the center of town on Route 51 to a point 12 miles west of the center of town on Route 82. Before a contractor can give an estimate for the cost of the job, she must determine the length of the new road. Since the roads form a right triangle, the Pythagorean Theorem can be used to find the length of the new road.

$$\begin{aligned} c^2 &= a^2 + b^2 \\ c^2 &= 16^2 + 12^2 \\ c^2 &= 256 + 144 \\ c^2 &= 400 \\ c &= 20 \end{aligned}$$

The new road will be 20 miles long.

Solve.

1. Suppose a person wishes to get from point S to point W, how many less miles will the person travel by taking the new road instead of Routes 51 and 82? **8 miles**
2. If the new road is built from a point 14 miles south of the center of town to a point 10 miles west of the center of town, what is the length of the new road? **about 17.2 miles**
3. How much shorter would the road in Exercise 2 be than the road in the example? **about 2.8 miles**

Science and Math Lab Manual, pp. 57–62

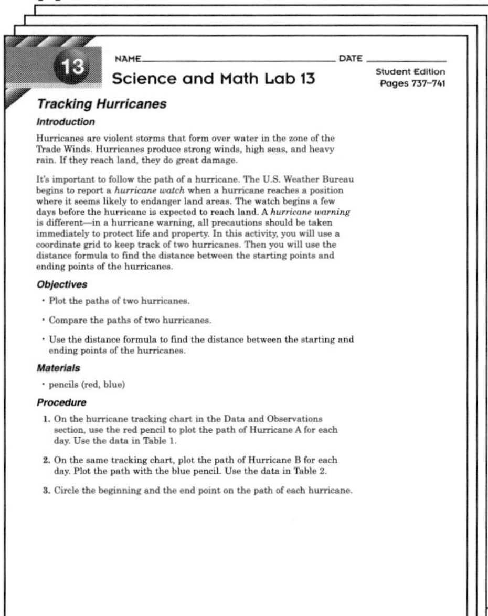

13 NAME ______ DATE ______

Science and Math Lab 13 — Student Edition Pages 737–741

Tracking Hurricanes

Introduction

Hurricanes are violent storms that form over water in the zone of the Trade Winds. Hurricanes produce strong winds, high seas, and heavy rain. If they reach land, they do great damage.

It's important to follow the path of a hurricane. The U.S. Weather Bureau begins to report a *hurricane watch* when a hurricane reaches a position where it seems likely to endanger land areas. The watch begins a few days before the hurricane is expected to reach land. A *hurricane warning* is different—in a hurricane warning, all precautions should be taken immediately to protect life and property. In this activity, you will use a coordinate grid to keep track of two hurricanes. Then you will use the distance formula to find the distance between the starting points and ending points of the hurricanes.

Objectives

- Plot the paths of two hurricanes.
- Compare the paths of two hurricanes.
- Use the distance formula to find the distance between the starting and ending points of the hurricanes.

Materials

- pencils (red, blue)

Procedure

1. On the hurricane tracking chart in the Data and Observations section, use the red pencil to plot the path of Hurricane A for each day. Use the data in Table 1.
2. On the same tracking chart, plot the path of Hurricane B for each day. Plot the path with the blue pencil. Use the data in Table 2.
3. Circle the beginning and the end point on the path of each hurricane.

Problem of the Week Cards, 35, 36

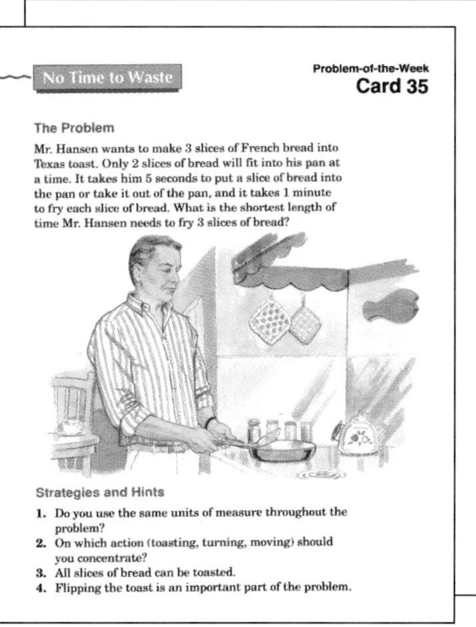

No Time to Waste

Problem-of-the-Week **Card 35**

The Problem

Mr. Hansen wants to make 3 slices of French bread into Texas toast. Only 2 slices of bread will fit into his pan at a time. It takes him 5 seconds to put a slice of bread into the pan or take it out of the pan, and it takes 1 minute to fry each slice of bread. What is the shortest length of time Mr. Hansen needs to fry 3 slices of bread?

Strategies and Hints

1. Do you use the same units of measure throughout the problem?
2. On which action (toasting, turning, moving) should you concentrate?
3. All slices of bread can be toasted.
4. Flipping the toast is an important part of the problem.

MAKING MATHEMATICS RELEVANT

This two-page introduction to the chapter provides students with an opportunity to explore contemporary topics and their applications to mathematics.

Background Information

Teens Talk to Parents Students will notice certain trends in the data. Young teenagers need help with their homework and need to feel loved. As they grow older, they need these less. Perhaps surprisingly, teenagers do not think it is important that their parents stay at home instead of working. But one concern is strong at all ages. Teenagers want their parents to teach them about sex.

CHAPTER 13

Exploring Radical Expressions and Equations

Objectives

In this chapter, you will:

- use the Pythagorean theorem to solve problems,
- simplify radical expressions,
- solve problems involving radical equations,
- solve quadratic equations by completing the square, and
- solve problems by identifying subgoals.

Teens Talk to Parents

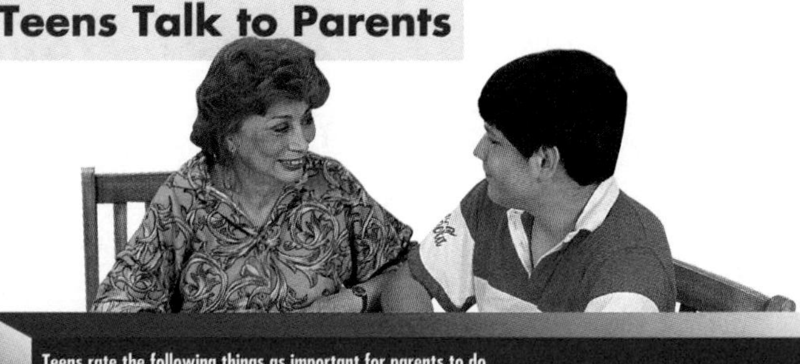

Teens rate the following things as important for parents to do for children who are 12 years old or younger: ○
For people their own age: ■

		Very important	Somewhat important	Not too important			Very important	Somewhat important	Not too important
Help them with their homework	○	74%	22%	4%	Talk with their teachers regularly	○	43%	44%	13%
	■	42%	38%	20%		■	32%	42%	25%
Give them a hug at least once a week	○	70%	34%	8%	Be active in school activities like the PTA	○	39%	43%	17%
	■	46%	34%	19%		■	24%	42%	34%

Source: *Oakland Press*, 1995

Almost everyone agrees that people who can communicate, who really listen to one another, have the healthiest relationships. This is certainly true when talking about teenagers and their parents. Kids can learn much from the experiences of their parents, and adults can benefit from the fresh ideas they share with their teens. What is the communication like in your home? What topics do you think are important to discuss with your parents?

TIME Line

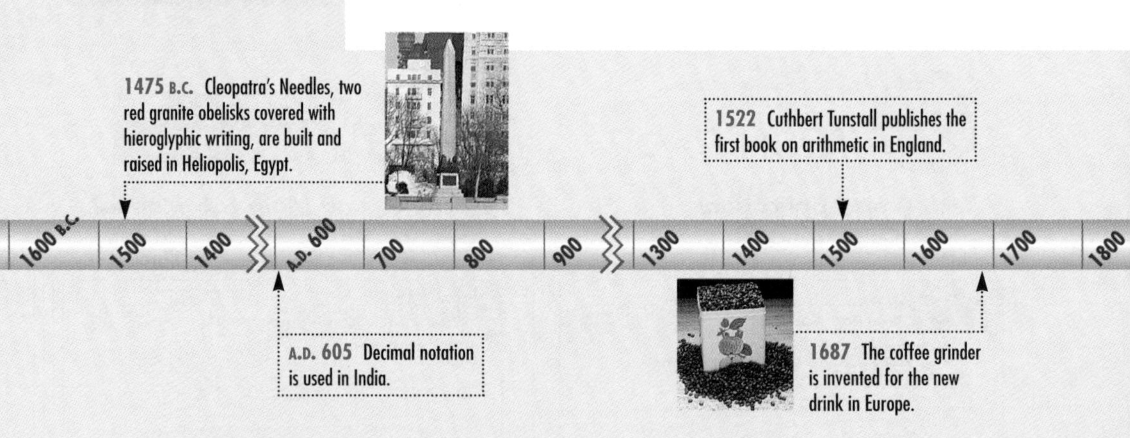

1475 B.C. Cleopatra's Needles, two red granite obelisks covered with hieroglyphic writing, are built and raised in Heliopolis, Egypt.
1522 Cuthbert Tunstall publishes the first book on arithmetic in England.
1600 B.C.
1500
1400
A.D. 600
700
800
900
1300
1400
1500
1600
1700
1800
A.D. 605 Decimal notation is used in India.
1687 The coffee grinder is invented for the new drink in Europe.

TIME Line

Students might find it interesting to do a class presentation on the history of number systems. Today we take zero, negative numbers, functions, and decimals for granted. But, as the time line indicates, decimal notation was not used until A.D. 605. Students might want to track down the invention of zero and the negative numbers.

interNET CONNECTION

Information about children and divorce, alcohol and other drugs, and children and TV violence is available on the Internet.

World Wide Web
http://www.psych.med.umich.edu/web/aacap/factsFam/

PEOPLE IN THE NEWS

At age 12, **Ryan Holladay** of Arlington, Virginia, is already a published author. His book *What Preteens Want Their Parents to Know* contains about 180 tips for parents on subjects from discipline and homework to love and respect. "It's tough growing up from being a little kid to a teenager," says Holladay. "I noticed lots of advice books written by adults, but none by kids. So I began writing one myself." Holladay surveyed children aged 9 to 12 for more ideas for his book, which took two years to complete. Since his book was published, Holladay has been on TV and radio shows to talk about it. His best advice to parents and children: "Be flexible."

Chapter Project

Form teams and conduct your own survey of what teens want from their parents or guardians. You can use the questions from the chart on page 710 or add your own. Keep track of the age, gender, and number of people questioned and their responses. See if you can make conclusions about the attitude of teens toward their parents' communication with them. Create charts or graphs using your data and report to the class on your findings.

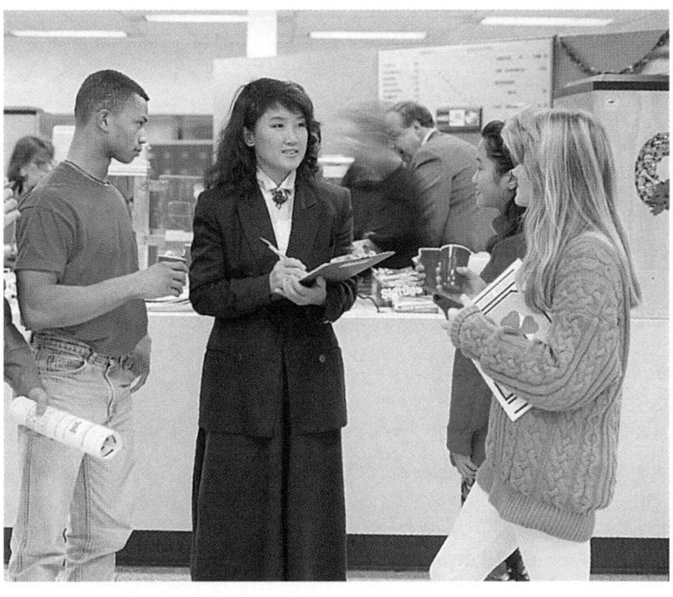

1968 Luis W. Alvarez wins the Nobel Prize for Physics for his discovery of resonance particles.
1994 Kim Campbell becomes Canada's first female prime minister.
1890
1900
1910
1920
1930
1940
1950
1960
1970
1980
1990
2000
1963 Charles Moore takes his famous photograph for *Life* magazine, *Birmingham Riots*, during the civil rights struggle in Alabama.
1987 The African-American experience is portrayed in the photograph *Basket of Millet* by artist Elisabeth Sunday.

Alternative Chapter Projects

Two other chapter projects are included in the *Investigations and Projects Masters*. In Chapter 13 Project A, pp. 73–74, students extend the topic in the chapter opener. In Chapter 13 Project B, pp. 75–76, students construct spirals.

PEOPLE IN THE NEWS

Ryan's book is available in many libraries. It was published in 1994 by McCracken Press, New York, NY.

Chapter Project

Cooperative Learning Groups should discuss what type of chart or graph would best represent the data collected. They might consider bar graphs, line graphs, or circle graphs. They should debate the pros and cons of each. They should conclude that bar graphs are the most descriptive.

Investigations and Projects Masters, p. 73

NAME________________ DATE________

13 Chapter 13 Project A — Student Edition Pages 712–748

Radical Survey

1. Although you may not have thought about it, your parents and other adults you know probably spent time as teenagers studying math just as you are doing now. For this project, you will work in a small group to survey adults about their high school years and their experiences with math during and after high school. Begin by brainstorming a list of questions. Some possibilities are shown below.
   - What was high school like in your day?
   - What issues concerned you in high school?
   - What is your view of today's high schools?
   - Were your experiences with math in high school positive or negative? Explain your answer.
   - How do you or someone you work with use math on the job?
2. Design a questionnaire containing the five best questions from exercise 1. Distribute copies of your questionnaire to at least 20 adults. Be sure to explain the procedure for returning the completed questionnaires.
3. Make charts and graphs to organize and explain the data you and your group collected in exercise 2.
4. Use the data your group collected to prepare a newspaper article or a radio or TV report about adults and their experiences in high school.

13-1A LESSON NOTES

NCTM Standards: 1–5, 7

Objective
Use a geoboard to find areas by using the Pythagorean theorem.

Recommended Time
Demonstration and discussion: 15 minutes; Exercises: 30 minutes

Instructional Resources
For each student or group of students
Student Manipulative Kit
- geoboard

Modeling Mathematics Masters
- p. 6 (geoboard pattern)
- p. 34 (worksheet)

For teacher demonstration
Algebra and Geometry Overhead Manipulative Resources

1 FOCUS

Motivating the Lesson
Discuss how triangles and squares relate to each other. Remind students of the meaning of *hypotenuse* and other terms related to triangles.

2 TEACH

Teaching Tip Students may be more willing to believe that the area of the square on the hypotenuse is 2 square units if they separate the square into right triangles as shown below.

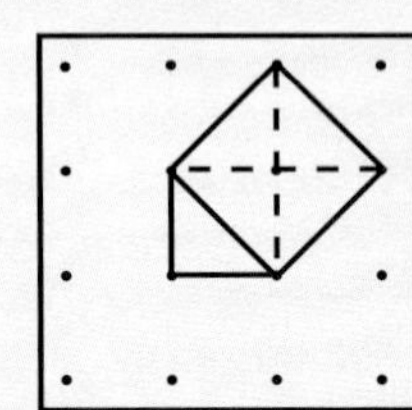

3 PRACTICE/APPLY

Assignment Guide

Core: 1–10
Enriched: 1–10

MODELING MATHEMATICS

13–1A The Pythagorean Theorem

A Preview of Lesson 13–1

Materials:  geoboard  dot paper

In this activity, you will use the Pythagorean theorem to build squares on a geoboard or dot paper.

Activity **Make a square with an area of 2 square units.**

Step 1 Start with a right triangle like the one shown below.

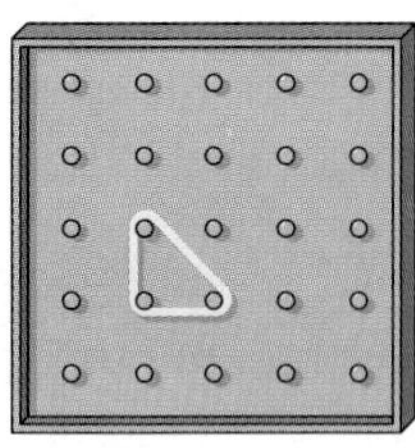

Step 2 Build squares on the two legs. Each square has an area of 1 square unit.

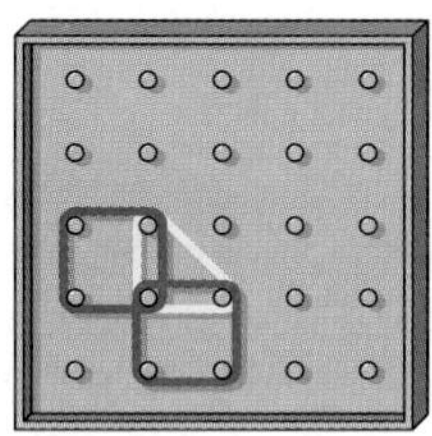

Step 3 Now build a square on the hypotenuse.

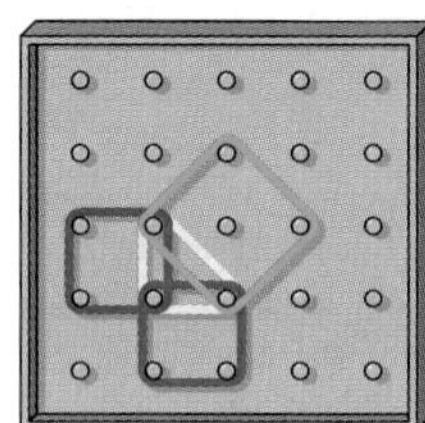

You can find the area of the square on the hypotenuse by using the Pythagorean theorem. Let c represent the measure of the hypotenuse and a and b represent the measures of the legs.

$$c^2 = a^2 + b^2$$
$$= 1^2 + 1^2 \quad \text{Replace } a \text{ with 1 and } b \text{ with 1.}$$
$$= 1 + 1 \text{ or } 2$$

The area of the square on the hypotenuse is 2 square units.

10. The total area of the 4 triangles equals the area of the large square minus the area of the triangle in each corner minus the area of the small square, 49 − 4(6) − 16 or 9 square units.

Model **Build squares on each side of the triangles shown below using a geoboard or dot paper. Record the area of each square.**

1. 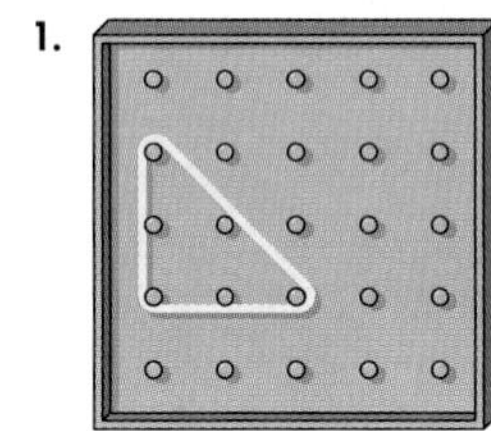
4 + 4 = 8

2. 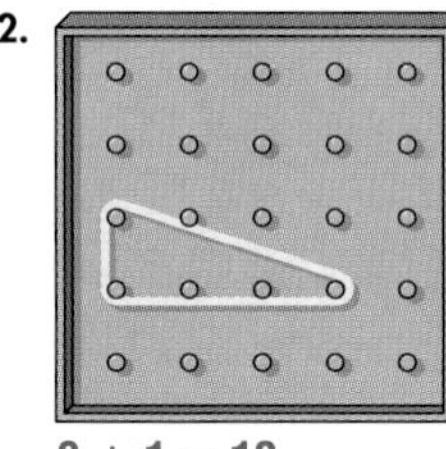
9 + 1 = 10

3. 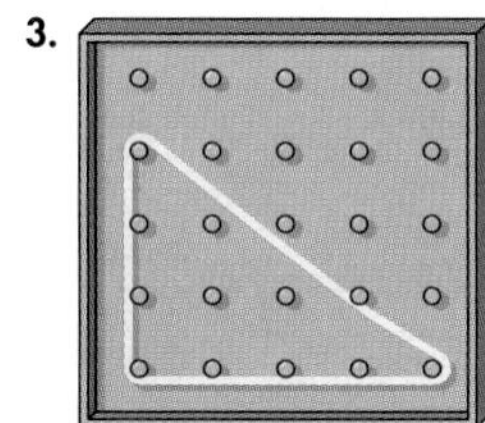
16 + 9 = 25

Draw **Draw a square on dot paper having each area.**

4–9. See Solutions Manual.

4. 4 square units
5. 9 square units
6. 8 square units
7. 13 square units
8. 17 square units
9. 32 square units

Write

10. Write a paragraph explaining how to find the total area of the shaded triangles in the drawing at the right.

4 ASSESS

Observing students working in cooperative groups is an excellent method of assessment.

GLENCOE Technology

Interactive Mathematics Tools Software

This multimedia software provides an interactive lesson that has students manipulate a triangle and chart the observations of the relationship of the sides of the right angle. A **Computer Journal** gives students an opportunity to write about what they have learned.

For Windows & Macintosh

13-1 Integration: Geometry The Pythagorean Theorem

- To use the Pythagorean theorem to solve problems.

You can use the Pythagorean theorem to solve problems involving sailing and travel.

FYI

Nearly 700 women applied for positions on the *America*[3] team. The final crew consisted of 28 women of various backgrounds, including an areospace engineer, a body builder, and a student.

Sailing

One of the most prestigious sailboat races in the world is the America's Cup. In 1995, for the first time in the 144-year history of the race, one of the sailboats, *America*[3], had an all-woman crew. Theirs was one of three U.S. boats vying for the most prized trophy in sailing.

A sailboat's *mast* and *boom* form a right angle. The sail itself, called a *mainsail*, is in the shape of a right triangle.

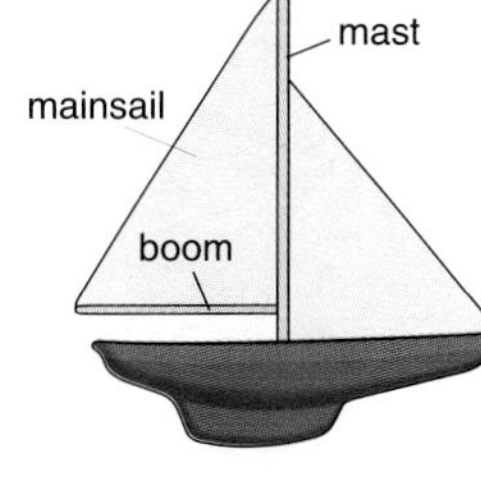

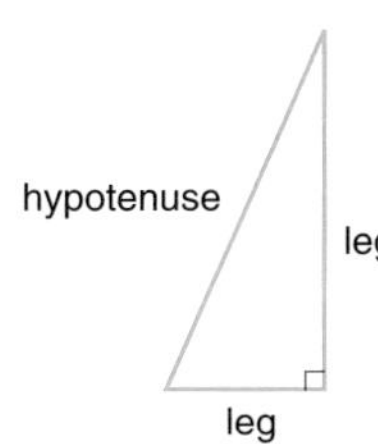

Recall that the side opposite the right angle in a right triangle is called the *hypotenuse*. This side is always the longest side of a right triangle. The other two sides are called the *legs* of the triangle.

To find the length of any side of a right triangle when the lengths of the other two are known, you can use a formula named for the Greek mathematician Pythagoras.

The Pythagorean Theorem	**If a and b are the measures of the legs of a right triangle and c is the measure of the hypotenuse, then $c^2 = a^2 + b^2$.**

You can use the Pythagorean theorem to find the length of the hypotenuse of a right triangle when the lengths of the legs are known.

LOOK BACK

Refer to Lesson 2-8 to review square roots.

Example 1 **Find the length of the hypotenuse of a right triangle if $a = 12$ and $b = 5$.**

$c^2 = a^2 + b^2$ *Pythagorean theorem*

$c^2 = 12^2 + 5^2$ *$a = 12$ and $b = 5$*

$c^2 = 144 + 25$

$c^2 = 169$

$c = \pm\sqrt{169}$

$c = \pm 13$ *Disregard -13. Why?*

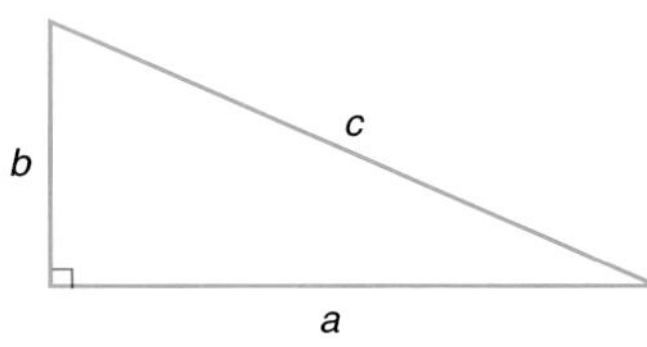

The length of the hypotenuse is 13 units.

FYI

The America's Cup was first run in 1851 in Great Britain as the Hundred Guinea Cup. It was won by a New York schooner, the *America*. In 1857 the New York Yacht Club began running the competition. In 1987 the San Diego Yacht Club took over.

13-1 LESSON NOTES

NCTM Standards: 1–5, 7

Instructional Resources

- Study Guide Master 13-1
- Practice Master 13-1
- Enrichment Master 13-1
- Modeling Mathematics Masters, pp. 69–71
- Multicultural Activity Masters, p. 25
- Real-World Applications, 31
- Tech Prep Applications Masters, p. 25

Transparency 13-1A contains the 5-Minute Check for this lesson; **Transparency 13-1B** contains a teaching aid for this lesson.

Recommended Pacing	
Honors Pacing	Day 2 of 11
Block Scheduling*	Day 1 of 7
Alg. 1 in Two Years*	Days 2 & 3 of 22

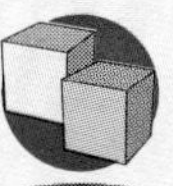

*For more information on pacing and possible lesson plans, refer to the *Block Scheduling Booklet* and *Algebra 1 in Two Years.*

1 FOCUS

5-Minute Check
(over Chapter 12)

1. Simplify and state the excluded values for $\frac{x^2-16}{x^2+6x+8} \cdot \frac{x-4}{x+2}$; **−4, −2**

Find each quotient.

2. $\frac{2a+4}{a+1} \div \frac{3a^2+8a+4}{2a^2+5a+3}$ **$\frac{4a+12}{3a+2}$**
3. $\frac{3b^2+5b-2}{b-1}$ **$3b+8+\frac{6}{b-1}$**

Find each difference.

4. $\frac{5x}{x-2} - \frac{2x}{2-x}$ **$\frac{7x}{x-2}$**
5. $\frac{3}{a+1} - \frac{2a}{2a+3}$ **$\frac{-2a^2+4a+9}{(a+1)(2a+3)}$**

Motivating the Lesson

Hands-On Activity Open this lesson by measuring the length and width of the classroom. Then have students measure the length of a diagonal that cuts across the room. Write the three measurements on the board. Challenge students to determine a relationship among the numbers.

2 TEACH

In-Class Examples

For Example 1
Find the length of the hypotenuse of each triangle if the measures of the legs are given below.

a. $a = 12, b = 16$ **20**
b. $a = 11, b = \sqrt{23}$ **12**
c. $a = 5, b = 7$ **$\sqrt{74}$ or 8.60**

For Example 2
If c is the measure of the hypotenuse of a right triangle, find each missing measure.

a. $a = 4, c = 10, b = ?$
$\sqrt{84}$ or 9.17
b. $a = 5, c = 6, b = ?$
$\sqrt{11}$ or 3.32
c. $c = 10, b = 7, a = ?$
$\sqrt{51}$ or 7.14

For Example 3
Determine whether the following side measures would form right triangles.

a. $a = 3, b = 8, c = \sqrt{73}$ **yes**
b. $a = 8, b = 9, c = 11$ **no**
c. $a = 13, b = 84, c = 85$ **yes**

For Example 4
Find the length of the diagonal of a rectangle whose length is 8 meters and whose width is 5 meters.
$\sqrt{89}$ or 9.43 meters

GLOBAL CONNECTIONS

Sails were originally made of animal hide. Then woven reeds were used. Sailcloth can be made of flax fiber, cotton, or other fabrics.

Example  **Find the length of side a if $b = 9$ and $c = 21$. Round to the nearest hundredth.**

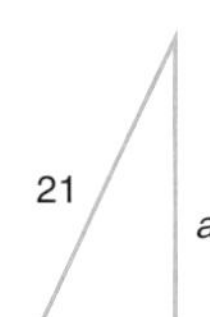

$c^2 = a^2 + b^2$ *Pythagorean theorem*
$21^2 = a^2 + 9^2$ *$b = 9$ and $c = 21$*
$441 = a^2 + 81$
$360 = a^2$ *Use a calculator to approximate $\sqrt{360}$ to the nearest hundredth.*
$\pm\sqrt{360} = a$
$18.97 \approx a$ *Only the positive value of a has meaning in this situation.*

The length of the leg, to the nearest hundredth, is 18.97 units.

The following corollary, based on the Pythagorean theorem, can be used to determine whether a triangle is a right triangle.

Corollary to the Pythagorean Theorem — ***If c is the measure of the longest side of a triangle and $c^2 \neq a^2 + b^2$, then the triangle is not a right triangle.***

Example  **Determine whether the following side measures would form right triangles.**

a. 6, 8, 10

Since the measure of the longest side is 10, let $c = 10$, $a = 6$, and $b = 8$. Then determine whether $c^2 = a^2 + b^2$.

$10^2 \stackrel{?}{=} 6^2 + 8^2$
$100 \stackrel{?}{=} 36 + 64$
$100 = 100$ ✓

Since $c^2 = a^2 + b^2$, the triangle is a right triangle.

b. 7, 9, 12

Since the measure of the longest side is 12, let $c = 12$, $a = 7$, and $b = 9$. Then determine whether $c^2 = a^2 + b^2$.

$12^2 \stackrel{?}{=} 7^2 + 9^2$
$144 \stackrel{?}{=} 49 + 81$
$144 \neq 130$

Since $c^2 \neq a^2 + b^2$, the triangle is not a right triangle.

GLOBAL CONNECTIONS

One of the earliest inventions was the sail, and it was the most important means of powering boats for most of maritime history. Boats with sails appear in Egyptian rock carvings near the Nile, dating from 3300 B.C.

Example 4

World Cultures

Agriculture was very important in ancient Aztec culture. Aztec farmers kept records of their farms including calculations of the dimensions and area. Because the terrain was so rough, very few of the farms were rectangular. Yet, by using measuring ropes that measured length with a unit called a *quahuitl* (about 2.5 meters), they were able to make accurate calculations. In the farm shown at the right, the farmer measured three sides of his farm. He had trouble measuring the fourth side because it was located in a dense forest. Find the measure of the fourth side.

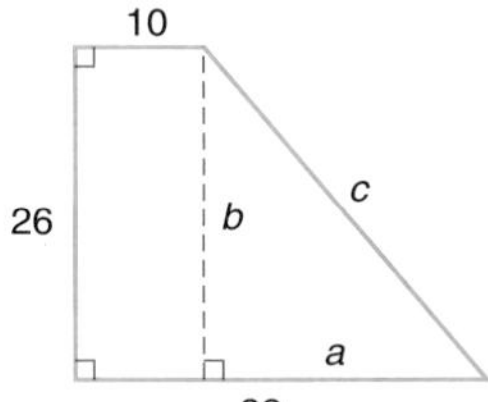

Alternative Learning Styles

Visual Challenge students to illustrate the Pythagorean theorem geometrically using other figures such as semicircles. Since $A = \pi r^2$ is the formula for the area of a circle, $A = \frac{1}{2}(\pi r^2)$ or $A = \frac{1}{2}\pi r^2$ is the formula for the area of a semicircle. Use this formula to illustrate the Pythagorean theorem using semicircles. The sides of the right triangle have lengths 3, 4, and 5 units.

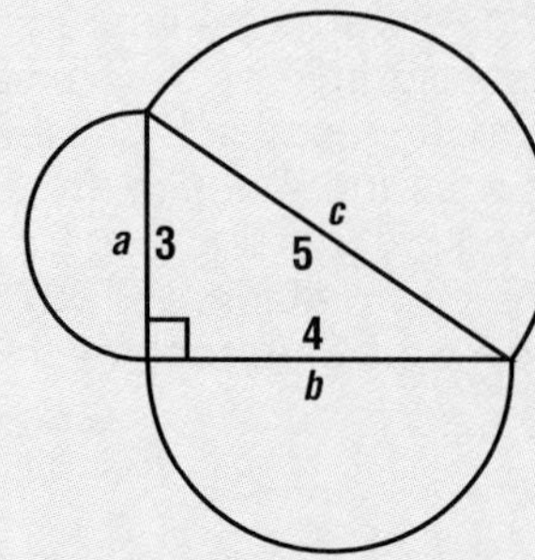

Explore Let c represent the length of the fourth side of the farm. Note that it is the hypotenuse of a right triangle.

Plan Use the Pythagorean theorem to find c. Let $a = 32 - 10$ or 22 and $b = 26$. Then solve the resulting equation.

Solve
$$c^2 = (22)^2 + (26)^2$$
$$c^2 = 484 + 676$$
$$c^2 = 1160$$
$$c \approx 34.06$$

The length of the forest side of the farm is approximately 34.06 quahuitls.

Examine Check the solution by substituting 34.06 for c in the Pythagorean theorem.

$$c^2 = a^2 + b^2$$
$$(34.06)^2 \stackrel{?}{=} (22)^2 + (26)^2$$
$$1160 = 1160 \checkmark$$

CHECK FOR UNDERSTANDING

Communicating Mathematics

3b. $3^2 + 4^2 \stackrel{?}{=} 5^2$

$9 + 16 \stackrel{?}{=} 25$

$25 = 25 \checkmark$

Study the lesson. Then complete the following. 1–2. See margin.

1. **Draw** a right triangle and label each side with a letter.
2. **Explain** how you can determine whether a triangle is a right triangle if you know the lengths of the three sides.
3. In 1955, Greece issued the stamp shown at the left to honor the 2500th anniversary of the Pythagorean School. Notice that there is a triangle bordered on each side by a checkerboard pattern.
   a. Count the number of squares along each side of the triangle. 3, 4, 5
   b. Use the Pythagorean theorem to show that it is a right triangle.
4. When taking the square root of a number, you can get a positive and a negative number. Why then, when using the Pythagorean theorem, is only the positive value used? Explain. See margin.

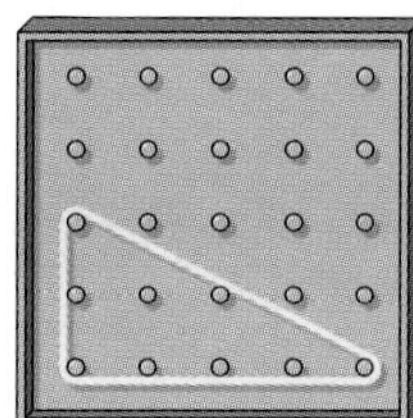

MODELING MATHEMATICS

5. Use a geoboard or dot paper to build squares on each side of the triangle at the right. Record the areas of the squares. 16 + 4 = 20

Guided Practice

Solve each equation. Assume each variable represents a positive number.

6. $5^2 + 12^2 = c^2$ 13
7. $a^2 + 24^2 = 25^2$ 7
8. $16^2 + b^2 = 20^2$ 12

Find the length of each missing side. Round to the nearest hundredth.

9. 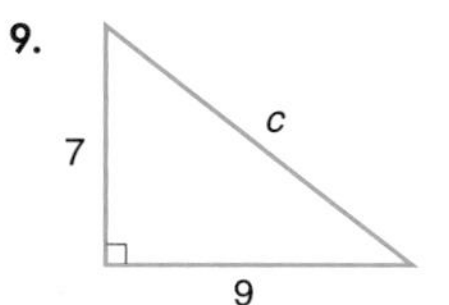

11.40

10. 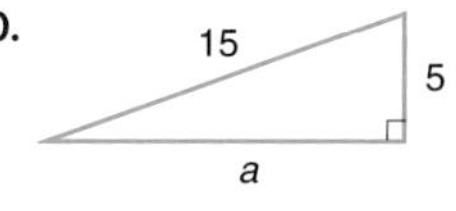

14.14

3 PRACTICE/APPLY

Check for Understanding

Exercises 1–17 are designed to help you assess your students' understanding through reading, writing, speaking, and modeling. You should work through Exercises 1–5 with your students and then monitor their work on Exercises 6–17.

Additional Answers

1. 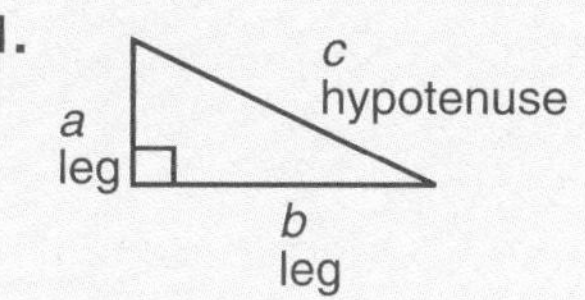

2. **By squaring the two shorter sides and adding them; then, comparing their sum to the square of the third or longer side. If they are equal, it is a right triangle.**
4. **Since the Pythagorean theorem finds the length of a side, it has to be positive. Distance is not negative.**

Reteaching

Using Manipulatives Supply a geoboard to each pair of students. Stretch rubber bands over the pegs to form right triangles. Find the lengths of all sides. Demonstrate how to find the lengths $\sqrt{2}$, $\sqrt{3}$, $\sqrt{5}$, $\sqrt{10}$, $\sqrt{13}$, and $3\sqrt{2}$. Form a square with an area of 2 square units.

Assignment Guide

Core: 19–47 odd, 49–56
Enriched: 18–44 even, 45–56

For **Extra Practice,** see p. 785.

The red A, B, and C flags, printed only in the Teacher's Wraparound Edition, indicate the level of difficulty of the exercises.

If *c* is the measure of the hypotenuse of a right triangle, find each missing measure. Round answers to the nearest hundredth, if necessary.

11. $a = 9, b = 12, c = ?$ **15**
12. $a = \sqrt{11}, c = 6, b = ?$ **5**
13. $b = \sqrt{30}, c = \sqrt{34}, a = ?$ **2**
14. $a = 7, b = 4, c = ?$ **$\sqrt{65} \approx 8.06$**

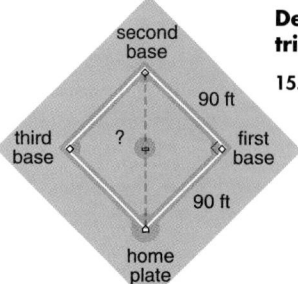

Determine whether the following side measures would form right triangles. Explain why or why not.

15. 12, 16, 20 **yes**
16. 2, 8, 8 **no; $2^2 + 8^2 \neq 8^2$**

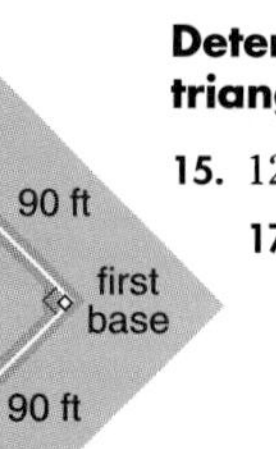

17. Baseball A baseball scout uses many different tests to determine whether or not to draft a particular player. One test for catchers is to see how quickly they can throw a ball from home plate to second base. On a baseball diamond, the distance from one base to the next is 90 feet. What is the distance from home plate to second base? **127.28 ft**

EXERCISES

Practice

Find the length of each missing side. Round to the nearest hundredth.

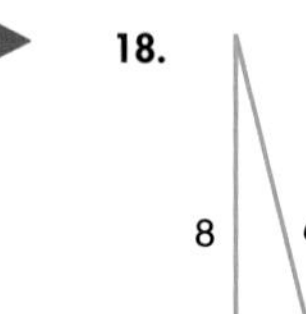

18.
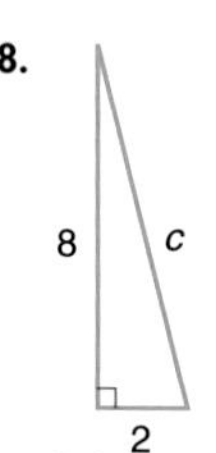

8.25

19.
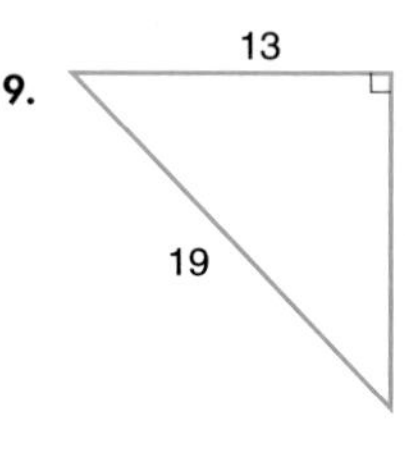

13.86

20.
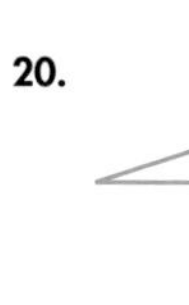

15.20

21.
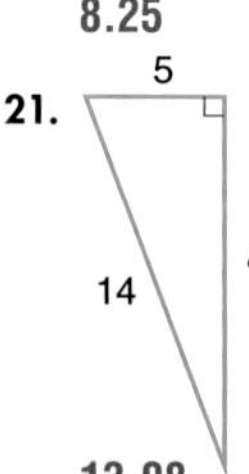

13.08

22.
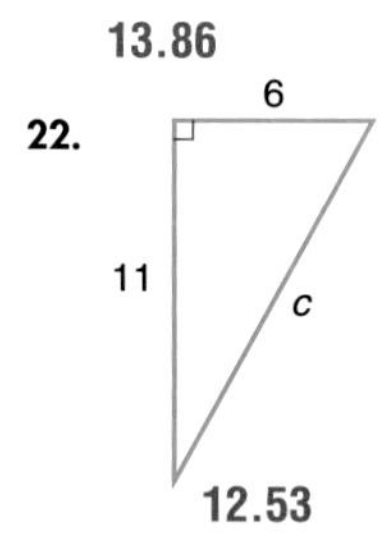

12.53

23.
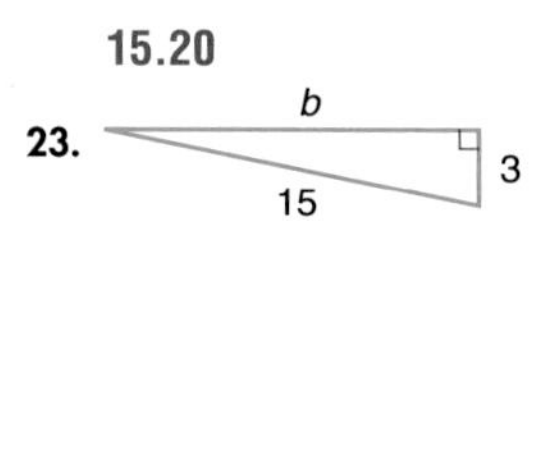

14.70

If *c* is the measure of the hypotenuse of a right triangle, find each missing measure. Round answers to the nearest hundredth.

24. $a = 16, b = 30, c = ?$ **34**
25. $a = 11, c = 61, b = ?$ **60**
26. $b = 13, c = \sqrt{233}, a = ?$ **8**
27. $a = \sqrt{7}, b = \sqrt{9}, c = ?$ **4**
28. $a = 6, b = 3, c = ?$ **$\sqrt{45} \approx 6.71$**
29. $b = \sqrt{77}, c = 12, a = ?$
30. $b = 10, c = 11, a = ?$ **$\sqrt{21} \approx 4.58$**
31. $a = 4, b = \sqrt{11}, c = ?$
32. $a = 15, b = \sqrt{28}, c = ?$
33. $a = 12, c = 17, b = ?$

Determine whether the following side measures would form right triangles. Explain why or why not.

34. 6, 9, 12
35. 45, 60, 75 **yes**
36. 30, 40, 50 **yes**
37. 11, 12, 15
38. 16, $\sqrt{32}$, 20
39. 15, $\sqrt{31}$, 16 **yes**

29. $\sqrt{67} \approx 8.19$
31. $\sqrt{27} \approx 5.20$
32. $\sqrt{253} \approx 15.91$
33. $\sqrt{145} \approx 12.04$
34. no; $6^2 + 9^2 \neq 12^2$
37. no; $11^2 + 12^2 \neq 15^2$
38. no; $16^2 + (\sqrt{32})^2 \neq 20^2$

Study Guide Masters, p. 88

NAME ______________________ DATE __________

Study Guide

Student Edition Pages 713–718

Integration: Geometry
The Pythagorean Theorem

The side opposite the right angle in a right triangle is called the **hypotenuse.** This side is always the longest side of a right triangle. The other two sides are called the **legs** of the triangle. To find the length of any side of a right triangle, given the lengths of the other two sides, you can use the *Pythagorean theorem.*

The Pythagorean Theorem
In a right triangle, if a and b are the measures of the legs and c is the measure of the hypotenuse, then $c^2 = a^2 + b^2$.

Example: Find the length of a leg of a right triangle if $a = 8$ and $c = 10$.

$$c^2 = a^2 + b^2$$
$$10^2 = 8^2 + b^2$$
$$100 = 64 + b^2$$
$$b^2 = 36$$
$$b = 6$$

The length of the leg is 6 units.

If c is the measure of the hypotenuse of a right triangle, find each missing measure. Round answers to the nearest hundredth.

1. $a = 10, b = 12, c = \underline{?}$ 15.62
2. $a = 9, b = 12, c = \underline{?}$ 15
3. $a = 12, b = \underline{?}, c = 16$ 10.58
4. $a = \underline{?}, b = 6, c = 8$ 5.29
5. $a = \sqrt{5}, b = \sqrt{10}, c = \underline{?}$ 3.87
6. $a = \underline{?}, b = \sqrt{8}, c = \sqrt{18}$ 3.16

For each problem, make a drawing. Then use an equation to solve the problem. Round answers to the nearest hundredth.

7. How long is a pipeline that runs diagonally across a square field 6 kilometers on a side? 8.49 km

8. A support wire on a television tower is 90 meters long and meets the ground 35 meters from the center of the base of the tower. What is the height of the tower? 89.92 m

Alternative Teaching Strategies

Student Diversity The following is a proof of the Pythagorean theorem discovered by James Garfield in 1876.

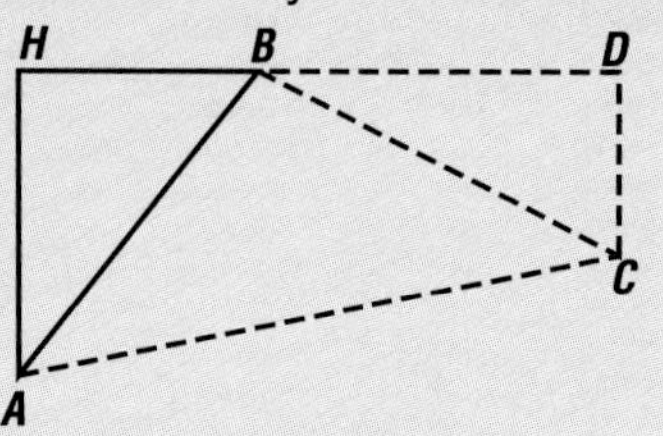

In the figure, $BD = AH$, and $DC \parallel AH$ and equal to BH. Area of trapezoid $CDHA$ = area of ACB + 2 area of ABH.
$\frac{1}{2}(AH + CD) \times HD = \frac{1}{2}AB^2 + 2 \times \frac{1}{2}AH \times HB$ or $(AH + HB)^2 = AB^2 + 2AH \times HB$, so $AB^2 = BH^2 + AH^2$.
Therefore, $h^2 = a^2 + b^2$.

Geometry

Use an equation to solve each problem. Round answers to the nearest hundredth.

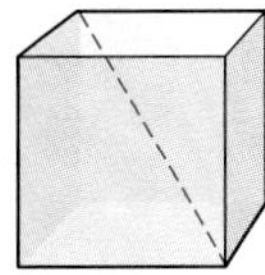

40. Find the length of the diagonal of a square if its area is 128 cm^2. **16 cm**

41. $\sqrt{75}$ **in. or about 8.66 in.**

41. Find the length of the diagonal of a cube if each side of the cube is 5 inches long.

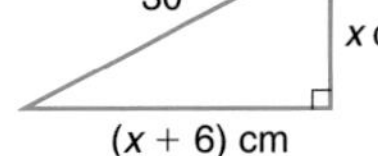

42. A right triangle has one leg that is 6 centimeters longer than the other. The hypotenuse is 30 centimeters long.
   a. Write an equation to find the length of the legs.
   b. Find the length of each leg of the triangle.

42a. $x^2 + (x+6)^2 = (30)^2$
42b. 18 cm, 24 cm

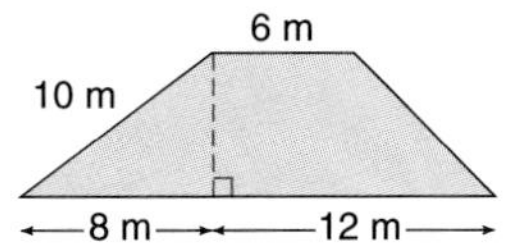

43. Look at the trapezoid at the right.
   a. Find the perimeter. (*Hint:* Drawing a second height will be helpful.) **44.49 m**
   b. Find the area. **78 m^2**

Programming

44. The graphing calculator program at the right calculates the hypotenuse of a right triangle and displays the right triangle being measured.

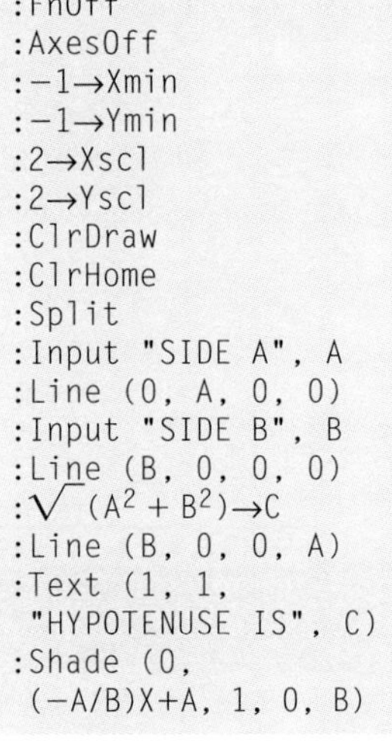

```
PROGRAM:PYTH
:FnOff
:AxesOff
:−1→Xmin
:−1→Ymin
:2→Xscl
:2→Yscl
:ClrDraw
:ClrHome
:Split
:Input "SIDE A", A
:Line (0, A, 0, 0)
:Input "SIDE B", B
:Line (B, 0, 0, 0)
:√ (A² + B²)→C
:Line (B, 0, 0, A)
:Text (1, 1,
 "HYPOTENUSE IS", C)
:Shade (0,
 (−A/B)X+A, 1, 0, B)
```

Find the hypotenuse c of each right triangle below, given the measures of sides a and b. Round to the nearest hundredth.

a. $a = 2, b = 7$ **7.28**
b. $a = 9, b = 12$ **15**
c. $a = 13, b = 15$ **19.85**
d. $a = 6, b = 9$ **10.82**
e. $a = 12, b = 16$ **20**

Critical Thinking

45. The window in Julia's attic was a square with an area of 1 square foot. After she remodeled her home, the width and height of the new attic window were the same as the original, but its area was half that of the original window. If the new window is also a square, explain how this is possible. Include a drawing. **See margin.**

Applications and Problem Solving

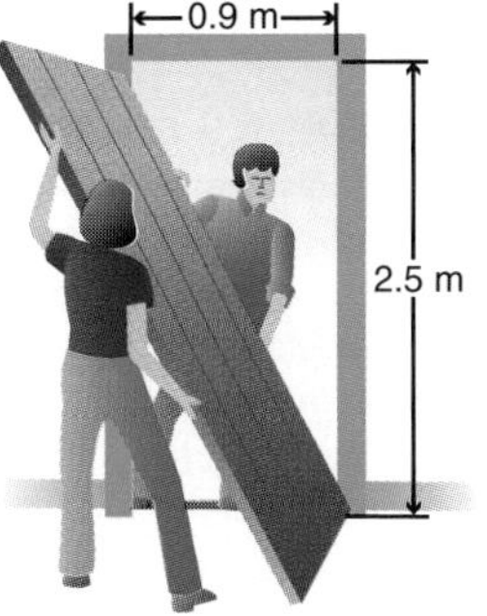
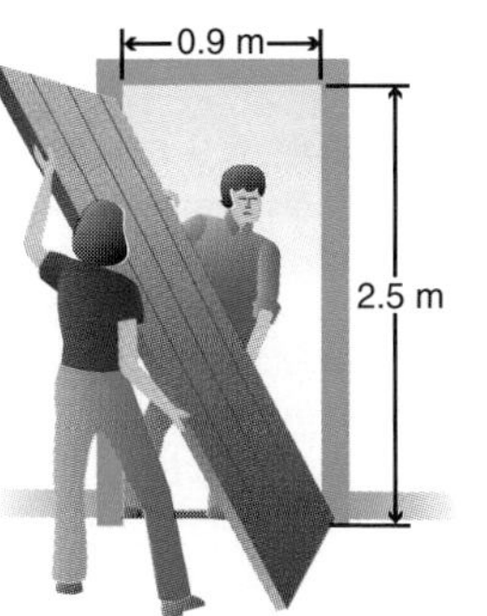

46. **Sailing** Refer to the application at the beginning of the lesson. If the edge of the mainsail that is attached to the mast is 100 feet long and the edge of the mainsail that is attached to the boom is 60 feet long, what is the length of the longest edge of the mainsail? **about 116.62 ft**

47. about 2.66 m

47. **Construction** The walls of the Downtown Recreation Center are being covered with paneling. The doorway into one room is 0.9 meters wide and 2.5 meters high. What is the length of the longest rectangular panel that can be taken through this doorway diagonally?

Using the Programming Exercises The program given in Exercise 44 is for use with a TI-82 graphing calculator. For other programmable calculators, have students consult their owner's manual for commands similar to those presented here.

Additional Answer

45.

1 ft
1 ft
Area = 1 ft^2 or 144 in^2

6 in.
6 in.
$(6\sqrt{2})$in.
Area = $(6\sqrt{2})^2$ or 72 in^2

Practice Masters, p. 88

13-1 NAME______ DATE______
Practice
Student Edition Pages 713–718

Integration: Geometry
The Pythagorean Theorem

If c is the measure of the hypotenuse of a right triangle, find each missing measure. Round answers to the nearest hundredth.

1. $a = 3, b = 4, c =$ 5
2. $a = 6, c = 10, b =$ 8
3. $b = 12, c = 13, a =$ 5
4. $a = 6, c = 12, b =$ $\sqrt{108} \approx 10.39$
5. $a = 8, b = 6, c =$ 10
6. $a = 5, c = 13, b =$ 12
7. $b = 0.8, c = 1.0, a =$ 0.6
8. $a = 11, b = 4, c =$ $\sqrt{137} \approx 11.70$
9. $a = \sqrt{12}, b = 6, c =$ $\sqrt{48} \approx 6.93$
10. $b = 11, c = \sqrt{289}, a =$ $\sqrt{168} \approx 12.96$
11. $a = 19, b = \sqrt{39}, c =$ 20
12. $a = \sqrt{6}, b = \sqrt{19}, c =$ 5

Determine whether the following side measures would form right triangles. Explain why or why not.

13. 20, 21, 29 yes
14. 15, 30, 34 no
15. 9, $\sqrt{40}$, 11 yes
16. 21, 72, 75 yes

4 ASSESS

Closing Activity

Writing At the end of class, have students write down the key points to consider when using the Pythagorean theorem.

Additional Answer

50.

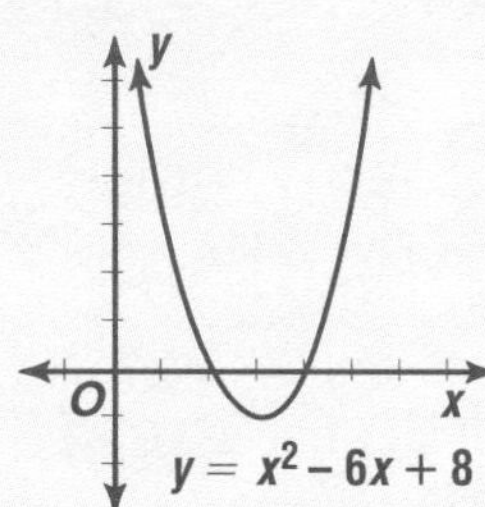

48. **Travel** The cruise ship M.S. Starward has a right triangle-shaped dance floor on the cabaret deck. The lengths of the two shortest sides of the dance floor are equal, and the longest side next to the stage is 36 feet long. How long is one side of the dance floor? **about 25.5 ft**

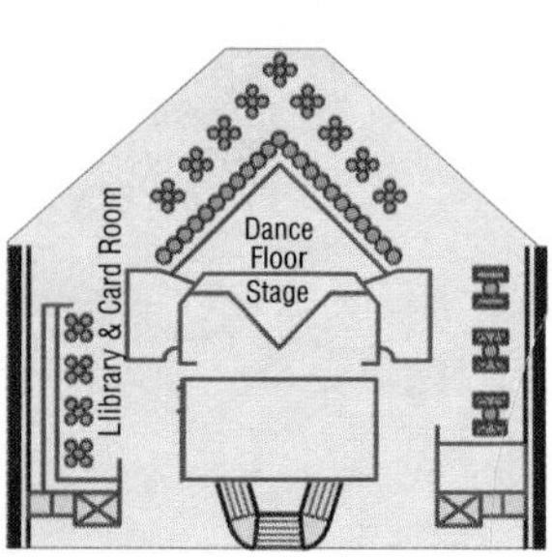

Mixed Review

49. Solve $\frac{a+2}{a} + \frac{a+5}{a} = 1$. (Lesson 12–8) **−7**

50. Find an equation of the axis of symmetry and the coordinates of the vertex of the graph of $y = x^2 - 6x + 8$. Then draw the graph. (Lesson 11–1) **50. $x = 3$; $(3, -1)$; See margin for graph.**

51. Simplify $(4r^5)(3r^2)$. (Lesson 9–1) **$12r^7$**

52. **Number Theory** The sum of two numbers is 38. Their difference is 6. Find the numbers. (Lesson 8–3) **16, 22**

53. Solve $5c - 2 \geq c$. (Lesson 7–2) $\left\{c \mid c \geq \frac{1}{2}\right\}$

54. State the slope and y-intercept of the line graphed at the right. Then write an equation of the line in slope-intercept form. (Lesson 6–4) **$-2, 3, y = -2x + 3$**

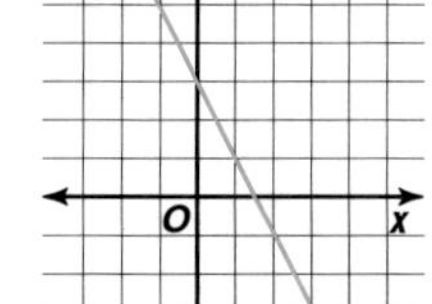

55. Given $f(x) = 2x^2 - 5x + 8$, find $f(-3)$. (Lesson 5–5) **55. 41**

56. Solve $-\frac{5}{6}y = 15$. (Lesson 3–2) **−18**

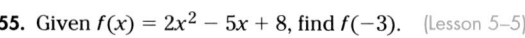

WORKING ON THE

In·ves·ti·ga·tion

Refer to the Investigation on pages 656–657.

A Growing Concern

1 Develop a detailed plan for planting the lawn, plants, and trees for the Sanchez family's yard. Indicate the dimensions of the lawn area as well as the location of the plants and trees you selected. Mark the length of each property line on the drawing.

2 Research the plants and trees that you feel would be good specimens for the areas in which you indicated that they would be planted. Explain why you chose those plants and trees. Think about all four seasons and which plants best suit each season. Also consider the amount of sunlight each area receives during the day.

3 Calculate the cost of materials, labor, and profit margin of planting these items and laying the sprinkler system. Justify your costs and make a case for the quantity and location of the plants, trees, and lawn area in your design.

4 Be sure to add the plants, trees, and sprinkler system to your overall design.

Add the results of your work to your Investigation Folder.

Enrichment Masters, p. 88

13-1 NAME ______________ DATE ______

Enrichment

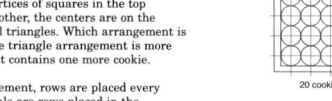

Student Edition Pages 713–718

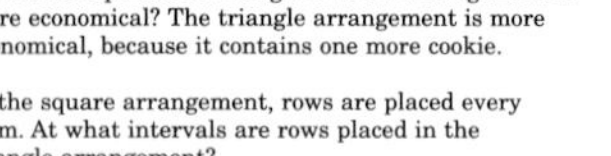

A Space-Saving Method

Two arrangements for cookies on a 32 cm by 40 cm cookie sheet are shown at the right. The cookies have 8-cm diameters after they are baked. The centers of the cookies are on the vertices of squares in the top arrangement. In the other, the centers are on the vertices of equilateral triangles. Which arrangement is more economical? The triangle arrangement is more economical, because it contains one more cookie.

8 cm

20 cookies

In the square arrangement, rows are placed every 8 cm. At what intervals are rows placed in the triangle arrangement?

Look at the right triangle labeled a, b, and c. A leg a of the triangle is the radius of a cookie, or 4 cm. The hypotenuse c is the sum of two radii, or 8 cm. Use the Pythagorean theorem to find b, the interval of the rows.

b = ?

20 cookies

$$c^2 = a^2 + b^2$$
$$8^2 = 4^2 + b^2$$
$$64 - 16 = b^2$$
$$\sqrt{48} = b$$
$$4\sqrt{3} = b$$
$$b = 4\sqrt{3} \approx 6.93$$

The rows are placed approximately every 6.93 cm.

Solve each problem.

1. Suppose cookies with 10-cm diameters are arranged in the triangular pattern shown above. What is the interval b of the rows? **8.66 cm**

2. Find the diameter of a cookie if the rows are placed in the triangular pattern every $3\sqrt{3}$ cm. **6 cm**

3. Describe other practical applications in which this kind of triangular pattern can be used to economize on space. **Answers will vary.**

Extension

Connections The length of a rectangle is 2 feet less than twice its width. Each diagonal is 34 feet. What are the dimensions of the rectangle? **16 ft, 30 ft**

In·ves·ti·ga·tion

Working on the Investigation

The Investigation on pages 656–657 is designed to be a long-term project that is completed over several days or weeks. Encourage students to keep their materials in their Investigation Folder as they work on the Investigation.

13-2 Simplifying Radical Expressions

- To simplify square roots, and
- to simplify radical expressions.

Why IT'S IMPORTANT

You can use radical expressions to solve problems involving physics and racing.

Physics

The period of a pendulum is the time in seconds that it takes the pendulum to make one complete swing back and forth. The formula for the period P of a pendulum is $P = 2\pi\sqrt{\frac{\ell}{32}}$, where ℓ is the length of the pendulum in feet. Suppose a clock makes one "tick" after each complete swing back and forth of a 2-foot-long pendulum. How many ticks would the clock make in one minute? *This problem will be solved in Example 4.*

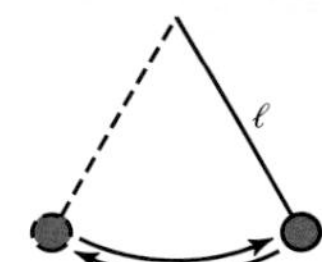

Can $2\pi\sqrt{\frac{\ell}{32}}$ be simplified? One rule used for simplifying radical expressions is that a radical expression is in *simplest form* if the **radicand,** the expression under the radical sign, contains no perfect square factors other than one. The following property can be used to simplify square roots.

Product Property of Square Roots	**For any numbers *a* and *b*, where $a \geq 0$ and $b \geq 0$, $\sqrt{ab} = \sqrt{a} \cdot \sqrt{b}$.**

The product property of square roots and prime factorization can be used to simplify radical expressions in which the radicand is not a perfect square.

Example  **Simplify.**

a. $\sqrt{18}$

$\sqrt{18} = \sqrt{3 \cdot 3 \cdot 2}$ *Prime factorization of 18*

$= \sqrt{3^2} \cdot \sqrt{2}$ *Product property of square roots*

$= 3\sqrt{2}$

b. $\sqrt{140}$

$140 = \sqrt{2 \cdot 2 \cdot 5 \cdot 7}$ *Prime factorization of 140*

$= \sqrt{2^2} \cdot \sqrt{5 \cdot 7}$ *Product property of square roots*

$= 2\sqrt{35}$

When finding the principal square root of an expression containing variables, be sure that the result is not negative. Consider the expression $\sqrt{x^2}$. Its simplest form is not x since, for example, $\sqrt{(-4)^2} \neq -4$. For radical expressions like $\sqrt{x^2}$, use absolute value to ensure nonnegative results.

$\sqrt{x^2} = |x|$ $\sqrt{x^3} = |x|\sqrt{x}$ $\sqrt{x^4} = x^2$ $\sqrt{x^5} = x^2\sqrt{x}$ $\sqrt{x^6} = |x^3|$

13-2 LESSON NOTES

NCTM Standards: 1–5

Instructional Resources

- Study Guide Master 13-2
- Practice Master 13-2
- Enrichment Master 13-2
- Assessment and Evaluation Masters, p. 352
- Tech Prep Applications Masters, p. 26

Transparency 13-2A contains the 5-Minute Check for this lesson; **Transparency 13-2B** contains a teaching aid for this lesson.

Recommended Pacing	
Honors Pacing	Day 3 of 11
Block Scheduling*	Day 2 of 7
Alg. 1 in Two Years*	Days 4, 5, & 6 of 22

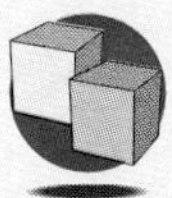

*For more information on pacing and possible lesson plans, refer to the *Block Scheduling Booklet* and *Algebra 1 in Two Years.*

1 FOCUS

5-Minute Check
(over Lesson 13-1)

1. Find the length of the hypotenuse of a right triangle if the legs have measures of 7 and 11. $\sqrt{170}$
2. Find the measure of the other leg of a right triangle if one leg is 24 and the hypotenuse is 25. 7

The measures of three sides of a triangle are given. Determine whether each triangle is a right triangle.

3. 12, 13, 15 no
4. 6, 8, 10 yes
5. 9, 16, 20 no

For $\sqrt{x^3}$, absolute value is not necessary. If x were negative, then x^3 would be negative, and $\sqrt{x^3}$ would not be defined as a real number. *Why is absolute value not necessary for $\sqrt{x^4}$?*

Example  **Simplify $\sqrt{72x^3y^4z^5}$.**

$$\begin{aligned}\sqrt{72x^3y^4z^5} &= \sqrt{2^3 \cdot 3^2 \cdot x^3 \cdot y^4 \cdot z^5} && \textit{Prime factorization}\\ &= \sqrt{2^2} \cdot \sqrt{2} \cdot \sqrt{3^2} \cdot \sqrt{x^2} \cdot \sqrt{x} \cdot \sqrt{y^4} \cdot \sqrt{z^4} \cdot \sqrt{z} && \textit{Product property}\\ &= 2 \cdot 3 \cdot |x| \cdot y^2 \cdot z^2 \cdot \sqrt{2xz} && \textit{Simplify.}\\ &= 6|x|y^2z^2\sqrt{2xz} && \textit{The absolute value of x ensures a nonegative result.}\end{aligned}$$

The product property can also be used to multiply square roots.

Example  **Simplify $\sqrt{5} \cdot \sqrt{35}$.**

$$\begin{aligned}\sqrt{5} \cdot \sqrt{35} &= \sqrt{5} \cdot \sqrt{5} \cdot \sqrt{7} && \textit{Product property of square roots}\\ &= \sqrt{5^2 \cdot 7}\\ &= 5\sqrt{7}\end{aligned}$$

You can use a graphing calculator to explore and analyze radical expressions.

EXPLORATION — GRAPHING CALCULATORS

The formula $Y = 91.4 - (91.4 - T)[0.478 + 0.301(\sqrt{x} - 0.02x)]$ can be used to calculate the windchill factor. In this formula, Y represents the windchill, T represents the outside Fahrenheit temperature, and x represents the wind speed in miles per hour. When a meteorologist says that the temperature is 12 degrees, but it feels like 18 degrees below zero because of the wind, you can use the formula to find how fast the wind is blowing. Set the range at [0, 40] by 2 and [−50, 40] by 5.

Enter: [Y=] 91.4 [−] [(] 91.4 [−] [ALPHA] [T] [)] [(] .478 [+] .301 [(] [2nd] [√] [X,T,θ] [−] .02 [X,T,θ] [)] [)] [ENTER] [2nd] [QUIT]

Store 12 into T by pressing 12 [STO▸] [ALPHA] [T] [ENTER]. Graph the function and trace along the graph until Y is about −18. The value of X will be about 10.2. So the wind is blowing at approximately 10 miles per hour.

Your Turn **a. about 7.2 mph**

a. Use the graph to find the wind speed if the temperature feels like −7°.

b. Enter the formula into Y1, Y2, and Y3 using different variables for T. Store three different temperatures for these variables and graph them simultaneously. **b–c. See students' work.**

c. Analyze and compare the graphs.

CAREER CHOICES

A **meteorologist** studies the atmosphere, its physical characteristics, motions, processes, and effects on the environment.

A bachelor's degree in meteorology is required, but a graduate degree is needed for advancement.

For more information, contact:

American Meteorological Society
45 Beacon St.
Boston, MA 02108-3693

Motivating the Lesson

Questioning You may want to open this lesson with a review of prime numbers and what it means to write the prime factorization of a number. You may want to use the Sieve of Eratostenes to develop primes less than 100. Absolute value should also be reviewed.

2 TEACH

In-Class Examples

For Example 1
Simplify.

a. $\sqrt{147}$ $7\sqrt{3}$
b. $\sqrt{605}$ $11\sqrt{5}$

For Example 2
Simplify. Use absolute value symbols when necessary.

a. $\sqrt{x^6}$ $|x^3|$
b. $\sqrt{49x^2}$ $7|x|$

For Example 3
Simplify.

a. $\sqrt{6} \cdot \sqrt{10}$ $2\sqrt{15}$
b. $2\sqrt{14} \cdot \sqrt{21}$ $14\sqrt{6}$

Teaching Tip Point out that to simplify a square root with a variable, absolute value symbols are necessary when the variable has an even exponent and the exponent of the square root is odd. For example, in $\sqrt{x^4} = x^2$, since x is squared in the answer, it will automatically be positive. In $\sqrt{x^6} = x^3$, in order to guarantee that x^3 is positive, $|x^3|$ is necessary instead of x^3.

EXPLORATION

This activity gives students experience with real-world applications using radicals.

Alternative Learning Styles

Auditory Write $\sqrt{x^a} = |x^b|$ on the board. Have students come to the board one at a time and guess values for a and b that might work. When a student is wrong, the next student must explain why. When a student is right, the next student must try to guess the pattern and predict another correct answer. **$a = 2, b = 1$; $a = 6, b = 3$; $a = 10, b = 5$; and so on**

CAREER CHOICES

Meteorologists study variations in temperature, humidity, air pressure, wind, cloud cover, and precipitation. Dynamic meteorology deals with motions of the atmosphere and airflow. Synoptic meteorology deals with the weather.

Example 4 Refer to the connection at the beginning of the lesson. How many ticks would the clock make in one minute? Use 3.14 for π and round to the nearest whole number.

Physics

Explore The clock makes one "tick" after each complete swing back and forth of its 2-foot-long pendulum. The formula for the period P of a pendulum is $P = 2\pi\sqrt{\frac{\ell}{32}}$.

Plan First find P, the number of seconds it takes for the pendulum to go back and forth one time. Then find $\frac{60}{P}$, the number of times the clock's pendulum swings back and forth in one minute.

Solve

$$P = 2\pi\sqrt{\frac{\ell}{32}}$$

$$\frac{2}{32} \approx 2(3.14)\sqrt{} \quad \textit{\pi \approx 3.14}$$

$$\approx 6.28 \cdot \sqrt{\frac{1}{16}}$$

$$\approx 6.28 \cdot \frac{1}{4} \text{ or } 1.57$$

So it takes about 1.57 seconds for the pendulum to go back and forth once.

$$\frac{60}{1.57} \approx 38.22$$

Thus, the clock makes about 38 ticks per minute.

Examine Since 38.22×1.57 is about 40×1.5 or 60, the answer seems reasonable.

You can divide square roots and simplify radical expressions that involve division by using the quotient property of square roots.

Quotient Property of Square Roots

For any numbers a and b, where $a \geq 0$ and $b > 0$, $\sqrt{\frac{a}{b}} = \frac{\sqrt{a}}{\sqrt{b}}$.

A fraction containing radicals is in simplest form if no radicals are left in the denominator.

Example 5

a. Simplify $\frac{\sqrt{56}}{\sqrt{7}}$ and $\sqrt{\frac{34}{25}}$.

b. Compare the expressions using $<$, $>$, or $=$.

FYI

The world's largest pendulum clock is in Tokyo, Japan. It is part of a water-mill clock and is 73 feet 9.75 inches long.

a. $\frac{\sqrt{56}}{\sqrt{7}} = \sqrt{\frac{56}{7}}$ *Quotient property of square roots* $\qquad \sqrt{\frac{34}{25}} = \frac{\sqrt{34}}{\sqrt{25}}$

$= \sqrt{8}$ $\qquad = \frac{\sqrt{34}}{5}$

$= \sqrt{4} \cdot \sqrt{2}$

$= 2\sqrt{2}$

In-Class Examples

For Example 4

If the clock pendulum were 1.5 feet long, how many ticks would the clock sound in one minute? $\frac{60}{1.36}$ **or 44.12**

For Example 5

a. Simplify $\frac{\sqrt{324}}{\sqrt{3}}$ and $\sqrt{\frac{98}{4}}$. $\mathbf{6\sqrt{3}, \frac{7\sqrt{2}}{2}}$

b. Compare the expressions using $>$, $<$, or $=$. $\frac{\sqrt{324}}{\sqrt{3}} > \sqrt{\frac{98}{4}}$

GLENCOE Technology

Interactive Mathematics Tools Software

This multimedia software provides an interactive lesson that explores traveling upstream and downstream in a boat by manipulating the current and distance traveled. A **Computer Journal** gives students an opportunity to write about what they have learned.

For Windows & Macintosh

FYI

Galileo determined that the period of a pendulum is constant by timing a swinging cathedral lamp using his pulse. Changing the length of the pendulum changes the period; changing the mass does not.

b. You can compare these expressions by estimating their values and then using a scientific calculator to find approximations for each simplified expression.

Estimate: $2\sqrt{2} \rightarrow$ *Since 2 is a little more than 1, $2\sqrt{2}$ will be a little more than 2.*

$\frac{\sqrt{34}}{5} \rightarrow \frac{\sqrt{36}}{5} = \frac{6}{5}$ This will be a little more than 1. So, $2\sqrt{2} > \frac{\sqrt{34}}{5}$.

Verify by using a calculator.

Enter: 2 [X] 2 [$\sqrt{x}$] [=] 2.828427125

Enter: 34 [$\sqrt{x}$] [÷] 5 [=] 1.166190379

Since $2.8 > 1.2$, then $\frac{\sqrt{56}}{\sqrt{7}} > \sqrt{\frac{34}{25}}$.

PEANUTS®

In the cartoon above, Woodstock simplified the radical expression by **rationalizing the denominator.** This method may be used to remove or eliminate radicals from the denominator of a fraction.

Example  **Simplify.**

a. $\frac{\sqrt{5}}{\sqrt{3}}$

$$\frac{\sqrt{5}}{\sqrt{3}} = \frac{\sqrt{5}}{\sqrt{3}} \cdot \frac{\sqrt{3}}{\sqrt{3}} \quad \textit{Note that } \frac{\sqrt{3}}{\sqrt{3}} = 1.$$
$$= \frac{\sqrt{15}}{3}$$

b. $\frac{\sqrt{7}}{\sqrt{12}}$

$$\frac{\sqrt{7}}{\sqrt{12}} = \frac{\sqrt{7}}{\sqrt{2 \cdot 2 \cdot 3}}$$
$$= \frac{\sqrt{7}}{\sqrt{2 \cdot 2 \cdot 3}} \cdot \frac{\sqrt{3}}{\sqrt{3}}$$
$$= \frac{\sqrt{7}\sqrt{3}}{2 \cdot 3}$$
$$= \frac{\sqrt{21}}{6}$$

Binomials of the form $a\sqrt{b} + c\sqrt{d}$ and $a\sqrt{b} - c\sqrt{d}$ are called **conjugates** of each other. For example, $6 + \sqrt{2}$ and $6 - \sqrt{2}$ are conjugates. Conjugates are useful when simplifying radical expressions because their product is always a rational number with no radicals.

$$(6 + \sqrt{2})(6 - \sqrt{2}) = 6^2 - (\sqrt{2})^2$$
$$= 36 - 2$$
$$= 34$$

Use the pattern $(a - b)(a + b) = a^2 - b^2$ to simplify the product.

This is true because of the following.

$$(\sqrt{2})^2 = \sqrt{2} \cdot \sqrt{2}$$
$$= \sqrt{2 \cdot 2}$$
$$= \sqrt{2^2} \text{ or } 2$$

Teaching Tip Remind students that $\frac{\sqrt{3}}{\sqrt{3}} = 1$, $\frac{\sqrt{2}}{\sqrt{2}} = 1$, and so on. Multiplying an expression by 1 never affects the value of an expression. The goal is to transform the denominator into an integer.

In-Class Examples

For Example 6
Simplify.

a. $\frac{\sqrt{50}}{\sqrt{5}}$ $\sqrt{10}$

b. $\frac{2\sqrt{7}}{\sqrt{2}}$ $\sqrt{14}$

For Example 7
Simplify.

a. $\frac{2}{3 + \sqrt{5}}$ $\frac{3 - \sqrt{5}}{2}$

b. $\frac{1}{4 - \sqrt{2}}$ $\frac{4 + \sqrt{2}}{14}$

c. $\frac{3}{5 - \sqrt{3}}$ $\frac{15 + 3\sqrt{3}}{22}$

Teaching Tip In Example 7, stress that the product of the binomials $4 - \sqrt{3}$ and $4 + \sqrt{3}$ is the difference of squares.

Classroom Vignette

"In expressions such as the one in Example 7, I remind my students that $4 - \sqrt{3}$ is a binomial and they can't simplify it as follows. $\frac{\not{4}}{\not{4} - \sqrt{3}} = \frac{1}{-\sqrt{3}}$"

Barbara S Owens

Barbara Owens
Davidson Middle School
Davidson, North Carolina

Conjugates are often used to rationalize the denominators of fractions containing square roots.

Example  **Simplify** $\frac{4}{4-\sqrt{3}}$.

To rationalize the denominator, multiply both the numerator and denominator by $4+\sqrt{3}$, which is the conjugate of $4-\sqrt{3}$.

$$\frac{4}{4-\sqrt{3}} = \frac{4}{4-\sqrt{3}} \cdot \frac{4+\sqrt{3}}{4+\sqrt{3}} \quad \textit{Notice that } \frac{4+\sqrt{3}}{4+\sqrt{3}} = 1.$$

$$= \frac{4(4)+4\sqrt{3}}{4^2-\left(\sqrt{3}\right)^2}$$

Use the distributive property to multiply numerators.
Use the pattern $(a-b)(a+b) = a^2-b^2$ to multiply denominators.

$$= \frac{16+4\sqrt{3}}{16-3}$$

$$= \frac{16+4\sqrt{3}}{13}$$

When simplifying radical expressions, check the following conditions to determine if the expression is in simplest form.

Simplest Radical Form

A radical expression is in simplest form when the following three conditions have been met.
1. **No radicands have perfect square factors other than 1.**
2. **No radicands contain fractions.**
3. **No radicals appear in the denominator of a fraction.**

CHECK FOR UNDERSTANDING

Communicating Mathematics

Study the lesson. Then complete the following. **2–4. See margin.**

1. **Explain** why absolute values are sometimes needed when simplifying radical expressions containing variables. **to ensure nonnegative results**

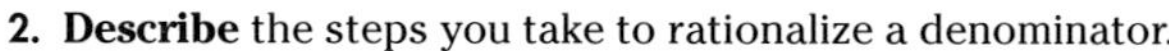

2. **Describe** the steps you take to rationalize a denominator.

3. **You Decide** Niara showed the following equations to her friend Melanie and said, "I know that 6 can't equal 10, but all these steps make sense!" Melanie said, "One of the steps must be wrong." Who is correct? Can you explain the mistake?

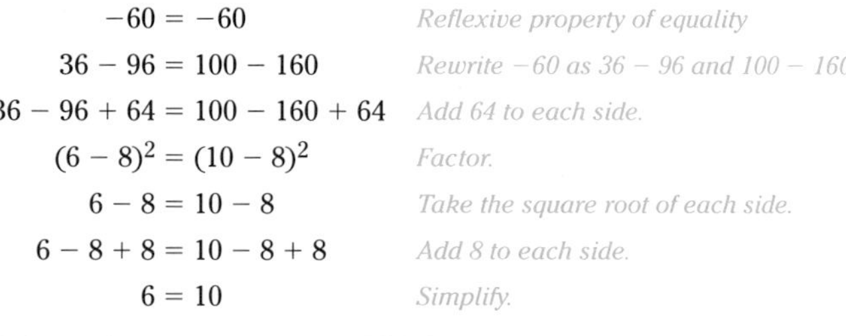

$-60 = -60$	*Reflexive property of equality*
$36 - 96 = 100 - 160$	*Rewrite −60 as 36 − 96 and 100 − 160.*
$36 - 96 + 64 = 100 - 160 + 64$	*Add 64 to each side.*
$(6-8)^2 = (10-8)^2$	*Factor.*
$6 - 8 = 10 - 8$	*Take the square root of each side.*
$6 - 8 + 8 = 10 - 8 + 8$	*Add 8 to each side.*
$6 = 10$	*Simplify.*

4. Refer to the cartoon on page 722. Obviously, Woodstock realized that rationalizing the denominator is important. What are other steps that you may have to use to simplify a radical expression?

3 PRACTICE/APPLY

Check for Understanding

Exercises 1–19 are designed to help you assess your students' understanding through reading, writing, speaking, and modeling. You should work through Exercises 1–4 with your students and then monitor their work on Exercises 5–19.

Additional Answers

2. **Multiply the numerator and denominator of a fraction by the same number so that a radical is not left in the denominator.**
3. **Melanie is correct. In step 4, the square root of a negative number was taken. This value is not defined in the set of real numbers.**
4. **Sample answer: You may have to take steps so that the radicand contains no perfect squares or fractions.**

Study Guide Masters, p. 89

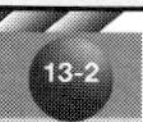

NAME________________ DATE ________

13-2 Study Guide — Student Edition Pages 719–725

Simplifying Radical Expressions

The product property of square roots and prime factorization can be used to simplify irrational square roots. When you simplify radical expressions with variables, use absolute values to ensure nonnegative results.

Example 1: Simplify $\sqrt{180}$.

$$\sqrt{180} = \sqrt{2\cdot2\cdot3\cdot3\cdot5} = \sqrt{2\cdot2}\cdot\sqrt{3\cdot3}\cdot\sqrt{5} = 2\cdot3\cdot\sqrt{5} = 6\sqrt{5}$$

Example 2: Simplify $\sqrt{100a^2}$.

$$\sqrt{100a^2} = \sqrt{100}\cdot\sqrt{a^2} = 10|a|$$

Use the quotient property of square roots and a method called **rationalizing the denominator** when simplifying radical expressions involving division. Study the example below.

Example 3: Simplify $\sqrt{\frac{56}{45}}$.

$$\sqrt{\frac{56}{45}} = \frac{\sqrt{56}}{\sqrt{45}} = \frac{2\cdot\sqrt{14}}{3\cdot\sqrt{5}} = \frac{2\sqrt{14}}{3\sqrt{5}}\cdot\frac{\sqrt{5}}{\sqrt{5}} = \frac{2\sqrt{70}}{15}$$

Simplify. Leave in radical form and use absolute value symbols when necessary.

1. $\sqrt{18}$ **$3\sqrt{2}$**
2. $\sqrt{68}$ **$2\sqrt{17}$**
3. $\sqrt{60}$ **$2\sqrt{15}$**
4. $\sqrt{75}$ **$5\sqrt{3}$**
5. $\sqrt{162}$ **$9\sqrt{2}$**
6. $\sqrt{4a^2}$ **$2|a|$**
7. $\sqrt{9x^4}$ **$3x^2$**
8. $\sqrt{300a^4}$ **$10a^2\sqrt{3}$**
9. $\sqrt{128c^6}$ **$8|c^3|\sqrt{2}$**
10. $\sqrt{5}\cdot\sqrt{10}$ **$5\sqrt{2}$**
11. $\sqrt{3x^2}\cdot3\sqrt{3x^4}$ **$9|x^3|$**
12. $4\sqrt{10}\cdot3\sqrt{6}$ **$24\sqrt{15}$**
13. $\frac{\sqrt{9}}{\sqrt{18}}$ **$\frac{\sqrt{2}}{2}$**
14. $\frac{\sqrt{8}}{\sqrt{24}}$ **$\frac{\sqrt{3}}{3}$**
15. $\frac{\sqrt{x^6}}{\sqrt{y^4}}$ **$\frac{|x^3|}{y^2}$**
16. $\frac{\sqrt{100}}{\sqrt{121}}$ **$\frac{10}{11}$**
17. $\frac{\sqrt{75}}{\sqrt{3}}$ **25**
18. $\frac{8\sqrt{2}}{2\sqrt{8}}$ **2**
19. $\frac{\sqrt{4}}{3-\sqrt{5}}$ **$\frac{3+\sqrt{5}}{2}$**
20. $\frac{\sqrt{8}}{2\sqrt{7}+4\sqrt{10}}$ **$\frac{4\sqrt{5}-\sqrt{14}}{33}$**

Reteaching

Using Problem Solving To convince students that rationalizing the denominator produces equivalent results, have them use a calculator to find approximate values of the following expressions.

$$\frac{2}{\sqrt{3}} \qquad \frac{2\cdot\sqrt{3}}{\sqrt{3}\cdot\sqrt{3}} \qquad \frac{2\sqrt{3}}{3}$$

The values all should be about 1.1547.

Alternative Teaching Strategies

Student Diversity Discuss the difference between $2\sqrt{2}\cdot5$, $2\sqrt{2}\cdot\sqrt{5}$, and $2\sqrt{2}\cdot5\sqrt{5}$. Have students write steps for each operation, trade with another student, and check the answers.

Error Analysis

Students may make errors when squaring the product of a rational and an irrational number. For example, they may incorrectly evaluate $(3\sqrt{2})(3\sqrt{2})$ as $9 \cdot 4$ or $3 \cdot 2$. Point out that the rational numbers and the irrational numbers should be squared and their product should be found.

Assignment Guide

Core: 21–53 odd, 54, 55, 57–66
Enriched: 20–52 even, 54–66

For **Extra Practice,** see p. 785.

The red A, B, and C flags, printed only in the Teacher's Wraparound Edition, indicate the level of difficulty of the exercises.

Practice Masters, p. 89

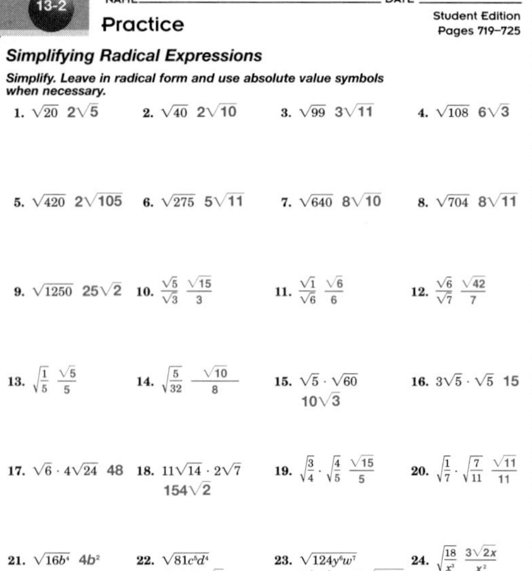

13-2 NAME______________________ DATE __________

Practice — Student Edition Pages 719–725

Simplifying Radical Expressions

Simplify. Leave in radical form and use absolute value symbols when necessary.

1. $\sqrt{20}$ $2\sqrt{5}$
2. $\sqrt{40}$ $2\sqrt{10}$
3. $\sqrt{99}$ $3\sqrt{11}$
4. $\sqrt{108}$ $6\sqrt{3}$
5. $\sqrt{420}$ $2\sqrt{105}$
6. $\sqrt{275}$ $5\sqrt{11}$
7. $\sqrt{640}$ $8\sqrt{10}$
8. $\sqrt{704}$ $8\sqrt{11}$
9. $\sqrt{1250}$ $25\sqrt{2}$
10. $\frac{\sqrt{5}}{\sqrt{3}}$ $\frac{\sqrt{15}}{3}$
11. $\frac{\sqrt{1}}{\sqrt{6}}$ $\frac{\sqrt{6}}{6}$
12. $\frac{\sqrt{6}}{\sqrt{7}}$ $\frac{\sqrt{42}}{7}$
13. $\sqrt{\frac{1}{5}}$ $\frac{\sqrt{5}}{5}$
14. $\sqrt{\frac{5}{32}}$ $\frac{\sqrt{10}}{8}$
15. $\sqrt{5} \cdot \sqrt{60}$ $10\sqrt{3}$
16. $3\sqrt{5} \cdot \sqrt{5}$ 15
17. $\sqrt{6} \cdot 4\sqrt{24}$ 48
18. $11\sqrt{14} \cdot 2\sqrt{7}$ $154\sqrt{2}$
19. $\sqrt{\frac{3}{4}} \cdot \sqrt{\frac{4}{5}}$ $\frac{\sqrt{15}}{5}$
20. $\sqrt{\frac{1}{7}} \cdot \sqrt{\frac{7}{11}}$ $\frac{\sqrt{11}}{11}$
21. $\sqrt{16b^4}$ $4b^2$
22. $\sqrt{81c^5d^4}$ $9c^2d^2\sqrt{c}$
23. $\sqrt{124y^6w^7}$ $2|y^3| \cdot w^3\sqrt{31w}$
24. $\sqrt{\frac{18}{x^3}}$ $\frac{3\sqrt{2x}}{x^2}$
25. $\frac{3}{5-\sqrt{2}}$ $\frac{15+3\sqrt{2}}{23}$
26. $\frac{5}{\sqrt{7}+\sqrt{3}}$ $\frac{5\sqrt{7}-5\sqrt{3}}{4}$
27. $\frac{6x}{5+\sqrt{x}}$ $\frac{30x-6x\sqrt{x}}{25-x}$
28. $\frac{3\sqrt{7}}{-1-\sqrt{27}}$ $\frac{3\sqrt{7}-9\sqrt{21}}{26}$

Guided Practice

State the conjugate of each expression. Then multiply the expression by its conjugate.

5. $5 + \sqrt{2}$ **$5 - \sqrt{2}$; 23**
6. $\sqrt{3} - \sqrt{7}$ **$\sqrt{3} + \sqrt{7}$; -4**

State the fraction by which each expression should be multiplied to rationalize the denominator.

7. $\frac{4}{\sqrt{7}}$ **$\frac{\sqrt{7}}{\sqrt{7}}$**
8. $\frac{2\sqrt{5}}{4-\sqrt{3}}$ **$\frac{4+\sqrt{3}}{4+\sqrt{3}}$**

Simplify. Leave in radical form and use absolute value symbols when necessary.

9. $\sqrt{18}$ **$3\sqrt{2}$**
10. $\frac{\sqrt{20}}{\sqrt{5}}$ **2**
11. $\sqrt{\frac{3}{7}}$ **$\frac{\sqrt{21}}{7}$**
12. $\sqrt{\frac{2}{3}} \cdot \sqrt{\frac{5}{2}}$ **$\frac{\sqrt{15}}{3}$**
13. $(\sqrt{2} + 4)(\sqrt{2} + 6)$ **$10\sqrt{2} + 26$**
14. $(y - \sqrt{5})(y + \sqrt{5})$ **$y^2 - 5$**
15. $\frac{6}{3-\sqrt{2}}$ **$\frac{18+6\sqrt{2}}{7}$**
16. $\sqrt{80a^2b^3}$ **$4|a|b\sqrt{5b}$**

Compare each pair of expressions using <, >, or =.

17. $4\sqrt{3} \cdot \sqrt{3}$, $\sqrt{48} + \sqrt{8}$ **>**
18. $\sqrt{\frac{12}{7}}$, $\frac{\sqrt{18}\cdot\sqrt{2}}{\sqrt{7}\cdot\sqrt{3}}$ **=**

19. **Water Supply** There is a relationship between a city's capacity to supply water to its citizens and the city's size. Suppose a city has a population P (in thousands). Then the number of gallons per minute that are required to assure water adequacy is given by the expression $1020\sqrt{P}(1 - 0.01\sqrt{P})$. If a city has a population of 55,000 people, how many gallons per minute must the city's pumping stations be able to supply? **about 7003.5 gal/min**

EXERCISES

Practice

Simplify. Leave in radical form and use absolute value symbols when necessary.

A
20. $\sqrt{75}$ **$5\sqrt{3}$**
21. $\sqrt{80}$ **$4\sqrt{5}$**
22. $\sqrt{280}$ **$2\sqrt{70}$**
23. $\sqrt{500}$ **$10\sqrt{5}$**
24. $\frac{\sqrt{7}}{\sqrt{3}}$ **$\frac{\sqrt{21}}{3}$**
25. $\frac{\sqrt{5}}{\sqrt{10}}$ **$\frac{\sqrt{2}}{2}$**
26. $\sqrt{\frac{2}{7}}$ **$\frac{\sqrt{14}}{7}$**
27. $\sqrt{\frac{11}{32}}$ **$\frac{\sqrt{22}}{8}$**
28. $5\sqrt{10} \cdot 3\sqrt{10}$
29. $7\sqrt{30} \cdot 2\sqrt{6}$
30. $\sqrt{\frac{3}{5}} \cdot \sqrt{\frac{7}{3}}$
31. $\sqrt{\frac{1}{6}} \cdot \sqrt{\frac{6}{11}}$ **$\frac{\sqrt{11}}{11}$**
32. $\sqrt{40b^4}$
33. $\sqrt{54a^2b^2}$
34. $\sqrt{60m^2y^4}$
35. $\sqrt{147x^5y^7}$

B
36. $\sqrt{\frac{t}{8}}$ **$\frac{\sqrt{2t}}{4}$**
37. $\sqrt{\frac{27}{p^2}}$ **$\frac{3\sqrt{3}}{|p|}$**
38. $\sqrt{\frac{5n^5}{4m^5}}$
39. $\frac{\sqrt{9x^5y}}{\sqrt{12x^2y^6}}$
40. $(1 + 2\sqrt{5})^2$
41. $(y - \sqrt{7})^2$
42. $(\sqrt{m} + \sqrt{20})^2$
43. $\frac{14}{\sqrt{8}-\sqrt{5}}$
44. $\frac{9a}{6+\sqrt{a}}$
45. $\frac{2\sqrt{5}}{-4+\sqrt{8}}$
46. $\frac{3\sqrt{7}}{5\sqrt{3}+3\sqrt{5}}$
47. $\frac{\sqrt{c}-\sqrt{d}}{\sqrt{c}+\sqrt{d}}$
48. $(\sqrt{2x} - \sqrt{6})(\sqrt{2x} + \sqrt{6})$
49. $(x - 4\sqrt{3})(x - \sqrt{3})$

28. 150
29. $84\sqrt{5}$
30. $\frac{\sqrt{35}}{5}$
32. $2b^2\sqrt{10}$
33. $3|ab|\sqrt{6}$
34. $2|m|y^2\sqrt{15}$
35. $7x^2y^3\sqrt{3xy}$
38. $\frac{n^2\sqrt{5mn}}{2|m^3|}$
39. $\frac{x\sqrt{3xy}}{2y^3}$
40–47. See margin.
48. $2x - 6$
49. $x^2 - 5x\sqrt{3} + 12$

40. $21 + 4\sqrt{5}$
41. $y^2 - 2y\sqrt{7} + 7$
42. $m + 4\sqrt{5m} + 20$
43. $\frac{28\sqrt{2}+14\sqrt{5}}{3}$
44. $\frac{54a - 9a\sqrt{a}}{36 - a}$
45. $\frac{-2\sqrt{5}-\sqrt{10}}{2}$
46. $\frac{5\sqrt{21}-3\sqrt{35}}{10}$
47. $\frac{c - 2\sqrt{cd} + d}{c - d}$

Compare each pair of expressions using <, >, or =.

50. $\sqrt{\frac{8}{9}} \cdot \frac{2}{\sqrt{8}}, \frac{2}{\sqrt{51}} \cdot \sqrt{\frac{17}{3}}$ **=**

51. $\sqrt{10} \cdot \sqrt{30}, \frac{10}{\sqrt{5}+9}$ **>**

52. $\frac{2}{\sqrt{6}-\sqrt{5}}, \frac{20}{6+\sqrt{3}}$ **>**

53. $\frac{3\sqrt{2}-\sqrt{7}}{2\sqrt{3}-5\sqrt{2}}, \frac{4\sqrt{5}-3\sqrt{7}}{\sqrt{6}}$ **<**

Critical Thinking

54. Determine whether $\sqrt{a \cdot b} = \sqrt{a} \cdot \sqrt{b}$ is true for negative real numbers. Give examples to support your answer. **No, because square roots of negative numbers are not defined in the set of real numbers; $\sqrt{(-2)\cdot(-3)} \neq \sqrt{-2}\cdot\sqrt{-3}$**

Applications and Problem Solving

55. **Racing** In yacht racing from 1958 to 1987, 12-meter boats were not really 12 meters long, but the formula that governed their design contained numbers that equaled 12. In the expression $\frac{\sqrt{S}+L-F}{2.37}$, S is the area of the sails, L is the waterline length, and F is the distance from the deck of the boat to the waterline. The result must be less than or equal to 12 for a boat to be classified as a 12-meter boat. Determine if a boat for which $S = 158\text{ m}^2$, $L = 17.5$ m, and $F = 2$ m could be classified as a 12-meter boat.

56. **Electricity** The voltage V required for a circuit is given by $V = \sqrt{PR}$ where P is the power in watts and R is the resistance in ohms. Find the volts needed to light a 75-watt bulb with a resistance of 110 ohms. **about 90.83 volts**

Mixed Review

55. Yes, the result is about 11.84.

57. 18.44 cm

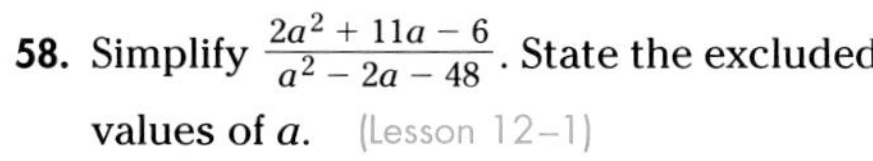

58. $\frac{2a-1}{a-8}$; $a \neq 8, -6$

59. 1, $\frac{2}{3}$

61. 2.7×10^9 acres

63. $y = \frac{1}{5}x + 6$

64. {(−1, 3), (−1, 4), (1, 4), (1, −3), (3, 5)}; {−1, 1, 3}; {3, 4, −3, 5}

57. **Geometry** Find the length of the missing side of the triangle shown at the right. Round to the nearest hundredth. (Lesson 13–1)

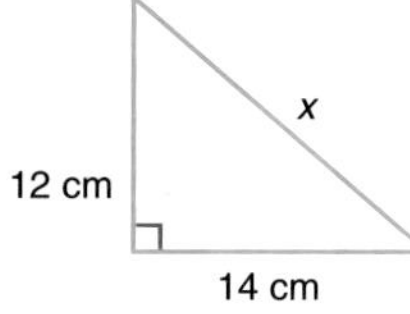

58. Simplify $\frac{2a^2+11a-6}{a^2-2a-48}$. State the excluded values of a. (Lesson 12–1)

59. Solve $3x^2 - 5x + 2 = 0$ by using the quadratic formula. (Lesson 11–3)

60. Factor $12a^2b^3 - 28ab^2c^2$. (Lesson 10–2) **$4ab^2(3ab - 7c^2)$**

61. **Forests** The largest forested areas in the world are located in northern Russia. They cover 2,700,000,000 acres, and they make up 25% of the world's forests. Express the number of acres in scientific notation. (Lesson 9–3)

62. Solve $4 - 2.3t < 17.8$. (Lesson 7–3) **$\{t \mid t > -6\}$**

63. **Geometry** Write an equation of the line that is perpendicular to the graph of $y = -5x + 2$ and passes through the point at (0, 6). (Lesson 6–6)

64. Express the relation shown in the graph at the right as a set of ordered pairs. Then state the domain and range of the relation. (Lesson 5–2)

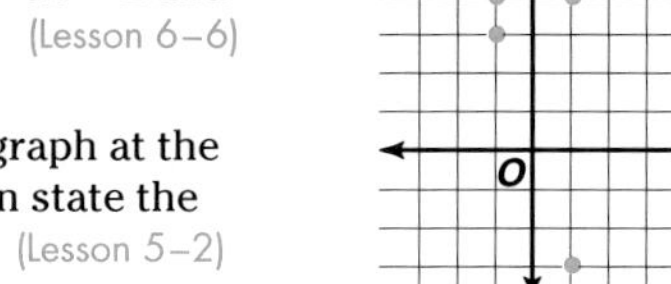

65. Solve $6(x + 3) = 3x$. (Lesson 3–5) **−6**

66. Write an algebraic expression for the verbal expression *one-fourth the square of a number.* (Lesson 1–1) **$\frac{1}{4}x^2$**

Extension

Problem Solving Simplify $\frac{2}{\sqrt{3}+\sqrt{5}+\sqrt{6}}$. $\frac{4\sqrt{3}+2\sqrt{5}+\sqrt{6}-3\sqrt{10}}{14}$

4 ASSESS

Closing Activity

Writing Have students summarize this section by writing a list of the key points that they learned.

Chapter 13, Quiz A (Lessons 13-1 and 13-2), is available in the *Assessment and Evaluation Masters,* p. 352.

Enrichment Masters, p. 89

13–2 NAME ______ DATE ______

Enrichment

Student Edition Pages 719–725

Rational Exponents

You have developed the following properties of powers when a is a positive real number and m and n are integers.

$a^m \cdot a^n = a^{m+n}$	$(ab)^m = a^m b^m$	$a^0 = 1$
$(a^m)^n = a^{mn}$	$\frac{a^m}{a^n} = a^{m-n}$	$a^{-m} = \frac{1}{a^m}$

Exponents need not be restricted to integers. We can define rational exponents so that operations involving them will be governed by the properties for integer exponents.

$\left(a^{\frac{1}{2}}\right)^2 = a^{\frac{1}{2}\cdot 2} = a$	$\left(a^{\frac{1}{3}}\right)^3 = a^{\frac{1}{3}\cdot 3}$	$\left(a^{\frac{1}{n}}\right)^n = a^{\frac{1}{n}\cdot n} = a$
$a^{\frac{1}{2}}$ squared is a.	$a^{\frac{1}{3}}$ cubed is a.	$a^{\frac{1}{n}}$ to the n power is a.
$a^{\frac{1}{2}}$ is a square root of a.	$a^{\frac{1}{3}}$ is a cube root of a.	$a^{\frac{1}{n}}$ is an nth root of a.
$a^{\frac{1}{2}} = \sqrt{a}$	$a^{\frac{1}{3}} = \sqrt[3]{a}$	$a^{\frac{1}{n}} = \sqrt[n]{a}$

Now let us investigate the meaning of $a^{\frac{m}{n}}$.

$a^{\frac{m}{n}} = a^{m\cdot\frac{1}{n}}(a^m)^{\frac{1}{n}} = \sqrt[n]{a^m}$ $\qquad a^{\frac{m}{n}} = a^{\frac{1}{n}\cdot m} = \left(a^{\frac{1}{n}}\right)^m = (\sqrt[n]{a})^m$

Therefore, $a^{\frac{m}{n}} = \sqrt[n]{a^m}$ or $(\sqrt[n]{a})^m$.

Example 1: Write $\sqrt[4]{a^3}$ in exponential form.
$\sqrt[4]{a^3} = a^{\frac{3}{4}}$

Example 2: Write $a^{\frac{2}{5}}$ in radical form.
$a^{\frac{2}{5}} = \sqrt[5]{a^2}$

Example 3: Find $\frac{a^{\frac{2}{3}}}{a^{\frac{1}{2}}}$.

$$\frac{a^{\frac{2}{3}}}{a^{\frac{1}{2}}} = a^{\frac{2}{3}-\frac{1}{2}} = a^{\frac{4}{6}-\frac{3}{6}} = a^{\frac{1}{6}} \text{ or } \sqrt[6]{a}$$

Write each expression in radical form.

1. $b^{\frac{3}{2}}$ **$\sqrt{b^3}$**
2. $3c^{\frac{1}{2}}$ **$3\sqrt{c}$**
3. $(3c)^{\frac{1}{2}}$ **$\sqrt{3c}$**

Write each expression in exponential form.

4. $\sqrt[3]{b^4}$ **$b^{\frac{4}{3}}$**
5. $\sqrt{4a^3}$ **$(4a^3)^{\frac{1}{2}} = 2a^{\frac{3}{2}}$**
6. $2 \cdot \sqrt[3]{b^2}$ **$2b^{\frac{2}{3}}$**

Perform the operation indicated. Answers should show positive exponents only.

7. $\left(a^3b^{\frac{1}{4}}\right)^2$ **$a^6b^{\frac{1}{2}}$**
8. $\frac{-8a^{\frac{3}{4}}}{2a^{\frac{1}{2}}}$ **$-4a^{\frac{1}{4}}$**
9. $\left(\frac{b^{\frac{1}{2}}}{b^{-\frac{2}{3}}}\right)^3$ **$b^{\frac{7}{2}}$**
10. $\sqrt{a^3} \cdot \sqrt{a}$ **a^2**
11. $\left(a^2b^{-\frac{1}{3}}\right)^{-\frac{1}{2}}$ **$\frac{b^{\frac{1}{6}}}{a}$**
12. $-2a^{\frac{1}{3}}b^0\left(5a^{\frac{1}{2}}b^{-\frac{2}{3}}\right)$ **$\frac{-10a^{\frac{5}{6}}}{b^{\frac{2}{3}}}$**

2 TEACH

Teaching Tip If students can use the distributive property with ease, they should have no trouble adding and subtracting expressions with like radicands.

In-Class Examples

For Example 1

Simplify $6\sqrt{7} - 2\sqrt{7} - \sqrt{3} + 4\sqrt{3}$. $\mathbf{4\sqrt{7} + 3\sqrt{3}}$

For Example 2

Simplify $3\sqrt{45} + 4\sqrt{80} - 2\sqrt{125}$. $\mathbf{15\sqrt{5}}$

For Example 3

Find the exact measure of the perimeter of a rectangle whose width is $4\sqrt{6} + \sqrt{3}$ units and whose length is $2\sqrt{3} - 4$ units. $\mathbf{8\sqrt{6} + 6\sqrt{3} - 8}$ **sq units**

Teaching Tip In Example 3, remind students that only when the radicands are alike can terms be added or subtracted.

In Example 1b, the expression $3\sqrt{6} + 7\sqrt{7}$ cannot be simplified further because the radicands are different. There are no common factors, and each radicand is in simplest form.

If the radicals in a radical expression are not in simplest form, simplify them first. Then use the distributive property wherever possible to further simplify the expression.

Example 2 **Simplify $4\sqrt{27} + 5\sqrt{12} + 8\sqrt{75}$. Then use a scientific calculator to verify your answer.**

$$\begin{aligned} 4\sqrt{27} + 5\sqrt{12} + 8\sqrt{75} &= 4\sqrt{3^2 \cdot 3} + 5\sqrt{2^2 \cdot 3} + 8\sqrt{5^2 \cdot 3} \\ &= 4(\sqrt{3^2} \cdot \sqrt{3}) + 5(\sqrt{2^2} \cdot \sqrt{3}) + 8(\sqrt{5^2} \cdot \sqrt{3}) \\ &= 4(3\sqrt{3}) + 5(2\sqrt{3}) + 8(5\sqrt{3}) \\ &= 12\sqrt{3} + 10\sqrt{3} + 40\sqrt{3} \\ &= 62\sqrt{3} \end{aligned}$$

The exact answer is $62\sqrt{3}$. Now, use a calculator to verify.

First, find a decimal approximation for the original expression.

Enter: 4 [×] 27 [√x] [+] 5 [×] 12 [√x] [+] 8 [×] 75 [√x] [=] *107.3871501*

Next, find a decimal approximation for the simplified expression.

Enter: 62 [×] 3 [√x] [=] *107.3871501*

Since the approximations are equal, the results have been verified.

Example 3

APPLICATION

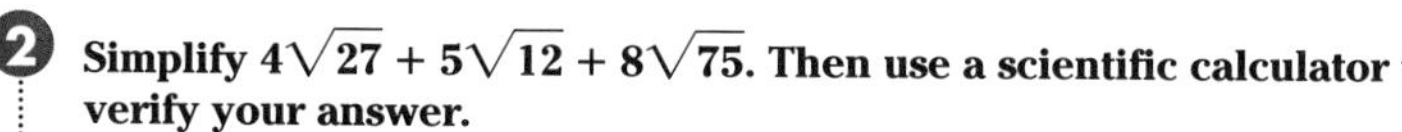

Travel

Refer to the application at the beginning of the lesson. How much farther is the captain able to see than the passengers on the pool deck?

$$\begin{aligned} d &= \sqrt{\frac{3(120)}{2}} - \sqrt{\frac{3(72)}{2}} \\ &= \sqrt{\frac{360}{2}} - \sqrt{\frac{216}{2}} \\ &= \sqrt{180} - \sqrt{108} \\ &= \sqrt{6^2 \cdot 5} - \sqrt{6^2 \cdot 3} \\ &= 6\sqrt{5} - 6\sqrt{3} \\ &\approx 3.02 \end{aligned}$$

The captain can see about 3 miles farther.

In the last lesson, you multiplied conjugates and expressions with like radicands. Multiplying two radical expressions with different radicands is similar to multiplying two binomials.

Example  **Simplify $(2\sqrt{3} - \sqrt{5})(\sqrt{10} + 4\sqrt{6})$.**

LOOK BACK

Refer to Lesson 9-7 to review multiplying polynomials.

$$(2\sqrt{3} - \sqrt{5})(\sqrt{10} + 4\sqrt{6})$$

$$= \underbrace{(2\sqrt{3})(\sqrt{10})}_{\text{First terms}} + \underbrace{(2\sqrt{3})(4\sqrt{6})}_{\text{Outer terms}} + \underbrace{(-\sqrt{5})(\sqrt{10})}_{\text{Inner terms}} + \underbrace{(-\sqrt{5})(4\sqrt{6})}_{\text{Last terms}}$$

$= 2\sqrt{30} + 8\sqrt{18} - \sqrt{50} - 4\sqrt{30}$ *Multiply.*

$= 2\sqrt{30} + 24\sqrt{2} - 5\sqrt{2} - 4\sqrt{30}$ *Simplify each term.*

$= -2\sqrt{30} + 19\sqrt{2}$ *Combine like terms.*

In-Class Example

For Example 4

Simplify $\left(5\sqrt{10} + 4\sqrt{2}\right)\left(2\sqrt{5} - 5\right)$. $30\sqrt{2} - 17\sqrt{10}$

CHECK FOR UNDERSTANDING

Communicating Mathematics

Study the lesson. Then complete the following.

1. **Write** three radical expressions that have the same radicand. 1. Sample answer: $4\sqrt{3}$, $2\sqrt{3}$, $6\sqrt{3}$
2. **Explain** why you should simplify each radical in a radical expression before adding or subtracting. **to determine if there are any like radicands**
3. **Explain** why $\sqrt{x} + \sqrt{y} \neq \sqrt{x + y}$. Give an example using numbers. **See margin.**

4. Explain how you use the distributive property to simplify like radicands that are added or subtracted. **See margin.**

Guided Practice

Name the expressions in each group that will have the same radicand after each expression is written in simplest form.

5. $\underline{3\sqrt{5}}, 5\sqrt{6}, \underline{3\sqrt{20}}$
6. $\underline{-5\sqrt{7}}, \underline{2\sqrt{28}}, 6\sqrt{14}$
7. $\sqrt{24}, \sqrt{12}, \sqrt{18}, \sqrt{28}$ **none**
8. $\underline{9\sqrt{32}}, \underline{2\sqrt{50}}, \sqrt{48}, \underline{3\sqrt{200}}$

Simplify. **9. $13\sqrt{6}$ 10. in simplest form 11. $12\sqrt{7x}$**

9. $3\sqrt{6} + 10\sqrt{6}$
10. $2\sqrt{5} - 5\sqrt{2}$
11. $8\sqrt{7x} + 4\sqrt{7x}$

Simplify. Then use a calculator to verify your answer.

12. $8\sqrt{5} + 3\sqrt{5}$ **$11\sqrt{5}$; 24.60**
13. $8\sqrt{3} - 2\sqrt{2} + 3\sqrt{2} + 5\sqrt{3}$ **13. $13\sqrt{3} + \sqrt{2}$; 23.93**
14. $2\sqrt{3} + \sqrt{12}$ **$4\sqrt{3}$; 6.93**
15. $\sqrt{7} + \sqrt{\frac{1}{7}}$ **$\frac{8}{7}\sqrt{7}$; 3.02**

Simplify.

16. $\sqrt{2}(\sqrt{18} + 4\sqrt{3})$ **$6 + 4\sqrt{6}$**
17. $(4 + \sqrt{5})(4 - \sqrt{5})$ **11**
18. **Geometry** Find the exact measures of the perimeter and area in simplest form for the rectangle at the right. **$8\sqrt{7} - 6\sqrt{3}$; $4\sqrt{21} - 12$**

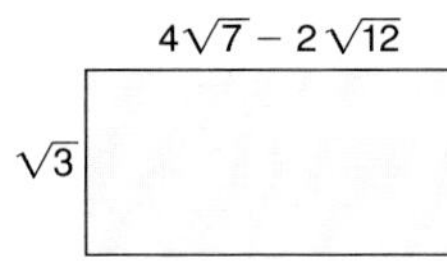

3 PRACTICE/APPLY

Check for Understanding

Exercises 1–18 are designed to help you assess your students' understanding through reading, writing, speaking, and modeling. You should work through Exercises 1–4 with your students and then monitor their work on Exercises 5–18.

Additional Answers

3. Sample answer:
$\sqrt{x} + \sqrt{y} \stackrel{?}{=} \sqrt{x + y}$
$\sqrt{3} + \sqrt{6} \stackrel{?}{=} \sqrt{9}$
$4.1815 \neq 3$
4. The distributive property allows you to add like terms. Radicals with like radicands can be added or subtracted.

Study Guide Masters, p. 90

NAME ______ DATE ______

Study Guide — Student Edition Pages 727–731

Operations with Radical Expressions

When adding or subtracting radical expressions, use the distributive and commutative properties to simplify the expressions. If radical expressions are not in simplest form, first simplify.

Example 1: Simplify $10\sqrt{6} - 5\sqrt{3} + 6\sqrt{3} - 4\sqrt{6}$.

$$\begin{aligned} 10\sqrt{6} - 5\sqrt{3} + 6\sqrt{3} - 4\sqrt{6} &= 10\sqrt{6} - 4\sqrt{6} - 5\sqrt{3} + 6\sqrt{3} \\ &= (10 - 4)\sqrt{6} + (-5 + 6)\sqrt{3} \\ &= 6\sqrt{6} + \sqrt{3} \end{aligned}$$

Example 2: Simplify $3\sqrt{12} + 5\sqrt{75}$.

$$\begin{aligned} 3\sqrt{12} + 5\sqrt{75} &= 3\sqrt{2^2 \cdot 3} + 5\sqrt{5^2 \cdot 3} \\ &= 3 \cdot 2\sqrt{3} + 5 \cdot 5\sqrt{3} \\ &= 6\sqrt{3} + 25\sqrt{3} \\ &= (6 + 25)\sqrt{3} \\ &= 31\sqrt{3} \end{aligned}$$

Simplify. Then use a calculator to verify your answer.

1. $8 + 3\sqrt{2}$ — $8 + 3\sqrt{2}$; $8 + 4.24 = 12.24$
2. $\sqrt{12} - \sqrt{27}$ — $-\sqrt{3}$; $3.46 - 5.20 \approx -1.73$
3. $\sqrt{27} - 2\sqrt{3}$ — $\sqrt{3}$; $5.20 - 3.46 \approx 1.73$
4. $\sqrt{20} + 2\sqrt{5} - 3\sqrt{5}$ — $\sqrt{5}$; $4.47 + 4.47 - 6.71 \approx 2.24$
5. $-5\sqrt{6} + 8\sqrt{6}$ — $3\sqrt{6}$; $-12.25 + 19.60 = 7.35$
6. $\sqrt{200} - 3\sqrt{2}$ — $7\sqrt{2}$; $14.14 - 4.24 = 9.90$
7. $\sqrt{54} + \sqrt{24}$ — $5\sqrt{6}$; $7.35 + 4.90 = 12.25$
8. $\sqrt{18} - 3\sqrt{8} + \sqrt{50}$ — $2\sqrt{2}$; $4.24 - 8.49 + 7.07 \approx 2.83$
9. $\sqrt{80} - \sqrt{20} + \sqrt{180}$ — $8\sqrt{5}$; $8.94 - 4.47 + 13.42 = 17.89$
10. $2\sqrt{28} + 3\sqrt{63} - \sqrt{\frac{54}{3}}$ — $13\sqrt{7} - 3\sqrt{2}$; $10.58 + 23.81 - 4.24 = 30.15 = 34.39 - 4.24$
11. $\sqrt{12} + \sqrt{\frac{1}{3}}$ — $\frac{7\sqrt{3}}{3}$; $3.46 + 0.58 = 4.04$
12. $\sqrt{54} - \sqrt{\frac{1}{6}} + \sqrt{24}$ — $\frac{29}{6}\sqrt{6}$; $7.35 - 0.41 + 4.90 = 11.84$

Reteaching

Logical Thinking Have students verify that the simplest form of a radical expression is equivalent to the radical expression for several problems. Use a calculator and round to 4 places. For example:

$$\sqrt{18} + \sqrt{50} = 8\sqrt{2}$$
$$4.243 + 7.071 = 8(1.414)$$
$$11.314 = 11.314$$

Assignment Guide

Core: 19–45 odd, 46–53
Enriched: 18–42 even, 43–53
All: Self Test, 1–10

For **Extra Practice,** see p. 785.

The red A, B, and C flags, printed only in the Teacher's Wraparound Edition, indicate the level of difficulty of the exercises.

EXERCISES

Practice

Simplify.

A

19. $25\sqrt{13} + \sqrt{13}$ **$26\sqrt{13}$**
20. $7\sqrt{2} - 15\sqrt{2} + 8\sqrt{2}$ **0**
21. $2\sqrt{6} - 8\sqrt{3}$ **in simplest form**
22. $2\sqrt{11} - 6\sqrt{11} - 3\sqrt{11}$ **$-7\sqrt{11}$**
23. $18\sqrt{2x} + 3\sqrt{2x}$ **$21\sqrt{2x}$**
24. $3\sqrt{5m} - 5\sqrt{5m}$ **$-2\sqrt{5m}$**

Simplify. Then use a calculator to verify your answer.

25. $4\sqrt{3} + 7\sqrt{3} - 2\sqrt{3}$ **$9\sqrt{3}$; 15.59**
26. $5\sqrt{5} + 3\sqrt{5} - 18\sqrt{5}$

B

27. $\sqrt{6} + 2\sqrt{2} + \sqrt{10}$
28. $4\sqrt{6} + \sqrt{7} - 6\sqrt{2} + 4\sqrt{7}$
29. $3\sqrt{7} - 2\sqrt{28}$ **$-\sqrt{7}$; −2.65**
30. $2\sqrt{50} - 3\sqrt{32}$ **$-2\sqrt{2}$; −2.83**
31. $3\sqrt{27} + 5\sqrt{48}$ **$29\sqrt{3}$; 50.23**
32. $2\sqrt{20} - 3\sqrt{24} - \sqrt{180}$
33. $\sqrt{80} + \sqrt{98} + \sqrt{128}$
34. $\sqrt{10} - \sqrt{\frac{2}{5}}$ **$\frac{4}{5}\sqrt{10}$; 2.53**
35. $3\sqrt{3} - \sqrt{45} + 3\sqrt{\frac{1}{3}}$
36. $6\sqrt{\frac{7}{4}} + 3\sqrt{28} - 10\sqrt{\frac{1}{7}}$

Simplify. **37. $10\sqrt{2} + 3\sqrt{10}$**

C

37. $\sqrt{5}(2\sqrt{10} + 3\sqrt{2})$
38. $\sqrt{6}(\sqrt{3} + 5\sqrt{2})$ **$3\sqrt{2} + 10\sqrt{3}$**
39. $(2\sqrt{10} + 3\sqrt{15})(3\sqrt{3} - 2\sqrt{2})$
40. $(\sqrt{5} - \sqrt{2})(\sqrt{14} + \sqrt{35})$ **$3\sqrt{7}$**
41. $(\sqrt{6} + \sqrt{8})(\sqrt{24} + \sqrt{2})$
42. $(5\sqrt{2} + 3\sqrt{5})(2\sqrt{10} - 3)$

26. $-10\sqrt{5}$; −22.36
27. $\sqrt{6} + 2\sqrt{2} + \sqrt{10}$; 8.44
28. $4\sqrt{6} - 6\sqrt{2} + 5\sqrt{7}$; 14.54
32. $-2\sqrt{5} - 6\sqrt{6}$; −19.17
33. $4\sqrt{5} + 15\sqrt{2}$; 30.16
35. $4\sqrt{3} - 3\sqrt{5}$; 0.22
36. $\frac{53\sqrt{7}}{7}$; 20.03
39. $19\sqrt{5}$
41. $10\sqrt{3} + 16$
42. $15\sqrt{2} + 11\sqrt{5}$

Critical Thinking

43. Explain why the simplified form of $\sqrt{(x-5)^2}$ must have an absolute value sign, but $\sqrt{(x-5)^4}$ does not need one. **Both $(x-5)^2$ and $(x-5)^4$ must be nonnegative, but $x - 5$ may be negative.**

Applications and Problem Solving

44. **Construction** *Slip forming* is the fastest method of erecting tall concrete buildings. With this method, the 1815-foot CN Tower in Toronto, Canada, was built at an average speed of 20 feet per day. At the beginning of the week, construction workers were 530 feet above the ground. After one week of construction, they were 670 feet above the ground. How many more miles could they see from the top of the building at the end of the week than at the beginning? Write your answer in exact form and as an approximation to the nearest hundredth. (*Hint:* Use the formula from the application at the beginning of the lesson.) **$\sqrt{1005} - \sqrt{795}$; 3.51 mi**

45. **Construction** A wire is stretched from the top of a 12-foot pole to a stake in the ground and then to the base of the pole. If a total of 20 feet of wire is needed, how far is the stake from the pole? (*Hint:* In the figure, $a + b = 20$.) **$6\frac{2}{5}$ ft**

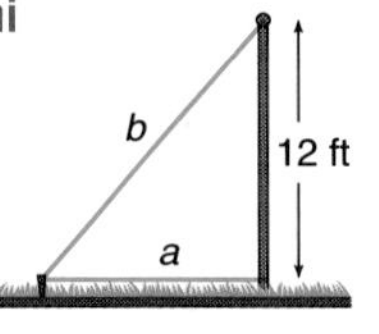

Practice Masters, p. 90

NAME______________________ DATE ____________

Practice

Student Edition
Pages 727–731

Operations with Radical Expressions

Simplify. Then use a calculator to verify your answer.
Exact and approximate values are given.

1. $6\sqrt{5} - 2\sqrt{5} + 8\sqrt{5}$ **$12\sqrt{5}$; 13.42 − 4.47 + 17.89 = 26.84**
2. $\sqrt{15} + 8\sqrt{15} - 12\sqrt{15}$ **$-3\sqrt{15}$; 3.87 + 30.98 − 46.48 = −11.63**
3. $4\sqrt{3} + 2\sqrt{12}$ **$8\sqrt{3}$; 6.93 + 6.93 = 13.86**
4. $8\sqrt{54} - 4\sqrt{6}$ **$20\sqrt{6}$; 58.79 − 9.80 = 48.99**
5. $8\sqrt{32} - 4\sqrt{8}$ **$24\sqrt{2}$; 45.25 − 11.31 = 33.94**
6. $2\sqrt{45} + 4\sqrt{20}$ **$14\sqrt{5}$; 13.42 + 17.89 = 31.31**
7. $5\sqrt{128} + 2\sqrt{18}$ **$46\sqrt{2}$; 56.57 + 8.49 = 65.06**
8. $3\sqrt{75} - \sqrt{243}$ **$6\sqrt{3}$; 25.98 − 15.59 = 10.39**
9. $\sqrt{28} + \sqrt{63}$ **$5\sqrt{7}$; 5.29 + 7.94 = 13.23**
10. $\sqrt{44} - \sqrt{11}$ **$\sqrt{11}$; 6.63 − 3.32 = 3.31**
11. $\sqrt{27} + \sqrt{48} + \sqrt{12}$ **$9\sqrt{3}$; 5.20 + 6.93 + 3.46 = 15.59**
12. $\sqrt{72} + \sqrt{50} - \sqrt{8}$ **$9\sqrt{2}$; 8.49 + 7.07 − 2.83 = 12.73**
13. $\sqrt{180} - 5\sqrt{5} + \sqrt{20}$ **$3\sqrt{5}$; 13.42 − 11.18 + 4.47 = 6.71**
14. $\sqrt{40} - \sqrt{10} + \sqrt{90}$ **$4\sqrt{10}$; 6.32 − 3.16 + 9.49 = 12.65**
15. $2\sqrt{32} + 3\sqrt{50} - 3\sqrt{18}$ **$14\sqrt{2}$; 11.31 + 21.21 − 12.73 = 19.79**
16. $\sqrt{27} + \sqrt{18} + \sqrt{300}$ **$3\sqrt{2} + 13\sqrt{3}$; 5.20 + 4.24 + 17.32 = 26.76**
17. $\sqrt{14} - \sqrt{\frac{2}{7}}$ **$\frac{6}{7}\sqrt{14}$; 3.74 − 0.53 = 3.21**
18. $\sqrt{50} + \sqrt{32} - \sqrt{\frac{1}{2}}$ **$\frac{17}{2}\sqrt{2}$; 7.07 + 5.66 − 0.71 = 12.02**

Alternative Teaching Strategies

Reading Algebra Have a student read the expressions $2(\sqrt{6} + \sqrt{7})$ and $2\sqrt{6} + \sqrt{7}$ to the class. The student should read the first one as "2 times the quantity radical 6 plus radical 7." The student should read the second one as "2 times radical 6 plus radical 7."

Mixed Review

46. Simplify $\frac{\sqrt{3}}{\sqrt{6}}$. (Lesson 13–2) $\frac{\sqrt{2}}{2}$

47. Find $\frac{x^2 - y^2}{3} \cdot \frac{9}{x + y}$. (Lesson 12–2) $3x - 3y$

48. Factor $3a^2 + 19a - 14$. (Lesson 10–3) $(3a - 2)(a + 7)$

49. Find the degree of $16s^3t^2 + 3s^2t + 7s^6t$. (Lesson 9–4) 7

50. **Statistics** Tim's scores on the first four of five 50-point quizzes were 47, 45, 48, and 45. What score must he receive on the fifth quiz to have an average of at least 46 points for all the quizzes? (Lesson 7–3) **45**

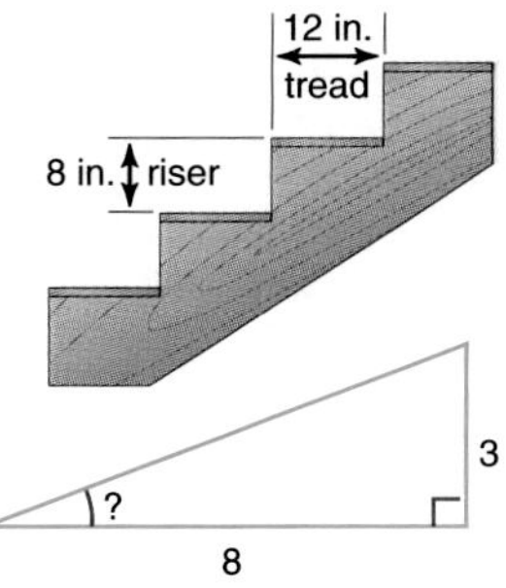

51. **Carpentry** When building a stairway, a carpenter considers the ratio of riser to tread. Write a ratio to describe the steepness of the stairs. (Lesson 6–1) $\frac{2}{3}$

52. **Geometry** Find the measure of the marked acute angle to the nearest degree. (Lesson 4–3) **21°**

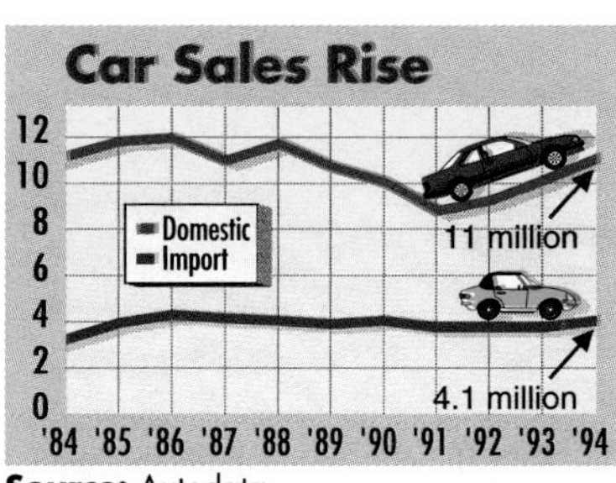

Source: Autodata

53. **Interpret Graphs** The graph at the left compares domestic and import brand auto sales from 1984 to 1994. (Lesson 1–9)
   a. During the ten-year period, when were domestic car sales the lowest? How many domestic cars were sold during that year? **1991; about 9 million**
   b. In 1990, how many more domestic cars were sold than imports? **6 million**

SELF TEST

If c is the measure of the hypotenuse of a right triangle, find each missing measure. Round answers to the nearest hundredth. (Lesson 13–1)

1. $a = 21, b = 28, c = ?$ **35**
2. $a = \sqrt{41}, c = 8, b = ?$ **4.80**
3. $b = 28, c = 54, a = ?$ **46.17**

Simplify. Leave in radical form and use absolute value symbols when necessary. (Lesson 13–2)

4. $\sqrt{20}$ $2\sqrt{5}$
5. $2\sqrt{5} \cdot \sqrt{5}$ **10**
6. $\frac{\sqrt{42x^2}}{\sqrt{6y^3}}$ $\frac{|x|\sqrt{7y}}{y^2}$

Simplify. (Lesson 13–3)

7. $8\sqrt{6} + 3\sqrt{6}$ $11\sqrt{6}$
8. $10\sqrt{17} + 9\sqrt{7} - 8\sqrt{17} + 6\sqrt{7}$ $2\sqrt{17} + 15\sqrt{7}$
9. $(6 + \sqrt{3})(2\sqrt{5} - \sqrt{3})$ $12\sqrt{5} - 6\sqrt{3} + 2\sqrt{15} - 3$

10. **Geometry** Find the perimeter and area of the figure at the right. Round answers to the nearest hundredth. (Lesson 13–1) **53.37 cm, 138 cm²**

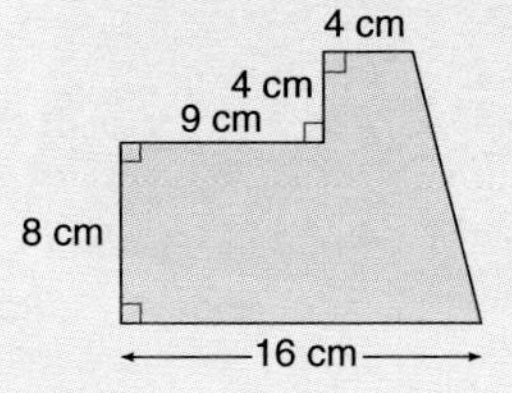

4 ASSESS

Closing Activity

Speaking Have students explain how addition and subtraction of monomials is similar to that of radical expressions.

Chapter 13, Quiz B (Lesson 13-3), is available in the *Assessment and Evaluation Masters*, p. 352.

Mid-Chapter Test (Lessons 13-1 through 13-3) is available in the *Assessment and Evaluation Masters*, p. 351.

Extension

Connections Simplify $\frac{\sqrt{a}}{\sqrt{b}} + \frac{\sqrt{b}}{\sqrt{a}}$. $(a + b)\frac{\sqrt{ab}}{ab}$

SELF TEST

The Self Test provides students with a brief review of the concepts and skills in Lessons 13-1 through 13-3. Lesson numbers are given to the right of exercises or instruction lines so students can review concepts not yet mastered.

Enrichment Masters, p. 90

13–3 NAME ______ DATE ______

Enrichment — Student Edition Pages 727–731

Pythagorean Triples

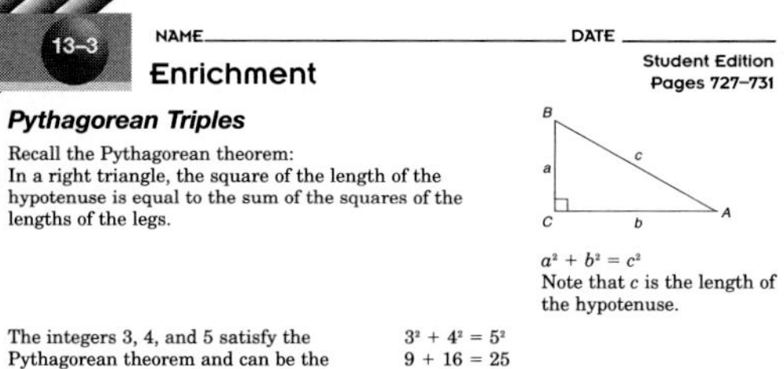

Recall the Pythagorean theorem:
In a right triangle, the square of the length of the hypotenuse is equal to the sum of the squares of the lengths of the legs.

$a^2 + b^2 = c^2$
Note that c is the length of the hypotenuse.

The integers 3, 4, and 5 satisfy the Pythagorean theorem and can be the lengths of the sides of a right triangle. $3^2 + 4^2 = 5^2$; $9 + 16 = 25$; $25 = 25$

Furthermore, for any positive integer n, the numbers $3n$, $4n$, and $5n$ satisfy the Pythagorean theorem. For $n = 2$: $6^2 + 8^2 = 10^2$; $36 + 64 = 100$; $100 = 100$

If three numbers satisfy the Pythagorean theorem, they are called a **Pythagorean triple.** Here is an easy way to find other Pythagorean triples.

The numbers a, b, and c are a Pythagorean triple if $a = m^2 - n^2$, $b = 2mn$, and $c = m^2 + n^2$, where m and n are relatively prime positive integers and $m > n$.

Example: Choose $m = 5$ and $n = 2$.
$a = m^2 - n^2 = 5^2 - 2^2 = 25 - 4 = 21$
$b = 2mn = 2(5)(2) = 20$
$c = m^2 + n^2 = 5^2 + 2^2 = 25 + 4 = 29$
Check: $20^2 + 21^2 = 29^2$; $400 + 441 = 841$; $841 = 841$

Use the following values of m and n to find Pythagorean triples.

1. $m = 3$ and $n = 2$ 12, 5, 13
2. $m = 4$ and $n = 1$ 8, 15, 17
3. $m = 5$ and $n = 3$ 30, 16, 34
4. $m = 6$ and $n = 5$ 60, 11, 61
5. $m = 10$ and $n = 7$ 140, 51, 149
6. $m = 8$ and $n = 5$ 80, 39, 89

13-4 LESSON NOTES

NCTM Standards: 1–5

Instructional Resources

- Study Guide Master 13-4
- Practice Master 13-4
- Enrichment Master 13-4

Transparency 13-4A contains the 5-Minute Check for this lesson; **Transparency 13-4B** contains a teaching aid for this lesson.

Recommended Pacing	
Honors Pacing	Day 6 of 11
Block Scheduling*	Day 4 of 7
Alg. 1 in Two Years*	Days 11, 12, & 13 of 22

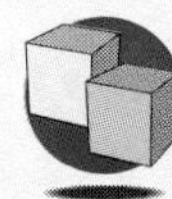

*For more information on pacing and possible lesson plans, refer to the *Block Scheduling Booklet* and *Algebra 1 in Two Years.*

1 FOCUS

5-Minute Check
(over Lesson 13-3)

Simplify. Use absolute value symbols when necessary.

1. $4\sqrt{2} + 5\sqrt{3} - \sqrt{2} + 3\sqrt{3}$ **$3\sqrt{2} + 8\sqrt{3}$**
2. $8\sqrt{11} + 5\sqrt{11} - 16\sqrt{11}$ **$-3\sqrt{11}$**
3. $3\sqrt{5} + 2\sqrt{20}$ **$7\sqrt{5}$**
4. $(3\sqrt{2} - 4\sqrt{5})(5\sqrt{2} + 2\sqrt{5})$ **$-10 - 14\sqrt{10}$**
5. $\sqrt{3}(\sqrt{6} + 2\sqrt{3})$ **$3\sqrt{2} + 6$**

13-4 Radical Equations

What YOU'LL LEARN

- To solve radical equations.

Why IT'S IMPORTANT

You can use radical equations to solve problems involving oceanography and recreation.

FYI

A tsunami may begin as a 2-foot high wave. After traveling hundreds of miles across the ocean at speeds of 450 to 500 miles per hour, it could approach shallow coastal waters as a towering 50-foot wall of water, capable of destroying anything in its path.

APPLICATION

Oceanography

Topographical map of the ocean surface

The Tonga Trench in the Pacific Ocean is a potential source for a *tsunami* (su-nom'-ee), a large ocean wave generated by an undersea earthquake. The formula for a tsunami's speed s in meters per second is $s = 3.1\sqrt{d}$, where d is the depth of the ocean in meters.

Equations like $s = 3.1\sqrt{d}$ that contain radicals with variables in the radicand are called **radical equations**. To solve these equations, first isolate the radical on one side of the equation. Then square each side of the equation to eliminate the radical.

Find the depth of the Tonga Trench if a tsunami's speed is 322 meters per second.

$322 = 3.1\sqrt{d}$ *Replace s with 322.*

$\frac{322}{3.1} = \frac{3.1\sqrt{d}}{3.1}$ *Divide each side by 3.1.*

$\left(\frac{322}{3.1}\right)^2 = \left(\sqrt{d}\right)^2$ *Square each side of the equation.*

$\left(\frac{322}{3.1}\right)^2 = d$ Use a scientific calculator to simplify $\left(\frac{322}{3.1}\right)^2$.

Enter: (322 ÷ 3.1) x^2 10789.17794

The depth of the Tonga Trench is approximately 10,789 meters. *Check this result by substituting 10,789 for d into the original formula.*

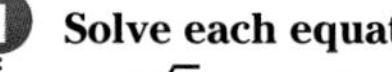

Example 1 **Solve each equation.**

a. $\sqrt{x} + 4 = 7$

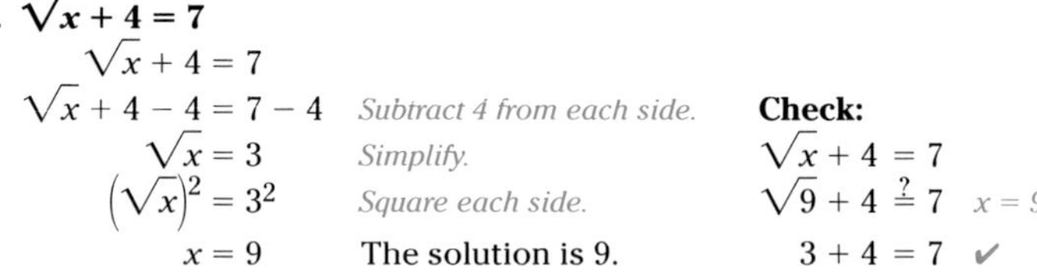

$\sqrt{x} + 4 = 7$

$\sqrt{x} + 4 - 4 = 7 - 4$ *Subtract 4 from each side.*

$\sqrt{x} = 3$ *Simplify.*

$(\sqrt{x})^2 = 3^2$ *Square each side.*

$x = 9$ The solution is 9.

Check:
$\sqrt{x} + 4 = 7$
$\sqrt{9} + 4 \stackrel{?}{=} 7$ *x = 9*
$3 + 4 = 7$ ✔

b. $\sqrt{x + 3} + 5 = 9$

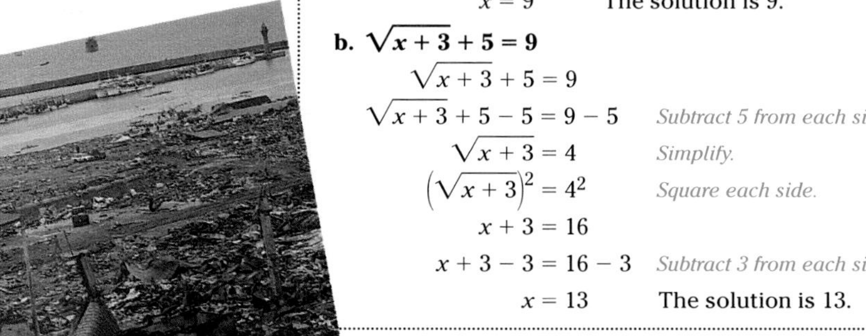

$\sqrt{x + 3} + 5 = 9$

$\sqrt{x + 3} + 5 - 5 = 9 - 5$ *Subtract 5 from each side.*

$\sqrt{x + 3} = 4$ *Simplify.*

$(\sqrt{x + 3})^2 = 4^2$ *Square each side.*

$x + 3 = 16$

$x + 3 - 3 = 16 - 3$ *Subtract 3 from each side.*

$x = 13$ The solution is 13. *Check this result.*

Squaring each side of an equation does not necessarily produce results that satisfy the original equation. Therefore, you must check all solutions when you solve radical equations.

FYI

Tsunamis are usually caused by earthquakes or volcanic eruptions. When the ocean floor is quickly raised or lowered, it generates a seismic wave.

GLENCOE Technology

CD-ROM Interaction

A multimedia simulation links radical expressions and the Pythagorean theorem with the installation of an electricity-generating windmill. A blackline master activity with teacher's notes provides a follow-up to the CD-ROM simulation.

For Windows & Macintosh

Example 2 **Solve $\sqrt{3x-5} = x - 5$.**

$$\sqrt{3x-5} = x - 5$$
$$(\sqrt{3x-5})^2 = (x-5)^2 \quad \textit{Square each side.}$$
$$3x - 5 = x^2 - 10x + 25 \quad \textit{Simplify.}$$
$$3x - 3x - 5 + 5 = x^2 - 10x + 25 - 3x + 5 \quad \textit{Add } -3x \textit{ and 5 to each side.}$$
$$0 = x^2 - 13x + 30 \quad \textit{Simplify.}$$
$$0 = (x-10)(x-3) \quad \textit{Factor.}$$
$$x - 10 = 0 \text{ or } x - 3 = 0 \quad \textit{Use the zero product property.}$$
$$x = 10 \qquad x = 3$$

Check:

$$\sqrt{3x-5} = x - 5 \qquad \sqrt{3x-5} = x - 5$$
$$\sqrt{3(10)-5} \stackrel{?}{=} 10 - 5 \qquad \sqrt{3(3)-5} \stackrel{?}{=} 3 - 5$$
$$\sqrt{30-5} \stackrel{?}{=} 5 \qquad \sqrt{9-5} \stackrel{?}{=} -2$$
$$\sqrt{25} \stackrel{?}{=} 5 \qquad \sqrt{4} \stackrel{?}{=} -2$$
$$5 = 5 \ \checkmark \qquad 2 \neq -2$$

Since 3 does not satisfy the original equation, 10 is the only solution.

You can use the *Mathematics Exploration Toolkit* (*MET*) to solve equations involving square roots.

EXPLORATION GRAPHING SOFTWARE

The CALC commands below will be used.

ADD (add)	SUBTRACT (sub)	MULTIPLY (mult)
DIVIDE (div)	FACTOR (fac)	RAISETO (rai)
SIMPLIFY (simp)	STORE (sto)	SUBSTITUTE (subs)

To enter the square root symbol, type &.

Solve $\sqrt{x-2} = x - 4$.

Enter:	**Result:**
&(x − 2) = x − 4	$\sqrt{x-2} = x - 4$
sto *a*	Saves the equation as *a*.
rai 2	$(\sqrt{x-2})^2 = (x-4)^2$
simp	$x - 2 = x^2 - 8x + 16$
sub x − 2	$x - 2 - (x-2) = x^2 - 8x + 16 - (x-2)$
simp	$0 = x^2 - 9x + 18$
fac	$0 = (x-6)(x-3)$

By inspection, the solutions are $x = 6$ or $x = 3$. However, 3 does not satisfy the original equation. Therefore, 6 is the only solution.

Your Turn
Use CALC to solve each equation.

a. $3 + \sqrt{2x} = 7$ **8**
b. $\sqrt{x+1} = x - 1$ **3**
c. $\sqrt{x} + 6 = 1$ **no solution**
d. $\sqrt{3x-8} = 5$ **11**
e. $x + \sqrt{6-x} = 4$ **2**
f. $\sqrt{3x-9} = 2x + 6$ **15**

Motivating the Lesson

Questioning Open the lesson by asking students to square each side of the following equations.

1. $\sqrt{x} = 12$
2. $\sqrt{n} = 8$
3. $\sqrt{2y} = -11$

Ask students to describe the results. Develop the idea that squaring each side of the equation eliminates the radical.

2 TEACH

Teaching Tip Emphasize the importance of isolating the radical by showing an example of what happens if it is not isolated.

$$\sqrt{y} + 3 = 0$$
$$(\sqrt{y} + 3)^2 = 0$$
$$y + 6\sqrt{y} + 9 = 0$$

The radical still remains.

Teaching Tip Stress that squaring each side of an equation often produces results that do not check. Thus, it is important to check all possible solutions. A result that does not check is called an *extraneous value.*

In-Class Examples

For Example 1
Solve each equation. Check your solution.

a. $3 + \sqrt{m} = 13$ **$m = 100$**
b. $\sqrt{(4m+1)} - 5 = 0$ **$m = 6$**
c. $\sqrt{(6m-3)} - 3 = 0$ **$m = 2$**

For Example 2
Solve each equation. Check your solution.

a. $\sqrt{(y+2)} = y - 4$ **7**
b. $\sqrt{(3x-5)} = x - 5$ **10**

Cooperative Learning

Brainstorming Have students work in small groups to brainstorm answers to this question: For what values of *a* is $\sqrt{|a|} = |a|$? **0, 1, −1**
Have one student test the other students' guesses.

For more information on the brainstorming strategy, see *Cooperative Learning in the Mathematics Classroom*, one of the titles in the Glencoe Mathematics Professional Series, page 30.

EXPLORATION

This activity makes use of the *MET* software. It may seem to relieve students of thinking. But, in fact, it makes them think about the role of each step in the program. Encourage students to explain in their own words what each built-in function does.

4 ASSESS

Closing Activity

Writing Have students develop new equations to solve. Collect the papers and then distribute randomly. Have students solve the new equations and return to the maker to be graded.

Additional Answer

50.

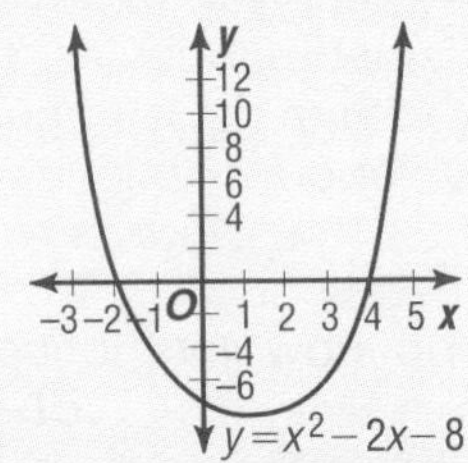

Mathematics and SOCIETY

Discuss the fact that in some problems, students were given data and then used graphing calculators to find a curve of best-fit. The *r* value was almost never 1. Did they settle for a poor model that they understood well? Why do models rarely fit the data perfectly? **Some of the factors include overly simple models, inaccuracy of measurements of data, other variables not occurring in the equation, and so on.**

Enrichment Masters, p. 91

NAME__________________ DATE________

13-4 **Enrichment** — Student Edition Pages 732–736

Using Radical Equations

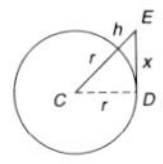

The circle with center C and radius r represents Earth. If your eye is at point E, the distance to the horizon is x, or the length of tangent segment ED. The segment drawn from C to D forms a right angle with ED. Thus, $\triangle EDC$ is a right triangle. Apply the Pythagorean theorem.

$$(\text{length of } CD)^2 + (\text{length of } ED)^2 = (\text{length of } EC)^2$$
$$r^2 + x^2 = (r + h)^2$$
$$x^2 = (r + h)^2 - r^2$$
$$x = \sqrt{(r + h)^2 - r^2}$$

1. Show that this equation is equivalent to $x = \sqrt{2r + h} \cdot \sqrt{h}$
   $\sqrt{(r + h)^2 - r^2} = \sqrt{r^2 + 2rh + h^2 - r^2} =$
   $\sqrt{2rh + h^2} = \sqrt{h(2r + h)} = \sqrt{2r + h} \cdot \sqrt{h}$

If the distance h is very small compared to r, $\sqrt{2r + h}$ is close to $\sqrt{2r}$.

$$x \approx \sqrt{2r} \cdot \sqrt{h}$$

The radius of Earth is about 20,900,000 feet.

$\sqrt{2r} \approx \sqrt{2(20{,}900{,}000)} = \sqrt{41{,}800{,}000} \approx 6465$ feet. Thus, $x \approx 6465\sqrt{h}$.

If you are h feet above Earth, you are about $6465\sqrt{h}$ feet from the horizon. Since there are 5280 feet in one mile,

$$6465\sqrt{h} \text{ feet} = \frac{5465\sqrt{h}}{5280} \text{ miles} \approx 1.22\sqrt{h} \text{ miles}.$$

Thus, if you are h feet above Earth's surface, you can see $1.22\sqrt{h}$ miles in any direction.

2. How far can you see to the nearest mile if your eye is:
   a. 1454 feet above the ground (the height of the Sears Tower)? **about 47 miles**
   b. 30,000 feet above the ground (altitude for a commercial airliner)? **about 211 miles**
   c. $5\frac{1}{2}$ feet from the ground? **about 2.9 miles**

A strong wind can severely alter the effect of an actual temperature on the human body. For example, a temperature of 32°F is much more dangerous on a windy day than on a still day. Windchill is a temperature value assigned to a particular combination of wind speed and temperature. If w is the speed of the wind in miles per hour and t is the actual temperature in degrees Fahrenheit, then the approximate windchill temperature in °F is given by the formula

$$\text{Windchill temperature} = 92.4 - \frac{(6.91\sqrt{w} + 10.45 - 0.477w)(91.4 - t)}{22.1}.$$

This formula gives reasonably accurate results for the windchill temperature when $5 \le w \le 30$ and $-30 \le t \le 50$.

3. Find the windchill temperature to the nearest degree when the actual temperature is −15°F and the wind speed is
   a. 10 mi/h b. 20 mi/h c. 30 mi/h. **−40°F −61°F −71°F**
4. Find the windchill temperature to the nearest degree when the wind speed is 20 mi/h and the actual temperature is
   a. 30°F b. 0°F c. −30°F **4°F −39°F −82°F**

47. **Travel** The speed s that a car is traveling in miles per hour, and the distance d in feet that it will skid when the brakes are applied, are related by the formula $s = \sqrt{30fd}$. In this formula, f is the coefficient of friction, which depends on the type and condition of the road. Sylvia Kwan told police she was traveling at about 30 miles per hour when she applied the brakes and skidded on a wet concrete road. The length of her skid marks was measured at 110 feet.
   a. If $f = 0.4$ for a wet concrete road, should Ms. Kwan's car have skidded that far when she applied the brakes?
   b. How fast was she traveling? **about 36 miles per hour**
   47a. No, at 30 mph, her car should have skidded 75 feet after the brakes were applied and not 110 feet.

Mixed Review

48. Simplify $5\sqrt{6} - 11\sqrt{3} - 8\sqrt{6} + \sqrt{27}$. (Lesson 13–3) $\mathbf{-3\sqrt{6} - 8\sqrt{3}}$
49. Find $(6b^2 + 4b + 20) \div (b + 5)$. (Lesson 12–4)
50. Find the roots of $x^2 - 2x - 8 = 0$ by graphing its related function. (Lesson 11–2) **−2, 4; See margin for graph.**
51. Find the GCF of $12x^2y^3$ and $42xy^4$. (Lesson 10–1) $\mathbf{6xy^3}$
52. Find $(6a - 2m) - (4a + 7m)$. (Lesson 9–5) $\mathbf{2a - 9m}$
53. Use substitution to solve the system of equations. (Lesson 8–2) **(−4, 5)**
    $x + 4y = 16$
    $3x + 6y = 18$
54. Write an equation in slope-intercept form of the line that passes through the points at (6, −1) and (3, 2). (Lesson 6–4) $\mathbf{y = -x + 5}$
55. **Probability** If the odds that an event will occur are 8:5, what is the probability that the event will occur? (Lesson 4–6)
56. Find $\frac{3}{7} + \left(-\frac{4}{9}\right)$. (Lesson 2–5) $\mathbf{-\frac{1}{63}}$

49. $\mathbf{6b - 26 + \frac{150}{b + 5}}$

55. $\mathbf{\frac{8}{13}}$

Mathematics and SOCIETY

Nonlinear Math

The excerpt below appeared in an article in *Business Week* on September 5, 1994.

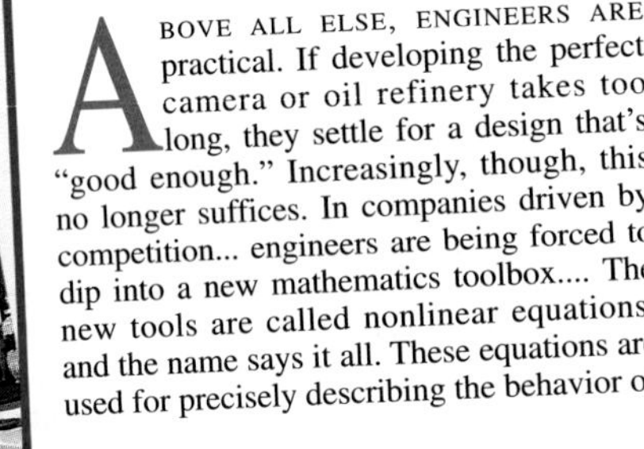

Above all else, engineers are practical. If developing the perfect camera or oil refinery takes too long, they settle for a design that's "good enough." Increasingly, though, this no longer suffices. In companies driven by competition... engineers are being forced to dip into a new mathematics toolbox.... The new tools are called nonlinear equations, and the name says it all. These equations are used for precisely describing the behavior of things with an unpredictable facet. That's nearly everything—from the workings of car engines to the actions of DNA molecules. Even baking a cake is nonlinear: turning up the oven's temperature twice as high won't bake the cake twice as fast. And with some industrial recipes, such as those for making drugs and plastics, a tiny change in ingredients or processing conditions can mean a huge difference in the finished product. Nonlinear math can help explain such lopsided effects. ■

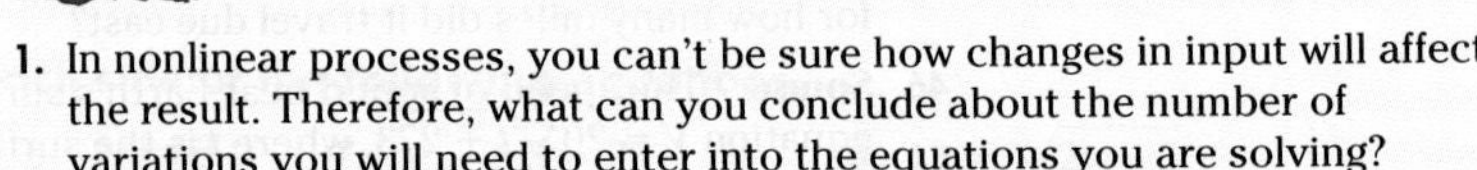

1. In nonlinear processes, you can't be sure how changes in input will affect the result. Therefore, what can you conclude about the number of variations you will need to enter into the equations you are solving?
2. Because of the huge number of variables that can be involved, solving nonlinear equations can require many millions of calculations. Why do you think the use of these equations has only recently begun to expand into many industries? **1–2. See margin.**

Answers for Mathematics and Society

1. Sample answer: You need to enter all the possible variations. This could involve entering huge numbers of variables.
2. Sample answer: Advances in computing power and the widespread use of computers have enabled more organizations to solve nonlinear equations.

13-5 Integration: Geometry The Distance Formula

What YOU'LL LEARN

- To find the distance between two points in the coordinate plane.

Why IT'S IMPORTANT

You can use the distance formula to solve problems involving art and communication.

INTEGRATION Geometry

In a coordinate plane, consider the points $A(-2, 6)$ and $B(5, 3)$. These two points do not lie on the same horizontal or vertical line. Therefore, you cannot find the distance between them by simply subtracting the x- or y-coordinates. A different method must be used.

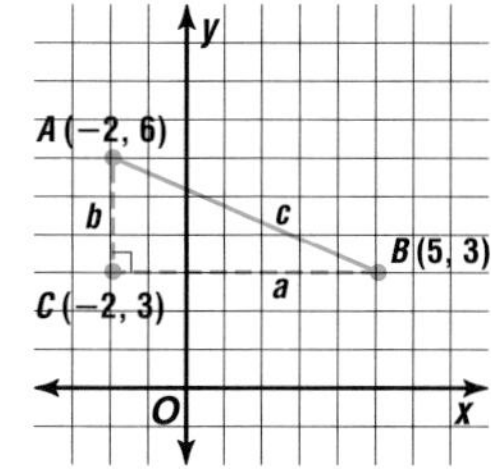

Notice that a right triangle can be formed by drawing lines parallel to the axes through points at $(-2, 6)$ and $(5, 3)$. These lines intersect at $C(-2, 3)$. The measure of side b is the difference of the y-coordinates of the endpoints, $6 - 3$ or 3. The measure of side a is the difference of the x-coordinates of the endpoints, $5 - (-2)$ or 7.

Now the Pythagorean theorem can be used to find c, the distance between $A(-2, 6)$ and $B(5, 3)$.

$c^2 = a^2 + b^2$ *Pythagorean theorem*

$c^2 = 7^2 + 3^2$ *Replace a with 7 and b with 3.*

$c^2 = 49 + 9$

$c^2 = 58$

$c = \sqrt{58}$

$c \approx 7.62$

The distance between points A and B is approximately 7.62 units.

The method used for finding the distance between $A(-2, 6)$ and $B(5, 3)$ can also be used to find the distance between any two points in the coordinate plane. This method can be described by the following formula.

The Distance Formula

The distance d between any two points with coordinates (x_1, y_1) and (x_2, y_2) is given by the following formula.

$$d = \sqrt{(x_2 - x_1)^2 + (y_2 - y_1)^2}$$

Example  **Find the distance between the points with coordinates (3, 5) and (6, 4).**

$d = \sqrt{(x_2 - x_1)^2 + (y_2 - y_1)^2}$

$= \sqrt{(6 - 3)^2 + (4 - 5)^2}$ *$(x_1, y_1) = (3, 5)$ and $(x_2, y_2) = (6, 4)$*

$= \sqrt{3^2 + (-1)^2}$

$= \sqrt{9 + 1}$

$= \sqrt{10}$ or about 3.16 units

13-5 LESSON NOTES

NCTM Standards: 1–5, 8

Instructional Resources

- Study Guide Master 13-5
- Practice Master 13-5
- Enrichment Master 13-5
- Assessment and Evaluation Masters, p. 353
- Graphing Calculator Masters, p. 13
- Multicultural Activity Masters, p. 26
- Science and Math Lab Manual, pp. 57–62

Transparency 13-5A contains the 5-Minute Check for this lesson; **Transparency 13-5B** contains a teaching aid for this lesson.

Recommended Pacing	
Honors Pacing	Day 7 of 11
Block Scheduling*	Day 5 of 7
Alg. 1 in Two Years*	Days 14, 15, & 16 of 22

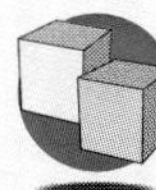

*For more information on pacing and possible lesson plans, refer to the *Block Scheduling Booklet* and *Algebra 1 in Two Years.*

1 FOCUS

5-Minute Check
(over Lesson 13-4)

Simplify.

1. $\sqrt{x} = 4\sqrt{3}$ 48
2. $5 - \sqrt{4a} = 2$ $\frac{9}{4}$
3. $\frac{\sqrt{3m}}{\sqrt{2}} - 7 = 1$ $\frac{128}{3}$
4. $\sqrt{4x^2} - 5 = 3x$ 1
5. Find two numbers that have a geometric mean of $10\sqrt{2}$ given that one number is 30 more than the other. 5, 35

Motivating the Lesson

Hands-On Activity Have students use a geoboard to construct several right triangles with one horizontal leg and one vertical leg. Measure the length of the hypotenuse and each leg. Confirm that those measurements fit the Pythagorean theorem.

2 TEACH

In-Class Examples

For Example 1
Find the distance between each pair of points whose coordinates are given.

a. (2, 0), (3, 5) $\sqrt{26}$ or 5.099
b. (1, 1), (4, 4) $3\sqrt{2}$ or 4.243
c. (−1, −2), (−3, 0) $2\sqrt{2}$ or 2.828

For Example 2
Determine if a triangle, *ABC*, with vertices *A*(2, 1), *B*(3, −2), and *C*(−3, −1) is an isosceles triangle. **no**

For Example 3
Find the value of *a* if the distance between the points at (*a*, 5) and (5, 1) is 5 units. **2**

Teaching Tip Encourage students to check the results in Example 3. Point out that the negative result does not need to be omitted. It represents the *y*-coordinate of a point in the plane, not a distance.

Additional Answer

1. **The values that are subtracted are squared before being added, the square of a negative is always positive, and distances are never negative numbers.**

Example 2

Geometry

Determine if triangle *ABC* with vertices *A*(−3, 4), *B*(5, 2), and *C*(−1, −5) is an isosceles triangle.

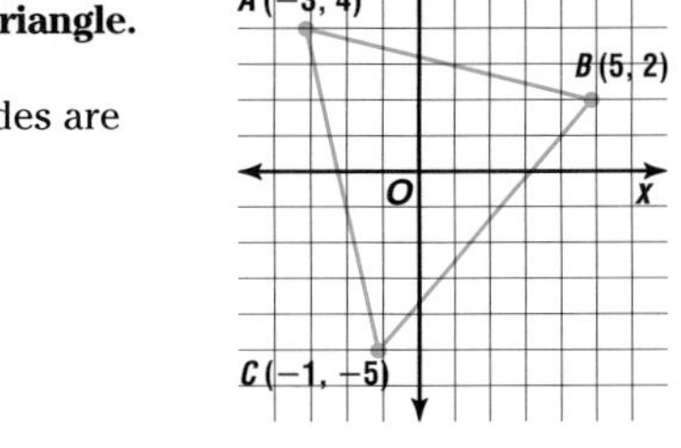

A triangle is isosceles if at least two sides are congruent. Find AB, BC, and AC.

$$AB = \sqrt{[5-(-3)]^2 + (2-4)^2} = \sqrt{8^2 + (-2)^2} \text{ or } \sqrt{68}$$

$$BC = \sqrt{(-1-5)^2 + (-5-2)^2} = \sqrt{(-6)^2 + (-7)^2} \text{ or } \sqrt{85}$$

$$AC = \sqrt{[-1-(-3)]^2 + (-5-4)^2} = \sqrt{2^2 + (-9)^2} \text{ or } \sqrt{85}$$

Since $\overline{BC}$ and $\overline{AC}$ have the same length, $\sqrt{85}$, they are congruent. So, triangle *ABC* is an isosceles triangle.

Suppose you know the coordinates of a point, one coordinate of another point, and the distance between the two points. You can use the distance formula to find the missing coordinate.

Example 3 **Find the value of *a* if the distance between the points with coordinates (−3, −2) and (*a*, −5) is 5 units.**

$$d = \sqrt{(x_2 - x_1)^2 + (y_2 - y_1)^2}$$
$$5 = \sqrt{[a-(-3)]^2 + [-5-(-2)]^2} \quad \text{Let } x_2 = a,\ x_1 = -3,\ y_2 = -5,\ y_1 = -2, \text{ and } d = 5.$$
$$5 = \sqrt{(a+3)^2 + (-3)^2}$$
$$5 = \sqrt{a^2 + 6a + 9 + 9}$$
$$5 = \sqrt{a^2 + 6a + 18}$$
$$(5)^2 = \left(\sqrt{a^2 + 6a + 18}\right)^2 \quad \text{Square each side.}$$
$$25 = a^2 + 6a + 18$$
$$0 = a^2 + 6a - 7$$
$$0 = (a+7)(a-1) \quad \text{Factor.}$$
$$a + 7 = 0 \text{ or } a - 1 = 0 \quad \text{Zero product property}$$
$$a = -7 \qquad a = 1$$

The value of *a* is −7 or 1.

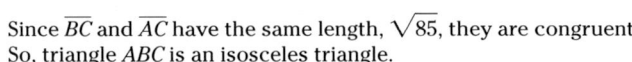

CHECK FOR UNDERSTANDING

Communicating Mathematics

Study the lesson. Then complete the following.

1. **Explain** why the value calculated under the radical sign in the distance formula will never be negative. **See margin.**
2. a. **Write** two ordered pairs and label them $A(x_1, y_1)$ and $B(x_2, y_2)$. Does it matter which ordered pair is first when using the distance formula? Explain. **a–b. See students' work.**
   b. Find the distance between *A* and *B*.

Reteaching

Using Alternative Methods

1. Show that a triangle with vertices *A*(−3, 4), *B*(5, 2), and *C*(−1, −5) is isosceles but not equilateral.
2. Show that a quadrilateral with vertices *A*(1, 1), *B*(4, −2), *C*(1, −3), and *D*(−2, 0) is a parallelogram.
3. Show that (2, 3) is the center of a circle through (5, 7), (−1, −1), and (−2, 0).

GLENCOE Technology

Interactive Mathematics Tools Software

This multimedia software provides an interactive lesson that allows students to derive the distance formula from the Pythagorean theorem and to practice the skill by substituting coordinates into the formula. A **Computer Journal** gives students an opportunity to write about what they have learned.

For Windows & Macintosh

3. a. **Explain** how you can find the distance between $X(12, 4)$ and $Y(3, 4)$ without using the distance formula. **a–b. See margin.**
   b. **Explain** how you can find the distance between $S(-2, 7)$ and $T(-2, -5)$ without using the distance formula.
4. Refer to Example 3. Check your answer by using the distance formula. **See students' work.**

Guided Practice

Find the distance between each pair of points whose coordinates are given. Express answers in simplest radical form and as decimal approximations rounded to the nearest hundredth if necessary.

5. $(6, 8), (3, 4)$ **5**
6. $(3, 7), (-2, -5)$ **13**
7. $(2, 2), (5, -1)$ **$3\sqrt{2}$ or 4.24**
8. $(2, 7), (10, -4)$ **$\sqrt{185}$ or 13.60**

Find the value of *a* if the points with the given coordinates are the indicated distance apart.

9. $(4, 7), (a, 3); d = 5$ **7 or 1**
10. $(5, a), (6, 1); d = \sqrt{10}$ **−2 or 4**

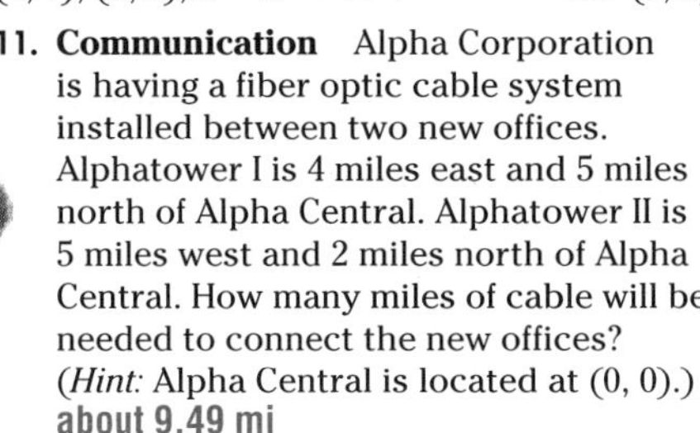

11. **Communication** Alpha Corporation is having a fiber optic cable system installed between two new offices. Alphatower I is 4 miles east and 5 miles north of Alpha Central. Alphatower II is 5 miles west and 2 miles north of Alpha Central. How many miles of cable will be needed to connect the new offices? (*Hint:* Alpha Central is located at (0, 0).) **about 9.49 mi**

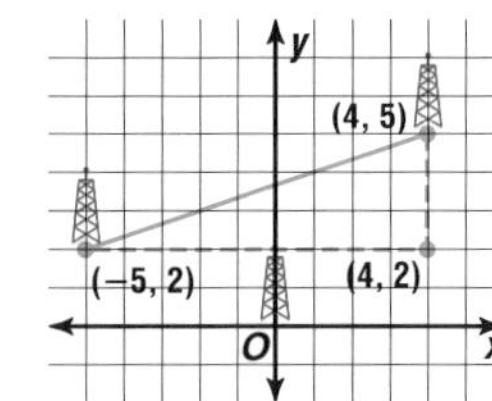

EXERCISES

Practice

Find the distance between each pair of points whose coordinates are given. Express answers in simplest radical form and as decimal approximations rounded to the nearest hundredth if necessary.

12. $(5, -1), (11, 7)$ **10**
13. $(-4, 2), (4, 17)$ **17**
14. $(-3, 8), (5, 4)$ **$4\sqrt{5}$ or 8.94**
15. $(-8, -4), (-3, -8)$ **$\sqrt{41}$ or 6.40**

16. $(9, -2), (3, -6)$ **$2\sqrt{13}$ or 7.21**
17. $(4, 2), \left(6, -\frac{2}{3}\right)$ **$\frac{10}{3}$ or 3.33**
18. $\left(3, \frac{3}{7}\right), \left(4, -\frac{2}{7}\right)$ **$\frac{\sqrt{74}}{7}$ or 1.23**
19. $\left(\frac{4}{5}, -1\right), \left(2, -\frac{1}{2}\right)$ **$\frac{13}{10}$ or 1.30**
20. $(4\sqrt{5}, 7), (6\sqrt{5}, 1)$ **$2\sqrt{14}$ or 7.48**
21. $(5\sqrt{2}, 8), (7\sqrt{2}, 10)$ **$2\sqrt{3}$ or 3.46**

Find the value of *a* if the points with the given coordinates are the indicated distance apart.

24. −2 or −12
25. −10 or 4
27. 8 or 32

22. $(3, -1), (a, 7); d = 10$ **9 or −3**
23. $(-4, a), (4, 2); d = 17$ **17 or −13**
24. $(a, 5), (-7, 3); d = \sqrt{29}$
25. $(6, -3), (-3, a); d = \sqrt{130}$
26. $(10, a), (1, -6); d = \sqrt{145}$ **2 or −14**
27. $(20, -5), (a, 9); d = \sqrt{340}$

Geometry

Determine if the triangles with the following vertices are isosceles triangles.

28. $L(7, -4), M(-1, 2), N(5, -6)$ **yes**
29. $T(1, -8), U(3, 5), V(-1, 7)$ **no**

GLENCOE Technology

Interactive Mathematics Tools Software

This multimedia software provides an interactive lesson that uses the Pythagorean theorem to find the distance of a plane, if given the horizontal and vertical distances from the tracking station. A **Computer Journal** gives students an opportunity to write about what they have learned.

For Windows & Macintosh

Additional Answers

3a. The distance between them is the absolute value of the difference of their *x*-coordinates, $|12 - 3|$ or 9 units.

3b. The distance between them is the absolute value of the difference of their *y*-coordinates, $|7 - (-5)|$ or 12 units.

3 PRACTICE/APPLY

Check for Understanding

Exercises 1–11 are designed to help you assess your students' understanding through reading, writing, speaking, and modeling. You should work through Exercises 1–4 with your students and then monitor their work on Exercises 5–11.

Error Analysis

Make sure that students subtract the coordinates in corresponding order when using the distance formula. For example, some students may substitute as follows.

$$d = \sqrt{(x_2 - x_1)^2 + (y_1 - y_2)^2}, \text{ or}$$

$$d = \sqrt{(x_1 - x_2)^2 + (y_2 - y_1)^2}$$

Assignment Guide

Core: 13–35 odd, 36–41
Enriched: 12–32 even, 33–41

For **Extra Practice,** see p. 786.

The red A, B, and C flags, printed only in the Teacher's Wraparound Edition, indicate the level of difficulty of the exercises.

Study Guide Masters, p. 92

NAME ______________________ DATE ________

Study Guide

Student Edition Pages 737–741

Integration: Geometry
The Distance Formula

The Pythagorean theorem can be used to derive the distance formula shown at the right. The distance formula can then be used to find the distance between any two points in the coordinate plane.

The Distance Formula
The distance between any two points with coordinates (x_1, y_1) and (x_2, y_2) is given by the following formula. $d = \sqrt{(x_2 - x_1)^2 + (y_2 - y_1)^2}$

Example: Find the distance between $(-5, 2)$ and $(4, 5)$.

In the diagram at the right below, the distance to be found is d. Use the distance formula.

$$\begin{aligned} d &= \sqrt{(x_2 - x_1)^2 + (y_2 - y_1)^2} \\ &= \sqrt{[4 - (-5)]^2 + (5 - 2)^2} \\ &= \sqrt{9^2 + 3^2} \\ &= \sqrt{81 + 9} \\ &= \sqrt{90} = 3\sqrt{10} \end{aligned}$$

The distance is $3\sqrt{10}$ units.

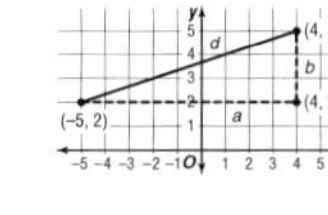

Find the distance between each pair of points whose coordinates are given. Express answers in simplest radical form and as decimal approximations rounded to the nearest hundredth.

1. $(1, 5), (3, 1)$ — $2\sqrt{5}$; 4.47
2. $(0, 0), (6, 8)$ — 10
3. $(-2, -8), (7, -3)$ — $\sqrt{106}$; 10.30
4. $(6, -7), (-2, 8)$ — 17
5. $(1, 5), (-8, 4)$ — $\sqrt{82}$; 9.06
6. $(3, -4), (-4, -4)$ — 7

Find the value of a if the points with the given coordinates are the indicated distance apart.

7. $(-3, -2), (a, -5); d = 5$ — −7 or 1
8. $(2, -1), (a, 3); d = 5$ — −1 or 5
9. $(a, 3), (2, -4); d = \sqrt{74}$ — −3 or 7

Using the Programming Exercises The program given in Exercise 32 is for use with a TI-82 graphing calculator. For other programmable calculators, have students consult their owner's manual for commands similar to those presented here.

Additional Answer

33. The distance between (3, −2) and (−3, 7) is $3\sqrt{13}$ units. The distance between (−3, 7) and (−9, 3) is $2\sqrt{13}$ units. The distance between (3, −2) and (−9, 3) is 13 units. Since $\left(3\sqrt{13}\right)^2 + \left(2\sqrt{13}\right)^2 = 13^2$, the triangle is a right triangle.

fabulous FIRSTS

The patent for Alexander Graham Bell's telephone was granted in 1876. By 1887 there were more than 150,000 phones in the U.S., 26,000 in the United Kingdom, 9000 in France, and 7000 in Russia. By 1979 the U.S. had 153,000,000 telephones.

Practice Masters, p. 92

13-5 NAME________ DATE________

Practice — Student Edition Pages 737–741

Integration: Geometry
The Distance Formula

Find the distance between each pair of points whose coordinates are given. Express answers in simplest radical form and as decimal approximations rounded to the nearest hundredth.

1. (9, 7), (1, 1) 10
2. (5, 2), (8, −2) 5
3. (1, −3), (1, 4) 7
4. (7, 2), (−5, 7) 13
5. (5, 2), (3, 10) $2\sqrt{17}$ or 8.25
6. (−1, −4), (−6, 0) $\sqrt{41}$ or 6.40
7. (−3, −1), (−11, 3) $4\sqrt{5}$ or 8.94
8. (−3, −8), (−7, 2) $2\sqrt{29}$ or 10.77
9. (0, −4), (3, 2) $3\sqrt{5}$ or 6.71
10. (−6, 3), (10, 3) 16
11. $\left(2, -\frac{1}{2}\right), \left(1, \frac{1}{2}\right)$ $\sqrt{2}$ or 1.41
12. $\left(\frac{2}{3}, -1\right), \left(2, \frac{1}{3}\right)$ $\frac{4\sqrt{2}}{3}$ or 1.89
13. $(\sqrt{3}, 3), (2\sqrt{3}, 5)$ $\sqrt{7}$ or 2.65
14. $(2\sqrt{2}, -1), (3\sqrt{2}, 4)$ $3\sqrt{3}$ or 5.20

Find the value of a if the points with the given coordinates are the indicated distance apart.

15. (−2, −5), (a, 7); d = 13 a = 3, −7
16. (8, −2), (5, a); d = 3 a = −2
17. (4, a), (1, 6); d = 5 a = 2, 10
18. (a, 4), (−3, −2); $d = \sqrt{61}$ a = 2, −8

30. $12\sqrt{10}$ or 37.9 units
31. $\sqrt{157} \neq \sqrt{101}$; Trapezoid is not isosceles.

30. Find the perimeter of square *QRST* if two of the vertices are *Q*(6, 7) and *R*(−3, 4).

31. If the diagonals of a trapezoid have the same length, then the trapezoid is isosceles. Find the lengths of the diagonals of the trapezoid with vertices *A*(−2, 2), *B*(10, 6), *C*(9, 8), and *D*(0, 5) to determine if it is isosceles.

Programming

32. The program at the right calculates the distance between a pair of points whose coordinates are given.

Find the distance between each pair of points.

a. *A*(6, −3), *B*(12, 5) 10
b. *M*(−3, 5), *N*(12, −2) ≈16.55
c. *S*(6.8, 9.9), *T*(−5.9, 4.3) ≈13.88

```
PROGRAM: DISTANCE
:ClrDraw
:Input "X1=", Q
:Input "Y1=", R
:Input "X2=", S
:Input "Y2=", T
:Line (Q, R, S, T)
:√((Q−S)²+(R−T)²)→D
:Text (5, 50, "DIST=", D)
```

Critical Thinking

33. Use the distance formula to show that the triangle with vertices at (3, −2), (−3, 7), and (−9, 3) is a right triangle. See margin.

Applications and Problem Solving

34. Art Egyptian artists about 5000 years ago decorated tombs of pharaohs by painting their pictures on the walls. The artists used small sketches on grids as a reference. In Egypt, the main standard of length was the cubit. It was the length of a man's forearm from the elbow to the tip of the outstretched fingers. Use the grid at the right to find the length of a cubit to the nearest inch if each unit represents 3.3 inches. 18 in.

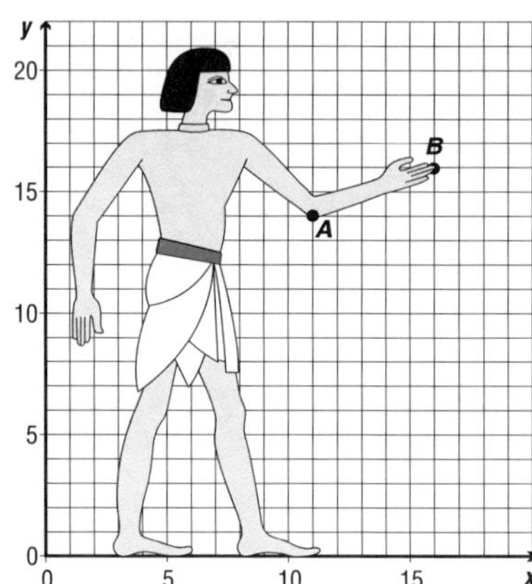

35. 109 miles

35. Telecommunications In order to set long distance rates, phone companies first superimpose an imaginary coordinate grid over the United States. Then the location of each exchange is represented by an ordered pair on the grid. The units of this grid are approximately equal to 0.316 miles. So, a distance of 3 units on the grid equals an actual distance of about 3(0.316) or 0.948 miles. Suppose the exchanges in two cities are at (132, 428) and (254, 105). Find the actual distance between these cities to the nearest mile.

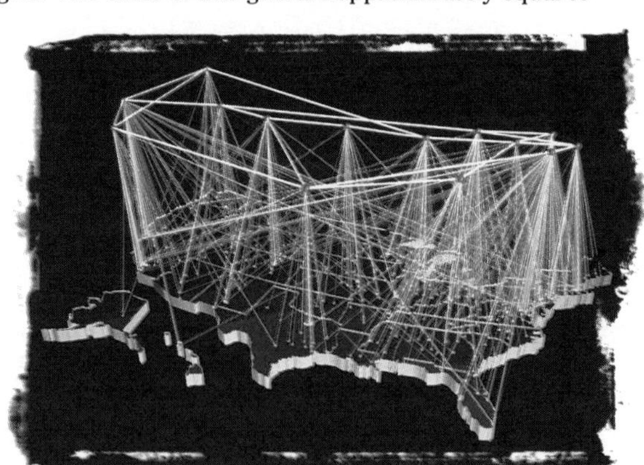

fabulous FIRSTS

Granville T. Woods (1856–1910)

Granville T. Woods patented an improved telephone transmitter in 1884. This was the first of more than twenty patents that he received for his inventions.

Extension

Connections The endpoints of $\overline{AB}$ are *A*(−2, 6) and *B*(4, 2). The endpoints of $\overline{CD}$ are *C*(3, 7) and *D*(1, 4). Show that $\overline{CD}$ is a perpendicular bisector of $\overline{AB}$.

a. Find the midpoint of *AB*. (1, 4); Since the midpoint of $\overline{AB}$ is point *D*, $\overline{CD}$ is a bisector of $\overline{AB}$.

b. Find the slopes of $\overline{AB}$ and $\overline{CD}$. slope of $\overline{AB} = -\frac{2}{3}$; slope of $\overline{CD} = \frac{3}{2}$; Since $\left(-\frac{2}{3}\right)\left(\frac{3}{2}\right) = -1$, $\overline{AB}$ and $\overline{CD}$ are perpendicular. Therefore, $\overline{CD}$ is a perpendicular bisector of $\overline{AB}$.

Mixed Review

36. Physics The time t, in seconds, it takes an object to drop d feet is given by the formula $4t = \sqrt{d}$. Jessica and Lu-Chan each dropped a stone at the same time, but Jessica dropped hers from a spot higher than Lu-Chan's. Lu-Chan's stone hit the ground 1 second before Jessica's. If Jessica's stone dropped 112 feet farther than Lu-Chan's, how long did it take her stone to hit the ground? (Lesson 13–4) **4 seconds**

37. Simplify $\sqrt{\frac{8}{9}}$. (Lesson 13–2) $\frac{2\sqrt{2}}{3}$

38. Find $\frac{4p^3}{p-1} \div \frac{p^2}{p-1}$. (Lesson 12–3) **4p**

39. No solution; see margin for graph.

39. Graph the system of equations. Then determine whether the system has *one* solution, *no* solution, or *infinitely many* solutions. (Lesson 8–1)
$4x - y = 2$
$y - 4x = 4$

40. Consumerism Jackie wanted to buy a new coat that cost $145. If she waited until the coat went on sale for 30% off the original price, how much money did Jackie save? (Lesson 4–5) **$43.50**

41. Air Conditioning The formula for determining the BTU (British Thermal Units) rating of the air conditioner necessary to cool a room is BTU = Area (sq ft) × Exposure Factor × Climate Factor. Use this formula to determine the BTU necessary to cool each of the rooms described in the following chart. (Lesson 2–6)

	Room Dimensions (feet)	Exposure Factor	Climate Factor
a.	22 by 16	North: 20	Buffalo: 1.05
b.	13 by 12	West: 25	Portland: 0.95
c.	17 by 14	East: 25	Topeka: 1.05
d.	26 by 18	South: 30	San Diego: 1.00
e.	23.5 by 15.3	North: 20	Tacoma: 0.95

41a. 7392 BTU
41b. 3705 BTU
41c. 6247.5 BTU
41d. 14,040 BTU
41e. 6831.45 BTU

WORKING ON THE In·ves·ti·ga·tion

Refer to the Investigation on pages 656–657.

A Growing Concern

1 Review your scale drawing. Make sure you have included everything that you think the Sanchez family wanted or will want in the future. Make any changes that you feel need to be made now that the plan is complete.

2 The Sanchez family had asked for a 3-dimensional model of the design. On a piece of cardboard, draw the boundary lines of the Sanchez property. Use modeling clay, construction paper, paint, markers, and whatever else you need to create a 3-dimensional model of your design.

3 Use a flashlight to model the sun's movement during the day. Note the patterns and the length of time certain areas are shaded.

Add the results of your work to your Investigation Folder.

In·ves·ti·ga·tion

Working on the Investigation

The Investigation on pages 656–657 is designed to be a long-term project that is completed over several days or weeks. Encourage students to keep their materials in their Investigation Folder as they work on the Investigation.

4 ASSESS

Closing Activity

Modeling Draw several right triangles on a plain sheet of paper. Place graph paper under the paper and assign coordinates to all three vertices. Find the length of the hypotenuse.

Chapter 13, Quiz C (Lessons 13-4 and 13-5), is available in the *Assessment and Evaluation Masters*, p. 353.

Additional Answer

39.

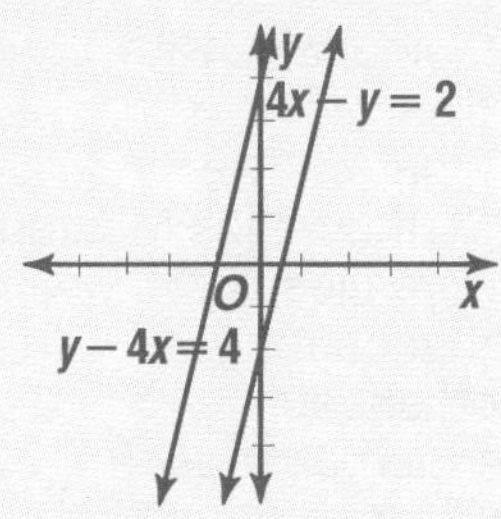

Enrichment Masters, p. 92

NAME ____ DATE ____

13-5 Enrichment

Student Edition Pages 737–741

The Wheel of Theodorus

The Greek mathematicians were intrigued by problems of representing different numbers and expressions using geometric constructions.

Theodorus, a Greek philosopher who lived about 425 B.C., is said to have discovered a way to construct the sequence $\sqrt{1}, \sqrt{2}, \sqrt{3}, \sqrt{4}, \cdots$.

The beginning of his construction is shown. You start with an isosceles right triangle with sides 1 unit long.

Use the figure above. Write each length as a radical expression in simplest form.

1. line segment AO $\sqrt{1}$
2. line segment BO $\sqrt{2}$
3. line segment CO $\sqrt{3}$
4. line segment DO $\sqrt{4}$
5. Describe how each new triangle is added to the figure. Draw a new side of length 1 at right angles to the last hypotenuse. Then draw the new hypotenuse.
6. The length of the hypotenuse of the first triangle is $\sqrt{2}$. For the second triangle, the length is $\sqrt{3}$. Write an expression for the length of the hypotenuse of the nth triangle. $\sqrt{n+1}$
7. Show that the method of construction will always produce the next number in the sequence. (*Hint:* Find an expression for the hypotenuse of the $(n + 1)$th triangle.) $\sqrt{(\sqrt{n})^2 + (1)^2} = \sqrt{n+1}$
8. In the space below, construct a Wheel of Theodorus. Start with a line segment 1 centimeter long. When does the Wheel start to overlap? after length $\sqrt{18}$

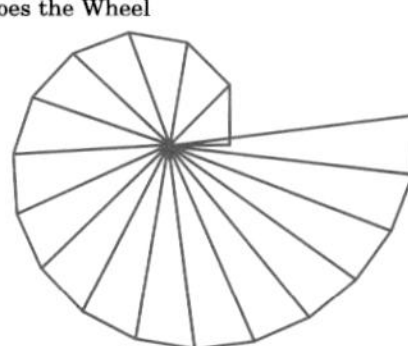

13-6A LESSON NOTES

NCTM Standards: 1–5

Objective
Use algebra tiles as a model for completing the square.

Recommended Time
Demonstration and discussion: 15 minutes; Exercises: 30 minutes

Instructional Resources
For each student or group of students
Student Manipulative Kit
- algebra tiles
- equation mat

Modeling Mathematics Masters
- p. 1 (algebra tiles)
- p. 14 (equation mat)
- p. 35 (worksheet)

For teacher demonstration
Algebra and Geometry Overhead Manipulative Resources

1 FOCUS

Motivating the Lesson
This modeling activity makes it clear that the phrase "completing the square" is literally true. The result of each modeling activity is a geometric square.

2 TEACH

Teaching Tip Point out that when completing the square, you can only add 1-tiles to each side of the mat.

3 PRACTICE/APPLY

Assignment Guide

Core: 1–8
Enriched: 1–8

4 ASSESS

Observing students working in cooperative groups is an excellent method of assessment.

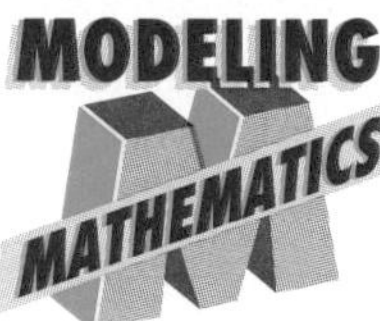

A Preview of Lesson 13–6

13–6A Completing the Square

Materials: algebra tiles, equation mat

One way to solve a quadratic equation is by **completing the square.** To use this method, the quadratic expression on one side of the equation must be a perfect square. You can use algebra tiles as a model for completing the square.

Activity **Use algebra tiles to complete the square for the equation $x^2 + 4x + 1 = 0$.**

Step 1 Subtract 1 from each side of the equation.

$$x^2 + 4x + 1 - 1 = 0 - 1$$
$$x^2 + 4x = -1$$

Then model the equation $x^2 + 4x = -1$.

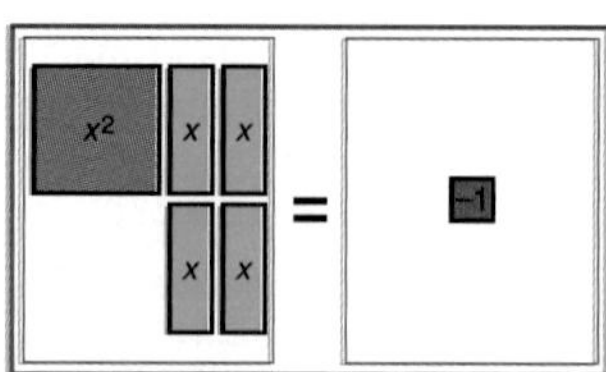

Step 2 Begin to arrange the x^2-tile and the x-tiles into a square.

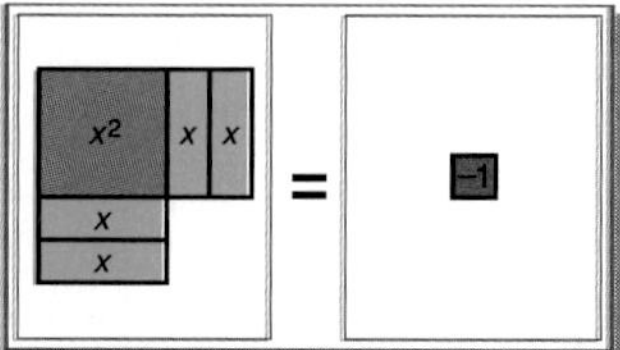

Step 3 In order to complete the square, you need to add 4 1-tiles to the left side of the mat. Since you are modeling an equation, add 4 1-tiles to the right side of the mat.

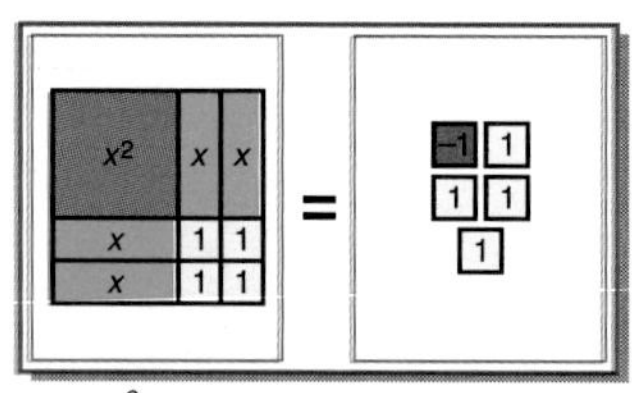

$x^2 + 4x + 4 = -1 + 4$

Step 4 Remove the zero pair on the right side of the mat. You have completed the square, and the equation is $x^2 + 4x + 4 = 3$ or $(x + 2)^2 = 3$.

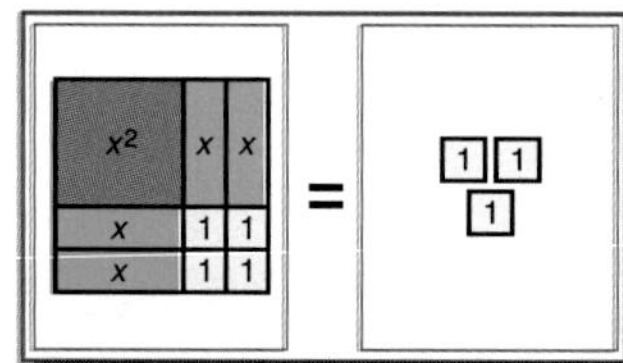

1. $(x + 2)^2 = 1$ **2.** $(x - 3)^2 = 4$ **3.** $(x + 2)^2 = 5$

Model **Use algebra tiles to complete the square for each equation.**

1. $x^2 + 4x + 3 = 0$
2. $x^2 - 6x + 5 = 0$
3. $x^2 + 4x - 1 = 0$
4. $x^2 - 2x + 5 = 3$ **$(x - 1)^2 = -1$**
5. $x^2 - 4x + 7 = 8$ **$(x - 2)^2 = 5$**
6. $0 = x^2 + 8x - 3$ **$(x + 4)^2 = 19$**

Draw

7. In the equations shown above, the coefficient of x was always an even number. Sometimes you have an equation like $x^2 + 3x - 1 = 0$ in which the coefficient of x is an odd number. Complete the square by making a drawing. **See margin.**

Write

8. Write a paragraph explaining how you could complete the square with models without first rewriting the equation. Include a drawing. **See margin.**

Additional Answers

7. $(x + 1.5)^2 = 3.25$

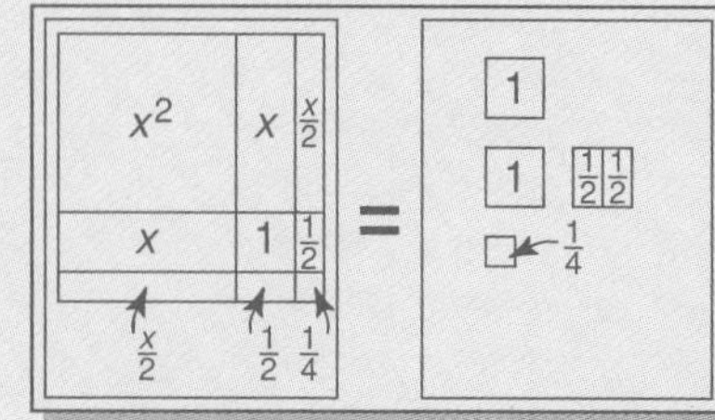

8. **Sample answer: You could model the expression on one side of the mat. Then add the appropriate 1-tiles needed to complete the square to each side of the mat.**

13-6 Solving Quadratic Equations by Completing the Square

What YOU'LL LEARN

- To solve quadratic equations by completing the square, and
- to solve problems by identifying subgoals.

Why IT'S IMPORTANT

You can solve quadratic equations to solve problems involving geography and construction.

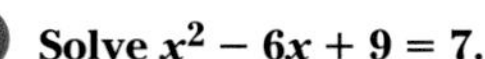

CONNECTION

Geography

The greatest flow of any river in the world is that of the Amazon, which discharges an average of 4.2 million cubic feet of water per second into the Atlantic Ocean. The rate at which water flows in a river varies depending on the distance from the shore.

Amazon River

Suppose the Castillos have a home on the bank of a river that is 40 yards wide. The rate of the river is given by the equation $y = -0.01x^2 + 0.4x$. Mr. and Mrs. Castillo do not want their children to wade in the water if the current is greater than 3 miles per hour. Find how many yards from shore the water flows at 3 miles per hour. *This problem will be solved in Example 4.*

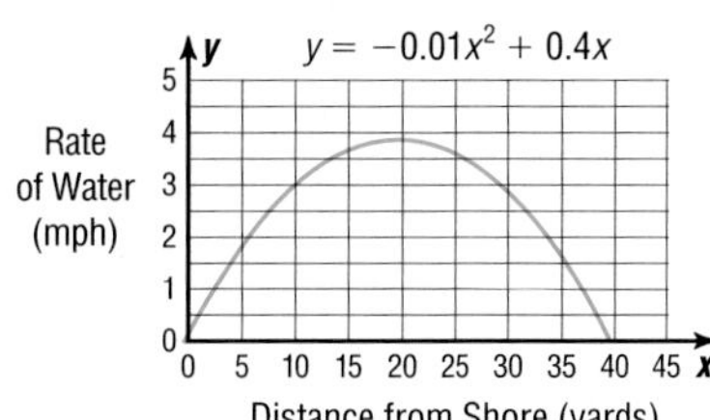

You can solve some quadratic equations by taking the square root of each side.

Example 1 Solve $x^2 - 6x + 9 = 7$.

$x^2 - 6x + 9 = 7$

$(x - 3)^2 = 7$ *$x^2 - 6x + 9$ is a perfect square trinomial.*

$\sqrt{(x - 3)^2} = \sqrt{7}$ *Take the square root of each side.*

$|x - 3| = \sqrt{7}$

$x - 3 = \pm\sqrt{7}$ *Why is this the case?*

$x = 3 \pm \sqrt{7}$ *Add 3 to each side.*

The solution set is $\{3 + \sqrt{7}, 3 - \sqrt{7}\}$.

To use the method shown in Example 1, the quadratic expression on one side of the equation must be a perfect square. However, few quadratic expressions are perfect squares. To make any quadratic expression a perfect square, a method called **completing the square** may be used.

Consider the pattern for squaring a binomial such as $x + 5$.

$$(x + 5)^2 = x^2 + 2(5)x + 5^2$$
$$= x^2 + 10x + 25$$
$$\left(\frac{10}{2}\right)^2 \rightarrow 5^2$$ *Notice that one half of 10 is 5 and 5^2 is 25.*

Alternative Learning Styles

Kinesthetic Ask students to list important ideas, terminology, and problems that have been covered in the chapter. Have them write each item on the front of an index card. On the back, have them explain or define the term in their own words using complete sentences. They can use these cards to study alone or in pairs.

13-6 LESSON NOTES

NCTM Standards: 1–5

Instructional Resources

- Study Guide Master 13-6
- Practice Master 13-6
- Enrichment Master 13-6
- Assessment and Evaluation Masters, p. 353
- Real-World Applications, 32

Transparency 13-6A contains the 5-Minute Check for this lesson; **Transparency 13-6B** contains a teaching aid for this lesson.

Recommended Pacing	
Standard Pacing	Day 9 of 11
Honors Pacing	Day 6 of 7
Block Scheduling*	Days 18 & 19 of 22

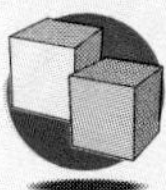

*For more information on pacing and possible lesson plans, refer to the *Block Scheduling Booklet* and *Algebra 1 in Two Years.*

1 FOCUS

5-Minute Check
(over Lesson 13-5)

Find the distance between each pair of points whose coordinates are given.

1. (6, 0)(12, 8) 10
2. (−6, −4)(2, 15) $5\sqrt{17}$
3. (0, −11)(−6, −15) $2\sqrt{13}$

Determine if the triangles with the following vertices are isosceles triangles.

4. $A(0, 7)$, $B(8, 5)$, $C(2, -2)$ yes
5. $X(1, -3)$, $Y(3, 10)$, $Z(-1, 12)$ no

Motivating the Lesson

Questioning Introduce the lesson with a review of factoring. Ask students to factor each of the following.

1. $x^2 + 4x + 4$ $(x + 2)^2$
2. $x^2 - 6x + 9$ $(x - 3^2)$
3. $n^2 + 10n + 25$ $(n + 5)^2$
4. $y^2 - 2y + 1$ $(y - 1)^2$

2 TEACH

Teaching Tip While students learned other ways of solving quadratic equations in Chapter 11, they needed to learn to work with radicals before being introduced to completing the square.

In-Class Examples

For Example 1
Solve $x^2 + 6x + 9 = 5$.
$\{-3 + \sqrt{5}, -3 - \sqrt{5}\}$

For Example 2
Find the value of *c* that makes each trinomial a perfect square.

a. $x^2 - 20x + c$ 100
b. $y^2 + y + c$ $\frac{1}{4}$
c. $m^2 - 3m + c$ $\frac{9}{4}$

For Example 3
Solve each equation by completing the square.

a. $x^2 - 6x + 8 = 0$ $\{2, 4\}$
b. $x^2 - 6x + 5 = 0$ $\{1, 5\}$
c. $m^2 + 4m - 9 = 0$
$\{-2 + \sqrt{13}, -2 - \sqrt{13}\}$
d. $x^2 - 6x + 7 = 0$
$\{3 + \sqrt{2}, 3 - \sqrt{2}\}$

For Example 4
How far from the shore is the river current 3.5 miles per hour?
20 ± 7.07

Teaching Tip A calculator may be used to aid in checking the roots of an equation.

To complete the square for a quadratic expression of the form $x^2 + bx$, you can follow the steps below.

Step 1 Find $\frac{1}{2}$ of b, the coefficient of x.

Step 2 Square the result of Step 1.

Step 3 Add the result of Step 2 to $x^2 + bx$, the original expression.

Example 2 **Find the value of *c* that makes each trinomial a perfect square.**

a. $x^2 + 20x + c$

Step 1	Find $\frac{1}{2}$ of 20.	$\frac{20}{2} = 10$
Step 2	Square the result of Step 1.	$10^2 = 100$
Step 3	Add the result of Step 2 to $x^2 + 20x$.	$x^2 + 20x + 100$

Thus, $c = 100$. Notice that $x^2 + 20x + 100 = (x + 10)^2$.

b. $x^2 - 15x + c$

Step 1	Find $\frac{1}{2}$ of -15.	$\frac{-15}{2} = -7.5$
Step 2	Square the result of Step 1.	$(-7.5)^2 = 56.25$
Step 3	Add the result of Step 2 to $x^2 - 15x$.	$x^2 - 15x + 56.25$

Thus, $c = 56.25$. Notice that $x^2 - 15x + 56.25 = (x - 7.5)^2$.

Example 3 **Solve $x^2 + 8x - 18 = 0$ by completing the square.**

$x^2 + 8x - 18 = 0$ *Notice that $x^2 + 8x - 18$ is not a perfect square.*
$x^2 + 8x = 18$ *Add 18 to each side. Then complete the square.*
$x^2 + 8x + 16 = 18 + 16$ *Since $\left(\frac{8}{2}\right)^2 = 16$, add 16 to each side.*
$(x + 4)^2 = 34$ *Factor $x^2 + 8x + 16$.*
$x + 4 = \pm\sqrt{34}$ *Take the square root of each side.*
$x = -4 \pm \sqrt{34}$ *Subtract 4 from each side.*
$x = -4 + \sqrt{34}$ or $x = -4 - \sqrt{34}$

The solution set is $\{-4 + \sqrt{34}, -4 - \sqrt{34}\}$. *Check this result.*

The method for solving quadratic equations cannot be used unless the coefficient of the first term is 1. To solve a quadratic equation in which the leading coefficient is not 1, divide each term by the coefficient.

Example 4 **Refer to the application at the beginning of the lesson. How far from shore will the rate of the current be 3 miles per hour?**

Geography

$y = -0.01x^2 + 0.4x$
$3 = -0.01x^2 + 0.4x$ *Replace y with 3 since the rate is 3 mph.*
$-300 = x^2 - 40x$ *Divide each side by −0.01.*
$-300 + 400 = x^2 - 40x + 400$ *Complete the square. $\left(\frac{-40}{2}\right)^2 = 400$*
$100 = (x - 20)^2$ *Factor $x^2 - 40x + 400$.*
$\pm 10 = x - 20$ *Take the square root of each side.*
$20 \pm 10 = x$ *Add 20 to each side.*

Cooperative Learning

Pairs Check Have students work in pairs. One student will work an example of completing the square and then give it to the other student to check. The student doing the checking must either say what is wrong or, if correct, explain each step. Then switch roles. For more information on the pairs check strategy, see *Cooperative Learning in the Mathematics Classroom,* one of the titles in the Glencoe Mathematics Professional Series, pages 12–13.

The solutions are 20 + 10 or 30 and 20 − 10 or 10. Thus, the children should not be allowed to wade more than 10 yards from the shore; between 10 and 30 yards from the shore the water is flowing too fast.

Check this result with a graphing calculator by graphing the equation $y = -0.01x^2 + 0.4x$, and then estimating.

Sometimes finding the solution to a problem requires several steps. An important strategy for solving such problems is to **identify subgoals.** This strategy involves taking steps that will either produce part of the solution or make the problem easier to solve.

Example 5

PROBLEM SOLVING
Identify Subgoals

A square is extended in one direction by 14 centimeters. The resulting rectangle has an area of 51 square centimeters. What is the length of each side of the original square?

Finding an equation to represent this problem will be easier if you develop the equation in steps rather than trying to write one directly from the given information.

Step 1 First, let x be the length of each side of the original square. Then the area of the square is x^2 cm.

Step 2 The extension is 14 cm long and x cm wide, so its area is $14x$ cm^2.

Step 3 Add the measures of the areas and set them equal to 51.
$x^2 + 14x = 51$

Step 4 Complete the square to find the value of x.

$$x^2 + 14x = 51$$
$$x^2 + 14x + 49 = 51 + 49 \quad \textit{Since } \left(\tfrac{14}{2}\right)^2 = 49\textit{, add 49 to each side.}$$
$$(x + 7)^2 = 100 \quad \textit{Factor } x^2 + 14x + 49.$$
$$x + 7 = \pm 10 \quad \textit{Find the square root of each side.}$$
$$x = -7 \pm 10$$
$$x = -7 + 10 \qquad x = -7 - 10$$
$$= 3 \qquad = -17$$

Since lengths cannot be negative, the length of each side of the original square is 3 centimeters. *Check this result.*

CHECK FOR UNDERSTANDING

Communicating Mathematics

Study the lesson. Then complete the following.

1. **Explain** which method for solving a quadratic equation always produces an exact solution, graphing or completing the square. **completing the square**
2. **Explain** the three steps used to complete the square for the expression $x^2 + bx$.
3. **Write** a quadratic equation that has no real solutions. After completing the square, explain how you could tell it had no real solutions.

2–3. See margin.

In-Class Example

For Example 5
A square is extended in one direction 15 centimeters. The resulting rectangle has an area of 250 square centimeters. What is the length of each side of the original square? **10 cm**

3 PRACTICE/APPLY

Check for Understanding

Exercises 1–13 are designed to help you assess your students' understanding through reading, writing, speaking, and modeling. You should work through Exercises 1–4 with your students and then monitor their work on Exercises 5–13.

Error Analysis
Students need to first look at the coefficient of the first term to be sure it is 1. Make sure they divide *all* terms by that coefficient if it is not 1.
Example: $3x^2 - 7x = 3$

$x^2 - \frac{7x}{3} = 3$ no $\qquad x^2 - \frac{7x}{3} = 1$ yes

Study Guide Masters, p. 93

13-6 NAME ____ DATE ____
Study Guide — Student Edition Pages 743–747

Solving Quadratic Equations by Completing the Square

Equations containing quadratic expressions that are not perfect squares can be solved by a method called *completing the square.* To complete the square for an expression of the form $x^2 + bx$, follow the steps listed below.

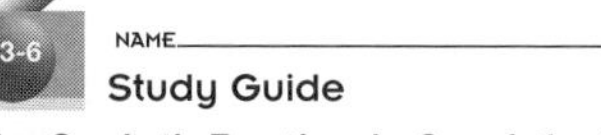

Completing the Square for an Expression of the Form $x^2 + bx$
Step 1 Find one-half of b, the coefficient of x.
Step 2 Square the result of **Step 1**.
Step 3 Add the result of **Step 2** to $x^2 + bx$.

Example: Solve $x^2 + 2x - 15 = 0$ by completing the square.

$x^2 + 2x - 15 = 0$	$x^2 + 2x - 15$ is not a perfect square.
$x^2 + 2x = 15$	Add 15 to each side.
$x^2 + 2x + 1 = 15 + 1$	Complete the square for $x^2 + 2x$.
$(x + 1)^2 = 16$	Factor $x^2 + 2x + 1$.
$x + 1 = \pm 4$	Find the square root of each side.
$x = \pm 4 - 1$	Subtract 1 from each side.
$x = -4 - 1$ or $x = 4 - 1$	
$x = -5$ $\quad x = 3$	

The solution set is {−5, 3}.

Solve each equation by completing the square. Leave irrational roots in simplest radical form.

1. $y^2 + 6y + 5 = 0$ **−1, −5**
2. $r^2 + 4r - 21 = 0$ **−7, 3**
3. $y^2 + 8y + 15 = 0$ **−5, −3**
4. $x^2 - 12x + 35 = 0$ **5, 7**
5. $3y^2 - 2y - 1 = 0$ **$-\frac{1}{3}$, 1**
6. $d^2 + 3d - 18 = 0$ **−6, 3**
7. $n^2 + 16n + 14 = 0$ **$-8 \pm 5\sqrt{2}$**
8. $t^2 - 2t = 15$ **−3, 5**
9. $-2x^2 + 10x + 22 = 0$ **$\frac{5 \pm \sqrt{69}}{2}$**

Reteaching

Using Discussion Discuss with students how to take half of a fractional or decimal coefficient. Does it change the method of completing the square? How do you square a fraction or a decimal?

Additional Answers

2. Step 1: Find one-half of b.
   Step 2: Square the result of Step 1.
   Step 3: Add the result of Step 2 to $x^2 + bx$.
3. Sample answer: $x^2 + 4x + 12 = 0$, $(x + 2)^2 = -8$; Since the number on the right side is negative, there are no real solutions.

Assignment Guide

Core: 15–41 odd, 42, 43, 45–52
Enriched: 14–40 even, 42–52

For **Extra Practice,** see p. 786.

The red A, B, and C flags, printed only in the Teacher's Wraparound Edition, indicate the level of difficulty of the exercises.

Additional Answer

42b.

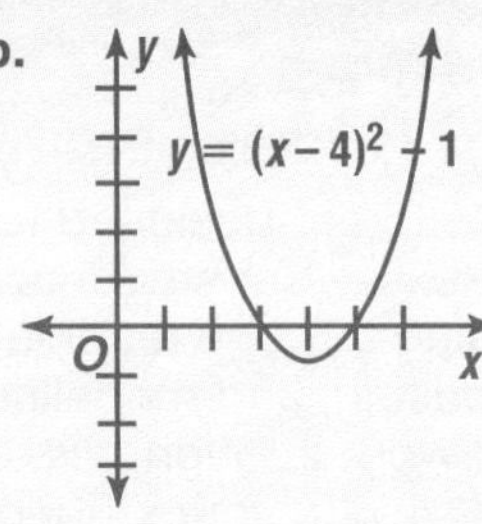

Practice Masters, p. 93

13-6 NAME________________ DATE________
Practice — Student Edition Pages 743–748

Solving Quadratic Equations by Completing the Square

Find the value of c that makes each trinomial a perfect square.

1. $g^2 + 6g + c$ 9
2. $m^2 - 5m + c$ $\frac{25}{4}$
3. $y^2 + 4y + c$ 4

Solve each equation by completing the square. Leave irrational roots in simplest radical form.

4. $x^2 + 4x - 12 = 0$ 2, −6
5. $y^2 - 6x + 4 = 0$ $3 + \sqrt{5}, 3 - \sqrt{5}$
6. $e^2 - 8e + 15 = 0$ 5, 3
7. $2p^2 - 3p = 0$ $\frac{3}{2}, 0$
8. $z^2 - 4z + 1 = 0$ $2 + \sqrt{3}, 2 - \sqrt{3}$
9. $r^2 + 8r + 10 = 0$ $4 + \sqrt{6}, 4 - \sqrt{6}$
10. $\frac{1}{2}x^2 - \frac{1}{2}x - 10 = 0$ 5, −4
11. $0.9a^2 + 5.4a - 4 = 0$ $\frac{2}{3}, -\frac{20}{3}$

4. Use algebra tiles to complete the square for the equation $x^2 + 6x + 2 = 0$. $(x + 3)^2 = 7$

Guided Practice

Find the value of c that makes each trinomial a perfect square.

5. $x^2 + 16x + c$ **64**
6. $a^2 - 7a + c$ $\frac{49}{4}$

7. −1, −3
10. $\frac{5}{2}$ 12. $\frac{5}{2}$, −4
16. $\frac{25}{4}$ 17. $\frac{121}{4}$

Solve each equation by completing the square. Leave irrational roots in simplest radical form.

7. $x^2 + 4x + 3 = 0$
8. $d^2 - 8d + 7 = 0$ **1, 7**
9. $a^2 - 4a = 21$ **7, −3**
10. $4x^2 - 20x + 25 = 0$
11. $r^2 - 4r = 2$ $2 \pm \sqrt{6}$
12. $2t^2 + 3t - 20 = 0$

13. **Sports** The dimensions of a regulation high school basketball court are 50 feet by 84 feet. The builders of an indoor sports arena can afford to construct an arena of 5600 square feet. They want it to have a regulation basketball court and walkways the same width around the court. Find the dimensions of the walkway. **4.9 ft**

EXERCISES

Practice

23. $\frac{7}{3}$
24. $12 \pm 3\sqrt{15}$
26. $4 \pm 2\sqrt{5}$
27. $5 \pm 4\sqrt{3}$
28. 3, $\frac{1}{2}$

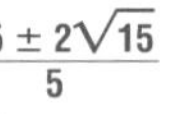

29. $\frac{-5 \pm 2\sqrt{15}}{5}$

30. $-\frac{3}{2}$, 4 31. $\frac{2}{3}$, −1
32. $-0.125 \pm \sqrt{0.515625}$
33. $\frac{7 \pm \sqrt{85}}{6}$
34. $\frac{5 \pm \sqrt{17}}{4}$
39. $2 \pm \sqrt{4 - c}$

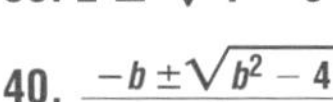

40. $\frac{-b \pm \sqrt{b^2 - 4c}}{2}$
41. $b(-2 \pm \sqrt{3})$

Find the value of c that makes each trinomial a perfect square.

A

14. $x^2 - 6x + c$ **9**
15. $b^2 + 8b + c$ **16**
16. $m^2 - 5m + c$
17. $a^2 + 11a + c$
18. $9t^2 - 18t + c$ **9**
19. $\frac{1}{2}x^2 - 4x + c$ **8**

Solve each equation by completing the square. Leave irrational roots in simplest radical form. 20. −3, −4 21. 4, 1 22. 1, −15

20. $x^2 + 7x + 10 = -2$
21. $a^2 - 5a + 2 = -2$
22. $r^2 + 14r - 9 = 6$
23. $9b^2 - 42b + 49 = 0$
24. $x^2 - 24x + 9 = 0$
25. $t^2 + 4 = 6t$ $3 \pm \sqrt{5}$

B

26. $m^2 - 8m = 4$
27. $p^2 - 10p = 23$
28. $x^2 - \frac{7}{2}x + \frac{3}{2} = 0$
29. $5x^2 + 10x - 7 = 0$
30. $\frac{1}{2}d^2 - \frac{5}{4}d - 3 = 0$
31. $0.3t^2 + 0.1t = 0.2$
32. $b^2 + 0.25b = 0.5$
33. $3p^2 - 7p - 3 = 0$
34. $2r^2 - 5r + 8 = 7$

Find the value of c that makes each trinomial a perfect square.

35. $x^2 + cx + 81$ **18, −18**
36. $4x^2 + cx + 225$ **60, −60**

C

37. $cx^2 + 30x + 75$ **3**
38. $cx^2 - 18x + 36$ $\frac{9}{4}$

Solve each equation by completing the square. Leave irrational roots in simplest radical form.

39. $x^2 - 4x + c = 0$
40. $x^2 + bx + c = 0$
41. $x^2 + 4bx + b^2 = 0$

Critical Thinking

42. **Geometry** Consider the quadratic function $y = x^2 - 8x + 15$.
   a. Write the function in the form $y = (x - h)^2 + k$. $y = (x - 4)^2 - 1$
   b. Graph the function. **See margin**
   c. What is the relationship of the point (h, k) to the graph? **It is the vertex of the graph.**

Tech Prep

Sportscaster Students who are interested in sportscasting may wish to do further research on the use of mathematics in that occupation, as mentioned in Exercise 13. For more information in tech prep, see the *Teacher's Handbook*.

Applications and Problem Solving

43. **Identify Subgoals** Two trains left the same station at the same time. One was traveling due north at a speed that was 10 mph faster than the other train, which was traveling due east. After one hour, the trains were 71 miles apart. How fast, to the nearest mile per hour, was each train traveling? (*Hint:* Use the Pythagorean theorem.) **45 mph, 55 mph**

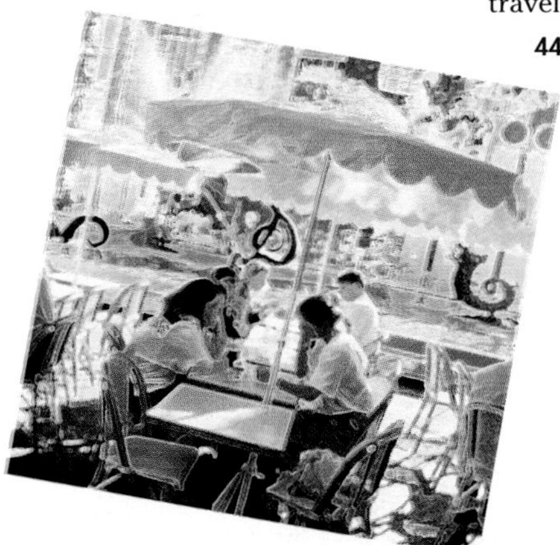

44. **Construction** Arlando's Restaurant wants to add an outdoor café on the side of the restaurant. They are having a special water fountain shipped in that is 10 by 15 feet, and Arlando can afford to buy 1800 square feet of space next to his restaurant. He wishes to have a dining area around the fountain, of equal width all around.

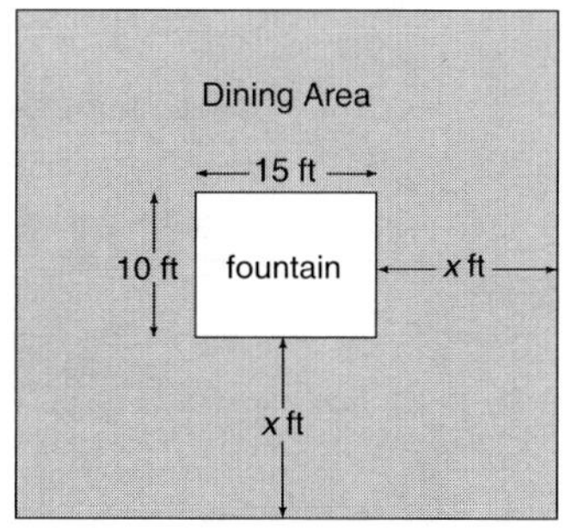

a. Write an equation for x, the width of the dining area around the fountain and solve it. **$(15 + 2x)(10 + 2x) = 1800$; 15**

b. What should be the length and width of the piece of land Arlando buys for the café? **45 ft by 40 ft**

Mixed Review

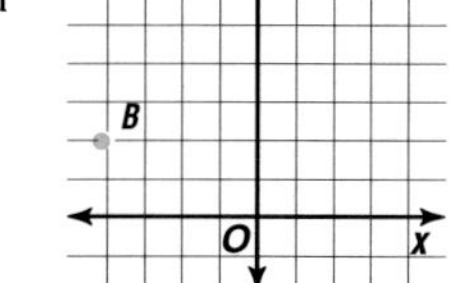

45. **Geometry** Find the distance between points A and B graphed at the right. Round your answer to the nearest hundredth. (Lesson 13–5) **8.06**

46. **Geometry** The measures of the sides of a triangle are 5, 7, and 9. Determine whether this triangle is a right triangle. (Lesson 13–1) **no**

47. **Astronomy** Earth, Jupiter, and Saturn revolve around the Sun about once every 1, 12, and 30 years, respectively. The last time Jupiter and Saturn appeared close to each other in Earth's night sky was in 1982. When will this happen again? (Lesson 12–6) **2042**

48. Find $(5y - 3)(y + 2)$. (Lesson 9–7) **$5y^2 + 7y - 6$**

49. Use elimination to solve the system of equations. (Lesson 8–4) **(2, 2)**

$5y - 4x = 2$

$2y + x = 6$

50. If the graph of $P(x, y)$ satisfies the given conditions, name the quadrant in which point P is located. (Lesson 5–1)

a. $x > 0, y < 0$ **IV** b. $x < 0, y = 3$ **II** c. $x = -1, y < 0$ **III**

51. **Food** The chart at the right compares the Calorie content of regular tacos and the light tacos at Taco Bell®. (Lesson 3–7)

a. Find the mean number of Calories for the regular tacos. **224.6**

b. Find the mean number of Calories for the light tacos. **172**

Item	Calories in Regular Tacos	Calories in Light Tacos
Taco	180	140
Soft Taco	220	180
Taco Supreme™	230	160
Soft Taco Supreme®	270	200
Chicken Soft Taco	223	180

52. Evaluate $|a| + |4b|$ if $a = -3$ and $b = -6$. (Lesson 2–3) **27**

4 ASSESS

Closing Activity

Speaking Have students explain how to use the completing the square method to solve $x^2 - 36 = 0$.

Chapter 13, Quiz D (Lessons 13-5 and 13-6), is available in the *Assessment and Evaluation Masters*, p. 353.

Enrichment Masters, p. 93

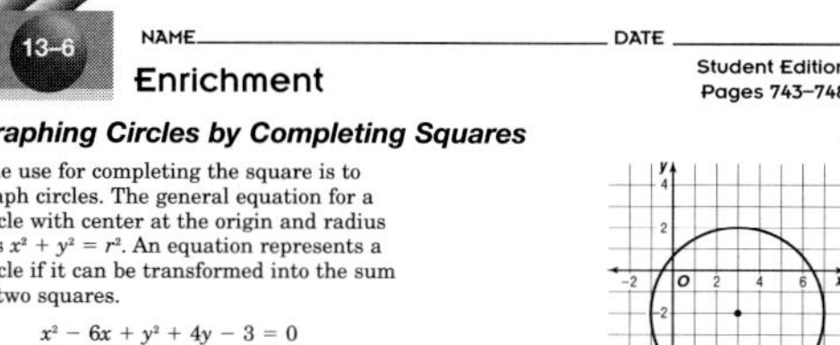

13–6 NAME__________ DATE ______

Enrichment

Student Edition Pages 743–748

Graphing Circles by Completing Squares

One use for completing the square is to graph circles. The general equation for a circle with center at the origin and radius r is $x^2 + y^2 = r^2$. An equation represents a circle if it can be transformed into the sum of two squares.

$$x^2 - 6x + y^2 + 4y - 3 = 0$$
$$(x^2 - 6x + \quad) + (y^2 + 4y + \quad) = 0$$
$$(x^2 - 6x + 9) + (y^2 + 4y + 4) = 3 + 9 + 4$$
$$(x - 3)^2 + (y + 2)^2 = 4^2$$

Notice that the center of the circle is at the point $(3, -2)$.

Transform each equation into the sum of two squares. Then graph the circle represented by the equation. Use the coordinate plane provided at the bottom of the page.

1. $x^2 - 14x + y^2 + 6y + 49 = 0$
$(x - 7)^2 + (y + 3)^2 = 3^2$
2. $x^2 + y^2 - 8y - 9 = 0$
$x^2 + (y - 4)^2 = 5^2$
3. $x^2 + 10x + y^2 + 21 = 0$
$(x + 5)^2 + y^2 = 2^2$
4. $x^2 + y^2 + 10y + 16 = 0$
$x^2 + (y + 5)^2 = 3^2$
5. $x^2 - 30x + y^2 + 209 = 0$
$(x - 15)^2 + y^2 = 4^2$
6. $x^2 - 18x + y^2 - 12y + 116 = 0$
$(x^2 - 9)^2 + (y - 6)^2 = 1^2$
7. $x^2 + 30x + y^2 - 4y + 193 = 0$
$(x + 15)^2 + (y - 2)^2 = 6^2$
8. $x^2 + 38x + y^2 - 12y + 393 = 0$
$(x + 19)^2 + (y - 6)^2 = 2^2$

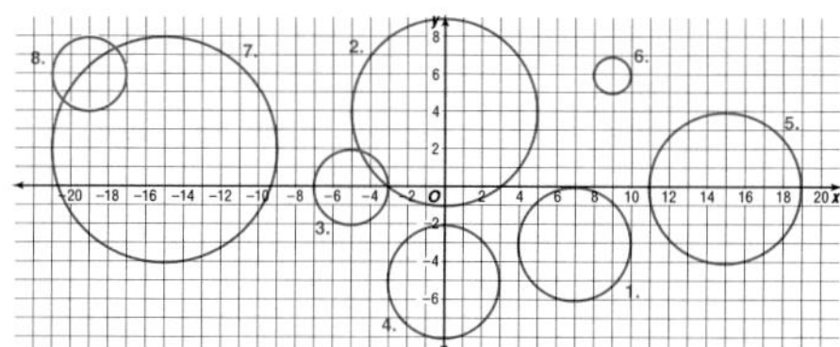

Extension

Problem Solving Solve the equation for x in terms of a by completing the square. State the restrictions on a so that the equation will have real-number roots.

$ax^2 - 3x + 4 = 0$

$x = \dfrac{3 \pm \sqrt{9 - 16a}}{2a}$; $a \le \dfrac{9}{16}$, $a \neq 0$

CLOSING THE

In•ves•ti•ga•tion

A Growing Concern

Refer to the Investigation on pages 656–657.

When landscape businesses prepare a proposal for a client, they prepare all the specifications for the bid and include a sketch of the proposal and a photo display of other work they have done for individuals or companies. They also include a list of references that they give to prospective clients. These clients can then call these people and see other work the company has done in order to verify their credentials. What else might be good to include in a sales presentation to a prospective client?

Analyze

You have made a scale drawing of your design and organized your computations in various ways. It is now time to analyze your design and verify your conclusions.

PORTFOLIO ASSESSMENT

You may want to keep your work on this Investigation in your portfolio.

1 Look over your scale drawing of the landscape design and verify the dimensions and placement of each item.

2 Refer to your data on the shade patterns in the yard. Review your knowledge of the plants and trees that you suggested and verify the amount of sun and/or shade that they will get or that they need. Make any necessary changes.

3 Organize a detailed financial bid for the Sanchez family's backyard design. The bid must include the cost of supplies, materials, labor costs, and a profit margin for each phase of the project.

Present

Select members of your class to represent the Sanchez family. Present your proposal, scale drawing, model, and written report as you would for the real Sanchez family.

4 Begin the presentation by stating the requirements that the Sanchez family had specified that they wanted or needed for their backyard design.

5 Explain the process you used to develop your plan. Justify your design and fully explain your bid costs.

6 Prepare a detailed plan for the Sanchez family. Submit in this plan one scale drawing, the model, a written description of the backyard design, and a detailed financial bid.

7 Make sure that each aspect of the design is justified in writing. Include any options that may negate the terms of the proposal, such as a change in the types of plants used.

8 Summarize your plan with a sales statement that details why your plan is superior.

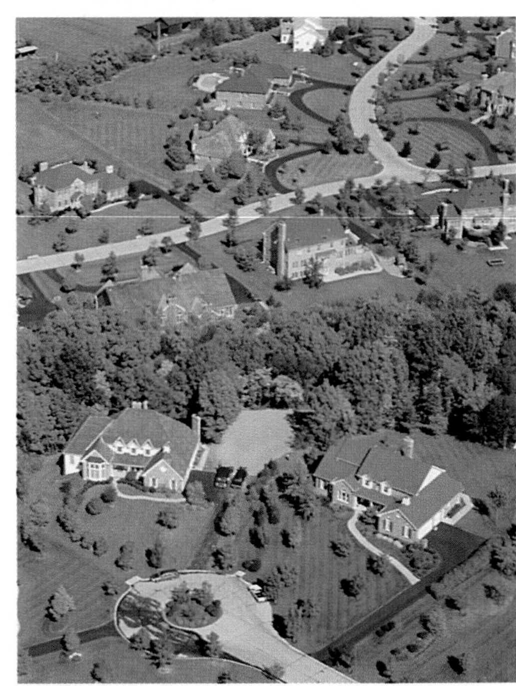

In•ves•ti•ga•tion

TEACHER NOTES

Closing the Investigation

This activity provides students an opportunity to bring their work on the Investigation to a close. For each Investigation, students should present their findings to the class. Here are some ways students can display their work.

- Conduct and report on an interview or survey.
- Write a letter, proposal, or report.
- Write an article for the school or local paper.
- Make a display, including graphs and/or charts.
- Plan an activity.

Assessment

To assess students' understanding of the concepts and topics explored in the Investigation and its follow-up activities, you may wish to examine students' Investigation Folders.

The scoring guide provided in the *Investigations and Projects Masters,* p. 23, provides a means for you to score students' work on the Investigation.

Investigations and Projects Masters, p. 23

Scoring Guide
Chapters 12 and 13
Investigation

Level	Specific Criteria
3 Superior	• Shows thorough understanding of the concepts of *landscaping plans, profit margins, percents, scale drawings, models, calculating dimensions,* and *using a price chart to calculate costs.* • Uses appropriate strategies to solve problems. • Computations are correct. • Written explanations are exemplary. • Drawings, model, report, and presentation are appropriate and sensible. • Goes beyond the requirements of the investigation.
2 Satisfactory, with Minor Flaws	• Shows understanding of the concepts of *landscaping plans, profit margins, percents, scale drawings, models, calculating dimensions,* and *using a price chart to calculate costs.* • Uses appropriate strategies to solve problems. • Computations are mostly correct. • Written explanations are effective. • Drawings, model, report, and presentation are appropriate and sensible. • Satisfies the requirements of the investigation.
1 Nearly Satisfactory, with Obvious Flaws	• Shows understanding of most of the concepts of *landscaping plans, profit margins, percents, scale drawings, models, calculating dimensions,* and *using a price chart to calculate costs.* • May not use appropriate strategies to solve problems. • Computations are mostly correct. • Written explanations are satisfactory. • Drawings, model, report, and presentation are appropriate and sensible. • Satisfies the requirements of the investigation.
0 Unsatisfactory	• Shows little or no understanding of the concepts of *landscaping plans, profit margins, percents, scale drawings, models, calculating dimensions,* and *using a price chart to calculate costs.* • Does not use appropriate strategies to solve problems. • Computations are incorrect. • Written explanations are not satisfactory. • Drawings, model, report, and presentation are not appropriate or sensible. • Does not satisfy the requirements of the investigation.

CHAPTER 13 HIGHLIGHTS

VOCABULARY

After completing this chapter, you should be able to define each term, property, or phrase and give an example or two of each.

Algebra

completing the square (pp. 742, 743)
conjugate (p. 722)
distance formula (p. 737)
product property of square roots (p. 719)
quotient property of square roots (p. 721)
radical equations (p. 732)
radicand (p. 719)
rationalizing the denominator (p. 722)
simplest radical form (p. 723)

Geometry

Pythagorean theorem (p. 713)

Problem Solving

identify subgoals (p. 745)

UNDERSTANDING AND USING THE VOCABULARY

State whether each sentence is *true* or *false*. If false, replace the underlined word or number to make a true sentence.

1. The binomials $-3 + \sqrt{7}$ and $\underline{3 - \sqrt{7}}$ are conjugates. **false, $-3 - \sqrt{7}$**
2. In the expression $-4\sqrt{5}$, the radicand is $\underline{5}$. **true**
3. The rational expression $\frac{1 + \sqrt{3}}{2 - \sqrt{5}}$ becomes $\underline{2 + 2\sqrt{3} + \sqrt{5} + \sqrt{15}}$ when the denominator is rationalized. **false, $-2 - 2\sqrt{3} - \sqrt{5} - \sqrt{15}$**
4. The value of c that makes the trinomial $t^2 - 3t + c$ a perfect square is $\underline{9}$. **false, $\frac{9}{4}$**
5. The $\underline{\text{longest}}$ side of a right triangle is the hypotenuse. **true**
6. The distance formula can be expressed using the equation $\underline{d^2 = (x_2 - x_1)^2 + (y_2 - y_1)^2}$. **true**
7. After the first step in solving the rational equation $\sqrt{3x + 19} = x + 3$, you would have the equation $\underline{3x + 19 = x^2 + 9}$. **false, $3x + 19 = x^2 + 6x + 9$**
8. The two sides that form the right angle in a right triangle are called the $\underline{\text{legs}}$ of the triangle. **true**
9. The expression $\underline{\frac{2x\sqrt{3x}}{\sqrt{6y}}}$ is in simplest radical form. **false, $\frac{x\sqrt{2xy}}{y}$**
10. To verify whether a triangle with sides having lengths of 25, 20, and 15 is a right triangle, the Pythagorean theorem would be used: $\underline{15^2 = 25^2 + 20^2}$. **false, $25^2 = 15^2 + 20^2$**

Instructional Resources

Three multiple-choice tests and three free-response tests are provided in the *Assessment and Evaluation Masters*. Forms 1A and 2A are for honors pacing, and Forms 1B, 1C, 2B, and 2C are for average pacing. Chapter 13 Test, Form 1B is shown at the right. Chapter 13 Test, Form 2B is shown on the next page.

Using the CHAPTER HIGHLIGHTS

The Chapter Highlights begins with an alphabetical listing of the new terms, properties, and phrases that were introduced in this chapter. Have students define each term and provide an example or two of it, if appropriate.

Assessment and Evaluation Masters, pp. 339–340

13 NAME ________ DATE ________

Chapter 13 Test, Form 1B

Write the letter for the correct answer in the blank at the right of each problem.

1. What is the measure of the hypotenuse of a right triangle with side $a = \sqrt{6}$ and side $b = 8$? — 1. B
   A. $\sqrt{58}$ B. $\sqrt{70}$ C. $\sqrt{14}$ D. 2
2. Which of the following are the measures of three sides of a right triangle? — 2. A
   A. 6, 8, 10 B. 5, 9, 11 C. 11, 13, 16 D. 3, 8, 12
3. What is the length of the diagonal of a rectangle with a length of 12 m and width of 5 m? — 3. B
   A. 169 m B. 13 m C. $\sqrt{17}$ m D. $\sqrt{119}$ m
4. Simplify $\sqrt{96}$. — 4. D
   A. $6\sqrt{4}$ B. $6\sqrt{16}$ C. $16\sqrt{6}$ D. $4\sqrt{6}$
5. Simplify $\sqrt{28x^2y^3}$. — 5. C
   A. $7x|y|\sqrt{2y}$ B. $2x|y|\sqrt{7y}$ C. $2|x|y\sqrt{7y}$ D. $7|x|y\sqrt{2y}$
6. Simplify $\sqrt{\frac{18}{40}}$. — 6. C
   A. $\frac{\sqrt{3}}{2}$ B. $\frac{3\sqrt{20}}{20}$ C. $\frac{3\sqrt{5}}{10}$ D. $\frac{\sqrt{45}}{10}$
7. Simplify $\frac{5}{4 + \sqrt{6}}$. — 7. A
   A. $\frac{4 - \sqrt{6}}{2}$ B. $\frac{20 - 5\sqrt{6}}{22}$ C. $\frac{20 + 5\sqrt{6}}{22}$ D. $\frac{-\sqrt{6}}{2}$
8. Simplify $5\sqrt{7} - \sqrt{7}$. — 8. A
   A. $4\sqrt{7}$ B. -35 C. 5 D. $-5\sqrt{7}$
9. Which expression cannot be simplified? — 9. B
   A. $5\sqrt{8} + 6\sqrt{2}$ B. $5\sqrt{10} + 3\sqrt{5}$
   C. $7\sqrt{48} + \sqrt{108}$ D. $\sqrt{52} + \sqrt{13}$
10. Simplify $\sqrt{18} - \sqrt{54} + 2\sqrt{50}$. — 10. A
    A. $13\sqrt{2} - 3\sqrt{6}$ B. $-4\sqrt{3} + 4\sqrt{5}$
    C. $-4\sqrt{3} - 4\sqrt{5}$ D. $8\sqrt{2} - 3\sqrt{6}$

13 NAME ________ DATE ________

Chapter 13 Test, Form 1B (continued)

11. Simplify $\sqrt{15} + \sqrt{\frac{3}{5}}$. — 11. B
    A. $\frac{2\sqrt{15}}{5}$ B. $\frac{6\sqrt{15}}{5}$ C. $2\sqrt{3}$ D. $\frac{\sqrt{15}}{5}$
12. Solve $\sqrt{3x - 2} = 4$. — 12. B
    A. $\{12\}$ B. $\{6\}$ C. $\left\{\frac{2}{3}\right\}$ D. $\left\{\frac{3}{2}\right\}$
13. Solve $\sqrt{3a + 28} = a$. — 13. B
    A. $\{-4\}$ B. $\{7\}$ C. $\{7, -4\}$ D. no solution
14. Solve $\sqrt{5n - 1} - n = 1$. — 14. C
    A. $\{1, 2\}$ B. $\{-1, -2\}$ C. $\left\{\frac{1}{4}\right\}$ D. $\{1\}$
15. What is the distance between $(-3, 4)$ and $(2, 7)$? — 15. A
    A. $\sqrt{34}$ B. $\sqrt{74}$ C. $2\sqrt{30}$ D. $\sqrt{10}$
16. The distance between $(n, 4)$ and $(-2, 7)$ is $\sqrt{73}$. What is one possible value for n? — 16. D
    A. -6 B. 10 C. 8 D. -10
17. What is the distance between the origin and (6, 8)? — 17. A
    A. 10 B. 2 C. 14 D. $\sqrt{28}$
18. What value of c makes $y^2 - y + c$ a perfect square? — 18. C
    A. $\frac{1}{2}$ B. $-\frac{1}{2}$ C. $\frac{1}{4}$ D. $-\frac{1}{4}$
19. If $b^2 + 6b - 10 = 0$ is to be solved by completing the square, what would be the best thing to do first? — 19. B
    A. Divide each side by 8. B. Add 10 to each side.
    C. Subtract 8 from each side. D. Add 2 to each side.
20. Solve $d^2 - 6d + 4 = 0$. — 20. A
    A. $3 \pm \sqrt{5}$ B. $6 \pm 2\sqrt{5}$ C. $\frac{-6 \pm \sqrt{5}}{2}$ D. $\frac{3 \pm 2\sqrt{5}}{2}$

Bonus Simplify $\sqrt{4x^2 + 12x + 9}$. — Bonus $|2x + 3|$

CHAPTER 13 STUDY GUIDE AND ASSESSMENT

SKILLS AND CONCEPTS

OBJECTIVES AND EXAMPLES

Upon completing this chapter, you should be able to:

- use the Pythagorean theorem to solve problems (Lesson 13–1)

Find the length of the missing side.

$c^2 = a^2 + b^2$

$25^2 = 15^2 + b^2$

$625 = 225 + b^2$

$400 = b^2$

$20 = b$

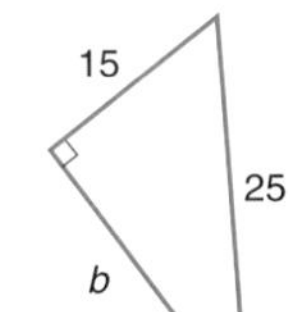

REVIEW EXERCISES

Use these exercises to review and prepare for the chapter test.

If c is the measure of the hypotenuse of a right triangle, find each missing measure. Round answers to the nearest hundredth.

11. $a = 30, b = 16, c = ?$ **34**
12. $a = 6, b = 10, c = ?$ **$2\sqrt{34} \approx 11.66$**
13. $a = 10, c = 15, b = ?$ **$5\sqrt{5} \approx 11.18$**
14. $b = 4, c = 56, a = ?$ **$4\sqrt{195} \approx 55.86$**
15. $a = 18, c = 30, b = ?$ **24**
16. $a = 1.2, b = 1.6, c = ?$ **2**

Determine whether the following side measures would form right triangles. Explain why or why not.

17. 9, 16, 20 **no; $9^2 + 16^2 \neq 20^2$**
18. 20, 21, 29 **yes**
19. 9, 40, 41 **yes**
20. 18, $\sqrt{24}$, 30 **no; $18^2 + (\sqrt{24})^2 \neq 30^2$**

OBJECTIVES AND EXAMPLES

- simplify radical expressions (Lesson 13–2)

$$\sqrt{343x^2y^3} = \sqrt{7 \cdot 7^2 \cdot x^2 \cdot y \cdot y^2}$$
$$= \sqrt{7} \cdot \sqrt{7^2} \cdot \sqrt{x^2} \cdot \sqrt{y} \cdot \sqrt{y^2}$$
$$= 7|x|y\sqrt{7y}$$

$$\frac{3}{5-\sqrt{2}} = \frac{3}{5-\sqrt{2}} \cdot \frac{5+\sqrt{2}}{5+\sqrt{2}}$$
$$= \frac{3(5) + 3\sqrt{2}}{5^2 - (\sqrt{2})^2}$$
$$= \frac{15 + 3\sqrt{2}}{25 - 2}$$
$$= \frac{15 + 3\sqrt{2}}{23}$$

REVIEW EXERCISES

Simplify. Leave in radical form and use absolute value symbols when necessary.

21. $\sqrt{480}$ **$4\sqrt{30}$**
22. $\sqrt{\frac{60}{y^2}}$ **$\frac{2\sqrt{15}}{|y|}$**
23. $\sqrt{44a^2b^5}$ **$2|a|b^2\sqrt{11b}$**
24. $\sqrt{96x^4}$ **$4x^2\sqrt{6}$**
25. $(3 - 2\sqrt{12})^2$ **$57 - 24\sqrt{3}$**
26. $\frac{9}{3+\sqrt{2}}$ **$\frac{27 - 9\sqrt{2}}{7}$**
27. $\frac{2\sqrt{7}}{3\sqrt{5} + 5\sqrt{3}}$ **$\frac{3\sqrt{35} - 5\sqrt{21}}{-15}$**
28. $\frac{\sqrt{3a^3b^4}}{\sqrt{8ab^{10}}}$ **$\frac{|a|\sqrt{6}}{4|b^3|}$**

Using the STUDY GUIDE AND ASSESSMENT

Skills and Concepts Encourage students to refer to the objectives and examples on the left as they complete the review exercises on the right.

Assessment and Evaluation Masters, pp. 345–346

13 NAME______ DATE______

Chapter 13 Test, Form 2B

1. Use the Pythagorean theorem to find the hypotenuse of a right triangle if side $a = 6$ and side $b = 10$. — 1. $2\sqrt{34}$, or 11.66
2. Find the width of a rectangle with a diagonal of 25 cm and a length of 24 cm. — 2. 7 cm
3. A rope from the top of a mast on a sailboat is attached to a point on the deck 5 feet from the base of the mast. If the rope is 13 feet long, how high is the mast? — 3. 12 ft

Simplify. Leave in radical form and use absolute value symbols when necessary.

4. $\sqrt{24}$ — 4. $2\sqrt{6}$
5. $\sqrt{75y^4w^3}$ — 5. $5y^2w\sqrt{3w}$
6. $\frac{\sqrt{14}}{\sqrt{45}}$ — 6. $\frac{\sqrt{70}}{15}$
7. $\frac{6}{6-\sqrt{11}}$ — 7. $\frac{36 + 6\sqrt{11}}{25}$
8. $3\sqrt{11} + 2\sqrt{11}$ — 8. $5\sqrt{11}$
9. $\sqrt{20} + 2\sqrt{45}$ — 9. $8\sqrt{5}$
10. $\sqrt{14} + \sqrt{\frac{2}{7}}$ — 10. $\frac{8}{7}\sqrt{14}$
11. $\sqrt{6}(\sqrt{30} + 4\sqrt{10})$ — 11. $6\sqrt{5} + 8\sqrt{15}$

13 NAME______ DATE______

Chapter 13 Test, Form 2B (continued)

Solve each equation. Check the solutions.

12. $\sqrt{x-8} = x - 10$ — 12. 12
13. $\sqrt{\frac{9a}{3}} - 4 = 0$ — 13. $\frac{16}{3}$
14. $\sqrt{m} = 2\sqrt{3}$ — 14. 12
15. Find the distance between $(0, -4)$ and $(5, 2)$. — 15. $\sqrt{61}$, or 7.81
16. The distance between $(3, 5)$ and $(7, a)$ is 5. Find the two possible values for a. — 16. 2, 8
17. Find the distance from the origin to $(5, -4)$. — 17. $\sqrt{41}$

Solve by completing the square.

18. $r^2 + 18r + 9 = 0$ — 18. $-9 \pm 6\sqrt{2}$
19. $b^2 + 8b + 10 = 0$ — 19. $-4 \pm \sqrt{6}$
20. Find the value for c that makes $x^2 + 3x + c$ a perfect square. — 20. $\frac{9}{4}$

Bonus For what values of x is $\sqrt{7x-4}$ a real number? — Bonus $x \geq \frac{4}{7}$

GLENCOE Technology

Test and Review Software

You may use this software, a combination of an item generator and item bank, to create your own tests or worksheets. Types of items include free response, multiple choice, short answer, and open ended.

For IBM & Macintosh

OBJECTIVES AND EXAMPLES	REVIEW EXERCISES

- simplify radical expressions involving addition, subtraction, and multiplication (Lesson 13–3)

$\sqrt{6} - \sqrt{54} + 3\sqrt{12} + 5\sqrt{3}$
$= \sqrt{6} - \sqrt{3^2 \cdot 6} + 3\sqrt{2^2 \cdot 3} + 5\sqrt{3}$
$= \sqrt{6} - (\sqrt{3^2} \cdot \sqrt{6}) + 3(\sqrt{2^2} \cdot \sqrt{3}) + 5\sqrt{3}$
$= \sqrt{6} - 3\sqrt{6} + 3(2\sqrt{3}) + 5\sqrt{3}$
$= \sqrt{6} - 3\sqrt{6} + 6\sqrt{3} + 5\sqrt{3}$
$= -2\sqrt{6} + 11\sqrt{3}$

30. $5\sqrt{13} + 5\sqrt{15}$

Simplify. Then use a calculator to verify your answer.

29. $2\sqrt{6} - \sqrt{48}$ **$2\sqrt{6} - 4\sqrt{3}$**
30. $2\sqrt{13} + 8\sqrt{15} - 3\sqrt{15} + 3\sqrt{13}$
31. $4\sqrt{27} + 6\sqrt{48}$ **$36\sqrt{3}$**
32. $5\sqrt{18} - 3\sqrt{112} - 3\sqrt{98}$ **$-6\sqrt{2} - 12\sqrt{7}$**
33. $\sqrt{8} + \sqrt{\frac{1}{8}}$ **$\frac{9\sqrt{2}}{4}$**
34. $4\sqrt{7k} - 7\sqrt{7k} + 2\sqrt{7k}$ **$-\sqrt{7k}$**

- solve radical equations (Lesson 13–4)

Solve $\sqrt{5 - 4x} - 6 = 7$.

$\sqrt{5 - 4x} - 6 = 7$
$\sqrt{5 - 4x} = 13$
$(\sqrt{5 - 4x})^2 = (13)^2$
$5 - 4x = 169$
$-4x = 164$
$x = -41$

Solve each equation. Check your solution.

35. $\sqrt{3x} = 6$ **12**
36. $\sqrt{t} = 2\sqrt{6}$ **24**
37. $\sqrt{7x - 1} = 5$ **$\frac{26}{7}$**
38. $\sqrt{x + 4} = x - 8$ **12**
39. $\sqrt{r} = 3\sqrt{5}$ **45**
40. $\sqrt{3x - 14} + x = 6$ **5**
41. $\sqrt{\frac{4a}{3}} - 2 = 0$ **3**
42. $9 = \sqrt{\frac{5n}{4}} - 1$ **80**
43. $10 + 2\sqrt{b} = 0$ **no solution**
44. $\sqrt{a + 4} = 6$ **32**

- find the distance between two points in the coordinate plane (Lesson 13–5)

Find the distance between the pair of points with coordinates $(-5, 1)$ and $(1, 5)$.

$d = \sqrt{(x_2 - x_1)^2 + (y_2 - y_1)^2}$
$= \sqrt{(1 - (-5))^2 + (5 - 1)^2}$
$= \sqrt{6^2 + 4^2}$
$= \sqrt{36 + 16}$ or $\sqrt{52} \approx 7.21$

Find the distance between each pair of points whose coordinates are given.

45. $(9, -2), (1, 13)$ **17**
46. $(4, 2), (7, -9)$ **$\sqrt{130} \approx 11.40$**
47. $(4, -6), (-2, 7)$ **$\sqrt{205} \approx 14.32$**
48. $(2\sqrt{5}, 9), (4\sqrt{5}, 3)$ **$2\sqrt{14} \approx 7.48$**

Find the value of a if the points with the given coordinates are the indicated distance apart.

49. $(-3, 2), (1, a); d = 5$ **5 or −1**
50. $(5, -2), (a, -3); d = \sqrt{170}$ **18 or −8**
51. $(1, 1), (4, a); d = 5$ **5 or −3**

Applications and Problem Solving Encourage students to work through the exercises in the Applications and Problem Solving section to strengthen their problem-solving skills.

OBJECTIVES AND EXAMPLES

- solve quadratic equations by completing the square (Lesson 13–6)

Solve $y^2 + 6y + 2 = 0$ by completing the square.

$$y^2 + 6y + 2 = 0$$
$$y^2 + 6y = -2$$
$$y^2 + 6y + 9 = -2 + 9 \quad \text{Complete the square. Since } \left(\frac{6}{2}\right)^2 = 9\text{, add 9 to each side}$$
$$(y + 3)^2 = 7$$
$$y + 3 = \pm\sqrt{7}$$
$$y = -3 \pm\sqrt{7}$$

The solution set is $-3 + \sqrt{7}$ and $-3 - \sqrt{7}$.

53. $\frac{49}{4}$ 55. $\frac{1}{9}$ 57. $\frac{7 \pm \sqrt{69}}{2}$ 58. $-\frac{3}{2}, -\frac{5}{2}$

REVIEW EXERCISES

Find the value of c that makes each trinomial a perfect square.

52. $y^2 - 12y + c$ **36**
53. $m^2 + 7m + c$
54. $b^2 + 18b + c$ **81**
55. $p^2 - \frac{2}{3}p + c$

Solve each equation by completing the square. Leave irrational roots in simplest radical form.

56. $x^2 - 16x + 32 = 0$ **$8 \pm 4\sqrt{2}$**
57. $m^2 - 7m = 5$
58. $4a^2 + 16a + 15 = 0$
59. $\frac{1}{2}y^2 + 2y - 1 = 0$ **$-2 \pm \sqrt{6}$**
60. $n^2 - 3n + \frac{5}{4} = 0$ **$\frac{1}{2}, \frac{5}{2}$**

APPLICATIONS AND PROBLEM SOLVING

61. **Geometry** The sides of a triangle measure $4\sqrt{24}$ cm, $5\sqrt{6}$ cm, and $3\sqrt{54}$ cm. What is the perimeter of the triangle? (Lesson 13–3) **$22\sqrt{6} \approx 53.9$ cm**

62. **Sight Distance** In the movie *Angels in the Outfield*, Roger decides to climb a tree in order to be able to see the Angels game better. The formula $V = 3.5\sqrt{h}$ relates height and distance, where h is your height in meters above the ground and V is the distance in kilometers that you can see. (Lesson 13–4) **62b. 256 m**
    a. How far could Roger see if he had climbed 9 meters to the top of a tree? **10.5 km**
    b. How high would someone have to be if she wanted to be able to see 56 kilometers?

63. **Geometry** What kind of triangle has vertices at (4, 2), (−3, 1), and (5, −4)? (Lesson 13–5) **scalene triangle**

64. **Number Theory** When 6 is subtracted from 10 times a number, the result is equal to 4 times the square of the number. Find the numbers. (Lesson 13–6) **1, 1.5**

65. **Nature** An 18-foot tall tree is broken by the wind. The top of the tree falls and touches the ground 12 feet from its base. How many feet from the base of the tree did the break occur? (Lesson 13–1) **5 ft**

66. **Physics** The time T (in seconds) required for a pendulum of length L (in feet) to make one complete swing back and forth is given by the formula $T = 2\pi\sqrt{\frac{L}{32}}$. How long does it take a pendulum 4 feet long to make one complete swing? (Lesson 13–2) **$\frac{\sqrt{2}}{2}\pi$ or about 2.22 s**

67. **Law Enforcement** Lina told the police officer that she was traveling at 55 mph when she applied the brakes and skidded. The skid marks at the scene were 240 feet long. Should Lina's car have skidded that far if it was traveling at 55 mph? Use the formula $s = \sqrt{15d}$. (Lesson 13–4) **No, it should skid about 201.7 ft.**

A practice test for Chapter 13 is provided on page 799.

ALTERNATIVE ASSESSMENT

COOPERATIVE LEARNING PROJECT

Battleship In this project, you will determine the placement of ships for a game. A game that Ryan and Nicholas enjoy playing is a form of the game Battleship®. They determine the lengths of five ships that they will draw on their grids. Each time a ship intersects a grid coordinate, that is one of the points where the ship can be attacked and hit. A ship can be placed on the grid horizontally, vertically, or diagonally. When all of the grid coordinates for a ship are hit, the ship is sunk. The goal is to sink all of the other person's ships first.

For their first game, Ryan and Nicholas decided on $2\sqrt{5}$ units, 3 units, $\sqrt{13}$ units, $\sqrt{2}$ units, and 2 units for their ship lengths. Place these five ships on the grid below making sure that they have the above stated lengths. How many hits would another player have to get in order to sink all of the ships on your grid? **3 + 4 + 2 + 2 + 3 = 14**

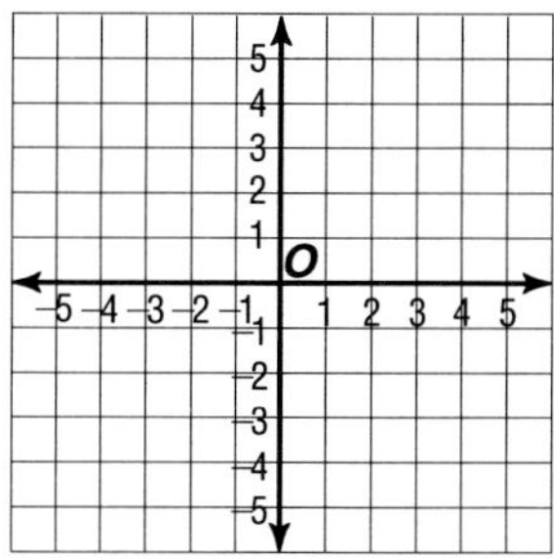

Follow these guidelines for your Battleship game.

- Determine whether each of the ships with the specified lengths have to be drawn horizontally, vertically, or diagonally.
- Devise a chart that will be helpful in organizing the data you will need; for example, length of ship, length of legs of triangle formed by diagonal ships, and number of grid coordinates for each ship.
- Create a grid that will incorporate the above specifications for the ships.
- Write a summary of the above game.
- Play the game with a fellow classmate.

THINKING CRITICALLY

- Is the sum of two irrational numbers always an irrational number? Explain.
- For any real number n, $\sqrt{n^2} = |n|$. What rule would be true for $\sqrt{n^t}$ if t were odd?

$\sqrt{n^t} = n^{\frac{t-1}{2}}\sqrt{n}$

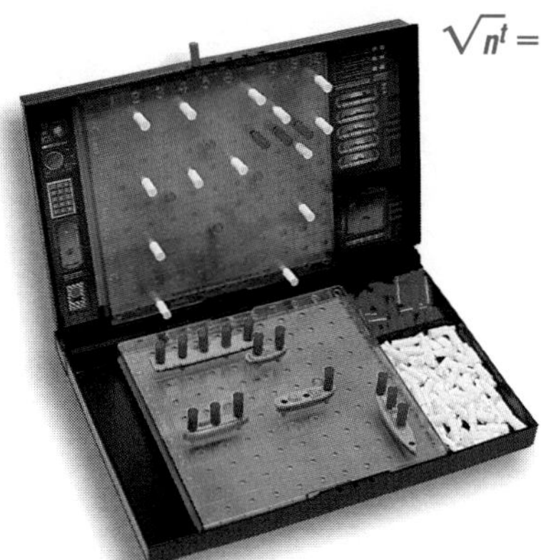

PORTFOLIO

Visualizing a concept or skill is a good way to help understand a concept. Using colored pencils is one way to visualize steps that need to be remembered or aid in "seeing" items that are easy to forget. When simplifying a rational expression, write the perfect square number or variables using colored pencils. This will serve to remind you what needs to be written outside the radical sign and what needs to stay inside the radical sign. Use one of the exercises in this chapter and use colored pencils to simplify it. Write how you chose what to put in color and how it aided you in the problem-solving process. Place this in your portfolio.

SELF EVALUATION

When solving a problem, one may look for alternative methods to use. The same method may not work in every situation. Being open-minded about a different process is a positive value. Comparing methods can also be beneficial.

Assess yourself. Are you a person that sticks to the same method for a solution or do you also look for other means? Do you compare methods or not? Think of two problems that you have had, one from your daily life and one from your mathematical experiences, where you used two different methods to solve each of them.

Alternative Assessment

The Alternative Assessment section provides students with the opportunity to assess their own work by thinking critically, working with others, keeping a portfolio, and honestly evaluating their own progress. For more information on alternative forms of assessment, see *Alternative Assessment in the Mathematics Classroom,* one of the titles in the Glencoe Mathematics Professional Series.

Performance Assessment

Performance Assessment tasks for this chapter are included in the *Assessment and Evaluation Masters.* A scoring guide is also provided.

Assessment and Evaluation Masters, pp. 350, 361

NAME ________ DATE ________

Chapter 13 Performance Assessment

Instructions: *Demonstrate your knowledge by giving a clear, concise solution to each problem. Be sure to include all relevant drawings and justify your answers. You may show your solution in more than one way or investigate beyond the requirements of the problem.*

1. Mrs. Raines is having a sprinkler system installed in her flower bed. The diagram below shows the location of sprinklers A, B, C, and D. The water source S is in the center of the bed.

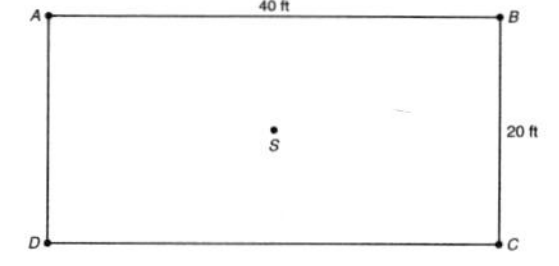

a. How much pipe would be required to hook each sprinkler directly to the water source? Explain your answer.

b. Do you think she could save pipe by running pipe from S to A, from A to B, from B to C, and from C to D? Explain your answer.

c. Would running pipe from A to B and from C to D before connecting the midpoint of each line to the water source S save pipe? Justify your answer.

d. If A is connected to D and B is connected to C and the midpoints of these two lines connected to the water source, would Mrs. Raines save pipe over the arrangement in part c? Why?

e. Would the arrangement in part d save pipe over the arrangement in part a? Justify your answer.

f. What other considerations might Mrs. Raines consider in choosing the hook-up arrangement?

g. Find the pipe arrangement that would use the least pipe for a square of side a feet. Justify your answer.

2. Solve $8\sqrt{4z^2 - 43} = 40$. Justify each step in your solution.

Scoring Guide
Chapter 13
Performance Assessment

Level	Specific Criteria
3 Superior	• Shows thorough understanding of the concepts of *finding square roots, using the Pythagorean theorem, simplifying radical expressions,* and *solving radical equations.* • Uses appropriate strategies to solve problems. • Computations are correct. • Written explanations are exemplary. • Diagrams are accurate and appropriate. • Goes beyond requirements of some or all problems.
2 Satisfactory, with Minor Flaws	• Shows understanding of the concepts of *finding square roots, using the Pythagorean theorem, simplifying radical expressions,* and *solving radical equations.* • Uses appropriate strategies to solve problems. • Computations are mostly correct. • Written explanations are effective. • Diagrams are mostly accurate and appropriate. • Satisfies all requirements of problems.
1 Nearly Satisfactory, with Serious Flaws	• Shows understanding of most of the concepts of *finding square roots, using the Pythagorean theorem, simplifying radical expressions,* and *solving radical equations.* • May not use appropriate strategies to solve problems. • Computations are mostly correct. • Written explanations are satisfactory. • Diagrams are mostly accurate and appropriate. • Satisfies most requirements of problems.
0 Unsatisfactory	• Shows little or no understanding of the concepts of *finding square roots, using the Pythagorean theorem, simplifying radical expressions,* and *solving radical equations.* • May not use appropriate strategies to solve problems. • Computations are incorrect. • Written explanations are not satisfactory. • Diagrams are not accurate or appropriate. • Does not satisfy requirements of problems.

For Additional Assessment

For Reference

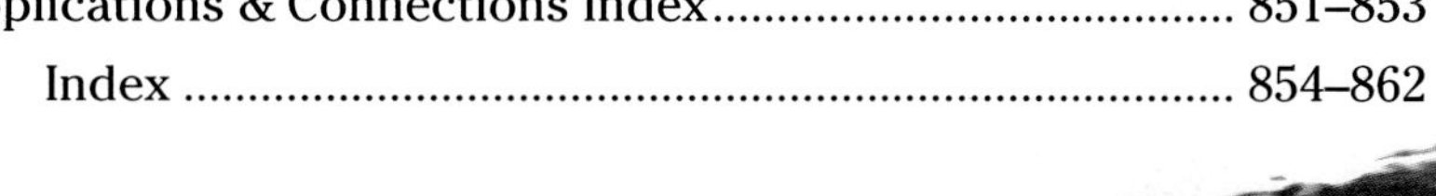

EXTRA PRACTICE

Lesson 1-1 Write an algebraic expression for each verbal expression.

1. the product of x and 7 **$7x$**
2. the quotient of r and s **$\frac{r}{s}$**
3. the sum of b and 21 **$b + 21$**
4. a number t decreased by 6 **$t - 6$**
5. a number a to the third power **a^3**
6. sixteen squared **16^2**

Write a verbal expression for each algebraic expression.

7. $n - 7$ **n minus 7**
8. xy **the product of x and y**
9. m^5 **m to the fifth power**
10. 8^4 **8 to the fourth power**
11. $6r^2$ **6 times r squared**
12. $z^7 + 2$ **z to the seventh power plus 2**

Write each expression as an expression with exponents.

13. $5 \cdot 5 \cdot 5$ **5^3**
14. $7 \cdot a \cdot a \cdot a \cdot a$ **$7a^4$**
15. $2(m)(m)(m)$ **$2m^3$**
16. $5 \cdot 5 \cdot 5 \cdot x \cdot x \cdot y$ **5^3x^2y**
17. $p \cdot p \cdot p \cdot p \cdot p \cdot p$ **p^6**
18. $4 \cdot 4 \cdot 4 \cdot t \cdot t$ **4^3t^2**

Evaluate each expression.

19. 2^4 **16**
20. 8^2 **64**
21. 7^3 **343**
22. 10^4 **10,000**
23. 3^6 **729**
24. 4^5 **1024**

Lesson 1-2 Give the next two items for each pattern.

1.

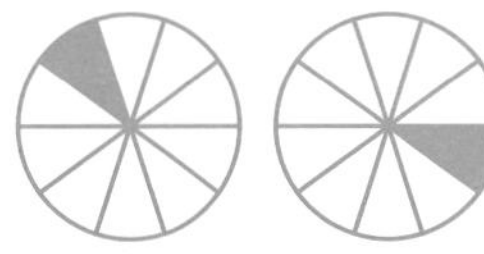
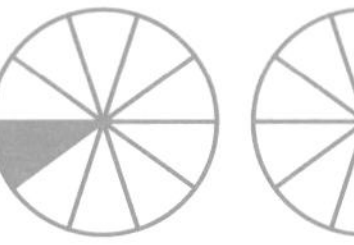

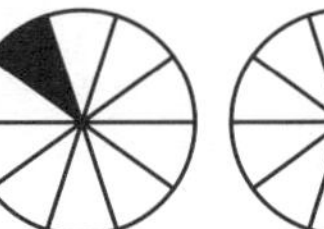
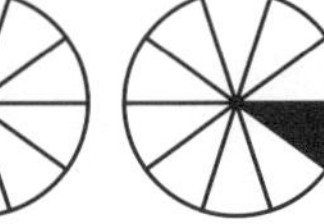

2.

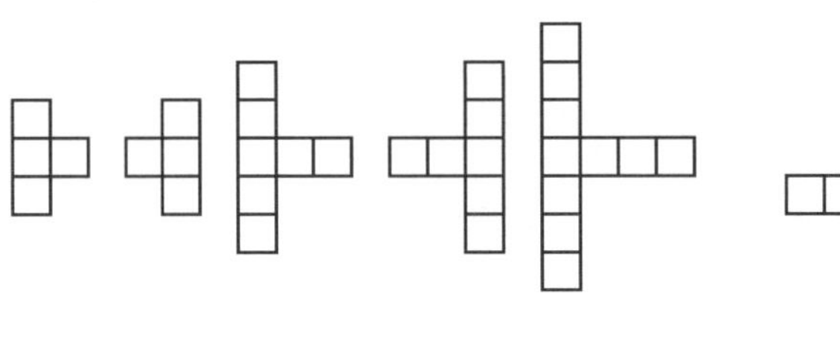

3. 12, 23, 34, 45, ... **56, 67**
4. 39, 33, 27, 21, ... **15, 9**
5. 6, 7.2, 8.4, 9.6, ... **10.8, 12**
6. 86, 81.5, 77, 72.5, ... **68, 63.5**
7. 4, 8, 16, 32, ... **64, 128**
8. 3125, 625, 125, 25, ... **5, 1**
9. 15, 16, 18, 21, 25, 30, ... **36, 43**
10. $w - 2, w - 4, w - 6, w - 8, \ldots$ **$w - 10, w - 12$**
11. 13, 10, 11, 8, 9, 6, ... **7, 4**

Lesson 1-3 Evaluate each expression.

1. $3 + 8 \div 2 - 5$ **2**
2. $4 + 7 \cdot 2 + 8$ **26**
3. $5(9 + 3) - 3 \cdot 4$ **48**
4. $4(11 + 7) - 9 \cdot 8$ **0**
5. $5^3 + 6^3 - 5^2$ **316**
6. $16 \div 2 \cdot 5 \cdot 3 \div 6$ **20**
7. $7(5^3 + 3^2)$ **938**
8. $\frac{9 \cdot 4 + 2 \cdot 6}{7 \cdot 7}$ **$\frac{48}{49}$**
9. $25 - \frac{1}{3}(18 + 9)$ **16**

Evaluate each expression when $a = 2$, $b = 5$, $x = 4$, and $n = 10$.

10. $8a + b$ **21**
11. $12x + ab$ **58**
12. $a(6 - 3n)$ **−48**
13. $bx + an$ **40**
14. $x^2 - 4n$ **−24**
15. $3b + 16a - 9n$ **−43**
16. $n^2 + 3(a + 4)$ **118**
17. $(2x)^2 + an - 5b$ **59**
18. $[a + 8(b - 2)]^2 \div 4$ **169**

Lesson 1-4 **Suppose the number 16,782 is rounded to 16,800 and plotted using stem 16 and leaf 8. Write the stem and leaf for each number below if the numbers are part of the same set of data.**

1. 24,640 stem 24, leaf 6
2. 35,788 stem 35, leaf 8
3. 4239 stem 4, leaf 2
4. 5865 stem 5, leaf 9
5. 611 stem 0, leaf 6
6. 17,903 stem 17, leaf 9
7. The stem-and-leaf plot at the right gives the average weekly earnings for various occupations in 1993. Use the plot to answer each question.
   a. What were the highest weekly earnings? $740–$749
   b. What were the lowest weekly earnings? $260–$269
   c. How many occupations have weekly earnings of at least $500? 6
   d. What does 5 | 1 represent? weekly earnings of $510–$519

Average Weekly Earnings

Stem	Leaf
2	6 7
3	1 2 8 9
4	1 3 6 8 8 8 9
5	0 1 6
6	
7	1 2 4

3 | 8 = $380–$389

Lesson 1-5 **State whether each equation is *true* or *false* for the value of the variable given.**

1. $b + \frac{2}{3} = \frac{3}{4} + \frac{1}{3}, b = \frac{1}{2}$ false
2. $\frac{2 + 13}{y} = \frac{3}{5}y, y = 5$ true
3. $x^8 = 9^4, x = 3$ true
4. $4t^2 - 5(3) = 9, t = 7$ false
5. $\frac{3^2 - 5x}{3^2 - 1} \leq 2, x = 4$ true
6. $a^6 \div 4 \div a^3 \div a < 3, a = 2$ true

Find the solution set for each inequality if the replacement set for *x* is {4, 5, 6, 7, 8} and for *y* is {10, 12, 14, 16}.

7. $x + 2 > 7$ {6, 7, 8}
8. $x - 1 > 3$ {5, 6, 7, 8}
9. $2y - 15 \leq 17$ {10, 12, 14, 16}
10. $y + 12 < 25$ {10, 12}
11. $\frac{y + 12}{7} \geq 4$ {16}
12. $\frac{2(x - 2)}{3} < \frac{4}{7 - 5}$ {4}
13. $x - 4 > \frac{x + 2}{3}$ {8}
14. $y^2 - 100 \geq 4y$ {14, 16}
15. $9x - 20 \geq x^2$ {4, 5}
16. $0.3(x + 4) \leq 0.4(2x + 3)$ {4, 5, 6, 7, 8}
17. $1.3x - 12 < 0.9x + 4$ {4, 5, 6, 7, 8}
18. $1.2y - 8 \leq 0.7y - 3$ {10}

Solve each equation.

19. $x = \frac{17 + 9}{2}$ 13
20. $3(8) + 4 = b$ 28
21. $\frac{18 - 7}{13 - 2} = y$ 1
22. $28 - (-14) = z$ 42
23. $20.4 - 5.67 = t$ 14.73
24. $t = 91.8 \div 27$ 3.4
25. $-\frac{5}{8}\left(-\frac{4}{5}\right) = c$ $\frac{1}{2}$
26. $8\frac{1}{12} - 5\frac{5}{12} = e$ $2\frac{2}{3}$
27. $\frac{3}{4} - \frac{9}{16} = s$ $\frac{3}{16}$
28. $\frac{5}{8} + \frac{1}{4} = y$ $\frac{7}{8}$
29. $n = \frac{84 \div 7}{18 \div 9}$ 6
30. $d = 3\frac{1}{2} \div 2$ $\frac{7}{4}$

Lesson 1-6 **Name the property or properties illustrated by each statement.**

1. If $8 \cdot 3 = 24$, then $24 = 8 \cdot 3$. symmetric (=)
2. $6 + (4 + 1) = 6 + 5$ substitution (=)
3. $(12 - 3)(7) = 9(7)$ substitution (=)
4. $qrs = 1qrs$ multiplicative identity
5. $\left(\frac{8}{9}\right)\left(\frac{9}{8}\right) = 1$ multiplicative inverse
6. $4\left(6^2 \cdot \frac{1}{36}\right) = 4$ multiplicative inverse, multiplicative identity
7. $0 + 45 = 45$ additive identity
8. If $5 = 9 - 4$, then $9 - 4 = 5$. symmetric (=)
9. $2(0) = 0$ multiplicative property of 0
10. $16 + 37 = 16 + 37$ reflexive (=)
11. $1(57) = 57$ multiplicative identity
12. $0 + h = 0$ additive identity
13. If $9 + 1 = 10$ and $10 = 5(2)$, then $9 + 1 = 5(2)$. transitive (=)

Additional Answers for Lesson 1-8

1. multiplicative identity
2. commutative property (+)
3. commutative property (×)
4. additive identity
5. distributive property
6. commutative property (+)
7. associative property (×)
8. multiplicative property of zero
9. substitution (=)
10. associative property (+)
11. associative property (×)
12. associative property (+)

EXTRA PRACTICE

Lesson 1-7 **Use the distributive property to rewrite each expression without parentheses.**

1. $3(5 + w)$ **$15 + 3w$**
2. $(h - 8)7$ **$7h - 56$**
3. $6(y + 4)$ **$6y + 24$**
4. $9(3n + 5)$ **$27n + 45$**
5. $32\left(x - \frac{1}{8}\right)$ **$32x - 4$**
6. $c(7 - d)$ **$7c - cd$**

Use the distributive property to find each product.

7. $6 \cdot 55$ **330**
8. $\left(4\frac{1}{18}\right) \times 18$ **73**
9. $15(108)$ **1620**
10. $14(3.7)$ **51.8**
11. $689 \cdot 5$ **3445**
12. 7×314 **2198**

Simplify each expression, if possible. If not possible, write *in simplest form*.

13. $13a + 5a$ **$18a$**
14. $21x - 10x$ **$11x$**
15. $8(3x + 7)$ **$24x + 56$**
16. $4m - 4n$ **in simplest form**
17. $3(5am - 4)$ **$15am - 12$**
18. $15x^2 + 7x^2$ **$22x^2$**
19. $9y^2 + 13y^2 + 3$ **$22y^2 + 3$**
20. $11a^2 - 11a^2 + 12a^2$ **$12a^2$**
21. $6a + 7a + 12b + 8b$ **$13a + 20b$**

Lesson 1-8 **Name the property illustrated by each statement.** **1–12. See margin.**

1. $1 \cdot a^2 = a^2$
2. $x^2 + (y + z) = x^2 + (z + y)$
3. $ax + 2b = xa + 2b$
4. $29 + 0 = 29$
5. $5(a + 3b) = 5a + 15b$
6. $5a + 3b = 3b + 5a$
7. $(4 \cdot c) \cdot d = 4 \cdot (c \cdot d)$
8. $(6x^3) \cdot 0 = 0$
9. $(4 + 1)x + 2 = 5x + 2$
10. $(a + b) + 3 = a + (b + 3)$
11. $5(ab) = (5a)b$
12. $5a + \left(\frac{1}{2}b + c\right) = \left(5a + \frac{1}{2}b\right) + c$

Simplify. **15. $3a + 13b + 2c$ 17. $-4p - 2q$ 22. $-11 + 3uv + u$ 23. $15x + 10y$ 24. $11.8a + 8.8b$**

13. $5a + 6b + 7a$ **$12a + 6b$**
14. $8x + 4y + 9x$ **$17x + 4y$**
15. $3a + 5b + 2c + 8b$
16. $\frac{2}{3}x^2 + 5x + x^2$ **$\frac{5}{3}x^2 + 5x$**
17. $(4p - 7q) + (5q - 8p)$
18. $8q + 5r - 7q - 6r$ **$q - r$**
19. $4(2x + y) + 5x$ **$13x + 4y$**
20. $9r^5 + 2r^2 + r^5$ **$10r^5 + 2r^2$**
21. $12b^3 + 12 + 12b^3$ **$24b^3 + 12$**
22. $7 + 3(uv - 6) + u$
23. $3(x + 2y) + 4(3x + y)$
24. $6.2(a + b) + 2.6(a + b) + 3a$
25. $3 + 8(st + 3w) + 3st$ **$3 + 11st + 24w$**
26. $5.4(s - 3t) + 3.6(s - 4)$ **$9s - 16.2t - 14.4$**
27. $3[4 + 5(2x + 3y)]$ **$12 + 30x + 45y$**

Lesson 1-9 **Match each description with the most appropriate graph.**

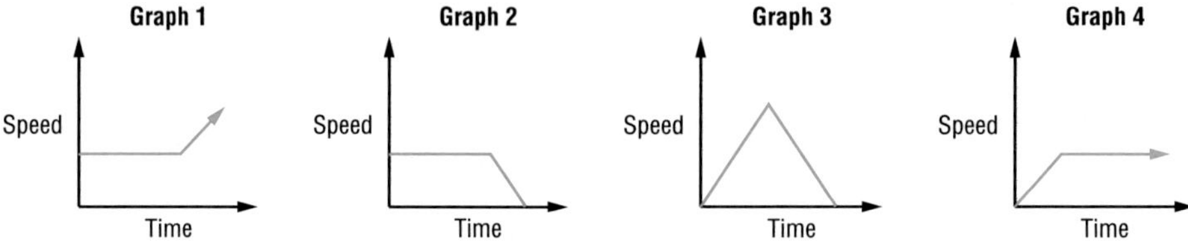

1. Jeremy picks up his speed while jogging down a hill. Then he slows down until he comes to a stop. **Graph 3**
2. Ilene jogs up a hill at a steady speed. Then she runs down the hill and picks up her speed. **Graph 1**
3. Casey runs down a hill and picks up his speed. Then he continues to jog at a steady pace. **Graph 4**
4. Luisa jogs along a road at a steady speed. Then she slows down until she comes to a stop. **Graph 2**

Lesson 2–1 Name the set of numbers graphed. **1–6. See margin.**

1. −6 −5 −4 −3 −2 −1 0 1 2 3 4 5 6
2. −4 −3 −2 −1 0 1 2 3 4 5 6 7 8
3. −3 −2 −1 0 1 2 3 4 5 6 7 8 9
4. 4 5 6 7 8 9 10 11 12 13 14 15 16
5. −10 −9 −8 −7 −6 −5 −4 −3 −2 −1 0 1 2
6. −5 −4 −3 −2 −1 0 1 2 3 4 5 6 7

Graph each set of numbers on a number line. 7–12. See margin.

7. {−2, −4, −6}
8. {. . . , −3, −2, −1, 0}
9. {integers greater than −1}
10. {integers less than −5 and greater than −10}
11. {integers less than or equal to 3}
12. {integers less than 0 and greater than or equal to −6}

Lesson 2–2 Use the line plot below to answer each question.

20 22 24 26 28 30 32 34 36 38 40 42 44 46 48 50 52 54

1. What was the highest score on the test? **50**
2. What was the lowest score on the test? **22**
3. How many students took the test? **31**
4. How many students scored in the 40s? **8**
5. What score was received by the most students? **26**

Make a line plot for each set of data. 6–9. See margin.

6. 134, 167, 137, 138, 120, 134, 145, 155, 152, 159, 164, 135, 144, 156
7. 19, 12, 11, 11, 7, 7, 8, 13, 12, 12, 9, 9, 8, 15, 11, 4, 12, 7, 7, 6
8. 66, 74, 72, 78, 68, 75, 80, 69, 62, 65, 63, 78, 81, 78, 76, 87, 80, 69, 81, 76, 79, 70, 62, 73, 85, 87, 70
9. 152, 156, 133, 154, 129, 146, 174, 138, 185, 141, 169, 176, 179, 168, 185, 154, 199, 200

Lesson 2–3 Find each sum or difference.

1. $-3 + 16$ **13**
2. $27 - 19$ **8**
3. $8 - 13$ **−5**
4. $14 + (-9)$ **5**
5. $-18 + (-11)$ **−29**
6. $-25 + 47$ **22**
7. $19m - 12m$ **$7m$**
8. $8h - 23h$ **$-15h$**
9. $24b - (-9b)$ **$33b$**
10. $97 + (-79)$ **18**
11. $4 + (-12) + (-18)$ **−26**
12. $7 + (-11) + 32$ **28**
13. $|-28 + (-67)|$ **95**
14. $|-89 + 46|$ **43**
15. $|-285 + (-641)|$ **926**
16. $-35 - (-12)$ **−23**
17. $24 + (-15)$ **9**
18. $-15 + (-13)$ **−28**
19. $-7 + (-21)$ **−28**
20. $8 - 17 + (-3)$ **−12**
21. $27 - 14 - (-19)$ **32**
22. $|-9 + 15|$ **6**
23. $\begin{bmatrix} 5 & -4 \\ 0 & 3 \end{bmatrix} + \begin{bmatrix} -3 & 4 \\ -2 & -4 \end{bmatrix}$ $\begin{bmatrix} 2 & 0 \\ -2 & -1 \end{bmatrix}$
24. $\begin{bmatrix} 1 & -4 \\ 5 & -6 \end{bmatrix} - \begin{bmatrix} 4 & -3 \\ 7 & -1 \end{bmatrix}$ $\begin{bmatrix} -3 & -1 \\ -2 & -5 \end{bmatrix}$

Additional Answers for Lesson 2-1

1. {−3, −2, −1, 0, 1, 2, 3, 4}
2. {−2, 0, 2, 3, 6}
3. {2, 3, 4}
4. {7, 8, 9, 10, 11, 12, 13}
5. {−8, −6, −4, −2, 0}
6. {−2, −1, 0, 1, 2, 3, 4}
7. −8 −7 −6 −5 −4 −3 −2 −1 0
8. −4 −3 −2 −1 0 1 2 3 4
9. −4 −3 −2 −1 0 1 2 3 4
10. −12 −11 −10 −9 −8 −7 −6 −5 −4
11. −1 0 1 2 3 4 5 6 7
12. −7 −6 −5 −4 −3 −2 −1 0 1

Additional Answers for Lesson 2-2

6. 118 124 130 136 142 148 154 160 166

7.

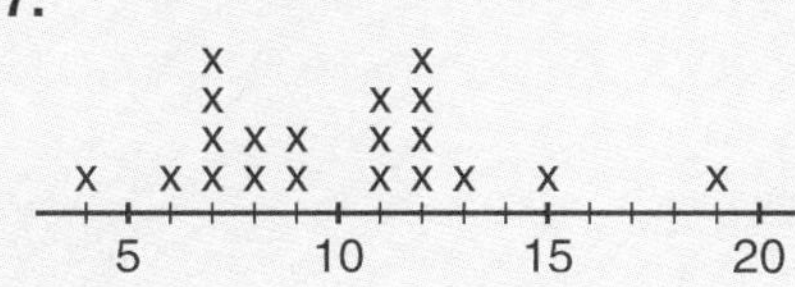

8.

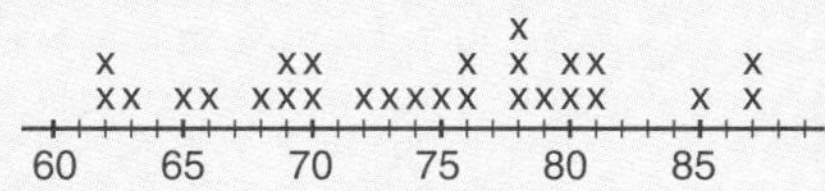

9.

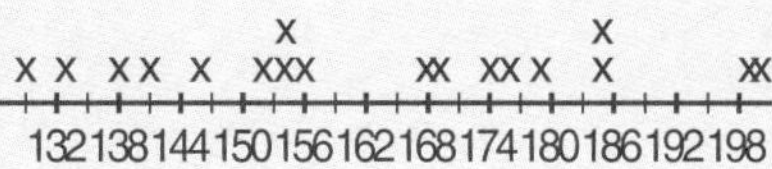

EXTRA PRACTICE

Lesson 2-4 **Replace each __?__ with <, >, or = to make each sentence true.**

1. $6 \underline{\ ?\ } -4$ $>$
2. $12 \underline{\ ?\ } -21$ $>$
3. $-4 \underline{\ ?\ } -10$ $>$
4. $4 \underline{\ ?\ } 14$ $<$
5. $-13 \underline{\ ?\ } -8$ $<$
6. $-5 + 2 \underline{\ ?\ } -3$ $=$
7. $7 \underline{\ ?\ } 13 - (-6)$ $<$
8. $3.4 - 5.7 \underline{\ ?\ } -2$ $<$
9. $\frac{18}{-6} \underline{\ ?\ } -3$ $=$
10. $\frac{8}{13} \underline{\ ?\ } \frac{9}{14}$ $<$
11. $\frac{25}{-8} \underline{\ ?\ } \frac{-28}{7}$ $>$
12. $6\left(\frac{5}{3}\right) \underline{\ ?\ } \left(\frac{3}{2}\right)6$ $>$
13. $\frac{0.6}{7} \underline{\ ?\ } \frac{1.8}{12}$ $<$
14. $24.6 \underline{\ ?\ } 13.8 - (-12.8)$ $<$
15. $-54 + 26.5 \underline{\ ?\ } 27.5$ $<$
16. $\frac{5.4}{18} \underline{\ ?\ } -4 + 1$ $>$
17. $(4.1)(0.2) \underline{\ ?\ } 8.4$ $<$
18. $-\frac{12}{17} \underline{\ ?\ } -\frac{9}{14}$ $<$

Lesson 2-5 **Find each sum or difference.**

1. $-\frac{11}{9} + \left(-\frac{7}{9}\right)$ **-2**
2. $\frac{5}{11} - \frac{6}{11}$ **$-\frac{1}{11}$**
3. $\frac{2}{7} - \frac{3}{14}$ **$\frac{1}{14}$**
4. $-4.8 + 3.2$ **-1.6**
5. $-1.7 - 3.9$ **-5.6**
6. $-72.5 - 81.3$ **-153.8**
7. $-\frac{3}{5} + \frac{5}{6}$ **$\frac{7}{30}$**
8. $\frac{3}{8} + \left(-\frac{7}{12}\right)$ **$-\frac{5}{24}$**
9. $-\frac{7}{15} + \left(-\frac{5}{12}\right)$ **$-\frac{53}{60}$**
10. $-4.5 - 8.6$ **-13.1**
11. $89.3 - (-14.2)$ **103.5**
12. $-0.007 + 0.06$ **0.053**
13. $-\frac{2}{7} + \frac{3}{14} + \frac{3}{7}$ **$\frac{5}{14}$**
14. $-\frac{3}{5} + \frac{6}{7} + \left(-\frac{2}{35}\right)$ **$\frac{1}{5}$**
15. $\frac{7}{3} + \left(-\frac{5}{6}\right) + \left(-\frac{2}{3}\right)$ **$\frac{5}{6}$**
16. $-4.13 + (-5.18) + 9.63$ **0.32**
17. $6.7 + (-8.1) + (-7.3)$ **-8.7**
18. $\frac{3}{4} + \left(-\frac{5}{8}\right) + \frac{3}{32}$ **$\frac{7}{32}$**
19. $1.9 - (-7)$ **8.9**
20. $-1.8 - 3.7$ **-5.5**
21. $-18 - (-1.3)$ **-16.7**

Lesson 2-6 **Find each product.**

1. $5(12)$ **60**
2. $(-6)(11)$ **-66**
3. $(-7)(-5)$ **35**
4. $\left(-\frac{7}{8}\right)\left(-\frac{1}{3}\right)$ **$\frac{7}{24}$**
5. $(-5)\left(-\frac{2}{5}\right)$ **2**
6. $(-6)(4)(-3)$ **72**
7. $(4)(-2)(-1)(-3)$ **-24**
8. $(-6.8)(-5.415)(3.1)$ **114.1482**
9. $(-5.34)(3.2)$ **-17.088**
10. $\left(\frac{3}{5}\right)\left(-\frac{5}{7}\right)$ **$-\frac{3}{7}$**
11. $-\frac{7}{15}\left(\frac{9}{14}\right)$ **$-\frac{3}{10}$**
12. $(4.2)(-5.1)(3.6)$ **-77.112**
13. $-6\left(\frac{5}{3}\right)\left(\frac{9}{10}\right)$ **-9**
14. $(3)(-6)(0)(-1)$ **0**
15. $(-21)(-2)(-1)$ **-42**

Lesson 2-7 Simplify.

1. $\frac{-48}{8}$ **−6**
2. $-49 \div (-7)$ **7**
3. $-64 \div 8$ **−8**
4. $-\frac{3}{4} \div 9$ $-\frac{1}{12}$
5. $-9 \div \left(-\frac{10}{17}\right)$ $\frac{153}{10} = 15\frac{3}{10}$
6. $\frac{-450n}{10}$ **−45*n***
7. $\frac{-36a}{-6}$ **6*a***
8. $\frac{63a}{-9}$ **−7*a***
9. $8 \div \left(-\frac{5}{4}\right)$ $-\frac{32}{5} = -6\frac{2}{5}$
10. $\frac{\frac{7}{8}}{-10}$ $-\frac{7}{80}$
11. $\frac{12}{-8}$ $-\frac{15}{2} = -7\frac{1}{2}$
12. $\frac{6a + 24}{6}$ $a + 4$
13. $\frac{20a + 30b}{-2}$ $-10a - 15b$
14. $\frac{\frac{5}{11}}{-6}$ $-\frac{11}{30}$
15. $\frac{70a - 42b}{-14}$ $-5a + 3b$
16. $\frac{-32x + 12y}{-4}$ $8x - 3y$
17. $-\frac{7}{12} \div \frac{1}{18}$ $-\frac{21}{2} = -10\frac{1}{2}$
18. $\frac{5}{\frac{15}{-7}}$ $-\frac{7}{3} = -2\frac{1}{3}$

Lesson 2-8 Find each square root. Use a calculator if necessary. Round to the nearest hundredth if the result is not a whole number.

1. $-\sqrt{81}$ **−9**
2. $\sqrt{0.0016}$ **0.04**
3. $\pm\sqrt{206}$ **±14.35**
4. $\pm\sqrt{\frac{81}{64}}$ $\pm\frac{9}{8}$
5. $\sqrt{85}$ **9.22**
6. $-\sqrt{\frac{36}{196}}$ $-\frac{6}{14}$ or $-\frac{3}{7}$
7. $-\sqrt{149}$ **−12.21**
8. $\pm\sqrt{961}$ **±31**
9. $\sqrt{10.24}$ **3.2**

Evaluate each expression. Use a calculator if necessary. Round to the nearest hundredth if the result is not a whole number.

10. $\sqrt{m}$, if $m = 529$ **23**
11. $-\sqrt{c - d}$, if $c = 1.097$ and $d = 1.0171$ **−0.28**
12. $-\sqrt{ab}$, if $a = 1.2$ and $b = 2.7$ **−1.8**
13. $\pm\sqrt{\frac{x}{y}}$, if $x = 144$ and $y = 1521$ **±0.31**

Lesson 2-9 Translate each sentence into an equation, inequality, or formula.

1. The square of a decreased by the cube of b is equal to c. $a^2 - b^3 = c$
2. Twenty-nine decreased by the product of x and y is less than z. $29 - xy < z$
3. The perimeter P of a parallelogram is twice the sum of the lengths of two adjacent sides a and b. $P = 2(a + b)$
4. Four-fifths of the product of m, n, and the square of p is greater than 26. $\frac{4}{5}mnp^2 > 26$
5. Thirty increased by the quotient of s and t is equal to v. $30 + \frac{s}{t} = v$
6. The area A of a trapezoid is half the product of the height h and the sum of the two parallel bases a and b. $A = \frac{1}{2}h(a + b)$

EXTRA PRACTICE

Lesson 3-1 **Solve each equation. Then check your solution.**

1. $-2 + g = 7$ **9**
2. $9 + s = -5$ **−14**
3. $-4 + y = -9$ **−5**
4. $m + 6 = 2$ **−4**
5. $t + (-4) = 10$ **14**
6. $v - 7 = -4$ **3**
7. $a - (-6) = -5$ **−11**
8. $-2 - x = -8$ **6**
9. $d + (-44) = -61$ **−17**
10. $e - (-26) = 41$ **15**
11. $p - 47 = 22$ **69**
12. $-63 - f = -82$ **19**
13. $c + 5.4 = -11.33$ **−16.73**
14. $-6.11 + b = 14.321$ **20.431**
15. $-5 = y - 22.7$ **17.7**
16. $-5 - q = 1.19$ **−6.19**
17. $n + (-4.361) = 59.78$ **64.141**
18. $t - (-46.1) = -3.673$ **−49.773**
19. $\frac{7}{10} - a = \frac{1}{2}$ $\mathbf{\frac{1}{5}}$
20. $f - \left(-\frac{1}{8}\right) = \frac{3}{10}$ $\mathbf{\frac{7}{40}}$
21. $-4\frac{5}{12} = t - \left(-10\frac{1}{36}\right)$ $\mathbf{-14\frac{4}{9}}$
22. $x + \frac{3}{8} = \frac{1}{4}$ $\mathbf{-\frac{1}{8}}$
23. $1\frac{7}{16} + s = \frac{9}{8}$ $\mathbf{-\frac{5}{16}}$
24. $17\frac{8}{9} = d + \left(-2\frac{5}{6}\right)$ $\mathbf{20\frac{13}{18}}$

Lesson 3-2 **Solve each equation. Then check your solution.**

1. $-5p = 35$ **−7**
2. $-3x = -24$ **8**
3. $62y = -2356$ **−38**
4. $\frac{a}{-6} = -2$ **12**
5. $\frac{c}{-59} = -7$ **413**
6. $\frac{f}{14} = -63$ **−882**
7. $84 = \frac{x}{97}$ **8148**
8. $\frac{w}{5} = 3$ **15**
9. $\frac{q}{9} = -3$ **−27**
10. $\frac{2}{5}x = \frac{4}{7}$ $\mathbf{\frac{10}{7}}$ **or** $\mathbf{1\frac{3}{7}}$
11. $\frac{z}{6} = -\frac{5}{12}$ $\mathbf{-\frac{5}{2}}$
12. $-\frac{5}{9}r = 7\frac{1}{2}$ $\mathbf{-13\frac{1}{2}}$
13. $2\frac{1}{6}j = 5\frac{1}{5}$ $\mathbf{2\frac{2}{5}}$
14. $3 = 1\frac{7}{11}q$ $\mathbf{1\frac{5}{6}}$
15. $-1\frac{3}{4}p = -\frac{5}{8}$ $\mathbf{\frac{5}{14}}$
16. $57k = 0.1824$ **0.0032**
17. $0.0022b = 0.1958$ **89**
18. $5j = -32.15$ **−6.43**
19. $\frac{w}{-2} = -2.48$ **4.96**
20. $\frac{z}{2.8} = -6.2$ **−17.36**
21. $\frac{x}{-0.063} = 0.015$ **−0.000945**
22. $15\frac{3}{8} = -5.125p$ **−3**
23. $-7.25 = -3\frac{5}{8}g$ **2**
24. $-18\frac{1}{4} = 2.50x$ **−7.3**

Lesson 3-3 **Solve each equation. Then check your solution.**

1. $2x - 5 = 3$ **4**
2. $4t + 5 = 37$ **8**
3. $7a + 6 = -36$ **−6**
4. $47 = -8g + 7$ **−5**
5. $-3c - 9 = -24$ **5**
6. $5k - 7 = -52$ **−9**
7. $5s + 4s = -72$ **−8**
8. $3x - 7 = 2$ **3**
9. $8 + 3x = 5$ **−1**
10. $-3y + 7.569 = 24.069$ **−5.5**
11. $7 - 9.1f = 137.585$ **−14.35**
12. $6.5 = 2.4m - 4.9$ **4.75**
13. $\frac{e}{5} + 6 = -2$ **−40**
14. $\frac{d}{4} - 8 = -5$ **12**
15. $-\frac{4}{13}y - 7 = 6$ $\mathbf{-42\frac{1}{4}}$
16. $\frac{p + 10}{3} = 4$ **2**
17. $\frac{h - 7}{6} = 1$ **13**
18. $\frac{5f + 1}{8} = -3$ **−5**
19. $\frac{4n - 8}{-2} = 12$ **−4**
20. $\frac{2a}{7} + 9 = 3$ **−21**
21. $\frac{-3t - 4}{2} = 8$ $\mathbf{-6\frac{2}{3}}$

762

Lesson 3-4 Find the complement of each angle measure.

1. 15° **75°**
2. 79° **11°**
3. 88° **2°**
4. a° **$(90 - a)^\circ$**
5. $(3c)^\circ$ **$(90 - 3c)^\circ$**
6. $(b - 15)^\circ$ **$(105 - b)^\circ$**

Find the supplement of each angle measure.

7. 156° **24°**
8. 94° **86°**
9. 21° **159°**
10. a° **$(180 - a)^\circ$**
11. $(3c)^\circ$ **$(180 - 3c)^\circ$**
12. $(b - 15)^\circ$ **$(195 - b)^\circ$**

Find the measure of the third angle of each triangle in which the measures of two angles of the triangle are given.

13. $90^\circ, 2^\circ$ **88°**
14. $34^\circ, 132^\circ$ **14°**
15. $111^\circ, 28^\circ$ **41°**
16. a°, b° **$(180 - a - b)^\circ$**
17. $a^\circ, (a - 15)^\circ$ **$(195 - 2a)^\circ$**
18. $b^\circ, (3b - 2)^\circ$ **$(182 - 4b)^\circ$**

Lesson 3-5 Solve each equation. Then check your solution.

1. $6(y - 5) = 18$ **8**
2. $-21 = 7(p - 10)$ **7**
3. $3(h + 2) = 12$ **2**
4. $-3(x + 2) = -18$ **4**
5. $11.2n + 6 = 5.2n$ **−1**
6. $2m + 5 - 6m = 25$ **−5**
7. $3z - 1 = 23 - 3z$ **4**
8. $5a - 5 = 7a - 19$ **7**
9. $5b + 12 = 3b - 6$ **−9**
10. $3x - 5 = 7x + 7$ **−3**
11. $1.9s + 6 = 3.1 - s$ **−1**
12. $2.85y - 7 = 12.85y - 2$ **$-\frac{1}{2}$**
13. $2.9m + 1.7 = 3.5 + 2.3m$ **3**
14. $3(x + 1) - 5 = 3x - 2$ **identity**
15. $4(2y - 1) = -10(y - 5)$ **3**
16. $\frac{6v - 9}{3} = v$ **3**
17. $\frac{3t + 1}{4} = \frac{3}{4}t - 5$ **no solution**
18. $\frac{2}{5}y + \frac{y}{2} = 9$ **10**
19. $3y - \frac{4}{5} = \frac{1}{3}y$ **$\frac{3}{10}$**
20. $\frac{3}{4}x - 4 = 7 + \frac{1}{2}x$ **44**
21. $\frac{x}{2} - \frac{1}{3} = \frac{x}{3} - \frac{1}{2}$ **−1**

Lesson 3-6 Solve each equation for *x*.

1. $x + r = q$ **$q - r$**
2. $ax + 4 = 7$ **$\frac{3}{a}$**
3. $2bx - b = -5$ **$\frac{-5 + b}{2b}$**
4. $\frac{x - c}{c + a} = a$ **$a^2 + ca + c$**
5. $\frac{x + y}{c} = d$ **$cd - y$**
6. $\frac{ax + 1}{2} = b$ **$\frac{2b - 1}{a}$**
7. $\frac{x + t}{4} = d$ **$4d - t$**
8. $6x - 7 = -r$ **$\frac{7 - r}{6}$**
9. $kx + 4y = 5z$ **$\frac{5z - 4y}{k}$**
10. $ax - 6 = t$ **$\frac{t + 6}{a}$**
11. $\frac{2}{3}x + a = b$ **$\frac{3}{2}(b - a)$**
12. $q(x + 1) = 5$ **$\frac{5}{q} - 1$**
13. $\frac{x - y}{z} = 8$ **$8z + y$**
14. $\frac{7a + b}{x} = 1$ **$7a + b$**
15. $\frac{4cx + t}{7} = 2$ **$\frac{14 - t}{4c}$**
16. $\frac{9x - 4c}{z} = z$ **$\frac{z^2 + 4c}{9}$**
17. $cx + a = bx$ **$\frac{-a}{c - b}$**
18. $\frac{12q - x}{5} = t$ **$12q - 5t$**

EXTRA PRACTICE

EXTRA PRACTICE

Lesson 3-7

1. **Geography** The areas in square miles of the 20 largest natural U.S. lakes are given below. Find the mean, median, and mode of the areas. **5166.6; 541.5; none**

   31,700 1697 242 700 22,300 451 374 432 23,000 207
   1000 1361 315 625 9910 215 458 360 7550 435

2. **Work** Each number below represents the number of days that each employee of the Cole Corporation was absent during 1996. Find the mean, median, and mode of the number of days absent. **4.45; 4; 3, 5**

   0 10 8 5 8 9 3 3 2 9 7 0 4 2 4 6 2
   9 13 3 1 5 5 7 2 6 5 3 4 7 1 1 5 3
   3 1 4 5 1 2

3. **Football** Tailback Michael Anderson, of the West High Bears, averaged 137.6 yards rushing per game for the first five games of the season. He rushes for 155 yards in the sixth game. What is his new rushing average? **140.5**

4. **Olympic Games** One of the events in the Winter Olympics is the men's 500-meter speed skating. The winning times for this event are shown at the right. Find the mean, median, and mode of the times. **40.164; 40.15; 40.2, 43.4**

Year	Time(s)	Year	Time(s)
1932	43.4	1968	40.3
1936	43.4	1972	39.4
1948	43.1	1976	39.2
1952	43.2	1980	38.0
1956	40.2	1984	38.2
1960	40.2	1988	36.5
1964	40.1	1992	37.1

Lesson 4-1 Solve each proportion.

1. $\frac{4}{5} = \frac{x}{20}$ **16**
2. $\frac{b}{63} = \frac{3}{7}$ **27**
3. $\frac{y}{5} = \frac{3}{4}$ **$\frac{15}{4} = 3.75$**
4. $\frac{7}{4} = \frac{3}{a}$ **$\frac{12}{7} \approx 1.71$**
5. $\frac{t-5}{4} = \frac{3}{2}$ **11**
6. $\frac{x}{9} = \frac{0.24}{3}$ **0.72**
7. $\frac{n}{3} = \frac{n+4}{7}$ **3**
8. $\frac{12q}{-7} = \frac{30}{14}$ **$-\frac{5}{4} = -1.25$**
9. $\frac{1}{y-3} = \frac{3}{y-5}$ **2**
10. $\frac{r-1}{r+1} = \frac{3}{5}$ **4**
11. $\frac{a-3}{8} = \frac{3}{4}$ **9**
12. $\frac{6p-2}{7} = \frac{5p+7}{8}$ **5**
13. $\frac{2}{9} = \frac{k+3}{2}$ **$-\frac{23}{9} \approx -2.56$**
14. $\frac{5m-3}{4} = \frac{5m+3}{6}$ **3**
15. $\frac{w-5}{4} = \frac{w+3}{3}$ **−27**
16. $\frac{96.8}{t} = \frac{12.1}{7}$ **56**
17. $\frac{x}{6.03} = \frac{4}{17.42}$ **1.385**
18. $\frac{4n+5}{5} = \frac{2n+7}{7}$ **0**

Lesson 4-2 Determine whether each pair of triangles is similar. Justify your answer.

1. **no**

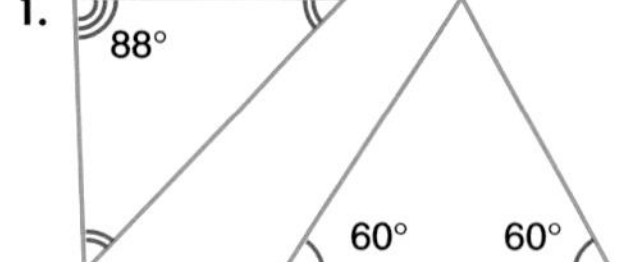

2. **yes**

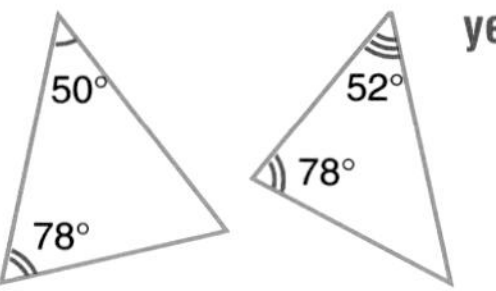

3. **no**

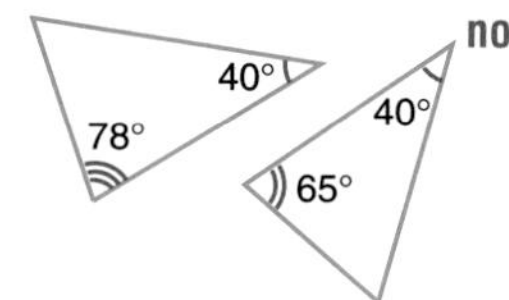

4. **no**

5. **yes**

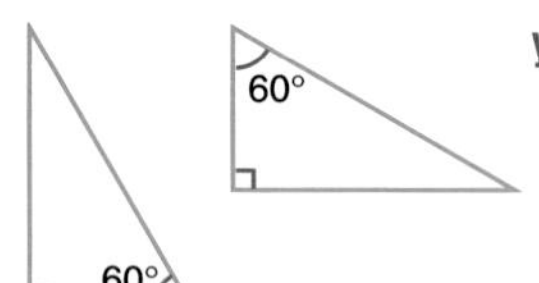

6. **no**

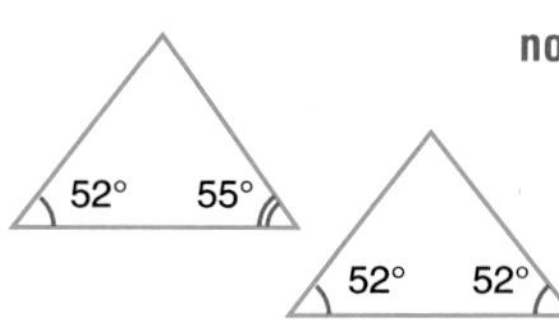

Lesson 4-3 For each triangle, find sin *N*, cos *N*, and tan *N* to the nearest thousandth. **1–3. See margin.**

1. Triangle MNP: MN = 39, NP = 15, MP = 36, right angle at P.
2. Triangle SRN: SR = 14, SN = 50, RN = 48, right angle at R.
3. Triangle NXY: NY = 37, XY = 35, NX = 12, right angle at X.

Use a calculator to find the value of each trigonometric ratio to the nearest ten thousandth.

4. cos 25° **0.9063**
5. tan 31° **0.6009**
6. sin 71° **0.9455**
7. cos 64° **0.4384**
8. tan 9° **0.1584**
9. sin 2° **0.0349**

Use a calculator to find the measure of each angle to the nearest degree.

10. tan $B = 0.5427$ **28°**
11. cos $A = 0.8480$ **32°**
12. sin $J = 0.9654$ **75°**
13. cos $Q = 0.3645$ **69°**
14. sin $R = 0.2104$ **12°**
15. tan $V = 0.956$ **44°**

Lesson 4-4 Write each ratio as a percent and then as a decimal.

1. $\frac{1}{4}$ **25%, 0.25**
2. $\frac{34}{100}$ **34%, 0.34**
3. $\frac{4}{25}$ **16%, 0.16**
4. $\frac{3}{20}$ **15%, 0.15**
5. $\frac{7}{8}$ **$87\frac{1}{2}$%, 0.875**
6. $\frac{9}{10}$ **90%, 0.9**
7. $\frac{24}{40}$ **60%, 0.6**
8. $\frac{4}{50}$ **8%, 0.08**
9. $\frac{7}{15}$ **$46\frac{2}{3}$%, $0.4\overline{6}$**
10. $\frac{4}{9}$ **$44\frac{4}{9}$%, $0.\overline{4}$**
11. $\frac{36}{15}$ **240%, 2.4**
12. $\frac{18}{4}$ **450%, 4.5**

Use a proportion to answer each question.

13. Twenty-four is what percent of 48? **50%**
14. What percent of 70 is 14? **20%**
15. Nine is what percent of 72? **$12\frac{1}{2}$%; 12.5%**
16. Fourteen is 17.5% of what number? **80**
17. What percent of 16 is 5.12? **32%**
18. What number is 25% of 64? **16**
19. What percent of 80 is 2? **2.5%**
20. Forty-five is what percent of 112.5? **40%**

Lesson 4-5 State whether each percent of change is a percent of increase or a percent of decrease. Then find the percent of increase or decrease. Round to the nearest whole percent.

1. original: $100
new: $67 **D; 33%**
2. original: 62 acres
new: 98 acres **I, 58%**
3. original: 322 people
new: 289 people **D, 10%**
4. original: 78 pennies
new: 36 pennies **D, 54%**
5. original: $212
new: $230 **I, 8%**
6. original: 35 mph
new: 65 mph **I, 86%**

Find the final price of each item. When there is a discount and sales tax, first compute the discount price and then compute the sales tax and final price.

7. television: $299
sales tax: 4.5%
$312.46
8. boots: $49.99
discount: 15%
sales tax: 3.5% **$43.98**
9. backpack: $28.95
discount: 10%
sales tax: 5% **$27.35**
10. software: $36.99
sales tax: 6.25%
$39.30
11. jacket: $65
discount: 30%
sales tax: 4% **$47.32**
12. book: $15.95
sales tax: 7%
$17.07

EXTRA PRACTICE

Additional Answers for Lesson 4-3

1. sin $N = 0.923$, cos $N = 0.385$, tan $N = 2.400$
2. sin $N = 0.280$, cos $N = 0.960$, tan $N = 0.292$
3. sin $N = 0.946$, cos $N = 0.324$, tan $N = 2.917$

Lesson 4–6 **Determine the probability of each event.** 2. 1

1. a coin will land tails up $\frac{1}{2}$
2. There is a December 1 this year.
3. A baby will be a girl. $\frac{1}{2}$
4. Next year will have 400 days. 0
5. This is an algebra book. 1
6. Today is Wednesday. $\frac{1}{7}$

Find the probability of each outcome if a computer randomly chooses a letter in the word "success."

7. the letter e $\frac{1}{7}$
8. P(not c) $\frac{5}{7}$
9. the letter s $\frac{3}{7}$
10. the letter b 0
11. P(vowel) $\frac{2}{7}$
12. the letters u or c $\frac{3}{7}$

Find the odds of each outcome if a die is rolled.

13. a 4 1:5
14. a number greater than 3 1:1
15. a multiple of 3 1:2
16. a number less than 5 2:1
17. an odd number 1:1
18. not a 6 5:1

Lesson 4–7

1. **Advertising** An advertisement for an orange drink claims that the drink contains 10% orange juice. How much pure orange juice would have to be added to 5 quarts of the drink to obtain a mixture containing 40% orange juice? 2.5 quarts
2. **Finance** Jane Pham is investing $6000 in two accounts, part at 4.5% and the remainder at 6%. If the total annual interest earned from the two accounts is $279, how much did Jane deposit at each rate? $5400 at 4.5%, $600 at 6%
3. **Entertainment** At the Golden Oldies Theater, tickets for adults cost $5.50 and tickets for children cost $3.50. How many of each kind of ticket was purchased if 21 tickets were bought for $83.50? 5 adults, 16 children
4. **Automotives** A car radiator has a capacity of 14 quarts and is filled with a 20% antifreeze solution. How much must be drained off and replaced with pure antifreeze to obtain a 40% antifreeze solution? 3.5 quarts
5. **Nutrition** A liter of cream has 9.2% butterfat. How much skim milk containing 2% butterfat should be added to the cream to obtain a mixture with 6.4% butterfat? $0.\overline{63}$ L

Lesson 4–8 **Determine which equations represent inverse variations and which represent direct variations. Then find the constant of variation.**

1. $ab = 6$ I, 6
2. $\frac{50}{y} = x$ I, 50
3. $\frac{1}{5}a = d$ D, $\frac{1}{5}$
4. $s = 3t$ D, 3
5. $14 = cd$ I, 14
6. $2x = y$ D, 2

Solve. Assume that y varies directly as x. 13. −12 16. −6

7. If $y = 45$ when $x = 9$, find y when $x = 7$. 35
8. If $y = 18$ when $x = 27$, find x when $y = 8$. 12
9. If $y = 450$ when $x = 6$, find y when $x = 10$. 750
10. If $y = 6$ when $x = 48$, find y when $x = 20$. 2.5
11. If $y = 25$ when $x = 20$, find x when $y = 35$. 28
12. If $y = 100$ when $x = 40$, find y when $x = 16$. 40
13. If $y = -7$ when $x = -1$, find x when $y = -84$.
14. If $y = 5$ when $x = -10$, find y when $x = 50$. −25
15. If $y = 24$ when $x = 6$, find y when $x = 14$. 56
16. If $y = -10$ when $x = -4$, find x when $y = -15$.

Solve. Assume that y varies inversely as x. 25. 11.7 26. 2.5

17. If $y = 54$ when $x = 4$, find x when $y = 27$. 8
18. If $y = 18$ when $x = 6$, find x when $y = 12$. 9
19. If $y = 2$ when $x = 26$, find y when $x = 4$. 13
20. If $y = 3$ when $x = 8$, find x when $y = 4$. 6
21. If $y = 12$ when $x = 24$, find x when $y = 9$. 32
22. If $y = 8$ when $x = -8$, find y when $x = -16$. 4
23. If $y = 3$ when $x = -8$, find y when $x = 4$. −6
24. If $y = 27$ when $x = \frac{1}{3}$, find y when $x = \frac{3}{4}$. 12
25. If $y = 19.5$ when $x = 6.3$, find x when $y = 10.5$.
26. If $y = 4.8$ when $x = 10$, find y when $x = 19.2$.

766

Lesson 5-1 Refer to the coordinate plane below. Write the ordered pair for each point. Name the quadrant in which the point is located.

1. B (1, 2), I
2. T (−5, 0), none
3. P (6, −2), IV
4. Q (0, 6), none
5. A (−2, −2), III
6. K (4, 5), I
7. J (2, −5), IV
8. L (4, 0), none
9. S (−3, 5), II
10. D (7, 3), I
11. M (−7, −5), III
12. N (−6, 4), II

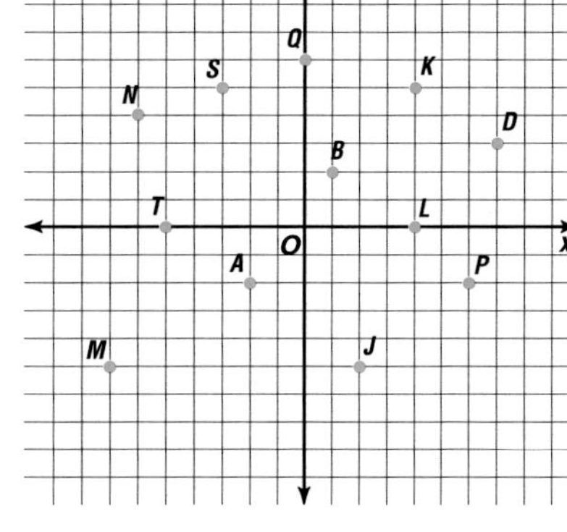

Graph each point. 13–20. See margin.

13. $A(2, -4)$
14. $B(3, 5)$
15. $C(-4, 0)$
16. $D(-4, 3)$
17. $E(-5, -5)$
18. $F(-1, 1)$
19. $G(0, -3)$
20. $H(2, 3)$

Lesson 5-2 State the domain and range of each relation.

1. {(5, 2), (0, 0), (−9, −1)}
{5, 0, −9}; {2, 0, −1}
2. {(−4, 2), (−2, 0), (0, 2), (2, 4)}
{−4, −2, 0, 2}; {2, 0, 4}
3. {(7, 5), (−2, −3), (4, 0), (5, −7), (−9, 2)}
{7, −2, 4, 5, −9}; {5, −3, 0, −7, 2}
4. {(3.1, −1), (−4.7, 3.9), (2.4, −3.6), (−9, 12.12)}
{3.1, −4.7, 2.4, −9}; {−1, 3.9, −3.6, 12.12}

Express the relation shown in each table, mapping, or graph as a set of ordered pairs. Then state the domain, range, and inverse of the relation. 5–10. See margin.

5.

x	y
1	3
2	4
3	5
4	6
5	7

6.

x	y
−4	1
−2	3
0	1
2	3
4	1

7. X: −1, −2, −6 → Y: 5, 4, 1

8.

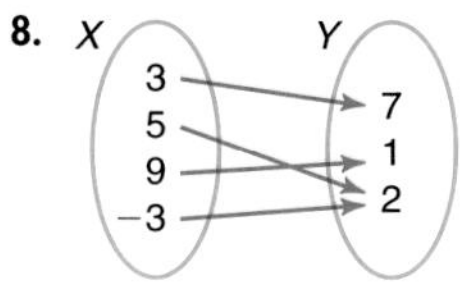

9.

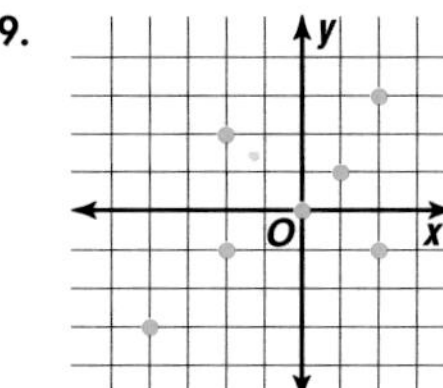

10.

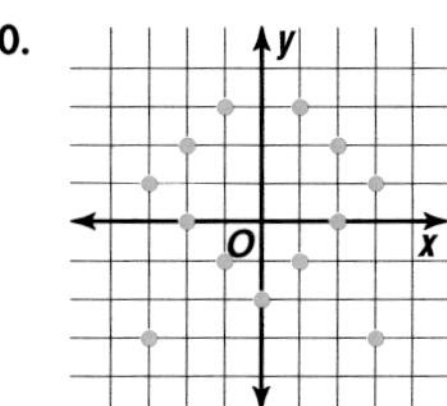

Lesson 5-3 Which ordered pairs are solutions of each equation?

		a.	b.	c.	d.
1. $3r = 8s - 4$	a, b, c	$\left(\frac{2}{3}, \frac{3}{4}\right)$	$\left(0, \frac{1}{2}\right)$	(4, 2)	(2, 4)
2. $3y = x + 7$	c, d	(2, 4)	(2, −1)	(2, 3)	(−1, 2)
3. $4x = 8 - 2y$	a	(2, 0)	(0, 2)	(0.5, −3)	(1, −2)
4. $3n = 10 - 4m$	b, c	(0, 3)	(−2, 6)	(1, 2)	(2, 1)

Solve each equation if the range is {−3, −1, 0, 2, 3}. 5–10. See margin.

5. $y = 2x$
6. $y = 5x + 1$
7. $2a + b = 4$
8. $4r + 3s = 13$
9. $5b = 8 - 4a$
10. $6m - n = -3$

EXTRA PRACTICE

Additional Answers for Lesson 5-1

13–20.

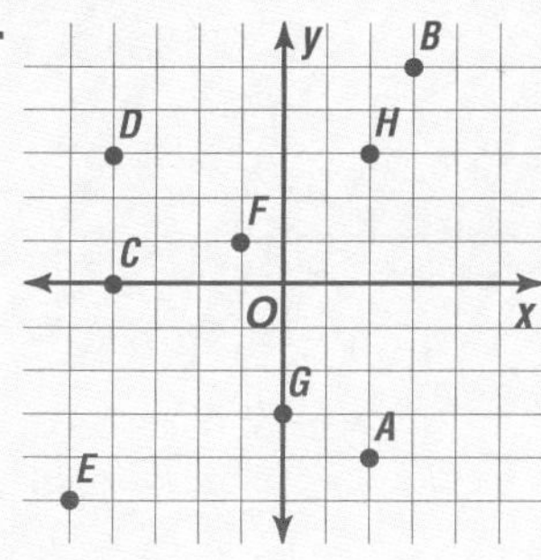

Additional Answers for Lesson 5-2

5. {(1, 3), (2, 4), (3, 5), (4, 6), (5, 7)}; {1, 2, 3, 4, 5}; {3, 4, 5, 6, 7}; {(3, 1), (4, 2), (5, 3), (6, 4), (7, 5)}
6. {(−4, 1), (−2, 3), (0, 1), (2, 3), (4, 1)}; {−4, −2, 0, 2, 4}; {1, 3}; {(1, −4), (3, −2), (1, 0), (3, 2), (1, 4)}
7. {(−1, 5), (−2, 5), (−2, 4), (−2, 1), (−6, 1)}; {−1, −2, −6}; {5, 4, 1}; {(5, −1), (5, −2), (4, −2), (1, −2), (1, −6)}
8. {(3, 7), (5, 2), (9, 1), (−3, 2)}; {3, 5, 9, −3}; {7, 1, 2}; {(7, 3), (2, 5), (1, 9), (2, −3)}
9. {(0, 0), (1, 1), (2, 3), (2, −1), (−2, 2), (−2, −1), (−4, −3)}; {0, 1, 2, −2, −4}; {0, 1, 3, −1, 2, −3}; {(0, 0), (1, 1), (3, 2), (−1, 2), (2, −2), (−1, −2), (−3, −4)}
10. {(−3, 1), (−3, −3), (−2, 0), (−2, 2), (−1, 3), (−1, −1), (0, −2), (1, 3), (1, −1), (2, 0), (2, 2), (3, 1), (3, −3)}; {−3, −2, −1, 0, 1, 2, 3}; {1, −3, 0, 2, 3, −1, −2}; {(1, −3), (−3, −3), (0, −2), (2, −2), (3, −1), (−1, −1), (−2, 0), (3, 1), (−1, 1), (0, 2), (2, 2), (1, 3), (−3, 3)}

Additional Answers for Lesson 5-3

5. $\left\{\left(-\frac{3}{2}, -3\right), \left(-\frac{1}{2}, -1\right), (0, 0), (1, 2), \left(\frac{3}{2}, 3\right)\right\}$
6. $\left\{\left(-\frac{4}{5}, -3\right), \left(-\frac{2}{5}, -1\right), \left(-\frac{1}{5}, 0\right), \left(\frac{1}{5}, 2\right), \left(\frac{2}{5}, 3\right)\right\}$
7. $\left\{\left(\frac{7}{2}, -3\right), \left(\frac{5}{2}, -1\right), (2, 0), (1, 2), \left(\frac{1}{2}, 3\right)\right\}$
8. $\left\{\left(\frac{11}{2}, -3\right), (4, -1), \left(\frac{13}{4}, 0\right), \left(\frac{7}{4}, 2\right), (1, 3)\right\}$
9. $\left\{\left(\frac{23}{4}, -3\right), \left(\frac{13}{4}, -1\right), (2, 0), \left(-\frac{1}{2}, 2\right), \left(-\frac{7}{4}, 3\right)\right\}$
10. $\left\{(-1, -3), \left(-\frac{2}{3}, -1\right), \left(-\frac{1}{2}, 0\right), \left(-\frac{1}{6}, 2\right), (0, 3)\right\}$

Additional Answers for Lesson 5-4

7.

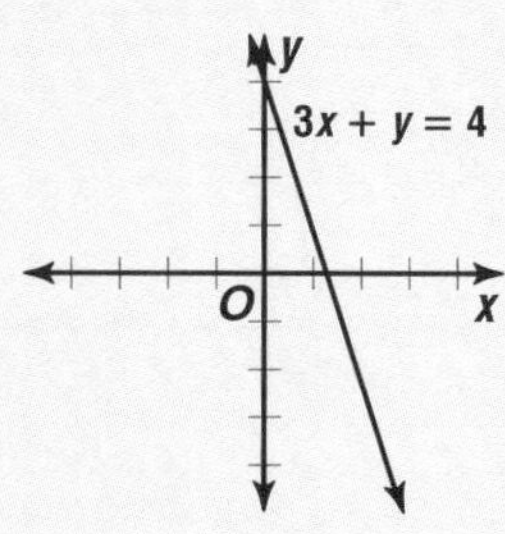

8.

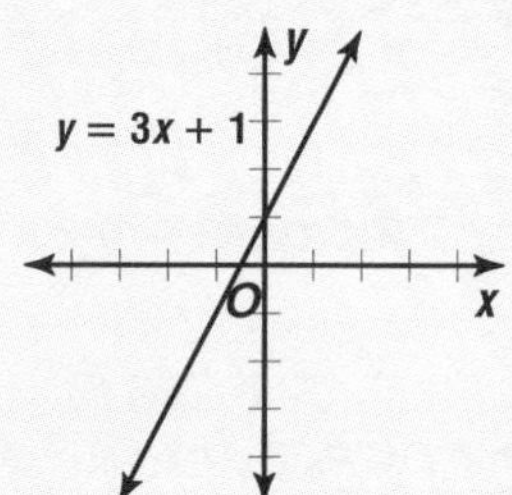

9.

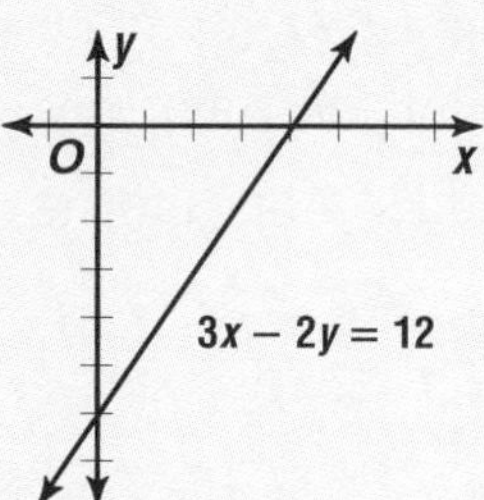

10.

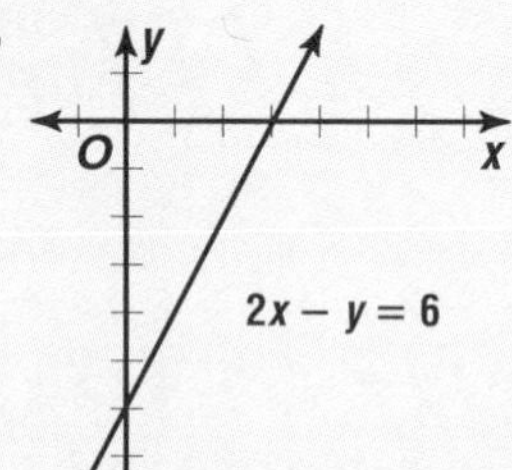

11.

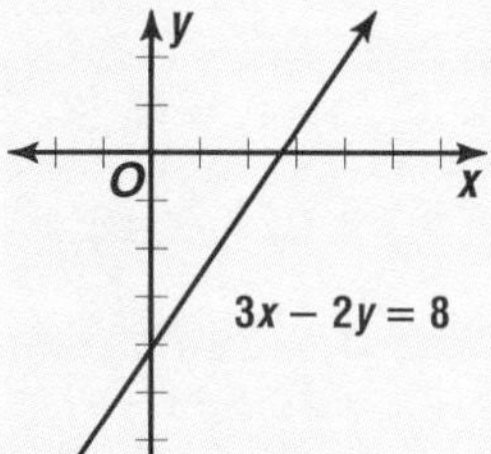

12.

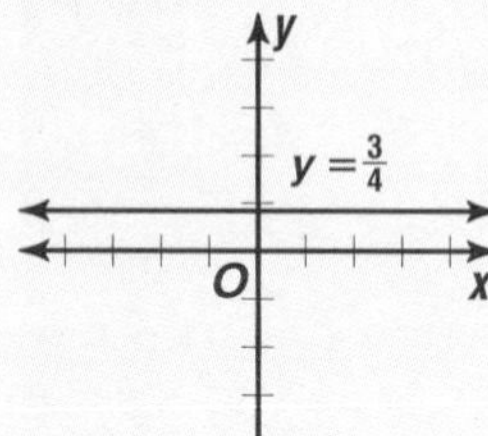

13.

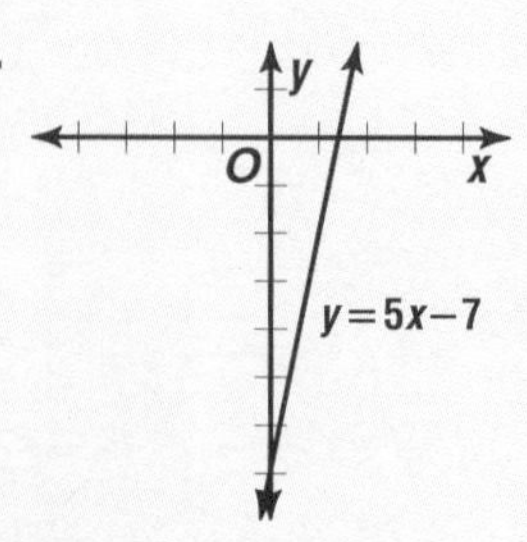

Lesson 5-4 Determine whether each equation is a linear equation. If an equation is linear, rewrite it in the form $Ax + By = C$. 4. yes; $3x - 7y = -7$

1. $3x = 2y$ yes; $3x - 2y = 0$
2. $2x - 3 = y^2$ no
3. $3x - 2y = 8$ yes
4. $5x - 7y = 2x - 7$
5. $2x + 5x = 7y$ yes; $7x - 7y = 0$
6. $\frac{1}{x} + \frac{5}{y} = -4$ no

Graph each equation. 7–18. See margin.

7. $3x + y = 4$
8. $y = 3x + 1$
9. $3x - 2y = 12$
10. $2x - y = 6$
11. $3x - 2y = 8$
12. $y = \frac{3}{4}$
13. $y = 5x - 7$
14. $x + \frac{1}{3}y = 6$
15. $x = -\frac{5}{2}$
16. $5x - 2y = 8$
17. $4x + 2y = 9$
18. $4x + 3y = 12$

Lesson 5-5 Determine whether each relation is a function.

1. no

x	y
1	3
2	5
1	−7
2	9
3	3

2. no

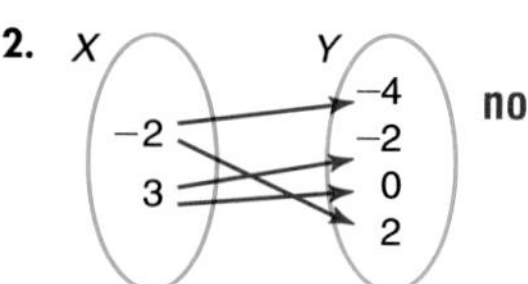

3. yes

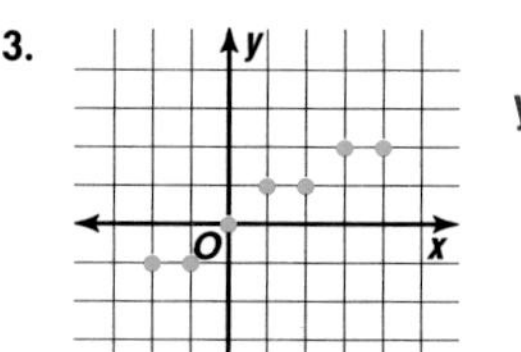

4. no

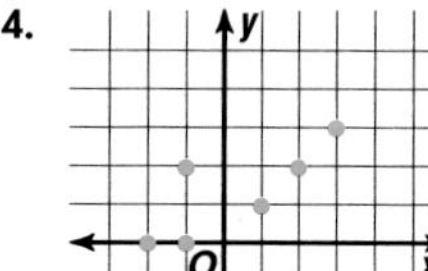

5. yes

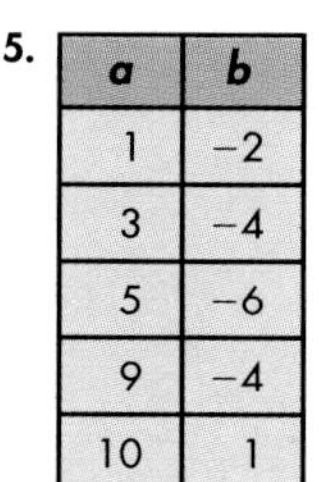

a	b
1	−2
3	−4
5	−6
9	−4
10	1

6. yes

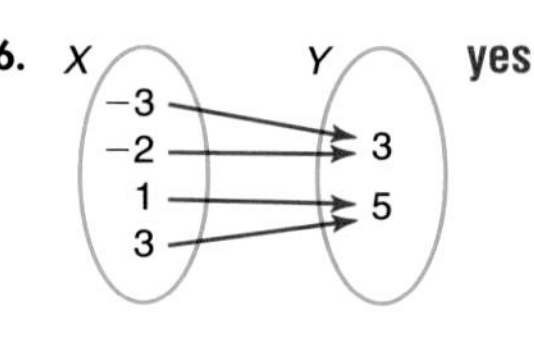

7. $\{(-2, 4), (1, 3), (5, 2), (1, 4)\}$ no
8. $\{(5, 4), (-6, 5), (4, 5), (0, 4)\}$ yes
9. $\{(3, 1), (5, 1), (7, 1)\}$ yes
10. $\{(3, -2), (4, 7), (-2, 7), (4, 5)\}$ no
11. $y = 2$ yes
12. $x^2 + y = 11$ yes

Lesson 5-6 Write an equation for each relation. 9. $y = \frac{2}{3}x - 4$ 10. $y = -2x + 3$

1. $f(x) = \frac{1}{2}x - 5$

x	2	4	6	8	10
f(x)	−4	−3	−2	−1	0

2. $f(x) = x^2 - 4$

x	−2	−1	0	1	2
f(x)	0	−3	−4	−3	0

3. $f(x) = 4x + 3$

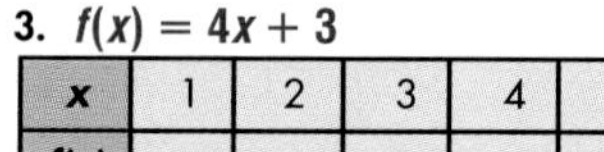

x	1	2	3	4	5
f(x)	7	11	15	19	23

4.

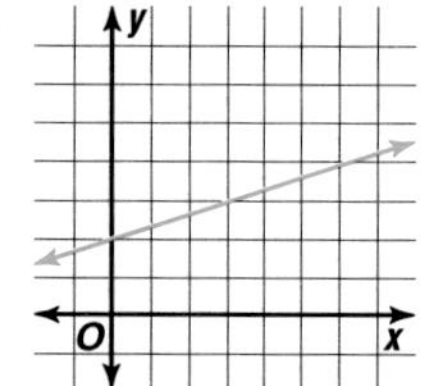

$y = \frac{1}{3}x + 2$

5.

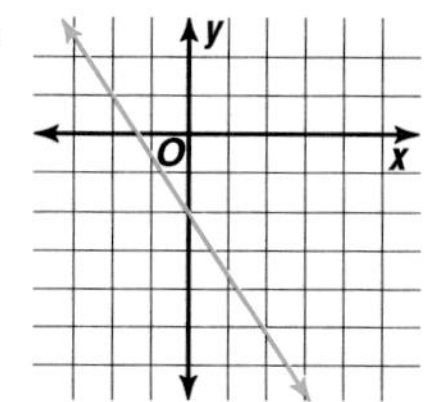

$y = -\frac{3}{2}x - 2$

6.

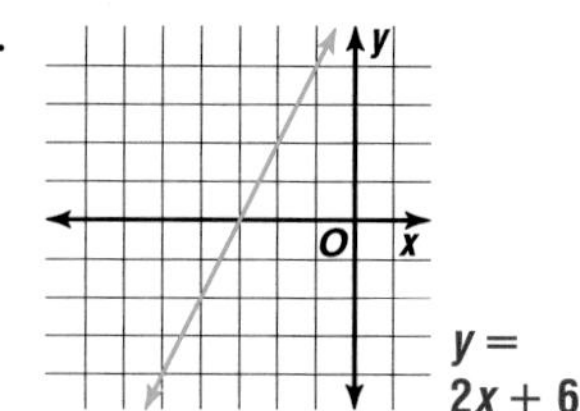

$y = 2x + 6$

7. $\{(3, 12), (4, 14), (5, 16), (6, 18), (7, 20)\}$ $y = 2x + 6$
8. $\{(1, 4), (2, 8), (3, 12), (-2, -8), (-3, -12)\}$ $y = 4x$
9. $\{(-6, -8), (-3, -6), (0, -4), (3, -2), (6, 0)\}$
10. $\{(-3, 9), (-2, 7), (1, 1), (4, -5), (6, -9)\}$

14.

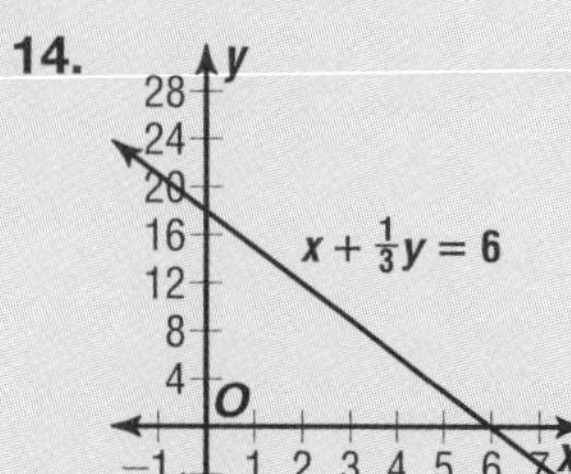

15.

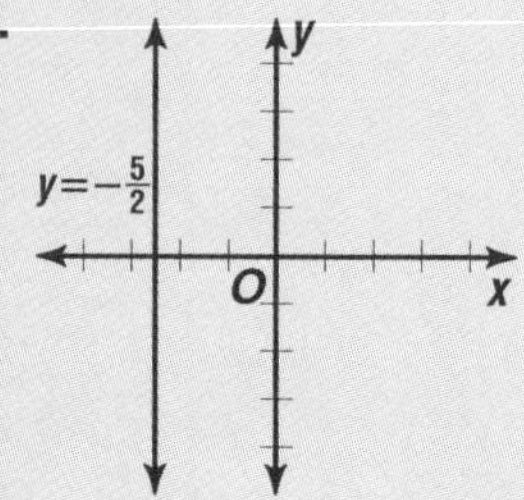

16.

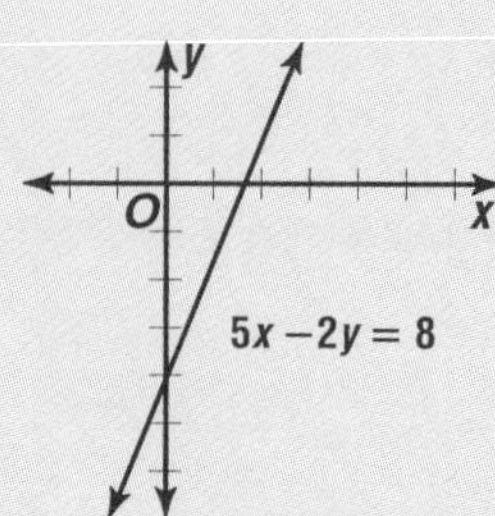

Lesson 5-7 Find the range, median, upper quartile, lower quartile, and interquartile range for each set of data. 4. 8; 4; 7.1; 2.4; 4.7

1. 56, 45, 37, 43, 10, 34 **46; 40; 45; 34; 11**
2. 77, 78, 68, 96, 99, 84, 65 **34; 78; 96; 68; 28**
3. 30, 90, 40, 70, 50, 100, 80, 60 **70; 65; 85; 45; 40**
4. 4, 5.2, 1, 3, 2.4, 6, 3.7, 8, 1.3, 7.1, 9
5. 25°, 56°, 13°, 44°, 0°, 31°, 73°, 66°, 4°, 29°, 37° **73°; 31°; 56°; 13°; 43°**
6. 234, 648, 369, 112, 527, 775, 406, 268, 400 **663; 400; 587.5; 251; 336.5**

7.

Stem	Leaf
0	0 2 3
1	1 7 9
2	2 3 5 6
3	3 4 4 5 9
4	0 7 8 8

$2|2 = 22$

48; 26; 39; 17; 22

8.

Stem	Leaf
7	3 4 7 8
8	0 0 3 5 7
9	4 6 8
10	0 1 8
11	1 9

$9|4 = 9.4$

4.6; 8.7; 10.05; 7.9; 2.15

9.

Stem	Leaf
25	0 3 7 9
26	1 3 4 5 5 6
27	1 5 6 6 9
28	1 2 3 5 8
29	2 5 6 9

$27|5 = 2750$

490; 2755; 2840; 2635; 205

Lesson 6-1 Determine the slope of each line.

1. a **3**
2. b **−1**
3. c **$\frac{3}{2}$**
4. d **0**
5. e **undefined**
6. f **$-\frac{1}{5}$**
7. g **$\frac{1}{3}$**
8. h **−2**

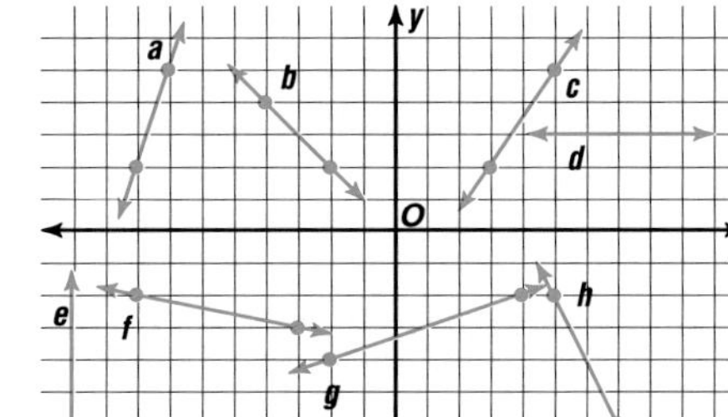

Determine the slope of the line that passes through each pair of points.

9. (−2, 2), (3, −3) **−1**
10. (−2, −8), (1, 4) **4**
11. (3, 4), (4, 6) **2**
12. (−5, 4), (−1, 11) **$\frac{7}{4}$**
13. (18, −4), (6, −10) **$\frac{1}{2}$**
14. (−4, −6), (−4, −8) **undefined**

Determine the value of r so the line that passes through each pair of points has the given slope.

15. $(-1, r), (1, -4), m = -5$ **6**
16. $(-2, 1), (r, 4), m = \frac{3}{5}$ **3**
17. $(-1, 3), (-3, r), m = -3$ **9**
18. $(3, r), (7, -2), m = \frac{1}{2}$ **−4**
19. $(r, -2), (-7, -1), m = -\frac{1}{4}$ **−3**
20. $(-3, 2), (7, r), m = \frac{2}{3}$ **$\frac{26}{3}$**

Lesson 6-2 Write the point-slope form of an equation of the line that passes through the given point and has the given slope. 2. $y - 4 = -5(x - 4)$ 3. $y + 4 = \frac{3}{4}(x + 2)$

1. $(5, -2), m = 3$ **$y + 2 = 3(x - 5)$**
2. $(5, 4), m = -5$
3. $(-2, -4), m = \frac{3}{4}$
4. $(-3, 1), m = 0$ **$y - 1 = 0$**
5. $(-1, 0), m = \frac{2}{3}$ **$y = \frac{2}{3}(x + 1)$**
6. $(0, 6), m = -2$ **$y - 6 = -2x$**

Write the standard form of an equation of the line that passes through the given point and has the given slope. 7. $x + 2y = -12$

7. $(-6, -3), m = -\frac{1}{2}$
8. $(4, -3), m = 2$ **$2x - y = 11$**
9. $(5, 4), m = -\frac{2}{3}$ **$2x + 3y = 22$**
10. $(1, 3), m =$ undefined **$x = 1$**
11. $(-2, 6), m = 0$ **$y = 6$**
12. $(6, -2), m = \frac{4}{3}$ **$4x - 3y = 30$**

Additional Answers for Lesson 5-4

17.

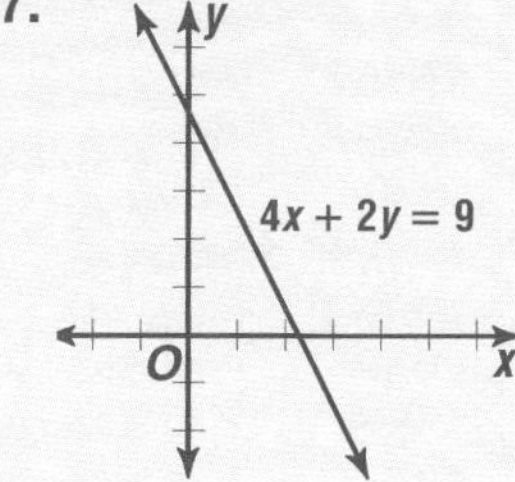

18.

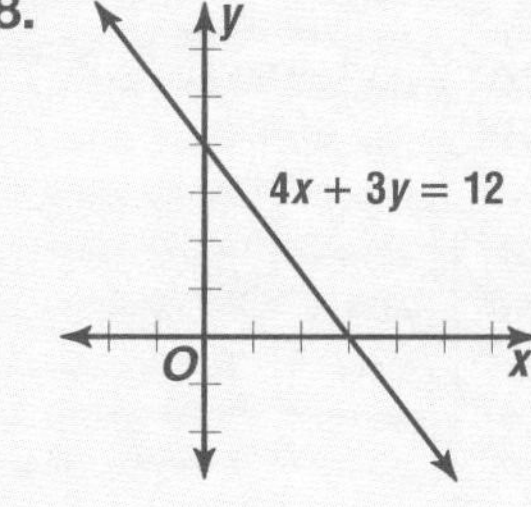

Additional Answers for Lesson 6-4

7. $y = -\frac{2}{5}x + 2$; $2x + 5y = 10$
8. $y = 5x - 15$; $5x - y = 15$
9. $y = -\frac{7}{4}x + 2$; $7x + 4y = 8$
10. $y = -\frac{4}{3}x + \frac{5}{3}$; $4x + 3y = 5$
11. $y = -6x + 15$; $6x + y = 15$
12. $y = 12x - 24$; $12x - y = 24$

Additional Answers for Lesson 6-5

1.
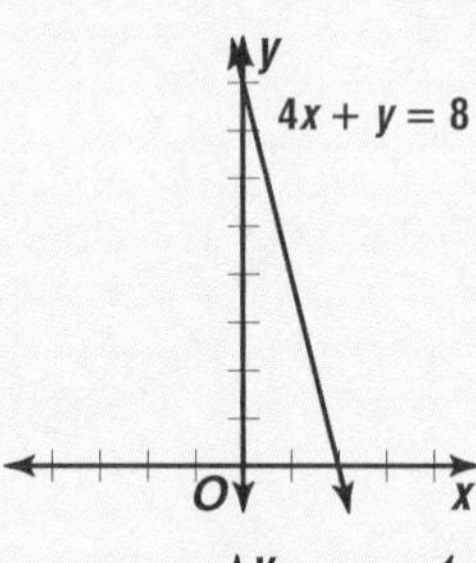

2.
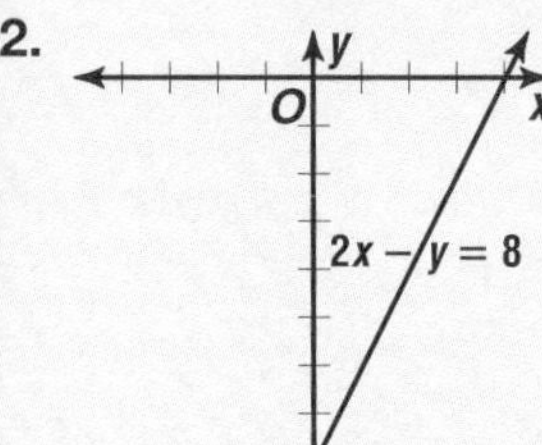

3.
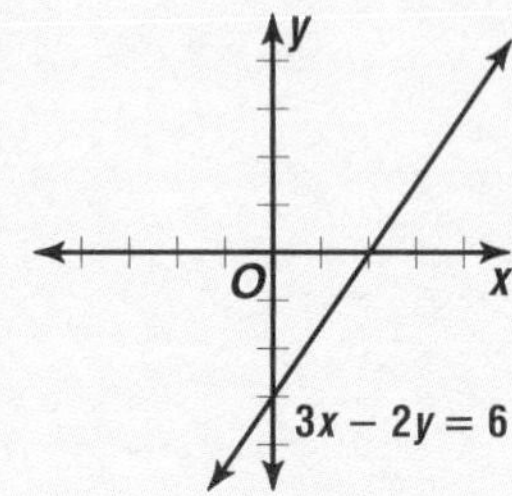

4.
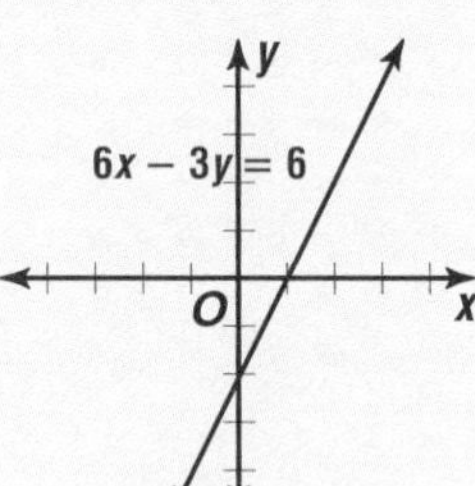

5.
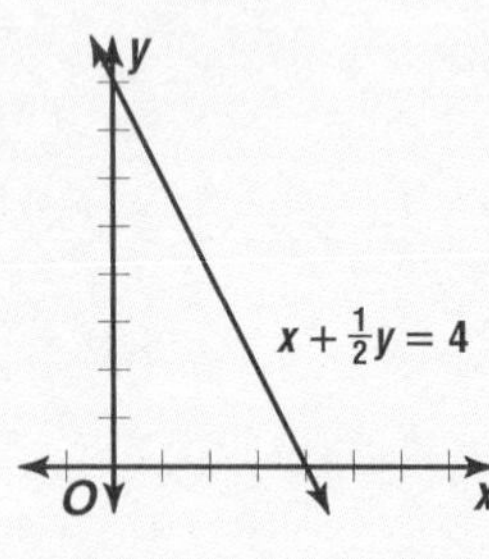

EXTRA PRACTICE

Lesson 6-3 **Explain whether a scatter plot for each pair of variables would probably show a *positive*, *negative*, or *no* correlation between the variables.**

1. playing time of a basketball game and points scored **positive**
2. age of a car and its value **negative**
3. heights of mothers and sons **positive**
4. temperature of the water in an outdoor swimming pool and the temperature outside **positive**
5. the weight of a person's car and the amount they pay to use a turnpike **no**
6. the number of guests at a birthday party and the amount of food remaining after the party **negative**
7. The scatter plot at the right compares the number of hours per week people watched television to the number of hours per week they spent doing some physical activity.
   a. As people watched more television, what happened to the number of hours they spend doing some physical activity? **It decreases.**
   b. Is there a correlation between the variables? Is it positive or negative? **yes, negative**

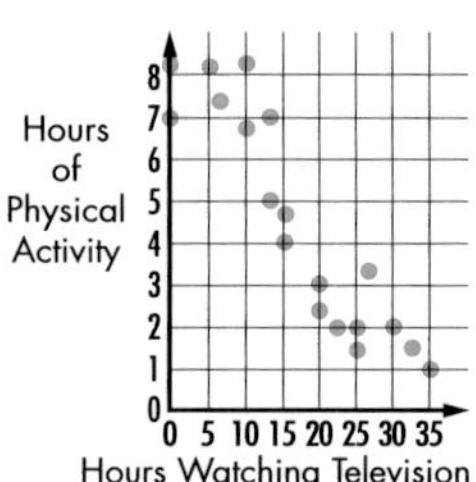

Lesson 6-4 **Find the x- and y-intercepts of the graph of each equation.**

1. $3x + 2y = 6$ **2, 3**
2. $5x + y = 10$ **2, 10**
3. $2x + 5y = -11$ $-\frac{11}{2}, -\frac{11}{5}$
4. $3y = 12$ **none, 4**
5. $y - 6x = 5$ $-\frac{5}{6}, 5$
6. $x = -2$ **−2, none**

Write an equation in slope-intercept form of a line with the given slope and y-intercept. Then write the equation in standard form. **7–12. See margin.**

7. $m = -\frac{2}{5}, b = 2$
8. $m = 5, b = -15$
9. $m = -\frac{7}{4}, b = 2$
10. $m = -\frac{4}{3}, b = \frac{5}{3}$
11. $m = -6, b = 15$
12. $m = 12, b = -24$

Find the slope and y-intercept of the graph of each equation.

13. $y - \frac{3}{5}x = -\frac{1}{4}$ $\frac{3}{5}, -\frac{1}{4}$
14. $y = 3x - 7$ **3, −7**
15. $\frac{2}{3}x + \frac{1}{6}y = 2$ **−4, 12**
16. $2x + 3y = 5$ $-\frac{2}{3}, \frac{5}{3}$
17. $3y = 8x + 2$ $\frac{8}{3}, \frac{2}{3}$
18. $5y = -8x - 2$ $-\frac{8}{5}, -\frac{2}{5}$

Write an equation in standard form for a line that passes through each pair of points.

19. (−1, 7), (8, −2) $x + y = 6$
20. (6, 0), (0, 4) $2x + 3y = 12$
21. (8, −1), (7, −1) $y = -1$
22. (1, 0), (0, 1) $x + y = 1$
23. (5, 7), (−1, 6) $x - 6y = -37$
24. (−3, −5), (3, −15) $5x + 3y = -30$

Lesson 6-5 **Graph each equation.** **1–18. See Solutions Manual.**

1. $4x + y = 8$
2. $2x - y = 8$
3. $3x - 2y = 6$
4. $6x - 3y = 6$
5. $x + \frac{1}{2}y = 4$
6. $4x + 5y = 20$
7. $y + 3 = -2(x + 4)$
8. $y - 1 = 3(x - 5)$
9. $y + 6 = -\frac{2}{3}(x + 1)$
10. $y - 5 = 4(x + 6)$
11. $y - 2 = (x + 7)$
12. $3(x - 1) = y + \frac{4}{5}$
13. $y = \frac{3}{4}x + 4$
14. $y = 4x - 1$
15. $-4x + y = 6$
16. $-2x + y = 3$
17. $5y - 6 = 3x$
18. $y = \frac{3}{2}x - 5$

6.
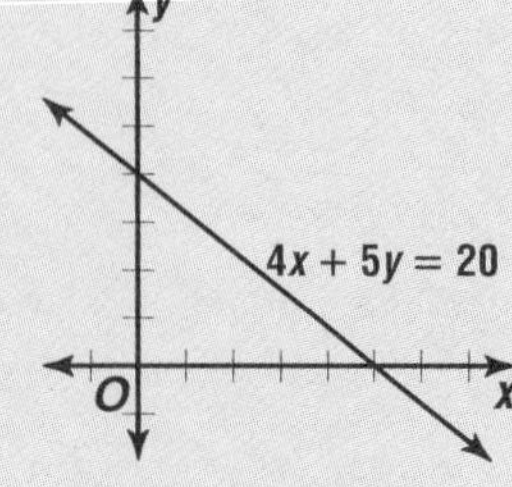

7.
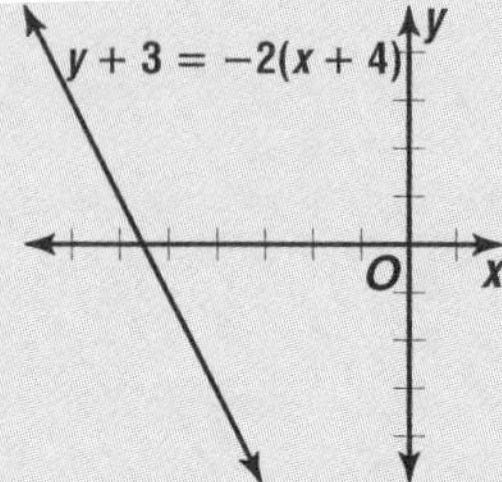

8.
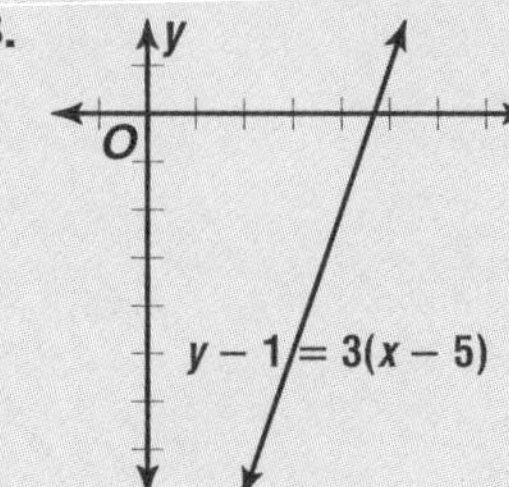

Lesson 6–6 **Determine whether the graphs of each pair of equations are *parallel, perpendicular,* or *neither.***

1. $2x + 3y = -12$
   $2x + 3y = 6$ parallel
2. $-4x + 3y = 12$
   $x + 3y = 12$ neither
3. $y = -3x + 9$
   $y + 3x = 14$ parallel
4. $y = 0.5x + 8$
   $2y = -8x - 3$ neither
5. $y = 7x + 2$
   $y = 2x + 7$ neither
6. $y + 5 = -9$
   $y + x = y - 6$ perpendicular

Write an equation in slope-intercept form of the line that passes through the given point and is parallel to the graph of each equation. 9. $y = \frac{2}{3}x + 2$ 11. $y = -\frac{3}{8}x + 4$

7. $(1, 6), y = 4x - 2$ $y = 4x + 2$
8. $(4, 6), y = 2x - 7$ $y = 2x - 2$
9. $(-3, 0), y = \frac{2}{3}x + 1$
10. $(2, 3), x - 5y = 7$ $y = \frac{1}{5}x + \frac{13}{5}$
11. $(0, 4), 3x + 8y = 4$
12. $(5, -2), y = -3x - 7$ $y = -3x + 13$

Write an equation in slope-intercept form of the line that passes through the given point and is perpendicular to the graph of each equation. 13. $y = \frac{5}{3}x - 1$ 14. $y = \frac{1}{6}x + \frac{10}{3}$ 15. $y = -\frac{4}{3}x$

13. $(0, -1), y = -\frac{3}{5}x + 4$
14. $(-2, 3), 6x + y = 4$
15. $(0, 0), y = \frac{3}{4}x - 1$
16. $(4, 0), 4x - 3y = 2$ $y = -\frac{3}{4}x + 3$
17. $(6, 7), 3x - 5y = 1$ $y = -\frac{5}{3}x + 17$
18. $(5, -1), 8x + 4y = 15$ $y = \frac{1}{2}x - \frac{7}{2}$

Lesson 6–7 **Find the coordinates of the midpoint of a segment with each pair of endpoints.**

1. $L(12, 2), M(8, 4)$ $(10, 3)$
2. $S(9, 5), T(17, 3)$ $(13, 4)$
3. $D(17, 9), E(11, -3)$ $(14, 3)$
4. $F(4, 2), G(8, -6)$ $(6, -2)$
5. $M(19, -3), N(11, 5)$ $(15, 1)$
6. $B(-6, 5), C(8, -11)$ $(1, -3)$
7. $T(-11, 6), U(13, 4)$ $(1, 5)$
8. $A(-6, 1), B(8, 9)$ $(1, 5)$
9. $J(6.4, -3), K(1.8, -3)$
10. $R(19, 5), S(7, 4)$ $\left(13, \frac{9}{2}\right)$
11. $G(8, 10), H(16, -6)$ $(12, 2)$
12. $C(7.6, 8.3), D(-5, 6.1)$ $(1.3, 7.2)$

9. $(4.1, -3)$

Find the coordinates of the other endpoint of a segment given one endpoint and the midpoint *M*.

13. $P(9, 3), M(1, 2)$ $(-7, 1)$
14. $C(3, 5), M(5, -7)$ $(7, -19)$
15. $G(5, -9), M\left(8, -\frac{15}{2}\right)$ $(11, -6)$
16. $J(4, -7), M(-2, -3)$ $(-8, 1)$
17. $A(-3, 8), M(3, -5)$ $(9, -18)$
18. $F(5, 7), M(5, 6)$ $(5, 5)$
19. $T(-6, 12), M(4, 1)$ $(14, -10)$
20. $D(-8, 5), M\left(-\frac{1}{2}, 2\right)$ $(7, -1)$
21. $S(16, -9), M\left(\frac{3}{2}, -\frac{13}{2}\right)$
22. $U(-9, 14), M(0, 3)$ $(9, -8)$
23. $F(21, 18), M(19, 11)$ $(17, 4)$
24. $X\left(\frac{1}{4}, \frac{1}{3}\right), M\left(\frac{3}{16}, \frac{1}{3}\right)$ $\left(\frac{1}{8}, \frac{1}{3}\right)$

21. $(-13, -4)$

Lesson 7–1 **Solve each inequality. Then check your solution.** 11. $\{b \mid b \geq -1\}$ 12. $\{s \mid s \geq -4.9\}$

1. $c + 9 \leq 3$ $\{c \mid c \leq -6\}$
2. $d - (-3) < 13$ $\{d \mid d < 10\}$
3. $z - 4 > 20$ $\{z \mid z > 24\}$
4. $h - (-7) > -2$ $\{h \mid h > -9\}$
5. $-11 > d - 4$ $\{d \mid d < -7\}$
6. $2x > x - 3$ $\{x \mid x > -3\}$
7. $2x - 3 \geq x$ $\{x \mid x \geq 3\}$
8. $16 + w < -20$ $\{w \mid w < -36\}$
9. $14p > 5 + 13p$ $\{p \mid p > 5\}$
10. $-7 < 16 - z$ $\{z \mid z < 23\}$
11. $-5 + 14b \leq -4 + 15b$
12. $2s - 6.5 \geq -11.4 + s$
13. $1.1v - 1 > 2.1v - 3$ $\{v \mid v < 2\}$
14. $\frac{1}{2}t + \frac{1}{4} \geq \frac{3}{2}t - \frac{2}{3}$ $\left\{t \mid t \leq \frac{11}{12}\right\}$
15. $9x < 8x - 2$ $\{x \mid x < -2\}$
16. $-2 + 9n \leq 10n$ $\{n \mid n \geq -2\}$
17. $a - 2.3 \geq -7.8$ $\{a \mid a \geq -5.5\}$
18. $5z - 6 > 4z$ $\{z \mid z > 6\}$

EXTRA PRACTICE

Additional Answers for Lesson 6-5

9.
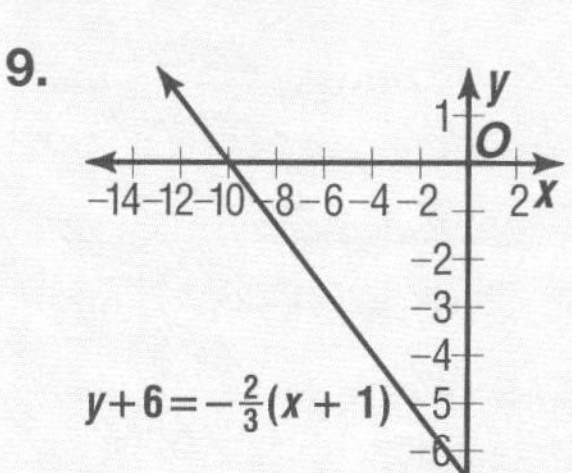

10.
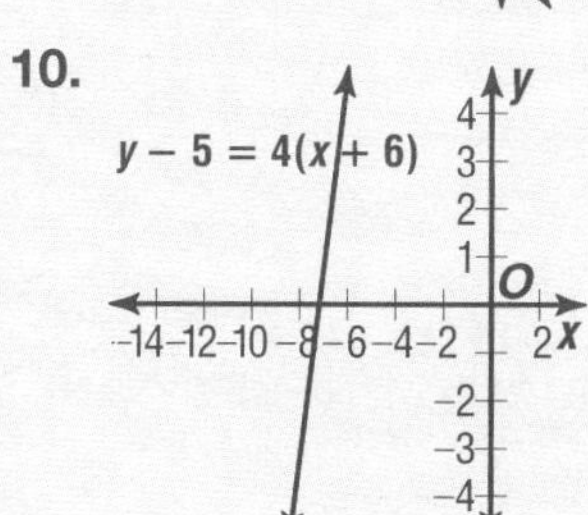

11.
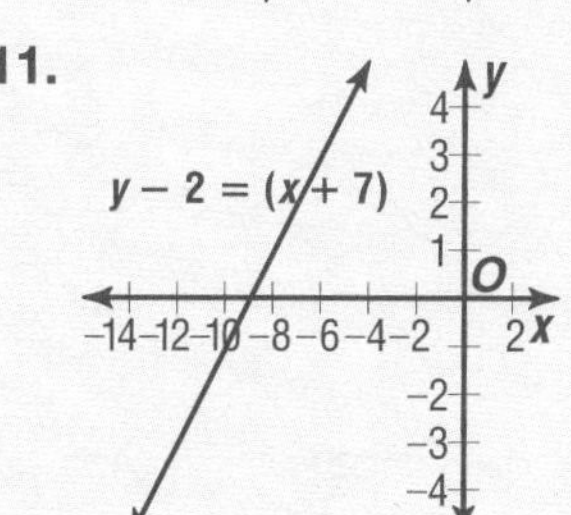

12.
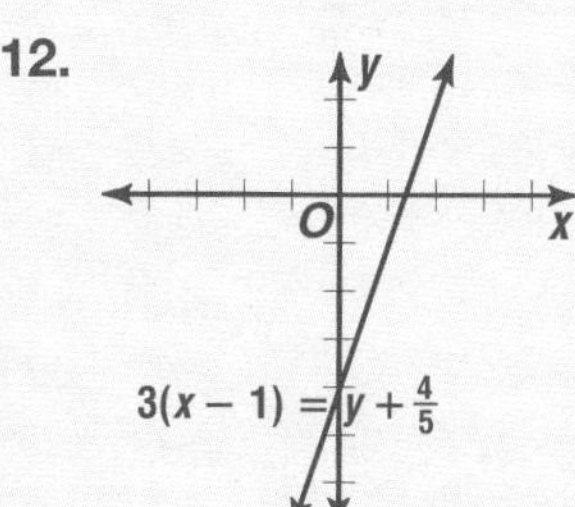

13.
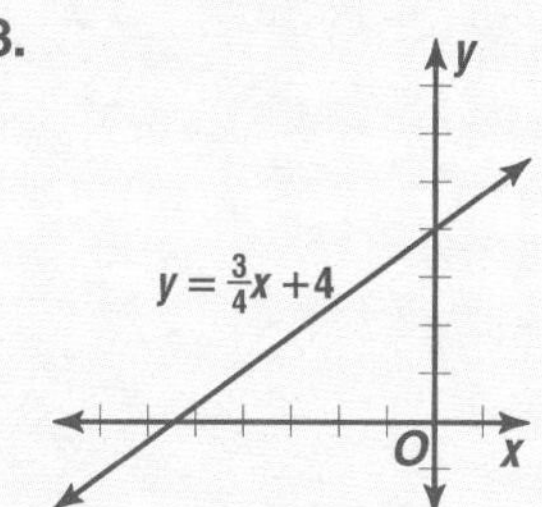

14.
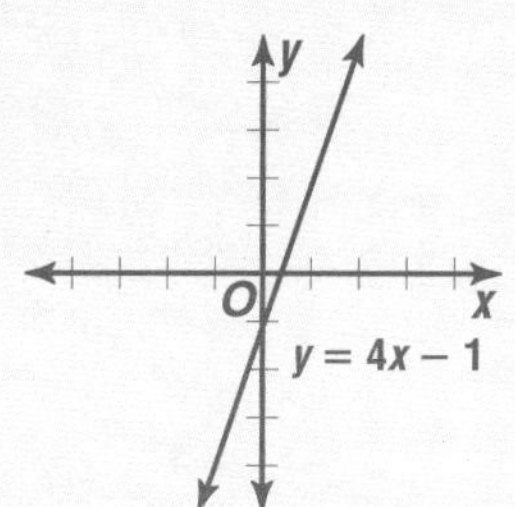

15.
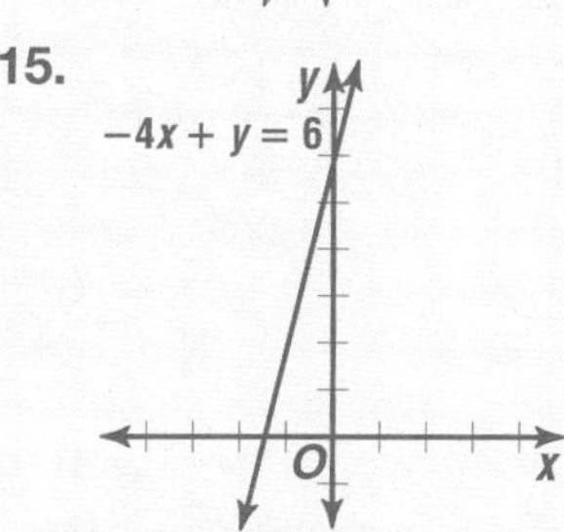

16.
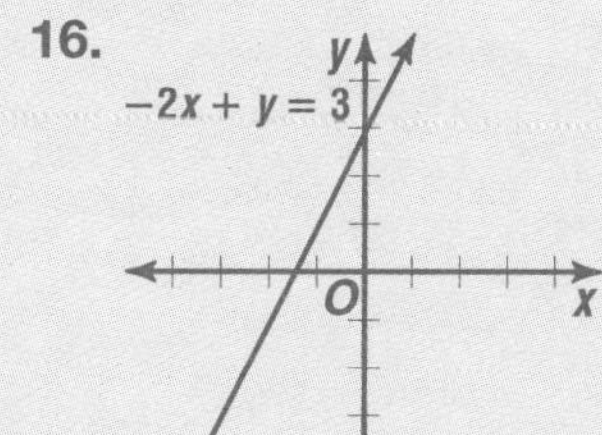

17.
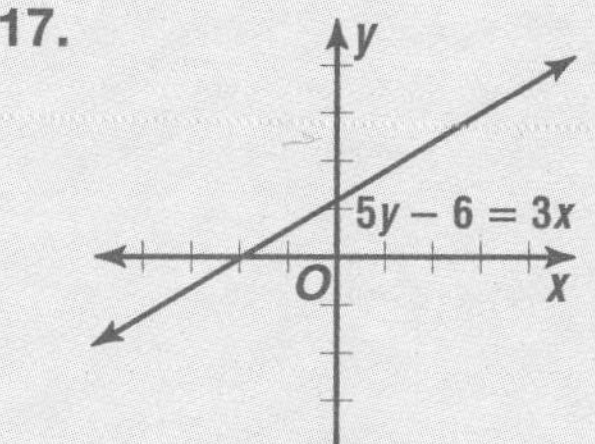

18.
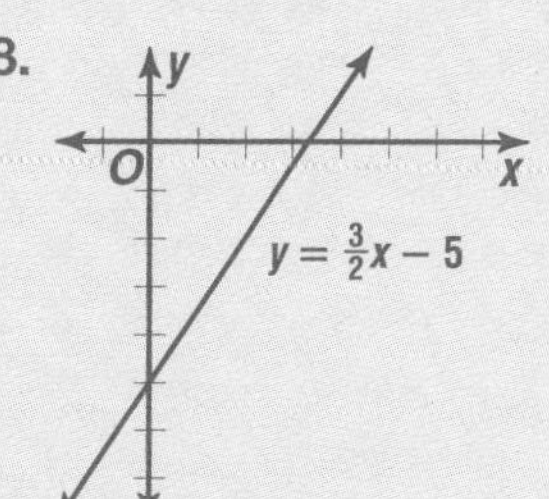

Additional Answers for Lesson 7-4

1. −8 −7 −6 −5 −4 −3 −2 −1 0 1 2 3 4

2. −4 −3 −2 −1 0 1 2 3 4

3. −3 −2 −1 0 1 2 3 4 5

4. −3 −2 −1 0 1 2 3 4 5

5.

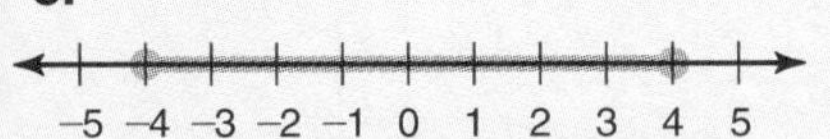

6. −9 −8 −7 −6 −5 −4 −3 −2 −1 0 1 2

7. −5 −4 −3 −2 −1 0 1 2 3 4

8. −4 −3 −2 −1 0 1 2 3 4

9. −3 −2 −1 0 1 2 3 4 5

10. 2 1 0 1 2

13. −4 −3 −2 −1 0 1 2 3 4

EXTRA PRACTICE

Lesson 7-2 **Solve each inequality. Then check your solution.**

1. $7b \geq -49$ $\{b \mid b \geq -7\}$
2. $-5j < -60$ $\{j \mid j > 12\}$
3. $\frac{w}{3} > -12$ $\{w \mid w > -36\}$
4. $\frac{p}{5} < 8$ $\{p \mid p < 40\}$
5. $-8f < 48$ $\{f \mid f > -6\}$
6. $\frac{t}{-4} \geq -10$ $\{t \mid t \leq 40\}$
7. $\frac{128}{-g} < 4$ $\{g \mid g > -32\}$
8. $-4.3x < -2.58$ $\{x \mid x > 0.6\}$
9. $4c \geq -6$ $\left\{c \mid c \geq -\frac{3}{2}\right\}$
10. $6 \leq 0.8n$ $\{n \mid n \geq 7.5\}$
11. $\frac{2}{3}m \geq -22$ $\{m \mid m \geq -33\}$
12. $-25 > \frac{a}{-6}$ $\{a \mid a > 150\}$
13. $-15a < -28$ $\left\{a \mid a > \frac{28}{15}\right\}$
14. $-\frac{7}{9}x < 42$ $\{x \mid x > -54\}$
15. $\frac{3y}{8} \leq 32$ $\left\{y \mid y \leq \frac{256}{3}\right\}$
16. $-7y \geq 91$ $\{y \mid y \leq -13\}$
17. $0.8t > 0.96$ $\{t \mid t > 1.2\}$
18. $\frac{4}{7}z \leq -\frac{2}{5}$ $\left\{z \mid z \leq -\frac{7}{10}\right\}$

Lesson 7-3 **Solve each inequality. Then check your solution.** 18. $\left\{y \mid y > \frac{11}{4}\right\}$

1. $3y - 4 > -37$ $\{y \mid y > -11\}$
2. $7s - 12 < 13$ $\left\{s \mid s < \frac{25}{7}\right\}$
3. $-5e + 9 > 24$ $\{e \mid e < -3\}$
4. $-6v - 3 \geq -33$ $\{v \mid v \leq 5\}$
5. $-2k + 12 < 30$ $\{k \mid k > -9\}$
6. $-2x + 1 < 16 - x$ $\{x \mid x > -15\}$
7. $15t - 4 > 11t - 16$ $\{t \mid t > -3\}$
8. $13 - y \leq 29 + 2y$ $\left\{y \mid y \geq -\frac{16}{3}\right\}$
9. $5q + 7 \leq 3(q + 1)$ $\{q \mid q \leq -2\}$
10. $2(w + 4) \geq 7(w - 1)$ $\{w \mid w \leq 3\}$
11. $-4t - 5 > 2t + 13$ $\{t \mid t < -3\}$
12. $\frac{2t + 5}{3} < -9$ $\{t \mid t < -16\}$
13. $\frac{z}{4} + 7 \geq -5$ $\{z \mid z \geq -48\}$
14. $13r - 11 > 7r + 37$ $\{r \mid r > 8\}$
15. $8c - (c - 5) > c + 17$ $\{c \mid c > 2\}$
16. $-5(k + 4) \geq 3(k - 4)$ $\{k \mid k \leq -1\}$
17. $9m + 7 < 2(4m - 1)$ $\{m \mid m < -9\}$
18. $3(3y + 1) < 13y - 8$
19. $5x \leq 10(3x + 4)$ $\left\{x \mid x \geq -\frac{8}{5}\right\}$
20. $3\left(a + \frac{2}{3}\right) \geq a - 1$ $\left\{a \mid a \geq -\frac{3}{2}\right\}$
21. $0.7(n - 3) \leq n - 0.6(n + 5)$ $\{n \mid n \leq -3\}$

1–14. See margin for graphs.

Lesson 7-4 **Solve each compound inequality. Then graph the solution set.**

1. $2 + x < -5$ or $2 + x > 5$ $\{x \mid x < -7$ or $x > 3\}$
2. $-4 + t > -5$ or $-4 + t < 7$ {all numbers}
3. $3 \leq 2g + 7$ and $2g + 7 \leq 15$ $\{g \mid -2 \leq g \leq 4\}$
4. $2v - 2 \leq 3v$ and $4v - 1 \geq 3v$ $\{v \mid v \geq 1\}$
5. $3b - 4 \leq 7b + 12$ and $8b - 7 \leq 25$ $\{b \mid -4 \leq b \leq 4\}$
6. $-9 < 2z + 7 < 10$ $\left\{z \mid -8 < z < \frac{3}{2}\right\}$
7. $5m - 8 \geq 10 - m$ or $5m + 11 < -9$ $\{m \mid m \geq 3$ or $m < -4\}$
8. $12c - 4 \leq 5c + 10$ or $-4c - 1 \leq c + 24$ {all numbers}
9. $2h - 2 \leq 3h \leq 4h - 1$ $\{h \mid h \geq 1\}$
10. $3p + 6 < 8 - p$ and $5p + 8 \geq p + 6$ $\left\{p \mid -\frac{1}{2} \leq p < \frac{1}{2}\right\}$
11. $2r + 8 > 16 - 2r$ and $7r + 21 < r - 9$ $\varnothing$
12. $4j + 3 < j + 22$ and $j - 3 < 2j - 15$ $\varnothing$
13. $2(q - 4) \leq 3(q + 2)$ or $q - 8 \leq 4 - q$ {all numbers}
14. $\frac{1}{2}w + 5 \geq w + 2 \geq \frac{1}{2}w + 9$ $\varnothing$

Lesson 7-5

1. **Food** For breakfast at Paul's Place, you can select one item from each of the following categories for \$1.99.

Meat	Potato	Bread	Beverage
ham sausage bacon	hash browns country potatoes	toast muffin bagel biscuit	juice coffee

   a. What is the probability that a customer will have ham for breakfast? $\frac{1}{3}$

   b. What is the probability of selecting a biscuit and hash browns? $\frac{1}{8}$

   c. What is the probability of having a bagel, country potatoes, and coffee? $\frac{1}{16}$

2. **Music** In order to raise money for a trip to the opera, the music club has set up a lottery using two-digit numbers. The first digit will be a numeral from 1 to 4. The second digit will be a numeral from 3 to 8. The first digit in Trudy's lottery number is 2, but she can't remember the second digit. If only one two-digit lottery number is drawn, and that number has 2 as the first digit, what is the probability that Trudy will win? $\frac{1}{6} = 0.1\overline{6}$

3. **Law** A three-judge panel is being used to settle a dispute. Both sides in the dispute have decided that a majority decision will be upheld. If each judge will render a favorable decision based on the evidence presented two-thirds of the time, what is the probability that the correct side will win the dispute? $\frac{20}{27} \approx 0.741$

Lesson 7-6 Solve each open sentence. Then graph the solution set. 2, 5–15. See margin.

1. $|a - 5| = -3$ $\varnothing$
2. $|g + 6| > 8$
3. $|t - 5| \leq 3$ $\{t \mid 2 \leq t \leq 8\}$
4. $|a + 5| \geq 0$ {all numbers}
5. $|14 - 2z| = 16$
6. $|y - 9| < 19$
7. $|2m - 5| > 13$
8. $|14 - w| \geq 20$
9. $|13 - 5y| = 8$
10. $|3p + 5| \leq 23$
11. $|6b - 12| \leq 36$
12. $|25 - 3x| < 5$
13. $|7 + 8x| > 39$
14. $|4c + 5| \geq 25$
15. $|4 - 5s| > 46$

Lesson 7-7

1. **Travel** Speeds of the fastest train runs in the U.S. and Canada are given below in miles per hour. Make a box-and-whisker plot of the data. See margin.

93.5	82.5	89.3	83.8	81.8	86.8
90.8	84.9	95.0	83.1	83.2	88.2

2. **Basketball** The numbers below represent the 20 highest points-scored-per-game averages for a season in the NBA from 1947 to 1990. Make a box-and-whisker plot of this data. See margin.

35.0	33.5	32.5	37.9	31.2	38.4	34.5	34.7	32.9	44.8
31.7	37.1	36.5	34.0	50.4	32.3	33.6	34.8	33.1	35.6

3. **Baseball** The stem-and-leaf plot at the right shows the number of home runs hit by the home run leaders in the National League in 1990.

   a. Find the median, upper quartile, lower quartile, and interquartile range. 26.5, 32.5, 24, 8.5

   b. Are there any outliers? If so, name them. no

   c. Draw a box-and-whisker plot of the data. See margin.

Stem	Leaf
2	2 3 3 4 4 4 4 5 5 6 7 7 8
3	2 2 3 3 5 7
4	0

$3|3 = 33$

Additional Answers for Lesson 7-6

2. $\{g \mid g < -14 \text{ or } g > 2\}$

−22 −18 −14 −10 −6 −2 2 6 10

3. $\{t \mid 2 \leq t \leq 8\}$

1 2 3 4 5 6 7 8 9

4. {all numbers}

−4 −3 −2 −1 0 1 2 3 4

5. $\{z \mid z = -1 \text{ or } z = 15\}$

−2 0 2 4 6 8 10 12 14

6. $\{y \mid -10 < y < 28\}$

−10 0 10 20 30

7. $\{m \mid m < -4 \text{ or } m > 9\}$

−6 −2 2 6 10

8. $\{w \mid w \leq -6 \text{ or } w \geq 34\}$

−12 −6 0 6 12 18 24 30 36

9. $\left\{y \mid y = 1 \text{ or } y = \frac{21}{5}\right\}$

−2 −1 0 1 2 3 4 5 6

10. $\left\{p \mid -\frac{28}{3} \leq p \leq 6\right\}$

−10 −8 −6 −4 −2 0 2 4 6 8

11. $\{b \mid -4 \leq b \leq 8\}$

−6 −4 −2 0 2 4 6 8 10

12. $\left\{x \mid \frac{20}{3} < x < 10\right\}$

4 5 6 7 8 9 10 11 12

13. $\left\{x \mid x < -\frac{23}{4} \text{ or } x > 4\right\}$

−10 −8 −6 −4 −2 0 2 4 6

14. $\left\{c \mid c \leq -\frac{30}{4} \text{ or } c \geq 5\right\}$

−8 −6 −4 −2 0 2 4 6 8

15. $\left\{s \mid s < -\frac{42}{5} \text{ or } s > 10\right\}$

−14 −10 −6 −2 2 6 10 14 18

Additional Answers for Lesson 7-7

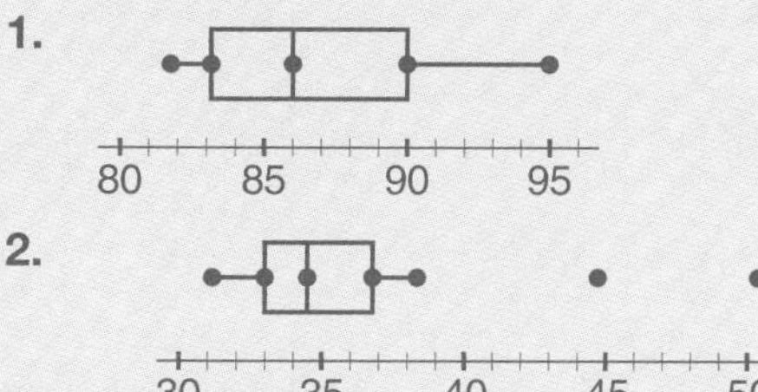

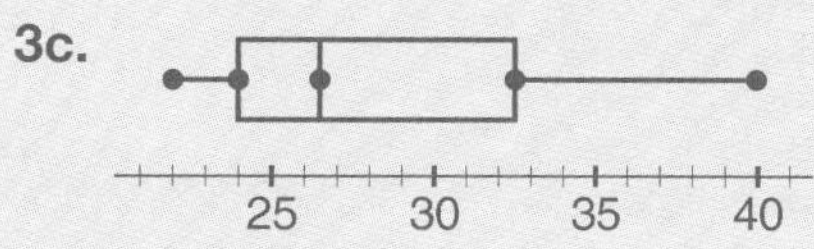

EXTRA PRACTICE

Lesson 7-8 Graph each inequality. **1–15. See Solutions Manual.**

1. $y \leq -2$
2. $x < 4$
3. $x + y < -2$
4. $x + y > -4$
5. $y > 4x - 1$
6. $3x + y > 1$
7. $3y - 2x \leq 2$
8. $x < y$
9. $3x + y > 4$
10. $5x - y < 5$
11. $-4x + 3y \geq 12$
12. $-x + 3y \leq 9$
13. $y > -3x + 7$
14. $3x + 8y \leq 4$
15. $5x - 2y \geq 6$

1–15. See Solutions Manual for graphs.

Lesson 8-1 Graph each system of equations. Then determine whether the system has *one* solution, *no* solution, or *infinitely many* solutions. If the system has one solution, name it.

1. $y = 3x$; $4x + 2y = 30$ **(3, 9)**
2. $x = -2y$; $3x + 5y = 21$ **(42, −21)**
3. $y = x + 4$; $3x + 2y = 18$ **(2, 6)**
4. $x + y = 6$; $x - y = 2$ **(4, 2)**
5. $x + y = 6$; $3x + 3y = 3$ **no solution**
6. $y = -3x$; $4x + y = 2$ **(2, −6)**
7. $x + y = 8$; $x - y = 2$ **(5, 3)**
8. $\frac{1}{5}x - y = \frac{12}{5}$; $3x - 5y = 6$ **(−3, −3)**
9. $x + 2y = 0$; $y + 3 = -x$ **(−6, 3)**
10. $x + 2y = -9$; $x - y = 6$ **(1, −5)**
11. $x + \frac{1}{2}y = 3$; $y = 3x - 4$ **(2, 2)**
12. $\frac{2}{3}x + \frac{1}{2}y = 2$; $4x + 3y = 12$ **infinitely many**
13. $y = x - 4$; $x + \frac{1}{2}y = \frac{5}{2}$ **(3, −1)**
14. $2x + y = 3$; $4x + 2y = 6$ **infinitely many**
15. $12x - y = -21$; $\frac{1}{2}x + \frac{2}{3}y = -3$ **(−2, −3)**

Lesson 8-2 Use substitution to solve each system of equations. If the system *does not* have exactly one solution, state whether it has *no* solution or *infinitely many* solutions.

1. $y = x$; $5x = 12y$ **(0, 0)**
2. $y = 7 - x$; $2x - y = 8$ **(5, 2)**
3. $x = 5 - y$; $3y = 3x + 1$ $\left(\frac{7}{3}, \frac{8}{3}\right)$
4. $3x + y = 6$; $y + 2 = x$ **(2, 0)**
5. $x - 3y = 3$; $2x + 9y = 11$ $\left(4, \frac{1}{3}\right)$
6. $3x = -18 + 2y$; $x + 3y = 4$ $\left(-\frac{46}{11}, \frac{30}{11}\right)$
7. $x + 2y = 10$; $-x + y = 2$ **(2, 4)**
8. $2x = 3 - y$; $2y = 12 - x$ **(−2, 7)**
9. $6y - x = -36$; $y = -3x$ $\left(\frac{36}{19}, \frac{-108}{19}\right)$
10. $\frac{3}{4}x + \frac{1}{3}y = 1$; $x - y = 10$ **(4, −6)**
11. $x + 6y = 1$; $3x - 10y = 31$ **(7, −1)**
12. $3x - 2y = 12$; $\frac{3}{2}x - y = 3$ **no solution**
13. $2x + 3y = 5$; $4x - 9y = 9$ $\left(\frac{12}{5}, \frac{1}{15}\right)$
14. $x = 4 - 8y$; $3x + 24y = 12$ **infinitely many**
15. $3x - 2y = -3$; $25x + 10y = 215$ **(5, 9)**

Lesson 8–3 **State whether addition, subtraction, or substitution would be most convenient to solve each system of equations. Then solve the system.**

1. $x + y = 7$
   $x - y = 9$ +; (8, −1)
2. $2x - y = 32$
   $2x + y = 60$ +; (23, 14)
3. $-y + x = 6$
   $y + x = 5$ +; $\left(\frac{11}{2}, -\frac{1}{2}\right)$
4. $s + 2t = 6$
   $3s - 2t = 2$ +; (2, 2)
5. $x = y - 7$
   $2x - 5y = -2$ sub; (−11, −4)
6. $3x + 5y = -16$
   $3x - 2y = -2$ −; (−2, −2)
7. $x - y = 3$
   $x + y = 3$ +; (3, 0)
8. $x + y = 8$
   $2x - y = 6$ +; $\left(\frac{14}{3}, \frac{10}{3}\right)$
9. $2s - 3t = -4$
   $s = 7 - 3t$ sub; (1, 2)
10. $-6x + 16y = -8$
    $6x - 42 = 16y$ +; ∅
11. $3x + 0.2y = 7$
    $3x = 0.4y + 4$ −; (2, 5)
12. $9x + 2y = 26$
    $1.5x - 2y = 13$
13. $\frac{2}{3}x - \frac{1}{2}y = 14$
    $\frac{5}{6}x - \frac{1}{2}y = 18$ −; (24, 4)
14. $4x - \frac{1}{3}y = 8$
    $5x + \frac{1}{3}y = 6$ +; $\left(\frac{14}{9}, -\frac{16}{3}\right)$
15. $2x - y = 3$
    $\frac{2}{3}x - y = -1$ −; (3, 3)

12. +; $\left(\frac{26}{7}, -\frac{26}{7}\right)$

Lesson 8–4 **Use elimination to solve each system of equations.**

1. $x + 8y = 3$
   $4x - 2y = 7$ $\left(\frac{31}{17}, \frac{5}{34}\right)$
2. $4x - y = 4$
   $x + 2y = 3$ $\left(\frac{11}{9}, \frac{8}{9}\right)$
3. $3y - 8x = 9$
   $y - x = 2$ $\left(-\frac{3}{5}, \frac{7}{5}\right)$
4. $x + 4y = 30$
   $2x - y = -6$ $\left(\frac{2}{3}, \frac{22}{3}\right)$
5. $3x - 2y = 0$
   $4x + 4y = 5$ $\left(\frac{1}{2}, \frac{3}{4}\right)$
6. $9x - 3y = 5$
   $x + y = 1$ $\left(\frac{2}{3}, \frac{1}{3}\right)$
7. $-3x + 2y = 10$
   $-2x - y = -5$ (0, 5)
8. $2x + 5y = 13$
   $4x - 3y = -13$ (−1, 3)
9. $5x + 3y = 4$
   $-4x + 5y = -18$ (2, −2)
10. $2x - 7y = 9$
    $-3x + 4y = 6$ (−6, −3)
11. $2x - 6y = -16$
    $5x + 7y = -18$ (−5, 1)
12. $6x - 3y = -9$
    $-8x + 2y = 4$ $\left(\frac{1}{2}, 4\right)$
13. $\frac{1}{3}x - y = -1$
    $\frac{1}{5}x - \frac{2}{5}y = -1$ (−9, −2)
14. $3x - 5y = 8$
    $4x - 7y = 10$ (6, 2)
15. $x - 0.5y = 1$
    $0.4x + y = -2$ (0, −2)

Lesson 8–5 **Solve each system of inequalities by graphing.** 1–12. See Solutions Manual.

1. $x > 3$
   $y < 6$
2. $y > 2$
   $y > -x + 2$
3. $x \leq 2$
   $y - 3 \geq 5$
4. $x + y \leq -1$
   $2x + y \leq 2$
5. $y \geq 2x + 2$
   $y \geq -x - 1$
6. $y \leq x + 3$
   $y \geq x + 2$
7. $x + 3y \geq 4$
   $2x - y < 5$
8. $y - x > 1$
   $y + 2x \leq 10$
9. $5x - 2y > 15$
   $2x - 3y < 6$
10. $4x + 3y > 4$
    $2x - y < 0$
11. $4x + 5y \geq 20$
    $y \geq x + 1$
12. $-4x + 10y \leq 5$
    $-2x + 5y < -1$

EXTRA PRACTICE

775

Lesson 9-1 Simplify.

1. $a^5(a)(a^7)$ a^{13}
2. $(r^3t^4)(r^4t^4)$ r^7t^8
3. $(x^3y^4)(xy^3)$ x^4y^7
4. $(bc^3)(b^4c^3)$ b^5c^6
5. $(-3mn^2)(5m^3n^2)$ $-15m^4n^4$
6. $[(3^3)^2]^2$ $531{,}441$
7. $(3s^3t^2)(-4s^3t^2)$ $-12s^6t^4$
8. $x^3(x^4y^3)$ x^7y^3
9. $(1.1g^2h^4)^3$ $1.331g^6h^{12}$
10. $-\frac{3}{4}a(a^2b^3c^4)$ $-\frac{3}{4}a^3b^3c^4$
11. $\left(\frac{1}{2}w^3\right)^2(w^4)^2$ $\frac{1}{4}w^{14}$
12. $\left(\frac{2}{3}y^3\right)(3y^2)^3$ $18y^9$
13. $[(-2^3)^3]^2$ $262{,}144$
14. $(10s^3t)(-2s^2t^2)^3$ $-80s^9t^7$
15. $(-0.2u^3w^4)^3$ $-0.008u^9w^{12}$

Lesson 9-2 Simplify. Assume no denominator is equal to zero.

1. $\frac{b^6c^5}{b^3c^2}$ b^3c^3
2. $\frac{(-a)^4b^8}{a^4b^7}$ b
3. $\frac{(-x)^3y^3}{x^3y^6}$ $-\frac{1}{y^3}$
4. $\frac{12ab^5}{4a^4b^3}$ $\frac{3b^2}{a^3}$
5. $\frac{24x^5}{-8x^2}$ $-3x^3$
6. $\frac{-9h^2k^4}{18h^5j^3k^4}$ $\frac{-1}{2h^3j^3}$
7. $\frac{a^0}{2a^{-3}}$ $\frac{a^3}{2}$
8. $\frac{9a^2b^7c^3}{2a^5b^4c}$ $\frac{9b^3c^2}{2a^3}$
9. $\frac{-15xy^5z^7}{-10x^4y^6z^4}$ $\frac{3z^3}{2x^3y}$
10. $\frac{(u^{-3}v^3)^2}{(u^3v)^{-3}}$ u^3v^9
11. $\frac{(-r)s^5}{r^{-3}s^{-4}}$ $-r^4s^9$
12. $\frac{28a^{-4}b^0}{14a^3b^{-1}}$ $\frac{2b}{a^7}$
13. $\frac{(j^2k^3l)^4}{(jk^4)^{-1}}$ $j^9k^{16}l^4$
14. $\left(\frac{-2x^4y}{4y^2}\right)^0$ 1
15. $\frac{3m^7n^2p^4}{9m^2np^3}$ $\frac{m^5np}{3}$

Lesson 9-3 Express each number in scientific notation.

1. 6500 6.5×10^3
2. 953.56 9.5356×10^2
3. 0.697 6.97×10^{-1}
4. 843.5 8.435×10^2
5. 568,000 5.68×10^5
6. 0.0000269 2.69×10^{-5}
7. 0.121212 1.21212×10^{-1}
8. 543×10^4 5.43×10^6
9. 739.9×10^{-5} 7.399×10^{-3}
10. 6480×10^{-2} 6.48×10
11. 0.366×10^{-7} 3.66×10^{-8}
12. 167×10^3 1.67×10^5

19. 1.302×10^2; 130.2 20. 7.8×10^8; 780,000,000

Evaluate. Express each result in scientific and standard notation.

13. $(2 \times 10^5)(3 \times 10^{-8})$ 6×10^{-3}; 0.006
14. $\frac{4.8 \times 10^3}{1.6 \times 10^1}$ 3×10^2; 300
15. $(4 \times 10^2)(1.5 \times 10^6)$ 6×10^8; 600,000,000
16. $\frac{8.1 \times 10^2}{2.7 \times 10^{-3}}$ 3×10^5; 300,000
17. $\frac{7.8 \times 10^{-5}}{1.3 \times 10^{-7}}$ 6×10^2; 600
18. $(2.2 \times 10^{-2})(3.2 \times 10^5)$ 7.04×10^3; 7040

6×10^{-3}; 0.006

19. $(3.1 \times 10^4)(4.2 \times 10^{-3})$
20. $(78 \times 10^6)(0.01 \times 10^3)$
21. $\frac{2.31 \times 10^{-2}}{3.3 \times 10^{-3}}$ 7×10^0; 7

Lesson 9-4 State whether each expression is a polynomial. If the expression is a polynomial, identify it as a *monomial*, a *binomial*, or a *trinomial* and find the degree of the polynomial.

1. $5x^2y + 3xy + 7$ **yes, trinomial, 3**
2. 0 **yes, monomial, 0**
3. $\frac{5}{k} - k^2y$ **no**
4. $3a^2x - 5a$ **yes, binomial, 3**
5. $a + \frac{5}{c}$ **no**
6. $14abcd - 6d^3$ **yes, binomial, 4**
7. $\frac{a^3}{3}$ **yes, monomial, 3**
8. $-4h^3$ **yes, monomial, 3**
9. $x^2 - \frac{x}{2} + \frac{1}{3}$ **yes, trinomial, 2**

Arrange the terms of each polynomial so that the powers of x are in descending order. **10–15. See margin.**

10. $5x^2 - 3x^3 + 7 + 2x$
11. $-6x + x^5 + 4x^3 - 20$
12. $5b + b^3x^2 + \frac{2}{3}bx$
13. $21p^2x + 3px^3 + p^4$
14. $3ax^2 - 6a^2x^3 + 7a^3 - 8x$
15. $\frac{1}{3}s^2x^3 + 4x^4 - \frac{2}{5}s^4x^2 + \frac{1}{4}x$

Lesson 9-5 Find each sum or difference.

1. $-7t^2 + 4ts - 6s^2$ (+) $-5t^2 - 12ts + 3s^2$ **$-12t^2 - 8ts - 3s^2$**
2. $6a^2 - 7ab - 4b^2$ (−) $2a^2 + 5ab + 6b^2$ **$4a^2 - 12ab - 10b^2$**
3. $4a^2 - 10b^2 + 7c^2$; $-5a^2 + 2c^2 + 2b$; (+) $7b^2 - 7c^2 + 7a$ **$-a^2 - 3b^2 + 2c^2 + 7a + 2b$**
4. $z^2 + 6z - 8$ (−) $4z^2 - 7z - 5$ **$-3z^2 + 13z - 3$**
5. $(4d + 3e - 8f) - (-3d + 10e - 5f + 6)$ **$7d - 7e - 3f - 6$**
6. $(7g + 8h - 9) + (-g - 3h - 6k)$ **$6g + 5h - 6k - 9$**
7. $(9x^2 - 11xy - 3y^2) - (x^2 - 16xy + 12y^2)$ **$8x^2 + 5xy - 15y^2$**
8. $(-3m + 9mn - 5n) + (14m - 5mn - 2n)$ **$11m + 4mn - 7n$**
9. $(4x^2 - 8y^2 - 3z^2) - (7x^2 - 14z^2 - 12)$ **$-3x^2 - 8y^2 + 11z^2 + 12$**
10. $(17z^4 - 5z^2 + 3z) - (4z^4 + 2z^3 + 3z)$ **$13z^4 - 2z^3 - 5z^2$**
11. $(6 - 7y + 3y^2) + (3 - 5y - 2y^2) + (-12 - 8y + y^2)$ **$2y^2 - 20y - 3$**
12. $(-3x^2 + 2x - 5) + (2x - 6) + (5x^2 + 3) + (-9x^2 - 7x + 4)$ **$-7x^2 - 3x - 4$**

8. $36m^4n + 4m^3n - 20m^2n^2$ 9. $-16s^3t^5 + 28s^6t^2 - 12s^2t^5$

Lesson 9-6 Find each product.

1. $-3(8x + 5)$ **$-24x - 15$**
2. $3b(5b + 8)$ **$15b^2 + 24b$**
3. $1.1a(2a + 7)$ **$2.2a^2 + 7.7a$**
4. $\frac{1}{2}x(8x - 6)$ **$4x^2 - 3x$**
5. $7xy(5x^2 - y^2)$ **$35x^3y - 7xy^3$**
6. $5y(y^2 - 3y + 6)$ **$5y^3 - 15y^2 + 30y$**
7. $-ab(3b^2 + 4ab - 6a^2)$ **$-3ab^3 - 4a^2b^2 + 6a^3b$**
8. $4m^2(9m^2n + mn - 5n^2)$
9. $4st^2(-4s^2t^3 + 7s^5 - 3st^3)$
10. $-\frac{1}{3}x(9x^2 + x - 5)$ **$-3x^3 - \frac{1}{3}x^2 + \frac{5}{3}x$**
11. $-2mn(8m^2 - 3mn + n^2)$ **$-16m^3n + 6m^2n^2 - 2mn^3$**
12. $-\frac{3}{4}ab^2\left(\frac{1}{3}b^2 - \frac{4}{9}b + 1\right)$ **$-\frac{1}{4}ab^4 + \frac{1}{3}ab^3 - \frac{3}{4}ab^2$**

Solve.

13. $-3(2a - 12) + 48 = 3a - 3$ **$\frac{29}{3}$**
14. $-6(12 - 2w) = 7(-2 - 3w)$ **$\frac{58}{33}$**
15. $a(a - 6) + 2a = 3 + a(a - 2)$ **$-\frac{3}{2}$**
16. $11(a - 3) + 5 = 2a + 44$ **8**
17. $q(2q + 3) + 20 = 2q(q - 3)$ **$-\frac{20}{9}$**
18. $w(w + 12) = w(w + 14) + 12$ **-6**
19. $x(x + 8) - x(x + 3) - 23 = 3x + 11$ **17**
20. $y(y - 12) + y(y + 2) + 25 = 2y(y + 5) - 15$ **2**
21. $x(x - 3) + 4x - 3 = 8x + 4 + x(3 + x)$ **$-\frac{7}{10}$**
22. $c(c - 3) + 4(c - 2) = 12 - 2(4 + c) - c(1 - c)$ **3**

EXTRA PRACTICE

Additional Answers for Lesson 9-4

10. $-3x^3 + 5x^2 + 2x + 7$
11. $x^5 + 4x^3 - 6x - 20$
12. $b^3x^2 + \frac{2}{3}bx + 5b$
13. $3px^3 + 21p^2x + p^4$
14. $-6a^2x^3 + 3ax^2 - 8x + 7a^3$
15. $4x^4 + \frac{1}{3}s^2x^3 - \frac{2}{5}s^4x^2 + \frac{1}{4}x$

EXTRA PRACTICE

Lesson 9–7 Find each product.

1. $(d + 2)(d + 3)$ $d^2 + 5d + 6$
2. $(z + 7)(z - 4)$ $z^2 + 3z - 28$
3. $(m - 8)(m - 5)$
4. $(2x - 5)(x + 6)$ $2x^2 + 7x - 30$
5. $(7a - 4)(2a - 5)$
6. $(4x + y)(2x - 3y)$
7. $(7v + 3)(v + 4)$ $7v^2 + 31v + 12$
8. $(7s - 8)(3s - 2)$
9. $(4g + 3h)(2g - 5h)$
10. $(4a + 3)(2a - 1)$ $8a^2 + 2a - 3$
11. $(7y - 1)(2y - 3)$ $14y^2 - 23y + 3$
12. $(2x + 3y)(5x + 2y)$ $10x^2 + 19xy + 6y^2$
13. $(12r - 4s)(5r + 8s)$ $60r^2 + 76rs - 32s^2$
14. $(x - 2)(x^2 + 2x + 4)$ $x^3 - 8$
15. $(3x + 5)(2x^2 - 5x + 11)$ $6x^3 - 5x^2 + 8x + 55$
16. $(4s + 5)(3s^2 + 8s - 9)$ $12s^3 + 47s^2 + 4s - 45$
17. $(3a + 5)(-8a^2 + 2a + 3)$
18. $(5x - 2)(-5x^2 + 2x + 7)$
19. $(x^2 - 7x + 4)(2x^2 - 3x - 6)$
20. $(a^2 + 2a + 5)(a^2 - 3a - 7)$
21. $(5x^4 - 2x^2 + 1)(x^2 - 5x + 3)$

3. $m^2 - 13m + 40$ 5. $14a^2 - 43a + 20$ 6. $8x^2 - 10xy - 3y^2$ 8. $21s^2 - 38s + 16$ 9. $8g^2 - 14gh - 15h^2$
17. $-24a^3 - 34a^2 + 19a + 15$ 18. $-25x^3 + 20x^2 + 31x - 14$ 19. $2x^4 - 17x^3 + 23x^2 + 30x - 24$
20. $a^4 - a^3 - 8a^2 - 29a - 35$ 21. $5x^6 - 25x^5 + 13x^4 + 10x^3 - 5x^2 - 5x + 3$

Lesson 9–8 Find each product.

1. $(t + 7)^2$ $t^2 + 14t + 49$
2. $(w - 12)(w + 12)$ $w^2 - 144$
3. $(q - 4h)^2$ $q^2 - 8qh + 16h^2$
4. $(10x + 11y)(10x - 11y)$
5. $(4e + 3)^2$ $16e^2 + 24e + 9$
6. $(2b - 4d)(2b + 4d)$
7. $(a + 2b)^2$ $a^2 + 4ab + 4b^2$
8. $(4x + y)^2$ $16x^2 + 8xy + y^2$
9. $(6m + 2n)^2$
10. $(5c - 2d)^2$ $25c^2 - 20cd + 4d^2$
11. $(5b - 6)(5b + 6)$ $25b^2 - 36$
12. $(1 + x)^2$ $1 + 2x + x^2$
13. $(4x - 9y)^2$ $16x^2 - 72xy + 81y^2$
14. $(8a - 2b)(8a + 2b)$ $64a^2 - 4b^2$
15. $\left(\frac{1}{2}a + b\right)^2$ $\frac{1}{4}a^2 + ab + b^2$
16. $(5a - 12b)^2$
17. $(a - 3b)^2$ $a^2 - 6ab + 9b^2$
18. $(7a^2 + b)(7a^2 - b)$ $49a^4 - b^2$
19. $(x + 2)(x - 2)(2x + 5)$ $2x^3 + 5x^2 - 8x - 20$
20. $(4x - 1)(4x + 1)(x - 4)$ $16x^3 - 64x^2 - x + 4$
21. $(x - 3)(x + 3)(x - 4)(x + 4)$ $x^4 - 25x^2 + 144$

4. $100x^2 - 121y^2$ 6. $4b^2 - 16d^2$ 9. $36m^2 + 24mn + 4n^2$ 16. $25a^2 - 120ab + 144b^2$

Lesson 10–1 Find the factors of each number.

1. 17 1, 17
2. 21 1, 3, 7, 21
3. 81 1, 3, 9, 27, 81
4. 24 1, 2, 3, 4, 6, 8, 12, 24
5. 18 1, 2, 3, 6, 9, 18
6. 22 1, 2, 11, 22

State whether each number is *prime* or *composite*. If the number is composite, find its prime factorization.

7. 39 composite; $3 \cdot 13$
8. 89 prime
9. 72 composite; $2^3 \cdot 3^2$
10. 41 prime
11. 57 composite; $3 \cdot 19$
12. 60 composite; $2^2 \cdot 3 \cdot 5$

Factor each expression completely. Do not use exponents.

13. -64 $-1 \cdot 2 \cdot 2 \cdot 2 \cdot 2 \cdot 2 \cdot 2$
14. -26 $-1 \cdot 2 \cdot 13$
15. -240 $-1 \cdot 2 \cdot 2 \cdot 2 \cdot 2 \cdot 3 \cdot 5$
16. -231 $-1 \cdot 3 \cdot 7 \cdot 11$
17. $44rs^2t^3$ $2 \cdot 2 \cdot 11 \cdot r \cdot s \cdot s \cdot t \cdot t \cdot t$
18. $756(mn)^2$ $2 \cdot 2 \cdot 3 \cdot 3 \cdot 3 \cdot 7 \cdot m \cdot m \cdot n \cdot n$

Find the GCF of the given monomials.

19. 16, 60 4
20. 15, 50 5
21. -80, 45 5
22. 29, -58 29
23. 305, 55 5
24. 252, 126 126
25. 128, 245 1
26. $7y^2$, $14y^2$ $7y^2$
27. $4xy$, $-6x$ $2x$
28. $35t^2$, $7t$ $7t$
29. $16pq^2$, $12p^2q$ $4pq$
30. 5, 15, 10 5
31. $12mn$, $10mn$, $15mn$ mn
32. 14, 12, 20 2
33. $26jk^4$, $16jk^3$, $8j^2$ $2j$

Lesson 10-2 **Complete. In exercises with two blanks, both blanks represent the same expression.**

1. $6x + 3y = 3(\underline{?} + y)$ **$2x$**
2. $8x^2 - 4x = 4x(2x - \underline{?})$ **1**
3. $12a^2b + 6a = 6a(\underline{?} + 1)$ **$2ab$**
4. $14r^2t - 42t = 14t(\underline{?} - 3)$ **r^2**
5. $24x^2 + 12y^2 = 12(\underline{?} + y^2)$ **$2x^2$**
6. $12xy + 12x^2 = \underline{?}\,(y + x)$ **$12x$**
7. $(bx + by) + (3ax + 3ay) = b(\underline{?}) + 3a(\underline{?})$ **$x + y$**
8. $(10x^2 - 6xy) + (15x - 9y) = 2x(\underline{?}) + 3(\underline{?})$ **$5x - 3$**
9. $(6x^3 + 6x) + (7x^2y + 7y) = 6x(\underline{?}) + 7y(\underline{?})$ **$x^2 + 1$**

Factor each polynomial.

10. $10a^2 + 40a$ **$10a(a + 4)$**
11. $15wx - 35wx^2$ **$5wx(3 - 7x)$**
12. $27a^2b + 9b^3$ **$9b(3a^2 + b^2)$**
13. $11x + 44x^2y$ **$11x(1 + 4xy)$**
14. $16y^2 + 8y$ **$8y(2y + 1)$**
15. $14mn^2 + 2mn$ **$2mn(7n + 1)$**
16. $25a^2b^2 + 30ab^3$ **$5ab^2(5a + 6b)$**
17. $2m^3n^2 - 16m^2n^3 + 8mn$ **$2mn(m^2n - 8mn^2 + 4)$**
18. $2ax + 6xc + ba + 3bc$ **$(2x + b)(a + 3c)$**
19. $6mx - 4m + 3rx - 2r$ **$(2m + r)(3x - 2)$**
20. $3ax - 6bx + 8b - 4a$ **$(a - 2b)(3x - 4)$**
21. $a^2 - 2ab + a - 2b$ **$(a + 1)(a - 2b)$**
22. $8ac - 2ad + 4bc - bd$ **$(2a + b)(4c - d)$**
23. $2e^2g + 2fg + 4e^2h + 4fh$ **$2(e^2 + f)(g + 2h)$**

Lesson 10-3 **Complete.**

1. $p^2 + 9p - 10 = (p + \underline{?})(p - 1)$ **10**
2. $y^2 - 2y - 35 = (y + 5)(y - \underline{?})$ **7**
3. $4a^2 + 4a - 63 = (2a - 7)(2a \underline{?} 9)$ **$+$**
4. $4r^2 - 25r + 6 = (r - 6)(\underline{?} - 1)$ **$4r$**
5. $b^2 + 12b + 35 = (b + 5)(b + \underline{?})$ **7**
6. $3x^2 - 7x - 6 = (3x + 2)(x \underline{?} 7)$ **$-$**
7. $3a^2 - 2a - 21 = (a \underline{?} 3)(3a + 7)$ **$-$**
8. $4y^2 + 11y + 6 = (\underline{?} + 3)(y + 2)$ **$4y$**
9. $2z^2 - 11z + 15 = (\underline{?} - 5)(z - 3)$ **$2z$**
10. $6n^2 + 7n - 3 = (2n + \underline{?})(3n - 1)$ **3**

Factor each trinomial, if possible. If the trinomial cannot be factored using integers, write *prime*.

11. $5x^2 - 17x + 14$ **$(5x - 7)(x - 2)$**
12. $a^2 - 9a - 36$ **$(a - 12)(a + 3)$**
13. $x^2 + 2x - 15$ **$(x + 5)(x - 3)$**
14. $n^2 - 8n + 15$ **$(n - 3)(n - 5)$**
15. $b^2 + 22b + 21$ **$(b + 21)(b + 1)$**
16. $c^2 + 2c - 3$ **$(c + 3)(c - 1)$**
17. $x^2 - 5x - 24$ **$(x - 8)(x + 3)$**
18. $2n^2 - 11n + 7$ **prime**
19. $8m^2 - 10m + 3$
20. $z^2 + 15z + 36$ **$(z + 3)(z + 12)$**
21. $s^2 - 13st - 30t^2$
22. $6y^2 + 2y - 2$ **$2(3y^2 + y - 1)$**
23. $2r^2 + 3r - 14$ **$(2r + 7)(r - 2)$**
24. $5x - 6 + x^2$ **$(x - 1)(x + 6)$**
25. $x^2 - 4xy - 5y^2$ **$(x - 5y)(x + y)$**
26. $5r^2 - 3r + 15$ **prime**
27. $18v^2 + 42v + 12$
28. $4k^2 + 2k - 12$ **$2(2k - 3)(k + 2)$**

19. $(4m - 3)(2m - 1)$ 21. $(s + 2t)(s - 15t)$ 27. $6(3v + 1)(v + 2)$

Lesson 10-4 **Factor each polynomial, if possible. If the polynomial cannot be factored, write *prime*.** **24, 26–30. See margin. 14. $(3x - 10y)(3x + 10y)$ 19. $-5(3m + 1)(3m - 1)$**

1. $x^2 - 9$ **$(x - 3)(x + 3)$**
2. $a^2 - 64$ **$(a - 8)(a + 8)$**
3. $t^2 - 49$ **$(t - 7)(t + 7)$**
4. $4x^2 - 9y^2$ **$(2x - 3y)(2x + 3y)$**
5. $1 - 9z^2$ **$(1 - 3z)(1 + 3z)$**
6. $16a^2 - 9b^2$ **$(4a - 3b)(4a + 3b)$**
7. $8x^2 - 12y^2$ **$4(2x^2 - 3y^2)$**
8. $a^2 - 4b^2$ **$(a - 2b)(a + 2b)$**
9. $x^2 - y^2$ **$(x - y)(x + y)$**
10. $75r^2 - 48$ **$3(5r + 4)(5r - 4)$**
11. $x^2 - 36y^2$ **$(x - 6y)(x + 6y)$**
12. $3a^2 - 16$ **prime**
13. $12t^2 - 75$ **$3(2t - 5)(2t + 5)$**
14. $9x^2 - 100y^2$
15. $49 - a^2b^2$ **$(7 - ab)(7 + ab)$**
16. $12a^2 - 48$ **$12(a - 2)(a + 2)$**
17. $169 - 16t^2$ **$(13 - 4t)(13 + 4t)$**
18. $8r^2 - 4$ **$4(2r^2 - 1)$**
19. $-45m^2 + 5$
20. $9x^4 - 16y^2$ **$(3x^2 - 4y)(3x^2 + 4y)$**
21. $36b^2 - 64$ **$4(3b - 4)(3b + 4)$**
22. $5g^2 - 20h^2$ **$5(g - 2h)(g + 2h)$**
23. $\frac{1}{4}n^2 - 16$ **$\left(\frac{1}{2}n - 4\right)\left(\frac{1}{2}n + 4\right)$**
24. $\frac{1}{4}t^2 - \frac{4}{9}p^2$
25. $(r - t)^2 + t^2$ **prime**
26. $12x^3 - 27xy^2$
27. $0.01n^2 - 1.69r^2$
28. $0.04m^2 - 0.09n^2$
29. $(x - y)^2 - y^2$
30. $162m^4 - 32n^8$

EXTRA PRACTICE

Additional Answers for Lesson 10-4

24. $\left(\frac{1}{2}t - \frac{2}{3}p\right)\left(\frac{1}{2}t + \frac{2}{3}p\right)$
26. $3x(2x - 3y)(2x + 3y)$
27. $(0.1n - 1.3r)(0.1n + 1.3r)$
28. $(0.2m - 0.3n)(0.2m + 0.3n)$
29. $x(x - 2y)$
30. $2(3m - 2n^2)(3m + 2n^2)(9m^2 + 4n^4)$

EXTRA PRACTICE

Lesson 10-5 Determine whether each trinomial is a perfect square trinomial. If so, factor it.

1. $x^2 + 12x + 36$ yes; $(x + 6)^2$
2. $n^2 - 13n + 36$ no
3. $a^2 + 4a + 4$ yes; $(a + 2)^2$
4. $b^2 - 14b + 49$ yes; $(b - 7)^2$
5. $x^2 + 20x - 100$ no
6. $y^2 - 10y + 100$ no
7. $9b^2 - 6b + 1$ yes; $(3b - 1)^2$
8. $4x^2 + 4x + 1$ yes; $(2x + 1)^2$
9. $2n^2 + 17n + 21$ no
10. $9x^2 - 10x + 4$ no
11. $9y^2 + 8y - 16$ no
12. $4a^2 - 20a + 25$ yes; $(2a - 5)^2$

Factor each polynomial, if possible. If the polynomial cannot be factored, write *prime*.

13. $n^2 - 8n + 16$ $(n - 4)^2$
14. $4k^2 - 4k + 1$ $(2k - 1)^2$
15. $x^2 + 16x + 64$ $(x + 8)^2$
16. $t^2 - 4t + 1$ prime
17. $x^2 + 22x + 121$ $(x + 11)^2$
18. $s^2 + 30s + 225$ $(s + 15)^2$
19. $1 - 10z + 25z^2$ $(1 - 5z)^2$
20. $9p^2 - 56p + 49$ prime
21. $9n^2 - 36nm + 36m^2$
22. $16a^2 + 81 - 72a$ $(4a - 9)^2$
23. $9x^2 + 12xy + 4y^2$ $(3x + 2y)^2$
24. $m^2 + 16mn + 64n^2$ $(m + 8n)^2$
25. $8t^4 + 56t^3 + 98t^2$ $2t^2(2t + 7)^2$
26. $4p^2 + 12pr + 9r^2$ $(2p + 3r)^2$
27. $16m^4 - 72m^2n^2 + 81n^4$ $(2m - 3n)^2(2m + 3n)^2$

21. $9(n - 2m)^2$

Lesson 10-6 Solve each equation. Check your solutions.

12. $\{0, -2\}$

1. $y(y - 12) = 0$ $\{0, 12\}$
2. $2x(5x - 10) = 0$ $\{0, 2\}$
3. $7a(a + 6) = 0$ $\{0, -6\}$
4. $(b - 3)(b - 5) = 0$ $\{3, 5\}$
5. $(p - 5)(p + 5) = 0$ $\{5, -5\}$
6. $(4t + 4)(2t + 6) = 0$ $\{-1, -3\}$
7. $(3x - 5)^2 = 0$ $\left\{\frac{5}{3}\right\}$
8. $x^2 - 6x = 0$ $\{0, 6\}$
9. $n^2 + 36n = 0$ $\{0, -36\}$
10. $2x^2 + 4x = 0$ $\{0, -2\}$
11. $2x^2 = x^2 - 8x$ $\{0, -8\}$
12. $7y - 1 = -3y^2 + y - 1$
13. $\frac{1}{2}y^2 - \frac{1}{4}y = 0$ $\left\{0, \frac{1}{2}\right\}$
14. $\frac{5}{6}x^2 - \frac{1}{3}x = \frac{1}{3}x$ $\left\{0, \frac{4}{5}\right\}$
15. $\frac{2}{3}x = \frac{1}{3}x^2$ $\{0, 2\}$
16. $\frac{3}{4}a^2 + \frac{7}{8}a = a$ $\left\{0, \frac{1}{6}\right\}$
17. $n^2 - 3n = 0$ $\{0, 3\}$
18. $3x^2 - \frac{3}{4}x = 0$ $\left\{0, \frac{1}{4}\right\}$
19. $8a^2 = -4a$ $\left\{-\frac{1}{2}, 0\right\}$
20. $(2y + 8)(3y + 24) = 0$ $\{-4, -8\}$
21. $(4x - 7)(3x + 5) = 0$ $\left\{\frac{7}{4}, -\frac{5}{3}\right\}$

Lesson 11-1 Write the equation of the axis of symmetry and find the coordinates of the vertex of the graph of each equation. State if the vertex is a maximum or minimum. Then graph the equation.

1–24. See Solutions Manual.

1. $y = x^2 + 6x + 8$
2. $y = -x^2 + 3x$
3. $y = -x^2 + 7$
4. $y = x^2 + x + 3$
5. $y = -x^2 + 4x + 5$
6. $y = 3x^2 + 6x + 16$
7. $y = -x^2 + 2x - 3$
8. $y = 3x^2 + 24x + 80$
9. $y = x^2 - 4x - 4$
10. $y = 5x^2 - 20x + 37$
11. $y = 3x^2 + 6x + 3$
12. $y = 2x^2 + 12x$
13. $y = x^2 - 6x + 5$
14. $y = \frac{1}{2}x^2 + 3x + \frac{9}{2}$
15. $y = \frac{1}{4}x^2 - 4x + \frac{15}{4}$
16. $y = 4x^2 - 1$
17. $y = -2x^2 - 2x + 4$
18. $y = 6x^2 - 12x - 4$
19. $y = x^2 - 1$
20. $y = -x^2 + x + 1$
21. $y = -5x^2 - 3x + 2$
22. $y = x^2 - x - 6$
23. $y = 2x^2 + 5x - 2$
24. $y = -3x^2 - 18x - 15$

Lesson 11–2 State the real roots of each quadratic equation whose related function is graphed below.

1. 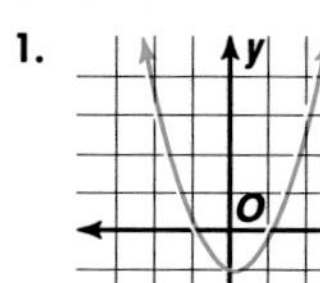 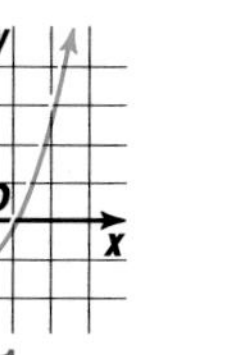 $-1, 1$

2. $0, 2$

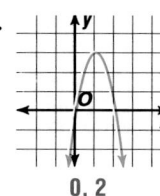

3. 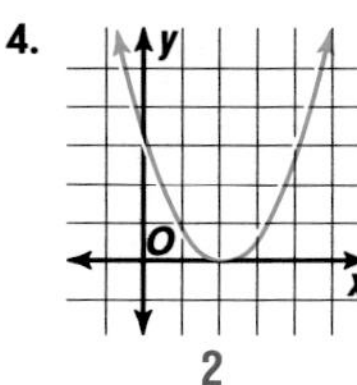 $\varnothing$

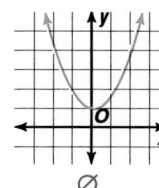

4. 2

Solve each equation by graphing. If exact roots cannot be found, state the consecutive integers between which the roots lie. 5–19. See Solutions Manual for graphs. 8. $0 < r < 1, 3 < r < 4$

5. $x^2 + 2x - 3 = 0$ $-3, 1$
6. $-x^2 + 6x - 5 = 0$ $1, 5$
7. $-a^2 - 2a + 3 = 0$ $-3, 1$
8. $2r^2 - 8r + 5 = 0$
9. $-3x^2 + 6x - 9 = 0$ $\varnothing$
10. $c^2 + c = 0$ $-1, 0$
11. $3t^2 + 2 = 0$ $\varnothing$
12. $-b^2 + 5b + 2 = 0$
13. $3x^2 + 7x = 1$
14. $x^2 + 5x - 24 = 0$ $-8, 3$
15. $8 - k^2 = 0$
16. $x^2 - 7x = 18$ $-2, 9$
17. $a^2 + 12a + 36 = 0$ -6
18. $64 - x^2 = 0$ $-8, 8$
19. $-4x^2 + 2x = -1$

12. $-1<b<0, 5<b<6$ 13. $-3<x<-2, 0<x<1$ 15. $-3 < k < -2, 2 < k < 3$ 19. $-1<x<0, 0<x<1$

Lesson 11–3 Solve each equation by using the quadratic formula. Approximate irrational roots to the nearest hundredth.

1. $x^2 - 8x - 4 = 0$ $8.47, -0.47$
2. $x^2 + 7x + 6 = 0$ $-6, -1$
3. $x^2 + 5x - 6 = 0$ $-6, 1$
4. $y^2 - 7y - 8 = 0$ $-1, 8$
5. $m^2 - 2m = 35$ $-5, 7$
6. $4n^2 - 20n = 0$ $0, 5$
7. $m^2 + 4m + 2 = 0$ $-0.59, -3.41$
8. $2t^2 - t - 15 = 0$ $-2.5, 3$
9. $5t^2 = 125$ $-5, 5$
10. $t^2 + 16 = 0$ $\varnothing$
11. $-4x^2 + 8x = -3$ $2.32, -0.32$
12. $3k^2 + 2 = -8k$ $-2.39, -0.28$
13. $8t^2 + 10t + 3 = 0$ $-\frac{1}{2}, -\frac{3}{4}$
14. $3x^2 - \frac{5}{4}x - \frac{1}{2} = 0$ $\frac{2}{3}, -\frac{1}{4}$
15. $-5b^2 + 3b - 1 = 0$ $\varnothing$
16. $s^2 + 8s + 7 = 0$ $-7, -1$
17. $d^2 - 14d + 24 = 0$ $2, 12$
18. $3k^2 + 11k = 4$ $-4, \frac{1}{3}$
19. $n^2 - 3n + 1 = 0$ $2.62, 0.38$
20. $2z^2 + 5z - 1 = 0$ $0.19, -2.69$
21. $3h^2 = 27$ $3, -3$
22. $3f^2 + 2f = 6$ $1.12, -1.79$
23. $2x^2 = 0.7x + 0.3$ $0.6, -0.25$
24. $3w^2 - 8w + 2 = 0$ $2.39, 0.28$
25. $2r^2 - r - 3 = 0$ $-1, \frac{3}{2}$
26. $x^2 - 9x = 5$ $9.52, -0.52$
27. $6t^2 - 4t - 9 = 0$ $1.60, -0.94$

Lesson 11–4 Use a calculator to determine the approximate value of each expression to the nearest hundredth.

1. $3^{1.6}$ 5.80
2. $10^{-0.2}$ 0.63
3. $\left(\frac{1}{3}\right)^{-1.4}$ 4.66
4. $\left(\frac{2}{3}\right)^{5.1}$ 0.13
5. $40(2^{-0.5})$ 28.28
6. $10(2^{-1.6})$ 3.30
7. $0.3(4^{0.8})$ 0.91
8. $30(0.75^{-3.6})$ 84.51
9. $5^{1.75}$ 16.72

Graph each function. State the y-intercept. 10–21. See Solutions Manual for graphs.

10. $y = 3^x + 1$ 2
11. $y = 2^x - 5$ -4
12. $y = 2^{x+3}$ 8
13. $y = 3^{x+1}$ 3
14. $y = \left(\frac{1}{4}\right)^x$ 1
15. $y = 5\left(\frac{2}{5}\right)^x$ 5
16. $y = 3 \cdot 2^x$ 3
17. $y = 4 \cdot 5^x$ 4
18. $y = 6^x$ 1
19. $y = 3^x$ 1
20. $y = \left(\frac{1}{8}\right)^x$ 1
21. $y = \left(\frac{3}{4}\right)^x$ 1

Solve each equation.

22. $6^{3x-4} = 6^x$ 2
23. $3^4 = 3^{2x+2}$ 1
24. $4^x = 4^{5x+8}$ -2
25. $2^x = 4^{x+1}$ -2
26. $5^{4x} = 5^{-4}$ -1
27. $2^{x+3} = 2^{-5}$ -8

EXTRA PRACTICE

Lesson 11-5 **Determine whether each exponential equation represents growth or decay.**

1. $y = 3.89(1.05)^x$ growth
2. $y = 476(0.35)^x$ decay
3. $y = 19{,}520(0.98)^x$ decay
4. $y = 16(1.0432)^x$ growth
5. $y = 1.01(1.099)^x$ growth
6. $y = 84(0.03)^x$ decay

7. **Education** Marco withdrew all of the $2500 in his savings account to pay the tuition for his first semester at college. The account had earned 12% interest compounded monthly, and no withdrawals or additional deposits were made.
   a. If Marco's original deposit was $1250, how long ago did he open the account? **about 5 years, 10 months ago**
   b. If Marco's original deposit was $1500, how long ago did he open the account? **about 4 years, 3 months ago**

8. **Finance** Erin saved $500 of the money she earned working at the Dairy Dream last summer. She deposited the money in a certificate of deposit that earns 8.75% interest compounded monthly. If she rolls over the CD at the same rate each year, when will Erin's CD have a balance of $800? **5 years, 5 months**

9. **Demographics** In 1994, the metropolitan area of Pensacola, Florida, had a population of 371,000. The growth rate from 1990 to 1994 was 7.7%.
   a. Write an exponential equation for the area's growth. $y = 371{,}000(1.077)^t$
   b. Compute the estimated population for Pensacola in the year 2000. **578,986**

Lesson 12-1 **Simplify each rational expression. State the excluded values of the variables.**

1. $\frac{13a}{39a^2}$ $\frac{1}{3a}$; $a \neq 0$
2. $\frac{38x^2}{42xy}$ $\frac{19x}{21y}$; $x, y \neq 0$
3. $\frac{14y^2z}{49yz^3}$ $\frac{2y}{7z^2}$; $y, z \neq 0$
4. $\frac{p+5}{2(p+5)}$ $\frac{1}{2}$; $p \neq -5$
5. $\frac{79a^2b}{158a^3bc}$ $\frac{1}{2ac}$; $a, b, c \neq 0$
6. $\frac{a+b}{a^2-b^2}$ $\frac{1}{a-b}$; $a \neq \pm b$
7. $\frac{y+4}{(y-4)(y+4)}$ $\frac{1}{y-4}$; $y \neq 4, -4$
8. $\frac{c^2-4}{(c+2)^2}$ $\frac{c-2}{c+2}$; $c \neq -2$
9. $\frac{a^2-a}{a-1}$ a; $a \neq 1$
10. $\frac{(w-4)(w+4)}{(w-2)(w-4)}$ $\frac{w+4}{w-2}$; $w \neq 2, 4$
11. $\frac{m^2-2m}{m-2}$ m; $m \neq 2$
12. $\frac{x^2+4}{x^4-16}$ $\frac{1}{x^2-4}$; $x \neq \pm 2$
13. $\frac{r^3-r^2}{r-1}$ r^2; $r \neq 1$
14. $\frac{3m^3}{6m^2-3m}$ $\frac{m^2}{2m-1}$; $m \neq 0, \frac{1}{2}$
15. $\frac{4t^2-8}{4t-4}$ $\frac{t^2-2}{t-1}$; $t \neq 1$
16. $\frac{6y^3-12y^2}{12y^2-18}$ $\frac{y^2(y-2)}{2y^2-3}$; $y \neq \pm\frac{\sqrt{6}}{2}$
17. $\frac{x-3}{x^2+x-12}$ $\frac{1}{x+4}$; $x \neq -4, 3$
18. $\frac{5x^2+10x+5}{3x^2+6x+3}$ $\frac{5}{3}$; $x \neq -1$

Lesson 12-2 **Find each product. Assume that no denominator has a value of 0.**

1. $\frac{a^2b}{b^2c} \cdot \frac{c}{d}$ $\frac{a^2}{bd}$
2. $\frac{6a^2n}{8n^2} \cdot \frac{12n}{9a}$ a
3. $\frac{2a^2d}{3bc} \cdot \frac{9b^2c}{16ad^2}$ $\frac{3ab}{8d}$
4. $\frac{10n^3}{6x^3} \cdot \frac{12n^2x^4}{25n^2x^2}$ $\frac{4n^3}{5x}$
5. $\left(\frac{2a}{b}\right)^2 \cdot \frac{5c}{6a}$ $\frac{10ac}{3b^2}$
6. $\frac{6m^3n}{10a^2} \cdot \frac{4a^2m}{9n^3}$ $\frac{4m^4}{15n^2}$
7. $\frac{5n-5}{3} \cdot \frac{9}{n-1}$ 15
8. $\frac{a^2}{a-b} \cdot \frac{3a-3b}{a}$ $3a$
9. $\frac{2a+4b}{5} \cdot \frac{25}{6a+8b}$ $\frac{5a+10b}{3a+4b}$
10. $\frac{4t}{4t+40} \cdot \frac{3t+30}{2t}$ $\frac{3}{2}$
11. $\frac{3k+9}{k} \cdot \frac{k^2}{k^2-9}$ $\frac{3k}{k-3}$
12. $\frac{7xy^3}{11z^2} \cdot \frac{44z^3}{21x^2y}$ $\frac{4y^2z}{3x}$
13. $\frac{3}{x-y} \cdot \frac{(x-y)^2}{6}$ $\frac{x-y}{2}$
14. $\frac{x+5}{3x} \cdot \frac{12x^2}{x^2+7x+10}$ $\frac{4x}{x+2}$
15. $\frac{a^2-b^2}{4} \cdot \frac{16}{a+b}$ $4a-4b$
16. $\frac{4a+8}{a^2-25} \cdot \frac{a-5}{5a+10}$ $\frac{4}{5a+25}$
17. $\frac{r^2}{r-s} \cdot \frac{r^2-s^2}{s^2}$ $\frac{r^3+r^2s}{s^2}$
18. $\frac{a^2-b^2}{a-b} \cdot \frac{7}{a+b}$ 7

Lesson 12-3 Find each quotient. Assume that no denominator has a value of 0.

1. $\frac{5m^2n}{12a^2} \div \frac{30m^4}{18an}$ $\frac{n^2}{4am^2}$
2. $\frac{25g^7h}{28t^3} \div \frac{5g^5h^2}{42s^2t^3}$ $\frac{15g^2s^2}{2h}$
3. $\frac{6a+3b}{36} \div \frac{3a+2b}{45}$ $\frac{5}{2}$
4. $\frac{x^2y}{18z} \div \frac{2yz}{3x^2}$ $\frac{x^4}{12z^2}$
5. $\frac{p^2}{14qr^3} \div \frac{2r^2p}{7q}$ $\frac{p}{4r^5}$
6. $\frac{5e-f}{5e+f} \div (25e^2 - f^2)$ $\frac{1}{(5e+f)^2}$
7. $\frac{t^2-2t-15}{t-5} \div \frac{t+3}{t+5}$ $t+5$
8. $\frac{5x+10}{x+2} \div (x+2)$ $\frac{5}{x+2}$
9. $\frac{3d}{2d^2-3d} \div \frac{9}{2d-3}$ $\frac{1}{3}$
10. $\frac{3v^2-27}{15v} \div \frac{v+3}{v^2}$ $\frac{v(v-3)}{5}$
11. $\frac{3g^2+15g}{4} \div \frac{g+5}{g^2}$ $\frac{3g^3}{4}$
12. $\frac{b^2-9}{4b} \div (b-3)$ $\frac{b+3}{4b}$
13. $\frac{p^2}{y^2-4} \div \frac{p}{2-y}$ $-\frac{p}{y+2}$
14. $\frac{k^2-81}{k^2-36} \div \frac{k-9}{k+6}$ $\frac{k+9}{k-6}$
15. $\frac{2a^3}{a+1} \div \frac{a^2}{a+1}$ $2a$
16. $\frac{x^2-16}{16-x^2} \div \frac{7}{x}$ $-\frac{x}{7}$
17. $\frac{y}{5} \div \frac{y^2-25}{5-y}$ $\frac{-y}{5y+25}$
18. $\frac{3m}{m+1} \div (m-2)$ $\frac{3m}{m^2-m-2}$

Lesson 12-4 Find each quotient.

1. $(2x^2 - 11x - 20) \div (2x+3)$ $x - 7 + \frac{1}{2x+3}$
2. $(a^2 + 7a + 12) \div (a+3)$ $a+4$
3. $(m^2 + 9m + 20) \div (m+5)$ $m+4$
4. $(x^2 - 2x - 35) \div (x-7)$ $x+5$
5. $(c^2 + 12c + 36) \div (c+9)$ $c + 3 + \frac{9}{c+9}$
6. $(y^2 - 2y - 30) \div (y+7)$ $y - 9 + \frac{33}{y+7}$
7. $(3t^2 - 14t - 24) \div (3t+4)$ $t-6$
8. $(2r^2 - 3r - 35) \div (2r+7)$ $r-5$
9. $\frac{12n^2+36n+15}{6n+3}$ $2n+5$
10. $\frac{10x^2+29x+21}{5x+7}$ $2x+3$
11. $\frac{4t^3+17t^2-1}{4t+1}$ t^2+4t-1
12. $\frac{2a^3+9a^2+5a-12}{a+3}$ $2a^2+3a-4$
13. $\frac{4m^3+5m-21}{2m-3}$ $2m^2+3m+7$
14. $\frac{6t^3+5t^2+12}{2t+3}$ $3t^2-2t+3+\frac{3}{2t+3}$
15. $\frac{27c^2-24c+8}{9c-2}$ $3c-2+\frac{4}{9c-2}$
16. $\frac{3b^3+8b^2+b-7}{b+2}$ $3b^2+2b-3-\frac{1}{b+2}$
17. $\frac{t^3-19t+9}{t-4}$ $t^2+4t-3-\frac{3}{t-4}$
18. $\frac{9d^3+5d-8}{3d-2}$ $3d^2+2d+3-\frac{2}{3d-2}$

Lesson 12-5 Find each sum or difference. Express in simplest form.

1. $\frac{4}{z}+\frac{3}{z}$ $\frac{7}{z}$
2. $\frac{a}{12}+\frac{2a}{12}$ $\frac{a}{4}$
3. $\frac{5}{2t}+\frac{-7}{2t}$ $-\frac{1}{t}$
4. $\frac{y}{2}+\frac{y}{2}$ y
5. $\frac{b}{x}+\frac{2}{x}$ $\frac{b+2}{x}$
6. $\frac{5x}{24}-\frac{3x}{24}$ $\frac{x}{12}$
7. $\frac{7p}{p}-\frac{8p}{p}$ -1
8. $\frac{8k}{5m}-\frac{3k}{5m}$ $\frac{k}{m}$
9. $\frac{y}{2}+\frac{y-6}{2}$ $y-3$
10. $\frac{a+2}{6}-\frac{a+3}{6}$ $-\frac{1}{6}$
11. $\frac{8}{m-2}-\frac{6}{m-2}$ $\frac{2}{m-2}$
12. $\frac{x}{x+1}+\frac{1}{x+1}$ 1
13. $\frac{2n}{2n-5}+\frac{5}{5-2n}$ 1
14. $\frac{y}{b+6}-\frac{2y}{b+6}$ $\frac{-y}{b+6}$
15. $\frac{x-y}{2-y}+\frac{x+y}{y-2}$ $\frac{2y}{y-2}$
16. $\frac{r^2}{r-s}+\frac{s^2}{r-s}$ $\frac{r^2+s^2}{r-s}$
17. $\frac{12n}{3n+2}+\frac{8}{3n+2}$ 4
18. $\frac{6x}{x+y}+\frac{6y}{x+y}$ 6

EXTRA PRACTICE

EXTRA PRACTICE

Lesson 12-6 Find each sum or difference.

1. $\frac{s}{3} + \frac{2s}{7}$ $\frac{13s}{21}$
2. $\frac{5}{2a} + \frac{-3}{6a}$ $\frac{2}{a}$
3. $\frac{2n}{5} - \frac{3m}{4}$ $\frac{8n - 15m}{20}$
4. $\frac{6}{5x} + \frac{7}{10x^2}$ $\frac{12x + 7}{10x^2}$
5. $\frac{3z}{7w^2} - \frac{2z}{w}$ $\frac{3z - 14wz}{7w^2}$
6. $\frac{s}{t^2} - \frac{r}{3t}$ $\frac{3s - rt}{3t^2}$
7. $\frac{5}{xy} + \frac{6}{yz}$ $\frac{5z + 6x}{xyz}$
8. $\frac{2}{t} + \frac{t + 3}{s}$ $\frac{2s + t^2 + 3t}{st}$
9. $\frac{a}{a - b} + \frac{b}{2b + 3a}$ $\frac{3a^2 + 3ab - b^2}{3a^2 - ab - 2b^2}$
10. $\frac{a}{a^2 - 4} - \frac{4}{a + 2}$ $\frac{-3a + 8}{a^2 - 4}$
11. $\frac{4a}{2a + 6} + \frac{3}{a + 3}$ $\frac{2a + 3}{a + 3}$
12. $\frac{m}{1(m - n)} - \frac{5}{m}$ $\frac{m^2 - 5m + 5n}{m(m - n)}$
13. $\frac{-3}{a - 5} + \frac{-6}{a^2 - 5a}$ $\frac{-3a - 6}{a^2 - 5a}$
14. $\frac{3t + 2}{3t - 6} - \frac{t + 2}{t^2 - 4}$ $\frac{3t - 1}{3t - 6}$
15. $\frac{y + 5}{y - 5} + \frac{2y}{y^2 - 25}$ $\frac{y^2 + 12y + 25}{y^2 - 25}$
16. $\frac{-18}{y^2 - 9} + \frac{7}{3 - y}$ $\frac{-7y - 39}{y^2 - 9}$
17. $\frac{c}{c^2 - 4c} - \frac{5c}{c - 4}$ $\frac{1 - 5c}{c - 4}$
18. $\frac{t + 10}{t^2 - 100} + \frac{1}{t - 10}$ $\frac{2}{t - 10}$

Lesson 12-7 Write each mixed expression as a rational expression.

1. $4 + \frac{2}{x}$ $\frac{4x + 2}{x}$
2. $8 + \frac{5}{3t}$ $\frac{24t + 5}{3t}$
3. $3b + \frac{b + 1}{2b}$ $\frac{6b^2 + b + 1}{2b}$
4. $2n + \frac{4 + n}{n}$ $\frac{2n^2 + n + 4}{n}$
5. $a^2 + \frac{2}{a - 2}$ $\frac{a^3 - 2a^2 + 2}{a - 2}$
6. $3r^2 + \frac{4}{2r + 1}$ $\frac{6r^3 + 3r^2 + 4}{2r + 1}$

Simplify.

7. $\frac{3\frac{1}{2}}{4\frac{3}{4}}$ $\frac{14}{19}$
8. $\frac{\frac{x^2}{y}}{\frac{y}{x^3}}$ $\frac{x^5}{y^2}$
9. $\frac{\frac{t^4}{u}}{\frac{t^3}{u^2}}$ tu
10. $\frac{\frac{x^3}{y^2}}{\frac{x + y}{x - y}}$ $\frac{x^3(x - y)}{y^2(x + y)}$
11. $\frac{\frac{y}{3} + \frac{5}{6}}{2 + \frac{5}{y}}$ $\frac{y}{6}$
12. $\frac{\frac{1}{x} + \frac{1}{y}}{\frac{1}{y} - \frac{1}{x}}$ $\frac{x + y}{x - y}$
13. $\frac{t - 2}{\frac{t^2 - 4}{t^2 + 5t + 6}}$ $t + 3$
14. $\frac{\frac{y^2 - 1}{y^2 + 3y - 4}}{y + 1}$ $\frac{1}{y + 4}$

Lesson 12-8 Solve each equation.

1. $\frac{k}{6} + \frac{2k}{3} = -\frac{5}{2}$ -3
2. $\frac{3x}{5} + \frac{3}{2} = \frac{7x}{10}$ 15
3. $\frac{18}{b} = \frac{3}{b} + 3$ 5
4. $\frac{3}{5x} + \frac{7}{2x} = 1$ $\frac{41}{10}$
5. $\frac{2a - 3}{6} = \frac{2a}{3} + \frac{1}{2}$ -3
6. $\frac{x + 1}{x} + \frac{x + 4}{x} = 6$ $\frac{5}{4}$
7. $\frac{2b - 3}{7} - \frac{b}{2} = \frac{b + 3}{14}$ $-\frac{9}{4}$
8. $\frac{2y}{y - 4} - \frac{3}{5} = 3$ 9
9. $\frac{2t}{t + 3} + \frac{3}{t} = 2$ 3
10. $\frac{5x}{x + 1} + \frac{1}{x} = 5$ $\frac{1}{4}$
11. $\frac{r - 1}{r + 1} - \frac{2r}{r - 1} = -1$ 0
12. $\frac{m}{m + 1} + \frac{5}{m - 1} = 1$ $-\frac{3}{2}$
13. $\frac{5}{5 - p} - \frac{p^2}{5 - p} = -2$ -5 or 3
14. $\frac{14}{b - 6} = \frac{1}{2} + \frac{6}{b - 8}$ 10 or 20
15. $\frac{r}{3r + 6} - \frac{r}{5r + 10} = \frac{2}{5}$ -3
16. $\frac{4x}{2x + 3} - \frac{2x}{2x - 3} = 1$ $\frac{1}{2}$
17. $\frac{2a - 3}{a - 3} - 2 = \frac{12}{a + 2}$ $\frac{14}{3}$
18. $\frac{z + 3}{z - 1} + \frac{z + 1}{z - 3} = 2$ 2

Lesson 13-1 **If c is the measure of the hypotenuse of a right triangle, find each missing measure. Round answers to the nearest hundredth.**

1. $b = 20, c = 29, a = ?$ 21
2. $a = 7, b = 24, c = ?$ 25
3. $a = 2, b = 6, c = ?$ $\sqrt{40} \approx 6.32$
4. $b = 10, c = \sqrt{200}, a = ?$ 10
5. $a = 3, c = 3\sqrt{2}, b = ?$ 3
6. $a = 6, c = 14, b = ?$ $\sqrt{160} \approx 12.65$
7. $a = \sqrt{11}, c = \sqrt{47}, b = ?$ 6
8. $a = \sqrt{13}, b = 6, c = ?$ 7
9. $a = \sqrt{6}, b = 3, c = ?$ $\sqrt{15} \approx 3.87$
10. $b = \sqrt{75}, c = 10, a = ?$ 5
11. $b = 9, c = \sqrt{130}, a = ?$ 7
12. $a = 9, c = 15, b = ?$ 12
13. $b = 5, c = 11, a = ?$ $\sqrt{96} \approx 9.80$
14. $a = \sqrt{33}, b = 4, c = ?$ 7

Determine whether the following side measures would form right triangles.

15. 14, 48, 50 yes
16. 20, 30, 40 no
17. 21, 72, 75 yes
18. 5, 12, $\sqrt{119}$ yes
19. 15, 39, 36 yes
20. $\sqrt{5}$, 12, 13 no
21. 10, 12, $\sqrt{22}$ no
22. 2, 3, 4 no
23. $\sqrt{7}$, 8, $\sqrt{71}$ yes

Lesson 13-2 **Simplify. Leave in radical form and use absolute value symbols when necessary.**

1. $\sqrt{50}$ $5\sqrt{2}$
2. $\sqrt{20}$ $2\sqrt{5}$
3. $\sqrt{162}$ $9\sqrt{2}$
4. $\sqrt{700}$ $10\sqrt{7}$
5. $\frac{\sqrt{3}}{\sqrt{5}}$ $\frac{\sqrt{15}}{5}$
6. $\frac{\sqrt{72}}{\sqrt{6}}$ $2\sqrt{3}$
7. $\sqrt{\frac{8}{7}}$ $\frac{2\sqrt{14}}{7}$
8. $\sqrt{\frac{7}{32}}$ $\frac{\sqrt{14}}{8}$
9. $\sqrt{10} \cdot \sqrt{20}$ $10\sqrt{2}$
10. $\sqrt{7} \cdot \sqrt{3}$ $\sqrt{21}$
11. $6\sqrt{2} \cdot \sqrt{3}$ $6\sqrt{6}$
12. $5\sqrt{6} \cdot 2\sqrt{3}$ $30\sqrt{2}$
13. $\sqrt{4x^4y^3}$ $2x^2y\sqrt{y}$
14. $\sqrt{200m^2y^3}$
15. $\sqrt{12ts^3}$ $2s\sqrt{3ts}$
16. $\sqrt{175a^4b^6}$
17. $\sqrt{\frac{54}{g^2}}$ $\frac{3\sqrt{6}}{|g|}$
18. $\sqrt{99x^3y^7}$
19. $\sqrt{\frac{32c^5}{9d^2}}$ $\frac{4c^2\sqrt{2c}}{3|d|}$
20. $\sqrt{\frac{27p^4}{3p^2}}$ $3|p|$
21. $\frac{1}{3+\sqrt{5}}$ $\frac{3-\sqrt{5}}{4}$
22. $\frac{2}{\sqrt{3}-5}$ $\frac{\sqrt{3}+5}{-11}$
23. $\frac{\sqrt{3}}{\sqrt{3}-5}$ $\frac{3+5\sqrt{3}}{-22}$
24. $\frac{\sqrt{6}}{7-2\sqrt{3}}$ $\frac{7\sqrt{6}+6\sqrt{2}}{37}$
25. $(\sqrt{p}+\sqrt{10})^2$
26. $(2\sqrt{5}+\sqrt{7})(2\sqrt{5}-\sqrt{7})$ 13
27. $(t-2\sqrt{3})(t-\sqrt{3})$

14. $10|m|y\sqrt{2y}$ 16. $5a^2|b^3|\sqrt{7}$ 18. $3xy^3\sqrt{11xy}$ 25. $p + 2\sqrt{10p} + 10$ 27. $t^2 - 3t\sqrt{3} + 6$

Lesson 13-3 **Simplify.** 4. $14\sqrt{7} - \sqrt{2}$ 10. $-6\sqrt{7} + 24\sqrt{5}$ 15. $69\sqrt{2} - 10\sqrt{3}$

1. $3\sqrt{11} + 6\sqrt{11} - 2\sqrt{11}$ $7\sqrt{11}$
2. $6\sqrt{13} + 7\sqrt{13}$ $13\sqrt{13}$
3. $2\sqrt{12} + 5\sqrt{3}$ $9\sqrt{3}$
4. $9\sqrt{7} - 4\sqrt{2} + 3\sqrt{2} + 5\sqrt{7}$
5. $3\sqrt{5} - 5\sqrt{3}$ in simplest form
6. $4\sqrt{8} - 3\sqrt{5}$ $8\sqrt{2} - 3\sqrt{5}$
7. $2\sqrt{27} - 4\sqrt{12}$ $-2\sqrt{3}$
8. $8\sqrt{32} + 4\sqrt{50}$ $52\sqrt{2}$
9. $\sqrt{45} + 6\sqrt{20}$ $15\sqrt{5}$
10. $2\sqrt{63} - 6\sqrt{28} + 8\sqrt{45}$
11. $14\sqrt{3t} + 8\sqrt{3t}$ $22\sqrt{3t}$
12. $7\sqrt{6x} - 12\sqrt{6x}$ $-5\sqrt{6x}$
13. $5\sqrt{7} - 3\sqrt{28}$ $-\sqrt{7}$
14. $7\sqrt{8} - \sqrt{18}$ $11\sqrt{2}$
15. $7\sqrt{98} + 5\sqrt{32} - 2\sqrt{75}$
16. $4\sqrt{6} + 3\sqrt{2} - 2\sqrt{5}$ in simplest form
17. $-3\sqrt{20} + 2\sqrt{45} - \sqrt{7}$ $-\sqrt{7}$
18. $4\sqrt{75} + 6\sqrt{27}$ $38\sqrt{3}$
19. $10\sqrt{\frac{1}{5}} - \sqrt{45} - 12\sqrt{\frac{5}{9}}$ $-5\sqrt{5}$
20. $\sqrt{15} - \sqrt{\frac{3}{5}}$ $\frac{4\sqrt{15}}{5}$
21. $3\sqrt{\frac{1}{3}} - 9\sqrt{\frac{1}{12}} + \sqrt{243}$ $\frac{17\sqrt{3}}{2}$

EXTRA PRACTICE

EXTRA PRACTICE

Lesson 13-4 Solve each equation. Check your solution.

1. $\sqrt{5x} = 5$ **5**
2. $4\sqrt{7} = \sqrt{-m}$ **−112**
3. $\sqrt{t} - 5 = 0$ **25**
4. $\sqrt{3b} + 2 = 0$ **no real solution**
5. $\sqrt{x - 3} = 6$ **39**
6. $5 - \sqrt{3x} = 1$ $\frac{16}{3}$
7. $2 + 3\sqrt{y} = 13$ $\frac{121}{9}$
8. $\sqrt{3g} = 6$ **12**
9. $\sqrt{a} - 2 = 0$ **4**
10. $\sqrt{2j} - 4 = 8$ **72**
11. $5 + \sqrt{x} = 9$ **16**
12. $\sqrt{5y + 4} = 7$ **9**
13. $7 + \sqrt{5c} = 9$ $\frac{4}{5}$
14. $2\sqrt{5t} = 10$ **5**
15. $\sqrt{44} = 2\sqrt{p}$ **11**
16. $4\sqrt{x - 5} = 15$ $\frac{305}{16}$
17. $4 - \sqrt{x - 3} = 9$ **no real solution**
18. $\sqrt{10x^2 - 5} = 3x$ $\sqrt{5}$
19. $\sqrt{2a^2 - 144} = a$ **12**
20. $\sqrt{3y + 1} = y - 3$ **8**
21. $\sqrt{2x^2 - 12} = x$ $2\sqrt{3}$
22. $\sqrt{b^2 + 16} + 2b = 5b$ $\sqrt{2}$
23. $\sqrt{m + 2} + m = 4$ **2**
24. $\sqrt{3 - 2c} + 3 = 2c$ $\frac{3}{2}$

Lesson 13-5 Find the distance between each pair of points whose coordinates are given. Express answers in simplest radical form and as decimal approximations rounded to the nearest hundredth.

1. (4, 2), (−2, 10) **10**
2. (−5, 1), (7, 6) **13**
3. (4, −2), (1, 2) **5**
4. (−2, 4), (4, −2) $6\sqrt{2}$ **or 8.49**
5. (3, 1), (−2, −1) $\sqrt{29}$ **or 5.39**
6. (−2, 4), (7, −8) **15**
7. (−5, 0), (−9, 6) $2\sqrt{13}$ **or 7.21**
8. (5, −1), (5, 13) **14**
9. (2, −3), (10, 8) $\sqrt{185}$ **or 13.60**
10. (−7, 5), (2, −7) **15**
11. (−6, −2), (−5, 4) $\sqrt{37}$ **or 6.08**
12. (8, −10), (3, 2) **13**
13. (4, −3), (7, −9) $3\sqrt{5}$ **or 6.71**
14. (6, 3), (9, 7) **5**
15. (10, 0), (9, 7) $5\sqrt{2}$ **or 7.07**
16. (2, −1), (−3, 3) $\sqrt{41}$ **or 6.40**
17. (−5, 4), (3, −2) **10**
18. (0, −9), (0, 7) **16**
19. (−1, 7), (8, 4) $3\sqrt{10}$ **or 9.49**
20. (−9, 2), (3, −3) **13**
21. $(3\sqrt{2}, 7), (5\sqrt{2}, 9)$
22. (6, 3), (10, 0) **5**
23. (3, 6), (5, −5) $5\sqrt{5}$ **or 11.18**
24. (−4, 2), (5, 4) $\sqrt{85}$ **or 9.22**

21. $2\sqrt{3}$ **or 3.46**

Lesson 13-6 Find the value of c that makes each trinomial a perfect square.

1. $a^2 + 6a + c$ **9**
2. $x^2 + 10x + c$ **25**
3. $t^2 + 12t + c$ **36**
4. $y^2 - 9y + c$ $\frac{81}{4}$
5. $p^2 - 14p + c$ **49**
6. $b^2 + 5b + c$ $\frac{25}{4}$

Solve each equation by completing the square. Leave irrational roots in simplest radical form.

7. $x^2 - 4x = 5$ **−1, 5**
8. $t^2 + 12t - 45 = 0$ **−15, 3**
9. $b^2 + 4b - 12 = 0$ **2, −6**
10. $a^2 - 8a - 84 = 0$ **14, −6**
11. $c^2 + 6 = -5c$ **−3, −2**
12. $t^2 - 7t = -10$ **2, 5**
13. $p^2 - 8p + 5 = 0$ $4 \pm \sqrt{11}$
14. $a^2 + 4a + 2 = 0$ $-2 \pm \sqrt{2}$
15. $2y^2 + 7y - 4 = 0$ $-4, \frac{1}{2}$
16. $t^2 + 3t = 40$ **5, −8**
17. $x^2 + 8x - 9 = 0$ **−9, 1**
18. $y^2 + 5y - 84 = 0$ **−12, 7**
19. $x^2 + 2x - 6 = 0$ $-1 \pm \sqrt{7}$
20. $t^2 + 12t + 32 = 0$ **−4, −8**
21. $2x - 3x^2 = -8$ $2, -\frac{4}{3}$
22. $2y^2 - y - 9 = 0$ $\frac{1 \pm \sqrt{73}}{4}$
23. $2z^2 - 5z - 4 = 0$ $\frac{5 \pm \sqrt{57}}{4}$
24. $4t^2 - 6t - \frac{1}{2} = 0$ $\frac{3 \pm \sqrt{11}}{4}$

Write an algebraic expression for each verbal expression.

1. the sum of a number x and 13 $x + 13$
2. the reciprocal of a number x squared $\frac{1}{x^2}$
3. the cube of a number x decreased by 7 $x^3 - 7$
4. the product of 5 and a number x squared $5x^2$

Find the next two items for each pattern. 5. See Solutions Manual.

5.   
6. 4, 7, 10, 13, . . . 16, 19
7. 2, 5, 10, 17, . . . 26, 37

Evaluate each expression.

8. $5^2 - 12$ 13
9. $(0.5)^3 + 2 \cdot 7$ 14.125
10. $\frac{2}{5}(16 - 9)$ $2\frac{4}{5}$

Evaluate each expression when $a = 2$, $b = 0.5$, $c = 3$, and $d = \frac{4}{3}$.

11. $a^2b + c$ 5
12. $(cd)^3$ 64
13. $(a + d)c$ 10

Solve each equation.

14. $y = (4.5 + 0.8) - 3.2$ 2.1
15. $4^2 - 3(4 - 2) = y$ 10
16. $\frac{2^3 - 1^3}{2 + 1} = y$ $\frac{7}{3}$

Name the property illustrated by each statement.

17. $a = a + 0$ additive identity
18. $\frac{1}{a} \cdot a = 1$ multiplicative inverse
19. If $a = b$, and $b = c$, then $a = c$. transitive
20. $a(bc) = (ab)c$ associative (×)
21. $8(st) = 8(ts)$ commutative (×)
22. $7y + 5x - 4y = 5x + 7y - 4y$ commutative (+)

Simplify each expression.

23. $2m + 3m$ $5m$
24. $4x + 2y - 2x + y$ $2x + 3y$
25. $3(2a + b) - 1.5a - 1.5b$ $4.5a + 1.5b$

Use the stem-and-leaf plot below to complete Exercises 26–27.

Stem	Leaf
1	1 4 6 8 8
2	0 3 3 3 5 7 9
3	0 0 2 6

$2|4 = 24$

26. List the set of data represented by the plot on the left. 11, 14, 16, 18, 18, 20, 23, 23, 23, 25, 27, 29, 30, 30, 32, 36
27. Which number is used most frequently? 23

Sketch a reasonable graph for each situation. 28–30. See Solutions Manual.

28. A basketball is shot from the free throw line and falls through the net.
29. A slammer is dropped on a stack of pogs and bounces off.
30. An infant grows through adulthood.

Solve. 31. 250 miles

31. **Travel** If a car travels at an average speed of 50 miles per hour, how far can the car travel in 5 hours? Use the distance formula $d = r \cdot t$.
32. **Geometry** If the area of a circle is given by the equation $A = \pi r^2$, find the area if the radius is 4 in. (Use 3.14 for π.) 50.24 in^2
33. If a farmer wants to put a fence around his rectangular garden that measures 11 yards by 14 yards, how many yards of fencing will he need? 50 yd

CHAPTER 2 TEST

CHAPTER TEST

Find each sum or difference.

1. $12 - 19$ -7
2. $-21 + (-34)$ -55
3. $1.654 + (-2.367)$ -0.713
4. $-\frac{7}{16} - \frac{3}{8}$ $-\frac{13}{16}$
5. $18b + 13xy - 46b$ $-28b + 13xy$
6. $6.32 - (-7.41)$ 13.73
7. $\frac{5}{8} + \left(-\frac{3}{16}\right) + \left(-\frac{3}{4}\right)$ $-\frac{5}{16}$
8. $32y + (-73y)$ $-41y$
9. $|-28 + (-13)|$ 41
10. $\begin{bmatrix} -8 & 5 \\ 2 & -3 \end{bmatrix} + \begin{bmatrix} 3 & 6 \\ -4 & -7 \end{bmatrix}$ $\begin{bmatrix} -5 & 11 \\ -2 & -10 \end{bmatrix}$
11. $\begin{bmatrix} 1 & 0 \\ -9 & 3 \end{bmatrix} - \begin{bmatrix} 5 & 8 \\ -4 & 2 \end{bmatrix}$ $\begin{bmatrix} -4 & -8 \\ -5 & 1 \end{bmatrix}$

Evaluate each expression.

12. $-x - 38$, if $x = -2$ -36
13. $\left|-\frac{1}{2} + z\right|$, if $z = \frac{1}{4}$ $\frac{1}{4}$
14. $mp - k$, if $m = -12$, $p = 1.5$, and $k = -8$ -10
15. $w^2 - 15$, if $w = 5$ 10

Replace each __?__ with <, >, or = to make each sentence true.

16. -14 __?__ -15 $>$
17. $\frac{9}{20}$ __?__ $\frac{7}{15}$ $<$
18. -4.65 __?__ -4.45 $<$

Find a number between the given numbers. Answers will vary; sample answers given.

19. $-\frac{2}{3}$ and $-\frac{9}{14}$ $-\frac{55}{84}$
20. $\frac{4}{7}$ and $\frac{9}{4}$ $\frac{79}{56}$
21. $\frac{12}{7}$ and $\frac{15}{8}$ $\frac{99}{56}$

Simplify.

22. $\frac{8(-3)}{2}$ -12
23. $(-5)(-2)(-2) - (-6)(-3)$ -38
24. $\frac{2}{3}\left(\frac{1}{2}\right) - \left(-\frac{3}{2}\right)\left(-\frac{2}{3}\right)$ $-\frac{2}{3}$
25. $\frac{70x - 30y}{-5}$ $-14x + 6y$
26. $\frac{7}{\frac{-2}{5}}$ $-\frac{35}{2}$
27. $\frac{3}{4}(8x + 12y) - \frac{5}{7}(21x - 35y)$ $-9x + 34y$

Find each square root. Use a calculator if necessary. Round to the nearest hundredth if the result is not a whole number.

28. $\pm\sqrt{\frac{16}{81}}$ $\pm\frac{4}{9} \approx 0.44$
29. $\sqrt{40}$ 6.32
30. $\sqrt{2.89}$ 1.7

31. **Statistics** The height, in inches, of the students in a health class are 65, 63, 68, 66, 72, 61, 62, 63, 59, 58, 61, 74, 65, 63, 71, 60, 62, 63, 71, 70, 59, 66, 61, 62, 68, 69, 64, 63, 70, 61, 68, and 67.
    a. Make a line plot of the data on the heights of the students in health class. See Solutions Manual.
    b. What was the most common height for the students in this class? 63

32. Define a variable and write an equation for the following problem. Do *not* solve.
    Each week for several weeks, Save-a-Buck stores reduced the price of a sofa by \$18.25. The original price was \$380.25. The final reduced price was \$252.50. For how many weeks was the sofa on sale? Let w = number of weeks on sale; $380.25 - 18.25w = 252.50$.

33. Translate $r = (a - b)^3$ into a verbal sentence. The number r equals the cube of the difference of a and b.

Solve each equation. Then check your solution.

1. $-15 - k = 8$ **−23**
2. $-1.2x = 7.2$ **−6**
3. $\frac{3}{4}y = -27$ **−36**
4. $\frac{t-7}{4} = 11$ **51**
5. $-12 = 7 - \frac{y}{3}$ **57**
6. $k - 16 = -21$ **−5**
7. $t - (-3.4) = -5.3$ **−8.7**
8. $-3(x + 5) = 8x + 18$ **−3**
9. $2 - \frac{1}{4}(b - 12) = 9$ **−16**
10. $\frac{r}{5} - 3 = \frac{2r}{5} + 16$ **−95**
11. $25 - 7w = 46$ **−3**
12. $-w + 11 = 4.6$ **6.4**

Find the mean, median, and mode for each set of data.

13. 67, 31, 15, 49, 31, 35, 42, 27
    37.125, 33, 31

14.

Stem	Leaf
18	0 5 8
19	3 3 4 4
20	8 8 9
21	4 5 5 5 9

$19|4 = 194$

202, 208, 215

Define a variable, write an equation, and solve each problem.

15. The sum of two integers is −23. One integer is −84. Find the other integer.
    $-84 + x = -23$; **61**
16. Negative two thirds of a number is eight fifths. What is the number? $-\frac{2}{3}x = \frac{8}{5}$; $-\frac{12}{5}$
17. What number decreased by 37 is −65?
    $x - 37 = -65$; **−28**
18. The measures of two angles of a triangle are 23° and 121°. Find the measure of the third angle. $180 = 23 + 121 + x$; **36°**
19. Find two consecutive odd integers whose sum is 172. $x + (x + 2) = 172$; **85, 87**

Solve each equation or formula for the variable specified.

20. $h = at - 0.25vt^2$, for a $a = \frac{h + 0.25vt^2}{t}$
21. $A = \frac{1}{2}(b + B)h$, for h $h = \frac{2A}{b + B}$

Solve.

22. **Geometry** One of two supplementary angles measures 37° less than four times the other. Find the measure of each angle. **43.4°, 136.6°**
23. **Grades** James had a cumulative score of 338 points after his first four tests, each worth 100 points. What is the least that he would need to score on his next chapter test in order to have a mean score of 82%? **72%**
24. **Merchandise Stock** A store has 49 cartons of yogurt, some plain, some with blueberry flavoring. There are six times as many cartons of plain yogurt as flavored yogurt. How many cartons of plain yogurt are there? **42 cartons**
25. **Consumerism** Teri went to SuperValue to buy some groceries. Her total bill was \$8.51. She bought 2 pounds of grapes at \$0.99 per pound, a gallon of milk for \$2.59, and three cans of orange juice. She used a coupon worth 50¢ for the three cans of orange juice. How much was each can of orange juice? **\$1.48**

CHAPTER 4 TEST

CHAPTER TEST

Solve each proportion.

1. $\frac{2}{5} = \frac{x-3}{-2}$ **$\frac{11}{5}$**
2. $\frac{n}{4} = \frac{3.25}{52}$ **$\frac{1}{4}$**
3. $\frac{x-3}{x+5} = \frac{9}{11}$ **39**
4. $\frac{x+1}{-3} = \frac{x-4}{5}$ **$\frac{7}{8}$**

Solve.

5. Find 6.5% of 80. **5.2**
6. 42 is what percent of 126? **$33\frac{1}{3}\%$**
7. 84 is 60% of what number? **140**
8. What number decreased by 20% is 16? **20**
9. 24 is what percent of 8? **300%**
10. 54 is 20% more than what number? **45**
11. A price decreased from $60 to $45. Find the percent of decrease. **25%**
12. The price in dollars p minus a 15% discount is $3.40. Find p. **$4**

ΔABC and ΔJKH are similar. For each set of measures given, find the measures of the remaining sides.

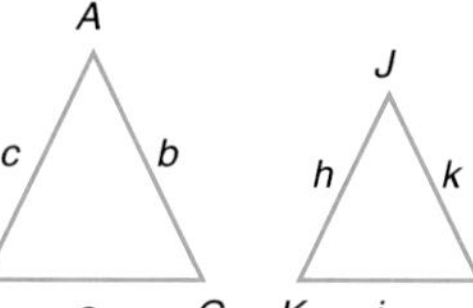

13. $c = 20, h = 15, k = 16, j = 12$
    $a = 16, b = \frac{64}{3}$
14. $c = 12, b = 13, a = 6, h = 10$
    $j = 5, k = \frac{65}{6}$
15. $k = 5, c = 6.5, b = 7.5, a = 4.5$
    $h = 4.\overline{3}, j = 3$
16. $h = 1\frac{1}{2}, c = 4\frac{1}{2}, k = 2\frac{1}{4}, a = 3$
    $b = 6\frac{3}{4}, j = 1$

Solve each right triangle. State the side lengths to the nearest tenth and the angle measures to the nearest degree.

17. 18.

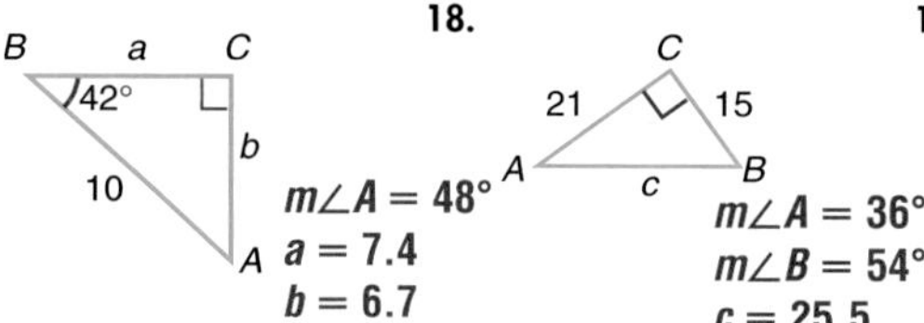

17. **$m\angle A = 48°$, $a = 7.4$, $b = 6.7$**

18. **$m\angle A = 36°$, $m\angle B = 54°$, $c = 25.5$**

19. **$m\angle A = 44°$, $m\angle B = 46°$, $a = 20$**

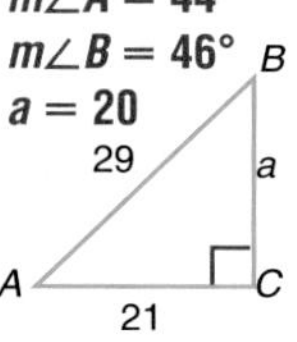

20. **$a = 29.9$, $c = 30.9$, $m\angle A = 75°$**

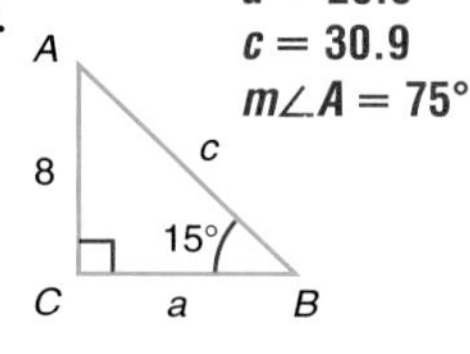

Solve.

21. **Music** During a 20-song sequence on a popular radio station, 8 soft-rock, 7 hard-rock, and 5 rap songs are played at random. You tune in to the station. **b. 1:3**
    a. What is the probability that a hard-rock song is playing? **$\frac{7}{20}$**
    b. What are the odds that a rap song is playing?
    c. What is the probability that a soft-rock song or a hard-rock song is playing? **$\frac{3}{4}$**
22. Two buildings are separated by an alley. Joe is looking out of a window 60 feet above the ground in one building. He observes the measurement of the angle of depression of the base of the second building to be 50° and the angle of elevation of the top to be 40°. How high is the second building? **about 102 feet**
23. **Finance** Gladys deposited an amount of money in the bank at 6.5% annual interest. After 6 months, she received $7.80 interest. How much money had Gladys deposited in the bank? **$240**
24. **Stereo** Larry is buying a stereo that costs $399. Since he is an employee of the store, he receives a 15% discount. He also has to pay 6% sales tax. If the discount is computed first, what is the total cost of Larry's stereo? **$359.50**
25. **Travel** At the same time Kris leaves Washington, D.C., for Detroit, Michigan, Amy leaves Detroit for Washington, D.C. The distance between the cities is 510 miles. Amy's average speed is 5 miles per hour faster than Kris's. How fast is Kris's average speed if they pass each other in 6 hours? **40 mph**

CHAPTER 5 TEST

1. Graph $K(0, -5)$, $M(3, -5)$, and $N(-2, -3)$. **See margin.**

2. Name the quadrant in which $P(-5, 1)$ is located. **II**

Express the relations shown in each table, mapping, or graph as a set of ordered pairs. Then state the domain, range, and inverse of the relation. **3–5. See margin.**

3.

x	f(x)
0	−1
2	4
4	5
6	10

4. 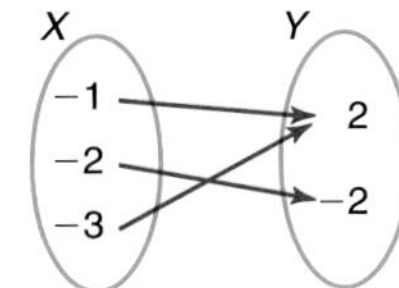

5. 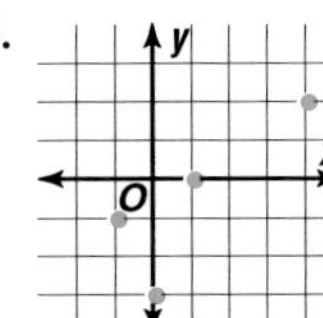

Solve each equation if the domain is {−2, −1, 0, 2, 4}.

6. $y = -4x + 10$
7. $4 - 2x = 5y$
8. $-x + 3y = 1$

6. $\{(-2, 18), (-1, 14), (0, 10), (2, 2), (4, -6)\}$

7. $\left\{\left(-2, \frac{8}{5}\right), \left(-1, \frac{6}{5}\right), \left(0, \frac{4}{5}\right), (2, 0), \left(4, \frac{-4}{5}\right)\right\}$

8. $\left\{\left(-2, \frac{-1}{3}\right), (-1, 0), \left(0, \frac{1}{3}\right), (2, 1), \left(4, \frac{5}{3}\right)\right\}$

Graph each equation. **9–11. See margin.**

9. $x + 2y = -1$
10. $-3x = 5 - y$
11. $-4 = x - \frac{1}{2}y$

Determine whether each relation is a function.

12. $\{(2, 4), (3, 2), (4, 6), (5, 4)\}$ **yes**
13. $8y = 7 + 3x$ **yes**

If $f(x) = -2x + 5$ and $g(x) = x^2 - 4x + 1$, find each value.

14. $f\left(\frac{1}{2}\right)$ **4**
15. $g(-2)$ **13**
16. $-2g(3)$ **4**

Write an equation for each relation.

17.

x	1	2	3	4	7
y	3	8	13	18	33

$y = 5x - 2$

18.

x	1	3	5	7	9	11	13
y	5	17	29	41	53	65	77

$y = 6x - 1$

Solve. **19c. Gina had more consistent scores since she had the smaller interquartile range.**

19. **School** Art and Gina's test scores in algebra are given below.
Art: 87, 54, 78, 97, 65, 82, 75, 68, 82, 73, 66, 75
Gina: 70, 80, 57, 100, 73, 74, 65, 77, 91, 69, 71, 76
 a. Find the range and interquartile range for each set of scores.
 b. Identify any outliers. **Gina's score of 100**
 c. Which one had the more consistent scores?

20. **Sales** When you use Jay's Taxi Service, a two-mile trip costs \$6.30, a five-mile trip costs \$11.25, and a ten-mile trip costs \$19.50. Write an equation to describe this relationship and use it to find the cost of a one-mile trip. **$y = 1.65x + 3$; \$4.65**

19a. Art: range, 43, interquartile range, 15; Gina: range, 43, interquartile range, 9

CHAPTER 6 TEST

CHAPTER TEST

Determine the slope of the line that passes through each pair of points.

1. (5, 8), (−3, 7) $\frac{1}{8}$
2. (−2, 5), (2, 9) 1

Find the slope and x- and y-intercepts of the graph of each equation.

3. $x - 8y = 3$ $\frac{1}{8}, 3, -\frac{3}{8}$
4. $3x - 2y = 9$ $\frac{3}{2}, 3, -\frac{9}{2}$
5. $y = 7$ 0, none, 7

Graph each equation. 6–8. See Solutions Manual.

6. $4x - 3y = 24$
7. $2x + 7y = 16$
8. $y = \frac{2}{3}x + 3$

Determine whether the graphs of each pair of equations are *parallel, perpendicular,* or *neither.*

9. $y = 4x - 11$ parallel
   $2y + 1 = 8x$
10. $-7y = 4x + 14$ perpendicular
    $7x - 4y = -12$

Write an equation in standard form of the line that satisfies the given conditions.

11. passes through (2, 5) and (8, −3)
    $4x + 3y = 23$
12. passes through (−2, −1) and (6, −4)
    $3x + 8y = -14$
13. has slope of 2 and y-intercept = 3
    $2x - y = -3$
14. has y-intercept = −4 and passes through (5, −3)
    $x - 5y = 20$
15. slope = $\frac{3}{4}$ and passes through (6, −2)
    $3x - 4y = 26$
16. parallel to $6x - y = 7$ and passes through (−2, 8)
    $6x - y = -20$

Write an equation in slope-intercept form of the line that satisfies the given conditions. 18. $y = \frac{1}{5}x - \frac{23}{5}$

17. passes through (4, −2) and the origin $y = -\frac{1}{2}x$
18. passes through (−2, −5) and (8, −3)
19. passes through (6, 4) with y-intercept = −2
20. slope = $-\frac{2}{3}$ and y-intercept = 5 $y = -\frac{2}{3}x + 5$
21. slope = 6 and passes through (−3, −4)
    $y = 6x + 14$ 19. $y = x - 2$
22. perpendicular to $5x - 3y = 9$ and passes through the origin $y = -\frac{3}{5}x$
23. parallel to $3x + 7y = 4$ and passes through (5, −2)
    $y = -\frac{3}{7}x + \frac{1}{7}$
24. perpendicular to $x + 3y = 7$ and passes through (5, 2) $y = 3x - 13$
25. Find the coordinates of the other endpoint of segment *AB* given *A*(−2, −7) and midpoint *M*(6, −5). *B*(14, −3)
26. The table below shows the number of students per computer in American classrooms since 1983.

School Year	Students per Computer	School Year	Students per Computer
1983–84	125	1989–90	22
1984–85	75	1990–91	20
1985–86	50	1991–92	18
1986–87	37	1992–93	16
1987–88	32	1993–94	14
1988–89	25	1994–95	12

a. Make a scatter plot of the data. See Solutions Manual.
b. Describe the correlation between the variables. negative
c. Are these data best modeled by a line? Explain your answer. No, they are not linear.

CHAPTER 7 TEST

Solve each inequality. Then check your solution.

1. $-12 \le d + 7$ $\{d \mid d \ge -19\}$
2. $7x < 6x - 11$ $\{x \mid x < -11\}$
3. $z - 1 \ge 2z - 3$ $\{z \mid z \le 2\}$
4. $5 - 4b > -23$ $\{b \mid b < 7\}$
5. $-\frac{2}{3}r \le \frac{7}{12}$ $\left\{r \mid r \ge -\frac{7}{8}\right\}$
6. $8y + 3 < 13y - 9$ $\left\{y \mid y > \frac{12}{5}\right\}$
7. $8(1 - 2z) \le 25 + z$ $\{z \mid z \ge -1\}$
8. $0.3(m + 4) > 0.5(m - 4)$ $\{m \mid m < 16\}$
9. $\frac{2n - 3}{-7} \le 5$ $\{n \mid n \ge -16\}$
10. $y + \frac{5}{8} > \frac{11}{24}$ $\left\{y \mid y > -\frac{1}{6}\right\}$

Solve each compound inequality. Then graph the solution set. 11–16. See Solutions Manual for graphs.

11. $x + 1 > -2$ and $3x < 6$ $\{x \mid -3 < x < 2\}$
12. $2n + 1 \ge 15$ or $2n + 1 \le -1$ $\{n \mid n \ge 7 \text{ or } n \le -1\}$
13. $8 + 3t > 2$ and $-12 > 11t - 1$ $\{t \mid -2 < t < -1\}$
14. $|2x - 1| < 5$ $\{x \mid -2 < x < 3\}$
15. $|5 - 3b| \ge 1$ $\left\{b \mid b \ge 2 \text{ or } b \le \frac{4}{3}\right\}$
16. $|3 - 5y| < 8$ $\left\{y \mid -1 < y < \frac{11}{5}\right\}$

Define a variable, write an inequality, and solve each problem. Then check your solution.

17. Twice a number subtracted from 12 is no less than the number increased by 27. $\{n \mid n \le -5\}$
18. Seven less than twice a number is between 71 and 83. $\{n \mid 39 < n < 45\}$
19. The product of two integers is no less than 30. One of the integers is 6. What is the other integer? 5 or more
20. The average of four consecutive odd integers is less than 20. What are the greatest integers that satisfy this condition? 15, 17, 19, 21

Graph each inequality. 21–23. See Solutions Manual for graphs.

21. $y \ge 5x + 1$
22. $x - 2y > 8$
23. $3x - 2y < 6$

Solve.

24. **Business** Two men and three women are each waiting for a job interview. There is only enough time to interview two people before lunch. Two people are chosen at random.
    a. What is the probability that both people are women? $\frac{3}{10}$ or 0.3
    b. What is the probability that at least one person is a woman? $\frac{9}{10}$ or 0.9
    c. Which is more likely, one of the people is a woman and the other is a man, or both people are either men or women? man and woman
25. **Fire Safety** The city council of McBride is investigating the efficiency of the fire department. The time taken by the fire department to respond to a fire alarm was surveyed. It was found that the response times in minutes for 17 alarms were as follows.

    1, 3, 2, 2, 1, 9, 4, 6, 1, 10, 1, 4, 5, 10, 1, 3, 6

    a. Draw a box-and-whisker plot of the data. See Solutions Manual.
    b. Between what two values of the data is the middle 50% of the data? 1 and 6

CHAPTER 8 TEST

CHAPTER TEST

Graph each system of equations. Then determine whether the system has *one* solution, *no* solution, or *infinitely many* solutions. If the system has one solution, name it. 1–6. See Solutions Manual for graphs.

1. $y = x + 2$
   $y = 2x + 7$ one; $(-5, -3)$

2. $x + 2y = 11$
   $x = 14 - 2y$ no solution

3. $2x + 5y = 16$
   $5x - 2y = 11$ one; $(3, 2)$

4. $3x + y = 5$
   $2y - 10 = -6x$ infinitely many

5. $y + 2x = -1$
   $y - 4 = -2x$ no solution

6. $2x + y = -4$
   $5x + 3y = -6$ one; $(-6, 8)$

Use substitution or elimination to solve each system of equations.

7. $y = 7 - x$
   $x - y = -3$ $(2, 5)$

8. $x = 2y - 7$
   $y - 3x = -9$ $(5, 6)$

9. $x + y = 8$
   $x - y = 2$ $(5, 3)$

10. $3x - y = 11$
    $x + 2y = -36$ $(-2, -17)$

11. $3x + y = 10$
    $3x - 2y = 16$ $(4, -2)$

12. $5x - 3y = 12$
    $-2x + 3y = -3$ $(3, 1)$

13. $2x + 5y = 12$
    $x - 6y = -11$ $(1, 2)$

14. $x + y = 6$
    $3x - 3y = 13$ $\left(\frac{31}{6}, \frac{5}{6}\right)$

15. $3x + \frac{1}{3}y = 10$
    $2x - \frac{5}{3}y = 35$ $(5, -15)$

16. $8x - 6y = 14$
    $6x - 9y = 15$ $(1, -1)$

17. $5x - y = 1$
    $y = -3x + 1$ $\left(\frac{1}{4}, \frac{1}{4}\right)$

18. $7x + 3y = 13$
    $3x - 2y = -1$ $(1, 2)$

Solve each system of inequalities by graphing. 19–21. See Solutions Manual for graphs.

19. $y \leq 3$
    $y > -x + 2$

20. $x \leq 2y$
    $2x + 3y \leq 7$

21. $x > y + 1$
    $2x + y \geq -4$

Solve.

22. **Number Theory** The units digit of a two-digit number exceeds twice the tens digit by 1. Find the number if the sum of its digits is 10. 37

23. **Geometry** The difference between the length and width of a rectangle is 7 cm. Find the dimensions of the rectangle if its perimeter is 50 cm. 16 cm by 9 cm

24. **Finance** Last year, Jodi invested \$10,000, part at 6% annual interest and the rest at 8% annual interest. If she received \$760 in interest at the end of the year, how much did she invest at each rate? \$2000 at 6%, \$8000 at 8%

25. **Organize Data** Joey sold 30 peaches from his fruit stand for a total of \$7.50. He sold small ones for 20 cents each and large ones for 35 cents each. How many of each kind did he sell? 20 small and 10 large

CHAPTER 9 TEST

Simplify. Assume that no denominator is equal to zero.

1. $(a^2b^4)(a^3b^5)$ **a^5b^9**
2. $(-12abc)(4a^2b^4)$ **$-48a^3b^5c$**
3. $\left(\frac{3}{5}m\right)^2$ **$\frac{9}{25}m^2$**
4. $(-3a)^4(a^5b)^2$ **$81a^{14}b^2$**
5. $(-5a^2)(-6b^3)^2$ **$-180a^2b^6$**
6. $(5a)^2b + 7a^2b$ **$32a^2b$**
7. $\frac{y^{11}}{y^6}$ **y^5**
8. $\frac{mn^4}{m^3n^2}$ **$\frac{n^2}{m^2}$**
9. $\frac{9a^2bc^2}{63a^4bc}$ **$\frac{c}{7a^2}$**
10. $\frac{48a^2bc^5}{(3ab^3c^2)^2}$ **$\frac{16c}{3b^5}$**
11. $\frac{14ab^{-3}}{21a^2b^{-5}}$ **$\frac{2b^2}{3a}$**
12. $\frac{(10a^2bc^4)^{-2}}{(5^{-1}a^{-1}b^{-5})^2}$ **$\frac{b^8}{4a^2c^8}$**

Express each number in scientific notation.

13. 46,300 **4.63×10^4**
14. 0.003892 **3.892×10^{-3}**
15. 284×10^3 **2.84×10^5**
16. 0.0031×10^4 **3.1×10**

Evaluate. Express each result in scientific notation.

17. $(3 \times 10^3)(2 \times 10^4)$ **6×10^7**
18. $\frac{2.5 \times 10^3}{5 \times 10^{-3}}$ **5×10^5**
19. $\frac{14.72 \times 10^{-4}}{3.2 \times 10^{-3}}$ **4.6×10^{-1}**
20. $(15 \times 10^{-7})(3.1 \times 10^4)$ **4.65×10^{-2}**
21. Find the degree of $5ya^3 - 7 - y^2a^2 + 2y^3a$ and arrange the terms so that the powers of y are in descending order. **4; $2y^3a - y^2a^2 + 5ya^3 - 7$**

Find each sum or difference.

22.
$$\begin{array}{rl} & 5ax^2 + 3a^2x - 7a^3 \\ (+) & 2ax^2 - 8a^2x \qquad + 4 \\ \hline & 7ax^2 - 5a^2x - 7a^3 + 4 \end{array}$$

23.
$$\begin{array}{rl} & x^3 - 3x^2y + 4xy^2 + y^3 \\ (-) & 7x^3 + x^2y - 9xy^2 + y^3 \\ \hline & -6x^3 - 4x^2y + 13xy^2 \end{array}$$

24. $(n^2 - 5n + 4) - (5n^2 + 3n - 1)$ **$-4n^2 - 8n + 5$**
25. $(ab^3 - 4a^2b^2 + ab - 7) + (-2ab^3 + 4ab^2 + 3ab + 2)$ **$-ab^3 - 4a^2b^2 + 4ab^2 + 4ab - 5$**

Simplify. **32. $3x^3 - 17x^2 - 9x$**

26. $(h - 5)^2$ **$h^2 - 10h + 25$**
27. $(2x - 5)(7x + 3)$ **$14x^2 - 29x - 15$**
28. $(4x - y)(4x + y)$ **$16x^2 - y^2$**
29. $(2a^2b + b^2)^2$ **$4a^4b^2 + 4a^2b^3 + b^4$**
30. $3x^2y^3(2x - xy^2)$ **$6x^3y^3 - 3x^3y^5$**
31. $(4m + 3n)(2m - 5n)$ **$8m^2 - 14mn - 15n^2$**
32. $x^2(x - 8) - 3x(x^2 - 7x + 3) + 5(x^3 - 6x^2)$
33. $(x - 6)(x^2 - 4x + 5)$ **$x^3 - 10x^2 + 29x - 30$**

CHAPTER 10 TEST

CHAPTER TEST

Find the GCF of the given monomials.

1. 48, 64 **16**
2. $18a^2b$, $28a^3b^2$ **$2a^2b$**
3. $6x^2y^3$, $12x^2y^2z$, $15x^2y$ **$3x^2y$**

Factor each polynomial, if possible. If the polynomial cannot be factored using integers, write *prime*.

4. $25y^2 - 49w^2$ **$(5y - 7w)(5y + 7w)$**
5. $t^2 - 16t + 64$ **$(t - 8)^2$**
6. $x^2 + 14x + 24$ **$(x + 12)(x + 2)$**
7. $28m^2 + 18m$ **$2m(14m + 9)$**
8. $a^2 - 11ab + 18b^2$ **$(a - 2b)(a - 9b)$**
9. $12x^2 + 23x - 24$ **$(3x + 8)(4x - 3)$**
10. $2h^2 - 3h - 18$ **prime**
11. $6x^3 + 15x^2 - 9x$ **$3x(x + 3)(2x - 1)$**
12. $4my - 20m + 3py - 15p$ **$(y - 5)(4m + 3p)$**
13. $x^3 - 4x^2 - 9x + 36$ **$(x - 4)(x - 3)(x + 3)$**
14. $36a^2b^3 - 45ab^4$ **$9ab^3(4a - 5b)$**
15. $36m^2 + 60mn + 25n^2$ **$(6m + 5n)^2$**
16. $\frac{1}{4}a^2 - \frac{4}{9}$ **$\left(\frac{1}{2}a - \frac{2}{3}\right)\left(\frac{1}{2}a + \frac{2}{3}\right)$**
17. $64p^2 - 63p + 16$ **prime**
18. $15a^2b + 5a^2 - 10a$ **$5a(3ab + a - 2)$**
19. $6y^2 - 5y - 6$ **$(2y - 3)(3y + 2)$**
20. $4s^2 - 100t^2$ **$4(s - 5t)(s + 5t)$**
21. $2d^2 + d - 1$ **$(2d - 1)(d + 1)$**
22. $3g^2 + g + 1$ **prime**
23. $2xz + 2yz - x - y$ **$(x + y)(2z - 1)$**

Solve each equation. Check your solutions.

24. $(4x - 3)(3x + 2) = 0$ **$\left\{\frac{3}{4}, -\frac{2}{3}\right\}$**
25. $18s^2 + 72s = 0$ **$\{0, -4\}$**
26. $4x^2 = 36$ **$\{-3, 3\}$**
27. $t^2 + 25 = 10t$ **$\{5\}$**
28. $a^2 - 9a - 52 = 0$ **$\{-4, 13\}$**
29. $x^3 - 5x^2 - 66x = 0$ **$\{-6, 0, 11\}$**
30. $2x^2 = 9x + 5$ **$\left\{-\frac{1}{2}, 5\right\}$**
31. $3b^2 + 6 = 11b$ **$\left\{\frac{2}{3}, 3\right\}$**

Solve.

32. **Geometry** A rectangle is 4 inches wide by 7 inches long. When the length and width are increased by the same amount, the area is increased by 26 square inches. What are the dimensions of the new rectangle? **6 in. by 9 in.**

33. **Construction** A rectangular lawn is 24 feet wide by 32 feet long. A sidewalk will be built along the inside edges of all four sides. The remaining lawn will have an area of 425 square feet. How wide will the walk be? **3.5 feet**

CHAPTER 11 TEST

Write the equation of the axis of symmetry and find the coordinates of the vertex of the graph of each equation. State if the vertex is a maximum or minimum. Then graph the equation.

1. $y = x^2 - 4x + 13$ $x = 2$; (2, 9); minimum
2. $y = -3x^2 - 6x + 4$ $x = -1$; (−1, 7); maximum
3. $y = 2x^2 + 3$ $x = 0$; (0, 3); minimum
4. $y = -1(x - 2)^2 + 1$ $x = 2$; (2, 1); maximum

1–4. See Solutions Manual for graphs.

Solve each equation by graphing. If exact roots cannot be found, state the consecutive integers between which the roots lie.

5. $x^2 - 2x + 2 = 0$ $\varnothing$
6. $x^2 + 6x = -7$ $-5 < x < -4, -2 < x < -1$
7. $x^2 + 24x + 144 = 0$ -12
8. $2x^2 - 8x = 42$ $7, -3$

Solve each equation.

9. $x^2 + 7x + 6 = 0$ $-1, -6$
10. $2x^2 - 5x - 12 = 0$ $4, -\frac{3}{2}$
11. $6n^2 + 7n = 20$ $\frac{4}{3}, -\frac{5}{2}$
12. $3k^2 + 2k = 5$ $1, -\frac{5}{3}$
13. $y^2 - \frac{3y}{5} + \frac{2}{25} = 0$ $\frac{2}{5}, \frac{1}{5}$
14. $-3x^2 + 5 = 14x$ $\frac{1}{3}, -5$
15. $4^{x-2} = 16^{2x+5}$ -4
16. $1000^x = 10{,}000^{6x+4}$ $-\frac{16}{21}$
17. $5^{x^2} = 5^{15-2x}$ $-5, 3$
18. $\left(\frac{1}{2}\right)^{x-2} = 4^{5x}$ $\frac{2}{11}$

Graph each function. State the *y*-intercept. 19–21. See Solutions Manual for graphs.

19. $y = \left(\frac{1}{2}\right)^x$ 1
20. $y = 4 \cdot 2^x$ 4
21. $y = \left(\frac{1}{3}\right)^x - 3$ -2

Solve.

22. **Automobile** Adina Ley needs to replace her car. If she leases a car, she will pay $410 a month for 2 years and then has the option to buy the car for $14, 458. The price of the car now is $17,369. If the car depreciates at 16% per year, how will the depreciated price compare with the buyout price of the lease? The depreciated value is $2202 less than the buyout price.

23. **Geometry** The area of a certain square is one-half the area of the rectangle formed if the length of one side of the square is increased by 2 cm and the length of an adjacent side is increased by 3 cm. What are the dimensions of the square? 6 cm by 6 cm

24. **Number Theory** Find two integers whose sum is 21 and whose product is 90. 6, 15

25. **Investment** After 6 years, a certain investment is worth $8479. If the money was invested at 9% interest compounded semiannually, find the original amount that was invested. $5000

CHAPTER 12 TEST

CHAPTER TEST

Simplify each rational expression. State the excluded values of the variables.

1. $\frac{5-2m}{6m-15}$ $-\frac{1}{3}, m \neq \frac{5}{2}$
2. $\frac{3+x}{2x^2+5x-3}$ $\frac{1}{2x-1}, x \neq \frac{1}{2}, -3$
3. $\frac{4c^2+12c+9}{2c^2-11c-21}$ $\frac{2c+3}{c-7}, c \neq -\frac{3}{2}, 7$

Simplify each expression.

4. $\frac{1-\frac{9}{t}}{1-\frac{81}{t^2}}$ $\frac{t}{t+9}$
5. $\frac{\frac{5}{6}+\frac{u}{t}}{\frac{2u}{t}-3}$ $\frac{6u+5t}{12u-18t}$
6. $\frac{x+4+\frac{5}{x-2}}{x+6+\frac{15}{x-2}}$ $\frac{x-1}{x+1}$

Perform the indicated operations.

7. $\frac{2x}{x-7}-\frac{14}{x-7}$ 2
8. $\frac{n+3}{2n-8} \cdot \frac{6n-24}{2n+1}$ $\frac{3n+9}{2n+1}$
9. $(10m^2+9m-36) \div (2m-3)$ $5m+12$
10. $\frac{x^2+4x-32}{x+5} \cdot \frac{x-3}{x^2-7x+12}$ $\frac{x+8}{x+5}$
11. $\frac{z^2+2z-15}{z^2+9z+20} \div (z-3)$ $\frac{1}{z+4}$
12. $\frac{4x^2+11x+6}{x^2-x-6} \div \frac{x^2+8x+16}{x^2+x-12}$ $\frac{4x+3}{x+4}$
13. $(10z^4+5z^3-z^2) \div 5z^3$ $2z+1-\frac{1}{5z}$
14. $\frac{y}{7y+14}+\frac{6}{3y+6}$ $\frac{y+14}{7y+14}$
15. $\frac{x+5}{x+2}+6$ $\frac{7x+17}{x+2}$
16. $\frac{x^2-1}{x+1}-\frac{x^2+1}{x-1}$ $\frac{-2x}{x-1}$
17. $\frac{-3}{a-5}+\frac{15}{a^2-5a}$ $\frac{-3}{a}$
18. $\frac{8}{m^2} \cdot \left(\frac{m^2}{2c}\right)^2$ $\frac{2m^2}{c^2}$

Solve each equation.

19. $\frac{2}{3t}+\frac{1}{2}=\frac{3}{4t}$ $\frac{1}{6}$
20. $\frac{2e}{e-4}-2=\frac{4}{e+5}$ -14
21. $\frac{4}{h-4}=\frac{3h}{h+3}$ $6, -\frac{2}{3}$

Solve each formula for the variable indicated.

22. $F=G\left(\frac{Mm}{d^2}\right)$, for G $G=\frac{Fd^2}{Mm}$
23. $\frac{1}{R_T}=\frac{1}{R_1}+\frac{1}{R_2}$, for R_2 $R_2=\frac{R_TR_1}{R_1-R_T}$

Solve. 25. $7\frac{1}{17}$ ohms

24. **Keyboarding** Willie can type a 200 word essay in 6 hours. Myra can type the same essay in $4\frac{1}{2}$ hours. If they work together, how long will it take them to type the essay? $2\frac{4}{7}$ hours

25. **Electronics** Three appliances are connected in parallel: a lamp of resistance 120 ohms, a toaster of resistance 20 ohms, and an iron of resistance 12 ohms. Find the total resistance.

CHAPTER 13 TEST

Simplify. Leave in radical form and use absolute value symbols when necessary.

1. $\sqrt{480}$ $4\sqrt{30}$
2. $\sqrt{72} \cdot \sqrt{48}$ $24\sqrt{6}$
3. $\sqrt{54x^4y}$ $3x^2\sqrt{6y}$
4. $\sqrt{\frac{32}{25}}$ $\frac{4\sqrt{2}}{5}$
5. $\sqrt{\frac{3x^2}{4n^3}}$ $\frac{|x|\sqrt{3n}}{2n^2}$
6. $\sqrt{6} + \sqrt{\frac{2}{3}}$ $\frac{4\sqrt{6}}{3}$
7. $\left(x + \sqrt{3}\right)^2$ $x^2 + 2x\sqrt{3} + 3$
8. $\frac{7}{7 + \sqrt{5}}$ $\frac{49 - 7\sqrt{5}}{44}$
9. $3\sqrt{50} - 2\sqrt{8}$ $11\sqrt{2}$
10. $\sqrt{\frac{10}{3}} \cdot \sqrt{\frac{4}{30}}$ $\frac{2}{3}$
11. $2\sqrt{27} + \sqrt{63} - 4\sqrt{3}$ $2\sqrt{3} + 3\sqrt{7}$
12. $\left(1 - \sqrt{3}\right)\left(3 + \sqrt{2}\right)$ $3 + \sqrt{2} - 3\sqrt{3} - \sqrt{6}$

Find the distance between each pair of points whose coordinates are given. Express answers in simplest radical form.

13. $(4, 7), (4, -2)$ 9
14. $(-9, 2), \left(\frac{2}{3}, \frac{1}{2}\right)$ $\frac{\sqrt{3445}}{6}$
15. $(-1, 1), (1, -5)$ $2\sqrt{10}$

Find the length of each missing side. Round to the nearest hundredth.

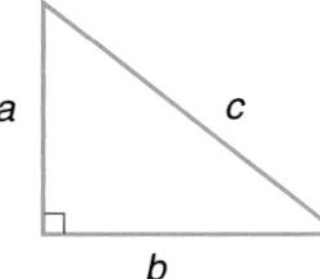

16. $a = 8, b = 10, c = ?$ 12.81
17. $a = 12, c = 20, b = ?$ 16
18. $a = 6\sqrt{2}, c = 12, b = ?$ 8.49
19. $b = 13, c = 17, a = ?$ 10.95

Solve each equation. Check your solution.

20. $\sqrt{4x + 1} = 5$ 6
21. $\sqrt{4x - 3} = 6 - x$ 3
22. $y^2 - 5 = -8y$ $-4 \pm \sqrt{21}$
23. $2x^2 - 10x - 3 = 0$ $\frac{5}{2} \pm \frac{\sqrt{31}}{2}$

Solve.

24. **Geometry** Find the measures of the perimeter and area, in simplest form, for the rectangle shown at the right. $16\sqrt{2} - 4\sqrt{6}$, $16\sqrt{3} - 18$

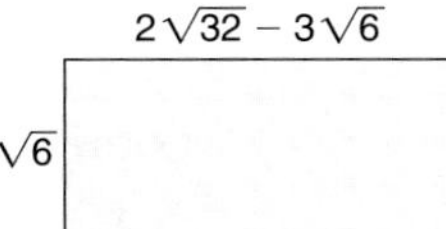

25. **Sports** A hiker leaves her camp in the morning. How far is she from camp after walking 9 miles due west and then 12 miles due north? 15 miles

GLOSSARY

GLOSSARY

Glossary

A

absolute value (85) The absolute value of a number is its distance from zero on a number line.

acute triangle (165) In an acute triangle, all of the angles measure less than 90 degrees.

adding integers (86)

1. To add integers with the *same* sign, add their absolute values. Give the result the same sign as the integers.
2. To add integers with *different* signs, subtract the lesser absolute value from the greater absolute value. Give the result the same sign as the integer with the greater absolute value.

addition property for inequality (385) For all numbers a, b, and c, the following are true:
1. if $a > b$, then $a + c > b + c$.
2. if $a < b$, then $a + c < b + c$.

addition property of equality (144) For any numbers a, b, and c, if $a = b$, then $a + c = b + c$.

additive identity (37) For any number a, $a + 0 = 0 + a = a$.

additive inverse property (87) For any number a, $a + (-a) = 0$.

algebraic expression (6) An expression consisting of one or more numbers and variables along with one or more arithmetic operations.

angle of depression (208) An angle of depression is formed by a horizontal line and a line of sight below it.

angle of elevation (208) An angle of elevation is formed by a horizontal line and a line of sight above it.

associative property (51) For any numbers a, b, and c, $(a + b) + c = a + (b + c)$ and $(ab)c = a(bc)$.

axes (254) Two perpendicular number lines that are used to locate points on a coordinate plane.

axis of symmetry (612) The equation of the axis of symmetry for the graph of $y = ax^2 + bx + c$, where $a \neq 0$, is $x = -\frac{b}{2a}$.

B

back-to-back stem-and-leaf plot (27) A back-to-back stem-and-leaf plot is used to compare two sets of data. The same stem is used for the leaves of both plots.

base **1.** (7) In an expression of the form x^n, the base is x. **2.** (215) The number that is divided into the percentage in the percent proportion.

best-fit line (341) A line drawn on a scatter plot that passes close to most of the data points.

binomial (514) The sum of two monomials.

boundary (437) A boundary of an inequality is a line that separates the coordinate plane into half-planes.

box-and-whisker plot (427) A type of diagram or graph that shows the quartiles and extreme values of data.

C

coefficient (47) The numerical factor in a term.

commutative property (51) For any numbers a and b, $a + b = b + a$ and $ab = ba$.

comparison property (94) For any two numbers a and b, exactly one of the following sentences is true.

$$a < b \qquad a = b \qquad a > b$$

complementary angles (163) Two angles are complementary if the sum of their measures is 90 degrees.

complete graph (278) A complete graph shows the origin, the points at which the graph crosses the x- and y-axes, and other important characteristics of the graph.

completeness property (121) Each real number corresponds to exactly one point on the number line. Each point on the number line corresponds to exactly one real number.

completeness property for points in the plane (256)

1. Exactly one point in the plane is named by a given ordered pair of numbers.

2. Exactly one ordered pair of numbers names a given point in the plane.

completing the square (742, 743) To add a constant term to a binomial of the form $x^2 + bx$ so that the resulting trinomial is a perfect square.

complex fraction (114) If a fraction has one or more fractions in the numerator or denominator, it is called a complex fraction.

composite numbers (558) A whole number, greater than 1, that is not prime.

compound event (414) A compound event consists of two or more simple events.

compound inequality (405) Two inequalities connected by *and* or *or.*

congruent angles (164) Angles that have the same measure.

conjugates (722) Two binomials of the form $a\sqrt{b} + c\sqrt{d}$ and $a\sqrt{b} - c\sqrt{d}$.

consecutive integers (158) Consecutive integers are integers in counting order.

consistent (455) A system of equations is said to be consistent when it has at least one ordered pair that satisfies both equations.

constant of variation (239) The number k in equations of the form $y = kx$ and $xy = k$.

constants (496) Monomials that are real numbers.

coordinate (73) The number that corresponds to a point on a number line.

coordinate plane (254) The plane containing the x- and y-axes.

corresponding angles (201) Matching angles in similar triangles, which have equal measures.

corresponding sides (201) The sides opposite the corresponding angles in similar triangles.

cosine (206) In a right triangle with acute angle A, the cosine of angle $A = \frac{\text{measure of leg adjacent to angle } A}{\text{measure of hypotenuse}}$.

cross products (94) When two fractions are compared, the cross products are the products of the terms on the diagonals.

D

data (25) Numerical information.

defining the variable (127) Choosing a variable to represent one of the unspecified numbers in a problem.

degree 1. (515) The degree of a monomial is the sum of the exponents of its variables.
2. (516) The degree of a polynomial is the degree of the term of the greatest degree.

density property (96) Between every pair of distinct rational numbers, there are infinitely many rational numbers.

dependent (456) A system of equations that has an infinite number of solutions.

dependent variable (58) The variable in a function whose value is determined by the independent variable.

difference of squares (545) Two perfect squares separated by a subtraction sign, $a^2 - b^2 = (a + b)(a - b)$.

dimensional analysis (174) The process of carrying units throughout a computation.

direct variation (239) A direct variation is described by an equation of the form $y = kx$, where $k \neq 0$.

discrete mathematics (88) A branch of mathematics that deals with finite or discontinuous quantities.

discriminant (632) In the quadratic formula, the expression $b^2 - 4ac$.

distance formula (737) The distance d between any two points with coordinates (x_1, y_1) and (x_2, y_2) is given by the following formula.

$$d = \sqrt{(x_2 - x_1)^2 + (y_2 - y_1)^2}$$

distributive property (46) For any numbers a, b, and c:

1. $a(b + c) = ab + ac$ and $(b + c)a = ba + ca$.
2. $a(b - c) = ab - ac$ and $(b - c)a = ba - ca$.

dividing rational numbers (112) The quotient of two rational numbers having the same sign is positive. The quotient of two rational numbers having different signs is negative.

division property for inequality (393) For all numbers a, b, and c, the following are true:

1. If c is positive and $a < b$, then $\frac{a}{c} < \frac{b}{c}$, and if c is positive and $a > b$, then $\frac{a}{c} > \frac{b}{c}$.
2. If c is negative and $a < b$, then $\frac{a}{c} > \frac{b}{c}$, and if c is negative and $a > b$, then $\frac{a}{c} < \frac{b}{c}$.

division property of equality (151) For any numbers a, b, and c, with $c \neq 0$, if $a = b$, then $\frac{a}{c} = \frac{b}{c}$.

domain (263) The set of all first coordinates from the ordered pairs in a relation.

draw a diagram (406) A problem-solving strategy that is often used as an organizational tool.

E

element (33) A member of a set.

elimination (469) The elimination method of solving a system of equations is a method that uses addition or subtraction to eliminate one of the variables to solve for the other variable.

equally likely (229) Outcomes that have an equal chance of occurring.

equation (33) A mathematical sentence that contains an equals sign, =.

equation in two variables (271) An equation in two variables contains two unknown values.

equilateral triangle (164) A triangle in which all the sides have the same length and all the angles have the same measure.

equivalent equation (144) Equations that have the same solution.

equivalent expressions (47) Expressions that denote the same number.

evaluate (8) To find the value of an expression when the values of the variables are known.

excluded value (660) A value is excluded from the domain of a variable because if that value were substituted for the variable, the result would have a denominator of zero.

exponent (7) In an expression of the form x^n, the exponent is n.

exponential function (634, 635) A function that can be described by an equation of the form $y = a^x$, where $a > 0$ and $a \neq 1$.

extraneous solutions (697) Solutions derived from an equation that are not solutions of the original equation.

extreme values (427) The least value and the greatest value in a set of data.

extremes (196) *See* proportion.

F

factored form (559) A monomial is written in factored form when it is expressed as the product of prime numbers and variables where no variable has an exponent greater than 1.

factoring (564) To express a polynomial as the product of monomials and polynomials.

factoring by grouping (567) A method of factoring polynomials with four or more terms.

factors (6) In a multiplication expression, the quantities being multiplied are called factors.

family of graphs (354) A family of graphs includes graphs and equations of graphs that have at least one characteristic in common.

FOIL method (537) To multiply two binomials, find the sum of the products of

F the first terms,
O the outside terms,
I the inside terms, and
L the last terms.

formula (128) An equation that states a rule for the relationship between certain quantities.

function **1.** (56) A relationship between input and output in which the output depends on the input. **2.** (287) A relation in which each element of the domain is paired with exactly one element of the range.

functional notation (289) In functional notation, the equation $y = x + 5$ is written as $f(x) = x + 5$.

G

general equation for exponential decay (644) The general equation for exponential decay is represented by the formula $A = C(1 - r)^t$.

general equation for exponential growth (643) The general equation for exponential growth is represented by the formula $A = C(1 + r)^t$.

graph (73, 255) To draw, or plot, the points named by certain numbers or ordered pairs on a number line or coordinate plane, respectively.

greatest common factor (GCF) (559) The greatest common factor of two or more integers is the greatest number that is a factor of all the integers.

guess and check (574) A problem-solving strategy in which several values or combinations of values are tried in order to find a solution to a problem.

H

half-life (638) The half-life of an element is defined as the time it takes for one-half a quantity of a radioactive element to decay.

half-plane (437) The region of a graph on one side of a boundary is called a half-plane.

horizontal axis (56) The horizontal line in a graph that represents the independent variable.

hypotenuse (206) The side of a right triangle opposite the right angle.

I

identify subgoals (745) A problem-solving strategy that uses a series of small steps, or subgoals.

identity (170) An equation that is true for every value of the variable.

inconsistent (456) A system of equations is said to be inconsistent when it has no ordered pair that satisfies both equations.

independent (456) A system of equations is said to be independent if the system has exactly one solution.

independent variable (58) The variable in a function whose value is subject to choice is the independent variable. The independent variable affects the value of the dependent variable.

inequality (33) A mathematical sentence having the symbols $<$, $\leq$, $>$, or $\geq$.

integers (73) The set of numbers represented as $\{..., -3, -2, -1, 0, 1, 2, 3, ...\}$.

interquartile range (304, 306) The difference between the upper quartile and the lower quartile of a set of data. It represents the middle half, or 50%, of the data in the set.

intersection (406) The intersection of two sets A and B is the set of elements common to both A and B.

inverse of a relation (264) The inverse of any relation is obtained by switching the coordinates in each ordered pair.

inverse variation (241) An inverse variation is described by an equation of the form $xy = k$, where $k \neq 0$.

irrational numbers (120) A number that cannot be expressed in the form $\frac{a}{b}$, where a and b are integers and $b \neq 0$.

isosceles triangle (164) In an isosceles triangle, at least two angles have the same measure and at least two sides have the same length.

L

least common denominator (LCD) (686) The least common denominator is the least common multiple of the denominators of two or more fractions.

least common multiple (LCM) (685) The least common multiple of two or more integers is the least positive integer that is divisible by each of the integers.

legs (206) The sides of a right triangle that are not the hypotenuse.

like terms (47) Terms that contain the same variables, with corresponding variables with the same power.

linear equation (280) An equation whose graph is a line.

linear function (278) An equation whose graph is a nonvertical line.

line plot (78) Numerical data displayed on a number line.

look for a pattern (13, 497, 637) A problem-solving strategy often involving the use of tables to organize information so that a pattern may be determined.

lower quartile (306) The lower quartile divides the lower half of a set of data into two equal parts.

M

make an organized list (685) A problem-solving strategy that uses an organized list to arrange and evaluate data in order to determine a solution.

mapping (263) A mapping pairs one element in the domain with one element in the range.

matrix (88) A matrix is a rectangular arrangement of elements in rows and columns.

maximum (611) The highest point on the graph of a curve, such as a the vertex of parabola that opens downward.

mean (178) The mean of a set of data is the sum of the numbers in the set divided by the number of numbers in the set.

means (196) *See* proportion.

measures of central tendency (178) Numbers known as measures of central tendency are often used to describe sets of data because they represent a centralized, or middle, value.

measures of variation (306) Measures of variation are used to describe the distribution of data.

median (178) The median is the middle number of a set of data when the numbers are arranged in numerical order.

midpoint (369) A point that is halfway between the endpoints of a segment.

minimum (611) The lowest point on the graph of a curve, such as a the vertex of parabola that opens upward.

mixed expression (690) An algebraic expression that contains a monomial and a rational expression.

mode (178) The mode of a set of data is the number that occurs most often in the set.

monomial (496) A monomial is a number, a variable, or a product of a number and one or more variables.

multi-step equations (157) Multi-step equations are equations that need more than one operation to solve them.

multiplicative identity (38) For any number a, $a \cdot 1 = 1 \cdot a = a$.

multiplicative inverse (38) For every nonzero number $\frac{a}{b}$, where $a, b \neq 0$, there is exactly one number $\frac{b}{a}$ such that $a \cdot b = 1$.

multiplication property for inequality (393) For all numbers a, b, and c, the following are true.

1. If c is positive and $a < b$, then $ac < bc$, $c \neq 0$, and if c is positive and $a > b$, then $ac > bc$, $c \neq 0$.
2. If c is negative and $a < b$, then $ac > bc$, $c \neq 0$, and if c is negative and $a > b$, then $ac < bc$, $c \neq 0$.

multiplicative property of -1 (107) The product of any number and -1 is its additive inverse.

$$-1(a) = -a \text{ and } a(-1) = -a$$

multiplicative property of equality (150) For any numbers a, b, and c, if $a = b$, then $a \cdot c = b \cdot c$.

multiplicative property of zero (38) For any number a, $a \cdot 0 = 0 \cdot a = 0$.

N

negative correlation (340) There is a negative correlation between x and y if the values are related in opposite ways.

negative exponent (503) For any nonzero number a and any integer n, $a^{-n} = \frac{1}{a^n}$.

negative number (73) Any number that is less than zero.

number line (72) A line with equal distances marked off to represent numbers.

number theory (158) The study of numbers and the relationships between them.

O

obtuse triangle (165) An obtuse triangle has one angle with measure greater than 90 degrees.

odds (229) The odds of an event occurring is the ratio of the number of ways the event can occur (successes) to the number of ways the event cannot occur (failures).

open sentences (32) Mathematical statements with one or more variables, or unknown numbers.

opposites (87) The opposite of a number is its additive inverse.

order of operations (19)

1. Simplify the expressions inside grouping symbols, such as parentheses, brackets, and braces, and as indicated by fraction bars.
2. Evaluate all powers.
3. Do all multiplications and divisions from left to right.
4. Do all additions and subtractions from left to right.

ordered pair (57) Pairs of numbers used to locate points in the coordinate plane.

organize data (464) Organizing data is useful before solving a problem. Some ways to organize data are to use tables, charts, different types of graphs, or diagrams.

origin (254) The point of intersection of the two axes in the coordinate plane.

outcomes (413) Outcomes are all possible combinations of a counting problem.

outlier (307) In a set of data, a value that is much greater or much less than the rest of the data can be called a outlier.

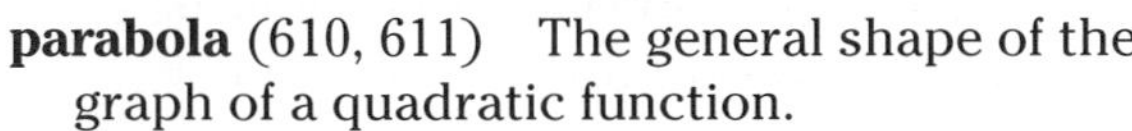

parabola (610, 611) The general shape of the graph of a quadratic function.

parallel lines (362) Lines in the plane that never intersect. Nonvertical parallel lines have the same slope.

parallelogram (362) A quadrilateral in which opposite sides are parallel.

parent graph (354) The simplest of the graphs in a family of graphs.

percent (215) A percent is a ratio that compares a number to 100.

percentage (215) The number that is divided by the base in a percent proportion.

percent of decrease (222) The ratio of an amount of decrease to the previous amount, expressed as a percent.

percent of increase (222) The ratio of an amount of increase to the previous amount, expressed as a percent.

percent proportion (215) $\frac{\text{Percentage}}{\text{Base}} = \frac{r}{100}$

perfect square (119) A rational number whose square root is a rational number.

perfect square trinomial (587) A trinomial which, when factored, has the form $(a + b)^2 = (a + b)(a + b)$ or $(a - b)^2 = (a - b)(a - b)$.

perpendicular lines (362) Lines that meet to form right angles.

point-slope form (333) For any point (x_1, y_1) on a nonvertical line having slope m, the point-slope form of a linear equation is as follows:

$$y - y_1 = m(x - x_1).$$

polynomial (513, 514) A polynomial is a monomial or a sum of monomials.

positive correlation (340) There is a positive correlation between x and y if the values are related in the same way.

power (7) An expression of the form x^n is known as a power.

power of a monomial (498) For any numbers a and b, and any integers m, n, and p,

$$(a^m b^n)^p = a^{mp} b^{np}.$$

power of a power (498) For any number a, and all integers m and n, $(a^m)^n = a^{mn}$.

power of a product (498) For all numbers a and b, and any integer m, $(ab)^m = a^m b^m$.

prime factorization (558) A whole number expressed as a product of factors that are all prime numbers.

prime number (558) A prime number is a whole number, greater than 1, whose only factors are 1 and itself.

prime polynomial (577) A polynomial that cannot be written as a product of two polynomials with integral coefficients is called a prime polynomial.

principal square root (119) The nonnegative square root of an expression.

probability (228) The ratio that tells how likely it is that an event will take place.

$$P(\text{event}) = \frac{\text{number of favorable outcomes}}{\text{total number of possible outcomes}}$$

problem-solving plan (126)

1. Explore the problem.
2. Plan the solution.
3. Solve the problem.
4. Examine the solution.

product (6) The result of multiplication.

product of powers (497) For any number a, and all integers m and n, $a^m \cdot a^n = a^{m+n}$.

product property of square roots (719) For any number a and b, where $a \geq 0$ and $b \geq 0$, $\sqrt{ab} = \sqrt{a} \cdot \sqrt{b}$.

proportion (195) In a proportion, the product of the extremes is equal to the product of the means. If $\frac{a}{b} = \frac{c}{d}$, then $ad = bc$.

Pythagorean theorem (713) If a and b are the measures of the legs of a right triangle and c is the measure of the hypotenuse, then $c^2 = a^2 + b^2$.

Pythagorean triple (583) Three whole numbers a, b, and c such that $a^2 + b^2 = c^2$.

quadrant (254) One of the four regions into which the x- and y-axes separate the coordinate plane.

quadratic equation (620) A quadratic equation is one in which the value of the related quadratic function is 0.

quadratic formula (628) The roots of a quadratic equation in the form $ax^2 + bx + c = 0$, where $a \neq 0$, are given by the formula $x = \frac{-b \pm \sqrt{b^2 - 4ac}}{2a}$.

quadratic function (610, 611) A quadratic function is a function that can be described by an equation of the form $y = ax^2 + bx + c$, where $a \neq 0$.

quartiles (304, 306) In a set of data, the quartiles are values that divide the data into four equal parts.

quotient of powers (501) For all integers m and n and any nonzero number a, $\frac{a^m}{a^n} = a^{m-n}$.

quotient property of square roots (721) For any numbers a and b, where $a > 0$ and $b > 0$, $\sqrt{\frac{a}{b}} = \frac{\sqrt{a}}{\sqrt{b}}$.

radical equations (732) Equations that contain radicals with variables in the radicand.

radical sign (119) The symbol $\sqrt{}$, indicating the principal or nonnegative root of an expression.

radicand (719) The radicand is the expression under the radical sign.

random (229) When an outcome is chosen without any preference, the outcome occurs at random.

range **1.** (263) The set of all second coordinates from the ordered pairs in the relation. **2.** (306) The difference between the greatest and the least values of a set of data.

rate **1.** (197) The ratio of two measurements having different units of measure. **2.** (215) In the percent proportion, the rate is the fraction with a denominator of 100.

ratio (195) A ratio is a comparison of two numbers by division.

rational equation (696) A rational equation is an equation that contains rational expressions.

rational expression (660) A rational expression is an algebraic fraction whose numerator and denominator are polynomials.

rational numbers (93) A rational number is a number that can be expressed in the form $\frac{a}{b}$, where a and b are integers and $b \neq 0$.

rationalizing the denominator (722) Rationalizing the denominator of a radical expression is a method used to remove or eliminate the radicals from the denominator of a fraction.

real numbers (121) The set of rational numbers and the set of irrational numbers together form the set of real numbers.

reciprocal (38) The multiplicative inverse of a number.

reflexive property of equality (39) For any number a, $a = a$.

regression line (342) The most accurate best-fit line for a set of data, and can be determined with a graphing calculator or computer.

relation (260, 263) A relation is a set of ordered pairs.

replacement set (33) A set of numbers from which replacements for a variable may be chosen.

right triangle (165) A right triangle has one angle with a measure of 90 degrees.

rise (325) The vertical change in a line.

roots (620) The solutions of a quadratic equation.

run (325) The horizontal change in a line.

scalar multiplication (108) In scalar multiplication, each element of a matrix is multiplied by a constant.

scale (197) A ratio called a scale is used when making a model to represent something that is too large or too small to be conveniently drawn at actual size.

scatter plot (339) In a scatter plot, the two sets of data are plotted as ordered pairs in the coordinate plane.

scientific notation (506) A number is expressed in scientific notation when it is in the form $a \times 10^n$, where $1 \leq a < 10$ and n is an integer.

set (33) A collection of objects or numbers.

set-builder notation (385) A notation used to describe the members of a set. For example, $\{y \mid y < 17\}$ represents the set of all numbers y such that y is less than 17.

sequence (13) A set of numbers in a specific order.

similar triangles (201) If two triangles are similar, the measures of their corresponding sides are proportional, and the measures of their corresponding angles are equal.

simple events (414) A single event in a probability problem.

simple interest (217) The amount paid or earned for the use of money. The formula $I = prt$ is used to solve simple interest problems.

simplest form (47) An expression is in simplest form when it is replaced by an equivalent expression having no like terms and no parentheses.

simplest radical form (723) A radical expression is in simplest radical form when the following three conditions have been met.

1. No radicands have perfect square factors other than one.
2. No radicands contain fractions.
3. No radicals appear in the denominator of a fraction.

sine (206) In a right triangle with acute angle A, sine of angle $A = \frac{\text{measure of leg opposite angle } A}{\text{measure of the hypotenuse}}$.

slope (324, 325) The ratio of the rise to the run as you move from one point to another along a line.

slope-intercept form (347) An equation of the form $y = mx + b$, where m is the slope and b is the y-intercept of a given line.

solution (32) A replacement for the variable in an open sentence that results in a true sentence.

solution of an equation in two variables (271) If a true statement results when the numbers in an ordered pair are substituted into an equation in two variables, then the ordered pair is a solution of the equation.

solution set (33) The set of all replacements for the variable in an open sentence that result in a true sentence.

solve an equation (145) To solve an equation means to isolate the variable having a coefficient of 1 on one side of the equation.

solving an open sentence (32) Finding a replacement for the variable that results in a true sentence.

solving a triangle (208) Finding the measures of all sides and angles of a right triangle.

square of a difference (544) If a and b are any numbers, $(a - b)^2 = (a - b)(a - b) = a^2 - 2ab + b^2$.

square of a sum (543) If a and b are any numbers, $(a + b)^2 = (a + b)(a + b) = a^2 + 2ab + b^2$.

square root (118, 119) One of two identical factors of a number.

standard form (333) The standard form of a linear equation is $Ax + By = C$, where A, B, and C are integers, $A \geq 0$, and A and B are not both zero.

statistics (25) A branch of mathematics concerned with methods of collecting, organizing, and interpreting data.

stem-and-leaf plot (26) In a stem-and-leaf plot, each piece of data is separated into two numbers that are used to form a stem and a leaf. The data are organized into two columns. The column on the left contains the stem and the column on the right contains the leaves.

substitution (462) The substitution method of solving a system of equations is a method that uses substitution of one equation into the other equation to solve for the other variable.

substitution property of equality (39) If $a = b$, then a may be replaced by b in any expression.

subtracting integers (87) To subtract a number, add its additive inverse. For any numbers a and b, $a - b = a + (-b)$.

subtraction property for inequality (385) For all numbers a, b, and c, the following are true:

1. if $a > b$, then $a - c > b - c$.
2. if $a < b$, then $a - c < b - c$.

subtraction property of equality (146) For any numbers a, b, and c, if $a = b$, then $a - c = b - c$.

supplementary angles (162) Two angles are supplementary if the sum of their measures is 180 degrees.

symmetric property of equality (39) For any numbers a and b, if $a = b$, then $b = a$.

symmetry (612) Symmetrical figures are those in which the figure can be folded and each half matches the other exactly.

system of equations (455) A set of equations with the same variables.

system of inequalities (482) A set of inequalities with the same variables.

T

tangent (206) In a right triangle, the tangent of angle $A = \frac{\text{measure of leg opposite angle } A}{\text{measure of leg adjacent to angle } A}$.

term **1.** (13) A number in a sequence. **2.** (47) A number, a variable, or a product or quotient of numbers and variables.

transitive property of equality (39) For any numbers a, b, and c, if $a = b$ and $b = c$, then $a = c$.

tree diagram (413) A tree diagram is a diagram used to show the total number of possible outcomes.

triangle (163) A triangle is a polygon with three sides and three angles.

trigonometric ratios (206) $m \angle C = 90$

$\sin A = \frac{a}{c}$ $\cos A = \frac{b}{c}$ $\tan A = \frac{a}{b}$

trinomials (514) A trinomial is the sum of three monomials.

uniform motion (235) When an object moves at a constant speed, or rate, it is said to be in uniform motion.

union (408) The union of two sets A and B is the set of elements contained in both A or B.

unique factorization theorem (558) The prime factorization of every number is unique except for the order in which the factors are written.

unit cost (95) The cost of one unit of something.

upper quartile (306) The upper quartile divides the upper half of a set of data into two equal parts.

use a model (339) A problem-solving strategy that uses models, or simulations, of mathematical situations that are difficult to solve directly.

use a table (255) A problem-solving strategy that uses tables to organize and solve problems.

variable (6) Variables are symbols that are used to represent unspecified numbers.

Venn diagrams (73) Venn diagrams are diagrams that use circles or ovals inside a rectangle to show relationships of sets.

vertex (610, 611) The maximum or minimum point of a parabola.

vertical axis (56) The vertical line in a graph that represents the dependent variable.

vertical line test (289) If any vertical line passes through no more than one point of the graph of a relation, then the relation is a function.

weighted average (233) The weighted average M of a set of data is the sum of the product of each number in the set and its weight divided by the sum of all the weights.

whiskers (428) The whiskers of a box-and-whisker plot are the segments that are drawn from the lower quartile to the least value and from the upper quartile to the greatest value.

whole numbers (72) The set of whole numbers is represented by {0, 1, 2, 3, ...}.

work backward (156) A problem-solving strategy that uses inverse operations to determine an original value.

***x*-axis** (254) The horizontal number line.

***x*-coordinate** (254) The first number in an ordered pair.

***x*-intercept** (346) The coordinate at which a graph intersects the x-axis.

***y*-axis** (254) The vertical number line.

***y*-coordinate** (254) The second number in an ordered pair.

***y*-intercept** (346) The coordinate at which a graph intersects the y-axis.

zero exponent (502) For any nonzero number a, $a^0 = 1$.

zero product property (594) For all numbers a and b, if $ab = 0$, then $a = 0$, $b = 0$, or both a and b equal 0.

zeros (620) The zeros of a function are the roots, or x-intercepts, of the function.

SPANISH GLOSSARY

A

absolute value/valor absoluto (85) El valor absoluto de un número equivale al número de unidades que dicho número dista de cero en la recta numérica.

acute triangle/triángulo agudo (165) En un triángulo agudo, todos los ángulos miden menos de 90°.

adding integers/suma de enteros (86)

1. Para sumar enteros del *mismo* signo, suma los valores absolutos de los números. Da al resultado el mismo signo de los números.
2. Para sumar enteros de *distinto* signo, resta el valor menor del valor mayor. Da al resultado el mismo signo que el número con el mayor valor absoluto.

addition property of equality/propiedad de adición de la igualdad (144) Para cualquiera de los números a, b y c, si $a = b$, entonces $a + c = b + c$.

addition property of inequality/propiedad de adición de la desigualdad (385) Para todos los números a, b y c:

1. si $a > b$, entonces $a + c > b + c$;
2. si $a < b$, entonces $a + c < b + c$.

additive identity/identidad aditiva (37) Para cualquier número a, $a + 0 = 0 + a = a$.

additive inverse property/propiedad del inverso de la adición (87) Para cualquier número a, $a + (-a) = 0$.

algebraic expression/expresión algebraica (6) Una expresión que consiste de uno o más números y variables, además de una o más operaciones aritméticas.

angle of depression/ángulo de depresión (208) El ángulo de depresión se forma por una línea horizontal y una línea visual por debajo de la misma.

angle of elevation/ángulo de elevación (208) Un ángulo de elevación se forma por una línea horizontal y una línea visual por encima de la misma.

associative property/propiedad asociativa (51) Para cualquiera de los números a, b, y c, $(a + b) + c = a + (b + c)$ y $(ab)c = a(bc)$.

axes/ejes (254) Dos rectas numéricas perpendiculares que se usan para ubicar puntos en un plano de coordenadas.

axis of symmetry/eje de simetría (612) La ecuación para el eje de simetría de la gráfica $y = ax^2 + bx + c$, en la cual $a \neq 0$ y $x = -\frac{b}{2a}$.

B

back-to-back stem-and-leaf plot/diagrama de tallo y hojas consecutivo (27) Un diagrama de tallo y hojas consecutivo se usa para comparar dos conjuntos de datos. El mismo tallo se usa para las hojas de ambos diagramas.

base/base **1.** (7) En una expresión de la forma x^n, la base es x. **2.** (215) El número que se divide entre el porcentaje en una proporción de porcentaje.

best-fit line/línea de mejor encaje (341) Es una línea que se dibuja en un diagrama de dispersión y que pasa cerca de la mayoría de los puntos de datos.

binomial/binomio (514) La suma de dos monomios.

boundary/frontera (437) La frontera de una desigualdad es una línea que separa el plano de coordenadas en dos mitades de planos.

box-and-whisker plot/diagrama de caja y patillas (427) Un tipo de diagrama o gráfica que muestra los valores cuartílicos y los extremos de los datos.

C

coefficient/coeficiente (47) El factor numérico de un término.

commutative property/propiedad conmutativa (51) Para cualquiera de los números a y b, $a + b = b + a$ y $ab = ba$.

comparison property/propiedad de comparación (94) Para cualquier par de números a y b, exactamente una de las siguientes operaciones es válida.

$$a < b \qquad a = b \qquad a > b$$

complementary angles/ángulos complementarios (163) Dos ángulos son complementarios si la suma de sus medidas es 90 grados.

complete graph/gráfica completa (278) Una gráfica completa muestra el origen, los puntos en donde la gráfica cruza el eje de coordenadas x y el eje de coordenadas y, además de otras características importantes en la gráfica.

completeness property/propiedad del completo (121) Cada número real corresponde exactamente a un punto en la recta numérica. Cada punto en la recta numérica corresponde exactamente a un número real.

completeness property for points in the plane/propiedad del completo de los puntos en el plano (256)

1. Exactamente un punto en el plano es nombrado por un par ordenado de números dado.

2. Exactamente un par ordenado de números nombra un punto dado en el plano.

completing the square/completar el cuadrado (742, 743) Sumar un término constante a un binomio de la forma $x^2 = bx$, de modo que el trinomio resultante sea un cuadrado perfecto.

complex fraction/fracción compleja (114) Si una fracción tiene uno o más fracciones en el numerador o denominador, entonces se llama una fracción compleja.

composite number/número compuesto (558) Un número entero, mayor de 1, que no es un número primo.

compound event/evento compuesto (414) Un evento compuesto consiste de dos o más eventos simples.

compound inequality/desigualdad compuesta (405) Dos desigualdades conectadas por *y* u *o*.

congruent angles/ángulos congruentes (164) Ángulos que tienen la misma medida.

conjugates/conjugados (722) Dos binomios de la forma $a\sqrt{b} + c\sqrt{d}$ y $a\sqrt{b} - c\sqrt{d}$.

consecutive integers/números enteros consecutivos (158) Son los números enteros en el orden de contar.

consistent/consistente (455) Se dice que un sistema de ecuaciones es consistente cuando tiene por lo menos un par ordenado que satisface ambas ecuaciones.

constant of variation/constante de variación (239) El número k en ecuaciones de la forma $y = kx$ y $xy = k$.

constants/constantes (469) Los monomios que son números reales.

coordinate/coordenada (73) El número que corresponde a un punto en una recta numérica.

coordinate plane/plano de coordenadas (254) El plano que contiene el eje de coordenadas x y el eje de coordenadas y.

corresponding angles/ángulos correspondientes (201) Ángulos que encajan en triángulos semejantes y que tienen medidas iguales.

corresponding sides/lados correspondientes (201) Los lados opuestos a los ángulos correspondientes en triángulo semejantes.

cosine/coseno (206) En un triángulo rectángulo con ángulo agudo A, el coseno del

ángulo $A = \frac{\text{medida del cateto adyacente al ángulo } A}{\text{medida de la hipotenusa}}$.

cross products/productos cruzados (94) Cuando se comparan dos fracciones, los productos cruzados son los productos de los términos en diagonales.

D

data/datos (25) La información numérica.

defining the variable/definir la variable (127) El escoger una variable para representar uno de los números no especificados en un problema.

degree/grado **1.** (515) El grado de un monomio es la suma de los exponentes de sus variables. **2.** (516) El grado de un polinomio es el grado del término con el grado más alto.

density property/propiedad de la densidad (96) Entre cada par de números racionales distintos, existe una infinidad de números racionales.

dependent/dependiente (456) Un sistema de ecuaciones que tiene un número infinito de soluciones.

dependent variable/variable dependiente (58) La variable en una función cuyo valor lo determina la variable independiente.

difference of squares/diferencia de cuadrados (545) Dos cuadrados perfectos separados por un signo de sustracción, $a^2 - b^2 = (a + b)(a - b)$.

dimensional analysis/análisis dimensional (174) El proceso de llevar unidades a lo largo de un cómputo.

direct variation/variación directa (239) Una función lineal descrita por una ecuación de la forma $y = kx$, en que $k \neq 0$.

discrete mathematics/matemáticas de números discretos (88) Una rama de las matemáticas que estudia los conjuntos de números finitos o interrumpidos.

discriminant/discriminante (632) En la fórmula cuadrática, la expresión $b^2 - 4ac$ se denomina la discriminante.

distance formula/fórmula de distancia (737) La distancia d entre cualquier par de puntos con coordenadas (x_1, y_1) y (x_2, y_2) es dada por la siguiente fórmula.

$$d = \sqrt{(x_2 - x_1)^2 + (y_2 - y_1)^2}.$$

distributive property/propiedad distributiva (46) Para cualquiera de los números a, b y c:

1. $a(b + c) = ab + ac$ y $(b + c)a = ba + ca$.

2. $a(b - c) = ab - ac$ y $(b - c)a = ba - ca$.

dividing rational numbers/división de números racionales (112) El cociente de dos números racionales que tienen el mismo signo es positivo. El cociente de dos números racionales que tienen diferente signo es negativo.

division property of equality/propiedad de división de la igualdad (151) Para cualquiera de los números a, b y c, en que $c \neq 0$, si $a = b$, entonces $\frac{a}{c} = \frac{b}{c}$.

division property of inequality/propiedad de división de la desigualdad (393) Para cualquiera de los números reales a, b y c:

1. si c es positivo y $a < b$, entonces $\frac{a}{c} < \frac{b}{c}$, y

   si c es positivo y $a > b$, entonces $\frac{a}{c} > \frac{b}{c}$.

2. si c es negativo y $a < b$, entonces $\frac{a}{c} > \frac{b}{c}$, y

   si c es negativo y $a > b$, entonces $\frac{a}{c} < \frac{b}{c}$.

domain/dominio (263) El conjunto de todas las primeras coordenadas de los pares ordenados de una relación.

draw a diagram/trazar un diagrama (406) Una estrategia para resolver problemas que a menudo se usa como una herramienta organizadora.

E

element/elemento (33) Un miembro de un conjunto.

elimination/eliminación (469) El método de eliminación para resolver un sistema de ecuaciones es un método que usa adición o sustracción para eliminar una de las variables y así despejar la otra variable.

equally likely/igualmente verosímil (229) Respuestas que tienen igual posibilidad de ocurrir.

equation/ecuación (33) Un enunciado matemático que contiene un signo de igualdad, =.

equation in two variables/ecuación en dos variables (271) Una ecuación en dos variables contiene dos valores desconocidos.

equilateral triangle/triángulo equilátero (164) Un triángulo en el cual todos los lados y todos los ángulos tienen la misma medida.

equivalent equation/ecuación equivalente (144) Ecuaciones que tienen la misma solución.

equivalente expressions/expresiones equivalentes (47) Expresiones que denotan el mismo número.

evaluate/evaluar (8) El método de hallar el valor de una expresión cuando se conocen los valores de las variables.

excluded value/valor excluido (660) Se excluye un valor del dominio de una variable porque si se sustituyera ese valor por la variable, el resultado tendría un cero en el denominador.

exponent/exponente (7) En una expresión de la forma x^n, el exponente es n.

exponential function/función exponencial (634, 635) Una función que se puede describir por una ecuación de la forma $y = a^x$, en que $a > 0$ y $a \neq 1$.

extraneous solutions/soluciones extrañas (697) Soluciones obtenidas de una ecuación y las cuales no son soluciones aceptables para la ecuación original.

extreme values/valores extremos (427) Los valores extremos son el menor valor y el mayor valor en un conjunto de datos.

extremes/extremos (196) *Ver* proporción.

factored form/forma factorial (559) Un monomio está escrito en forma factorial cuando está expresado como el producto de números primos y variables y las variables no tienen un exponente mayor de 1.

factoring/factorizar (564) Expresar un polinomio como el producto de monomios o polinomios.

factoring by grouping/factorizando por grupos (567) Un método de factorización de polinomios de cuatro o más términos.

factors/factores (6) En una expresión de multiplicación, los factores son las cantidades que se multiplican.

family of graphs/familia de gráficas (354) Una familia de gráficas incluye gráficas y ecuaciones de gráficas que tienen por lo menos una característica en común.

FOIL method/método FOIL (537) Para multiplicar dos binomios, halla la suma de los productos de los primeros términos, los términos de afuera, los términos de adentro y los últimos términos.

formula/fórmula (128) Una ecuación que enuncia una regla para la relación entre ciertas cantidades.

function/función **1.** (56) Una relación entre los datos de entrada y de salida, en que los datos de salida dependen de los datos de entrada. **2.** (287) Una relación en que cada elemento del dominio se aparea exactamente con un elemento de la amplitud.

functional notation/notación funcional (289) En notación funcional, la ecuación $y = x + 5$ se escribe $f(x) = x + 5$.

general equation for exponential decay/ ecuación general para la disminución exponencial (644) La ecuación general para la disminución exponencial está representada por $A = C(1 - r)^t$.

general equation for exponential growth/ ecuación general para el crecimiento exponencial (643) La ecuación general para el crecimiento exponencial está representada por $A = C(1 + r)^t$.

graph/gráfica (73, 255) Consiste en dibujar, o trazar sobre una recta numérica o plano de coordenadas, los puntos nombrados por ciertos números o pares ordenados.

greatest common factor (GCF)/máximo común divisor (MCD) (559) El máximo común divisor de dos o más números enteros es el número mayor que es un factor de todos los números.

guess and check/conjetura y cotejo (574) Una estrategia para resolver problemas en la cual se prueban varios valores o combinaciones de valores para hallar una solución al problema.

half-life/media vida (638) La media vida de un elemento radioactivo es el tiempo que tarda en desintegrarse la mitad del elemento.

half-plane/medio plano (437) La región de un plano en un lado de una recta en el plano.

horizontal axis/eje horizontal (56) La recta horizontal en una gráfica que representa la variable independiente.

hypotenuse/hipotenusa (206) El lado de un triángulo rectángulo opuesto al ángulo recto.

I

identify subgoals/identificación de submetas (745) Una estrategia para resolver problemas que utiliza una serie de pasos secundarios o submetas.

identity/identidad (170) Una ecuación que es cierta para cualquier valor de la variable.

inconsistent/inconsistente (456) Se dice que un sistema de ecuaciones es inconsistente cuando no tiene pares ordenados que satisfacen ambas ecuaciones.

independent/independiente (456) Se dice que un sistema de ecuaciones es independiente si el sistema tiene exactamente una solución.

independent variable/variable independiente (58) La variable en una función cuyo valor está sujeto a elección se dice que es una variable independiente. La variable independiente determina el valor de la variable dependiente.

inequality/desigualdad (33) Un enunciado matemático que contiene los símbolos $<$, $\leq$, $>$, o $\geq$.

integers/enteros (73) El conjunto de enteros representados por $\{\ldots, -3, -2, -1, 0, 1, 2, 3, \ldots\}$.

interquartile range/amplitud intercuartílica (304, 306) La diferencia entre el cuartil superior y el inferior de un conjunto de datos. Representa la mitad inferior, ó 50%, de los datos en un conjunto.

intersection/intersección (406) Para dos conjuntos A y B, es el conjunto de elementos comunes a ambos A y B.

inverse of a relation/inverso de una relación (264) El inverso de cualquier relación se obtiene intercambiando las coordenadas en cada par ordenado.

inverse variation/variación inversa (241) Una variación inversa se describe por una ecuación de la forma $xy = k$, en que $k \neq 0$.

irrational number/números irracional (120) Un número que no se puede expresar en la forma $\frac{a}{b}$, en que a y b son números enteros y $b \neq 0$.

isosceles triangle/triángulo isósceles (164) En un triángulo isósceles, por lo menos dos ángulos tienen la misma medida y por lo menos dos lados tienen la misma longitud.

least common denominator (LCD)/mínimo común denominador (MCD) (686) El mínimo común múltiplo de los denominadores de dos o más fracciones.

least common multiple (LCM)/mínimo común múltiplo (MCM) (685) Para dos o más enteros, el MCM es el menor número entero positivo divisible entre cada uno de los enteros.

legs/catetos (206) Los lados de un triángulo rectángulo que forman el ángulo recto.

like terms/términos semejantes (47) Términos que contienen las mismas variables, con variables correspondientes que tienen la misma potencia.

line plot/esquema lineal (78) Datos numéricos desplegados sobre una recta numérica.

linear equation/ecuación lineal (280) Una ecuación que puede escribirse de la forma $Ax + By = C$, en que A y B no son ambos iguales a cero. La gráfica de una ecuación lineal es una recta.

linear function/función lineal (278) Ecuación cuya gráfica es una recta no vertical.

look for a pattern/busca un patrón (13, 497, 637) Una estrategia para resolver problemas que a menudo involucra el uso de tablas para organizar la información de manera que se pueda determinar un patrón.

lower quartile/cuartil inferior (306) El cuartil inferior divide la mitad inferior de un conjunto de datos en dos partes iguales.

make an organized list/haz una lista organizada (685) Una estrategia para resolver problemas que utiliza una lista organizada para arreglar y evaluar los datos y determinar una solución.

mapping/relación (263) Una relación aparea un elemento en el dominio con un elemento en la amplitud.

matrix/matriz (88) Una matriz es un arreglo rectangular de elementos en hileras y columnas.

maximum/máximo (611) El punto más alto en la gráfica de una curva, tal como el vértice de la parábola que se abre hacia abajo.

mean/media (178) La media de un conjunto de datos es la suma de los números en el conjunto dividida entre el número de números en el conjunto.

means/media proporcional (196) *Ver* proporción.

measures of central tendency/medidas de tendencia central (178) Los números que se usan a menudo para describir conjuntos de datos porque estos representan un valor centralizado o en el medio.

measures of variation/medidas de variación (306) Números usados para describir la amplitud o distribución de los datos.

median/mediana (178) La mediana es el número en el centro de un conjunto de datos cuando los números se organizan en orden numérico.

midpoint/punto medio (369) El punto medio de un segmento es el punto equidistante entre los extremos del segmento.

minimum/mínimo (611) El punto más bajo en la gráfica de una curva, tal como el vértice de una parábola que se abre hacia arriba.

mixed expression/expresión mixta (690) Una expresión algebraica que contiene un monomio y una expresión racional.

mode/modal (178) El número que ocurre con más frecuencia en un conjunto.

monomial/monomio (496) Un número, una variable o el producto de un número y una o más variables.

multi-step equations/ecuaciones múltiples (157) Ecuaciones que requieren más de una operación para resolverlas.

multiplicative identity/identidad de multiplicación (38) Para cualquier número a, $a \cdot 1 = 1 \cdot a = a$.

multiplicative inverse/inverso multiplicativo (38) Para cualquier número no cero a, hay exactamente un número $\frac{1}{a}$, tal que $a \cdot \frac{1}{a} = \frac{1}{a} \cdot a = 1$.

multiplicative property of −1/propiedad multiplicativa de −1 (107) El producto de cualquier número y −1 es el inverso aditivo del número.

$$-1(a) = -a \text{ y } a(-1) = -a.$$

multiplicative property of equality/propiedad multiplicativa de la igualdad (150) Para cualquiera de los números a, b y c, si $a = b$, entonces $a \cdot c = b \cdot c$.

multiplication property of inequality/propiedad de multiplicación de la desigualdad (393) Para todos los números a, b y c, lo siguiente es cierto:

1. Si c es positivo y $a < b$, entonces $ac < bc$, y si c es positivo y $a > b$, entonces $ac > bc$.
2. Si c es negativo y $a < b$, entonces $ac > bc$, y si c es negativo y $a > b$, entonces $ac < bc$.

multiplicative property of zero/propiedad multiplicativa de cero (38) Para cualquier número a, $a \cdot 0 = 0 \cdot a = 0$.

negative correlation/correlación negativa (340) Existe una correlación negativa entre x y y si los valores están relacionados de maneras opuestas.

negative exponent/exponente negativo (503) Para cualquiera de los números no cero a y cualquier número entero n, $a^{-n} = \frac{1}{a^n}$

negative number/número negativo (73) Cualquier número que es menos de cero.

number line/recta numérica (72) Una recta con marcas equidistantes que se usa para representar números.

number theory/teoría de números (158) El estudio de los números y de las relaciones entre los mismos.

obtuse triangle/triángulo obtuso (165) Un triángulo obtuso tiene un ángulo cuya medida es mayor de 90 grados.

odds/posibilidades (229) Las posibilidades de que un evento ocurra son la proporción del número de formas en que el evento puede ocurrir (éxitos) comparada con el número de formas en que puede no ocurrir (fracasos).

open sentences/enunciados abiertos (32) Enunciado matemático que contiene una o más variables, o incógnitas.

opposites/opuestos (87) El opuesto de un número es su inverso aditivo.

order of operations/orden de operaciones (19)

1. Simplifica las expresiones dentro de los símbolos de agrupación, tales como paréntesis, corchetes y como lo indiquen las barras de fracción.
2. Evalúa todas las potencias.
3. Realiza todas las multiplicaciones y las divisiones de izquierda a derecha.
4. Realiza todas las sumas y las restas de izquierda a derecha.

ordered pair/par ordenado (57) Pares de números que se usan para ubicar puntos en el plano de coordenadas.

organize data/organizar datos (464) Una estrategia útil antes de resolver un problema. Algunas formas de organizar los datos son el uso de tablas, esquemas, diferentes tipos de gráficas o diagramas.

origin/origen (254) El punto de intersección de los dos ejes en el plano de coordenadas.

outcomes/resultados (413) Todas las maneras en que puede ocurrir un evento.

outlier/valor atípico (307) Cualquier elemento en un conjunto de datos que es por lo menos 1.5 veces el valor de la amplitud intercuartílica mayor que el cuartil superior, o menor que el cuartil inferior.

parabola/parábola (610, 611) La forma general de la gráfica de una función cuadrática.

SPANISH GLOSSARY

parallel lines/rectas paralelas (362) Rectas en el plano que nunca se intersecan. Las rectas paralelas no verticales tienen la misma pendiente.

parallelogram/paralelogramo (362) Un cuadrilátero cuyos lados opuestos son paralelos.

parent graph/gráfica principal (354) La gráfica más simple en una familia de gráficas.

percent/por ciento (215) Un por ciento es una proporción que compara un número con 100.

percentage/porcentaje (215) El número que se divide entre la base en un por ciento de proporción.

percent of decrease/porcentaje de disminución (222) La proporción de una cantidad de disminución comparada con una cantidad previa, expresada en forma de por ciento.

percent of increase/porcentaje de aumento (222) La proporción de una cantidad de aumento comparada con una cantidad previa, expresada en forma de por ciento.

percent proportion/proporción de porcentaje (215)

$$\frac{\text{Percentaje}}{\text{Base}} = \frac{r}{100}$$

perfect square/cuadrado perfecto (119) Un número racional cuya raíz cuadrada es un número racional.

perfect square trinomial/cuadrado perfecto trinómico (587) Un trinomio que al factorizarse tiene la forma $(a + b)^2 = (a + b)(a + b)$ o $(a - b)^2 = (a - b)(a - b)$.

perpendicular lines/rectas perpendiculares (362) Rectas que se encuentran para formar ángulos rectos.

point-slope form/forma punto–pendiente (333) Para cualquier punto (x_1, y_1) sobre una recta no vertical cuya pendiente es m, la forma punto–pendiente de una ecuación lineal es la siguiente:

$$y - y_1 = m(x - x_1).$$

polynomial/polinomio (513, 514) Un polinomio es un monomio o la suma de monomios.

positive correlation/correlación positiva (340) Existe una correlación positiva entre x y y si los valores están relacionados de la misma forma.

power/potencia (7) Una expresión de la forma x^n se conoce como una potencia.

power of a monomial/potencia de un monomio (498) Para cualquiera de los números a y b, y para todos los números enteros m, n y p, $(a^m b^n)^p = a^{mp} b^{np}$.

power of a power/potencia de una potencia (498) Para cualquier número a y todos los números enteros m y n, $(a^m)^n = a^{mn}$.

power of a product/potencia de un producto (498) Para todos los números a y b y cualquier número entero m, $(ab)^m = a^m b^m$.

prime factorization/factorización prima (558) Un número entero expresado como un producto de factores que son todos números primos.

prime number/número primo (558) Un número primo es un número entero mayor que 1 cuyos únicos factores son 1 y el número mismo.

prime polynomial/polinomio primo (577) Un polinomio que no se puede escribir como el producto de dos polinomios con coeficientes integrales se llama un polinomio primo.

principal square root/raíz cuadrada principal (119) La raíz cuadrada no negativa de una expresión.

probability/probabilidad (228) Una razón que expresa la posibilidad de que algún evento suceda.

$$P(\text{evento}) = \frac{\text{número de resultados favorables}}{\text{número de resultados posibles}}$$

problem-solving plan/plan para solucionar problemas (126)

1. Explorar el problema.
2. Planificar la solución.
3. Resolver el problema.
4. Examinar la solución.

product/producto (6) El resultado de la multiplicación.

product of powers/producto de potencias (497) Para cualquiera de los números a y todos los números enteros m y n, $a^m \cdot a^n = a^{m+n}$.

product property of square roots/propiedad del producto de raíces cuadradas (719) Para cualquiera de los números a y b, en que $a \geq 0$ y $b \geq 0$, $\sqrt{ab} = \sqrt{a} \cdot \sqrt{b}$.

proportion/proporción (195) En una proporción, el producto de los extremos es igual al producto de las medias. Si $\frac{a}{b} = \frac{c}{d}$, entonces $ad = bc$.

Pythagorean theorem/teorema de Pitágoras (713) Si a y b son las medidas de los catetos de un triángulo rectángulo y c es la medida de la hipotenusa, entonces, $c^2 = a^2 + b^2$.

Pythagorean triple/triplete de Pitágoras (583) Tres números enteros a, b, y c, tales que $a^2 + b^2 = c^2$.

quadrant/cuadrante (254) Una de las cuatro regiones en que el eje x y el eje y separan el plano de coordenadas.

quadratic equation/ecuación cuadrática (620) Una ecuación cuadrática es una en que el valor de la función cuadrática relacionada es 0.

quadratic formula/fórmula cuadrática (628) Las raíces de una ecuación cuadrática en la forma $ax^2 + bx + c = 0$, en la cual $a \neq 0$, son dadas por la fórmula

$$x = \frac{-b \pm \sqrt{b^2 - 4ac}}{2a}.$$

quadratic function/función cuadrática (610, 611) Función que se puede describir con una equación de la forma $ax^2 + bx + c = 0$, en la cual $a \neq 0$.

quartiles/cuartiles (304, 306) En un conjunto de datos, los cuartiles son valores que dividen los datos en cuatro partes iguales.

quotient of powers/cociente de potencias (501) Para todos los números enteros m y n y todo número no cero a, $\frac{a^m}{a^n} = a^{m-n}$.

quotient property of square roots/propiedad del cociente de raíces cuadradas (721) Para cualquiera de los números a y b, en que $a > 0$ y $b > 0$, $\sqrt{\frac{a}{b}} = \frac{\sqrt{a}}{\sqrt{b}}$.

R

radical equations/ecuaciones radicales (732) Ecuaciones que contienen radicales con variables en el radicando.

radical sign/signo radical (119) El símbolo $\sqrt{\ }$ que se usa para indicar la raíz cuadrada principal no negativa de una expresión.

radicand/radicando (719) El radicando es la expresión debajo del signo radical.

random/al azar (229) Cuando se escoge un resultado sin ninguna preferencia, el resultado ocurre al azar.

range/amplitud **1.** (263) El conjunto de todas las segundas coordenadas de los pares ordenados de una relación. **2.** (306) La diferencia entre los valores mayor y menor en un conjunto de datos.

rate/razón (197) La proporción de dos medidas que se dan en diferentes unidades de medida.

rate/tasa (215) En una proporción de por ciento, la tasa es la fracción con 100 como denominador.

ratio/razón (195) Una razón es una comparación de dos números mediante división.

rational equation/ecuación racional (696) Una ecuación racional es una ecuación que contiene expresiones racionales.

rational expression/expresión racional (660) Una expresión racional es una fracción algebraica cuyo numerador y cuyo denominador son polinomios.

rational number/número racional (93) Un número racional es un número que se puede expresar en la forma $\frac{a}{b}$, en que a y b son números enteros y $b \neq 0$.

rationalizing the denominator/racionalizando el denominador (722) Un proceso que se usa para quitar o eliminar radicales del denominador de una fracción, en una expresión radical.

real numbers/números reales (121) El conjunto de números irracionales junto con los números racionales.

reciprocal/recíproco (38) El inverso multiplicativo de un número.

reflexive property of equality/propiedad reflexiva de la igualdad (39) Para cualquier número a, $a = a$.

regression line/línea de regresión (342) La línea más exacta de mejor ajuste para un conjunto de datos. Se puede determinar con una calculadora de gráficar o una computadora.

relation/relación (260, 263) Una relación es un conjunto de pares ordenados.

replacement set/conjunto de substitución (33) Un conjunto de números de los cuales se pueden escoger números para reemplazar una variable.

right triangle/triángulo rectángulo (165) Un triángulo que tiene un ángulo con una medida de 90 grados.

rise/altura (325) El cambio vertical en una recta.

roots/raíces (620) Las soluciones para una ecuación cuadrática.

run/carrera (325) El cambio horizontal en una recta.

S

scalar multiplication/multiplicación escalar (108) En multiplicación escalar, cada elemento de una matriz se multiplica por una constante.

scale/escala (197) Una razón llamada una escala se usa en la construcción de un modelo para representar algo que es muy grande o muy pequeño para ser dibujado a su tamaño real.

scatter plot/diagrama de dispersión (339) Gráfica que muestra dos conjuntos de datos trazados como puntos (pares ordenados) en el plano de coordenadas.

scientific notation/notación científica (506) Un número está expresado en notación científica cuando está en la forma de $a \times 10^n$, en que $1 \leq a < 10$ y n es un número entero.

set/conjunto (33) Una colección de objetos o números.

set-builder notation/notación de construcción de conjuntos (385) Una notación que se usa para describir los miembros de un conjunto. Por ejemplo, $\{y \mid y < 17\}$ representa el conjunto de todos los números y de modo que y es menor que 17.

sequence/sucesión (13) Un conjunto de números en un orden específico.

similar triangles/triángulos semejantes (201) Si dos triángulos son semejantes, las medidas de sus lados correspondientes son proporcionales y las medidas de sus ángulos correspondientes son iguales.

simple event/evento simple (414) Evento sencillo en un problema de probabilidad.

simple interest/interés simple (217) La cantidad pagada o ganada por el uso de una cantidad de dinero. La fórmula $I = prt$ se usa para resolver problemas de interés simple.

simplest form/forma reducida (47) Una expresión está en su forma reducida cuando ha sido reemplazada por una expresión similar que no tiene términos semejantes ni paréntesis.

simplest radical form/forma radical reducida (723) Una expresión radical se encuentra en forma reducida cuando se satisfacen las siguientes condiciones:

1. Ningún radicando tiene factores cuadrados perfectos además de uno.
2. Ningún radicando contiene fracciones.
3. No aparece ningún radical en el denominador de una fracción.

sine/seno (206) En un triángulo rectángulo con ángulo agudo A, el seno del ángulo $A = \frac{\text{medida del cateto opuesto al ángulo } A}{\text{medida de la hipotenusa}}$.

slope/pendiente (324, 325) El cambio vertical (altura) al cambio horizontal (carrera) a medida que te mueves de un punto a otro a lo largo de la recta.

slope-intercept form/forma pendiente-intersección (347) Una ecuación de la forma $y = mx + b$, en que m es la pendiente y b es la intersección en y de una recta dada.

solution/solución (32) Una sustitución por una variable en una ecuación que resulta en una ecuación válida.

solution of an equation in two variables/ solución de una ecuación de dos variables (271) Si al sustituir los números en un par ordenado se satisface una ecuación de dos variables, entonces el par ordenado es una solución de la ecuación.

solution set/conjunto de solución (33) El conjunto de todos los sustitutos para una variable en un enunciado abierto que satisfacen el enunciado.

solve an equation/resuelve una ecuación (145) Resolver una ecuación quiere decir aislar la variable cuyo coeficiente es 1, en un lado de la ecuación.

solving an open sentence/resolviendo un enunciado abierto (32) Hallar un sustituto que satisface la variable.

solving a triangle/resolviendo un triángulo (208) El proceso de hallar las medidas de todos los lados y ángulos de un triángulo rectángulo.

square of a difference/cuadrado de una diferencia (544) Si a y b son cualquier par de números, $(a - b)^2 = (a - b)(a - b) = a^2 - 2ab + b^2$.

square of a sum/cuadrado de una suma (543) Si a y b son cualquier par de números, $(a + b)^2 = (a + b)(a + b) = a^2 + 2ab + b^2$.

square root/raíz cuadrada (118, 119) Uno de los dos factores idénticos de un número.

standard form/forma estándar (333) La forma estándar de una ecuación lineal es $Ax + By = C$, en la cual A, B y C son números enteros, $A \geq 0$ y A y B no son ceros ambos.

statistics/estadística (25) Una rama de las matemáticas que tiene que ver con los métodos de recolección, organización e interpretación de datos.

stem-and-leaf plot/gráfica de tallo y hojas (26) En una gráfica de tallo y hojas, cada dato se separa en dos números que se usan para formar un tallo y las hojas. Los datos se organizan en dos columnas. La columna de la izquierda contiene el tallo y la columna de la derecha las hojas.

substitution/sustitución (462) El método de sustitución para resolver un sistema de ecuaciones es un método que sustituye una ecuación en la otra ecuación para despejar la otra variable.

substitution property of equality/propiedad de sustitución de la igualdad (39) Si $a = b$, entonces, a se puede reemplazar por b en cualquier expresión.

subtracting integers/sustracción de enteros (87) Para restar un entero, suma su inverso aditivo. Para cualquiera de los enteros a y b, $a - b = a + (-b)$.

subtraction property of equality/propiedad de sustracción de la igualdad (146) Para cualquiera de los números a, b y c, si $a = b$, entonces $a - c = b - c$.

subtraction property of inequality/propiedad de sustracción de la desigualdad (385) Para todos los números a, b y c los siguientes son ciertos:

1. si $a > b$, entonces $a - c > b - c$;
2. si $a < b$, entonces $a - c < b - c$.

supplementary angles/ángulos suplementarios (162) Dos ángulos son suplementarios si la suma de sus medidas es 180 grados.

symmetric property of equality/propiedad simétrica de la igualdad (39) Para cualquiera de los números a y b, si $a = b$, entonces $b = a$.

symmetry/simetría (612) Las figuras simétricas son aquellas en que la figura se puede doblar y cada mitad es exactamente igual a la otra.

system of equations/sistema de ecuaciones (455) Un conjunto de ecuaciones con las mismas variables.

system of inequalities/sistema de desigualdades (482) Un conjunto de desigualdades con las mismas variables.

T

tangent/tangente (206) En un triángulo rectángulo con ángulo agudo A, la tangente del ángulo $A = \frac{\text{medida del cateto opuesto al ángulo } A}{\text{medida del cateto adyacente al ángulo } A}$.

term/término **1.** (13) Un número en una sucesión. **2.** (47) Un número, una variable o un producto o cociente de números y variables.

transitive property of equality/propiedad transitiva de la igualdad (39) Para cualquiera de los números a, b y c, si $a = b$ y $b = c$, entonces $a = c$.

tree diagram/diagrama de árbol (413) Un diagrama de árbol es un diagrama que se usa para mostrar el número total de posibles resultados.

triangle/triángulo (163) Un triángulo es un polígono con tres lados y tres ángulos.

trigonometric ratios/razones trigonométricas (206) Para el triángulo rectángulo ABC con ángulo agudo A, el seno $A = \frac{a}{c}$, coseno $A = \frac{b}{c}$, tangente $A = \frac{a}{b}$.

trinomial/trinomio (514) La suma de tres monomios.

uniform motion/movimiento uniforme (235) Cuando un objeto se mueve a una velocidad, o a un ritmo constante, se dice que se mueve en movimiento uniforme.

union/unión (408) Para dos conjuntos A y B, el conjunto de los elementos contenidos en ambos A o B o A y B.

unique factorization theorem/teorema de la factorización única (558) La factorización prima de cada número es única excepto por el orden en que se escriben los factores.

unit cost/costo unitario (95) El costo de una unidad de un artículo.

upper quartile/cuartil superior (306) El cuartil superior divide la parte superior de un conjunto de datos en dos partes iguales.

use a model/usa un modelo (339) Una estrategia para resolver problemas que usa modelos, o simulaciones, de situaciones matemáticas difíciles de resolver directamente.

use a table/usa una tabla (255) Una estrategia para resolver problemas que usa tablas para organizar y resolver problemas.

variable/variable (6) Las variables son símbolos que se usan para representar números desconocidos.

Venn diagrams/diagramas de Venn (73) Los diagramas de Venn son diagramas que usan círculos u óvalos dentro de un rectángulo para mostrar relaciones de conjuntos.

vertex/vértice (610, 611) El punto máximo o mínimo de una parábola.

vertical axis/eje vertical (56) La recta vertical en una gráfica que representa la variable dependiente.

vertical line test/prueba de recta vertical (289) Si cualquier recta vertical pasa por un solo punto de la gráfica de una relación, entonces la relación es una función.

weighted average/promedio ponderado (233) El promedio ponderado M de un conjunto de datos es la suma del producto de cada número en el conjunto y su peso divididos entre la suma de todos los pesos.

whiskers/patillas (428) Las patillas de un diagrama de caja y patillas son los segmentos que se dibujan desde el cuartil inferior hasta el mínimo valor y desde el cuartil superior hasta el máximo valor.

whole numbers/números enteros (72) El conjunto de números enteros se representa por $\{0,1,2,3,...\}$.

work backward/trabaja al revés (156) Una estrategia para resolver problemas que usa operaciones inversas para determinar un valor inicial.

x*-axis/eje *x (254) La recta numérica horizontal en el plano de coordenadas.

x*-coordinate/coordenada *x (254) El primer número en un par ordenado.

x*-intercept/intersección con el eje *x (346) La coordenada x de un punto donde una gráfica interseca el eje x.

y*-axis/eje *y (254) La recta numérica vertical en el plano de coordenadas.

y*-coordinate/coordenada *y (254) El segundo número en un par ordenado.

y*-intercept/intersección con el eje *y (346) La coordenada y de un punto donde una gráfica interseca el eje y.

Z

zero exponent/exponente cero (502) Para cualquier número no cero a, $a^0 = 1$.

zero product property/propiedad del producto cero (594) Para todos los números a y b, si $ab = 0$, entonces $a = 0$, $b = 0$, o ambos a y b son iguales a cero.

zeros/ceros (620) Las raíces, o intersecciones con el eje x de la gráfica de una función.

SELECTED ANSWERS

CHAPTER 1 EXPLORING EXPRESSIONS, EQUATIONS, AND FUNCTIONS

Pages 9–11 Lesson 1-1
7. $3y^2 - 6$ **9.** 3 times x squared increased by 4 **11.** a^7 **13.** 32 **15.** $k + 20$ **17.** a^7 **19.** $\frac{2x^2}{3}$ or $\frac{2}{3}x^2$ **21.** $b + 8$ **23.** 4 times m to the fifth power **25.** c squared plus 23 **27.** 2 times 4 times 5 squared **29.** 8^2 **31.** 4^7 **33.** z^5 **35.** 49 **37.** 64 **39.** 16 **41.** $3(55 - w^3)$ **43.** $a + b + \frac{a}{b}$ **45.** $y + 10x$ **47.** 16; 16 **47a.** They are equal; no; no. **47b.** no; for example, $2^3 \neq 3^2$ **49a.** $3.5x$ **49b.** $3.5y$ **49c.** $3.5x + 3.5y$

Pages 15–18 Lesson 1-2
5.

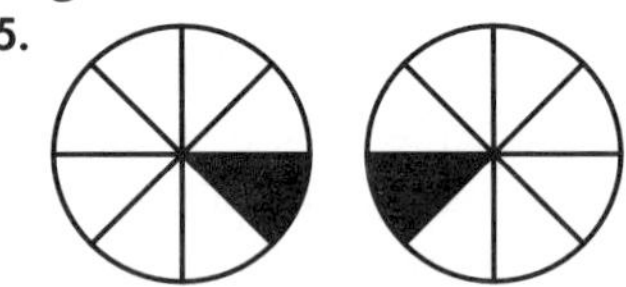

7. $5x + 1$, $6x + 1$

9a.

4^1	4^2	4^3	4^4	4^5
4	16	64	256	1024

9b. 6; $4^6 = 4096$ **9c.** 4; When the exponent is odd, 4 is in the ones place. **11.**

13. 48, 96 **15.** 25, 36 **17.** $a + 7$, $a + 9$

19a.

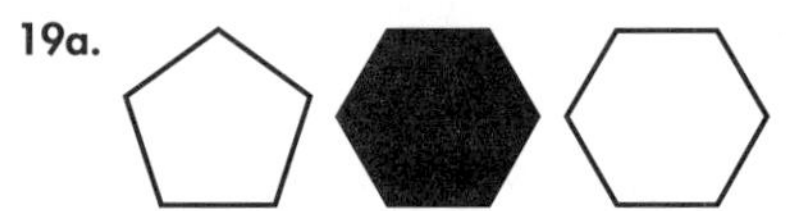

19b. White; even-numbered figures are white. **19c.** 12 sides; The shapes come in pairs. Since $19 \div 2 = 9.5$, the 19th figure will be part of the 10th pair. The 10th pair will have 10 + 2, or 12 sides. **21a.** 4; 9; 16; 25 **21b.** The sums are perfect square numbers; that is 1^2, 2^2, 3^2, 4^2, 5^2, ... **21c.** 10,000 **21d.** x^2 **23a.** 1,999,998; 2,999,997; 3,999,996; 4,999,995 **23b.** 8,999,991 **25.** 4, 7, 10, 13, 16 **27.** 1:13 P.M. **29a.** 100 cards **31.** x cubed divided by 9 **33.** 4^3; 64 cubes **35.** $x + \frac{1}{11}x$

Pages 22–24 Lesson 1-3
5. 173 **7.** 4 **9.** 25 **11.** $a^2 - a$ **13.** 14 **15.** 14 **17.** 9 **19.** 60 **21.** $\frac{11}{18}$ **23.** 6 **25.** 126 **27.** 147 **29.** 126 **31.** $r^2 + 3s$; 19 **33.** $(r + s)t^2$; $\frac{7}{4}$ **35.** 19.5 mm **37.** 22 in. **39.** 16.62 **41.** 3.77 **43a.** Sample answer: $(4 - 2)5 \div (3 + 2)$ **43b.** $4 \times 2 \times 5 \times 3 \times 2$ **45a.** $\frac{1}{3}Bh$ **45b.** 198,450 m^3 **47.** 16, 19.5 **49.** June 18 **51.** $t + 3$ **53.** 9 more than two times y

Pages 28–31 Lesson 1-4
5. stem 12, leaf 2 **7.** stem 126, leaf 9 **9a.** 35 tickets sold one day **9b.** 31 tickets **9c.** 75 tickets **9d.** 1043 tickets **11.** stem 13, leaf 3 **13.** stem 44, leaf 3 **15.** stem 111, leaf 3 **17.** stem 14, leaf 3 **19.** stem 111, leaf 4 **21.** rounded to the nearest hundred: 9, 12, 24, 27, 38, 39, 40 **23a.** Possible answer: Both teens and young adults had similar distribution of responses. **23b.** Sample answer: The market research shows that the new game is equally appealing to teens and young adults. Therefore, we should concentrate our marketing efforts towards both teens and young adults.

25a.

1980	Stem	1993
4	1	3
6	2	4
8 5 1	3	3 7 7
	4	6
1	5	

3|7 = 3700

25b. 3000 to 3900; 3000 to 3900 **25c.** Sample answer: Farm sizes seem to be shrinking. **27.** 15 **29a.** $\frac{s}{5}$ **29b.** 2 miles; yes **31.**

Pages 34–36 Lesson 1-5
7. false **9.** {1, 3, 5} **11.** 75 **13.** false **15.** true **17.** true **19.** {5} **21.** $\left\{\frac{1}{2}, \frac{3}{4}\right\}$ **23.** {10, 15, 20} **25.** 3 **27.** 2 **29.** $4\frac{5}{6}$ **31.** Sample answer: $p = 1$ and $q = 2$, $p = 2$ and $q = 10$, $p = 3$ and $q = 8$, $p = 4$ and $q = 20$, and $p = 5$ and $q = 15$ **33a.** $C = \frac{3500 \cdot 4}{14}$ **33b.** 1000 Calories **35.** 2, 5, 6, 7, 9 **37.** 8 **39.** 5 less than x to the fifth power

Page 36 Self Test
1. $3a + b^2$ **3.** 11:04, 11:08, 11:51, 11:55 **5.** 408

7.

Stem	Leaf
4	8
5	4
6	7
7	7
8	5 9

6|7 = 67

9. {6, 7, 8}

Pages 41–43 Lesson 1-6
7. $\frac{2}{9}$ **9.** c **11.** e **13.** d **15.** f

17. $6(12 - 48 \div 4) + 9 \cdot 1$
$= 6(12 - 12) + 9 \cdot 1$ *Substitution (=)*
$= 6(0) + 9 \cdot 1$ *Substitution (=)*
$= 0 + 9 \cdot 1$ *Multiplicative property of 0*
$= 0 + 9$ *Multiplicative identity*
$= 9$ *Additive identity*

19a. $4(20) + 7$
19b. $4(20) + 7$
$= 80 + 7$ *Substitution (=)*
$= 87$ *Substitution (=)*

19c. 87 years **21.** 9 **23.** $\frac{1}{p}$ **25.** $\frac{2}{3}$ **27.** substitution (=) **29.** multiplicative identity **31.** multiplicative inverse, multiplicative identity **33.** symmetric (=) **35.** reflexive (=) **37.** substitution (=); substitution (=); multiplicative identity; multiplicative inverse; substitution (=) **39.** substitution (=); substitution (=); substitution (=); multiplicative identity

41. $(15 - 8) \div 7 \cdot 25$
$= 7 \div 7 \cdot 25$ *Substitution(=)*
$= 1 \cdot 25$ *Substitution (=)*
$= 25$ *Multiplicative identity*

43. $(2^5 - 5^2) + (4^2 - 2^4)$
$= (32 - 25) + (16 - 16)$ *Substitution (=)*
$= 7 + 0$ *Substitution (=)*
$= 7$ *Additive identity*

45. $5^3 + 9\left(\frac{1}{3}\right)^2$
$= 125 + \frac{1}{9}9(\quad)$ *Substitution (=)*
$= 125 + 1$ *Multiplicative inverse*
$= 126$ *Substitution (=)*

47a. $[21(12 \cdot 2)] + [23(15 \cdot 2)] + [67(10 \cdot 2)]$
47b. $[21(12 \cdot 2)] + [23(15 \cdot 2)] + [67(10 \cdot 2)]$
$= [21(24)] + [23(30)] + 67(20)]$ *Substitution (=)*
$= 504 + 690 + 1340$ *Substitution (=)*
$= 2534$ *Substitution (=)*
47c. \$25.34 **49.** true **51.** false **55.** 36 **57.** $12y$

Page 44 Lesson 1–7A

1. $2x + 2$ **3.** $4x + 2$ **5.** false;

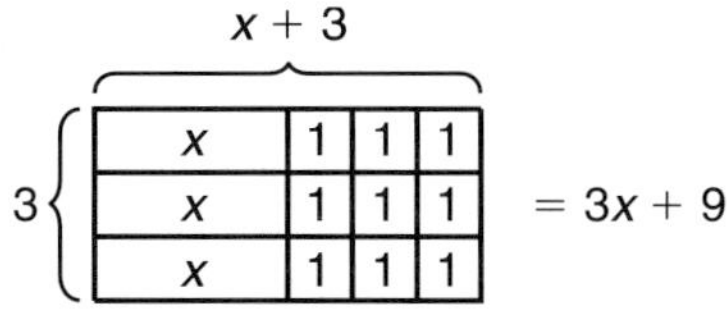

7a. Adita **7b.** Answers will vary. Answers should include the concept that multiplication by 3 is distributed over both terms in parentheses.

Pages 48–50 Lesson 1–7

5. b **7.** a **9.** c **11.** $2a - 2b$ **13.** 60 **15.** 7 **17.** $4y^4$, y^4 **19.** $3t^2 + 4t$ **21.** $23a^2b + 3ab^2$ **23.** $5.35(24) + 5.35(32)$; $5.35(24 + 32)$ **25.** $5g - 45$ **27.** $24m + 48$ **29.** $5a - ab$ **31.** 52 **33.** 60 **35.** 645 **37.** in simplest form **39.** $12a + 15b$ **41.** in simplest form **43.** $3x + 4y$ **45.** $1\frac{3}{5}a$ **47.** $9x + 5y$ **49.** no; sample counterexample: $2 + (4 \cdot 5) \neq (2 + 4)(2 + 5)$ **51a.** $2[x + (x + 14)] = 4x + 28$ **51b.** 96 **51c.** 527 ft^2 **53.** multiplicative property of 0 **55a.** $d = (1129)(2)$ **55b.** 2258 ft **57.** 2 **59.** 8 years

Pages 53–55 Lesson 1–8

7. commutative (+) **9.** associative (×) **11.** $5a + 2b$ **13.** $14x + 3y$

15. $6z^2 + (7 + z^2 + 6)$
$= 6z^2 + (z^2 + 7 + 6)$ *Commutative (+)*
$= (6z^2 + z^2) + (7 + 6)$ *Associative (+)*
$= (6 + 1)z^2 + (7 + 6)$ *Distributive property*
$= 7z^2 + 13$ *Substitution (=)*

17. multiplicative identity **19.** associative (×) **21.** distributive property **23.** commutative (×) **25.** associative (+) **27.** $10x + 5y$ **29.** $7x + 10y$ **31.** $10x + 2y$ **33.** $32a^2 + 16$ **35.** $5x$ **37.** $\frac{3}{4} + \frac{5}{3}m + \frac{4}{3}n$

39. $2(s + t) - s$
$= 2s + 2t - s$ *Distributive property*
$= 2t + 2s - s$ *Commutative (+)*
$= 2t + (2s - s)$ *Associative (+)*
$= 2t + (2s - 1s)$ *Multiplicative identity*
$= 2t + (2 - 1)s$ *Distributive property*
$= 2t + 1s$ *Substitution (=)*
$= 2t + s$ *Multiplicative identity*

41. $5xy + 3xy$
$= (5 + 3)xy$ *Distributive property*
$= 8xy$ *Substitution (=)*

43. $\frac{1}{100}$; Each denominator and the following numerator represent the number 1. The resulting expression is 1 in the numerator and 100 in the denominator multiplied by numerous 1s. Since 1 is the multiplicative identity, the product is $\frac{1}{100}$. **45.** no; Sample example: Let $a = 1$ and b = 2, then 1 * 2 = 1 + 2(2) or 5 and 2 * 1 = 2 + 2(1) or 4. **47a.** $G = 3.73$ **49.** $100d + 80d + 8d$, $188d$ **51.** $\frac{3}{2}$

53.

Stem	Leaf
10	0 0 0 1 1 1 1 3
9	5 7 7 9 9 9

10|3 = 103

55. ▼▲

Pages 59–62 Lesson 1–9

5a. Graph 3 **5b.** Graph 4 **5c.** Graph 2 **5d.** Graph 1 **7a.** False, *A* is the younger player, but *B* runs the mile in less time. **7b.** False, *A* made more 3-point shots, but *B* made more 2-point shots. **7c.** True, *B* is the older player and *B* made more 2-point shots. **7d.** True, *A* is the younger player and *A* made more free-throw shots. **9.** Graph a; An average person makes no money as a child, then his or her income rises for several years, and finally levels off. **11a.** Graph 5 **11b.** Graph 3 **11c.** Graph 1 **11d.** Graph 4 **11e.** Graph 6 **11f.** Graph 2

13.

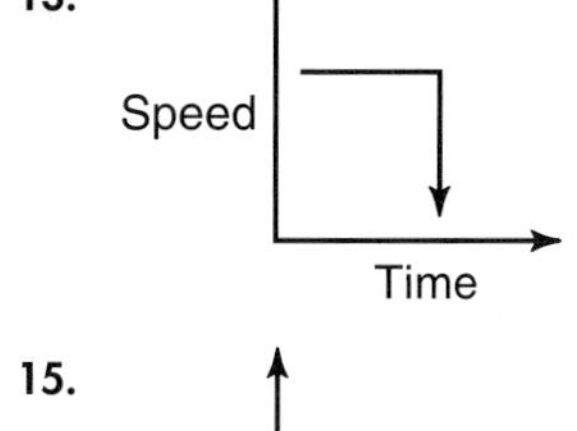

15.

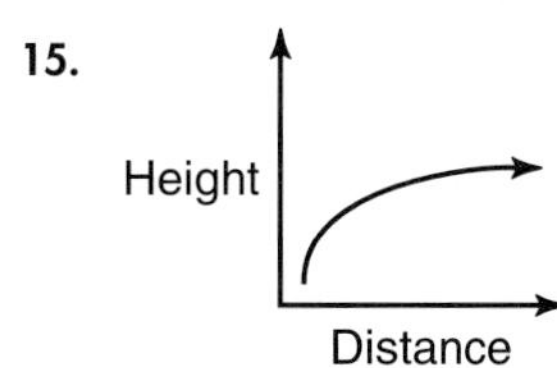

17a. horizontal axis: time in years; vertical axis: millions of dollars **17b.** In 1991 the telethon raised 45 million dollars. **17c.** Money raised in the telethon decreased, then increased steadily. **17d.** The money raised increased. **19.** $7p + 11q + 9$ **21.** 14 **23.** $\frac{1}{3}Bh$

Page 63 Chapter 1 Highlights

1. a **3.** e **5.** f **7.** g **9.** b

Pages 64–66 Chapter 1 Study Guide and Assessment

11. x^5 **13.** $5x^2$ **15.** three times a number *m* to the fifth power **17.** the difference of four times *m* squared and twice *m* **19.** 32, 64 **21.** $4x + y$, $5x + y$ **23.** 10^1, 10^2, 10^3, 10^4 **25.** 11 **27.** 9 **29.** 26 **31.** 2.4 **33.** 9.2 **35.** 90

37. No; some former presidents are still living. **39.** false **41.** true **43.** $13\frac{1}{2}$ **45.** $\frac{1}{3}$ **47.** 5 **49.** additive identity **51.** $2 \times 4 + 2 \times 7$ **53.** $1 - 3p$ **55.** 294 **57.** $8m + 8n$ **59.** commutative (+) **61.** commutative (×) **63.** $5x + 5y$ **65.** $3pq$ **67.** Graph c **69a.** $80s$ **69b.** 320 **69c.** 640 **71a.** $3 + 4 > 5, 4 + 5 > 3, 3 + 5 > 4$ **71b.** $x < 9$ ft and $x >$ 3 ft

CHAPTER 2 EXPLORING RATIONAL NUMBERS

Pages 75–77 Lesson 2–1

7. $\{-1, 0, 1, 2, \ldots\}$

9.

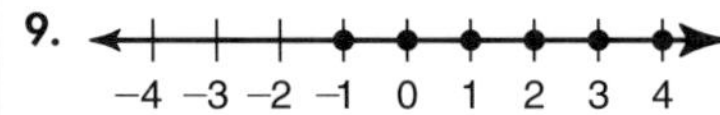

11. $-4 + (-3) = -7$ **13.** -5 **15.** 9-yard gain **17.** $\{-4, -3, -2, -1\}$ **19.** $\{-7, -3\}$ **21.** $\{\ldots, -5, -4, -3, -2, -1, 0\}$

23. 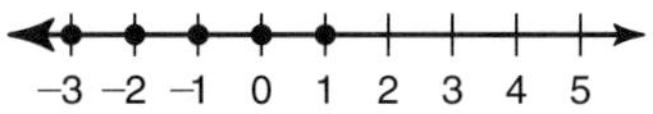 –3 –2 –1 0 1 2 3 4 5

25. 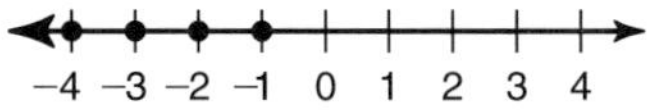 –4 –3 –2 –1 0 1 2 3 4

27. 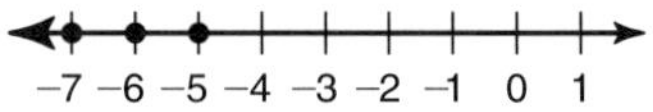 –7 –6 –5 –4 –3 –2 –1 0 1

29. 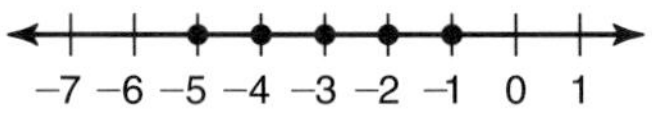 –7 –6 –5 –4 –3 –2 –1 0 1

31. 13 **33.** -5 **35.** -12 **37.** 6 **39.** -23 **41.** $-17 + 82 = 65$; 65°F **43a.** 4°F **43b.** -31°F **43c.** -15°F **43d.** 9°F

45.

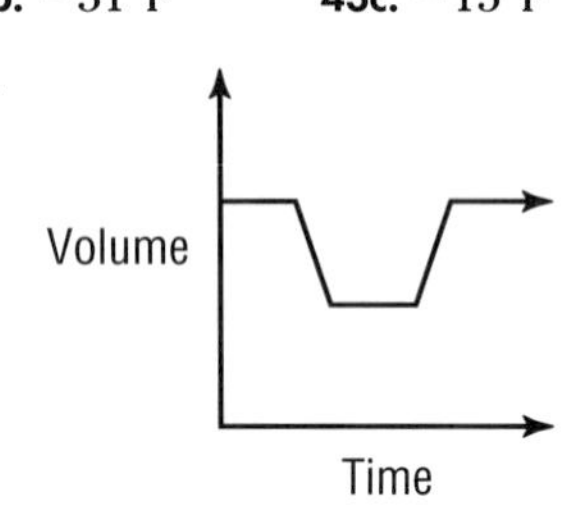

47. commutative (×) **49.** substitution (=) **51.** 13.9

Pages 80–83 Lesson 2–2

5. from 35 to 80;

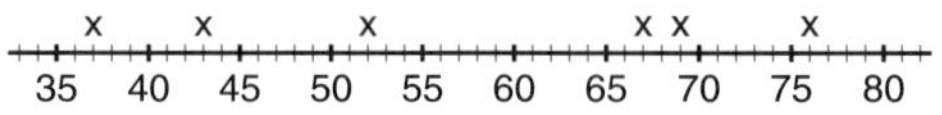

35 40 45 50 55 60 65 70 75 80

7a.

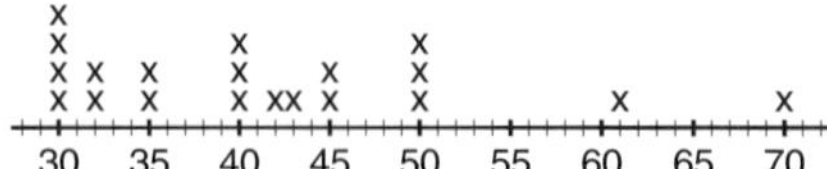

7b. 70 mph; cheetah **7c.** 30 mph **7d.** 30 mph **7e.** 12 **7f.** 4 **9a.** from 20 to 75 by 5;

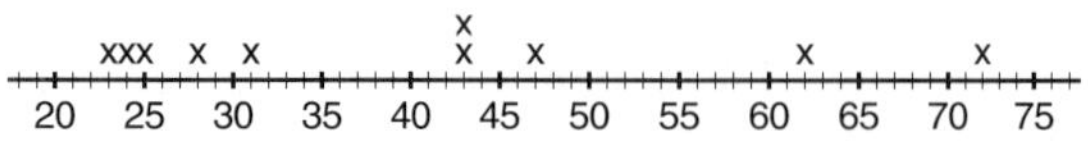

20 25 30 35 40 45 50 55 60 65 70 75

9b. yes; 25 and 42 **11a.** Ms. Martinez's **11b.** No; in Ms. Martinez's class, the hours that the students talked on the phone were close to each other (4 ± 2 hours). In Mr. Thomas' class, the students had a wider range of hours spent talking on the phone. The range was from 0 to 8 hours, with no hours clustered around a certain number.

13. Mr. Thomas', 3.6 hours; Ms. Martinez's, 3.7 hours; yes **15.** associative (×) **17.** 2

Page 84 Lesson 2–3A

1. 6 **3.** -2 **5.** 2 **7.** 6 **9.** false **11.** true **13.** Answers should include using the number line.

Pages 90–92 Lesson 2–3

7. $-7, 7$ **9.** 0, 0 **11.** -4 **13.** -11 **15.** 0 **17.** $40c$ **19.** -4 **21.** 3 **23.** $\begin{bmatrix} -1 & 4 \\ 6 & -6 \end{bmatrix}$ **25.** $-12, 12$ **27.** 302, 302 **29.** 0 **31.** -32 **33.** -16 **35.** -70 **37.** -15 **39.** -5 **41.** $40b$ **43.** $-29p$ **45.** $26d$ **47.** -4 **49.** -15 **51.** 5 **53.** 8 **55.** 99 **57.** -59 **59.** $\begin{bmatrix} 0 & 3 \\ 2 & 2 \end{bmatrix}$ **61.** $\begin{bmatrix} -4 & -5 \\ -4 & -5 \\ 10 & -3 \end{bmatrix}$ **63.** 24th floor

65a.

Monday	*Sesame*	*Poppy*	*Blue*	*Plain*
East Store	120	80	64	75
West Store	65	105	77	53

Tuesday	*Sesame*	*Poppy*	*Blue*	*Plain*
East Store	112	76	56	74
West Store	69	95	82	50

65b.

Monday + Tuesday	*Sesame*	*Poppy*	*Blue*	*Plain*
East Store	232	156	120	149
West Store	134	200	159	103

65c.

Monday − Tuesday	*Sesame*	*Poppy*	*Blue*	*Plain*
East Store	8	4	8	1
West Store	−4	10	−5	3

This matrix represents the difference between Monday's and Tuesday's sales in each category. **67a.** from 100 to 146

67b. 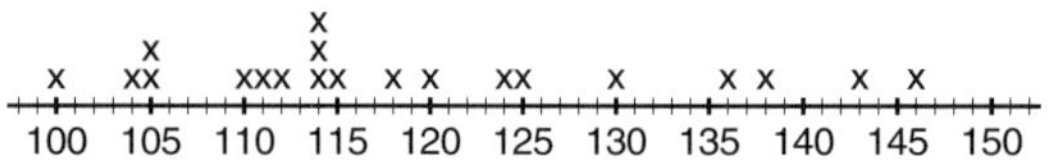 100 105 110 115 120 125 130 135 140 145 150

67c. yes; 114 **67d.** Siberian Elm; American Elm

69.

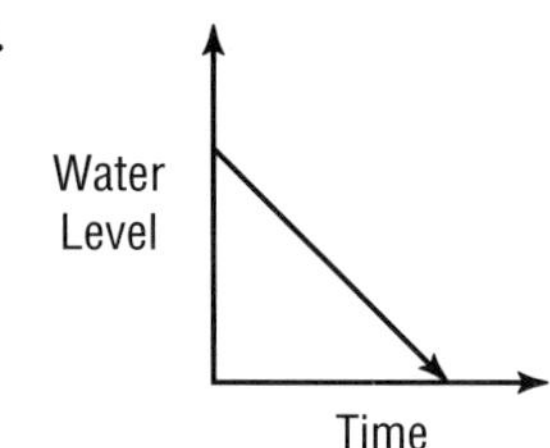

71. $55y^2$ **73.** symmetric (=) **75.** 14, 20, 29, 34, 37, 38, 43, 59, 64, 74, 84 **77.** 24

Pages 97–99 Lesson 2–4

7. $<$ **9.** $=$ **11.** $-0.5, \frac{3}{4}, \frac{7}{8}, 2.5$ **15.** $>$ **17.** $<$ **19.** $<$ **21.** $>$ **23.** $\frac{3}{8}, \frac{2}{3}, \frac{6}{7}$ **25.** $\frac{3}{23}, \frac{8}{42}, \frac{4}{14}$ **27.** $-\frac{2}{5}, -0.2, 0.2$ **29.** a 16-ounce drink for \$0.59 **31.** a package of 75 paper plates for \$3.29 **33.** Sample answer: $\frac{2}{3}$ **35.** Sample answer: $\frac{7}{20}$ **37.** Sample answer: $\frac{1}{6}$ **39.** $E = \frac{4}{14}$ or $\frac{2}{7}$; $G = \frac{10}{14}$ or $\frac{5}{7}$; $H = \frac{13}{14}$ **41.** 0.375 inch **43.** $-9, 9$

45. 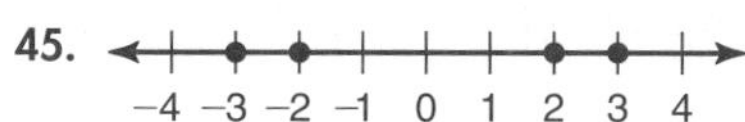

47. $m + 2n + \frac{3}{2}$ **49.** two times x squared plus six

Page 99 Self Test

1. 

3. -17 **5.** 55

7a. 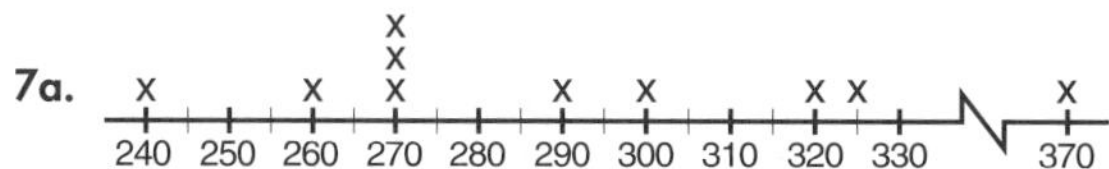

7b. yes; \$270 **9.** $<$

Pages 102–104 Lesson 2–5

5. $-\frac{1}{9}$ **7.** $-2\frac{5}{8}$ **9.** 0.88 **11.** 5.75 **13a.** $+\frac{1}{2}$

13b. 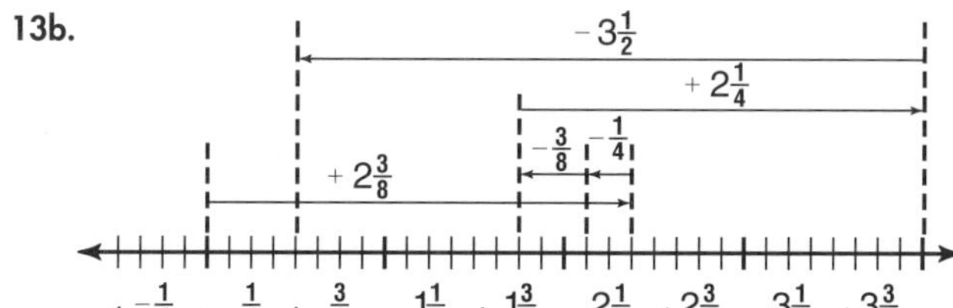

15. $-\frac{11}{16}$ **17.** -1.3 **19.** $\frac{4}{9}$ **21.** -0.2007 **23.** $-8\frac{5}{8}$

25. 0.0485 **27.** -22.94 **29.** -2.17 **31.** 1 **33.** -3.5

35. -16.7 **37.** $-\frac{52}{21}$ **39.** $\begin{bmatrix} -3.8 & -0.1 \\ 1.7 & 2.9 \end{bmatrix}$ **41.** $\begin{bmatrix} \frac{1}{4} & 11 \\ -5 & 8 \\ \frac{1}{2} & -12 \end{bmatrix}$

43. Answers will vary. Sample answer: $\begin{bmatrix} 5 & 3 \\ 1 & 7 \end{bmatrix}, \begin{bmatrix} 4 & -3 \\ 1 & 2 \end{bmatrix}, \begin{bmatrix} 9 & 0 \\ 2 & 9 \end{bmatrix}, \begin{bmatrix} 18 & 0 \\ 4 & 18 \end{bmatrix}; \begin{bmatrix} 5 & 3 \\ 1 & 7 \end{bmatrix} + \begin{bmatrix} 4 & -3 \\ 1 & 2 \end{bmatrix} = \begin{bmatrix} 9 & 0 \\ 2 & 9 \end{bmatrix} = \begin{bmatrix} 4 & -3 \\ 1 & 2 \end{bmatrix} + \begin{bmatrix} 5 & 3 \\ 1 & 7 \end{bmatrix}; \left(\begin{bmatrix} 5 & 3 \\ 1 & 7 \end{bmatrix} + \begin{bmatrix} 4 & -3 \\ 1 & 2 \end{bmatrix}\right) + \begin{bmatrix} 9 & 0 \\ 2 & 9 \end{bmatrix} = \begin{bmatrix} 18 & 0 \\ 4 & 18 \end{bmatrix} = \begin{bmatrix} 5 & 3 \\ 1 & 7 \end{bmatrix} + \left(\begin{bmatrix} 4 & -3 \\ 1 & 2 \end{bmatrix} + \begin{bmatrix} 9 & 0 \\ 2 & 9 \end{bmatrix}\right)$ **45a.** 9.01, 10.13, 11.25

45b. 1.12 **45c.** $-2, -\frac{5}{4}, -\frac{1}{2}, \frac{1}{4}, 1; -2\frac{1}{2}$ **47.** $<$

49a.

10 15 20 25 30 35 40 45 50 55 60

49b. no **51.** multiplicative property of 0 **53.** $\frac{7}{24}$

Page 105 Lesson 2–6A

3. -10 **5.** -10 **7.** 10

Pages 109–111 Lesson 2–6

5. -18 **7.** 24 **9.** $-\frac{4}{5}$ **11.** -4 **13.** $-46st$ **15.** $\begin{bmatrix} -6 & 12 \\ -3 & 15 \end{bmatrix}$

17. \$28.65 **19.** -60 **21.** -1 **23.** 2 **25.** 0.00879 **27.** $-\frac{6}{5}$

29. $-\frac{9}{17}$ **31.** 85.7095 **33.** 3 **35.** -6 **37.** $\frac{13}{12}$ **39.** $-\frac{179}{24}$

41. $\frac{25}{24}$ **43.** $-30rt + 4s$ **45.** $21x$ **47.** $16.48x - 5.3y$

49. $\begin{bmatrix} 2 & 6 & 3 \\ \frac{5}{2} & 5 & 1 \end{bmatrix}$ **51.** $\begin{bmatrix} -9 & 22.68 \\ -22.4 & -10 \\ 28.8 & 11.12 \end{bmatrix}$ **53.** $\begin{bmatrix} 6 & 18 & 4 \\ 0 & 2 & \frac{8}{3} \end{bmatrix}$

55. It is negative. **57a.** No; it meets the requirement that no dimensions can be less than 20 feet, but the minimum of 1250 square feet is not met because this yard would have 1216 square feet. **57b.** No; it meets the requirement that yards must have a minimum of 1250 square feet because it would have 1330 square feet, but the requirement that no dimension can be less than 20 feet is not met because it has a side of length 19 feet. **57c.** Answers will vary. Sample answer: 30 feet.
59. -2.2 **61.** -20 **63.** -7 **65.** $\{6, 7\}$ **67.** 408

Pages 115–117 Lesson 2–7

5. -4 **7.** $-\frac{3}{32}$ **9.** $-\frac{5}{48}$ **11.** $-6x$ **13.** about 2 minutes

15. 6 **17.** $-\frac{1}{18}$ **19.** $-\frac{1}{16}$ **21.** -6 **23.** $-\frac{1}{12}$ **25.** $\frac{243}{10}$

27. $-\frac{35}{3}$ **29.** $-65m$ **31.** $r + 3$ **33.** $-20a - 25b$

35. $-14c + 6d$ **37.** $-a - 4b$ **39.** -1.2 **41.** $-0.8\overline{3}$ **43.** 4

45a. 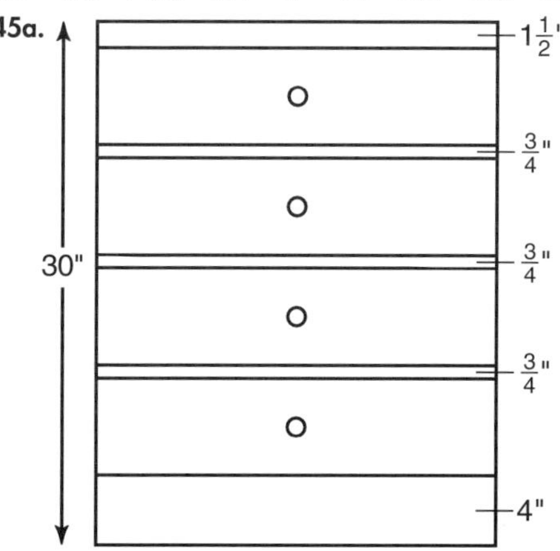

45b. $5\frac{9}{16}$ inches **47a.** 17.4 ± 0.161 cm **47b.** 0.161 cm

49. $\frac{13}{12}$ **51a.** $\begin{bmatrix} -2 & 4 \\ 3 & 12 \end{bmatrix}$ **51b.** $\begin{bmatrix} 4 & 4 \\ 7 & 2 \end{bmatrix}$ **53.** $20b + 24$

55a.

Stem	Leaf
13	0
15	4
17	2 3
19	4
20	0
22	0 2
24	5
25	1 4
27	9
29	0
30	4 7
33	0
34	6
52	5

$13|0 = 13{,}000$

55b. \$13,000 **55c.** \$52,500 **55d.** 9

Page 118 Lesson 2–8A

1. 4–5 **3.** 13–14 **5.** 1–2

Pages 123–125 Lesson 2–8

7. 8 **9.** 11.05 **11.** 16 **13.** Q **15.** Q

17. 2 3 4 5 6 7 8 9 10

19. −2 −1 0 1 2 3 4 5 6

21. 13 **23.** $\frac{2}{3}$ **25.** 20.49 **27.** 15 **29.** −25 **31.** $\frac{3}{5}$ **33.** 9.33 **35.** −47 **37.** −20 **39.** Z, Q **41.** Q **43.** Q **45.** Q **47.** I **49.** Q **51.**

−6 −5 −4 −3 −2 −1 0 1 2

53. −7 −6 −5 −4 −3 −2 −1 0 1

55. −15 −14 −13 −12 −11 −10 −9 −8 −7

57. −5.6 −5.5 −5.4 −5.3 −5.2 −5.1 −5.0 −4.9 −4.8

59. $4\frac{1}{2}$ $4\frac{3}{4}$ 5 $5\frac{1}{4}$ $5\frac{1}{2}$ $5\frac{3}{4}$ 6 $6\frac{1}{4}$ $6\frac{1}{2}$

61. Yes; it lies between 27 and 28. **63a.** $d = 1.4\sqrt{1275}$; Estimates should be close to 50. **63b.** 49.989999 **65.** Fibonacci **67.** −12 **69.** a 1-pound package of lunch meat for $1.95 **71.** from 500 to 520 by 5s;

500 505 510 515 520

73. b **75a.** to the nearest ten thousand **75b.** 10 **75c.** $690,000 **77.** 26, 32

Pages 130–132 Lesson 2–9

5a. 4 points **5b.** 15 questions **5c.** 86 **5d.** 6 points **5e.** $15 - n$ **5f.** $4n$ **7.** $n + 5 \geq 48$ **9.** Let s = average speed; $s = \frac{189}{3} = 63$. **11.** The product of a times the sum of y and 1 is b. **13a.** 640 feet **13b.** $y - 80$ **13c.** $\ell + \ell + w + w$ or $2\ell + 2w$ **13d.** 200 ft **13e.** 24,000 ft^2 **13f.** Is there enough space to build the playground as designed? **15.** $a^2 + b^3 = 25$ **17.** $x < (a - 4)^2$ **19.** $\frac{7}{8}(a + b + c^2) = 48$

21. Sample answer: x = first number
$x + 25$ = second number
$25 + 2x = 106$

23. Sample answer:
y = how much money Hector saves
$y = 2(7)(50)$

25. V is equal to the quotient of a times h and three. **27.** Sample answer: Three times the number of miles home is from school is 36. How many miles away from home is the school? **29.** Sample answer: The number of hours spent waiting for a doctor in the waiting room is 5 minutes more than 6 times the number of appointments she has. How long would you have to wait for a doctor with d appointments?

31. $ab + cd + 2d - 8$

33. Sample answer: $4.5 + x + x = 24 + x$
$x = 19.5$

35a. $A = \frac{1}{2}h(a + b)$ **35b.** $A = 108\frac{1}{2}$ ft^2 **37.** −8

39. $\begin{bmatrix} 8 & -1.6 \\ 12 & -20 \end{bmatrix}$ **41.** −25, 25 **43.** {2, 3, 4, 5}

Page 133 Chapter 2 Highlights

1. true **3.** true **5.** false, sample answer: $\frac{\frac{1}{2}}{5}$ **7.** true **9.** true

Pages 134–136 Chapter 2 Study Guide and Assessment

11. −4 −3 −2 −1 0 1 2 3 4 5 6

13. −3 −2 −1 0 1 2 3 4 5

15. −5 **17.** −4 **19.** −5 **21.** 56% **23.** 8 **25.** −2 **27.** 4 **29.** −5 **31.** > **33.** < **35.** Sample answer: 0 **37.** −1 **39.** 12.37 **41.** −13.26 **43.** −99 **45.** $-\frac{3}{7}$

47. −90 **49.** −9 **51.** $\frac{-4}{35}$ **53.** −109 **55.** 14 **57.** −30

59a.

5 10 15 20 25 30

59b. 14 detergents **61.** 1.25 liters of soda for $1.31 **63a.** They both saw a gain of $\frac{1}{4}$ for the week. **63b.** CompNet **63c.** It changed $-\frac{3}{4}$ from Wednesday to Thursday. **65.** Let n = the number; $3n - 21 = 57$.

CHAPTER 3 SOLVING LINEAR EQUATIONS

Page 143 Lesson 3–1A

1. 1 **3.** −11 **5.** 8 **7.** 4 **9.** no **11.** yes

Pages 148–149 Lesson 3–1

7. −17 **9.** −12 **11.** −14 **13.** $n + (-56) = -82$; −26 **15.** −9 **17.** −1.3 **19.** 1.4 **21.** −3 **23.** −6 **25.** −7 **27.** 24 **29.** $1\frac{1}{4}$ **31.** $-\frac{1}{10}$ **33.** $n + 5 = 34$; 29 **35.** $n - (-23) = 35$; 12 **37.** $n + (-35) = 98$; 133 **39a.** $203{,}600 + s = 19{,}300{,}000$; 19,096,400 subscribers **39b.** Sample answer: 25 million **41a.** 24 years old **41b.** 24 years **41c.** 78 years **41d.** 34 years old **41e.** 24 years **43.** −5 **45.** $-0.73x$ **47.** 48

Pages 152–154 Lesson 3–2

7. 17 **9.** 48 **11.** $\frac{11}{15}$ **13.** $-7n = 1.477$; −0.211 **15.** −9 **17.** −17.33 or $-17\frac{1}{3}$ **19.** −14 **21.** 12.81 **23.** 600 **25.** −275 **27.** −25 **29.** $8\frac{6}{13}$ **31.** −8 **33.** $-12n = -156$; 13 **35.** $\frac{4}{3}n = 4.82$; 3.615 **37.** 45 **39.** 30 **41.** 12 **43a.** $7\ell = 350$, 50 people; $7\ell = 583$, ≈ 83 people **43b.** $\frac{p}{7} = 65$; 455 people **45a.** $0.10m = 2.30$; 23 minutes **45b.** $0.10(18) = c$; $1.80 **47.** $17 + 1\frac{1}{2}y = 33\frac{1}{2}$; 11 years **49.** $\frac{19}{28}$ **51a.** The number of bags increased during the day. **51b.** The machine was refilled. **53.** 31

Page 155 Lesson 3–3A

1. 5 **3.** −2 **5.** 2 **7.** 4 **9.** −2

Pages 159–161 Lesson 3–3

9. −6 **11.** 35 **13.** −17 **15.** $n + (n + 2) + (n + 4) = 21$;

5, 7, 9 **17.** −1 **19.** −2 **21.** −42.72 **23.** $12\frac{2}{3}$ **25.** −6 **27.** −48 **29.** 30 **31.** 57 **33.** $4\frac{3}{7}$ **35.** $12 - 2n = -7$; $9\frac{1}{2}$ **37.** $n + (n + 1) + (n + 2) + (n + 3) = 86$; 20, 21, 22, 23 **39.** $n + (n + 2) + (n + 4) = 39$; 11 m, 13 m, 15 m **41.** 28 **43.** −38 **45.** $(3n - 1) + (3n + 1) + (3n + 3)$ **47.** $2a + 2 = 26$; 12 letters **51.** I **53.** −5 **55.** 5

Page 161 Self Test

1. 44 **3.** 6 **5.** $-8\frac{1}{3}$ **7.** $23 - n = 42$; −19 **9.** $n + (n + 2) = 126$; 62, 64

Pages 165–167 Lesson 3–4

7. 79°, 169° **9.** $(90 - 3x)°$, $(180 - 3x)°$ **11.** $(110 - x)°$, $(200 - x)°$ **13.** 85° **15.** $(c - 38) + c = 90$; 26°, 64° **17.** 48°, 138° **19.** none, 55° **21.** 69°, 159° **23.** none, 81° **25.** $(90 - 3a)°$, $(180 - 3a)°$ **27.** $(128 - b)°$, $(218 - b)°$ **29.** 70° **31.** 105° **33.** 138° **35.** $(190 - 2p)°$ **37.** 45° **39.** $x + (x - 30) = 180$; 75° **41.** $x + (30 + 3x) = 90$; 15°, 75° **43.** $x + 3x + 4x = 180$; 22.5°, 67.5°, 90° **45.** 60° **47.** 40° **49a.** \$80 **49b.** \$74.75 **51.** $7t$

53a.

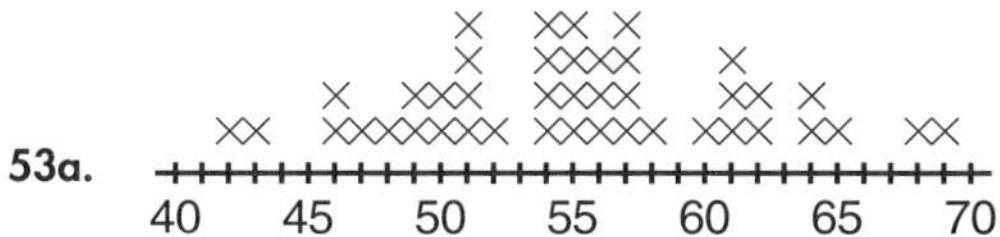

53b. yes; 51, 54, 55, and 57

55.

Stem	Leaf
4	2 3 6 6 7 8 9 9
5	0 0 1 1 1 1 2 4
	4 4 4 5 5 5 5 6
	6 6 7 7 7 7 8
6	0 1 1 1 2 2 4 4
	5 8 9

5 | 2 = 52

Pages 170–172 Lesson 3–5

5. $-\frac{1}{2}$; Add $4x$ to each side, subtract 10 from each side, then divide each side by 14. **7.** −25; Subtract $\frac{1}{5}$ from each side, subtract 3 from each side, then multiply by $\frac{5}{2}$. **9.** all numbers **11.** $2(n + 2) = 3n - 13$; 17, 19 **13.** $\frac{1}{2}n + 16 = \frac{2}{3}n - 4$; 120 **15.** no solution **17.** −2 **19.** 5.6 **21.** −16 **23.** no solution **25.** 4 **27.** 2.6 or $2\frac{3}{5}$ **29.** 8 **31.** 2 **33.** all numbers **35.** $\frac{1}{5}n + 5n = 7n - 18$; 10 **37.** $n + (n + 2) + (n + 4) = 180$; 58°, 60°, 62° **39a.** $4x + 6 = 4x + 6$; identity **39b.** $5x - 7 = x + 3$; 2.5 **39c.** $-3x + 6 = 3x - 6$; 2 **39d.** $5.4x + 6.8 = 4.6x + 2.8$; −5 **39e.** $2x - 8 = 2x - 6$; no solution **41a.** 6.827 years **41b.** Air conditioner sales are decreasing while fan sales are increasing. **43.** 148° **45.** −2.7 **47.** $\frac{11}{9}$ **49.** 11.05

Pages 175–177 Lesson 3–6

5. $\frac{3x - 7}{4}$ **7.** $\frac{c - b}{2}$ **9.** $\frac{2S - nt}{n}$ **11.** $\frac{b}{9}$ **13.** $3c - a$ **15.** $\frac{v - r}{t}$ **17.** $\frac{I}{pt}$ **19.** $\frac{H}{0.24I^2t}$ **21.** $\frac{-t + 5}{4}$ **23.** $\frac{4}{3}(c - b)$ **25.** $\frac{5x + y}{2}$ **27.** $2x + 12 = 3y - 31$; $\frac{3y - 43}{2}$ **29.** $\frac{2}{3}x + 5 = \frac{1}{2}y - 3$; $\frac{3}{4}y - 12$ **31.** \$900 **33a.** 36 words per minute **33b.** Clarence, with 74 words per minute **35.** 5° **37.** $x \neq 2$ **39.** {8}

Pages 181–183 Lesson 3–7

11. 8.2; 8; 8 **13.** Sample answer: 20, 20, 30, 50, 90, 90 **15a.** 3.857; 3; 3 **15b.** mean **17.** 96.8; 50; none **19.** 9; 9; none **21.** 5.69; 4.56; none **23.** 212.94; 218; 219 **25.** 14 **27a.** mean **27b.** Yes, because the article indicates that the median value is in the middle of the data set. **27c.** No, he should have used the term *mean* since the mean can be affected by extremely high values, causing the majority to be below the mean value. **29.** $1660\frac{3}{5}$; $1076\frac{1}{2}$; none **31.** 37.5 mph **33.** 16 **35.** −20 **37.** $5 + 9ac + 14b$

Page 185 Chapter 3 Highlights

1. b **3.** i **5.** d **7.** k **9.** a **11.** g

Pages 186–188 Chapter 3 Study Guide and Assessment

13. −16 **15.** 21 **17.** −10 **19.** −32 **21.** −116 **23.** 7 **25.** $83\frac{1}{3}$ **27.** 10 **29.** −16 **31.** 3 **33.** −153 **35.** 11 **37.** 50 **39.** 16, 17 **41.** 21°; 111° **43.** $(70 - y)°$; $(160 - y)°$ **45.** 30° **47.** 15° **49.** $180° - (z + z - 30°)$ or $(210° - 2z)$ **51.** −30 **53.** 2 **55.** $\frac{y}{5}$ **57.** $\frac{a}{y - c}$ **59.** 21; 21; 21 **61a.** 25.5 years **63.** $13\frac{3}{4}$ years

CHAPTER 4 USING PROPORTIONAL REASONING

Pages 198–200 Lesson 4–1

5. = **7.** 12 **9.** 5 **11.** 4.62 **13.** 9.5 gallons **15.** = **17.** ≠ **19.** = **21.** 9 **23.** $\frac{8}{5}$; 1.6 **25.** 0.84 **27.** $-\frac{149}{6}$; −24.8 **29.** 11 **31.** 2.28 **33.** 1.251

35a.

Louis' age	1	2	3	6	10	20	30
Mariah's age	9	**10**	**11**	**14**	**18**	**28**	**38**

35b. 9, 5, $3.\overline{6}$, $2.\overline{3}$, 1.8, 1.4, $1.2\overline{6}$ **35c.** $r = \frac{y + 8}{y}$ **35d.** The ratio gets smaller. **35e.** No; if the ratio equaled 1, Mariah and Louis would be the same age. **37.** 85 movies **39.** 23; 19; 18 **41.** $4x - 2x = 100$; 50 **43a.** 93 **43b.** \$9695 **43c.** more than 40 **43d.** $93 - p$ **43e.** yes **45.** 10

Pages 203–205 Lesson 4–2

5. ΔDEF **7.** yes **9.** $\ell = 12$, $m = 6$ **11.** 27 feet high **13.** ΔDFE **15.** no **17.** yes **19.** no **21.** $a = \frac{55}{6}$, $b = \frac{22}{3}$ **23.** $d = \frac{51}{5}$, $c = 9$ **25.** $a = 2.78$, $c = 4.24$ **27.** $b = 16.2$, $d = 6.3$ **29.** $c = \frac{7}{2}$, $d = \frac{17}{8}$ **31.** Sample answer:

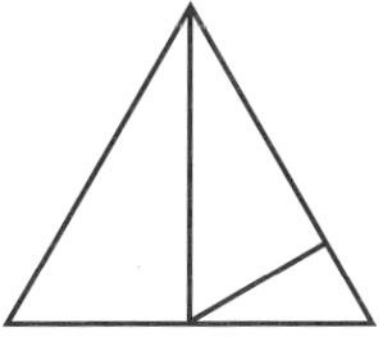

33a. $\frac{3}{9} = \frac{d}{d + 39}$; $d = 19.5$ ft; yes **33b.** When the ball is

served from 8 feet, $d = 23.4$ ft, and the serve is a fault. When the ball is served from 10 feet, $d = 16.7$ ft. Therefore, tall players have an easier time serving. **35.** $16\frac{1}{2}$ feet by 21 feet **37.** 160 **39.** 60 million; 420 million
41. 23;
$(19 - 12) \div 7 \cdot 23$
$= 7 \div 7 \cdot 23$ *Substitution (=)*
$= 1 \cdot 23$ *Substitution (=)*
$= 23$ *Multiplicative identity*

Pages 211–214 Lesson 4–3

7. $\sin Y = 0.600$, $\cos Y = 0.800$, $\tan Y = 0.750$ **9.** 0.8192 **11.** 0.1228 **13.** 46° **15.** 36° **17.** $m\angle B = 50°$, $AC = 12.3$ m, $BC = 10.3$ m **19.** $m\angle A = 30°$, $AC = 13.9$ m, $BC = 8$ m **21.** $\sin G = 0.6$, $\cos G = 0.8$, $\tan G = 0.75$ **23.** $\sin G = 0.471$, $\cos G = 0.882$, $\tan G = 0.533$ **25.** $\sin G = 0.923$, $\cos G = 0.385$, $\tan G = 2.4$ **27.** 0.3584 **29.** 0.9703 **31.** 0.3746 **33.** 33° **35.** 22° **37.** 62° **39.** 77° **41.** 58° **43.** 18° **45.** $m\angle B = 60°$, $AC = 12.1$ m, $BC = 7$ m **47.** $m\angle A = 45°$, $AC = 6$ ft, $AB = 8.5$ ft **49.** $m\angle A = 63°$, $AC = 9.1$ in., $BC = 17.8$ in. **51.** $m\angle A = 23°$, $m\angle B = 67°$, $AB \approx 12.8$ ft **53.** $m\angle A = 30°$, $m\angle B = 60°$, $AC = 5.2$ cm **55a.** true; $\sin C = \frac{c}{a}$, $\cos B = \frac{c}{a}$ **55b.** $\cos B = \frac{c}{a}$, $\frac{1}{\sin B} = \frac{a}{b}$ **55c.** $\tan C = \frac{c}{b}$, $\frac{\cos C}{\sin C} = \frac{b}{a} \div \frac{c}{a} = \frac{b}{c}$ **55d.** true; $\tan C = \frac{c}{b}$, $\frac{\sin C}{\cos C} = \frac{c}{a} \div \frac{b}{a} = \frac{c}{b}$ **55e.** true; $\sin B = \frac{b}{a}$, $(\tan B)(\cos B) = \frac{b}{c} \cdot \frac{c}{a} = \frac{b}{a}$ **57.** 2229 ft **59.** 3 **61.** 10 **63.** N, W, Z, Q **65.** $\frac{2}{15}$ **67.** $9x + 2y$ **69.** 125

Pages 218–221 Lesson 4–4

7. 43%, 0.43 **9.** 55% **11.** 15 **13.** \$52.50 **15.** \$1200 at 10%, \$6000 at 14% **17.** 30%, 0.30 **19.** 70%, 0.70 **21.** 180%, 1.8 **23.** $62\frac{1}{2}$%; 0.625 **25.** 50% **27.** $12\frac{1}{2}$%, 12.5% **29.** 400% **31.** 2.5% **33.** 72.3 **35.** 702.4 **37.** 28% **39.** 25.92 **41.** 12% **43.** \$3125 **45.** \$2400 **47.** 39%, 34%, 15%, 12% **49.** 11.5% **51.** \$5.45 **53a.** about 54 people; about 23 people **53b.** Respondents could choose more than one entree. **55.** about 277 feet **57.** $\frac{77}{4}$; 19.25 **59.** −4.7 **61.** $-28y$ **63.** 2

Page 221 Self Test

1. 5 **3.** 5 **5.** $m\angle B = 34°$, $b \approx 11.5$, $c \approx 20.5$ **7.** no **9.** $62\frac{1}{2}$%, 62.5%

Pages 225–227 Lesson 4–5

5. I; 40% **7.** D; 50% **9.** \$85.85 **11.** \$196 **13.** I; 69% **15.** D; 55% **17.** D; 27% **19.** I; 2% **21.** \$233.24 **23.** \$19.90 **25.** \$85.39 **27.** \$16.59 **29.** \$30.09 **31.** no **33.** 133% **35.** 8.6% **37.** an increase of more than 22% **39.** 11% **41.** 65 ft **43.** 11 **45.** heavy, 571; light, 286

Pages 230–232 Lesson 4–6

7. $\frac{1}{2}$ **9.** $\frac{2}{3}$ **11.** 5:1 **13.** $\frac{1}{2}$ **15.** $\frac{1}{2}$ **17.** $\frac{1}{2}$ **19.** $\frac{2}{11}$ **21.** 0 **23.** $\frac{8}{11}$ **25.** 1:5 **27.** 4:11 **29.** 23:7 **31.** $\frac{1}{2}$ **33.** 1:3 **35.** $\frac{8}{13}$ **37.** $\frac{1}{7}$ **39a.** 1:14 **39b.** $\frac{1}{27}$ **41a.** $\frac{3}{26}$ **41b.** $\frac{4}{13}$ **41c.** 3:23 **43.** 9° **45.** \$50 **47.** $\frac{4}{21}$ **49.** 75

Pages 236–238 Lesson 4–7

5. 226 dozen chocolate chip; 311 dozen peanut butter **7.** 11:30 A.M. **9.** 5 quarters **11.** 3 hours **13.** 2.5 hours **15.** 46 mph **17.** 67 mph **19a.** 7 **19b.** 16 **21.** Sample answer: How much pure antifreeze must be mixed with a 20% solution to produce 40 quarts of a 28% solution? **23.** 0 **25.** \$11,000 **27.** 49 **29.** associative (+)

Pages 243–244 Lesson 4–8

5. I, 5 **7.** 10 **9.** 99 **11.** 20 inches from the 8-ounce weight, 16 inches from the 10-ounce weight **13.** I, 15 **15.** I, 9 **17.** D, 4 **19.** $26\frac{1}{4}$ **21.** −8 **23.** 12 **25.** 6.075 **27.** 8.3875 **29.** $\frac{1}{4}$ **31.** 18 pounds **33.** 1.6 feet **35.** 2:1 **37.** \$4300 at 5%, \$7400 at 7% **39.** −12

Page 245 Chapter 4 Highlights

1. angle of elevation **3.** equal, proportional **5.** legs **7.** odds **9.** inverse variation

Pages 246–248 Chapter 4 Study Guide and Assessment

11. = **13.** ≠ **15.** 18 **17.** 16 **19.** $d = \frac{45}{8}$, $e = \frac{27}{4}$ **21.** $b = \frac{44}{3}$, $d = 6$ **23.** 0.528 **25.** 0.849 **27.** 1.607 **29.** 39° **31.** 80° **33.** $m\angle B = 45°$, $AB = 8.5$, $BC = 6$ **35.** $m\angle A \approx 66°$, $AB \approx 9.8$, $m\angle B \approx 24°$ **37.** 48 **39.** 87.5% **41.** 0.1881 **43.** decrease, 13% **45.** increase, 6% **47.** decrease, 10% **49.** \$85.66 **51.** \$9272.23 **53.** $\frac{1}{12}$ **55.** $\frac{5}{6}$ **57.** 10:39 **59.** 34:15 **61.** 21 **63.** 6 **65.** 21 **67.** 48 **69.** 450 mph and 530 mph

CHAPTER 5 GRAPHING RELATIONS AND FUNCTIONS

Pages 257–259 Lesson 5–1

5. (−3, −1); III **7.** (0, −2); none
8–11.

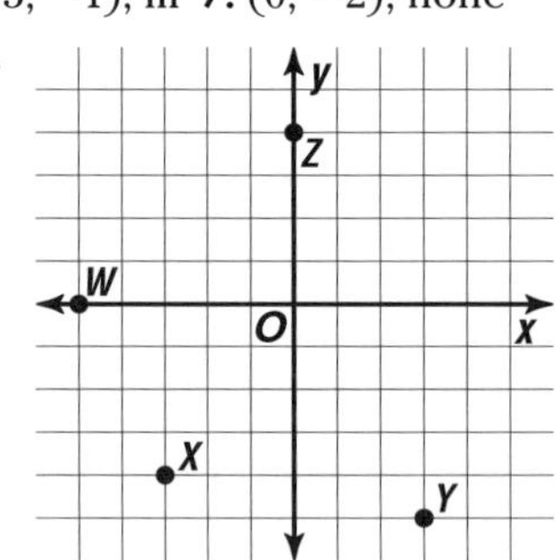

13. (−1, −3); III **15.** (0, 3); none **17.** (0, 0); none **19.** (3, −2); IV **21.** (2, 2); I **23.** (−5, 4); II
25–36.

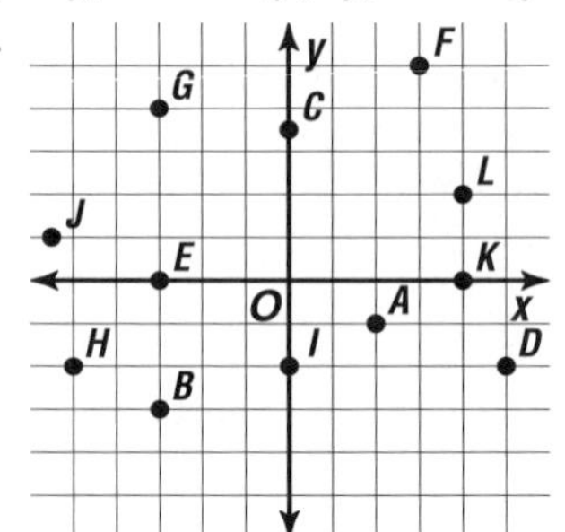

37.

41a. New Orleans **41b.** Oregon **41c.** Sample answer: (75°, 40°) **41d.** Honolulu, Hawaii **41f.** Sample answer: The longitude lines are not the same distance apart; they meet at the poles. **45.** $3\frac{1}{2}$ hours **47.** 4; 3; none **49.** $-a - 5$ **51.** $31a + 21b$

Page 261 Lesson 5–2A

1–6.

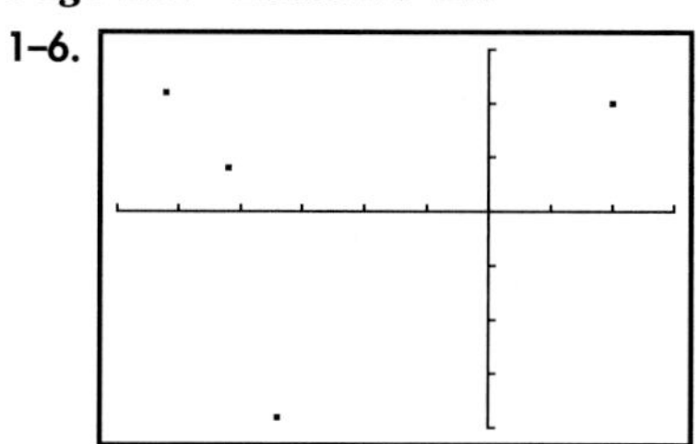

7. They are points on the axes.

Pages 266–269 Lesson 5–2

5. D = {0, 1, 2}; R = {2, −2, 4} **7.** {(1, 3), (2, 4), (3, 5), (5, 7)}; D = {1, 2, 3, 5}; R = {3, 4, 5, 7}; Inv = {(3, 1), (4, 2), (5, 3), (7, 5)} **9.** {(1, 3), (2, 2), (4, 9), (6, 5)}; D = {1, 2, 4, 6}; R = {2, 3, 5, 9}; Inv = {(3, 1), (2, 2), (9, 4), (5, 6)} **11.** {(−2, 2), (−1, 1), (0, 1), (1, 1), (1, −1), (2, −1), (3, 1)}; D = {−2, −1, 0, 1, 2, 3}; R = {−1, 1, 2}; Inv = {(2, −2), (1, −1), (1, 0), (1, 1), (−1, 1), (−1, 2), (1, 3)}

13.

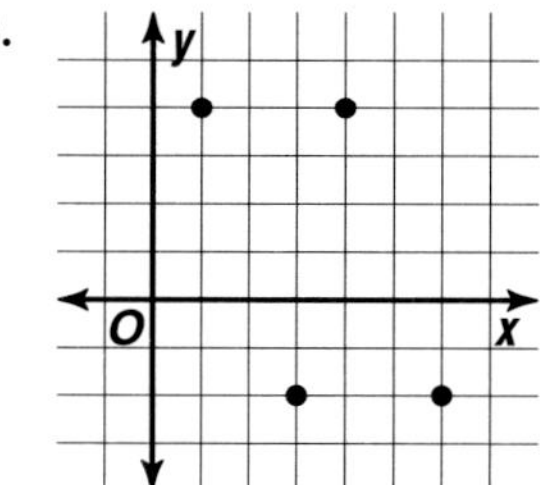

15b. 5.6% **15c.** Except for the first 6 months of 1992, the unemployment rate seems to be decreasing. **17.** D = {−5, −2, 1, 3}; R = {7} **19.** D = $\left\{-5\frac{1}{4}, -3, \frac{1}{2},\right.$

(1, −2)}; D = {1, 3, 4, 6}; R ={−2, 4}; Inv = {(4, 6), (−2, 4), (4, 3), (−2, 1)} **23.** {(6, 0), (−3, 5), (2, −2), (−3, 3)}; D = {−3, 2, 6}, R = {−2, 0, 3, 5}; Inv = {(0, 6), (5, −3), (−2, 2), (3, −3)} **25.** (3, 4), (3, 2), (2, 9), (5, 4), (5, 8), (−7, 2)}; D = {−7, 2, 3, 5}; R = {2, 4, 8, 9}; Inv = {(4, 3), (2, 3), (9, 2), (4, 5), (8, 5), (2, −7)} **27.** {(0, 25), (1, 50), (2, 75), (3, 100)}; D = {0, 1, 2, 3}; R = {25, 50, 75, 100}; Inv = {(25, 0), (50, 1), (75, 2), (100, 3)} **29.** {(−3, 4), (−2, 2), (−1, −2), (2, 2)}; D = {−3, −2, −1, 2}; R = {−2, 2, 4}; Inv = {(4, −3), (2, −2), (−2, −1), (2, 2)} **31.** {(−3, 3), (−3, −3), (3, 3), (3, −3), (0, 0)}; D = {−3, 0, 3}; R = {−3, 0, 3}; Inv = {(3, −3), (−3, −3), (3, 3), (−3, 3), (0, 0)} **33.** {(−3, 1), (−1, 1), (2, 1), (3, 1), (4, 1)}; D = {−3, −1, 2, 3, 4}; R = {1}; Inv = {(1, −3), (1, −1), (1, 2), (1, 3), (1, 4)}

35.

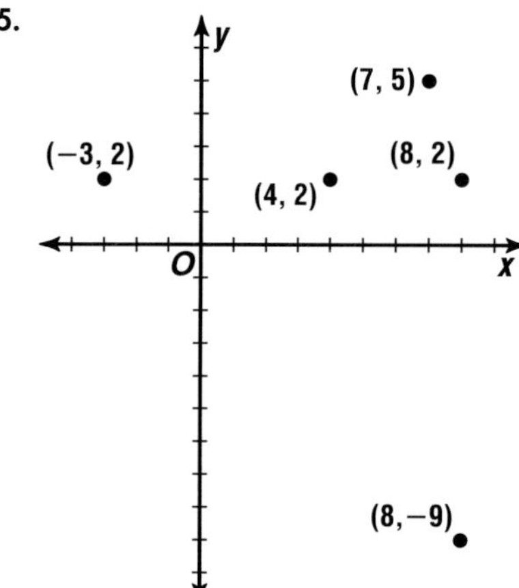

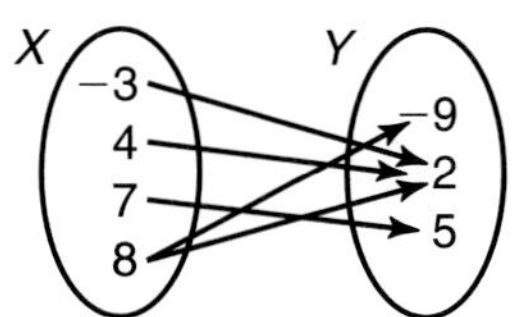

37.

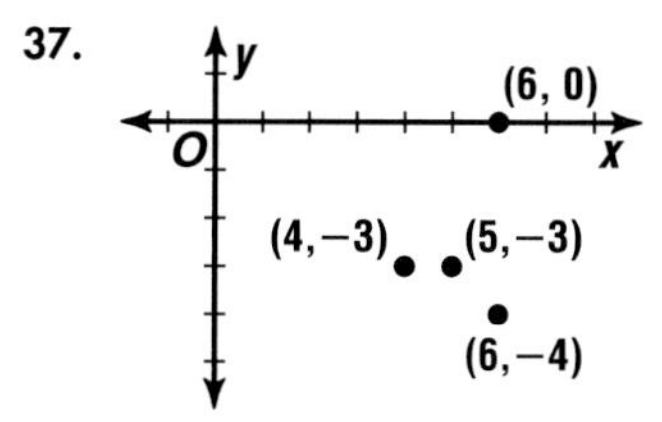

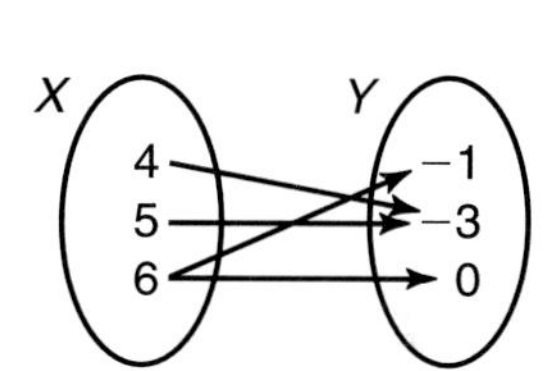

39a. [−10, 10] by [−10, 10]

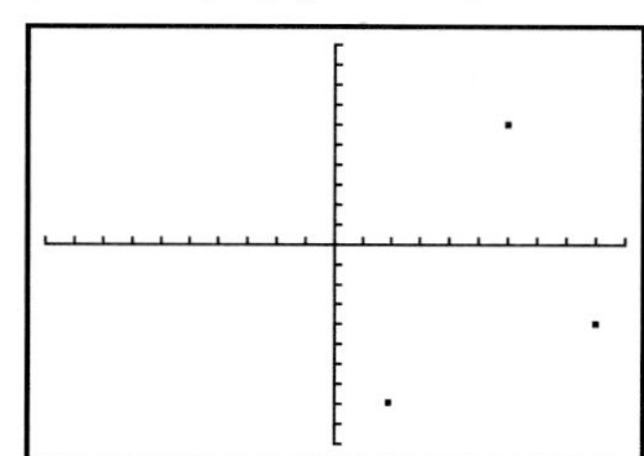

39b. {(10, 0), (−8, 2), (6, 6), (−4, 9)}

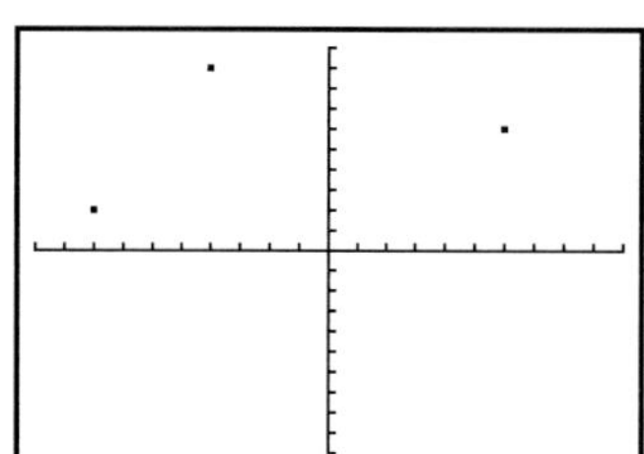

39c.

(x, y)	Quadrant	Inverse's Quadrant
(0, 10)	I	I
(2, −8)	IV	II
(6, 6)	I	I
(9, −4)	IV	II

41. If a point lies in Quadrants I or III, its inverse will lie in the same quadrant as the point. If a point lies in Quadrant II, its inverse lies in Quadrant IV, and vice versa. If a point lies on the *x*-axis, its inverse lies on the *y*-axis and vice versa. **43.** Sample answers: **a.** 157 billion, 191 billion

b. Retail sales have increased from 1992 to 1994.
c. As unemployment decreases, retail sales increase, because people have more money to spend.

45a.

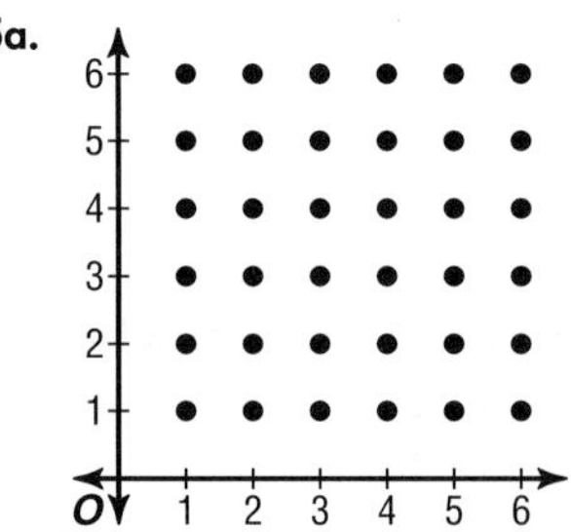

D = R = {1, 2, 3, 4, 5, 6} **45b.** D = R = {1, 2, 3, 4, 5, 6}; relation = inverse
45c. 11 possible sums **45d.**

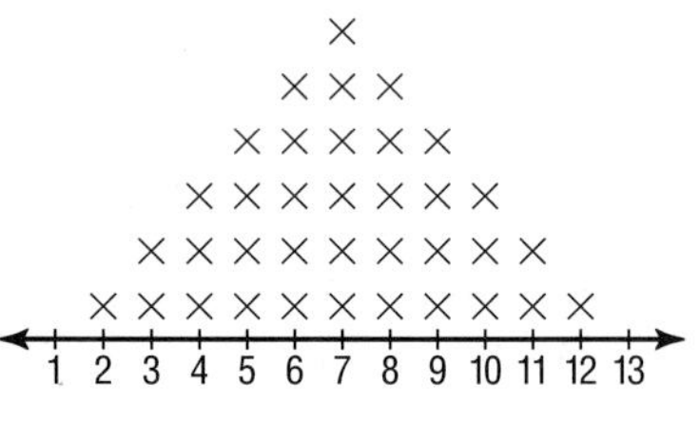

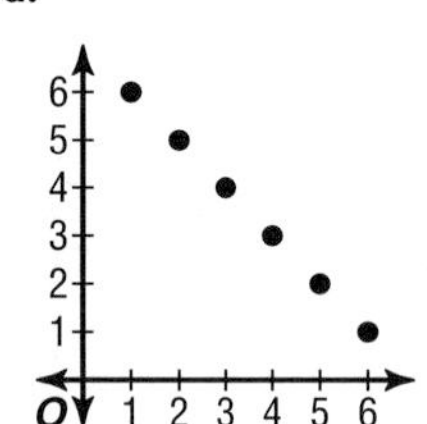

45e. $\frac{6}{36}$ or $\frac{1}{6}$; There are 6 out of 36 ways to roll a sum of 7.

47. 32 **49.** \$480 **51.** 2214, 2290 **53.** $4 + 80x + 32y$

Page 270 Lesson 5–3A
1. {(−3, −19), (−2, −15), (−1, −11), (0, −7), (1, −3), (2, −1), (3, 5)} **3.** {(−3, −10.4), (−2, −9.2), (−1, −8), (0, −6.8), (1, −5.6), (2, −4.4), (3, −3.2)}

Pages 274–277 Lesson 5–3
7. {(−2, −1), (−1, 1), (0, 3), (1, 5), (2, 7), (3, 9)} **9.** a, d
11. {(−2, 8), (−1, 7.5), (0, 7), (1, 6.5), (2, 6)}

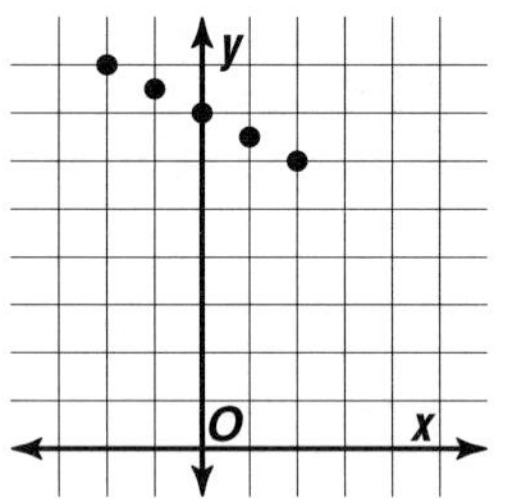

13. a, c **15.** a, b **17.** c **19.** {(−3, −12), (−2, −8), (0, 0), (3, 12), (6, 24)} **21.** {(−3, 10), (−2, 9), (0, 7), (3, 4), (6, 1)} **23.** {(−3, 11), (−2, 9.5), (0, 6.5), (3, 2), (6, −2.5)} **25.** {(−3, −12), (−2, −7), (0, 3), (3, 18), (6, 33)} **27.** {(−3, 1.8), (−2, 1.4), (0, 0.6), (3, −0.6), (6, −1.8)}

29.

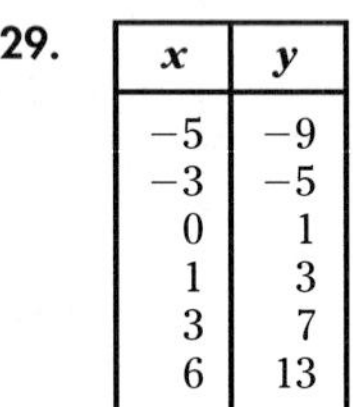

x	y
−5	−9
−3	−5
0	1
1	3
3	7
6	13

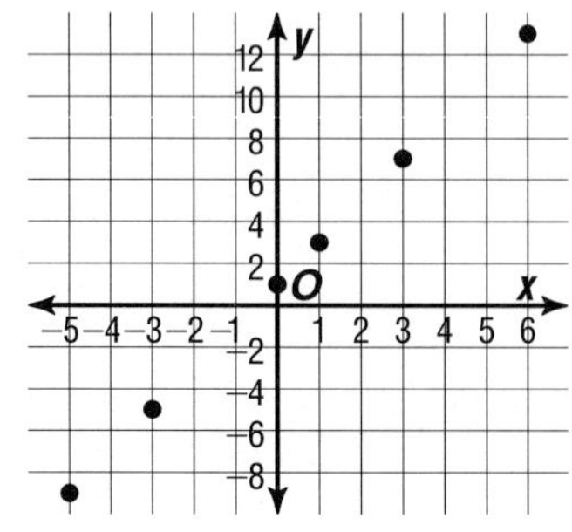

31.

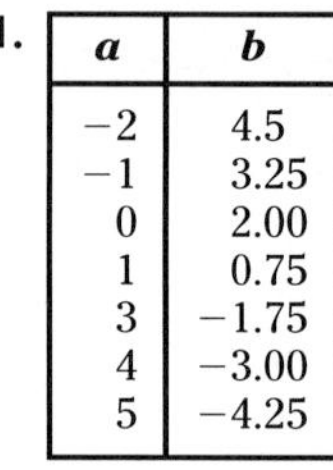

a	b
−2	4.5
−1	3.25
0	2.00
1	0.75
3	−1.75
4	−3.00
5	−4.25

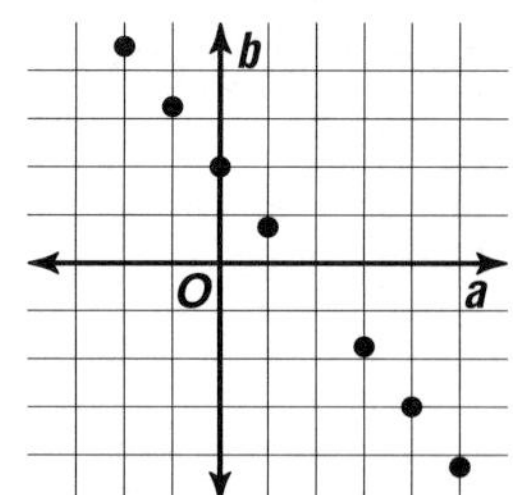

33a. $3x + 4y = 180$ **33b.** $y = 45 - \frac{3}{4}x$ **33c.** Sample answer: (1, 44.25), (2, 43.5), (3, 42.75), (4, 42), (5, 41.25)
35. $\left\{-\frac{2}{3}, -\frac{1}{3}, 0, \frac{2}{3}, 1\right\}$ **37.** $\left\{-\frac{7}{4}, -\frac{1}{2}, 2, \frac{13}{4}, \frac{9}{2}\right\}$

39.

x	y
−2	−8
−1	−1
0	0
1	1
2	8

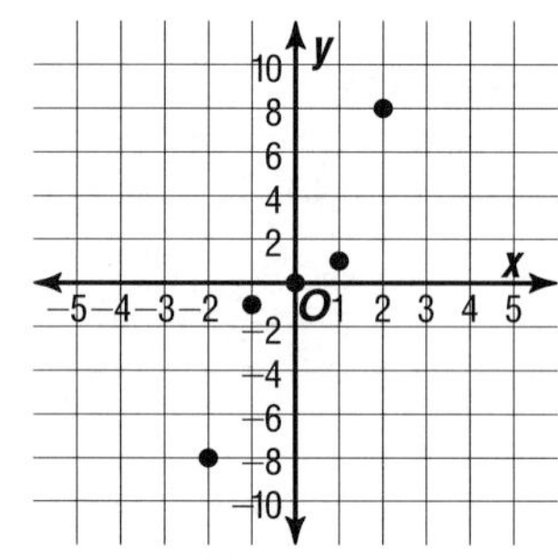

41. {(−2.5, −4.26), (−1.75, −3.21), (0, −0.76), (1.25, 0.99), (3.333, 3.902)} **43.** [(−100, 350), (−30, $116.\overline{6}$), (0, $16.\overline{6}$), (120, $-383.\overline{3}$), (360, $-1183.\overline{3}$), (720, $-2383.\overline{3}$)}
45a. {−6, −4, 0, 4, 6} **45b.** {−13, −8, −4, 4, 8, 13}
45c. {−5, 0, 4, 8, 13} **47a.** $D = \frac{m}{V}$ **47b.** silver and gasoline

49. Women

L	S	(L, S)
$9\frac{1}{3}$	6	$\left(9\frac{1}{3}, 6\right)$
$9\frac{5}{6}$	$7\frac{1}{2}$	$\left(9\frac{5}{6}, 7\frac{1}{2}\right)$
$10\frac{1}{6}$	$8\frac{1}{2}$	$\left(10\frac{1}{6}, 8\frac{1}{2}\right)$
$10\frac{2}{3}$	10	$\left(10\frac{2}{3}, 10\right)$

Men

L	S	(L, S)
$11\frac{1}{3}$	8	$\left(11\frac{1}{3}, 8\right)$
$11\frac{5}{6}$	$9\frac{1}{2}$	$\left(11\frac{5}{6}, 9\frac{1}{2}\right)$
$12\frac{1}{3}$	11	$\left(12\frac{1}{3}, 11\right)$
$12\frac{5}{6}$	$12\frac{1}{2}$	$\left(12\frac{5}{6}, 12\frac{1}{2}\right)$

51a.

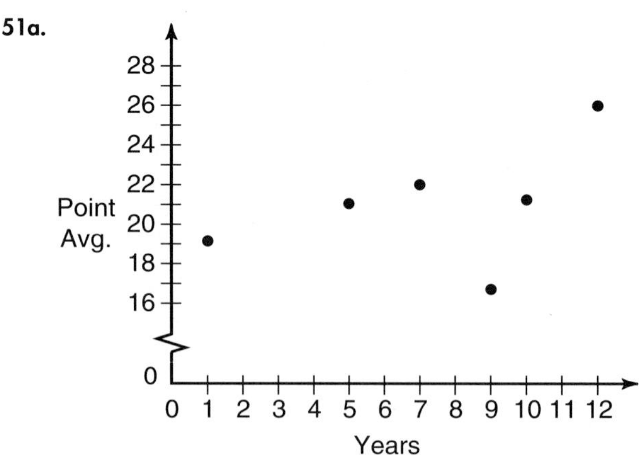

51b. The more years played, the higher the point per game average of the player. **53.** \$467.50 **55.** \$18,000
57a. $2w + 2\ell = 148$ **57b.** w, 14.25 in.; ℓ, 59.75 in.
59. −7.976

Page 280 Lesson 5–4A

1. 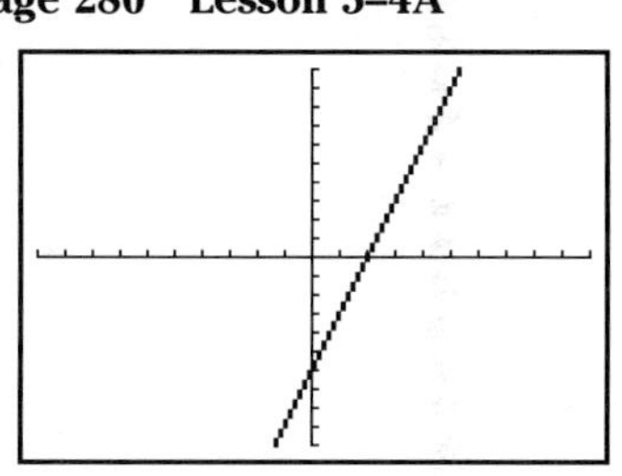

3. 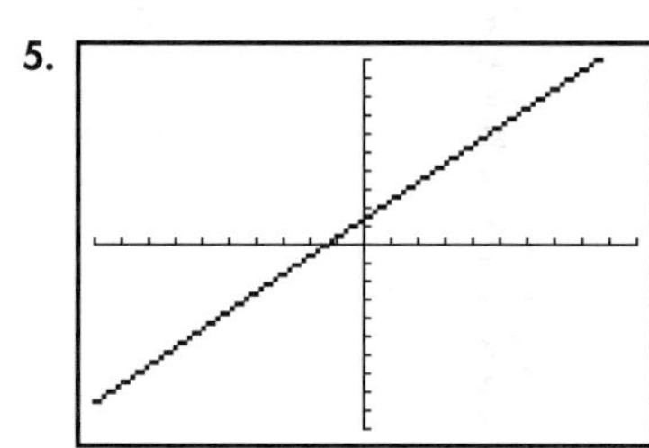

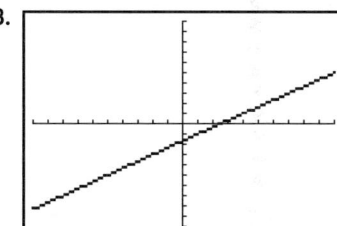

5. 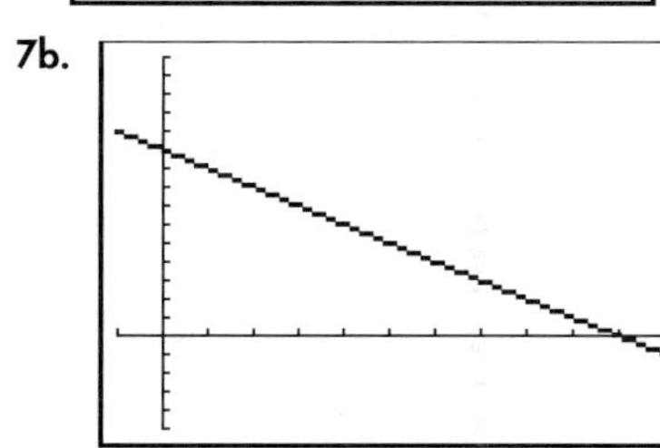

7a. Sample answer: [−10, 110] by [−5, 15]

7b. 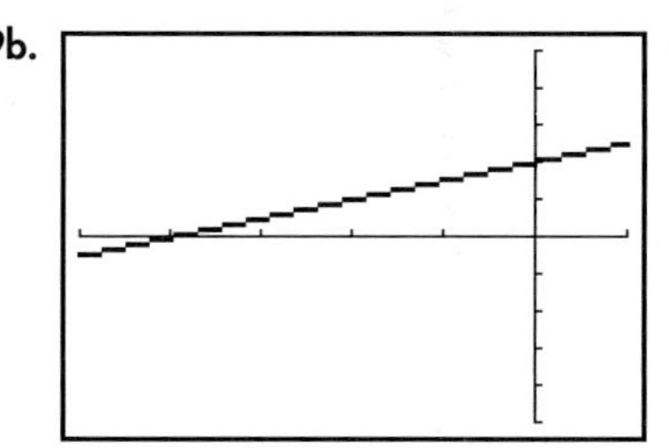

7c. Sample answer: (0, 10), (100, 0), (10, 9)

9a. Sample answer: [−2.5, 0.5] by [−0.05, 0.05]

9b.

9c. Sample answer: (−2, 0), (0, 0.02), (1, 0.03)

Pages 283–286 Lesson 5–4

7. yes; $2x + y = 6$ **9.** yes; $3x + 2y = 7$

11. 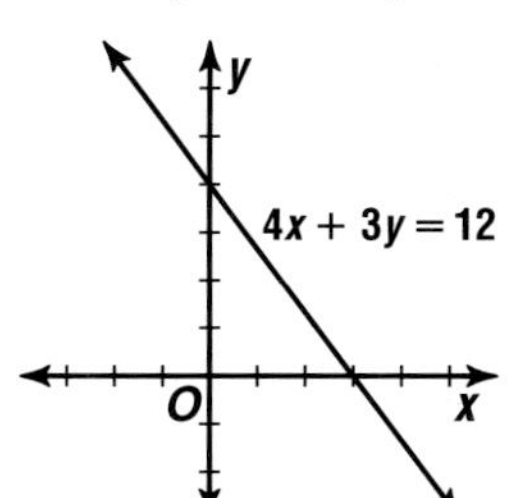

13. 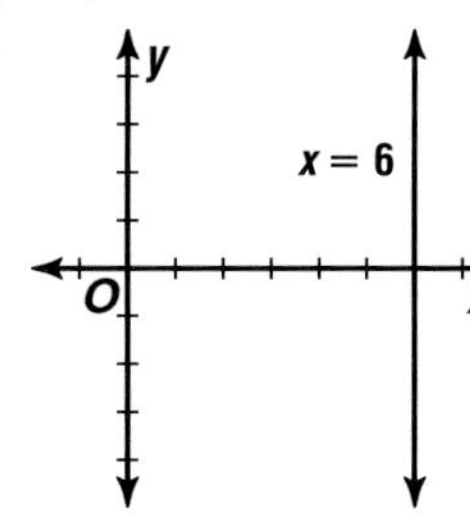

15. 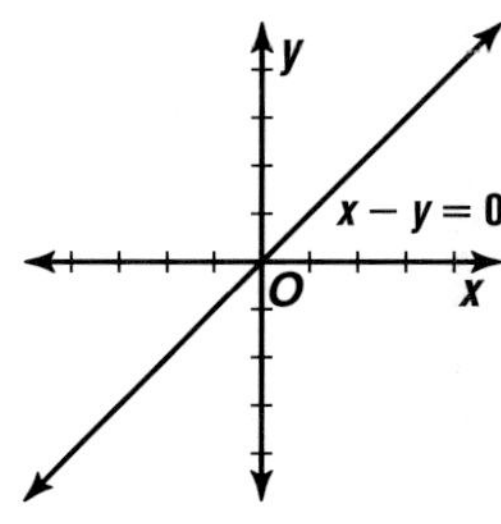

17. yes; $\frac{3}{5}x - \frac{2}{3}y = 5$ **19.** no **21.** yes; $3x - 2y = 8$ **23.** yes; $7x - 7y = 0$ **25.** yes; $3m - 2n = 0$ **27.** yes; $6a - 7b = -5$

29. 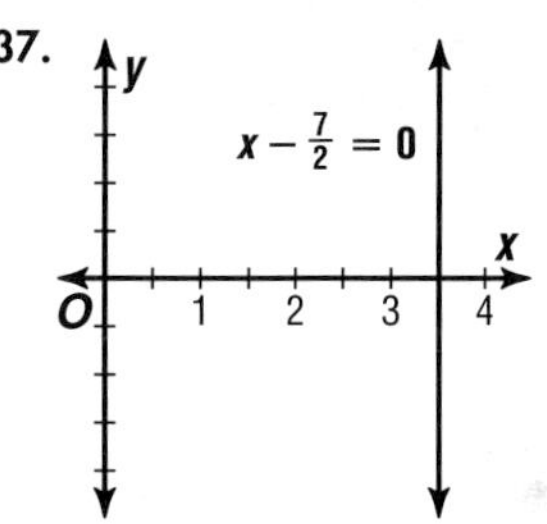

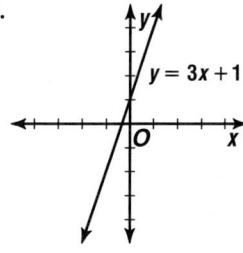

33. 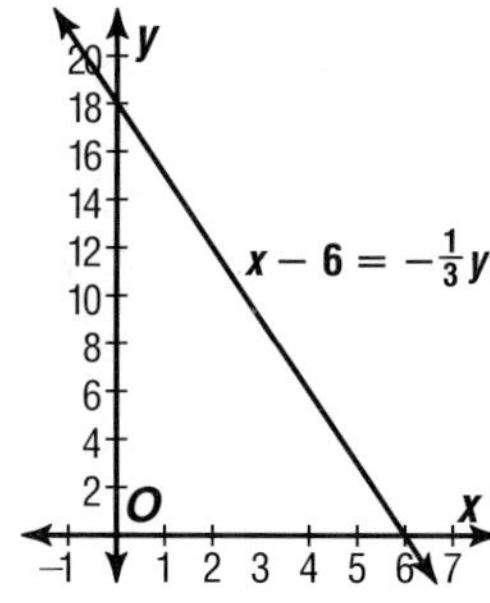

37. 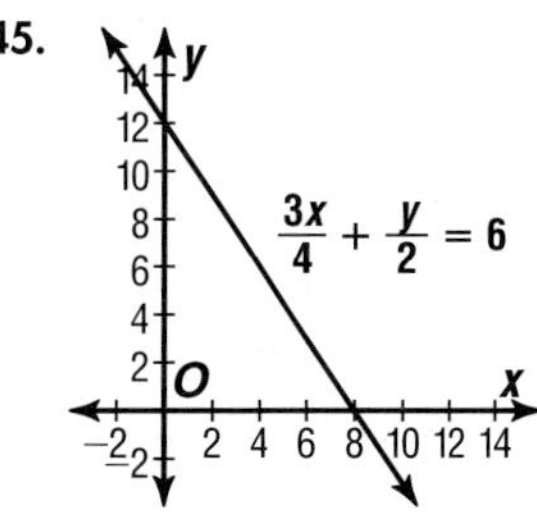

41. 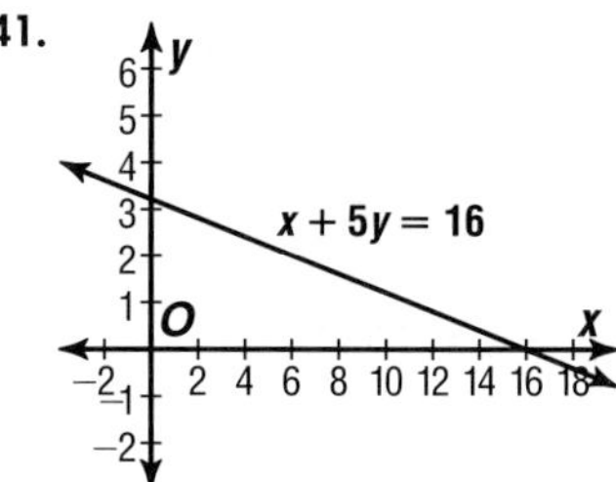

45.

47. −5, 7.5 **49.** 6, 9, 10

51. [−5, 15] by [−10, 10], Xscl: 1, Yscl: 1

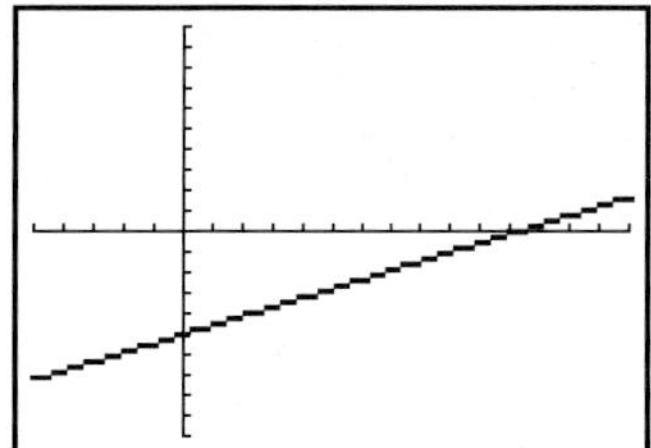

53. [−10, 10] by [−2, 2], Xscl: 1, Yscl: 0.25

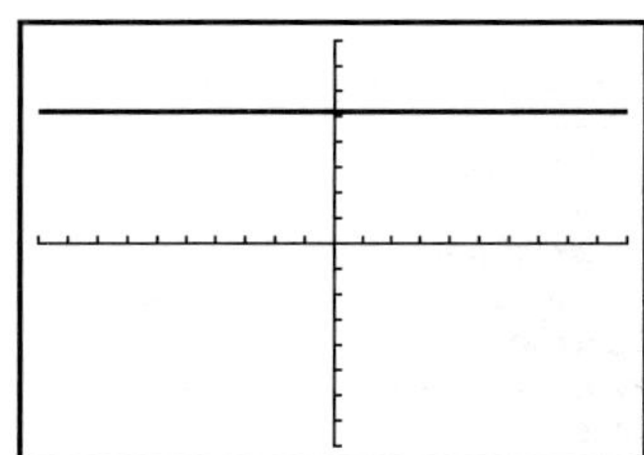

55. [−5, 25] by [−20, 5], Xscl: 5, Yscl: 5

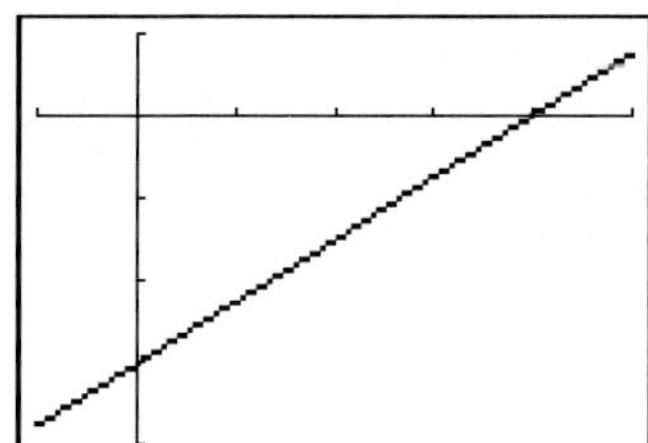

57a. Sample answer: Parallel lines that slant upward and intersect the x-axis at -7, -2.5, 0, and 4.5.

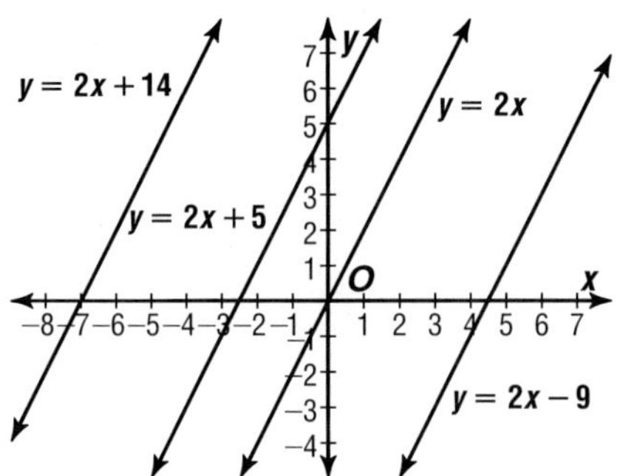

57b. Sample answer: Parallel lines that slant downward and intersect the x-axis at 0, $1\frac{1}{3}$, $2\frac{1}{3}$, and $-3\frac{1}{3}$.

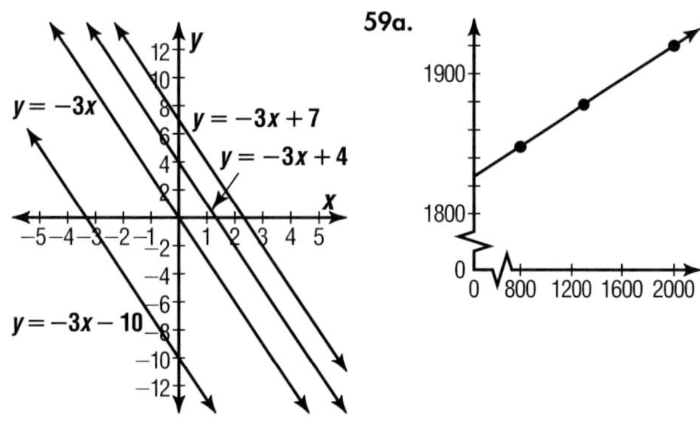

59b. yes, but only if her sales are \$1300 or \$2000 over target **61.** $y = 3 - 4x$

63a.

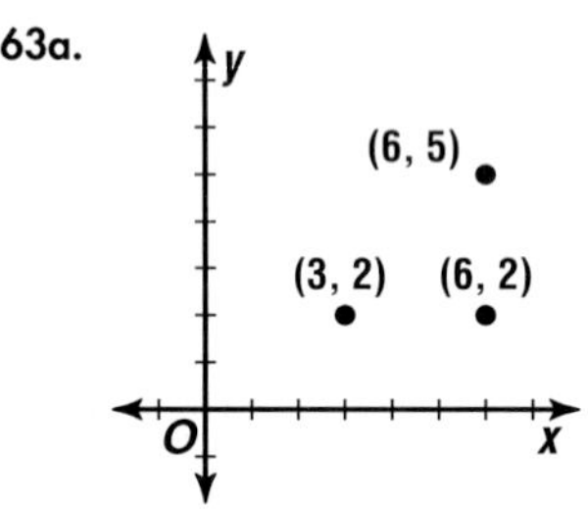

63b. (3, 5) **63c.** 12 inches **65.** 15% **67.** 120° **69.** $\frac{1}{48}$

Page 286 Self Test

1.

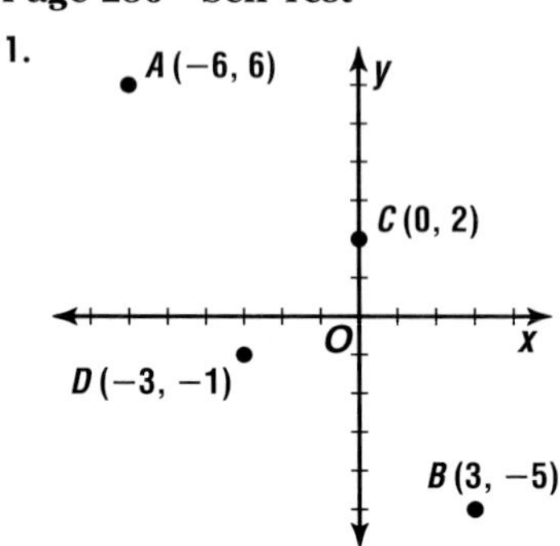

3. $\{(-5, -3), (-1, 4), (4, 4), (4, -3)\}$, D = $\{-5, -1, 4\}$, R = $\{-3, 4\}$, Inv = $\{(-3, -5), (4, -1), (4, 4), (-3, 4)\}$

5. $b = 3 - \frac{2}{3}a$: $\left\{\left(-2, 4\frac{1}{3}\right), \left(-1, 3\frac{2}{3}\right), (0, 3), \left(1, 2\frac{1}{3}\right), (3, 1)\right\}$

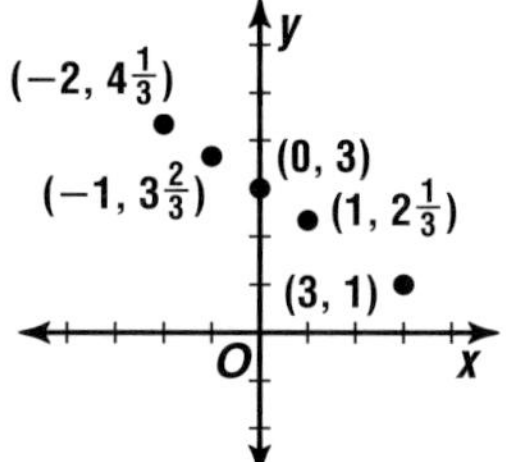

7.

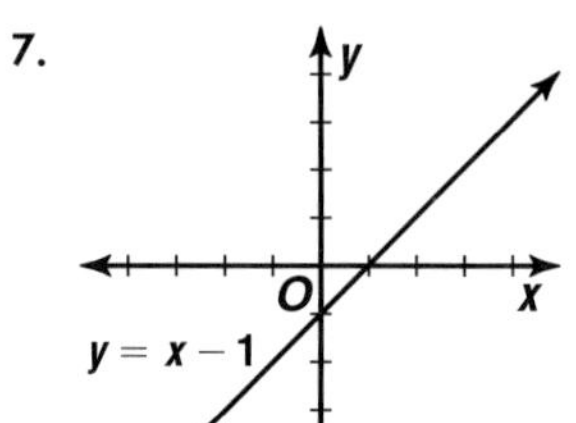

9.

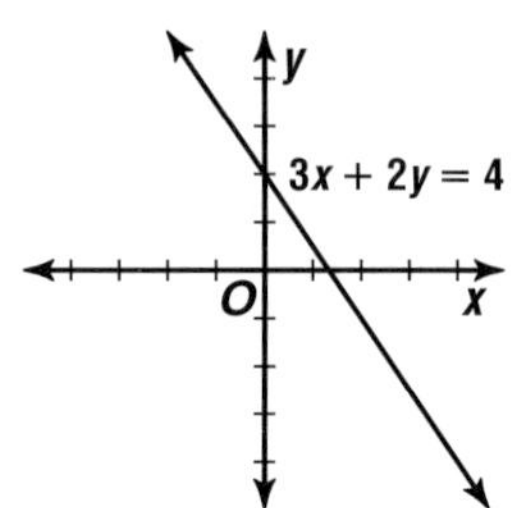

Pages 291–294 Lesson 5–5

7. no **9.** yes **11.** yes **13.** no **15.** -10 **17.** $3w + 2$ **19.** yes **21.** no **23.** yes **25.** yes **27.** yes **29.** yes **31.** yes **33.** no **35.** -14 **37.** $-\frac{9}{25}$ **39.** 5.25 **41.** -18 **43.** $9b^2 - 6b$ **45.** 3 **47.** $5a^4 - 10a^2$ **49.** $24p - 36$ **51a.** Sample answer: $f(x) = x$. **51b.** Sample answer: $f(x) = x^2$ **53.** a; sample answer: because they have a single price per shirt, and since you cannot have a fraction of a shirt, you must use points instead of lines. It is a function.

55a. D: $0 \le k \le 24$; R: $0 \le g \le 100$

55b.

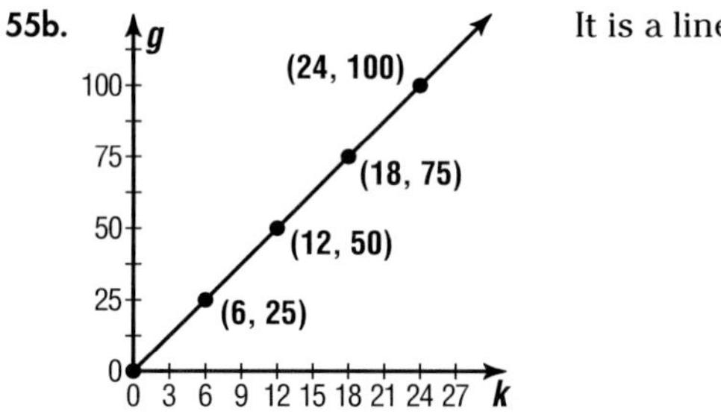

It is a line.

55c. 24 karats **57a.** \$260.87 **57b.** Sample equation: $B = \frac{3}{23}P$ **59.** $\{-16, -13, 5, 11\}$ **61.** $\frac{3}{13}$ **63.** 45° **65.** -48

67a.

91 105 110 115 120 125 130 135 149

67b. 149 **67c.** 91 **67d.** 121 and 125 **67e.** 13 players

Pages 299–302 Lesson 5–6

5. $f(x) = 2x + 6$ **7.** $y = \frac{1}{2}x - \frac{3}{2}$ **9.** -6 **11.** 48, 60 **13.** 8, -2 **15.** $f(x) = 5x$ **17.** $g(x) = 11 - x$ **19.** $h(x) = \frac{1}{3}x - 2$ **21.** $y = -3x$ **23.** $y = \frac{1}{2}x$ **25.** $y = 2x - 10$ **27.** $xy = -24$ **29.** $y = x^3$ **31.** y-intercept: $f(0)$, x-intercept: $f(x) = 0$ **33a.** $f(x) = 34x - 34$ **33b.** You must go deeper in fresh water to get the same pressure as in ocean water. **35a.** 1514 C **35b.** 11.04 C/min

35c.

Minutes	Calories
1	11.04
2	22.08
3	33.12
4	44.16
5	55.2
6	66.24
7	77.28
8	88.32
9	99.36
10	110.4

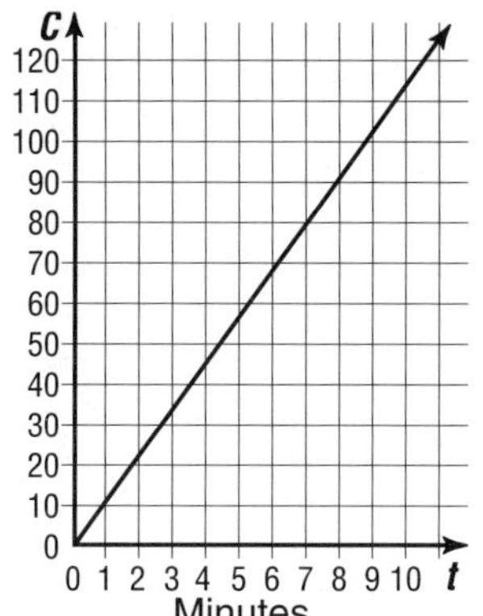

35d. $C(t) = 11.04t$; yes **35e.** 1324.8 C burned; 189.2 C remaining **37.** D = {1, 3, 5}; R = {2, 4, 5} **39.** 5

41.

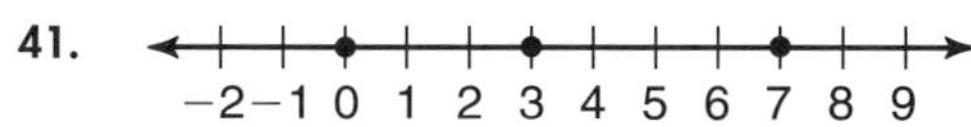

Page 304 Lesson 5–7A

1. Q1, 14; Med, 17; Q3, 20.5; R, 11; IQR, 6.5
3. Q1, 68; Med. 78; Q3, 96; R, 34; IQR, 28
5. Q1, 3.4; Med, 5.3, Q3, 21; R, 77; IQR, 17.6
7. At least 50% of the data is clustered around the median.

Pages 309–313 Lesson 5–7

7. 45, 40, 45, 34, 11; 11 **9.** 48, 26, 39, 17, 22 **11.** 34, 78, 96, 68, 28 **13.** 10, 5, 8, 2, 6 **15.** 1.1, 30.6, 30.9, 30.05, 0.85 **17.** 340, 1075, 1125, 1025, 100 **19.** 39, 218, 221, 202, 19 **21a.** 9,198,630; 11,750,000; 5,700,000; 24,000,000; 6,050,000 **21b.** No; the libraries would have accumulated books throughout the years. **23.** 3760, 3224, 4201.5, 2708.5, 1421 **25.** 21,674; 9790; 12,194; 5475; 6719 **27a.** males: 36, 29, 37.5, 44, 15; females: 23, 29, 33, 39, 10 **27b.** There are no outliers. **27c.** The ages of the top female golfers are less varied than those of the top male golfers. **31.** c, d **33a.** $6.2 + p = 9.4$; about 3.2 million people **33b.** $6.0 + p = 6.9$; about 0.9 million people

Page 315 Chapter 5 Highlights

1. e **3.** d **5.** j **7.** c **9.** b

Pages 316–318 Chapter 5 Study Guide and Assessment

10–13.

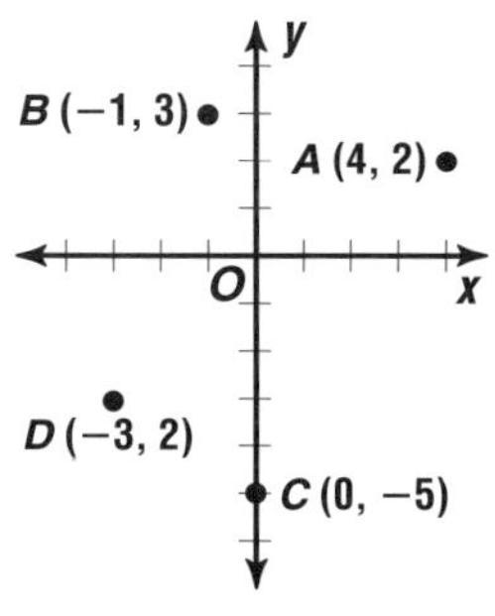

15. (2, −1); IV **17.** (1, 1); I **19.** D = {−3, 4}, R = {5, 6} **21.** D = {−3, −2, −1, 0}, R = {0, 1, 2} **23.** {(2, 0), (−1, 3), (2, 2), (−1, −2)} **25.** {(−4, −13), (−2, −11), (0, −9), (2, −7), (4, −5)} **27.** $\left\{\left(-4, -5\frac{1}{3}\right), \left(-2, -2\frac{2}{3}\right), (0, 0), \left(2, 2\frac{2}{3}\right), \left(4, 5\frac{1}{3}\right)\right\}$

29.

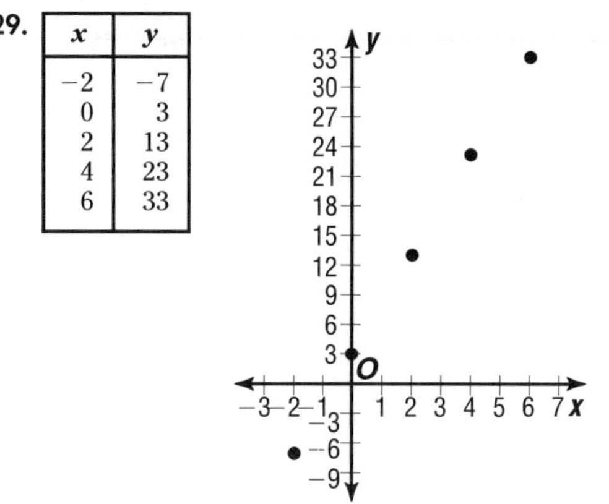

x	y
−2	−7
0	3
2	13
4	23
6	33

31.

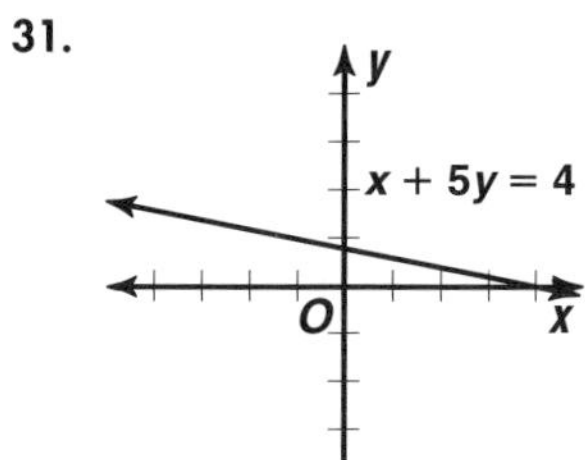

33.

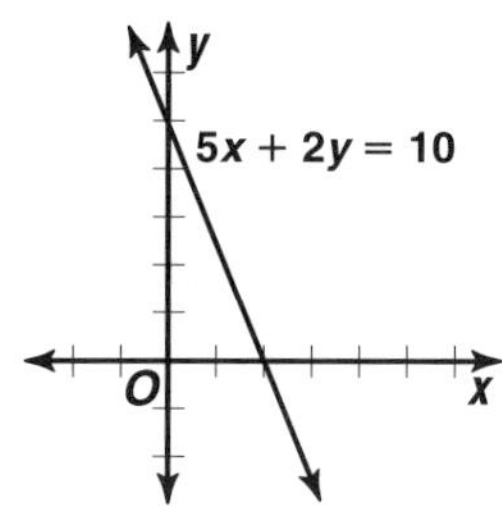

35.

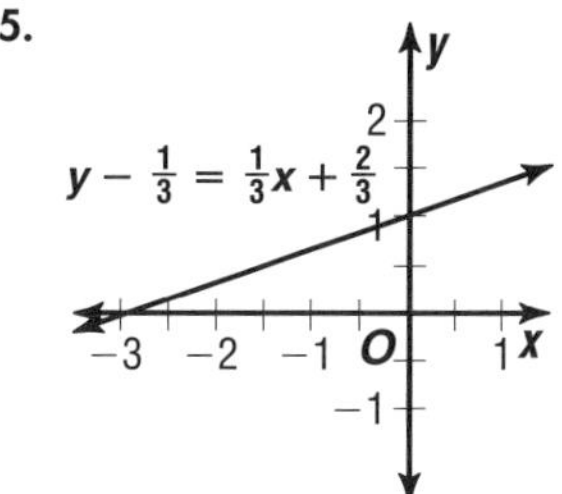

37. yes **39.** no **41.** yes **43.** 3 **45.** $a^2 + a + 1$ **47.** $2a^2 - 14a + 26$ **49.** $y = x - 4$ **51.** 70, 65, 85, 45, 40 **53.** 37, 73, 77, 62, 15

55a.

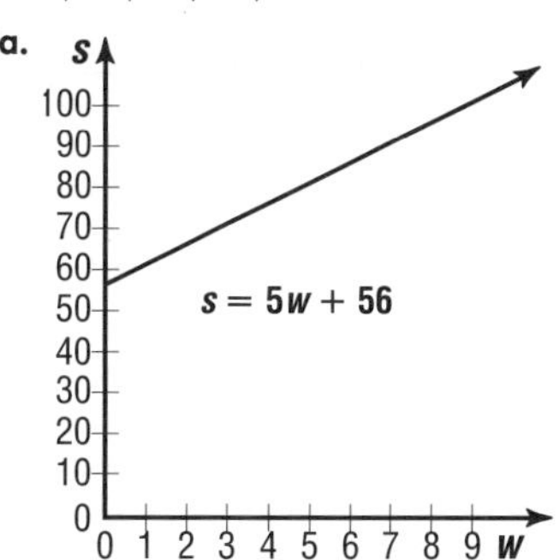

55b. \$56 **57a.** 9.6, 6, 7.05, 8.3, 2.3 **57b.** 12.6, 14.7

CHAPTER 6 ANALYZING LINEAR EQUATIONS

Page 324 Lesson 6–1A

1. $-\frac{1}{2}$ **3.** Sample answer: (3, 3) **5.** Yes; sample answer: suppose the endpoints are $C(-4, -1)$ and $D(-2, -2)$. Let the upper right peg represent (0, 0). To go from C to D, the y valuc decreases by 1 and the x value increased by 2. The ratio is $\frac{-1}{2}$. If you use the rule from Exercise 4, $\frac{-2-(-1)}{-2-(-4)} = \frac{-1}{2}$. The results are the same.

Pages 329–331 Lesson 6–1

7. $\frac{5}{3}$ **9.** $\frac{3}{2}$ **11.** undefined **13.** 2 **15.** $-\frac{1}{5}$ **17.** 0 **19.** 1

21. $\frac{4}{7}$ **23.** $-\frac{2}{3}$ **25.** undefined **27.** undefined **29.** $\frac{9}{5}$
31. -5 **33.** -1 **35.** 7
37.

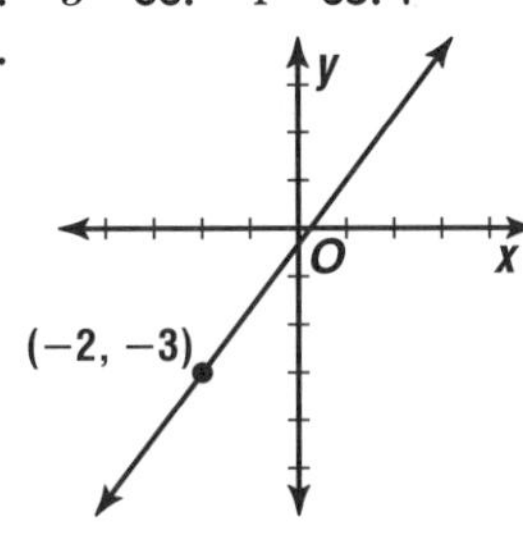

39. Sample answer: (7, 6). The slope of the line containing A and B is -2. Use the slope to go 2 units down and 1 unit to the right of either point.
41. about 1478 feet
43. 11,160 feet **45.** 77; 5; 8; 3.2; 4.8 **47.** 12% **49.** -6

51a. 12 animals
51b.

51c. 15 years **51d.** 4 animals

Pages 336–338 Lesson 6–2
5. 4; $(2, -3)$ **7.** 3; $(-7, 1)$ **9.** $y - 4 = -3(x + 2)$
11. $3x + 4y = -9$ **13.** $2x - y = -6$ **15.** $y + 2 = -\frac{4}{7}(x + 1)$ or $y - 2 = -\frac{4}{7}(x + 8)$; $4x + 7y = -18$ **17.** $3x - 5y = -2$
19. $y - 5 = 3(x - 4)$ **21.** $y - 1 = -4(x + 6)$ **23.** $y - 3 = -2(x - 1)$ **25.** $y + 3 = \frac{3}{4}(x - 8)$ **27.** $4x - y = -5$
29. $3x - 2y = -24$ **31.** $2x + 5y = 26$ **33.** $y - 2 = -\frac{1}{3}(x + 5)$ or $y + 1 = -\frac{1}{3}(x - 4)$ **35.** $y + 1 = \frac{3}{7}(x + 8)$ or $y - 5 = \frac{3}{7}(x - 6)$ **37.** $y + 2 = 0(x - 4)$ or $y + 2 = 0(x - 8)$
39. $4x + 3y = 39$ **41.** $11x + 8y = 17$ **43.** $36x - 102y = 61$
45. $\frac{5-1}{5-9} = \frac{4}{-4}$ or -1; An equation of the line is $(y - 1) = -1(x - 9)$. Let $y = 0$ in the equation and see if $x = 10$.

$$0 - 1 = -x + 9$$
$$-10 = -x$$
$$10 = x$$

(10, 0) lies on the line. Since (10, 0) is a point on the x-axis, the line intersects the x-axis at (10, 0).
47a. No; for a rise of 30 inches the ramp must be 30 feet long, but there only 18 feet available.
47b.

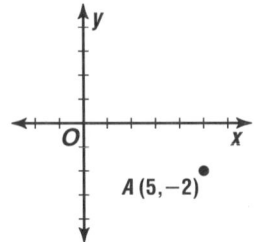

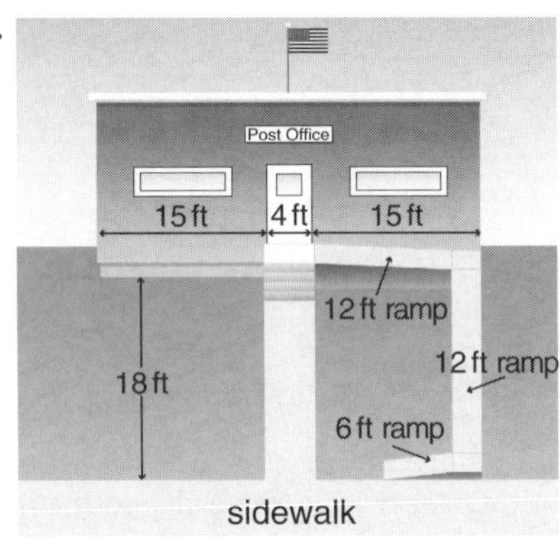

49a. \$0.26, \$0.09 **49b.** \$0.32 **51.** 28 **53.** No; because there could be 2 pink and 1 white or 1 pink and 2 white.
55. associative property of addition

Pages 343–345 Lesson 6–3
7. positive **9a.** It gets better. **9b.** yes; negative
9c. (7, 18) and (8, 16) **11.** positive **13.** no **15.** no
17. positive **19.** negative **21.** Yes; sample reason: Dots are grouped in an upward diagonal pattern. **23.** c; If 1 is correct then 19 are wrong, if 2 are correct, then 18 are wrong, and so on. Graph c shows these pairs of numbers. **25a.** Sample answer: The more taxation increases, the more in debt the government becomes.
25b. Sample answer: You work harder and your grades go up. **25c.** Sample answer: As more money is spent on research, fewer people die of cancer. **25d.** Sample answer: Comparing the number of professional golfers with the number of holes-in-one. **27a.** The correlation shown by the graph shows a slightly positive correlation between SAT scores and graduation rate. **27b.** Sample equation: $y = 0.89x + 1142.17$

29a., c.

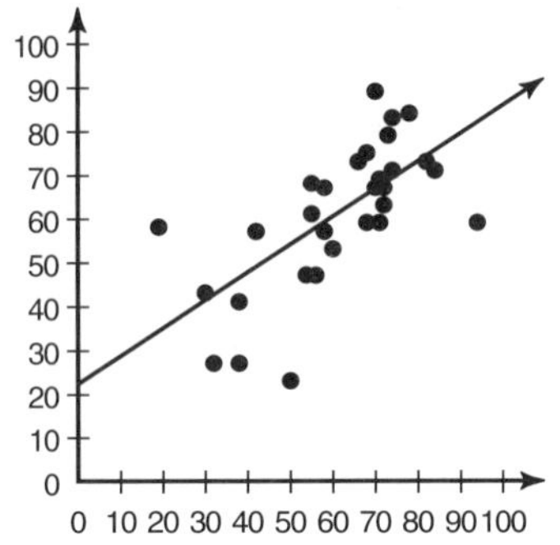

29b. positive **29c.** Sample answer: $15x - 13y = 129$
31a. $7500x - y = 120{,}000$ **31b.** 15,000 feet **31c.** No; it only describes the plane's path in that part of the flight.
33. 75° **35.** 1.45

Pages 350–353 Lesson 6–4
7a. 2 **7b.** 2, -4 **7c.** $y = 2x - 4$ **9.** $-\frac{28}{3}, \frac{7}{2}$
11. $y = \frac{2}{3}x - 10$, $2x - 3y = 30$ **13.** $-2, -4$ **15.** $x + 5y = 13$
17. $y = \frac{11}{3}x$; 44 **19.** $-\frac{3}{2}$; 0, 0; $y = -\frac{3}{2}x$ **21.** $\frac{2}{5}$, 5, -2, $y = \frac{2}{5}x - 2$ **23.** 2, $\frac{8}{7}$ **25.** $-\frac{5}{6}$, 5 **27.** none, 6
29. $y = 7x - 2$, $7x - y = 2$ **31.** $y = -1.5x + 3.75$, $6x + 4y = 15$ **33.** $y = -7$, $y = -7$ **35.** $-\frac{5}{4}, \frac{5}{2}$ **37.** -4, 12
39. $\frac{4}{5}, \frac{4}{15}$ **41.** $y = -2$ **43.** $x = 3$ **45.** $y = 9$
47. $y = \frac{11}{24}x$, $\frac{33}{2}$ **49.** $y = \frac{2}{3}x - \frac{8}{3}$ **51.** $(-3, -1)$
53a. $y = 2.04x - 21.32$ **55a.** 42.24 ft^3 **55b.** 270 K
57a. $y = 0.1x + 3$ **57b.** Go to the other bank, since this one would charge you \$5.50. **59.** 0
61.

63. 13 ft 11 in.
65. -9
67. $14x + 14$

Page 353 Self Test
1. -1 **3.** undefined **5.** $y - 4 = \frac{1}{2}(x + 6)$ **7.** $x + 5y = 17$
9. $-6, -14$

Page 355 Lesson 6–5A

1. All graphs are of the family $y = -ax + 0$, where a represents different negative slopes.

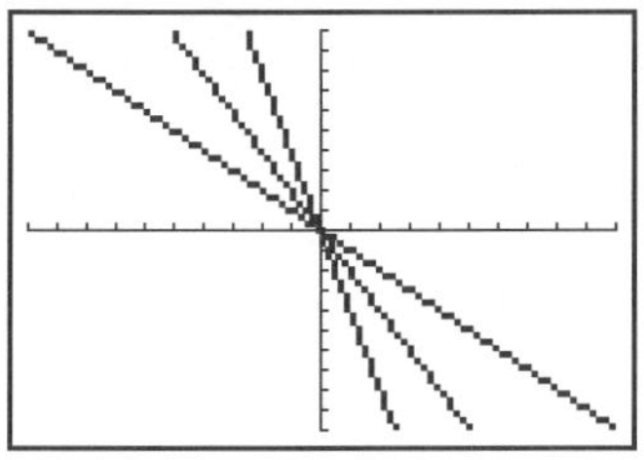

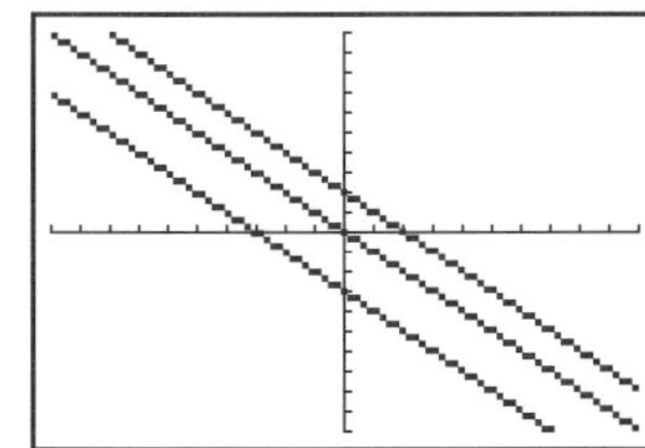

3. All graphs have the same slope, -1, but have different y-intercepts.

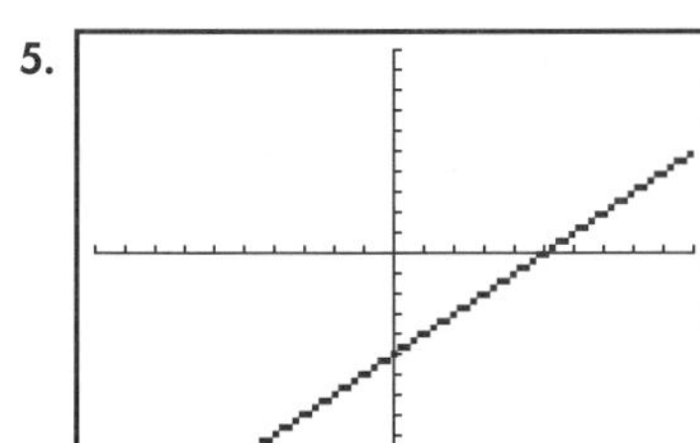

5.

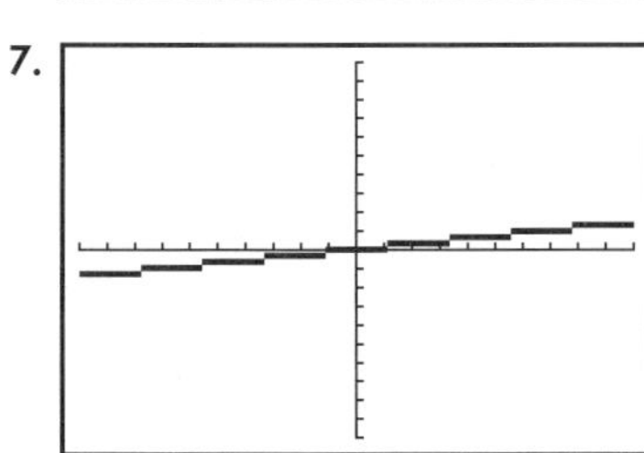

7.

9. $y = -x + 2.5$

Pages 359–360 Lesson 6–5

7.

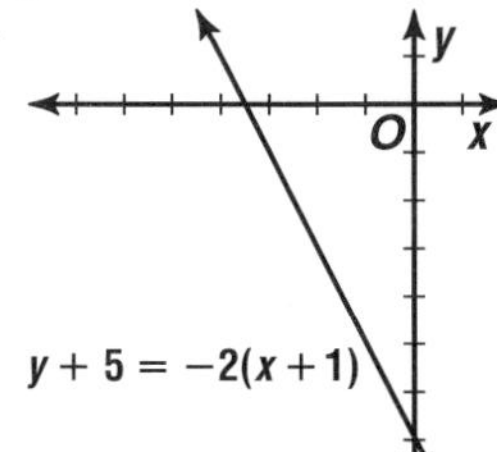

9.

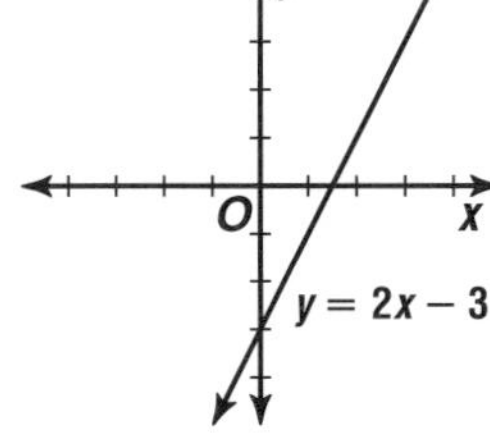

11.

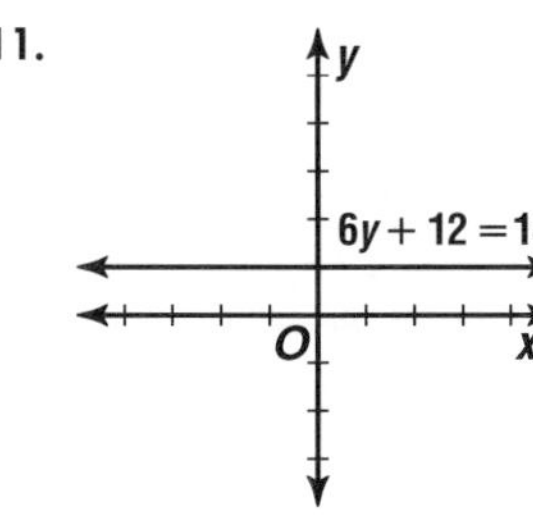

13a.

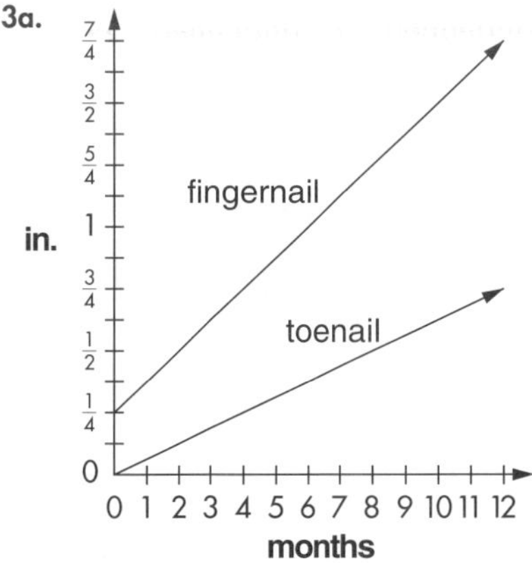

13b. $1\frac{3}{4}$ inches **13c.** $\frac{1}{16}$ inch **13d.** See 13a.

13e. the rate of growth per month

15.

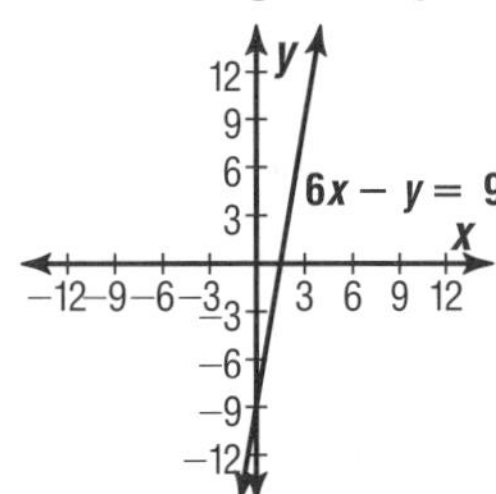

19.

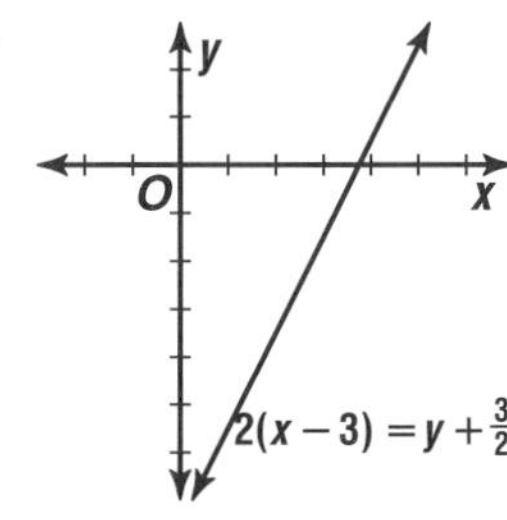

23.

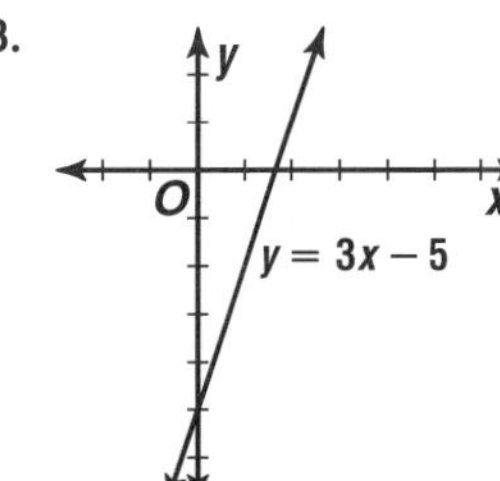

27.

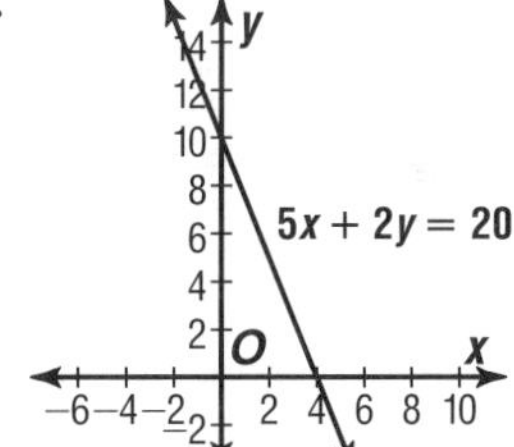

31.

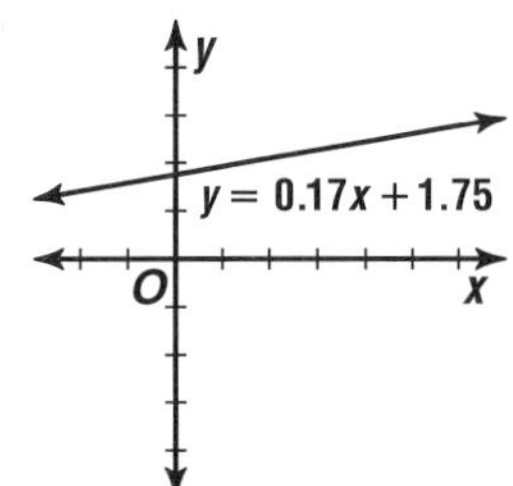

35.

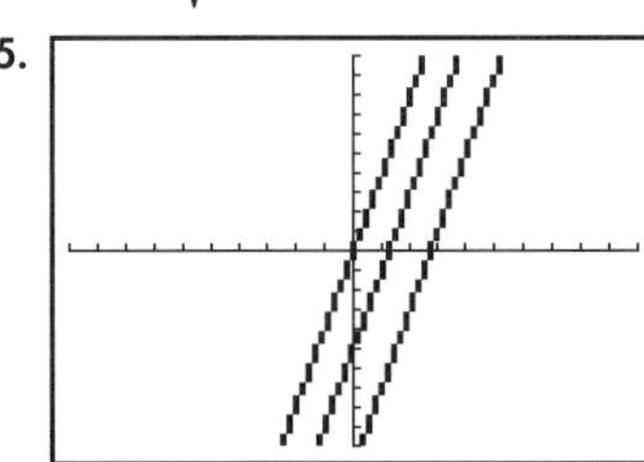

All lines have the same slope 4.

37a.

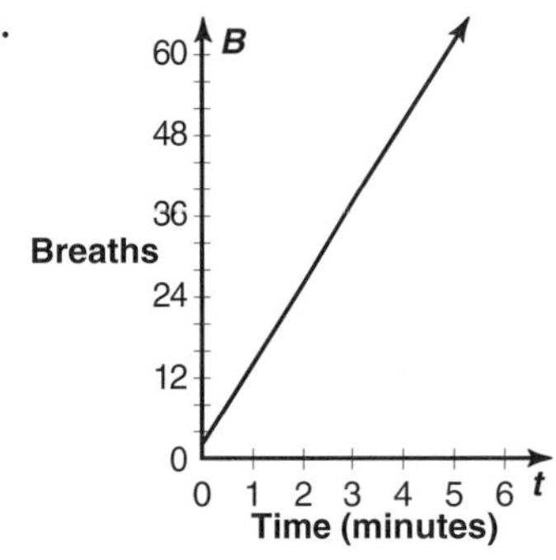

37b. 122 breaths

39. $y = \frac{2}{5}x + 12$

41. 4:30 P.M. **43.** 1.8

45. a number m minus 1

Pages 366–368 Lesson 6–6
7. $\frac{2}{3}, -\frac{3}{2}$ **9.** perpendicular **11.** $y = \frac{5}{6}x - \frac{21}{2}$
13. $y = x - 9$ **15.** $y = \frac{9}{5}x$ **17.** perpendicular
19. parallel **21.** perpendicular **23.** perpendicular
25. $y = x + 1$ **27.** $y = \frac{8}{7}x + \frac{2}{7}$ **29.** $y = 2.5x - 5$
31. $y = \frac{2}{3}x + 4$ **33.** $y = -\frac{9}{2}x + 14$ **35.** $y = 3x - 19$
37. $y = -\frac{1}{5}x - 1$ **39.** $y = -\frac{7}{2}x + 11$ **41.** $y = -3$
43. $y = \frac{5}{4}x$ **45.** $y = \frac{1}{3}x - 6$ **47.** No, because the slope of $\overline{AC}$ is $\frac{6}{7}$ and the slope of $\overline{BC}$ is $-\frac{2}{3}$. These slopes are not negative reciprocals of each other, so the lines are not perpendicular and the figure is not a rhombus.
49a. $y = \frac{1}{2}x + \frac{7}{2}, y = -2x + 11$ **49b.** right or 90° angle
51.

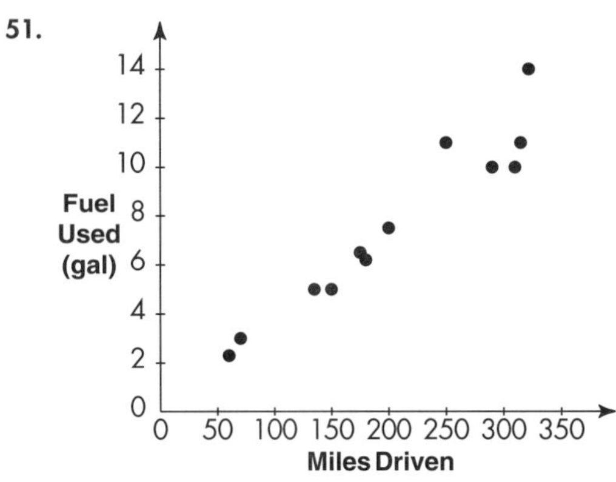

53. \$242.80 **55a.** $\frac{5}{12}; \frac{7}{12}$ **55b.** 16-karat gold
57. multiplicative identity

Pages 372–374 Lesson 6–7
5. (1.5, 6) **7.** (−6, 1) **9.** (−11, 7) **11.** (8, 9.8)
13. (12.5, 6) **15.** (1, 5) **17.** (3, 2) **19.** $\left(\frac{1}{2}, 1\right)$
21. (0.8, 2.7) **23.** $(4x, 9y)$ **25.** (−7, 0) **27.** (21, −6)
29. (9, 10) **31.** $\left(\frac{5}{6}, \frac{1}{3}\right)$ **33.** $B(2.3, 6.8)$ **35.** $P(6.65, -1.85)$
37. (−1, 5) **39.** $\left(1, \frac{5}{2}\right)$ **41a.** $N(6, 3), M(10, 3)$
41b. parallel, $MN = \frac{1}{2}AB$ **43a.** $P(-4, 1), Q(10, -1), R(2, 9)$ **43b.** 62 square units; Sample answer: The area of the smaller triangle is $\frac{1}{2}bh$. Since the base of the larger triangle is twice that of the smaller one and the height is also twice the length of the small one, the area of the larger is $\frac{1}{2}(2b)(2h)$, or $2bh$. This is 4 times the area of the small one. **45.** $y = -\frac{7}{9}x - \frac{8}{3}$ **47.** yes; $9x - 6y = 7$
49. −20 **51.** $3.1x + 1.54$

Page 375 Chapter 6 Highlights
1. parallel **3.** midpoint **5.** perpendicular
7. slope-intercept **9.** slope

Pages 376–378 Chapter 6 Study Guide and Assessment
11. $-\frac{1}{3}$ **13.** $\frac{2}{5}$ **15.** $\frac{25}{3}$ **17.** $y + 3 = -2(x - 4)$
19. $y - 7 = 0$ **21.** $y - 3 = \frac{3}{5}x$ **23.** $y - 1 = -\frac{6}{7}(x - 4)$
25. $3x - y = 18$ **27.** $3x - 4y = 22$ **29.** $y = 5$ **31.** $x = -2$
33a. Yes; it is positive **33b.** Sample answer: 35 stories
33c. $y = 0.03x + 21$

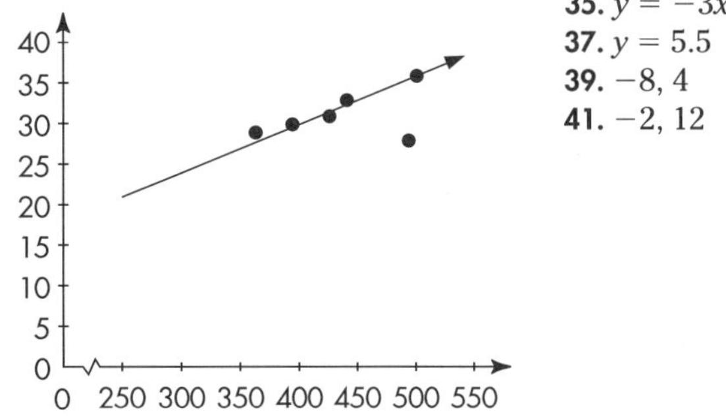

35. $y = -3x$
37. $y = 5.5$
39. −8, 4
41. −2, 12

43. **45.**

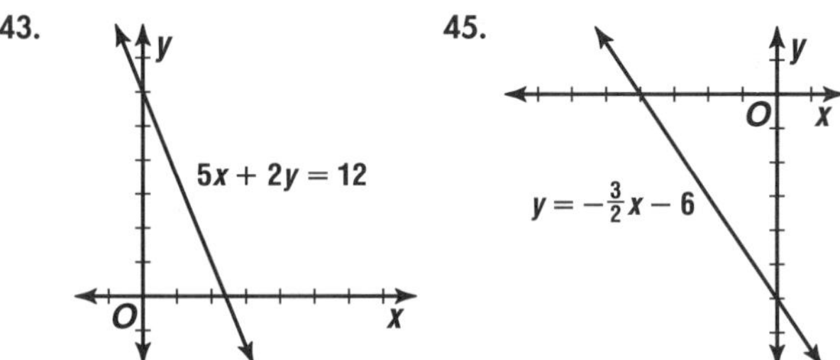

47. **49.**

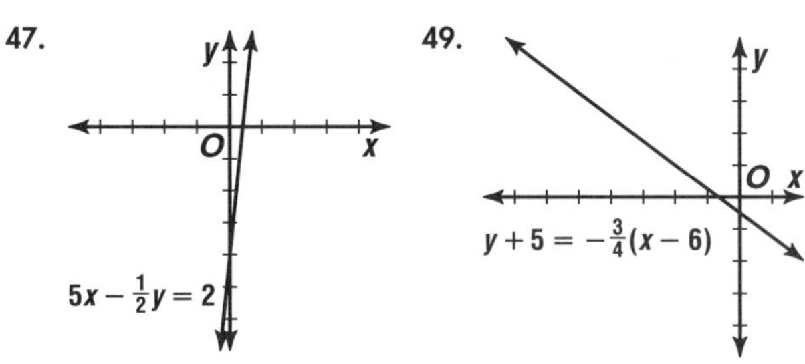

51. $y = -\frac{7}{2}x - 14$ **53.** $y = -\frac{3}{8}x + \frac{13}{2}$ **55.** $y = 5x - 15$
57. $\left(1, -\frac{5}{2}\right)$ **59.** $\left(5, \frac{11}{2}\right)$ **61.** $\left(\frac{7}{2}, -\frac{3}{2}\right)$ **63.** (13, 11)
65. (−5, 20) **67.** $d = 45t - 10$

CHAPTER 7 SOLVING LINEAR EQUATIONS

Pages 388–390 Lesson 7–1
7. c **9.** d **11.** $\{x \mid x > -5\}$ **13.** $\{y \mid y < -5\}$
15. $x - 17 < -13, \{x \mid x < 4\}$
17. $\{a \mid a < 18\}$ 13 14 15 16 17 18 19 20 21
19. $\{x \mid x \leq 1\}$ −4 −3 −2 −1 0 1 2 3 4
21. $\left\{x \mid x > \frac{11}{3}\right\}$ 1 2 3 $\frac{11}{3}$ 5 6 7 8 9
23. $\{x \mid x > 2\}$ −1 0 1 2 3 4 5 6 7
25. $\left\{x \mid x < \frac{3}{8}\right\}$ **27.** $\{x \mid x \leq 15\}$ **29.** $\{x \mid x < 0.98\}$
31. $\{r \mid r < 10\}$ **33.** $x - (-4) \geq 9, \{x \mid x \geq 5\}$ **35.** $3x < 2x + 8, \{x \mid x < 8\}$ **37.** $20 + x < 53, \{x \mid x < 33\}$ **39.** $2x > x - 6, \{x \mid x > -6\}$ **41.** 12 **43.** −2 **45a.** no **45b.** yes **45c.** yes **45d.** yes **47.** The value of x falls between −2.4 and 3.6. **49a.** $x \leq \$12.88$ **49b.** Sample answer: There may be sales tax on his purchases. **51.** $y = -3x + 3$
53. 42, 131, 145, 159, 28 **55.** 12 **57.** <

Page 391 Lesson 7–2A

1. Sample answer: The variable remains on the left, but the inequality symbol is reversed. **3.** When the coefficient of x is positive, you can solve the inequality like an equation and retain the same inequality symbol. If the coefficient of x is negative, you can solve like an equation, but the symbol must be reversed.

Pages 396–398 Lesson 7–2

9. multiply by $-\frac{1}{6}$ or divide by -6; yes; $\{y \mid y \le 4\}$
11. multiply by 4; no; $\{x \mid x < -20\}$ **13.** $\{x \mid x < 30\}$
15. $\{t \mid t \le -30\}$ **17.** $\frac{1}{5}x \le 4.025$; $\{x \mid x \le 20.125\}$ **19.** $s \ge 12$
21. $\{b \mid b > -12\}$ **23.** $\{x \mid x \ge -44\}$ **25.** $\{r \mid r < -6\}$
27. $\{t \mid t < 169\}$ **29.** $\{g \mid g \ge 7.5\}$ **31.** $\{x \mid x \ge -0.7\}$
33. $\left\{r \mid r < -\frac{1}{20}\right\}$ **35.** $\{x \mid x < -27\}$ **37.** $\{m \mid m \ge -24\}$
39. $36 \ge \frac{1}{2}x$; $\{x \mid x \le 72\}$ **41.** $\frac{3}{4}x \le -24$; $\{x \mid x \le -32\}$
43. $-8x \le 144$; -18 or greater **45.** $y < 7.14$ meters **47.** $\ge$
49. $<$ **51.** up to 416 miles **53.** at least 5883 signatures
55. $(-1, 1)$ **57.** -2 **59.** \$155.64 **61.** 65 yd by 120 yd

Pages 402–404 Lesson 7–3

7. c **9.** $\{x \mid x > 2\}$ **11.** $\{d \mid d > -125\}$ **13.** $\{2, 3\}$
15a. $x + (x + 2) > 75$ **15b.** $x > 36.5$ **15c.** Sample answer: 38 and 40. **17.** $\{-10, -9, \ldots, 2, 3\}$ **19.** $\{-10, -9, \ldots, -5, -4\}$ **21.** $\{t \mid t > 3\}$ **23.** $\{w \mid w \le 15\}$ **25.** $\{n \mid n > -9\}$ **27.** $\{m \mid m < 15\}$ **29.** $\{x \mid x < -15\}$ **31.** $\left\{p \mid p \le \frac{14}{3}\right\}$ **33.** $\{x \mid x > -10\}$ **35.** $\{k \mid k \le -1\}$ **37.** $\{y \mid y < -1\}$ **39.** $3(x + 7) > 5x - 13$; $\{x \mid x < 17\}$ **41.** $2x + 2 \le 18$ for $x > 0$; 7 and 9; 5 and 7; 3 and 5; 1 and 3 **43.** no solution $\{\varnothing\}$ **45a.** $x \le -8$ **45b.** $x > 8$ **45c.** $x > 2$ **45d.** $x \le -1$
47. $x + 0.04x + 0.15(x + 0.04x) \le \$50$, $x \le \$41.80$ **49.** at least \$571,428.57 **51a.** at most 2.9 weeks **51b.** no change **51c.** at most 4.1 weeks **53.** $\{y \mid y > 10\}$ **55.** $3x + 2y = 14$
57. $\{-5, -3, -2, 4, 16\}$ **59.** 25.1; 23.5; no mode

Pages 409–412 Lesson 7–4

7. $0 \le x \le 9$

−1 0 1 2 3 4 5 6 7 8 9 10

9. $-3 < x \le 1$ **11.** The solution is the empty set. There are no numbers greater than 5 but less than -3.
13. $\{h \mid h \le -7$ or $h \ge 1\}$

15. $\{w \mid 1 > w \ge -5\}$

−7 −6 −5 −4 −3 −2 −1 0 1 2 3

17. Drawings will vary; 16 pieces. **19.** $\varnothing$

21. −7 −6 −5 −4 −3 −2 −1 0 1 **23.** −10 −8 −6 −4 −2 0 2 4

25. $-4 \le x \le 5$ **27.** $x \le -2$ or $x > 1$
29. $\{x \mid -1 < x < 5\}$

−3 −2 −1 0 1 2 3 4 5 6 7

31. $\{x \mid x < -2$ or $x > 3\}$

−4 −3 −2 −1 0 1 2 3 4 5

33. $\{c \mid c < 7\}$

1 2 3 4 5 6 7 8 9

35. $\varnothing$

37. $\{x \mid x$ is a real number.$\}$

0 1 2 3 4 5 6 7 8

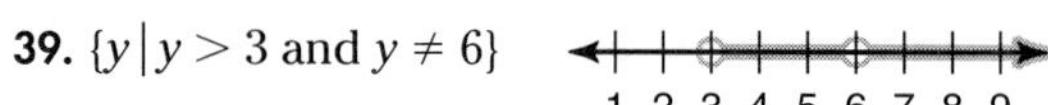

39. $\{y \mid y > 3$ and $y \ne 6\}$

41. $\{x \mid x$ is a real number.$\}$

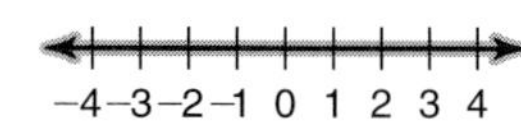

43. $\{w \mid w < 4\}$

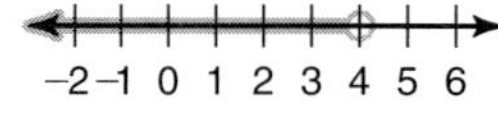

45. Sample answer: $x > 5$ and $x < -4$ **47.** $n + 2 \le 6$ or $n + 2 \ge 10$; $\{n \mid n \le 4$ or $n \ge 8\}$ **49.** $31 \le 6n - 5 \le 37$; $\{n \mid 6 \le n \le 7\}$
51. $\{m \mid -4 < m < 1\}$

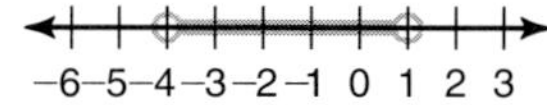

53a. $\{x \mid x < -7$ or $x > 1\}$ **53b.** $\{x \mid -5 \le x < 1\}$
55. $-4 \le x \le -1.5$ or $x \ge 2$ **57.** $4.4 < x < 6.7$
59. $\left\{m \mid m \ge \frac{44}{3}\right\}$ **61.** -5

63.

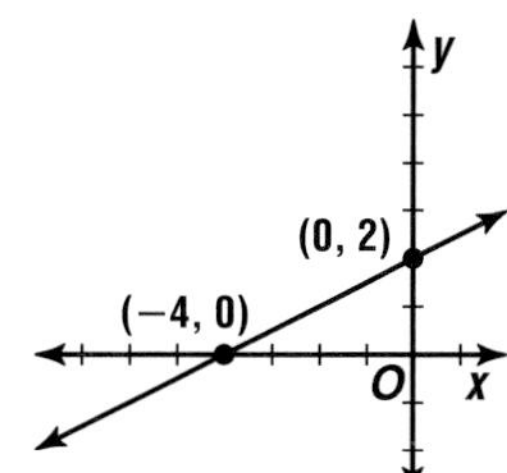

65. a little more than half a mile **67.** $18px - 15bg$

Page 412 Self Test

1. $\{y \mid y \ge -17\}$ **3.** $\{n \mid n < 4\}$ **5.** $\{g \mid g < -5\}$
7. c **9.** more than 17 points

Pages 415–419 Lesson 7–5

7a. outcomes from tree diagram:
burger, soup, lemonade; burger, soup, soft drink; burger, salad, lemonade; burger, salad, soft drink; burger, french fries, lemonade; burger, french fries, soft drink; sandwich, soup, lemonade; sandwich, soup, soft drink; sandwich, salad, lemonade; sandwich, salad, soft drink; sandwich, french fries, lemonade; sandwich, french fries, soft drink; taco, soup, lemonade; taco, soup, soft drink; taco, salad, lemonade; taco, salad, soft drink; taco, french fries, lemonade; taco, french fries, soft drink; pizza, soup, lemonade; pizza, soup, soft drink; pizza, salad, lemonade; pizza, salad, soft drink; pizza, french fries, lemonade; pizza, french fries, soft drink **7b.** $\frac{1}{3}$ or $0.\overline{3}$ **7c.** $\frac{1}{12}$ or $0.08\overline{3}$ **7d.** $\frac{1}{24}$ or $0.041\overline{6}$ **9a.** 15 **9b.** $\frac{1}{5}$ or 0.2 **11.** $\frac{1}{3}$ or $0.\overline{3}$
13a. R3-G5, R3-R10, R3-B10, R3-G1, R3-Y14, B3-G5, B3-R10, B3-B10, B3-G1, B3-Y14, R5-G5, R5-R10, R5-B10, R5-G1, R5-Y14, R14-G5, R14-R10, R14-B10, R14-G1, R14-Y14, Y10-G5, Y10-R10, Y10-B10, Y10-G1, Y10-Y14 **13b.** $\frac{3}{25}$ or 0.12
13c. $\frac{2}{25}$ or 0.08 **13d.** 0 **13e.** $\frac{14}{25}$ or 0.56 **15a.** about 5.6%
15b. about 26.3% **17.** 32% **19.** between 83 and 99, inclusive **21.** 3; -9 **23a.** 50° **23b.** 130° **23c.** yes

Pages 423–426 Lesson 7–6

5. c **7.** c **9.** d
11. $\{m \mid m \le -5$ or $m \ge 5\}$

−8 −6 −4 −2 0 2 4 6 8

13. $\{r \mid -9 < r < 3\}$

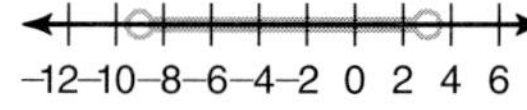

15. $|x| = 2$

SELECTED ANSWERS

17. $\{-2, 6\}$ −6 −4 −2 0 2 4 6 8 10

19. $\varnothing$

21. $\{y \mid 1 \leq y \leq 3\}$ −2 −1 0 1 2 3 4 5 6

23. $\varnothing$

25. $\left\{e \mid \frac{5}{3} < e < 3\right\}$ 1 2 3 4

27. $\{y \mid y$ is a real number.$\}$ −4 −3 −2 −1 0 1 2 3 4

29. $\{w \mid 0 \leq w \leq 18\}$ 0 2 4 6 8 10 12 14 16 18

31. $\{-2, 3\}$ −4 −3 −2 −1 0 1 2 3 4

33. $\left\{x \mid x \leq -\frac{8}{3} \text{ or } x \geq 4\right\}$ −4 −3 −2 −1 0 1 2 3 4 5 6

35. $|p - 1| \leq 0.01$ **37.** $|t - 50| > 50$ **39.** $|x + 1| = 3$ **41.** $|x - 1| \leq 1$ **43.** $|x - 8| \geq 3$ **45.** $\{-2, -1, 0, 1, 2\}$ **47.** $2a + 1$ **49.** $a \neq 0$; never **51.** $\frac{8}{13}$ or 0.61 **53.** no; $52 \leq s \leq 66$ **55.** $\$16{,}500 \leq p \leq \$18{,}000$ **57a.** Outcomes from tree diagram: BBBB, BBBG, BBGB, BBGG, BGBB, BGBG, BGGB, BGGG, GBBB, GBBG, GBGB, GBGG, GGBB, GGBG, GGGB, GGGG **57b.** $\frac{1}{16}$ or 0.0625, regardless of gender **57c.** $\frac{3}{8}$ or 0.375 **59.** $\{x \mid x \geq -1\}$ **61.** $\{k \mid k \geq -15\}$

63.

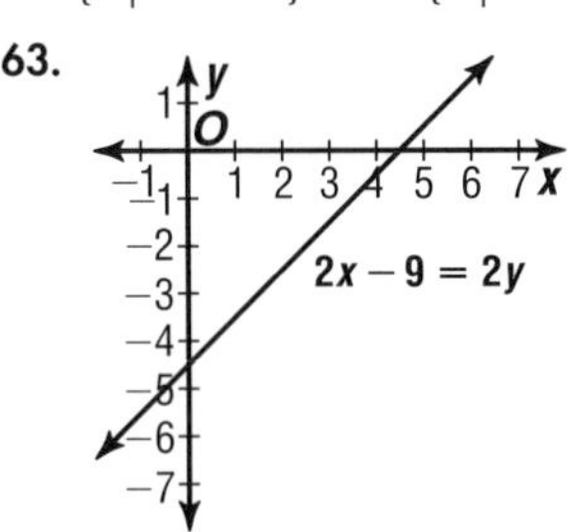

65. $m = -6 - \frac{n}{2}$

Pages 430–432 Lesson 7–7

7a. A; 25, 65, 30, 60, 40; B; 20, 70, 40, 60, 45 **7b.** B **7c.** A **7d.** B **9a.** Q2 = 6.5, Q3 = 16, Q1 = 5, IQR = 11 **9b.** no

9c.

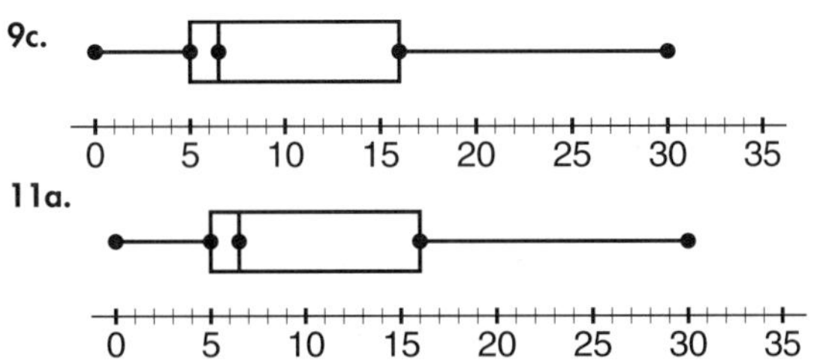

11b. clustered with lots of outliers **11c.** There are four western states that have more American Indian people than other states. **11d.** It is greater than the median.

13a.

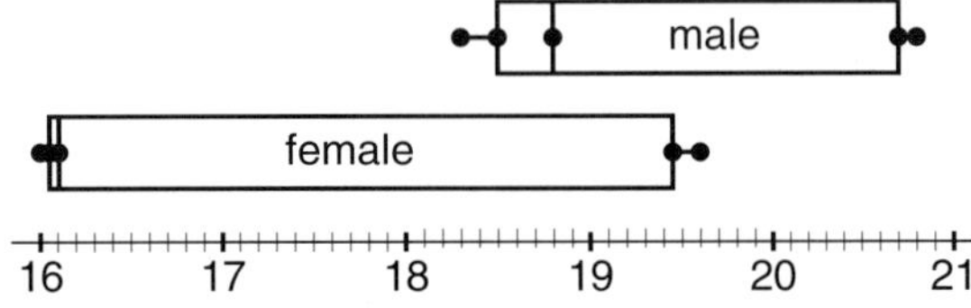

13b. 1990 **15.** Sample answer: Class A appears to be a more difficult class than B because the students don't do as well. **17.** $\{m \mid m > 1\}$

19.

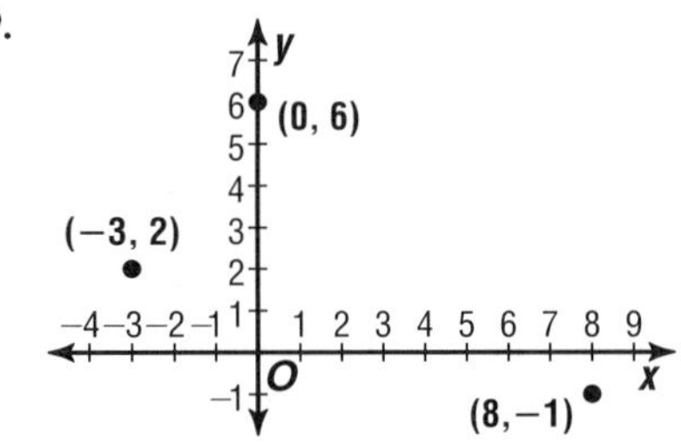

Page 434 Lesson 7–7B

1. women identifying objects with their left hands **3.** The left hand data are more clustered. **5.** males

Page 435 Lesson 7–8A

1.

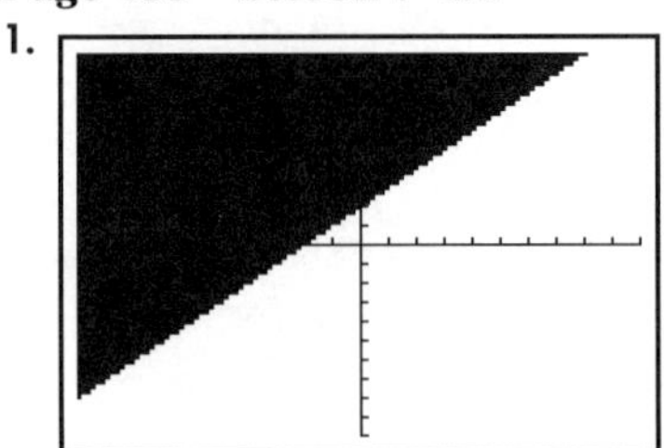

3.

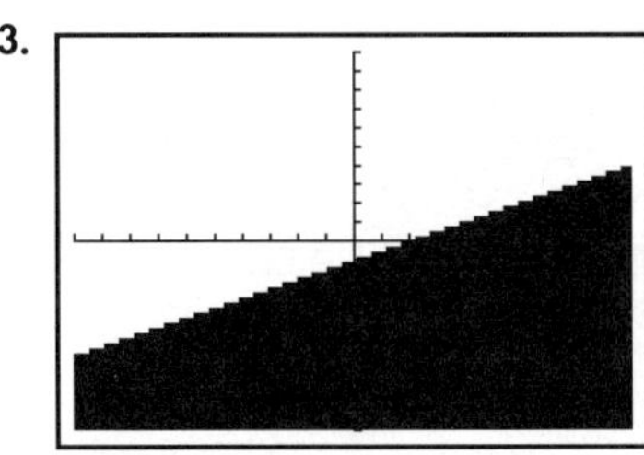

5.

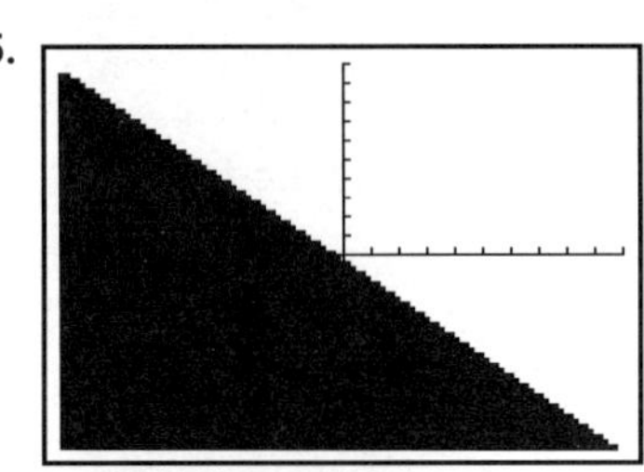

7.

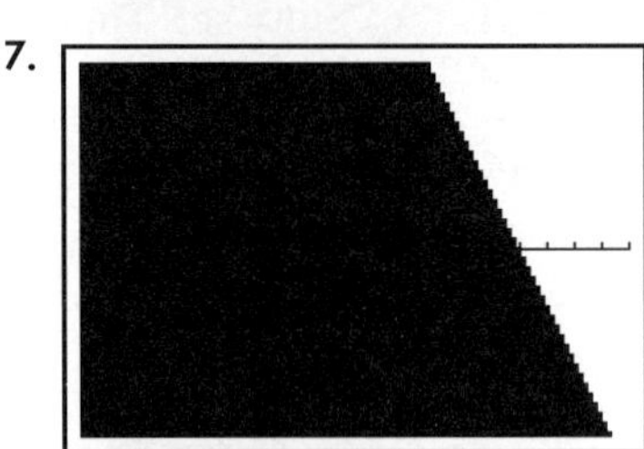

9.

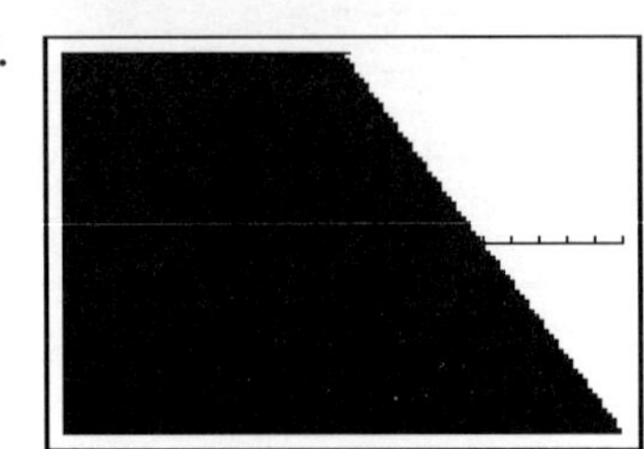

Page 439–441 Lesson 7–8

5. a **7.** b **9.** a, c; no

11.

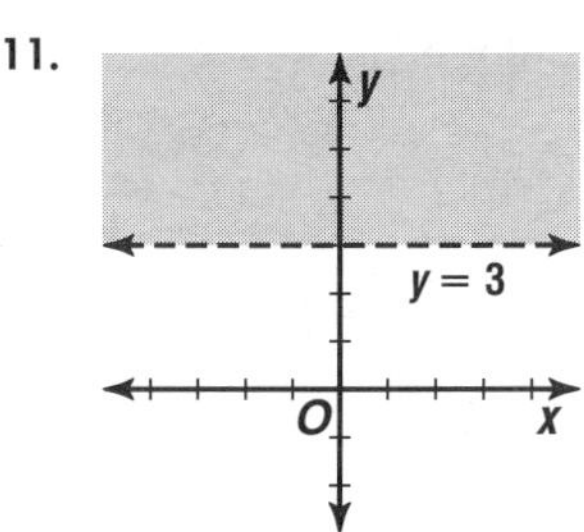

13.

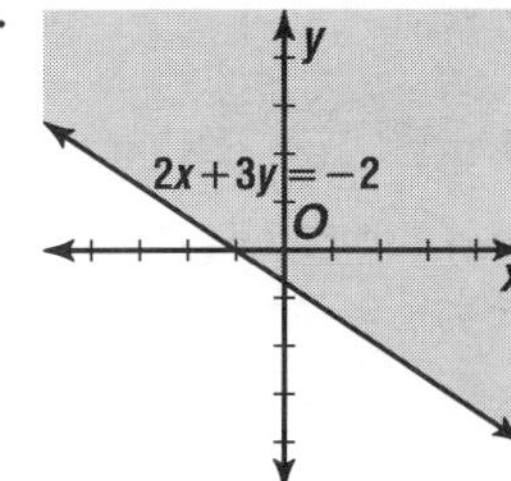

15.

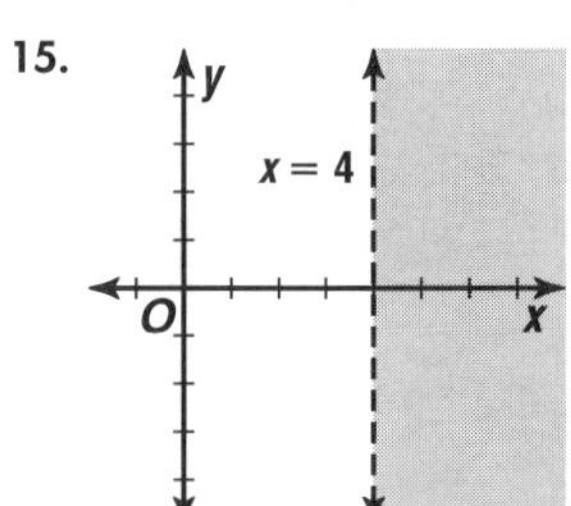

17.

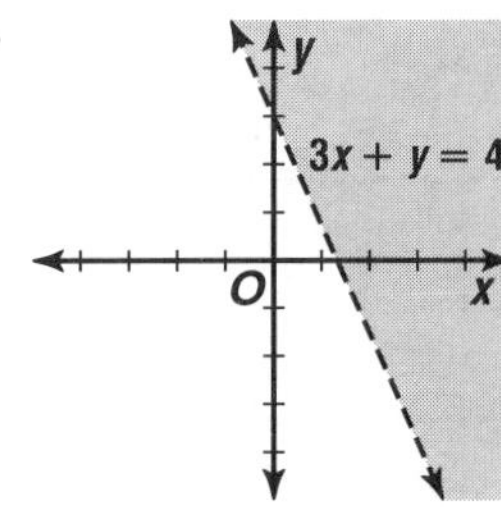

19. {(1, 1), (1, 2)} **21.** ∅

23. y = −3

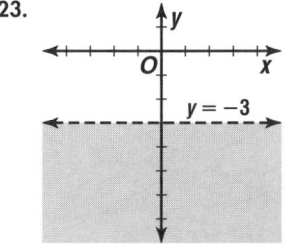

25.

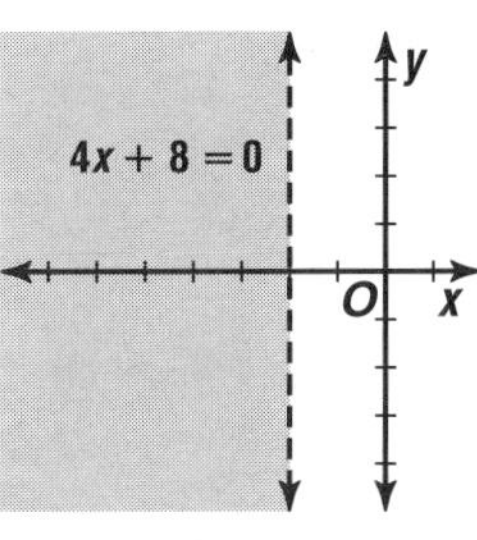

27.

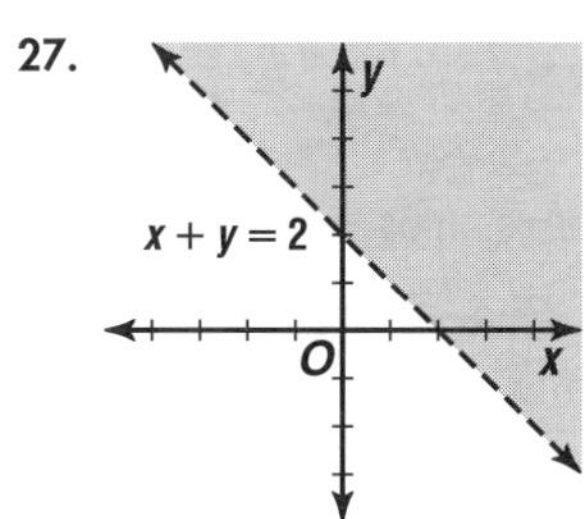

29.

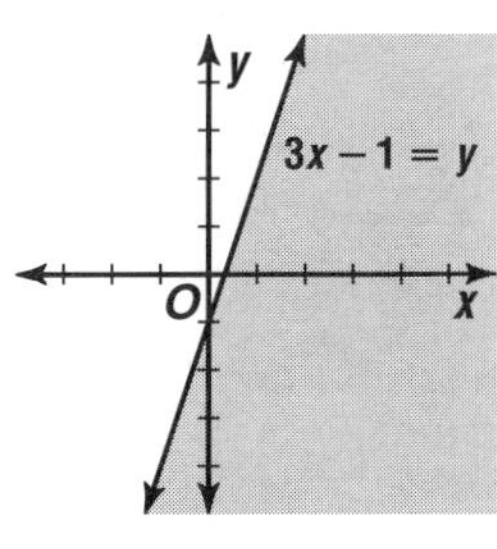

31.

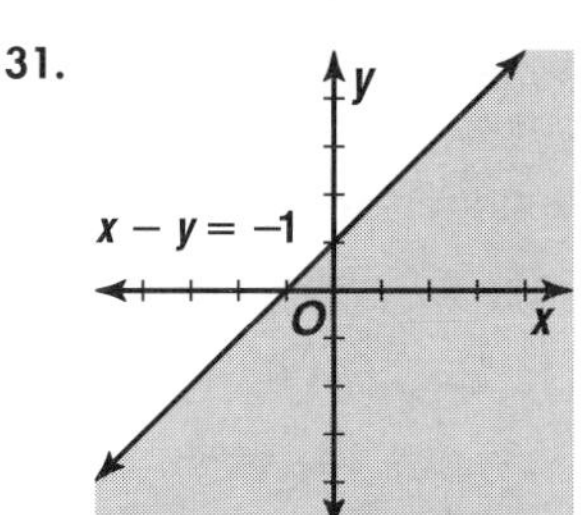

33.

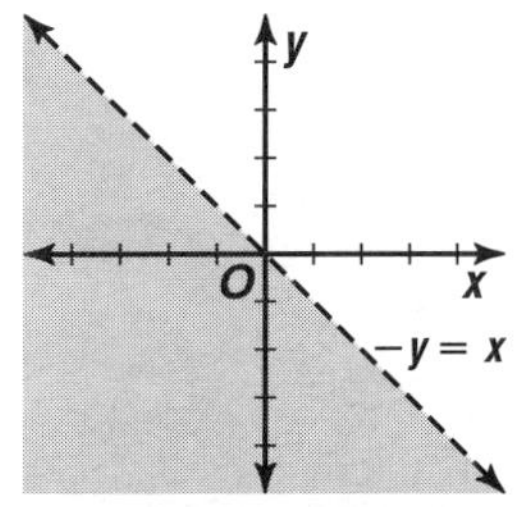

35.

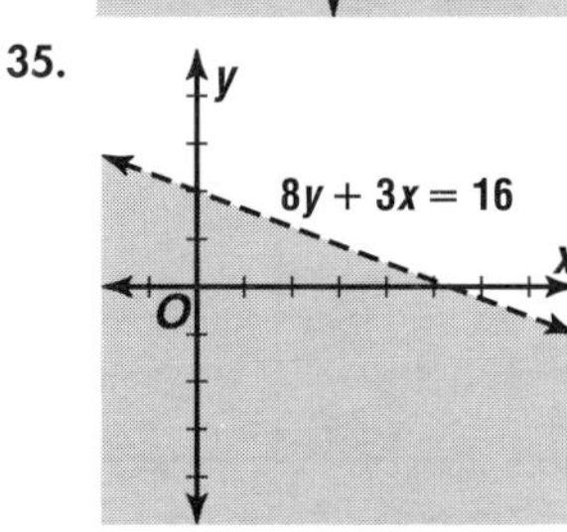

37.

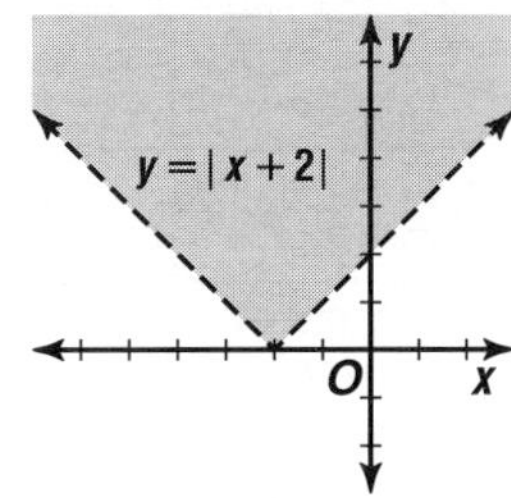

39.

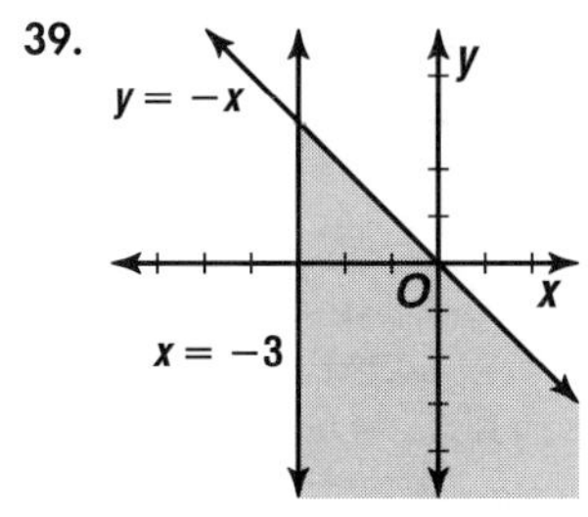

41.

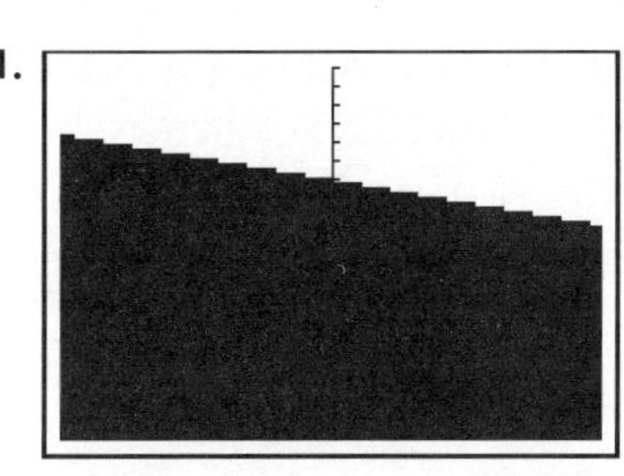

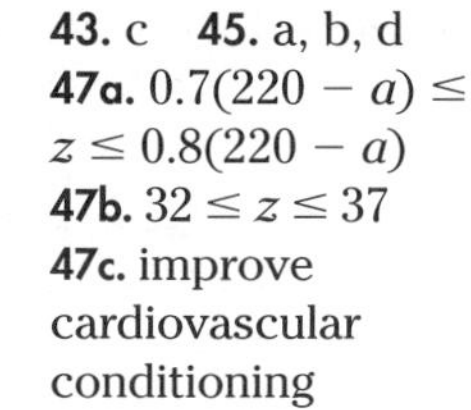

43. c **45.** a, b, d
47a. $0.7(220 - a) \leq z \leq 0.8(220 - a)$
47b. $32 \leq z \leq 37$
47c. improve cardiovascular conditioning

49.

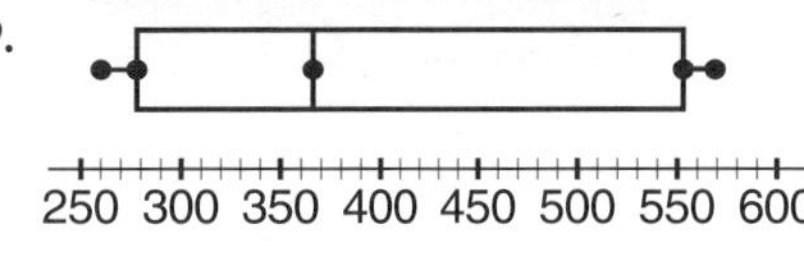

51. $y = 4x - 2$

53.

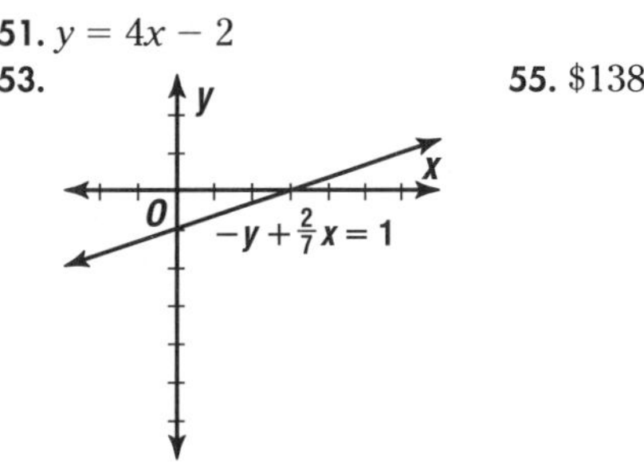

55. \$138.22

Page 443 Chapter 7 Highlights

1. h **3.** i **5.** c, f **7.** d **9.** g **11.** a

Pages 444–446 Chapter 7 Study Guide and Assessment

13. $\{n \mid n > -35\}$ **15.** $\{p \mid p \leq -18\}$ **17.** $3n > 4n - 8$, $\{n \mid n < 8\}$ **19.** $\{w \mid w \leq -15\}$ **21.** $\{x \mid x > 32\}$ **23.** $-\frac{3}{4}n \leq 30$, $\{n \mid n \geq -40\}$ **25.** $\{-5, -4, \ldots 0, 1\}$

27. $\left\{y \mid y \leq -\frac{9}{2}\right\}$ **29.** $\left\{x \mid x > -\frac{5}{2}\right\}$ **31.** $\{z \mid z \leq 20\}$

33. $\{a \mid a$ is a real number$\}$ −4 −3 −2 −1 0 1 2 3 4

35. $\{b \mid b \leq 5\}$ −1 0 1 2 3 4 5 6 7

37a. $\frac{2}{7}$ **37b.** $\frac{1}{2}$ **37c.** $\frac{2}{7}$ **37d.** 0 **37e.** $\frac{3}{14}$ **37f.** $\frac{1}{14}$

39. $\{y \mid y > -5$ or $y < -5\}$ −9 −8 −7 −6 −5 −4 −3 −2 −1

41. $\{k \mid 3 \geq k \geq -4\}$ −5 −4 −3 −2 −1 0 1 2 3 4 5

43. $\left\{y \mid y \geq \frac{21}{5} \text{ or } y \leq 1\right\}$ −2 −1 0 1 2 3 4 5 6

45.

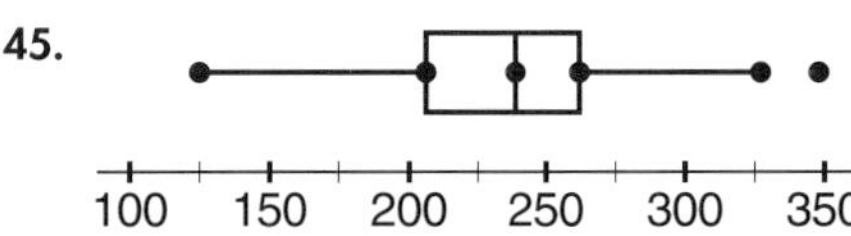

47. $\{(2, -1), (-1, 1)\}$ **49.** $\{(5, 10), (3, 6)\}$

51.

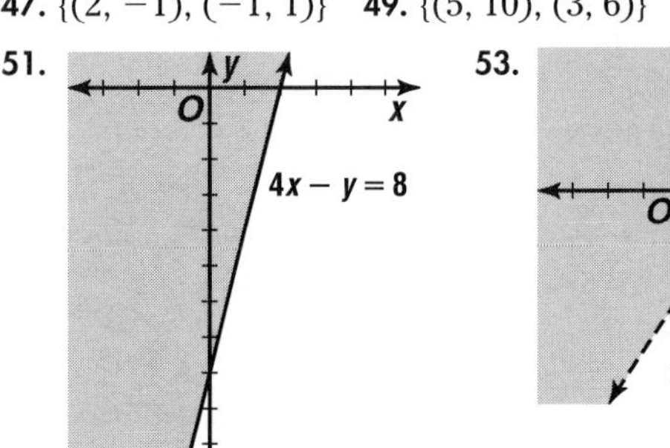

53.

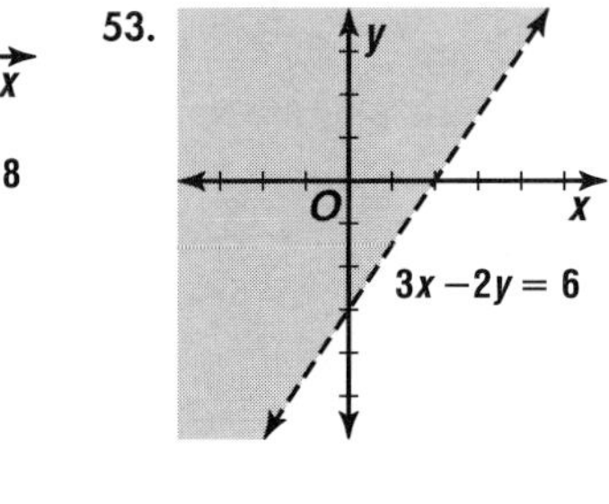

55. 17 to 20 books

CHAPTER 8 SYSTEMS OF LINEAR EQUATIONS AND INEQUALITIES

Page 453 Lesson 8–1A

1. (1, 8) **3.** (2.86, 4.57) **5.** (2.28, 3.08) **7.** (−2.9, 5.6) **9.** (1.14, −3.29)

Pages 458–461 Lesson 8-1

9. no solution **11.** one; (−6, 2) **13.** yes **15.** (3, 5) **17.** (2, −6)

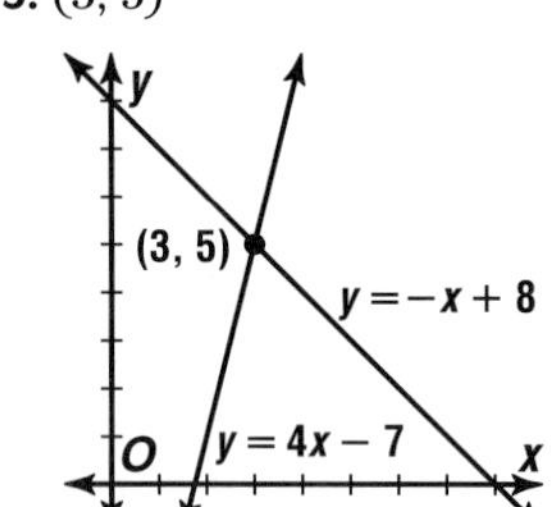

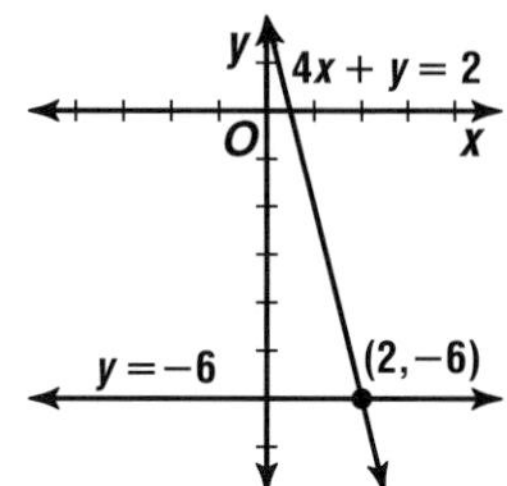

19. (−6, 8)

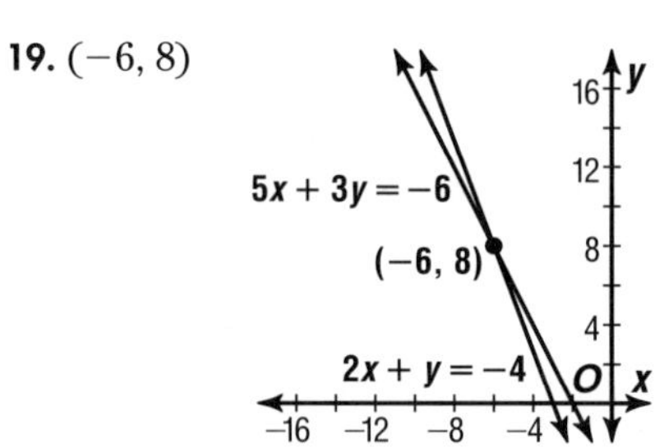

21. one, (3, −1) **23.** no solution **25.** one, (3, 3)

27. (2, −2) **29.** (−1, 3)

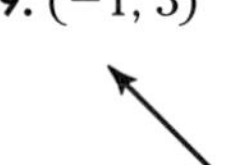

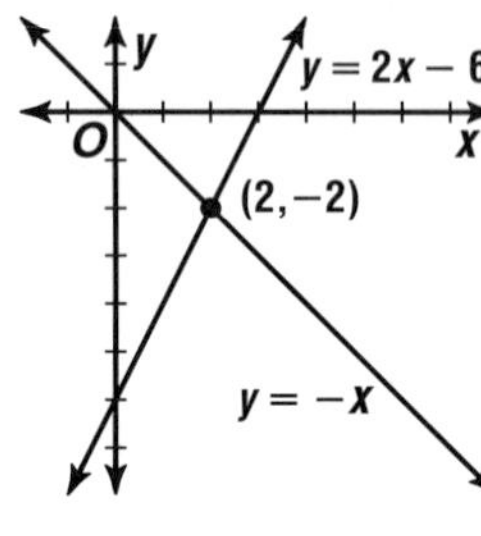

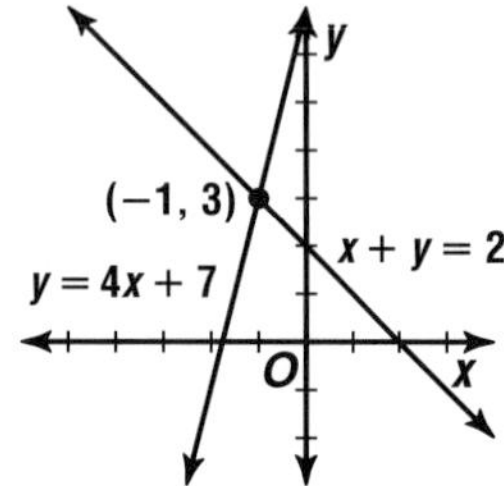

31. (−2, 4) **33.** (2, 0)

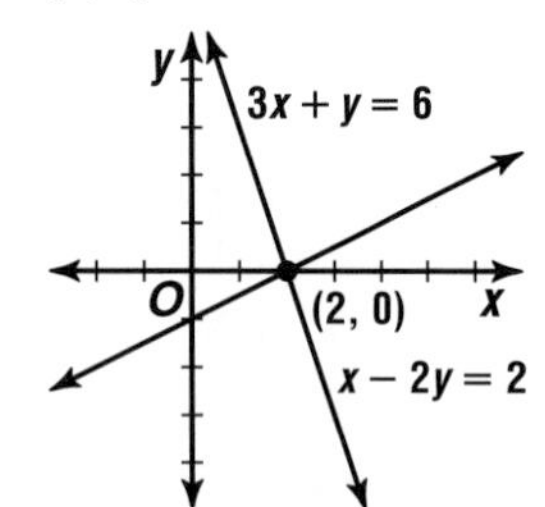

35. infinitely many **37.** no solution

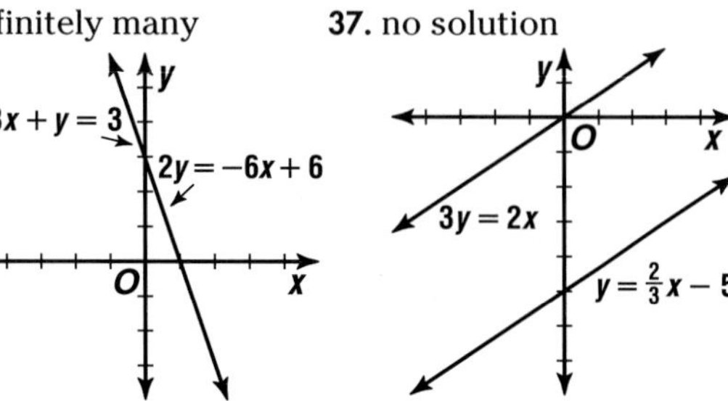

39. (8, 6) **41.** infinitely many

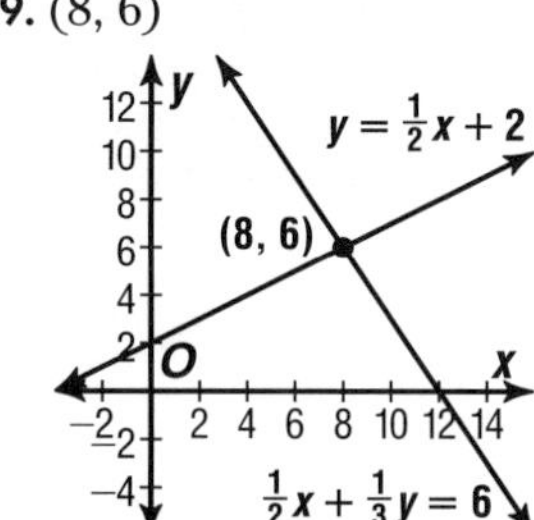

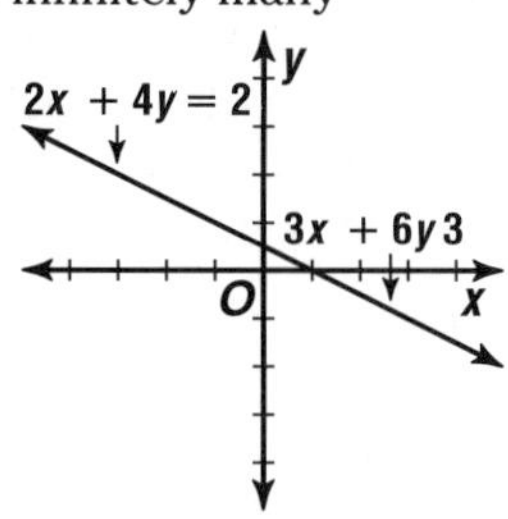

43. 15 square units **45.** (1.71, −2.57) **47.** (−0.25, −3.25) **49.** $A = -3, B = 2$

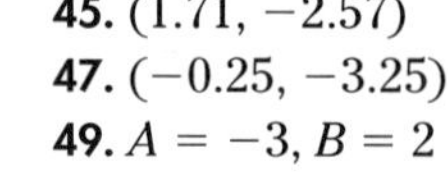

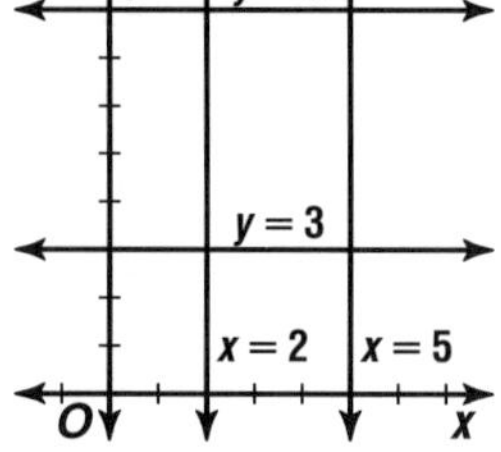

51. A.D. 376

53. −1.5, −13.5 **55.** $y = -2x + 2$ **57.** $7\frac{1}{2}\%$ **59a.** how many three- and four-bedroom homes will be built **59b.** $100 - h$ **59c.** 80 homes

Pages 466–468 Lesson 8–2

7. $x = 8 - 4y; y = 2 - \frac{1}{4}x$ **9.** $x = -\frac{0.75}{0.8}y - 7.5; y = -\frac{0.8}{0.75}x - 8$ **11.** $\left(3, \frac{3}{2}\right)$ **13.** no solution **15.** infinitely many **17.** (3, 1) **19.** (−4, 4) **21.** (4, −1) **23.** (2, 0) **25.** (9, 1) **27.** (2, 5) **29.** (4, 2) **31.** (5, 2) **33.** $\left(\frac{8}{3}, \frac{13}{3}\right)$ **35.** (36, −6, −84) **37.** (14, 27, −6) **39a.** $y = 1000 + 5x$, $y = 13x$ **39b.** 125 tickets **41a.** 26.5 years **41b.** 33.8 seconds

43a.

	75% Gold (18-carat)	50% Gold (12-carat)	58% Gold (14-carat)
Total Grams	x	y	300
Grams of Pure Gold	$0.75x$	$0.50y$	0.58(300)

43b. $x + y = 300$; $0.75x + 0.50y = 0.58(300)$ **43c.** 96 grams of 18-carat gold, 204 grams of 12-carat gold **45.** 37 shares **47.** {(−1, −7), (4, 8), (7, 17), (13, 35)} **49.** −3 **51.** $m - 12$

Pages 472–474 Lesson 8–3

5. addition, (1, 0) **7.** subtraction, $\left(-\frac{5}{2}, -2\right)$ **9.** substitution, (1, 4) **11.** 8, 48 **13.** (1, −4), (1.29, −4.05) **15.** +; (6, 2) 17. −; (4, −1) **19.** −; (4, −7) **21.** +; (5, 1) **23.** sub; infinitely many **25.** −; (−2, 3) **27.** −; $\left(\frac{1}{2}, 1\right)$

29. +; (10, −15) **31.** +; (1.75, 2.5) **33.** 11, 53 **35.** 5, 8 **37.** (2, 3, 7) **39.** (14, 27, −6) **41.** Ling, 1.45 hours or 1 hour, 27 minutes; José, 1.15 hours, or 1 hour, 9 minutes **43.** 320 gal of 25% and 180 gal of 50% **45.** −3 **47.** $-\frac{7}{6}$ **49.** substitution (=)

Page 474 Self Test

1. (1, −2)

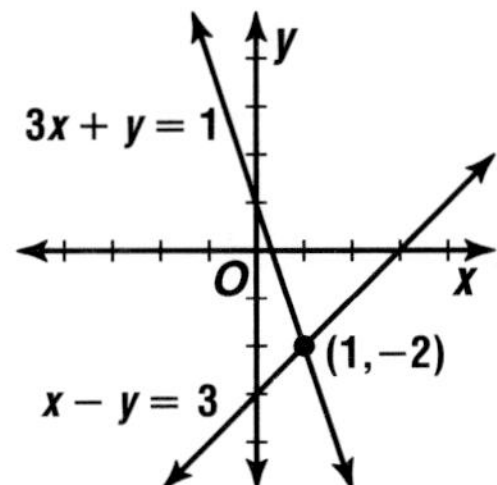

3. infinitely many

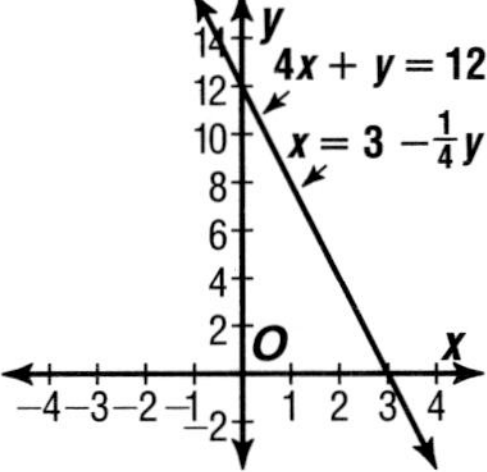

5. (−9, −7) **7.** (10, 15) **9.** (4, −2)

Pages 478–481 Lesson 8–4

5. (−1, 1); Multiply the first equation by −3, then add. **7.** (−9, −13); Multiply the second equation by 5, then add. **9.** (−1, −2); Multiply the first equation by 5, multiply the second equation by −8, then add. **11.** b; (2, 0) **13.** c; (4, 1) **15.** (2, 1) **17.** (5, −2) **19.** (2, −5) **21.** (−4, −7) **23.** (−1, −2) **25.** (4, −6) **27.** (13, −2) **29.** (10, 12) **31.** 6, 9 **33.** elimination, addition; $\left(2, \frac{1}{8}\right)$ **35.** substitution or elimination, multiplication; infinitely many **37.** elimination, subtraction; (24, 4) **39.** (11, 12) **41.** $\left(\frac{1}{3}, \frac{1}{6}\right)$ **43.** (−2, 7), (2, 2), (7, 5) **45.** 6 2-seat tables, 11 4-seat tables **47.** (3, −4) **49.** $\frac{1}{3}$ **51.** 165 yd **53.** 2 cups

Pages 485–486 Lesson 8–5

7.

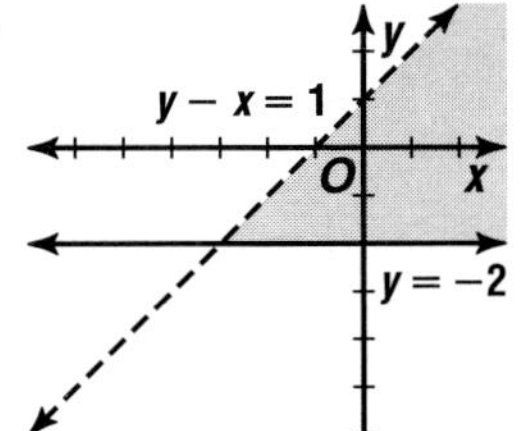

9.

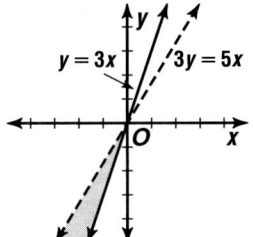

11.

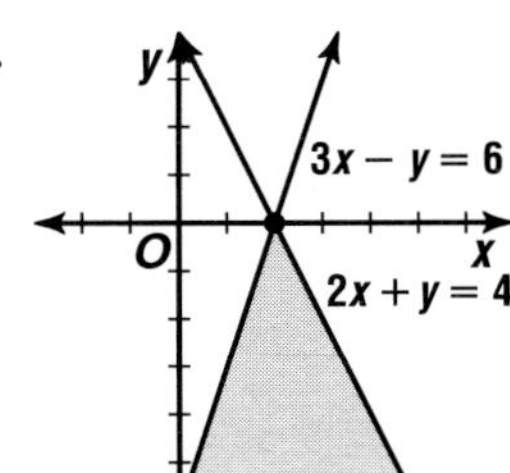

13. $y > x, y \leq x + 4$

15.

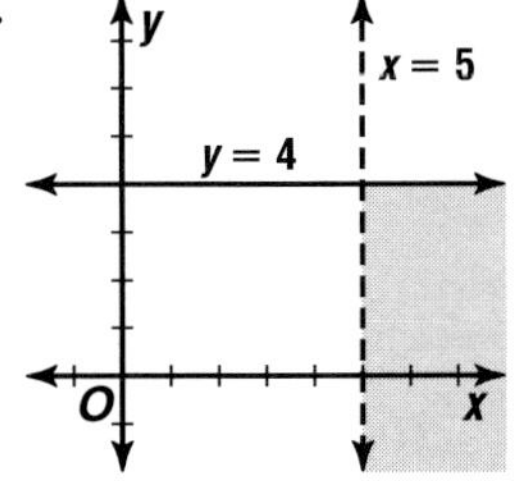

17.

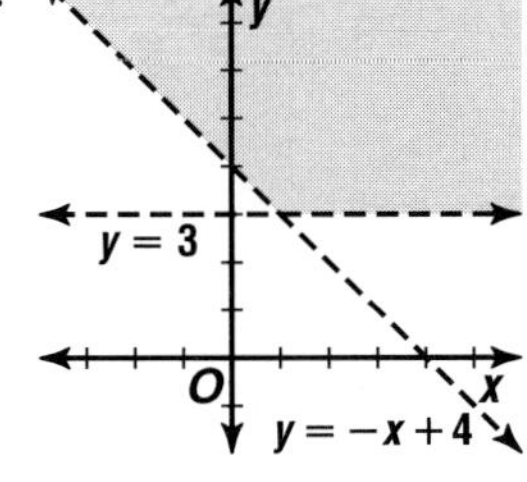

19.

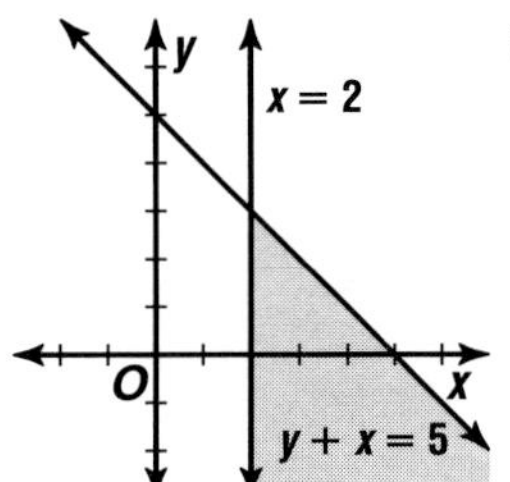

21.

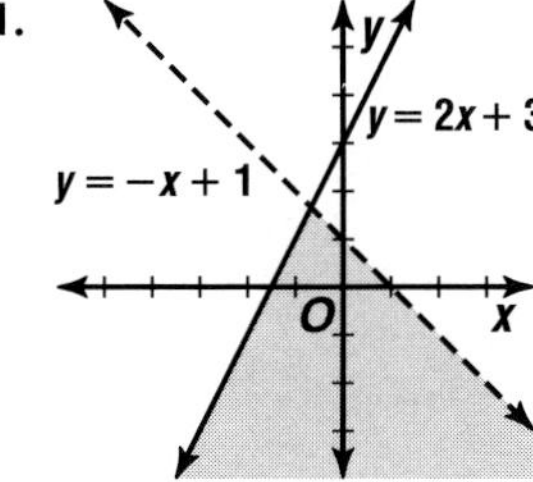

23.

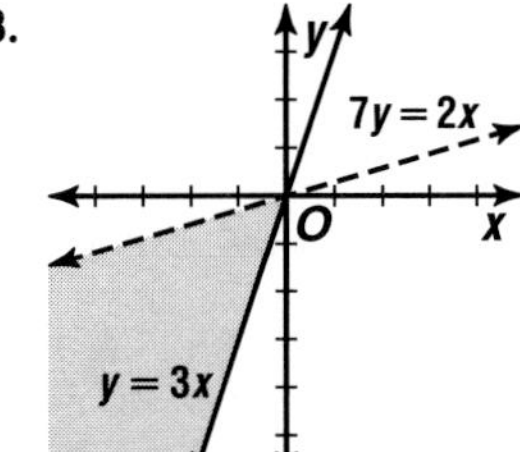

25.

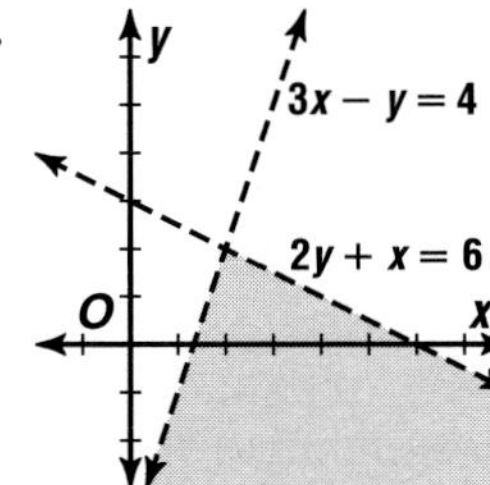

27.

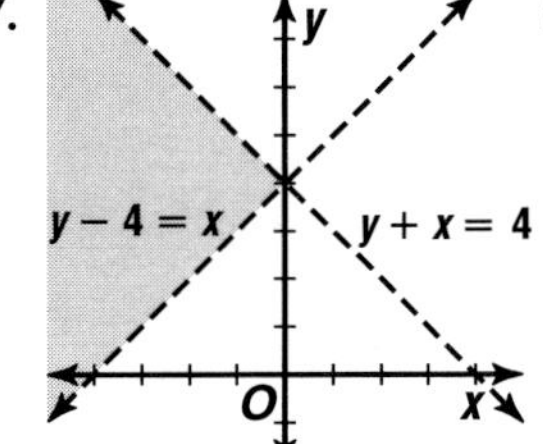

29.

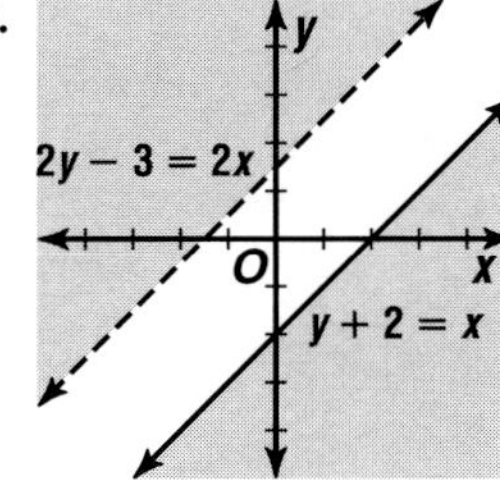

31.

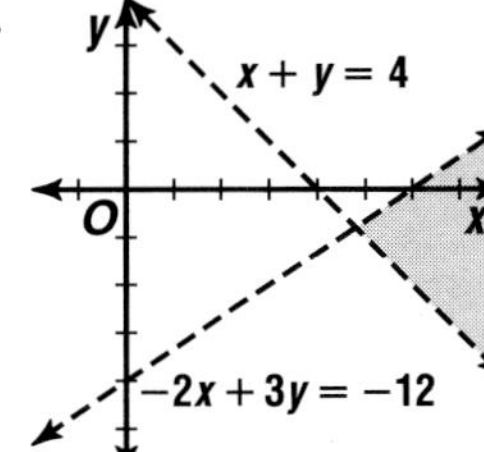

33. $y > -1, x \geq -2$ **35.** $y \leq x, y > x - 3$ **37.** $x \geq 0, y \geq 0, x + 2y \leq 6$

39.

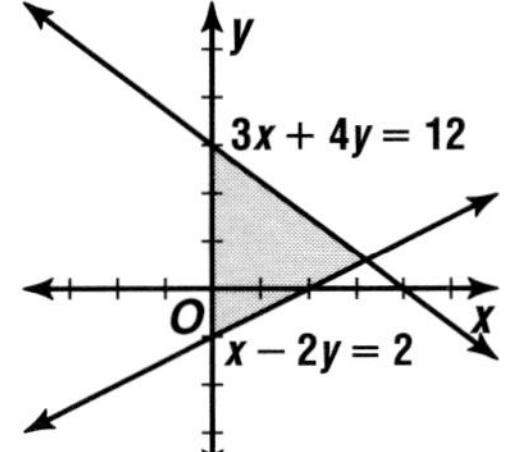

41.

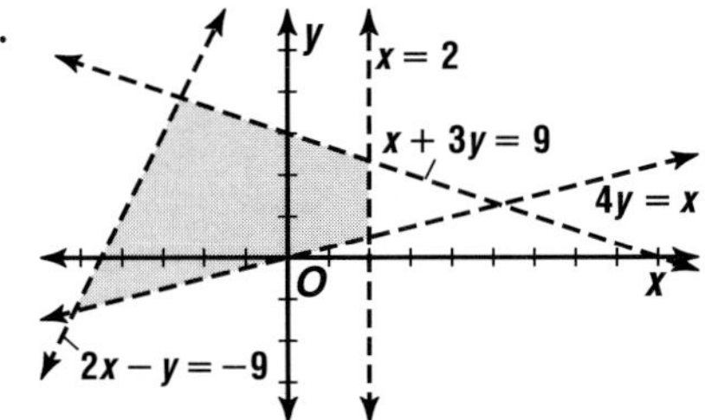

43.

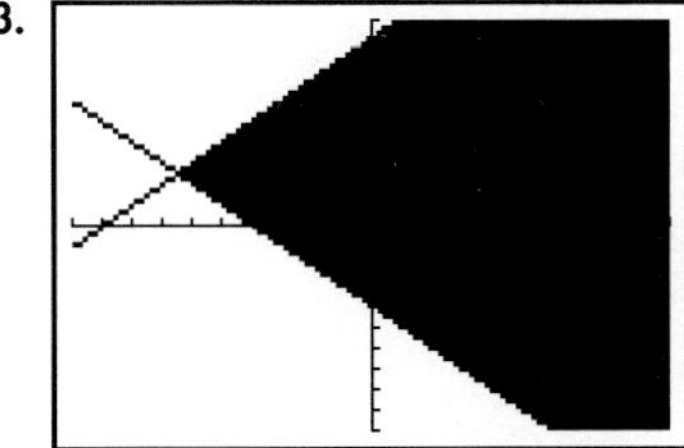

45.

47. Sample answer: walk, 15 min, jog, 15 min; walk, 10 min, jog, 20 min; walk, 5 min, jog, 25 min. **49.** 4 \$5 bills, 8 \$20 bills **51.** $y = -\frac{1}{2}x + \frac{9}{2}$ **53.** 10 **55.** Let y = the number of yards gained in both games; $y = 134 + (134 - 17)$

Page 487 Chapter 8 Highlights

1. substitution **3.** inconsistent **5.** elimination **7.** infinitely many **9.** no **11.** second

Pages 488–490 Chapter 8 Study Guide and Assessment

13. no solution **15.** $(-2, -7)$

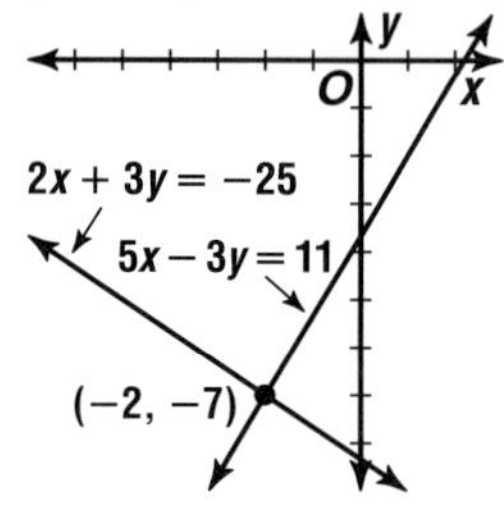

17. no solution

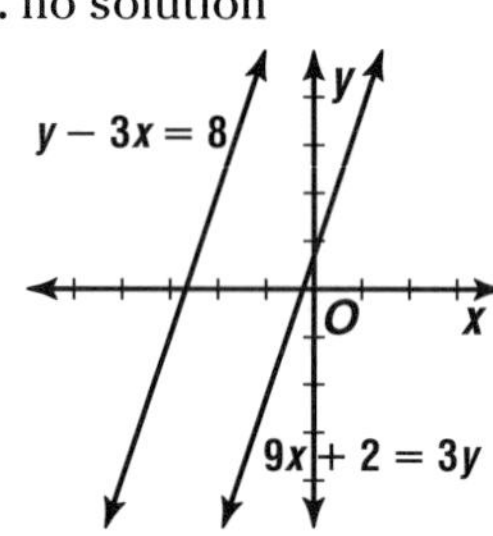

19. one, $(2, -2)$

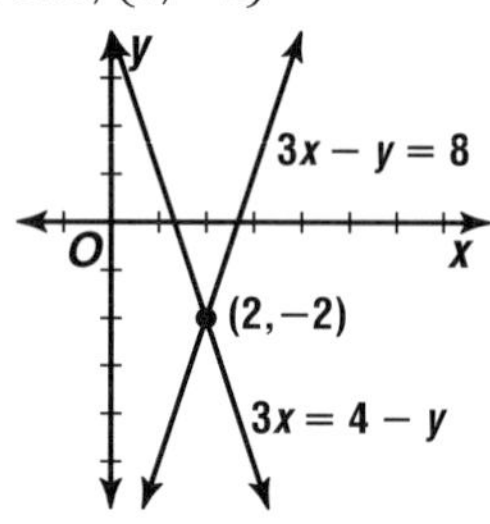

21. $(0, 2)$ **23.** $\left(\frac{1}{2}, \frac{1}{2}\right)$ **25.** $(2, -1)$ **27.** $(5, 1)$ **29.** $(8, -2)$ **31.** $(-9, -7)$ **33.** $\left(\frac{3}{5}, 3\right)$ **35.** $(2, -1)$ **37.** $\left(\frac{7}{9}, 0\right)$ **39.** $(0, 0)$ **41.** $(13, -2)$

43.

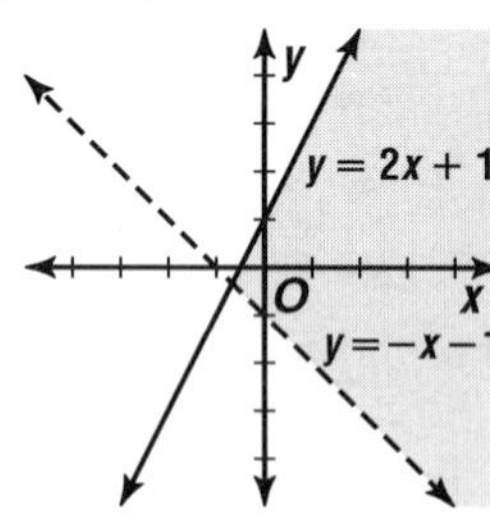

45.

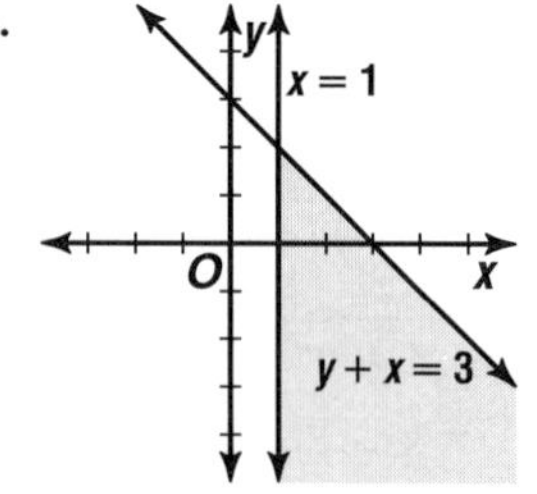

47.

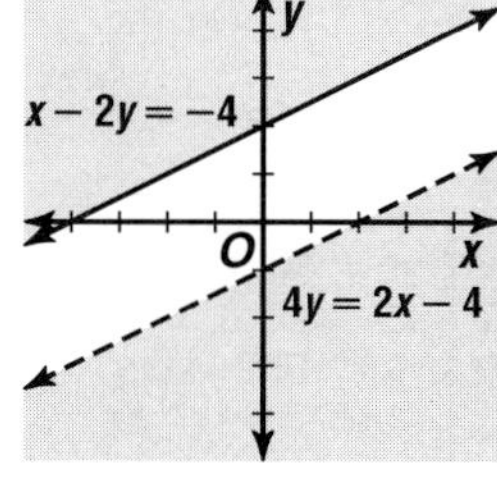

49. 35 **51a.** \$75 **51b.** \$15

CHAPTER 9 EXPLORING POLYNOMIALS

Pages 499–500 Lesson 9–1

5. no **7.** no **9.** a^{12} **11.** 3^{16} or 43,046,721 **13.** $9a^2y^6$ **15.** $15a^4b^3$ **17.** m^4n^3 **19.** 2^{12} or 4096 **21.** $a^{12}x^8$ **23.** $6x^4y^4z^4$ **25.** $a^2b^2c^2$ **27.** $\frac{4}{25}d^2$ **29.** $0.09x^6y^4$ **31.** $90y^{10}$ **33.** $4a^3b^5$ **35.** $-520x^9$ **37.** -2^4 equals $-(2)(2)(2)(2)$ or -16 and $(-2)^4$ equals $(-2)(-2)(-2)(-2)$ or 16. **39.** 301 parts **41.** one; $(-6, 3)$

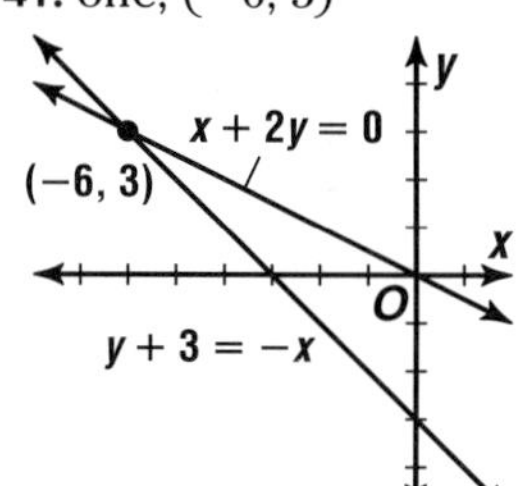

43. $n = 2m + 1$

m	−3	−2	−1	0	1
n	−5	−3	−1	1	3

45. 136° **47.** $1.11y + 0.06$

Pages 504–505 Lesson 9–2

7. $\frac{1}{121}$ **9.** 36 **11.** $\frac{6}{r^4}$ **13.** $\frac{\pi}{4}$ **15.** a^2 **17.** m^6 **19.** $\frac{m^3}{3}$ **21.** $\frac{c^3}{b^8}$ **23.** b^8 **25.** $-s^6$ **27.** $-\frac{4b^3}{c^3}$ **29.** $\frac{1}{64a^6}$ **31.** $\frac{s^3}{r^3}$ **33.** $\frac{a}{4c^2}$ **35.** m^{3+n} **37.** 3^{4x-6} **39.** $\frac{1}{q^{18}}$ **41.** \$1257.14 **43.** $98a^5b^4$ **45.** $|x + 1| < 3$ **47.** $(3, -5)$ **49.** $\frac{52}{41}$

Pages 509–512 Lesson 9–3

5. 43,400,000; 1.515×10^3 **7.** 507,000,000; 4.4419×10^4 **9.** 4,551,400,000; 7.14×10^2 **11.** 1.672×10^{-21} mg **13.** 4×10^{-6} in. **15.** 6.2×10^{-7}; 0.00000062 **17.** 6×10^7; 60,000,000 **19.** 9.5×10^{-3} **21.** 8.76×10^{10} **23.** 3.1272×10^8 **25.** 9.0909×10^{-2} **27.** 7.86×10^4 **29.** 7×10^{-10} **31.** 9.9×10^{-6} **33.** 6×10^{-3}; 0.006 **35.** 8.992×10^{-7}; 0.0000008992 **37.** 4×10^{-2}; 0.04 **39.** 6.5×10^{-6}; 0.0000065 **41.** 6.6×10^{-6}; 0.0000066 **43.** 1.2×10^{-4}; 0.00012 **45.** 2.4336×10^{-1} **47.** 2.8×10^5 **49a.** Sample answer: overflow **49b.** Multiply 3.7 and 5.6 and multiply 10^{112} and 10^{10}. Then write the product in scientific notation. **49c.** 2.072×10^{123} **51.** 1,000,000,000,001 **53.** 6.75×10^{18} molecules **55a.** about 90.1 kg

57.

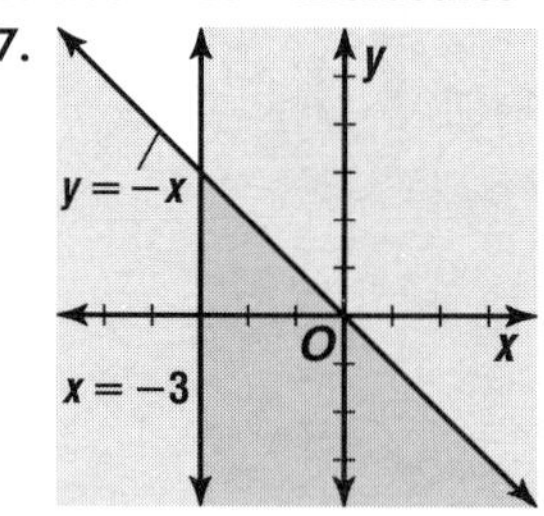

59. 25% **61.** $y = x + 5$ **63.** yes **65.** 27° **67.** −163

Page 513 Lesson 9–4A

1.

3.

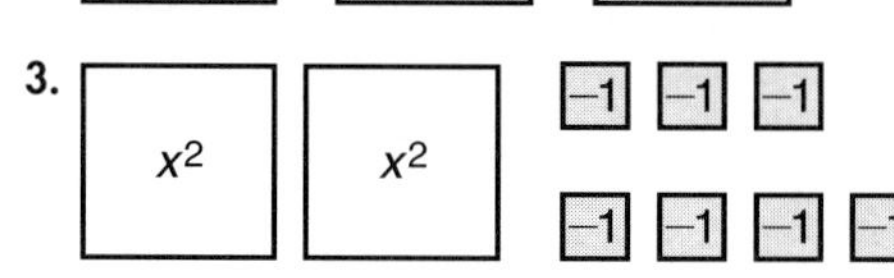

5. $x^2 - 3x + 2$ **7.** $2x^2 - x - 5$ **9.** x^2, x, and 1 represent the areas of the tiles.

Pages 517–519 Lesson 9–4

5. yes; trinomial **7.** yes; binomial **9.** 0 **11.** $x^8 - 12x^6 + 5x^3 - 11x$ **13a.** $\ell w - \pi r^2 - s^2$ **13b.** about 91.43 square units **15.** yes; trinomial **17.** yes; binomial **19.** yes; trinomial **21.** 5 **23.** 3 **25.** 9 **27.** 4 **29.** $x^5 + 3x^3 + 5$ **31.** $-x^7 + abx^2 - bcx + 34$ **33.** $1 + x^2 + x^3 + x^5$ **35.** $7a^3x + \frac{2}{3}x^2 - 8a^3x^3 + \frac{1}{5}x^5$ **37.** $2ab + \pi b^2$; about 353.10 square units **39.** $ab - 4x^2$; 116 square units **41b.** $8a^4 + 9a^3 + 4a^2 + 3a^1 + 5a^0$ **43.** about 153 eggs **45.** 4.235×10^4 **47.** 7, 8, 9 **49.** $\left\{(-3, 8), \left(1, \frac{8}{3}\right), (3, 0), (9, -8)\right\}$ **51.** \$38.75

Page 521 Lesson 9–5A

1. $-x^2 + 6$ **3.** $-2x^2 - 6x$ **5.** $3x^2 - x + 4$

7. true

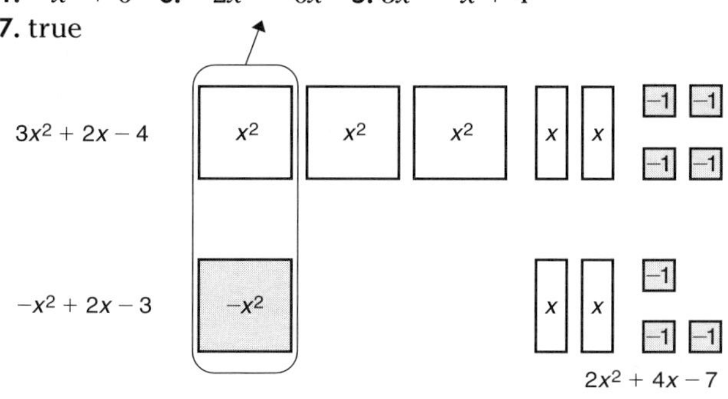

9. Method from Activity 2:

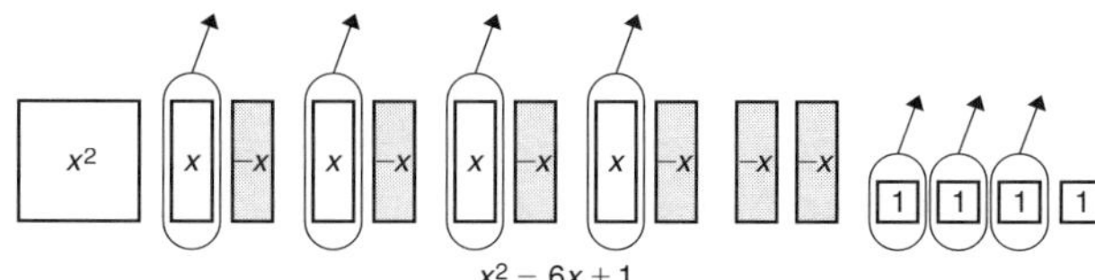

You need to add zero-pairs so that you can remove 4 green x-tiles.

Method from Activity 4:

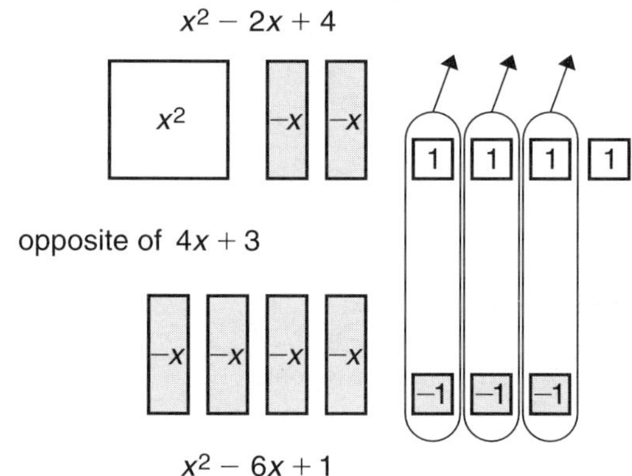

You remove all zero-pairs to find the difference in simplest form.

Pages 524–527 Lesson 9–5

7. $6a^2 - 3$ **9.** $4x^2 + 3y^2 - 8y - 7x$ **11.** $-8y^2$ and $3y^2$; $2x$ and $4x$ **13.** $3p^3q$ and $10p^3q$; $-2p$ and $-p$ **15.** $6m^2n^2 + 8mn - 28$ **17.** $-4y^2 + 5y + 3$ **19.** $7p^3 - 3p^2 - 2p - 7$ **21.** $10x^2 + 13xy$ **23.** $4a^3 + 2a^2b - b^2 + b^3$ **25.** $-4a + 6b - 5c$ **27.** $3a - 11m$ **29.** $-2n^2 + 7n + 5$ **31.** $13x - 2y$ **33.** $-y^3 + 3y + 3$ **35.** $4z^3 - 2z^2 + z$ **37.** $2x + 3y$ **39.** $353 - 18x$ **41.** $-2n^2 - n + 4$ **43.** $719x^2$ cubic stories **45.** $-3x - 2x^3 + 4x^5$

47. (3, 1)

y
(3, 1)
O
x
2x – 5y = 1
5x – 3y = 12

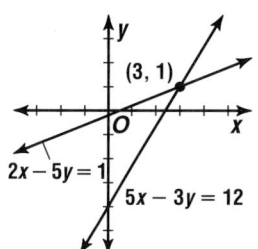

49. Sample answer: Let h = the number of hours for the repair and let c = the total charge; $c = 34h + 15$.

51. $-\frac{50}{3}$

Page 527 Self Test

1. $-6n^5y^7$ **3.** $-12ab^4$ **5.** 5.67×10^6 **7.** about 1.53×10^4 seconds or 4.25 hours **9.** $5x^2 - 4x - 14$

Page 528 Lesson 9–6A

1. $x^2 + 2x$ **3.** $2x^2 + 2x$ **5.** $2x^2 + x$

7. true

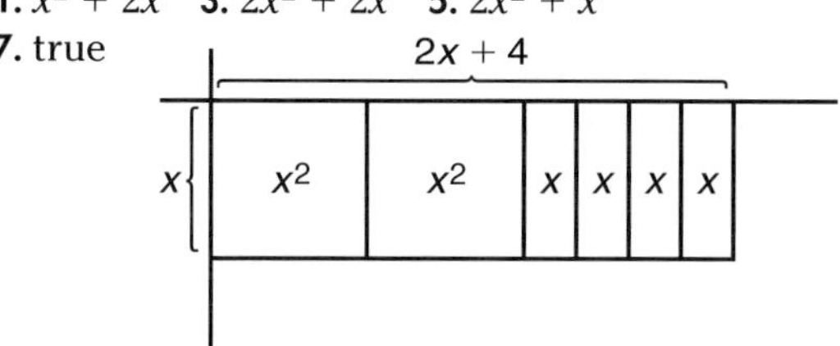

9a. $3x$ and $x + 15$

9b. $(3x^2 + 45x)$ square feet

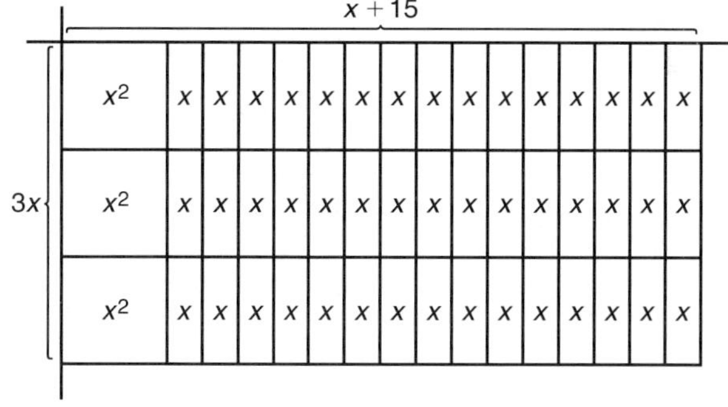

Pages 531–533 Lesson 9–6

5. $-63b^4c - 7b$ **7.** $10y^2 - 26y$ **9.** $3w^2 - 2w$ **11.** $\frac{103}{19}$ **13a.** $a^2 + a$ **13b.** $a^2 + 2a$ **15.** $\frac{1}{3}x^2 - 9x$ **17.** $-20m^5 - 8m^4$ **19.** $30m^5 - 40m^4n + 60m^3n^3$ **21.** $-28d^3 + 16d^2 - 12d$ **23.** $-32r^2s^2 - 56r^2s + 112rs^3$ **25.** $\frac{36}{5}x^3y + x^3 - 24x^2y$ **27.** $36t^2 - 42$ **29.** $61y^3 - 16y^2 + 167y - 18$ **31.** $53a^3 - 57a^2 + 7a$ **33.** $-\frac{77}{8}$ **35.** 0 **37.** $\frac{23}{24}$ **39.** 2 **41.** $15p^2 + 32p$ **43.** Sample answer: $1(8a^2b + 18ab)$, $a(8ab + 18b)$, $b(8a^2 + 18a)$, $2(4a^2b + 9ab)$, $(2a)(4ab + 9b)$, $(2b)(4a^2 + 9a)$, $(ab)(8a + 18)$, $(2ab)(4a + 9)$ **45.** $1.50t + 1.25mt$

47a.

0 diagonals

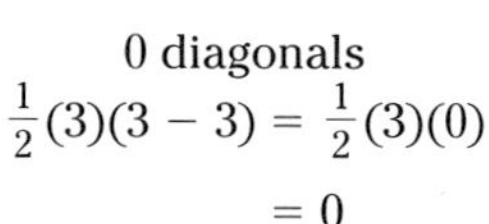

$\frac{1}{2}(3)(3 - 3) = \frac{1}{2}(3)(0)$

$= 0$

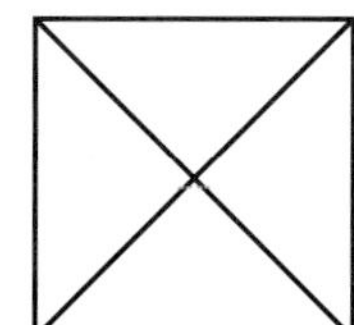

2 diagonals

$\frac{1}{2}(4)(4 - 3) = \frac{1}{2}(4)(1)$

$= 2$

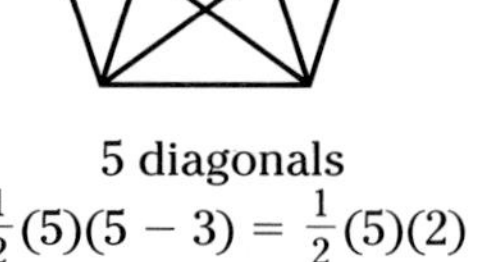

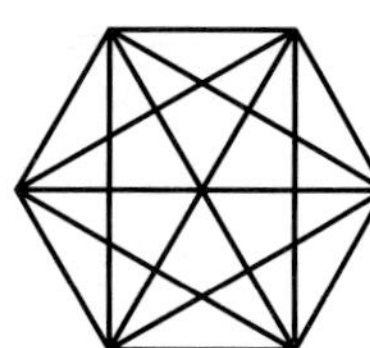

5 diagonals

$\frac{1}{2}(5)(5-3) = \frac{1}{2}(5)(2)$
$= 5$

9 diagonals

$\frac{1}{2}(6)(6-3) = \frac{1}{2}(6)(3)$
$= 9$

47b. $\frac{1}{2}n^2 - \frac{3}{2}n$ **47c.** 90 diagonals **49.** 75 gal of 50%, 25 gal of 30% **51.** $a^2 - 1$ **53.** 11 days **55.** 25.7

Page 535 Lesson 9–7A

1. $x^2 + 3x + 2$ **3.** $x^2 - 6x + 8$ **5.** $2x^2 - 2$

7. false

$x + 6$ (top); $x + 4$ (side)

x^2	x	x	x	x	x	x
x	1	1	1	1	1	1
x	1	1	1	1	1	1
x	1	1	1	1	1	1
x	1	1	1	1	1	1

9. false

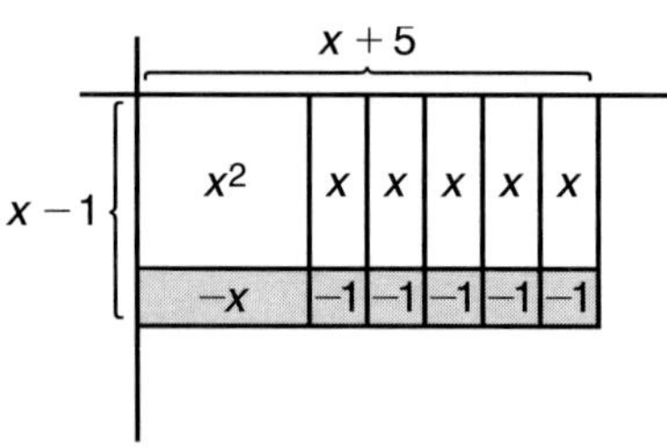

11. By the distributive property, $(x + 3)(x + 2) = x(x + 2) + 3(x + 2)$. The top row represents $x(x + 2)$ or $x^2 + 2x$. The bottom row represents $3(x + 2)$ or $3x + 6$.

Pages 539–541 Lesson 9–7

5. $d^2 + 10d + 16$ **7.** $y^2 - 4y - 21$ **9.** $2x^2 + 9x - 5$ **11.** $10a^2 + 11ab - 6b^2$ **13a.** $a^3 + 3a^2 + 2a$ **13c.** The result is the same as the product in part b. **15.** $c^2 - 10c + 21$ **17.** $w^2 - 6w - 27$ **19.** $10b^2 - b - 3$ **21.** $169x^2 - 9$ **23.** $0.15v^2 - 2.9v - 14$ **25.** $\frac{2}{3}a^2 + \frac{1}{18}ab - \frac{1}{3}b^2$ **27.** $0.63p^2 + 3.9pq + 6q^2$ **29.** $6x^3 + 11x^2 - 68x + 55$ **31.** $9x^3 - 45x^2 + 62x - 16$ **33.** $20d^4 - 9d^3 + 73d^2 - 39d + 99$ **35.** $10x^4 + 3x^3 + 51x^2 - 16x - 48$ **37.** $a^4 - a^3 - 8a^2 - 29a - 35$ **39.** $63y^3 - 57y^2 - 36y$ **41a.** 28 cm, 20 cm, 16 cm **41b.** 8960 cm^3 **41c.** 8960 **41d.** They are the same measure. **43.** $-8x^4 - 6x^3 + 24x^2 - 12x + 80$ **45.** $-3x^3 - 5x^2 - 6x + 24$ **47a.** Sample answer: $x - 2$, $x + 3$ **47b.** Sample answer: $x^2 + x - 6$ **47c.** larger, 2 sq ft **49.** $\{y \mid y < -6\}$ **51.** $y = 2x + 1$ **53.** $a = 4, y = 9$ **55.** 15°C

Pages 546–547 Lesson 9–8

7. $m^2 - 6mn + 9n^2$ **9.** $m^4 + 8m^2n + 16n^2$ **11.** $25 - 10x + x^2$ **13.** $x^2 + 8xy + 16y^2$ **15.** $9b^2 - 6ab + a^2$ **17.** $81p^2 - 4q^2$ **19.** $25b^2 - 120ab + 144a^2$ **21.** $x^6 + 2x^3a^2 + a^4$ **23.** $64x^4 - 9y^2$ **25.** $1.21g^2 + 2.2gh^5 + h^{10}$ **27.** $\frac{16}{9}x^4 - y^2$ **29.** $9x^3 - 45x^2 - x + 5$ **31.** $a^3 + 9a^2b + 27ab^2 + 27b^3$ **33.** $x^2 + y^2 + z^2 + 2xy + 2yz + 2xz$

	x	y	z
x	x^2	xy	xz
y	xy	y^2	yz
z	xz	yz	z^2

35a. $2\pi s + 7\pi$ square meters **35b.** about 28.27 square meters **35c.** about 40.84 square meters **37.** $6t^2 - 3t - 3$

39a.

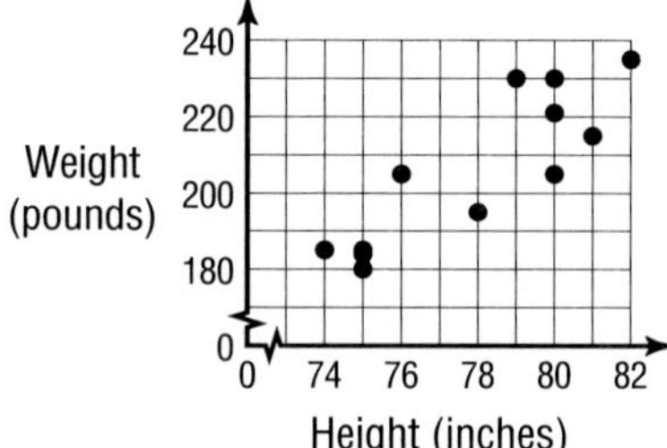

39b. the taller the player, the greater the weight

41.

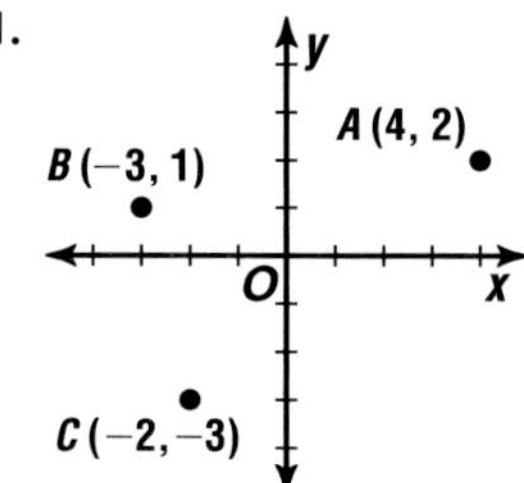

43. 5

Page 549 Chapter 9 Highlights

1. e **3.** h **5.** i **7.** f **9.** a

Pages 550–552 Chapter 9 Study Guide and Assessment

11. y^7 **13.** $20a^5x^5$ **15.** $576x^5y^2$ **17.** $-\frac{1}{2}m^4n^8$ **19.** y^4 **21.** $3b^3$ **23.** $\frac{a^4}{2b}$ **25.** 2.4×10^5 **27.** 4.88×10^9 **29.** 7.96×10^5 **31.** 6×10^{11} **33.** 6×10^{-1} **35.** 1.68×10^{-1} **37.** 2 **39.** 5 **41.** 4 **43.** $3x^4 + x^2 - x - 5$ **45.** $-3x^3 + x^2 - 5x + 5$ **47.** $16m^2n^2 - 2mn + 11$ **49.** $21m^4 - 10m - 1$ **51.** $12a^3b - 28ab^3$ **53.** $8x^5y - 12x^4y^3 + 4x^2y^5$ **55.** $2x^2 - 17xy^2 + 10x + 10y^2$ **57.** $r^2 + 4r - 12$ **59.** $4x^2 + 13x - 12$ **61.** $18x^2 - 0.125$ **63.** $2x^3 + 15x^2 - 11x - 9$ **65.** $x^2 - 36$ **67.** $16x^2 + 56x + 49$ **69.** $25x^2 - 9y^2$ **71.** $36a^2 - 60ab + 25b^2$ **73.** \$305.26 **75.** no; after 6 years

CHAPTER 10 USING FACTORING

Pages 561–563 Lesson 10–1

5. 1, 2, 4 **7.** prime **9.** $-1 \cdot 2 \cdot 3 \cdot 5$ **11.** 4 **13.** $6d$ **15.** $4gh$ **17.** 40 in. **19.** 1, 67 **21.** 1, 2, 4, 5, 8, 10, 16, 20, 40, 80 **23.** 1, 5, 10, 19, 25, 38, 50, 95, 190, 950 **25.** composite; $3^2 \cdot 7$ **27.** prime **29.** composite; $2^2 \cdot 5 \cdot 7 \cdot 11$ **31.** $-1 \cdot 3 \cdot 3 \cdot 13$ **33.** $2 \cdot 2 \cdot b \cdot b \cdot b \cdot d \cdot d$ **35.** $-1 \cdot 2 \cdot 7 \cdot 7 \cdot a \cdot a \cdot b$ **37.** 9 **39.** 1 **41.** 19 **43.** $7pq$ **45.** $15r^2t^2$ **47.** 12 **49.** 6 **51.** $8m^2n$ **53.** $-12x^3yz^2$ **55.** $3m^2n$ **57.** 29 cm by 47 cm **59.** 3, 5; 5, 7; 11, 13; 17, 19; 29, 31; 41, 43; 59, 61; 71, 73

Selected Answers

61a. $2b^3 \times 1 \times 1$, $2b^2 \times b \times 1$, $2b \times b \times b$, $b^3 \times 2 \times 1$, $b^2 \times 2b \times 1$, and $b^2 \times 2 \times b$ **61c.** $4b^3 + 1$ or 865, $2b^3 + 2b^2 + 1$ or 505, $5b^2$ or 180, $3b^3 + 2$ or 650, $2b^3 + b^2 + b$ or 474, and $b^3 + 2b^2 + 2b$ or 300, respectively **61d.** Though the volume remains constant, the surface areas vary greatly.
63. 1500 squares of sod **65.** $3b$
67.

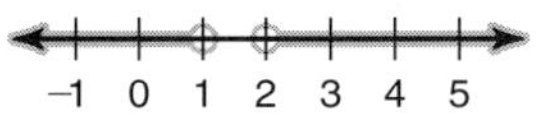

69. $x = 4$ **71.** $9\frac{3}{5}$ ft **73.** -4

Page 564 Lesson 10–2A
1. $3(x + 3)$ **3.** $x(3x + 4)$
5. no

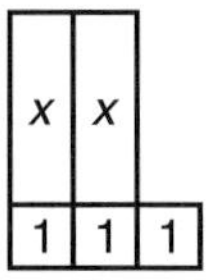

7. yes

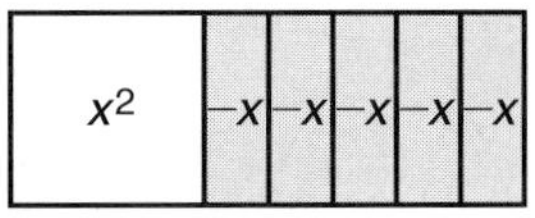

9. Binomials can be factored if they can be represented by a rectangle. Examples: $4x + 4$ can be factored and $4x + 3$ cannot be factored.

Pages 569–571 Lesson 10–2
7. 3 **9.** 1 **11.** $4xy^2$ **13.** $(a + b)(x + y)$ **15.** $(x - y)(3a + 4b)$ **17.** $3t$ **19.** $x(29y - 3)$ **21.** $3c^2d(1 - 2d)$
23. $(r + k)(x + 2y)$ **25a.** $g = \frac{1}{2}n(n - 1)$ **25b.** 91 games
25c. 63 games **27.** $8rs$ **29.** $2y - 5$ **31.** $5k - 7p$
33. $2xz(7 - 9z)$ **35.** $a(17 - 41ab)$ **37.** $(m + x)(2y + 7)$
39. $3xy(x^2 - 3y + 12)$ **41.** $(2x^2 - 5y^2)(x - y)$ **43.** $(2x - 5y)(2a - 7b)$ **45.** $7abc(4abc + 3ac - 2)$ **47.** $2(2m + r)(3x - 2)$ **49.** $(7x + 3t - 4)(a + b)$ **51.** $8a - 4b + 8c + 16d + ab + 64$ **53.** $4r^2(4 - \pi)$ **55.** $(4z + 3m)$ cm by $(z - 6)$ cm **57.** Sample answer: $(3a + 2b)(ab + 6)$, $(3a + ab)(2b + 6)$, $(2b + ab)(3a + 6)$ **59.** $(2s - 3)(s + 8)$ **61.** A and C sharp **63.** 9 **65.** $\{z \mid z \geq -1.654\}$ **67.** $\left\{(-3, 10), \left(0, \frac{11}{2}\right), (1, 4), \left(2, \frac{5}{2}\right), (5, -2)\right\}$ **69.** $y = -\frac{4}{3}x + \frac{7}{3}$ **71.** 7

Page 573 Lesson 10–3A
1. $(x + 1)(x + 5)$ **3.** $(x + 3)(x + 4)$ **5.** $(x - 1)(x - 2)$
7. $(x - 1)(x + 5)$
9. yes

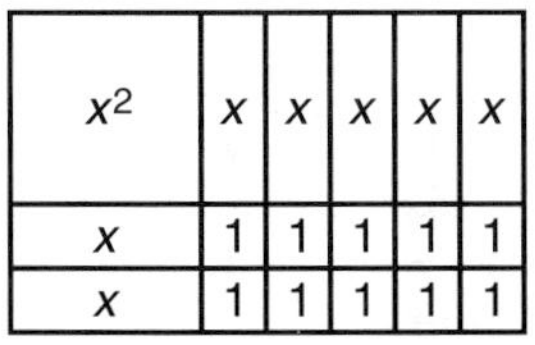

11. no

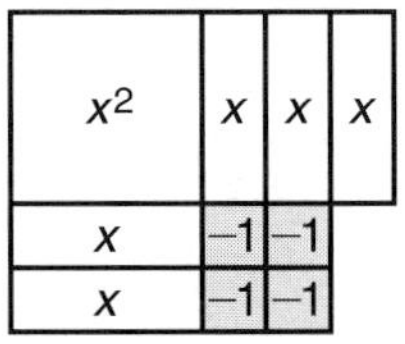

13. Trinomials can be factored if they can be represented by a rectangle. Sample answers: $x^2 + 4x + 4$ can be factored and $x^2 + 6x + 4$ cannot be factored.

Pages 578–580 Lesson 10–3
5. 3, 8 **7.** 8, 5 **9.** -2, -6 **11.** – **13.** $(t + 3)(t + 4)$
15. $2(y + 2)(y - 3)$ **17.** prime **19.** 9, -9, 15, -15
21. $(3x^2 + 2x)$ m^2 **23.** – **25.** 5 **27.** 5 **29.** $(m - 4)(m - 10)$
31. prime **33.** $(2x + 7)(x - 3)$ **35.** $(2x + 3)(x - 4)$
37. $(2n - 7)(2n + 5)$ **39.** $(2 + 3m)(5 + 2m)$ **41.** prime
43. $2x(3x + 8)(2x - 5)$ **45.** $2a^2b(5a - 7b)(2a - 3b)$
47. 7, -7, 11, -11 **49.** 1, -1, 11, -11, 19, -19, 41, -41

51. 6, 4 **53.** r cm, $(5r + 6)$ cm, $(3r - 7)$ cm **55.** no; $(2x + 3)(x - 1)$ **57.** no; $(x - 3)(x - 3)$ **59.** 27 ft^3
61. $(3x - 10)$ shares **63.** $(5, -2)$ **65.** $y = -\frac{2}{3}x + \frac{14}{3}$
67. D = {0, 1, 2}; R = {2, −2, 4} **69.** 90° **71.** $3x + 4y$

Page 580 Self Test
1. $10n^2$ **3.** $6xy(3y - 4x)$ **5.** $(2q + 3)(q - 6)$ **7.** $(3y - 5)(y - 1)$ **9.** 41,312,432 or 23,421,314

Pages 584–586 Lesson 10–4
7. yes **9.** no **11.** d **13.** a **15.** $(1 - 4g)(1 + 4g)$
17. $5(2m - 3n)(2m + 3n)$ **19.** $(x - y)(x + y)(x^2 + y^2)$
21. 5 **23.** $(2 - v)(2 + v)$ **25.** $(10d - 1)(10d + 1)$
27. $2(z - 7)(z + 7)$ **29.** prime **31.** $17(1 - 2k)(1 + 2k)$
33. prime **35.** prime **37.** $(ax - 0.8y)(ax + 0.8y)$
39. $\frac{1}{2}(3a - 7b)(3a + 7b)$ **41.** $(a + b - c - d)(a + b + c + d)$ **43.** $(x^2 - 2y)(x^2 + 2y)(x^4 + 4y^2)$ **45.** $(a^2 + 5b^2)(a - 2b)(a + 2b)$ **47.** 624 **49.** $(2a - b)$ in., $(2a + b)$ in.
51. $(x - 2)$ ft, $(x + 4)$ ft **53.** $(a - 5b)$ in., $(a + 5b)$ in., $(5a + 3b)$ in. **55a.** square **55b.** 25 cm^2 **57.** 9, 12, 15
59. \$3.16, \$1.50, \$1.25, \$1.20 **61.** (4, 16)
63.

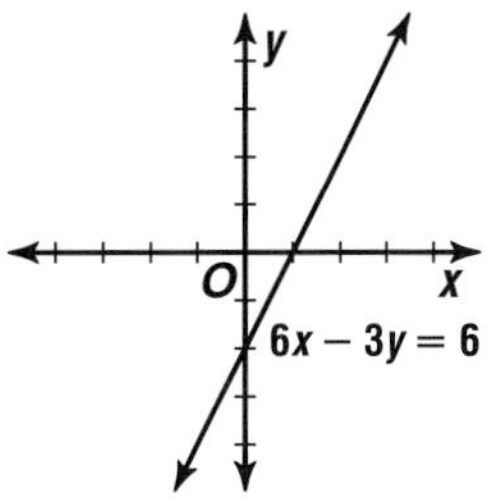

65.

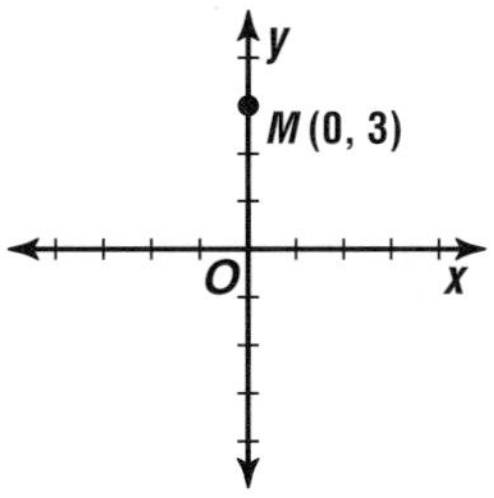

67. 5.14, 3.6, 0.6

Pages 591–593 Lesson 10–5
7. 8a **9.** 6c **11.** yes; $(2n - 7)^2$ **13.** yes; $(4b - 7c)^2$
15. $4(a - 3b)(a + 3b)$ **17.** $2(5g + 2)^2$ **19.** $(2a - 3b)(2a + 3b)(5x - y)$ **21.** yes; $(r - 4)^2$ **23.** yes; $(7p - 2)^2$ **25.** no
27. yes; $(2m + n)^2$ **29.** no **31.** no **33.** yes; $\left(\frac{1}{2}a + 3\right)^2$
35. $a(45a - 32b)$ **37.** $(v - 15)^2$ **39.** prime **41.** $3(y - 7)(y + 7)$ **43.** $2(3a - 4)^2$ **45.** $(y^2 + z^2)(x - 1)(x + 1)$
47. $(a^2 + 2)(4a + 3b^2)$ **49.** $0.7(p - 3q)(p - 2q)$
51. $(g^2 - 3h)(g + 3)^2$ **53.** -110, 110 **55.** 9 **57.** $(6y + 26)$ cm **59.** $(8x^2 - 22x + 14)$ cm^2 **61a.** $a \geq b$ **61b.** $a \leq b$
61c. $a = b$ **63a.** \$1166.40 **63b.** $p(1 + r)^2$ **63c.** \$1144.90
65. $50.6t^2 + 21t - 102$ **67.** 20 hours
69a.

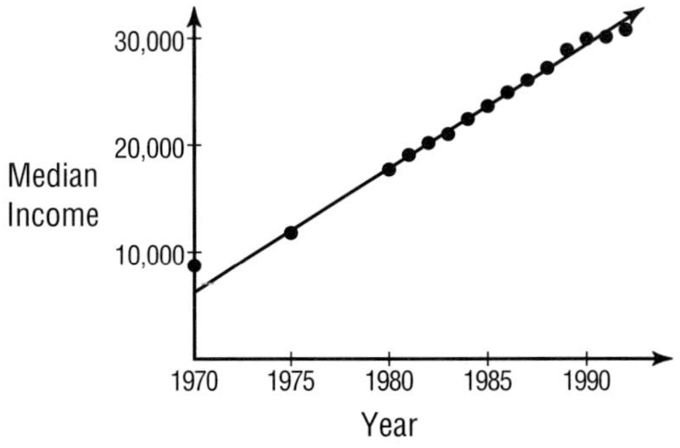

69b. Sample answer: Yes; $I = 1200y + 6000$, where I is the median income and y is the number of years since 1970.
71. 1:3 **73.** 152 ft

Pages 598–600 Lesson 10-6

5. {0, −5} **7.** $\left\{0, \frac{5}{3}\right\}$ **9.** {5} **11.** 6 **13.** {0, 24} **15.** $\left\{\frac{3}{2}, \frac{8}{3}\right\}$ **17.** {−9, −4} **19.** {−8, 8} **21.** {0, 4} **23.** $\left\{-\frac{1}{3}, -\frac{5}{2}\right\}$ **25.** {12, −4} **27.** {−9, 0, 9} **29.** {−5, 7} **31.** −14 and −12 or 12 and 14 **33.** 5 cm **35a.** {−6, 1}; {−6, 1} **35b.** They are equivalent; they have the same solution. **37a.** about 3.35 s **37b.** His ideas about falling objects differed from what most people thought to be true. He believed that Earth is a moving planet and that the sun and planets do not revolve around Earth. **39.** about 90,180 ft or 17 mi **41.** 0.5 km **43.** $40q^2 + rq - 6r^2$ **45.** 14; −4 **47.** 4.355 minutes or about 4 minutes 21 seconds

Page 601 Chapter 10 Highlights

1. false; composite **3.** false; sample answer: 64 **5.** false; $2^4 \cdot 3$ **7.** true **9.** false; zero product property

Pages 602–604 Chapter 10 Study Guide and Assessment

11. composite, $2^2 \cdot 7$ **13.** composite, $2 \cdot 3 \cdot 5^2$ **15.** prime **17.** 5 **19.** $4ab$ **21.** $5n$ **23.** $7x^2$ **25.** $13(x + 2y)$ **27.** $6ab(4ab - 3)$ **29.** $12pq(3pq - 1)$ **31.** $(a - 4c)(a + b)$ **33.** $(4k - p^2)(4k^2 - 7p)$ **35.** $(8m - 3n)(3a + 5b)$ **37.** $(y + 3)(y + 4)$ **39.** prime **41.** $(r - 4)(2r + 5)$ **43.** $(b - 4)(b + 4)$ **45.** $(4a - 9b^2)(4a + 9b^2)$ **47.** prime **49.** $(a + 9)^2$ **51.** $(2 - 7r)^2$ **53.** $6b(b - 2g)^2$ **55.** $(5x - 12)^2$ **57.** $\left\{\frac{2}{3}, -\frac{7}{4}\right\}$ **59.** {0, −17} **61.** {−5, −8} **63.** $\left\{-\frac{2}{5}\right\}$ **65.** 384 **67.**

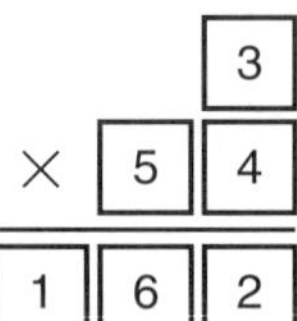

69. 9, 11; −11, −9

CHAPTER 11 EXPLORING QUADRATIC AND EXPONENTIAL FUNCTIONS

Page 610 Lesson 11-1A

1. (−8, −5)

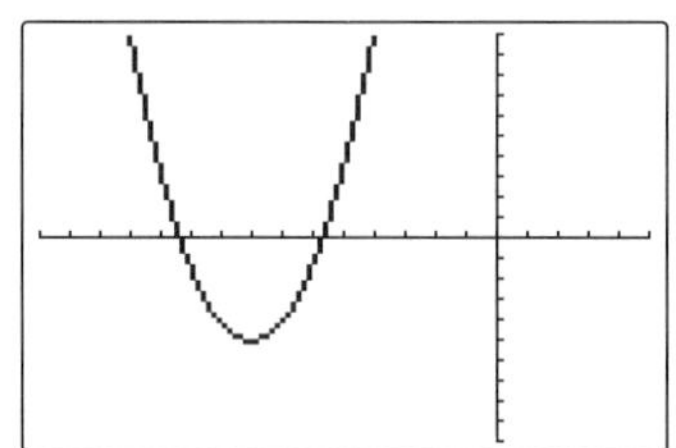

3. (5, 0)

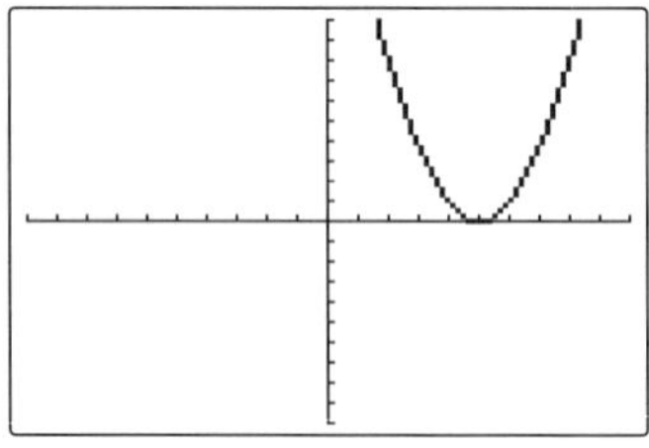

5. (10, 14)

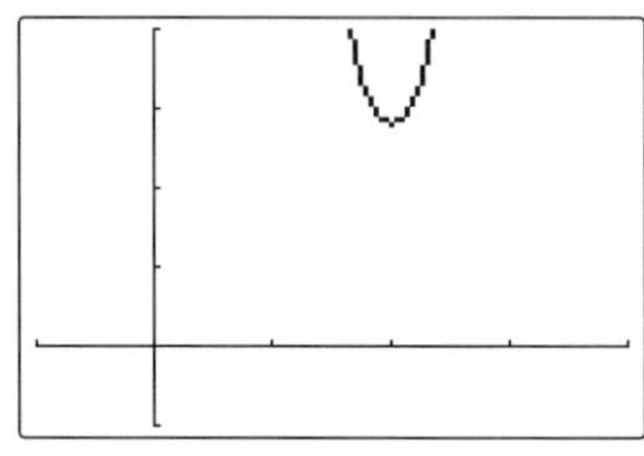

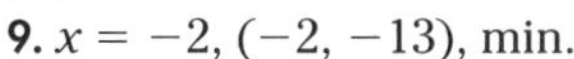

Pages 615–617 Lesson 11-1

7. $x = 0$, (0, 2), min.

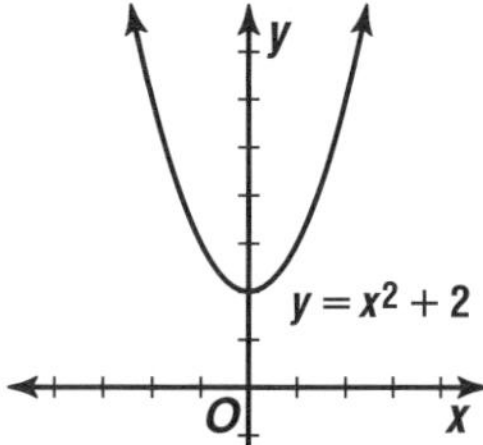

9. $x = -2$, (−2, −13), min.

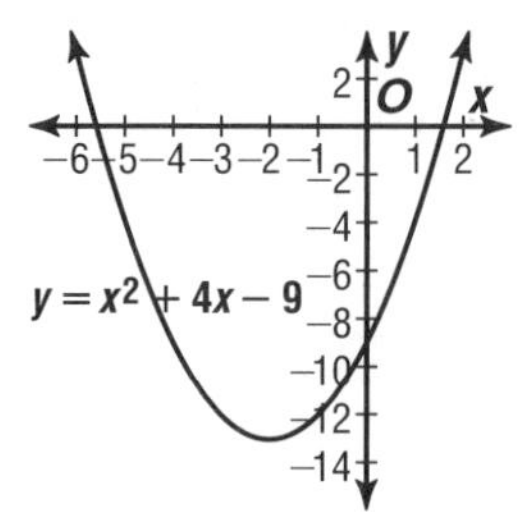

11. $x = 2.5$, (2.5, 12.25), max. **13.** c

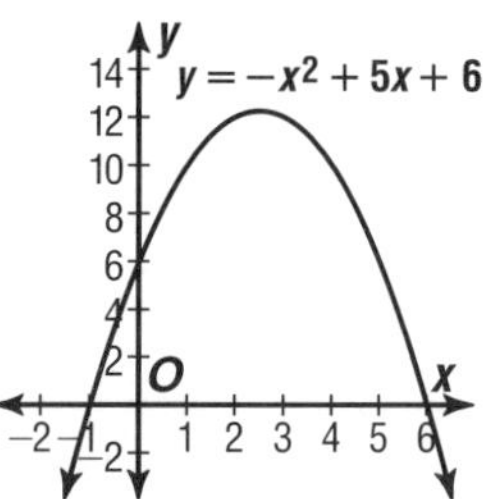

15. $x = 0$, (0, 0), min.

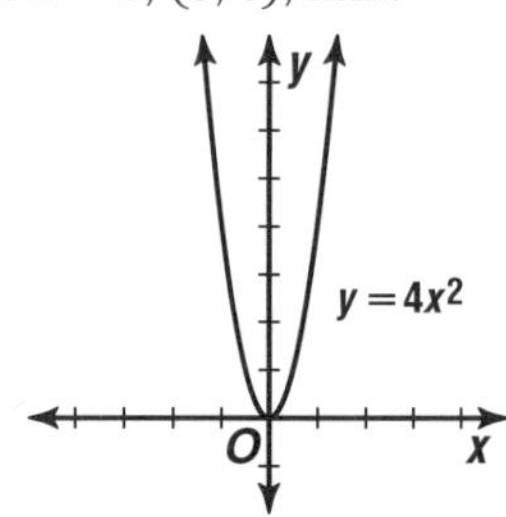

17. $x = -1$, (−1, 17), min.

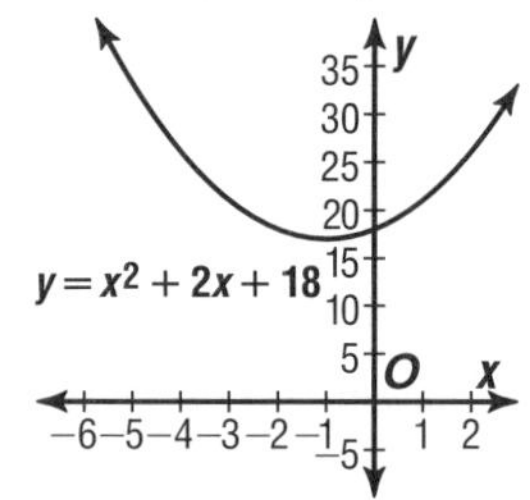

19. $x = 0$, (0, −5), min.

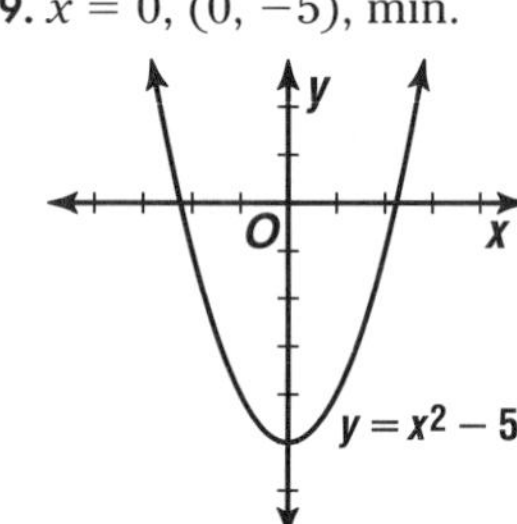

21. $x = -3$, (−3, −29), min.

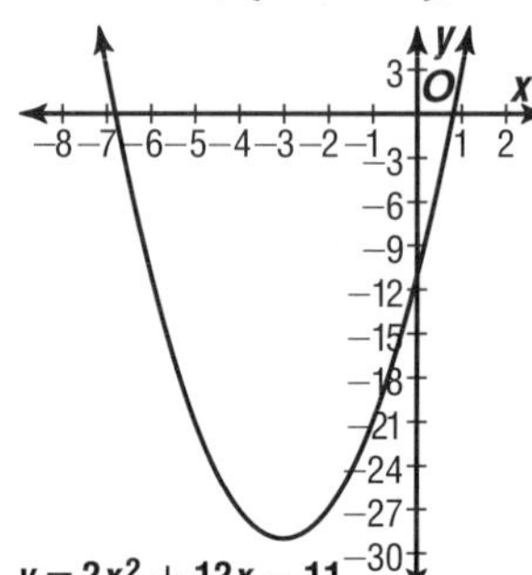

23. $x = 0$, (0, −25), min.

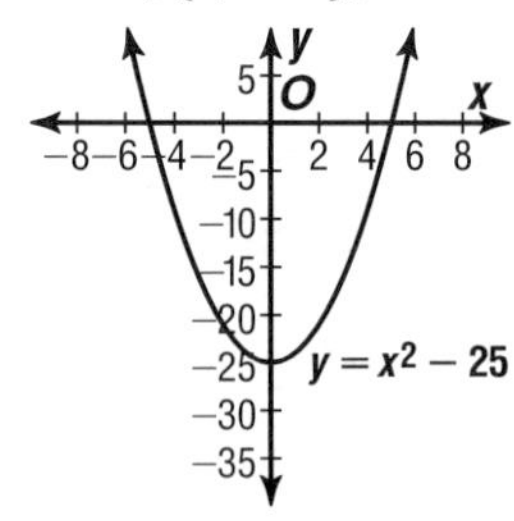

25. $x = -1$, (−1, 7), max.

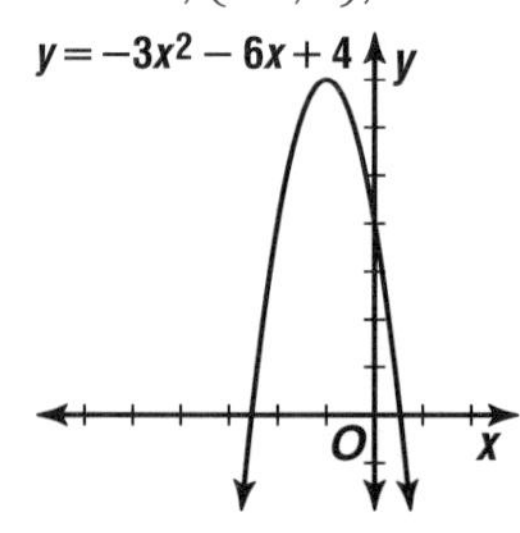

27. $x = -1$, (−1, −20), min.

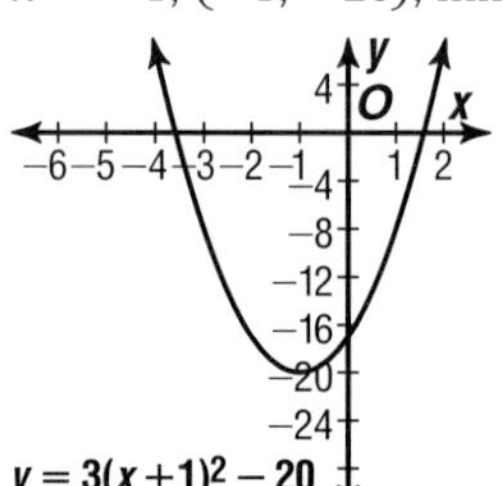

29. $x = -1$, (−1, −1), min.

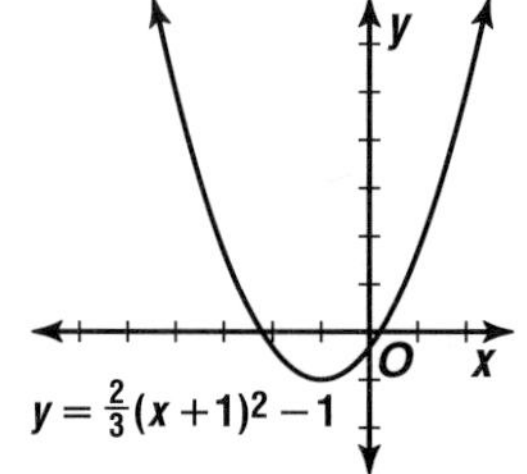

31. *c* **33.**

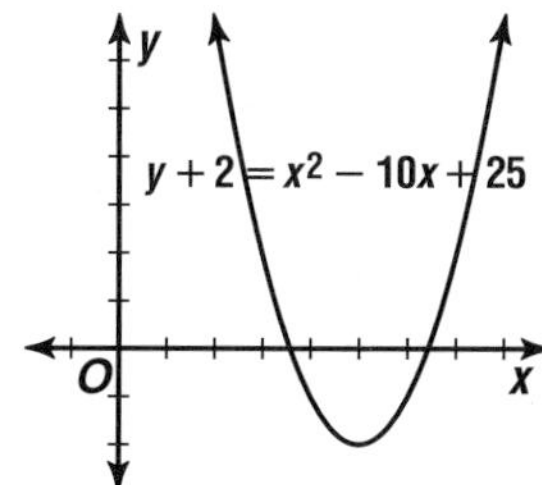

35.

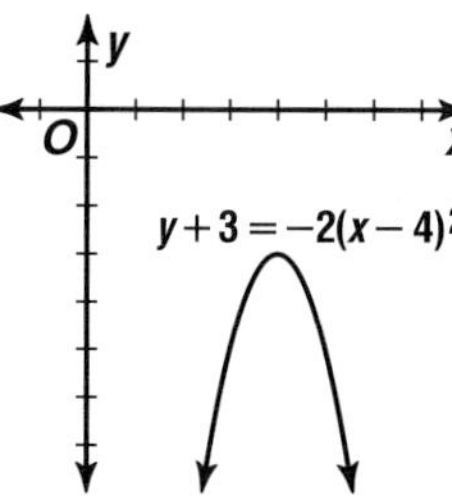

37. $x = -1$ **39.** $x = 2$

41. $(-1.10, 125.8)$

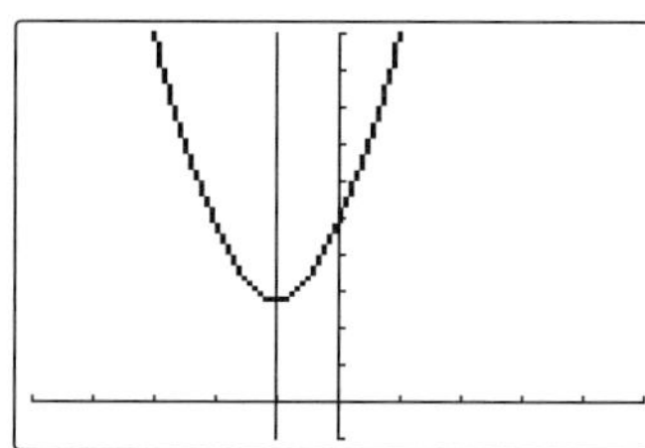

43. $(-0.15, 90.14)$

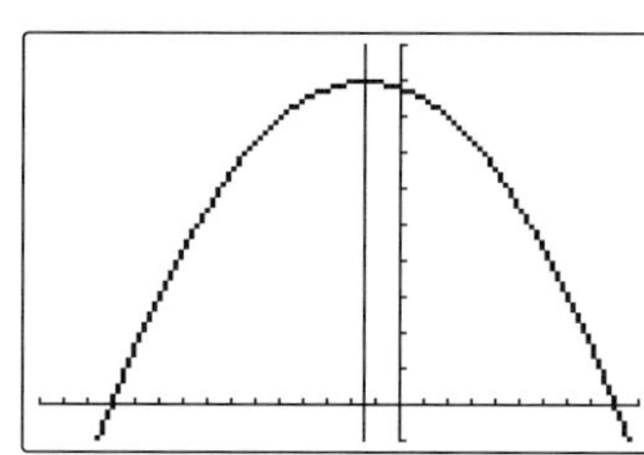

45a.

Year	t	$U(t)$
1970	0	329.96
1975	5	370.31
1980	10	559.16
1985	15	896.51
1990	20	1382.36
1993	23	1745.15

45b. D: $0 \le t \le 23$; R: $326 < U(t) < 1746$

45c.

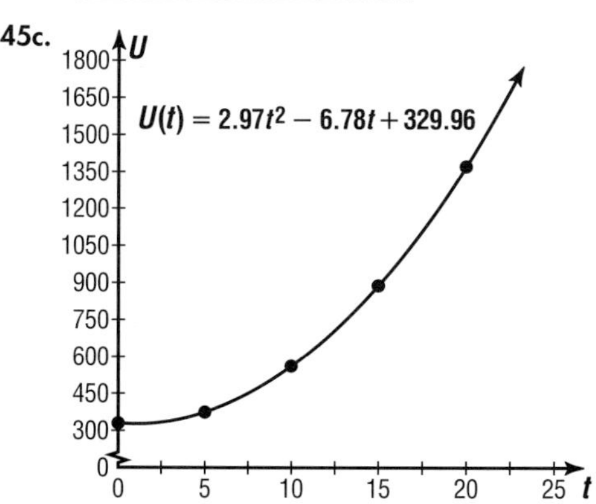

47. 0.15 km **49a.** 6790 km **49b.** 2.2792×10^8 km

51a.

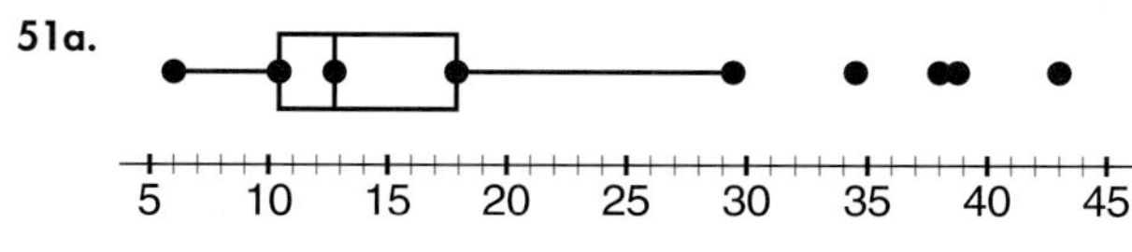

51b. 34.5, 38, 38.7, 43 **53.** -5

Page 619 Lesson 11–1B

1. All the graphs open downward from the origin. $y = -2x^2$ is narrower than $y = -x^2$ and $y = -5x^2$ is the narrowest.

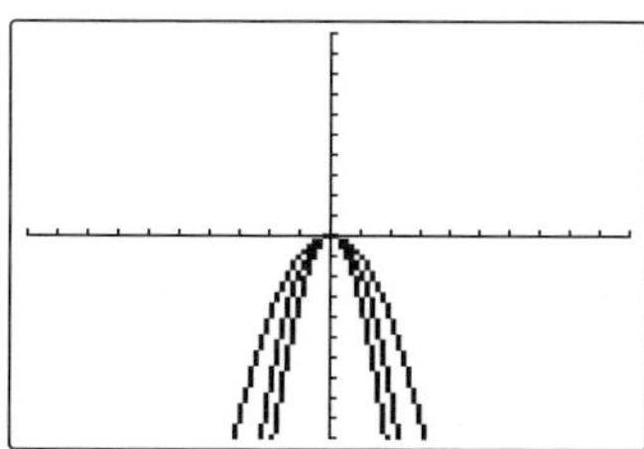

3. All open downward, have the same shape, and have vertices along the x-axis. However, each vertex is different.

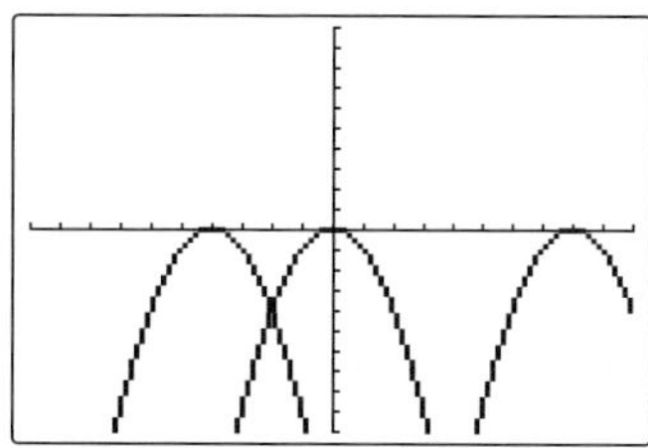

5. It will open upward, have a vertex at the origin, and be narrower than $y = x^2$.

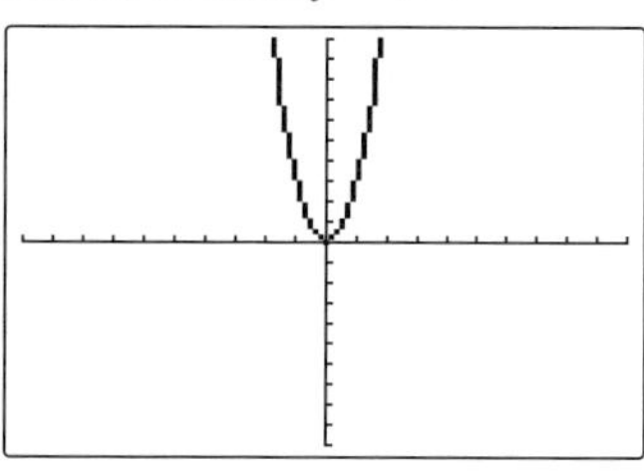

7. It will have vertex at the origin, open downward, and be wider than $y = x^2$.

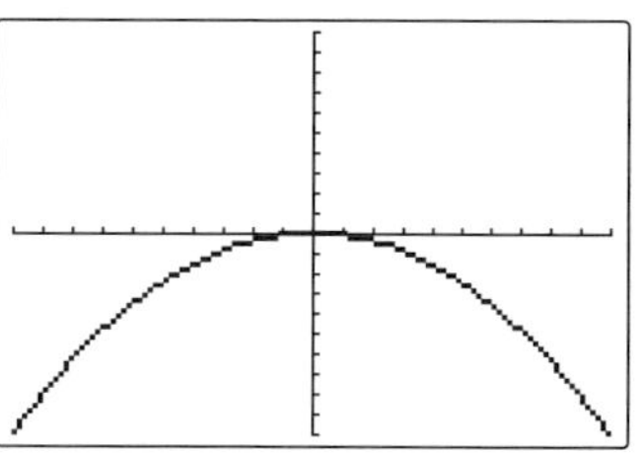

9a. If $|a| > 1$, the graph is narrower than the graph of $y = x^2$. If $0 < |a| < 1$, the graph is wider than the graph of $y = x^2$. If $a < 0$, it opens downward; if $a > 0$, it opens upward. **9b.** The graph has the same shape as $y = x^2$, but is shifted a units (up if $a > 0$, down if $a < 0$). **9c.** The graph has the same shape as $y = x^2$, but is shifted a units (left if $a > 0$, right if $a < 0$). **9d.** The graph has the same shape as $y = x^2$ but is shifted a units left or right and b units up or down as prescribed in 9b and 9c.

Pages 624–627 Lesson 11–2

7. 2 real roots **9.** $-1, 2$

11. $-3, 8$

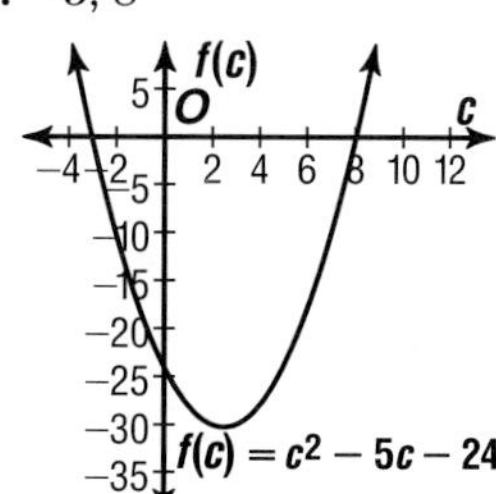

13. $-2 < w < -1, 4 < w < 5$

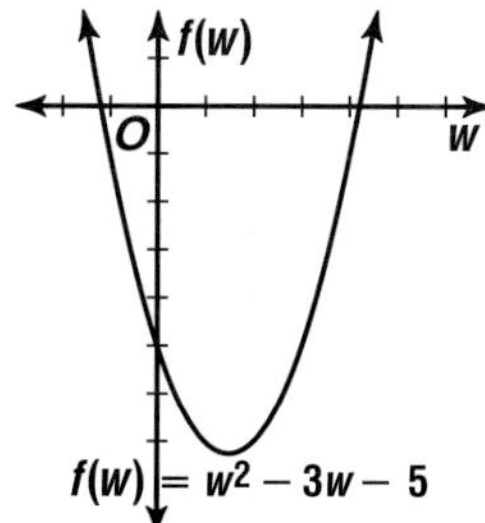

SELECTED ANSWERS

Selected Answers

15. 5

17. $-2, -6$ **19.** 2 **21.** $-4, 4$ **23.** -3 **25.** $\varnothing$ **27.** $-4, 1$ **29.** 4 **31.** $8 < a < 9, -1 < a < 0$ **33.** $3, -1 < n < 0$

35.

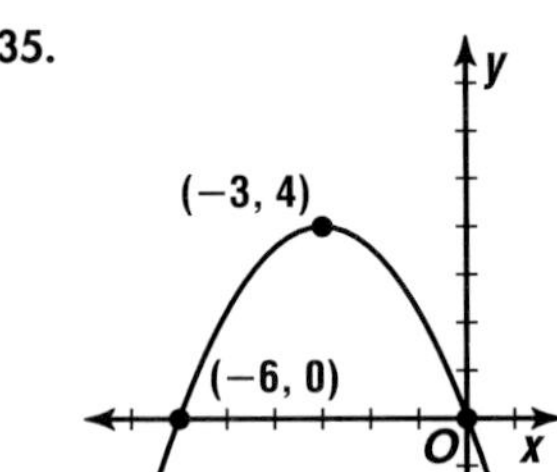

37.

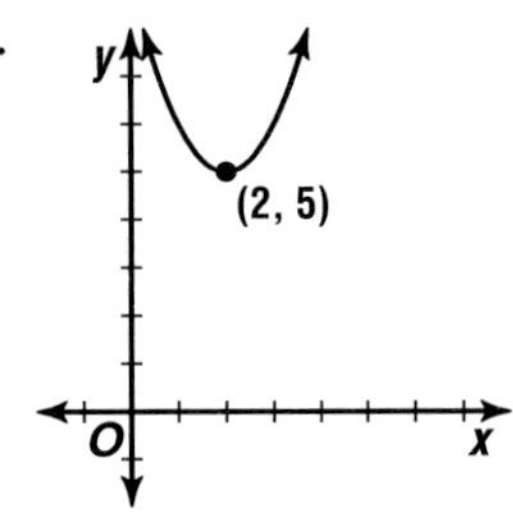

39. $-4, -8$ or $4, 8$ **41.** no solution **43.** $-2, -6$ **45.** no y-intercepts **47.** no real roots **49.** 0.29, 1.71 **51.** $-0.79, 2.54$ **53.** $-2, k > -2, k < -2$

55a.

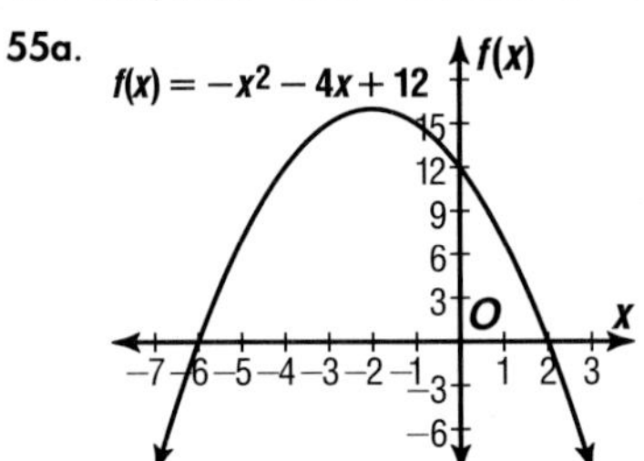

55b. 8 feet **55c.** 16 feet **55d.** $85\frac{1}{3}$ ft^2 **55e.** \$297

57. $x = 0$; $(0, 4)$

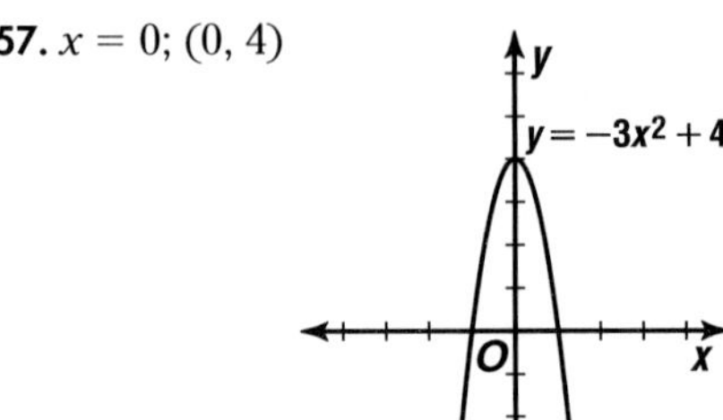

59. $(4x - 3)$ m by $(2x - 1)$ m **61.** 6 **63.** $-4 < x < \frac{4}{3}$ **65.** \$24.75

Pages 631–633 Lesson 11–3

5. $1, 3, -18; 3, -6$ **7.** $4, -2, 15$; no real roots **9.** 1.83, -2.33 **11.** $-1, -6$ **13.** about 29.4 ft/s **15.** $-8.61, -1.39$ **17.** $-\frac{4}{5}, 1$ **19.** $-\frac{5}{3}, 4$ **21.** $-2.91, 2.41$ **23.** 7 **25.** $-\frac{3}{2}, -\frac{5}{3}$ **27.** $-6, 6$ **29.** $-\frac{1}{3}, 1$ **31.** $-3, -\frac{1}{2}$ **33.** $0.60, -0.25$ **35.** 0.5, 2 **37.** $-0.2, 1.4$ **39.** Sample answer: 4, 20, 23; $4x^2 + 20x + 23 = 0$ **41a.** 10, 1 **41b.** none **41c.** -0.5 **41d.** $-1, 0.7142857143$ **43a.** Discriminant is not a perfect square. **43b.** Discriminant is a perfect square but the expression is not an integer **43c.** Discriminant is a perfect square and the equation can be factored. **43d.** Discriminant is 0 and the equation is a perfect square. **45a.** Y, 10 mg; C, 8.75 mg **45b.** $0 = a^2 - 11a + 12$; 1.2 yr, 9.8 yr **47.** $\{0, -9\}$ **49.** 24, 18 **51.** yes **53.** -4°F

Page 633 Self Test

1. $x = \frac{1}{2}, \left(\frac{1}{2}, -\frac{25}{4}\right)$

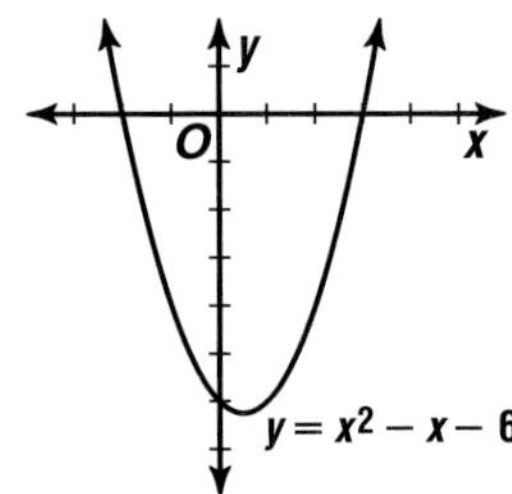

3. $x = 0$, $(0, 7)$

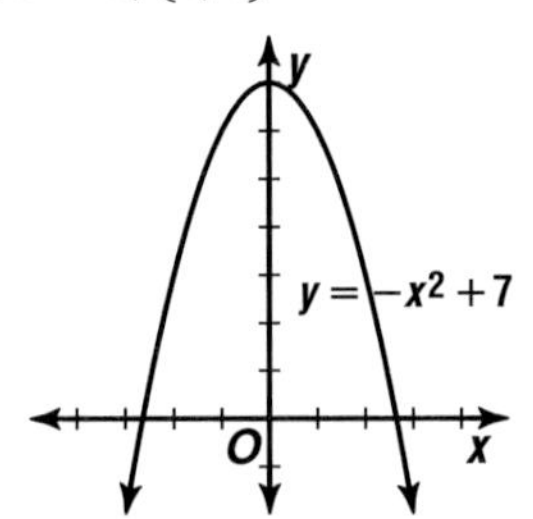

5. $-9, 9$

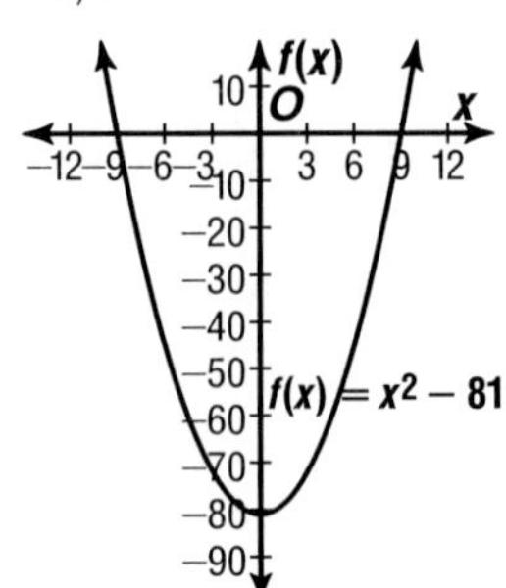

7. $\varnothing$

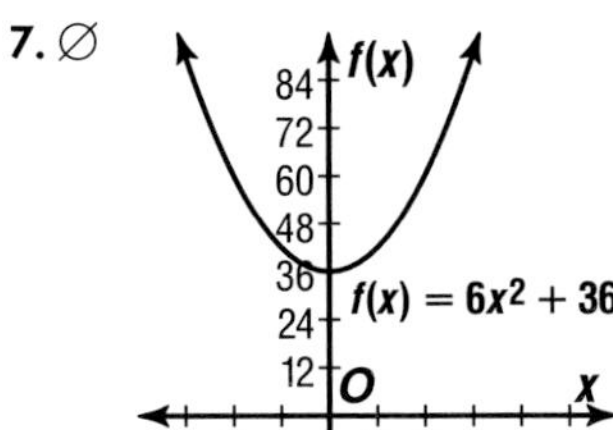

9. $\varnothing$

Page 634 Lesson 11–4A

1.

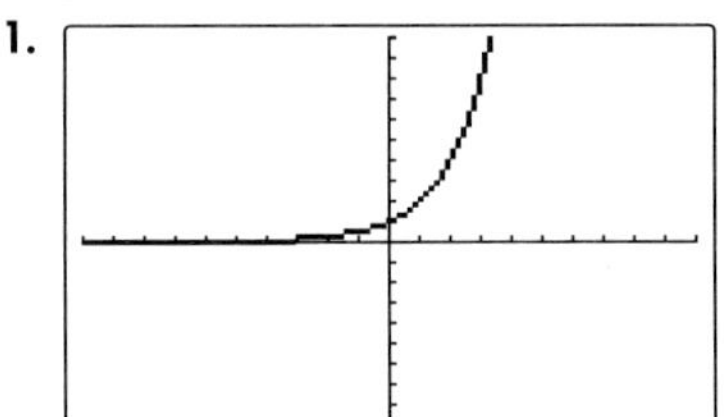

3.

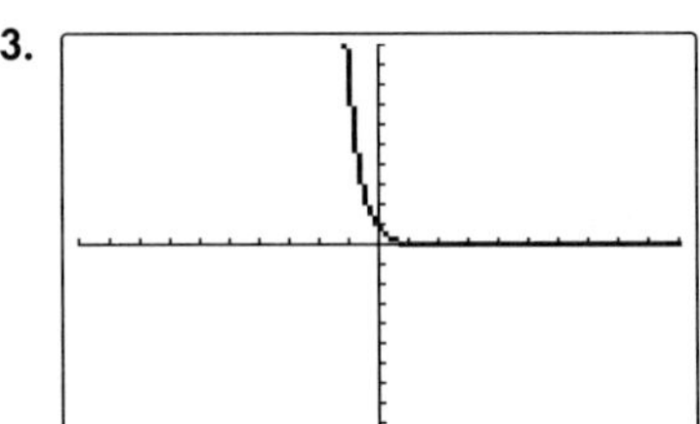

5.

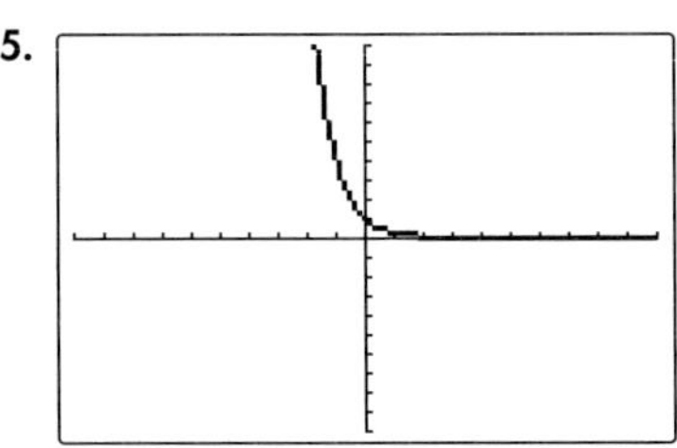

7.

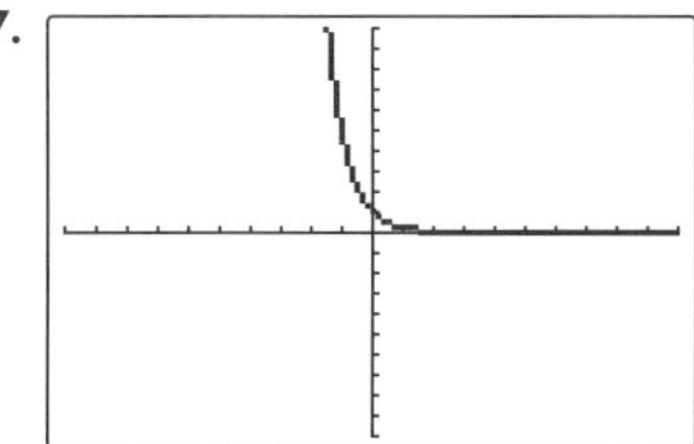

9.

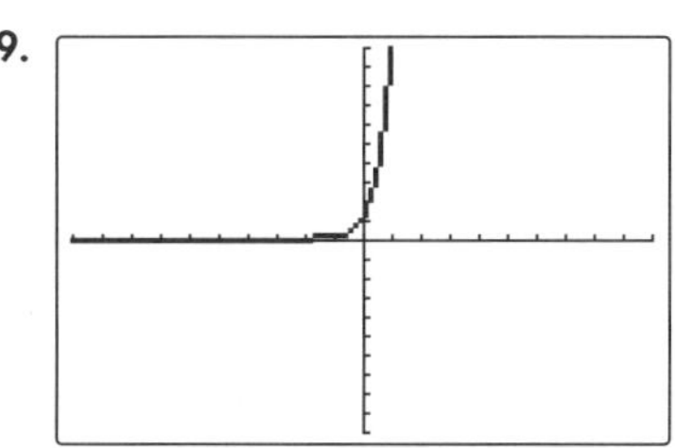

Pages 640–642 Lesson 11–4

7. 5.20 **9.** 12.51

11. 7

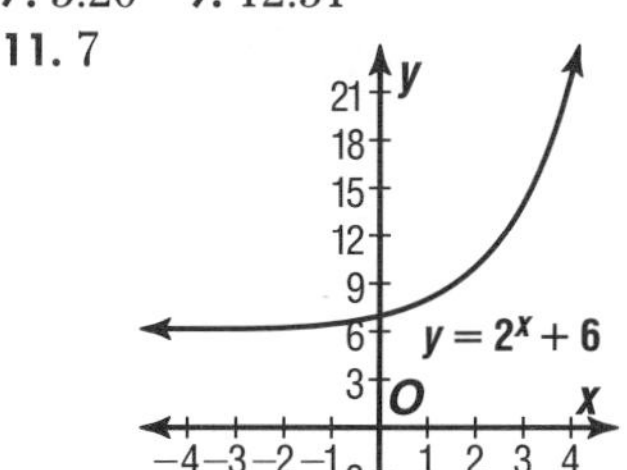

13. −2 **15.** −2 **17.** 10.56 **19.** 1.63 **21.** 25.86 **23.** 0.86

25. 5

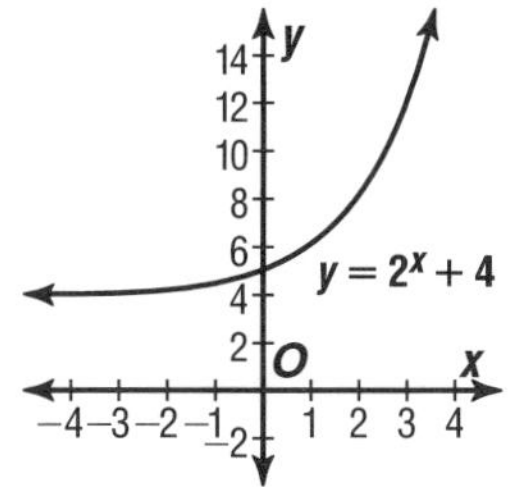

27. 3

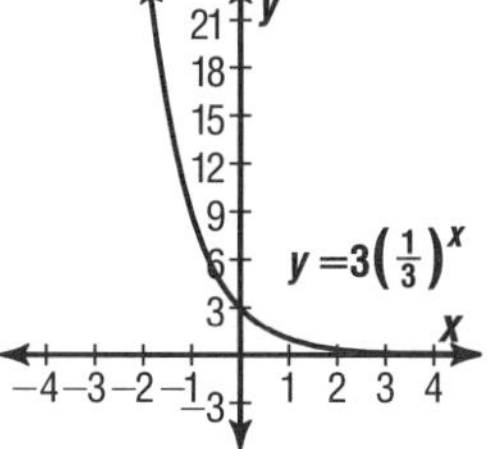

29. 1

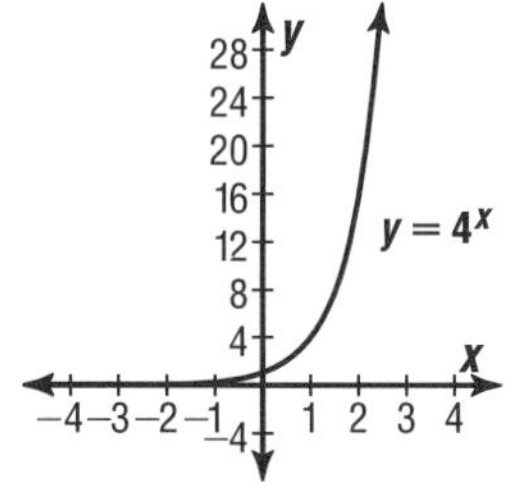

31. no, linear **33.** no, no pattern **35.** −7 **37.** −3 **39.** 3 **41.** 3, −1

43. All have the same shape but different *y*-intercepts.

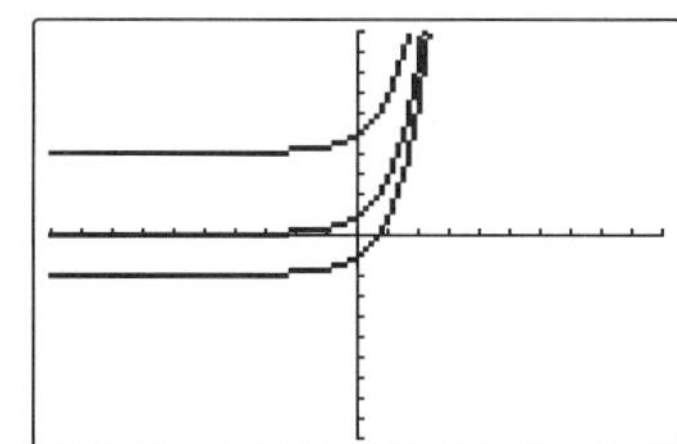

45. All have the same shape but are positioned at different places along the *x*-axis.

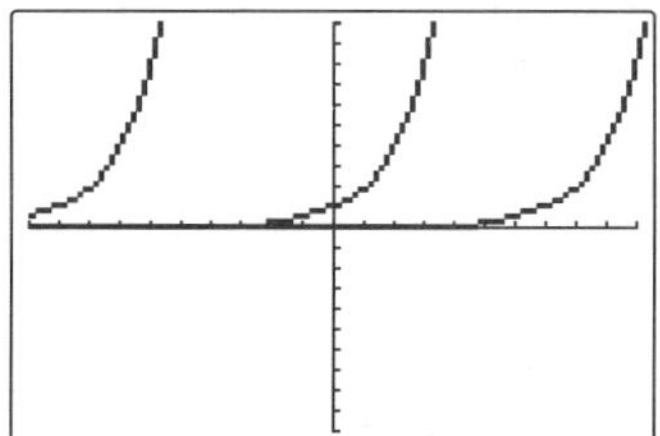

47a. 2.51 **47b.** You cannot write 10 as a power or 2.5. **49.** 1600 bacteria **51.** $-3, -\frac{1}{2}$ **53.** $\frac{1}{5}ab(4a - 3b - 1)$ **55.** 184

57a–b.

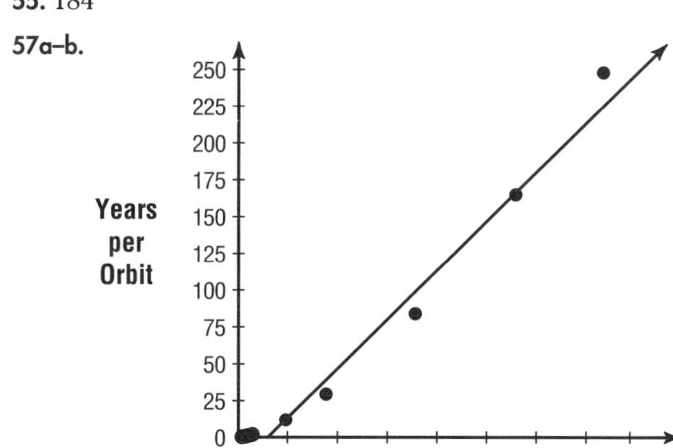

Sample answer: $y = 0.07x - 12$ **57c.** 275 years **59.** 60

Pages 646–649 Lesson 11–5

3a. 1 **3b.** 2 **3c.** 4 **3d.** 365 **5.** decay **7.** about $70,000,000,000 **9.** $661.44 **11.** $10,962.19 **13a.** Each equation represents growth and *t* is the number of years since 1994. $y = 58.7(1.025)^t$, $y = 919.9(1.019)^t$, $y = 35.6(1.021)^t$, $y = 2.9(1.013)^t$ **13b.** 68.1 million, 1029.9 million, 40.3 million, 3.1 million

13c.

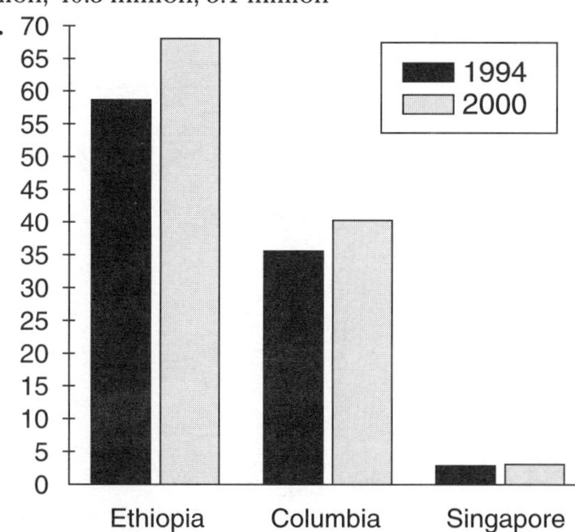

15a. $y = 231.5(1.0963)^t$, t = years since 1950, growth **15b.** 57.6 trillion **17a.** $6931.53 **17b.** $6954.58 **17c.** $23.05 **17d.** daily **21.** $a > 1$, growth; $0 < a < 1$, decay; *x* represents time **23.** $-\frac{1}{2}$ **25.** $-\frac{r^{10}t^2}{3}$ **27.** $y = -2.5x + 2$

Page 651 Chapter 11 Highlights

1. d **3.** i **5.** c **7.** b **9.** f

Pages 652–654 Chapter 11 Study Guide and Assessment

11. $x = 0$; (0, 4) **13.** $x = -1$; (−1, −20) **15.** $x = -1$; (−1, −1)

17. $y = x^2 - 3x - 4$

19. $y = 3(x + 1)^2 - 20$

21. −3, 4 **23.** $-5 < x < -4, 0 < x < 1$ **25.** 3, 7 **27.** 10, −2 **29.** $\frac{3}{2}, -\frac{5}{2}$ **31.** $1, \frac{4}{9}$ **33.** 2, 3 **35.** 0.47, −0.71

37. 7 $y = 3^x + 6$

39. 1

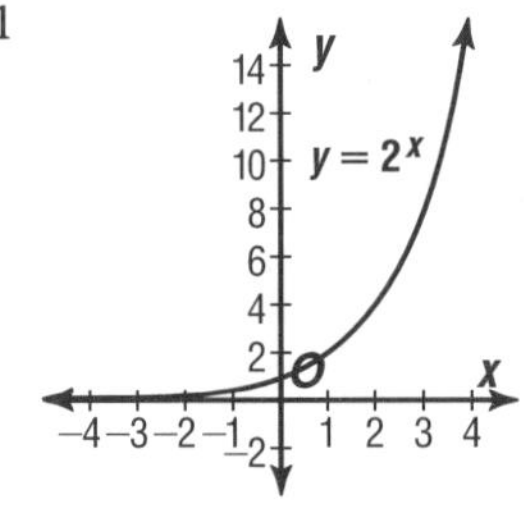

41. −3 **43.** −6 **45.** $-3, \frac{3}{2}$ **47.** \$12,067.68 **49.** \$24,688.37 **51.** 1.3 seconds and 4.7 seconds **53.** between 4 meters and 5 meters **55.** CD, which yields \$541.50 vs. savings at \$530.92

CHAPTER 12 EXPLORING RATIONAL EXPRESSIONS AND EQUATIONS

Pages 663–665 Lesson 12–1

7. $7a^2b$; $\frac{-b^2}{3a^3}$; $a \neq 0, b \neq 0$ **9.** b; $\frac{3}{b+5}$; $b \neq 0$, $b \neq -5$ **11.** $m - 3$; $\frac{1}{m+3}$; $m \neq \pm 3$ **13a.** 5 **13b.** 750 lb **15.** $\frac{5y}{2z}$; $y \neq 0, z \neq 0$ **17.** $\frac{4}{5x}$; $x \neq 0, y \neq 0$ **19.** y; $y \neq -\frac{1}{3}$ **21.** $y + 7x$; $y \neq 7x$ **23.** $\frac{1}{a-4}$; $a \neq 4, -3$ **25.** $\frac{1}{x+4}$; $x \neq -4$ **27.** $\frac{a+6}{a+4}$; $a \neq -4, 2$ **29.** $\frac{b+1}{b-9}$; $b \neq 9, 4$ **31.** $\frac{4}{5}$; $x \neq -1$ **33.** $-\frac{5}{3}$ **35.** not possible, $y \neq -2$ **37a.** 192 **37b.** 3840 lb **39a.** 9.5°/mile **39b.** $-\frac{95}{10} = -\frac{19}{2}$

39c.

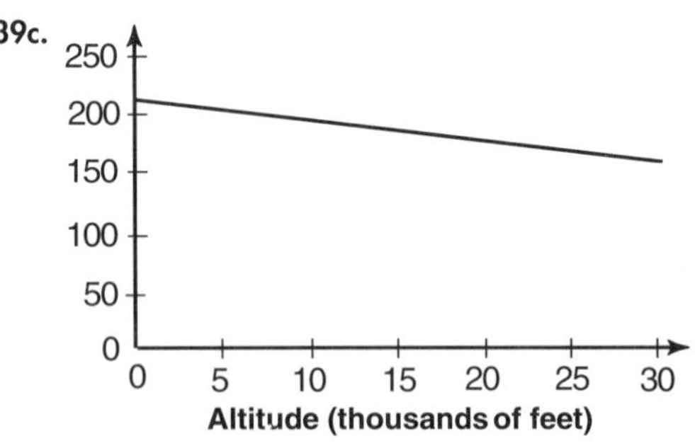

41. ± 11

43. $y = 2x - 1$, $y = x - 2$

45. $y = 3x + 2$

Page 666 Lesson 12–1B

1. $\frac{x-5}{x+5}$; −5 **3.** $\frac{1}{2x}$; 0, 4.5 **5.** $\frac{x-1}{x-8}$; 8 **7.** −1; 3 **9.** $-\frac{x+5}{x+6}$; −6, 5

Pages 669–670 Lesson 12–2

5. $\frac{9m}{2xy}$ **7.** $\frac{4}{x+3}$ **9.** $\frac{4}{3}$ **11a.** 91.44 cm/yd **11b.** changing centimeters to yards **13.** $\frac{3a}{2}$ **15.** $\frac{2}{x}$ **17.** $8y$ **19.** $\frac{4m}{3(m+5)}$ **21.** $\frac{2m}{a+9}$ **23.** 2 **25.** $3(x - y)$ **27.** 36 **29.** $\frac{4x}{3x-5}$ **31.** $\frac{4}{5(x+5)}$ **33.** 21.8 mph; converts ft/s to mph **35.** 10,800 Calories **37.** Sample answer: $\frac{3(x+2)}{x+7} \cdot \frac{2(x-3)}{x-4}$; $\frac{6}{x+7} \cdot \frac{x^2-x-1}{x-4}$ **39a.** $\frac{1 \text{ franc}}{0.1981 \text{ dollars}}$; $\frac{1 \text{ dollar}}{5.05 \text{ francs}}$ **39b.** 12,500 pesos $\cdot \frac{0.1597 \text{ dollars}}{1 \text{ peso}} \cdot \frac{1 \text{ franc}}{0.1981 \text{ dollars}}$; about 10,077 francs **39c.** Convert to American dollars and then to Hong Kong's dollar. **41.** 729 **43.** 4 **45.** 0

Pages 672–674 Lesson 12–3

5. $\frac{5}{x}$ **7.** $\frac{y+3}{x^2-9}$ **9.** $\frac{1}{x^2+2x+5}$ **11.** $\frac{(m+2)^2}{4}$ **13.** $\frac{x+4}{x+6}$ **15.** 0.5 ft^3 **17.** $\frac{a}{a+11}$ **19.** $\frac{b^2m}{c^2}$ **21.** $\frac{m}{m+5}$ **23.** $\frac{4}{z+3}$ **25.** $\frac{2(x+5)}{x+1}$ **27.** $x + 3$ **29.** $x + 5$ **31.** $(x + 5)(x + 3)$ **33.** 88 ft/s; sample answer: changes 60 mph to ft/s **35.** 27.5 yd^3; changes ft^3 to yd^3 **37.** $\frac{x-y}{4}$ **39a.** 5500 miles **39b.** 41.3 trains **39c.** 10,618.2 miles; 27.3 trains **41.** $\frac{8}{m-5}$ **43.** $(4a - 3b^2)^2$ **45.** $x < -2$ **47.** $7 + 2 = 2 + 7$

Pages 678–680 Lesson 12–4

5. $3b^2 - 5$ **7.** $t - 1$ **9.** $2m + 3 + \frac{-3}{m+2}$ **11.** $(2x + 3)$ m **13.** $\frac{b}{3} + 3 - \frac{7}{3b}$ **15.** $\frac{m}{7} + 1 - \frac{4}{m}$ **17.** $a^2 + 2a + 12 + \frac{3}{a-2}$ **19.** $m - 3 - \frac{2}{m+7}$ **21.** $2x + 6 + \frac{1}{x-7}$ **23.** $3m^2 + 4k^2 - 2$ **25.** $a + 7 - \frac{1}{a+3}$ **27.** $3x - 2 + \frac{21}{2x+3}$ **29.** $t + 4 - \frac{5}{4t+1}$ **31.** $8x^2 + \frac{32x}{7} - 9$ **33a.** $x^2 + 4x + 4$ **33b.** $5t^2 - 3t - 2$ **33c.** $2a^2 + 3a - 4$ **35.** −15 **37.** \$480,000 **39.** $-\frac{x}{7}$ **41.** $3(x - 7)(x + 5)$ **43.** (1, 3)

45. −3 −2 −1 0 1 2 3 4 5 6

Page 680 Self Test

1. $\frac{25x}{36y}$ **3.** $\frac{(x+4)(4x^2+2x-3)}{(x+3)(x+2)(x-1)}$ **5.** $\frac{x-5}{x+4}$ **7.** $2x - 5$ **9.** $4(3x - 1)$ or $(12x - 4)$ units

Pages 683–684 Lesson 12–5

5. $\frac{10}{x}$ **7.** $\frac{10}{a+2}$ **9.** 3 **11.** m **13.** $-\frac{a}{6}$ **15.** 2 **17.** $\frac{7}{x+7}$ **19.** 1 **21.** 1 **23.** $\frac{4x}{x+2}$ **25.** $\frac{2y-6}{y+3}$ **27.** 2 **29.** $\frac{25x-93}{2x+5}$ **31.** $\frac{24x+26y}{7x-2y}$ **33.** b **35.** 442 days **37.** $\pm\frac{6}{5}$ **39.** −3, 4 **41.** $\frac{1}{216}$

Pages 687–689 Lesson 12–6

3. x^2 **5.** $90m^2b^2$ **7.** $(x-3)(x+3)$ **9.** $\frac{7+9m}{15m^2}$

11. $\frac{-20}{3(x+2)}$ **13.** $\frac{22-7y}{6y^2}$ **15.** 168 members **17.** $\frac{-5x}{63}$

19. $\frac{7yz+3}{xyz}$ **21.** $\frac{2s-3t}{s^2t^2}$ **23.** $\frac{7z+8}{(z+5)(z-4)}$ **25.** $\frac{k^2+k-10}{(k+5)(k+3)}$

27. $\frac{-17r-32}{(3r-2)(r-5)}$ **29.** $\frac{10+m}{2(2m-3)}$ **31.** $\frac{3w-4}{3(5w+2)}$

33. $\frac{-5n-4}{(n+3)(n+1)}$ **35.** $\frac{-17t+27}{(t+3)(10t-9)}$ **37.** $\frac{v^2+11v-2}{(v+4)(v-1)}$

39a. 360, 3, 120, 360 **39b.** 540, 6, 90, 540 **39c.** The GCF times the LCM of two numbers is equal to the product of the two numbers. **39d.** Divide the GCF into the product of the two numbers to find the LCM. **41.** 15 months later or July 20, 1997 **43.** 2 **45.** $\frac{3 \pm \sqrt{41}}{4} \approx 2.35$ or -0.85 **47.** $-15a^6b^3$ **49.** 4

Pages 693–695 Lesson 12–7

3. $\frac{8x+3}{x}$ **5.** $\frac{6m^2+m+1}{2m}$ **7.** $\frac{9}{5}$ **9.** $\frac{y}{2}$ **11.** $\frac{a-b}{3}$

13. $\frac{3x+15}{x+3}$ **15.** $\frac{6x-1}{2x+1}$ **17.** $\frac{5r^2+r-48}{r^2-9}$ **19.** $\frac{4}{3}$ **21.** ab

23. $\frac{6}{x}$ **25.** $\frac{a-5}{a-2}$ **27.** $\frac{1}{x+3}$ **29.** $\frac{m+6}{m+2}$ **31.** $\frac{y-3}{y+2}$

33. $\frac{t^2+2t+2}{-t^2}$ **35.** $\frac{32b^2}{35}$ **37a.** $x \neq 0, x \neq -3$ **37b.** $x \neq 0, x \neq -3$ **37c.** yes **39a.** $\frac{fs}{s-v}$ **39b.** 413.5 **39c.** 2 **39d.** 6

41. $\frac{2x+3}{x+3}$ **43.** $(2x+1)(2x-1)$ **45.** $(-3, 4)$

47. $w = \frac{P-2\ell}{2}$

Pages 700–702 Lesson 12–8

5. -12 **7.** $-\frac{1}{2}$ **9.** 0 **11a.** $\frac{1}{5}$ **11b.** $\frac{3}{5}$ **11c.** $\frac{x}{5}$ **13.** 3.429 ohms **15.** 8 ohms, 4 ohms **17.** $\frac{50}{3}$ **19.** $\frac{5}{4}$ **21.** 0

23. 2 or $-\frac{1}{3}$ **25.** 4 or $-\frac{1}{3}$ **27.** -13 **29.** $\frac{3}{5}$ **31.** 3 **33.** 6

35. 4 ohms **37.** $R_1 = \frac{R_2R_T}{R_2 - R_T}$ **39.** $n = \frac{IR}{E - Ir}$ **41.** 10

43. 96 ohms **45a.** 2.5 hours **45b.** 2.19 hours **45c.** after 923 miles **47a.** $w = \frac{c\ell}{100}$ **47b.** $\ell = \frac{100w}{c}$ **49.** 10

51. $\frac{b+4}{4(b+2)^2}$

53.

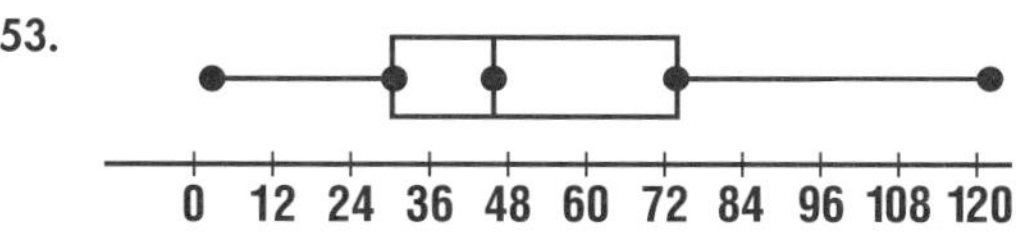

55. $\frac{48}{25}$ **57.** 36

Page 703 Chapter 12 Highlights

1. false, rational **3.** true **5.** false, $x^2 - 144$ **7.** false, least common multiple

Pages 704–706 Chapter 12 Study Guide and Assessment

9. $z, z \neq 3$ **11.** $\frac{a}{a+2}, a \neq 0, -2$ **13.** $\frac{1}{(b+3)(b+2)}$, $b \neq \pm 2, \pm 3$ **15.** $\frac{3axy}{10}$ **17.** $\frac{3}{a^2-3a}$ **19.** $b + 7$

21. $\frac{y^2-4y}{3}$ **23.** $\frac{2m-3}{3m-2}$ **25.** x^2+4x-2 **27.** $x^3+5x^2+12x+23+\frac{52}{x-2}$ **29.** $\frac{2a}{m^2}$ **31.** $\frac{2m+3}{5}$ **33.** $a+b$

35. $\frac{22n}{21}$ **37.** $\frac{4c^2+9d}{6cd^2}$ **39.** $\frac{4r}{r-3}$ **41.** $\frac{5m-8}{m-2}$ **43.** $\frac{3x}{y}$

45. $\frac{x^2+8x-65}{x^2+8x+12}$ **47.** -5 **49.** $-\frac{1}{4}$ **51.** 5, $-\frac{1}{4}$ **53.** 6 bicycles, 24 tricycles, and 24 wagons **55.** $1\frac{7}{8}$ hours

CHAPTER 13 EXPLORING RADICAL EXPRESSIONS AND EQUATIONS

Page 712 Lesson 13–1A

1. $4 + 4 = 8$ **3.** $16 + 9 = 25$

5.

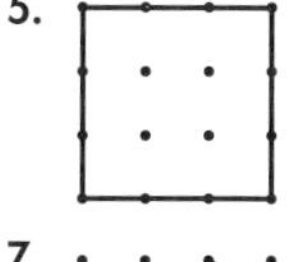

7. **9.**

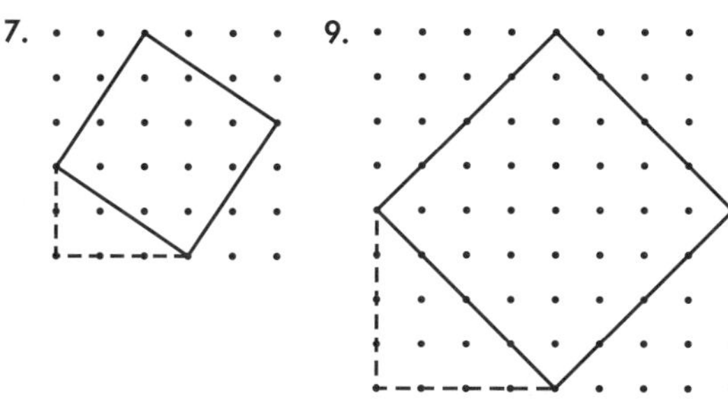

Pages 715–718 Lesson 13–1

7. 7 **9.** 11.40 **11.** 15 **13.** 2 **15.** yes **17.** 127.28 ft **19.** 13.86 **21.** 13.08 **23.** 14.70 **25.** 60 **27.** 4 **29.** $\sqrt{67} \approx 8.19$ **31.** $\sqrt{27} \approx 5.20$ **33.** $\sqrt{145} \approx 12.04$ **35.** yes **37.** no; $11^2 + 12^2 \neq 15^2$ **39.** yes **41.** $\sqrt{75}$ in. or about 8.66 in. **43a.** 44.49 m **43b.** 78 m^2

45.

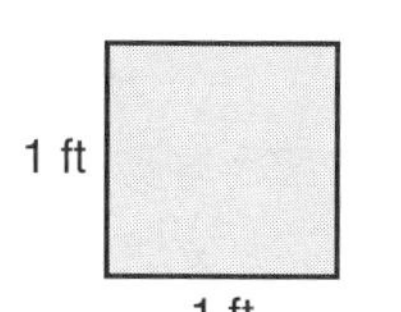

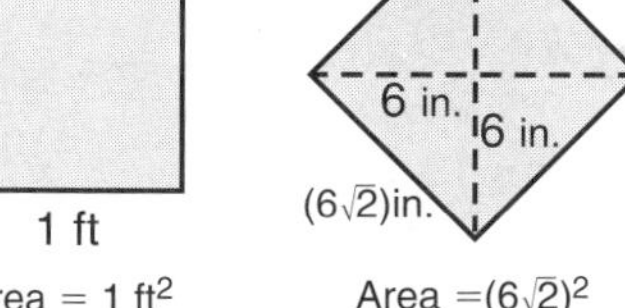

47. about 2.66 m **49.** -7 **51.** $12r^7$ **53.** $\left\{c \mid c \geq \frac{1}{2}\right\}$ **55.** 41

Pages 724–725 Lesson 13–2

5. $5 - \sqrt{2}$; 23 **7.** $\frac{\sqrt{7}}{\sqrt{7}}$ **9.** $3\sqrt{2}$ **11.** $\frac{\sqrt{21}}{7}$ **13.** $10\sqrt{2} + 26$ **15.** $\frac{18+6\sqrt{2}}{7}$ **17.** > **19.** about 7003.5 gal/min

21. $4\sqrt{5}$ **23.** $10\sqrt{5}$ **25.** $\frac{\sqrt{2}}{2}$ **27.** $\frac{\sqrt{22}}{8}$ **29.** $84\sqrt{5}$

31. $\frac{\sqrt{11}}{11}$ **33.** $3|ab|\sqrt{6}$ **35.** $7x^2y^3\sqrt{3xy}$ **37.** $\frac{3\sqrt{3}}{|p|}$

39. $\frac{x\sqrt{3xy}}{2y^3}$ **41.** $y^2 - 2y\sqrt{7} + 7$ **43.** $\frac{28\sqrt{2}+14\sqrt{5}}{3}$

45. $\frac{-2\sqrt{5}-\sqrt{10}}{2}$ **47.** $\frac{c-2\sqrt{cd}+d}{c-d}$ **49.** $x^2 - 5x\sqrt{3} + 12$

51. > **53.** < **55.** Yes; the result is about 11.84. **57.** 18.44 cm **59.** 1, $\frac{2}{3}$ **61.** 2.7×10^9 acres

63. $y = \frac{1}{5}x + 6$ **65.** -6

Page 726 Lesson 13–2B
1. $14\sqrt{7}$ or 37.04 **3.** $2\sqrt{3} + 3\sqrt{2}$ or 7.71 **5.** $\frac{4\sqrt{7}}{7}$ or 1.51
7. $3\sqrt{3}$ or 5.20 **9.** $\frac{\sqrt{3}}{2} - \frac{\sqrt{5}}{3} + 3\sqrt{2}$ or 4.36
11. $\frac{-16\sqrt{6}}{3}$ or -13.06

Pages 729–731 Lesson 13–3
5. $3\sqrt{5}, 3\sqrt{20}$ **7.** none **9.** $13\sqrt{6}$ **11.** $12\sqrt{7x}$
13. $13\sqrt{3} + \sqrt{2}$; 23.93 **15.** $\frac{8}{7}\sqrt{7}$; 3.02 **17.** 11
19. $26\sqrt{13}$ **21.** in simplest form **23.** $21\sqrt{2x}$ **25.** $9\sqrt{3}$; 15.59 **27.** $\sqrt{6} + 2\sqrt{2} + \sqrt{10}$; 8.44 **29.** $-\sqrt{7}$; -2.65 **31.** $29\sqrt{3}$; 50.23 **33.** $4\sqrt{5} + 15\sqrt{2}$; 30.16 **35.** $4\sqrt{3} - 3\sqrt{5}$; 0.22 **37.** $10\sqrt{2} + 3\sqrt{10}$ **39.** $19\sqrt{5}$ **41.** $10\sqrt{3} + 16$ **43.** Both $(x-5)^2$ and $(x-5)^4$ must be nonnegative, but $x - 5$ may be negative. **45.** $6\frac{2}{5}$ ft **47.** $3x - 3y$
49. 7 **51.** $\frac{2}{3}$ **53a.** 1991; about 9 million **53b.** 6 million

Page 731 Self Test
1. 35 **3.** 46.17 **5.** 10 **7.** $11\sqrt{6}$ **9.** $12\sqrt{5} - 6\sqrt{3} + 2\sqrt{15} - 3$

Pages 734–736 Lesson 13–4
7. $a + 3 = 4$ **9.** 16 **11.** 36 **13.** -12 **15.** 3.9 mph
17. -63 **19.** no real solution **21.** $\frac{81}{25}$ **23.** 2 **25.** 24
27. $\frac{3}{25}$ **29.** $\pm\frac{4}{3}\sqrt{3}$ **31.** $\sqrt{7}$ **33.** 6 **35.** $-3, -10$ or 3, 10
37. 16 **39.** no real solution **41.** $\left(\frac{1}{4}, \frac{25}{36}\right)$ **43.** $\sqrt{3} + 1$
45. $54\frac{2}{3}$ mi **47a.** No, at 30 mph, her car should have skidded 75 feet after the breaks were applied and not 110 feet. **47b.** about 36 miles per hour **49.** $6b - 26 + \frac{150}{b+5}$ **51.** $6xy^3$ **53.** $(-4, 5)$ **55.** $\frac{8}{13}$

Pages 739–741 Lesson 13–5
5. 5 **7.** $3\sqrt{2}$ or 4.24 **9.** 7 or 1 **11.** about 9.49 mi **13.** 17
15. $\sqrt{41}$ or 6.40 **17.** $\frac{10}{3}$ or 3.33 **19.** $\frac{13}{10}$ or 1.30
21. $2\sqrt{3}$ or 3.46 **23.** 17 or -13 **25.** -10 or 4 **27.** 8 or 32 **29.** no **31.** $\sqrt{157} \neq \sqrt{101}$; Trapezoid is not isosceles. **33.** The distance between $(3, -2)$ and $(-3, 7)$ is $3\sqrt{13}$ units. The distance between $(-3, 7)$ and $(-9, 3)$ is $2\sqrt{13}$ units. The distance between $(3, -2)$ and $(-9, 3)$ is 13 units. Since $(3\sqrt{13})^2 + (2\sqrt{13})^2 = 13^2$, the triangle is a right triangle. **35.** 109 miles **37.** $\frac{2\sqrt{2}}{3}$

39. no solution

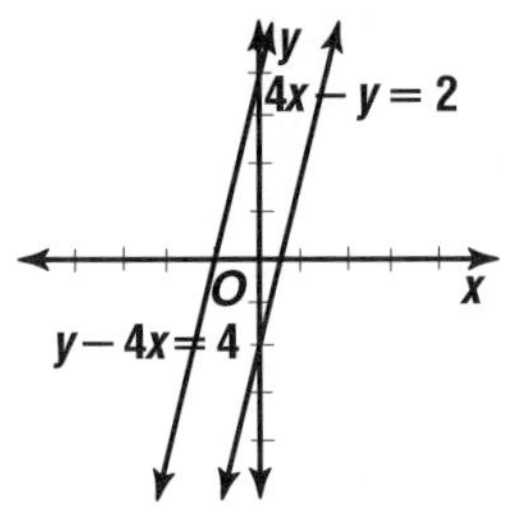

41a. 7392 BTU **41b.** 3705 BTU **41c.** 6247.5 BTU
41d. 14,040 BTU **41e.** 6831.45 BTU

Page 742 Lesson 13–6A
1. $(x+2)^2 = 1$ **3.** $(x+2)^2 = 5$ **5.** $(x-2)^2 = 5$
7. $(x + 1.5)^2 = 3.25$

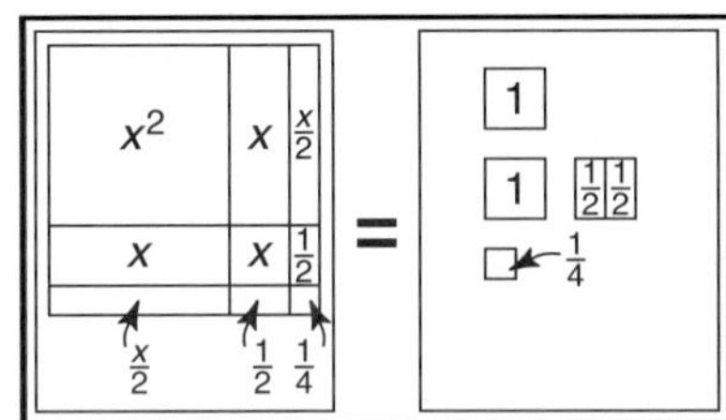

Pages 746–747 Lesson 13–6
5. 64 **7.** $-1, -3$ **9.** 7, -3 **11.** $2 \pm \sqrt{6}$ **13.** 4.9 ft
15. 16 **17.** $\frac{121}{4}$ **19.** 8 **21.** 4, 1 **23.** $\frac{7}{3}$ **25.** $3 \pm \sqrt{5}$
27. $5 \pm 4\sqrt{3}$ **29.** $\frac{-5 \pm 2\sqrt{15}}{5}$ **31.** $\frac{2}{3}, -1$ **33.** $\frac{7 \pm \sqrt{85}}{6}$
35. 18, -18 **37.** 3 **39.** $2 \pm \sqrt{4 - c}$ **41.** $b(-2 \pm \sqrt{3})$
43. 45 mph, 55 mph **45.** 8.06 **47.** 2042 **49.** (2, 2)
51a. 224.6 **51b.** 172

Page 749 Chapter 13 Highlights
1. false, $-3 - \sqrt{7}$ **3.** false, $-2 - 2\sqrt{3} - \sqrt{5} - \sqrt{15}$
5. true **7.** false, $3x + 19 = x^2 + 6x + 9$ **9.** false, $\frac{x\sqrt{2xy}}{y}$

Pages 750–752 Chapter 13 Study Guide and Assessment
11. 34 **13.** $5\sqrt{5} \approx 11.18$ **15.** 24 **17.** no; $9^2 + 16^2 \neq 20^2$
19. yes **21.** $4\sqrt{30}$ **23.** $2|a|b^2\sqrt{11b}$ **25.** $57 - 24\sqrt{3}$
27. $\frac{3\sqrt{35} - 5\sqrt{21}}{-15}$ **29.** $2\sqrt{6} - 4\sqrt{3}$ **31.** $36\sqrt{3}$ **33.** $\frac{9\sqrt{2}}{4}$
35. 12 **37.** $\frac{26}{7}$ **39.** 45 **41.** 3 **43.** no solution **45.** 17
47. $\sqrt{205} \approx 14.32$ **49.** 5 or -1 **51.** 5 or -3 **53.** $\frac{49}{4}$
55. $\frac{1}{9}$ **57.** $\frac{7 \pm \sqrt{69}}{2}$ **59.** $-2 \pm \sqrt{6}$ **61.** $22\sqrt{6} \approx 53.9$
63. scalene triangle **65.** 5 ft **67.** No, it should skid about 201.7 ft.

PHOTO CREDITS

Cover (t)FPG, (b)Nawrocki Stock, (bkgd)Superstock; **iii** (t)courtesy Home of the Hamburger, Inc., (b)Aaron Haupt; **viii** (l)Terry Eiler, (r)Paul L. Ruben **ix** (t)Geoff Butler, (b)Skip Comer; **x** (t)Stephen Webster, (bl)Aaron Haupt, (br)Mark Peterson/Tony Stone Images; **xi** Tom Tietz/Tony Stone Images; **xii** (t)Stan Fellerman/The Stock Market, (b)Aaron Haupt; **xiii** (t)Tom McCarthy/The Stock Market, (c)Geoff Butler, (b)Charles H. Kuhn Collection, The Ohio State University Cartoon,Graphic and Photgraphic Arts Research Library/Reprinted with Special Permission of King Features Syndicate; **xiv** (t)Clark James Mishler/The Stock Market, (b)Julian Baum/Science Photo Library/Photo Researchers; **xv** (t)Jonathan Daniel/Allsport USA, (c)Steve Niedorf/The Image Bank, (b)Jody Dole/The Image Bank; **xvi** Tate Gallery, London/Art Resource, NY; **xvii** (t)Morton & White, (b)NCSA, University of IL/Science Photo Library/Photo Researchers, **4** (l)E. Hobson/Ancient Art & Architecture Collection, (r)David Ball/The Stock Market; **5** (tl)John Neubauer, (tr)Scott Cunningham, (bl)Ron Sheridan/Ancient Art & Architecture Collection, (br)Ulf Anderson/Gamma Liaison; **6** Doug Martin; **10** (l)Aaron Haupt, (r)KS Studio; **11** (t)Ted Rice, (b)courtesy Math Association Of America; **12** Terry Eiler; **18** Aaron Haupt; **19** Comnet/Westlight; **24** (t)Stephen Studd/Tony Stone Images, (b)Ken Reid/FPG; **27** (t)Cynthia Johnson/Gamma Liaison, (bl br)KS Studio; **30** (t)KS Studio, (b)Bruce Hands/Tony Stone Images; **31** Phil Degginger/Tony Stone Images; **32** courtesy Home of the Hamburger, Inc.; **36** Gail Shumwat/FPG; **37** Duomo/Steven E. Sutton; **40** Doug Martin; **41** Mary Ann Evans; **43, 45** Aaron Haupt; **46** David Madison; **50** Eugene G. Schulz; **52** Kenji Kerins; **54** Doug Martin; **55** (l)KS Studio, (r)Paul L. Ruben; **57** Ken Huang/The Image Bank; **58** Aaron Haupt; **59** HERMAN ©1975 Jim Unger. Reprinted with permission of UNIVERSAL PRESS SYNDICATE. All rights reserved. **60** J. Parker/Westlight; **61** Aaron Haupt; **62** Sonderegger/Westlight; **67** (l)Aaron Haupt, (r)Eclipse Studios; **70** (t)Geoff Butler, (b)Frank Lerner; **71** (tl)James Wasserman, (tr)Doug Martin, (cl)Archive Photos, (cr)courtesy Gae Veit, (b)courtesy Nestle USA; **72-73** Geoff Butler **75** KS Studio; **76** Geoff Butler; **77** Kenji Kerins; **78** James Avery/Everett Collection; **79** Geoff Butler; **80** (l)Mark Petersen/Tony Stone Images, (r)R.B. Sanchez/The Stock Market; **82** Geoff Butler; **83** Focus on Sports; **85** AP/NASA/Wide World Photos; **88** Doug Martin; **89** Geoff Butler; **91** (t)American Photo Bank, (b)Geoff Butler; **92** William Clark/Tony Stone Images; **95** (t)Duomo/Paul Sutton, (b)Aaron Haupt; **97** Aaron Haupt; **98** (l)UPI/Corbis-Bettmann, (r)Warren Morgan/Westlight; **99** Doug Martin; **100** (t)Westlight, (b)Stewart Cohen/Tony Stone Images; **102** Archive Photos; **103** James Marshall/The Stock Market; **104** (t)Aaron Haupt, (c)Gary Mortimore/Allsport USA, (b)Geoff Butler; **106** R. Foulds/Washington Stock Photos; **108** Geoff Butler; **110** P. Saloutos/The Stock Market, (inset)Aaron Haupt; **112** (t)Holmes-Lebel/FPG, (b)Corbis-Bettmann; **113** Kenji Kerins, (inset)Aaron Haupt; **116** Aaron Haupt; **117** (l)David Stoecklein/The Stock Market, (r)David Woods/The Stock Market; **119** Johnny Johnson/Tony Stone Images; **122** CLOSE TO HOME ©1994 John McPherson/Dist. of UNIVERSAL PRESS SYNDICATE. Reprinted with permission. All rights reserved. **124** (t)Joe Towers/The Stock Market, (c)Nick Nicholson/The Image Bank, (b)Aaron Haupt; **125** (l)Aaron Haupt, (r)David Cannon/Allsport USA; **126** (t)Ian R. Howarth, (cl)Charles W. Melton, (cr)Roy Morsch/The Stock Market, (bl)Meryl Joseph/The Stock Market (br)William J. Weber; **127** (t)Elaine Shay, (b)John Neubauer; **129, 130** Aaron Haupt; **131** Bob Daemmrich; **132** Doug Martin; **136** Mark Burnett; **140** (t,chart)Mark Scott/FPG, (c b)Ron Sheridan/Ancient Art & Architecture Collection, (b,chart)Frank Cezus; **141** (tl)courtesy Marianne Ragins, (tr)David Woods/The Stock Market, (bl)Ron Sheridan/Ancient Art & Architecture Collection, (bc)Corbis-Bettmann, (br)Gamma Liaison; **145** Michael Burr/NFLP; **149** Rick Weber; **150** AP/Wide World Photos; **153** Rick Weber; **154** Dave Lawrence/The Stock Market; **156** (t)Wide World Photos, (b)H.G. Ross/FPG; **157** Archive Photos; **160** Benn Mitchell/The Image Bank; **161** Skip Comer; **162** Ralph Cowan/Tony Stone Images; **163** Allsport USA **167** (t)StudiOhio, (b)Rick Weber; **168** (l)UPI/Corbis-Bettmann, (r)LPI/M. Yada/FPG; **170** E. Alan McGee/FPG; **172** Paul Barton/The Stock Market; **174** Rick Weber; **176** Aaron Haupt; **177** Bruce Ayres/Tony Stone Images; **178** David M. Grossman/Photo Researchers; **180** Archive Photos/Martin Keene; **181** Focus On Sports; **183** Aaron Haupt; **184** (l)David M. Dennis, (r)NASA; **189** Rick Weber; **190** James Meyer/The Image Bank; **191** Morton & White, (bkgd)James Meyer/The Image Bank; **192** (t)Tim Courlas, (cl)Corbis-Bettmann, (cr)Glencoe photo, (b)Archive Photos; **193** (t)John Karl Breun, (c)KS Studio, (bl)Corbis-Bettmann, (br)UPI/Corbis-Bettmann **195** Morton & White; **196** Larry Hamill; **197** (t)Universal Pictures/Shooting Star, (b)Phil Schofield/Tony Stone Images; **198** (l)David R. Frazier Photolibrary, (r)Glencoe photo; **199** (t)Biophoto Associates/Photo Researchers, (b)Morton & White; **200** Kenji Kerins; **201** (t)Gus Chan, (c)TJ Collection/Shooting Star, (b)AP/Wide World Photos; **202** Aaron Haupt; **203** James N. Westwater; **205** (t)Mike Yamashita/Westlight, (b)Scott Padgett/Tradd Street Stock; **206** (t)John Elk III, (b)Jurgen Vogt/The Image Bank; **209** AP Photo/Luc Novovitch/Wide World Photos; **213** (t)Ulf Wallin/The Image Bank, (b)Doug Martin; **214** (t)Focus on Sports, (b)courtesy Ryan Morgan; **215** (t)Doug Martin, (b)Morton & White; **217** (t)Everett Collection, (bl br)Morton & White; **218** Leo Nason/The Image Bank; **219** KS Studio; **220** (t)Elaine Shay, (tc tb)Morton & White, (b)John Mead/Science Photo Library/Photo Researchers; **222** (bkgd)Morton & White, (t)Dr. E.R. Degginger/Color-Pic, (b)Charles Cangialosi/ Westlight; **223, 224** Morton & White; **226** (t)Morton & White, (b)KS Studios; **227** Kenji Kerins; **228** (t)Roger Dons/People Weekly, (b)Morton & White; **229, 230, 231** Morton & White; **232** (t)Everett Collection, (b)Morton & White, (c)Chris Hackett/The Image Bank; **233** Everett Collection; **234** (t)Don Mason/The Stock Market, (b)Morton & White; **236** Skip Comer; **237** (t)Moron & White Photgraphic, (b)Chuck Bankuti/Shooting Star; **238** (t)Kenji Kerins, (b)Morton & White; **239** KS Studios; **240** Matt Meadows; **241** Morton & White; **242** Photo Researchers; **244** NASA; **248-249** KS Studio; **252** (t)Terry Vine/Tony Stone Images, (c)Ronald Sheridan/Ancient Art & Architecture Collection, (b)Corbis-Bettmann; **253** (tl)Weldon McDougal, (tr)David W. Hamiliton/The Image Bank, (cl)Jim Cornfield/Westlight, (cr)Glencoe photo, (br)courtesy Council for the Traditional Arts; **254** (t)Mimi Ostendorf-Smith, (b)James Blank/The Stock Market; **262** (t)Tom Tietz/Tony Stone Images, (c)Daniel J. Cox/Tony Stone Iamges, (b)Tim Davis/Tony Stone Images; **264** James P. Rowan/Tony Stone Images; **266** Stephen Marks/The Image Bank; **269** Stephen Webster; **271** (t)Doug Martin, (c)Brent Turner/BLT Productions, (b)Aaron Haupt; **273, 277** Aaron Haupt; **280** Eclipse Studios; **281** (t)Doug Martin, (b)Tom McGuire; **282** KS Studio; **284** Ruth Dixon; **285** (t)KS Studio, (b)Aaron Haupt; **286** Bob Daemmrich; **287** David O. Hill/Photo Researchers; **290** KS Studio; **293** Stephen Marks/The Image Bank; **294** (l)Duomo/Al Tielemans, (r)Duomo/Bryan Yablonsky; **295** Alese & Mort Pecher/The Stock Market; **299** 1995 Watterman/Dist. by Universal Press Syndicate; **301** (t)Smithsonian Institution, (b)David Madison; **302** MAK-1; **304** KS Studio; **306** (t)Aaron Haupt, (b)Doug Martin; **308** David Barnes/The Stock Market; **309** KS Studio; **310** Aaron Haupt; **311** (t)Joan Marcus/Photofest, (c bl)Photofest, (br)Aaron Haupt; **313** (t)Aaron Haupt, (b)Harald Sund The Image Bank; **314** (t)Eric Lars Bakke/Allsport USA, (c b)In Fisherman Magazine; **318** Aaron Haupt; **319** Denis Valentine/The Stock Market; **321** (t)Matt Meadows, (b)Morton & White; **322** (t)Corbis-Bettmann, (b)Culver Pictures; **323** (tl)courtesy Jorge Arturo Pineda Aguilar, (tr)Morton & White. (cl)National Portrait Gallery, Washington, D.C./Art Resource, NY, (cr)Archive Photos/Consolidated News, (b)Alan S. Weiner/People Weekly; **325, 328** Morton & White; **330** J. Coolidge/The Image Bank; **331** (tl)John M. Roberts/The Stock Market, (tc)Andrea Pistolesi/The Image Bank, (tr)Dallas & John Heaton/Westlight, (cl)Tom Walker/Tony Stone Images, (cr)Art Wolfe/Tony Stone Images,(bl)UPI/Corbis-Bettmann, (br)Gail Shumway/FPG; **332** (t)Grant V. Faint/The Image Bank, (c)Scott Markewitz/FPG, (b)Van Bucher/Photo Researchers; **335, 337** Doug Martin; **338, 339** Morton & White; **340** (t)Tom Stack/Tom Stack & Associates, (c)Bob Daemmrich, (b)Tracy I. Borland; **342** Stan Fellerman/The Stock Market; **343** (t)Doug Martin, (b)Morton & White; **344-345** Morton & White Photographic; **349** Corbis-Bettmann; **350** John McDermott/Tony Stone Images; **353** Morton & White; **354** Dick Luria/FPG; **356** Morton & White; **358** E.R. Degginger/Color-Pic; **360** Gabe Palmer/The Stock Market; **362** (t)E. Alan McGee/FPG, (b)Bob Taylor/FPG; **364** Terry Qing/FPG; **368** Bryan Peterson/The Stock Market; **369, 372** Morton & White; **374** Rainer Grosskopf/Tony Stone Images; **37** Life Images; **382** (t)Bob Daemmrich/Tony Stone Images, (b)Ron Sheridan/Ancient Art & Architecture Collection; **383** (tl)Phil Matt, (tr)Ken Edward/Science Source/Photo Researchers, (cl)John Berry/courtesy Center for Forensic Science, Univ. of New Haven, CT, (cr)courtesy University of TX at Brownsville, (bl)Culver Pictures, (br)US Justice Department; **384, 386** Aaron Haupt; **390** (bkgd)KS Studio, (t)Doug Martin; **392** The Gordon Hart Collection, Bluffton, IN; **394** Life Images; **395** Kenji Kerins; **397** (l)courtesy the office of Congresswoman Patsy Mink, (r)Doug Martin; **398** (l)Duomo/Paul Sutton, (r)Mark Lawrence/The Stock Market; **399** Aaron Haupt; **404** (t)Kenji Kerins, (b)Doug Martin; **405** (t)Alain Evrard/Photo

Researchers, (c)Nikolas/Konstantinou/Tony Stone Images, (b)Norbert Wu/Tony Stone Images; **407** Aaron Haupt; **411** Pal Hermansen/Tony Stone Images; **413** (t)Suzanne Murphy-Larronde/FPG (b)Bill Ross/Westlight; **415** (bkgd)Duomo/William Sallaz,(t)KS Studio; **416** Travelpix/FPG; **417** (t)KS Studio, (b)Matt Meadows; 418 (t)Eclipse Studios, (b)Skip Comer; **419** Aaron Haupt; **420** (t)Joseph Drivas/The Image Bank, (bl)Michael Salas/The Image Bank; **422** Aaron Haupt; **423** Scott Cunningham; **425** (t)David R. Frazier/Photo Researchers, (b)MAK-1; **429** C.T. Tracy/FPG; **430** Gary Buss/FPG; **431** Lawrence Migdale; **432** courtesy The White House; **433** KS Studio; **436** Lori Adamski Peek/Tony Stone Images; **438** Court Mast/FPG; **440** Kenji Kerins; **442** (bkgd)Jane Shemilt/Science Photo Library/Photo Researchers, (t)Aaron Haupt, (b)Photo Researchers; **449** Steve Lissau; **450** (t)David Madison, (bl)Ronald Sheridan/Ancient Art & Architecture Collection, (br)Corbis-Bettmann; **451** (tl)Tom Lynn/Sports Illustrated, (tr)AP Photo/Elise Amendola/Wide World Photos, (cl)Corbis-Bettmann, (cr)Reuters/Corbis-Bettmann, (b)Corbis-Bettmann; **455** AP Photo/Mark Lenniham/Wide World Photos; **460** (t)Dinodia/Jagdish Agarwal, (b)Dinodia/Jagdish Agarwal; **461** Matt Meadows; **462** Clark James Mishler/The Stock Market; **465** (t)Bob Daemmrich/Tony Stone Images , (b)Aaron Haupt; **467** (l)Aaron Haupt, (r)PEANUTS reprinted by permission of United Feature Syndicate, Inc.; **468** (t)Ronald Johnson/The Image Bank, (b)Brian Brake/Photo Researchers; **469** (t)Archive Photos, (l)Charles H. Kuhn Collection, The Ohio State University Cartoon,Graphic, and Photographic Arts Research Library/Reprinted with Special Permission of King Features Syndicate; **471** Doug Martin; **473** M. Gerber/LGI Photo Agency; **475** J-L Charmet/Science Photo Library/Photo Researchers; **477** (t)Bob Daemmrich, (b)AP/Wide World Photos; **481** (t)Aaron Haupt, (b)Tim Courlas, **482** Aaron Haupt; **483** James Blank/FPG; **485** Nino Mascardi/The Image Bank; **486** Elaine Shay **490** Tracy Borland; **494** (t)Reprinted with special permission of King Features Syndicate, (c)Corbis-Bettmann, (b)Corbis-Bettmann; **495** (t)Richard Lee/Detroit Free Press, (c)Gary Buss/FPG, (b)Culver Pictures; **496** (t)People Weekly ©1995 Keith Philpott, (b)AP/Wide World Photos; **500** (t)Will Crocker/The Image Bank, (b)Doug Martin; **501** (t)Aaron Haupt; **505** (t,inset)Aaron Haupt, (t)Julie Marcotte/Westlight, (b)Kenji Kerins; **506** (t)Aaron Haupt, (b)Mason Morfit/FPG; **507** David R. Frazier Photolibrary; **508** (t)Julian Baum/Science Photo Library/Photo Researchers, (b)Archive Photos/PA News; **509** Centers For Disease Control; **510** (l)Demmie Todd/Shooting Star, (r)Adam Hart-Davis/Science Photo Library/Photo Researchers; **511** (t)Prof. Aaron Pollack/Science Photo Library/Photo Researchers, (b)Doug Martin; **512** (t)Co Rentmeester/The Image Bank, (b)Garry Gay/The Image Bank; **514** (t)Tate Gallery, London/Art Resource, N.Y., (bl)The Metropolitan Museum of Art, Gift of Adele R. Levy, 1958/Malcolm Varon, (br)Erich Lessing/Art Resource, NY; **515** David R. Frazier/Photo Researchers; **518** Gene Frazier; **519** Scott Cunningham; **522** (t)Aaron Haupt, (b)Doug Martin; **526** Bill Ross/Westlight; **527** (r)Aaron Haupt, (l)Richard Hutchings/Photo Researchers; **529** (t)Aaron Haupt, (b)Jeff Spielman/The Image Bank; **531** (t)Robert Tringali/Sportschrome East/West, (b)Mark Gibson/The Stock Market; **533** (t)Eric A. Wessman/Viesti Associates, (b)Charlotte Raymond/Photo Researchers; **536** (t)Jon Blumb, (b)Mark Burnett/Photo Researchers; **540** Steve Elmore/The Stock Market; **541** Aaron Haupt; **542** (t)Larry Lefever from Grant Heilman, (b)Aaron Haupt; **547** Aaron Haupt; **548** Steve Lissau; **553** (l)Geoff Butler, (r)Morton & White; **555** (l)Ira Block/The Image Bank, (r)David R. Frazier Photolibrary; **556** Corbis-Bettmann; **557** (t)Everett Collection, (ct)Morton & White, (cb)NASA, (b)UPI/Corbis-Bettmann; **558** Morton & White; **559** National Archives; **560** (t)Archive Photos, (b)Morton & White; **561-562** Morton & White; **563** (t)Morton & White, (b)Jeff Shallit; **565** (t)Barros & Barros/The Image Bank, (b)KS Studio; **566** Kenji Kerins; **569** (r)Larry Hamill, (l)Otto Greule/Allsport USA; **570** Kenji Kerins; **571** (l)National Portrait Gallery, Smithsonian Institution, Gift of TIME, Inc./Art Resource, NY; **579** Derek Berwin/The Image Bank; **583** (t)Corbis-Bettmann, (b)Ron Sheridan/Ancient Art & Architecture Collection; **586** Brent Turner/BLT Productions; **593** L.D.Gordon/The Image Bank; **594** Piffret/Photo Researchers; **596** Janeart LTD./The Image Bank; **597** (t)Joe Towers/The Stock Market, (b)Roy Marsch/The Stock Market; **598** (l)Steve Niedorf/The Image Bank; **599** (l)Chuck Doyle, (r)Archive Photos; **600** Kenji Kerins; **608** (t)Tim Courlas, (c)North Wind Picture Archive, (b)Corbis-Bettmann; **609** (t)Kathleen Cabble/St. Petersburg Times, (c bl)Archive Photos, (br)Archive Photos/Frank Edwards; **611** Arteaga Photo Ltd.; **613** (l)American Photo Library, (r)Ed & Chris Kumler; **615** Morton & White; **616** (t)Mark Burnett, (b)Morton & White; **620** Aaron Haupt; **626** David R. Frazier Photolibrary; **628** (t)Library of Congress, (b)Aaron Rezny/The Stock Market; **631** Harald Sund/The Image Bank; **633** (t)Marc Grimberg/The Image Bank, (b)Morton & White; **635** Morton & White; **638** David M. Dennis; **641** Howard Sochurek/The Stock Market; **642** (t)Jody Dole/The Image Bank, (b)courtesy Bob Paz/Caltech; **645** (t)Michael Melford/The Image Bank, (c)James N. Westwater, (b)Gary Cralle/The Image Bank; **646** Culver Pictures; **647** Elaine Shay; **648** (t)Herbert Schaible/The Image Bank, (b)Hans Wol/The Image Bank; **650** (t)Morton & White, (bl)John Phelan/The Stock Market, (br)John Maher/The Stock Market; **654** (l)Doug Martin, (r)Co Rentmeester/The Image Bank; **655** (l)Mark Burnett, (c)G. Petrov/Washington Stock Photo, (r)M. Winn/Washington Stock Photo; **657** Tony Freeman/PhotoEdit; **658** (t)Lou Jacobs Jr. from Grant Heilman, (b)Ron Sheridan/Ancient Art & Architecture Collection; **659** (t)courtesy Joe Maktima, (cl)Lawrence Migdale, (cr)M. Miller/H. Armstrong Roberts, (b)NASA; **660** (t)Morton & White, (b)MAK-1; **662, 663** Morton & White; **664** (t)Ron Watts/Westlight, (b)Doug Martin; **665** (t)Ned Gillette/The Stock Market, (b)Lynn Campbell; **667** Benjamin Rondel/The Stock Market; **669** (t)KS Studio, (b)Morton & White; **670** (t)Morton & White, (b)Bob Mullenix; **671** (l)Stephen Dunn/Allsport USA, (r)Brett Palmer/The Stock Market; **672** (l)Mark Gibson/The Stock Market, (r)Larry Hamill; **674** Aaron Haupt; **675, 677** KS Studio; **680** (t)Ward's Natural Science, (bl br)Aaron Haupt; **681** Corbis-Bettmann; **684** Dave Merrill & Co./National Geographic Society; **685** (br)CalTech & Carnegie Institution, (others)NASA; **687** Roy Morsch/The Stock Market; **688** (t)UPI/Corbis-Bettmann, (b)Morton & White; **689** (t)GK & Vikki Hart/The Image Bank, (b)Morton & White; **690** Morton & White; **694** Andrew Unangst/The Image Bank; **696** Kenji Kerins; **697** (l)Patrick W. Stoll/Westlight, (r)Morton & White; **698** Doug Martin; **701** (l)Morton & White, (r)Guido Alberto Rossi/The Image Bank; **702** Archive Photos; **706** "Dennis The Menace" used by permission of Hank Ketcham and © by North American Syndicate; **707** Morton & White; **710** (t)Glencoe photo, (c)Ron Sheridan/Ancient Art & Architecture Collection, (b)Matt Meadows; **711** (tl)Annie Griffiths Belt, (tr)Doug Martin, (cl)University of California/Lawrence Berkeley Laboratory, (cr)Archive Photos/Reuters/Peter Jones, (b)Lucille Day/Black Star; **713** Stock Newport/Daniel Forster; **715** (t)Ulrike Welsch/Photo Researchers, (b)Glencoe photo; **717** David R. Frazier Photolibrary; **718** Archive Photos/David Lees; **719** W. Warren/Westlight; **720** Doug Martin; **721** Gerard Photography; **722** PEANUTS reprinted by permission of United Features Syndicate, Inc. **723** Bob Daemmrich; **724** (l)Flip Chalfant/The Image Bank, (r)Glencoe photo; **725** KS Studio; **727** (t)Tom Bean/The Stock Market, (b)D. Logan/H. Armstrong Roberts; **728** William Kennedy/The Image Bank; **730** K. Scholz/H. Armstrong Roberts; **732** (t)Kaz Mori/The Image Bank, (c)DLR, ESA/Science Photo Library/Photo Researchers, (b)Greg Davis/The Stock Market; **736** (t)Joe Azzara/The Image Bank, (b)KS Studio; **739** Ken Davies/Masterfile; **740** (l)Glencoe photo, (r)NCSA, University of IL/Science Photo Library/Photo Researchers; **743** (t)David R. Frazier Photolibrary, (c)Dallas & John Heaton/Westlight, (b)Wong How Man/Westlight; **746** Doug Martin; **747** Sylvain Grandadam/Photo Researchers; **748** David R. Frazier Photolibrary; **752** The Walt Disney Co./Shooting Star; **753** Aaron Haupt; **755** (tl)Ken Edwards/Science Source/Photo Researchers, (tr)Kaz Mori/The Image Bank, (bl)David Ball/The Stock Market, (br)E. Alan McGee/FPG.

APPLICATIONS & CONNECTIONS INDEX

A

B

C

D

E

INDEX

Red type denotes items only in the Teacher's Wraparound Edition.

 A

 B

 C

G

H

I

J

K

L

M

T